VOLUME 2

Teacher's Edition

Geometry

Common Core

Randall I. Charles
Basia Hall
Dan Kennedy
Laurie E. Bass
Art Johnson
Stuart J. Murphy
Grant Wiggins

PEARSON

Boston, Massachusetts • Chandler, Arizona • Glenview, Illinois • Hoboken, New Jersey

Acknowledgments appear on page T949, which constitutes an extension of this copyright page.

PEARSON

ISBN-13: 978-0-13-328122-4
ISBN-10: 0-13-328122-1
3 4 5 6 7 8 9 10 V057 18 17 16 15 14

VOLUME 2

Geometry *Teacher's Edition Contents*

Common Core State Standards
for Mathematics High School

The following shows the High School Standards for Mathematical Content that are taught in *Pearson Geometry Common Core Edition* ©2015. Included are all of the standards that make up Achieve's Pathway for Geometry and PARCC's Model Content Frameworks. Standards that are part of Achieve's Pathway are indicated with an (A) on the standard code, while those that are part of the PARCC Model Content Frameworks are indicated with a (P) on the standard code. Standards that begin with (+) indicate additional mathematics that students should learn in order to take advanced courses such as calculus, advanced statistics, or discrete mathematics.

Geometry		Where to Find
Congruence		
Experiment with transformations in the plane		
G-CO.A.1(A)(P)	Know precise definitions of angle, circle, perpendicular line, parallel line, and line segment, based on the undefined notions of point, line, distance along a line, and distance around a circular arc.	1-2, 1-3, 1-4, 1-6, 3-1, 10-6
G-CO.A.2(A)(P)	Represent transformations in the plane using, e.g., transparencies and geometry software; describe transformations as functions that take points in the plane as inputs and give other points as outputs. Compare transformations that preserve distance and angle to those that do not (e.g., translation versus horizontal stretch).	9-1, CB 9-1, 9-2, 9-3, 9-4, 9-6
G-CO.A.3(A)(P)	Given a rectangle, parallelogram, trapezoid, or regular polygon, describe the rotations and reflections that carry it onto itself.	CB* 9-3
G-CO.A.4(A)(P)	Develop definitions of rotations, reflections, and translations in terms of angles, circles, perpendicular lines, parallel lines, and line segments.	9-1, 9-2, 9-3
G-CO.A.5(A)(P)	Given a geometric figure and a rotation, reflection, or translation, draw the transformed figure using, e.g., graph paper, tracing paper, or geometry software. Specify a sequence of transformations that will carry a given figure onto another.	9-1, 9-2, CB 9-2, 9-3, 9-4

*CB = Concept Byte

Geometry		Where to Find
Understand congruence in terms of rigid motions		
G-CO.B.6(A)(P)	Use geometric descriptions of rigid motions to transform figures and to predict the effect of a given rigid motion on a given figure; given two figures, use the definition of congruence in terms of rigid motions to decide if they are congruent.	9-1, 9-2, 9-3, 9-4, 9-5
G-CO.B.7(A)(P)	Use the definition of congruence in terms of rigid motions to show that two triangles are congruent if and only if corresponding pairs of sides and corresponding pairs of angles are congruent.	9-5
G-CO.B.8(A)(P)	Explain how the criteria for triangle congruence (ASA, SAS, and SSS) follow from the definition of congruence in terms of rigid motions.	9-5
Prove geometric theorems		
G-CO.C.9(A)(P)	Prove theorems about lines and angles. *Theorems include: vertical angles are congruent; when a transversal crosses parallel lines, alternate interior angles are congruent and corresponding angles are congruent; points on a perpendicular bisector of a line segment are exactly those equidistant from the segment's endpoints.*	2-6, 3-2, 5-2
G-CO.C.10(A)(P)	Prove theorems about triangles. *Theorems include: measures of interior angles of a triangle sum to 180°; base angles of isosceles triangles are congruent; the segment joining midpoints of two sides of a triangle is parallel to the third side and half the length; the medians of a triangle meet at a point.*	3-5, 4-5, 5-1, 5-4
G-CO.C.11(A)(P)	Prove theorems about parallelograms. *Theorems include: opposite sides are congruent, opposite angles are congruent, the diagonals of a parallelogram bisect each other and its converse, rectangles are parallelograms with congruent diagonals.*	6-2, 6-3, 6-4, 6-5
Make geometric constructions		
G-CO.D.12(A)(P)	Make formal geometric constructions with a variety of tools and methods (compass and straightedge, string, reflective devices, paper folding, dynamic geometric software, etc.). *Copying a segment; copying an angle; bisecting a segment; bisecting an angle; constructing perpendicular lines, including the perpendicular bisector of a line segment; and constructing a line parallel to a given line through a point not on the line.*	1-6, CB 1-6b, CB 3-2, 3-6, 4-4, CB 4-5, 5-2, CB 6-9, CB 7-5
G-CO.D.13(A)(P)	Construct an equilateral triangle, a square, and a regular hexagon inscribed in a circle.	3-6, 4-5, 10-3

Geometry		Where to Find
Similarity, Right Triangles, and Trigonometry		
Understand similarity in terms of similarity transformations		
G-SRT.A.1(A)(P)	Verify experimentally the properties of dilations given by a center and a scale factor.	9-6, CB 9-6
G-SRT.A.1a(A)(P)	A dilation takes a line not passing through the center of the dilation to a parallel line, and leaves a line passing through the center unchanged.	9-6, CB 9-6
G-SRT.A.1b(A)(P)	The dilation of a line segment is longer or shorter in the ratio given by the scale factor.	9-6, CB 9-6
G-SRT.A.2(A)(P)	Given two figures, use the definition of similarity in terms of similarity transformations to decide if they are similar; explain using similarity transformations the meaning of similarity for triangles as the equality of all corresponding pairs of angles and the proportionality of all corresponding pairs of sides.	9-7
G-SRT.A.3(A)(P)	Use the properties of similarity transformations to establish the AA criterion for two triangles to be similar.	9-7
Prove theorems involving similarity		
G-SRT.B.4(A)(P)	Prove theorems about triangles. *Theorems include: a line parallel to one side of a triangle divides the other two proportionally, and conversely the Pythagorean Theorem proved using triangle similarity.*	7-5, 8-1
G-SRT.B.5(A)(P)	Use congruence and similarity criteria for triangles to solve problems and to prove relationships in geometric figures.	4-2, 4-3, 4-4, 4-5, 4-6, 4-7, 5-1, 5-2, 5-4, 6-1, 6-2, 6-3, 6-4, 6-5, 6-6, 7-2, 7-3, 7-4
Define trigonometric ratios and solve problems involving right triangles		
G-SRT.C.6(A)(P)	Understand that by similarity, side ratios in right triangles are properties of the angles in the triangle, leading to definitions of trigonometric ratios for acute angles.	CB 8-3, 8-3
G-SRT.C.7(A)(P)	Explain and use the relationship between the sine and cosine of complementary angles.	8-3, CB 8-4
G-SRT.C.8(A)(P)	Use trigonometric ratios and the Pythagorean Theorem to solve right triangles in applied problems.	8-1, 8-2, 8-3, 8-4
Apply trigonometry to general triangles		
G-SRT.D.9(A)	(+) Derive the formula $A = \frac{1}{2}ab\sin(C)$ for the area of a triangle by drawing an auxiliary line from a vertex perpendicular to the opposite side.	10-5
G-SRT.D.10(A)	(+) Prove the Laws of Sines and Cosines and use them to solve problems.	8-5, 8-6
G-SRT.D.11(A)	(+) Understand and apply the Law of Sines and the Law of Cosines to find unknown measurements in right and non-right triangles (e.g., surveying problems, resultant forces).	8-5, 8-6

Geometry		Where to Find
Circles		
Understand and apply theorems about circles		
G-C.A.1(A)(P)	Prove that all circles are similar.	10-6
G-C.A.2(A)(P)	Identify and describe relationships among inscribed angles, radii, and chords. *Include the relationship between central, inscribed, and circumscribed angles; inscribed angles on a diameter are right angles; the radius of a circle is perpendicular to the tangent where the radius intersects the circle.*	10-6, 12-1, 12-2, 12-3
G-C.A.3(A)(P)	Construct the inscribed and circumscribed circles of a triangle, and prove properties of angles for a quadrilateral inscribed in a circle.	5-3, 12-3
G-C.A.4(A)	(+) Construct a tangent line from a point outside a given circle to the circle.	12-3
Find arc lengths and areas of sectors of circles		
G-C.B.5(A)(P)	Derive using similarity the fact that the length of the arc intercepted by an angle is proportional to the radius, and define the radian measure of the angle as the constant of proportionality; derive the formula for the area of a sector.	10-6, CB 10-6, 10-7
Expressing Geometric Properties with Equations		
Translate between the geometric description and the equation for a conic section		
G-GPE.A.1(A)(P)	Derive the equation of a circle of given center and radius using the Pythagorean Theorem; complete the square to find the center and radius of a circle given by an equation.	12-5
G-GPE.A.2(A)	Derive the equation of a parabola given a focus and directrix.	CB 12-5
Use coordinates to prove simple geometric theorems algebraically		
G-GPE.B.4(A)(P)	Use coordinates to prove simple geometric theorems algebraically.	6-9
G-GPE.B.5(A)(P)	Prove the slope criteria for parallel and perpendicular lines and use them to solve geometric problems (e.g., find the equation of a line parallel or perpendicular to a given line that passes through a given point).	3-8, 7-3, 7-4
G-GPE.B.6(A)(P)	Find the point on a directed line segment between two given points that partitions the segment in a given ratio.	1-3, 1-7, CB 1-7
G-GPE.B.7(A)(P)	Use coordinates to compute perimeters of polygons and areas of triangles and rectangles, e.g., using the distance formula.	6-7, 10-1, CB 10-7

Geometry		Where to Find
Geometric Measurement and Dimension		
Explain volume formulas and use them to solve problems		
G-GMD.A.1(A)(P)	Give an informal argument for the formulas for the circumference of a circle, area of a circle, volume of a cylinder, pyramid, and cone. *Use dissection arguments, Cavalieri's principle, and informal limit arguments.*	11-4, CB 10-7
G-GMD.A.3(A)(P)	Use volume formulas for cylinders, pyramids, cones, and spheres to solve problems.	11-4, 11-5, 11-6
Visualize relationships between two-dimensional and three-dimensional objects		
G-GMD.B.4(A)(P)	Identify the shapes of two-dimensional cross-sections of three-dimensional objects, and identify three-dimensional objects generated by rotations of two-dimensional objects.	11-1, 12-6
Modeling with Geometry		
Apply geometric concepts in modeling situations		
G-MG.A.1(A)(P)	Use geometric shapes, their measures, and their properties to describe objects (e.g., modeling a tree trunk or a human torso as a cylinder).	8-3, 10-1, 10-2, 10-3, 11-2, 11-3, 11-4, 11-5, 11-6, 11-7
G-MG.A.2(A)(P)	Apply concepts of density based on area and volume in modeling situations (e.g., persons per square mile, BTUs per cubic foot).	11-7, CB 11-7
G-MG.A.3(A)(P)	Apply geometric methods to solve design problems (e.g., designing an object or structure to satisfy physical constraints or minimize cost; working with typographic grid systems based on ratios).	3-4

Statistics and Probability		Where to Find
Conditional Probability and the Rules of Probability		
Understand independence and conditional probability and use them to interpret data		
S-CP.A.1(A)	Describe events as subsets of a sample space (the set of outcomes) using characteristics (or categories) of the outcomes, or as unions, intersections, or complements of other events ("or," "and," "not").	13-1
S-CP.A.2(A)	Understand that two events *A* and *B* are independent if the probability of *A* and *B* occurring together is the product of their probabilities, and use this characterization to determine if they are independent.	13-6
S-CP.A.3(A)	Understand the conditional probability of *A* given *B* as *P*(*A* and *B*)/*P*(*B*), and interpret independence of *A* and *B* as saying that the conditional probability of *A* given *B* is the same as the probability of *A*, and the conditional probability of *B* given *A* is the same as the probability of *B*.	13-6
S-CP.A.4(A)	Construct and interpret two-way frequency tables of data when two categories are associated with each object being classified. Use the two way table as a sample space to decide if events are independent and to approximate conditional probabilities.	13-1, 13-2, 13-5
S-CP.A.5(A)	Recognize and explain the concepts of conditional probability and independence in everyday language and everyday situations.	13-2, 13-6
Use the rules of probability to compute probabilities of compound events in a uniform probability model		
S-CP.B.6(A)	Find the conditional probability of *A* given *B* as the fraction of *B*'s outcomes that also belong to *A*, and interpret the answer in terms of the model.	13-6
S-CP.B.7(A)	Apply the Addition Rule, $P(A \text{ or } B) = P(A) + P(B) - P(A \text{ and } B)$, and interpret the answer in terms of the model.	13-4
S-CP.B.8(A)	(+) Apply the general Multiplication Rule in a uniform probability model, $P(A \text{ and } B) = P(A)P(B\|A) = P(B)P(A\|B)$, and interpret the answer in terms of the model.	13-4
S-CP.B.9(A)	(+) Use permutations and combinations to compute probabilities of compound events and solve problems.	13-4
Using Probability to Make Decisions		
Use probability to evaluate outcomes of decisions		
S-MD.B.6(A)	(+) Use probabilities to make fair decisions (e.g., drawing by lots, using a random number generator).	13-7, CB 13-7
S-MD.B.7(A)	(+) Analyze decisions and strategies using probability concepts (e.g., product testing, medical testing, pulling a hockey goalie at the end of a game).	13-7, CB 13-7

Geometry *Pacing Guide*

The development of Pearson Geometry Common Core Edition was informed by Appendix A and the PARCC Model Content Frameworks as well as considerations about the shifts in content expectations of the Common Core State Standards. The result is a course that offers flexibility of implementation. Teachers can teach a curriculum that addresses all of the standards in the model courses in Appendix A or one that aligns to the PARCC Model Content Frameworks. This Pacing Guide identifies lessons that have content that aligns with content expectations for Appendix A and with content expectations for PARCC, and also indicates lessons that can be used for enrichment or extension.

The suggested number of days for each chapter is based on a traditional 45-minute class period and on a 90-minute block period. The total of 160 days of instruction leaves time for assessments or other special days that vary from school to school.

KEY
✓= Geometry Content
❍= Prerequisite Content
❑= Content for Enrichment

	Common Core State Standards	PARCC	Appendix A
Chapter 1 Tools of Geometry		**Traditional 10**	**Block 5**
1-1 Nets and Drawings for Visualizing Geometry	Prepares for G-CO.A.1	❍	❍
1-2 Points, Lines, and Planes	G-CO.A.1	✓	✓
1-3 Measuring Segments	G-CO.A.1, G-GPE.B.6	✓	✓
1-4 Measuring Angles	G-CO.A.1	✓	✓
1-5 Exploring Angle Pairs	Prepares for G-CO.A.1 and G-CO.C.9	❍	❍
Concept Byte: Compass Designs	Prepares for G-CO.D.12	❑	❑
1-6 Basic Constructions	G-CO.A.1, G-CO.D.12	✓	✓
Concept Byte: Exploring Constructions	G-CO.D.12	✓	✓
1-7 Midpoint and Distance in the Coordinate Plane	Prepares for G-GPE.B.4 and G-GPE.B.7; also G-GPE.B.6	✓	✓
Concept Byte: Partitioning a Line Segment	G-GPE.B.6	✓	✓
Review: Classifying Polygons	Prepares for G-MG.A.1	❍	❍
1-8 Perimeter, Circumference, and Area	N-Q.A.1	❍	❍
Concept Byte: Comparing Perimeters and Areas	Prepares for G-MG.A.2	❍	❍

	Common Core State Standards	PARCC	Appendix A
Chapter 2 Reasoning and Proof		**Traditional 12**	**Block 6**
2-1 Patterns and Inductive Reasoning	Prepares for G-CO.C.9, G-CO.C.10, and G-CO.C.11	❍	❍
2-2 Conditional Statements	Prepares for G-CO.C.9, G-CO.C.10, and G-CO.C.11	❍	❍
Concept Byte: Logic and Truth Tables	Prepares for G-CO.C.9, G-CO.C.10, and G-CO.C.11	❑	❑
2-3 Biconditionals and Definitions	Prepares for G-CO.C.9, G-CO.C.10, and G-CO.C.11	❍	❍
2-4 Deductive Reasoning	Prepares for G-CO.C.9, G-CO.C.10, and G-CO.C.11	❍	❍
2-5 Reasoning in Algebra and Geometry	Prepares for G-CO.C.9, G-CO.C.10, and G-CO.C.11	❍	❍
2-6 Proving Angles Congruent	G-CO.C.9	✓	✓
Chapter 3 Parallel and Perpendicular Lines		**Traditional 12**	**Block 6**
3-1 Lines and Angles	G-CO.A.1, Prepares for G-CO.C.9	✓	✓
Concept Byte: Parallel Lines and Related Angles	G-CO.D.12, Prepares for G-CO.C.9	✓	✓
3-2 Properties of Parallel Lines	G-CO.C.9	✓	✓
3-3 Proving Lines Parallel	Extends G-CO.C.9	❑	❑
3-4 Parallel and Perpendicular Lines	G-MG.A.3	✓	✓
Concept Byte: Perpendicular Lines and Planes	Extends G-CO.A.1	❑	❑
3-5 Parallel Lines and Triangles	G-CO.C.10	✓	✓
Concept Byte: Exploring Spherical Geometry	Extends G-CO.A.1	❑	❑
3-6 Constructing Parallel and Perpendicular Lines	G-CO.D.12, G-CO.D.13	✓	✓
3-7 Equations of Lines in the Coordinate Plane	Prepares for G-GPE.B.5	❍	❍
3-8 Slopes of Parallel and Perpendicular Lines	G-GPE.B.5	✓	✓
Chapter 4 Congruent Triangles		**Traditional 16**	**Block 8**
4-1 Congruent Figures	Prepares for G-SRT.B.5	❍	❍
Concept Byte: Building Congruent Triangles	Prepares for G-SRT.B.5	❑	❑
4-2 Triangle Congruence by SSS and SAS	G-SRT.B.5	✓	✓
4-3 Triangle Congruence by ASA and AAS	G-SRT.B.5	✓	✓
Concept Byte: Exploring AAA and SSA	Extends G-SRT.B.5	❑	❑
4-4 Using Corresponding Parts of Congruent Triangles	G-CO.D.12, G-SRT.B.5	✓	✓
Concept Byte: Paper-Folding Conjectures	G-CO.D.12	✓	✓
4-5 Isosceles and Equilateral Triangles	G-CO.C.10, G-CO.D.13, G-SRT.B.5	✓	✓
Algebra Review: Systems of Linear Equations	Reviews A-REI.C.6	❍	❍
4-6 Congruence in Right Triangles	G-SRT.B.5	✓	✓
4-7 Congruence in Overlapping Triangles	G-SRT.B.5	✓	✓

	Common Core State Standards	PARCC	Appendix A
Chapter 5 Relationships Within Triangles		**Traditional 14**	**Block 7**
Concept Byte: Investigating Midsegments	Prepares for G-CO.C.10	❍	❍
5-1 Midsegments of Triangles	G-CO.C.10, G-CO.D.12, G-SRT.B.5	✓	✓
5-2 Perpendicular and Angle Bisectors	G-CO.C.9, G-CO.D.12, G-SRT.B.5	✓	✓
Concept Byte: Paper Folding Bisectors	Prepares for G-C.A.3	❍	❍
5-3 Bisectors in Triangles	G-C.A.3	✓	✓
Concept Byte: Special Segments in Triangles	Prepares for G-CO.C.9	❍	❍
5-4 Medians and Altitudes	G-CO.C.10, G-SRT.B.5	✓	✓
5-5 Indirect Proof	Extends G-CO.C.10	❑	❑
Algebra Review: Solving Inequalities	Reviews A-CED.A.1	❍	❍
5-6 Inequalities in One Triangle	Extends G-CO.C.10	❑	❑
5-7 Inequalities in Two Triangles	Extends G-CO.C.10	❑	❑
Chapter 6 Polygons and Quadrilaterals		**Traditional 18**	**Block 9**
Concept Byte: Exterior Angles of Polygons	Prepares for G-SRT.B.5	❑	❑
6-1 The Polygon-Angle Sum Theorems	G-SRT.B.5	✓	✓
6-2 Properties of Parallelograms	G-CO.C.11, G-SRT.B.5	✓	✓
6-3 Proving That a Quadrilateral Is a Parallelogram	G-CO.C.11, G-SRT.B.5	✓	✓
6-4 Properties of Rhombuses, Rectangles, and Squares	G-CO.C.11, G-SRT.B.5	✓	✓
6-5 Conditions for Rhombuses, Rectangles, and Squares	G-CO.C.11, G-SRT.B.5	✓	✓
6-6 Trapezoids and Kites	G-SRT.B.5	✓	✓
6-7 Polygons in the Coordinate Plane	G-GPE.B.7	✓	✓
6-8 Applying Coordinate Geometry	Prepares for G-GPE.B.4	❍	❍
Concept Byte: Quadrilaterals in Quadrilaterals	G-CO.D.12	✓	✓
6-9 Proofs Using Coordinate Geometry	G-GPE.B.4	✓	✓

	Common Core State Standards	PARCC	Appendix A
Chapter 7 Similarity		**Traditional 10**	**Block 5**
7-1 Ratios and Proportions	Prepares for G-SRT.B.5	❍	❍
Algebra Review: Solving Quadratic Equations	Reviews A-CED.A.1	❍	❍
7-2 Similar Polygons	G-SRT.B.5	✓	✓
Concept Byte: Fractals	Extends G-SRT.B.5	❑	❑
7-3 Proving Triangles Similar	G-SRT.B.5, G-GPE.B.5	✓	✓
7-4 Similarity in Right Triangles	G-SRT.B.5, G-GPE.B.5	✓	✓
Concept Byte: The Golden Ratio	Extends G-SRT.B.5	❑	❑
Concept Byte: Exploring Proportions in Triangles	G-CO.D.12	✓	✓
7-5 Proportions in Triangles	G-SRT.B.4	✓	✓
Chapter 8 Right Triangles and Trigonometry		**Traditional 10**	**Block 6**
Concept Byte: The Pythagorean Theorem	Prepares for G-SRT.B.4	❍	❍
8-1 The Pythagorean Theorem and Its Converse	G-SRT.B.4, G-SRT.C.8	✓	✓
8-2 Special Right Triangles	G-SRT.C.8	✓	✓
Concept Byte: Exploring Trigonometric Ratios	G-SRT.C.6	❑	❑
8-3 Trigonometry	G-SRT.C.6, G-SRT.C.7, G-SRT.C.8, G-MG.A.1	✓	✓
Concept Byte: Complementary Angles and Trigonometric Ratios	G-SRT.C.7	✓	✓
8-4 Angles of Elevation and Depression	G-SRT.C.8	✓	✓
8-5 Law of Sines	G-SRT.D.10, G-SRT.D.11		✓
8-6 Law of Cosines	G-SRT.D.10, G-SRT.D.11		✓

	Common Core State Standards	PARCC	Appendix A
Chapter 9 Transformations		**Traditional 14**	**Block 7**
Concept Byte: Tracing Paper Transformations	G-CO.A.2	✓	✓
9-1 Translations	G-CO.A.2, G-CO.A.4, G-CO.A.5, G-CO.B.6	✓	✓
Concept Byte: Paper Folding and Reflections	G-CO.A.5	✓	✓
9-2 Reflections	G-CO.A.2, G-CO.A.4, G-CO.A.5, G-CO.B.6	✓	✓
9-3 Rotations	G-CO.A.2, G-CO.A.4, G-CO.A.5, G-CO.B.6	✓	✓
Concept Byte: Symmetry	G-CO.A.3	✓	✓
9-4 Compositions of Isometries	G-CO.A.2, G-CO.A.5, G-CO.B.6	✓	✓
9-5 Triangle Congruence	G-CO.B.6, G-CO.B.7, G-CO.B.8	✓	✓
Concept Byte: Exploring Dilations	G-SRT.A.1a, G-SRT.A.1b	✓	✓
9-6 Dilations	G-CO.A.2, G-SRT.A.1a, G-SRT.A.1b, G-SRT.A.2	✓	✓
9-7 Similarity Transformations	G-SRT.A.2, G-SRT.A.3	✓	✓
Chapter 10 Area		**Traditional 12**	**Block 6**
Concept Byte: Transforming to Find Area	Prepares for G-GMD.A.3	❑	❑
10-1 Areas of Parallelograms and Triangles	G-GPE.B.7, G-MG.A.1	✓	✓
10-2 Areas of Trapezoids, Rhombuses, and Kites	G-MG.A.1	✓	✓
10-3 Areas of Regular Polygons	G-CO.D.13, G-MG.A.1	✓	✓
10-4 Perimeters and Areas of Similar Figures	Prepares for G-GMD.A.3	❑	❑
10-5 Trigonometry and Area	G-SRT.D.9	✓	✓
10-6 Circles and Arcs	G-CO.A.1, G-C.A.1, G-C.A.2, G-C.B.5	✓	✓
Concept Byte: Radian Measure	G-C.B.5	✓	✓
Concept Byte: Exploring the Area of a Circle	G-GMD.A.1	✓	✓
10-7 Areas of Circles and Sectors	G-C.B.5	✓	✓
Concept Byte: Inscribed and Circumscribed Figures	G-GPE.B.7	❑	❑
10-8 Geometric Probability	Prepares for S-CP.A.1		✓

	Common Core State Standards	PARCC	Appendix A
Chapter 11 Surface Area and Volume		**Traditional 12**	**Block 6**
11-1 Space Figures and Cross Sections	G-GMD.B.4	✓	✓
Concept Byte: Perspective Drawing	Extends G-GMD.B.4	❑	❑
Concept Byte: Literal Equations	Reviews A-CED.A.4	❍	❍
11-2 Surface Areas of Prisms and Cylinders	G-MG.A.1	✓	✓
11-3 Surface Areas of Pyramids and Cones	G-MG.A.1	✓	✓
11-4 Volumes of Prisms and Cylinders	G-GMD.A.1, G-GMD.A.2, G-GMD.A.3, G-MG.A.1	✓	✓
Concept Byte: Finding Volume	Prepares for G-GMD.A.1	❍	❍
11-5 Volumes of Pyramids and Cones	G-GMD.A.3, G-MG.A.1	✓	✓
11-6 Surface Areas and Volumes of Spheres	G-GMD.A.3, G-MG.A.1	✓	✓
Concept Byte: Exploring Similar Solids	Prepares for G-MG.A.2	❑	❑
11-7 Areas and Volumes of Similar Solids	G-MG.A.1, G-MG.A.2	❑	❑
Chapter 12 Circles		**Traditional 10**	**Block 5**
12-1 Tangent Lines	G-C.A.2	✓	✓
Concept Byte: Paper Folding With Circles	Prepares for G-C.A.2	❍	❍
12-2 Chords and Arcs	G-C.A.2	✓	✓
12-3 Inscribed Angles	G-C.A.2, G-C.A.3, G-C.A.4	✓	✓
Concept Byte: Exploring Chords and Secants	Extends G-C.A.2	❑	❑
12-4 Angle Measures and Segment Lengths	Extends G-C.A.2	❑	❑
12-5 Circles in the Coordinate Plane	G-GPE.A.1	✓	✓
Concept Byte: Equation of a Parabola	G-GPE.A.2		✓
12-6 Locus: A Set of Points	G-GMD.B.4	✓	✓
Chapter 13 Probability		**Traditional 10**	**Block 5**
13-1 Experimental and Theoretical Probability	S-CP.A.1, S-CP.A.4		✓
13-2 Probability Distributions and Frequency Tables	S-CP.A.4, S-CP.A.5		✓
13-3 Permutations and Combinations	Prepares for S-CP.B.9		✓
13-4 Compound Probability and Probability of Multiple Events	S-CP.B.7, S-CP.B.8, S-CP.B.9		✓
13-5 Probability Models	S-CP.A.4		✓
13-6 Conditional Probability Formulas	S-CP.A.2, S-CP.A.3, S-CP.A.5, S-CP.B.6		✓
13-7 Modeling Randomness	S-MD.B.6, S-MD.B.7		✓
Concept Byte: Probability and Decision Making	S-MD.B.6, S-MD.B.7		✓

1

Tools of Geometry

Chapters 1 & 2

Numbers
Quantities
Reason quantitatively and use units to solve problems

Geometry
Congruence
Experiment with transformations in the plane
Prove geometric theorems
Make geometric constructions

2 Reasoning and Proof

Visual See It!

Reasoning Try It!

Practice Do It!

3

Parallel and Perpendicular Lines

Chapters 3 & 4

Geometry

Congruence

- Experiment with transformations in the plane
- Prove geometric theorems
- Make geometric constructions

Similarity, Right Triangles, and Trigonometry

- Prove theorems involving similarity

Geometry continued

Expressing Geometric Properties with Equations

- Use coordinates to prove simple geometric theorems algebraically

Modeling with Geometry

- Apply geometric concepts in modeling situations

4 Congruent Triangles

Visual See It!

Reasoning Try It!

Practice Do It!

5

Relationships Within Triangles

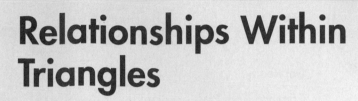

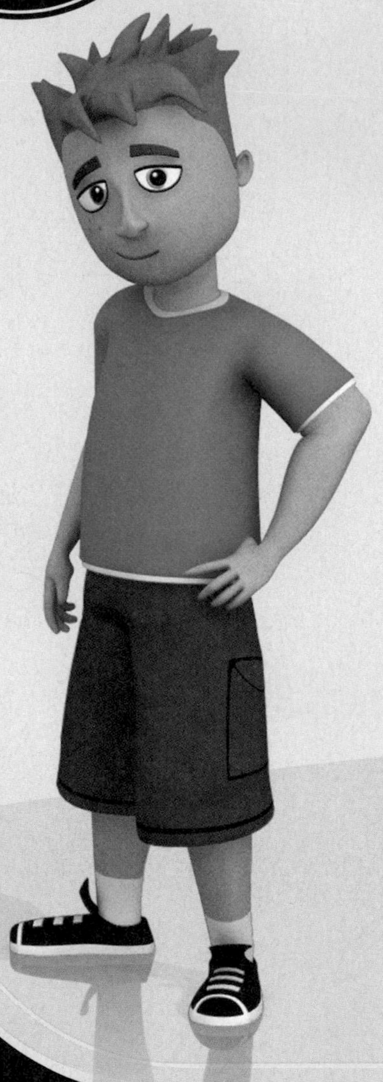

Chapters 5 & 6

Geometry

Congruence

Prove geometric theorems

Make geometric constructions

Similarity, Right Triangles, and Trigonometry

Prove theorems involving similarity

Geometry continued

Circles

Understand and apply theorems about circles

Expressing Geometric Properties with Equations

Use coordinates to prove simple geometric theorems algebraically

6 Polygons and Quadrilaterals

Visual See It!

Reasoning Try It!

Practice Do It!

7

Similarity

Chapters 7 & 8

Geometry

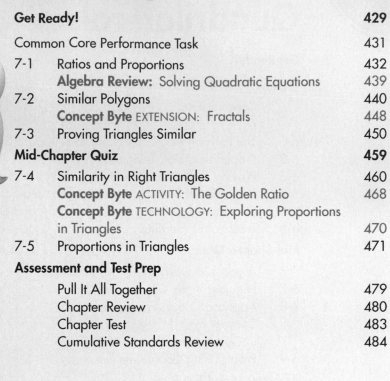

Similarity, Right Triangles, and Trigonometry

Prove theorems involving similarity

Define trigonometric ratios and solve problems involving right triangles

Apply trigonometry to general triangles

Geometry continued

Expressing Geometric Properties with Equations

Use coordinates to prove simple geometric theorems algebraically

Modeling with Geometry

Apply geometric concepts in modeling situations

8 Right Triangles and Trigonometry

9

Transformations

Chapters 9 & 10

Geometry

Congruence

- Experiment with transformations in the plane
- Understand congruence in terms of rigid motions
- Make geometric constructions

Similarity, Right Triangles, and Trigonometry

- Understand similarity in terms of similarity transformations
- Apply trigonometry to general triangles

Geometry continued

Circles

- Understand and apply theorems about circles
- Find arc lengths and areas of sectors of circles

Expressing Geometric Properties with Equations

- Use coordinates to prove simple geometric theorems algebraically

Modeling with Geometry

- Apply geometric concepts in modeling situations

10

Area

Visual See It!

Reasoning Try It!

Practice Do It!

11

Surface Area and Volume

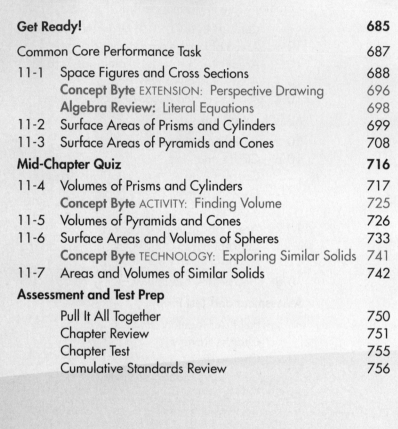

Chapters 11 & 12

Geometry

Circles

Understand and apply theorems about circles

Expressing Geometric Properties with Equations

Translate between the geometric description and the equation for a conic section

Geometry continued

Geometric Measurement and Dimension

Explain volume formulas and use them to solve problems

Visualize relationships between two-dimensional and three-dimensional objects

Modeling with Geometry

Apply geometric concepts in modeling situations

12 Circles

Visual **See It!**

Reasoning **Try It!**

Practice **Do It!**

COMMON CORE COMMON CORE COMMON CORE

13

Probability

Chapter 13

Probability

Conditional Probability and the Rules of Probability

Understand independence and conditional probability and use them to interpret data

Use the rules of probability to compute probabilities of compound events in a uniform probability model

Using Probability to Make Decisions

Use probability to evaluate outcomes of decisions

CHAPTER 7

Get Ready!

Lesson 3-2

Properties of Parallel Lines

Use the diagram at the right. Find the measure of each angle. Justify your answer.

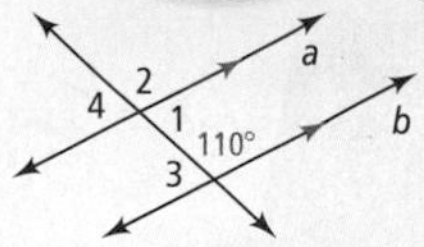

1. ∠1 **2.** ∠2 **3.** ∠3 **4.** ∠4

Lesson 4-1

Naming Congruent Parts

△PAC ≅ △DHL. Complete each congruence statement.

5. $\overline{PC}$ ≅ ? **6.** ∠H ≅ ? **7.** ∠PCA ≅ ? **8.** △HDL ≅ ?

Lessons 4-2 and 4-3

Triangle Congruence

Write a congruence statement for each pair of triangles. Explain why the triangles are congruent.

9.

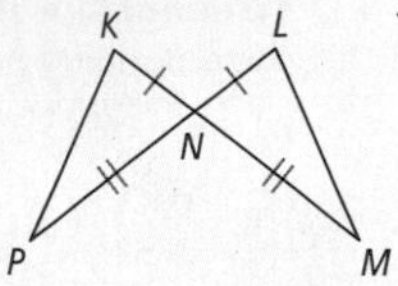

10.

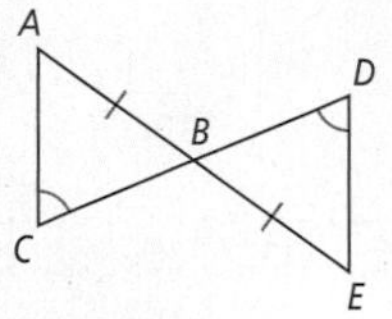

11.

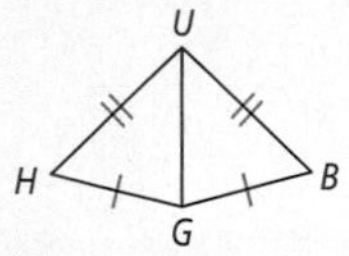

Lesson 5-1

Midsegments of Triangles

Use the diagram at the right for Exercises 12–13.

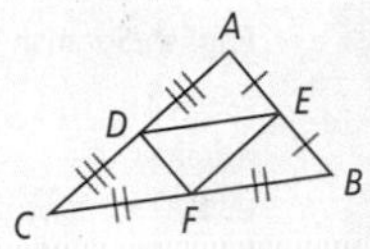

12. If BC = 12, then BF = ? and DE = ?.

13. If EF = 4.7, then AD = ? and AC = ?.

Looking Ahead Vocabulary

14. An artist sketches a person. She is careful to draw the different parts of the person's body in *proportion*. What does *proportion* mean in this situation?

15. Siblings often look *similar* to each other. How might two geometric figures be *similar?*

16. A road map has a *scale* on it that tells you how many miles are equivalent to a distance of 1 inch on the map. How would you use the scale to estimate the distance between two cities on the map?

Get Ready!

Assign this diagnostic assessment to determine if students have the prerequisite skills for Chapter 7.

Lesson	Skill
3-2	Properties of Parallel Lines
4-1	Naming Congruent Parts
4-2 and 4-3	Triangle Congruence
5-1	Midsegments of Triangles

To remediate students, select from these resources (available for every lesson).

- Online Problems (PowerGeometry.com)
- Reteaching (All-in-One Teaching Resources)
- Practice (All-in-One Teaching Resources)

Why Students Need These Skills

PROPERTIES OF PARALLEL LINES Angle relationships will be extended to similar triangles.

NAMING CONGRUENT PARTS The order in which vertices are listed is important when identifying similar triangles.

TRIANGLE CONGRUENCE Congruence statements are formed by postulates. Likewise, similarity statements will be formed from the AA ~ postulate.

MIDSEGMENTS OF TRIANGLES Properties of proportionality within triangles will be explored.

Looking Ahead Vocabulary

PROPORTION Ask students to identify other real-world objects that are in proportion to one another.

SIMILAR Using the example of a golf ball and basketball, have students identify characteristics that define similarity.

SCALE Show examples of maps and blueprint drawings that use scales.

Answers

Get Ready!

1. 70; if lines are ∥, same-side int. ∠s are suppl.

2. 110; if lines are ∥, corresponding ∠s are ≅.

3. 70; adjacent angles forming a straight ∠ are suppl.

4. 70; it is a vert. ∠ with ∠1; vert. ∠s are ≅.

5. $\overline{DL}$

6. ∠A

7. ∠DLH

8. △APC

9. △KNP ≅ △LNM by SAS.

10. △BAC ≅ △BED by AAS.

11. △UGH ≅ △UGB by SSS.

12. 6, 6

13. 4.7, 9.4

14. Answers may vary. Sample: The relative sizes of the body parts in the drawing are the same as those of a real person.

15. Answers may vary. Sample: They might be similar if they have the same shape.

16. Answers may vary. Sample: Measure the number of inches on the map between the two cities, and multiply that number of inches by the number of miles represented by 1 in.

Chapter 7 Overview

Chapter 7 expands on students' understandings and skills related to similarity. In this chapter, students will develop the answers to the Essential Questions as they learn the concepts and skills bulleted below.

BIG idea Similarity

ESSENTIAL QUESTION How do you use proportions to find side lengths in similar polygons?

- Students will form proportions based on known lengths of corresponding sides.

BIG idea Reasoning and Proof

ESSENTIAL QUESTION How do you show two triangles are similar?

- Students will use the Angle-Angle Similarity Postulate.
- Students will use the Side-Angle-Side Similarity Theorem.
- Students will use the Side-Side-Side Similarity Theorem.

BIG idea Visualization

ESSENTIAL QUESTION How do you identify corresponding parts of similar triangles?

- A key to understanding corresponding parts of similar triangles is to show the triangles in like orientations.

Content Standards

Following are the standards covered in this chapter. Modeling standards are indicated by a star symbol (★).

CONCEPTUAL CATEGORY Geometry

Domain Similarity, Right Triangles, and Trigonometry G-SRT

Cluster Prove theorems involving similarity (Standards G-SRT.B.4, G-SRT.B.5)
LESSONS 7-2, 7-3, 7-4, 7-5

Domain Expressing Geometric Properties with Equations G-GPE

Cluster Use coordinates to prove simple geometric theorems algebraically (Standard G-GPE.B.4)
LESSON 7-4

Domain Congruence G-CO

Cluster Understand congruence in terms of rigid motions (Standard G-CO.D.12)
LESSON 7-5 CB

CHAPTER 7

Similarity

Download videos connecting math to your world.

Interactive! Vary numbers, graphs, and figures to explore math concepts.

The online Solve It will get you in gear for each lesson.

Math definitions in English and Spanish

Online access to stepped-out problems aligned to Common Core

Get and view your assignments online.

Extra practice and review online

Virtual Nerd™ tutorials with built-in support

Chapter Preview

BIG ideas

1 **Similarity**
Essential Question How do you use proportions to find side lengths in sim[ilar] polygons?

2 **Reasoning and Proof**
Essential Question How do you sho[w] two triangles are similar?

3 **Visualization**
Essential Question How do you ider[ntify] corresponding parts of similar triangle[s]

Vocabulary

English/Spanish Vocabulary Audio Online:

English	Spanish
extremes of a proportion, p. 434	valores extremos de una proporción
geometric mean, p. 462	media geométrica
indirect measurement, p. 454	medición indirecta
means of a proportion, p. 434	valores medios de una proporción
proportion, p. 434	proporción
ratio, p. 432	razón
scale drawing, p. 443	dibujo a escala
scale factor, p. 440	factor de escala
similar, p. 440	semejante
similar polygons, p. 440	polígonos semejantes

DOMAINS

- Similarity, Right Triangles, and Trigonom[etry]
- Expressing Geometric Properties with Equations
- Mathematical Practice: Construct viable arguments

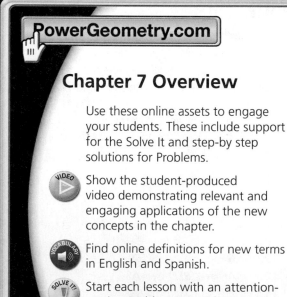

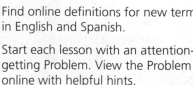

PowerGeometry.com

Chapter 7 Overview

Use these online assets to engage your students. These include support for the Solve It and step-by step solutions for Problems.

Show the student-produced video demonstrating relevant and engaging applications of the new concepts in the chapter.

Find online definitions for new terms in English and Spanish.

Start each lesson with an attention-getting Problem. View the Problem online with helpful hints.

Common Core Performance Task

Adjusting a Graphing Calculator Window

Lillian graphs the functions $y = 2x + 3$ and $y = -\frac{1}{2}x + 1$ on her graphing calculator. She knows the lines are perpendicular because the product of the slopes of the lines is -1. However, the lines do not appear to be perpendicular on the screen.

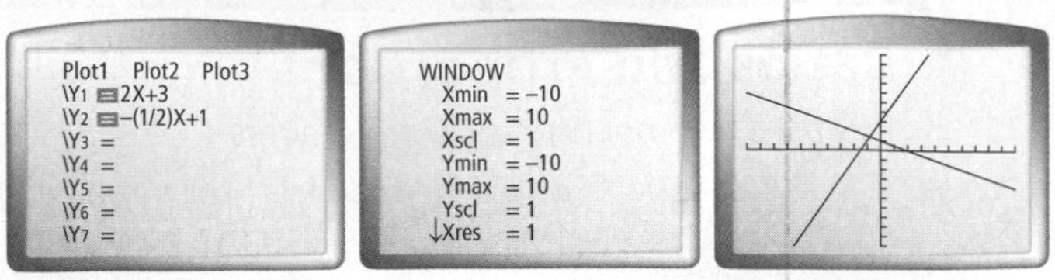

The WINDOW screen shows the intervals of the x-axis (from Xmin to Xmax) and the y-axis (from Ymin to Ymax) that are visible in the graph. Lillian wants to adjust the values of Xmin and Xmax that she entered so that the graph of the lines is not distorted. She measures her calculator's rectangular screen and finds the width is 5.1 cm and the height is 3.4 cm.

Task Description

Determine the values Lillian should enter for Xmin and Xmax so that the graph is not distorted.

Connecting the Task to the Math Practices

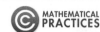

MATHEMATICAL PRACTICES

As you complete the task, you'll apply these Standards for Mathematical Practice.

- You'll use ratios to help you understand the problem. (MP 1)
- You'll determine whether two rectangles are similar and explain your reasoning. (MP 3)

Increase students' depth of knowledge with interactive online activities.

Show Problems from each lesson solved step by step. Instant replay allows students to go at their own pace when studying online.

Assign homework to individual students or to an entire class.

Prepare students for the Mid-Chapter Quiz and Chapter Test with online practice and review.

Virtual Nerd™ Access Virtual Nerd student-centered math tutorials that directly relate to the content of the lesson.

Overview of the Performance Task

Students will use ratios of side lengths in similar figures to adjust the window on a graphing calculator so that a graph is not distorted.

Students will work on the Performance Task in the following places in the chapter.

- Lesson 7-1 (p. 438)
- Lesson 7-2 (p. 447)
- Pull It All Together (p. 479)

Introducing the Performance Task

Tell students to read the problem on this page. Do not have them start work on the problem at this time, but ask them the following questions.

Q How can you use a graphing calculator to explore this problem situation? **[Sample: I can graph the lines on a graphing calculator using various values for Xmin, Xmax, Ymin, and Ymax, and observe whether the lines appear to be perpendicular.]**

Q What do you notice about the number of units that are displayed along the x-axis and along the y-axis in the graph? **[Sample: 20 units (from −10 to 10) are displayed along each axis.]**

PARCC CLAIMS

Sub-Claim A: Major Content with Connections to Practices
Sub-Claim C: Highlighted Practices MP 3, 6 with Connections to Content

SBAC CLAIMS

Claim 2: Problem Solving
Claim 3: Communicating Reasoning

CHAPTER 7

SIMILARITY

Math Background

PROFESSIONAL DEVELOPMENT

The Understanding by Design® methodology was central to the development of the Big Ideas and the Essential Understandings. These will help your students build a structure on which to make connections to prior learning.

Similarity

BIG idea Two geometric figures are similar when corresponding lengths are proportional and corresponding angles are congruent.

ESSENTIAL UNDERSTANDINGS

7-1 An equation can be written stating that two ratios are equal, and if the equation contains a variable, it can be solved to find the value of the variable.

7-2 Ratios and proportions can be used to decide whether two polygons are similar and to find unknown side lengths of similar figures.

7-4 Drawing in the altitude to the hypotenuse of a right triangle forms three pairs of similar right triangles.

7-5 When two or more parallel lines intersect other lines, proportional segments are formed.

Reasoning and Proof

BIG idea Definitions establish meanings and remove possible misunderstanding. Other truths are more complex and difficult to see. It is often possible to verify complex truths by reasoning from simpler ones by using deductive reasoning.

ESSENTIAL UNDERSTANDINGS

7-2 to 7-3 Ratios and proportions can be used to prove whether two polygons are similar and to find unknown side lengths. Triangles can be shown to be similar based on the relationship of two or three pairs of corresponding parts.

7-4 It can be proven that the three pairs of right triangles formed by drawing in the altitude to the hypotenuse are similar.

Visualization

BIG idea Visualization can help you see the relationships between two figures and help you connect the properties of real objects with two-dimensional drawings of these objects.

ESSENTIAL UNDERSTANDINGS

7-3 to 7-4 Two triangles can be shown to be similar. Drawing in the altitude to the hypotenuse of a right triangle forms three pairs of similar right triangles.

7-5 When two or more parallel lines intersect other lines, proportional segments are formed.

Ratios and Proportions

A *ratio* is a comparison of two numbers.

Cereal is on sale: 3 boxes for 5 dollars. This ratio can be written $\frac{5 \text{ dollars}}{3 \text{ boxes}}$.

You can make a table to show the cost of different numbers of boxes of cereal. Simple tables like this can help students see equivalent ratios.

Dollars	5	10	15	20
Boxes of Cereal	3	6	9	12

The two numbers in each column can be expressed as ratios:

$\frac{5}{3}, \frac{10}{6}, \frac{15}{9}, \frac{20}{12}$

A proportion is a statement equating two ratios and can be used to solve problems. To find the cost of 5 boxes of cereal, use the proportion

$$\begin{matrix} \text{extreme} & \frac{3}{5} = \frac{5}{x} & \text{mean} \\ \text{mean} & & \text{extreme} \end{matrix}$$

The cross-products of proportions are equal. "The product of the means equals the product of the extremes." One way to find x is

$3x = 25$

$x = 8.33$

The cost of 5 boxes equals $8.33.

Common Errors With Ratios and Proportions

If two ratios are equal, $\frac{a}{b} = \frac{c}{d}$, their inverses are also equal, $\frac{b}{a} = \frac{d}{c}$. However, $\frac{a}{b} \neq \frac{d}{c}$. Cereal boxes must be the numerator in both ratios or the denominator in both ratios.

Mathematical Practices

Reason abstractly and quantitatively Students decontextualize to represent the problem in terms of ratios, solve it, and then contextualize to make sense of their solution.

Look for and make use of structure Students examine patterns in equivalent ratios, seeing the structure as shown in the table at the top of this column, as numerators increase by multiples of 5 and denominators increase by multiples of 3.

Proving Similar Triangles

Two triangles are similar if and only if corresponding angles are congruent and corresponding sides are proportional.

Recall from Chapter 4 that SSS, SAS, ASA, AAS, and HL postulates could each be used to prove that triangles are congruent.

In the case of similarity, you can prove two triangles are similar if two angles are congruent, all three sides are proportional, or if one angle is congruent and the two adjacent sides are proportional

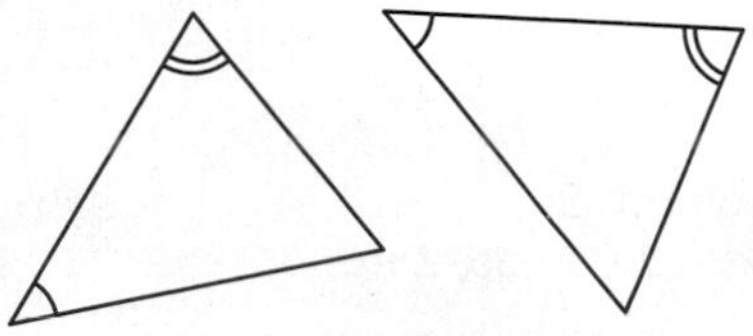

Common Errors With Proving Similar Triangles

If students are given similar triangles that are oriented in such a way that their corresponding sides are oriented differently, students may incorrectly conclude that the triangles are not similar based on incorrectly setting up the proportion.

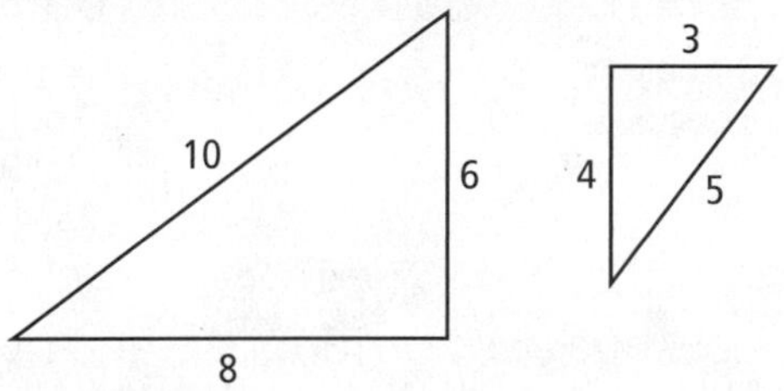

The two triangles above are similar. However, the student may set up the proportion, $\frac{4}{6} = \frac{3}{8}$ and wrongly conclude that the two triangles are not similar. Students should always compare the longest side to the longest, the shortest side to the shortest, and the remaining sides.

Mathematical Practices

Construct viable arguments and critique the reasoning of others In working with congruent triangles, students build on deductive arguments about congruent triangles to establish new results.

Look for and make use of structure Students learn to see patterns in noncongruent triangles, matching corresponding sides or angles to make comparisons.

Proportions in Triangles

Special segments in triangles divide the sides proportionally.

Angle Bisectors

If $\overrightarrow{AD}$ bisects $\angle CAB$ then $\frac{CD}{DB} = \frac{CA}{BA}$.

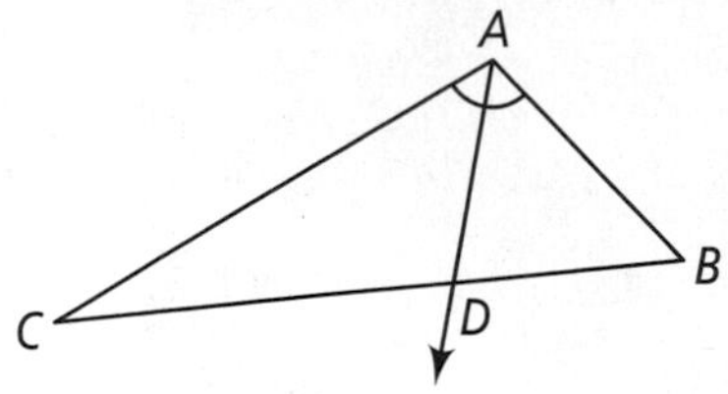

Altitudes in Right Triangles

If $\overline{CD}$ is the altitude drawn to the hypotenuse in a right triangle, then

1. $\frac{AD}{CD} = \frac{CD}{DB}$
2. $\frac{AB}{AC} = \frac{AC}{AD}$
3. $\frac{AB}{CB} = \frac{CB}{DB}$
4. $\triangle ABC \sim \triangle CBD \sim \triangle ACD$

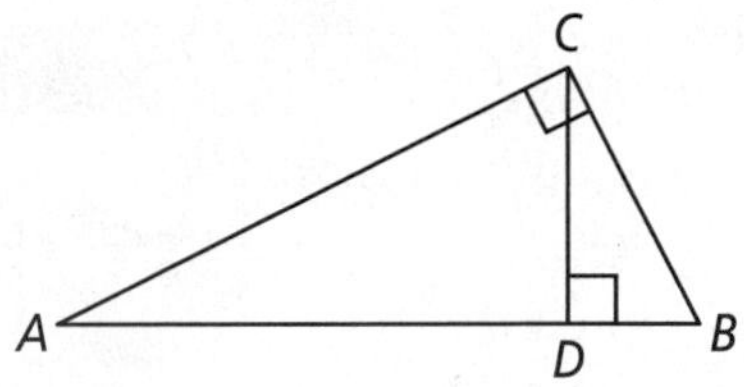

Side-Splitter Theorem

If $\overline{DE} \parallel \overline{AC}$, then $\frac{AD}{DB} = \frac{CE}{EB}$.

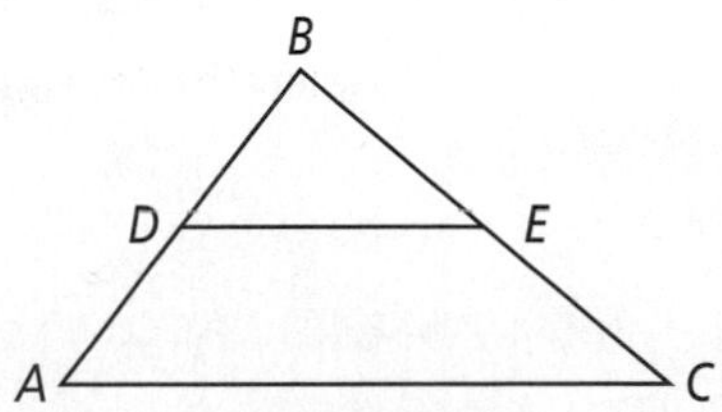

Common Errors With Bisectors

In the first triangle above, students might incorrectly assume that the angle bisector of $\angle CAB$ in $\triangle CAB$ divides $\overline{CB}$ into two equal parts. In fact, the parts are proportional to the sides $\overline{CA}$ and $\overline{AB}$.

Mathematical Practices

Construct viable arguments and critique the reasoning of others Students use auxiliary lines to establish the Triangle Angle Bisector Theorem, building on a previously established result (the Side-Splitter Theorem) to make sense of an otherwise nonintuitive conclusion.

CHAPTER 7

SIMILARITY

Pacing and Assignment Guide

Lesson	Teaching Day(s)	Basic (TRADITIONAL)	Average (TRADITIONAL)	Advanced (TRADITIONAL)	Block (BLOCK)
7-1	1	Problems 1–3 Exs. 9–16	Problems 1–3 Exs. 9–15 odd, 33–34	Problems 1–3 Exs. 9–15 odd, 33–34	**Day 1** Problems 1–5 Exs. 9–31 odd, 33–54
	2	Problems 4–5 Exs. 17–32, 37–38, 40–41, 46, 48–51	Problems 4–5 Exs. 17–31 odd, 35–54	Problems 4–5 Exs. 17–31 odd, 35–60	
7-2	1	Problems 1–2 Exs. 9–17	Problems 1–2 Exs. 9–17 odd	Problems 1–2 Exs. 9–17 odd	**Day 2** Problems 1–5 Exs. 9–23 odd, 25–47
	2	Problems 3–5 Exs. 18–27, 32, 34, 37, 39, 47	Problems 3–5 Exs. 19–23 odd, 25–47	Problems 3–5 Exs. 19–23 odd, 25–50	
7-3	1	Problems 1–2 Exs. 7–12, 37–52	Problems 1–2 Exs. 7–11 odd, 37–52	Problems 1–4 Exs. 7–17 odd, 18–52	**Day 3** Problems 1–4 Exs. 7–17 odd, 18–34, 37–52
	2	Problems 3–4 Exs. 13–19, 22–25, 32	Problems 3–4 Exs. 13–17 odd, 18–34		
7-4	1	Problems 1–4 Exs. 9–22, 24–25, 30, 37–39, 48–56	Problems 1–4 Exs. 9–21 odd, 23–44, 48–56	Problems 1–4 Exs. 9–21 odd, 23–56	**Day 4** Problems 1–4 Exs. 9–21 odd, 23–44, 48–56
7-5	1	Problems 1–3 Exs. 9–26, 33, 36, 38–39, 41, 43, 51–64	Problems 1–3 Exs. 9–23 odd, 25–47, 51–64	Problems 1–3 Exs. 9–23 odd, 25–64	Problems 1–3 Exs. 9–23 odd, 25–47, 51–64
Review	1	Chapter 7 Review	Chapter 7 Review	Chapter 7 Review	**Day 5** Chapter 7 Review
Assess	1	Chapter 7 Test	Chapter 7 Test	Chapter 7 Test	Chapter 7 Test
Total		**10 Days**	**10 Days**	**9 Days**	**5 Days**

Note: Pacing does not include Concept Bytes and other feature pages.

Resources

KEY

- I = Interactive asset at PowerGeometry.com
- E = Editable master at PowerGeometry.com
- P = Available in Print
- M = Master at PowerGeometry.com
- ✓= CD-ROM

	For the Chapter	7-1	7-2	7-3	7-4	7-5
Planning						
Teacher Center Online Planner & Grade Book	I	I	I	I	I	I
Interactive Learning & Guided Instruction						
My Math Video	I					
Solve It!		I M	I M	I M	I M	I M
Student Companion		P M	P M	P M	P M	P M
Vocabulary Support		I P M	I P M	I P M	I P M	I P M
Got It? Support		I P	I P	I P	I P	I P
Online Problems		I	I	I	I	I
Additional Problems		M	M	M	M	M
English Language Learner Support (TR)		E P M	E P M	E P M	E P M	E P M
Activities, Games, and Puzzles		E M	E M	E M	E M	E M
Teaching With TI Technology With CD-ROM						✓ P
TI-Nspire™ Support CD-ROM		✓	✓	✓	✓	✓
Lesson Check & Practice						
Student Companion		P M	P M	P M	P M	P M
Lesson Check Support		I P	I P	I P	I P	I P
Practice and Problem Solving Workbook		P	P	P	P	P
Think About a Plan (TR)		E P M	E P M	E P M	E P M	E P M
Practice Form G (TR)		E P M	E P M	E P M	E P M	E P M
Standardized Test Prep (TR)		P M	P M	P M	P M	P M
Practice *Form K* (TR)		E P M	E P M	E P M	E P M	E P M
Extra Practice	E M					
Find the Errors!	M					
Enrichment (TR)		E P M	E P M	E P M	E P M	E P M
Answers and Solutions CD-ROM	✓	✓	✓	✓	✓	✓
Assess & Remediate						
ExamView CD-ROM	✓	✓	✓	✓	✓	✓
Lesson Quiz		I M	I M	I M	I M	I M
Quizzes and Tests *Form G* (TR)	E P M			E P		E P
Quizzes and Tests *Form K* (TR)	E P M			E P		E P
Reteaching (TR)		E P M	E P M	E P M	E P M	E P M
Performance Tasks (TR)	P M					
Cumulative Review (TR)	P M					
Progress Monitoring Assessments	I P M					

(TR) Available in All-In-One Teaching Resources

1 Interactive Learning

Solve It!

PURPOSE To use ratios to represent quantities and find equivalent ratios

PROCESS Students may find the ratios of wins to total games played in simplest form for each team or convert ratios to percents to compare each team's record.

FACILITATE

Q How can each year's record be expressed as a ratio? **[Sample: It can be expressed as a ratio of wins to losses such as 60 : 24.]**

Q Is there an equivalent ratio for 60 : 24 for the year 1890? Explain. **[Yes; 60 : 24 can be divided by 12 on both sides to equal 5 : 2.]**

ANSWER See Solve It in Answers on next page.

CONNECT THE MATH Students should realize that even though the four teams did not play the same number of games, their win-to-loss ratios can be equivalent. To determine equivalent ratios, the ratios should be written in simplest form.

2 Guided Instruction

Problem 1

Q What relationship does the conversion factor used in the problem describe? **[It shows the relationship 12 in. = 1 ft]**

Got It? **ERROR PREVENTION**

Be sure that students remember to express the width and the height in the same unit.

7-1 Ratios and Proportions

Common Core State Standards

Prepares for G-SRT.B.5 Use . . . similarity criteria for triangles to solve problems and to prove relationships in geometric figures.

MP 1, MP 3, MP 4, MP 6, MP 7

Objective To write ratios and solve proportions

SOLVE IT! Getting Ready!

The table at the right gives the wins and losses of a baseball team. In which year(s) did the team have the best record? Explain.

Year	Wins	Losses
1890	60	24
1930	110	44
1970	110	52
2010	108	54

The year the team had the most wins is not necessarily the year in which it had the best record.

MATHEMATICAL PRACTICES In the Solve It, you compared two quantities for four years.

Essential Understanding You can write a *ratio* to compare two quantities.

Lesson Vocabulary
- ratio
- extended ratio
- proportion
- extremes
- means
- Cross Products Property

A **ratio** is a comparison of two quantities by division. You can write the ratio of two numbers a and b, where $b \neq 0$, in three ways: $\frac{a}{b}$, $a : b$, and a to b. You usually express a and b in the same unit and write the ratio in simplest form.

Problem 1 Writing a Ratio

Bonsai Trees **The bonsai bald cypress tree is a small version of a full-size tree. A Florida bald cypress tree called the Senator stands 118 ft tall. What is the ratio of the height of the bonsai to the height of the Senator?**

Think

How can you write the heights using the same unit?

You can convert the height of the Senator to inches or the height of the bonsai tree to feet.

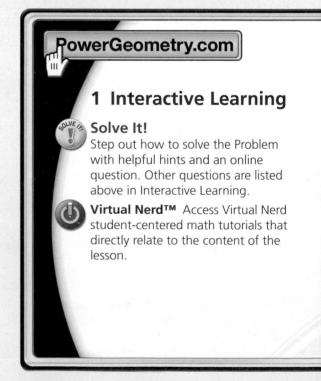

Express both heights in the same unit. To convert 118 ft to inches, multiply by the conversion factor $\frac{12 \text{ in.}}{1 \text{ ft}}$.

$118 \text{ ft} = \frac{118 \text{ ft}}{1} \cdot \frac{12 \text{ in.}}{1 \text{ ft}} = (118 \cdot 12) \text{ in.} = 1416 \text{ in.}$

Write the ratio as a fraction in simplest form.

$\frac{\text{height of bonsai} \rightarrow}{\text{height of Senator} \rightarrow} \frac{15 \text{ in.}}{118 \text{ ft}} = \frac{15 \text{ in.}}{1416 \text{ in.}} = \frac{(3 \cdot 5) \text{ in.}}{(3 \cdot 472) \text{ in.}} = \frac{5}{472}$

The ratio of the height of the bonsai to the height of the Senator is $\frac{5}{472}$ or 5 : 472.

Got It? 1. A bonsai tree is 18 in. wide and stands 2 ft tall. What is the ratio of the width of the bonsai to its height?

7-1 Preparing to Teach

BIG idea Proportionality

ESSENTIAL UNDERSTANDINGS
- A ratio can be written to compare two quantities.
- An equation can be written stating that two ratios are equal.
- If the equation contains a variable, it can be solved to find the value of the variable.

Math Background

The topic of ratios is an important foundation that leads to solving problems that involve scale drawings and similar figures. Students need to become comfortable with ratios written as fractions, as well as decimals and percents. They will need to identify equivalent ratios in order to determine when figures are similar. Students should not only be able to write proportions to show the relationship of similar figures and scale drawings, but should be able to solve proportions for a missing term.

Mathematical Practices

Attend to precision. Students will use colons and division to represent ratios and proportions. They will also set up algebraic expressions using ratios and proportions.

PowerGeometry.com

1 Interactive Learning

Solve It! Step out how to solve the Problem with helpful hints and an online question. Other questions are listed above in Interactive Learning.

Virtual Nerd™ Access Virtual Nerd student-centered math tutorials that directly relate to the content of the lesson.

Problem 2 Dividing a Quantity Into a Given Ratio

Fundraising **Members of the school band are buying pots of tulips and pots of daffodils to sell at their fundraiser. They plan to buy 120 pots of flowers. The ratio $\frac{\text{number of tulip pots}}{\text{number of daffodil pots}}$ will be $\frac{2}{3}$. How many pots of each type of flower should they buy?**

Plan

How do you write expressions for the numbers of pots? Multiply the numerator 2 and the denominator 3 by the factor x. $\frac{2x}{3x} = \frac{2}{3}$.

Think	Write
If the ratio $\frac{\text{number of tulip pots}}{\text{number of daffodil pots}}$ is $\frac{2}{3}$, it must be in the form $\frac{2x}{3x}$.	Let $2x$ = the number of tulip pots. Let $3x$ = the number of daffodil pots.
The total number of flower pots is 120. Use this fact to write an equation. Then solve for x.	$2x + 3x = 120$ $5x = 120$ $x = 24$
Substitute 24 for x in the expressions for the numbers of pots.	$2x = 2(24) = 48$ $3x = 3(24) = 72$
Write the answer in words.	The band members should buy 48 tulip pots and 72 daffodil pots.

Got It? 2. The measures of two supplementary angles are in the ratio 1 : 4. What are the measures of the angles?

An **extended ratio** compares three (or more) numbers. In the extended ratio $a : b : c$, the ratio of the first two numbers is $a : b$, the ratio of the last two numbers is $b : c$, and the ratio of the first and last numbers is $a : c$.

Problem 3 Using an Extended Ratio

The lengths of the sides of a triangle are in the extended ratio 3 : 5 : 6. The perimeter of the triangle is 98 in. What is the length of the longest side?

Think

How do you use the solution of the equation to answer the question? Substitute the value for x in the expression for the length of the longest side.

Sketch the triangle. Use the extended ratio to label the sides with expressions for their lengths.

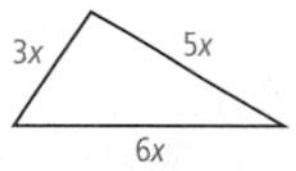

$3x + 5x + 6x = 98$	The perimeter is 98 in.
$14x = 98$	Simplify.
$x = 7$	Divide each side by 14.

The expression that represents the length of the longest side is $6x$. $6(7) = 42$, so the length of the longest side is 42 in.

Got It? 3. The lengths of the sides of a triangle are in the extended ratio 4 : 7 : 9. The perimeter is 60 cm. What are the lengths of the sides?

Problem 2

Students use ratios to write equations that show the relationship between two quantities.

Q In the ratio $\frac{2x}{3x}$, what does x represent? **[the common factor of the numerator and denominator]**

Q How many ratios can be written that equal $\frac{2}{3}$? Of these, how many meet the criteria of the number of pots bought? **[Infinitely many; only one ratio has a numerator and denominator that have a sum of 120.]**

Got It? **VISUAL LEARNERS**

Have students attempt to draw a pair of supplementary angles that fit this description before calculating the answer algebraically.

Problem 3

Q What ratios are described by the extended ratio in Problem 3? **[$\frac{3x}{5x}$, $\frac{5x}{6x}$, and $\frac{3x}{6x}$]**

Q What steps are used to check the answer to this problem? **[Find the lengths of all three sides. Check that the sum is 98 and verify that the side lengths are in the proportion given by the extended table.]**

Got It? **VISUAL LEARNERS**

Have students sketch the triangle and label the side lengths using the ratio.

Q What equation can you write using the ratios and the perimeter? **[$4x + 7x + 9x = 60$]**

2 Guided Instruction

Each Problem is worked out and supported online.

Problem 1
Writing a Ratio
Animated

Problem 2
Dividing a Quantity Into a Given Ratio

Problem 3
Using an Extended Ratio

Alternative Problem 3
Using an Extended Ratio
Animated

Problem 4
Solving a Proportion
Animated

Problem 5
Writing Equivalent Proportions

Alternative Problem 5
Writing Equivalent Proportions

Answers

Solve It!

1890 and 1930; explanations may vary. Sample: In both years the team won $\frac{5}{7}$, or about 71.4% of the games they played.

Got It?

1. 3 : 4
2. 36, 144
3. 12 cm, 21 cm, 27 cm

Take Note

Have students write their own examples of proportions and verify that their examples are proportions using the Cross Products Property.

Here's Why It Works

Q What relationship does *bd* have with the proportion? **[*bd* is a common multiple for the two denominators.]**

Q Why is that relationship important? **[Multiplying both ratios by a common multiple eliminates the denominators.]**

Problem 4

Q Why is the Cross Products Property used to solve these proportions? **[The products of the means and extremes are used to write a linear equation.]**

Q In 4B, what property is used to find the product of the extremes? **[The extremes are 3 and $y + 4$. The Distributive Property is used to find the product.]**

Got It?

ERROR PREVENTION

For 4b, if students give an answer of $\frac{1}{12}$, remind them to use the Distributive Property with 3 and $m + 1$. Have students check their answers by substituting their solutions back into the original proportion.

Essential Understanding If two ratios are equivalent, you can write an equation stating that the ratios are equal. If the equation contains a variable, you can solve the equation to find the value of the variable.

An equation that states that two ratios are equal is called a **proportion.** The first and last numbers in a proportion are the **extremes.** The middle two numbers are the **means.**

extremes: $2 : 3 = 4 : 6$:means

extremes → $\frac{2}{3} = \frac{4}{6}$ ← means

take note

Key Concept Cross Products Property

Words	Symbols	Example
In a proportion, the product of the extremes equals the product of the means.	If $\frac{a}{b} = \frac{c}{d}$, where $b \neq 0$ and $d \neq 0$, then $ad = bc$.	$\frac{2}{3} = \frac{4}{6}$ $2 \cdot 6 = 3 \cdot 4$ $12 = 12$

Here's Why It Works Begin with $\frac{a}{b} = \frac{c}{d}$, where $b \neq 0$ and $d \neq 0$.

$bd \cdot \frac{a}{b} = \frac{c}{d} \cdot bd$ Multiply each side of the proportion by *bd*.

$\frac{\cancel{b}d}{1} \cdot \frac{a}{\cancel{b}} = \frac{c}{\cancel{d}} \cdot \frac{b\cancel{d}}{1}$ Divide the common factors.

$ad = bc$ Simplify.

Problem 4 Solving a Proportion

Algebra What is the solution of each proportion?

A $\frac{6}{x} = \frac{5}{4}$

$6(4) = 5x$ Cross Products Property

$24 = 5x$ Simplify.

$x = \frac{24}{5}$ Solve for the variable.

The solution is $\frac{24}{5}$ or 4.8.

Think

Does the solution check?

$\frac{6}{\frac{24}{5}} \stackrel{?}{=} \frac{5}{4}$

$6 \cdot 4 \stackrel{?}{=} \frac{24}{5} \cdot 5$

$24 = 24$ ✓

B $\frac{y+4}{9} = \frac{y}{3}$

$3(y + 4) = 9y$

$3y + 12 = 9y$

$12 = 6y$

$y = 2$

The solution is 2.

Got It? 4. What is the solution of each proportion?

a. $\frac{9}{2} = \frac{a}{14}$ b. $\frac{15}{m+1} = \frac{3}{m}$

Additional Problems

1. A pigmy rattlesnake has an average length of 18 in., while a Western diamondback rattlesnake averages a length of 5 ft 6 in. What is the ratio of the length of a pigmy rattlesnake to the length of a Western diamondback rattlesnake?

ANSWER 3 : 11

2. The measures of two complementary angles are in the ratio 1 : 3. What are the measures of the angles?

ANSWER 22.5°, 67.5°

3. The lengths of the sides of a triangle are in the extended ratio 2 : 4 : 5. The perimeter of the triangle is 77 in. What is the length of the longest side?

ANSWER 35 in.

4. What is the solution of each proportion?

a. $\frac{3}{8} = \frac{5}{x}$

b. $\frac{y+3}{12} = \frac{5}{6}$

ANSWER a. $\frac{40}{3}$ **b.** 7

5. In the diagram, $\frac{x}{10} = \frac{y}{9}$. What ratio completes the equivalent proportion $\frac{x}{y} = \frac{■}{■}$? Use one of the Properties of Proportions to justify your answer.

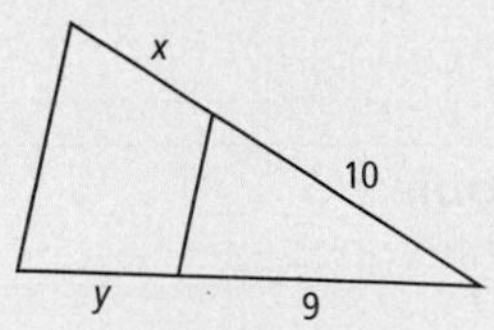

ANSWER $\frac{x}{y} = \frac{10}{9}$

Using the Properties of Equality, you can rewrite proportions in equivalent forms.

take note

Key Concept Properties of Proportions

a, b, c, and d do not equal zero.

Property	How to Apply It
(1) $\frac{a}{b}=\frac{c}{d}$ is equivalent to $\frac{b}{a}=\frac{d}{c}$.	Write the reciprocal of each ratio. $\frac{2}{3}=\frac{4}{6}$ becomes $\frac{3}{2}=\frac{6}{4}$.
(2) $\frac{a}{b}=\frac{c}{d}$ is equivalent to $\frac{a}{c}=\frac{b}{d}$.	Switch the means. $\frac{2}{3}=\frac{4}{6}$ becomes $\frac{2}{4}=\frac{3}{6}$.
(3) $\frac{a}{b}=\frac{c}{d}$ is equivalent to $\frac{a+b}{b}=\frac{c+d}{d}$.	In each ratio, add the denominator to the numerator. $\frac{2}{3}=\frac{4}{6}$ becomes $\frac{2+3}{3}=\frac{4+6}{6}$.

Plan

How do you decide which property of proportions applies?
Look at how the positions of the known parts of the incomplete proportion relate to their positions in the original proportion.

Problem 5 Writing Equivalent Proportions

In the diagram, $\frac{x}{6}=\frac{y}{7}$. What ratio completes the equivalent proportion $\frac{x}{y}=\frac{\blacksquare}{\blacksquare}$? Justify your answer.

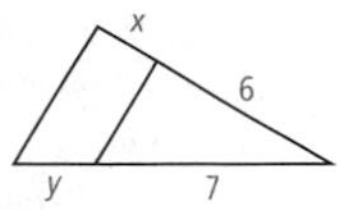

Method 1

$\frac{x}{6}=\frac{y}{7}$

$\frac{x}{y}=\frac{6}{7}$ Property of Proportions (2)

Method 2

$\frac{x}{6}=\frac{y}{7}$

$7x=6y$ Cross Products Property

$\frac{7x}{7y}=\frac{6y}{7y}$ To solve for $\frac{x}{y}$, divide each side by $7y$.

$\frac{x}{y}=\frac{6}{7}$ Simplify.

The ratio that completes the proportion is $\frac{6}{7}$.

Got It? 5. For parts (a) and (b), use the proportion $\frac{x}{6}=\frac{y}{7}$. What ratio completes the equivalent proportion? Justify your answer.

a. $\frac{6}{x}=\frac{\blacksquare}{\blacksquare}$

b. $\frac{\blacksquare}{\blacksquare}=\frac{y+7}{7}$

c. **Reasoning** Explain why $\frac{6}{x-6}=\frac{7}{y-7}$ is an equivalent proportion to $\frac{x}{6}=\frac{y}{7}$.

Take Note

Have students write their true proportions and then carry out the actions described in the properties.

Problem 5

Q In Method 1, why is Property (2) used? **[It shows the means in the two ratios changing places.]**

Got It?

Q Which property is applied in 5a? Explain. **[Property (1) because it shows the proportion is preserved by taking the reciprocal of each ratio.]**

Q How is the expression $\frac{y+7}{7}$ written as the sum of two ratios? How can the sum of these ratios clarify property (3)? **[$\frac{y}{7}+\frac{7}{7}$; It shows that 1 (expressed as a fraction) is added to each side of the proportion.]**

Answers

Got It? (continued)

4a. 63

b. 0.25

5a. $\frac{7}{y}$; Prop. of Proportions (1)

b. $\frac{x+6}{6}$; Prop. of Proportions (3)

c. The proportion is equivalent to $\frac{x-6}{6}=\frac{y-7}{7}$ by Prop. of Proportions (1). Then by Prop. of Proportions (3), $\frac{x-6+6}{6}=\frac{x-7+7}{7}$, which simplifies to $\frac{x}{6}=\frac{y}{7}$.

3 Lesson Check

Do you know HOW?

- If students have difficulty with Exercise 4, then have them describe the change made to $\frac{a}{7}$ and then make the same change to $\frac{13}{b}$.

Do you UNDERSTAND?

- If students have difficulty with Exercise 6, then remind them to rewrite the extended ratio as $3x : 6x : 7x$ and then choose a value for x.

Close

Q When two ratios are written as a proportion, what relationship must be true? **[The product of the means equals the product of the extremes.]**

Q When a proportion contains a variable, what property can be used to solve for the unknown quantity? **[Cross Products Property]**

Lesson Check

Do you know HOW?

1. To the nearest millimeter, a cell phone is 84 mm long and 46 mm wide. What is the ratio of the width to the length?

2. Two angle measures are in the ratio 5 : 9. Write expressions for the two angle measures in terms of the variable x.

3. What is the solution of the proportion $\frac{20}{z} = \frac{5}{3}$?

4. For $\frac{a}{7} = \frac{13}{b}$ complete each equivalent proportion.

a. $\frac{a}{■} = \frac{7}{■}$ **b.** $\frac{a-7}{7} = \frac{■}{■}$ **c.** $\frac{7}{a} = \frac{■}{■}$

Do you UNDERSTAND? MATHEMATICAL PRACTICES

5. Vocabulary What is the difference between a ratio and a proportion?

6. Open-Ended The lengths of the sides of a triangle are in the extended ratio 3 : 6 : 7. What are two possible sets of side lengths, in inches, for the triangle?

7. Error Analysis What is the error in the solution of the proportion shown at the right?

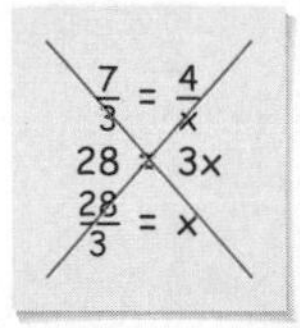

8. What is a proportion that has means 6 and 18 and extremes 9 and 12?

Practice and Problem-Solving Exercises MATHEMATICAL PRACTICES

A Practice **Write the ratio of the first measurement to the second measurement.** See Problem 1.

9. length of a tennis racket: 2 ft 4 in.
length of a table tennis paddle: 10 in.

10. height of a table tennis net: 6 in.
height of a tennis net: 3 ft

11. diameter of a table tennis ball: 40 mm
diameter of a tennis ball: 6.8 cm

12. length of a tennis court: 26 yd
length of a table tennis table: 9 ft

13. Baseball A baseball team played 154 regular season games. The ratio of the number of games they won to the number of games they lost was $\frac{5}{2}$. How many games did they win? How many games did they lose? See Problem 2.

14. The measures of two supplementary angles are in the ratio 5 : 7. What is the measure of the larger angle?

15. The lengths of the sides of a triangle are in the extended ratio 6 : 7 : 9. The perimeter of the triangle is 88 cm. What are the lengths of the sides? See Problem 3.

16. The measures of the angles of a triangle are in the extended ratio 4 : 3 : 2. What is the measure of the largest angle?

Algebra **Solve each proportion.** See Problem 4.

17. $\frac{1}{3} = \frac{x}{12}$ **18.** $\frac{9}{5} = \frac{3}{x}$ **19.** $\frac{4}{x} = \frac{5}{9}$ **20.** $\frac{y}{10} = \frac{15}{25}$ **21.** $\frac{9}{24} = \frac{12}{n}$

22. $\frac{11}{14} = \frac{b}{21}$ **23.** $\frac{3}{5} = \frac{6}{x+3}$ **24.** $\frac{y+7}{9} = \frac{8}{5}$ **25.** $\frac{5}{x-3} = \frac{10}{x}$ **26.** $\frac{n+4}{8} = \frac{n}{4}$

PowerGeometry.com

3 Lesson Check

For a digital lesson check, use the Got It questions.

Support in Geometry Companion

- Lesson Check

4 Practice

Assign homework to individual students or to an entire class.

Answers

Lesson Check

1. 23 : 42
2. $5x$, $9x$
3. 12
4a. $\frac{a}{13} = \frac{7}{b}$
b. $\frac{a-7}{7} = \frac{13-b}{b}$
c. $\frac{7}{a} = \frac{b}{13}$
5. A ratio is a single comparison, while a proportion is a statement that two ratios are equal.
6. Answers may vary. Sample: 3 in., 6 in., 7 in.; or 6 in., 12 in., 14 in.
7. The second line should equate the product of the means and the product of the extremes: $7x = 12$. Then the third line would be $x = \frac{12}{7}$.
8. $\frac{9}{6} = \frac{18}{12}$, $\frac{9}{18} = \frac{6}{12}$, $\frac{12}{6} = \frac{18}{9}$, or $\frac{12}{18} = \frac{6}{9}$

Practice and Problem-Solving Exercises

9. $\frac{14}{5}$ or 14 : 5
10. $\frac{1}{6}$ or 1 : 6
11. $\frac{10}{17}$ or 10 : 17
12. $\frac{26}{3}$ or 26 : 3
13. won 110, lost 44
14. 105
15. 24 cm, 28 cm, 36 cm
16. 80
17. 4
18. $\frac{5}{3}$
19. $\frac{36}{5}$
20. 6
21. 32
22. 16.5
23. 7
24. 7.4
25. 6
26. 4

In the diagram, $\frac{a}{b} = \frac{3}{4}$. Complete each statement. Justify your answer. See Problem 5.

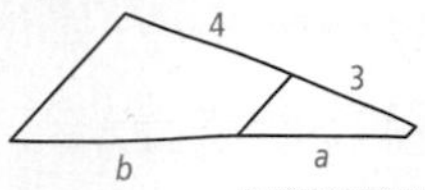

27. $\frac{b}{a} = \frac{■}{■}$ 28. $4a = ■$ 29. $\frac{■}{■} = \frac{b}{4}$

30. $\frac{■}{■} = \frac{7}{4}$ 31. $\frac{a+b}{b} = \frac{■}{■}$ 32. $\frac{b}{■} = \frac{4}{■}$

B Apply

Coordinate Geometry Use the graph. Write each ratio in simplest form.

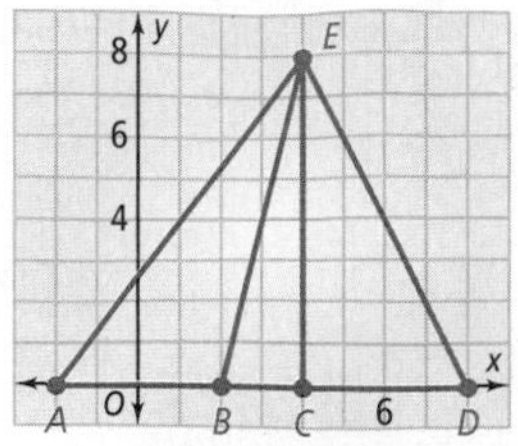

33. $\frac{AC}{BD}$ 34. $\frac{AE}{EC}$ 35. slope of $\overline{EB}$ 36. slope of $\overline{ED}$

37. **Think About a Plan** The area of a rectangle is 150 in.^{2}. The ratio of the length to the width is 3 : 2. Find the length and the width.
- What is the formula for the area of a rectangle?
- How can you use the given ratio to write expressions for the length and width?

Art To draw a face, you can sketch the head as an oval and then lightly draw horizontal lines to help locate the eyes, nose, and mouth. You can use the extended ratios shown in the diagrams to help you place the lines for an adult's face or for a baby's face.

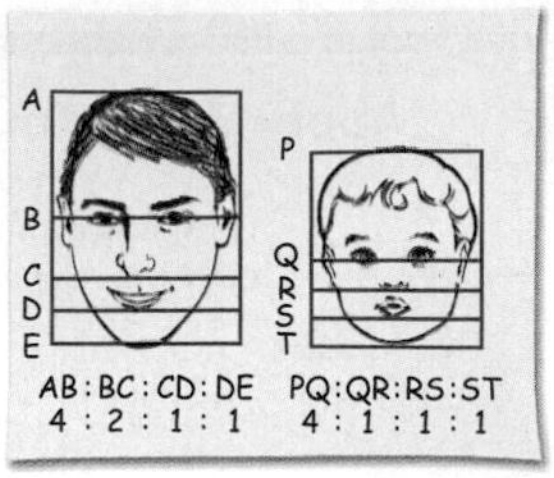

38. If $AE = 72$ cm in the diagram, find AB, BC, CD, and DE.

39. You draw a baby's head as an oval that is 21 in. from top to bottom.
   a. How far from the top should you place the line for the eyes?
   b. Suppose you decide to make the head an adult's head. How far up should you move the line for the eyes?

Algebra Solve each proportion.

40. $\frac{1}{7y-5} = \frac{2}{9y}$ 41. $\frac{4a+1}{7} = \frac{2a}{3}$ 42. $\frac{5}{x+2} = \frac{3}{x+1}$ 43. $\frac{2b-1}{4} = \frac{b-2}{12}$

44. The ratio of the length to the width of a rectangle is 9 : 4. The width of the rectangle is 52 mm. Write and solve a proportion to find the length.

45. **Open-Ended** Draw a quadrilateral that satisfies this condition: The measures of the consecutive angles are in the extended ratio 4 : 5 : 4 : 7.

46. **Reasoning** The means of a proportion are 4 and 15. List all possible pairs of positive integers that could be the extremes of the proportion.

47. **Writing** Describe how to use the Cross Products Property to determine whether $\frac{10}{26} = \frac{16}{42}$ is a true proportion.

48. **Reasoning** Explain how to use two different properties of proportions to change the proportion $\frac{3}{4} = \frac{12}{16}$ into the proportion $\frac{12}{3} = \frac{16}{4}$.

4 Practice

ASSIGNMENT GUIDE
Basic: 9–32, 37, 38, 40, 41, 46, 48–51
Average: 9–31 odd, 33–54
Advanced: 9–31 odd, 33–60

Mathematical Practices are supported by exercises with red headings. Here are the Practices supported in this lesson:

MP 1: Make Sense of Problems Ex. 6, 37, 45
MP 1: Persevere in Solving Problems Ex. 46
MP 3: Communicate Ex. 47
MP 3: Critique the Reasoning of Others Ex. 7
MP 7: Use Structure Ex. 48

Applications exercises have blue headings. Exercises 13, 38, and 39 support MP 4: Model.

EXERCISE 46: Use the Think About a Plan worksheet in the **Practice and Problem Solving Workbook** (also available in the Teaching Resources in print and online) to further support students' development in becoming independent learners.

HOMEWORK QUICK CHECK
To check students' understanding of key skills and concepts, go over Exercises 13, 25, 37, 46, and 48.

27. $\frac{4}{3}$; Prop. of Proportions (1)
28. $3b$; Cross Products Prop.
29. $\frac{a}{3}$; Prop. of Proportions (2)
30. $\frac{a+b}{b}$; Prop. of Proportions (3)
31. $\frac{7}{4}$; Prop. of Proportions (3)
32. $\frac{b}{a} = \frac{4}{3}$; Prop. of Proportions (1)
33. 1
34. $\frac{5}{4}$
35. 4
36. −2
37. length: 15 in.; width: 10 in.
38. $AB = 36$ cm, $BC = 18$ cm; $CD = 9$ cm, $DE = 9$ cm
39a. 12 in.
b. 1.5 in.
40. 2
41. 1.5
42. 0.5
43. 0.2
44. $\frac{9}{4} = \frac{x}{52}$; 117 mm
45.

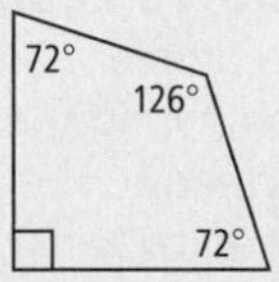

46. 1 and 60, 2 and 30, 3 and 20, 4 and 15, 5 and 12, 6 and 10
47. The product of the means is $26 \cdot 16 = 416$, and the product of the extremes is $10 \cdot 42 = 420$. Since $416 \neq 420$, it is not a valid proportion.
48. $\frac{3}{4} = \frac{12}{16}$ is equivalent to $\frac{3}{12} = \frac{4}{16}$ by Prop. of Proportions (2). $\frac{3}{12} = \frac{4}{16}$ is equivalent to $\frac{12}{3} = \frac{16}{4}$ by Prop. of Proportions (1).

Answers

Practice and Problem-Solving Exercises (continued)

49. $\frac{9}{4}$; divide each side by $4n$.

50. $\frac{30}{18}$; mult. each side by $\frac{t}{18}$.

51. $\frac{b}{2}$; Prop. of Proportions (3)

52. $\frac{b}{d}$; Prop. of Proportions (3) and (2)

53. $\frac{c}{d}$; Prop. of Proportions (2), then (3), then (2)

54. $\frac{c+2d}{d}$; apply Prop. of Proportions (3) twice.

55. $\frac{a}{b} = \frac{c}{d}$ (given); $\frac{a}{b}(bd) = \frac{c}{d}(bd)$ (Mult. Prop. of =); $ad = bc$ (simplify and Commutative Prop. of Mult.); $bc = ad$ (Sym. Prop. of =); $\frac{bc}{ac} = \frac{ad}{ac}$ (Div. Prop. of =); $\frac{b}{a} = \frac{d}{c}$ (simplify)

56. $\frac{a}{b} = \frac{c}{d}$ (given); $\frac{a}{b}(bd) = \frac{c}{d}(bd)$ (Mult. Prop. of =); $ad = bc$ (simplify and Commutative Prop. of Mult.); $\frac{ad}{cd} = \frac{bc}{cd}$ (Div. Prop. of =); $\frac{a}{c} = \frac{b}{d}$ (simplify)

57. $\frac{a}{b} = \frac{c}{d}$ (given); $\frac{a}{b} + 1 = \frac{c}{d} + 1$ (Add. Prop. of Eq.); $\frac{a}{b} + \frac{b}{b} = \frac{c}{d} + \frac{d}{d}$ (Subst. Prop. of Eq.); $\frac{a+b}{b} = \frac{c+d}{d}$ (simplify)

58. -3 or 4

59. $-\frac{5}{6}, \frac{1}{2}$

60. $x = 5, y = 24$

Complete each statement. Justify your answer.

49. If $4m = 9n$, then $\frac{m}{n} = \frac{\blacksquare}{\blacksquare}$.

50. If $\frac{30}{t} = \frac{18}{r}$, then $\frac{t}{r} = \frac{\blacksquare}{\blacksquare}$.

51. If $\frac{a+5}{5} = \frac{b+2}{2}$, then $\frac{a}{5} = \frac{\blacksquare}{\blacksquare}$.

52. If $\frac{a}{b} = \frac{c}{d}$, then $\frac{a+b}{c+d} = \frac{\blacksquare}{\blacksquare}$.

53. If $\frac{a}{b} = \frac{c}{d}$, then $\frac{a+c}{b+d} = \frac{\blacksquare}{\blacksquare}$.

54. If $\frac{a}{b} = \frac{c}{d}$, then $\frac{a+2b}{b} = \frac{\blacksquare}{\blacksquare}$.

Challenge

Algebra **Use properties of equality to justify each property of proportions.**

55. $\frac{a}{b} = \frac{c}{d}$ is equivalent to $\frac{b}{a} = \frac{d}{c}$.

56. $\frac{a}{b} = \frac{c}{d}$ is equivalent to $\frac{a}{c} = \frac{b}{d}$.

57. $\frac{a}{b} = \frac{c}{d}$ is equivalent to $\frac{a+b}{b} = \frac{c+d}{d}$.

Algebra **Solve each proportion for the variable(s).**

58. $\frac{x-3}{3} = \frac{2}{x+2}$

59. $\frac{3-4x}{1+5x} = \frac{1}{2+3x}$

60. $\frac{x}{6} = \frac{x+10}{18} = \frac{4x}{y}$

Apply What You've Learned

MATHEMATICAL PRACTICES
MP 1

Look back at the information on page 431 about the dimensions of Lillian's graphing calculator screen. Select all of the following that are true. Explain your reasoning.

A. The perimeter of the calculator screen is 10 cm.

B. The ratio of the width of the screen to the height of the screen is 3 : 2.

C. The ratio of the width of the screen to the height of the screen is 15 : 9.

D. The ratio of the width of the screen to the height of the screen is the same whether you measure the dimensions in centimeters or millimeters.

E. A rectangle with a width of 2 ft and a height of 16 in. has the same ratio of width to height as Lillian's calculator screen.

F. An extended ratio that compares the dimensions of the four sides of Lillian's calculator screen is 3 : 2 : 3 : 2.

G. The extended ratio that compares the dimensions of the four sides of Lillian's calculator screen is 2 : 1 : 2 : 1.

Apply What You've Learned

The solution of the problem on page 431 depends on comparing the shape of the calculator screen to the shape of the rectangle determined by Xmin, Xmax, Ymin, and Ymax. Here students use ratios to investigate the shape of the calculator screen.

Mathematical Practices

Students begin to **make sense of the problem** by finding the ratio of the length to the width for Lillian's calculator screen. (MP 1)

ANSWERS

Choices B, D, E, and F are all true.

7-1 Lesson Resources

Differentiated Remediation

Instructional Support

Geometry Companion

Students can use the **Geometry Companion** worktext (4 pages) . . .

- New Vocabulary
- Key Concepts
- Got It for each Problem
- Lesson Check

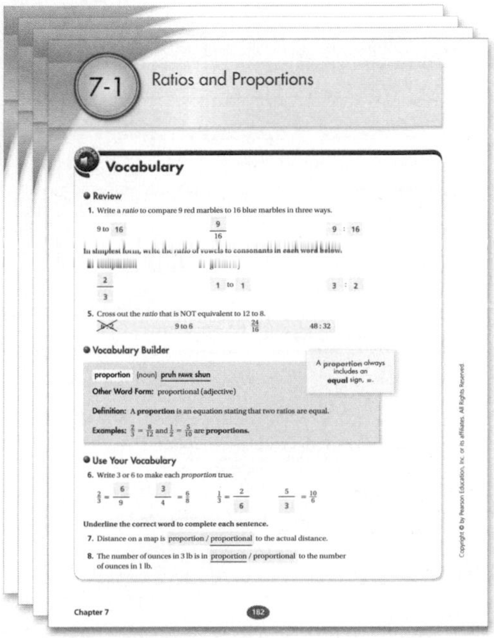

ELL Support

Use Graphic Organizers Tell students to make a 3-column KWL table. The columns are labeled "know," "want to know," and "learned." In the first column, have students write a declarative sentence about each of the following words: *ratio*, *proportion*, *extreme*, *mean*, and *cross product*. In the second column, have them write a question about each word.

Give the students an example to help them get started. Here is an example of what a student might write for *ratio*:
K: A ratio compares two numbers.
W: What ratios do triangles have?

After the lesson, ask students to write what they have learned about each word in the third column.

5 Assess & Remediate

Lesson Quiz

1. **Do you UNDERSTAND?** Jessica is making a scale model of the Empire State Building. Her model will have a height of 16 cm. The actual height of the building is about 448 m. What is the ratio of the height of Jessica's model to the height of the building?
2. The measures of two supplementary angles are in the ratio 4 : 5. What are the measures of the angles?
3. In the diagram, $\frac{x}{15} = \frac{y}{6}$. What ratio completes the equivalent proportion $\frac{x}{y} = \frac{\blacksquare}{\blacksquare}$? Use one of the Properties of Proportions to justify your answer.

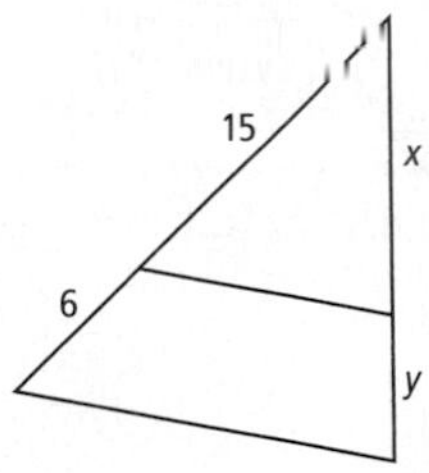

ANSWERS TO LESSON QUIZ

1. 1:2800
2. 80, 100
3. $\frac{x}{y} = \frac{15}{6}$

PRESCRIPTION FOR REMEDIATION

Use the student work on the Lesson Quiz to prescribe a differentiated review assignment.

Points	Differentiated Remediation
0–1	Intervention
2	On-level
3	Extension

PowerGeometry.com

5 Assess & Remediate

Assign the Lesson Quiz. Appropriate intervention, practice, or enrichment is automatically generated based on student performance.

Intervention

- **Reteaching** (2 pages) Provides reteaching and practice exercises for the key lesson concepts. Use with struggling students or absent students.
- **English Language Learner Support** Helps students develop and reinforce mathematical vocabulary and key concepts.

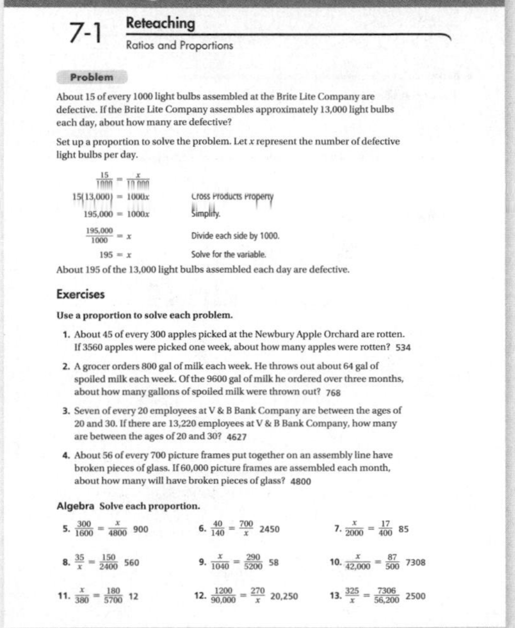

All-in-One Resources/Online

Reteaching

7-1 **Reteaching**
Ratios and Proportions

Problem

About 15 of every 1000 light bulbs assembled at the Brite Lite Company are defective. If the Brite Lite Company assembles approximately 13,000 light bulbs each day, about how many are defective?

Set up a proportion to solve the problem. Let x represent the number of defective light bulbs per day.

$\frac{15}{1000} = \frac{x}{13{,}000}$

$15(13{,}000) = 1000x$ Cross Products Property

$195{,}000 = 1000x$ Simplify.

$\frac{195{,}000}{1000} = x$ Divide each side by 1000.

$195 = x$ Solve for the variable.

About 195 of the 13,000 light bulbs assembled each day are defective.

Exercises

Use a proportion to solve each problem.

1. About 45 of every 300 apples picked at the Newbury Apple Orchard are rotten. If 3560 apples were picked one week, about how many apples were rotten? 534
2. A grocer orders 800 gal of milk each week. He throws out about 64 gal of spoiled milk each week. Of the 9600 gal of milk he ordered over three months, about how many gallons of spoiled milk were thrown out? 768
3. Seven of every 20 employees at V & B Bank Company are between the ages of 20 and 30. If there are 13,220 employees at V & B Bank Company, how many are between the ages of 20 and 30? 4627
4. About 56 of every 700 picture frames put together on an assembly line have broken pieces of glass. If 60,000 picture frames are assembled each month, about how many will have broken pieces of glass? 4800

Algebra **Solve each proportion.**

5. $\frac{300}{1600} = \frac{x}{4800}$ 900
6. $\frac{40}{140} = \frac{700}{x}$ 2450
7. $\frac{x}{2000} = \frac{17}{400}$ 85
8. $\frac{35}{x} = \frac{150}{2400}$ 560
9. $\frac{x}{1040} = \frac{290}{5200}$ 58
10. $\frac{x}{42{,}000} = \frac{87}{500}$ 7308
11. $\frac{x}{380} = \frac{180}{5700}$ 12
12. $\frac{1200}{90{,}000} = \frac{270}{x}$ 20,250
13. $\frac{325}{x} = \frac{7306}{56{,}200}$ 2500

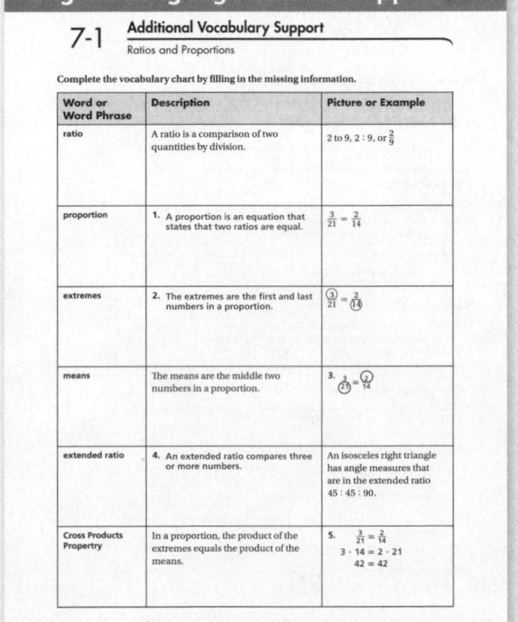

All-in-One Resources/Online

English Language Learner Support

7-1 **Additional Vocabulary Support**
Ratios and Proportions

Complete the vocabulary chart by filling in the missing information.

Word or Word Phrase	Description	Picture or Example
ratio	A ratio is a comparison of two quantities by division.	2 to 9, 2 : 9, or $\frac{2}{9}$
proportion	1. A proportion is an equation that states that two ratios are equal.	$\frac{3}{21} = \frac{2}{14}$
extremes	2. The extremes are the first and last numbers in a proportion.	$\frac{(3)}{21} = \frac{2}{(14)}$
means	The means are the middle two numbers in a proportion.	3. $\frac{3}{(21)} = \frac{(2)}{14}$
extended ratio	4. An extended ratio compares three or more numbers.	An isosceles right triangle has angle measures that are in the extended ratio 45 : 45 : 90.
Cross Products Property	In a proportion, the product of the extremes equals the product of the means.	5. $\frac{3}{21} = \frac{2}{14}$ $3 \cdot 14 = 2 \cdot 21$ $42 = 42$

Differentiated Remediation *continued*

On-Level

- **Practice** (2 pages) Provides extra practice for each lesson. For simpler practice exercises, use the Form K Practice pages found in the All-in-One Teaching Resources and online.
- **Think About a Plan** Helps students develop specific problem-solving skills and strategies by providing scaffolded guiding questions.
- **Standardized Test Prep** Focuses on all major exercises, all major question types, and helps students prepare for the high-stakes assessments.

Extension

- **Enrichment** Provides students with interesting problems and activities that extend the concepts of the lesson.
- **Activities, Games, and Puzzles** Worksheets that can be used for concepts development, enrichment, and for fun!

Practice and Problem Solving Wkbk/ All-in-One Resources/Online

Practice page 1

7-1 **Practice** *Form G*
Ratios and Proportions

Write the ratio of the first measurement to the second measurement.

1. diameter of a salad plate: 8 in. — diameter of a dinner plate: 1 ft $\frac{2}{3}$
2. weight of a cupcake: 2 oz — weight of a cake: 2 lb 2 oz $\frac{1}{17}$
3. garden container width: 2 ft 6 in. — garden container length: 8 ft $\frac{5}{16}$
4. width of a canoe: 28 in. — length of a canoe: 12 ft 6 in. $\frac{14}{75}$
5. height of a book: 11 in. — height of a bookshelf: 3 ft 3 in. $\frac{11}{39}$
6. The perimeter of a rectangle is 280 cm. The ratio of the width to the length is 3 : 4. What is the length of the rectangle? **80 cm**
7. The ratio of country albums to jazz albums in a music collection is 2 : 3. If the music collection has 45 albums, how many are country albums? **18**
8. The lengths of the sides of a triangle are in the extended ratio 3 : 6 : 8. The triangle's perimeter is 510 cm. What are the lengths of the sides? **90 cm, 180 cm, 240 cm**

Algebra **Solve each proportion.**

9. $\frac{x}{4} = \frac{13}{52}$ **1**
10. $\frac{x}{2x+1} = \frac{16}{40}$ **2**
11. $\frac{9}{10} = \frac{9x}{70}$ **7**
12. $\frac{2}{7} = \frac{b+1}{56}$ **15**
13. $\frac{11}{y} = \frac{9}{27}$ **33**
14. $\frac{3}{34} = \frac{m}{51}$ **4.5**

Use the proportion $\frac{x}{z} = \frac{6}{5}$. Complete each statement. Justify your answer.

15. $\frac{x}{6} = \frac{\square}{\square}$ $\frac{z}{5}$; Prop. of Proportions (2)
16. $\frac{x+z}{z} = \frac{\square}{\square}$ $\frac{11}{5}$; Prop. of Proportions (3)
17. $\frac{z}{x} = \frac{\square}{\square}$ $\frac{5}{6}$; Prop. of Proportions (1)
18. $5x = \square$ $6z$; Cross Products Property
19. The measures of two consecutive angles in a parallelogram are in the ratio 4 : 11. What are the measures of the four angles of the parallelogram? **48, 48, 132, 132**

Practice and Problem Solving Wkbk/ All-in-One Resources/Online

Practice page 2

7-1 **Practice** (continued) *Form G*
Ratios and Proportions

Coordinate Geometry **Use the graph. Write each ratio in simplest form.**

20. $\frac{AB}{BD}$ $\frac{4}{7}$
21. $\frac{AE}{EC}$ $\frac{5}{3}$
22. $\frac{EC}{BC}$ $\frac{3}{2}$
23. $\frac{\text{slope of } \overline{BE}}{\text{slope of } \overline{AE}}$ $\frac{2}{1}$ or 2

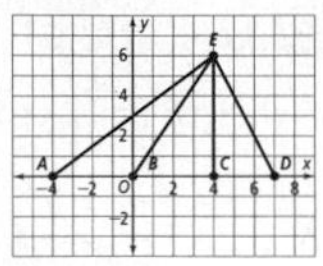

24. A band director needs to purchase new uniforms. The ratio of small to medium to large uniforms is 3 : 4 : 6.
    a. If there are 260 total uniforms to purchase, how many will be small? **60**
    b. How many of these uniforms will be medium? **80**
    c. How many of these uniforms will be large? **120**
25. The measures of two complementary angles are in the ratio 2 : 3. What is the measure of the smaller angle? **36**
26. The measures of two supplementary angles are in the ratio 4 : 11. What is the measure of the larger angle? **132**
27. The means of a proportion are 4 and 17. List all possible pairs of positive integers that could be the extremes of the proportion. **1 and 68, 2 and 34, 4 and 17**
28. The extremes of a proportion are 5 and 14. List all possible pairs of positive integers that could be the means of the proportion. **1 and 70, 2 and 35, 5 and 14, 7 and 10**

Algebra **Solve each proportion.**

29. $\frac{(x-1)}{(x+1)} = \frac{10}{14}$ **6**
30. $\frac{7}{50} = \frac{x}{30}$ **4.2**

31. **Writing** Explain why solving proportions is an important skill for solving geometry problems. **Answers may vary. Sample: Many geometric properties involve ratios. You can use proportions to model them and solve problems.**
32. Draw a triangle that satisfies this condition: The ratio of the interior angles is 7 : 11 : 12. **Triangle should have angles that measure 42, 66, and 72.**

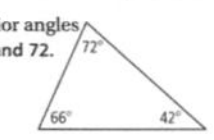

All-in-One Resources/Online

Enrichment

7-1 **Enrichment**
Ratios and Proportions

Ratios and proportions occur frequently in everyday situations. Some involve linear equations, such as those concerning menu planning and recipes, whereas others, often involving geometry, require quadratic equations.

Use ratios and proportions to solve each problem.

1. A meatloaf recipe uses 4 lb of hamburger to feed 6 people. How many pounds of hamburger will be used to feed 15 people? **10 lb**
2. The tenth grade at Milford High School has a dance every year. Last year there were 80 students in the tenth grade, and the party cost \$200. This year there are 100 students in the tenth grade. How much should they plan to spend? **\$250**
3. If it costs \$200,000 to build a sidewalk around a rectangular field whose dimensions are 200 yd by 800 yd, how much will it cost to build a sidewalk around a rectangular field whose dimensions are 300 yd by 900 yd? **\$240,000**
4. The cost of buying a plot of land in Happy Valley depends on the area of the plot. If a rectangular plot of land whose dimensions are 200 yd by 800 yd costs \$100,000, what is the cost of a rectangular plot of land whose dimensions are 300 yd by 900 yd? **\$168,750**
5. If it costs \$5980 to have a picket fence installed around a rectangular lot that is 110 ft by 150 ft, how much will it cost to have a picket fence installed around a rectangular lot that is 125 ft by 170 ft? **\$6785**
6. At Pools-a-Plenty it costs \$165 for a swimming pool cover for a round, aboveground pool that is 30 ft in diameter. How much will a cover for a pool that is 24 ft in diameter cost at Pools-a-Plenty? Use 3.14 for π. **\$105.60**
7. Fencing was purchased for two rectangular plots of land. The first plot measured 80 yd by 100 yd, and the cost of the fencing was \$900. The cost of the fencing for the second plot was \$2400, and one of the dimensions of the plot was 120 yd. What was the other dimension? **360 yd**
8. If 3 hens lay 10 eggs in 5 days, how many eggs will 3 hens lay in 20 days? **40 eggs**
9. If 8 hens lay 14 eggs in 6 days, how many eggs will 24 hens lay in 6 days? **42 eggs**
10. If 4 hens lay 7 eggs in 3 days, how many eggs will 12 hens lay in 9 days? **63 eggs**
11. If Jenna can walk 6 mi in 2 h, how many miles could she walk in 2.5 h, assuming she keeps the same pace? **7.5 mi**
12. Suppose 2.5 lb of grass seed can cover a plot of land that is 30 ft by 30 ft. How much grass seed is needed to cover a plot of land 45 ft by 60 ft? **7.5 lb**

Practice and Problem Solving Wkbk/ All-in-One Resources/Online

Think About a Plan

7-1 **Think About a Plan**
Ratios and Proportions

Reasoning The means of a proportion are 4 and 15. List all possible pairs of positive integers that could be the extremes of the proportion.

Understanding the Problem

1. What is a proportion? **an equation that states two ratios are equal**
2. What are some of the forms in which a proportion can be written? **Answers may vary. Samples: $a : b = c : d$ or $\frac{a}{b} = \frac{c}{d}$**
3. Explain the difference between the means and the extremes of a proportion. Use an example in your explanation. **When you write a proportion in the form $a : b = c : d$, the first and last numbers are the extremes and the middle numbers are the means. In this example a and d are the extremes and b and c are the means.**

Planning the Solution

4. How can you write the proportion described in the problem, using variables for the extremes? Should you use the same variable for the extremes or different variables? **$a : 4 = 15 : b$; use different variables because the extremes may be different numbers.**
5. How can you rewrite the proportion as equivalent fractions? **$\frac{a}{4} = \frac{15}{b}$**
6. How do you solve for variables in a proportion? Apply this to the proportion you wrote in Step 5. **Find the cross-products; $ab = 60$.**

Getting an Answer

7. Look at the equation you wrote in Step 6. How do the two variables on the one side of the equation relate to the value on the other side? **The value is the product of the two variables.**
8. How can you use factoring to find all the positive integers that could represent the values of the variables? **List all factor pairs for 60. These are the possible values for the extremes.**
9. Find the solution to the problem. **1 and 60; 2 and 30; 3 and 20; 4 and 15; 5 and 12; 6 and 10**

Practice and Problem Solving Wkbk/ All-in-One Resources/Online

Standardized Test Prep

7-1 **Standardized Test Prep**
Ratios and Proportions

Gridded Response

Solve each exercise and enter your answer on the grid provided.

Use the graph at the right for Exercises 1 and 2.

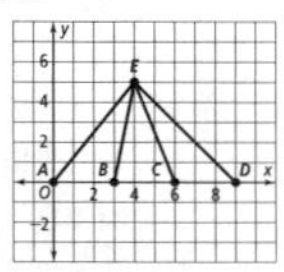

1. What is $\frac{AD}{AB}$ in simplest form?
2. What is $\frac{\text{slope of } \overline{BE}}{\text{slope of } \overline{AE}}$ in simplest form?
3. What is the value of x in the proportion $\frac{(x-1)}{5} = \frac{(4x+2)}{35}$?
4. What is the value of x in the proportion $\frac{x+1}{x+3} = \frac{15}{21}$?
5. The lengths of the sides of a triangle are in the extended ratio 3 : 10 : 12. The perimeter is 400 cm. What is the length of the longest side in centimeters?

Answers

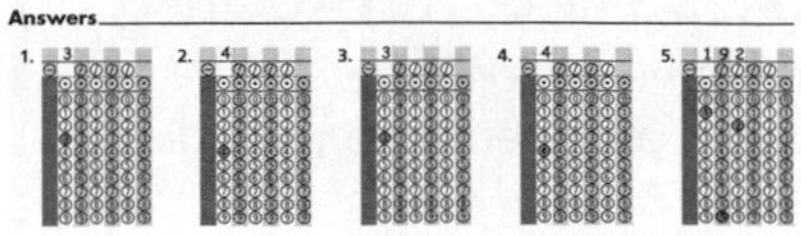

Online Teacher Resource Center

Activities, Games, and Puzzles

7-1 **Puzzle: Crossword**
Ratios and Proportions

Solve each problem. Write your answer in the crossword puzzle below. Each numerical answer should be written in word form. (Don't forget hyphens where appropriate!)

1S	I	X	2T	Y	-	T	H	3R	E	4E
			H					E		X
5R	A	T	I	O				C		T
			R			6T		I		R
7S	I	X	T	Y		W		P		E
			Y		8F	O	U	R		M
			-					O		E
			N					C		S
		9F	I	V	E			A		
			N					L		
	10E	L	E	V	E	N		S		

Across

1. The ratio of two complementary angles is 3 to 7. Find the measure of the larger one.
5. A __?__ is a comparison of two quantities by division.
7. The measures of the angles of a triangle are in the ratio 1 : 3 : 5. How many degrees is the middle-sized angle?
8. Find the value of x: $\frac{6}{x+5} = \frac{2}{3}$
9. Find the value of y: $\frac{y}{6} = \frac{7.5}{9}$
10. If $\frac{a}{3} = \frac{c}{11}$, then $\frac{a+3}{3} = \frac{c + \underline{?}}{11}$.

Down

2. Mary is making a party mix to bring to school. She believes that the perfect ratio of pretzels to peanuts is 3 to 5. If Mary knows that she is going to use 65 peanuts in the mix, how many pretzels will she need?
3. $\frac{a}{b} = \frac{c}{d}$ is equivalent to $\frac{b}{a} = \frac{d}{c}$. In other words, if two fractions are in proportion, their __?__ are also in proportion.
4. The first and last numbers in a proportion are known as the __?__.
6. Find the value of z: $\frac{4z-2}{3} = \frac{2z+6}{5}$

Algebra Review

Use With Lesson 7-2

Solving Quadratic Equations

Common Core State Standards
Reviews A-CED.A.1 Create equations and inequalities in one variable and use them to solve problems.

Equations in the form $ax^2 + bx + c = 0$, where $a \neq 0$, are quadratic equations in standard form. You can solve some quadratic equations in standard form by factoring and using the Zero-Product Property:

If $ab = 0$, then $a = 0$ or $b = 0$.

You can solve all quadratic equations in standard form using the quadratic formula:

If $ax^2 + bx + c = 0$, where $a \neq 0$, then $x = \frac{-b \pm \sqrt{b^2 - 4ac}}{2a}$.

Example

Algebra **Solve for *x*. For irrational solutions, give both the exact answer and the answer rounded to the nearest hundredth.**

A

$7x^2 + 6x - 1 = 0$	The equation is in standard form.
$(7x - 1)(x + 1) = 0$	Factor.
$7x - 1 = 0$ or $x + 1 = 0$	Use the Zero-Product Property.
$x = \frac{1}{7}$ or $x = -1$	Solve for *x*.

B

$-3x^2 - 5x + 1 = 0$	The equation is in standard form.
$a = -3, b = -5, c = 1$	Identify *a*, *b*, and *c*.
$x = \frac{-(-5) \pm \sqrt{(-5)^2 - 4(-3)(1)}}{2(-3)}$	Substitute in the quadratic formula.
$x = \frac{5 \pm \sqrt{37}}{-6}$	Simplify.
$x = -$ [illegible] or $x = -$ [illegible]	Write the two solutions separately.
$x \approx -1.85$ or $x \approx 0.18$	Use a calculator and round to the nearest hundredth.

Exercises

Algebra **Solve for *x*. For irrational solutions, give both the exact answer and the answer rounded to the nearest hundredth.**

1. $x^2 + 5x - 14 = 0$
2. $4x^2 - 13x + 3 = 0$
3. $2x^2 + 7x + 3 = 0$
4. $5x^2 + 2x - 2 = 0$
5. $2x^2 - 10x + 11 = 0$
6. $8x^2 - 2x - 3 = 0$
7. $2x^2 + 3x - 20 = 0$
8. $x^2 - x - 210 = 0$
9. $x^2 - 4x = 0$
10. $x^2 - 25 = 0$
11. $6x^2 + 10x = 5$
12. $1 = 2x^2 - 6x$

Guided Instruction

PURPOSE To review solving quadratic equations using the Zero Product Property and the quadratic formula

PROCESS Students will
- write a quadratic equation in standard form.
- solve quadratic equations by factoring and applying the Zero-Product Property.
- solve quadratic equations by using the quadratic formula.

DISCUSS The examples and exercises focus on the steps taken to solve quadratic equations by factoring and by using the quadratic formula.

Example

In this Example students solve two quadratic equations, one by factoring and one by using the quadratic formula.

Q Can you use the quadratic formula to solve every quadratic equation? Explain. **[Yes. Every quadratic equation can be written in standard form and values of *a*, *b*, and *c* can be substituted into the formula.]**

Q When you are using the quadratic formula, how can you tell if there is only one solution? **[the expression under the radical simplifies to zero]**

Q What conclusion can you draw about a quadratic equation when its solutions are irrational? **[The quadratic equation cannot be factored using integers.]**

Answers

Exercises

1. −7, 2
2. $\frac{1}{4}$, 3
3. −3, $-\frac{1}{2}$
4. $\frac{-1 + \sqrt{11}}{5}$, $\frac{-1 - \sqrt{11}}{5}$; 0.46, −0.86
5. $\frac{5 + \sqrt{3}}{2}$, $\frac{5 - \sqrt{3}}{2}$; 3.37, 1.63
6. $-\frac{1}{2}$, $\frac{3}{4}$
7. −4, $\frac{5}{2}$
8. −14, 15
9. 0, 4
10. −5, 5
11. $\frac{-5 + \sqrt{55}}{6}$, $\frac{-5 - \sqrt{55}}{6}$; 0.40, −2.07
12. $\frac{3 + \sqrt{11}}{2}$, $\frac{3 - \sqrt{11}}{2}$; 3.16, −0.16

1 Interactive Learning

Solve It!

PURPOSE To determine properties of similar polygons

PROCESS Students may find the ratios of heights and ratios of widths to see which screens have proportional sides or find the height-to-width ratios of all three screens.

FACILITATE

Q If a video that is filmed for a movie theater screen is shown on a television that does not have the same height-to-width ratio, how might the images appear on the screen? **[You can display part of the theater image and fill the TV screen. The complete theater image cannot be made to fill a TV screen without distorting the image. You can scale the theater image to fit, but part of the TV screen will not be filled.]**

ANSWER See Solve It in Answers on next page.

CONNECT THE MATH Students should recognize that the relationship between similar figures lies in the proportionality of their sides. This allows a movie filmed for a theater screen to be shrunk to fit on a television screen. "Letterbox" format is a video format that preserves the film's original aspect ratio.

2 Guided Instruction

TAKE NOTE Ask students to describe the two conditions necessary for figures to be similar. Draw similar figures on the board and ask students to identify the conditions that show the figures are similar.

7-2 Similar Polygons

Common Core State Standards

G-SRT.B.5 Use . . . similarity criteria for triangles to solve problems and to prove relationships in geometric figures.

MP 1, MP 3, MP 4, MP 6

Objective To identify and apply similar polygons

SOLVE IT! Getting Ready!

A movie theater screen is in the shape of a rectangle 45 ft wide by 25 ft high. Which of the TV screen formats at the right do you think would show the most complete scene from a movie shown on the theater screen? Explain.

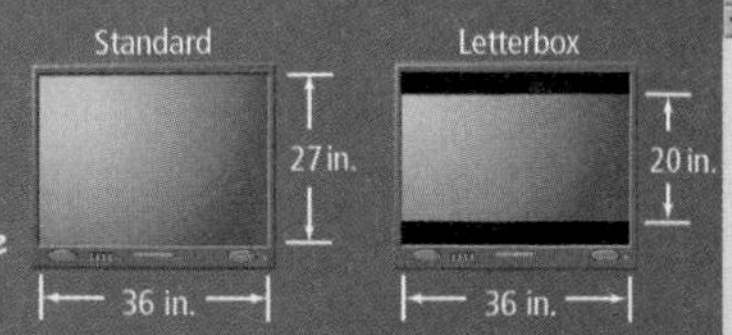

Can you use what you've learned before about ratios to help you solve this problem?

MATHEMATICAL PRACTICES

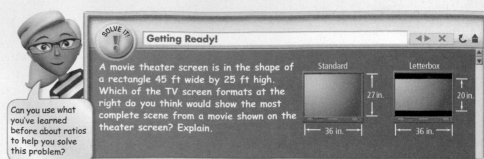

Similar figures have the same shape but not necessarily the same size. You can abbreviate *is similar to* with the symbol ~.

Essential Understanding You can use ratios and proportions to decide whether two polygons are similar and to find unknown side lengths of similar figures.

Lesson Vocabulary
- similar figures
- similar polygons
- extended proportion
- scale factor
- scale drawing
- scale

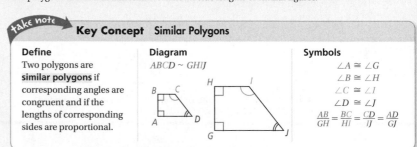

take note **Key Concept** Similar Polygons

Define	Diagram	Symbols
Two polygons are **similar polygons** if corresponding angles are congruent and if the lengths of corresponding sides are proportional.	$ABCD \sim GHIJ$ 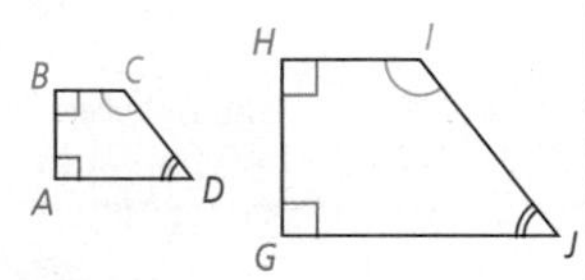	$\angle A \cong \angle G$ $\angle B \cong \angle H$ $\angle C \cong \angle I$ $\angle D \cong \angle J$ $\frac{AB}{GH} = \frac{BC}{HI} = \frac{CD}{IJ} = \frac{AD}{GJ}$

You write a similarity statement with corresponding vertices in order, just as you write a congruence statement. When three or more ratios are equal, you can write an **extended proportion.** The proportion $\frac{AB}{GH} = \frac{BC}{HI} = \frac{CD}{IJ} = \frac{AD}{GJ}$ is an extended proportion.

A **scale factor** is the ratio of corresponding linear measurements of two similar figures. The ratio of the lengths of corresponding sides $\overline{BC}$ and $\overline{YZ}$, or more simply stated, the ratio of corresponding sides, is $\frac{BC}{YZ} = \frac{20}{8} = \frac{5}{2}$. So the scale factor of $\triangle ABC$ to $\triangle XYZ$ is $\frac{5}{2}$ or 5 : 2.

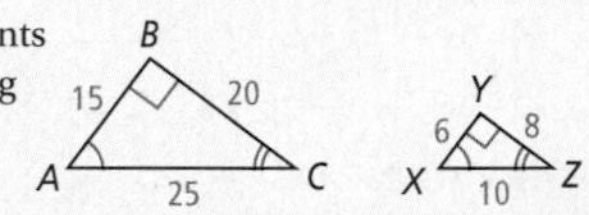

$\triangle ABC \sim \triangle XYZ$

7-2 Preparing to Teach

BIG ideas **Proportionality**
Reasoning and Proof

ESSENTIAL UNDERSTANDINGS
- Ratios and proportions can be used to decide whether two polygons are similar and to find unknown side lengths of similar figures.
- All lengths in a scale drawing are proportional to their corresponding actual lengths.

Math Background

Scale drawings and similar figures appear in many real-world situations. Similar figures are used to create scale drawings, produce reductions or enlargements of existing figures, and conduct indirect measurement. Students should be comfortable with writing and simplifying ratios and using the Cross Products Property. Students will use these skills to solve problems involving unknown side lengths and to prove that figures are similar. Students should understand that proving figures similar is the necessary first step before using proportions to find missing dimensions. Any proportion can be set up in more than one correct way—and more than one incorrect way. Students may need extra guidance to be sure their ratios are written correctly.

Mathematical Practices

Attend to precision. Using a knowledge of extended proportions, students will define and make explicit use of the term "scale factor" between similar polygons. They will also solve situations involving scale factor.

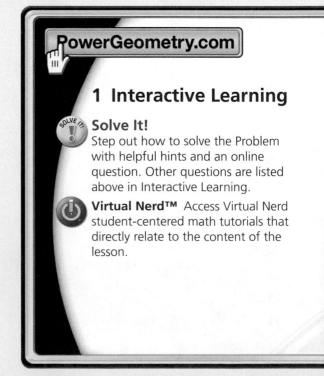

PowerGeometry.com

1 Interactive Learning

Solve It!
Step out how to solve the Problem with helpful hints and an online question. Other questions are listed above in Interactive Learning.

Virtual Nerd™ Access Virtual Nerd student-centered math tutorials that directly relate to the content of the lesson.

Problem 1 Understanding Similarity

$\triangle MNP \sim \triangle SRT$

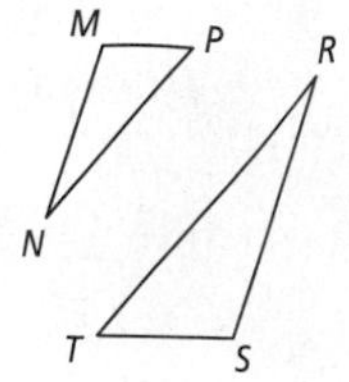

A **What are the pairs of congruent angles?**

$\angle M \cong \angle S$, $\angle N \cong \angle R$, and $\angle P \cong \angle T$

B **What is the extended proportion for the ratios of corresponding sides?**

$\frac{MN}{SR} = \frac{NP}{RT} = \frac{MP}{ST}$

Think
How can you use the similarity statement to write ratios of corresponding sides?
Use the order of the sides in the similarity statement. $\overline{MN}$ corresponds to $\overline{SR}$, so $\frac{MN}{SR}$ is a ratio of corresponding sides.

Got It? 1. $DEFG \sim HJKL$.
a. What are the pairs of congruent angles?
b. What is the extended proportion for the ratios of the lengths of corresponding sides?

Problem 2 Determining Similarity

Are the polygons similar? If they are, write a similarity statement and give the scale factor.

A ***JKLM*** **and** ***TUVW***

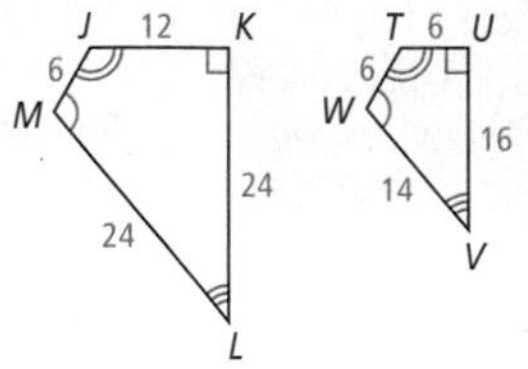

Step 1 Identify pairs of congruent angles.

$\angle J \cong \angle T$, $\angle K \cong \angle U$, $\angle L \cong \angle V$, and $\angle M \cong \angle W$

Step 2 Compare the ratios of corresponding sides.

$\frac{JK}{TU} = \frac{12}{6} = \frac{2}{1}$ $\frac{KL}{UV} = \frac{24}{16} = \frac{3}{2}$

$\frac{LM}{VW} = \frac{24}{14} = \frac{12}{7}$ $\frac{JM}{TW} = \frac{6}{6} = \frac{1}{1}$

Corresponding sides are not proportional, so the polygons are not similar.

Think
How do you identify corresponding sides?
The included side between a pair of angles of one polygon corresponds to the included side between the corresponding pair of congruent angles of another polygon.

B $\triangle ABC$ **and** $\triangle EFD$

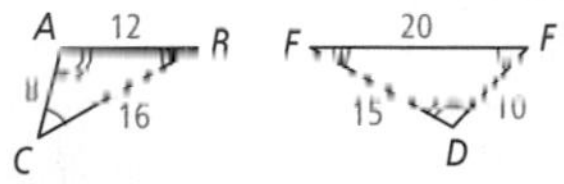

Step 1 Identify pairs of congruent angles.

$\angle A \cong \angle D$, $\angle B \cong \angle E$, and $\angle C \cong \angle F$

Step 2 Compare the ratios of corresponding sides.

$\frac{AB}{DE} = \frac{12}{15} = \frac{4}{5}$ $\frac{BC}{EF} = \frac{16}{20} = \frac{4}{5}$ $\frac{AC}{DF} = \frac{8}{10} = \frac{4}{5}$

Yes; $\triangle ABC \sim \triangle DEF$ and the scale factor is $\frac{4}{5}$ or 4 : 5.

Got It? 2. Are the polygons similar? If they are, write a similarity statement and give the scale factor.

a.
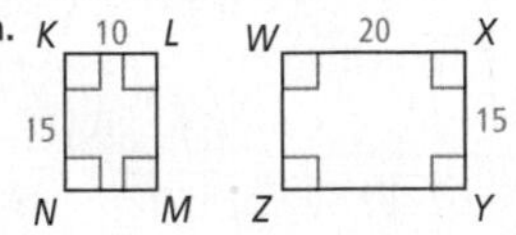

b.
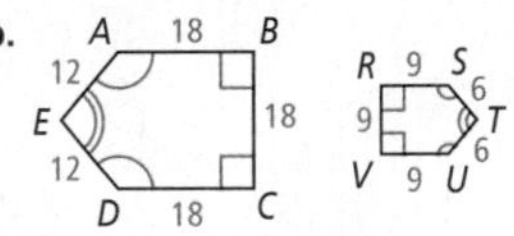

Problem 1

Q How do you know which angles are corresponding? **[The similarity statement shows the order in which the vertices correspond.]**

Q What is another way to write the extended ratio of corresponding sides? Explain. **[$\frac{RS}{NM} = \frac{RT}{NP} = \frac{ST}{MP}$; the sides of the larger triangle are in the numerators of these ratios.]**

Got It? ERROR PREVENTION

Remind students that there will be four pairs of corresponding sides and corresponding angles, because the figures are quadrilaterals.

Problem 2

Use ratios to write equations that show the relationship between two quantities.

Q In 2A, for the figures to be similar, what angle must be congruent to $\angle L$? **[$\angle V$]**

Q For 2B, what triangle is similar to $\triangle CAB$? **[$\triangle FDE$]**

Q What other scale factor may be used in 2B? Explain. **[A scale factor of 5 : 4 may be used if the larger triangle is listed first in the similarity statement.]**

Got It? VISUAL LEARNERS

Remind students to determine which sides are proportional before writing the similarity statement. The vertices should be listed in corresponding order.

2 Guided Instruction

Each Problem is worked out and supported online.

Problem 1
Understanding Similarity
Animated

Problem 2
Determining Similarity
Animated

Problem 3
Using Similar Polygons
Animated

Problem 4
Using Similarity

Problem 5
Using a Scale Drawing

Support in Geometry Companion
- Vocabulary
- Key Concepts
- Got It?

Answers

Solve It!

The letterbox screen has the same ratio of width to height as the movie screen, 9 : 5 in both cases.

Got It?

1a. $\angle D \cong \angle H$, $\angle E \cong \angle J$, $\angle F \cong \angle K$, $\angle G \cong \angle L$

b. $\frac{DE}{HJ} = \frac{EF}{JK} = \frac{FG}{KL} = \frac{GD}{LH}$

2a. not similar

b. $ABCDE \sim SRVUT$ or $ABCDE \sim UVRST$; 2 : 1

Problem 3

Q What four sides should be included in the proportion to solve for *x* and why? **[*FG, BC, ED, AD*; The unknown segment length *x* is the length of $\overline{FG}$, which corresponds to $\overline{BC}$. The only known corresponding side lengths are *ED* and *AD*.]**

Q Can you check that your answer is correct? Explain. **[Yes; find the ratio of *FG* to *BC* and check that it is equivalent to the ratio of *ED* to *AD*.]**

Got It?

Point out to students there are two ways to find the value of *y*: Set up a proportion as in Problem 3 or multiply the scale factor by the corresponding side length.

Problem 4

Q Which happens if the width of the poster is maximized instead of the height? **[The new poster would have a height of 57.6 in. (4.8 ft), which would not fit the wall height of 48 in. (4 ft).]**

Got It?

Q By what scale factor can the poster be enlarged? **[4.8]**

Plan

Can you rely on the diagram alone to set up the proportion? No, you need to use the similarity statement to identify corresponding sides in order to write ratios that are equal.

Problem 3 Using Similar Polygons

Algebra ***ABCD* ~ *EFGD*. What is the value of *x*?**

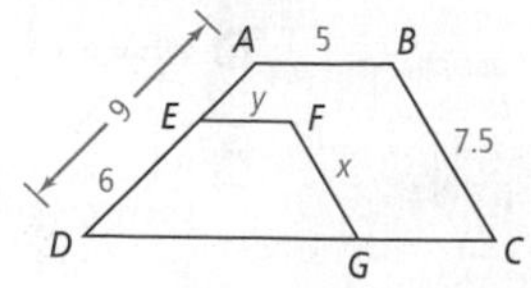

Ⓐ 4.5 Ⓒ 7.2

Ⓑ 5 Ⓓ 11.25

$\frac{FG}{BC} = \frac{ED}{AD}$ Corresponding sides of similar polygons are proportional.

$\frac{x}{7.5} = \frac{6}{9}$ Substitute.

$9x = 45$ Cross Products Property

$x = 5$ Divide each side by 9.

The value of *x* is 5. The correct answer is B.

Got It? 3. Use the diagram in Problem 3. What is the value of *y*?

Think

You can't solve the problem until you know which dimension fills the space first.

Problem 4 Using Similarity

Design Your class is making a rectangular poster for a rally. The poster's design is 6 in. high by 10 in. wide. The space allowed for the poster is 4 ft high by 8 ft wide. What are the dimensions of the largest poster that will fit in the space?

Step 1 Determine whether the height or width will fill the space first.

Height: 4 ft = 48 in. Width: 8 ft = 96 in.

48 in. ÷ 6 in. = 8 96 in. ÷ 10 in. = 9.6

The design can be enlarged at most 8 times.

Step 2 The greatest height is 48 in., so find the width.

$\frac{6}{48} = \frac{10}{x}$ Corresponding sides of similar polygons are proportional.

$6x = 480$ Cross Products Property

$x = 80$ Divide each side by 6.

The largest poster is 48 in. by 80 in. or 4 ft by $6\frac{2}{3}$ ft.

Got It? 4. Use the same poster design in Problem 4. What are the dimensions of the largest complete poster that will fit in a space 3 ft high by 4 ft wide?

Additional Problems

1. △*QRS* ~ △*DEF*.
   a. What are the pairs of congruent angles?
   b. What is the extended proportion for the ratios of corresponding side lengths?

   ANSWER a. ∠*Q* ≅ ∠*D*, ∠*R* ≅ ∠*E*, ∠*S* ≅ ∠*F*
   b. $\frac{QR}{DE} = \frac{RS}{EF} = \frac{QS}{DF}$

2. Are the polygons similar? If they are, write a similarity statement and give the scale factor.
   a. *ABC* and *FED*

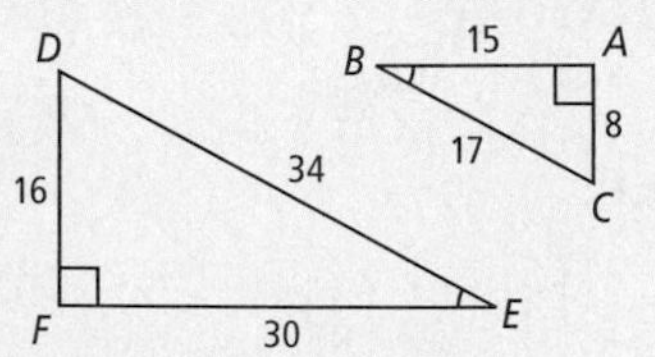

   b. *LMNO* and *QRST*

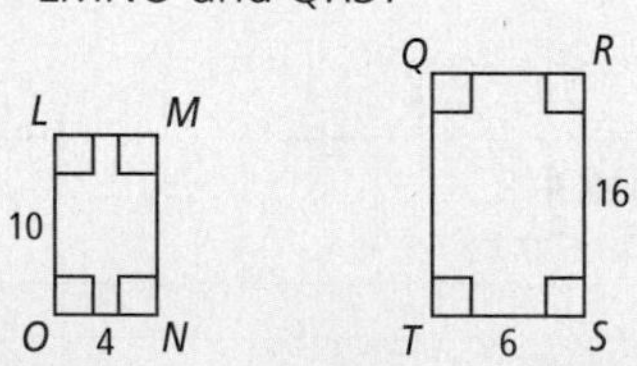

   ANSWER a. yes; △*ABC* ~ △*FED*, scale factor 1 : 2 **b.** not similar

3. *WXYZ* ~ *RSTZ*. What is the value of *x*?

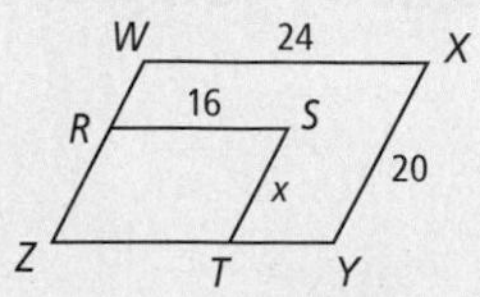

   A. 12 **C.** 14
   B. $13\frac{1}{3}$ **D.** $14\frac{1}{2}$

   ANSWER B

4. Jan uses an overhead projector to enlarge a picture 5 in. high and 7 in. wide. She projects the picture on a blackboard 4 ft 2 in. high and 12 ft wide. What are the dimensions of the largest picture that can be projected on the blackboard?

   ANSWER 4 ft 2 in. by 5 ft 10 in.

In a **scale drawing,** all lengths are proportional to their corresponding actual lengths. The **scale** is the ratio that compares each length in the scale drawing to the actual length. The lengths used in a scale can be in different units. For example, a scale might be written as 1 cm to 50 km, 1 in. = 100 mi, or 1 in. : 10 ft.

You can use proportions to find the actual dimensions represented in a scale drawing.

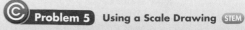

Problem 5 Using a Scale Drawing STEM

Design **The diagram shows a scale drawing of the Golden Gate Bridge in San Francisco. The distance between the two towers is the main span. What is the actual length of the main span of the bridge?**

Scale: 1 cm = 200 m

Think

Why is it helpful to use a scale in different units?
1 cm : 200 m in the same units would be 1 cm : 20,000 cm. When solving the problem, $\frac{1}{200}$ is easier to work with than $\frac{1}{20{,}000}$.

The length of the main span in the scale drawing is 6.4 cm. Let *s* represent the main span of the bridge. Use the scale to set up a proportion.

$\frac{1}{200} = \frac{6.4}{s}$ $\frac{\text{length in drawing (cm)}}{\text{actual length (m)}}$

$s = 1280$ Cross Products Property

The actual length of the main span of the bridge is 1280 m.

Got It? **5. a.** Use the scale drawing in Problem 5. What is the actual height of the towers above the roadway?

b. Reasoning The Space Needle in Seattle is 605 ft tall. A classmate wants to make a scale drawing of the Space Needle on an $8\frac{1}{2}$ in.-by-11 in. sheet of paper. He decides to use the scale 1 in. = 50 ft. Is this a reasonable scale? Explain.

PowerGeometry.com | Lesson 7-2 Similar Polygons | 440

Problem 5

Q How can the scale factor in the drawing be written as a ratio? **[$\frac{1\text{ cm}}{200\text{ m}}$]**

Q How does this ratio compare to the ratios used in Lesson 7-1? How do they differ? **[In Lesson 7-1, the units in the numerator and denominator were the same. In this ratio, the units are different.]**

Q How are units used when writing a proportion for this problem? **[Both ratios have the same units in the numerators and the same units in the denominators.]**

Q Is there another method you can use to solve this problem? Explain. **[Yes, you could multiply the length on the drawing by the scale factor of the drawing.]**

Got It?

For 5b, to solve the problem, students must decide whether to scale the height of the Space Needle to the width of the paper or to its length. Have students calculate the scale using first the length, then the width.

Q If the drawing is placed on the paper so that the height of the Space Needle fills the $8\frac{1}{2}$ in. of the width of the paper, what is the maximum height, drawn to scale, that will fit on the page? **[425 ft]**

Q If the drawing is placed on the paper so that the height of the Space Needle fills the 11 in. length of the paper, what is the maximum height, drawn to scale, that will fit on the page? **[550 ft]**

Answers

Got It? (continued)

3. $\frac{10}{3}$

4. 28.8 in. high by 48 in. wide

5a. Using 0.8 cm as the height of the towers, then $\frac{1}{200} = \frac{0.8}{h}$ and $h = 160$ m.

b. No; using a scale of 1 in. = 50 ft, the paper must be more than 12 in. long.

3 Lesson Check

Do you know HOW?

- If students have difficulty with Exercise 3, then have them write and simplify the ratios of corresponding sides.

Do you UNDERSTAND?

- If students have difficulty with Exercise 7, then have them draw examples of each property.

Close

Q What two conditions must be true for similar figures? **[Corresponding angles are congruent, and corresponding sides are proportional.]**

Q How can you use the scale factor between two similar figures to find an unknown side length? **[Set up a proportion between the scale factor and the corresponding side lengths or multiply the known side length by the scale factor.]**

Lesson Check

Do you know HOW?

JDRT ~ *WHYX*. Complete each statement.

1. $\angle D \cong$?

2. $\frac{RT}{YX} = \frac{■}{WX}$

3. Are the polygons similar? If they are, write a similarity statement and give the scale factor.

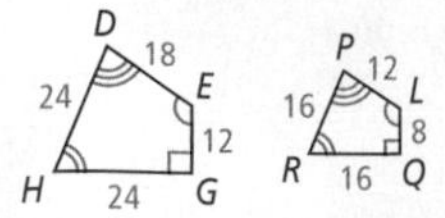

4. $\triangle FGH \sim \triangle MNP$. What is the value of x?

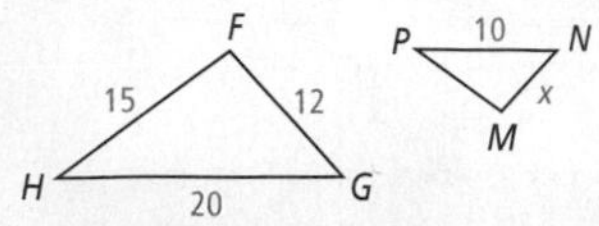

Do you UNDERSTAND? MATHEMATICAL PRACTICES

5. **Vocabulary** What does the scale on a scale drawing indicate?

6. **Error Analysis** The polygons at the right are similar. Which similarity statement is *not* correct? Explain.

   A. *TRUV* ~ *NPQU*
   B. *RUVT* ~ *QUNP*

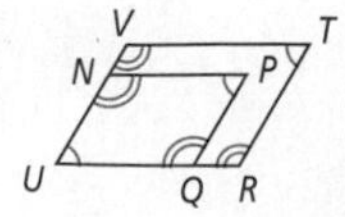

7. **Reasoning** Is similarity reflexive? Transitive? Symmetric? Justify your reasoning.

8. The triangles at the right are similar. What are three similarity statements for the triangles?

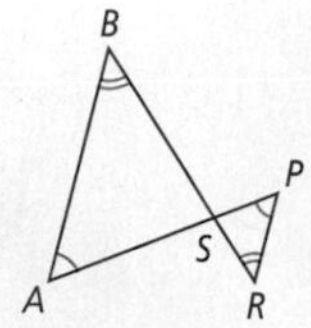

Practice and Problem-Solving Exercises MATHEMATICAL PRACTICES

A Practice

List the pairs of congruent angles and the extended proportion that relates the corresponding sides for the similar polygons.

See Problem 1.

9. *RSTV* ~ *DEFG*

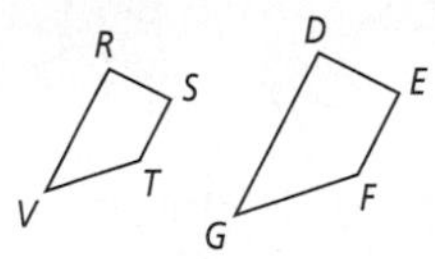

10. $\triangle CAB \sim \triangle WVT$

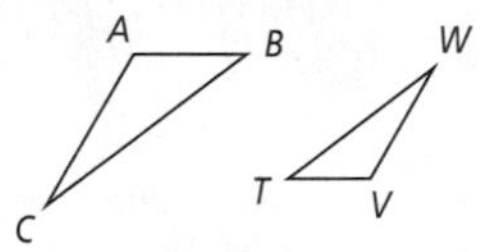

11. *KLMNP* ~ *HGFDC*

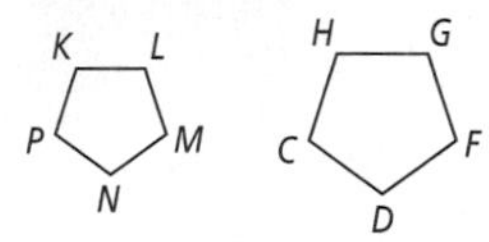

Determine whether the polygons are similar. If so, write a similarity statement and give the scale factor. If not, explain.

See Problem 2.

12.

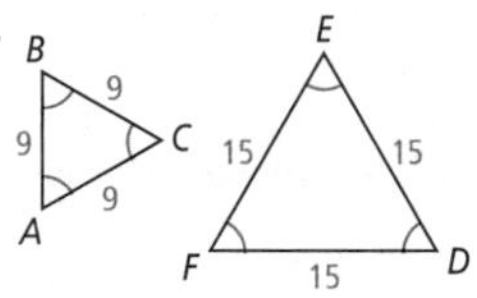

13.

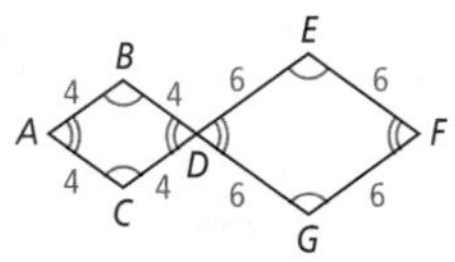

14.

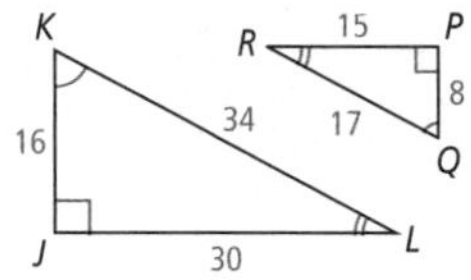

PowerGeometry.com

3 Lesson Check

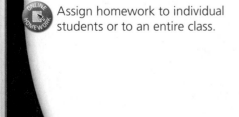

For a digital lesson check, use the Got It questions.

Support in Geometry Companion

- Lesson Check

4 Practice

Assign homework to individual students or to an entire class.

Answers

Lesson Check

1. $\angle H$
2. *JT*
3. yes; *DEGH* ~ *PLQR*; 3 : 2
4. 6
5. Answers may vary. Sample: The scale indicates how many units of length of the actual object are represented by each unit of length in the drawing.
6. A is incorrect. Sample explanation: In the diagram, $\angle T$ corresp. to $\angle P$ (or to $\angle U$), but in the similarity statement *TRUV* ~ *NPQV*, $\angle T$ corresp. to $\angle N$.
7. Every figure is ~ to itself, so similarity is reflexive. If figure 1 ~ figure 2 and figure 2 ~ figure 3, then figure 1 ~ figure 3, so similarity is transitive. If figure 1 ~ figure 2, then figure 2 ~ figure 1, so the similarity is symmetric.
8. any three of the following: $\triangle ABS \sim \triangle PRS$, $\triangle ASB \sim \triangle PSR$, $\triangle SAB \sim \triangle SPR$, $\triangle SBA \sim \triangle SRP$, $\triangle BAS \sim \triangle RPS$, $\triangle BSA \sim \triangle RSP$

Practice and Problem-Solving Exercises

9. $\angle R \cong \angle D$, $\angle S \cong \angle E$, $\angle T \cong \angle F$, $\angle V \cong \angle G$; $\frac{RS}{DE} = \frac{ST}{EF} = \frac{TV}{FG} = \frac{VR}{GD}$
10. $\angle C \cong \angle W$, $\angle A \cong \angle V$, $\angle B \cong \angle T$; $\frac{CA}{WV} = \frac{AB}{VT} = \frac{BC}{TW}$
11. $\angle K \cong \angle H$, $\angle L \cong \angle G$, $\angle M \cong \angle F$, $\angle N \cong \angle D$, $\angle P \cong \angle C$; $\frac{KL}{HG} = \frac{LM}{GF} = \frac{MN}{FD} = \frac{NP}{DC} = \frac{PK}{CH}$
12. $\triangle ABC \sim \triangle DEF$ (in any order); scale factor is 3 : 5.
13. *ABDC* ~ *FEDG* (or *ABDC* ~ *FGDE*, *ABDC* ~ *DEFG*, *ABDC* ~ *DGFE*); scale factor is 2 : 3.
14. $\triangle JKL \sim \triangle PQR$; scale factor is 2 : 1.

15. 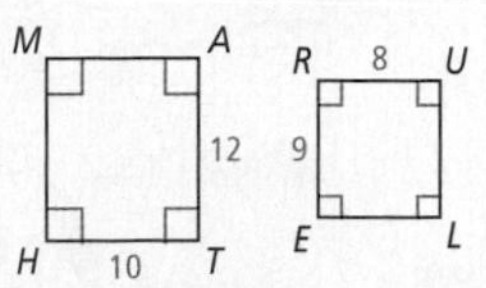

16. 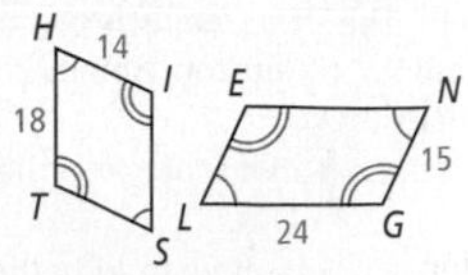

17. 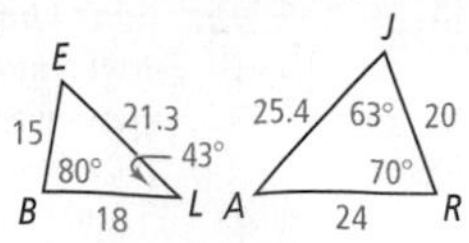

Algebra **The polygons are similar. Find the value of each variable.** See Problem 3.

18. 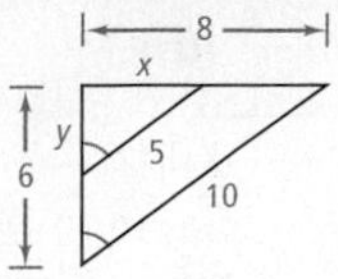

19. 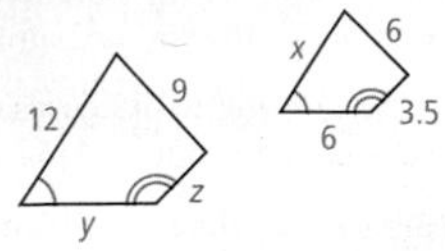

20. 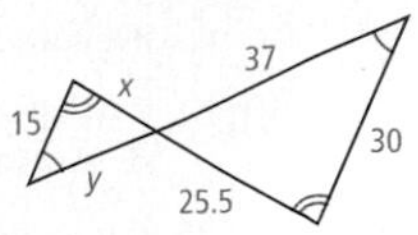

STEM **21. Web Page Design** The space allowed for the mascot on a school's Web page is 120 pixels wide by 90 pixels high. Its digital image is 500 pixels wide by 375 pixels high. What is the largest image of the mascot that will fit on the Web page? See Problem 4.

22. Art The design for a mural is 16 in. wide and 9 in. high. What are the dimensions of the largest possible complete mural that can be painted on a wall 24 ft wide by 14 ft high?

STEM **23. Architecture** You want to make a scale drawing of New York City's Empire State Building using the scale 1 in. = 250 ft. If the building is 1250 ft tall, how tall should you make the building in your scale drawing? See Problem 5.

24. Cartography A cartographer is making a map of Pennsylvania. She uses the scale 1 in. = 10 mi. The actual distance between Harrisburg and Philadelphia is about 95 mi. How far apart should she place the two cities on the map?

B Apply

In the diagram below, $\triangle DFG \sim \triangle HKM$. Find each of the following.

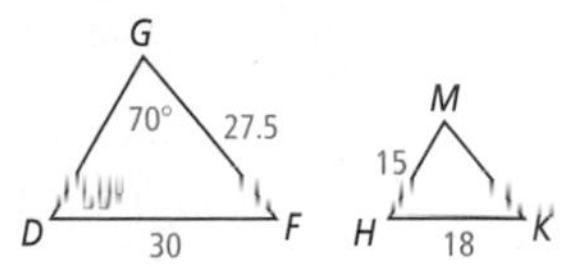

25. the scale factor of $\triangle HKM$ to $\triangle DFG$

26. $m\angle K$

27. $\frac{GD}{MH}$

28. MK

29. GD

30. Flags A company produces a standard-size U.S. flag that is 3 ft by 5 ft. The company also produces a giant-size flag that is similar to the standard-size flag. If the shorter side of the giant-size flag is 36 ft, what is the length of its longer side?

31. a. Coordinate Geometry What are the measures of $\angle A$, $\angle ABC$, $\angle BCD$, $\angle CDA$, $\angle E$, $\angle F$, and $\angle G$? Explain.

b. What are the lengths of $\overline{AB}$, $\overline{BC}$, $\overline{CD}$, $\overline{DA}$, $\overline{AE}$, $\overline{EF}$, $\overline{FG}$, and $\overline{AG}$?

c. Is $ABCD$ similar to $AEFG$? Justify your answer.

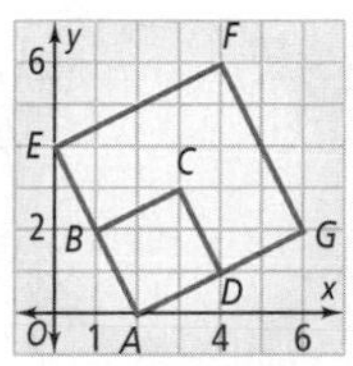

4 Practice

ASSIGNMENT GUIDE

Basic: 9–27, 32, 34, 37, 39, 47

Average: 9–23 odd, 25–47

Advanced: 9–23 odd, 25–50

Mathematical Practices are supported by exercises with red headings. Here are the Practices supported in this lesson:

MP 1: Make Sense of Problems Ex. 32, 36

MP 3: Construct Arguments Ex. 7, 33, 34

MP 3: Critique the Reasoning of Others Ex. 6

Applications exercises have blue headings. Exercises 21–24, 30, and 47 support MP 4: Model.

STEM exercises focus on science or engineering applications.

EXERCISE 39: Use the Think About a Plan worksheet in the **Practice and Problem Solving Workbook** (also available in the Teaching Resources in print and online) to further support students' development in becoming independent learners.

HOMEWORK QUICK CHECK

To check students' understanding of key skills and concepts, go over Exercises 13, 23, 32, 34, and 39.

15. Not similar; sample explanation: The ratio of the longer sides is $\frac{12}{9}$ or $\frac{4}{3}$, and the ratio of the shorter sides is $\frac{10}{8}$ or $\frac{5}{4}$. Since $\frac{4}{3} \neq \frac{5}{4}$, the corresp. sides are not proportional and the figures are not ~.

16. Not similar; sample explanation: The ratio of the longer sides is $\frac{18}{24}$ or $\frac{3}{4}$, and the ratio of the shorter sides is $\frac{14}{15}$. Since $\frac{3}{4} \neq \frac{14}{15}$, the corresp. sides are not proportional and the figures are not ~.

17. Not similar; sample explanation: The ∠ measures are not the same.

18. $x = 4$, $y = 3$

19. $x = 8$, $y = 9$, $z = 5.25$

20. $x = 12.75$, $y = 18.5$

21. 120 pixels wide by 90 pixels high

22. 288 in. wide, 162 in. high; or 24 ft wide, 13.5 ft high

23. 5 in. **24.** 9.5 in. **25.** 3 : 5

26. 51 **27.** 5 : 3 **28.** 16.5

29. 25 **30.** 60 ft

31a. The slope of $\overline{AB}$, $\overline{CD}$, $\overline{AE}$, and $\overline{FG}$ is −2. The slope of $\overline{BC}$, $\overline{AD}$, $\overline{EF}$, and $\overline{AG}$ is $\frac{1}{2}$. For each pair of consecutive sides of $ABCD$, the slopes are negative reciprocals, so $ABCD$ has four rt. ∠s. Similarly, $AEFG$ has four rt. ∠s. The measure of $\angle A$, $\angle ABC$, $\angle BCD$, $\angle CDA$, $\angle E$, $\angle F$, and $\angle G$, is 90.

b. By the Distance Formula, $AB = BC = CD = AD = \sqrt{5}$ and $AE = EF = FG = AG = 2\sqrt{5}$.

c. All the angles of $AEFG$ and $ABCD$ are ≅.
$\frac{AB}{AE} = \frac{BC}{EF} = \frac{CD}{FG} = \frac{AD}{AG} = \frac{\sqrt{5}}{2\sqrt{5}} = \frac{1}{2}$
The corresp. sides are proportional, so $AEFG \sim ABCD$.

Answers

Practice and Problem-Solving Exercises (continued)

32. The distance on the map is about 2.8 cm, so the actual distance is (2.8)(112), or about 314 km.

33. No; for polygons with more than 3 sides, you also need to know that corresp. ⦞ are ≅ in order to state that the polygons are ~.

34. Answers may vary. Sample: If two figures are ≅, then corresp. ⦞ are ≅ and corresp. sides are ≅. Therefore the ratio of each pair of corresp. sides is 1, so the sides are proportional with a scale factor of 1 : 1.

35. 1 : 3

36. Check students' work.

37. $x = 10$; 2 : 1

38. $x = 4.4$; 4 : 3

39–42. Check students' work.

43. always

44. never

45. sometimes

46. sometimes

47. 21 ft by 40 ft

48a. 24 in., 32 in.

b. Ratio of perimeters: $\frac{54}{72} = \frac{3}{4}$, scale factor is $\frac{12}{16} = \frac{3}{4}$; they are the same.

32. Think About a Plan The Davis family is planning to drive from San Antonio to Houston. About how far will they have to drive?

- How can you find the distance between the two cities on the map?
- What proportion can you set up to solve the problem?

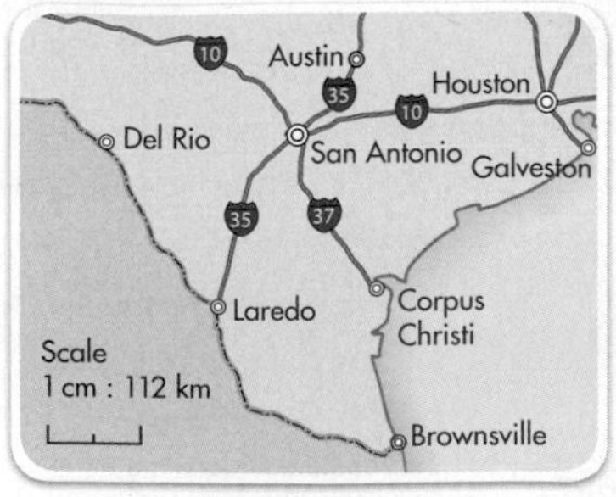

33. Reasoning Two polygons have corresponding side lengths that are proportional. Can you conclude that the polygons are similar? Justify your reasoning.

34. Writing Explain why two congruent figures must also be similar. Include scale factor in your explanation.

35. $\triangle JLK$ and $\triangle RTS$ are similar. The scale factor of $\triangle JLK$ to $\triangle RTS$ is 3 : 1. What is the scale factor of $\triangle RTS$ to $\triangle JLK$?

36. Open-Ended Draw and label two different similar quadrilaterals. Write a similarity statement for each and give the scale factor.

Algebra Find the value of x. Give the scale factor of the polygons.

37. $\triangle WLJ \sim \triangle QBV$

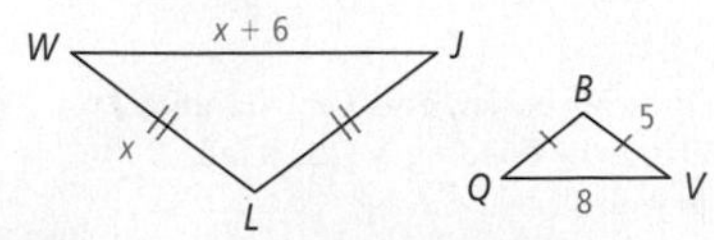

38. $GKNM \sim VRPT$

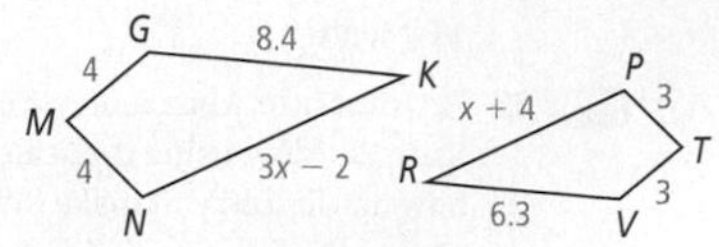

Sports Choose a scale and make a scale drawing of each rectangular playing surface.

39. A soccer field is 110 yd by 60 yd.

40. A volleyball court is 60 ft by 30 ft.

41. A tennis court is 78 ft by 36 ft.

42. A football field is 360 ft by 160 ft.

Determine whether each statement is *always, sometimes,* or *never* true.

43. Any two regular pentagons are similar.

44. A hexagon and a triangle are similar.

45. A square and a rhombus are similar.

46. Two similar rectangles are congruent.

STEM **47. Architecture** The scale drawing at the right is part of a floor plan for a home. The scale is 1 cm = 10 ft. What are the actual dimensions of the family room?

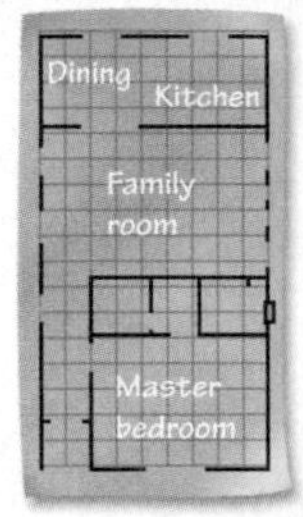

Challenge

48. The lengths of the sides of a triangle are in the extended ratio 2 : 3 : 4. The perimeter of the triangle is 54 in.

a. The length of the shortest side of a similar triangle is 16 in. What are the lengths of the other two sides of this triangle?

b. Compare the ratio of the perimeters of the two triangles to their scale factor. What do you notice?

49. In rectangle $BCEG$, $BC : CE = 2 : 3$. In rectangle $LJAW$, $LJ : JA = 2 : 3$. Show that $BCEG \sim LJAW$.

50. Prove the following statement: If $\triangle ABC \sim \triangle DEF$ and $\triangle DEF \sim \triangle GHK$, then $\triangle ABC \sim \triangle GHK$.

Apply What You've Learned

MATHEMATICAL PRACTICES MP 3

Look back at the information on page 431 about the graph Lillian wants to adjust. The screens on pages 431 are shown again below. In the Apply What You've Learned in Lesson 7-1, you determined the ratio of the width to the height of Lillian's calculator screen.

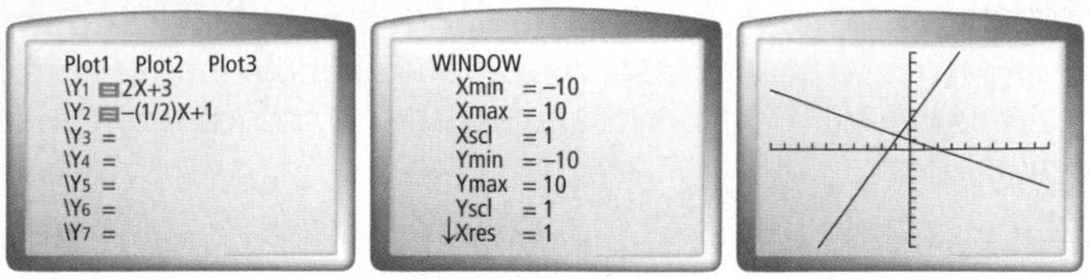

a. Consider the "viewing rectangle" determined by the values of Xmin, Xmax, Ymin, and Ymax. While the width and height of the calculator screen do not change, the width and height of the viewing rectangle depend on the values entered for Xmin, Xmax, Ymin, and Ymax. What are the width and height of the viewing rectangle for Lillian's graph shown above?

b. Is the viewing rectangle similar to Lillian's rectangular calculator screen? Explain.

c. If the intersecting lines are to be graphed without distortion, what must be true about the viewing rectangle?

d. If Lillian uses Xmin = −30 and Xmax = 30, will she be able to graph the lines without distortion? Use similarity to explain why or why not.

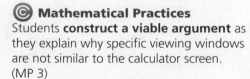

49. All ∠s in any rectangle are right ∠s, so all corresp. ∠s are ≅. The ratio of two pair of consecutive sides for each rectangle is the same. Since opposite sides of a parallelogram are equal, the other two pair of sides will also have the same ratio. So corresp. sides form equal ratios and are proportional. So $BCEG \sim LJAW$.

50. If $\triangle ABC \sim \triangle DEF$ then $\angle A \cong \angle D$, $\angle B \cong \angle E$, $\angle C \cong \angle F$, and $\frac{AB}{DE} = \frac{BC}{EF} = \frac{AC}{DF}$. If $\triangle DEF \sim \triangle GHK$, then $\angle D \cong \angle G$, $\angle E \cong \angle H$, $\angle F \cong \angle K$, and $\frac{DE}{GH} = \frac{EF}{HK} = \frac{DF}{GK}$. Using Prop. of Proportions (2), you can write $\frac{AB}{BC} = \frac{DE}{EF}$ and $\frac{AB}{AC} = \frac{DE}{DF}$ from the first extended proportion and $\frac{DE}{EF} = \frac{GH}{HK}$ and $\frac{DE}{DF} = \frac{GH}{GK}$ from the second extended proportion. Then $\frac{AB}{BC} = \frac{GH}{HK}$ and $\frac{AB}{AC} = \frac{GH}{GK}$ by the Transitive Prop. of Equality; applying Prop. of Proportions (2) again gives $\frac{AB}{GH} = \frac{BC}{HK}$ and $\frac{AB}{GH} = \frac{AC}{GK}$. Since $\angle A \cong \angle G$, $\angle B \cong \angle H$, and $\angle C \cong \angle K$ by the Transitive Prop. of ≅, and $\frac{AB}{GH} = \frac{BC}{HK} = \frac{AC}{GK}$ by the Transitive Prop. of Equality, then $\triangle ABC \sim \triangle GHK$.

Apply What You've Learned

In the Apply What You've Learned for Lesson 7-1, students used the dimensions given on page 431 to determine the ratio of the width to the height of Lillian's calculator screen. Now students consider the ratio of the width to the height for the "viewing rectangle" determined by Xmin, Xmax, Ymin, and Ymax.

Mathematical Practices

Students **construct a viable argument** as they explain why specific viewing windows are not similar to the calculator screen. (MP 3)

ANSWERS

a. Width: 20 units; height: 20 units

b. No; although corresponding angles are congruent, the lengths of corresponding sides are not proportional. The ratio of the width of the calculator screen to the width of the viewing rectangle is 0.255, and the ratio of the height of the calculator screen to the height of the viewing rectangle is 0.17.

c. The viewing rectangle must be similar to the calculator screen.

d. No; the width of the viewing rectangle will be 60 units and the height is 20 units. This rectangle is not similar to the calculator screen because the lengths of corresponding sides are still not proportional.

7-2 Lesson Resources

Differentiated Remediation

Instructional Support

Geometry Companion

Students can use the **Geometry Companion** worktext (4 pages) . . .

- New Vocabulary
- Key Concepts
- Got It for each Problem
- Lesson Check

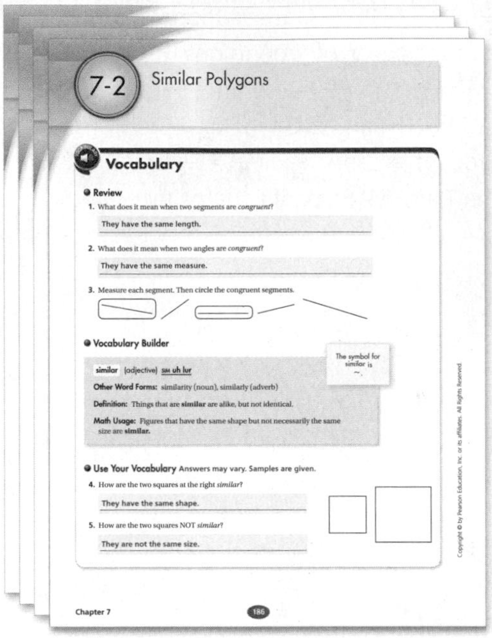

7-2 Similar Polygons

Vocabulary

Review

1. What does it mean when two segments are *congruent*?

They have the same length.

2. What does it mean when two angles are *congruent*?

They have the same measure.

3. Measure each segment. Then circle the congruent segments.

Vocabulary Builder

similar (adjective) sım uh lur

Other Word Forms: similarity (noun), similarly (adverb)

Definition: Things that are **similar** are alike, but not identical.

Math Usage: Figures that have the same shape but not necessarily the same size are **similar**.

The symbol for similar is ~

Use Your Vocabulary Answers may vary. Samples are given.

4. How are the two squares at the right *similar*?

They have the same shape.

5. How are the two squares NOT *similar*?

They are not the same size.

Chapter 7 186

ELL Support

Focus on Language Have groups of students list ways they have heard the word *similar* used. Tell each group: Use familiar words to write a definition for *similar*. Have students compare their list and definition with another group.

Say: In your group, list five examples of things that are similar. You may also draw pictures of similar things. Give examples, such as twins and two pencils, to help students get started.

Have a volunteer draw *similar* triangles on the board. Have each group write out answers to the following questions: Do the triangles fit your definition of similar? Ask: Is a large triangle similar to a small triangle?

5 Assess & Remediate

Lesson Quiz

1. $\triangle HJK \sim \triangle CND$.

  a. What are the pairs of congruent angles?

  b. What is the extended proportion for the ratios of corresponding sides?

2. Are the polygons similar? If they are, write a similarity statement and give the scale factor.

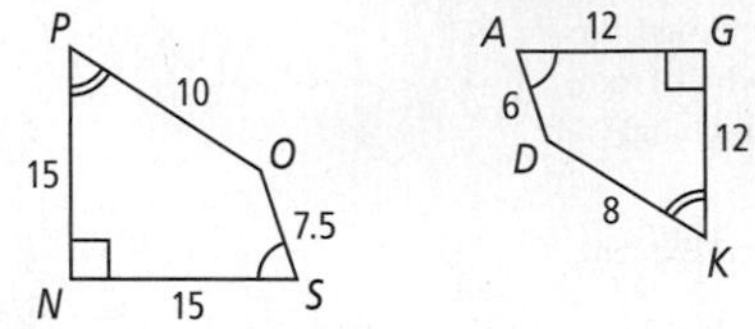

3. Do you UNDERSTAND? Harold has a photograph that is 8 in. × 10 in. that he wants to print at a reduced size to fit into a frame that is 6 in. tall. What will be the dimensions of the new photograph?

ANSWERS TO LESSON QUIZ

1. a. $\angle H \cong \angle C$, $\angle J \cong \angle N$, $\angle K \cong \angle D$;

  b. $\frac{HJ}{CN} = \frac{JK}{ND} = \frac{HK}{CD}$

2. yes; $POSN \sim KDAG$, scale factor: 5 : 4

3. 4.8 in. × 6 in.

PRESCRIPTION FOR REMEDIATION

Use the student work on the Lesson Quiz to prescribe a differentiated review assignment.

Points	Differentiated Remediation
0–1	Intervention
2	On-level
3	Extension

PowerGeometry.com

5 Assess & Remediate

Assign the Lesson Quiz. Appropriate intervention, practice, or enrichment is automatically generated based on student performance.

Intervention

- **Reteaching** (2 pages) Provides reteaching and practice exercises for the key lesson concepts. Use with struggling students or absent students.
- **English Language Learner Support** Helps students develop and reinforce mathematical vocabulary and key concepts.

All-in-One Resources/Online

Reteaching

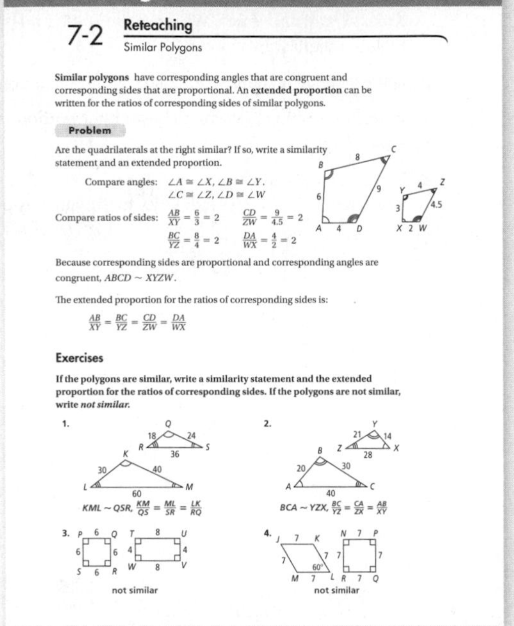

7-2 Reteaching

Similar Polygons

Similar polygons have corresponding angles that are congruent and corresponding sides that are proportional. An **extended proportion** can be written for the ratios of corresponding sides of similar polygons.

Problem

Are the quadrilaterals at the right similar? If so, write a similarity statement and an extended proportion.

Compare angles: $\angle A \cong \angle X$, $\angle B \cong \angle Y$, $\angle C \cong \angle Z$, $\angle D \cong \angle W$

Compare ratios of sides: $\frac{AB}{XY} = \frac{6}{3} = 2$, $\frac{CD}{ZW} = \frac{9}{4.5} = 2$, $\frac{BC}{YZ} = \frac{8}{4} = 2$, $\frac{DA}{WX} = \frac{4}{2} = 2$

Because corresponding sides are proportional and corresponding angles are congruent, $ABCD \sim XYZW$.

The extended proportion for the ratios of corresponding sides is:

$\frac{AB}{XY} = \frac{BC}{YZ} = \frac{CD}{ZW} = \frac{DA}{WX}$

Exercises

If the polygons are similar, write a similarity statement and the extended proportion for the ratios of corresponding sides. If the polygons are not similar, write *not similar*.

1. $KML \sim QSR$, $\frac{KM}{QS} = \frac{ML}{SR} = \frac{LK}{RQ}$

2. $BCA \sim YZX$, $\frac{BC}{YZ} = \frac{CA}{ZX} = \frac{AB}{XY}$

3. not similar

4. not similar

All-in-One Resources/Online

English Language Learner Support

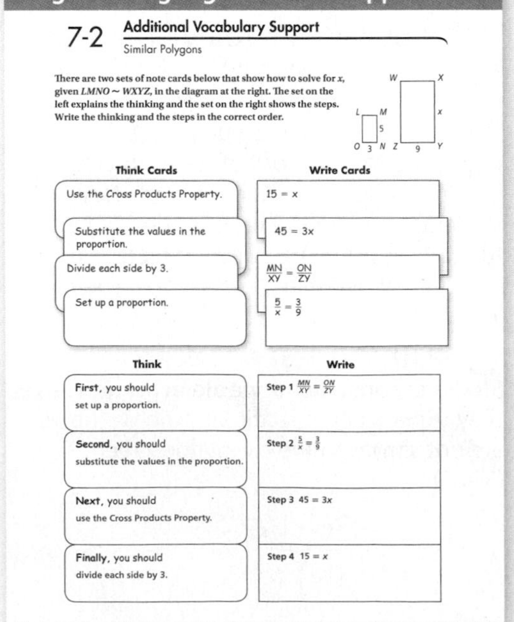

7-2 Additional Vocabulary Support

Similar Polygons

There are two sets of note cards below that show how to solve for *x*, given $LMNO \sim WXYZ$, in the diagram at the right. The set on the left explains the thinking and the set on the right shows the steps. Write the thinking and the steps in the correct order.

Think Cards	Write Cards
Use the Cross Products Property.	$15 = x$
Substitute the values in the proportion.	$45 = 3x$
Divide each side by 3.	$\frac{MN}{XY} = \frac{ON}{ZY}$
Set up a proportion.	$\frac{5}{x} = \frac{3}{9}$

Think	Write
First, you should set up a proportion.	Step 1 $\frac{MN}{XY} = \frac{ON}{ZY}$
Second, you should substitute the values in the proportion.	Step 2 $\frac{5}{x} = \frac{3}{9}$
Next, you should use the Cross Products Property.	Step 3 $45 = 3x$
Finally, you should divide each side by 3.	Step 4 $15 = x$

Differentiated Remediation *continued*

On-Level

- **Practice** (2 pages) Provides extra practice for each lesson. For simpler practice exercises, use the Form K Practice pages found in the All-in-One Teaching Resources and online.
- **Think About a Plan** Helps students develop specific problem-solving skills and strategies by providing scaffolded guiding questions.
- **Standardized Test Prep** Focuses on all major exercises, all major question types, and helps students prepare for the high-stakes assessments.

Extension

- **Enrichment** Provides students with interesting problems and activities that extend the concepts of the lesson.
- **Activities, Games, and Puzzles** Worksheets that can be used for concepts development, enrichment, and for fun!

Practice and Problem Solving Wkbk/ All-in-One Resources/Online

Practice page 1

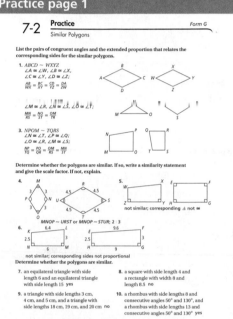

7-2 **Practice** Form G
Similar Polygons

List the pairs of congruent angles and the extended proportion that relates the corresponding sides for the similar polygons.

1. $ABCD \sim WXYZ$
$\angle A \cong \angle W$, $\angle B \cong \angle X$, $\angle C \cong \angle Y$, $\angle D \cong \angle Z$; $\frac{AB}{WX} = \frac{BC}{XY} = \frac{CD}{YZ} = \frac{DA}{ZW}$

2. $\angle M \cong \angle R$, $\angle N \cong \angle S$, $\angle O \cong \angle T$; $\frac{MN}{RS} = \frac{NO}{ST} = \frac{OM}{TR}$

3. $NPOM \sim TQRS$
$\angle N \cong \angle T$, $\angle P \cong \angle Q$; $\angle O \cong \angle R$, $\angle M \cong \angle S$; $\frac{NP}{TQ} = \frac{PO}{QR} = \frac{OM}{RS} = \frac{MN}{ST}$

Determine whether the polygons are similar. If so, write a similarity statement and give the scale factor. If not, explain.

4. $MNOP \sim URST$ or $MNOP \sim STUR$; 2 : 3
5. not similar; corresponding $\angle$ not $\cong$
6. not similar; corresponding sides not proportional

Determine whether the polygons are similar.

7. an equilateral triangle with side length 6 and an equilateral triangle with side length 15 yes
8. a square with side length 4 and a rectangle with width 8 and length 8.5 no
9. a triangle with side lengths 3 cm, 4 cm, and 5 cm, and a triangle with side lengths 18 cm, 19 cm, and 20 cm no
10. a rhombus with side lengths 8 and consecutive angles 50° and 130°, and a rhombus with side lengths 13 and consecutive angles 50° and 130° yes

Practice and Problem Solving Wkbk/ All-in-One Resources/Online

Practice page 2

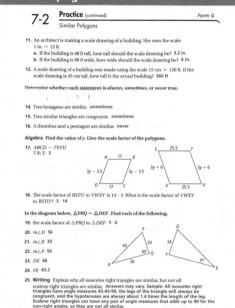

7-2 **Practice** (continued) Form G
Similar Polygons

11. An architect is making a scale drawing of a building. She uses the scale 1 in. = 15 ft.
a. If the building is 48 ft tall, how tall should the scale drawing be? 3.2 in.
b. If the building is 90 ft wide, how wide should the scale drawing be? 6 in.
12. A scale drawing of a building was made using the scale 15 cm = 120 ft. If the scale drawing is 45 cm tall, how tall is the actual building? 360 ft

Determine whether each statement is *always*, *sometimes*, or *never* true.

14. Two hexagons are similar. sometimes
15. Two similar triangles are congruent. sometimes
16. A rhombus and a pentagon are similar. never

Algebra Find the value of *y*. Give the scale factor of the polygons.

17. $ABCD \sim TSVU$ 7.5; 2 : 3
18. The scale factor of $RSTU$ to $VWXY$ is 14 : 3. What is the scale factor of $VWXY$ to $RSTU$? 3 : 14

In the diagram below, $\triangle PRQ \sim \triangle DEF$. Find each of the following.

19. the scale factor of $\triangle PRQ$ to $\triangle DEF$ 5 : 6
20. $m\angle D$ 56
21. $m\angle R$ 35
22. $m\angle P$ 56
23. DE 48
24. FE 43.2
25. **Writing** Explain why all isosceles right triangles are similar, but not all scalene right triangles are similar. Answers may vary. Sample: All isosceles right triangles have angle measures 45-45-90, the legs of the triangle will always be congruent, and the hypotenuses are always about 1.4 times the length of the leg. Scalene right triangles can have any pair of angle measures that adds up to 90 for the non-right angles, so they are not all similar.

All-in-One Resources/Online

Enrichment

7-2 **Enrichment**
Similar Polygons

Floor Plans

Architects, engineers, and other professionals make scale drawings to design or present building plans. A floor plan of the second floor of a house is shown below. Use the scale to find the actual dimensions of each room.

1. playroom 18 ft by 10 ft
2. library 18 ft by 14 ft
3. master bedroom 18 ft by 16 ft
5. closet 3 ft by 10 ft

Scale: 1 in. = 16 ft

Someone who wants to rearrange a room can make use of a scale drawing of the room that includes furniture. Two-dimensional shapes can represent the objects that sit on the floor in the room.

Make a scale drawing of a room in which you spend a lot of time, such as your classroom or bedroom, including any objects that take up floor space.

6. Choose an appropriate scale so the drawing covers most of an 8.5 in.-by-11 in. piece of paper. What scale did you choose? Answers may vary. Sample: 1 in. = 3 ft
7. What shape is the room? Measure the dimensions of the room and draw the shape to represent the room's outline. Answers may vary. Sample: rectangle; 15 ft by 24 ft
8. List three objects that take up floor space. Measure the dimensions of each object, then determine their dimensions in the scale drawing. You can round to the nearest millimeter or quarter of an inch.

Object	Actual Dimensions	Scale Factor	Dimensions on Drawing
Sample: table	Sample: 4 ft by 8 ft	192 : 1	Sample: 0.25 in. by 0.50 in.

9. Complete the scale drawing. Remember to measure the distance between objects so that this is accurately represented in the drawing. Check students' work.

Practice and Problem Solving Wkbk/ All-in-One Resources/Online

Think About a Plan

7-2 **Think About a Plan**
Similar Polygons

Sports Choose a scale and make a scale drawing of a rectangular soccer field that is 110 yd by 60 yd.

1. What is a scale drawing? How does a figure in a scale drawing relate to an actual figure?
Answers may vary. Sample: A scale drawing is enlarged or reduced proportionally to the actual figure. A figure in a scale drawing and the actual figure are similar figures.
2. What is a scale? What will the scale of your drawing compare? Write a ratio to represent this.
Answers may vary. Sample: a ratio of the actual size to the size in the drawing; the soccer field's actual length to the length in the drawing; actual length : length of drawing
3. To select a scale you need to choose a unit for the drawing. Assuming you are going to make your drawing on a typical sheet of paper, which customary unit of length should you use? inches
4. You have to choose how many yards each unit you chose in Step 3 will represent. The soccer field is 110 yd long. What is the least number of yards each unit can represent and still fit on an 8.5 in.-by-11 in. sheet of paper? Explain. Does this scale make sense for your scale drawing?
The least number of yards each inch can represent is 10 yd. If the scale is 1 in. = 10 yd, the scale drawing will be 11 in. long, which is the length of the paper. It might make sense to use a scale that makes the drawing smaller.
5. Choose the scale of your drawing. Answers may vary. Sample: 1 in. = 20 yd
6. How can you use the scale to write a proportion to find the length of the field in the scale drawing? Write and solve a proportion to find the length of the soccer field in the scale drawing.
Answers may vary. Sample: Make a proportion using the actual length of the soccer field, the length in the drawing, and the scale factor. $\frac{110 \text{ yd}}{\ell \text{ in.}} = \frac{20 \text{ yd}}{1 \text{ in.}}$; 5.5 in.
7. Write and solve a proportion to find the width of the soccer field in the scale drawing. Answers may vary. Sample: $\frac{60 \text{ yd}}{w \text{ in.}} = \frac{20 \text{ yd}}{1 \text{ in.}}$; 3 in.
8. Use a ruler to create the scale drawing on a separate piece of paper. Check students' work.

Practice and Problem Solving Wkbk/ All-in-One Resources/Online

Standardized Test Prep

7-2 **Standardized Test Prep**
Similar Polygons

Multiple Choice

For Exercises 1–5, choose the correct letter.

1. You make a scale drawing of a tree using the scale 5 in. = 27 ft. If the tree is 67.5 ft tall, how tall is the scale drawing? D
(A) 10 in. (B) 11.5 in. (C) 12 in. (D) 12.5 in.
2. You make a scale drawing of a garden plot using the scale 2 in. = 17 ft. If the length of a row of vegetables on the drawing is 3 in., how long is the actual row? G
(F) 17 ft (G) 25.5 ft (H) 34 ft (I) 42.5 ft
3. The scale factor of $\triangle RST$ to $\triangle DEC$ is 3 : 13. What is the scale factor of $\triangle DEC$ to $\triangle RST$? D
(A) 3 : 13 (B) 1 : 39 (C) 39 : 1 (D) 13 : 3
4. $\triangle ACB \sim \triangle FED$. What is the value of *x*? I
(F) 4 (G) 4.2 (H) 4.5 (I) 5
5. $MNOP \sim QRST$ with a scale factor of 5 : 4. MP = 85 mm. What is the value of QT? B
(A) 60 mm (B) 68 mm (C) 84 mm (D) 106.25 mm

Short Response

6. Are the triangles at the right similar? Explain.
[2] Yes; corresponding angles are congruent and lengths of corresponding sides are proportional. [1] recognition that corresponding angles are congruent or corresponding side lengths are proportional. [0] No explanation given.

Online Teacher Resource Center

Activities, Games, and Puzzles

7-2 **Activity: Similarity Investigation**
Similar Polygons

Construct

Construct parallelogram $ABCD$ whose diagonals intersect at E. Measure its sides and angles. Construct the midpoints of $\overline{AE}$, $\overline{BE}$, $\overline{CE}$, and $\overline{DE}$ called P, Q, R, and S, respectively.

Construct a quadrilateral $PQRS$ and measure its sides and angles.

Investigate

1. Drag the vertices of $ABCD$ and observe the effect on $PQRS$. Classify $PQRS$ as specifically as possible. parallelogram
2. Explain why this classification holds. The sides of $PQRS$ are mid segments, and therefore parallel, to the sides of $ABCD$.
3. Comparing corresponding angles and sides, verify that $ABCD$ and $PQRS$ are similar. corresponding angles are congruent; corresponding sides are proportional
4. Find the similarity ratio. Without measuring, make a conjecture about the ratio of the areas of the two figures. Test your conjecture by measuring. 2 : 1, 4 : 1
5. Do you think the results in Exercises 1 through 4 are true for quadrilaterals that are not parallelograms? Test your answer by constructing and measuring. yes

Extend

Step 1
Construct quadrilateral $ABCD$ with point V not on $ABCD$. Construct line segments from each vertex of $ABCD$ to V. Construct the midpoints of $\overline{AV}$, $\overline{BV}$, $\overline{CV}$, and $\overline{DV}$, called M, N, O, and P, respectively. Point V is called the vanishing point.

Step 2
Draw segments connecting the corresponding vertices of $ABCD$ to $MNOP$. Then hide the segments joining the vertices of $ABCD$ to V. The three dimensional object that results is an example of a drawing in one point perspective.

6. Measure the sides and angles of $ABCD$ and $MNOP$ and verify that the quadrilaterals are similar. corresponding sides are in a ratio of 2 : 1; corresponding angles are congruent

Guided Instruction

PURPOSE To introduce students to fractals and their properties

PROCESS Students will
- draw fractals by iteration.
- draw a Koch Curve.
- draw a Koch Snowflake.

DISCUSS The examples and exercises focus on drawing fractals and portions of fractals.

Q Where might you see fractals in everyday life? **[Answers will vary. Samples: crystals, rivers, broccoli, blood vessels, lightning, mountain ranges, coastlines, clouds, bark]**

Example 1

This Example demonstrates the steps to drawing a fractal tree.

Q What geometric figure is the beginning or Stage 0 of this fractal? **[a line segment with length 1 unit]**

Q Because the lengths of the segments in Stage 1 are $\frac{1}{3}$ unit, and those of Stage 2 are $\frac{1}{9}$ unit, what will be the length of the segments in Stage 3? **[$\frac{1}{27}$ unit]**

Example 2

The Koch Curve is a famous fractal made in 1904 by Swedish mathematician Helge von Koch. The Koch Curve is used to model coastlines.

Q Does the total length of all the segments in the Koch Curve fractal stay the same, get shorter, or get longer as you proceed incrementally from stage to stage? **[It gets longer each time.]**

Mathematical Practices This Concept Byte supports students in becoming proficient in looking for patterns, Mathematical Practice 7.

Concept Byte — Fractals

Use With Lesson 7-2

EXTENSION

Common Core State Standards

Extends G-SRT.B.5 Use . . . similarity criteria for triangles to solve problems and to prove relationships in geometric figures.

MP 7

Fractals are objects that have three important properties:

- You can form fractals by repeating steps. This process is called *iteration*.
- They require infinitely many iterations. In practice, you can continue until the objects become too small to draw. Even then the steps could continue in your mind.
- At each stage, a portion of the object is a reduced copy of the entire object at the previous stage. This property is called *self-similarity*.

Example 1

The segment at the right of length 1 unit is Stage 0 of a fractal tree. Draw Stage 1 and Stage 2 of the tree. For each stage, draw two branches from the top third of each segment.

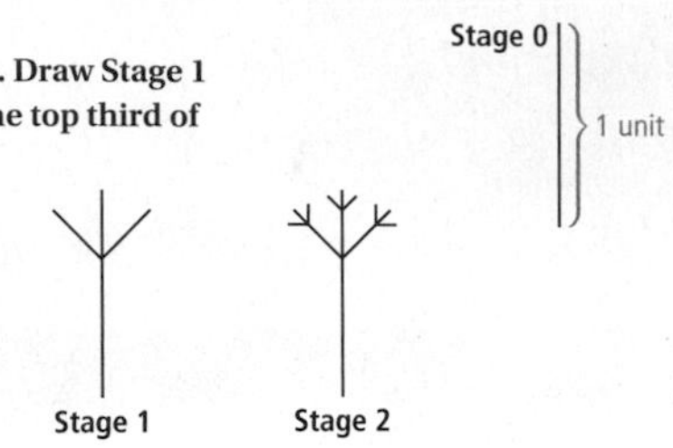

- To draw Stage 1, find the point that is $\frac{1}{3}$ unit from the top of the segment. From this point, draw two segments of length $\frac{1}{3}$ unit.
- To draw Stage 2, find the point that is $\frac{1}{9}$ unit from the top of each branch of Stage 1. From each of these points, draw two segments of length $\frac{1}{9}$ unit. The length of each new branch is $\frac{1}{3}$ of $\frac{1}{3}$ unit which is $\frac{1}{9}$ unit.

Amazingly, some fractals are used to describe natural formations such as mountain ranges and clouds. In 1904, Swedish mathematician Helge von Koch made the Koch Curve, a fractal that is used to model coastlines.

Example 2

The segment at the right of length 1 unit is Stage 0 of a Koch Curve. Draw Stages 1–4 of the curve. For each stage, replace the middle third of each segment with two segments, both equal in length to the middle third.

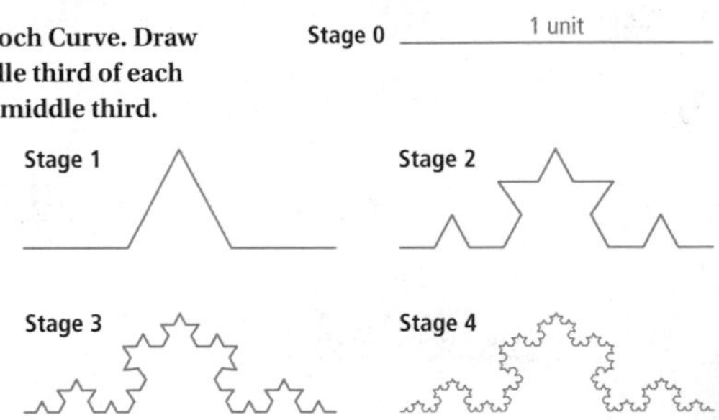

- For Stage 1, replace the middle third with two segments that are each $\frac{1}{3}$ unit long.
- For Stage 2, replace the middle third of each segment of Stage 1 with two segments that are each $\frac{1}{9}$ unit long.
- Continue with a third and fourth iteration.

Example 3

The equilateral triangle at the right is Stage 0 of a Koch Snowflake. Draw Stage 1 of the snowflake by first drawing an equilateral triangle on the middle third of each side. Then erase the middle third of each side of the original triangle.

- To draw an equilateral triangle on the middle third of a side, find the two points that are $\frac{1}{3}$ unit from an endpoint of the side. From each point, draw a segment of length $\frac{1}{3}$ unit. Each segment must make a 60° angle with the side of the original triangle.

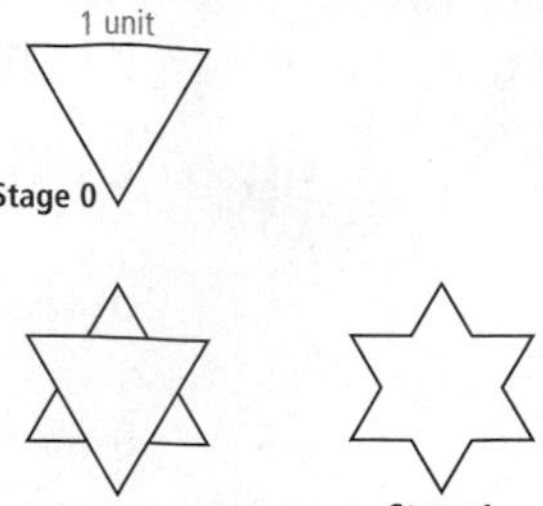

Exercises

1. Draw Stage 3 of the fractal tree in Example 1.

Use the Koch Curve in Example 2 for Exercises 2–4.

2. Complete the table to find the length of the Koch Curve at each stage.

Stage	0	1	2
Length	1	■	■

3. Examine the results of Exercise 2 and look for a pattern. Use this pattern to predict the length of the Koch Curve at Stage 3 and at Stage 4.
4. Suppose you are able to complete a Koch Curve to Stage *n*.
   a. Write an expression for the length of the curve.
   b. What happens to the length of the curve as *n* increases?
5. Draw Stage 2 of the Koch Snowflake in Example 3.

Stage 3 of the Koch Snowflake is shown at the right. Use it and the earlier stages to answer Exercises 6–8.

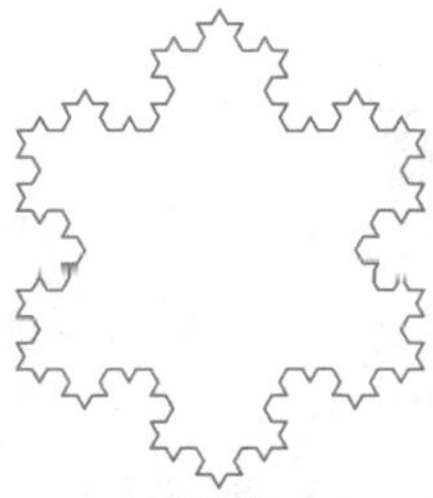
Stage 3

6. At each stage, is the snowflake equilateral?
7. a. Complete the table to find the perimeter at each stage.

Stage	Number of Sides	Length of a Side	Perimeter
0	3	1	3
1	■	$\frac{1}{3}$	■
2	48	■	■
3	■	■	■

   b. Predict the perimeter at Stage 4.
   c. Will there be a stage at which the perimeter is greater than 100 units? Explain.
8. Exercises 4 and 7 suggest that there is no bound on the perimeter of the Koch Snowflake. Is this true about the area of the Koch Snowflake? Explain.

Example 3

This Example shows the steps to draw the Koch Snowflake — a popular "line bender" fractal.

Q What is the difference in Stage 0 of the Koch Curve and the Koch Snowflake? **[The Koch Curve begins with a line segment, and the Koch Snowflake begins with an equilateral triangle.]**

ERROR PREVENTION

Provide graphing paper to make it easier to keep track of the stages iterations increase.

SYNTHESIZING

Students adept at using graphing calculators may enjoy programming their calculator to draw fractals. Programs are also available online at websites for TI calculators.

EXTENSION

An interesting property of fractals is their fractional dimension. Students are familiar with two and three-dimensional objects, but fractals have dimensions anywhere between one and three dimensions. The Koch Snowflake drawn in this activity has a dimension of 1.262. Another famous fractal, the Sierpinski Triangle, has a dimension of 1.585. Students can research this topic and find the dimensions of other fractals.

Answers

Exercises

1.

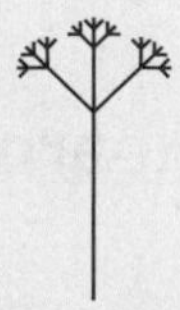

2. $\frac{4}{3}$, $\frac{16}{9}$

3. $\frac{64}{27}$, $\frac{256}{81}$

4a. $\frac{4^n}{3^n}$

b. It gets greater and greater, with no limit.

5.

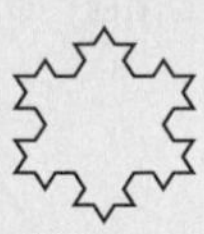

6. yes

7a. 12, 4; $\frac{1}{9}$, $\frac{16}{3}$; 192, $\frac{1}{27}$, $\frac{64}{9}$

b. $\frac{256}{27}$ or $9.\overline{481}$

c. Yes; the perimeter increases by a factor of $\frac{4}{3}$ at each stage, so it expands without limit.

8. No; a circle drawn around Stage 0 would also contain every subsequent stage.

1 Interactive Learning

Solve It!

PURPOSE To determine whether two triangles are similar

PROCESS Students may measure the side lengths and show that they are proportional or use the Triangle Sum Theorem to find the missing angle measures.

FACILITATE

- **Q** How can you determine whether the triangles are similar? **[Polygons are similar if corresponding angles are congruent and pairs of corresponding side lengths are proportional.]**
- **Q** Are the unmarked angles in the triangles congruent? How do you know? **[Yes; by the Triangle Sum Theorem, both angle measures are 70°.]**
- **Q** What must be true about the side lengths in similar triangles? **[Corresponding side lengths must be proportional.]**

ANSWER See Solve It in Answers on next page.

CONNECT THE MATH Emphasize that proving triangles similar with the definition requires showing three pairs of angles are congruent, and that an extended proportion is true for corresponding side lengths. Learning the triangle similarity theorems will shorten this process.

2 Guided Instruction

TAKE NOTE Emphasize that the Angle-Angle Similarity Postulate is stated in the form of an If-Then statement. Have students identify the hypothesis and conclusion of the postulate. Discuss the logic behind the postulate.

7-3 Proving Triangles Similar

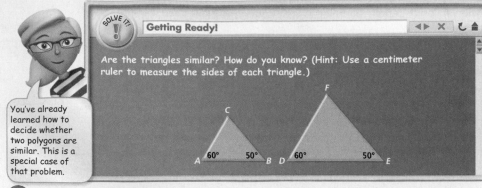

Common Core State Standards

G-SRT.B.5 Use . . . similarity criteria for triangles to solve problems and to prove relationships in geometric figures. **Also G-GPE.B.5**

MP 1, MP 3, MP 4

Objectives To use the AA ~ Postulate and the SAS ~ and SSS ~ Theorems
To use similarity to find indirect measurements

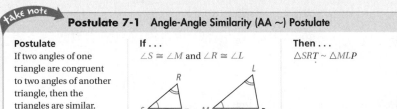

Solve It! Getting Ready!

Are the triangles similar? How do you know? (Hint: Use a centimeter ruler to measure the sides of each triangle.)

You've already learned how to decide whether two polygons are similar. This is a special case of that problem.

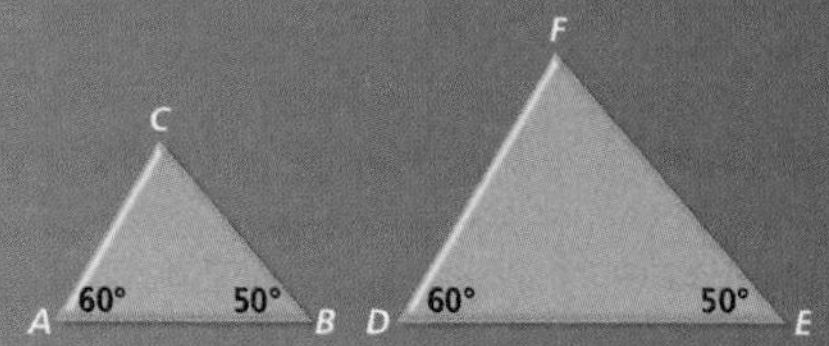

MATHEMATICAL PRACTICES In the Solve It, you determined whether the two triangles are similar. That is, you needed information about all three pairs of angles and all three pairs of sides. In this lesson, you'll learn an easier way to determine whether two triangles are similar.

Lesson Vocabulary
- indirect measurement

Essential Understanding You can show that two triangles are similar when you know the relationships between only two or three pairs of corresponding parts.

Take note — Postulate 7-1 Angle-Angle Similarity (AA ~) Postulate

Postulate	If . . .	Then . . .
If two angles of one triangle are congruent to two angles of another triangle, then the triangles are similar.	$\angle S \cong \angle M$ and $\angle R \cong \angle L$ 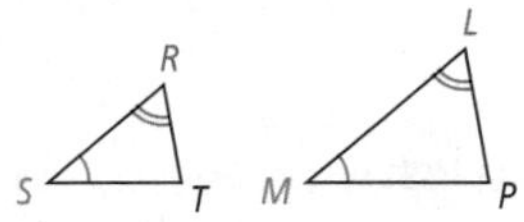	$\triangle SRT \sim \triangle MLP$

7-3 Preparing to Teach

BIG ideas **Reasoning and Proof**
Visualization

ESSENTIAL UNDERSTANDINGS
- Triangles can be shown to be similar based on the relationship of two or three pairs of corresponding parts.
- Similar triangles can be used to find unknown measurements.

Math Background

The definition of similar triangles involves a complicated set of conditions. Postulates and theorems about similarity can provide a shortcut to proving triangles similar. Similar triangles can be used for indirect measurement in a variety of circumstances. Similar triangles are also part of the basic idea of proportionality, which extends through all areas of geometry. Proving statements in geometry helps students understand how to develop logical arguments in other areas of their lives.

Mathematical Practices

Construct viable arguments and critique the reasoning of others. Students will prove the similarity of triangles using the stated assumptions provided by the AA Postulate and the SAS ~ and SSS ~ Theorems.

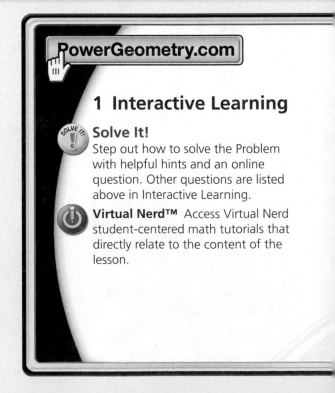

PowerGeometry.com

1 Interactive Learning

Solve It! Step out how to solve the Problem with helpful hints and an online question. Other questions are listed above in Interactive Learning.

Virtual Nerd™ Access Virtual Nerd student-centered math tutorials that directly relate to the content of the lesson.

Plan

What do you need to show that the triangles are similar? To use the AA ~ Postulate, you need to prove that two pairs of angles are congruent.

Problem 1 Using the AA ~ Postulate

Are the two triangles similar? How do you know?

A $\triangle RSW$ **and** $\triangle VSB$

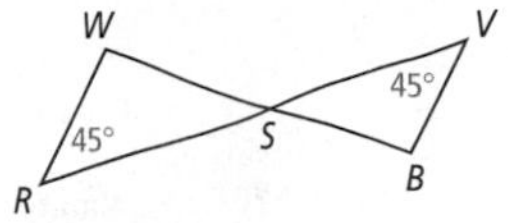

$\angle R \cong \angle V$ because both angles measure 45°.
$\angle RSW \cong \angle VSB$ because vertical angles are congruent.
So, $\triangle RSW \sim \triangle VSB$ by the AA ~ Postulate.

B $\triangle JKL$ **and** $\triangle PQR$

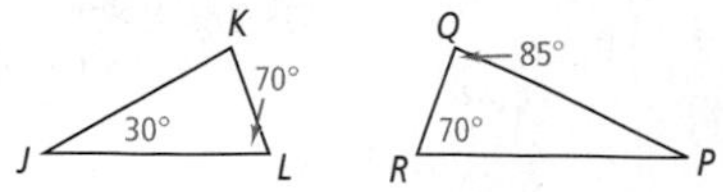

$\angle L \cong \angle R$ because both angles measure 70°.
By the Triangle Angle-Sum Theorem,
$m\angle K = 180 - 30 - 70 = 80$ and
$m\angle P = 180 - 85 - 70 = 25$. Only one pair of angles is congruent. So, $\triangle JKL$ and $\triangle PQR$ are *not* similar.

Got It? 1. Are the two triangles similar? How do you know?

a.

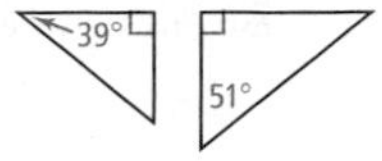

b.

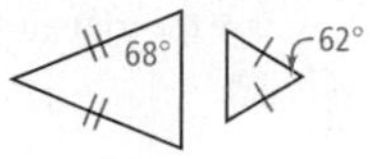

Here are two other ways to determine whether two triangles are similar.

take note

Theorem 7-1 Side-Angle-Side Similarity (SAS ~) Theorem

Theorem
If an angle of one triangle is congruent to an angle of a second triangle, and the sides that include the two angles are proportional, then the triangles are similar.

If . . .
$\frac{AB}{QR} = \frac{AC}{QS}$ and $\angle A \cong \angle Q$

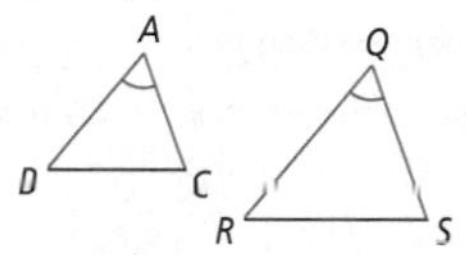

Then . . .
$\triangle ABC \sim \triangle QRS$

You will prove Theorem 7-1 in Exercise 35.

Theorem 7-2 Side-Side-Side Similarity (SSS ~) Theorem

Theorem
If the corresponding sides of two triangles are proportional, then the triangles are similar.

If . . .
$\frac{AB}{QR} = \frac{AC}{QS} = \frac{BC}{RS}$

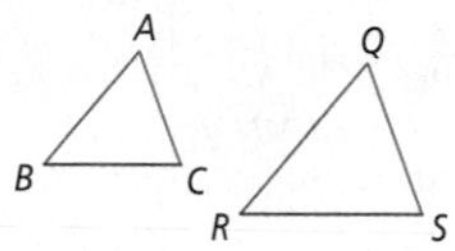

Then . . .
$\triangle ABC \sim \triangle QRS$

You will prove Theorem 7-2 in Exercise 36.

Problem 1

Q How many pairs of angles are labeled with the same measure? What does this mean about the angles? **[1; The angles are congruent.]**

Q How many pairs of angles are needed to use the AA~ Postulate? **[2]**

Q What theorem relates the measures of the angles with vertex 5? **[The Vertical Angles Theorem states that vertical angles are congruent.]**

Got It? SYNTHESIZING

Ask students to describe the types of triangles in each diagram. Have them list the properties of the triangles. Students should use the properties of these triangles to determine similarity.

Take Note

Review the concept of included angles. Ask students to explain in their own words why each theorem is true. Focus the discussion on how only three conditions can satisfy the full definition of similar triangles.

2 Guided Instruction

Each Problem is worked out and supported online.

Problem 1
Using the AA ~ Postulate
Animated

Problem 2
Verifying Triangle Similarity
Animated

Problem 3
Proving Triangles Similar
Animated

Problem 4
Finding Lengths in Similar Triangles

Support in Geometry Companion

- Vocabulary
- Key Concepts
- Got It?

Answers

Solve It!

Yes; corresp. ∡ are ≅ and corresp. sides are proportional.

Got It?

1a. The measures of the two acute ∡ in each △ are 39 and 51, so the ⧍ are ~ by the AA ~ Post.

b. Each of the base ∡ in the △ at the left measures 68, while each of the base ∡ in the △ at the right measures $\frac{1}{2}(180 - 62) = 59$; the ⧍ are not ~.

Problem 2

Q In 2A, what parts of the triangles are given? **[All six side measures are given.]**

Q What similarity postulate uses these measures? **[SSS ~ Theorem]**

Q In order to prove the triangles similar, what computation must be performed? Why? **[The ratios of corresponding side lengths must be simplified to show that the ratios are equivalent.]**

Q In 2B, what measurements of the triangles are given? **[Two sets of side lengths are given.]**

Q What further information do you need to prove the triangles are similar in 2B? **[One pair of congruent angles is needed to use the SAS ~ Theorem.]**

Q Is there a pair of congruent angles in the diagram? What theorem, postulate, or property confirms their congruence? **[Yes, ∠K is contained in both triangles. By the Reflexive Property of Congruence, ∠K ≅ ∠K.]**

Proof Proof of Theorem 7-1: Side-Angle-Side Similarity Theorem

Given: $\frac{AB}{QR} = \frac{AC}{QS}$, $\angle A \cong \angle Q$

Prove: $\triangle ABC \sim \triangle QRS$

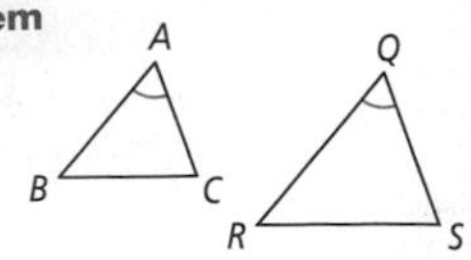

Plan for Proof: Choose X on $\overline{RQ}$ so that $QX = AB$. Draw $\overleftrightarrow{XY} \parallel \overline{RS}$. Show that $\triangle QXY \sim \triangle QRS$ by the AA ~ Postulate. Then use the proportion $\frac{QX}{QR} = \frac{QY}{QS}$ and the given proportion $\frac{AB}{QR} = \frac{AC}{QS}$ to show that $AC = QY$. Then prove that $\triangle ABC \cong \triangle QXY$. Finally, prove that $\triangle ABC \sim \triangle QRS$ by the AA ~ Postulate.

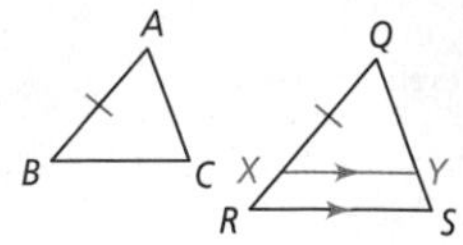

Problem 2 Verifying Triangle Similarity

Are the triangles similar? If so, write a similarity statement for the triangles.

A

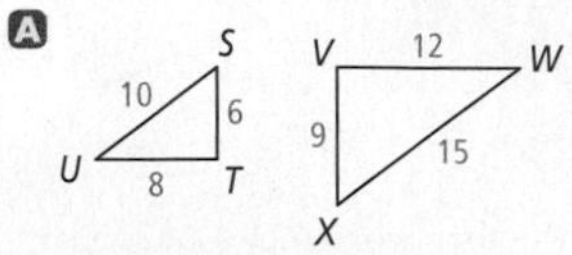

Use the side lengths to identify corresponding sides. Then set up ratios for each pair of corresponding sides.

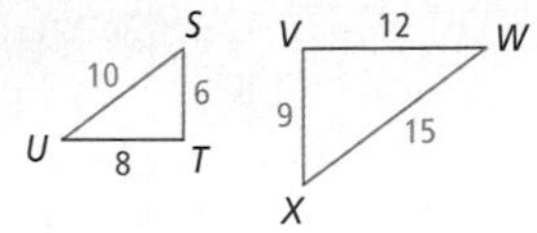

Shortest sides	$\frac{ST}{XV} = \frac{6}{9} = \frac{2}{3}$
Longest sides	$\frac{US}{WX} = \frac{10}{15} = \frac{2}{3}$
Remaining sides	$\frac{TU}{VW} = \frac{8}{12} = \frac{2}{3}$

All three ratios are equal, so corresponding sides are proportional. $\triangle STU \sim \triangle XVW$ by the SSS ~ Theorem.

Plan

How can you make it easier to identify corresponding sides and angles?
Sketch and label two separate triangles.

B

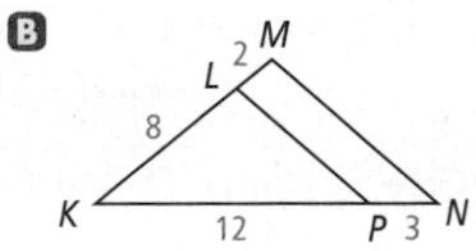

$\angle K \cong \angle K$ by the Reflexive Property of Congruence.

$\frac{KL}{KM} = \frac{8}{10} = \frac{4}{5}$ and $\frac{KP}{KN} = \frac{12}{15} = \frac{4}{5}$.

So, $\triangle KLP \sim \triangle KMN$ by the SAS ~ Theorem.

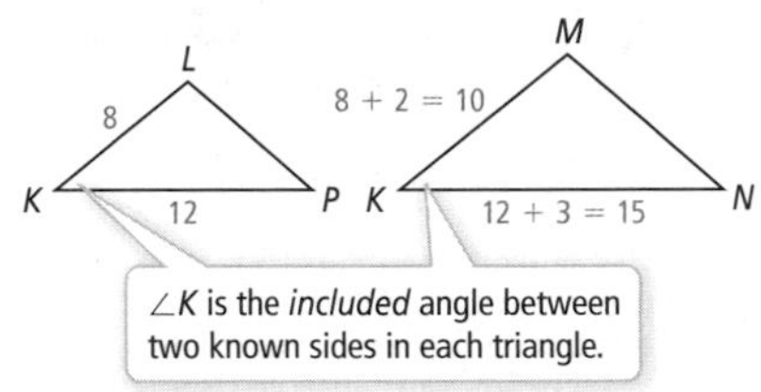

Additional Problems

1. Are the two triangles similar? How do you know?

a. △RST and △NOP

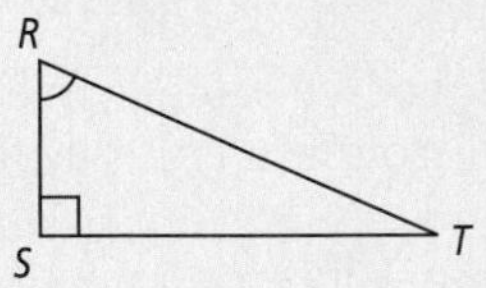

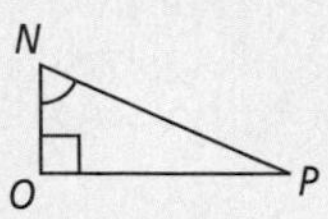

b. △GHT and △DEF

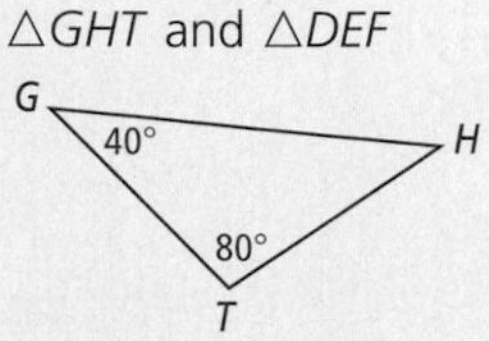

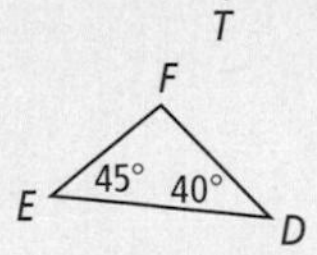

ANSWER a. yes, by the AA Similarity Postulate **b.** not similar

2. Are the triangles similar? If so, write a similarity statement.

a.

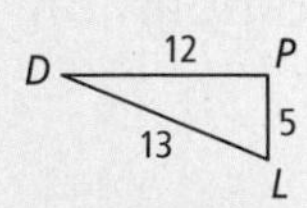

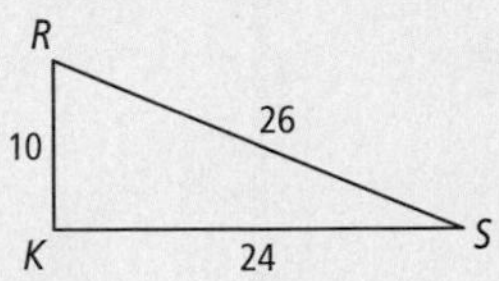

b.

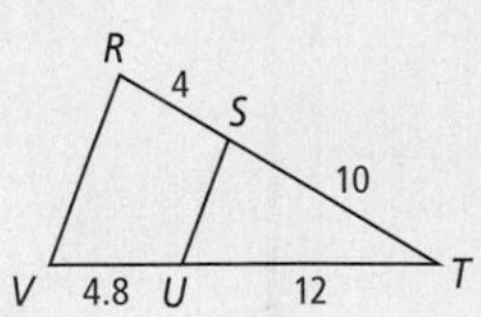

ANSWER a. $\triangle DPL \sim \triangle SKR$ by the SSS Similarity Theorem **b.** $\triangle TSU \sim \triangle TRV$ by the SAS Similarity Theorem

3. Given: $m\angle S = 35$, $m\angle LTC = 55$, right angles STL and TLC

Prove: $\triangle STL \sim \triangle CLT$

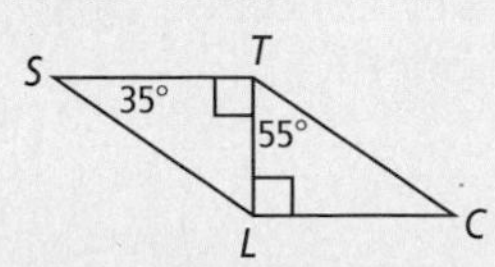

ANSWER It is given that $m\angle S = 35°$, $m\angle LTC = 55°$, and angles STL and TLC are right angles. Since the sum of the interior angles of a triangle is 180°, $m\angle C = 180° - 90° - 55°$, or 35°. So, two angles of triangle STL are congruent to two angles of triangle CLT. So, $\triangle STL \sim \triangle CLT$ by the AA Similarity Postulate.

4. Benjamin places a mirror 40 ft from the base of an oak tree. When he stands at a distance of 5 ft from the mirror, he can see the top of the tree in the reflection. If Benjamin is 5 ft 8 in. tall, what is the height of the oak tree?

ANSWER 45 ft 4 in.

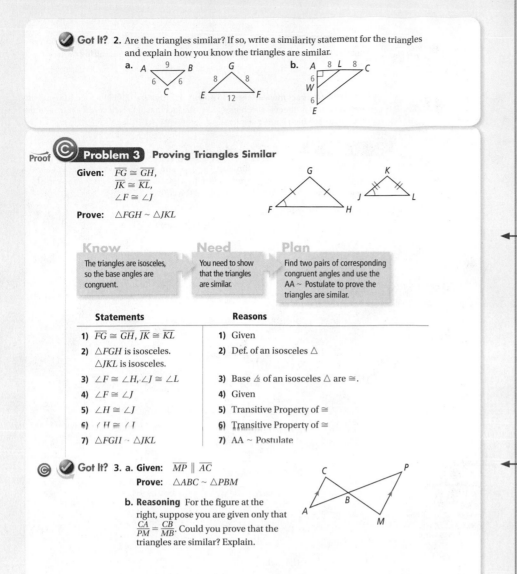

Got It? 2. Are the triangles similar? If so, write a similarity statement for the triangles and explain how you know the triangles are similar.

a.

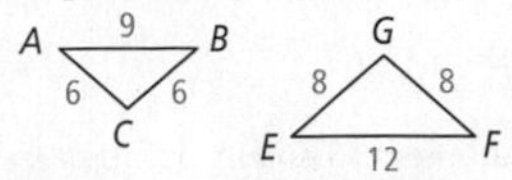

b.

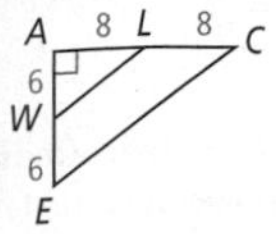

Proof

Problem 3 Proving Triangles Similar

Given: $\overline{FG} \cong \overline{GH}$, $\overline{JK} \cong \overline{KL}$, $\angle F \cong \angle J$

Prove: $\triangle FGH \sim \triangle JKL$

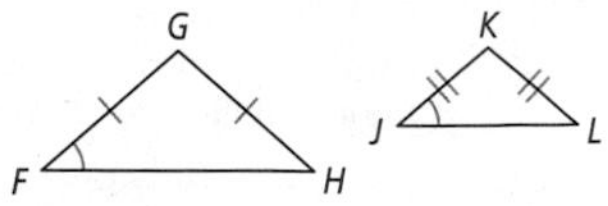

Know
The triangles are isosceles, so the base angles are congruent.

Need
You need to show that the triangles are similar.

Plan
Find two pairs of corresponding congruent angles and use the AA ~ Postulate to prove the triangles are similar.

Statements	Reasons
1) $\overline{FG} \cong \overline{GH}$, $\overline{JK} \cong \overline{KL}$	1) Given
2) $\triangle FGH$ is isosceles. $\triangle JKL$ is isosceles.	2) Def. of an isosceles $\triangle$
3) $\angle F \cong \angle H$, $\angle J \cong \angle L$	3) Base $\angle$s of an isosceles $\triangle$ are $\cong$.
4) $\angle F \cong \angle J$	4) Given
5) $\angle H \cong \angle J$	5) Transitive Property of $\cong$
6) $\angle H \cong \angle L$	6) Transitive Property of $\cong$
7) $\triangle FGH \sim \triangle JKL$	7) AA ~ Postulate

Got It? 3. a. Given: $\overline{MP} \parallel \overline{AC}$

Prove: $\triangle ABC \sim \triangle PBM$

b. Reasoning For the figure at the right, suppose you are given only that $\frac{CA}{PM} = \frac{CB}{MB}$. Could you prove that the triangles are similar? Explain.

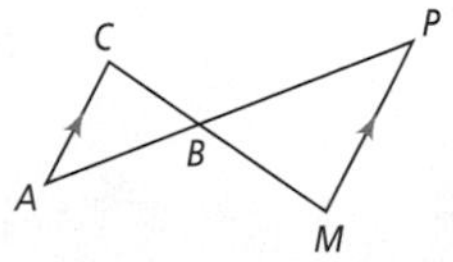

Got It? VISUAL LEARNERS

For 2b, have students draw each triangle separately to confirm which angles and sides are corresponding.

Problem 3

When writing a proof, students may find it easier to make a plan that starts at the conclusion and work backward based on the information they need.

Q What information do you need to prove the triangles are similar? **[two pairs of congruent angles for the AA ~ Postulate, two pairs of proportional sides and the congruent included angles for the SAS ~ Theorem, or three pairs of proportional sides for the SSS ~ Theorem]**

Q What properties of an isosceles triangle help you prove that the triangles are similar? **[Isosceles triangles have one vertex angle and two congruent base angles.]**

Q Which pairs of angles are congruent within each triangle? **[$\angle F \cong \angle H$ and $\angle J \cong \angle L$]**

Got It?

Q What theorem about parallel lines may be useful in proving the triangles similar? **[When two parallel lines are cut by a transversal, alternate interior angles are congruent.]**

Q Are there any special pairs of angles present in the diagram? Explain. **[Yes; there are two pairs of alternate interior angles and one pair of vertical angles.]**

Answers

Got It? (continued)

2a. The ratio for each of the three pairs of corresp. sides is 3 : 4, so $\triangle ABC \sim \triangle EFG$ by SSS ~.

b. $\angle A$ is in each $\triangle$ and $\frac{AL}{AC} = \frac{AW}{AE} = \frac{1}{2}$, so $\triangle ALW \sim \triangle ACE$ by SAS~.

3a. $\overline{MP} \parallel \overline{AC}$ (given), so $\angle A \cong \angle P$ and $\angle C \cong \angle M$ because if two lines are $\parallel$, then alt. int. $\angle$s are $\cong$. So $\triangle ABC \sim \triangle PBM$ by AA~.

b. No; the $\cong$ vertical angles are not included by the proportional sides, so it is not possible to prove that the triangles are similar.

Problem 4

Q How can you tell the two triangles in the diagram are similar? **[They contain two pairs of congruent angles: the right angles formed at the ground and the angles at the mirror.]**

Q What postulate verifies that the triangles are similar? **[AA ~ Postulate]**

Q How do you know which sides are corresponding to set up a proportion? **[The corresponding sides are included between congruent angles.]**

Got It?

Q What characteristic needed to prove the triangles similar is not given directly by the problem? **[The angles of light entering and exiting the mirror are congruent.]**

Essential Understanding Sometimes you can use similar triangles to find lengths that cannot be measured easily using a ruler or other measuring device.

You can use **indirect measurement** to find lengths that are difficult to measure directly. One method of indirect measurement uses the fact that light reflects off a mirror at the same angle at which it hits the mirror.

Problem 4 Finding Lengths in Similar Triangles

Rock Climbing **Before rock climbing, Darius wants to know how high he will climb. He places a mirror on the ground and walks backward until he can see the top of the cliff in the mirror. What is the height of the cliff?**

Plan

Before solving for x, verify that the triangles are similar. $\triangle HTV \sim \triangle JSV$ by the AA~ Postulate because $\angle T \cong \angle S$ and $\angle HVT \cong \angle JVS$.

$\triangle HTV \sim \triangle JSV$	AA ~ Postulate
$\frac{HT}{JS} = \frac{TV}{SV}$	Corresponding sides of ~ triangles are proportional.
$\frac{5.5}{x} = \frac{6}{34}$	Substitute.
$187 = 6x$	Cross Products Property
$31.2 \approx x$	Solve for x.

The cliff is about 31 ft high.

Got It? 4. Reasoning Why is it important that the ground be flat to use the method of indirect measurement illustrated in Problem 4? Explain.

Answers

Got It? (continued)

4. The triangles formed will not be similar unless both Darius and the cliff form right angles with the ground.

Lesson Check

Do you know HOW?

Are the triangles similar? If yes, write a similarity statement and explain how you know they are similar.

1.

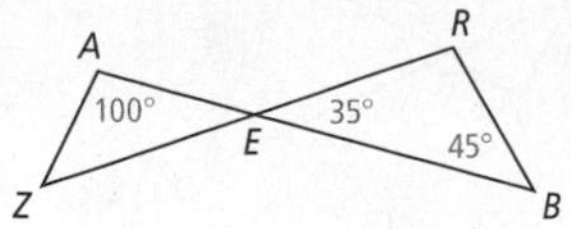

2.

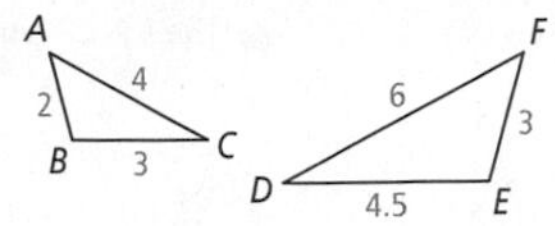

3.

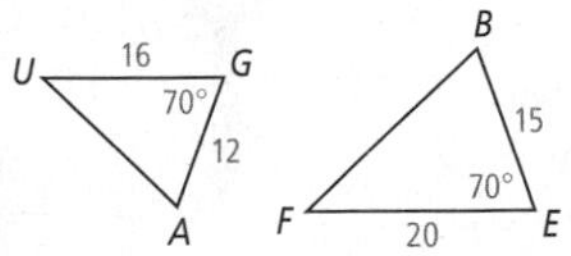

Do you UNDERSTAND? MATHEMATICAL PRACTICES

4. Vocabulary How could you use indirect measurement to find the height of the flagpole at your school?

5. Error Analysis Which solution for the value of *x* in the figure at the right is *not* correct? Explain.

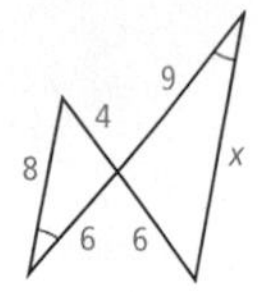

A.
$\frac{4}{8} = \frac{8}{x}$
$4x = 72$
$x = 18$

B.
$\frac{8}{x} = \frac{4}{6}$
$48 = 4x$
$12 = x$

6. a. Compare and Contrast How are the SAS Similarity Theorem and the SAS Congruence Postulate alike? How are they different?

b. How are the SSS Similarity Theorem and the SSS Congruence Postulate alike? How are they different?

Practice and Problem-Solving Exercises MATHEMATICAL PRACTICES

A Practice Determine whether the triangles are similar. If so, write a similarity statement and name the postulate or theorem you used. If not, explain.

See Problems 1 and 2.

7.

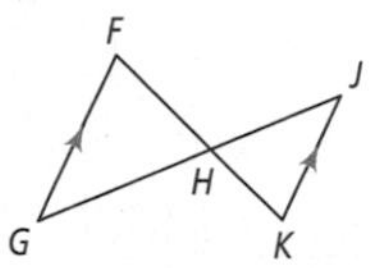

8.

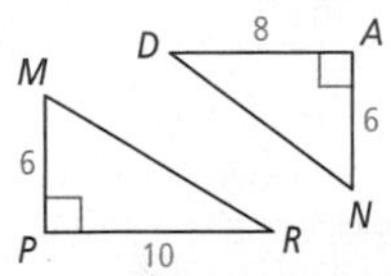

9.

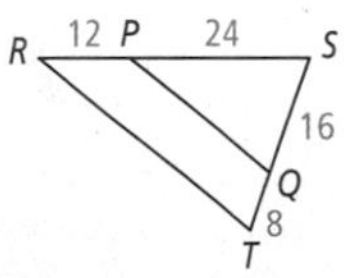

10.

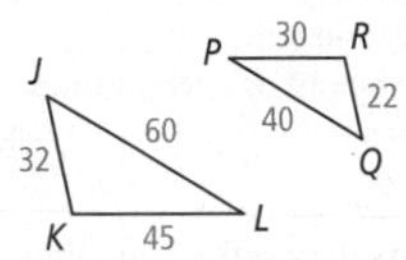

11.

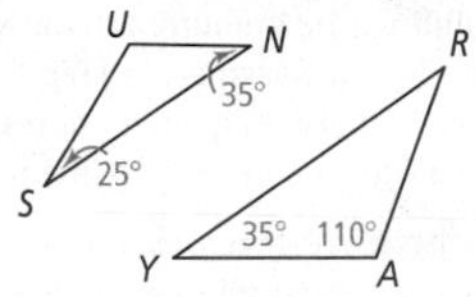

12.

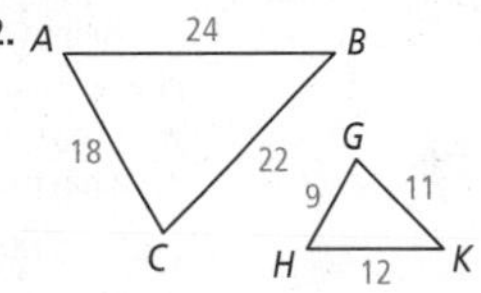

3 Lesson Check

Do you know HOW?

- If students have difficulty with Exercise 3, then have them redraw the triangles so that they are oriented the same way.

Do you UNDERSTAND?

- If students have difficulty with Exercise 4, then have them review Problem 4.

Close

Q What is the minimum number of conditions necessary to prove two triangles similar? What are the conditions? **[Two pairs of congruent angles can prove two triangles similar.]**

Q What other sets of conditions will prove two triangles are similar? **[two pairs of proportional side lengths and the included angle or three pairs of proportional side lengths]**

Lesson Check

1. Yes; $m\angle R = 180 - (35 + 45) = 100$, and $\angle AEZ \cong \angle REB$ (Vert. ∠s are ≅.), so $\triangle AEZ \sim \triangle REB$ by AA~.
2. Yes; the ratios of corresp. sides are all 2 : 3, so $\triangle ABC \sim \triangle FED$ by SSS~.
3. Yes; $\angle G \cong \angle E$ and $\frac{UG}{FE} = \frac{AG}{BE} = \frac{4}{5}$, so $\triangle GUA \sim \triangle EFB$ by SAS~.
4. Answers may vary. Sample: Measure your shadow and the flagpole's shadow. Use the proportion $\frac{\text{your shadow}}{\text{flagpole's shadow}} = \frac{\text{your height}}{\text{flagpole's height}}$.
5. Method A is not correct because the ratio, $\frac{4}{8}$ does not use corresp. sides.

6a. Answers may vary. Sample: Both use two pairs of corresp. sides and the ∠s included by those sides, but SAS~ uses pairs of equal ratios, while SAS ≅ uses pairs of ≅ sides.

b. Both involve all three sides of a △, but corresp. sides are proportional for SSS ~ and ≅ for SSS ≅.

Practice and Problem-Solving Exercises

7. $\triangle FGH \sim \triangle KJH$; AA~
8. Not ~; using the sides that contain the rt. ∠s, the ratio of the shorter sides is 1 : 1, while the ratio of the longer sides is 5 : 4.
9. $\triangle RST \sim \triangle PSQ$; SAS~
10. Not ~; $\frac{JL}{PQ} = \frac{KL}{PR} = \frac{3}{2}$, but $\frac{JK}{RQ} = \frac{16}{11}$.
11. Not ~; $m\angle U = 180 - (25 + 35) = 120$, while $m\angle A = 110$.
12. $\triangle ABC \sim \triangle HKG$; SSS~

PowerGeometry.com

3 Lesson Check

For a digital lesson check, use the Got It questions.

Support in Geometry Companion

- Lesson Check

4 Practice

Assign homework to individual students or to an entire class.

4 Practice

ASSIGNMENT GUIDE

Basic: 7–19, 22–25, 32
Average: 7–17 odd, 18–34
Advanced: 7–17 odd, 18–36
Standardized Test Prep: 37–40
Mixed Review: 41–52

Mathematical Practices are supported by exercises with red headings. Here are the Practices supported in this lesson:

MP 1: Make Sense of Problems Ex. 6, 22
MP 3: Construct Arguments Ex. 29, 31
MP 3: Critique the Reasoning of Others Ex. 5

Applications exercises have blue headings. Exercises 15–17 and 23 support MP 4: Model.

EXERCISE 23: Use the Think About a Plan worksheet in the **Practice and Problem Solving Workbook** (also available in the Teaching Resources in print and online) to further support students' development in becoming independent learners.

HOMEWORK QUICK CHECK

To check students' understanding of key skills and concepts, go over Exercises 11, 17, 22, 23, and 32.

13. Proof **Given:** $\angle ABC \cong \angle ACD$
Prove: $\triangle ABC \sim \triangle ACD$

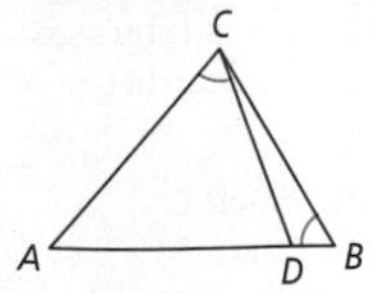

14. Proof **Given:** $PR = 2NP$, $PQ = 2MP$
Prove: $\triangle MNP \sim \triangle QRP$

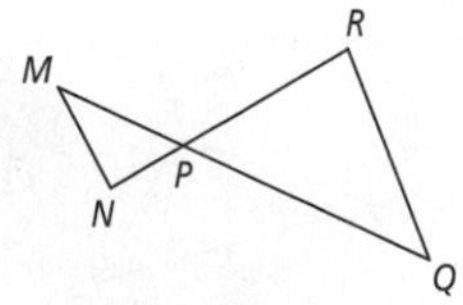

See Problem 3.

Indirect Measurement **Explain why the triangles are similar. Then find the distance represented by x.**

See Problem 4.

15.

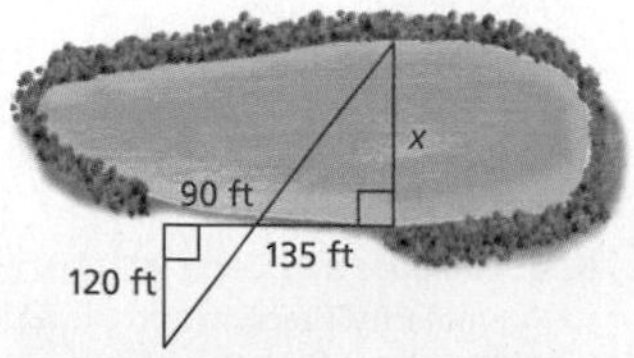

16.

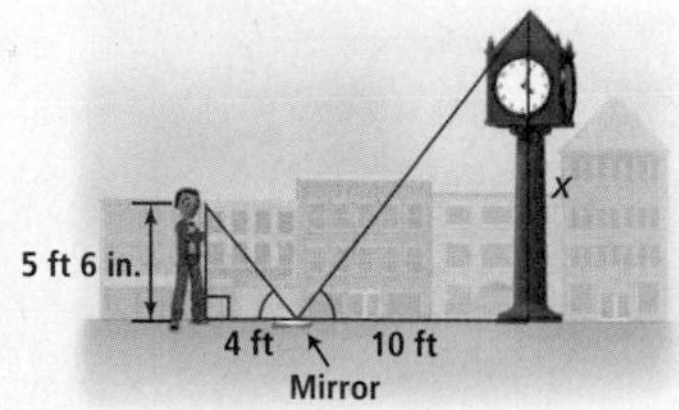

17. Washington Monument At a certain time of day, a 1.8-m-tall person standing next to the Washington Monument casts a 0.7-m shadow. At the same time, the Washington Monument casts a 65.8-m shadow. How tall is the Washington Monument?

B Apply

Can you conclude that the triangles are similar? If so, state the postulate or theorem you used and write a similarity statement. If not, explain.

18.

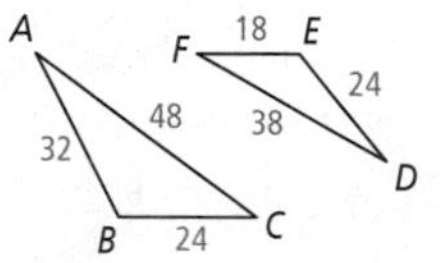

19.

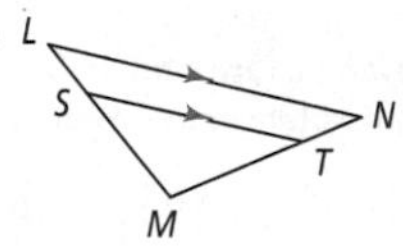

20.

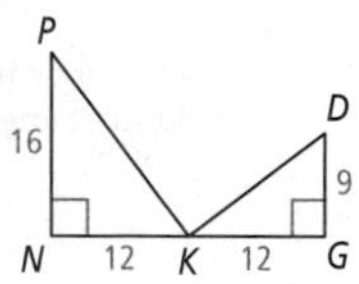

21. a. Are two isosceles triangles always similar? Explain.
b. Are two right isosceles triangles always similar? Explain.

22. Think About a Plan On a sunny day, a classmate uses indirect measurement to find the height of a building. The building's shadow is 12 ft long and your classmate's shadow is 4 ft long. If your classmate is 5 ft tall, what is the height of the building?
- Can you draw and label a diagram to represent the situation?
- What proportion can you use to solve the problem?

23. Indirect Measurement A 2-ft vertical post casts a 16-in. shadow at the same time a nearby cell phone tower casts a 120-ft shadow. How tall is the cell phone tower?

Answers

Practice and Problem-Solving Exercises (continued)

13. $\angle A \cong \angle A$ (Refl. Prop. of $\cong$) and $\angle ABC \cong \angle ACD$ (given), so $\triangle ABC \sim \triangle ACD$ by AA~.

14. $\angle MPN \cong \angle QPR$ (Vert. ∠s are $\cong$.), and the given information tells us $\frac{PR}{PN} = \frac{PQ}{PM} = \frac{2}{1}$. So $\triangle MNP \sim \triangle QRP$ by SAS~.

15. There are a pair of $\cong$ vert. ∠s and a pair of $\cong$ rt. ∠s, so the △s are ~ by AA~; 180 ft

16. A pair of $\cong$ ∠s are given and the two rt. ∠s are $\cong$, so the △s are ~ by AA~; 13.75 ft or 13 ft 9 in.

17. about 169.2 m

18. Not ~; $\frac{AB}{DF} = \frac{BC}{EF} = \frac{4}{3}$, $\frac{AC}{FD} = \frac{48}{38} = \frac{24}{19}$.

19. $\triangle LMN \sim \triangle SMT$ by AA~.

20. $\frac{NK}{NP} = \frac{GD}{GK}$ and $\angle N \cong \angle G$, so $\triangle NKP \sim \triangle GDK$ by SAS~.

21a. No; the ratios of the sides that form the vertex ∠s are =, but the vertex ∠s may not be $\cong$.

b. Yes; sample explanation: An isosc. rt. △ has two ∠s 45°, so any two isosc. rt. △s are ~ by AA~.

22. 15 ft **23.** 180 ft

Algebra For each pair of similar triangles, find the value of x.

24.

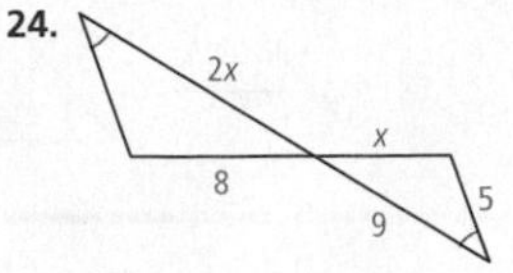

25.

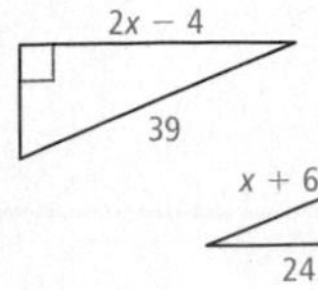

26.

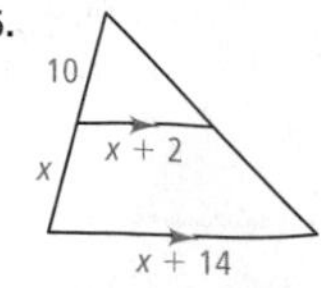

27. Given: $\overline{PQ} \perp \overline{QT}$, $\overline{ST} \perp \overline{TQ}$, $\frac{PQ}{ST} = \frac{QR}{TV}$
Proof **Prove:** $\triangle VKR$ is isosceles.

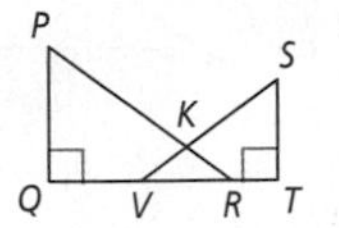

28. Given: $\overline{AB} \parallel \overline{CD}$, $\overline{BC} \parallel \overline{DG}$
Proof **Prove:** $AB \cdot CG = CD \cdot AC$

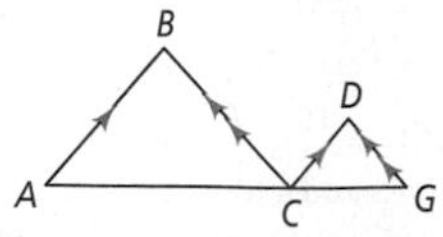

29. Reasoning Does any line that intersects two sides of a triangle and is parallel to the third side of the triangle form two similar triangles? Justify your reasoning.

30. Constructions Draw any $\triangle ABC$ with $m\angle C = 30$. Use a straightedge and compass to construct $\triangle LKJ$ so that $\triangle LKJ \sim \triangle ABC$.

31. Reasoning In the diagram at the right, $\triangle PMN \sim \triangle SRW$. $\overline{MQ}$ and $\overline{RT}$ are altitudes. The scale factor of $\triangle PMN$ to $\triangle SRW$ is 4 : 3. What is the ratio of $\overline{MQ}$ to $\overline{RT}$? Explain how you know.

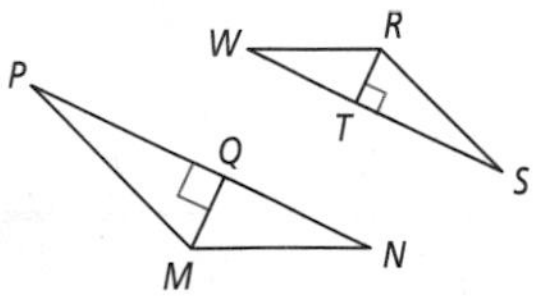

32. Coordinate Geometry $\triangle ABC$ has vertices $A(0, 0)$, $B(2, 4)$, and
Proof $C(4, 2)$. $\triangle RST$ has vertices $R(0, 3)$, $S(-1, 5)$, and $T(-2, 4)$. Prove that $\triangle ABC \sim \triangle RST$. (*Hint:* Graph $\triangle ABC$ and $\triangle RST$ in the coordinate plane.)

33. Write a proof of the following: Any two nonvertical parallel
Proof lines have equal slopes.

Given: Nonvertical lines ℓ_1 and ℓ_2, $\ell_1 \parallel \ell_2$, $\overline{EF}$ and $\overline{BC}$ are $\perp$ to the x-axis

Prove: $\frac{BC}{AC} = \frac{EF}{DF}$

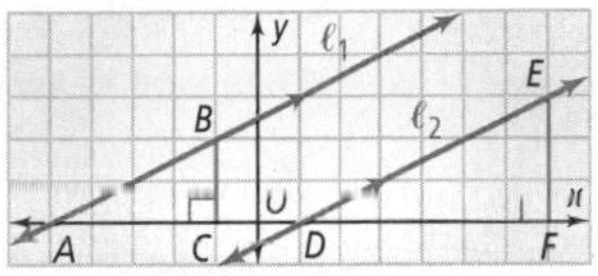

34. Use the diagram in Exercise 33. Prove: Any two nonvertical lines with equal slopes
Proof are parallel.

Challenge **35.** Prove the Side-Angle-Side Similarity Theorem (Theorem 7-1).
Proof **Given:** $\frac{AB}{QR} = \frac{AC}{QS}$, $\angle A \cong \angle Q$
Prove: $\triangle ABC \sim \triangle QRS$

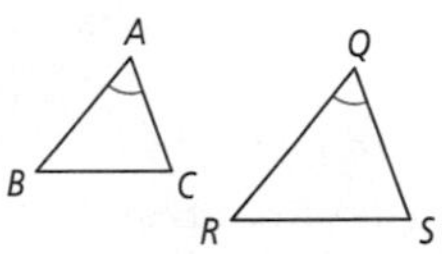

24. 6 **25.** 20 **26.** 10

27. In $\triangle PQR$ and $\triangle STV$, $\angle Q \cong \angle T$ because $\perp$ lines form rt. ∠s, which are $\cong$. The sides that contain the ∠s are proportional (given). So $\triangle PQR \sim \triangle STV$ by SAS~, and $\angle KRV \cong \angle KVR$ because corresp. ∠s of ~ △s are $\cong$. Thus $\triangle VKR$ is isosc. by the Converse of Isosc. △ Thm.

28. $\angle A \cong \angle DCG$ and $\angle ACB \cong \angle G$ (If lines are $\parallel$ then corresp. ∠s are $\cong$.). So $\triangle ABC \sim \triangle CDG$ by AA~. Then $\frac{AB}{CD} = \frac{AC}{CG}$ because corresp. sides of ~ △s are proportional, and $AB \cdot CG = CD \cdot AC$ by the Cross Products Prop.

29. Yes; the two $\parallel$ lines and the two sides determine two pair of $\cong$ corr. ∠s, so two △s are ~ by AA~.

30.

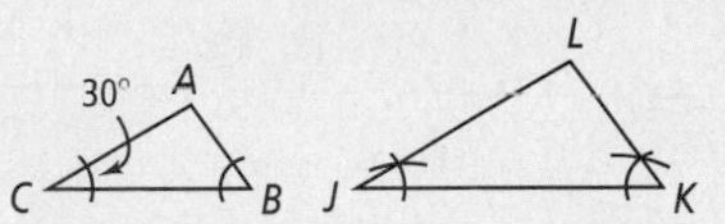

Draw a line segment $\overline{JK}$ of any length. At J, construct an angle $\cong$ to $\angle C$. At K, construct an angle $\cong$ to $\angle B$. Extend the sides of the angles till they intersect. Label the intersection point L. $\triangle LKJ \sim \triangle ABC$ by AA~.

31. 4 : 3; sample explanation: Since $\angle P \cong \angle S$ and $\angle PQM \cong \angle STR$, $\triangle PQM \sim \triangle STR$ by AA~. So the ratio $\frac{MQ}{RT} = \frac{PM}{SR}$ = the ratio of corresp. sides in $\triangle PMN$ and $\triangle SRW$ namely, 4 : 3.

32. Use the Distance Formula: $AB = AC = 2\sqrt{5}$ and $BC = 2\sqrt{2}$, while $RS = RT = \sqrt{5}$ and $ST = \sqrt{2}$. $\triangle ABC \sim \triangle RST$ (SSS~) because $\frac{AB}{RS} = \frac{BC}{ST} = \frac{AC}{RT} = 2$.

33. It is given that $\ell_1 \parallel \ell_2$, so $\angle BAC \cong \angle EDF$ because if lines are $\parallel$, then corresponding ∠s are $\cong$. The given $\perp$ lines mean $\angle ACB \cong \angle DFE$ because $\perp$ lines form rt. ∠s, which are $\cong$. So $\triangle ABC \sim \triangle DEF$ by AA~, and $\frac{BC}{EF} = \frac{AC}{DF}$ because corresp. sides of ~△s are proportional. Then Prop. of Proportions (2) lets us conclude that $\frac{BC}{AC} = \frac{EF}{DF}$.

34. $\frac{BC}{AC} = \frac{EF}{DE}$, $\overline{EF} \perp \overline{AF}$, $\overline{BC} \perp \overline{AF}$ (Given); $\angle ACB$ and $\angle DFE$ are rt. ∠s. (Def. of $\perp$); $\angle ACB \cong \angle DFE$ (All rt. ∠s are $\cong$.); $\triangle ABC \sim \triangle DEF$ (SAS ~); $\angle BAC \cong \angle EDF$ (Def. of similar); $\ell_1 \parallel \ell_2$ (If corr. ∠s are $\cong$, then $\parallel$ lines.)

35.

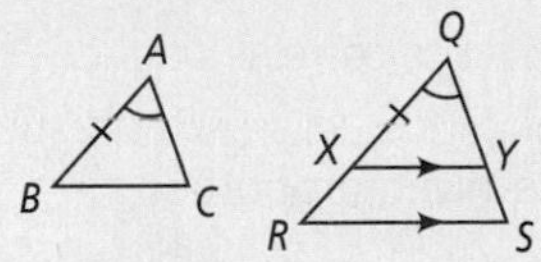

Choose point X on $\overline{QR}$ so that $QX = AB$. Then draw $\overline{XY} \parallel \overline{RS}$ (Through a point not on a line, there is exactly one line $\parallel$ to the given line.). $\angle A \cong \angle Q$ (Given) and $\angle QXY \cong \angle R$ (If two lines are $\parallel$, then corresp. ∠s are $\cong$.), so $\triangle QXY \sim \triangle QRS$ by AA~. Therefore, $\frac{QX}{QR} = \frac{XY}{RS} = \frac{QY}{QS}$ because corresp. sides of ~ △s are proportional. Since $QX = AB$, substitute QX for AB in the given proportion $\frac{AB}{QR} = \frac{AC}{QS}$ to get $\frac{QX}{QR} = \frac{AC}{QS}$. Therefore, $\frac{QX}{QR} = \frac{QY}{QS} = \frac{AC}{QS}$, and $QY = AC$. So $\triangle ABC \cong \triangle QXY$ by SAS. $\angle B \cong \angle QXY$ (Corresp. parts of $\cong$ △s are $\cong$.) and $\angle B \cong \angle R$ by the Transitive Prop. of $\cong$. Therefore, $\triangle ABC \sim \triangle QRS$ by AA~.

Answers

Practice and Problem-Solving Exercises (continued)

36.

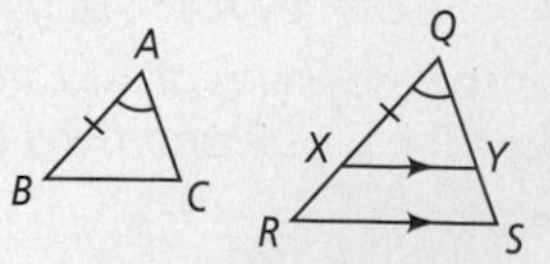

Choose point X on $\overline{QR}$ so that $QX = AB$. Then draw $\overline{XY} \parallel \overline{RS}$ (Through a point not on a line, there is exactly one line $\parallel$ to the given line.). $\angle A \cong \angle Q$ and $\angle QXY \cong \angle R$ (If lines are $\parallel$, then corresp. ∡ are ≅.), so $\triangle QXY \sim \triangle QRS$ by AA~. Therefore, $\frac{QX}{QR} = \frac{XY}{RS} = \frac{QY}{QS}$ because corresp. sides of ~ ⧍ are proportional. Since $QX = AB$, substitute QX for AB in the given proportion $\frac{AB}{QR} = \frac{AC}{QS} = \frac{BC}{RS}$ to get $\frac{QX}{QR} = \frac{AC}{QS} = \frac{BC}{RS}$. Therefore, $\frac{AC}{QS} = \frac{QY}{QS}$ and $\frac{BC}{RS} = \frac{XY}{RS}$. So $BC = XY$ and $AC = QY$. Then $\triangle ABC \cong QXY$ by SSS ≅. $\angle B \cong \angle QXY$ and $\angle C \cong \angle QYX$ (Corresp. parts of ≅ ⧍ are ≅.). Since $\overline{XY} \parallel \overline{RS}$, $\angle QXY \cong \angle R$ and $\angle QYX \cong \angle S$ (If lines are $\parallel$, then corresp. ∡ are ≅.). $\angle B \cong \angle R$ and $\angle C \cong \angle S$ (Transitive Prop.) Therefore, $\triangle ABC \sim \triangle QRS$ by AA~.

37. C **38.** G **39.** C

40. [4] Answers may vary. Sample:

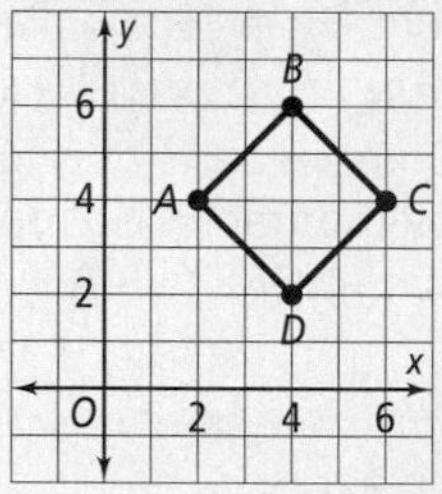

Slope of $\overline{AB} = 1$, slope of $\overline{BC} = -1$, slope of $\overline{CD} = 1$, slope of $\overline{AD} = -1$; for each pair of consecutive sides the slopes are negative reciprocals, so the figure has four rt. ∡. By the Distance Formula, $AB = BC = CD = AD = \sqrt{8}$, so all four sides are ≅. Since $ABCD$ is a quadrilateral with four rt. ∡ and four ≅ sides, it is a square.

[3] minor error in a calculation

[2] includes some, but not all, of the conditions necessary for a figure to be a square

[1] incomplete OR incorrect explanation

41. 2 : 3 **42.** 135

43. 12 **44.** $\frac{3}{2}$

45. 125; obtuse **46.** 88; acute

47. 180; straight **48.** 110; obtuse

49. 8, 18; x, 24; 6

50. m, 18; 12, 20; $\frac{40}{3}$ or $13\frac{1}{3}$

51. $x + 2$, 9; 15, x; 3

52. $x + 4$, 5; $x - 3$, 9; $\frac{47}{4}$ or 11.75

36. Prove the Side-Side-Side Similarity Theorem (Theorem 7-2).
Proof
Given: $\frac{AB}{QR} = \frac{AC}{QS} = \frac{BC}{RS}$
Prove: $\triangle ABC \sim \triangle QRS$

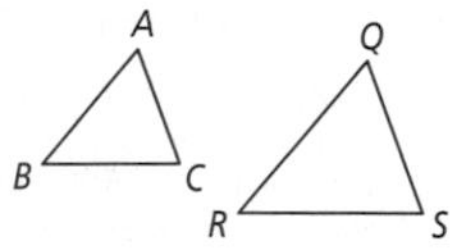

Standardized Test Prep

SAT/ACT

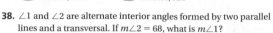

37. Complete the statement $\triangle ABC \sim$ __?__. By which postulate or theorem are the triangles similar?

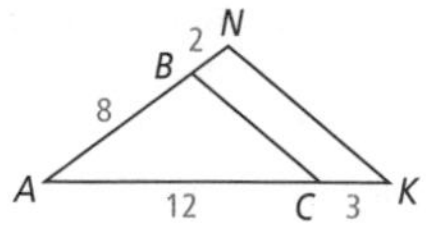

(A) $\triangle AKN$; SSS ~ (B) $\triangle AKN$; SAS ~ (C) $\triangle ANK$; SAS ~ (D) $\triangle ANK$; AA ~

38. $\angle 1$ and $\angle 2$ are alternate interior angles formed by two parallel lines and a transversal. If $m\angle 2 = 68$, what is $m\angle 1$?

(F) 22 (G) 68 (H) 112 (I) 122

39. The length of a rectangle is twice its width. If the perimeter of the rectangle is 72 in., what is the length of the rectangle?

(A) 12 in. (B) 18 in. (C) 24 in. (D) 36 in.

Extended Response

40. Graph $A(2, 4)$, $B(4, 6)$, $C(6, 4)$, and $D(4, 2)$. What type of polygon is $ABCD$? Justify your answer.

Mixed Review

***TRAP* ~ *EZYD*. Use the diagram at the right to find the following.** See Lesson 7-2.

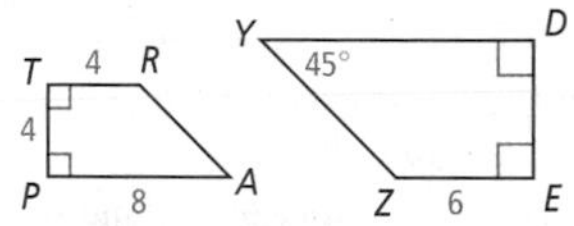

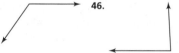

41. the scale factor of *TRAP* to *EZYD*

42. $m\angle R$ **43.** DY **44.** $\frac{DE}{PT}$

Use a protractor to find the measure of each angle. Classify the angle as *acute*, *right*, *obtuse*, or *straight*. See Lesson 1-4.

45.

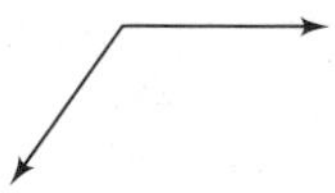

46.

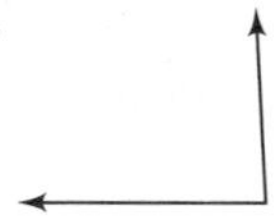

47.

48.

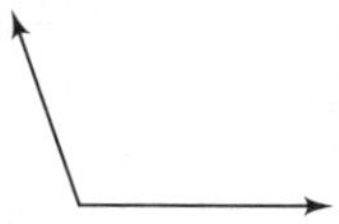

Get Ready! **To prepare for Lesson 7-4, do Exercises 49–52.**

Algebra **Identify the means and extremes of each proportion. Then solve for *x*.** See Lesson 7-1.

49. $\frac{x}{8} = \frac{18}{24}$ **50.** $\frac{12}{m} = \frac{18}{20}$ **51.** $\frac{15}{x+2} = \frac{9}{x}$ **52.** $\frac{x-3}{x+4} = \frac{5}{9}$

7-3 Lesson Resources

Differentiated Remediation

Instructional Support

Geometry Companion

Students can use the **Geometry Companion** worktext (4 pages) . . .

- New Vocabulary
- Key Concepts
- Got It for each Problem
- Lesson Check

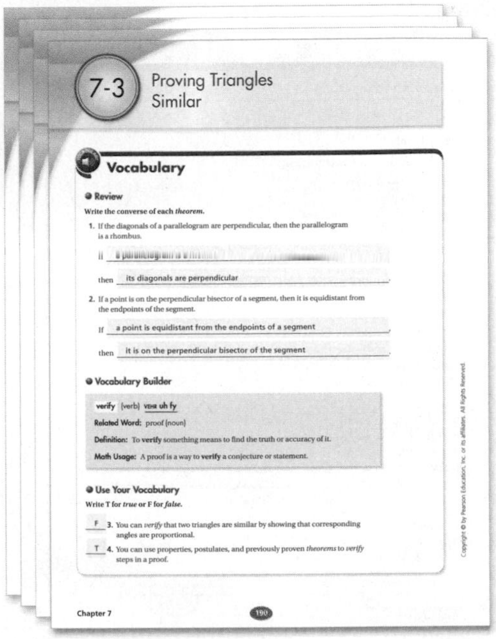
7-3 Proving Triangles Similar

Vocabulary

Review

Write the converse of each *theorem*.

1. If the diagonals of a parallelogram are perpendicular, then the parallelogram is a rhombus.

If [illegible]

then its diagonals are perpendicular.

2. If a point is on the perpendicular bisector of a segment, then it is equidistant from the endpoints of the segment.

If a point is equidistant from the endpoints of a segment,

then it is on the perpendicular bisector of the segment.

Vocabulary Builder

verify (verb) VEHR uh fy

Related Word: proof (noun)

Definition: To **verify** something means to find the truth or accuracy of it.

Math Usage: A proof is a way to **verify** a conjecture or statement.

Use Your Vocabulary

Write T for *true* or F for *false*.

F 3. You can *verify* that two triangles are similar by showing that corresponding angles are proportional.

T 4. You can use properties, postulates, and previously proven *theorems* to *verify* steps in a proof.

Copyright © by Pearson Education, Inc. or its affiliates. All Rights Reserved.

Chapter 7 190

ELL Support

Use Role Playing Divide students into small groups. Say: Draw two triangles. Make the length of each side of the second triangle one-and-a-half times the length of a side of the first triangle. Use one group's triangles for the next activity.

Hold up the two triangles and say, "I accuse these two triangles of being similar. We will hold a trial to see if the triangles are guilty of being similar." Assign students the roles of prosecutor, defense lawyer, witnesses, and members of the jury. Set a chair in front of the classroom for witnesses to sit. Have the prosecutor and defense lawyer question the witnesses. Then have the jury vote to determine whether the triangles are similar or not.

5 Assess & Remediate

Lesson Quiz

1. Are $\triangle KRA$ and $\triangle FLN$ similar? How do you know?

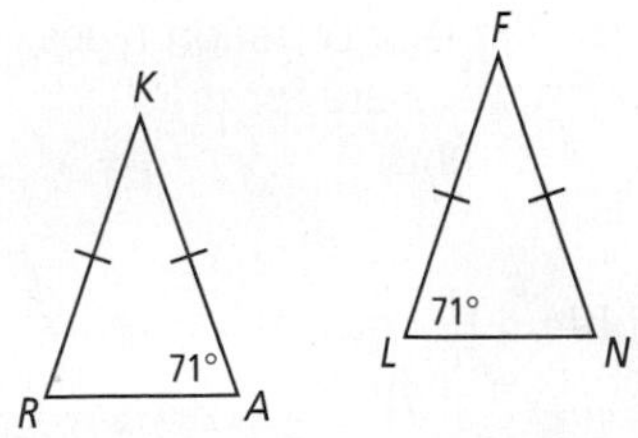

2. Do you UNDERSTAND? A flagpole casts a shadow 18 ft long. At the same time, Rachael casts a shadow that is 4 ft long. If Rachael is 5 ft tall, what is the height of the flagpole?

3. Are the triangles similar? If so, what is the similarity statement?

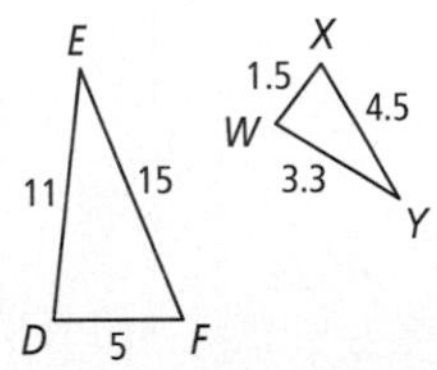

ANSWERS TO LESSON QUIZ

1. yes; by the AA Similarity Postulate

2. 22 ft 6 in.

3. yes; $\triangle DEF \sim \triangle WYX$

PRESCRIPTION FOR REMEDIATION

Use the student work on the Lesson Quiz to prescribe a differentiated review assignment.

Points	Differentiated Remediation
0–1	Intervention
2	On-level
3	Extension

PowerGeometry.com

5 Assess & Remediate

Assign the Lesson Quiz. Appropriate intervention, practice, or enrichment is automatically generated based on student performance.

Intervention

- **Reteaching** (2 pages) Provides reteaching and practice exercises for the key lesson concepts. Use with struggling students or absent students.
- **English Language Learner Support** Helps students develop and reinforce mathematical vocabulary and key concepts.

All-in-One Resources/Online

Reteaching

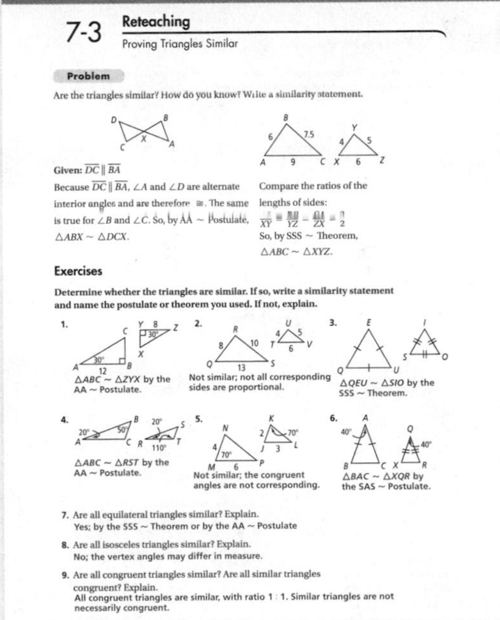
7-3 Reteaching
Proving Triangles Similar

Problem

Are the triangles similar? How do you know? Write a similarity statement.

Given: $\overline{DC} \parallel \overline{BA}$

Because $\overline{DC} \parallel \overline{BA}$, $\angle A$ and $\angle D$ are alternate interior angles and are therefore ≅. The same is true for $\angle B$ and $\angle C$. So, by AA ~ Postulate, $\triangle ABX \sim \triangle DCX$.

Compare the ratios of the lengths of sides: [illegible] So, by SSS ~ Theorem, $\triangle ABC \sim \triangle XYZ$.

Exercises

Determine whether the triangles are similar. If so, write a similarity statement and name the postulate or theorem you used. If not, explain.

1. $\triangle ABC \sim \triangle ZYX$ by the AA ~ Postulate.
2. Not similar; not all corresponding sides are proportional.
3. $\triangle QEU \sim \triangle SIO$ by the SSS ~ Theorem.
4. $\triangle ABC \sim \triangle RST$ by the AA ~ Postulate.
5. Not similar; the congruent angles are not corresponding.
6. $\triangle BAC \sim \triangle XQR$ by the SAS ~ Postulate.
7. Are all equilateral triangles similar? Explain.
Yes; by the SSS ~ Theorem or by the AA ~ Postulate
8. Are all isosceles triangles similar? Explain.
No; the vertex angles may differ in measure.
9. Are all congruent triangles similar? Are all similar triangles congruent? Explain.
All congruent triangles are similar, with ratio 1 : 1. Similar triangles are not necessarily congruent.

All-in-One Resources/Online

English Language Learner Support

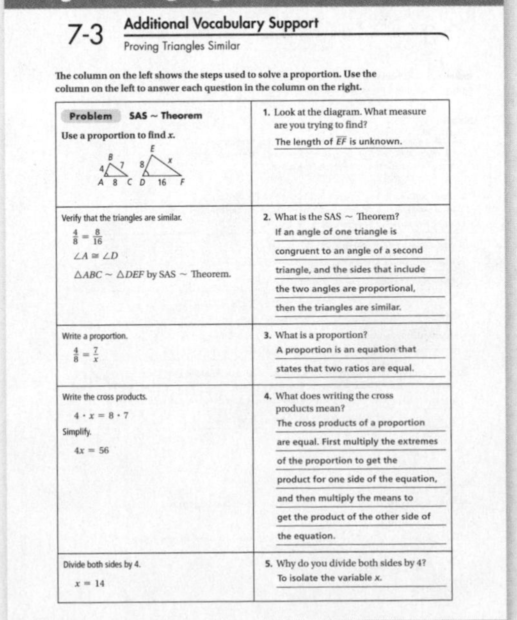
7-3 Additional Vocabulary Support
Proving Triangles Similar

The column on the left shows the steps used to solve a proportion. Use the column on the left to answer each question in the column on the right.

Problem SAS ~ Theorem **Use a proportion to find x.**	1. Look at the diagram. What measure are you trying to find? The length of $\overline{EF}$ is unknown.
Verify that the triangles are similar. $\frac{4}{8} = \frac{8}{16}$ $\angle A \cong \angle D$ $\triangle ABC \sim \triangle DEF$ by SAS ~ Theorem.	2. What is the SAS ~ Theorem? If an angle of one triangle is congruent to an angle of a second triangle, and the sides that include the two angles are proportional, then the triangles are similar.
Write a proportion. $\frac{4}{8} = \frac{7}{x}$	3. What is a proportion? A proportion is an equation that states that two ratios are equal.
Write the cross products. $4 \cdot x = 8 \cdot 7$ Simplify. $4x = 56$	4. What does writing the cross products mean? The cross products of a proportion are equal. First multiply the extremes of the proportion to get the product for one side of the equation, and then multiply the means to get the product of the other side of the equation.
Divide both sides by 4. $x = 14$	5. Why do you divide both sides by 4? To isolate the variable x.

Differentiated Remediation *continued*

On-Level

- **Practice** (2 pages) Provides extra practice for each lesson. For simpler practice exercises, use the Form K Practice pages found in the All-in-One Teaching Resources and online.
- **Think About a Plan** Helps students develop specific problem-solving skills and strategies by providing scaffolded guiding questions.
- **Standardized Test Prep** Focuses on all major exercises, all major question types, and helps students prepare for the high-stakes assessments.

Extension

- **Enrichment** Provides students with interesting problems and activities that extend the concepts of the lesson.
- **Activities, Games, and Puzzles** Worksheets that can be used for concepts development, enrichment, and for fun!

Practice and Problem Solving Wkbk/ All-in-One Resources/Online

Practice page 1

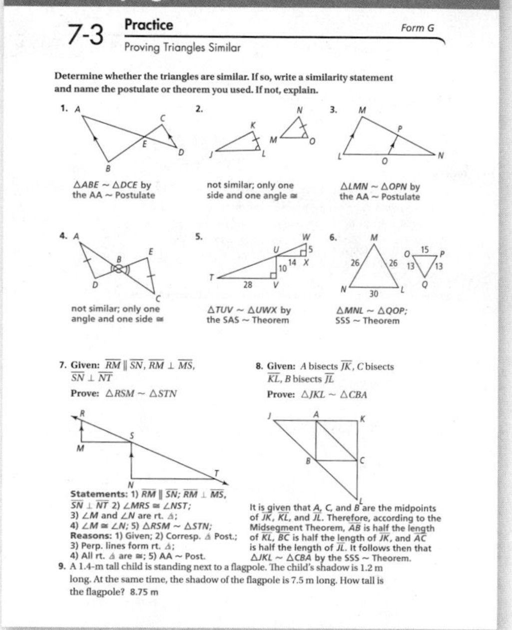

7-3 **Practice** — Proving Triangles Similar — *Form G*

Determine whether the triangles are similar. If so, write a similarity statement and name the postulate or theorem you used. If not, explain.

1. $\triangle ABE \sim \triangle DCE$ by the AA ~ Postulate
2. not similar; only one side and one angle ≅
3. $\triangle LMN \sim \triangle OPN$ by the AA ~ Postulate
4. not similar; only one angle and one side ≅
5. $\triangle TUV \sim \triangle UWX$ by the SAS ~ Theorem
6. $\triangle MNL \sim \triangle QOP$; SSS ~ Theorem
7. **Given:** $\overline{RM} \parallel \overline{SN}$, $\overline{RM} \perp \overline{MS}$, $\overline{SN} \perp \overline{NT}$ **Prove:** $\triangle RSM \sim \triangle STN$
   Statements: 1) $\overline{RM} \parallel \overline{SN}$; $\overline{RM} \perp \overline{MS}$, $\overline{SN} \perp \overline{NT}$ 2) $\angle MRS \cong \angle NST$; 3) $\angle M$ and $\angle N$ are rt. ∡; 4) $\angle M \cong \angle N$; 5) $\triangle RSM \sim \triangle STN$; Reasons: 1) Given; 2) Corresp. ∡ Post.; 3) Perp. lines form rt. ∡; 4) All rt. ∡ are ≅; 5) AA ~ Post.
8. **Given:** A bisects $\overline{JK}$, C bisects $\overline{KL}$, B bisects $\overline{JL}$ **Prove:** $\triangle JKL \sim \triangle CBA$
   It is given that A, C, and B are the midpoints of $\overline{JK}$, $\overline{KL}$, and $\overline{JL}$. Therefore, according to the Midsegment Theorem, $\overline{AB}$ is half the length of $\overline{KL}$, $\overline{BC}$ is half the length of $\overline{JK}$, and $\overline{AC}$ is half the length of $\overline{JL}$. It follows then that $\triangle JKL \sim \triangle CBA$ by the SSS ~ Theorem.
9. A 1.4-m tall child is standing next to a flagpole. The child's shadow is 1.2 m long. At the same time, the shadow of the flagpole is 7.5 m long. How tall is the flagpole? 8.75 m

Practice and Problem Solving Wkbk/ All-in-One Resources/Online

Practice page 2

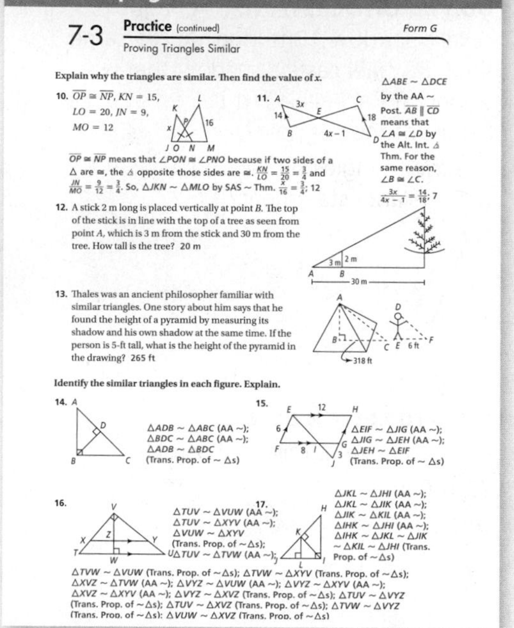

7-3 **Practice** (continued) — Proving Triangles Similar — *Form G*

Explain why the triangles are similar. Then find the value of x.

10. $\overline{OP} \cong \overline{NP}$, $KN = 15$, $LO = 20$, $JN = 9$, $MO = 12$
    $\overline{OP} \cong \overline{NP}$ means that $\angle PON \cong \angle PNO$ because if two sides of a △ are ≅, the ∡ opposite those sides are ≅. $\frac{KN}{LO} = \frac{15}{20} = \frac{3}{4}$ and $\frac{JN}{MO} = \frac{9}{12} = \frac{3}{4}$. So, $\triangle JKN \sim \triangle MLO$ by SAS ~ Thm. $\frac{x}{16} = \frac{3}{4}$; 12
11. $\triangle ABE \sim \triangle DCE$ by the AA ~ Post. $\overline{AB} \parallel \overline{CD}$ means that $\angle A \cong \angle D$ by the Alt. Int. ∡ Thm. For the same reason, $\angle B \cong \angle C$. $\frac{3x}{4x-1} = \frac{14}{18}$; 7
12. A stick 2 m long is placed vertically at point B. The top of the stick is in line with the top of a tree as seen from point A, which is 3 m from the stick and 30 m from the tree. How tall is the tree? 20 m
13. Thales was an ancient philosopher familiar with similar triangles. One story about him says that he found the height of a pyramid by measuring its shadow and his own shadow at the same time. If the person is 5-ft tall, what is the height of the pyramid in the drawing? 265 ft

Identify the similar triangles in each figure. Explain.

14. $\triangle ADB \sim \triangle ABC$ (AA ~); $\triangle BDC \sim \triangle ABC$ (AA ~); $\triangle ADB \sim \triangle BDC$ (Trans. Prop. of ~ △s)
15. $\triangle EIF \sim \triangle JIG$ (AA ~); $\triangle JIG \sim \triangle JEH$ (AA ~); $\triangle JEH \sim \triangle EIF$ (Trans. Prop. of ~ △s)
16. $\triangle TUV \sim \triangle VUW$ (AA ~); $\triangle TUV \sim \triangle XYV$ (AA ~); $\triangle VUW \sim \triangle XYV$ (Trans. Prop. of ~△s); $\triangle TUV \sim \triangle TVW$ (AA ~); $\triangle TVW \sim \triangle VUW$ (Trans. Prop. of ~△s); $\triangle TVW \sim \triangle XYV$ (Trans. Prop. of ~△s); $\triangle XVZ \sim \triangle TVW$ (AA ~); $\triangle VYZ \sim \triangle VUW$ (AA ~); $\triangle VYZ \sim \triangle XYV$ (AA ~); $\triangle XVZ \sim \triangle XYV$ (AA ~); $\triangle VYZ \sim \triangle XVZ$ (Trans. Prop. of ~△s); $\triangle TUV \sim \triangle VYZ$ (Trans. Prop. of ~△s); $\triangle TUV \sim \triangle XVZ$ (Trans. Prop. of ~△s); $\triangle TVW \sim \triangle VYZ$ (Trans. Prop. of ~△s); $\triangle VUW \sim \triangle XVZ$ (Trans. Prop. of ~△s)
17. $\triangle JKL \sim \triangle JHI$ (AA ~); $\triangle JKL \sim \triangle JIK$ (AA ~); $\triangle JIK \sim \triangle KIL$ (AA ~); $\triangle IHK \sim \triangle JHI$ (AA ~); $\triangle IHK \sim \triangle JKL \sim \triangle JIK \sim \triangle KIL \sim \triangle JHI$ (Trans. Prop. of ~△s)

All-in-One Resources/Online

Enrichment

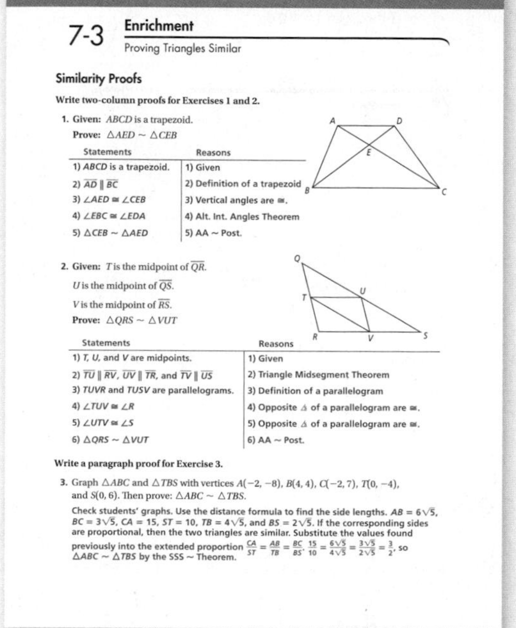

7-3 **Enrichment** — Proving Triangles Similar

Similarity Proofs

Write two-column proofs for Exercises 1 and 2.

1. **Given:** ABCD is a trapezoid. **Prove:** $\triangle AED \sim \triangle CEB$

Statements	Reasons
1) ABCD is a trapezoid.	1) Given
2) $\overline{AD} \parallel \overline{BC}$	2) Definition of a trapezoid
3) $\angle AED \cong \angle CEB$	3) Vertical angles are ≅.
4) $\angle EBC \cong \angle EDA$	4) Alt. Int. Angles Theorem
5) $\triangle CEB \sim \triangle AED$	5) AA ~ Post.

2. **Given:** T is the midpoint of $\overline{QR}$. U is the midpoint of $\overline{QS}$. V is the midpoint of $\overline{RS}$. **Prove:** $\triangle QRS \sim \triangle VUT$

Statements	Reasons
1) T, U, and V are midpoints.	1) Given
2) $\overline{TU} \parallel \overline{RV}$, $\overline{UV} \parallel \overline{TR}$, and $\overline{TV} \parallel \overline{US}$	2) Triangle Midsegment Theorem
3) TUVR and TUSV are parallelograms.	3) Definition of a parallelogram
4) $\angle TUV \cong \angle R$	4) Opposite ∡ of a parallelogram are ≅.
5) $\angle UTV \cong \angle S$	5) Opposite ∡ of a parallelogram are ≅.
6) $\triangle QRS \sim \triangle VUT$	6) AA ~ Post.

Write a paragraph proof for Exercise 3.

3. Graph $\triangle ABC$ and $\triangle TBS$ with vertices $A(-2, -8)$, $B(4, 4)$, $C(-2, 7)$, $T(0, -4)$, and $S(0, 6)$. Then prove: $\triangle ABC \sim \triangle TBS$.
   Check students' graphs. Use the distance formula to find the side lengths. $AB = 6\sqrt{5}$, $BC = 3\sqrt{5}$, $CA = 15$, $ST = 10$, $TB = 4\sqrt{5}$, and $BS = 2\sqrt{5}$. If the corresponding sides are proportional, then the two triangles are similar. Substitute the values found previously into the extended proportion $\frac{CA}{ST} = \frac{AB}{TB} = \frac{BC}{BS}$. $\frac{15}{10} = \frac{6\sqrt{5}}{4\sqrt{5}} = \frac{3\sqrt{5}}{2\sqrt{5}} = \frac{3}{2}$, so $\triangle ABC \sim \triangle TBS$ by the SSS ~ Theorem.

Practice and Problem Solving Wkbk/ All-in-One Resources/Online

Think About a Plan

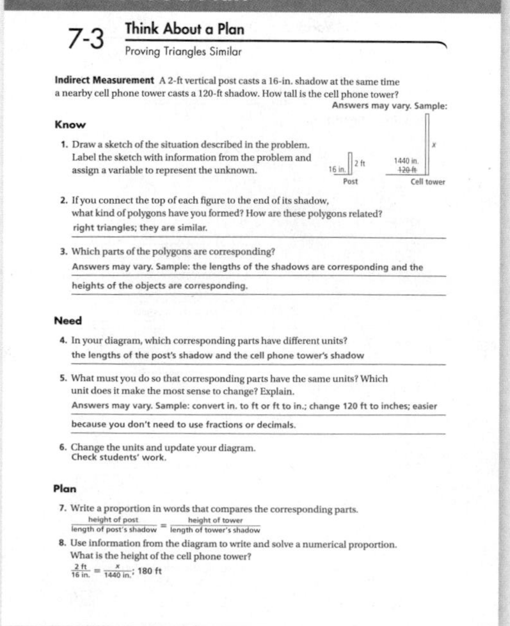

7-3 **Think About a Plan** — Proving Triangles Similar

Indirect Measurement A 2-ft vertical post casts a 16-in. shadow at the same time a nearby cell phone tower casts a 120-ft shadow. How tall is the cell phone tower?

Know

1. Draw a sketch of the situation described in the problem. Label the sketch with information from the problem and assign a variable to represent the unknown. Answers may vary. Sample:
2. If you connect the top of each figure to the end of its shadow, what kind of polygons have you formed? How are these polygons related? right triangles; they are similar.
3. Which parts of the polygons are corresponding? Answers may vary. Sample: the lengths of the shadows are corresponding and the heights of the objects are corresponding.

Need

4. In your diagram, which corresponding parts have different units? the lengths of the post's shadow and the cell phone tower's shadow
5. What must you do so that corresponding parts have the same units? Which unit does it make the most sense to change? Explain. Answers may vary. Sample: convert in. to ft or ft to in.; change 120 ft to inches; easier because you don't need to use fractions or decimals.
6. Change the units and update your diagram. Check students' work.

Plan

7. Write a proportion in words that compares the corresponding parts. $\frac{\text{height of post}}{\text{length of post's shadow}} = \frac{\text{height of tower}}{\text{length of tower's shadow}}$
8. Use information from the diagram to write and solve a numerical proportion. What is the height of the cell phone tower? $\frac{2 \text{ ft}}{16 \text{ in.}} = \frac{x}{1440 \text{ in.}}$; 180 ft

Practice and Problem Solving Wkbk/ All-in-One Resources/Online

Standardized Test Prep

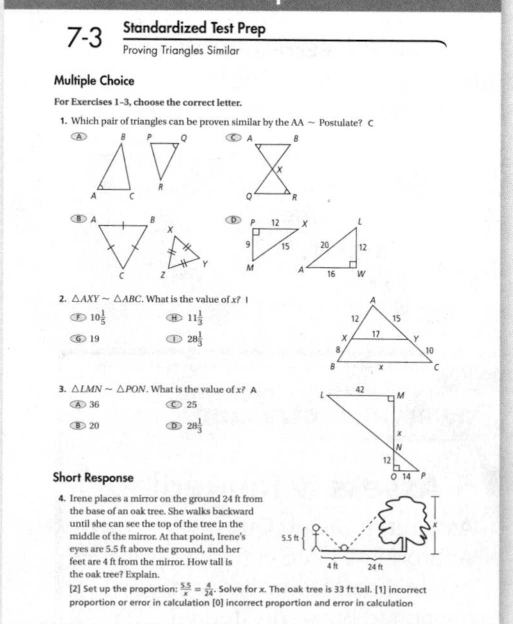

7-3 **Standardized Test Prep** — Proving Triangles Similar

Multiple Choice

For Exercises 1–3, choose the correct letter.

1. Which pair of triangles can be proven similar by the AA ~ Postulate? C
2. $\triangle AXY \sim \triangle ABC$. What is the value of x? I
   F $10\frac{1}{5}$ H $11\frac{1}{3}$ G 19 I $28\frac{1}{3}$
3. $\triangle LMN \sim \triangle PON$. What is the value of x? A
   A 36 C 25 B 20 D $28\frac{1}{3}$

Short Response

4. Irene places a mirror on the ground 24 ft from the base of an oak tree. She walks backward until she can see the top of the tree in the middle of the mirror. At that point, Irene's eyes are 5.5 ft above the ground, and her feet are 4 ft from the mirror. How tall is the oak tree? Explain.
   [2] Set up the proportion: $\frac{5.5}{x} = \frac{4}{24}$. Solve for x. The oak tree is 33 ft tall. [1] incorrect proportion or error in calculation [0] incorrect proportion and error in calculation

All-in-One Resources/Online

Activities, Games, and Puzzles

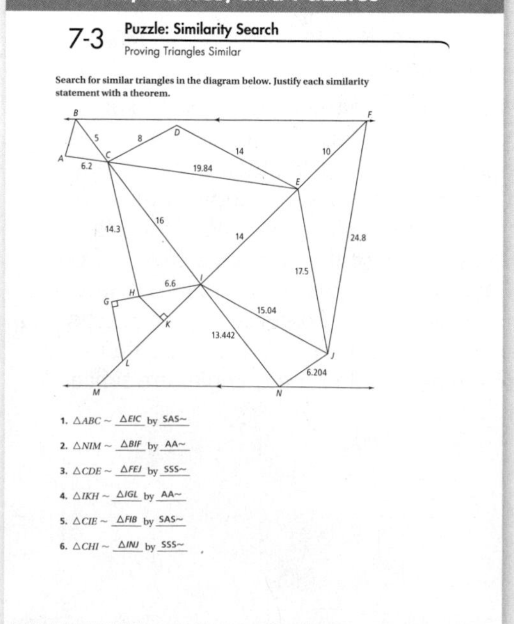

7-3 **Puzzle: Similarity Search** — Proving Triangles Similar

Search for similar triangles in the diagram below. Justify each similarity statement with a theorem.

1. $\triangle ABC \sim$ $\triangle EIC$ by SAS~
2. $\triangle NIM \sim$ $\triangle BIF$ by AA~
3. $\triangle CDE \sim$ $\triangle FEJ$ by SSS~
4. $\triangle IKH \sim$ $\triangle IGL$ by AA~
5. $\triangle CIE \sim$ $\triangle FIB$ by SAS~
6. $\triangle CHI \sim$ $\triangle INJ$ by SSS~

7 Mid-Chapter Quiz

MathXL® for School Go to PowerGeometry.com

Do you know HOW?

1. A bookcase is 4 ft tall. A model of the bookcase is 6 in. tall. What is the ratio of the height of the model bookcase to the height of the real bookcase?

2. If $\frac{a}{b} = \frac{9}{10}$, complete this statement: $\frac{a}{9} = \frac{\blacksquare}{\blacksquare}$.

3. Are the two polygons shown below similar? If so, give the similarity ratio of the first polygon to the second. If not, explain.

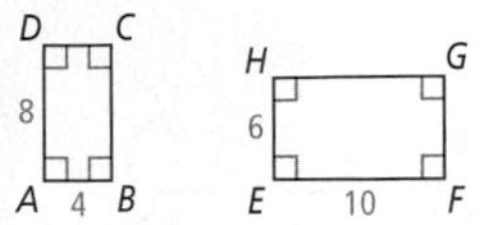

Solve each proportion.

4. $\frac{y}{6} = \frac{18}{54}$

5. $\frac{5}{7} = \frac{x-2}{4}$

6. On the scale drawing of a floor plan 2 in. = 5 ft. A room is 7 in. long on the scale drawing. Find the actual length of the room.

$\triangle ABC \sim \triangle DBF$. Complete each statement.

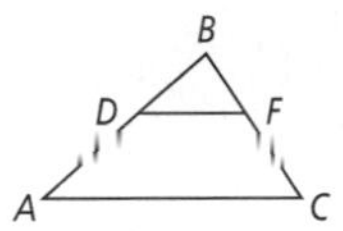

7. $m\angle A = m\angle$ __?__

8. $\frac{AB}{DB} = \frac{BC}{\blacksquare}$

9. A postcard is 6 in. by 4 in. A printing shop will enlarge it so that the longer side is any length up to 3 ft. Find the dimensions of the biggest enlargement.

10. **Algebra** Find the value of x.

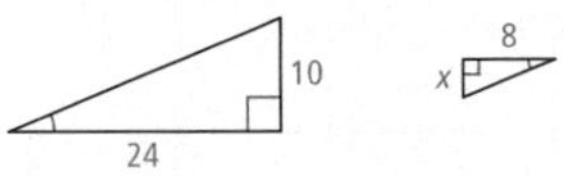

Are the triangles similar? If so, write a similarity statement and name the postulate or theorem you used. If not, explain.

11.

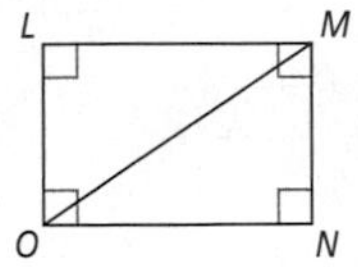

12.

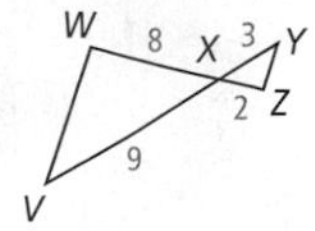

13.

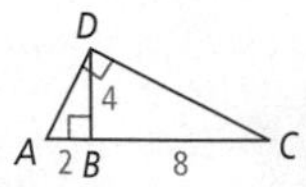

14.

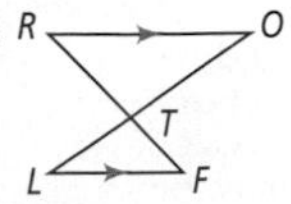

Algebra **Explain why the triangles are similar. Then find the value of x.**

15.

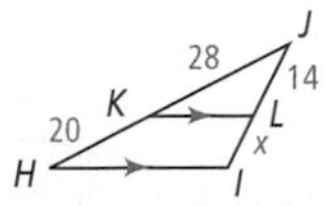

16.

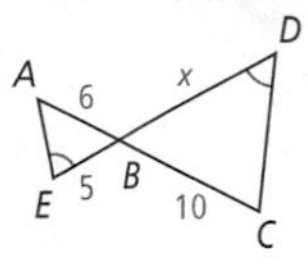

17. In a garden, a birdbath 2 ft 6 in. tall casts an 18-in. shadow at the same time an oak tree casts a 90-ft shadow. How tall is the oak tree?

Do you UNDERSTAND?

18. **Writing** You find an old scale drawing of your home, but the scale has faded and you cannot read it. How can you find the scale of the drawing?

19. **Reasoning** The sides of one triangle are twice as long as the corresponding sides of a second triangle. What is the relationship between the angles?

20. **Error Analysis** Your classmate says that since all congruent polygons are similar, all similar polygons must be congruent. Is he right? Explain.

Answers

Mid-Chapter Quiz

1. 1 : 8
2. $\frac{b}{10}$
3. Not ~; the ratio of the shorter sides is $\frac{2}{3}$ and the ratio of the longer sides is $\frac{4}{5}$. Since $\frac{2}{3} \neq \frac{4}{5}$, the polygons are not ~.
4. 2
5. $\frac{34}{7}$
6. 17.5 ft
7. *BDF*
8. *BF*
9. 3 ft by 2 ft (or 36 in. by 24 in.)
10. $\frac{10}{3}$
11. $\triangle LOM \sim \triangle NMO$; SAS~, AA~, or SSS~
12. No; corresp. sides are not proportional.
13. $\triangle ABD \sim \triangle DBC \sim \triangle ADC$: SSS ~, AA ~, or SAS ~
14. $\triangle TOR \sim \triangle TLF$; AA~
15. AA~; if lines are ∥, corresp. ∠s are ≅; 10
16. AA~; vert. ∠s are ≅; $\frac{25}{3}$
17. 150 ft
18. Answers may vary. Sample: Find a length *h* that you can measure on your house, and then find the length *d* on the drawing that represents *h*. The scale of the drawing is *d* : *h*.
19. The two △s are ~ by the SSS~ Theorem. Since the △s are ~, the corresp. ∠s are ≅.
20. No; two ≅ polygons have a 1 : 1 scale factor, but two ~ polygons can have any scale factor, so corresp. sides in ~ polygons are proportional but not necessarily ≅.

1 Interactive Learning

Solve It!

PURPOSE To show that the triangles created by drawing altitudes in a right triangle are similar

PROCESS Students may align the right angles of each triangle to see which angles are corresponding, redraw the triangles so they are oriented the same way, or use properties of complementary angles to show that the acute angles are congruent.

FACILITATE

Q What is the relationship of the angles that are produced by cutting apart the corners of the paper? **[They are complementary.]**

Q How are the opposite sides of the paper related? **[They are parallel.]**

Q What type of line is formed by the first cut in relationship to the parallel lines? **[It is a transversal.]**

ANSWER See Solve It in Answers on next page.

CONNECT THE MATH Ask students to identify the type of line that was drawn to separate the upper right-hand triangle into two similar triangles. Review the definition of an altitude. They will be using properties of this line to show that triangles are similar.

2 Guided Instruction

TAKE NOTE Have students use the congruence statements to write angle-congruence and side-proportionality statements. Be sure that students understand how each triangle is related to the other triangles.

7-4 Similarity in Right Triangles

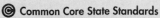

Common Core State Standards
G-SRT.B.5 Use . . . similarity criteria for triangles to solve problems and to prove relationships in geometric figures. **Also G-GPE.B.5**
MP 1, MP 3, MP 4

Objective To find and use relationships in similar right triangles

Analyze the situation first. Think about how you will match angles.

MATHEMATICAL PRACTICES

SOLVE IT! Getting Ready!

Draw a diagonal of a rectangular piece of paper to form two right triangles. In one triangle, draw the altitude from the right angle to the hypotenuse. Number the angles as shown. Cut out the three triangles. How can you match the angles of the triangles to show that all three triangles are similar? Explain how you know the matching angles are congruent.

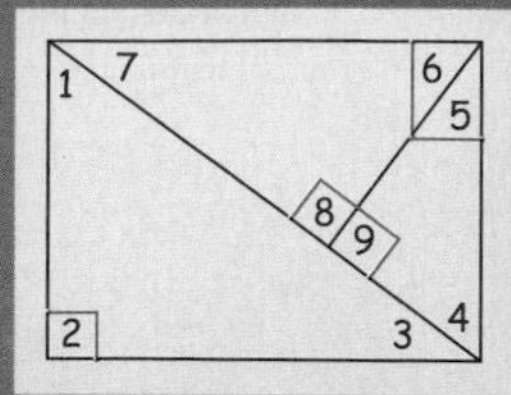

In the Solve It, you looked at three similar right triangles. In this lesson, you will learn new ways to think about the proportions that come from these similar triangles. You began with three separate, nonoverlapping triangles in the Solve It. Now you will see the two smaller right triangles fitting side-by-side to form the largest right triangle.

Essential Understanding When you draw the *altitude to the hypotenuse* of a right triangle, you form three pairs of similar right triangles.

Lesson Vocabulary
- geometric mean

take note

Theorem 7-3

Theorem
The altitude to the hypotenuse of a right triangle divides the triangle into two triangles that are similar to the original triangle and to each other.

If . . .
$\triangle ABC$ is a right triangle with right $\angle ACB$, and $\overline{CD}$ is the altitude to the hypotenuse

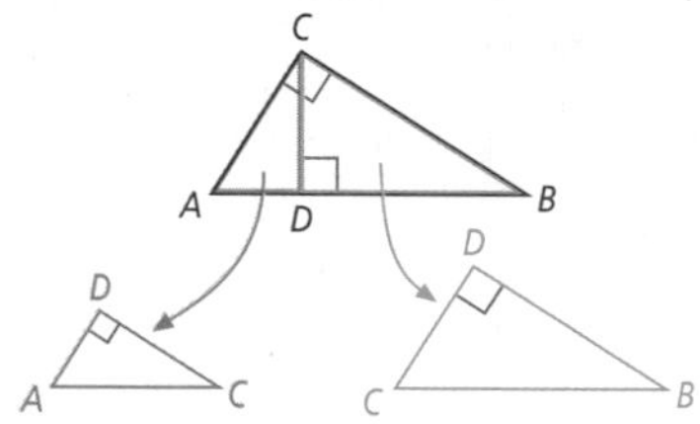

Then . . .
$\triangle ABC \sim \triangle ACD$
$\triangle ABC \sim \triangle CBD$
$\triangle ACD \sim \triangle CBD$

7-4 Preparing to Teach

BIG ideas **Reasoning and Proof**
Visualization
Proportionality

ESSENTIAL UNDERSTANDINGS
- Drawing in the altitude to the hypotenuse of a right triangle forms three pairs of similar right triangles.
- The altitude to the hypotenuse of a right triangle, the segments formed by the altitude, and the sides of the right triangle have lengths that are related using geometric means.

Math Background

Similar triangles are created when the altitude of a right triangle is drawn to the hypotenuse. The segments created in and existing in these triangles are related to the concept of geometric mean. A geometric mean is an average of factors that contribute to a specific product. It is used throughout geometry and mathematics to average quantities that are multiplied together. The relationship between the segments in right triangles with an altitude can be used for indirect measurement as well.

Mathematical Practices

Make sense of problems and persevere in solving them. Students will use the special case of drawing an altitude in right triangles to find similar triangles. They will also define and make explicit use of the term "geometric mean" based on this altitude.

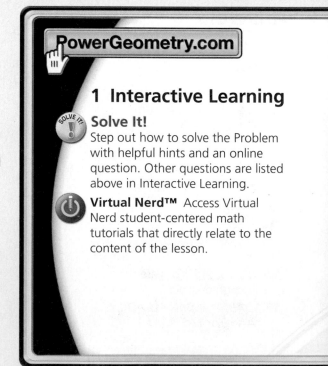

PowerGeometry.com

1 Interactive Learning

Solve It!
Step out how to solve the Problem with helpful hints and an online question. Other questions are listed above in Interactive Learning.

Virtual Nerd™ Access Virtual Nerd student-centered math tutorials that directly relate to the content of the lesson.

Proof **Proof of Theorem 7-3**

Given: Right $\triangle ABC$ with right $\angle ACB$ and altitude $\overline{CD}$

Prove: $\triangle ACD \sim \triangle ABC$, $\triangle CBD \sim \triangle ABC$, $\triangle ACD \sim \triangle CBD$

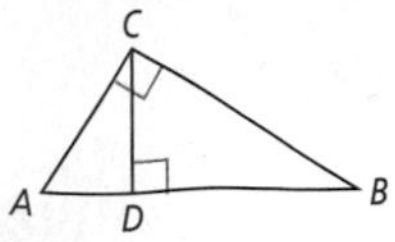

Statements	Reasons
1) $\angle ACB$ is a right angle.	1) Given
2) $\overline{CD}$ is an altitude.	2) Given
3) $\overline{CD} \perp \overline{AB}$	3) Definition of altitude
4) $\angle ADC$ and $\angle CDB$ are right angles.	4) Definition of $\perp$
5) $\angle ADC \cong \angle ACB$, $\angle CDB \cong \angle ACB$	5) All right ⦞ are $\cong$.
6) $\angle A \cong \angle A$, $\angle B \cong \angle B$	6) Reflexive Property of $\cong$
7) $\triangle ACD \sim \triangle ABC$, $\triangle CBD \sim \triangle ABC$	7) AA ~ Postulate
8) $\angle ACD \cong \angle B$	8) Corresponding ⦞ of ~ △s are $\cong$.
9) $\angle ADC \cong \angle CDB$	9) All right ⦞ are $\cong$.
10) $\triangle ACD \sim \triangle CBD$	10) AA ~ Postulate

Problem 1 **Identifying Similar Triangles**

What similarity statement can you write relating the three triangles in the diagram?

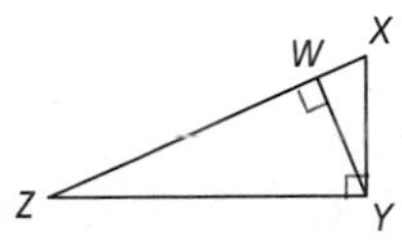

Plan

What will help you see the corresponding vertices?
Sketch the triangles separately in the same orientation.

$\overline{YW}$ is the altitude to the hypotenuse of right $\triangle XYZ$, so you can use Theorem 7-3. There are three similar triangles.

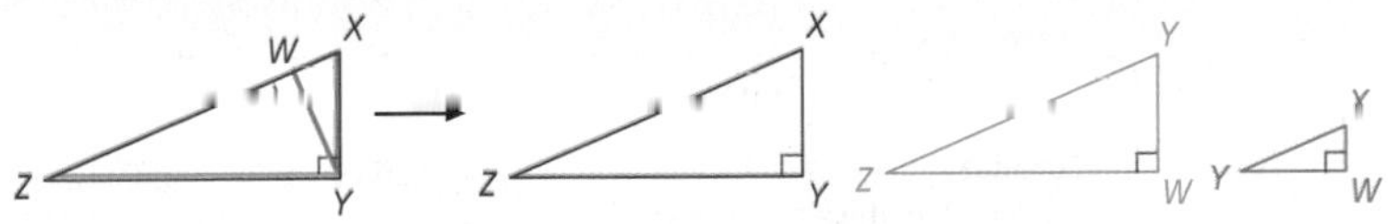

$\triangle XYZ \sim \triangle YWZ \sim \triangle XWY$

Got It? 1. a. What similarity statement can you write relating the three triangles in the diagram?
b. **Reasoning** From the similarity statement in part (a), write two different proportions using the ratio $\frac{SR}{SP}$.

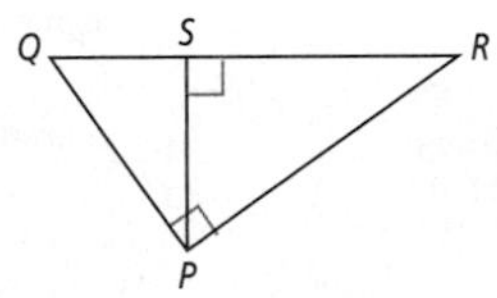

It may help students to reorganize the statements to address one set of triangles at a time. Begin by having students draw each triangle separately with the right angle pointing upward. They should first show that $\triangle ACD \sim \triangle ABC$ by finding two angles in those triangles congruent and proving the triangles similar by AA~. Next, they can show that $\triangle CBD \sim \triangle ABC$. Finally, have students show $\triangle ACD \sim \triangle CDB$ by AA~. Note that in the proof $\angle B$ is only shown as $\angle B$. As students examine their separated triangles, make sure they know $\angle CBA$ and $\angle CBD$ are the same angle.

Problem 1

Q Which angles are congruent based on the diagram? Why? **[$\angle XYZ$, $\angle XWY$, and $\angle YWZ$; all three are right angles.]**

Q According to the Reflexive Property, which pairs of angles are congruent? **[$\angle Z \cong \angle Z$, $\angle X \cong \angle X$]**

Q How do you know the order in which to list the vertices? **[When there is a correspondence between figures, corresponding vertices should be listed in the same order.]**

Got It?

Remind students to draw each triangle separately with the same orientation before they attempt to write a similarity statement or answer 1b.

2 Guided Instruction

Each Problem is worked out and supported online.

Problem 1
Identifying Similar Triangles
Animated

Problem 2
Finding the Geometric Mean
Animated

Problem 3
Using the Corollaries
Animated

Problem 4
Finding a Distance

Support in Geometry Companion

- Vocabulary
- Key Concepts
- Got It?

Answers

Solve It?

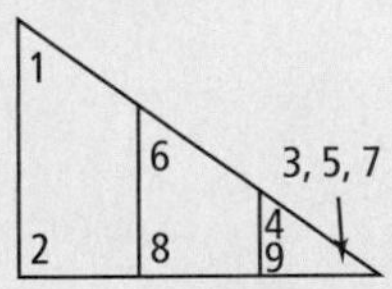

$\angle 2 \cong \angle 8 \cong \angle 9$ because all rt. ⦞ are congruent. In the original diagram, $\angle 1 \cong \angle 4$ and $\angle 3 \cong \angle 7$ because they are alt. int. ⦞ of $\parallel$ lines. By the Triangle-Angle-Sum Thm. you can show that $\angle 1 \cong \angle 6$ and $\angle 3 \cong \angle 5$.

Got It?

1a. $\triangle PRQ \sim \triangle SPQ \sim \triangle SRP$

b. $\frac{SR}{SP} = \frac{SP}{SQ}$, $\frac{SR}{SP} = \frac{PR}{QP}$

Problem 2

Q How would you state the definition of geometric mean in your own words? **[Answers will vary. Sample: The geometric mean of two numbers has the same ratio to the first number that the second number has to the mean.]**

Q How do you find the geometric mean of two numbers? **[Set up a proportion where the means are unknown and the given numbers are the extremes.]**

Got It? SYNTHESIZING

Remind students to set up a proportion with the unknowns in the mean positions. Prompt students to extend the definition of geometric mean to include the geometric mean of three numbers. **[The geometric mean of three numbers is the cube root of the product of the numbers.]**

Take Note

Ask students to find a set of integers that satisfy Corollary 1 to Theorem 7-3. Have them sketch the triangle and label each segment accordingly. Then they should show that the numbers satisfy Corollary 1. Students could also use the corollary to find the numbers by finding equivalent ratios.

Proportions in which the means are equal occur frequently in geometry. For any two positive numbers a and b, the **geometric mean** of a and b is the positive number x such that $\frac{a}{x} = \frac{x}{b}$.

Problem 2 Finding the Geometric Mean

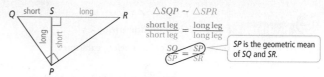

Think
How do you use the definition of geometric mean?
Set up a proportion with x in both means positions. The numbers 6 and 15 go into the extremes positions.

Multiple Choice What is the geometric mean of 6 and 15?

(A) 90 (B) $3\sqrt{10}$ (C) $9\sqrt{10}$ (D) 30

$\frac{6}{x} = \frac{x}{15}$ Definition of geometric mean

$x^2 = 90$ Cross Products Property

$x = \sqrt{90}$ Take the positive square root of each side.

$x = 3\sqrt{10}$ Write in simplest radical form.

The geometric mean of 6 and 15 is $3\sqrt{10}$. The correct answer is B.

Got It? 2. What is the geometric mean of 4 and 18?

In Got It 1 part (b), you used a pair of similar triangles to write a proportion with a geometric mean.

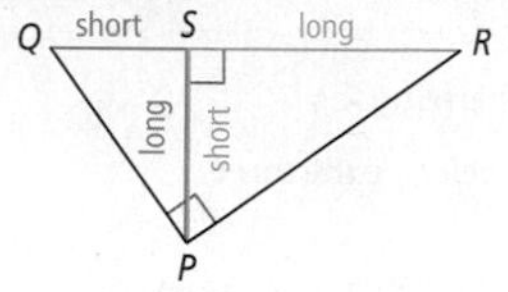

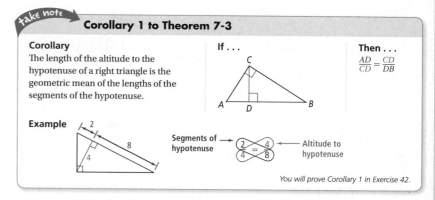

$\triangle SQP \sim \triangle SPR$

$\frac{\text{short leg}}{\text{short leg}} = \frac{\text{long leg}}{\text{long leg}}$

$\frac{SQ}{SP} = \frac{SP}{SR}$ — *SP* is the geometric mean of *SQ* and *SR*.

This illustrates the first of two important corollaries of Theorem 7-3.

take note — Corollary 1 to Theorem 7-3

Corollary
The length of the altitude to the hypotenuse of a right triangle is the geometric mean of the lengths of the segments of the hypotenuse.

If . . .

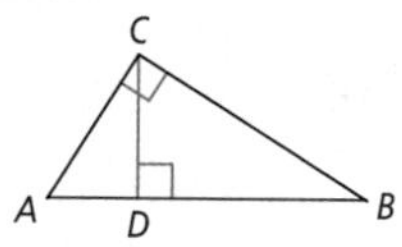

Then . . .
$\frac{AD}{CD} = \frac{CD}{DB}$

Example

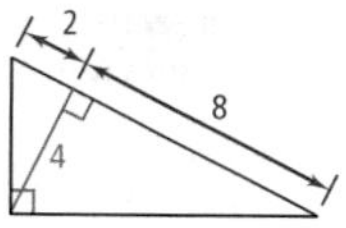

Segments of hypotenuse → $\frac{2}{4} = \frac{4}{8}$ ← Altitude to hypotenuse

You will prove Corollary 1 in Exercise 42.

Additional Problems

1. What similarity statement can you write relating the three triangles in the diagram?

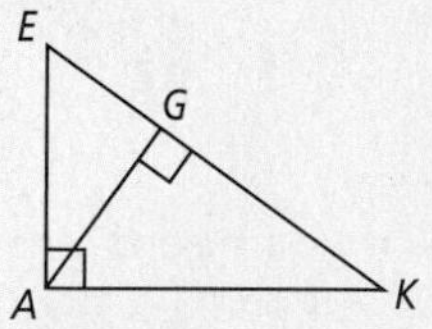

ANSWER $\triangle EAK \sim \triangle AGK \sim \triangle EGA$

2. What is the geometric mean of 5 and 12?

A. 60

B. $12\sqrt{3}$

C. $4\sqrt{15}$

D. $2\sqrt{15}$

ANSWER D

3. What are the values of x and y?

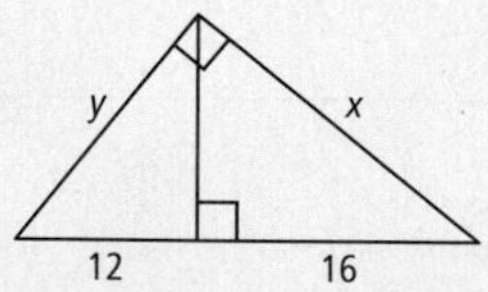

ANSWER $x = 8\sqrt{7}, y = 4\sqrt{21}$

4. Maggie has a kite with the dimensions shown below. What is the width of the kite?

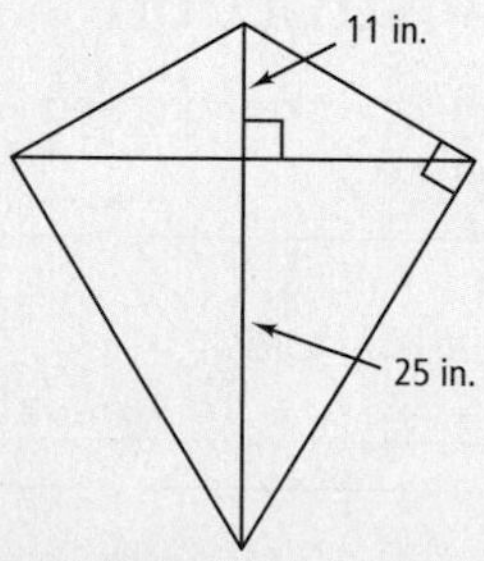

ANSWER $10\sqrt{11}$ in.

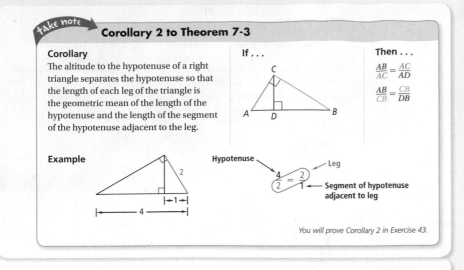

Take Note — Corollary 2 to Theorem 7-3

Corollary	If . . .	Then . . .
The altitude to the hypotenuse of a right triangle separates the hypotenuse so that the length of each leg of the triangle is the geometric mean of the length of the hypotenuse and the length of the segment of the hypotenuse adjacent to the leg.	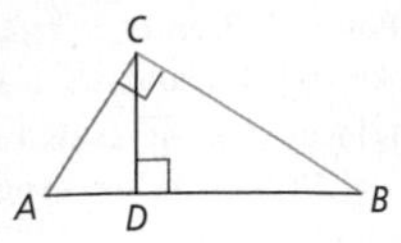	$\frac{AB}{AC} = \frac{AC}{AD}$ $\frac{AB}{CB} = \frac{CB}{DB}$

Example

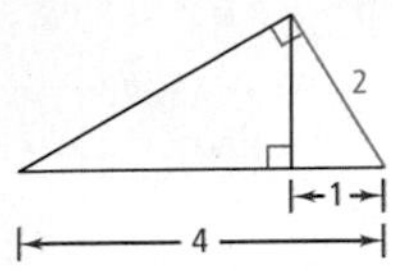

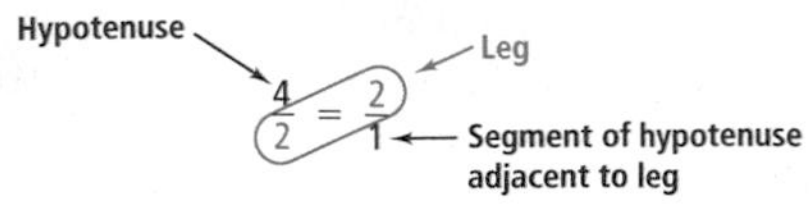

You will prove Corollary 2 in Exercise 43.

The corollaries to Theorem 7-3 give you ways to write proportions using lengths in right triangles without thinking through the similar triangles. To help remember these corollaries, consider the diagram and these properties.

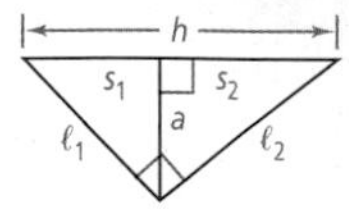

Corollary 1

$\frac{s_1}{a} = \frac{a}{s_2}$

Corollary 2

$\frac{h}{\ell_1} = \frac{\ell_1}{s_1}, \frac{h}{\ell_2} = \frac{\ell_2}{s_2}$

Problem 3 Using the Corollaries

Algebra What are the values of x and y?

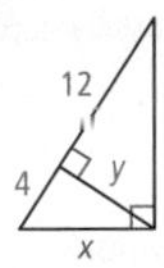

Plan

How do you decide which corollary to use?
If you are using or finding an altitude, use Corollary 1. If you are using or finding a leg or hypotenuse, use Corollary 2.

Use Corollary 2.

$\frac{4 + 12}{x} = \frac{x}{4}$	Write a proportion.	$\frac{4}{y} = \frac{y}{12}$
$x^2 = 64$	Cross Products Property	$y^2 = 48$
$x = \sqrt{64}$	Take the positive square root.	$y = \sqrt{48}$
$x = 8$	Simplify.	$y = 4\sqrt{3}$

Use Corollary 1.

Got It? 3. What are the values of x and y?

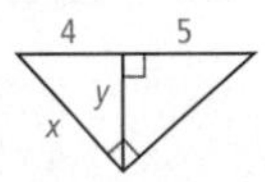

Take Note

Have students compare and contrast the corollaries to Theorem 7-3. The first corollary relates the segments of the hypotenuse created by the altitude. The second corollary relates the entire hypotenuse to the length of a leg and the segment of the hypotenuse adjacent to that leg. Also, have students read the proportions using analogy language: "*AB* is to *AC* as *AC* is to *AD*."

Problem 3

Have students redraw each triangle in the same orientation. This helps them to write proportions based on the similarity of the triangles.

Q Which corollary relates segments in the triangle to the segment marked *x*? Explain. **[The segment marked as *x* is a leg of the largest right triangle. Corollary 2 relates this segment to the entire hypotenuse and the adjacent segment of the hypotenuse.]**

Got It?

Q What proportion do you write to solve for *x*? Which corollary did you use? **[$\frac{4}{x} = \frac{x}{9}$; Corollary 2]**

Q What proportion do you write to solve for *y*? Which corollary did you use? **[$\frac{4}{y} = \frac{y}{5}$; Corollary [illegible]]**

Answers

Got It? (continued)

2. $6\sqrt{2}$

3. $x = 6, y = 2\sqrt{5}$

Problem 4

Q What must you do in order to find the entire length of the hypotenuse? **[Add the lengths of the two segments of the hypotenuse.]**

Got It?

Q Would it have been possible to find the distance from *B* to *D* on the larger triangle without solving for *x* first? Explain. **[No, the corollary relates the length of the altitude to the lengths of the segments of the hypotenuse. You can only have one unknown in a proportion if you are to solve it.]**

3 Lesson Check

Do you know HOW?

- If students have difficulty with Exercises 3-6, then have them redraw the triangles so that they are oriented the same way.

Do you UNDERSTAND?

- If students have difficulty with Exercise 8, then have them review Problem 3.

Close

Q How is the geometric mean used in right triangles? **[The altitude of a right triangle to the hypotenuse is the geometric mean of the segments of the hypotenuse it creates. A leg of a right triangle is the geometric mean of the hypotenuse and the segment of the hypotenuse created by the altitude, adjacent to the leg.]**

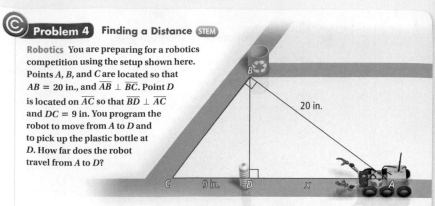

Problem 4 Finding a Distance STEM

Robotics You are preparing for a robotics competition using the setup shown here. Points *A*, *B*, and *C* are located so that $AB = 20$ in., and $\overline{AB} \perp \overline{BC}$. Point *D* is located on $\overline{AC}$ so that $\overline{BD} \perp \overline{AC}$ and $DC = 9$ in. You program the robot to move from *A* to *D* and to pick up the plastic bottle at *D*. How far does the robot travel from *A* to *D*?

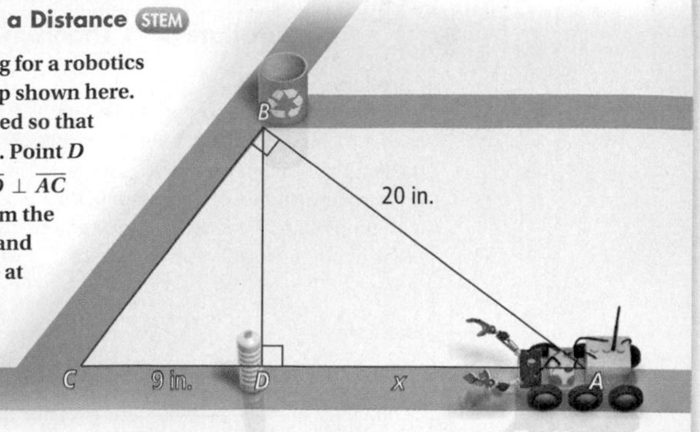

Think

You can't solve this equation by taking the square root. What do you do?
Write the quadratic equation in the standard form $ax^2 + bx + c = 0$. Then solve by factoring or use the quadratic formula.

$\frac{x+9}{20} = \frac{20}{x}$	Corollary 2
$x^2 + 9x = 400$	Cross Products Property
$x^2 + 9x - 400 = 0$	Subtract 400 from each side.
$(x - 16)(x + 25) = 0$	Factor.
$x - 16 = 0$ or $(x + 25) = 0$	Zero-Product Property
$x = 16$ or $x = -25$	Solve for *x*.

Only the positive solution makes sense in this situation. The robot travels 16 in.

Got It? 4. From point *D*, the robot must turn right and move to point *B* to put the bottle in the recycling bin. How far does the robot travel from *D* to *B*?

Lesson Check

Do you know HOW?

Find the geometric mean of each pair of numbers.

1. 4 and 9 **2.** 4 and 12

Use the figure to complete each proportion.

3. $\frac{g}{e} = \frac{e}{\blacksquare}$ **4.** $\frac{j}{d} = \frac{d}{\blacksquare}$

5. $\frac{\blacksquare}{f} = \frac{f}{\blacksquare}$ **6.** $\frac{j}{\blacksquare} = \frac{\blacksquare}{g}$

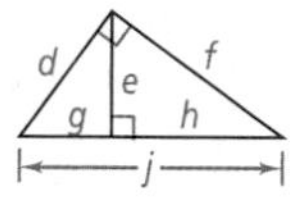

Do you UNDERSTAND? MATHEMATICAL PRACTICES

7. Vocabulary Identify the following in $\triangle RST$.
a. the hypotenuse
b. the segments of the hypotenuse
c. the segment of the hypotenuse adjacent to leg $\overline{ST}$

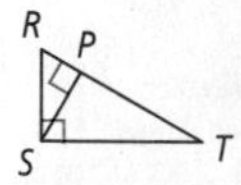

8. Error Analysis A classmate wrote an incorrect proportion to find *x*. Explain and correct the error.

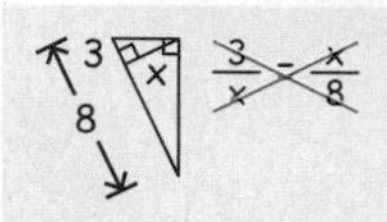

PowerGeometry.com

3 Lesson Check

For a digital lesson check, use the Got It questions.

Support in Geometry Companion
- Lesson Check

4 Practice

Assign homework to individual students or to an entire class.

Answers

Got It? (continued)

4. 12 in.

Lesson Check

1. 6
2. $\sqrt{48}$ or $4\sqrt{3}$
3. *h*
4. *g*
5. *j*, *h* or *h*, *j*
6. *d*, *d*
7a. $\overline{RT}$
b. $\overline{RP}$, $\overline{PT}$
c. $\overline{PT}$
8. The length 8 is the entire hypotenuse, so the segments of the hypotenuse have lengths 3 and 5. The correct proportion is $\frac{3}{x} = \frac{x}{5}$.

Practice and Problem-Solving Exercises

MATHEMATICAL PRACTICES

A Practice

Write a similarity statement relating the three triangles in each diagram. See Problem 1.

9. 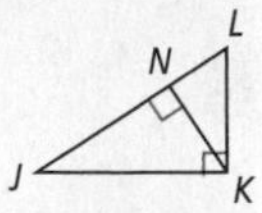

10. 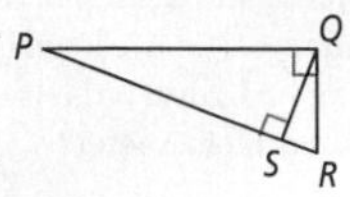

11. 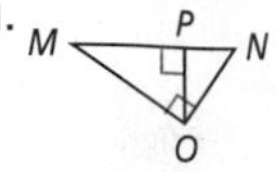

Algebra Find the geometric mean of each pair of numbers. See Problem 2.

12. 4 and 10 **13.** 3 and 48 **14.** 5 and 125

15. 7 and 9 **16.** 3 and 16 **17.** 4 and 49

Algebra Solve for x and y. See Problems 3 and 4.

18. 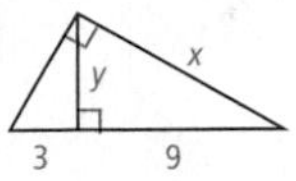

19. 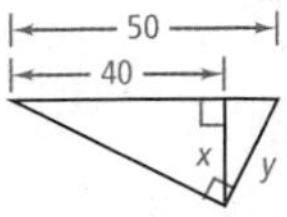

20. 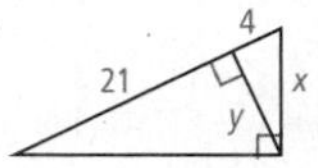

21. 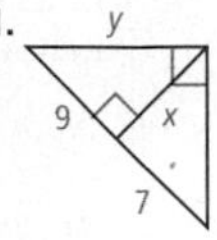

STEM **22. Architecture** The architect's side-view drawing of a saltbox-style house shows a post that supports the roof ridge. The support post is 10 ft tall. How far from the front of the house is the support post positioned?

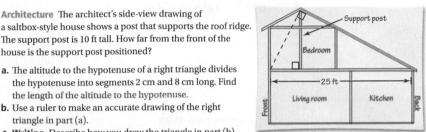

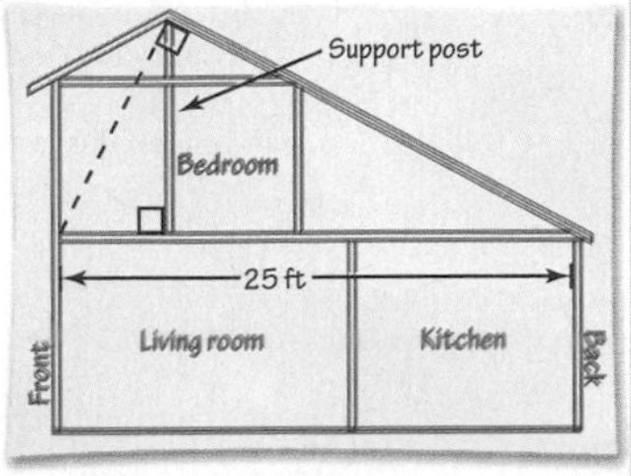

B Apply

23. a. The altitude to the hypotenuse of a right triangle divides the hypotenuse into segments 2 cm and 8 cm long. Find the length of the altitude to the hypotenuse.

b. Use a ruler to make an accurate drawing of the right triangle in part (a).

c. Writing Describe how you drew the triangle in part (b).

Algebra Find the geometric mean of each pair of numbers.

24. 1 and 1000 **25.** 5 and 1.25 **26.** $\sqrt{8}$ and $\sqrt{2}$ **27.** $\frac{1}{2}$ and 2 **28.** $\sqrt{28}$ and $\sqrt{7}$

29. Reasoning A classmate says the following statement is true: The geometric mean of positive numbers a and b is $\sqrt{ab}$. Do you agree? Explain.

30. Think About a Plan The altitude to the hypotenuse of a right triangle divides the hypotenuse into segments with lengths in the ratio 1 : 2. The length of the altitude is 8. How long is the hypotenuse?

- How can you use the given ratio to help you draw a sketch of the triangle?
- How can you use the given ratio to write expressions for the lengths of the segments of the hypotenuse?
- Which corollary to Theorem 7-3 applies to this situation?

4 Practice

ASSIGNMENT GUIDE

Basic: 9–22, 24–25, 30, 37–39
Average: 9–21 odd, 23–44
Advanced: 9–21 odd, 23–47
Standardized Test Prep: 48–50
Mixed Review: 51–56

Mathematical Practices are supported by exercises with red headings. Here are the Practices supported in this lesson:

MP 1: Make Sense of Problems Ex. 30
MP 3: Construct Arguments Ex. 32
MP 3: Communicate Ex. 23c
MP 3: Critique the Reasoning of Others Ex. 8, 29

Applications exercises have blue headings. Exercises 22 and 31 support MP 4: Model.

STEM exercises focus on science or engineering applications.

EXERCISE 37: Use the Think About a Plan worksheet in the **Practice and Problem Solving Workbook** (also available in the Teaching Resources in print and online) to further support students' development in becoming independent learners.

HOMEWORK QUICK CHECK

To check students' understanding of key skills and concepts, go over Exercises 9, 19, 25, 30, and 37.

Practice and Problem-Solving Exercises

9. Answers may vary. Sample: $\triangle KJL \sim \triangle NJK \sim \triangle NKL$

10. Answers may vary. Sample: $\triangle QPR \sim \triangle SPQ \sim \triangle SQR$

11. Answers may vary. Sample: $\triangle OMN \sim \triangle PMO \sim \triangle PON$

12. $\sqrt{40}$ or $2\sqrt{10}$

13. 12

14. 25

15. $\sqrt{63}$ or $3\sqrt{7}$

16. $\sqrt{48}$ or $4\sqrt{3}$

17. 14

18. $x = 6\sqrt{3}, y = 3\sqrt{3}$

19. $x = 20, y = 10\sqrt{5}$

20. $x = 10, y = 2\sqrt{21}$

21. $x = 3\sqrt{7}, y = 12$

22. 5 ft

23a. 4 cm

b. 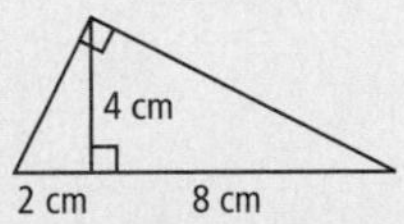

c. Answers may vary. Sample: Draw a 10-cm segment. Construct a $\perp$ of length 4 cm that is 2 cm from one endpoint; connect to form a $\triangle$.

24. $10\sqrt{10}$

25. 2.5

26. 2

27. 1

28. $\sqrt{14}$

29. Yes; the proportion $\frac{a}{\sqrt{ab}} = \frac{\sqrt{ab}}{b}$ is true by the Cross Products Prop. and satisfies the definition of the geometric mean.

30. $12\sqrt{2}$ units

Answers

Practice and Problem-Solving Exercises (continued)

31. 8.50 m

32. They are equal. Sample explanation: Let a = length of the altitude and $2x$ = the length of the hypotenuse. Then $\frac{a}{x} = \frac{x}{a}$, so $a = x$.

33. $\ell_1 = \sqrt{2}$, $\ell_2 = \sqrt{2}$, $a = 1$, $s_2 = 1$

34. $\ell_1 = 6\sqrt{2}$, $\ell_2 = 6\sqrt{2}$, $h = 12$, $s_2 = 6$

35. $\ell_2 = 2\sqrt{3}$, $h = 4$, $a = \sqrt{3}$, $s_1 = 1$

36. $\ell_1 = 6$, $h = 12$, $a = 3\sqrt{3}$, $s_2 = 9$

37. (−2, 6), (10, 6)

38. 3 **39.** 4 **40.** 6 **41.** 5

42. $\triangle ACD \sim \triangle CBD$ by Thm. 7-3, so $\frac{AD}{CD} = \frac{CD}{BD}$ because corresp. sides of ~ △ are proportional.

43. $\triangle ABC \sim \triangle ACD$ and $\triangle ABC \sim \triangle CBD$ by Thm. 7-3. Then $\frac{AB}{AC} = \frac{AC}{AD}$ and $\frac{AB}{CB} = \frac{BC}{BD}$ because corresp. sides of ~ △ are proportional.

44. Rt. $\triangle ABC$ with alt. to the hypotenuse $\overline{AB}$ (given); $\frac{a}{b} = \frac{b}{c}$ (Corollary 1 to Thm. 7-3); Slope of $\overleftrightarrow{AC} = \frac{b}{a}$ and slope of $\overleftrightarrow{BC} = -\frac{b}{c}$. Since $\frac{a}{b} = \frac{b}{c}$, $-\frac{a}{b} = -\frac{b}{c}$. Therefore the product of the slopes, $\frac{b}{a} \cdot -\frac{a}{b} = -1$.

45a.

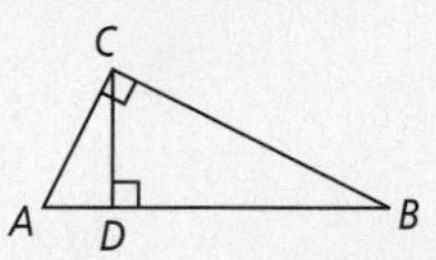

Given: $\overline{AC} \perp \overline{BC}$, $\overline{AB} \perp \overline{CD}$

Prove: $AC \cdot BC = AB \cdot CD$

b. The conjecture is true. You can express the area of $\triangle ABC$ as $\frac{1}{2}(AC)(BC)$ or as $\frac{1}{2}(AB)(CD)$, so $AC \cdot BC = AB \cdot CD$.

---

31. Archaeology To estimate the height of a stone figure, Anya holds a small square up to her eyes and walks backward from the figure. She stops when the bottom of the figure aligns with the bottom edge of the square and the top of the figure aligns with the top edge of the square. Her eye level is 1.84 m from the ground. She is 3.50 m from the figure. What is the height of the figure to the nearest hundredth of a meter?

32. Reasoning Suppose the altitude to the hypotenuse of a right triangle bisects the hypotenuse. How does the length of the altitude compare with the lengths of the segments of the hypotenuse? Explain.

The diagram shows the parts of a right triangle with an altitude to the hypotenuse. For the two given measures, find the other four. Use simplest radical form.

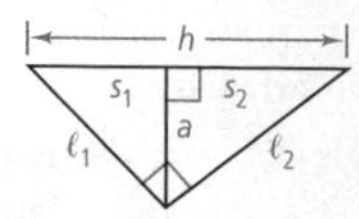

33. $h = 2$, $s_1 = 1$ **34.** $a = 6$, $s_1 = 6$ **35.** $\ell_1 = 2$, $s_2 = 3$ **36.** $s_1 = 3$, $\ell_2 = 6\sqrt{3}$

37. Coordinate Geometry $\overline{CD}$ is the altitude to the hypotenuse of right $\triangle ABC$. The coordinates of A, D, and B are (4, 2), (4, 6), and (4, 15), respectively. Find all possible coordinates of point C.

Algebra **Find the value of x.**

38.

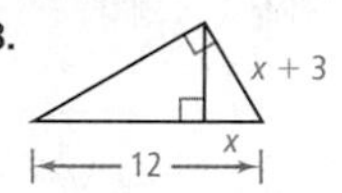

39.

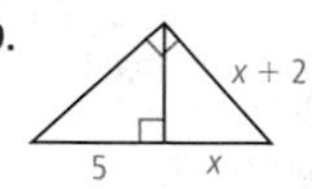

40.

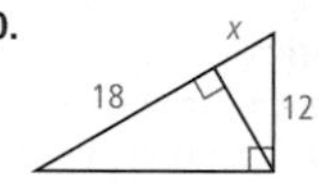

41.

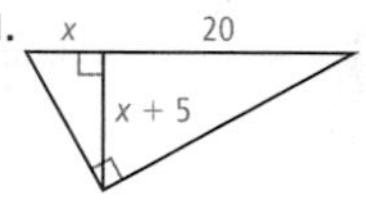

Use the figure at the right for Exercises 42–43.

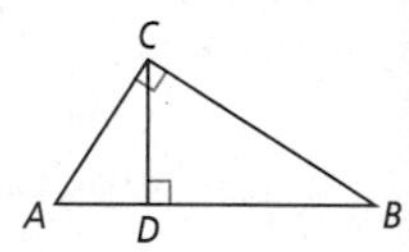

42. Prove Corollary 1 to Theorem 7-3.
Proof
Given: Right $\triangle ABC$ with altitude to the hypotenuse $\overline{CD}$
Prove: $\frac{AD}{CD} = \frac{CD}{DB}$

43. Prove Corollary 2 to Theorem 7-3.
Proof
Given: Right $\triangle ABC$ with altitude to the hypotenuse $\overline{CD}$
Prove: $\frac{AB}{AC} = \frac{AC}{AD}$, $\frac{AB}{BC} = \frac{BC}{DB}$

44. **Given:** Right $\triangle ABC$ with altitude $\overline{CD}$ to the hypotenuse $\overline{AB}$
Proof
Prove: The product of the slopes of perpendicular lines is −1.

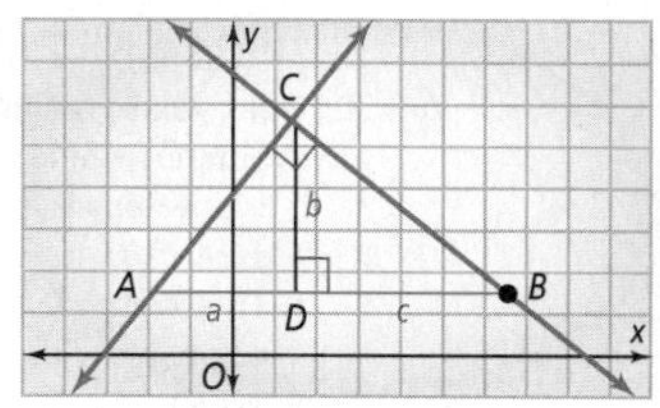

Challenge

45. a. Consider the following conjecture: The product of the lengths of the two legs of a right triangle is equal to the product of the lengths of the hypotenuse and the altitude to the hypotenuse. Draw a figure for the conjecture. Write the *Given* information and what you are to *Prove*.

b. Reasoning Is the conjecture true? Explain.

46. a. In the diagram, $c = x + y$. Use Corollary 2 to Theorem 7-3 to write two more equations involving a, b, c, x, and y.
b. The equations in part (a) form a system of three equations in five variables. Reduce the system to one equation in three variables by eliminating x and y.
c. State in words what the one resulting equation tells you.

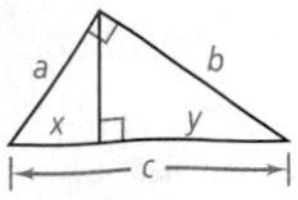

47. Given: In right $\triangle ABC$, $\overline{BD} \perp \overline{AC}$, and $\overline{DE} \perp \overline{BC}$.
Proof **Prove:** $\frac{AD}{DC} = \frac{BE}{EC}$

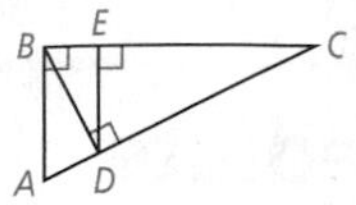

Standardized Test Prep

SAT/ACT

48. The altitude to the hypotenuse of a right triangle divides the hypotenuse into segments of lengths 5 and 15. What is the length of the altitude?

Ⓐ 3 Ⓑ $5\sqrt{3}$ Ⓒ 10 Ⓓ $5\sqrt{5}$

49. A triangle has side lengths 3 in., 4 in., and 6 in. The longest side of a similar triangle is 15 in. What is the length of the shortest side of the similar triangle?

Ⓕ 1 in. Ⓖ 1.2 in. Ⓗ 7.5 in. Ⓘ 10 in.

Short Response

50. Two students disagree about the measures of angles in a kite. They know that two angles measure 124 and 38. But they get different answers for the other two angles. Can they both be correct? Explain.

Mixed Review

51. Write a similarity statement for the two triangles. How do you know they are similar?

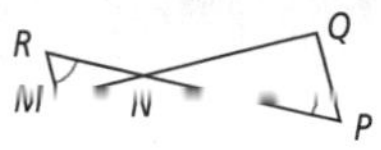

See Lesson 7-3.

Algebra Find the values of x and y in $\square RSTV$.

See Lesson 6-2.

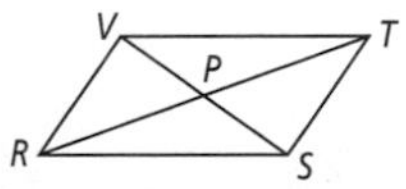

52. $RP = 2x$, $PT = y + 2$, $VP = y$, $PS = x + 3$

53. $RV = 2x + 3$, $VT = 5x$, $TS = y + 5$, $SR = 4y - 1$

Get Ready! **To prepare for Lesson 7-5, do Exercises 54–56.**

The two triangles in each diagram are similar. Find the value of x in each.

See Lesson 7-2.

54.

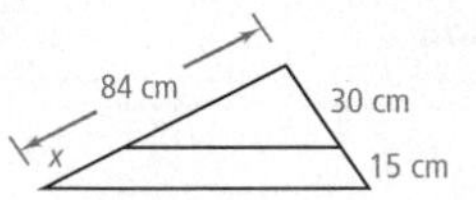

55.

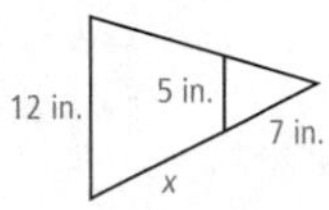

56.

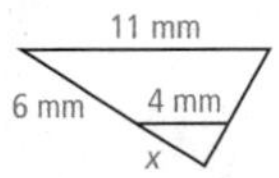

46a. $\frac{x}{a} = \frac{a}{c}$, $\frac{y}{b} = \frac{b}{c}$
b. $c^2 = a^2 + b^2$
c. The square of the hypotenuse equals the sum of the squares of the legs.

47. Answers may vary. Sample: $\triangle ABC \sim \triangle DEC$ (AA ~ Post.), so $\frac{AC}{DC} = \frac{BC}{EC}$ (Corr. sides of ~ ▲s are in prop.). By the Subtraction Property of =, $\frac{AC - DC}{DC} = \frac{BC - EC}{EC}$, or $\frac{AD}{DC} = \frac{BE}{EC}$.

48. B **49.** H

50. [2] Yes; they can both be correct. The three possibilities for the measures of the four angles can be 38, 124, 124, 74; 38, 124, 38, 160; and 38, 124, 99, 99.
[1] incomplete OR incorrect answer

51. $\angle R \cong \angle P$ (given) and $\angle RNM \cong \angle PNQ$ (Vert. ∠s are ≅.), so $\triangle NRM \sim \triangle NPQ$ by AA ~.

52. $x = 5$, $y = 8$

53. $x = 3$, $y = 4$

54. 28 cm **55.** 9.8 in.

56. $\frac{24}{7}$ mm or $3\frac{3}{7}$ mm

7-4 Lesson Resources

Differentiated Remediation

Instructional Support

Geometry Companion

Students can use the **Geometry Companion** worktext (4 pages) . . .

- New Vocabulary
- Key Concepts
- Got It for each Problem
- Lesson Check

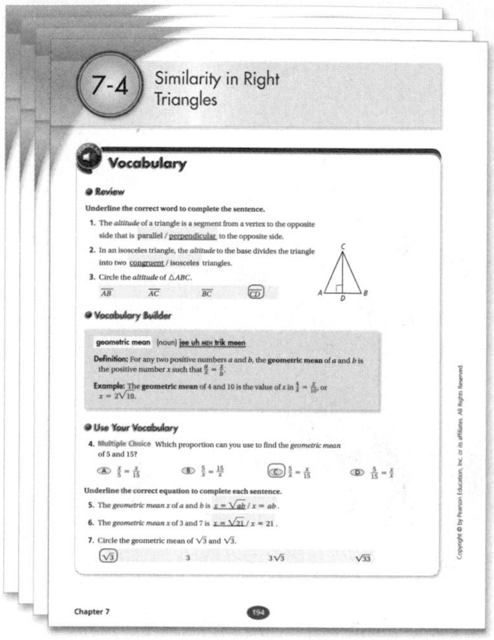

7-4 Similarity in Right Triangles

Vocabulary

Review

Underline the correct word to complete the sentence.

1. The *altitude* of a triangle is a segment from a vertex to the opposite side that is parallel / perpendicular to the opposite side.
2. In an isosceles triangle, the *altitude* to the base divides the triangle into two congruent / isosceles triangles.
3. Circle the *altitude* of $\triangle ABC$.

$\overline{AB}$ $\overline{AC}$ $\overline{BC}$ $\overline{CD}$

Vocabulary Builder

geometric mean (noun) jee uh MEH trik meen

Definition: For any two positive numbers a and b, the **geometric mean** of a and b is the positive number x such that $\frac{a}{x} = \frac{x}{b}$.

Example: The **geometric mean** of 4 and 10 is the value of x in $\frac{4}{x} = \frac{x}{10}$, or $x = 2\sqrt{10}$.

Use Your Vocabulary

4. Multiple Choice Which proportion can you use to find the *geometric mean* of 5 and 15?

A $\frac{x}{5} = \frac{x}{15}$ B $\frac{5}{x} = \frac{15}{x}$ C $\frac{5}{x} = \frac{x}{15}$ D $\frac{5}{15} = \frac{x}{x}$

Underline the correct equation to complete each sentence.

5. The *geometric mean* x of a and b is $x = \sqrt{ab}$ / $x = ab$.
6. The *geometric mean* x of 3 and 7 is $x = \sqrt{21}$ / $x = 21$.
7. Circle the geometric mean of $\sqrt{3}$ and $\sqrt{3}$.

$\sqrt{3}$ 3 $3\sqrt{3}$ $\sqrt{33}$

Chapter 7 194

ELL Support

Focus on Communication Have a volunteer draw a right triangle on the board. Point to a leg of the triangle. Ask: What is the name of this side? Point to the hypotenuse and repeat the question. Draw the altitude perpendicular to the hypotenuse and touching the corner between the two legs. Ask: What is the name of this line? [the altitude to the hypotenuse] Ask volunteers to point to the two right triangles that the altitude forms.

Have small groups draw a right triangle like the one on the board. Say: Label your triangle, then list the similar triangles. Pair groups together. Have one group describe its triangle and the other group draw a copy based on the description.

5 Assess & Remediate

Lesson Quiz

1. What similarity statement can you write relating the three triangles in the diagram?

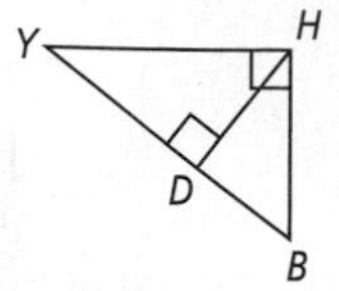

2. What is the geometric mean of 6 and 16?
3. **Do you UNDERSTAND?** What are the values of x and y?

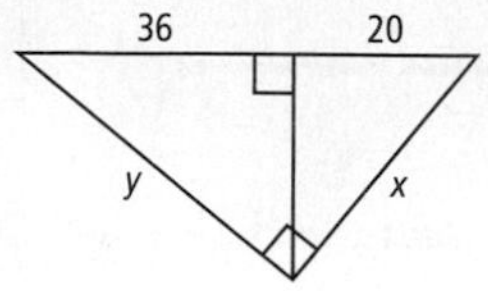

ANSWERS TO LESSON QUIZ

1. $\triangle YHB \sim \triangle YDH \sim \triangle HDB$
2. $4\sqrt{6}$
3. $x = 4\sqrt{70}$, $y = 12\sqrt{14}$

PRESCRIPTION FOR REMEDIATION

Use the student work on the Lesson Quiz to prescribe a differentiated review assignment.

Points	Differentiated Remediation
0–1	Intervention
2	On-level
3	Extension

PowerGeometry.com

5 Assess & Remediate

Assign the Lesson Quiz. Appropriate intervention, practice, or enrichment is automatically generated based on student performance.

Intervention

- **Reteaching** (2 pages) Provides reteaching and practice exercises for the key lesson concepts. Use with struggling students or absent students.
- **English Language Learner Support** Helps students develop and reinforce mathematical vocabulary and key concepts.

All-in-One Resources/Online

Reteaching

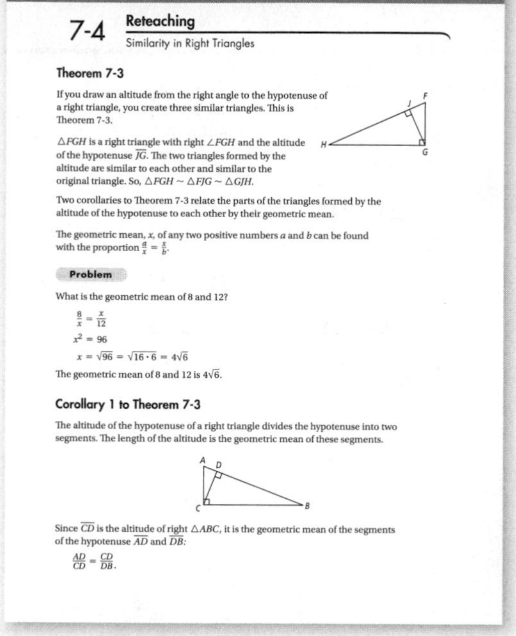

7-4 Reteaching

Similarity in Right Triangles

Theorem 7-3

If you draw an altitude from the right angle to the hypotenuse of a right triangle, you create three similar triangles. This is Theorem 7-3.

$\triangle FGH$ is a right triangle with right $\angle FGH$ and the altitude of the hypotenuse $\overline{JG}$. The two triangles formed by the altitude are similar to each other and similar to the original triangle. So, $\triangle FGH \sim \triangle FJG \sim \triangle GJH$.

Two corollaries to Theorem 7-3 relate the parts of the triangles formed by the altitude of the hypotenuse to each other by their geometric mean.

The geometric mean, x, of any two positive numbers a and b can be found with the proportion $\frac{a}{x} = \frac{x}{b}$.

Problem

What is the geometric mean of 8 and 12?

$\frac{8}{x} = \frac{x}{12}$

$x^2 = 96$

$x = \sqrt{96} = \sqrt{16 \cdot 6} = 4\sqrt{6}$

The geometric mean of 8 and 12 is $4\sqrt{6}$.

Corollary 1 to Theorem 7-3

The altitude of the hypotenuse of a right triangle divides the hypotenuse into two segments. The length of the altitude is the geometric mean of these segments.

Since $\overline{CD}$ is the altitude of right $\triangle ABC$, it is the geometric mean of the segments of the hypotenuse $\overline{AD}$ and $\overline{DB}$:

$\frac{AD}{CD} = \frac{CD}{DB}$.

All-in-One Resources/Online

English Language Learner Support

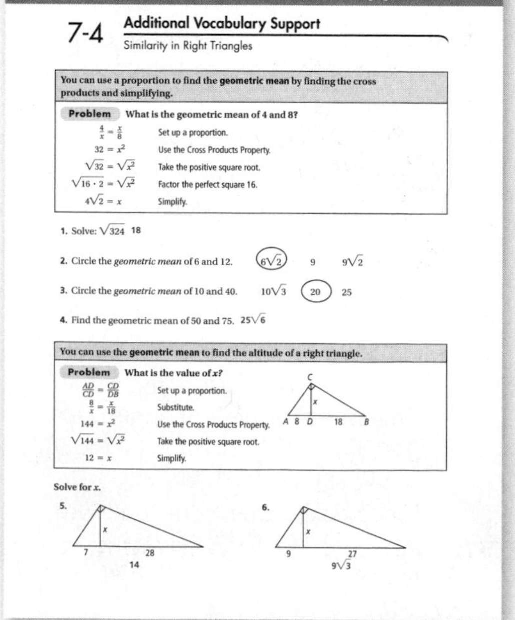

7-4 Additional Vocabulary Support

Similarity in Right Triangles

You can use a proportion to find the **geometric mean** by finding the cross products and simplifying.

Problem What is the geometric mean of 4 and 8?

$\frac{4}{x} = \frac{x}{8}$	Set up a proportion.
$32 = x^2$	Use the Cross Products Property.
$\sqrt{32} = \sqrt{x^2}$	Take the positive square root.
$\sqrt{16 \cdot 2} = \sqrt{x^2}$	Factor the perfect square 16.
$4\sqrt{2} = x$	Simplify.

1. Solve: $\sqrt{324}$ 18
2. Circle the *geometric mean* of 6 and 12. $6\sqrt{2}$ 9 $9\sqrt{2}$
3. Circle the *geometric mean* of 10 and 40. $10\sqrt{3}$ 20 25
4. Find the geometric mean of 50 and 75. $25\sqrt{6}$

You can use the **geometric mean** to find the altitude of a right triangle.

Problem What is the value of x?

$\frac{AD}{CD} = \frac{CD}{DB}$	Set up a proportion.
$\frac{8}{x} = \frac{x}{18}$	Substitute.
$144 = x^2$	Use the Cross Products Property.
$\sqrt{144} = \sqrt{x^2}$	Take the positive square root.
$12 = x$	Simplify.

Solve for x.

5. 7, 28 — 14
6. 9, 27 — $9\sqrt{3}$

Differentiated Remediation *continued*

On-Level

- **Practice** (2 pages) Provides extra practice for each lesson. For simpler practice exercises, use the Form K Practice pages found in the All-in-One Teaching Resources and online.
- **Think About a Plan** Helps students develop specific problem-solving skills and strategies by providing scaffolded guiding questions.
- **Standardized Test Prep** Focuses on all major exercises, all major question types, and helps students prepare for the high-stakes assessments.

Extension

- **Enrichment** Provides students with interesting problems and activities that extend the concepts of the lesson.
- **Activities, Games, and Puzzles** Worksheets that can be used for concepts development, enrichment, and for fun!

Practice and Problem Solving Wkbk/ All-in-One Resources/Online

Practice page 1

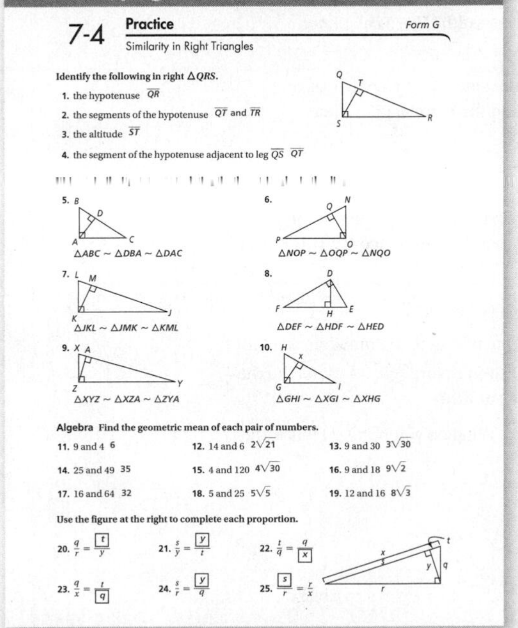

7-4 Practice — Form G
Similarity in Right Triangles

Identify the following in right $\triangle QRS$.
1. the hypotenuse $\overline{QR}$
2. the segments of the hypotenuse $\overline{QT}$ and $\overline{TR}$
3. the altitude $\overline{ST}$
4. the segment of the hypotenuse adjacent to leg $\overline{QS}$ $\overline{QT}$

5. $\triangle ABC \sim \triangle DBA \sim \triangle DAC$
6. $\triangle NOP \sim \triangle OQP \sim \triangle NQO$
7. $\triangle JKL \sim \triangle JMK \sim \triangle KML$
8. $\triangle DEF \sim \triangle HDF \sim \triangle HED$
9. $\triangle XYZ \sim \triangle XZA \sim \triangle ZYA$
10. $\triangle GHI \sim \triangle XGI \sim \triangle XHG$

Algebra Find the geometric mean of each pair of numbers.
11. 9 and 4 6
12. 14 and 6 $2\sqrt{21}$
13. 9 and 30 $3\sqrt{30}$
14. 25 and 49 35
15. 4 and 120 $4\sqrt{30}$
16. 9 and 18 $9\sqrt{2}$
17. 16 and 64 32
18. 5 and 25 $5\sqrt{5}$
19. 12 and 16 $8\sqrt{3}$

Use the figure at the right to complete each proportion.
20. $\frac{q}{r} = \frac{t}{y}$
21. $\frac{s}{y} = \frac{y}{t}$
22. $\frac{t}{q} = \frac{q}{x}$
23. $\frac{q}{x} = \frac{t}{q}$
24. $\frac{s}{r} = \frac{y}{q}$
25. $\frac{s}{r} = \frac{r}{x}$

Practice and Problem Solving Wkbk/ All-in-One Resources/Online

Practice page 2

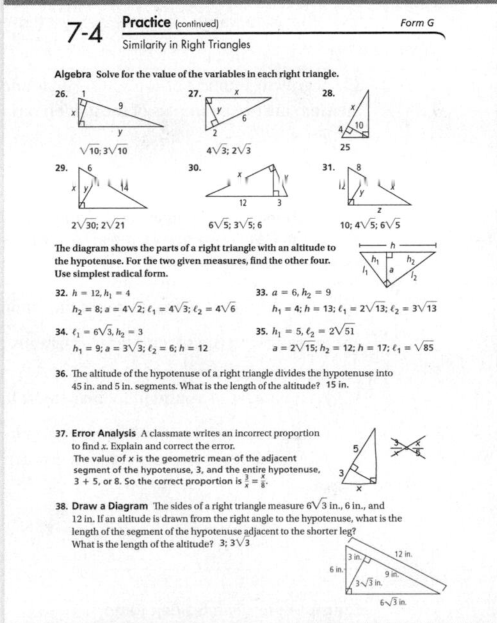

7-4 Practice (continued) — Form G
Similarity in Right Triangles

Algebra Solve for the value of the variables in each right triangle.
26. $\sqrt{10}$; $3\sqrt{10}$
27. $4\sqrt{3}$; $2\sqrt{3}$
28. 25
29. $2\sqrt{30}$; $2\sqrt{21}$
30. $6\sqrt{5}$; $3\sqrt{5}$; 6
31. 10; $4\sqrt{5}$; $6\sqrt{5}$

The diagram shows the parts of a right triangle with an altitude to the hypotenuse. For the two given measures, find the other four. Use simplest radical form.
32. $h = 12, h_1 = 4$ — $h_2 = 8; a = 4\sqrt{2}; \ell_1 = 4\sqrt{3}; \ell_2 = 4\sqrt{6}$
33. $a = 6, h_2 = 9$ — $h_1 = 4; h = 13; \ell_1 = 2\sqrt{13}; \ell_2 = 3\sqrt{13}$
34. $\ell_1 = 6\sqrt{3}, h_2 = 3$ — $h_1 = 9; a = 3\sqrt{3}; \ell_2 = 6; h = 12$
35. $h_1 = 5, \ell_2 = 2\sqrt{51}$ — $a = 2\sqrt{15}; h_2 = 12; h = 17; \ell_1 = \sqrt{85}$
36. The altitude of the hypotenuse of a right triangle divides the hypotenuse into 45 in. and 5 in. segments. What is the length of the altitude? 15 in.
37. **Error Analysis** A classmate writes an incorrect proportion to find x. Explain and correct the error. The value of x is the geometric mean of the adjacent segment of the hypotenuse, 3, and the entire hypotenuse, 3 + 5, or 8. So the correct proportion is $\frac{3}{x} = \frac{x}{8}$.
38. **Draw a Diagram** The sides of a right triangle measure $6\sqrt{3}$ in., 6 in., and 12 in. If an altitude is drawn from the right angle to the hypotenuse, what is the length of the segment of the hypotenuse adjacent to the shorter leg? What is the length of the altitude? 3; $3\sqrt{3}$

All-in-One Resources/Online

Enrichment

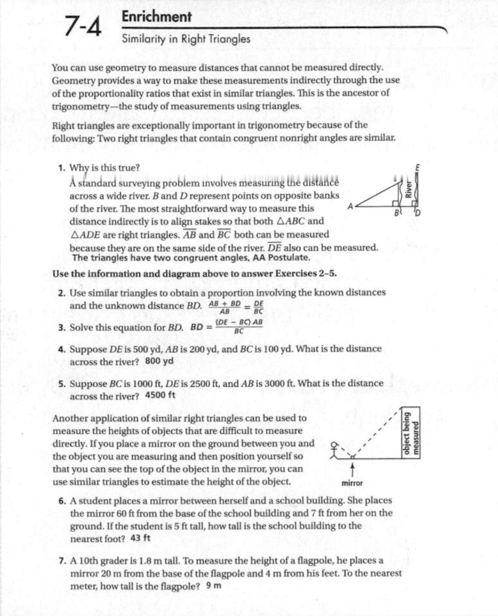

7-4 Enrichment
Similarity in Right Triangles

You can use geometry to measure distances that cannot be measured directly. Geometry provides a way to make these measurements indirectly through the use of the proportionality ratios that exist in similar triangles. This is the ancestor of trigonometry—the study of measurements using triangles.

Right triangles are exceptionally important in trigonometry because of the following: Two right triangles that contain congruent nonright angles are similar.

1. Why is this true?
A standard surveying problem involves measuring the distance across a wide river. B and D represent points on opposite banks of the river. The most straightforward way to measure this distance indirectly is to align stakes so that both $\triangle ABC$ and $\triangle ADE$ are right triangles. $\overline{AB}$ and $\overline{BC}$ both can be measured because they are on the same side of the river. $\overline{DE}$ also can be measured. The triangles have two congruent angles, AA Postulate.

Use the information and diagram above to answer Exercises 2–5.
2. Use similar triangles to obtain a proportion involving the known distances and the unknown distance BD. $\frac{AB + BD}{AB} = \frac{DE}{BC}$
3. Solve this equation for BD. $BD = \frac{(DE - BC)AB}{BC}$
4. Suppose DE is 500 yd, AB is 200 yd, and BC is 100 yd. What is the distance across the river? 800 yd
5. Suppose BC is 1000 ft, DE is 2500 ft, and AB is 3000 ft. What is the distance across the river? 4500 ft

Another application of similar right triangles can be used to measure the heights of objects that are difficult to measure directly. If you place a mirror on the ground between you and the object you are measuring and then position yourself so that you can see the top of the object in the mirror, you can use similar triangles to estimate the height of the object.
6. A student places a mirror between herself and a school building. She places the mirror 60 ft from the base of the school building and 7 ft from her on the ground. If the student is 5 ft tall, how tall is the school building to the nearest foot? 43 ft
7. A 10th grader is 1.8 m tall. To measure the height of a flagpole, he places a mirror 20 m from the base of the flagpole and 4 m from his feet. To the nearest meter, how tall is the flagpole? 9 m

Practice and Problem Solving Wkbk/ All-in-One Resources/Online

Think About a Plan

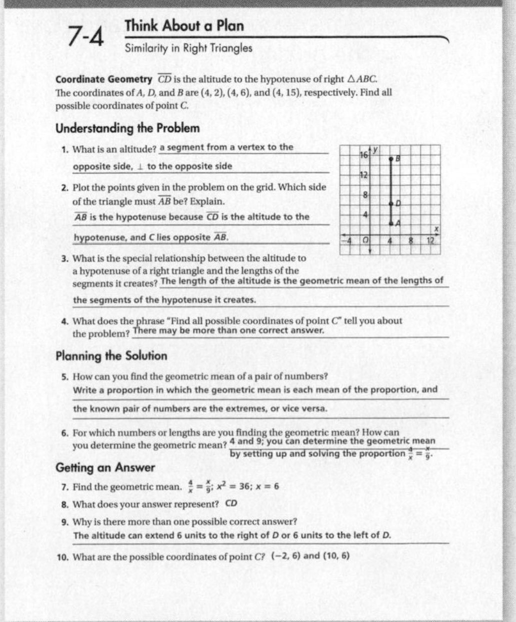

7-4 Think About a Plan
Similarity in Right Triangles

Coordinate Geometry $\overline{CD}$ is the altitude to the hypotenuse of right $\triangle ABC$. The coordinates of A, D, and B are (4, 2), (4, 6), and (4, 15), respectively. Find all possible coordinates of point C.

Understanding the Problem
1. What is an altitude? a segment from a vertex to the opposite side, ⊥ to the opposite side
2. Plot the points given in the problem on the grid. Which side of the triangle must $\overline{AB}$ be? Explain. $\overline{AB}$ is the hypotenuse because $\overline{CD}$ is the altitude to the hypotenuse, and C lies opposite $\overline{AB}$.
3. What is the special relationship between the altitude to a hypotenuse of a right triangle and the lengths of the segments it creates? The length of the altitude is the geometric mean of the lengths of the segments of the hypotenuse it creates.
4. What does the phrase "Find all possible coordinates of point C" tell you about the problem? There may be more than one correct answer.

Planning the Solution
5. How can you find the geometric mean of a pair of numbers? Write a proportion in which the geometric mean is each mean of the proportion, and the known pair of numbers are the extremes, or vice versa.
6. For which numbers or lengths are you finding the geometric mean? How can you determine the geometric mean? 4 and 9; you can determine the geometric mean by setting up and solving the proportion $\frac{4}{x} = \frac{x}{9}$.

Getting an Answer
7. Find the geometric mean. $\frac{4}{x} = \frac{x}{9}$; $x^2 = 36$; $x = 6$
8. What does your answer represent? CD
9. Why is there more than one possible correct answer? The altitude can extend 6 units to the right of D or 6 units to the left of D.
10. What are the possible coordinates of point C? (−2, 6) and (10, 6)

Practice and Problem Solving Wkbk/ All-in-One Resources/Online

Standardized Test Prep

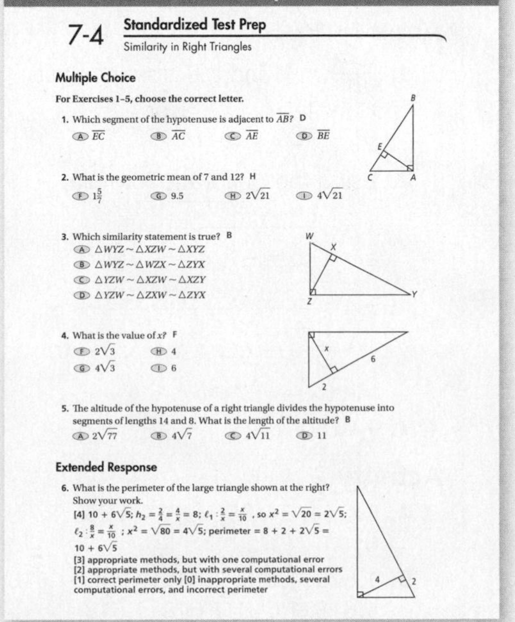

7-4 Standardized Test Prep
Similarity in Right Triangles

Multiple Choice
For Exercises 1–5, choose the correct letter.
1. Which segment of the hypotenuse is adjacent to $\overline{AB}$? D
(A) $\overline{EC}$ (B) $\overline{AC}$ (C) $\overline{AE}$ (D) $\overline{BE}$
2. What is the geometric mean of 7 and 12? H
(F) $1\frac{5}{7}$ (G) 9.5 (H) $2\sqrt{21}$ (I) $4\sqrt{21}$
3. Which similarity statement is true? B
(A) $\triangle WYZ \sim \triangle XZW \sim \triangle XYZ$
(B) $\triangle WYZ \sim \triangle WZX \sim \triangle ZYX$
(C) $\triangle YZW \sim \triangle XZW \sim \triangle XZY$
(D) $\triangle YZW \sim \triangle ZXW \sim \triangle ZYX$
4. What is the value of x? F
(F) $2\sqrt{3}$ (G) $4\sqrt{3}$ (H) 4 (I) 6
5. The altitude of the hypotenuse of a right triangle divides the hypotenuse into segments of lengths 14 and 8. What is the length of the altitude? B
(A) $2\sqrt{77}$ (B) $4\sqrt{7}$ (C) $4\sqrt{11}$ (D) 11

Extended Response
6. What is the perimeter of the large triangle shown at the right? Show your work.
[4] $10 + 6\sqrt{5}$; $h_2 = \frac{2}{4} = \frac{4}{h_2} = 8$; $\ell_1 : \frac{2}{\ell_1} = \frac{\ell_1}{10}$, so $x^2 = \sqrt{20} = 2\sqrt{5}$; $\ell_2 : \frac{8}{\ell_2} = \frac{\ell_2}{10}$; $x^2 = \sqrt{80} = 4\sqrt{5}$; perimeter $= 8 + 2 + 2\sqrt{5} =$ $10 + 6\sqrt{5}$
[3] appropriate methods, but with one computational error
[2] appropriate methods, but with several computational errors
[1] correct perimeter only [0] inappropriate methods, several computational errors, and incorrect perimeter

All-in-One Resources/Online

Activities, Games, and Puzzles

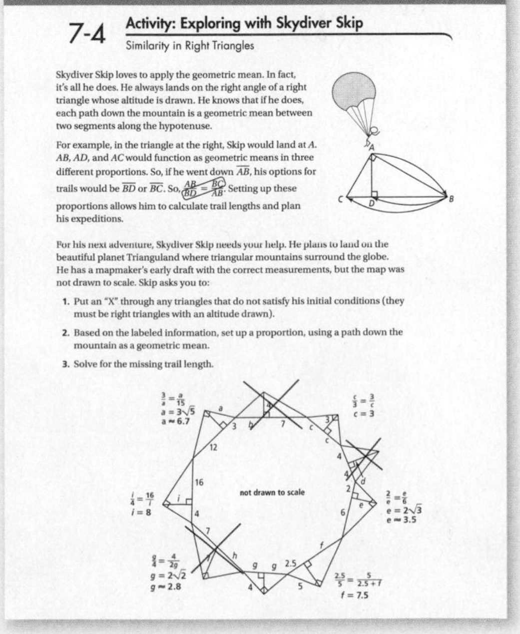

7-4 Activity: Exploring with Skydiver Skip
Similarity in Right Triangles

Skydiver Skip loves to apply the geometric mean. In fact, it's all he does. He always lands on the right angle of a right triangle whose altitude is drawn. He knows that if he does, each path down the mountain is a geometric mean between two segments along the hypotenuse.

For example, in the triangle at the right, Skip would land at A. AB, AD, and AC would function as geometric means in three different proportions. So, if he went down $\overline{AB}$, his options for trails would be $\overline{BD}$ or $\overline{BC}$. So, $\frac{AB}{BD} = \frac{BC}{AB}$. Setting up these proportions allows him to calculate trail lengths and plan his expeditions.

For his next adventure, Skydiver Skip needs your help. He plans to land on the beautiful planet Trianguland where triangular mountains surround the globe. He has a mapmaker's early draft with the correct measurements, but the map was not drawn to scale. Skip asks you to:
1. Put an "X" through any triangles that do not satisfy his initial conditions (they must be right triangles with an altitude drawn).
2. Based on the labeled information, set up a proportion, using a path down the mountain as a geometric mean.
3. Solve for the missing trail length.

Guided Instruction

PURPOSE To explore the golden ratio and how it relates to the Fibonacci sequence

PROCESS Students will
- derive the golden ratio.
- extend the Fibonacci sequence.

DISCUSS Mathematicians have studied the Golden Ratio for centuries because of its prevalence in art and nature and how it relates to the Fibonacci sequence.

Activity 1

In this Activity, students derive the golden ratio.

Q Because $AC = x$ and $CB = 1$, what is the length of $\overline{AB}$? **[$x + 1$]**

Q How is the quadratic equation in Question 2 obtained? **[After known values are substituted into the proportion, the Cross Products Property is used and the variable term is isolated on one side.]**

Activity 2

Examples of the Fibonacci sequence are found in nature's spiral growth patterns, including sunflower seeds and the veins on the plant leaves on a stem.

ERROR PREVENTION

Caution students not to rush when continuing the terms of the Fibonacci sequence. If one number is calculated incorrectly, then every successive number will be incorrect.

Mathematical Practices This Concept Byte supports students in becoming proficient in looking for patterns, Mathematical Practice 7.

Concept Byte: The Golden Ratio

Use With Lesson 7-4

ACTIVITY

Common Core State Standards

Extends G-SRT.B.5 Use . . . similarity criteria for triangles to solve problems and to prove relationships in geometric figures.

MP 7

In his book *Elements*, Euclid defined the *extreme and mean ratio* using a proportion formed by dividing a line segment at a particular point, as shown at the right. In the diagram, C divides $\overline{AB}$ so that the length of $\overline{AC}$ is the geometric mean of the lengths of $\overline{AB}$ and $\overline{CB}$. That is, $\frac{AB}{AC} = \frac{AC}{CB}$. The ratio $\frac{AC}{CB}$ is known today as the *golden ratio*, which is about 1.618 : 1.

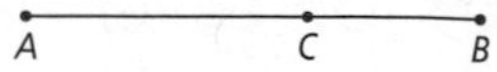

Rectangles in which the ratio of the length to the width is the golden ratio are *golden rectangles*. A golden rectangle can be divided into a square and a rectangle that is similar to the original rectangle. A pattern of golden rectangles is shown at the right.

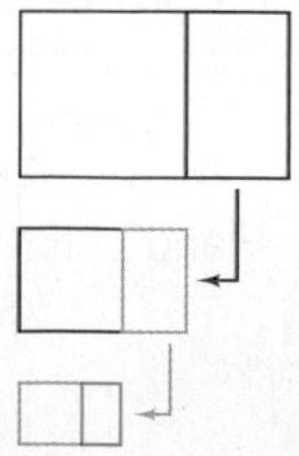

Activity 1

To derive the golden ratio, consider $\overline{AB}$ divided by C so that $\frac{AB}{AC} = \frac{AC}{CB}$.

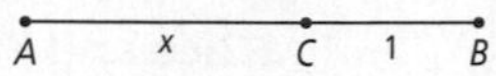

1. Use the diagram at the right to write a proportion that relates the lengths of the segments. How can you rewrite the proportion as a quadratic equation?
2. Use the quadratic formula to solve the quadratic equation in Question 1. Why does only one solution makes sense in this situation?
3. What is the value of x to the nearest ten-thousandth? Use a calculator.

Spiral growth patterns of sunflower seeds and the spacing of plant leaves on the stem are two examples of the golden ratio and the Fibonacci sequence in nature.

Activity 2

In the Fibonacci sequence, each term after the first two terms is the sum of the preceding two terms. The first six terms of the Fibonacci sequence are 1, 1, 2, 3, 5, and 8.

4. What are the next nine terms of the Fibonacci sequence?
5. Starting with the second term, the ratios of each term to the previous term for the first six terms are $\frac{1}{1} = 1$, $\frac{2}{1} = 2$, $\frac{3}{2} = 1.5$, $\frac{5}{3} = 1.666\ldots$, and $\frac{8}{5} = 1.6$. What are the next nine ratios rounded to the nearest thousandth?
6. Compare the ratios you found in Question 5. What do you notice? How is the Fibonacci sequence related to the golden ratio?

Answers

Activity 1

1. $\frac{x+1}{x} = \frac{x}{1}$; find the cross products to write $x^2 = x + 1$ or $x^2 - x - 1 = 0$.
2. Using the Quadratic Formula, $x = \frac{-b \pm \sqrt{b^2 - 4ac}}{2a}$ with $a = 1$, $b = -1$, $c = -1$ gives $x = \frac{-(-1) \pm \sqrt{(-1)^2 - 4(1)(-1)}}{2(1)} = \frac{1 \pm \sqrt{1+4}}{2}$; since $1 - \sqrt{5} < 0$, then $x = \frac{1 + \sqrt{5}}{2}$.
3. 1.6180

Activity 2

4. 13, 21, 34, 55, 89, 144, 233, 377, 610
5. 1.625, 1.615, 1.619, 1.618, 1.618, 1.618, 1.618, 1.618, 1.618
6. For terms in the Fibonacci sequence, the ratio of each term to the previous term gets closer and closer to the golden ratio.

Exercises

7. The golden rectangle is considered to be pleasing to the human eye. Of the following rectangles, which do you prefer? Is it a golden rectangle?

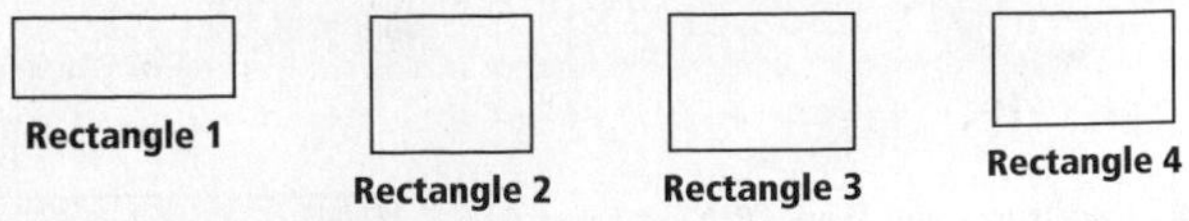

8. A drone is a male honeybee. Drones have only one parent, a queen. Workers and queens are female honeybees. Females have two parents, a drone and a queen. Part of the family tree showing the ancestors of a drone is shown below, where D represents a drone and Q represents a queen.

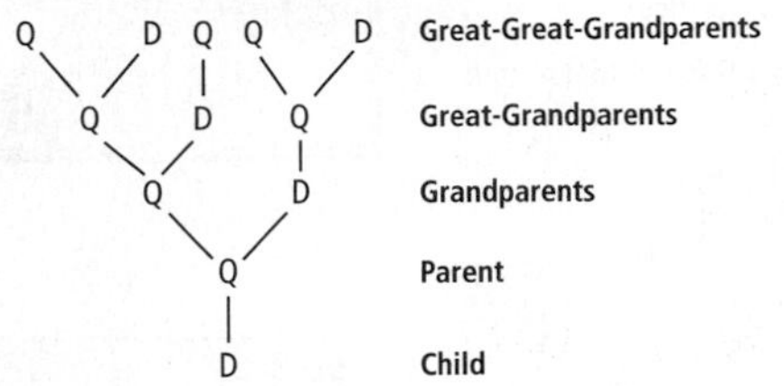

a. Continue the family tree for three more generations of ancestors.

b. Count the number of honeybees in each generation. What pattern do you notice?

9. What is the relationship between the flowers and the Fibonacci sequence?

10. In $\triangle ABC$, point D divides the hypotenuse into the golden ratio. That is, $AD : DB$ is about $1.618 : 1$. $\overline{CD}$ is an altitude. Using the value 1.618 for AD and the value 1 for DB, solve for x. What do you notice?

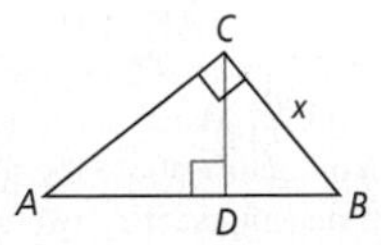

Exercises

7. Answers may vary. Students who select Rectangle 4 have selected a rectangle whose dimensions are close to the golden ratio.

8a.

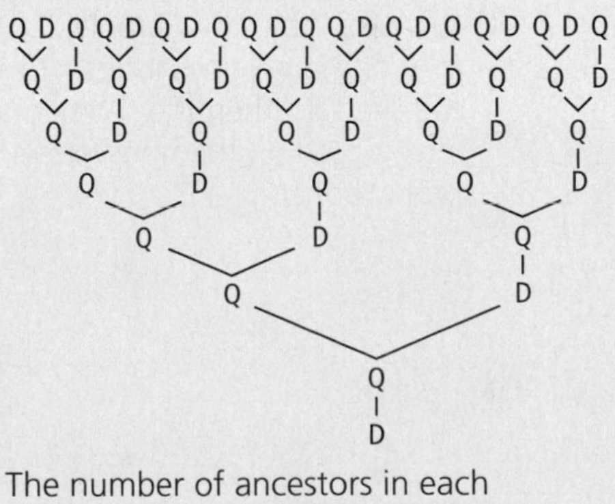

b. The number of ancestors in each generation is a Fibonacci number.

9. The number of petals is a Fibonacci number.

10. If $AD = 1.618$ and $DB = 1$, then $AB = 2.618$. So $\frac{1}{x} = \frac{x}{2.618}$ and $x^2 = 2.618$. So $x = \sqrt{2.618} = 1.618$.

Guided Instruction

PURPOSE To use geometry software to investigate proportions in triangles

PROCESS Students will

- construct a triangle.
- construct a line parallel to a given side of a triangle.

DISCUSS Students make constructions using geometry software that will enable them to explore and make conjectures regarding proportions that exist within triangles.

Activity 1

In this Activity, students draw a triangle and construct a line parallel to one side of the triangle.

Q What do you notice about the ratios of the segment lengths? **[They are equal.]**

Q What conjecture can be made about triangles *ABC* and *DBE*? **[They are similar.]**

Q If $BD = 8$, $DA = 12$, and $BE = 10$, what is *EC*? **[15]**

Activity 2

In this Activity, students construct a triangle in the coordinate plane and locate points that satisfy certain conditions.

Q After locating points *B* and *C*, measure *AB* and *AC*. What do you notice about the ratio of *AC* to *AE* and the ratio of *AB* to *AD*? **[The ratios are the same.]**

Mathematical Practices This Concept Byte supports students in becoming proficient in using appropriate tools, Mathematical Practice 5.

Concept Byte
Use With Lesson 7-5
TECHNOLOGY

Exploring Proportions in Triangles

Common Core State Standards
G-CO.D.12 Make formal geometric construc[tions] with a variety of tools and methods . . .
MP 5

Activity 1

Use geometry software to draw $\triangle ABC$. Construct point *D* on $\overline{AB}$. Next, construct a line through *D* parallel to $\overline{AC}$. Then construct the intersection *E* of the parallel line with $\overline{BC}$.

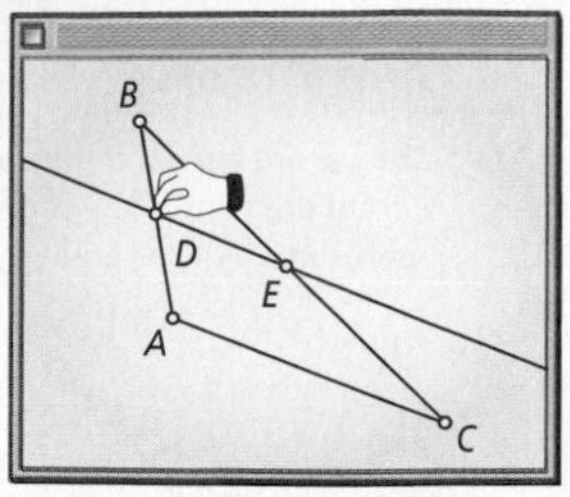

1. Measure $\overline{BD}$, $\overline{DA}$, $\overline{BE}$, and $\overline{EC}$. Calculate the ratios $\frac{BD}{DA}$ and $\frac{BE}{EC}$.

2. Manipulate $\triangle ABC$ and observe $\frac{BD}{DA}$ and $\frac{BE}{EC}$. What do you notice?

3. Make a conjecture about the four segments formed by a line parallel to one side of a triangle intersecting the other two sides.

Activity 2

Use geometry software to construct $\triangle ADE$ with vertices $A(3, 3)$, $D(-2, 0)$, and $E(5, 1)$.

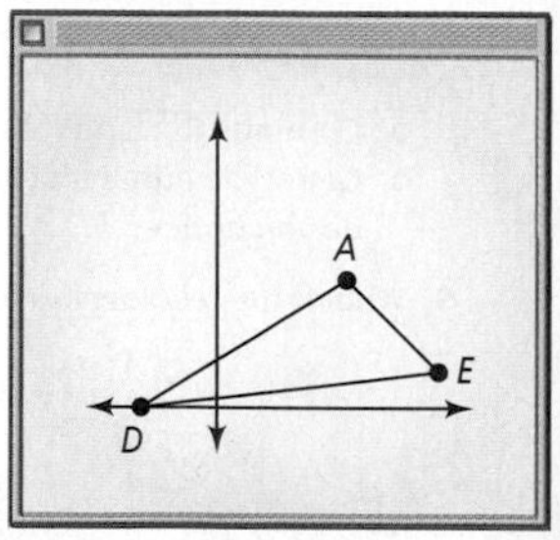

4. Measure *AD*, *AE*, and *DE*. Give your answers to the nearest tenth.

5. Suppose *B* is the point on $\overline{AD}$ such that $AB = \frac{2}{3}AD$, and *C* is the point on $\overline{AE}$ such that $\overline{BC} \parallel \overline{DE}$. Describe how you could approximate the coordinates of point *B*.

6. Now use the geometry software to draw $\overline{BC}$ and manipulate the segment so that it satisfies the conditions given in Exercise 5. What are the coordinates of points *B* and *C*?

Exercises

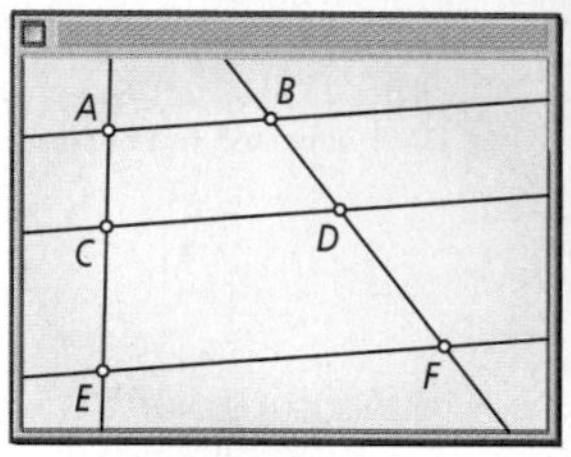

7. Construct $\overleftrightarrow{AB} \parallel \overleftrightarrow{CD} \parallel \overleftrightarrow{EF}$. Then construct two transversals that intersect all three parallel lines. Measure $\overline{AC}$, $\overline{CE}$, $\overline{BD}$, and $\overline{DF}$. Calculate the ratios $\frac{AC}{CE}$ and $\frac{BD}{DF}$. Manipulate the locations of *A* and *B* and observe $\frac{AC}{CE}$ and $\frac{BD}{DF}$. Make a conjecture about the segments of the transversals formed by the three parallel lines intersecting two transversals.

8. Suppose four or more parallel lines intersect two transversals. Make a conjecture about the segments of the transversals.

Answers

Activity 1

1. Students' values for *BD*, *DA*, *BE*, and *BC* will vary, but $\frac{BD}{DA} = \frac{BE}{EC}$.

2. For each location of *D*, $\frac{BD}{DA} = \frac{BC}{EC}$.

3. The $\parallel$ line divides the sides into proportional segments.

Activity 2

4. $AD = 5.8$, $AE = 2.8$, $DE = 7.1$

5. Sample: You could divide $\overline{AD}$ into thirds, and estimate the coordinates of the point at $\frac{2}{3}$ the distance from *A* to *D*.

6. $B\left(-\frac{1}{3}, 1\right)$, $C\left(4\frac{1}{3}, 1\frac{2}{3}\right)$

Exercises

7. Answers may vary. Sample: If three $\parallel$ lines intersect two transversals, then the segments intercepted on the transversals are proportional.

8. Answers may vary. Sample: If four or more $\parallel$ lines intersect two transversals, then the segments intercepted on the transversals are proportional.

7-5 Proportions in Triangles

Common Core State Standards
G-SRT.B.4 Prove theorems about triangles . . . a line parallel to one side of a triangle divides the other two proportionally . . .
MP 1, MP 3, MP 4

Objective To use the Side-Splitter Theorem and the Triangle-Angle-Bisector Theorem

Use what you know about similar triangles to plan a pathway to a solution.

MATHEMATICAL PRACTICES

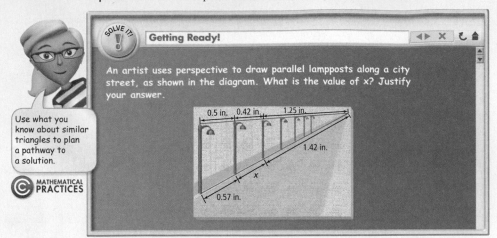

Getting Ready!

An artist uses perspective to draw parallel lampposts along a city street, as shown in the diagram. What is the value of *x*? Justify your answer.

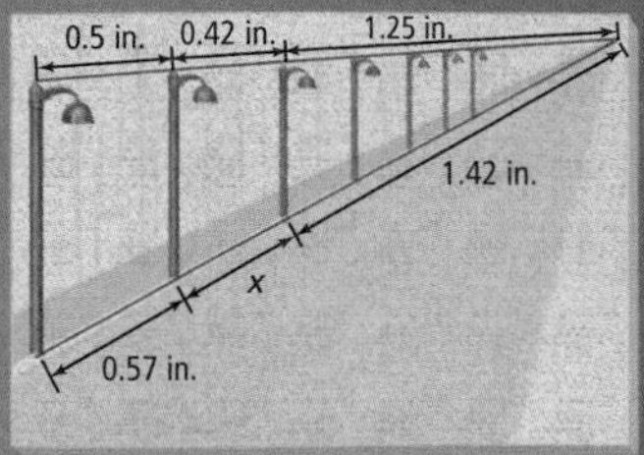

The Solve It involves parallel lines cut by two transversals that intersect. In this lesson, you will learn how to use proportions to find lengths of segments formed by parallel lines that intersect two or more transversals.

Essential Understanding When two or more parallel lines intersect other lines, proportional segments are formed.

take note

Theorem 7-4 Side-Splitter Theorem

Theorem
If a line is parallel to one side of a triangle and intersects the other two sides, then it divides those sides proportionally.

If . . .
$\overleftrightarrow{RS} \parallel \overleftrightarrow{XY}$

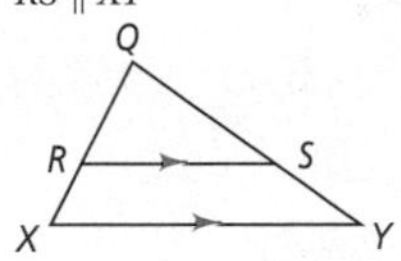

Then . . .
$\frac{XR}{RQ} = \frac{YS}{SQ}$

1 Interactive Learning

Solve It!

PURPOSE To use similar triangles to show that parallel lines divide segments proportionally

PROCESS Students may prove that all triangles in the diagram are similar by AA~ or create proportions involving the side lengths of similar triangles to solve for the unknown.

FACILITATE

Q Are there any similar triangles in the diagram? How do you know the triangles are similar? **[All the triangles that include the far right vertex and one lamp post as a side are similar.]**

Q How can you use these angles to make a statement about the triangles in the drawing? **[By AA~, all triangles in the diagram are similar.]**

ANSWER See Solve It in Answers on next page.

CONNECT THE MATH Students should recognize that the parallel lines (lampposts in the Solve It) separate one large triangle into smaller similar triangles. By the properties of similar triangles, the lengths of the sides are proportional. This leads to the Side-Splitter Theorem, which states that parallel lines divide segments proportionally.

2 Guided Instruction

TAKE NOTE Have students draw each triangle separately. Then write proportions that relate the sides of the triangles.

7-5 Preparing to Teach

BIG ideas **Reasoning and Proof**
Visualization
Proportionality

ESSENTIAL UNDERSTANDINGS

- When two or more parallel lines intersect other lines, proportional segments are formed.
- The bisector of an angle of a triangle divides the opposite side into two segments with lengths proportional to the sides of the triangle that form the angle.

Math Background

Similar triangles are used to prove the theorems and corollaries presented in this chapter. Without an understanding of proportionality, students will not be able to reason through these theorems. The Side-Splitter Theorem and its corollary lead to solving problems with indirect measurement and splitting figures into similar figures.

Mathematical Practices

Construct viable arguments and critique the reasoning of others. Students will use their knowledge of similarity in triangles to prove the Triangle-Angle-Bisector Theorem in Exercise 47.

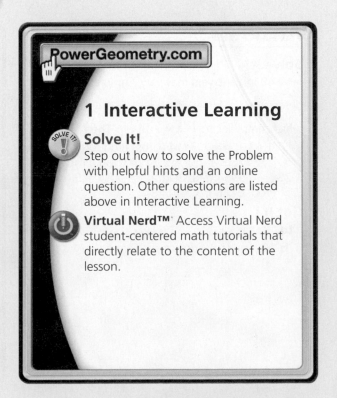

PowerGeometry.com

1 Interactive Learning

Solve It!
Step out how to solve the Problem with helpful hints and an online question. Other questions are listed above in Interactive Learning.

Virtual Nerd™ Access Virtual Nerd student-centered math tutorials that directly relate to the content of the lesson.

Q How can you use the fact that $\overleftrightarrow{RS} \parallel \overleftrightarrow{XY}$ in the proof? **[Because the lines are parallel, the corresponding angles are congruent.]**

Q What two triangles are in the diagram? **[$\triangle XQY$ and $\triangle RQS$]**

Q How are the two triangles related? **[They are similar.]**

Q If the triangles are similar, what statement can you make about their side lengths? **[They are proportional.]**

Q How can you use the Segment Addition Postulate to write the side lengths? **[$XQ = XR + RQ$ and $YQ = YS + SQ$]**

Proof of Theorem 7-4: Side-Splitter Theorem

Given: $\triangle QXY$ with $\overleftrightarrow{RS} \parallel \overleftrightarrow{XY}$

Prove: $\frac{XR}{RQ} = \frac{YS}{SQ}$

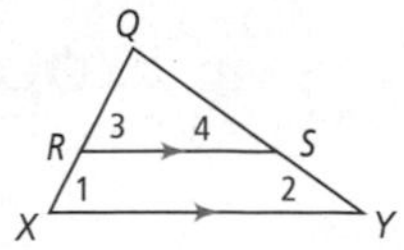

Statements	Reasons
1) $\overleftrightarrow{RS} \parallel \overleftrightarrow{XY}$	1) Given
2) $\angle 1 \cong \angle 3$, $\angle 2 \cong \angle 4$	2) If lines are $\parallel$, then corresponding ∠s are $\cong$.
3) $\triangle QXY \sim \triangle QRS$	3) AA ~ Postulate
4) $\frac{XQ}{RQ} = \frac{YQ}{SQ}$	4) Corresponding sides of ~ △s are proportional.
5) $XQ = XR + RQ$, $YQ = YS + SQ$	5) Segment Addition Postulate
6) $\frac{XR + RQ}{RQ} = \frac{YS + SQ}{SQ}$	6) Substitution Property
7) $\frac{XR}{RQ} = \frac{YS}{SQ}$	7) Property of Proportions (3)

Problem 1

Q What condition of the Side-Splitter Theorem is marked in the diagram? **[$\overline{KL}$ is parallel to $\overline{PN}$.]**

Q What conclusion can you draw using the Side-Splitter Theorem? **[The segments of the sides of the triangle are proportional.]**

Problem 1 Using the Side-Splitter Theorem

GRIDDED RESPONSE

Plan

How can you use the parallel lines in the diagram?

$\overline{KL}$ is parallel to one side of $\triangle MNP$. Use the Side-Splitter Theorem to set up a proportion.

What is the value of x in the diagram at the right?

$\frac{PK}{KM} = \frac{NL}{LM}$ Side-Splitter Theorem

$\frac{x+1}{12} = \frac{x}{9}$ Substitute.

$9x + 9 = 12x$ Cross Products Property

$9 = 3x$ Subtract $9x$ from each side.

$3 = x$ Divide each side by 3.

Grid in the number 3.

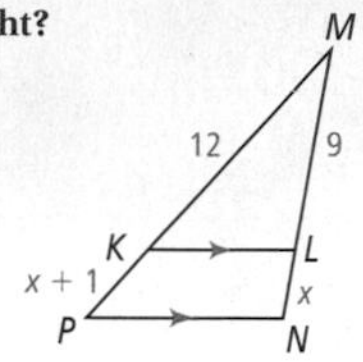

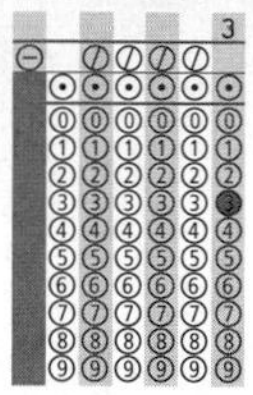

Got It? 1. a. What is the value of a in the diagram at the right?

b. Reasoning In $\triangle XYZ$, $\overline{RS}$ joins $\overline{XY}$ and $\overline{YZ}$ with R on $\overline{XY}$ and S on $\overline{YZ}$, and $\overline{RS} \parallel \overline{XZ}$. If $\frac{YR}{RX} = \frac{YS}{SZ} = 1$, what must be true about RS? Justify your reasoning.

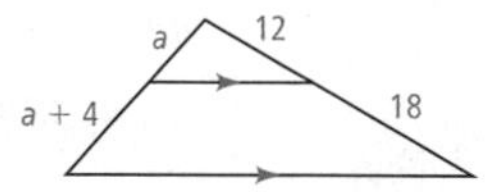

Got It?

For 1b, have students create a diagram of the figure described. Have students substitute numbers for the lengths of segments that satisfy the conditions in the problem. Then, ask students to identify the type of special segment that $\overline{RS}$ represents.

Answers

Solve It!

≈ 0.48 in.; answers may vary. Sample: The $\parallel$ lines determine similar △s, so $\frac{1.25}{1.25 + 0.42} = \frac{1.42}{1.42 + x}$, which simplifies to $\frac{1.25}{1.67} = \frac{1.42}{1.42 + x}$. Then $1.25(1.42 + x) = 2.3714$ (Cross Products Prop.); $1.775 + 1.25x = 2.3714$ (Distr. Prop.); $1.25x = 0.5964$ (Subst. Prop. of =); $x = 0.47712$ (Div. Prop. of Eq.); $x \approx 0.48$.

Got It?

1a. 8

b. $RS = \frac{1}{2}XZ$ (Midsegment Thm.)

PowerGeometry.com

2 Guided Instruction

Each Problem is worked out and supported online.

Problem 1
Using the Side-Splitter Theorem
Animated

Problem 2
Finding a Length
Animated

Problem 3
Using the Triangle-Angle-Bisector Theorem
Animated

Support in Geometry Companion
- Vocabulary
- Key Concepts
- Got It?

Plan

What information does the diagram give you?
The lines separating the campsites are parallel. Think of the river and the edge of the road as transversals. Then the boundaries along the road and river for each campsite are proportional.

take note

Corollary — Corollary to the Side-Splitter Theorem

Corollary
If three parallel lines intersect two transversals, then the segments intercepted on the transversals are proportional.

If . . .
$a \parallel b \parallel c$

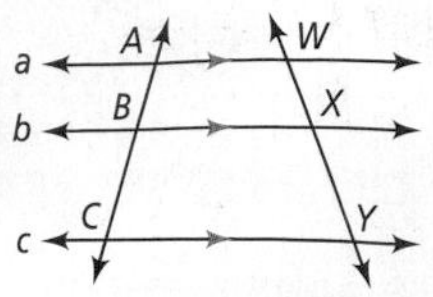

Then . . .
$\frac{AB}{BC} = \frac{WX}{XY}$

You will prove the Corollary to Theorem 7-4 in Exercise 46.

Problem 2 — Finding a Length

Camping **Three campsites are shown in the diagram. What is the length of Site A along the river?**

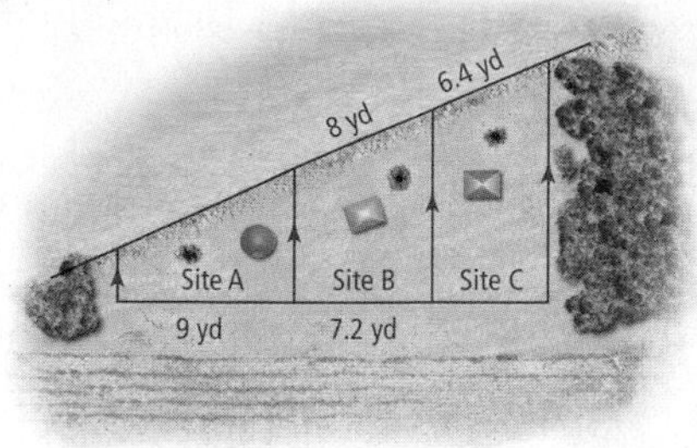

Let x be the length of Site A along the river.

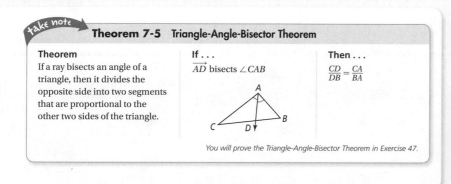

$\frac{x}{8} = \frac{9}{7.2}$ Corollary to the Side-Splitter Theorem

$7.2x = 72$ Cross Products Property

$x = 10$ Divide each side by 7.2.

The length of Site A along the river is 10 yd.

Got It? **2.** What is the length of Site C along the road?

Essential Understanding The bisector of an angle of a triangle divides the opposite side into two segments with lengths proportional to the sides of the triangle that form the angle.

take note

Theorem 7-5 — Triangle-Angle-Bisector Theorem

Theorem
If a ray bisects an angle of a triangle, then it divides the opposite side into two segments that are proportional to the other two sides of the triangle.

If . . .
$\overrightarrow{AD}$ bisects $\angle CAB$

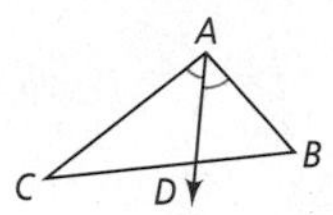

Then . . .
$\frac{CD}{DB} = \frac{CA}{BA}$

You will prove the Triangle-Angle-Bisector Theorem in Exercise 47.

Take Note
To see how the Corollary to the Side-Splitter Theorem connects to the theorem itself, continue the transversal lines until they intersect. Then students can identify similar triangles in the diagram and prove the corollary.

Problem 2
Q For which proportional sides have both measurements been given? **[Both sides of Site B have been given.]**

Q Which side should be labeled as the unknown? **[the length of Site A along the river]**

Q If the river side of Site B is in the numerator of the first ratio, what measurement should be in the denominator of the second ratio? **[the length of Site A along the road]**

Got It?
Remind students that the numerators of both ratios must be corresponding sides, and so must the two denominators.

Take Note
Have students redraw the triangles so that they have the same orientation. Then, ask students to make a similarity statement that shows how the two triangles and their sides are related.

Additional Problems

1. What is the value of x in the diagram?

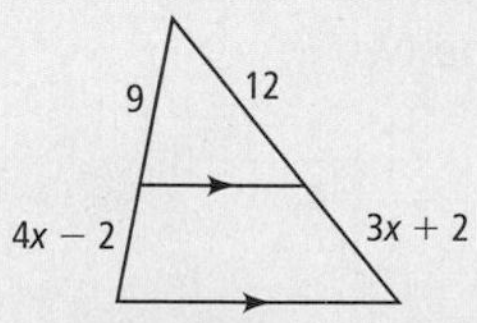

ANSWER 2

2. Two plots of land are shown below. What is the unknown length, x?

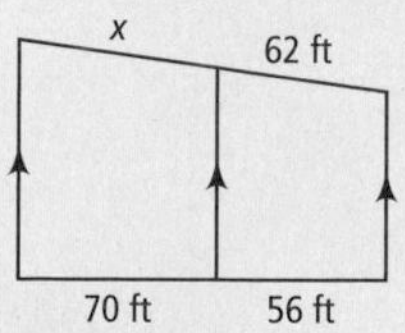

ANSWER 77.5 ft

3. What is the value of x in the diagram?

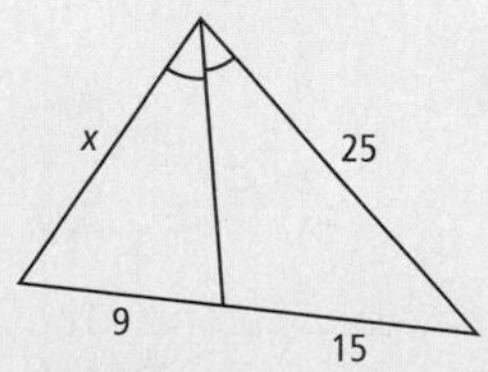

ANSWER 15

Answers

Got It? (continued)

2. 5.76 yd

Problem 3

Q What type of segment is drawn inside the triangle? How do you know? **[angle bisector; the angles it creates are congruent.]**

Q Using the properties of proportions, how can the proportion be rewritten so that the x is in a numerator? **[Property 1 of proportions allows the proportion to be written as $\frac{18}{10} = \frac{x}{12}$.]**

Got It?
If students struggle with this problem, have them label each point in the diagram and write the proportion based on the Triangle-Angle-Bisector Theorem. Then students can substitute values into the proportion.

3 Lesson Check

Do you know HOW?
- If students have difficulty with Exercises 1-3, then have them review Problem 2.

Do you UNDERSTAND?
- If students have difficulty with Exercise 7, then have them review Problem 3.

Close

Q When parallel lines intersect two or more segments, what is the relationship between the segments formed? **[The segments formed between the parallel lines are proportional.]**

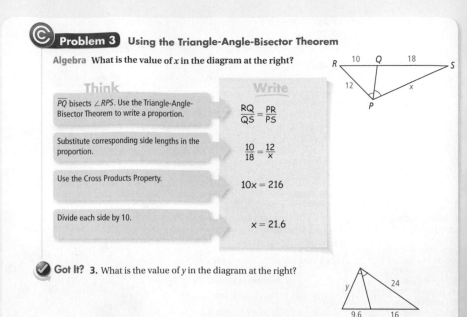

Problem 3 Using the Triangle-Angle-Bisector Theorem

Algebra What is the value of x in the diagram at the right?

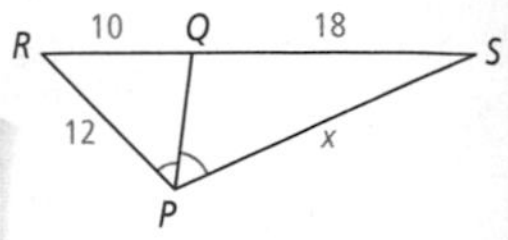

Think	Write
$\overline{PQ}$ bisects $\angle RPS$. Use the Triangle-Angle-Bisector Theorem to write a proportion.	$\frac{RQ}{QS} = \frac{PR}{PS}$
Substitute corresponding side lengths in the proportion.	$\frac{10}{18} = \frac{12}{x}$
Use the Cross Products Property.	$10x = 216$
Divide each side by 10.	$x = 21.6$

Got It? 3. What is the value of y in the diagram at the right?

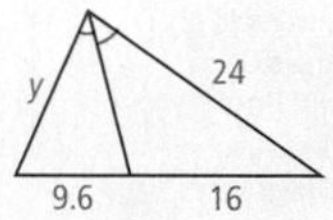

Lesson Check

Do you know HOW?

Use the figure to complete each proportion.

1. $\frac{a}{b} = \frac{\blacksquare}{e}$
2. $\frac{b}{\blacksquare} = \frac{e}{f}$
3. $\frac{a}{b+c} = \frac{\blacksquare}{e+f}$

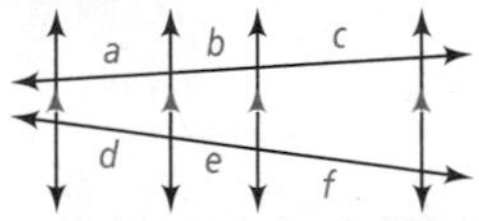

What is the value of x in each figure?

4.
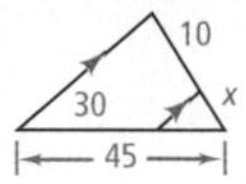

5.
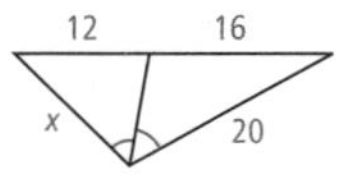

Do you UNDERSTAND? MATHEMATICAL PRACTICES

6. **Compare and Contrast** How is the Corollary to the Side-Splitter Theorem related to Theorem 6-7: If three (or more) parallel lines cut off congruent segments on one transversal, then they cut off congruent segments on every transversal?

7. **Compare and Contrast** How are the Triangle-Angle-Bisector Theorem and Corollary 1 to Theorem 7-3 alike? How are they different?

8. **Error Analysis** A classmate says you can use the Side-Splitter Theorem to find both x and y in the diagram. Explain what is wrong with your classmate's statement.

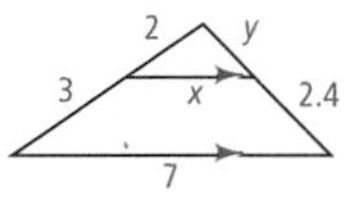

PowerGeometry.com

3 Lesson Check
For a digital lesson check, use the Got It questions.

Support in Geometry Companion
- Lesson Check

4 Practice
Assign homework to individual students or to an entire class.

Answers

Got It? (continued)
3. 14.4

Lesson Check
1. *d*
2. *c*
3. *d*
4. 5
5. 15
6. Answers may vary. Sample: The Corollary to the Side-Splitter Thm. takes the same three (or more) ∥ lines as in Thm. 6-7, but instead of cutting off ≅ segments it allows the segments to be proportional.
7. Answers may vary. Sample: Alike: Both involve a △ and a seg. from one vertex to the opposite side of the △. Different: In Corollary 1 to Thm. 7-3, the △ is a rt. △ and the seg. is an alt., while in the △-∠-Bis. Thm. the △ does not have to be a rt. △ and the seg. is an ∠ bis.
8. The Side-Splitter Thm. involves only the segments formed on the two sides intersected by the ∥ line. (To find x, you can use a proportion involving the two ~ △s.)

Practice and Problem-Solving Exercises

A Practice

Algebra Solve for x. See Problem 1.

9.
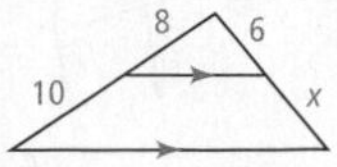

10.
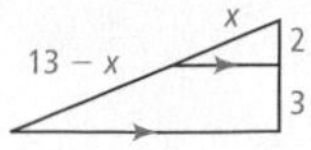

11.
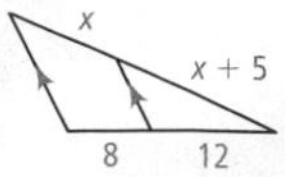

12.
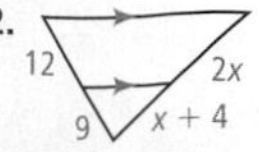

Marine Biology Use the information shown on the auger shell. See Problem 2.

13. What is the value of x?

14. What is the value of y?

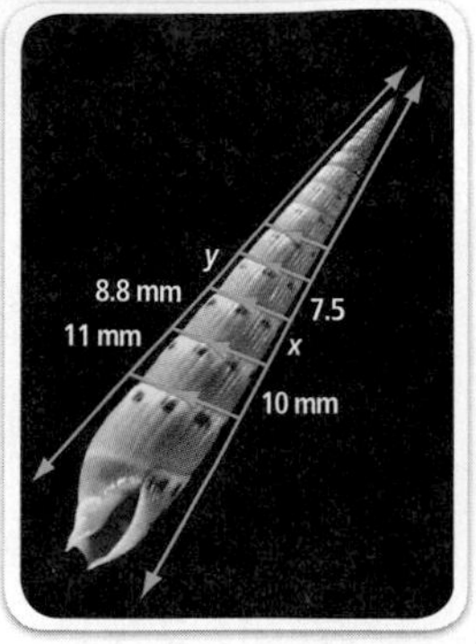

Algebra Solve for x.

15.
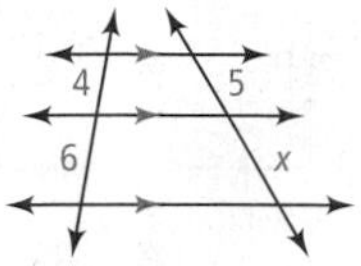

16.
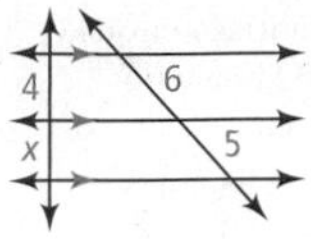

17.
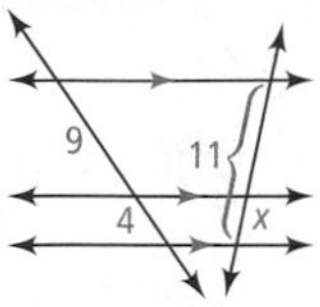

18.
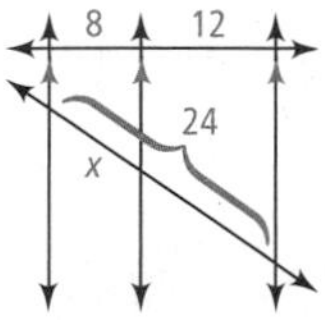

Algebra Solve for x. See Problem 3.

19.
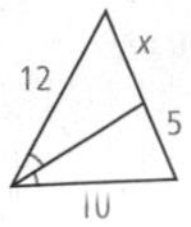

20.
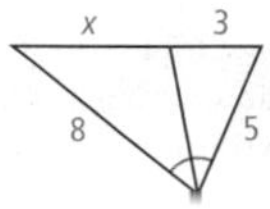

21.
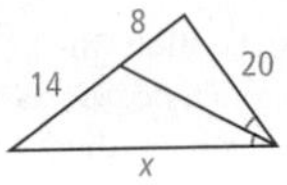

22.
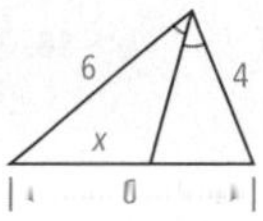

23. Writing The size of an oil spill on the open ocean is difficult to measure directly. Use the figure at the right to describe how you could find the length of the oil spill indirectly. What measurements and calculations would you use?

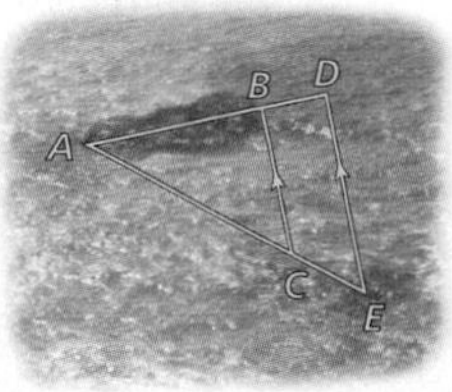

24. The lengths of the sides of a triangle are 5 cm, 12 cm, and 13 cm. Find the lengths, to the nearest tenth, of the segments into which the bisector of each angle divides the opposite side.

Practice and Problem-Solving Exercises

9. 7.5 **10.** 5.2
11. 10 **12.** 8
13. 8 mm **14.** 8.25 mm
15. 7.5 **16.** $3\frac{1}{3}$
17. $3\frac{5}{13}$ **18.** 9.6
19. 6 **20.** 4.8
21. 35 **22.** 3.6
23. Use the Side-Splitter Thm. to write the proportion $\frac{AB}{BD} = \frac{AC}{CE}$, then find the values of BD, AC, and CE to calculate the unknown length AB.
24. 5-cm side: 2.4 cm, 2.6 cm
12-cm side: $3\frac{1}{3}$ cm, $8\frac{2}{3}$ cm
13-cm side: about 3.8 cm, about 9.2 cm

4 Practice

ASSIGNMENT GUIDE
Basic: 9–26, 33, 36, 38–41, 43
Average: 9–23 odd, 25–47
Advanced: 9–23 odd, 25–50
Standardized Test Prep: 51–54
Mixed Review: 55–64

Mathematical Practices are supported by exercises with red headings. Here are the Practices supported in this lesson:

MP 1: Make Sense of Problems Ex. 6, 7, 36
MP 3: Construct Arguments Ex. 42, 43
MP 3: Communicate Ex. 23
MP 3: Critique the Reasoning of Others Ex. 8

Applications exercises have blue headings. Exercises 31 and 32 support MP 4: Model.

STEM exercises focus on science or engineering applications.

EXERCISE 41: Use the Think About a Plan worksheet in the **Practice and Problem Solving Workbook** (also available in the Teaching Resources in print and online) to further support students' development in becoming independent learners.

HOMEWORK QUICK CHECK
To check students' understanding of key skills and concepts, go over Exercises 15, 23, 26, 41, and 43.

Answers

Practice and Problem-Solving Exercises (continued)

25. *KS* by the △-∠-Bis. Thm.
26. *SQ* by the Side-Splitter Thm.
27. *JP* by the Side-Splitter Thm.
28. *KP* by the △-∠-Bis. Thm.
29. *KM* by the △-∠-Bis. Thm.
30. *MP* by the Corollary to the Side-Splitter Thm.
31. 575 ft **32.** 750 ft
33. 20 **34.** 2.5 **35.** $\frac{2}{7}$ or 3
36. $x = 18$ m, $y = 12$ m
37. $\frac{XR}{RQ} = \frac{YS}{SQ}$ (Given); $\frac{XR + RQ}{RQ} = \frac{YS + SQ}{SQ}$ (Prop. of Proportions (3)); $XQ = XR + RQ$, $YQ = YS + SQ$ (Seg. Add. Post.); $\frac{XQ}{RQ} = \frac{YQ}{SQ}$ (Subst.); $\angle Q \cong \angle Q$ (Refl. Prop. of ≅); $\triangle XQY \sim \triangle RQS$ (SAS ~ Post.); $\angle 1 \cong \angle 2$ (Corresp. ∡ of ~ ▵ are ≅.); $\overline{RS} \parallel \overline{XY}$ (If corresp. ∡ are ≅, the lines are ∥.)

B Apply

Use the figure at the right to complete each proportion. Justify your answer.

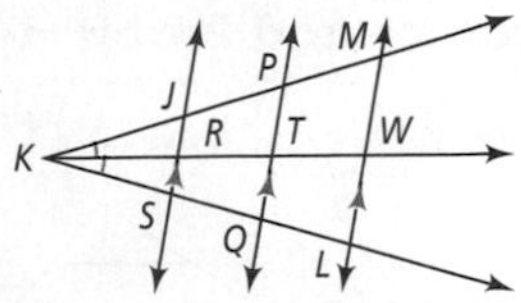

25. $\frac{RS}{■} = \frac{JR}{KJ}$ **26.** $\frac{KJ}{JP} = \frac{KS}{■}$

27. $\frac{QL}{PM} = \frac{SQ}{■}$ **28.** $\frac{PT}{■} = \frac{TQ}{KQ}$

29. $\frac{KL}{LW} = \frac{■}{MW}$ **30.** $\frac{■}{KP} = \frac{LQ}{KQ}$

STEM **Urban Design** In Washington, D.C., E. Capitol Street, Independence Avenue, C Street, and D Street are parallel streets that intersect Kentucky Avenue and 12th Street.

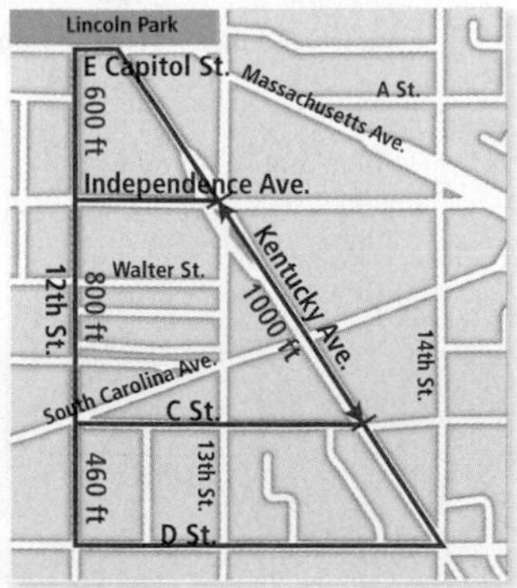

31. How long (to the nearest foot) is Kentucky Avenue between C Street and D Street?

32. How long (to the nearest foot) is Kentucky Avenue between E. Capitol Street and Independence Avenue?

Algebra Solve for *x*.

33.

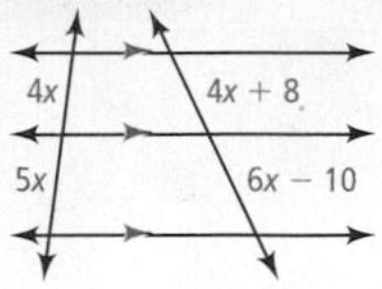

34.

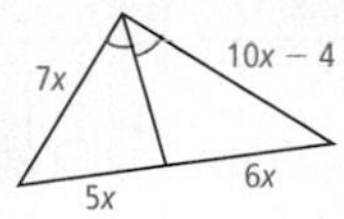

35.

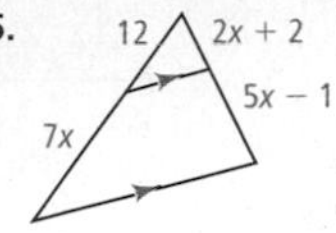

C 36. Think About a Plan The perimeter of the triangular lot at the right is 50 m. The surveyor's tape bisects an angle. Find the lengths *x* and *y*.
- How can you use the perimeter to write an equation in *x* and *y*?
- What other relationship do you know between *x* and *y*?

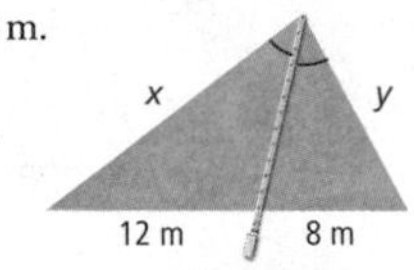

37. Proof Prove the Converse of the Side-Splitter Theorem: If a line divides two sides of a triangle proportionally, then it is parallel to the third side.

Given: $\frac{XR}{RQ} = \frac{YS}{SQ}$

Prove: $\overline{RS} \parallel \overline{XY}$

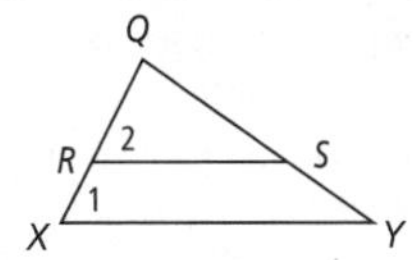

Determine whether the red segments are parallel. Explain each answer. You can use the theorem proved in Exercise 37.

38.

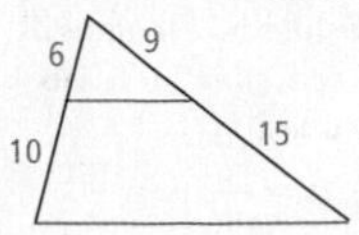

39.

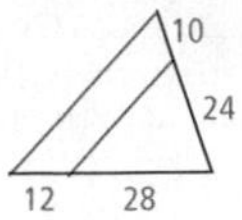

40.

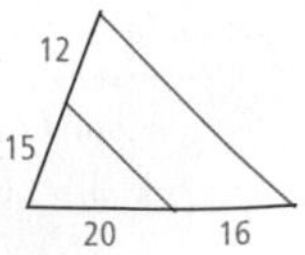

41. An angle bisector of a triangle divides the opposite side of the triangle into segments 5 cm and 3 cm long. A second side of the triangle is 7.5 cm long. Find all possible lengths for the third side of the triangle.

42. Open-Ended In a triangle, the bisector of an angle divides the opposite side into two segments with lengths 6 cm and 9 cm. How long could the other two sides of the triangle be? (*Hint:* Make sure the three sides satisfy the Triangle Inequality Theorem.)

43. Reasoning In $\triangle ABC$, the bisector of $\angle C$ bisects the opposite side. What type of triangle is $\triangle ABC$? Explain your reasoning.

Algebra Solve for x.

44.

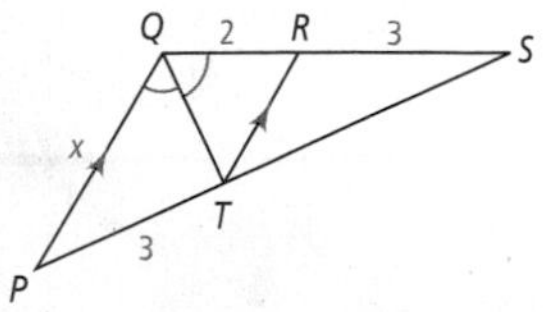

45.

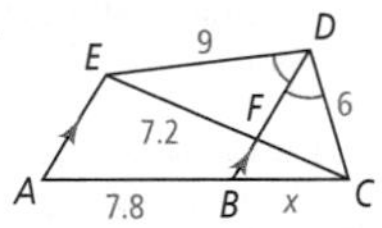

46. Prove the Corollary to the Side-Splitter Theorem. In the diagram from page 473, draw the auxiliary line $\overleftrightarrow{CW}$ and label its intersection with line b as point P.
Proof

Given: $a \parallel b \parallel c$

Prove: $\frac{AB}{BC} = \frac{WX}{XY}$

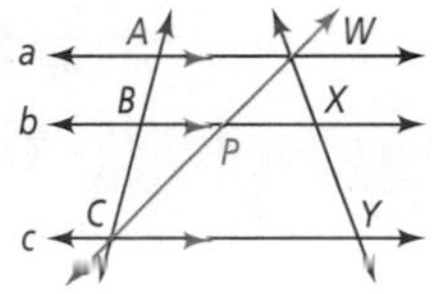

47. Prove the Triangle-Angle-Bisector Theorem. In the diagram from page 473, draw the auxiliary line $\overleftrightarrow{BE}$ so that $\overleftrightarrow{BE} \parallel \overleftrightarrow{DA}$. Extend $\overline{CA}$ to meet $\overleftrightarrow{BE}$ at point F.
Proof

Given: $\overrightarrow{AD}$ bisects $\angle CAB$.

Prove: $\frac{CD}{DB} = \frac{CA}{BA}$

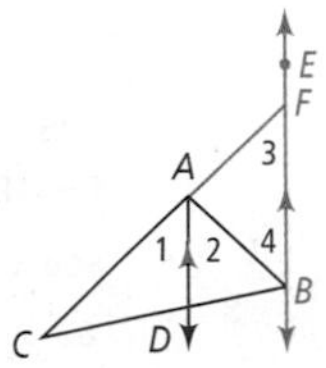

Challenge

48. Use the definition in part (a) to prove the statements in parts (b) and (c).
a. Write a definition for a midsegment of a parallelogram.
b. A parallelogram midsegment is parallel to two sides of the parallelogram.
c. A parallelogram midsegment bisects the diagonals of a parallelogram.

38. yes; $\frac{6}{10} = \frac{9}{15}$ (Converse of Side-Splitter Thm.)

39. no; $\frac{28}{12} \neq \frac{24}{10}$

40. yes; $\frac{15}{12} = \frac{20}{16}$ (Converse of Side-Splitter Thm.)

41. 12.5 cm or 4.5 cm

42. Answers may vary. Sample: 10 and 15, 8 and 12, or any two sides in the ratio 2 : 3 where the shorter side > 6 cm and < 15 cm.

43. Isosc.; $AC : BC$ is 1 : 1 by the $\triangle$-$\angle$-Bis. Thm.

44. $\frac{10}{3}$

45. 5.2

46. $\frac{AB}{BC} = \frac{WP}{PC}$ by the Side-Splitter Thm., and $\frac{WP}{PC} = \frac{WX}{XY}$. Therefore $\frac{AB}{BC} = \frac{WX}{XY}$ by the Transitive Prop. of =.

47. By the Side-Splitter Thm., $\frac{CD}{DB} = \frac{CA}{AF}$. By the Corresp. $\angle$s Post., $\angle 3 \cong \angle 1$. Since $\overrightarrow{AD}$ bisects $\angle CAB$, $\angle 1 \cong \angle 2$. By the Alt. Int. $\angle$s Thm., $\angle 2 \cong \angle 4$. So, $\angle 3 \cong \angle 4$ by the Trans. Prop. of $\cong$. By the Converse of the Isosc. $\triangle$ Thm., $BA = AF$. Substituting BA for AF, $\frac{CD}{DB} = \frac{CA}{BA}$.

48a. Answers may vary. Sample: A midsegment of a parallelogram connects the midpts. of two opposite sides of the parallelogram.

b.

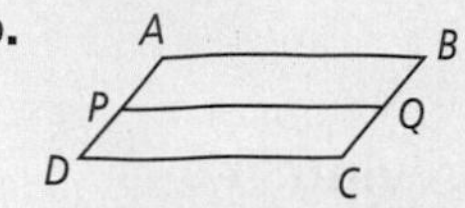

Given: $\overline{PQ}$ is a midsegment of parallelogram $ABCD$.

Prove: $\overline{PQ} \parallel \overline{AB}$, $\overline{PQ} \parallel \overline{DC}$

$\overline{AD} \cong \overline{BC}$ and $\overline{AD} \parallel \overline{BC}$ (properties of parallelograms), so $PD = \frac{1}{2}(AD) = \frac{1}{2}(BC) = QC$, and $PA = \frac{1}{2}(AD) = \frac{1}{2}(BC) = BQ$, both by the def. of midpt. Therefore, $ABQP$ and $PQCD$ are parallelograms because each has a pair of opposite sides that are $\cong$ and $\parallel$. So $\overline{PQ} \parallel \overline{AB}$ and $\overline{PQ} \parallel \overline{DC}$ because opposite sides of a parallelogram are $\parallel$.

c.

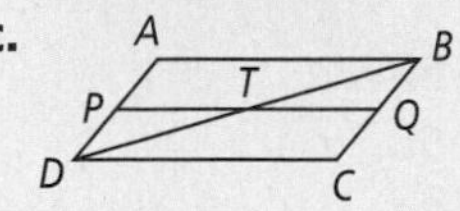

Given: Parallelogram $ABCD$ and midsegment $\overline{PQ}$

Prove: $\overline{PQ}$ bisects $\overline{BD}$.

From part (b) of this exercise, $\overline{AB} \parallel \overline{PQ} \parallel \overline{DC}$. Since $\overline{AP} \cong \overline{PD}$ by the def. of midsegment, $\overline{DT} \cong \overline{TB}$ because if $\parallel$ lines cut off $\cong$ segments on one transversal, they cut off $\cong$ segments on every transversal (Thm. 6-7). Since $\overline{PQ}$ contains the midpt. of BD, then PQ bisects BD by the def. of bisect. Also, point T is the midpt. of both diagonals (because the diagonals of a parallelogram have the same midpt.), so $\overline{PQ}$ bisects both diagonals of the parallelogram.

Answers

Practice and Problem-Solving Exercises (continued)

49. Use the diagram with Ex. 47, with $\overline{AD} \parallel \overline{EB}$. It is given that $\frac{CD}{DB} = \frac{CA}{BA}$, and you want to prove that $\angle 1 \cong \angle 2$. By the Side-Splitter Thm., $\frac{CA}{AF} = \frac{CD}{DB}$. So $\frac{CA}{BA} = \frac{CA}{AF}$ by the Transitive Prop. of $=$, and $BA = AF$. Therefore, $\angle 3 \cong \angle 4$ by the Isosc. $\triangle$Thm. Using properties of $\parallel$ lines, $\angle 1 \cong \angle 3$ and $\angle 2 \cong \angle 4$. So $\angle 1 \cong \angle 2$ by the Transitive Prop. of $\cong$, and $\overline{AD}$ bisects $\angle CAB$ by the def. of $\angle$ bis.

50a. 90 units **b.** 14 units

51. 20 **52.** 52

53. 118 **54.** 66 in. **55.** m

56. m **57.** c **58.** h

59. (3, −3) **60.** (0, 2) **61.** (1.5, 2.5)

62. $(3\text{ m})^2 = 9\text{ m}^2$, $(4\text{ m})^2 = 16\text{ m}^2$, $(5\text{ m})^2 = 25\text{ m}^2$

63. $(5\text{ in.})^2 = 25\text{ in.}^2$, $(12\text{ in.})^2 = 144\text{ in.}^2$, $(13\text{ in.})^2 = 169\text{ in.}^2$

64. $(4\text{ m})^2 = 16\text{ m}^2$, $(4\sqrt{2}\text{ m})^2 = 32\text{ m}^2$

49. State the converse of the Triangle-Angle-Bisector Theorem. Give a convincing argument that the converse is true or a counterexample to prove that it is false.

50. In $\triangle ABC$, the bisectors of $\angle A$, $\angle B$, and $\angle C$ cut the opposite sides into lengths a_1 and a_2, b_1 and b_2, and c_1 and c_2, respectively, labeled in order counterclockwise around $\triangle ABC$. Find the perimeter of $\triangle ABC$ for each set of values.

a. $b_1 = 16, b_2 = 20, c_1 = 18$ **b.** $a_1 = \frac{5}{3}, a_2 = \frac{10}{3}, b_1 = \frac{15}{4}$

Standardized Test Prep

GRIDDED RESPONSE

SAT/ACT

51. What is the value of x in the figure at the right?

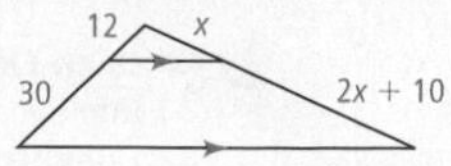

52. Suppose $\triangle VLQ \sim \triangle PSX$. If $m\angle V = 48$ and $m\angle L = 80$, what is $m\angle X$?

53. In the diagram at the right, $\overline{PR} \cong \overline{QR}$. For what value of x is $\overline{TS}$ parallel to $\overline{QP}$?

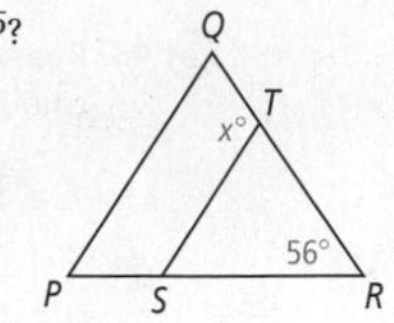

54. Leah is playing basketball on an outdoor basketball court. The 10-ft pole supporting the basketball net casts a 15-ft shadow. At the same time, the length of Leah's shadow is 8 ft 3 in. What is Leah's height in inches? You can assume both Leah and the pole supporting the net are perpendicular to the ground.

Mixed Review

Use the figure to complete each proportion. See Lesson 7-4.

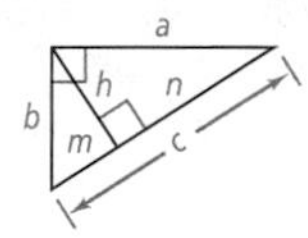

55. $\frac{n}{h} = \frac{h}{\blacksquare}$ **56.** $\frac{\blacksquare}{b} = \frac{b}{c}$

57. $\frac{n}{a} = \frac{a}{\blacksquare}$ **58.** $\frac{m}{h} = \frac{\blacksquare}{n}$

Find the center of the circle that you can circumscribe about each $\triangle ABC$. See Lesson 5-3.

59. $A(0, 0)$, $B(6, 0)$, $C(0, -6)$ **60.** $A(2, 5)$, $B(-2, 5)$, $C(-2, -1)$ **61.** $A(-2, 0)$, $B(5, 5)$, $C(-2, 5)$

Get Ready! To prepare for Lesson 8-1, do Exercises 62–64.

Square the lengths of the sides of each triangle. See p. 829.

62.

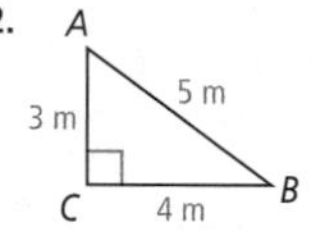

63.

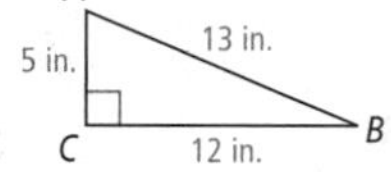

64.

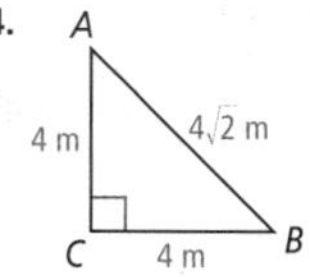

7-5 Lesson Resources

Differentiated Remediation

Instructional Support

Geometry Companion

Students can use the **Geometry Companion** worktext (4 pages) . . .

- New Vocabulary
- Key Concepts
- Got It for each Problem
- Lesson Check

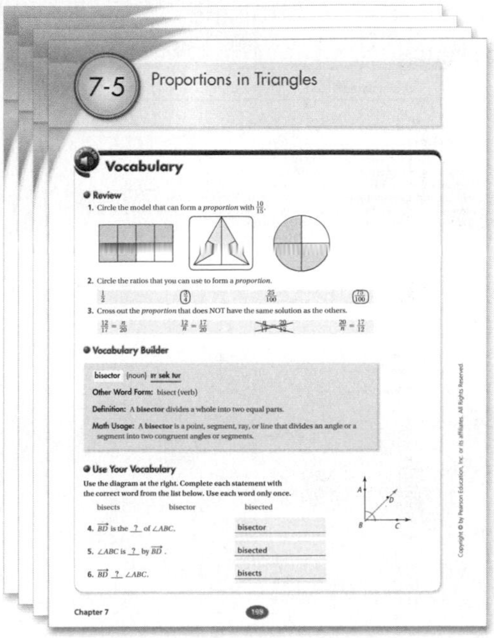

ELL Support

Use Graphic Organizers Ask: What is the central idea of this chapter? [Similar triangles or Similarity] Have groups of students make a graphic organizer for the chapter. Say: Use the Essential Understandings for each lesson in the chapter. Encourage the groups to use examples, illustrations, diagrams, and familiar words in their graphic organizer.

Assign each group one lesson from the chapter. Have the group make a more detailed graphic organizer for that lesson. Say: Explain the lesson vocabulary words and each of the problems in your organizers. Have each group show their graphic organizers to the class.

5 Assess & Remediate

Lesson Quiz

1. What is the value of x in the diagram?

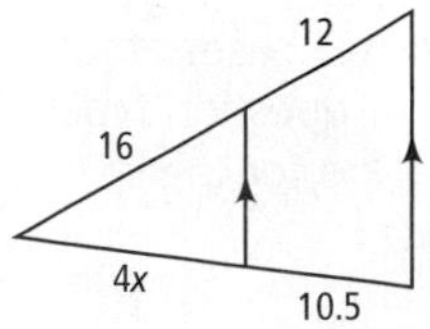

2. What is the value of x in the diagram?

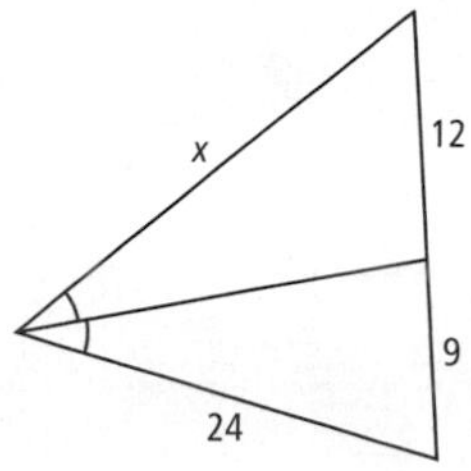

3. Do you UNDERSTAND? What is the length of Site A along $\overline{QZ}$?

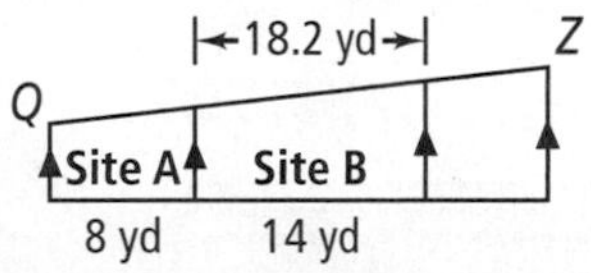

ANSWERS TO LESSON QUIZ

1. 3.5

2. 32

3. 10.4 yd

PRESCRIPTION FOR REMEDIATION

Use the student work on the Lesson Quiz to prescribe a differentiated review assignment.

Points	Differentiated Remediation
0–1	Intervention
2	On-level
3	Extension

PowerGeometry.com

5 Assess & Remediate

Assign the Lesson Quiz. Appropriate intervention, practice, or enrichment is automatically generated based on student performance.

Intervention

- **Reteaching** (2 pages) Provides reteaching and practice exercises for the key lesson concepts. Use with struggling students or absent students.
- **English Language Learner Support** Helps students develop and reinforce mathematical vocabulary and key concepts.

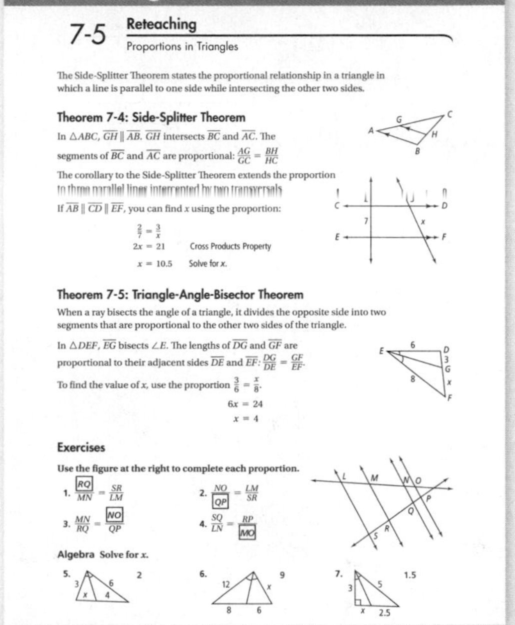

All-in-One Resources/Online

Reteaching

7-5 Reteaching

Proportions in Triangles

The Side-Splitter Theorem states the proportional relationship in a triangle in which a line is parallel to one side while intersecting the other two sides.

Theorem 7-4: Side-Splitter Theorem

In $\triangle ABC$, $\overline{GH} \parallel \overline{AB}$. $\overline{GH}$ intersects $\overline{BC}$ and $\overline{AC}$. The segments of $\overline{BC}$ and $\overline{AC}$ are proportional: $\frac{AG}{GC} = \frac{BH}{HC}$

The corollary to the Side-Splitter Theorem extends the proportion to three parallel lines intercepted by two transversals.

If $\overline{AB} \parallel \overline{CD} \parallel \overline{EF}$, you can find x using the proportion:

$\frac{2}{7} = \frac{3}{x}$

$2x = 21$ Cross Products Property

$x = 10.5$ Solve for x.

Theorem 7-5: Triangle-Angle-Bisector Theorem

When a ray bisects the angle of a triangle, it divides the opposite side into two segments that are proportional to the other two sides of the triangle.

In $\triangle DEF$, $\overline{EG}$ bisects $\angle E$. The lengths of $\overline{DG}$ and $\overline{GF}$ are proportional to their adjacent sides $\overline{DE}$ and $\overline{EF}$: $\frac{DG}{DE} = \frac{GF}{EF}$.

To find the value of x, use the proportion $\frac{3}{6} = \frac{x}{8}$.

$6x = 24$

$x = 4$

Exercises

Use the figure at the right to complete each proportion.

1. $\frac{RQ}{MN} = \frac{SR}{LM}$
2. $\frac{NO}{QP} = \frac{LM}{SR}$
3. $\frac{MN}{RQ} = \frac{NO}{QP}$
4. $\frac{SQ}{LN} = \frac{RP}{MO}$

Algebra Solve for x.

5. 2
6. 9
7. 1.5

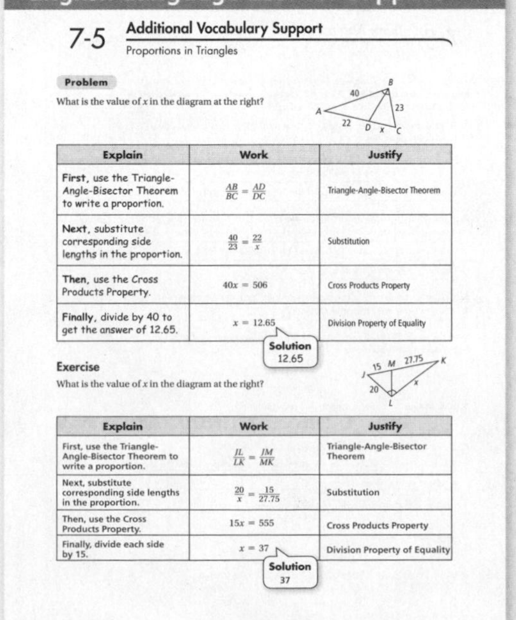

All-in-One Resources/Online

English Language Learner Support

7-5 Additional Vocabulary Support

Proportions in Triangles

Problem

What is the value of x in the diagram at the right?

Explain	Work	Justify
First, use the Triangle-Angle-Bisector Theorem to write a proportion.	$\frac{AB}{BC} = \frac{AD}{DC}$	Triangle-Angle-Bisector Theorem
Next, substitute corresponding side lengths in the proportion.	$\frac{40}{23} = \frac{22}{x}$	Substitution
Then, use the Cross Products Property.	$40x = 506$	Cross Products Property
Finally, divide by 40 to get the answer of 12.65.	$x = 12.65$	Division Property of Equality

Solution 12.65

Exercise

What is the value of x in the diagram at the right?

Explain	Work	Justify
First, use the Triangle-Angle-Bisector Theorem to write a proportion.	$\frac{JL}{LK} = \frac{JM}{MK}$	Triangle-Angle-Bisector Theorem
Next, substitute corresponding side lengths in the proportion.	$\frac{20}{x} = \frac{15}{27.75}$	Substitution
Then, use the Cross Products Property.	$15x = 555$	Cross Products Property
Finally, divide each side by 15.	$x = 37$	Division Property of Equality

Solution 37

Differentiated Remediation *continued*

On-Level

- **Practice** (2 pages) Provides extra practice for each lesson. For simpler practice exercises, use the Form K Practice pages found in the All-in-One Teaching Resources and online.
- **Think About a Plan** Helps students develop specific problem-solving skills and strategies by providing scaffolded guiding questions.
- **Standardized Test Prep** Focuses on all major exercises, all major question types, and helps students prepare for the high-stakes assessments.

Extension

- **Enrichment** Provides students with interesting problems and activities that extend the concepts of the lesson.
- **Activities, Games, and Puzzles** Worksheets that can be used for concepts development, enrichment, and for fun!

Practice and Problem Solving Wkbk/ All-in-One Resources/Online

Practice page 1

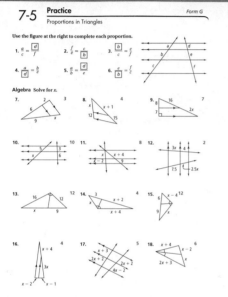

7-5 **Practice** *Form G*
Proportions in Triangles

Use the figure at the right to complete each proportion.

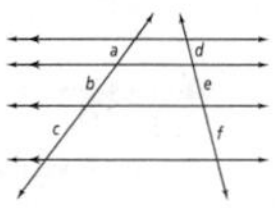

1. $\frac{a}{c} = \frac{\boxed{d}}{f}$ 2. $\frac{f}{e} = \frac{c}{\boxed{b}}$ 3. $\frac{\boxed{b}}{c} = \frac{e}{f}$
4. $\frac{a}{\boxed{d}} = \frac{b}{e}$ 5. $\frac{a}{b} = \frac{\boxed{d}}{e}$ 6. $\frac{e}{\boxed{b}} = \frac{f}{c}$

Algebra Solve for *x*.

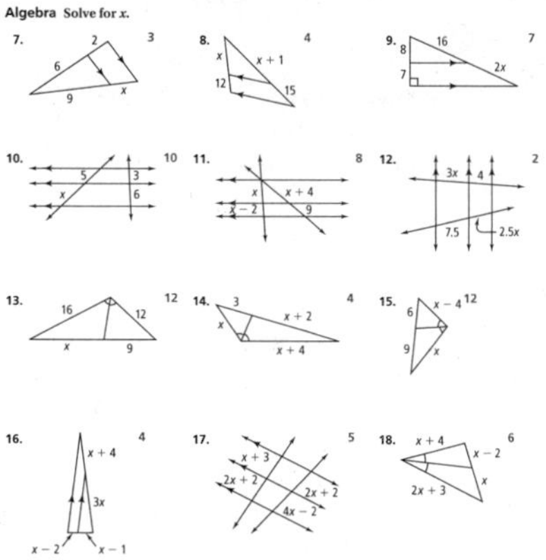

Practice page 2

7-5 **Practice** (continued) *Form G*
Proportions in Triangles

19. **Compare and Contrast** How is the Triangle-Angle-Bisector Theorem similar to Corollary 2 of Theorem 7-3? How is it different?
Answers may vary. Sample: Both relate to a line that intercepts an angle of a triangle and its opposite side. In both, the segments created by the intersecting line are related proportionally to the sides of the triangle. Corollary 2 of Theorem 7-3 is only true of right triangles with an altitude to the hypotenuse. The Triangle-Angle-Bisector Theorem relates to all triangles that contain an angle bisector that intersects the opposite side.
20. **Reasoning** In $\triangle FGH$, the bisector of $\angle F$ also bisects the opposite side. The ratio of each half of the bisected side to each of the other sides is 1 : 2. What type of triangle is $\triangle FGH$? Explain.
$\triangle FGH$ is an equilateral triangle. Because the side has been bisected, each segment is the same length. So, their sum is: $x + x = 2x$. This is the same as the length of a side.
21. **Error Analysis** Your classmate says you can use the Triangle-Angle-Bisector Theorem to find the value of *x* in the diagram. Explain what is wrong with your classmate's statement.

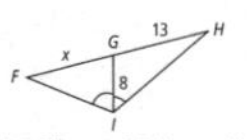

The classmate is confusing this Theorem with Corollary 1 to Theorem 7-3. You could only find the value of *x* if $\triangle FHI$ were a right triangle with right $\angle I$, and $\overline{IG}$ were an altitude to the hypotenuse.
22. **Reasoning** An angle bisector of a triangle divides the opposite side of the triangle into segments 3 in. and 6 in. long. A second side of the triangle is 5 in. long. Find the length of the third side of the triangle. Explain how you arrived at the correct length.
10 in.; The other possible side length is 2.5 in., but because 2.5 in. + 5 in. < 9 in., it violates the Triangle Inequality Theorem.
23. The flag of Antigua and Barbuda is like the image at the right. In the image, $\overline{DE} \parallel \overline{CF} \parallel \overline{BG}$.

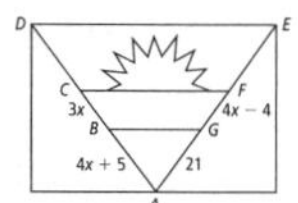

a. An artist has made a sketch of the flag for a mural. The measures indicate the length of the lines in feet. What is the value of *x*? 4
b. What type of triangle is $\triangle ACF$? Explain.
$\triangle ACF$ is isosceles. Because $x = 4$, $\overline{CB} \cong \overline{FG}$ and $\overline{BA} \cong \overline{GA}$. Because $CA = CB + BA$ and $FA = FG + GA$, by substitution $\overline{CA} \cong \overline{FA}$.
c. **Given:** $\overline{DE} \parallel \overline{CF} \parallel \overline{BG}$
Prove: $\triangle ABG \sim \triangle ACF \sim \triangle ADE$
Statements: 1) $\overline{DE} \parallel \overline{CF} \parallel \overline{BG}$; 2) $\angle EDC \cong \angle FCB \cong \angle GBA$; 3) $\angle DEF \cong \angle CFG \cong \angle BGA$; 4) $\triangle ABG \sim \triangle ACF \sim \triangle ADE$;
Reasons: 1) Given; 2) If lines are ∥, corresponding ∠s are ≅; 3) If lines are ∥, corresponding ∠s are ≅; 4) AA ~

All-in-One Resources/Online

Enrichment

7-5 **Enrichment**
Proportions in Triangles

Great Geometers

One of the first great women mathematicians was born in ancient Greece in the middle of the 4th century CE. Taught by her father, she was also a preeminent astronomer and philosopher. She became the head of the Neoplatonist school of philosophy and was widely known. She wrote commentaries on several mathematical works and was consulted in the construction of an astrolabe and hydroscope. She was eventually murdered by religious zealots who disagreed with her philosophical teachings.

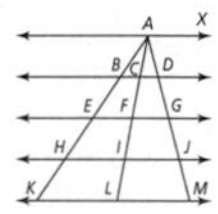

To find the name of this mathematician and astronomer, consider the diagram, and use the information given below to find the lengths of all the segments.

$\overline{AX} \parallel \overline{BC} \parallel \overline{EF} \parallel \overline{HI} \parallel \overline{KL}$

$\overline{AL}$ bisects $\angle KAM$

$AK = 40$, $EF = 12$, $BE = 16$, $HK = 4$, $AD = 4$, $AC = 6$, and $AE = 24$.

Compute the lengths associated with the various letters, and place the letters in the appropriate spaces below.

Letter	Segment Length	Numerical length	Letter	Segment Length	Numerical length
1. I	*HI*	?	2. H	*IL* − *JM*	?
3. A	*EK* + *IL*	?	4. P	*IL*	?
5. O	*DG*	?	6. A	*FI* − *CD*	?
7. A	*LM* + *BC*	?	8. F	*IJ*	?
9. X	*FI* + *HK*	?	10. A	*BC*	?
11. T	*IL* + *JM*	?	12. I	*GJ*	?
13. E	*CF*	?	14. R	*IJ* + *DG*	?
15. Y	*JM*	?	16. A	*LM*	?
17. L	*IL* + *JM* + *GJ*	?	18. N	*FG* + *FI*	?
19. D	*DM*	?			

H Y P A T I A O F
1 2 3 4 5 6 7 8 9

A L E X A N D R I A
10 11 12 13 14 15 16 17 18 19

Practice and Problem Solving Wkbk/ All-in-One Resources/Online

Think About a Plan

7-5 **Think About a Plan**
Proportions in Triangles

An angle bisector of a triangle divides the opposite side of the triangle into segments 5 cm and 3 cm long. A second side of the triangle is 7.5 cm long. Find all possible lengths for the third side of the triangle.

1. What is the Triangle-Angle-Bisector Theorem? What relationships does it specifically describe?
It states that the angle bisector of a triangle divides the opposite side of the triangle into proportional segments. So the ratio of one segment and its adjacent side is proportional to the ratio of the second segment and its adjacent side.
2. What information is given in this problem? What information is not given?
You know that the lengths of the two segments created by the angle bisector are 5 cm and 3 cm long. You know the length of the second side is 7.5 cm, but you don't know to which segment this side is adjacent.
3. What does the phrase "all possible lengths" tell you about the problem?
There may be more than one answer.
4. In the space below, draw all the possible representations of the triangle described in the problem.

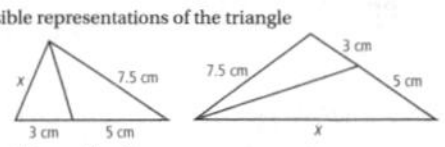

5. How can proportions be used to solve this problem?
Write a proportion comparing the ratio of each segment to its adjacent side.
6. How many proportions will you need to set up? Explain.
You will need to set up two proportions because you don't know the segment that is adjacent to the 7.5 cm side.
7. Use the space below to write and solve the proportions.
$\frac{3}{7.5} = \frac{5}{x}$, $3x = 37.5$, $x = 12.5$ $\frac{3}{x} = \frac{5}{7.5}$, $5x = 22.5$, $x = 4.5$
8. What are the possible lengths for the third side of the triangle?
12.5 cm or 4.5 cm

Practice and Problem Solving Wkbk/ All-in-One Resources/Online

Standardized Test Prep

7-5 **Standardized Test Prep**
Proportions in Triangles

Multiple Choice

For Exercises 1–5, choose the correct letter.

For Exercises 1 and 2, use the diagram at the right.

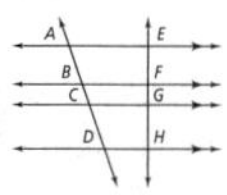

1. Which makes the proportion true? $\frac{AB}{\square} = \frac{EF}{GH}$ C
Ⓐ *AD* Ⓒ *CD*
Ⓑ *DH* Ⓓ *BC*
2. Which proportion is *not* true? I
Ⓕ $\frac{BC}{CD} = \frac{FG}{GH}$ Ⓖ $\frac{AC}{CD} = \frac{EG}{GH}$ Ⓗ $\frac{BD}{FH} = \frac{AD}{EH}$ Ⓘ $\frac{AB}{AE} = \frac{EF}{BF}$
3. What is the value of *y*? A
Ⓐ 2 Ⓒ 3
Ⓑ 4 Ⓓ 6

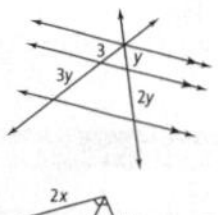

4. What is the value of *x*? G
Ⓕ 3 Ⓗ 6
Ⓖ 8 Ⓘ 12

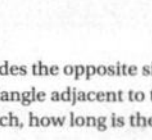

5. In $\triangle DEF$, the bisector of $\angle F$ divides the opposite sides into segments that are 4 and 9 in. long. The side of the triangle adjacent to the 4 in. segment is 6 in. long. To the nearest tenth of an inch, how long is the third side of the triangle? D
Ⓐ 2.7 in. Ⓑ 6 in. Ⓒ 13 in. Ⓓ 13.5 in.

Short Response

6. In $\triangle QRS$, $\overline{XY} \parallel \overline{SR}$. $\overline{XY}$ divides $\overline{QR}$ and $\overline{QS}$ into segments as follows: $\overline{SX} = 3$, $\overline{XQ} = 2x$, $\overline{RY} = 4.5$, and $\overline{YQ} = 7.5$. Write a proportion to find *x*. What is the length of $\overline{QS}$? [2] $\frac{7.5}{4.5} = \frac{2x}{3}$; $x = 2.5$; 8 [1] incorrect proportion or error in calculation [0] incorrect proportion and error in calculation

Online Teacher Resource Center

Activities, Games, and Puzzles

7-5 **Game: Proportions Relay**
Proportions in Triangles

Setup

Your teacher will divide the class into teams of four. Sit in a row with your teammates. Each team member will receive a problem on a slip of paper. Leave the problem face down until your teacher instructs you to begin. The first person in your team will have enough information to solve their problem. Subsequent players will have problems that contain a measurement labeled (TNWYWR) for "The Number Which You Will Receive."

Game Play

When your teacher tells you to begin, turn your paper over. The first person on each team completes the problem, writes the *answer only* on the first line of your team's answer slip, and passes the slip to the second person. The second person replaces the (TNWYWR) in his/her problem with the answer from the first person. Then, the second problem can be solved and the answer passed to the third person, and so on. When the fourth person has completed his/her problem, he/she raises the answer slip in the air.

If you pass an answer and then later realize it was incorrect, you may write the corrected answer on a new answer slip and pass it. Then, subsequent players must revise their answers based on the new information. While you wait for the answer from the player(s) before you, set up the problem.

Ending the Game

Each round lasts a maximum of six minutes. When your teacher indicates that time is up, your team must submit its answer slip immediately. Your team earns a point for each correct answer. If your team correctly answers all four problems in under 3 minutes, you earn 2 bonus points. After 3 rounds, the team with the most points wins.

Team: Round: A	Team: Round: B	Team: Round: C
1. 7.5	1. 4	1. 10
2. 10	2. 6	2. 20
3. 12.8	3. 12.5	3. 10
4. 8	4. 9.28	4. 8
Points earned:	Points earned:	Points earned:

7 Pull It All Together

Completing the Performance Task

Look back at your results from the Apply What You've Learned sections in Lessons 7-1 and 7-2. Use the work you did to complete the following.

To solve these problems, you will pull together many concepts and skills that you have studied about similarity.

1. Solve the problem in the Task Description on page 431 by determining the values Lillian should enter for Xmin and Xmax so that the graph is not distorted. Show all your work and explain each step of your solution.
2. **Reflect** Choose one of the Mathematical Practices below and explain how you applied it in your work.
   MP 1: Make sense of problems and persevere in solving them.
   MP 3: Construct viable arguments and critique the reasoning of others.

On Your Own

Lillian graphs the function $y = x$ on her brother's graphing calculator. She knows that the slope of the line is 1 and that the line should bisect two of the right angles formed by the two axes. However, the line does not appear to bisect the right angles on the screen.

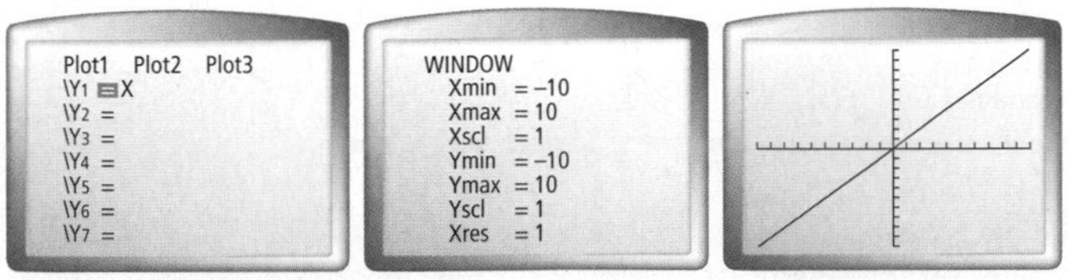

This time, Lillian wants to adjust the values of Ymin and Ymax so that the graph of the line is not distorted. In the calculator's manual, Lillian reads that the screen is 96 pixels wide by 64 pixels high.

Determine the values Lillian should enter for Ymin and Ymax Lillian so that the graph is not distorted.

Completing the Performance Task

In the Apply What You've Learned sections in Lessons 7-1 and 7-2, students compared the shape of Lillian's calculator screen to the shape of the rectangle determined by Xmin, Xmax, Ymin, and Ymax. Here, students use their findings and what they have learned about proportions and similarity to complete the Performance Task. Ask students the following questions as they work toward solving the problem.

Q How can you set up and solve a proportion to complete the Performance Task? **[Sample: Let *a* be the value of Xmax. Then the width of the viewing rectangle is 2*a*. Solve the proportion $\frac{2a}{20} = \frac{3}{2}$ for *a*.]**

Q How can you use a graphing calculator to check that you found the correct values of Xmin and Xmax? **[Sample: Using a graphing calculator with a screen for which the ratio of the width to height is 3 : 2, enter your values for Xmin and Xmax in the WINDOW screen (with Ymin = −10 and Ymax = 10), and then graph the lines. The graphs of the lines should now appear to be perpendicular.]**

FOSTERING MATHEMATICAL DISCOURSE

Not all graphing calculator screens have a width-to-height ratio of 3 : 2. Have students discuss how the values they found for Xmin and Xmax in the Performance Task would change for various calculator screen dimensions.

ANSWERS

1. Xmax = −15, Xmax = 15
2. Check students' work.

On Your Own

This problem is similar to the problem posed on page 431, but now students must adjust the values of Ymin and Ymax in order to create a graph that is not distorted. Students should strive to solve this problem independently.

ANSWER

Ymin = $-6\frac{2}{3}$, Ymax = $6\frac{2}{3}$

Essential Questions

BIG idea Similarity

ESSENTIAL QUESTION How do you use proportions to find side lengths in similar polygons?
ANSWER You can set up and solve proportions using corresponding sides of similar polygons.

BIG ideas Reasoning and Proof

ESSENTIAL QUESTION How do you show two triangles are similar?
ANSWER Two triangles are similar if certain relationships exist between two or three pairs of corresponding parts.

BIG idea Visualization

ESSENTIAL QUESTION How do you identify corresponding parts of similar triangles?
ANSWER Sketch and label triangles separately in the same orientation to see how the sides and vertices correspond.

7 Chapter Review

Connecting BIG ideas and Answering the Essential Questions

1 Similarity
You can set up and solve proportions using corresponding sides of similar polygons.

2 Reasoning and Proof
Two triangles are similar if certain relationships exist between two or three pairs of corresponding parts.

3 Visualization
Sketch and label triangles separately in the same orientation to see how the vertices correspond.

Ratios and Proportions (Lesson 7-1)
The Cross Products Property states that if $\frac{a}{b} = \frac{c}{d}$, then $ad = bc$.

Similar Polygons (Lesson 7-2)
Corresponding angles of similar polygons are congruent, and corresponding sides of similar polygons are proportional.

Proving Triangles Similar (Lesson 7-3)
Angle-Angle Similarity (AA ~) Postulate
Side-Angle-Side Similarity (SAS ~) Theorem
Side-Side-Side Similarity (SSS ~) Theorem

Seeing Similar Triangles (Lessons 7-3 and 7-4)

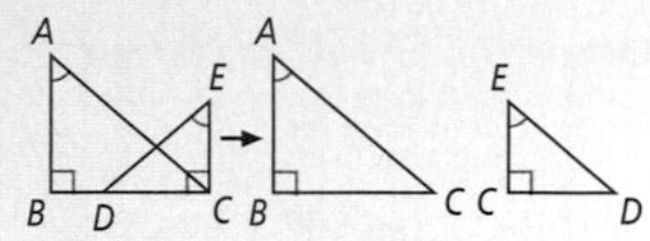

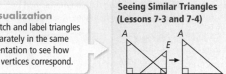

$\triangle ABC \sim \triangle ECD$

Proportions in Triangles (Lessons 7-4 and 7-5)

Geometric Means in Right Triangles

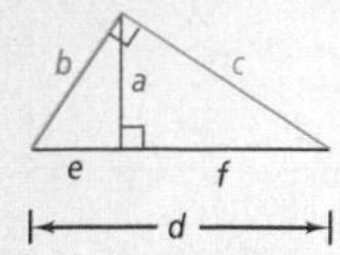

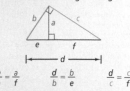

$\frac{e}{a} = \frac{a}{f}$ $\quad \frac{d}{b} = \frac{b}{e}$ $\quad \frac{d}{c} = \frac{c}{f}$

Side-Splitter Theorem

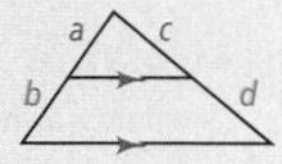

$\frac{a}{b} = \frac{c}{d}$

Triangle-Angle-Bisector Theorem

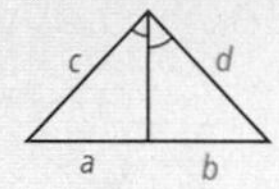

$\frac{a}{b} = \frac{c}{d}$

Chapter Vocabulary

- extended proportion (p. 440)
- extended ratio (p. 433)
- extremes (p. 434)
- geometric mean (p. 462)
- indirect measurement (p. 454)
- means (p. 434)
- proportion (p. 434)
- ratio (p. 432)
- scale drawing (p. 443)
- scale factor (p. 440)
- similar figures (p. 440)
- similar polygons (p. 440)

Choose the correct term to complete each sentence.

1. Two polygons are _?_ if their corresponding angles are congruent and corresponding sides are proportional.

2. A(n) _?_ is a statement that two ratios are equal.

3. The ratio of the lengths of corresponding sides of two similar polygons is the _?_.

4. The Cross Products Property states that the product of the _?_ is equal to the product of the _?_.

Summative Questions

Use the following prompts as you review this chapter with your students. The prompts are designed to help you assess your students' understanding of the Big Ideas they have studied.

- How do you form a proportion?
- How do you solve a proportion?
- What conditions must be true for two polygons to be similar?
- When a figure is made up of more than one polygon, how can you visualize the vertices and sides of each polygon?

Answers

Chapter Review

1. similar
2. proportion
3. scale factor
4. means, extremes (in either order)

7-1 Ratios and Proportions

Quick Review

A **ratio** is a comparison of two quantities by division. A **proportion** is a statement that two ratios are equal. The **Cross Products Property** states that if $\frac{a}{b} = \frac{c}{d}$, where $b \neq 0$ and $d \neq 0$, then $ad = bc$.

Example

What is the solution of $\frac{x}{x+3} = \frac{4}{6}$?

$6x = 4(x + 3)$	Cross Products Property
$6x = 4x + 12$	Distributive Property
$2x = 12$	Subtract $4x$ from each side.
$x = 6$	Divide each side by 2.

Exercises

5. A high school has 16 math teachers for 1856 math students. What is the ratio of math teachers to math students?

6. The measures of two complementary angles are in the ratio 2 : 3. What is the measure of the smaller angle?

Algebra **Solve each proportion.**

7. $\frac{x}{7} = \frac{18}{21}$

8. $\frac{6}{11} = \frac{15}{2x}$

9. $\frac{x}{3} = \frac{x+4}{5}$

10. $\frac{8}{x+9} = \frac{2}{x-3}$

7-2 and 7-3 Similar Polygons and Proving Triangles Similar

Quick Review

Similar polygons have congruent corresponding angles and proportional corresponding sides. You can prove triangles similar with limited information about congruent corresponding angles and proportional corresponding sides.

Postulate or Theorem	What You Need
Angle-Angle (AA ~)	two pairs of ≅ angles
Side-Angle-Side (SAS ~)	two pairs of proportional sides and the included angles ≅
Side-Side-Side (SSS ~)	three pairs of proportional sides

Example

Is $\triangle ABC$ similar to $\triangle RQP$? How do you know?

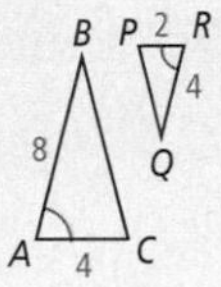

You know that $\angle A \cong \angle R$.

$\frac{AB}{RQ} = \frac{AC}{RP} = \frac{2}{1}$, so the triangles are similar by the SAS ~ Theorem.

Exercises

The polygons are similar. Write a similarity statement and give the scale factor.

11.

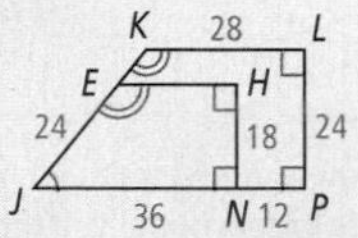

12.

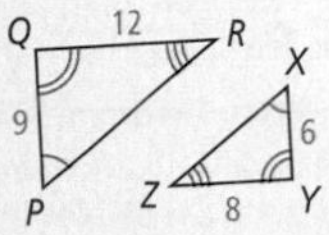

13. **City Planning** The length of a rectangular playground in a scale drawing is 12 in. If the scale is 1 in. = 10 ft, what is the actual length?

14. **Indirect Measurement** A 3-ft vertical post casts a 24-in. shadow at the same time a pine tree casts a 30-ft shadow. How tall is the pine tree?

Are the triangles similar? How do you know?

15.

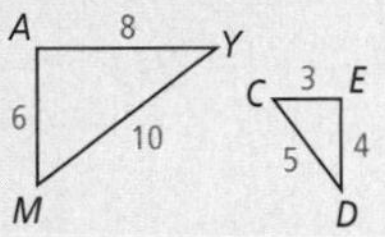

16.

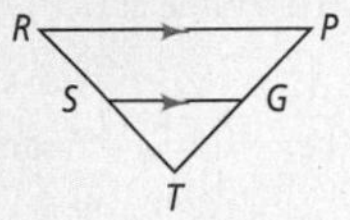

5. 1 : 116 or $\frac{1}{116}$ **6.** 36
7. 6 **8.** $\frac{55}{4}$ or $13\frac{3}{4}$
9. 6 **10.** 7
11. *JEHN* ~ *JKLP*; 3 : 4
12. $\triangle PQR \sim \triangle XYZ$; 3 : 2
13. 120 ft **14.** 45 ft
15. The ratio of each pair of corresp. sides is 2 : 1, so $\triangle AMY \sim \triangle ECD$ by SSS~.
16. If lines are ∥, then corresp. ∠s are ≅, so $\triangle RPT \sim \triangle SGT$ by AA~.

Answers

Chapter Review (continued)

17. 12 **18.** $2\sqrt{15}$

19. $x = 6\sqrt{2}$, $y = 6\sqrt{6}$

20. $\sqrt{35}$

21. $x = 2\sqrt{21}$; $y = 4\sqrt{3}$

22. $x = 12$, $y = 4\sqrt{5}$

23. 7.5 **24.** 3.6 **25.** 22.5

26. 12 **27.** 17.5 **28.** 77

7-4 Similarity in Right Triangles

Quick Review

$\overline{CD}$ is the altitude to the hypotenuse of right $\triangle ABC$.

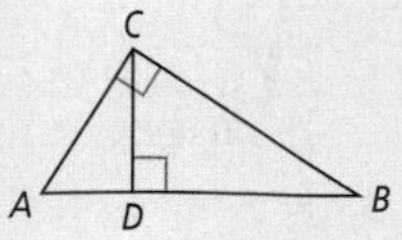

- $\triangle ABC \sim \triangle ACD$, $\triangle ABC \sim \triangle CBD$, and $\triangle ACD \sim \triangle CBD$
- $\frac{AD}{CD} = \frac{CD}{DB}$, $\frac{AB}{AC} = \frac{AC}{AD}$, and $\frac{AB}{CB} = \frac{CB}{DB}$

Example

What is the value of x?

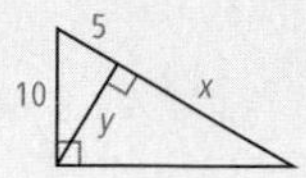

$\frac{5 + x}{10} = \frac{10}{5}$ Write a proportion.

$5(5 + x) = 100$ Cross Products Property

$25 + 5x = 100$ Distributive Property

$5x = 75$ Subtract 25 from each side.

$x = 15$ Divide each side by 5.

Exercises

Find the geometric mean of each pair of numbers.

17. 9 and 16 **18.** 5 and 12

Algebra **Find the value of each variable. Write your answer in simplest radical form.**

19.

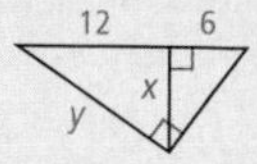

20.

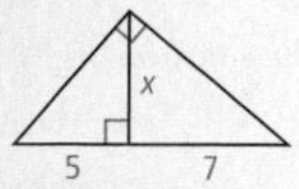

21.

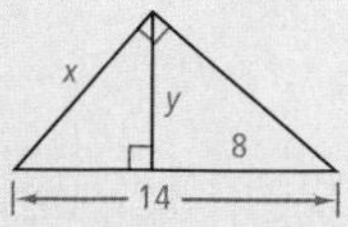

22.

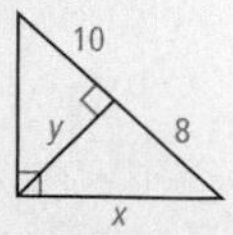

7-5 Proportions in Triangles

Quick Review

Side-Splitter Theorem and Corollary

If a line parallel to one side of a triangle intersects the other two sides, then it divides those sides proportionally. If three parallel lines intersect two transversals, then the segments intercepted on the transversals are proportional.

Triangle-Angle-Bisector Theorem

If a ray bisects an angle of a triangle, then it divides the opposite side into two segments that are proportional to the other two sides of the triangle.

Example

What is the value of x?

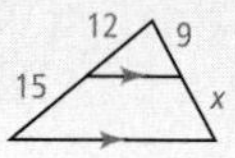

$\frac{12}{15} = \frac{9}{x}$ Write a proportion.

$12x = 135$ Cross Products Property

$x = 11.25$ Divide each side by 12.

Exercises

Algebra **Find the value of x.**

23.

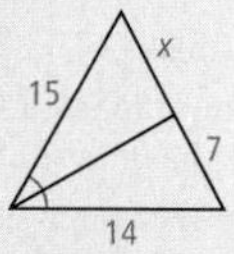

24.

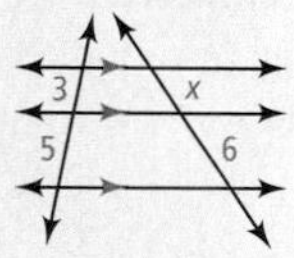

25.

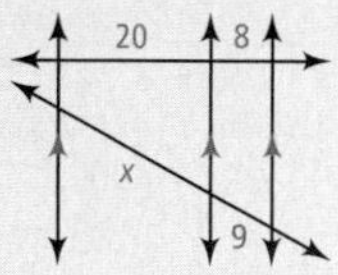

26.

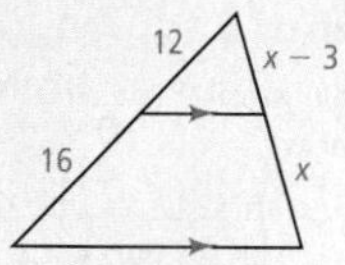

27.

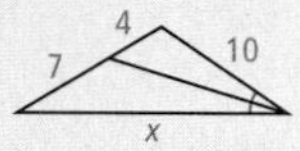

28.

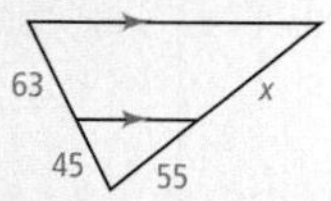

Chapter Test

Do you know HOW?

Algebra **Solve each proportion.**

1. $\frac{x}{3} = \frac{8}{12}$

2. $\frac{4}{x+2} = \frac{16}{9}$

3. Are the polygons below similar? If they are, write a similarity statement and give the scale factor.

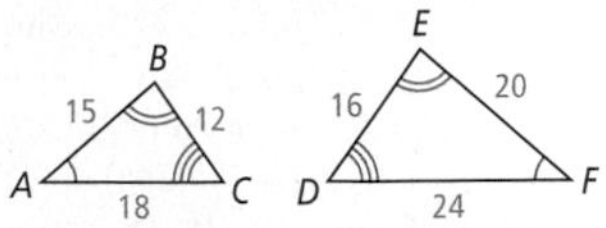

Algebra **The figures in each pair are similar. Find the value of each variable.**

4.

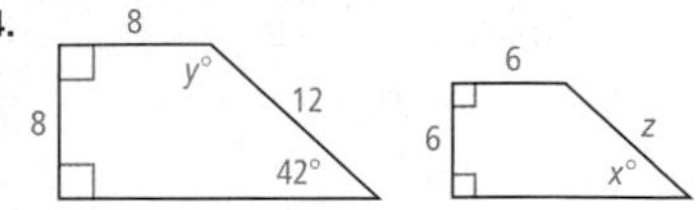

5.

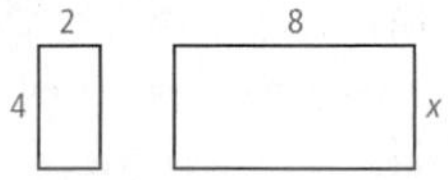

6.

Determine whether the triangles are similar. If so, write a similarity statement and name the postulate or the theorem you used. If not, explain.

7.

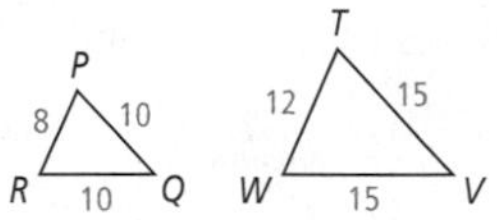

8.

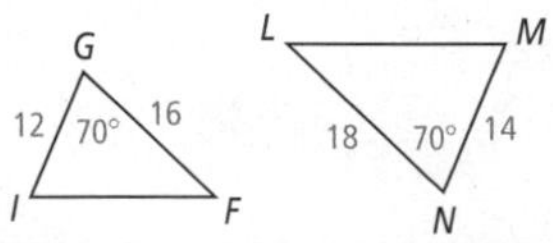

9. **Indirect Measurement** A meterstick perpendicular to the ground casts a 1.5-m shadow. At the same time, a telephone pole casts a shadow that is 9 m. How tall is the telephone pole?

10. **Photography** A photographic negative is 3 cm by 2 cm. A similar print from the negative is 9 cm long on its shorter side. What is the length of the longer side?

11. What is the geometric mean of 10 and 15?

Algebra **Find the value of *x*.**

12.

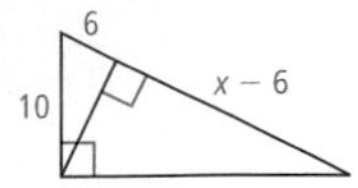

13.

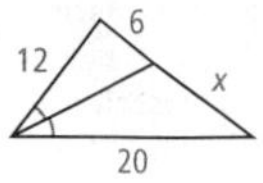

14.

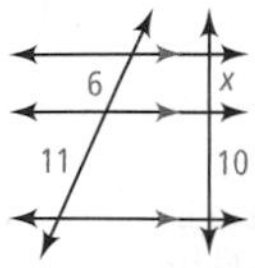

15.

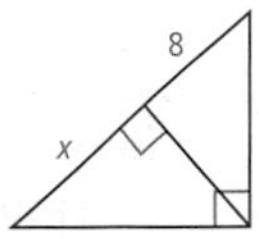

Do you UNDERSTAND?

16. **Reasoning** In the diagram, $\overline{MN} \parallel \overline{LK}$. Is $\frac{JM}{ML}$ equal to $\frac{MN}{LK}$? Explain.

17. **Reasoning** ▱ABCD ~ ▱PQRS. $\overline{BD}$ is a diagonal of ▱ABCD and $\overline{SQ}$ is a diagonal of ▱PQRS. Is △BCD similar to △QRS? Justify your reasoning.

Determine whether each statement is *always*, *sometimes*, or *never* true.

18. A parallelogram is similar to a trapezoid.

19. Two rectangles are similar.

20. If the vertex angles of two isosceles triangles are congruent, then the triangles are similar.

Answers

Chapter Test

1. 2
2. $\frac{1}{4}$
3. △ABC ~ △FED; 3 : 4 or $\frac{3}{4}$
4. $x = 42$, $y = 138$, $z = 9$
5. $x = 4$
6. $x = 63$, $y = 8$
7. △QRP ~ △VWT by SSS~.
8. No; the corresp. sides are not proportional.
9. 6 m
10. 13.5 cm
11. $5\sqrt{6}$
12. $\frac{50}{3}$ or $16\frac{2}{3}$
13. 10
14. $\frac{60}{11}$ or $5\frac{5}{11}$
15. 10
16. No. Explanations may vary. Sample: △JMN ~ △JLK by AA~. The ratio $\frac{MN}{LK}$ is a ratio of corresp. sides in the two △s, but the ratio $\frac{JM}{ML}$ is NOT a ratio of corresp. sides. A correct ratio would be $\frac{JM}{JL} = \frac{MN}{LK}$.
17. If ABCD ~ PQRS, then ∠C ≅ ∠R and $\frac{BC}{QR} = \frac{CD}{RS}$. So △BCD ~ △QRS by SAS~.
18. never
19. sometimes
20. always

Item Number	Lesson	Content Standard
1	6-4	*G-CO.C.11
2	5-4	*G-CO.C.10
3	6-6	G-CO.C.11
4	7-3	G-CO.C.10
5	1-7	*G-GPE.B.7
6	7-5	G-GPE.B.6
7	7-4	G-SRT.C.6
8	5-1	G-SRT.B.5
9	7-2	G-SRT.B.5
10	1-6	G-CO.D.12
11	6-1	G-SRT.B.5
12	7-2	G-SRT.B.5
13	3-2	G-CO.C.11
14	6-1	G-MG.A.3
15	7-2	G-SRT.B.5
16	6-2	G-CO.C.11
17	4-5	G-CO.C.10
18	7-5	G-GPE.B.6
19	6-4	G-CO.C.11
20	3-6	G-CO.D.12
21	7-2	G-SRT.B.5
22	4-2	G-CO.C.10
23	7-4	G-SRT.B.5

*Prepares for standard

7 Common Core Cumulative Standards Review

ASSESSMENT

TIPS FOR SUCCESS

Some test questions ask you to find the measure of an interior or exterior angle of a polygon. Read the sample question at the right. Then follow the tips to answer it.

In the figure below, *ABCDE* is a regular pentagon. What is the measure, in degrees, of $\angle ABE$?

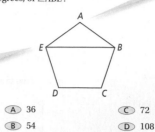

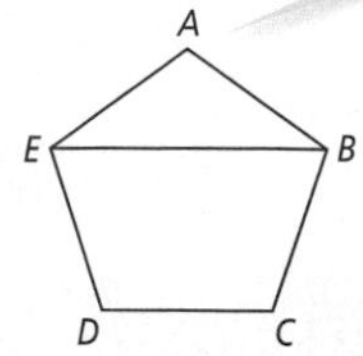

Ⓐ 36 Ⓑ 54 Ⓒ 72 Ⓓ 108

TIP 1

List what you know about $\triangle ABE$.
- $\overline{AE} \cong \overline{AB}$ because the pentagon is regular.
- $\angle ABE \cong \angle AEB$ because $\triangle ABE$ is isosceles.
- $m\angle A + m\angle ABE + m\angle AEB = 180$

TIP 2

Find $m\angle A$ and then use it to find $m\angle ABE$.

Think It Through

By the Polygon Angle-Sum Theorem, the sum of the interior angle measures of *ABCDE* is $(5-2)180 = 3(180) = 540$. So $m\angle A = \frac{540}{5} = 108$. Then $108 + m\angle ABE + m\angle AEB = 180$. Since $\angle ABE \cong \angle AEB$, $108 + 2 \cdot m\angle ABE = 180$. So $m\angle ABE = \frac{180-108}{2} = \frac{72}{2} = 36$. The correct answer is A.

Vocabulary Builder

As you solve test items, you must understand the meanings of mathematical terms. Match each term with its mathematical meaning.

A. corollary
B. geometric mean
C. midsegment
D. scale

I. the ratio of a length in a scale drawing to the actual length
II. a segment connecting the midpoints of two sides of a triangle
III. a statement that follows immediately from a theorem
IV. for positive numbers *a* and *b*, the positive number *x* such that $\frac{a}{x} = \frac{x}{b}$

Selected Response

Read each question. Then write the letter of the correct answer on your paper.

1. What is a name for the quadrilateral below?

I. square
II. rectangle
III. rhombus
IV. parallelogram

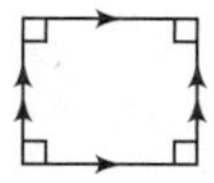

Ⓐ I only
Ⓑ IV only
Ⓒ II and IV
Ⓓ I, II, and IV

2. In which point do the bisectors of the angles of a triangle meet?

Ⓕ centroid
Ⓖ circumcenter
Ⓗ incenter
Ⓘ orthocenter

Answers

Common Core Cumulative Standards Review

A. III
B. IV
C. II
D. I
1. C
2. H

3. Which quadrilateral does NOT always have perpendicular diagonals?

Ⓐ square
Ⓑ rhombus
Ⓒ kite
Ⓓ isosceles trapezoid

4. Which of the following facts would be sufficient to prove $\triangle ACE \sim \triangle BCD$?

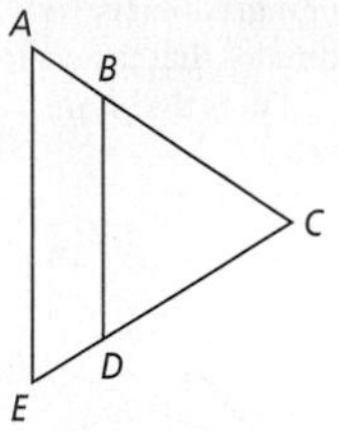

Ⓕ $\triangle BCD$ is a right triangle.
Ⓖ $\overline{AB} \cong \overline{ED}$
Ⓗ $m\angle A = m\angle E$
Ⓘ $\overline{AE} \parallel \overline{BD}$

5. What is the midpoint of the segment whose endpoints are $M(6, -11)$ and $N(-18, 7)$?

Ⓐ $(-6, -2)$
Ⓑ $(6, 2)$
Ⓒ $(-12, 9)$
Ⓓ $(12, -9)$

6. Use the figure below. By which theorem or postulate does $x = 3$?

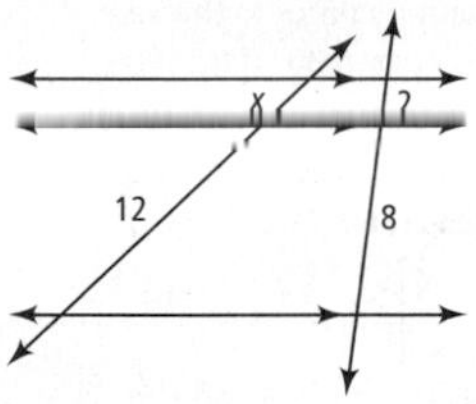

Ⓕ SAS Postulate
Ⓖ If three parallel lines intersect two transversals, then the segments intercepted on the transversals are proportional.
Ⓗ Opposite sides of a parallelogram are congruent.
Ⓘ If two lines are parallel to the same line, then they are parallel to each other.

7. Which angle is congruent to $\angle DCB$?

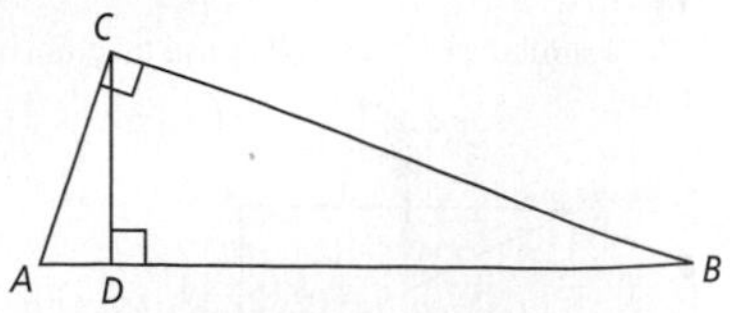

Ⓐ $\angle B$
Ⓑ $\angle CDB$
Ⓒ $\angle A$
Ⓓ $\angle ACD$

8. In the figure below, $\overline{EF}$ is a midsegment of $\triangle ABC$ and $ADGC$ is a rectangle. What is the area of $\triangle EDA$?

Ⓕ 49 cm^2
Ⓖ 98 cm^2
Ⓗ 294 cm^2
Ⓘ 588 cm^2

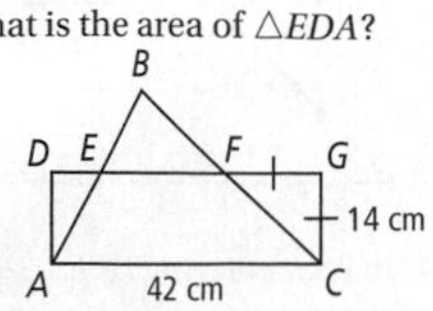

9. Andrew is looking at a map that uses the scale 1 in. = 5 mi. On the map, the distance from Westville to Allentown is 9 in. Which proportion CANNOT be used to find the actual distance?

Ⓐ $\frac{1\text{ in.}}{5\text{ mi}} = \frac{9\text{ in.}}{d}$
Ⓑ $\frac{5\text{ mi}}{d} = \frac{9\text{ in.}}{1\text{ in.}}$
Ⓒ $\frac{d}{9\text{ in.}} = \frac{5\text{ mi}}{1\text{ in.}}$
Ⓓ $\frac{5\text{ mi}}{1\text{ in.}} = \frac{d}{9\text{ in.}}$

10. What type of construction is shown below?

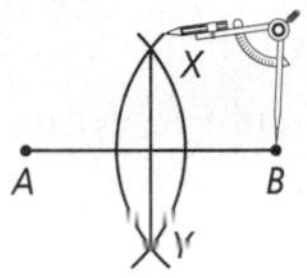

Ⓕ angle bisector
Ⓖ perpendicular bisector
Ⓗ congruent angles
Ⓘ congruent triangles

11. A student is sketching an 11-sided regular polygon. What is the sum of the measures of the polygon's first five angles to the nearest degree?

Ⓐ 147
Ⓑ 720
Ⓒ 736
Ⓓ 1620

3. D
4. I
5. A
6. G
7. C
8. F
9. B
10. G
11. C

Answers

Common Core Cumulative Standards Review (continued)

12. 138.72 **13.** 30

14. 135 **15.** 80

16. 7.5 **17.** 34 **18.** 3.75

19. [2] $x = 8$; the diagonals of a rectangle are $\cong$.

[1] incorrect answer OR incorrect explanation

20. [2]

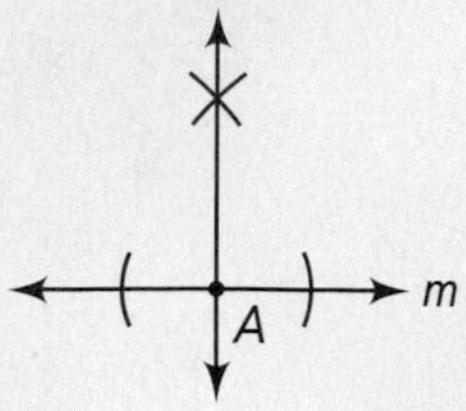

Draw arcs from A intersecting line m at two points. Open the compass wider and draw an arc above m from each of the two points. Draw a line from A to the intersection of arcs above A.

[1] incorrect drawing OR incomplete/incorrect steps

21. [2] Answers may vary. Sample: No; the model would be more than 10 ft tall.

[1] incomplete OR incorrect explanation

22. [4] **a.** No; only one pair of $\cong$ sides and $\cong$ vert. $\angle$s is not enough information.

b. Any of the following: $\overline{BD} \cong \overline{EC}$; $\angle BCD \cong \angle EDC$; $\angle CBG \cong \angle DEG$; $\angle BCG \cong \angle EDG$

c. $\overline{AF} \cong \overline{FA}$ by the Refl. Prop. of $\cong$.

d. SSS; we have information about the three sides of the two $\triangle$s, but we have no information about any $\angle$s of the $\triangle$s.

[3] minor error in one explanation

[2] incorrect or missing explanations in one or two parts

[1] incomplete OR incorrect answers in three or more parts

23. [4] By the Pythagorean Theorem, $x = 40$. Because the two $\triangle$s are $\sim$, $\frac{40}{30} = \frac{30}{y}$ and $y = 22.5$. Total distance $= 40 + 22.5 = 62.5$ yd.

[3] complete explanation; minor error in a calculation, but the rest of the calculations are consistent with that result

[2] complete explanation with two or more errors in calculation

[1] correct calculation but no explanation

Constructed Response

12. Triangle ABC is similar to triangle HIJ. Find the area of rectangle $HIJK$.

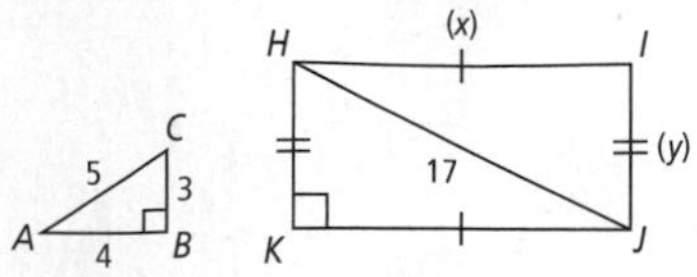

13. What is the value of x in the figure below?

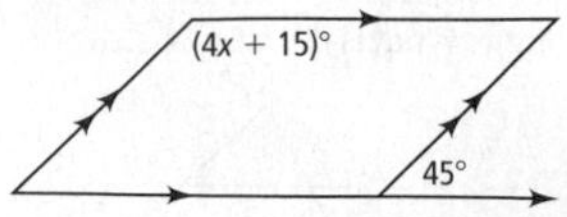

14. In hexagon $ABCDEF$, $\angle A$ and $\angle B$ are right angles. If $\angle C \cong \angle D \cong \angle E \cong \angle F$, what is the measure of $\angle F$ in degrees?

15. A scale drawing of a swimming pool and deck is shown below. Use the scale 1 in. = 2 m. What is the area of the deck in square meters?

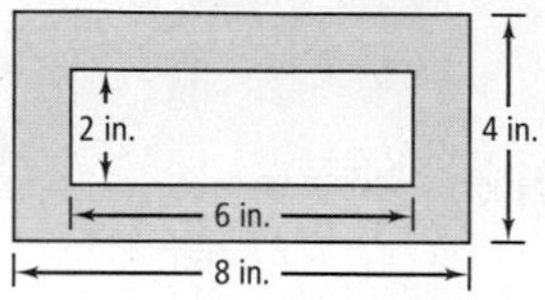

16. In parallelogram $ABCD$ below, DB is 15. What is DE?

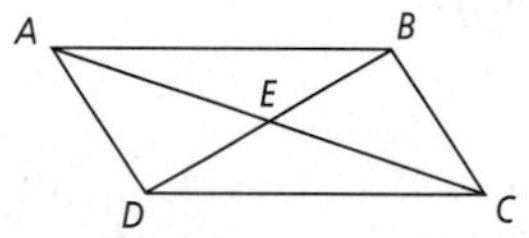

17. The measure of the vertex angle of an isosceles triangle is 112. What is the measure of a base angle?

18. What is the value of x in the figure below?

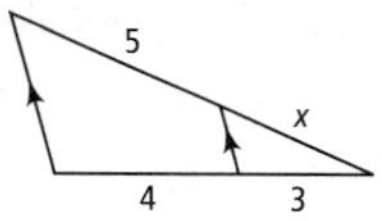

19. In rectangle $ABCD$, $AC = 5(x - 2)$ and $BD = 3(x + 2)$. What is the value of x? Justify your answer.

20. Draw line m with point A on it. Construct a line perpendicular to m at A. What steps did you take to perform the construction?

21. Petra visited the Empire State Building, which is approximately 1454 ft tall. She estimates that the scale of the model she bought is 1 in. = 12 ft. Is this scale reasonable? Explain.

Extended Response

22. In the diagram, $AB = FE$, $BC = ED$, and $AE = FB$.

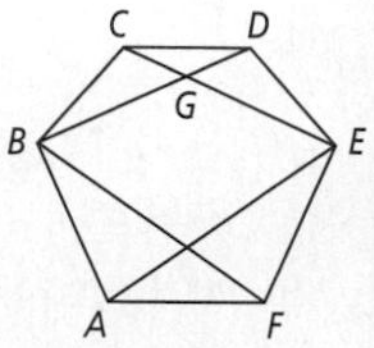

a. Is there enough information to prove $\triangle BCG \cong \triangle EDG$? Explain.

b. What one additional piece of information would allow you to prove $\triangle BCD \cong \triangle EDC$? Explain.

c. What can you conclude from the diagram that would help you prove $\triangle BAF \cong \triangle EFA$?

d. In part (c), is $\triangle BAF \cong \triangle EFA$ by SAS or SSS? Explain.

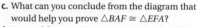

23. At a campground, the 50-yd path from your campsite to the information center forms a right angle with the path from the information center to the lake. The information center is located 30 yd from the bathhouse. How far is your campsite from the lake? Show your work.

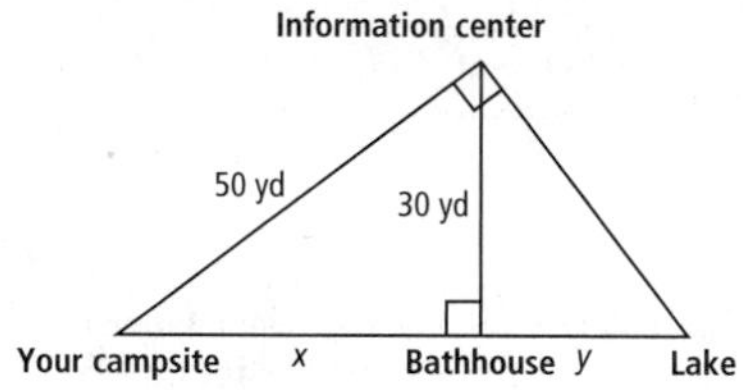

Get Ready!

CHAPTER 8

Lesson 7-1 **Solving Proportions**

Algebra Solve for x. If necessary, round answers to the nearest thousandth.

1. $0.2734 = \frac{x}{17}$ **2.** $0.5858 = \frac{24}{x}$ **3.** $0.8572 = \frac{5271}{x}$ **4.** $0.5 = \frac{x}{3x + 5}$

Lesson 7-3 **Proving Triangles Similar**

Name the postulate or theorem that proves each pair of triangles similar.

5. $\overline{CD} \parallel \overline{AB}$

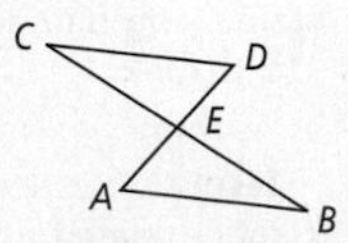

6.

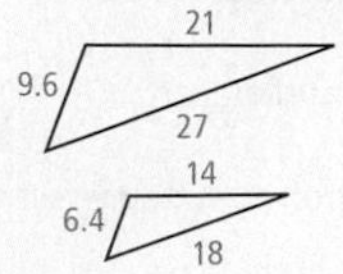

7. $\overline{JK} \perp \overline{ML}$

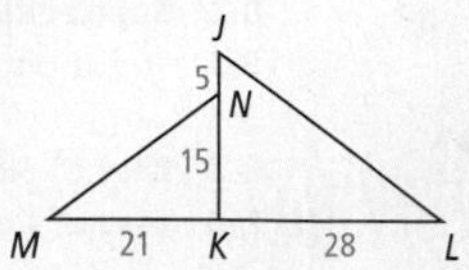

Lesson 7-4 **Similarity in Right Triangles**

Algebra Find the value of x in $\triangle ABC$ with right $\angle C$ and altitude $\overline{CD}$.

8.

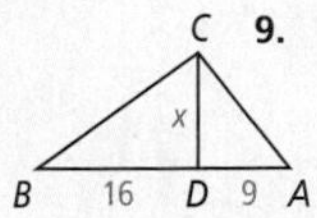

9.

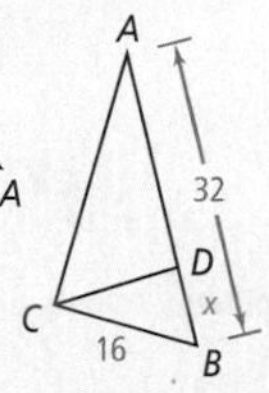

10.

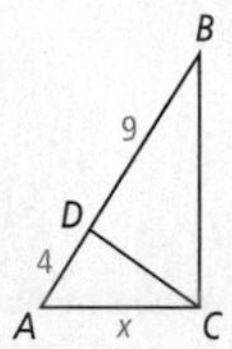

11.

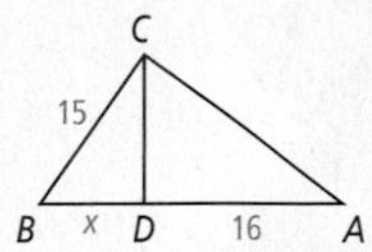

Looking Ahead Vocabulary

12. People often describe the height of a mountain as its *elevation*. How might you describe an *angle of elevation* in geometry?

13. You see the prefix *tri-* in many words, such as *triad, triathlon, trilogy*, and *trimester.* What does the prefix indicate in these words? What geometric figure do you think is associated with the phrase *trigonometric ratio*? Explain.

14. You can use the Pythagorean Theorem to derive the *Law of Cosines*. What do you think you can find by using the *Law of Cosines*?

Answers

Get Ready!

1. 4.648
2. 40.970
3. 6149.090
4. −5
5. AA ~
6. SSS ~
7. SAS ~
8. 12
9. 8
10. $2\sqrt{13}$
11. 9
12. Answers may vary. Sample: When something is "elevated" you look up to see it, so an ∠ of elevation is formed by a horizontal line and the line of sight.
13. Answers may vary. Sample: The prefix *tri-* means 3; triangles are associated with trigonometric ratios.
14. Answers may vary. Sample: Because the Law of Cosines can be derived from the Pythagorean Theorem, and you can use the Law of Cosines to find angle measures, the Law of Cosines can probably be used to find side lengths and angle measures.

Get Ready!

Assign this diagnostic assessment to determine if students have the prerequisite skills for Chapter 8.

Lesson	Skill
7-1	Solving Proportions
7-3	Proving Triangles Similar
7-4	Similarity in Right Triangles

To remediate students, select from these resources (available for every lesson).

- Online Problems (PowerGeometry.com)
- Reteaching (All-in-One Teaching Resources)
- Practice (All-in-One Teaching Resources)

Why Students Need These Skills

SOLVING PROPORTIONS Students will use proportions in conjunction with trigonometric ratios to make indirect measurements.

PROVING TRIANGLES SIMILAR Students will use relationships of similar triangles when solving problems related to right triangle trigonometry.

SIMILARITY IN RIGHT TRIANGLES Concepts of similar right triangles will help students understand properties of special right triangles.

Looking Ahead Vocabulary

ANGLE OF ELEVATION Have students identify objects in the classroom that they are using an angle of elevation to view.

LAW OF COSINES Have students draw a triangle and label the vertices *A, B,* and *C*, and the sides opposite those vertices *a, b,* and *c*, respectively. Have them list the included angle for each pair of sides.

TRIGONOMETRIC RATIO Have students name other words that use the prefix *tri-*.

Chapter 8 Overview

In Chapter 8 students explore concepts related to right triangles, including trigonometry. Students will develop the answers to the Essential Questions as they learn the concepts and skills shown below.

BIG idea Measurement

ESSENTIAL QUESTION How do you find a side length or angle measure in a right triangle?

- Students will use the Pythagorean Theorem.
- Students will use concepts of 30-60-90 and 45-45-90 triangles.
- Students will use trigonometric ratios to form proportions.

BIG idea Similarity

ESSENTIAL QUESTION How do trigonometric ratios relate to similar right triangles?

- Students will examine the sine ratio.
- Students will examine the cosine ratio.
- Students will examine the tangent ratio.

Content Standards

Following are the standards covered in this chapter. Modeling standards are indicated by a star symbol (★).

CONCEPTUAL CATEGORY Geometry

Domain Similarity, Right Triangles, and Trigonometry G-SRT

Cluster Prove theorems involving similarity (Standard G-SRT.B.4)
LESSON 8-1

Cluster Define trigonometric ratios and solve problems involving right triangles (Standards G-SRT.C.7, G-SRT.C.8★)
LESSONS 8-1, 8-2, 8-3, 8-4, 8-5

Cluster Apply trigonometry to general triangles (Standards G-SRT.D.10, G-SRT.D.11)
LESSON 8-5

Domain Modeling with Geometry G-MG

Cluster Apply geometric concepts in modeling situations (Standard G-MG.A.1★)
LESSON 8-3

CHAPTER 8

Right Triangles and Trigonometry

Download videos connecting math to your world.

Interactive! Vary numbers, graphs, and figures to explore math concepts.

The online Solve It will get you in gear for each lesson.

Math definitions in English and Spanish

Online access to stepped-out problems aligned to Common Core

Get and view your assignments online.

Extra practice and review online

Virtual Nerd™ tutorials with built-in support

Chapter Preview

BIG ideas

1 **Measurement**
Essential Question How do you find a side length or angle measure in a right triangle?

2 **Similarity**
Essential Question How do trigonometric ratios relate to similar right triangles?

Vocabulary

English/Spanish Vocabulary Audio Online:

English	Spanish
angle of depression, *p. 516*	ángulo de depresión
angle of elevation, *p. 516*	ángulo de elevación
cosine, *p. 507*	coseno
Law of Cosines, *p. 526*	Ley de cosenos
Law of Sines, *p. 522*	Ley de senos
Pythagorean triple, *p. 492*	tripleta de Pitágoras
sine, *p. 507*	seno
tangent, *p. 507*	tangente

DOMAINS

- Similarity, Right Triangles, and Trigonometry
- Modeling with Geometry

PowerGeometry.com

Chapter 8 Overview

Use these online assets to engage your students. These include support for the Solve It and step-by-step solutions for Problems.

Show the student-produced video demonstrating relevant and engaging applications of the new concepts in the chapter.

Find online definitions for new terms in English and Spanish.

Start each lesson with an attention-getting problem. View the Problem online with helpful hints.

Common Core Performance Task

Locating a Forest Fire

Rangers in the two lookout towers at a state forest notice a plume of smoke, shown at point *C* in the diagram. The towers are 2000 m apart. One ranger observes the smoke at an angle of 54°. The other ranger observes it at an angle of 30°. Both angles are measured from the line that connects the two towers. When the rangers call to report the fire, they must state the location of the fire using distances to the north and west of Lookout Tower B.

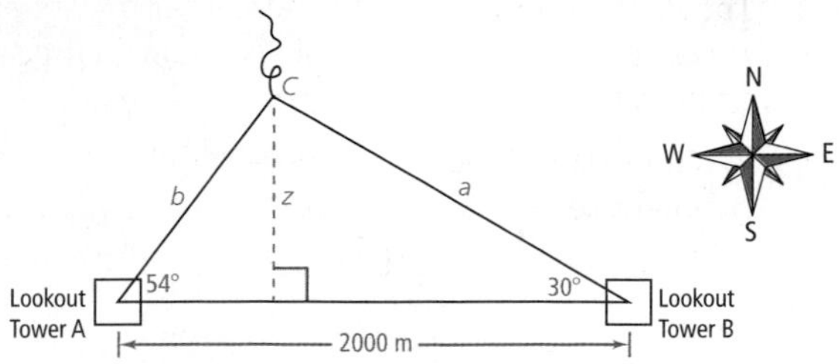

Task Description

Find how far to the north and west of Lookout Tower B the fire is located. Round your answers to the nearest tenth of a meter.

Connecting the Task to the Math Practices

MATHEMATICAL PRACTICES

As you complete the task, you'll apply several Standards for Mathematical Practice.

- You'll use trigonometric ratios to relate lengths in the diagram. (MP 1)
- You'll solve trigonometric equations to find unknown distances. (MP 6)
- You'll verify results by another method and reason abstractly to write an equation. (MP 1, MP 2)

Increase students' depth of knowledge with interactive online activities.

Show Problems from each lesson solved step by step. Instant replay allows students to go at their own pace when studying online.

Assign homework to individual students or to an entire class.

Prepare students for the Mid-Chapter Quiz and Chapter Test with online practice and review.

Virtual Nerd™ Access Virtual Nerd student-centered math tutorials that directly relate to the content of the lesson.

Overview of the Performance Task

Students will use trigonometry to write equations relating various distances determined by the positions of two lookout towers and a fire. They will use their equations to find how far the fire is to the north and west of Lookout Tower B.

Students will work on the Performance Task in the following places in the chapter.

- Lesson 8-3 (p. 513)
- Lesson 8-5 (p. 526)
- Lesson 8-6 (p. 532)
- Pull It All Together (p. 533)

Introducing the Performance Task

Tell students to read the problem on this page. Do not have them start work on the problem at this time, but ask them the following questions.

Q What is a strategy you could try in order to solve the problem? **[Sample: I could use relationships within right triangles to find the distances.]**

Q Can you use the Pythagorean Theorem to relate the two distances you are asked to find? Explain. **[Yes; the two distances are the lengths of the legs of a right triangle with hypotenuse of length *a*.]**

PARCC CLAIMS

Sub-Claim A: Major Content with Connections to Practices
Sub-Claim D: Highlighted Practices MP 4 with Connections to Content

SBAC CLAIMS

Claim 2: Problem Solving
Claim 3: Communicating Reasoning

CHAPTER 8

RIGHT TRIANGLES AND TRIGONOMETRY

Math Background

PROFESSIONAL DEVELOPMENT

The Understanding by Design® methodology was central to the development of the Big Ideas and the Essential Understandings. These will help your students build a structure on which to make connections to prior learning.

Measurement

BIG idea Some attributes of geometric figures, such as length, area, volume, and angle measure, are measurable. Units are used to describe these attributes.

ESSENTIAL UNDERSTANDINGS

8-1 If the lengths of any two sides of a right triangle are known, the length of the third side can be found by using the Pythagorean Theorem.

8-2 Certain right triangles have properties that allow their side lengths to be determined without using the Pythagorean Theorem.

8-3 If certain combinations of side lengths and angle measures of a right triangle are known, ratios can be used to find other side lengths and angle measures.

8-5 If you know the measures of two angles and any side (AAS or ASA), or two sides and a nonincluded angle (SSA), you can use the Law of Sines to find the other measures of the triangle.

8-6 If you know the measures of two sides and the included angle (SAS), or the measures of all three sides (SSS), you can use the Law of Cosines to find the other measures of the triangle.

Similarity

BIG idea Two geometric figures are similar when corresponding lengths are proportional and corresponding angles are congruent.

ESSENTIAL UNDERSTANDINGS

8-3 Ratios can be used to find side lengths and angle measures of a right triangle when certain combinations of side lengths and angles measures are known.

8-4 The angles of elevation and depression are the acute angles of right triangles formed by a horizontal distance and a vertical height.

8-5 If you know the measures of two angles and any side (AAS or ASA), or two sides and a nonincluded angle (SSA), you can use the Law of Sines to find the other measures of the triangle.

8-6 If you know the measures of two sides and the included angle (SAS), or the measures of all three sides (SSS), you can use the Law of Cosines to find the other measures of the triangle.

Pythagorean Theorem

The Pythagorean Theorem states that the square of the hypotenuse of a right triangle is equal to the sum of the squares of the other two sides.

In algebraic notation, $a^2 + b^2 = c^2$, where c is the hypotenuse, and a and b are the legs.

The hypotenuse c is the longest side of the triangle.

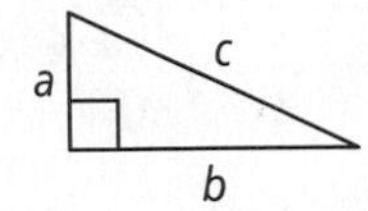

The sets of whole numbers which satisfy the Pythagorean Theorem are known as Pythagorean triples. One example is the lengths 3, 4, and 5.

The Pythagorean Theorem can also be used to determine if a triangle is acute, right, or obtuse.

If $a^2 + b^2 > c^2$, then the triangle is acute.

If $a^2 + b^2 = c^2$, then the triangle is right.

If $a^2 + b^2 < c^2$, then the triangle is obtuse.

Common Errors With Pythagorean Theorem

When using the Pythagorean Theorem to classify a triangle as right, obtuse, or acute, conditions for obtuse and acute can be confused because they seem to be counterintuitive.

For example:

A triangle has side lengths 7, 15, and 18. Is it acute, obtuse, or right?

$18^2 \overset{?}{>} 7^2 + 15^2$

$324 \overset{?}{>} 49 + 225$

$324 > 274$

The student might conclude that the triangle is acute instead of obtuse.

Mathematical Practices

Construct viable arguments and critique the reasoning of others Students explore a famous result, the Pythagorean Theorem, and discover not only Pythagorean triples but their multiples (6-8-10 is mathematically similar to 3-4-5).

Look for and make use of structure The proof in Exercise 49 builds on work from Chapter 7 on altitudes in right triangles.

30°-60°-90° and 45°-45°-90° Triangles

You can use the ratios of the side lengths of 30°-60°-90° and 45°-45°-90° triangles to set up proportions and solve for unknown side lengths.

45°-45°-90° 30°-60°-90°

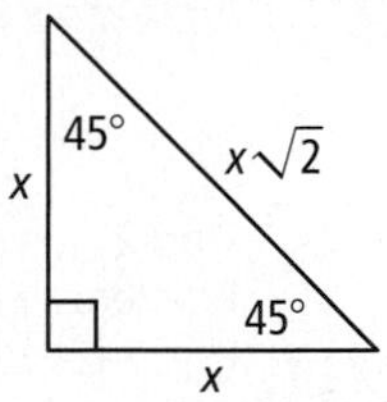

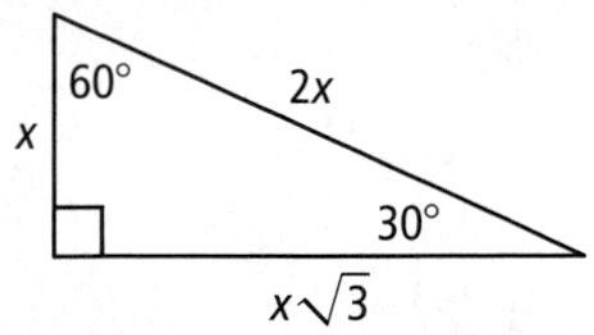

To find the unknown length in a 45°-45°-90° triangle:

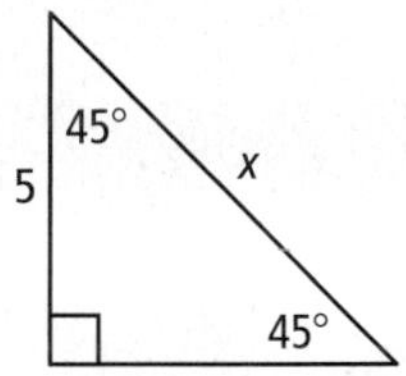

$\frac{x}{5} = \frac{\sqrt{2}}{1}$

$x = 5\sqrt{2}$

To find the unknown length in a 30°-60°-90° triangle:

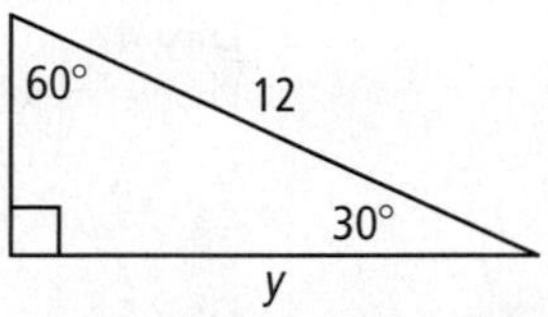

$\frac{y}{12} = \frac{\sqrt{3}}{2}$

$2y = 12\sqrt{3}$

$y = 6\sqrt{3}$

Common Errors With 30°-60°-90° and 45°-45°-90° Triangles

Often students are unsure of how to solve for a side length when they are not given the length of the shortest side in a 30°-60°-90° triangle. Instruct students to set up an equation with *x* and solve.

For example:

$6 = x\sqrt{3}$

$\frac{6}{\sqrt{3}} = \frac{x\sqrt{3}}{\sqrt{3}}$

$\frac{\sqrt{3}}{\sqrt{3}} \cdot \frac{6}{\sqrt{3}} = x$

$x = 2\sqrt{3}$

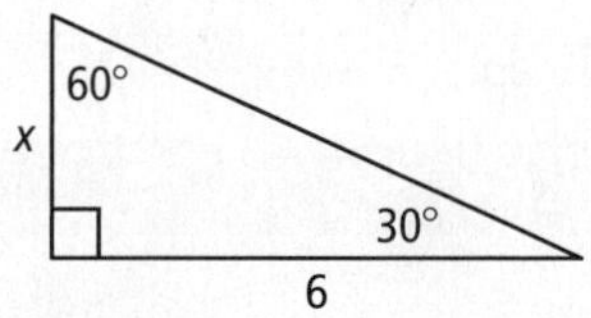

Mathematical Practices

Model with Mathematics Students combine algebraic and geometric thinking to develop the special triangles and their side ratios. They see patterns in the lengths of sides and can solve for nonintegral as well as integral sides in figures.

Look for and express regularity in repeated reasoning They apply these ideas to figures other than right triangles in the exercises and look at some applied cases where these triangles might be used.

Trigonometric Ratios

With respect to $\angle A$, there are six trigonometric ratios:

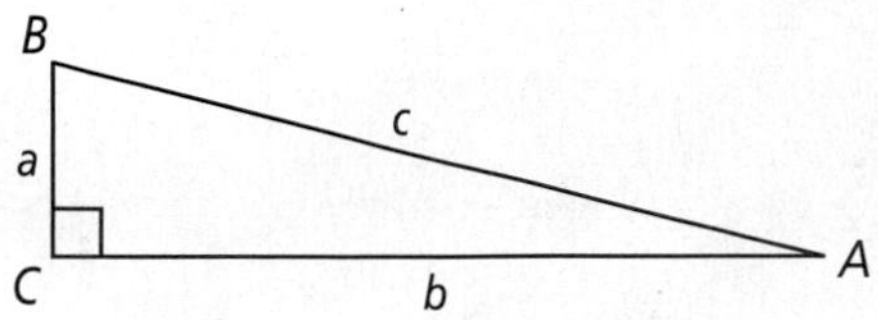

$\sin A = \frac{a}{c} = \frac{\text{opposite}}{\text{hypotenuse}}$ $\csc A = \frac{c}{a} = \frac{\text{hypotenuse}}{\text{opposite}}$

$\cos A = \frac{b}{c} = \frac{\text{adjacent}}{\text{hypotenuse}}$ $\sec A = \frac{c}{b} = \frac{\text{hypotenuse}}{\text{adjacent}}$

$\tan A = \frac{a}{b} = \frac{\text{opposite}}{\text{adjacent}}$ $\cot A = \frac{b}{a} = \frac{\text{adjacent}}{\text{opposite}}$

Trigonometric ratios provide the tools needed to find unknown right triangle measurements when certain combinations of side lengths and angle measures are given. When finding an unknown side length, use the direct ratio. When finding an unknown angle measure, use the inverse ratio.

Example 1: $a = 8$, $A = 43$

$\sin A = \frac{a}{c} = \frac{\text{opposite}}{\text{hypotenuse}}$

$\sin 43 = \frac{8}{c}$

$c = \frac{8}{\sin 43} \approx 11.7$

Example 2: $b = 6$, $c = 8$

$\cos A = \frac{b}{c} = \frac{\text{adjacent}}{\text{hypotenuse}}$

$\cos A = \frac{6}{8} = \frac{3}{4}$

$\cos^{-1}(\cos A) = \cos^{-1}\left(\frac{3}{4}\right)$

$A = \cos^{-1}\left(\frac{3}{4}\right) \approx 41.4$

Common Errors With Trigonometric Ratios

Students may get confused about when to use the trigonometric functions and when to use the inverse trigonometric functions. Instruct students always to write the equations setting the trigonometric term equal to a ratio. Then they will see if they need to use the inverse.

Mathematical Practices

Make sense of problems and persevere in solving them Students learn trigonometric ratios and apply them to specific situations.

Model with Mathematics They decontextualize a problem, which they represent with a trigonometric equation, and then contextualize it, to make sense of the results.

Reason abstractly and quantitatively They look at exercises that build on these skills, first considering simpler cases and then working up to the more complex problem at hand.

CHAPTER 8

RIGHT TRIANGLES AND TRIGONOMETRY

Pacing and Assignment Guide

		TRADITIONAL			BLOCK
Lesson	**Teaching Day(s)**	**Basic**	**Average**	**Advanced**	**Block**
8-1	**1**	Problems 1–3 Exs. 7–23, 55–62	Problems 1–3 Exs. 7–23 odd, 55–62	Problems 1–5 Exs. 7–31 odd, 33–62	**Day 1** Problems 1–5 Exs. 7–31 odd, 33–51, 55–62
	2	Problems 4–5 Exs. 24–34, 36–37, 43–45, 50	Problems 4–5 Exs. 25–31 odd, 33–51		
8-2	**1**	Problems 1–3 Exs. 7–14 all	Problems 1–3 Exs. 7–13 odd	Problems 1–3 Exs. 7–13 odd	**Day 2** Problems 1–5 Exs. 7–21 odd, 23–32
	2	Problems 4–5 Exs. 15–22 all, 23–27 odd, 29–31	Problems 4–5 Exs. 15–21 odd, 23–32	Problems 4–5 Exs. 15–21 odd, 23–33	
8-3	**1**	Problems 1–2 Exs. 11–21, 54–62	Problems 1–2 Exs. 11–21 odd, 54–62	Problems 1–3 Exs. 11–27 odd, 28–62	**Day 3** Problems 1–3 Exs. 11–27 odd, 28–47, 54–62
	2	Problem 3 Exs. 22–27, 29, 35, 39–42	Problem 3 Exs. 23–27 odd, 28–47		
8-4	**1**	Problems 1–3 Exs. 9–22, 23–26, 29–31, 36–44	Problems 1–3 Exs. 9–23 odd, 24–33, 36–44	Problems 1–3 Exs. 9–22 odd, 23–44	**Day 4** Problems 1–3 Exs. 9–23 odd, 24–33, 36–44
8-5	**1**	Problems 1–3 Exs. 6–13, 14, 16	Problems 1–3 Exs. 7–13 odd, 14–16	Problems 1–3 Exs. 7–13 odd, 14–18	**Day 5** Problems 1–3 Exs. 7–13 odd, 14–16
8-6	**1**	Problems 1–3 Exs. 7–17, 18–22 even, 23, 26	Problems 1–3 Exs. 7–17 odd, 18–26	Problems 1–3 Exs. 7–17 odd, 18–30	Problems 1–3 Exs. 7–17 odd, 18–26
Review	**1**	Chapter 8 Review	Chapter 8 Review	Chapter 8 Review	**Day 6** Chapter 8 Review Chapter 8 Test
Assess	**1**	Chapter 8 Test	Chapter 8 Test	Chapter 8 Test	
Total		**11 Days**	**11 Days**	**9 Days**	**6 Days**

Note: Pacing does not include Concept Bytes and other feature pages.

Resources

KEY
I = Interactive asset at PowerGeometry.com
E = Editable master at PowerGeometry.com
P = Available in Print
M = Master at PowerGeometry.com
✓ = CD-ROM

	For the Chapter	8-1	8-2	8-3	8-4	8-5	8-6
Planning							
Teacher Center Online Planner & Grade Book	I	I	I	I	I	I	I
Interactive Learning & Guided Instruction							
My Math Video	I						
Solve It!		I M	I M	I M	I M	I M	I M
Student Companion		P M	P M	P M	P M	P M	P M
Vocabulary Support		I P M	I P M	I P M	I P M	I P M	I P M
Got It? Support		I P	I P	I P	I P	I P	I P
Dynamic Activity	I						
Online Problems		I	I	I	I	I	I
Additional Problems		M	M	M	M	M	M
English Language Learner Support (TR)		E P M	E P M	E P M	E P M	E P M	E P M
Activities, Games, and Puzzles		E M	E M	E M	E M	E M	E M
Teaching With TI Technology With CD-ROM		✓ P					
TI-Nspire™ Support CD-ROM		✓	✓	✓	✓		
Lesson Check & Practice							
Student Companion		P M	P M	P M	P M	P M	P M
Lesson Check Support		I P	I P	I P	I P	I P	I P
Practice and Problem Solving Workbook		P	P	P	P	P	P
Think About a Plan (TR)		E P M	E P M	E P M	E P M	E P M	E P M
Practice Form G (TR)		E P M	E P M	E P M	E P M	E P M	E P M
Standardized Test Prep (TR)		P M	P M	P M	P M	P M	P M
Practice *Form K* (TR)		E P M	E P M	E P M	E P M	E P M	E P M
Extra Practice	E M						
Find the Errors!	M						
Enrichment (TR)		E P M	E P M	E P M	E P M	E P M	E P M
Answers and Solutions CD-ROM	✓	✓	✓	✓	✓	✓	✓
Assess & Remediate							
ExamView CD-ROM	✓	✓	✓	✓	✓	✓	✓
Lesson Quiz		I M	I M	I M	I M	I M	I M
Quizzes and Tests *Form G* (TR)	E P M			E P M			E P M
Quizzes and Tests *Form K* (TR)	E P M			E P M			E P M
Reteaching (TR)		E P M	E P M	E P M	E P M	E P M	E P M
Performance Tasks (TR)	P M						
Cumulative Review (TR)	P M						
Progress Monitoring Assessments	I P M						

(TR) Available in All-In-One Teaching Resources

Guided Instruction

PURPOSE To help students understand why the Pythagorean Theorem works

PROCESS Students should work in pairs dividing the work equally.

DISCUSS This activity is geared for tactile learners. It is a great way for the students to actually get their hands on why the Pythagorean Theorem works.

Activity

This Activity allows students to demonstrate visually how the side lengths of three squares assimilate to form a right triangle.

Q Does it matter which side of your rectangle you label *a* and which side you label *b*? Explain. **[No; sides *a* and *b* are interchangeable.]**

Q Do the dimensions of your original triangle affect the results of this Activity? Explain. **[No; you can make any rectangle. The largest square will always be the same size as the other two squares combined.]**

Mathematical Practices This Concept Byte supports students in becoming proficient in shifting perspective, Mathematical Practice 7.

Concept Byte
Use With Lesson 8-1
ACTIVITY

The Pythagorean Theorem

Common Core State Standards
Prepares for G-SRT.B.4 Prove theorems about triangles . . . the Pythagorean Theorem . . .
MP 7

You will learn the Pythagorean Theorem in Lesson 8-1. The activity below will help you understand why the theorem is true.

Activity

Step 1 Using graph paper, draw any rectangle and label the width *a* and the length *b*.

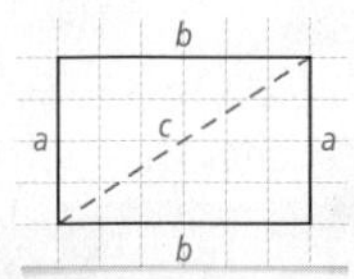

Step 2 Cut four rectangles with width *a* and length *b* from the graph paper. Then cut each rectangle on its diagonal, *c*, forming eight congruent triangles.

Step 3 Cut three squares from colored paper, one with sides of length *a*, one with sides of length *b*, and one with sides of length *c*.

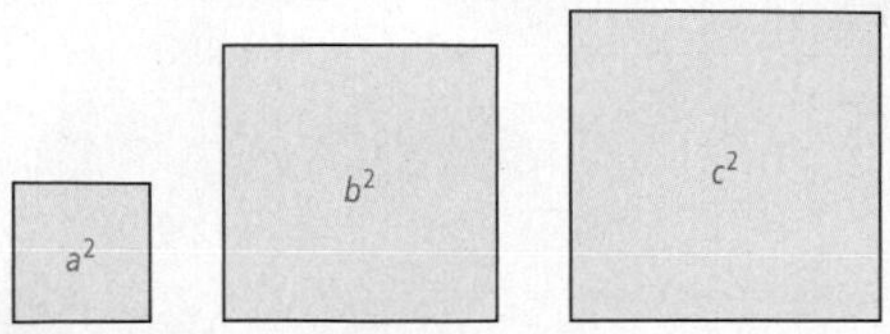

Step 4 Separate the 11 pieces into two groups.

Group 1: four triangles and the two smaller squares
Group 2: four triangles and the largest square

Step 5 Arrange the pieces of each group to form a square.

1. a. How do the areas of the two squares you formed in Step 5 compare?
b. Write an algebraic expression for the area of each of these squares.
c. What can you conclude about the areas of the three squares you cut from colored paper? Explain.
d. Repeat the activity using a new rectangle and different *a* and *b* values. What do you notice?

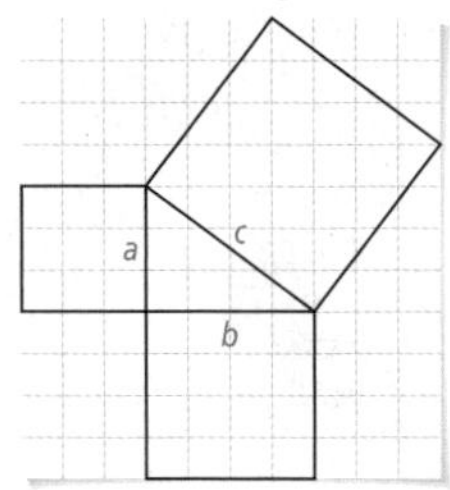

2. a. Express your conclusion as an algebraic equation.
b. Use a ruler with any rectangle to find actual measures for *a*, *b*, and *c*. Do these measures confirm your equation in part (a)?

3. Explain how the diagram at the right represents your equation in Question 2.

4. Does your equation work for nonright triangles? Explore and explain.

Answers

Activity

1a. The areas are equal.

b. $a^2 + b^2 + 2ab$, $c^2 + 2ab$

c. The sum of the areas of the two smaller squares = the area of the larger square.

d. The same relationship occurs.

2a. $a^2 + b^2 = c^2$

b. Yes; check students' work.

3. The sum of the squares of the lengths of the two legs = the square of the length of the hypotenuse.

4. No; $c^2 > a^2 + b^2$ for an obtuse △, and $c^2 < a^2 + b^2$ for an acute △.

8-1 The Pythagorean Theorem and Its Converse

Common Core State Standards
G-SRT.C.8 Use . . . the Pythagorean Theorem to solve right triangles in applied problems. **Also G-SRT.B.4**
MP 1, MP 3, MP 4, MP 8

Objective To use the Pythagorean Theorem and its converse

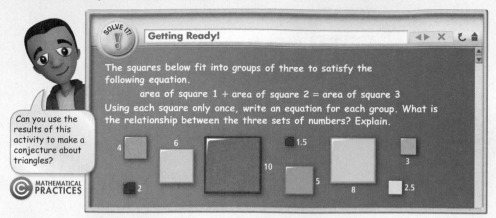

Can you use the results of this activity to make a conjecture about triangles?

MATHEMATICAL PRACTICES

SOLVE IT! Getting Ready!

The squares below fit into groups of three to satisfy the following equation.

area of square 1 + area of square 2 = area of square 3

Using each square only once, write an equation for each group. What is the relationship between the three sets of numbers? Explain.

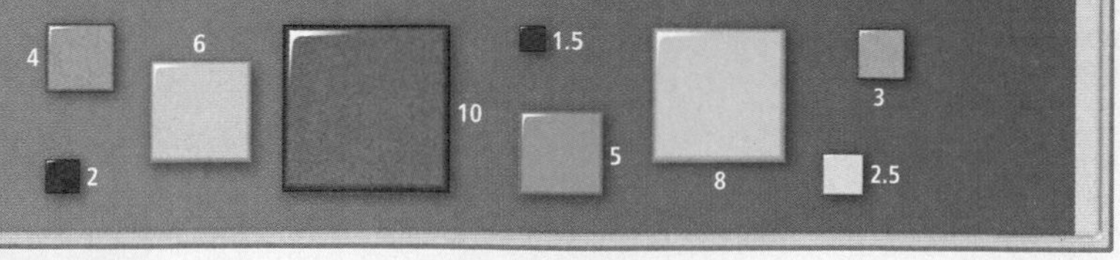

The equations in the Solve It demonstrate an important relationship in right triangles called the Pythagorean Theorem. This theorem is named for Pythagoras, a Greek mathematician who lived in the 500s B.C. We now know that the Babylonians, Egyptians, and Chinese were aware of this relationship before its discovery by Pythagoras. There are many proofs of the Pythagorean Theorem. You will see one proof in this lesson and others later in the book.

Lesson Vocabulary
- Pythagorean triple

Essential Understanding If you know the lengths of any two sides of a right triangle, you can find the length of the third side by using the Pythagorean Theorem.

take note **Theorem 8-1 Pythagorean Theorem**

Theorem	If . . .	Then . . .
If a triangle is a right triangle, then the sum of the squares of the lengths of the legs is equal to the square of the length of the hypotenuse.	$\triangle ABC$ is a right triangle 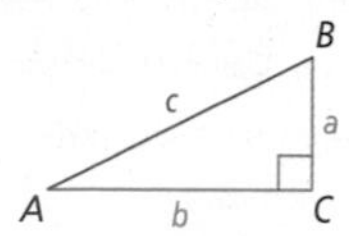	$(\text{leg}_1)^2 + (\text{leg}_2)^2 = (\text{hypotenuse})^2$ $a^2 + b^2 = c^2$

You will prove Theorem 8-1 in Exercise 49.

1 Interactive Learning

Solve It!

PURPOSE To explore the relationship of Pythagorean triples

PROCESS Students may
- use the formula for the area of a square.
- use trial and error to write the equations.
- use algebraic properties.

FACILITATE

Q How do you determine the area of a square? **[Square the side length.]**

Q Which property states that you can multiply each side of an equation by the same number and create an equivalent equation? **[the Multiplication Property of Equality]**

ANSWER See Solve It in Answers on next page.

CONNECT THE MATH In the Solve It, students discover the relationship of the Pythagorean Theorem. In the lesson, students learn the Pythagorean Theorem and how to use it to find unknown side lengths in a right triangle.

2 Guided Instruction

Take Note

Use properties of real numbers to change the form of the Pythagorean Theorem so that it is solved for a^2 and then for b^2.

8-1 Preparing to Teach

BIG ideas **Measurement**
Reasoning and Proof

ESSENTIAL UNDERSTANDINGS
- If the lengths of any two sides of a right triangle are known, the length of the third side can be found using the Pythagorean Theorem.
- If the lengths of all sides of a triangle are known, it can be determined whether the triangle is acute, right, or obtuse.

Math Background

The Pythagorean Theorem was one of the first theorems used by mathematicians in ancient civilizations. Although named after and credited to the Greek mathematician Pythagoras because of his proof of the theorem, the notion of the theorem actually dates back to a millennium earlier, when it was first used by the Babylonians.

The distance formula, used in coordinate geometry, is a derivative of the Pythagorean Theorem and is the foundation on which much of trigonometry is based.

There are numerous proofs of the Pythagorean Theorem, including those alluded to in the Concept Byte preceding this lesson. As an extension of this lesson, have students research various examples of proofs of the theorem and present their findings to the class.

Mathematical Practices

Make sense of problems and persevere in solving them. Students will analyze the relationship between sides in a right triangle through the Pythagorean Theorem.

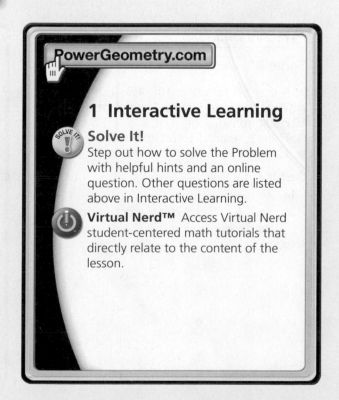

PowerGeometry.com

1 Interactive Learning

Solve It!
Step out how to solve the Problem with helpful hints and an online question. Other questions are listed above in Interactive Learning.

Virtual Nerd™ Access Virtual Nerd student-centered math tutorials that directly relate to the content of the lesson.

Problem 1

Q What are the lengths of the legs in the right triangle? **[20 and 21]**

Q Does it matter which of these side lengths is assigned to *a* and which to *b*? Explain. **[No, addition is commutative, so the squared values can be added in any order.]**

Q Why is only the principal (positive) square root found when solving the equation $c^2 = 841$? **[The length of a side cannot be negative.]**

Got It? VISUAL LEARNERS

Even though the problem can be solved by substituting the given values into the formula and solving for *c*, students benefit from drawing a sketch of the triangle. This action reinforces identifying the legs and the hypotenuse.

Problem 2

Q How can you decide which side of the triangle is the hypotenuse? **[The hypotenuse is the side opposite the right angle.]**

Q What are the lengths of the legs of the right triangle? **[*x* and 8]**

Got It?

Q What are the lengths of the legs of the right triangle? **[6 units and *x* units]**

Q What is the length of the hypotenuse? **[12 units]**

A **Pythagorean triple** is a set of nonzero whole numbers *a*, *b*, and *c* that satisfy the equation $a^2 + b^2 = c^2$. Below are some common Pythagorean triples.

3, 4, 5   5, 12, 13   8, 15, 17   7, 24, 25

If you multiply each number in a Pythagorean triple by the same whole number, the three numbers that result also form a Pythagorean triple. For example, the Pythagorean triples 6, 8, 10, and 9, 12, 15 each result from multiplying the numbers in the triple 3, 4, 5 by a whole number.

Problem 1 Finding the Length of the Hypotenuse

What is the length of the hypotenuse of $\triangle ABC$? Do the side lengths of $\triangle ABC$ form a Pythagorean triple? Explain.

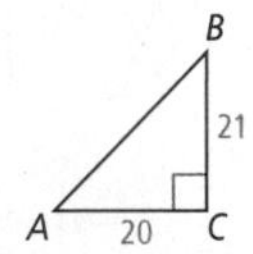

$(\text{leg}_1)^2 + (\text{leg}_2)^2 = (\text{hypotenuse})^2$ Pythagorean Theorem

$a^2 + b^2 = c^2$

$21^2 + 20^2 = c^2$ Substitute 21 for a and 20 for b.

$441 + 400 = c^2$ Simplify.

$841 = c^2$

$c = 29$ Take the positive square root.

Think
Is the answer reasonable?
Yes. The hypotenuse is the longest side of a right triangle. The value for c, 29, is greater than 20 and 21.

The length of the hypotenuse is 29. The side lengths 20, 21, and 29 form a Pythagorean triple because they are whole numbers that satisfy $a^2 + b^2 = c^2$.

Got It? 1. a. The legs of a right triangle have lengths 10 and 24. What is the length of the hypotenuse?
b. Do the side lengths in part (a) form a Pythagorean triple? Explain.

Problem 2 Finding the Length of a Leg

Algebra **What is the value of *x*? Express your answer in simplest radical form.**

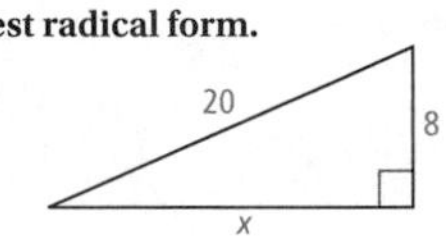

$a^2 + b^2 = c^2$ Pythagorean Theorem

$8^2 + x^2 = 20^2$ Substitute.

$64 + x^2 = 400$ Simplify.

$x^2 = 336$ Subtract 64 from each side.

$x = \sqrt{336}$ Take the positive square root.

$x = \sqrt{16(21)}$ Factor out a perfect square.

$x = 4\sqrt{21}$ Simplify.

Plan
Which side lengths do you have?
Remember from Chapter 4 that the side opposite the 90° angle is always the hypotenuse. So you have the lengths of the hypotenuse and one leg.

Got It? 2. The hypotenuse of a right triangle has length 12. One leg has length 6. What is the length of the other leg? Express your answer in simplest radical form.

Answers

Solve It!

$3^2 + 4^2 = 5^2$, $6^2 + 8^2 = 10^2$, $1.5^2 + 2^2 = 2.5^2$; answers may vary. Sample: The numbers 6, 8, and 10 result from multiplying 3, 4, and 5 by 2. The numbers 3, 4, and 5 result from multiplying 1.5, 2, and 2.5 by 2.

Got It?

1a. 26

b. Yes; 10, 24, and 26 are whole numbers that satisfy $a^2 + b^2 = c^2$.

2. $6\sqrt{3}$

PowerGeometry.com

2 Guided Instruction

Each Problem is worked out and supported online.

Problem 1
Finding the Length of the Hypotenuse
Animated

Problem 2
Finding the Length of a Leg
Animated

Problem 3
Finding Distance

Problem 4
Identifying a Right Triangle

Problem 5
Classifying a Triangle
Animated

Support in Geometry Companion
- Vocabulary
- Key Concepts
- Got It?

Problem 3 Finding Distance

Dog Agility Dog agility courses often contain a seesaw obstacle, as shown below. To the nearest inch, how far above the ground are the dog's paws when the seesaw is parallel to the ground?

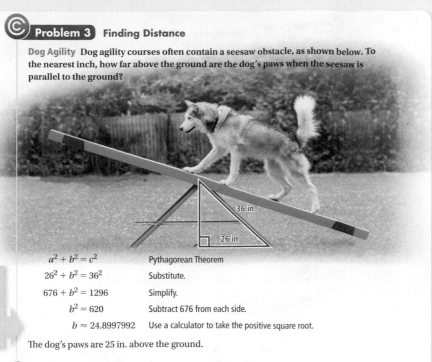

$a^2 + b^2 = c^2$	Pythagorean Theorem
$26^2 + b^2 = 36^2$	Substitute.
$676 + b^2 = 1296$	Simplify.
$b^2 = 620$	Subtract 676 from each side.
$b \approx 24.8997992$	Use a calculator to take the positive square root.

The dog's paws are 25 in. above the ground.

Think
How do you know when to use a calculator?
This is a real-world situation. Real-world distances are not usually expressed in radical form.

Got It? 3. The size of a computer monitor is the length of its diagonal. You want to buy a 19-in. monitor that has a height of 11 in. What is the width of the monitor? Round to the nearest tenth of an inch.

You can use the Converse of the Pythagorean Theorem to determine whether a triangle is a right triangle.

Take note

Theorem 8-2 Converse of the Pythagorean Theorem

Theorem	If . . .	Then . . .
If the sum of the squares of the lengths of two sides of a triangle is equal to the square of the length of the third side, then the triangle is a right triangle.	$a^2 + b^2 = c^2$	$\triangle ABC$ is a right triangle

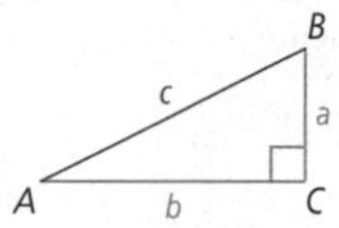

You will prove Theorem 8-2 in Exercise 52.

Problem 3

Q What formula can be used to find the distance between two points? **[$d = \sqrt{(x_2 - x_1)^2 + (y_2 - y_1)^2}$]**

Q Can you use the distance formula to complete this problem? Why or why not? **[Answers may vary. Sample: No, because the diagram is not in the coordinate plane.]**

Q Which side of the right triangle is unknown? **[a leg]**

Q Is the square of 24.8997992 exactly 620? Explain. **[No, it is an approximation of $\sqrt{620}$.]**

Got It? VISUAL LEARNERS

Encourage students to make a sketch of the computer monitor and label its parts prior to completing the problem.

Take Note

Because this theorem is the converse of Theorem 8-1, the Pythagorean Theorem can be written as a biconditional statement. Ask students to write such a statement.

Additional Problems

1. What is the length of the hypotenuse of $\triangle LMN$? Do the side lengths of $\triangle LMN$ form a Pythagorean triple? Explain.

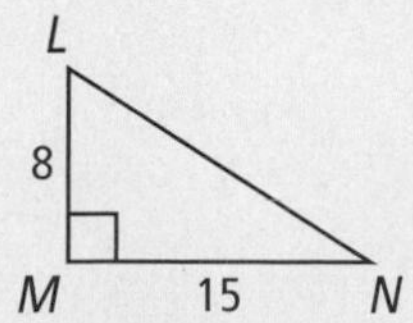

ANSWER 17, yes, all three side lengths are nonzero whole numbers.

2. What is the value of *x*? Express your answer in simplest radical form.

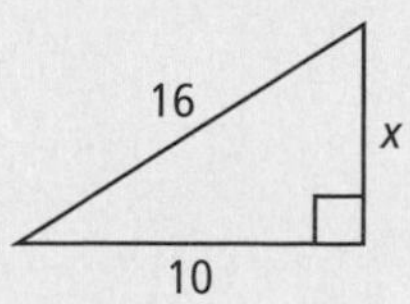

ANSWER $2\sqrt{39}$

3. Jamal leans a 12-ft-long ladder against the side of a house. The base of the ladder is 4 ft from the house. To the nearest tenth of a foot, how high on the house does the ladder reach?

ANSWER 11.3 ft

4. A triangle has side lengths 12.5, 30, and 32.5. Is it a right triangle? Explain.

ANSWER Yes, the triangle is a right triangle because $12.5^2 + 30^2 = 32.5^2$.

5. A triangle has side lengths 4, 9, and 12. Is it acute, obtuse, or right? Explain.

ANSWER obtuse; $4^2 + 9^2 = 97$ and $12^2 = 144$, so the triangle is obtuse by Theorem 8-3.

Answers

Got It? (continued)

3. 15.5 in.

Problem 4

Q How would the third line of the solution differ if you assigned the side length 84 to *a*? **[The left side of the equation would be 7056 + 169.]**

Q Are the numbers 13, 84, and 85 a Pythagorean triple? Explain. **[Yes; all three lengths are nonzero whole numbers, and $13^2 + 84^2 = 85^2$.]**

Got It? ERROR PREVENTION

If students are unsure of the answer for 4b, encourage them to experiment with the numbers in 4a as a way to check.

Take Note

Students can use geometry software to explore these two theorems. They can begin by constructing a right triangle with sides *a*, *b*, and *c*, and recording the values of $a^2 + b^2$ and c^2. Next, students can manipulate the measure of angle *C*, the right angle, so that the triangle becomes acute and then obtuse. Students should record the angle measures of each triangle for classification reasons as well as the values of $a^2 + b^2$ and c^2.

Plan

How do you know where each of the side lengths goes in the equation?
Work backward. If the triangle is a right triangle, then the hypotenuse is the longest side. So use the greatest number for *c*.

Problem 4 Identifying a Right Triangle

A triangle has side lengths 85, 84, and 13. Is the triangle a right triangle? Explain.

$a^2 + b^2 \stackrel{?}{=} c^2$	Pythagorean Theorem
$13^2 + 84^2 \stackrel{?}{=} 85^2$	Substitute 13 for *a*, 84 for *b*, and 85 for *c*.
$169 + 7056 \stackrel{?}{=} 7225$	Simplify.
$7225 = 7225$ ✓	

Yes, the triangle is a right triangle because $13^2 + 84^2 = 85^2$.

Got It? **4. a.** A triangle has side lengths 16, 48, and 50. Is the triangle a right triangle? Explain.

b. Reasoning Once you know which length represents the hypotenuse, does it matter which length you substitute for *a* and which length you substitute for *b*? Explain.

The theorems below allow you to determine whether a triangle is acute or obtuse. These theorems relate to the Hinge Theorem, which states that the longer side is opposite the larger angle and the shorter side is opposite the smaller angle.

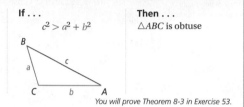

take note

Theorem 8-3

Theorem	If . . .	Then . . .
If the square of the length of the longest side of a triangle is greater than the sum of the squares of the lengths of the other two sides, then the triangle is obtuse.	$c^2 > a^2 + b^2$ 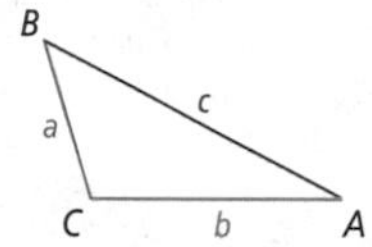	$\triangle ABC$ is obtuse

You will prove Theorem 8-3 in Exercise 53.

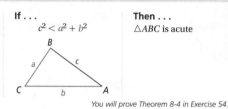

Theorem 8-4

Theorem	If . . .	Then . . .
If the square of the length of the longest side of a triangle is less than the sum of the squares of the lengths of the other two sides, then the triangle is acute.	$c^2 < a^2 + b^2$ 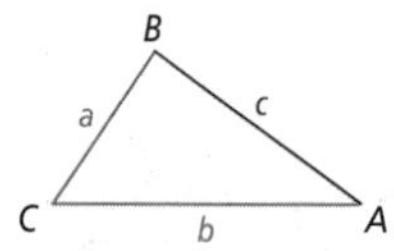	$\triangle ABC$ is acute

You will prove Theorem 8-4 in Exercise 54.

Answers

Got It? (continued)

4a. No; $16^2 + 48^2 \neq 50^2$.

b. No; $a^2 + b^2 = b^2 + a^2$ for any values of *a* and *b*.

Plan

What information do you need?
You need to know how the square of the longest side compares to the sum of the squares of the other two sides.

Problem 5 Classifying a Triangle

A triangle has side lengths 6, 11, and 14. Is it *acute, obtuse,* or *right*?

$c^2 \blacksquare a^2 + b^2$	Compare c^2 to $a^2 + b^2$.
$14^2 \blacksquare 6^2 + 11^2$	Substitute the greatest value for c.
$196 \blacksquare 36 + 121$	Simplify.
$196 > 157$	

Since $c^2 > a^2 + b^2$, the triangle is obtuse.

Got It? 5. Is a triangle with side lengths 7, 8, and 9 *acute, obtuse,* or *right*?

Lesson Check

Do you know HOW?

What is the value of x in simplest radical form?

1. 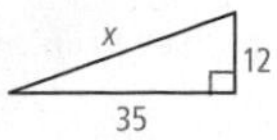

2. 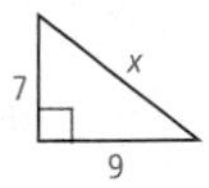

3. 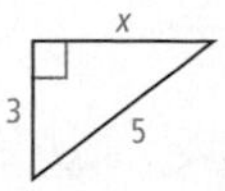

4. 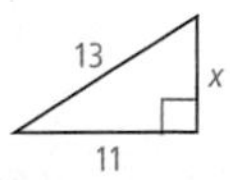

Do you UNDERSTAND? MATHEMATICAL PRACTICES

5. **Vocabulary** Describe the conditions that a set of three numbers must meet in order to form a Pythagorean triple.

6. **Error Analysis** A triangle has side lengths 16, 34, and 30. Your friend says it is not a right triangle. Look at your friend's work and describe the error.

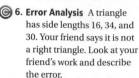

~~$16^2 + 34^2 \stackrel{?}{=} 30^2$~~
~~$256 + 1156 \stackrel{?}{=} 900$~~
~~$1412 \neq 900$~~

Practice and Problem-Solving Exercises MATHEMATICAL PRACTICES

A Practice **Algebra** **Find the value of x.** See Problem 1.

7. 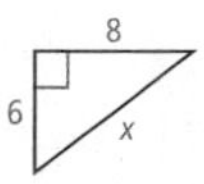

8. 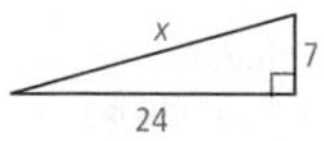

9. 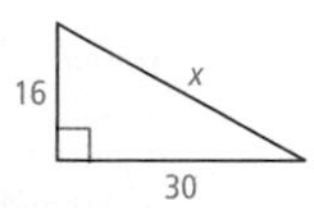

10. 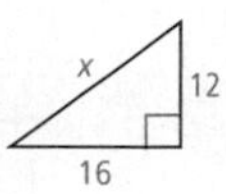

11. 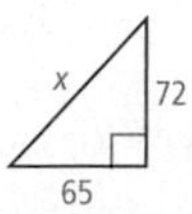

12. 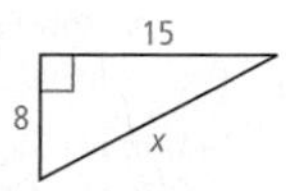

Does each set of numbers form a Pythagorean triple? Explain.

13. 4, 5, 6
14. 10, 24, 26
15. 15, 20, 25

Problem 5

Q Which of the given numbers do you know are the side lengths *a* and *b*? Why? **[6 and 11 because 14 is the longest side.]**

Q What is the square of the longest side? **[196]**

Q What sentence describes the relationship between the sum of the squares of the lengths of the two shorter sides and the square of the length of the longest side? **[The sum of the squares of the lengths of the two shorter sides is less than the square of the length of the longest side.]**

Got It? ERROR PREVENTION

Q What is the sum of the squares of the lengths of the shorter sides? **[113]**

Q What is the square of the length of the longest side? **[81]**

3 Lesson Check

Do you know HOW?

- If students have difficulty with Exercise 3, then have them decide if the *x* is substituted into the side of the equation that contains addition.

Do you UNDERSTAND?

- If students have difficulty with Exercise 5, then have them review the definition of a Pythagorean triple on page 492.

Close

Q What is the difference between how the Pythagorean Theorem and its converse are used? **[The Pythagorean Theorem is used to determine the length of the third side of a right triangle given two of the sides. The converse is used to determine if three given side lengths form a right triangle.]**

Answers

Got It? (continued)

5. acute

Lesson Check

1. 37
2. $\sqrt{130}$
3. 4
4. $4\sqrt{3}$
5. The three numbers *a*, *b*, and *c* must be whole numbers that satisfy $a^2 + b^2 = c^2$.
6. The longest side is 34, so the student should have tested $16^2 + 30^2 \stackrel{?}{=} 34^2$.

Practice and Problem-Solving Exercises

7. 10
8. 25
9. 34
10. 20
11. 97
12. 17
13. no; $4^2 + 5^2 \neq 6^2$
14. yes; $10^2 + 24^2 = 26^2$
15. yes; $15^2 + 20^2 = 25^2$

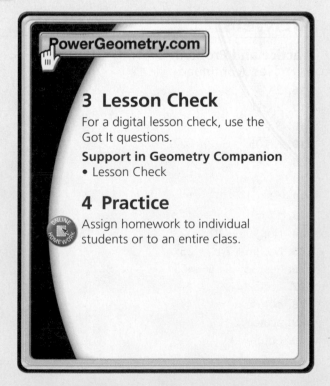

PowerGeometry.com

3 Lesson Check

For a digital lesson check, use the Got It questions.

Support in Geometry Companion
- Lesson Check

4 Practice

Assign homework to individual students or to an entire class.

4 Practice

ASSIGNMENT GUIDE

Basic: 7–34, 36–37, 43–45, 50
Average: 7–31 odd, 33–51
Advanced: 7–31 odd, 33–54
Standardized Test Prep: 55–58
Mixed Review: 59–62

Mathematical Practices are supported by exercises with red headings. Here are the Practices supported in this lesson:

MP 1: Make Sense of Problems Ex. 33
MP 3: Critique the Reasoning of Others Ex. 6
MP 8: Repeated Reasoning Ex. 43–48

Applications exercises have blue headings. Exercise 50 supports MP 4: Model.

STEM exercises focus on science or engineering applications.

EXERCISE 50: Use the Think About a Plan worksheet in the **Practice and Problem Solving Workbook** (also available in the Teaching Resources in print and online) to further support students' development in becoming independent learners.

HOMEWORK QUICK CHECK

To check students' understanding of key skills and concepts, go over Exercises 17, 29, 33, 43, and 50.

Algebra **Find the value of x. Express your answer in simplest radical form.**

See Problem 2.

16.
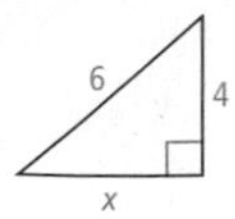

17.
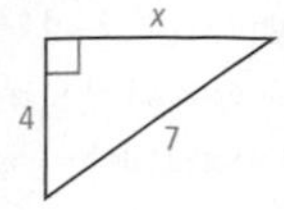

18.
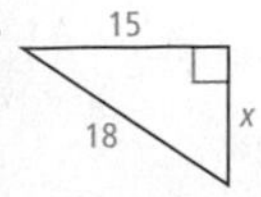

19.
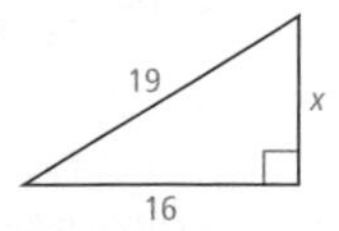

20.
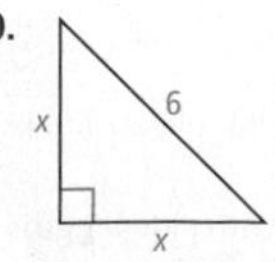

21.
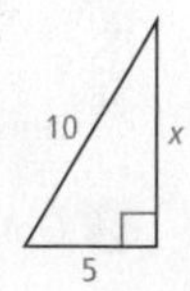

22. Home Maintenance A painter leans a 15-ft ladder against a house. The base of the ladder is 5 ft from the house. To the nearest tenth of a foot, how high on the house does the ladder reach?

See Problem 3.

23. A walkway forms one diagonal of a square playground. The walkway is 24 m long. To the nearest meter, how long is a side of the playground?

Is each triangle a right triangle? Explain.

See Problem 4.

24.
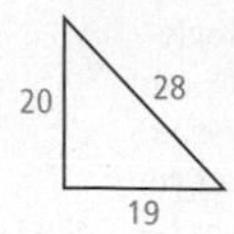

25.
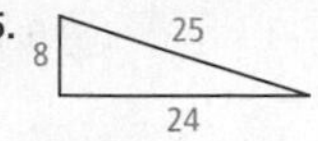

26.
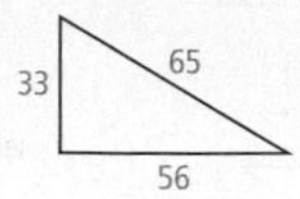

The lengths of the sides of a triangle are given. Classify each triangle as *acute*, *right*, or *obtuse*.

See Problem 5.

27. 4, 5, 6
28. 0.3, 0.4, 0.6
29. 11, 12, 15
30. $\sqrt{3}$, 2, 3
31. 30, 40, 50
32. $\sqrt{11}$, $\sqrt{7}$, 4

B Apply

33. Think About a Plan You want to embroider a square design. You have an embroidery hoop with a 6-in. diameter. Find the largest value of x so that the entire square will fit in the hoop. Round to the nearest tenth.

- What does the diameter of the circle represent in the square?
- What do you know about the sides of a square?
- How do the side lengths of the square relate to the length of the diameter?

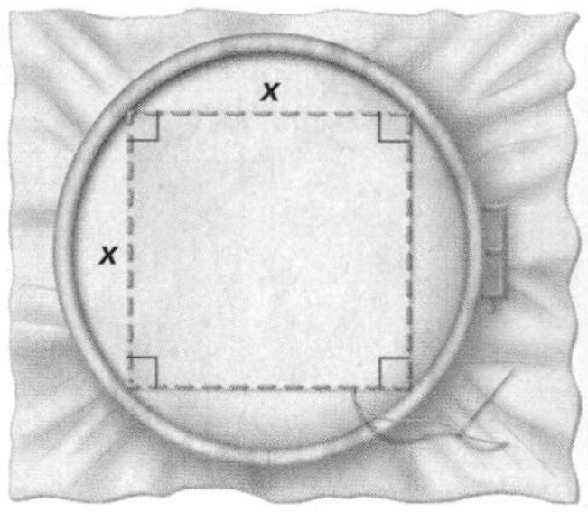

34. In parallelogram $RSTW$, $RS = 7$, $ST = 24$, and $RT = 25$. Is $RSTW$ a rectangle? Explain.

Answers

Practice and Problem-Solving Exercises (continued)

16. $2\sqrt{5}$
17. $\sqrt{33}$
18. $3\sqrt{11}$
19. $\sqrt{105}$
20. $3\sqrt{2}$
21. $5\sqrt{3}$
22. 14.1 ft
23. 17 m
24. No; $19^2 + 20^2 \neq 28^2$.
25. No; $8^2 + 24^2 \neq 25^2$.
26. Yes; $33^2 + 56^2 = 65^2$.
27. acute
28. obtuse
29. acute
30. obtuse
31. right
32. acute
33. 4.2 in.
34. Yes; $7^2 + 24^2 = 25^2$, so $\angle S$ is a rt. $\angle$.

35. Coordinate Geometry You can use the Pythagorean Theorem to prove the Distance Formula. Let points $P(x_1, y_1)$ and $Q(x_2, y_2)$ be the endpoints of the hypotenuse of a right triangle.

Proof

a. Write an algebraic expression to complete each of the following: $PR =$? and $QR =$? .

b. By the Pythagorean Theorem, $PQ^2 = PR^2 + QR^2$. Rewrite this statement by substituting the algebraic expressions you found for PR and QR in part (a).

c. Complete the proof by taking the square root of each side of the equation that you wrote in part (b).

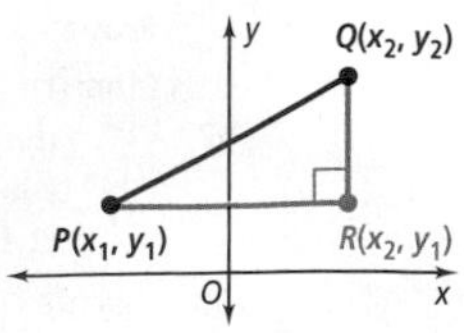

Algebra **Find the value of *x*. If your answer is not an integer, express it in simplest radical form.**

36.

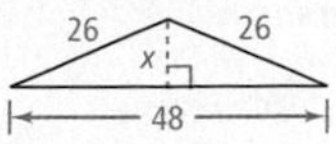

37.

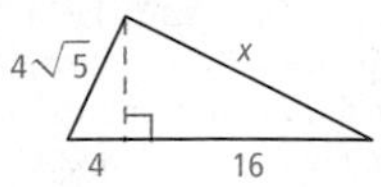

38.

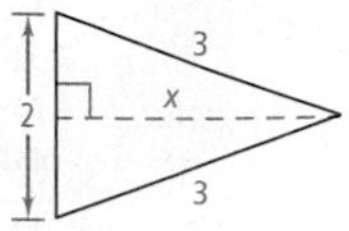

For each pair of numbers, find a third whole number such that the three numbers form a Pythagorean triple.

39. 20, 21 **40.** 14, 48 **41.** 13, 85 **42.** 12, 37

Open-Ended **Find integers *j* and *k* such that (a) the two given integers and *j* represent the side lengths of an acute triangle and (b) the two given integers and *k* represent the side lengths of an obtuse triangle.**

43. 4, 5 **44.** 2, 4 **45.** 6, 9 **46.** 5, 10 **47.** 6, 7 **48.** 9, 12

49. Prove the Pythagorean Theorem.

Proof

Given: $\triangle ABC$ is a right triangle.

Prove: $a^2 + b^2 = c^2$

(*Hint:* Begin with proportions suggested by Theorem 7-3 or its corollaries.)

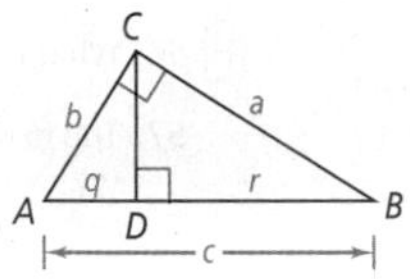

STEM **50. Astronomy** The Hubble Space Telescope orbits 600 km above Earth's surface. Earth's radius is about 6370 km. Use the Pythagorean Theorem to find the distance *x* from the telescope to Earth's horizon. Round your answer to the nearest ten kilometers. (Diagram is not to scale.)

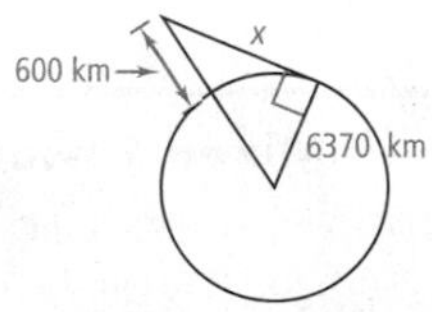

51. Prove that if the slopes of two lines have product -1, then the lines are perpendicular. Use parts (a)–(c) to write a coordinate proof.

a. First, argue that neither line can be horizontal or vertical.

b. Then, tell why the lines must intersect. (*Hint:* Use indirect reasoning.)

c. Place the lines in the coordinate plane. Choose a point on ℓ_1 and find a related point on ℓ_2. Complete the proof.

35a. $|x_2 - x_1|$; $|y_2 - y_1|$

b. $PQ^2 = (x_2 - x_1)^2 + (y_2 - y_1)^2$

c. $PQ = \sqrt{(x_2 - x_1)^2 + (y_2 - y_1)^2}$

36. 10

37. $8\sqrt{5}$

38. $2\sqrt{2}$

39. 29

40. 50

41. 84

42. 35

43–48. Answers may vary. Samples are given.

43a. 6

b. 7

44a. 4

b. 5

45a. 8

b. 11

46a. 11

b. 12

47a. 8

b. 10

48a. 14

b. 16

49. $\frac{q}{b} = \frac{b}{c}$ and $\frac{r}{a} = \frac{a}{c}$ because each leg is the geometric mean of the adj. hypotenuse segment and the hypotenuse. By the Cross Products Property, $b^2 = qc$ and $a^2 = rc$. Then $a^2 + b^2 = qc + rc = c(q + r)$. Substituting c for $q + r$ gives $a^2 + b^2 = c^2$.

50. 2830 km

51a. Horiz. lines have slope 0, and vert. lines have undef. slope. Neither could be mult. to get -1.

b. Assume the lines do not intersect. Then they have the same slope m. Then $m \cdot m = m^2 = -1$, which is impossible. So the lines must intersect.

c. Let ℓ_1 be $y = \frac{b}{a}x$ and ℓ_2 be $y = -\frac{a}{b}x$. Define $C(a, b)$, $A(0, 0)$, and $B(a, -\frac{a^2}{b})$.

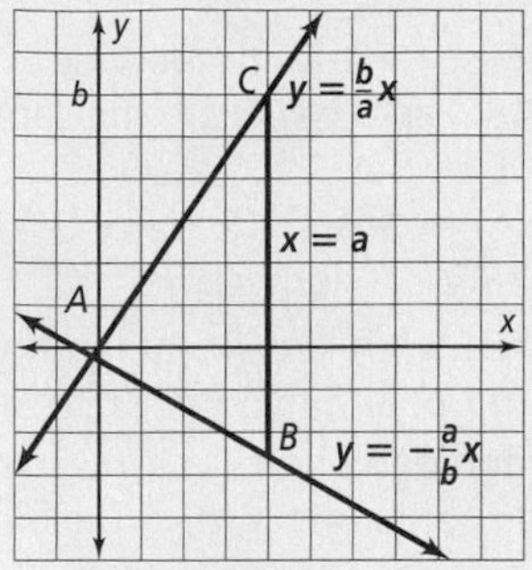

Using the Distance Formula, $AC = \sqrt{a^2 + b^2}$, $BA = \sqrt{a^2 + \frac{a^4}{b^2}}$, and $CB = b + \frac{a^2}{b}$. Then $AC^2 + BA^2 = CB^2$ and $m\angle A = 90$ by the Conv. of the Pythagorean Thm. So $\ell_1 \perp \ell_2$.

Note: the choice of the coordinates of B is challenging.

Answers

Practice and Problem-Solving Exercises (continued)

52. Draw right $\triangle FDE$ with legs $\overline{DE}$ of length a and $\overline{EF}$ of length b, and hypotenuse of length x. Then $a^2 + b^2 = x^2$ by the Pythagorean Thm. You are given $\triangle ABC$ with sides of length a, b, and c, and $a^2 + b^2 = c^2$. By subst., $c^2 = x^2$, so $c = x$. Since all side lengths of $\triangle ABC$ and $\triangle FDE$ are the same, $\triangle ABC \cong \triangle FDE$ by SSS. $\angle C \cong \angle E$ because corresp. parts of $\cong$ ▲ are $\cong$, so $m\angle C = 90$. Therefore, $\triangle ABC$ is a right $\triangle$.

53. Draw right $\triangle FDE$ with legs $\overline{DE}$ of length a and $\overline{EF}$ of length b, and hypotenuse of length x. By the Pythagorean Thm., $a^2 + b^2 = x^2$. $\triangle ABC$ has sides of length a, b, and c, where $c^2 > a^2 + b^2$. $c^2 > x^2$ and $c > x$ by Prop. of Inequalities. If $c > x$, then $m\angle C > m\angle E$ by the Converse of the Hinge Thm. An angle with measure > 90 is obtuse, so $\triangle ABC$ is an obtuse $\triangle$.

54. Draw right $\triangle FDE$ with legs $\overline{DE}$ of length a and $\overline{EF}$ of length b, and hypotenuse of length x. By the Pythagorean Thm., $a^2 + b^2 = x^2$. $\triangle ABC$ has sides of length a, b, and c, where $c^2 < a^2 + b^2$. $c^2 < x^2$ and $c < x$ by Prop. of Inequalities. If $c < x$, then $m\angle C < m\angle E$ by the Converse of the Hinge Thm. An angle with measure < 90 is acute, so $\triangle ABC$ is an acute $\triangle$.

55. 4 **56.** 23

57. 61 **58.** 2.25

59. 4, 5 **60.** $\sqrt{3}$

61. $15\sqrt{2}$ **62.** $\frac{16\sqrt{3}}{3}$

Challenge

52. Use the plan and write a proof of Theorem 8-2 (Converse of the Pythagorean Theorem). *Proof*

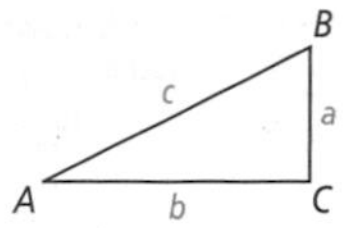

Given: $\triangle ABC$ with sides of length a, b, and c, where $a^2 + b^2 = c^2$

Prove: $\triangle ABC$ is a right triangle.

Plan: Draw a right triangle (not $\triangle ABC$) with legs of lengths a and b. Label the hypotenuse x. By the Pythagorean Theorem, $a^2 + b^2 = x^2$. Use substitution to compare the lengths of the sides of your triangle and $\triangle ABC$. Then prove the triangles congruent.

53. Use the plan and write a proof of Theorem 8-3. *Proof*

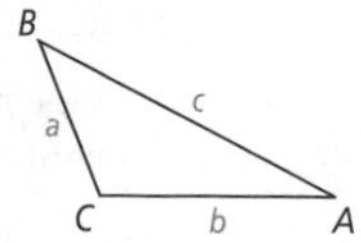

Given: $\triangle ABC$ with sides of length a, b, and c, where $c^2 > a^2 + b^2$

Prove: $\triangle ABC$ is an obtuse triangle.

Plan: Draw a right triangle (not $\triangle ABC$) with legs of lengths a and b. Label the hypotenuse x. By the Pythagorean Theorem, $a^2 + b^2 = x^2$. Use substitution to compare lengths c and x. Then use the Converse of the Hinge Theorem to compare $\angle C$ to the right angle.

54. Prove Theorem 8-4. *Proof*

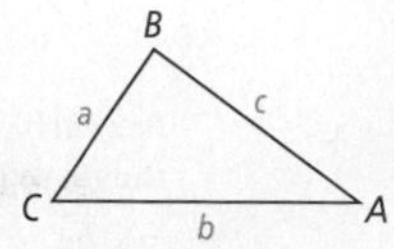

Given: $\triangle ABC$ with sides of length a, b, and c, where $c^2 < a^2 + b^2$

Prove: $\triangle ABC$ is an acute triangle.

Standardized Test Prep

GRIDDED RESPONSE

SAT/ACT

55. A 16-ft ladder leans against a building, as shown. To the nearest foot, how far is the base of the ladder from the building?

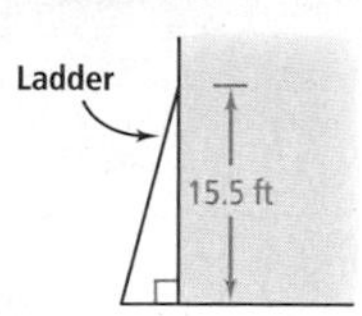

56. What is the measure of the complement of a 67° angle?

57. The measure of the vertex angle of an isosceles triangle is 58. What is the measure of one of the base angles?

58. The length of rectangle $ABCD$ is 4 in. The length of similar rectangle $DEFG$ is 6 in. How many times greater than the area of $ABCD$ is the area of $DEFG$?

Mixed Review

59. $\triangle ABC$ has side lengths $AB = 8$, $BC = 9$, and $AC = 10$. Find the lengths of the segments formed on $\overline{BC}$ by the bisector of $\angle A$. See Lesson 7-5.

Get Ready! **To prepare for Lesson 8-2, do Exercises 60–62.**

Simplify each expression. See Review, p. 399.

60. $\sqrt{9} \div \sqrt{3}$ **61.** $30 \div \sqrt{2}$ **62.** $\frac{16}{\sqrt{3}}$

8-1 Lesson Resources

Differentiated Remediation

Instructional Support

Geometry Companion

Students can use the **Geometry Companion** worktext (4 pages) . . .

- New Vocabulary
- Key Concepts
- Got It for each Problem
- Lesson Check

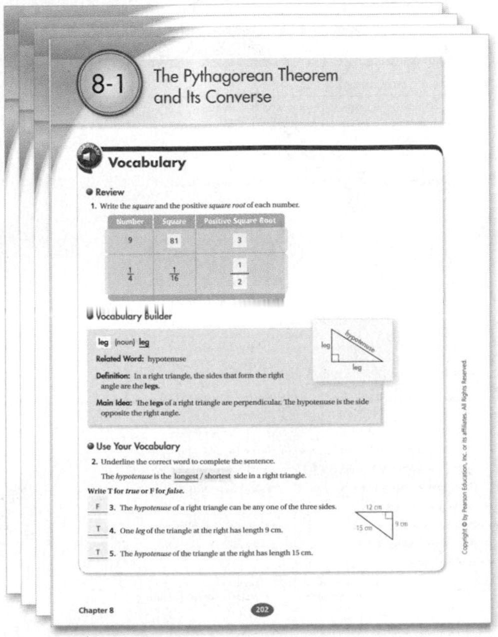

ELL Support

Use Graphic Organizers Have students make an organizer to organize the lesson concepts. At the top, have them draw a right triangle. Do the same on the board. Trace the legs and hypotenuse as you say their name. Label the legs *a* and *b* and the hypotenuse *c*. Write the Pythagorean Theorem and the following underneath: $(\text{leg})^2 + (\text{leg})^2 = (\text{hypotenuse})^2$.

Ask: If you have two side lengths of a right triangle, how can you use the Pythagorean Theorem to find an unknown length? Then draw two lines to two examples: one which asks how to find the hypotenuse and one that asks how to find a missing leg measure. Then ask: How can you use the Pythagorean Theorem to prove this triangle is a right triangle? Complete the organizer after the lesson.

5 Assess & Remediate

Lesson Quiz

1. What is the length of the hypotenuse of $\triangle RST$? Do the side lengths of $\triangle RST$ form a Pythagorean triple? Explain.

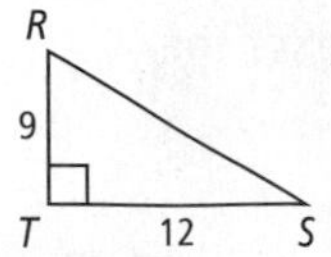

2. Cassie's computer monitor is in the shape of a rectangle. The screen on the monitor is 11.5 in. high and 18.5 in. wide. What is the length of the diagonal? Round to the nearest tenth of an inch.

3. A triangle has side lengths 24, 32, and 42. Is it a right triangle? Explain.

4. A triangle has side lengths 9, 10, and 12. Is it acute, obtuse, or right? Explain.

5. Do you UNDERSTAND? Can three segments with lengths 4 cm, 6 cm, and 11 cm be assembled to form an acute triangle, a right triangle, or an obtuse triangle? Explain.

ANSWERS TO LESSON QUIZ

1. 15; Yes, because all three side lengths are whole numbers.

2. 21.8 in.

3. No, the side lengths do not satisfy the Pythagorean Theorem.

4. Acute; $9^2 + 10^2 = 181$ and $12^2 = 144$, so the triangle is acute by Theorem 8-4.

5. Because $4 + 6 < 11$, the three lengths cannot form a triangle of any kind.

PRESCRIPTION FOR REMEDIATION

Use the student work on the Lesson Quiz to prescribe a differentiated review assignment.

Points	Differentiated Remediation
0–2	Intervention
3–4	On-level
5	Extension

PowerGeometry.com

5 Assess & Remediate

Assign the Lesson Quiz. Appropriate intervention, practice, or enrichment is automatically generated based on student performance.

Intervention

- **Reteaching** (2 pages) Provides reteaching and practice exercises for the key lesson concepts. Use with struggling students or absent students.
- **English Language Learner Support** Helps students develop and reinforce mathematical vocabulary and key concepts.

All-in-One Resources/Online

Reteaching

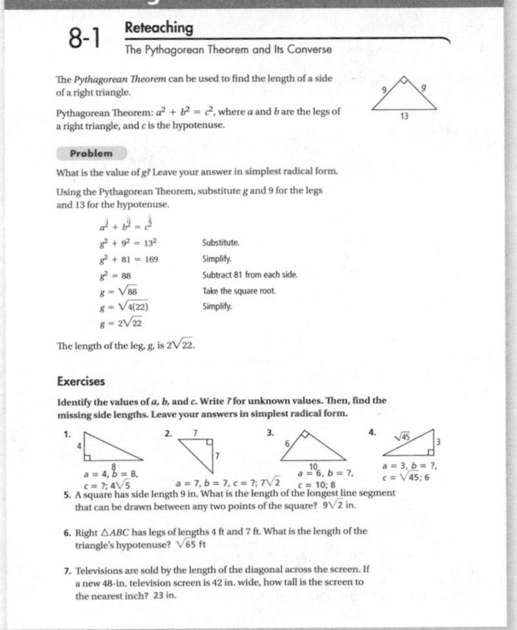

All-in-One Resources/Online

English Language Learner Support

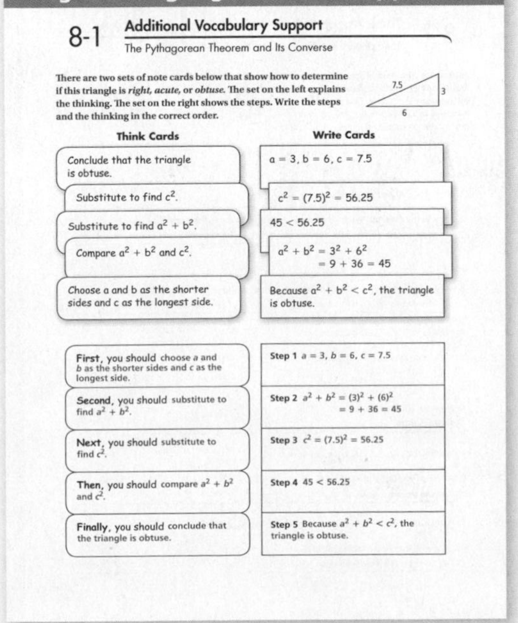

Differentiated Remediation *continued*

On-Level

- **Practice** (2 pages) Provides extra practice for each lesson. For simpler practice exercises, use the Form K Practice pages found in the All-in-One Teaching Resources and online.
- **Think About a Plan** Helps students develop specific problem-solving skills and strategies by providing scaffolded guiding questions.
- **Standardized Test Prep** Focuses on all major exercises, all major question types, and helps students prepare for the high-stakes assessments.

Extension

- **Enrichment** Provides students with interesting problems and activities that extend the concepts of the lesson.
- **Activities, Games, and Puzzles** Worksheets that can be used for concepts development, enrichment, and for fun!

Practice and Problem Solving Wkbk/ All-in-One Resources/Online

Practice page 1

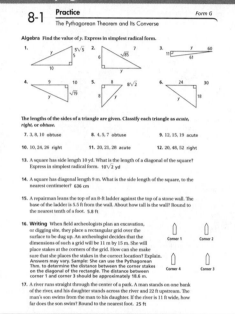

8-1 **Practice** Form G
The Pythagorean Theorem and Its Converse

Algebra Find the value of *y*. Express in simplest radical form.

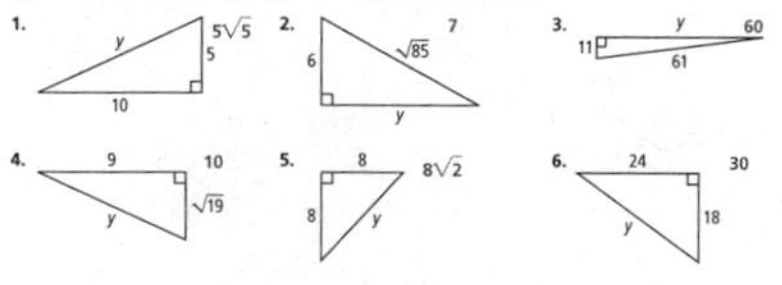

The lengths of the sides of a triangle are given. Classify each triangle as *acute*, *right*, or *obtuse*.

7. 3, 8, 10 obtuse
8. 4, 5, 7 obtuse
9. 12, 15, 19 acute
10. 10, 24, 26 right
11. 20, 21, 28 acute
12. 20, 48, 52 right
13. A square has side length 10 yd. What is the length of a diagonal of the square? Express in simplest radical form. $10\sqrt{2}$ yd
14. A square has diagonal length 9 m. What is the side length of the square, to the nearest centimeter? 636 cm
15. A repairman leans the top of an 8-ft ladder against the top of a stone wall. The base of the ladder is 5.5 ft from the wall. About how tall is the wall? Round to the nearest tenth of a foot. 5.8 ft
16. **Writing** When field archeologists plan an excavation, or digging site, they place a rectangular grid over the surface to be dug up. An archeologist decides that the dimensions of such a grid will be 11 m by 15 m. She will place stakes at the corners of the grid. How can she make sure that she places the stakes in the correct location? Explain. Answers may vary. Sample: She can use the Pythagorean Thm. to determine the distance between the corner stakes on the diagonal of the rectangle. The distance between corner 1 and corner 3 should be approximately 18.6 m.
17. A river runs straight through the center of a park. A man stands on one bank of the river, and his daughter stands across the river and 22 ft upstream. The man's son swims from the man to his daughter. If the river is 11 ft wide, how far does the son swim? Round to the nearest foot. 25 ft

Practice and Problem Solving Wkbk/ All-in-One Resources/Online

Practice page 2

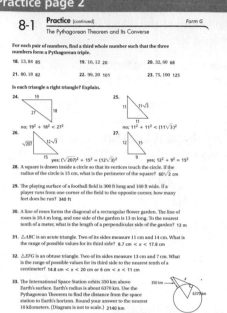

8-1 **Practice** (continued) Form G
The Pythagorean Theorem and Its Converse

For each pair of numbers, find a third whole number such that the three numbers form a Pythagorean triple.

18. 13, 84 85
19. 16, 12 20
20. 32, 60 68
21. 80, 18 82
22. 99, 20 101
23. 75, 100 125

Is each triangle a right triangle? Explain.

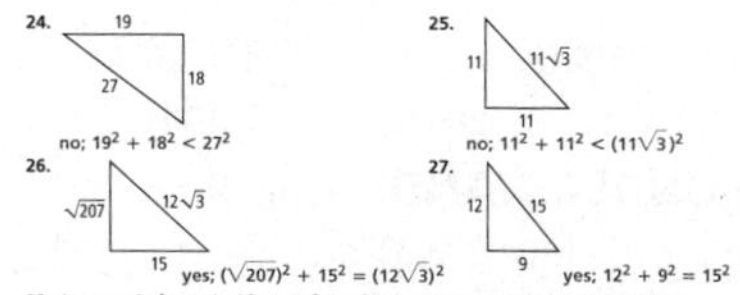

24. no; $19^2 + 18^2 < 27^2$
25. no; $11^2 + 11^2 < (11\sqrt{3})^2$
26. yes; $(\sqrt{207})^2 + 15^2 = (12\sqrt{3})^2$
27. yes; $12^2 + 9^2 = 15^2$
28. A square is drawn inside a circle so that its vertices touch the circle. If the radius of the circle is 15 cm, what is the perimeter of the square? $60\sqrt{2}$ cm
29. The playing surface of a football field is 300 ft long and 160 ft wide. If a player runs from one corner of the field to the opposite corner, how many feet does he run? 340 ft
30. A line of roses forms the diagonal of a rectangular flower garden. The line of roses is 18.4 m long, and one side of the garden is 13 m long. To the nearest tenth of a meter, what is the length of a perpendicular side of the garden? 13 m
31. $\triangle ABC$ is an acute triangle. Two of its sides measure 11 cm and 14 cm. What is the range of possible values for its third side? 8.7 cm < *x* < 17.8 cm
32. $\triangle EFG$ is an obtuse triangle. Two of its sides measure 13 cm and 7 cm. What is the range of possible values for its third side to the nearest tenth of a centimeter? 14.8 cm < *x* < 20 cm or 6 cm < *x* < 11 cm
33. The International Space Station orbits 350 km above Earth's surface. Earth's radius is about 6370 km. Use the Pythagorean Theorem to find the distance from the space station to Earth's horizon. Round your answer to the nearest 10 kilometers. (Diagram is not to scale.) 2140 km

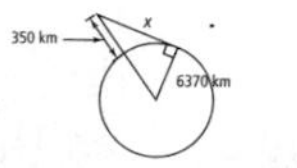

All-in-One Resources/Online

Enrichment

8-1 **Enrichment**
The Pythagorean Theorem and Its Converse

Art and Design

An artist plans to use a pattern of triangles made from variously colored strings in her work. She takes a 28-ft piece of wood and marks off points every 2 ft along the length. At the midpoint, she connects a 9-ft piece of wire at a right angle.

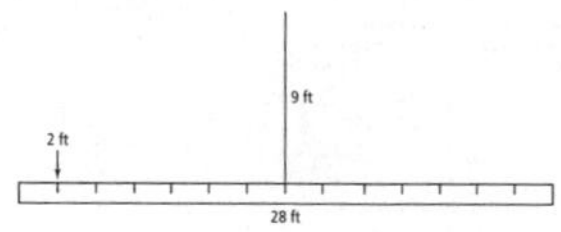

She plans to connect strings from the ends of the wood and from every marked-off point to the top of the wire, including the ends of the wood and the center. From left to right, she plans on connecting purple, red, yellow, dark blue, white, mint green, gray, rose, tan, dark green, brown, light blue, mauve, cream, and burgundy strings.

Use the Pythagorean Theorem and the sketch above to determine the following lengths, to the nearest inch:

1. the purple string 16 ft 8 in.
2. the red string 15 ft
3. the yellow string 13 ft 5 in.
4. the dark blue string 12 ft
5. the white string 10 ft 10 in.
6. the mint green string 9 ft 10 in.
7. the gray string 9 ft 3 in.
8. the rose string 9 ft
9. the tan string 9 ft 3 in.
10. the dark green string 9 ft 10 in.
11. the brown string 10 ft 10 in.
12. the light blue string 12 ft
13. the mauve string 13 ft 5 in.
14. the cream string 15 ft
15. the burgundy string 16 ft 8 in.
16. Develop plans for your own art project involving right triangles, taking inspiration from this piece. Calculate all of the lengths of sides of the right triangles you use. Check students' work.

Practice and Problem Solving Wkbk/ All-in-One Resources/Online

Think About a Plan

8-1 **Think About a Plan**
The Pythagorean Theorem and Its Converse

Astronomy The Hubble Space Telescope orbits 600 km above Earth's surface. Earth's radius is about 6370 km. Use the Pythagorean Theorem to find the distance *x* from the telescope to Earth's horizon. Round your answer to the nearest ten kilometers. (Diagram is not to scale.)

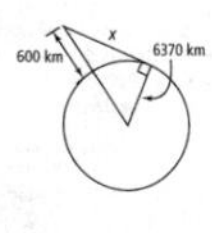

Know

1. Write the Pythagorean Theorem. $a^2 + b^2 = c^2$
2. Look at the diagram. What could replace *a* and *b* in the Pythagorean Theorem? *x*, the distance from the telescope to the horizon, and the radius of Earth, 6370 km

Need

3. How can you find the value of *c*, the hypotenuse of the right triangle? Add the radius, 6370 km, and the distance from the telescope to Earth's surface, 600 km.
4. What is the value of *c*? 6970 km
5. Substitute the known and unknown values for *a*, *b*, and *c*. $6370^2 + x^2 = 6970^2$

Plan

6. How can you find the value of x^2? Subtract 6370^2 from 6970^2.
7. What is the value of x^2? 8,004,000 km^2
8. How can you find the value of *x*? Find the square root of 8,004,000 km^2.
9. What is the value of *x*, to the nearest 10 kilometers? 2830 km
10. Is your answer reasonable? Explain. Yes; Sample: The distance makes sense in this situation.

Practice and Problem Solving Wkbk/ All-in-One Resources/Online

Standardized Test Prep

8-1 **Standardized Test Prep**
The Pythagorean Theorem and Its Converse

Gridded Response

Solve each exercise and enter your answer on the grid provided.
What is the value of *x*?

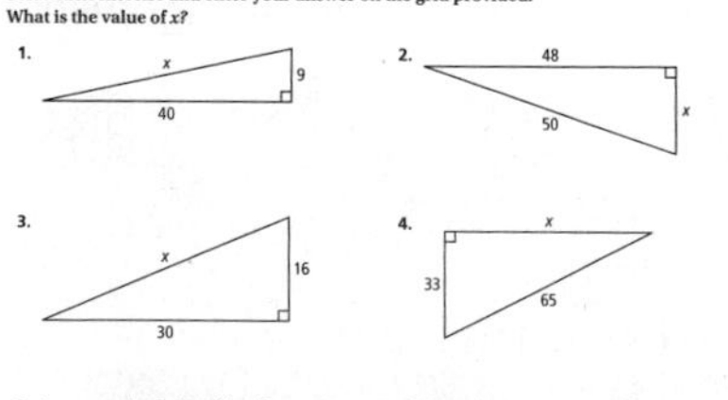

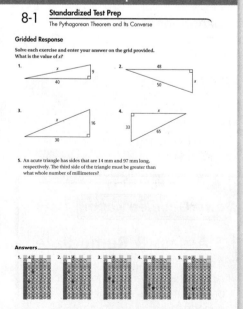

5. An acute triangle has sides that are 14 mm and 97 mm long, respectively. The third side of the triangle must be greater than what whole number of millimeters?

Answers

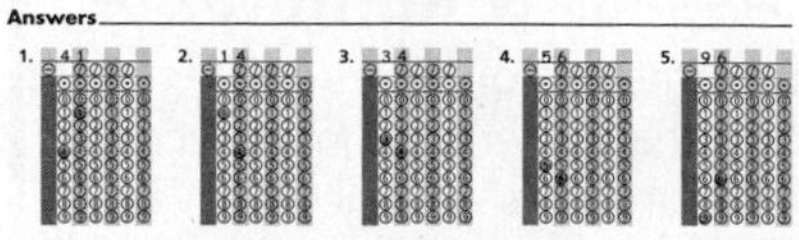

Online Teacher Resource Center

Activities, Games, and Puzzles

8-1 **Puzzle: Crossword**
The Pythagorean Theorem and Its Converse

Solve each problem. Write your answer in the crossword puzzle below. Each numerical answer should be written in word form. (Include hyphens where appropriate.)

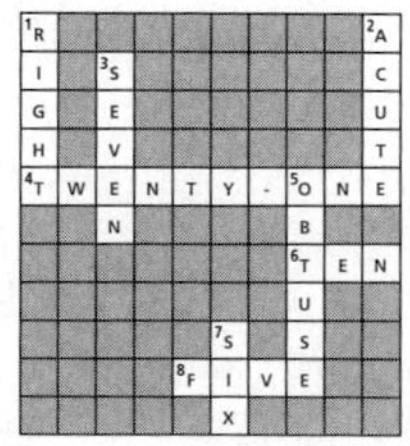

Across

4. You are buying a new TV. The 29-in. TV you are interested in is 20 in. high. How many inches wide is the TV? (Hint: a 29-in. TV measures 29 in. along its *diagonal*.)
6. Given $\triangle ABC$ with coordinates $A(-3, 2)$, $B(-3, -4)$, and $C(5, -4)$, what is the length of $\overline{AC}$?
8. A worker needs to wash the windows of a house. In order to reach the windows on the second story, his 13-ft ladder must reach 12 ft up the side of his house. How many feet from the house should he place the base of the ladder?

Down

1. Give the angle classification for the triangle with side lengths 16 cm, 63 cm, and 65 cm.
2. An artist cuts wood to make frames for her artwork. When she is done, she has scrap pieces that measure 11 in., 5 in., and 10 in. She connects these pieces at their endpoints. Classify the triangle they make by angle.
3. 25, 24, and __?__ are a Pythagorean triple.
5. Give the angle classification of the triangle with side lengths 3 cm, 5 cm, and 7 cm.
7. After finding the value of *k* as a radical in simplest form, what is the integer inside the square root?

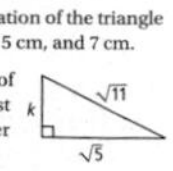

8-2 Special Right Triangles

Common Core State Standards
G-SRT.C.8 Use . . . the Pythagorean Theorem to solve right triangles in applied problems.
MP 1, MP 3, MP 4

Objective To use the properties of 45°-45°-90° and 30°-60°-90° triangles

There are a lot of similar right triangles here. In the lesson, you'll learn a shortcut for finding some of these distances.

SOLVE IT! Getting Ready!

This map of part of a college campus shows a square "quad" area with walking paths. The distance from the dorm to the dining hall is 150 yd.

Suppose you go from your dorm to the dining hall, to the science lab, to your dorm, to the student center, to the library, and finally back to your dorm. To the nearest tenth, how far do you walk? Justify your answer. (Assume you always take the most direct routes and stay on the paths.)

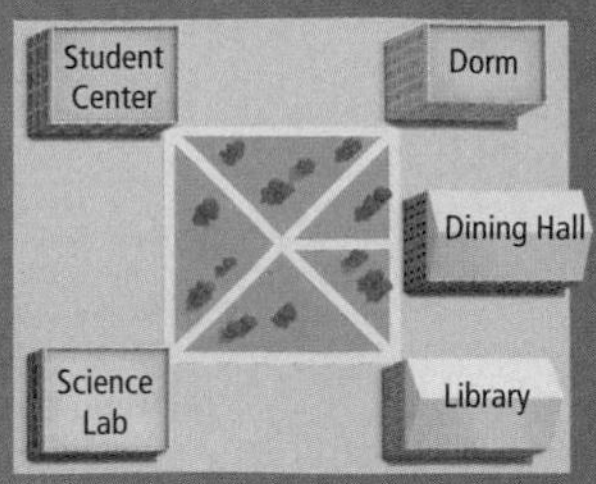

MATHEMATICAL PRACTICES

The Solve It involves triangles with angles 45°, 45°, and 90°.

Essential Understanding Certain right triangles have properties that allow you to use shortcuts to determine side lengths without using the Pythagorean Theorem.

The acute angles of a right isosceles triangle are both 45° angles. Another name for an isosceles right triangle is a 45°-45°-90° triangle. If each leg has length x and the hypotenuse has length y, you can solve for y in terms of x.

$x^2 + x^2 = y^2$ Use the Pythagorean Theorem.

$2x^2 = y^2$ Simplify.

$x\sqrt{2} = y$ Take the positive square root of each side.

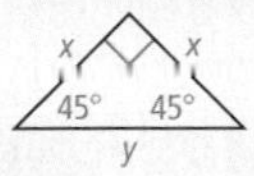

You have just proved the following theorem.

Take note

Theorem 8-5 45°-45°-90° Triangle Theorem

In a 45°-45°-90° triangle, both legs are congruent and the length of the hypotenuse is $\sqrt{2}$ times the length of a leg.

hypotenuse $= \sqrt{2} \cdot$ leg

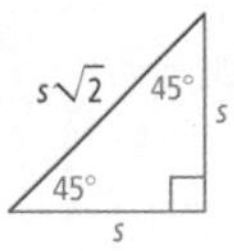

1 Interactive Learning

Solve It!

PURPOSE To use the Pythagorean Theorem to explore the lengths of the sides of 45°-45°-90° triangles

PROCESS Students may use congruent triangle theorems, the Pythagorean Theorem, or algebraic properties.

FACILITATE

Q What is the distance from the dining hall to the library? Explain. **[150 yd, because of congruent triangles]**

Q What is the length of each side of the quad? Explain. **[300 yd by the Segment Addition Postulate]**

Q What is the distance from the dorm to the science lab? Explain. **[Using the Pythagorean Theorem, it is 424.3 yd.]**

Q What is the distance from the dining hall to the intersection of the diagonals of the quad? **[Answers may vary. Sample: using the Pythagorean Theorem, it is 150 yd.]**

ANSWER See Solve It in Answers on next page.

CONNECT THE MATH Students use elements on a diagram to form a right triangle. In this lesson, students learn about a special right triangle with angles measuring 45°, 45°, and 90°.

2 Guided Instruction

Take Note

Using algebraic properties and rules for simplifying radicals, show students how to solve the equation so that the length of the leg is given in terms of the length of the hypotenuse.

8-2 Preparing to Teach

BIG ideas **Measurement**
Reasoning and Proof

ESSENTIAL UNDERSTANDING

- Certain right triangles have properties that allow their side lengths to be determined without using the Pythagorean Theorem.

Math Background

The study of the relationship of the lengths of the sides of the special right triangles provides a bridge between the study of the Pythagorean Theorem and the study of trigonometry. The properties of right triangles and isosceles triangles, along with the Pythagorean Theorem, can be easily used to determine the ratios of the sides of triangles that contain acute angles of 45° and 45° or of 30° and 60°. It is helpful to students to present these triangles in different orientations so that students can internalize the importance of the relative positions within the triangles. Students should memorize the ratios of the sides in the special triangles because they provide shortcuts for finding lengths of sides.

In coming lessons, student will learn that all similar right triangles have constant ratios of side lengths which are referred to as trigonometric functions.

Mathematical Practices

Make sense of problems and persevere in solving them. Students will compute the relationships between sides in right triangles in the special cases of 45-degree and 30-60-degree right triangles.

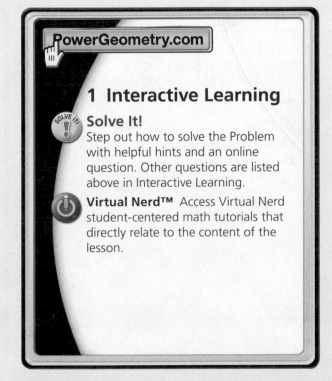

PowerGeometry.com

1 Interactive Learning

Solve It! Step out how to solve the Problem with helpful hints and an online question. Other questions are listed above in Interactive Learning.

Virtual Nerd™ Access Virtual Nerd student-centered math tutorials that directly relate to the content of the lesson.

Problem 1

Q How do you simplify the expression $2\sqrt{2} \cdot \sqrt{2}$? **[$2\sqrt{2} \cdot \sqrt{2} = 2 \cdot 2 = 4$]**

Q Can you use the Pythagorean Theorem to determine the length of the hypotenuse in 1A? Explain. **[Yes; because it is an isosceles triangle, you know the lengths of both legs.]**

Q How can you use the Pythagorean Theorem to check your work? **[You can substitute the lengths of all three sides into the Pythagorean Theorem.]**

Got It? SYNTHESIZING

Q According to Theorem 8-5, what product gives the length of the hypotenuse? **[$5\sqrt{3}(\sqrt{2})$]**

Q How do you multiply terms that contain radicals? **[Multiply the whole numbers together and multiply the radicals by finding the product of the numbers under the radicals.]**

Problem 2

Q Why is the expression $\frac{6}{\sqrt{2}}$ not in simplified form? **[In simplest form, you cannot have a radical in the denominator of a fraction.]**

Q What does it mean to rationalize a denominator? **[to change the form of the fraction so that it does not have a radical in the denominator]**

Q How can you work backward to solve this problem? **[You can multiply each answer choice by $\sqrt{2}$ to see if you get 6.]**

Got It?

Encourage students to check that the leg length is accurate by multiplying it by $\sqrt{2}$ to make sure that they get the length of the hypotenuse.

Problem 1 Finding the Length of the Hypotenuse

Think
Why is only one leg labeled?
A 45°-45°-90° triangle is a right isosceles triangle, so the legs have equal lengths.

What is the value of each variable?

A

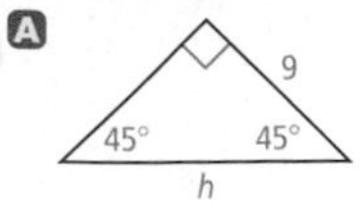

hypotenuse $= \sqrt{2} \cdot$ leg	45°-45°-90° △ Theorem
$h = \sqrt{2} \cdot 9$	Substitute.
$h = 9\sqrt{2}$	Simplify.

B

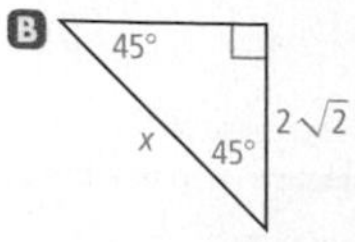

hypotenuse $= \sqrt{2} \cdot$ leg

$x = \sqrt{2} \cdot 2\sqrt{2}$

$x = 4$

Got It? 1. What is the length of the hypotenuse of a 45°-45°-90° triangle with leg length $5\sqrt{3}$?

Problem 2 Finding the Length of a Leg

Think
Can you eliminate any of the choices?
The variable x represents the length of a leg. Since the hypotenuse is the longest side of a right triangle, $x < 6$. You can eliminate choices C and D.

Multiple Choice What is the value of x?

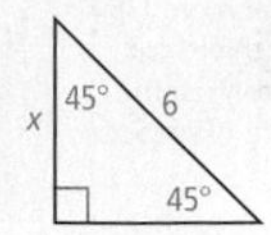

Ⓐ 3　　Ⓒ 6

Ⓑ $3\sqrt{2}$　　Ⓓ $6\sqrt{2}$

hypotenuse $= \sqrt{2} \cdot$ leg	45°-45°-90° Triangle Theorem
$6 = \sqrt{2} \cdot x$	Substitute.
$x = \frac{6}{\sqrt{2}}$	Divide each side by $\sqrt{2}$.
$x = \frac{6}{\sqrt{2}} \cdot \frac{\sqrt{2}}{\sqrt{2}}$	Multiply by a form of 1 to rationalize the denominator.
$x = \frac{6\sqrt{2}}{2}$	Simplify.
$x = 3\sqrt{2}$	Simplify.

The correct answer is B.

Got It? 2. a. The length of the hypotenuse of a 45°-45°-90° triangle is 10. What is the length of one leg?

b. Reasoning In Problem 2, why can you multiply $\frac{6}{\sqrt{2}}$ by $\frac{\sqrt{2}}{\sqrt{2}}$?

Answers

Solve It!

1960.7 yd; $150 + 150 + 150\sqrt{2} + 300\sqrt{2} + 300 + 300\sqrt{2} + 300$

Total: $900 + 750\sqrt{2} \approx 1960.7$ yd

Got It?

1. $5\sqrt{6}$

2a. $5\sqrt{2}$

b. $\frac{\sqrt{2}}{\sqrt{2}} = 1$, so multiplying by $\frac{\sqrt{2}}{\sqrt{2}}$ is the same as multiplying by 1.

3. 141 ft

PowerGeometry.com

2 Guided Instruction

Each Problem is worked out and supported online.

Problem 1
Finding the Length of the Hypotenuse
Animated

Problem 2
Finding the Length of a Leg
Animated

Problem 3
Finding Distance

Problem 4
Using the Length of One Side

Problem 5
Applying the 30°-60°-90° Triangle Theorem
Animated

Support in Geometry Companion
- Vocabulary
- Key Concepts
- Got It?

When you apply the 45°-45°-90° Triangle Theorem to a real-life example, you can use a calculator to evaluate square roots.

Think

How do you know that *d* is a hypotenuse?
The diagonal *d* is part of two right triangles. The hypotenuse of a right triangle is always opposite the 90° angle. So *d* must be a hypotenuse.

Problem 3 Finding Distance

Softball **A high school softball diamond is a square. The distance from base to base is 60 ft. To the nearest foot, how far does a catcher throw the ball from home plate to second base?**

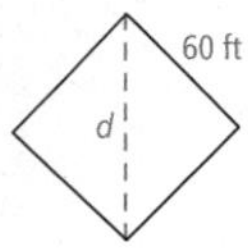

The distance d is the length of the hypotenuse of a 45°-45°-90° triangle.

$d = 60\sqrt{2}$ hypotenuse $= \sqrt{2} \cdot$ leg

$d \approx 84.85281374$ Use a calculator.

The catcher throws the ball about 85 ft from home plate to second base.

Got It? 3. You plan to build a path along one diagonal of a 100 ft-by-100 ft square garden. To the nearest foot, how long will the path be?

Another type of special right triangle is a 30°-60°-90° triangle.

take note

Theorem 8-6 30°-60°-90° Triangle Theorem

In a 30°-60°-90° triangle, the length of the hypotenuse is twice the length of the shorter leg. The length of the longer leg is $\sqrt{3}$ times the length of the shorter leg.

hypotenuse = 2 • shorter leg

longer leg = $\sqrt{3}$ • shorter leg

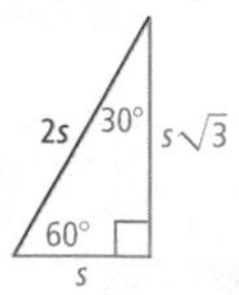

Proof Proof of Theorem 8-6: 30°-60°-90° Triangle Theorem

For equilateral $\triangle WXZ$, altitude $\overline{WY}$ bisects $\angle W$ and is the perpendicular bisector of $\overline{XZ}$. So, $\overline{WY}$ divides $\triangle WXZ$ into two congruent 30°-60°-90° triangles.

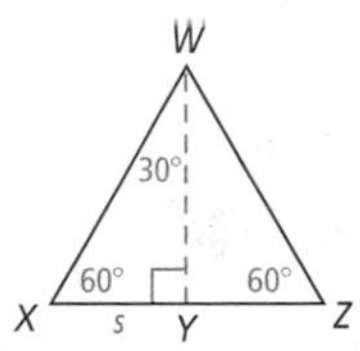

Thus, $XY = \frac{1}{2}XZ = \frac{1}{2}XW$, or $XW = 2XY = 2s$.

$XY^2 + YW^2 = XW^2$	Use the Pythagorean Theorem.
$s^2 + YW^2 = (2s)^2$	Substitute *s* for *XY* and 2*s* for *XW*.
$YW^2 = 4s^2 - s^2$	Subtract s^2 from each side.
$YW^2 = 3s^2$	Simplify.
$YW = s\sqrt{3}$	Take the positive square root of each side.

Problem 3

Q How do you know that the triangles formed in the square are 45°-45°-90° triangles? **[Answers may vary. Sample: Diagonals bisect the angles of a square, and each angle of a square is a right angle.]**

Q What part of the right triangle does the 60 ft measurement represent? **[leg, because the hypotenuse is opposite the right angle]**

Q Is 85 ft the longest throwing distance from one player to another when both players are in the infield of the softball diamond? Explain. **[Yes, the diagonal of the square is the longest measurement.]**

Got It? VISUAL LEARNERS

Q Can you use Theorem 8-5 to determine the length of the diagonal of a garden that is not a square? Explain. **[No, because the diagonal would not form 45°-45°-90° triangles.]**

Take Note

Ask students to determine the coordinates of point *C* in the following diagram such that $AC = AB = BC$. Students will need to set up an equation using the distance formula in order to determine the *y*-coordinate.

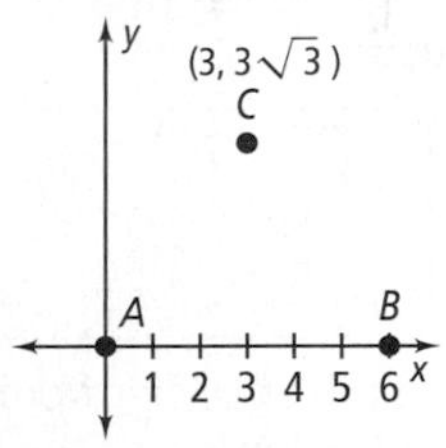

Additional Problems

1. What is the value of each variable?

a.

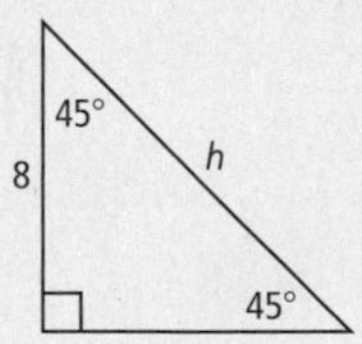

b.

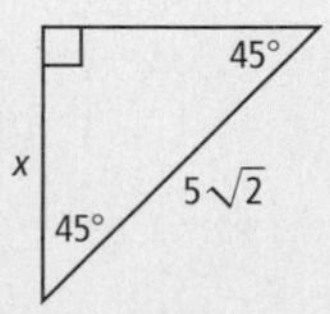

ANSWER **a.** $8\sqrt{2}$ **b.** 5

2. What is the value of *x*?

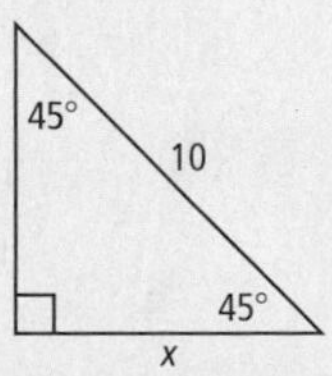

A. $5\sqrt{2}$
B. $10\sqrt{2}$
C. 5
D. 10

ANSWER A

3. A courtyard is shaped like a square with 250-ft-long sides. What is the distance from one corner of the courtyard to the opposite corner? Round to the nearest tenth.

ANSWER 353.6 ft

4. What is the value of *x*?

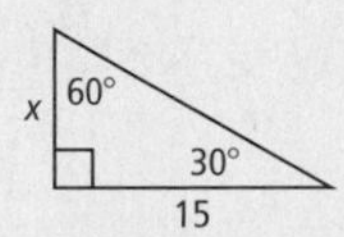

ANSWER $5\sqrt{3}$

5. What is the height of an equilateral triangle with sides that are 12 cm long? Round to the nearest tenth.

ANSWER 10.4 cm

Problem 4

Q Is the leg whose length is given the shorter leg, the longer leg, or the hypotenuse? How can you tell? **[longer leg; it is across from the 60° angle.]**

Q How is the length of the shorter leg of a 30°-60°-90° triangle related to the length of the longer leg? **[The longer leg is $\sqrt{3}$ times as long as the shorter leg.]**

Got It? ERROR PREVENTION

Q How is the length of the hypotenuse of a 30°-60°-90° triangle related to the length of the shorter leg? **[The hypotenuse is twice the length of the shorter leg.]**

Problem 5

Q Is the leg whose length is given the shorter leg, the longer leg, or the hypotenuse? How can you tell? **[Longer leg; it is across from the 60° angle.]**

Q Does Theorem 8-6 express a relationship between the length of the longer leg and the length of the hypotenuse? Explain. **[No, the length of the hypotenuse is expressed in terms of the shorter leg.]**

Q How could you solve this problem using two steps? **[Determine the length of the shorter leg, and then double the length to get the length of the hypotenuse.]**

Got It?

Q How does this problem differ from Problem 5? **[You are given the length of the hypotenuse of a 30°-60°-90° triangle instead of the length of the longer leg.]**

You can also use the 30°-60°-90° Triangle Theorem to find side lengths.

Problem 4 Using the Length of One Side

Algebra **What is the value of d in simplest radical form?**

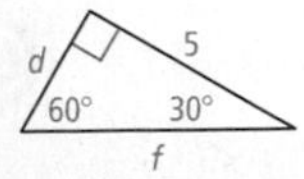

Think	Write
In a 30°-60°-90° triangle, the leg opposite the 60° angle is the longer leg. So d represents the length of the shorter leg. Write an equation relating the legs.	longer leg = $\sqrt{3}$ • shorter leg $5 = d\sqrt{3}$
Divide each side by $\sqrt{3}$ to solve for d.	$d = \frac{5}{\sqrt{3}}$
The value of d is not in simplest radical form because there is a radical in the denominator. Multiply d by a form of 1.	$\frac{5}{\sqrt{3}} \cdot \frac{\sqrt{3}}{\sqrt{3}} = \frac{5\sqrt{3}}{3}$ So $d = \frac{5\sqrt{3}}{3}$.

Got It? 4. In Problem 4, what is the value of f in simplest radical form?

Problem 5 Applying the 30°-60°-90° Triangle Theorem

Plan

How does knowing the shape of the pendants help? Since the triangle is equilateral, you know that an altitude divides the triangle into two congruent 30°-60°-90° triangles.

Jewelry Making **An artisan makes pendants in the shape of equilateral triangles. The height of each pendant is 18 mm. What is the length s of each side of a pendant to the nearest tenth of a millimeter?**

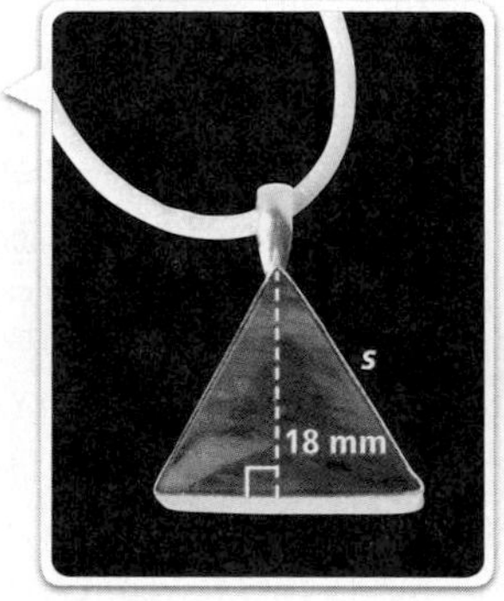

The hypotenuse of each 30°-60°-90° triangle is s. The shorter leg is $\frac{1}{2}s$.

$18 = \sqrt{3}\left(\frac{1}{2}s\right)$ longer leg = $\sqrt{3}$ • shorter leg

$18 = \frac{\sqrt{3}}{2}s$ Simplify.

$\frac{2}{\sqrt{3}} \cdot 18 = s$ Multiply each side by $\frac{2}{\sqrt{3}}$.

$s \approx 20.78460969$ Use a calculator.

Each side of a pendant is about 20.8 mm long.

Got It? 5. Suppose the sides of a pendant are 18 mm long. What is the height of the pendant to the nearest tenth of a millimeter?

Answers

Got It? (continued)

4. $\frac{10\sqrt{3}}{3}$
5. 15.6 mm

Lesson Check

Do you know HOW?

What is the value of *x*? If your answer is not an integer, express it in simplest radical form.

1.

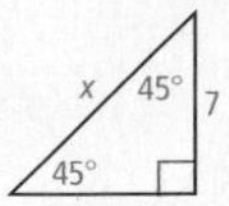

2.

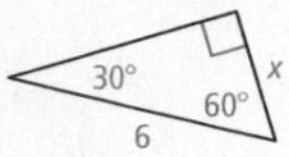

3.

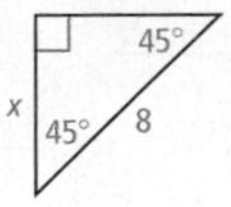

4.

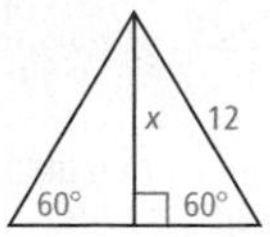

Do you UNDERSTAND? MATHEMATICAL PRACTICES

5. Error Analysis Sandra drew the triangle below. Rika said that the labeled lengths are not possible. With which student do you agree? Explain.

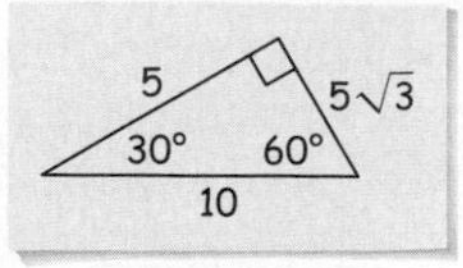

6. Reasoning A test question asks you to find two side lengths of a 45°-45°-90° triangle. You know that the length of one leg is 6, but you forgot the special formula for 45°-45°-90° triangles. Explain how you can still determine the other side lengths. What are the other side lengths?

Practice and Problem-Solving Exercises MATHEMATICAL PRACTICES

A Practice Find the value of each variable. If your answer is not an integer, express it in simplest radical form. See Problems 1 and 2.

7.

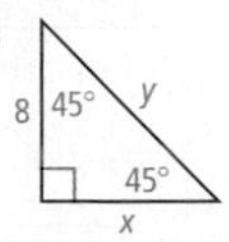

8.

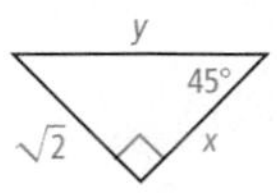

9.

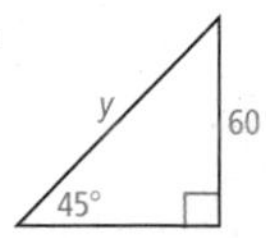

10.

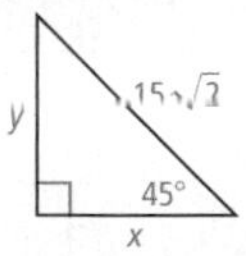

11.

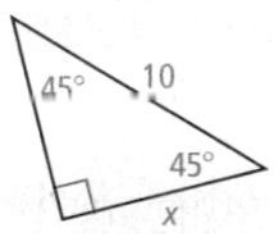

12.

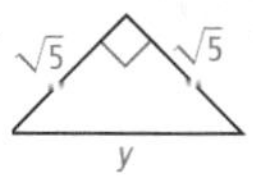

13. Dinnerware Design What is the side length of the smallest square plate on which a 20-cm chopstick can fit along a diagonal without any overhang? Round your answer to the nearest tenth of a centimeter. See Problem 3.

14. Aviation The four blades of a helicopter meet at right angles and are all the same length. The distance between the tips of two adjacent blades is 36 ft. How long is each blade? Round your answer to the nearest tenth of a foot.

3 Lesson Check

Do you know HOW?

- If students have difficulty with Exercise 3, then have them review Problem 2 to understand how to rationalize a denominator.

Do you UNDERSTAND?

- If students have difficulty with Exercise 5, then remind them that 30°-60°-90° triangles are a subset of right triangles, and that satisfying the Pythagorean Theorem alone does not guarantee that it is a 30°-60°-90° triangle.

Close

Q What are special right triangles? Why are they studied? **[They are 30°-60°-90° triangles and 45°-45°-90° triangles. They are studied because their special properties can be used as shortcuts for finding the lengths of the sides of these triangles.]**

Lesson Check

1. $7\sqrt{2}$
2. 3
3. $4\sqrt{2}$
4. $6\sqrt{3}$
5. Rika; 5 should be opposite the 30° ∠ and $5\sqrt{3}$ should be opposite the 60° ∠.
6. Answers may vary. Sample: The △ is isosc. The length of each leg is the same. Use the Pythagorean Thm. to find the hypotenuse; 6, $6\sqrt{2}$.

Practice and Problem-Solving Exercises

7. $x = 8$, $y = 8\sqrt{2}$
8. $x = \sqrt{2}$, $y = 2$
9. $60\sqrt{2}$
10. $x = 15$, $y = 15$.
11. $5\sqrt{2}$
12. $\sqrt{10}$
13. 14.1 cm
14. 25.5 ft

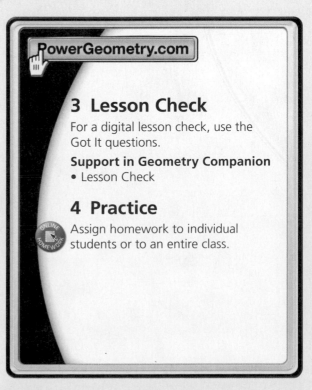

PowerGeometry.com

3 Lesson Check

For a digital lesson check, use the Got It questions.

Support in Geometry Companion

- Lesson Check

4 Practice

Assign homework to individual students or to an entire class.

4 Practice

ASSIGNMENT GUIDE

Basic: 7–22 all, 23–27 odd, 29–31
Average: 7–21 odd, 23–32
Advanced: 7–21 odd, 23–33
Standardized Test Prep: 34–37
Mixed Review: 38–43

Mathematical Practices are supported by exercises with red headings. Here are the Practices supported in this lesson:

MP 3: Construct Arguments Ex. 6
MP 3: Critique the Reasoning of Others Ex. 5
MP 4: Model with Mathematics Ex. 29, 31

Applications exercises have blue headings. Exercises 13, 14, 21, 22, and 30 support MP 4: Model.

STEM exercises focus on science or engineering applications.

EXERCISE 30: Use the Think About a Plan worksheet in the **Practice and Problem Solving Workbook** (also available in the Teaching Resources in print and online) to further support students' development in becoming independent learners.

HOMEWORK QUICK CHECK

To check students' understanding of key skills and concepts, go over Exercises 7, 21, 29, 30, and 31.

Algebra **Find the value of each variable. If your answer is not an integer, express it in simplest radical form.** See Problems 4 and 5.

15.

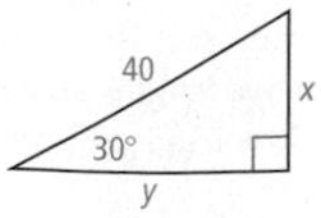

16.

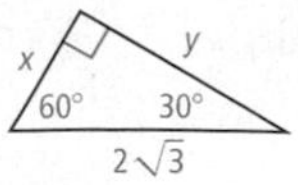

17.

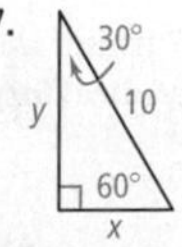

18.

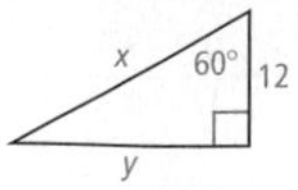

19.

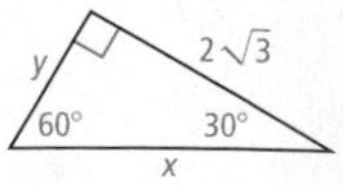

20.

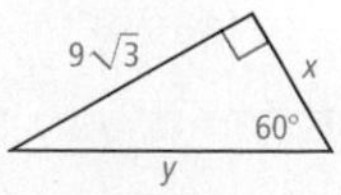

STEM **21. Architecture** An escalator lifts people to the second floor of a building, 25 ft above the first floor. The escalator rises at a 30° angle. To the nearest foot, how far does a person travel from the bottom to the top of the escalator?

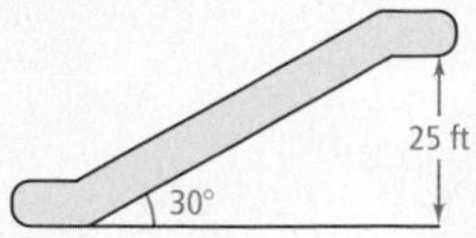

STEM **22. City Planning** Jefferson Park sits on one square city block 300 ft on each side. Sidewalks across the park join opposite corners. To the nearest foot, how long is each diagonal sidewalk?

B Apply

Algebra **Find the value of each variable. If your answer is not an integer, express it in simplest radical form.**

23.

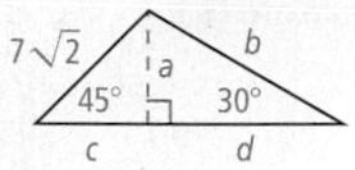

24.

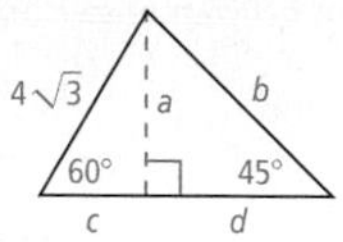

25.

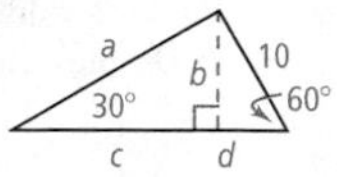

26.

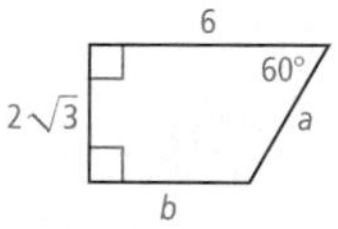

27.

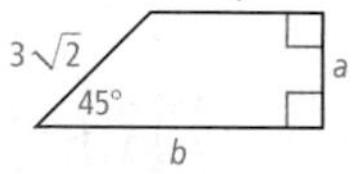

28.

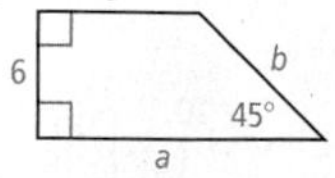

29. Think About a Plan A farmer's conveyor belt carries bales of hay from the ground to the barn loft. The conveyor belt moves at 100 ft/min. How many seconds does it take for a bale of hay to go from the ground to the barn loft?

- Which part of a right triangle does the conveyor belt represent?
- You know the speed. What other information do you need to find time?
- How are minutes and seconds related?

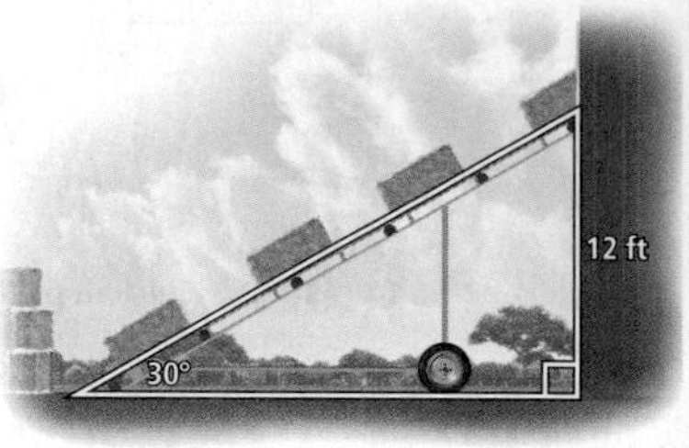

Answers

Practice and Problem-Solving Exercises (continued)

15. $x = 20$, $y = 20\sqrt{3}$
16. $x = \sqrt{3}$, $y = 3$
17. $x = 5$, $y = 5\sqrt{3}$
18. $x = 24$, $y = 12\sqrt{3}$
19. $x = 4$, $y = 2$
20. $x = 9$, $y = 18$
21. 50 ft
22. 424 ft
23. $a = 7$, $b = 14$, $c = 7$, $d = 7\sqrt{3}$
24. $a = 6$, $b = 6\sqrt{2}$, $c = 2\sqrt{3}$, $d = 6$
25. $a = 10\sqrt{3}$, $b = 5\sqrt{3}$, $c = 15$, $d = 5$
26. $a = 4$, $b = 4$
27. $a = 3$, $b = 7$
28. $a = 14$, $b = 6\sqrt{2}$
29. 14.4 s

30. **House Repair** After heavy winds damaged a house, workers placed a 6-m brace against its side at a 45° angle. Then, at the same spot on the ground, they placed a second, longer brace to make a 30° angle with the side of the house.
   a. How long is the longer brace? Round to the nearest tenth of a meter.
   b. About how much higher does the longer brace reach than the shorter brace?

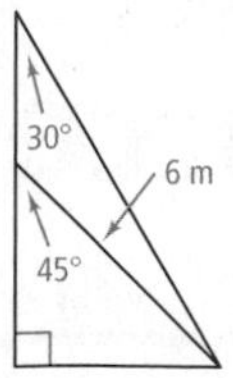

31. **Open-Ended** Write a real-life problem that you can solve using a 30°-60°-90° triangle with a 12-ft hypotenuse. Show your solution.

32. **Constructions** Construct a 30°-60°-90° triangle using a segment that is the given side.
   a. the shorter leg   b. the hypotenuse   c. the longer leg

Challenge

33. **Geometry in 3 Dimensions** Find the length d, in simplest radical form, of the diagonal of a cube with edges of the given length.
   a. 1 unit   b. 2 units   c. s units

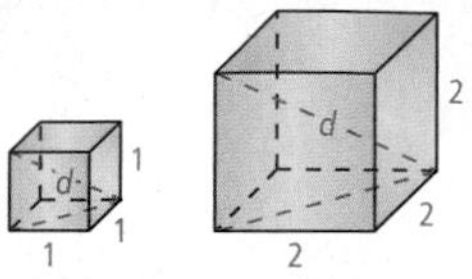

Standardized Test Prep

SAT/ACT

34. The longer leg of a 30°-60°-90° triangle is 6. What is the length of the hypotenuse?
   (A) $2\sqrt{3}$ (B) $3\sqrt{2}$ (C) $4\sqrt{3}$ (D) 12

35. Which triangle is NOT a right triangle?

(F)
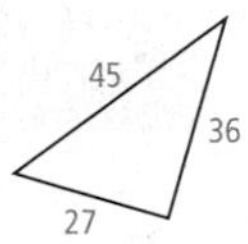

(G)
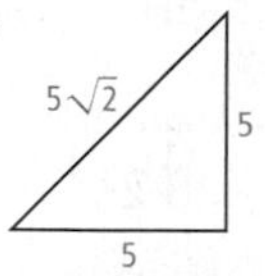

(H)
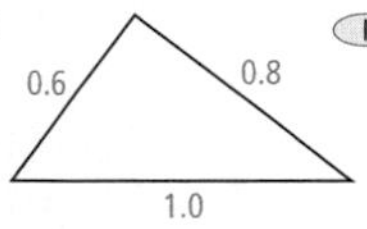

(I)
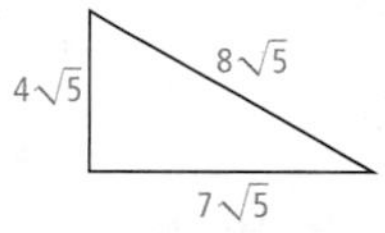

36. Suppose p is false and q is true. Which statement is NOT true?
   (A) $p \rightarrow q$ (B) $\sim q \vee (p \wedge q)$ (C) $p \vee q$ (D) $(p \vee q) \wedge \sim p$

Short Response

37. In right $\triangle ABC$, $\angle C$ is the right angle and $\overline{CD}$ is the altitude drawn to the hypotenuse. If $AD = 3$ and $DB = 9$, what is AC? Show your work.

Mixed Review

38. A right triangle has a 6-in. hypotenuse and a 5-in. leg. Find the length of the other leg in simplest radical form.

See Lesson 8-1.

39. An isosceles triangle has 20-cm legs and a 16-cm base. Find the length of the altitude to the base in simplest radical form.

Get Ready! **To prepare for Lesson 8-3, do Exercises 40–43.**

Algebra Solve each proportion.

See Lesson 7-1.

40. $\frac{x}{3} = \frac{4}{7}$ 41. $\frac{6}{11} = \frac{x}{9}$ 42. $\frac{8}{15} = \frac{4}{x}$ 43. $\frac{5}{x} = \frac{7}{12}$

38. $\sqrt{11}$ in.
39. $4\sqrt{21}$ cm
40. $\frac{12}{7}$
41. $\frac{54}{11}$
42. $\frac{15}{2}$
43. $\frac{60}{7}$

30a. 8.5 m
b. 3.1 m
31. Answers may vary. Sample: A ramp up to a door is 12 ft long. The ramp forms a 30° ∠ with the ground. How high off the ground is the door? 6 ft
32. Answers may vary. Samples using the following segment are given.

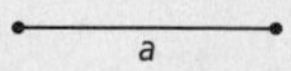

a.
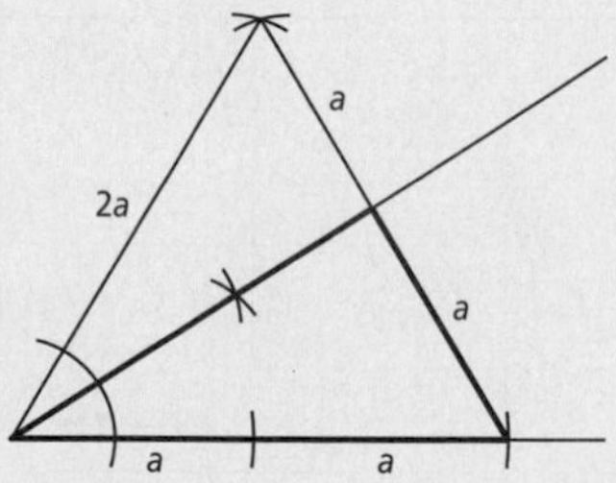

b.
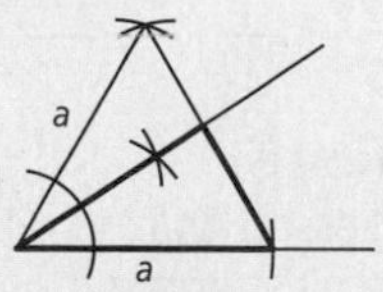

c.
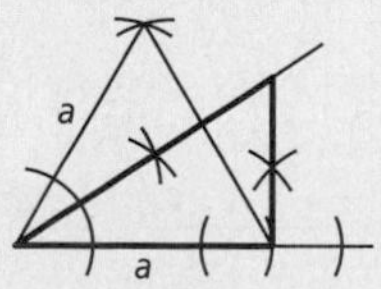

33a. $\sqrt{3}$ units
b. $2\sqrt{3}$ units
c. $s\sqrt{3}$ units
34. C
35. I
36. B
37. [2] $AC = 6$; $\frac{3}{AC} = \frac{AC}{12}$, $AC^2 = 36$
[1] correct proportion, but minor computational error

8-2 Lesson Resources

Differentiated Remediation

Instructional Support

Geometry Companion

Students can use the **Geometry Companion** worktext (4 pages) . . .

- New Vocabulary
- Key Concepts
- Got It for each Problem
- Lesson Check

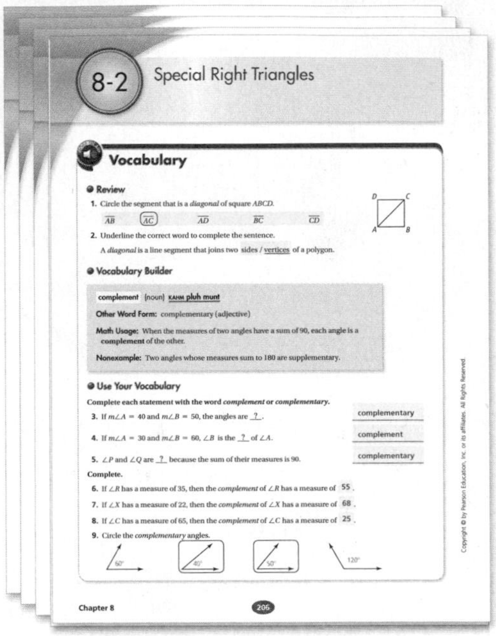
8-2 Special Right Triangles

Vocabulary

Review

1. Circle the segment that is a *diagonal* of square *ABCD*.

2. Underline the correct word to complete the sentence.

A diagonal is a line segment that joins two sides / vertices of a polygon.

Vocabulary Builder

complement (noun) KAHM pluh munt

Other Word Form: complementary (adjective)

Math Usage: When the measures of two angles have a sum of 90, each angle is a complement of the other.

Nonexample: Two angles whose measures sum to 180 are supplementary.

Use Your Vocabulary

Complete each statement with the word *complement* or *complementary*.

3. complementary

4. complement

5. complementary

Complete.

6. 55

7. 68

8. 25

9. Circle the *complementary* angles.

Chapter 8

ELL Support

Use Manipulatives Have students work in small groups. On graph paper, have each group of students draw a square of a different size with a diagonal. Ask them to cut out the square and fold it along the diagonal. Ask: How would you classify this triangle? Then have them measure the two acute angles and the length of each leg. Ask: How can you find the length of the hypotenuse without measuring? Have them multiply the length of one leg by $\sqrt{2}$. Students should compare their results with one another and make a conjecture.

5 Assess & Remediate

Lesson Quiz

1. What is the value of *h*?

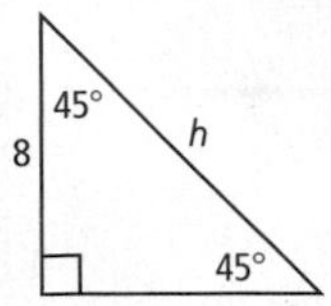

2. What is the value of *x*?

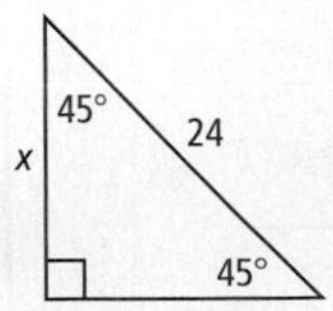

3. Do you UNDERSTAND? A company logo is shaped like an equilateral triangle with 2-in.-long sides. What is the height of the logo? Round to the nearest tenth.

4. What is the value of *a*?

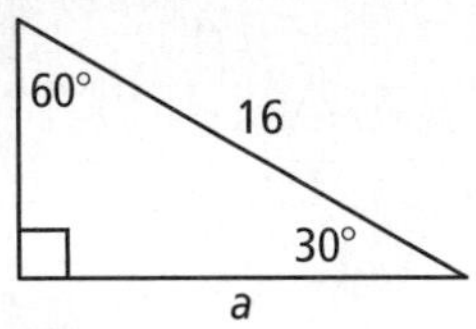

ANSWERS TO LESSON QUIZ

1. $8\sqrt{2}$

2. $12\sqrt{2}$

3. 1.7 in.

4. $8\sqrt{3}$

PRESCRIPTION FOR REMEDIATION

Use the student work on the Lesson Quiz to prescribe a differentiated review assignment.

Points	Differentiated Remediation
0–2	Intervention
3	On-level
4	Extension

PowerGeometry.com

5 Assess & Remediate

Assign the Lesson Quiz. Appropriate intervention, practice, or enrichment is automatically generated based on student performance.

Intervention

- **Reteaching** (2 pages) Provides reteaching and practice exercises for the key lesson concepts. Use with struggling students or absent students.
- **English Language Learner Support** Helps students develop and reinforce mathematical vocabulary and key concepts.

All-in-One Resources/Online

Reteaching

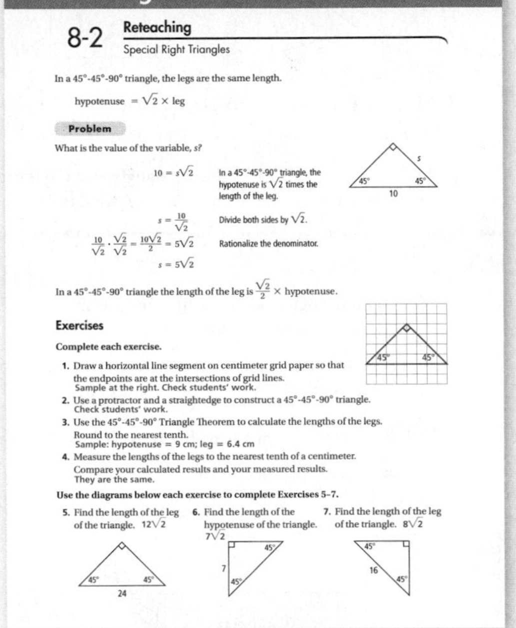
8-2 Reteaching

Special Right Triangles

In a 45°-45°-90° triangle, the legs are the same length.

hypotenuse = $\sqrt{2}$ × leg

Problem

What is the value of the variable, *s*?

$10 = s\sqrt{2}$ In a 45°-45°-90° triangle, the hypotenuse is $\sqrt{2}$ times the length of the leg.

$s = \frac{10}{\sqrt{2}}$ Divide both sides by $\sqrt{2}$.

$\frac{10}{\sqrt{2}} \cdot \frac{\sqrt{2}}{\sqrt{2}} = \frac{10\sqrt{2}}{2} = 5\sqrt{2}$ Rationalize the denominator.

$s = 5\sqrt{2}$

In a 45°-45°-90° triangle the length of the leg is $\frac{\sqrt{2}}{2}$ × hypotenuse.

Exercises

Complete each exercise.

1. Draw a horizontal line segment on centimeter grid paper so that the endpoints are at the intersections of grid lines. Sample at the right. Check students' work.

2. Use a protractor and a straightedge to construct a 45°-45°-90° triangle. Check students' work.

3. Use the 45°-45°-90° Triangle Theorem to calculate the lengths of the legs. Round to the nearest tenth. Sample: hypotenuse = 9 cm; leg ≈ 6.4 cm

4. Measure the lengths of the legs to the nearest tenth of a centimeter. Compare your calculated results and your measured results. They are the same.

Use the diagrams below each exercise to complete Exercises 5–7.

5. Find the length of the leg of the triangle. $12\sqrt{2}$

6. Find the length of the hypotenuse of the triangle. $7\sqrt{2}$

7. Find the length of the leg of the triangle. $8\sqrt{2}$

All-in-One Resources/Online

English Language Learner Support

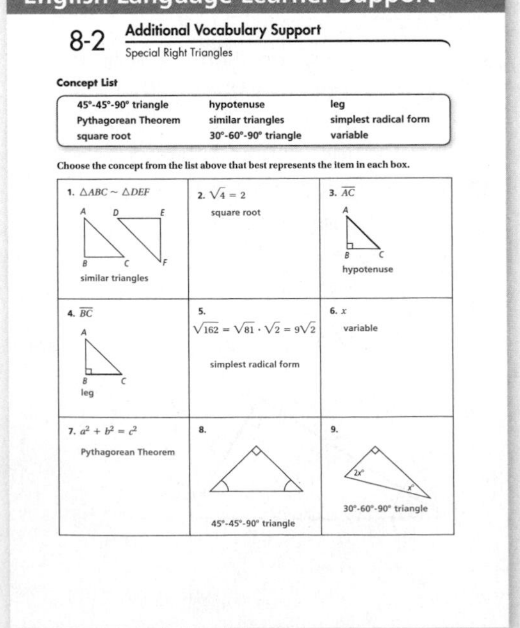
8-2 Additional Vocabulary Support

Special Right Triangles

Concept List

45°-45°-90° triangle	hypotenuse	leg
Pythagorean Theorem	similar triangles	simplest radical form
square root	30°-60°-90° triangle	variable

Choose the concept from the list above that best represents the item in each box.

1. $\triangle ABC \sim \triangle DEF$ similar triangles	2. $\sqrt{4} = 2$ square root	3. $\overline{AC}$ hypotenuse
4. $\overline{BC}$ leg	5. $\sqrt{162} = \sqrt{81} \cdot \sqrt{2} = 9\sqrt{2}$ simplest radical form	6. *x* variable
7. $a^2 + b^2 = c^2$ Pythagorean Theorem	8. 45°-45°-90° triangle	9. 30°-60°-90° triangle

Differentiated Remediation *continued*

On-Level

- **Practice** (2 pages) Provides extra practice for each lesson. For simpler practice exercises, use the Form K Practice pages found in the All-in-One Teaching Resources and online.
- **Think About a Plan** Helps students develop specific problem-solving skills and strategies by providing scaffolded guiding questions.
- **Standardized Test Prep** Focuses on all major exercises, all major question types, and helps students prepare for the high-stakes assessments.

Extension

- **Enrichment** Provides students with interesting problems and activities that extend the concepts of the lesson.
- **Activities, Games, and Puzzles** Worksheets that can be used for concepts development, enrichment, and for fun!

Practice and Problem Solving Wkbk/ All-in-One Resources/Online

Practice page 1

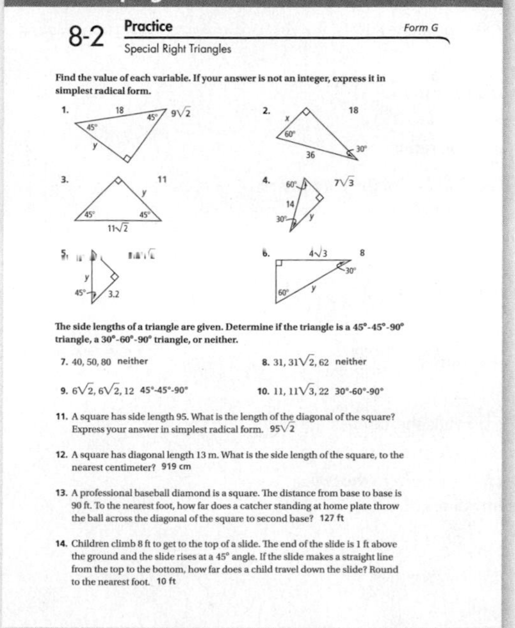
8-2 Practice — Special Right Triangles — Form G

Find the value of each variable. If your answer is not an integer, express it in simplest radical form.

The side lengths of a triangle are given. Determine if the triangle is a 45°-45°-90° triangle, a 30°-60°-90° triangle, or neither.

7. 40, 50, 80 neither
8. 31, $31\sqrt{2}$, 62 neither
9. $6\sqrt{2}$, $6\sqrt{2}$, 12 45°-45°-90°
10. 11, $11\sqrt{3}$, 22 30°-60°-90°
11. A square has side length 95. What is the length of the diagonal of the square? Express your answer in simplest radical form. $95\sqrt{2}$
12. A square has diagonal length 13 m. What is the side length of the square, to the nearest centimeter? 919 cm
13. A professional baseball diamond is a square. The distance from base to base is 90 ft. To the nearest foot, how far does a catcher standing at home plate throw the ball across the diagonal of the square to second base? 127 ft
14. Children climb 8 ft to get to the top of a slide. The end of the slide is 1 ft above the ground and the slide rises at a 45° angle. If the slide makes a straight line from the top to the bottom, how far does a child travel down the slide? Round to the nearest foot. 10 ft

Practice and Problem Solving Wkbk/ All-in-One Resources/Online

Practice page 2

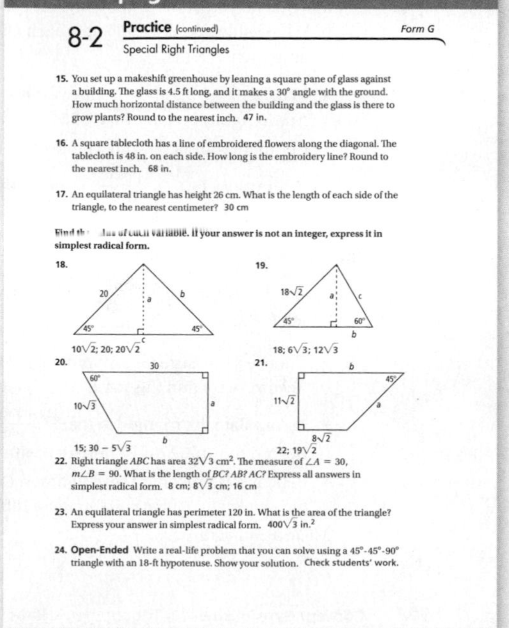
8-2 Practice (continued) — Special Right Triangles — Form G

15. You set up a makeshift greenhouse by leaning a square pane of glass against a building. The glass is 4.5 ft long, and it makes a 30° angle with the ground. How much horizontal distance between the building and the glass is there to grow plants? Round to the nearest inch. 47 in.
16. A square tablecloth has a line of embroidered flowers along the diagonal. The tablecloth is 48 in. on each side. How long is the embroidery line? Round to the nearest inch. 68 in.
17. An equilateral triangle has height 26 cm. What is the length of each side of the triangle, to the nearest centimeter? 30 cm

Find the value of each variable. If your answer is not an integer, express it in simplest radical form.

18. $10\sqrt{2}$; 20; $20\sqrt{2}$
19. 18; $6\sqrt{3}$; $12\sqrt{3}$
20. 15; $30 - 5\sqrt{3}$
21. 22; $19\sqrt{2}$
22. Right triangle ABC has area $32\sqrt{3}$ cm². The measure of $\angle A = 30$, $m\angle B = 90$. What is the length of $\overline{BC}$? $\overline{AB}$? $\overline{AC}$? Express all answers in simplest radical form. 8 cm; $8\sqrt{3}$ cm; 16 cm
23. An equilateral triangle has perimeter 120 in. What is the area of the triangle? Express your answer in simplest radical form. $400\sqrt{3}$ in.²
24. Open-Ended Write a real-life problem that you can solve using a 45°-45°-90° triangle with an 18-ft hypotenuse. Show your solution. Check students' work.

All-in-One Resources/Online

Enrichment

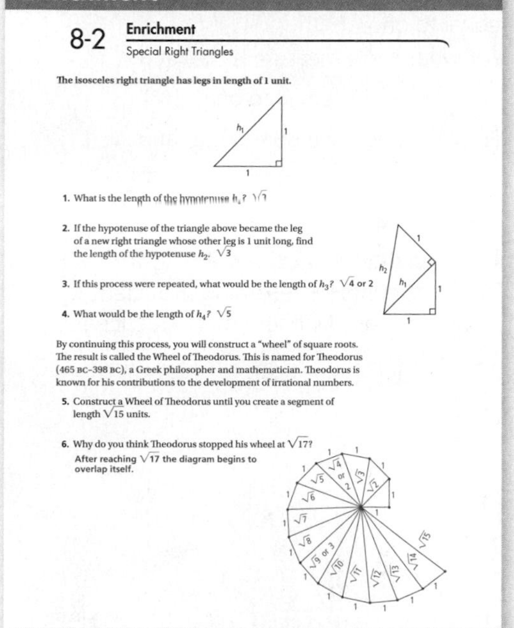
8-2 Enrichment — Special Right Triangles

The isosceles right triangle has legs in length of 1 unit.

1. What is the length of the hypotenuse h_1? $\sqrt{2}$
2. If the hypotenuse of the triangle above became the leg of a new right triangle whose other leg is 1 unit long, find the length of the hypotenuse h_2. $\sqrt{3}$
3. If this process were repeated, what would be the length of h_3? $\sqrt{4}$ or 2
4. What would be the length of h_4? $\sqrt{5}$

By continuing this process, you will construct a "wheel" of square roots. The result is called the Wheel of Theodorus. This is named for Theodorus (465 BC–398 BC), a Greek philosopher and mathematician. Theodorus is known for his contributions to the development of irrational numbers.

5. Construct a Wheel of Theodorus until you create a segment of length $\sqrt{15}$ units.
6. Why do you think Theodorus stopped his wheel at $\sqrt{17}$? After reaching $\sqrt{17}$ the diagram begins to overlap itself.

Practice and Problem Solving Wkbk/ All-in-One Resources/Online

Think About a Plan

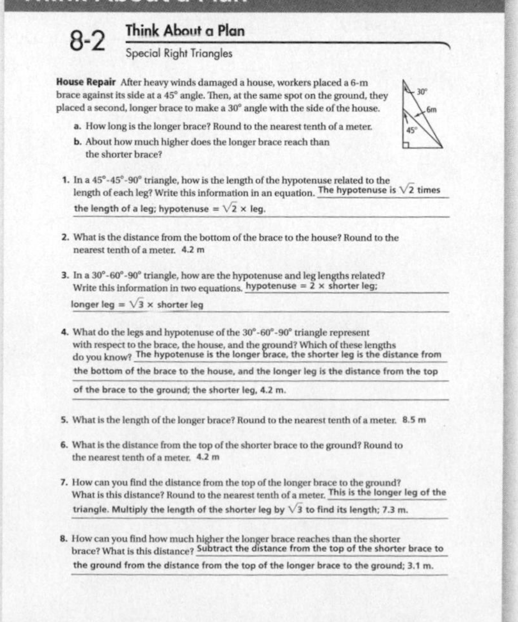
8-2 Think About a Plan — Special Right Triangles

House Repair After heavy winds damaged a house, workers placed a 6-m brace against its side at a 45° angle. Then, at the same spot on the ground, they placed a second, longer brace to make a 30° angle with the side of the house.
a. How long is the longer brace? Round to the nearest tenth of a meter.
b. About how much higher does the longer brace reach than the shorter brace?

1. In a 45°-45°-90° triangle, how is the length of the hypotenuse related to the length of each leg? Write this information in an equation. The hypotenuse is $\sqrt{2}$ times the length of a leg; hypotenuse = $\sqrt{2}$ × leg.
2. What is the distance from the bottom of the brace to the house? Round to the nearest tenth of a meter. 4.2 m
3. In a 30°-60°-90° triangle, how are the hypotenuse and leg lengths related? Write this information in two equations. hypotenuse = 2 × shorter leg; longer leg = $\sqrt{3}$ × shorter leg
4. What do the legs and hypotenuse of the 30°-60°-90° triangle represent with respect to the brace, the house, and the ground? Which of these lengths do you know? The hypotenuse is the longer brace, the shorter leg is the distance from the bottom of the brace to the house, and the longer leg is the distance from the top of the brace to the ground; the shorter leg, 4.2 m.
5. What is the length of the longer brace? Round to the nearest tenth of a meter. 8.5 m
6. What is the distance from the top of the shorter brace to the ground? Round to the nearest tenth of a meter. 4.2 m
7. How can you find the distance from the top of the longer brace to the ground? What is this distance? Round to the nearest tenth of a meter. This is the longer leg of the triangle. Multiply the length of the shorter leg by $\sqrt{3}$ to find its length; 7.3 m.
8. How can you find how much higher the longer brace reaches than the shorter brace? What is this distance? Subtract the distance from the top of the shorter brace to the ground from the distance from the top of the longer brace to the ground; 3.1 m.

Practice and Problem Solving Wkbk/ All-in-One Resources/Online

Standardized Test Prep

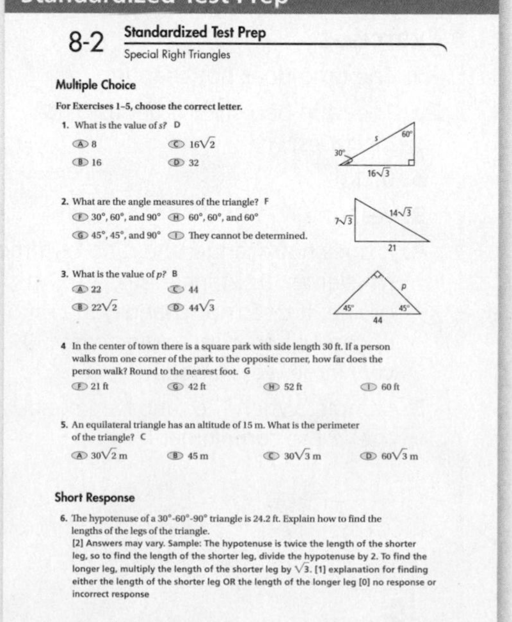
8-2 Standardized Test Prep — Special Right Triangles

Multiple Choice

For Exercises 1–5, choose the correct letter.

1. What is the value of s? D
A 8 B 16 C $16\sqrt{2}$ D 32
2. What are the angle measures of the triangle? F
F 30°, 60°, and 90° G 45°, 45°, and 90° H 60°, 60°, and 60° I They cannot be determined.
3. What is the value of p? B
A 22 B $22\sqrt{2}$ C 44 D $44\sqrt{3}$
4. In the center of town there is a square park with side length 30 ft. If a person walks from one corner of the park to the opposite corner, how far does the person walk? Round to the nearest foot. G
F 21 ft G 42 ft H 52 ft I 60 ft
5. An equilateral triangle has an altitude of 15 m. What is the perimeter of the triangle? C
A $30\sqrt{2}$ m B 45 m C $30\sqrt{3}$ m D $60\sqrt{3}$ m

Short Response

6. The hypotenuse of a 30°-60°-90° triangle is 24.2 ft. Explain how to find the lengths of the legs of the triangle.
[2] Answers may vary. Sample: The hypotenuse is twice the length of the shorter leg, so to find the length of the shorter leg, divide the hypotenuse by 2. To find the longer leg, multiply the length of the shorter leg by $\sqrt{3}$. [1] explanation for finding either the length of the shorter leg OR the length of the longer leg [0] no response or incorrect response

Online Teacher Resource Center

Activities, Games, and Puzzles

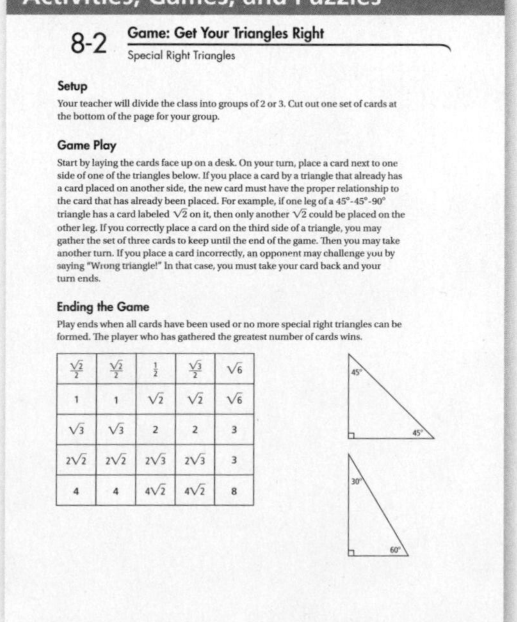
8-2 Game: Get Your Triangles Right — Special Right Triangles

Setup

Your teacher will divide the class into groups of 2 or 3. Cut out one set of cards at the bottom of the page for your group.

Game Play

Start by laying the cards face up on a desk. On your turn, place a card next to one side of one of the triangles below. If you place a card by a triangle that already has a card placed on another side, the new card must have the proper relationship to the card that has already been placed. For example, if one leg of a 45°-45°-90° triangle has a card labeled $\sqrt{2}$ on it, then only another $\sqrt{2}$ could be placed on the other leg. If you correctly place a card on the third side of a triangle, you may gather the set of three cards to keep until the end of the game. Then you may take another turn. If you place a card incorrectly, an opponent may challenge you by saying "Wrong triangle!" In that case, you must take your card back and your turn ends.

Ending the Game

Play ends when all cards have been used or no more special right triangles can be formed. The player who has gathered the greatest number of cards wins.

$\frac{\sqrt{2}}{2}$	$\frac{\sqrt{2}}{2}$	$\frac{1}{2}$	$\frac{\sqrt{3}}{2}$	$\sqrt{6}$
1	1	$\sqrt{2}$	$\sqrt{2}$	$\sqrt{6}$
$\sqrt{3}$	$\sqrt{3}$	2	2	3
$2\sqrt{2}$	$2\sqrt{2}$	$2\sqrt{3}$	$2\sqrt{3}$	3
4	4	$4\sqrt{2}$	$4\sqrt{2}$	8

Guided Instruction

PURPOSE To use geometry software to explore the trigonometric ratios of sine, cosine, and tangent
PROCESS Students will construct a triangle using geometry software. They will manipulate a point on the triangle to see how moving the point impacts the three trigonometric ratios.
DISCUSS Students have not yet been introduced to the trigonometric ratios; in this activity they are exploring ratios of side lengths to one another.

Activity

This Activity allows students to see how the ratio of side lengths of a right triangle changes as an acute angle in the triangle gets larger or smaller.

Q Why is the ratio $\frac{ED}{AE}$ a function of $\angle A$? **[The length of $\overline{ED}$ changes as the size of $\angle A$ changes.]**

Q For what angle measure is the ratio $\frac{\text{leg opposite } \angle A}{\text{hypotenuse}}$ equal to one? **[90°]**

Q For what angle measure is the ratio $\frac{\text{leg adjacent to } \angle A}{\text{hypotenuse}}$ equal to one? **[0°]**

Q For what angle measure is the ratio $\frac{\text{leg opposite } \angle A}{\text{leg adjacent to } \angle A}$ equal to one? **[45°]**

Mathematical Practices This Concept Byte supports students in becoming proficient in using appropriate tools, Mathematical Practice 5.

Concept Byte
Use With Lesson 8-3
TECHNOLOGY

Exploring Trigonometric Ratios

Common Core State Standards
G-SRT.C.6 Understand that by similarity, side ratios in right triangles are properties of the angles in the triangle, leading to definitions of trigonometric ratios for acute angles.
MP 5

Construct

Use geometry software to construct $\overrightarrow{AB}$ and $\overrightarrow{AC}$ so that $\angle A$ is acute. Through a point D on $\overrightarrow{AB}$, construct a line perpendicular to $\overrightarrow{AB}$ that intersects $\overrightarrow{AC}$ in point E.

Moving point D changes the size of $\triangle ADE$. Moving point C changes the size of $\angle A$.

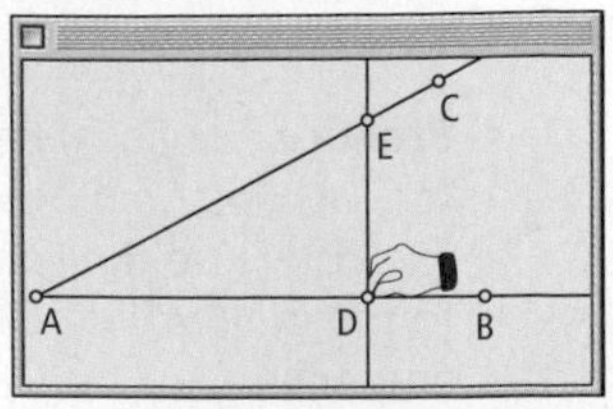

Exercises

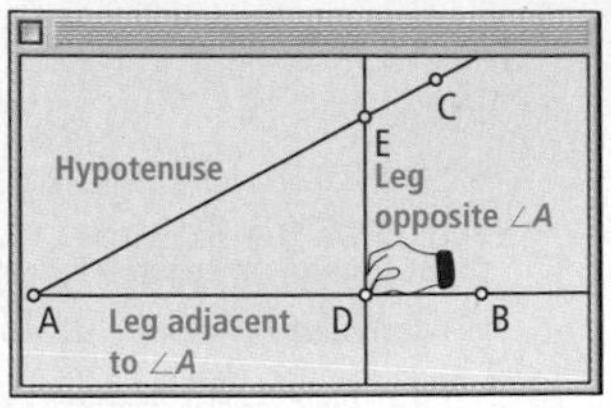

1. • Measure $\angle A$ and find the lengths of the sides of $\triangle ADE$.
   • Calculate the ratio $\frac{\text{leg opposite } \angle A}{\text{hypotenuse}}$, which is $\frac{ED}{AE}$.
   • Move point D to change the size of $\triangle ADE$ without changing $m\angle A$.
   What do you observe about the ratio as the size of $\triangle ADE$ changes?

2. • Move point C to change $m\angle A$.
   a. What do you observe about the ratio as $m\angle A$ changes?
   b. What value does the ratio approach as $m\angle A$ approaches 0? As $m\angle A$ approaches 90?

3. • Make a table that shows values for $m\angle A$ and the ratio $\frac{\text{leg opposite } \angle A}{\text{hypotenuse}}$. In your table, include 10, 20, 30, . . . , 80 for $m\angle A$.
   • Compare your table with a table of trigonometric ratios.
   Do your values for $\frac{\text{leg opposite } \angle A}{\text{hypotenuse}}$ match the values in one of the columns of the table? What is the name of this ratio in the table?

Extend

4. Repeat Exercises 1–3 for $\frac{\text{leg adjacent to } \angle A}{\text{hypotenuse}}$, which is $\frac{AD}{AE}$, and $\frac{\text{leg opposite} \angle A}{\text{leg adjacent to} \angle A}$, which is $\frac{ED}{AD}$.

5. • Choose a measure for $\angle A$ and determine the ratio $r = \frac{\text{leg opposite } \angle A}{\text{hypotenuse}}$. Record $m\angle A$ and this ratio.
   • Manipulate the triangle so that $\frac{\text{leg adjacent to } \angle A}{\text{hypotenuse}}$ has the same value r. Record this $m\angle A$ and compare it with your first value of $m\angle A$.
   • Repeat this procedure several times. Look for a pattern in the two measures of $\angle A$ that you found for different values of r.
   Make a conjecture.

Answers

Exercises

1. The ratio does not change.
2a. The ratio becomes larger as $m\angle A$ increases.
b. 0; 1
3. yes; sine
4. It does not change; the ratio becomes smaller as $m\angle A$ increases; 1, 0; yes; cosine; it does not change; the ratio becomes larger as $\angle A$ increases; 0; very large; yes; tangent.
5. Sample: When the ratios are equal, the ∠s are complements.

8-3 Trigonometry

Common Core State Standards
G-SRT.C.8 Use trigonometric ratios and the Pythagorean Theorem to solve right triangles in applied problems. **Also, G-SRT.C.7, G-MG.A.1**
MP 1, MP 3, MP 4, MP 6

Objective To use the sine, cosine, and tangent ratios to determine side lengths and angle measures in right triangles

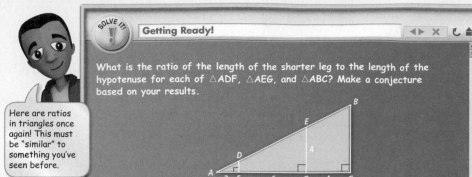

Here are ratios in triangles once again! This must be "similar" to something you've seen before.

MATHEMATICAL PRACTICES

Getting Ready!

What is the ratio of the length of the shorter leg to the length of the hypotenuse for each of △ADF, △AEG, and △ABC? Make a conjecture based on your results.

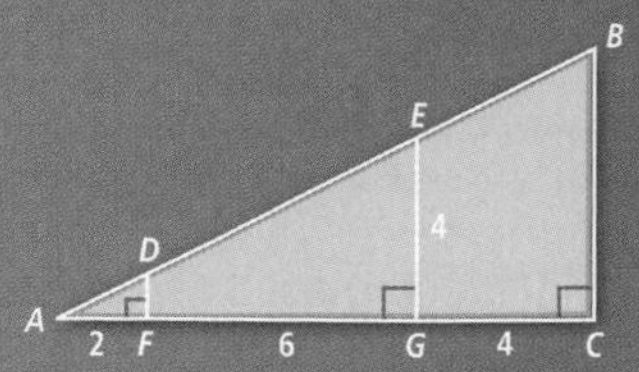

Essential Understanding If you know certain combinations of side lengths and angle measures of a right triangle, you can use ratios to find other side lengths and angle measures.

Any two right triangles that have a pair of congruent acute angles are similar by the AA Similarity Postulate. Similar right triangles have equivalent ratios for their corresponding sides called **trigonometric ratios.**

Lesson Vocabulary
- trigonometric ratios
- sine
- cosine
- tangent

take note

Key Concept Trigonometric Ratios

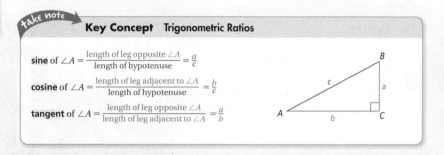

sine of $\angle A = \dfrac{\text{length of leg opposite } \angle A}{\text{length of hypotenuse}} = \dfrac{a}{c}$

cosine of $\angle A = \dfrac{\text{length of leg adjacent to } \angle A}{\text{length of hypotenuse}} = \dfrac{b}{c}$

tangent of $\angle A = \dfrac{\text{length of leg opposite } \angle A}{\text{length of leg adjacent to } \angle A} = \dfrac{a}{b}$

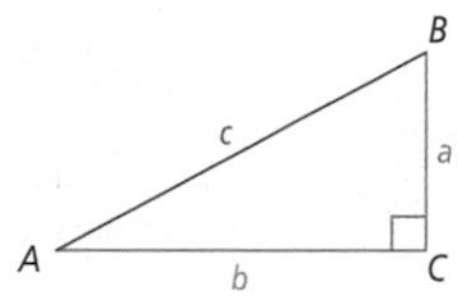

1 Interactive Learning

Solve It!

PURPOSE To make a conjecture about the ratios of the lengths of the corresponding sides of similar triangles

PROCESS Students may use similar triangle postulates, the Pythagorean Theorem, or logical reasoning.

FACILITATE

Q What is *DF*? Explain. **[*DF* = 1; ∠*ADF* is similar to ∠*AEG*.]**

Q What is the length of the hypotenuse of △*AEG*? **[$4\sqrt{5}$]**

Q What is the ratio of the length of the shortest leg of △*AEG* to the length of the hypotenuse of △*AEG*? **[The ratio is $\frac{\sqrt{5}}{5}$.]**

ANSWER See Solve It in Answers on next page.

CONNECT THE MATH In the Solve It, students find the ratio of the short leg to the hypotenuse in three triangles. In this lesson, students learn the ratios of the side lengths to the hypotenuse and to each other.

2 Guided Instruction

Take Note

Some students have difficulty correctly identifying the opposite and adjacent sides for a given angle. Provide practice for students by drawing right triangles in different orientations and using different combinations of variables to represent the length of each side.

8-3 Preparing to Teach

BIG ideas **Reasoning and Proof**
Measurement

ESSENTIAL UNDERSTANDING
- If certain combinations of side lengths and angle measures of a right triangle are known, ratios can be used to find other side lengths and angle measures.

Math Background

From the study of similar triangles and right triangles, students learn that the lengths of corresponding sides in similar right triangles have constant ratios.

Right triangles and special right triangles lead into the study of right triangle trigonometry. Right triangle trigonometry is the foundation for the study of unit circle trigonometry. When using right triangles, as in this lesson, only the ratios of angles measuring between zero and ninety degrees can be considered. Once unit circle trigonometry is introduced, ratios of angles with any real number measure can be considered.

Three additional right triangle trigonometric ratios exist but are not covered in this lesson. Those ratios are cotangent (the reciprocal of the tangent ratio), secant (the reciprocal of the cosine ratio), and cosecant (the reciprocal of the sine ratio).

Mathematical Practices

Attend to precision. Students will define sine, cosine, and tangent, as well as compute their values in right triangles.

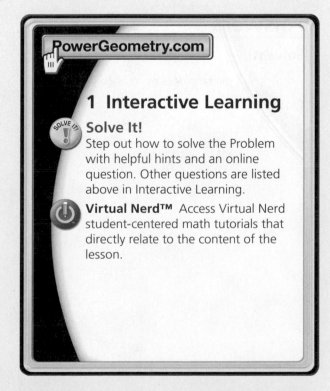

PowerGeometry.com

1 Interactive Learning

Solve It!
Step out how to solve the Problem with helpful hints and an online question. Other questions are listed above in Interactive Learning.

Virtual Nerd™ Access Virtual Nerd student-centered math tutorials that directly relate to the content of the lesson.

Problem 1

Q How can you identify the hypotenuse of a right triangle? **[It is the side opposite the right angle.]**

Q How can you identify the adjacent leg of an angle? **[It is the side that is one of the rays of the angle, but is not the hypotenuse.]**

Q If in a new right triangle $\triangle ABC$, $\sin A = \frac{8}{17}$, what do you know about $\triangle ABC$ and $\triangle TGR$? **[They are similar triangles.]**

Got It?

ERROR PREVENTION

Q How is tan *T* related to tan *G*? **[They are reciprocals.]**

Q How is cos *T* related to sin *G*? **[They are the same.]**

Q How is sin *T* related to cos *G*? **[They are the same.]**

Problem 2

Q In relation to the 5° angle, what do the sides in the diagram represent? **[The side representing the 150-ft drop is the adjacent side and the side representing the distance from the base of the tower is the opposite side.]**

Q Which trigonometric ratio involves the lengths of the adjacent and opposite sides? **[tan]**

Q What is the measure of the angle formed by the ground and the tower? Explain. **[85; 180 − 90 − 5 = 85.]**

Q What trigonometric ratio of the 85° angle could you use to determine *x*? **[tan]**

You can abbreviate the ratios as

$$\sin A = \frac{\text{opposite}}{\text{hypotenuse}},\ \cos A = \frac{\text{adjacent}}{\text{hypotenuse}},\ \text{and } \tan A = \frac{\text{opposite}}{\text{adjacent}}.$$

Problem 1 Writing Trigonometric Ratios

Think

How do the sides relate to $\angle T$?
$\overline{GR}$ is across from, or *opposite*, $\angle T$. $\overline{TR}$ is next to, or *adjacent* to, $\angle T$. $\overline{TG}$ is the *hypotenuse* because it is opposite the 90° angle.

What are the sine, cosine, and tangent ratios for $\angle T$?

$$\sin T = \frac{\text{opposite}}{\text{hypotenuse}} = \frac{8}{17}$$

$$\cos T = \frac{\text{adjacent}}{\text{hypotenuse}} = \frac{15}{17}$$

$$\tan T = \frac{\text{opposite}}{\text{adjacent}} = \frac{8}{15}$$

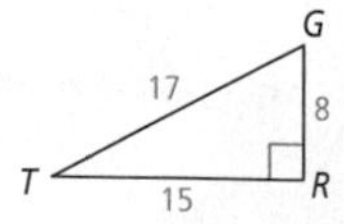

Got It? 1. Use the triangle in Problem 1. What are the sine, cosine, and tangent ratios for $\angle G$?

In Chapter 7, you used similar triangles to measure distances indirectly. You can also use trigonometry for indirect measurement.

Problem 2 Using a Trigonometric Ratio to Find Distance

Landmarks **In 1990, the Leaning Tower of Pisa was closed to the public due to safety concerns. The tower reopened in 2001 after a 10-year project to reduce its tilt from vertical. Engineers' efforts were successful and resulted in a tilt of 5°, reduced from 5.5°. Suppose someone drops an object from the tower at a height of 150 ft. How far from the base of the tower will the object land? Round to the nearest foot.**

Plan

What is the first step?
Look at the triangle and determine how the sides of the triangle relate to the given angle.

The given side is adjacent to the given angle. The side you want to find is opposite the given angle.

$\tan 5° = \frac{x}{150}$	Use the tangent ratio.
$x = 150(\tan 5°)$	Multiply each side by 150.
150 tan 5 enter	Use a calculator.
$x \approx 13.12329953$	

The object will land about 13 ft from the base of the tower.

Answers

Solve It!

$\frac{\sqrt{5}}{5}, \frac{\sqrt{5}}{5}, \frac{\sqrt{5}}{5}$; the ratio does not change for similar △.

Got It?

1. $\frac{15}{17}, \frac{8}{17}, \frac{15}{8}$

PowerGeometry.com

2 Guided Instruction

Each Problem is worked out and supported online.

Problem 1
Writing Trigonometric Ratios
Animated

Problem 2
Using a Trigonometric Ratio to Find Distance
Animated

Problem 3
Using Inverses
Animated

Support in Geometry Companion
- Vocabulary
- Key Concepts
- Got It?

Got It? 2. For parts (a)–(c), find the value of w to the nearest tenth.

a. 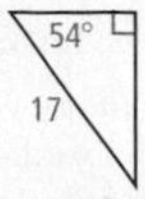

b. 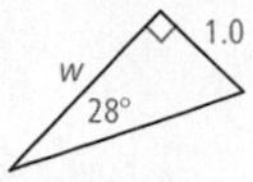

c. 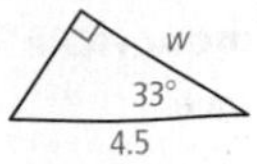

d. A section of Filbert Street in San Francisco rises at an angle of about 17°. If you walk 150 ft up this section, what is your vertical rise? Round to the nearest foot.

If you know the sine, cosine, or tangent ratio for an angle, you can use an inverse ($\sin^{-1}$, $\cos^{-1}$, or $\tan^{-1}$) to find the measure of the angle.

Problem 3 Using Inverses

What is $m\angle X$ to the nearest degree?

A 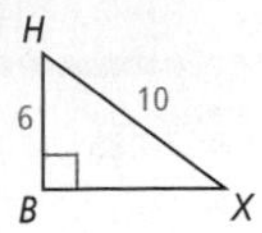

You know the lengths of the hypotenuse and the side opposite $\angle X$.

Use the sine ratio.

B 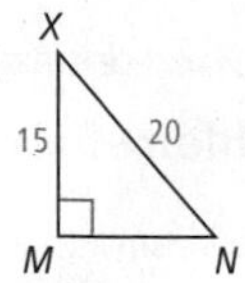

You know the lengths of the hypotenuse and the side adjacent to $\angle X$.

Use the cosine ratio.

A		B
$\sin X = \frac{6}{10}$	Write the ratio.	$\cos X = \frac{15}{20}$
$m\angle X = \sin^{-1}\left(\frac{6}{10}\right)$	Use the inverse.	$m\angle X = \cos^{-1}\left(\frac{15}{20}\right)$
sin-1 6 ÷ 10 enter	Use a calculator.	cos-1 15 ÷ 20 enter
$m\angle X \approx 36.86989765$		$m\angle X \approx 41.40962211$
≈ 37		≈ 41

Think

When should you use an inverse?
Use an inverse when you know two side lengths of a right triangle and you want to find the measure of one of the acute angles.

Got It? 3. a. Use the figure at the right. What is $m\angle Y$ to the nearest degree?

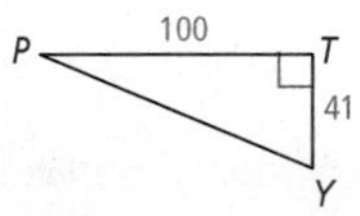

b. Reasoning Suppose you know the lengths of all three sides of a right triangle. Does it matter which trigonometric ratio you use to find the measure of any of the three angles? Explain.

Got It? VISUAL LEARNERS

Encourage students to sketch a diagram of 2d before solving the problem. Make sure that they label the hypotenuse of the triangle 150 ft.

Problem 3

Q How could you use the tangent ratio to determine $m\angle X$? **[You could first determine the third side of the triangle using the Pythagorean Theorem. Then you can compute the tangent ratio.]**

Q How could you use $\angle H$ to determine $m\angle X$? **[You could first determine $m\angle H$ using the cosine ratio, and then use the Triangle Angle-Sum Theorem to determine $m\angle X$.]**

Got It? ERROR PREVENTION

If students do not get the correct angle measurements using their calculators, make certain that their calculators are in degree mode rather than in radian mode. Explain to students that radians are another unit used to measure angles.

Additional Problems

1. What are the sine, cosine, and tangent ratios for $\angle P$?

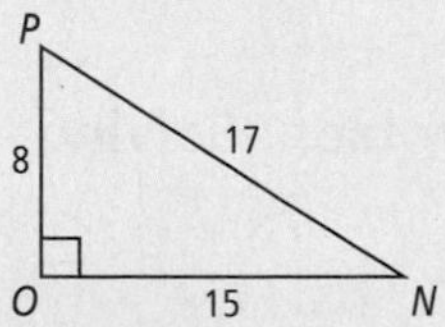

ANSWER $\sin P = \frac{15}{17}$, $\cos P = \frac{8}{17}$, $\tan P = \frac{15}{8}$

2. A cliff forms an angle of 84° with a lake. If you stand on the edge of a cliff and drop a rock at a height of 48 ft above the water, how far from the cliff will the rock strike the water? Round to the nearest whole foot.

ANSWER about 5 ft

3. What is $m\angle T$ to the nearest degree?

a. 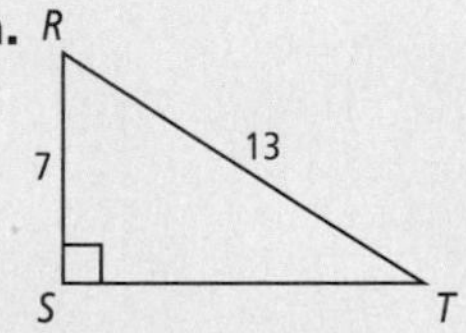

b. 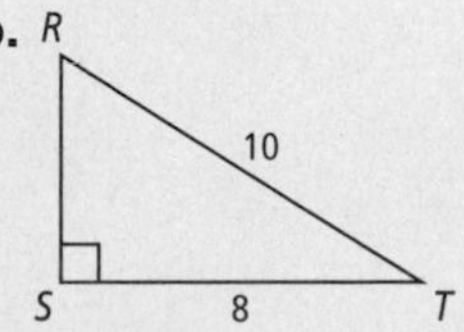

ANSWER a. about 33° **b.** about 37°

Answers

Got It? (continued)

2a. 13.8
b. 1.9
c. 3.8
d. 44 ft
3a. 68
b. No; you can use any of the three trigonometric ratios as long as you identify the appropriate leg that is opp. or adj. to each acute ∠.

3 Lesson Check

Do you know HOW?

- If students have difficulty with Exercises 1-6, then have them review Problem 1. They should write the ratio using the words *opposite, adjacent*, and *hypotenuse* for each before substituting the values for *a*, *b*, and *c*.

Do you UNDERSTAND?

- If students have difficulty with Exercise 10, then have them refer to the triangles used in the Solve It.

Close

Q How could you determine the value of sin 35° without using a calculator? **[Draw a right triangle with one acute angle of 35°. Measure the length of the opposite side and the hypotenuse. Then find the ratio of the length of the opposite side to the length of the hypotenuse.]**

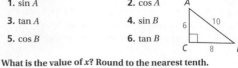

Lesson Check

Do you know HOW?

Write each ratio.

1. sin *A* **2.** cos *A*

3. tan *A* **4.** sin *B*

5. cos *B* **6.** tan *B*

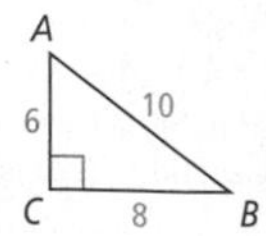

What is the value of *x*? Round to the nearest tenth.

7.

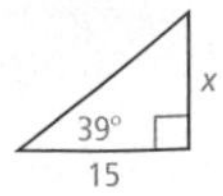

8.

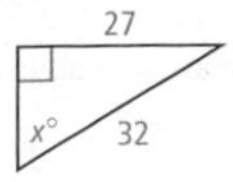

Do you UNDERSTAND? MATHEMATICAL PRACTICES

9. Vocabulary Some people use SOH-CAH-TOA to remember the trigonometric ratios for sine, cosine, and tangent. Why do you think that word might help? (*Hint*: Think of the first letters of the ratios.)

10. Error Analysis A student states that sin *A* > sin *X* because the lengths of the sides of △*ABC* are greater than the lengths of the sides of △*XYZ*. What is the student's error? Explain.

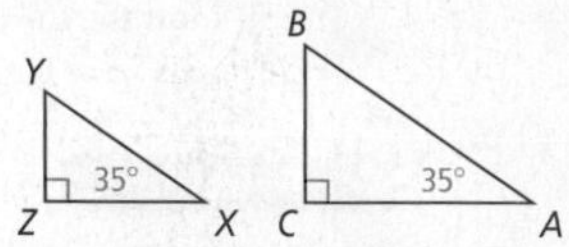

Practice and Problem-Solving Exercises MATHEMATICAL PRACTICES

A Practice

Write the ratios for sin *M*, cos *M*, and tan *M*. See Problem 1.

11.

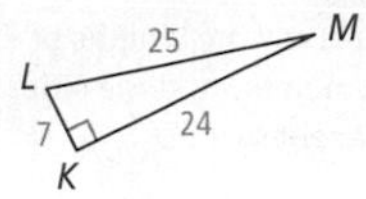

12.

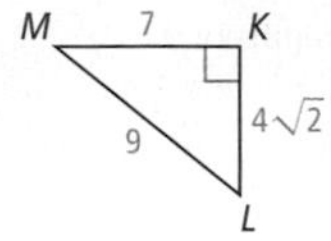

13.

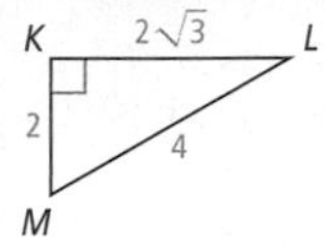

Find the value of *x*. Round to the nearest tenth. See Problem 2.

14.

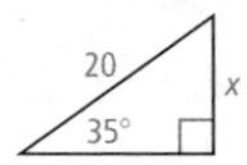

15.

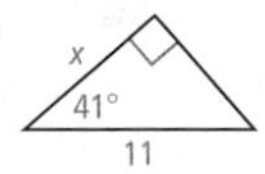

16.

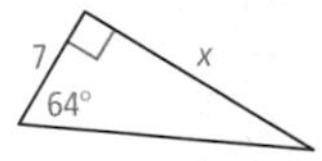

17.

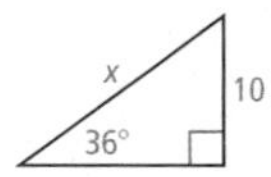

18.

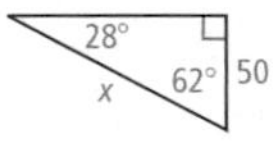

19.

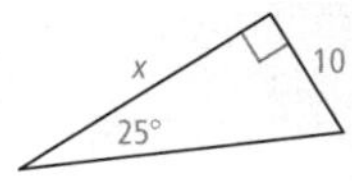

20. Recreation A skateboarding ramp is 12 in. high and rises at an angle of 17°. How long is the base of the ramp? Round to the nearest inch.

21. Public Transportation An escalator in the subway station has a vertical rise of 195 ft 9.5 in., and rises at an angle of 10.4°. How long is the escalator? Round to the nearest foot.

PowerGeometry.com

3 Lesson Check

For a digital lesson check, use the Got It questions.

Support in Geometry Companion

- Lesson Check

4 Practice

Assign homework to individual students or to an entire class.

Answers

Lesson Check

1. $\frac{8}{10}$ or $\frac{4}{5}$

2. $\frac{6}{10}$ or $\frac{3}{5}$

3. $\frac{8}{6}$ or $\frac{4}{3}$

4. $\frac{6}{10}$ or $\frac{3}{5}$

5. $\frac{8}{10}$ or $\frac{4}{5}$

6. $\frac{6}{8}$ or $\frac{3}{4}$

7. 12.1

8. 57.5

9. The word is made up of the first letters of each ratio: $S = \frac{O}{H}$, $C = \frac{A}{H}$, and $T = \frac{O}{A}$.

10. No; $\sin X = \frac{YZ}{YX}$, $\sin A = \frac{BC}{BA}$, and $\triangle XYZ \sim \triangle ABC$ by AA ~, so $\frac{YZ}{YX} = \frac{BC}{BA}$ because corresp. sides of ~ ▲s are proportional. Therefore, $\sin X = \sin A$.

Practice and Problem-Solving Exercises

11. $\frac{7}{25}$, $\frac{24}{25}$, $\frac{7}{24}$

12. $\frac{4\sqrt{2}}{9}$, $\frac{7}{9}$, $\frac{4\sqrt{2}}{7}$

13. $\frac{\sqrt{3}}{2}$, $\frac{1}{2}$, $\sqrt{3}$

14. 11.5

15. 8.3

16. 14.4

17. 17.0

18. 106.5

19. 21.4

20. 39 in.

21. 1085 ft

Find the value of x. Round to the nearest degree.

See Problem 3.

22. 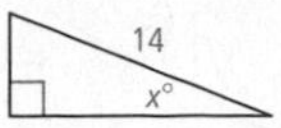

23. 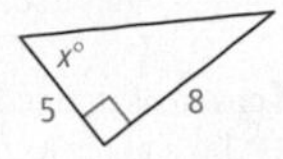

24. 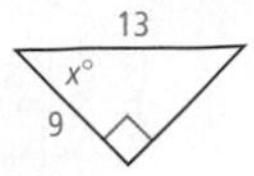

25. 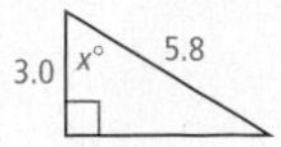

26. 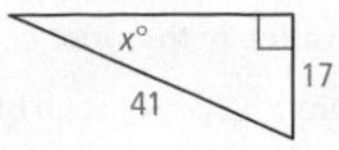

27. 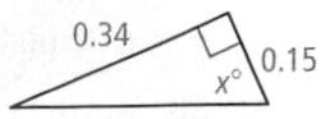

B Apply

28. The lengths of the diagonals of a rhombus are 2 in. and 5 in. Find the measures of the angles of the rhombus to the nearest degree.

29. **Think About a Plan** Carlos plans to build a grain bin with a radius of 15 ft. The recommended slant of the roof is 25°. He wants the roof to overhang the edge of the bin by 1 ft. What should the length x be? Give your answer in feet and inches.

- What is the position of the side of length x in relation to the given angle?
- What information do you need to find a side length of a right triangle?
- Which trigonometric ratio could you use?

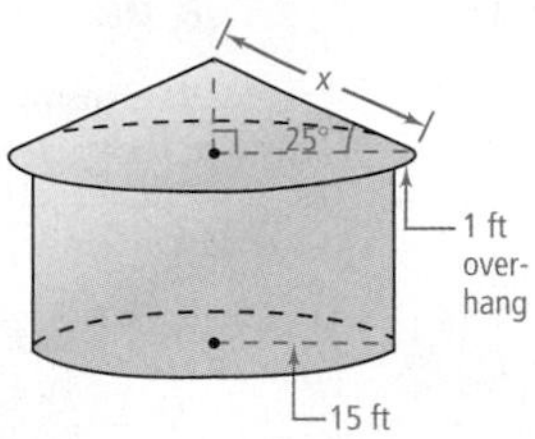

An *identity* is an equation that is true for all the allowed values of the variable. Use what you know about trigonometric ratios to show that each equation is an identity.

30. $\tan X = \frac{\sin X}{\cos X}$

31. $\sin X = \cos X \cdot \tan X$

32. $\cos X = \frac{\sin X}{\tan X}$

Find the values of w and then x. Round lengths to the nearest tenth and angle measures to the nearest degree.

33. 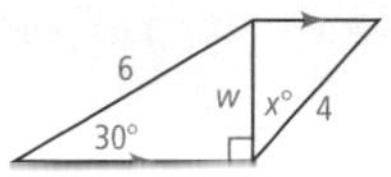

34. 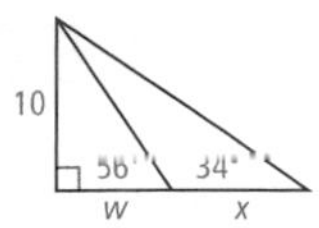

35. 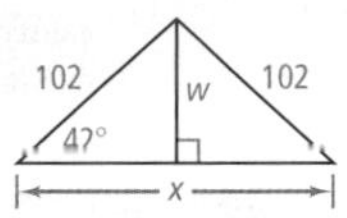

STEM 36. **Pyramids** All but two of the pyramids built by the ancient Egyptians have faces inclined at 52° angles. Suppose an archaeologist discovers the ruins of a pyramid. Most of the pyramid has eroded, but the archaeologist is able to determine that the length of a side of the square base is 82 m. How tall was the pyramid, assuming its faces were inclined at 52°? Round your answer to the nearest meter.

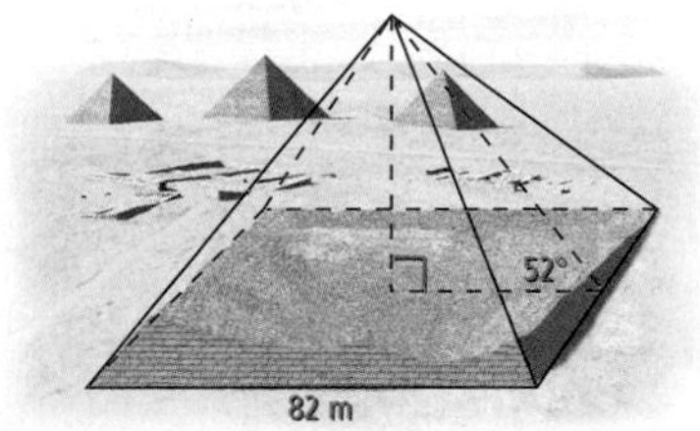

4 Practice

ASSIGNMENT GUIDE

Basic: 11–27, 29, 35, 39–42
Average: 11–27 odd, 28–47
Advanced: 11–27 odd, 28–53

Mathematical Practices are supported by exercises with red headings. Here are the Practices supported in this lesson:

MP 1: Make Sense of Problems Ex. 29
MP 3: Construct Arguments Ex. 47
MP 3: Communicate Ex. 37b, 39
MP 3: Critique the Reasoning of Others Ex. 10

Applications exercises have blue headings. Exercises 20, 21, 36, and 53 support MP 4: Model.

STEM exercises focus on science or engineering applications.

EXERCISE 35: Use the Think About a Plan worksheet in the **Practice and Problem Solving Workbook** (also available in the Teaching Resources in print and online) to further support students' development in becoming independent learners.

HOMEWORK QUICK CHECK

To check students' understanding of key skills and concepts, go over Exercises 17, 25, 29, 35, and 39.

22. 21 **23.** 58
24. 46 **25.** 59
26. 24 **27.** 66
28. 44 and 136
29. about 17 ft 8 in.
30. $\frac{\sin X}{\cos X} = \sin X \cdot \frac{1}{\cos X} = \frac{\text{opposite}}{\text{hypotenuse}} \cdot \frac{\text{hypotenuse}}{\text{adjacent}} = \frac{\text{opposite}}{\text{adjacent}} = \tan X$
31. $\cos X \cdot \tan X = \frac{\text{adjacent}}{\text{hypotenuse}} \cdot \frac{\text{opposite}}{\text{adjacent}} = \frac{\text{opposite}}{\text{hypotenuse}} = \sin X$
32. $\sin X \cdot \frac{1}{\tan X} = \frac{\text{opposite}}{\text{hypotenuse}} \cdot \frac{\text{adjacent}}{\text{opposite}} = \frac{\text{adjacent}}{\text{hypotenuse}} = \cos X$
33. $w = 3$, $x \approx 41$
34. $w \approx 6.7$, $x \approx 8.1$
35. $w \approx 68.3$, $x \approx 151.6$
36. 52 m

Answers

Practice and Problem-Solving Exercises (continued)

37a. They are equal; yes; sine and cosine of compl. ∠s are =.

b. ∠B; ∠A

c. Sample: The cosine is the complement's sine.

38. Answers may vary. Samples are given.

a. $\sin A = \frac{\text{opposite}}{\text{hypotenuse}}$, and the hypotenuse of a right △ is always the longest side, so $\sin A$ is a proper fraction, and $\sin A < 1$.

b. $\cos A = \frac{\text{adjacent}}{\text{hypotenuse}}$, and the hypotenuse of a rt. △ is always the longest side, so $\cos A$ is a proper fraction, and $\cos A < 1$.

39a.

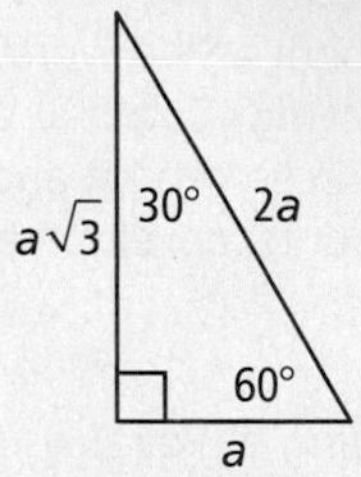

Using the ratio of sides $1 : \sqrt{3} : 2$ for a 30°-60°-90° △, $\tan 60° = \frac{\sqrt{3}}{1} = \sqrt{3}$.

b. Answers may vary. Sample: $\sin 60° = \sqrt{3} \cdot \cos 60°$

40. $\frac{15}{12}$ or $\frac{5}{4}$

41. $\frac{15}{9}$ or $\frac{5}{3}$

42. $\frac{9}{12}$ or $\frac{3}{4}$

43. $\frac{15}{9}$ or $\frac{5}{3}$

44. $\frac{15}{12}$ or $\frac{5}{4}$

45. $\frac{12}{9}$ or $\frac{4}{3}$

46a. 0.99985

b. 1

c. 1; 89.9; yes, $\sin X = 1$ and is not < 1.

d. For ∠s with measures that approach 90, the opposite side and hypotenuse are almost the same length, and $\sin X$ approaches 1.

47a. No; answers may vary. Sample: $\tan 45° + \tan 30° = 1 + \frac{\sqrt{3}}{3} \approx 1.6$, but $\tan 75° \approx 3.7$.

b. No; assume $\tan A - \tan B = \tan(A - B)$; $\tan A = \tan B + \tan(A - B)$ by the Add. Prop. of =; let $A = B + C$, then $\tan(B + C) = \tan B + \tan C$ by the Subst. Prop.; part (a) proved this false; this contradicts the assumption, so $\tan A - \tan B \neq \tan(A - B)$.

48. $(\sin A)^2 + (\cos A)^2 = \left(\frac{a}{c}\right)^2 + \left(\frac{b}{c}\right)^2 =$
$\frac{a^2}{c^2} + \frac{b^2}{c^2} = \frac{a^2 + b^2}{c^2} = \frac{c^2}{c^2} = 1$

---

37. a. In △ABC at the right, how does sin A compare to cos B? Is this true for the acute angles of other right triangles?

b. Reading Math The word cosine is derived from the words *complement's sine.* Which angle in △ABC is the complement of ∠A? Of ∠B?

c. Explain why the derivation of the word cosine makes sense.

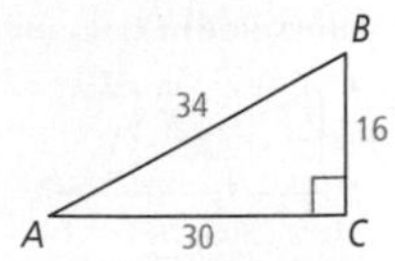

38. Proof For right △ABC with right ∠C, prove each of the following.

a. $\sin A < 1$

b. $\cos A < 1$

39. a. Writing Explain why $\tan 60° = \sqrt{3}$. Include a diagram with your explanation.

b. Make a Conjecture How are the sine and cosine of a 60° angle related? Explain.

The sine, cosine, and tangent ratios each have a reciprocal ratio. The reciprocal ratios are cosecant (csc), secant (sec), and cotangent (cot). Use △ABC and the definitions below to write each ratio.

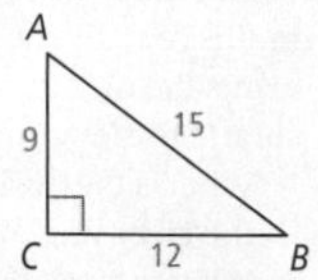

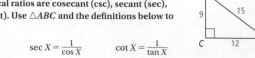

$\csc X = \frac{1}{\sin X}$ $\sec X = \frac{1}{\cos X}$ $\cot X = \frac{1}{\tan X}$

40. csc A **41.** sec A **42.** cot A

43. csc B **44.** sec B **45.** cot B

46. Graphing Calculator Use the table feature of your graphing calculator to study sin X as X gets close to (but not equal to) 90. In the y= screen, enter Y1 = sin X.

a. Use the tblset feature so that X starts at 80 and changes by 1. Access the table. From the table, what is sin X for X = 89?

b. Perform a "numerical zoom-in." Use the tblset feature, so that X starts with 89 and changes by 0.1. What is sin X for X = 89.9?

c. Continue to zoom-in numerically on values close to 90. What is the greatest value you can get for sin X on your calculator? How close is X to 90? Does your result contradict what you are asked to prove in Exercise 38a?

d. Use right triangles to explain the behavior of sin X found above.

47. a. Reasoning Does $\tan A + \tan B = \tan(A + B)$ when $A + B < 90$? Explain.

b. Does $\tan A - \tan B = \tan(A - B)$ when $A - B > 0$? Use part (a) and indirect reasoning to explain.

Challenge **Verify that each equation is an identity by showing that each expression on the left simplifies to 1.**

48. $(\sin A)^2 + (\cos A)^2 = 1$ **49.** $(\sin B)^2 + (\cos B)^2 = 1$

50. $\frac{1}{(\cos A)^2} - (\tan A)^2 = 1$ **51.** $\frac{1}{(\sin A)^2} - \frac{1}{(\tan A)^2} = 1$

52. Show that $(\tan A)^2 - (\sin A)^2 = (\tan A)^2 \cdot (\sin A)^2$ is an identity.

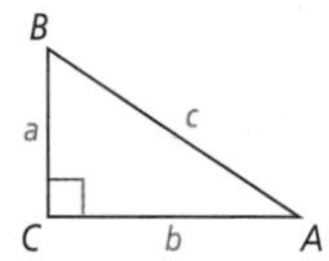

---

49. $(\sin B)^2 + (\cos B)^2 =$
$\left(\frac{b}{c}\right)^2 + \left(\frac{a}{c}\right)^2 =$
$\frac{b^2}{c^2} + \frac{a^2}{c^2} =$
$\frac{b^2 + a^2}{c^2} = \frac{c^2}{c^2} = 1$

50. $\frac{1}{(\cos A)^2} - (\tan A)^2 =$
$\left(1 \div \frac{b^2}{c^2}\right) - \frac{a^2}{b^2} = \frac{c^2}{b^2} - \frac{a^2}{b^2} =$
$\frac{c^2 - a^2}{b^2} = \frac{b^2}{b^2} = 1$

51. $\frac{1}{(\sin A)^2} - \frac{1}{(\tan A)^2} = \frac{1}{\left(\frac{a}{c}\right)^2} - \frac{1}{\left(\frac{a}{b}\right)^2} =$
$\frac{c^2}{a^2} - \frac{b^2}{a^2} =$
$\frac{c^2 - b^2}{a^2} = \frac{a^2}{a^2} = 1$

52. $(\tan A)^2 - (\sin A)^2 =$
$\left(\frac{a}{b}\right)^2 - \left(\frac{a}{c}\right)^2 = \frac{a^2}{b^2} - \frac{a^2}{c^2} =$
$\frac{a^2 c^2}{b^2 c^2} - \frac{a^2 b^2}{b^2 c^2} = \frac{a^2 c^2 - a^2 b^2}{b^2 c^2} =$
$\frac{a^2(c^2 - b^2)}{b^2 c^2} =$
$\frac{a^2 \cdot a^2}{b^2 c^2} = \left(\frac{a}{b}\right)^2 \left(\frac{a}{c}\right)^2 =$
$(\tan A)^2 (\sin A)^2$

STEM 53. Astronomy The Polish astronomer Nicolaus Copernicus devised a method for determining the sizes of the orbits of planets farther from the sun than Earth. His method involved noting the number of days between the times that a planet was in the positions labeled A and B in the diagram. Using this time and the number of days in each planet's year, he calculated c and d.

a. For Mars, $c = 55.2$ and $d = 103.8$. How far is Mars from the sun in astronomical units (AU)? One astronomical unit is defined as the average distance from Earth to the center of the sun, about 93 million miles.

b. For Jupiter, $c = 21.9$ and $d = 100.8$. How far is Jupiter from the sun in astronomical units?

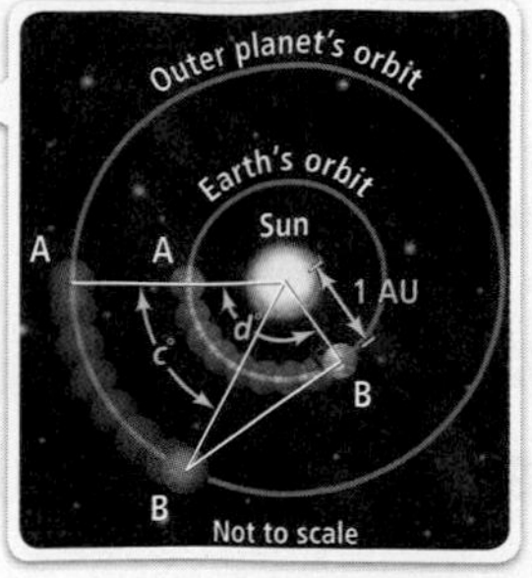

53a. 1.5 AU
b. 5.2 AU

Apply What You've Learned

MATHEMATICAL PRACTICES MP 1

Look back at the information on page 489 about the fire in a state forest. The diagram is shown again below.

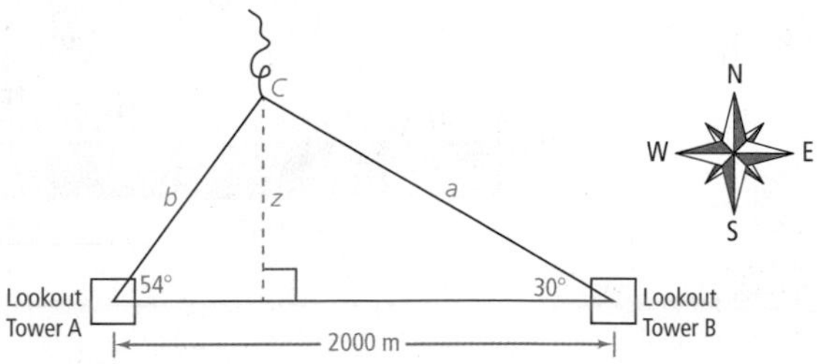

Select all of the following that are true. Explain your reasoning.

A. $\sin 54° = \frac{z}{b}$

B. $\cos 30° = \frac{z}{2000}$

C. $\tan 30° = \frac{z}{2000}$

D. $\sin 30° = \frac{z}{a}$

E. $\tan 54° = \frac{z}{2000}$

F. $\cos 54° = \frac{z}{b}$

Apply What You've Learned

Here students use trigonometric ratios to relate unknown distances in the diagram from page 489. Later in the chapter, they will use the Law of Sines to find the distances a and b. If you choose not to cover Lessons 8-5 and 8-6, have students represent the parts of the 2000 m distance between the towers in terms of a variable x. Students can write additional trigonometric ratios using pairs of the distances a, b, z, x, and $2000 - x$.

Mathematical Practices

Students use trigonometric ratios to **make sense** of relationships in the right triangles in the diagram. (MP 1)

ANSWERS

Choices A and D are both true.

8-3 Lesson Resources

Differentiated Remediation

Instructional Support

Geometry Companion

Students can use the **Geometry Companion** worktext (4 pages) . . .

- New Vocabulary
- Key Concepts
- Got It for each Problem
- Lesson Check

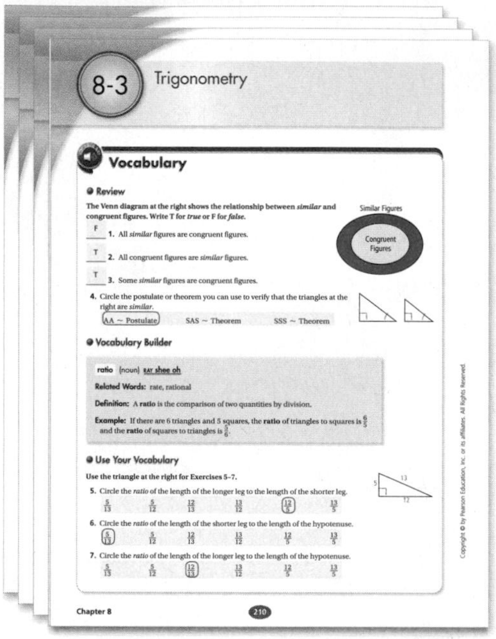
8-3 Trigonometry

Vocabulary

Review

The Venn diagram at the right shows the relationship between *similar* and congruent figures. Write T for *true* or F for *false*.

F 1. All *similar* figures are congruent figures.

T 2. All congruent figures are *similar* figures.

T 3. Some *similar* figures are congruent figures.

4. Circle the postulate or theorem you can use to verify that the triangles at the right are *similar*.

AA ~ Postulate SAS ~ Theorem SSS ~ Theorem

Similar Figures / Congruent Figures

Vocabulary Builder

ratio (noun) RAY shee oh

Related Words: rate, rational

Definition: A **ratio** is the comparison of two quantities by division.

Example: If there are 6 triangles and 5 squares, the **ratio** of triangles to squares is $\frac{6}{5}$ and the **ratio** of squares to triangles is $\frac{5}{6}$.

Use Your Vocabulary

Use the triangle at the right for Exercises 5–7.

5. Circle the *ratio* of the length of the longer leg to the length of the shorter leg.

$\frac{5}{13}$ $\frac{5}{12}$ $\frac{12}{13}$ $\frac{13}{12}$ $\frac{12}{5}$ $\frac{13}{5}$

6. Circle the *ratio* of the length of the shorter leg to the length of the hypotenuse.

$\frac{5}{13}$ $\frac{5}{12}$ $\frac{12}{13}$ $\frac{13}{12}$ $\frac{12}{5}$ $\frac{13}{5}$

7. Circle the *ratio* of the length of the longer leg to the length of the hypotenuse.

$\frac{5}{13}$ $\frac{5}{12}$ $\frac{12}{13}$ $\frac{13}{12}$ $\frac{12}{5}$ $\frac{13}{5}$

Copyright © by Pearson Education, Inc. or its affiliates. All Rights Reserved.

Chapter 8 210

ELL Support

Focus on Language Display a word wall with vocabulary words and other key words and pictures. For example, write *isosceles right triangle* and *ratio*, with their definitions from the chapter and a picture of each. It may help students if pictures and words are colored by topic. For example, triangles may be one color and quadrilaterals another.

Use Manipulatives A bike ramp has a height of 30 in. and a base length of 48 in. Draw a picture to model the situation. Calculate the angle of elevation of the ramp.

5 Assess & Remediate

Lesson Quiz

1. What are the sine, cosine, and tangent ratios for $\angle B$?

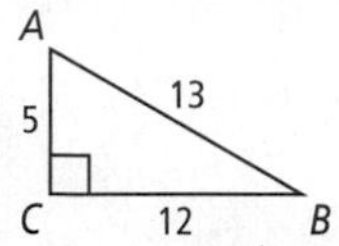

2. What is the value of x? Round to the nearest tenth.

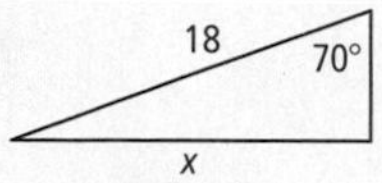

3. What is $m\angle X$ to the nearest degree?

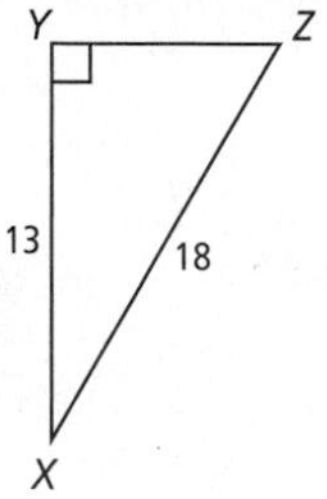

4. **Do you UNDERSTAND?** Can a sine be greater than 1? Explain.

ANSWERS TO LESSON QUIZ

1. $\sin B = \frac{5}{13}$, $\cos B = \frac{12}{13}$, $\tan B = \frac{5}{12}$
2. 16.9
3. 44
4. No; a leg of a right triangle cannot be longer than the hypotenuse.

PRESCRIPTION FOR REMEDIATION

Use the student work on the Lesson Quiz to prescribe a differentiated review assignment.

Points	Differentiated Remediation
0–2	Intervention
3	On-level
4	Extension

PowerGeometry.com

5 Assess & Remediate

Assign the Lesson Quiz. Appropriate intervention, practice, or enrichment is automatically generated based on student performance.

Intervention

- **Reteaching** (2 pages) Provides reteaching and practice exercises for the key lesson concepts. Use with struggling students or absent students.
- **English Language Learner Support** Helps students develop and reinforce mathematical vocabulary and key concepts.

All-in-One Resources/Online

Reteaching

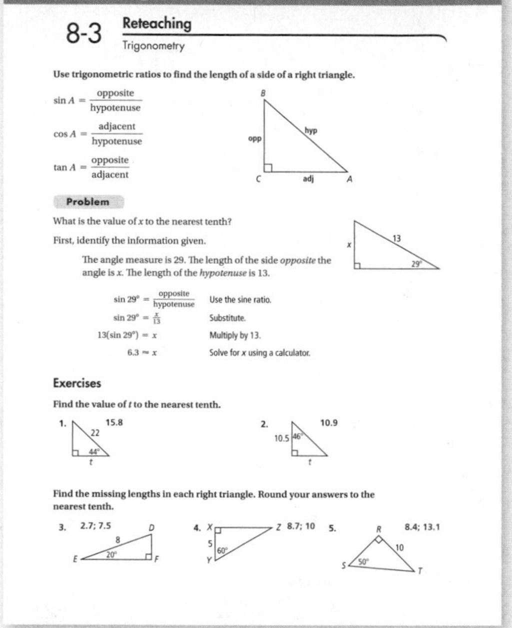
8-3 Reteaching

Trigonometry

Use trigonometric ratios to find the length of a side of a right triangle.

$\sin A = \frac{\text{opposite}}{\text{hypotenuse}}$

$\cos A = \frac{\text{adjacent}}{\text{hypotenuse}}$

$\tan A = \frac{\text{opposite}}{\text{adjacent}}$

B, C, A, opp, hyp, adj

Problem

What is the value of x to the nearest tenth?

First, identify the information given.

The angle measure is 29. The length of the side *opposite* the angle is x. The length of the *hypotenuse* is 13.

$\sin 29° = \frac{\text{opposite}}{\text{hypotenuse}}$ Use the sine ratio.

$\sin 29° = \frac{x}{13}$ Substitute.

$13(\sin 29°) = x$ Multiply by 13.

$6.3 \approx x$ Solve for x using a calculator.

Exercises

Find the value of t to the nearest tenth.

1. 15.8 (22, 44°, t)

2. 10.9 (10.5, 46°, t)

Find the missing lengths in each right triangle. Round your answers to the nearest tenth.

3. 2.7; 7.5 (D, 8, E, 20°, F)

4. 8.7; 10 (X, Z, 5, 60°, Y)

5. 8.4; 13.1 (R, 10, S, 50°, T)

All-in-One Resources/Online

English Language Learner Support

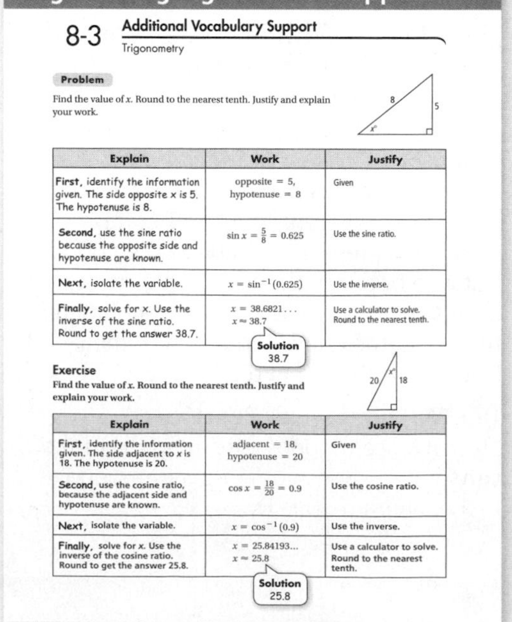
8-3 Additional Vocabulary Support

Trigonometry

Problem

Find the value of x. Round to the nearest tenth. Justify and explain your work.

(8, 5, x°)

Explain	Work	Justify
First, identify the information given. The side opposite x is 5. The hypotenuse is 8.	opposite = 5, hypotenuse = 8	Given
Second, use the sine ratio because the opposite side and hypotenuse are known.	$\sin x = \frac{5}{8} = 0.625$	Use the sine ratio.
Next, isolate the variable.	$x = \sin^{-1}(0.625)$	Use the inverse.
Finally, solve for x. Use the inverse of the sine ratio. Round to get the answer 38.7.	$x = 38.6821\ldots$ $x \approx 38.7$	Use a calculator to solve. Round to the nearest tenth.

Solution 38.7

Exercise

Find the value of x. Round to the nearest tenth. Justify and explain your work.

(20, 18, x°)

Explain	Work	Justify
First, identify the information given. The side adjacent to x is 18. The hypotenuse is 20.	adjacent = 18, hypotenuse = 20	Given
Second, use the cosine ratio, because the adjacent side and hypotenuse are known.	$\cos x = \frac{18}{20} = 0.9$	Use the cosine ratio.
Next, isolate the variable.	$x = \cos^{-1}(0.9)$	Use the inverse.
Finally, solve for x. Use the inverse of the cosine ratio. Round to get the answer 25.8.	$x = 25.84193\ldots$ $x \approx 25.8$	Use a calculator to solve. Round to the nearest tenth.

Solution 25.8

Differentiated Remediation *continued*

On-Level

- **Practice** (2 pages) Provides extra practice for each lesson. For simpler practice exercises, use the Form K Practice pages found in the All-in-One Teaching Resources and online.
- **Think About a Plan** Helps students develop specific problem-solving skills and strategies by providing scaffolded guiding questions.
- **Standardized Test Prep** Focuses on all major exercises, all major question types, and helps students prepare for the high-stakes assessments.

Extension

- **Enrichment** Provides students with interesting problems and activities that extend the concepts of the lesson.
- **Activities, Games, and Puzzles** Worksheets that can be used for concepts development, enrichment, and for fun!

Practice and Problem Solving Wkbk/ All-in-One Resources/Online

Practice page 1

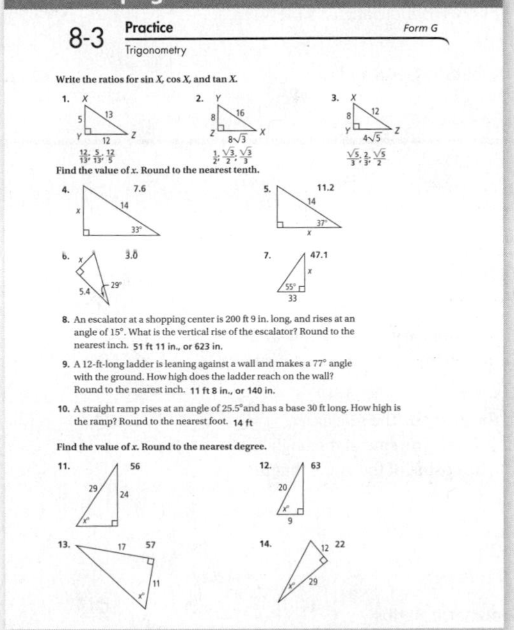

8-3 Practice — Trigonometry — Form G

Write the ratios for sin X, cos X, and tan X.

Find the value of x. Round to the nearest tenth.

4. 7.6 5. 11.2 6. 3.0 7. 47.1

8. An escalator at a shopping center is 200 ft 9 in. long, and rises at an angle of 15°. What is the vertical rise of the escalator? Round to the nearest inch. 51 ft 11 in., or 623 in.

9. A 12-ft-long ladder is leaning against a wall and makes a 77° angle with the ground. How high does the ladder reach on the wall? Round to the nearest inch. 11 ft 8 in., or 140 in.

10. A straight ramp rises at an angle of 25.5° and has a base 30 ft long. How high is the ramp? Round to the nearest foot. 14 ft

Find the value of x. Round to the nearest degree.

11. 56 12. 63 13. 57 14. 22

Practice and Problem Solving Wkbk/ All-in-One Resources/Online

Practice page 2

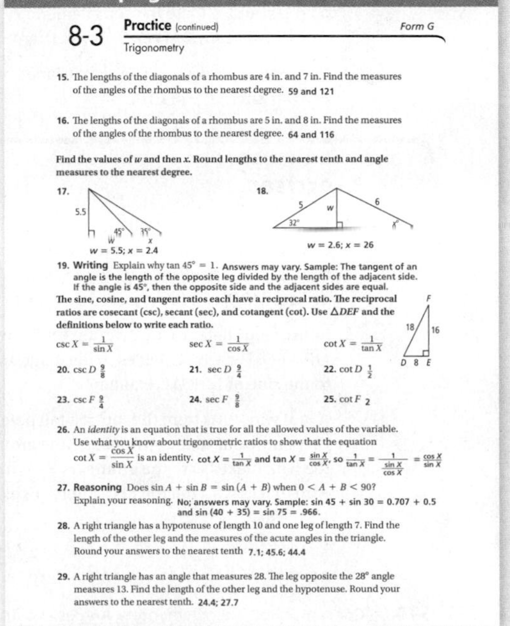

8-3 Practice (continued) — Trigonometry — Form G

15. The lengths of the diagonals of a rhombus are 4 in. and 7 in. Find the measures of the angles of the rhombus to the nearest degree. 59 and 121

16. The lengths of the diagonals of a rhombus are 5 in. and 8 in. Find the measures of the angles of the rhombus to the nearest degree. 64 and 116

Find the values of w and then x. Round lengths to the nearest tenth and angle measures to the nearest degree.

17. w = 5.5; x = 2.4 18. w = 2.6; x = 26

19. Writing Explain why tan 45° = 1. Answers may vary. Sample: The tangent of an angle is the length of the opposite leg divided by the length of the adjacent side. If the angle is 45°, then the opposite side and the adjacent sides are equal.

The sine, cosine, and tangent ratios each have a reciprocal ratio. The reciprocal ratios are cosecant (csc), secant (sec), and cotangent (cot). Use △DEF and the definitions below to write each ratio.

26. An identity is an equation that is true for all the allowed values of the variable. Use what you know about trigonometric ratios to show that the equation $\cot X = \frac{\cos X}{\sin X}$ is an identity.

27. Reasoning Does sin A + sin B = sin (A + B) when 0 < A + B < 90? Explain your reasoning. No; answers may vary. Sample: sin 45 + sin 30 = 0.707 + 0.5 and sin (40 + 35) = sin 75 = .966.

28. A right triangle has a hypotenuse of length 10 and one leg of length 7. Find the length of the other leg and the measures of the acute angles in the triangle. Round your answers to the nearest tenth. 7.1; 45.6; 44.4

29. A right triangle has an angle that measures 28. The leg opposite the 28° angle measures 13. Find the length of the other leg and the hypotenuse. Round your answers to the nearest tenth. 24.4; 27.7

All-in-One Resources/Online

Enrichment

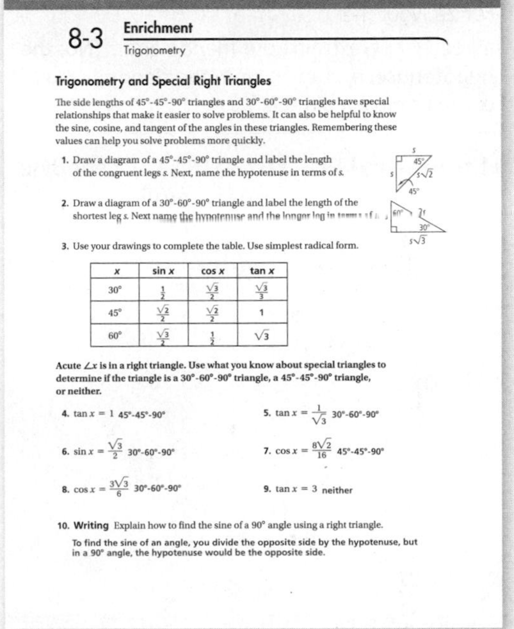

8-3 Enrichment — Trigonometry

Trigonometry and Special Right Triangles

The side lengths of 45°-45°-90° triangles and 30°-60°-90° triangles have special relationships that make it easier to solve problems. It can also be helpful to know the sine, cosine, and tangent of the angles in these triangles. Remembering these values can help you solve problems more quickly.

1. Draw a diagram of a 45°-45°-90° triangle and label the length of the congruent legs s. Next, name the hypotenuse in terms of s.

2. Draw a diagram of a 30°-60°-90° triangle and label the length of the shortest leg s.

3. Use your drawings to complete the table. Use simplest radical form.

x	sin x	cos x	tan x
30°	$\frac{1}{2}$	$\frac{\sqrt{3}}{2}$	$\frac{\sqrt{3}}{3}$
45°	$\frac{\sqrt{2}}{2}$	$\frac{\sqrt{2}}{2}$	1
60°	$\frac{\sqrt{3}}{2}$	$\frac{1}{2}$	$\sqrt{3}$

Acute ∠x is in a right triangle. Use what you know about special triangles to determine if the triangle is a 30°-60°-90° triangle, a 45°-45°-90° triangle, or neither.

4. tan x = 1 45°-45°-90° 5. $\tan x = \frac{1}{\sqrt{3}}$ 30°-60°-90°

6. $\sin x = \frac{\sqrt{3}}{2}$ 30°-60°-90° 7. $\cos x = \frac{8\sqrt{2}}{16}$ 45°-45°-90°

8. $\cos x = \frac{3\sqrt{3}}{6}$ 30°-60°-90° 9. tan x = 3 neither

10. Writing Explain how to find the sine of a 90° angle using a right triangle. To find the sine of an angle, you divide the opposite side by the hypotenuse, but in a 90° angle, the hypotenuse would be the opposite side.

Practice and Problem Solving Wkbk/ All-in-One Resources/Online

Think About a Plan

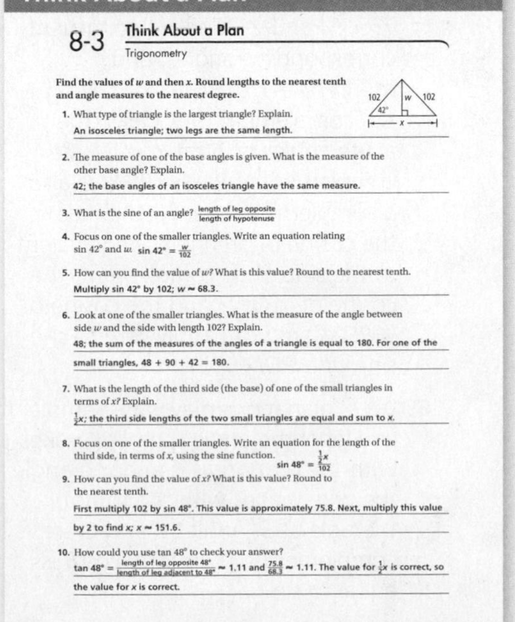

8-3 Think About a Plan — Trigonometry

Find the values of w and then x. Round lengths to the nearest tenth and angle measures to the nearest degree.

1. What type of triangle is the largest triangle? Explain. An isosceles triangle; two legs are the same length.

2. The measure of one of the base angles is given. What is the measure of the other base angle? Explain. 42; the base angles of an isosceles triangle have the same measure.

3. What is the sine of an angle? length of leg opposite / length of hypotenuse

4. Focus on one of the smaller triangles. Write an equation relating sin 42° and w. $\sin 42° = \frac{w}{102}$

5. How can you find the value of w? What is this value? Round to the nearest tenth. Multiply sin 42° by 102; w ≈ 68.3.

6. Look at one of the smaller triangles. What is the measure of the angle between side w and the side with length 102? Explain. 48; the sum of the measures of the angles of a triangle is equal to 180. For one of the small triangles, 48 + 90 + 42 = 180.

7. What is the length of the third side (the base) of one of the small triangles in terms of x? Explain. $\frac{1}{2}x$; the third side lengths of the two small triangles are equal and sum to x.

8. Focus on one of the smaller triangles. Write an equation for the length of the third side, in terms of x, using the sine function. $\sin 48° = \frac{\frac{1}{2}x}{102}$

9. How can you find the value of x? What is this value? Round to the nearest tenth. First multiply 102 by sin 48°. This value is approximately 75.8. Next, multiply this value by 2 to find x; x ≈ 151.6.

10. How could you use tan 48° to check your answer? tan 48° = length of leg opposite 48° / length of leg adjacent to 48° ≈ 1.11 and $\frac{75.8}{68.3}$ ≈ 1.11. The value for $\frac{1}{2}x$ is correct, so the value for x is correct.

Practice and Problem Solving Wkbk/ All-in-One Resources/Online

Standardized Test Prep

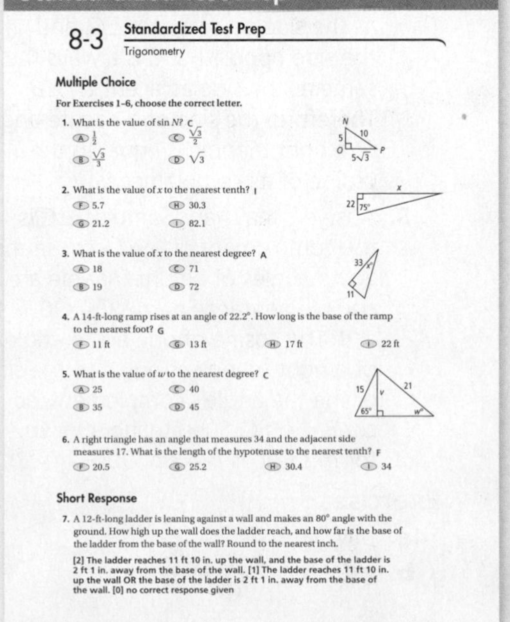

8-3 Standardized Test Prep — Trigonometry

Multiple Choice

For Exercises 1–6, choose the correct letter.

1. What is the value of sin N? C

2. What is the value of x to the nearest tenth? I (F) 5.7 (H) 30.3 (G) 21.2 (I) 82.1

3. What is the value of x to the nearest degree? A (A) 18 (C) 71 (B) 19 (D) 72

4. A 14-ft-long ramp rises at an angle of 22.2°. How long is the base of the ramp to the nearest foot? G (F) 11 ft (G) 13 ft (H) 17 ft (I) 22 ft

5. What is the value of w to the nearest degree? C (A) 25 (C) 40 (B) 35 (D) 45

6. A right triangle has an angle that measures 34 and the adjacent side measures 17. What is the length of the hypotenuse to the nearest tenth? F (F) 20.5 (G) 25.2 (H) 30.4 (I) 34

Short Response

7. A 12-ft-long ladder is leaning against a wall and makes an 80° angle with the ground. How high up the wall does the ladder reach, and how far is the base of the ladder from the base of the wall? Round to the nearest inch.

[2] The ladder reaches 11 ft 10 in. up the wall, and the base of the ladder is 2 ft 1 in. away from the base of the wall. [1] The ladder reaches 11 ft 10 in. up the wall OR the base of the ladder is 2 ft 1 in. away from the base of the wall. [0] no correct response given

Online Teacher Resource Center

Activities, Games, and Puzzles

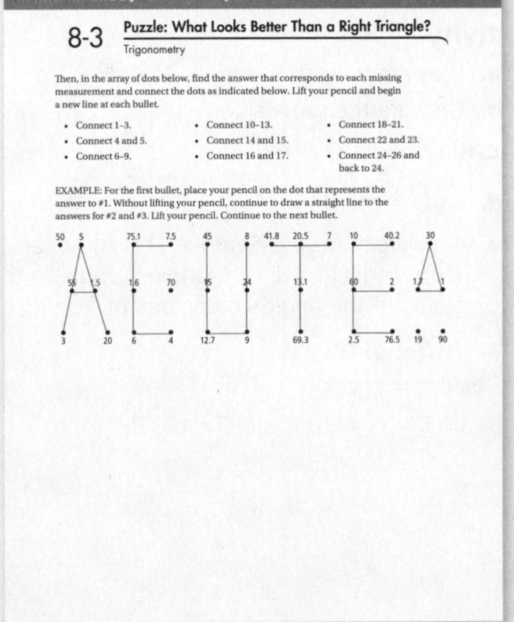

8-3 Puzzle: What Looks Better Than a Right Triangle? — Trigonometry

Then, in the array of dots below, find the answer that corresponds to each missing measurement and connect the dots as indicated below. Lift your pencil and begin a new line at each bullet.

- Connect 1–3.
- Connect 4 and 5.
- Connect 6–9.
- Connect 10–13.
- Connect 14 and 15.
- Connect 16 and 17.
- Connect 18–21.
- Connect 22 and 23.
- Connect 24–26 and back to 24.

EXAMPLE: For the first bullet, place your pencil on the dot that represents the answer to #1. Without lifting your pencil, continue to draw a straight line to the answers for #2 and #3. Lift your pencil. Continue to the next bullet.

Guided Instruction

PURPOSE To explain and use the relationship between the sine and cosine of complementary angles

PROCESS Students will

- discover and apply the relationship between the sine and cosine of complementary angles.

DISCUSS A right triangle has one right angle and two acute angles that are complementary, which means the sum of their measures is 90.

Have students determine the sine and cosine of the acute angles for several different right triangles.

Activity

Q For 1a, there are two sides of the triangle adjacent to $\angle R$. How do you determine which side is the adjacent side? **[The adjacent side is the leg that forms the right angle. The other side that is adjacent to $\angle R$ is the hypotenuse.]**

Q For 2a, your friend says that $\sin B = \frac{a}{b}$. Do you agree? If not, explain your friend's error. **[No; the hypotenuse, *a*, should be in the denominator. The correct ratio for sin *B* is $\frac{b}{a}$.]**

Mathematical Practices This Concept Byte supports students in becoming proficient in constructing arguments, Mathematical Practice 3.

Concept Byte

Use With Lesson 8-3

ACTIVITY

Complementary Angles and Trigonometric Ratios

Common Core State Standards

G-SRT.C.7 Explain and use the relationship between the sine and cosine of complementary angles.

MP 3

The acute angles of a right triangle are complementary because the sum of their measures is 90. There is a relationship between the sine and cosine of complementary angles.

Activity

What is the relationship between the sine and cosine of complementary angles?

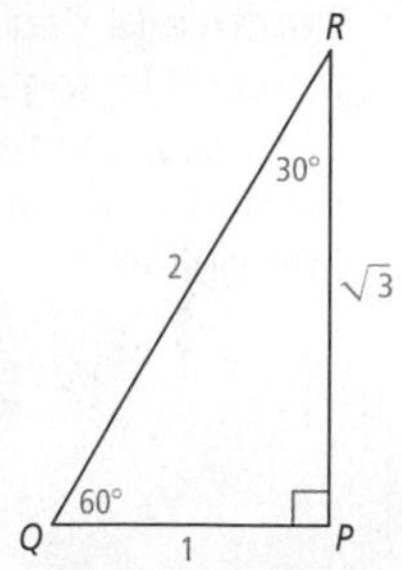

1. Refer to right triangle *PQR*, shown at the right.
  a. How is the side opposite $\angle R$ related to the side adjacent to $\angle Q$?
  b. How is the side adjacent to $\angle R$ related to the side opposite $\angle Q$?
  c. Compare sin *R* and cos *Q*.
  d. Compare sin *Q* and cos *R*.
  e. Make a conjecture based on your results from parts (c) and (d).

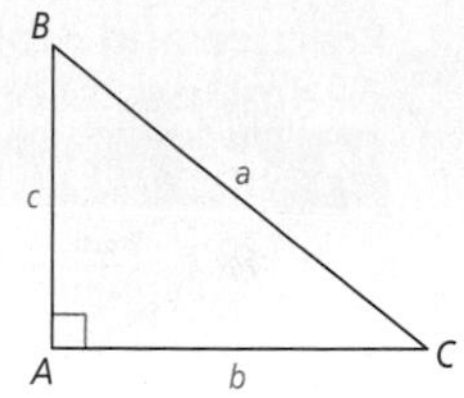

2. Refer to right triangle *ABC*, shown at the right.
  a. Write sine and cosine ratios for angles *B* and *C*. What do you notice about the relationship between sin *B* and cos *C*? Between sin *C* and cos *B*?
  b. If $\angle B$ and $\angle C$ are the complementary angles of a right triangle, is the statement $\sin B = \cos C$ always true? Explain.

3. Reasoning Prove that in a right triangle with acute angles *B* and *C*, $\cos B = \sin (90 - m\angle B)$.

Exercises

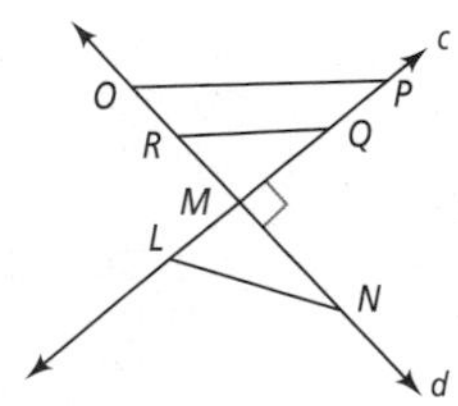

4. In the diagram at the right, line *c* is perpendicular to line *d*.
  a. In $\triangle LMN$, cos *L* is equal to the sine of which angle?
  b. In $\triangle MOP$, sin *P* is equal to the cosine of which angle?
  c. In $\triangle MRQ$, cos *Q* is equal to the sine of which angle?

5. Reasoning Refer to the diagram for Exercise 4. Suppose you also know that $\overline{OP} \parallel \overline{RQ}$. For which angles in the diagram is the cosine of the angle equivalent to the sine of $\angle POM$? Explain.

6. A taut wire runs from the top of a tall pole to the ground. The pole is perpendicular to the ground and the ground is level. The sine of the angle that the wire makes with the ground is $\frac{12}{13}$. What is the cosine of the angle that the wire makes with the top of the pole? Explain.

Answers

Activity

1a. They are the same side.
b. They are the same side.
c. $\sin R = \cos Q = \frac{1}{2}$
d. $\sin Q = \cos R = \frac{\sqrt{3}}{2}$
e. Answers may vary. Sample: The sine of an acute angle in a right triangle is equal to the cosine of the angle's complement.

2a. $\sin B = \frac{b}{a}$, $\cos B = \frac{c}{a}$, $\sin C = \frac{c}{a}$, $\cos C = \frac{b}{a}$; $\sin B = \cos C$ and $\sin C = \cos B$

b. Yes answers may vary. Sample: The side opposite $\angle B$ is always the same as the side adjacent to $\angle C$, and the side opposite $\angle C$ is always the same as the side adjacent to $\angle B$. Therefore, the sine of an acute angle of a right triangle is equal to the cosine of its complement.

3. Answers may vary. Sample: $\angle C$ is the complement of $\angle B$ because the acute angles of a right triangle are complementary. So, $m\angle C = 90 - m \angle B$. The cosine of one acute angle of a right triangle is equal to the sine of the the angle's complement, so $\cos B = \sin C$. Substituting $90 - m\angle B$ for $m\angle C$, $\cos B = \sin (90 - m\angle B)$.

Exercises

4a. $\angle N$
b. $\angle O$
c. $\angle R$

5. $\angle OPM$ and $\angle RQM$; answers may vary. Sample: Because $\overline{OP} \parallel \overline{RQ}$, $\angle OPM$ and $\angle RQM$ are congruent corresponding angles, and $m\angle OPM = m\angle RQM$. $\angle POM$ is a complement of $\angle OPM$ (the acute angles of right $\triangle POM$ are complementary), so $\angle POM$ is also a complement of $\angle RQM$. Since the cosine of an angle is equivalent to the sine of its complement, the cosine of $\angle OPM$ and the cosine of $\angle RQM$ are both equivalent to the sine of $\angle POM$.

6. $\frac{12}{13}$; explanations may vary. Sample: If you label the angle the wire makes with the ground as *C* and the angle the wire makes with the top of the pole as *B*, angles *B* and *C* are complements. Therefore, cos *B* is equal to sin *C*.

8 Mid-Chapter Quiz

MathXL® for School
Go to PowerGeometry.com

Do you know HOW?

Algebra Find the value of each variable. Express your answer in simplest radical form.

1.

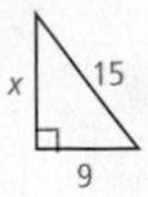

2.

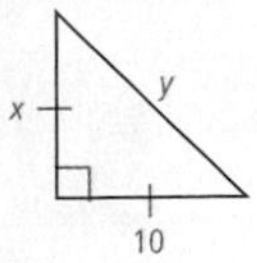

3.

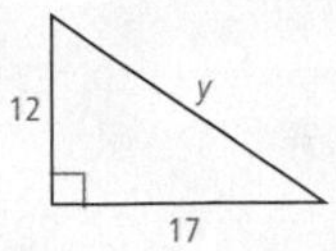

4.

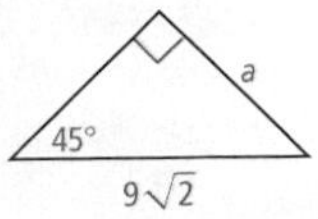

5.

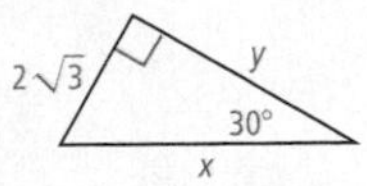

6.

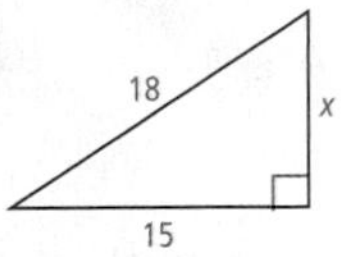

Given the following triangle side lengths, identify the triangle as *acute*, *right*, or *obtuse*.

7. 7, 8, 9

8. 15, 36, 39

9. 10, 12, 16

10. A square has a 40-cm diagonal. How long is each side of the square? Round to the nearest tenth of a centimeter.

Write the sine, cosine, and tangent ratios for ∠A and ∠B.

11.

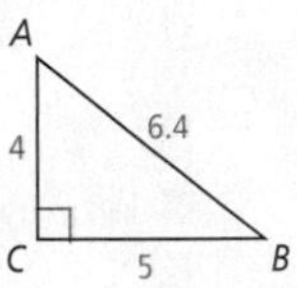

12.

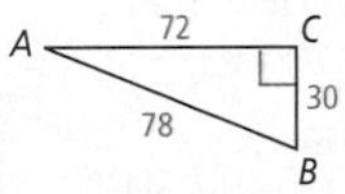

Does each set of numbers form a Pythagorean triple? Explain.

13. 32, 60, 68

14. 1, 2, 3

15. 2.5, 6, 6.5

16. Landscaping A landscaper uses a 13-ft wire to brace a tree. The wire is attached to a protective collar around the trunk of the tree. If the wire makes a 60° angle with the ground, how far up the tree is the protective collar located? Round to the nearest tenth of a foot.

Algebra Find the value of *x*. Round to the nearest tenth.

17.

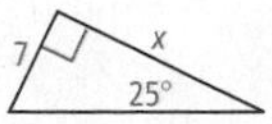

18.

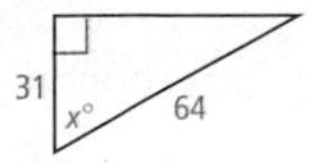

19.

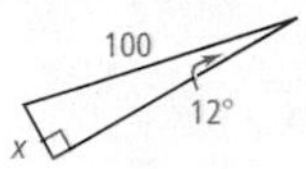

20.

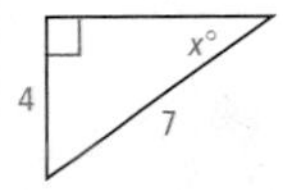

Do you UNDERSTAND?

21. Compare and Contrast What are the similarities between the methods you use to determine whether a triangle is acute, obtuse, or right? What are the differences?

22. In the figure below, which angle has the greater sine value? The greater cosine value? Explain.

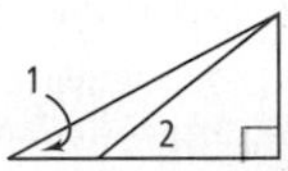

23. Reasoning What angle has a tangent of 1? Explain. (Do not use a calculator or a table.)

Answers

Mid-Chapter Quiz

1. 12

2. $x = 10$, $y = 10\sqrt{2}$

3. $\sqrt{433}$

4. 9

5. $x = 4\sqrt{3}$, $y = 6$

6. $3\sqrt{11}$

7. acute

8. right

9. obtuse

10. 28.3 cm

11. $\sin A = \frac{5}{6.4}$ or $\frac{25}{32}$; $\cos A = \frac{4}{6.4}$ or $\frac{5}{8}$; $\tan A = \frac{5}{4}$; $\sin B = \frac{4}{6.4}$ or $\frac{5}{8}$; $\cos B = \frac{5}{6.4}$ or $\frac{25}{32}$; $\tan B = \frac{4}{5}$

12. $\sin A = \frac{30}{78}$ or $\frac{5}{13}$; $\cos A = \frac{72}{78}$ or $\frac{12}{13}$; $\tan A = \frac{30}{72}$ or $\frac{5}{12}$; $\sin B = \frac{72}{78}$ or $\frac{12}{13}$; $\cos B = \frac{30}{78}$ or $\frac{5}{13}$; $\tan B = \frac{72}{30}$ or $\frac{12}{5}$

13. yes; $32^2 + 60^2 = 68^2$

14. no; $1^2 + 2^2 \neq 3^2$

15. No; $2.5^2 + 6^2 = 6.5^2$, but they are not whole numbers.

16. 11.3 ft

17. 15.0

18. 61.0

19. 20.8

20. 34.8

21. Answers may vary. Sample: Using *a*, *b*, and *c* as the lengths of the sides, with *c* as the longest length, compare $a^2 + b^2$ with c^2. If $a^2 + b^2 < c^2$, the △ is obtuse, if $a^2 + b^2 = c^2$, the △ is rt., and if $a^2 + b^2 > c^2$, the △ is acute.

22. ∠2; ∠1; answers may vary. Sample: sin ∠2 > sin ∠1 because they have the same opposite side and ∠2 has a shorter hypotenuse; cos ∠1 > cos ∠2 because the side adjacent to ∠1 is longer than the side adjacent to ∠2.

23. 45°; the legs of a rt. isosc. △ have the same length. So, $\tan 45° = \frac{s}{s} = 1$.

PowerGeometry.com

MathXL for School

Prepare students for the Mid-Chapter Quiz and Chapter Test with online practice and review.

1 Interactive Learning

Solve It!

PURPOSE To determine an angle of depression using inverse trigonometric ratios and geometric reasoning
PROCESS Students may use trigonometric ratios, knowledge of complementary angles, or properties of parallel lines.

FACILITATE

- **Q** What is the measure of the angle in the upper corner of the triangle formed by the lead, spotlight B, and the vertical line from the spotlight to the ground? Explain. **[≈22°, $\tan^{-1}\frac{10}{25} = 21.801$.]**
- **Q** What is the measure of the angle in the upper corner of the triangle formed by the lead, spotlight A, and the vertical line from the spotlight to the ground? Explain. **[≈22°, the triangles are congruent.]**
- **Q** How are the acute angles in the lower corners of the triangles related to the angles below horizontal that the lamps are set at? **[They are congruent, because they are alternate interior angles.]**
- **Q** At what angle above a horizontal line does the female lead see the spotlights in the original diagram? **[68.2°]**

ANSWER See Solve It in Answers on next page.
CONNECT THE MATH The Solve It situation illustrates an angle of depression in context. In the lesson, students distinguish between angles of elevation and depression and how to calculate the measure of these angles.

8-4 Angles of Elevation and Depression

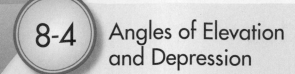

Common Core State Standards
G-SRT.C.8 Use trigonometric ratios . . . to solve right triangles in applied problems.
MP 1, MP 3, MP 4, MP 6

Objective To use angles of elevation and depression to solve problems

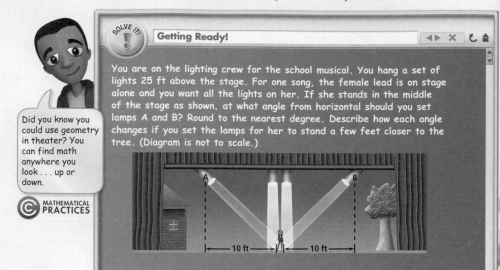

Did you know you could use geometry in theater? You can find math anywhere you look . . . up or down.

MATHEMATICAL PRACTICES

SOLVE IT! Getting Ready!

You are on the lighting crew for the school musical. You hang a set of lights 25 ft above the stage. For one song, the female lead is on stage alone and you want all the lights on her. If she stands in the middle of the stage as shown, at what angle from horizontal should you set lamps A and B? Round to the nearest degree. Describe how each angle changes if you set the lamps for her to stand a few feet closer to the tree. (Diagram is not to scale.)

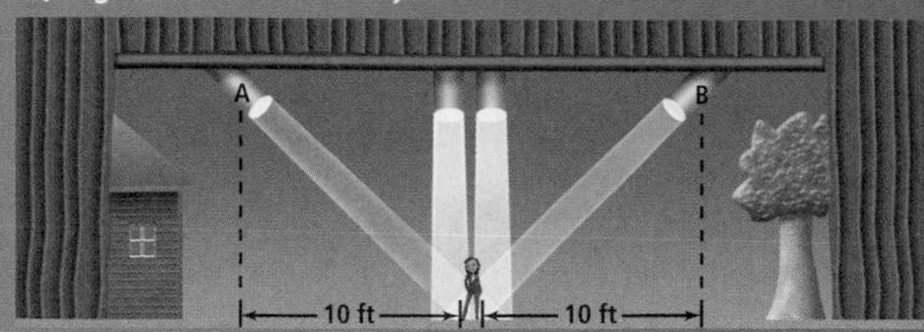

Lesson Vocabulary
- angle of elevation
- angle of depression

The angles in the Solve It are formed below the horizontal black pipe. Angles formed above and below a horizontal line have specific names.

Suppose a person on the ground sees a hang glider at a 38° angle above a horizontal line. This angle is the **angle of elevation.**

At the same time, a person in the hang glider sees the person on the ground at a 38° angle below a horizontal line. This angle is the **angle of depression.**

Notice that the angle of elevation is congruent to the angle of depression because they are alternate interior angles.

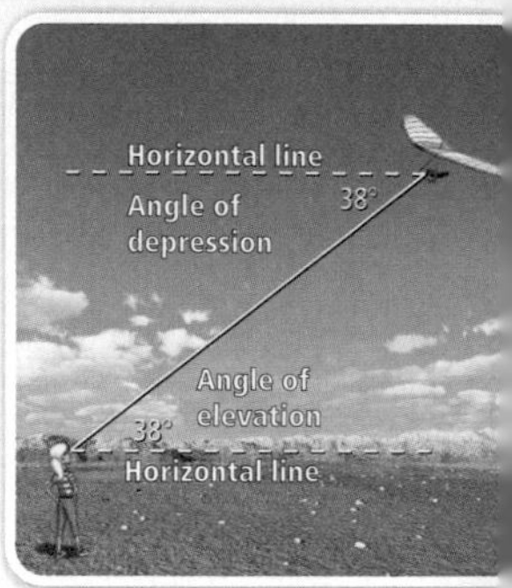

Essential Understanding You can use the angles of elevation and depression as the acute angles of right triangles formed by a horizontal distance and a vertical height.

8-4 Preparing to Teach

BIG ideas **Reasoning and Proof**
Coordinate Geometry

ESSENTIAL UNDERSTANDING
- The angles of elevation and depression are the acute angles of right triangles formed by a horizontal distance and a vertical height.

Math Background

Many practical applications of trigonometry in surveying, construction, aeronautics and other fields involve indirect measurement. Indirect measurement is a technique used to measure something when the use of a measuring device is either impractical or impossible.

In order to see an object above your eye level, you must raise your line of sight. The angle formed from the horizontal at your eye level to the raised or elevated line of sight is called the angle of elevation. An angle of elevation is measured from a horizontal below an object to the line of sight of an object.

Similarly, an angle of depression describes the angle from a horizontal at the top of an object to the line of sight below.

The vertex of an angle of elevation or depression is always at the endpoint of a horizontal ray and a ray along the line of sight to the object. Emphasize that in any given situation, these angles are always congruent. After presenting the real-world examples from the text, encourage students to write examples of their own.

Mathematical Practices

Attend to precision. Students will make explicit use of the terms "angle of elevation" and "angle of depression" and will solve for them.

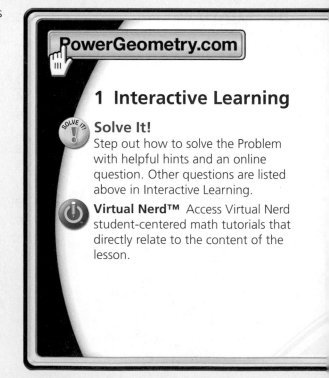

PowerGeometry.com

1 Interactive Learning

Solve It!
Step out how to solve the Problem with helpful hints and an online question. Other questions are listed above in Interactive Learning.

Virtual Nerd™ Access Virtual Nerd student-centered math tutorials that directly relate to the content of the lesson.

Plan

How can you tell if it is an angle of elevation or depression?
Place your finger on the vertex of the angle. Trace along the nonhorizontal side of the angle. See if your finger is above (elevation) or below (depression) the vertex.

Problem 1 Identifying Angles of Elevation and Depression

What is a description of the angle as it relates to the situation shown?

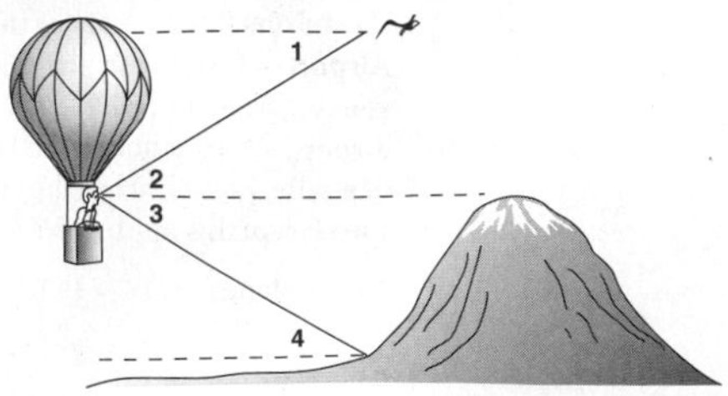

A ∠1

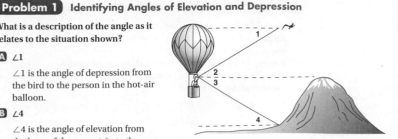

∠1 is the angle of depression from the bird to the person in the hot-air balloon.

B ∠4

∠4 is the angle of elevation from the base of the mountain to the person in the hot-air balloon.

Got It? 1. Use the diagram in Problem 1. What is a description of the angle as it relates to the situation shown?

a. ∠2 **b.** ∠3

Problem 2 Using the Angle of Elevation

Wind Farm Suppose you stand 53 ft from a wind farm turbine. Your angle of elevation to the hub of the turbine is 56.5°. Your eye level is 5.5 ft above the ground. Approximately how tall is the turbine from the ground to its hub?

$\tan 56.5^\circ = \frac{x}{53}$ Use the tangent ratio.

$x = 53(\tan 56.5^\circ)$ Solve for x.

53 [tan] 56.5 [enter] Use a calculator.

80.07426526

Think

Why does your eye level matter here?
Your normal line of sight is a horizontal line. The angle of elevation starts from this eye level, not from the ground.

So $x \approx 80$, which is the height from your eye level to the hub of the turbine. To find the total height of the turbine, add the height from the ground to your eyes. Since $80 + 5.5 = 85.5$, the wind turbine is about 85.5 ft tall from the ground to its hub.

Got It? 2. You sight a rock climber on a cliff at a 32° angle of elevation. Your eye level is 6 ft above the ground and you are 1000 ft from the base of the cliff. What is the approximate height of the rock climber from the ground?

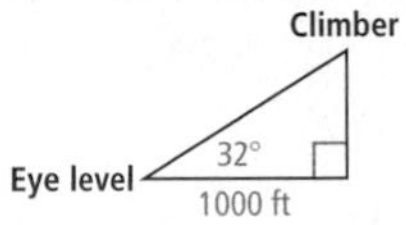

2 Guided Instruction

Problem 1

Q How is ∠1 related to ∠2? Explain. **[They are alternate interior angles, so they are congruent.]**

Q Using the diagram, can you determine the angle of elevation from the base of the mountain to the bird? **[No, there is not enough information.]**

Got It? VISUAL LEARNERS

Q What are the two pairs of congruent angles shown in the diagram in Problem 1? **[∠1 and ∠2; ∠3 and ∠4]**

Problem 2

Q In relation to the angle of elevation in the diagram, which two sides of the triangle are labeled? **[The opposite side is *x*, and the adjacent side is 53 ft.]**

Q Which trigonometric ratio involves the lengths of the adjacent and opposite sides? **[tangent]**

Q What are the measures of the other two angles in the triangle? Explain. **[90°, since it is a right angle and 33.5°, since 180 − 90 − 56.5 = 33.5.]**

Q What trigonometric ratio of the 33.5° angle could you use to determine *x*? **[tangent]**

Got It?

Ask students to identify the angle of depression from the line of sight from the climber to the person on the ground. Make certain that students answer 32° rather than 58°.

2 Guided Instruction

Each Problem is worked out and supported online.

Problem 1
Identifying Angles of Elevation and Depression

Problem 2
Using the Angle of Elevation

Problem 3
Using the Angle of Depression

Support in Geometry Companion
- Vocabulary
- Key Concepts
- Got It?

Answers

Solve It!

Lamp A: 68°; Lamp B: 68°; The measure of the angle of Lamp A decreases, and the measure of the angle of Lamp B increases.

Got It?

1a. ∠ of elevation from the person in the hot-air balloon to bird

b. ∠of depression from the person in the hot-air balloon to base of mountain

2. about 631 ft

Problem 3

Q Is the airplane 2714 ft above the runway as he begins his descent? Explain. **[No, 2714 ft is his altitude above sea level, not his altitude above the runway.]**

Q If the angle of descent is 3°, what is the angle of elevation from the point of contact on the runway to the airplane? Explain. **[The angle of elevation is congruent to the angle of depression because they are alternate interior angles, so it is also 3°.]**

Q To the nearest tenth of a mile, how much horizontal distance does the airplane cover as it makes its descent? **[6.2 mi]**

Got It?
It is implied that the life raft is at sea level and therefore has an altitude of zero.

3 Lesson Check

Do you know HOW?
- If students have difficulty with Exercises 1-5, then have them review Problem 1.

Do you UNDERSTAND?
- If students have difficulty with Exercise 8, then remind them that an angle of depression is the angle below a horizontal line.

Close

Q If two buildings are 30 ft apart and the angle of elevation from the top of the first to the top of the second is 19°, what is the angle of depression from the top of the second to the top of the first? What is the difference in their heights to the nearest tenth of a foot? **[19°; 10.3 ft]**

Problem 3 Using the Angle of Depression

To approach runway 17 of the Ponca City Municipal Airport in Oklahoma, the pilot must begin a 3° descent starting from a height of 2714 ft above sea level. The airport is 1007 ft above sea level. To the nearest tenth of a mile, how far from the runway is the airplane at the start of this approach?

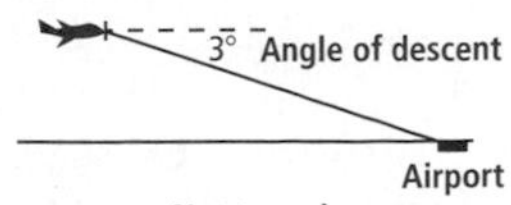

The airplane is 2714 − 1007, or 1707 ft, above the level of the airport.

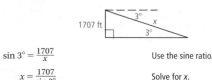

Think
Why is the angle of elevation also 3°?
The path of the airplane before descent is parallel to the ground. So the angles formed by the path of descent are congruent alternate interior angles.

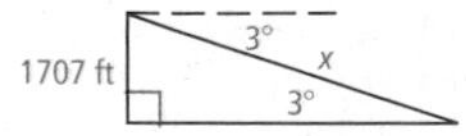

$\sin 3° = \frac{1707}{x}$ Use the sine ratio.

$x = \frac{1707}{\sin 3°}$ Solve for x.

1707 ÷ sin 3 enter 32616.19969 Use a calculator.

÷ 5280 enter 6.177310548 Divide by 5280 to convert feet to miles.

The airplane is about 6.2 mi from the runway.

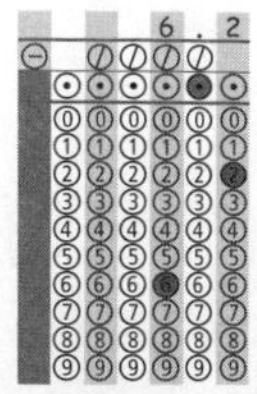

Got It? 3. An airplane pilot sights a life raft at a 26° angle of depression. The airplane's altitude is 3 km. What is the airplane's horizontal distance d from the raft?

Lesson Check

Do you know HOW?
What is a description of each angle as it relates to the diagram?

1. ∠1
2. ∠2
3. ∠3
4. ∠4
5. ∠5

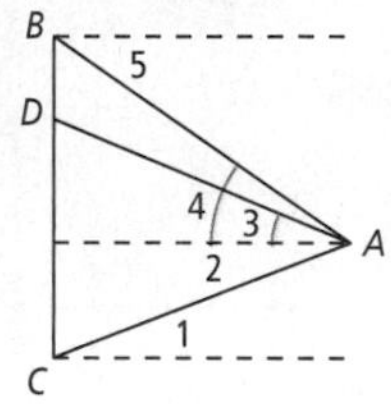

6. What are two pairs of congruent angles in the diagram above? Explain why they are congruent.

Do you UNDERSTAND? MATHEMATICAL PRACTICES

7. **Vobabulary** How is an angle of elevation formed?

8. **Error Analysis** A homework question says that the angle of depression from the bottom of a house window to a ball on the ground is 20°. Below is your friend's sketch of the situation. Describe your friend's error.

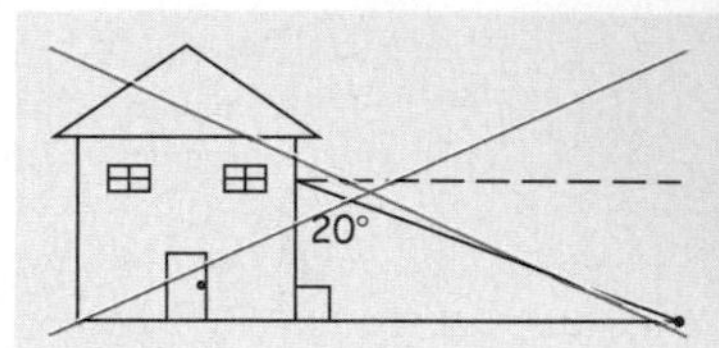

PowerGeometry.com

3 Lesson Check
For a digital lesson check, use the Got It questions.

Support in Geometry Companion
- Lesson Check

4 Practice
Assign homework to individual students or to an entire class.

Additional Problems

1. What is a description of the angle as it relates to the situation shown?

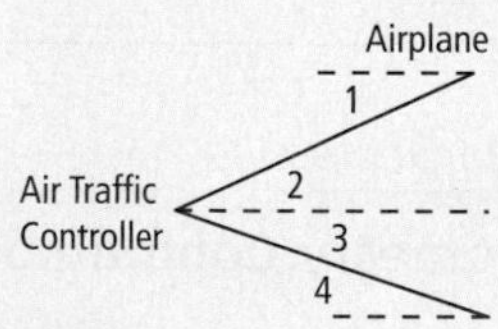

a. ∠1
b. ∠4

ANSWER a. ∠1 is the angle of depression from the airplane to the air traffic controller **b.** ∠4 is the angle of elevation from the ground crew to the air traffic controller.

2. Kyle stands 120 ft from the base of a tree. The angle of elevation from eye level to the top of the tree is 40°. What is the height of the tree to the nearest foot? Kyle's height to eye level is 5 ft.

ANSWER 106 ft

3. A rescue worker is located 175 ft above the ground in a lighthouse. He spots a ship on the water at an angle of depression of 62°. How far from the base of the lighthouse is the ship? Round to the nearest foot.

ANSWER 93 ft

Practice and Problem-Solving Exercises — MATHEMATICAL PRACTICES

A Practice

Describe each angle as it relates to the situation in the diagram.

See Problem 1.

9. ∠1 **10.** ∠2 **11.** ∠3 **12.** ∠4

13. ∠5 **14.** ∠6 **15.** ∠7 **16.** ∠8

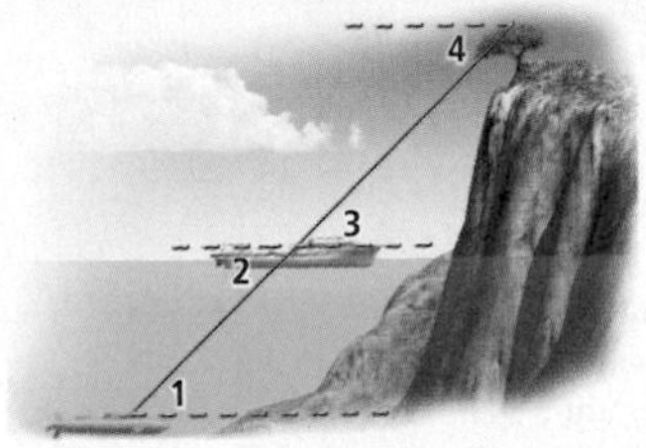

Find the value of x. Round to the nearest tenth of a unit.

See Problem 2.

17.

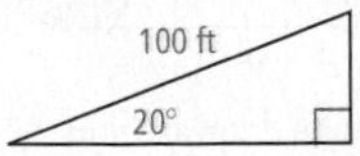

18.

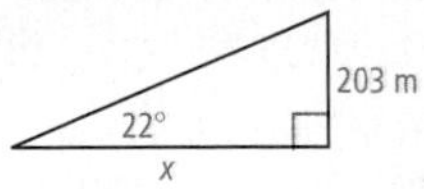

STEM **19. Meteorology** A meteorologist measures the angle of elevation of a weather balloon as 41°. A radio signal from the balloon indicates that it is 1503 m from his location. To the nearest meter, how high above the ground is the balloon?

Find the value of x. Round to the nearest tenth of a unit.

See Problem 3.

20.

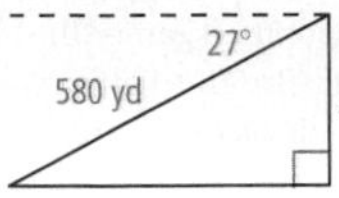

21.

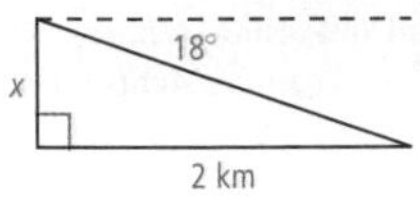

22. Indirect Measurement A tourist looks out from the crown of the Statue of Liberty, approximately 250 ft above ground. The tourist sees a ship coming into the harbor and measures the angle of depression as 18°. Find the distance from the base of the statue to the ship to the nearest foot.

B Apply

23. Flagpole The world's tallest unsupported flagpole is a 282-ft-tall steel pole in Surrey, British Columbia. The shortest shadow cast by the pole during the year is 137 ft long. To the nearest degree, what is the angle of elevation of the sun when casting the flagpole's shortest shadow?

4 Practice

ASSIGNMENT GUIDE

Basic: 9–22, 23–26, 29–31
Average: 9–23 odd, 24–33
Advanced: 9–23 odd, 24–35
Standardized Test Prep: 36–38
Mixed Review: 39–44

Mathematical Practices are supported by exercises with red headings. Here are the Practices supported in this lesson:

MP 1: Make Sense of Problems Ex. 24
MP 3: Communicate Ex. 29
MP 3: Critique the Reasoning of Others Ex. 8

Applications exercises have blue headings. Exercises 22, 23, and 30–35 support MP 4: Model.

STEM exercises focus on science or engineering applications.

EXERCISE 23: Use the Think About a Plan worksheet in the **Practice and Problem Solving Workbook** (also available in the Teaching Resources in print and online) to further support students' development in becoming independent learners.

HOMEWORK QUICK CHECK

To check students' understanding of key skills and concepts, go over Exercises 11, 21, 23, 24, and 29.

Answers

Got It? (continued)

3. about 6.2 km

Lesson Check

1. ∠ of elevation from *C* to *A*

2. ∠ of depression from *A* to *C*

3. ∠ of elevation from *A* to *D*

4. ∠ of elevation from *A* to *B*

5. ∠ of depression from *B* to *A*

6. ∠1 ≅ ∠2 (alt. int. ∠s); ∠4 ≅ ∠5 (alt. int. ∠s)

7. Answers may vary. Sample: An ∠ of elevation is formed by two rays with a common endpoint when one ray is horizontal and the other ray is above the horizontal ray.

8. Answers may vary. Sample: The ∠ labeled in the sketch is the complement of the ∠ of depression.

Practice and Problem-Solving Exercises

9. ∠ of elevation from sub to boat

10. ∠ of depression from boat to sub

11. ∠ of elevation from boat to tree

12. ∠ of depression from tree to boat

13. ∠ of elevation from Max to top of waterfall

14. ∠ of elevation from Maya to top of waterfall

15. ∠ of depression from top of waterfall to Max

16. ∠ of depression from top of waterfall to Maya

17. 34.2 ft

18. 502.4 m

19. 986 m

20. 263.3 yd

21. 0.6 km

22. 769 ft

23. 64°

Answers

Practice and Problem-Solving Exercises (continued)

24. 193 m
25. 72, 72
26. 46, 46
27. 27, 27
28. 20, 20
29a. length of any guy wire = distance on the ground from the tower to the guy wire div. by the cosine of the ∠ formed by the guy wire and the ground
b. height of attachment = distance on the ground from the tower to the guy wire times the tangent of the ∠ formed by the guy wire and the ground
30. 5
31. about 2.8
32. 0.5; about 85
33. 3300 m

24. Think About a Plan Two office buildings are 51 m apart. The height of the taller building is 207 m. The angle of depression from the top of the taller building to the top of the shorter building is 15°. Find the height of the shorter building to the nearest meter.
- How can a diagram help you?
- How does the angle of depression from the top of the taller building relate to the angle of elevation from the top of the shorter building?

Algebra The angle of elevation *e* from *A* to *B* and the angle of depression *d* from *B* to *A* are given. Find the measure of each angle.

25. e: $(7x - 5)°$, d: $4(x + 7)°$

26. e: $(3x + 1)°$, d: $2(x + 8)°$

27. e: $(x + 21)°$, d: $3(x + 3)°$

28. e: $5(x - 2)°$, d: $(x + 14)°$

29. Writing A communications tower is located on a plot of flat land. The tower is supported by several guy wires. Assume that you are able to measure distances along the ground, as well as angles formed by the guy wires and the ground. Explain how you could estimate each of the following measurements.
a. the length of any guy wire
b. how high on the tower each wire is attached

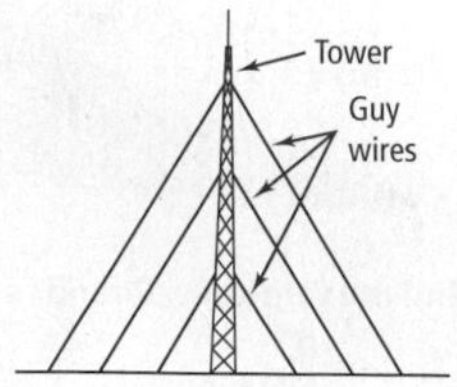

Flying An airplane at a constant altitude *a* flies a horizontal distance *d* toward you at velocity *v*. You observe for time *t* and measure its angles of elevation $\angle E_1$ and $\angle E_2$ at the start and end of your observation. Find the missing information.

30. $a = ■$ mi, $v = 5$ mi/min, $t = 1$ min, $m\angle E_1 = 45$, $m\angle E_2 = 90$

31. $a = 2$ mi, $v = ■$ mi/min, $t = 15$ s, $m\angle E_1 = 40$, $m\angle E_2 = 50$

32. $a = 4$ mi, $d = 3$ mi, $v = 6$ mi/min, $t = ■$ min, $m\angle E_1 = 50$, $m\angle E_2 = ■$

33. Aerial Television A blimp provides aerial television views of a football game. The television camera sights the stadium at a 7° angle of depression. The altitude of the blimp is 400 m. What is the line-of-sight distance from the television camera to the base of the stadium? Round to the nearest hundred meters.

Challenge

34. **Firefighting** A firefighter on the ground sees fire break through a window near the top of the building. The angle of elevation to the windowsill is 28°. The angle of elevation to the top of the building is 42°. The firefighter is 75 ft from the building and her eyes are 5 ft above the ground. What roof-to-windowsill distance can she report by radio to firefighters on the roof?

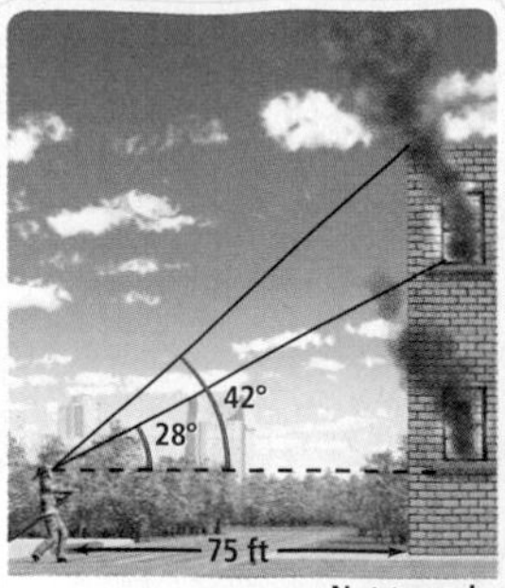

Not to scale

35. **Geography** For locations in the United States, the relationship between the latitude ℓ and the greatest angle of elevation a of the sun at noon on the first day of summer is $a = 90° - \ell + 23.5°$. Find the latitude of your town. Then determine the greatest angle of elevation of the sun for your town on the first day of summer.

Standardized Test Prep

SAT/ACT

36. A 107-ft-tall building casts a shadow of 90 ft. To the nearest whole degree, what is the angle of elevation of the sun?

(A) 33° (B) 40° (C) 50° (D) 57°

37. Which assumption should you make to prove indirectly that the sum of the measures of the angles of a parallelogram is 360?

(F) The sum of the measures of the angles of a parallelogram is 360.
(G) The sum of the measures of the angles of a parallelogram is not 360.
(H) The sum of the measures of consecutive angles of a parallelogram is 180.
(I) The sum of the measures of the angles of a parallelogram is 180.

Extended Response

38. A parallelogram has four congruent sides.
a. Name the types of parallelograms that have this property.
b. What is the most precise name for the figure, based only on the given description? Explain.
c. Draw a diagram to show the categorization of parallelograms.

Mixed Review

Find the value of x. Round to the nearest tenth of a unit. See Lesson 8-3.

39. 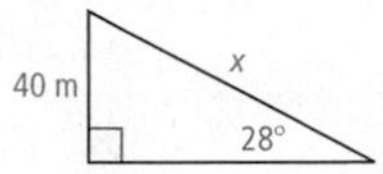

40. 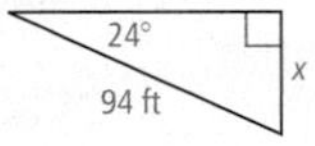

41. 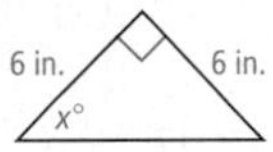

Get Ready! **To prepare for Lesson 8-5, do Exercises 42–44.**

Find the distance between each pair of points. See Lesson 1-7.

42. (0, 0) and (8, 2)
43. (−15, −2) and (0, 0)
44. (−2, 12) and (0, 0)

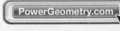

34. about 27.7 ft
35. Check students' work.
36. C
37. G
38. [4] **a.** rhombus and square
b. Rhombus; no information is given about a rt. ∠.
c. Answers may vary. Sample:

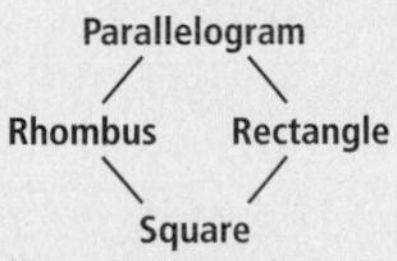

[3] parts (a) and (b) correct, but incomplete categorization of ▱ in part (c)
[2] two parts correct
[1] one part correct
39. 85.2 m
40. 38.2 ft
41. 45
42. $2\sqrt{17} \approx 8.2$
43. $\sqrt{229} \approx 15.1$
44. $2\sqrt{37} \approx 12.2$

8-4 Lesson Resources

Differentiated Remediation

Instructional Support

Geometry Companion

Students can use the **Geometry Companion** worktext (4 pages) . . .

- New Vocabulary
- Key Concepts
- Got It for each Problem
- Lesson Check

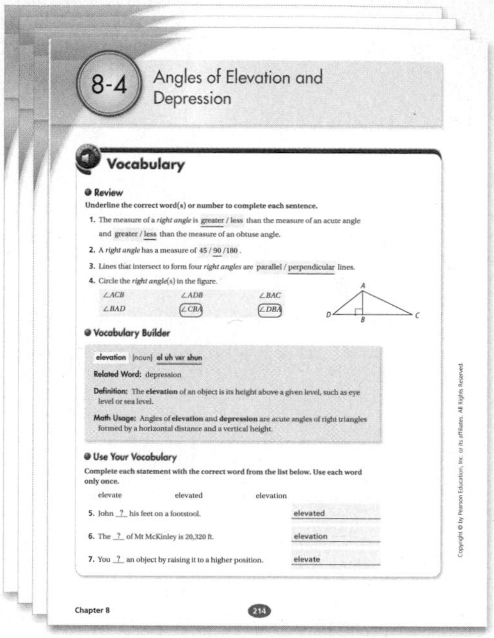
8-4 Angles of Elevation and Depression

Vocabulary

Review

Underline the correct word(s) or number to complete each sentence.

1. The measure of a *right angle* is greater / less than the measure of an acute angle and greater / less than the measure of an obtuse angle.
2. A *right angle* has a measure of 45 / 90 / 180.
3. Lines that intersect to form four *right angles* are parallel / perpendicular lines.
4. Circle the *right angle(s)* in the figure.

∠ACB ∠ADB ∠BAC ∠BAD ∠CBA ∠DBA

Vocabulary Builder

elevation (noun) el uh vay shun

Related Word: depression

Definition: The elevation of an object is its height above a given level, such as eye level or sea level.

Math Usage: Angles of elevation and depression are acute angles of right triangles formed by a horizontal distance and a vertical height.

Use Your Vocabulary

Complete each statement with the correct word from the list below. Use each word only once.

elevate elevated elevation

5. John _?_ his feet on a footstool. elevated
6. The _?_ of Mt McKinley is 20,320 ft. elevation
7. You _?_ an object by raising it to a higher position. elevate

Chapter 8 214

ELL Support

Focus on Language Focus on *elevation* and *depression*. What are synonyms and antonyms of elevation and depression? **[Synonyms of *elevation* may include *altitude*, *mountain*, or *roof*. Synonyms of *depression* may include *sag*, *crater*, or *dent*. *Elevation* and *depression* are antonyms.]**

Use Multiple Representation Read through the examples of elevation and depression in this lesson, such as the balloon, geography, and windmill problems. Draw pictures on the board and trace the angles of elevation and depression. Divide students into heterogeneous pairs. Invite students to think of their own examples with angles of elevation and depression. Students can draw their ideas on the board to share with the class.

5 Assess & Remediate

Lesson Quiz

1. Missy stands at a horizontal distance of 45 ft from the base of a building. The angle of elevation from eye level to the top of the building is 48°. Missy's height to eye level is 5 ft. What is the height of the building to the nearest foot?

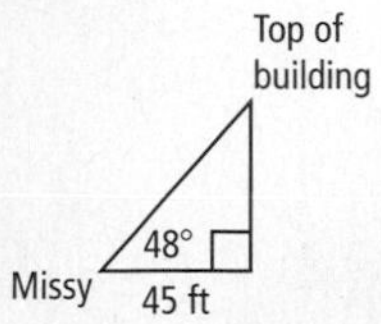

2. The flagpole in Terry's schoolyard is 42 ft tall. On a sunny day, the flagpole casts a shadow 20 ft long. What is the angle of elevation of the sun at that moment? Round to the nearest tenth of a degree.

3. Kurt leans a 20-ft long ladder against the side of his house. The ladder reaches to a height of 18.9 feet up the side of the house. What is the angle of elevation of the ladder to the nearest tenth of a degree?

4. Do you UNDERSTAND? You sight the top of a 50-ft tree from a point on the ground 50 ft from the base of the tree. What is your angle of elevation?

ANSWERS TO LESSON QUIZ

1. 55 ft
2. 64.5°
3. 70.9°
4. 45°

PRESCRIPTION FOR REMEDIATION

Use the student work on the Lesson Quiz to prescribe a differentiated review assignment.

Points	Differentiated Remediation
0–2	Intervention
3	On-level
4	Extension

PowerGeometry.com

5 Assess & Remediate

Assign the Lesson Quiz. Appropriate intervention, practice, or enrichment is automatically generated based on student performance.

Intervention

- **Reteaching** (2 pages) Provides reteaching and practice exercises for the key lesson concepts. Use with struggling students or absent students.
- **English Language Learner Support** Helps students develop and reinforce mathematical vocabulary and key concepts.

All-in-One Resources/Online

Reteaching

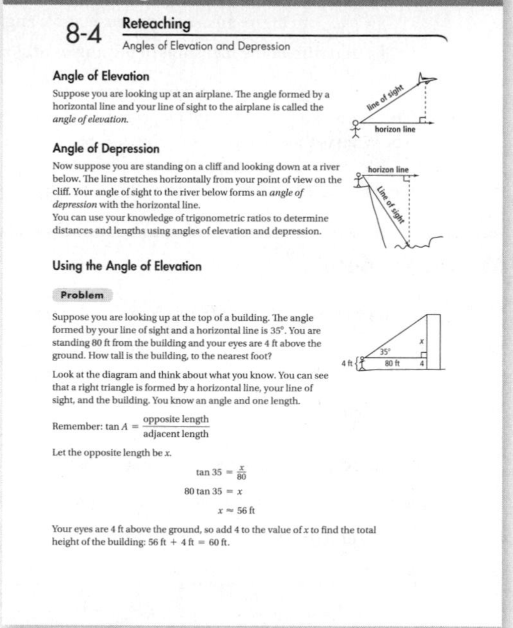
8-4 Reteaching
Angles of Elevation and Depression

Angle of Elevation

Suppose you are looking up at an airplane. The angle formed by a horizontal line and your line of sight to the airplane is called the *angle of elevation.*

Angle of Depression

Now suppose you are standing on a cliff and looking down at a river below. The line stretches horizontally from your point of view on the cliff. Your angle of sight to the river below forms an *angle of depression* with the horizontal line.

You can use your knowledge of trigonometric ratios to determine distances and lengths using angles of elevation and depression.

Using the Angle of Elevation

Problem

Suppose you are looking up at the top of a building. The angle formed by your line of sight and a horizontal line is 35°. You are standing 80 ft from the building and your eyes are 4 ft above the ground. How tall is the building, to the nearest foot?

Look at the diagram and think about what you know. You can see that a right triangle is formed by a horizontal line, your line of sight, and the building. You know an angle and one length.

Remember: $\tan A = \frac{\text{opposite length}}{\text{adjacent length}}$

Let the opposite length be x.

$\tan 35 = \frac{x}{80}$

$80 \tan 35 = x$

$x \approx 56$ ft

Your eyes are 4 ft above the ground, so add 4 to the value of x to find the total height of the building: 56 ft + 4 ft = 60 ft.

All-in-One Resources/Online

English Language Learner Support

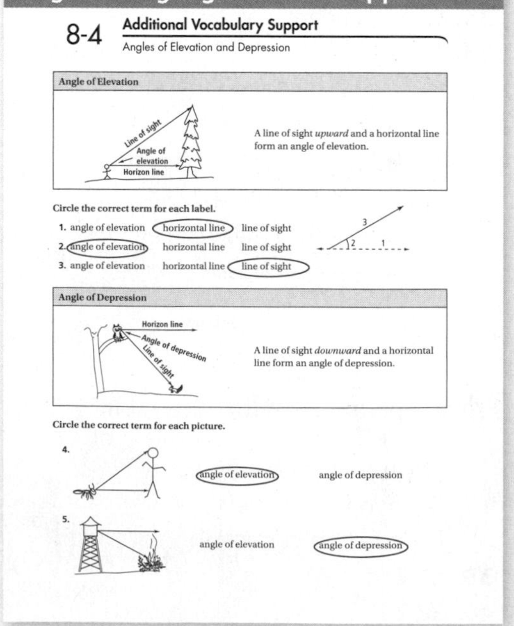
8-4 Additional Vocabulary Support
Angles of Elevation and Depression

Angle of Elevation

A line of sight *upward* and a horizontal line form an angle of elevation.

Circle the correct term for each label.

1. angle of elevation / horizontal line / line of sight
2. angle of elevation / horizontal line / line of sight
3. angle of elevation / horizontal line / line of sight

Angle of Depression

A line of sight *downward* and a horizontal line form an angle of depression.

Circle the correct term for each picture.

4. angle of elevation / angle of depression
5. angle of elevation / angle of depression

Differentiated Remediation *continued*

On-Level

- **Practice** (2 pages) Provides extra practice for each lesson. For simpler practice exercises, use the Form K Practice pages found in the All-in-One Teaching Resources and online.
- **Think About a Plan** Helps students develop specific problem-solving skills and strategies by providing scaffolded guiding questions.
- **Standardized Test Prep** Focuses on all major exercises, all major question types, and helps students prepare for the high-stakes assessments.

Extension

- **Enrichment** Provides students with interesting problems and activities that extend the concepts of the lesson.
- **Activities, Games, and Puzzles** Worksheets that can be used for concepts development, enrichment, and for fun!

Practice and Problem Solving Wkbk/ All-in-One Resources/Online

Practice page 1

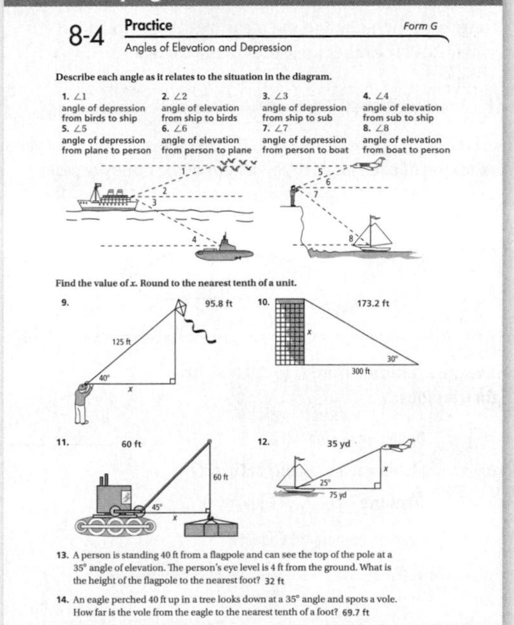

8-4 Practice — Form G
Angles of Elevation and Depression

Describe each angle as it relates to the situation in the diagram.

1. ∠1 angle of depression from birds to ship
2. ∠2 angle of elevation from ship to birds
3. ∠3 angle of depression from ship to sub
4. ∠4 angle of elevation from sub to ship
5. ∠5 angle of depression from plane to person
6. ∠6 angle of elevation from person to plane
7. ∠7 angle of depression from person to boat
8. ∠8 angle of elevation from boat to person

Find the value of x. Round to the nearest tenth of a unit.

9. 95.8 ft
10. 173.2 ft
11. 60 ft
12. 35 yd

13. A person is standing 40 ft from a flagpole and can see the top of the pole at a 35° angle of elevation. The person's eye level is 4 ft from the ground. What is the height of the flagpole to the nearest foot? 32 ft

14. An eagle perched 40 ft up in a tree looks down at a 35° angle and spots a vole. How far is the vole from the eagle to the nearest tenth of a foot? 69.7 ft

Practice and Problem Solving Wkbk/ All-in-One Resources/Online

Practice page 2

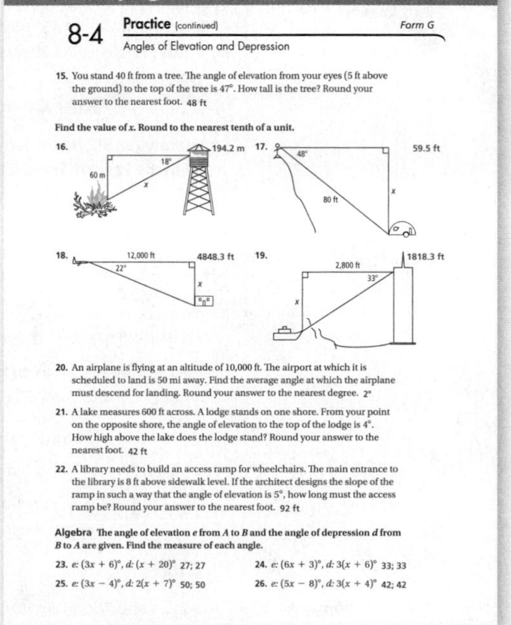

8-4 Practice (continued) — Form G
Angles of Elevation and Depression

15. You stand 40 ft from a tree. The angle of elevation from your eyes (5 ft above the ground) to the top of the tree is 47°. How tall is the tree? Round your answer to the nearest foot. 48 ft

Find the value of x. Round to the nearest tenth of a unit.

16. 194.2 m
17. 59.5 ft
18. 4848.3 ft
19. 1818.3 ft

20. An airplane is flying at an altitude of 10,000 ft. The airport at which it is scheduled to land is 50 mi away. Find the average angle at which the airplane must descend for landing. Round your answer to the nearest degree. 2°

21. A lake measures 600 ft across. A lodge stands on one shore. From your point on the opposite shore, the angle of elevation to the top of the lodge is 4°. How high above the lake does the lodge stand? Round your answer to the nearest foot. 42 ft

22. A library needs to build an access ramp for wheelchairs. The main entrance to the library is 8 ft above sidewalk level. If the architect designs the slope of the ramp in such a way that the angle of elevation is 5°, how long must the access ramp be? Round your answer to the nearest foot. 92 ft

Algebra The angle of elevation e from A to B and the angle of depression d from B to A are given. Find the measure of each angle.

23. e: (3x + 6)°, d: (x + 20)° 27; 27
24. e: (6x + 3)°, d: 3(x + 6)° 33; 33
25. e: (3x − 4)°, d: 2(x + 7)° 50; 50
26. e: (5x − 8)°, d: 3(x + 4)° 42; 42

All-in-One Resources/Online

Enrichment

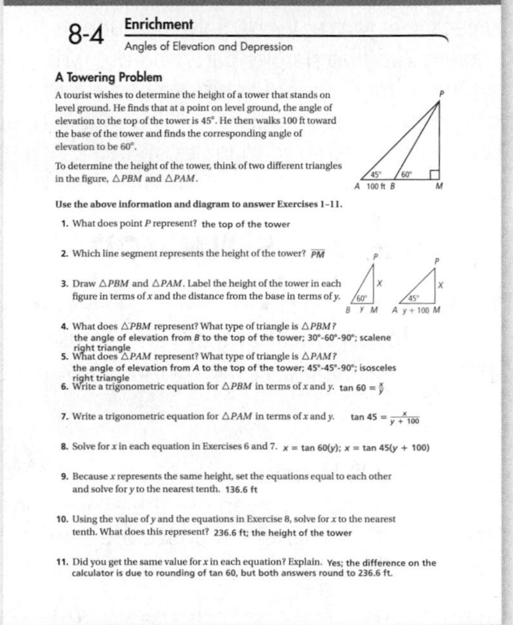

8-4 Enrichment
Angles of Elevation and Depression

A Towering Problem

A tourist wishes to determine the height of a tower that stands on level ground. He finds that at a point on level ground, the angle of elevation to the top of the tower is 45°. He then walks 100 ft toward the base of the tower and finds the corresponding angle of elevation to be 60°.

To determine the height of the tower, think of two different triangles in the figure, △PBM and △PAM.

Use the above information and diagram to answer Exercises 1–11.

1. What does point P represent? the top of the tower
2. Which line segment represents the height of the tower? $\overline{PM}$
3. Draw △PBM and △PAM. Label the height of the tower in each figure in terms of x and the distance from the base in terms of y.
4. What does △PBM represent? What type of triangle is △PBM? the angle of elevation from B to the top of the tower; 30°-60°-90°; scalene right triangle
5. What does △PAM represent? What type of triangle is △PAM? the angle of elevation from A to the top of the tower; 45°-45°-90°; isosceles right triangle
6. Write a trigonometric equation for △PBM in terms of x and y. $\tan 60 = \frac{x}{y}$
7. Write a trigonometric equation for △PAM in terms of x and y. $\tan 45 = \frac{x}{y + 100}$
8. Solve for x in each equation in Exercises 6 and 7. $x = \tan 60(y)$; $x = \tan 45(y + 100)$
9. Because x represents the same height, set the equations equal to each other and solve for y to the nearest tenth. 136.6 ft
10. Using the value of y and the equations in Exercise 8, solve for x to the nearest tenth. What does this represent? 236.6 ft; the height of the tower
11. Did you get the same value for x in each equation? Explain. Yes; the difference on the calculator is due to rounding of tan 60, but both answers round to 236.6 ft.

Practice and Problem Solving Wkbk/ All-in-One Resources/Online

Think About a Plan

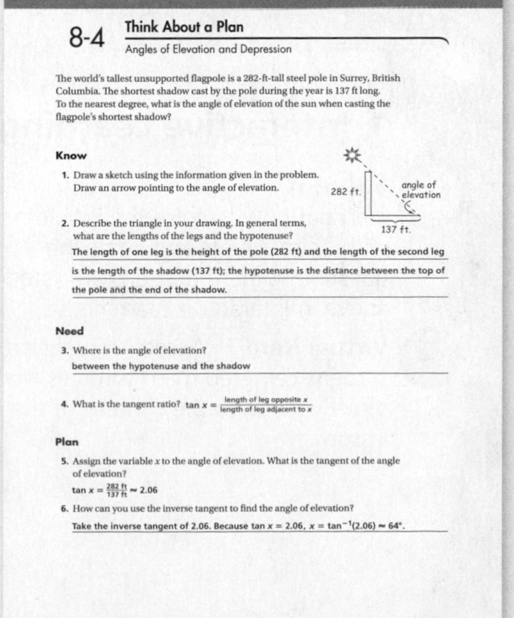

8-4 Think About a Plan
Angles of Elevation and Depression

The world's tallest unsupported flagpole is a 282-ft-tall steel pole in Surrey, British Columbia. The shortest shadow cast by the pole during the year is 137 ft long. To the nearest degree, what is the angle of elevation of the sun when casting the flagpole's shortest shadow?

Know

1. Draw a sketch using the information given in the problem. Draw an arrow pointing to the angle of elevation.
2. Describe the triangle in your drawing. In general terms, what are the lengths of the legs and the hypotenuse? The length of one leg is the height of the pole (282 ft) and the length of the second leg is the length of the shadow (137 ft); the hypotenuse is the distance between the top of the pole and the end of the shadow.

Need

3. Where is the angle of elevation? between the hypotenuse and the shadow
4. What is the tangent ratio? $\tan x = \frac{\text{length of leg opposite } x}{\text{length of leg adjacent to } x}$

Plan

5. Assign the variable x to the angle of elevation. What is the tangent of the angle of elevation? $\tan x = \frac{282 \text{ ft}}{137 \text{ ft}} \approx 2.06$
6. How can you use the inverse tangent to find the angle of elevation? Take the inverse tangent of 2.06. Because $\tan x = 2.06$, $x = \tan^{-1}(2.06) \approx 64°$.

Practice and Problem Solving Wkbk/ All-in-One Resources/Online

Standardized Test Prep

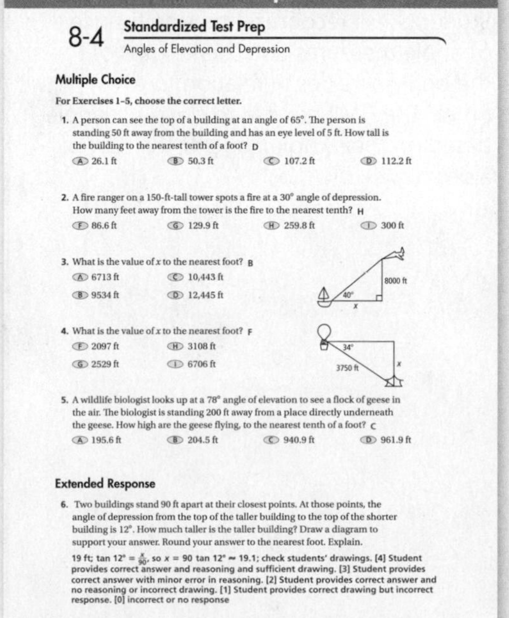

8-4 Standardized Test Prep
Angles of Elevation and Depression

Multiple Choice

For Exercises 1–5, choose the correct letter.

1. A person can see the top of a building at an angle of 65°. The person is standing 50 ft away from the building and has an eye level of 5 ft. How tall is the building to the nearest tenth of a foot? D
 A 26.1 ft B 50.3 ft C 107.2 ft D 112.2 ft
2. A fire ranger on a 150-ft-tall tower spots a fire at a 30° angle of depression. How many feet away from the tower is the fire to the nearest foot? H
 F 86.6 ft G 129.9 ft H 259.8 ft I 300 ft
3. What is the value of x to the nearest foot? B
 A 6713 ft B 9534 ft C 10,443 ft D 12,445 ft
4. What is the value of x to the nearest foot? F
 F 2097 ft G 2529 ft H 3108 ft I 6706 ft
5. A wildlife biologist looks up at a 78° angle of elevation to see a flock of geese in the air. The biologist is standing 200 ft away from a place directly underneath the geese. How high are the geese flying, to the nearest tenth of a foot? C
 A 195.6 ft B 204.5 ft C 940.9 ft D 961.9 ft

Extended Response

6. Two buildings stand 90 ft apart at their closest points. At those points, the angle of depression from the top of the taller building to the top of the shorter building is 12°. How much taller is the taller building? Draw a diagram to support your answer. Round your answer to the nearest foot. Explain.
 19 ft; $\tan 12° = \frac{x}{90}$, so $x = 90 \tan 12° \approx 19.1$; check students' drawings. [4] Student provides correct answer and reasoning and sufficient drawing. [3] Student provides correct answer with minor error in reasoning. [2] Student provides correct answer and no reasoning or incorrect drawing. [1] Student provides correct drawing but incorrect response. [0] incorrect or no response

Online Teacher Resource Center

Activities, Games, and Puzzles

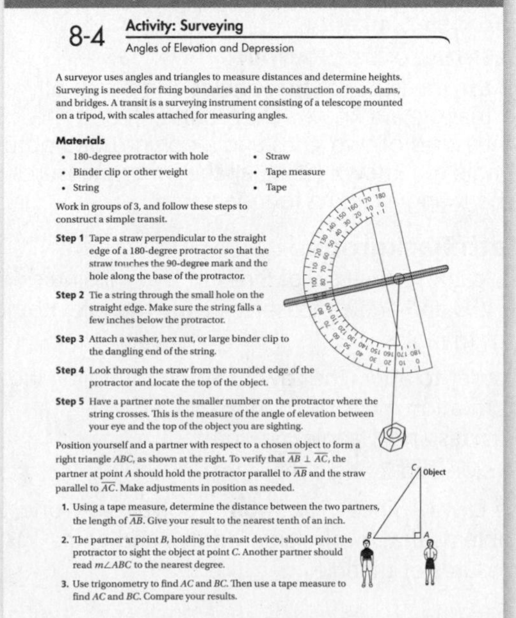

8-4 Activity: Surveying
Angles of Elevation and Depression

A surveyor uses angles and triangles to measure distances and determine heights. Surveying is needed for fixing boundaries and in the construction of roads, dams, and bridges. A transit is a surveying instrument consisting of a telescope mounted on a tripod, with scales attached for measuring angles.

Materials
- 180-degree protractor with hole
- Binder clip or other weight
- String
- Straw
- Tape measure
- Tape

Work in groups of 3, and follow these steps to construct a simple transit.

Step 1 Tape a straw perpendicular to the straight edge of a 180-degree protractor so that the straw touches the 90-degree mark and the hole along the base of the protractor.
Step 2 Tie a string through the small hole on the straight edge. Make sure the string falls a few inches below the protractor.
Step 3 Attach a washer, hex nut, or large binder clip to the dangling end of the string.
Step 4 Look through the straw from the rounded edge of the protractor and locate the top of the object.
Step 5 Have a partner note the smaller number on the protractor where the string crosses. This is the measure of the angle of elevation between your eye and the top of the object you are sighting.

Position yourself and a partner with respect to a chosen object to form a right triangle ABC, as shown at the right. To verify that $\overline{AB} \perp \overline{AC}$, the partner at point A should hold the protractor parallel to $\overline{AB}$ and the straw parallel to $\overline{AC}$. Make adjustments in position as needed.

1. Using a tape measure, determine the distance between the two partners, the length of $\overline{AB}$. Give your result to the nearest tenth of an inch.
2. The partner at point B, holding the transit device, should pivot the protractor to sight the object at point C. Another partner should read m∠ABC to the nearest degree.
3. Use trigonometry to find AC and BC. Then use a tape measure to find AC and BC. Compare your results.

1 Interactive Learning

Solve It!

PURPOSE To use triangles to find missing lengths.

PROCESS Students may see that they can draw an altitude from the edge of the rock shelf to the water to divide the figure into two right triangles. This will allow them to use the sine ratio to find the length of the altitude. Then they can use the sine ratio and the altitude to solve for x.

FACILITATE

Q How can you break the figure into two right triangles? **[Draw an altitude from the edge of the rock shelf to the water.]**

Q Which trigonometric ratio can you use to find the length of the altitude? How will this help you solve for x? **[Sine; use the altitude and the sine ratio to find x.]**

ANSWER See Solve It in Answers on next page.

CONNECT THE MATH In the Solve It, students break a problem into two simpler parts and use what they know about trigonometric ratios to find a missing length. In this lesson, students will learn how to use the Law of Sines to solve triangles.

2 Guided Instruction

Take Note

EXTENSION

Emphasize to students that they can also use reciprocals when applying the Law of Sines: $\frac{a}{\sin A} = \frac{b}{\sin B} = \frac{c}{\sin C}$.

Here's Why It Works

Show students how they can use the expressions for the area of the triangle to derive the Law of Sines: $\frac{1}{2}bc \sin A = \frac{1}{2}ac \sin B = \frac{1}{2}ab \sin C$.

8-5 Law of Sines

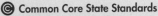

Common Core State Standards
G-SRT.D.11 Understand and apply the Law of Sines . . . to find unknown measurements in right and non-right triangles . . . **Also G-SRT.D.10**
MP 1, MP 3, MP 4, MP 7

Objectives To apply the Law of Sines

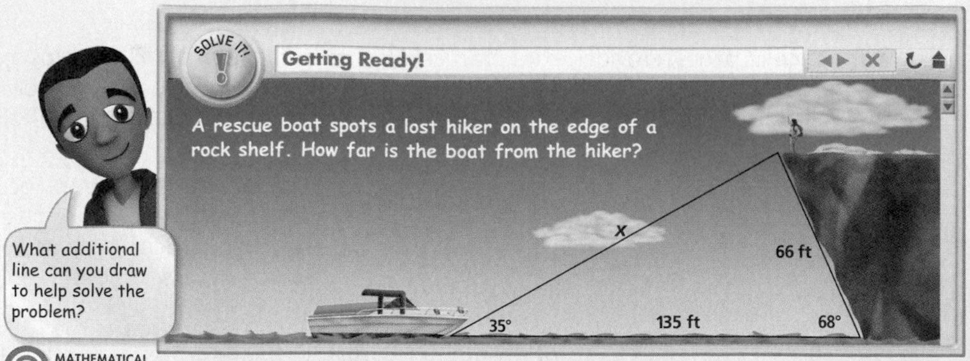

In the Solve It, you used what you know about triangles to find missing lengths.

Lesson Vocabulary
- Law of Sines

Essential Understanding If you know the measures of two angles and the length of a side (AAS or ASA), or two side lengths and the measure of a nonincluded obtuse angle (SSA), then you can find all the other measures of the triangle.

take note

Key Concept Law of Sines

For any $\triangle ABC$, let the lengths of the sides opposite angles A, B, and C be a, b, and c, respectively. Then the **Law of Sines** relates the sine of each angle to the length of the opposite side.

$$\frac{\sin A}{a} = \frac{\sin B}{b} = \frac{\sin C}{c}$$

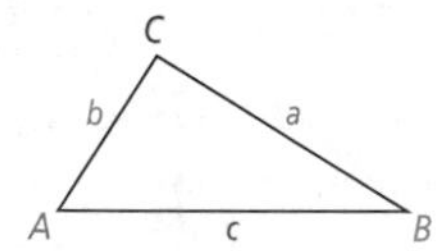

Here's Why It Works Draw the altitude from C to $\overline{AB}$ and label it h. $\triangle ACD$ and $\triangle BCD$ are right triangles.

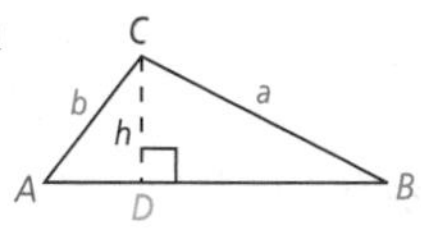

$\sin A = \frac{h}{b}$ and $\sin B = \frac{h}{a}$	Definition of sine
$b \sin A = h$ and $a \sin B = h$	Multiplication Property of Equality
$b \sin A = a \sin B$	Transitive Property of Equality
$\frac{\sin A}{a} = \frac{\sin B}{b}$	Division Property of Equality

8-5 Preparing to Teach

BIG idea Similarity

ESSENTIAL UNDERSTANDING

- If the measures of two angles and a side of a triangle are known (AAS or ASA), or if the measures of two sides and a nonincluded obtuse angle are known (SSA), all the other measures of the triangle can be found.

Math Background

You can use the Law of Sines to find missing side lengths and angle measures in any triangle, not just right triangles.

In order to apply the Law of Sines, the following information must be known:

- 2 sides and 1 angle measure.
- 1 side and 2 angle measures.

The Law of Sines allows you to solve real-world problems that involve side lengths and angle measures of triangles.

Mathematical Practices

Look for and make use of structure. Students will recognize the significance of angle measures and the lengths of the opposite sides in relation to each other. They will be able to solve a triangle based on their knowledge of the Law of Sines.

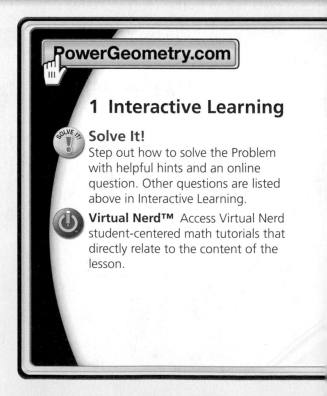

PowerGeometry.com

1 Interactive Learning

Solve It! Step out how to solve the Problem with helpful hints and an online question. Other questions are listed above in Interactive Learning.

Virtual Nerd™ Access Virtual Nerd student-centered math tutorials that directly relate to the content of the lesson.

Problem 1 Using the Law of Sines (AAS)

In $\triangle ABC$, $m\angle A = 48$, $m\angle B = 93$, and $AC = 15$. To the nearest tenth, what is the length of $\overline{BC}$?

Plan

How will drawing a diagram help you solve the problem? Drawing a diagram will help you to visualize the problem. Carefully draw a diagram and label it with all of the given information.

Think	Write
Draw and label $\triangle ABC$. You are given two angle measures and the length of a nonincluded side (AAS).	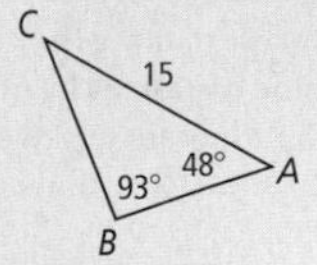
Use the Law of Sines to write an equation.	$\frac{\sin 93°}{15} = \frac{\sin 48°}{BC}$
Solve for BC.	$BC = \frac{15 \sin 48°}{\sin 93°}$
Use a calculator to find BC.	$BC \approx 11.16247016$

The length of $\overline{BC}$ is about 11.2.

Got It? 1. In $\triangle ABC$ above, what is AB to the nearest tenth?

You can also use the Law of Sines to find missing angle measures.

Problem 2 Using the Law of Sines (SSA)

In $\triangle RST$, $RT = 11$, $ST = 18$, and $m\angle R = 120$. To the nearest tenth, what is $m\angle S$?

Think

What unknown angle should you use? Use $\angle S$ because it is opposite a known side length.

Step 1 Draw and label a diagram.

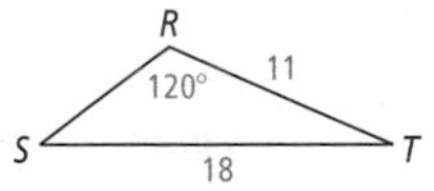

Step 2 Use the Law of Sines to set up an equation.

$$\frac{\sin 120°}{18} = \frac{\sin S}{11}$$

Step 3 Find $m\angle S$.

$$\sin S = \frac{11 \sin 120°}{18} \qquad \text{Solve for } \sin S.$$

$$m\angle S = \sin^{-1}\left(\frac{11 \sin 120°}{18}\right) \approx 31.95396690 \qquad \text{Use the inverse.}$$

$m\angle S$ is about 32.0.

Got It? 2. In $\triangle KLM$, $LM = 9$, $KM = 14$, and $m\angle L = 105$. To the nearest tenth, what is $m\angle K$?

Problem 1 — ERROR PREVENTION

Emphasize to students how drawing a diagram will help them set up the problem correctly.

Q How does drawing a diagram help you to avoid making mistakes? **[The diagram makes it easier to see which angles are opposite which sides and set up the proportion correctly.]**

Got It?

Q Which angle is opposite $\overline{AB}$? **[C]**

Q Do you need to find the measure of the third angle to use the Law of Sines? **[Yes, in order to find *AB* you need to know the measure of angle opposite it.]**

Problem 2

In this problem, students will use the Law of Sines to find a missing angle given two sides and the nonincluded angle.

Q How do you know that the answer should be an acute angle? **[Because the measure of angle *R* is an obtuse angle, and a triangle can have only one obtuse angle.]**

Got It?

Q In $\triangle KLM$, why can you use the Law of Sines to solve directly for angle K? **[Because you know the measure of the side opposite *K* and another corresponding side and angle ratio.]**

2 Guided Instruction

Each problem is worked out and supported online.

Problem 1
Using the Law of Sines (AAS)

Problem 2
Using the Law of Sines (SSA)

Problem 3
Using the Law of Sines to Solve a Problem

Support in Geometry Companion
- Vocabulary
- Key Concepts
- Got It?

Answers

Solve It!

106.7 ft; draw an altitude from the edge of the rock shelf to the water to form two right triangles. Use the sine ratio to solve for the altitude, 61.2 ft. Then use the sine ratio to find x, 106.7 ft.

Got It?

1. 9.5 units
2. 38.4
3. 40.6 ft

Problem 3 **SYNTHESIZING**

In this problem, students will apply the Law of Sines to solve a real world problem.

Q Would you be able to solve this problem without using the Law of Sines? Explain. **[Yes, you could break the figure into two right triangles and use trigonometric ratios.]**

Q Why do you need to find the third angle before solving the problem? **[Because this is the angle opposite the known side length.]**

Got It?

Q Do you need to first find the angle formed by 2nd base, right-fielder, and 1st base? Explain. **[Yes, this is the angle opposite the known side length.]**

Q How would you find the distance from the right-fielder to 1st base? **[Replace 40° with 68° and use the Law of Sines to solve for the side opposite the 68° angle.]**

You can apply the Law of Sines to real-world problems involving triangles.

Problem 3 **Using the Law of Sines to Solve a Problem**

A ship has been at sea longer than expected and has only enough fuel to safely sail another 42 miles. Port City Lighthouse and Cove Town Lighthouse are located 40 miles apart along the coast. At sea, the captain cannot determine distances by observation. The triangle formed by the lighthouses and the ship is shown. Can the ship sail safely to either lighthouse?

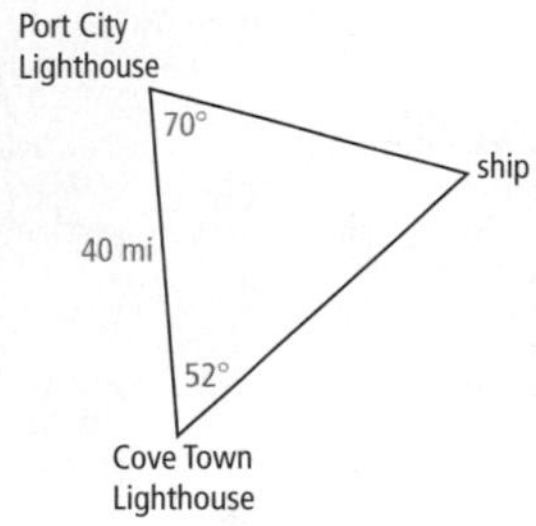

Think

What other information do you need to know to use the Law of Sines?
You can use the Law of Sines to find the distances from the ship to each lighthouse if you can find the angle with the ship as its vertex.

Step 1 Find the measure of the angle formed by Port City, the cruise ship, and Cove Town.

The sum of the measures of the angles of a triangle is 180°. Subtract the given angle measures from 180.

$180 - 70 - 52 = 58$

Step 2 Use the Law of Sines to find the distances from the ship to each lighthouse.

Port City		Cove Town
$\frac{\sin 58^\circ}{40} = \frac{\sin 52^\circ}{d}$	Law of Sines	$\frac{\sin 58^\circ}{40} = \frac{\sin 70^\circ}{d}$
$d \cdot \sin 58^\circ = 40 \cdot \sin 52^\circ$	Find the cross products.	$d \cdot \sin 58^\circ = 40 \cdot \sin 70^\circ$
$d = \frac{40 \cdot \sin 52^\circ}{\sin 58^\circ}$	Divide each side by sin 58°.	$d = \frac{40 \cdot \sin 70^\circ}{\sin 58^\circ}$
$d \approx 37.16821049$	Use a calculator.	$d \approx 44.32260977$

The distance from the ship to Port City Lighthouse is about 37.2 miles, and the distance to Cove Town Lighthouse is about 44.3 miles.

The ship can sail safely to Port City Lighthouse but not to Cove Town Lighthouse.

Got It? **3.** The right-fielder fields a softball between first base and second base as shown in the figure. If the right-fielder throws the ball to second base, how far does she throw the ball?

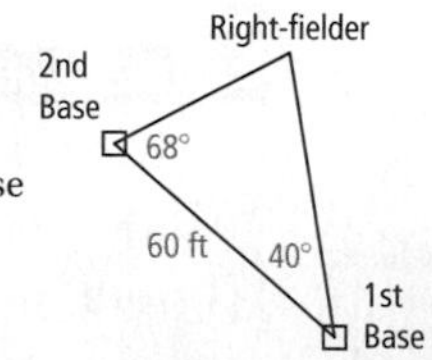

Lesson Check

Do you know HOW?

1. In $\triangle ABC$, $AB = 7$, $BC = 10$, and $m\angle A = 80$. To the nearest tenth, what is $m\angle C$?

2. What is x?

3. What is y?

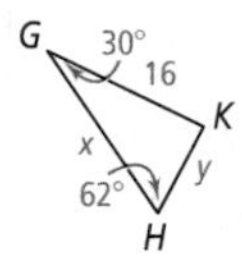

Do you UNDERSTAND? MATHEMATICAL PRACTICES

4. Reasoning If you know the three side lengths of a triangle, can you use the Law of Sines to find the missing angle measures? Explain.

5. Error Analysis In $\triangle PQR$, $PQ = 4$ cm, $QR = 3$ cm, and $m\angle R = 75$. Your friend uses the Law of Sines to write $\frac{\sin 75^\circ}{3} = \frac{\sin P}{4}$ to find $m\angle P$. Explain the error.

Additional Problems

1. In $\triangle LMN$, $m\angle N = 28°$, $m\angle L = 125°$, and $LM = 11.4$. To the nearest tenth, what is the length of $\overline{MN}$?
ANSWER 19.9 units

2. In $\triangle RST$, $RS = 9$, $ST = 13$, and $m\angle T = 43°$. To the nearest tenth, what is $m\angle R$?
ANSWER 80.1°

3. A quarterback (QB), linebacker (LB), and wide receiver (WR) are positioned on a football field as shown in the figure. If the linebacker sprints to tackle the quarterback, how far must he run?

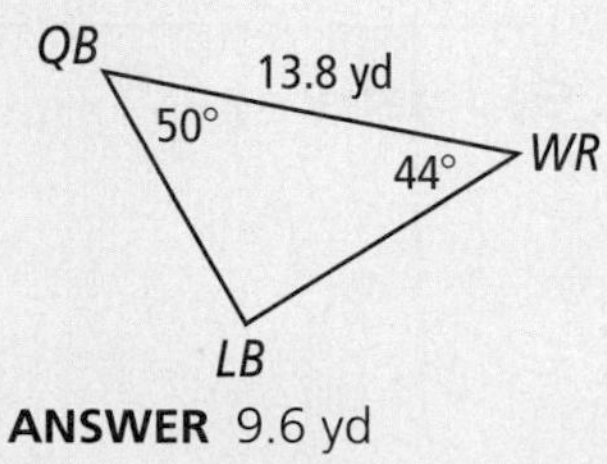

ANSWER 9.6 yd

Practice and Problem-Solving Exercises

MATHEMATICAL PRACTICES

A Practice **Use the information given to solve.**

See Problems 1 and 2.

6. In $\triangle ABC$, $m\angle A = 70$, $m\angle C = 62$, and $BC = 7.3$. To the nearest tenth, what is AB?

7. In $\triangle XYZ$, $m\angle Y = 80$, $XY = 14$, and $XZ = 17$. To the nearest tenth, what is $m\angle Z$?

Use the Law of Sines to find the values of x and y. Round to the nearest tenth.

8.

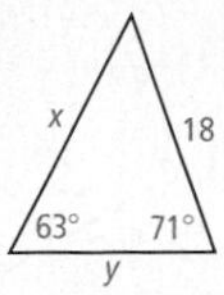

9.

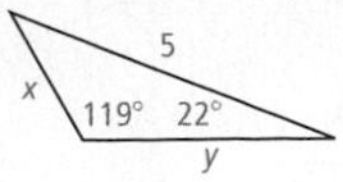

10.

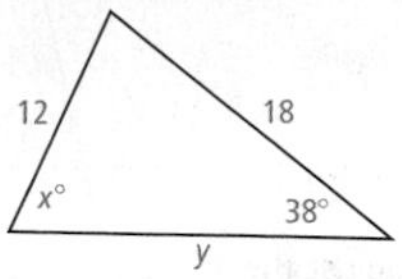

11.

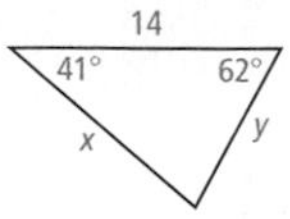

See Problem 3.

12. The main sail of a sailboat has the dimensions shown in the figure at the right. To the nearest tenth of a foot, what is the height of the main sail?

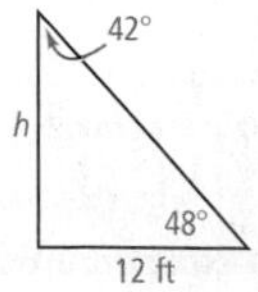

B Apply

13. A portion of a city map is shown in the figure at the right. If you walk along Maple Street between 2nd Street and Elm Grove Lane, how far do you walk? Round your answer to the nearest tenth of a yard.

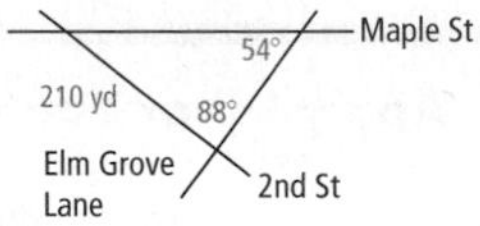

14. Navigation The Bermuda Triangle is a historically famous region of the Atlantic Ocean. The vertices of the triangle are formed by Miami, FL; Bermuda; and San Juan, Puerto Rico. The approximate dimensions of the Bermuda Triangle are shown in the figure at the right. Explain how you would find the distance from Bermuda to Miami. What is this distance to the nearest mile?

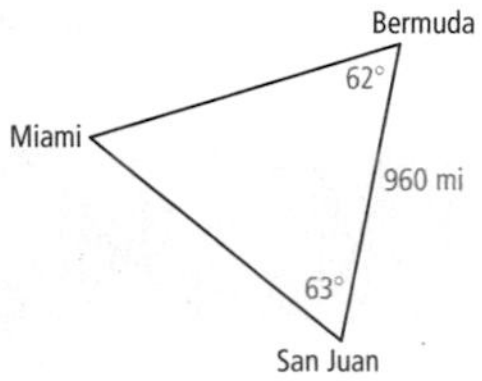

15. Think About a Plan An airplane took off from an airport and started flying toward its destination 210 miles due east. After flying 80 miles east, it encountered a storm and altered its course by turning left 22°. When it was past the storm, it turned right 30° and flew in a straight line until it reached its destination. How far was the plane from its destination when it made the 30° turn?
- Would drawing a diagram help you visualize the problem situation?
- What are you being asked to find?
- What measures do you know?

3 Lesson Check

Do you know HOW? ERROR PREVENTION

- If students have trouble solving Exercise 1, then have them draw a diagram and label it using the given information.
- If students have difficulty setting up a proportion in Exercise 2, then remind them that they need to solve for the measure of angle K first.

Do you UNDERSTAND?

- For Exercise 4, have students draw a triangle and label the side lengths to see that they need at least one angle measure to apply the Law of Sines.
- If students have difficulty identifying the error in Exercise 5, then have them draw a diagram and label it using the given information.

Close

Q The Law of Sines relates equivalent ratios of a triangle to form a proportion. What are the ratios used in the Law of Sines? **[the ratios of the sines of angle measures to the lengths of their opposite sides]**

Q How can you solve a triangle if you know two angle measures and a side length or two side lengths and an angle measure? **[If necessary, solve for the third angle measure. Then write and solve a proportion using equivalent ratios of the sines of angle measures and the lengths of their opposite sides.]**

Answers

Lesson Check

1. 43.6

2. about 18.1

3. about 9.1

4. No, you need to know at least one angle measure to use the Law of Sines.

5. Angle R is opposite side PQ, so the proportion should be written as $\frac{\sin 75°}{4} = \frac{\sin P}{3}$.

Practice and Problem-Solving Exercises

6. 6.9

7. 54.2

8. $x \approx 19.1$; $y \approx 14.5$

9. $x \approx 2.1$; $y \approx 3.6$

10. $x \approx 67.4$; $y \approx 18.8$

11. $x \approx 12.7$; $y \approx 9.4$

12. 13.3 ft

13. 259.4 yd

14. Sample: Find the angle measure for Miami (55°). Then set up and solve the proportion $\frac{\sin 55°}{960} = \frac{\sin 63°}{d}$. The distance is about 1044 miles.

15. about 97.4 mi

4 Practice

ASSIGNMENT GUIDE
Basic: 6–13, 14, 16
Average: 7–13 odd, 14–16
Advanced: 7–13 odd, 14–18

Mathematical Practices are supported by exercises with red headings. Here are the Practices supported in this lesson:

MP 1: Make Sense of Problems Ex. 15
MP 3: Construct Arguments Ex. 4
MP 3: Critique the Reasoning of Others Ex. 5

Applications exercises have blue headings. Exercises 14, 16, and 22 support MP 4: Model.

EXERCISE 14: Use the Think About a Plan worksheet in the **Practice and Problem Solving Workbook** (also available in the Teaching Resources in print and online) to further support students' development in becoming independent learners.

HOMEWORK QUICK CHECK
To check students' understanding of key skills and concepts, go over Exercises 9, 12, 14, 15, and 16.

Answers

Practice and Problem-Solving Exercises (continued)

16. 284.3 ft

17. 36.9°

18. Since $\sin 90° = 1$, the formula $\text{Area} = \frac{1}{2}bc \sin A$ becomes $\text{Area} = \frac{1}{2}bc$, which is the more familiar formula for the area of a triangle with height b and base c.

16. Zipline A zipline is constructed over a ravine as shown in the diagram at the right. What is the horizontal distance from the bottom of the ladder to the platform where the zipline ends? Round your answer to the nearest tenth of a foot.

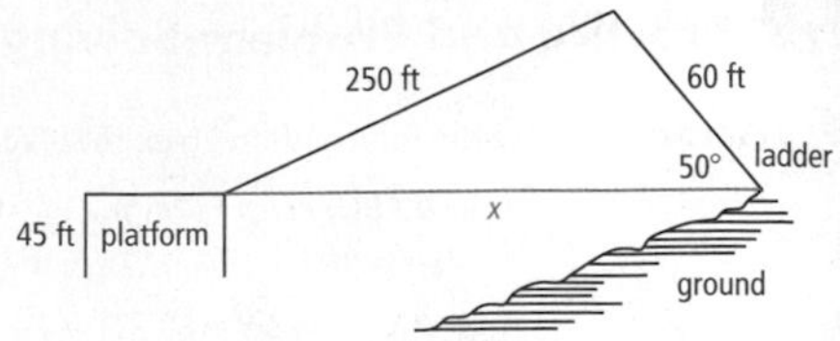

17. If $m\angle DEG = m\angle D + m\angle G + 43$, what is $m\angle EFG$?

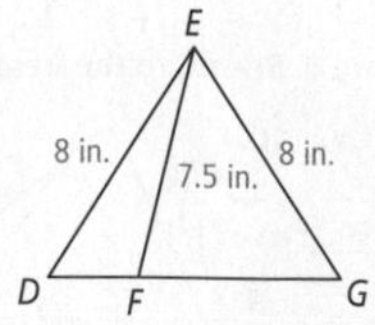

Challenge

18. You can use the formula $\text{Area} = \frac{1}{2}bc \sin A$ to find the area of the triangle shown at the right. Show how this formula becomes the more familiar formula for the area of a triangle if $\triangle ABC$ is a right triangle and $m\angle A = 90$.

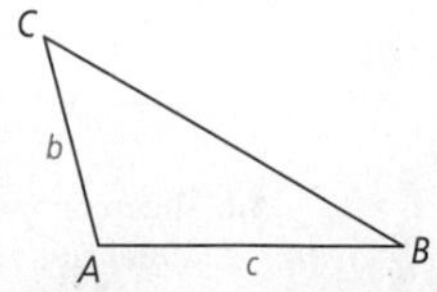

Apply What You've Learned

MATHEMATICAL PRACTICES
MP 6

Look back at the information on page 489 about the fire in a state forest. The diagram is shown again below.

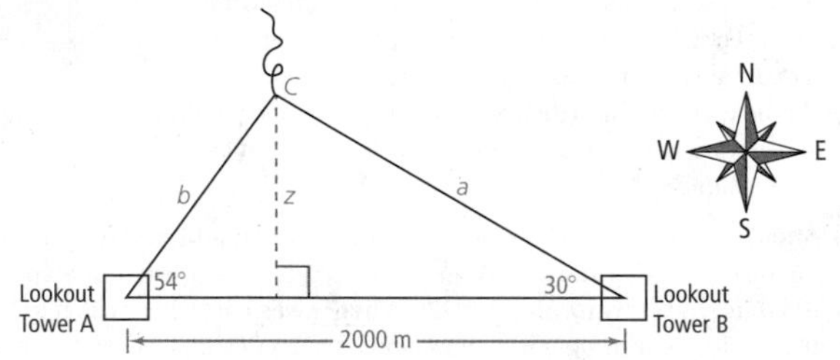

a. What is the measure of the angle between the distances labeled a and b?

b. Write and solve an equation using the Law of Sines to find the distance between Lookout Tower A and the fire. Round to the nearest tenth.

c. Write and solve an equation using the Law of Sines to find the distance between Lookout Tower B and the fire. Round to the nearest tenth.

Apply What You've Learned

In the Apply What You've Learned for Lesson 8-3, students identified valid trigonometric relationships among the unknown lengths a, b, and z in the diagram from page 489. Here, they use the Law of Sines to find approximate values of a and b.

Mathematical Practices
Students **attend to precision** as they write and solve proportions to find unknown distances. (MP 6)

ANSWERS

a. 96°

b. $\frac{\sin 96°}{2000} = \frac{\sin 30°}{b}$; $b \approx 1005.5$ m

c. $\frac{\sin 96°}{2000} = \frac{\sin 54°}{a}$; $a \approx 1626.9$ m

8-5 Lesson Resources

Differentiated Remediation

Additional Instructional Support

Geometry Companion

Students can use the **Geometry Companion** worktext (4 pages) as you teach the lesson. Use the Companion to support

- Solve It!
- New Vocabulary
- Key Concepts

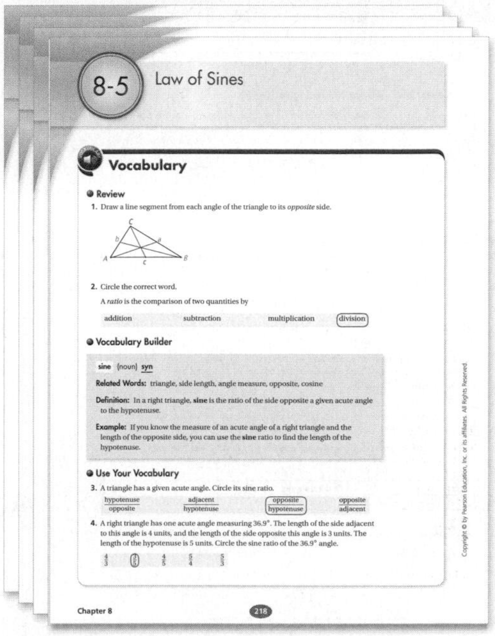

8-5 Law of Sines

Vocabulary

Review

1. Draw a line segment from each angle of the triangle to its *opposite* side.

2. Circle the correct word.

A *ratio* is the comparison of two quantities by

addition   subtraction   multiplication   (division)

Vocabulary Builder

sine (noun) syn

Related Words: triangle, side length, angle measure, opposite, cosine

Definition: In a right triangle, **sine** is the ratio of the side opposite a given acute angle to the hypotenuse.

Example: If you know the measure of an acute angle of a right triangle and the length of the opposite side, you can use the **sine** ratio to find the length of the hypotenuse.

Use Your Vocabulary

3. A triangle has a given acute angle. Circle its sine ratio.

$\frac{\text{hypotenuse}}{\text{opposite}}$ $\frac{\text{adjacent}}{\text{hypotenuse}}$ $\frac{\text{opposite}}{\text{hypotenuse}}$ $\frac{\text{opposite}}{\text{adjacent}}$

4. A right triangle has one acute angle measuring 36.9°. The length of the side adjacent to this angle is 4 units, and the length of the side opposite this angle is 3 units. The length of the hypotenuse is 5 units. Circle the sine ratio of the 36.9° angle.

$\frac{4}{3}$ $\frac{3}{5}$ $\frac{4}{5}$ $\frac{5}{4}$ $\frac{5}{3}$

Copyright © by Pearson Education, Inc. or its affiliates. All Rights Reserved.

Chapter 8   218

ELL Support

Have Students Report Back Orally To deepen students' understanding of the Law of Sines, have them describe orally how to apply the Law of Sines to solve a triangle. Draw a triangle on the board and label the vertices and sides. Be sure students are able to identify the correct ratios of angles to their opposite sides needed to apply the Law of Sines. Then draw several triangles on the board with three measurements given and have students explain whether they can use the Law of Sines to solve the triangles. Ask them to describe the ratios they would use as they tell you how to set up the proportion.

5 Assess & Remediate

Lesson Quiz

Use the Lesson Quiz to assess students' mastery of the skills and concepts of this lesson.

1. In $\triangle ABC$, $m\angle B = 55$, $m\angle C = 100$, and $AC = 6$. To the nearest tenth, what is the length of $\overline{AB}$?
2. In $\triangle GHI$, $GH = 6$, $GI = 7$, and $m\angle H = 64$. To the nearest tenth, what is $m\angle I$?
3. **Do you UNDERSTAND?** Two dogs are swimming after a ball that was thrown in a pond as shown in the figure. About how far from the ball is Molly?

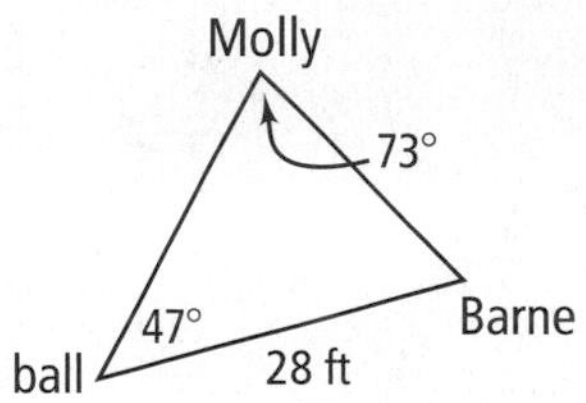

ANSWERS TO LESSON QUIZ

1. 7.2 units
2. 50.4
3. about 25.4 ft

PRESCRIPTION FOR REMEDIATION

Use the student work on the Lesson Quiz to prescribe a differentiated review assignment.

Points	Differentiated Remediation
0–1	Intervention
2	On-level
3	Extension

PowerGeometry.com

5 Assess & Remediate

Assign the Lesson Quiz. Appropriate intervention, practice, or enrichment is automatically generated based on student performance.

Intervention

- **Reteaching** (2 pages) Provides reteaching and practice exercises for the key lesson concepts. Use with struggling students or absent students.
- **English Language Learner Support** Helps students develop and reinforce mathematical vocabulary and key concepts.

All-in-One Resources/Online

Reteaching

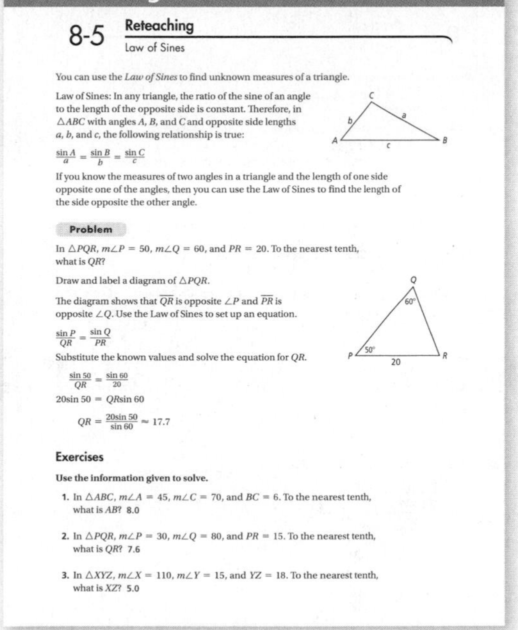

8-5 Reteaching

Law of Sines

You can use the *Law of Sines* to find unknown measures of a triangle.

Law of Sines: In any triangle, the ratio of the sine of an angle to the length of the opposite side is constant. Therefore, in $\triangle ABC$ with angles A, B, and C and opposite side lengths a, b, and c, the following relationship is true:

$$\frac{\sin A}{a} = \frac{\sin B}{b} = \frac{\sin C}{c}$$

If you know the measures of two angles in a triangle and the length of one side opposite one of the angles, then you can use the Law of Sines to find the length of the side opposite the other angle.

Problem

In $\triangle PQR$, $m\angle P = 50$, $m\angle Q = 60$, and $PR = 20$. To the nearest tenth, what is QR?

Draw and label a diagram of $\triangle PQR$.

The diagram shows that $\overline{QR}$ is opposite $\angle P$ and $\overline{PR}$ is opposite $\angle Q$. Use the Law of Sines to set up an equation.

$$\frac{\sin P}{QR} = \frac{\sin Q}{PR}$$

Substitute the known values and solve the equation for QR.

$$\frac{\sin 50}{QR} = \frac{\sin 60}{20}$$

$$20\sin 50 = QR\sin 60$$

$$QR = \frac{20\sin 50}{\sin 60} \approx 17.7$$

Exercises

Use the information given to solve.

1. In $\triangle ABC$, $m\angle A = 45$, $m\angle C = 70$, and $BC = 6$. To the nearest tenth, what is AB? **8.0**
2. In $\triangle PQR$, $m\angle P = 30$, $m\angle Q = 80$, and $PR = 15$. To the nearest tenth, what is QR? **7.6**
3. In $\triangle XYZ$, $m\angle X = 110$, $m\angle Y = 15$, and $YZ = 18$. To the nearest tenth, what is XZ? **5.0**

All-in-One Resources/Online

English Language Learner Support

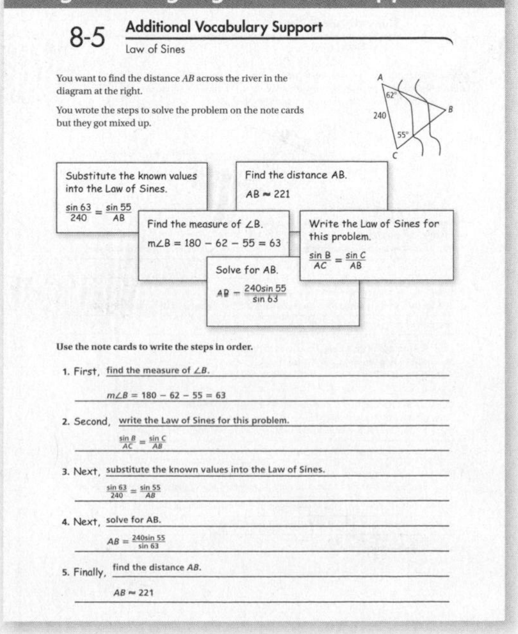

8-5 Additional Vocabulary Support

Law of Sines

You want to find the distance AB across the river in the diagram at the right.

You wrote the steps to solve the problem on the note cards but they got mixed up.

Use the note cards to write the steps in order.

1. First, find the measure of $\angle B$.
   $m\angle B = 180 - 62 - 55 = 63$
2. Second, write the Law of Sines for this problem.
   $\frac{\sin B}{AC} = \frac{\sin C}{AB}$
3. Next, substitute the known values into the Law of Sines.
   $\frac{\sin 63}{240} = \frac{\sin 55}{AB}$
4. Next, solve for AB.
   $AB = \frac{240\sin 55}{\sin 63}$
5. Finally, find the distance AB.
   $AB \approx 221$

Differentiated Remediation *continued*

On-Level

- **Practice** (2 pages) Provides extra practice for each lesson. For simpler practice exercises, use the Form K Practice pages found in the All-in-One Teaching Resources and online.
- **Think About a Plan** Helps students develop specific problem-solving skills and strategies by providing scaffolded guiding questions.
- **Standardized Test Prep** Focuses on all major exercises, all major question types, and helps students prepare for the high-stakes assessments.

Extension

- **Enrichment** Provides students with interesting problems and activities that extend the concepts of the lesson.
- **Activities, Games, and Puzzles** Worksheets that can be used for concepts development, enrichment, and for fun!

Practice and Problem Solving Wkbk/ All-in-One Resources/Online

Practice page 1

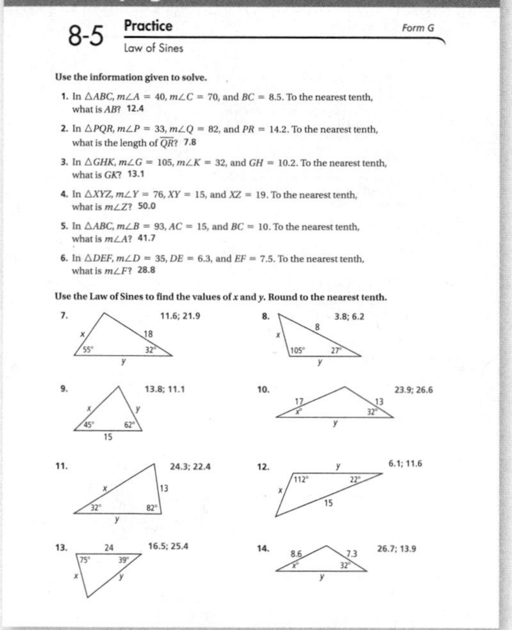

8-5 Practice — Form G
Law of Sines

Use the information given to solve.

1. In $\triangle ABC$, $m\angle A = 40$, $m\angle C = 70$, and $BC = 8.5$. To the nearest tenth, what is AB? 12.4
2. In $\triangle PQR$, $m\angle P = 33$, $m\angle Q = 82$, and $PR = 14.2$. To the nearest tenth, what is the length of $\overline{QR}$? 7.8
3. In $\triangle GHK$, $m\angle G = 105$, $m\angle K = 32$, and $GH = 10.2$. To the nearest tenth, what is GK? 13.1
4. In $\triangle XYZ$, $m\angle Y = 76$, $XY = 15$, and $XZ = 19$. To the nearest tenth, what is $m\angle Z$? 50.0
5. In $\triangle ABC$, $m\angle B = 93$, $AC = 15$, and $BC = 10$. To the nearest tenth, what is $m\angle A$? 41.7
6. In $\triangle DEF$, $m\angle D = 35$, $DE = 6.3$, and $EF = 7.5$. To the nearest tenth, what is $m\angle F$? 28.8

Use the Law of Sines to find the values of x and y. Round to the nearest tenth.

7. 11.6; 21.9
8. 3.8; 6.2
9. 13.8; 11.1
10. 23.9; 26.6
11. 24.3; 22.4
12. 6.1; 11.6
13. 16.5; 25.4
14. 26.7; 13.9

Practice and Problem Solving Wkbk/ All-in-One Resources/Online

Practice page 2

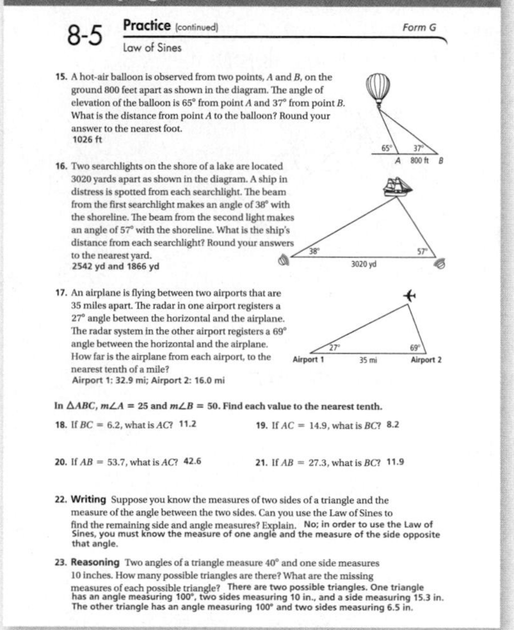

8-5 Practice (continued) — Form G
Law of Sines

15. A hot-air balloon is observed from two points, A and B, on the ground 800 feet apart as shown in the diagram. The angle of elevation of the balloon is 65° from point A and 37° from point B. What is the distance from point A to the balloon? Round your answer to the nearest foot. 1026 ft
16. Two searchlights on the shore of a lake are located 3020 yards apart as shown in the diagram. A ship in distress is spotted from each searchlight. The beam from the first searchlight makes an angle of 38° with the shoreline. The beam from the second light makes an angle of 57° with the shoreline. What is the ship's distance from each searchlight? Round your answers to the nearest yard. 2542 yd and 1866 yd
17. An airplane is flying between two airports that are 35 miles apart. The radar in one airport registers a 27° angle between the horizontal and the airplane. The radar system in the other airport registers a 69° angle between the horizontal and the airplane. How far is the airplane from each airport, to the nearest tenth of a mile? Airport 1: 32.9 mi; Airport 2: 16.0 mi

In $\triangle ABC$, $m\angle A = 25$ and $m\angle B = 50$. Find each value to the nearest tenth.

18. If $BC = 6.2$, what is AC? 11.2
19. If $AC = 14.9$, what is BC? 8.2
20. If $AB = 53.7$, what is AC? 42.6
21. If $AB = 27.3$, what is BC? 11.9
22. **Writing** Suppose you know the measures of two sides of a triangle and the measure of the angle between the two sides. Can you use the Law of Sines to find the remaining side and angle measures? Explain. No; in order to use the Law of Sines, you must know the measure of one angle and the measure of the side opposite that angle.
23. **Reasoning** Two angles of a triangle measure 40° and one side measures 10 inches. How many possible triangles are there? What are the missing measures of each possible triangle? There are two possible triangles. One triangle has an angle measuring 100°, two sides measuring 10 in., and a side measuring 15.3 in. The other triangle has an angle measuring 100° and two sides measuring 6.5 in.

All-in-One Resources/Online

Enrichment

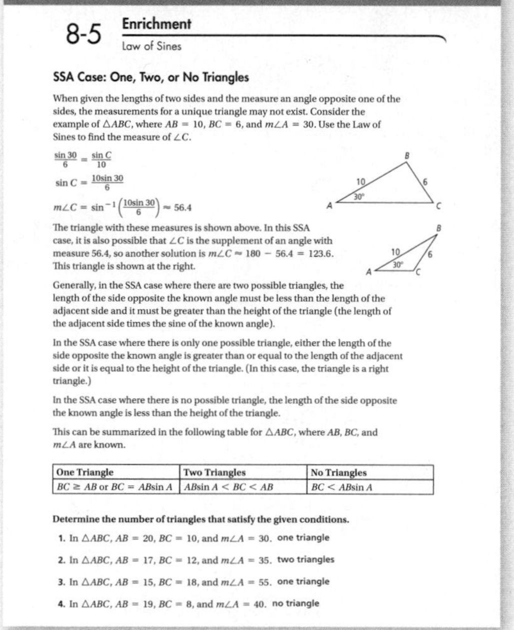

8-5 Enrichment
Law of Sines

SSA Case: One, Two, or No Triangles

When given the lengths of two sides and the measure an angle opposite one of the sides, the measurements for a unique triangle may not exist. Consider the example of $\triangle ABC$, where $AB = 10$, $BC = 6$, and $m\angle A = 30$. Use the Law of Sines to find the measure of $\angle C$.

$\frac{\sin 30}{6} = \frac{\sin C}{10}$

$\sin C = \frac{10 \sin 30}{6}$

$m\angle C = \sin^{-1}\left(\frac{10 \sin 30}{6}\right) \approx 56.4$

The triangle with these measures is shown above. In this SSA case, it is also possible that $\angle C$ is the supplement of an angle with measure 56.4, so another solution is $m\angle C \approx 180 - 56.4 = 123.6$. This triangle is shown at the right.

Generally, in the SSA case where there are two possible triangles, the length of the side opposite the known angle must be less than the length of the adjacent side and it must be greater than the height of the triangle (the length of the adjacent side times the sine of the known angle).

In the SSA case where there is only one possible triangle, either the length of the side opposite the known angle is greater than or equal to the length of the adjacent side or it is equal to the height of the triangle. (In this case, the triangle is a right triangle.)

In the SSA case where there is no possible triangle, the length of the side opposite the known angle is less than the height of the triangle.

This can be summarized in the following table for $\triangle ABC$, where AB, BC, and $m\angle A$ are known.

One Triangle	Two Triangles	No Triangles
$BC \geq AB$ or $BC = AB \sin A$	$AB \sin A < BC < AB$	$BC < AB \sin A$

Determine the number of triangles that satisfy the given conditions.

1. In $\triangle ABC$, $AB = 20$, $BC = 10$, and $m\angle A = 30$. one triangle
2. In $\triangle ABC$, $AB = 17$, $BC = 12$, and $m\angle A = 35$. two triangles
3. In $\triangle ABC$, $AB = 15$, $BC = 18$, and $m\angle A = 55$. one triangle
4. In $\triangle ABC$, $AB = 19$, $BC = 8$, and $m\angle A = 40$. no triangle

Practice and Problem Solving Wkbk/ All-in-One Resources/Online

Think About a Plan

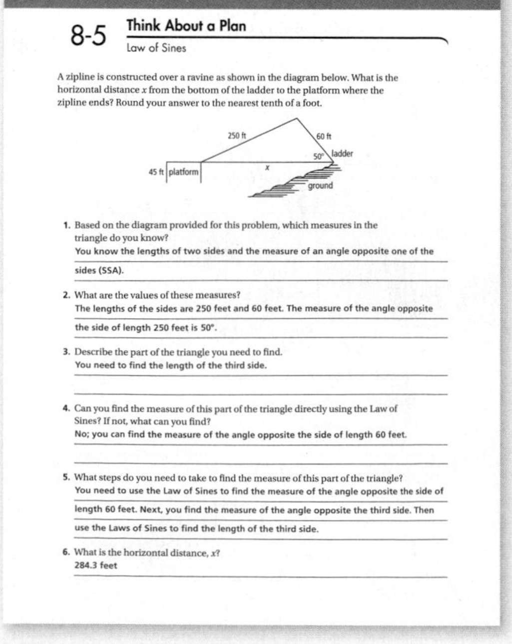

8-5 Think About a Plan
Law of Sines

A zipline is constructed over a ravine as shown in the diagram below. What is the horizontal distance x from the bottom of the ladder to the platform where the zipline ends? Round your answer to the nearest tenth of a foot.

1. Based on the diagram provided for this problem, which measures in the triangle do you know? You know the lengths of two sides and the measure of an angle opposite one of the sides (SSA).
2. What are the values of these measures? The lengths of the sides are 250 feet and 60 feet. The measure of the angle opposite the side of length 250 feet is 50°.
3. Describe the part of the triangle you need to find. You need to find the length of the third side.
4. Can you find the measure of this part of the triangle directly using the Law of Sines? If not, what can you find? No; you can find the measure of the angle opposite the side of length 60 feet.
5. What steps do you need to take to find the measure of this part of the triangle? You need to use the Law of Sines to find the measure of the angle opposite the side of length 60 feet. Next, you find the measure of the angle opposite the third side. Then use the Laws of Sines to find the length of the third side.
6. What is the horizontal distance, x? 284.3 feet

Practice and Problem Solving Wkbk/ All-in-One Resources/Online

Standardized Test Prep

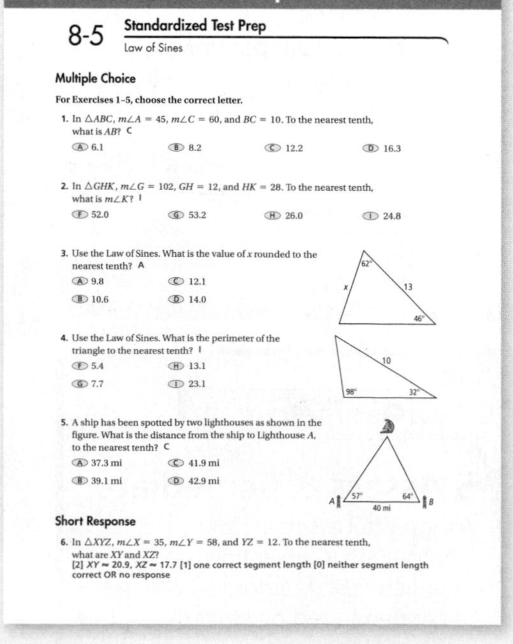

8-5 Standardized Test Prep
Law of Sines

Multiple Choice

For Exercises 1–5, choose the correct letter.

1. In $\triangle ABC$, $m\angle A = 45$, $m\angle C = 60$, and $BC = 10$. To the nearest tenth, what is AB? C
   A. 6.1 B. 8.2 C. 12.2 D. 16.3
2. In $\triangle GHK$, $m\angle G = 102$, $GH = 12$, and $HK = 28$. To the nearest tenth, what is $m\angle K$? I
   F. 52.0 G. 53.2 H. 26.0 I. 24.8
3. Use the Law of Sines. What is the value of x rounded to the nearest tenth? A
   A. 9.8 B. 10.6 C. 12.1 D. 14.0
4. Use the Law of Sines. What is the perimeter of the triangle to the nearest tenth? I
   F. 5.4 G. 7.7 H. 13.1 I. 23.1
5. A ship has been spotted by two lighthouses as shown in the figure. What is the distance from the ship to Lighthouse A, to the nearest tenth? C
   A. 37.3 mi B. 39.1 mi C. 41.9 mi D. 42.9 mi

Short Response

6. In $\triangle XYZ$, $m\angle X = 35$, $m\angle Y = 58$, and $YZ = 12$. To the nearest tenth, what are XY and XZ? [2] $XY \approx 20.9$, $XZ \approx 17.7$ [1] one correct segment length [0] neither segment length correct OR no response

Online Teacher Resource Center

Activities, Games, and Puzzles

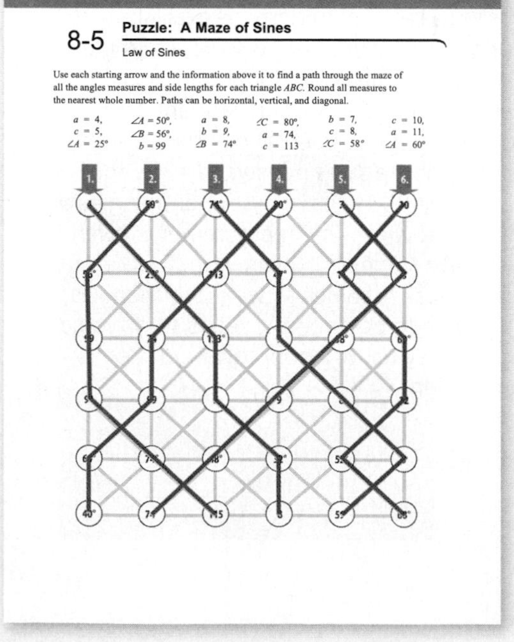

8-5 Puzzle: A Maze of Sines
Law of Sines

Use each starting arrow and the information above it to find a path through the maze of all the angles measures and side lengths for each triangle ABC. Round all measures to the nearest whole number. Paths can be horizontal, vertical, and diagonal.

1. $a = 4$, $c = 5$, $\angle A = 25°$
2. $\angle A = 50°$, $\angle B = 56°$, $b = 99$
3. $a = 8$, $b = 9$, $\angle B = 74°$
4. $\angle C = 80°$, $a = 74$, $c = 113$
5. $b = 7$, $c = 8$, $\angle C = 58°$
6. $c = 10$, $a = 11$, $\angle A = 60°$

8-6 Law of Cosines

Common Core State Standards
G-SRT.D.11 Understand and apply the . . . Law of Cosines . . . **Also G-SRT.D.10**
MP 1, MP 3, MP 4, MP 7

Objective To apply the Law of Cosines

Think about how the areas of the squares are related to the side lengths of the triangle. What does this remind you of?

Getting Ready!

In the diagram, $\triangle ABC$ is an acute triangle. Use what you know about right triangle trigonometry to write an expression for the area of the shaded region that uses a, b, and C.

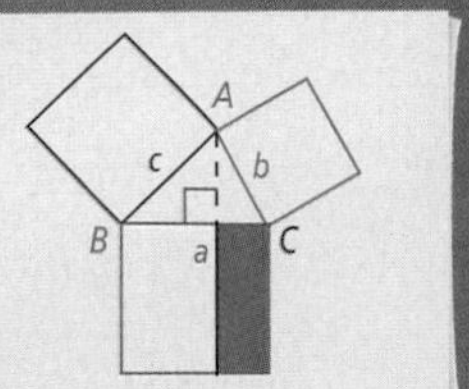

MATHEMATICAL PRACTICES In the Solve It, you used right triangle trigonometry to write an expression to describe a side length. You can also find relationships between the angle measures and the side lengths of nonright triangles.

Essential Understanding If you know the measures of two side lengths and the measure of the included angle (SAS), or all three side lengths (SSS), then you can find all the other measures of the triangle.

Lesson Vocabulary
- Law of Cosines

Key Concept Law of Cosines

For any $\triangle ABC$, the **Law of Cosines** relates the cosine of each angle to the side lengths of the triangle.

$a^2 = b^2 + c^2 - 2bc \cos A$
$b^2 = a^2 + c^2 - 2ac \cos B$
$c^2 = a^2 + b^2 - 2ab \cos C$

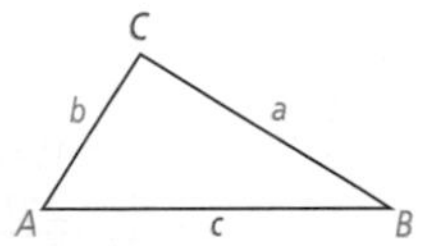

Here's Why It Works Note that $b^2 = x^2 + h^2$ and $x = b \cos A$. Use the Pythagorean Theorem with $\triangle BCD$ and simplify.

$a^2 = (c - x)^2 + h^2$	Pythagorean Theorem
$a^2 = c^2 - 2cx + x^2 + h^2$	Simplify.
$a^2 = c^2 - 2cb \cos A + b^2$	Substitute b^2 for $x^2 + h^2$ and $b \cos A$ for x.
$a^2 = b^2 + c^2 - 2bc \cos A$	Commutative Property

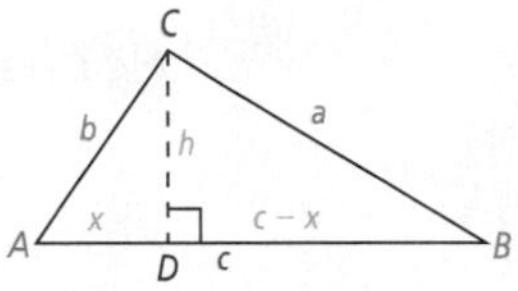

1 Interactive Learning

Solve It!

PURPOSE To use right triangle trigonometry to represent an area.

PROCESS Students may see that they need to find an expression for the width of the rectangular shaded region. The cosine function relates angle C, a, and b.

FACILITATE

Q What kinds of triangles are formed when the dashed altitude is drawn from vertex A to side BC of triangle ABC? Why is this important? **[Right triangles; this allows you to use trigonometric ratios.]**

Q Which trigonometric ratio relates the adjacent side to the hypotenuse b? **[cosine ratio]**

ANSWER See Solve It in Answers on next page.

CONNECT THE MATH In the Solve It, students use trigonometry to find relationships between angles and side lengths of a right triangle. In this lesson, they will use the Law of Cosines to find relationships between sides and angles of any triangle.

2 Guided Instruction

Take Note

Q What is the relationship between the side lengths and the angle on the right side? **[The angle is the included angle for the sides.]**

Here's Why It Works

The Law of Cosines is also known as the Generalized Pythagorean Theorem because it applies to any triangle. In the equation $c^2 = a^2 + b^2 - 2ab \cos C$, what is the result when C is a right angle? **[The equation becomes $c^2 - a^2 + b^2$.]**

8-6 Preparing to Teach

BIG idea Similarity

ESSENTIAL UNDERSTANDING

- If the measures of two side lengths of a triangle and their included angle (SAS) are known, or all three side lengths are known (SSS), all the other measures of the triangle can be found.

Math Background

In order to use the Law of Sines to solve a triangle, you must know two sides and a nonincluded angle measure, or one side and two angle measures. Explain to students that the Law of Cosines can be used to solve triangles for which the Law of Sines cannot be used, such as

- when three side lengths are known (SSS)
- when two sides and the included angle are known (SAS)

The formula for the Law of Sines is simpler than the Law of Cosines. Many times when solving a triangle, the Law of Cosines can be applied to find a missing measure, and then the Law of Sines can be used to find the remaining unknown measures.

Mathematical Practices

Look for and make use of structure. With the Law of Cosines, students will recognize the significance of angle measures and the lengths of adjacent sides in relation to each other. With the Law of Cosines, they will be able to solve a triangle using SSS and SAS.

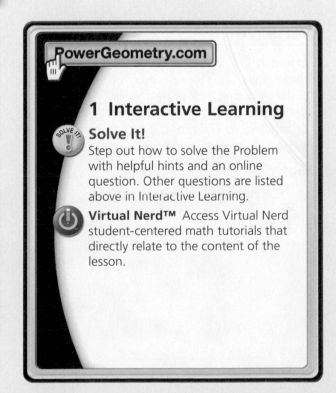

PowerGeometry.com

1 Interactive Learning

Solve It! Step out how to solve the Problem with helpful hints and an online question. Other questions are listed above in Interactive Learning.

Virtual Nerd™ Access Virtual Nerd student-centered math tutorials that directly relate to the content of the lesson.

Problem 1 EXTENSION

Q Is it possible to solve for b using the Law of Sines? Explain. **[No, you need to know an angle measure and an opposite side length to use the Law of Sines.]**

Q Would the solution method be any different if the vertices of $\triangle ABC$ were different letters, say *WCP*? Explain. **[No, the triangle is a SAS case regardless of the labels on the vertices, so the solution method will be the same.]**

Got It?

Q Does it make sense for a triangle to have a negative side length or a negative angle measure? **[No, side lengths and angles must be positive.]**

Q For what kinds of angles is the cosine function negative? What do you think happens in the formula for the Law of Cosines in these cases? **[Obtuse angles; the negative signs cancel to result in positive values.]**

Problem 2

In this problem, students will apply the Law of Cosines to a triangle when all side lengths are known.

Q What is the smallest angle measure in $\triangle TUV$? What is the largest? How can you use this as a simple check of your answer? **[V, T; the result should be a small acute angle.]**

Got It?

Q How could you use the result from Problem 1 and the Law of Sines to find $m\angle T$? **[Solve $\frac{\sin 37.0^\circ}{4.4} = \frac{\sin T}{7.1}$ for T.]**

Problem 1 Using the Law of Cosines (SAS)

Find b to the nearest tenth.

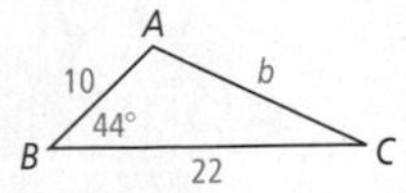

Know	Need	Plan
$\overline{AB}$ is opposite $\angle C$, so $AB = c = 10$. $\overline{BC}$ is opposite $\angle A$, so $BC = a = 22$. $m\angle B = 44$	the length of $\overline{AC}$	Because you know $m\angle B$ and need b, substitute the angle measure and the two side lengths into $b^2 = a^2 + c^2 - 2ac\cos B$ and solve for b.

$b^2 = a^2 + c^2 - 2ac\cos B$ Law of Cosines

$b^2 = 22^2 + 10^2 - 2(22)(10)\cos 44^\circ$ Substitute.

$b \approx 16.35513644$ Use a calculator.

The value of b is about 16.4.

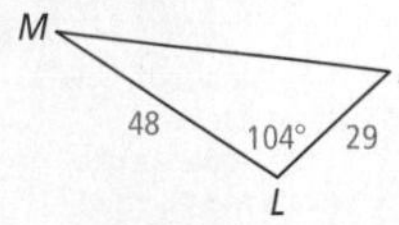

Got It? 1. Find *MN* to the nearest tenth.

Problem 2 Using the Law of Cosines (SSS)

In $\triangle TUV$, $TU = 4.4$, $UV = 7.1$, and $TV = 6.7$. Find $m\angle V$ to the nearest tenth of a degree.

Plan
How can you use what you know to find $m\angle V$?
You know the three side lengths (SSS) so you can use the Law of Cosines to find $m\angle V$.

Step 1 Draw and label a diagram.

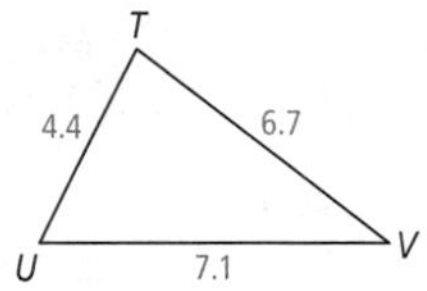

Step 2 Use the Law of Cosines to set up an equation.

$TU^2 = UV^2 + TV^2 - 2(UV)(TV)\cos V$

$4.4^2 = 7.1^2 + 6.7^2 - 2(7.1)(6.7)\cos V$

Step 3 Solve for $m\angle V$.

$19.36 = 50.41 + 44.89 - 95.14\cos V$ Simplify.

$\frac{-75.94}{-95.14} = \cos V$ Solve for cos V.

$V = \cos^{-1}\left(\frac{-75.94}{-95.14}\right)$ Solve for $m\angle V$.

$m\angle V \approx 37.04219062$ Use a caclulator.

The measure of $\angle V$ is about 37.0.

Got It? 2. In $\triangle TUV$ above, find $m\angle T$ to the nearest tenth degree.

Answers

Solve It!

Area of the shaded region = $ab\cos C$; the height of the shaded region is a, and the width is x. Use the cosine function to solve for x in terms of b and C: $\cos C = \frac{x}{b}$ so $x = b\cos C$. So, the area of the rectangular shaded region is $ab\cos C$.

Got It?

1. 61.8
2. 76.4°
3. 2.1 mi

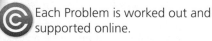

PowerGeometry.com

2 Guided Instruction

Each Problem is worked out and supported online.

Problem 1
Using the Law of Cosines (SAS)

Problem 2
Using the Law of Cosines (SSS)

Problem 3
Using the Law of Cosines to Solve a Problem

Support in Geometry Companion
- Vocabulary
- Key Concepts
- Got It?

You can use the Law of Cosines to solve real world problems involving triangles.

Problem 3 Using the Law of Cosines to Solve a Problem

Think

What do you need to find before you can use the law of Cosines?

You need to find the measure of the angle opposite d in the triangle before you can apply the Law of Cosines.

An air traffic controller is tracking a plane 2.1 kilometers due south of the radar tower. A second plane is located 3.5 kilometers from the tower at a heading of N 75° E (75° east of north). To the nearest tenth of a kilometer, how far apart are the two planes?

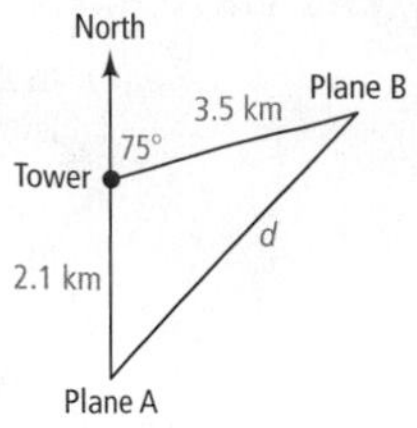

The north-south line in the figure represents a straight angle. Let the angle opposite d be $\angle D$. Use supplementary angles to find $m\angle D$. Use supplementary angles to find the measure of the angle opposite d.

$m\angle D = 180 - 75 = 105$ Supplementary angles

Use the Law of Cosines to solve for d.

$d^2 = a^2 + b^2 - 2ab\cos D$ Law of Cosines

$d^2 = 3.5^2 + 2.1^2 - 2(3.5)(2.1)\cos 105°$ Substitute.

$d \approx 4.52378602$ Use a calculator.

The distance between the two planes is about 4.5 kilometers.

Got It? 3. You and a friend hike 1.4 miles due west from a campsite. At the same time, two other friends hike 1.9 miles at a heading of S 11° W (11° west of south) from the campsite. To the nearest tenth of a mile, how far apart are the two groups?

Lesson Check

Do you know HOW?

1. In $\triangle ABC$, $AB = 7$, $BC = 10$, and $m\angle B = 80$. To the nearest tenth, what is b?
2. In $\triangle QRS$, $QR = 31.9$, $RS = 25.2$, and $QS = 37.6$. To the nearest tenth, what is $m\angle R$?
3. In $\triangle LMN$, $LN = 7$, $MN = 10$, and $m\angle N = 48$. To the nearest tenth, what is the area of $\triangle LMN$?
4. What are $m\angle X$, $m\angle Y$, and $m\angle Z$?

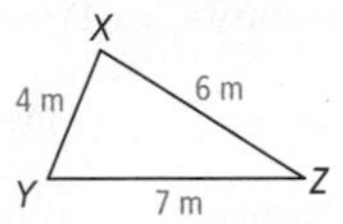

Do you UNDERSTAND? MATHEMATICAL PRACTICES

5. **Error Analysis** In $\triangle ABC$, $AC = 15$ ft, $BC = 12$ ft, and $m\angle C = 32$. A student solved for c for $a = 12$ ft, $b = 15$ ft, and $m\angle C = 32$. What was the error?

~~$C = 12^2 + 15^2 - 2(12)(15)\cos 32°$~~
~~$C = 369 - 360\cos 32°$~~
~~$C = 63.7$~~

6. **Reasoning** Explain how you would find the measure of the largest angle of a triangle if given the measures of the three side lengths.

Problem 3

In this problem, students will use the Law of Cosines to solve a real-world application problem.

Q Is it possible to use the Law of Sines to solve for d? Explain. **[No, you do not know an angle measure and opposite side length.]**

Q Why do you think N 75° E is interpreted as 75° east of north? **[To find the heading, you begin by pointing north and then rotate 75° clockwise to the east.]**

Got It? VISUAL LEARNERS

Suggest to students that they draw a diagram of the problem situation to help them see how to solve it.

Q How do you draw a ray to represent a heading of S 11° W? **[Begin by pointing due south and then rotate 11° clockwise to the west.]**

3 Lesson Check

Do you know HOW? ERROR INTERVENTION

- If students have trouble solving Exercises 1–3, then have them draw diagrams and label them using the given information.

Do you UNDERSTAND?

- In Exercise 6, remind students that the largest angle of a triangle will be opposite the longest side.

Close

Q What is an advantage of using the Law of Cosines over the Law of Sines to solve a triangle? What is a disadvantage? **[Advantage: the Law of Cosines can be used to solve SSS and SAS triangles; disadvantage: the Law of Cosines is more computationally involved.]**

Additional Problems

1. Find YT to the nearest tenth.

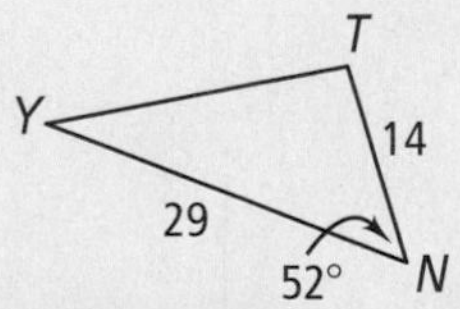

ANSWER 23.2

2. In $\triangle QFA$, $QF = 9$, $FA = 5$, and $QA = 8$. Find $m\angle F$ to the nearest tenth of a degree.
ANSWER 62.2°

3. A boat leaves a port and travels 1.3 kilometers due east. A second boat travels 1.8 kilometers from the port at a heading of S 22° E (22° east of south). To the nearest tenth of a mile, how far apart are the two boats?
ANSWER 1.8 km

Answers

Lesson Check

1. 11.2
2. 81.5
3. 26.0 sq. units
4. $m\angle X \approx 86.4$, $m\angle Y \approx 58.8$, $m\angle Z \approx 34.8$
5. The variable c should be squared, so the actual value of c is about 7.98.
6. Use the Law of Cosines to solve for the measure of the angle that is opposite the longest side of the triangle.

4 Practice

ASSIGNMENT GUIDE

Basic: 7–17, 18–22 even, 23, 26
Average: 7–17 odd, 18–26
Advanced: 7–17 odd, 18–28

Mathematical Practices are supported by exercises with red headings. Here are the Practices supported in this lesson:

MP 1: Persevere in Solving Problems Ex. 28
MP 3: Construct Arguments Ex. 6
MP 3: Critique the Reasoning of Others Ex. 5

Applications exercises have blue headings. Exercises 15, 16, 17, 22, 24, and 26 support MP 4: Model.

EXERCISE 22: Use the Think About a Plan worksheet in the **Practice and Problem Solving Workbook** (also available in the Teaching Resources in print and online) to further support students' development in becoming independent learners.

HOMEWORK QUICK CHECK

To check students' understanding of key skills and concepts, go over Exercises 13, 18, 20, 23, and 26.

Practice and Problem-Solving Exercises

MATHEMATICAL PRACTICES

A Practice Use the information given to solve. See Problems 1 and 2.

7. In $\triangle QRS$, $m\angle R = 38$, $QR = 11$, and $RS = 16$. To the nearest tenth, what is the length of $\overline{QS}$?

8. In $\triangle WXY$, $WX = 20.4$, $XY = 16.4$, and $WY = 25.3$. To the nearest tenth, what is $m\angle W$?

9. In $\triangle JKL$, $JK = 2.6$, $KL = 6.4$, and $m\angle K = 10.5$. To the nearest tenth, what is the length of $\overline{JL}$?

10. In $\triangle DEF$, $DE = 13$, $EF = 24$, and $FD = 27$. To the nearest tenth, what is $m\angle E$?

Use the Law of Cosines to find the values of x and y. Round to the nearest tenth.

11.

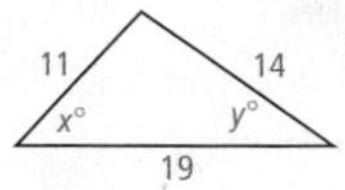

12.

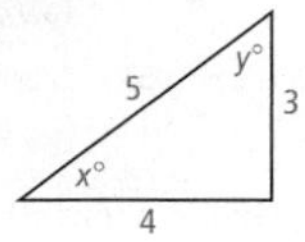

13.

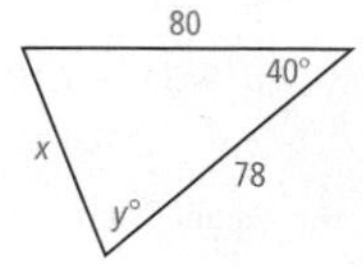

14.

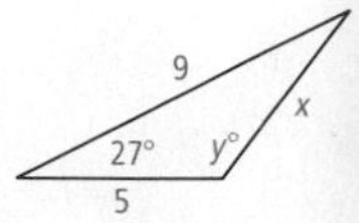

Use the Law of Cosines to solve each problem. See Problem 3.

15. Baseball After fielding a ground ball, a pitcher is located 110 feet from first base and 57 feet from home plate as shown in the figure at the right. To the nearest tenth, what is the measure of the angle with its vertex at the pitcher?

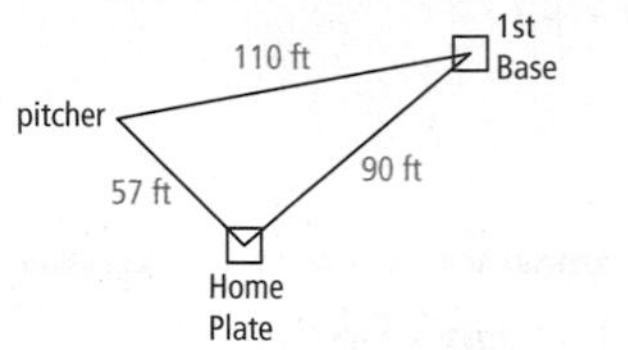

16. Zipline One side of a ravine is 14 ft long. The other side is 12 ft long. A 20 ft zipline runs from the top of one side of the ravine to the other. To the nearest tenth, at what angle do the sides of the ravine meet?

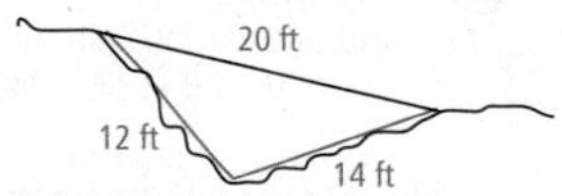

B Apply

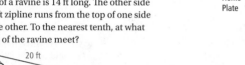

17. Think About a Plan A walking path around the outside of a garden is shaped like a triangle. Two sides of the path that measure 32 ft and 39 ft form a 76° angle. If you walk around the entire path one time, how far have you walked? Write your answer to the nearest foot.

- What information do you need to find before you can solve this problem?
- How can you find the information you need?
- Can drawing a diagram help you solve this problem?

Answers

Practice and Problem-Solving Exercises

7. 10.0
8. 40.3
9. 3.9
10. 88.5
11. $x \approx 46.8$, $y \approx 35.0$
12. $x \approx 36.9$, $y \approx 53.1$
13. $x \approx 54.1$, $y \approx 72.0$
14. $x \approx 5.1$, $y \approx 126.5$
15. 54.7
16. 100.3
17. 115 ft

18. **Airplane** A commuter plane flies from City A to City B, a distance of 90 mi due north. Due to bad weather, the plane is redirected at take-off to a heading N 60° W (60° west of north). After flying 57 mi, the plane is directed to turn northeast and fly directly toward City B. To the nearest tenth, how many miles did the plane fly on the last leg of the trip?

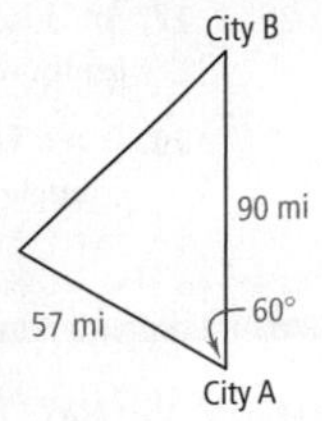

For each triangle shown below, determine whether you would use the Law of Sines or Law of Cosines to find the value of *x*. Then find the value of *x* to the nearest tenth.

19.

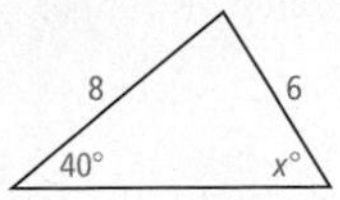

20.

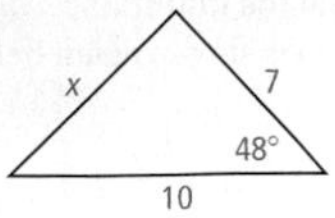

21.

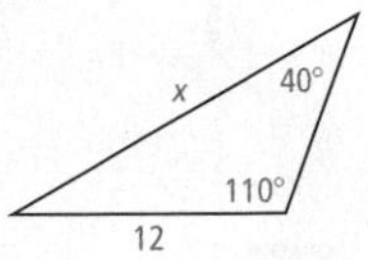

22.

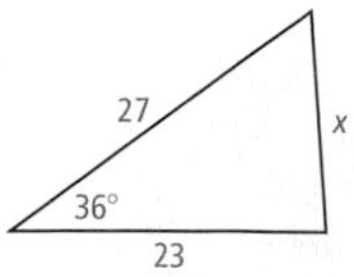

23. A 15-ft water slide has a 9.5-ft ladder which meets the slide at a 95° angle. To the nearest tenth, what is the distance between the end of the slide and the bottom of the ladder?

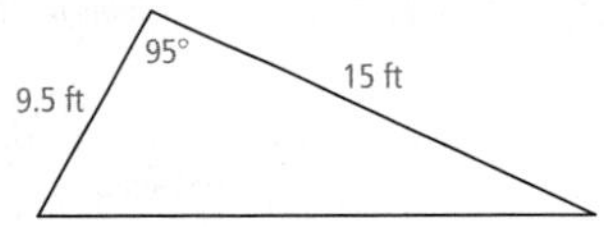

24. **Flags** The dimensions of a triangular flag are 18 ft by 25 ft by 27 ft. To the nearest tenth, what is the measure of the angle formed by the two shorter sides?

Challenge

25. Parallelogram *QRST* has a perimeter of 62 mm. To the nearest tenth, what is the length of $\overline{TR}$?

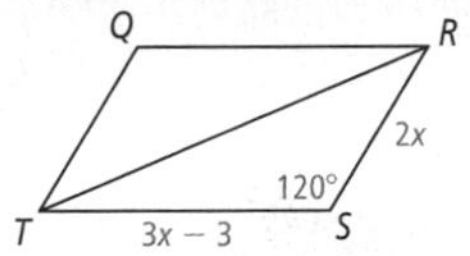

STEM 26. **Surveying** A surveyor measures the distance to the base of a monument to be 12.4 meters at an angle of elevation of 11°. At an angle of elevation of 26°, the distance to the top of the monument is 13.3 meters. What is the height of the monument to the nearest tenth?

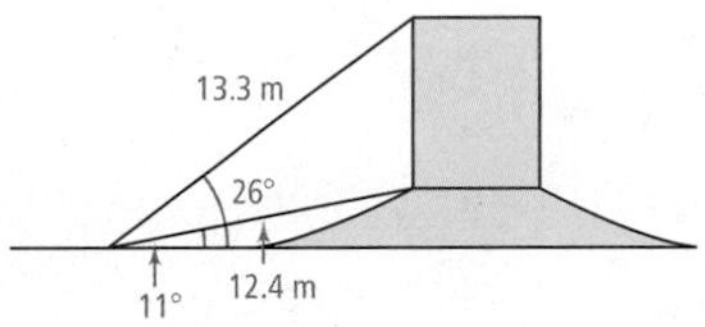

18. 78.9 mi
19. Law of Sines; 59.0
20. Law of Cosines; 7.4
21. Law of Sines; 17.5
22. Law of Cosines; 15.9
23. 18.4 ft
24. 75.9
25. 26.9
26. 3.5 m

Answers

Practice and Problem-Solving Exercises (continued)

27. 73.7, 53.1, 53.1

28. Answers may vary. Sample: If you are given the measure of each angle of a triangle, then you cannot solve for the side lengths of the triangle.

27. An isosceles triangle XYZ has a base of 12 in. and a height of 8 in. To the nearest tenth, what are the measures of the angles?

28. Open-Ended Describe a situation in which you are given three measures of a triangle but are unable to solve the triangle for the other three measures.

Apply What You've Learned

MATHEMATICAL PRACTICES
MP 1, MP 2

Look back at the information on page 489 about the fire in a state forest. The diagram is shown again below.

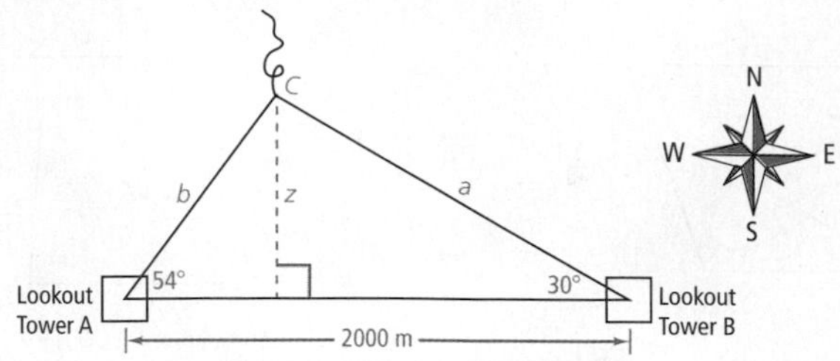

For parts (a) and (b), refer to the distances you found in the Apply What You Learned section in Lesson 8-5.

a. Use the Law of Cosines to verify the distance between Lookout Tower A and the fire.

b. Use the Law of Cosines to verify the distance between Lookout Tower B and the fire.

c. Write an equation that can be used to find how far west of Lookout Tower B the fire is located.

Apply What You've Learned

In the Apply What You've Learned for Lesson 8-5, students found the values of a and b in the diagram from page 489. Here, they verify those measures using the Law of Cosines, and write an equation that can be used to find how far west of Lookout Tower B the fire is located.

Mathematical Practices

In parts (a) and (b), students **persevere in solving the problem** as they verify previous results using another method. (MP 1)

In part (c), students **reason abstractly** to write a trigonometric equation based on the situation. (MP 2)

ANSWERS

a. $1005.5^2 \stackrel{?}{=} 1626.9^2 + 2000^2 - 2(1626.9)(2000)(\cos 30°)$; $1{,}011{,}030.3 \approx 1{,}011{,}056.7$

b. $1626.9^2 \stackrel{?}{=} 1005.5^2 + 2000^2 - 2(1005.5)(2000)(\cos 54°)$; $2{,}646{,}803.6 \approx 2{,}646{,}958.0$

c. Answers may vary. Sample: $\cos 30° = \frac{x}{1626.9}$ where x represents the fire's distance to the west of Lookout Tower B.

8-6 Lesson Resources

Differentiated Remediation

Additional Instructional Support

Geometry Companion

Students can use the **Geometry Companion** worktext (4 pages) as you teach the lesson. Use the Companion to support

- Solve It!
- New Vocabulary
- Key Concepts

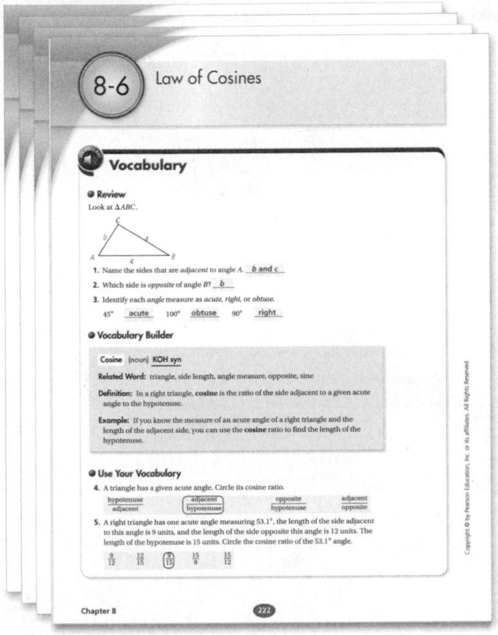

8-6 Law of Cosines

Vocabulary

Review

Look at $\triangle ABC$.

1. Name the sides that are *adjacent* to angle *A*. b and c
2. Which side is *opposite* of angle *B*? b
3. Identify each *angle* measure as *acute, right,* or *obtuse.*

45° acute 100° obtuse 90° right

Vocabulary Builder

Cosine (noun) KOH syn

Related Word: triangle, side length, angle measure, opposite, sine

Definition: In a right triangle, **cosine** is the ratio of the side adjacent to a given acute angle to the hypotenuse.

Example: If you know the measure of an acute angle of a right triangle and the length of the adjacent side, you can use the **cosine** ratio to find the length of the hypotenuse.

Use Your Vocabulary

4. A triangle has a given acute angle. Circle its cosine ratio.

$\frac{\text{hypotenuse}}{\text{adjacent}}$ $\frac{\text{adjacent}}{\text{hypotenuse}}$ $\frac{\text{opposite}}{\text{hypotenuse}}$ $\frac{\text{adjacent}}{\text{opposite}}$

5. A right triangle has one acute angle measuring 53.1°, the length of the side adjacent to this angle is 9 units, and the length of the side opposite this angle is 12 units. The length of the hypotenuse is 15 units. Circle the cosine ratio of the 53.1° angle.

$\frac{9}{12}$ $\frac{12}{15}$ $\frac{9}{15}$ $\frac{15}{9}$ $\frac{15}{12}$

Chapter 8 222

ELL Support

Connect to Prior Knowledge of Math

Students should compare and contrast the processes of solving triangles using the Law of Cosines and the Law of Sines. Ask students to name advantages and disadvantages of each method.

- Ask: If a triangle can be solved using either method, which one would you use and why?

Draw several triangles on the board with different combinations of angles and side lengths given such as SSS, SAS, SSA, AAS, and AAA. Ask students to identify which triangles can be solved using either method, which triangles can be solved using only one method, and which triangles cannot be solved.

5 Assess & Remediate

Lesson Quiz

Use the Lesson Quiz to assess students' mastery of the skills and concepts of this lesson.

1. In $\triangle ABC$, $AC = 15$, $BC = 11$, and $m\angle C = 75$. To the nearest tenth, what is the length of $\overline{AB}$?
2. In $\triangle LMN$, $LM = 6$, $MN = 9$, and $LN = 7$. Find $m\angle M$ to the nearest tenth of a degree.
3. **Do you UNDERSTAND?** A quarterback looks downfield at a receiver who is 14 yards away. The receiver is covered, so the quarterback turns 30° to his right and sees another receiver who is 19 yards away. To the nearest tenth of a yard, how far apart are the two receivers?

ANSWERS TO LESSON QUIZ

1. 16.1 units
2. 51.0
3. 9.8 yd

PRESCRIPTION FOR REMEDIATION

Use the student work on the Lesson Quiz to prescribe a differentiated review assignment.

Points	Differentiated Remediation
0–1	Intervention
2	On-level
3	Extension

PowerGeometry.com

5 Assess & Remediate

Assign the Lesson Quiz. Appropriate intervention, practice, or enrichment is automatically generated based on student performance.

Intervention

- **Reteaching** (2 pages) Provides reteaching and practice exercises for the key lesson concepts. Use with struggling students or absent students.
- **English Language Learner Support** Helps students develop and reinforce mathematical vocabulary and key concepts.

All-in-One Resources/Online

Reteaching

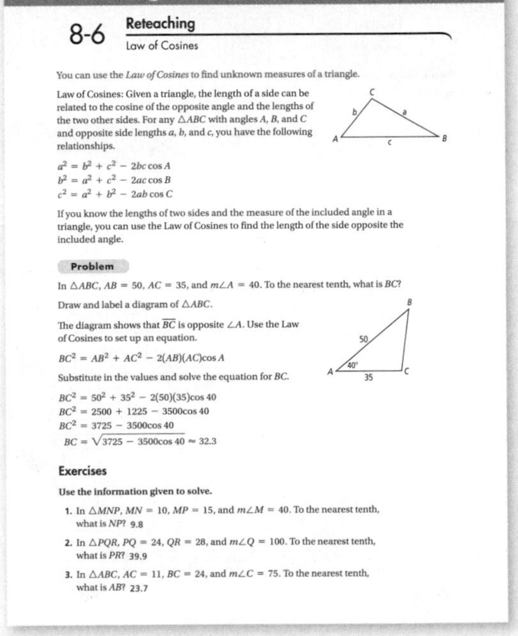

8-6 Reteaching

Law of Cosines

You can use the *Law of Cosines* to find unknown measures of a triangle.

Law of Cosines: Given a triangle, the length of a side can be related to the cosine of the opposite angle and the lengths of the two other sides. For any $\triangle ABC$ with angles *A*, *B*, and *C* and opposite side lengths *a*, *b*, and *c*, you have the following relationships.

$a^2 = b^2 + c^2 - 2bc\cos A$
$b^2 = a^2 + c^2 - 2ac\cos B$
$c^2 = a^2 + b^2 - 2ab\cos C$

If you know the lengths of two sides and the measure of the included angle in a triangle, you can use the Law of Cosines to find the length of the side opposite the included angle.

Problem

In $\triangle ABC$, $AB = 50$, $AC = 35$, and $m\angle A = 40$. To the nearest tenth, what is BC?

Draw and label a diagram of $\triangle ABC$.

The diagram shows that $\overline{BC}$ is opposite $\angle A$. Use the Law of Cosines to set up an equation.

$BC^2 = AB^2 + AC^2 - 2(AB)(AC)\cos A$

Substitute in the values and solve the equation for BC.

$BC^2 = 50^2 + 35^2 - 2(50)(35)\cos 40$
$BC^2 = 2500 + 1225 - 3500\cos 40$
$BC^2 = 3725 - 3500\cos 40$
$BC = \sqrt{3725 - 3500\cos 40} \approx 32.3$

Exercises

Use the information given to solve.

1. In $\triangle MNP$, $MN = 10$, $MP = 15$, and $m\angle M = 40$. To the nearest tenth, what is NP? 9.8
2. In $\triangle PQR$, $PQ = 24$, $QR = 28$, and $m\angle Q = 100$. To the nearest tenth, what is PR? 39.9
3. In $\triangle ABC$, $AC = 11$, $BC = 24$, and $m\angle C = 75$. To the nearest tenth, what is AB? 23.7

All-in-One Resources/Online

English Language Learner Support

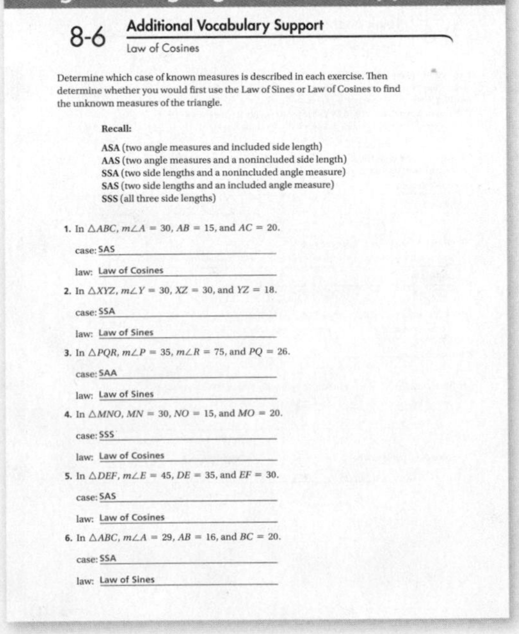

8-6 Additional Vocabulary Support

Law of Cosines

Determine which case of known measures is described in each exercise. Then determine whether you would first use the Law of Sines or Law of Cosines to find the unknown measures of the triangle.

Recall:

ASA (two angle measures and included side length)
AAS (two angle measures and a nonincluded side length)
SSA (two side lengths and a nonincluded angle measure)
SAS (two side lengths and an included angle measure)
SSS (all three side lengths)

1. In $\triangle ABC$, $m\angle A = 30$, $AB = 15$, and $AC = 20$.
case: SAS
law: Law of Cosines
2. In $\triangle XYZ$, $m\angle Y = 30$, $XZ = 30$, and $YZ = 18$.
case: SSA
law: Law of Sines
3. In $\triangle PQR$, $m\angle P = 35$, $m\angle R = 75$, and $PQ = 26$.
case: SAA
law: Law of Sines
4. In $\triangle MNO$, $MN = 30$, $NO = 15$, and $MO = 20$.
case: SSS
law: Law of Cosines
5. In $\triangle DEF$, $m\angle E = 45$, $DE = 35$, and $EF = 30$.
case: SAS
law: Law of Cosines
6. In $\triangle ABC$, $m\angle A = 29$, $AB = 16$, and $BC = 20$.
case: SSA
law: Law of Sines

Differentiated Remediation *continued*

On-Level

- **Practice** (2 pages) Provides extra practice for each lesson. For simpler practice exercises, use the Form K Practice pages found in the All-in-One Teaching Resources and online.
- **Think About a Plan** Helps students develop specific problem-solving skills and strategies by providing scaffolded guiding questions.
- **Standardized Test Prep** Focuses on all major exercises, all major question types, and helps students prepare for the high-stakes assessments.

Practice and Problem Solving Wkbk/ All-in-One Resources/Online

Practice page 1

8-6 Practice *Form G*
Law of Cosines

Use the information given to solve.

1. In $\triangle ABC$, $m\angle A = 40$, $AB = 9.2$, and $AC = 8.5$. To the nearest tenth, what is BC? 6.1
2. In $\triangle PQR$, $m\angle Q = 112$, $PQ = 12.5$, and $QR = 14.2$. To the nearest tenth, what is the length of $\overline{PR}$? 22.2
3. In $\triangle GHK$, $GH = 11$, $HK = 21$, and $GK = 15$. To the nearest tenth, what is $m\angle K$? 30.1
4. In $\triangle XYZ$, $XY = 23$, $YZ = 15$, and $XZ = 19$. To the nearest tenth, what is $m\angle Z$? 84.3
5. In $\triangle ABC$, $m\angle B = 53$, $AB = 9.2$, and $BC = 7.3$. To the nearest tenth, what is AC? 7.6
6. In $\triangle DEF$, $DE = 12.1$, $EF = 5.8$, and $DF = 10.2$. To the nearest tenth, what is $m\angle F$? 94.2

Use the Law of Cosines to find the values of *x* and *y*. Round to the nearest tenth.

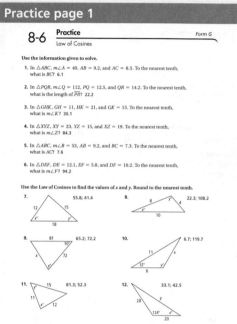

7. 55.8; 41.4
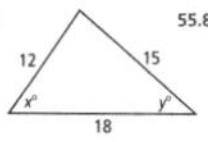

8. 22.3; 108.2
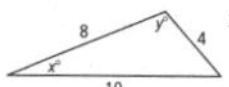

9. 65.2; 72.2
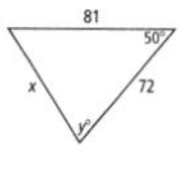

10. 6.7; 119.7
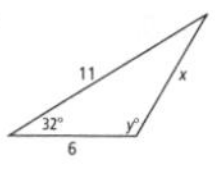

11. 81.3; 52.3
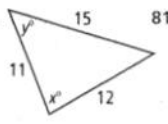

12. 33.1; 42.5
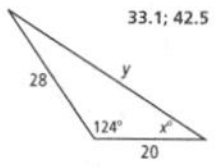

Practice and Problem Solving Wkbk/ All-in-One Resources/Online

Practice page 2

8-6 Practice (continued) *Form G*
Law of Cosines

13. One airplane is 78 miles due south of a control tower. Another airplane is 52 miles from the tower at a heading of N 38° E (38° east of north). To the nearest tenth of a mile, how far apart are the two airplanes?
123.2 mi

14. A coach sets up a triangular race course. One corner is 100 feet from the start/finish and the other corner is 85 feet from the start/finish. If the angle at the start/finish measures 55°, what is the total length of the course? Round to the nearest tenth of a foot.
271.5 ft

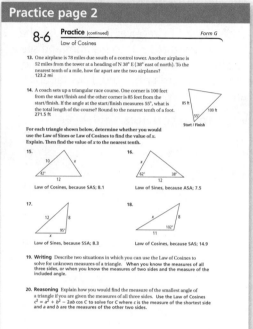

For each triangle shown below, determine whether you would use the Law of Sines or Law of Cosines to find the value of *x*. Explain. Then find the value of *x* to the nearest tenth.

15.
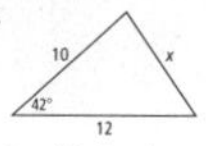

Law of Cosines, because SAS; 8.1

16.
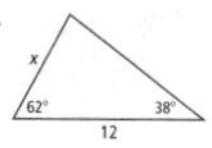

Law of Sines, because ASA; 7.5

17.
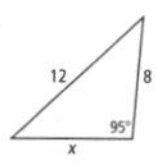

Law of Sines, because SSA; 8.3

18.
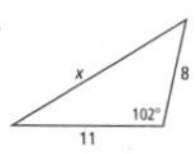

Law of Cosines, because SAS; 14.9

19. **Writing** Describe two situations in which you can use the Law of Cosines to solve for unknown measures of a triangle. When you know the measures of all three sides, or when you know the measures of two sides and the measure of the included angle.

20. **Reasoning** Explain how you would find the measure of the smallest angle of a triangle if you are given the measures of all three sides. Use the Law of Cosines $c^2 = a^2 + b^2 - 2ab \cos C$ to solve for *C* where *c* is the measure of the shortest side and *a* and *b* are the measures of the other two sides.

Practice and Problem Solving Wkbk/ All-in-One Resources/Online

Think About a Plan

8-6 Think About a Plan
Law of Cosines

A commuter plane flies from City A to City B, a distance of 90 mi due north. Due to bad weather, the plane is redirected at take-off to a heading N 60° W (60° west of north). After flying 57 mi, the plane is directed to turn northeast and fly directly toward City B. To the nearest tenth, how many miles did the plane fly on the last leg of the trip?

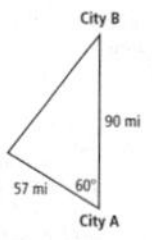

1. Based on the diagram provided for this problem, which measures of the triangle do you know?
You know the lengths of two sides and the measure of the included angle (SAS).
2. What are the values of these measures?
The lengths of the sides are 90 miles and 57 miles. The measure of the included angle is 60°.
3. Describe the part of the triangle you need to find.
You need to find the length of the side opposite the included angle measuring 60°.
4. What concept will you use to write an equation? What is the equation?
Law of Cosines; $a^2 = 90^2 + 57^2 - 2(90)(57)\cos 60$
5. Solve the equation. What is the distance of the last leg of the trip?
$a = \sqrt{6219} \approx 78.9$; 78.9 miles

Practice and Problem Solving Wkbk/ All-in-One Resources/Online

Standardized Test Prep

8-6 Standardized Test Prep
Law of Cosines

Multiple Choice

For Exercises 1–5, choose the correct letter.

1. To the nearest tenth, what is the value of *x*? A
Ⓐ 10.7 Ⓑ 21.9 Ⓒ 39.8 Ⓓ 113.9

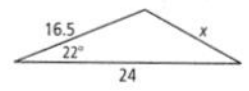

2. To the nearest tenth, what is the value of *x*? H
Ⓕ 7.7 Ⓖ 50.1 Ⓗ 59.9 Ⓘ 70.0

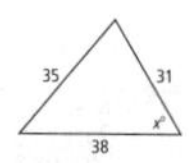

3. To the nearest tenth, what is the value of *x*? D
Ⓐ 27.2 Ⓑ 17.3 Ⓒ 13.4 Ⓓ 12.3

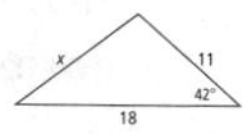

4. To the nearest tenth, what is the value of *x*? I
Ⓕ 36.5 Ⓖ 49.1 Ⓗ 56.3 Ⓘ 74.6

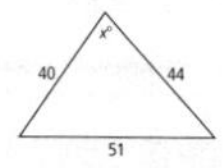

5. In $\triangle ABC$, $a = 20$, $b = 12$, and $c = 30$. What is the measure of the largest angle to the nearest tenth? B
Ⓐ 15.6° Ⓑ 26.6° Ⓒ 42.1° Ⓓ 137.9°

Short Response

6. To the nearest tenth, what are the measures of the remaining side and angles in the triangle?
[2] $r \approx 15.3$, $m\angle P \approx 55.7$, $m\angle Q \approx 76.3$ [1] correct length for *r* OR correct measures for $\angle P$ and $\angle Q$ [0] no correct measures given OR no response

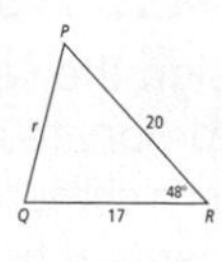

Extension

- **Enrichment** Provides students with interesting problems and activities that extend the concepts of the lesson.
- **Activities, Games, and Puzzles** Worksheets that can be used for concepts development, enrichment, and for fun!

All-in-One Resources/Online

Enrichment

8-6 Enrichment
Law of Cosines

Flight Paths

By using degree measurements to represent compass directions, you can describe the heading, or direction, in which a plane is traveling. In this system, 0° (360°) corresponds to due north, 90° corresponds to due east, 180° corresponds to due south, and 270° corresponds to due west.

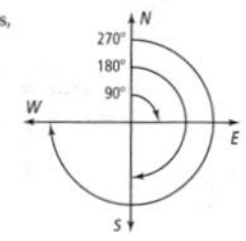

Angles are measured in a *clockwise* direction. This is different from measuring angles in standard position on a coordinate system.

With this method you can describe a flight path in terms of distances and headings. For example, suppose a plane flies 300 mi on a heading of 45°; then the plane changes course and flies 200 mi on a heading of 150°.

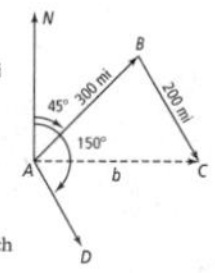

If you could determine the angle between the two "legs" of the trip ($\angle B$), you could then use the Law of Cosines to find how far the plane has traveled from its point of departure (b).

Because $\overline{BC}$ is parallel to $\overline{AD}$, $\angle B$ is supplementary to $\angle BAD$, which is 105°. Thus $\angle B = 180° - 105° = 75°$. Using the law of cosines, $b^2 = 300^2 + 200^2 - 2(300)(200)\cos 75°$, and $b = 314.6$ mi.

Determine the heading on which the plane would have to travel to return to point *A*.

1. First use the Law of Sines to find $\angle BAC$, and then add 45°.
$m\angle BAC = 37.9°$; $m\angle BAC + 45° = 82.9°$
2. Add 180° to find the return heading.
return heading is 262.9°

Suppose a plane flies 240 mi on a heading of 35°. Then the plane changes course and flies 160 mi on a heading of 160°.

3. Determine the heading on which the plane would have to travel to return to its point of origin.
return heading is 256.5°

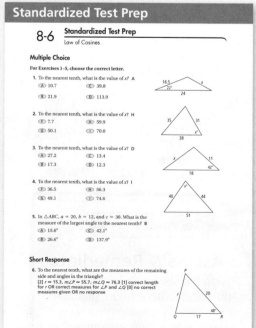

Online Teacher Resource Center

Activities, Games, and Puzzles

8-6 Game: Who can make the longest side?
Law of Cosines

Materials: two cubes numbered 1 to 6

Number of Players: 2 to 3

Number of Rounds: 10

Action of each Round: Roll a pair of cubes three times. Place numbers as lengths of two sides of a triangle and the included angle. Use the Law of Cosines to find the length of the third side.

Goal for Each Round: Have the longest third side of a triangle

Score: In each round, 1 point is earned by the player with a longer side from each player with a shorter side length. If there are three players, it is possible to earn 1 or 2 points.

Play: For each roll, a player can choose to do one the following:

- Use the numbers that land face up on the cubes as the lengths of two sides of a triangle.
- Use the numbers that land face up on the cubes as the measure of the included angle.
- Pass on the numbers of the roll. Each player is allowed to pass on only once per round.

An example for one round of play:

First roll is 3 and 4. Players can choose to use 3 and 4 for side lengths or make the measure of the included angle 34° or 43°. Players can also choose to pass on these numbers.

Second roll is 1 and 5. Players can choose to use 1 and 5 as side lengths or make the measure of the included angle 15° or 51°. If a player did not pass on the first roll, he or she can pass on this roll. If the player chooses not to pass, he or she must use these numbers for the measurement(s) not yet labeled.

Third roll is 2 and 4. If a player has not passed on either the first or second roll, this is the roll that must be passed. If a player has passed one of the previous rolls this round, these numbers for the measurement(s) not yet labeled.

For the above rolls, below is an example of triangles for Players 1, 2, and 3. These triangles are only three of many options. Player 1 scores 1 point for beating Player 3. Player 2 scores 2 points for beating both Players 1 and 2. Player 3 scores 0 points.

Player 1:
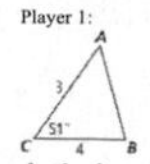

$c^2 = 3^2 + 4^2 - 2(3)(4)\cos 51°$
$c^2 = 25 - 24\cos 51°$
$c^2 \approx 9.90$
$c \approx 3.15$

Player 2:
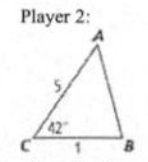

$c^2 = 5^2 + 1^2 - 2(5)(1)\cos 42°$
$c^2 = 26 - 10\cos 42°$
$c^2 \approx 18.57$
$c \approx 4.31$

Player 3:
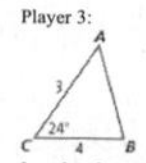

$c^2 = 4^2 + 3^2 - 2(4)(3)\cos 24°$
$c^2 = 25 - 24\cos 24°$
$c^2 \approx 3.07$
$c \approx 1.75$

8 Pull It All Together

Completing the Performance Task

Look back at your results from the Apply What You've Learned sections in Lessons 8-3, 8-5, and 8-6. Use the work you did to complete the following.

To solve these problems, you will pull together many concepts and skills that you have studied about right triangles and trigonometry.

1. Solve the problem in the Task Description on page 489 by finding how far to the north and west of Lookout Tower B the fire is located. Round your answers to the nearest tenth of a meter. Show all your work and explain each step of your solution.

2. **Reflect** Choose one of the Mathematical Practices below and explain how you applied it in your work on the Performance Task.
   - MP 1: Make sense of problems and persevere in solving them.
   - MP 2: Reason abstractly and quantitatively.
   - MP 6: Attend to precision.

On Your Own

Another fire is sighted by the rangers at the two lookout towers. The ranger at Lookout Tower A observes the smoke at an angle of 65°. The ranger at Lookout Tower B observes the smoke at an angle of 45°. Both angles are measured from the line that connects the two lookout towers.

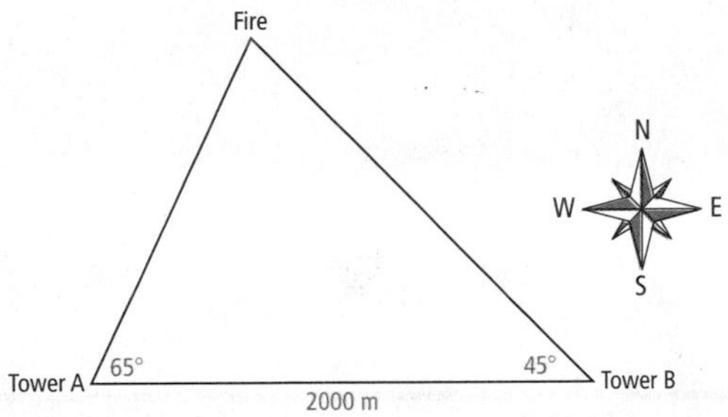

How far north of Lookout Tower A is this fire located? Round your answer to the nearest tenth of a meter.

Completing the Performance Task

In the Apply What You've Learned sections in Lessons 8-3, 8-5, and 8-6, students found equations that can be used to find the required distances in the diagram on page 489, and found the distance from each lookout tower to the fire. Here, students use this information to complete the Performance Task. Ask students the following questions as they work toward solving the problem.

Q How can you use the work you have done in the chapter to solve the problem? **[Sample: I can use trigonometric equations I wrote in Lessons 8-3 and 8-6 and the values of *a* and *b* that I found in Lesson 8-5 to solve for *z* and for the distance west of Lookout Tower B.]**

Q How can you check that your answer is reasonable? **[Sample: I can use the Pythagorean Theorem to check that the distances I found make a right triangle.]**

FOSTERING MATHEMATICAL DISCOURSE

Have students discuss alternative methods of solving the problem in the Performance Task, including methods that do not use the Law of Sines or the Law of Cosines.

ANSWERS

1. about 813.5 m to the north and 1408.9 m to the west
2. Check students' work.

On Your Own

This problem is similar to the problem posed on page 489, but now students must draw an auxiliary line. Students should strive to solve this problem independently.

ANSWER

about 1364.0 m

Essential Questions

BIG idea Measurement

ESSENTIAL QUESTION How do you find a side length or angle measure in a right triangle?

ANSWER Use the Pythagorean Theorem or trigonometric ratios to find a side length or angle measure of a right triangle. The Law of Sines and the Law of Cosines can be used to find missing side lengths and angle measures of any triangle.

BIG idea Similarity

ESSENTIAL QUESTION How do trigonometric ratios relate to similar right triangles?

ANSWER A trigonometric ratio compares the lengths of two sides of a right triangle. The ratios remain constant within a group of similar right triangles.

Summative Questions

Use the following prompts as you review this chapter with your students. The prompts are designed to help you assess your students' understanding of the BIG Ideas they have studied.

- What is the Pythagorean Theorem? When is it used?
- What are the trigonometric ratios? What are they used for?
- What is a vector? What are vectors used for?

8 Chapter Review

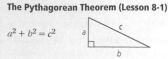

Connecting BIG ideas and Answering the Essential Questions

1 Measurement
Use the Pythagorean Theorem or trigonometric ratios to find a side length or angle measure of a right triangle. The Law of Sines and the Law of Cosines can be used to find missing side lengths and angle measures of any triangle.

2 Similarity
A trigonometric ratio compares the lengths of two sides of a right triangle. The ratios remain constant within a group of similar right triangles.

The Pythagorean Theorem (Lesson 8-1)

$a^2 + b^2 = c^2$

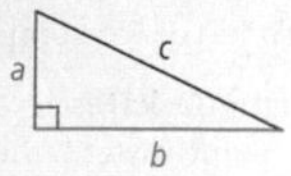

Trigonometry (Lesson 8-3)

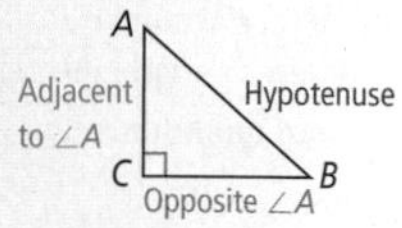

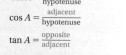

$\sin A = \frac{\text{opposite}}{\text{hypotenuse}}$

$\cos A = \frac{\text{adjacent}}{\text{hypotenuse}}$

$\tan A = \frac{\text{opposite}}{\text{adjacent}}$

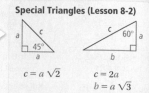

Special Triangles (Lesson 8-2)

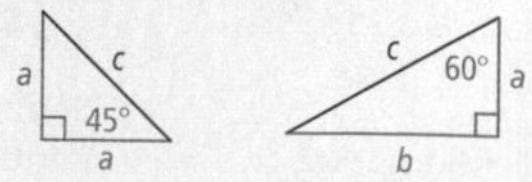

$c = a\sqrt{2}$

$c = 2a$
$b = a\sqrt{3}$

Angles of Elevation and Depression (Lesson 8-4)

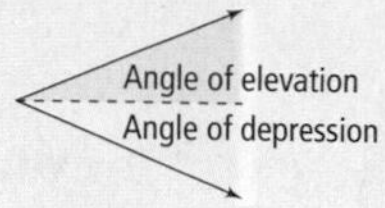

Law of Sines and Law of Cosines (Lessons 8-5 and 8-6)

$\frac{\sin A}{a} = \frac{\sin B}{b} = \frac{\sin C}{c}$

$a^2 = b^2 + c^2 - 2bc \cos A$
$b^2 = a^2 + c^2 - 2ac \cos B$
$c^2 = a^2 + b^2 - 2ab \cos C$

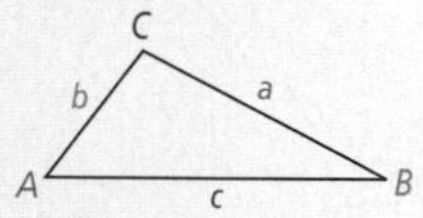

Chapter Vocabulary

- angle of depression (p. 516)
- angle of elevation (p. 516)
- cosine (p. 507)
- Law of Cosines (p. 526)
- Law of Sines (p. 522)
- Pythagorean triple (p. 492)
- sine (p. 507)
- tangent (p. 507)
- trigonometric ratios (p. 507)

Choose the correct term to complete each sentence.

1. _?_ are equivalent ratios for the corresponding sides of two triangles.
2. A(n) _?_ is formed by a horizontal line and the line of sight above that line.
3. A set of three nonzero whole numbers that satisfy $a^2 + b^2 = c^2$ form a(n) _?_.

Answers

Chapter Review

1. Trigonometric ratios
2. ∠ of elevation
3. Pythagorean triple

8-1 The Pythagorean Theorem and Its Converse

Quick Review

The **Pythagorean Theorem** holds true for any right triangle.

$$(\text{leg}_1)^2 + (\text{leg}_2)^2 = (\text{hypotenuse})^2$$

$$a^2 + b^2 = c^2$$

The Converse of the Pythagorean Theorem states that if $a^2 + b^2 = c^2$, where c is the greatest side length of a triangle, then the triangle is a right triangle.

Example

What is the value of x?

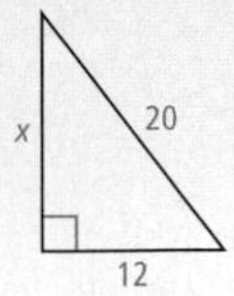

$a^2 + b^2 = c^2$	Pythagorean Theorem
$x^2 + 12^2 = 20^2$	Substitute.
$x^2 = 256$	Simplify.
$x = 16$	Take the square root.

Exercises

Find the value of x. If your answer is not an integer, express it in simplest radical form.

4.

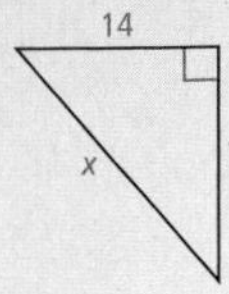

5.

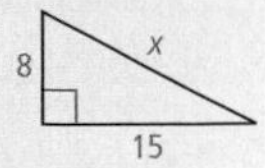

6.

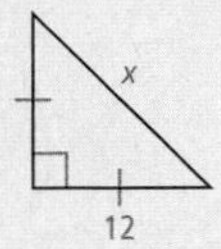

7.

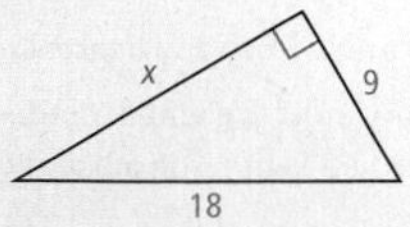

8-2 Special Right Triangles

Quick Review

45°-45°-90° Triangle

hypotenuse = $\sqrt{2}$ • leg

30°-60°-90° Triangle

hypotenuse = 2 • shorter leg
longer leg = $\sqrt{3}$ • shorter leg

Example

What is the value of x?

The triangle is a 30°-60°-90° triangle, and x represents the length of the longer leg.

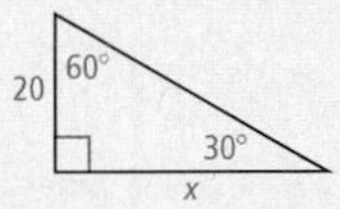

longer leg = $\sqrt{3}$ • shorter leg

$x = 20\sqrt{3}$

Exercises

Find the value of each variable. If your answer is not an integer, express it in simplest radical form.

8.

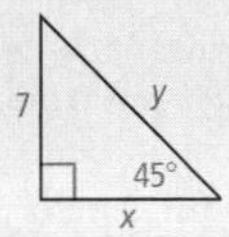

9.

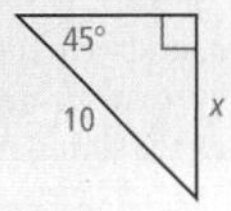

10.

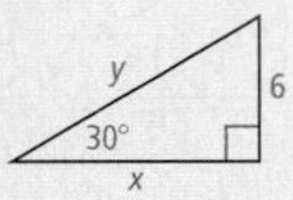

11.

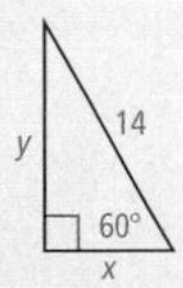

12. A square garden has sides 50 ft long. You stretch a hose from one corner of the garden to another corner along the garden's diagonal. To the nearest tenth, how long is the hose?

4. $2\sqrt{113}$
5. 17
6. $12\sqrt{2}$
7. $9\sqrt{3}$
8. $x = 7,\ y = 7\sqrt{2}$
9. $5\sqrt{2}$
10. $x = 6\sqrt{3},\ y = 12$
11. $x = 7,\ y = 7\sqrt{3}$
12. 70.7 ft

Answers

Chapter Review (continued)

13. $\frac{2\sqrt{19}}{20}$ or $\frac{\sqrt{19}}{10}$; $\frac{18}{20}$ or $\frac{9}{10}$; $\frac{2\sqrt{19}}{18}$ or $\frac{\sqrt{19}}{9}$

14. $\frac{16}{20}$ or $\frac{4}{5}$; $\frac{12}{20}$ or $\frac{3}{5}$; $\frac{16}{12}$ or $\frac{4}{3}$

15. 16.5

16. 33.1

17. 38.2 ft

18. $x = 14.7$ cm

19. $x = 29.9°$

20. $m\angle D = 82.1$

21. $m\angle N = 86.0$

8-3 and 8-4 Trigonometry and Angles of Elevation and Depression

Quick Review

In right $\triangle ABC$, C is the right angle.

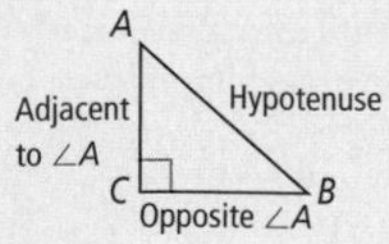

$\sin \angle A = \frac{\text{leg opposite } \angle A}{\text{hypotenuse}}$

$\cos \angle A = \frac{\text{leg adjacent to } \angle A}{\text{hypotenuse}}$

$\tan \angle A = \frac{\text{leg opposite } \angle A}{\text{leg adjacent to } \angle A}$

Example

What is FE to the nearest tenth?

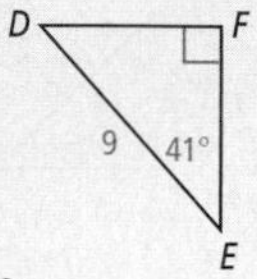

You know the length of the hypotenuse, and $\overline{FE}$ is the side adjacent to $\angle E$.

$\cos 41° = \frac{FE}{9}$ Use cosine.

$FE = 9(\cos 41°)$ Multiply each side by 9.

$FE \approx 6.8$ Use a calculator.

Exercises

Express sin A, cos A, and tan A as ratios.

13.

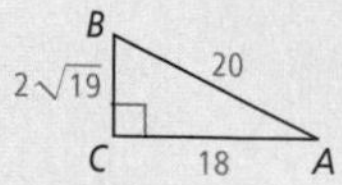

14.

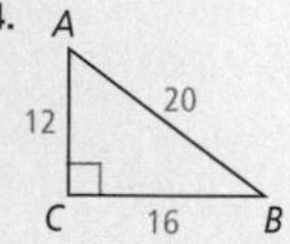

Find the value of x to the nearest tenth.

15.

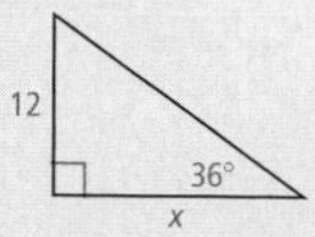

16.

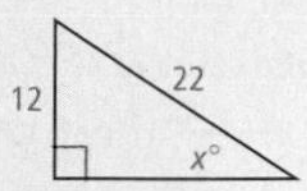

17. While flying a kite, Linda lets out 45 ft of string and anchors it to the ground. She determines that the angle of elevation of the kite is 58°. What is the height of the kite from the ground? Round to the nearest tenth.

8-5 and 8-6 Law of Sines and Law of Cosines

Quick Review

In $\triangle ABC$, a, b, and c are the lengths of the sides opposite $\angle A$, $\angle B$, and $\angle C$, respectively. The Law of Sines and the Law of Cosines are summarized below.

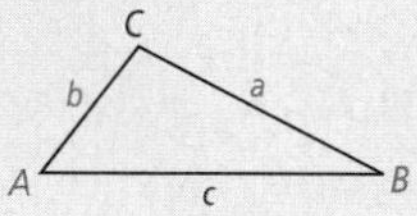

$\frac{\sin A}{a} = \frac{\sin B}{b} = \frac{\sin C}{c}$

$a^2 = b^2 + c^2 - 2bc \cos A$

$b^2 = a^2 + c^2 - 2ac \cos B$

$c^2 = a^2 + b^2 - 2ab \cos C$

Example

What is GH?

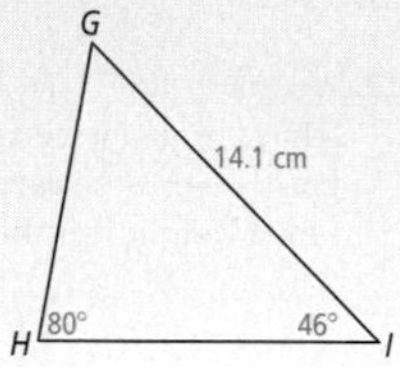

Use the Law of Sines to find GH.

$\frac{\sin 46°}{GH} = \frac{\sin 80°}{14.1}$

$GH \sin 80° = 14.1 \sin 46°$

$0.9848\,GH = 10.1427$

$GH \approx 10.3$

Exercises

Find the value of x to the nearest tenth.

18.

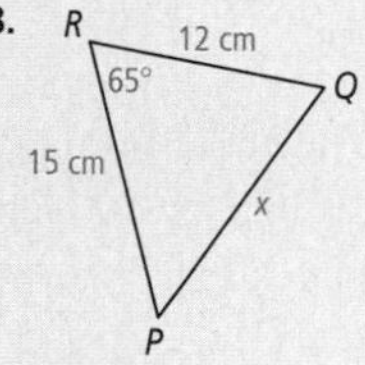

19.

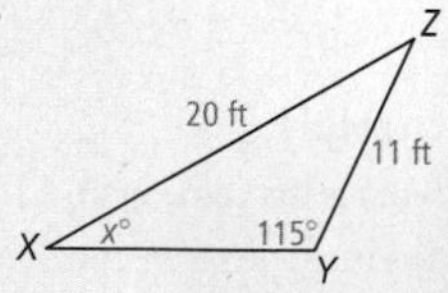

20. In $\triangle DEF$, sides d, e, and f are opposite $\angle D$, $\angle E$, and $\angle F$ respectively. The side lengths are $d = 25$ in., $e = 18$ in., and $f = 20$ in. Find the $m\angle D$ to the nearest tenth.

21. In $\triangle LMN$, sides ℓ, m, and n are opposite $\angle L$, $\angle M$, and $\angle N$ respectively. You know that $m = 3$ cm, $n = 8$ cm, and $m\angle L = 72°$. Find the $m\angle N$ to the nearest tenth.

8 Chapter Test

MathXL® for School
Go to PowerGeometry.com

Do you know HOW?

Algebra Find the value of each variable. Express your answer in simplest radical form.

1.
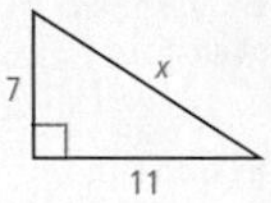

2.
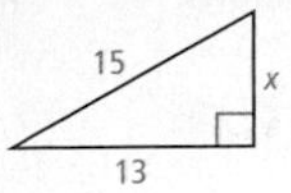

3.
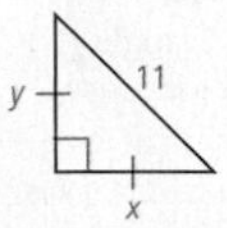

4.
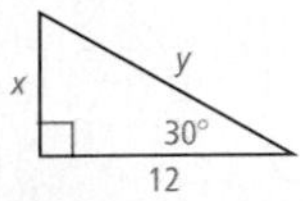

Given the following triangle side lengths, identify the triangle as *acute, right,* or *obtuse*.

5. 9 cm, 10, cm, 12, cm

6. 8 m, 15 m, 17 m

7. 5 in., 6 in., 10 in.

Express sin *B*, cos *B*, and tan *B* as ratios.

8.
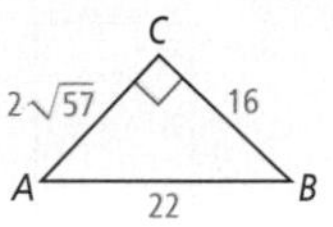

9.
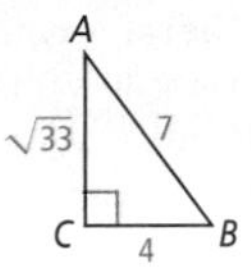

Find each missing value to the nearest tenth.

10. $\tan \blacksquare^\circ = 1.11$

11. $\sin 34^\circ = \frac{5}{\blacksquare}$

12. $\cos \blacksquare^\circ = \frac{12}{15}$

13. A woman stands 15 ft from a statue. She looks up at an angle of 60° to see the top of the statue. Her eye level is 5 ft above the ground. How tall is the statue to the nearest foot?

Find the value of *x*. Round lengths to the nearest tenth and angle measures to the nearest degree.

14.
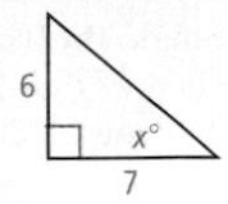

15.
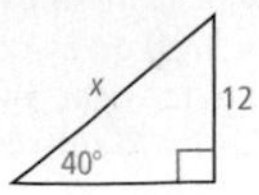

16.
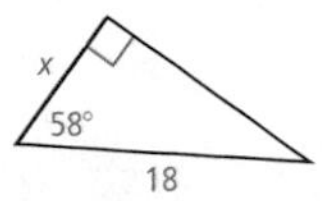

17.
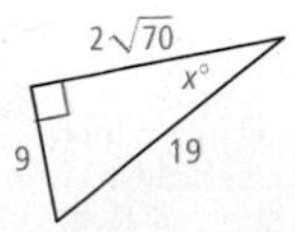

18. Find the $m\angle A$ to the nearest tenth.

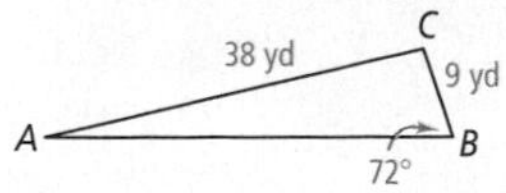

19. Find *TU* to the nearest tenth.

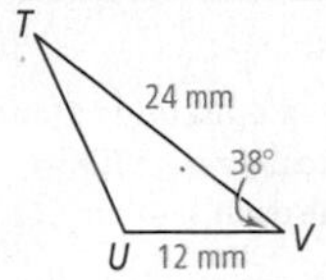

20. In $\triangle KLP$, $k = 13$ mi, $\ell = 10$ mi, and $p = 8$ mi. Find $m\angle K$ to the nearest tenth.

21. In $\triangle ABC$, $a = 8$, $b = 10$, and $m\angle B = 120$. Find the $m\angle C$ to the nearest tenth.

Do you UNDERSTAND?

22. Writing Explain why $\sin x^\circ = \cos (90 - x)^\circ$. Include a diagram with your explanation.

23. Reasoning Suppose that you know all three angle measures of a triangle. Can you use Law of Sines or Law of Cosines to find the side lengths? Explain.

24. Reasoning If you know the measures of both acute angles of a right triangle, can you determine the lengths of the sides? Explain.

Chapter Test

1. $\sqrt{170}$

2. $2\sqrt{14}$

3. $x = y = \frac{11\sqrt{2}}{2}$

4. $x = 4\sqrt{3}$, $y = 8\sqrt{3}$

5. acute

6. right

7. obtuse

8. $\frac{2\sqrt{57}}{22}$ or $\frac{\sqrt{57}}{11}$; $\frac{16}{22}$ or $\frac{8}{11}$; $\frac{2\sqrt{57}}{16}$ or $\frac{\sqrt{57}}{8}$

9. $\frac{\sqrt{33}}{7}$; $\frac{4}{7}$; $\frac{\sqrt{33}}{4}$

10. 48.0

11. 8.9

12. 36.9

13. 31 ft

14. 41

15. 18.7

16. 9.5

17. 28

18. 13.0

19. 16.3 mm

20. 91.8

21. 16.1

22.

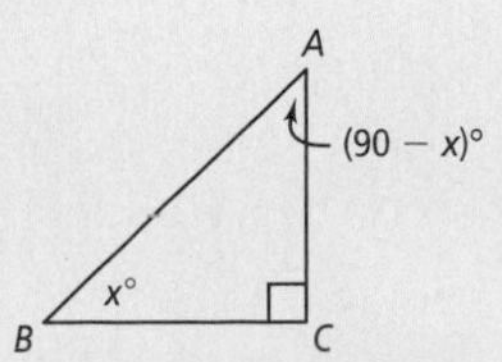

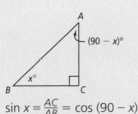

$\sin x = \frac{AC}{AB} = \cos (90 - x)$

23. No; you need at least one side to use the Law of Sines and two sides to use the Law of Cosines. You can find an infinite number of similar triangles that have congruent corresponding angles, therefore at least one side length must be given.

24. No; you need at least one side length to determine the lengths of the sides.

PowerGeometry.com

MathXL for School

Prepare students for the Mid-Chapter Quiz and Chapter Test with online practice and review.

Item Number	Lesson	Content Standard
1	8-3	G-SRT.C.8
2	4-5	*G-CO.C.10
3	8-4	G-SRT.C.8
4	2-2	MP 3
5	6-2	G-CO.A.4
6	5-6	G-CO.C.10
7	3-3	G-CO.C.9
8	8-1	G-MG.A.3
9	8-1	G-SRT.D.10
10	8-2	G-SRT.C.8
11	8-1	G-SRT.C.8
12	8-3	G-SRT.C.8
13	7-3	G-SRT.B.5
14	6-1	G-SRT.D.10
15	7-2	G-SRT.B.5
16	8-2	G-SRT.C.8
17	1-5	*G-C.A.2
18	6-6	G-SRT.B.5
19	7-4	G-SRT.D.10
20	6-7	*G-GPE.B.7
21	7-3	G-MG.A.3

*Prepares for standard

8 Common Core Cumulative Standards Review

ASSESSMENT

TIPS FOR SUCCESS

Some test questions require you to use relationships in right triangles. Read the sample question at the right. Then follow the tips to answer it.

Every day, Michael goes for a run through the park. The park is shaped like a square. He follows the path shown by the arrows. About how far does Michael run?

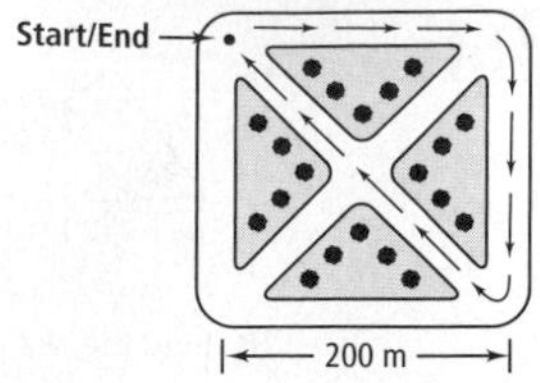

A 282.8 m
B 600 m
C 682.8 m
D 800 m

TIP 1

Michael's path is an isosceles right triangle with each leg about 200 m long. This means that he runs more than 600 m. You can eliminate choices A and B.

TIP 2

The length of the hypotenuse must be less than 400 m, so the path must be less than 800 m. You can eliminate choice D.

Think It Through

In a right isosceles triangle, the length of the hypotenuse is $\sqrt{2}$ times the length of a leg. So the distance Michael runs is $200 + 200 + \sqrt{2}(200) \approx 682.8$.

The correct answer is C.

Vocabulary Builder

As you solve test items, you must understand the meanings of mathematical terms. Choose the correct term to complete each sentence.

I. In a right triangle, the (*sine*, *cosine*) of an acute angle is the ratio of the length of the side opposite the angle to the length of the hypotenuse.

II. Polygons that have congruent corresponding angles and corresponding sides that are proportional are (*similar*, *congruent*) polygons.

III. Angles of a polygon that share a side are (*adjacent*, *consecutive*) angles.

IV. A (*proportion*, *ratio*) is a comparison of two numbers using division.

Selected Response

Read each question. Then write the letter of the correct answer on your paper.

1. What is the approximate area of the rectangle at the right?

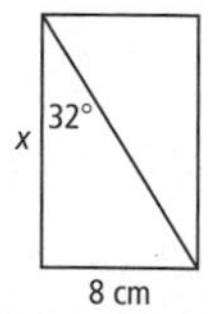

A 102 cm^2
B 75 cm^2
C 63 cm^2
D 45 cm^2

2. $\triangle ABC$ has $AB = 7$, $BC = 24$, and $CA = 24$. Which statement is true?

F $\triangle ABC$ is an equilateral triangle.
G $\triangle ABC$ is an isosceles triangle.
H $\angle C$ is the largest angle.
I $\angle B$ is the smallest angle.

Answers

Common Core Cumulative Standards Review

I. sine
II. similar
III. consecutive
IV. ratio
1. A
2. G

3. From the top of a 45-ft-tall building, the angle of depression to the edge of a parking lot is 48°. About how many feet is the base of the building from the edge of the parking lot?

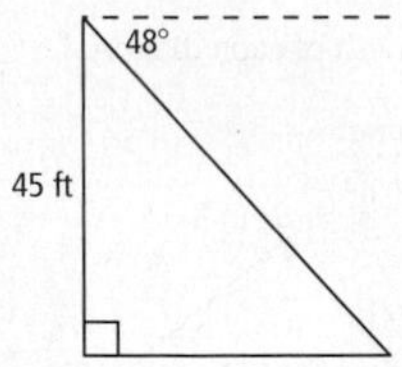

Ⓐ 50 ft
Ⓑ 40.5 ft
Ⓒ 20.7 ft
Ⓓ $13\sqrt{2}$ ft

4. What is the converse of the following statement?

If you study in front of the television, then you do not score well on exams.

Ⓕ If you do not study in front of the television, then you score well on exams.
Ⓖ If you score well on exams, then you do not study in front of the television.
Ⓗ If you do not score well on exams, then you study in front of the television.
Ⓘ If you study in front of the television, then you score well on exams.

5. What are the values of x and of y in the parallelogram below?

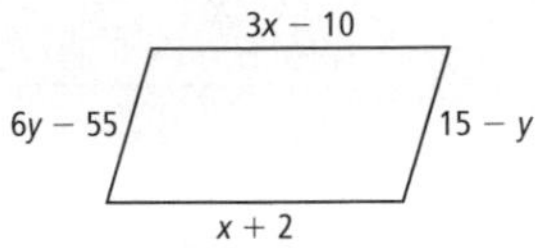

Ⓐ $x = 3, y = 5$
Ⓑ $x = 6, y = 10$
Ⓒ $x = 6, y = 14$
Ⓓ $x = 2, y = 2$

6. In $\triangle HTQ$, if $m\angle H = 72$ and $m\angle Q = 55$, what is the correct order of the lengths of the sides from least to greatest?

Ⓕ TQ, HQ, HT
Ⓖ TQ, HT, HQ
Ⓗ HQ, HT, TQ
Ⓘ HQ, TQ, HT

7. If $m\angle 2 = m\angle 3$, which statement must be true?

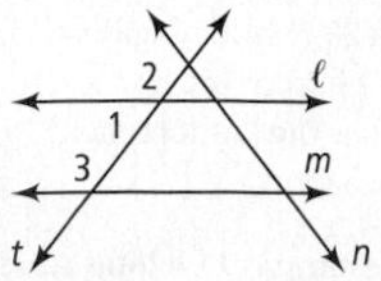

Ⓐ $\ell \parallel m$
Ⓑ $t \perp m$
Ⓒ $t \perp \ell$
Ⓓ $m\angle 1 = m\angle 2$

8. A bike messenger has just been asked to make an additional stop. Now, instead of biking straight from the law office to the court, she is going to stop at City Hall in between. Approximately how many additional miles will she bike?

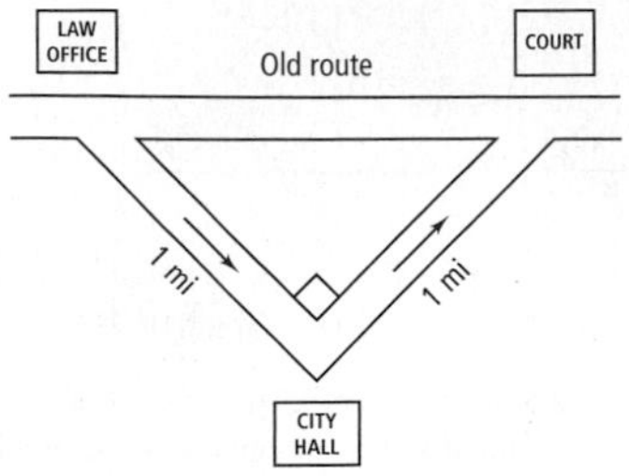

Ⓕ 1.4 mi
Ⓖ 0.6 mi
Ⓗ 2 mi
Ⓘ 3.4 mi

9. In $\triangle XYZ$, $XY = 12$, $YZ = 10$, and $ZX = 8$. To the nearest tenth of a degree, what is the $m\angle Z$?

Ⓐ 12.8
Ⓑ 55.8
Ⓒ 41.4
Ⓓ 82.8

10. What is the value of y?

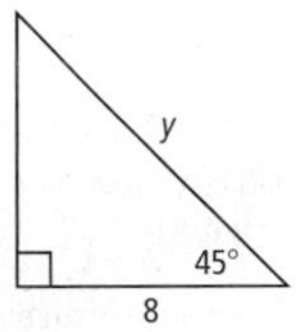

Ⓕ 16
Ⓖ $8\sqrt{2}$
Ⓗ 8
Ⓘ $8\sqrt{3}$

3. B
4. H
5. B
6. H
7. A
8. G
9. D
10. G

Answers

Common Core Cumulative Standards Review (continued)

11. 18.7
12. 4.6
13. 28
14. 42.26
15. 14
16. 112
17. 77
18. [2] **a.** $BD = AC$; $2x - 8 + x - 4 = x + 2$; $3x - 12 = x + 2$; $2x = 14$; $x = 7$
b. 9
[1] minor computational error
19. [2] **a.** Law of Sines; the measures of two sides and a nonincluded angle are given.
b. 49.9
[1] one part correct
20. [4] No; the distance from (1, 7) to (4, 3) is 5, but the distance from (1, 7) to (6.5, 11) is about 6.8. A regular polygon must have all sides ≅.
[3] correct answer, insufficient explanation
[2] correct answer, incorrect explanation
[1] correct answer, no explanation
21. [4] **a.** $\frac{ED}{CB} = \frac{DF}{BA}$, $\frac{20}{25} = \frac{DF}{40}$, $DF = 32$ in.
b. Area $\triangle EDF = \frac{1}{2} DF \cdot DE = \frac{1}{2}(32)(20)$ $= 320$ in.^{2}, area $\triangle ABC = \frac{1}{2} AB \cdot BC$ $= \frac{1}{2}(40)(25) = 500$ in.^{2}, so dark area = 820 in.^{2}; light area = area of rectangle $ABCD$ − dark area = (40)(25) in.^{2} − 820 in.^{2} = 180 in.^{2}
c. Yes, since $\triangle EDF \sim \triangle CBA$, $\angle DEF \cong \angle BCA$. Since complements of ≅ ∠s are ≅, $\angle DFE \cong \angle ACF$. Therefore, $\overline{EF} \parallel \overline{AC}$ because if ≅ corresp. ∠s, lines are ∥.
[3] correct method, one computational error
[2] parts (a) and (b) correct, incomplete proof of part (c)
[1] correct answers, no work shown

Constructed Response

11. A roofer leans a 20-ft ladder against a house. The base of the ladder is 7 ft from the house. How high, in feet, on the house does the ladder reach? Round to the nearest tenth of a foot.

12. A ship's loading ramp is 15 ft long and makes an angle of 18° with the dock. How many feet above the dock, to the nearest tenth of a foot, is the ship's deck?

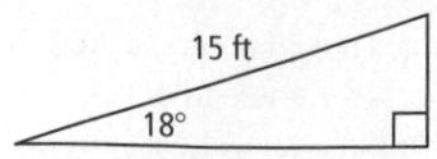

13. What is the value of x in the figure below?

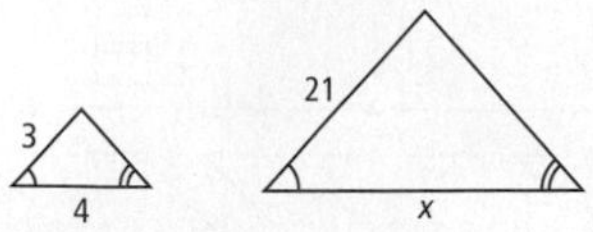

14. In $\triangle ABC$, $AC = 32$, $BC = 15$, and $m\angle C = 124°$. To the nearest hundredth, what is the length of $\overline{AB}$?

15. A front view of a barn is shown. The doorway is a square. Using a scale of 1 in. : 7 ft, what is the height, in feet, of the barn's doorway?

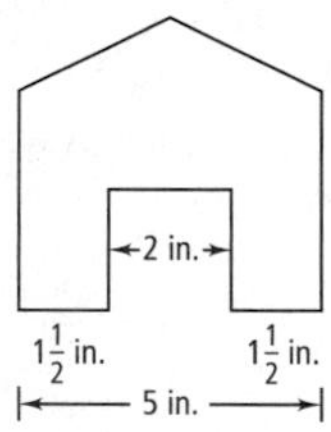

16. In a 30°-60°-90° triangle, the longer leg measures $56\sqrt{3}$ cm. How many centimeters long is the hypotenuse?

17. The measure of an angle is 12 more than 5 times its complement. What is the measure of the angle?

18. In the trapezoid at the right, $BE = 2x - 8$, $DE = x - 4$, and $AC = x + 2$.

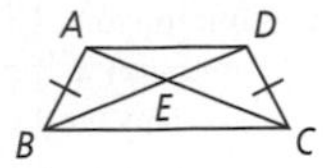

a. Write and solve an equation for x.
b. Find the length of each diagonal.

19. Use the diagram below.

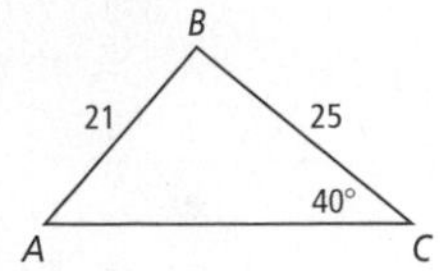

a. Explain whether you would use Law of Sines or Law of Cosines to find $m\angle A$.
b. To the nearest tenth, what is $m\angle A$?

Extended Response

20. A clothing store window designer is preparing a new window display. Using the lower left-hand corner of the window as the origin, she marks points at (1, 7), (4, 3), (9, 3), (12, 7), and (6.5, 11). The designer uses tape to connect the points to form a polygon. Is the polygon an equilateral pentagon? Justify your answer.

21. A youth organization is designing a 25 in.-by-40 in. rectangular flag, as shown below. The designers want the shaded triangles to be similar.

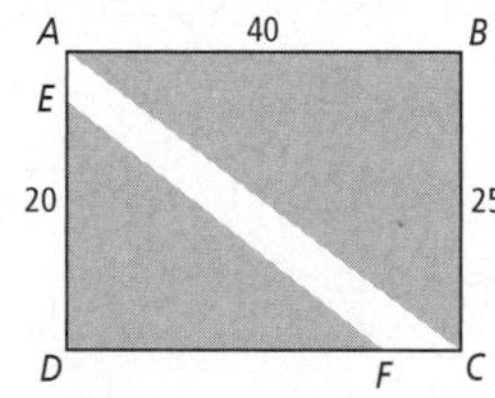

a. What should DF be in order to make $\triangle EDF \sim \triangle CBA$? Explain your reasoning.
b. In square inches, how much dark fabric do the designers need? How much light fabric?
c. Are the hypotenuses of the two triangles parallel? How do you know? (*Hint:* Use corresponding angles.)

CHAPTER 9

Get Ready!

Lesson 4-1

Congruent Figures

The triangles in each exercise are congruent. For each pair, complete the congruence statement $\triangle ABC \cong$? .

1.
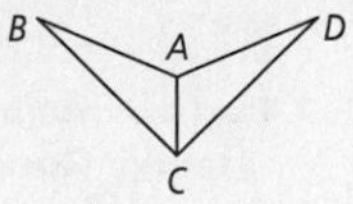

2.
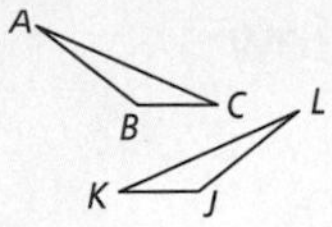

3.
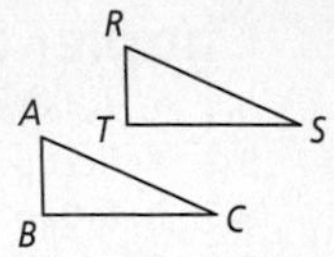

4.
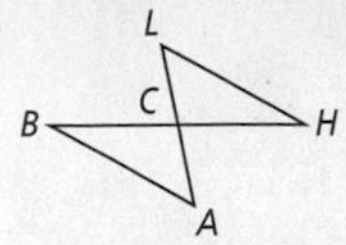

Lesson 6-1

Regular Polygons

Determine the measure of an angle of the given regular polygon. Draw a figure to help explain your reasoning.

5. pentagon **6.** octagon **7.** decagon **8.** 18-gon

Lessons 6-2, 6-4, 6-5, and 6-6

Quadrilaterals

Determine whether a diagonal of the given quadrilateral *always, sometimes,* or *never* produces congruent triangles.

9. rectangle **10.** isosceles trapezoid **11.** kite **12.** parallelogram

Lesson 7-2

Scale Drawing

The scale of a blueprint is 1 in. = 20 ft.

13. The length of a wall in the blueprint is 2.5 in. What is the length of the actual wall? Explain your reasoning.

14. What is the blueprint length of an entrance that is 5 ft wide? Explain your reasoning.

Looking Ahead Vocabulary

15. Think about your *reflection* in a mirror. If you raise your right hand, which hand appears to be raised in your reflection? If you are standing 2 ft from the mirror, how far away from you does your reflection appear to be?

16. The minute hand of a clock *rotates* as the minutes go by. What part of the minute hand stays fixed as the hand rotates?

17. The pupils in your eyes *dilate* in the dark. What do you think it means to *dilate* a geometric figure?

Answers

Get Ready!

1. $\triangle ADC$
2. $\triangle LJK$
3. $\triangle RTS$
4. $\triangle LHC$
5. 108
6. 135
7. 144
8. 160
9. always
10. never
11. sometimes
12. always
13. 50 ft; Because 1 in. = 20 ft, 2.5 in. = 2.5(20 ft) = 50 ft.
14. 0.25 in.; Because 20 ft = 1 in. and 5 ft = $\frac{1}{4}$(20 ft), by substitution, 5 ft = $\frac{1}{4}$(1 in.) = 0.25 in.
15. left hand; 4 ft
16. the point at the center of the clock
17. Answers may vary. Sample: When you dilate a geometric figure, you change its size.

Get Ready!

Assign this diagnostic assessment to determine if students have the prerequisite skills for Chapter 9.

Lesson	Skill
4-1	Congruent Figures
6-1	Regular Polygons
6-2, 6-4, 6-5, and 6-6	Quadrilaterals
7-2	Scale Drawing

To remediate students, select from these resources (available for every lesson).

- Online Problems (PowerGeometry.com)
- Reteaching (All-in-One Teaching Resources)
- Practice (All-in-One Teaching Resources)

Why Students Need These Skills

CONGRUENT FIGURES Students will identify whether figures remain congruent after being transformed.

REGULAR POLYGONS Students will determine which transformations map a regular polygon onto itself.

QUADRILATERALS The results of transformations on quadrilaterals, as well as other polygons, will be explored.

SCALE DRAWINGS Understanding scale drawings will help students comprehend the concept of dilations.

Looking Ahead Vocabulary

REFLECTION Have students describe what happens when they look at a mirror.

ROTATION Have students identify other real-world objects that rotate.

DILATE Ask students to conjecture about why an eye doctor may dilate the pupils of patients.

TRANSLATION Have students examine other uses of the word *translation*.

Chapter 9 Overview

In Chapter 9 students explore concepts related to transformations. Students will develop the answers to the Essential Questions as they learn the concepts and skills shown below.

BIG idea Transformations

ESSENTIAL QUESTION How can you change a figure's position without changing its size and shape? How can you change a figure's size without changing its shape?

- Students will explore translations, reflections, and rotations.
- Students will explore dilations.

BIG idea Coordinate Geometry

ESSENTIAL QUESTION How can you represent a transformation in the coordinate plane?

- Transformations will be conducted both on and off a coordinate plane.
- Students will determine the new coordinates of a polygon after any given transformation.

BIG idea Visualization

ESSENTIAL QUESTION How do you recognize congruence and similarity in figures?

- Students will identify congruence transformations and prove congruence using isometries.
- Students will identify similarity transformations and verify properties of similarity.

Content Standards

Following are the standards covered in this chapter.

CONCEPTUAL CATEGORY Geometry

Domain Congruence G-CO

Cluster Experiment with transformations in the plane (Standards G-CO.A.2, G-CO.A.4, G-CO.A.5)

LESSONS 9-1, 9-2, 9-3, 9-4, 9-6

Cluster Understand congruence in terms of rigid motions (Standards G-CO.B.6, G-CO.B.7, G-CO.B.8)

LESSONS 9-1, 9-2, 9-3, 9-4, 9-5

Domain Similarity, Right Triangles, and Trigonometry G-SRT

Cluster Understand similarity in terms of similarity transformations (Standards G-SRT.A.2, G-SRT.A.3)

LESSON 9-7

CHAPTER 9

Transformations

Download videos connecting math to your world.

Interactive! Vary numbers, graphs, and figures to explore math concepts.

The online Solve It will get you in gear for each lesson.

Math definitions in English and Spanish

Online access to stepped-out problems aligned to Common Core

Get and view your assignments online.

Extra practice and review online

Virtual Nerd™ tutorials with built-in support

Chapter Preview

BIG ideas

1 **Transformations**
Essential Questions How can you change a figure's position without changing its size and shape? How can you change a figure's size without changin
shape?

2 **Coordinate Geometry**
Essential Question How can you represent a transformation in the coordinate plane?

3 **Visualization**
Essential Question How do you reco
congruence and similarity in figures?

DOMAINS
- Congruence
- Similarity, Right Triangles, and Trigonom

Vocabulary

English/Spanish Vocabulary Audio Online:

English	Spanish
congruence transformation, *p. 580*	transformación de congruencia
dilation, *p. 587*	dilatación
image, *p. 545*	imagen
isometry, *p. 570*	isometría
preimage, *p. 545*	preimagen
reflection, *p. 554*	reflexión
rigid motion, *p. 545*	movimiento rígido
rotation, *p. 561*	rotación
similarity transformation, *p. 596*	transformación de semejanza
translation, *p. 547*	traslación

PowerGeometry.com

Chapter 9 Overview

Use these online assets to engage your students. These include support for the Solve It and step-by step solutions for Problems.

Show the student-produced video demonstrating relevant and engaging applications of the new concepts in this chapter.

Find online definitions for new terms in English and Spanish.

Start each lesson with an attention-getting Problem. View the Problem online with helpful hints.

Common Core Performance Task

Programming a Video Game

Alicia is a computer programmer working on an interactive puzzle for a company that makes video games. She is writing a program that uses transformations to move the puzzle piece shown by $\triangle ABC$ into the target area shown by $\triangle DEF$.

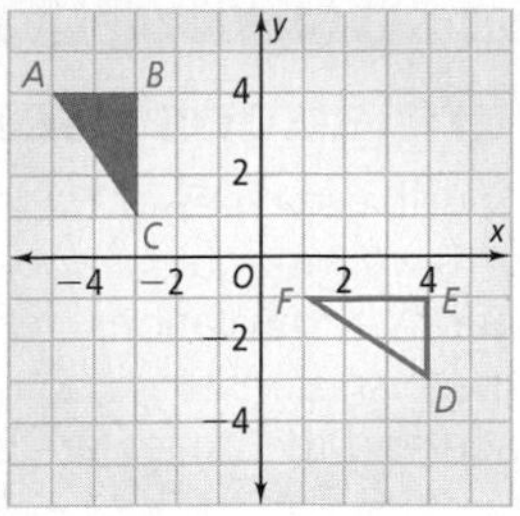

Alicia's boss asks her to write programs for two different cases. She must convince her boss that in each case the program uses as few moves as possible.

- Case 1: Reflections are allowed only across vertical and horizontal lines.
- Case 2: Reflections across vertical and horizontal lines are not allowed.

Task Description

Find a sequence of transformations that moves the puzzle piece to the target area for Case 1 and for Case 2. Use as few moves as possible in each case.

Connecting the Task to the Math Practices

MATHEMATICAL PRACTICES

As you complete the task, you'll apply several Standards for Mathematical Practice.

- You'll make sense of the problem by exploring single transformations of the puzzle piece. (MP 1)
- You'll use pencil and paper to look at how the transformations move the puzzle piece. (MP 5)
- You'll use properties of transformations to identify and explain characteristics of solutions to the problem. (MP 3)

Increase students' depth of knowledge with interactive online activities.

Show Problems from each lesson solved step by step. Instant replay allows students to go at their own pace when studying online.

Assign homework to individual students or to an entire class.

Prepare students for the Mid-Chapter Quiz and Chapter Test with online practice and review.

Virtual Nerd™ Access Virtual Nerd student-centered math tutorials that directly relate to the content of the lesson.

Overview of the Performance Task

Students will use properties of transformations to look for a sequence of transformations that will move a puzzle piece to a target area in as few moves as possible.

Students will work on the Performance Task in the following places in the chapter.

- Lesson 9-2 (p. 560)
- Lesson 9-3 (p. 567)
- Lesson 9-4 (p. 576)
- Pull It All Together (p. 601)

Introducing the Performance Task

Tell students to read the problem on this page. Do not have them start work on the problem at this time, but ask them the following questions.

Q What is a strategy you could try in order to solve the problem? **[Sample: I can trace the puzzle piece and cut it out. Then I can try different ways of moving it to the target area.]**

Q What are some different ways you can move the puzzle piece? **[Sample: slide the puzzle piece in one direction, turn the puzzle piece over, rotate the puzzle piece]**

PARCC CLAIMS

Sub-Claim A: Major Content with Connections to Practices
Sub-Claim C: Highlighted Practices MP 3, 6 with Connections to Content

SBAC CLAIMS

Claim 1: Concepts and Procedures
Claim 3: Communicating Reasoning

TRANSFORMATIONS
Math Background

PROFESSIONAL DEVELOPMENT

The Understanding by Design® methodology was central to the development of the Big Ideas and the Essential Understandings. These will help your students build a structure on which to make connections to prior learning.

Transformations

BIG idea Transformations may be described geometrically or by coordinates. Symmetries of figures may be defined and classified by transformations.

ESSENTIAL UNDERSTANDINGS

9-1 to 9-3 The distance between any two points and the angles in a geometric figure stay the same when (1) its location and orientation changes, (2) it is flipped across a line, or (3) it is turned about a point.

9-4 One of two congruent figures in a plane can be mapped onto the other by a single reflection, translation, rotation, or glide reflection.

9-6 A scale factor can be used to make a larger or smaller copy of a figure that is similar to the original figure.

Coordinate Geometry

BIG idea A coordinate system in a plane is formed by two perpendicular number lines, called the *x*- and *y*- axes, and the quadrants they form. It is possible to verify some complex truths using deductive reasoning in combination with Distance, Midpoint, and Slope formulas.

ESSENTIAL UNDERSTANDINGS

9-1 to 9-3 The distance between any two points and the angles in a geometric figure stay the same when (1) its location and orientation changes, (2) it is flipped across a line, or (3) it is turned about a point.

9-6 A scale factor can be used to make a larger or smaller copy of a figure that is also similar to the original figure.

9-5 and 9-7 You can use coordinate geometry to prove triangle congruence and verify properties of similarity.

Visualization

BIG idea Visualization can help you see the relationships between two figures and help you connect properties of real objects with two-dimensional drawings of these objects.

ESSENTIAL UNDERSTANDINGS

9-5 If two figures can be mapped to each other by a sequence of rigid motions, then the figures are congruent.

9-7 Two figures are similar if there is a similarity transformation that maps one to the other.

Transformations (Isometries)

There are four basic types of transformations: *translations*, *reflections*, *rotations* and *dilations*. Three of these transformations—translations, reflections and rotations—are isometries. An *isometry* is a transformation in which the preimage and the image are congruent.

Translations

A translation is a slide of a figure.

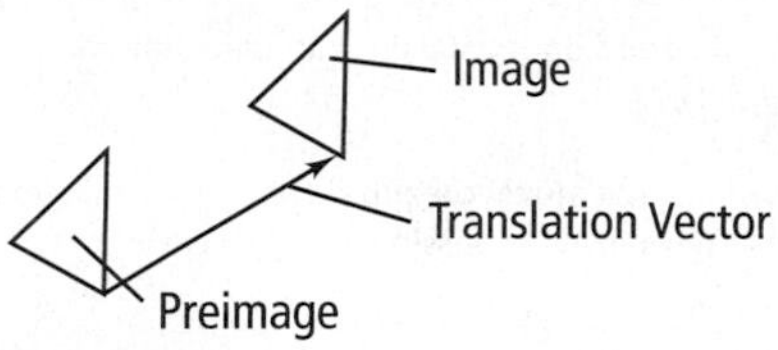

Reflections

A reflection flips a figure over a line of reflection.

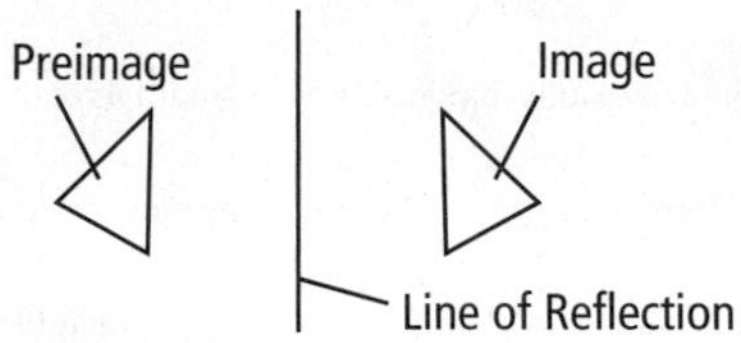

Rotations

A rotation turns a figure around a point by a given angle.

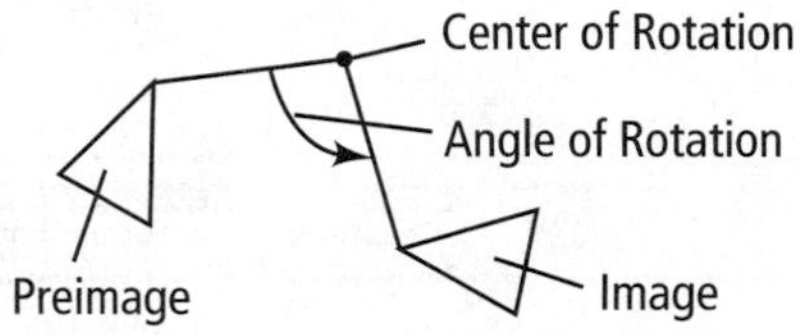

Common Errors With Transformations

Translations have the same size, shape, and orientation. Reflections, however, can have different orientations. Students might not think that a reflection is an isometry because the two figures are oriented differently.

Mathematical Practices

Attend to precision. Students define transformations and their properties. They distinguish between rotations, reflections, and translations in terms of orientation but see them all as rigid motions.

Symmetry

A figure is *symmetric* if there is an isometry that maps the figure onto itself.

A figure that can be a reflection of itself has *line symmetry*. A figure that can rotate less than 180° onto itself has *rotational symmetry*. A figure that can rotate 180° onto itself has *point symmetry*.

A figure can have multiple *lines of symmetry*. A figure can also rotate multiple times onto itself during a rotation of 360°. The number of times it rotates onto itself in a 360° rotation is its *order* of symmetry.

Regular polygons are symmetrical figures. Not all regular polygons have point symmetry, however. Consider the following regular polygons:

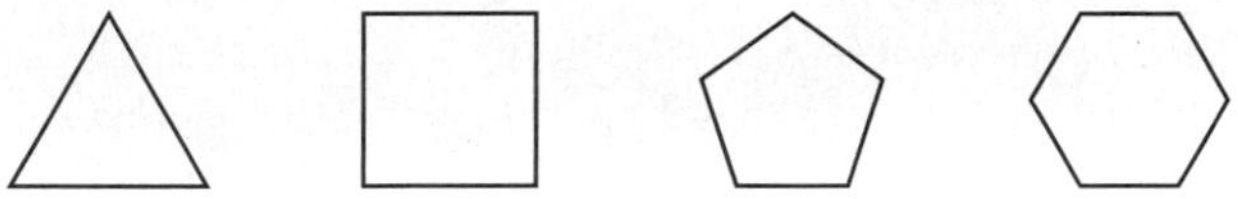

Regular Polygon	Symmetries	Lines of Symmetry	Order of Symmetry
Triangle	line rotational	3	3
Square	line rotational point	4	4
Pentagon	line rotational	5	5
Hexagon	line rotational point	6	6

Common Errors With Symmetry

A regular hexagon has 6 lines of symmetry.

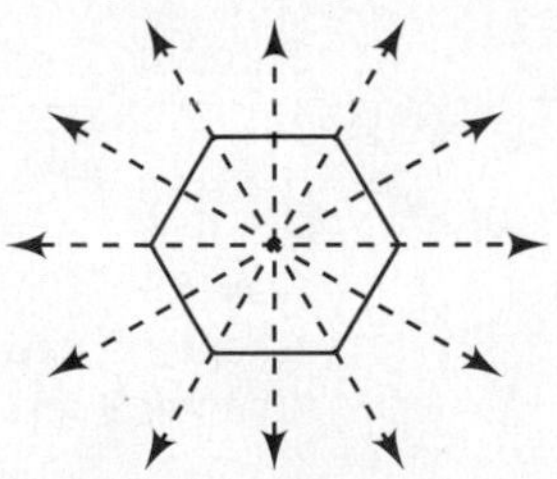

Students might count 12 lines of symmetry not realizing that they are counting the same lines twice.

Mathematical Practices

Look for and make use of structure. Students see the importance of symmetry in transformations, such as those in reflections, which guarantee two congruent angles in an isosceles triangle, and rotational symmetry, which explains the congruent sides and angles of a regular polygon.

Dilations

Dilations are another type of transformation. However, unlike translations, reflections, and rotations, dilations are not isometries. This is because the image resulting from dilation is similar to the original figure but not congruent.

Recall from Chapter 7 that corresponding sides of similar figures are proportional.

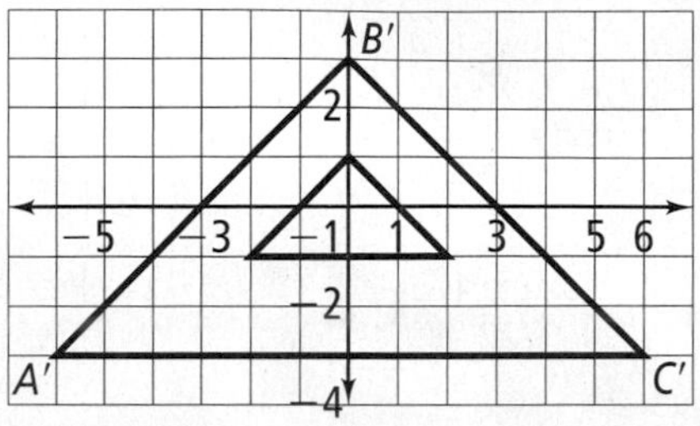

$\triangle A'B'C'$ is a dilation of $\triangle ABC$. So, corresponding sides are proportional. The *scale factor* of a dilation tells the amount by which the figure was enlarged or reduced. Dilations with a scale factor greater than 1 are *enlargements*. Dilations with a scale factor less than one are *reductions*.

$A'C' = 6 - (-6) = 12$

$AC = 2 - (-2) = 4$

$\frac{A'C'}{AC} = \frac{12}{4} = 3$

The scale factor of the dilation is 3. The dilation is an enlargement.

Common Errors With Dilations

Given that the larger figure is a dilation image of the smaller figure, students might make an error when finding the scale factor:

Incorrect: $\frac{8}{16} = \frac{1}{2}$ Correct: $\frac{16}{8} = 2$

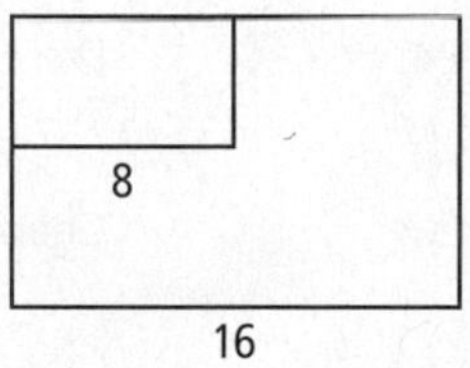

Mathematical Practices

Look for and make use of structure. Students examine dilations in which the center of the dilation is on the figure, on one vertex, in its interior and in its exterior and can see that for a given scale factor each of these images are congruent to each other but still similar to the preimage.

CHAPTER 9

TRANSFORMATIONS

Pacing and Assignment Guide

Lesson	Teaching Day(s)	Basic (Traditional)	Average (Traditional)	Advanced (Traditional)	Block
9-1	1	Problems 1–3 Exs. 7–15 all	Problems 1–3 Exs. 7–15 odd	Problems 1–3 Exs. 7–15 odd	**Day 1** Problems 1–5 Exs. 7–19 odd, 21–33
	2	Problems 4–5 Exs. 16–23 all, 24–28 even	Problems 4–5 Exs. 17, 19, 21–33	Problems 4–5 Exs. 17, 19, 21–35	
9-2	1	Problems 1–4 Exs. 7–20 all, 24, 27, 30–36, 45–57	Problems 1–4 Exs. 7–19 odd, 21–38, 45–57	Problems 1–4 Exs. 7–19 odd, 21–57	**Day 2** Problems 1–3 Exs. 7–19 odd, 21–38, 49–57
9-3	1	Problems 1–2 Exs. 9–22 all, 50–53	Problems 1–2 Exs. 9–21 odd, 50–53	Problems 1–2 Exs. 9–21 odd, 50–53	Problems 1–2 Exs. 9–21 odd, 50–53
	2	Problem 3 Exs. 23–25, 28–37, 38–46 even	Problem 3 Exs. 23, 25, 26–47	Problem 3 Exs. 23, 25, 26–49	**Day 3** Problem 3 Exs. 23, 25, 26–47
9-4	1	Problems 1–3 Exs. 6–15 all, 16–26 even, 30–38 even	Problems 1–3 Exs. 7–15 odd, 16–38	Problems 1–3 Exs. 7–15 odd, 16–40	Problems 1–3 Exs. 7–15 odd, 16–38
9-5	1	Problems 1–3 Exs. 6–11, 21, 29–40	Problems 1–3 Exs. 7–11 odd, 29–40	Problems 1–3 Exs. 7–11 odd, 29–40	**Day 4** Problems 1–5 Exs. 7–15 odd, 17–27, 29–40
	2	Problems 4–5 Exs. 12–16, 18–28 even	Problems 4–5 Exs. 13, 15, 17–28	Problems 4–5 Exs. 13, 15, 17–28	
9-6	1	Problems 1–3 Exs. 7–22 all, 24–34 even, 35, 38, 42–52 even, 56–63	Problems 1–3 Exs. 7–21 odd, 23–52, 56–63	Problems 1–3 Exs. 7–21 odd, 23–63	**Day 5** Problems 1–3 Exs. 7–21 odd, 23–52, 56–63
9-7	1	Problems 1–4 Exs. 5–17, 22, 24, 30–39	Problems 1–4 Exs. 5–15 odd, 16–24, 30–39	Problems 1–4 Exs. 5–15 odd, 16–39	Problems 1–4 Exs. 5–15 odd, 16–24, 30–39
Review	1	Chapter 9 Review	Chapter 9 Review	Chapter 9 Review	**Day 6** Chapter 9 Review Chapter 9 Test
Assess	1	Chapter 9 Test	Chapter 9 Test	Chapter 9 Test	
Total		**12 Days**	**12 Days**	**12 Days**	**6 Days**

Note: Pacing does not include Concept Bytes and other feature pages.

Resources

KEY
I = Interactive asset at PowerGeometry.com
E = Editable master at PowerGeometry.com
P = Available in Print
M = Master at PowerGeometry.com
✓ = CD-ROM

	For the Chapter	9-1	9-2	9-3	9-4	9-5	9-6	9-7
Planning								
Teacher Center Online Planner & Grade Book	I	I	I	I	I	I	I	I
Interactive Learning & Guided Instruction								
My Math Video	I							
Solve It!		I M	I M	I M	I M	I M	I M	I M
Student Companion		P M	P M	P M	P M	P M	P M	
Vocabulary Support		I P M	I P M	I P M	I P M	I P M	I P M	I P M
Got It? Support		I P	I P	I P	I P	I P	I P	I P
Dynamic Activity	I							
Online Problems		I	I	I	I	I	I	I
Additional Problems		M	M	M	M	M	M	M
English Language Learner Support (TR)		E P M	E P M	E P M	E P M	E P M	E P M	E P M
Activities, Games, and Puzzles		E M	E M	E M	E M	E M	E M	E M
Teaching With TI Technology With CD-ROM			✓ P					
TI-Nspire™ Support CD-ROM		✓	✓	✓	✓		✓	
Lesson Check & Practice								
Student Companion		P M	P M	P M	P M	P M	P M	P M
Lesson Check Support		I P	I P	I P	I P	I P	I P	I P
Practice and Problem Solving Workbook		P	P	P	P	P	P	P
Think About a Plan (TR)		E P M	E P M	E P M	E P M	E P M	E P M	E P M
Practice Form G (TR)		E P M	E P M	E P M	E P M	E P M	E P M	E P M
Standardized Test Prep (TR)		P M	P M	P M	P M	P M	P M	P M
Practice *Form K* (TR)		E P M	E P M	E P M	E P M	E P M	E P M	E P M
Extra Practice	E M							
Find the Errors!	M							
Enrichment (TR)		E P M	E P M	E P M	E P M	E P M	E P M	E P M
Answers and Solutions CD-ROM	✓	✓	✓	✓	✓	✓	✓	✓
Assess & Remediate								
ExamView CD-ROM	✓	✓	✓	✓	✓	✓	✓	✓
Lesson Quiz		I M	I M	I M	I M	I M	I M	I M
Quizzes and Tests *Form G* (TR)	E P M						E P M	E P M
Quizzes and Tests *Form K* (TR)	E P M						E P M	E P M
Reteaching (TR)		E P M	E P M	E P M	E P M	E P M	E P M	E P M
Performance Tasks (TR)	P M							
Cumulative Review (TR)	P M							
Progress Monitoring Assessments	I P M							

(TR) Available in All-In-One Teaching Resources

Guided Instruction

PURPOSE To perform translations, rotations, and reflections of geometric figures on coordinate planes by using tracings.

PROCESS Students will

- translate a triangle on a coordinate plane, determine the coordinates of the new triangle, and compare the new triangle to the original triangle.
- rotate a triangle on a coordinate plane 90° about the origin, determine the coordinates of the new triangle, and compare the coordinates of the new triangle to those of the original triangle.
- reflect a triangle on a coordinate plane across the *x*-axis and compare the orientation of the new triangle to that of the original triangle.

DISCUSS Be sure students have a solid grasp on the three different types of transformations. Ask, "What information is needed when performing a translation? What information is needed when performing a rotation? What information is needed when performing a reflection?"

Activity

In this Activity students translate, rotate, and reflect a triangle on a coordinate plane.

Q If the point *A*(–3, 3) is moved up 4 units and to the right 2 units, what is the location of the new point *A*′? **[(–1, 7)]**

Q If the point *B*(–1, 1) is moved up 4 units and to the right 2 units, what is the location to the new point *B*′? **[(1, 5]**

Q If the point *C*(1, 4) is moved up 4 units and to the right 2 units, what is the location of the new point *C*′? **[(3, 8]**

TACTILE LEARNERS

Using tracings to translate, rotate, and reflect geometric figures will help tactile learners. Encourage students to use tracings to transform geometric figures on coordinate planes until they feel comfortable doing the transformations without the tracings.

Mathematical Practices This Concept Byte supports students in becoming proficient in using appropriate tools, Mathematical Practice 5.

Concept Byte — Tracing Paper Transformations

Use With Lesson 9-1

ACTIVITY

Common Core State Standards
G-CO.A.2 Represent transformations in the plane . . .
MP 5

In this activity, you will use tracing paper to perform translations, rotations, and reflections.

Activity

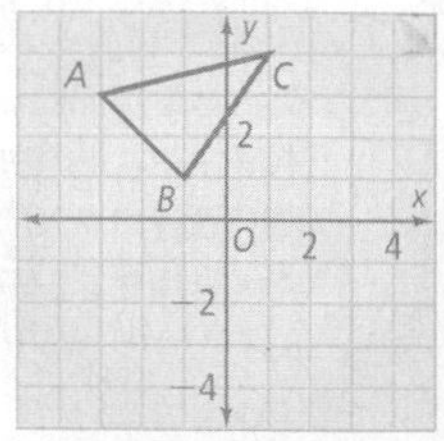

Step 1 Copy $\triangle ABC$ and the *x*- and *y*-axis on graph paper. Trace the copy of $\triangle ABC$ on tracing paper.

Step 2 Translate $\triangle ABC$ up 4 units and to the right 2 units by sliding the tracing paper. Draw the new triangle on the graph paper and label it $\triangle A'B'C'$ so that the original vertices *A*, *B*, and *C* correspond to the vertices *A*′, *B*′, and *C*′ of the new triangle. What are the coordinates of the vertices of $\triangle A'B'C'$? What is the same about the triangles? What is different?

Step 3 Align your tracing of $\triangle ABC$ with the original and then trace the positive *x*-axis and the origin.

Step 4 Rotate $\triangle ABC$ 90° about the origin by keeping the origin in place and aligning the traced axis with the positive *y*-axis. You can use the point of your pencil to hold the origin in place as you rotate the triangle. Draw the image of $\triangle ABC$ after the rotation on the graph paper and label it $\triangle A''B''C''$. Compare the coordinates of the vertices of $\triangle ABC$ with the coordinates of the vertices of $\triangle A''B''C''$. Describe the pattern.

Step 5 Flip your tracing of $\triangle ABC$ over and align the origin and the traced positive *x*-axis to reflect $\triangle ABC$ across the *x*-axis. Draw and label the reflected triangle $\triangle A'''B'''C'''$ on the graph paper. What do you notice about the orientations of the triangles?

Exercises

Use tracing paper. Find the images of each triangle for a translation 3 units left and 5 units down, a 90° rotation about the origin, and a reflection across the *x*-axis.

1.
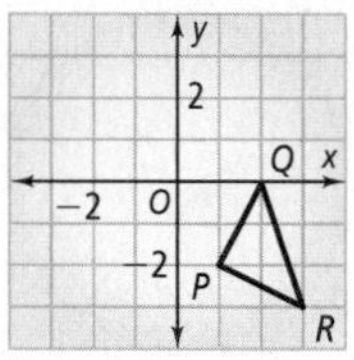

2.
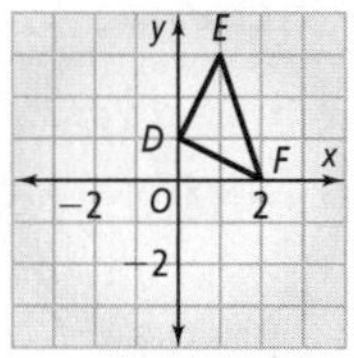

3.
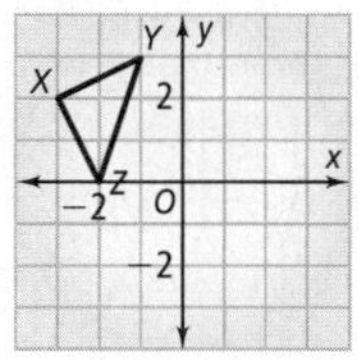

Answers

Exercises

1. translation: *P*′(−2, −7), *Q*′(−1, 5), *R*′(0, −8); rotation: *P*′(2, 1), *Q*′(0, 2), *R*′(3, 3); reflection: *P*′(1, 2), *Q*′(2, 0), *R*′(3, 3)

2. translation: *D*′(−3, −4), *E*′(−2, −2), *F*′(−1, −5); rotation: *D*′(−1, 0), *E*′(−3, 1), *F*′(0, 2); reflection: *D*′(0, −1), *E*′(1, −3), *F*′(2, 0)

3. translation: *X*′(−6, −3), *Y*′(−4, −2), *Z*′(−5, −5); rotation: *X*′(−3, −2), *Y*′(−3, −1), *Z*′(0, −2); reflection: *X*′(−3, −2), *Y*′(−1, −3), *Z*′(0, −2)

9-1 Translations

Common Core State Standards
G-CO.A.2 Represent transformations in the plane . . . describe transformations as functions that take points in the plane as inputs and give other points as outputs . . . **Also G-CO.A.4, G-CO.B.6**
MP 1, MP 3, MP 4, MP 7

Objectives To identify rigid motions
To find translation images of figures

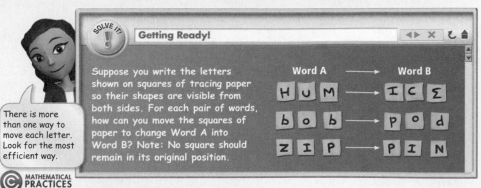

Solve It! Getting Ready!

Suppose you write the letters shown on squares of tracing paper so their shapes are visible from both sides. For each pair of words, how can you move the squares of paper to change Word A into Word B? Note: No square should remain in its original position.

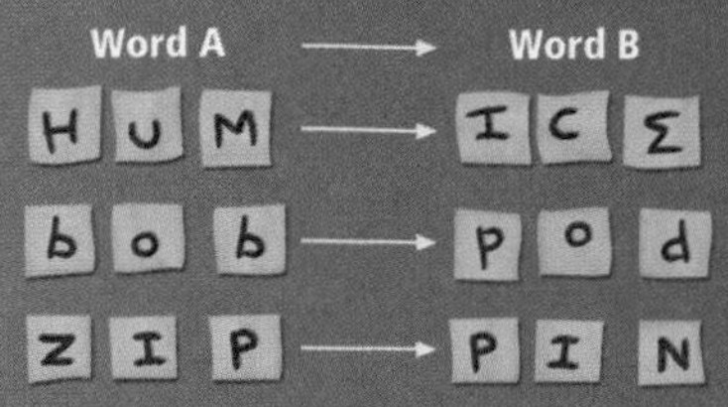

There is more than one way to move each letter. Look for the most efficient way.

MATHEMATICAL PRACTICES

In the Solve It, you described changes in positions of letters. In this lesson, you will learn some of the mathematical language used to describe changes in positions of geometric figures.

Lesson Vocabulary
- transformation
- preimage
- image
- rigid motion
- translation
- composition of transformations

Essential Understanding You can change the position of a geometric figure so that the angle measures and the distance between any two points of a figure stay the same.

A **transformation** of a geometric figure is a function, or *mapping* that results in a change in the position, shape, or size of the figure. When you play dominoes, you often move the dominoes by flipping them, sliding them, or turning them. Each move is a type of transformation. The diagrams below illustrate some basic transformations that you will study.

The domino flips. The domino slides. The domino turns.

In a transformation, the original figure is the **preimage.** The resulting figure is the **image.** Some transformations, like those shown by the dominoes, preserve distance and angle measures. To preserve distance means that the distance between any two points of the image is the same as the distance between the corresponding points of the preimage. To preserve angles means that the angles of the image have the same angle measure as the corresponding angles of the preimage. A transformation that preserves distance and angle measures is called a **rigid motion.**

1 Interactive Learning

Solve It!

PURPOSE To describe transformations

PROCESS Students may
- use visual judgment.
- write the letters on tracing paper and physically transform them.

FACILITATE

Q How can you change the letter H into I? **[Turn the square 90° either clockwise or counterclockwise.]**

Q How can you change the letter b into d? **[Flip the square over.]**

Q How can you move the letter P into its new location? **[Slide it to the front of the word.]**

ANSWER See Solve It in Answers on next page.

CONNECT THE MATH In the Solve It, students describe transformations to each letter to achieve the new word. In the lesson, students learn about these types of transformations, specifically translations.

2 Guided Instruction

Have students hold up their left hand, palms outward, with their thumbs forming a 90° angle with their index fingers. Have each student model each type of transformation with the right hand. They hold up their left hands to represent the preimages and then position their right hands to represent each image as you call out one of the transformations.

9-1 Preparing to Teach

BIG ideas **Transformations**
Coordinate Geometry

ESSENTIAL UNDERSTANDINGS
- The location and orientation of a geometric figure can be changed while preserving distance and angle measures.
- The distance between any two points, angle measures, and orientation of a geometric figure remain the same when the figure is translated in one direction.

Math Background

Transformations describe the movement of geometric figures. Any figure can be transformed by sliding, rotating, flipping, or using a combination of these movements. Slides, or translations, are simple linear movements. They can represent the movement of a car along a driveway or the repeating of a symbol in a pattern. A translation preserves distance and angle measures; so it is a rigid motion. In future lessons, students will learn about other rigid motions and a type of transformation that is not a rigid motion.

Mathematical Practices

Look for and make use of structure. In translations of geometric figures, students will see a geometric figure as a whole and shift the figure on a coordinate plane.

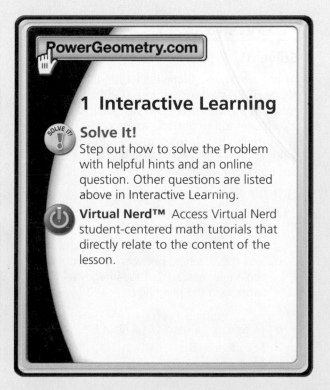

PowerGeometry.com

1 Interactive Learning

Solve It! Step out how to solve the Problem with helpful hints and an online question. Other questions are listed above in Interactive Learning.

Virtual Nerd™ Access Virtual Nerd student-centered math tutorials that directly relate to the content of the lesson.

Problem 1

Q What is the relationship between the image and the preimage? **[They appear to be similar figures.]**

Q Is the distance between any two points in the image the same as the corresponding distance in the preimage? **[No]**

Got It? VISUAL LEARNERS

Have students state whether the two figures appear to be congruent. Ask students to describe the transformation performed on each set of figures.

Problem 2

Q How can you use the transformation statement to identify pairs of corresponding parts? **[The transformation statement lists the vertices in order for both figures. Corresponding points are in corresponding locations in the names of the figures.]**

Q If you could move the image on top of the preimage, what movement would you need to do? **[slide]**

Got It? VISUAL LEARNERS

Have students visualize lining up the triangles one on top of the other so that the first vertex named in each part of the transformation statement are aligned. This will help them identify which points are corresponding.

Think

What must be true about a rigid motion?
In a rigid motion, the image and the preimage must preserve distance and angle measures.

Problem 1 Identifying a Rigid Motion

Does the transformation at the right appear to be a rigid motion? Explain.

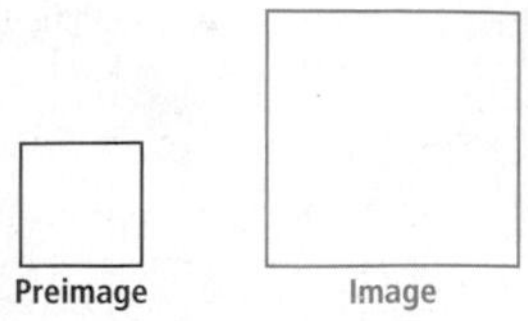

No, a rigid motion preserves both distance and angle measure. In this transformation, the distances between the vertices of the image are not the same as the corresponding distances in the preimage.

Got It? 1. Does the transformation appear to be a rigid motion? Explain.

a.

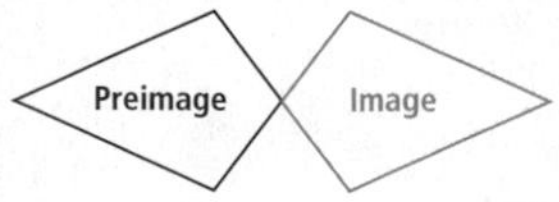

b.

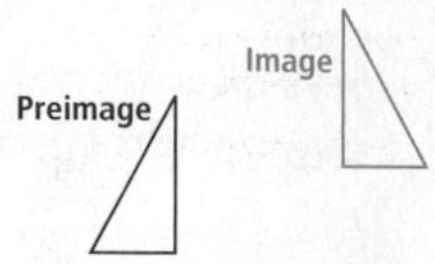

A transformation maps every point of a figure onto its image and may be described with arrow notation (→). Prime notation (′) is sometimes used to identify image points. In the diagram below, K' is the image of K.

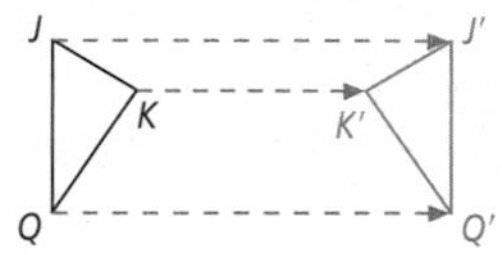

$\triangle JKQ \to \triangle J'K'Q'$
$\triangle JKQ$ maps onto $\triangle J'K'Q'$.

Notice that you list corresponding points of the preimage and image in the same order, as you do for corresponding points of congruent or similar figures.

Plan

How do you identify corresponding points?
Corresponding points have the same position in the names of the preimage and image. You can use the statement $EFGH \to E'F'G'H'$.

Problem 2 Naming Images and Corresponding Parts

In the diagram, $EFGH \to E'F'G'H'$.

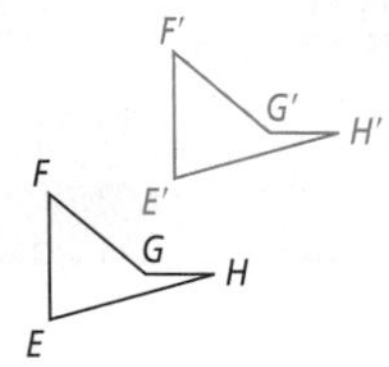

$EFGH \to E'F'G'H'$

A What are the images of $\angle F$ and $\angle H$?

$\angle F'$ is the image of $\angle F$. $\angle H'$ is the image of $\angle H$.

B What are the pairs of corresponding sides?

$\overline{EF}$ and $\overline{E'F'}$ $\quad$ $\overline{FG}$ and $\overline{F'G'}$

$\overline{EH}$ and $\overline{E'H'}$ $\quad$ $\overline{GH}$ and $\overline{G'H'}$

Got It? 2. In the diagram, $\triangle NID \to \triangle SUP$.

a. What are the images of $\angle I$ and point D?

b. What are the pairs of corresponding sides?

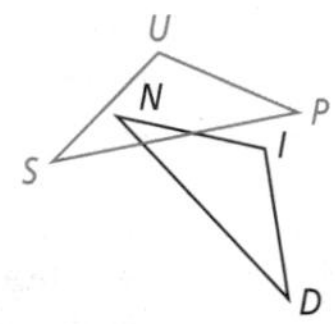

Answers

Solve It!

Answers may vary. Sample: "HUM" to "ICE": rotate H 90° clockwise, rotate U 90° clockwise, and rotate M 90° counterclockwise; "bob" to "pod": turn b over top to bottom, rotate o 180° clockwise, and turn b over left to right; "ZIP" to "PIN": rotate Z 90° counterclockwise and slide it to the third position, turn I over, and slide P to the first position.

Got It?

1a. Yes; the distances between the vertices and the angle measures of the image are the same as in the preimage.

b. Yes; the distances between the vertices and the angle measures of the image are the same as in the preimage.

2a. $\angle U$; P

b. $\overline{NI}$ and $\overline{SU}$, $\overline{ID}$ and $\overline{UP}$, $\overline{DN}$ and $\overline{PS}$

3. See page 548.

PowerGeometry.com

2 Guided Instruction

Each Problem is worked out and supported online.

Problem 1
Identifying a Rigid Motion

Problem 2
Naming Images and Corresponding Parts

Problem 3
Finding the Image of a Translation
Animated

Problem 4
Writing a Rule to Describe a Translation
Animated

Problem 5
Composing Translations
Animated

Support in Geometry Companion
- Vocabulary
- Key Concepts
- Got It?

take note

Key Concept Translation

A **translation** is a transformation that maps all points of a figure the same distance in the same direction.

You write the translation that maps $\triangle ABC$ onto $\triangle A'B'C'$ as $T(\triangle ABC) = \triangle A'B'C'$. A translation is a rigid motion with the following properties.

If $T(\triangle ABC) = \triangle A'B'C'$, then

- $AA' = BB' = CC'$
- $AB = A'B'$, $BC = B'C'$, $AC = A'C'$
- $m\angle A = m\angle A'$, $m\angle B = m\angle B'$, $m\angle C = m\angle C'$

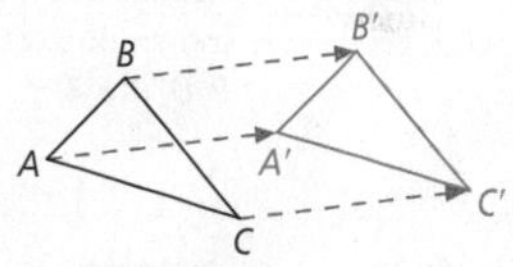

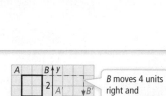

The diagram at the right shows a translation in the coordinate plane. Each point of $ABCD$ is translated 4 units right and 2 units down. So each (x, y) pair in $ABCD$ is mapped to $(x + 4, y - 2)$. You can use the function notation $T_{<4,-2>}(ABCD) = A'B'C'D'$ to describe this translation, where 4 represents the translation of each point of the figure along the x-axis and -2 represents the translation along the y-axis.

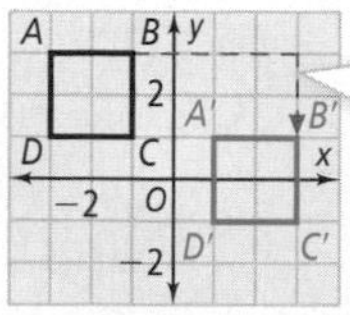

Problem 3 Finding the Image of a Translation

What are the vertices of $T_{<-2,-5>}(\triangle PQR)$? Graph the image of $\triangle PQR$.

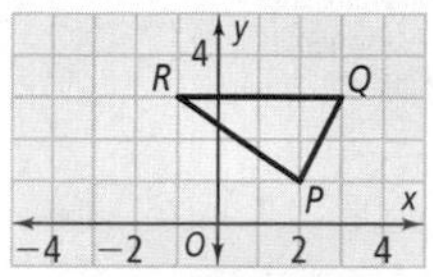

Think
What does the rule tell you about the direction each point moves?
-2 means that each point moves 2 units left. -5 means that each point moves 5 units down.

Identify the coordinates of each vertex. Use the translation rule to find the coordinates of each vertex of the image.

$T_{<-2,-5>}(P) = (2 - 2, 1 - 5)$, or $P'(0, -4)$.

$T_{<-2,-5>}(Q) = (3 - 2, 3 - 5)$, or $Q'(1, -2)$.

$T_{<-2,-5>}(R) = (-1 - 2, 3 - 5)$, or $R'(-3, -2)$.

To graph the image of $\triangle PQR$, first graph P', Q', and R'. Then draw $\overline{P'Q'}$, $\overline{Q'R'}$, and $\overline{R'P'}$.

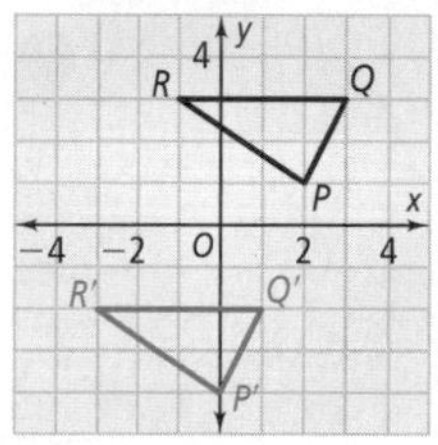

Got It? 3. a. What are the vertices of $T_{<1,-4>}(\triangle ABC)$? Copy $\triangle ABC$ and graph its image.

b. Reasoning Draw $\overline{AA'}$, $\overline{BB'}$, and $\overline{CC'}$. What relationships exist among these three segments? How do you know?

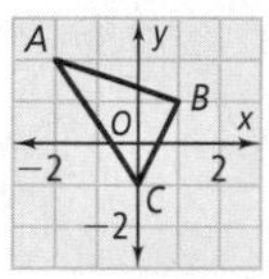

Take Note

Focus students on the segments connecting the image and preimage. Emphasize that each point is moved the same distance in the same direction. Have students study the translation rule. To practice writing these rules, give students translation rules and ask them to describe the movements needed.

Problem 3

Q How can you describe the translation given by the rule? **[Slide the triangle 2 units left and 5 units down.]**

Q What should be added to the x-coordinate of each vertex? **[-2]**

Q What should be added to the y-coordinate of each vertex? **[-5]**

Point out that the image can be obtained in two different ways.

- 1: Make the translation from each vertex and graph the new point. Then write the new ordered pairs.
- 2: Add -2 to each x-coordinate and -5 to each y-coordinate to find the new ordered pairs of the image. Then graph the ordered pairs.

Students should select one method to find the image and use the other to check their answer.

Got It? ERROR PREVENTION

Have students describe the transformation before graphing the image. Be sure that students check their work by verifying that each coordinate satisfies the translation rule. Ask them to use the Distance Formula to verify that the segments connecting a point to its image are congruent.

Additional Problems

1. Does the transformation below appear to be a rigid motion? Explain.

ANSWER Yes, the distances between the vertices and the angle measures of the image are the same as in the preimage.

2. In the diagram, $\triangle A'B'C'$ is an image of $\triangle ABC$ after a translation.

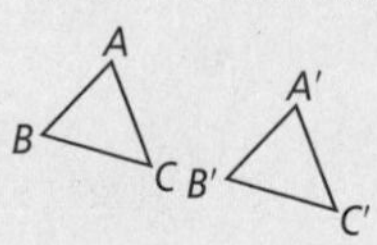

a. What are the images of $\angle A$ and $\angle B$?

b. What are the pairs of corresponding sides?

ANSWER a. $\angle A'$ and $\angle B'$ **b.** $\overline{AB}$ and $\overline{A'B'}$, $\overline{BC}$ and $\overline{B'C'}$, $\overline{AC}$ and $\overline{A'C'}$

3. What are the vertices of $T_{<-3,4>}(\triangle RST)$? Graph the image.

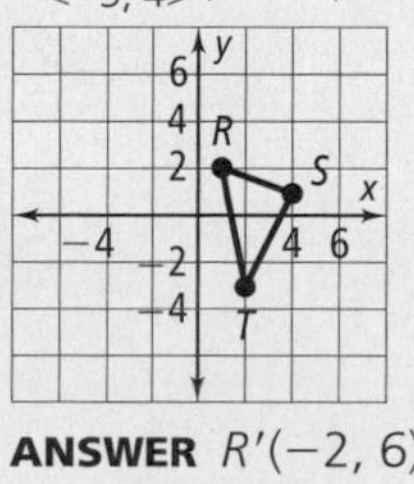

ANSWER $R'(-2, 6)$, $S'(1, 5)$, $T'(-1, 1)$

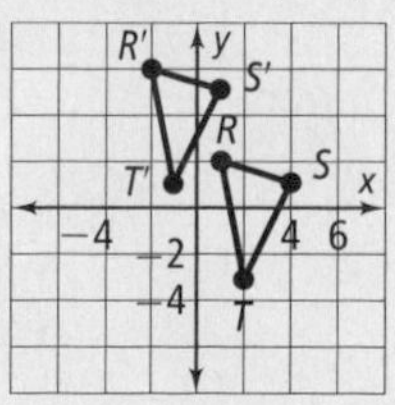

4. What is a rule that describes the translation that maps $\triangle LMN$ to $\triangle L'M'N'$?

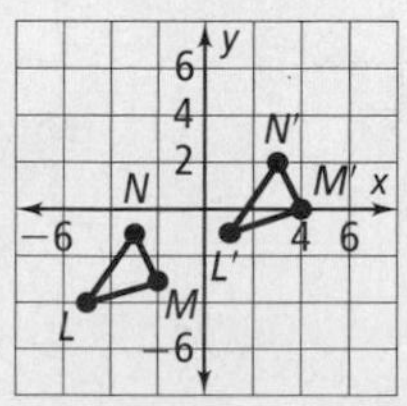

ANSWER $T_{<6,3>}(\triangle LMN)$

5. Carmen leaves her apartment and travels 2 blocks east and 5 blocks north to the library. Then she travels 4 blocks west and 1 block south to the grocery store. Where is Carmen in relation to her apartment?

ANSWER 2 blocks west and 4 blocks north of her apartment

Problem 4

Q How many units has the figure been shifted along the *x*-axis? **[+8]**

Q How many units has the figure been shifted along the *y*-axis? **[−2]**

Q How can you verify that your rule is correct? **[Write the ordered pairs of the vertices of the preimage. Add 8 to each *x*-coordinate. Subtract 2 from each *y*-coordinate. Write the new ordered pairs and verify that they match the location of the graphed image.]**

Got It?

Have students graph the preimage and image on the same coordinate grid. They should be able to count the number of units that the figure has been translated in each direction. Have students verify their answers by showing that the coordinates of corresponding points satisfy the translation rule that was produced.

Composition is a word that is used in many disciplines. Discuss the different uses of the word *composition* and its meanings in various situations. Some examples include: in language arts, composition is the arrangement of words; in chemistry, composition is the qualitative and quantitative makeup of a chemical compound; in music, composition is the arrangement of notes. Guide students to realize that in each instance a composition is the bringing together of multiple elements to form a new entity. With these analogies, students should be able to better understand that a composition of transformations consists of multiple transformations.

Problem 4 Writing a Rule to Describe a Translation

What is a rule that describes the translation that maps *PQRS* onto *P′Q′R′S′*?

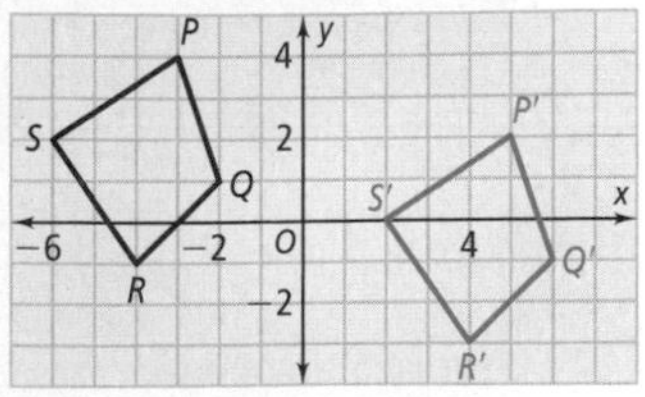

Know
The coordinates of the vertices of both figures

Need
An algebraic relationship that maps each point of *PQRS* onto *P′Q′R′S′*

Plan
Use one pair of corresponding vertices to find the change in the horizontal direction *x* and the change in the vertical direction *y*. Then use the other vertices to verify.

Think
How do you know which pair of corresponding vertices to use?
A translation moves all points the same distance and the same direction. You can use any pair of corresponding vertices.

Use $P(-3, 4)$ and its image $P'(5, 2)$.

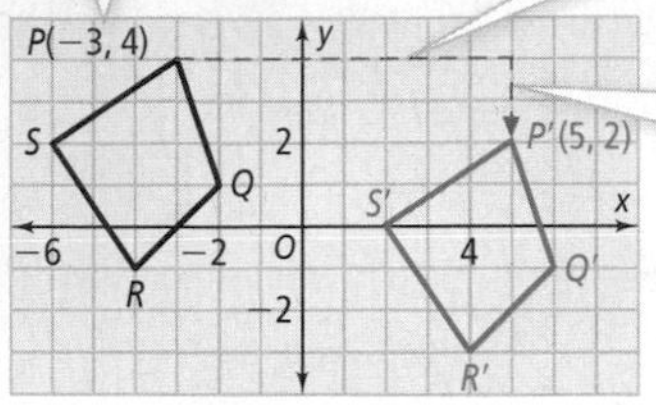

Horizontal change: $5 - (-3) = 8$
$x \rightarrow x + 8$

Vertical change: $2 - 4 = -2$
$y \rightarrow y - 2$

The translation maps each (x, y) to $(x + 8, y - 2)$. The translation rule is $T_{<8, -2>}(PQRS)$.

Got It? 4. The translation image of $\triangle LMN$ is $\triangle L'M'N'$ with $L'(1, -2)$, $M'(3, -4)$, and $N'(6, -2)$. What is a rule that describes the translation?

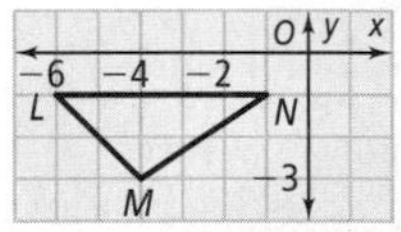

A **composition of transformations** is a combination of two or more transformations. In a composition, you perform each transformation on the image of the preceding transformation.

In the diagram at the right, the field hockey ball can move from Player 3 to Player 5 by a direct pass. This translation is represented by the blue arrow. The ball can also be passed from Player 3 to Player 9, and then from Player 9 to Player 5. The two red arrows represent this composition of translations.

In general, the composition of any two translations is another translation.

Answers

Got It? (continued)

3a–b. Graph:

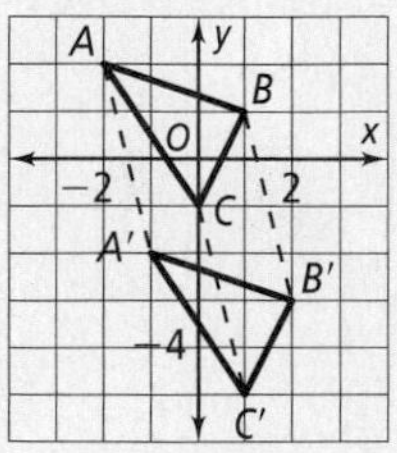

a. $A'(-1, -2)$, $B'(2, -3)$, $C'(1, -5)$

b. $AA' = BB' = CC'$ and $\overline{AA'} \parallel \overline{BB'} \parallel \overline{CC'}$ because rigid motions preserve distance, and the slope of each segment is −4.

4. $T_{<7, -1>}(\triangle LMN)$

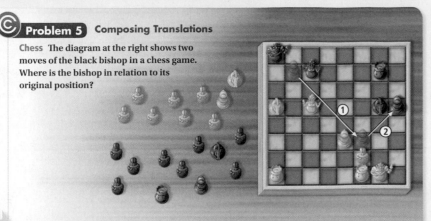

Problem 5 Composing Translations

Chess **The diagram at the right shows two moves of the black bishop in a chess game. Where is the bishop in relation to its original position?**

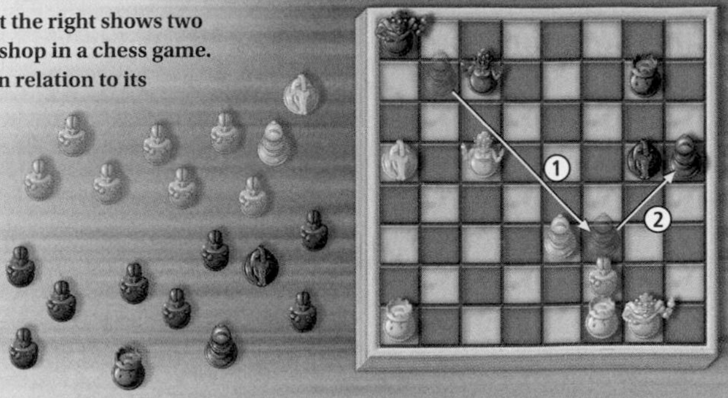

Think

How can you define the bishop's original position?
You can think of the chessboard as a coordinate plane with the bishop's original position at the origin.

Use (0, 0) to represent the bishop's original position. Write translation rules to represent each move.

$T_{<4,-4>}(x, y) = (x + 4, y - 4)$ The bishop moves 4 squares right and 4 squares down.

$T_{<2,2>}(x, y) = (x + 2, y + 2)$ The bishop moves 2 squares right and 2 squares up.

The bishop's current position is the composition of the two translations.

First, $T_{<4,-4>}(0, 0) = (0 + 4, 0 - 4)$, or $(4, -4)$.

Then, $T_{<2,2>}(4, -4) = (4 + 2, -4 + 2)$, or $(6, -2)$.

The bishop is 6 squares right and 2 squares down from its original position.

Got It? **5.** The bishop next moves 3 squares left and 3 squares down. Where is the bishop in relation to its original position?

Lesson Check

Do you know HOW?

1. If $\triangle JPT \rightarrow \triangle J'P'T'$, what are the images of P and $\overline{TJ}$?

2. Copy the graph at the right. Graph $T_{<-3,-4>}(NILE)$.

3. Point $H(x, y)$ moves 12 units left and 4 units up. What is a rule that describes this translation?

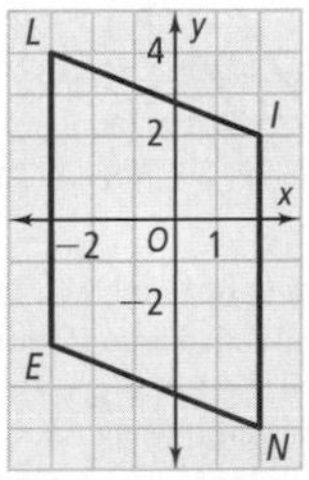

Do you UNDERSTAND? MATHEMATICAL PRACTICES

4. Vocabulary What is true about a transformation that is not a rigid motion? Include a sketch of an example.

5. Error Analysis Your friend says the transformation $\triangle ABC \rightarrow \triangle PQR$ is a translation. Explain and correct her error.

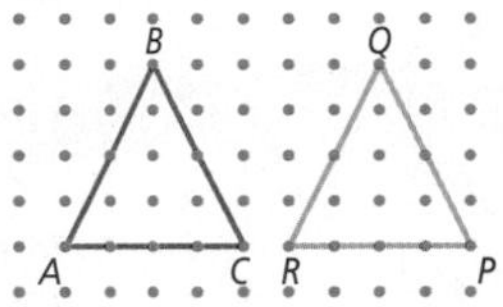

6. Reasoning Write the translation $T_{<1,-3>}(x, y)$ as a composition of a horizontal translation and a vertical translation.

Problem 5

Q How can you describe the first move of the bishop? **[The piece moved right 4 squares and down 4 squares.]**

Q What is the bishop's final position in relation to its starting point? **[The piece is 6 squares to the right and 2 squares down from its starting position.]**

Got It?

Have students mark the square on the chess board that corresponds to this new move. Then, have students work backward by counting the squares from the starting position to the final position.

3 Lesson Check

Do you know HOW?

- If students have difficulty with Exercise 2, then have them review Problem 3 to determine how movements left and movements up are written.

Do you UNDERSTAND?

- If students have difficulty with Exercise 6, then have them draw a graph of the translation and identify the vertical and horizontal shifts separately.

Close

Q What is a transformation that preserves distance and angle measures? **[a rigid motion]**

Q How can you write a rule for a translation? **[Write the coordinates of the points as an algebraic expression that gives the number of units shifted in each direction.]**

5. 3 squares right and 5 squares down

Lesson Check

1. P'; $\overline{T'J'}$

2.

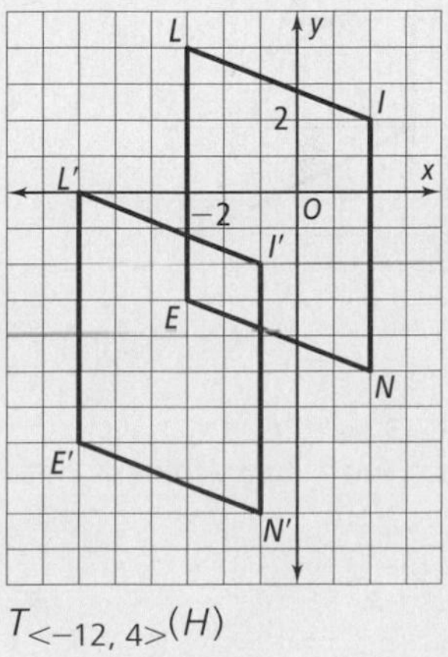

3. $T_{<-12,4>}(H)$

4. Answers may vary. Sample: There are points in the image that are not the same distance from each other as the corresponding points in the preimage.

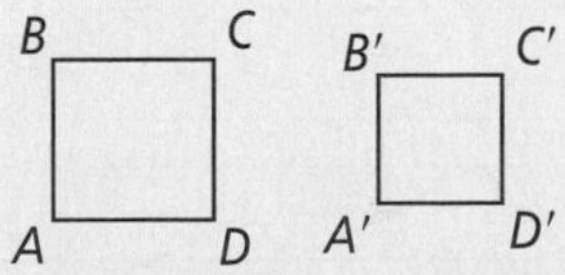

5. The transformation that maps $\triangle ABC$ to $\triangle PQR$ maps A to P and C to R, so it is a reflection, not a translation. The transformation that maps $\triangle ABC$ onto $\triangle RQP$ is a translation.

6. $T_{<1,0>}(x, y)$ followed by $T_{<0,-3>}(x, y)$

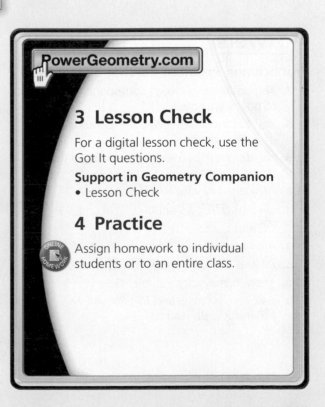

PowerGeometry.com

3 Lesson Check

For a digital lesson check, use the Got It questions.

Support in Geometry Companion
- Lesson Check

4 Practice

Assign homework to individual students or to an entire class.

4 Practice

ASSIGNMENT GUIDE

Basic: 7–23 all, 24–28 even
Average: 7–19 odd, 21–33
Advanced: 7–19 odd, 21–35
Standardized Test Prep: 36–39
Mixed Review: 40–44

Mathematical Practices are supported by exercises with red headings. Here are the Practices supported in this lesson:

MP 1: Make Sense of Problems Ex. 6, 22
MP 3: Construct Arguments Ex. 28
MP 3: Communicate Ex. 35
MP 3: Critique the Reasoning of Others Ex. 5

Applications exercises have blue headings. Exercises 19, 20, 27, 29, 30, and 40 support MP 4: Model.

STEM exercises focus on science or engineering applications.

EXERCISE 23: Use the Think About a Plan worksheet in the **Practice and Problem Solving Workbook** (also available in the Teaching Resources in print and online) to further support students' development in becoming independent learners.

HOMEWORK QUICK CHECK

To check students' understanding of key skills and concepts, go over Exercises 11, 15, 22, 23, and 28.

Practice and Problem-Solving Exercises

MATHEMATICAL PRACTICES

A Practice

Tell whether the transformation appears to be a rigid motion. Explain. See Problem 1.

7.

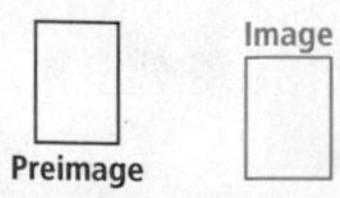

8.

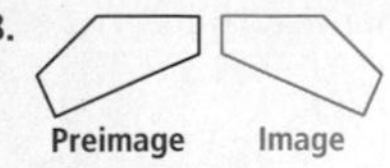

9.

In each diagram, the blue figure is an image of the black figure. See Problem 2.
(a) Choose an angle or point from the preimage and name its image.
(b) List all pairs of corresponding sides.

10.

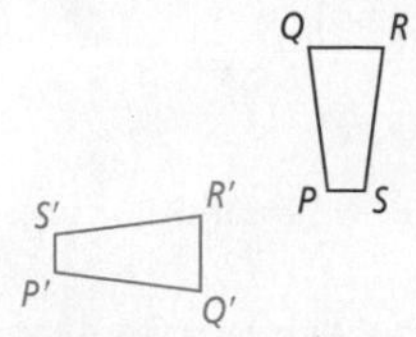

11.

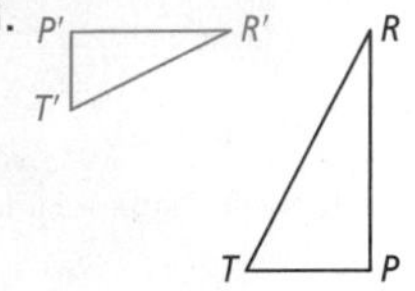

12.

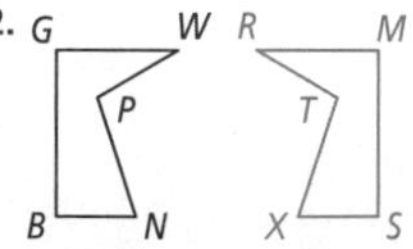

Copy each graph. Graph the image of each figure under the given translation. See Problem 3.

13. $T_{<3, 2>}(x, y)$

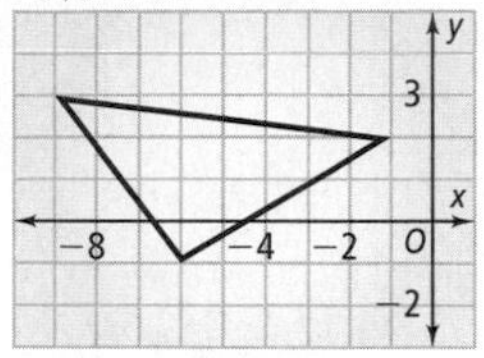

14. $T_{<5, -1>}(x, y)$

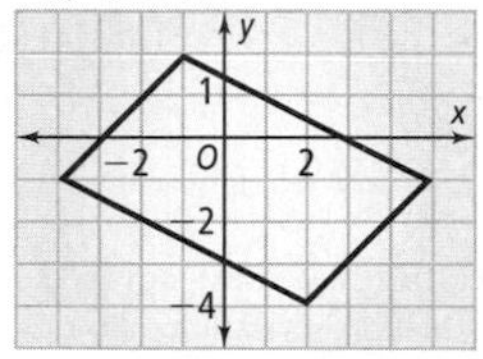

15. $T_{<-2, 5>}(x, y)$

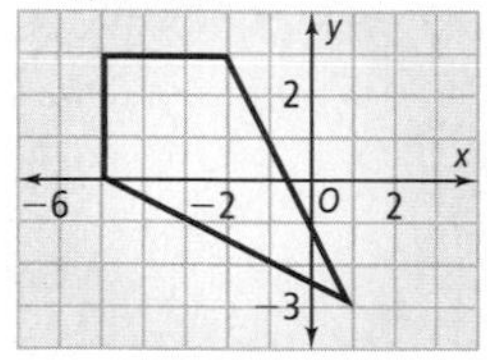

The blue figure is a translation image of the black figure. Write a rule to describe each translation. See Problem 4.

16.

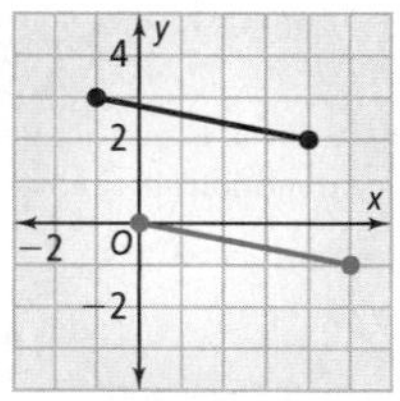

17.

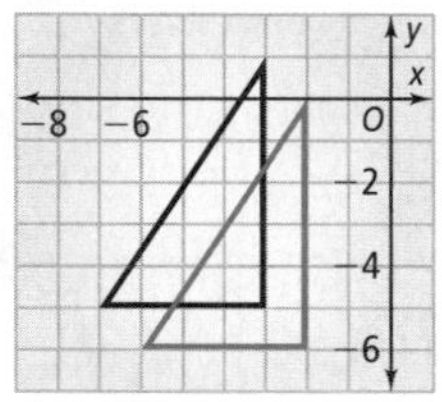

18.

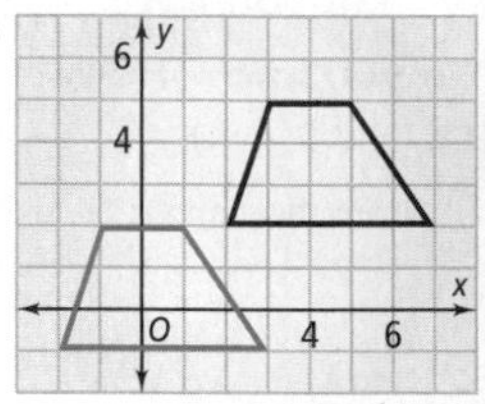

19. Travel You are visiting San Francisco. From your hotel near Union Square, you walk 4 blocks east and 4 blocks north to the Wells Fargo History Museum. Then you walk 5 blocks west and 3 blocks north to the Cable Car Barn Museum. Where is the Cable Car Barn Museum in relation to your hotel? See Problem 5.

Answers

Practice and Problem-Solving Exercises

7. Yes; distances between corresponding pairs of points are equal.

8. Yes; distances between corresponding pairs of points are equal.

9. No; distances between corresponding pairs of points are not equal.

10a. Answers may vary. Sample: $\angle Q \rightarrow \angle Q'$

b. $\overline{QR}$ and $\overline{Q'R'}$; $\overline{RS}$ and $\overline{R'S'}$; $\overline{SP}$ and $\overline{S'P'}$; $\overline{QP}$ and $\overline{Q'P'}$

11a. Answers may vary. Sample: $\angle R \rightarrow \angle R'$

b. $\overline{RP}$ and $\overline{R'P'}$; $\overline{PT}$ and $\overline{P'T'}$; $\overline{RT}$ and $\overline{R'T'}$

12a. Answers may vary. Sample: $G \rightarrow M$

b. $\overline{GW}$ and $\overline{MR}$; $\overline{WP}$ and $\overline{RT}$; $\overline{PN}$ and $\overline{TX}$; $\overline{NB}$ and $\overline{XS}$; $\overline{BG}$ and $\overline{SM}$

13.

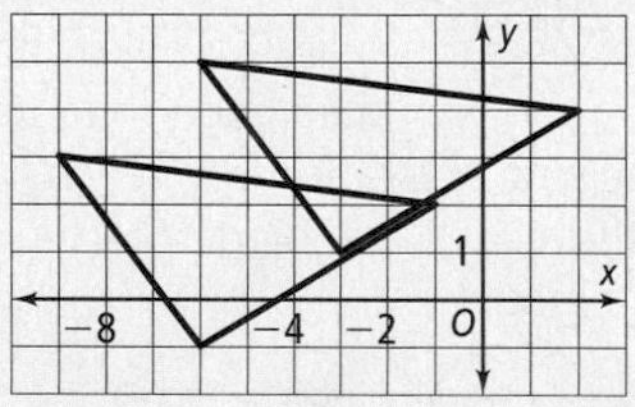

14.

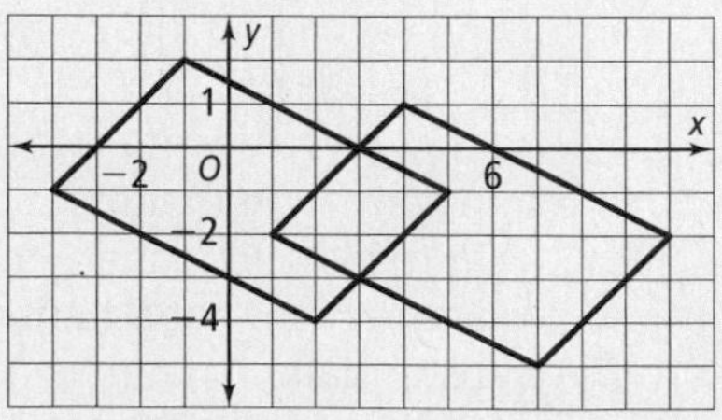

15.

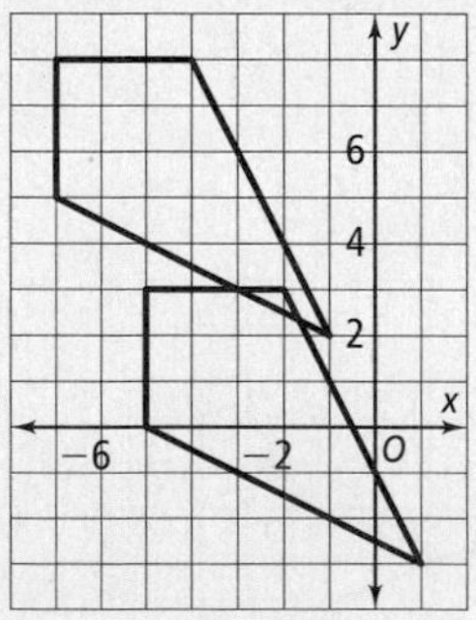

16. $T_{<1, -3>}(x, y)$

17. $T_{<1, -1>}(x, y)$

18. $T_{<-4, -3>}(x, y)$

19. 1 block west and 7 blocks north

20. **Travel** Your friend and her parents are visiting colleges. They leave their home in Enid, Oklahoma, and drive to Tulsa, which is 107 mi east and 18 mi south of Enid. From Tulsa, they go to Norman, 83 mi west and 63 mi south of Tulsa. Where is Norman in relation to Enid?

B Apply

21. In the diagram at the right, the orange figure is a translation image of the red figure. Write a rule that describes the translation.

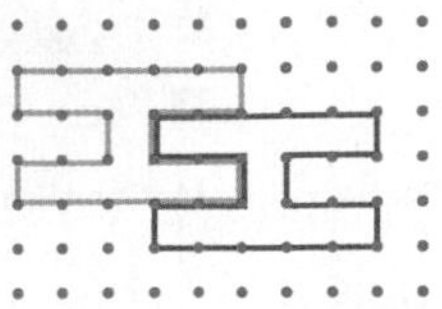

22. **Think About a Plan** $\triangle MUG$ has coordinates $M(2, -4)$, $U(6, 6)$, and $G(7, 2)$. A translation maps point M to $M'(-3, 6)$. What are the coordinates of U' and G' for this translation?
   - How can you use a graph to help you visualize the problem?
   - How can you find a rule that describes the translation?

23. **Coordinate Geometry** $PLAT$ has vertices $P(-2, 0)$, $L(-1, 1)$, $A(0, 1)$, and $T(-1, 0)$. The translation $T_{<2, -3>}(PLAT) = P'L'A'T'$. Show that $\overline{PP'}$, $\overline{LL'}$, $\overline{AA'}$, and $\overline{TT'}$ are all parallel.

Geometry in 3 Dimensions Follow the sample at the right. Use each figure, graph paper, and the given translation to draw a three-dimensional figure.

SAMPLE Use the rectangle and the translation $T_{<3, 1>}(x, y)$ to draw a box.

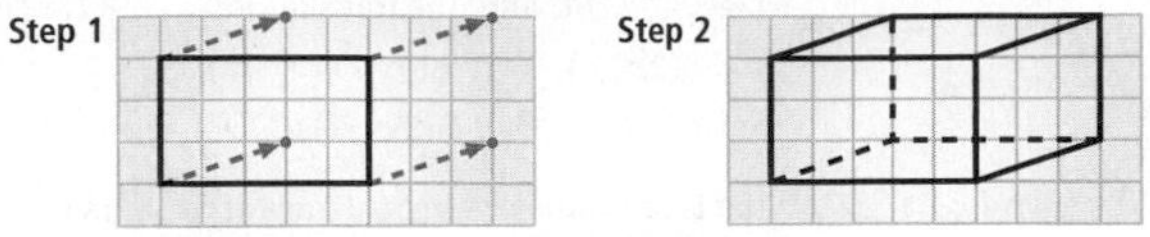

24. $T_{<2, -1>}(x, y)$

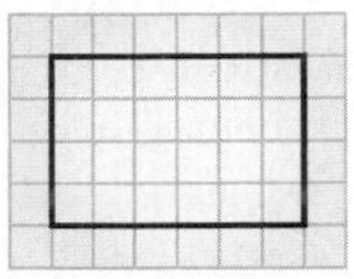

25. $T_{<-2, 2>}(x, y)$

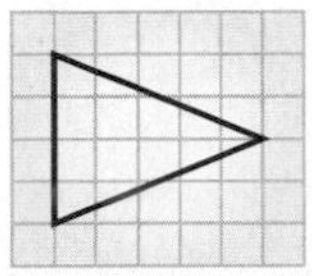

26. $T_{<-3, -5>}(x, y)$

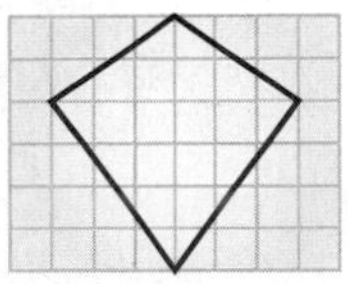

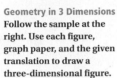

27. **Open-Ended** You are a graphic designer for a company that manufactures wrapping paper. Make a design for wrapping paper that involves translations.

28. **Reasoning** If $T_{<5, 7>}(\triangle MNO) = \triangle M'N'O'$, what translation rule maps $\triangle M'N'O'$ onto $\triangle MNO$?

29. **Landscaping** The diagram at the right shows the site plan for a backyard storage shed. Local law, however, requires the shed to sit at least 15 ft from property lines. Describe how to move the shed to comply with the law.

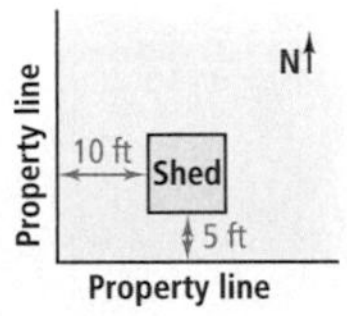

STEM 30. **Computer Animation** You write a computer animation program to help young children learn the alphabet. The program draws a letter, erases the letter, and makes it reappear in a new location two times. The program uses the following composition of translations to move the letter.

$T_{<5, 7>}(x, y)$ followed by $T_{<-9, -2>}(x, y)$

Suppose the program makes the letter W by connecting the points (1, 2), (2, 0), (3, 2), (4, 0) and (5, 2). What points does the program connect to make the last W?

20. 24 mi east and 81 mi south
21. $T_{<-3, 1>}(x, y)$
22. $U'(1, 16)$, $G'(2, 12)$
23. The vertices of $P'L'A'T'$ are $P'(0, -3)$, $L'(1, -2)$, $A'(2, -2)$, and $T'(1, -3)$. Slope of $\overline{PP'}$ = slope of $\overline{LL'}$ = slope of $\overline{AA'}$ = slope of $\overline{TT'} = -\frac{3}{2}$, so $\overline{PP'} \parallel \overline{LL'} \parallel \overline{AA'} \parallel \overline{TT'}$.
24.

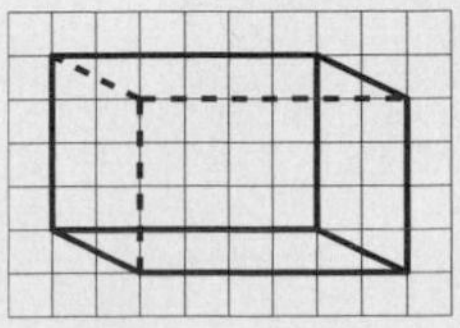

25.

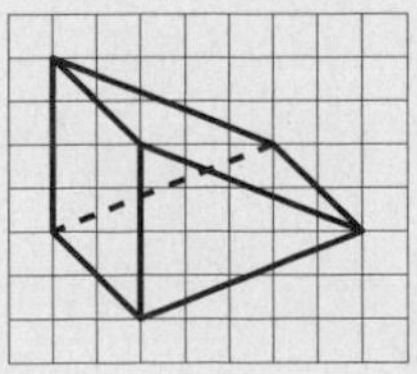

26.

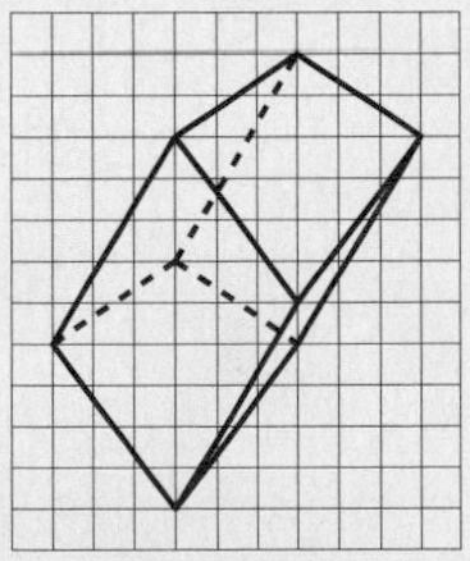

27. Check students' work.
28. $T_{<-5, -7>}(\triangle M'N'O')$
29. at least 5 ft east and 10 ft north
30. (−3, 7), (−2, 5), (−1, 7), (0, 5), (1, 7)

Answers

Practice and Problem-Solving Exercises (continued)

31. Answers may vary. Sample:
$T_{<4, -1>}(\triangle JKL)$
$T_{<2, -1>}(\triangle JKL)$
$T_{<4, -4>}(\triangle JKL)$

32. $T_{<-2, 14>}(x, y)$

33. $T_{<13, -2.5>}(x, y)$

34. The midpts. of $\overline{AB}$, $\overline{BC}$, and $\overline{AC}$ are $(-3, 2)$, $(-1, -2)$, and $(0, 1)$, respectively. The translation that maps (x, y) onto $(x + 4, y + 2)$ translates those midpts. to $(1, 4)$, $(3, 0)$, and $(4, 3)$, respectively. The same translation moves A, B, and C to $A'(2, 7)$, $B'(0, 1)$, and $C'(6, -1)$, so the midpts. of $\overline{A'B'}$, $\overline{B'C'}$, and $\overline{A'C'}$ are $(1, 4)$, $(3, 0)$, and $(4, 3)$, respectively.

35. Translate a line segment in some other direction than along the segment. Then connect the endpoints of the line segment and its image to form a parallelogram.

36. A

37. F

38. B

39. [2] **a.** $(-5, 2)$
b. Yes; answers may vary. Sample: The slope of $\overline{DB} = 1$ and the slope of $\overline{AC} = -1$, so $\overline{DB} \perp \overline{AC}$. Since $ABCD$ is a ▱ with ⊥ diagonals, $ABCD$ is a rhombus.
[1] one part incorrect or missing OR incorrect explanation

40. about 431.7 km

41. $\overline{BC} \cong \overline{EF}$ and $\overline{BC} \parallel \overline{EF}$ (given), so $\angle BCA \cong \angle F$ (corresp. ∡ of ∥ lines are ≅). $\overline{AD} \cong \overline{DC} \cong \overline{CF}$ (given), so $AC = AD + DC = DC + CF = DF$ (Segment Addition Post., Trans. Prop. of Equality). So $\triangle BCA \cong \triangle EFD$ by SAS, and $\overline{AB} \cong \overline{DE}$ (corresp. parts of ≅ ▵ are ≅).

42. $y = -2$

43. $x = -1$

44. $y = -x + 1$

31. Use the graph at the right. Write three different translation rules for which the image of $\triangle JKL$ has a vertex at the origin.

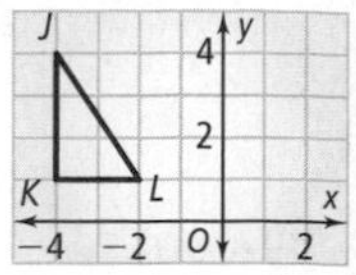

Find a translation that has the same effect as each composition of translations.

32. $T_{<2, 5>}(x, y)$ followed by $T_{<-4, 9>}(x, y)$

33. $T_{<12, 0.5>}(x, y)$ followed by $T_{<1, -3>}(x, y)$

Challenge

34. Coordinate Geometry $\triangle ABC$ has vertices $A(-2, 5)$, $B(-4, -1)$, and $C(2, -3)$. If $T_{<4, 2>}(\triangle ABC) = \triangle A'B'C'$, show that the images of the midpoints of the sides of $\triangle ABC$ are the midpoints of the sides of $\triangle A'B'C'$.

35. Writing Explain how to use translations to draw a parallelogram.

Standardized Test Prep

SAT/ACT

36. $\triangle ABC$ has vertices $A(-5, 2)$, $B(0, -4)$, and $C(3, 3)$. What are the vertices of the image of $\triangle ABC$ after the translation $T_{<7, -5>}(\triangle ABC)$?

(A) $A'(2, -3)$, $B'(7, -9)$, $C'(10, -2)$
(B) $A'(-12, -3)$, $B'(-7, -9)$, $C'(-4, -2)$
(C) $A'(-12, 7)$, $B'(-7, 1)$, $C'(-4, 8)$
(D) $A'(2, -3)$, $B'(10, -2)$, $C'(7, -9)$

37. What is the value of x in the figure at the right?

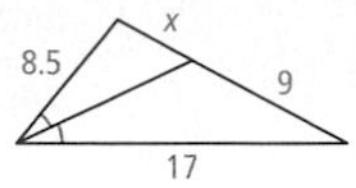

(F) 4.5
(G) 16
(H) 18
(I) 18.5

38. In $\triangle PQR$, $PQ = 4.5$, $QR = 4.4$, and $RP = 4.6$. Which statement is true?

(A) $m\angle P + m\angle Q < m\angle R$
(B) $\angle Q$ is the largest angle.
(C) $\angle R$ is the largest angle.
(D) $m\angle R < m\angle P$

Short Response

39. ▱$ABCD$ has vertices $A(0, -3)$, $B(-4, -2)$, and $D(-1, 1)$.
a. What are the coordinates of C?
b. Is ▱$ABCD$ a rhombus? Explain.

Mixed Review

40. Navigation An airplane landed at a point 100 km east and 420 km south from where it took off. If the airplane flew in a straight line from where it took off to where it lands, how far did it fly? See Lesson 8-1.

41. Given: $\overline{BC} \cong \overline{EF}$, $\overline{BC} \parallel \overline{EF}$, $\overline{AD} \cong \overline{DC} \cong \overline{CF}$ See Lesson 4-7.

Prove: $\overline{AB} \cong \overline{DE}$

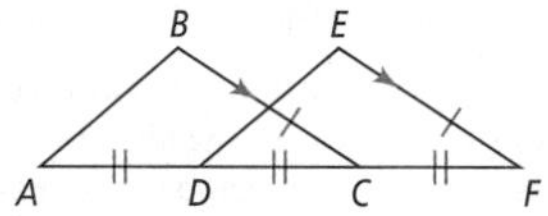

Get Ready! **To prepare for Lesson 9-2, do Exercises 42–44.**

Write an equation for the line through A perpendicular to the given line. See Lesson 3-8.

42. $A(1, -2)$; $x = -2$

43. $A(-1, -1)$; $y = 1$

44. $A(-1, 2)$; $y = x$

9-1 Lesson Resources

Differentiated Remediation

Instructional Support

Geometry Companion

Students can use the **Geometry Companion** worktext (4 pages) . . .

- New Vocabulary
- Key Concepts
- Got It for each Problem
- Lesson Check

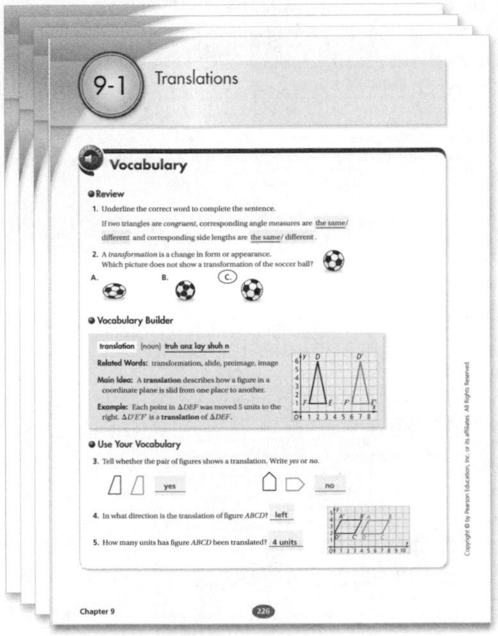
9-1 Translations

Vocabulary

Review

1. Underline the correct word to complete the sentence.
If two triangles are congruent, corresponding angle measures are the same/ different and corresponding side lengths are the same/ different.

2. A transformation is a change in form or appearance. Which picture does not show a transformation of the soccer ball?

Vocabulary Builder

translation (noun) truh anz lay shuh n

Related Words: transformation, slide, preimage, image

Main Idea: A translation describes how a figure in a coordinate plane is slid from one place to another.

Example: Each point in △DEF was moved 5 units to the right. △D′E′F′ is a translation of △DEF.

Use Your Vocabulary

3. Tell whether the pair of figures shows a translation. Write yes or no.
yes no

4. In what direction is the translation of figure ABCD? left

5. How many units has figure ABCD been translated? 4 units

Chapter 9 226

ELL Support

Use Manipulatives Assign groups of students the following activity to support the development of geometrical language. Have each group either make polygons out of construction paper or use polygonal manipulatives. Have students point out the corresponding parts of congruent polygons. Have groups discuss what attributes of congruent polygons can be different. **[Although the size and shape are the same, the color and position can be different.]**

Next have groups move one of a pair of congruent polygons so that it is on top of the other. Have members of the group describe how the polygon was moved. Have each group describe how the movement is like a "slide" or like a "translation." Have students compare this meaning of translation with other meanings of translation that they know.

5 Assess & Remediate

Lesson Quiz

1. Does the transformation below appear to be a rigid motion? Explain.

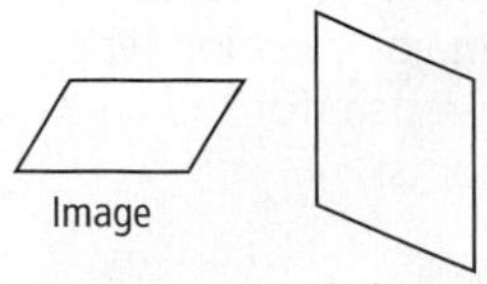

2. What is a rule that describes the translation that maps *QRST* onto *Q′R′S′T′*?

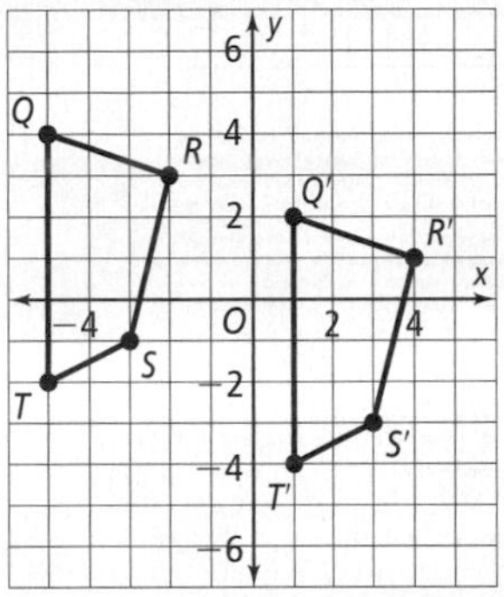

3. **Do you UNDERSTAND?** $\triangle ABC$ has vertices $A(5, 6)$, $B(6, -3)$, and $C(7, 2)$. $T_{<-2, 3>}(\triangle ABC = \triangle A'B'C')$. What are the coordinates of the vertices of $\triangle A'B'C'$?

ANSWERS TO LESSON QUIZ

1. No, the distances between the vertices of the image are not the same as the corresponding distances in the preimage.
2. $T_{<6, -2>}(QRST)$
3. $A'(3, 9)$, $B'(4, 0)$, $C'(5, 5)$

PRESCRIPTION FOR REMEDIATION

Use the student work on the Lesson Quiz to prescribe a differentiated review assignment.

Points	Differentiated Remediation
0–1	Intervention
2	On-level
3	Extension

PowerGeometry.com

5 Assess & Remediate

Assign the Lesson Quiz. Appropriate intervention, practice, or enrichment is automatically generated based on student performance.

Intervention

- **Reteaching** (2 pages) Provides reteaching and practice exercises for the key lesson concepts. Use with struggling students or absent students.
- **English Language Learner Support** Helps students develop and reinforce mathematical vocabulary and key concepts.

All-in-One Resources/Online

Reteaching

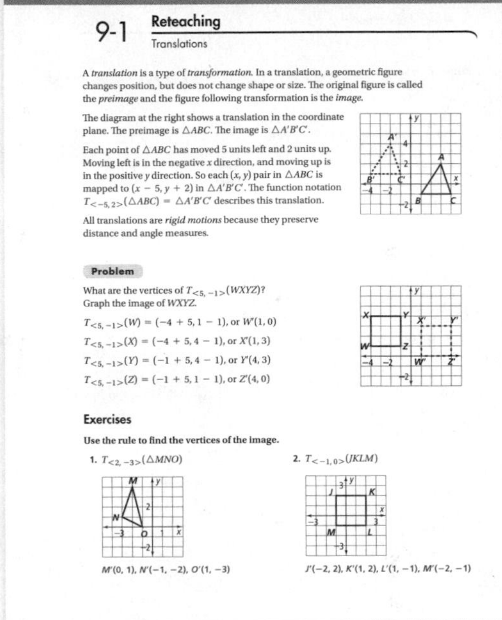
9-1 Reteaching
Translations

A *translation* is a type of *transformation*. In a translation, a geometric figure changes position, but does not change shape or size. The original figure is called the *preimage* and the figure following transformation is the *image*.

The diagram at the right shows a translation in the coordinate plane. The preimage is △ABC. The image is △A′B′C′.

Each point of △ABC has moved 5 units left and 2 units up. Moving left is in the negative x direction, and moving up is in the positive y direction. So each (x, y) pair in △ABC is mapped to (x − 5, y + 2) in △A′B′C′. The function notation $T_{<-5, 2>}(\triangle ABC) = \triangle A'B'C'$ describes this translation.

All translations are *rigid motions* because they preserve distance and angle measures.

Problem

What are the vertices of $T_{<5, -1>}(WXYZ)$? Graph the image of WXYZ.

$T_{<5, -1>}(W) = (-4 + 5, 1 - 1)$, or $W'(1, 0)$
$T_{<5, -1>}(X) = (-4 + 5, 4 - 1)$, or $X'(1, 3)$
$T_{<5, -1>}(Y) = (-1 + 5, 4 - 1)$, or $Y'(4, 3)$
$T_{<5, -1>}(Z) = (-1 + 5, 1 - 1)$, or $Z'(4, 0)$

Exercises

Use the rule to find the vertices of the image.

1. $T_{<2, -3>}(\triangle MNO)$
2. $T_{<-1, 0>}(JKLM)$

$M'(0, 1), N'(-1, -2), O'(1, -3)$
$J'(-2, 2), K'(1, 2), L'(1, -1), M'(-2, -1)$

All-in-One Resources/Online

English Language Learner Support

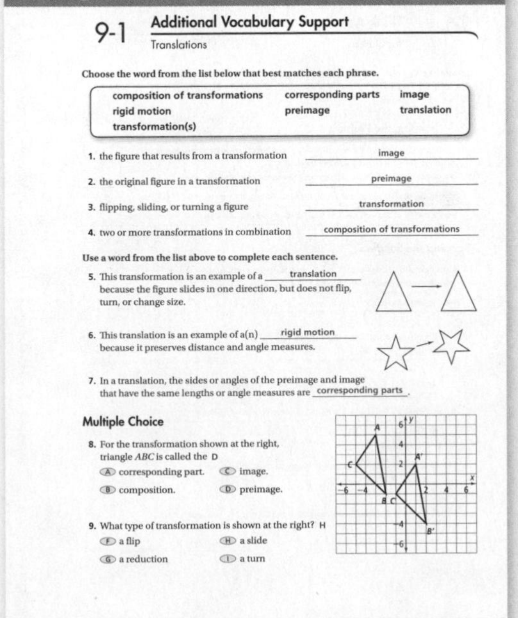
9-1 Additional Vocabulary Support
Translations

Choose the word from the list below that best matches each phrase.

composition of transformations, corresponding parts, image, rigid motion, preimage, translation, transformation(s)

1. the figure that results from a transformation — image
2. the original figure in a transformation — preimage
3. flipping, sliding, or turning a figure — transformation
4. two or more transformations in combination — composition of transformations

Use a word from the list above to complete each sentence.

5. This transformation is an example of a translation because the figure slides in one direction, but does not flip, turn, or change size.
6. This translation is an example of a(n) rigid motion because it preserves distance and angle measures.
7. In a translation, the sides or angles of the preimage and image that have the same lengths or angle measures are corresponding parts.

Multiple Choice

8. For the transformation shown at the right, triangle ABC is called the D
A corresponding part. C image.
B composition. D preimage.

9. What type of transformation is shown at the right? H
F a flip H a slide
G a reduction I a turn

Differentiated Remediation *continued*

On-Level

- **Practice** (2 pages) Provides extra practice for each lesson. For simpler practice exercises, use the Form K Practice pages found in the All-in-One Teaching Resources and online.
- **Think About a Plan** Helps students develop specific problem-solving skills and strategies by providing scaffolded guiding questions.
- **Standardized Test Prep** Focuses on all major exercises, all major question types, and helps students prepare for the high-stakes assessments.

Extension

- **Enrichment** Provides students with interesting problems and activities that extend the concepts of the lesson.
- **Activities, Games, and Puzzles** Worksheets that can be used for concepts development, enrichment, and for fun!

Practice and Problem Solving Wkbk/ All-in-One Resources/Online

Practice page 1

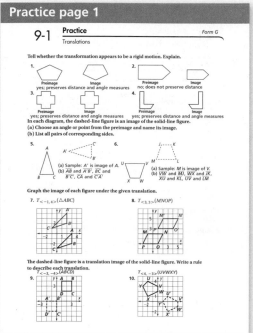

9-1 **Practice** *Form G*
Translations

Tell whether the transformation appears to be a rigid motion. Explain.

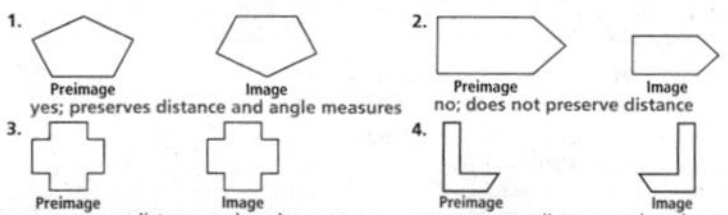

1. yes; preserves distance and angle measures
2. no; does not preserve distance
3. yes; preserves distance and angle measures
4. yes; preserves distance and angle measures

In each diagram, the dashed-line figure is an image of the solid-line figure.
(a) Choose an angle or point from the preimage and name its image.
(b) List all pairs of corresponding sides.

5. (a) Sample: A' is image of A. (b) $\overline{AB}$ and $\overline{A'B'}$, $\overline{BC}$ and $\overline{B'C'}$, $\overline{CA}$ and $\overline{C'A'}$
6. (a) Sample: M is image of V. (b) $\overline{VW}$ and $\overline{MJ}$, $\overline{WX}$ and $\overline{JK}$, $\overline{XU}$ and $\overline{KL}$, $\overline{UV}$ and $\overline{LM}$

Graph the image of each figure under the given translation.

7. $T_{<-1, 4>}(\triangle ABC)$
8. $T_{<3, 3>}(MNOP)$

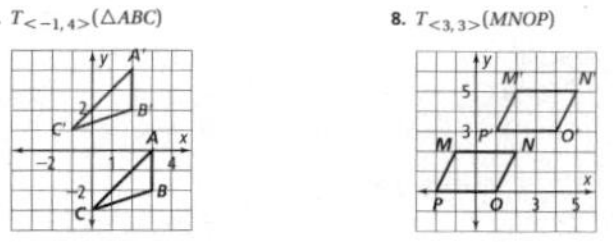

The dashed-line figure is a translation image of the solid-line figure. Write a rule to describe each translation.

9. $T_{<-3, -4>}(ABCD)$
10. $T_{<4, -3>}(UVWXY)$

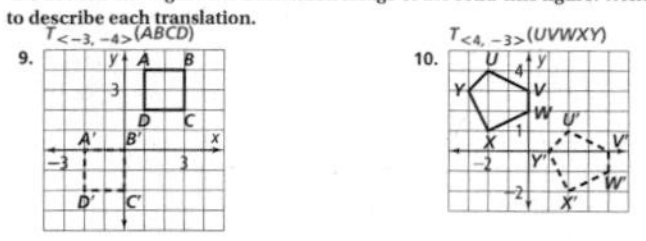

Practice and Problem Solving Wkbk/ All-in-One Resources/Online

Practice page 2

9-1 **Practice** (continued) *Form G*
Translations

11. You are visiting Washington, D.C. From the American History Museum you walk 5 blocks east and 1 block south to the Air and Space Museum. Then you walk 8 blocks west to the Washington Monument. Where is the Washington Monument in relation to the American History Museum? **1 block south and 3 blocks west**
12. You and some friends go to a book fair where booths are set out in rows. You buy drinks at the refreshment stand and then walk 8 rows north and 2 rows east to the science fiction booth. Then you walk 1 row south and 2 rows west to the children's book booth. Where is the children's book booth in relation to the refreshment stand? **7 rows north**
13. **Reasoning** If $T_{<10, 15>}(PQRS) = P'Q'R'S'$, what translation maps $P'Q'R'S'$ onto $PQRS$? $T_{<-10, -15>}(P'Q'R'S') = PQRS$
14. $\triangle XYZ$ has coordinates $X(2, 3)$, $Y(1, 4)$, and $Z(8, 9)$. A translation maps X to $X'(4, 7)$. What are the coordinates for Y' and Z' for this translation? $Y'(3, 8)$; $Z'(10, 13)$
15. Use the graph at the right. Write three different translation rules for which the image of $\triangle RST$ has a vertex at the origin.
$T_{<-1, -3>}(\triangle RST)$; $T_{<-5, -1>}(\triangle RST)$; $T_{<-2, 2>}(\triangle RST)$

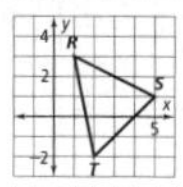

16. Use the graph at the right. Write three different translation rules for which the image of $\triangle BCD$ has a vertex at the origin.
$T_{<3, -3>}(\triangle BCD)$; $T_{<-4, -1>}(\triangle BCD)$; $T_{<0, 2>}(\triangle BCD)$

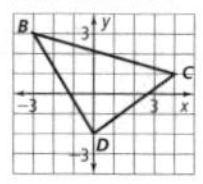

Graph the image of each figure under the given translation.

17. $T_{<-3, 4>}(\triangle DEF)$
18. $T_{<-5, 1>}(KLMN)$

All-in-One Resources/Online

Enrichment

9-1 **Enrichment**
Translations

Soccer Field Translations

Below is a diagram of half a soccer field with seven offensive players. By passing and dribbling the ball, the players try to score by getting the ball into the goal. When a player passes, dribbles, or kicks the ball, the ball is being translated on the soccer field.

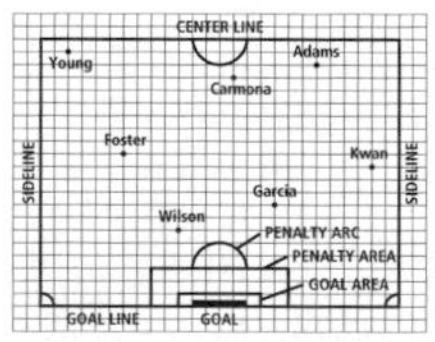

Use the diagram to complete each Exercise. Each unit on the grid represents 1 unit on a coordinate grid. Assume the ball's starting position is (x, y).

1. Garcia passes the ball to Wilson. What is the rule for this translation? $T_{<-7, -2>}(x, y)$
2. Young passes the ball to Foster. What is the rule for this translation? $T_{<4, -8>}(x, y)$
3. Kwan passes the ball to Carmona. What is the rule for this translation? $T_{<-10, 7>}(x, y)$
4. Adams passes the ball to Kwan, who then passes the ball to Garcia. Write two rules to describe these translations. $T_{<4, -8>}(x, y)$ and $T_{<-7, -3>}(x, y)$
5. Carmona passes the ball according to the rule $T_{<-8, -6>}(x, y)$. Who receives Carmona's pass? **Foster**
6. Wilson kicks the ball according to the rule $T_{<3, -6>}(x, y)$. Is a goal scored if the goalie does not block the ball? **yes**
7. Young dribbles the ball according to the rule $T_{<0, -10>}(x, y)$, and then passes the ball according to the rule $T_{<8, -4>}(x, y)$. Who receives Young's pass? **Wilson**
8. If Carmona kicks the ball toward the goal and the goalie stops it in the penalty area, which of the following rules could represent Carmona's kick? **B**
   - A $T_{<-6, -12>}(x, y)$
   - B $T_{<-3, -15>}(x, y)$
   - C $T_{<0, -12>}(x, y)$
   - D $T_{<5, -15>}(x, y)$

Practice and Problem Solving Wkbk/ All-in-One Resources/Online

Think About a Plan

9-1 **Think About a Plan**
Translations

Coordinate Geometry Quadrilateral $PLAT$ has vertices $P(-2, 0)$, $L(-1, 1)$, $A(0, 1)$, and $T(-1, 0)$. $T_{<2, -3>}(PLAT) = P'L'A'T'$. Show that $\overline{PP'}$, $\overline{LL'}$, $\overline{AA'}$, and $\overline{TT'}$ are all parallel.

Understanding the Problem

1. How can you use the three sentences of the problem to help you break the problem down into manageable pieces?
Answers may vary. Sample: The first sentence contains the given information. The second gives the translation rule I can use to obtain the coordinates of the image. The third directs me toward what I need to prove.

Planning the Solution

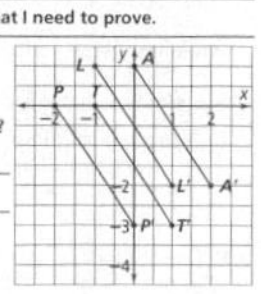

2. How can you use a diagram to help you solve the problem?
Plot the given points and the points of the image on a coordinate grid.
3. Use the given translation to find the image points.
$P'(0, -3)$, $L'(1, -2)$, $A'(2, -2)$, $T'(1, -3)$
4. Graph $\overline{PP'}$, $\overline{LL'}$, $\overline{AA'}$, and $\overline{TT'}$. What theorems about parallel lines come to mind when you look at the graph?
If the slopes of two lines are equal, then the lines are parallel.

Getting an Answer

5. Find the slopes of the lines that connect each original point to its image.
$-\frac{3}{2}$; $-\frac{3}{2}$; $-\frac{3}{2}$; $-\frac{3}{2}$
6. How can you use the slopes to establish that $\overline{PP'}$, $\overline{LL'}$, $\overline{AA'}$, and $\overline{TT'}$ are all parallel?
The lines are parallel because the lines all have the same slope.

Practice and Problem Solving Wkbk/ All-in-One Resources/Online

Standardized Test Prep

9-1 **Standardized Test Prep**
Translations

Multiple Choice

For Exercises 1–4, choose the correct letter.

1. In the diagram, $\triangle A'B'C'$ is an image of $\triangle ABC$. Which rule describes this translation? **B**
   - A $T_{<-5, -3>}(\triangle ABC)$
   - B $T_{<5, 3>}(\triangle ABC)$
   - C $T_{<-3, -5>}(\triangle ABC)$
   - D $T_{<3, 5>}(\triangle ABC)$

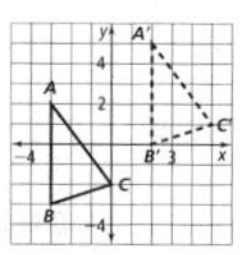

2. If $T_{<3, -7>}(TUVW) = T'U'V'W'$, what translation maps $T'U'V'W'$ onto $TUVW$? **I**
   - F $T_{<3, -7>}(T'U'V'W')$
   - G $T_{<-7, 3>}(T'U'V'W')$
   - H $T_{<7, -3>}(T'U'V'W')$
   - I $T_{<-3, 7>}(T'U'V'W')$
3. Which of the following is true for a rigid motion? **A**
   - A The preimage and the image have the same measurements.
   - B The preimage is larger than the image.
   - C The preimage is smaller than the image.
   - D The preimage is in the same position as the image.
4. $\triangle RSV$ has coordinates $R(2, 1)$, $S(3, 2)$, and $V(2, 6)$. A translation maps point R to R' at $(-4, 8)$. What are the coordinates for S' for this translation? **H**
   - F $(-6, -4)$
   - G $(-3, 2)$
   - H $(-3, 9)$
   - I $(-4, 13)$

Short Response

5. $\triangle LMP$ has coordinates $L(3, 4)$, $M(6, 6)$, and $P(5, 5)$. A translation maps point L to L' at $(7, -4)$. What are the coordinates for M' and for P' for this translation?
[2] $M'(10, -2)$ and $P'(9, -3)$ [1] one correct coordinate given
[0] no correct coordinates given

Online Teacher Resource Center

Activities, Games, and Puzzles

9-1 **Game: Finders Keepers**
Translations

Setup

You and a partner should choose one of the two game boards on which to play. Cut out the black and gray figures to the right of the game board to use as markers. Decide which of you will use each color and who will go first.

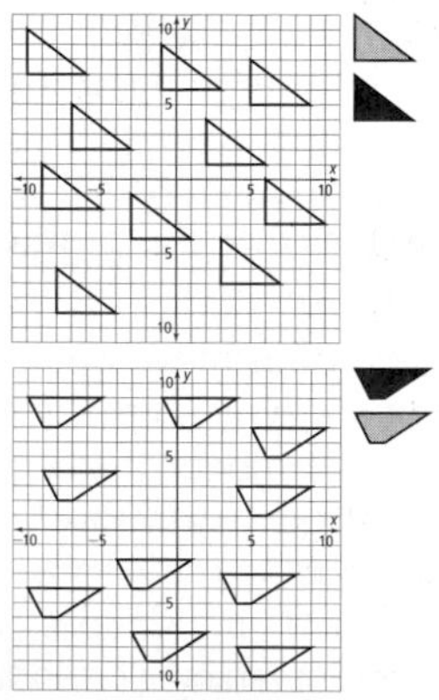

Game Play

Place your game pieces with one vertex at the origin. Be sure that your piece is placed in the same direction as the figures on the board so that a translation would map it onto one of the existing figures. The object of the game is to claim each figure on the game board by moving your piece directly on top of it. On your turn, write a translation that you believe will translate your piece onto one of the existing figures. Show the translation to your opponent. Then, he or she must move your figure according to the translation you wrote. If your piece lands exactly on a figure, you may claim the figure by writing your initials on it, and your turn is over. If your piece does not land exactly on a figure, you may not claim a figure, but your turn is still over. Either way, your piece remains in the location to which it was mapped by your translation.

Ending the Game

Play ends when all of the figures have been claimed or when the teacher indicates that time has run out. Whoever has claimed the most figures wins.

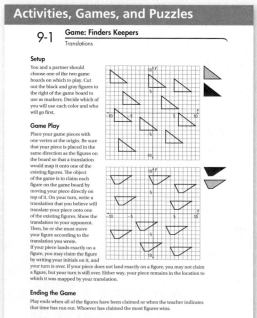

Concept Byte
Use With Lesson 9-2

ACTIVITY

Paper Folding and Reflections

Common Core State Standards
G-CO.A.5 Given a geometric figure and a rotation, reflection, or translation, draw the transformed figure . . .
MP 5

In Activity 1, you will see how a figure and its *reflection* image are related. In Activity 2, you will use these relationships to construct a reflection image.

Activity 1

Step 1 Use a piece of tracing paper and a straightedge. Using less than half the page, draw a large, scalene triangle. Label its vertices *A*, *B*, and *C*.

Step 2 Fold the paper so that your triangle is covered. Trace $\triangle ABC$ using a straightedge.

Step 3 Unfold the paper. Label the traced points corresponding to *A*, *B*, and *C* as *A′*, *B′*, and *C′*, respectively. $\triangle A'B'C'$ is a reflection image of $\triangle ABC$. The fold is the reflection line.

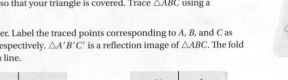

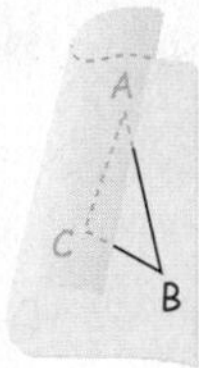

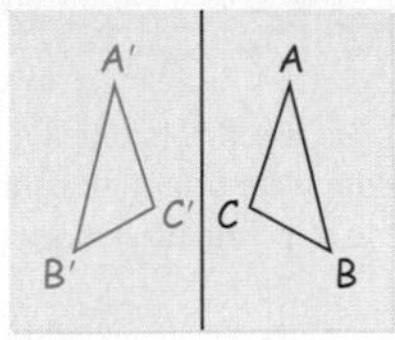

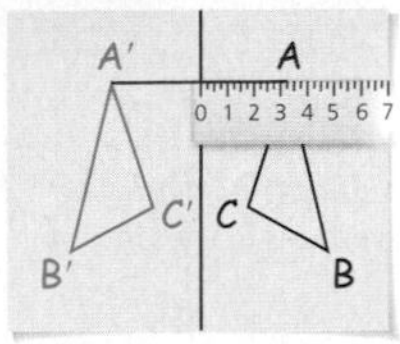

1. Use a ruler to draw $\overline{AA'}$. Measure the perpendicular distances from *A* to the fold and from *A′* to the fold. What do you notice?
2. Measure the angles formed by the fold and $\overline{AA'}$. What are the angle measures?
3. Repeat Exercises 1 and 2 for *B* and *B′* and for *C* and *C′*. Then, make a conjecture: How is the reflection line related to the segment joining a point and its image?

Activity 2

Step 1 On regular paper, draw a simple shape or design made of segments. Use less than half the page. Draw a reflection line near your figure.

Step 2 Use a compass and straightedge to construct a perpendicular to the reflection line through one point of your drawing.

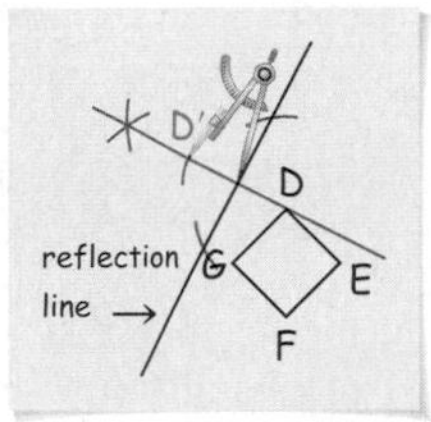

4. Explain how you can use a compass and the perpendicular you drew to find the reflection image of the point you chose.
5. Connect the reflection images for several points of your shape and complete the image. Check the accuracy of the reflection image by folding the paper along the reflection line and holding it up to a light source.

Answers

Activity 1

1. The distances are equal.
2. 90°
3. The results are the same; the reflection line is the ⊥ bis. of the seg. joining a pt. and its image.

Activity 2

4. Open the compass to the length of the segment from the line of reflection to the point you want to reflect. Using the same compass setting, draw an arc that intersects the perpendicular on the opposite side of the line of reflection.
5. Check students' work.

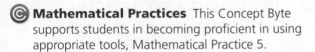

Guided Instruction

PURPOSE To use paper-folding to see how a figure and its *reflection* image are related, and to construct a reflection image using a compass and straightedge

PROCESS Students will

- create a reflection of a scalene triangle using tracing paper and paper-folding and make a conjecture about the distances and angle measures formed by the line of reflection and the segments which join the vertices of their original triangle with their images.
- draw a shape and a nearby reflection line, and construct the reflected image of the shape

DISCUSS Discuss the concept of a mirror image. Have students predict which vertex of the triangle will be closest to the line of reflection after the triangle is reflected over the line.

Activity 1

This Activity uses paper folding to generate an image over a line of reflection.

Q What can you conclude about the perpendicular distances from the vertices of the original triangle to the fold line, and the perpendicular distances from the vertices of the reflection image of their original triangle to the fold line? **[They are equal.]**

Q What is another name for the fold line (reflection line) with reference to the segments that connect the points of the original shape with their reflective images? **[perpendicular bisector]**

Activity 2

This Activity uses a compass and straightedge to generate an image over a line of reflection.

Q What geometric construction is used in Step 2? **[Given a point outside a line, construct the perpendicular to the line from the given point.]**

Q What geometric construction is used in question #4? **[Given a segment, construct a segment congruent to the given segment.]**

Mathematical Practices This Concept Byte supports students in becoming proficient in using appropriate tools, Mathematical Practice 5.

1 Interactive Learning

Solve It!

PURPOSE To describe reflections

PROCESS Students may
- use visual judgment.
- cut out the shapes and trace their reflections.

FACILITATE

Q Which part of each shape stays in the same location? **[The side that is on the yellow line stays in the same location.]**

Q Does the size of each shape change when you flip it? **[no]**

Q If a point on a shape is 3 mm from the yellow line, how far from the yellow line is it after it has been flipped? **[3 mm]**

Q What changes about a shape after you flip it? **[The shape is oriented in a different way.]**

ANSWER See Solve It in Answers on next page.

CONNECT THE MATH Students describe the characteristics of a reflection as a rigid motion that changes orientation. In the lesson, students study reflections across a line and graph reflected images.

2 Guided Instruction

Take Note

Have students demonstrate the second property of the definition using congruent triangles and the definition of a perpendicular bisector. Review the definition of rigid motion and ask students to verify that a reflection is a rigid motion.

Q What must be true for a transformation to be a rigid motion? **[Corresponding distances and angle measures must be preserved.]**

9-2 Reflections

Common Core State Standards

G-CO.A.5 Given a geometric figure and a rotation, reflection, or translation, draw the transformed figure . . . . **Also G-CO.A.2, G-CO.A.4, G-CO.B.6**

MP 1, MP 3, MP 4

Objective To find reflection images of figures

If it's hard to visualize, try using grid paper to draw the shapes.

MATHEMATICAL PRACTICES

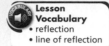

Solve It! Getting Ready!

Look at the shapes at the right. Visualize flipping each shape across its yellow line. What word do the images of the shapes form? Copy the shapes as they are shown and sketch the results of flipping them.

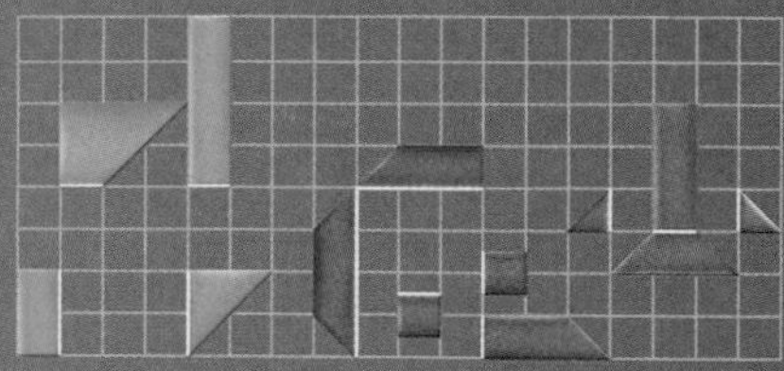

In the Solve It, you reflected shapes across lines. Notice that when you reflect a figure, the shapes have *opposite orientations*. Two figures have opposite orientations if the corresponding vertices of the preimage and image read in opposite directions.

Lesson Vocabulary
- reflection
- line of reflection

The vertices of $\triangle BUG$ read clockwise.

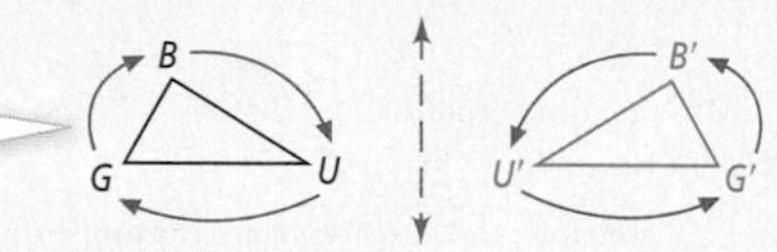

The vertices of $\triangle B'U'G'$ read counterclockwise.

Essential Understanding When you reflect a figure across a line, each point of the figure maps to another point the same distance from the line but on the other side. The orientation of the figure reverses.

take note **Key Concept** Reflection Across a Line

A **reflection** across a line m, called the **line of reflection,** is a transformation with the following properties:
- If a point A is on line m, then the image of A is itself (that is, $A' = A$).
- If a point B is not on line m, then m is the perpendicular bisector of $\overline{BB'}$.

You write the reflecion across m that takes P to P' as $R_m(P) = P'$.

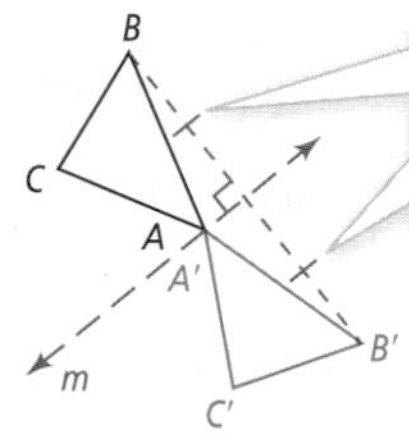

The preimage B and its image B' are equidistant from the line of reflection.

9-2 Preparing to Teach

BIG ideas **Transformations**
Coordinate Geometry

ESSENTIAL UNDERSTANDING
- When you reflect a figure across a line, each point of the figure goes to another point the same distance from the line, but on the other side.

Math Background

A reflection is another type of transformation that is a rigid motion. Students can connect their intuitive understanding of mirrors with the geometric properties of reflections. If you want to see yourself closer in the mirror, you move closer to the mirror because the image is the same distance from the mirror as the preimage.

Reflections may be difficult to graph when the line of reflection is not vertical or horizontal. Encourage students to use the Distance Formula or right triangles to graph these reflections. In later lessons, student will learn about another rigid motion called a rotation, as well as types of transformations that do not preserve distance.

Mathematical Practices

Make sense of problems and persevere in solving them. Students will plan a solution pathway for drawing reflections of a figure in a coordinate plane that includes: identifying the vertices of the preimage; identifying the line of reflection; and determining the distance each vertex lies from the line of reflection.

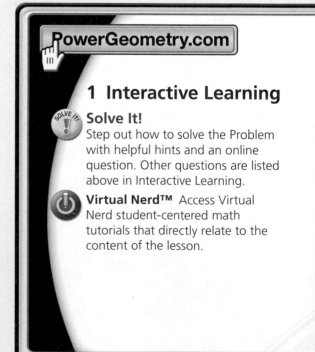

PowerGeometry.com

1 Interactive Learning

Solve It! Step out how to solve the Problem with helpful hints and an online question. Other questions are listed above in Interactive Learning.

Virtual Nerd™ Access Virtual Nerd student-centered math tutorials that directly relate to the content of the lesson.

You can use the equation of a line of reflection in the function notation. For example, $R_{y=x}$ describes the reflection across the line $y = x$.

Problem 1 Reflecting a Point Across a Line

Multiple Choice Point P has coordinates (3, 4). What are the coordinates of $R_{y=1}(P)$?

Ⓐ (3, −4) Ⓑ (0, 4) Ⓒ (3, −2) Ⓓ (−3, −2)

Graph point P and the line of reflection $y = 1$. P and its reflection image across the line must be equidistant from the line of reflection.

Think

How does a graph help you visualize the problem?
A graph shows that $y = 1$ is a horizontal line, so the line through P that is perpendicular to the line of reflection is a vertical line.

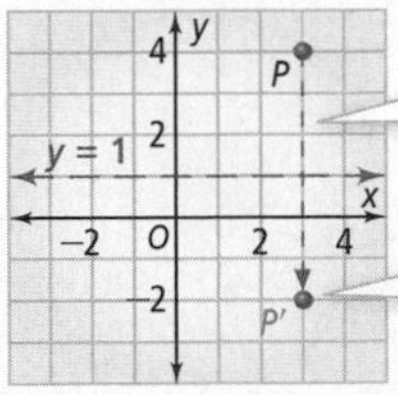

P is 3 units above the line $y = 1$, so P' must be 3 units below the line $y = 1$. The line $y = 1$ is the perpendicular bisector of $\overline{PP'}$ if P' is (3, −2). The correct answer is C.

Got It? 1. $R_{x=1}(P) = P'$. What are the coordinates of P'?

You can also use the notation R_m to describe reflections of figures. The diagram below shows $R_m(\triangle ABC)$, and function notation is used to describe some of the properties of reflections.

Take note — **Property** Properties of Reflections

- Reflections preserve distance.
  If $R_m(A) = A'$, and $R_m(B) = B'$, then $AB = A'B'$.
- Reflections preserve angle measure.
  If $R_m(\angle ABC) = \angle A'B'C'$, then $m\angle ABC = m\angle A'B'C'$.
- Reflections map each point of the preimage to one and only one corresponding point of its image.
  $R_m(A) = A'$ if and only if $R_m(A') = A$.

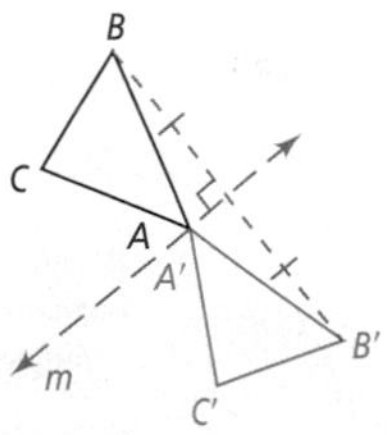

Observe that the above properties mean that reflections are rigid motions, which you learned about in Lesson 9-1.

Problem 1

Q What type of line is the line of reflection? **[horizontal]**

Q How far is P from the line of reflection? Specify above or below the line. **[3 units above]**

Q In what quadrant will the reflection of P appear? **[Quadrant IV]**

Q How far from the line of reflection is the image? Specify above or below. **[3 units below]**

Got It? VISUAL LEARNERS
Have students draw a graph with the reflection line and the preimage.

Take Note

Discuss with students the similarities and differences between reflections and translations. Both are rigid motions. Translations preserve orientation while reflections reverse the orientation of the figure.

Q Is it possible to write a translation rule that maps a figure to the same position as a reflection of the figure over a line m? **[No; since translations preserve orientation of figures and reflections reverse orientation, it is not possible.]**

2 Guided Instruction

Each Problem is worked out and supported online.

Problem 1
Reflecting a Point Across a Line
Animated

Problem 2
Graphing a Reflection Image
Animated

Problem 3
Writing a Reflection Rule

Problem 4
Using Properties of Reflections

Support in Geometry Companion

- Vocabulary
- Key Concepts
- Got It?

Answers

Solve It!

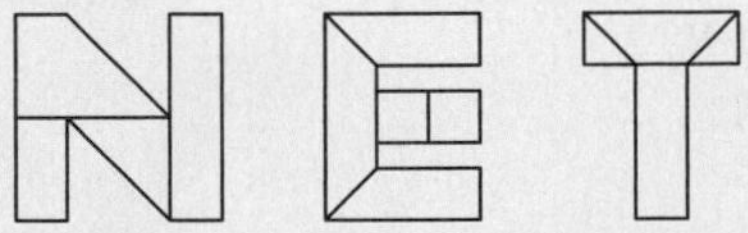

Got It?

1. (−1, 4)

Problem 2

Q Where is point *B* in relation to the line of reflection? **[Point B is on the line of reflection.]**

Q How does the location of point *B* dictate where the image of *B* will be located? **[If a point is on the line of reflection, then the point is its own image.]**

Q How far should the images of points *A* and *C* be from the line of reflection? Explain. **[The images will be the same distance from the line of reflection as their preimages. Point A is 3 units from the line. Point C is 4 units from the line.]**

Got It? ERROR PREVENTION

Have students draw a diagram of the preimage and image and verify by checking distances.

Problem 3

Q How can you easily identify the corresponding vertices of each triangle? **[Since the triangles appear to be right triangles, identify the right angle and hypotenuse in each triangle.]**

Q What are two possible lines of reflection in the diagram? **[lines *m* and *k*]**

Got It? VISUAL LEARNERS

Have students draw the reflection of the vertex of the right angle in Triangle 1 across line *m* and across line *k*. Once they find the triangle that has the correct image for the vertex, they can check the other vertices.

Problem 2 Graphing a Reflection Image

Coordinate Geometry **Graph points $A(-3, 4)$, $B(0, 1)$, and $C(4, 2)$. Graph and label $R_{y\text{-axis}}(\triangle ABC)$.**

Think
$\triangle ABC$ intersects the line of reflection. How will the image relate to the line of reflection?
The image will also intersect the line of reflection.

Step 1
Graph $\triangle ABC$. Show the *y*-axis as the dashed line of reflection.

Step 2
Find A', B', and C'. B' is in the same position as B because B is on the line of reflection. Locate A' and C' so that the *y*-axis is the perpendicular bisector of $\overline{AA'}$ and $\overline{CC'}$.

Step 3
Draw $\triangle A'B'C'$.

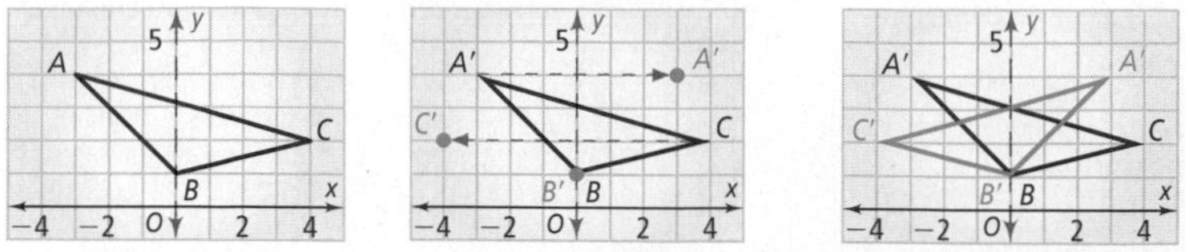

Got It? 2. Graph $\triangle ABC$ from Problem 2. Graph and label $R_{x\text{-axis}}(\triangle ABC)$.

Problem 3 Writing a Reflection Rule

Each triangle in the diagram is a reflection of another triangle across one of the given lines. How can you describe Triangle 2 by using a reflection rule?

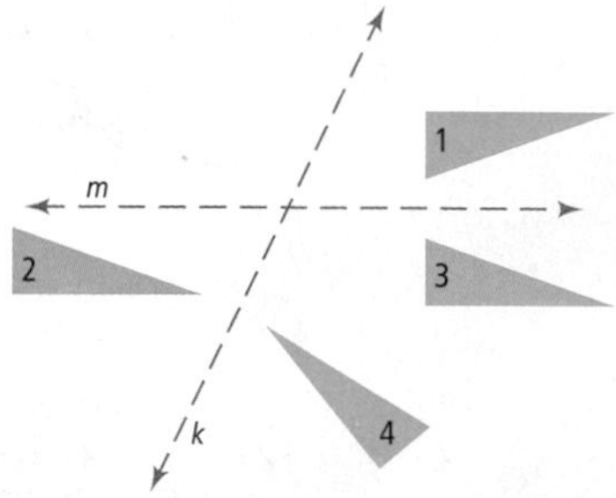

Plan
If Triangle 2 is the image of a reflection, what do you know about the preimage?
The preimage has opposite orientation, and lies on the opposite side of the line of reflection.

Triangle 2 is the image of a reflection, so find the preimage and the line of reflection to write a rule.

The preimage cannot be Triangle 3 because Triangle 2 and Triangle 3 have the same orientation and reflections reverse orientation.

Check Triangles 1 and 4 by drawing line segments that connect the corresponding vertices of Triangle 2. Because neither line *k* nor line *m* is the perpendicular bisector of the segment drawn from Triangle 1 to Triangle 2, Triangle 1 is not the preimage.

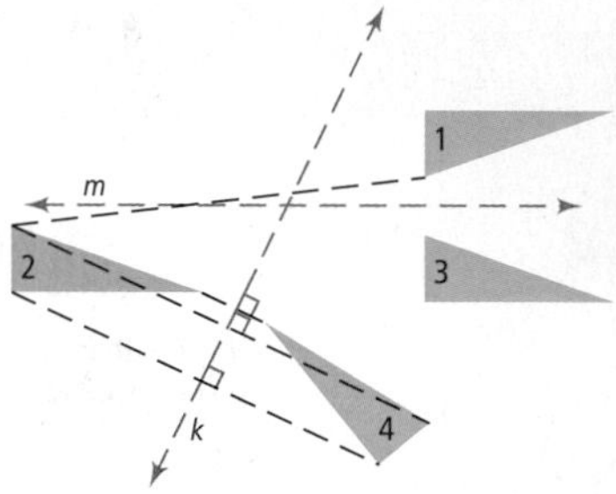

Line *k* is the perpendicular bisector of the segments joining corresponding vertices of Triangle 2 and Triangle 4. So, Triangle 2 = R_k(Triangle 4).

Got It? 3. How can you use a reflection rule to describe Triangle 1? Explain.

Additional Problems

1. Point *F* has coordinates (2, 6). What are the coordinates of $R_{y=-3}(F)$?

A. (−5, 6)
B. (2, −6)
C. (−5, −12)
D. (2, −12)

ANSWER D

2. Use points $X(-4, 3)$, $Y(2, 6)$, and $Z(-1, -8)$. What are the coordinates of $R_{x\text{-axis}}(\triangle XYZ)$?

ANSWER $X'(-4, -3)$, $Y'(2, -6)$, $Z'(-1, 8)$

3. Trapezoid 2 is a reflection of another trapezoid across one of the given lines. How can you describe Trapezoid 2 by using a reflection rule?

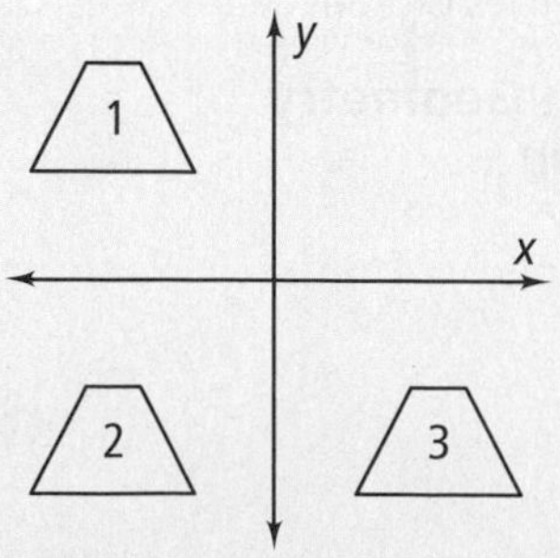

ANSWER Trapezoid 2 = $R_{y\text{-axis}}$ (Trapezoid 3)

4. In the diagram, $R_t(A) = A$, $R_t(B) = D$, and $R_t(C) = C$. Use the properties of reflections to describe how you know that *ABCD* is a kite.

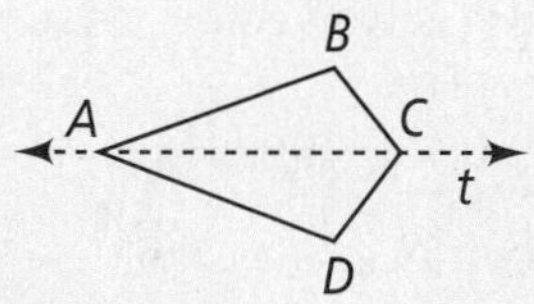

ANSWER Since $R_t(A) = A$, $R_t(B) = D$, and reflections preserve distance, $R_t(\overline{AB}) = \overline{AD}$. So, $AB = AD$. Since $R_t(C) = C$ and $R_t(B) = D$, $R_t(\overline{CB}) = \overline{CD}$. So, $CB = CD$. Therefore, *ABCD* has two pairs of consecutive sides congruent, so by definition, *ABCD* is a kite.

You can use the properties of reflections to prove statements about figures.

Problem 4 Using Properties of Reflections

Plan

What do you have to know about $\triangle GHJ$ to show that it is an isosceles triangle? Isosceles triangles have at least two congruent sides.

In the diagram, $R_t(G) = G$, $R_t(H) = J$, and $R_t(D) = D$. Use the properties of reflections to describe how you know that $\triangle GHJ$ is an isosceles triangle.

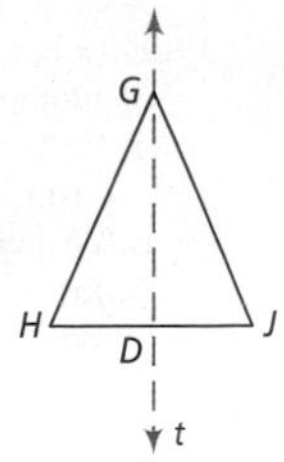

Since $R_t(G) = G$, $R_t(H) = J$, and reflections preserve distance, $R_t(\overline{GH}) = \overline{GJ}$. So, $GH = GJ$ and, by definition, $\triangle GHJ$ is an isosceles triangle.

Got It? **4.** Can you use properties of reflections to prove that $\triangle GHJ$ is equilateral? Explain.

Lesson Check

Do you know HOW?

Use the graph of $\triangle FGH$.

1. What are the coordinates of $R_{y\text{-axis}}(H)$?

2. What are the coordinates of $R_{x=3}(G)$?

3. Graph and label $R_{y=4}(\triangle FGH)$.

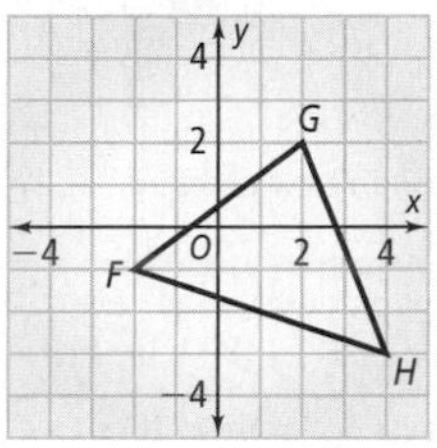

Do you UNDERSTAND? MATHEMATICAL PRACTICES

4. Vocabulary What is the relationship between a line of reflection and a segment joining corresponding points of the preimage and image?

5. Error Analysis A classmate sketched $R_s(A) = A'$ as shown in the diagram.

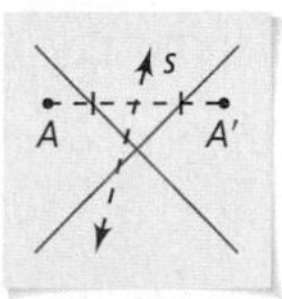

a. Explain your classmate's error.

b. Copy point A and line s and show the correct location of A'.

6. What are the coordinates of a point $P(x, y)$ reflected across the y-axis? Across the x-axis? Use reflection notation to write your answer.

Practice and Problem-Solving Exercises MATHEMATICAL PRACTICES

A Practice Find the coordinates of each image.

See Problem 1.

7. $R_{x=1}(Q)$ **8.** $R_{y=-1}(P)$

9. $R_{y\text{-axis}}(S)$ **10.** $R_{y=0.5}(T)$

11. $R_{x=-3}(U)$ **12.** $R_{x\text{-axis}}(V)$

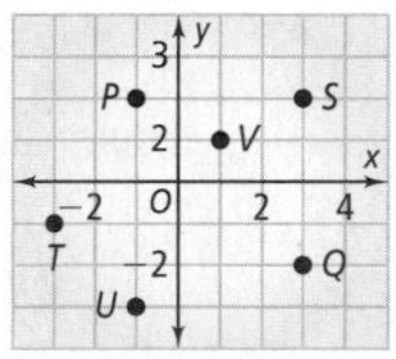

Problem 4

Ask students if there is another way to prove that $\triangle GHJ$ is an isosceles triangle using properties of reflections. Remind them that the base angles of an isosceles triangle are congruent and that reflections preserve angle measurements.

Got It?

Q What do you need to know about $\triangle GHJ$ to show that it is equilateral? **[Equilateral triangles have three congruent sides.]**

Q What do you need to prove to show $\triangle GHJ$ is equilateral? **[$HJ = HG$ or $HJ = GJ$]**

3 Lesson Check

Do you know HOW?

- If students have difficulty with Exercise 1, then have them review Problem 1 to find out how a reflection across the y-axis changes an ordered pair.

Do you UNDERSTAND?

- If students have difficulty with Exercise 5, then have them review Problem 2.

Close

Q What type of transformation flips a figure over a line? **[a reflection]**

Q How is a line of reflection related to the segment that connects the points of a preimage to its image? **[The line of reflection is the perpendicular bisector of the segment.]**

Answers

Got It? (continued)

2.

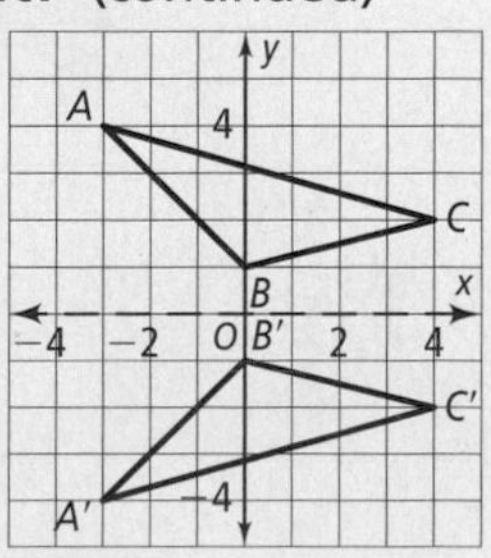

3. Triangle 1 = R_m(Triangle 3)

4. No; it is not possible to prove that $\triangle GHJ$ is equilateral since you cannot prove that $HJ = HG$ or $HJ = GJ$.

Lesson Check

1. $(-4, -3)$

2. $(4, 2)$

3.

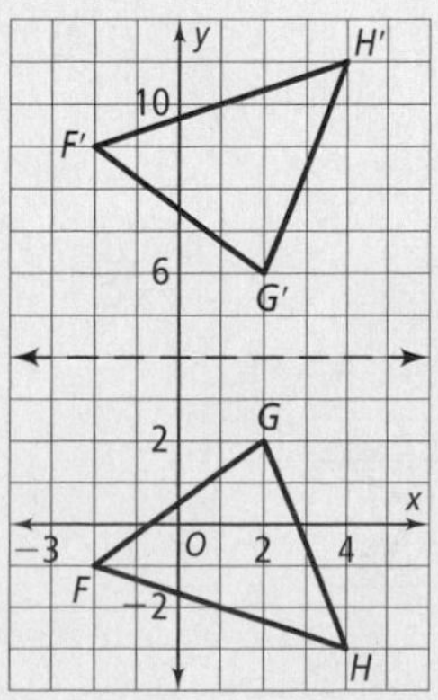

PowerGeometry.com

3 Lesson Check

For a digital lesson check, use the Got It questions.

Support in Geometry Companion

- Lesson Check

4 Practice

Assign homework to individual students or to an entire class.

4 Practice

ASSIGNMENT GUIDE
Basic: 7–20 all, 24, 27, 30–36
Average: 7–19 odd, 21–38
Advanced: 7–19 odd, 21–44

Mathematical Practices are supported by exercises with red headings. Here are the Practices supported in this lesson:

MP 1: Make Sense of Problems Ex. 26
MP 3: Construct Arguments Ex. 35, 38–43
MP 3: Critique the Reasoning of Others Ex. 5

Applications exercises have blue headings. Exercises 25, 30, and 36 support MP 4: Model.

EXERCISE 29: Use the Think About a Plan worksheet in the **Practice and Problem Solving Workbook** (also available in the Teaching Resources in print and online) to further support students' development in becoming independent learners.

HOMEWORK QUICK CHECK
To check students' understanding of key skills and concepts, go over Exercises 7, 15, 26, 30, and 36.

Coordinate Geometry Given points $J(1, 4)$, $A(3, 5)$, and $G(2, 1)$, graph $\triangle JAG$ and its reflection image as indicated. See Problem 2.

13. $R_{x\text{-axis}}$ **14.** $R_{y\text{-axis}}$ **15.** $R_{y=2}$ **16.** $R_{y=5}$ **17.** $R_{x=-1}$ **18.** $R_{x=2}$

19. Each figure in the diagram at the right is a reflection of another figure across one of the reflection lines. See Problem 3.
 a. Write a reflection rule to describe Figure 3. Justify your answer.
 b. Write a reflection rule to describe Figure 2. Justify your answer.
 c. Write a reflection rule to describe Figure 4. Justify your answer.

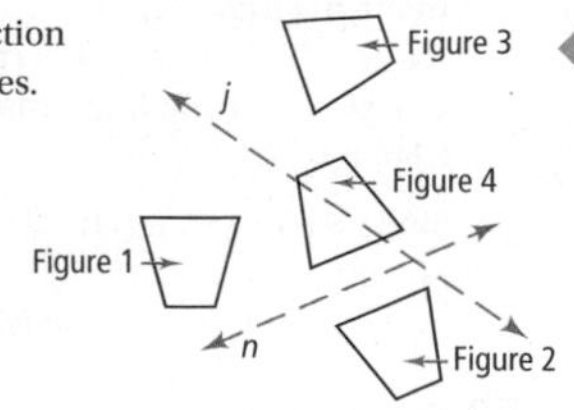

20. In the diagram at the right, $LMNP$ is a rectangle with $LM = 2MN$. See Problem 4.
 a. Copy the diagram. Then sketch $R_{\overline{LM}}(LMNP)$.
 b. What figure results from the reflection? Use properties of reflections to justify your solution.

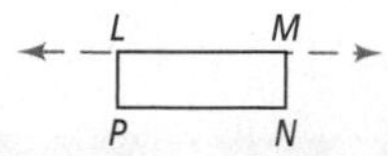

B Apply

Copy each figure and line ℓ. Draw each figure's reflection image across line ℓ.

21.

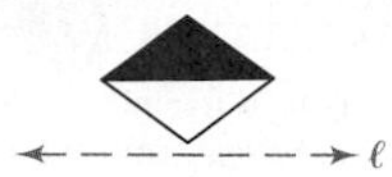

22.

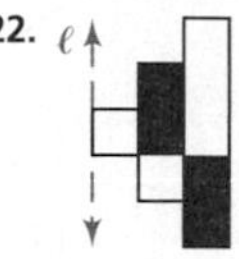

23. Coordinate Geometry The following steps explain how to reflect point A across the line $y = x$.

Step 1 Draw line ℓ through $A(5, 1)$ perpendicular to the line $y = x$. The slope of $y = x$ is 1, so the slope of line ℓ is $1 \cdot (-1)$, or -1.

Step 2 From A, move two units left and two units up to $y = x$. Then move two more units left and two more units up to find the location of A' on line ℓ. The coordinates of A' are $(1, 5)$.

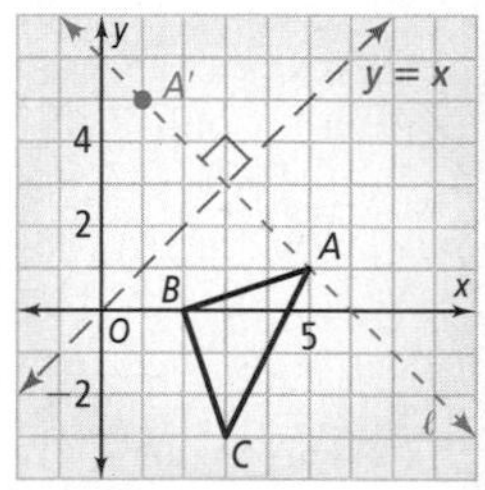

 a. Copy the diagram. Then draw the lines through B and C that are perpendicular to the line $y = x$. What is the slope of each line?
 b. $R_{y=x}(B) = B'$ and $R_{y=x}(C) = C'$. What are the coordinates of B' and C'?
 c. Graph $\triangle A'B'C'$.
 d. Make a Conjecture Compare the coordinates of the vertices of $\triangle ABC$ and $\triangle A'B'C'$. Make a conjecture about the coordinates of the point $P(a, b)$ reflected across the line $y = x$.

24. Coordinate Geometry $\triangle ABC$ has vertices $A(-3, 5)$, $B(-2, -1)$, and $C(0, 3)$. Graph $R_{y=-x}(\triangle ABC)$ and label it. (*Hint:* See Exercise 23.)

Answers

Lesson Check (continued)

4. The line of reflection is the ⊥ bis. of any seg. whose endpts. are corresp. pts. of the preimage and image.

5. $\overline{AA'}$ should be ⊥ to r.

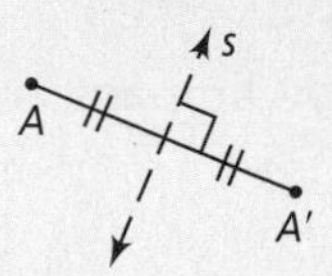

6. $R_{y\text{-axis}}(x, y) = (-x, y)$; $R_{x\text{-axis}}(x, y) = (x, -y)$
7. $(-1, -2)$
8. $(-1, -4)$
9. $(-3, 2)$
10. $(-3, 2)$
11. $(-5, -3)$
12. $(1, -1)$

Practice and Problem-Solving Exercises

13. $J'(1, -4)$, $A'(3, -5)$, $G'(2, -1)$

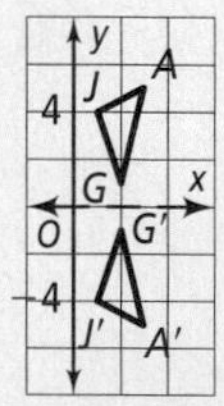

14. $J'(-1, 4)$, $A'(-3, 5)$, $G'(-2, 1)$

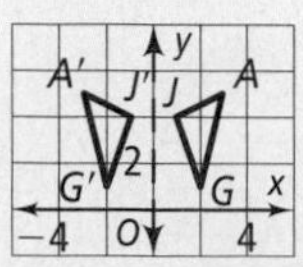

15. $J'(1, 0)$, $A'(3, -1)$, $G'(2, 3)$

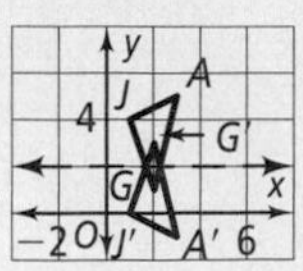

16. $J'(1, 6)$, $A'(3, 5)$, $G'(2, 9)$

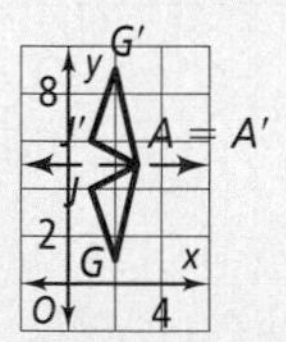

17. $J'(-3, 4)$, $A'(-5, 5)$, $G'(-4, 1)$

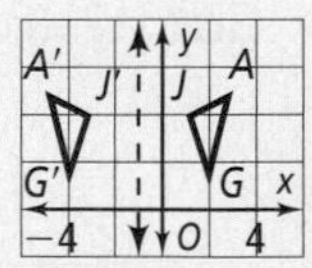

18. $J'(3, 4)$, $A'(1, 5)$, $G'(2, 1)$

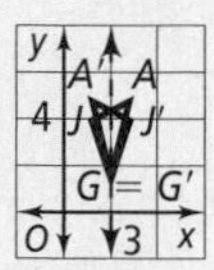

19a. Figure 3 = R_j (Figure 1) because line j is the perpendicular bisector of the line segments between corresponding vertices of Figures 1 and 3.
 b. Figure 2 = R_n (Figure 4) because line n is the perpendicular bisector of the line segments between corresponding vertices of Figures 2 and 4.
 c. Figure 4 = R_n (Figure 2) because line n is the perpendicular bisector of the line segments between corresponding vertices of Figures 4 and 2.

25. **Recreation** When you play pool, you can use the fact that the ball bounces off the side of the pool table at the same angle at which it hits the side. Suppose you want to put the ball at point B into the pocket at point P by bouncing it off side $\overline{RS}$. Off what point on $\overline{RS}$ should the ball bounce? Draw a diagram and explain your reasoning.

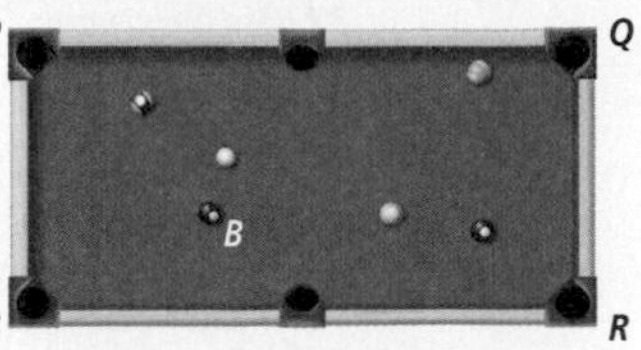

26. **Think About a Plan** The coordinates of the vertices of $\triangle FGH$ are $F(2, -1)$, $G(-2, -2)$, and $H(-4, 3)$. Graph $\triangle FGH$ and $R_{y=x-3}(\triangle FGH)$.
- What is the relationship between the line $y = x - 3$ and $\overline{FF'}$, $\overline{GG'}$, and $\overline{HH'}$?
- How can you use slope to find the image of each vertex?

27. In the diagram $R(ABCDE) = A'B'C'D'E'$.
  a. What are the midpoints of $\overline{AA'}$ and $\overline{DD'}$?
  b. What is the equation of the line of reflection?
  c. Write a rule that describes this reflection.

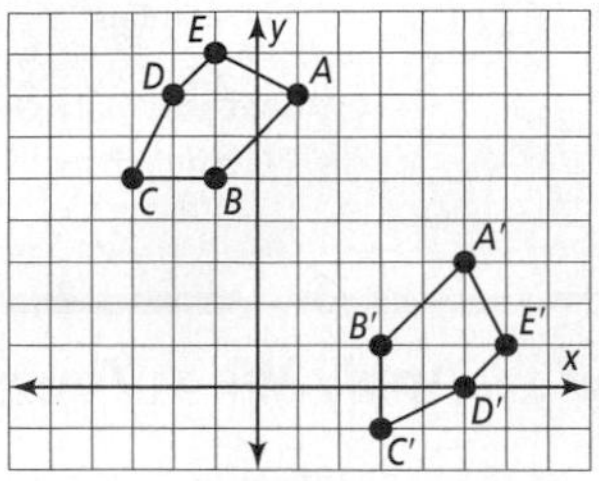

Copy each pair of figures. Then draw the line of reflection you can use to map one figure onto the other.

28. 

29. 

30. **History** The work of artist and scientist Leonardo da Vinci (1452–1519) has an unusual characteristic. His handwriting is a mirror image of normal handwriting.
  a. Write the mirror image of the sentence, "Leonardo da Vinci was left-handed." Use a mirror to check how well you did.
  b. Explain why the fact about da Vinci in part (a) might have made mirror writing seem natural to him.

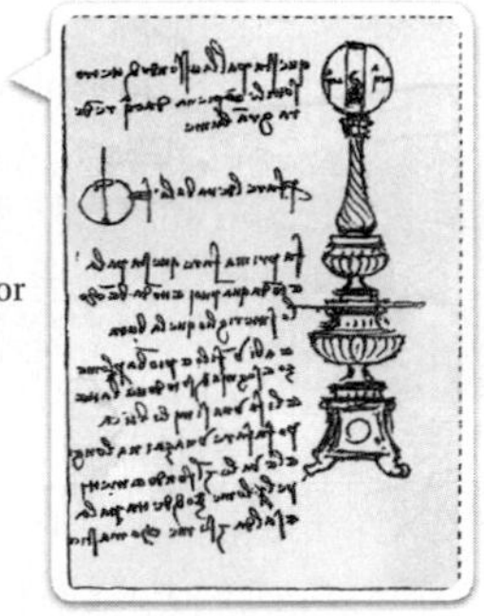

31. **Open-Ended** Give three examples from everyday life of objects or situations that show or use reflections.

Find the image of $O(0, 0)$ after two reflections, first across line ℓ_1 and then across line ℓ_2.

32. $\ell_1: y = 3$, ℓ_2: x-axis

33. $\ell_1: x = -2$, ℓ_2: y-axis

34. ℓ_1: x-axis, ℓ_2: y-axis

35. **Reasoning** When you reflect a figure across a line, does every point on the preimage move the same distance? Explain.

36. **Security** Recall that when a ray of light hits a mirror, it bounces off the mirror at the same angle at which it hits the mirror. You are installing a security camera. At what point on the mirrored wall should you aim the camera at C in order to view the door at D? Draw a diagram and explain your reasoning.

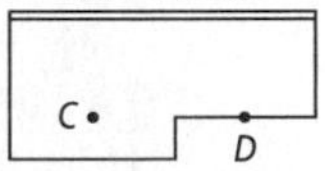

20a.

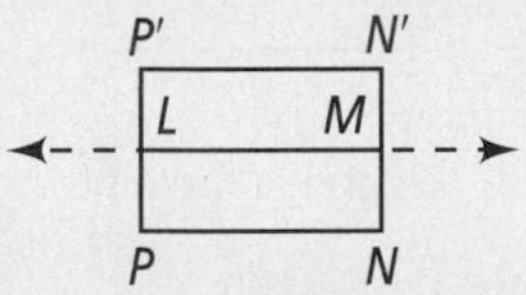

b. square; Since $R_{LM}(M) = M$, $R_{LM}(N) = N'$, and reflections preserve distance, $R_{LM}(\overline{MN}) = \overline{MN'}$. So, $MN = MN'$ and $NN' = 2MN$. Since $LM = 2MN$ and $NN' = 2MN$, by substitution $LM = NN'$. Therefore, in the new figure $PNN'P'$ the length equals the width, so the figure is a square.

21.

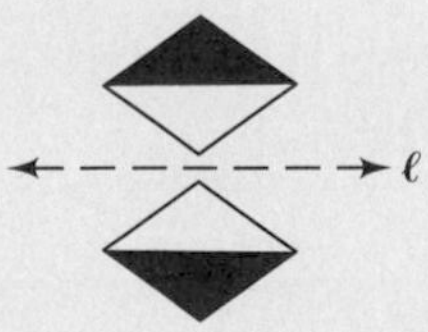

22.

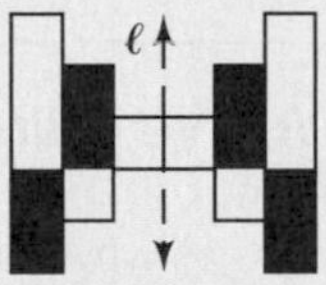

23a. -1

b. $B'(0, 2)$; $C'(-3, 3)$

c.

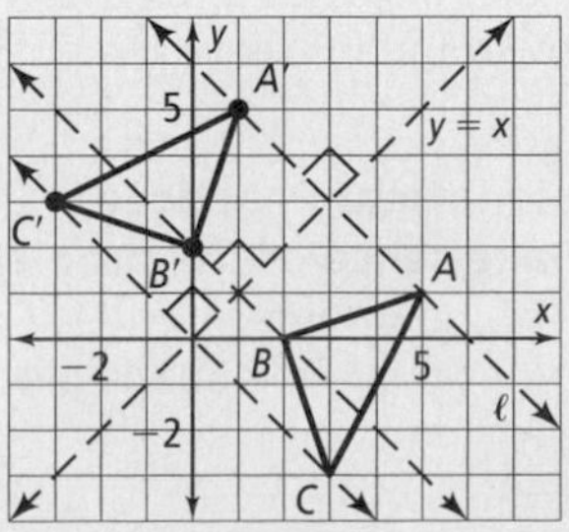

d. The coordinates of P' will be (b, a); the x- and y-coordinates will switch.

24.

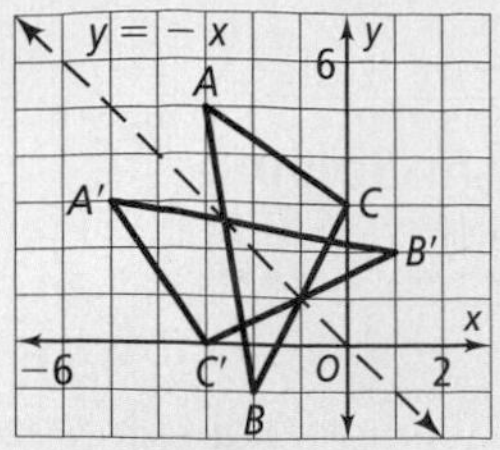

25. Reflect P across $\overline{SR}$ to P'. Because the pool table is a rectangle, $\overline{PS} \perp \overline{SR}$, and thus P' is collinear with S and P. The ball should bounce off the point T that is the intersection of $\overline{BP'}$ and $\overline{SR}$. Let A be the point on $\overline{SP}$ that the ball rolls to after it bounces off $\overline{SR}$. To see why A is the same point as P, look at $\triangle AST$ and $\triangle P'ST$.

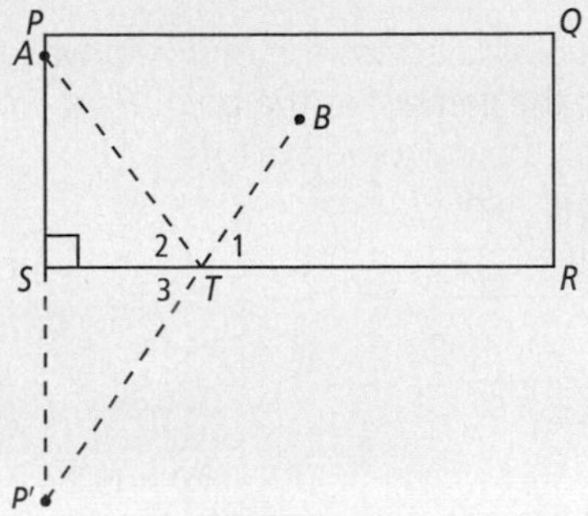

Since the ball bounces off $\overline{SR}$ so that $\angle 1 \cong \angle 2$ and $\angle 1 \cong \angle 3$ (vertical ∠s), $\angle 2 \cong \angle 3$ by the Trans. Prop. of $\cong$. Right △s AST and $P'ST$ are $\cong$ and $\overline{TS} \cong \overline{TS}$, so $\triangle ATS \cong \triangle P'TS$ by ASA. Then $\overline{AS} \cong \overline{P'S}$ because corresp. parts of $\cong$ △s are $\cong$. But $\overline{P'S} \cong \overline{PS}$ by the definition of reflection across a line, so A and P must be the same point.

26.

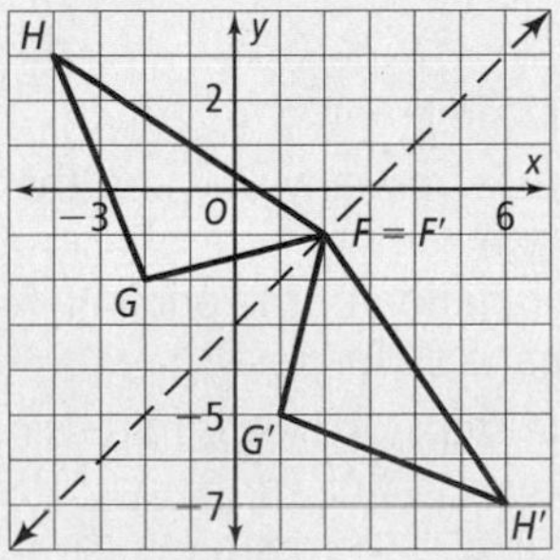

27a. (3, 5), (1.5, 3.5)

b. $y = x + 2$

c. $R_{y=x+2}(ABCDE) = A'B'C'D'E'$

28.

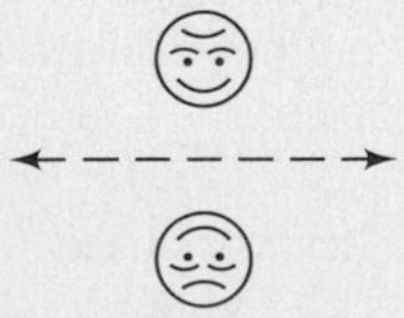

29.

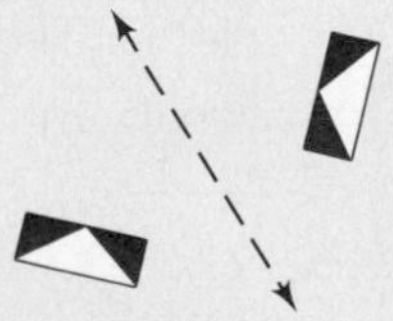

30–42. See next page.

Answers

Practice and Problem-Solving Exercises (continued)

30a. Leonardo da Vinci was left-handed. (printed mirrored)

b. Answers may vary. Sample: For mirror writing, he would write from left to right. Because he was left-handed, his writing hand would not cover up the words he had just written.

31. Answers may vary. Sample: scissors, baseball glove, golf clubs

32. (0, −6) **33.** (4, 0) **34.** (0, 0)

35. No; each point moves a distance equal to twice the point's distance from the line of reflection.

36. Reflect point D across the mirrored wall to D'. Aim the camera at the point P where $\overline{CD'}$ intersects the mirrored wall.

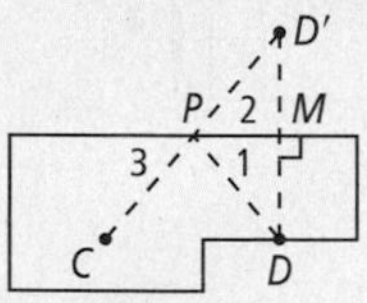

To show that a ray of light traveling from D to P will bounce off P and into the camera at C, show that $\angle 1 \cong \angle 3$. By the definition of reflection across a line, the mirrored wall is the ⊥ bis. of $\overline{DD'}$, so $\overline{MD} \cong \overline{MD'}$ and $\angle DMP \cong \angle D'MP$. $\overline{PM} \cong \overline{PM}$, so $\triangle DMP \cong \triangle D'MP$ by SAS, and $\angle 1 \cong \angle 2$ because corresp. parts of ≅ △s are ≅. Then, since $\angle 2 \cong \angle 3$ (vert. ∠s), $\angle 1 \cong \angle 3$ by the Trans. Prop. of ≅.

37a. (3, 1) **b.** (−1, −3) **c.** (−3, −1) **d.** (1, 3)

e. They are the same point.

38. Yes; reflect a △ across any side and then reflect the image across the ⊥ bis. of that side. The combination of the original △ and the second image △ forms a ▱.

39. Yes; follow the steps of Exercise 38 using one leg of an isosc. △ to first form a ▱. Then reflect the original △ across the ⊥ bis. of the base of the second △ to form an isosc. trapezoid.

40. Yes; reflect an acute scalene △ across any side, an obtuse scalene △ across its longest side, a nonright isosc. (but not equilateral) △ across either leg, or a nonisosc. rt. △ across its hyp.

41. Yes; reflect an isosc. △ across its base.

42. Yes; follow the steps of Exercise 44 using a rt. △ and the hyp. as the first reflection line.

43. Yes; reflect an isosc. rt. △ across its hyp.

44. The slope of $\overline{AB}$ is $\frac{a-b}{b-a} = \frac{a-b}{-1(a-b)} = -1$. The slope of $y = x$ is 1. Since $(1)(-1) = -1$, the lines are ⊥. The midpt. of $\overline{AB}$ is $\left(\frac{b+a}{2}, \frac{a+b}{2}\right)$, which is a point on the line $y = x$.

37. Use the diagram at the right. Find the coordinates of each image point.
a. $R_{y=x}(A) = A'$
b. $R_{y=-x}(A') = A''$
c. $R_{y=x}(A'') = A'''$
d. $R_{y=-x}(A''') = A''''$
e. How are A and A'''' related?

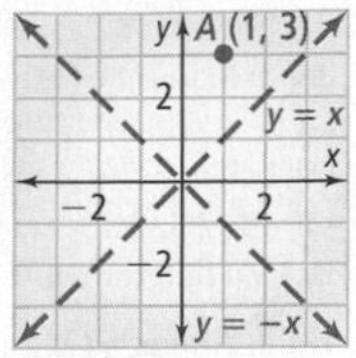

Challenge

Reasoning **Can you form the given type of quadrilateral by drawing a triangle and then reflecting one or more times? Explain.**

38. parallelogram **39.** isosceles trapezoid **40.** kite
41. rhombus **42.** rectangle **43.** square

44. **Coordinate Geometry** Show that $R_{y=x}(A) = B$ for point $A(a, b)$ and $B(b, a)$. (*Hint:* Show that $y = x$ is the perpendicular bisector of $\overline{AB}$.)

Apply What You've Learned

MATHEMATICAL PRACTICES MP 1

Look back at the information about the video game on page 543, and review the requirements for Case 1 and Case 2 of the video game program. The graph of the puzzle piece and target area is shown again below.

Select all of the following that are true. Explain your reasoning.

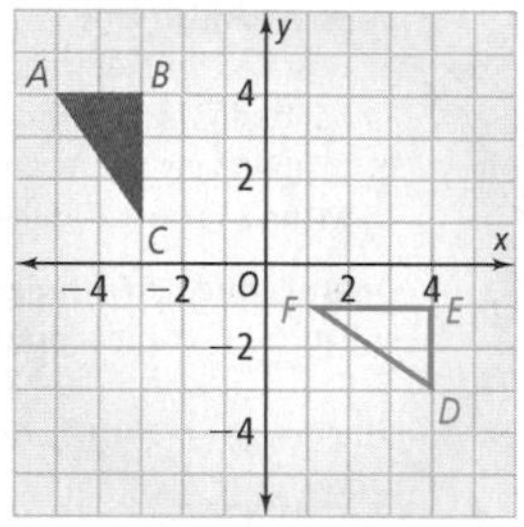

A. $T_{<2,\,-2>}(x, y)$ moves $\triangle ABC$ so that the image of C is the reflection of F across the y-axis.
B. $T_{<2,\,-2>}(x, y)$ moves $\triangle ABC$ so that the image of $\triangle ABC$ is the reflection of $\triangle DEF$ across the y-axis.
C. $R_{y=x}(C) = F$
D. F is the reflection of $(-1, 1)$ across the line $y = x$.
E. F is the reflection of C across the line $y = 2x + 2$.
F. E is the reflection of B across the line $y = 2x + 2$.
G. The image of the puzzle piece after a reflection across a line has the same orientation as the target area.
H. The problem cannot be solved for Case 1 using only reflections.

Apply What You've Learned

Here students will look at translations and reflections as ways to move the puzzle piece toward the target area in the video game described on page 543. Later in the chapter, they will consider how rotations move the puzzle piece.

Mathematical Practices

Students **make sense of the problem** by considering how some individual transformations move the puzzle piece. (MP 1)

ANSWERS

Choices A, D, E, G, and H are all true.

9-2 Lesson Resources

Differentiated Remediation

Instructional Support

Geometry Companion

Students can use the **Geometry Companion** worktext (4 pages) . . .

- New Vocabulary
- Key Concepts
- Got It for each Problem
- Lesson Check

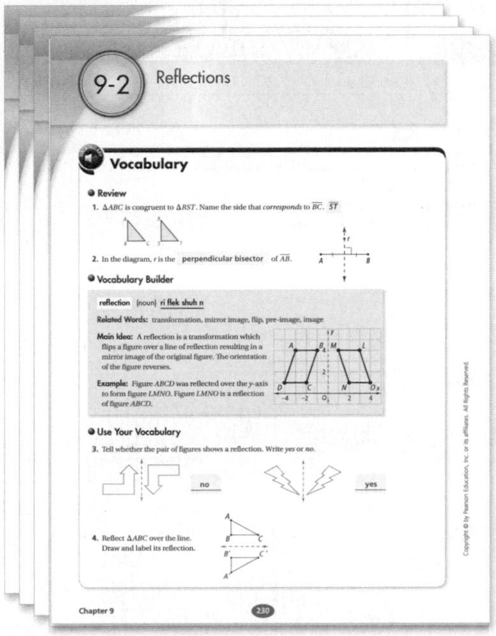

9-2 Reflections

Vocabulary

Review

1. $\triangle ABC$ is congruent to $\triangle RST$. Name the side that *corresponds* to $\overline{BC}$. $\overline{ST}$

2. In the diagram, *r* is the perpendicular bisector of $\overline{AB}$.

Vocabulary Builder

reflection (noun) ri flek shuh n

Related Words: transformation, mirror image, flip, pre-image, image

Main Idea: A reflection is a transformation which flips a figure over a line of reflection resulting in a mirror image of the original figure. The orientation of the figure reverses.

Example: Figure *ABCD* was reflected over the *y*-axis to form figure *LMNO*. Figure *LMNO* is a reflection of figure *ABCD*.

Use Your Vocabulary

3. Tell whether the pair of figures shows a reflection. Write *yes* or *no*. no yes

4. Reflect $\triangle ABC$ over the line. Draw and label its reflection.

Chapter 9 230

ELL Support

Connect to Prior Knowledge Have groups of students work with a mirror and describe how objects relate to their mirror images or reflections. Let each group demonstrate how the orientation of the reflection is opposite to the original object, e.g., a person's right hand matches the left hand of the mirror image.

Next have each group draw a polygon on a coordinate grid and discuss questions like the following:

- What would a mirror show if you set it on the coordinate grid next to the polygon? **[a reflection of the polygon]**
- Would the reflection have the same orientation as the polygon? **[No. The orientation will be opposite.]**
- How far will the reflection seem to be from the mirror? **[The same distance from the mirror as the polygon is.]**

5 Assess & Remediate

Lesson Quiz

1. Point *D* has coordinates (4, 1). What are the coordinates of $R_{x=2}(D)$?
2. A design for the math club logo reflects a triangle across the *y*-axis. Graph the reflection.

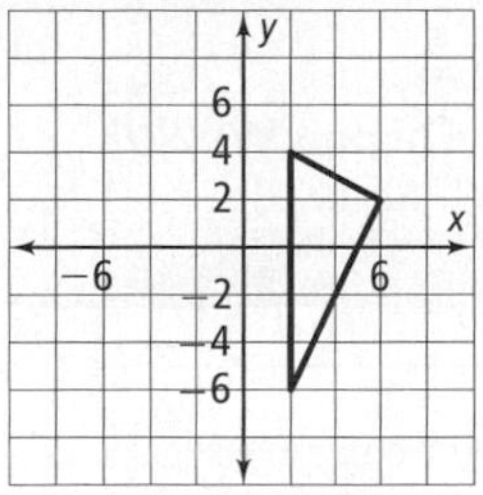

3. $\triangle ABC$ has vertices $A(-2, 0)$, $B(2, 5)$, $C(3, 1)$. The reflection of image of $\triangle ABC$ is $\triangle A'B'C'$ with vertices $A'(-2, 0)$, $B'(2, -5)$, $C'(3, -1)$. What is a rule that describes the reflection?

ANSWERS TO LESSON QUIZ

1. (0, 1)
2.

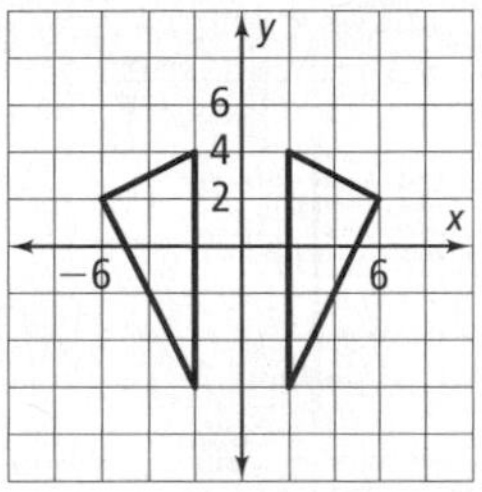

3. $R_{x\text{-axis}}(\triangle ABC) = \triangle A'B'C'$

PRESCRIPTION FOR REMEDIATION

Use the student work on the Lesson Quiz to prescribe a differentiated review assignment.

Points	Differentiated Remediation
0–1	Intervention
2	On-level
3	Extension

PowerGeometry.com

5 Assess & Remediate

Assign the Lesson Quiz. Appropriate intervention, practice, or enrichment is automatically generated based on student performance.

Intervention

- **Reteaching** (2 pages) Provides reteaching and practice exercises for the key lesson concepts. Use with struggling students or absent students.
- **English Language Learner Support** Helps students develop and reinforce mathematical vocabulary and key concepts.

All-in-One Resources/Online

Reteaching

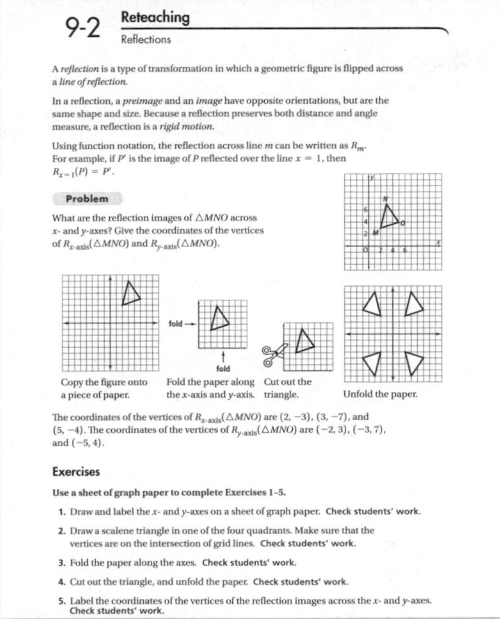

9-2 Reteaching

Reflections

A *reflection* is a type of transformation in which a geometric figure is flipped across a *line of reflection*.

In a reflection, a *preimage* and an *image* have opposite orientations, but are the same shape and size. Because a reflection preserves both distance and angle measure, a reflection is a *rigid motion*.

Using function notation, the reflection across line *m* can be written as R_m. For example, if P' is the image of *P* reflected over the line $x = 1$, then $R_{x=1}(P) = P'$.

Problem

What are the reflection images of $\triangle MNO$ across *x*- and *y*-axes? Give the coordinates of the vertices of $R_{x\text{-axis}}(\triangle MNO)$ and $R_{y\text{-axis}}(\triangle MNO)$.

Copy the figure onto a piece of paper. Fold the paper along the *x*-axis and *y*-axis. Cut out the triangle. Unfold the paper.

The coordinates of the vertices of $R_{x\text{-axis}}(\triangle MNO)$ are $(2, -3)$, $(3, -7)$, and $(5, -4)$. The coordinates of the vertices of $R_{y\text{-axis}}(\triangle MNO)$ are $(-2, 3)$, $(-3, 7)$, and $(-5, 4)$.

Exercises

Use a sheet of graph paper to complete Exercises 1–5.

1. Draw and label the *x*- and *y*-axes on a sheet of graph paper. **Check students' work.**
2. Draw a scalene triangle in one of the four quadrants. Make sure that the vertices are on the intersection of grid lines. **Check students' work.**
3. Fold the paper along the axes. **Check students' work.**
4. Cut out the triangle, and unfold the paper. **Check students' work.**
5. Label the coordinates of the vertices of the reflection images across the *x*- and *y*-axes. **Check students' work.**

All-in-One Resources/Online

English Language Learner Support

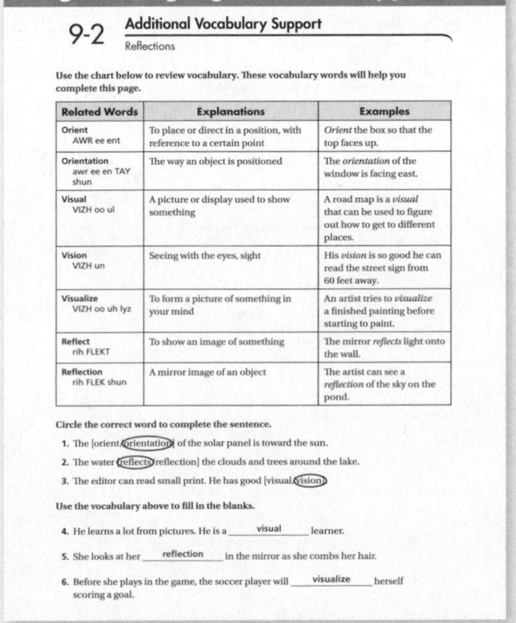

9-2 Additional Vocabulary Support

Reflections

Use the chart below to review vocabulary. These vocabulary words will help you complete this page.

Related Words	Explanations	Examples
Orient AWR ee ent	To place or direct in a position, with reference to a certain point	*Orient* the box so that the top faces up.
Orientation awr ee en TAY shun	The way an object is positioned	The *orientation* of the window is facing east.
Visual VIZH oo ul	A picture or display used to show something	A road map is a *visual* that can be used to figure out how to get to different places.
Vision VIZH un	Seeing with the eyes, sight	His *vision* is so good he can read the street sign from 60 feet away.
Visualize VIZH oo uh lyz	To form a picture of something in your mind	An artist tries to *visualize* a finished painting before starting to paint.
Reflect rih FLEKT	To show an image of something	The mirror *reflects* light onto the wall.
Reflection rih FLEK shun	A mirror image of an object	The artist can see a *reflection* of the sky on the pond.

Circle the correct word to complete the sentence.

1. The [orient/orientation] of the solar panel is toward the sun.
2. The water [reflects/reflection] the clouds and trees around the lake.
3. The editor can read small print. He has good [visual/vision].

Use the vocabulary above to fill in the blanks.

4. He learns a lot from pictures. He is a ___visual___ learner.
5. She looks at her ___reflection___ in the mirror as she combs her hair.
6. Before she plays in the game, the soccer player will ___visualize___ herself scoring a goal.

Differentiated Remediation *continued*

On-Level

- **Practice** (2 pages) Provides extra practice for each lesson. For simpler practice exercises, use the Form K Practice pages found in the All-in-One Teaching Resources and online.
- **Think About a Plan** Helps students develop specific problem-solving skills and strategies by providing scaffolded guiding questions.
- **Standardized Test Prep** Focuses on all major exercises, all major question types, and helps students prepare for the high-stakes assessments.

Extension

- **Enrichment** Provides students with interesting problems and activities that extend the concepts of the lesson.
- **Activities, Games, and Puzzles** Worksheets that can be used for concepts development, enrichment, and for fun!

Practice and Problem Solving Wkbk/ All-in-One Resources/Online

Practice page 1

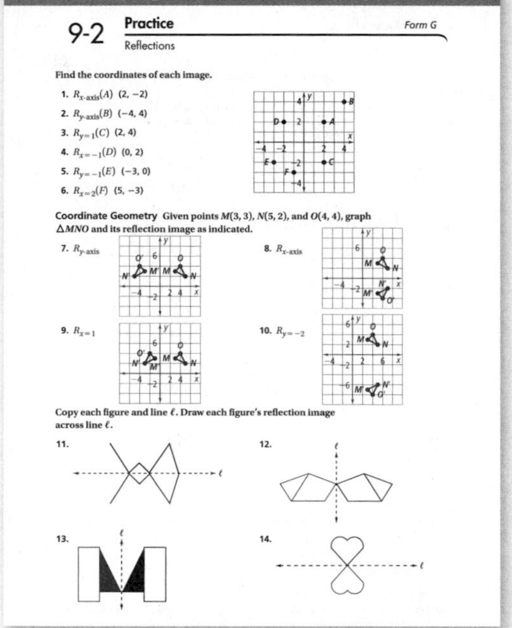
9-2 Practice Form G
Reflections

Find the coordinates of each image.

1. $R_{x\text{-axis}}(A)$ (2, −2)
2. $R_{y\text{-axis}}(B)$ (−4, 4)
3. $R_{y=1}(C)$ (2, 4)
4. $R_{x=-1}(D)$ (0, 2)
5. $R_{y=-1}(E)$ (−3, 0)
6. $R_{x=2}(F)$ (5, −3)

Coordinate Geometry Given points M(3, 3), N(5, 2), and O(4, 4), graph △MNO and its reflection image as indicated.

7. $R_{y\text{-axis}}$
8. $R_{x\text{-axis}}$
9. $R_{x=1}$
10. $R_{y=-2}$

Copy each figure and line ℓ. Draw each figure's reflection image across line ℓ.

11.
12.
13.
14.

Practice and Problem Solving Wkbk/ All-in-One Resources/Online

Practice page 2

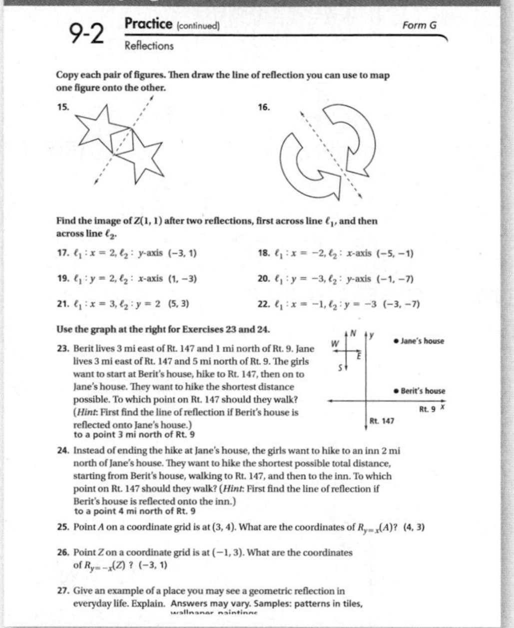
9-2 Practice (continued) Form G
Reflections

Copy each pair of figures. Then draw the line of reflection you can use to map one figure onto the other.

15.
16.

Find the image of Z(1, 1) after two reflections, first across line ℓ_1, and then across line ℓ_2.

17. $\ell_1: x = 2, \ell_2$: y-axis (−3, 1)
18. $\ell_1: x = -2, \ell_2$: x-axis (−5, −1)
19. $\ell_1: y = 2, \ell_2$: x-axis (1, −3)
20. $\ell_1: y = -3, \ell_2$: y-axis (−1, −7)
21. $\ell_1: x = 3, \ell_2: y = 2$ (5, 3)
22. $\ell_1: x = -1, \ell_2: y = -3$ (−3, −7)

Use the graph at the right for Exercises 23 and 24.

23. Berit lives 3 mi east of Rt. 147 and 1 mi north of Rt. 9. Jane lives 3 mi east of Rt. 147 and 5 mi north of Rt. 9. The girls want to start at Berit's house, hike to Rt. 147, then on to Jane's house. They want to hike the shortest distance possible. To which point on Rt. 147 should they walk? (*Hint:* First find the line of reflection if Berit's house is reflected onto Jane's house.) to a point 3 mi north of Rt. 9
24. Instead of ending the hike at Jane's house, the girls want to hike to an inn 2 mi north of Jane's house. They want to hike the shortest possible total distance, starting from Berit's house, walking to Rt. 147, and then to the inn. To which point on Rt. 147 should they walk? (*Hint:* First find the line of reflection if Berit's house is reflected onto the inn.) to a point 4 mi north of Rt. 9
25. Point A on a coordinate grid is at (3, 4). What are the coordinates of $R_{y=x}(A)$? (4, 3)
26. Point Z on a coordinate grid is at (−1, 3). What are the coordinates of $R_{y=-x}(Z)$? (−3, 1)
27. Give an example of a place you may see a geometric reflection in everyday life. Explain. Answers may vary. Samples: patterns in tiles, wallpaper, paintings

All-in-One Resources/Online

Enrichment

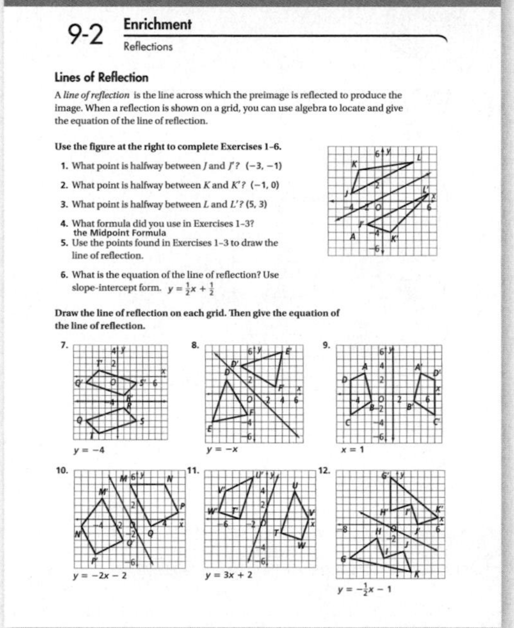
9-2 Enrichment
Reflections

Lines of Reflection

A *line of reflection* is the line across which the preimage is reflected to produce the image. When a reflection is shown on a grid, you can use algebra to locate and give the equation of the line of reflection.

Use the figure at the right to complete Exercises 1–6.

1. What point is halfway between J and J′? (−3, −1)
2. What point is halfway between K and K′? (−1, 0)
3. What point is halfway between L and L′? (5, 3)
4. What formula did you use in Exercises 1–3? the Midpoint Formula
5. Use the points found in Exercises 1–3 to draw the line of reflection.
6. What is the equation of the line of reflection? Use slope-intercept form. $y = \frac{1}{2}x + \frac{1}{2}$

Draw the line of reflection on each grid. Then give the equation of the line of reflection.

7. $y = -4$
8. $y = -x$
9. $x = 1$
10. $y = -2x - 2$
11. $y = 3x + 2$
12. $y = -\frac{1}{2}x - 1$

Practice and Problem Solving Wkbk/ All-in-One Resources/Online

Think About a Plan

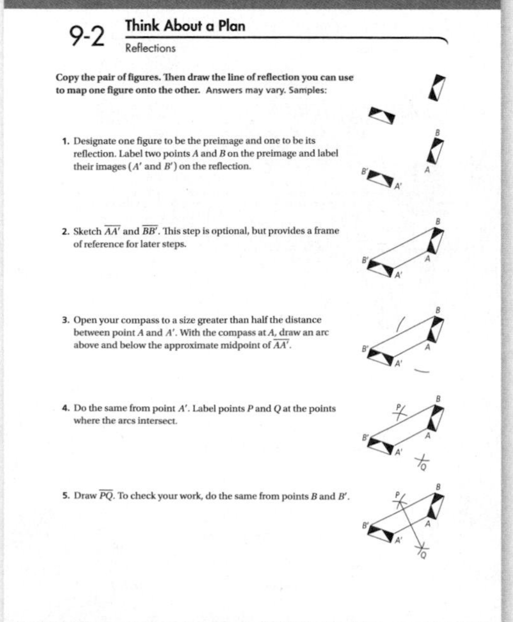
9-2 Think About a Plan
Reflections

Copy the pair of figures. Then draw the line of reflection you can use to map one figure onto the other. Answers may vary. Samples:

1. Designate one figure to be the preimage and one to be its reflection. Label two points A and B on the preimage and label their images (A′ and B′) on the reflection.
2. Sketch $\overline{AA'}$ and $\overline{BB'}$. This step is optional, but provides a frame of reference for later steps.
3. Open your compass to a size greater than half the distance between point A and A′. With the compass at A, draw an arc above and below the approximate midpoint of $\overline{AA'}$.
4. Do the same from point A′. Label points P and Q at the points where the arcs intersect.
5. Draw $\overline{PQ}$. To check your work, do the same from points B and B′.

Practice and Problem Solving Wkbk/ All-in-One Resources/Online

Standardized Test Prep

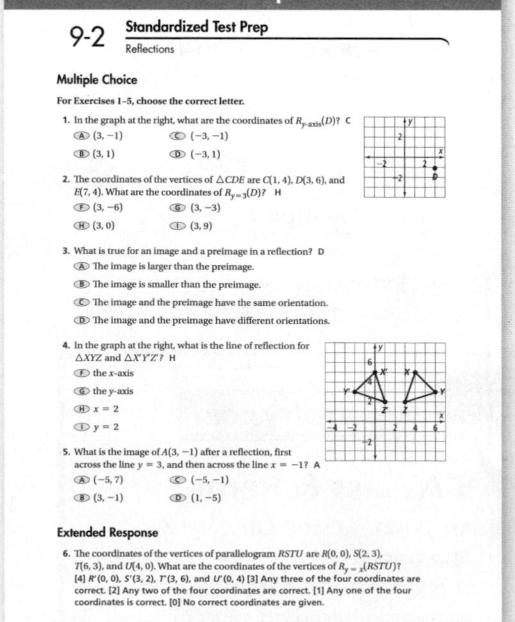
9-2 Standardized Test Prep
Reflections

Multiple Choice

For Exercises 1–5, choose the correct letter.

1. In the graph at the right, what are the coordinates of $R_{x\text{-axis}}(D)$? C
   A (3, −1) C (−3, −1)
   B (3, 1) D (−3, 1)
2. The coordinates of the vertices of △CDE are C(1, 4), D(6, 3), and E(7, 4). What are the coordinates of $R_{y=3}(D)$? H
   F (3, −6) G (3, −3)
   H (3, 0) I (3, 9)
3. What is true for an image and a preimage in a reflection? D
   A The image is larger than the preimage.
   B The image is smaller than the preimage.
   C The image and the preimage have the same orientation.
   D The image and the preimage have different orientations.
4. In the graph at the right, what is the line of reflection for △XYZ and △X′Y′Z′? H
   F the x-axis
   G the y-axis
   H x = 2
   I y = 2
5. What is the image of A(3, −1) after a reflection, first across the line y = 3, and then across the line x = −1? A
   A (−5, 7) C (−5, −1)
   B (3, −1) D (1, −5)

Extended Response

6. The coordinates of the vertices of parallelogram RSTU are R(0, 0), S(2, 3), T(6, 3), and U(4, 0). What are the coordinates of the vertices of $R_{y=x}(RSTU)$? [4] R′(0, 0), S′(3, 2), T′(3, 6), and U′(0, 4) [3] Any three of the four coordinates are correct. [2] Any two of the four coordinates are correct. [1] Any one of the four coordinates is correct. [0] No correct coordinates are given.

Online Teacher Resource Center

Activities, Games, and Puzzles

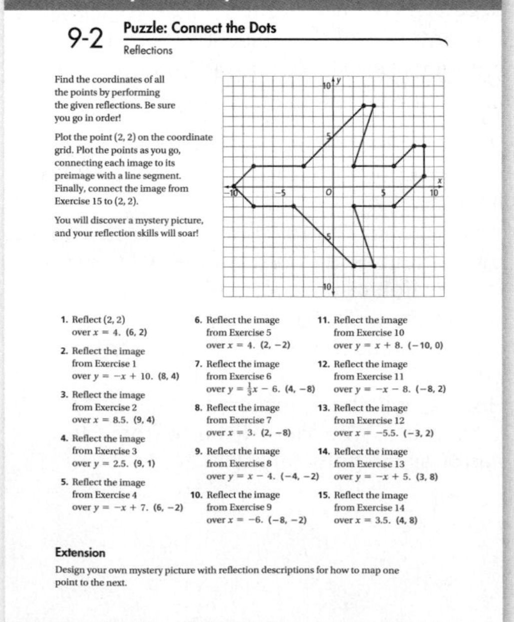
9-2 Puzzle: Connect the Dots
Reflections

Find the coordinates of all the points by performing the given reflections. Be sure you go in order!

Plot the point (2, 2) on the coordinate grid. Plot the points as you go, connecting each image to its preimage with a line segment. Finally, connect the image from Exercise 15 to (2, 2).

You will discover a mystery picture, and your reflection skills will soar!

1. Reflect (2, 2) over x = 4. (6, 2)
2. Reflect the image from Exercise 1 over y = −x + 10. (8, 4)
3. Reflect the image from Exercise 2 over x = 8.5. (9, 4)
4. Reflect the image from Exercise 3 over y = 2.5. (9, 1)
5. Reflect the image from Exercise 4 over y = −x + 7. (6, −2)
6. Reflect the image from Exercise 5 over x = 4. (2, −2)
7. Reflect the image from Exercise 6 over $y = \frac{1}{3}x - 6$. (4, −8)
8. Reflect the image from Exercise 7 over x = 3. (2, −8)
9. Reflect the image from Exercise 8 over y = x − 4. (−4, −2)
10. Reflect the image from Exercise 9 over x = −6. (−8, −2)
11. Reflect the image from Exercise 10 over y = x + 8. (−10, 0)
12. Reflect the image from Exercise 11 over y = −x − 8. (−8, 2)
13. Reflect the image from Exercise 12 over x = −5.5. (−3, 2)
14. Reflect the image from Exercise 13 over y = −x + 5. (3, 8)
15. Reflect the image from Exercise 14 over x = 3.5. (4, 8)

Extension

Design your own mystery picture with reflection descriptions for how to map one point to the next.

9-3 Rotations

Common Core State Standards
G-CO.A.4 Develop definitions of rotations . . . in terms of angles, circles, perpendicular lines, parallel lines, and line segments. **Also G-CO.A.2, G-CO.B.6**
MP 1, MP 3, MP 4

Objective To draw and identify rotation images of figures

Notice the position of the point, in relation to the x- and y-axis, as it rotates around the origin.

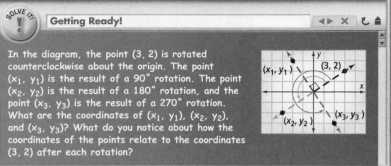

Getting Ready!

In the diagram, the point (3, 2) is rotated counterclockwise about the origin. The point (x_1, y_1) is the result of a 90° rotation. The point (x_2, y_2) is the result of a 180° rotation, and the point (x_3, y_3) is the result of a 270° rotation. What are the coordinates of (x_1, y_1), (x_2, y_2), and (x_3, y_3)? What do you notice about how the coordinates of the points relate to the coordinates (3, 2) after each rotation?

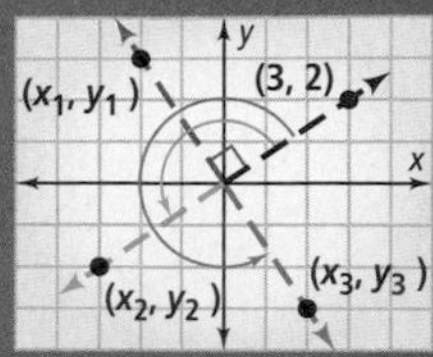

MATHEMATICAL PRACTICES

In the Solve It, you thought about how the coordinates of a point change as it turns, or *rotates*, about the origin on a coordinate grid. In this lesson, you will learn how to recognize and construct rotations of geometric figures.

Essential Understanding Rotations preserve distance, angle measures, and orientation of figures.

Lesson Vocabulary
- rotation
- center of rotation
- angle of rotation

Key Concept Rotation About a Point

A **rotation** of $x°$ about a point Q, called the **center of rotation,** is a transformation with these two properties:

- The image of Q is itself (that is, $Q' = Q$).
- For any other point V, $QV' = QV$ and $m\angle VQV' = x$.

The number of degrees a figure rotates is the **angle of rotation.**

A rotation about a point is a rigid motion. You write the $x°$ rotation of $\triangle UVW$ about point Q as $r_{(x°, Q)}(\triangle UVW) = \triangle U'V'W'$.

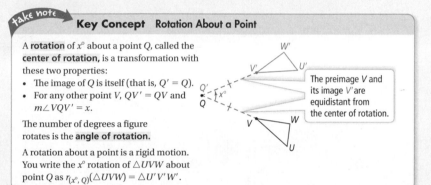

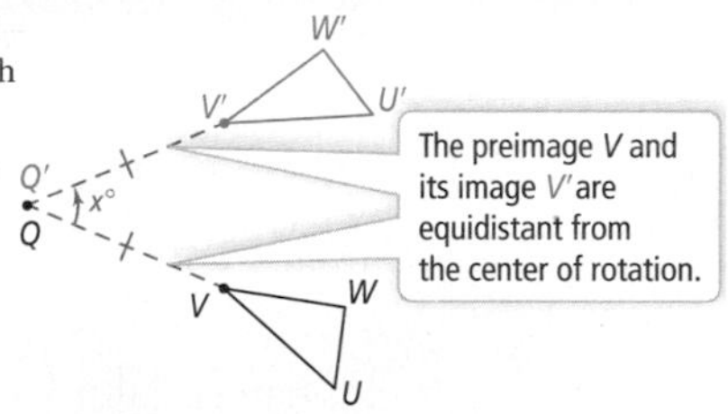

Unless stated otherwise, rotations in this book are counterclockwise.

1 Interactive Learning

Solve It!

PURPOSE To determine the image of a point after several rotations about the origin.

PROCESS Students may
- use visual judgment.
- analyze the coordinates of the point after each rotation.

FACILITATE

Q What are the coordinates of (x_1, y_1)? **[(−2, 3)]**

Q How can you describe a general rule that relates the coordinates of (x_1, y_1) to the coordinates of (x, y)? **[Multiply the *y*-coordinate in (*x*, *y*) by −1, and then switch the *x*- and *y*-coordinates.]**

ANSWER See Solve It in Answers on next page.

CONNECT THE MATH Students will visualize the results of rotations about a center point. In the lesson, students make rotations about a center of rotation and identify the angle of rotation.

2 Guided Instruction

Take Note

Focus on the angle of rotation with students. Emphasize that an angle is typically measured counterclockwise. This is counterintuitive and may confuse many students. Practice drawing simple figures rotated at given angles counterclockwise.

9-3 Preparing to Teach

BIG ideas **Transformations**
Coordinate Geometry

ESSENTIAL UNDERSTANDING
- Distances, angle measures, and orientation of a geometric figure stay the same when a figure is rotated about a center of rotation.

Math Background

Students will use their knowledge of angles and skills with constructions to learn how to sketch or construct the rotation of a figure. They will learn the relationship between a rotation's preimage and image. To identify the angle of rotation, students can measure the angle formed by a point, the center of rotation, and the point's image. A rotation is a rigid motion that preserves orientation. Students will combine this isometry with others to form compositions of transformations in the next lesson. They will also learn about a type of transformation that is not a rigid motion.

Mathematical Practices

Make sense of problems and persevere in solving them. In drawing rotations of geometric figures, students will plan a solution pathway that includes: identifying the vertices of the preimage; identifying the center and angle of rotation; and determining the vertices of the image.

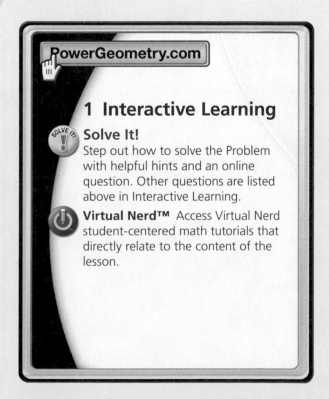

PowerGeometry.com

1 Interactive Learning

Solve It! Step out how to solve the Problem with helpful hints and an online question. Other questions are listed above in Interactive Learning.

Virtual Nerd™ Access Virtual Nerd student-centered math tutorials that directly relate to the content of the lesson.

Problem 1

It may help students to trace each angle of rotation ($\angle OCO'$, $\angle LCL'$, and $\angle BCB'$).

Q What should be the measure of the angle between a point, the center of rotation, and the point's image? **[100°]**

Q How do you know how far from C to sketch each point? **[Each image point should be the same distance from C as its preimage.]**

Got It? (VISUAL LEARNERS)

Be sure that students understand how to construct a copy of the figure. Emphasize that the center of rotation is now point B.

Take Note (EXTENSION)

Have students make a conjecture about the rules for *clockwise* rotations. A 90° clockwise rotation is the same as a 270° counterclockwise rotation. A 180° clockwise rotation is the same as a 180° counterclockwise rotation. A 270° clockwise rotation is the same as a 90° counterclockwise rotation.

Point out to students that when a figure is rotated 360° about the origin, the figure returns to its original position. The coordinates of the image is (x, y), which is the same as the preimage.

Problem 1 Drawing a Rotation Image

What is the image of $r_{(100°, C)}(\triangle LOB)$?

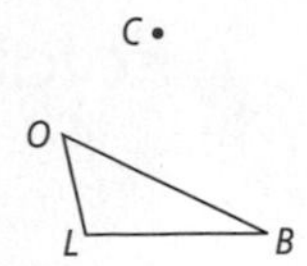

Plan

How do you use the definition of rotation about a point to help you get started?
You know that O and O' must be equidistant from C and that $m\angle OCO'$ must be 100.

Step 1 Draw $\overline{CO}$. Use a protractor to draw a 100° angle with vertex C and side $\overline{CO}$.

Step 2 Use a compass to construct $\overline{CO'} \cong \overline{CO}$.

Step 3 Locate B' and L' in a similar manner.

Step 4 Draw $\triangle L'O'B'$.

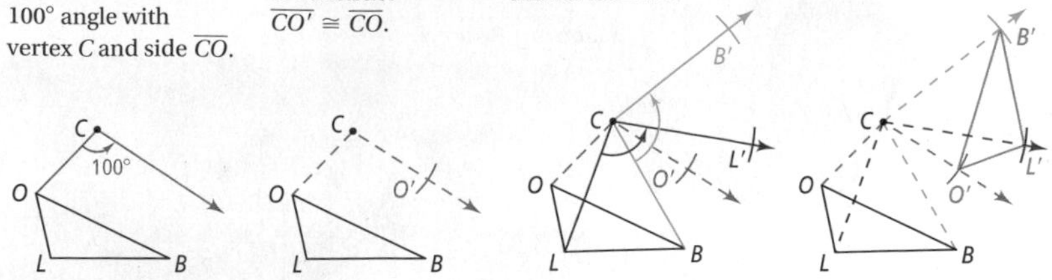

Got It? 1. Copy $\triangle LOB$ from Problem 1. What is the image of $\triangle LOB$ for a 50° rotation about B?

When a figure is rotated 90°, 180°, or 270° about the origin O in a coordinate plane, you can use the following rules.

take note — Key Concept Rotation in the Coordinate Plane

$r_{(90°, O)}(x, y) = (-y, x)$

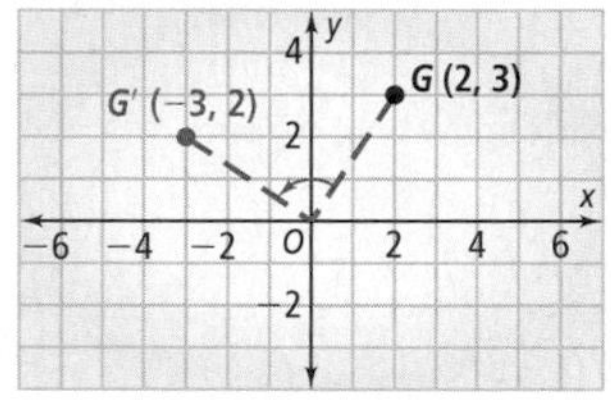

$r_{(180°, O)}(x, y) = (-x, -y)$

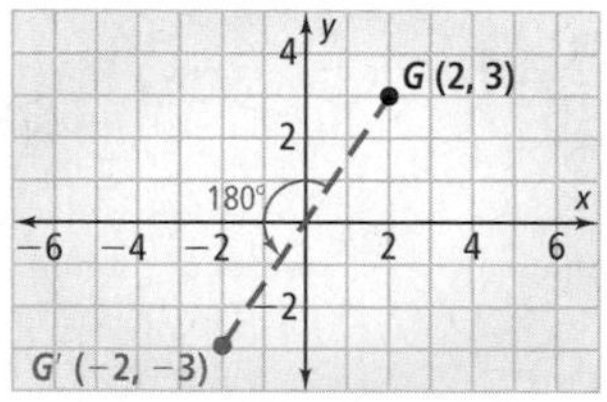

$r_{(270°, O)}(x, y) = (y, -x)$

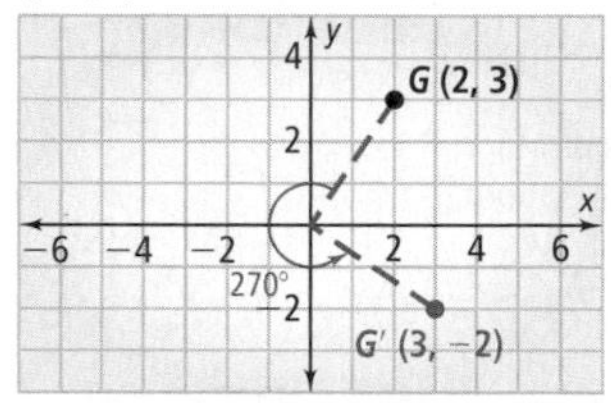

$r_{(360°, O)}(x, y) = (x, y)$

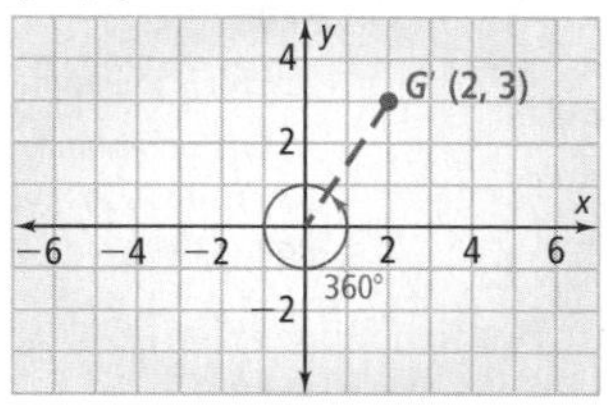

Answers

Solve It!

$(x_1, y_1) = (-2, 3)$, $(x_2, y_2) = (-3, -2)$, $(x_3, y_3) = (2, -3)$;
90° rotation: $(x_1, y_1) = (-y, x)$
180° rotation: $(x_2, y_2) = (-x, -y)$
270° rotation: $(x_3, y_3) = (y, -x)$

Got It?

1.

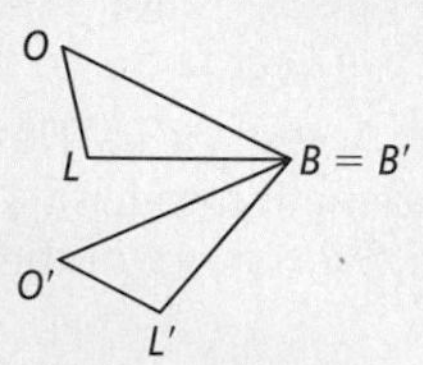

2–3. See page 564.

PowerGeometry.com

2 Guided Instruction

Each Problem is worked out and supported online.

Problem 1
Drawing a Rotation Image
Animated

Problem 2
Drawing Rotations in a Coordinate Plane

Problem 3
Using Properties of Rotations

Support in Geometry Companion
- Vocabulary
- Key Concepts
- Got It?

Problem 2 Drawing Rotations in a Coordinate Plane

Plan

How do you know where to draw the vertices on the coordinate plane?
Use the rules for rotating a point and apply them to each vertex of the figure. Then graph the points and connect them to draw the image.

PQRS **has vertices** $P(1, 1)$, $Q(3, 3)$, $R(4, 1)$, **and** $S(3, 0)$.
What is the graph of $r_{(90°, O)}(PQRS)$.

First, graph the images of each vertex.

$P' = r_{(90°, O)}(1, 1) = (-1, 1)$

$Q' = r_{(90°, O)}(3, 3) = (-3, 3)$

$R' = r_{(90°, O)}(4, 1) = (-1, 4)$

$S' = r_{(90°, O)}(3, 0) = (0, 3)$

Next, connect the vertices to graph $P'Q'R'S'$.

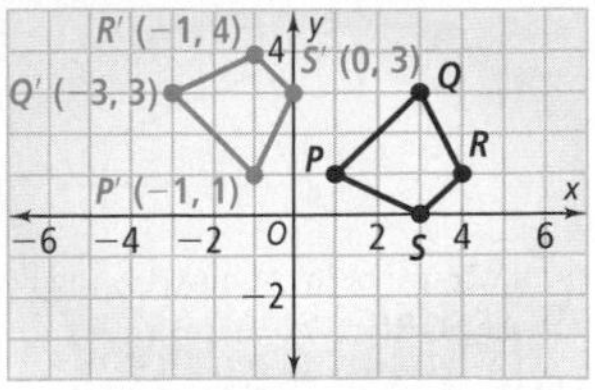

Got It? 2. Graph $r_{(270°, O)}(FGHI)$.

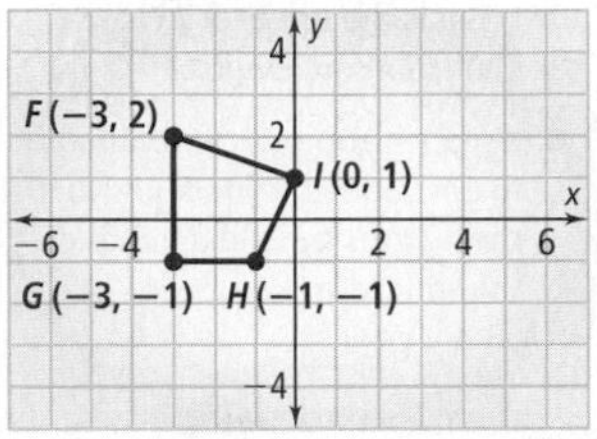

You can use the properties of rotations to solve problems.

Problem 3 Using Properties of Rotations

Think

What do you know about rotations that can help you show that opposite sides of the parallelogram are equal?
You know that rotations are rigid motions, so if you show that the opposite sides can be mapped to each other, then the side lengths must be equal.

In the diagram, ***WXYZ*** **is a parallelogram, and** ***T*** **is the midpoint of the diagonals. How can you use the properties of rotations to show that the lengths of the opposite sides of the parallelogram are equal?**

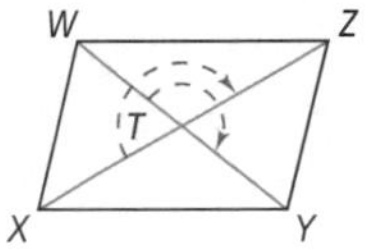

Because T is the midpoint of the diagonals, $XT = ZT$ and $WT = YT$. Since W and Y are equidistant from T, and the measure of $\angle WTY = 180$, you know that $r_{(180°, T)}(W) = Y$. Similarly, $r_{(180°, T)}(X) = Z$.

You can rotate every point on $\overline{WX}$ in this same way, so $r_{(180°, T)}(\overline{WX}) = \overline{YZ}$. Likewise, you can map $\overline{WZ}$ to $\overline{YX}$ with $r_{(180°, T)}(\overline{WZ}) = \overline{YX}$.

Because rotations are rigid motions and preserve distance, $WX = YZ$ and $WZ = YX$.

Got It? 3. Can you use the properties of rotations to prove that $WXYZ$ is a rhombus? Explain.

Problem 2

Have students visually check the graph of the image by looking at the distance of each vertex from the origin. The distance should be the same for corresponding vertices. Students can also draw a line segment from the origin to P and a line segment from the origin to P'. The two line segments should form a 90° angle.

Q What is a rule, in words, for a rotation of 90° counterclockwise about the origin? **[multiply the *y*-coordinate by −1 and switch the *x*- and *y*-coordinates.]**

Got It?

Encourage students to graph the preimage so they can visually check that the graph of the image rotation make sense.

Q What is a rule, in words, for a rotation of 270° counterclockwise about the origin? **[multiply the *x*-coordinate by −1 and switch the *x*- and *y*-coordinates.]**

Problem 3

Review the properties of parallelograms.

Q What do you need to show? **[$WX = YZ$ and $WZ = YX$]**

Q Do you know any of the angle measures in the parallelogram? **[Yes; $m\angle XTZ = m\angle WTY = 180°$]**

Q What is $r_{(180°, T)}(W)$? **[Y]**

Got It?

Q What would you need to show to prove $WXYZ$ is a rhombus? **[$WX = XY = YZ = ZW$]**

Additional Problems

1. What is the image of $\triangle DEF$ after a 75° rotation about point P?

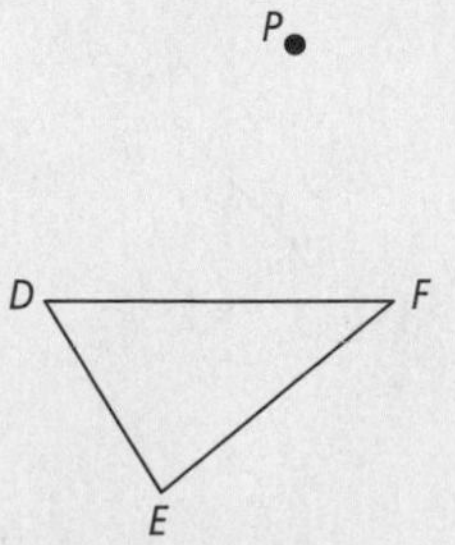

ANSWER

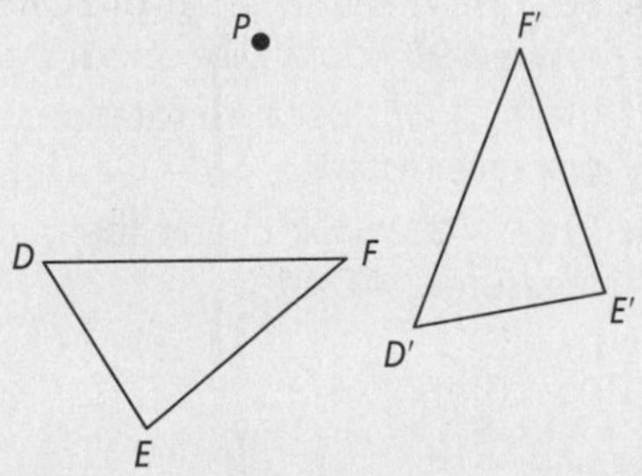

2. Triangle ABC has vertices $A(1, 1)$, $B(4, 2)$, and $C(0, 4)$. Graph $\triangle ABC$ and $r_{(180°, 0)}(\triangle ABC)$.

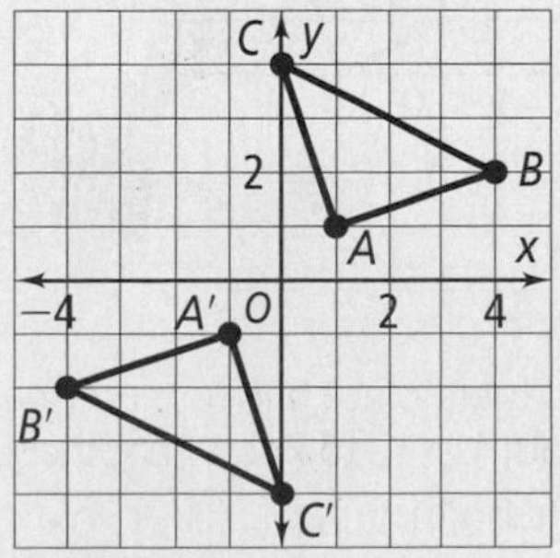

ANSWER

3. In the diagram, $WXYZ$ is a parallelogram, and T is the midpoint of the diagonals. Can you use the properties of rotations to prove that the lengths of the diagonals are equal?

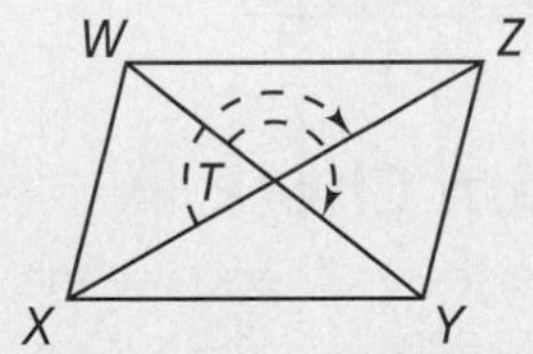

ANSWER No; There is no rotation that maps X onto W and Z to Y, so you cannot show that $XZ = WY$.

3 Lesson Check

Do you know HOW?

- If students have difficulty with Exercise 4, then have them review Problem 3 to see how to use properties of rotations to solve problems.

Do you UNDERSTAND?

- If students have difficulty with Exercise 8, then suggest that they choose an ordered pair for point *P* and perform the rotations to determine the image of point *P*. Then they can identify what happens to the elements of the ordered pair of the image.

Close

Q What two things do you need to find the image of a figure after a rotation? **[the center of rotation and the angle of rotation]**

Q How are the image and preimage related in a rotation? **[The angle between the rays connecting image and preimage with the center of rotation is equal to the angle of rotation.]**

Lesson Check

Do you know HOW?

1. Copy the figure and point *P*. Draw $r_{(70°, P)}(\triangle ABC)$.

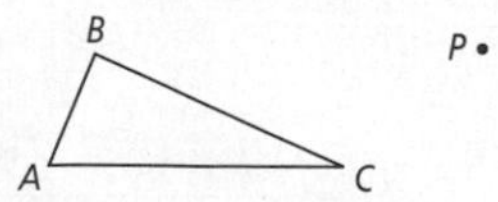

In the figure below, point *A* is the center of square *SQRE*.

2. What is $r_{(90°, A)}(E)$?

3. What is the image of $\overline{RQ}$ for a 180° rotation about *A*?

4. Use the properties of rotations to describe how you know that the lengths of the diagonals of the square are equal.

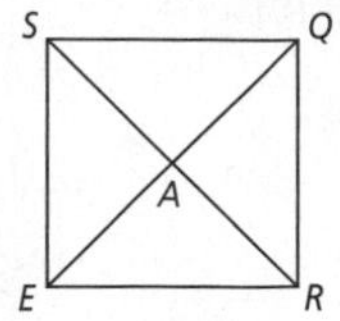

Do you UNDERSTAND? MATHEMATICAL PRACTICES

5. Vocabulary $\triangle A'B'C'$ is a rotation image of $\triangle ABC$ about point *O*. Describe how to find the angle of rotation.

6. Error Analysis A classmate drew a 115° rotation of $\triangle PQR$ about point *P*, as shown at the right. Explain and correct your classmate's error.

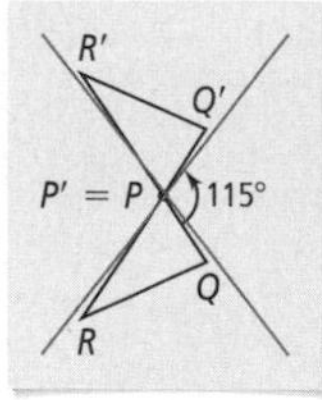

7. Compare and Contrast Compare rotating a figure about a point to reflecting the figure across a line. How are the transformations alike? How are they different?

8. Reasoning Point *P*(*x*, *y*) is rotated about the origin by 135° and then by 45°. What are the coordinates of the image of point *P*? Explain

Practice and Problem-Solving Exercises MATHEMATICAL PRACTICES

A Practice

Copy each figure and point *P*. Draw the image of each figure for the given rotation about *P*. Use prime notation to label the vertices of the image.

See Problem 1.

9. 60°

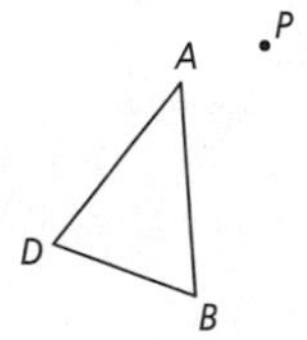

10. 90°

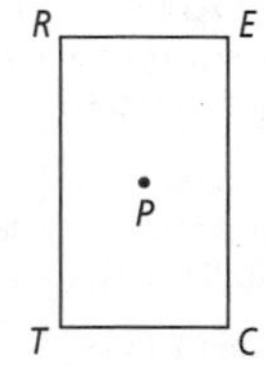

11. 180°

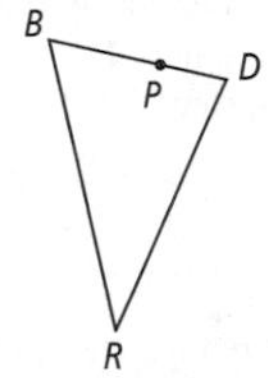

12. 90°

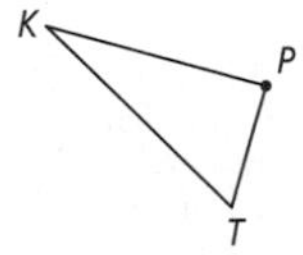

Copy each figure and point *P*. Then draw the image of $\overline{JK}$ for a 180° rotation about *P*. Use prime notation to label the vertices of the image.

13.

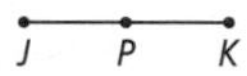

14.

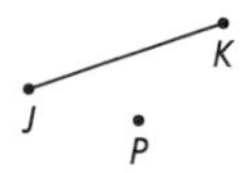

15.

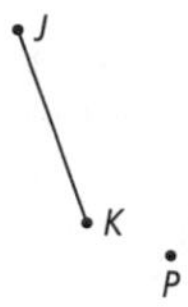

16.

PowerGeometry.com

3 Lesson Check

For a digital lesson check, use the Got It questions.

Support in Geometry Companion

- Lesson Check

4 Practice

Assign homework to individual students or to an entire class.

Answers

Got It? (continued)

2.

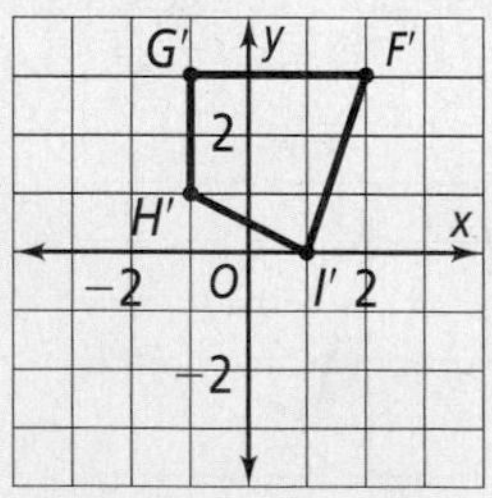

3. No; there is not enough information about *WXYZ* to know that there is a rotation that maps $\overline{XW}$ to $\overline{WZ}$.

Lesson Check

1.

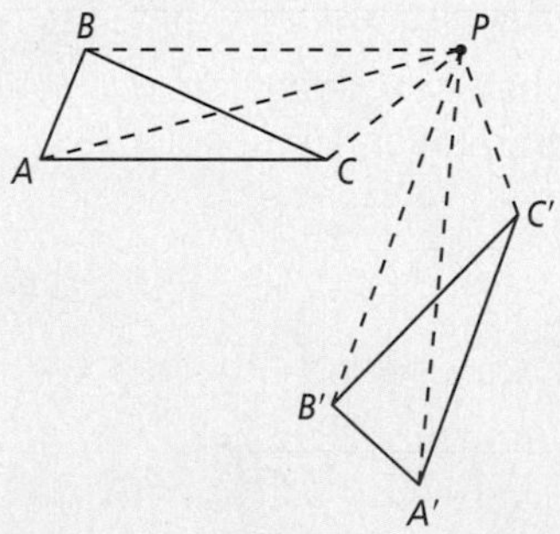

2. *R*

3. $\overline{SE}$

4. Because *A* is the center of *SQRE*, $\overline{SR}$ rotated 90° clockwise about point *A* maps to $\overline{QE}$. Because rotations preserve distance, *SR* = *QE*.

5. Draw $\overline{AO}$ and $\overline{A'O}$ and then measure $\angle AOA'$.

For Exercises 17–19, use the graph at the right.

See Problem 2.

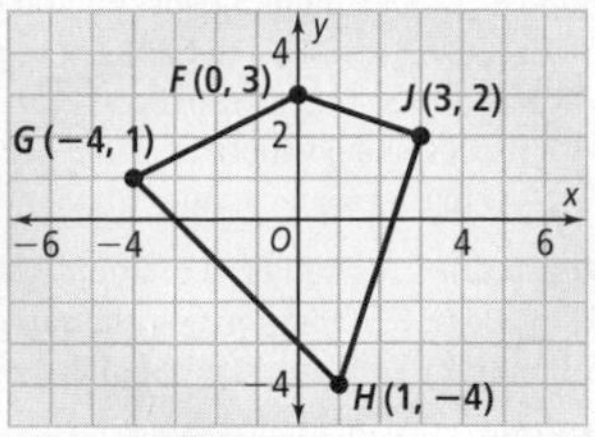

17. Graph $r_{(90°, O)}(FGHJ)$.

18. Graph $r_{(180°, O)}(FGHJ)$.

19. Graph $r_{(270°, O)}(FGHJ)$.

20. The coordinates of $\triangle PRS$ are $P(-3, 2)$, $R(2, 5)$, and $S(0, 0)$. What are the coordinates of the vertices of $r_{(270°, O)}(\triangle PRS)$?

21. $V'W'X'Y'$ has vertices $V'(-3, 2)$, $W'(5, 1)$, $X'(0, 4)$, and $Y'(-2, 0)$. If $r_{(90°, O)}(VWXY) = V'W'X'Y'$, what are the coordinates of $VWXY$?

22. Ferris Wheel A Ferris wheel is drawn on a coordinate plane so that the first car is located at the point (30, 0). What are the coordinates of the first car after a rotation of 270° about the origin?

For Exercises 23–25, use the diagram at the right. *TQNV* is a rectangle. *M* is the midpoint of the diagonals.

See Problem 3.

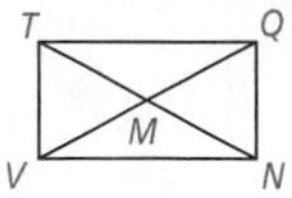

23. Use the properties of rotations to show that the measures of both pairs of opposite sides are equal in length.

24. Reasoning Can you use the properties of rotations to show that the measures of the lengths of the diagonals are equal?

25. Reasoning Can you use properties of rotations to conclude that the diagonals of *TQNV* bisect the angles of *TQNV*? Explain.

B Apply

26. In the diagram at the right, $\overline{M'N'}$ is the rotation image of $\overline{MN}$ about point *E*. Name all pairs of angles and all pairs of segments that have equal measures in the diagram.

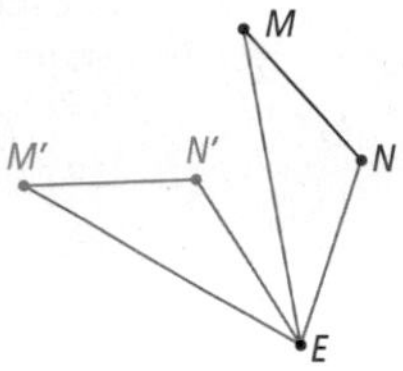

27. Language Arts Symbols are used in dictionaries to help users pronounce words correctly. The symbol ə is called a *schwa*. It is used in dictionaries to represent neutral vowel sounds such as *a* in *ago*, *i* in *sanity*, and *u* in *focus*. What transformation maps a ə to a lowercase e?

Find the angle of rotation about *C* that maps the black figure to the blue figure.

28.

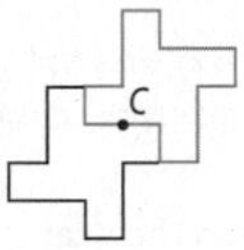

29.

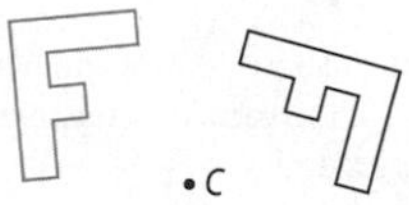

30.

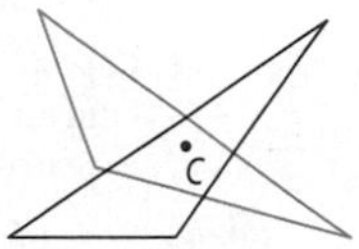

4 Practice

ASSIGNMENT GUIDE

Basic: 9–25 all, 28–37, 38–46 even
Average: 9–25 odd, 26–47
Advanced: 9–25 odd, 26–49
Standardized Test Prep: 50–53

Mathematical Practices are supported by exercises with red headings. Here are the Practices supported in this lesson:

MP 1: Make Sense of Problems Ex. 7, 31
MP 3: Construct Arguments Ex. 8, 24, 25, 32, 49
MP 3: Communicate Ex. 33, 36
MP 3: Critique the Reasoning of Others Ex. 5, 37

Applications exercises have blue headings. Exercise 22 supports MP 4: Model.

EXERCISE 34: Use the Think About a Plan worksheet in the **Practice and Problem Solving Workbook** (also available in the Teaching Resources in print and online) to further support students' development in becoming independent learners.

HOMEWORK QUICK CHECK

To check students' understanding of key skills and concepts, go over Exercises 11, 21, 31, 34, and 37.

6. The diagram shows a reflection, not a rotation. *R′* is a 115° clockwise rotation of *R*. All points of $\triangle PQR$ must be rotated counterclockwise.

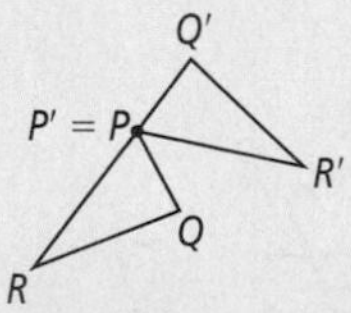

7. Both are rigid motions. A reflection reverses orientation. A rotation has the same orientation.

8. $(-x, -y)$; Sample: The coordinates are the same as a single rotation of 180° since $135° + 45° = 180°$.

Practice and Problem-Solving Exercises

9.

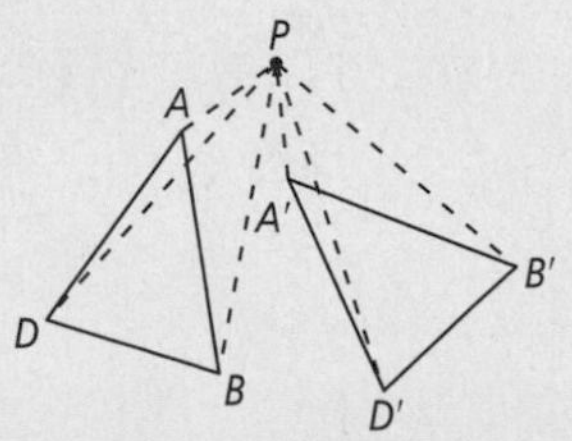

10.

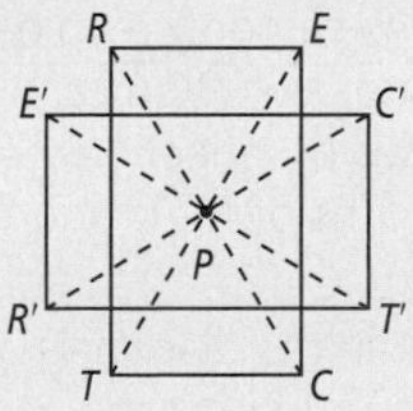

11.

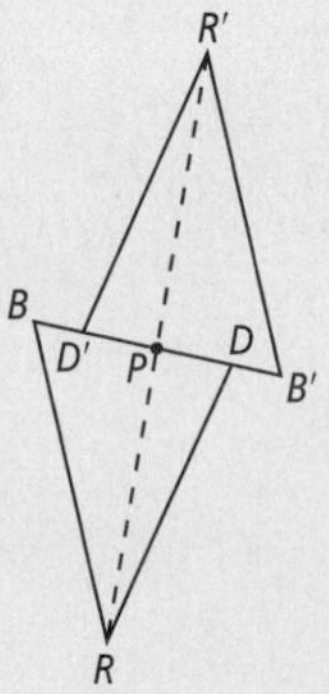

12.

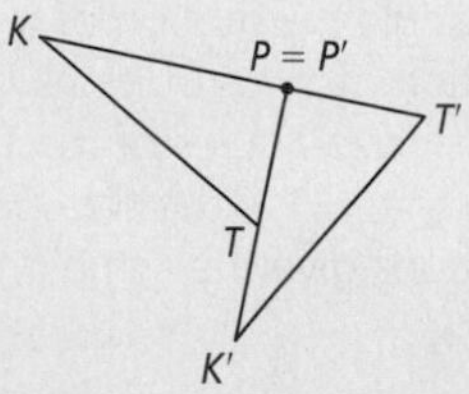

13.

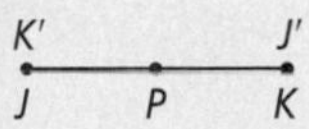

14.

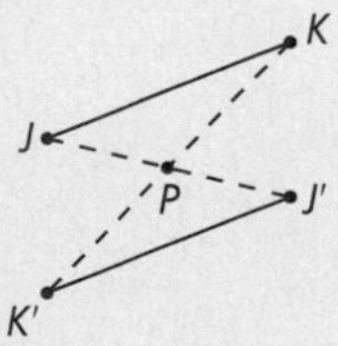

15.

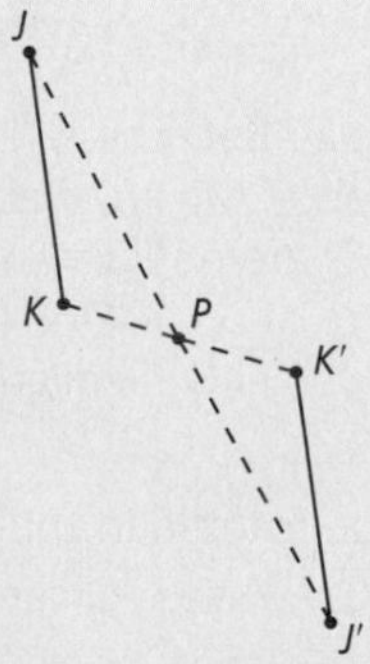

16–30. See next page.

Answers

Practice and Problem-Solving Exercises (continued)

16.

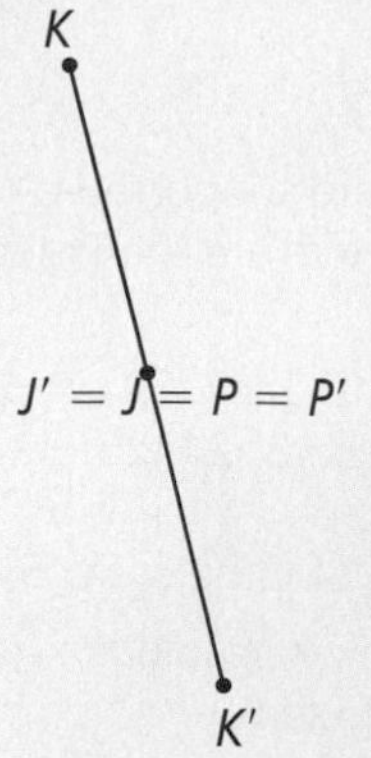

17.

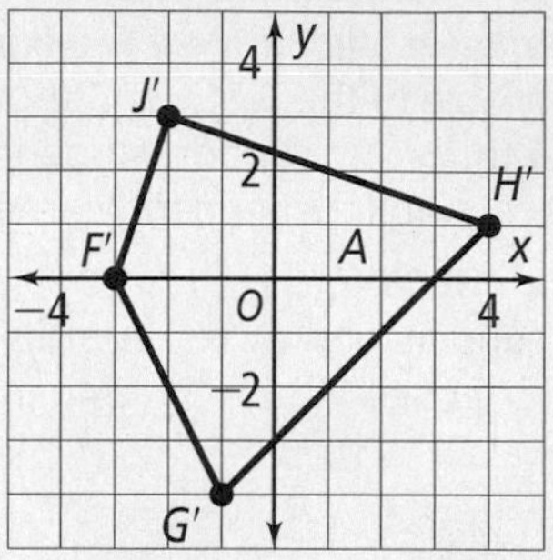

18.

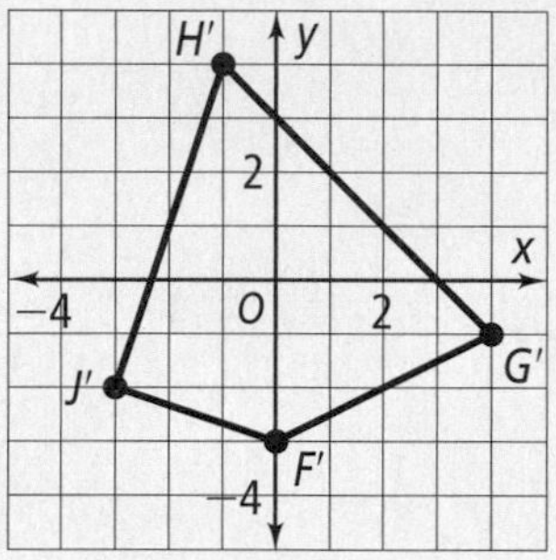

19.

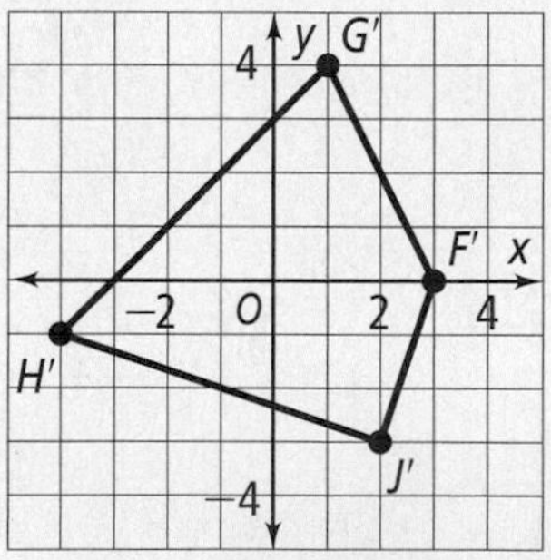

31. Think About a Plan The Millenium Wheel, also known as the London Eye, contains 32 observation cars. Determine the angle of rotation that will bring Car 3 to the position of Car 18.
- How do you find the angle of rotation that a car travels when it moves one position counterclockwise?
- How many positions does Car 3 move?

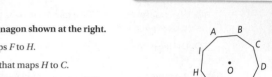

32. Reasoning For center of rotation P, does an $x°$ rotation followed by a $y°$ rotation give the same image as a $y°$ rotation followed by an $x°$ rotation? Explain.

33. Writing Describe how a series of rotations can have the same effect as a 360° rotation about a point X.

34. Coordinate Geometry Graph $A(5, 2)$. Graph B, the image of A for a 90° rotation about the origin O. Graph C, the image of A for a 180° rotation about O. Graph D, the image of A for a 270° rotation about O. What type of quadrilateral is $ABCD$? Explain.

Point O is the center of the regular nonagon shown at the right.

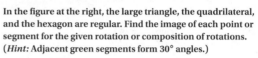

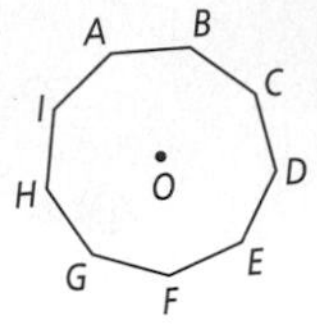

35. Find the angle of rotation that maps F to H.

36. Open-Ended Describe a rotation that maps H to C.

37. Error Analysis Your friend says that $\overline{AB}$ is the image of $\overline{ED}$ for a 120° rotation about O. What is wrong with your friend's statement?

In the figure at the right, the large triangle, the quadrilateral, and the hexagon are regular. Find the image of each point or segment for the given rotation or composition of rotations. (*Hint:* Adjacent green segments form 30° angles.)

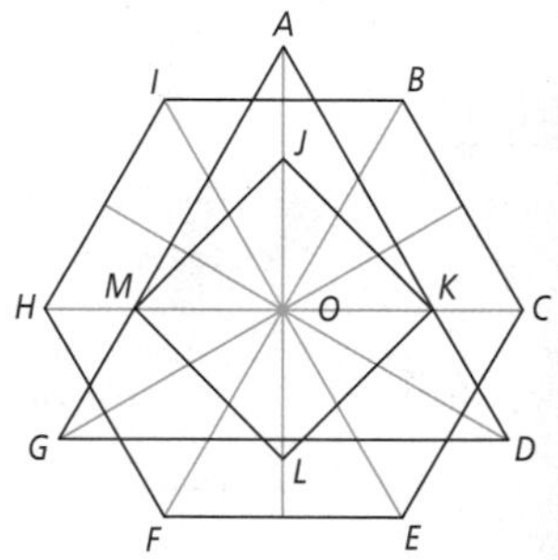

38. $r_{(120°, O)}(B)$ **39.** $r_{(270°, O)}(L)$

40. $r_{(300°, O)}(\overline{IB})$ **41.** $r_{(60°, O)}(E)$

42. $r_{(180°, O)}(\overline{JK})$ **43.** $r_{(240°, O)}(G)$

44. $r_{(120°, H)}(F)$ **45.** $r_{(270°, L)}(M)$

46. $r_{(180°, O)}(I)$ **47.** $r_{(270°, O)}(M)$

Challenge

48. Coordinate Geometry Draw $\triangle LMN$ with vertices $L(2, -1)$, $M(6, -2)$, and $N(4, 2)$. Find the coordinates of the vertices after a 90° rotation about the origin and about each of the points L, M, and N.

49. Reasoning If you are given a figure and a rotation image of the figure, how can you find the center and angle of rotation?

20. $P'(2, 3)$, $R'(5, -2)$, $S'(0, 0)$

21. $V(2, 3)$, $W(1, -5)$, $X(4, 0)$, $Y(0, 2)$

22. (0, –30)

23. Answers may vary. Sample: Because M is the midpoint of the diagonals, $VM = QM$ and $TM = NM$. Since V and Q are equidistant from M, and the measure of $\angle VMQ = 180$, you know that $r_{(180°, M)}(V) = Q$. Similarly, $r_{(180°, M)}(T) = N$.

Every point on $\overline{VT}$ can be rotated in this same way, so $r_{(180°, T)}(\overline{VT}) = \overline{QN}$. Also, $\overline{TQ}$ can be mapped to $\overline{NV}$, so $r_{(180°, M)}(\overline{TQ}) = \overline{NV}$.

Because rotations are rigid motions and preserve distance, $TV = NQ$ and $TQ = VN$.

24. No. Rotations can only map each diagonal onto itself. You can use the properties of reflections to show that the lengths of the diagonals are equal.

25. No. In general, the diagonals of a rectangle do not bisect the angles of a rectangle.

26. $\overline{MN} \cong \overline{M'N'}$; $\overline{EN} \cong \overline{EN'}$; $\overline{ME} \cong \overline{M'E}$; $\angle M \cong \angle M'$; $\angle N \cong \angle N'$; $\angle MEN \cong \angle M'EN'$; $\angle MEM' \cong \angle NEN'$

27. a 180° rotation

28. 180°

29. 110°

30. 290°

31. 168.75°

32. Yes; the angle of rotation of a composition of two rotations is the sum of the two angles of rotations. Since $x + y = y + x$, the two compositions give the same image.

33. any two rotations of $a°$ and $b°$ if $a > 0$, $b > 0$, and $a + b = 360$

34. Square; all sides are ≅ and all ∠s are 90°.

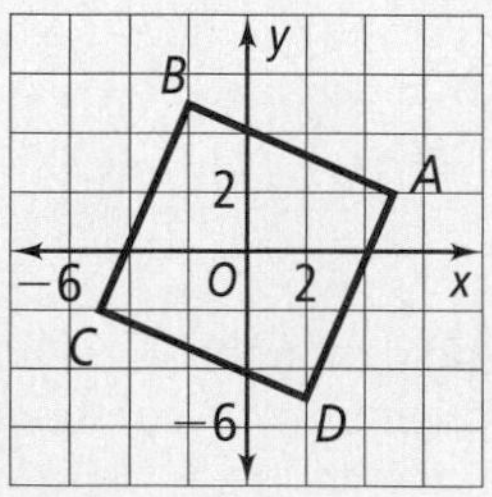

35. 280°

36. Answers may vary. Sample: 40° and 160° rotations about point O

37. The image of $\overline{ED}$ is $\overline{BA}$, not $\overline{AB}$.

38. H **39.** M

40. $\overline{BC}$ **41.** C

42. $\overline{LM}$ **43.** A

44. I **45.** K

46. E **47.** J

Standardized Test Prep

SAT/ACT

50. What is the image of $(1, -6)$ for a 90° counterclockwise rotation about the origin?

Ⓐ $(6, 1)$ Ⓑ $(-1, 6)$ Ⓒ $(-6, -1)$ Ⓓ $(-1, -6)$

51. The costume crew for your school musical makes aprons like the one shown. If blue ribbon costs \$1.50 per foot, what is the cost of ribbon for six aprons?

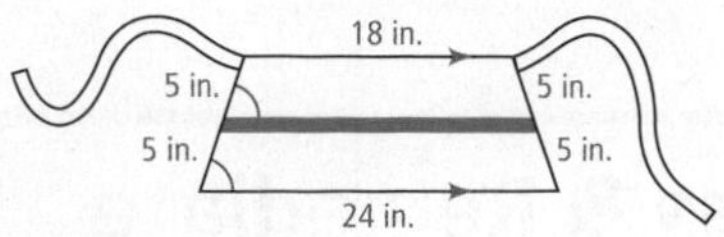

Ⓕ \$15.75 Ⓗ \$42.00

Ⓖ \$31.50 Ⓘ \$63.00

52. In $\triangle ABC$, $m\angle A + m\angle B = 84$. Which statement must be true?

Ⓐ $BC > AC$ Ⓑ $AC > BC$ Ⓒ $AB > BC$ Ⓓ $BC > AB$

Short Response

53. Use the following statement: If two lines are parallel, then the lines do not intersect.

a. What are the converse, inverse, and contrapositive of the statement?

b. What is the truth value of each statement you wrote in part (a)? If a statement is false, give a counterexample.

Apply What You've Learned

MATHEMATICAL PRACTICES MP 5

Look back at the information about the video game on page 543. The graph of the puzzle piece and target area is shown again below.

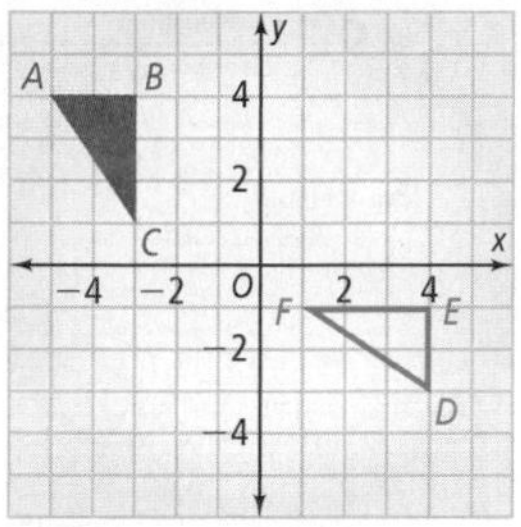

In the Apply What You've Learned in Lesson 9-2, you looked at how some translations and reflections move the puzzle piece. Now you will look at how rotations move the puzzle piece.

a. How does the orientation of the puzzle piece compare to the orientation of the target area?

b. Can you move the puzzle piece to the target area using only rotations? Explain.

For parts (c)–(e), copy the graph and then graph the image of $\triangle ABC$ for the given rotation.

c. $r_{(90^\circ, O)}(x, y)$ **d.** $r_{(180^\circ, O)}(x, y)$ **e.** $r_{(270^\circ, O)}(x, y)$

48. about the origin: $L'(1, 2)$, $M'(2, 6)$, $N'(-2, 4)$

about L: $L'(2, -1)$, $M'(3, 3)$, $N'(-1, 1)$

about M: $L'(5, -6)$, $M'(6, -2)$, $N'(2, -4)$

about N: $L'(7, 0)$, $M'(8, 4)$, $N'(4, 2)$

49. Draw two segments connecting preimage points A and B to image points A' and B'. If $\overline{AA'}$ and $\overline{BB'}$ are not collinear, construct the $\perp$ bis. of $\overline{AA'}$ and $\overline{BB'}$ to find C, the center of rotation. $m\angle ACA'$ is the $\angle$ of rotation. If $\overline{AA'}$ and $\overline{BB'}$ are collinear, then the midpoint of $\overline{AA'}$ is the center of rotation, and 180° is the angle of rotation.

50. A

51. F

52. C

53. [2] **a.** Converse: If two lines do not intersect, then they are $\parallel$; inverse: If two lines are not $\parallel$, then they intersect; contrapositive: If two lines intersect, then they are not $\parallel$.

b. Converse: false; a counterexample is two skew lines; inverse: false; a counterexample is two skew lines; contrapositive: true.

[1] one part incorrect OR incomplete.

Apply What You've Learned

PERFORMANCE TASK

In the Apply What You've Learned for Lesson 9-2, students explored how translations and reflections move the puzzle piece in the graph shown on page 543. Now, they will look at how rotations move the puzzle piece.

Mathematical Practices

Students **use appropriate tools** (paper and pencil) to graph rotations of the puzzle piece and look at how the image relates to the target area. (MP 5)

ANSWERS

a. The orientation of the puzzle piece and the target area are opposites.

b. No; rotations do not change the orientation of a figure. Because the orientations of the target area and puzzle piece are opposites, you cannot move the puzzle piece to the target area using only rotations.

c.

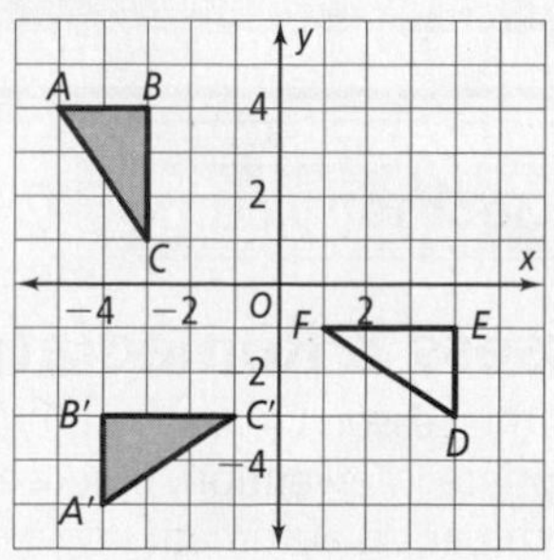

d.

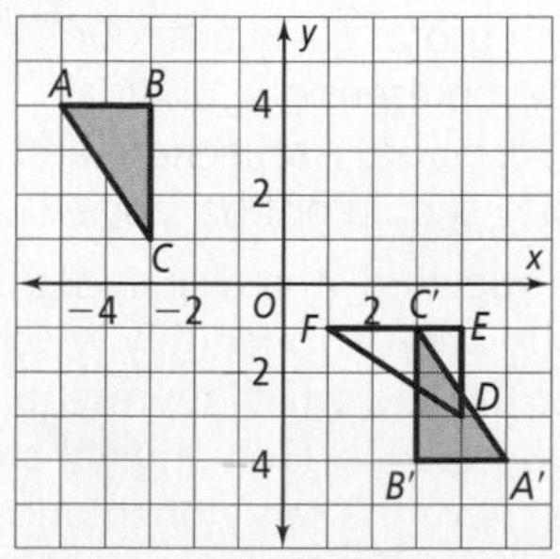

e.

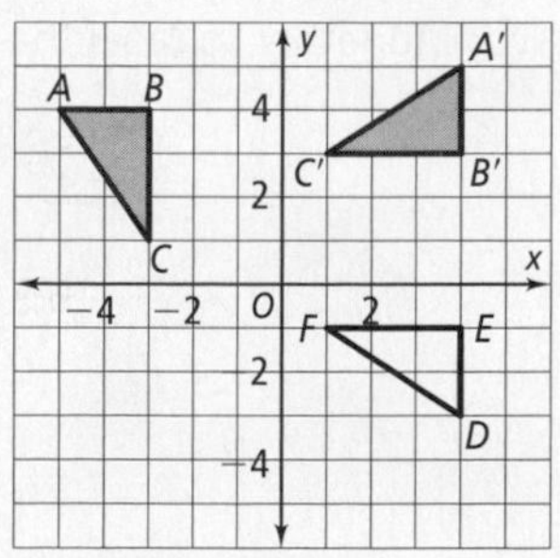

9-3 Lesson Resources

Differentiated Remediation

Instructional Support

Geometry Companion

Students can use the **Geometry Companion** worktext (4 pages) . . .

- New Vocabulary
- Key Concepts
- Got It for each Problem
- Lesson Check

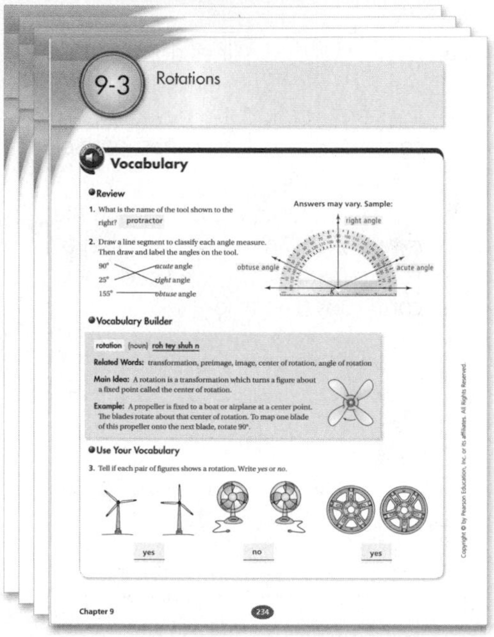

9-3 Rotations
Vocabulary
Review
1. What is the name of the tool shown to the right? protractor
Answers may vary. Sample:
2. Draw a line segment to classify each angle measure. Then draw and label the angles on the tool.
90° acute angle
25° right angle
135° obtuse angle
Vocabulary Builder
rotation (noun) roh tay shuh n
Related Words: transformation, preimage, image, center of rotation, angle of rotation
Main Idea: A rotation is a transformation which turns a figure about a fixed point called the center of rotation.
Example: A propeller is fixed to a boat or airplane at a center point. The blades rotate about that center of rotation. To map one blade of this propeller onto the next blade, rotate 90°.
Use Your Vocabulary
3. Tell if each pair of figures shows a rotation. Write yes or no.
yes
no
yes
Chapter 9
234

ELL Support

Focus on Communication Use the following activity to strengthen students' abilities to discuss transformations. Have small groups of students make four identical copies of a polygon. Have each group place the four copies so that they represent one original polygon, and a translation, a reflection, and a rotation. The display should not be labeled.

Next have each group study the display set up by a different group and try to identify which polygon is the original and which represents each of the transformations. (The original and the translation will have the same orientation.) Have the group that made the display and the group studying it work together to label the display properly.

5 Assess & Remediate

Lesson Quiz

1. Point *I* is the center of regular octagon *ABCDEFGH*. What is the image of the given point or segment for the given rotation?

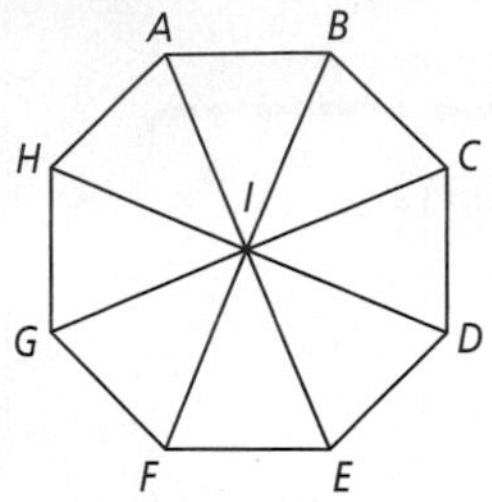

a. 225° rotation of *B* about *I*

b. 135° rotation of $\overline{CD}$ about *I*

2. Triangle *RST* has vertices *R*(–2, –2), *S*(1, 1), and *T*(–1, 3). Find the coordinates of the image of each vertex for the given rotation.

a. $r_{(90°,\,O)}(\triangle RST)$

b. $r_{(180°,\,O)}(\triangle RST)$

c. $r_{(270°,\,O)}(\triangle RST)$

ANSWERS TO LESSON QUIZ

1. a. point *E*

b. $\overline{AH}$

2. a. $R'(2, -2)$, $S'(-1, 1)$, and $T'(-3, -1)$.

b. $R'(2, 2)$, $S'(-1, -1)$, and $T'(1, -3)$.

c. $R'(-2, 2)$, $S'(1, -1)$, and $T'(3, 1)$.

PRESCRIPTION FOR REMEDIATION

Use the student work on the Lesson Quiz to prescribe a differentiated review assignment.

Points	Differentiated Remediation
0	Intervention
1	On-level
2	Extension

PowerGeometry.com

5 Assess & Remediate

Assign the Lesson Quiz. Appropriate intervention, practice, or enrichment is automatically generated based on student performance.

Intervention

- **Reteaching** (2 pages) Provides reteaching and practice exercises for the key lesson concepts. Use with struggling students or absent students.
- **English Language Learner Support** Helps students develop and reinforce mathematical vocabulary and key concepts.

All-in-One Resources/Online

Reteaching

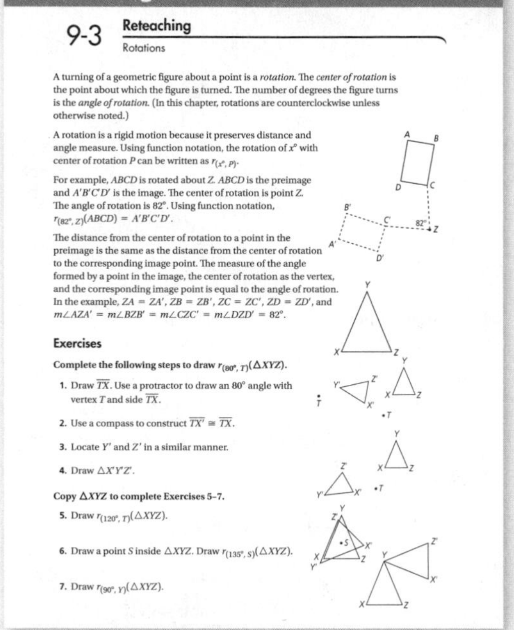

9-3 Reteaching
Rotations
A turning of a geometric figure about a point is a rotation. The center of rotation is the point about which the figure is turned. The number of degrees the figure turns is the angle of rotation. (In this chapter, rotations are counterclockwise unless otherwise noted.)
A rotation is a rigid motion because it preserves distance and angle measure. Using function notation, the rotation of x° with center of rotation P can be written as r(x°, P).
For example, ABCD is rotated about Z. ABCD is the preimage and A'B'C'D' is the image. The center of rotation is point Z. The angle of rotation is 82°. Using function notation, r(82°, Z)(ABCD) = A'B'C'D'.
The distance from the center of rotation to a point in the preimage is the same as the distance from the center of rotation to the corresponding image point. The measure of the angle formed by a point in the image, the center of rotation as the vertex, and the corresponding image point is equal to the angle of rotation. In the example, ZA = ZA', ZB = ZB', ZC = ZC', ZD = ZD', and m∠AZA' = m∠BZB' = m∠CZC' = m∠DZD' = 82°.
Exercises
Complete the following steps to draw r(80°, T)(△XYZ).
1. Draw TX. Use a protractor to draw an 80° angle with vertex T and side TX.
2. Use a compass to construct TX' ≅ TX.
3. Locate Y' and Z' in a similar manner.
4. Draw △X'Y'Z'.
Copy △XYZ to complete Exercises 5–7.
5. Draw r(120°, T)(△XYZ).
6. Draw a point S inside △XYZ. Draw r(135°, S)(△XYZ).
7. Draw r(90°, Y)(△XYZ).

All-in-One Resources/Online

English Language Learner Support

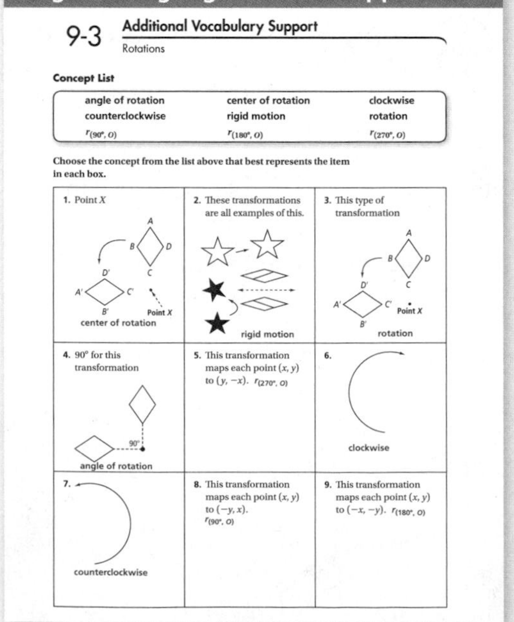

9-3 Additional Vocabulary Support
Rotations
Concept List
angle of rotation
center of rotation
clockwise
counterclockwise
rigid motion
rotation
r(90°, O)
r(180°, O)
r(270°, O)
Choose the concept from the list above that best represents the item in each box.
1. Point X
center of rotation
2. These transformations are all examples of this.
rigid motion
3. This type of transformation
rotation
4. 90° for this transformation
angle of rotation
5. This transformation maps each point (x, y) to (y, −x). r(270°, O)
6.
clockwise
7.
counterclockwise
8. This transformation maps each point (x, y) to (−y, x). r(90°, O)
9. This transformation maps each point (x, y) to (−x, −y). r(180°, O)

Differentiated Remediation *continued*

On-Level

- **Practice** (2 pages) Provides extra practice for each lesson. For simpler practice exercises, use the Form K Practice pages found in the All-in-One Teaching Resources and online.
- **Think About a Plan** Helps students develop specific problem-solving skills and strategies by providing scaffolded guiding questions.
- **Standardized Test Prep** Focuses on all major exercises, all major question types, and helps students prepare for the high-stakes assessments.

Extension

- **Enrichment** Provides students with interesting problems and activities that extend the concepts of the lesson.
- **Activities, Games, and Puzzles** Worksheets that can be used for concepts development, enrichment, and for fun!

Practice and Problem Solving Wkbk/ All-in-One Resources/Online

Practice page 1

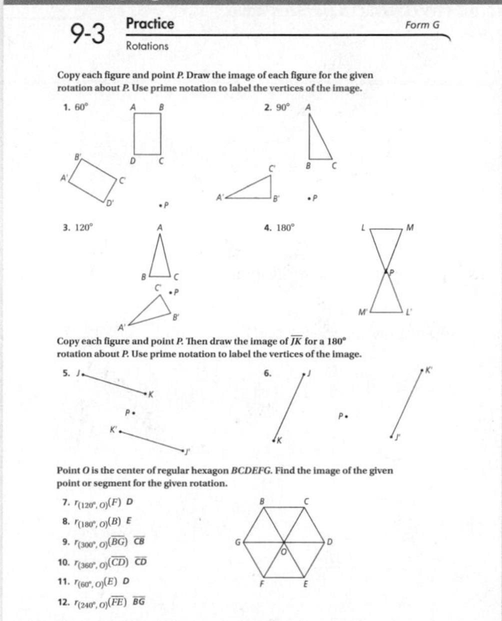

9-3 **Practice** Form G

Rotations

Copy each figure and point *P*. Draw the image of each figure for the given rotation about *P*. Use prime notation to label the vertices of the image.

1. 60°
2. 90°
3. 120°
4. 180°

Copy each figure and point *P*. Then draw the image of $\overline{JK}$ for a 180° rotation about *P*. Use prime notation to label the vertices of the image.

5.
6.

Point *O* is the center of regular hexagon *BCDEFG*. Find the image of the given point or segment for the given rotation.

7. $r_{(120°, O)}(F)$ D
8. $r_{(180°, O)}(B)$ E
9. $r_{(300°, O)}(\overline{BG})$ $\overline{CB}$
10. $r_{(360°, O)}(\overline{CD})$ $\overline{CD}$
11. $r_{(60°, O)}(E)$ D
12. $r_{(240°, O)}(\overline{FE})$ $\overline{BG}$

Practice and Problem Solving Wkbk/ All-in-One Resources/Online

Practice page 2

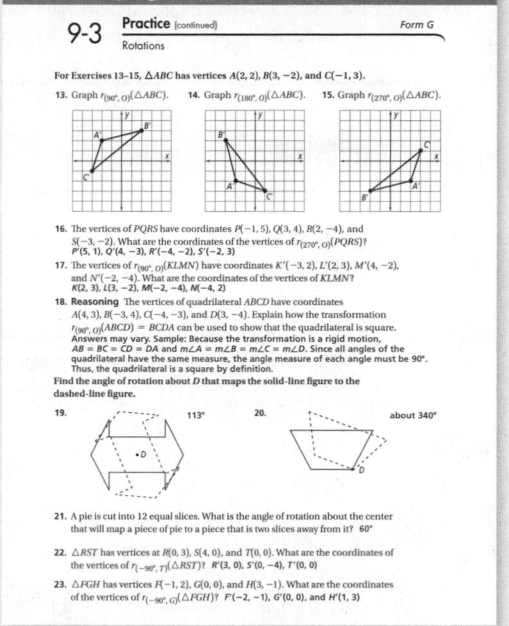

9-3 **Practice** (continued) Form G

Rotations

For Exercises 13–15, $\triangle ABC$ has vertices $A(2, 2)$, $B(3, -2)$, and $C(-1, 3)$.

13. Graph $r_{(90°, O)}(\triangle ABC)$.
14. Graph $r_{(180°, O)}(\triangle ABC)$.
15. Graph $r_{(270°, O)}(\triangle ABC)$.

16. The vertices of *PQRS* have coordinates $P(-1, 5)$, $Q(3, 4)$, $R(2, -4)$, and $S(-3, -2)$. What are the coordinates of the vertices of $r_{(270°, O)}(PQRS)$? **$P'(5, 1)$, $Q'(4, -3)$, $R'(-4, -2)$, $S'(-2, 3)$**
17. The vertices of $r_{(90°, O)}(KLMN)$ have coordinates $K'(-3, 2)$, $L'(2, 3)$, $M'(4, -2)$, and $N'(-2, -4)$. What are the coordinates of the vertices of *KLMN*? **$K(2, 3)$, $L(3, -2)$, $M(-2, -4)$, $N(-4, 2)$**
18. **Reasoning** The vertices of quadrilateral *ABCD* have coordinates $A(4, 3)$, $B(-3, 4)$, $C(-4, -3)$, and $D(3, -4)$. Explain how the transformation $r_{(90°, O)}(ABCD) = BCDA$ can be used to show that the quadrilateral is square. **Answers may vary. Sample: Because the transformation is a rigid motion, $AB = BC = CD = DA$ and $m\angle A = m\angle B = m\angle C = m\angle D$. Since all angles of the quadrilateral have the same measure, the angle measure of each angle must be 90°. Thus, the quadrilateral is a square by definition.**

Find the angle of rotation about *D* that maps the solid-line figure to the dashed-line figure.

19. **113°**
20. **about 340°**

21. A pie is cut into 12 equal slices. What is the angle of rotation about the center that will map a piece of pie to a piece that is two slices away from it? **60°**
22. $\triangle RST$ has vertices at $R(0, 3)$, $S(4, 0)$, and $T(0, 0)$. What are the coordinates of the vertices of $r_{(-90°, T)}(\triangle RST)$? **$R'(3, 0)$, $S'(0, -4)$, $T'(0, 0)$**
23. $\triangle FGH$ has vertices $F(-1, 2)$, $G(0, 0)$, and $H(3, -1)$. What are the coordinates of the vertices of $r_{(-90°, G)}(\triangle FGH)$? **$F'(-2, -1)$, $G'(0, 0)$, and $H'(1, 3)$**

All-in-One Resources/Online

Enrichment

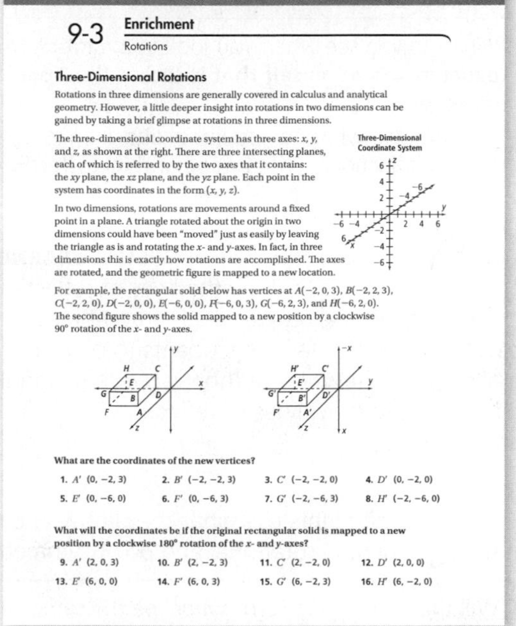

9-3 **Enrichment**

Rotations

Three-Dimensional Rotations

Rotations in three dimensions are generally covered in calculus and analytical geometry. However, a little deeper insight into rotations in two dimensions can be gained by taking a brief glimpse at rotations in three dimensions.

The three-dimensional coordinate system has three axes: *x*, *y*, and *z*, as shown at the right. There are three intersecting planes, each of which is referred to by the two axes that it contains: the *xy* plane, the *xz* plane, and the *yz* plane. Each point in the system has coordinates in the form (x, y, z).

Three-Dimensional Coordinate System

In two dimensions, rotations are movements around a fixed point in a plane. A triangle rotated about the origin in two dimensions could have been "moved" just as easily by leaving the triangle as is and rotating the *x*- and *y*-axes. In fact, in three dimensions this is exactly how rotations are accomplished. The axes are rotated, and the geometric figure is mapped to a new location.

For example, the rectangular solid below has vertices at $A(-2, 0, 3)$, $B(-2, 2, 3)$, $C(-2, 2, 0)$, $D(-2, 0, 0)$, $E(-6, 0, 0)$, $F(-6, 0, 3)$, $G(-6, 2, 3)$, and $H(-6, 2, 0)$. The second figure shows the solid mapped to a new position by a clockwise 90° rotation of the *x*- and *y*-axes.

What are the coordinates of the new vertices?

1. A' **(0, −2, 3)**
2. B' **(−2, −2, 3)**
3. C' **(−2, −2, 0)**
4. D' **(0, −2, 0)**
5. E' **(0, −6, 0)**
6. F' **(0, −6, 3)**
7. G' **(−2, −6, 3)**
8. H' **(−2, −6, 0)**

What will the coordinates be if the original rectangular solid is mapped to a new position by a clockwise 180° rotation of the *x*- and *y*-axes?

9. A' **(2, 0, 3)**
10. B' **(2, −2, 3)**
11. C' **(2, −2, 0)**
12. D' **(2, 0, 0)**
13. E' **(6, 0, 0)**
14. F' **(6, 0, 3)**
15. G' **(6, −2, 3)**
16. H' **(6, −2, 0)**

Practice and Problem Solving Wkbk/ All-in-One Resources/Online

Think About a Plan

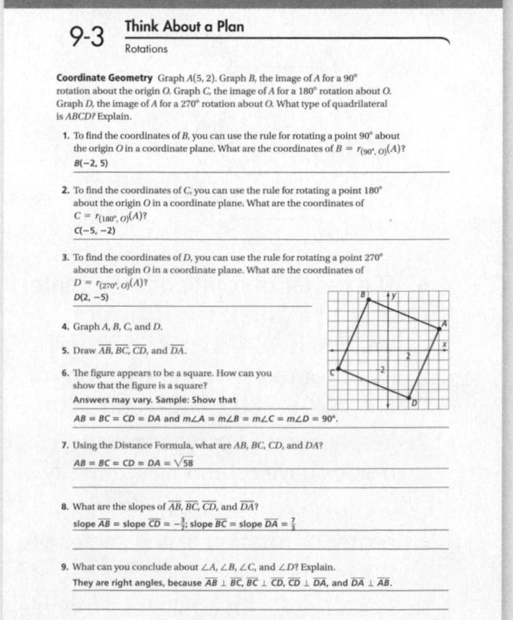

9-3 **Think About a Plan**

Rotations

Coordinate Geometry Graph $A(5, 2)$. Graph *B*, the image of *A* for a 90° rotation about the origin *O*. Graph *C*, the image of *A* for a 180° rotation about *O*. Graph *D*, the image of *A* for a 270° rotation about *O*. What type of quadrilateral is *ABCD*? Explain.

1. To find the coordinates of *B*, you can use the rule for rotating a point 90° about the origin *O* in a coordinate plane. What are the coordinates of $B = r_{(90°, O)}(A)$? **B(−2, 5)**
2. To find the coordinates of *C*, you can use the rule for rotating a point 180° about the origin *O* in a coordinate plane. What are the coordinates of $C = r_{(180°, O)}(A)$? **C(−5, −2)**
3. To find the coordinates of *D*, you can use the rule for rotating a point 270° about the origin *O* in a coordinate plane. What are the coordinates of $D = r_{(270°, O)}(A)$? **D(2, −5)**
4. Graph *A*, *B*, *C*, and *D*.
5. Draw $\overline{AB}$, $\overline{BC}$, $\overline{CD}$, and $\overline{DA}$.
6. The figure appears to be a square. How can you show that the figure is a square? **Answers may vary. Sample: Show that $AB = BC = CD = DA$ and $m\angle A = m\angle B = m\angle C = m\angle D = 90°$.**
7. Using the Distance Formula, what are *AB*, *BC*, *CD*, and *DA*? **$AB = BC = CD = DA = \sqrt{58}$**
8. What are the slopes of $\overline{AB}$, $\overline{BC}$, $\overline{CD}$, and $\overline{DA}$? **slope $\overline{AB}$ = slope $\overline{CD} = -\frac{3}{7}$; slope $\overline{BC}$ = slope $\overline{DA} = \frac{7}{3}$**
9. What can you conclude about $\angle A$, $\angle B$, $\angle C$, and $\angle D$? Explain. **They are right angles, because $\overline{AB} \perp \overline{BC}$, $\overline{BC} \perp \overline{CD}$, $\overline{CD} \perp \overline{DA}$, and $\overline{DA} \perp \overline{AB}$.**

Practice and Problem Solving Wkbk/ All-in-One Resources/Online

Standardized Test Prep

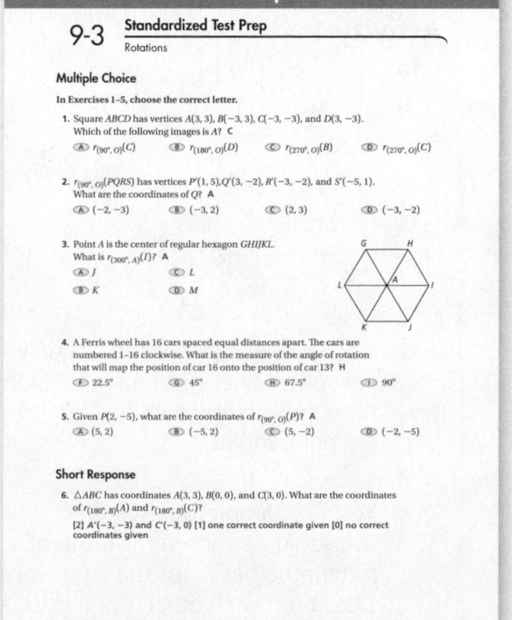

9-3 **Standardized Test Prep**

Rotations

Multiple Choice

In Exercises 1–5, choose the correct letter.

1. Square *ABCD* has vertices $A(3, 3)$, $B(-3, 3)$, $C(-3, -3)$, and $D(3, -3)$. Which of the following images is *A*? **C**
   A $r_{(90°, O)}(C)$ B $r_{(180°, O)}(D)$ C $r_{(270°, O)}(B)$ D $r_{(270°, O)}(C)$
2. $r_{(90°, O)}(PQRS)$ has vertices $P'(1, 5)$, $Q'(3, -2)$, $R'(-3, -2)$, and $S'(-5, 1)$. What are the coordinates of *Q*? **A**
   A (−2, −3) B (−3, 2) C (2, 3) D (−3, −2)
3. Point *A* is the center of regular hexagon *GHIJKL*. What is $r_{(300°, A)}(I)$? **A**
   A *J* B *K* C *L* D *M*
4. A Ferris wheel has 16 cars spaced equal distances apart. The cars are numbered 1–16 clockwise. What is the measure of the angle of rotation that will map the position of car 16 onto the position of car 13? **H**
   F 22.5° G 45° H 67.5° I 90°
5. Given $P(2, -5)$, what are the coordinates of $r_{(90°, O)}(P)$? **A**
   A (5, 2) B (−5, 2) C (5, −2) D (−2, −5)

Short Response

6. $\triangle ABC$ has coordinates $A(3, 3)$, $B(0, 0)$, and $C(3, 0)$. What are the coordinates of $r_{(180°, B)}(A)$ and $r_{(180°, B)}(C)$? **[2] $A'(-3, -3)$ and $C'(-3, 0)$ [1] one correct coordinate given [0] no correct coordinates given**

Online Teacher Resource Center

Activities, Games, and Puzzles

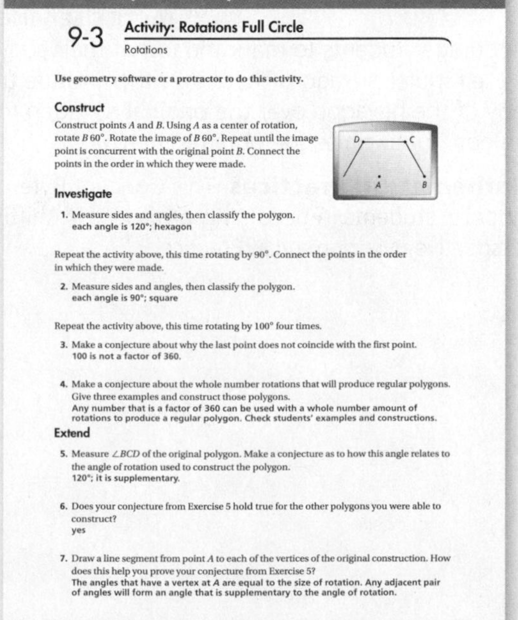

9-3 **Activity: Rotations Full Circle**

Rotations

Use geometry software or a protractor to do this activity.

Construct

Construct points *A* and *B*. Using *A* as a center of rotation, rotate *B* 60°. Rotate the image of *B* 60°. Repeat until the image point is concurrent with the original point *B*. Connect the points in the order in which they were made.

Investigate

1. Measure sides and angles, then classify the polygon. **each angle is 120°; hexagon**

Repeat the activity above, this time rotating by 90°. Connect the points in the order in which they were made.

2. Measure sides and angles, then classify the polygon. **each angle is 90°; square**

Repeat the activity above, this time rotating by 100° four times.

3. Make a conjecture about why the last point does not coincide with the first point. **100 is not a factor of 360.**
4. Make a conjecture about the whole number rotations that will produce regular polygons. Give three examples and construct those polygons. **Any number that is a factor of 360 can be used with a whole number amount of rotations to produce a regular polygon. Check students' examples and constructions.**

Extend

5. Measure $\angle BCD$ of the original polygon. Make a conjecture as to how this angle relates to the angle of rotation used to construct the polygon. **120°; it is supplementary.**
6. Does your conjecture from Exercise 5 hold true for the other polygons you were able to construct? **yes**
7. Draw a line segment from point *A* to each of the vertices of the original construction. How does this help you prove your conjecture from Exercise 5? **The angles that have a vertex at *A* are equal to the size of rotation. Any adjacent pair of angles will form an angle that is supplementary to the angle of rotation.**

Guided Instruction

PURPOSE To identify reflectional (line), rotational, and point symmetry using reflections and rotations.

PROCESS Students will
- use reflections to identify line or reflectional symmetry.
- use rotations to identify rotational and point symmetry.
- design figures with reflectional and rotational symmetry.

DISCUSS Students use what they have learned about reflections and rotations to identify reflectional (line), rotational, and point symmetry. Then students make a design with reflectional and rotational symmetry.

Activity 1

In this activity students use what they know about reflections to identify reflectional symmetry and lines of symmetry.

Q What do you see when you look in a mirror? **[An exact image of myself that is facing the opposite direction.]**

Q In geometry, what do you know about an image and its reflection? **[The image and its reflection are identical except for their orientations.]**

VISUAL LEARNERS

Encourage students to draw the lines of reflection or symmetry on their copies of the rhombus and isosceles trapezoid. Tell the students to copy a parallelogram that is not a rhombus or a rectangle and attempt to draw lines of reflection for their parallelogram.

Activity 2

In this activity students use what they know about rotations to identify rotational and point symmetry.

Q Will the outline of a Ferris wheel be the same each time it stops to let people on and off? **[yes]**

Q Will all objects have the same outline if they are rotated 360°? **[yes]**

TACTILE LEARNERS

Encourage students to mark the top of their copy of the regular hexagon. Then have them rotate the copy of the hexagon over the original hexagon to see how many times the outlines match.

Mathematical Practices This Concept Byte supports students in becoming proficient in shifting perspective, Mathematical Practice 7.

Concept Byte — Symmetry

Use With Lesson 9-3

ACTIVITY

Common Core State Standards
G-CO.A.3 Given a rectangle, parallelogram, trapezoid, or regular polygon, describe the rotations and reflections that carry it onto itself.
MP 7

You can use what you know about reflections and rotations to identify types of **symmetry**. A figure has symmetry if there is a rigid motion that maps the figure onto itself.

A figure has **line symmetry**, or **reflectional symmetry**, if there is a reflection for which the figure is its own image. The line of reflection is called the **line of symmetry**.

A figure has a **rotational symmetry**, if its image, after a rotation of less than 360°, is exactly the same as the original figure. A figure has **point symmetry** if a 180° rotation about a center of rotation maps the figure onto itself.

Activity 1

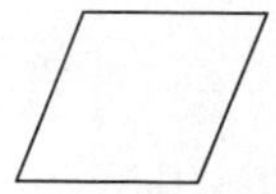

1. Use a straightedge to copy the rhombus at the right.
   a. How many lines of reflection, or lines of symmetry, does the rhombus have?
   b. Draw all of the lines of symmetry.
2. Do all parallelograms have reflectional symmetry? Explain your reasoning.
3. The isosceles trapezoid at the right has only 1 pair of parallel sides. How many lines of symmetry does the trapezoid have?
4. Do all isosceles trapezoids have reflectional symmetry? Do all trapezoids have reflectional symmetry? Explain.

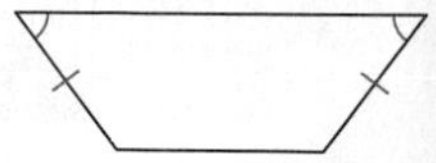

Activity 2

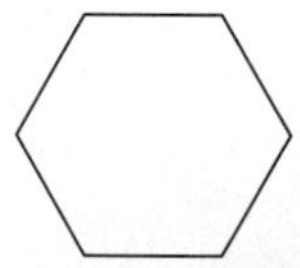

5. Use a straightedge to copy the regular hexagon at the right.
   a. How many lines of symmetry does a regular hexagon have?
   b. Draw all of the lines of symmetry.
6. What are the center and angle(s) of the rotations that map the regular hexagon onto itself?
7. Do all regular polygons have rotational symmetry? Explain your reasoning.
8. Do all regular polygons have point symmetry? Explain.

Answers

Activity 1

1a. 2

b.

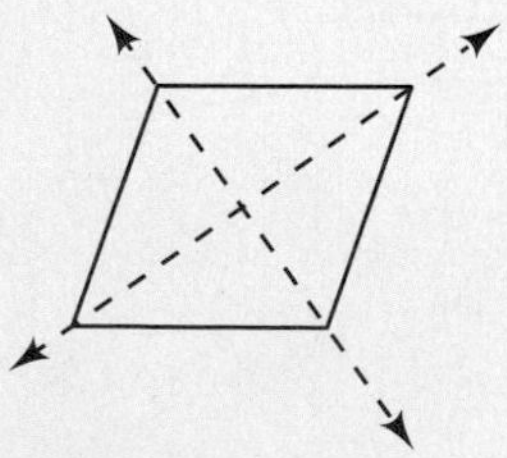

2. No; Sample: A parallelogram only has reflectional symmetry if all sides have equal lengths or if all angles are right angles.

3. 1

4. Yes; no; Sample: An isosceles trapezoid has one vertical line of symmetry that maps the congruent sides onto each other. If the sides are not congruent, then it will not have reflectional symmetry.

Activity 2

5a. 6

b.

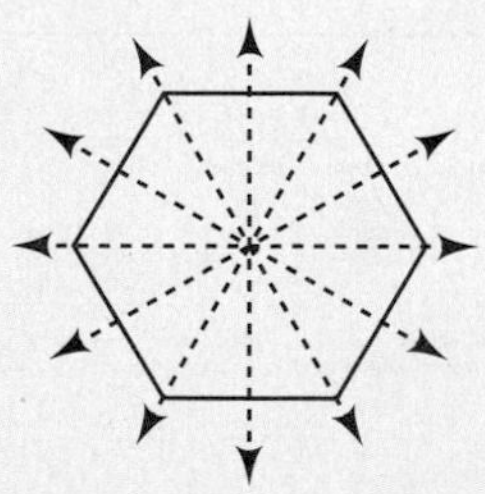

6. The center of rotation is the center of the hexagon. The angles of rotation is 60°, so the figure will rotate onto itself after rotations of 60°, 120°, 180°, 240°, and 300°.

7. Yes; all regular polygons with n sides have rotational symmetry with angle of rotation $\frac{360°}{n}$ and center of rotation at the center of the polygon.

8. No; Sample: An equilateral triangle does not have point symmetry.

Activity 3

Copy and cut out the shapes below. Shade $\frac{1}{2}$ of each square to represent the orange sections. Arrange the shapes to make a design that has both reflectional symmetry and rotational symmetry.

9. Draw the design you made.
10. How many lines of symmetry does your design have? Sketch each line of symmetry.
11. Why are the colors of the tiles important to the symmetry?
12. Does your design have more than one of angle of rotation that maps it onto itself? If so, what are they?
13. Can you change the center of rotation and still map the figure onto itself? Explain.

Exercises

Tell what type(s) of symmetry each figure has. Sketch the figure and the line(s) of symmetry, and give the angle(s) of rotation when appropriate.

14. 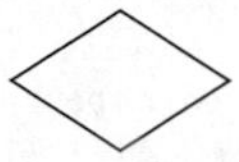

15. 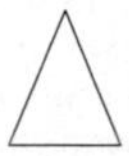

16. 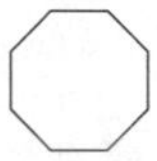

17. **Vocabulary** If a figure has point symmetry, must it also have rotational symmetry? Explain.
18. **Writing** A quadrilateral with vertices (1, 5) and (−2, −3) has point symmetry about the origin.
    a. Show that the quadrilateral is a parallelogram.
    b. How can you use point symmetry to find the other vertices?
19. **Error Analysis** Your friend thinks that the regular pentagon in the diagram has 10 lines of symmetry. Explain and correct your friend's error.

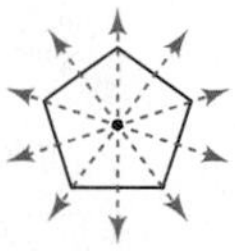

Activity 3

In this Activity students will make a design that has both reflectional and rotational symmetry.

Q How do you know if a figure has reflectional symmetry? **[If every point on one side of a line has a corresponding point on the other side of the line that is the same distance from the line, the figure has reflectional symmetry.]**

Q What items do you see in this classroom that have reflectional symmetry? **[Answers will vary. Possible answers are a chalkboard or the teacher's desk.]**

Q How do you know if a figure has rotational symmetry? **[If the image of the figure after it is rotated less than 360° is exactly the same as the figure, the figure has rotational symmetry.]**

Q What items do you see in this classroom that have rotational symmetry? **[Answers will vary. Possible answers are a ceiling light or a tile on the floor.]**

TACTILE LEARNERS

Encourage students to move their copies of the squares around until they have a design that has both reflectional and rotational symmetry. Challenge students to see how many different designs with the both symmetries they can form.

Activity 3

9–13. Check students' work.

Exercises

14. reflectional; rotational; point; 180°

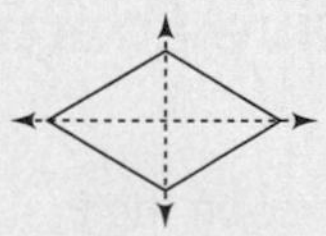

15. reflectional

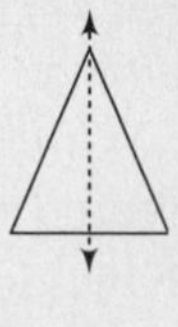

16. reflectional; rotational; point; 45°, 90°, 135°, 180°, 225°, 270°, 315°

17. Yes; point symmetry means it is its own image for a 180° rotation, and that satisfies the definition of rotational symmetry.

18a. Rotate the vertices 180° about the origin to find the other vertices. The other two vertices are (−1, −5) and (2, 3).

b. The slopes of two opposite sides are −2 and the slopes of the other two opposite sides are $\frac{8}{3}$, so the quadrilateral has two pairs of opposite sides parallel.

19. Your friend counted the arrowheads instead of the lines; there are 5 lines of symmetry.

1 Interactive Learning

Solve It!

PURPOSE To describe a translation as a composition of reflections

PROCESS Students may use visual reasoning to draw the two lines of reflection.

FACILITATE

Q If you want the translation to be horizontal, what should be true about the line of reflection? **[It should be a vertical line.]**

Q What will the letter look like after a single reflection? **[The letter will be backwards.]**

Q After the first reflection, where should you place the second line of reflection? **[The second line should be halfway between the first image and the final image.]**

ANSWER See Solve It in Answers on next page.

CONNECT THE MATH In Solve It, students see that a single transformation can be written as a composition of translations. The lesson summarizes compositions of which isometries result in a single transformation.

2 Guided Instruction

Take Note

For each composition taught in the lesson, have students verify that the composition is an isometry using coordinate geometry and simple figures.

9-4 Compositions of Isometries

Common Core State Standards

G-CO.B.6 Use geometric descriptions of rigid motions to transform figures and to predict the effect of a given rigid motion on a given figure . . . **Also G-CO.A.5**

MP 1, MP 3, MP 6

Objectives To find compositions of isometries, including glide reflections
To classify isometries

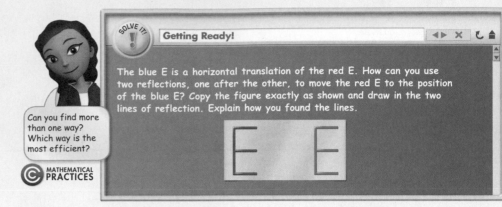

Can you find more than one way? Which way is the most efficient?

MATHEMATICAL PRACTICES

Getting Ready!

The blue E is a horizontal translation of the red E. How can you use two reflections, one after the other, to move the red E to the position of the blue E? Copy the figure exactly as shown and draw in the two lines of reflection. Explain how you found the lines.

Lesson Vocabulary
- glide reflection
- isometry

In the Solve It, you looked for a way to use two reflections to produce the same image as a given horizontal translation. In this lesson, you will learn that any rigid motion can be expressed as a composition of reflections.

The term *isometry* means same distance. An **isometry** is a transformation that preserves distance, or length. So, translations, reflections, and rotations are isometries.

Essential Understanding You can express all isometries as compositions of reflections.

Expressing isometries as compositions of reflections depends on the following theorem.

take note — Theorem 9-1

The composition of two or more isometries is an isometry.

There are only four kinds of isometries.

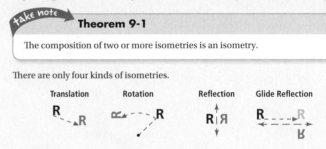

You will learn about *glide reflections* later in the lesson.

9-4 Preparing to Teach

BIG ideas **Transformations**
Coordinate Geometry

ESSENTIAL UNDERSTANDINGS

- If there is an isometry that maps a figure to another, then you can map one onto the other by using a composition of reflections.

Math Background

Students will apply their knowledge of transformations to compositions of transformations. Two congruent figures can be mapped onto each other using no more than three reflections. Students will learn how to write a transformation as a composition. They will also compare the properties preserved under each type of transformation. Translations and rotations preserve orientation. Reflections and glide reflections do not.

The theorems in this lesson are sophisticated, and their proofs are too complicated for students at this point in their studies. Nevertheless, their inclusion is warranted because the theorems themselves are easily understood and applied.

Mathematical Practices

Attend to precision. Students will make explicit use of the terms "reflectional symmetry" and "rotational symmetry" and will identify geometric figures that have one, both, or neither.

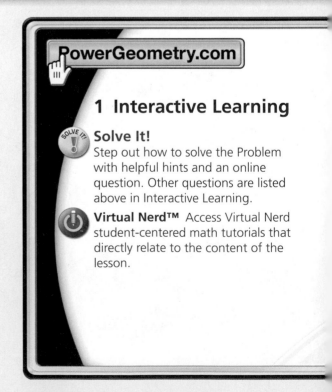

PowerGeometry.com

1 Interactive Learning

Solve It! Step out how to solve the Problem with helpful hints and an online question. Other questions are listed above in Interactive Learning.

Virtual Nerd™ Access Virtual Nerd student-centered math tutorials that directly relate to the content of the lesson.

In Lesson 9-1, you learned that a composition of transformations is a combination of two or more transformations, one performed after the other.

Take note

Theorem 9-2 Reflections Across Parallel Lines

A composition of reflections across two parallel lines is a translation.
You can write this composition as
$(R_m \circ R_\ell)(\triangle ABC) = \triangle A''B''C''$
or $R_m(R_\ell(\triangle ABC)) = \triangle A''B''C''$.
$\overline{AA''}$, $\overline{BB''}$, and $\overline{CC''}$ are all perpendicular to lines ℓ and m.

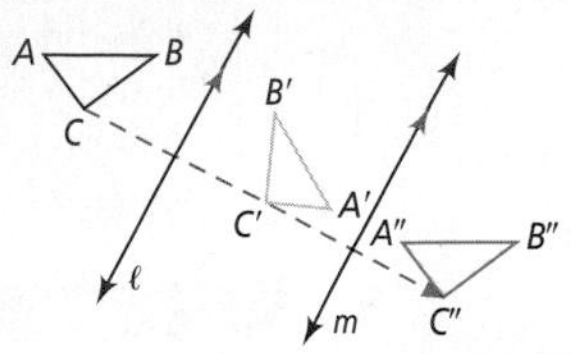

Problem 1 Composing Reflections Across Parallel Lines

What is $(R_m \circ R_\ell)(J)$? What is the distance of the resulting translation?

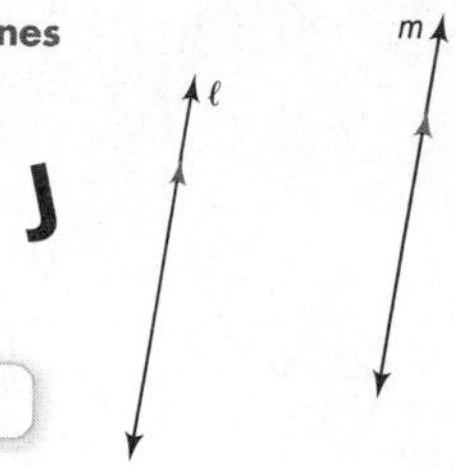

As you do the two reflections, keep track of the distance moved by a point P of the preimage.

Think
How do you know that $PA = AP'$, $P'B = BP''$, and $\overleftrightarrow{AB} \perp \ell$?
All three statements are true by the definition of reflection across a line.

Step 1 Reflect J across ℓ. $PA = AP'$, so $PP' = 2AP'$.

Step 2 Reflect the image across m. $P'B = BP''$, so $P'P'' = 2P'B$.

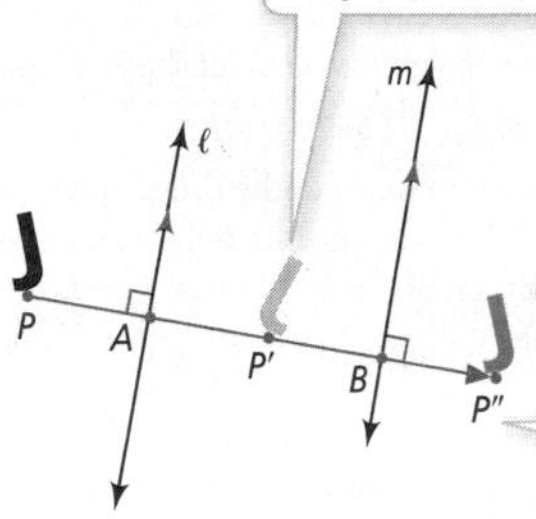

P moved a total distance of $2AP' + 2P'B$, or $2AB$.

The red arrow shows the translation. The total distance P moved is $2 \cdot AB$. Because $\overleftrightarrow{AB} \perp \ell$, AB is the distance between ℓ and m. The distance of the translation is twice the distance between ℓ and m.

Got It? **1. a.** Draw parallel lines ℓ and m as in Problem 1. Draw J between ℓ and m. What is the image of $(R_m \circ R_\ell)(J)$? What is the distance of the resulting translation?
b. Reasoning Use the results of part (a) and Problem 1. Make a conjecture about the distance of any translation that is the result of a composition of reflections across two parallel lines.

Take Note

Remind students that $R_m \circ R_\ell(\triangle ABC)$ means to reflect $\triangle ABC$ over line ℓ first, and then line m. Students should notice that the transformation that maps the original figure onto the final figure can be written as a single translation. Ask students to make a conjecture about how they could write a single transformation if the figure was reflected over a third parallel line (reflection).

Problem 1

Q What will the letter look like after a reflection across line ℓ? **[The letter will be backwards and the same distance from line ℓ as the preimage.]**

Q How is the letter moving compared to the two lines of reflection? **[The letter is moving on a line perpendicular to the lines of reflection.]**

Q What will the letter look like after a reflection across line m? **[The letter will be oriented the same as the preimage and the same distance from m as the first image.]**

Q What is the total distance that the letter moved? **[The letter moved twice the distance between the two lines.]**

Got It?

For 1a, students should perform the two reflections and then compare their final diagrams with the diagram given in Problem 1. Students should compare and contrast the diagrams to help them answer the question.

For 1b, have students use grid paper to perform the composition. Using grid paper will allow them to plot their points accurately. Students can test their conjectures by starting at different points on the grid.

2 Guided Instruction

Each Problem is worked out and supported online.

Problem 1
Composing Reflections Across Parallel Lines

Problem 2
Composing Reflections Across Intersecting Lines
Animated

Problem 3
Finding a Glide Reflection Image
Animated

Support in Geometry Companion
- Vocabulary
- Key Concepts
- Got It?

Answers

Solve It!

Answers may vary. Sample:

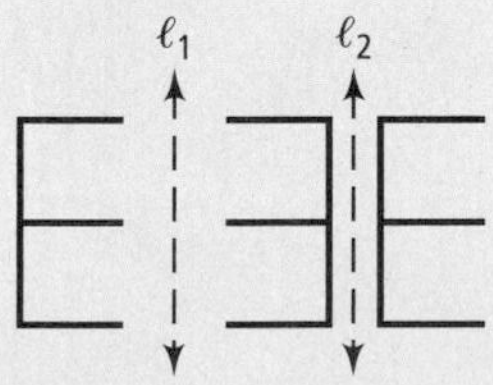

Draw a vertical line, ℓ_1, and reflect the red E across it (any vertical line may be used as the first line of reflection). Join one point, P, of the image to its corresponding point, P', on the blue E. Find the midpoint of $\overline{PP'}$ and draw the line perpendicular to $\overline{PP'}$ that passes through the midpoint of $\overline{PP'}$. This line is the second line of reflection, ℓ_2.

Got It?

1. See back of book.

Take Note

See if students can predict the relationship between the angle formed by the lines of reflection and the angle of rotation. Students can sketch a composition and measure angles. They should recognize that the angle of rotation is twice the angle formed by the lines of reflection.

Problem 2

Q What will the letter look like after the first reflection over line ℓ? **[The letter will be backwards and at the same distance from ℓ as the preimage.]**

Q How is the angle formed by the lines related to the angle formed by the preimage, point of intersection, and image? **[The angles are congruent.]**

Q About what point does the letter appear to be rotated? **[point C]**

Got It? — ERROR PREVENTION

Have students copy the figure onto paper and sketch each reflection. Then, students can measure the angle of rotation formed by the composition. Have them check their conjectures by starting the preimage J at several different locations.

---

take note

Theorem 9-3 Reflections Across Intersecting Lines

A composition of reflections across two intersecting lines is a rotation.

You can write this composition as $(R_m \circ R_\ell)(\triangle ABC) = \triangle A''B''C''$ or $R_m(R_\ell(\triangle ABC)) = \triangle A''B''C''$.

The figure is rotated about the point where the two lines intersect. In this case, point Q.

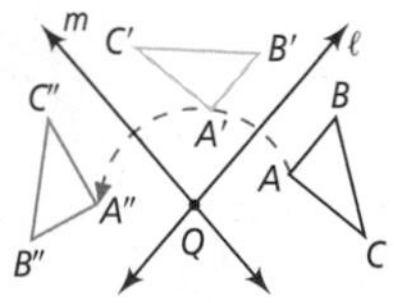

Problem 2 Composing Reflections Across Intersecting Lines

Lines ℓ and m intersect at point C and form a 70° angle. What is $(R_m \circ R_\ell)(\text{J})$? What are the center of rotation and the angle of rotation for the resulting rotation?

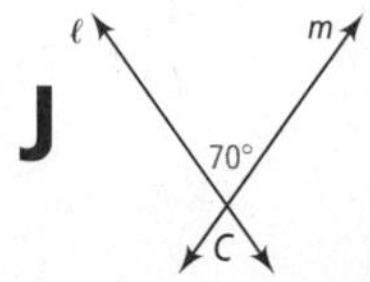

After you do the reflections, follow the path of a point P of the preimage.

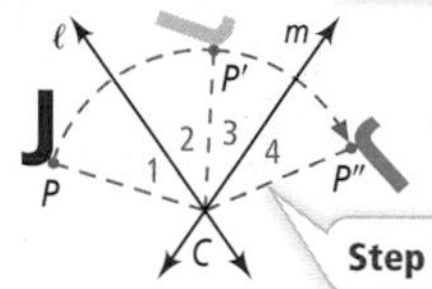

Step 1 Reflect J across ℓ.

Step 2 Reflect the image across m.

Step 3 Draw the angles formed by joining P, P', and P'' to C.

Think

How do you show that $m\angle1 = m\angle2$?
If you draw $\overline{PP'}$ and label its intersection point with line ℓ as A, then $PA = P'A$ and $\overline{PP'} \perp \ell$. So, by the Converse of the Angle Bisector Theorem, $m\angle1 = m\angle2$.

J is rotated clockwise about the intersection point of the lines. The center of rotation is C. You know that $m\angle2 + m\angle3 = 70$. You can use the definition of reflection to show that $m\angle1 = m\angle2$ and $m\angle3 = m\angle4$. So, $m\angle1 + m\angle2 + m\angle3 + m\angle4 = 140$. The angle of rotation is 140° clockwise.

Got It? 2. a. Use the diagram at the right. What is $(R_b \circ R_a)(\text{J})$? What are the center and the angle of rotation for the resulting rotation?

b. Reasoning Use the results of part (a) and Problem 2. Make a conjecture about the center of rotation and the angle of rotation for any rotation that is the result of any composition of reflections across two intersecting lines.

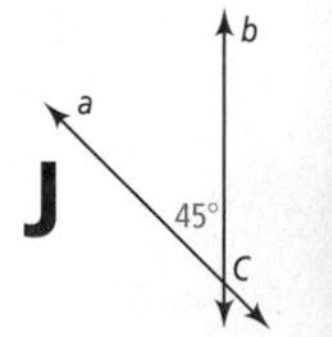

Any composition of isometries can be represented by either a reflection, translation, rotation, or glide reflection. A **glide reflection** is the composition of a translation (a glide) and a reflection across a line parallel to the direction of translation. You can map a left paw print onto a right paw print with a glide reflection.

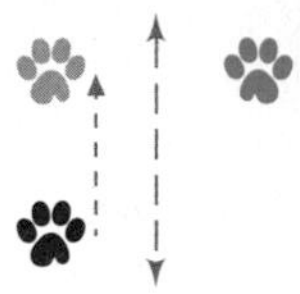

Additional Problems

1. What is $R_m \circ R_\ell$? What is the distance of the resulting translation?

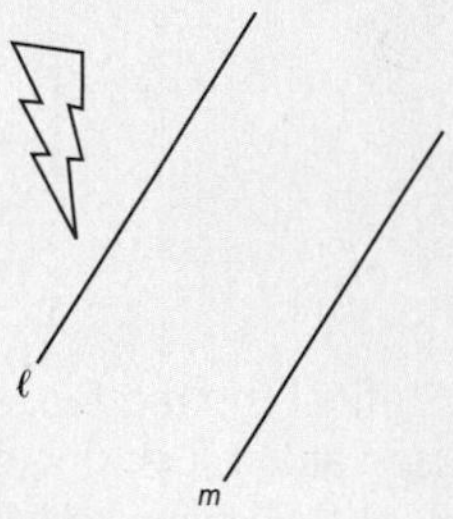

ANSWER The distance of the resulting translation is twice the perpendicular distance between lines ℓ and m.

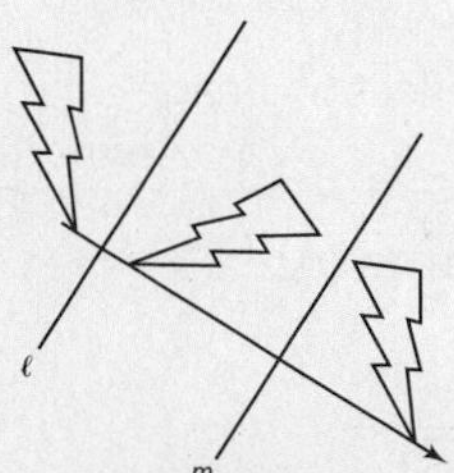

2. What is $R_m \circ (R_\ell(F))$? What are the center of rotation and the angle of rotation for the resulting rotation?

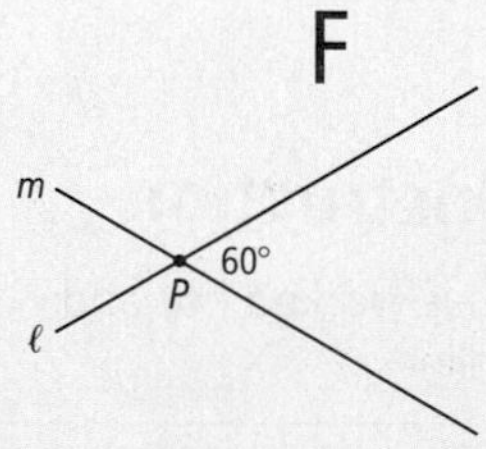

ANSWER a 120° clockwise rotation about point P

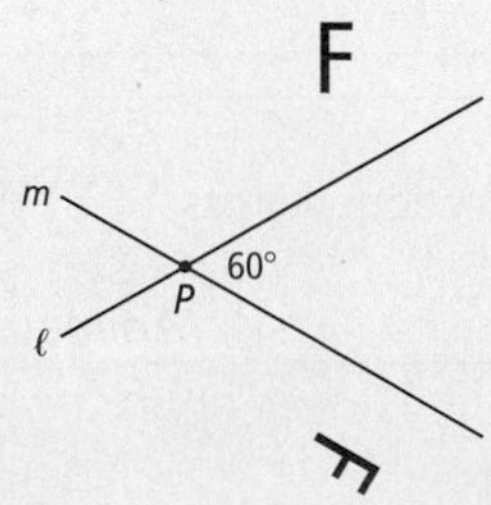

Problem 3 Finding a Glide Reflection Image

Coordinate Geometry What is $(R_{x=0} \circ T_{<0,\,-5>})(\triangle TEX)$?

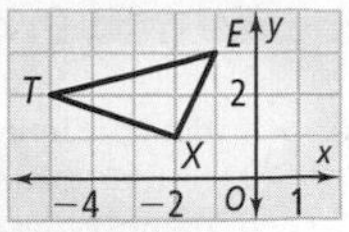

Know	Need	Plan
• The vertices of $\triangle TEX$ • The translation rule • The line of reflection	The image of $\triangle TEX$ for the glide reflection	First use the translation rule to translate $\triangle TEX$. Then reflect the translation image of each vertex across the line of reflection.

Use the translation rule $T_{<0,\,-5>}(\triangle TEX)$ to move $\triangle TEX$ down 5 units.

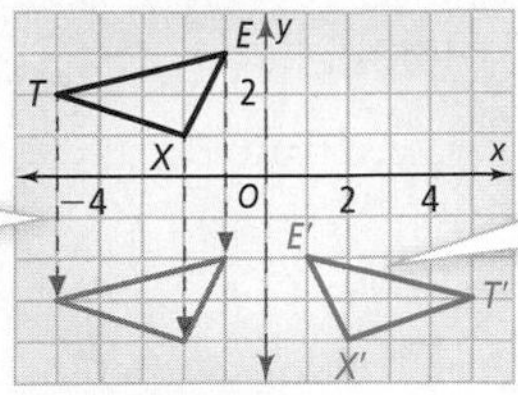

Reflect the image of $\triangle TEX$ across the line $x = 0$.

Got It? 3. Graph $\triangle TEX$ from Problem 3. What is the image of $\triangle TEX$ for the glide reflection $(R_{y=-2} \circ T_{<1,\,0>})(\triangle TEX)$?

Lesson Check

Do you know HOW?

Copy the diagrams below. Sketch the image of Z reflected across line *a*, then across line *b*.

1.

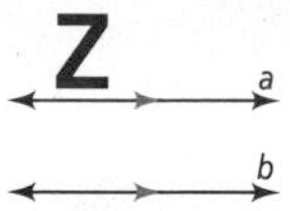

2.

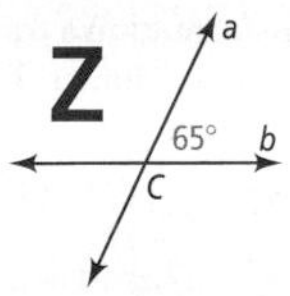

3. $\triangle PQR$ has vertices $P(0, 5)$, $Q(5, 3)$, and $R(3, 1)$. What are the vertices of the image of $\triangle PQR$ for the glide reflection $(R_{y=-2} \circ T_{<3,\,-1>})(\triangle PQR)$?

Do you UNDERSTAND? MATHEMATICAL PRACTICES

4. **Vocabulary** In a glide reflection, what is the relationship between the direction of the translation and the line of reflection?

5. **Error Analysis** You reflect $\triangle DEF$ first across line m and then across line n. Your friend says you can get the same result by reflecting $\triangle DEF$ first across line n and then across line m. Explain your friend's error.

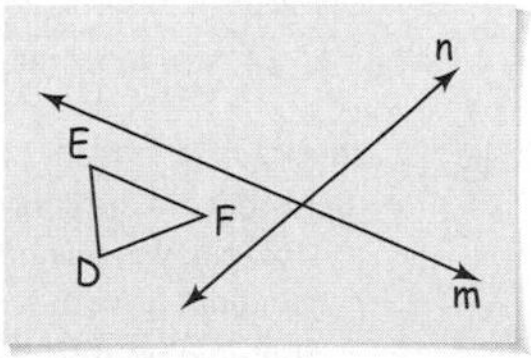

Problem 3

Q Which direction is the translation? Explain. **[The translation is down because 5 is subtracted from the *y*-coordinates.]**

Q After the translation, in what quadrant will the triangle be located? **[Quadrant III]**

Q After the reflection, in what quadrant will the triangle be located? **[Quadrant IV]**

Got It? SYNTHESIZING

Have students draw a diagram of the situation. Before they perform each transformation, have them predict where the image will be.

3 Lesson Check

Do you know HOW?

- If students have difficulty with Exercise 2, then have them review Problem 2 to see what a reflection across two intersecting lines looks like.

Do you UNDERSTAND?

- If students have difficulty with Exercise 4, then have them review Problem 3 to see how the line of reflection fits in a glide reflection.

Close

Q How many reflections are necessary at a maximum to map one congruent figure onto another? **[3]**

Q Which isometries do not change the orientation of a figure? **[translations and rotations]**

3. Graph $\triangle RST$ with vertices $R(-5, 4)$, $S(-2, 3)$, and $T(-3, 1)$. What is the image of $\triangle RST$ for the glide reflection $(R_{y=0} \circ T_{<8,\,0>})(\triangle RST)$?

ANSWER

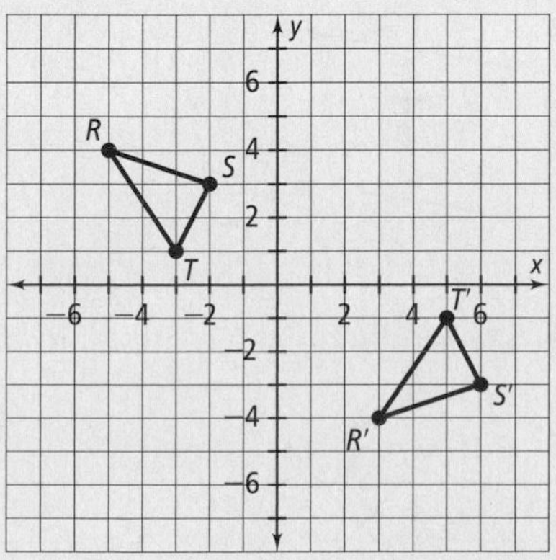

Answers

Got It? (continued)

2a.

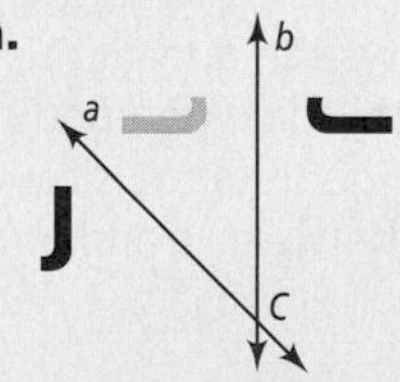

The center of rotation is *C*. The angle of rotation is 90° clockwise.

b. The center of rotation is the intersection of the lines of reflection; the ∠ of rotation is two times the measure of the acute or right ∠ formed by the lines of reflection.

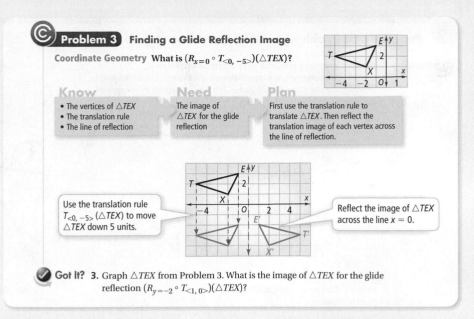

3.

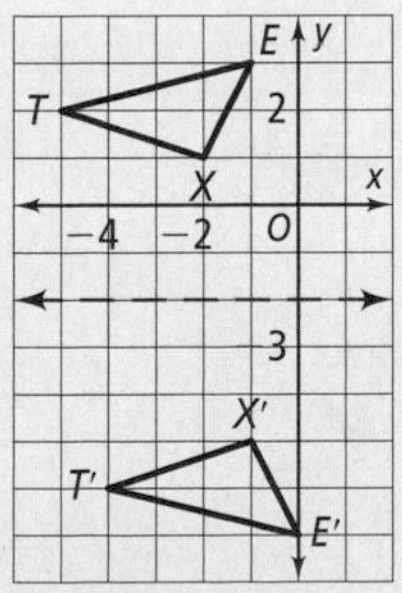

Lesson Check

1.

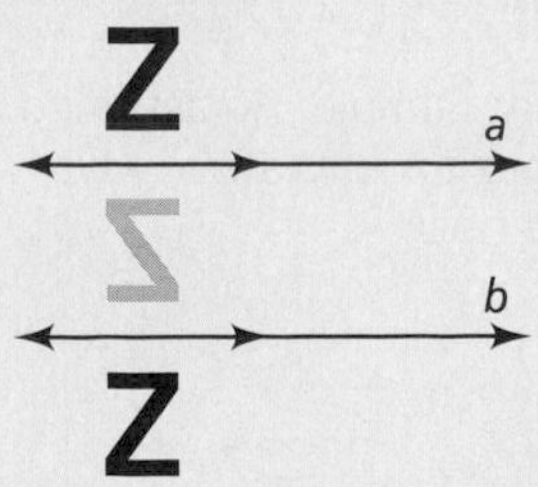

2–5. See next page.

4 Practice

ASSIGNMENT GUIDE

Basic: 6–15 all, 16–26 even, 30–38 even

Average: 7–15 odd, 16–38

Advanced: 7–15 odd, 16–40

Mathematical Practices are supported by exercises with red headings. Here are the Practices supported in this lesson:

MP 1: Make Sense of Problems Ex. 22

MP 3: Construct Arguments Ex. 28, 40

MP 3: Communicate Ex. 26, 27

MP 3: Critique the Reasoning of Others Ex. 5

Applications exercises have blue headings.

EXERCISE 30: Use the Think About a Plan worksheet in the **Practice and Problem Solving Workbook** (also available in the Teaching Resources in print and online) to further support students' development in becoming independent learners.

HOMEWORK QUICK CHECK

To check students' understanding of key skills and concepts, go over Exercises 13, 16, 22, 26, and 30.

Practice and Problem-Solving Exercises

MATHEMATICAL PRACTICES

A Practice

Find the image of each letter after the transformation $R_m \circ R_\ell$. Is the resulting transformation a translation or a rotation? For a translation, describe the direction and distance. For a rotation, tell the center of rotation and the angle of rotation.

See Problems 1 and 2.

6.

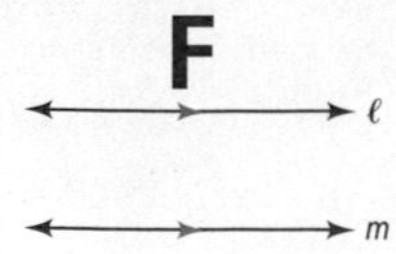

7.

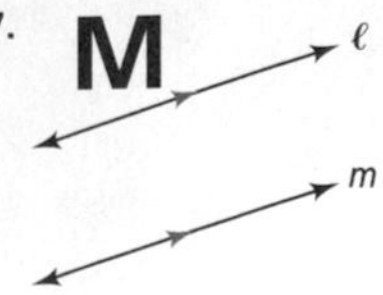

8.

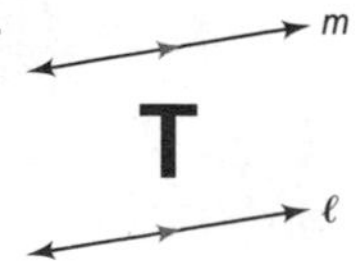

9.

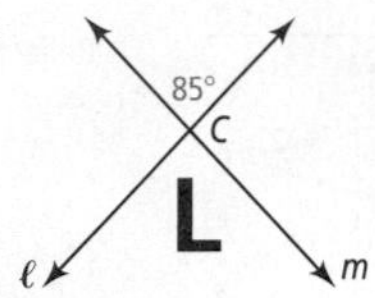

10.

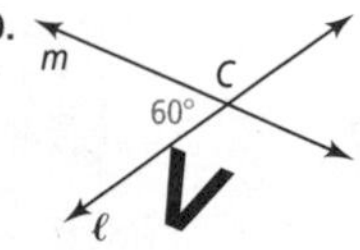

11.

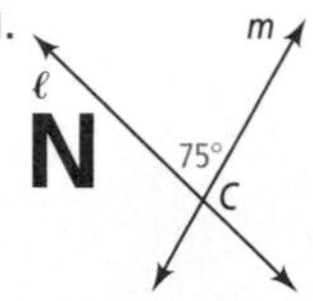

Graph $\triangle PNB$ and its image after the given transformation.

See Problem 3.

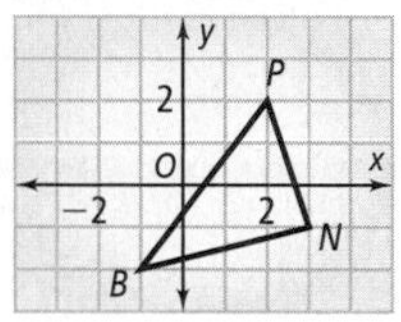

12. $(R_{y=3} \circ T_{<2,\,0>})(\triangle PNB)$

13. $(R_{x=0} \circ T_{<0,\,-3>})(\triangle PNB)$

14. $(R_{y=0} \circ T_{<2,\,2>})(\triangle PNB)$

15. $(R_{y=x} \circ T_{<-1,\,1>})(\triangle PNB)$

B Apply

Use the given points and lines. Graph $\overline{AB}$ and its image $\overline{A''B''}$ after a reflection first across ℓ_1 and then across ℓ_2. Is the resulting transformation a translation or a rotation? For a translation, describe the direction and distance. For a rotation, tell the center of rotation and the angle of rotation.

16. $A(1, 5)$ and $B(2, 1)$; $\ell_1: x = 3$; $\ell_2: x = 7$

17. $A(2, 4)$ and $B(3, 1)$; ℓ_1: x-axis; ℓ_2: y-axis

18. $A(-4, -3)$ and $B(-4, 0)$; $\ell_1: y = x$; $\ell_2: y = -x$

19. $A(2, -5)$ and $B(-1, -3)$; $\ell_1: y = 0$; $\ell_2: y = 2$

20. $A(6, -4)$ and $B(5, 0)$; $\ell_1: x = 6$; $\ell_2: x = 4$

21. $A(-1, 0)$ and $B(0, -2)$; $\ell_1: y = -1$; $\ell_2: y = 1$

22. Think About a Plan Let A' be the point $(1, 5)$. If $(R_{y=1} \circ T_{<3,\,0>})(A) = A'$, then what are the coordinates of A?

- How can you *work backwards* to find the coordinates of A?
- Should A be to the left or to the right of A'?
- Should A be above or below A'?

Answers

Lesson Check (continued)

2.

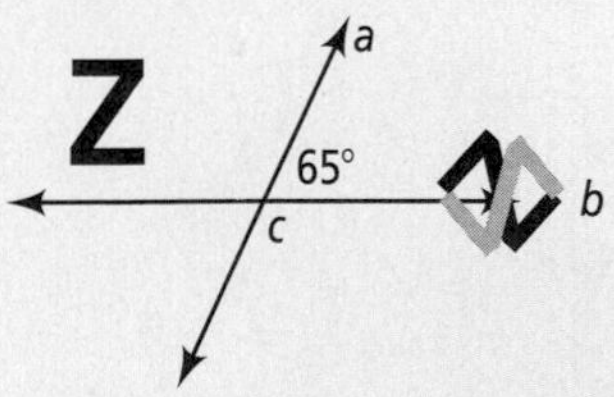

3. $P'(3, -8)$, $Q'(8, -6)$, $R'(6, -4)$

4. parallel

5. Answers may vary. Sample: He assumed that reflections are commutative, when, in general, they are not.

Practice and Problem-Solving Exercises

6–8. A translation; the arrow in the diagram shows the direction, determined by a line perpendicular to ℓ and m. The distance is twice the distance between ℓ and m.

6.

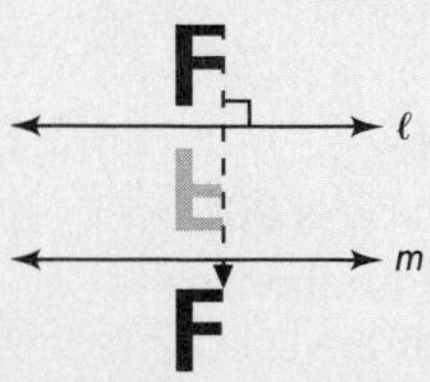

7.

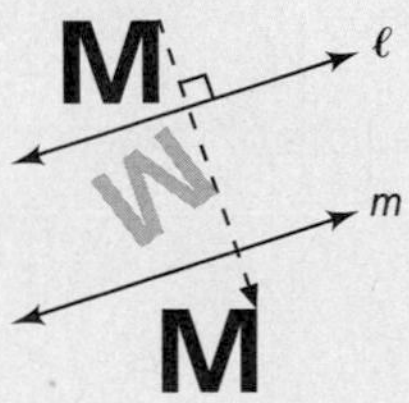

8.

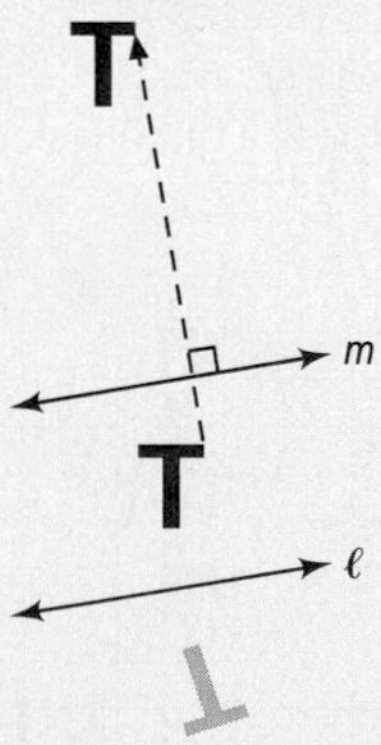

9–11. A rotation; the center of rotation is C.

9.

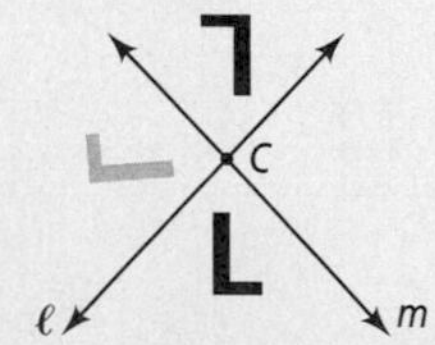

The $\angle$ of rotation is 170° clockwise.

Describe the isometry that maps the black figure onto the blue figure.

23.

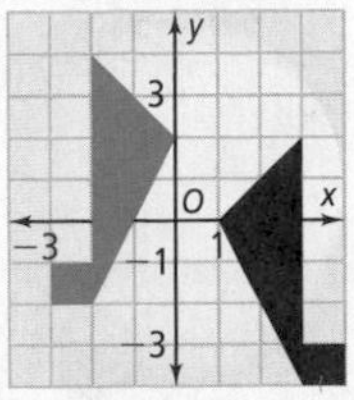

24.

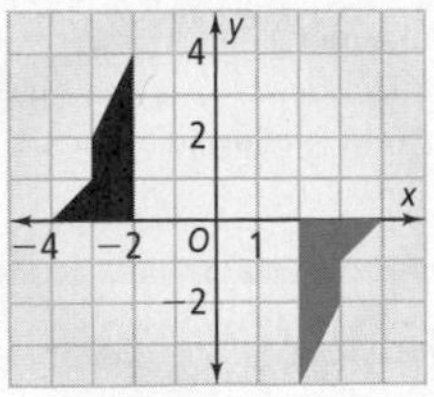

25. Which transformation maps the black triangle onto the blue triangle?

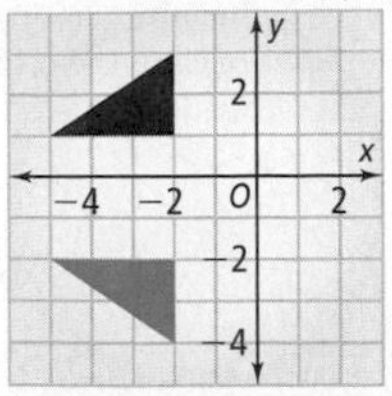

A $R_{x=2} \circ T_{<0,-3>}$

B $r_{(180°, O)}$

C $R_{y=-\frac{1}{2}}$

D $r_{(180°, O)} \circ R_{x\text{-axis}}$

26. Writing Reflections and glide reflections are *odd isometries*, while translations and rotations are *even isometries*. Use what you have learned in this lesson to explain why these categories make sense.

27. Open-Ended Draw $\triangle ABC$. Describe a reflection, a translation, a rotation, and a glide reflection. Then draw the image of $\triangle ABC$ for each transformation.

28. Reasoning The definition states that a glide reflection is the composition of a translation and a reflection. Explain why these can occur in either order.

Identify each mapping as a translation, reflection, rotation, or glide reflection. Write the rule for each translation, reflection, rotation, or glide reflection. For glide reflections, write the rule as a composition of a translation and a reflection.

29. $\triangle ABC \rightarrow \triangle EDC$

30. $\triangle EDC \rightarrow \triangle PQM$

31. $\triangle MNJ \rightarrow \triangle EDC$

32. $\triangle HIF \rightarrow \triangle HGF$

33. $\triangle PQM \rightarrow \triangle JLM$

34. $\triangle MNP \rightarrow \triangle EDC$

35. $\triangle JLM \rightarrow \triangle MNJ$

36. $\triangle PQM \rightarrow \triangle KJN$

37. $\triangle KJN \rightarrow \triangle ABC$

38. $\triangle HGF \rightarrow \triangle KJN$

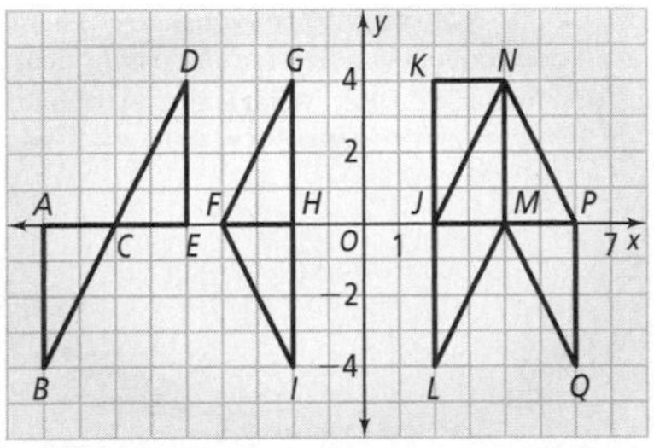

10.

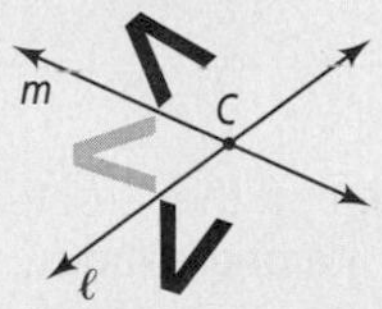

The ∠ of rotation is 120° clockwise.

11.

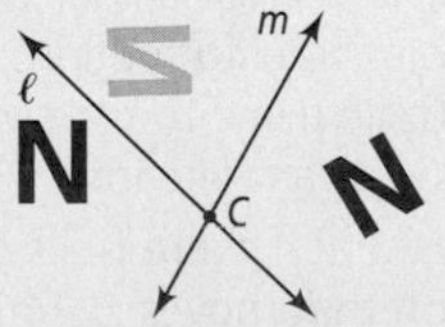

The ∠ of rotation is 150° clockwise.

12.

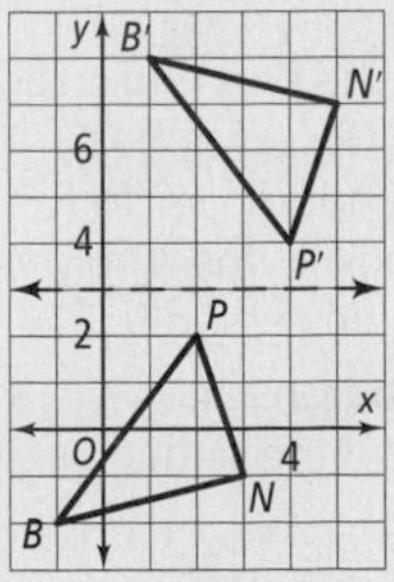

13.

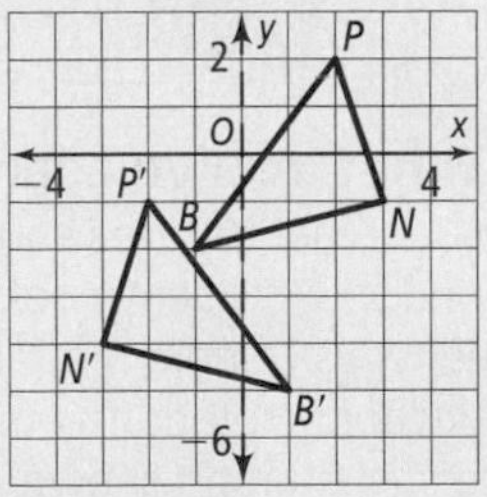

14.

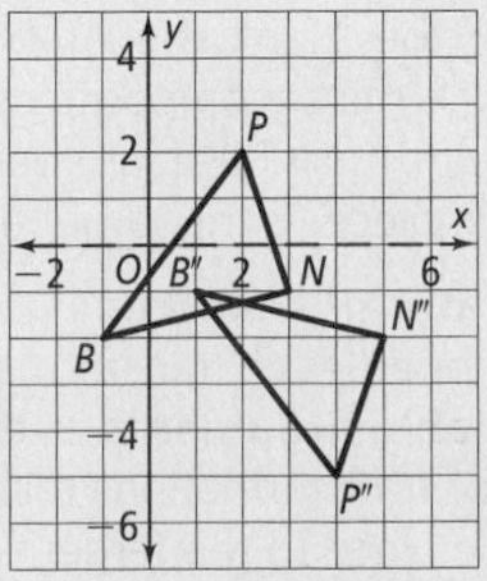

15.

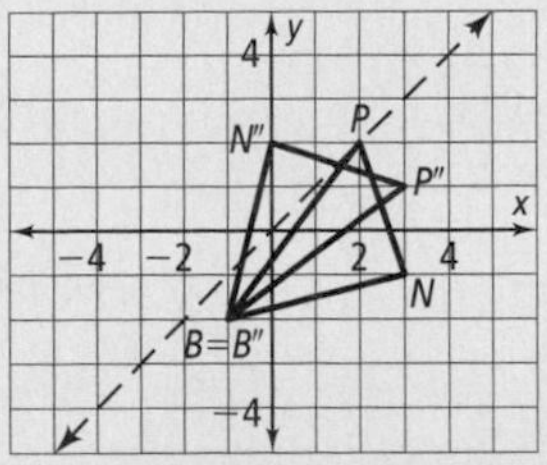

16.

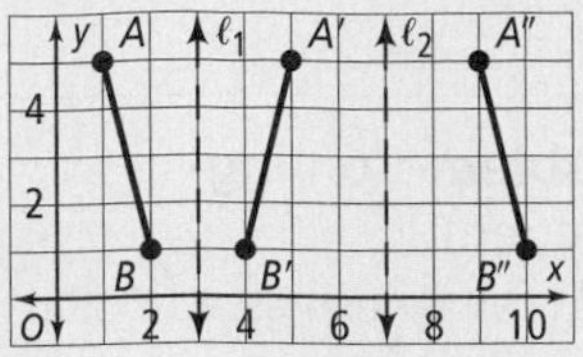

a translation 8 units to the right

17.

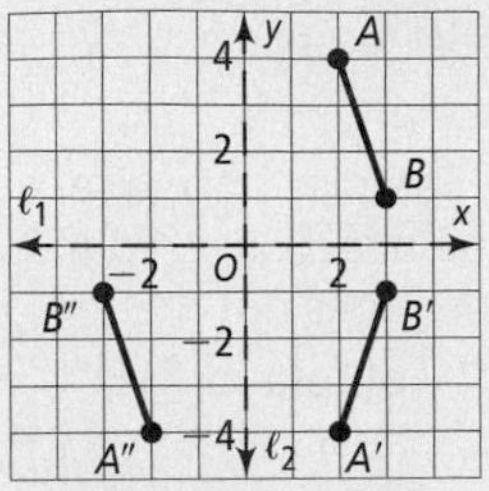

a 180° rotation about (0, 0)

18.

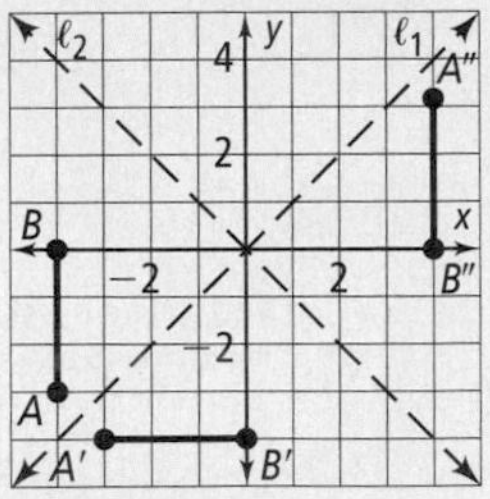

a 180° rotation about (0, 0)

19.

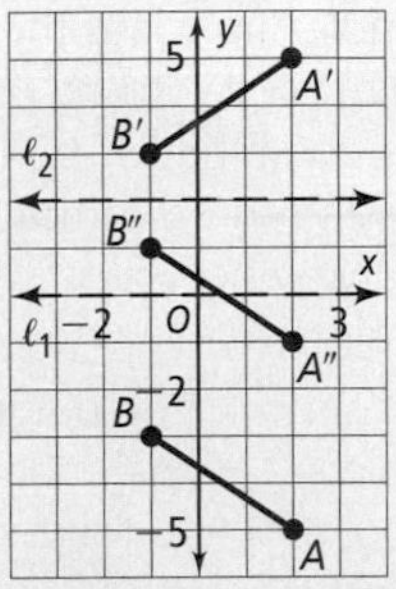

a translation 4 units up

20.

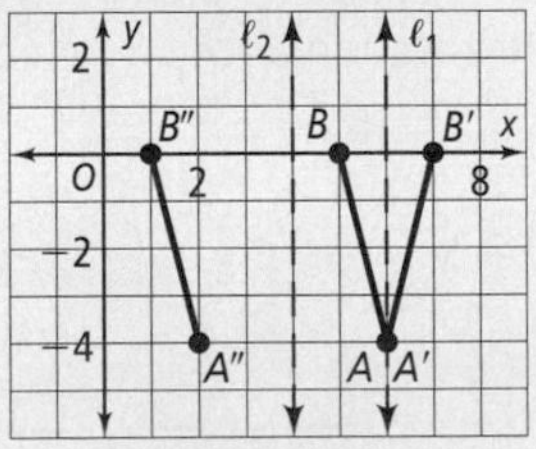

a translation 4 units left

21.

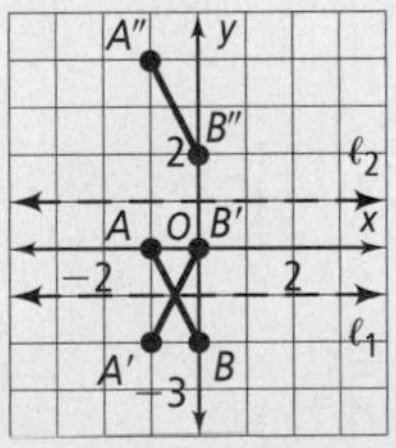

a translation 4 units up

22. $B(-2, -3)$

23. $R_{x=\frac{1}{2}} \circ T_{<0, 2>}$

24. $r_{(180°, O)}$

25. C

26–38. See next page.

Answers

Practice and Problem-Solving Exercises (continued)

26. Odd isometries can be expressed as the composition of an odd number of reflections. Even isometries are the composition of an even number of reflections.

27. Check students' work.

28. Answers may vary. Sample: Since a reflection moves a point in the direction ⊥ to the translation, the order does not matter.

29. rotation; center C, ∠ of rotation 180°

30. glide reflection; $(R_{y=0} \circ T_{<11, 0>})(x, y)$

31. translation; $T_{<-9, 0>}(x, y)$

32. reflection; $y = 0$

33. reflection; $x = 4$

34. reflection; $x = -\frac{1}{2}$

35. rotation; center (3, 0), ∠ of rotation 180°

36. glide reflection; $(R_{x=4} \circ T_{<0, 4>})(x, y)$

37. translation; $T_{<-11, -4>}(x, y)$

38. rotation; center (0, 2), ∠ of rotation 180°

39. Answers may vary. Sample: Translate the black R so that one point moves to its corresponding point on the blue R. Then reflect across a line passing through that point and the point halfway between two other corresponding points.

40. No; answers may vary. Sample: The diagram shows a counterexample. A'' is the image of A for a 40° rotation about P (A'), followed by a reflection across line ℓ, while A'''' is the image of A for a reflection across line ℓ (A'''), followed by a 40° rotation about P.

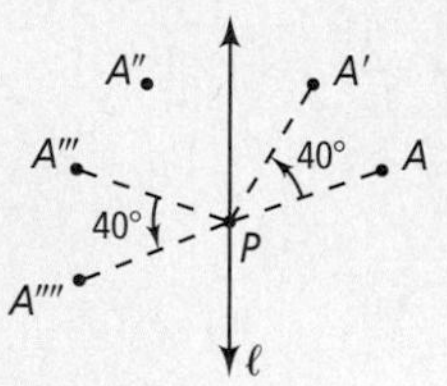

Challenge **39.** Describe a glide reflection that maps the black R to the blue R.

40. Reasoning Does an $x°$ rotation about a point P followed by a reflection across a line ℓ give the same image as a reflection across ℓ followed by an $x°$ rotation about P? Explain.

Apply What You've Learned

MATHEMATICAL PRACTICES MP 3

Look back at the information about the video game on page 543 and review the requirements of Case 1 and Case 2. The graph of the puzzle piece and target area is shown again below.

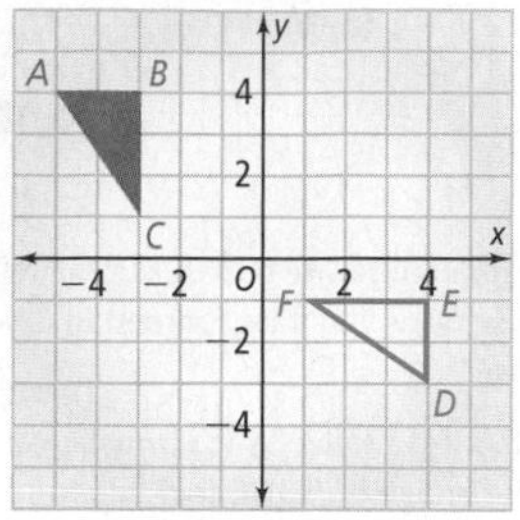

In the Apply What You've Learned sections in Lessons 9-2 and 9-3, you explored how various translations, reflections, and rotations move the puzzle piece. Now you will think about how to use a composition of transformations to move the puzzle piece to the target area.

a. Explain why any composition of transformations that moves the puzzle piece to the target area must include a reflection.

b. Explain why you must use at least two transformations to move the puzzle piece to the target area.

c. Case 2 does not allow reflections across vertical or horizontal lines. Consider the position of $R_{y=x}(\triangle DEF)$. What is the relationship of $R_{y=x}(\triangle DEF)$ to $\triangle ABC$?

Apply What You've Learned

In the Apply What You've Learned for Lesson 9-2 and 9-3, students looked at how the three types of rigid motions—translations, reflections, rotations—move the puzzle piece in the graph shown on page 543. Now, they think about what characteristics their final compositions of transformations must have in order to move the puzzle piece to the target area.

Mathematical Practices

Students use properties of transformations to **construct viable arguments** to explain characteristics of transformations that will move the puzzle piece to the target area. (MP 3)

ANSWERS

a. The puzzle piece and the target area have opposite orientations. A figure's orientation is changed only by reflection across a line, so any sequence of transformations that moves the puzzle piece to the target area must include a reflection.

b. Part (a) shows that if there is a single transformation that moves $\triangle ABC$ to $\triangle DEF$, it must be a reflection. The reflection that maps C to F is across the line that is the perpendicular bisector of $\overline{CF}$. However, this reflection does not map B to E, as shown by part (F) of the Apply What You've Learned in Lesson 9-2.

c. $R_{y=x}(\triangle DEF)$ is a translation of $\triangle ABC$ 2 units to the right.

9-4 Lesson Resources

Differentiated Remediation

Instructional Support

Geometry Companion

Students can use the **Geometry Companion** worktext (4 pages) . . .

- New Vocabulary
- Key Concepts
- Got It for each Problem
- Lesson Check

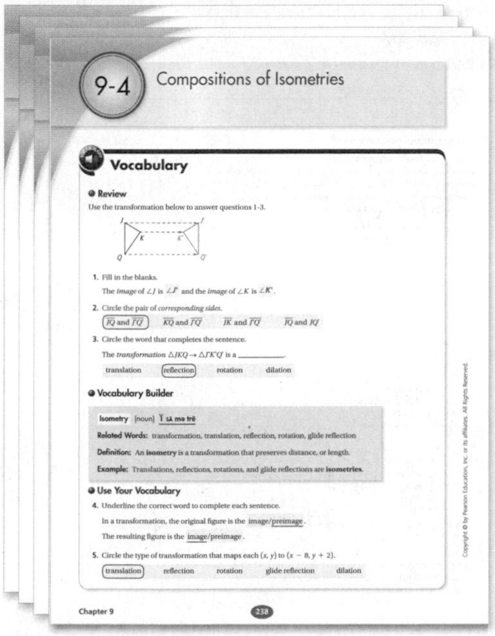
9-4 Compositions of Isometries

Vocabulary

Review

Use the transformation below to answer questions 1-3.

1. Fill in the blanks.
   The *image* of ∠J is ∠J′ and the *image* of ∠K is ∠K′.
2. Circle the pair of *corresponding sides*.
   $\overline{JQ}$ and $\overline{J'Q'}$ $\overline{KQ}$ and $\overline{J'Q'}$ $\overline{JK}$ and $\overline{J'Q'}$ $\overline{JQ}$ and JQ'
3. Circle the word that completes the sentence.
   The *transformation* △JKQ → △J′K′Q′ is a ________.
   translation reflection rotation dilation

Vocabulary Builder

Isometry (noun) Ī SÄ mə trē

Related Words: transformation, translation, reflection, rotation, glide reflection

Definition: An **isometry** is a transformation that preserves distance, or length.

Example: Translations, reflections, rotations, and glide reflections are **isometries**.

Use Your Vocabulary

4. Underline the correct word to complete each sentence.
   In a transformation, the original figure is the image/preimage.
   The resulting figure is the image/preimage.
5. Circle the type of transformation that maps each (x, y) to $(x - 8, y + 2)$.
   translation reflection rotation glide reflection dilation

Copyright © by Pearson Education, Inc. or its affiliates. All Rights Reserved.

Chapter 9 238

ELL Support

Use Graphic Organizers Use the following activity to help students integrate the vocabulary learned through the chapter. Have small groups of students work together to make a graphic organizer for the concepts in this chapter. Suggest that students use transformation as a central concept for their organizer. Symmetry, isometry, and the different types of transformations can be placed in various sections. Groups should use familiar words and illustrations to explain the concepts in their graphic organizer.

Have groups compare other graphic organizers with their own and revise their graphic organizer to incorporate insights that they have learned from the other organizers.

5 Assess & Remediate

Lesson Quiz

1. **Do you UNDERSTAND?** What is $(R_m \circ R_\ell)$(basketball)? What is the distance of the resulting translation?

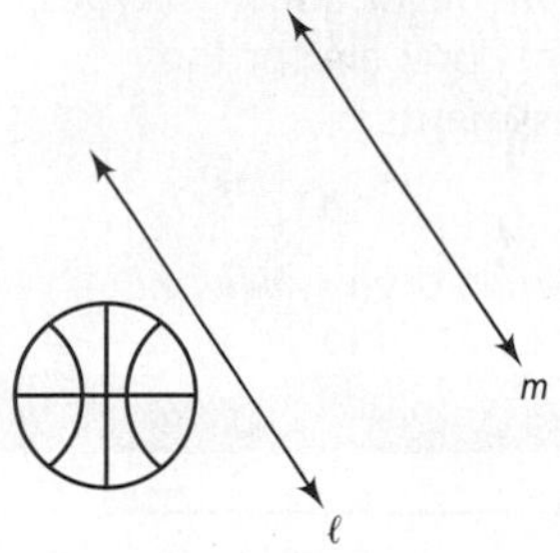

2. What are the vertices of the image of $\triangle ABC$ for the glide reflection $(R_{y=0} \circ T_{<4,\,0>})(\triangle ABC)$? Triangle ABC has vertices $A(-5, 6)$, $B(-3, 2)$, and $C(-1, 2)$.

ANSWERS TO LESSON QUIZ

1. The direction is shown by the arrow. The distance is twice the perpendicular distance between lines ℓ and m.

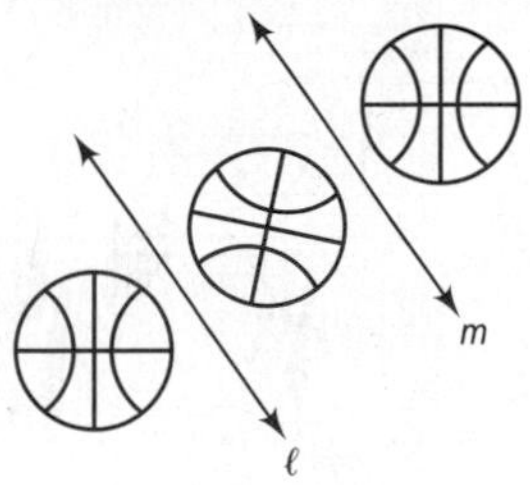

2. $A'(-1, -6)$, $B'(1, -2)$, $C'(3, -2)$

PRESCRIPTION FOR REMEDIATION

Use the student work on the Lesson Quiz to prescribe a differentiated review assignment.

Points	Differentiated Remediation
0	Intervention
1	On-level
2	Extension

PowerGeometry.com

5 Assess & Remediate

Assign the Lesson Quiz. Appropriate intervention, practice, or enrichment is automatically generated based on student performance.

Intervention

- **Reteaching** (2 pages) Provides reteaching and practice exercises for the key lesson concepts. Use with struggling students or absent students.
- **English Language Learner Support** Helps students develop and reinforce mathematical vocabulary and key concepts.

All-in-One Resources/Online

Reteaching

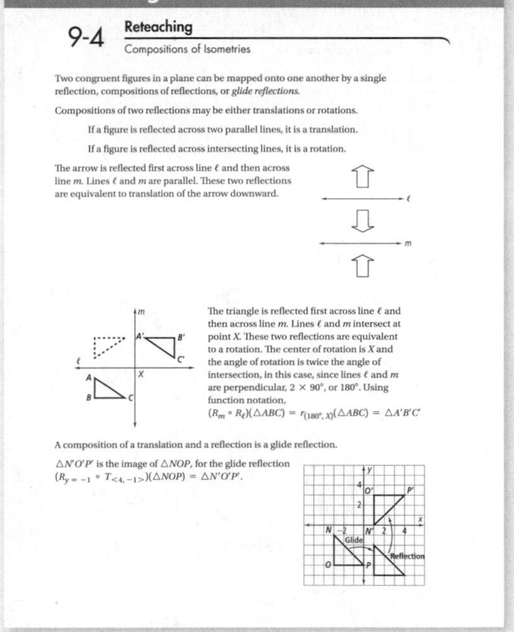
9-4 Reteaching
Compositions of Isometries

Two congruent figures in a plane can be mapped onto one another by a single reflection, compositions of reflections, or *glide reflections*.

Compositions of two reflections may be either translations or rotations.

If a figure is reflected across two parallel lines, it is a translation.

If a figure is reflected across intersecting lines, it is a rotation.

The arrow is reflected first across line ℓ and then across line m. Lines ℓ and m are parallel. These two reflections are equivalent to translation of the arrow downward.

The triangle is reflected first across line ℓ and then across line m. Lines ℓ and m intersect at point X. These two reflections are equivalent to a rotation. The center of rotation is X and the angle of rotation is twice the angle of intersection, in this case, since lines ℓ and m are perpendicular, $2 \times 90°$, or $180°$. Using function notation, $(R_m \circ R_\ell)(\triangle ABC) = r_{(180°,\,X)}(\triangle ABC) = \triangle A'B'C'$

A composition of a translation and a reflection is a glide reflection.

$\triangle N'O'P'$ is the image of $\triangle NOP$, for the glide reflection $(R_{y=-1} \circ T_{<4,\,-1>})(\triangle NOP) = \triangle N'O'P'$.

All-in-One Resources/Online

English Language Learner Support

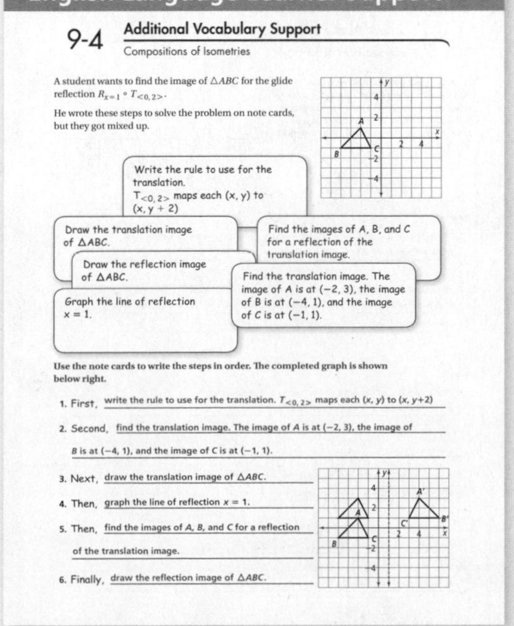
9-4 Additional Vocabulary Support
Compositions of Isometries

A student wants to find the image of $\triangle ABC$ for the glide reflection $R_{x=1} \circ T_{<0,\,2>}$.

He wrote these steps to solve the problem on note cards, but they got mixed up.

Write the rule to use for the translation. $T_{<0,\,2>}$ maps each (x, y) to (x, y + 2)

Draw the translation image of △ABC.

Find the images of A, B, and C for a reflection of the translation image.

Draw the reflection image of △ABC.

Find the translation image. The image of A is at (−2, 3), the image of B is at (−4, 1), and the image of C is at (−1, 1).

Graph the line of reflection x = 1.

Use the note cards to write the steps in order. The completed graph is shown below right.

1. First, write the rule to use for the translation. $T_{<0,\,2>}$ maps each (x, y) to $(x, y+2)$
2. Second, find the translation image. The image of A is at $(-2, 3)$, the image of B is at $(-4, 1)$, and the image of C is at $(-1, 1)$.
3. Next, draw the translation image of $\triangle ABC$.
4. Then, graph the line of reflection $x = 1$.
5. Then, find the images of A, B, and C for a reflection of the translation image.
6. Finally, draw the reflection image of $\triangle ABC$.

Differentiated Remediation *continued*

On-Level

- **Practice** (2 pages) Provides extra practice for each lesson. For simpler practice exercises, use the Form K Practice pages found in the All-in-One Teaching Resources and online.
- **Think About a Plan** Helps students develop specific problem-solving skills and strategies by providing scaffolded guiding questions.
- **Standardized Test Prep** Focuses on all major exercises, all major question types, and helps students prepare for the high-stakes assessments.

Extension

- **Enrichment** Provides students with interesting problems and activities that extend the concepts of the lesson.
- **Activities, Games, and Puzzles** Worksheets that can be used for concepts development, enrichment, and for fun!

Practice and Problem Solving Wkbk/ All-in-One Resources/Online

Practice page 1

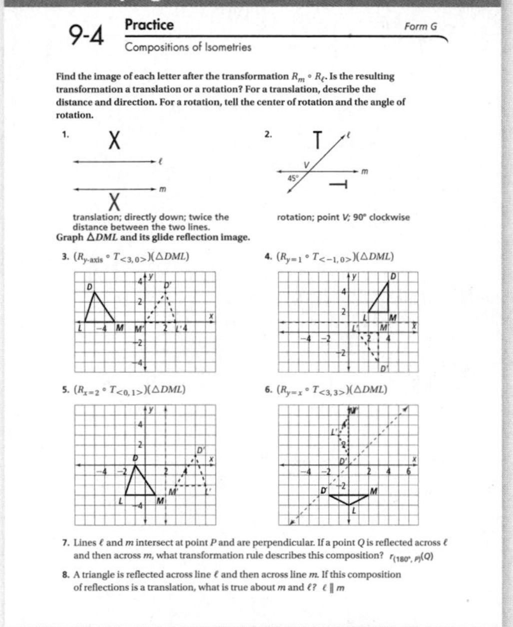

9-4 **Practice** *Form G*
Compositions of Isometries

Find the image of each letter after the transformation $R_m \circ R_\ell$. Is the resulting transformation a translation or a rotation? For a translation, describe the distance and direction. For a rotation, tell the center of rotation and the angle of rotation.

1. translation; directly down; twice the distance between the two lines.

2. rotation; point V; 90° clockwise

Graph $\triangle DML$ and its glide reflection image.

3. $(R_{y\text{-axis}} \circ T_{<3, 0>})(\triangle DML)$

4. $(R_{y=1} \circ T_{<-1, 0>})(\triangle DML)$

5. $(R_{x=2} \circ T_{<0, 1>})(\triangle DML)$

6. $(R_{y=x} \circ T_{<3, 3>})(\triangle DML)$

7. Lines ℓ and m intersect at point P and are perpendicular. If a point Q is reflected across ℓ and then across m, what transformation rule describes this composition? $r_{(180°, P)}(Q)$

8. A triangle is reflected across line ℓ and then across line m. If this composition of reflections is a translation, what is true about m and ℓ? $\ell \parallel m$

Practice and Problem Solving Wkbk/ All-in-One Resources/Online

Practice page 2

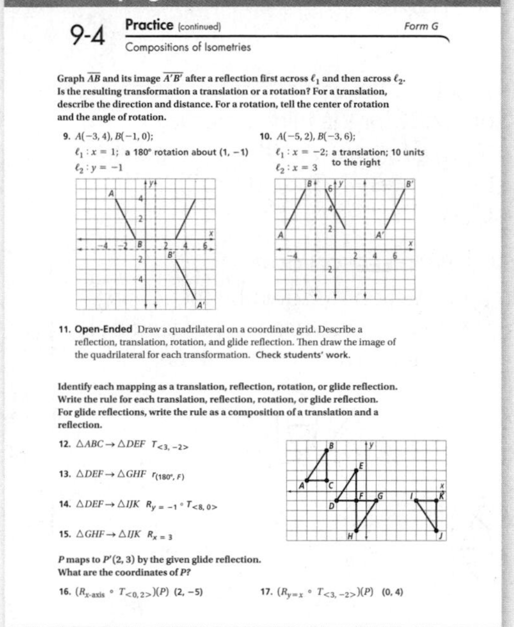

9-4 **Practice** (continued) *Form G*
Compositions of Isometries

Graph $\overline{AB}$ and its image $\overline{A'B'}$ after a reflection first across ℓ_1 and then across ℓ_2. Is the resulting transformation a translation or a rotation? For a translation, describe the direction and distance. For a rotation, tell the center of rotation and the angle of rotation.

9. $A(-3, 4), B(-1, 0)$;
$\ell_1 : x = 1$; a 180° rotation about $(1, -1)$
$\ell_2 : y = -1$

10. $A(-5, 2), B(-3, 6)$;
$\ell_1 : x = -2$; a translation; 10 units to the right
$\ell_2 : x = 3$

11. **Open-Ended** Draw a quadrilateral on a coordinate grid. Describe a reflection, translation, rotation, and glide reflection. Then draw the image of the quadrilateral for each transformation. Check students' work.

Identify each mapping as a translation, reflection, rotation, or glide reflection. Write the rule for each translation, reflection, rotation, or glide reflection. For glide reflections, write the rule as a composition of a translation and a reflection.

12. $\triangle ABC \to \triangle DEF$ $T_{<3, -2>}$

13. $\triangle DEF \to \triangle GHF$ $r_{(180°, F)}$

14. $\triangle DEF \to \triangle IJK$ $R_{y=-1} \circ T_{<8, 0>}$

15. $\triangle GHF \to \triangle IJK$ $R_{x=3}$

P maps to $P'(2, 3)$ by the given glide reflection. What are the coordinates of P?

16. $(R_{x\text{-axis}} \circ T_{<0, 2>})(P)$ (2, −5)

17. $(R_{y=x} \circ T_{<3, -2>})(P)$ (0, 4)

All-in-One Resources/Online

Enrichment

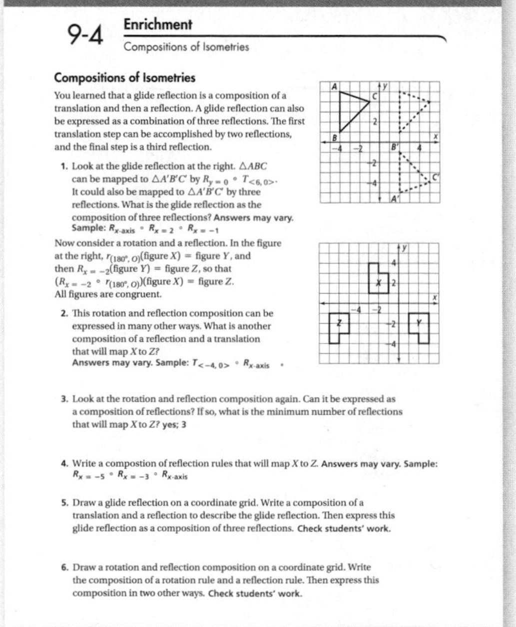

9-4 **Enrichment**
Compositions of Isometries

Compositions of Isometries

You learned that a glide reflection is a composition of a translation and then a reflection. A glide reflection can also be expressed as a combination of three reflections. The first translation step can be accomplished by two reflections, and the final step is a third reflection.

1. Look at the glide reflection at the right. $\triangle ABC$ can be mapped to $\triangle A'B'C'$ by $R_{y=0} \circ T_{<6, 0>}$. It could also be mapped to $\triangle A'B'C'$ by three reflections. What is the glide reflection as the composition of three reflections? Answers may vary. Sample: $R_{x\text{-axis}} \circ R_{x=2} \circ R_{x=-1}$

Now consider a rotation and a reflection. In the figure at the right, $r_{(180°, O)}$(figure X) = figure Y, and then $R_{x=-2}$(figure Y) = figure Z, so that $(R_{x=-2} \circ r_{(180°, O)})$(figure X) = figure Z. All figures are congruent.

2. This rotation and reflection composition can be expressed in many other ways. What is another composition of a reflection and a translation that will map X to Z? Answers may vary. Sample: $T_{<-4, 0>} \circ R_{x\text{-axis}}$

3. Look at the rotation and reflection composition again. Can it be expressed as a composition of reflections? If so, what is the minimum number of reflections that will map X to Z? yes; 3

4. Write a compostion of reflection rules that will map X to Z. Answers may vary. Sample: $R_{x=-5} \circ R_{x=-3} \circ R_{x\text{-axis}}$

5. Draw a glide reflection on a coordinate grid. Write a composition of a translation and a reflection to describe the glide reflection. Then express this glide reflection as a composition of three reflections. Check students' work.

6. Draw a rotation and reflection composition on a coordinate grid. Write the composition of a rotation rule and a reflection rule. Then express this composition in two other ways. Check students' work.

Practice and Problem Solving Wkbk/ All-in-One Resources/Online

Think About a Plan

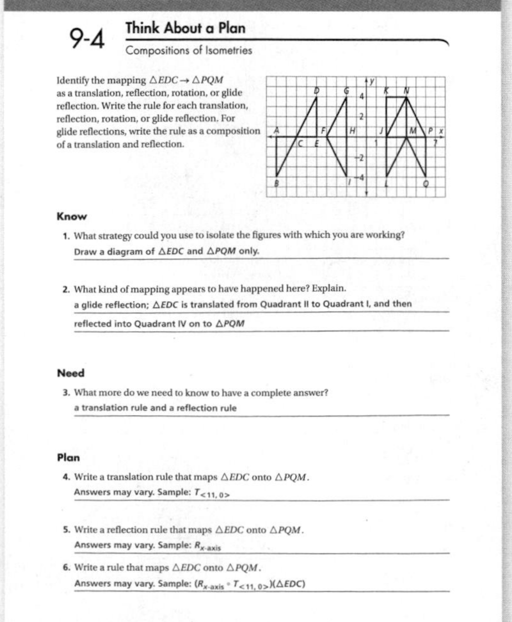

9-4 **Think About a Plan**
Compositions of Isometries

Identify the mapping $\triangle EDC \to \triangle PQM$ as a translation, reflection, rotation, or glide reflection. Write the rule for each translation, reflection, rotation, or glide reflection. For glide reflections, write the rule as a composition of a translation and reflection.

Know

1. What strategy could you use to isolate the figures with which you are working?
Draw a diagram of $\triangle EDC$ and $\triangle PQM$ only.

2. What kind of mapping appears to have happened here? Explain.
a glide reflection; $\triangle EDC$ is translated from Quadrant II to Quadrant I, and then reflected into Quadrant IV on to $\triangle PQM$

Need

3. What more do we need to know to have a complete answer?
a translation rule and a reflection rule

Plan

4. Write a translation rule that maps $\triangle EDC$ onto $\triangle PQM$.
Answers may vary. Sample: $T_{<11, 0>}$

5. Write a reflection rule that maps $\triangle EDC$ onto $\triangle PQM$.
Answers may vary. Sample: $R_{x\text{-axis}}$

6. Write a rule that maps $\triangle EDC$ onto $\triangle PQM$.
Answers may vary. Sample: $(R_{x\text{-axis}} \circ T_{<11, 0>})(\triangle EDC)$

Practice and Problem Solving Wkbk/ All-in-One Resources/Online

Standardized Test Prep

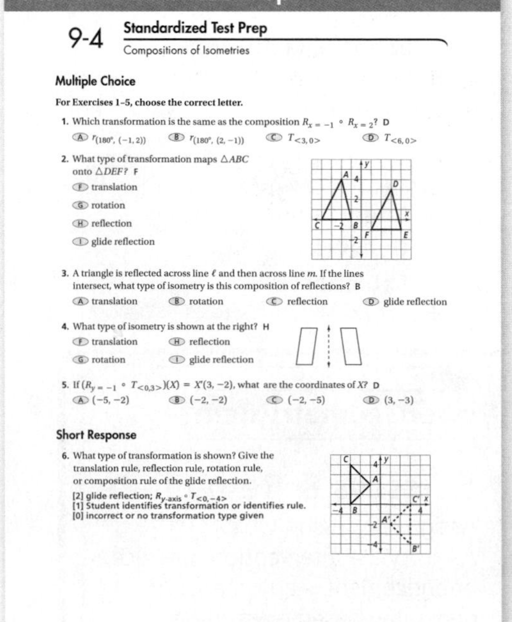

9-4 **Standardized Test Prep**
Compositions of Isometries

Multiple Choice

For Exercises 1–5, choose the correct letter.

1. Which transformation is the same as the composition $R_{x=-1} \circ R_{x=2}$? D
Ⓐ $r_{(180°, (-1, 2))}$ Ⓑ $r_{(180°, (2, -1))}$ Ⓒ $T_{<3, 0>}$ Ⓓ $T_{<6, 0>}$

2. What type of transformation maps $\triangle ABC$ onto $\triangle DEF$? F
Ⓕ translation
Ⓖ rotation
Ⓗ reflection
Ⓘ glide reflection

3. A triangle is reflected across line ℓ and then across line m. If the lines intersect, what type of isometry is this composition of reflections? B
Ⓐ translation Ⓑ rotation Ⓒ reflection Ⓓ glide reflection

4. What type of isometry is shown at the right? H
Ⓕ translation Ⓗ reflection
Ⓖ rotation Ⓘ glide reflection

5. If $(R_{y=-1} \circ T_{<0,3>})(X) = X'(3, -2)$, what are the coordinates of X? D
Ⓐ (−5, −2) Ⓑ (−2, −2) Ⓒ (−2, −5) Ⓓ (3, −3)

Short Response

6. What type of transformation is shown? Give the translation rule, reflection rule, rotation rule, or composition rule of the glide reflection.
[2] glide reflection; $R_{y\text{-axis}} \circ T_{<0, -4>}$
[1] Student identifies transformation or identifies rule.
[0] incorrect or no transformation type given

Online Teacher Resource Center

Activities, Games, and Puzzles

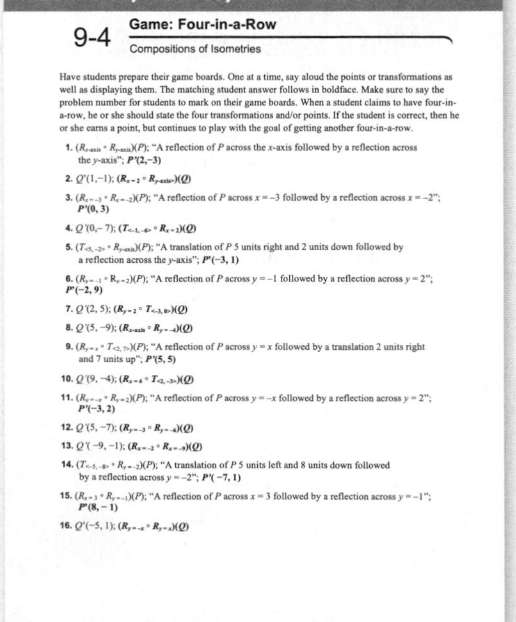

9-4 **Game: Four-in-a-Row**
Compositions of Isometries

Have students prepare their game boards. One at a time, say aloud the points or transformations as well as displaying them. The matching student answer follows in boldface. Make sure to say the problem number for students to mark on their game boards. When a student claims to have four-in-a-row, he or she should state the four transformations and/or points. If the student is correct, then he or she earns a point, but continues to play with the goal of getting another four-in-a-row.

1. $(R_{x\text{-axis}} \circ R_{y\text{-axis}})(P)$; "A reflection of P across the x-axis followed by a reflection across the y-axis"; **$P'(2, -3)$**
2. $Q'(1, -1)$; **$(R_{x=2} \circ R_{y\text{-axis}})(Q)$**
3. $(R_{x=-3} \circ R_{x=-2})(P)$; "A reflection of P across $x = -3$ followed by a reflection across $x = -2$"; **$P'(0, 3)$**
4. $Q'(0, -7)$; **$(T_{<-1, -6>} \circ R_{x=2})(Q)$**
5. $(T_{<5, -2>} \circ R_{y\text{-axis}})(P)$; "A translation of P 5 units right and 2 units down followed by a reflection across the y-axis"; **$P'(-3, 1)$**
6. $(R_{y=-1} \circ R_{y=2})(P)$; "A reflection of P across $y = -1$ followed by a reflection across $y = 2$"; **$P'(-2, 9)$**
7. $Q'(2, 5)$; **$(R_{y=2} \circ T_{<-3, 0>})(Q)$**
8. $Q'(5, -9)$; **$(R_{x\text{-axis}} \circ R_{y=-4})(Q)$**
9. $(R_{y=x} \circ T_{<2, 7>})(P)$; "A reflection of P across $y = x$ followed by a translation 2 units right and 7 units up"; **$P'(5, 5)$**
10. $Q'(9, -4)$; **$(R_{x=6} \circ T_{<2, -3>})(Q)$**
11. $(R_{y=-x} \circ R_{y=2})(P)$; "A reflection of P across $y = -x$ followed by a reflection across $y = 2$"; **$P'(-3, 2)$**
12. $Q'(5, -7)$; **$(R_{y=-3} \circ R_{y=-6})(Q)$**
13. $Q'(-9, -1)$; **$(R_{x=-2} \circ R_{x=-9})(Q)$**
14. $(T_{<-5, -8>} \circ R_{y=-2})(P)$; "A translation of P 5 units left and 8 units down followed by a reflection across $y = -2$"; **$P'(-7, 1)$**
15. $(R_{x=3} \circ R_{y=-1})(P)$; "A reflection of P across $x = 3$ followed by a reflection across $y = -1$"; **$P'(8, -1)$**
16. $Q'(-5, 1)$; **$(R_{y=-x} \circ R_{y=x})(Q)$**

Mid-Chapter Quiz

Do you know HOW?

Tell whether the transformation appears to be an isometry. Explain.

1. 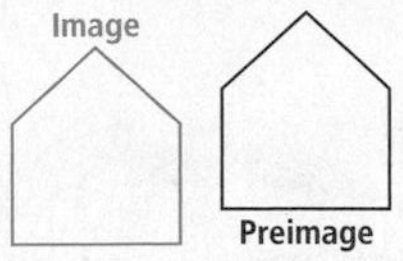

2. 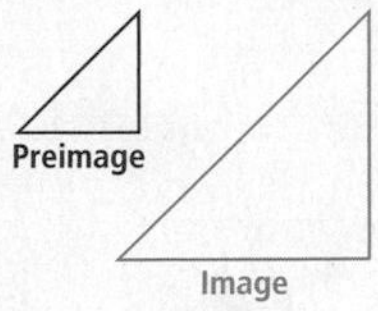

3. What rule describes the translation 5 units left and 10 units up?

4. Describe the translation $T_{<-3,5>}(x, y)$ in words.

5. Find a translation that has the same effect as the composition $T_{<-3,2>} \circ T_{<7,-2>}$.

Find each reflection image.

6. $R_{x\text{-axis}}(5, -3)$

7. $R_{y\text{-axis}}(5, -3)$

8. $\triangle WXY$ has vertices $W(-4, 1)$, $X(2, -7)$, and $Y(0, -3)$. What are the vertices of $T_{<-3,5>}(\triangle WXY)$?

9. $\triangle ABC$ has vertices $A(-1, 4)$, $B(2, 0)$, and $C(4, 3)$. Graph the reflection image of $\triangle ABC$ across the line $x = -1$.

Copy each figure and point *A*. Draw the image of each figure for the given angle of rotation about *A*. Use prime notation to label the vertices of the image.

10. 40°

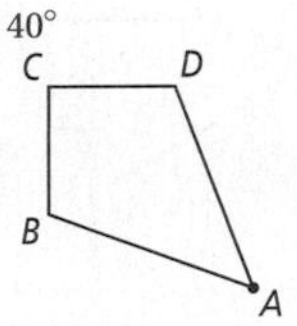

11. 90°

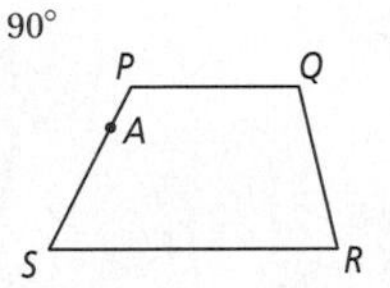

Graph $\triangle JKL$ and its image after each composition of transformations.

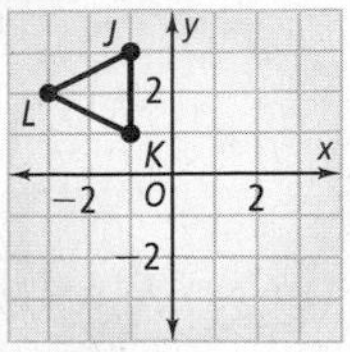

12. $(T_{<0,-4>} \circ R_{y\text{-axis}})(\triangle JKL)$

13. $(R_{x\text{-axis}} \circ r_{(180°,O)})(\triangle JKL)$

14. $(r_{(90°,O)} \circ T_{<-1,-3>})(\triangle JKL)$

Do you UNDERSTAND?

15. **Reasoning** The point $(5, -9)$ is the image under the translation $T_{<-3,2>}(x, y) = (5, -9)$. What is its preimage?

16. **Reasoning** The point $T(5, -1)$ is reflected first across the *x*-axis and then across the *y*-axis. What is the distance between the image and the preimage? Explain.

17. **Coordinate Geometry** The point $L(a, b)$ is in Quadrant I. What are the coordinates of the image of *L* after the composition of transformations $R_{y\text{-axis}} \circ r_{(90°,O)}$?

The two letters in each pair can be mapped onto each other. Does one figure appear to be a translation image, a reflection image, or a rotation image of the other?

18. 

19. 

20. **Error Analysis** Your friend draws the diagram at the right to show the reflection of ▱*PQRS* across the *x*-axis. Explain and correct your friend's error.

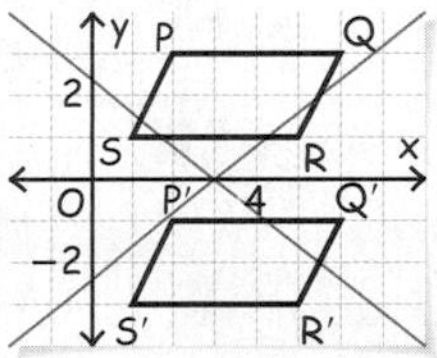

Answers

Mid-Chapter Quiz

1. Yes; the transformation is a translation.
2. No; the transformation is a dilation.
3. $T_{<-5,10>}(x, y)$
4. a translation of 3 units left and 5 units up
5. $T_{<4,0>}(x, y)$
6. (5, 3)
7. (−5, −3)
8. $W'(-7, 6)$, $X'(-1, -2)$, $Y'(-3, 2)$
9. 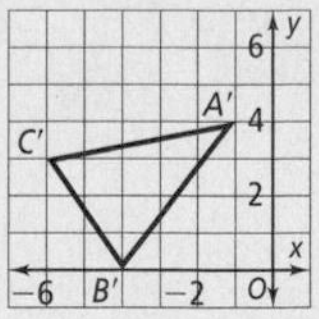

10. 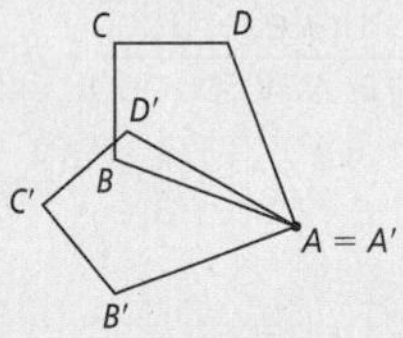

11. 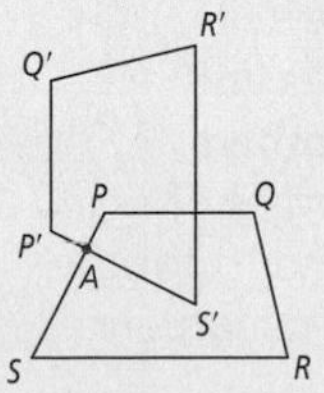

12. 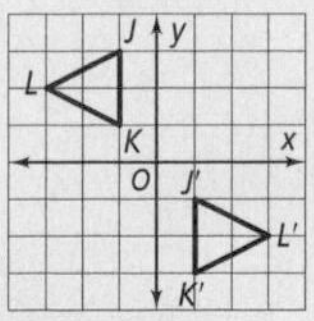

13. 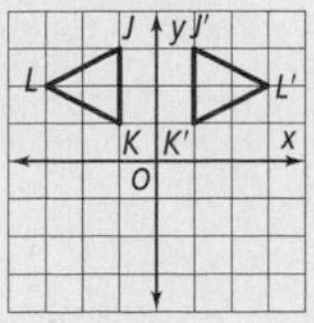

14. 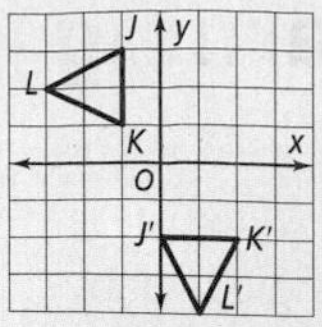

15. (8, −11)
16. $2\sqrt{26}$; The image is (−5, 1), and the distance between (5, −1) and (−5, 1) is $\sqrt{10^2 + (-2)^2} = \sqrt{104} = 2\sqrt{26}$.
17. (*b*, *a*)
18. rotation image
19. reflection image
20. Your friend translated *PQRS* down 4 units instead of reflecting it; the vertices of *PQRS* reflected across the *x*-axis are $P'(2, -3)$, $Q'(6, -3)$, $R'(5, -1)$, and $S'(1, -1)$.

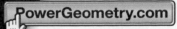

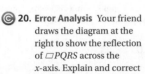

1 Interactive Learning

Solve It!

PURPOSE To develop an intuitive sense of congruence when working with rigid motions

PROCESS Students may determine that the two wings are identical because one was formed by tracing the other so they overlap exactly.

FACILITATE

Q Which angles do you think have the same measures in the two figures? Explain. **[corresponding angles; the figures are congruent]**

Q Which sides do you think have the same lengths in the two figures? Explain. **[corresponding sides; the figures are congruent]**

Problem 1

In this problem, students will identify corresponding angles and side lengths after a composition of rigid motions.

Q Would the answer change if trapezoid *LMNO* were rotated first and then reflected to create trapezoid *GHJK*? Explain. **[No, the corresponding angles and sides would still be the same.]**

Got It?

Q How can you use the order of the vertices of the preimage and image to solve the problem? **[The corresponding vertices are listed in order in the composition statement.]**

9-5 Congruence Transformations

Common Core State Standards

G-CO.B.7 Use the definition of congruence in terms of rigid motions to show that two triangles are congruent . . .
Also G-CO.B.6, G-CO.B.8
MP 1, MP 3, MP 4

Objective To identify congruence transformations
To prove triangle congruence using isometries

Are there other methods you could use to create two identical wings?

Getting Ready!

Suppose that you want to create two identical wings for a model airplane. You draw one wing on a large sheet of tracing paper, fold it along the dashed line, and then trace the first wing. How do you know that the two wings are identical?

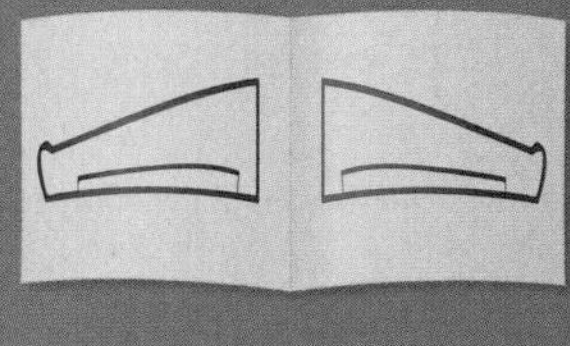

MATHEMATICAL PRACTICES

In the Solve It, you may have used the properties of rigid motions to describe why the wings are identical.

Lesson Vocabulary
- congruent
- congruence transformation

Essential Understanding You can use compositions of rigid motions to understand congruence.

Plan

How can you use the properties of isometries to find equal angle measures and equal side lengths?
Isometries preserve angle measure and distance, so identify corresponding angles and corresponding side lengths.

Problem 1 Identifying Equal Measures

The composition $(R_n \circ r_{(90°, P)})(LMNO) = GHJK$ is shown at the right.

A Which angle pairs have equal measures?

Because compositions of isometries preserve angle measure, corresponding angles have equal measures.

$m\angle L = m\angle G$, $m\angle M = m\angle H$, $m\angle N = m\angle J$, and $m\angle O = m\angle K$.

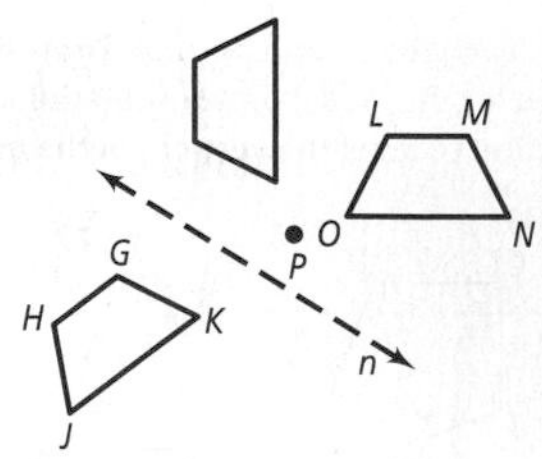

B Which sides have equal lengths?

By definition, isometries preserve distance. So, corresponding side lengths have equal measures.

$LM = GH$, $MN = HJ$, $NO = JK$, and $LO = GK$.

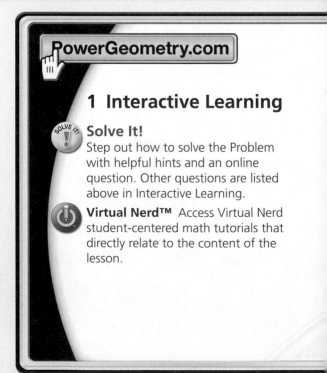

Got It? 1. The composition $(R_t \circ T_{<2, 3>})(\triangle ABC) = \triangle XYZ$. List all of the pairs of angles and sides with equal measures.

9-5 Preparing to Teach

BIG ideas Transformations
Visualization

ESSENTIAL UNDERSTANDING
- If two figures can be mapped to each other by a sequence of rigid motions, then the figures are congruent.

Math Background

In previous lessons, students have learned that two figures are congruent if and only if corresponding sides have the same length and corresponding angles have the same measure. In this lesson, students explore the concept of congruence transformations. Two figures are congruent if and only if there is a rigid motion or composition of rigid motions that maps one figure to the other.

This new approach to determining congruence can be used to verify postulates such as the SAS Postulate or the SSS Postulate. Students will learn that if there is a way to map one figure onto another through a series of rigid motions, then the figures are congruent.

Mathematical Practices

Construct viable arguments and critique the reasoning of others. Students will use the properties of rigid motions to construct arguments for the validity of the SAS and SSS congruence postulates.

PowerGeometry.com

1 Interactive Learning

Solve It! Step out how to solve the Problem with helpful hints and an online question. Other questions are listed above in Interactive Learning.

Virtual Nerd™ Access Virtual Nerd student-centered math tutorials that directly relate to the content of the lesson.

In Problem 1 you saw that compositions of rigid motions preserve corresponding side lengths and angle measures. This suggests another way to define congruence.

take note **Key Concept** Congruent Figures

Two figures are **congruent** if and only if there is a sequence of one or more rigid motions that maps one figure onto the other.

Problem 2 Identifying Congruent Figures

Think

Does one rigid motion count as a sequence?
Yes. It is a sequence of length 1.

Which pairs of figures in the grid are congruent? For each pair, what is a sequence of rigid motions that maps one figure to the other?

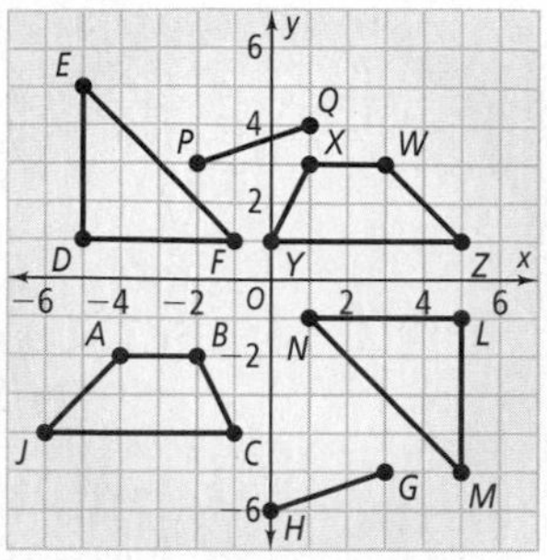

Figures are congruent if and only if there is a sequence of rigid motions that maps one figure to the other. So, to find congruent figures, look for sequences of translations, rotations, and reflections that map one figure to another.

Because $r_{(180°, O)}(\triangle DEF) = \triangle LMN$, the triangles are congruent.

Because $(T_{<-1, 5>} \circ R_{y\text{-axis}})(ABCJ) = WXYZ$, the trapezoids are congruent.

Because $T_{<-2, 9>}(\overline{HG}) = \overline{PQ}$, the line segments are congruent.

Got It? **2.** Which pairs of figures in the grid are congruent? For each pair, what is a sequence of rigid motions that map one figure to the other?

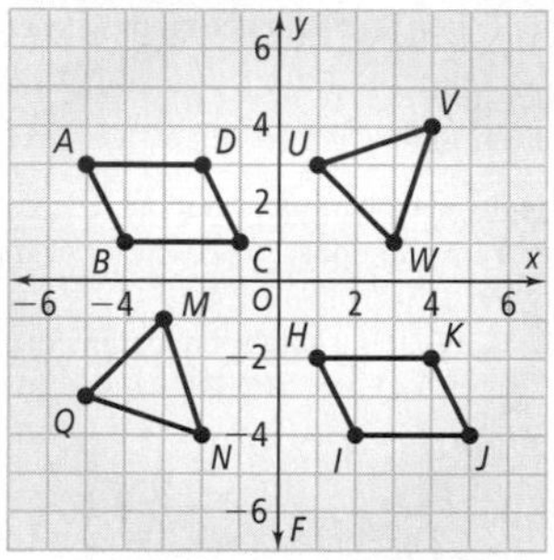

2 Guided Instruction

Take Note

EXTENSION

Discuss with students the differences between this definition and previously learned definitions of congruent figures.

Problem 2

VISUAL LEARNERS

Point out to students that the first step should always be to look for two figures that appear to have the same size and shape. Then to verify congruence, they must identify a rigid motion or composition of rigid motions that maps one figure to the other.

Q Is there another composition of rigid motions that you could have used to determine the congruence of $\triangle DEF$ and $\triangle LMN$? Explain. **[Yes. For example, a composition of reflections across both axes.]**

Got It?

Q How might you have verified the congruence of $\triangle UVW$ and $\triangle MNO$ prior to learning the methods of this lesson? Explain. **[Sample answer: Use the Distance Formula to show that the sides have the same lengths and then apply the SSS Postulate.]**

2 Guided Instruction

Each problem is worked out and supported online.

Problem 1
Identifying Equal Measures

Problem 2
Identifying Congruent Figures

Problem 3
Identifying Congruence Transformations

Problem 4
Verifying the SAS Postulate

Problem 5
Determining Congruence

Support in Geometry Companion
- Vocabulary
- Key Concepts
- Got It?

Answers

Solve It!

There is a rigid motion (reflection) from one wing onto the other. When you fold one side of the paper onto the other, the two wings coincide exactly.

Got It?

1. $m\angle A = m\angle X$, $m\angle B = m\angle Y$, $m\angle C = m\angle Z$; $AB = XY$, $BC = YZ$, $AC = XZ$
2. $\triangle UVW$ can be mapped to $\triangle QNM$ by a translation followed by a reflection. Parallelogram $ABCD$ can be mapped to $HIJK$ by a translation.

Problem 3

In this problem, students will identify a congruence transformation that maps one triangle onto another.

Q Is there a single rigid motion that can be used to map $\triangle JQV$ onto $\triangle EWT$? Explain. **[No, $\triangle JQV$ cannot be reflected, translated, or rotated to overlap $\triangle EWT$. A composition of rigid motions is necessary.]**

Got It?

Q What do you need to show to prove that the triangles are congruent? **[You need to show that there is a congruence transformation from one triangle onto the other]**

Q Looking at the orientation of the triangles, how do you think one triangle can be transformed to map it to the other triangle? **[Sample answer: translation and rotation]**

Because compositions of rigid motions take figures to congruent figures, they are also called **congruence transformations**.

Problem 3 Identifying Congruence Transformations

In the diagram at the right, $\triangle JQV \cong \triangle EWT$. What is a congruence transformation that maps $\triangle JQV$ onto $\triangle EWT$?

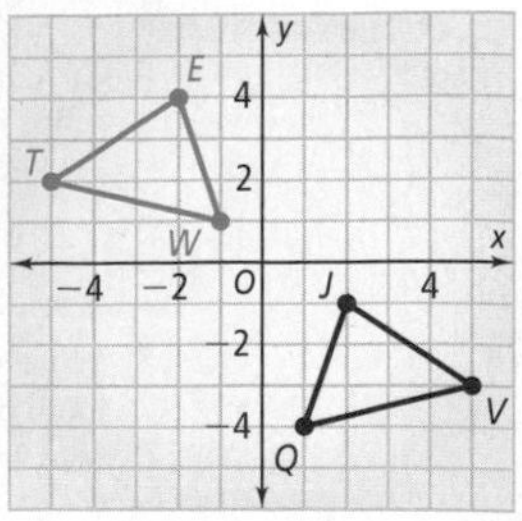

Know
The coordinates of the vertices of the triangles

Need
A sequence of rigid motions that maps $\triangle JQV$ onto $\triangle EWT$

Plan
Identify the corresponding parts and find a congruence transformation that maps the preimage to the image. Then use the vertices to verify the congruence transformation.

Because $\triangle EWT$ lies above $\triangle JQV$ on the plane, a translation can map $\triangle JQV$ up on the plane. Also, notice that $\triangle EWT$ is on the opposite side of the y-axis and has the opposite orientation of $\triangle JQV$. This suggests that the triangle is reflected across the y-axis.

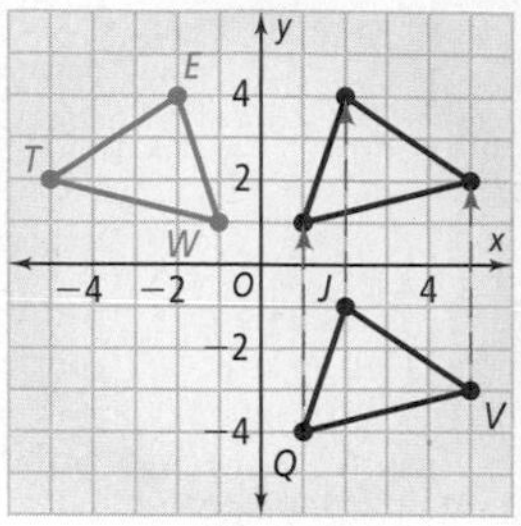

It appears that a translation of $\triangle JQV$ up 5 units, followed by a reflection across the y-axis maps $\triangle JQV$ to $\triangle EWT$. Verify by using the coordinates of the vertices.

$$T_{<0,\,5>}(x, y) = (x, y + 5)$$
$$T_{<0,\,5>}(J) = (2, 4)$$
$$R_{y\text{-axis}}(2, 4) = (-2, 4) = E$$

Next, verify that the sequence maps Q to W and V to T.

$$T_{<0,\,5>}(Q) = (1, 1) \qquad T_{<0,\,5>}(V) = (5, 2)$$
$$R_{y\text{-axis}}(1, 1) = (-1, 1) = W \qquad R_{y\text{-axis}}(5, 2) = (-5, 2) = T$$

So, the congruence transformation $R_{y\text{-axis}} \circ T_{<0,\,5>}$ maps $\triangle JQV$ onto $\triangle EWT$. Note that there are other possible congruence transformations that map $\triangle JQV$ onto $\triangle EWT$.

Got It? 3. What is a congruence transformation that maps $\triangle NAV$ to $\triangle BCY$?

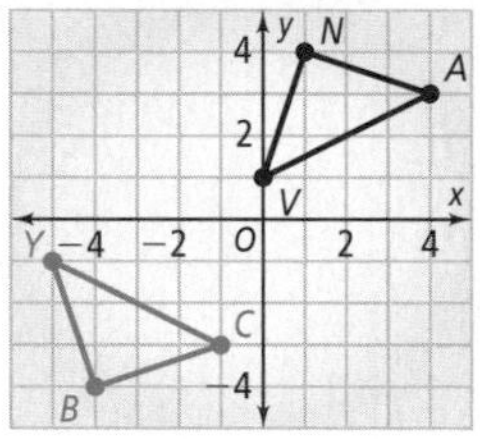

Additional Problems

1. Let $(r_{(180°,\,O)} \circ T_{<-5,\,6>})(\triangle QMC) = \triangle APT$. List all of the pairs of angles and sides with equal measures.
ANSWER $m\angle Q = m\angle A$, $m\angle M = m\angle P$, $m\angle C = m\angle T$; $QM = AP$, $MC = PT$, $QC = AT$

2. Identify a pair of congruent figures on the coordinate grid. Describe a sequence of rigid motions that maps one figure onto the other.

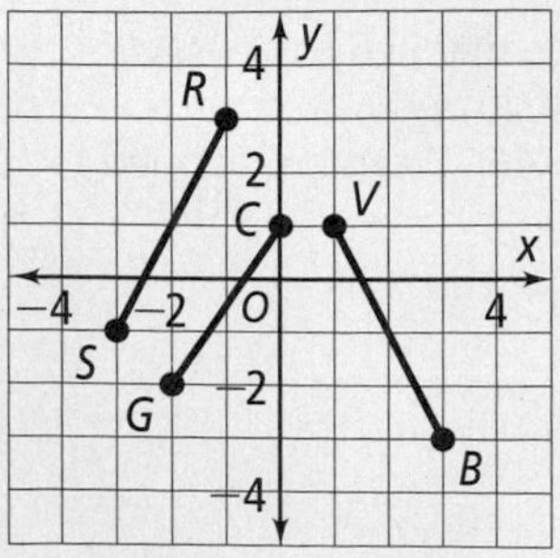

ANSWER $\overline{RS} \cong \overline{VB}$; $(R_{y\text{-axis}} \circ T_{<0,\,-2>})(\overline{RS}) = \overline{VB}$

3. Identify the congruence transformation that maps $\triangle WAN$ to $\triangle PQR$.

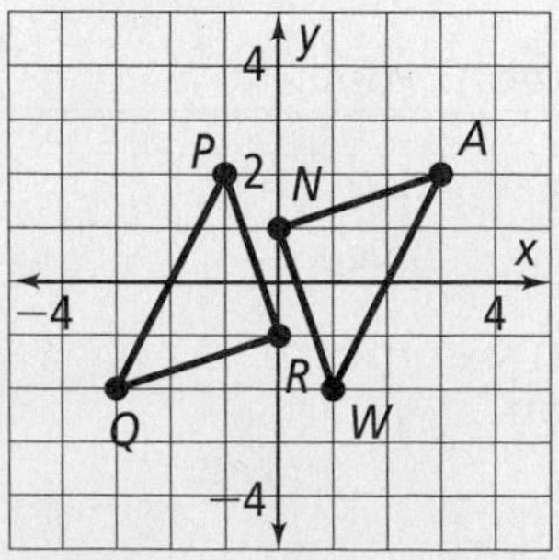

ANSWER $r_{(180°,\,O)}(\triangle WAN) = \triangle PQR$

4. Given: $\overline{BN} \cong \overline{YT}$, $\overline{NG} \cong \overline{TZ}$, $\overline{BG} \cong \overline{YZ}$
Prove: $\triangle BNG \cong \triangle YTZ$

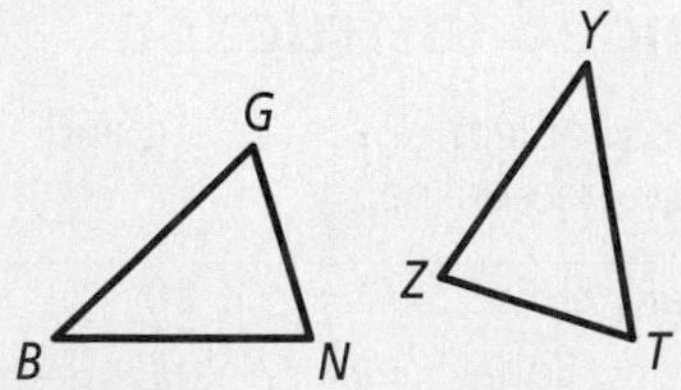

ANSWER Check students' work.

5. Are the figures below congruent? Explain.

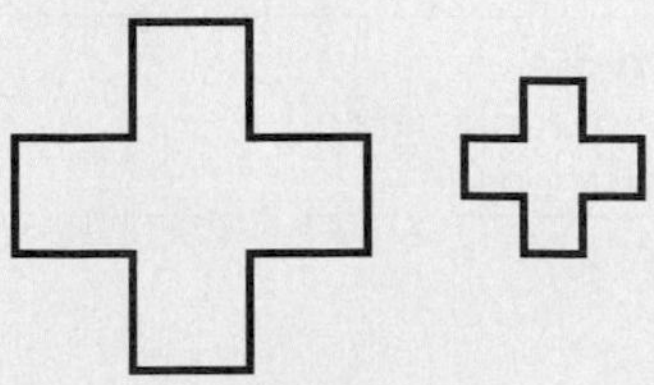

ANSWER No, there is no congruence transformation that maps one to the other.

In Chapter 4, you studied triangle congruence postulates and theorems. You can use congruence transformations to justify criteria for determining triangle congruence.

Proof

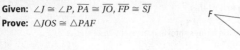

Problem 4 Verifying the SAS Postulate

Think

How do you show that the two triangles are congruent?
Find a congruence transformation that maps one onto the other.

Given: $\angle J \cong \angle P$, $\overline{PA} \cong \overline{JO}$, $\overline{FP} \cong \overline{SJ}$

Prove: $\triangle JOS \cong \triangle PAF$

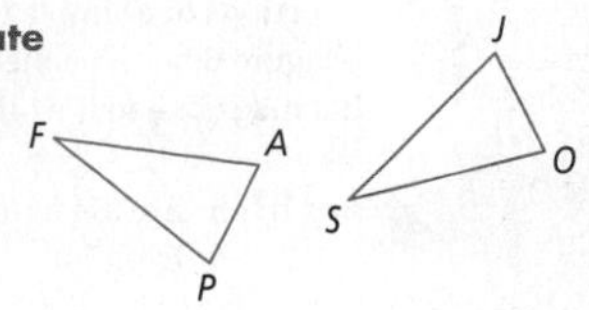

Step 1 Translate $\triangle PAF$ so that points A and O coincide.

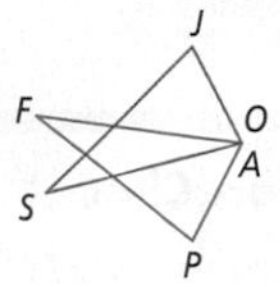

Step 2 Because $\overline{PA} \cong \overline{JO}$, you can rotate $\triangle PAF$ about point A so that $\overline{PA}$ and $\overline{JO}$ coincide.

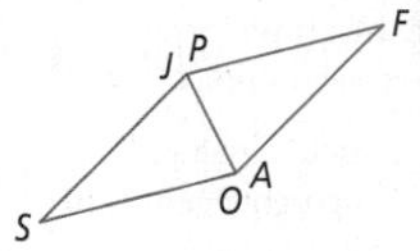

Step 3 Reflect $\triangle PAF$ across $\overline{PA}$. Because reflections preserve angle measure and distance, and because $\angle J \cong \angle P$ and $\overline{FP} \cong \overline{SJ}$, you know that the reflection maps $\angle P$ to $\angle J$ and $\overline{FP}$ to $\overline{SJ}$. Since points S and F coincide, $\triangle PAF$ coincides with $\triangle JOS$.

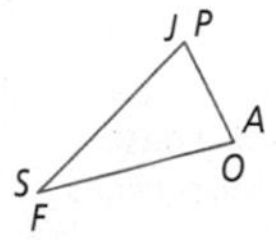

There is a congruence transformation that maps $\triangle PAF$ onto $\triangle JOS$, so $\triangle PAF \cong \triangle JOS$.

Got It? 4. Verify the SSS postulate.

Given: $\overline{TD} \cong \overline{EN}$, $\overline{YT} \cong \overline{SE}$, $\overline{YD} \cong \overline{SN}$

Prove: $\triangle YDT \cong \triangle SNE$

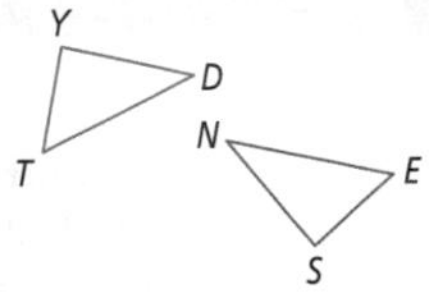

In Problem 4, you used the transformational approach to prove triangle congruence. Because this approach is more general, you can use what you know about congruence transformations to determine whether any two figures are congruent.

Problem 4 EXTENSION

In this problem, students will verify the SAS Postulate for proving triangle congruence by finding a congruence transformation that maps one triangle onto the other.

Q Do you think you could use similar methods to verify the SSS Postulate, the ASA Postulate, and the AAS Theorem? Explain. **[Yes, the only difference would be the given information. In all cases, you would simply identify a congruence transformation that maps one triangle to the other.]**

Got It?

Q If a friend is having difficulty seeing the congruence transformation in this problem, how might you help him or her? **[Sample answer: Suggest tracing the triangles on tracing paper and cutting them out. Then slide, flip, and turn the cut outs until they overlap.]**

Answers

Got It? (continued)

3. Answers may vary. Sample: translation 5 units left, reflection across the x-axis

4. Answers may vary. Sample: Translate $\triangle YDT$ so that point D and point N coincide. Since $\overline{TD} \cong \overline{EN}$, you can rotate $\triangle YDT$ so that $\overline{TD}$ and $\overline{EN}$ coincide. Since rotations preserve angle measure and distance, the other two pairs of sides will also coincide. Therefore, this composition of a translation followed by a rotation maps $\triangle YDT$ to $\triangle SNE$, and $\triangle YDT \cong \triangle SNE$.

Problem 5

In this problem, students will determine the congruence of two plane figures.

Challenge students to see how many different ways they can show that figures A and B are congruent.

Got It?

Q How can you show that two figures are not congruent? **[If corresponding distances are not equal, figures are not congruent.]**

3 Lesson Check

Do you know HOW? ERROR INTERVENTION

- If students have trouble solving Exercise 1, then ask them which triangles could be placed on top of each other so that they coincide.

Do you UNDERSTAND?

- Have students share their responses to Exercise 5 with the rest of the class so that everyone is exposed to different real world examples of congruence transformations.

Close

Q What do you know about corresponding sides and angles of figures that are mapped using a composition of rigid motions? **[They have equal measures.]**

Q How can you show that two figures are congruent? **[Show that there is a sequence of rigid motions that maps one figure onto the other.]**

Q Suppose two figures are congruent. What do you know about how the figures are related in the plane? **[There is a congruence transformation that maps one figure onto the other.]**

Plan

How can you determine whether the figures are congruent?
You can find a congruence transformation that maps Figure A onto Figure B.

Problem 5 Determining Congruence

Is Figure A congruent to Figure B? Explain how you know.

Figure A can be mapped to Figure B by a sequence of reflections or a simple translation. So, Figure A is congruent to Figure B because there is a congruence transformation that maps one to the other.

Figure A

Figure B

Got It? 5. Are the figures shown at the right congruent? Explain.

Lesson Check

Do you know HOW?

Use the graph for Exercises 1 and 2.

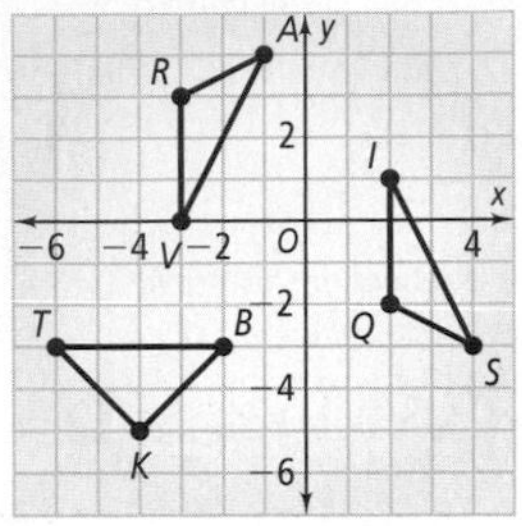

1. Identify a pair of congruent figures and write a congruence statement.
2. What is a congruence transformation that relates two congruent figures?

Do you UNDERSTAND? MATHEMATICAL PRACTICES

3. How can the definition of congruence in terms of rigid motions be more useful than a definition of congruence that relies on corresponding angles and sides?
4. **Reasoning** Is a composition of a rotation followed by a glide reflection a congruence transformation? Explain.
5. **Open Ended** What is an example of a board game in which a game piece is moved by using a congruence transformation?

Practice and Problem-Solving Exercises MATHEMATICAL PRACTICES

A Practice **For each coordinate grid, identify a pair of congruent figures. Then determine a congruence transformation that maps the preimage to the congruent image.**

See Problem 1 and 2.

6.
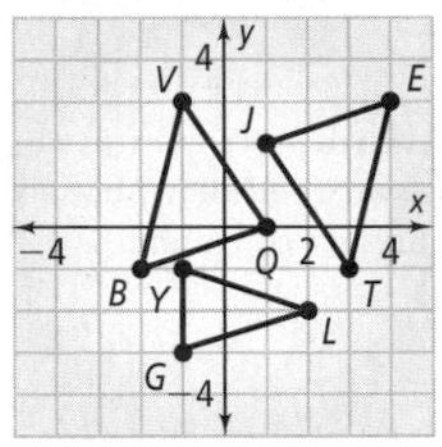

7.
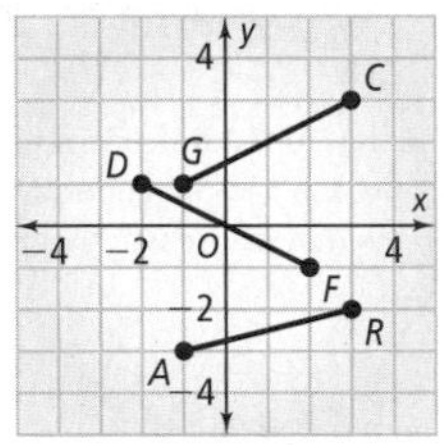

8.
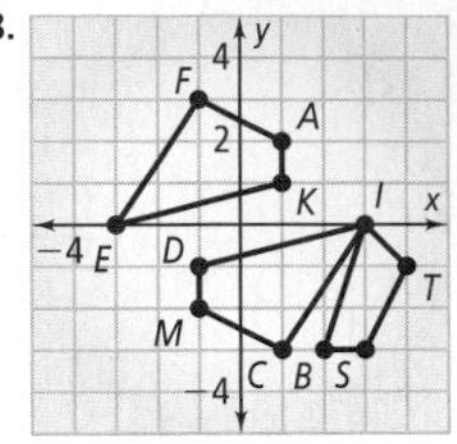

PowerGeometry.com

3 Lesson Check

For a digital lesson check, use the Got It questions.

Support in Geometry Companion
- Lesson Check

4 Practice

Assign homework to individual students or to an entire class.

Answers

Got It? (continued)

5. No, there is no congruence transformation from one figure onto the other.

Lesson Check

1. $\triangle RAV \cong \triangle QSI$
2. Answers may vary. Sample answer: Translate triangle *RAV* 5 units right and 1 unit down; then reflect across the *x*-axis.
3. Answers may vary. Sample answer: Using transformations, you can define congruence of figures other than polygons.
4. Yes, because a rotation is a rigid motion and a glide reflection is a composition of a translation and reflection, so a rotation followed by a glide reflection is a congruence transformation.
5. Sample answer: The game of chess requires that the chess pieces move on the board by using congruence transformations.

Practice and Problem-Solving Exercises

6. $\triangle BVQ \cong \triangle ETJ$; Sample answer: Rotate triangle *BVQ* 180° about the origin; then translate 2 units right and 2 units up.
7. $\overline{GC} \cong \overline{FD}$; Sample answer: Reflect segment *GC* over the *y*-axis; then translate 1 unit right and 2 units down.
8. *FAKE* ≅ *CMDI*; Sample answer: Rotate *FAKE* 180° about the origin.

In Exercises 9–11, find a congruence transformation that maps $\triangle LMN$ to $\triangle RST$. See Problem 3.

9.

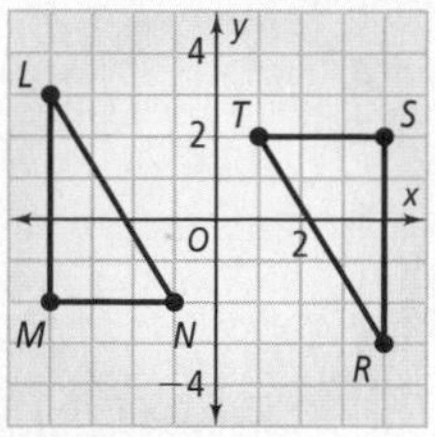

10.

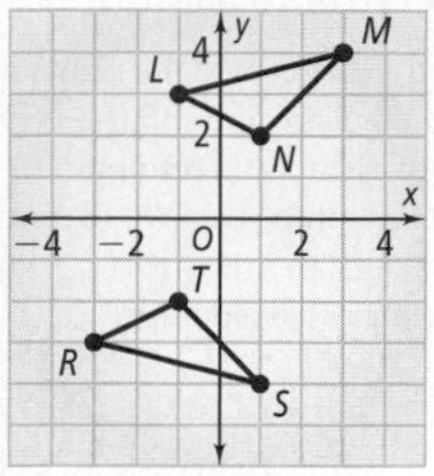

11.

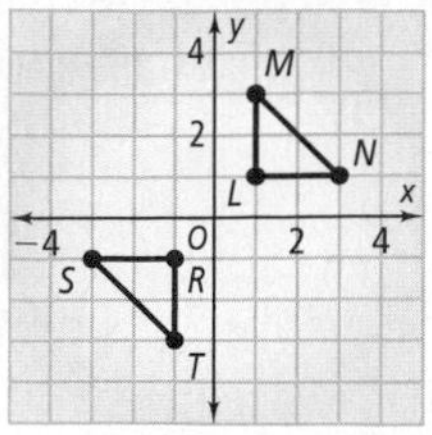

12. Verify the ASA Postulate for triangle congruence by using congruence transformations. See Problem 4.

Proof

Given: $\overline{EK} \cong \overline{LH}$ **Prove:** $\triangle EKS \cong \triangle HLA$

$\angle E \cong \angle H$

$\angle K \cong \angle L$

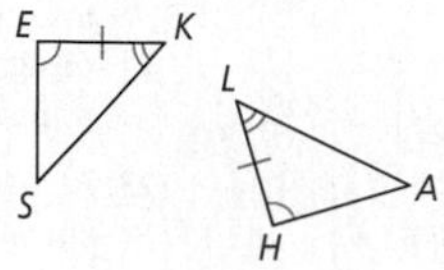

13. Verify the AAS Postulate for triangle congruence by using congruence transformations.

Proof

Given: $\angle I \cong \angle V$ **Prove:** $\triangle NVZ \cong \triangle CIQ$

$\angle C \cong \angle N$

$\overline{QC} \cong \overline{NZ}$

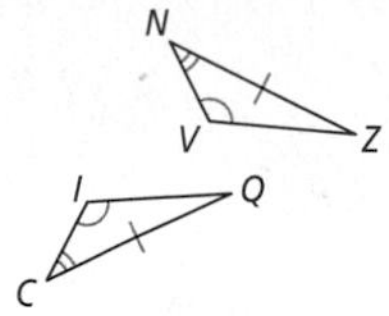

In Exercises 14–16, determine whether the figures are congruent. If so, describe a congruence transformation that maps one to the other. If not, explain. See Problem 5.

14.

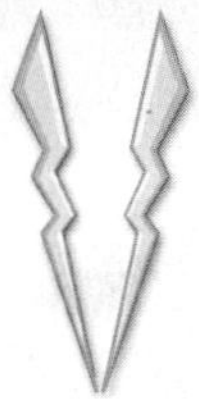

15.

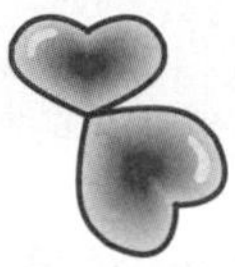

16.

B Apply

Construction **The figure at the right shows a roof truss of a new building. Identify an isometry or composition of isometries to justify each of the following statements.**

17. Triangle 1 is congruent to triangle 3.

18. Triangle 1 is congruent to triangle 4.

19. Triangle 2 is congruent to triangle 5.

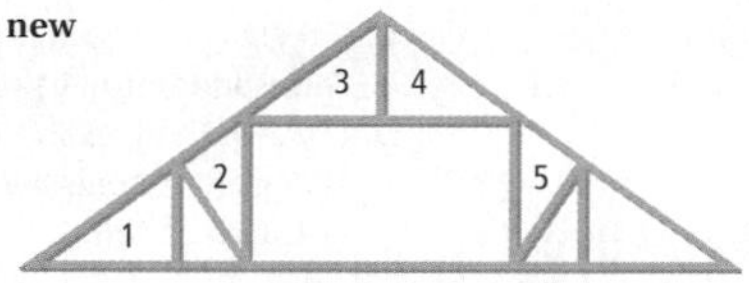

4 Practice

ASSIGNMENT GUIDE

Basic: 6–16, 18–20 even, 21, 22–26 even

Average: 7–15 odd, 17–27

Advanced: 7–15 odd, 17–28

Standardized Test Prep: 29–32

Mixed Review: 33–40

Mathematical Practices are supported by exercises with red headings. Here are the Practices supported in this lesson:

MP 1: Make sense of Problems Ex. 21

MP 3: Construct Arguments Ex. 4, 28

MP 4: Model Ex. 5

Applications exercises have blue headings. Exercises 22 and 23 support MP 4: Model.

EXERCISE 27: Use the Think About a Plan worksheet in the **Practice and Problem Solving Workbook** (also available in the Teaching Resources in print and online) to further support students' development in becoming independent learners.

HOMEWORK QUICK CHECK

To check students' understanding of key skills and concepts, go over Exercises 11, 18, 20, 21, and 25.

9. Sample answer: Rotate $\triangle LMN$ 180° about the origin.

10. Sample answer: Reflect $\triangle LMN$ 180° over the x-axis; then translate 2 units to the left.

11. Sample answer: Rotate $\triangle LMN$ 180° about the origin.

12. Answers will vary. Sample answer: Rotate $\triangle LAH$ so that side HA is parallel to side ES. Translate $\triangle LAH$ so that points E and H coincide. $\triangle EKS \cong \triangle HLA$

13. Answers will vary. Sample answer: Translate $\triangle IQC$ so that points Q and Z coincide. Rotate $\triangle IQC$ so that sides QC and NZ coincide. Reflect $\triangle IQC$ across side NZ so that the triangles overlap. $\triangle IQC \cong \triangle VZN$

14. yes, reflection

15. no, there is no congruence transformation

16. no, there is no congruence transformation

17. translation

18. translation and reflection

19. reflection

Answers

Practice and Problem-Solving Exercises (continued)

20. Answers may vary. Sample: Translate and rotate $\triangle ABC$ so that $\overline{AB}$ coincides with $\overline{DE}$. Because rigid motions preserve angle measures, $\overline{BC}$ and $\overline{EF}$ lie along the same line. If points B and E are on opposite sides of $\overline{AB}$, reflect $\triangle ABC$ across $\overline{AB}$. Suppose that points E and F do not coincide. Then consider $\triangle ACF$ By assumption $\overline{DF} \cong \overline{AC}$, so $\overline{AF} \cong \overline{AC}$. By the Isosceles Triangle Theorem, $\angle ACF \cong \angle AFC$. One of these angles must be obtuse since it is the exterior angle of a right triangle. But $\angle ACF$ and $\angle AFC$ cannot both be obtuse, so the assumption that points C and F do not coincide must be wrong. So points C and F coincide, and therefore is a sequence of rigid transformations that maps $\triangle ABC \cong \triangle DEF$.

21. Answers may vary. Sample: Translate the top triangle down 6 units; reflect across the x-axis; rotate the bottom triangle 180° about the point $(-3, 0)$, then reflect across the line $x = -3$; reflect the bottom triangle across the line $x = -4$, then rotate 180° about the point $(-4, 0)$.

22. reflection or rotation

23a. rotations and glide reflections

b. translations and glide reflections

24. glide reflection

25a. congruence transformations preserve distances and angle measures

b. Use SAS, proven in Problem 4.

20. Let $\triangle ABC$ and $\triangle DEF$ be right triangles with $\overline{AB} \cong \overline{DE}$, $\overline{AC} \cong \overline{DF}$, and $m\angle B = m\angle E = 90°$. Prove the Hypotenuse-Leg Theorem by showing that there is a sequence of rigid motions that maps $\triangle ABC$ to $\triangle DEF$.

21. Think About a Plan The figure at the right shows two congruent, isosceles triangles. What are four different isometries that map the top triangle onto the bottom triangle?

- How can you use the three basic rigid motions to map the top triangle onto the bottom triangle?
- What other isometries can you use?

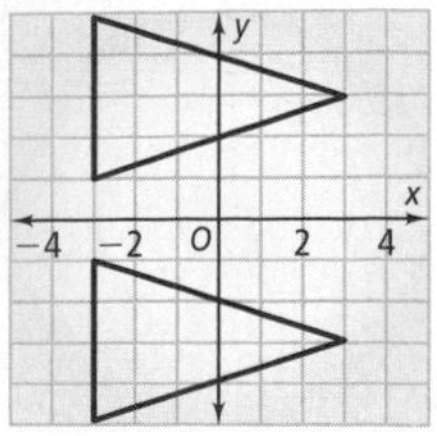

22. Graphic Design Most companies have a logo that is used on company letterhead and signs. A graphic designer sketches the logo at the right. What congruence transformations might she have used to draw this logo?

23. Art Artists frequently use congruence transformations in their work. The artworks shown below are called *tessellations*. What types of congruence transformations can you identify in the tessellations?

a.

b.

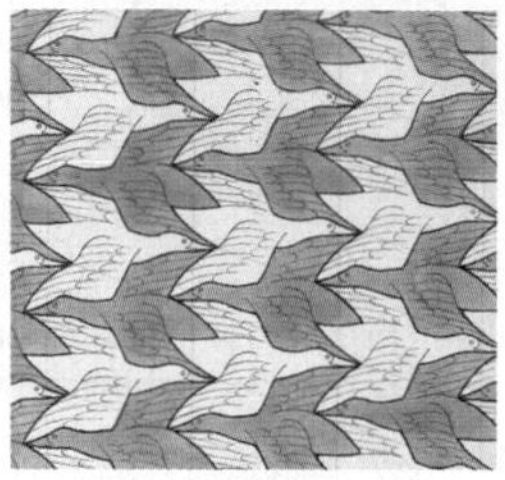

24. In the footprints shown below, what congruence transformations can you use to extend the footsteps?

25. Proof Prove the statements in parts (a) and (b) to show congruence in terms of transformations is equivalent to the criteria for triangle congruence you learned in Chapter 4.

a. If there is a congruence transformation that maps $\triangle ABC$ to $\triangle DEF$ then corresponding pairs of sides and corresponding pairs of angles are congruent.

b. In $\triangle ABC$ and $\triangle DEF$, if corresponding pairs of sides and corresponding pairs of angles are congruent, then there is a congruence transformation that maps $\triangle ABC$ to $\triangle DEF$.

26. **Baking** Cookie makers often use a cookie press so that the cookies all look the same. The baker fills a cookie sheet for baking in the pattern shown. What types of congruence transformations are being used to set each cookie on the sheet?

27. Use congruence transformations to prove the Isosceles Triangle Theorem.
Proof

Given: $\overline{FG} \cong \overline{FH}$

Prove: $\angle G \cong \angle H$

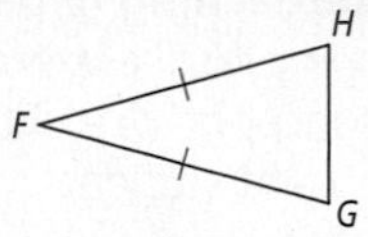

Challenge

28. **Reasoning** You project an image for viewing in a large classroom. Is the projection of the image an example of a congruence transformation? Explain your reasoning.

Standardized Test Prep

GRIDDED RESPONSE

SAT/ACT

29. To the nearest hundredth, what is the value of x in the diagram at the right?

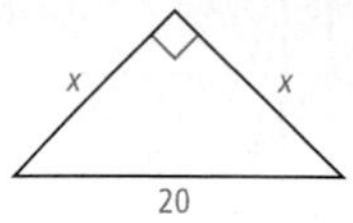

30. In $\triangle FGH$ and $\triangle XYZ$, $\angle G$ and $\angle Y$ are right angles. $\overline{FH} \cong \overline{XZ}$ and $\overline{GH} \cong \overline{YZ}$. If $GH = 7$ ft and $XY = 9$ ft, what is the area of $\triangle FGH$ in square inches?

31. $\triangle ACB$ is isosceles with base $\overline{AB}$. Point D is on $\overline{AB}$ and $\overline{CD}$ is the bisector of $\angle C$. If $CD = 5$ in. and $DB = 4$ in., what is BC to the nearest tenth of an inch?

32. Two angle measures of $\triangle JKL$ are 30 and 60. The shortest side measures 10 cm. What is the length, in centimeters, of the longest side of the triangle?

Mixed Review

33. A triangle has vertices $A(3, 2)$, $B(4, 1)$, and $C(4, 3)$. Find the coordinates of the images of A, B, and C for a glide reflection with translation $(x, y) \rightarrow (x, y + 1)$ and reflection line $x = 0$.

See Lesson 9-4.

The lengths of two sides of a triangle are given. What are the possible lengths for the third side?

See Lesson 5-6.

34. 16 in., 26 in.
35. 19.5 ft, 20.5 ft
36. 9 m, 9 m
37. $4\frac{1}{2}$ yd, 8 yd

Get Ready! **To prepare for Lessons 9-6, do Exercises 38–40.**

Determine the scale drawing dimensions of each room using a scale of $\frac{1}{4}$ in. = 1 ft.

See Lesson 7-2.

38. kitchen: 12 ft by 16 ft
39. bedroom: 8 ft by 10 ft
40. laundry room: 6 ft by 9 ft

26. Sample answer: reflections, translations, rotations, glide reflections

27. Sample answer: Draw and label the midpoint of $\overline{GH}$ as point M. Draw $\overline{FM}$. Using the identity mapping, $\overline{FM} \cong \overline{FM}$. It is given that $\overline{FG} \cong \overline{FH}$, and $\overline{GM} \cong \overline{MH}$ by the definition of midpoint. Therefore, if triangle FHM is reflected across $\overline{FM}$, it will overlap triangle FGM. Because there is an isometry mapping triangle FHM onto triangle FGM, $\triangle FHM \cong \triangle FGM$. Therefore, $\angle G \cong \angle H$ because corresp. parts of $\cong$ ⚠ are $\cong$.

28. Sample answer: No, the image is likely enlarged to be seen more easily so it is not congruent to the preimage.

29. 14.14

30. 31.5

31. 6.4

32. 20

33. $A'(-3, 3)$, $B'(-4, 2)$, $C'(-4, 4)$

34. greater than 10 in. and less than 42 in.

35. greater than 1 ft and less than 40 ft

36. greater than 0 m and less than 18 m

37. greater than 3.5 yd and less than 12.5 yd

38. 3 in. by 4 in.

39. 2 in. by 2.5 in.

40. 1.5 in. by 2.25 in.

9-5 Lesson Resources

Additional Instructional Support

Geometry Companion

Students can use the **Geometry Companion** worktext (4 pages) as you teach the lesson. Use the Companion to support

- Solve It!
- New Vocabulary
- Key Concepts

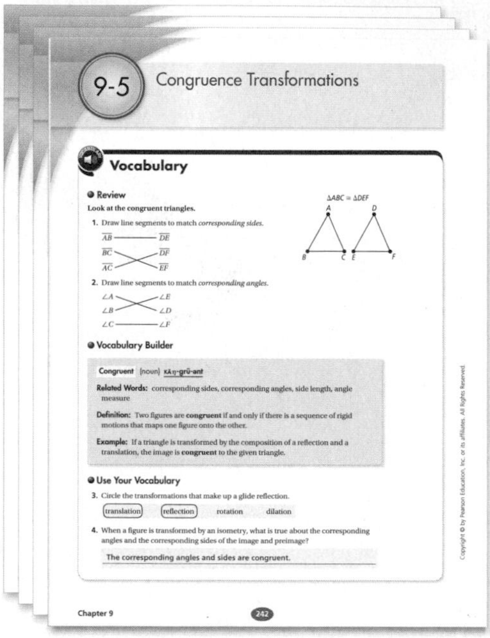

9-5 Congruence Transformations

Vocabulary

Review

Look at the congruent triangles.

$\triangle ABC \cong \triangle DEF$

1. Draw line segments to match *corresponding sides*.

$\overline{AB}$ — $\overline{DE}$
$\overline{BC}$ — $\overline{DF}$
$\overline{AC}$ — $\overline{EF}$

2. Draw line segments to match *corresponding angles*.

$\angle A$ — $\angle E$
$\angle B$ — $\angle D$
$\angle C$ — $\angle F$

Vocabulary Builder

Congruent (noun) KAŋ-grū-ant

Related Words: corresponding sides, corresponding angles, side length, angle measure

Definition: Two figures are **congruent** if and only if there is a sequence of rigid motions that maps one figure onto the other.

Example: If a triangle is transformed by the composition of a reflection and a translation, the image is **congruent** to the given triangle.

Use Your Vocabulary

3. Circle the transformations that make up a glide reflection.

translation reflection rotation dilation

4. When a figure is transformed by an isometry, what is true about the corresponding angles and the corresponding sides of the image and preimage?

The corresponding angles and sides are congruent.

Chapter 9 242

ELL Support

Connect to Prior Knowledge of Math

Review with students the Reflexive, Symmetric, and Transitive Properties of Congruence. Discuss with students how you can use congruence transformations to justify these properties.

- Reflexive Property – The identity mapping from a figure onto itself can be used to demonstrate the Reflexive Property of Congruence.
- Symmetric Property – If an isometry is performed on a figure, then the inverse isometry can be performed to map the image back onto the preimage.
- Transitive Property – In a composition of two isometries, the intermediate image (Figure A') is congruent to the preimage (Figure A) and the final image (Figure A'') is congruent to the intermediate image (Figure A'). Because there is a congruence transformation that maps Figure A onto Figure A'', all three figures are congruent.

5 Assess & Remediate

Lesson Quiz

Use the Lesson Quiz to assess students' mastery of the skills and concepts of this lesson.

1. What is a congruence transformation that maps $\triangle KNO$ to $\triangle AWN$?

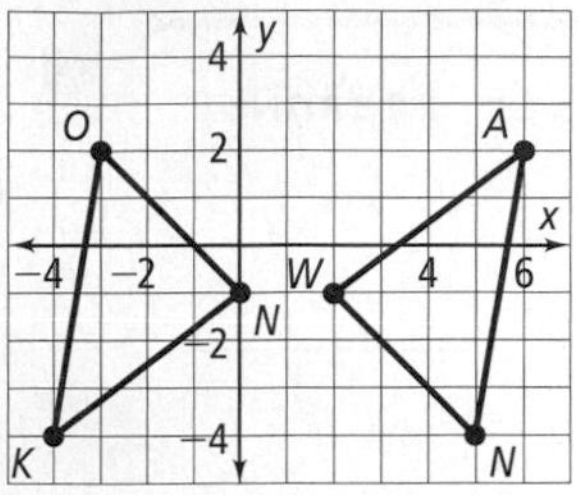

2. **Do you UNDERSTAND?** Is there only one congruence transformation that maps one figure to another figure? Explain.
3. Are the figures below congruent? Explain.

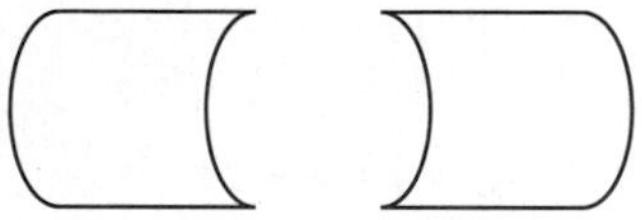

ANSWERS TO LESSON QUIZ

1. Sample answer: $(T_{<2,\,-2>} \circ r_{(180°,\,O)})(\triangle KNO) = \triangle AWN$.
2. Sample answer: No, there are typically other congruence transformations that map one figure to another. For example, any translation or rotation can be described as a composition of reflections.
3. Yes, there is a congruence transformation (reflection) from one figure to the other.

PRESCRIPTION FOR REMEDIATION

Use the student work on the Lesson Quiz to prescribe a differentiated review assignment.

Points	Differentiated Remediation
0–1	Intervention
2	On-level
3	Extension

PowerGeometry.com

5 Assess & Remediate

Assign the Lesson Quiz. Appropriate intervention, practice, or enrichment is automatically generated based on student performance.

Differentiated Remediation

Intervention

- **Reteaching** (2 pages) Provides reteaching and practice exercises for the key lesson concepts. Use with struggling students or absent students.
- **English Language Learner Support** Helps students develop and reinforce mathematical vocabulary and key concepts.

All-in-One Resources/Online

Reteaching

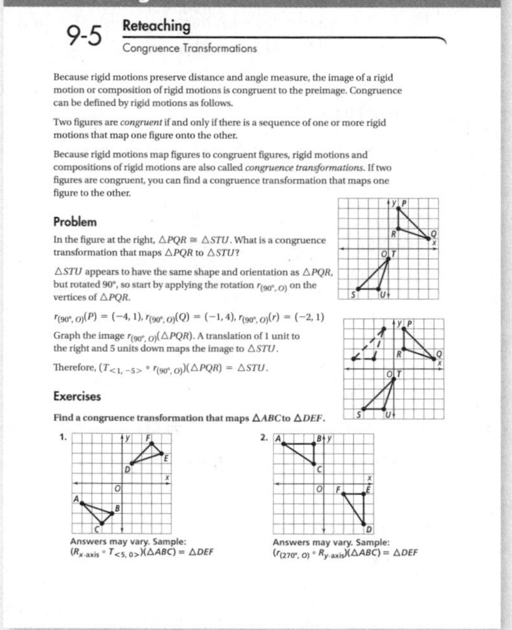

9-5 Reteaching
Congruence Transformations

Because rigid motions preserve distance and angle measure, the image of a rigid motion or composition of rigid motions is congruent to the preimage. Congruence can be defined by rigid motions as follows.

Two figures are *congruent* if and only if there is a sequence of one or more rigid motions that map one figure onto the other.

Because rigid motions map figures to congruent figures, rigid motions and compositions of rigid motions are also called *congruence transformations*. If two figures are congruent, you can find a congruence transformation that maps one figure to the other.

Problem

In the figure at the right, $\triangle PQR \cong \triangle STU$. What is a congruence transformation that maps $\triangle PQR$ to $\triangle STU$?

$\triangle STU$ appears to have the same shape and orientation as $\triangle PQR$, but rotated 90°, so start by applying the rotation $r_{(90°,\,O)}$ on the vertices of $\triangle PQR$.

$r_{(90°,\,O)}(P) = (-4, 1)$, $r_{(90°,\,O)}(Q) = (-1, 4)$, $r_{(90°,\,O)}(r) = (-2, 1)$

Graph the image $r_{(90°,\,O)}(\triangle PQR)$. A translation of 1 unit to the right and 5 units down maps the image to $\triangle STU$.

Therefore, $(T_{<1,\,-5>} \circ r_{(90°,\,O)})(\triangle PQR) = \triangle STU$.

Exercises

Find a congruence transformation that maps $\triangle ABC$ to $\triangle DEF$.

1. Answers may vary. Sample: $(R_{x\text{-axis}} \circ T_{<5,\,0>})(\triangle ABC) = \triangle DEF$

2. Answers may vary. Sample: $(r_{(270°,\,O)} \circ R_{y\text{-axis}})(\triangle ABC) = \triangle DEF$

All-in-One Resources/Online

English Language Learner Support

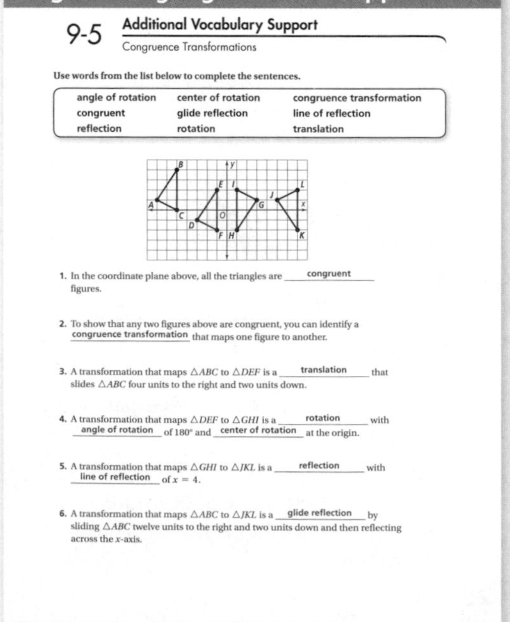

9-5 Additional Vocabulary Support
Congruence Transformations

Use words from the list below to complete the sentences.

angle of rotation	center of rotation	congruence transformation
congruent	glide reflection	line of reflection
reflection	rotation	translation

1. In the coordinate plane above, all the triangles are congruent figures.
2. To show that any two figures above are congruent, you can identify a congruence transformation that maps one figure to another.
3. A transformation that maps $\triangle ABC$ to $\triangle DEF$ is a translation that slides $\triangle ABC$ four units to the right and two units down.
4. A transformation that maps $\triangle DEF$ to $\triangle GHI$ is a rotation with angle of rotation of 180° and center of rotation at the origin.
5. A transformation that maps $\triangle GHI$ to $\triangle JKL$ is a reflection with line of reflection of $x = 4$.
6. A transformation that maps $\triangle ABC$ to $\triangle JKL$ is a glide reflection by sliding $\triangle ABC$ twelve units to the right and two units down and then reflecting across the x-axis.

Differentiated Remediation *continued*

On-Level

- **Practice** (2 pages) Provides extra practice for each lesson. For simpler practice exercises, use the Form K Practice pages found in the All-in-One Teaching Resources and online.
- **Think About a Plan** Helps students develop specific problem-solving skills and strategies by providing scaffolded guiding questions.
- **Standardized Test Prep** Focuses on all major exercises, all major question types, and helps students prepare for the high-stakes assessments.

Extension

- **Enrichment** Provides students with interesting problems and activities that extend the concepts of the lesson.
- **Activities, Games, and Puzzles** Worksheets that can be used for concepts development, enrichment, and for fun!

Practice and Problem Solving Wkbk/ All-in-One Resources/Online

Practice page 1

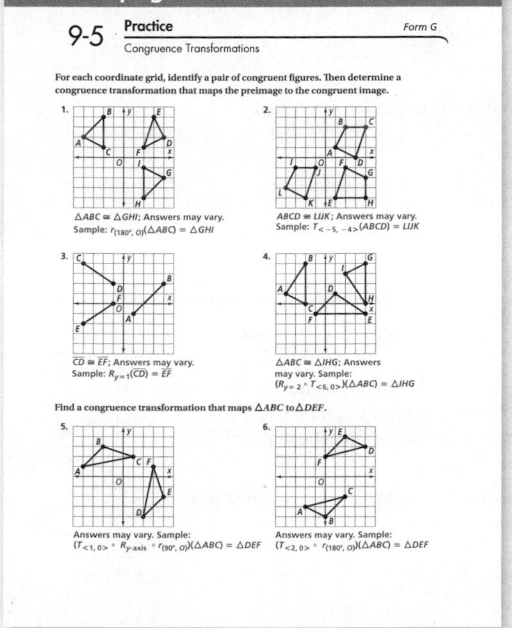

Practice and Problem Solving Wkbk/ All-in-One Resources/Online

Practice page 2

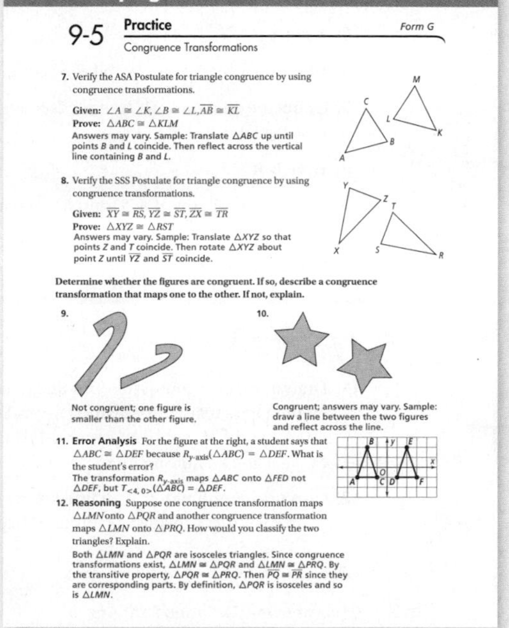

All-in-One Resources/Online

Enrichment

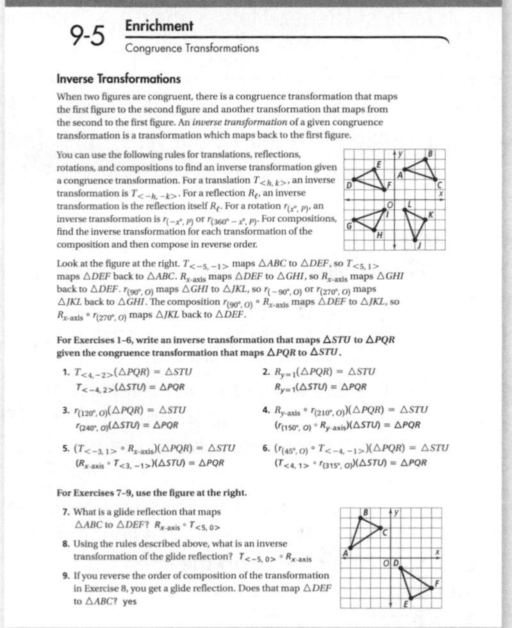

Practice and Problem Solving Wkbk/ All-in-One Resources/Online

Think About a Plan

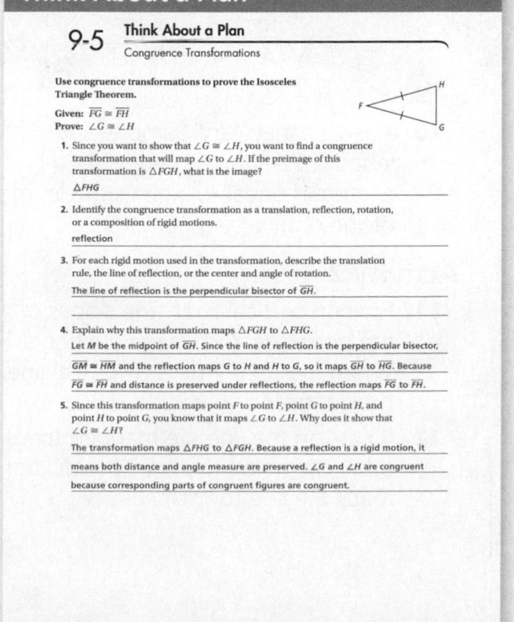

Practice and Problem Solving Wkbk/ All-in-One Resources/Online

Standardized Test Prep

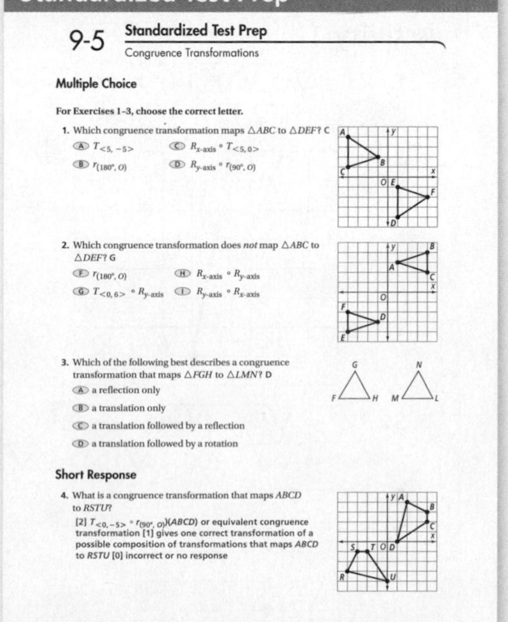

Online Teacher Resource Center

Activities, Games, and Puzzles

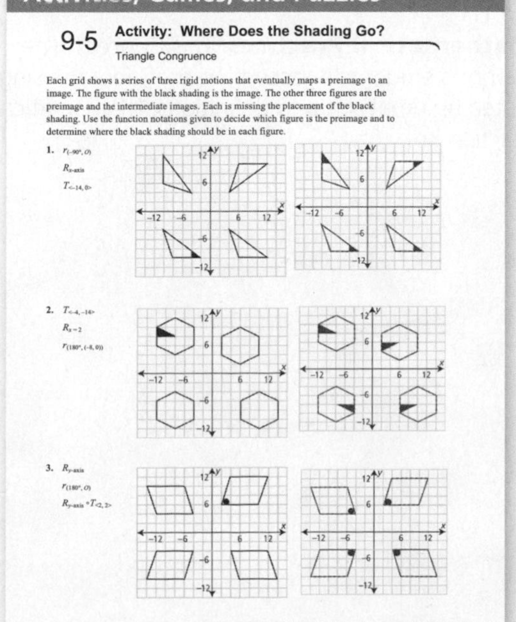

Guided Instruction

PURPOSE To dilate lines and geometric figures and make conjectures about properties of dilations.

PROCESS Students will

- graph dilations of a line with scale factors of 2 and $\frac{1}{2}$ and make conjectures about how dilations change the length of lines.
- graph dilations of lines through the origin and not through the origin and make conjectures about relationships between the preimage and image that do not involve length.

DISCUSS Students first graph dilations of a line and make a conjecture about how the scale factor affects the length of the image. Then they make a conjecture about how the slopes of the image and pre-image are related. They also observe the effect of a dilation of a line that passes through the origin when the center of dilation is at the origin.

Activity 1

In this activity students dilate a line on a coordinate plane.

Q Does a translation of a line change its length? **[no]**
Q Does a rotation of a line change its length? **[no]**
Q Does a reflection of a line change its length? **[no]**

Activity 2

In this activity students observe that dilations map lines through the origin to themselves and lines not through the origin to lines parallel to the preimage.

Q If a line and a dilation of the line intersect at a point, what must be true about the dilation? **[The center of the dilation must be at the point of intersection.]**
Q In Exercise 12, what happens when you draw rays from the origin through points on a line through the origin? **[The rays lie on the line through the origin.]**

VISUAL LEARNERS

Encourage students to make dilations of other geometric figures on a coordinate plane. Ask them if their conjectures still seem valid.

Mathematical Practices This Concept Byte supports students in becoming proficient in using stated assumptions and definitions, Mathematical Practice 3.

Concept Byte: Exploring Dilations

Use With Lesson 9-6

ACTIVITY

Common Core State Standards
G-SRT.A.1b The dilation of a line segment is longer or shorter in the ratio given by the scale factor. **Also G-SRT.A.1a**
MP 3

In this activity, you will explore the properties of dilations. A dilation is defined by a center of dilation and a scale factor.

Activity 1

To dilate a segment by a scale factor n with center of dilation at the origin, you measure the distance from the origin to each point on the segment. The diagram at the right shows the dilation of $\overline{GH}$ by the scale factor 3 with center of dilation at the origin. To locate the dilation image of $\overline{GH}$, draw rays from the origin through points G and H. Then, measure the distance from the origin to G. Next, find the point along the same ray that is 3 times that distance. Label the point G'. Now dilate the endpoint H similarly. Draw $\overline{G'H'}$.

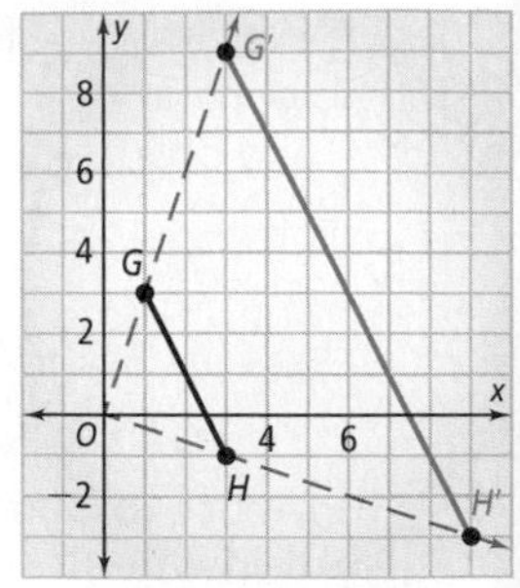

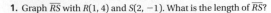

1. Graph $\overline{RS}$ with $R(1, 4)$ and $S(2, -1)$. What is the length of $\overline{RS}$?
2. Graph the dilations of the endpoint of $\overline{RS}$ by scale factor 2 and center of dilation at the origin. Label the dilated endpoints R' and S'.
3. What are the coordinates of R' and S'?
4. Graph $\overline{R'S'}$.
5. What is $R'S'$?
6. How do the lengths of $\overline{RS}$ and $\overline{R'S'}$ compare?
7. Graph the dilation of $\overline{RS}$ by scale factor $\frac{1}{2}$ with center of dilation at the origin. Label the dilation $\overline{R''S''}$.
8. What is $R''S''$?
9. How do the lengths of $\overline{R'S'}$ and $\overline{R''S''}$ compare?
10. What can you conjecture about the length of a line segment that has been dilated by scale factor n?

Activity 2

11. Draw a line on a coordinate grid that does not pass through the origin. Use the method in Activity 1 to construct several dilations of the line you drew with different scale factors (not equal to 1). Make a conjecture relating the slopes of the original line and the dilations.
12. On a new coordinate grid, draw a line through the origin. What happens when you try to construct a dilation of this line? Explain.

Answers

Activity 1

1. $RS = \sqrt{(2-1)^2 + (-1-4)^2}$
$= \sqrt{1 + 25} = \sqrt{26}$

2–4.

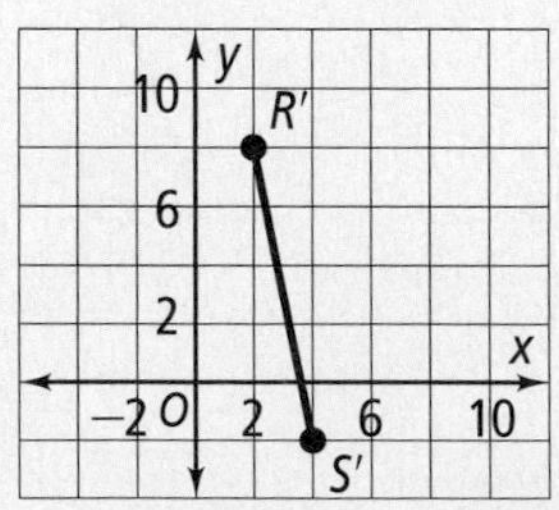

5. $R'S' = \sqrt{(4-2)^2 + (-2-8)^2}$
$= \sqrt{4 - 100} = \sqrt{104} = 2\sqrt{26}$

6. $R'S' = 2RS$

7.

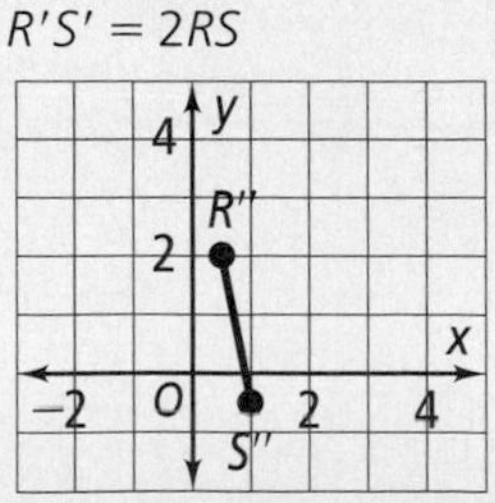

8. $R''S'' = \sqrt{\left(1 - \frac{1}{2}\right)^2 + \left(-\frac{1}{2} - 2\right)^2}$
$= \sqrt{\frac{1}{4} + \frac{25}{4}} = \sqrt{\frac{26}{4}} = \frac{1}{2}\sqrt{26}$

9. $R''S'' = \frac{1}{2}RS$

10. Answers may vary. Sample: The length of a line segment after a dilation is equal to r times the length of the original segment.

Activity 2

11. For non-vertical lines, the slopes of the preimage and image are equal. When the preimage is a vertical line, the image is also a vertical line.

12. Points on the line are mapped to other points on the line. The dilation maps the line to itself.

9-6 Dilations

Common Core State Standards
G-SRT.A.1a A dilation takes a line not passing through the center of the dilation to a parallel line, . . . **Also G-SRT.A.1b, G-CO.A.2, G-SRT.A.2**
MP 1, MP 3, MP 4, MP 7

Objective To understand dilation images of figures

Do you think you can model this using rigid motions?

MATHEMATICAL PRACTICES

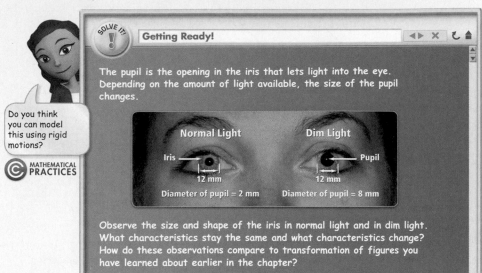

Solve It! Getting Ready!

The pupil is the opening in the iris that lets light into the eye. Depending on the amount of light available, the size of the pupil changes.

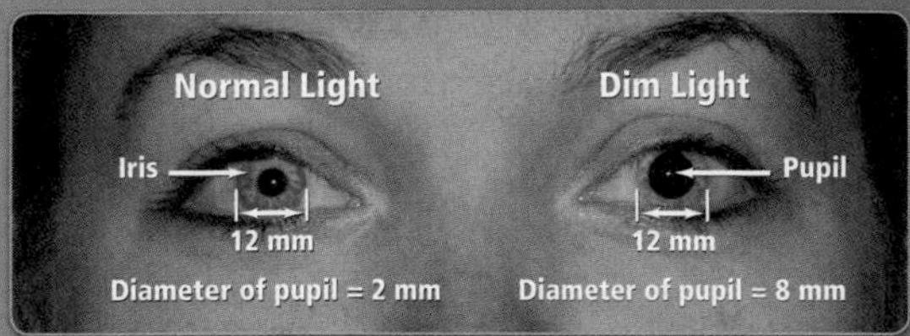

Observe the size and shape of the iris in normal light and in dim light. What characteristics stay the same and what characteristics change? How do these observations compare to transformation of figures you have learned about earlier in the chapter?

Lesson Vocabulary
- dilation
- center of dilation
- scale factor of a dilation
- enlargement
- reduction

In the Solve It, you looked at how the pupil of an eye changes in size, or *dilates*. In this lesson, you will learn how to dilate geometric figures.

Essential Understanding You can use a scale factor to make a larger or smaller copy of a figure that is also similar to the original figure.

take note **Key Concept** Dilation

A **dilation** with **center of dilation** C and **scale factor** n, $n > 0$, can be written as $D_{(n,\, C)}$. A dilation is a transformation with the following properties.
- The image of C is itself (that is, $C' = C$).
- For any other point R, R' is on $\overrightarrow{CR}$ and $CR' = n \cdot CR$, or $n = \frac{CR'}{CR}$.
- Dilations preserve angle measure.

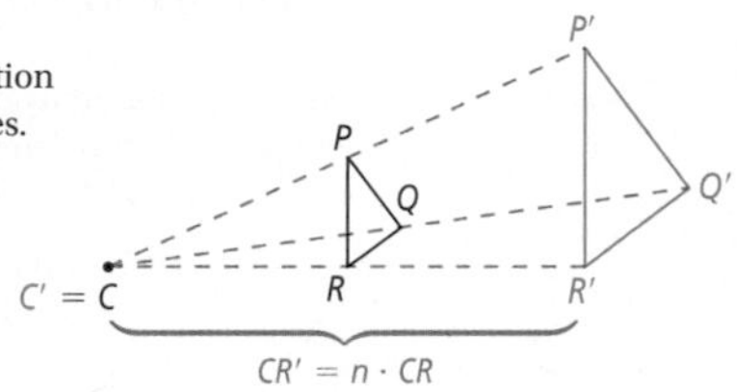

1 Interactive Learning

Solve It!

PURPOSE To identify a dilation as a similarity transformation

PROCESS Students may extend their knowledge of similar figures to describe a dilation.

FACILITATE

Q What is the definition of similar figures? **[Similar figures are figures of the same shape, but different sizes. They have the same angle measures and proportional sides.]**

Q How do circles fit into the definition of similar figures? **[All circles are the same shape, but can be different sizes. All circles are similar to each other.]**

Q Is each radii in the dilated pupil proportional to the corresponding radii in the dim light pupil? **[Yes]**

ANSWER See Solve It in Answers on next page.

CONNECT THE MATH Students review their understanding of similar figures to include circles and realize that similar figures can be reductions or enlargements of each other. In the lesson, students learn that a dilation is a transformation where the preimage and image are similar figures.

2 Guided Instruction

Take Note

Ask students to identify the scale factor between the image and the preimage. Be sure that students understand the difference between a reduction and an enlargement.

9-6 Preparing to Teach

BIG ideas **Transformations**
Coordinate Geometry

ESSENTIAL UNDERSTANDING
- A scale factor can be used to make a larger or smaller copy of a figure that is also similar to the original figure.

Math Background

The final type of transformation presented is a dilation. A dilation is not a rigid motion because it does not preserve the size of the figure. A dilation produces a pair of similar figures whose corresponding side lengths are proportional. The scale factor related to a dilation is equal to the ratio of distances between pairs of corresponding points.

Dilations are used in multiple real-world areas, such as science, technology, and photography. Students can connect the terms in this lesson with reducing and enlarging photos.

Mathematical Practices

Look for and make use of structure. Students will build on their knowledge of scale factors in Lesson 7-2 to construct an understanding of dilations. They will also see dilations as being defined by a center and a scale factor.

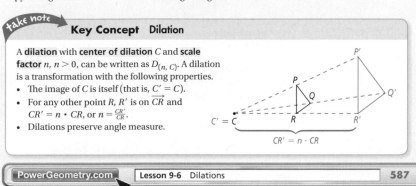

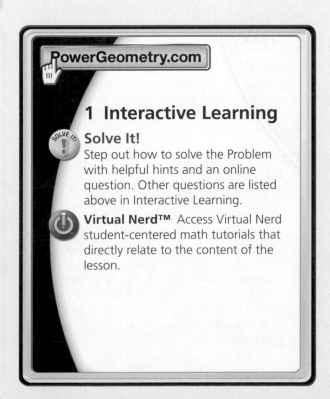

PowerGeometry.com

1 Interactive Learning

Solve It! Step out how to solve the Problem with helpful hints and an online question. Other questions are listed above in Interactive Learning.

Virtual Nerd™ Access Virtual Nerd student-centered math tutorials that directly relate to the content of the lesson.

Problem 1

Q Which triangle is larger? **[△X′T′R′]**
Q Which triangle is the preimage? **[△XTR]**
Q Which side corresponds to $\overline{XT}$? **[$\overline{X'T'}$]**
Q What is the ratio of corresponding sides? **[$\frac{X'T'}{XT} = \frac{4+8}{4} = \frac{12}{4} = 3$]**

Got It?

Q Is the dilation a reduction or an enlargement? Justify your answer. **[It is a reduction because *J′K′L′M′* is smaller than *JKLM*.]**
Q What is the advantage to having the figures on a coordinate grid? **[You can identify the ordered pairs of the vertices.]**
Q How can you find the length of corresponding sides? **[Use the Distance Formula.]**

Have students consider both a dilation that is a reduction of the preimage and an enlargement of the preimage when they read the statement about multiplying the coordinates of a point by the scale factor.

Q If the preimage and image are the same size, what would be the value of *n*? Explain. **[1; multiplying by 1 does not change a number.]**
Q If the image is an enlargement of the preimage, write an inequality that describes *n*? Explain. **[$n > 1$; multiplying by a number greater than 1 produces a greater number.]**
Q If the image is a reduction of the preimage, write an inequality that describes *n*. Explain. **[$n < 1$; multiplying by a number less than 1 produces a lesser number.]**

The scale factor *n* of a dilation is the ratio of a length of the image to the corresponding length in the preimage, with the image length always in the numerator. For the figure shown on page 587, $n = \frac{CR'}{CR} = \frac{R'P'}{RP} = \frac{P'Q'}{PQ} = \frac{Q'R'}{QR}$.

A dilation is an **enlargement** if the scale factor *n* is greater than 1. The dilation is a **reduction** if the scale factor *n* is between 0 and 1.

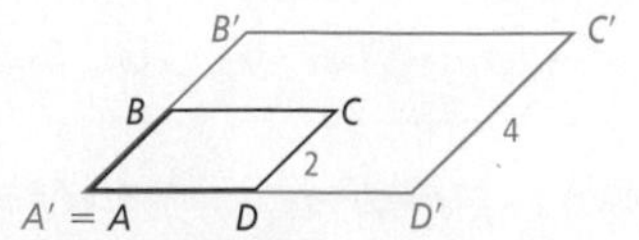

Enlargement
center *A*, scale factor 2

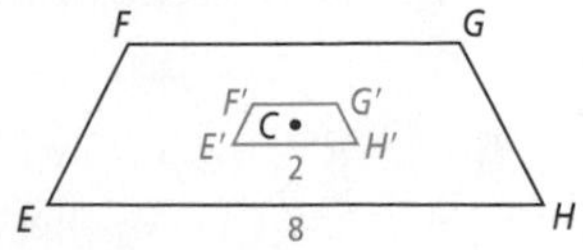

Reduction
center *C*, scale factor $\frac{1}{4}$

Problem 1 Finding a Scale Factor

Multiple Choice Is $D_{(n, X)}(\triangle XTR) = \triangle X'T'R'$ an enlargement or a reduction? What is the scale factor *n* of the dilation?

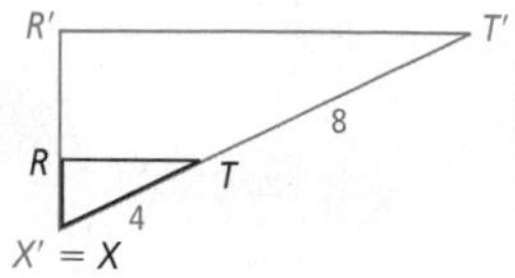

Ⓐ enlargement; $n = 2$　Ⓒ reduction; $n = \frac{1}{3}$
Ⓑ enlargement; $n = 3$　Ⓓ reduction; $n = 3$

The image is larger than the preimage, so the dilation is an enlargement.
Use the ratio of the lengths of corresponding sides to find the scale factor.

$$n = \frac{X'T'}{XT} = \frac{4+8}{4} = \frac{12}{4} = 3$$

△X′T′R′ is an enlargement of △XTR, with a scale factor of 3. The correct answer is B.

Think
Why is the scale factor not $\frac{4}{12}$, or $\frac{1}{3}$?
The scale factor of a dilation always has the image length (or the distance between a point on the image and the center of dilation) in the numerator.

Got It? 1. Is $D_{(n, O)}(JKLM) = J'K'L'M'$ an enlargement or a reduction? What is the scale factor *n* of the dilation?

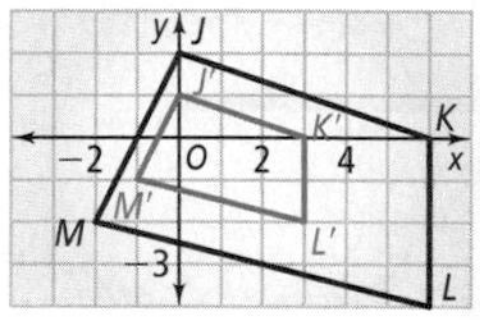

In Got It 1, you looked at a dilation of a figure drawn in the coordinate plane. In this book, all dilations of figures in the coordinate plane have the origin as the center of dilation. So you can find the dilation image of a point *P*(*x*, *y*) by multiplying the coordinates of *P* by the scale factor *n*. A dilation of scale factor *n* with center of dilation at the origin can be written as

$$D_n(x, y) = (nx, ny)$$

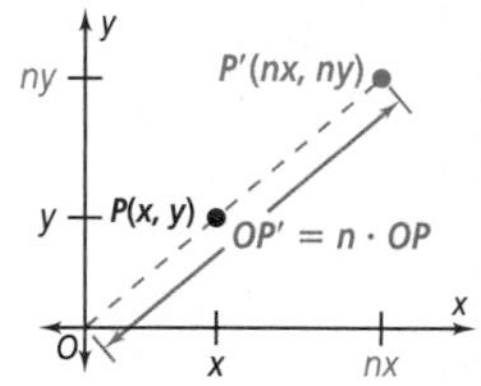

Answers

Solve It!

The shape stays the same, but the size changes. In transformations, the size and the shape of the figure stays the same. In this case, only the shape stays the same.

Got It?

1. reduction; $n = \frac{1}{2}$

2a. $P'(-0.5, 0)$, $Z'(-1, 0.5)$, $G'(0, -1)$

b. Sample answer: $\overline{PZ}$ and $\overline{P'Z'}$ lie on the same line. $\overline{PG}$ and $\overline{P'G'}$ are parallel, and so are $\overline{GZ}$ and $\overline{G'Z'}$. Since $\overline{PZ}$ and $\overline{P'Z'}$, you can conjecture that lines that pass through the center of dilation are left as is, and that a dilation of a line that does not pass through the center of dilation results in a line parallel to the preimage.

3. 5.1 cm

PowerGeometry.com

2 Guided Instruction

Each Problem is worked out and supported online.

Problem 1
Finding a Scale Factor

Problem 2
Finding a Dilation Image
Animated

Problem 3
Using a Scale Factor to Find a Length
Animated

Support in Geometry Companion
- Vocabulary
- Key Concepts
- Got It?

Will the vertices of the triangle move closer to (0, 0) or farther from (0, 0)?
The scale factor is 2, so the dilation is an enlargement. The vertices will move farther from (0, 0).

Problem 2 Finding a Dilation Image

What are the coordinates of the vertices of $D_2(\triangle PZG)$? Graph the image of $\triangle PZG$.

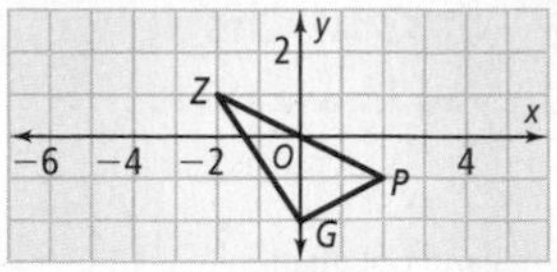

Identify the coordinates of each vertex. The center of dilation is the origin and the scale factor is 2, so use the dilation rule $D_2(x, y) = (2x, 2y)$.

$D_2(P) = (2 \cdot 2, 2 \cdot (-1))$, or $P'(4, -2)$.

$D_2(Z) = (2 \cdot (-2), 2 \cdot 1)$, or $Z'(-4, 2)$.

$D_2(G) = (2 \cdot 0, 2 \cdot (-2))$, or $G'(0, -4)$.

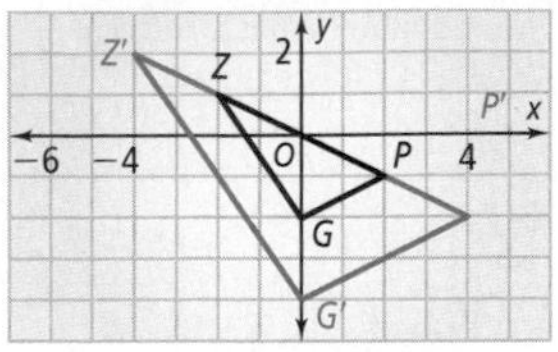

To graph the image of $\triangle PZG$, graph P', Z', and G'. Then draw $\triangle P'Z'G'$.

Got It? 2. a. What are the coordinates of the vertices of $D_{\frac{1}{2}}(\triangle PZG)$?

b. Reasoning How are $\overline{PZ}$ and $\overline{P'Z'}$ related? How are $\overline{PG}$ and $\overline{P'G'}$, and $\overline{GZ}$ and $\overline{G'Z'}$ related? Use these relationships to make a conjecture about the effects of dilations on lines.

Dilations and scale factors help you understand real-world enlargements and reductions, such as images seen through a microscope or on a computer screen.

What does a scale factor of 7 tell you?
A scale factor of 7 tells you that the ratio of the image length to the actual length is 7, or $\frac{\text{image length}}{\text{actual length}} = 7$.

Problem 3 Using a Scale Factor to Find a Length

Biology **A magnifying glass shows you an image of an object that is 7 times the object's actual size. So the scale factor of the enlargement is 7. The photo shows an apple seed under this magnifying glass. What is the actual length of the apple seed?**

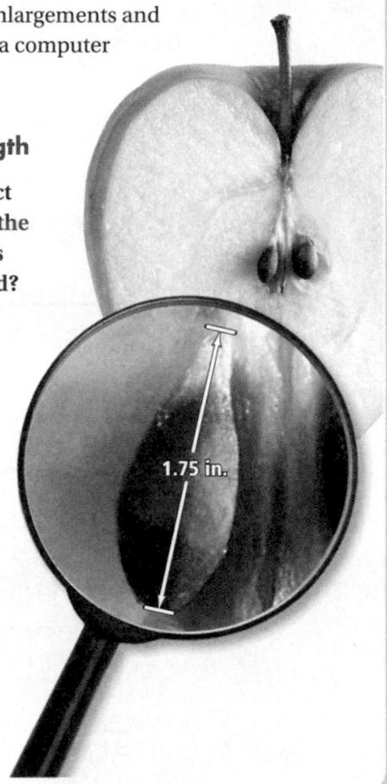

$1.75 = 7 \cdot p$ image length = scale factor • actual length

$0.25 = p$ Divide each side by 7.

The actual length of the apple seed is 0.25 in.

Got It? 3. The height of a document on your computer screen is 20.4 cm. When you change the zoom setting on your screen from 100% to 25%, the new image of your document is a dilation of the previous image with scale factor 0.25. What is the height of the new image?

Problem 2

Q Is the dilation a reduction or an enlargement? Justify your answer. **[The dilation is an enlargement because the scale factor is greater than 1.]**

Q How do you find the coordinates of points of a dilation centered at the origin? **[Multiply each element of ordered pairs of the preimage by the scale factor.]**

Got It? **ERROR PREVENTION**
Have students identify the dilation as a reduction or an enlargement. Make sure that students show their work for calculating the new coordinates for each point. Students should multiply each *x*- and *y*-coordinate by the scale factor.

Problem 3

Q Is the magnifying glass reducing or enlarging the image of the seed? **[enlarging]**

Q How can you find the actual length of the seed? **[Divide the length in the magnification by the magnification factor.]**

Got It? **ERROR PREVENTION**
Have students write a proportion that relates the corresponding heights of the documents and the scale factor. Use properties of proportions to solve for the new height.

Additional Problems

1. Is $D_n(\triangle ABC) = \triangle A'B'C'$ an enlargement or a reduction? What is the scale factor n of the dilation?

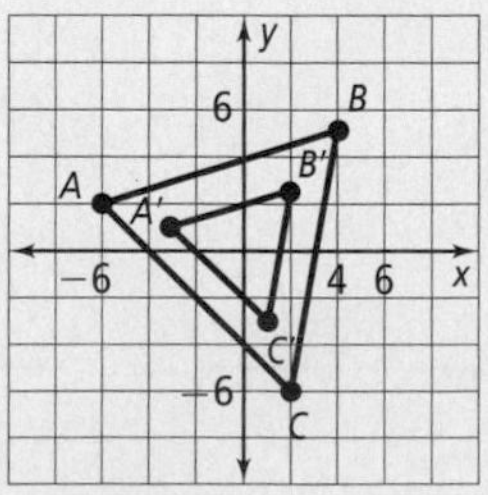

ANSWER reduction, $n = \frac{1}{2}$

2. What are the coordinates of the vertices of $D_2(\triangle LMN)$? Graph the image of $\triangle LMN$.

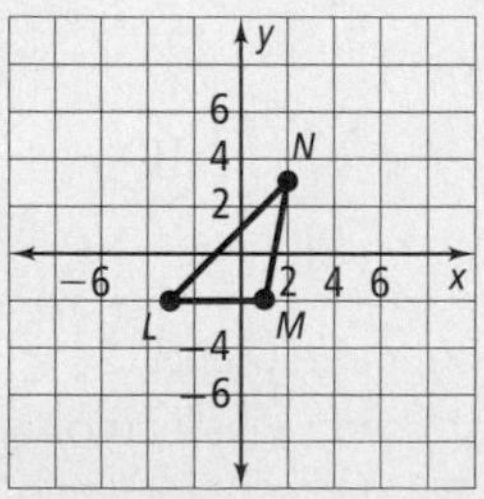

ANSWER $L'(-6, -4)$, $M'(2, -4)$, $N'(4, 6)$

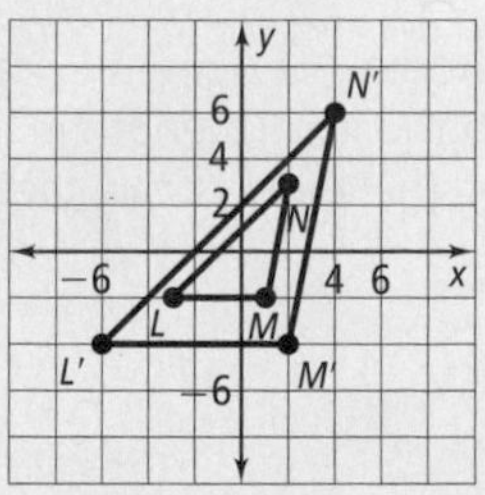

3. Jessica has a photograph that is 4 inches high and 6 inches wide. She wants to enlarge the picture by a scale factor of 1.25. What are the new dimensions of the photograph?

ANSWER 5 inches high and 7.5 inches wide

3 Lesson Check

Do you know HOW?

- If students have difficulty with Exercises 2–4, then have them review Problem 2 to see which factors are to be multiplied.

Do you UNDERSTAND?

- If students have difficulty with Exercise 6, then have them review Problem 1 to determine if the process is the same or the opposite of what is shown as the solution.

Close

Q How are the image and preimage related in a dilation? **[The ratio of corresponding side lengths in the two figures is equal to the scale factor of the dilation.]**

Q What is the difference between a reduction and an enlargement? **[An enlargement makes the figure bigger and has a scale factor greater than 1. A reduction makes the figure smaller and has a scale factor between 0 and 1.]**

Lesson Check

Do you know HOW?

1. The blue figure is a dilation image of the black figure with center of dilation C. Is the dilation an enlargement or a reduction? What is the scale factor of the dilation?

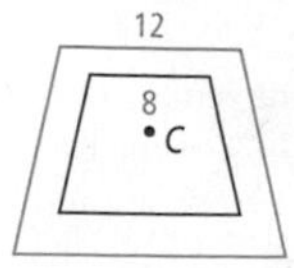

Find the image of each point.

2. $D_2(1, -5)$ **3.** $D_{\frac{1}{2}}(0, 6)$ **4.** $D_{10}(0, 0)$

Do you UNDERSTAND? MATHEMATICAL PRACTICES

5. Vocabulary Describe the scale factor of a reduction.

6. Error Analysis The blue figure is a dilation image of the black figure for a dilation with center A.

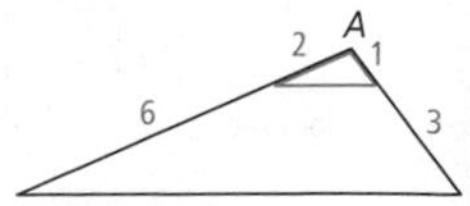

Two students made errors when asked to find the scale factor. Explain and correct their errors.

A.

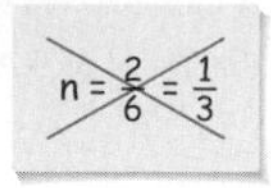

B.

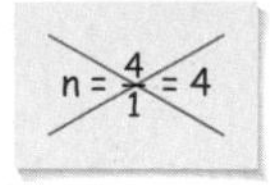

Practice and Problem-Solving Exercises MATHEMATICAL PRACTICES

A Practice

The blue figure is a dilation image of the black figure. The labeled point is the center of dilation. Tell whether the dilation is an enlargement or a reduction. Then find the scale factor of the dilation.

See Problem 1.

7.

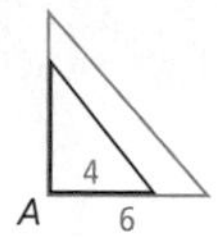

8.

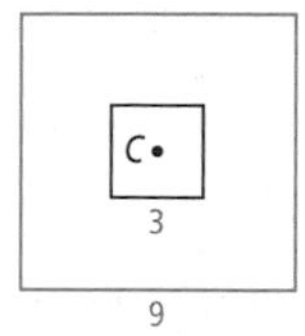

9.

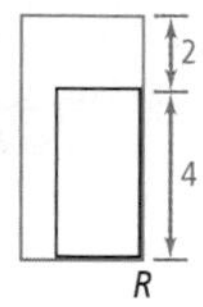

10.

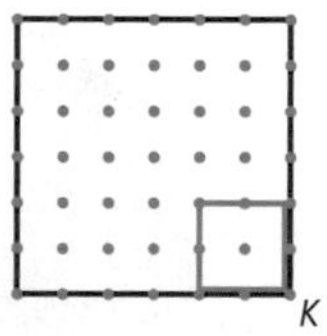

11.

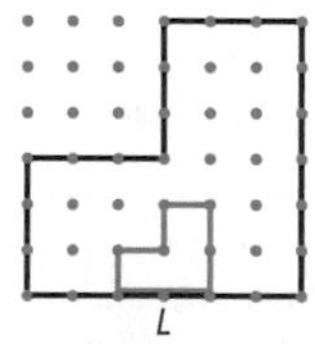

12.

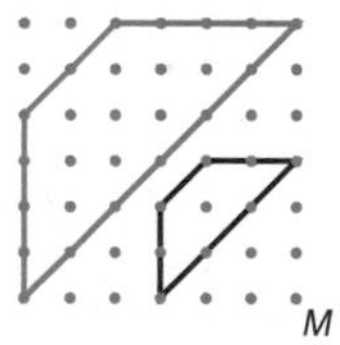

13.

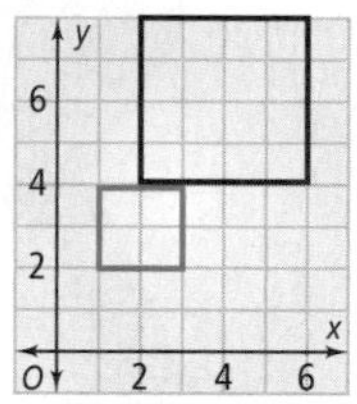

14.

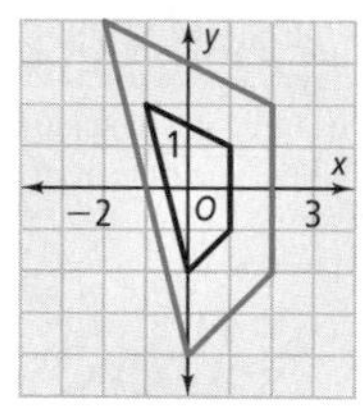

15.

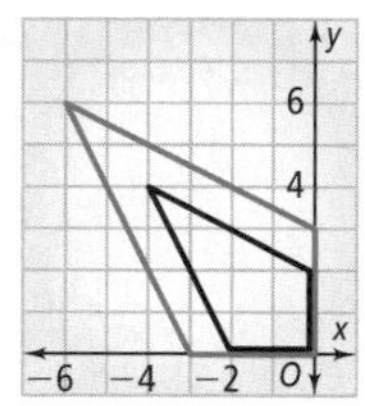

PowerGeometry.com

3 Lesson Check

For a digital lesson check, use the Got It questions.

Support in Geometry Companion

- Lesson Check

4 Practice

Assign homework to individual students or to an entire class.

Answers

Lesson Check

1. enlargement; 1.5

2. $D'(2, -10)$

3. $T'(0, 3)$

4. $M'(0, 0)$

5. a number between 0 and 1

6a. The student used 6, instead of $2 + 6 = 8$, as the preimage length in the denominator; the correct scale factor is $n = \frac{2}{2+6} = \frac{1}{4}$.

b. The student did not write the scale factor with the image length in the numerator; the correct scale factor is $n = \frac{1}{4}$.

Practice and Problem-Solving Exercises

7. enlargement; $\frac{3}{2}$

8. enlargement; 3

9. enlargement; $\frac{3}{2}$

10. reduction; $\frac{1}{3}$

11. reduction; $\frac{1}{3}$

12. enlargement; 2

13. reduction; $\frac{1}{2}$

14. enlargement; 2

15. enlargement; $\frac{3}{2}$

Find the images of the vertices of $\triangle PQR$ for each dilation. Graph the image.

See Problem 2.

16. $D_3(\triangle PQR)$

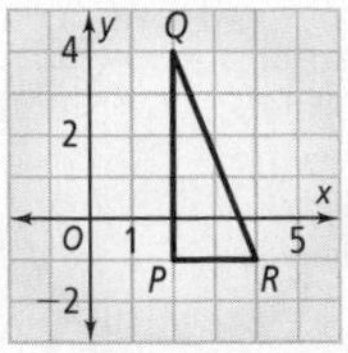

17. $D_{10}(\triangle PQR)$

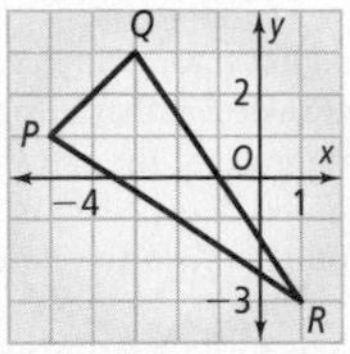

18. $D_{\frac{3}{4}}(\triangle PQR)$

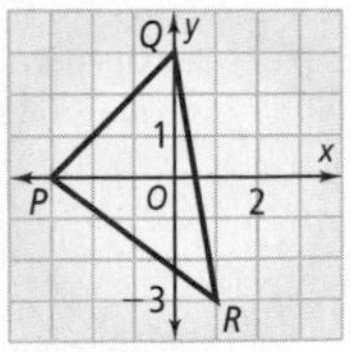

Magnification You look at each object described in Exercises 19–22 under a magnifying glass. Find the actual dimension of each object.

See Problem 3.

19. The image of a button is 5 times the button's actual size and has a diameter of 6 cm.

20. The image of a pinhead is 8 times the pinhead's actual size and has a width of 1.36 cm.

21. The image of an ant is 7 times the ant's actual size and has a length of 1.4 cm.

22. The image of a capital letter N is 6 times the letter's actual size and has a height of 1.68 cm.

B Apply

Find the image of each point for the given scale factor.

23. $L(-3, 0)$; $D_5(L)$

24. $N(-4, 7)$; $D_{0.2}(N)$

25. $A(-6, 2)$; $D_{1.5}(A)$

26. $F(3, -2)$; $D_{\frac{1}{3}}(F)$

27. $B\left(\frac{5}{4}, -\frac{3}{2}\right)$; $D_{\frac{1}{10}}(B)$

28. $Q\left(6, \frac{\sqrt{3}}{2}\right)$; $D_{\sqrt{6}}(Q)$

Use the graph at the right. Find the vertices of the image of $QRTW$ for a dilation with center (0, 0) and the given scale factor.

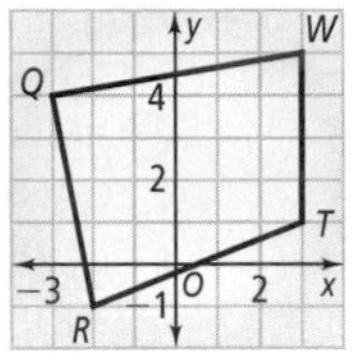

29. $\frac{1}{4}$

30. 0.6

31. 0.9

32. 10

33. 100

34. Compare and Contrast Compare the definition of scale factor of a dilation to the definition of scale factor of two similar polygons. How are they alike? How are they different?

35. Think About a Plan The diagram at the right shows $\triangle LMN$ and its image $\triangle L'M'N'$ for a dilation with center P. Find the values of x and y. Explain your reasoning.

- What is the relationship between $\triangle LMN$ and $\triangle L'M'N'$?
- What is the scale factor of the dilation?
- Which variable can you find using the scale factor?

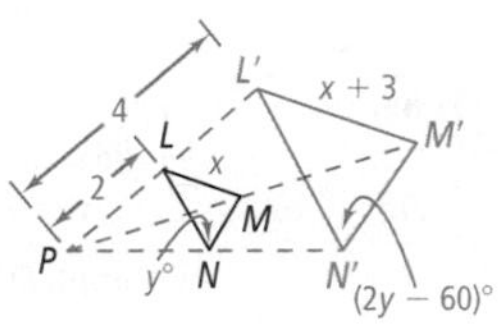

36. Writing An equilateral triangle has 4-in. sides. Describe its image for a dilation with center at one of the triangle's vertices and scale factor 2.5.

4 Practice

ASSIGNMENT GUIDE

Basic: 7–22 all, 24–34 even, 35, 38, 40–48 even
Average: 7–21 odd, 23–48
Advanced: 7–21 odd, 23–51
Standardized Test Prep: 52–55
Mixed Review: 56–59

Mathematical Practices are supported by exercises with red headings. Here are the Practices supported in this lesson:

MP 1: Make Sense of Problems Ex. 35
MP 3: Communicate Ex. 34, 36, 39
MP 3: Construct Arguments Ex. 44
MP 3: Critique the Reasoning of Others Ex. 6, 45–48

Applications exercises have blue headings. Exercises 19–22, and 51 support MP 4: Model.

EXERCISE 44: Use the Think About a Plan worksheet in the **Practice and Problem Solving Workbook** (also available in the Teaching Resources in print and online) to further support students' development in becoming independent learners.

HOMEWORK QUICK CHECK

To check students' understanding of key skills and concepts, go over Exercises 15, 17, 34, 35, and 44.

16. $P'(6, -3)$, $Q'(6, 12)$, $R'(12, -3)$

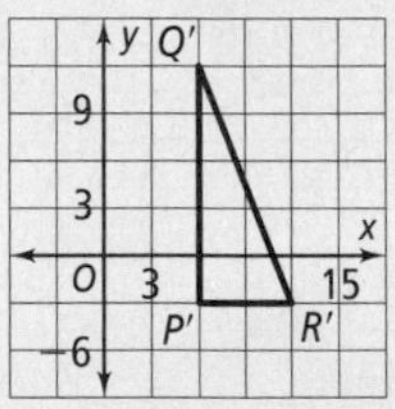

17. $P'(-50, 10)$, $Q'(-30, 30)$, $R'(10, -30)$

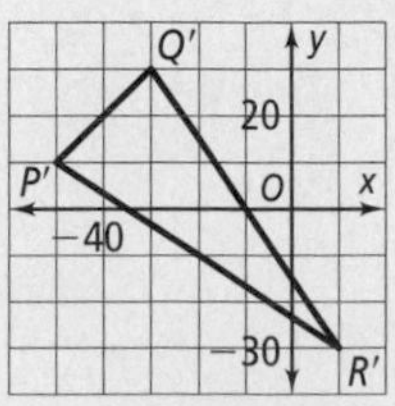

18. $P'\left(-\frac{9}{4}, 0\right)$, $Q'\left(0, \frac{9}{4}\right)$, $R'\left(\frac{3}{4}, -\frac{9}{4}\right)$

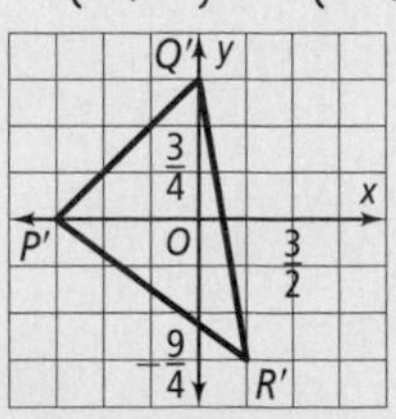

19. 1.2 cm

20. 0.17 cm

21. 0.2 cm

22. 0.28 cm

23. $L'(-15, 0)$

24. $N'(-0.8, 1.4)$

25. $A'(-9, 3)$

26. $F'\left(1, -\frac{2}{3}\right)$

27. $B'\left(\frac{1}{8}, -\frac{3}{20}\right)$

28. $Q'\left(6\sqrt{6}, \frac{3\sqrt{2}}{2}\right)$

29. $Q'\left(-\frac{3}{4}, 1\right)$, $R'\left(-\frac{1}{2}, -\frac{1}{4}\right)$, $T'\left(\frac{3}{4}, \frac{1}{4}\right)$, $W'\left(\frac{3}{4}, \frac{5}{4}\right)$

30. $Q'(-1.8, 2.4)$, $R'(-1.2, -0.6)$, $T'(1.8, 0.6)$, $W'(1.8, 3)$

31. $Q'(-2.7, 3.6)$, $R'(-1.8, -0.9)$, $T'(2.7, 0.9)$, $W'(2.7, 4.5)$

32. $Q'(-30, 40)$, $R'(-20, -10)$, $T'(30, 10)$, $W'(30, 50)$

33. $Q'(-300, 400)$, $R'(-200, -100)$, $T'(300, 100)$, $W'(300, 500)$

34. Answers may vary. Sample: Each type of scale factor is a constant ratio of corresp. lengths. For a dilation, the scale factor is always the ratio of an image length to a corresp. preimage length, while for similar figures, the scale factor ratio can relate the two figures in either order. The scale factor of two similar figures is always the ratio of the lengths of two corresponding sides, while the scale factor of a dilation is also the ratio of the distances of corresponding points from the center of dilation. If the center is not on the preimage, then these distances are not lengths of corresponding sides of the image and preimage.

35–36. See next page.

Answers

Practice and Problem-Solving Exercises (continued)

35. $x = 3$, $y = 60$; the image of a dilation is similar to the preimage, so $\triangle L'N'M' \sim \triangle LNM$. The ratio of the corresp. sides is the same as the scale factor of the dilation, which is 4 : 2, or 2 : 1. To find x, solve the proportion $\frac{x+3}{x} = \frac{2}{1}$. $y = 60$ because corresponding angles of $\sim$ figures are $\cong$.

36. Answers may vary. Sample: The image is an equilateral $\triangle$ with sides 10 in. long. For two of the pairs of corresp. sides, the corresp. sides lie on the same line. The sides of the third pair of corresp. sides are $\parallel$.

37.

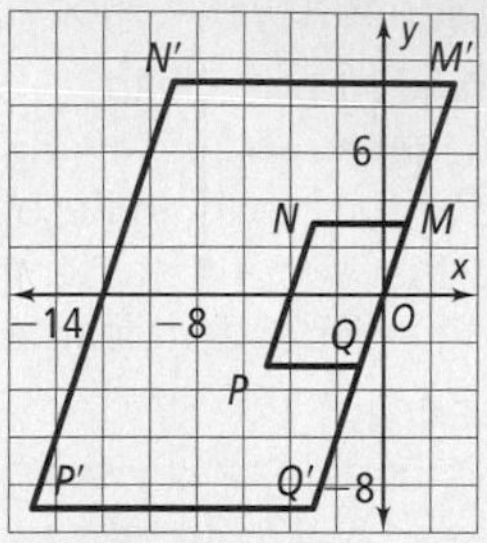

38.

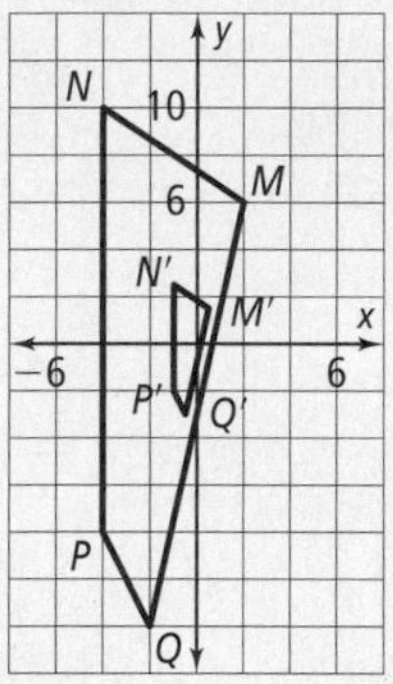

39. Check students' work.

40. $I'J' = 10$ in.; $H'J' = 12$ in.

41. $HI = 32$ ft; $I'J' = 7.5$ ft

42. Let ℓ be given by the equation $y = ax$ for some real number a. If $C = (c_1, c_2)$ is on ℓ, then $c_2 = ac_1$, and $kc_2 = k(ac_1) = a(kc_1)$, so $D_k(C) = (kc_1, kc_2)$ is also on ℓ.

43a. Since $\overleftrightarrow{AB}$ does not pass through the origin, it has x- and y-intercepts $(a, 0)$ and $(0, b)$ for some $a, b \neq 0$. So $D_k(a, 0) = (ka, 0)$ and $D_k(0, b) = (0, kb)$. So $\overleftrightarrow{A'B'}$ has different x- and y-intercepts than $\overleftrightarrow{AB}$ and thus $\overleftrightarrow{AB} \neq \overleftrightarrow{A'B'}$.

b. The slope of $\overleftrightarrow{AB}$ is $\frac{b_2 - a_2}{b_1 - a_1}$. Since $A' = (ka_1, ka_2)$ and $B' = (kb_1, kb_2)$, the slope of $\overleftrightarrow{A'B'}$ is $\frac{kb_2 - ka_2}{kb_1 - ka_1} = \frac{b_2 - a_2}{b_1 - a_1}$.

c. If $a_1 = b_1$ then $\overleftrightarrow{AB}$ is vertical. Also, $a_1 = b_1$ implies $ka_1 = kb_1$, so $\overleftrightarrow{AB}$ is also vertical.

44. Connect corresp. points A and A', and B and B'. Extend $\overline{AA'}$ and $\overline{BB'}$ until they intersect. The intersection point is the center of dilation. The scale factor is the length of $\overline{A'B'}$ divided by the length of $\overline{AB}$.

45. False; a dilation does not map a segment to a $\cong$ segment unless the scale factor is 1.

46. False; a dilation with a scale factor greater than 1 is an enlargement.

47. True; the image and preimage are $\sim$, so the corresp. $\angle$s are $\cong$.

48. False; for example, if the center of dilation is on the preimage, then it is also on the image.

49.

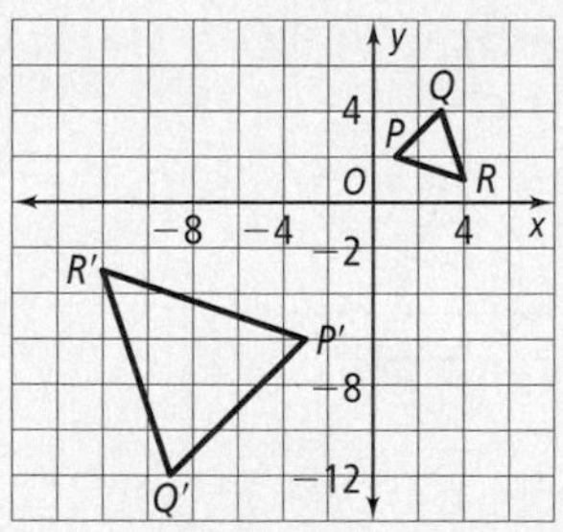

Coordinate Geometry Graph *MNPQ* and its image *M′N′P′Q′* for a dilation with center (0, 0) and the given scale factor.

37. $M(1, 3)$, $N(-3, 3)$, $P(-5, -3)$, $Q(-1, -3)$; 3

38. $M(2, 6)$, $N(-4, 10)$, $P(-4, -8)$, $Q(-2, -12)$; $\frac{1}{4}$

39. Open-Ended Use the dilation command in geometry software or drawing software to create a design that involves repeated dilations, such as the one shown at the right. The software will prompt you to specify a center of dilation and a scale factor. Print your design and color it. Feel free to use other transformations along with dilations.

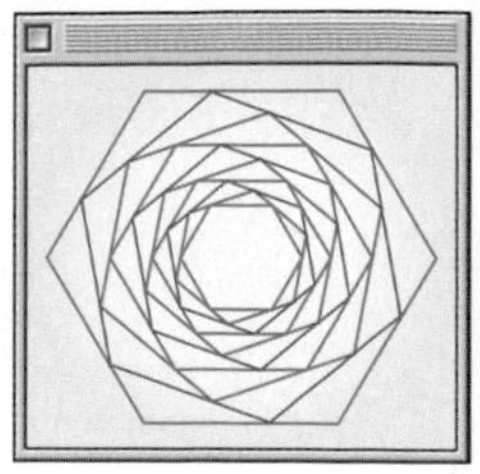

A dilation maps $\triangle HIJ$ onto $\triangle H'I'J'$. Find the missing values.

40. $HI = 8$ in. $H'I' = 16$ in.
$IJ = 5$ in. $I'J' = ■$ in.
$HJ = 6$ in. $H'J' = ■$ in.

41. $HI = ■$ ft $H'I' = 8$ ft
$IJ = 30$ ft $I'J' = ■$ ft
$HJ = 24$ ft $H'J' = 6$ ft

42. Let ℓ be a line through the origin. Show that $D_k(\ell) = \ell$ by showing that if $C = (c_1, c_2)$ is on ℓ, then $D_k(C)$ is also on ℓ.

43. Let $A = (a_1, a_2)$ and $B = (b_1, b_2)$, let $A' = D_k(A)$ and $B' = D_k(B)$ with $k \neq 1$, and suppose that $\overleftrightarrow{AB}$ does not pass through the origin.

a. Show that $\overleftrightarrow{AB} \neq \overleftrightarrow{A'B'}$ (*Hint:* What happens to the x- and y-intercepts of $\overleftrightarrow{AB}$ under the dilation D_k?)

b. Suppose that $a_1 \neq b_1$. Show that $\overleftrightarrow{AB}$ is parallel to $\overleftrightarrow{A'B'}$ by showing that they have the same slope.

c. Show that $\overleftrightarrow{AB} \parallel \overleftrightarrow{A'B'}$ if $a_1 = b_1$.

44. Reasoning You are given $\overline{AB}$ and its dilation image $\overline{A'B'}$ with A, B, A', and B' noncollinear. Explain how to find the center of dilation and scale factor.

Reasoning Write *true* or *false* for Exercises 45–48. Explain your answers.

45. A dilation is an isometry.

46. A dilation with a scale factor greater than 1 is a reduction.

47. For a dilation, corresponding angles of the image and preimage are congruent.

48. A dilation image cannot have any points in common with its preimage.

Challenge

Coordinate Geometry In the coordinate plane, you can extend dilations to include scale factors that are negative numbers. For Exercises 49 and 50, use $\triangle PQR$ with vertices $P(1, 2)$, $Q(3, 4)$, and $R(4, 1)$.

49. Graph $D_{-3}(\triangle PQR)$.

50. a. Graph $D_{-1}(\triangle PQR)$.

b. Explain why the dilation in part (a) may be called a *reflection through a point*. Extend your explanation to a new definition of point symmetry.

51. **Shadows** A flashlight projects an image of rectangle $ABCD$ on a wall so that each vertex of $ABCD$ is 3 ft away from the corresponding vertex of $A'B'C'D'$. The length of $\overline{AB}$ is 3 in. The length of $\overline{A'B'}$ is 1 ft. How far from each vertex of $ABCD$ is the light?

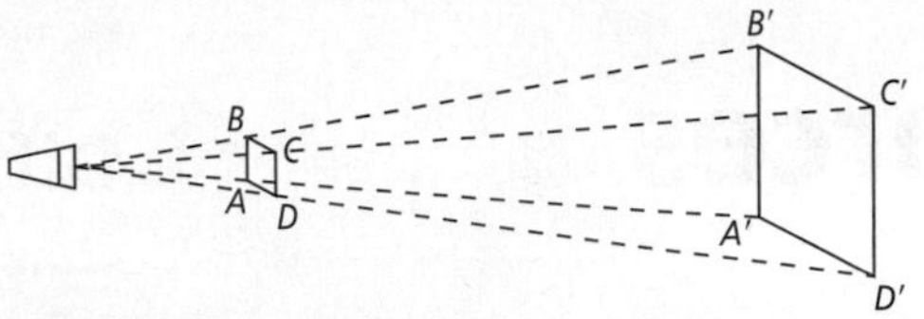

Standardized Test Prep

SAT/ACT

52. A dilation maps $\triangle CDE$ onto $\triangle C'D'E'$. If $CD = 7.5$ ft, $CE = 15$ ft, $D'E' = 3.25$ ft, and $C'D' = 2.5$ ft, what is DE?

Ⓐ 1.08 ft Ⓑ 5 ft Ⓒ 9.75 ft Ⓓ 19 ft

53. You want to prove indirectly that the diagonals of a rectangle are congruent. As the first step of your proof, what should you assume?

Ⓕ A quadrilateral is not a rectangle.

Ⓖ The diagonals of a rectangle are not congruent.

Ⓗ A quadrilateral has no diagonals.

Ⓘ The diagonals of a rectangle are congruent.

54. Which word can describe a kite?

Ⓐ equilateral Ⓑ equiangular Ⓒ convex Ⓓ scalene

Short Response

55. Use the figure at the right to answer the questions below.
   a. Does the figure have rotational symmetry? If so, identify the angle of rotation.
   b. Does the figure have reflectional symmetry? If so, how many lines of symmetry does it have?

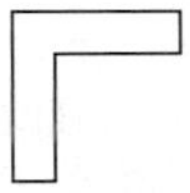

Mixed Review

56. $\triangle JKL$ has vertices $J(23, 2)$, $K(4, 1)$, and $L(1, 23)$. What are the coordinates of J', K', and L' if $(R_{x\text{-axis}} \circ T_{<2,-3>})(\triangle JKL) = \triangle J'K'L'$?

See Lesson 9-5.

Get Ready! **To prepare for Lesson 9-7, do Exercises 55–57.**

Algebra $TRSU \sim NMYZ$. Find the value of each variable.

See Lesson 7-2.

57. a 58. b 59. c

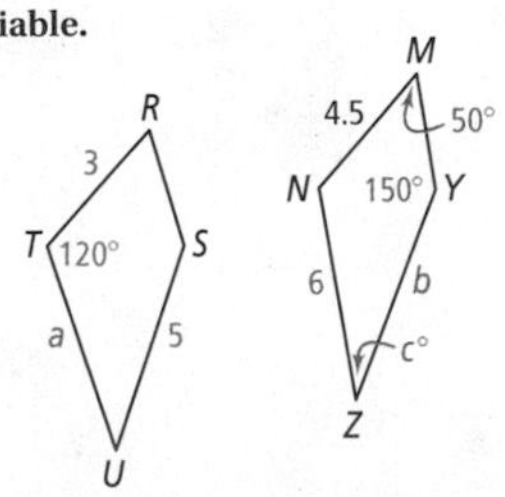

50a.

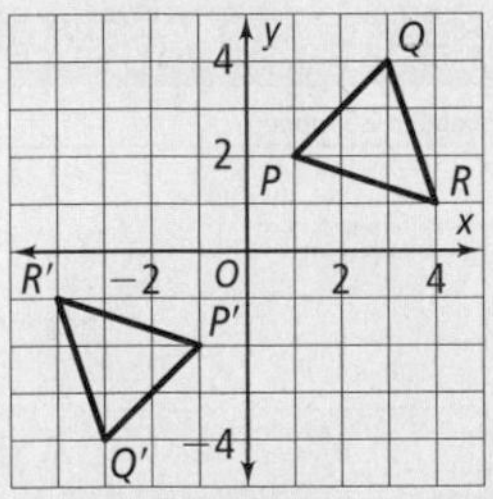

b. Answers may vary. Sample: The origin is the midpt. of each segment joining an image point to its preimage. A figure has point symmetry if there is a point P through which the figure reflects onto itself.

51. 1 ft
52. C
53. G
54. C
55. [2] **a.** no
b. yes; 1
[1] one part incorrect
56. $J'(25, 1)$, $K'(6, 2)$, $L'(3, -20)$
57. 4
58. 7.5
59. 40

9-6 Lesson Resources

Differentiated Remediation

Instructional Support

Geometry Companion

Students can use the **Geometry Companion** worktext (4 pages) . . .

- New Vocabulary
- Key Concepts
- Got It for each Problem
- Lesson Check

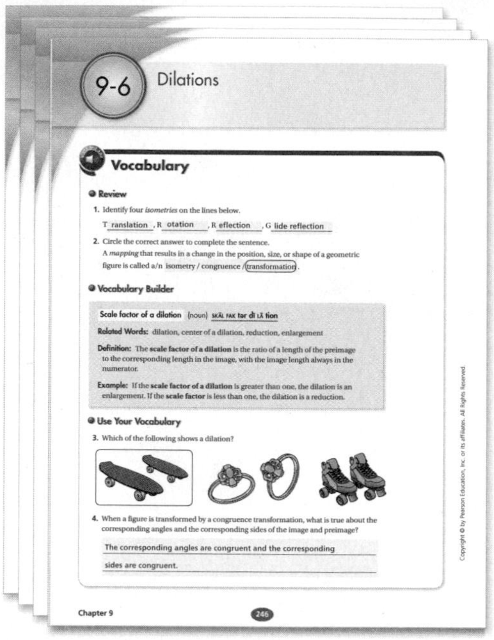
9-6 Dilations

Vocabulary

Review

1. Identify four *isometries* on the lines below.
T ranslation , R otation , R eflection , G lide reflection

2. Circle the correct answer to complete the sentence.
A *mapping* that results in a change in the position, size, or shape of a geometric figure is called a/n isometry / congruence / transformation.

Vocabulary Builder

Scale factor of a dilation (noun) SKĀL FAK tor of dī LĀ tion

Related Words: dilation, center of a dilation, reduction, enlargement

Definition: The **scale factor of a dilation** is the ratio of a length of the preimage to the corresponding length in the image, with the image length always in the numerator.

Example: If the **scale factor of a dilation** is greater than one, the dilation is an enlargement. If the **scale factor** is less than one, the dilation is a reduction.

Use Your Vocabulary

3. Which of the following shows a dilation?

4. When a figure is transformed by a congruence transformation, what is true about the corresponding angles and the corresponding sides of the image and preimage?
The corresponding angles are congruent and the corresponding sides are congruent.

Chapter 9 246

ELL Support

Focus on Language Use the following activity to familiarize students with the lesson vocabulary. Ask if students have ever heard of *shrink* or *growth rays.* Discuss how fictional *shrink rays* make an object smaller without changing its shape. Ask what real machine makes things larger or smaller. **[a copy machine]**

Next ask questions like the following:

- Does part of the word *enlargement* suggest its meaning? **[*Large* suggests enlargement makes things bigger.]**
- Does part of the word *reduction* suggest its meaning? **[*Reduce* suggests reduction makes things smaller.]**

Have students integrate the concepts from this activity by working in small groups to make graphic organizers for the lesson vocabulary.

5 Assess & Remediate

Lesson Quiz

1. Is $D_n(\triangle DEF) = \triangle D'E'F'$ an enlargement or a reduction? What is the scale factor n of the dilation?

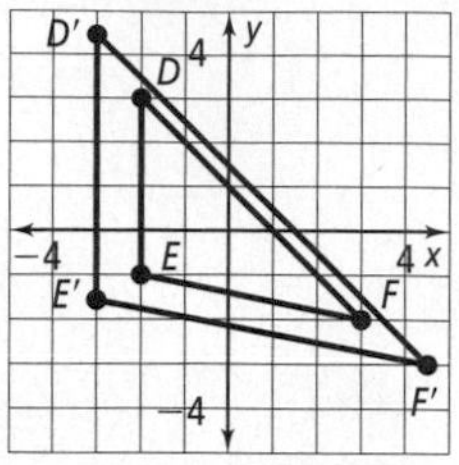

2. **Do you UNDERSTAND?** Quadrilateral $WXYZ$ has coordinates $W(-3, 2)$, $X(1, 3)$, $Y(2, -2)$, and $Z(-1, -2)$. What are the coordinates of the vertices of $D_2(WXYZ)$? Graph the preimage and image on a coordinate grid.

ANSWERS TO LESSON QUIZ

1. enlargement, 1.5
2. $W'(-6, 4)$, $X'(2, 6)$, $Y'(4, -4)$, $Z'(-2, -4)$.

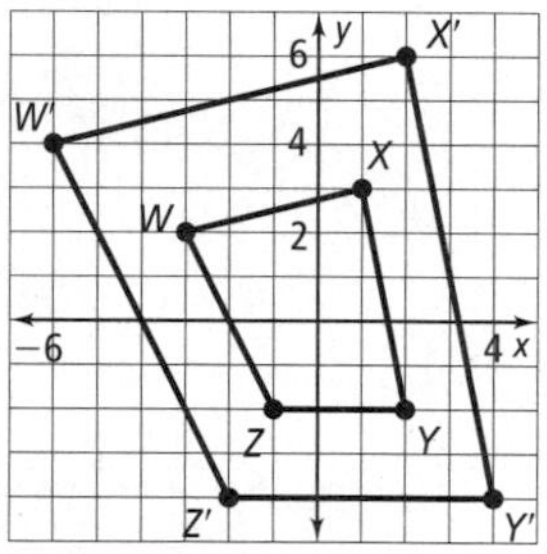

PRESCRIPTION FOR REMEDIATION

Use the student work on the Lesson Quiz to prescribe a differentiated review assignment.

Points	Differentiated Remediation
0	Intervention
1	On-level
2	Extension

PowerGeometry.com

5 Assess & Remediate

Assign the Lesson Quiz. Appropriate intervention, practice, or enrichment is automatically generated based on student performance.

Intervention

- **Reteaching** (2 pages) Provides reteaching and practice exercises for the key lesson concepts. Use with struggling students or absent students.
- **English Language Learner Support** Helps students develop and reinforce mathematical vocabulary and key concepts.

All-in-One Resources/Online

Reteaching

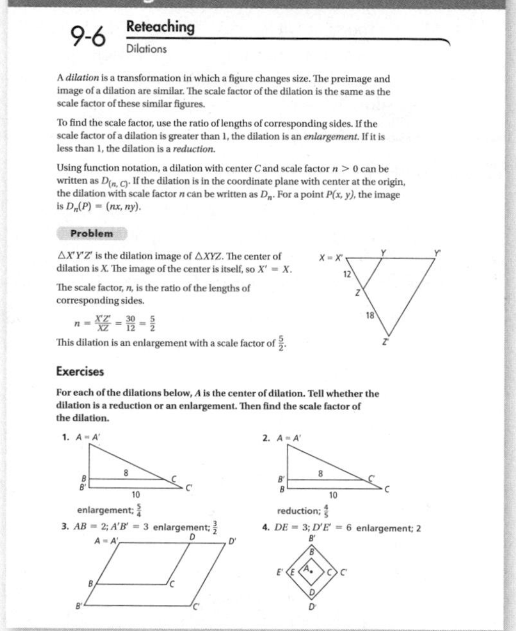
9-6 Reteaching
Dilations

A *dilation* is a transformation in which a figure changes size. The preimage and image of a dilation are similar. The scale factor of the dilation is the same as the scale factor of these similar figures.

To find the scale factor, use the ratio of lengths of corresponding sides. If the scale factor of a dilation is greater than 1, the dilation is an *enlargement*. If it is less than 1, the dilation is a *reduction*.

Using function notation, a dilation with center C and scale factor $n > 0$ can be written as $D_{(n, C)}$. If the dilation is in the coordinate plane with center at the origin, the dilation with scale factor n can be written as D_n. For a point $P(x, y)$, the image is $D_n(P) = (nx, ny)$.

Problem

$\triangle X'Y'Z'$ is the dilation image of $\triangle XYZ$. The center of dilation is X. The image of the center is itself, so $X' = X$.

The scale factor, n, is the ratio of the lengths of corresponding sides.

$n = \frac{X'Z'}{XZ} = \frac{30}{12} = \frac{5}{2}$

This dilation is an enlargement with a scale factor of $\frac{5}{2}$.

Exercises

For each of the dilations below, A is the center of dilation. Tell whether the dilation is a reduction or an enlargement. Then find the scale factor of the dilation.

1. $A = A'$ enlargement; $\frac{5}{4}$
2. $A = A'$ reduction; $\frac{4}{5}$
3. $AB = 2$; $A'B' = 3$ enlargement; $\frac{3}{2}$
4. $DE = 3$; $D'E' = 6$ enlargement; 2

All-in-One Resources/Online

English Language Learner Support

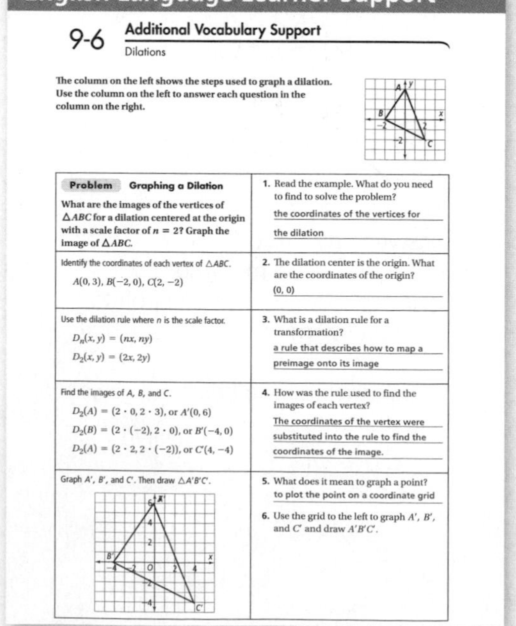
9-6 Additional Vocabulary Support
Dilations

The column on the left shows the steps used to graph a dilation. Use the column on the left to answer each question in the column on the right.

Problem Graphing a Dilation What are the images of the vertices of $\triangle ABC$ for a dilation centered at the origin with a scale factor of $n = 2$? Graph the image of $\triangle ABC$.	1. Read the example. What do you need to find to solve the problem? the coordinates of the vertices for the dilation
Identify the coordinates of each vertex of $\triangle ABC$. $A(0, 3)$, $B(-2, 0)$, $C(2, -2)$	2. The dilation center is the origin. What are the coordinates of the origin? (0, 0)
Use the dilation rule where n is the scale factor. $D_n(x, y) = (nx, ny)$ $D_2(x, y) = (2x, 2y)$	3. What is a dilation rule for a transformation? a rule that describes how to map a preimage onto its image
Find the images of A, B, and C. $D_2(A) = (2 \cdot 0, 2 \cdot 3)$, or $A'(0, 6)$ $D_2(B) = (2 \cdot (-2), 2 \cdot 0)$, or $B'(-4, 0)$ $D_2(A) = (2 \cdot 2, 2 \cdot (-2))$, or $C'(4, -4)$	4. How was the rule used to find the images of each vertex? The coordinates of the vertex were substituted into the rule to find the coordinates of the image.
Graph A', B', and C'. Then draw $\triangle A'B'C'$.	5. What does it mean to graph a point? to plot the point on a coordinate grid 6. Use the grid to the left to graph A', B', and C' and draw $A'B'C'$.

Differentiated Remediation *continued*

On-Level

- **Practice** (2 pages) Provides extra practice for each lesson. For simpler practice exercises, use the Form K Practice pages found in the All-in-One Teaching Resources and online.
- **Think About a Plan** Helps students develop specific problem-solving skills and strategies by providing scaffolded guiding questions.
- **Standardized Test Prep** Focuses on all major exercises, all major question types, and helps students prepare for the high-stakes assessments.

Extension

- **Enrichment** Provides students with interesting problems and activities that extend the concepts of the lesson.
- **Activities, Games, and Puzzles** Worksheets that can be used for concepts development, enrichment, and for fun!

Practice and Problem Solving Wkbk/ All-in-One Resources/Online

Practice page 1

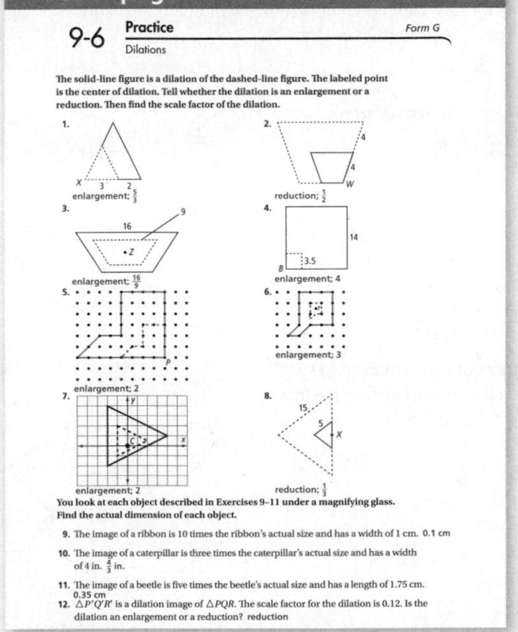

9-6 Practice — Dilations — Form G

The solid-line figure is a dilation of the dashed-line figure. The labeled point is the center of dilation. Tell whether the dilation is an enlargement or a reduction. Then find the scale factor of the dilation.

You look at each object described in Exercises 9–11 under a magnifying glass. Find the actual dimension of each object.

9. The image of a ribbon is 10 times the ribbon's actual size and has a width of 1 cm. 0.1 cm
10. The image of a caterpillar is three times the caterpillar's actual size and has a width of 4 in. $\frac{4}{3}$ in.
11. The image of a beetle is five times the beetle's actual size and has a length of 1.75 cm. 0.35 cm
12. $\triangle P'Q'R'$ is a dilation image of $\triangle PQR$. The scale factor for the dilation is 0.12. Is the dilation an enlargement or a reduction? reduction

Practice and Problem Solving Wkbk/ All-in-One Resources/Online

Practice page 2

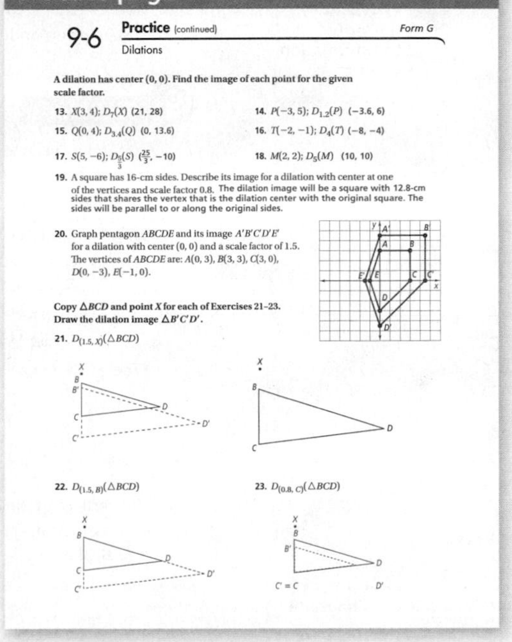

9-6 Practice (continued) — Dilations — Form G

A dilation has center (0, 0). Find the image of each point for the given scale factor.

13. $X(3, 4)$; $D_7(X)$ (21, 28)
14. $P(-3, 5)$; $D_{1.2}(P)$ (−3.6, 6)
15. $Q(0, 4)$; $D_{3.4}(Q)$ (0, 13.6)
16. $T(-2, -1)$; $D_4(T)$ (−8, −4)
17. $S(5, -6)$; $D_{\frac{5}{3}}(S)$ $(\frac{25}{3}, -10)$
18. $M(2, 2)$; $D_5(M)$ (10, 10)
19. A square has 16-cm sides. Describe its image for a dilation with center at one of the vertices and scale factor 0.8. The dilation image will be a square with 12.8-cm sides that shares the vertex that is the center with the original square. The sides will be parallel to or along the original sides.
20. Graph pentagon $ABCDE$ and its image $A'B'C'D'E'$ for a dilation with center (0, 0) and a scale factor of 1.5. The vertices of $ABCDE$ are: $A(0, 3)$, $B(3, 3)$, $C(3, 0)$, $D(0, -3)$, $E(-1, 0)$.

Copy $\triangle BCD$ and point X for each of Exercises 21–23. Draw the dilation image $\triangle B'C'D'$.

21. $D_{(1.5,\,X)}(\triangle BCD)$
22. $D_{(1.5,\,B)}(\triangle BCD)$
23. $D_{(0.8,\,C)}(\triangle BCD)$

All-in-One Resources/Online

Enrichment

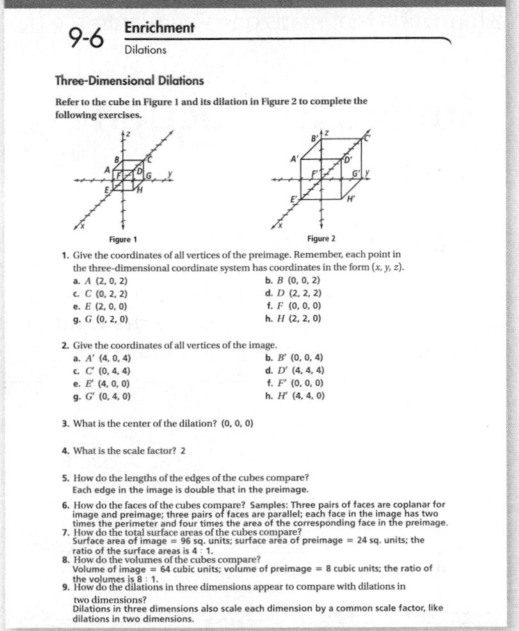

9-6 Enrichment — Dilations

Three-Dimensional Dilations

Refer to the cube in Figure 1 and its dilation in Figure 2 to complete the following exercises.

1. Give the coordinates of all vertices of the preimage. Remember, each point in the three-dimensional coordinate system has coordinates in the form (x, y, z).
   a. A (2, 0, 2) b. B (0, 0, 2) c. C (0, 2, 2) d. D (2, 2, 2) e. E (2, 0, 0) f. F (0, 0, 0) g. G (0, 2, 0) h. H (2, 2, 0)
2. Give the coordinates of all vertices of the image.
   a. A' (4, 0, 4) b. B' (0, 0, 4) c. C' (0, 4, 4) d. D' (4, 4, 4) e. E' (4, 0, 0) f. F' (0, 0, 0) g. G' (0, 4, 0) h. H' (4, 4, 0)
3. What is the center of the dilation? (0, 0, 0)
4. What is the scale factor? 2
5. How do the lengths of the edges of the cubes compare? Each edge in the image is double that in the preimage.
6. How do the faces of the cubes compare? Samples: Three pairs of faces are coplanar for image and preimage; three pairs of faces are parallel; each face in the image has two times the perimeter and four times the area of the corresponding face in the preimage.
7. How do the total surface areas of the cubes compare? Surface area of image = 96 sq. units; surface area of preimage = 24 sq. units; the ratio of the surface areas is 4 : 1.
8. How do the volumes of the cubes compare? Volume of image = 64 cubic units; volume of preimage = 8 cubic units; the ratio of the volumes is 8 : 1.
9. How do the dilations in three dimensions appear to compare with dilations in two dimensions? Dilations in three dimensions also scale each dimension by a common scale factor, like dilations in two dimensions.

Practice and Problem Solving Wkbk/ All-in-One Resources/Online

Think About a Plan

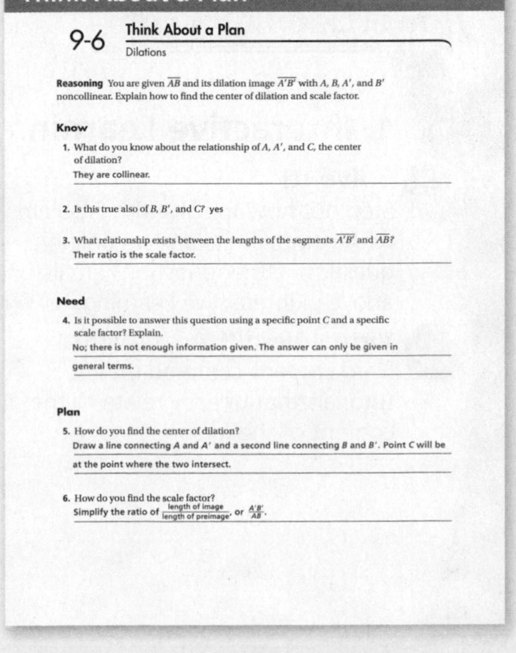

9-6 Think About a Plan — Dilations

Reasoning You are given $\overline{AB}$ and its dilation image $\overline{A'B'}$ with A, B, A', and B' noncollinear. Explain how to find the center of dilation and scale factor.

Know

1. What do you know about the relationship of A, A', and C, the center of dilation? They are collinear.
2. Is this true also of B, B', and C? yes
3. What relationship exists between the lengths of the segments $\overline{A'B'}$ and $\overline{AB}$? Their ratio is the scale factor.

Need

4. Is it possible to answer this question using a specific point C and a specific scale factor? Explain. No; there is not enough information given. The answer can only be given in general terms.

Plan

5. How do you find the center of dilation? Draw a line connecting A and A' and a second line connecting B and B'. Point C will be at the point where the two intersect.
6. How do you find the scale factor? Simplify the ratio of $\frac{\text{length of image}}{\text{length of preimage}}$, or $\frac{A'B'}{AB}$.

Practice and Problem Solving Wkbk/ All-in-One Resources/Online

Standardized Test Prep

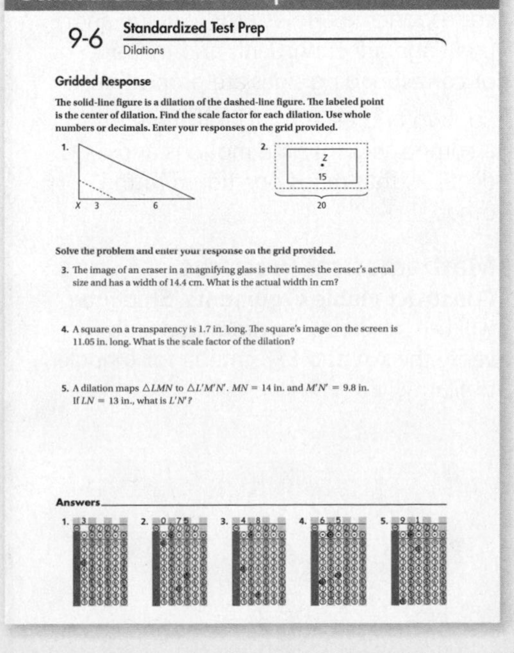

9-6 Standardized Test Prep — Dilations

Gridded Response

The solid-line figure is a dilation of the dashed-line figure. The labeled point is the center of dilation. Find the scale factor for each dilation. Use whole numbers or decimals. Enter your responses on the grid provided.

Solve the problem and enter your response on the grid provided.

3. The image of an eraser in a magnifying glass is three times the eraser's actual size and has a width of 14.4 cm. What is the actual width in cm?
4. A square on a transparency is 1.7 in. long. The square's image on the screen is 11.05 in. long. What is the scale factor of the dilation?
5. A dilation maps $\triangle LMN$ to $\triangle L'M'N'$. $MN = 14$ in. and $M'N' = 9.8$ in. If $LN = 13$ in., what is $L'N'$?

Answers

Online Teacher Resource Center

Activities, Games, and Puzzles

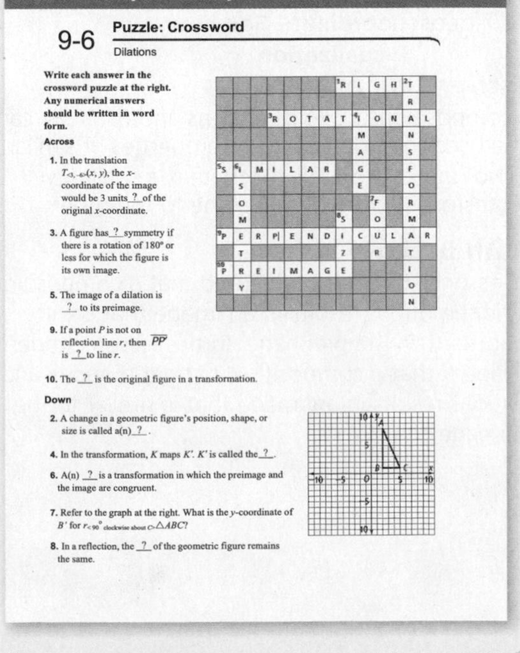

9-6 Puzzle: Crossword — Dilations

Write each answer in the crossword puzzle at the right. Any numerical answers should be written in word form.

Across

1. In the translation $T_{\langle -3,-6\rangle}(x, y)$, the x-coordinate of the image would be 3 units _?_ of the original x-coordinate.
3. A figure has _?_ symmetry if there is a rotation of 180° or less for which the figure is its own image.
5. The image of a dilation is _?_ to its preimage.
9. If a point P is not on reflection line r, then $\overline{PP'}$ is _?_ to line r.
10. The _?_ is the original figure in a transformation.

Down

2. A change in a geometric figure's position, shape, or size is called a(n) _?_.
4. In the transformation, K maps K'. K' is called the _?_.
6. A(n) _?_ is a transformation in which the preimage and the image are congruent.
7. Refer to the graph at the right. What is the y-coordinate of B' for $r_{(90°\ \text{clockwise about } C)}\triangle ABC$?
8. In a reflection, the _?_ of the geometric figure remains the same.

1 Interactive Learning

Solve It!

PURPOSE To describe the composition of transformations in a mapping of two triangles.
PROCESS Students should recognize that because the image is larger than the preimage, a dilation is one of the transformations used.

FACILITATE

Q Do the figures appear to have the same shape? Do they appear to have the same size? Explain. **[yes; no; The triangles appear to be similar.]**

ANSWER See Solve It in Answers on next page.
CONNECT THE MATH In the Solve It, students describe how to map from one triangle to a similar triangle through a composition of transformations. In this lesson, students will learn that two figures are similar if and only if there is a similarity transformation that maps one figure onto the other.

2 Guided Instruction

Problem 1

In this problem, students will draw the image of a similarity transformation.

Q Why is the reflection performed before the dilation? **[The transformation to the right of the ∘ symbol is always performed first.]**

9-7 Similarity Transformations

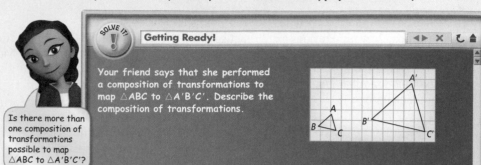

Common Core State Standards
G-SRT.A.2 Given two figures, use the definition of similarity in terms of similarity transformations to decide if they are similar . . . **Also G-SRT.A.3**
MP 1, MP 2, MP 3, MP 4

Objectives To identify similarity transformations and verify properties of similarity

Is there more than one composition of transformations possible to map △ABC to △A′B′C′?

Getting Ready!

Your friend says that she performed a composition of transformations to map △ABC to △A′B′C′. Describe the composition of transformations.

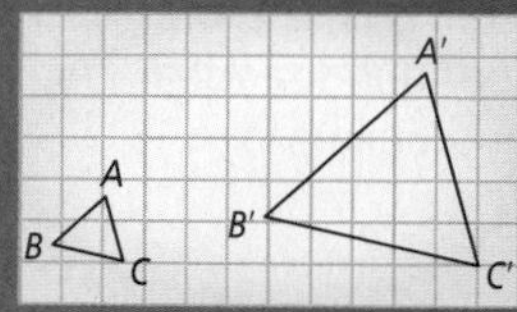

MATHEMATICAL PRACTICES

In the Solve It, you used a composition of a rigid motion and a dilation to describe the mapping from △ABC to △A′B′C′.

Lesson Vocabulary
- similarity transformation
- similar

Essential Understanding You can use compositions of rigid motions and dilations to help you understand the properties of similarity.

Problem 1 Drawing Transformations

△DEF has vertices D(2, 0), E(1, 4), and F(4, 2). What is the image of △DEF when you apply the composition $D_{1.5} \circ R_{y\text{-axis}}$?

Step 1 Find the vertices of $R_{y\text{-axis}}(\triangle DEF)$. Then connect the vertices to draw the image.
$R_{y\text{-axis}}(D) = D'(-2, 0)$
$R_{y\text{-axis}}(E) = E'(-1, 4)$
$R_{y\text{-axis}}(F) = F'(-4, 2)$

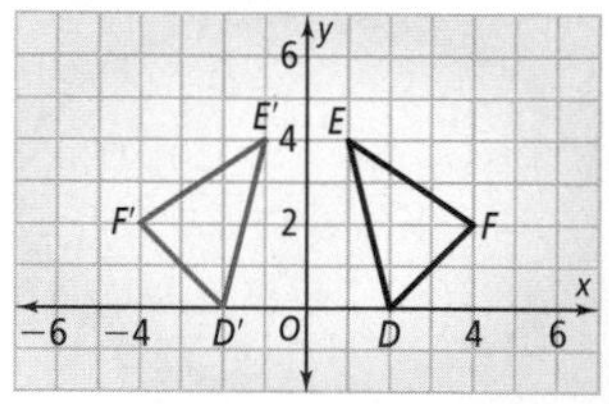

Step 2 Find the vertices of the dilation of △D′E′F′. Then connect the vertices to draw the image.
$D_{1.5}(D') = D''(-3, 0)$
$D_{1.5}(E') = E''(-1.5, 6)$
$D_{1.5}(F') = F''(-6, 3)$

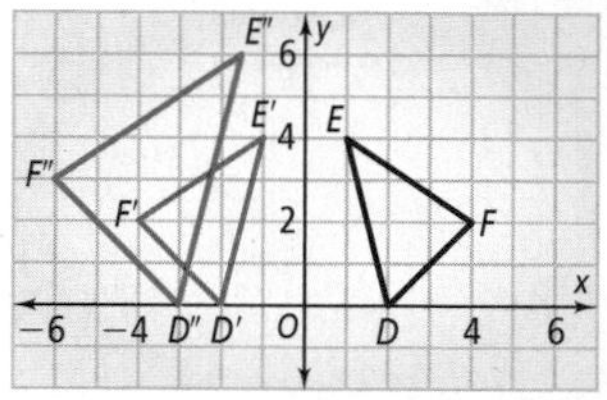

The vertices of the image after the composition of transformations are $D''(-3, 0)$, $E''(-1.5, 6)$, and $F''(-6, 3)$.

9-7 Preparing to Teach

BIG ideas Coordinate Geometry
Visualization

ESSENTIAL UNDERSTANDINGS
- Compositions of rigid motions and dilations can be used to understand the properties of similarity.
- Two figures are similar if there is a similarity transformation that maps one to the other.

Math Background

In Lesson 9-5, students learned that a composition of rigid motions results in an image that is congruent to the preimage. In this lesson, students will learn that a composition of rigid motions and dilations results in an image that is similar to the preimage.

When a figure undergoes a composition of rigid motions and dilations, the corresponding angles of the image and preimage are congruent, and the ratios of corresponding sides are proportional.

So, two figures are similar if there is a composition of rigid motions and dilations that maps one figure to the other.

Mathematical Practices

Construct viable arguments. Students will use similarity transformations to verify the AA and SSS criteria for triangle similarity.

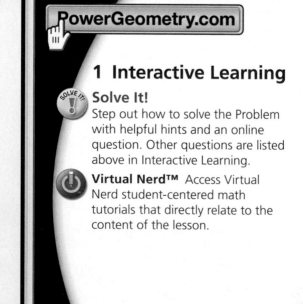

PowerGeometry.com

1 Interactive Learning

Solve It!
Step out how to solve the Problem with helpful hints and an online question. Other questions are listed above in Interactive Learning.

Virtual Nerd™ Access Virtual Nerd student-centered math tutorials that directly relate to the content of the lesson.

Got It? 1. Reasoning $\triangle LMN$ has vertices $L(-4, 2)$, $M(-3, -3)$, and $N(-1, 1)$. Suppose the triangle is translated 4 units right and 2 units up and then dilated by a scale factor of 0.5 with center of dilation at the origin. Sketch the resulting image of the composition of transformations.

Problem 2 Describing Transformations

What is a composition of rigid motions and a dilation that maps $\triangle RST$ to $\triangle PYZ$?

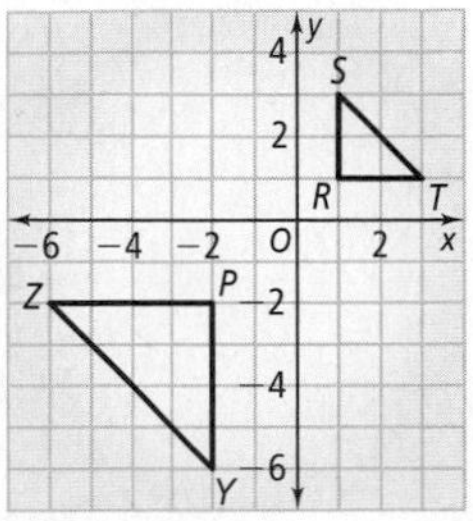

Know	Need	Plan
The vertices of the preimage and image	A composition of transformations that maps $\triangle RST$ to $\triangle PYZ$	Study the figures to determine how the image could have resulted from the preimage. Then use the vertices to verify the composition of transformations.

It appears that $\triangle RST$ was rotated and then enlarged to create $\triangle PYZ$. To verify the composition of transformations, begin by rotating the triangle 180° about the origin.

$r_{(180°, O)}(R) = R'(-1, -1)$ Use the rule $r_{(180°, O)}(x, y) = (-x, -y)$.

$r_{(180°, O)}(S) = S'(-1, -3)$

$r_{(180°, O)}(T) = T'(-3, -1)$

$\triangle PYZ$ appears to be about twice as large as $\triangle RST$. Scale the vertices of the intermediate image $R'S'T'$ to verify the composition.

$D_2(-1, -1) = P(-2, -2)$ Use the rule $D_2(x, y) = (2x, 2y)$.

$D_2(-1, -3) = Y(-2, -6)$

$D_2(-3, -1) = Z(-6, -2)$

The vertices of the dilation of $\triangle R'S'T'$ match the vertices of $\triangle PYZ$.

A rotation of 180° about the origin followed by a dilation with scale factor 2 maps $\triangle RST$ to $\triangle PYZ$.

Got It? 2. What is a composition of rigid motions and a dilation that maps trapezoid $ABCD$ to trapezoid $MNHP$?

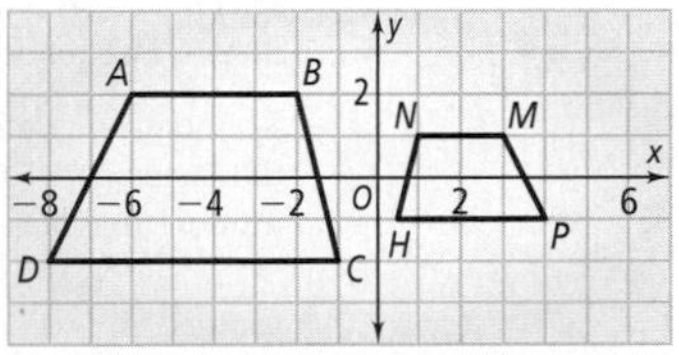

Got It?

Q How would you write this composition using function notation? **[$D_{0.5} \circ T_{<4, 2>}$]**

Problem 2

In this problem, students identify a sequence of transformations that maps a preimage onto an image.

Q In looking at the figures, what is the one type of transformation that was certainly used? Explain. **[dilation; The figures are different sizes.]**

Q Do you think there is more than one composition of transformations from $\triangle RST$ to $\triangle PYZ$? If so, give an example. **[Yes. Sample answer: Reflect $\triangle RST$ across the *y*-axis, then reflect across the *x*-axis, then dilate.]**

Got It?

Q What is a sequence of transformations that you could use to map from trapezoid *MNHP* back onto trapezoid *ABCD*? **[Sample answer: dilate by a scale factor of $\frac{4}{3}$, then reflect across the *y*-axis.]**

2 Guided Instruction

Each Problem is worked out and supported online.

Problem 1
Drawing Transformations

Problem 2
Describing Transformations

Problem 3
Finding Similarity Transformations

Problem 4
Determining Similarity

Support in Geometry Companion

- Vocabulary
- Key Concepts
- Got It?

Answers

Solve It!

Sample answer: translation followed by a dilation; Translate $\triangle ABC$ until points *B* and *B'* coincide. Then dilate by the appropriate scale factor until the two triangles overlap.

Got It?

1. $L''(0, 2)$, $M''(0.5, -0.5)$, $N''(1.5, 1.5)$
2. $D_{0.5} \circ R_{y\text{-axis}}$

Take Note

EXTENSION

Ask students what is needed in order to show that two figures are similar according to this Take Note definition.

Here's Why It Works

Students may have difficulty following the abstract algebraic proportions. Provide students with side lengths such as $PQ = 3$, $PR = 5$, $QR = 7$, $P'Q' = 6$, $P'R' = 10$, $Q'R' = 14$. Then have students calculate the scale factor k.

Problem 3

VISUAL LEARNERS

Some students may have difficulty seeing how a preimage was transformed to create the image. Have them simplify the process by only focusing on one pair of congruent angles or one pair of corresponding sides at a time.

Q $\overline{QA}$ corresponds to $\overline{ON}$. Which rigid motion(s) could have been performed to map $\overline{QA}$ to $\overline{ON}$? **[rotation and translation]**

Got It?

Pair students who have difficulty visualizing the similarity transformations with students who are more proficient to work together on the Got It problems. Ask them to share strategies on how to approach looking for similarity transformations.

Notice that the figures in Problems 1 and 2 appear to have the same shape but different sizes. Compositions of rigid motions and dilations map preimages to similar images. For this reason, they are called **similarity transformations**. Similarity transformations give you another way to think about similarity.

take note

Key Concept **Similar Figures**

Two figures are **similar** if and only if there is a similarity transformation that maps one figure onto the other.

Here's Why It Works Consider the composition of a rigid motion and a dilation shown at the right.

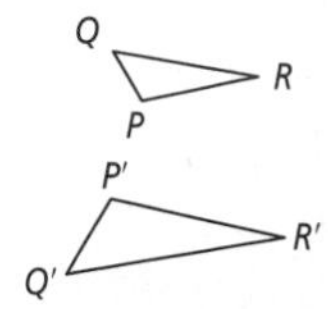

Because rigid motions and dilations preserve angle measure, $m\angle P = m\angle P'$, $m\angle Q = m\angle Q'$, and $m\angle R = m\angle R'$. So, corresponding angles are congruent.

Because there is a dilation, there is some scale factor k such that:

$$PQ = kP'Q' \quad QR = kQ'R' \quad PR = kP'R'$$

$$k = \frac{PQ}{P'Q'} \quad k = \frac{QR}{Q'R'} \quad k = \frac{PR}{P'R'}$$

So, $\frac{PQ}{P'Q'} = \frac{QR}{Q'R'} = \frac{PR}{P'R'}$.

Problem 3 Finding Similarity Transformations

Is there a similarity transformation that maps $\triangle PAQ$ to $\triangle TNO$? If so, identify the similarity transformation and write a similarity statement. If not, explain.

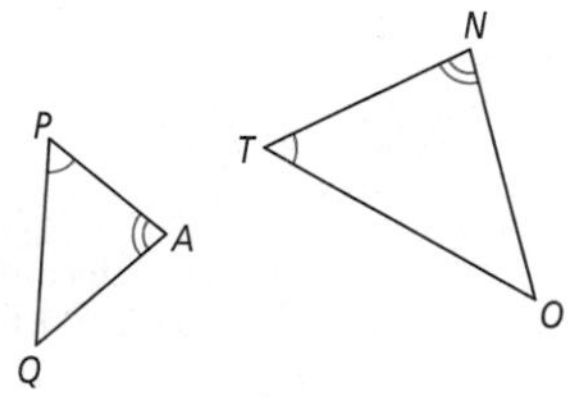

Think

Does it matter what the center of dilation is?

No. All that matters is that $k \cdot PA = TN$.

Although $PA \neq TN$, there is a scale factor k such that $k \cdot PA = TN$. Dilate $\triangle PAQ$ using this scale factor. Then $\overline{P'A'} \cong \overline{TN}$. Since dilations preserve angle measure, you also know that $\angle P' \cong \angle T$ and $\angle A' \cong \angle N$. Therefore, $\triangle P'A'Q' \cong \triangle TNO$ by ASA. This means that there is a sequence of rigid motions that maps $\triangle P'A'Q'$ onto $\triangle TNO$.

So, there is a dilation that maps $\triangle PAQ$ to $\triangle P'A'Q'$, and a sequence of rigid motions that maps $\triangle P'A'Q'$ to $\triangle TNO$. Therefore, there is a composition of a dilation and rigid motions that maps $\triangle PAQ$ onto $\triangle TNO$.

Got It? **3.** Is there a similarity transformation that maps $\triangle JKL$ to $\triangle RST$? If so, identify the similarity transformation and write a similarity statement. If not, explain.

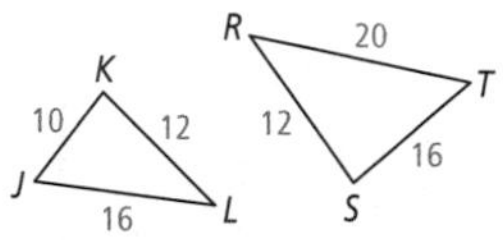

Additional Problems

1. Rectangle $HIJK$ has vertices $H(2, 3)$, $I(6, 3)$, $J(6, 5)$, and $K(2, 5)$. What is the image of $HIJK$ when you apply the composition $D_2 \circ r_{(180°, O)}$?

ANSWER $H''(-4, -6)$, $I''(-12, -6)$, $J''(-12, -10)$, $K''(-4, -10)$

2. What is a composition of rigid motions and a dilation that maps $\triangle WYB$ to $\triangle GCF$?

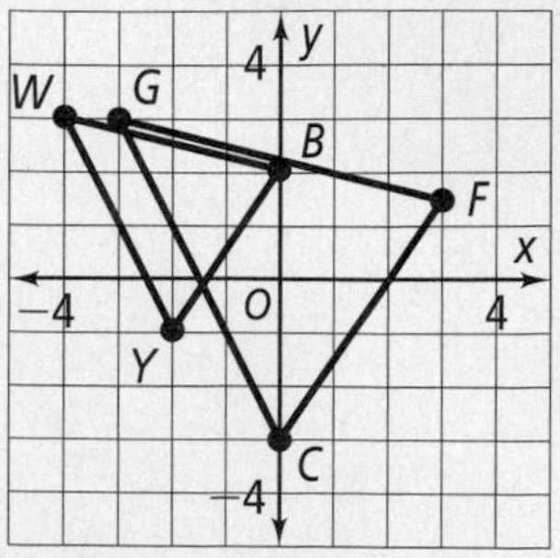

ANSWER $D_{1.5} \circ T_{<2, -1>}$

3. Is there a similarity transformation that maps $\triangle JAV$ to $\triangle NUR$? If so, identify the similarity transformation and write a similarity statement.

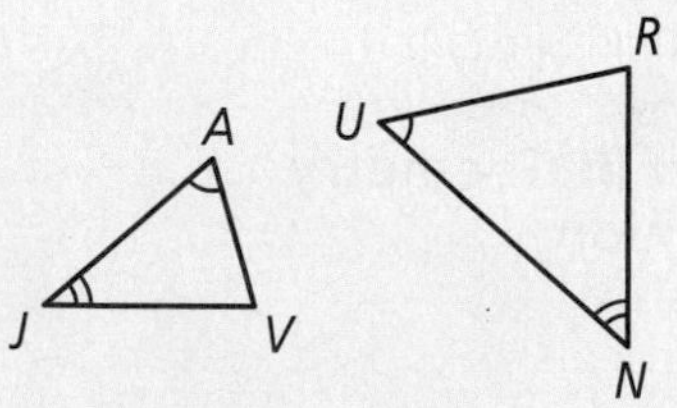

ANSWER Yes, a rotation then a dilation. $\triangle JAV \sim \triangle NUR$

4. Are the figures shown below similar? Explain.

ANSWER No, there is no similarity transformation that maps one figure to the other.

Similarity transformations provide a powerful general approach to similarity. In Problem 3, you used similarity transformations to verify the AA Postulate for triangle similarity. Another advantage to the transformational approach to similarity is that you can apply it to figures other than polygons.

Plan

How can you determine whether two figures are similar if you have no information about side lengths or angle measures?
Any two plane figures are similar if you can find a similarity transformation that maps one onto the other.

Problem 4 Determining Similarity

A new company is using a computer program to design its logo. Are the two figures used in the logo so far similar?

If you can find a similarity transformation between two figures, then you know they are similar. The smaller lightning bolt can be translated so that the tips coincide. Then it can be enlarged by some scale factor so that the two bolts overlap.

The figures are similar because there is a similarity transformation that maps one figure onto the other. The transformation is a translation followed by a dilation.

Got It? 4. Are the figures at the right similar? Explain.

Lesson Check

Do you know HOW?

Use the diagram below for Exercises 1 and 2.

1. What is a similarity transformation that maps $\triangle RST$ to $\triangle JKL$?

2. What are the coordinates of $(D_{\frac{1}{4}} \circ r_{(180°,\,O)})(\triangle RST)$?

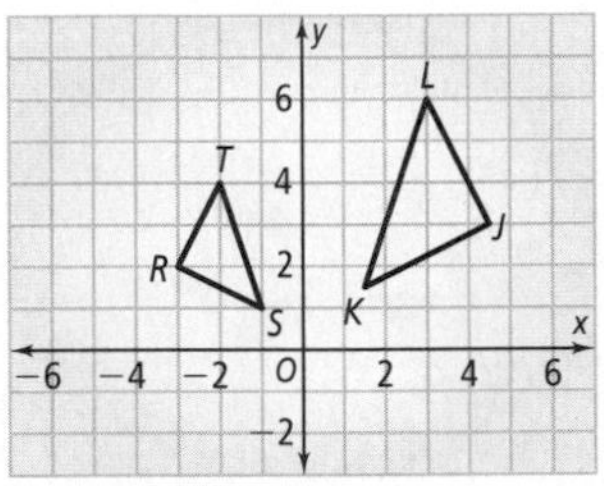

Do you UNDERSTAND? MATHEMATICAL PRACTICES

3. Vocabulary Describe how the word *dilation* is used in areas outside of mathematics. How do these applications relate the mathematical definition?

4. Open Ended For $\triangle TUV$ at the right, give the vertices of a similar triangle after a similarity transformation that uses at least 1 rigid motion.

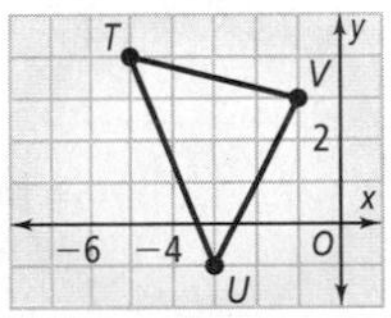

Problem 4

In this problem, students will determine if two figures are similar.

Q What do you know about the lengths of the segments used to create each lightning bolt? **[Corresponding segments are proportional.]**

Got It?

Q If you divide the height of the smaller face by the height of the larger face, what does the numerical result represent? **[the scale factor of the dilation]**

3 Lesson Check

Do you know HOW? ERROR INTERVENTION

- If students have trouble solving Exercise 1, then ask them which angles appear to be congruent.

Do you UNDERSTAND?

- In Exercise 3, ask students to share their examples with the rest of the class. Have students brainstorm different real world applications of similarity transformations.

Close

Q What do you know about corresponding sides and corresponding angles of figures that are mapped using a composition of rigid motions and dilations? **[corresponding angles are congruent; corresponding sides are proportional]**

Q How can you show that two figures are similar? **[Find a composition of rigid motions and dilations that maps one figure onto the other.]**

Answers

Got It? (continued)

3. No, there is no similarity transformation that maps one triangle to the other. The side lengths are not all proportional.

4. Yes, there is a similarity transformation: rotation, translation, and then dilation.

Lesson Check

1. Answers may vary. Sample: $D_{1.5} \circ R_{y\text{-axis}}$

2. $R''\left(\frac{3}{4}, -\frac{1}{2}\right)$, $S''\left(\frac{1}{4}, -\frac{1}{4}\right)$, $T''\left(\frac{1}{2}, -1\right)$

3. Sample answer: The pupils of your eyes dilate when you go from dark to bright locations or from bright to dark. The pupils are reduced or enlarged proportionally to form similar pupils.

4. Answers will vary. Check students' work.

4 Practice

ASSIGNMENT GUIDE

Basic: 5–17, 22, 24
Average: 5–15 odd, 16–24
Advanced: 5–15 odd, 16–29
Standardized Test Prep: 30–33
Mixed Review: 34–39

Mathematical Practices are supported by exercises with red headings. Here are the Practices supported in this lesson:

MP 1: Make Sense of Problems Ex. 17
MP 2: Reason Abstractly Ex. 4
MP 3: Construct Arguments Ex. 25, 29
MP 3: Communicate Ex. 16, 18

Applications exercises have blue headings. Exercises 23, 24, and 26 support MP 4: Model.

EXERCISE 23: Use the Think About a Plan worksheet in the **Practice and Problem Solving Workbook** (also available in the Teaching Resources in print and online) to further support students' development in becoming independent learners.

HOMEWORK QUICK CHECK

To check students' understanding of key skills and concepts, go over Exercises 9, 11, 17, 23, and 24.

Practice and Problem-Solving Exercises

MATHEMATICAL PRACTICES

A Practice

$\triangle MAT$ has vertices $M(6, -2)$, $A(4, -5)$, and $T(1, -2)$. For each of the following, sketch the image of the composition of transformations. See Problem 1.

5. reflection across the x-axis followed by a dilation by a scale factor of 0.5

6. rotation of 180° about the origin followed by a dilation by a scale factor of 1.5

7. translation 6 units up followed by a reflection across the y-axis and then a dilation by a scale factor of 2

For each graph, describe the composition of transformations that maps $\triangle FGH$ to $\triangle QRS$. See Problem 2.

8.
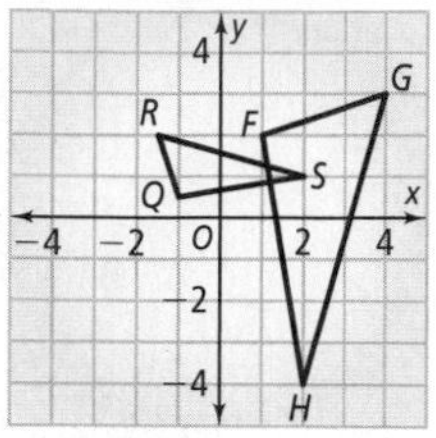

9.
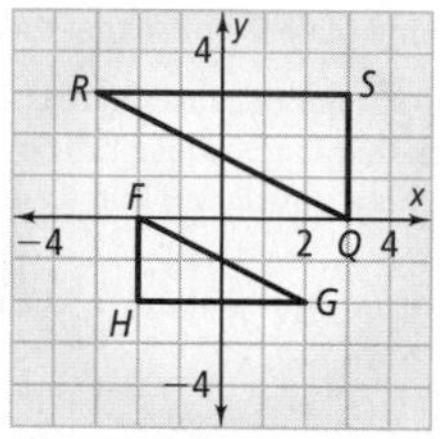

10.
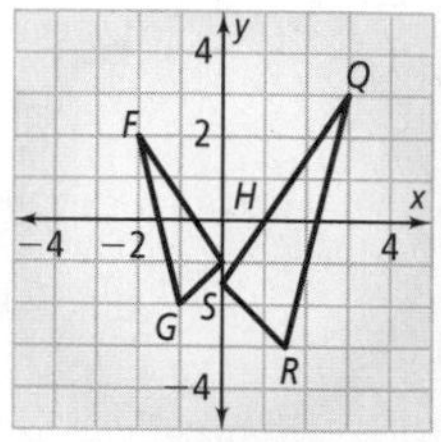

For each pair of figures, determine if there is a similarity transformation that maps one figure onto the other. If so, identify the similarity transformation and write a similarity statement. If not, explain. See Problem 3.

11.
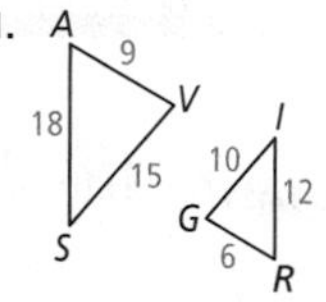

12.
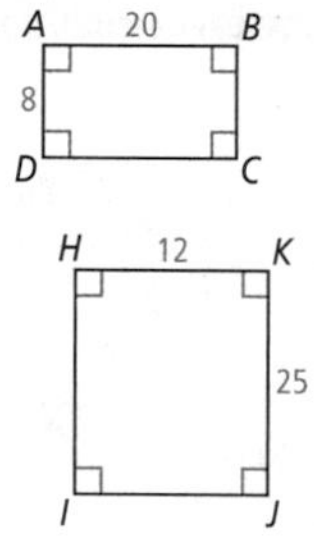

13.
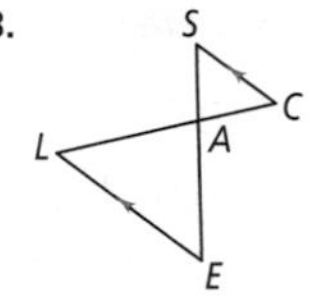

Determine whether or not each pair of figures below is similar. Explain your reasoning. See Problem 4.

14.

15.

Answers

Practice and Problem-Solving Exercises

5.
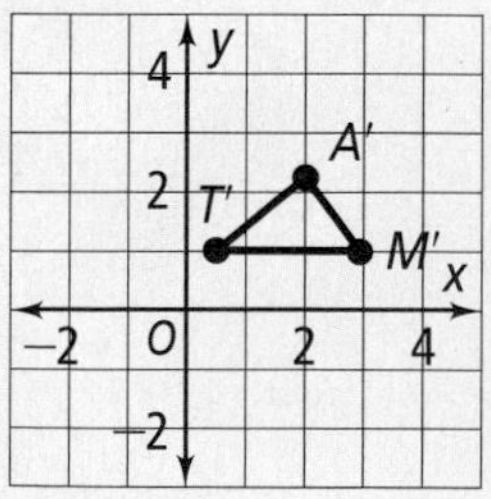

6.
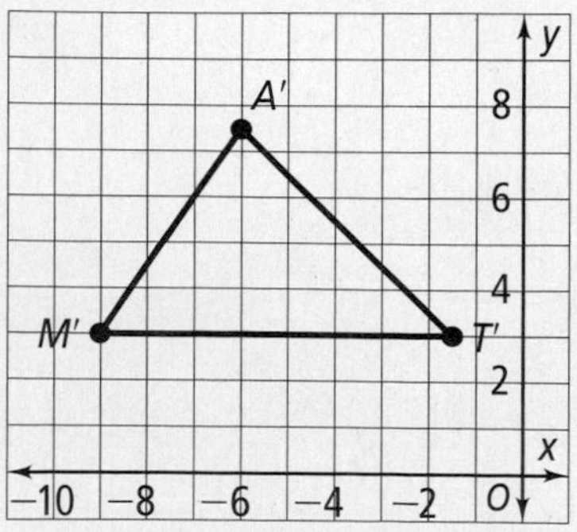

7.
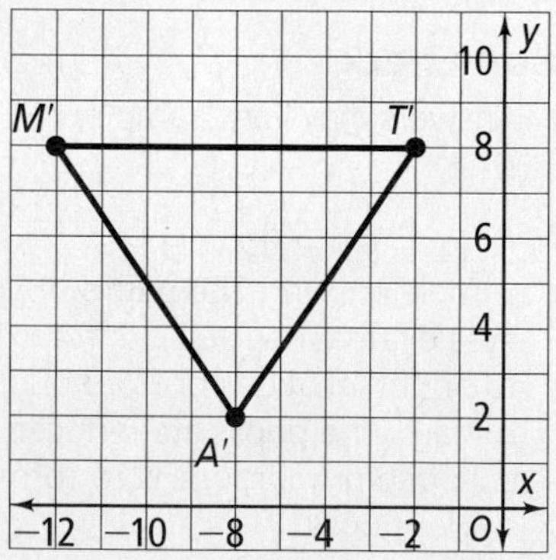

8. $D_{0.5} \circ r_{(90°, O)}$

9. $D_{1.5} \circ r_{(180°, O)}$

10. $D_{0.5} \circ R_{y\text{-axis}}$

11. Answers may vary. Sample answer: $\triangle AVS$ is similar to $\triangle RGI$. Translate $\triangle RGI$ so that points R and A coincide. Rotate by 180°. Then dilate with center A and scale factor 1.5.

12. Answers may vary. Sample answer: There is no similarity transformation. The figures are not similar because corresponding sides are not proportional.

13. Answers may vary. Sample answer: The similarity transformation is a rotation about point A followed by a dilation with respect to point A. $\triangle SCA$ is similar to $\triangle ELA$.

14. Answers may vary. Sample answer: Yes, there is a similarity transformation between the two figures: a translation and a rotation followed by a dilation.

15. Answers may vary. Sample answer: Yes, there is a similarity transformation between the two figures: a translation and a rotation followed by a dilation.

B Apply

16. **Writing** Your teacher uses geometry software program to plot $\triangle ABC$ with vertices $A(2, 1)$, $B(6, 1)$, and $C(6, 4)$. Then he used a similarity transformation to plot $\triangle DEF$ with vertices $D(-4, -2)$, $E(-12, -2)$, and $F(-12, -8)$. The corresponding angles of the two triangles are congruent. How can the Distance Formula be used to verify that the ratios of the corresponding sides are proportional? Verify that the figures are similar.

17. **Think About a Plan** Suppose that $\triangle JKL$ is formed by connecting the midpoints of $\triangle ABC$. Is $\triangle AJL$ similar to $\triangle ABC$? Explain.
    - How are the side lengths of $\triangle AJL$ related to the side lengths of $\triangle ABC$?
    - Can you find a similarity transformation that maps $\triangle AJL$ to $\triangle ABC$? Explain.

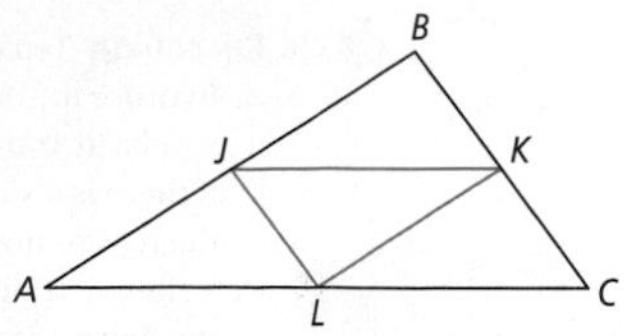

18. **Writing** What properties are preserved by rigid motions but not by similarity transformations?

Determine whether each statement is *always, sometimes,* or *never* true.

19. There is a similarity transformation between two rectangles.
20. There is a similarity transformation between two squares.
21. There is a similarity transformation between two circles.
22. There is a similarity transformation between a right triangle and an equilateral triangle.
23. **Indirect Measurement** A surveyor wants to use similar triangles to determine the distance across a lake as shown at the right.
    a. Are the two triangles in the figure similar? Justify your reasoning.
    b. What is the distance d across the lake?

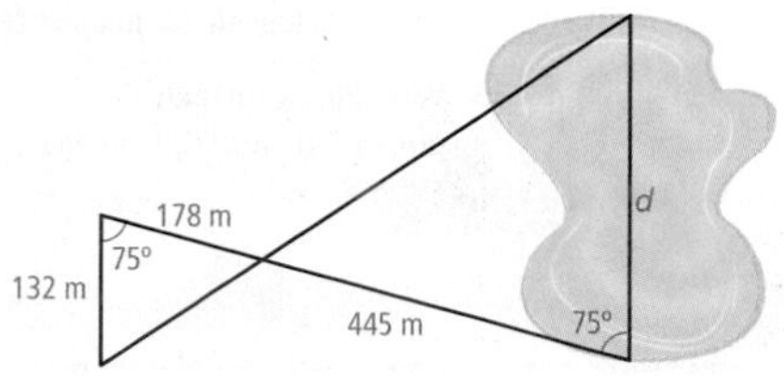

24. **Photography** A 4-inch by 6-inch rectangular photo is enlarged to fit an 8-inch by 10-inch frame. Are the two photographs similar? Explain.
25. **Reasoning** Is a rigid motion an example of a similarity transformation? Explain your reasoning and give an example.
26. **Art** A printing company enlarges a banner for a graduation party by a scale factor of 8.
    a. What are the dimensions of the larger banner?
    b. How can the printing company be sure that the enlarged banner is similar to the original?

16. Answers may vary. Sample answer: Use the Distance Formula to find the lengths of the sides. Then verify that the ratios of corresponding sides are proportional. $AB = 4$, $BC = 3$, $AC = 5$, $DE = 8$, $EF = 6$, and $DF = 10$. Since $\frac{AB}{DE} = \frac{BC}{EF} = \frac{AC}{DF} = 0.5$, the ratios of corresponding sides are proportional, and the figures are similar.
17. Yes, the triangles are similar because there is a similarity transformation that maps $\triangle AJL$ to $\triangle ABC$.
18. Distance is preserved by rigid motions, but not by similarity transformations.
19. sometimes
20. always
21. always
22. never
23. a. Yes, the triangles are similar because there is a similarity transformation between them: rotation followed by a dilation.;
    b. 330 m
24. No, the photos will not be similar. There is no similarity transformation that maps the smaller photo to the larger photo.
25. Answers may vary. Sample answer: Yes, a rigid motion is a similarity transformation with a scale factor of 1. The preimage and image of a rigid motion are congruent, so they are also similar.
26. a. 104 inches by 24 inches;
    b. Enlarging the banner is a dilation, which is an example of a similarity transformation.

Answers

Practice and Problem-Solving Exercises (continued)

27. Answers may vary. Sample answer: No, there is not a similarity transformation. To create $\triangle NOP$, $\triangle ABC$ is reflected across the x-axis. Then the x-coordinates are scaled by a factor of 5 and the y-coordinates are scaled by a factor of 4. Because the reflected triangle is 5 times as wide as $\triangle ABC$ but only 4 times as tall, the figures are not similar. Therefore, there is no similarity transformation between them.

28. Yes, the image of the transparency is rotated in space and dilated. The image on the wall is similar to the preimage on the transparency.

29. **a.** true; **b.** true; **c.** false

30. 2

31. 6

32. 38.9

33. 7.5

34. *H*, 180°; *I*, 180°; *N*, 180°; *O*, any rotation; *S*, 180°; *X*, 180°; *Z*, 180°

35. (–2, –2) or (7, 1)

36. 25 cm^2

37. 28 $in.^2$

38. 11.5 m^2

39. 1.5 ft^2

Challenge

27. If $\triangle ABC$ has vertices given by $A(u, v)$, $B(w, x)$, and $C(y, z)$, and $\triangle NOP$ has vertices given by $N(5u, -4v)$, $O(5w, -4x)$, and $P(5y, -4z)$, is there a similarity transformation that maps $\triangle ABC$ to $\triangle NOP$? Explain.

28. Overhead Projector When Mrs. Sheldon places a transparency on the screen of the overhead projector, the projector shows an enlargement of the transparency on the wall. Does this situation represent a similarity transformation? Explain.

29. Reasoning Tell whether each statement below is *true* or *false*.

a. In order to show that two figures are similar, it is sufficient to show that there is a similarity transformation that maps one figure to the other.

b. If there is a similarity transformation that maps one figure to another figure, then the figures are similar.

c. If there is a similarity transformation that maps one figure to another figure, then the figures are congruent.

Standardized Test Prep

GRIDDED RESPONSE

SAT/ACT

30. $\triangle STU$ has vertices $S(1, 2)$, $T(0, 5)$, and $U(-8, 0)$. What is the x-coordinate of S after a 270° rotation about the origin?

31. The diagonals of rectangle $PQRS$ intersect at O. $PO = 2x - 5$ and $OR = 7 - x$. What is the length of $\overline{QS}$?

32. The length of the hypotenuse of a 45°-45°-90° triangle is 55 in. What is the length of one of its legs to the nearest tenth of an inch?

33. You place a sprinkler so that it is equidistant from three rose bushes at points A, B, and C. How many feet is the sprinkler from A?

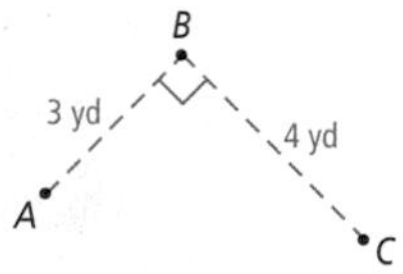

Mixed Review

34. Which capital letters of the alphabet are rotation images of themselves? Draw each letter and give an angle of rotation (< 360°). See Lesson 9-3.

35. Three vertices of an isosceles trapezoid are $(-2, 1)$, $(1, 4)$, and $(4, 4)$. Find all possible coordinates for the fourth vertex. See Lesson 6-7.

Get Ready! **To prepare for Lesson 10-1, do Exercises 34–37.**

Find the area of each figure. See Lesson 1-8.

36. a square with 5-cm sides

37. a rectangle with base 4 in. and height 7 in.

38. a 4.6 m-by-2.5 m rectangle

39. a rectangle with length 3 ft and width $\frac{1}{2}$ ft

9-7 Lesson Resources

Differentiated Remediation

Additional Instructional Support

Geometry Companion

Students can use the **Geometry Companion** worktext (4 pages)as you teach the lesson. Use the Companion to support

- Solve It!
- New Vocabulary
- Key Concepts

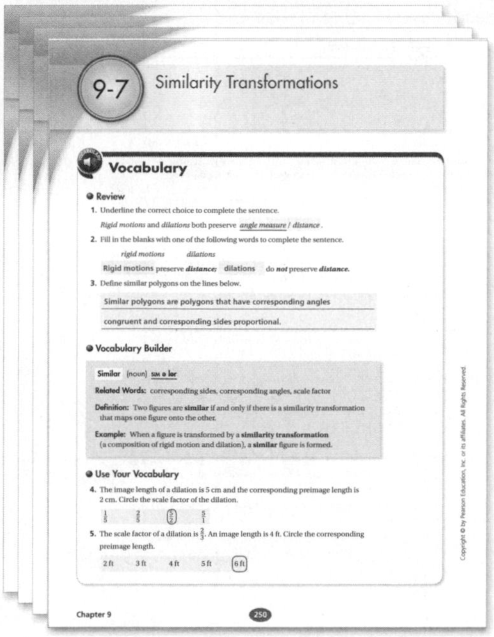

9-7 Similarity Transformations

Vocabulary

Review

1. Underline the correct choice to complete the sentence.
   Rigid motions and *dilations* both preserve *angle measure / distance*.
2. Fill in the blanks with one of the following words to complete the sentence.
   rigid motions *dilations*
   Rigid motions preserve ***distance;*** dilations do ***not*** preserve ***distance.***
3. Define similar polygons on the lines below.
   Similar polygons are polygons that have corresponding angles congruent and corresponding sides proportional.

Vocabulary Builder

Similar (noun) SIM ə lər

Related Words: corresponding sides, corresponding angles, scale factor

Definition: Two figures are **similar** if and only if there is a similarity transformation that maps one figure onto the other.

Example: When a figure is transformed by a **similarity transformation** (a composition of rigid motion and dilation), a **similar** figure is formed.

Use Your Vocabulary

4. The image length of a dilation is 5 cm and the corresponding preimage length is 2 cm. Circle the scale factor of the dilation.
   $\frac{1}{5}$ $\frac{2}{5}$ $\frac{5}{2}$ $\frac{5}{1}$
5. The scale factor of a dilation is $\frac{2}{3}$. An image length is 4 ft. Circle the corresponding preimage length.
   2 ft 3 ft 4 ft 5 ft 6 ft

Copyright © by Pearson Education, Inc. or its affiliates. All Rights Reserved.

Chapter 9 250

ELL Support

Use Manipulatives To help students who have difficulty seeing how a preimage is transformed to create an image in a similarity transformation, suggest that they follow the steps below.

- Use tracing paper to trace the image and preimage.
- Cut the drawings out and orient them the same way they appear in your book.
- Manipulate the preimage using flips, slides, and turns until it is orientated the same way as the image and two corresponding vertices are intersecting.
- Align the figures at each pair of corresponding vertices to be sure all pairs of corresponding angles are congruent.

Instruct students to keep track of the rigid motions that they used to get the first two corresponding vertices to intersect. Then write the composition of rigid motions and include the appropriate dilation.

5 Assess & Remediate

Lesson Quiz

Use the Lesson Quiz to assess students' mastery of the skills and concepts of this lesson.

1. $\triangle BCD$ has vertices $B(-8, 3)$, $C(-4, 6)$, and $D(2, 0)$. If the triangle is reflected across the x-axis and dilated by a scale factor of 0.5 with respect to the origin, what are the coordinates of the vertices of $\triangle B''C''D''$?
2. **Do you UNDERSTAND?** What is a sequence of transformations that maps $\triangle RST$ to $\triangle LMN$?

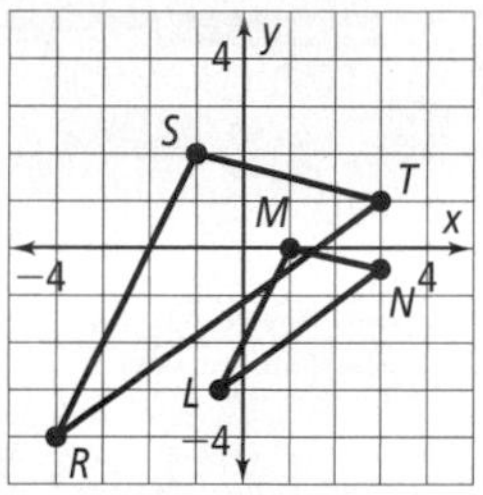

3. Are the figures below similar? Explain.

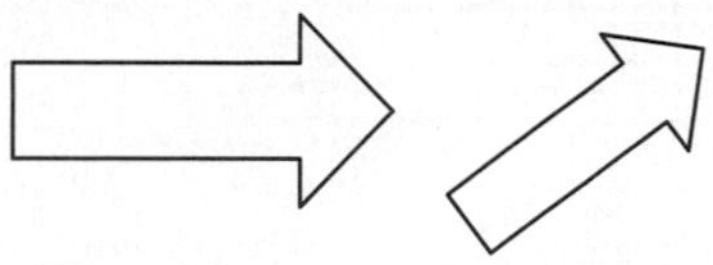

ANSWERS TO LESSON QUIZ

1. $B''(-4, -1.5)$, $C''(-2, -3)$, $D''(1, 0)$
2. $D_{0.5} \circ T_{<3, -2>}$
3. Yes, there is a similarity transformation: translation, rotation, then, dilation.

PRESCRIPTION FOR REMEDIATION

Use the student work on the Lesson Quiz to prescribe a differentiated review assignment.

Points	Differentiated Remediation
0–1	Intervention
2	On-level
3	Extension

PowerGeometry.com

5 Assess & Remediate

Assign the Lesson Quiz. Appropriate intervention, practice, or enrichment is automatically generated based on student performance.

Intervention

- **Reteaching** (2 pages) Provides reteaching and practice exercises for the key lesson concepts. Use with struggling students or absent students.
- **English Language Learner Support** Helps students develop and reinforce mathematical vocabulary and key concepts.

All-In-One Resources/Online

Reteaching

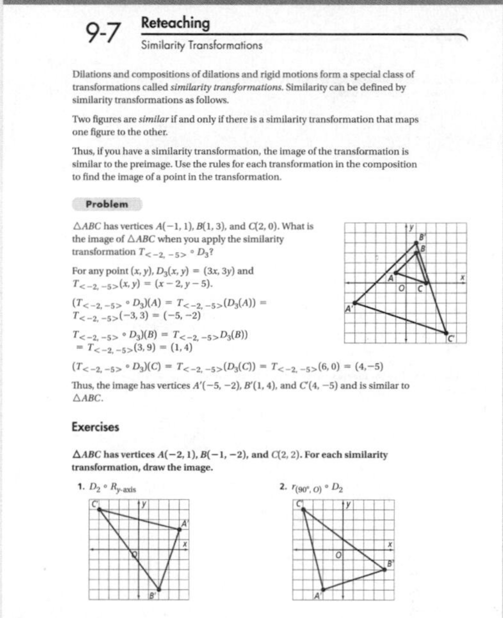

9-7 **Reteaching**
Similarity Transformations

Dilations and compositions of dilations and rigid motions form a special class of transformations called *similarity transformations.* Similarity can be defined by similarity transformations as follows.

Two figures are *similar* if and only if there is a similarity transformation that maps one figure to the other.

Thus, if you have a similarity transformation, the image of the transformation is similar to the preimage. Use the rules for each transformation in the composition to find the image of a point in the transformation.

Problem

$\triangle ABC$ has vertices $A(-1, 1)$, $B(1, 3)$, and $C(2, 0)$. What is the image of $\triangle ABC$ when you apply the similarity transformation $T_{<-2,-5>} \circ D_3$?

For any point (x, y), $D_3(x, y) = (3x, 3y)$ and $T_{<-2,-5>}(x, y) = (x - 2, y - 5)$.

$(T_{<-2,-5>} \circ D_3)(A) = T_{<-2,-5>}(D_3(A)) = T_{<-2,-5>}(-3, 3) = (-5, -2)$

$T_{<-2,-5>} \circ D_3)(B) = T_{<-2,-5>}D_3(B)) = T_{<-2,-5>}(3, 9) = (1, 4)$

$(T_{<-2,-5>} \circ D_3)(C) = T_{<-2,-5>}(D_3(C)) = T_{<-2,-5>}(6, 0) = (4, -5)$

Thus, the image has vertices $A'(-5, -2)$, $B'(1, 4)$, and $C'(4, -5)$ and is similar to $\triangle ABC$.

Exercises

$\triangle ABC$ has vertices $A(-2, 1)$, $B(-1, -2)$, and $C(2, 2)$. For each similarity transformation, draw the image.

1. $D_2 \circ R_{y\text{-axis}}$
2. $r_{(90°, O)} \circ D_2$

All-In-One Resources/Online

English Language Learner Support

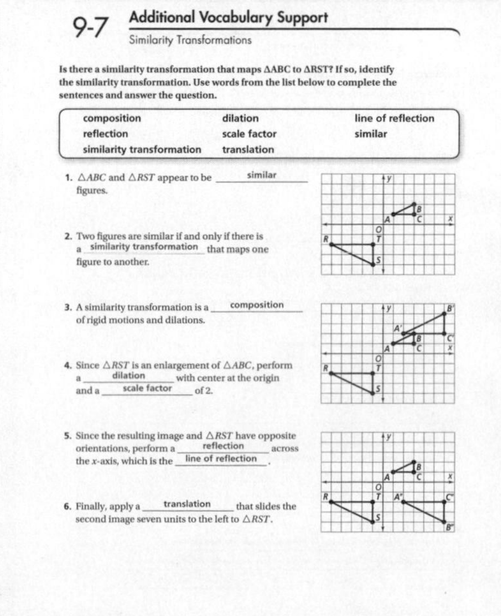

9-7 **Additional Vocabulary Support**
Similarity Transformations

Is there a similarity transformation that maps ΔABC to ΔRST? If so, identify the similarity transformation. Use words from the list below to complete the sentences and answer the question.

composition	dilation	line of reflection
reflection	scale factor	similar
similarity transformation	translation	

1. $\triangle ABC$ and $\triangle RST$ appear to be similar figures.
2. Two figures are similar if and only if there is a similarity transformation that maps one figure to another.
3. A similarity transformation is a composition of rigid motions and dilations.
4. Since $\triangle RST$ is an enlargement of $\triangle ABC$, perform a dilation with center at the origin and a scale factor of 2.
5. Since the resulting image and $\triangle RST$ have opposite orientations, perform a reflection across the x-axis, which is the line of reflection.
6. Finally, apply a translation that slides the second image seven units to the left to $\triangle RST$.

Differentiated Remediation *continued*

On-Level

- **Practice** (2 pages) Provides extra practice for each lesson. For simpler practice exercises, use the Form K Practice pages found in the All-in-One Teaching Resources and online.
- **Think About a Plan** Helps students develop specific problem-solving skills and strategies by providing scaffolded guiding questions.
- **Standardized Test Prep** Focuses on all major exercises, all major question types, and helps students prepare for the high-stakes assessments.

Extension

- **Enrichment** Provides students with interesting problems and activities that extend the concepts of the lesson.
- **Activities, Games, and Puzzles** Worksheets that can be used for concepts development, enrichment, and for fun!

Practice and Problem Solving Wkbk/ All-in-One Resources/Online

Practice page 1

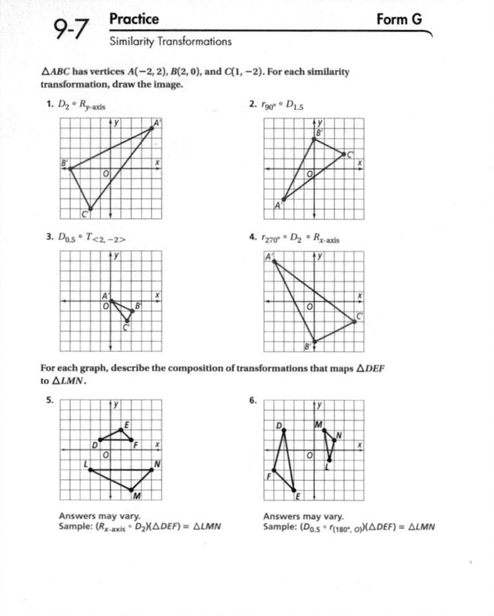

9-7 Practice — Form G
Similarity Transformations

△ABC has vertices A(−2, 2), B(2, 0), and C(1, −2). For each similarity transformation, draw the image.

1. $D_2 \circ R_{y\text{-axis}}$
2. $r_{(90°)} \circ D_{1.5}$
3. $D_{0.5} \circ T_{<2,\,-2>}$
4. $r_{270°} \circ D_2 \circ R_{x\text{-axis}}$

For each graph, describe the composition of transformations that maps △DEF to △LMN.

5. Answers may vary. Sample: $(R_{x\text{-axis}} \circ D_2)(\triangle DEF) = \triangle LMN$
6. Answers may vary. Sample: $(D_{0.5} \circ r_{(180°,\,O)})(\triangle DEF) = \triangle LMN$

Practice and Problem Solving Wkbk/ All-in-One Resources/Online

Practice page 2

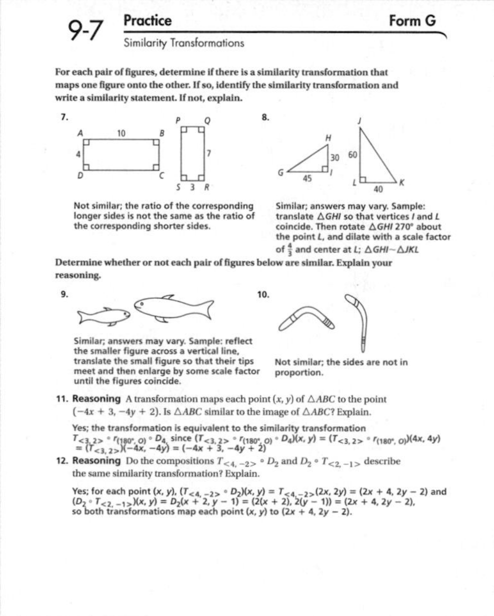

9-7 Practice — Form G
Similarity Transformations

For each pair of figures, determine if there is a similarity transformation that maps one figure onto the other. If so, identify the similarity transformation and write a similarity statement. If not, explain.

7. Not similar; the ratio of the corresponding longer sides is not the same as the ratio of the corresponding shorter sides.
8. Similar; answers may vary. Sample: translate △GHI so that vertices I and L coincide. Then rotate △GHI 270° about the point L, and dilate with a scale factor of $\frac{4}{3}$ and center at L; △GHI ~ △JKL

Determine whether or not each pair of figures below are similar. Explain your reasoning.

9. Similar; answers may vary. Sample: reflect the smaller figure across a vertical line, translate the small figure so that their tips meet and then enlarge by some scale factor until the figures coincide.
10. Not similar; the sides are not in proportion.
11. **Reasoning** A transformation maps each point (x, y) of △ABC to the point (−4x + 3, −4y + 2). Is △ABC similar to the image of △ABC? Explain.
Yes; the transformation is equivalent to the similarity transformation $T_{<3,2>} \circ r_{(180°,\,O)} \circ D_4$, since $(T_{<3,2>} \circ r_{(180°,\,O)} \circ D_4)(x, y) = (T_{<3,2>} \circ r_{(180°,\,O)})(4x, 4y) = (T_{<3,2>})(-4x, -4y) = (-4x + 3, -4y + 2)$
12. **Reasoning** Do the compositions $T_{<4,\,-2>} \circ D_2$ and $D_2 \circ T_{<2,\,-1>}$ describe the same similarity transformation? Explain.
Yes; for each point (x, y), $(T_{<4,-2>} \circ D_2)(x, y) = T_{<4,-2>}(2x, 2y) = (2x + 4, 2y - 2)$ and $(D_2 \circ T_{<2,-1>})(x, y) = D_2(x + 2, y - 1) = (2(x + 2), 2(y - 1)) = (2x + 4, 2y - 2)$, so both transformations map each point (x, y) to (2x + 4, 2y − 2).

All-in-One Resources/Online

Enrichment

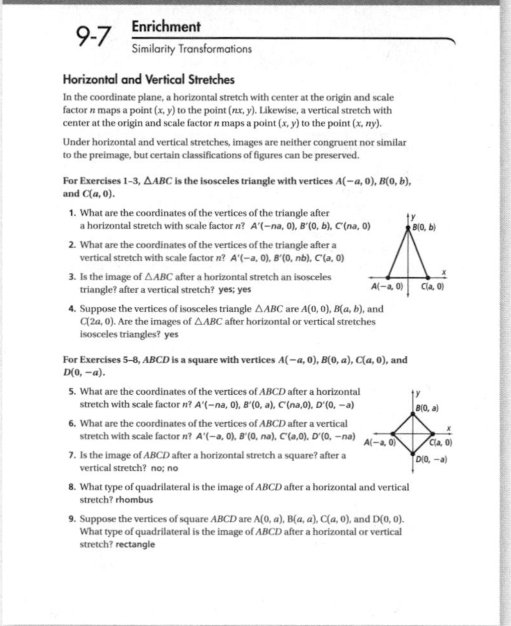

9-7 Enrichment
Similarity Transformations

Horizontal and Vertical Stretches

In the coordinate plane, a horizontal stretch with center at the origin and scale factor n maps a point (x, y) to the point (nx, y). Likewise, a vertical stretch with center at the origin and scale factor n maps a point (x, y) to the point (x, ny).

Under horizontal and vertical stretches, images are neither congruent nor similar to the preimage, but certain classifications of figures can be preserved.

For Exercises 1–3, △ABC is the isosceles triangle with vertices A(−a, 0), B(0, b), and C(a, 0).

1. What are the coordinates of the vertices of the triangle after a horizontal stretch with scale factor n? A′(−na, 0), B′(0, b), C′(na, 0)
2. What are the coordinates of the vertices of the triangle after a vertical stretch with scale factor n? A′(−a, 0), B′(0, nb), C′(a, 0)
3. Is the image of △ABC after a horizontal stretch an isosceles triangle? after a vertical stretch? yes; yes
4. Suppose the vertices of isosceles triangle △ABC are A(0, 0), B(a, b), and C(2a, 0). Are the images of △ABC after horizontal or vertical stretches isosceles triangles? yes

For Exercises 5–8, ABCD is a square with vertices A(−a, 0), B(0, a), C(a, 0), and D(0, −a).

5. What are the coordinates of the vertices of ABCD after a horizontal stretch with scale factor n? A′(−na, 0), B′(0, a), C′(na, 0), D′(0, −a)
6. What are the coordinates of the vertices of ABCD after a vertical stretch with scale factor n? A′(−a, 0), B′(0, na), C′(a, 0), D′(0, −na)
7. Is the image of ABCD after a horizontal stretch a square? after a vertical stretch? no; no
8. What type of quadrilateral is the image of ABCD after a horizontal and vertical stretch? rhombus
9. Suppose the vertices of square ABCD are A(0, a), B(a, a), C(a, 0), and D(0, 0). What type of quadrilateral is the image of ABCD after a horizontal or vertical stretch? rectangle

Practice and Problem Solving Wkbk/ All-in-One Resources/Online

Think About a Plan

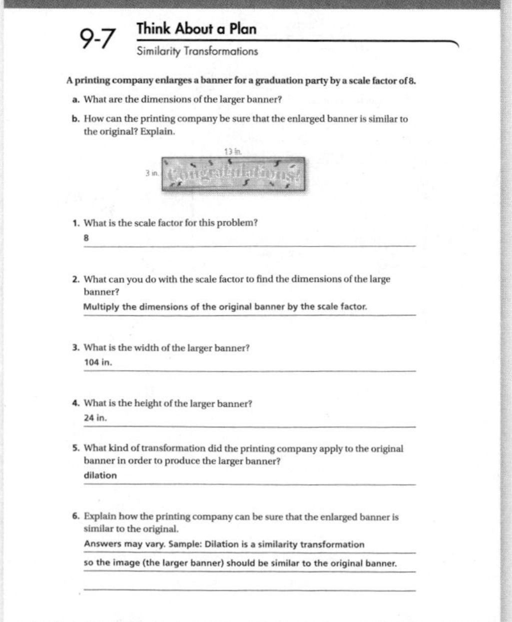

9-7 Think About a Plan
Similarity Transformations

A printing company enlarges a banner for a graduation party by a scale factor of 8.

a. What are the dimensions of the larger banner?
b. How can the printing company be sure that the enlarged banner is similar to the original? Explain.

1. What is the scale factor for this problem? 8
2. What can you do with the scale factor to find the dimensions of the large banner? Multiply the dimensions of the original banner by the scale factor.
3. What is the width of the larger banner? 104 in.
4. What is the height of the larger banner? 24 in.
5. What kind of transformation did the printing company apply to the original banner in order to produce the larger banner? dilation
6. Explain how the printing company can be sure that the enlarged banner is similar to the original. Answers may vary. Sample: Dilation is a similarity transformation so the image (the larger banner) should be similar to the original banner.

Practice and Problem Solving Wkbk/ All-in-One Resources/Online

Standardized Test Prep

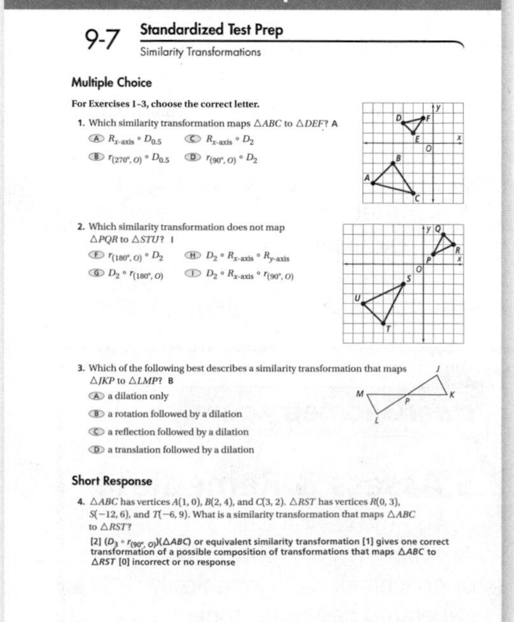

9-7 Standardized Test Prep
Similarity Transformations

Multiple Choice

For Exercises 1–3, choose the correct letter.

1. Which similarity transformation maps △ABC to △DEF? A
Ⓐ $R_{x\text{-axis}} \circ D_{0.5}$ Ⓒ $R_{x\text{-axis}} \circ D_2$
Ⓑ $r_{(270°,\,O)} \circ D_{0.5}$ Ⓓ $r_{(90°,\,O)} \circ D_2$
2. Which similarity transformation does not map △PQR to △STU? I
Ⓕ $r_{(180°,\,O)} \circ D_2$ Ⓗ $D_2 \circ R_{x\text{-axis}} \circ R_{y\text{-axis}}$
Ⓖ $D_2 \circ r_{(180°,\,O)}$ Ⓘ $D_2 \circ R_{x\text{-axis}} \circ r_{(90°,\,O)}$
3. Which of the following best describes a similarity transformation that maps △JKP to △LMP? B
Ⓐ a dilation only
Ⓑ a rotation followed by a dilation
Ⓒ a reflection followed by a dilation
Ⓓ a translation followed by a dilation

Short Response

4. △ABC has vertices A(1, 0), B(2, 4), and C(3, 2). △RST has vertices R(0, 3), S(−12, 6), and T(−6, 9). What is a similarity transformation that maps △ABC to △RST?
[2] $(D_3 \circ r_{(90°,\,O)})(\triangle ABC)$ or equivalent similarity transformation [1] gives one correct transformation of a possible composition of transformations that maps △ABC to △RST [0] incorrect or no response

Online Teacher Resource Center

Activities, Games, and Puzzles

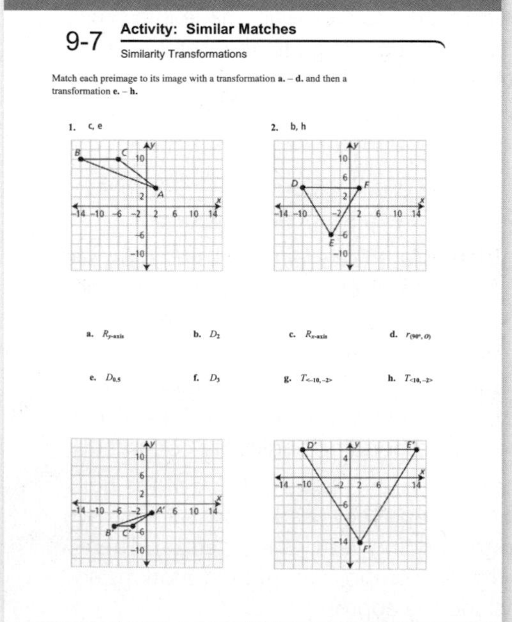

9-7 Activity: Similar Matches
Similarity Transformations

Match each preimage to its image with a transformation a. – d. and then a transformation e. – h.

1. c, e
2. b, h

a. $R_{y\text{-axis}}$ b. D_2 c. $R_{x\text{-axis}}$ d. $r_{(90°,\,O)}$
e. $D_{0.5}$ f. D_3 g. $T_{<-10,\,-2>}$ h. $T_{<10,\,-2>}$

Pull It All Together

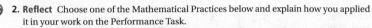

Completing the Performance Task

Look back at your results from the Apply What You've Learned sections in Lessons 9-2, 9-3, and 9-4. Use the work you did to complete the following.

To solve these problems you will pull together many concepts and skills that you have studied about transformations.

1. Solve the problem in the Task Description on page 543 by finding a sequence of transformations that uses as few moves as possible to move the puzzle piece to the target area for Case 1 and for Case 2. Show all your work and explain each step of your solution.
2. **Reflect** Choose one of the Mathematical Practices below and explain how you applied it in your work on the Performance Task.
   MP 1: Make sense of problems and persevere in solving them.
   MP 3: Construct viable arguments and critique the reasoning of others.
   MP 5: Use appropriate tools strategically.

On Your Own

Alicia must write a program that moves the puzzle piece $\triangle PQR$ to the target area $\triangle XYZ$, as shown in the graph below. Find a sequence of three different transformations (one translation, one reflection, and one rotation about the origin) that Alicia can use in her program.

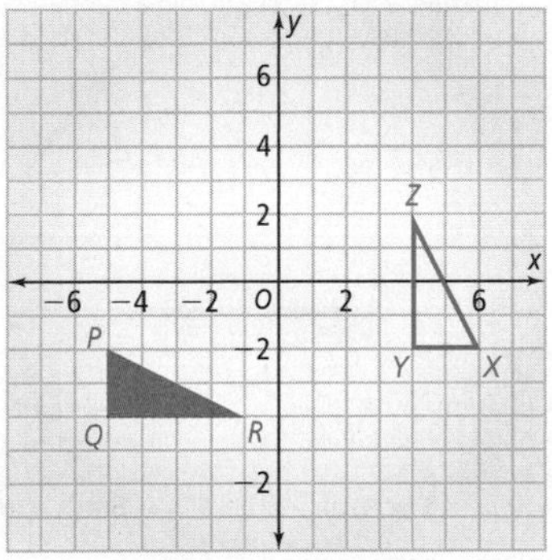

Completing the Performance Task

In the Apply What You've Learned sections in Lessons 9-2, 9-3, and 9-4, students looked at how various transformations move the puzzle piece in the graph shown on page 543, and thought about the characteristics of a composition of transformations that will move the puzzle piece to the target area. Now, they use their observations to complete the Performance Task. Ask students the following questions as they work toward solving the problem.

Q How can you use the work you have done in the chapter to solve the problem? **[Sample: I can use the transformations that I drew and look for transformation relationships between the images of the puzzle piece and the target area.]**

Q How can you check that your sequence of transformations works? **[Sample: I can graph $\triangle ABC$, carry out my transformations, and verify that the final image is $\triangle DEF$.]**

FOSTERING MATHEMATICAL DISCOURSE

Have students compare their sequences of transformations for each case and verify that different sequences successfully move the puzzle piece to the target area.

ANSWERS

1. Case 1: Answers may vary. Sample: Rotate the puzzle piece 270° about the origin and then reflect it across the line $y = 1$.
   Case 2: Answers may vary. Sample: Translate the puzzle piece two units right and then reflect it across the line $y = x$.
2. Check students' work.

On Your Own

This problem is similar to the problem posed on page 543, but now the student must find a sequence of three transformations that move the puzzle piece to the target area. Students should strive to solve this problem independently.

ANSWER

Answers may vary. Sample: Rotate the puzzle piece 90° about the origin, translate it up 3 units, and then reflect it across the line $x = 4$.

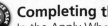

Essential Questions

BIG idea Transformations

ESSENTIAL QUESTION How can you change a figure's position without changing its size and shape? How can you change a figure's size without changing its shape?

ANSWER When you translate, reflect, or rotate a geometric figure, its size and shape stay the same. When you dilate a geometric figure, the figure is enlarged or reduced.

BIG idea Coordinate Geometry

ESSENTIAL QUESTION How can you represent a transformation in the coordinate plane?

ANSWER You can show a transformation in the coordinate plane by graphing a figure and its image.

BIG idea Visualization

ESSENTIAL QUESTION How do you recognize congruence and similarity in figures?

ANSWER If two figures are congruent, then you can visualize a congruence transformation that maps one figure to the other. If you can visualize a composition of rigid motions and dilations that map one figure to another, then the figures are similar.

9 Chapter Review

Connecting BIG ideas and Answering the Essential Questions

1 Transformations
When you translate, reflect, or rotate a geometric figure, its size and shape stay the same. When you dilate a geometric figure, the figure is enlarged or reduced.

2 Coordinate Geometry
You can show a transformation in the coordinate plane by graphing a figure and its image.

3 Visualization
If two figures are congruent, then you can visualize a congruence transformation that maps one figure to the other. If you can visualize a composition of rigid motions and dilations that map one figure to another, then the figures are similar.

Transformations (Lessons 9-1, 9-2, 9-3, and 9-6)
The black triangle is the preimage of each transformation.

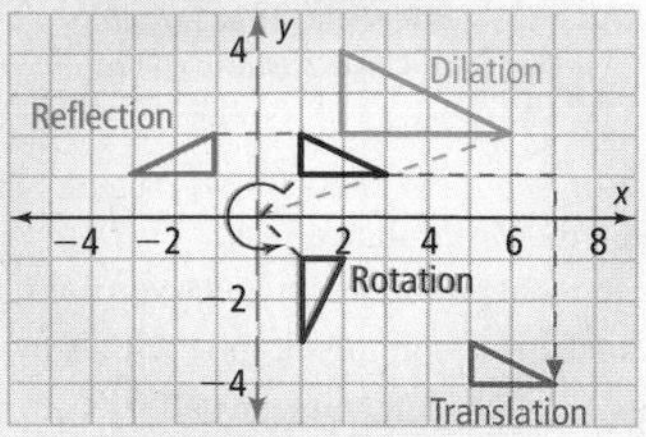

Congruence Transformations (Lesson 9-5)
Triangles are congruent if and only if there is a sequence of rigid motions that maps one triangle to the other.
Because $\triangle A''B''C''$ is the image of $\triangle ABC$ after a reflection and a translation, $\triangle ABC \cong \triangle A''B''C''$.

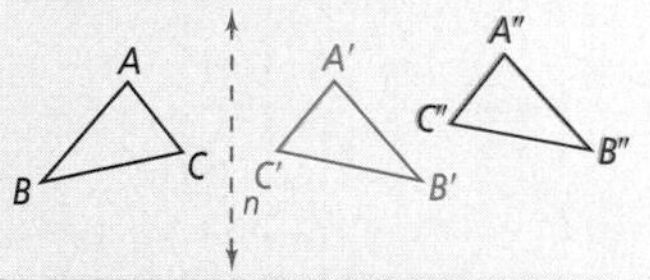

Composing Transformations (Lesson 9-4)
A glide reflection moves the black triangle down 3 units and then reflects it across the line $x = -2$.

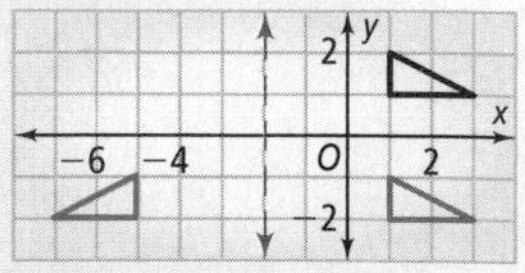

Similarity Transformations (Lesson 9-7)
Figures are similar if and only if there is a sequence of rigid motions and dilations that maps one figure onto the other.
Because $\triangle L''M''N''$ is the image of $\triangle LMN$ after a reflection and a dilation, $\triangle LMN$ is similar to $\triangle L''M''N''$.

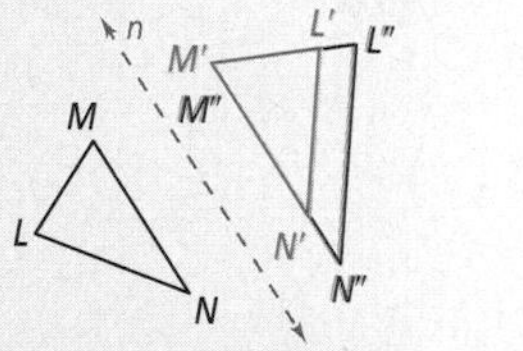

Chapter Vocabulary

- composition of transformations (p. 548)
- congruence transformation (p. 580)
- dilation (p. 587)
- glide reflection (p. 572)
- image (p. 545)
- isometry (p. 570)
- preimage (p. 545)
- reflection (p. 554)
- rigid motion (p. 545)
- rotation (p. 561)
- similarity transformation (p. 596)
- transformation (p. 545)
- translation (p. 547)

Choose the correct term to complete each sentence.

1. A(n) __?__ is a change in the position, shape, or size of a figure.

2. A(n) __?__ is a composition of rigid motions and dilations.

3. In a(n) __?__, all points of a figure move the same distance in the same direction.

4. A(n) __?__ is the result of a transformation.

602 **Chapter 9** Chapter Review

Summative Questions

Use the following prompts as you review this chapter with your students. The prompts are designed to help you assess your students' understanding of the Big Ideas they have studied.

- Which transformations are rigid motions?
- What happens in a dilation?
- What single transformation results from a composition of two reflections across parallel lines? intersecting lines?
- How can you verify that two figures are congruent or similar?

Answers

Chapter Review

1. transformation
2. similarity transformation
3. translation
4. image

9-1 Translations

Quick Review

A **transformation** of a geometric figure is a change in its position, shape, or size.

A **translation** is a rigid motion that maps all points of a figure the same distance in the same direction.

In a **composition of transformations,** each transformation is performed on the image of the preceding transformation.

Example

What are the coordinates of $T_{<-2, 3>}(5, -9)$?

Add -2 to the x-coordinate, and 3 to the y-coordinate.

$A(5, -9) \rightarrow (5 - 2, -9 + 3)$, or $A'(3, -6)$.

Exercises

5. a. A transformation maps *ZOWE* onto *LFMA*. Does the transformation appear to be a rigid motion? Explain.

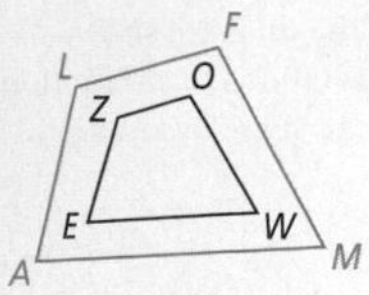

b. What is the image of $\overline{ZE}$? What is the preimage of *M*?

6. $\triangle RST$ has vertices $R(0, -4)$, $S(-2, -1)$, and $T(-6, 1)$. Graph $T_{<-4, 7>}(\triangle RST)$.

7. Write a rule to describe a translation 5 units left and 10 units up.

8. Find a single translation that has the same effect as the following composition of translations. $T_{<-4, 7>}$ followed by $T_{<3, 0>}$

9-2 Reflections

Quick Review

The diagram shows a **reflection** across line *r*. A reflection is rigid motion that preserves distance and angle measure. The image and preimage of a reflection have opposite orientations.

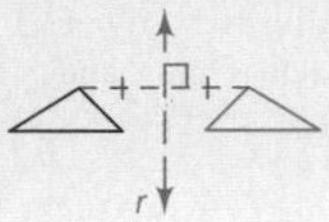

Example

Use points $P(1, 0)$, $Q(3, -2)$, and $R(4, 0)$. What is $R_{y\text{-axis}}(\triangle PQR)$?

Graph $\triangle PQR$. Find P', Q', and R' such that the y-axis is the perpendicular bisector of $\overline{PP'}$, $\overline{QQ'}$, and $\overline{RR'}$. Draw $\triangle P'Q'R'$.

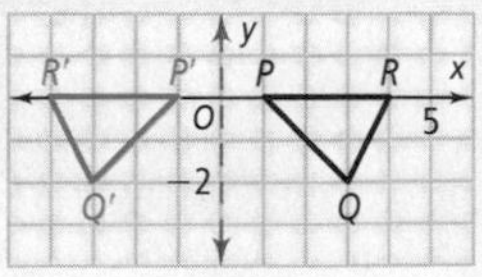

Exercises

Given points $A(6, 4)$, $B(-2, 1)$, and $C(5, 0)$, graph $\triangle ABC$ and each reflection image.

9. $R_{x\text{-axis}}(\triangle ABC)$

10. $R_{x=4}(\triangle ABC)$

11. $R_{y=x}(\triangle ABC)$

12. Copy the diagram. Then draw $R_{y\text{-axis}}(BGHT)$. Label the vertices of the image by using prime notation.

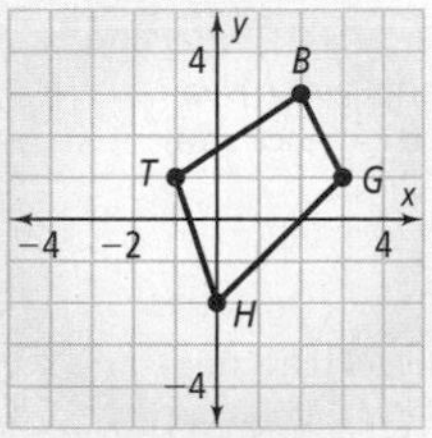

5a. No; the distances between corresponding points in the image and preimage are not the same.

b. $\overline{LA}$, *W*

6. $R'(-4, 3)$, $S'(-6, 6)$, $T'(-10, 8)$

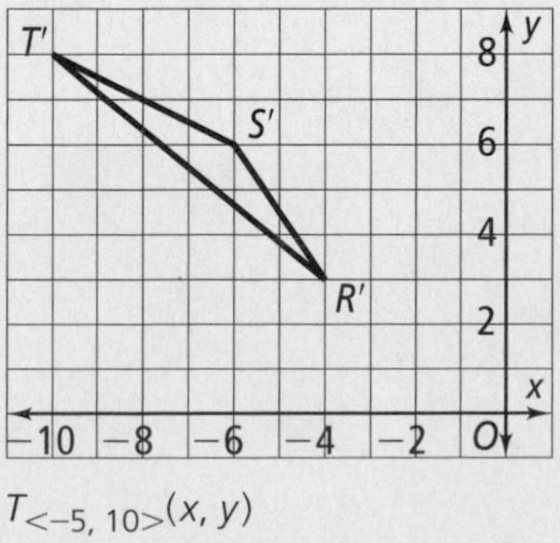

7. $T_{<-5, 10>}(x, y)$

8. $T_{<-2, 7>}(x, y)$

9. $A'(6, -4)$, $B'(-2, -1)$, $C'(5, 0)$

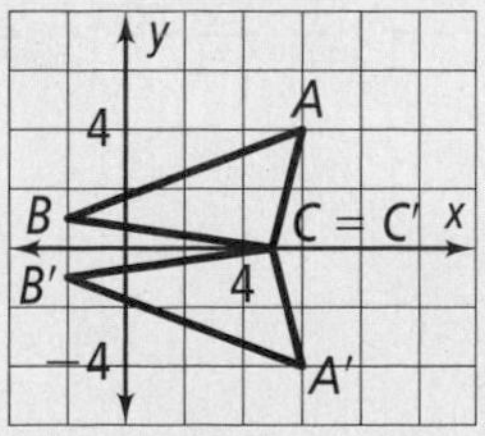

10. $A'(2, 4)$, $B'(10, 1)$, $C'(3, 0)$

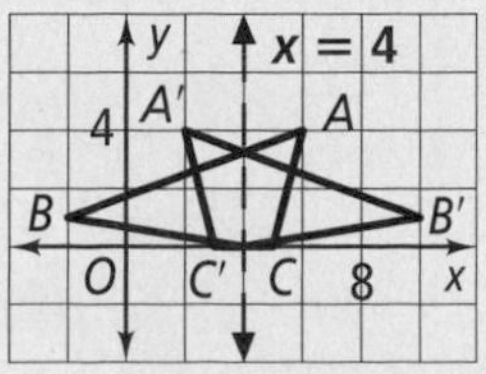

11. $A'(4, 6)$, $B'(1, -2)$, $C'(0, 5)$

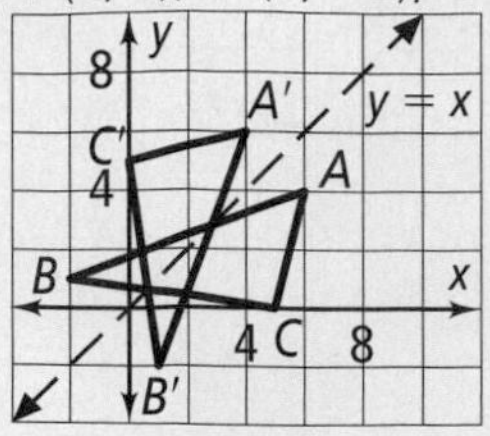

12.

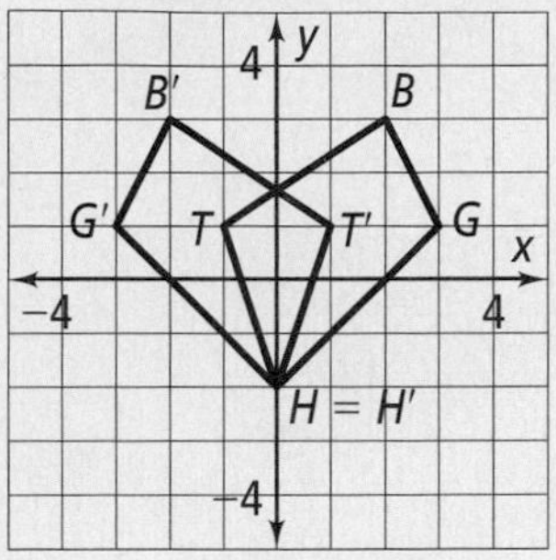

Answers

Chapter Review (continued)

13.

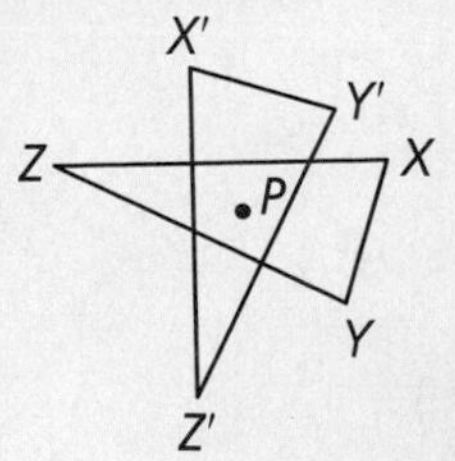

14. $P'(4, -1)$

15. Check students' graphs; $W'(-1, -3)$, $X'(2, -5)$, $Y'(8, 0)$, and $Z'(-1, -2)$.

16.

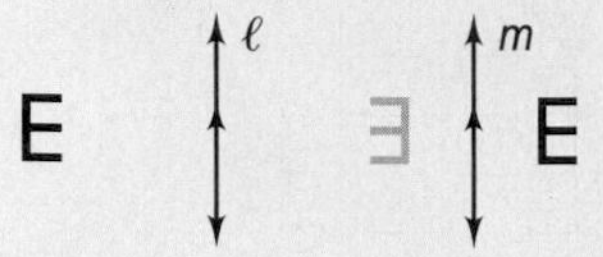

E is translated right, twice the distance between ℓ and *m*.

17. same; rotation

18. same; translation

19. opposite; glide reflection

20. $\triangle T'A'M'$ with vertices $T'(-4, -9)$, $A'(0, -5)$, $M'(-1, -10)$

9-3 Rotations

Quick Review

The diagram shows a **rotation** of $x°$ about point *R*. A rotation is rigid motion in which a figure and its image have the same orientation.

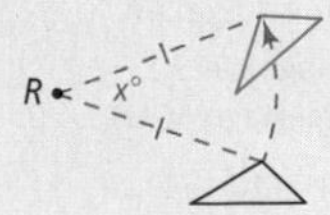

Example

***GHIJ* has vertices *G*(0, −3), *H*(4, 1), *I*(−1, 2), and *J*(−5, −2). What are the vertices of $r_{(90°, O)}(GHIJ)$?**

Use the rule $r_{(90°, O)}(x, y) = (-y, x)$.

$r_{(90°, O)}(G) = (3, 0)$

$r_{(90°, O)}(H) = (-1, 4)$

$r_{(90°, O)}(I) = (-2, -1)$

$r_{(90°, O)}(J) = (2, -5)$

Exercises

13. Copy the diagram below. Then draw $r_{(90°, P)}(\triangle ZXY)$. Label the vertices of the image by using prime notation.

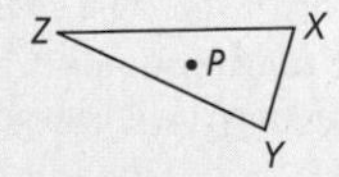

14. What are the coordinates of $r_{(180°, O)}(-4, 1)$?

15. *WXYZ* is a quadrilateral with vertices *W*(3, −1), *X*(5, 2), *Y*(0, 8), and *Z*(2, −1). Graph *WXYZ* and $r_{(270°, O)}(WXYZ)$.

9-4 Compositions of Isometries

Quick Review

An isometry is a transformation that preserves distance. All of the rigid motions, translations, reflections, and rotations, are isometries. A composition of isometries is also an isometry. All rigid motions can be expressed as a composition of reflections.

The diagram shows a **glide reflection** of N. A glide reflection is an isometry in which a figure and its image have opposite orientations.

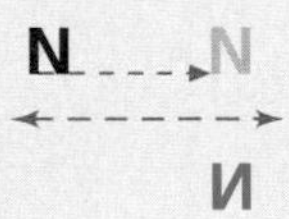

Example

Describe the result of reflecting P first across line ℓ and then across line *m*.

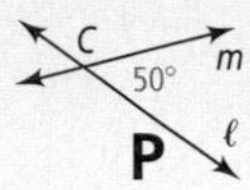

A composition of two reflections across intersecting lines is a rotation. The angle of rotation is twice the measure of the acute angle formed by the intersecting lines. P is rotated 100° about *C*.

Exercises

16. Sketch and describe the result of reflecting E first across line ℓ and then across line *m*.

Each figure is an isometry image of the figure at the right. Tell whether their orientations are the same or opposite. Then classify the isometry.

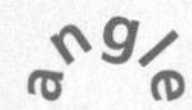

17.

18.

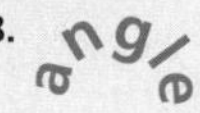

19.

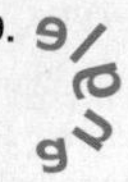

20. $\triangle TAM$ has vertices *T*(0, 5), *A*(4, 1), and *M*(3, 6). Find the image of $R_{y=-2} \circ T_{(-4, 0)}(\triangle TAM)$.

9-5 Congruence Transformations

Quick Review

Two figures are congruent if and only if there is a sequence of rigid motions that maps one figure onto the other.

Example

$R_{y\text{-axis}}(TGMB) = KWAV$.
What are all of the congruent angles and all of the congruent sides?

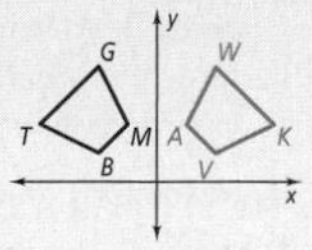

A reflection is a congruence transformation, so $TGMB \cong KWAV$, and corresponding angles and corresponding sides are congruent.

$\angle T \cong \angle K$, $\angle G \cong \angle W$, $\angle M \cong \angle A$, and $\angle B \cong \angle V$
$TG = KW$, $GM = WA$, $MB = AV$ and $TB = KV$

Exercises

21. In the diagram at the right, $\triangle LMN \cong \triangle XYZ$. Identify a congruence transformation that maps $\triangle LMN$ onto $\triangle XYZ$.

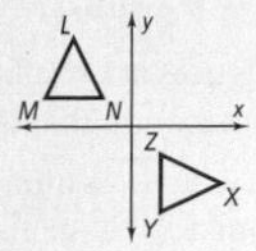

22. Fonts Graphic designers use some fonts because they have pleasing proportions or are easy to read from far away. The letters p and d to the right are used on a sign using a special font. Are the letters congruent? If so, describe a congruence transformation that maps one onto the other. If not, explain why not.

p d

9-6 Dilations

Quick Review

The diagram shows a **dilation** with center C and scale factor n. The preimage and image are similar.

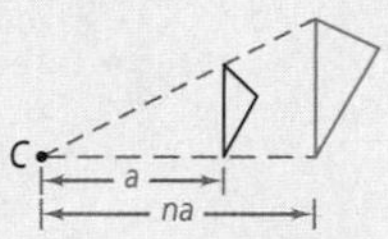

In the coordinate plane, if the origin is the center of a dilation with scale factor n, then $P(x, y) \rightarrow P'(nx, ny)$.

Example

The blue figure is a dilation image of the black figure. The center of dilation is A. Is the dilation an enlargement or a reduction? What is the scale factor?

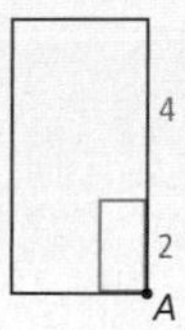

The image is smaller than the preimage, so the dilation is a reduction. The scale factor is $\frac{\text{image length}}{\text{original length}} = \frac{2}{2+4} = \frac{2}{6}$, or $\frac{1}{3}$.

Exercises

23. The blue figure is a dilation image of the black figure. The center of dilation is O. Tell whether the dilation is an enlargement or a reduction. Then find the scale factor.

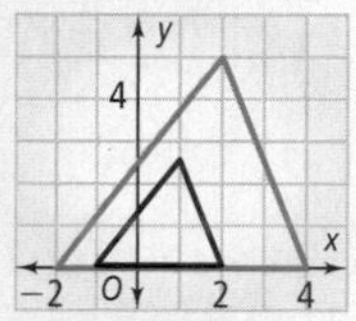

Graph the polygon with the given vertices. Then graph its image for a dilation with center (0, 0) and the given scale factor.

24. $M(-3, 4)$, $A(-6, -1)$, $T(0, 0)$, $H(3, 2)$; scale factor 5

25. $F(-4, 0)$, $U(5, 0)$, $N(-2, -5)$; scale factor $\frac{1}{2}$

26. A dilation maps $\triangle LMN$ onto $\triangle L'M'N'$. $LM = 36$ ft, $LN = 26$ ft, $MN = 45$ ft, and $L'M' = 9$ ft. Find $L'N'$ and $M'N'$.

21. $(r_{(90°, O)} \circ R_{x\text{-axis}})(\triangle XYZ)$

22. Answers may vary. Sample: Yes, the letters are congruent. The *p* can be mapped to the *d* with a composition of a translation followed by a rotation.

23. enlargement; 2

24. $M'(-15, 20)$, $A'(-30, -5)$, $T'(0, 0)$, $H'(15, 10)$; check students' graphs.

25.

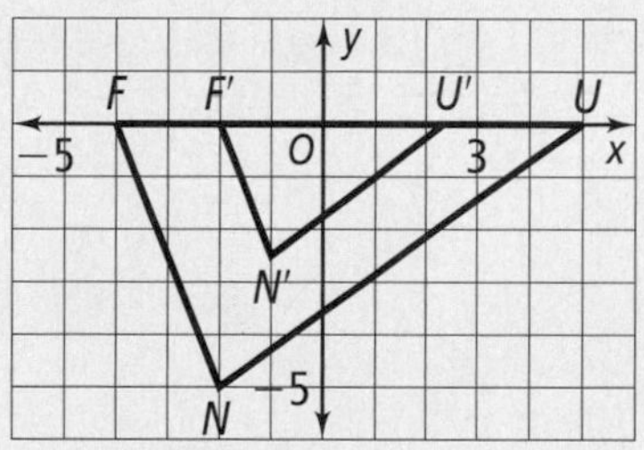

26. $L'N' = 6.5$ ft, $M'N' = 11.25$ ft

Answers

Chapter Review (continued)

27.

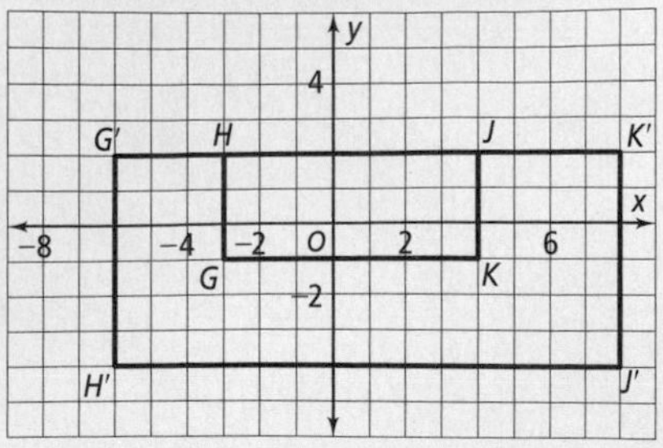

28. No. The side lengths are not proportional.

29. No, because all of the dimensions of the airplane must dilate by the same scale factor for the figures to be similar.

30. Answers may vary. Sample answer: The figures are similar because a composition of a translation, rotation, and a dilation maps *p* to *d*.

Lesson 9-7 **Similarity Transformations**

Quick Review

Two figures are similar if and only if there is a similarity transformation that maps one figure onto the other.

When a figure is transformed by a composition of rigid motions and dilations, the corresponding angles of the image and preimage are congruent, and the ratios of corresponding sides are proportional.

Example

Is $\triangle JKL$ similar to $\triangle DCX$? If so, write a similarity transformation rule. If not, explain why not.

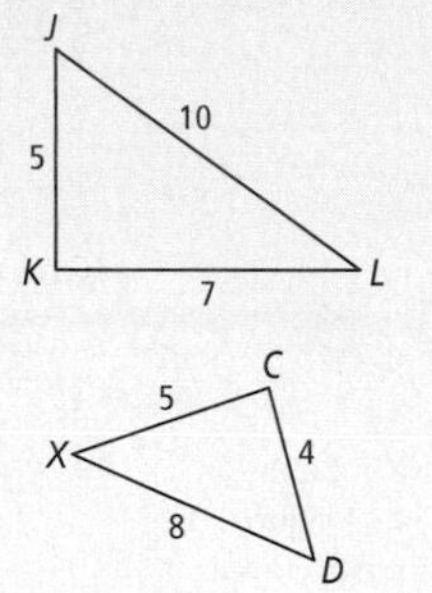

$\triangle JKL$ can be rotated and then translated so that J and D coincide and $\overline{JK}$ and $\overline{CD}$ are collinear. Then if $\triangle JKL$ is dilated by scale factor $\frac{4}{5}$, then $\triangle JKL$ will coincide with $\triangle DCX$. So, the $\triangle JKL$ is similar to $\triangle DCX$, and the similarity transformation is a rotation, followed by a translation, followed by a dilation of scale factor $\frac{4}{5}$.

Exercises

27. $\square GHJK$ has vertices $G(-3, -1)$, $H(-3, 2)$, $J(4, 2)$, and $K(4, -1)$. Draw $\square GHJK$ and its image when you apply the composition $D_2 \circ R_{x\text{-axis}}$.

28. Writing Suppose that you have an 8 in. by 12 in. photo of your friends and a 2 in. by 6 in. copy of the same picture. Are the two photos similar figures? How do you know?

29. Reasoning A model airplane has an overall length that is $\frac{1}{20}$ the actual plane's length, and an overall height that is $\frac{1}{18}$ the actual plane's height. Are the model airplane and the actual airplane similar figures? Explain.

30. Determine whether the figures below are similar. If so, write the similarity transformation rule. If not, explain.

9 Chapter Test

MathXL® for School
Go to PowerGeometry.com

Do you know HOW?

For Exercises 1–7, find the coordinates of the vertices of the image of *ABCD* for each transformation.

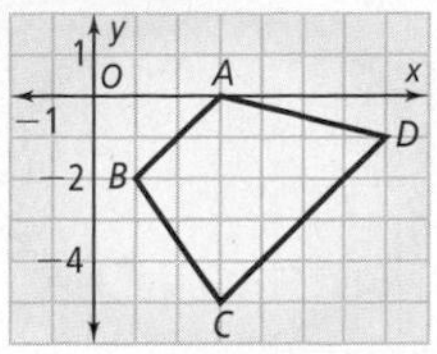

1. $R_{x=-4}(ABCD)$
2. $T_{<-6, 8>}(ABCD)$
3. $r_{(90°, O)}(ABCD)$
4. $D_{\frac{2}{3}}(ABCD)$
5. $(R_{x=0} \circ T_{<0, 5>})(ABCD)$
6. $R_{y=x}(ABCD)$
7. $D_3(ABCD)$
8. Write the translation rule that maps $P(-4, 2)$ to $P'(-1, -1)$.

What type of transformation has the same effect as each composition of transformations?

9. $R_{x=6} \circ T_{<0, -5>}$
10. $T_{<-3, 2>} \circ T_{<8, -4>}$
11. $R_{x=4} \circ R_{x=-2}$
12. $R_{y=x} \circ R_{y=-x}$

What type(s) of symmetry does each figure have?

13. 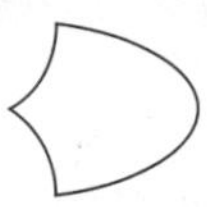

14. 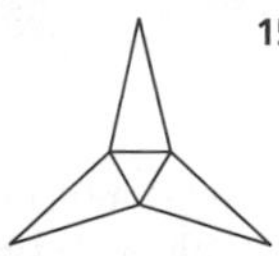

15. 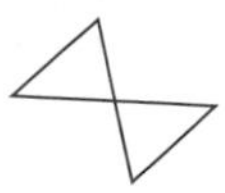

For Exercises 16 and 17, find the coordinates of the vertices of △*XYZ* with vertices *X*(3, 4), *Y*(2, 1), and *Z*(−2, 2) for each similarity transformation.

16. $(r_{(90°, O)} \circ R_{y=1})(\triangle XYZ)$
17. $(r_{(180°, O)} \circ T_{<2, -1>})(\triangle XYZ)$

Determine whether the figures are similar. If so, identify a similarity transformation that maps one to the other. If not, explain.

18. 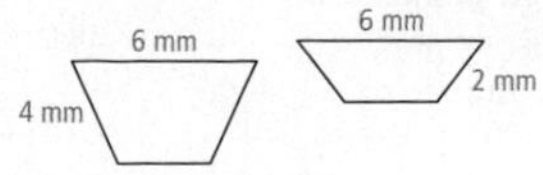

19. 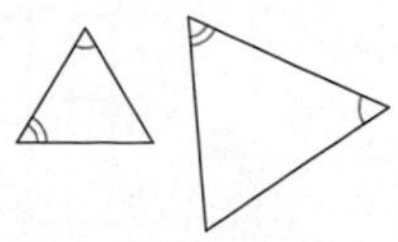

Identify the type of isometry that maps the black figure to the blue figure.

20. 

21. 

Do you UNDERSTAND?

22. **Vocabulary** Is a dilation an isometry? Explain.
23. **Writing** Line *m* intersects $\overline{UH}$ at *N*, and *UN* = *NH*. Must *H* be the reflection image of *U* across line *m*? Explain your reasoning.
24. **Coordinate Geometry** A dilation with center (0, 0) and scale factor 2.5 maps (*a*, *b*) to (10, −25). What are the values of *a* and *b*?
25. **Error Analysis** A classmate says that a certain regular polygon has 50° rotational symmetry. Explain your classmate's error.
26. **Reasoning** Choose points *A*, *B*, and *C* in the first quadrant. Find the coordinates of *A′*, *B′*, and *C′* by multiplying the coordinates of *A*, *B*, and *C* by −2. What composition of transformations maps △*ABC* onto △*A′B′C′*? Explain.

Chapter Test

1. $A'(-11, 0)$, $B'(-9, -2)$, $C'(-11, -5)$, $D'(-15, -1)$
2. $A'(-3, 8)$, $B'(-5, 6)$, $C'(-3, 3)$, $D'(1, 7)$
3. $A'(0, 3)$, $B'(2, 1)$, $C'(5, 3)$, $D'(1, 7)$
4. $A'(2, 0)$, $B'\left(\frac{2}{3}, -\frac{4}{3}\right)$, $C'\left(2, -\frac{10}{3}\right)$, $D'\left(\frac{14}{3}, -\frac{2}{3}\right)$
5. $A'(-3, 5)$, $B'(-1, 3)$, $C'(-3, 0)$, $D'(-7, 4)$
6. $A'(0, 3)$, $B'(-2, 1)$, $C'(-5, 3)$, $D'(-1, 7)$
7. $A'(9, 0)$, $B'(3, -6)$, $C'(9, -15)$, $D'(21, -3)$
8. $T_{<3, -3>}(P)$
9. glide reflection
10. translation
11. translation
12. rotation
13. line
14. line, rotational
15. rotational, point
16. $X'(2, 3)$, $Y'(-1, 2)$, $Z'(0, -2)$
17. $X'(-5, -3)$, $Y'(-4, 0)$, $Z'(0, -1)$
18. The figures as not similar because if the small trapezoid is translated so that the 6mm sides coincide, there is no dilation that maps the remaining sides to each other.
19. The figures are similar because you know that if two corresponding angles are congruent, then there is a similarity transformation that maps one triangle onto the other.
20. reflection
21. rotation
22. A dilation with scale factor $n \neq 1$ changes the size of the preimage, so it is not a rigid motion. A dilation with a scale factor 1 is a rigid motion.
23. No; line *m* does bis. $\overline{UH}$, but *H* is the reflection image of *U* if and only if *m* is also ⊥ to $\overline{UH}$.
24. $a = 4$, $b = -10$
25. Answers may vary. Sample: For a figure to have $x°$ rotational symmetry, *x* must divide into 360 with no remainder. Since 50 leaves a remainder when divided into 360, a figure cannot have 50° rotational symmetry.
26. Check students' points; answers may vary. Sample: a dilation with center (0, 0) and scale factor 2 followed by a 180° rotation about the origin; the transformation $(x, y) \rightarrow (2x, 2y)$ followed by the transformation $(x, y) \rightarrow (-x, -y)$ results in the transformation $(x, y) \rightarrow (-2x, -2y)$. If (*a*, *b*) is a point in the first quadrant, then the results of multiplying the coordinates by −2 are (−2*a*, −2*b*), which represents a point in the third quadrant. To map △*ABC* to △*A″B″C″*, one composition is a rotation by 180° (or $r_{(180°, O)}$) followed by a dilation by $n = 2$ (or $D_2(x, y)$).

PowerGeometry.com

MathXL for School

Prepare students for the Mid-Chapter Quiz and Chapter Test with online practice and review.

Item Number	Lesson	Content Standard
1	6-6	G-SRT.B.5
2	8-2	G-SRT.C.5
3	6-7	G-GPE.B.7
4	8-1	G-SRT.C.8
5	8-2	G-SRT.C.8
6	9-4	G-CO.A.3
7	6-3	G-CO.C.11
8	3-6	G-CO.D.12
9	5-4	G-CO.C.10
10	5-4	G-CO.C.10
11	6-1	G-SRT.B.5
12	6-4	G-CO.C.11
13	3-3	G-CO.C.9
14	4-4	G-SRT.B.5
15	1-8	G-GPE.B.7
16	6-2	G-CO.C.11
17	7-3	G-SRT.B.5
18	7-2	G-SRT.B.5
19	8-1	G-SRT.C.8
20	9-3	G-CO.A.4
21	8-2	G-SRT.C.8
22	8-5	G-SRT.D.11
23	4-3	G-SRT.B.5
24	6-7	G-GPE.B.4
25	9-4	G-CO.A.5

9 Common Core Cumulative Standards Review

ASSESSMENT

TIPS FOR SUCCESS

Some problems ask you to perform a transformation on a figure in the coordinate plane. Read the sample question at the right. Then follow the tips to answer it.

$\triangle G'H'K'$ is the image of $\triangle GHK$ for a dilation with center (0, 0) and $H'K' = 8$. What are the coordinates of H'?

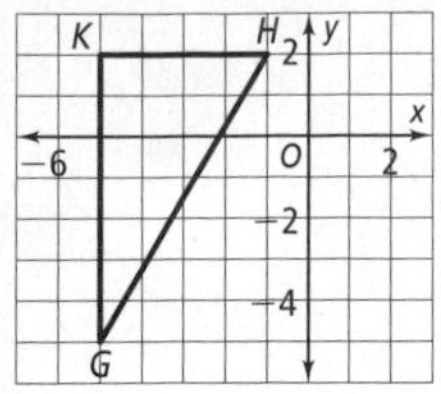

A (−2, 4)
B (−1, −2)
C (1, 2)
D (4, 2)

TIP 1

To calculate the scale factor of a dilation, you need to know the lengths of a pair of corresponding sides. You are given H'K'. You can use the graph to find HK.

TIP 2

Identify the coordinates of G, H, and K from the graph.

Think It Through

The scale factor is $\frac{H'K'}{HK} = \frac{8}{4} = 2$.

To find the coordinates of H', multiply the coordinates of H by the scale factor. H' is at $(2 \cdot (-1), 2 \cdot 2)$, or $(-2, 4)$. The correct answer is A.

Vocabulary Builder

As you solve problems, you must understand the meanings of mathematical terms. Match each term with its mathematical meaning.

A. centroid
B. circumcenter
C. isometry
D. similarity transformation
E. proportion
F. transformation

I. a transformation that preserves distance
II. the point of concurrency of the medians in a triangle
III. an equation that states that two ratios are equal
IV. a mapping that may result in a change in the position, shape, or size of a figure
V. a composition of rigid motions and dilations
VI. the point of concurrency of the perpendicular bisectors of a triangle

Selected Response

Read each question. Then write the letter of the correct answer on your paper.

1. Which quadrilateral must have congruent diagonals?

A kite
B rectangle
C parallelogram
D rhombus

2. In a 30°-60°-90° triangle, the shortest leg measures 13 in. What is the measure of the longer leg?

F 13 in.
G $13\sqrt{2}$ in.
H $13\sqrt{3}$ in.
I 26 in.

3. The vertices of $\square ABCD$ are $A(1, 7)$, $B(0, 0)$, $C(7, -1)$, and $D(8, 6)$. What is the perimeter of $\square ABCD$?

A 50
B 100
C $\sqrt{200}$
D $20\sqrt{2}$

Answers

Common Core Cumulative Standards Review

A. II
B. VI
C. I
D. V
E. III
F. IV
1. B
2. H
3. D

4. Mica and Joy are standing at corner A of the rectangular field shown below.

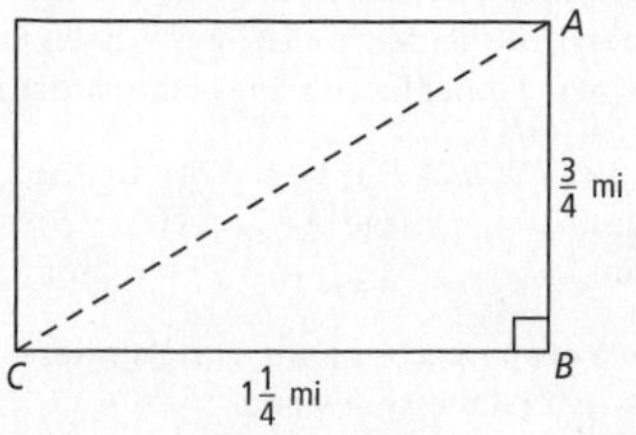

Mica walks diagonally across the field, from corner A to corner C. Joy walks from corner A to corner B, and then to corner C. To the nearest hundredth of a mile, how much farther did Joy walk than Mica?

(F) 0.54 mi (H) 1.46 mi

(G) 1 mi (I) 2 mi

5. What is the area of the square floor tile?

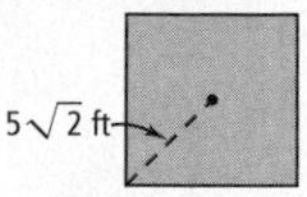

(A) 50 ft² (C) $100\sqrt{2}$ ft²

(B) 100 ft² (D) 150 ft²

6. What type of symmetry does the figure have?

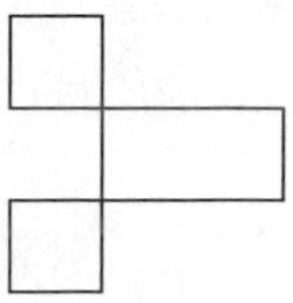

(F) 60° rotational symmetry

(G) 90° rotational symmetry

(H) line symmetry

(I) point symmetry

7. Which conditions allow you to conclude that a quadrilateral is a parallelogram?

(A) one pair of sides congruent, the other pair of sides parallel

(B) perpendicular, congruent diagonals

(C) diagonals that bisect each other

(D) one diagonal bisects opposite angles

8. If you are given a line and a point not on the line, what is the first step to construct the line parallel to the given line through the point?

(F) Construct an angle from a point on the line to the given point.

(G) Draw a straight line through the given point.

(H) Draw a ray from the given point that does not intersect the line.

(I) Label a point on the given line, and draw a line through that point and the given point

9. In a right triangle, which point lies on the hypotenuse?

(A) incenter (C) centroid

(B) orthocenter (D) circumcenter

10. In $\triangle LMN$, P is the centroid and $LE = 24$. What is PE?

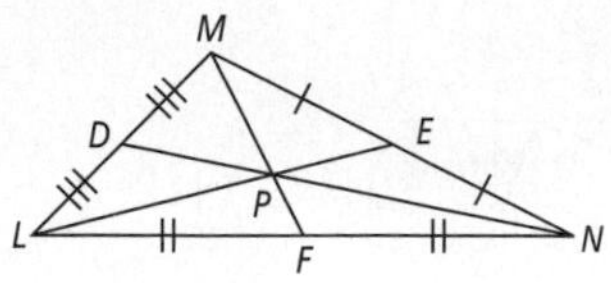

(F) 8 (H) 10

(G) 9 (I) 16

11. What is the sum of the angle measures of a 32-gon?

(A) 3200° (C) 5400°

(B) 3800° (D) 5580°

12. The diagonals of rectangle $PQRS$ intersect at H. What is the length of $\overline{QS}$?

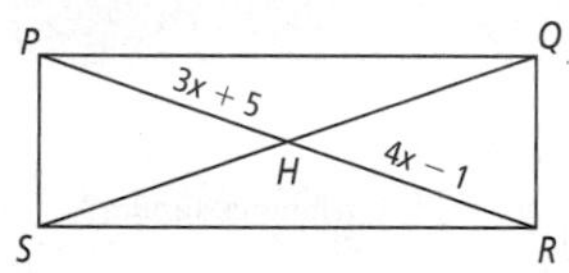

(F) 6 (H) 23

(G) 12 (I) 46

4. F
5. B
6. H
7. C
8. I
9. D
10. F
11. C
12. I

Answers

Common Core Cumulative Standards Review (continued)

13. 30
14. 95
15. 29
16. 62
17. 2
18. 27.5
19. 20
20. [2] Quadrants I and IV

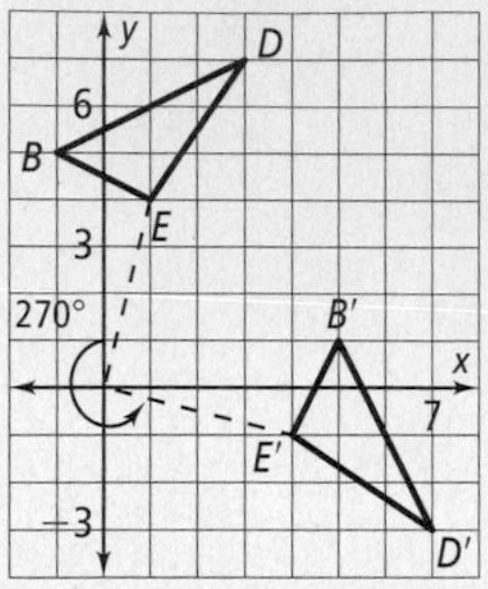

[1] incomplete OR incorrect diagram

21. [2] 24.5 units^2; a rt. isosc. △ is a 45°-45°-90° △, so the length of each leg is 7 units. Then the area of the △ is $\frac{1}{2} \cdot 7^2 = \frac{49}{2} = 24.5$ units^2 OR other appropriate method.
[1] appropriate method with one computational error

22. [2] 6.36
[1] one minor computational error

23. [2] $\overline{AB} \cong \overline{CB}$ (Given); $\angle A \cong \angle C$ (Isosc. △ Thm.); $\overline{BD} \perp \overline{AC}$ (Given); $\angle ADB$ and $\angle CDB$ are rt. ⦣ (Def. of ⊥); $\angle ADB \cong \angle CDB$ (All rt. ⦣ are ≅.); $\triangle ABD \cong \triangle CDB$ (AAS) OR other correct proof
[1] proof is incomplete OR contains an error

24. [4] Yes; the coordinates of the vertices are $A(1, 4)$, $B(2, -2)$, and $C(-4, -3)$. The slope of $\overline{AB}$ is $\frac{4-(-2)}{1-2} = -\frac{6}{1}$ and the slope of $\overline{BC}$ is $\frac{-2-(-3)}{2-(-4)} = \frac{1}{6}$. Since the product of their slopes is -1, $\overline{AB} \perp \overline{BC}$, so $\angle ABC$ is a rt. ∠. Thus $\triangle ABC$ is a rt. △ OR other appropriate method.
[3] appropriate method with one computational error
[2] appropriate method with two computational errors
[1] appropriate method with more than two computational errors

25. [4] Check students' graphs.
Yes, $LMNO \cong RSTV$ because
$(R_{x\text{-axis}} \circ T_{<5,\,-1>})(LMNO) = RSTV$
[3] student incorrectly wrote the congruence transformation rule
[2] student incorrectly answer two parts of the question
[1] correct answer; missing graph and congruence transformation rule

Constructed Response

13. What is the value of x for which $p \parallel q$?

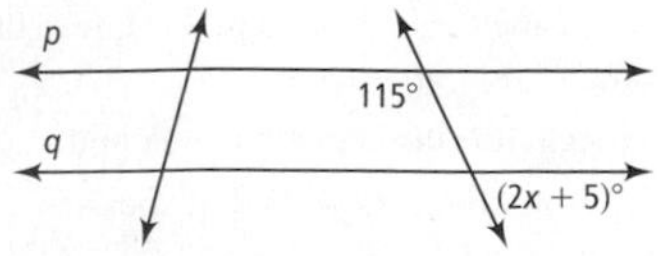

14. What is the measure of $\angle H$?

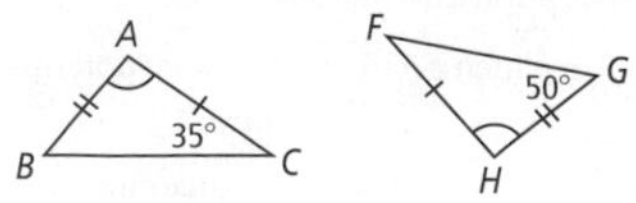

15. What is the area of the square, in square units, in the figure below?

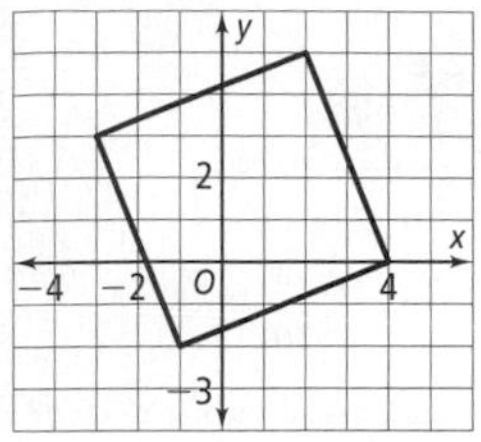

16. In ▱$PQRS$, what is the value of x?

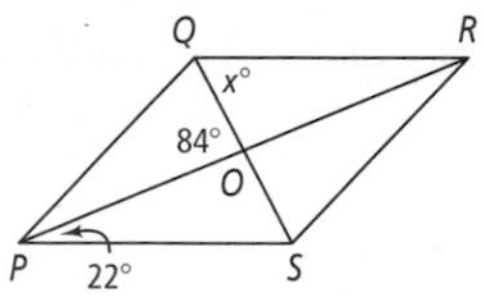

17. For what value of x are the two triangles similar?

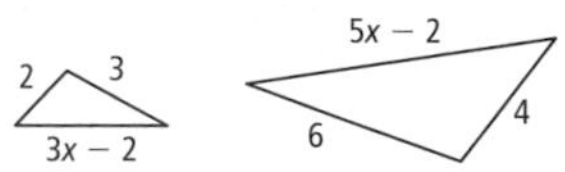

18. Your friend is 5 ft 6 in. tall. When your friend's shadow is 6 ft long, the shadow of a nearby sculpture is 30 ft long. What is the height, in feet, of the sculpture?

19. Lucia makes a triangular garden in one corner of her fenced rectangular backyard. She has 25 ft of edging to use along the unfenced side of the garden. One of the fenced sides of the garden is 15 ft long. What is the length, in feet, of the other fenced side of the garden?

20. $\triangle DEB$ has vertices $D(3, 7)$, $E(1, 4)$, and $B(-1, 5)$. In which quadrant(s) is the image of $r_{(270°,\,O)}(\triangle DEB)$? Draw a diagram.

21. What is the area of an isosceles right triangle whose hypotenuse is $7\sqrt{2}$? Show your work.

22. In $\triangle BGT$, $m\angle B = 48$, $m\angle G = 52$, and $GT = 6$ mm. What is BT? Write your answer to the nearest hundredth of a millimeter.

23. In $\triangle ABC$ below, $\overline{AB} \cong \overline{CB}$ and $\overline{BD} \perp \overline{AC}$. Prove that $\triangle ABD \cong \triangle CBD$.

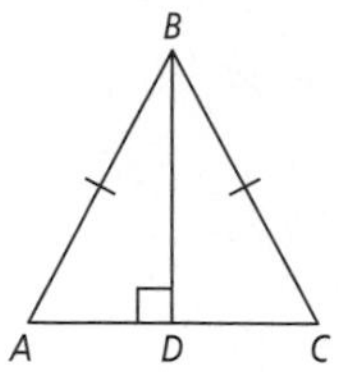

Extended Response

24. Is $\triangle ABC$ a right triangle? Justify your answer.

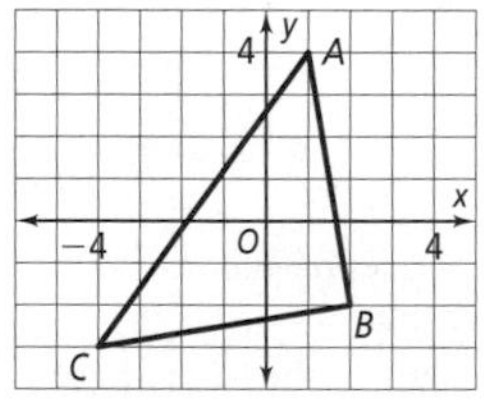

25. $LMNO$ has vertices $L(-4, 0)$, $M(-2, 3)$, $N(1, 1)$, and $O(-1, -2)$. $RSTV$ has vertices $R(1, 1)$, $S(3, -2)$, $T(6, 0)$, and $V(4, 3)$. Graph the two quadrilaterals. Is $LMNO \cong RSTV$? If so, write the rule for the congruence transformation that maps $LMNO$ to $RSTV$. If not, explain why not.

CHAPTER 10

Get Ready!

Skills Handbook, p. 888

Squaring Numbers and Finding Square Roots

Simplify.

1. 3^2 **2.** 8^2 **3.** 12^2 **4.** 15^2

5. $\sqrt{16}$ **6.** $\sqrt{64}$ **7.** $\sqrt{100}$ **8.** $\sqrt{169}$

Solve each quadratic equation.

9. $x^2 = 64$ **10.** $b^2 - 225 = 0$ **11.** $a^2 = 144$

Review, p. 399

Simplifying Radicals

Simplify. Leave your answer in simplest radical form.

12. $\sqrt{8}$ **13.** $\sqrt{27}$ **14.** $\sqrt{75}$ **15.** $4\sqrt{72}$

Lesson 1-8

Area

16. A garden that is 6 ft by 8 ft has a walkway that is 3 ft wide around it. What is the ratio of the area of the garden to the area of the garden and walkway? Write your answer in simplest form.

17. A rectangular rose garden is 8 m by 10 m. One bag of fertilizer can cover 16 m^2. How many bags of fertilizer will be needed to cover the entire garden?

Lessons 6-3 and 6-5

Classifying Quadrilaterals

Classify each quadrilateral as specifically as possible.

18.

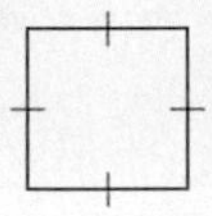

19.

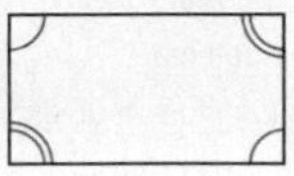

20.

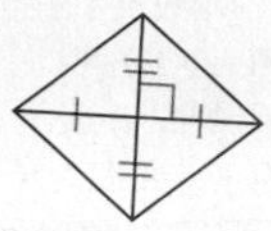

Looking Ahead Vocabulary

21. A *semi*annual school fundraiser is an event that occurs every half year. What might a *semi*circle look like in geometry?

22. A *major* skill is an important skill. How would you describe a *major* arc in geometry?

23. Two buildings are *adjacent* if they are next to each other. What do you think *adjacent* arcs on a geometric figure could be?

Answers

Get Ready!

1. 9
2. 64
3. 144
4. 225
5. 4
6. 8
7. 10
8. 13
9. ± 8
10. ± 15
11. ± 12
12. $2\sqrt{2}$
13. $3\sqrt{3}$
14. $5\sqrt{3}$
15. $24\sqrt{2}$
16. $\frac{2}{7}$
17. 5
18. rhombus
19. parallelogram
20. rhombus
21. Answers may vary. Sample: half of a circle
22. Answers may vary. Sample: more than half a circle
23. Answers may vary. Sample: arcs that are next to each other

Get Ready!

Assign this diagnostic assessment to determine if students have the prerequisite skills for Chapter 10.

Lesson	Skill
Skills Handbook, p. T889	Squaring Numbers and Finding Square Roots
Review, p. 399	Simplifying Radicals
1-8	Probability
6-3 and 6-5	Classifying Quadrilaterals

To remediate students, select from these resources (available for every lesson).

- Online Problems (PowerGeometry.com)
- Reteaching (All-in-One Teaching Resources)
- Practice (All-in-One Teaching Resources)

Why Students Need These Skills

SQUARING NUMBERS AND FINDING SQUARE ROOTS Squaring numbers and finding square roots will be used when calculating areas of similar figures.

SIMPLIFYING RADICALS Simplifying radicals allows students to identify and combine like terms, and will be applied to problems in geometry and algebra.

PROBABILITY The concept of probability will be extended to geometric probability.

CLASSIFYING QUADRILATERALS Area formulas for common quadrilaterals will be used. Students need to identify quadrilaterals in order to find their areas.

Looking Ahead Vocabulary

SEMICIRCLE Have students name other words that use the prefix *semi-*.

MAJOR Ask students whether a major arc or minor arc is larger.

ADJACENT Have students name items or persons that are adjacent to them in the classroom.

Chapter 10 Overview

In Chapter 10 students find areas of circles and polygons. Students will develop the answers to the Essential Questions as they learn the concepts and skills shown below.

BIG idea Measurement

ESSENTIAL QUESTION How do you find the area of a polygon or find the circumference and area of a circle?

- Students will use formulas to find areas of parallelograms, triangles, trapezoids, rhombuses, and kites.
- Students will explore area concepts related to regular polygons.
- Students will use trigonometry to find areas.
- Students will find circumferences and areas of circles.

BIG idea Similarity

ESSENTIAL QUESTION How do perimeters and areas of similar polygons compare?

- Students will examine ratios among similar figures.
- Given a figure and its area, students will be able to find the area of a figure similar to the original figure.

Content Standards

Following are the standards covered in this chapter. Modeling standards are indicated by a star symbol (★).

CONCEPTUAL CATEGORY Geometry

Domain Congruence G-CO

Cluster Experiment with transformations in the plane (Standard G-CO.A.1)
LESSON 10-6

Cluster Make Geometric Constructions (Standard G-CO.D.13)
LESSON 10-3

Domain Similarity, Right Triangles, and Trigonometry G-SRT

Cluster Apply trigonometry to general triangles (Standard G-SRT.D.9)
LESSON 10-5

Domain Circles G-C

Cluster Understand and apply theorems about circles (Standard G-C.A.1)
LESSON 10-6

Cluster Find arc lengths and areas of sectors of circles (Standard G-C.B.5)
LESSONS 10-6, 10-7

Domain Expressing Geometric Properties with Equations G-GPE

Cluster Use coordinates to prove simple geometric theorems algebraically (Standard G-GPE.B.7★)
LESSON 10-1

Domain Modeling with Geometry G-MG

Cluster Apply geometric concepts in modeling situations (Standard G-MG.A.1★)
LESSONS 10-1, 10-2, 10-3

CHAPTER 10

Area

Download videos connecting math to your world. (VIDEO)

Interactive! Vary numbers, graphs, and figures to explore math concepts. (DYNAMIC ACTIVITIES)

The online Solve It will get you in gear for each lesson. (SOLVE IT!)

Math definitions in English and Spanish (VOCABULARY)

Online access to stepped-out problems aligned to Common Core (ONLINE PROBLEMS)

Get and view your assignments online. (ONLINE HOMEWORK)

Extra practice and review online (MathXL FOR SCHOOL)

Virtual Nerd™ tutorials with built-in support

Chapter Preview

BIG ideas

1 **Measurement**
Essential Question How do you find the area of a polygon or find the circumference and area of a circle?

2 **Similarity**
Essential Question How do perimet and areas of similar polygons compar

Vocabulary

English/Spanish Vocabulary Audio Online:

English	Spanish
adjacent arcs, *p. 650*	arcos adyacentes
apothem, *p. 629*	apotema
arc length, *p. 653*	longitud de un arco
central angle, *p. 649*	ángulo central
concentric circles, *p. 651*	círculos concéntricos
congruent arcs, *p. 653*	arcos congruentes
diameter, *p. 649*	diámetro
major arc, *p. 649*	arco mayor
minor arc, *p. 649*	arco menor
radius, *pp. 629, 649*	radio
sector of a circle, *p. 661*	sector de un círculo
segment of a circle, *p. 662*	segmento de un círculo

DOMAINS

- Circles
- Expressing Geometric Properties with Equations
- Modeling with Geometry

PowerGeometry.com

Chapter 10 Overview

Use these online assets to engage your students. These include support for the Solve It and step-by step solutions for Problems.

VIDEO: Show the student-produced video demonstrating relevant and engaging applications of the new concepts in this chapter.

VOCABULARY: Find online definitions for new terms in English and Spanish.

SOLVE IT!: Start each lesson with an attention-getting Problem. View the Problem online with helpful hints.

Common Core Performance Task

Finding the Probability of Winning

The diagram below shows the target for a game at the school fair. The target is composed of a regular octagon inscribed in a circle with center *O*. The length of one side of the regular octagon is 9 in. Four vertices of the octagon are connected to form a quadrilateral, as shown.

To play the game, a person throws one dart at the target. The player wins a prize if the dart lands in any of the red or yellow regions.

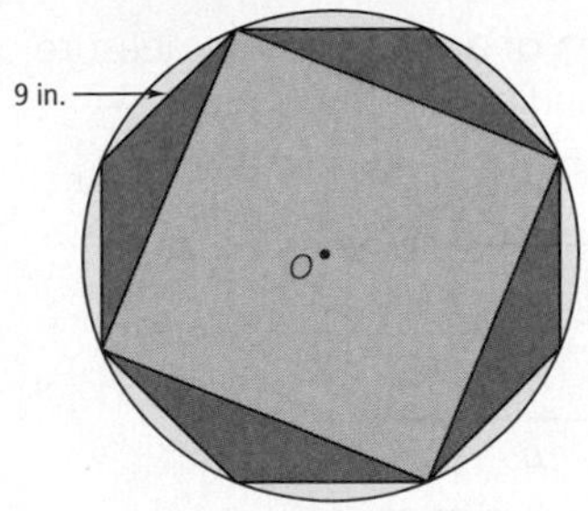

Task Description

Find the probability that a person who plays the game wins a prize. Round your answer to the nearest whole percent. Assume the player's dart is equally likely to land at any point of the target.

Connecting the Task to the Math Practices

MATHEMATICAL PRACTICES

As you complete the task, you'll apply several Standards for Mathematical Practice.

- You'll use trigonometric ratios and a calculator to find lengths needed to calculate areas. (MP 5)
- You'll draw in auxiliary line segments to help you find areas. (MP 7)
- You'll use more than one method to find areas and verify your results. (MP 1)

- Increase students' depth of knowledge with interactive online activities.
- Show Problems from each lesson solved step by step. Instant replay allows students to go at their own pace when studying online.
- Assign homework to individual students or to an entire class.
- Prepare students for the Mid-Chapter Quiz and Chapter Test with online practice and review.
- **Virtual Nerd™** Access Virtual Nerd student-centered math tutorials that directly relate to the content of the lesson.

Overview of the Performance Task

Students will find areas of a variety of shapes in a complex diagram. They will then use some of their results to calculate a geometric probability. Students may recognize that the area of the winning region can be found as the difference between the area of the circle and the area of the gray quadrilateral (which they can prove is a square). The area of the winning region can also be found as the sum of the areas of the red triangles and the yellow segments. The Apply What You've Learned sections use the latter approach, as a way to reinforce concepts and procedures students learn in the chapter. Both approaches require finding the length of the side of the square and the radius of the circle.

Students will work on the Performance Task in the following places in the chapter.

- Lesson 10-1 (p. 622)
- Lesson 10-5 (p. 648)
- Lesson 10-7 (p. 666)
- Pull It All Together (p. 675)

Introducing the Performance Task

Tell students to read the problem on this page. Do not have them start work on the problem at this time, but ask them the following questions.

Q What is a strategy you can try in order to solve the problem? **[Sample: I can determine the probability of winning a prize by finding the ratio of the total area of the red and yellow regions to the total area of the target.]**

Q All the line segments in the diagram are borders of regions of the target. What other line segments might you draw in to help you solve the problem? **[Sample: an altitude of a red triangle, a radius of the circle]**

PARCC CLAIMS

Sub-Claim A: Major Content with Connections to Practices

Sub-Claim B: Additional and Supporting Content with Connections to Practices

SBAC CLAIMS

Claim 1: Concepts and Procedures

Claim 2: Problem Solving

AREA

Math Background

PROFESSIONAL DEVELOPMENT

The Understanding by Design® methodology was central to the development of the Big Ideas and the Essential Understandings. These will help your students build a structure on which to make connections to prior learning.

Measurement

BIG idea Some attributes of geometric figures, such as length, area, volume, and angle measure, are measurable. Units are used to describe these attributes.

ESSENTIAL UNDERSTANDINGS

10–1 The area of a parallelogram or a triangle can be found when the length of its base and its height are known.

10–2 The area of a trapezoid can be found when the height and the lengths of its bases are known. The area of a rhombus or a kite can be found when the lengths of its diagonals are known.

10–3 The area of a regular polygon is a function of the distance from the center to a side and the perimeter.

10–5 Trigonometry can be used to find the area of a regular polygon when the length of a side, radius, or apothem is known or to find the area of a triangle when the length of two sides and the included angle are known.

10–6 The length of part of a circle's circumference can be found by relating it to an angle in the circle.

10–7 The area of parts of a circle formed by radii and arcs can be found when the circle's radius is known.

Similarity

BIG idea Two geometric figures are similar when corresponding lengths are proportional and corresponding angles are congruent. Areas of similar figures are proportional to the squares of their corresponding lengths.

ESSENTIAL UNDERSTANDING

10–4 Ratios can be used to compare the perimeters and areas of similar figures.

Area Formulas

The area of a figure is the measure of how much space is contained within the figure.

Rectangle

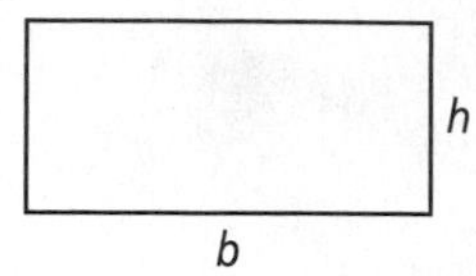

$A = bh$

Parallelogram

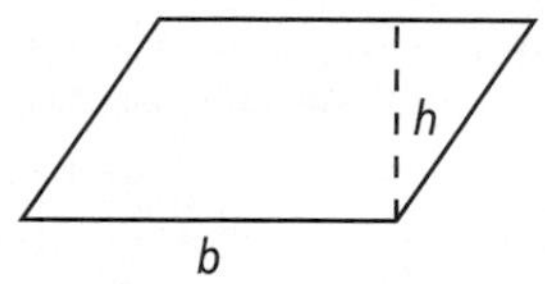

$A = bh$

Triangle

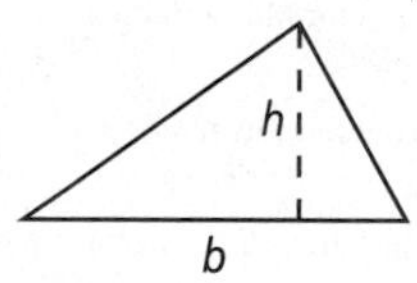

$A = \frac{1}{2}bh$

Trapezoid

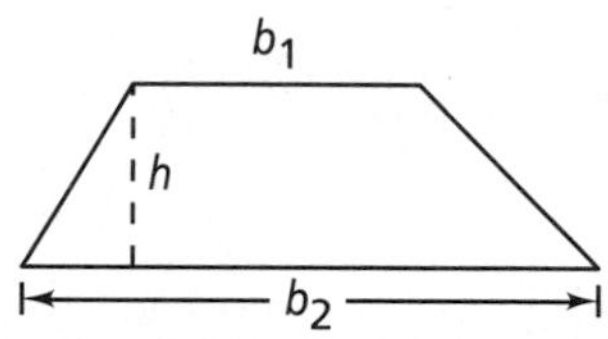

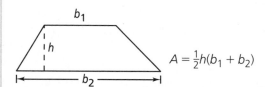

$A = \frac{1}{2}h(b_1 + b_2)$

Circle

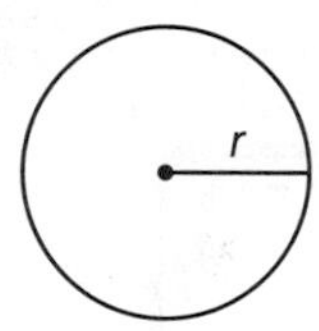

$A = \pi r^2$

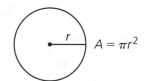

Common Errors With Area Formulas

When finding the area of a parallelogram, students might use the length of a side rather than the altitude.

Mathematical Practices

Look for and make use of structure Students build on their knowledge of a rectangle's area and develop area formulas for triangles, special quadrilaterals, and regular polygons. They base kites and rhombuses on their experience with right triangles and parallelograms and trapezoids on rectangles. They use isosceles triangles within a regular polygon.

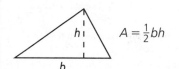

Similar Figures

Recall two properties of similar figures from Lesson 7-2:

1. Corresponding angles are congruent.
2. Corresponding sides are proportional.

There is also a relationship between the perimeters and areas of similar figures.

3. The ratio of the perimeters is the same as the scale factor.
4. The ratio of the areas is the square of the scale factor.

Consider the following similar triangles.

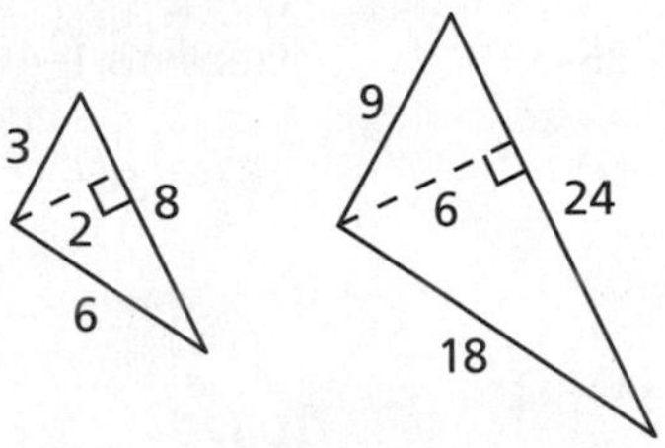

Scale factor $= \frac{3}{9} = \frac{8}{24} = \frac{6}{18} = \frac{1}{3}$

Ratio of perimeters $= \frac{3 + 8 + 6}{9 + 24 + 18} = \frac{17}{51} = \frac{1}{3}$

The ratio of their areas is the square of the ratio of their corresponding sides.

Ratio of areas $= \dfrac{\frac{1}{2}(8)(2)}{\frac{1}{2}(24)(6)} = \frac{8}{72} = \frac{1}{9} = \left(\frac{1}{3}\right)^2$

The ratio of their areas is the square of the ratio of their corresponding sides.

Common Errors With Similar Figures

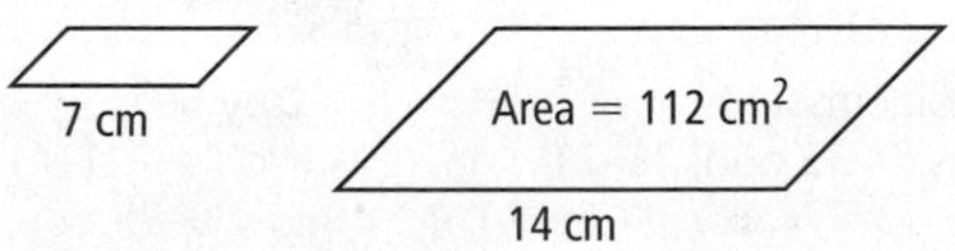

Scale factor $= \frac{14}{7} = \frac{2}{1}$

Students might forget to square the ratio when figuring the area of the second figure.

Scale factor $= \frac{2}{1} = \frac{112}{x}$ Scale factor $= \frac{4}{1} = \frac{112}{x}$

Mathematical Practices

Model with Mathematics/Look for and express regularity in repeated reasoning Students employ exploration to discover the pattern in the ratio of similar figures' areas. They examine specific applied situations where this concept is in use.

Geometric Probability

The probability of an event is equal to the ratio of the favorable outcomes to the possible outcomes. In geometry, probability can be represented using length or area models.

Length Models

Situations that involve linear measurements can be modeled using length.

Jenna is expecting a phone call sometime between 10 and 10:30. What is the probability of a phone call coming in the first 10 minutes?

Use a segment of length 30 to represent the window of time in which the phone call should happen. The segments from 0 to 10 represent the call coming in the first 10 minutes.

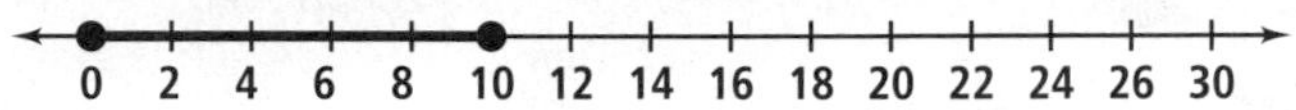

$p(\text{First 10 minutes}) = \frac{\text{length of favorable}}{\text{total length}} = \frac{10}{30} = \frac{1}{3}$

Area Models

The probability that a randomly selected point lies in a certain section of a figure can be found by comparing the areas of the section and the figure.

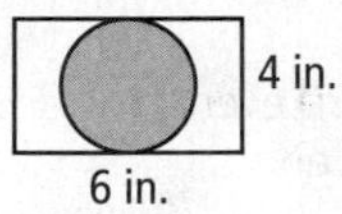

Find the probability that a randomly selected point will fall in the shaded region.

$p(\text{shaded}) = \frac{\text{area of circle}}{\text{area of rectangle}} = \frac{\pi \cdot 2^2}{6 \cdot 4} \approx 0.52$

Common Errors With Geometric Probability

Students may get confused about which length or area represents the favorable outcomes. To make sure that they calculate the correct probability, have them find each length or area separately before dividing to find the probability.

Mathematical Practices

Model with Mathematics/Make sense of problems and persevere in solving them In geometric probability, students model situations in time and space with line segments and regions, solving for length and area. They build on this thinking in later exercises to look at problems modeled with inequalities on both number lines and coordinate planes.

CHAPTER 10

AREA

Pacing and Assignment Guide

Lesson	Teaching Day(s)	Basic (Traditional)	Average (Traditional)	Advanced (Traditional)	Block
10-1	1	Problems 1–2 Exs. 8–13 all	Problems 1–2 Exs. 9–13 odd	Problems 1–4 Exs. 9–17 odd, 18–46	**Day 1** Problems 1–4 Exs. 9–17 odd, 18–43
	2	Problems 3–4 Exs. 14–17 all, 19, 22–23, 30–34 even, 37–38	Problems 3–4 Exs. 15–17 odd, 18–43		
10-2	1	Problems 1–2 Exs. 11–19 all, 45–53	Problems 1–2 Exs. 11–19 odd, 45–53	Problems 1–4 Exs. 11–25 odd, 26–53	**Day 2** Problems 1–4 Exs. 11–25 odd, 26–41, 45–53
	2	Problems 3–4 Exs. 20–25 all, 26–38 even	Problems 3–4 Exs. 21–25 odd, 26–41		
10-3	1	Problems 1–3 Exs. 8–25 all, 26-30 even, 31–33 all, 35, 44–52	Problems 1–3 Exs. 9–25 odd, 26–41, 44–52	Problems 1–3 Exs. 9–25 odd, 26–52	**Day 3** Problems 1–3 Exs. 9–25 odd, 26–41, 44–52
10-4	1	Problems 1–2 Exs. 9–16 all, 52–62	Problems 1–2 Exs. 9–15 odd, 52–62	Problems 1–4 Exs. 9–23 odd, 25–62	Problems 1–2 Exs. 9–15 odd, 52–62
	2	Problems 3–4 Exs. 17–24 all, 26–30 even, 31–33 all, 34–44 even	Problems 3–4 Exs. 17–23 odd, 25–47		**Day 4** Problems 3–4 Exs. 17–23 odd, 25–47
10-5	1	Problems 1–3 Exs. 6–22 all, 24–26 even, 27, 32–36 even	Problems 1–3 Exs. 7–19 odd, 20–37	Problems 1–3 Exs. 7–19 odd, 20–39	Problems 1–3 Exs. 7–19 odd, 20–37
10-6	1	Problems 1–4 Exs. 9–35 all, 36–50 even, 60–71	Problems 1–4 Exs. 9–35 odd, 36–56, 60–71	Problems 1–4 Exs. 9–35 odd, 36–71	**Day 5** Problems 1–4 Exs. 9–35 odd, 36–56, 60–71
10-7	1	Problems 1–3 Exs. 7–25 all, 26–34 even, 35–36	Problems 1–3 Exs. 7–25 odd, 26–44	Problems 1–3 Exs. 7–25 odd, 26–50	Problems 1–3 Exs. 7–25 odd, 26–44
10-8	1	Problems 1-2 Exs. 8–16 all, 46–62	Problems 1-2 Exs. 9–15 odd, 46–62	Problems 1-4 Exs. 9–23 odd, 25–62	**Day 6** Problems 1–4 Exs. 9–23 odd, 25–43, 46–62
	2	Problems 3–4 Exs. 17–24 all, 27–28, 32–40 even	Problems 3–4 Exs. 17–23 odd, 25–43		
Review	1	Chapter 10 Review	Chapter 10 Review	Chapter 10 Review	**Day 7** Chapter 10 Review
Assess	1	Chapter 10 Test	Chapter 10 Test	Chapter 10 Test	Chapter 10 Test
Total		**14 Days**	**14 Days**	**10 Days**	**7 Days**

Note: Pacing does not include Concept Bytes and other feature pages.

Resources

KEY
I = Interactive asset at PowerGeometry.com
E = Editable master at PowerGeometry.com
P = Available in Print
M = Master at PowerGeometry.com
✓ = CD-ROM

	For the Chapter	10-1	10-2	10-3	10-4	10-5	10-6	10-7	10-8
Planning									
Teacher Center Online Planner & Grade Book	I	I	I	I	I	I	I	I	I
Interactive Learning & Guided Instruction									
My Math Video	I								
Solve It!		I M	I M	I M	I M	I M	I M	I M	I M
Student Companion		P M	P M	P M	P M	P M	P M	P M	P M
Vocabulary Support		I P M	I P M	I P M	I P M	I P M	I P M	I P M	I P M
Got It? Support		I P	I P	I P	I P	I P	I P	I P	I P
Online Problems		I	I	I	I	I	I	I	I
Additional Problems		M	M	M	M	M	M	M	M
English Language Learner Support (TR)		E P M	E P M	E P M	E P M	E P M	E P M	E P M	E P M
Activities, Games, and Puzzles		E M	E M	E M	E M	E M	E M	E M	E M
Teaching With TI Technology With CD-ROM		✓ P	✓ P		✓ P				
TI-Nspire™ Support CD-ROM		✓	✓	✓	✓	✓	✓	✓	✓
Lesson Check & Practice									
Student Companion		P M	P M	P M	P M	P M	P M	P M	P M
Lesson Check Support		I P	I P	I P	I P	I P	I P	I P	I P
Practice and Problem Solving Workbook		P	P	P	P	P	P	P	P
Think About a Plan (TR)		E P M	E P M	E P M	E P M	E P M	E P M	E P M	E P M
Practice Form G (TR)		E P M	E P M	E P M	E P M	E P M	E P M	E P M	E P M
Standardized Test Prep (TR)		P M	P M	P M	P M	P M	P M	P M	P M
Practice *Form K* (TR)		E P M	E P M	E P M	E P M	E P M	E P M	E P M	E P M
Extra Practice	E M								
Find the Errors!	M								
Enrichment (TR)		E P M	E P M	E P M	E P M	E P M	E P M	E P M	E P M
Answers and Solutions CD-ROM	✓	✓	✓	✓	✓	✓	✓	✓	✓
Assess & Remediate									
ExamView CD-ROM	✓	✓	✓	✓	✓	✓	✓	✓	✓
Lesson Quiz		I M	I M	I M	I M	I M	I M	I M	I M
Quizzes and Tests *Form G* (TR)	E P M				E P M				E P M
Quizzes and Tests *Form K* (TR)	E P M				E P M				E P M
Reteaching (TR)		E P M	E P M	E P M	E P M	E P M	E P M	E P M	E P M
Performance Tasks (TR)	P M								
Cumulative Review (TR)	P M								
Progress Monitoring Assessments	I P M								

(TR) Available in All-In-One Teaching Resources

Guided Instruction

PURPOSE To derive formulas for the areas of polygons

PROCESS Students will derive formulas for the areas of polygons by comparing areas with those of a more recognizable, transformed polygon.

DISCUSS Explain the meaning of area. Show students how to count the areas of figures placed on grid paper.

Activity 1

This Activity has students derive the formula for the area of a parallelogram.

Q When you cut your parallelogram, what is the shape that you cut away from the parallelogram? **[a right triangle]**

Q Why is it important to make the cut perpendicular to the base of the parallelogram? **[to create a right angle that will be a vertex of the rectangle]**

Activity 2

This Activity has students derive the formula for the area of a triangle.

Q When you cut your triangle, what is the shape that you cut away from the triangle? **[a smaller triangle]**

Q What shape remains after you cut away the triangle off the top? **[a trapezoid]**

Q Why is it important to make the base of the triangle you cut away be the midsegment of the original triangle? **[so that when you rotate the small triangle, the side lengths will match]**

Mathematical Practices This Concept Byte supports students in becoming proficient in shifting perspective, Mathematical Practice 7.

Concept Byte

Use With Lessons 10-1 and 10-2

ACTIVITY

Transforming to Find Area

Common Core State Standards

Prepares for G-GMD.A.3 Use volume formulas for cylinders, pyramids, cones, and spheres to solve problems.

MP 7

You can use transformations to find formulas for the areas of polygons. In these activities, you will cut polygons into pieces and use the pieces to form different polygons.

Activity 1

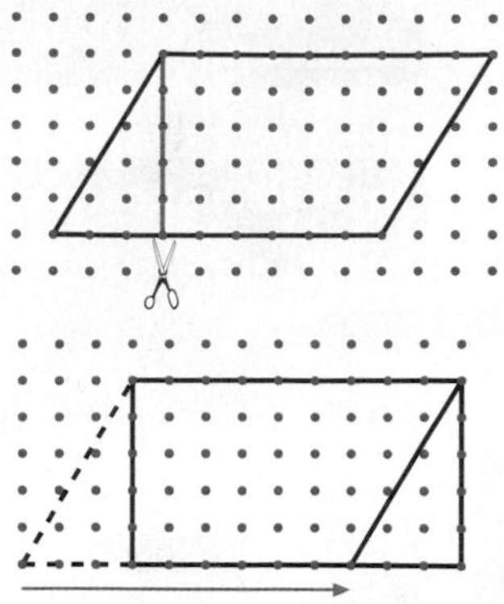

Step 1 Count and record the number of units in the base and the height of the parallelogram at the right.

Step 2 Copy the parallelogram onto grid paper.

Step 3 Cut out the parallelogram. Then cut it into two pieces as shown.

Step 4 Translate the triangle to the right through a distance equal to the base of the parallelogram.

The translation results in a rectangle. Since their pieces are congruent, the parallelogram and rectangle have the same area.

1. How many units are in the base of the rectangle? The height of the rectangle?
2. How do the base and height of the rectangle compare to the base and height of the parallelogram?
3. Write the formula for the area of the rectangle. Explain how you can use this formula to find the area of a parallelogram.

Activity 2

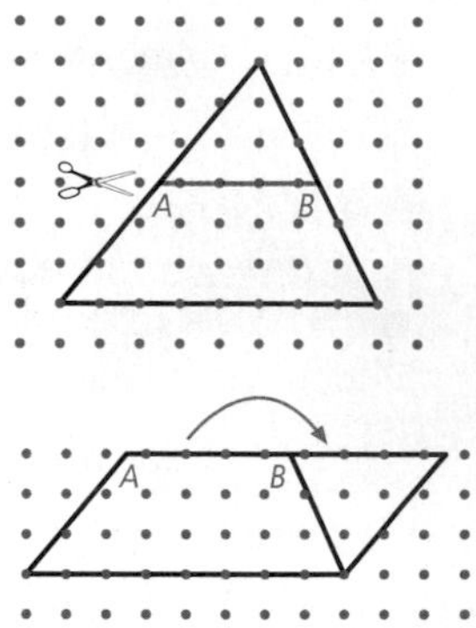

Step 1 Count and record the number of units in the base and the height of the triangle at the right.

Step 2 Copy the triangle onto grid paper. Mark the midpoints A and B and draw midsegment $\overline{AB}$.

Step 3 Cut out the triangle. Then cut it along $\overline{AB}$.

Step 4 Rotate the small triangle 180° about the point B.

The bottom part of the triangle and the image of the top part form a parallelogram.

4. How many units are in the base of the parallelogram? The height of the parallelogram?

Answers

Activity 1

1. 9; 5
2. They are the same.
3. $A = b \cdot h$; explanations may vary. Sample: A ▱ can be transformed into a rectangle with the same base and height.

Activity 2

4. 8; 3

5. How do the base and height of the parallelogram compare to the base and height of the original triangle? Write an expression for the height of the parallelogram in terms of the height h of the triangle.

6. Write your formula for the area of a parallelogram from Activity 1. Substitute the expression you wrote for the height of the parallelogram into this formula. You now have a formula for the area of a triangle.

Activity 3

Step 1 Count and record the bases and height of the trapezoid at the right.

Step 2 Copy the trapezoid. Mark the midpoints M and N, and draw midsegment $\overline{MN}$.

Step 3 Cut out the trapezoid. Then cut it along $\overline{MN}$.

Step 4 Transform the trapezoid into a parallelogram.

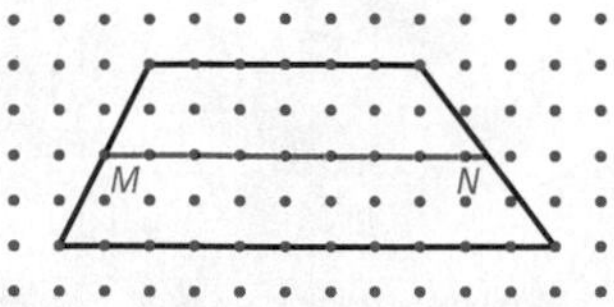

7. What transformation did you apply to form a parallelogram?

8. What is an expression for the base of the parallelogram in terms of the two bases, b_1 and b_2, of the trapezoid?

9. If h represents the height of the trapezoid, what is an expression in terms of h for the height of the parallelogram?

10. Substitute your expressions from Questions 8 and 9 into your area formula for a parallelogram. What is the formula for the area of a trapezoid?

Exercises

11. In Activity 2, can a different rotation of the small triangle form a parallelogram? If so, does using that rotation change your results? Explain.

12. Make another copy of the Activity 2 triangle. Find a rotation of the entire triangle so that the preimage and image together form a parallelogram. How can you use the parallelogram and your formula for the area of a parallelogram to find the formula for the area of a triangle?

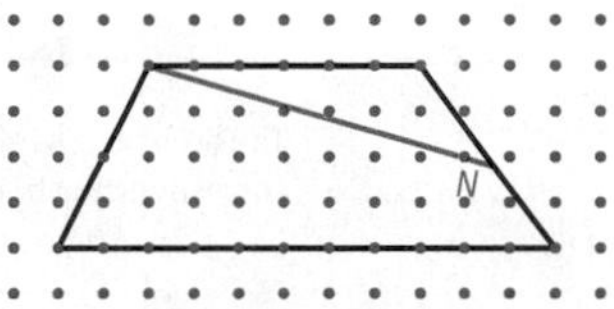

13. a. In the trapezoid at the right, a cut is shown from the midpoint of one leg to a vertex. What transformation can you apply to the top piece to form a triangle from the trapezoid?
    b. Use your formula for the area of a triangle to find a formula for the area of a trapezoid.

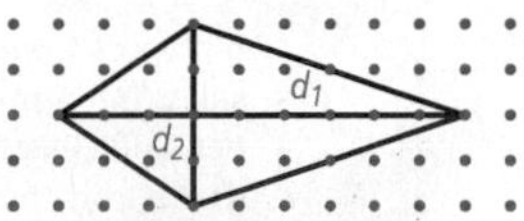

14. Count and record the lengths of the diagonals, d_1 and d_2, of the kite at the right. Copy and cut out the kite. Reflect half of the kite across the line of symmetry d_1 by folding the kite along d_1. Use your formula for the area of a triangle to find a formula for the area of a kite.

5. The bases are the same; the height of the ▱ is half the height of the △; $\frac{1}{2}h$

6. $A = b \cdot h$; $A = \frac{1}{2}b \cdot h$

Activity 3

7. a rotation of 180° around point M or point N

8. $b_1 + b_2$

9. $\frac{1}{2}h$

10. $\frac{1}{2}(b_1 + b_2)h$

Exercises

11. Yes; no; explanations may vary. Sample: If you rotate the small △ about point A, the image of the small △ and the bottom part of the original △ form a ▱.

12. Rotate the entire △ 180° about the midpt. of any side to form a ▱ that has the same base b and height h as the △. Since two △s form the ▱, its area is twice the area of a △. That is, bh = Area of ▱ = 2 • Area of △. So Area of △ $= \frac{1}{2}bh$.

13a. Rotate the △ by 180° about point N.

b. Area $= \frac{1}{2}(b_1 + b_2)h$

14. $d_1 = 9$ units, $d_2 = 4$ units; the area of each △ is half the area of the kite. Area of kite = 2 • Area of △ = $2 \cdot \left(\frac{1}{2}bh\right) = bh$, where $b = d_1$ and $h = \frac{1}{2}d_2$. So, Area of kite $= \frac{1}{2}d_1d_2$.

Activity 3

This Activity has students derive the formula for the area of a trapezoid.

Q When you cut your trapezoid, what is the shape that you cut away from the trapezoid? **[a smaller trapezoid]**

Q What shape remains after you cut away the trapezoid off the top? **[a trapezoid]**

1 Interactive Learning

Solve It!

PURPOSE To use the area of right isosceles triangles to find the area of a larger composite figure

PROCESS Students may

- divide the total area by the area of one block to find the upper and lower limits on the number of blocks used to build the stage.
- use the guess and check strategy to sketch diagrams of possible solutions.

FACILITATE

Q How many blocks will fill an area of approximately 1000 ft^2? **[approximately 31]**

Q How many blocks will fill an area of approximately 1400 ft^2? **[approximately 43]**

Q What are the length and width of the arrow stage with the largest possible area? **[56 ft by 64 ft]**

ANSWER See Solve It in Answers on next page.

CONNECT THE MATH In the Solve It, students use triangles to calculate areas. In the lesson, students learn formulas for areas of rectangles, parallelograms and triangles.

2 Guided Instruction

Take Note

Be sure that students understand the difference between the side length and height of a parallelogram. Focus on the concept that the distance from a point to a line is a perpendicular segment.

10-1 Areas of Parallelograms and Triangles

Common Core State Standards

G-MG.A.1 Use geometric shapes, their measures, and their properties to describe objects.
G-GPE.B.7 Use coordinates to compute perimeters of polygons and areas of triangles and rectangles.
MP 3, MP 4, MP 5, MP 6

Objective To find the area of parallelograms and triangles

You can combine triangles to make just about any shape!

MATHEMATICAL PRACTICES

Solve It! Getting Ready!

A stage is being set up for a concert at the arena. The stage is made up of blocks with tops that are congruent right triangles. The tops of two of the blocks, when put together, make an 8 ft-by-8 ft square. The band has requested that the stage be arranged to form the shape of an arrow. Draw a diagram that shows how the stage could be laid out in the shape of an arrow with an area of at least 1000 ft^2 but no more than 1400 ft^2.

Essential Understanding You can find the area of a parallelogram or a triangle when you know the length of its base and its height.

A parallelogram with the same base and height as a rectangle has the same area as the rectangle.

Lesson Vocabulary
- base of a parallelogram
- altitude of a parallelogram
- height of a parallelogram
- base of a triangle
- height of a triangle

take note

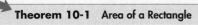

Theorem 10-1 Area of a Rectangle

The area of a rectangle is the product of its base and height.

$A = bh$

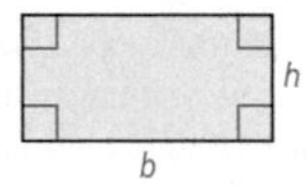

Theorem 10-2 Area of a Parallelogram

The area of a parallelogram is the product of a base and the corresponding height.

$A = bh$

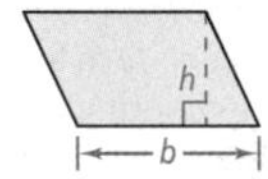

A **base of a parallelogram** can be any one of its sides. The corresponding **altitude** is a segment perpendicular to the line containing that base, drawn from the side opposite the base. The **height** is the length of an altitude.

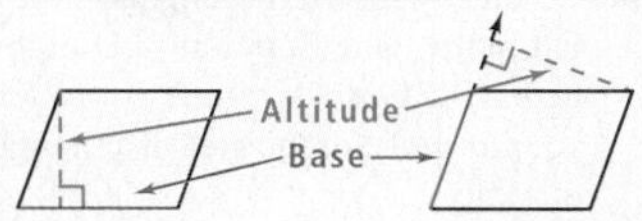

10-1 Preparing to Teach

BIG idea Measurement

ESSENTIAL UNDERSTANDING

- The area of a parallelogram or a triangle can be found when the length of its base and its height are known.

Math Background

Students will use what they know about the area of a square to find the area of rectangles. They will then find the area of a figure comprised of triangles. Ultimately, students will find the area of a parallelogram.

The area of a parallelogram is the product of the height and length of the base. The height of a parallelogram is any segment perpendicular to the line containing the base drawn from the side opposite the base. Students will learn the formula for the area of a triangle: $A = \frac{1}{2}bh$. They will also use these formulas to find the area of composite figures made of parallelograms and triangles.

Students often have difficulty understanding the relationship between the lengths of the sides of the rectangle and the area. The following model can help students understand this relationship. Take four strips of cardboard and tape them together to make a flexible parallelogram. By decreasing the size of the angle between the two sides, students will see that the area inside the parallelogram will be much less than when the sides form right angles. This demonstration illustrates that it is not the length of two sides that determines the area, but the length of one side and the perpendicular distance between parallel sides.

Mathematical Practices

Attend to precision. In calculating the areas of parallelograms and triangles, students will specify units of measure.

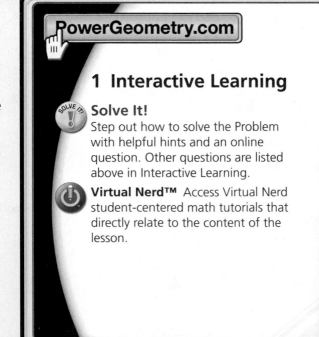

PowerGeometry.com

1 Interactive Learning

Solve It!
Step out how to solve the Problem with helpful hints and an online question. Other questions are listed above in Interactive Learning.

Virtual Nerd™ Access Virtual Nerd student-centered math tutorials that directly relate to the content of the lesson.

Problem 1 Finding the Area of a Parallelogram

Think

Why aren't the sides of the parallelogram considered altitudes?
Altitudes must be perpendicular to the bases. Unless the parallelogram is also a rectangle, the sides are not perpendicular to the bases.

What is the area of each parallelogram?

A

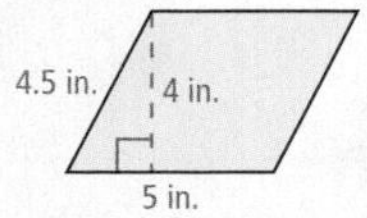

B

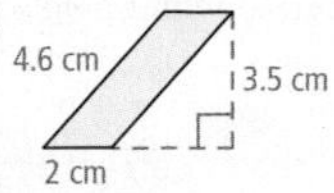

You are given each height. Choose the corresponding side to use as the base.

A	B
$A = bh$	$A = bh$
$= 5(4) = 20$ Substitute for b and h.	$= 2(3.5) = 7$
The area is 20 $in.^2$.	The area is 7 cm^2.

Got It? 1. What is the area of a parallelogram with base length 12 m and height 9 m?

Problem 2 Finding a Missing Dimension

Think

What does $\overline{CF}$ represent?
$\overline{CF}$ is an altitude of the parallelogram when $\overline{AD}$ and $\overline{BC}$ are used as bases.

For ▱ABCD, what is DE to the nearest tenth?

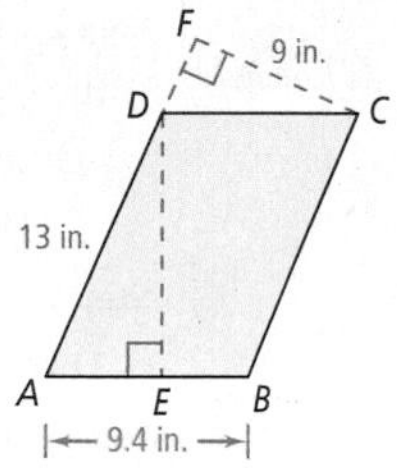

First, find the area of ▱ABCD. Then use the area formula a second time to find *DE*.

$A = bh$

$= 13(9) = 117$ Use base *AD* and height *CF*.

The area of ▱ABCD is 117 $in.^2$.

$A = bh$

$117 = 9.4(DE)$ Use base *AB* and height *DE*.

$DE = \frac{117}{9.4} \approx 12.4$

DE is about 12.4 in.

Got It? 2. A parallelogram has sides 15 cm and 18 cm. The height corresponding to a 15-cm base is 9 cm. What is the height corresponding to an 18-cm base?

You can rotate a triangle about the midpoint of a side to form a parallelogram.

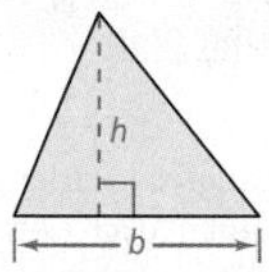

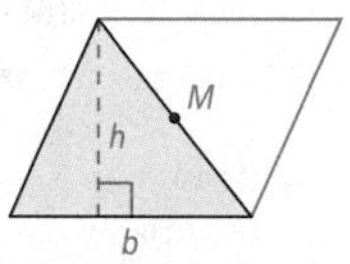

The area of the triangle is half the area of the parallelogram.

Problem 1

Q For 1A, what segment represents the height? Justify your answer. **[The 4 inch segment is the height of the parallelogram because it is perpendicular to one side.]**

Q In 2B, what information is extraneous? Explain. **[The 4.6 cm length of the side is extraneous information because only the length of the base and the height are needed to calculate the area.]**

Got It? ERROR PREVENTION

Have students draw the parallelogram and label the given lengths. Be sure they draw the height of the parallelogram correctly.

Problem 2

Q What is $\overline{CF}$? **[The height of the parallelogram with base $\overline{AD}$.]**

Q Can you find the area of the parallelogram using $\overline{AB}$ as the base? Explain. **[No, the corresponding height is not given.]**

Q Which measurements are used to calculate the area of the parallelogram? ***[AD* and *CF]***

Got It?

Have students draw a diagram similar to the one in Problem 2. They should be able to identify two sets of measurements that could be used to calculate the area of the parallelogram.

2 Guided Instruction

Each Problem is worked out and supported online.

Problem 1
Finding the Area of a Parallelogram
Animated

Problem 2
Finding a Missing Dimension

Problem 3
Finding the Area of a Triangle

Problem 4
Finding the Area of a Composite Figure

Support in Geometry Companion
- Vocabulary
- Key Concepts
- Got It?

Answers

Solve It!

Answers may vary. Sample:

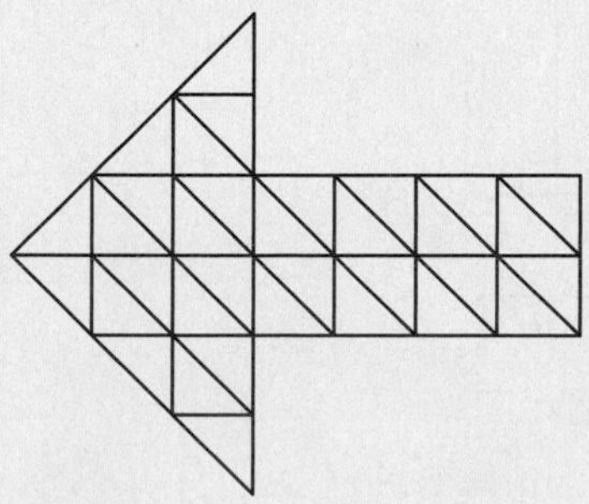

Got It?

1. 108 m^2

2. 7.5 cm

Take Note

Have students explain why the product of the height and base is multiplied by $\frac{1}{2}$. Emphasize the connection between the height of the triangle and an altitude of the triangle.

Problem 3

Q How can you find the amount of fabric needed for the sail? **[Find the area of the sail.]**

Q What measurements can be used to find the area of the sail? **[the height and base of the sail]**

Q In what units should your final answer be written? **[square feet]**

Got It?

Ask students to identify the height of the triangle. Be sure that they understand that either side length could be used as the height.

Problem 4

Q What two figures can you identify in the diagram? **[There is a triangle and a square.]**

Q How can you find the total area of the figure? **[Find the sum of the areas of the two figures.]**

Got It?

Challenge students to reason through this problem without calculating the new area. Have them verify their answers by calculating the new area and comparing the two answers.

take note

Theorem 10-3 Area of a Triangle

The area of a triangle is half the product of a base and the corresponding height.

$A = \frac{1}{2}bh$

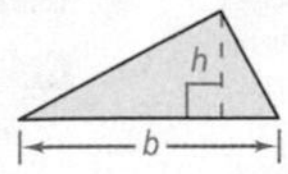

A **base of a triangle** can be any of its sides. The corresponding **height** is the length of the altitude to the line containing that base.

Problem 3 Finding the Area of a Triangle

Plan

Why do you need to convert the base and the height into inches?
You must convert them both because you can only multiply measurements with like units.

Sailing **You want to make a triangular sail like the one at the right. How many square feet of material do you need?**

Step 1 Convert the dimensions of the sail to inches.

$(12\text{ ft} \cdot \frac{12\text{ in.}}{1\text{ ft}}) + 2\text{ in.} = 146\text{ in.}$ Use a conversion factor.

$(13\text{ ft} \cdot \frac{12\text{ in.}}{1\text{ ft}}) + 4\text{ in.} = 160\text{ in.}$

Step 2 Find the area of the triangle.

$A = \frac{1}{2}bh$

$= \frac{1}{2}(160)(146)$ Substitute 160 for *b* and 146 for *h*.

$= 11{,}680$ Simplify.

Step 3 Convert 11,680 in.^{2} to square feet.

$11{,}680\text{ in.}^2 \cdot \frac{1\text{ ft}}{12\text{ in.}} \cdot \frac{1\text{ ft}}{12\text{ in.}} = 81\frac{1}{9}\text{ ft}^2$

You need $81\frac{1}{9}$ ft^2 of material.

Got It? 3. What is the area of the triangle?

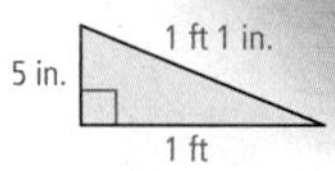

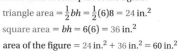

Problem 4 Finding the Area of an Irregular Figure

Plan

How do you know the length of the base of the triangle?
The lower part of the figure is a square. The base length of the triangle is the same as the base length of the square.

What is the area of the figure at the right?

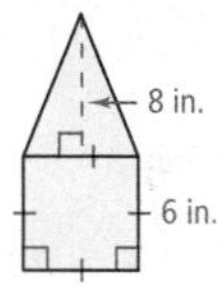

Find the area of each part of the figure.

triangle area $= \frac{1}{2}bh = \frac{1}{2}(6)8 = 24\text{ in.}^2$

square area $= bh = 6(6) = 36\text{ in.}^2$

area of the figure $= 24\text{ in.}^2 + 36\text{ in.}^2 = 60\text{ in.}^2$

Got It? 4. Reasoning Suppose the base lengths of the square and triangle in the figure above are doubled to 12 in., but the height of each polygon remains the same. How is the area of the figure affected?

Additional Problems

1. What is the area of each parallelogram?

a.

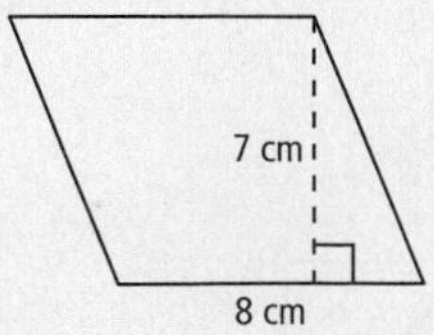

b.

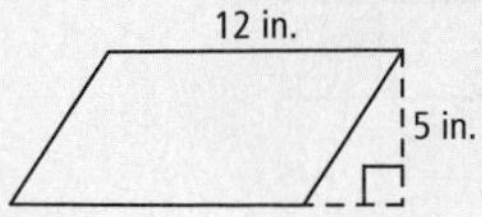

ANSWER a. 56 cm^2 **b.** 60 in.^{2}

2. In ▱$ABDE$, $AE = 12$ cm, $AB = 9$ cm, and $CB = 10$ cm. What is DF to the nearest tenth?

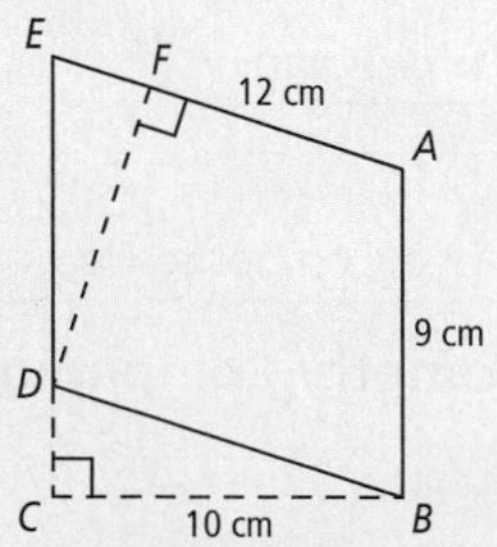

ANSWER 7.5 cm

3. A machinist is cutting out triangular metal braces to support the legs of a picnic table. Each triangular brace is 3 inches high and 2.2 inches wide. What is the area of each brace?

ANSWER 3.3 in.^{2}

4. What is the area of the figure below?

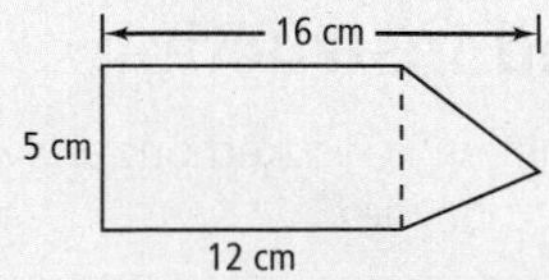

ANSWER 70 cm^2

Lesson Check

Do you know HOW?

Find the area of each parallelogram.

1.

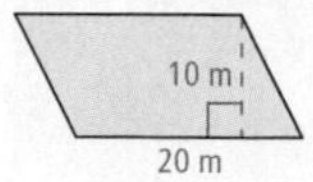

2.

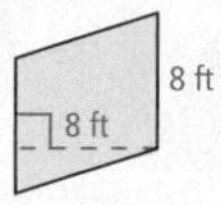

Find the area of each triangle.

3.

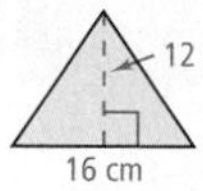

4.

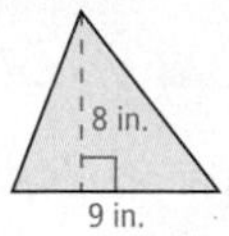

Do you UNDERSTAND? MATHEMATICAL PRACTICES

5. Vocabulary Does an altitude of a triangle have to lie inside the triangle? Explain.

6. Writing How can you show that a parallelogram and a rectangle with the same bases and heights have equal areas?

7. ▱*ABCD* is divided into two triangles along diagonal $\overline{AC}$. If you know the area of the parallelogram, how do you find the area of $\triangle ABC$?

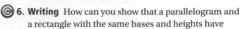

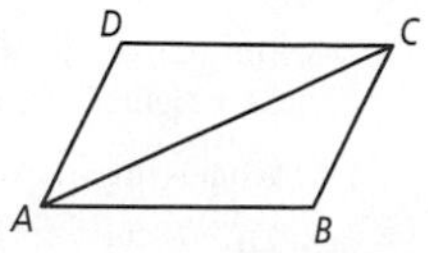

Practice and Problem-Solving Exercises MATHEMATICAL PRACTICES

A Practice

Find the area of each parallelogram.

See Problem 1.

8.

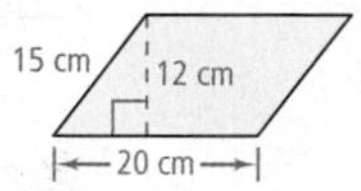

9.

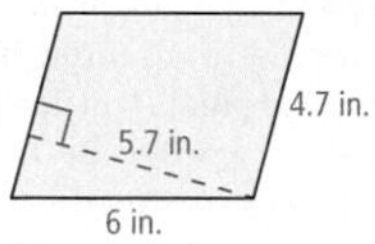

10.

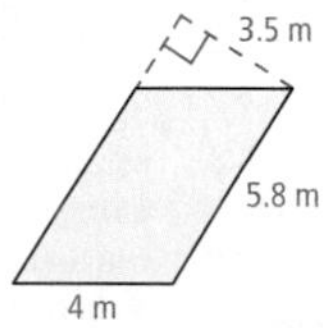

Find the value of *h* for each parallelogram.

See Problem 2.

11.

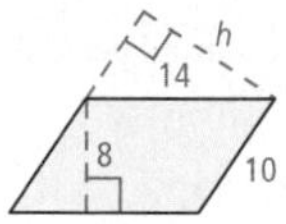

12.

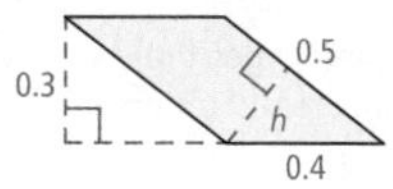

13.

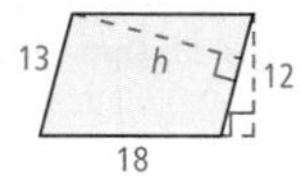

Find the area of each triangle.

See Problem 3.

14.

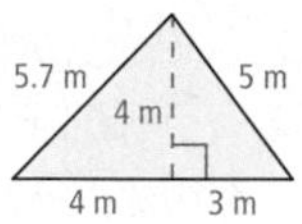

15.

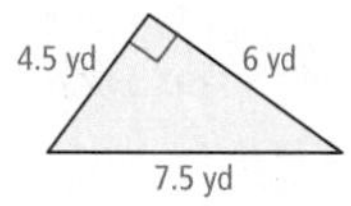

16.

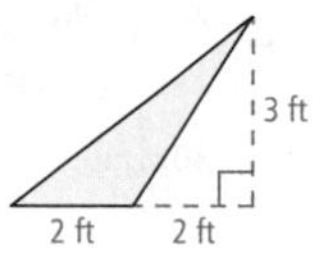

3 Lesson Check

Do you know HOW?

- If students have difficulty with Exercises 1-2, then have them review Problem 1 to write the formula and identify the values of the height and the base.

Do you UNDERSTAND?

- If students have difficulty with Exercise 6, then draw a diagram similar to the one in Problem 2, and ask students how they could rearrange the two figures formed by the dotted line to form a rectangle.

Close

Q How do you find the area of a parallelogram? **[Multiply the length of the base by the height of the parallelogram.]**

Q What is the formula for the area of a triangle? What does each variable represent? **[$A = \frac{1}{2}bh$, where *b* is the length of the base and *h* is the height of the triangle.]**

Answers

Got It! (continued)

3. 30 in.^{2} or $\frac{5}{24}$ ft^2

4. The area is doubled.

Lesson Check

1. 200 m^2

2. 64 ft^2

3. 96 cm^2

4. 36 in.^{2}

5. No; explanations may vary. Sample: two altitudes of an obtuse △ lie outside the △. The legs of a right △ are two altitudes of the △.

6. Answers may vary. Sample: You can cut and paste a section of the ▱ to make a rectangle that is ≅ to the given rectangle.

7. The area of $\triangle ABC$ is half the area of the ▱.

Practice and Problem-Solving Exercises

8. 240 cm^2

9. 26.79 in.^{2}

10. 20.3 m^2

11. 11.2 units

12. 0.24 unit

13. $16\frac{8}{13}$ units

14. 14 m^2

15. 13.5 yd^2

16. 3 ft^2

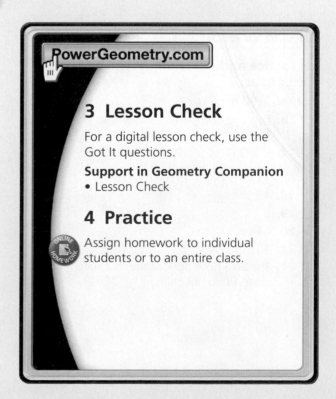

PowerGeometry.com

3 Lesson Check

For a digital lesson check, use the Got It questions.

Support in Geometry Companion

- Lesson Check

4 Practice

Assign homework to individual students or to an entire class.

4 Practice

ASSIGNMENT GUIDE

Basic: 8–17 all, 19, 22–23, 30–34 even, 37–38
Average: 9–17 odd, 18–43
Advanced: 9–17 odd, 18–46
Standardized Test Prep: 47–49

Mathematical Practices are supported by exercises with red headings. Here are the Practices supported in this lesson:

MP 3: Construct Arguments Ex. 6, 22, 31, 36a
MP 5: Use Appropriate Tools Ex. 23

Applications exercises have blue headings. Exercise 17 supports MP 4: Model.

EXERCISE 37: Use the Think About a Plan worksheet in the **Practice and Problem Solving Workbook** (also available in the Teaching Resources in print and online) to further support students' development in becoming independent learners.

HOMEWORK QUICK CHECK

To check students' understanding of key skills and concepts, go over Exercises 13, 17, 22, 23, and 37.

17. Urban Design A bakery has a 50 ft-by-31 ft parking lot. The four parking spaces are congruent parallelograms, the driving region is a rectangle, and the two areas for flowers are congruent triangles. See Problem 4.

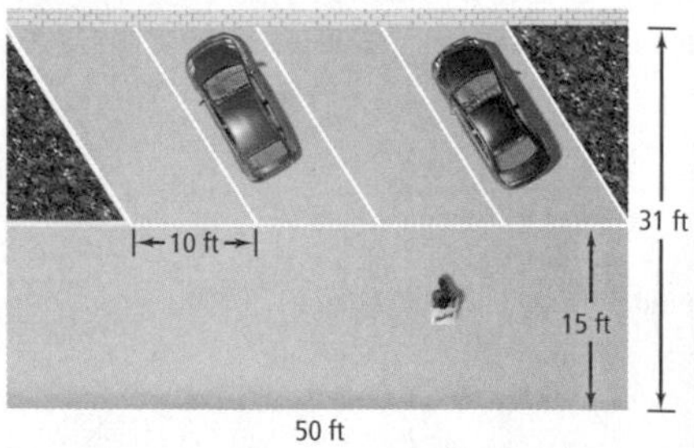

a. Find the area of the paved surface by adding the areas of the driving region and the four parking spaces.
b. Describe another method for finding the area of the paved surface.
c. Use your method from part (b) to find the area. Then compare answers from parts (a) and (b) to check your work.

B Apply

18. The area of a parallelogram is 24 in.^{2} and the height is 6 in. Find the length of the corresponding base.

19. What is the area of the figure at the right?

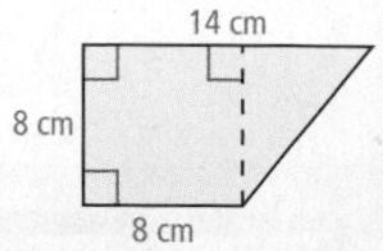

(A) 64 cm^2 (B) 88 cm^2 (C) 96 cm^2 (D) 112 cm^2

20. A right isosceles triangle has area 98 cm^2. Find the length of each leg.

21. Algebra The area of a triangle is 108 in.^{2}. A base and corresponding height are in the ratio 3 : 2. Find the length of the base and the corresponding height.

22. Think About a Plan Ki used geometry software to create the figure at the right. She constructed $\overleftrightarrow{AB}$ and a point C not on $\overleftrightarrow{AB}$. Then she constructed line k parallel to $\overleftrightarrow{AB}$ through point C. Next, Ki constructed point D on line k as well as $\overline{AD}$ and $\overline{BD}$. She dragged point D along line k to manipulate $\triangle ABD$. How does the area of $\triangle ABD$ change? Explain.

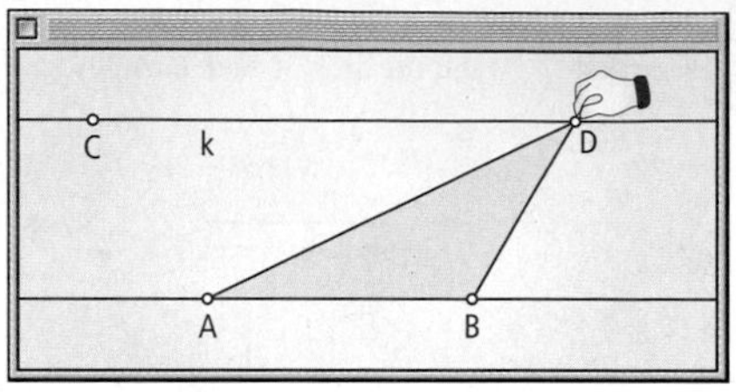

- Which dimensions of the triangle change when Ki drags point D?
- Do the lengths of AD and BD matter when calculating area?

23. Open-Ended Using graph paper, draw an acute triangle, an obtuse triangle, and a right triangle, each with area 12 units^2.

Find the area of each figure.

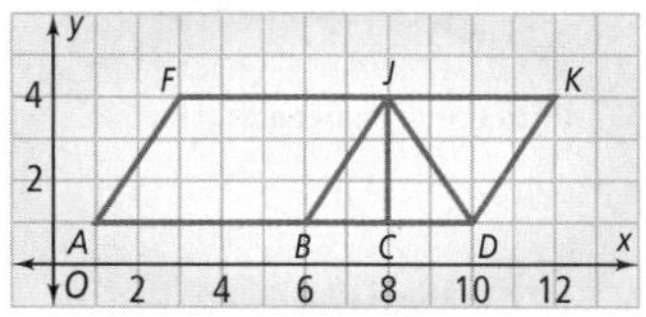

24. $\square ABJF$
25. $\triangle BDJ$
26. $\triangle DKJ$
27. $\square BDKJ$
28. $\square ADKF$
29. $\triangle BCJ$
30. trapezoid $ADJF$

31. Reasoning Suppose the height of a triangle is tripled. How does this affect the area of the triangle? Explain.

Answers

Practice and Problem-Solving Exercises (continued)

17a. 1390 ft^2
b. Find the entire area and subtract the areas for flowers.
c. $(50)(31) - 2\left[\frac{1}{2}(10)(16)\right] =$
$1550 - 160 = 1390$ ft^2
18. 4 in.
19. B
20. 14 cm
21. 18 in.; 12 in.
22. The area does not change; the height and base AB do not change.
23. Check students' work.
24. 15 units^2
25. 6 units^2
26. 6 units^2
27. 12 units^2
28. 27 units^2
29. 3 units^2
30. 21 units^2
31. The area is tripled; explanations may vary. Sample: If $A = \frac{1}{2}b \cdot h$, then $\frac{1}{2}(b \cdot 3h) = 3 \cdot \frac{1}{2}(b \cdot h) = 3A$.

For Exercises 32–35, (a) graph the lines and (b) find the area of the triangle enclosed by the lines.

32. $y = x, x = 0, y = 7$

33. $y = x + 2, y = 2, x = 6$

34. $y = -\frac{1}{2}x + 3, y = 0, x = -2$

35. $y = \frac{3}{4}x - 2, y = -2, x = 4$

36. Probability Your friend drew these three figures on a grid. A fly lands at random at a point on the grid.

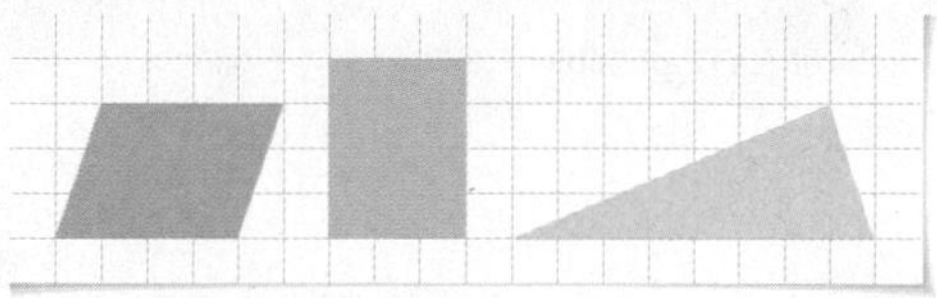

a. **Writing** Is the fly more likely to land on one of the figures or on the blank grid? Explain.

b. Suppose you know the fly lands on one of the figures. Is the fly more likely to land on one figure than on another? Explain.

Coordinate Geometry **Find the area of a polygon with the given vertices.**

37. $A(3, 9), B(8, 9), C(2, -3), D(-3, -3)$

38. $E(1, 1), F(4, 5), G(11, 5), H(8, 1)$

39. $D(0, 0), E(2, 4), F(6, 4), G(6, 0)$

40. $K(-7, -2), L(-7, 6), M(1, 6), N(7, -2)$

Find the area of each figure.

41.

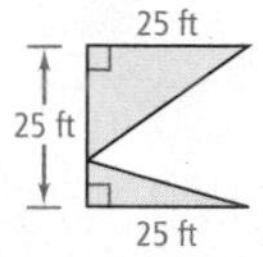

42.

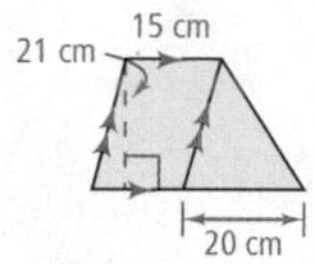

43.

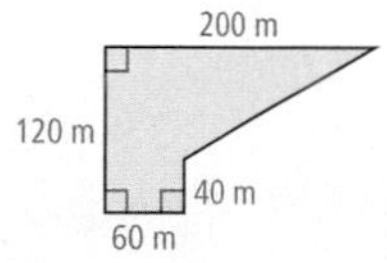

Challenge

History **The Greek mathematician Heron is most famous for this formula for the area of a triangle in terms of the lengths of its sides a, b, and c.**

$$A = \sqrt{s(s-a)(s-b)(s-c)}, \text{ where } s = \frac{1}{2}(a + b + c)$$

Use Heron's Formula and a calculator to find the area of each triangle. Round your answer to the nearest whole number.

44. $a = 8$ in., $b = 9$ in., $c = 10$ in.

45. $a = 15$ m, $b = 17$ m, $c = 21$ m

46. a. Use Heron's Formula to find the area of this triangle.

b. Verify your answer to part (a) by using the formula $A = \frac{1}{2}bh$.

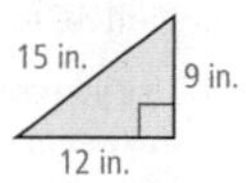

37. 60 units^2

38. 28 units^2

39. 20 units^2

40. 88 units^2

41. 312.5 ft^2

42. 525 cm^2

43. 12,800 m^2

44. 34 in.^{2}

45. 126 m^2

46a. 54 in.^{2}

b. 54 in.^{2}

32a.

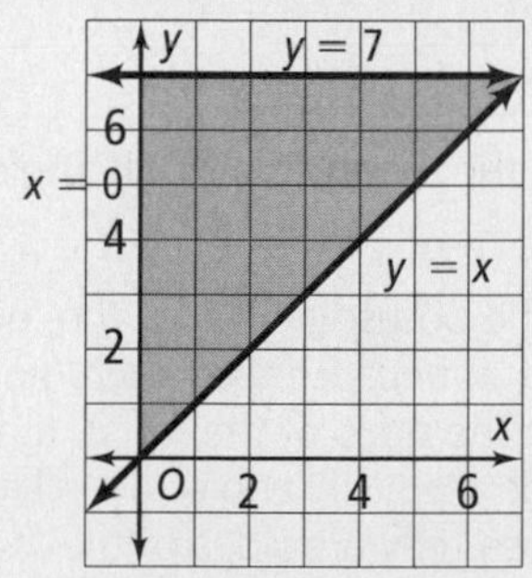

b. 24.5 units^2

33a.

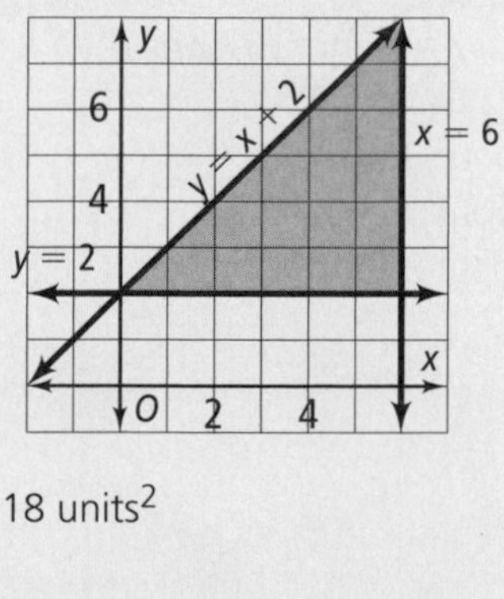

b. 18 units^2

34a.

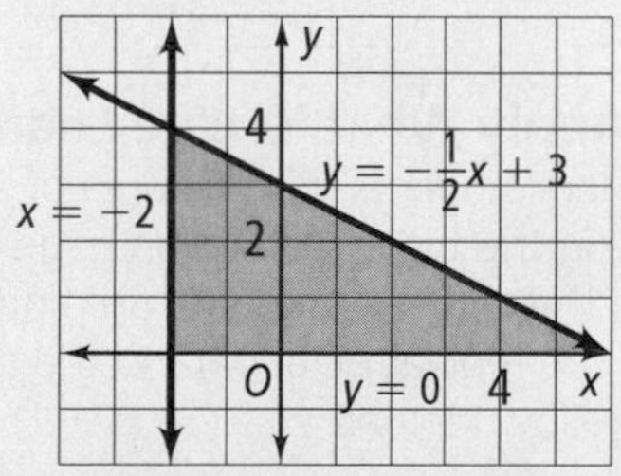

b. 16 units^2

35a.

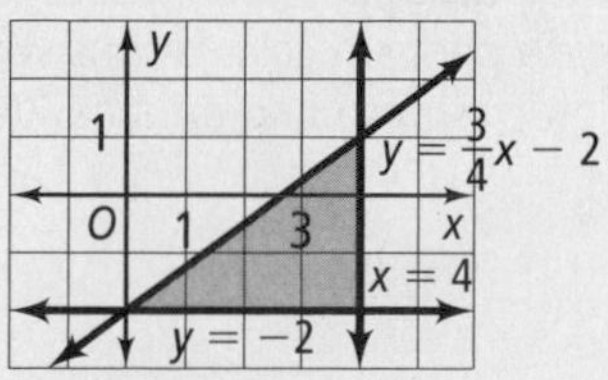

b. 6 units^2

36a. Blank grid; area of blank grid is 84 units^2, while the total area of the three figures is 36 units^2.

b. No; all three figures have the same area.

Answers

Practice and Problem-Solving Exercises (continued)

47. B

48. I

49. [2] No; the sum of the two shorter legs is $6 + 4$. By the △ Inequality Thm., that sum must be greater than the length of the third side of the △. Since $6 + 4 < 11$, a △ with sides 6, 4, and 11 is not possible.

[1] incomplete OR incorrect explanation

Standardized Test Prep

SAT/ACT

47. The lengths of the sides of a right triangle are 10 in., 24 in., and 26 in. What is the area of the triangle?

Ⓐ 116 in.^{2} Ⓑ 120 in.^{2} Ⓒ 130 in.^{2} Ⓓ 156 in.^{2}

48. In quadrilateral $ABCD$, $AB \cong BC \cong CD \cong DA$. Which type of quadrilateral could $ABCD$ never be classified as?

Ⓕ square Ⓖ rectangle Ⓗ rhombus Ⓘ kite

Short Response

49. Are the side lengths of $\triangle XYZ$ possible? Explain.

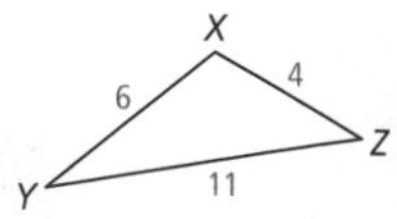

Apply What You've Learned

MATHEMATICAL PRACTICES MP 5

Look back at the information given about the target on page 613. The diagram of the target is shown again below, with three vertices of the regular octagon labeled A, B, and C. $\overline{BP}$ is drawn perpendicular to $\overline{AC}$.

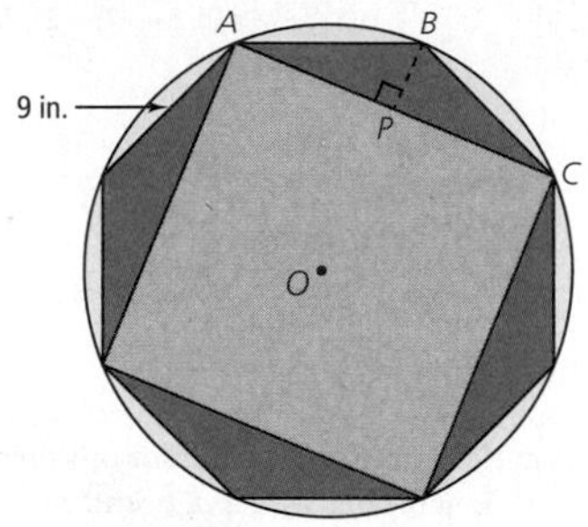

a. What is the measure of $\angle ABC$? Justify your answer.

b. Are the four red triangles congruent? Justify your answer.

c. What are the measures of the angles of $\triangle ABP$?

d. Use a trigonometric ratio to find BP to the nearest hundredth of an inch.

e. Find AC to the nearest hundredth of an inch.

f. Use your results from parts (d) and (e) to find the area of $\triangle ABC$. Round your answer to the nearest tenth of a square inch.

Apply What You've Learned

Here students find the area of a red triangle in the target shown on page 613. Later in the chapter, students will find the area of the regular octagon and the area of the yellow regions.

Mathematical Practices

Students **use an appropriate tool** (a calculator) to perform calculations with a trigonometric ratio to find an area. (MP 5)

ANSWERS

a. 135°; the measure of each angle of a regular n-gon is $\frac{(n-2)(180)}{n}$.

b. Yes; the obtuse angles of the red △s are all ≅ because they each measure 135°. The sides of the red △s that include the 135° angles are all ≅ because they are sides of the regular octagon. So, the four red △s are ≅ by SAS.

c. $m\angle APB = 90$; $m\angle ABP = 67.5$; $m\angle BAP = 22.5$

d. about 3.44 in.

e. about 16.63 in.

f. about 28.6 in.^{2}

10-1 Lesson Resources

Differentiated Remediation

Instructional Support

Geometry Companion

Students can use the **Geometry Companion** worktext (4 pages) . . .

- New Vocabulary
- Key Concepts
- Got It for each Problem
- Lesson Check

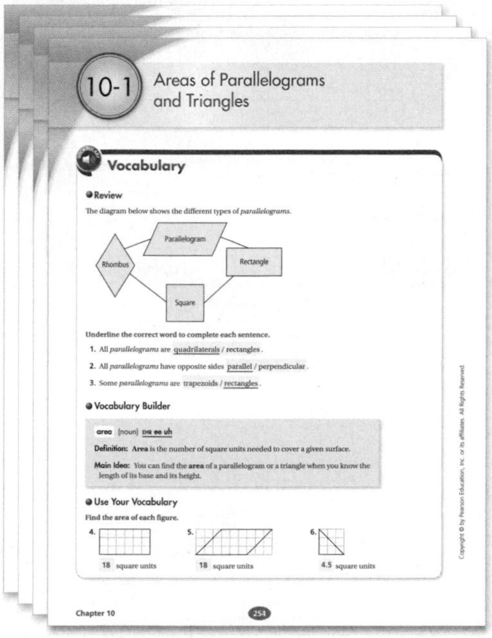

10-1 Areas of Parallelograms and Triangles

Vocabulary

Review

The diagram below shows the different types of *parallelograms.*

Underline the correct word to complete each sentence.

1. All *parallelograms* are quadrilaterals / rectangles .
2. All *parallelograms* have opposite sides parallel / perpendicular .
3. Some *parallelograms* are trapezoids / rectangles .

Vocabulary Builder

area (noun) EHR ee uh

Definition: **Area** is the number of square units needed to cover a given surface.

Main Idea: You can find the **area** of a parallelogram or a triangle when you know the length of its base and its height.

Use Your Vocabulary

Find the area of each figure.

4. 18 square units
5. 18 square units
6. 4.5 square units

Chapter 10 254

ELL Support

Connect to Prior Knowledge Draw an assortment of figures on the board that include triangles and quadrilaterals. Invite students to classify the figures. Have students contribute as you write a list on the board of the characteristics of polygons. Ask these questions:

How is a polygon named?
What is a regular polygon?
How are a rhombus and kite different?
How are they the same?

5 Assess & Remediate

Lesson Quiz

1. What is the area of the parallelogram?

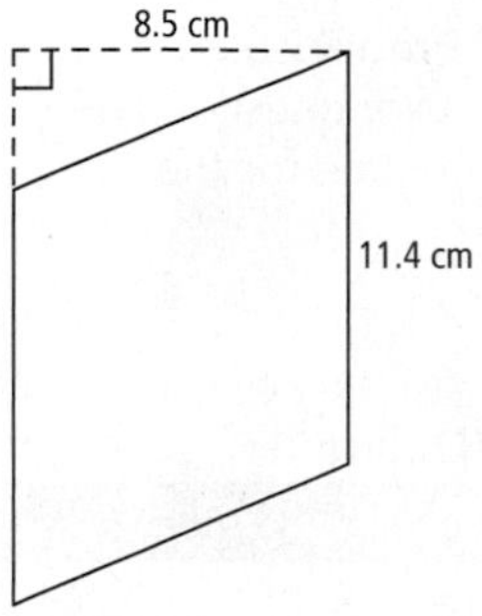

2. Do you UNDERSTAND? A triangular window pane is 4 feet 8 inches wide and 3 feet 6 inches high. What is the area of the window pane? Round to the nearest tenth square foot if necessary.

3. What is the area of the figure below?

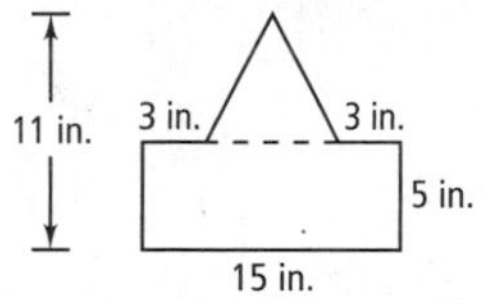

ANSWERS TO LESSON QUIZ

1. 96.9 cm^2
2. about 8.2 ft^2
3. 102 $in.^2$

PRESCRIPTION FOR REMEDIATION

Use the student work on the Lesson Quiz to prescribe a differentiated review assignment.

Points	Differentiated Remediation
0–1	Intervention
2	On-level
3	Extension

PowerGeometry.com

5 Assess & Remediate

Assign the Lesson Quiz. Appropriate intervention, practice, or enrichment is automatically generated based on student performance.

Intervention

- **Reteaching** (2 pages) Provides reteaching and practice exercises for the key lesson concepts. Use with struggling students or absent students.
- **English Language Learner Support** Helps students develop and reinforce mathematical vocabulary and key concepts.

All-in-One Resources/Online

Reteaching

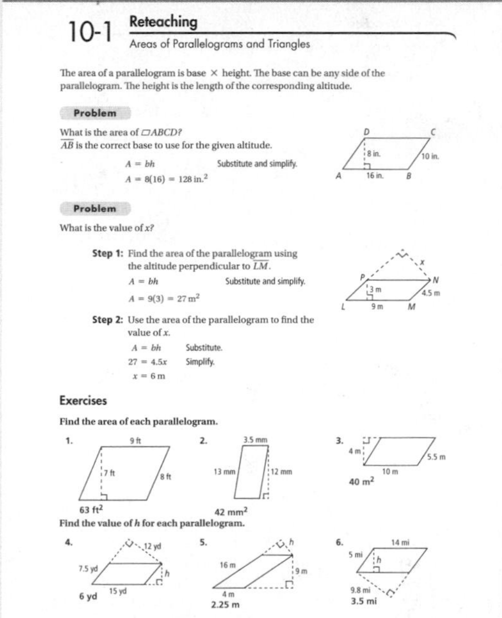

10-1 Reteaching
Areas of Parallelograms and Triangles

The area of a parallelogram is base × height. The base can be any side of the parallelogram. The height is the length of the corresponding altitude.

Problem

What is the area of ▱ABCD?
$\overline{AB}$ is the correct base to use for the given altitude.

$A = bh$ Substitute and simplify.
$A = 8(16) = 128 \text{ in.}^2$

Problem

What is the value of x?

Step 1: Find the area of the parallelogram using the altitude perpendicular to $\overline{LM}$.
$A = bh$ Substitute and simplify.
$A = 9(3) = 27 \text{ m}^2$

Step 2: Use the area of the parallelogram to find the value of x.
$A = bh$ Substitute.
$27 = 4.5x$ Simplify.
$x = 6 \text{ m}$

Exercises

Find the area of each parallelogram.

1. 9 ft, 7 ft, 8 ft — 63 ft^2
2. 3.5 mm, 13 mm, 12 mm — 42 mm^2
3. 4 m, 5.5 m, 10 m — 40 m^2

Find the value of h for each parallelogram.

4. 12 yd, 7.5 yd, h, 15 yd — 6 yd
5. h, 16 m, 9 m, 4 m — 2.25 m
6. 14 mi, 5 mi, h, 9.8 mi — 3.5 mi

All-in-One Resources/Online

English Language Learner Support

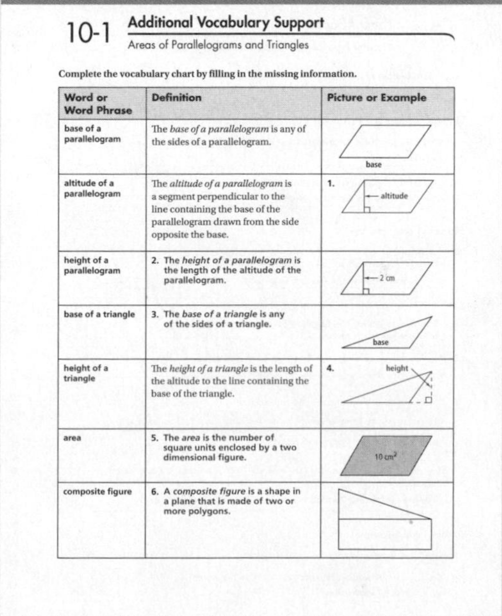

10-1 Additional Vocabulary Support
Areas of Parallelograms and Triangles

Complete the vocabulary chart by filling in the missing information.

Word or Word Phrase	Definition	Picture or Example
base of a parallelogram	The *base of a parallelogram* is any of the sides of a parallelogram.	base
altitude of a parallelogram	The *altitude of a parallelogram* is a segment perpendicular to the line containing the base of the parallelogram drawn from the side opposite the base.	1. altitude
height of a parallelogram	2. The *height of a parallelogram* is the length of the altitude of the parallelogram.	2 cm
base of a triangle	3. The *base of a triangle* is any of the sides of a triangle.	base
height of a triangle	The *height of a triangle* is the length of the altitude to the line containing the base of the triangle.	4. height
area	5. The *area* is the number of square units enclosed by a two dimensional figure.	10 cm^2
composite figure	6. A *composite figure* is a shape in a plane that is made of two or more polygons.	

Differentiated Remediation *continued*

On-Level

- **Practice** (2 pages) Provides extra practice for each lesson. For simpler practice exercises, use the Form K Practice pages found in the All-in-One Teaching Resources and online.
- **Think About a Plan** Helps students develop specific problem-solving skills and strategies by providing scaffolded guiding questions.
- **Standardized Test Prep** Focuses on all major exercises, all major question types, and helps students prepare for the high-stakes assessments.

Extension

- **Enrichment** Provides students with interesting problems and activities that extend the concepts of the lesson.
- **Activities, Games, and Puzzles** Worksheets that can be used for concepts development, enrichment, and for fun!

Practice and Problem Solving Wkbk/ All-in-One Resources/Online

Practice page 1

10-1 **Practice** *Form G*
Areas of Parallelograms and Triangles

Find the area of each parallelogram.

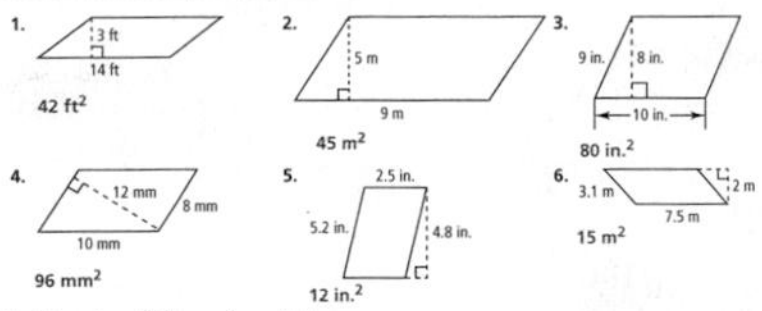

1. 42 ft²
2. 45 m²
3. 80 in.²
4. 96 mm²
5. 12 in.²
6. 15 m²

Find the value of *h* for each parallelogram.

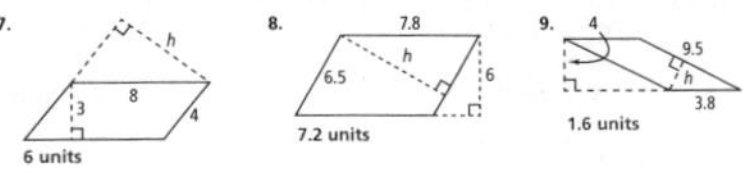

7. 6 units
8. 7.2 units
9. 1.6 units

Find the area of each triangle.

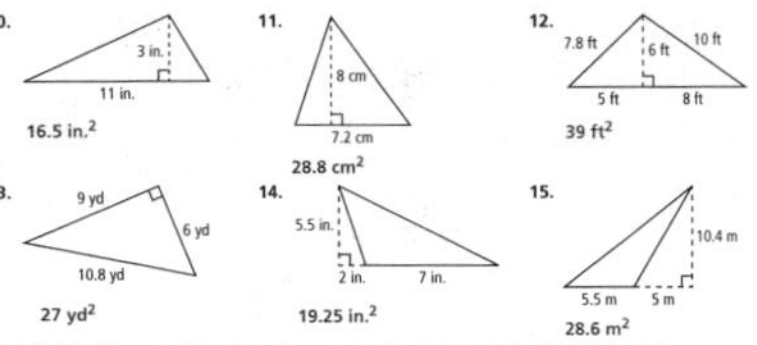

10. 16.5 in.²
11. 28.8 cm²
12. 39 ft²
13. 27 yd²
14. 19.25 in.²
15. 28.6 m²
16. **Algebra** In a parallelogram, a base, *b*, and its corresponding height, *h*, are in the ratio of 5 : 3. The area is 135 mm². Find *b* and *h*. 15 mm; 9 mm
17. **Reasoning** A triangle has an area of 18 ft². List all the possible positive integers that could represent its base and height.
1 ft and 36 ft; 2 ft and 18 ft; 3 ft and 12 ft; 4 ft and 9 ft; 6 ft and 6 ft; 9 ft and 4 ft; 12 ft and 3 ft; 18 ft and 2 ft; 36 ft and 1 ft

Practice and Problem Solving Wkbk/ All-in-One Resources/Online

Practice page 2

10-1 **Practice** (continued) *Form G*
Areas of Parallelograms and Triangles

18. A company wants to paint its logo on the side of a building. The entire area needs to be covered with a primer. The two triangular areas will be painted red, the rectangle containing the company's name will be white, and the rest of the parallelogram will be yellow.

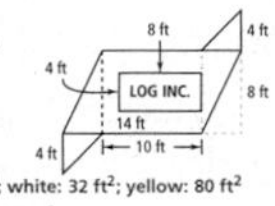

a. Find the area for each different color. red: 16 ft²; white: 32 ft²; yellow: 80 ft²
b. Find the area that must be painted with primer. 128 ft²

19. A scale drawing of the side view of a house is shown at the right. Find the total area (in square inches) of the side of the house in the drawing. 1.4 in.²

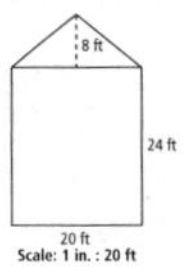

20. **Open-Ended** Using graph paper, draw a figure with area 42 units² made up of a parallelogram and a triangle. Check students' work.

Coordinate Geometry **Find the area of a polygon with the given vertices.**

21. *A*(2, 2), *B*(5, 2), *C*(3, −1), *D*(0, −1) 9 units²
22. *A*(1, 4), *B*(−2, −2), *C*(−7, −2), *D*(−4, 4) 30 units²
23. *A*(5, −3), *B*(−1, −3), *C*(−1, 2), *D*(5, 6) 42 units²
24. *A*(5, 0), *B*(5, 8), *C*(−1, 7), *D*(−1, −6) 63 units²

Find the area of each figure.

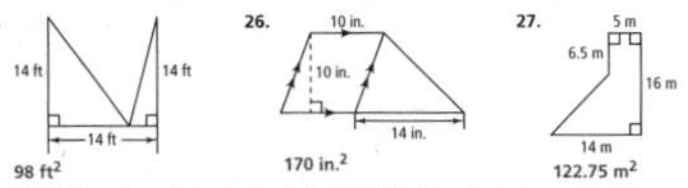

25. 98 ft²
26. 170 in.²
27. 122.75 m²
28. **Reasoning** A parallelogram has a height of 6 units and a corresponding base of 7 units. What is the area of each triangle formed when one diagonal of the parallelogram is drawn? What is the area of each small triangle formed when two diagonals are drawn? 21 unit²; 10.5 unit²
29. A parallelogram has sides 24 m and 5 m. The height corresponding to a 24-m base is 4 m. What it the height corresponding to a 5-m base? 19.2 m

All-in-One Resources/Online

Enrichment

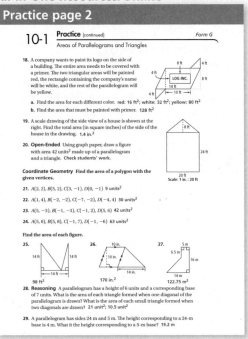

10-1 **Enrichment**
Areas of Parallelograms and Triangles

Unique Polygons

The area of some polygons uniquely determines their shape. In that case, any of the polygons with the same area must be congruent. Consider triangles.

1. How many possible triangles have an area of 20 units²? Draw and label three possible triangles.
Infinite number; check students' work.
2. How many possible right triangles have an area of 20 units²? Explain.
Infinite number; the ratio of the two legs has an infinite variation.
3. How many possible equilateral triangles have an area of 20 units²? Explain.
Only one; all equilateral triangles are similar. Only one unique size equilateral triangle will have an area of 20 units².
4. How many isosceles right triangles have an area of 20 units²? Explain.
Only one; all isosceles right triangles are similar. Only the isosceles right triangle with sides $2\sqrt{10}$ units long will have the area of 20 units².
5. Do any special parallelograms have shapes uniquely determined by their area? Explain.
Yes; squares are all similar.
6. Is the shape of a rhombus uniquely determined by its area? Explain.
No; The angles of a rhombus are uniquely determined by each ratio of length to height, so no two rhombuses with the same area but different ratios have the same shape.
7. The ratio of length to width for a golden rectangle is approximately 1.618 : 1. Is the shape of a golden rectangle uniquely determined by its area? Explain.
Yes, all golden rectangles are similar, so the area uniquely determines its shape.

Practice and Problem Solving Wkbk/ All-in-One Resources/Online

Think About a Plan

10-1 **Think About a Plan**
Areas of Parallelograms and Triangles

Coordinate Geometry Find the area of a polygon with the given vertices.
A(3, 9), *B*(8, 9), *C*(2, −3), *D*(−3, −3)

Understanding the Problem

1. What is the first thing you need to know about the polygon to find its area?
the type of polygon
2. What is the simplest way to determine this?
Draw the polygon on a coordinate grid.

Planning the Solution

3. Plot the points on the coordinate grid. What type of polygon is it? How can you tell?
Parallelogram; its opposite sides are congruent and parallel.

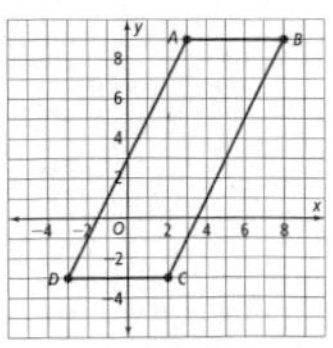

4. What is the formula for finding the area of this type of polygon? $A = bh$
5. Which segment will you use as the base?
Answers may vary. Sample: either $\overline{AB}$ or $\overline{DC}$
6. How can you add to the figure to find the height?
Answers may vary. Sample: Use $\overline{DC}$ as the base and extend $\overline{DC}$ so you can draw an altitude from *B* to the line containing $\overline{DC}$.
7. How can you find the measure of the base and height? What are the base and the height?
Either use the Distance Formula or just count on the grid; $b = 5$; $h = 12$.

Getting an Answer

8. Substitute the values for the base and height into the formula and solve. $A = 5 \times 12$; 60
9. What unit should you use for your answer? What is the area?
square units; 60 square units

Practice and Problem Solving Wkbk/ All-in-One Resources/Online

Standardized Test Prep

10-1 **Standardized Test Prep**
Areas of Parallelograms and Triangles

Multiple Choice

For Exercises 1–6, choose the correct letter.

1. What is the area of the parallelogram at the right? C
Ⓐ 18 in.² Ⓒ 36 in.²
Ⓑ 30 in.² Ⓓ 60 in.²

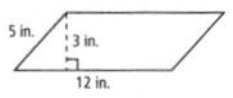

2. What is the area of the parallelogram at the right? G
Ⓕ 31.5 m² Ⓗ 84 m²
Ⓖ 63 m² Ⓘ 126 m²

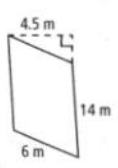

3. What is the value of *h* for the parallelogram at the right? A
Ⓐ 9.6 units Ⓒ 48 units
Ⓑ 26.7 units Ⓓ 96 units

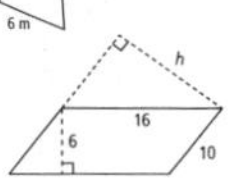

4. What is the area of the figure at the right? F
Ⓕ 26 in.² Ⓗ 52 in.²
Ⓖ 27 in.² Ⓘ 54 in.²

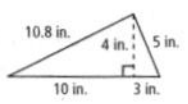

5. A parallelogram has sides 8 ft and 6 ft and an area of 54 ft². What is the length of the altitude to the 8-ft base? A
Ⓐ 6.75 ft Ⓑ 9 ft Ⓒ 24 ft Ⓓ 27 ft
6. What is the area of the figure at the right? I
Ⓕ 36 m² Ⓗ 72 m²
Ⓖ 60 m² Ⓘ 96 m²

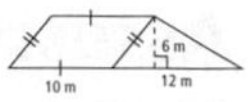

Short Response

7. In a triangle, a base and a corresponding height are in the ratio 5 : 2. The area is 80 ft². What is the base and the corresponding height? Show your work.
[2] $\frac{1}{2}5x(2x) = 80$; $5x^2 = 80$; $x = 4$ ft; base $= 5(4) = 20$ ft; height $= 2(4) = 8$ ft [1] error in calculation or incorrect work [0] error in calculation and no work provided

Online Teacher Resource Center

Activities, Games, and Puzzles

10-1 **Activity: Exploring Area**
Areas of Parallelograms and Triangles

Construct

Step 1 Use geometry software to construct $\overline{AD}$. Use 'Mark Center "A"' to rotate $\overline{AD}$ 90°. Label the new point *B*. Use a similar procedure to complete the construction of square *ABCD*.

Step 2 Draw two segments from vertex *A* to point *E* on $\overline{BC}$ and to point *F* on $\overline{CD}$.

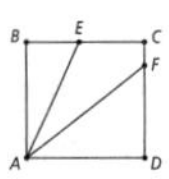

Investigate

Measure the areas of $\triangle ABE$, *AECF*, and $\triangle AFD$. Drag *E* and *F* until the three regions formed have equal area; that is, the area of the square has been trisected.

1. Calculate the ratios $\frac{BE}{EC}$ and $\frac{DF}{FC}$. Drag the vertices of the square and make a conjecture about the areas and their relationship to the ratios. both ratios are 2 : 1
2. Use the formula for the area of a triangle, $A = \frac{1}{2}bh$, to justify that the areas of the three regions are the same. (Hint: Label $\overline{AB}$ as 3*x*, then label each of the segments along $\overline{BC}$ and $\overline{CD}$ in terms of *x*. Find the areas of triangles *ABE* and *AFD* in terms of *x* as well as the area of *ABCD*.) The area of *ABCD* is $3x \cdot 3x = 9x^2$; the area of $\triangle ABE$ = area $\triangle ADF = \frac{1}{2} \cdot 2x \cdot 3x = 3x^2$; the area of $AECF = 9x^2 - 3x^2 - 3x^2 = 3x^2$; the areas are equal.

Extend

3. Construct square *ABCD* with $\overline{AE}$, $\overline{AF}$, and $\overline{AG}$ as shown in the diagram at right. Measure the areas of the four resulting regions. Drag *E*, *F*, and *G* so that the four regions have equal area.
Check students' work.

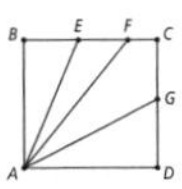

4. Measure the lengths of the segments along $\overline{BC}$ and $\overline{CD}$ and calculate their ratios with one another. Make a conjecture about the position of *E*, *F*, and *G*. *E* and *G* are midpoints of $\overline{BC}$ and $\overline{CD}$, respectively; *F* coincides with *C*.
5. Use the formula for the area of a triangle, $A = \frac{1}{2}bh$, to justify that the areas of the four regions are the same. Using 2*x* to represent the length of one side of *ABCD*, each triangle has area $\frac{1}{2} \cdot x \cdot 2x = x^2$. The areas are equal.
6. Repeat Exercises 3 through 5 by dividing square *ABCD* into five regions with equal area. The sides to which the segments are drawn are divided into segments in a ratio of 2 : 2 : 1.
7. Repeat Exercises 3 through 5 by dividing square *ABCD* into six regions with equal area. The sides to which the segments are drawn are divided into segments in a ratio of 1 : 1 : 1.
8. Make a conjecture as to how to position segments when dividing a square into an even number of regions with equal area. The sides to which the segments are drawn are divided into equal segments.
9. Make a conjecture as to how to position segments when dividing a square into an odd number of regions with equal area. The sides to which the segments are drawn are divided into segments in a ratio of 2 : 2 : … 2 : 1. The segment near the upper right-hand corner of the square is half the length of segments along either of divided sides.

10-2 Areas of Trapezoids, Rhombuses, and Kites

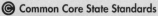

Common Core State Standards
G-MG.A.1 Use geometric shapes, their measures, and their properties to describe objects.
MP 1, MP 3, MP 4, MP 6

Objective To find the area of a trapezoid, rhombus, or kite

Rearranging figures into familiar shapes is an example of the Solve a Simpler Problem strategy.

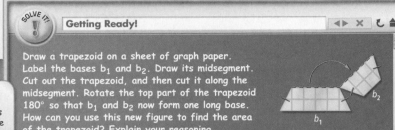

Solve It! Getting Ready!

Draw a trapezoid on a sheet of graph paper. Label the bases b_1 and b_2. Draw its midsegment. Cut out the trapezoid, and then cut it along the midsegment. Rotate the top part of the trapezoid 180° so that b_1 and b_2 now form one long base. How can you use this new figure to find the area of the trapezoid? Explain your reasoning.

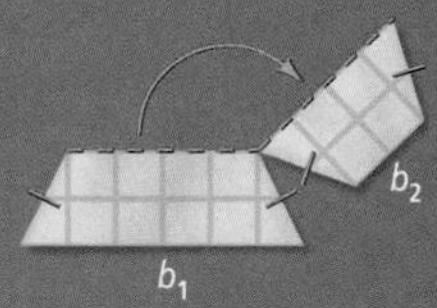

Mathematical Practices **Essential Understanding** You can find the area of a trapezoid when you know its height and the lengths of its bases.

The **height of a trapezoid** is the perpendicular distance between the bases.

Lesson Vocabulary
- height of a trapezoid

Take note

Theorem 10-4 Area of a Trapezoid

The area of a trapezoid is half the product of the height and the sum of the bases.

$A = \frac{1}{2}h(b_1 + b_2)$

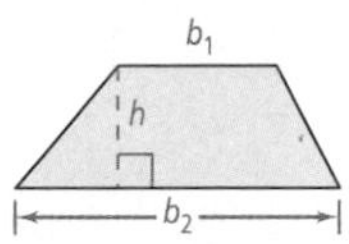

Problem 1 Area of a Trapezoid

Plan
Which borders of Nevada can you use as the bases of a trapezoid?
The two parallel sides of Nevada form the bases of a trapezoid.

Geography **What is the approximate area of Nevada?**

$A = \frac{1}{2}h(b_1 + b_2)$	Use the formula for area of a trapezoid.
$= \frac{1}{2}(309)(205 + 511)$	Substitute 309 for h, 205 for b_1, and 511 for b_2.
$= 110{,}622$	Simplify.

The area of Nevada is about 110,600 mi^2.

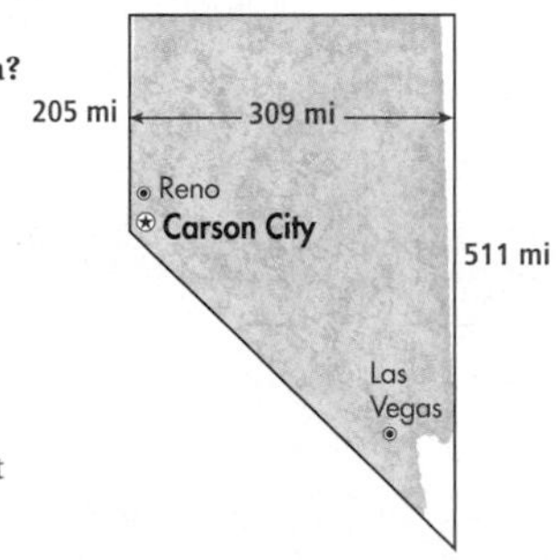

Got It? **1.** What is the area of a trapezoid with height 7 cm and bases 12 cm and 15 cm?

1 Interactive Learning

Solve It!

PURPOSE To use a parallelogram to find the area of a trapezoid

PROCESS Students may draw and label a diagram or cut out and transform a trapezoid as directed.

FACILITATE

Q What type of figure is created? **[a parallelogram]**

Q Let the height of the original trapezoid be *h*. What is the height of the parallelogram? **[$\frac{1}{2}h$]**

Q What is the length of the longest side of the parallelogram? **[$b_1 + b_2$]**

ANSWER See Solve It in Answers on next page.

CONNECT THE MATH In the Solve It, students manipulate a parallelogram and investigate the lengths of its shorter and longer bases. In the lesson, students derive the formula for the area of a trapezoid.

2 Guided Instruction

Take Note

Remind students that the height must be a segment from one base perpendicular to the other base.

Problem 1

Q Which sides are the bases of the trapezoid? **[The sides marked 205 mi and 511 mi are the bases.]**

Got It? VISUAL LEARNERS

Have students draw a diagram and label the known segments.

10-2 Preparing to Teach

BIG idea Measurement

ESSENTIAL UNDERSTANDINGS

- The area of a trapezoid can be found when the height and the lengths of its bases are known.
- The area of a rhombus or a kite can be found when the lengths of its diagonals are known.

Math Background

A diagonal of a trapezoid separates it into two triangles of the same height. Because students know the formula for the area of a triangle, students can derive the formula for the area of a trapezoid. Students may also find a way to derive the formula for the area of a trapezoid using the formula for the area of a parallelogram. Either way, students will derive and then use this formula to find the areas of given trapezoids. The area of a rhombus or a kite can be found by noting that the diagonals are perpendicular to each other so that the area will be half the product of the diagonals. Students will use this formula to find the areas of given figures.

Mathematical Practices

Attend to precision. In addition to specifying units of measure, students will also solve a real-life situation involving the area of Nevada in Problem 1.

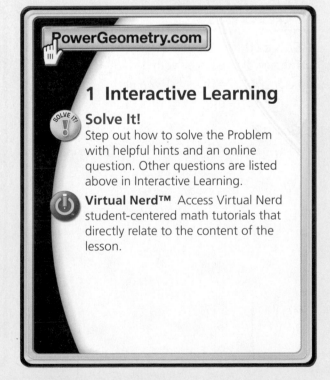

PowerGeometry.com

1 Interactive Learning

Solve It! Step out how to solve the Problem with helpful hints and an online question. Other questions are listed above in Interactive Learning.

Virtual Nerd™ Access Virtual Nerd student-centered math tutorials that directly relate to the content of the lesson.

Problem 2

Q How do you know that the figure created is a rectangle? **[All the angles are right angles.]**

Q Which side of the 30°-60°-90° triangle represents the height? What is its measure? **[The height is the longer leg that has a length equal to $\sqrt{3}$ times the length of the shorter leg.]**

Got It? ERROR PREVENTION

Ask students to identify what type of triangle will be formed by the height in the new trapezoid. A 45°-45°-90° triangle will have two legs of equal length.

Take Note

Review the definition and properties of the diagonals of a rhombus and a kite.

- A rhombus is a parallelogram with four congruent sides.
- A kite is a quadrilateral with two pairs of congruent adjacent sides and no opposite side congruent.

Problem 3

Q How can you find the length of the diagonals? **[Add the lengths of the segments that form the diagonals.]**

Got It?

Have students draw a diagram and label each diagonal with the given lengths.

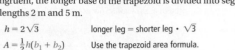

Problem 2 Finding Area Using a Right Triangle

Think

How are the sides related in a 30°-60°-90° triangle?
The length of the hypotenuse is 2 times the length of the shorter leg, and the longer leg is $\sqrt{3}$ times the length of the shorter leg.

What is the area of trapezoid *PQRS*?

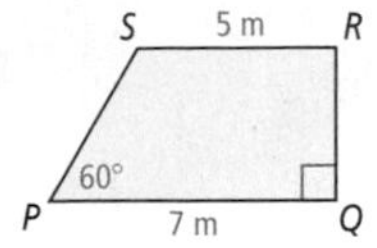

You can draw an altitude that divides the trapezoid into a rectangle and a 30°-60°-90° triangle. Since the opposite sides of a rectangle are congruent, the longer base of the trapezoid is divided into segments of lengths 2 m and 5 m.

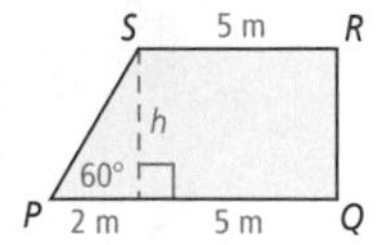

$h = 2\sqrt{3}$ longer leg = shorter leg • $\sqrt{3}$

$A = \frac{1}{2}h(b_1 + b_2)$ Use the trapezoid area formula.

$= \frac{1}{2}(2\sqrt{3})(7 + 5)$ Substitute $2\sqrt{3}$ for h, 7 for b_1, and 5 for b_2.

$= 12\sqrt{3}$ Simplify.

The area of trapezoid *PQRS* is $12\sqrt{3}$ m^2.

Got It? 2. Reasoning In Problem 2, suppose h decreases so that $m\angle P = 45$ while angles R and Q and the bases stay the same. What is the area of trapezoid *PQRS*?

Essential Understanding You can find the area of a rhombus or a kite when you know the lengths of its diagonals.

take note

Theorem 10-5 Area of a Rhombus or a Kite

The area of a rhombus or a kite is half the product of the lengths of its diagonals.

$A = \frac{1}{2}d_1d_2$

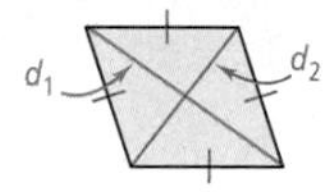

Rhombus

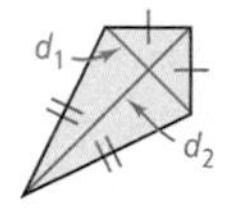

Kite

Problem 3 Finding the Area of a Kite

Think

Do you need to know the side lengths of the kite to find its area?
No. You only need the lengths of the diagonals.

What is the area of kite *KLMN*?

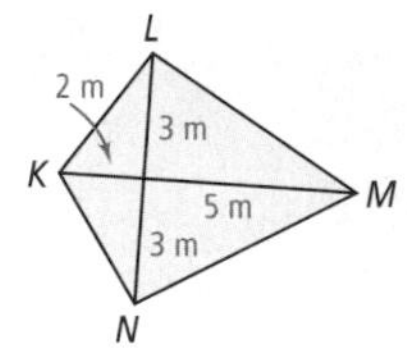

Find the lengths of the two diagonals:
$KM = 2 + 5 = 7$ m and $LN = 3 + 3 = 6$ m.

$A = \frac{1}{2}d_1d_2$ Use the formula for area of a kite.

$= \frac{1}{2}(7)(6)$ Substitute 7 for d_1 and 6 for d_2.

$= 21$ Simplify.

The area of kite *KLMN* is 21 m^2.

Got It? 3. What is the area of a kite with diagonals that are 12 in. and 9 in. long?

Answers

Solve It!

The new figure is a ▱ with base $b_1 + b_2$ and height $\frac{1}{2}h$. $A = (b_1 + b_2) \cdot \frac{1}{2}h$ or $A = \frac{1}{2}h(b_1 + b_2)$.

Got It?

1. 94.5 cm^2 **2.** 12 m^2 **3.** 54 in.^{2}

PowerGeometry.com

2 Guided Instruction

Each Problem is worked out and supported online.

Problem 1
Area of a Trapezoid

Problem 2
Finding Area Using a Right Triangle

Problem 3
Finding the Area of a Kite

Problem 4
Finding the Area of a Rhombus
Animated

Support in Geometry Companion
- Vocabulary
- Key Concepts
- Got It?

Problem 4 Finding the Area of a Rhombus

Think

How can you find the length of $\overline{AB}$?
$\overline{AB}$ is a leg of right $\triangle ABC$. You can use the Pythagorean Theorem, $a^2 + b^2 = c^2$, to find its length.

Car Pooling **The High Occupancy Vehicle (HOV) lane is marked by a series of "diamonds," or rhombuses painted on the pavement. What is the area of the HOV lane diamond shown at the right?**

$\triangle ABC$ is a right triangle. Using the Pythagorean Theorem, $AB = \sqrt{6.5^2 - 2.5^2} = 6$. Since the diagonals of a rhombus bisect each other, the diagonals of the HOV lane diamond are 5 ft and 12 ft.

$A = \frac{1}{2}d_1d_2$ Use the formula for area of a rhombus.

$= \frac{1}{2}(5)(12)$ Substitute 5 for d_1 and 12 for d_2.

$= 30$ Simplify.

The area of the HOV lane diamond is 30 ft^2.

Got It? 4. A rhombus has sides 10 cm long. If the longer diagonal is 16 cm, what is the area of the rhombus?

Lesson Check

Do you know HOW?

Find the area of each figure.

1.

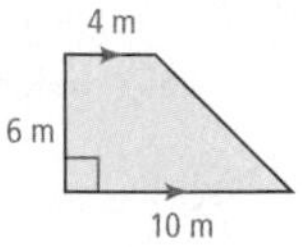

2.

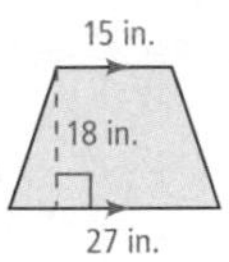

3.

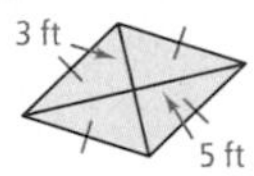

4.

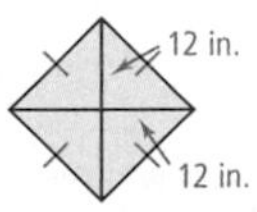

5.

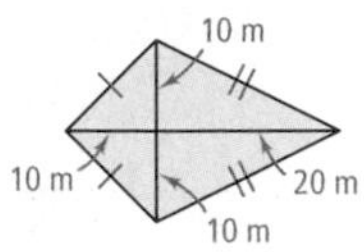

6.

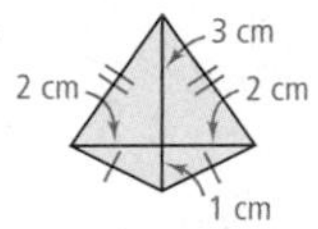

Do you UNDERSTAND? MATHEMATICAL PRACTICES

7. Vocabulary Can a trapezoid and a parallelogram with the same base and height have the same area? Explain.

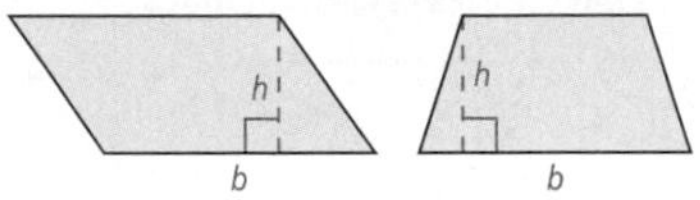

8. Reasoning Do you need to know all the side lengths to find the area of a trapezoid?

9. Reasoning Can you find the area of a rhombus if you only know the lengths of its sides? Explain.

10. Reasoning Do you need to know the lengths of the sides to find the area of a kite? Explain.

Problem 4

Q What do the dimensions given in the problem represent in a rhombus? **[They represent the diagonals of the rhombus.]**

Got It?

Students may be able to locate an actual image of an HOV lane on the Internet so that they might better picture the situation.

3 Lesson Check

Do you know HOW?

- If students have difficulty with Exercise 3, then have them review Problem 3 to get the proper formula.

Do you UNDERSTAND?

- If students have difficulty with Exercise 7, then have them write the formulas for the areas of each figure, using *x* for the length of the unknown shorter base. Set the expressions equal to each other and draw conclusions from the solution to your equation.

Close

Q What is the formula for the area of a trapezoid? What does each variable represent? **[$A = \frac{1}{2}h(b_1 + b_2)$, where h is the height, and b_1 and b_2 are the lengths of the bases.]**

Q What is the formula for the area of a rhombus or a kite? What does each variable represent? **[$A = \frac{1}{2}(d_1d_2)$, where d_1 and d_2 are the lengths of the diagonals.]**

Additional Problems

1. What is the area of the trapezoid below?

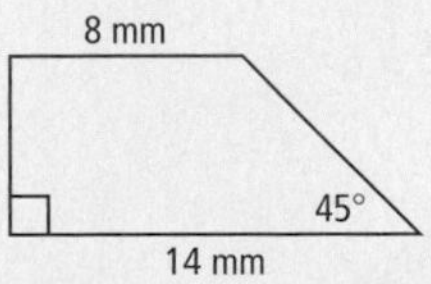

ANSWER 66 mm^2

2. What is the area of kite *ABCD*?

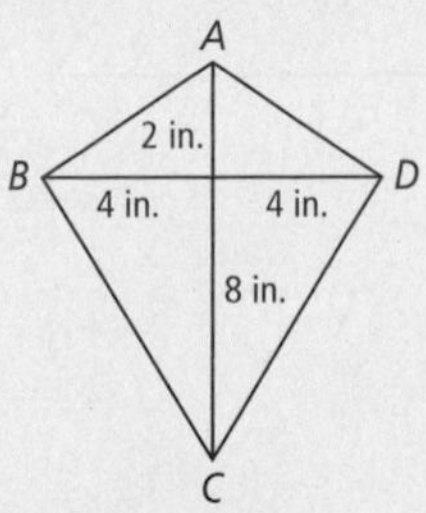

ANSWER 40 in.^{2}

3. What is the area of the rhombus below?

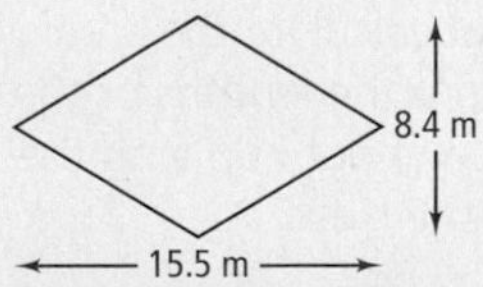

ANSWER 65.1 m^2

Answers

Got It? (continued)

4. 96 cm^2

Lesson Check

1–10. See next page.

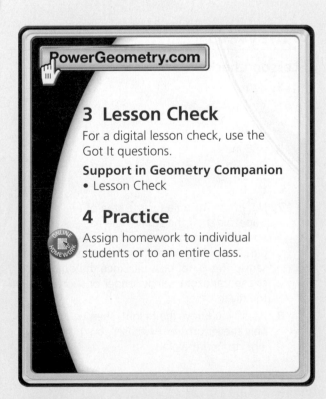

PowerGeometry.com

3 Lesson Check

For a digital lesson check, use the Got It questions.

Support in Geometry Companion

- Lesson Check

4 Practice

Assign homework to individual students or to an entire class.

4 Practice

ASSIGNMENT GUIDE
Basic: 11–25 all, 26–38 even
Average: 11–25 odd, 26–41
Advanced: 11–25 odd, 26–44
Standardized Test Prep: 45–47
Mixed Review: 48–53

Mathematical Practices are supported by exercises with red headings. Here are the Practices supported in this lesson:

MP 1: Make Sense of Problems Ex. 26
MP 3: Construct Arguments Ex. 8, 9, 10, 40
MP 3: Communicate Ex. 41b
MP 6: Attend to Precision Ex. 28

Applications exercises have blue headings. Exercises 16, 27, and 43 support MP 4: Model.

EXERCISE 36: Use the Think About a Plan worksheet in the **Practice and Problem Solving Workbook** (also available in the Teaching Resources in print and online) to further support students' development in becoming independent learners.

HOMEWORK QUICK CHECK
To check students' understanding of key skills and concepts, go over Exercises 13, 21, 26, 28, and 36.

Practice and Problem-Solving Exercises

MATHEMATICAL PRACTICES

A Practice

Find the area of each trapezoid See Problem 1.

11.
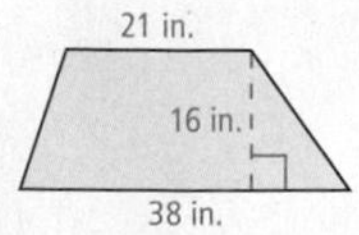

12.
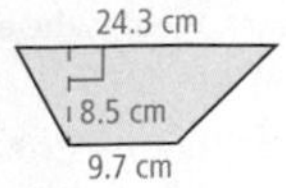

13.
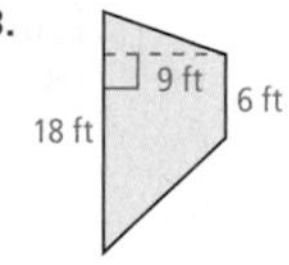

14. Find the area of a trapezoid with bases 12 cm and 18 cm and height 10 cm.

15. Find the area of a trapezoid with bases 2 ft and 3 ft and height $\frac{1}{3}$ ft.

16. Geography The border of Tennessee resembles a trapezoid with bases 340 mi and 440 mi and height 110 mi. Estimate the area of Tennessee by finding the area of the trapezoid.

Find the area of each trapezoid. If your answer is not an integer, leave it in simplest radical form. See Problem 2.

17.
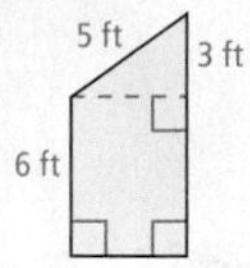

18.
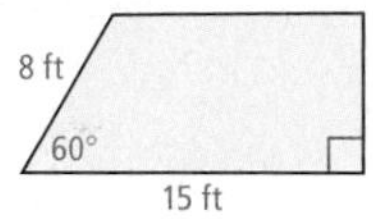

19.
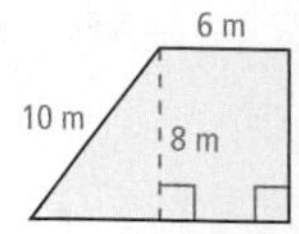

Find the area of each kite. See Problem 3.

20.
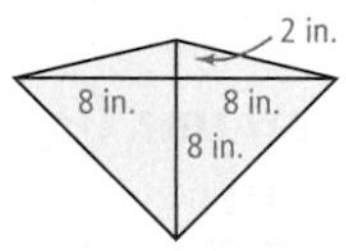

21.
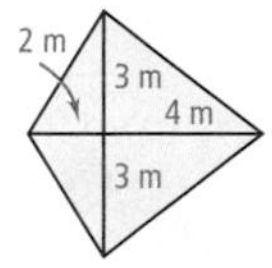

22.
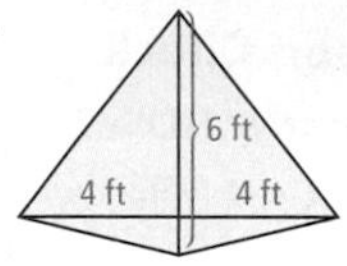

Find the area of each rhombus. See Problem 4.

23.
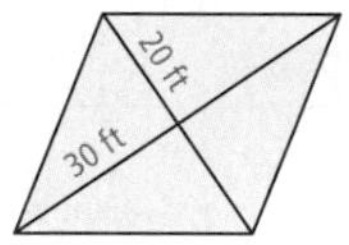

24.
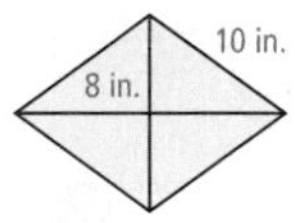

25.
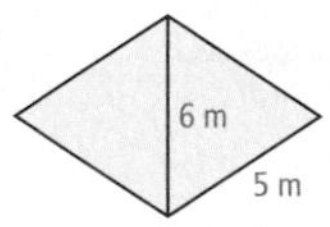

B Apply

26. Think About a Plan A trapezoid has two right angles, 12-m and 18-m bases, and an 8-m height. Sketch the trapezoid and find its perimeter and area.
- Are the right angles consecutive or opposite angles?
- How does knowing the height help you find the perimeter?

Answers

Lesson Check

1. 42 m^2
2. 378 $in.^2$
3. 30 ft^2
4. 288 $in.^2$
5. 300 m^2
6. 8 cm^2
7. No; in the formula for the area of a trapezoid, half the sum of the bases would have to equal the length of the base of the parallelogram in order for the areas to be the same. This is not possible since the other base of the trapezoid will be longer or shorter than the given base.
8. No; if you know the height, then you need only the lengths of the bases, but not the legs, to find the area.
9. No; unless the rhombus is a square, you cannot calculate the area without knowing the lengths of the diagonals.
10. No; you can calculate the area of a kite from the lengths of the diagonals, without knowing the lengths of the sides.

Practice and Problem-Solving Exercises

11. 472 $in.^2$
12. 144.5 cm^2
13. 108 ft^2
14. 150 cm^2
15. $\frac{5}{6}$ ft^2
16. 42,900 mi^2
17. 30 ft^2
18. $52\sqrt{3}$ ft^2
19. 72 m^2
20. 80 $in.^2$
21. 18 m^2
22. 24 ft^2
23. 1200 ft^2
24. 96 $in.^2$
25. 24 m^2
26.

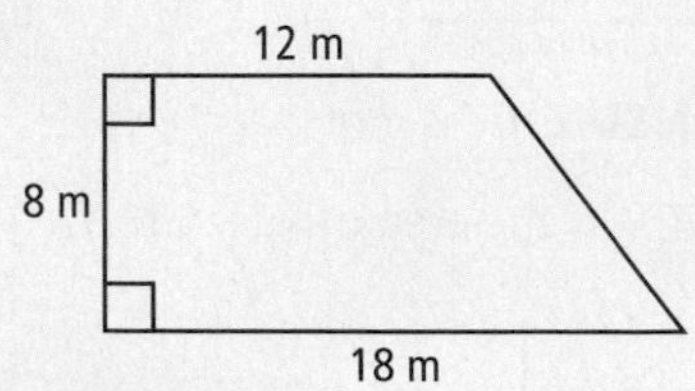

48 m; 120 m^2

27. Metallurgy The end of a gold bar has the shape of a trapezoid with the measurements shown. Find the area of the end.

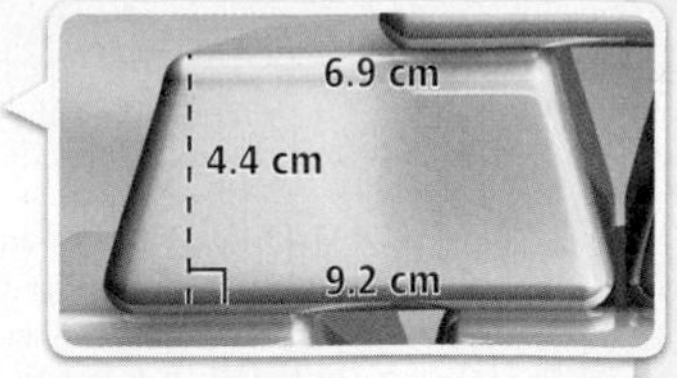

28. Open-Ended Draw a kite. Measure the lengths of its diagonals. Find its area.

Find the area of each trapezoid to the nearest tenth.

29.

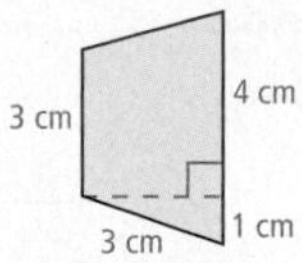

30.

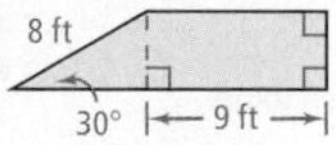

31.

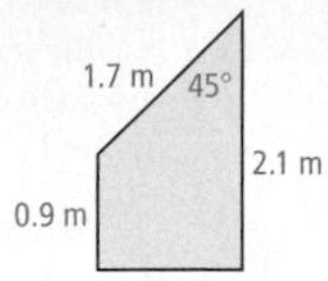

Coordinate Geometry **Find the area of quadrilateral *QRST*.**

32.

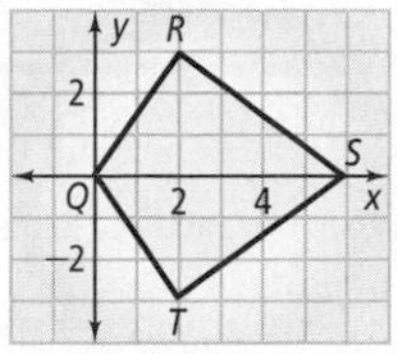

33.

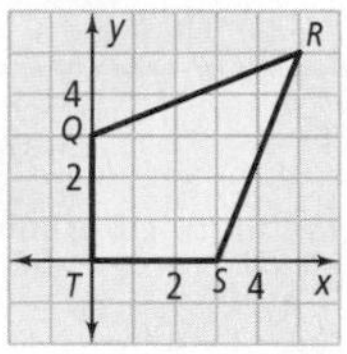

34.

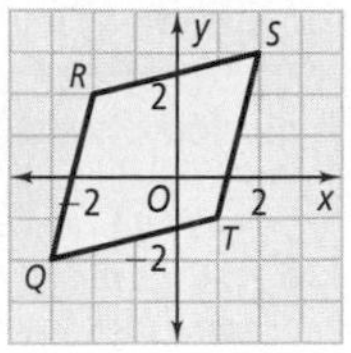

35. What is the area of the kite at the right?

Ⓐ 90 m^2 Ⓒ 135 m^2

Ⓑ 108 m^2 Ⓓ 216 m^2

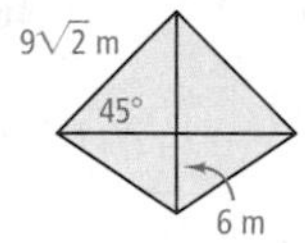

36. a. Coordinate Geometry Graph the lines $x = 0$, $x = 6$, $y = 0$, and $y = x + 4$.

b. What type of quadrilateral do the lines form?

c. Find the area of the quadrilateral.

Find the area of each rhombus. Leave your answer in simplest radical form.

37.

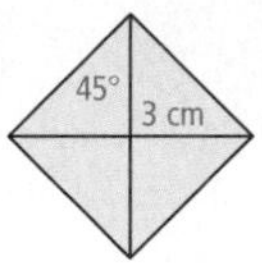

38.

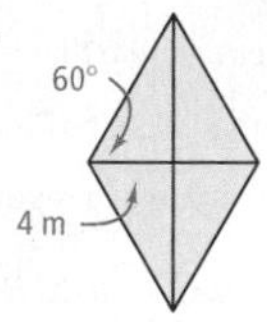

39.

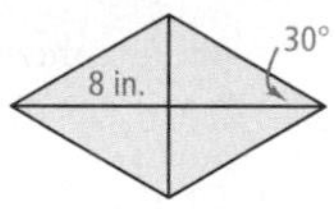

40. Visualization The kite has diagonals d_1 and d_2 congruent to the sides of the rectangle. Explain why the area of the kite is $\frac{1}{2}d_1d_2$.

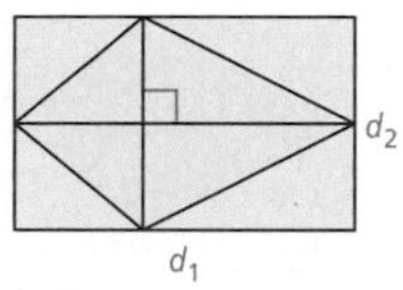

41. Draw a trapezoid. Label its bases b_1 and b_2 and its height h. Then draw a diagonal of the trapezoid.

a. Write equations for the area of each of the two triangles formed.

b. Writing Explain how you can justify the trapezoid area formula using the areas of the two triangles.

27. about 35.4 cm^2

28. Check students' work.

29. 11.3 cm^2

30. 49.9 ft^2

31. 1.8 m^2

32. 18 units^2

33. 15 units^2

34. 15 units^2

35. C

36a.

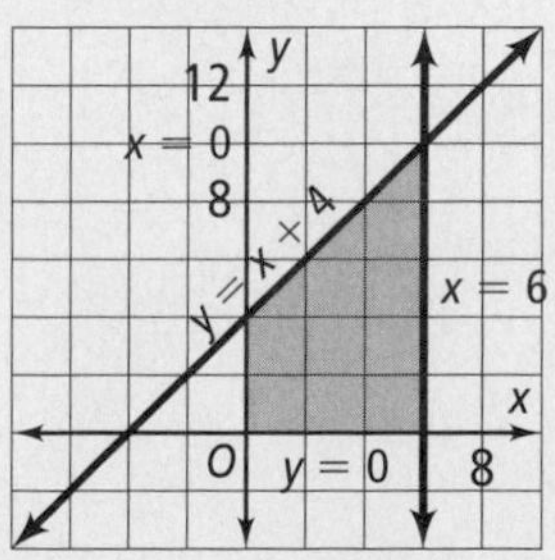

b. trapezoid

c. 42 units^2

37. 18 cm^2

38. $32\sqrt{3}\text{ m}^2$

39. $\frac{128\sqrt{3}}{3}\text{ in.}^2$

40. Explanations may vary. Sample: Each kite section is half of a corresp. rectangle section.

41a. $A = \frac{1}{2}b_1h$, $A = \frac{1}{2}b_2h$

b. Add the areas of the △s to get the area of the trapezoid: Area of trapezoid $= \frac{1}{2}b_1h + \frac{1}{2}b_2h = \frac{1}{2}h(b_1 + b_2)$.

Answers

Practice and Problem-Solving Exercises (continued)

42. height: 18 cm; bases: 12 cm and 24 cm

43. 1.5 m^2

44. $100 + 50\sqrt{3}$ or about 186.6 $in.^2$

45. A

46. H

47. [2]

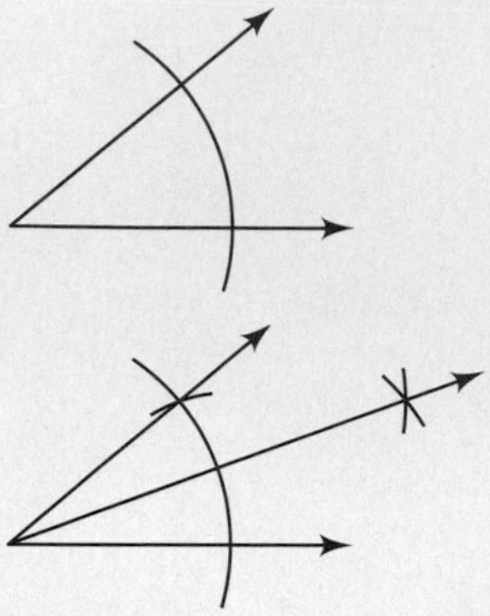

[1] sketch without construction marks

48. 72 cm^2

49. 15 ft

50. 140

51. $25\sqrt{3}$ cm^2

52. 50 ft^2

53. $\frac{100\sqrt{3}}{3}$ m^2

Challenge

42. **Algebra** One base of a trapezoid is twice the other. The height is the average of the two bases. The area is 324 cm^2. Find the height and the bases. (*Hint:* Let the smaller base be x.)

43. **Sports** Ty wants to paint one side of the skateboarding ramp he built. The ramp is 4 m wide. Its surface is modeled by the equation $y = 0.25x^2$. Use the trapezoids and triangles shown to estimate the area to be painted.

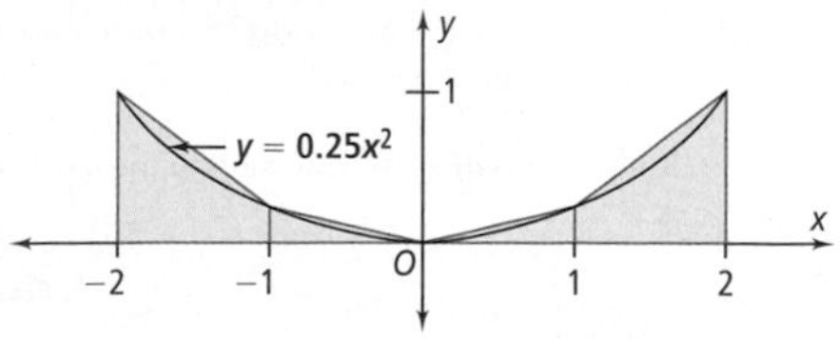

44. In trapezoid $ABCD$ at the right, $\overline{AB} \parallel \overline{DC}$. Find the area of $ABCD$.

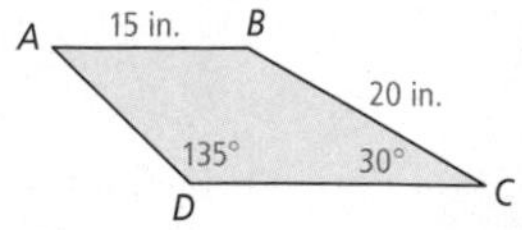

Standardized Test Prep

SAT/ACT

45. The area of a kite is 120 cm^2. The length of one diagonal is 20 cm. What is the length of the other diagonal?

(A) 12 cm (B) 20 cm (C) 24 cm (D) 48 cm

46. $\triangle ABC \sim \triangle XYZ$. $AB = 6$, $BC = 3$, and $CA = 7$. Which of the following are NOT possible dimensions of $\triangle XYZ$?

(F) $XY = 3$, $YZ = 1.5$, $ZX = 3.5$

(G) $XY = 9$, $YZ = 4.5$, $ZX = 10.5$

(H) $XY = 10$, $YZ = 7$, $ZX = 11$

(I) $XY = 18$, $YZ = 9$, $ZX = 21$

Short Response

47. Draw an angle. Construct a congruent angle and its bisector.

Mixed Review

48. Find the area of a right isosceles triangle that has one leg of length 12 cm. See Lesson 10-1.

49. A right isosceles triangle has area 112.5 ft^2. Find the length of each leg.

50. Find the measure of an interior angle of a regular nonagon. See Lesson 6-1.

Get Ready! **To prepare for Lesson 10-3, do Exercises 51–53.**

Find the area of each regular polygon. Leave radicals in simplest form. See Lesson 8-2.

51.

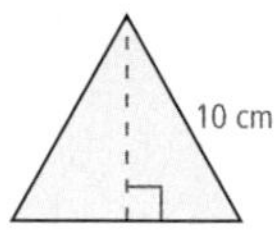

52.

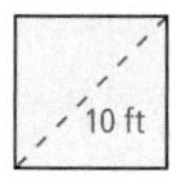

53.

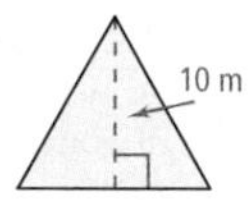

10-2 Lesson Resources

Differentiated Remediation

Instructional Support

Geometry Companion

Students can use the **Geometry Companion** worktext (4 pages) . . .

- New Vocabulary
- Key Concepts
- Got It for each Problem
- Lesson Check

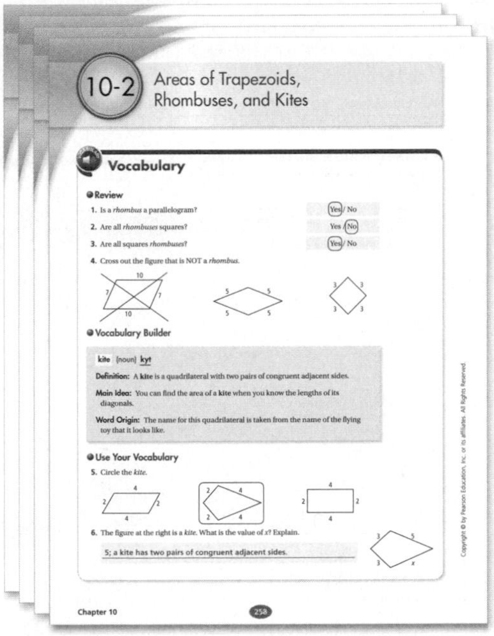
10-2 Areas of Trapezoids, Rhombuses, and Kites

Vocabulary

Review

1. Is a *rhombus* a parallelogram? Yes / No
2. Are all *rhombuses* squares? Yes / No
3. Are all squares *rhombuses*? Yes / No
4. Cross out the figure that is NOT a *rhombus*.

Vocabulary Builder

kite (noun) kyt

Definition: A kite is a quadrilateral with two pairs of congruent adjacent sides.

Main Idea: You can find the area of a kite when you know the lengths of its diagonals.

Word Origin: The name for this quadrilateral is taken from the name of the flying toy that it looks like.

Use Your Vocabulary

5. Circle the kite.
6. The figure at the right is a kite. What is the value of *x*? Explain.

5; a kite has two pairs of congruent adjacent sides.

Chapter 10 258

ELL Support

Use Manipulatives Arrange students into pairs of mixed abilities. Hand out grid paper. Use the Concept Byte at the beginning of the chapter as a guide. Tell students to trace a parallelogram, cut a triangle off the end, and then slide it to the other end to make a rectangle. Discuss the area of a parallelogram as it relates to the area of a rectangle. Ask: what is the formula for the area of a parallelogram? Challenge students to use their texts and grid paper to demonstrate the relationship of area between parallelograms and trapezoids, parallelograms/rectangles and triangles, and triangles and kites. Invite students to demonstrate how they can derive formulas from their manipulations.

5 Assess & Remediate

Lesson Quiz

1. What is the area of the trapezoid below?

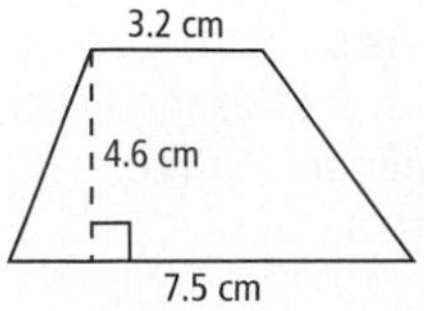

2. What is the area of kite *QRST*?

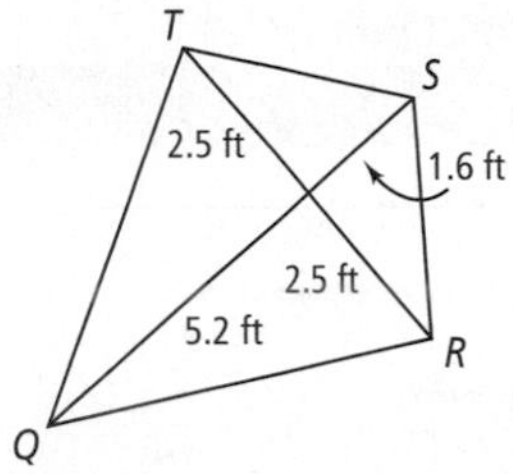

3. Suppose the area of a trapezoid is 126 yd^2. If the bases of the trapezoid are 17 yd and 11 yd long, what is the height?
4. **Do you UNDERSTAND?** Suppose a square has side length *s*. How could you use the formula for the area of a trapezoid to find the area of the square?

ANSWERS TO LESSON QUIZ

1. 24.61 cm^2
2. 17 ft^2
3. 9 yd
4. For a square, $h = b_1 = b_2 = s$. Substituting into the trapezoid area formula, you get $A = \frac{1}{2}s(s + s) = \frac{1}{2}s(2s) = s^2$. This result is consistent with the formula for the area of a square.

PRESCRIPTION FOR REMEDIATION

Use the student work on the Lesson Quiz to prescribe a differentiated review assignment.

Points	Differentiated Remediation
0–2	Intervention
3	On-level
4	Extension

PowerGeometry.com

5 Assess & Remediate

Assign the Lesson Quiz. Appropriate intervention, practice, or enrichment is automatically generated based on student performance.

Intervention

- **Reteaching** (2 pages) Provides reteaching and practice exercises for the key lesson concepts. Use with struggling students or absent students.
- **English Language Learner Support** Helps students develop and reinforce mathematical vocabulary and key concepts.

All-in-One Resources/Online

Reteaching

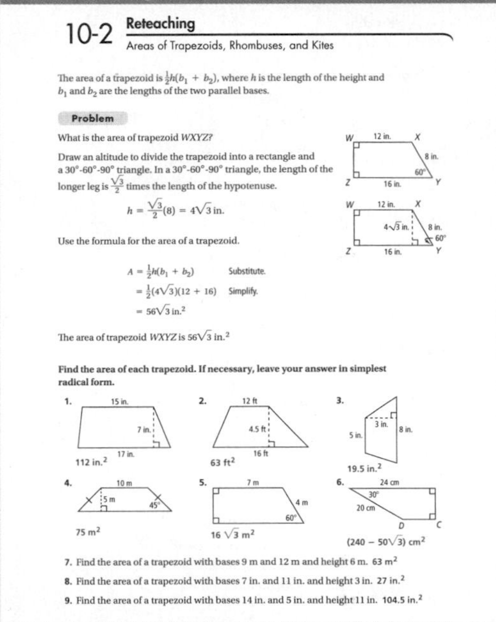
10-2 Reteaching

Areas of Trapezoids, Rhombuses, and Kites

The area of a trapezoid is $\frac{1}{2}h(b_1 + b_2)$, where *h* is the length of the height and b_1 and b_2 are the lengths of the two parallel bases.

Problem

What is the area of trapezoid *WXYZ*?

Draw an altitude to divide the trapezoid into a rectangle and a 30°-60°-90° triangle. In a 30°-60°-90° triangle, the length of the longer leg is $\frac{\sqrt{3}}{2}$ times the length of the hypotenuse.

$h = \frac{\sqrt{3}}{2}(8) = 4\sqrt{3}$ in.

Use the formula for the area of a trapezoid.

$A = \frac{1}{2}h(b_1 + b_2)$ Substitute.

$= \frac{1}{2}(4\sqrt{3})(12 + 16)$ Simplify.

$= 56\sqrt{3}$ in.^{2}

The area of trapezoid *WXYZ* is $56\sqrt{3}$ in.^{2}

Find the area of each trapezoid. If necessary, leave your answer in simplest radical form.

1. 112 in.^{2} 2. 63 ft^2 3. 19.5 in.^{2} 4. 75 m^2 5. $16\sqrt{3}$ m^2 6. $(240 - 50\sqrt{3})$ cm^2

7. Find the area of a trapezoid with bases 9 m and 12 m and height 6 m. 63 m^2
8. Find the area of a trapezoid with bases 7 in. and 11 in. and height 3 in. 27 in.^{2}
9. Find the area of a trapezoid with bases 14 in. and 5 in. and height 11 in. 104.5 in.^{2}

All-in-One Resources/Online

English Language Learner Support

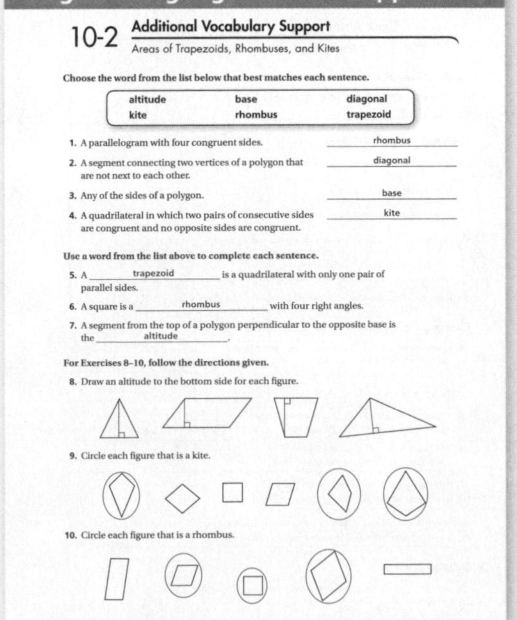
10-2 Additional Vocabulary Support

Areas of Trapezoids, Rhombuses, and Kites

Choose the word from the list below that best matches each sentence.

altitude, base, diagonal, kite, rhombus, trapezoid

1. A parallelogram with four congruent sides. rhombus
2. A segment connecting two vertices of a polygon that are not next to each other. diagonal
3. Any of the sides of a polygon. base
4. A quadrilateral in which two pairs of consecutive sides are congruent and no opposite sides are congruent. kite

Use a word from the list above to complete each sentence.

5. A trapezoid is a quadrilateral with only one pair of parallel sides.
6. A square is a rhombus with four right angles.
7. A segment from the top of a polygon perpendicular to the opposite base is the altitude.

For Exercises 8–10, follow the directions given.

8. Draw an altitude to the bottom side for each figure.
9. Circle each figure that is a kite.
10. Circle each figure that is a rhombus.

Differentiated Remediation *continued*

On-Level

- **Practice** (2 pages) Provides extra practice for each lesson. For simpler practice exercises, use the Form K Practice pages found in the All-in-One Teaching Resources and online.
- **Think About a Plan** Helps students develop specific problem-solving skills and strategies by providing scaffolded guiding questions.
- **Standardized Test Prep** Focuses on all major exercises, all major question types, and helps students prepare for the high-stakes assessments.

Extension

- **Enrichment** Provides students with interesting problems and activities that extend the concepts of the lesson.
- **Activities, Games, and Puzzles** Worksheets that can be used for concepts development, enrichment, and for fun!

Practice and Problem Solving Wkbk/ All-in-One Resources/Online

Practice page 1

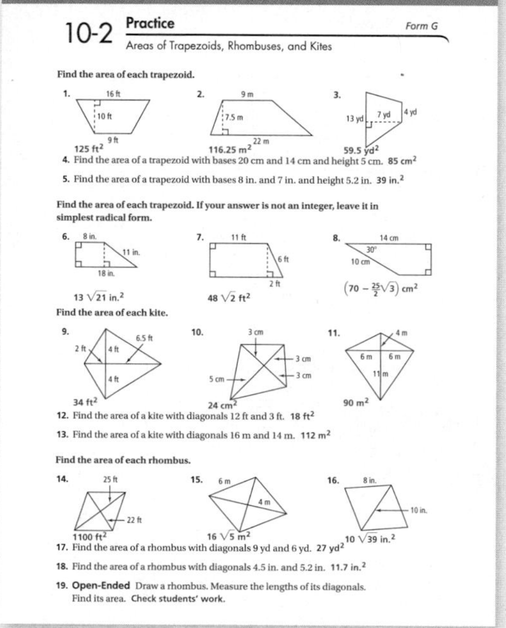

10-2 Practice — Form G
Areas of Trapezoids, Rhombuses, and Kites

Find the area of each trapezoid.

1. 125 ft² 2. 116.25 m² 3. 59.5 yd²
4. Find the area of a trapezoid with bases 20 cm and 14 cm and height 5 cm. 85 cm²
5. Find the area of a trapezoid with bases 8 in. and 7 in. and height 5.2 in. 39 in.²

Find the area of each trapezoid. If your answer is not an integer, leave it in simplest radical form.

6. $13\sqrt{21}$ in.² 7. $48\sqrt{2}$ ft² 8. $(70 - \frac{25}{2}\sqrt{3})$ cm²

Find the area of each kite.

9. 34 ft² 10. 24 cm² 11. 90 m²
12. Find the area of a kite with diagonals 12 ft and 3 ft. 18 ft²
13. Find the area of a kite with diagonals 16 m and 14 m. 112 m²

Find the area of each rhombus.

14. 1100 ft² 15. $16\sqrt{5}$ m² 16. $10\sqrt{39}$ in.²
17. Find the area of a rhombus with diagonals 9 yd and 6 yd. 27 yd²
18. Find the area of a rhombus with diagonals 4.5 in. and 5.2 in. 11.7 in.²
19. Open-Ended Draw a rhombus. Measure the lengths of its diagonals. Find its area. Check students' work.

Practice and Problem Solving Wkbk/ All-in-One Resources/Online

Practice page 2

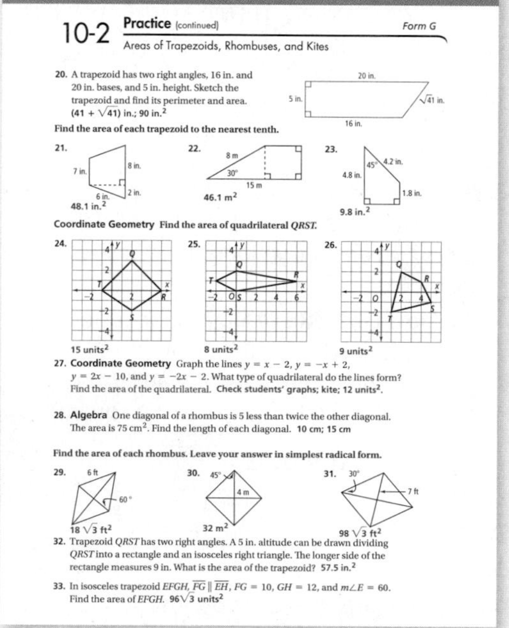

10-2 Practice (continued) — Form G
Areas of Trapezoids, Rhombuses, and Kites

20. A trapezoid has two right angles, 16 in. and 20 in. bases, and 5 in. height. Sketch the trapezoid and find its perimeter and area. $(41 + \sqrt{41})$ in.; 90 in.²

Find the area of each trapezoid to the nearest tenth.

21. 48.1 in.² 22. 46.1 m² 23. 9.8 in.²

Coordinate Geometry Find the area of quadrilateral *QRST*.

24. 15 units² 25. 8 units² 26. 9 units²
27. Coordinate Geometry Graph the lines $y = x - 2$, $y = -x + 2$, $y = 2x - 10$, and $y = -2x - 2$. What type of quadrilateral do the lines form? Find the area of the quadrilateral. Check students' graphs; kite; 12 units².
28. Algebra One diagonal of a rhombus is 5 less than twice the other diagonal. The area is 75 cm². Find the length of each diagonal. 10 cm; 15 cm

Find the area of each rhombus. Leave your answer in simplest radical form.

29. $18\sqrt{3}$ ft² 30. 32 m² 31. $98\sqrt{3}$ ft²
32. Trapezoid *QRST* has two right angles. A 5 in. altitude can be drawn dividing *QRST* into a rectangle and an isosceles right triangle. The longer side of the rectangle measures 9 in. What is the area of the trapezoid? 57.5 in.²
33. In isosceles trapezoid *EFGH*, $\overline{FG} \parallel \overline{EH}$, $FG = 10$, $GH = 12$, and $m\angle E = 60$. Find the area of *EFGH*. $96\sqrt{3}$ units²

All-in-One Resources/Online

Enrichment

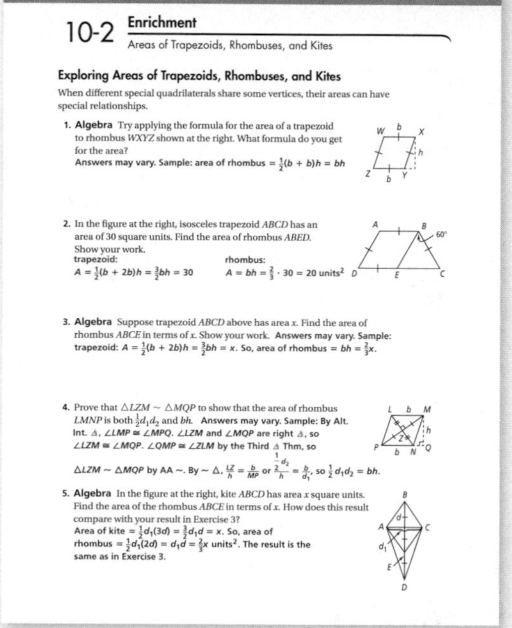

10-2 Enrichment
Areas of Trapezoids, Rhombuses, and Kites

Exploring Areas of Trapezoids, Rhombuses, and Kites

When different special quadrilaterals share some vertices, their areas can have special relationships.

1. Algebra Try applying the formula for the area of a trapezoid to rhombus *WXYZ* shown at the right. What formula do you get for the area? Answers may vary. Sample: area of rhombus $= \frac{1}{2}(b + b)h = bh$
2. In the figure at the right, isosceles trapezoid *ABCD* has an area of 30 square units. Find the area of rhombus *ABED*. Show your work. trapezoid: $A = \frac{1}{2}(b + 2b)h = \frac{3}{2}bh = 30$; rhombus: $A = bh = \frac{2}{3} \cdot 30 = 20$ units²
3. Algebra Suppose trapezoid *ABCD* above has area *x*. Find the area of rhombus *ABCE* in terms of *x*. Show your work. Answers may vary. Sample: trapezoid: $A = \frac{1}{2}(b + 2b)h = \frac{3}{2}bh = x$. So, area of rhombus $= bh = \frac{2}{3}x$.
4. Prove that $\triangle LZM \sim \triangle MQP$ to show that the area of rhombus *LMNP* is both $\frac{1}{2}d_1d_2$ and *bh*. Answers may vary. Sample: By Alt. Int. ∠s, ∠LMP ≅ ∠MPQ. ∠LZM and ∠MQP are right ∠s, so ∠LZM ≅ ∠MQP. ∠QMP ≅ ∠ZLM by the Third ∠s Thm, so △LZM ~ △MQP by AA ~. By ~ △, $\frac{LZ}{h} = \frac{LM}{MP}$ or $\frac{d_1/2}{h} = \frac{b}{d_2}$, so $\frac{1}{2}d_1d_2 = bh$.
5. Algebra In the figure at the right, kite *ABCD* has area *x* square units. Find the area of the rhombus *ABCE* in terms of *x*. How does this result compare with your result in Exercise 3? Area of kite $= \frac{1}{2}d_1(3d) = \frac{3}{2}d_1d = x$. So, area of rhombus $= \frac{1}{2}d_1(2d) = d_1d = \frac{2}{3}x$ units². The result is the same as in Exercise 3.

Practice and Problem Solving Wkbk/ All-in-One Resources/Online

Think About a Plan

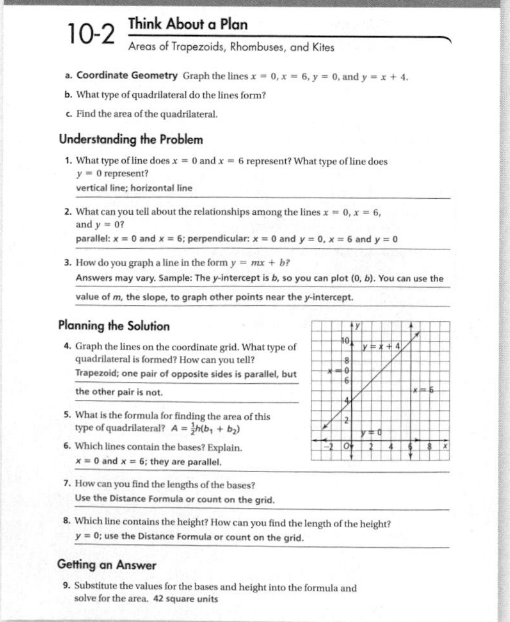

10-2 Think About a Plan
Areas of Trapezoids, Rhombuses, and Kites

a. Coordinate Geometry Graph the lines $x = 0$, $x = 6$, $y = 0$, and $y = x + 4$.
b. What type of quadrilateral do the lines form?
c. Find the area of the quadrilateral.

Understanding the Problem

1. What type of line does $x = 0$ and $x = 6$ represent? What type of line does $y = 0$ represent? vertical line; horizontal line
2. What can you tell about the relationships among the lines $x = 0$, $x = 6$, and $y = 0$? parallel: $x = 0$ and $x = 6$; perpendicular: $x = 0$ and $y = 0$, $x = 6$ and $y = 0$
3. How do you graph a line in the form $y = mx + b$? Answers may vary. Sample: The *y*-intercept is *b*, so you can plot (0, *b*). You can use the value of *m*, the slope, to graph other points near the *y*-intercept.

Planning the Solution

4. Graph the lines on the coordinate grid. What type of quadrilateral is formed? How can you tell? Trapezoid; one pair of opposite sides is parallel, but the other pair is not.
5. What is the formula for finding the area of this type of quadrilateral? $A = \frac{1}{2}h(b_1 + b_2)$
6. Which lines contain the bases? Explain. $x = 0$ and $x = 6$; they are parallel.
7. How can you find the lengths of the bases? Use the Distance Formula or count on the grid.
8. Which line contains the height? How can you find the length of the height? $y = 0$; use the Distance Formula or count on the grid.

Getting an Answer

9. Substitute the values for the bases and height into the formula and solve for the area. 42 square units

Practice and Problem Solving Wkbk/ All-in-One Resources/Online

Standardized Test Prep

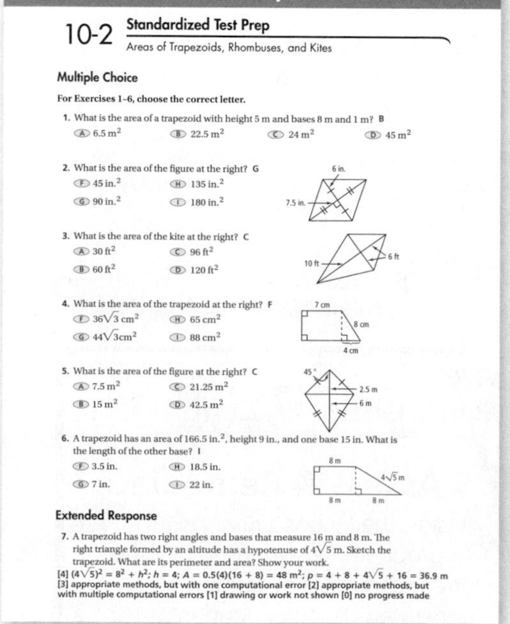

10-2 Standardized Test Prep
Areas of Trapezoids, Rhombuses, and Kites

Multiple Choice

For Exercises 1–6, choose the correct letter.

1. What is the area of a trapezoid with height 5 m and bases 8 m and 1 m? B
(A) 6.5 m² (B) 22.5 m² (C) 24 m² (D) 45 m²
2. What is the area of the figure at the right? G
(F) 45 in.² (G) 90 in.² (H) 135 in.² (I) 180 in.²
3. What is the area of the kite at the right? C
(A) 30 ft² (B) 60 ft² (C) 96 ft² (D) 120 ft²
4. What is the area of the trapezoid at the right? F
(F) $36\sqrt{3}$ cm² (G) $44\sqrt{3}$ cm² (H) 65 cm² (I) 88 cm²
5. What is the area of the figure at the right? C
(A) 7.5 m² (B) 15 m² (C) 21.25 m² (D) 42.5 m²
6. A trapezoid has an area of 166.5 in.², height 9 in., and one base 15 in. What is the length of the other base? I
(F) 3.5 in. (G) 7 in. (H) 18.5 in. (I) 22 in.

Extended Response

7. A trapezoid has two right angles and bases that measure 16 m and 8 m. The right triangle formed by an altitude has a hypotenuse of $4\sqrt{5}$ m. Sketch the trapezoid. What are its perimeter and area? Show your work.
[4] $(4\sqrt{5})^2 = 8^2 + h^2$; $h = 4$; $A = 0.5(4)(16 + 8) = 48$ m²; $p = 4 + 8 + 4\sqrt{5} + 16 = 36.9$ m [3] appropriate methods, but with one computational error [2] appropriate methods, but with multiple computational errors [1] drawing or work not shown [0] no progress made

Online Teacher Resource Center

Activities, Games, and Puzzles

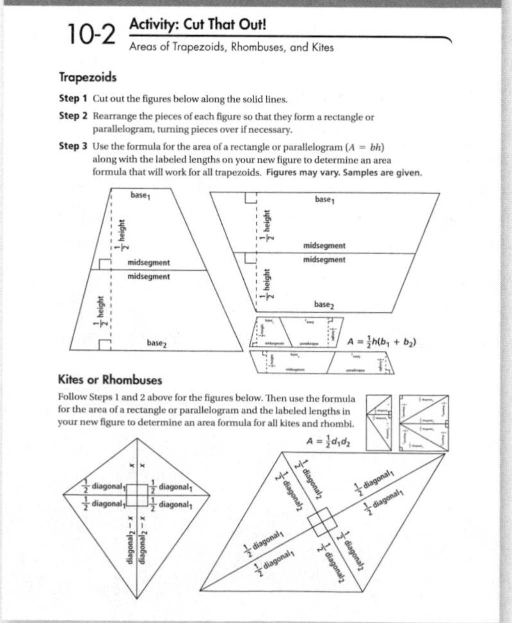

10-2 Activity: Cut That Out!
Areas of Trapezoids, Rhombuses, and Kites

Trapezoids

Step 1 Cut out the figures below along the solid lines.
Step 2 Rearrange the pieces of each figure so that they form a rectangle or parallelogram, turning pieces over if necessary.
Step 3 Use the formula for the area of a rectangle or parallelogram ($A = bh$) along with the labeled lengths on your new figure to determine an area formula that will work for all trapezoids. Figures may vary. Samples are given.

$A = \frac{1}{2}h(b_1 + b_2)$

Kites or Rhombuses

Follow Steps 1 and 2 above for the figures below. Then use the formula for the area of a rectangle or parallelogram and the labeled lengths in your new figure to determine an area formula for all kites and rhombi. $A = \frac{1}{2}d_1d_2$

10-3 Areas of Regular Polygons

Common Core State Standards
G-MG.A.1 Use geometric shapes, their measures, and their properties to describe objects. **Also G-CO.D.13**
MP 1, MP 3, MP 4, MP 6, MP 7

Objective To find the area of a regular polygon

Solve a simpler problem. Try using fewer sides to see what happens.

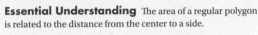

Solve It! Getting Ready!

You want to build a koi pond. For the border, you plan to use 3-ft-long pieces of wood. You have 12 pieces that you can connect together at any angle, including a straight angle. If you want to maximize the area of the pond, in what shape should you arrange the pieces? Explain your reasoning.

MATHEMATICAL PRACTICES

The Solve It involves the area of a polygon.

Lesson Vocabulary
- radius of a regular polygon
- apothem

Essential Understanding The area of a regular polygon is related to the distance from the center to a side.

You can circumscribe a circle about any regular polygon. The center of a regular polygon is the center of the circumscribed circle. The **radius of a regular polygon** is the distance from the center to a vertex. The **apothem** is the perpendicular distance from the center to a side.

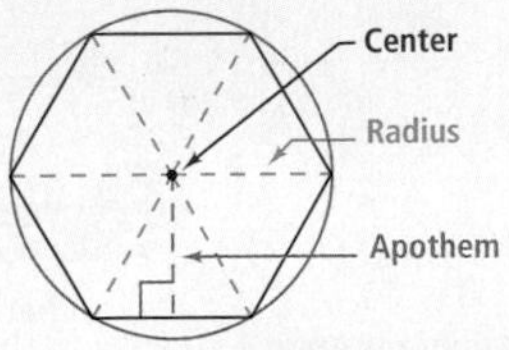

Problem 1 Finding Angle Measures

The figure at the right is a regular pentagon with radii and an apothem drawn. What is the measure of each numbered angle?

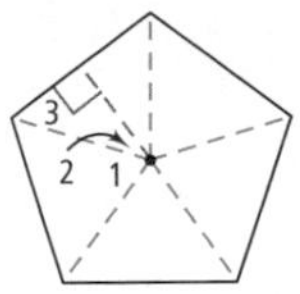

$m\angle 1 = \frac{360}{5} = 72$ — Divide 360 by the number of sides.

$m\angle 2 = \frac{1}{2}m\angle 1$ — The apothem bisects the vertex angle of the isosceles triangle formed by the radii.

$= \frac{1}{2}(72) = 36$

$90 + 36 + m\angle 3 = 180$ — The sum of the measures of the angles of a triangle is 180.

$m\angle 3 = 54$

$m\angle 1 = 72$, $m\angle 2 = 36$, and $m\angle 3 = 54$.

Think
How do you know the radii make isosceles triangles?
Since the pentagon is a regular polygon, the radii are congruent. So, the triangle made by two adjacent radii and a side of the polygon is an isosceles triangle.

Got It? 1. At the right, a portion of a regular octagon has radii and an apothem drawn. What is the measure of each numbered angle?

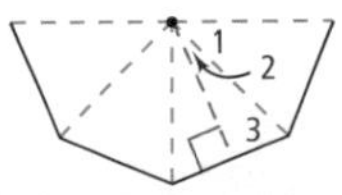

1 Interactive Learning

Solve It!

PURPOSE To realize that increasing the number of sides in a regular polygon increases its area

PROCESS Students may guess and check by drawing diagrams of the figures and calculate the area or solve a simpler problem by constructing regular polygons with fewer sides.

FACILITATE

Q Can you form a regular triangle with the pieces? If so, what is its area? **[Yes, the area is approximately 62.35 ft^2.]**

Q Can you form a regular quadrilateral with the pieces? If so, what is its area? **[Yes, the area is 81 ft^2.]**

Q Can you form a regular hexagon? If so, what is its area? (*Hint*: Use composite figures.) **[Yes, the area is approximately 93.53 ft^2.]**

ANSWER See Solve It in Answers on next page.

CONNECT THE MATH In the Solve It, students experiment with different regular polygons and investigate their areas. In the lesson, students learn that a regular *n*-gon contains the maximum area of any polygon with the same perimeter and number of sides.

2 Guided Instruction

Problem 1

Q What type of triangle is formed by the apothem and radius of the polygon? **[a right triangle]**

Got It? VISUAL LEARNERS

Have students draw a diagram of the right triangle and label each angle as they find its measure.

10-3 Preparing to Teach

BIG idea Measurement

ESSENTIAL UNDERSTANDING

- The area of a regular polygon is a function of the distance from the center to a side and the perimeter.

Math Background

The area of polygons can be found using the length of the apothem. The apothem is the perpendicular segment that connects the center of the polygon to the side. The apothem and radius create a right triangle within a polygon. Students will use their knowledge of special right triangles to solve for unknown side lengths in polygons. Note that students can find the area of a square in an easier way. Request that they use the new formula here and use previous knowledge to check their results.

Students often have difficulty understanding the relationship between a regular polygon and its inscribed and circumscribed circle. The relationship is critical to understanding how to determine the length of the apothem and the radius of the regular polygon.

Mathematical Practices

Look for and make use of structure.
In finding the area of a regular polygon, students will find a polygon of *n* sides to be composed of *n* congruent triangles and find the areas of those triangles. As a result, they will determine a formula for the area of a regular polygon.

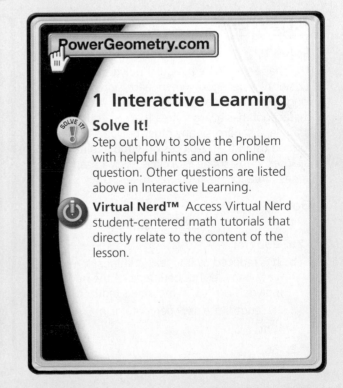

PowerGeometry.com

1 Interactive Learning

Solve It!
Step out how to solve the Problem with helpful hints and an online question. Other questions are listed above in Interactive Learning.

Virtual Nerd™ Access Virtual Nerd student-centered math tutorials that directly relate to the content of the lesson.

Take Note

Have students derive the formula for the area of a polygon by drawing a diagram. Lead them through the logic of adding the areas of all the isosceles triangles contained by the polygon.

Apothem may be a new vocabulary word for many students. Draw several diagrams on the board and verify that students can identify the apothems.

Problem 2

Q What values do you need to find the area of the polygon? **[the perimeter and the apothem length]**

Q How can you find the perimeter of the polygon? **[Find the sum of the side lengths or the product of the number of sides and the side length.]**

Got It?

ERROR PREVENTION

Have students draw a diagram of the pentagon and label the known segment lengths. Be sure that students perform both steps as in Problem 2. In 2b, use the perimeter formula, $n \cdot s$, to find the larger perimeter. Have students substitute $\frac{1}{2}s$ into the formula. Ask them to compare the results.

take note **Postulate 10-1**

If two figures are congruent, then their areas are equal.

Suppose you have a regular n-gon with side s. The radii divide the figure into n congruent isosceles triangles. By Postulate 10-1, the areas of the isosceles triangles are equal. Each triangle has a height of a and a base of length s, so the area of each triangle is $\frac{1}{2}as$.

Since there are n congruent triangles, the area of the n-gon is $A = n \cdot \frac{1}{2}as$. The perimeter p of the n-gon is the number of sides n times the length of a side s, or ns. By substitution, the area can be expressed as $A = \frac{1}{2}ap$.

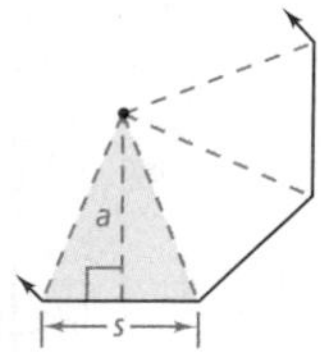

take note **Theorem 10-6 Area of a Regular Polygon**

The area of a regular polygon is half the product of the apothem and the perimeter.

$$A = \frac{1}{2}ap$$

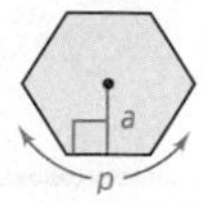

Plan

What do you know about the regular decagon?
A decagon has 10 sides, so $n = 10$. From the diagram, you know that the apothem a is 12.3 in., and the side length s is 8 in.

Problem 2 Finding the Area of a Regular Polygon

What is the area of the regular decagon at the right?

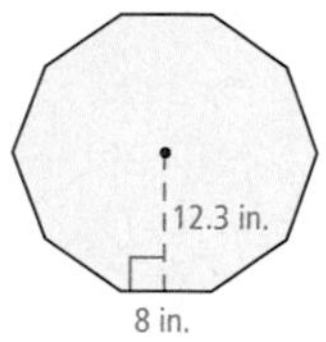

Step 1 Find the perimeter of the regular decagon.

$p = ns$	Use the formula for the perimeter of an n-gon.
$= 10(8)$	Substitute 10 for n and 8 for s.
$= 80$ in.	

Step 2 Find the area of the regular decagon.

$A = \frac{1}{2}ap$	Use the formula for the area of a regular polygon.
$= \frac{1}{2}(12.3)(80)$	Substitute 12.3 for a and 80 for p.
$= 492$	

The regular decagon has an area of 492 in.^{2}.

Got It? 2. a. What is the area of a regular pentagon with an 8-cm apothem and 11.6-cm sides?

b. Reasoning If the side of a regular polygon is reduced to half its length, how does the perimeter of the polygon change? Explain.

Answers

Solve It!

The shape should be a regular polygon with 12 sides (dodecagon), where each side is 3 ft long. Explanations may vary. Sample: A regular dodecagon has a larger area than an equilateral triangle, a square, or any other polygon that can be formed using the 12 pieces of wood and its shape is closer to the shape of a circle, which has the largest area for the fixed perimeter 36 ft.

Got It?

1. $m\angle 1 = 45$, $m\angle 2 = 22.5$, $m\angle 3 = 67.5$

2a. 232 cm^2

b. It is reduced by half; explanations may vary. Sample: The perimeter of the original polygon is $n \cdot s$. If the side is reduced to half its length, the new perimeter is $n \cdot \frac{1}{2}s$, or $\frac{1}{2}ns$.

3. 665 ft^2

PowerGeometry.com

2 Guided Instruction

Each Problem is worked out and supported online.

Problem 1
Finding Angle Measures

Problem 2
Finding the Area of a Regular Polygon
Animated

Problem 3
Using Special Triangles to Find Area

Support in Geometry Companion
- Vocabulary
- Key Concepts
- Got It?

Problem 3 Using Special Triangles to Find Area STEM

Zoology **A honeycomb is made up of regular hexagonal cells. The length of a side of a cell is 3 mm. What is the area of a cell?**

Know	Need	Plan
You know the length of a side, which you can use to find the perimeter.	The apothem	*Draw a diagram* to help find the apothem. Then use the area formula for a regular polygon.

Step 1 Find the apothem.

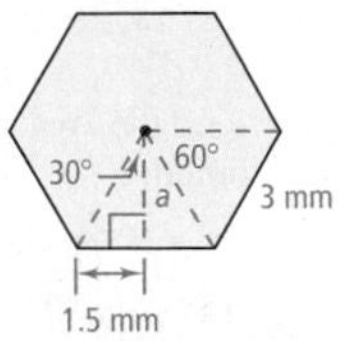

The radii form six 60° angles at the center, so you can use a 30°-60°-90° triangle to find the apothem.

$a = 1.5\sqrt{3}$ longer leg = $\sqrt{3}$ • shorter leg

Step 2 Find the perimeter.

$p = ns$ Use the formula for the perimeter of an n-gon.

$= 6(3)$ Substitute 6 for n and 3 for s.

$= 18$ mm

Step 3 Find the area.

$A = \frac{1}{2}ap$ Use the formula for the area of a regular polygon.

$= \frac{1}{2}(1.5\sqrt{3})(18)$ Substitute $1.5\sqrt{3}$ for a and 18 for p.

≈ 23.3826859 Use a calculator.

The area is about 23 mm^2.

Got It? **3.** The side of a regular hexagon is 16 ft. What is the area of the hexagon? Round your answer to the nearest square foot.

Lesson Check

Do you know HOW?

What is the area of each regular polygon? Round your answer to the nearest tenth.

1.

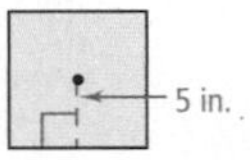

2.

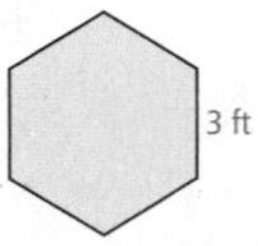

3.

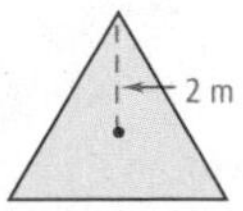

4.

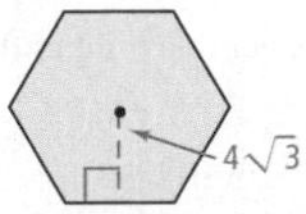

Do you UNDERSTAND? MATHEMATICAL PRACTICES

5. Vocabulary What is the difference between a radius and an apothem?

6. What is the relationship between the side length and the apothem in each figure?

a. a square

b. a regular hexagon

c. an equilateral triangle

7. Error Analysis Your friend says you can use special triangles to find the apothem of any regular polygon. What is your friend's error? Explain.

Problem 3

Q What type of triangle is created by the side of the hexagon, its apothem, and its radius? **[a 30°-60°-90° triangle]**

Q How can you find the length of the apothem of the hexagon? **[The apothem is the longer leg of the 30°-60°-90° triangle. Its length is $\sqrt{3}$ times the length of the shorter side.]**

Got It?

Have students draw a diagram and label the side length. They can draw the 30°-60°-90° triangle and determine the length of the apothem as in Problem 3.

3 Lesson Check

Do you know HOW?

- If students have difficulty with Exercise 1, then have them draw the right triangle created by the radius.

Do you UNDERSTAND?

- If students have difficulty with Exercise 6, then have them review Problem 3 and Exercise 1. Students can draw a diagram of each figure named and label side length and apothem for easier comparisons.

Close

Q What is the formula for the area of a regular polygon? What does each variable represent? **[$A = \frac{1}{2}ap$, where a is the length of the apothem and p is the perimeter of the polygon.]**

Additional Problems

1. The figure below is a regular hexagon with radii and an apothem drawn. What is the measure of each numbered angle?

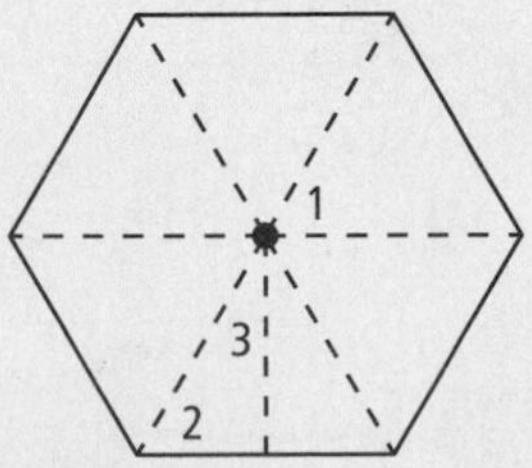

ANSWER $m\angle 1 = 60$, $m\angle 2 = 60$, $m\angle 3 = 30$

2. What is the area of the regular octagon below?

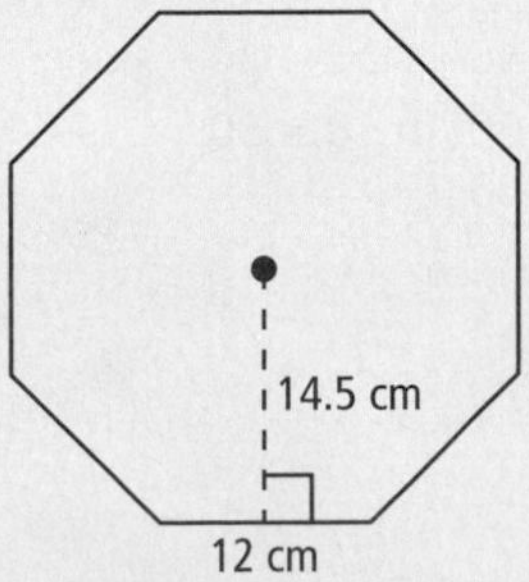

ANSWER 696 cm^2

3. Nancy has a gazebo shaped like a regular hexagon. Each side is 5 ft long. What is the area of the gazebo? Round to the nearest whole number.

ANSWER about 65 ft^2

Answers

Lesson Check

1–7. See next page.

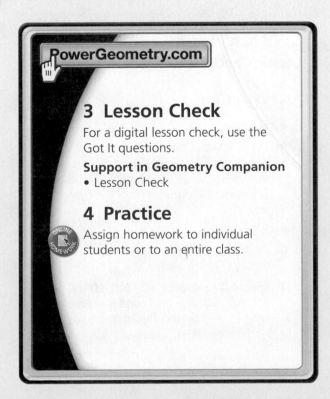

PowerGeometry.com

3 Lesson Check

For a digital lesson check, use the Got It questions.

Support in Geometry Companion

- Lesson Check

4 Practice

ONLINE HOMEWORK

Assign homework to individual students or to an entire class.

4 Practice

ASSIGNMENT GUIDE
Basic: 8–25 all, 26–30 even, 31–33 all, 35
Average: 9–25 odd, 26–41
Advanced: 9–25 odd, 26–42
Standardized Test Prep: 44–47
Mixed Review: 48–52

Mathematical Practices are supported by exercises with red headings. Here are the Practices supported in this lesson:

MP 1: Make Sense of Problems Ex. 36d
MP 3: Construct Arguments Ex. 35
MP 3: Critique the Reasoning of Others Ex. 7
MP 4: Model with Mathematics Ex. 31
MP 7: Look for Patterns Ex. 33d

Applications exercises have blue headings. Exercise 30 supports MP 4: Model.

STEM exercises focus on science or engineering applications.

EXERCISE 32: Use the Think About a Plan worksheet in the **Practice and Problem Solving Workbook** (also available in the Teaching Resources in print and online) to further support students' development in becoming independent learners.

HOMEWORK QUICK CHECK
To check students' understanding of key skills and concepts, go over Exercises 11, 23, 31, 32, and 35.

Practice and Problem-Solving Exercises

MATHEMATICAL PRACTICES

A Practice

Each regular polygon has radii and apothem as shown. Find the measure of each numbered angle. See Problem 1.

8.
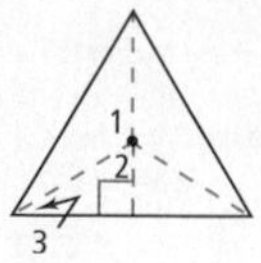

9.
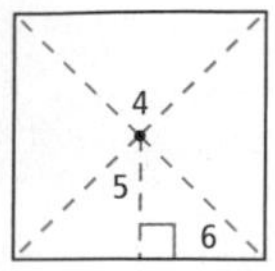

10.
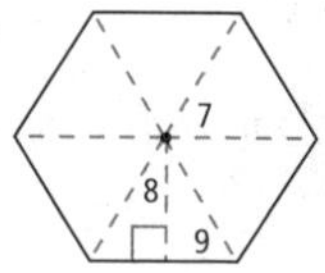

Find the area of each regular polygon with the given apothem a and side length s. See Problem 2.

11. pentagon, $a = 24.3$ cm, $s = 35.3$ cm
12. 7-gon, $a = 29.1$ ft, $s = 28$ ft
13. octagon, $a = 60.4$ in., $s = 50$ in.
14. nonagon, $a = 27.5$ in., $s = 20$ in.
15. decagon, $a = 19$ m, $s = 12.3$ m
16. dodecagon, $a = 26.1$ cm, $s = 14$ cm

Find the area of each regular polygon. Round your answer to the nearest tenth. See Problem 3.

17.
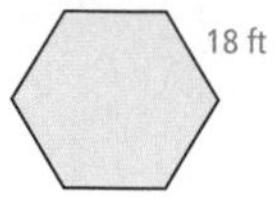

18.
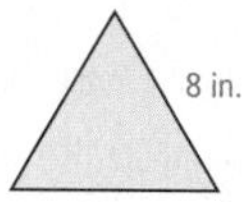

19.
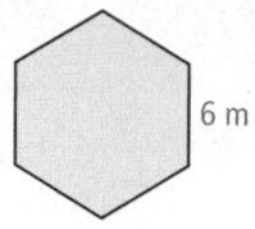

20. Art You are painting a mural of colored equilateral triangles. The radius of each triangle is 12.7 in. What is the area of each triangle to the nearest square inch?

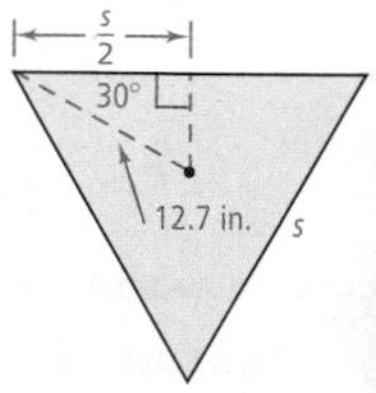

Find the area of each regular polygon with the given radius or apothem. If your answer is not an integer, leave it in simplest radical form.

21.
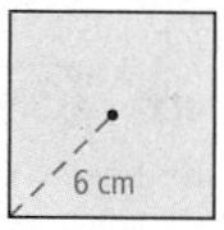

22.
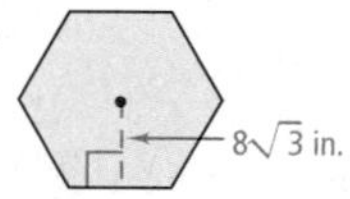

23.
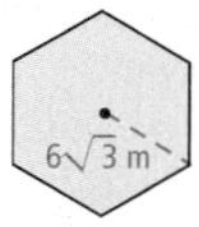

24.
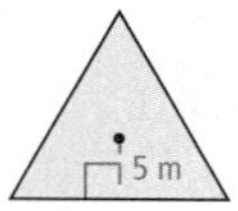

25.
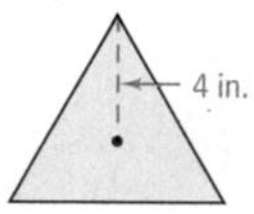

B Apply

Find the measures of the angles formed by (a) two consecutive radii and (b) a radius and a side of the given regular polygon.

26. pentagon **27.** octagon **28.** nonagon **29.** dodecagon

Answers

Lesson Check

1. 100.0 in.^2
2. 23.4 ft^2
3. 5.2 m^2
4. 166.3 units^2
5. A radius is the distance from the center to a vertex, while the apothem is the perpendicular distance from the center to a side.

6a. $s = 2a$
b. $s = \frac{2\sqrt{3}}{3}a$
c. $s = 2\sqrt{3}a$

7. Special ▲ have ∡ of 30°, 60°, 90° or 45°, 45°, 90° and are found in equilateral ▲, squares, and regular hexagons.

Practice and Problem-Solving Exercises

8. $m\angle 1 = 120$, $m\angle 2 = 60$, $m\angle 3 = 30$
9. $m\angle 4 = 90$, $m\angle 5 = 45$, $m\angle 6 = 45$
10. $m\angle 7 = 60$, $m\angle 8 = 30$, $m\angle 9 = 60$
11. 2144.475 cm^2
12. 2851.8 ft^2
13. $12{,}080 \text{ in.}^2$
14. 2475 in.^2
15. 1168.5 m^2
16. 2192.4 cm^2
17. 841.8 ft^2
18. 27.7 in.^2
19. 93.5 m^2
20. 210 in.^2
21. 72 cm^2
22. $384\sqrt{3} \text{ in.}^2$
23. $162\sqrt{3} \text{ m}^2$
24. $75\sqrt{3} \text{ m}^2$
25. $12\sqrt{3} \text{ in.}^2$
26a. 72
b. 54
27a. 45
b. 67.5
28a. 40
b. 70
29a. 30
b. 75

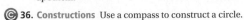

STEM **30. Satellites** One of the smallest space satellites ever developed has the shape of a pyramid. Each of the four faces of the pyramid is an equilateral triangle with sides about 13 cm long. What is the area of one equilateral triangular face of the satellite? Round your answer to the nearest whole number.

31. Think About a Plan The gazebo in the photo is built in the shape of a regular octagon. Each side is 8 ft long, and the enclosed area is 310.4 ft^2. What is the length of the apothem?
- How can you *draw a diagram* to help you solve the problem?
- How can you use the area of a regular polygon formula?

32. A regular hexagon has perimeter 120 m. Find its area.

33. The area of a regular polygon is 36 $in.^2$. Find the length of a side if the polygon has the given number of sides. Round your answer to the nearest tenth.
a. 3 **b.** 4 **c.** 6
d. Estimation Suppose the polygon is a pentagon. What would you expect the length of a side to be? Explain.

34. A portion of a regular decagon has radii and an apothem drawn. Find the measure of each numbered angle.

35. Writing Explain why the radius of a regular polygon is greater than the apothem.

36. Constructions Use a compass to construct a circle.
a. Construct two perpendicular diameters of the circle.
b. Construct diameters that bisect each of the four right angles.
c. Connect the consecutive points where the diameters intersect the circle. What regular polygon have you constructed?
d. Reasoning How can a circle help you construct a regular hexagon?

Find the perimeter and area of each regular polygon. Round to the nearest tenth, as necessary.

37. a square with vertices at $(-1, 0)$, $(2, 3)$, $(5, 0)$ and $(2, -3)$

38. an equilateral triangle with two vertices at $(-4, 1)$ and $(4, 7)$

39. a hexagon with two adjacent vertices at $(-2, 1)$ and $(1, 2)$

40. To find the area of an equilateral triangle, you can use the formula $A = \frac{1}{2}bh$ or $A = \frac{1}{2}ap$. A third way to find the area of an equilateral triangle is to use the formula $A = \frac{1}{4}s^2\sqrt{3}$. Verify the formula $A = \frac{1}{4}s^2\sqrt{3}$ in two ways as follows:
a. Find the area of Figure 1 using the formula $A = \frac{1}{2}bh$.
b. Find the area of Figure 2 using the formula $A = \frac{1}{2}ap$.

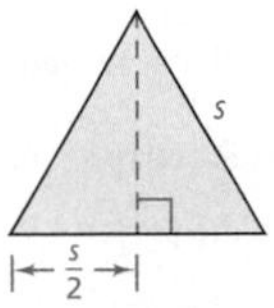

Figure 1

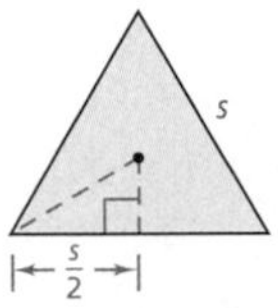

Figure 2

41. Proof For Problem 1 on page 629, write a proof that the apothem bisects the vertex angle of an isosceles triangle formed by two radii.

30. 73 cm^2
31. 9.7 ft
32. $600\sqrt{3}$ m^2 or about 1039.2 m^2
33a. 9.1 in.
b. 6 in.
c. 3.7 in.
d. Answers may vary. Sample: About 4 in.; the length of a side of a pentagon should be between 3.7 in. and 6 in.
34. $m\angle 1 = 36$, $m\angle 2 = 18$, $m\angle 3 = 72$
35. The apothem is one leg of a rt. △ and the radius is the hypotenuse.

36a–c.

regular octagon
d. Construct a 60° ∠ with the vertex at the center of the circle.
37. 17.0; 18
38. 30; 43.3
39. 19.0; 26.0

40a. $b = s$, $h = \frac{\sqrt{3}}{2}s$; $A = \frac{1}{2}bh = \frac{1}{2}s \cdot \frac{\sqrt{3}}{2s} = \frac{s^2\sqrt{3}}{4}$
b. $a = \frac{s\sqrt{3}}{6}$; $A = \frac{1}{2}ap = \frac{1}{2}\left(\frac{s\sqrt{3}}{6}\right)(3s) = \frac{s^2\sqrt{3}}{4}$
41. The apothem is ⊥ to a side of the pentagon. Two right ⚠s are formed with the radii of the pentagon. The ⚠s are ≅ by HL. So, the ⚠s formed by the apothem and radii are ≅ because corresp. parts of ≅ ⚠s are ≅. Therefore, the apothem bisects the vertex ∠.

Answers

Practice and Problem-Solving Exercises (continued)

42. For regular n-gon $ABCDE\ldots$, let P be the intersection of the bisectors of $\angle ABC$ and $\angle BCD$. $\overline{BC} \cong \overline{DC}$, $\angle BCP \cong \angle DCP$, and $\overline{CP} \cong \overline{CP}$, so $\triangle BCP \cong \triangle DCP$, and $\angle CBP \cong \angle CDP$ because corresp. parts of $\cong$ ⧌ are $\cong$. Since $\angle BCP$ is half the size of $\angle ABC$ and $\angle ABC \cong \angle CDE$, then $\angle CDP$ is half the size of $\angle CDE$. By a similar argument, P is on the bisector of each $\angle$ around the polygon. The smaller ⧌ formed by each of the $\angle$ bisectors are all $\cong$. By the Converse of the Isosc. $\triangle$ Thm., each of $\triangle APB$, $\triangle BPC$, $\triangle CPD$, etc., are isosc. with $\overline{AP} \cong \overline{BP} \cong \overline{CP}$, etc. Thus, P is equidistant from the polygon's vertices. So P is the center of the polygon and the $\angle$ bisectors are radii.

43a. (2.8, 2.8)

b. 5.6 units^2

c. 45 units^2

44. B

45. F

46. B

47. [2] $(2, 2\sqrt{3})$ and $(2, -2\sqrt{3})$, or equivalent decimal approximations (2, 3.464) and (2, −3.464); the length of each side of the $\triangle$ is 4 units. The third vertex must lie on the altitude of the triangle, which is a point on the line $x = 2$ and has x-coordinate 2. Using the Distance Formula, $\sqrt{(2-0)^2 + (y-0)^2} = 4$; $\sqrt{4 + y^2} = 4$; $4 + y^2 = 16$; $y^2 = 12$; $y = \pm\sqrt{12}$; $y = \pm 2\sqrt{3}$

[1] incomplete answer OR one incomplete statement

48. 46 m^2

49. 8 m

50. $P = 28$ in.; $A = 49$ in.^{2}

51. $P = 24$ m; $A = 32$ m^2

52. $P = 24$ cm; $A = 24$ cm^2

Challenge

42. (Proof) Prove that the bisectors of the angles of a regular polygon are concurrent and that they are, in fact, radii of the polygon. (*Hint:* For regular n-gon $ABCDE\ldots$, let P be the intersection of the bisectors of $\angle ABC$ and $\angle BCD$. Show that $\overline{DP}$ must be the bisector of $\angle CDE$.)

43. Coordinate Geometry A regular octagon with center at the origin and radius 4 is graphed in the coordinate plane.

a. Since V_2 lies on the line $y = x$, its x- and y-coordinates are equal. Use the Distance Formula to find the coordinates of V_2 to the nearest tenth.

b. Use the coordinates of V_2 and the formula $A = \frac{1}{2}bh$ to find the area of $\triangle V_1OV_2$ to the nearest tenth.

c. Use your answer to part (b) to find the area of the octagon to the nearest whole number.

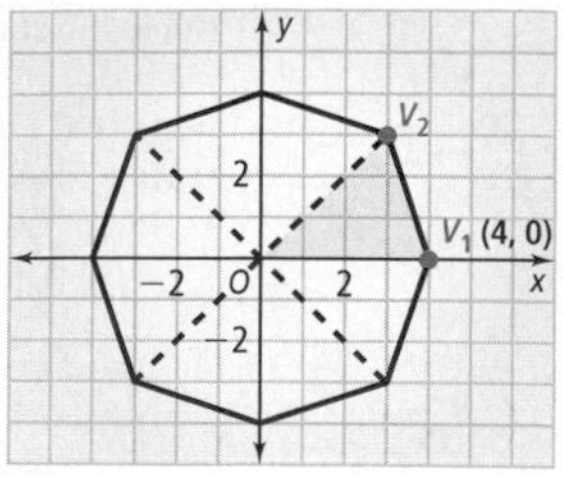

Standardized Test Prep

SAT/ACT

44. What is the area of a regular pentagon with an apothem of 25.1 mm and perimeter of 182 mm?

(A) 913.6 mm^2 (B) 2284.1 mm^2 (C) 3654.6 mm^2 (D) 4568.2 mm^2

45. What is the most precise name for a regular polygon with four right angles?

(F) square (G) parallelogram (H) trapezoid (I) rectangle

46. $\triangle ABC$ has coordinates $A(-2, 4)$, $B(3, 1)$, and $C(0, -2)$. If you reflect $\triangle ABC$ across the x-axis, what are the coordinates of the vertices of the image $\triangle A'B'C'$?

(A) $A'(2, 4)$, $B'(-3, 1)$, $C'(0, -2)$

(B) $A'(-2, -4)$, $B'(3, -1)$, $C'(0, 2)$

(C) $A'(4, -2)$, $B'(1, 3)$, $C'(-2, 0)$

(D) $A'(4, 2)$, $B'(1, -3)$, $C'(-2, 0)$

Short Response

47. An equilateral triangle on a coordinate grid has vertices at (0, 0) and (4, 0). What are the possible locations of the third vertex?

Mixed Review

48. What is the area of a kite with diagonals 8 m and 11.5 m? See Lesson 10-2.

49. The area of a trapezoid is 42 m^2. The trapezoid has a height of 7 m and one base of 4 m. What is the length of the other base?

Get Ready! To prepare for Lesson 10-4, do Exercises 50–52. See Lesson 1-8.

Find the perimeter and area of each figure.

50.

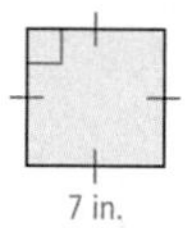

51.

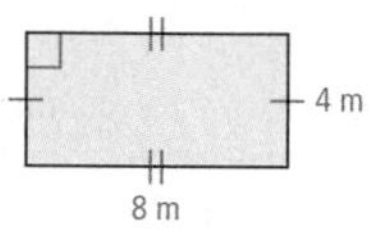

52.

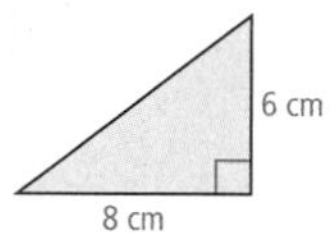

10-3 Lesson Resources

Differentiated Remediation

Instructional Support

Geometry Companion

Students can use the **Geometry Companion** worktext (4 pages) . . .

- New Vocabulary
- Key Concepts
- Got It for each Problem
- Lesson Check

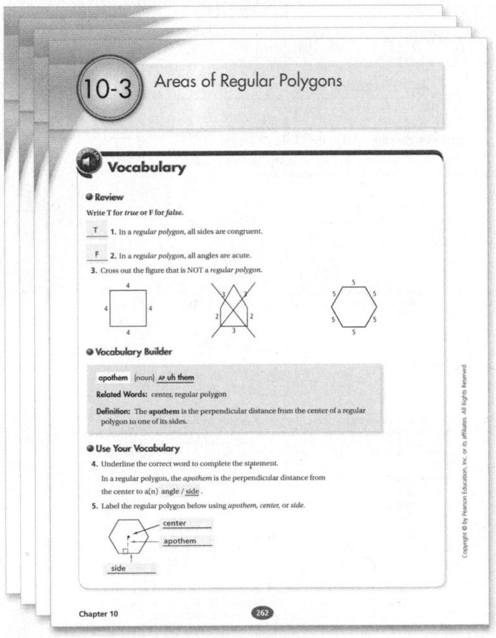

10-3 Areas of Regular Polygons

Vocabulary

Review

Write T for *true* or F for *false*.

T 1. In a *regular polygon*, all sides are congruent.

F 2. In a *regular polygon*, all angles are acute.

3. Cross out the figure that is NOT a *regular polygon*.

Vocabulary Builder

apothem (noun) AP uh them

Related Words: center, regular polygon

Definition: The **apothem** is the perpendicular distance from the center of a regular polygon to one of its sides.

Use Your Vocabulary

4. Underline the correct word to complete the statement.
In a regular polygon, the *apothem* is the perpendicular distance from the center to a(n) angle / side.

5. Label the regular polygon below using *apothem, center,* or *side*.

center
apothem
side

Copyright © by Pearson Education, Inc. or its affiliates. All Rights Reserved.

Chapter 10 262

ELL Support

Focus on Language Place the lesson on an overhead projector. Read aloud as students read along, pointing to each word as you read. As you come across vocabulary words or other key words such as circumscribe, regular, and apothem, think aloud as you determine their meanings. Invite students to rephrase sentences in their own words.

Write key words on the board while reading the lesson. Ask students for synonyms, or words with the same meaning. For example, a synonym for apothem may be inscribed radius. A synonym for perpendicular may be upright.

5 Assess & Remediate

Lesson Quiz

1. What is the area of the regular pentagon below?

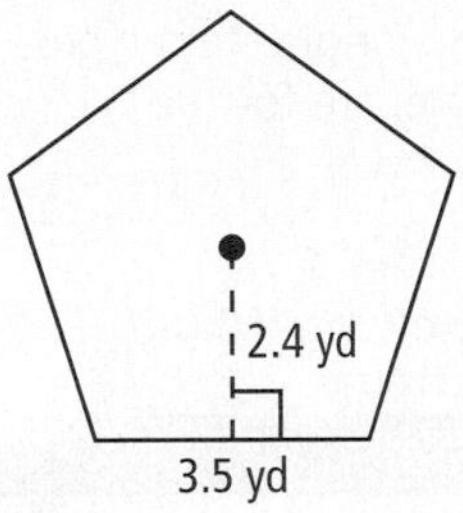

2. What is the length of the apothem of a regular hexagon with 10-cm sides? Round to the nearest tenth if necessary.

3. Do you UNDERSTAND? Geoff uses hexagonal tiles to create a tessellation pattern in his garden. What is the area of each tile? Round to the nearest whole number.

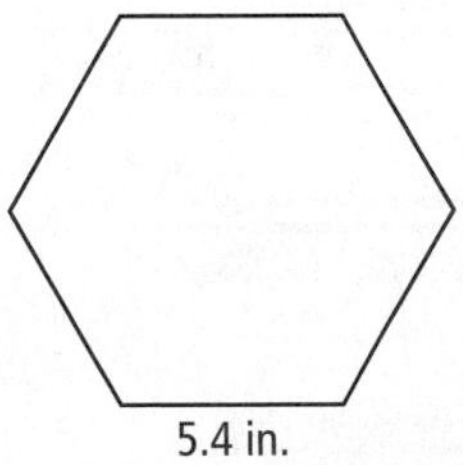

ANSWERS TO LESSON QUIZ

1. 21 yd^2

2. about 8.7 cm

3. about 76 in.^{2}

PRESCRIPTION FOR REMEDIATION

Use the student work on the Lesson Quiz to prescribe a differentiated review assignment.

Points	Differentiated Remediation
0–1	Intervention
2	On-level
3	Extension

PowerGeometry.com

5 Assess & Remediate

Assign the Lesson Quiz. Appropriate intervention, practice, or enrichment is automatically generated based on student performance.

Intervention

- **Reteaching** (2 pages) Provides reteaching and practice exercises for the key lesson concepts. Use with struggling students or absent students.
- **English Language Learner Support** Helps students develop and reinforce mathematical vocabulary and key concepts.

All-in-One Resources/Online

Reteaching

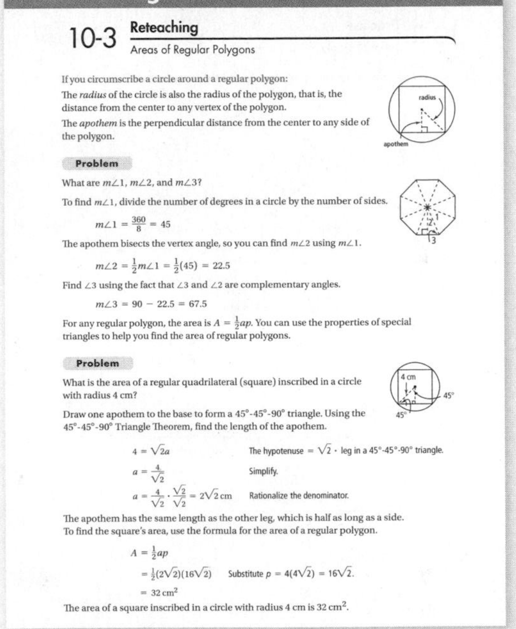

10-3 Reteaching
Areas of Regular Polygons

If you circumscribe a circle around a regular polygon:

The *radius* of the circle is also the radius of the polygon, that is, the distance from the center to any vertex of the polygon.

The *apothem* is the perpendicular distance from the center to any side of the polygon.

Problem

What are $m\angle 1$, $m\angle 2$, and $m\angle 3$?

To find $m\angle 1$, divide the number of degrees in a circle by the number of sides.

$m\angle 1 = \frac{360}{8} = 45$

The apothem bisects the vertex angle, so you can find $m\angle 2$ using $m\angle 1$.

$m\angle 2 = \frac{1}{2}m\angle 1 = \frac{1}{2}(45) = 22.5$

Find $\angle 3$ using the fact that $\angle 3$ and $\angle 2$ are complementary angles.

$m\angle 3 = 90 - 22.5 = 67.5$

For any regular polygon, the area is $A = \frac{1}{2}ap$. You can use the properties of special triangles to help you find the area of regular polygons.

Problem

What is the area of a regular quadrilateral (square) inscribed in a circle with radius 4 cm?

Draw one apothem to the base to form a 45°-45°-90° triangle. Using the 45°-45°-90° Triangle Theorem, find the length of the apothem.

$4 = \sqrt{2}a$ The hypotenuse $= \sqrt{2} \cdot$ leg in a 45°-45°-90° triangle.

$a = \frac{4}{\sqrt{2}}$ Simplify.

$a = \frac{4}{\sqrt{2}} \cdot \frac{\sqrt{2}}{\sqrt{2}} = 2\sqrt{2}$ cm Rationalize the denominator.

The apothem has the same length as the other leg, which is half as long as a side. To find the square's area, use the formula for the area of a regular polygon.

$A = \frac{1}{2}ap$

$= \frac{1}{2}(2\sqrt{2})(16\sqrt{2})$ Substitute $p = 4(4\sqrt{2}) = 16\sqrt{2}$.

$= 32\text{ cm}^2$

The area of a square inscribed in a circle with radius 4 cm is 32 cm^2.

All-in-One Resources/Online

English Language Learner Support

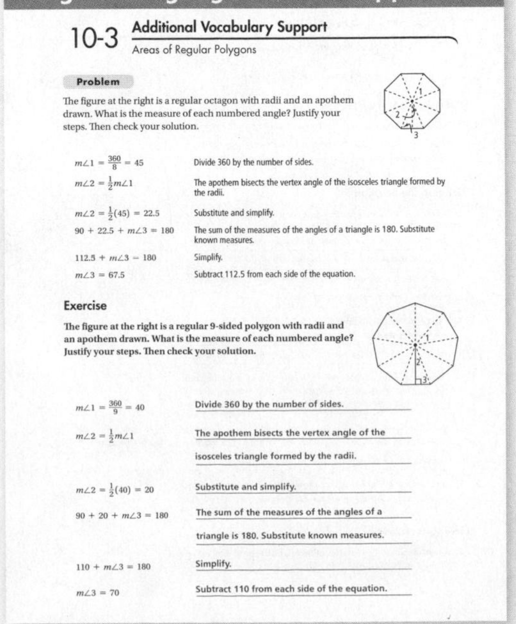

10-3 Additional Vocabulary Support
Areas of Regular Polygons

Problem

The figure at the right is a regular octagon with radii and an apothem drawn. What is the measure of each numbered angle? Justify your steps. Then check your solution.

$m\angle 1 = \frac{360}{8} = 45$	Divide 360 by the number of sides.
$m\angle 2 = \frac{1}{2}m\angle 1$	The apothem bisects the vertex angle of the isosceles triangle formed by the radii.
$m\angle 2 = \frac{1}{2}(45) = 22.5$	Substitute and simplify.
$90 + 22.5 + m\angle 3 = 180$	The sum of the measures of the angles of a triangle is 180. Substitute known measures.
$112.5 + m\angle 3 = 180$	Simplify.
$m\angle 3 = 67.5$	Subtract 112.5 from each side of the equation.

Exercise

The figure at the right is a regular 9-sided polygon with radii and an apothem drawn. What is the measure of each numbered angle? Justify your steps. Then check your solution.

$m\angle 1 = \frac{360}{9} = 40$	Divide 360 by the number of sides.
$m\angle 2 = \frac{1}{2}m\angle 1$	The apothem bisects the vertex angle of the isosceles triangle formed by the radii.
$m\angle 2 = \frac{1}{2}(40) = 20$	Substitute and simplify.
$90 + 20 + m\angle 3 = 180$	The sum of the measures of the angles of a triangle is 180. Substitute known measures.
$110 + m\angle 3 = 180$	Simplify.
$m\angle 3 = 70$	Subtract 110 from each side of the equation.

Differentiated Remediation *continued*

On-Level

- **Practice** (2 pages) Provides extra practice for each lesson. For simpler practice exercises, use the Form K Practice pages found in the All-in-One Teaching Resources and online.
- **Think About a Plan** Helps students develop specific problem-solving skills and strategies by providing scaffolded guiding questions.
- **Standardized Test Prep** Focuses on all major exercises, all major question types, and helps students prepare for the high-stakes assessments.

Extension

- **Enrichment** Provides students with interesting problems and activities that extend the concepts of the lesson.
- **Activities, Games, and Puzzles** Worksheets that can be used for concepts development, enrichment, and for fun!

Practice and Problem Solving Wkbk/ All-in-One Resources/Online

Practice page 1

10-3 **Practice** *Form G*
Areas of Regular Polygons

Each regular polygon has radii and apothem as shown. Find the measure of each numbered angle.

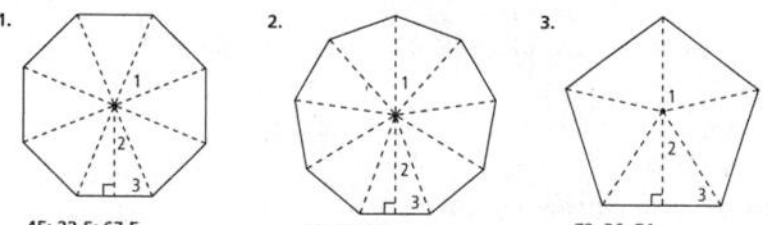

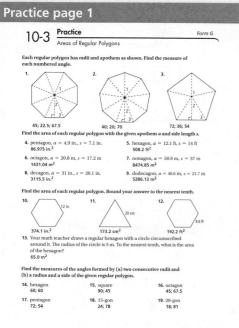

1. 45; 22.5; 67.5
2. 40; 20; 70
3. 72; 36; 54

Find the area of each regular polygon with the given apothem a and side length s.

4. pentagon, $a = 4.9$ in., $s = 7.1$ in. **86.975 in.^{2}**
5. hexagon, $a = 12.1$ ft, $s = 14$ ft **508.2 ft^2**
6. octagon, $a = 20.8$ m, $s = 17.2$ m **1431.04 m^2**
7. nonagon, $a = 50.9$ m, $s = 37$ m **8474.85 m^2**
8. decagon, $a = 31$ in., $s = 20.1$ in. **3115.5 in.^{2}**
9. dodecagon, $a = 40.6$ m, $s = 21.7$ m **5286.12 m^2**

Find the area of each regular polygon. Round your answer to the nearest tenth.

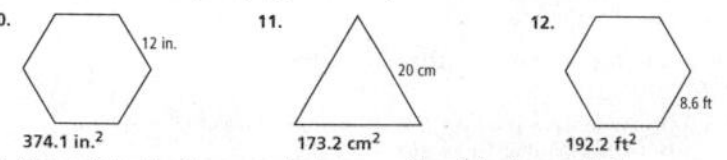

10. 374.1 in.^{2}
11. 173.2 cm^2
12. 192.2 ft^2

13. Your math teacher draws a regular hexagon with a circle circumscribed around it. The radius of the circle is 5 m. To the nearest tenth, what is the area of the hexagon? **65.0 m^2**

Find the measures of the angles formed by (a) two consecutive radii and (b) a radius and a side of the given regular polygon.

14. hexagon **60; 60**
15. square **90; 45**
16. octagon **45; 67.5**
17. pentagon **72; 54**
18. 15-gon **24; 78**
19. 20-gon **18; 81**

Practice and Problem Solving Wkbk/ All-in-One Resources/Online

Practice page 2

10-3 **Practice** (continued) *Form G*
Areas of Regular Polygons

Find the area of each regular polygon with the given radius or apothem. If your answer is not an integer, leave it in simplest radical form.

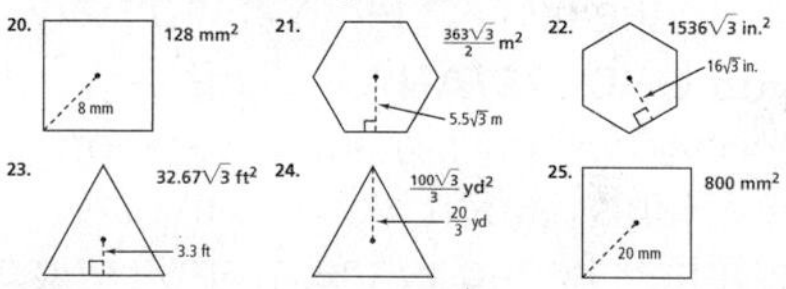

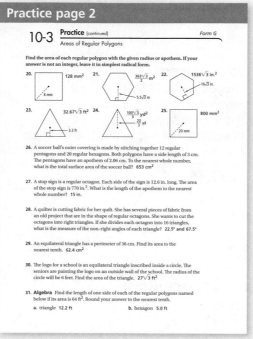

20. 128 mm^2
21. $\frac{363\sqrt{3}}{2}$ m^2
22. $1536\sqrt{3}$ in.^{2}
23. $32.67\sqrt{3}$ ft^2
24. $\frac{100\sqrt{3}}{3}$ yd^2
25. 800 mm^2

26. A soccer ball's outer covering is made by stitching together 12 regular pentagons and 20 regular hexagons. Both polygons have a side length of 3 cm. The pentagons have an apothem of 2.06 cm. To the nearest whole number, what is the total surface area of the soccer ball? **653 cm^2**
27. A stop sign is a regular octagon. Each side of the sign is 12.6 in. long. The area of the stop sign is 770 in.^{2}. What is the length of the apothem to the nearest whole number? **15 in.**
28. A quilter is cutting fabric for her quilt. She has several pieces of fabric from an old project that are in the shape of regular octagons. She wants to cut the octagons into right triangles. If she divides each octagon into 16 triangles, what is the measure of the non-right angles of each triangle? **22.5° and 67.5°**
29. An equilateral triangle has a perimeter of 36 cm. Find its area to the nearest tenth. **62.4 cm^2**
30. The logo for a school is an equilateral triangle inscribed inside a circle. The seniors are painting the logo on an outside wall of the school. The radius of the circle will be 6 feet. Find the area of the triangle. **$27\sqrt{3}$ ft^2**
31. **Algebra** Find the length of one side of each of the regular polygons named below if its area is 64 ft^2. Round your answer to the nearest tenth.
    a. triangle **12.2 ft**
    b. hexagon **5.0 ft**

All-in-One Resources/Online

Enrichment

10-3 **Enrichment**
Areas of Regular Polygons

Approximating the Area of the Circle

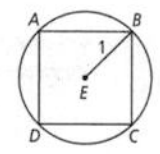

1. In the figure at the right, use square *ABCD* to approximate the area of ⊙*E*. Find the area using the formula $A = \frac{1}{2}ap$. **2 units^2**
2. Both the square and circle have radius 1. How much longer is the circumference of the circle than the perimeter of the square? How much greater is the area of the circle than the area of the square?
   $2\pi - 4\sqrt{2} \approx 0.626331$ unit; $\pi - 2 \approx 1.14159265$ units^2

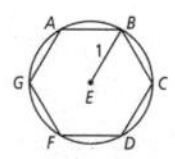

3. In the figure at the right, use hexagon *ABCDFG* to approximate the area of ⊙*E*. Find the area using the formula $A = \frac{1}{2}ap$. **2.598 units^2**
4. Both the hexagon and circle have radius 1. How much longer is the circumference of the circle than the perimeter of the hexagon? How much greater is the area of the circle than the area of the hexagon?
   $2\pi - 6 \approx 0.2831853$ unit; $\pi - \frac{3\sqrt{3}}{2} \approx 0.5435$ unit^2
5. Why are the perimeters and areas of both polygons less than the values for a circle with the same radius?
   Answers may vary. Sample: The polygons are inscribed inside the circle, so the perimeters must be less than the circumference and the areas must be less than the circle's area.
6. Predict what will happen if you use a 20-gon to approximate the area of a circle.
   Answers may vary. Sample: The area will be greater than 3 but less than π units^2.
7. Compare the area formula for polygons to the area formula for a circle, πr^2. What do the formulas have in common? How do they differ? Why does one formula have a factor of $\frac{1}{2}$? Why does one formula have a variable that is squared?
   Answers may vary. Sample: The variable a in the formula $\frac{1}{2}ap$ corresponds to r in the formula πr^2; half the circumference, πr, corresponds to $\frac{1}{2}p$, so the factor $\frac{1}{2}$ does not appear in the circle's area formula; the variable r is squared because it appears twice.

Practice and Problem Solving Wkbk/ All-in-One Resources/Online

Think About a Plan

10-3 **Think About a Plan**
Areas of Regular Polygons

A regular hexagon has perimeter 120 m. Find its area.

Understanding the Problem

1. What is the formula for the area of a regular polygon? **$A = \frac{1}{2}ap$**
2. What information is given? What information do you need?
   The perimeter is given; you need to find the length of the apothem.

Planning the Solution

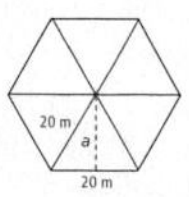

3. Divide the hexagon into six congruent triangles. What type of triangle are these? Explain how you know.
   Equilateral; answers may vary. Sample: In a regular hexagon, the central angle measures 360 ÷ 6 = 60. The base angles are congruent because the legs are congruent radii, so all three angles measure 60.
4. What is the length of the radius? Explain.
   20 m; each side of the hexagon is 120 ÷ 6 = 20 m. The sides of an equilateral triangle are congruent. Two sides are radii, so each radius is 20 m.
5. Draw an apothem. What type of triangle is formed by one radius, half of one side, and the apothem? What are the angles in this polygon?
   right triangle; 30°-60°-90°
6. What relationships exist among the sides of this type of triangle?
   The length of the shorter leg is half the length of the hypotenuse. The length of the longer leg is $\frac{\sqrt{3}}{2}$ times the length of the hypotenuse.
7. How can you find the length of the apothem? Find its length.
   Answers may vary. Sample: Multiply the radius by $\frac{\sqrt{3}}{2}$; $10\sqrt{3}$ m.

Getting an Answer

8. Substitute the values for the apothem and the perimeter into the formula and solve. **$A = \frac{1}{2}(10\sqrt{3})(120)$; $600\sqrt{3}$ m^2**

Practice and Problem Solving Wkbk/ All-in-One Resources/Online

Standardized Test Prep

10-3 **Standardized Test Prep**
Areas of Regular Polygons

Multiple Choice

For Exercises 1–6, choose the correct letter.

For Exercises 1 and 2, use the diagram at the right.

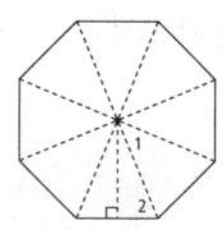

1. The figure at the right is a regular octagon with radii and an apothem drawn. What is $m\angle 1$? **B**
   Ⓐ 22.5 Ⓑ 45 Ⓒ 60 Ⓓ 67.5
2. What is $m\angle 2$? **I**
   Ⓕ 22.5 Ⓖ 45 Ⓗ 60 Ⓘ 67.5
3. A regular pentagon has an apothem of 3.2 m and an area of 37.2 m^2. What is the length of one side of the pentagon? **B**
   Ⓐ 3.96 m Ⓑ 4.65 m Ⓒ 11.875 m Ⓓ 23.75 m

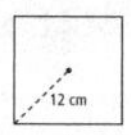

4. What is the area of the square at the right? **I**
   Ⓕ 16.97 cm^2 Ⓗ 144 cm^2 Ⓖ 72 cm^2 Ⓘ 288 cm^2
5. A regular hexagon has perimeter 60 in. What is the hexagon's area? **B**
   Ⓐ $75\sqrt{3}$ in.^{2} Ⓑ $150\sqrt{3}$ in.^{2} Ⓒ $300\sqrt{3}$ in.^{2} Ⓓ $600\sqrt{3}$ in.^{2}
6. For which regular polygon can you *not* use special triangles to find the apothem? **F**
   Ⓕ pentagon Ⓖ triangle Ⓗ square Ⓘ hexagon

Short Response

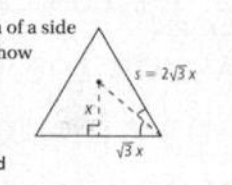

7. The area of an equilateral triangle is $108\sqrt{3}$ ft^2. What is the length of a side and the apothem in simplest radical form? Draw a diagram and show your work.
   [2] Substitute values into the area formula and solve: $A = \frac{1}{2}ap$; $108\sqrt{3} = \frac{1}{2}(x)(3)(2x\sqrt{3})$; $x = 6$; $s = 2\sqrt{3}(6) = 12\sqrt{3}$ ft; $a = 6$ ft; [1] correct procedure with some calculation errors [0] drawing and work not shown.

Online Teacher Resource Center

Activities, Games, and Puzzles

10-3 **Game: Pay Up!**
Areas of Regular Polygons

Materials

- Game board
- Colored squares, to use for game pieces
- Number cube
- Note pad, to keep score

Setup

Your teacher will divide the class into groups of 3 or 4 and will post answer cards around the room. Set your game piece on the square labeled START. Each player starts with 50 points and should roll the number cube to see who goes first. The player who rolls the greatest number goes first.

Game Play

On your turn, roll the number cube and move your game piece that number of spaces. If you land on an unclaimed figure, you and your opponents should each find the area of the polygon. Where possible, find the exact area. (Leave the radical in the answer.) When the apothem and a side of a figure are given, round your answers to the nearest tenth. Once everyone finds the area of the figure, locate the answer card that corresponds to the letter of your figure. Check the area of your figure on the back of the card. If you are correct, you may claim the figure by writing your initials *inside* it. If any of your opponents finds the correct answer, write his or her initials *above* the figure. Those opponents will not "pay rent" if they land on the figure in the future.

If you land on a figure that has been claimed by an opponent, you must "Pay Up!" The rent for each figure equals the number of sides it has. For example, if you land on a hexagon that is claimed by an opponent, you must subtract 6 from your score and add 6 to that opponent's score. If your initials are above the figure you land on, you do not pay. Either way, your turn is over.

If you land on a "Roll Again" space, roll the number cube again and proceed as you would on a regular turn. If you land on the "WIN 5" space, add 5 points to your score. If you land on "START," your turn is over.

Ending the Game

At the end of the game, complete the round so that each player has had the same number of turns. Find your point total. The player with the highest score wins.

10-4 Perimeters and Areas of Similar Figures

Common Core State Standards
Prepares for G-GMD.A.3 Use volume formulas for cylinders, pyramids, cones, and spheres to solve problems.
MP 1, MP 3, MP 4, MP 5, MP 7, MP 8

Objective To find the perimeters and areas of similar polygons

You already know that if you double the length and width of a rectangle, its area quadruples.

MATHEMATICAL PRACTICES

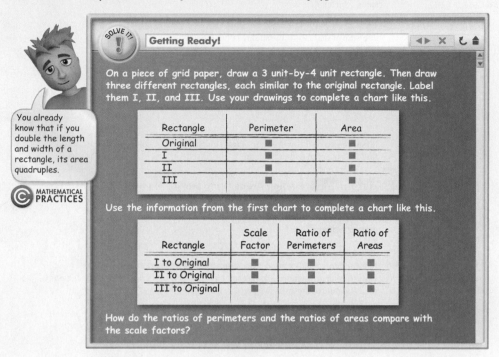

Solve It! Getting Ready!

On a piece of grid paper, draw a 3 unit-by-4 unit rectangle. Then draw three different rectangles, each similar to the original rectangle. Label them I, II, and III. Use your drawings to complete a chart like this.

Rectangle	Perimeter	Area
Original	■	■
I	■	■
II	■	■
III	■	■

Use the information from the first chart to complete a chart like this.

Rectangle	Scale Factor	Ratio of Perimeters	Ratio of Areas
I to Original	■	■	■
II to Original	■	■	■
III to Original	■	■	■

How do the ratios of perimeters and the ratios of areas compare with the scale factors?

In the Solve It, you compared the areas of similar figures.

Essential Understanding You can use ratios to compare the perimeters and areas of similar figures.

Take note

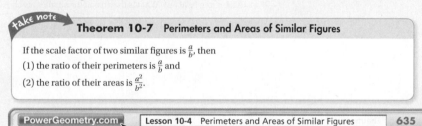

Theorem 10-7 Perimeters and Areas of Similar Figures

If the scale factor of two similar figures is $\frac{a}{b}$, then
(1) the ratio of their perimeters is $\frac{a}{b}$ and
(2) the ratio of their areas is $\frac{a^2}{b^2}$.

1 Interactive Learning

Solve It!

PURPOSE To determine the relationship between the perimeters and areas of similar figures
PROCESS Students may draw and calculate the perimeter and area of similar figures.

FACILITATE

Q How are the perimeters of the original rectangle and one that is enlarged by a scale factor of two related? **[The perimeter of the enlarged rectangle is 2 times the perimeter of the original rectangle.]**

Q How are the areas of the original rectangle and one that is enlarged by a scale factor of two related? **[The area of the enlarged rectangle is 4 times the area of the original rectangle.]**

Q Why do you think these numbers are related? **[The perimeter is a one-dimensional measurement so it increases by the same scale factor. The area is a two-dimensional measurement so it increases by the square of the scale factor.]**

ANSWER See Solve It in Answers on next page.
CONNECT THE MATH In Solve It, students should realize that the perimeters and areas of similar figures are related by the multiples of the scale factor. In the lesson, students find the perimeter and areas of similar figures and confirm the relationships revealed in Solve It.

2 Guided Instruction

Take Note

Have students check that the rectangles they drew in the Solve It satisfy this theorem.

10-4 Preparing to Teach

BIG ideas **Similarity**
Measurement

ESSENTIAL UNDERSTANDING

- Ratios can be used to compare the perimeters and areas of similar figures.

Math Background

Students should be able to answer the mathematical question: does perimeter always increase as the area increases? This question is typical of, and fundamental to, much of mathematics, in that it looks at relations between two quantities. It is valuable to know how one thing varies as the result of change in another. The study of functions examines this kind of relationship. In the field of fractals, the relationship between area and perimeter is less intuitive; it is possible for an infinitely large perimeter to surround a region of finite area.

Students know how to find the area of most types of geometric figures. In this lesson, students will apply these formulas to discover the relationship between the perimeters and areas of similar figures. If similar figures have a scale factor of $\frac{a}{b}$, then their perimeters are in the same ratio. The areas of the figures are in the ratio $\frac{a^2}{b^2}$.

Mathematical Practices

Look for and make use of structure. With knowledge of computing area, students will discern that both the ratio of perimeters and the ratio of areas of similar geometric figures are related to the scale factor.

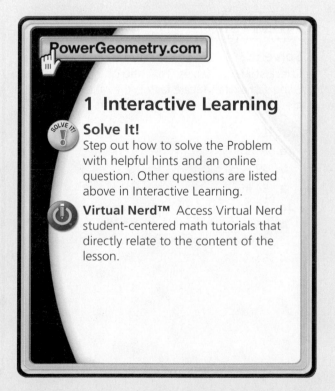

PowerGeometry.com

1 Interactive Learning

Solve It!
Step out how to solve the Problem with helpful hints and an online question. Other questions are listed above in Interactive Learning.

Virtual Nerd™ Access Virtual Nerd student-centered math tutorials that directly relate to the content of the lesson.

Problem 1

Q How is the ratio of the perimeters related to the ratio of the side lengths in similar figures? **[The ratios are equal.]**

Q How is the ratio of the areas related to the ratio of side lengths in similar figures? **[The ratio is the square of the ratio of side lengths.]**

Got It?

After students use the theorem to find the ratios in 1a and 1b, they can draw a diagram of any two similar polygons. Ask them to label the side lengths to represent the ratio 5 : 7. Students can calculate the perimeters and areas to check their work.

Problem 2

Q What is the ratio of the side lengths of the pentagons? $[\frac{4}{10} = \frac{2}{5}]$

Q What proportion can you write to find the area of the larger pentagon? $[\frac{4}{25} = \frac{27.5}{x}]$

Got It?

ERROR PREVENTION

Have students sketch the parallelograms. Have them write the proportion that they can use to find the area of the smaller parallelogram.

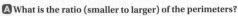

Plan

How do you find the scale factor?
Write the ratio of the lengths of two corresponding sides.

Problem 1 Finding Ratios in Similar Figures

The trapezoids at the right are similar. The ratio of the lengths of corresponding sides is $\frac{6}{9}$, or $\frac{2}{3}$.

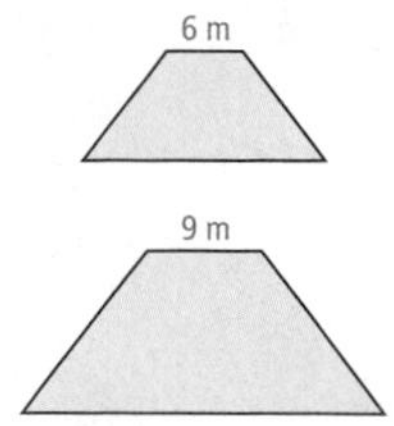

A What is the ratio (smaller to larger) of the perimeters?

The ratio of the perimeters is the same as the ratio of corresponding sides, which is $\frac{2}{3}$.

B What is the ratio (smaller to larger) of the areas?

The ratio of the areas is the square of the ratio of corresponding sides, which is $\frac{2^2}{3^2}$, or $\frac{4}{9}$.

Got It? 1. Two similar polygons have corresponding sides in the ratio 5 : 7.
a. What is the ratio (larger to smaller) of their perimeters?
b. What is the ratio (larger to smaller) of their areas?

When you know the area of one of two similar polygons, you can use a proportion to find the area of the other polygon.

Think

Can you eliminate any answer choices immediately?
Yes. Since the area of the smaller pentagon is 27.5 cm^2, you know that the area of the larger pentagon must be greater than that, so you can eliminate choice A.

Problem 2 Finding Areas Using Similar Figures

Multiple Choice The area of the smaller regular pentagon is about 27.5 cm^2. What is the best approximation for the area of the larger regular pentagon?

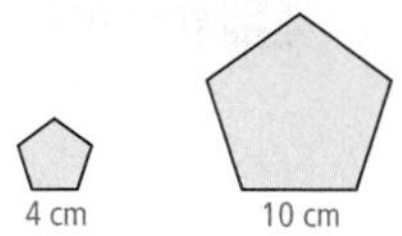

Ⓐ 11 cm^2 Ⓑ 69 cm^2 Ⓒ 172 cm^2 Ⓓ 275 cm^2

Regular pentagons are similar because all angles measure 108 and all sides in each pentagon are congruent. Here the ratio of corresponding side lengths is $\frac{4}{10}$, or $\frac{2}{5}$. The ratio of the areas is $\frac{2^2}{5^2}$, or $\frac{4}{25}$.

$\frac{4}{25} = \frac{27.5}{A}$	Write a proportion using the ratio of the areas.
$4A = 687.5$	Cross Products Property
$A = \frac{687.5}{4}$	Divide each side by 4.
$A = 171.875$	Simplify.

The area of the larger pentagon is about 172 cm^2. The correct answer is C.

Got It? 2. The scale factor of two similar parallelograms is $\frac{3}{4}$. The area of the larger parallelogram is 96 $in.^2$. What is the area of the smaller parallelogram?

Answers

Solve It!

Check students' tables. The ratio of the perimeters is the same as the scale factor; the ratio of the areas is the square of the scale factors.

Got It?

1a. 7 : 5
b. 49 : 25
2. 54 $in.^2$
3a. $6.94
b. In order for the two plots to be ~, the pairs of corresp. sides must have the same ratio.
4. $5\sqrt{5} : 3$

PowerGeometry.com

2 Guided Instruction

Each Problem is worked out and supported online.

Problem 1
Finding Ratios in Similar Figures

Problem 2
Finding Areas Using Similar Figures

Problem 3
Applying Area Ratios

Problem 4
Finding Perimeter Ratios

Support in Geometry Companion
- Vocabulary
- Key Concepts
- Got It?

Problem 3 Applying Area Ratios

Think
Do you need to know the shapes of the two plots of land?
No. As long as the plots are similar, you can compare their areas using their scale factor.

Agriculture During the summer, a group of high school students cultivated a plot of city land and harvested 13 bushels of vegetables that they donated to a food pantry. Next summer, the city will let them use a larger, similar plot of land. In the new plot, each dimension is 2.5 times the corresponding dimension of the original plot. How many bushels can the students expect to harvest next year?

The ratio of the dimensions is 2.5 : 1. So, the ratio of the areas is $(2.5)^2 : 1^2$, or 6.25 : 1. With 6.25 times as much land next year, the students can expect to harvest 6.25(13), or about 81, bushels.

Got It? 3. a. The scale factor of the dimensions of two similar pieces of window glass is 3 : 5. The smaller piece costs $2.50. How much should the larger piece cost?

b. Reasoning In Problem 3, why is it important that *each* dimension is 2.5 times the corresponding dimension of the original plot? Explain.

When you know the ratio of the areas of two similar figures, you can work backward to find the ratio of their perimeters.

Problem 4 Finding Perimeter Ratios

The triangles at the right are similar. What is the scale factor? What is the ratio of their perimeters?

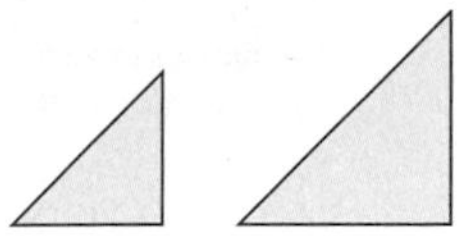

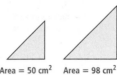

Area = 50 cm²   Area = 98 cm²

Know
The areas of the two similar triangles

Need
The scale factor

Plan
Write a proportion using the ratios of the areas.

$\frac{a^2}{b^2} = \frac{50}{98}$ Use $a^2 : b^2$ for the ratio of the areas.

$\frac{a^2}{b^2} = \frac{25}{49}$ Simplify.

$\frac{a}{b} = \frac{5}{7}$ Take the positive square root of each side.

The ratio of the perimeters equals the scale factor 5 : 7.

Got It? 4. The areas of two similar rectangles are 1875 ft^2 and 135 ft^2. What is the ratio of their perimeters?

Problem 3

Q What is the ratio of side lengths between the two plots? **[2.5 : 1]**

Q What is the ratio of the areas of the two plots? **[$2.5^2 : 1^2 = 6.25 : 1$]**

Q How can you use this to find the number of bushels of vegetables they can expect to harvest? **[Multiply this year's harvest by the ratio of areas.]**

Got It?

Make sure students realize that the cost of the glass would be based on the area of the piece. They should set up a proportion to find the price of the larger piece of glass.

Problem 4

Q What is the ratio of areas between the two triangles? **[$\frac{50}{98}$]**

Q How can you find the scale factor of the two triangles? **[Simplify and take the square root of the ratio of areas.]**

Q How is the ratio of the perimeters related to the scale factor? **[They are equal.]**

Got It?

Have students set up a ratio of the two areas.

Q What is the square root of the numerator? **[approximately 43.3 ft]**

Q What is the square root of the denominator? **[11.6 ft]**

Additional Problems

1. The rectangles below are similar. The ratio of the lengths of corresponding sides is $\frac{6}{8}$ or $\frac{3}{4}$.

a. What is the ratio (smaller to larger) of the perimeters?

b. What is the ratio (smaller to larger) of the areas?

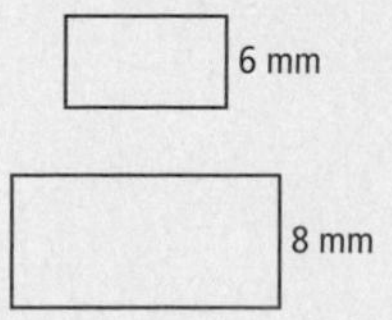

ANSWER a. $\frac{3}{4}$ **b.** $\frac{9}{16}$

2. The area of the smaller regular hexagon is about 127.3 ft^2. What is the best approximation for the area of the larger regular hexagon?

A. 181.9 ft^2
B. 224.7 ft^2
C. 259.8 ft^2
D. 278.5 ft^2

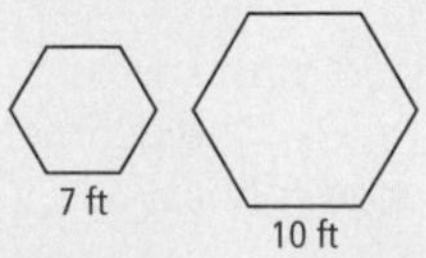

ANSWER C

3. It will cost Maria $150 to have carpet installed in a room that measures 12 ft by 10 ft. At this rate, how much would it cost her to have carpet installed in a similarly shaped family room with the larger dimension 18 ft?

ANSWER $337.50

4. The areas of two similar figures are 32 m^2 and 50 m^2. What is the scale factor? What is the ratio of their perimeters?

ANSWER 4 : 5; 4 : 5

3 Lesson Check

Do you know HOW?

- If students have difficulty with Exercise 2, then have them review Problem 2 to find out how to write the ratio of the side lengths, the ratio of the areas, and a proportion.

Do you UNDERSTAND?

- If students have difficulty with Exercise 6, then have them review Problem 4 to see the reasoning process of using the known information to find the unknown value.

Close

Q If the scale factor of two similar figures is $\frac{x}{y}$, what is the ratio of the perimeters of the figure? **[$\frac{x}{y}$]**

Q If the scale factor of two similar figures is $\frac{c}{d}$, what is the ratio of the areas of the figures? **[$\frac{c^2}{d^2}$]**

Lesson Check

Do you know HOW?

The figures in each pair are similar. What is the ratio of the perimeters and the ratio of the areas?

1.

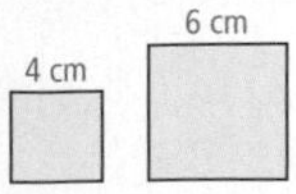

2.

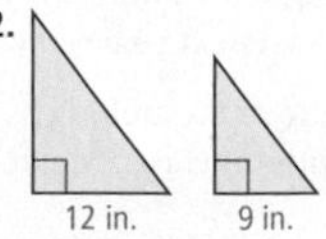

3. In Exercise 2, if the area of the smaller triangle is about 39 ft^2, what is the area of the larger triangle to the nearest tenth?

4. The areas of two similar rhombuses are 48 m^2 and 128 m^2. What is the ratio of their perimeters?

Do you UNDERSTAND? MATHEMATICAL PRACTICES

5. Reasoning How does the ratio of the areas of two similar figures compare to the ratio of their perimeters? Explain.

6. Reasoning The area of one rectangle is twice the area of another. What is the ratio of their perimeters? How do you know?

7. Error Analysis Your friend says that since the ratio of the perimeters of two polygons is $\frac{1}{2}$, the area of the smaller polygon must be one half the area of the larger polygon. What is wrong with this statement? Explain.

8. Compare and Contrast How is the relationship between the areas of two congruent figures different from the relationship between the areas of two similar figures?

Practice and Problem-Solving Exercises MATHEMATICAL PRACTICES

A Practice

The figures in each pair are similar. Compare the first figure to the second. Give the ratio of the perimeters and the ratio of the areas. See Problem 1.

9.

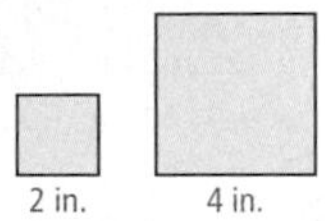

10.

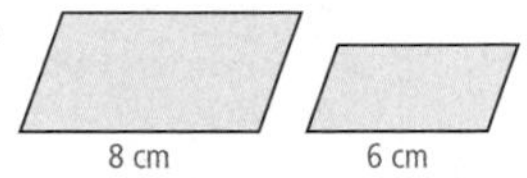

11.

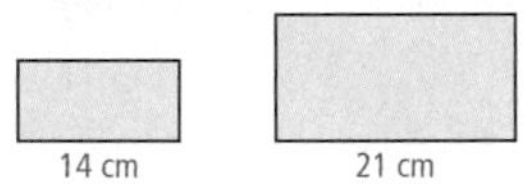

12.

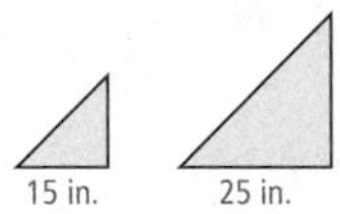

The figures in each pair are similar. The area of one figure is given. Find the area of the other figure to the nearest whole number. See Problem 2.

13.

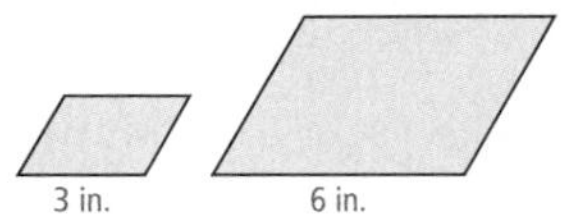

Area of smaller parallelogram = 6 in.^{2}

14.

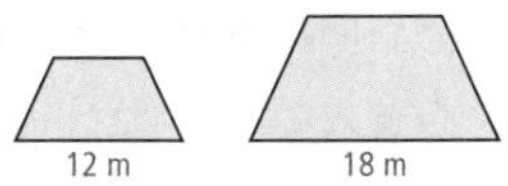

Area of larger trapezoid = 121 m^2

PowerGeometry.com

3 Lesson Check

For a digital lesson check, use the Got It questions.

Support in Geometry Companion

- Lesson Check

4 Practice

Assign homework to individual students or to an entire class.

Answers

Lesson Check

1. 2 : 3; 4 : 9

2. 4 : 3; 16 : 9

3. 69.3 ft^2

4. $\sqrt{6}$: 4

5. For two ~ figures, the ratio of their areas is the square of the ratio of the perimeters.

6. $\sqrt{2}$: 1; the ratio of the areas is 2 : 1, so the ratio of the perimeters is the square root of that ratio, which is $\sqrt{2}$: 1.

7. Answers may vary. Sample: The ratios of perimeters and areas of ~ figures are not = (unless the figures are ≅, in which case each ratio is 1).

8. The ratio of the areas of two ≅ figures is 1, while the ratio of the areas of two ~ figures is the square of the scale factor.

Practice and Problem-Solving Exercises

9. 1 : 2; 1 : 4

10. 4 : 3; 16 : 9

11. 2 : 3; 4 : 9

12. 3 : 5; 9 : 25

13. 24 in.^{2}

14. 54 m^2

15.

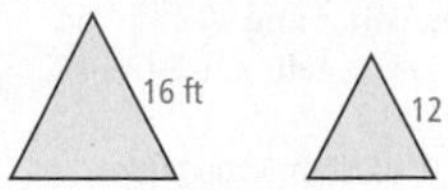

Area of larger triangle $= 105 \text{ ft}^2$

16.

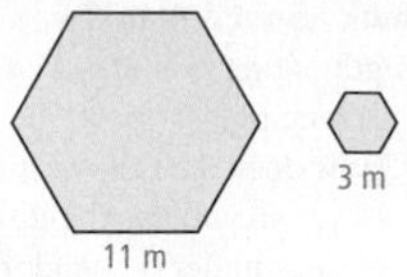

Area of smaller hexagon $= 23 \text{ m}^2$

17. Remodeling The scale factor of the dimensions of two similar wood floors is 4 : 3. It costs \$216 to refinish the smaller wood floor. At that rate, how much would it cost to refinish the larger wood floor?

See Problem 3.

18. Decorating An embroidered placemat costs \$3.95. An embroidered tablecloth is similar to the placemat, but four times as long and four times as wide. How much would you expect to pay for the tablecloth?

Find the scale factor and the ratio of perimeters for each pair of similar figures.

See Problem 4.

19. two regular octagons with areas 4 ft^2 and 16 ft^2

20. two triangles with areas 75 m^2 and 12 m^2

21. two trapezoids with areas 49 cm^2 and 9 cm^2

22. two parallelograms with areas 18 in.^2 and 32 in.^2

23. two equilateral triangles with areas $16\sqrt{3} \text{ ft}^2$ and $\sqrt{3} \text{ ft}^2$

24. two circles with areas $2\pi \text{ cm}^2$ and $200\pi \text{ cm}^2$

B Apply

The scale factor of two similar polygons is given. Find the ratio of their perimeters and the ratio of their areas.

25. 3 : 1 **26.** 2 : 5 **27.** $\frac{2}{3}$ **28.** $\frac{7}{4}$ **29.** 6 : 1

30. The area of a regular decagon is 50 cm^2. What is the area of a regular decagon with sides four times the sides of the smaller decagon?

(A) 200 cm^2 (B) 500 cm^2 (C) 800 cm^2 (D) 2000 cm^2

31. Error Analysis A reporter used the graphic below to show that the number of houses with more than two televisions had doubled in the past few years. Explain why this graphic is misleading.

4 Practice

ASSIGNMENT GUIDE

Basic: 9–24 all, 26–30 even, 31–33 all, 34–44 even
Average: 9–23 odd, 25–47
Advanced: 9–23 odd, 25–51
Standardized Test Prep: 52–55
Mixed Review: 56–62

Mathematical Practices are supported by exercises with red headings. Here are the Practices supported in this lesson:

MP 1: Make Sense of Problems Ex. 8
MP 1: Persevere in Solving Problems Ex. 32
MP 3: Construct Arguments Ex. 5, 6, 48–51
MP 3: Communicate Ex. 46
MP 3: Critique the Reasoning of Others Ex. 7, 31
MP 5: Use Appropriate Tools Ex. 41a
MP 8: Check for Reasonableness Ex. 41c, 47c

Applications exercises have blue headings. Exercises 17, 18, 40, and 47 support MP 4: Model.

STEM exercises focus on science or engineering applications.

EXERCISE 33: Use the Think About a Plan worksheet in the **Practice and Problem Solving Workbook** (also available in the Teaching Resources in print and online) to further support students' development in becoming independent learners.

HOMEWORK QUICK CHECK

To check students' understanding of key skills and concepts, go over Exercises 13, 17, 31, 32, and 33.

15. 59 ft^2
16. 309 m^2
17. \$384
18. \$63.20
19. 1 : 2; 1 : 2
20. 5 : 2; 5 : 2
21. 7 : 3; 7 : 3
22. 3 : 4; 3 : 4
23. 4 : 1; 4 : 1
24. 1 : 10; 1 : 10
25. 3 : 1; 9 : 1
26. 2 : 5; 4 : 25
27. 2 : 3; 4 : 9
28. 7 : 4; 49 : 16
29. 6 : 1; 36 : 1
30. C
31. While the ratio of lengths is 2 : 1, the ratio of areas is 4 : 1.

Answers

Practice and Problem-Solving Exercises (continued)

32. $2\frac{1}{4}$ in.-by-12 in.; 3 in.-by-16 in.

33. 252 m^2

34. $x = 2$ cm, $y = 3$ cm

35. $x = 2\sqrt{2}$ cm, $y = 3\sqrt{2}$ cm

36. $x = 4$ cm, $y = 6$ cm

37. $x = \frac{8\sqrt{3}}{3}$ cm, $y = 4\sqrt{3}$ cm

38. $x = 4\sqrt{2}$ cm, $y = 6\sqrt{2}$ cm

39. $x = 8$ cm, $y = 12$ cm

40. 0.3 cm^2

41a–b. Check students' work.

c. Estimates may vary. Sample: 205 m^2

42a. 5 : 2

b. 25 : 4

43a. 8 : 3

b. 64 : 9

44a. 2 : 1

b. 4 : 1

45a. $6\sqrt{3}$ cm^2

b. $54\sqrt{3}$ cm^2; $13.5\sqrt{3}$ cm^2; $96\sqrt{3}$ cm^2

46. Answers may vary. Sample: The proposed playground is more than adequate. The number of students has approximately doubled. The proposed playground would be four times larger than the original playground.

32. Think About a Plan Two similar rectangles have areas 27 in.^{2} and 48 in.^{2}. The length of one side of the larger rectangle is 16 in. What are the dimensions of both rectangles?

- How does the ratio of the similar rectangles compare to their scale factor?
- How can you use the dimensions of the larger rectangle to find the dimensions of the smaller rectangle?

33. The longer sides of a parallelogram are 5 m. The longer sides of a similar parallelogram are 15 m. The area of the smaller parallelogram is 28 m^2. What is the area of the larger parallelogram?

Algebra Find the values of x and y when the smaller triangle shown here has the given area.

34. 3 cm^2 **35.** 6 cm^2 **36.** 12 cm^2

37. 16 cm^2 **38.** 24 cm^2 **39.** 48 cm^2

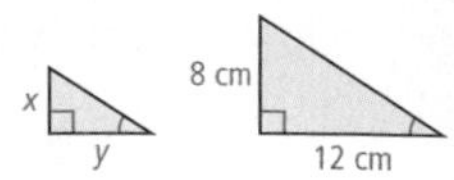

STEM **40. Medicine** For some medical imaging, the scale of the image is 3 : 1. That means that if an image is 3 cm long, the corresponding length on the person's body is 1 cm. Find the actual area of a lesion if its image has area 2.7 cm^2.

41. In $\triangle RST$, $RS = 20$ m, $ST = 25$ m, and $RT = 40$ m.

a. Open-Ended Choose a convenient scale. Then use a ruler and compass to draw $\triangle R'S'T' \sim \triangle RST$.

b. Constructions Construct an altitude of $\triangle R'S'T'$ and measure its length. Find the area of $\triangle R'S'T'$.

c. Estimation Estimate the area of $\triangle RST$.

Compare the blue figure to the red figure. Find the ratios of (a) their perimeters and (b) their areas.

42.

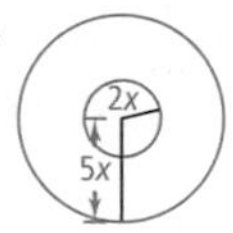

43.

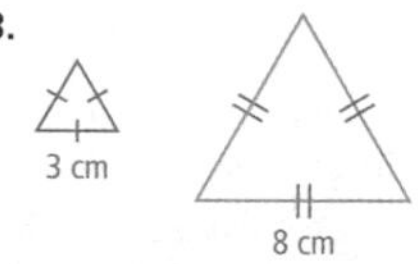

44.

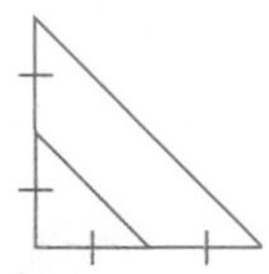

45. a. Find the area of a regular hexagon with sides 2 cm long. Leave your answer in simplest radical form.

b. Use your answer to part (a) and Theorem 10-7 to find the areas of the regular hexagons shown at the right.

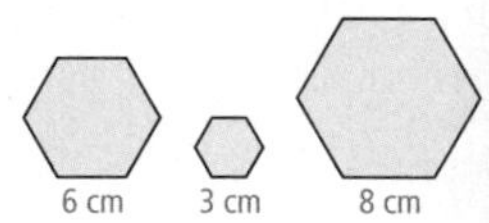

46. Writing The enrollment at an elementary school is going to increase from 200 students to 395 students. A parents' group is planning to increase the 100 ft-by-200 ft playground area to a larger area that is 200 ft by 400 ft. What would you tell the parents' group when they ask your opinion about whether the new playground will be large enough?

STEM **47. a. Surveying** A surveyor measured one side and two angles of a field, as shown in the diagram. Use a ruler and a protractor to draw a similar triangle.

b. Measure the sides and altitude of your triangle and find its perimeter and area.

c. Estimation Estimate the perimeter and area of the field.

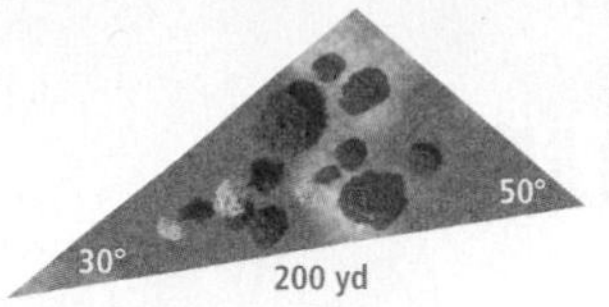

Challenge

Reasoning Complete each statement with *always, sometimes,* or *never*. Justify your answers.

48. Two similar rectangles with the same perimeter are __?__ congruent.

49. Two rectangles with the same area are __?__ similar.

50. Two rectangles with the same area and different perimeters are __?__ similar.

51. Similar figures __?__ have the same area.

Standardized Test Prep

GRIDDED RESPONSE

SAT/ACT

52. Two regular hexagons have sides in the ratio 3 : 5. The area of the smaller hexagon is 81 m^2. In square meters, what is the area of the larger hexagon?

53. What is the value of x in the diagram at the right?

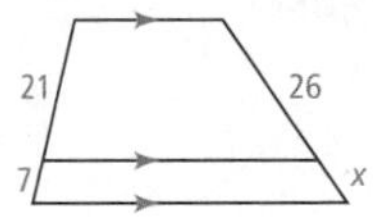

54. A trapezoid has base lengths of 9 in. and 4 in. and a height of 3 in. What is the area of the trapezoid in square inches?

55. In quadrilateral $ABCD$, $m\angle A = 62$, $m\angle B = 101$, and $m\angle C = 42$. What is $m\angle D$?

Mixed Review

Find the area of each regular polygon. See Lesson 10-3.

56. a square with a 5-cm radius

57. a pentagon with apothem 13.8 and side length 20

58. an octagon with apothem 12 and side length 10

59. An angle bisector divides the opposite side of a triangle into segments 4 cm and 6 cm long. A second side of the triangle is 8 cm long. What are all possible lengths for the third side of the triangle? See Lesson 7-5.

Get Ready! **To prepare for Lesson 10-5, do Exercises 60–62.**

Find the area of each regular polygon. See Lesson 10-3.

60.

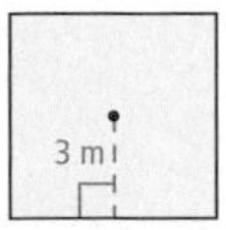

61.

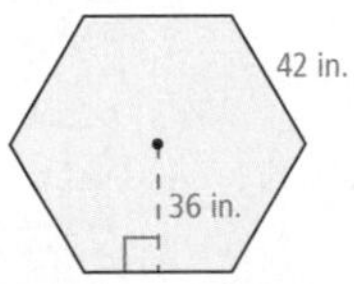

62.

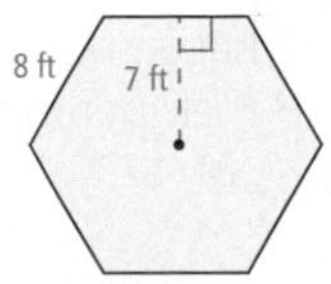

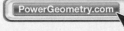

47a–c. Answers may vary. Samples are given.

a.

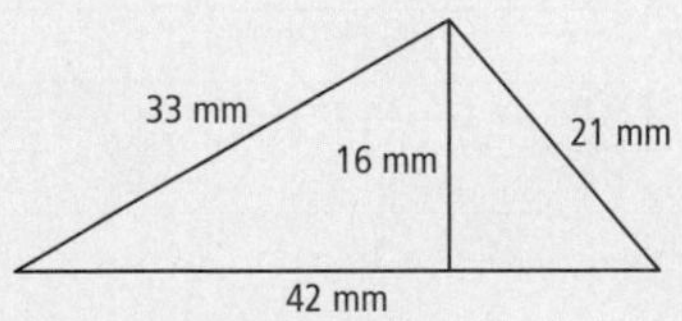

b. 96 mm; 336 mm^2

c. 457 yd; 7619 yd^2

48. Always; similar rectangles with = perimeters have a scale factor of 1 : 1, so they are ≅.

49. Sometimes; a 1 unit-by-8 unit rectangle and a 2 unit-by-4 unit rectangle have the same area, but they are not ~.

50. Never; if they were ≅ then both measures would be the same. If they were ~ but not ≅, their areas would not be =.

51. Sometimes; if they are ≅, they are ~ and have = areas.

52. 225

53. $\frac{26}{3}$

54. 19.5

55. 155

56. 50 cm^2

57. 690 $units^2$

58. 480 $units^2$

59. $5\frac{1}{3}$ cm, 12 cm

60. 36 m^2

61. 4536 $in.^2$

62. 168 ft^2

10-4 Lesson Resources

Differentiated Remediation

Instructional Support

Geometry Companion

Students can use the **Geometry Companion** worktext (4 pages) . . .

- New Vocabulary
- Key Concepts
- Got It for each Problem
- Lesson Check

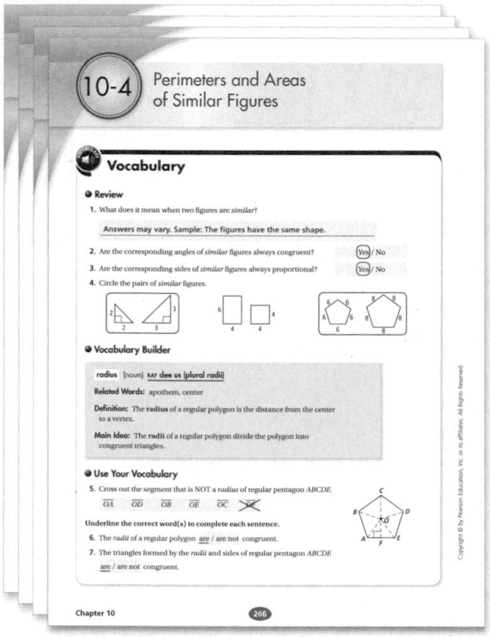

10-4 Perimeters and Areas of Similar Figures

Vocabulary

Review

1. What does it mean when two figures are *similar*?

Answers may vary. Sample: The figures have the same shape.

2. Are the corresponding angles of *similar* figures always congruent? Yes / No
3. Are the corresponding sides of *similar* figures always proportional? Yes / No
4. Circle the pairs of *similar* figures.

Vocabulary Builder

radius (noun) RAY dee us (plural radii)

Related Words: apothem, center

Definition: The **radius** of a regular polygon is the distance from the center to a vertex.

Main Idea: The **radii** of a regular polygon divide the polygon into congruent triangles.

Use Your Vocabulary

5. Cross out the segment that is NOT a *radius* of regular pentagon *ABCDE*.

$\overline{OA}$ $\overline{OD}$ $\overline{OB}$ $\overline{OE}$ $\overline{OC}$ $\overline{OF}$

Underline the correct word(s) to complete each sentence.

6. The *radii* of a regular polygon are / are not congruent.
7. The triangles formed by the *radii* and sides of regular pentagon *ABCDE* are / are not congruent.

Chapter 10 266

ELL Support

Assess Understanding Taneesha made an 80-in.-by-120-in. quilt by sewing together 54 pieces of fabric. Draw a picture to find about how many pieces of fabric she will need to make a quilt that measures 40 in. by 60 in.

Focus on Language Investigate the word *perimeter*. Analyze the word for the prefix and root. "Meter" is the measure. The prefix "peri-" means around. Ask students for the meaning (the outer boundary of a two-dimensional figure) and examples of objects in real life where perimeter may be used.

5 Assess & Remediate

Lesson Quiz

1. The triangles below are similar. The ratio of the lengths of corresponding sides is $\frac{4}{9}$.
   a. What is the ratio (smaller to larger) of the perimeters?
   b. What is the ratio (smaller to larger) of the areas?

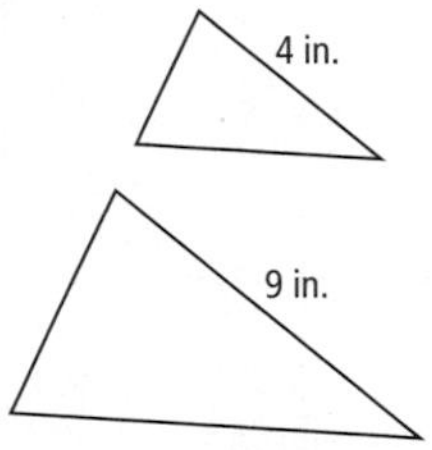

2. **Do you UNDERSTAND?** Mr. Williams is building a sandbox for his children. It costs \$228 for the sand if he builds a sandbox with dimensions 9 ft by 6 ft. How much will the sand cost if Mr. Williams decides to increase the size to $13\frac{1}{2}$ ft by 9 ft?
3. The areas of two similar figures are 20 cm^2 and 45 cm^2. What is the scale factor? What is the ratio of their perimeters?

ANSWERS TO LESSON QUIZ

1. a. $\frac{4}{9}$; b. $\frac{16}{81}$
2. \$512.73
3. 2 : 3; 2 : 3

PRESCRIPTION FOR REMEDIATION

Use the student work on the Lesson Quiz to prescribe a differentiated review assignment.

Points	Differentiated Remediation
0–1	Intervention
2	On-level
3	Extension

PowerGeometry.com

5 Assess & Remediate

Assign the Lesson Quiz. Appropriate intervention, practice, or enrichment is automatically generated based on student performance.

Intervention

- **Reteaching** (2 pages) Provides reteaching and practice exercises for the key lesson concepts. Use with struggling students or absent students.
- **English Language Learner Support** Helps students develop and reinforce mathematical vocabulary and key concepts.

All-in-One Resources/Online

Reteaching

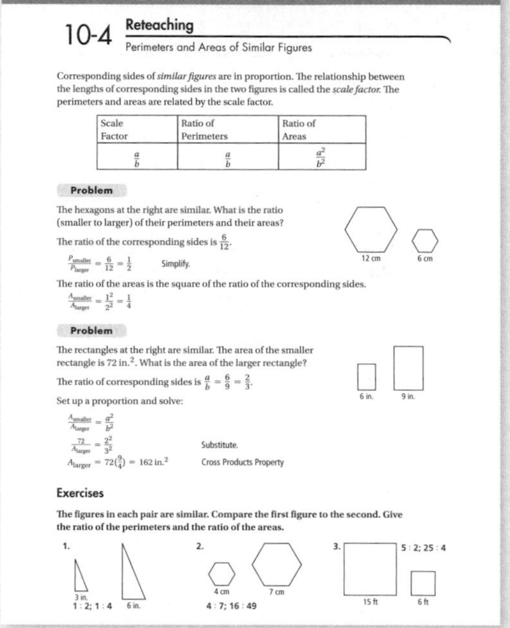

10-4 Reteaching

Perimeters and Areas of Similar Figures

Corresponding sides of *similar figures* are in proportion. The relationship between the lengths of corresponding sides in the two figures is called the *scale factor*. The perimeters and areas are related by the scale factor.

Scale Factor	Ratio of Perimeters	Ratio of Areas
$\frac{a}{b}$	$\frac{a}{b}$	$\frac{a^2}{b^2}$

Problem

The hexagons at the right are similar. What is the ratio (smaller to larger) of their perimeters and their areas?

The ratio of the corresponding sides is $\frac{6}{12}$.

$\frac{P_{smaller}}{P_{larger}} = \frac{6}{12} = \frac{1}{2}$ Simplify.

The ratio of the areas is the square of the ratio of the corresponding sides.

$\frac{A_{smaller}}{A_{larger}} = \frac{1^2}{2^2} = \frac{1}{4}$

12 cm 6 cm

Problem

The rectangles at the right are similar. The area of the smaller rectangle is 72 in.^{2}. What is the area of the larger rectangle?

The ratio of corresponding sides is $\frac{a}{b} = \frac{6}{9} = \frac{2}{3}$.

Set up a proportion and solve:

$\frac{A_{smaller}}{A_{larger}} = \frac{a^2}{b^2}$

$\frac{72}{A_{larger}} = \frac{2^2}{3^2}$ Substitute.

$A_{larger} = 72(\frac{9}{4}) = 162$ in.^{2} Cross Products Property

6 in. 9 in.

Exercises

The figures in each pair are similar. Compare the first figure to the second. Give the ratio of the perimeters and the ratio of the areas.

1. 3 in. 6 in. 1 : 2; 1 : 4
2. 4 cm 7 cm 4 : 7; 16 : 49
3. 15 ft 6 ft 5 : 2; 25 : 4

All-in-One Resources/Online

English Language Learner Support

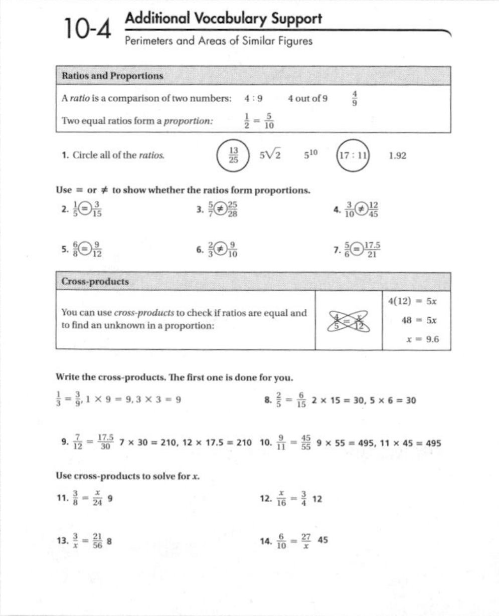

10-4 Additional Vocabulary Support

Perimeters and Areas of Similar Figures

Ratios and Proportions			
A *ratio* is a comparison of two numbers:	4 : 9	4 out of 9	$\frac{4}{9}$
Two equal ratios form a *proportion:*	$\frac{1}{2} = \frac{5}{10}$		

1. Circle all of the *ratios*. $\frac{13}{25}$ $5\sqrt{2}$ 5^{10} 17 : 11 1.92

Use = or ≠ to show whether the ratios form proportions.

2. $\frac{1}{5} = \frac{3}{15}$
3. $\frac{5}{7} \neq \frac{25}{28}$
4. $\frac{3}{10} \neq \frac{12}{45}$
5. $\frac{6}{8} = \frac{9}{12}$
6. $\frac{2}{3} \neq \frac{9}{10}$
7. $\frac{5}{6} = \frac{17.5}{21}$

Cross-products		
You can use *cross-products* to check if ratios are equal and to find an unknown in a proportion:	$\frac{4}{5} = \frac{x}{12}$	$4(12) = 5x$ $48 = 5x$ $x = 9.6$

Write the cross-products. The first one is done for you.

$\frac{1}{3} = \frac{3}{9}$, $1 \times 9 = 9$, $3 \times 3 = 9$

8. $\frac{2}{5} = \frac{6}{15}$ $2 \times 15 = 30$, $5 \times 6 = 30$
9. $\frac{7}{12} = \frac{17.5}{30}$ $7 \times 30 = 210$, $12 \times 17.5 = 210$
10. $\frac{9}{11} = \frac{45}{55}$ $9 \times 55 = 495$, $11 \times 45 = 495$

Use cross-products to solve for *x*.

11. $\frac{3}{8} = \frac{x}{24}$ 9
12. $\frac{x}{16} = \frac{3}{4}$ 12
13. $\frac{3}{x} = \frac{21}{56}$ 8
14. $\frac{6}{10} = \frac{27}{x}$ 45

Differentiated Remediation *continued*

On-Level

- **Practice** (2 pages) Provides extra practice for each lesson. For simpler practice exercises, use the Form K Practice pages found in the All-in-One Teaching Resources and online.
- **Think About a Plan** Helps students develop specific problem-solving skills and strategies by providing scaffolded guiding questions.
- **Standardized Test Prep** Focuses on all major exercises, all major question types, and helps students prepare for the high-stakes assessments.

Extension

- **Enrichment** Provides students with interesting problems and activities that extend the concepts of the lesson.
- **Activities, Games, and Puzzles** Worksheets that can be used for concepts development, enrichment, and for fun!

Practice and Problem Solving Wkbk/ All-in-One Resources/Online

Practice page 1

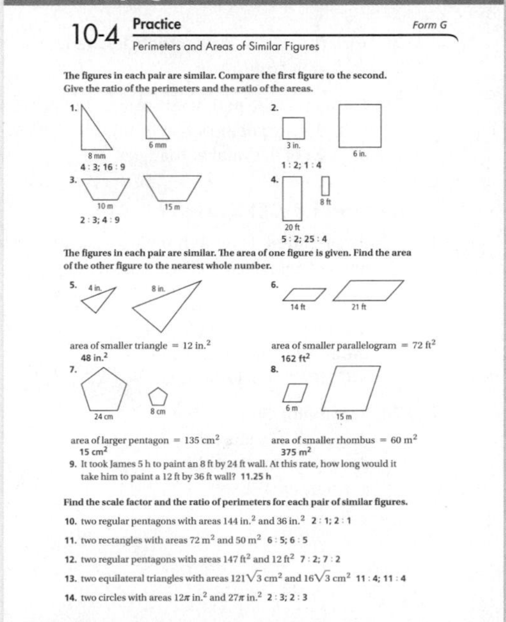

10-4 Practice — Form G
Perimeters and Areas of Similar Figures

The figures in each pair are similar. Compare the first figure to the second. Give the ratio of the perimeters and the ratio of the areas.

1. 4 : 3; 16 : 9
2. 1 : 2; 1 : 4
3. 2 : 3; 4 : 9
4. 5 : 2; 25 : 4

The figures in each pair are similar. The area of one figure is given. Find the area of the other figure to the nearest whole number.

5. area of smaller triangle = 12 in.² 48 in.²
6. area of smaller parallelogram = 72 ft² 162 ft²
7. area of larger pentagon = 135 cm² 15 cm²
8. area of smaller rhombus = 60 m² 375 m²
9. It took James 5 h to paint an 8 ft by 24 ft wall. At this rate, how long would it take him to paint a 12 ft by 36 ft wall? 11.25 h

Find the scale factor and the ratio of perimeters for each pair of similar figures.

10. two regular pentagons with areas 144 in.² and 36 in.² 2 : 1; 2 : 1
11. two rectangles with areas 72 m² and 50 m² 6 : 5; 6 : 5
12. two regular pentagons with areas 147 ft² and 12 ft² 7 : 2; 7 : 2
13. two equilateral triangles with areas $121\sqrt{3}$ cm² and $16\sqrt{3}$ cm² 11 : 4; 11 : 4
14. two circles with areas 12π in.² and 27π in.² 2 : 3; 2 : 3

Practice and Problem Solving Wkbk/ All-in-One Resources/Online

Practice page 2

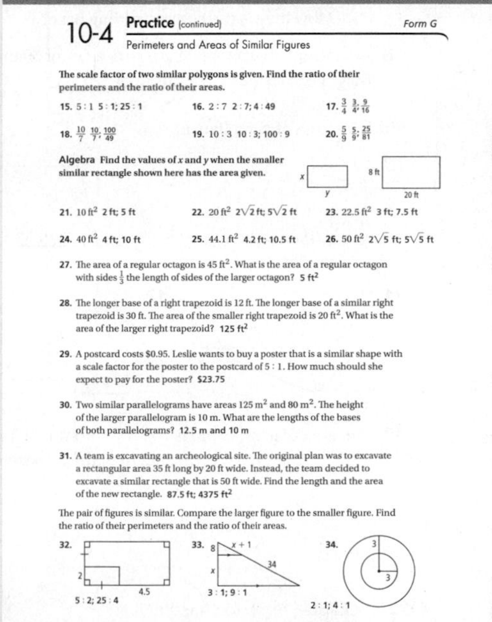

10-4 Practice (continued) — Form G
Perimeters and Areas of Similar Figures

The scale factor of two similar polygons is given. Find the ratio of their perimeters and the ratio of their areas.

15. 5 : 1 5 : 1; 25 : 1
16. 2 : 7 2 : 7; 4 : 49
17. $\frac{3}{4}$ $\frac{3}{4}$; $\frac{9}{16}$
18. $\frac{10}{7}$ $\frac{10}{7}$; $\frac{100}{49}$
19. 10 : 3 10 : 3; 100 : 9
20. $\frac{5}{9}$ $\frac{5}{9}$; $\frac{25}{81}$

Algebra Find the values of x and y when the smaller similar rectangle shown here has the area given.

21. 10 ft² 2 ft; 5 ft
22. 20 ft² $2\sqrt{2}$ ft; $5\sqrt{2}$ ft
23. 22.5 ft² 3 ft; 7.5 ft
24. 40 ft² 4 ft; 10 ft
25. 44.1 ft² 4.2 ft; 10.5 ft
26. 50 ft² $2\sqrt{5}$ ft; $5\sqrt{5}$ ft
27. The area of a regular octagon is 45 ft². What is the area of a regular octagon with sides $\frac{1}{3}$ the length of sides of the larger octagon? 5 ft²
28. The longer base of a right trapezoid is 12 ft. The longer base of a similar right trapezoid is 30 ft. The area of the smaller right trapezoid is 20 ft². What is the area of the larger right trapezoid? 125 ft²
29. A postcard costs $0.95. Leslie wants to buy a poster that is a similar shape with a scale factor for the poster to the postcard of 5 : 1. How much should she expect to pay for the poster? $23.75
30. Two similar parallelograms have areas 125 m² and 80 m². The height of the larger parallelogram is 10 m. What are the lengths of the bases of both parallelograms? 12.5 m and 10 m
31. A team is excavating an archeological site. The original plan was to excavate a rectangular area 35 ft long by 20 ft wide. Instead, the team decided to excavate a similar rectangle that is 50 ft wide. Find the length and the area of the new rectangle. 87.5 ft; 4375 ft²

The pair of figures is similar. Compare the larger figure to the smaller figure. Find the ratio of their perimeters and the ratio of their areas.

32. 5 : 2; 25 : 4
33. 3 : 1; 9 : 1
34. 2 : 1; 4 : 1

All-in-One Resources/Online

Enrichment

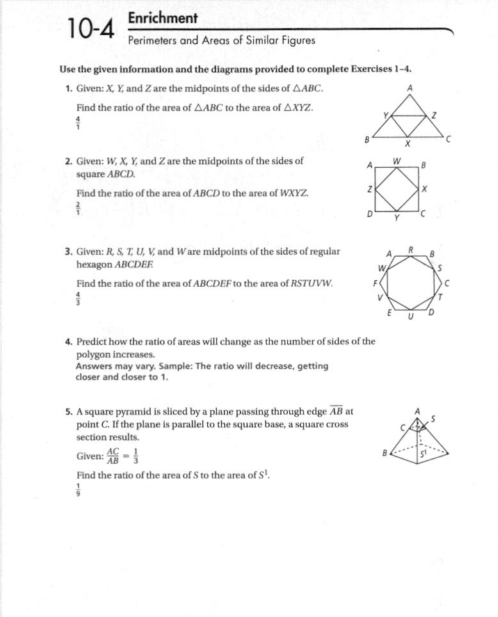

10-4 Enrichment
Perimeters and Areas of Similar Figures

Use the given information and the diagrams provided to complete Exercises 1–4.

1. Given: X, Y, and Z are the midpoints of the sides of △ABC. Find the ratio of the area of △ABC to the area of △XYZ. $\frac{4}{1}$
2. Given: W, X, Y, and Z are the midpoints of the sides of square ABCD. Find the ratio of the area of ABCD to the area of WXYZ. $\frac{2}{1}$
3. Given: R, S, T, U, V, and W are midpoints of the sides of regular hexagon ABCDEF. Find the ratio of the area of ABCDEF to the area of RSTUVW. $\frac{4}{3}$
4. Predict how the ratio of areas will change as the number of sides of the polygon increases. Answers may vary. Sample: The ratio will decrease, getting closer and closer to 1.
5. A square pyramid is sliced by a plane passing through edge $\overline{AB}$ at point C. If the plane is parallel to the square base, a square cross section results. Given: $\frac{AC}{AB} = \frac{1}{3}$ Find the ratio of the area of S to the area of S¹. $\frac{1}{9}$

Practice and Problem Solving Wkbk/ All-in-One Resources/Online

Think About a Plan

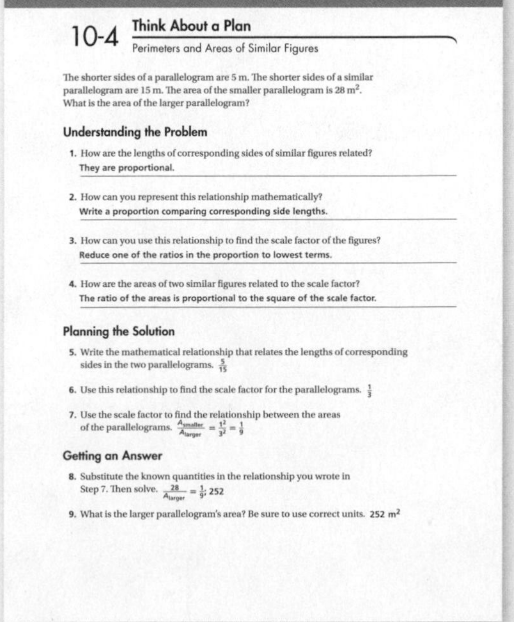

10-4 Think About a Plan
Perimeters and Areas of Similar Figures

The shorter sides of a parallelogram are 5 m. The shorter sides of a similar parallelogram are 15 m. The area of the smaller parallelogram is 28 m². What is the area of the larger parallelogram?

Understanding the Problem

1. How are the lengths of corresponding sides of similar figures related? They are proportional.
2. How can you represent this relationship mathematically? Write a proportion comparing corresponding side lengths.
3. How can you use this relationship to find the scale factor of the figures? Reduce one of the ratios in the proportion to lowest terms.
4. How are the areas of two similar figures related to the scale factor? The ratio of the areas is proportional to the square of the scale factor.

Planning the Solution

5. Write the mathematical relationship that relates the lengths of corresponding sides in the two parallelograms. $\frac{5}{15}$
6. Use this relationship to find the scale factor for the parallelograms. $\frac{1}{3}$
7. Use the scale factor to find the relationship between the areas of the parallelograms. $\frac{A_{smaller}}{A_{larger}} = \frac{1^2}{3^2} = \frac{1}{9}$

Getting an Answer

8. Substitute the known quantities in the relationship you wrote in Step 7. Then solve. $\frac{28}{A_{larger}} = \frac{1}{9}$; 252
9. What is the larger parallelogram's area? Be sure to use correct units. 252 m²

Practice and Problem Solving Wkbk/ All-in-One Resources/Online

Standardized Test Prep

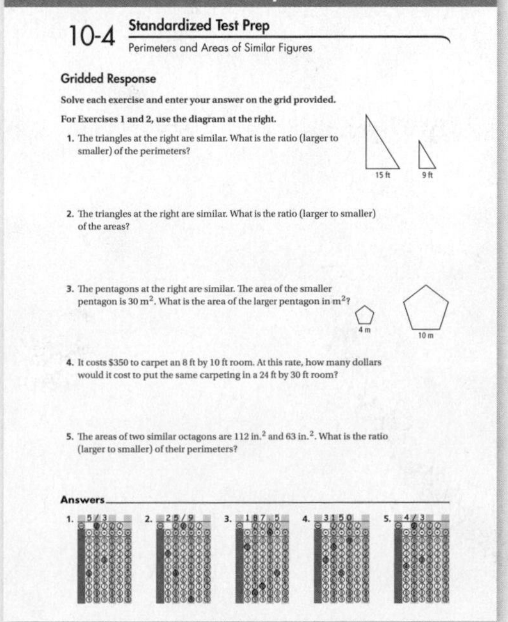

10-4 Standardized Test Prep
Perimeters and Areas of Similar Figures

Gridded Response

Solve each exercise and enter your answer on the grid provided.

For Exercises 1 and 2, use the diagram at the right.

1. The triangles at the right are similar. What is the ratio (larger to smaller) of the perimeters?
2. The triangles at the right are similar. What is the ratio (larger to smaller) of the areas?
3. The pentagons at the right are similar. The area of the smaller pentagon is 30 m². What is the area of the larger pentagon in m²?
4. It costs $350 to carpet an 8 ft by 10 ft room. At this rate, how many dollars would it cost to put the same carpeting in a 24 ft by 30 ft room?
5. The areas of two similar octagons are 112 in.² and 63 in.². What is the ratio (larger to smaller) of their perimeters?

Answers

1. 5/3 2. 25/9 3. 187.5 4. 3150 5. 4/3

Online Teacher Resource Center

Activities, Games, and Puzzles

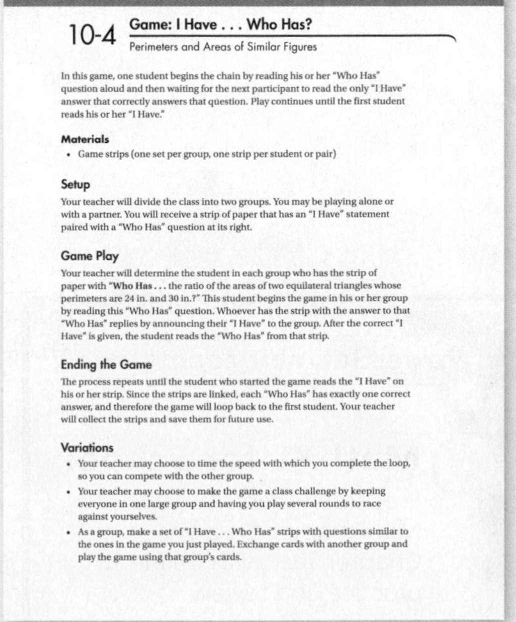

10-4 Game: I Have . . . Who Has?
Perimeters and Areas of Similar Figures

In this game, one student begins the chain by reading his or her "Who Has" question aloud and then waiting for the next participant to read the only "I Have" answer that correctly answers that question. Play continues until the first student reads his or her "I Have."

Materials

- Game strips (one set per group, one strip per student or pair)

Setup

Your teacher will divide the class into two groups. You may be playing alone or with a partner. You will receive a strip of paper that has an "I Have" statement paired with a "Who Has" question at its right.

Game Play

Your teacher will determine the student in each group who has the strip of paper with "Who Has . . . the ratio of the areas of two equilateral triangles whose perimeters are 24 in. and 30 in.?" This student begins the game in his or her group by reading this "Who Has" question. Whoever has the strip with the answer to that "Who Has" replies by announcing their "I Have" to the group. After the correct "I Have" is given, the student reads the "Who Has" from that strip.

Ending the Game

The process repeats until the student who started the game reads the "I Have" on his or her strip. Since the strips are linked, each "Who Has" has exactly one correct answer, and therefore the game will loop back to the first student. Your teacher will collect the strips and save them for future use.

Variations

- Your teacher may choose to time the speed with which you complete the loop, so you can compete with the other group.
- Your teacher may choose to make the game a class challenge by keeping everyone in one large group and having you play several rounds to race against yourselves.
- As a group, make a set of "I Have . . . Who Has" strips with questions similar to the ones in the game you just played. Exchange cards with another group and play the game using that group's cards.

Answers

Mid-Chapter Quiz

1. 84 $in.^2$
2. 112 cm^2
3. 48 m^2
4. 216 ft^2
5. 204 $in.^2$
6. 7 cm
7. 12 in.
8. 173 m^2
9. 135 $in.^2$
10. 54 m^2
11. 56 cm^2
12. 26 ft^2
13. 8 ft
14. 124.7 $in.^2$
15. 27.7 $in.^2$
16. 1110 cm^2
17. 65 ft^2
18. 60 in.
19. 45 $in.^2$
20. 6 : 5
21. 3 ft^2
22. Check students' work.
23. Method 1: Use the formula $A = \frac{1}{2}ap$. Method 2: Find the area of one equilateral △ and multiply it by 6; $162\sqrt{3}$ cm^2.
24. The area is quadrupled. If a kite has diagonals d_1 and d_2 then its area is $\frac{1}{2}d_1d_2$. If another kite has diagonals with lengths $2d_1$ and $2d_2$, then its area is $\frac{1}{2}(2d_1)(2d_2) = 4\left(\frac{1}{2}d_1d_2\right)$ or 4 times the area of the original kite.

10 Mid-Chapter Quiz

Do you know HOW?

Find the area of each figure.

1.
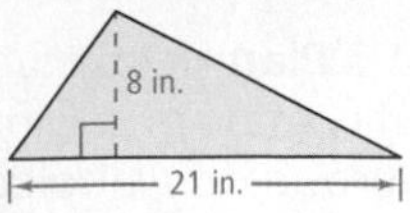

2.
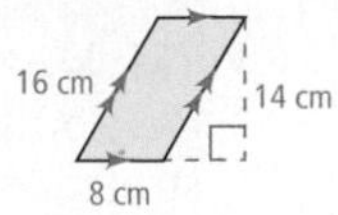

3.
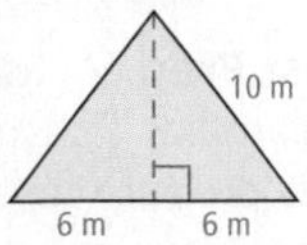

4.
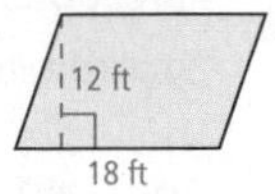

5. What is the area of a parallelogram with a base of 17 in. and a corresponding height of 12 in.?

6. If the base of a triangle is 10 cm, and its area is 35 cm^2, what is the height of the triangle?

7. The area of a parallelogram is 36 $in.^2$, and its height is 3 in. How long is the corresponding base?

8. An equilateral triangle has a perimeter of 60 m and a height of 17.3 m. What is its area?

Find the area of each figure.

9.
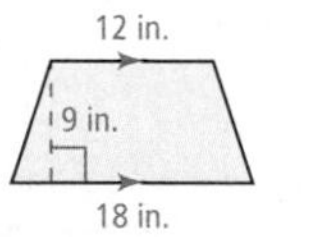

10.
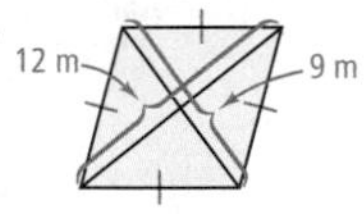

11.
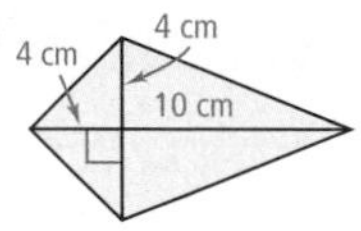

12.
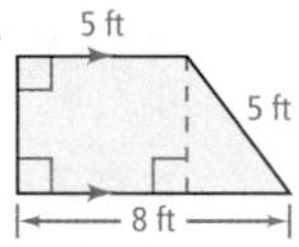

13. The area of a trapezoid is 100 ft^2. The sum of its two bases is 25 ft. What is the height of the trapezoid?

Find the area of each regular polygon. Round your answer to the nearest tenth.

14.
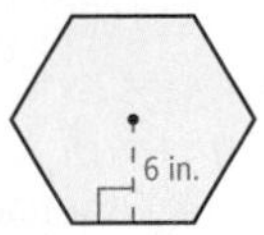

15.
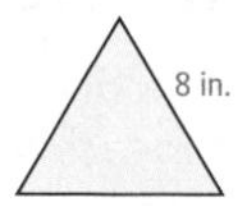

16. A regular octagon has sides 15 cm long. The apothem is 18.5 cm long. What is the area of the octagon?

17. The radius of a regular hexagon is 5 ft. What is the area of the hexagon to the nearest square foot?

The scale factor of $\triangle ABC$ to $\triangle DEF$ is 3 : 5. Fill in the missing information.

18. The perimeter of $\triangle ABC$ is 36 in. The perimeter of $\triangle DEF$ is __?__.

19. The area of $\triangle ABC$ is __?__. The area of $\triangle DEF$ is 125 $in.^2$.

20. The areas of two similar triangles are 1.44 and 1.00. Find their scale factor.

21. The ratio of the perimeters of two similar triangles is 1 : 3. The area of the larger triangle is 27 ft^2. What is the area of the smaller triangle?

Do you UNDERSTAND?

22. **Open-Ended** Draw a rhombus. Measure the lengths of the diagonals. What is the area?

23. **Writing** Describe two different methods for finding the area of regular hexagon *ABCDEF*. What is the area?

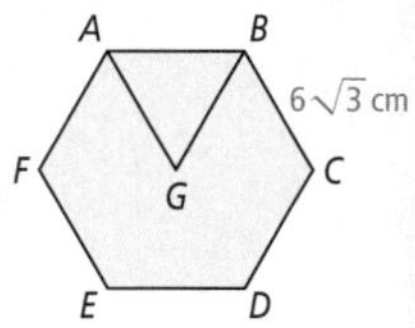

24. **Reasoning** Suppose the diagonals of a kite are doubled. How does this affect its area? Explain.

10-5 Trigonometry and Area

Common Core State Standards

G-SRT.D.9 Derive the formula $A = \frac{1}{2}ab \sin(C)$ for the area of a triangle by drawing an auxiliary line from a vertex perpendicular to the opposite side.

MP 1, MP 3, MP 4, MP 6, MP 8

Objective To find areas of regular polygons and triangles using trigonometry

Use techniques you've already learned to find the height of the triangle.

Solve It! — Getting Ready!

The pennant at the right is in the shape of an isosceles triangle. The measure of the vertex angle is 20. What is the area of the pennant? How do you know?

MATHEMATICAL PRACTICES

In this lesson you will use isosceles triangles and trigonometry to find the area of a regular polygon.

Essential Understanding You can use trigonometry to find the area of a regular polygon when you know the length of a side, radius, or apothem.

Problem 1 Finding Area

Think

What is the apothem in the diagram?
The apothem is the altitude of the isosceles triangle. The apothem bisects the central angle and the side of the polygon.

What is the area of a regular nonagon with 10-cm sides?

Draw a regular nonagon with center C. Draw $\overline{CP}$ and $\overline{CR}$ to form isosceles $\triangle PCR$. The measure of central $\angle PCR$ is $\frac{360}{9}$, or 40. The perimeter is $9 \cdot 10$, or 90 cm. Draw the apothem $\overline{CS}$. $m\angle PCS = \frac{1}{2}m\angle PCR = 20$ and $PS = \frac{1}{2}PR = 5$ cm.

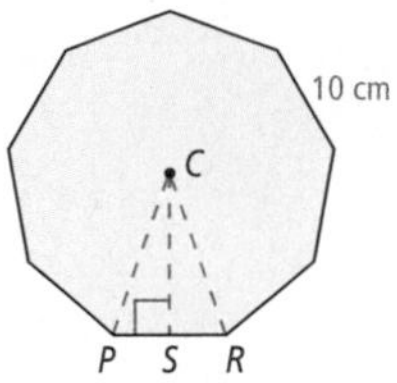

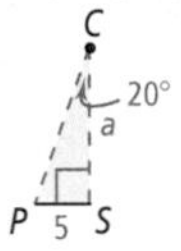

Let a represent CS. Find a and substitute into the area formula.

$\tan 20° = \frac{5}{a}$ Use the tangent ratio.

$a = \frac{5}{\tan 20°}$ Solve for a.

$A = \frac{1}{2}ap$

$= \frac{1}{2} \cdot \frac{5}{\tan 20°} \cdot 90$ Substitute $\frac{5}{\tan 20°}$ for a and 90 for p.

≈ 618.1824194 Use a calculator.

The area of the regular nonagon is about 618 cm^2.

Got It? 1. What is the area of a regular pentagon with 4-in. sides? Round your answer to the nearest square inch.

1 Interactive Learning

Solve It!

PURPOSE To use trigonometry to find the area of figures

PROCESS Students may draw a right triangle and use trigonometry to find the height of the pennant.

FACILITATE

Q What measurement do you need to find the area of the pennant? **[triangle height]**

Q What are the angle measures of each right triangle formed by drawing the height? Explain. **[The angles are 90°, 10°, and 80°. The vertex angle is bisected by the altitude to form the 10° angle.]**

Q What is the length of the short leg of the right triangle? Explain. **[Because the altitude is the perpendicular bisector in an isosceles triangle, the short leg has a measure of 5 in.]**

ANSWER See Solve It in Answers on next page.

CONNECT THE MATH Students use isosceles triangles and trigonometry to find the area of a regular polygon in the Solve It. In the lesson, students use trigonometry to find the area of different regular polygons and the area of a triangle.

2 Guided Instruction

Problem 1

Q How can you find the length of the apothem? **[Write a trigonometric equation with the tangent of half the central angle equal to the ratio of half the side length over *a*.]**

Got It? VISUAL LEARNERS

Have students draw the pentagon and the right triangle made by the radius and the apothem.

10-5 Preparing to Teach

BIG idea Measurement

ESSENTIAL UNDERSTANDINGS

- Trigonometry can be used to find the area of a regular polygon when the length of a side, radius, or apothem is known.
- Trigonometry can be used to find the area of a triangle when the lengths of two sides and the included angle are known.

Math Background

While finding the area of regular polygons is frequently necessary, you may not always know the necessary measurements to calculate it. Because a regular polygon can be subdivided into n congruent isosceles triangles, and each triangle subdivided into two congruent right triangles, you can use trigonometry to find unknown side lengths. Students will apply their knowledge of all three trigonometric ratios to solve for unknown measurements. In the case of regular polygons, students must know the number of sides and either the side length, the apothem length, or the radius of the polygon to find the area. In a triangle, students need the length of two adjacent sides and the measure of an included angle to find the area.

Mathematical Practices

Attend to precision. Using trigonometric functions, students will calculate the areas of triangles and regular polygons accurately, efficiently, and to a degree of precision appropriate to the context of the situation.

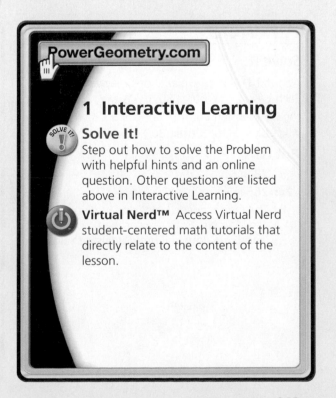

PowerGeometry.com

1 Interactive Learning

Solve It! Step out how to solve the Problem with helpful hints and an online question. Other questions are listed above in Interactive Learning.

Virtual Nerd™ Access Virtual Nerd student-centered math tutorials that directly relate to the content of the lesson.

Problem 2

Q What is the formula for the area of a regular octagon? **[$A = \frac{1}{2}ap$, where a is the length of the apothem, and p is the perimeter of the octagon.]**

Q Which trigonometric ratio can you use to find the length of the apothem? Explain. **[Because you know the hypotenuse and need the side adjacent to the known central angle, use the cosine ratio.]**

Q What trigonometric ratio can you use to find the length of half the side? Explain. **[Because you know the hypotenuse and need the side opposite the known central angle, use the sine ratio.]**

Q What other variable do you need to use the formula for the area of a regular polygon? **[the perimeter]**

Q What is the length of a side of the octagon? Show your work. **[$2x = 2 \cdot 16.2(\sin 22.5°) = 32.4(\sin 22.5°) = 12.4$ in.]**

Q What is the perimeter of the octagon? Show your work. **[$8 \cdot 32.4(\sin 22.5°) = 259.2(\sin 22.5°) = 99.2$ in.^{2}]**

Got It?

ERROR PREVENTION

For 2a, ask students to draw a diagram of the polygon and the right triangle that is formed. For 2b, have students identify the step in Problem 2 where the radius is doubled. Have them redo the problem with twice the radius and compare the final answers. They can identify the relationship. Challenge students to prove their conjectures using properties of algebra.

Problem 2 Finding Area

GRIDDED RESPONSE

Road Signs A stop sign is a regular octagon. The standard size has a 16.2-in. radius. What is the area of the stop sign to the nearest square inch?

Know
The radius and the number of sides of the octagon

Need
The apothem and the length of a side

Plan
Use trigonometric ratios to find the apothem and the length of a side

Step 1 Let a represent the apothem. Use the cosine ratio to find a. The measure of a central angle of the octagon is $\frac{360}{8}$, or 45. So $m\angle C = \frac{1}{2}(45) = 22.5$.

$\cos 22.5° = \frac{a}{16.2}$ — Use the cosine ratio.

$16.2(\cos 22.5°) = a$ — Multiply each side by 16.2.

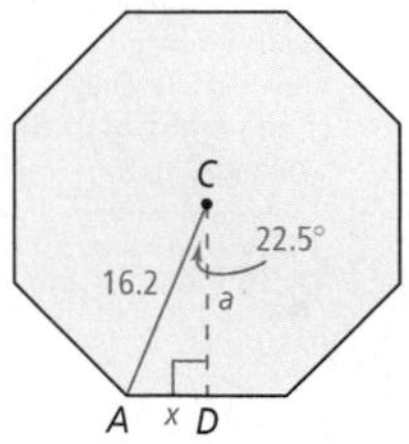

Step 2 Let x represent AD. Use the sine ratio to find x.

$\sin 22.5° = \frac{x}{16.2}$ — Use the sine ratio.

$16.2(\sin 22.5°) = x$ — Multiply each side by 16.2.

Step 3 Find the perimeter of the octagon.

$p = 8 \cdot$ length of a side

$= 8 \cdot 2x$ — The length of each side is $2x$.

$= 8 \cdot 2 \cdot 16.2(\sin 22.5°)$ — Substitute for x.

$= 259.2(\sin 22.5°)$ — Simplify.

Step 4 Substitute into the area formula.

$A = \frac{1}{2}ap$

$= \frac{1}{2} \cdot 16.2(\cos 22.5°) \cdot 259.2(\sin 22.5°)$ — Substitute for a and p.

≈ 742.2924146 — Use a calculator.

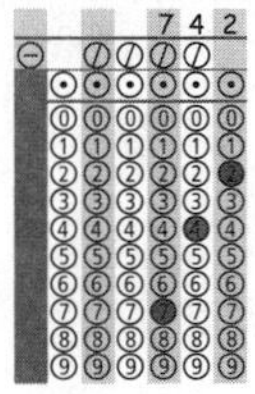

The area of the stop sign is about 742 in.^{2}.

Got It? **2. a.** A tabletop has the shape of a regular decagon with a radius of 9.5 in. What is the area of the tabletop to the nearest square inch?

b. Reasoning Suppose the radius of a regular polygon is doubled. How does the area of the polygon change? Explain.

Answers

Solve It!

About 141.8 in.^{2}; explanations may vary. Sample: Each base ∠ measures 80. If h is the height of the △, then $\tan 80° = \frac{h}{5}$. So, $h = 5 \cdot \tan 80°$. $A = \frac{1}{2}bh = \frac{1}{2}(10)(5 \cdot \tan 80°)$.

Got It?

1. 28 in.^{2}

2a. 265 in.^{2}

b. The area is quadrupled; explanations may vary. Sample: Both the apothem and the side length are doubled if the radius is doubled.

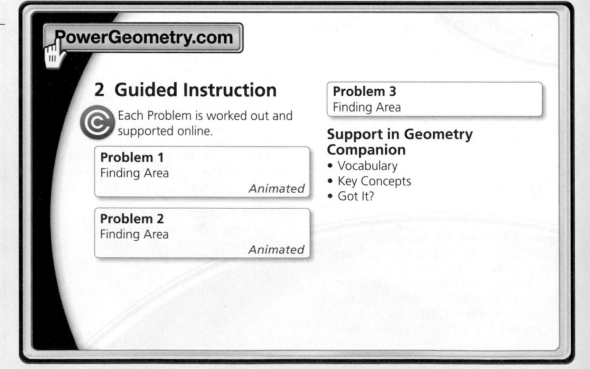

PowerGeometry.com

2 Guided Instruction

Each Problem is worked out and supported online.

Problem 1
Finding Area
Animated

Problem 2
Finding Area
Animated

Problem 3
Finding Area

Support in Geometry Companion
- Vocabulary
- Key Concepts
- Got It?

Essential Understanding You can use trigonometry to find the area of a triangle when you know the length of two sides and the included angle.

Suppose you want to find the area of $\triangle ABC$, but you know only $m\angle A$ and the lengths b and c. To use the formula $A = \frac{1}{2}bh$, you need to know the height. You can find the height by using the sine ratio.

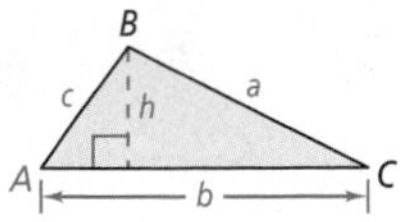

$\sin A = \frac{h}{c}$ Use the sine ratio.

$h = c(\sin A)$ Solve for h.

Now substitute for h in the formula Area $= \frac{1}{2}bh$.

Area $= \frac{1}{2}bc(\sin A)$

This completes the proof of the following theorem for the case in which $\angle A$ is acute.

take note

Theorem 10-8 Area of a Triangle Given SAS

The area of a triangle is half the product of the lengths of two sides and the sine of the included angle.

Area of $\triangle ABC = \frac{1}{2}bc(\sin A)$

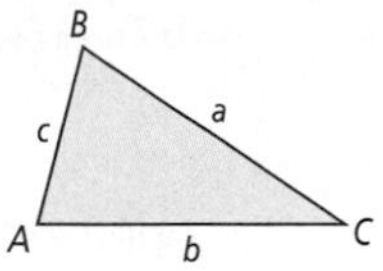

Plan

Which formula should you use?
The diagram gives the lengths of two sides and the measure of the included angle. Use the formula for the area of a triangle given SAS.

Problem 3 Finding Area

What is the area of the triangle?

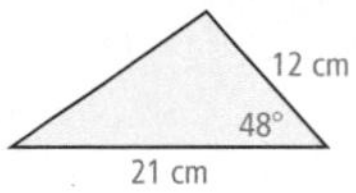

Area $= \frac{1}{2} \cdot$ side length $\cdot$ side length $\cdot$ sine of included angle

$= \frac{1}{2} \cdot 12 \cdot 21 \cdot \sin 48°$ Substitute.

≈ 93.63624801 Use a calculator.

The area of the triangle is about 94 cm^2.

Got It? 3. What is the area of the triangle? Round your answer to the nearest square inch.

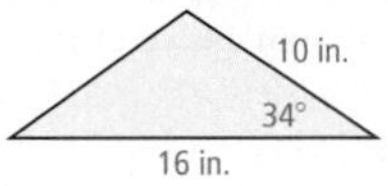

Draw the diagram at the top of page 645 on the board. Challenge students to write the height of the triangle in terms of the side lengths and measure of the angle. Then ask students to write a formula for the area of the triangle using their expression for the height. Have them compare and contrast their answers with other students.

Take Note

Ask students to write formulas for the area of the triangle given the other two sides and angle combinations. **[$A = \frac{1}{2}ab(\sin C)$, $A = \frac{1}{2}ac(\sin B)$]**

Problem 3

Q What information do you need to use the SAS area formula? **[the length of two sides and the measure of the included angle]**

Q Do you have the appropriate information to use the formula? Give the measurements you know. **[yes; one side = 12 cm; another side = 21 cm; the included angle = 48°]**

Got It?

Have students identify the two sides and the included angle that they need in order to use the SAS area formula.

Additional Problems

1. What is the area of a regular octagon with 7.5 cm sides?

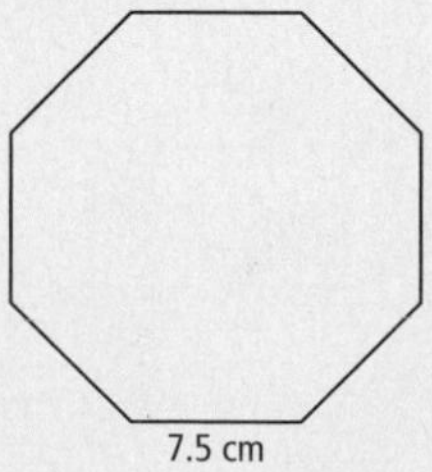

ANSWER about 271.6 cm^2

2. The gazebo at a park has a floor shaped like a regular hexagon with a radius of 6 ft. What is the area of the hexagonal floor to the nearest square foot?

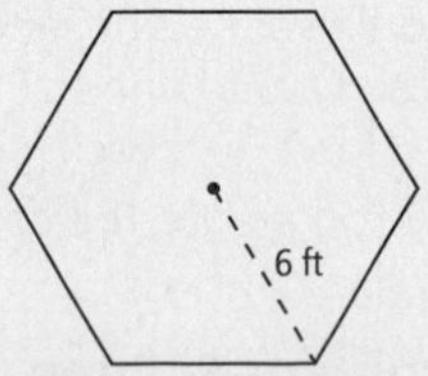

ANSWER about 94 ft^2

3. What is the area of the triangle?

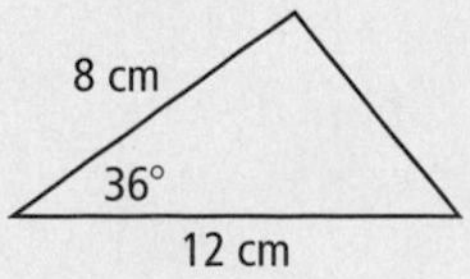

ANSWER about 28.2 cm^2

Answers

Got It? (continued)

3. 45 in.^{2}

3 Lesson Check

Do you know HOW?

- If students have difficulty with Exercise 2, then have them review Problem 1 to see how to draw a right triangle and find the length of the apothem.

Do you UNDERSTAND?

- If students have difficulty with Exercise 5, then have them review Problem 2 to compare their equations with those in the problem.

Close

Q How can you use trigonometry to help you find the area of a regular polygon? **[If you know what type of polygon you have, and the side length, radius, or apothem, you can use central angles and trigonometry to find the missing lengths in any regular polygon.]**

Q What information do you need to find the area of a triangle using trigonometry? **[the length of two sides and the measure of the included angle]**

Lesson Check

Do you know HOW?

What is the area of each regular polygon? Round your answers to the nearest tenth.

1.

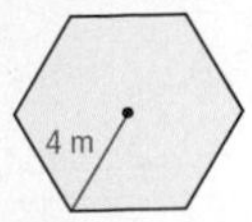

2.

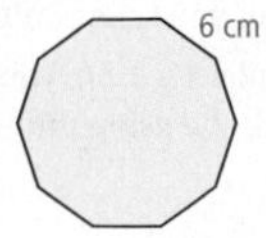

3. What is the area of the triangle at the right to the nearest square inch?

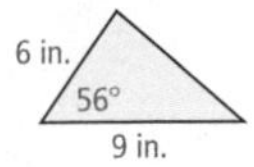

Do you UNDERSTAND? MATHEMATICAL PRACTICES

4. Reasoning A diagonal through the center of a regular hexagon is 12 cm long. Is it possible to find the area of this hexagon? Explain.

5. Error Analysis Your classmate needs to find the area of a regular pentagon with 8-cm sides. To find the apothem, he sets up and solves a trigonometric ratio. What error did he make? Explain.

$\frac{a}{4} = \tan 36°$
$a = 4 \cdot \tan 36°$

Practice and Problem-Solving Exercises MATHEMATICAL PRACTICES

A Practice

Find the area of each regular polygon. Round your answers to the nearest tenth. See Problems 1 and 2.

6. octagon with side length 6 cm
7. decagon with side length 4 yd
8. pentagon with radius 3 ft
9. nonagon with radius 7 in.
10. dodecagon with radius 20 cm
11. 20-gon with radius 2 mm
12. 18-gon with perimeter 72 mm
13. 15-gon with perimeter 180 cm

Find the area of each triangle. Round your answers to the nearest tenth. See Problem 3.

14.

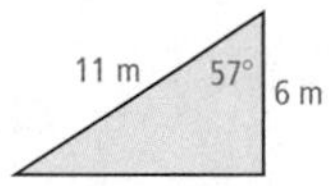

15.

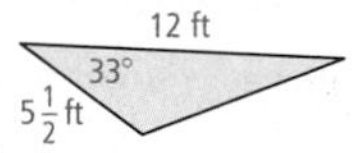

16.

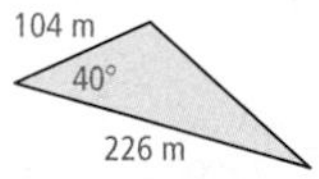

17.

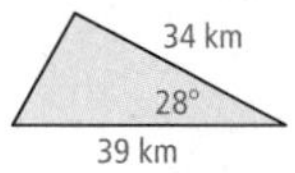

18.

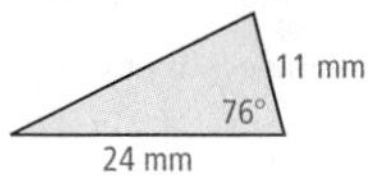

19.

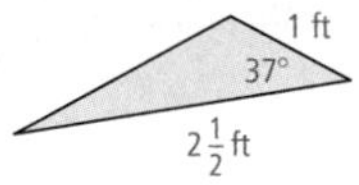

B Apply

20. *PQRST* is a regular pentagon with center *O* and radius 10 in. Find each measure. If necessary, round your answers to the nearest tenth.

a. $m\angle POQ$
b. $m\angle POX$
c. OX
d. PQ
e. perimeter of *PQRST*
f. area of *PQRST*

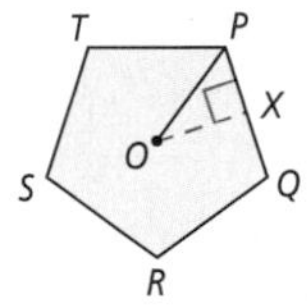

PowerGeometry.com

3 Lesson Check

For a digital lesson check, use the Got It questions.

Support in Geometry Companion

- Lesson Check

4 Practice

Assign homework to individual students or to an entire class.

Answers

Lesson Check

1. 41.6 m^2
2. 277.0 cm^2
3. 22 $in.^2$
4. Yes; the diagonal of a regular hexagon is two times the side, and you have several ways to find the area of a regular hexagon with 6-cm sides.
5. He set up the wrong ratio. The correct ratio is $\frac{4}{a} = \tan 36°$.

Practice and Problem-Solving Exercises

6. 173.8 cm^2
7. 123.1 yd^2
8. 21.4 ft^2
9. 141.7 $in.^2$
10. 1200 cm^2
11. 12.4 mm^2
12. 408.3 mm^2
13. 2540.5 cm^2
14. 27.7 m^2
15. 18.0 ft^2
16. 7554.0 m^2
17. 311.3 km^2
18. 128.1 mm^2
19. 0.8 ft^2
20a. 72
b. 36
c. 8.1 in.
d. 11.8 in.
e. 58.8 in.
f. 237.8 $in.^2$

21. Writing Describe three ways to find the area of a regular hexagon if you know only the length of a side.

22. Think About a Plan The surveyed lengths of two adjacent sides of a triangular plot of land are 80 yd and 150 yd. The angle between the sides is 67°. What is the area of the parcel of land to the nearest square yard?
- Can you *draw a diagram* to represent the situation?
- Which formula for the area of a triangle should you use?

Find the perimeter and area of each regular polygon to the nearest tenth.

23.

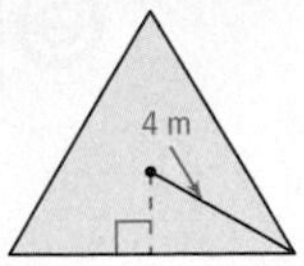

24.

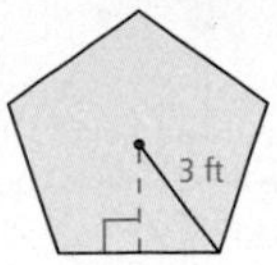

25.

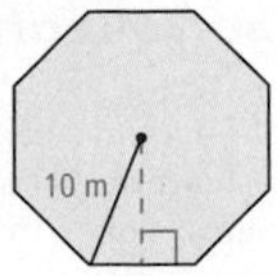

26.

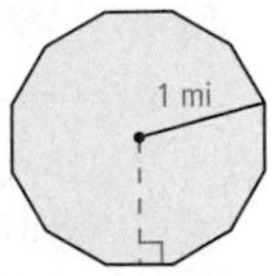

27. Architecture The Pentagon in Arlington, Virginia, is one of the world's largest office buildings. It is a regular pentagon, and the length of each of its sides is 921 ft. What is the area of land that the Pentagon covers to the nearest thousand square feet?

28. What is the area of the triangle shown at the right?

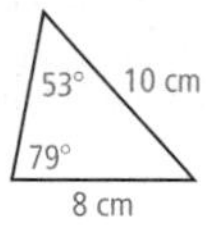

29. The central angle of a regular polygon is 10°. The perimeter of the polygon is 108 cm. What is the area of the polygon?

30. Replacement glass for energy-efficient windows costs $5/ft^2. About how much will you pay for replacement glass for a regular hexagonal window with a radius of 2 ft?

(A) $10.39 (B) $27.78 (C) $45.98 (D) $51.96

Regular polygons A and B are similar. Compare their areas.

31. The apothem of Pentagon A equals the radius of Pentagon B.

32. The length of a side of Hexagon A equals the radius of Hexagon B.

33. The radius of Octagon A equals the apothem of Octagon B.

34. The perimeter of Decagon A equals the length of a side of Decagon B.

The polygons are regular polygons. Find the area of the shaded region.

35.

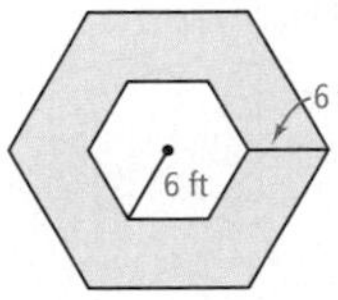

36.

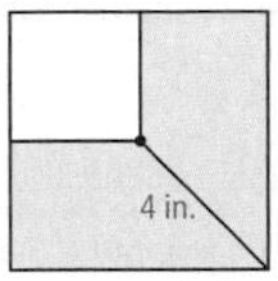

37.

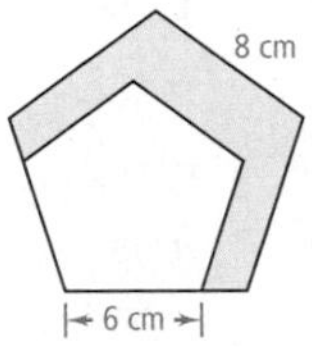

4 Practice

ASSIGNMENT GUIDE
Basic: 6–22 all, 24–26 even, 27, 32–36 even
Average: 7–19 odd, 20–37
Advanced: 7–19 odd, 20–39

Mathematical Practices are supported by exercises with red headings. Here are the Practices supported in this lesson:

MP 1: Make Sense of Problems Ex. 22
MP 3: Construct Arguments Ex. 4
MP 3: Critique the Reasoning of Others Ex. 5
MP 8: Repeated Reasoning 21

Applications exercises have blue headings. Exercises 27 and 38 support MP 4: Model.

STEM exercises focus on science or engineering applications.

EXERCISE 27: Use the Think About a Plan worksheet in the **Practice and Problem Solving Workbook** (also available in the Teaching Resources in print and online) to further support students' development in becoming independent learners.

HOMEWORK QUICK CHECK
To check students' understanding of key skills and concepts, go over Exercises 7, 17, 21, 22, and 27.

21. Multiply the formula for the area of an equilateral △, $A = \frac{s^2\sqrt{3}}{4}$, by 6 to get $\frac{3s^2\sqrt{3}}{2}$; use a 30°-60°-90° △ to find the height of one equilateral △ with side s, then multiply the area of that △ by 6; or use the tangent ratio to find the apothem and then use the formula $A = \frac{1}{2}ap$.
22. 5523 yd^2
23. 20.8 m, 20.8 m^2
24. 17.6 ft, 21.4 ft^2
25. 61.2 m, 282.8 m^2
26. 6.2 mi, 3 mi^2
27. 1,459,000 ft^2
28. about 29.7 cm^2
29. about 925.8 cm^2
30. D
31. area of Pentagon A ≈ 1.53 • (area of Pentagon B)
32. area of Hexagon A = area of Hexagon B
33. area of Octagon B ≈ 1.17 • (area of Octagon A)
34. area of Decagon A = 0.01 • (area of Decagon B)
35. $162\sqrt{3}$ ft^2 or about 280.6 ft^2
36. 24 in.^{2}
37. about 48.2 cm^2

Answers

Practice and Problem-Solving Exercises (continued)

38. 0.65

39. 320 ft

Challenge **38.** Segments are drawn between the midpoints of consecutive sides of a regular pentagon to form another regular pentagon. Find, to the nearest hundredth, the ratio of the area of the smaller pentagon to the area of the larger pentagon.

STEM **39. Surveying** A surveyor wants to mark off a triangular parcel with an area of 1 acre (1 acre = 43,560 ft^2). One side of the triangle extends 300 ft along a straight road. A second side extends at an angle of 65° from one end of the first side. What is the length of the second side to the nearest foot?

Apply What You've Learned

MATHEMATICAL PRACTICES MP 5, MP 7

Look back at the information given about the target on page 613. The diagram of the target is shown again below. In the Apply What You Learned in Lesson 10-1, you found the area of one of the red triangles.

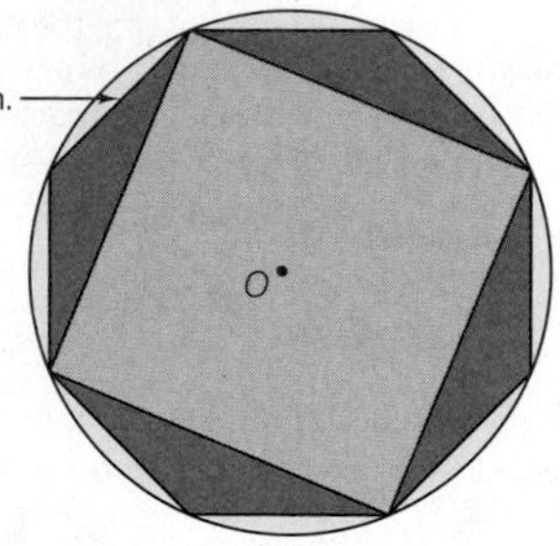

Find the area of the regular octagon in the target. Select all of the following that are true. Explain your reasoning.

A. The perimeter of the regular octagon is 72 in.

B. The apothem of the regular octagon is the same as the radius of the circle.

C. The grey quadrilateral is a rhombus, but not necessarily a square.

D. The apothem of the regular octagon, in inches, is equivalent to $\frac{9}{\tan 22.5°}$.

E. The apothem of the regular octagon, in inches, is equivalent to $\frac{4.5}{\tan 22.5°}$.

F. The apothem of the regular octagon, in inches, is equivalent to $\frac{9}{\tan 45°}$.

G. The area of the regular octagon is about 391.1 in.^{2}.

H. The area of the regular octagon is about 423.4 in.^{2}.

I. The area of the regular octagon is about 782.2 in.^{2}.

Apply What You've Learned

In the Apply What You've Learned for Lesson 10-1, students found the area of a red triangle in the target shown on page 613. Here, they find the area of the regular octagon. Later in the chapter, students will find the area of the yellow regions of the target.

Mathematical Practices

Students **look for and make use of structure** to draw in the auxiliary line segments that will allow them to find the area of the octagon. (MP 7)

Students **use an appropriate tool** (a calculator) to perform calculations with a trigonometric ratio to find an area. (MP 5)

ANSWERS

Choices A, E, and G are all true.

10-5 Lesson Resources

Differentiated Remediation

Instructional Support

Geometry Companion

Students can use the **Geometry Companion** worktext (4 pages) . . .

- New Vocabulary
- Key Concepts
- Got It for each Problem
- Lesson Check

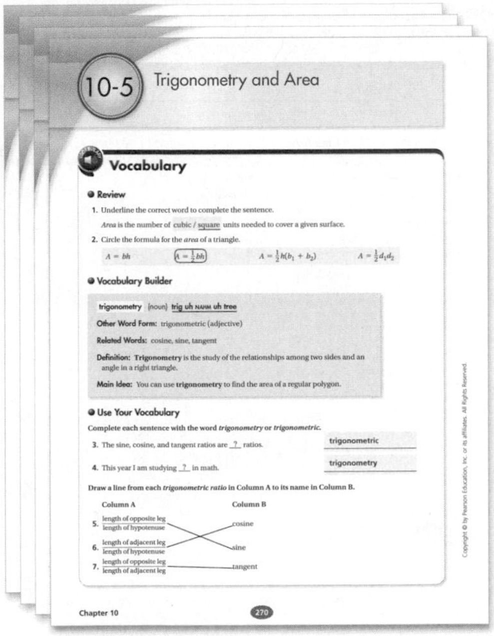

10-5 Trigonometry and Area

Vocabulary

Review

1. Underline the correct word to complete the sentence.

Area is the number of cubic / square units needed to cover a given surface.

2. Circle the formula for the *area* of a triangle.

$A = bh$ $\quad$ $A = \frac{1}{2}bh$ $\quad$ $A = \frac{1}{2}h(b_1 + b_2)$ $\quad$ $A = \frac{1}{2}d_1d_2$

Vocabulary Builder

trigonometry (noun) trig uh NAHM uh tree

Other Word Form: trigonometric (adjective)

Related Words: cosine, sine, tangent

Definition: **Trigonometry** is the study of the relationships among two sides and an angle in a right triangle.

Main Idea: You can use **trigonometry** to find the area of a regular polygon.

Use Your Vocabulary

Complete each sentence with the word *trigonometry* or *trigonometric*.

3. The sine, cosine, and tangent ratios are _?_ ratios. **trigonometric**

4. This year I am studying _?_ in math. **trigonometry**

Draw a line from each *trigonometric ratio* in Column A to its name in Column B.

Column A	Column B
5. $\frac{\text{length of opposite leg}}{\text{length of hypotenuse}}$	cosine
6. $\frac{\text{length of adjacent leg}}{\text{length of hypotenuse}}$	sine
7. $\frac{\text{length of opposite leg}}{\text{length of adjacent leg}}$	tangent

Chapter 10 $\quad$ 270

ELL Support

Use Manipulatives Arrange students into heterogeneous pairs. Hand out pattern blocks or a geometry template for the polygons covered in this chapter. Tell students to make a study guide for the lesson topic, including finding area of parallelograms, triangles, trapezoids, rhombi, kites, regular polygons, and similar figures. Their guides can include examples, definitions, formulas, and steps to find area. The blocks or templates can be used to draw examples. Students can share their work. Discuss the positive and negative attributes of each guide (Ask: what makes an effective study guide?) and encourage students to choose their favorites.

5 Assess & Remediate

Lesson Quiz

1. What is the area of a regular hexagon with 15 in. sides?

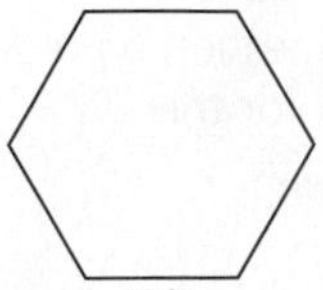

2. Do you UNDERSTAND? A regular decagon has sides that are 8 cm long. What is the area of the figure? Round to the nearest whole number if necessary?

3. What is the area of the triangle to the nearest tenth?

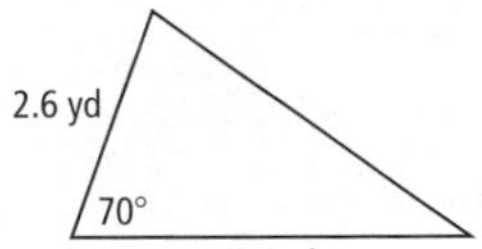

ANSWERS TO LESSON QUIZ

1. about 585 in.^{2}

2. about 492 cm^2

3. about 6.8 yd^2

PRESCRIPTION FOR REMEDIATION

Use the student work on the Lesson Quiz to prescribe a differentiated review assignment.

Points	Differentiated Remediation
0–1	Intervention
2	On-level
3	Extension

PowerGeometry.com

5 Assess & Remediate

Assign the Lesson Quiz. Appropriate intervention, practice, or enrichment is automatically generated based on student performance.

Intervention

- **Reteaching** (2 pages) Provides reteaching and practice exercises for the key lesson concepts. Use with struggling students or absent students.
- **English Language Learner Support** Helps students develop and reinforce mathematical vocabulary and key concepts.

All-in-One Resources/Online

Reteaching

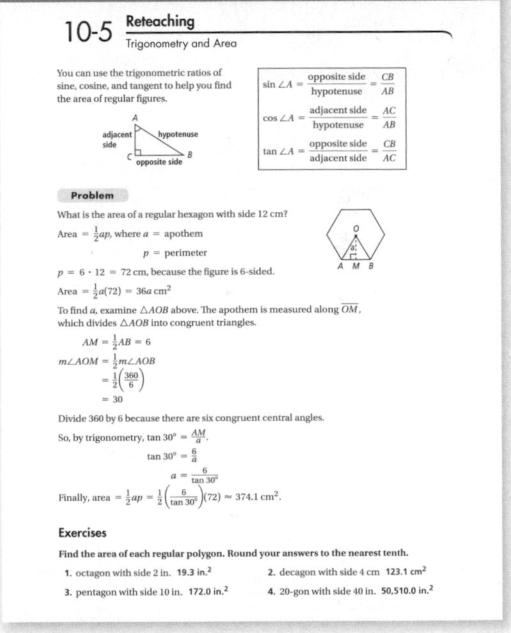

10-5 Reteaching

Trigonometry and Area

You can use the trigonometric ratios of sine, cosine, and tangent to help you find the area of regular figures.

$\sin \angle A = \frac{\text{opposite side}}{\text{hypotenuse}} = \frac{CB}{AB}$

$\cos \angle A = \frac{\text{adjacent side}}{\text{hypotenuse}} = \frac{AC}{AB}$

$\tan \angle A = \frac{\text{opposite side}}{\text{adjacent side}} = \frac{CB}{AC}$

Problem

What is the area of a regular hexagon with side 12 cm?

Area $= \frac{1}{2}ap$, where a = apothem

p = perimeter

$p = 6 \cdot 12 = 72$ cm, because the figure is 6-sided.

Area $= \frac{1}{2}a(72) = 36a$ cm^2

To find a, examine $\triangle AOB$ above. The apothem is measured along $\overline{OM}$, which divides $\triangle AOB$ into congruent triangles.

$AM = \frac{1}{2}AB = 6$

$m\angle AOM = \frac{1}{2}m\angle AOB = \frac{1}{2}\left(\frac{360}{6}\right) = 30$

Divide 360 by 6 because there are six congruent central angles.

So, by trigonometry, $\tan 30° = \frac{AM}{a}$.

$\tan 30° = \frac{6}{a}$

$a = \frac{6}{\tan 30°}$

Finally, area $= \frac{1}{2}ap = \frac{1}{2}\left(\frac{6}{\tan 30°}\right)(72) \approx 374.1$ cm^2.

Exercises

Find the area of each regular polygon. Round your answers to the nearest tenth.

1. octagon with side 2 in. **19.3 in.^{2}**
2. decagon with side 4 cm **123.1 cm^2**
3. pentagon with side 10 in. **172.0 in.^{2}**
4. 20-gon with side 40 in. **50,510.0 in.^{2}**

All-in-One Resources/Online

English Language Learner Support

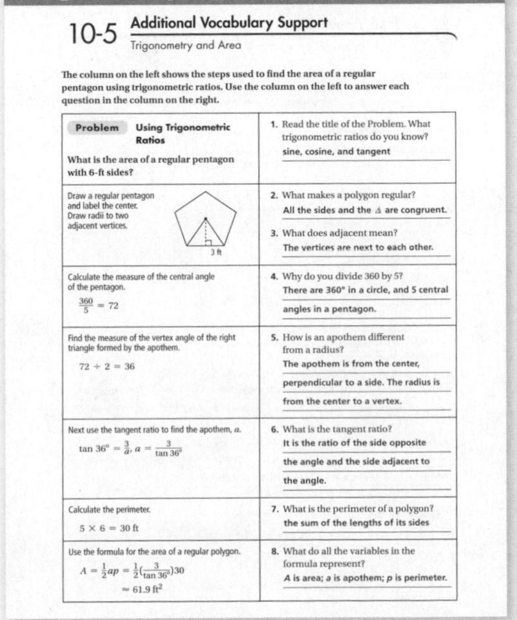

10-5 Additional Vocabulary Support

Trigonometry and Area

The column on the left shows the steps used to find the area of a regular pentagon using trigonometric ratios. Use the column on the left to answer each question in the column on the right.

Problem Using Trigonometric Ratios. **What is the area of a regular pentagon with 6-ft sides?**	1. Read the title of the Problem. What trigonometric ratios do you know? **sine, cosine, and tangent**
Draw a regular pentagon and label the center. Draw radii to two adjacent vertices. (3 ft)	2. What makes a polygon regular? **All the sides and the ∠ are congruent.** 3. What does adjacent mean? **The vertices are next to each other.**
Calculate the measure of the central angle of the pentagon. $\frac{360}{5} = 72$	4. Why do you divide 360 by 5? **There are 360° in a circle, and 5 central angles in a pentagon.**
Find the measure of the vertex angle of the right triangle formed by the apothem. $72 \div 2 = 36$	5. How is an apothem different from a radius? **The apothem is from the center, perpendicular to a side. The radius is from the center to a vertex.**
Next use the tangent ratio to find the apothem, a. $\tan 36° = \frac{3}{a}$, $a = \frac{3}{\tan 36°}$	6. What is the tangent ratio? **It is the ratio of the side opposite the angle and the side adjacent to the angle.**
Calculate the perimeter. $5 \times 6 = 30$ ft	7. What is the perimeter of a polygon? **the sum of the lengths of its sides**
Use the formula for the area of a regular polygon. $A = \frac{1}{2}ap = \frac{1}{2}\left(\frac{3}{\tan 36°}\right)30 \approx 61.9$ ft^2	8. What do all the variables in the formula represent? ***A* is area; *a* is apothem; *p* is perimeter.**

Differentiated Remediation *continued*

On-Level

- **Practice** (2 pages) Provides extra practice for each lesson. For simpler practice exercises, use the Form K Practice pages found in the All-in-One Teaching Resources and online.
- **Think About a Plan** Helps students develop specific problem-solving skills and strategies by providing scaffolded guiding questions.
- **Standardized Test Prep** Focuses on all major exercises, all major question types, and helps students prepare for the high-stakes assessments.

Extension

- **Enrichment** Provides students with interesting problems and activities that extend the concepts of the lesson.
- **Activities, Games, and Puzzles** Worksheets that can be used for concepts development, enrichment, and for fun!

Practice and Problem Solving Wkbk/ All-in-One Resources/Online

Practice page 1

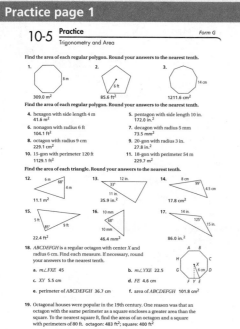

10-5 **Practice** — Form G
Trigonometry and Area

Find the area of each regular polygon. Round your answers to the nearest tenth.

1. 8 m — $309.0\ m^2$
2. 6 ft — $85.6\ ft^2$
3. 14 cm — $1211.6\ cm^2$

Find the area of each regular polygon. Round your answers to the nearest tenth.

4. hexagon with side length 4 m — $41.6\ m^2$
5. pentagon with side length 10 in. — $172.0\ in.^2$
6. nonagon with radius 6 ft — $104.1\ ft^2$
7. decagon with radius 5 mm — $73.5\ mm^2$
8. octagon with radius 9 cm — $229.1\ cm^2$
9. 20-gon with radius 3 in. — $27.8\ in.^2$
10. 15-gon with perimeter 120 ft — $1129.1\ ft^2$
11. 18-gon with perimeter 54 m — $229.7\ m^2$

Find the area of each triangle. Round your answers to the nearest tenth.

12. $11.1\ m^2$
13. $35.9\ in.^2$
14. $17.8\ cm^2$
15. $22.4\ ft^2$
16. $46.4\ mm^2$
17. $86.0\ in.^2$

18. *ABCDEFGH* is a regular octagon with center *X* and radius 6 cm. Find each measure. If necessary, round your answers to the nearest tenth.
   a. $m\angle FXE$ 45
   b. $m\angle YXE$ 22.5
   c. XY 5.5 cm
   d. FE 4.6 cm
   e. perimeter of *ABCDEFGH* 36.7 cm
   f. area of *ABCDEFGH* $101.8\ cm^2$

19. Octagonal houses were popular in the 19th century. One reason was that an octagon with the same perimeter as a square encloses a greater area than the square. To the nearest square ft, find the areas of an octagon and a square with perimeters of 80 ft. octagon: $483\ ft^2$; square: $400\ ft^2$

Practice and Problem Solving Wkbk/ All-in-One Resources/Online

Practice page 2

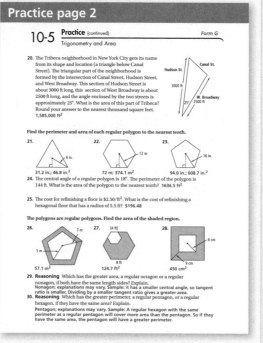

10-5 **Practice** (continued) — Form G
Trigonometry and Area

20. The Tribeca neighborhood in New York City gets its name from its shape and location (a triangle below Canal Street). The triangular part of the neighborhood is formed by the intersection of Canal Street, Hudson Street, and West Broadway. This section of Hudson Street is about 3000 ft long, this section of West Broadway is about 2500 ft long, and the angle enclosed by the two streets is approximately 25°. What is the area of this part of Tribeca? Round your answer to the nearest thousand square feet. $1{,}585{,}000\ ft^2$

Find the perimeter and area of each regular polygon to the nearest tenth.

21. 31.2 in.; $46.8\ in.^2$
22. 72 m; $374.1\ m^2$
23. 94.0 in.; $608.7\ in.^2$
24. The central angle of a regular polygon is 18°. The perimeter of the polygon is 144 ft. What is the area of the polygon to the nearest tenth? $1636.5\ ft^2$
25. The cost for refinishing a floor is $\$2.50/ft^2$. What is the cost of refinishing a hexagonal floor that has a radius of 5.5 ft? \$196.48

The polygons are regular polygons. Find the area of the shaded region.

26. $57.1\ m^2$
27. $124.7\ ft^2$
28. $450\ cm^2$

29. **Reasoning** Which has the greater area, a regular octagon or a regular nonagon, if both have the same length sides? Explain. Nonagon; explanations may vary. Sample: it has a smaller central angle, so tangent ratio is smaller. Dividing by a smaller tangent ratio gives a greater area.
30. **Reasoning** Which has the greater perimeter, a regular pentagon, or a regular hexagon, if they have the same area? Explain. Pentagon; explanations may vary. Sample: A regular hexagon with the same perimeter as a regular pentagon will cover more area than the pentagon. So if they have the same area, the pentagon will have a greater perimeter.

All-in-One Resources/Online

Enrichment

10-5 **Enrichment**
Trigonometry and Area

Using Triangles to Explore the Area of Polygons

The area of any polygon can be found by adding the areas of triangles inside the polygon. Use triangles to find each area. Show your work.

1. Find the area of the irregular quadrilateral *ABCD* at the right. If necessary, use fractions and square roots to express your answer. Then round to the nearest tenth. (*Hint:* Draw in diagonal $\overline{BD}$.) $\frac{1}{2}(3)(4) + \frac{1}{2}(5)(5)\sin 60 = 6 + \frac{25\sqrt{3}}{4} \approx 16.8$ units^2
2. Find the area of the irregular hexagon *ABCDEF* at the right. If necessary, use fractions and square roots to express your answers. Then round to the nearest tenth. (*Hint:* Draw diagonal $\overline{BD}$ and diagonal $\overline{AE}$.) $2[\frac{1}{2}(3)(3)\sin 60] + 4(3) = \frac{9\sqrt{3}}{2} + 12 \approx 19.8$ units^2
3. The area of the parallelogram at the right is $48\ cm^2$. Find *x*. $48 = (8)(x)\sin 60 = 8x\frac{\sqrt{3}}{2}$; 6.9 cm
4. **Algebra** Find the area of the irregular quadrilateral at the right. Express your answer in terms of *a*, *b*, *c*, and *d*. $\frac{1}{2}ab\sin 150 + \frac{1}{2}cd\sin 30 = \frac{(ab + cd)}{4}$ units^2
5. **Algebra** Find the area of the irregular hexagon at the right. Express your answer in terms of *x* and *y*. $2[\frac{1}{2}x(x)\sin 90] + (x\sqrt{2})y = x^2 + \sqrt{2}xy$ units^2

Practice and Problem Solving Wkbk/ All-in-One Resources/Online

Think About a Plan

10-5 **Think About a Plan**
Trigonometry and Area

Architecture The Pentagon in Arlington, Virginia, is one of the world's largest office buildings. It is a regular pentagon, and the length of each of its sides is 921 ft. What is the area of land the Pentagon covers to the nearest thousand square feet?

Understanding the Problem

1. The figure at the right is a regular pentagon. Label the side length with the information given in the problem.
2. The formula for the area of a regular polygon is $A = \frac{1}{2}ap$. Which measures do you need to solve the problem? perimeter and apothem length

Planning the Solution

3. How can you find the perimeter? Multiply the side length by the number of sides.
4. Draw an apothem and radius on the figure. Label the center *A*, and draw right $\triangle ABC$ formed by the apothem and radius, where $\overline{AB}$ is the hypotenuse. Check students' work.
5. How can you find $m\angle ABC$? Write this measure on the diagram. Answers may vary. Sample: Find the central angle: $360 \div 5 = 72$. Because the apothem bisects the central angle, $m\angle BAC = 36$. So, $m\angle ABC = 90 - m\angle BAC = 90 - 36 = 54$.
6. Which trigonometric ratio can you use to find the apothem? Explain. The tangent ratio; you know the length of the side adjacent to the 54° angle and you want to find the length of the side opposite that angle.

Getting an Answer

7. Complete the equation below. Then solve for *a*, the apothem. $\tan 54° = \frac{a}{460.5}$; $a \approx 633.8$ ft
8. Use the formula for the area of a regular polygon to find the area. $A \approx 1{,}459{,}000\ ft^2$

Practice and Problem Solving Wkbk/ All-in-One Resources/Online

Standardized Test Prep

10-5 **Standardized Test Prep**
Trigonometry and Area

Multiple Choice

For Exercises 1–6, choose the correct letter.

1. What is the area of the regular pentagon at the right? A
   (A) $688.2\ ft^2$ (B) $850.7\ ft^2$ (C) $951.1\ ft^2$ (D) $1376.4\ ft^2$
2. What is the area of the regular octagon at the right? G
   (F) $114.5\ m^2$ (G) $229.1\ m^2$ (H) $391.1\ m^2$ (I) $458.2\ m^2$
3. What is the area of a regular hexagon with sides 8 cm? B
   (A) $110.1\ cm^2$ (B) $166.3\ cm^2$ (C) $309.0\ cm^2$ (D) $332.6\ cm^2$
4. What is the area of a regular nonagon with a radius of 14 in.? H
   (F) $70.9\ in.^2$ (G) $141.7\ in.^2$ (H) $566.9\ in.^2$ (I) $1211.6\ in.^2$
5. What is the area of a 15-gon with a perimeter of 90 m? B
   (A) $528.2\ m^2$ (B) $635.1\ m^2$ (C) $1270.3\ m^2$ (D) $142{,}903.1\ m^2$
6. What is the area of the triangle at the right? G
   (F) $6.0\ ft^2$ (G) $8.6\ ft^2$ (H) $15.0\ ft^2$ (I) $17.2\ ft^2$

Short Response

7. What is the area of the triangle shown at the right? Show your work and explain how you got your answer.
   [2] $\frac{1}{2}(5)(5)(\sin 118) \approx 11.04\ in.^2$; because the base angles are equal, the triangle is isosceles and the sides opposite the base angles are congruent. The vertex angle is equal to $180 - 31 - 31 = 118$. To find the area, multiply the lengths of the congruent sides and the sine of the included 118° angle and divide by 2. [1] appropriate methods, but with one computational error or work not shown [0] incorrect answer with no work shown

Online Teacher Resource Center

Activities, Games, and Puzzles

10-5 **Game: Connect Three**
Trigonometry and Area

Materials

- Two types of place markers

Setup

Your teacher will divide the class into pairs. You and your partner will receive a set of cards. Mix the cards and place them figure-side-up next to the game board.

Game Play

Decide whether you or your partner will go first. The object of the game is to claim 3 squares in a row (horizontally, vertically, or diagonally). On your turn, choose a card and determine the area of the figure. Round your answer to the nearest tenth and include units. When you are done, flip the card over to see the answer. If you are correct, place one of your markers on a square on the 3-by-3 grid below, and your turn is over. If you are incorrect, return the card to the bottom of the pile, and your turn is over.

Ending the Game

The game ends when one player has three markers in a row.

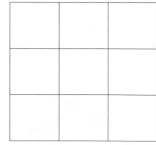

10-6 Circles and Arcs

Common Core State Standards
G-CO.A.1 Know precise definitions of . . . circle . . .
G-C.A.1 Prove that all circles are similar. **Also G-C.A.2, G-C.B.5**
MP 1, MP 3, MP 4, MP 6, MP 8

Objectives To find the measures of central angles and arcs
To find the circumference and arc length

Hm. Will the answer be more than 63 in. or less than 63 in.?

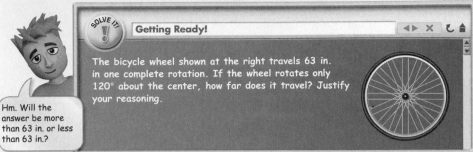

SOLVE IT! Getting Ready!

The bicycle wheel shown at the right travels 63 in. in one complete rotation. If the wheel rotates only 120° about the center, how far does it travel? Justify your reasoning.

MATHEMATICAL PRACTICES

Lesson Vocabulary
- circle
- center
- diameter
- radius
- congruent circles
- central angle
- semicircle
- minor arc
- major arc
- adjacent arcs
- circumference
- pi
- concentric circles
- arc length

In a plane, a **circle** is the set of all points equidistant from a given point called the **center.** You name a circle by its center. Circle P ($\odot P$) is shown below.

A **diameter** is a segment that contains the center of a circle and has both endpoints on the circle. A **radius** is a segment that has one endpoint at the center and the other endpoint on the circle. **Congruent circles** have congruent radii. A **central angle** is an angle whose vertex is the center of the circle.

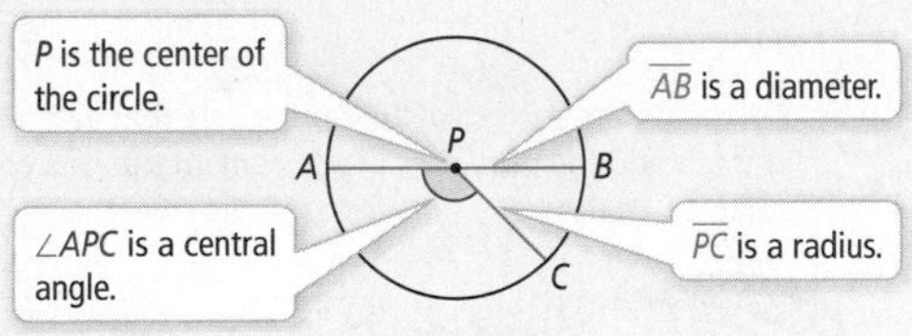

P is the center of the circle.
$\overline{AB}$ is a diameter.
A
P
B
C
∠APC is a central angle.
$\overline{PC}$ is a radius.

Essential Understanding You can find the length of part of a circle's circumference by relating it to an angle in the circle.

An arc is a part of a circle. One type of arc, a **semicircle,** is half of a circle. A **minor arc** is smaller than a semicircle. A **major arc** is larger than a semicircle. You name a minor arc by its endpoints and a major arc or a semicircle by its endpoints and another point on the arc.

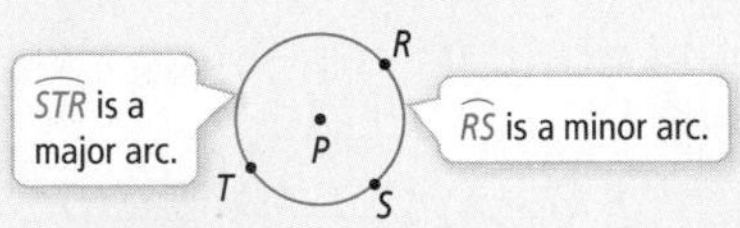

$\widehat{STR}$ is a major arc.
R
P
T
S
$\widehat{RS}$ is a minor arc.

1 Interactive Learning

Solve It!

PURPOSE To find the portion of the circumference represented by an arc

PROCESS Students may
- find the percentage of the circumference represented by the 120° angle.
- use a proportion to find the portion of the circumference represented by the 120° angle.

FACILITATE

Q What does the distance of one complete rotation represent? **[The distance represents the circumference of the wheel.]**

Q Will the wheel complete a full rotation? What does this mean in terms of the distance it travels? **[No, the wheel will not rotate completely around so it will not travel the full distance.]**

Q What portion of a rotation will the wheel complete? **[The wheel will complete $\frac{120}{360} = \frac{1}{3}$ of a rotation.]**

ANSWER See Solve It in Answers on next page.

CONNECT THE MATH In Solve It, students should see how an arc is related to its central angle. The lesson presents instruction about arc measurements, arc addition, and circumference.

2 Guided Instruction

Have students make a manipulative with diagrams of the vocabulary on this page. Allow them to refer to this manipulative throughout the lesson to reinforce definitions.

10-6 Preparing to Teach

BIG idea Measurement

ESSENTIAL UNDERSTANDING
- The length of part of a circle's circumference can be found by relating it to a central angle in the circle.

Math Background

In this lesson, students will learn important vocabulary related to circles. They will learn to identify major and minor arcs and their measures. An arc is measured by the central angle that defines it. Students will learn how to find the circumference of a circle. The ratio of the central angle of an arc to 360° can be used to find the length of the arc. Students will learn how to find distances along circular paths using circumference and arc length.

In 2002, Japanese mathematicians used a supercomputer to calculate the value of π to more than 1 trillion decimal places. In 1874, William Shanks set the record of 707 decimal places for the paper-and-pencil calculation of π. Today when a calculator is not available, most students use $\frac{22}{7}$ or 3.14 as an estimate for π. Estimates for π date back as far as 240 B.C.

Mathematical Practices

Attend to precision. Students will use clear definitions of both arc measure and arc length and compute both quantities.

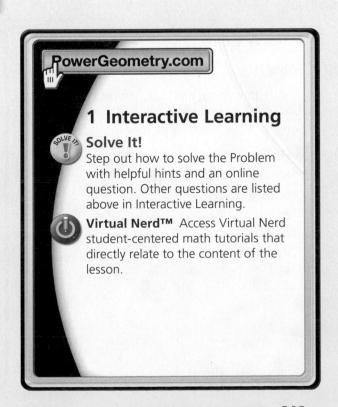

PowerGeometry.com

1 Interactive Learning

Solve It!
Step out how to solve the Problem with helpful hints and an online question. Other questions are listed above in Interactive Learning.

Virtual Nerd™ Access Virtual Nerd student-centered math tutorials that directly relate to the content of the lesson.

Problem 1

> **Q** How can you tell a minor arc in a circle? **[It is smaller than a semicircle.]**
>
> **Q** How can you indicate which direction you want the arc to go? **[List more than two points on the circle.]**
>
> **Q** What are major arcs? **[They are larger than a semicircle.]**

Got It? — ERROR PREVENTION

Students may benefit from tracing the figures with colored pencils. Have them redraw the circle three times and use four different colors on each part.

Take Note

Have students practice finding the measure of related arcs in a circle. Call out an arc measure and have them give the related measure.

Take Note

Discuss the similarities between Postulate 10-2 and the Segment and Angle Addition Postulates. Emphasize that they are called postulates because they are taken as self-evident.

Plan

How can you identify the minor arcs?
Since a minor arc contains all the points in the interior of a central angle, start by identifying the central angles in the diagram.

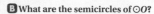

Problem 1 Naming Arcs

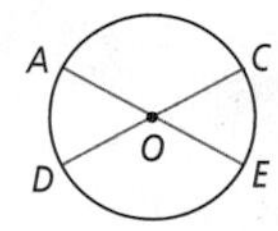

A What are the minor arcs of ⊙O?

The minor arcs are $\overset{\frown}{AD}$, $\overset{\frown}{CE}$, $\overset{\frown}{AC}$, and $\overset{\frown}{DE}$.

B What are the semicircles of ⊙O?

The semicircles are $\overset{\frown}{ACE}$, $\overset{\frown}{CED}$, $\overset{\frown}{EDA}$, and $\overset{\frown}{DAC}$.

C What are the major arcs of ⊙O that contain point A?

The major arcs that contain point A are $\overset{\frown}{ACD}$, $\overset{\frown}{CEA}$, $\overset{\frown}{EDC}$, and $\overset{\frown}{DAE}$.

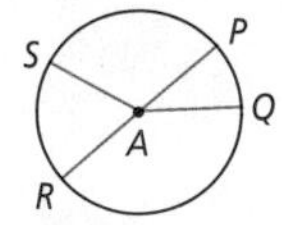

Got It? 1. a. What are the minor arcs of ⊙A?
b. What are the semicircles of ⊙A?
c. What are the major arcs of ⊙A that contain point Q?

take note — Key Concept Arc Measure

Arc Measure

The measure of a minor arc is equal to the measure of its corresponding central angle.

The measure of a major arc is the measure of the related minor arc subtracted from 360.

The measure of a semicircle is 180.

Example

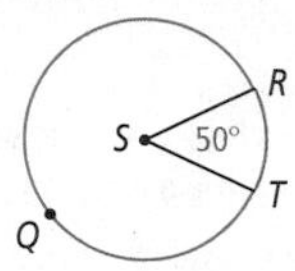

$m\overset{\frown}{RT} = m\angle RST = 50$
$m\overset{\frown}{TQR} = 360 - m\overset{\frown}{RT}$
$= 310$

Adjacent arcs are arcs of the same circle that have exactly one point in common. You can add the measures of adjacent arcs just as you can add the measures of adjacent angles.

take note — Postulate 10-2 Arc Addition Postulate

The measure of the arc formed by two adjacent arcs is the sum of the measures of the two arcs.

$$m\overset{\frown}{ABC} = m\overset{\frown}{AB} + m\overset{\frown}{BC}$$

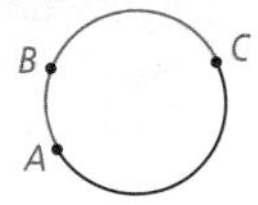

Answers

Solve It!

21 in.; explanations may vary. Sample: 120° is one third of a complete revolution, so the wheel will travel $\frac{1}{3} \cdot 63 = 21$ in. for a rotation of 120°.

Got It?

1a. $\overset{\frown}{SP}$, $\overset{\frown}{SQ}$, $\overset{\frown}{PQ}$, $\overset{\frown}{QR}$, $\overset{\frown}{RS}$
b. $\overset{\frown}{RSP}$, $\overset{\frown}{RQP}$
c. $\overset{\frown}{PQS}$, $\overset{\frown}{PSQ}$, $\overset{\frown}{SPR}$, $\overset{\frown}{QRS}$, $\overset{\frown}{RSQ}$
2a. 77
b. 103
c. 208
d. 283

PowerGeometry.com

2 Guided Instruction

Each Problem is worked out and supported online.

Problem 1
Naming Arcs
Animated

Problem 2
Finding the Measures of Arcs
Animated

Problem 3
Finding a Distance

Problem 4
Finding Arc Length

Support in Geometry Companion
- Vocabulary
- Key Concepts
- Got It?

Problem 2 Finding the Measures of Arcs

Think

How can you find $m\overset{\frown}{BD}$?

$\overline{BD}$ is formed by adjacent arcs $\overset{\frown}{BC}$ and $\overset{\frown}{CD}$. Use the Arc Addition Postulate.

What is the measure of each arc in $\odot O$?

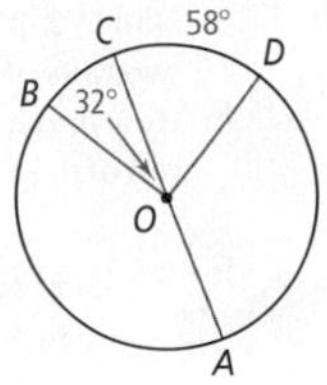

A $\overset{\frown}{BC}$ $\quad m\overset{\frown}{BC} = m\angle BOC = 32$

B $\overset{\frown}{BD}$ $\quad m\overset{\frown}{BD} = m\overset{\frown}{BC} + m\overset{\frown}{CD}$

$m\overset{\frown}{BD} = 32 + 58 = 90$

C $\overset{\frown}{ABC}$ $\quad \overset{\frown}{ABC}$ is a semicircle.

$m\overset{\frown}{ABC} = 180$

D $\overset{\frown}{AB}$ $\quad m\overset{\frown}{AB} = 180 - 32 = 148$

Got It? 2. What is the measure of each arc in $\odot C$?

a. $m\overset{\frown}{PR}$

b. $m\overset{\frown}{RS}$

c. $m\overset{\frown}{PRQ}$

d. $m\overset{\frown}{PQR}$

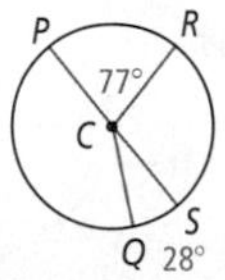

The **circumference** of a circle is the distance around the circle. The number **pi** (π) is the ratio of the circumference of a circle to its diameter.

take note

Theorem 10-9 Circumference of a Circle

The circumference of a circle is π times the diameter.

$C = \pi d$ or $C = 2\pi r$

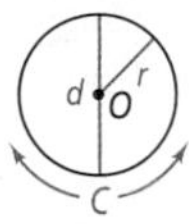

The number π is irrational, so you cannot write it as a terminating or repeating decimal. To approximate π, you can use 3.14, $\frac{22}{7}$, or the π key on your calculator.

Many properties of circles deal with ratios that stay the same no matter what size the circle is. This is because all circles are similar to each other. To see this, consider the circles at the right. There is a translation that maps circle O so that it shares the same center with circle P.

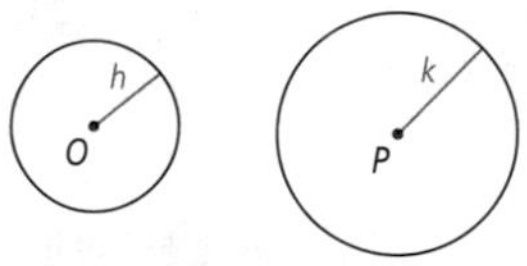

There also exists a dilation with scale factor $\frac{k}{h}$ that maps circle O to circle P. A translation followed by a dilation is a similarity transformation. Because a similarity transformation maps circle O to circle P, the two circles are similar.

Coplanar circles that have the same center are called **concentric circles**.

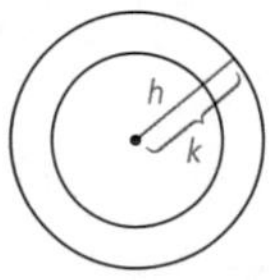

Concentric circles

Problem 2

Q How is the measure of an arc related to the measure of its central angle? **[The measures are the same.]**

Q How can you write $m\overset{\frown}{BD}$ as a sum of two other arc measures? **[$m\overset{\frown}{BC} + m\overset{\frown}{CD}$]**

Q How can you classify $\overset{\frown}{ABC}$? **[It is a semicircle.]**

Q What is $m\overset{\frown}{AB}$? **[$m\overset{\frown}{AB} = 180 - m\overset{\frown}{BC} = 180 - 32 = 148$]**

Got It?

ERROR PREVENTION

Have students calculate and label each central angle in the diagram.

Take Note

Review the relationship between radius and diameter. Be sure that students can identify both measurements in a circle. Have students research the history of pi (π). Emphasize that pi is an irrational number. Help students locate and understand how to use the pi (π) button on their calculators.

EXTENSION

Have students investigate the relationship between scale factor and circumference using several circles with different radii.

Additional Problems

1. a. What are the minor arcs of $\odot C$?

b. What are the semicircles of $\odot C$?

c. What are the major arcs of $\odot C$ that contain point B?

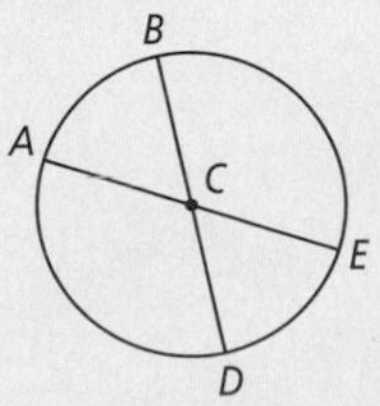

ANSWER a. arcs AB, BE, ED, and DA
b. arcs ABE, BED, EDA, and DAB
c. arcs BEA, EDB, DAE, and ABD

2. What is the measure of each arc in $\odot O$?

a. arc TOU

b. arc TUR

c. arc STU

d. arc ROU

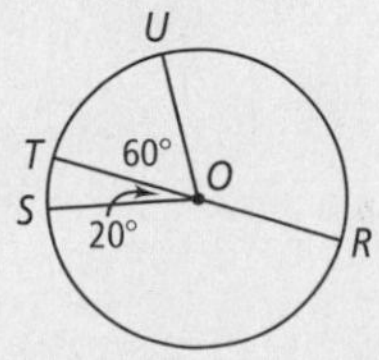

ANSWER a. 60 **b.** 180 **c.** 80 **d.** 120

3. A merry-go-round has seats that are 7 ft from the center of the ride and 10 ft from the center. How much farther does a child seated on the outside loop travel than a child seated on the inside loop in one complete revolution?

ANSWER about 18.8 ft

4. What is the length of each arc?

a. arc CD **b.** arc STR

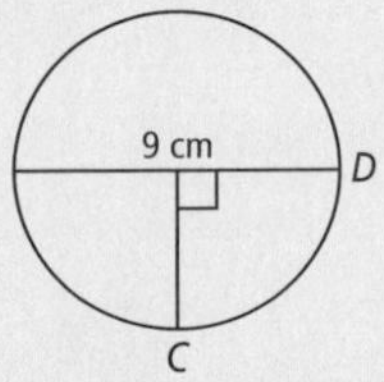

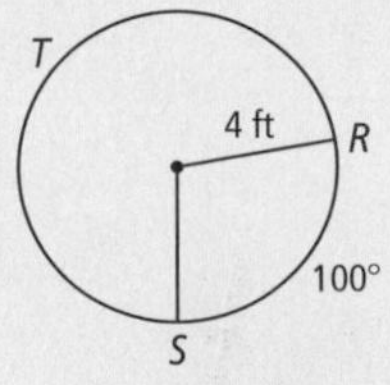

ANSWER a. about 7.1 cm **b.** about 18.2 ft

Problem 3

Q How can you find the distance traveled by wheels on the two tracks? **[Calculate the circumference of the two circles.]**

Q What is the radius of the outer circle? **[8 ft + 2 ft = 10 ft]**

Q How do you know that a wheel on the outer edge travels farther than a wheel on the inner edge? **[The radius of the outer edge is greater, so the circumference is greater.]**

Q How can you find the difference in the distances traveled? **[Calculate the difference between the two circumferences.]**

Got It?

Be sure that students find the radius of the entire outer circle. They should draw and label a diagram of the situation on their papers. In 3b, have students calculate the circumference of each circle using a common variable to relate the two radii. Have students write a ratio of the circumferences and simplify.

Plan

What do you need to find?
You need to find the distance around the track, which is the circumference of a circle.

Problem 3 Finding a Distance

Film **A 2-ft-wide circular track for a camera dolly is set up for a movie scene. The two rails of the track form concentric circles. The radius of the inner circle is 8 ft. How much farther does a wheel on the outer rail travel than a wheel on the inner rail of the track in one turn?**

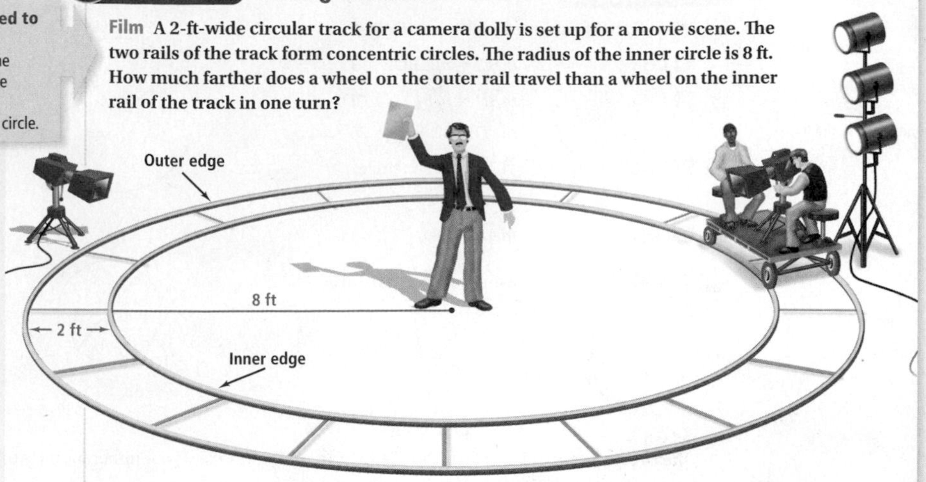

circumference of inner circle $= 2\pi r$ — Use the formula for the circumference of a circle.

$= 2\pi(8)$ — Substitute 8 for *r*.

$= 16\pi$ — Simplify.

The radius of the outer circle is the radius of the inner circle plus the width of the track.

radius of the outer circle $= 8 + 2 = 10$

circumference of outer circle $= 2\pi r$ — Use the formula for the circumference of a circle.

$= 2\pi(10)$ — Substitute 10 for *r*.

$= 20\pi$ — Simplify.

The difference in the two distances traveled is $20\pi - 16\pi$, or 4π ft.

$4\pi \approx 12.56637061$ — Use a calculator.

A wheel on the outer edge of the track travels about 13 ft farther than a wheel on the inner edge of the track.

Got It? **3. a.** A car has a circular turning radius of 16.1 ft. The distance between the two front tires is 4.7 ft. How much farther does a tire on the outside of the turn travel than a tire on the inside?

b. Reasoning Suppose the radius of $\odot A$ is equal to the diameter of $\odot B$. What is the ratio of the circumference of $\odot A$ to the circumference of $\odot B$? Explain.

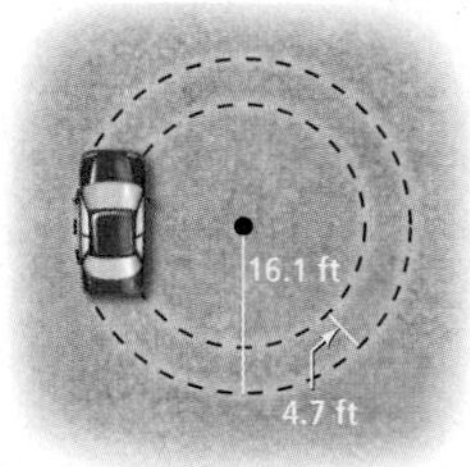

Answers

Got It? (continued)

3a. about 29.5 ft

b. 2 : 1; if the radius of $\odot A$ is *r*, then its circumference is $2\pi r$. $\odot B$ will have a circumference of πr. The ratio of their circumferences is $\frac{2\pi r}{\pi r} = \frac{2}{1}$, or 2 : 1.

The measure of an arc is in degrees, while the **arc length** is a fraction of the circumference.

Consider the arcs shown at the right. Since the circles are concentric, there is a dilation that maps C_1 to C_2. The same dilation maps the slice of the small circle to the slice of the large circle. Since corresponding lengths of similar figures are proportional,

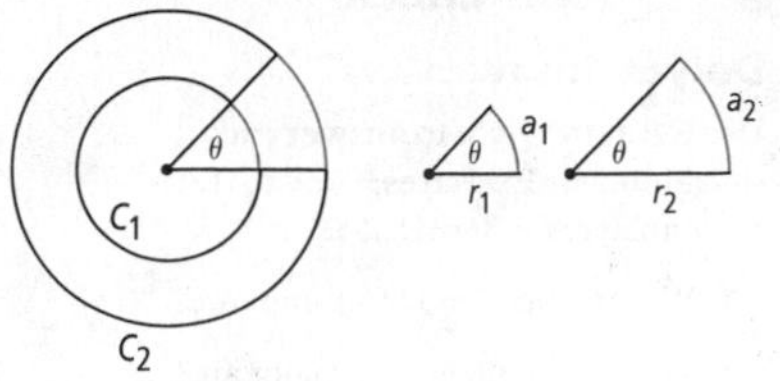

$$\frac{r_1}{r_2} = \frac{a_1}{a_2}$$

$$r_1 a_2 = r_2 a_1$$

$$a_1 = r_1 \frac{a_2}{r_2}$$

This means that the arc length a_1 is equal to the radius r_1 times some number. So for a given central angle, the length of the arc it intercepts depends only on the radius.

An arc of 60° represents $\frac{60}{360}$, or $\frac{1}{6}$, of the circle. So its arc length is $\frac{1}{6}$ of the circumference. This observation suggests the following theorem.

take note

Theorem 10-10 Arc Length

The length of an arc of a circle is the product of the ratio $\frac{\text{measure of the arc}}{360}$ and the circumference of the circle.

$$\text{length of } \widehat{AB} = \frac{m\widehat{AB}}{360} \cdot 2\pi r$$

$$= \frac{m\widehat{AB}}{360} \cdot \pi d$$

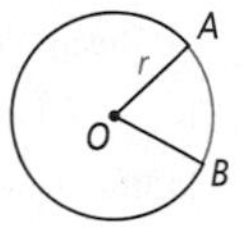

Problem 4 Finding Arc Length

What is the length of each arc shown in red? Leave your answer in terms of π.

Think

How do you know which formula to use?
It depends on whether the diameter is given or the radius is given.

A

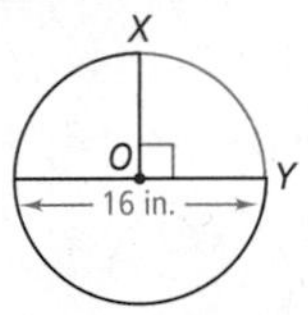

length of $\widehat{XY} = \frac{m\widehat{XY}}{360} \cdot \pi d$ Use a formula for arc length.

$= \frac{90}{360} \cdot \pi(16)$ Substitute.

$= 4\pi$ in. Simplify.

B

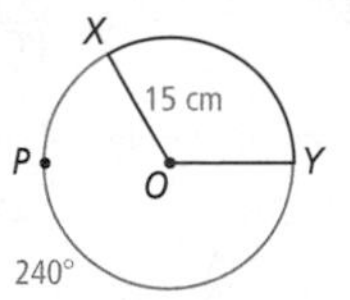

length of $\widehat{XPY} = \frac{m\widehat{XPY}}{360} \cdot 2\pi r$

$= \frac{240}{360} \cdot 2\pi(15)$

$= 20\pi$ cm

Got It? 4. What is the length of a semicircle with radius 1.3 m? Leave your answer in terms of π.

ERROR PREVENTION

Discuss the difference between arc measure and arc length. Arc measure is a degree measure similar to the measure of angles and is denoted by $m\widehat{AB}$. Arc length is the distance around the curve and is similar to the length of a line segment. Show students that it is possible for two arcs of different circles to have the same measure but different lengths. Similarly, it is possible for two arcs of different circles to have the same length but different measures.

Take Note

Students should begin to see that an arc is related to a circle by the central angle that defines it. Have them review their work in the Solve It. The ratio of the central angle to the total number of degrees in a circle will appear again when discussing the area of a sector. Be sure that students understand its significance.

Problem 4

Q In 4A, what fraction of the circle is represented by the central angle that created the arc? **[$\frac{1}{4}$]**

Q What fraction of the circle is represented by the highlighted arc in 4B? **[$\frac{2}{3}$]**

Got It?

Ask students to define a semicircle. They should be able to identify the fraction of a circle represented by a semicircle without first identifying the measure of its associated central angle.

4. 1.3π m

3 Lesson Check

Do you know HOW?

- If students have difficulty with Exercise 5, then have them review Theorem 10-9 to state the formula for the circumference.

Do you UNDERSTAND?

- If students have difficulty with Exercise 7, then have them review Problem 2 to find an arc and its arc length.

Close

Q What is the measure of an arc? **[It is equal to the measure of the central angle that defines it.]**

Q How do you find the circumference of a circle? **[Multiply 2π by the radius of the circle or multiply π by the diameter of the circle.]**

Q How can you find the length of an arc? **[Multiply the ratio of the arc measure to 360° by the circumference of the circle.]**

Lesson Check

Do you know HOW?

Use $\odot P$ at the right to answer each question. For Exercises 5 and 6, leave your answers in terms of π.

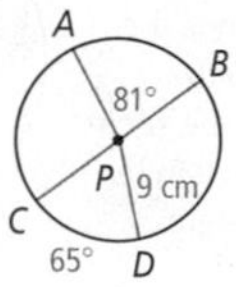

1. What is the name of a minor arc?
2. What is the name of a major arc?
3. What is the name of a semicircle?
4. What is $m\widehat{AB}$?
5. What is the circumference of $\odot P$?
6 What is the length of $\widehat{BD}$?

Do you UNDERSTAND? MATHEMATICAL PRACTICES

7. **Vocabulary** What is the difference between the measure of an arc and arc length? Explain.
8. **Error Analysis** Your class must find the length of $\widehat{AB}$. A classmate submits the following solution. What is the error?

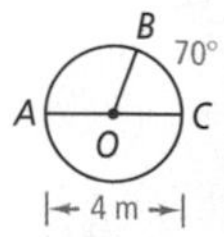

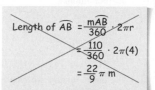

$$\text{Length of } \widehat{AB} = \frac{m\widehat{AB}}{360} \cdot 2\pi r = \frac{110}{360} \cdot 2\pi(4) = \frac{22}{9}\pi \text{ m}$$

Practice and Problem-Solving Exercises MATHEMATICAL PRACTICES

A Practice

Name the following in $\odot O$. See Problem 1.

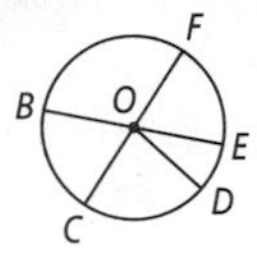

9. the minor arcs
10. the major arcs
11. the semicircles

Find the measure of each arc in $\odot P$. See Problem 2.

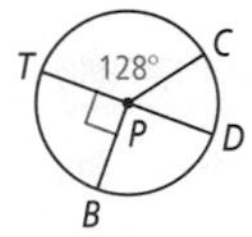

12. $\widehat{TC}$ 13. $\widehat{TBD}$ 14. $\widehat{BTC}$
15. $\widehat{TCB}$ 16. $\widehat{CD}$ 17. $\widehat{CBD}$
18. $\widehat{TCD}$ 19. $\widehat{DB}$ 20. $\widehat{TDC}$
21. $\widehat{TB}$ 22. $\widehat{BC}$ 23. $\widehat{BCD}$

Find the circumference of each circle. Leave your answer in terms of π. See Problem 3.

24.
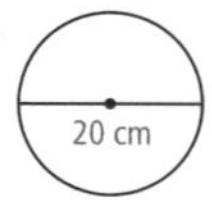

25.
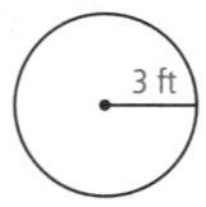

26.

27.
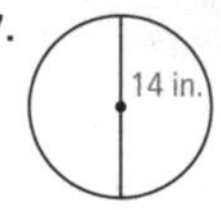

28. The camera dolly track in Problem 3 can be expanded so that the diameter of the outer circle is 70 ft. How much farther will a wheel on the outer rail travel during one turn around the track than a wheel on the inner rail?

PowerGeometry.com

3 Lesson Check

For a digital lesson check, use the Got It questions.

Support in Geometry Companion

- Lesson Check

4 Practice

Assign homework to individual students or to an entire class.

Answers

Lesson Check

1–3. Answers may vary. Samples are given.

1. $\widehat{AB}$
2. $\widehat{DAB}$
3. $\widehat{CAB}$
4. 81
5. 18π cm
6. $\frac{23\pi}{4}$ cm
7. The measure of an arc corresponds to the measure of a central angle; an arc length is a fraction of the circle's circumference.
8. The student substituted the diameter into the formula that requires the radius.

Practice and Problem-Solving Exercises

9. $\widehat{BC}$, $\widehat{BD}$, $\widehat{CD}$, $\widehat{CE}$, $\widehat{DE}$, $\widehat{DF}$, $\widehat{EF}$, $\widehat{FB}$
10. $\widehat{BDF}$, $\widehat{CDB}$, $\widehat{DEB}$, $\widehat{EFC}$, $\widehat{EFD}$, $\widehat{FBD}$, $\widehat{FBE}$, $\widehat{CFD}$
11. $\widehat{BCE}$, $\widehat{BFE}$, $\widehat{CBF}$, $\widehat{CDF}$
12. 128
13. 180
14. 218
15. 270
16. 52
17. 308
18. 180
19. 90
20. 232
21. 90
22. 142
23. 270
24. 20π cm
25. 6π ft
26. 8.4π m
27. 14π in.
28. About 13 ft

29. The wheel of a compact car has a 25-in. diameter. The wheel of a pickup truck has a 31-in. diameter. To the nearest inch, how much farther does the pickup truck wheel travel in one revolution than the compact car wheel?

Find the length of each arc shown in red. Leave your answer in terms of π.

See Problem 4.

30.

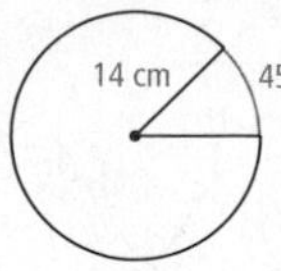

31.

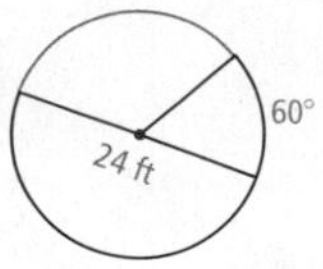

32.

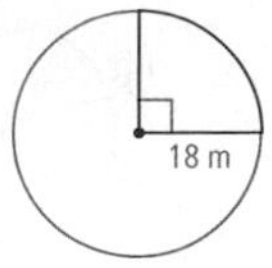

33.

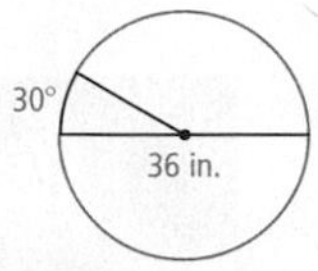

34.

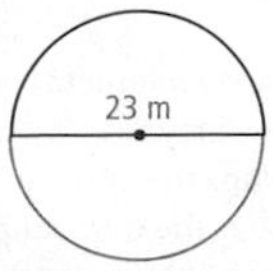

35.

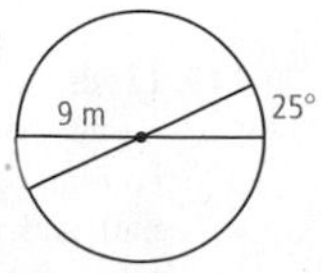

B Apply

36. **Think About a Plan** Nina designed a semicircular arch made of wrought iron for the top of a mall entrance. The nine segments between the two concentric semicircles are each 3 ft long. What is the total length of wrought iron used to make this structure? Round your answer to the nearest foot.
- What do you know from the diagram?
- What formula should you use to find the amount of wrought iron used in the semicircular arches?

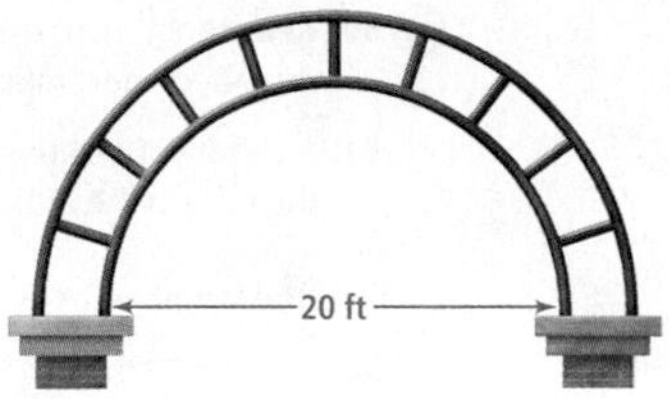

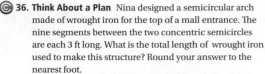

Find each indicated measure for ⊙O.

37. $m\angle EOF$	38. $m\overparen{EJH}$	39. $m\overparen{FH}$
40. $m\angle FOG$	41. $m\overparen{JEG}$	42. $m\overparen{HFJ}$

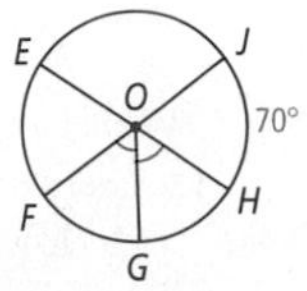

43. **Pets** A hamster wheel has a 7-in. diameter. How many feet will a hamster travel in 100 revolutions of the wheel?

STEM 44. **Traffic** Five streets come together at a traffic circle, as shown at the right. The diameter of the circle traveled by a car is 200 ft. If traffic travels counterclockwise, what is the approximate distance from East St. to Neponset St.?

Ⓐ 227 ft Ⓒ 454 ft

Ⓑ 244 ft Ⓓ 488 ft

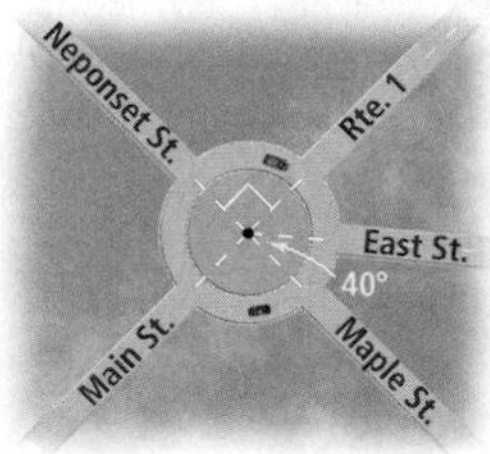

45. **Writing** Describe two ways to find the arc length of a major arc if you are given the measure of the corresponding minor arc and the radius of the circle.

29. 19 in.
30. $\frac{7\pi}{2}$ cm
31. 8π ft
32. 27π m
33. 33π in.
34. $\frac{23\pi}{2}$ m
35. $\frac{5\pi}{4}$ m
36. 99 ft
37. 70
38. 180
39. 110
40. 55
41. 235
42. 290
43. about 183.3 ft
44. B
45. Find the measure of the major arc, then use Thm. 10-10; or find the length of the minor arc using Thm. 10-10, then subtract that length from the circumference of the circle.

4 Practice

ASSIGNMENT GUIDE

Basic: 9–35 all, 36–50 even
Average: 9–35 odd, 36–56
Advanced: 9–35 odd, 36–59
Standardized Test Prep: 60–63
Mixed Review: 64–71

Mathematical Practices are supported by exercises with red headings. Here are the Practices supported in this lesson:

MP 1: Make Sense of Problems Ex. 36
MP 3: Construct Arguments Ex. 50
MP 3: Critique the Reasoning of Others Ex. 8
MP 8: Repeated Reasoning Ex. 45

Applications exercises have blue headings. Exercises 44, 49, and 59 support MP 4: Model.

STEM exercises focus on science or engineering applications.

EXERCISE 46: Use the Think About a Plan worksheet in the **Practice and Problem Solving Workbook** (also available in the Teaching Resources in print and online) to further support students' development in becoming independent learners.

HOMEWORK QUICK CHECK

To check students' understanding of key skills and concepts, go over Exercises 13, 31, 36, 46, and 50.

Answers

Practice and Problem-Solving Exercises (continued)

46a. 6
b. 30
c. 120
47. 38
48. 40
49. 31 m
50. The circumference is doubled; explanations may vary. Sample: Since $C = 2\pi r$, doubling the radius results in $2\pi(2r) = 2(2\pi r) = 2C$.
51. 3 : 4
52. 5.125π ft
53. 2.6π in.
54. 3π m
55. 7.9 units
56. 18 cm
57. Since $\overline{AR} \cong \overline{RW}$ and $AR + RW = AW$ by the Seg. Add. Post., $AW = 2 \cdot AR$. So the radius of the outer circle is twice the radius of the inner circle. Because $\angle QAR$ and $\angle SAU$ are vertical ⦣s, and $m\angle SAT = \frac{1}{2} m\angle SAU$, $m\angle QAR = 2 \cdot m\angle SAT$. The length of $\overset{\frown}{ST} = \frac{m\angle SAT}{360} \cdot 2\pi(2 \cdot AR) = \frac{m\angle SAT}{90} \cdot \pi(AR)$ and the length of $\overset{\frown}{QR} = \frac{m\angle QAR}{360} \cdot 2\pi(AR) = \frac{2 \cdot m\angle SAT}{360} \cdot 2\pi(AR) = \frac{m\angle SAT}{90} \cdot \pi(AR)$. Therefore the length of $\overset{\frown}{ST}$ = the length of $\overset{\frown}{QR}$ by the Trans. Prop. of Eq.
58. $\overline{AP} \cong \overline{BP}$ (Radii of a circle are ≅.); $\triangle APB$ is isosc. (def. of an isosc. △); $\angle A \cong \angle B$ (Isosc. △ Thm.); $\overline{AB} \parallel \overline{PC}$ (Given); $\angle B \cong \angle BPC$ (Alt. Int. ⦣ Thm.); $\angle A \cong \angle CPD$ (Corresp. ⦣ Post.); $\angle BPC \cong \angle CPD$ (Trans. Prop. of ≅); $m\angle BPC = m\overset{\frown}{BC}$ and $m\angle CPD = m\overset{\frown}{CD}$ (The measure of a minor arc is = to the measure of its corresp. central ∠.); $m\overset{\frown}{BC} = m\overset{\frown}{CD}$ (Trans. Prop. of =).

46. Time Hands of a clock suggest an angle whose measure is continually changing. How many degrees does a minute hand move through during each time interval?

a. 1 min **b.** 5 min **c.** 20 min

Algebra **Find the value of each variable.**

47.

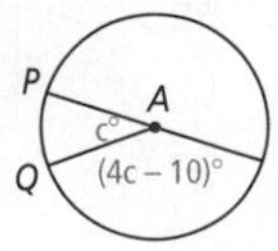

48.

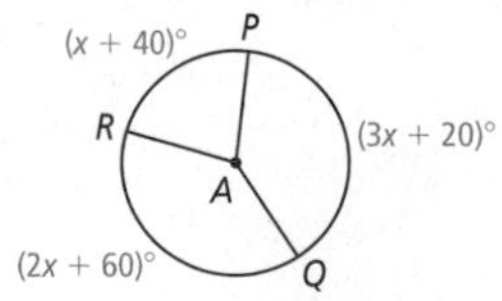

49. Landscape Design A landscape architect is constructing a curved path through a rectangular yard. The curved path consists of two 90° arcs. He plans to edge the two sides of the path with plastic edging. What is the total length of plastic edging he will need? Round your answer to the nearest meter.

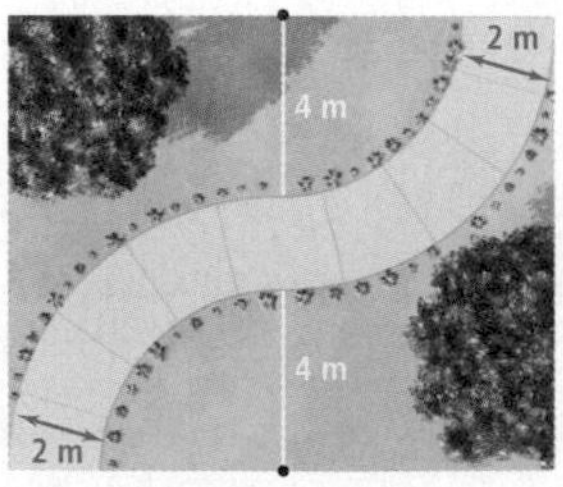

50. Reasoning Suppose the radius of a circle is doubled. How does this affect the circumference of the circle? Explain.

51. A 60° arc of ⊙A has the same length as a 45° arc of ⊙B. What is the ratio of the radius of ⊙A to the radius of ⊙B?

Find the length of each arc shown in red. Leave your answer in terms of π.

52.

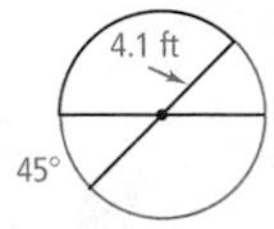

53.

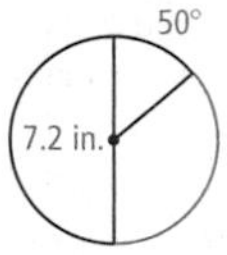

54.

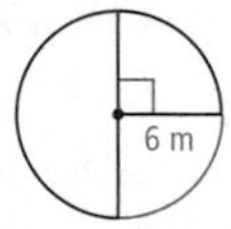

55. Coordinate Geometry Find the length of a semicircle with endpoints (1, 3) and (4, 7). Round your answer to the nearest tenth.

56. In ⊙O, the length of $\overset{\frown}{AB}$ is 6π cm and $m\overset{\frown}{AB}$ is 120. What is the diameter of ⊙O?

Challenge

57. The diagram below shows two concentric circles. $\overline{AR} \cong \overline{RW}$. Show that the length of $\overset{\frown}{ST}$ is equal to the length of $\overset{\frown}{QR}$.

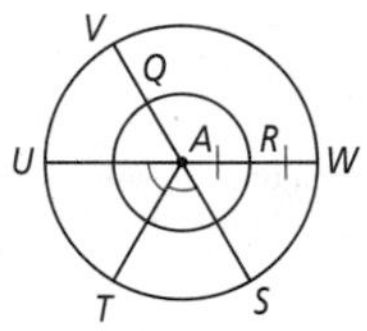

58. Given: ⊙P with $\overline{AB} \parallel \overline{PC}$
Proof **Prove:** $m\overset{\frown}{BC} = m\overset{\frown}{CD}$

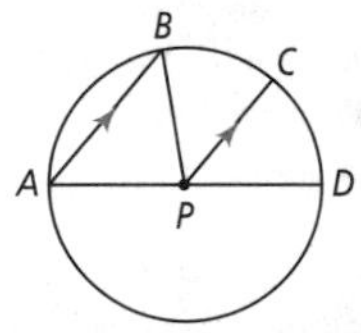

59. Sports An athletic field is a 100 yd-by-40 yd rectangle, with a semicircle at each of the short sides. A running track 10 yd wide surrounds the field. If the track is divided into eight lanes of equal width, what is the distance around the track along the inside edge of each lane?

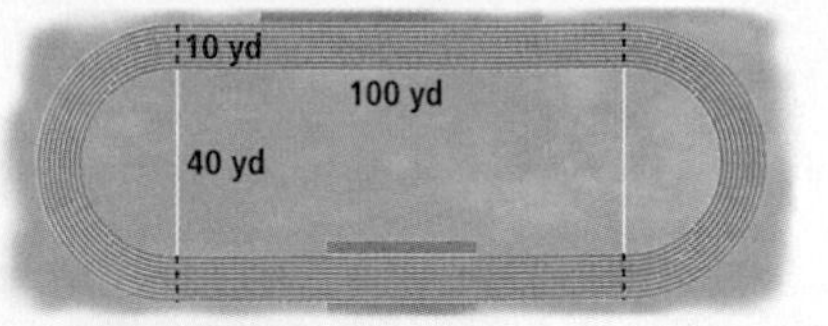

Standardized Test Prep

SAT/ACT

60. The radius of a circle is 12 cm. What is the length of a 60° arc?

(A) 3π cm (B) 4π cm (C) 5π cm (D) 6π cm

61. What is the image of *P* for a 135° clockwise rotation about the center of the regular octagon?

(F) *S* (G) *T* (H) *U* (I) *R*

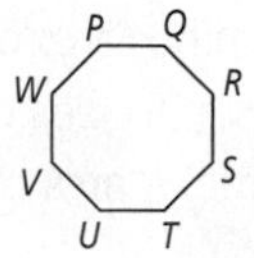

62. Which of the following are the sides of a right triangle?

(A) 6, 8, 12 (B) 8, 15, 17 (C) 9, 11, 23 (D) 5, 12, 15

Extended Response

63. Quadrilateral *ABCD* has vertices *A*(1, 1), *B*(4, 1), *C*(4, 6), and *D*(1, 6). Quadrilateral *RSTV* has vertices *R*(−3, 4), *S*(−3, −2), *T*(−13 −2), and *V*(−13, 4). Show that *ABCD* and *RSTV* are similar rectangles.

Mixed Review

Part of a regular dodecagon is shown at the right.

See Lesson 10-5.

64. What is the measure of each numbered angle?

65. The radius is 19.3 mm. What is the apothem?

66. What is the perimeter and area of the dodecagon to the nearest millimeter or square millimeter?

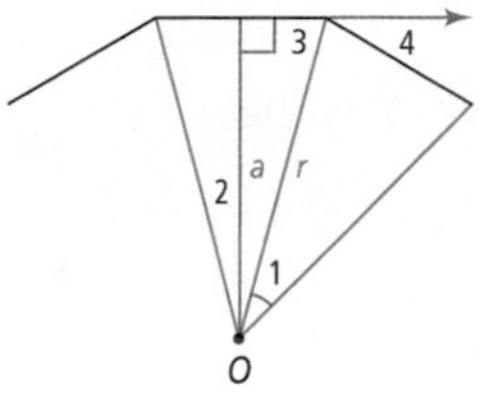

Can you conclude that the figure is a parallelogram? Explain.

See Lesson 6-3.

67.

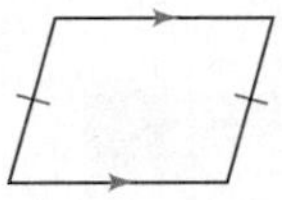

68.

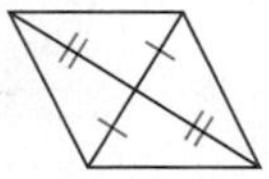

69.

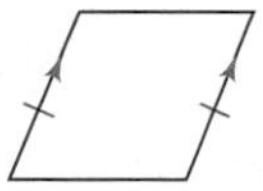

Get Ready! **To prepare for Lesson 10-7, do Exercises 70 and 71.**

See Lesson 10-6.

70. What is the circumference of a circle with diameter 17 in.?

71. What is the length of a 90° arc in a circle with radius 6 cm?

59. 325.7 yd, 333.5 yd, 341.4 yd, 349.2 yd, 357.1 yd, 365.0 yd, 372.8 yd, 380.6 yd

60. B **61.** F **62.** B

63. [4] Using the Distance Formula, $AB = CD = 3$, $BC = AD = 5$, $RS = TV = 6$, and $ST = RV = 10$. The slopes of $\overline{AB}$ and $\overline{CD} = 0$ and the slopes of $\overline{BC}$ and $\overline{AD}$ are undefined. So both $\overline{AB}$ and $\overline{CD}$ are ⊥ to $\overline{BC}$ and $\overline{AD}$. Therefore, *ABCD* is a rectangle and ∠s *A*, *B*, *C*, and *D* are rt. ∠s. The slopes of $\overline{RS}$ and $\overline{TV}$ are undefined and the slopes of $\overline{ST}$ and $\overline{RV} = 0$. So, *RSTV* is a rectangle and ∠s *R*, *S*, *T*, and *V* are rt. ∠s. Since all rt. ∠s are =, the pairs of corresponding ∠s are ≅. The short sides of the two rectangles are 3 and 6, and the long sides are 5 and 10. Since $\frac{3}{6} = \frac{5}{10} = \frac{1}{2}$, the corresp. sides are proportional. Therefore, $ABCD \sim RSTV$ by the def. of ~ polygons.

[3] one missing or incorrect step

[2] two missing or incorrect steps

[1] more than two missing or incorrect steps

64. $m\angle 1 = 30$, $m\angle 2 = 15$, $m\angle 3 = 75$, $m\angle 4 = 30$

65. 18.6 mm

66. Answers may vary slightly. Samples: 120 mm; 1116 mm^2

67. No; it could be an isosc. trapezoid.

68. Yes; the diagonals bis. each other, so it is a ▱.

69. Yes; one pair of sides is both ≅ and ∥, so it is a ▱.

70. 17π in. or about 53.4 in.

71. 3π cm or about 9.4 cm

10-6 Lesson Resources

Differentiated Remediation

Instructional Support

Geometry Companion

Students can use the **Geometry Companion** worktext (4 pages) . . .

- New Vocabulary
- Key Concepts
- Got It for each Problem
- Lesson Check

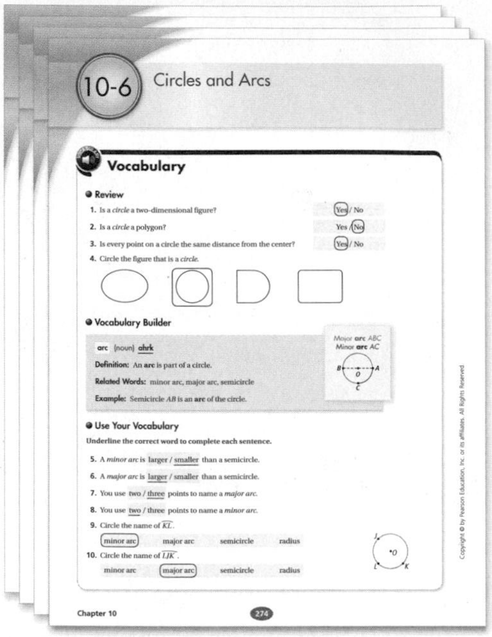
10-6 Circles and Arcs

Vocabulary

Review

1. Is a circle a two-dimensional figure? Yes / No
2. Is a circle a polygon? Yes / No
3. Is every point on a circle the same distance from the center? Yes / No
4. Circle the figure that is a circle.

Vocabulary Builder

arc (noun) ahrk

Definition: An arc is part of a circle.

Related Words: minor arc, major arc, semicircle

Example: Semicircle AB is an arc of the circle.

Use Your Vocabulary

Underline the correct word to complete each sentence.

5. A minor arc is larger / smaller than a semicircle.
6. A major arc is larger / smaller than a semicircle.
7. You use two / three points to name a major arc.
8. You use two / three points to name a minor arc.
9. Circle the name of $\overparen{KL}$. minor arc / major arc / semicircle / radius
10. Circle the name of $\overparen{LJK}$. minor arc / major arc / semicircle / radius

Chapter 10 274

ELL Support

Use Graphic Organizers Students can work in mixed pairs or small groups. Have students construct then cut out a circle to use as an organizer for key vocabulary. Have them draw and label the parts of a circle on their circle organizer using the vocabulary list from the lesson. Encourage students to use multiple colors to differentiate between parts. Ask students to add related words to the labels on their organizers.

5 Assess & Remediate

Lesson Quiz

1. Use the circle below for Questions 1–3.
   a. What are the minor arcs of ⊙*L*?
   b. What are the semicircles of ⊙*L*?
   c. What are the major arcs of ⊙*L* that contain point *K*?

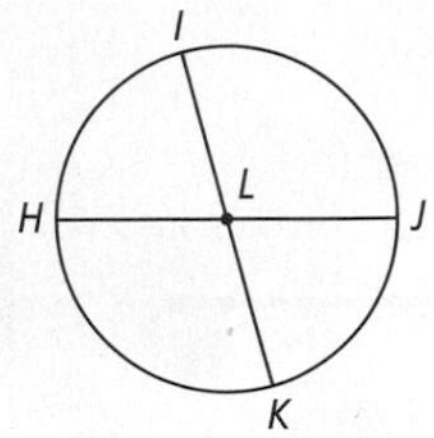

2. **Do you UNDERSTAND?** What is the measure of arc *WX* in ⊙*V*?

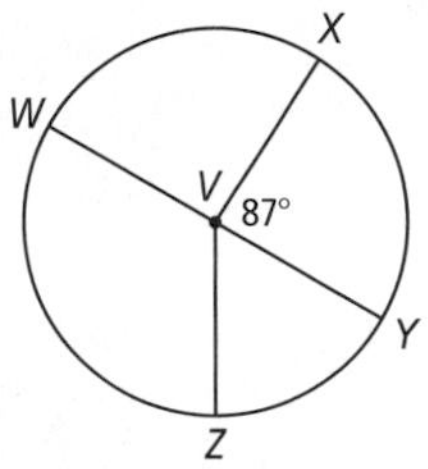

3. The radius of ⊙*C* is four times the radius of ⊙*D*. How many times greater is the circumference of ⊙*C* than ⊙*D*?

ANSWERS TO LESSON QUIZ

1. **a.** arcs *HI*, *IJ*, *JK*, and *KH* **b.** arcs *HIJ*, *IJK*, *JKH*, and *KHI* **c.** arcs *KHJ*, *HIK*, *IJH*, and *JKI*
2. 93°
3. four times

PRESCRIPTION FOR REMEDIATION

Use the student work on the Lesson Quiz to prescribe a differentiated review assignment.

Points	Differentiated Remediation
0–1	Intervention
2	On-level
3	Extension

PowerGeometry.com

5 Assess & Remediate

Assign the Lesson Quiz. Appropriate intervention, practice, or enrichment is automatically generated based on student performance.

Intervention

- **Reteaching** (2 pages) Provides reteaching and practice exercises for the key lesson concepts. Use with struggling students or absent students.
- **English Language Learner Support** Helps students develop and reinforce mathematical vocabulary and key concepts.

All-in-One Resources/Online

Reteaching

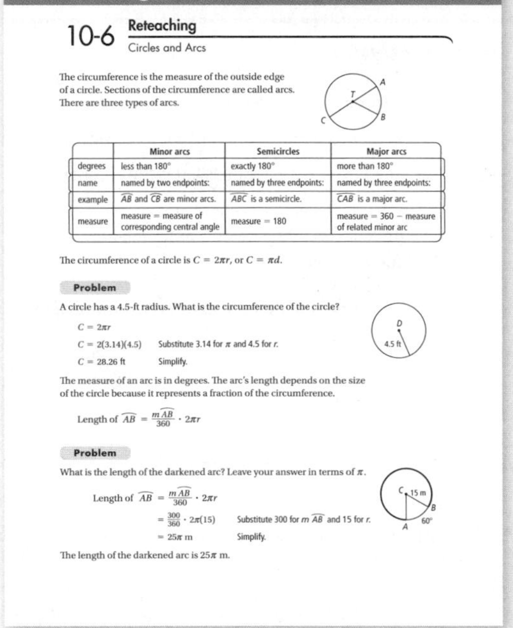
10-6 Reteaching

Circles and Arcs

The circumference is the measure of the outside edge of a circle. Sections of the circumference are called arcs. There are three types of arcs.

	Minor arcs	Semicircles	Major arcs
degrees	less than 180°	exactly 180°	more than 180°
name	named by two endpoints	named by three endpoints	named by three endpoints
example	$\overparen{AB}$ and $\overparen{CB}$ are minor arcs.	$\overparen{ABC}$ is a semicircle.	$\overparen{CAB}$ is a major arc.
measure	measure = measure of corresponding central angle	measure = 180	measure = 360 − measure of related minor arc

The circumference of a circle is $C = 2\pi r$, or $C = \pi d$.

Problem

A circle has a 4.5-ft radius. What is the circumference of the circle?

$C = 2\pi r$

$C = 2(3.14)(4.5)$ Substitute 3.14 for π and 4.5 for r.

$C = 28.26$ ft Simplify.

The measure of an arc is in degrees. The arc's length depends on the size of the circle because it represents a fraction of the circumference.

Length of $\overparen{AB} = \frac{m\overparen{AB}}{360} \cdot 2\pi r$

Problem

What is the length of the darkened arc? Leave your answer in terms of π.

Length of $\overparen{AB} = \frac{m\overparen{AB}}{360} \cdot 2\pi r$

$= \frac{300}{360} \cdot 2\pi(15)$ Substitute 300 for $m\overparen{AB}$ and 15 for r.

$= 25\pi$ m Simplify.

The length of the darkened arc is 25π m.

All-in-One Resources/Online

English Language Learner Support

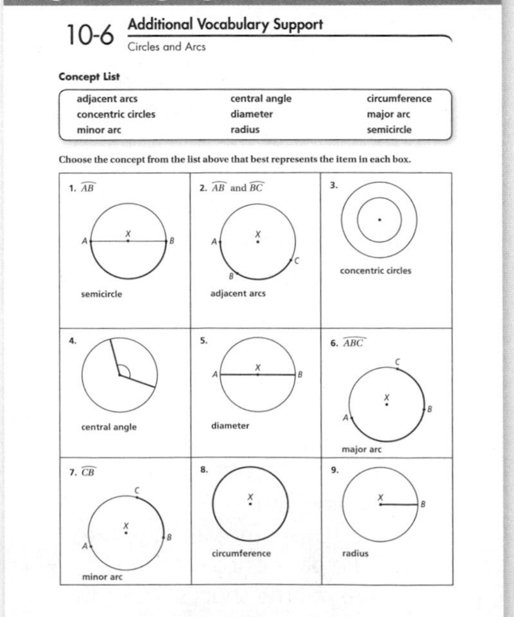
10-6 Additional Vocabulary Support

Circles and Arcs

Concept List

adjacent arcs	central angle	circumference
concentric circles	diameter	major arc
minor arc	radius	semicircle

Choose the concept from the list above that best represents the item in each box.

1. $\overparen{AB}$ semicircle
2. $\overparen{AB}$ and $\overparen{BC}$ adjacent arcs
3. concentric circles
4. central angle
5. diameter
6. $\overparen{ABC}$ major arc
7. $\overparen{CB}$ minor arc
8. circumference
9. radius

Differentiated Remediation *continued*

On-Level

- **Practice** (2 pages) Provides extra practice for each lesson. For simpler practice exercises, use the Form K Practice pages found in the All-in-One Teaching Resources and online.
- **Think About a Plan** Helps students develop specific problem-solving skills and strategies by providing scaffolded guiding questions.
- **Standardized Test Prep** Focuses on all major exercises, all major question types, and helps students prepare for the high-stakes assessments.

Extension

- **Enrichment** Provides students with interesting problems and activities that extend the concepts of the lesson.
- **Activities, Games, and Puzzles** Worksheets that can be used for concepts development, enrichment, and for fun!

Practice and Problem Solving Wkbk/ All-in-One Resources/Online

Practice page 1

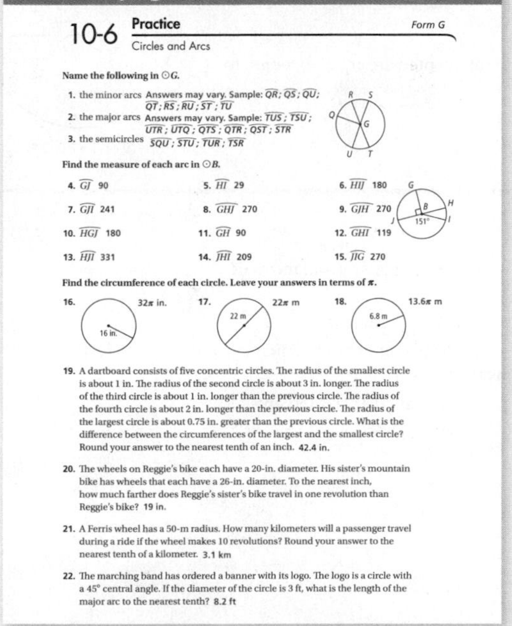
10-6 Practice — Form G
Circles and Arcs

Name the following in ⊙G.
1. the minor arcs Answers may vary. Sample: QR; QS; QU; QT; RS; RU; ST; TU
2. the major arcs Answers may vary. Sample: TUS; TSU; UTR; UTQ; QTS; QTR; QST; STR
3. the semicircles SQU; STU; TUR; TSR

Find the measure of each arc in ⊙B.
4. GJ 90 5. HI 29 6. HIJ 180
7. GJI 241 8. GHJ 270 9. GJH 270
10. HGJ 180 11. GH 90 12. GHI 119
13. HJI 331 14. JHI 209 15. JIG 270

Find the circumference of each circle. Leave your answers in terms of π.
16. 32π in. 17. 22π m 18. 13.6π m

19. A dartboard consists of five concentric circles. The radius of the smallest circle is about 1 in. The radius of the second circle is about 3 in. longer. The radius of the third circle is about 1 in. longer than the previous circle. The radius of the fourth circle is about 2 in. longer than the previous circle. The radius of the largest circle is about 0.75 in. greater than the previous circle. What is the difference between the circumferences of the largest and the smallest circle? Round your answer to the nearest tenth of an inch. 42.4 in.
20. The wheels on Reggie's bike each have a 20-in. diameter. His sister's mountain bike has wheels that each have a 26-in. diameter. To the nearest inch, how much farther does Reggie's sister's bike travel in one revolution than Reggie's bike? 19 in.
21. A Ferris wheel has a 50-m radius. How many kilometers will a passenger travel during a ride if the wheel makes 10 revolutions? Round your answer to the nearest tenth of a kilometer. 3.1 km
22. The marching band has ordered a banner with its logo. The logo is a circle with a 45° central angle. If the diameter of the circle is 3 ft, what is the length of the major arc to the nearest tenth? 8.2 ft

Practice and Problem Solving Wkbk/ All-in-One Resources/Online

Practice page 2

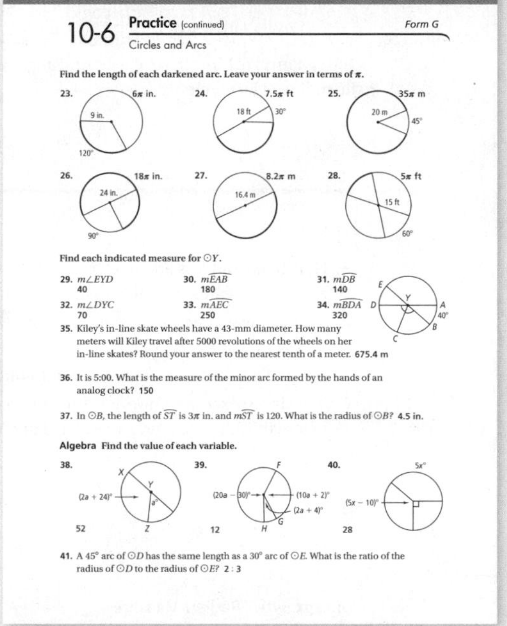
10-6 Practice (continued) — Form G
Circles and Arcs

Find the length of each darkened arc. Leave your answer in terms of π.
23. 6π in. 24. 7.5π ft 25. 35π m
26. 18π in. 27. 8.2π m 28. 5π ft

Find each indicated measure for ⊙Y.
29. m∠EYD 40 30. mEAB 180 31. mDB 140
32. m∠DYC 70 33. mAEC 250 34. mBDA 320
35. Kiley's in-line skate wheels have a 43-mm diameter. How many meters will Kiley travel after 5000 revolutions of the wheels on her in-line skates? Round your answer to the nearest tenth of a meter. 675.4 m
36. It is 5:00. What is the measure of the minor arc formed by the hands of an analog clock? 150
37. In ⊙B, the length of ST is 3π in. and mST is 120. What is the radius of ⊙B? 4.5 in.

Algebra Find the value of each variable.
38. 52 39. 12 40. 28
41. A 45° arc of ⊙D has the same length as a 30° arc of ⊙E. What is the ratio of the radius of ⊙D to the radius of ⊙E? 2 : 3

All-in-One Resources/Online

Enrichment

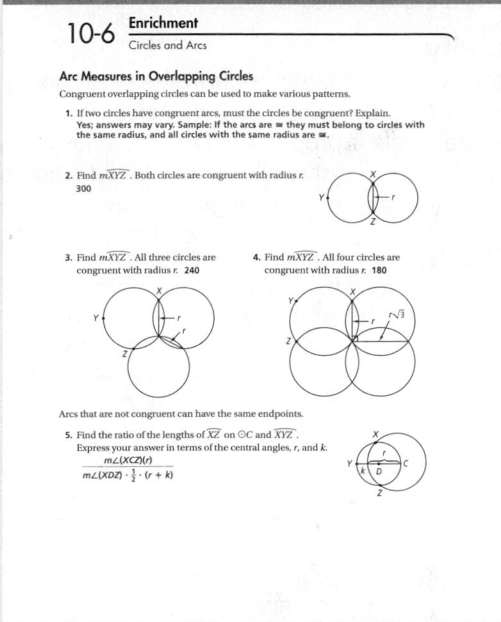
10-6 Enrichment
Circles and Arcs

Arc Measures in Overlapping Circles
Congruent overlapping circles can be used to make various patterns.
1. If two circles have congruent arcs, must the circles be congruent? Explain. Yes; answers may vary. Sample: If the arcs are ≅ they must belong to circles with the same radius, and all circles with the same radius are ≅.
2. Find mXYZ. Both circles are congruent with radius r. 300
3. Find mXYZ. All three circles are congruent with radius r. 240
4. Find mXYZ. All four circles are congruent with radius r. 180

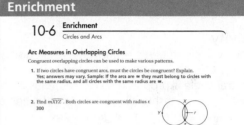

Arcs that are not congruent can have the same endpoints.
5. Find the ratio of the lengths of XZ on ⊙C and XYZ. Express your answer in terms of the central angles, r, and k.
m∠(XCZ)(r) / m∠(XDZ) · ½ · (r + k)

Practice and Problem Solving Wkbk/ All-in-One Resources/Online

Think About a Plan

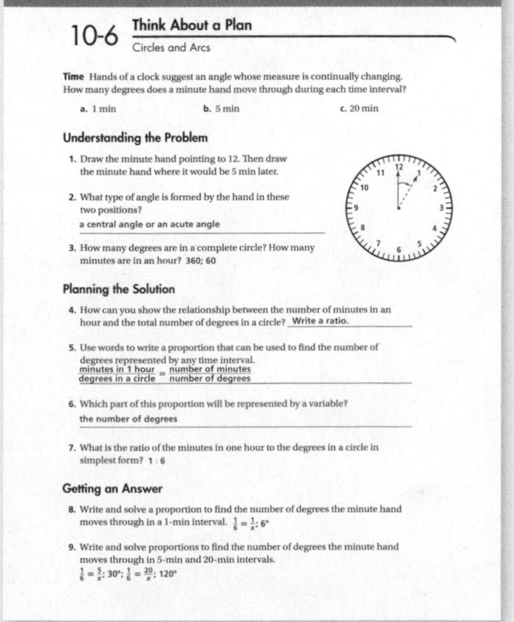
10-6 Think About a Plan
Circles and Arcs

Time Hands of a clock suggest an angle whose measure is continually changing. How many degrees does a minute hand move through during each time interval?
a. 1 min b. 5 min c. 20 min

Understanding the Problem
1. Draw the minute hand pointing to 12. Then draw the minute hand where it would be 5 min later.
2. What type of angle is formed by the hand in these two positions? a central angle or an acute angle
3. How many degrees are in a complete circle? How many minutes are in an hour? 360; 60

Planning the Solution
4. How can you show the relationship between the number of minutes in an hour and the total number of degrees in a circle? Write a ratio.
5. Use words to write a proportion that can be used to find the number of degrees represented by any time interval. minutes in 1 hour / degrees in a circle = number of minutes / number of degrees
6. Which part of this proportion will be represented by a variable? the number of degrees
7. What is the ratio of the minutes in one hour to the degrees in a circle in simplest form? 1 : 6

Getting an Answer
8. Write and solve a proportion to find the number of degrees the minute hand moves through in a 1-min interval. $\frac{1}{6} = \frac{1}{x}$; 6°
9. Write and solve proportions to find the number of degrees the minute hand moves through in 5-min and 20-min intervals. $\frac{1}{6} = \frac{5}{x}$; 30°; $\frac{1}{6} = \frac{20}{x}$; 120°

Practice and Problem Solving Wkbk/ All-in-One Resources/Online

Standardized Test Prep

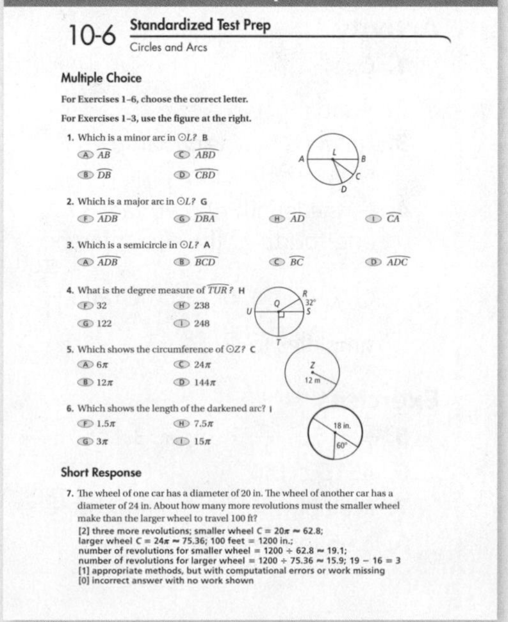
10-6 Standardized Test Prep
Circles and Arcs

Multiple Choice
For Exercises 1–6, choose the correct letter.
For Exercises 1–3, use the figure at the right.
1. Which is a minor arc in ⊙L? B
A AB C ABD B DB D CBD
2. Which is a major arc in ⊙L? G
F ADB G DBA H AD I CA
3. Which is a semicircle in ⊙L? A
A ADB B BCD C BC D ADC
4. What is the degree measure of TUR? H
F 32 H 238 G 122 I 248
5. Which shows the circumference of ⊙Z? C
A 6π C 24π B 12π D 144π
6. Which shows the length of the darkened arc? I
F 1.5π H 7.5π G 3π I 15π

Short Response
7. The wheel of one car has a diameter of 20 in. The wheel of another car has a diameter of 24 in. About how many more revolutions must the smaller wheel make than the larger wheel to travel 100 ft?
[2] three more revolutions; smaller wheel C = 20π ≈ 62.8; larger wheel C = 24π ≈ 75.36; 100 feet = 1200 in.; number of revolutions for smaller wheel = 1200 ÷ 62.8 ≈ 19.1; number of revolutions for larger wheel = 1200 ÷ 75.36 ≈ 15.9; 19 − 16 = 3
[1] appropriate methods, but with computational errors or work missing
[0] incorrect answer with no work shown

Online Teacher Resource Center

Activities, Games, and Puzzles

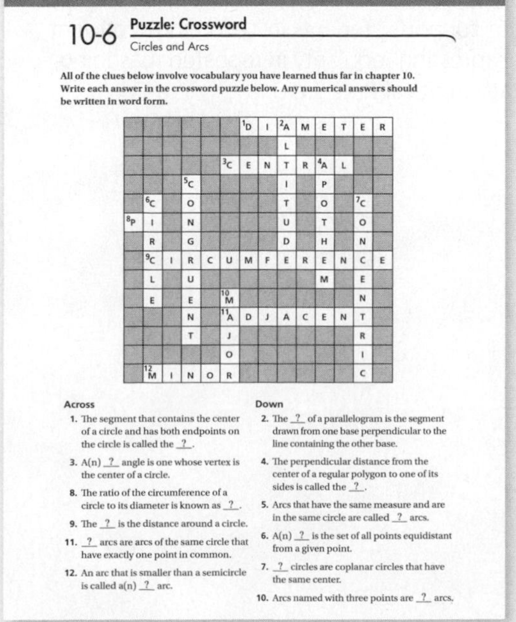
10-6 Puzzle: Crossword
Circles and Arcs

All of the clues below involve vocabulary you have learned thus far in chapter 10. Write each answer in the crossword puzzle below. Any numerical answers should be written in word form.

Across
1. The segment that contains the center of a circle and has both endpoints on the circle is called the _?_.
3. A(n) _?_ angle is one whose vertex is the center of a circle.
8. The ratio of the circumference of a circle to its diameter is known as _?_.
9. The _?_ is the distance around a circle.
11. _?_ arcs are arcs of the same circle that have exactly one point in common.
12. An arc that is smaller than a semicircle is called a(n) _?_ arc.

Down
2. The _?_ of a parallelogram is the segment drawn from one base perpendicular to the line containing the other base.
4. The perpendicular distance from the center of a regular polygon to one of its sides is called the _?_.
5. Arcs that have the same measure and are in the same circle are called _?_ arcs.
6. A(n) _?_ is the set of all points equidistant from a given point.
7. _?_ circles are coplanar circles that have the same center.
10. Arcs named with three points are _?_ arcs.

Guided Instruction

PURPOSE To explore radian measure and convert between radians and degrees

PROCESS Students will

- find the radian measure of a central angle as the ratio of the length of the intercepted arc the radius of the circle.
- convert angle measures between radians and degrees.

DISCUSS Talk to students about how degree measure is essentially an arbitrary scale. There are 360 degrees in a full circle, but you could just as easily use a division into 100 units, or 10, or any other convenient number. There is a unit of angle measure that arises naturally from the properties of a circle, namely, radian measure. Students may not be familiar with using the greek letter θ to represent an angle measure. Explain that in advanced courses using radian measure, greek letters are often used to represent angle measures.

Activity

In this Activity, students confirm the result discussed on page 653: the length of the intercepted arc of a central angle is proportional to the radius of the circle. This relationship allows for the definition of the radian measure of an angle.

Q Will the radian measure of a central angle of a circle change if the radius of the circle is increased or decreased? Explain. **[Sample answer: No, the radian measure will remain the same. The radius divides out in the ratio of arc length to radius.]**

Q What is the radian measure of a 360° central angle? **[2π radians]**

Q To the nearest tenth of a degree, what is the degree measure of a central angle of 1 radian? **[57.3]**

EXTENSION

Have students draw a circle and mark the degree measures at 30°, 45°, 60°, 90°, 120°, 135°, 150°, 180°, 210°, 225°, 240°, 270°, 300°, 315°, 330°, and 360°. Then mark the same angles using radians.

Mathematical Practices This Concept Byte supports students in becoming proficient in expressing regularity in repeated reasoning, Mathematical Practice 8.

Concept Byte — Radian Measure

Use With Lesson 10-6

ACTIVITY

Common Core State Standards

G-C.B.5 Derive using similarity the fact that the length of the arc intercepted by an angle is proportional to the radius, and define the radian measure of the angle as the constant of proportionality; derive the formula for the area of sector.

MP 8

Activity

Use the diagram of $\odot O$.

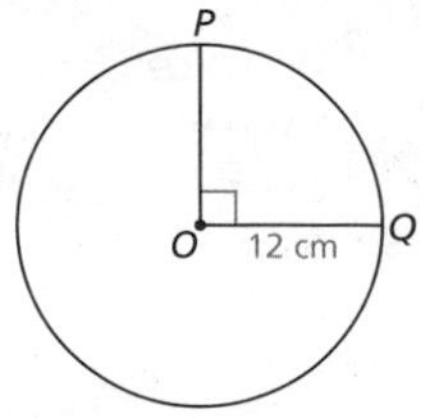

1. Find the length of $\widehat{PQ}$. Then find the ratio of the length of $\widehat{PQ}$ to the radius of $\odot O$. Leave your answers in terms of π.
2. Suppose the radius of $\odot O$ is 9 in., instead of 12 cm. Repeat Exercise 1 using this new value for the radius.
3. Choose another length for the radius of $\odot O$. Repeat Exercise 1 using your new value for the radius.
4. In any circle, what is the ratio of the length of an arc intercepted by a 90° central angle to the radius of the circle? Explain your reasoning.

On page 653, you saw that the length of an arc intercepted by a central angle is proportional to the radius of the circle. This relationship between arc length and radius is used to define a unit of angle measure called a *radian*.

The **radian measure** of a central angle of a circle is the ratio of the arc length of the intercepted arc to the radius of the circle.

$$\text{radian measure} = \frac{\text{arc length}}{\text{radius}}$$

One radian is the measure of a central angle with intercepted arc equal in length to the radius of the circle.

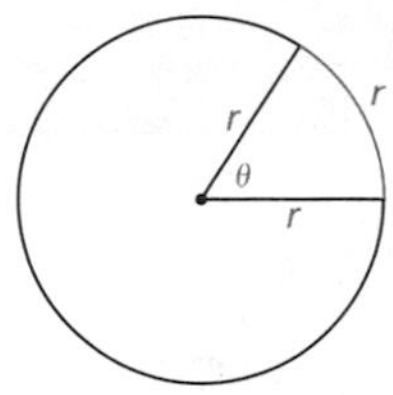

$\theta = 1$ radian

Exercises

5. How many radians are in 30°? In 135°? In 180°? Give your answers in terms of π and to the nearest hundredth of a radian.
6. How many degrees are in $\frac{\pi}{4}$ radians? In $\frac{\pi}{3}$ radians? In $\frac{5\pi}{6}$ radians?
7. **Reasoning** How can you convert angle measures from degrees to radians? How can you convert angle measures from radians to degrees?
8. In a circle of radius 8 mm, x is the measure of a central angle that intercepts an arc of length 15.6 mm. Find x in radians and to the nearest degree.

Answers

Activity

1. 6π; $\frac{\pi}{2}$
2. 4.5π; $\frac{\pi}{2}$
3. Answers may vary. Sample: radius = 10ft; 5π; $\frac{\pi}{2}$
4. $\frac{\pi}{2}$; the length of a 90° arc is one-fourth of the circumference of the circle, or $\frac{\pi r}{2}$. The ratio $\frac{\frac{\pi r}{2}}{r}$ simplifies to $\frac{\pi}{2}$.

Exercises

5. $\frac{\pi}{6}$, 0.52; $\frac{3\pi}{4}$, 2.36; π, 3.14
6. 45; 60; 150
7. Multiply by $\frac{\pi}{180}$; multiply by $\frac{180}{\pi}$.
8. 1.95 radians, 112°

Concept Byte

Use With Lesson 10-7

ACTIVITY

Exploring the Area of a Circle

Common Core State Standards

G-GMD.A.1 Give an informal argument for the formulas for the circumference of a circle, area of a circle, volume of a cylinder, pyramid, and cone.

MP 7

You can use transformations to find the formula for the area of a circle.

Activity

Step 1 Use a compass to draw a large circle. Fold the circle horizontally and then vertically. Cut the circle into four wedges on the fold lines.

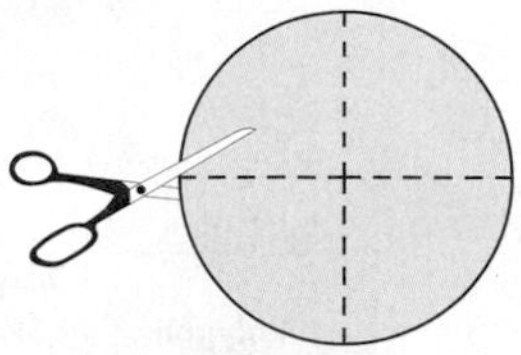

Step 2 Fold each wedge into quarters. Cut each wedge on the fold lines. You will have 16 wedges.

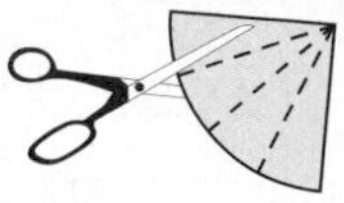

Step 3 Tape the wedges to a piece of paper to form the figure shown at the right. The figure resembles a parallelogram.

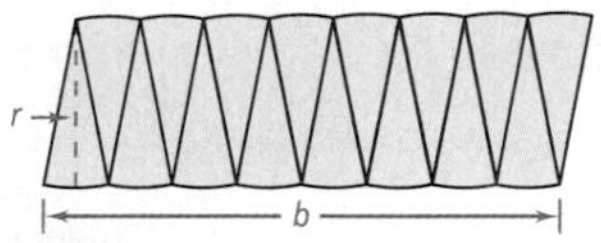

1. How does the area of the parallelogram compare with the area of the circle?
2. The base of the parallelogram is formed by arcs of the circle. Explain how the length b relates to the circumference C of the circle.
3. Explain how the length b relates to the radius r of the circle.
4. Write an expression for the area of the parallelogram in terms of r to write a formula for the area of a circle.

Exercises

Repeat Steps 1 and 2 from the activity. Tape the wedges to a piece of paper to form another figure that resembles a parallelogram, as shown at the right.

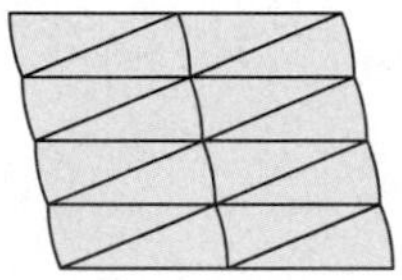

5. What are the base and height of the figure in terms of r?
6. Write an expression for the area of the figure to write a formula for the area of a circle. Is this expression the same as the one you wrote in the activity?

Answers

Activity

1. They are equal.
2. $b \approx \frac{1}{2}C$
3. $b \approx \frac{1}{2}C \approx \frac{1}{2}(2\pi r) \approx \pi r$
4. $A = \pi r \cdot r = \pi r^2$

Exercises

5. $b \approx 2r$, $h \approx \frac{1}{2}\pi r$
6. $A = b \cdot h = 2r \cdot \frac{1}{2}\pi r = \pi r^2$; yes

Guided Instruction

PURPOSE To derive the formula for area of a circle

PROCESS Students will

- transform a circle into a figure resembling a parallelogram.
- make a series of comparisons between their former circle and their transformed parallelogram.

DISCUSS Before having the students perform the Steps of the Activity and its Exercises, review the following formulas, as they will be required:

Area of a parallelogram: $A = bh$
Area of a circle: $A = \pi r^2$
Circumference of a circle: $C = 2\pi r$

Activity

In this Activity students explore the similarities between the area of a circle and the area of a parallelogram.

Q What benefit is there to cutting your circle into 16 sections, rather than just stopping at 4, or perhaps 8? **[Answers will vary. Sample: The smaller the pieces, the more the transformation resembles a parallelogram. If you stop at 4 or even 8 pieces, there are quite a few "bumps" along the edges.]**

Q What polygon is formed as the sections get smaller in size? **[rectangle]**

Mathematical Practices This Concept Byte supports students in becoming proficient in shifting perspective, Mathematical Practice 7.

1 Interactive Learning

Solve It!

PURPOSE To use the area of regular polygons to approximate the area and circumference of a circle with a radius of 1 unit

PROCESS Students may

- complete the table and use inductive reasoning to predict the area and circumference of a unit circle.
- reason that a polygon with a large number of sides approximates the area of a circle with the same radius.

FACILITATE

Q What is the pattern in the perimeter column of the table? **[The value is increasing.]**

Q What number does the value appear to be approaching? **[6.28 units]**

Q What is the pattern in the area column of the table? **[The value is increasing.]**

Q What number does the value appear to be approaching? **[3.14]**

ANSWER See Solve It in Answers on next page.

CONNECT THE MATH In the Solve It, students see that the area of a unit circle is π and the circumference is 2π. The lesson presents more instruction about circles, specifically about sectors and segments.

2 Guided Instruction

Take Note

Have students describe what they know about the value of π. Help students locate and use the pi key on their calculators. Discuss as a class the differences in answers when the π key is used on a calculator in place of one of the common estimates, 3.14 or $\frac{22}{7}$.

10-7 Areas of Circles and Sectors

Common Core State Standards
G-C.B.5 Derive . . . the formula for the area of a sector.
MP 1, MP 3, MP 4, MP 6, MP 8

Objective To find the areas of circles, sectors, and segments of circles

Try to find a pattern in these perimeters and areas to tell you what the circumference and area of a circle should be.

MATHEMATICAL PRACTICES

SOLVE IT! Getting Ready!

Each of the regular polygons in the table has radius 1. Use a calculator to complete the table for the perimeter and area of each polygon. Write out the first five decimal places.

Polygon	Number of Sides, n	Length of Side, s	Apothem, a	Perimeter ($P = ns$)	Area ($A = \frac{1}{2}ap$)
Decagon	10	2(sin 18°)	cos 18°	6.18033 . . .	2.93892 . . .
20-gon	20	2(sin 9°)	cos 9°	■	■
50-gon	50	2(sin 3.6°)	cos 3.6°	■	■
100-gon	100	2(sin 1.8°)	cos 1.8°	■	■
1000-gon	1000	2(sin 0.18°)	cos 0.18°	■	■

Look at the results in your table. Notice the perimeter and area of an n-gon as n gets very large. Now consider a circle with radius 1. What are the circumference and area of the circle? Explain your reasoning.

Lesson Vocabulary
- sector of a circle
- segment of a circle

In the Solve It, you explored the area of a circle.

Essential Understanding You can find the area of a circle when you know its radius. You can use the area of a circle to find the area of part of a circle formed by two radii and the arc the radii form when they intersect with the circle.

take note **Theorem 10-11** Area of a Circle

The area of a circle is the product of π and the square of the radius.

$A = \pi r^2$

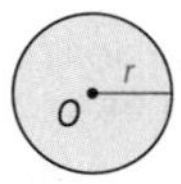

10-7 Preparing to Teach

BIG idea **Measurement**

ESSENTIAL UNDERSTANDINGS

- The area of a circle can be found when the circle's radius is known.
- The area of parts of a circle formed by radii and an arc can be found when the circle's radius and the arc's measure are known.

Math Background

Circumference and area formulas may be distinguished by using diameter to compute circumference ($C = \pi d$) and radius to compute area ($A = \pi r^2$).

Helping students to see that an arc is a fractional part of the circumference and that sector area is a fractional part of the circle's area facilitates understanding of these area concepts. Students may find it useful to repeat, "Segment area equals sector area minus triangle area."

Students need to realize the definition of segment that they have known up to this point in their study of mathematics relates to one-dimensional geometry. The definition students learn in this lesson is in regards to a circle, and therefore is a new concept related to two-dimensional geometry.

Mathematical Practices

Attend to precision. With the formula for the area of a circle, students will use clear definitions of both the area of a sector of a circle and the area of a segment of a circle, computing both quantities.

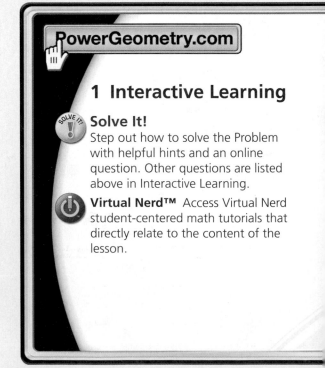

PowerGeometry.com

1 Interactive Learning

Solve It!
Step out how to solve the Problem with helpful hints and an online question. Other questions are listed above in Interactive Learning.

Virtual Nerd™ Access Virtual Nerd student-centered math tutorials that directly relate to the content of the lesson.

Problem 1 Finding the Area of a Circle

Think

What do you need in order to use the area formula?
You need the radius. The diameter is given, so you can find the radius by dividing the diameter by 2.

Sports **What is the area of the circular region on the wrestling mat?**

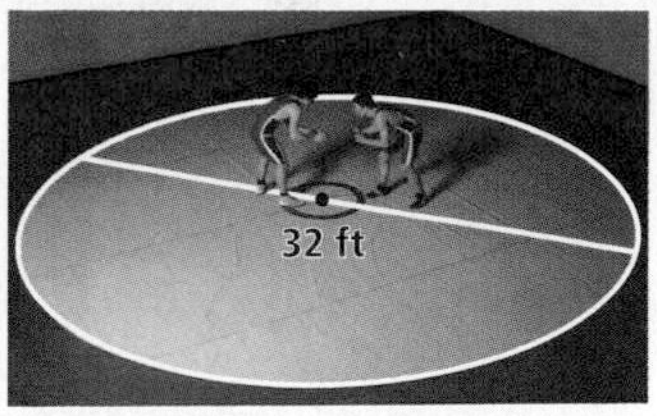

Since the diameter of the region is 32 ft, the radius is $\frac{32}{2}$, or 16 ft.

$A = \pi r^2$	Use the area formula.
$= \pi(16)^2$	Substitute 16 for r.
$= 256\pi$	Simplify.
≈ 804.2477193	Use a calculator.

The area of the wrestling region is about 804 ft^2.

Got It? **1. a.** What is the area of a circular wrestling region with a 42-ft diameter?
b. Reasoning If the radius of a circle is halved, how does its area change? Explain.

A **sector of a circle** is a region bounded by an arc of the circle and the two radii to the arc's endpoints. You name a sector using one arc endpoint, the center of the circle, and the other arc endpoint.

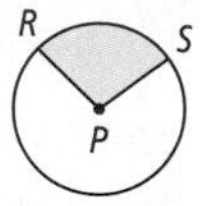

Sector *RPS*

The area of a sector is a fractional part of the area of a circle. The area of a sector formed by a 60° arc is $\frac{60}{360}$, or $\frac{1}{6}$, of the area of the circle.

take note

Theorem 10-12 Area of a Sector of a Circle

The area of a sector of a circle is the product of the ratio $\frac{\text{measure of the arc}}{360}$ and the area of the circle.

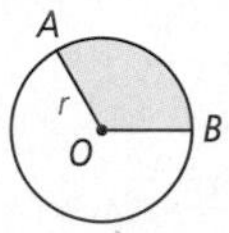

$$\text{Area of sector } AOB = \frac{m\overset{\frown}{AB}}{360} \cdot \pi r^2$$

Problem 2 Finding the Area of a Sector of a Circle

Think

What fraction of a circle's area is the area of a sector formed by a 72° arc?
The area of a sector formed by a 72° arc is $\frac{72}{360}$, or $\frac{1}{5}$, of the area of the circle.

What is the area of sector *GPH*? Leave your answer in terms of π.

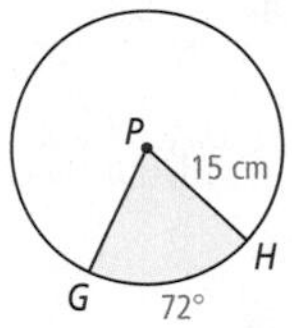

area of sector $GPH = \frac{m\overset{\frown}{GH}}{360} \cdot \pi r^2$	
$= \frac{72}{360} \cdot \pi(15)^2$	Substitute 72 for $m\overset{\frown}{GH}$ and 15 for r.
$= 45\pi$	Simplify.

The area of sector *GPH* is 45π cm^2.

Got It? **2.** A circle has a radius of 4 in. What is the area of a sector bounded by a 45° minor arc? Leave your answer in terms of π.

Problem 1

Q What measurement is given on the wrestling mat? **[the diameter of the wrestling region]**

Got It?

ERROR PREVENTION

For 1b, be sure that students verify their answer by substituting the expression for the radius $\frac{1}{2}r$, into the formula for the area of a circle and simplifying. Most students will think the area will be halved instead of quartered.

Take Note

Review the formula for finding the length of an arc from Lesson 10-6. Focus on the ratio of the central angle and 360°. Connect this to the formula for finding the area of a sector. Be sure that students understand that a sector, arc, and central angle are connected by the portion of the circle they represent.

Problem 2

Q What portion of the circle is represented by the shaded sector? **[$\frac{72}{360} = \frac{1}{5}$]**

Q How can you find the area of the sector? **[Multiply $\frac{1}{5}$ by the area of the circle.]**

Got It?

Have students sketch a diagram of the circle and sector. They should label each known measurement.

2 Guided Instruction

Each Problem is worked out and supported online.

Problem 1
Finding the Area of a Circle
Animated

Problem 2
Finding the Area of a Sector of a Circle

Problem 3
Finding the Area of a Segment of a Circle

Support in Geometry Companion
- Vocabulary
- Key Concepts
- Got It?

Answers

Solve It!

20-gon: 6.25737 . . . ; 3.09016 . . . ;
50-gon: 6.27905 . . . ; 3.13333 . . . ;
100-gon: 6.28215 . . . ; 3.13952 . . . ;
1000-gon: 6.28317 . . . ; 3.14157 . . .

About 6.28, or 2π units; about 3.14, or π units^2; explanations may vary. Sample: As the number of sides of a regular polygon with radius 1 increases, its shape gets closer and closer to the circumscribed circle of radius 1. The table shows that as the perimeter gets closer to 6.28, which $\approx 2\pi$ and the area gets closer to 3.14, which $\approx \pi$.

Got It?

1a. about 1385 ft^2

b. The area is $\frac{1}{4}$ the original area; explanations may vary. Sample: half the radius is $\frac{r}{2}$. So, if $A = \pi r^2$, then $\pi\left(\frac{r}{2}\right)^2 = \frac{1}{4}\pi r^2 = \frac{1}{4}A$.

2. 2π in.^{2}

Take Note

Have students identify the type of triangle in the second diagram. They should recognize the isosceles triangle formed by the radii and the segment of the circle. Remind students of the trigonometry they used to find the length of an apothem in regular polygons. They will be using the same methods to find the height of the triangle in the circle.

Problem 3

Q What two areas must you find to get the area of the shaded segment in the diagram? **[the area of the sector and the area of the isosceles triangle]**

Q What measurement must you find in order to find the area of the triangle? **[the height]**

Q What type of triangle is formed by the altitude of the isosceles triangle? **[30°-60°-90° triangle]**

Q What is the height of the triangle? **[$9\sqrt{3}$]**

Q How can you find the area of the shaded segment? **[Subtract the area of the triangle from the area of the sector.]**

Got It?

ERROR PREVENTION

Be sure that students complete each step carefully. Students should model Problem 3 in the book closely.

A part of a circle bounded by an arc and the segment joining its endpoints is a **segment of a circle.**

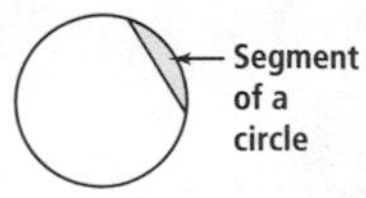

To find the area of a segment for a minor arc, draw radii to form a sector. The area of the segment equals the area of the sector minus the area of the triangle formed.

take note

Key Concept Area of a Segment

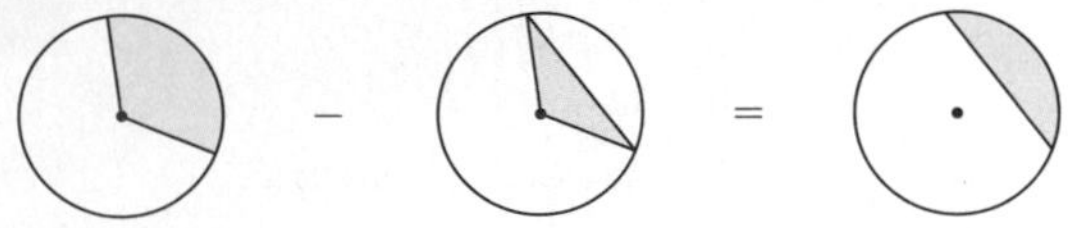

Area of sector − Area of triangle = Area of segment

Problem 3 Finding the Area of a Segment of a Circle

What is the area of the shaded segment shown at the right? Round your answer to the nearest tenth.

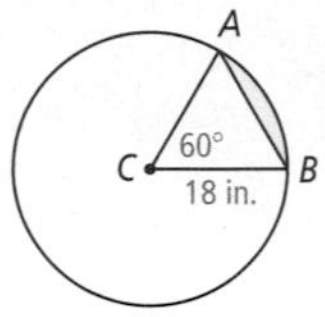

Know	Need	Plan
• The radius and $m\widehat{AB}$ • $\overline{CA} \cong \overline{CB}$ and $m\angle ACB$	The area of sector ACB and the area of $\triangle ACB$	Subtract the area of $\triangle ACB$ from the area of sector ACB.

area of sector $ACB = \frac{m\widehat{AB}}{360} \cdot \pi r^2$ Use the formula for area of a sector.

$= \frac{60}{360} \cdot \pi(18)^2$ Substitute 60 for $m\widehat{AB}$ and 18 for r.

$= 54\pi$ Simplify.

Think

What kind of triangle is $\triangle ACB$?
Since $\overline{CA} \cong \overline{CB}$, the base angles of $\triangle ACB$ are congruent. By the Triangle-Angle-Sum Theorem, $m\angle A = m\angle B = 60$. So, $\triangle ACB$ is equiangular, and therefore equilateral.

$\triangle ACB$ is equilateral. The altitude forms a 30°-60°-90° triangle.

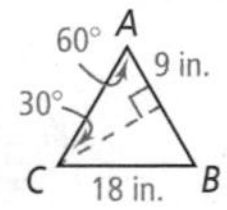

area of $\triangle ACB = \frac{1}{2}bh$ Use the formula for area of a triangle.

$= \frac{1}{2}(18)(9\sqrt{3})$ Substitute 18 for b and $9\sqrt{3}$ for h.

$= 81\sqrt{3}$ Simplify.

area of shaded segment $=$ area of sector $ACB -$ area of $\triangle ACB$

$= 54\pi - 81\sqrt{3}$ Substitute.

≈ 29.34988788 Use a calculator.

The area of the shaded segment is about 29.3 in.^{2}.

Got It? 3. What is the area of the shaded segment shown at the right? Round your answer to the nearest tenth.

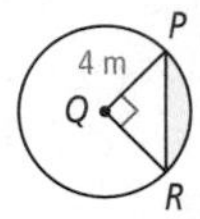

Additional Problems

1. What is the area of a circular ice skating rink with a diameter of 48 ft?

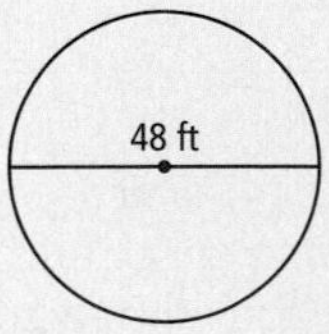

ANSWER about 1810 ft^2

2. What is the area of sector RST? Leave your answer in terms of π.

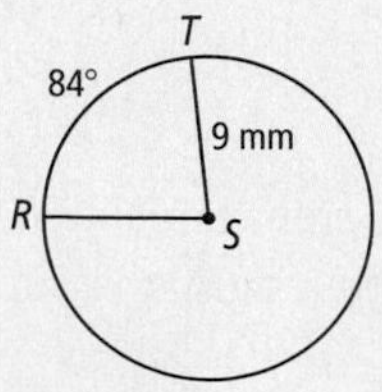

ANSWER 18.9π mm^2

3. What is the area of the shaded segment shown? Round your answer to the nearest whole number.

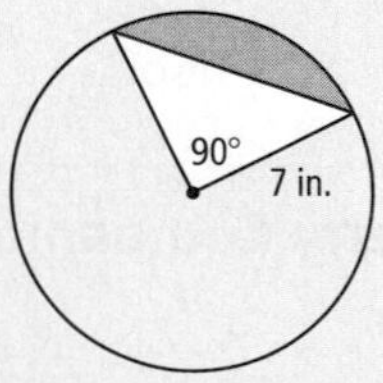

ANSWER about 14 in.^{2}

Answers

Got It? (continued)

3. 4.6 m^2

Lesson Check

Do you know HOW?

1. What is the area of a circle with diameter 16 in.? Leave your answer in terms of π.

Find the area of the shaded region of the circle. Leave your answer in terms of π.

2.

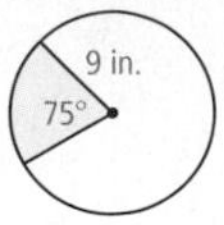

3.

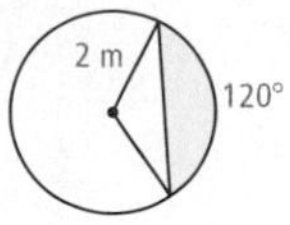

Do you UNDERSTAND? MATHEMATICAL PRACTICES

4. **Vocabulary** What is the difference between a sector of a circle and a segment of a circle?

5. **Reasoning** Suppose a sector of $\odot P$ has the same area as a sector of $\odot O$. Can you conclude that $\odot P$ and $\odot O$ have the same area? Explain.

6. **Error Analysis** Your class must find the area of a sector of a circle determined by a 150° arc. The radius of the circle is 6 cm. What is your classmate's error? Explain.

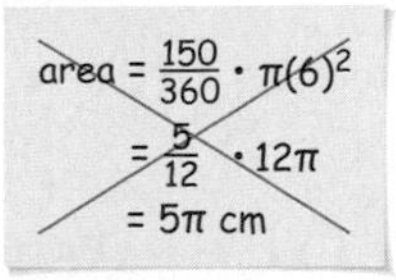

Practice and Problem-Solving Exercises MATHEMATICAL PRACTICES

A Practice Find the area of each circle. Leave your answer in terms of π. See Problem 1.

7.

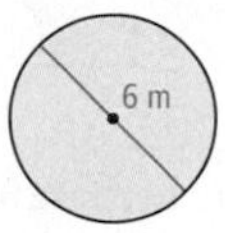

8.

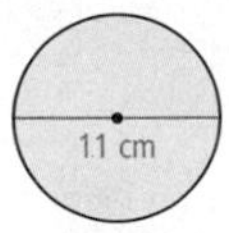

9.

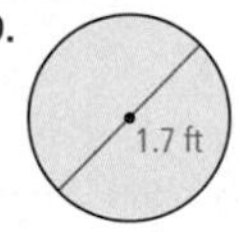

10.

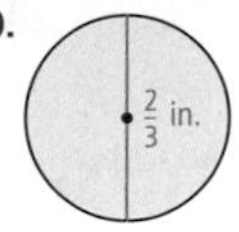

STEM 11. **Agriculture** Some farmers use a circular irrigation method. An irrigation arm acts as the radius of an irrigation circle. How much land is covered with an irrigation arm of 300 ft?

12. You use an online store locator to search for a store within a 5-mi radius of your home. What is the area of your search region?

Find the area of each shaded sector of a circle. Leave your answer in terms of π. See Problem 2.

13.

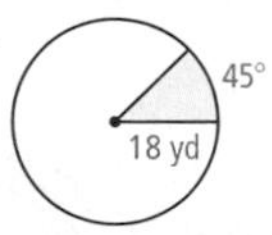

14.

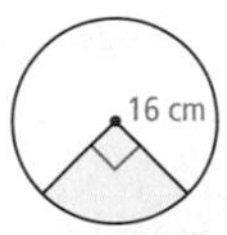

15.

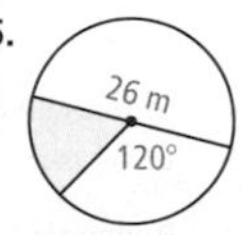

16.

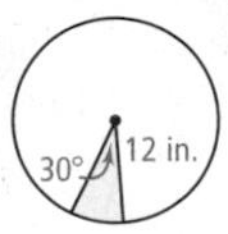

17.

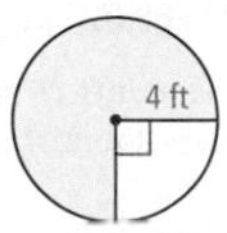

18.

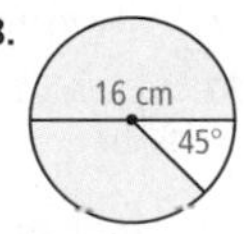

3 Lesson Check

Do you know HOW?

- If students have difficulty with Exercise 1, then have them review Problem 1 to write the formula for area and know how to find the value of r to substitute into the formula.

Do you UNDERSTAND?

- If students have difficulty with Exercise 5, then have them draw a diagram in which one circle is large with a small sector and another circle is small with a large sector.

Close

Q How do you find the area of a sector? **[Multiply the ratio of the arc measure and 360° by the area of the circle.]**

Q What is a segment of a circle? **[A segment of a circle is an area bounded by an arc and the line segment joining the endpoints of two radii in a circle.]**

Lesson Check

1. 64π in.^{2}
2. $\frac{135}{8}\pi$ in.^{2}, or 16.875π in.^{2}
3. $\left(\frac{4}{3}\pi - \sqrt{3}\right) m^2$
4. A sector of a circle is a region bounded by an arc and the two radii to the endpoints of the arc. A segment is a part of a circle bounded by an arc and the seg. joining the arc's endpoints.
5. No; the central ∠s corresponding to the arcs and the radii of the circles may be different. Circles with different radii do not have the same area.
6. 6^2 was incorrectly evaluated as $6 \cdot 2$.

Practice and Problem-Solving Exercises

7. 9π m^2
8. 30.25π cm^2
9. 0.7225π ft^2
10. $\frac{\pi}{9}$ in.^{2}
11. about 282,743 ft^2
12. about 78.5 mi^2
13. 40.5π yd^2
14. 64π cm^2
15. $\frac{169\pi}{6}$ m^2
16. 12π in.^{2}
17. 12π ft^2
18. 56π cm^2

PowerGeometry.com

3 Lesson Check

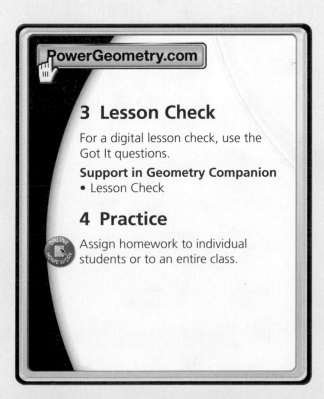

For a digital lesson check, use the Got It questions.

Support in Geometry Companion

- Lesson Check

4 Practice

Assign homework to individual students or to an entire class.

4 Practice

ASSIGNMENT GUIDE
Basic: 7–25 all, 26–34 even, 35–36
Average: 7–25 odd, 26–44
Advanced: 7–25 odd, 26–50

Mathematical Practices are supported by exercises with red headings. Here are the Practices supported in this lesson:

MP 1: Make Sense of Problems Ex. 34
MP 3: Construct Arguments Ex. 5, 36
MP 3: Critique the Reasoning of Others Ex. 6
MP 6: Attend to Precision Ex. 40
MP 8: Repeated Reasoning Ex. 41a

Applications exercises have blue headings. Exercises 11, 32, 33, 35, and 48 support MP 4: Model.

STEM exercises focus on science or engineering applications.

EXERCISE 35: Use the Think About a Plan worksheet in the **Practice and Problem Solving Workbook** (also available in the Teaching Resources in print and online) to further support students' development in becoming independent learners.

HOMEWORK QUICK CHECK
To check students' understanding of key skills and concepts, go over Exercises 9, 15, 34, 35, and 36.

Find the area of sector *TOP* in ⊙*O* using the given information. Leave your answer in terms of π.

19. $r = 5$ m, $m\overset{\frown}{TP} = 90$

20. $r = 6$ ft, $m\overset{\frown}{TP} = 15$

21. $d = 16$ in., $m\overset{\frown}{PT} = 135$

22. $d = 15$ cm, $m\overset{\frown}{POT} = 180$

Find the area of each shaded segment. Round your answer to the nearest tenth. See Problem 3.

23.
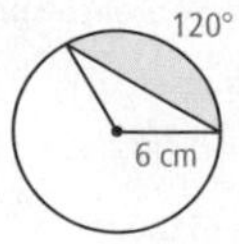

24.
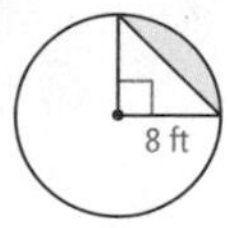

25.
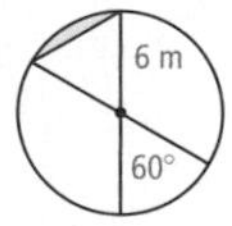

Find the area of the shaded region. Leave your answer in terms of π and in simplest radical form.

B Apply

26.
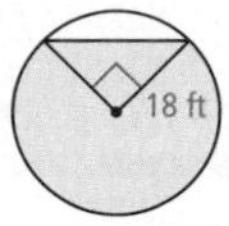

27.
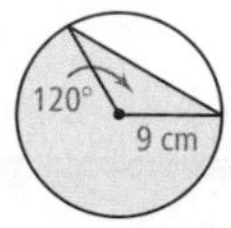

28.
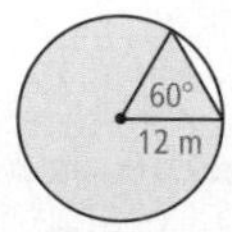

29.
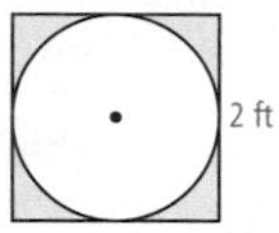

30.
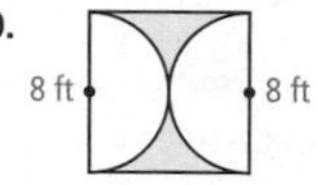

31.
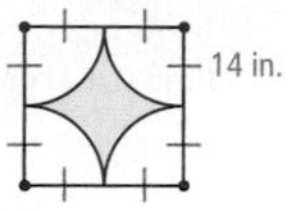

32. Transportation A town provides bus transportation to students living beyond 2 mi of the high school. What area of the town does *not* have the bus service? Round to the nearest tenth.

33. Design A homeowner wants to build a circular patio. If the diameter of the patio is 20 ft, what is its area to the nearest whole number?

34. Think About a Plan A circular mirror is 24 in. wide and has a 4-in. frame around it. What is the area of the frame?
- How can you *draw a diagram* to help solve the problem?
- What part of a circle is the width?
- Is there more than one area to consider?

STEM 35. Industrial Design Refer to the diagram of the regular hexagonal nut. What is the area of the hexagonal face to the nearest millimeter?

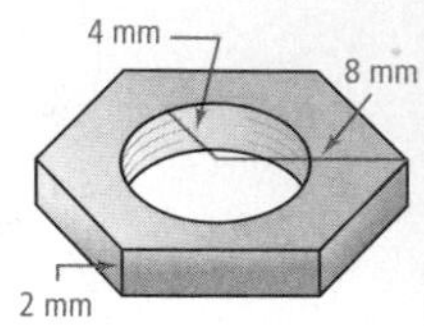

36. Reasoning $\overline{AB}$ and $\overline{CD}$ are diameters of ⊙*O*. Is the area of sector *AOC* equal to the area of sector *BOD*? Explain.

37. A circle with radius 12 mm is divided into 20 sectors of equal area. What is the area of one sector to the nearest tenth?

Answers

Practice and Problem-Solving Exercises (continued)

19. $\frac{25\pi}{4}$ m^2
20. $\frac{3\pi}{2}$ ft^2
21. 24π in.^{2}
22. 28.125π cm^2
23. 22.1 cm^2
24. 18.3 ft^2
25. 3.3 m^2
26. $(243\pi + 162)$ ft^2
27. $(54\pi + 20.25\sqrt{3})$ cm^2
28. $(120\pi + 36\sqrt{3})$ m^2
29. $(4 - \pi)$ ft^2
30. $(64 - 16\pi)$ ft^2
31. $(784 - 196\pi)$ in.^{2}
32. 12.6 mi^2
33. 314 ft^2
34. 112π in.^{2} or about 351.9 in.^{2}
35. 116 mm^2
36. Yes; $\angle AOC \cong \angle BOD$ (Vertical ∠s are ≅.), so the two sectors are ≅ and will have = areas.
37. 22.6 mm^2

38. The circumference of a circle is 26π in. What is its area? Leave your answer in terms of π.

39. In a circle, a 90° sector has area 36π in.^{2}. What is the radius of the circle?

40. Open-Ended Draw a circle and a sector so that the area of the sector is 16π cm^2. Give the radius of the circle and the measure of the sector's arc.

41. A method for finding the area of a segment determined by a minor arc is described in this lesson.

a. Writing Describe two ways to find the area of a segment determined by a major arc.

b. If $m\widehat{AB} = 90$ in a circle of radius 10 in., find the areas of the two segments determined by $\widehat{AB}$.

Find the area of the shaded segment to the nearest tenth.

42.

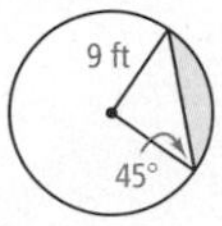

43.

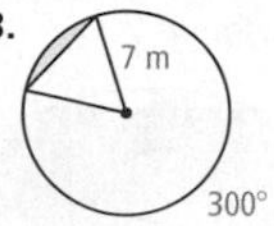

44.

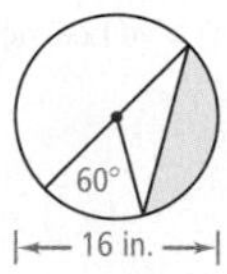

Challenge **Find the area of the shaded region. Leave your answer in terms of π.**

45.

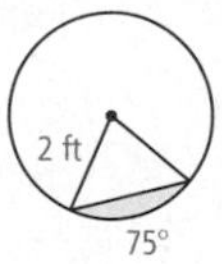

46.

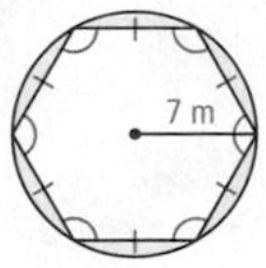

47.

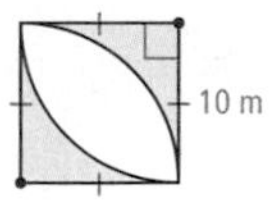

48. Recreation An 8 ft-by-10 ft floating dock is anchored in the middle of a pond. The bow of a canoe is tied to a corner of the dock with a 10-ft rope, as shown in the picture below.

a. Sketch a diagram of the region in which the bow of the canoe can travel.

b. What is the area of that region? Round your answer to the nearest square foot.

38. 169π in.^{2}

39. 12 in.

40. Check students' work.

41a. Answers may vary. Sample: Subtract the minor arc segment from the area of the circle; or add the areas of the major sector and the $\triangle$ that is part of the minor arc sector.

b. $(25\pi - 50)$ units^2; $(75\pi + 50)$ units^2

42. 23.1 ft^2

43. 4.4 m^2

44. 39.3 in.^{2}

45. $\left(\frac{5\pi}{6} - 2 \cdot \sin 75^\circ\right)$ ft^2, or $\left[\frac{5\pi}{6} - 4(\sin 37.5^\circ)(\cos 37.5^\circ)\right]$

46. $(49\pi - 73.5\sqrt{3})$ m^2

47. $(200 - 50\pi)$ m^2

48a.

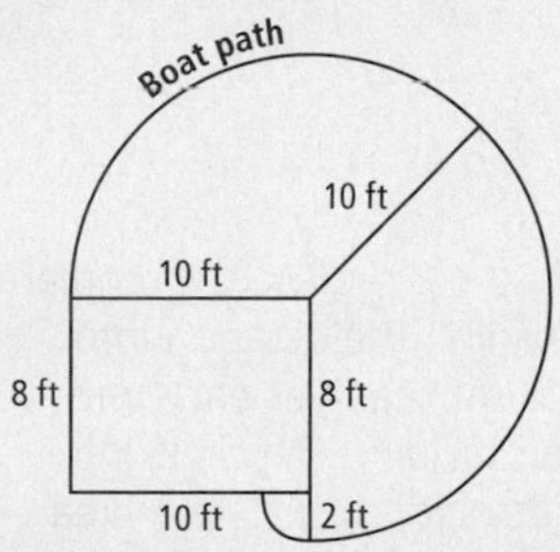

b. 239 ft^2

Answers

Practice and Problem-Solving Exercises (continued)

49. Blue region: Let $AB = 2$. Area of blue $= 4 - \pi$; area of yellow $= \pi - 2$, and $4 - \pi < \pi - 2$.

50. $\left(\frac{200\pi}{3} - 50\sqrt{3}\right)$ units^2

49. $\odot O$ at the right is inscribed in square $ABCD$ and circumscribed about square $PQRS$. Which is smaller, the blue region or the yellow region? Explain.

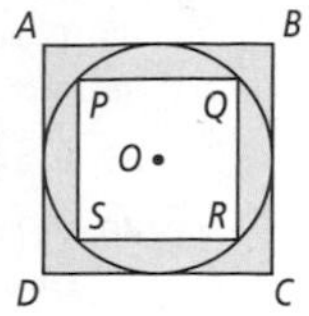

50. Circles T and U each have radius 10 and $TU = 10$. Find the area of the region that is contained inside both circles. (*Hint:* Think about where T and U must lie in a diagram of $\odot T$ and $\odot U$.)

Apply What You've Learned

MATHEMATICAL PRACTICES MP 1, MP 7

Look back at the information given about the target on page 613. The diagram of the target is shown again below. In the Apply What You Learned in Lesson 10-1, you found the area of one red triangle, and in the Apply What You've Learned in Lesson 10-5, you found the area of the regular octagon.

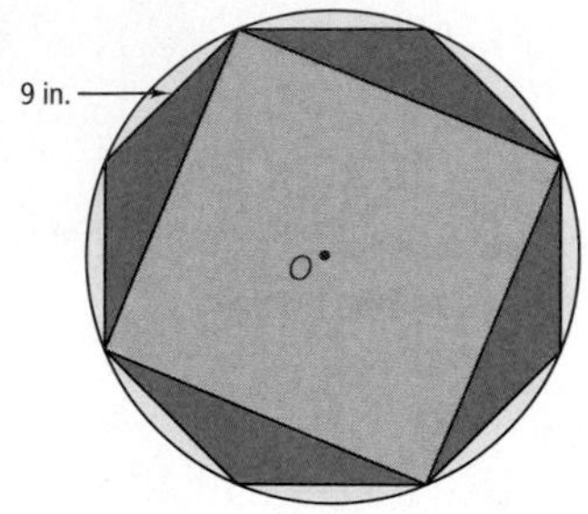

a. Is each yellow region of the target called a *segment* or a *sector* of $\odot O$?

b. Do the eight yellow regions all have the same area? Justify your answer.

c. What information do you need in order to find the area of the yellow regions of the target? Describe a method to find this information.

d. Describe a method to find the total area of the yellow regions of the target. Then find the total area of the yellow regions. Round your answer the nearest tenth of a square inch.

e. Use a different method to find the total area of the yellow regions of the target and check that you get the same result as in part (d).

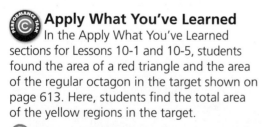

Apply What You've Learned

In the Apply What You've Learned sections for Lessons 10-1 and 10-5, students found the area of a red triangle and the area of the regular octagon in the target shown on page 613. Here, students find the total area of the yellow regions in the target.

Mathematical Practices

In part (c), students **look for and make use of structure** to draw in the auxiliary line segments that will allow them to find the radius of the circle. (MP 7)

In part (e), students **make sense of the problem** as they use a different method to check their result for the total area of the yellow regions of the target. (MP 1)

ANSWERS

a. a segment

b. Yes; draw in the 8 radii of the regular octagon, dividing the octagon into eight congruent triangles each with area A_1, and dividing the circle into 8 congruent sectors, each with area A_2. Then the area of each yellow segment is $A_2 - A_1$, so the areas of the 8 yellow segments are all the same.

c. The radius of the circle (which is also the radius of the octagon); methods may vary. Sample: draw two consecutive radii of the octagon and the apothem that lies between the two radii, forming two right triangles in which the hypotenuse is the radius of the circle. You can use a trigonometric ratio in either triangle to find the radius.

d. Methods may vary. Sample: Subtract the area of the regular octagon from the area of the circle; about 43.3 in.^{2}.

e. Methods may vary. Sample: Find the area of one yellow segment and multiply by 8.

10-7 Lesson Resources

Differentiated Remediation

Instructional Support

Geometry Companion

Students can use the **Geometry Companion** worktext (4 pages) . . .

- New Vocabulary
- Key Concepts
- Got It for each Problem
- Lesson Check

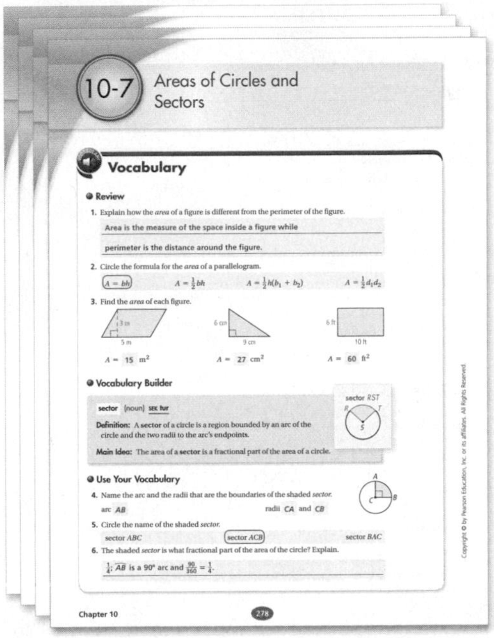
10-7 Areas of Circles and Sectors

Vocabulary

Review

1. Explain how the *area* of a figure is different from the perimeter of the figure.
Area is the measure of the space inside a figure while perimeter is the distance around the figure.

2. Circle the formula for the *area* of a parallelogram.
$A = bh$ (circled) $A = \frac{1}{2}bh$ $A = \frac{1}{2}h(b_1 + b_2)$ $A = \frac{1}{2}d_1d_2$

3. Find the *area* of each figure.
$A = 15$ m^2 $A = 27$ cm^2 $A = 60$ ft^2

Vocabulary Builder

sector (noun) SEK tur

Definition: A **sector** of a circle is a region bounded by an arc of the circle and the two radii to the arc's endpoints.

Main Idea: The area of a **sector** is a fractional part of the area of a circle.

Use Your Vocabulary

4. Name the arc and the radii that are the boundaries of the shaded *sector*.
arc AB radii CA and CB

5. Circle the name of the shaded *sector*.
sector ABC sector ACB (circled) sector BAC

6. The shaded *sector* is what fractional part of the area of the circle? Explain.
$\frac{1}{4}$; $\overset{\frown}{AB}$ is a 90° arc and $\frac{90}{360} = \frac{1}{4}$.

Chapter 10 278

Copyright © by Pearson Education, Inc. or its affiliates. All Rights Reserved.

ELL Support

Connect to Prior Knowledge Construct a circle on the board. Trace the circumference, diameter, and radius as you say the name of each part. Then model how to calculate *pi* from the ratio of the circumference to its diameter as you think aloud. Ask: do you predict this ratio to be true for all circles? Ask students to find and measure the circumference and diameter of circular shapes in the classroom to verify their predictions. Then hand out precut circles of various sizes. Have students measure the circumference and diameter. Tell them to use the ratio of the circumference to the diameter times the radius squared to calculate area. Discuss their methods and their results. The circles can then be used to find the area of a sector.

5 Assess & Remediate

Lesson Quiz

1. **Do you UNDERSTAND?** Suppose the landing pad for a helicopter is shaped like a circle with a 35-ft diameter. What is the area of the landing pad?

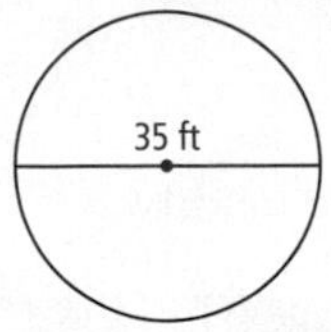

2. What is the area of sector *XYZ*? Leave your answer in terms of π.

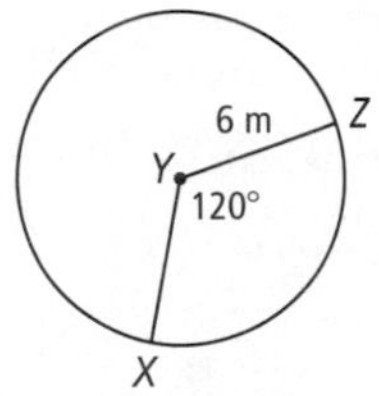

3. Suppose $\overline{XZ}$ is drawn in the circle from Question 2 above. What is the area of the segment between $\overline{XZ}$ and $\overset{\frown}{XZ}$ to the nearest tenth?

ANSWERS TO LESSON QUIZ

1. about 962 ft^2
2. 12π m^2
3. 22.1

PRESCRIPTION FOR REMEDIATION

Use the student work on the Lesson Quiz to prescribe a differentiated review assignment.

Points	Differentiated Remediation
0–1	Intervention
2	On-level
3	Extension

PowerGeometry.com

5 Assess & Remediate

Assign the Lesson Quiz. Appropriate intervention, practice, or enrichment is automatically generated based on student performance.

Intervention

- **Reteaching** (2 pages) Provides reteaching and practice exercises for the key lesson concepts. Use with struggling students or absent students.
- **English Language Learner Support** Helps students develop and reinforce mathematical vocabulary and key concepts.

All-in-One Resources/Online

Reteaching

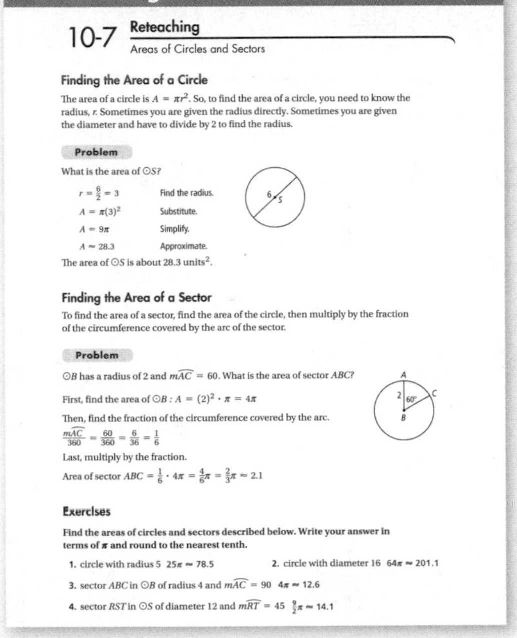
10-7 Reteaching
Areas of Circles and Sectors

Finding the Area of a Circle

The area of a circle is $A = \pi r^2$. So, to find the area of a circle, you need to know the radius, *r*. Sometimes you are given the radius directly. Sometimes you are given the diameter and have to divide by 2 to find the radius.

Problem

What is the area of ⊙*S*?

$r = \frac{6}{2} = 3$ Find the radius.
$A = \pi(3)^2$ Substitute.
$A = 9\pi$ Simplify.
$A \approx 28.3$ Approximate.

The area of ⊙*S* is about 28.3 units^2.

Finding the Area of a Sector

To find the area of a sector, find the area of the circle, then multiply by the fraction of the circumference covered by the arc of the sector.

Problem

⊙*B* has a radius of 2 and $m\overset{\frown}{AC} = 60$. What is the area of sector *ABC*?

First, find the area of ⊙*B* : $A = (2)^2 \cdot \pi = 4\pi$

Then, find the fraction of the circumference covered by the arc.

$\frac{m\overset{\frown}{AC}}{360} = \frac{60}{360} = \frac{6}{36} = \frac{1}{6}$

Last, multiply by the fraction.

Area of sector $ABC = \frac{1}{6} \cdot 4\pi = \frac{4}{6}\pi = \frac{2}{3}\pi \approx 2.1$

Exercises

Find the areas of circles and sectors described below. Write your answer in terms of π and round to the nearest tenth.

1. circle with radius 5 $25\pi \approx 78.5$
2. circle with diameter 16 $64\pi \approx 201.1$
3. sector *ABC* in ⊙*B* of radius 4 and $m\overset{\frown}{AC} = 90$ $4\pi \approx 12.6$
4. sector *RST* in ⊙*S* of diameter 12 and $m\overset{\frown}{RT} = 45$ $\frac{9}{2}\pi \approx 14.1$

All-in-One Resources/Online

English Language Learner Support

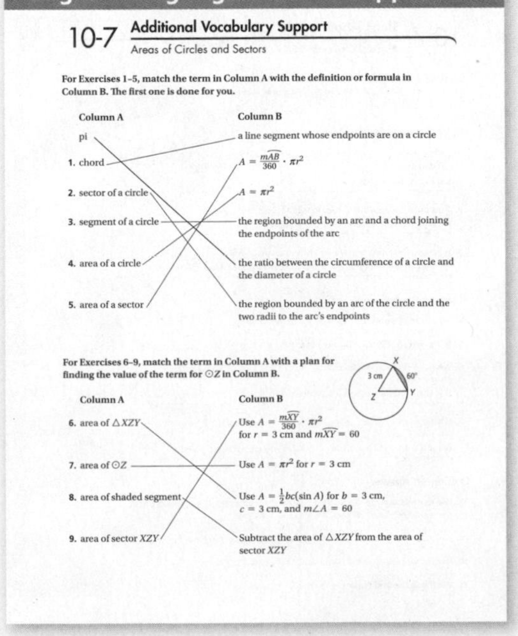
10-7 Additional Vocabulary Support
Areas of Circles and Sectors

For Exercises 1–5, match the term in Column A with the definition or formula in Column B. The first one is done for you.

Column A	Column B
pi	a line segment whose endpoints are on a circle
1. chord	$A = \frac{m\overset{\frown}{AB}}{360} \cdot \pi r^2$
2. sector of a circle	$A = \pi r^2$
3. segment of a circle	the region bounded by an arc and a chord joining the endpoints of the arc
4. area of a circle	the ratio between the circumference of a circle and the diameter of a circle
5. area of a sector	the region bounded by an arc of the circle and the two radii to the arc's endpoints

For Exercises 6–9, match the term in Column A with a plan for finding the value of the term for ⊙*Z* in Column B.

Column A	Column B
6. area of △*XZY*	Use $A = \frac{m\overset{\frown}{XY}}{360} \cdot \pi r^2$ for $r = 3$ cm and $m\overset{\frown}{XY} = 60$
7. area of ⊙*Z*	Use $A = \pi r^2$ for $r = 3$ cm
8. area of shaded segment	Use $A = \frac{1}{2}bc(\sin A)$ for $b = 3$ cm, $c = 3$ cm, and $m\angle A = 60$
9. area of sector *XZY*	Subtract the area of △*XZY* from the area of sector *XZY*

Differentiated Remediation *continued*

On-Level

- **Practice** (2 pages) Provides extra practice for each lesson. For simpler practice exercises, use the Form K Practice pages found in the All-in-One Teaching Resources and online.
- **Think About a Plan** Helps students develop specific problem-solving skills and strategies by providing scaffolded guiding questions.
- **Standardized Test Prep** Focuses on all major exercises, all major question types, and helps students prepare for the high-stakes assessments.

Extension

- **Enrichment** Provides students with interesting problems and activities that extend the concepts of the lesson.
- **Activities, Games, and Puzzles** Worksheets that can be used for concepts development, enrichment, and for fun!

Practice and Problem Solving Wkbk/ All-in-One Resources/Online

Practice page 1

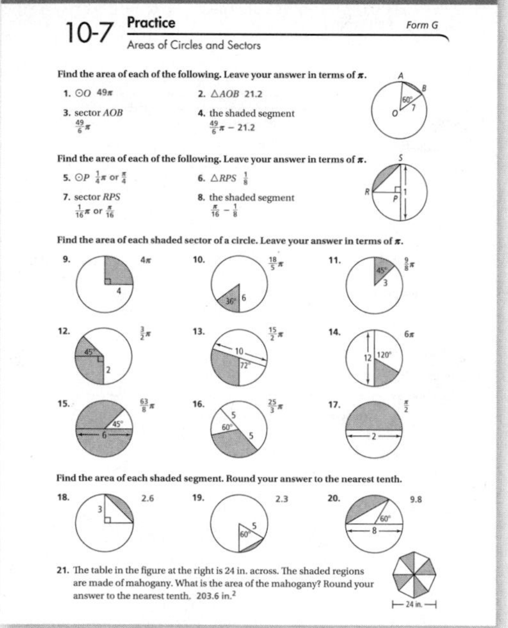

10-7 **Practice** *Form G*
Areas of Circles and Sectors

Find the area of each of the following. Leave your answer in terms of π.

1. $\odot O$ 49π
2. $\triangle AOB$ 21.2
3. sector AOB $\frac{49}{6}\pi$
4. the shaded segment $\frac{49}{6}\pi - 21.2$

Find the area of each of the following. Leave your answer in terms of π.

5. $\odot P$ $\frac{1}{4}\pi$ or $\frac{\pi}{4}$
6. $\triangle RPS$ $\frac{1}{8}$
7. sector RPS $\frac{1}{16}\pi$ or $\frac{\pi}{16}$
8. the shaded segment $\frac{\pi}{16} - \frac{1}{8}$

Find the area of each shaded sector of a circle. Leave your answer in terms of π.

9. 4π 10. $\frac{18}{5}\pi$ 11. $\frac{9}{8}\pi$
12. $\frac{3}{2}\pi$ 13. $\frac{15}{2}\pi$ 14. 6π
15. $\frac{63}{8}\pi$ 16. $\frac{25}{3}\pi$ 17. $\frac{\pi}{2}$

Find the area of each shaded segment. Round your answer to the nearest tenth.

18. 2.6 19. 2.3 20. 9.8

21. The table in the figure at the right is 24 in. across. The shaded regions are made of mahogany. What is the area of the mahogany? Round your answer to the nearest tenth. 203.6 in.^{2}

Practice and Problem Solving Wkbk/ All-in-One Resources/Online

Practice page 2

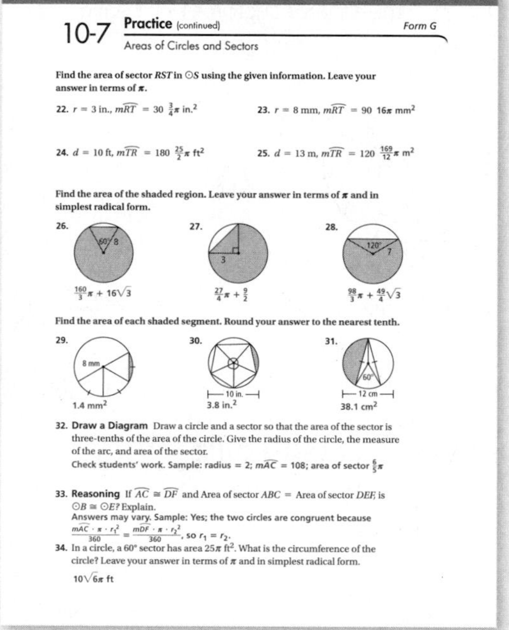

10-7 **Practice** (continued) *Form G*
Areas of Circles and Sectors

Find the area of sector RST in $\odot S$ using the given information. Leave your answer in terms of π.

22. $r = 3$ in., $m\widehat{RT} = 30$ $\frac{3}{4}\pi$ in.^{2}
23. $r = 8$ mm, $m\widehat{RT} = 90$ 16π mm^2
24. $d = 10$ ft, $m\widehat{TR} = 180$ $\frac{25}{2}\pi$ ft^2
25. $d = 13$ m, $m\widehat{TR} = 120$ $\frac{169}{12}\pi$ m^2

Find the area of the shaded region. Leave your answer in terms of π and in simplest radical form.

26. $\frac{160}{3}\pi + 16\sqrt{3}$ 27. $\frac{27}{4}\pi + \frac{9}{2}$ 28. $\frac{98}{3}\pi + \frac{49}{4}\sqrt{3}$

Find the area of each shaded segment. Round your answer to the nearest tenth.

29. 1.4 mm^2 30. 3.8 in.^{2} 31. 38.1 cm^2

32. **Draw a Diagram** Draw a circle and a sector so that the area of the sector is three-tenths of the area of the circle. Give the radius of the circle, the measure of the arc, and area of the sector.
Check students' work. Sample: radius = 2; $m\widehat{AC} = 108$; area of sector $\frac{6}{5}\pi$
33. **Reasoning** If $\widehat{AC} \cong \widehat{DF}$ and Area of sector ABC = Area of sector DEF, is $\odot B \cong \odot E$? Explain.
Answers may vary. Sample: Yes; the two circles are congruent because $\frac{m\widehat{AC} \cdot \pi \cdot r_1^2}{360} = \frac{m\widehat{DF} \cdot \pi \cdot r_2^2}{360}$, so $r_1 = r_2$.
34. In a circle, a 60° sector has area 25π ft^2. What is the circumference of the circle? Leave your answer in terms of π and in simplest radical form.
$10\sqrt{6}\pi$ ft

All-in-One Resources/Online

Enrichment

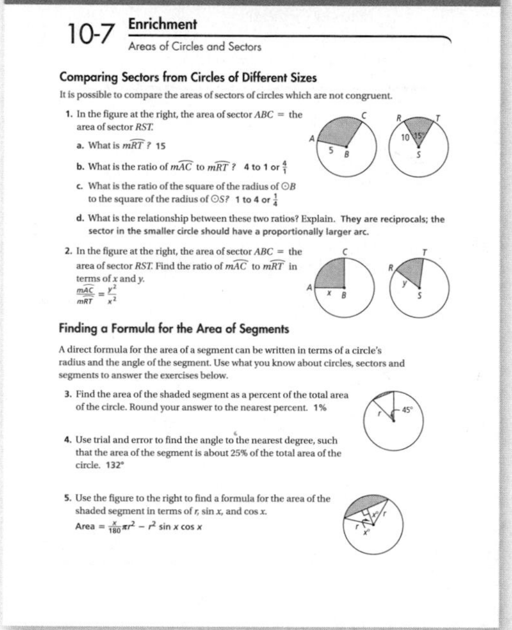

10-7 **Enrichment**
Areas of Circles and Sectors

Comparing Sectors from Circles of Different Sizes

It is possible to compare the areas of sectors of circles which are not congruent.

1. In the figure at the right, the area of sector ABC = the area of sector RST.
   a. What is $m\widehat{RT}$? 15
   b. What is the ratio of $m\widehat{AC}$ to $m\widehat{RT}$? 4 to 1 or $\frac{4}{1}$
   c. What is the ratio of the square of the radius of $\odot B$ to the square of the radius of $\odot S$? 1 to 4 or $\frac{1}{4}$
   d. What is the relationship between these two ratios? Explain. They are reciprocals; the sector in the smaller circle should have a proportionally larger arc.
2. In the figure at the right, the area of sector ABC = the area of sector RST. Find the ratio of $m\widehat{AC}$ to $m\widehat{RT}$ in terms of x and y.
$\frac{m\widehat{AC}}{m\widehat{RT}} = \frac{y^2}{x^2}$

Finding a Formula for the Area of Segments

A direct formula for the area of a segment can be written in terms of a circle's radius and the angle of the segment. Use what you know about circles, sectors and segments to answer the exercises below.

3. Find the area of the shaded segment as a percent of the total area of the circle. Round your answer to the nearest percent. 1%
4. Use trial and error to find the angle to the nearest degree, such that the area of the segment is about 25% of the total area of the circle. 132°
5. Use the figure to the right to find a formula for the area of the shaded segment in terms of r, $\sin x$, and $\cos x$.
Area $= \frac{x}{180}\pi r^2 - r^2 \sin x \cos x$

Practice and Problem Solving Wkbk/ All-in-One Resources/Online

Think About a Plan

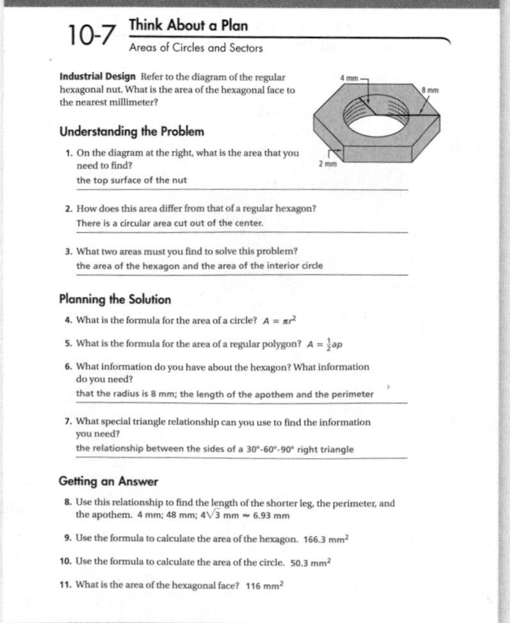

10-7 **Think About a Plan**
Areas of Circles and Sectors

Industrial Design Refer to the diagram of the regular hexagonal nut. What is the area of the hexagonal face to the nearest millimeter?

Understanding the Problem

1. On the diagram at the right, what is the area that you need to find?
the top surface of the nut
2. How does this area differ from that of a regular hexagon?
There is a circular area cut out of the center.
3. What two areas must you find to solve this problem?
the area of the hexagon and the area of the interior circle

Planning the Solution

4. What is the formula for the area of a circle? $A = \pi r^2$
5. What is the formula for the area of a regular polygon? $A = \frac{1}{2}ap$
6. What information do you have about the hexagon? What information do you need?
that the radius is 8 mm; the length of the apothem and the perimeter
7. What special triangle relationship can you use to find the information you need?
the relationship between the sides of a 30°-60°-90° right triangle

Getting an Answer

8. Use this relationship to find the length of the shorter leg, the perimeter, and the apothem. 4 mm; 48 mm; $4\sqrt{3}$ mm $\approx$ 6.93 mm
9. Use the formula to calculate the area of the hexagon. 166.3 mm^2
10. Use the formula to calculate the area of the circle. 50.3 mm^2
11. What is the area of the hexagonal face? 116 mm^2

Practice and Problem Solving Wkbk/ All-in-One Resources/Online

Standardized Test Prep

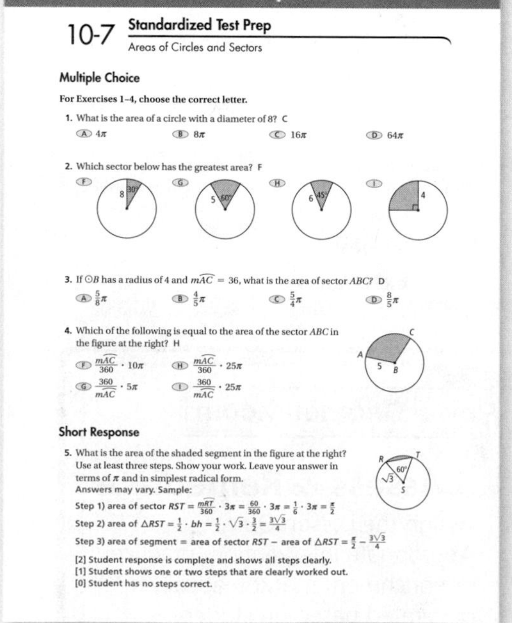

10-7 **Standardized Test Prep**
Areas of Circles and Sectors

Multiple Choice

For Exercises 1–4, choose the correct letter.

1. What is the area of a circle with a diameter of 8? C
   A 4π B 8π C 16π D 64π
2. Which sector below has the greatest area? F
   F G H I
3. If $\odot B$ has a radius of 4 and $m\widehat{AC} = 36$, what is the area of sector ABC? D
   A $\frac{5}{8}\pi$ B $\frac{4}{5}\pi$ C $\frac{5}{4}\pi$ D $\frac{8}{5}\pi$
4. Which of the following is equal to the area of the sector ABC in the figure at the right? H
   F $\frac{m\widehat{AC}}{360} \cdot 10\pi$ H $\frac{m\widehat{AC}}{360} \cdot 25\pi$
   G $\frac{360}{m\widehat{AC}} \cdot 5\pi$ I $\frac{360}{m\widehat{AC}} \cdot 25\pi$

Short Response

5. What is the area of the shaded segment in the figure at the right? Use at least three steps. Show your work. Leave your answer in terms of π and in simplest radical form.
Answers may vary. Sample:
Step 1) area of sector $RST = \frac{m\widehat{RT}}{360} \cdot 3\pi = \frac{60}{360} \cdot 3\pi = \frac{1}{6} \cdot 3\pi = \frac{\pi}{2}$
Step 2) area of $\triangle RST = \frac{1}{2} \cdot bh = \frac{1}{2} \cdot \sqrt{3} \cdot \frac{3}{2} = \frac{3\sqrt{3}}{4}$
Step 3) area of segment = area of sector RST − area of $\triangle RST = \frac{\pi}{2} - \frac{3\sqrt{3}}{4}$
[2] Student response is complete and shows all steps clearly.
[1] Student shows one or two steps that are clearly worked out.
[0] Student has no steps correct.

Online Teacher Resource Center

Activities, Games, and Puzzles

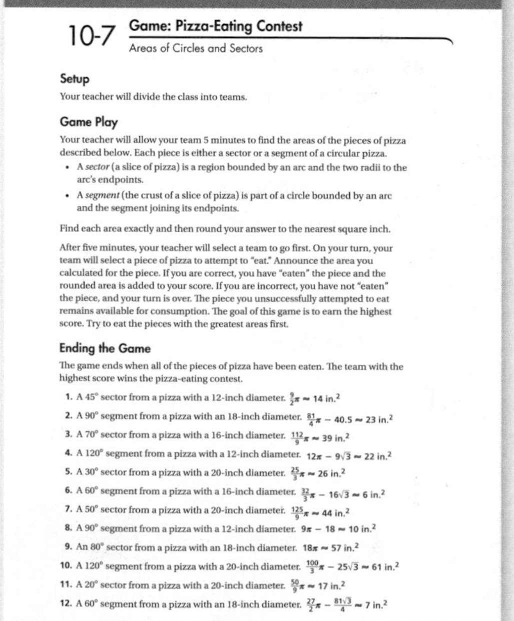

10-7 **Game: Pizza-Eating Contest**
Areas of Circles and Sectors

Setup

Your teacher will divide the class into teams.

Game Play

Your teacher will allow your team 5 minutes to find the areas of the pieces of pizza described below. Each piece is either a sector or a segment of a circular pizza.

- A *sector* (a slice of pizza) is a region bounded by an arc and the two radii to the arc's endpoints.
- A *segment* (the crust of a slice of pizza) is part of a circle bounded by an arc and the segment joining its endpoints.

Find each area exactly and then round your answer to the nearest square inch.

After five minutes, your teacher will select a team to go first. On your turn, your team will select a piece of pizza to attempt to "eat." Announce the area you calculated for the piece. If you are correct, you have "eaten" the piece and the rounded area is added to your score. If you are incorrect, you have not "eaten" the piece, and your turn is over. The piece you unsuccessfully attempted to eat remains available for consumption. The goal of this game is to earn the highest score. Try to eat the pieces with the greatest areas first.

Ending the Game

The game ends when all of the pieces of pizza have been eaten. The team with the highest score wins the pizza-eating contest.

1. A 45° sector from a pizza with a 12-inch diameter. $\frac{9}{2}\pi \approx 14$ in.^{2}
2. A 90° segment from a pizza with an 18-inch diameter. $\frac{81}{4}\pi - 40.5 \approx 23$ in.^{2}
3. A 70° sector from a pizza with a 16-inch diameter. $\frac{112}{9}\pi \approx 39$ in.^{2}
4. A 120° segment from a pizza with a 12-inch diameter. $12\pi - 9\sqrt{3} \approx 22$ in.^{2}
5. A 30° sector from a pizza with a 20-inch diameter. $\frac{25}{3}\pi \approx 26$ in.^{2}
6. A 60° segment from a pizza with a 16-inch diameter. $\frac{32}{3}\pi - 16\sqrt{3} \approx 6$ in.^{2}
7. A 50° sector from a pizza with a 20-inch diameter. $\frac{125}{9}\pi \approx 44$ in.^{2}
8. A 90° segment from a pizza with a 12-inch diameter. $9\pi - 18 \approx 10$ in.^{2}
9. An 80° sector from a pizza with an 18-inch diameter. $18\pi \approx 57$ in.^{2}
10. A 120° segment from a pizza with a 20-inch diameter. $\frac{100}{3}\pi - 25\sqrt{3} \approx 61$ in.^{2}
11. A 20° sector from a pizza with a 20-inch diameter. $\frac{50}{9}\pi \approx 17$ in.^{2}
12. A 60° segment from a pizza with an 18-inch diameter. $\frac{27}{2}\pi - \frac{81\sqrt{3}}{4} \approx 7$ in.^{2}

Concept Byte

Use With Lesson 10-7

ACTIVITY

Inscribed and Circumscribed Figures

Common Core State Standards

G-GPE.B.7 Use coordinates to compute perimeters of polygons and areas of triangles and rectangles, e.g., using the distance formula.
MP 7

In this Activity, you will compare the circumference and area of a circle with the perimeter and area of regular polygons inscribed in and circumscribed about the circle.

Activity 1

Write your answers as decimals rounded to the nearest tenth.

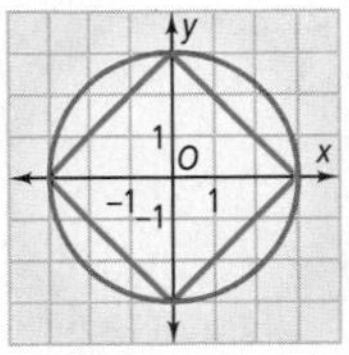

1. The square is inscribed in a circle.
a. What is the length of a side of the square?
b. What is the perimeter and area of the square?
c. What are the circumference and the area of the circle? Use 3.14 for π.
d. What is the ratio of the perimeter of the square to the circumference of the circle? What is the ratio of the area of the square to the area of the circle?

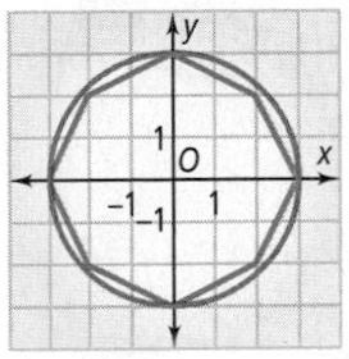

2. The regular octagon is inscribed in a circle.
a. What is the length of a side of the octagon?
b. What is the perimeter and area of the octagon?
c. The radius of the circle is the same as in Exercise 1. What is the ratio of the perimeter of the octagon to the circumference of the circle? What is the ratio of the area of the octagon to the area of the circle?

3. Make a Conjecture What will happen to the ratios as you increase the number of sides of the regular polygon?

Activity 2

Write your answers as decimals rounded to the nearest tenth, as necessary.

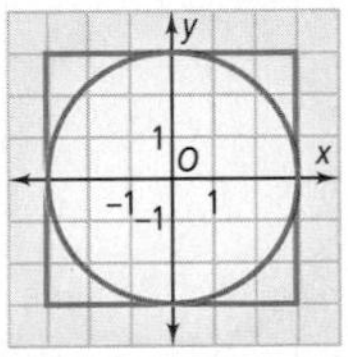

4. The square circumscribes the circle.
a. What is the perimeter and area of the square?
b. The radius of the circle is the same as in Exercise 1. What is the ratio of the circumference of the circle to the perimeter of the square? What is the ratio of the area of the circle to the area of the square?

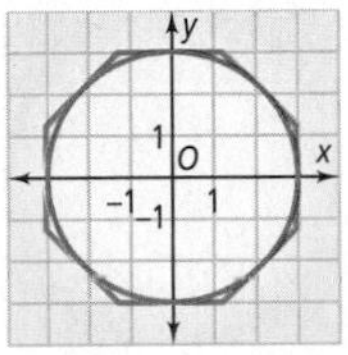

5. The octagon circumscribes the circle.
a. What is the perimeter and area of the octagon?
b. The radius of the circle is the same as in Exercise 1. What is the ratio of the circumference of the circle to the perimeter of the octagon? What is the ratio of the area of the circle to the area of the octagon?

6. Make a Conjecture What will happen to the ratios as you increase the number of sides of the regular polygon?

Answers

Activity 1

1a. 4.2
b. 16.8, 17.6
c. 18.8, 28.3
d. $\frac{16.8}{18.8} = \frac{42}{47}$; $\frac{176}{283}$
2a. 2.3
b. 18.4, 25.2
c. $\frac{46}{47}$; $\frac{252}{283}$
3. The ratios will increase and get closer to 1.

Activity 2

4a. 24; 36
b. $\frac{60}{47}$; $\frac{360}{283}$
5a. 20.0, 29.8
b. $\frac{50}{47}$; $\frac{298}{283}$
6. As you increase the number of sides, the ratios will approach 1.

Guided Instruction

PURPOSE To calculate, explore, and make conjectures about the circumference and area of a circle, and the perimeter and area of regular polygons that are inscribed in and circumscribed about the circle

PROCESS Students will

- find the circumference and area of a circle.
- find the perimeter and area of a square and an octagon inscribed in and circumscribed about the circle.
- use their calculations to explore ratios and make conjectures about those ratios as the number of sides of the regular polygon increases.

DISCUSS Have students review how to divide a regular polygon into congruent isosceles triangles, find the measure of the central angle, and then use trigonometric ratios to find missing measures such as the apothem and length of a side of the polygon.

Area of regular polygon: $A = \frac{1}{2}ap$, where a is the length of the apothem and p is perimeter

Activity 1

In this Activity students compare the circumference and area of a circle to the perimeter and area of a regular polygon that is inscribed in the circle.

Q How can you find the length of a side of the square? **[Use the Distance Formula or the Pythagorean Theorem.]**

Q How can you find the length of the apothem and one side of the octagon? **[Separate the octagon into 8 isosceles triangles. Then use one of the triangles and trigonometric ratios to find the apothem and length of a side.]**

Q Is the perimeter of the square or the octagon a better approximation for the circumference of the circle? **[octagon]**

Activity 2

In this Activity students compare the circumference and area of a circle to the perimeter and area of a regular polygon that circumscribes the circle.

Q How is the radius of the circle related to the apothem of the octagon? **[The lengths are the same.]**

EXTENSION

Have students find the perimeter of a regular polygon with 20 sides that is inscribed in and circumscribed about the same circle. Then find the ratio of the perimeter of the figure to the diameter of the circle. Ask them to make a conjecture about what value is being approximated as you increase the number of sides of the regular polygon. **[π]**

Mathematical Practices This Concept Byte supports students in becoming proficient in shifting perspective, Mathematical Practice 7.

1 Interactive Learning

Solve It!

PURPOSE To find a theoretical probability

PROCESS Students may list the possible results of the toss of three coins and use the formula for finding theoretical probability.

FACILITATE

Q How can you determine the possible outcomes? **[Make a tree diagram or list.]**

Q How many ways can the coin land heads up twice? **[3]**

Q How many possible outcomes are there? **[8]**

ANSWER See Solve It in Answers on next page.

CONNECT THE MATH Students review their knowledge of probability. In the lesson, students apply what they know about probability to situations where probability is calculated in regards to length and area.

2 Guided Instruction

Take Note

In geometric probability the sample space is represented by a literal space in the plane. Have students practice identifying subsets of linear figures in the classroom such as a portion of the number line or the side of one tile in a row of tiles.

10-8 Geometric Probability

Common Core State Standards

Prepares for S-CP.A.1 Describe events as subsets of a sample space . . . using characteristics . . . of the outcomes, or as unions, intersections, or complements of other events . . . .

MP 1, MP 3, MP 4

Objective To use segment and area models to find the probabilities of events

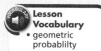

Solve It! **Getting Ready!**

A fair coin is equally likely to land heads up or tails up. Suppose you toss a fair coin three times. What is the probability that the coin will land tails up exactly twice? Explain your reasoning.

Try making a chart of all the possible outcomes to make sense of this problem.

MATHEMATICAL PRACTICES In the Solve It, you found a probability involving a coin. In this lesson you will find probabilities based on lengths and areas. The probability of an event, written *P*(event), is the likelihood that the event will occur.

When the possible outcomes are equally likely, the theoretical probability of an event is the ratio of the number of favorable outcomes to the number of possible outcomes.

$$P(\text{event}) = \frac{\text{number of favorable outcomes}}{\text{number of possible outcomes}}$$

Recall that a probability can be expressed as a fraction, a decimal, or a percent.

Lesson Vocabulary
- geometric probablility

Essential Understanding You can use geometric models to solve certain types of probability problems.

In **geometric probability,** points on a segment or in a region of a plane represent outcomes. The geometric probability of an event is a ratio that involves geometric measures such as length or area.

take note **Key Concept** Probability and Length

Point S on $\overline{AD}$ is chosen at random. The probability that S is on $\overline{BC}$ is the ratio of the length of $\overline{BC}$ to the length of $\overline{AD}$.

$$P(S \text{ on } \overline{BC}) = \frac{BC}{AD}$$

10-8 Preparing to Teach

ESSENTIAL UNDERSTANDING
- Certain problems in probability can be solved by modeling the situation with geometric measures.

Math Background

Probability, geometry and algebra all come together when investigating geometric probability. Students should have some knowledge of percentages and simple probability. Geometric probability is the study of probabilities involved in geometric problems under stated conditions, such as length, area, and volume for geometric objects.

Probability can be both applied to and modeled by geometry. Probability can be represented by linear models like line segments. The length of a part of the segment is compared to the total length of the segment. Area models represent probability in two-dimensions. Area probability is used in games and sports such as archery to determine point values.

Students will use composite figures to find areas of shaded regions and calculate the probability of a point falling in that region.

Mathematical Practices

Model with mathematics. Using a knowledge of length and area, students will compute the probability of a point ending up along a certain length or within a certain area. They will make an initial simplifying assumption about a probability ratio and then compute the ratio accurately to justify their conjecture.

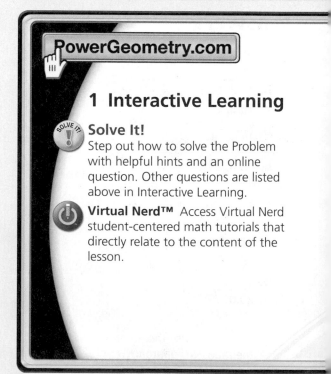

PowerGeometry.com

1 Interactive Learning

Solve It! Step out how to solve the Problem with helpful hints and an online question. Other questions are listed above in Interactive Learning.

Virtual Nerd™ Access Virtual Nerd student-centered math tutorials that directly relate to the content of the lesson.

Problem 1 Using Segments to Find Probability

Point K on $\overline{ST}$ is chosen at random. What is the probability that K lies on $\overline{QR}$?

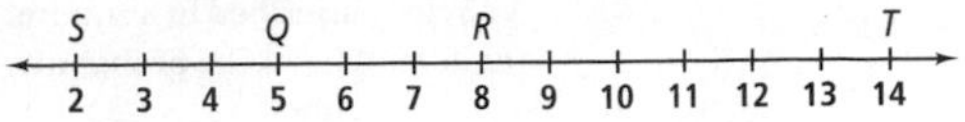

Think

How can you find the length of each segment?
You can use the Ruler Postulate to find the length of each segment.

$$P(K \text{ on } \overline{QR}) = \frac{\text{length of } \overline{QR}}{\text{length of } \overline{ST}} = \frac{|5-8|}{|2-14|} = \frac{3}{12}, \text{ or } \frac{1}{4}$$

The probability that K is on $\overline{QR}$ is $\frac{1}{4}$, or 25%.

Got It? 1. Use the diagram in Problem 1. Point H on $\overline{ST}$ is selected at random. What is the probability that H lies on $\overline{SR}$?

Problem 2 Using Segments to Find Probability

Transportation **A commuter train runs every 25 min. If a commuter arrives at the station at a random time, what is the probability that the commuter will have to wait at least 10 min for the train?**

Plan

How can you *draw a diagram* to model the situation?
Draw a line segment with endpoints 0 and 25 to represent the length of time between trains. Each point on the segment represents an arrival time.

Assume that a stop takes very little time. Draw a line segment to model the situation. The length of the entire segment represents the amount of time between trains. A commuter will have to wait at least 10 min for the train if the commuter arrives at any time between 0 and 15 min.

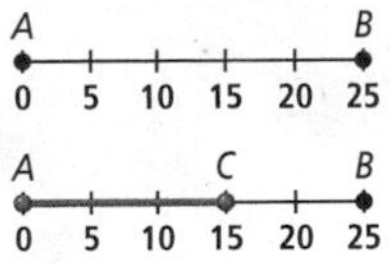

$$P(\text{waiting at least 10 min}) = \frac{\text{length of favorable segment}}{\text{length of entire segment}} = \frac{15}{25}, \text{ or } \frac{3}{5}$$

The probability that a commuter will have to wait at least 10 min for the train is $\frac{3}{5}$, or 60%.

Got It? 2. What is the probability that a commuter will have to wait no more than 5 min for the train?

When the points of a region represent equally likely outcomes, you can find probabilities by comparing areas.

Take note — Key Concept Probability and Area

Point S in region R is chosen at random. The probability that S is in region N is the ratio of the area of region N to the area of region R.

$$P(S \text{ in region } N) = \frac{\text{area of region } N}{\text{area of region } R}$$

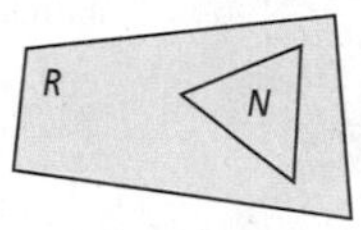

Problem 1

Q What measurements do you need to find the probability that K is on $\overline{QR}$? **[ST and QR]**

Q How do you find the length of a segment? **[Find the difference in the absolute values of the endpoints of the segment.]**

Got It?

Q What lengths are different compared to Problem 1? How? **[QR is replaced with SR.]**

Q What is the numerator used to calculate this probability? Show your computation. **[6; |8 − 2|]**

Problem 2

Q What are some examples of wait times that are "at least 10 minutes"? **[Sample: 10 min, 16 min, 24 min]**

Q What segment represents the interval between trains? **[the segment from 0 to 25 minutes]**

Q How can you find the probability of waiting at least 10 minutes for the train? **[Divide the length of the favorable segment by the total length.]**

Got It?

Q What segment represents a wait of 5 minutes or less? **[the segment from 20 to 25 minutes]**

Take Note

Have students practice identifying subsets of planes in the classroom such as a section of the blackboard, or color blocks on posters. Ask students to discuss similarities and differences between linear and area probability models.

2 Guided Instruction

Each Problem is worked out and supported online.

Problem 1
Using Segments to Find Probability

Problem 2
Using Segments to Find Probability

Problem 3
Using Area to Find Probability

Problem 4
Using Area to Find Probability

Support in Geometry Companion

- Vocabulary
- Key Concepts
- Got It?

Answers

Solve It!

$\frac{3}{8}$; explanations may vary. Sample: The possible outcomes for tossing a coin three times are (H, H, H), (H, H, T), (H, T, H), (H, T, T), (T, H, H), (T, H, T), (T, T, H), and (T, T, T). Three out of the eight possible outcomes have two tails.

Got It?

1. $\frac{1}{2}$ or 50%
2. $\frac{1}{5}$ or 20%

Problem 3

Q What two areas must you compare to find the probability? **[the area of the shaded area and the area of the square]**

Q What area should appear in the numerator of the ratio? **[the area of the square minus the area of the circle]**

Got It? **ERROR PREVENTION**

Have students identify and write a ratio in words before they perform any calculations. For the numerator of the ratio, they must remember to subtract the area of the white triangle from the area of the square.

Problem 4

Q How can you calculate the area of the blue zone? **[Subtract the area of the yellow and red circle from the area of the circle that includes the yellow, red, and blue zones.]**

Q What is the radius of the entire target? **[12.2(5) = 61 cm]**

Got It? **SYNTHESIZING**

Have students make a conjecture for 4b before they calculate the probability for each zone. Have them verify their conjecture. If the results do not support their conjecture, have students identify their errors.

Problem 3 Using Area to Find Probability

A circle is inscribed in a square. Point Q in the square is chosen at random. What is the probability that Q lies in the shaded region?

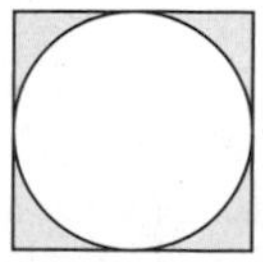

Know

The length of a side of the square, which is also the length of the diameter of the inscribed circle

Need

The areas of the square and the shaded region

Plan

Subtract the area of the circle from the area of the square to find the area of the shaded region. Then use it to find the probability.

$$\text{area of shaded region} = \text{area of square} - \text{area of circle}$$
$$= 6^2 - \pi(3)^2$$
$$= 36 - 9\pi$$

$$P(Q \text{ lies in shaded region}) = \frac{\text{area of shaded region}}{\text{area of square}}$$
$$= \frac{36 - 9\pi}{36} \approx 0.215$$

The probability that Q lies in the shaded region is about 0.215, or 21.5%.

Got It? **3.** A triangle is inscribed in a square. Point T in the square is selected at random. What is the probability that T lies in the shaded region?

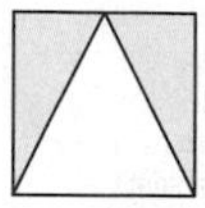

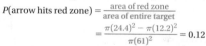

Problem 4 Using Area to Find Probability

Archery **An archery target has 5 colored scoring zones formed by concentric circles. The target's diameter is 122 cm. The radius of the yellow zone is 12.2 cm. The width of each of the other zones is also 12.2 cm. If an arrow hits the target at a random point, what is the probability that it hits the red zone?**

Plan

How can you find the area of the red zone?

The red zone lies between two concentric circles. To find the area of the red zone, subtract the areas of the two concentric circles.

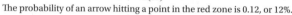

The red zone is the region between a circle with radius 12.2 + 12.2, or 24.4 cm and the yellow circle with radius 12.2 cm. The target is a circle with radius $\frac{122}{2}$, or 61 cm.

$$P(\text{arrow hits red zone}) = \frac{\text{area of red zone}}{\text{area of entire target}}$$
$$= \frac{\pi(24.4)^2 - \pi(12.2)^2}{\pi(61)^2} = 0.12$$

The probability of an arrow hitting a point in the red zone is 0.12, or 12%.

Got It? **4. a.** What is the probability that an arrow hits the yellow zone?

b. Reasoning If an arrow hits the target at a random point, is it more likely to hit the black zone or the red zone? Explain.

Additional Problems

1. Point P on $\overline{FJ}$ is chosen at random. What is the probability that P is on $\overline{GH}$?

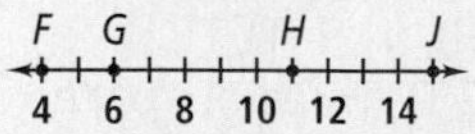

ANSWER $\frac{5}{11}$

2. A river ferry runs every 40 minutes. If a passenger arrives at the ferry station at a random time, what is the probability that he will have to wait at least 25 minutes for the ferry?

ANSWER $\frac{3}{8}$

3. A circle is inscribed in a square. A point N in the square is chosen at random. What is the probability that N lies in the shaded region?

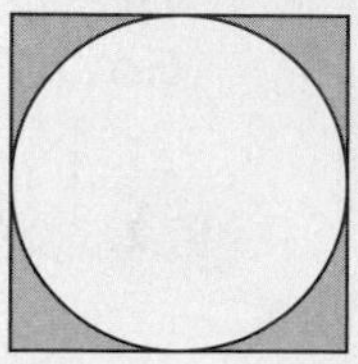

ANSWER about 21.5%

4. Suppose a dart lands randomly on the target below. What is the probability that the dart will land in the shaded region?

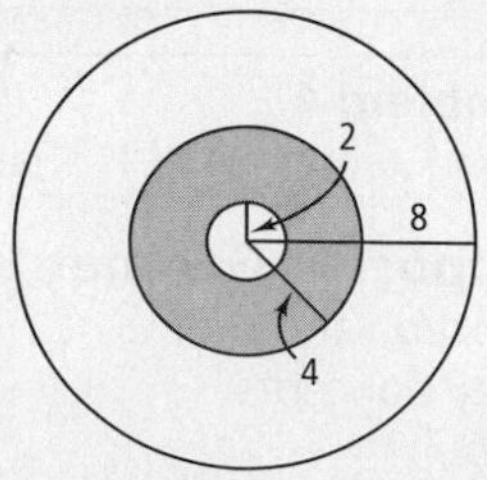

ANSWER 18.75%

Answers

Got It! (continued)

3. $\frac{1}{2}$ or 50%

4a. 0.04, or 4%

b. The black zone; the area of the black zone is greater than the area of the red zone, so $P(\text{black zone}) > P(\text{red zone})$.

Lesson Check

Do you know HOW?

Point T on $\overline{AD}$ is chosen at random. What is the probability that T lies on the given segment?

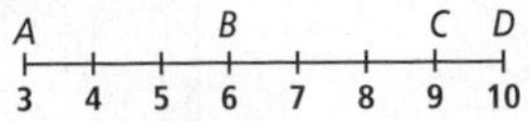

1. $\overline{AB}$ 2. $\overline{AC}$ 3. $\overline{BD}$ 4. $\overline{BC}$

5. A point K in the regular hexagon is chosen at random. What is the probability that K lies in the region that is *not* shaded?

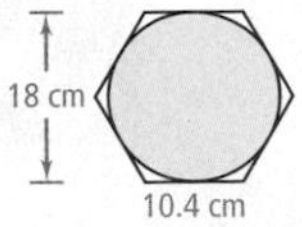

Do you UNDERSTAND? MATHEMATICAL PRACTICES

6. **Reasoning** In the figure at the right, $\frac{SQ}{QT} = \frac{1}{2}$. What is the probability that a point on $\overline{ST}$ chosen at random will lie on $\overline{QT}$? Explain.

7. **Error Analysis** Your class needs to find the probability that a point A in the square chosen at random lies in the shaded region. Your classmate's work is shown below. What is the error? Explain.

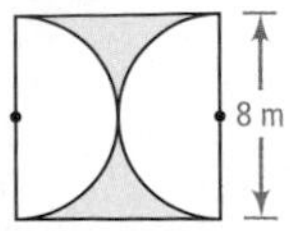

P(A in shaded region) = $\frac{\text{Area of semicircles}}{\text{Area of square}}$
$= \frac{16\pi}{64}$
$= 0.785$, or 79%

Practice and Problem-Solving Exercises MATHEMATICAL PRACTICES

A Practice A point on $\overline{AK}$ is chosen at random. Find the probability that the point lies on the given segment.

See Problem 1.

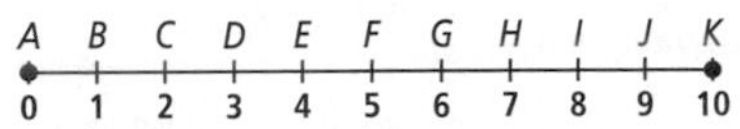

8. $\overline{CH}$ 9. $\overline{FG}$ 10. $\overline{DJ}$

11. $\overline{EI}$ 12. $\overline{AK}$ 13. $\overline{GK}$

See Problem 2.

14. **Transportation** At a given bus stop, a city bus stops every 16 min. If a student arrives at his bus stop at a random time, what is the probability that he will not have to wait more than 4 min for the bus?

15. **Traffic Lights** The cycle of the traffic light on Main Street at the intersection of Main Street and Commercial Street is 40 seconds green, 5 seconds yellow, and 30 seconds red. If you reach the intersection at a random time, what is the probability that the light is red?

16. **Communication** Your friend is supposed to call you between 3 P.M. and 4 P.M. At 3:20 P.M., you realize that your cell phone is off and you immediately turn it on. What is the probability that you missed your friend's call?

3 Lesson Check

Do you know HOW?

- If students have difficulty with Exercises 1-4, then have them review Problem 1 to determine how to find the length of the segments needed to calculate each probability.

Do you UNDERSTAND?

- If students have difficulty with Exercise 6, then have them represent SQ as x and ST as $3x$.

Close

Q How is theoretical probability calculated? **[the number of favorable outcomes divided by the number of possible outcomes]**

Q How can a geometric model represent probability? **[The length or area of a portion of a figure represents the favorable points while the entire figure represents all possible points.]**

Lesson Check

1. $\frac{3}{7}$
2. $\frac{6}{7}$
3. $\frac{4}{7}$
4. $\frac{3}{7}$
5. about 0.09, or 9%
6. $\frac{2}{3}$; explanations may vary. Sample: Since $\frac{SQ}{QT} = \frac{1}{2}$, you can let $SQ = x$ and $QT = 2x$, where x is not 0. Then $ST = 3x$ and the ratio $\frac{QT}{ST} = \frac{2x}{3x} = \frac{2}{3}$.
7. The numerator should be (area of square − area of semicircles); the favorable region is the shaded region and its area is the area left when the areas of the semicircles are subtracted from the area of the square.

Practice and Problem-Solving Exercises

8. $\frac{1}{2}$
9. $\frac{1}{10}$
10. $\frac{3}{5}$
11. $\frac{2}{5}$
12. 1
13. $\frac{2}{5}$
14. $\frac{1}{4}$, or 25%
15. $\frac{2}{5}$, or 40%
16. $\frac{1}{3}$, or about 33%

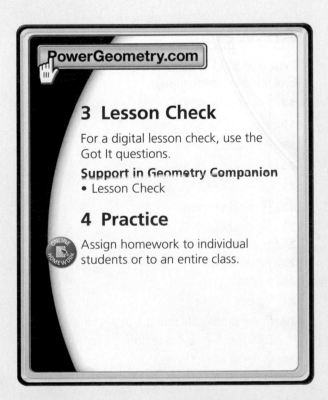

PowerGeometry.com

3 Lesson Check

For a digital lesson check, use the Got It questions.

Support in Geometry Companion

- Lesson Check

4 Practice

Assign homework to individual students or to an entire class.

4 Practice

ASSIGNMENT GUIDE
Basic: 8–24 all, 27–28, 32–40 even
Average: 9–23 odd, 25–43
Advanced: 9–23 odd, 25–45
Standardized Test Prep: 46–50
Mixed Review: 51–62

Mathematical Practices are supported by exercises with red headings. Here are the Practices supported in this lesson:

MP 1: Make Sense of Problems Ex. 27
MP 3: Construct Arguments Ex. 6, 28, 31, 40b
MP 3: Critique the Reasoning of Others Ex. 7
MP 4: Model with Mathematics Ex. 32

Applications exercises have blue headings. Exercises 14–16, 21–24, 29, 30, 40, 41, and 43 support MP 4: Model.

STEM exercises focus on science or engineering applications.

EXERCISE 40: Use the Think About a Plan worksheet in the **Practice and Problem Solving Workbook** (also available in the Teaching Resources in print and online) to further support students' development in becoming independent learners.

HOMEWORK QUICK CHECK
To check students' understanding of key skills and concepts, go over Exercises 11, 19, 27, 32, and 40.

A point in the figure is chosen at random. Find the probability that the point lies in the shaded region.

See Problems 3 and 4.

17.

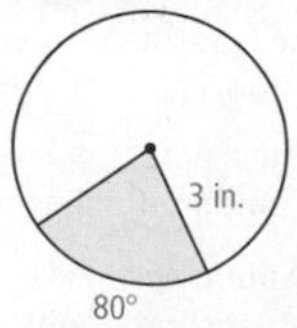

18.

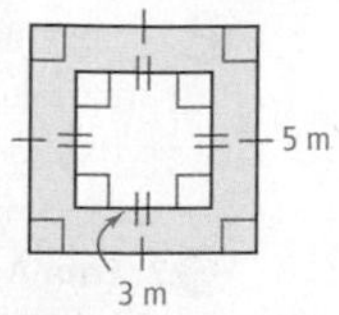

19.

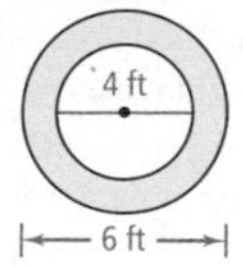

20.

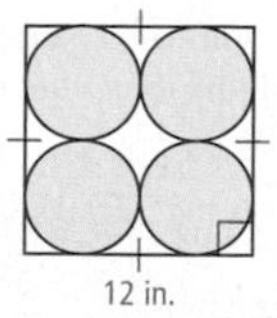

Target Game **A target with a diameter of 14 cm has 4 scoring zones formed by concentric circles. The diameter of the center circle is 2 cm. The width of each ring is 2 cm. A dart hits the target at a random point. Find the probability that it will hit a point in the indicated region.**

21. the center region
22. the blue region
23. either the blue or red region
24. any region

B Apply

25. Points M and N are on $\overline{ZB}$ with M between Z and N. $ZM = 5$, $NB = 9$, and $ZB = 20$. A point on $\overline{ZB}$ is chosen at random. What is the probability that the point is on $\overline{MN}$?

26. $\overline{BZ}$ contains $\overline{MN}$ and $BZ = 20$. A point on $\overline{BZ}$ is chosen at random. The probability that the point is also on $\overline{MN}$ is 0.3, or 30%. Find MN.

27. Think About a Plan Every 20 min from 4:00 P.M. to 7:00 P.M., a commuter train crosses Boston Road. For 3 min, a gate stops cars from crossing over the tracks as the train goes by. What is the probability that a motorist randomly arriving at the train crossing during this time interval will have to stop for a train?
- How can you represent the situation visually?
- What ratio can you use to solve the problem?

28. Reasoning Suppose a point in the regular pentagon is chosen at random. What is the probability that the point is *not* in the shaded region? Explain.

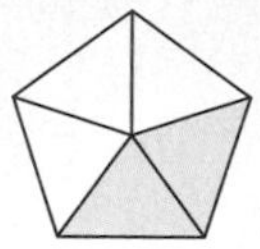

29. Commuting A bus arrives at a stop every 16 min and waits 3 min before leaving. What is the probability that a person arriving at the bus stop at a random time has to wait more than 10 min for a bus to leave?

STEM **30. Astronomy** Meteorites (mostly dust-particle size) are continually bombarding Earth. The surface area of Earth is about 65.7 million mi^2. The area of the United States is about 3.7 million mi^2. What is the probability that a meteorite landing on Earth will land in the United States?

31. Reasoning What is the probability that a point chosen at random on the circumference of $\odot C$ lies on $\widehat{AB}$? Explain how you know.

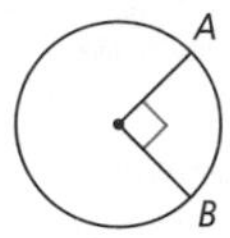

32. Writing Describe a real-life situation in which you would use geometric probability.

Answers

Practice and Problem-Solving Exercises (continued)

17. $\frac{2}{9}$, or about 22%
18. $\frac{16}{25}$, or 64%
19. $\frac{5}{9}$, or about 56%
20. $\frac{\pi}{4}$, or about 78.5%
21. $\frac{1}{49}$, or about 2%
22. $\frac{16}{49}$, or about 33%
23. $\frac{24}{49}$, or about 49%
24. 1, or 100%
25. $\frac{3}{10}$, or 30%
26. 6 units
27. $\frac{3}{20}$, or 15%
28. $\frac{3}{5}$; the probability that it is in any one of the 5 sectors is $\frac{1}{5}$, and 3 sectors are not shaded.
29. $\frac{9}{19}$, or about 47%
30. about 5.6%
31. $\frac{1}{4}$; $m\widehat{AB} = 90$, so the length of $\widehat{AB} = \frac{90}{360} \cdot 2\pi r = \frac{1}{4} \cdot 2\pi r$. The ratio of the length of $\widehat{AB}$ to the circumference is $\frac{1}{4}$.
32. Check students' work.

Algebra **Find the probability that coordinate x of a point chosen at random on $\overline{AK}$ satisfies the inequality.**

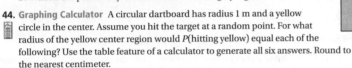

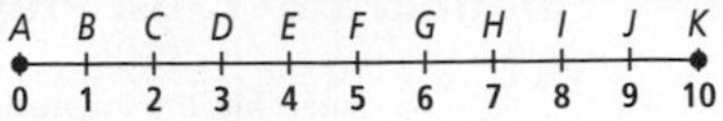

33. $2 \leq x \leq 8$

34. $2x \leq 8$

35. $5 \leq 11 - 6x$

36. $\frac{1}{2}x - 5 > 0$

37. $2 \leq 4x \leq 3$

38. $-7 \leq 1 - 2x \leq 1$

39. One type of dartboard is a square of radius 10 in. You throw a dart and hit the target. What is the probability that the dart lies within $\sqrt{10}$ in. of the square's center?

40. Games To win a prize at a carnival game, you must toss a quarter, so that it lands entirely within a circle as shown at the right. Assume that the center of a tossed quarter is equally likely to land at any point within the 8-in. square.

a. What is the probability that the quarter lands entirely in the circle in one toss?

b. Reasoning On average, how many coins must you toss to win a prize? Explain.

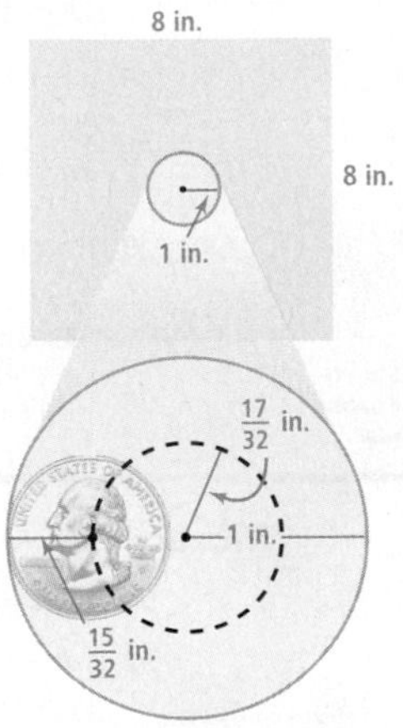

41. Traffic Patterns The traffic lights at Fourth and State Streets repeat themselves in 1-min cycles. A motorist will face a red light 60% of the time. Use this information to estimate how long the Fourth Street light is red during each 1-min cycle.

42. You have a 4-in. straw and a 6-in. straw. You want to cut the 6-in. straw into two pieces so that the three pieces form a triangle.

a. If you cut the straw to get two 3-in. pieces, can you form a triangle?

b. If the two pieces are 1 in. and 5 in., can you form a triangle?

c. If you cut the straw at a random point, what is the probability that you can form a triangle?

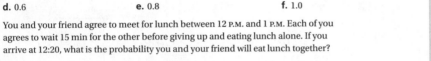

43. Target Game Assume that a dart you throw will land on the 12 in.-by-12 in. square dartboard and is equally likely to land at any point on the board. The diameter of the center circle is 2 in., and the width of each ring is 1 in.

a. What is the probability of hitting either the blue or yellow region?

b. What is the probability the dart will *not* hit the gray region?

Challenge

44. Graphing Calculator A circular dartboard has radius 1 m and a yellow circle in the center. Assume you hit the target at a random point. For what radius of the yellow center region would P(hitting yellow) equal each of the following? Use the table feature of a calculator to generate all six answers. Round to the nearest centimeter.

a. 0.2 **b.** 0.4 **c.** 0.5
d. 0.6 **e.** 0.8 **f.** 1.0

45. You and your friend agree to meet for lunch between 12 P.M. and 1 P.M. Each of you agrees to wait 15 min for the other before giving up and eating lunch alone. If you arrive at 12:20, what is the probability you and your friend will eat lunch together?

33. $\frac{3}{5}$
34. $\frac{2}{5}$
35. $\frac{1}{10}$
36. 0
37. $\frac{1}{40}$
38. $\frac{2}{5}$
39. $\frac{\pi}{20}$, or about 16%
40a. 1.4%
b. About 72 coins; the probability of winning on each toss is 0.014 and $(71)(0.014) < 1 < (72)(0.014)$.
41. 36 s
42a. yes
b. no
c. $\frac{2}{3}$
43a. about 8.7%
b. about 19.6%
44a. 45 cm
b. 63 cm
c. 71 cm
d. 77 cm
e. 89 cm
f. 100 cm
45. 50%

Answers

Practice and Problem-Solving Exercises (continued)

46. A

47. G

48. D

49. G

50. [2] $P = 4s = 24$, so s is 6 ft. The diagonal of a square is the hypotenuse of a 45°-45°-90° △ with leg s and its length is $s\sqrt{2}$. So the diagonal of the square is $6\sqrt{2}$ ft.
[1] no work shown OR missing units

51. 100π ft^2

52. 12π cm^2

53. $A'''(-2, -2)$; $B'''(1, -1)$; $C'''(-1, 1)$

54. $A'''(-5, 6)$; $B'''(-4, 3)$; $C'''(-2, 5)$

55. $A'''(-2, 1)$; $B'''(2, -1)$; $C'''(1, -2)$

56.

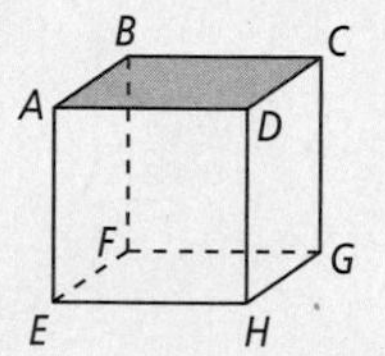

57.

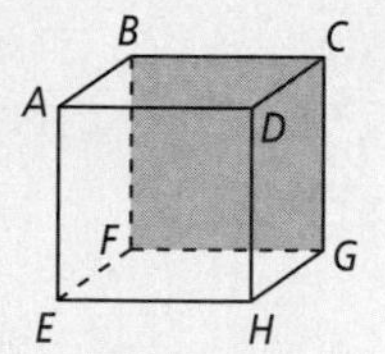

58.

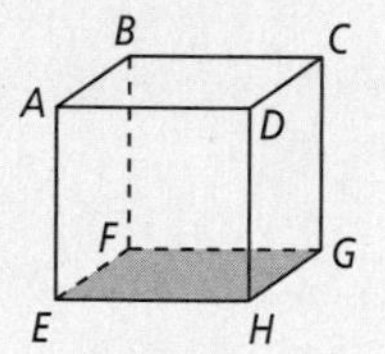

59.

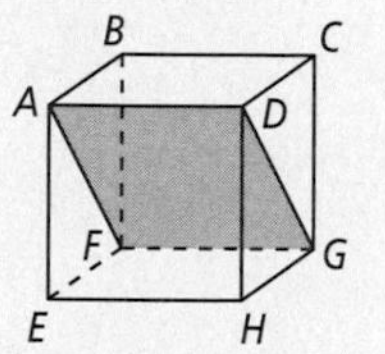

60.

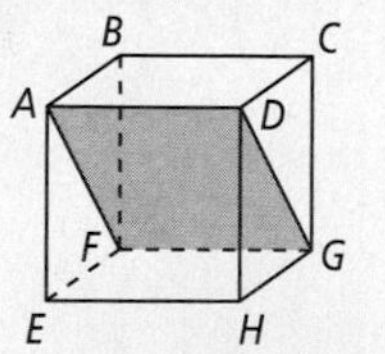

61.

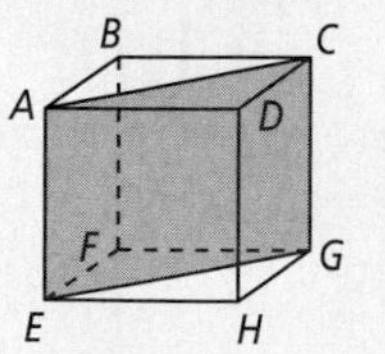

62. Sample:

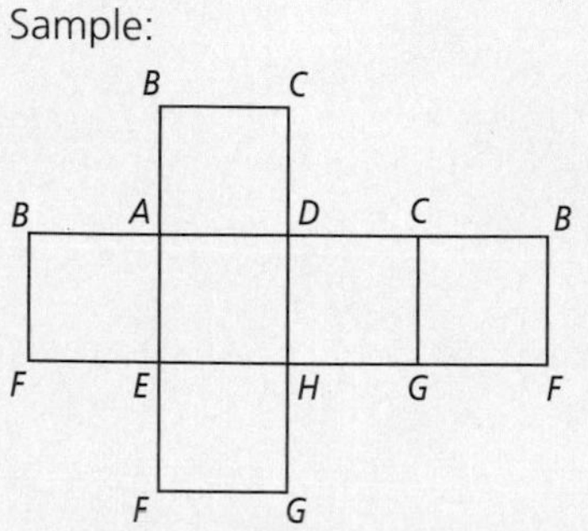

Standardized Test Prep

SAT/ACT

46. A dart hits the square dartboard shown. What is the probability that it lands in the shaded region?

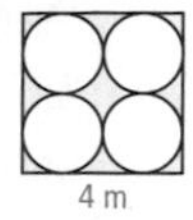

Ⓐ 21% Ⓑ 25% Ⓒ 50% Ⓓ 79%

47. A dilation maps $\triangle JKL$ onto $\triangle J'K'L'$ with a scale factor of 1.2. If $J'L' = 54$ cm, what is JL?

Ⓕ 43.2 cm Ⓖ 45 cm Ⓗ 54 cm Ⓘ 64.8 cm

48. What is the value of x in the figure at the right?

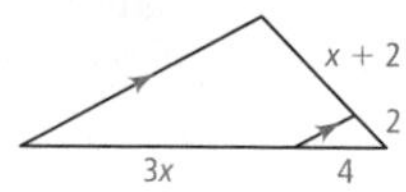

Ⓐ $\frac{1}{3}$ Ⓑ 2 Ⓒ 3 Ⓓ 4

49. The radius of $\odot P$ is 4.5 mm. The measure of central $\angle APB$ is 160. What is the length of $\overset{\frown}{AB}$?

Ⓕ 2π mm Ⓖ 4π mm Ⓗ 5π mm Ⓘ 9π mm

Short Response

50. The perimeter of a square is 24 ft. What is the length of the square's diagonal?

Mixed Review

51. A circle has circumference 20π ft. What is its area in terms of π? See Lesson 10-7.

52. A circle has radius 12 cm. What is the area of a sector of the circle with a 30° central angle in terms of π?

For Exercises 53–55, find the coordinates of the image, $\triangle A'''B'''C'''$, that result from the composition of transformations described. See Lesson 9-4.

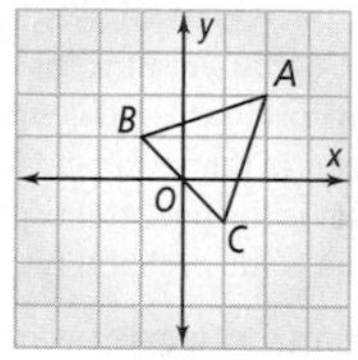

53. $R_{y\text{-axis}} \circ R_{x\text{-axis}}(\triangle ABC)$

54. $T_{<-3, 4>} \circ r_{(90°, O)}(\triangle ABC)$

55. $R_{y=1} \circ T_{<2, -1>}(\triangle ABC)$

Get Ready! To prepare for Lesson 11-1, do Exercises 56–62.

For each exercise, make a copy of the cube below. Shade the plane that contains the indicated points. See Lessons 1-1 and 1-2.

56. A, B, and C

57. B, F, and G

58. E, F, and H

59. A, D, and G

60. F, D, and G

61. A, C, and G

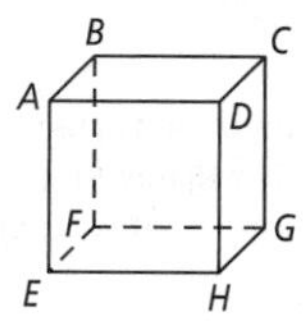

62. Draw a net for the cube at the right.

10-8 Lesson Resources

Differentiated Remediation

Instructional Support

Geometry Companion

Students can use the **Geometry Companion** worktext (4 pages) . . .

- New Vocabulary
- Key Concepts
- Got It for each Problem
- Lesson Check

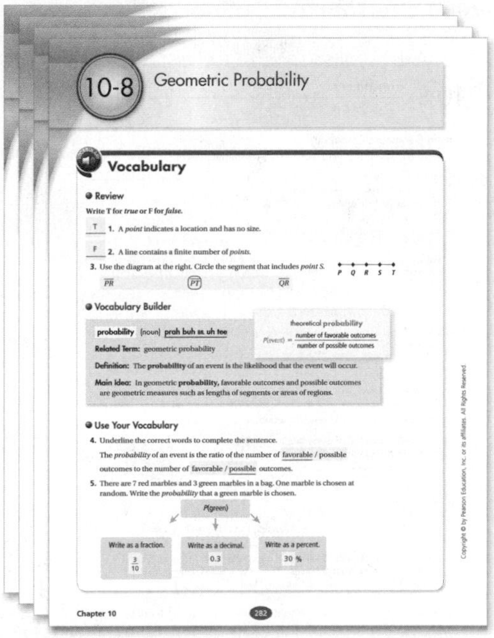
10-8 Geometric Probability

Vocabulary

Review

Write T for *true* or F for *false*.

T 1. A *point* indicates a location and has no size.

F 2. A line contains a finite number of points.

3. Use the diagram at the right. Circle the segment that includes *point S*.

Vocabulary Builder

probability (noun) prah buh BIL uh tee

Related Term: geometric probability

Definition: The probability of an event is the likelihood that the event will occur.

Main Idea: In geometric probability, favorable outcomes and possible outcomes are geometric measures such as lengths of segments or areas of regions.

Use Your Vocabulary

4. Underline the correct words to complete the sentence.

The probability of an event is the ratio of the number of favorable / possible outcomes to the number of favorable / possible outcomes.

5. There are 7 red marbles and 3 green marbles in a bag. One marble is chosen at random. Write the probability that a green marble is chosen.

Chapter 10

ELL Support

Assess Understanding Ask students to write about the similarities and differences between discrete probability and geometric probability in their own words. Ask the following questions:
What is discrete probability?
What is geometric probability?
In what ways are the two concepts related?

Focus on Cummunication Assign a student that is less proficient with another student that is more proficient. In pairs, ask students to read their paragraphs to each other. Encourage questions, clarifications, and oral explanations in familiar words.

5 Assess & Remediate

Lesson Quiz

1. Point O on $\overline{AD}$ is chosen at random. What is the probability that O is on $\overline{BC}$?

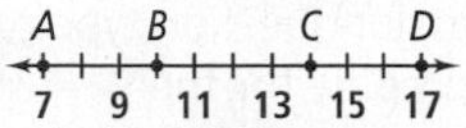

2. A bus picks up passengers at a bus stop every 12 minutes in the morning. Suppose Adrian arrives at the bus stop at a random time. What is the probability that he will have to wait 5 minutes or less for the next bus?

3. A regular octagon is inscribed in a square. A point P in the square is chosen at random. What is the probability that P lies in the shaded region?

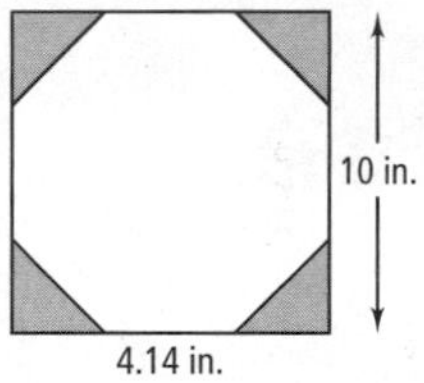

ANSWERS TO LESSON QUIZ

1. 0.4 or 40%

2. $\frac{5}{12}$

3. 17.2%

PRESCRIPTION FOR REMEDIATION

Use the student work on the Lesson Quiz to prescribe a differentiated review assignment.

Points	Differentiated Remediation
0–1	Intervention
2	On-level
3	Extension

PowerGeometry.com

5 Assess & Remediate

Assign the Lesson Quiz. Appropriate intervention, practice, or enrichment is automatically generated based on student performance.

Intervention

- **Reteaching** (2 pages) Provides reteaching and practice exercises for the key lesson concepts. Use with struggling students or absent students.
- **English Language Learner Support** Helps students develop and reinforce mathematical vocabulary and key concepts.

All-in-One Resources/Online

Reteaching

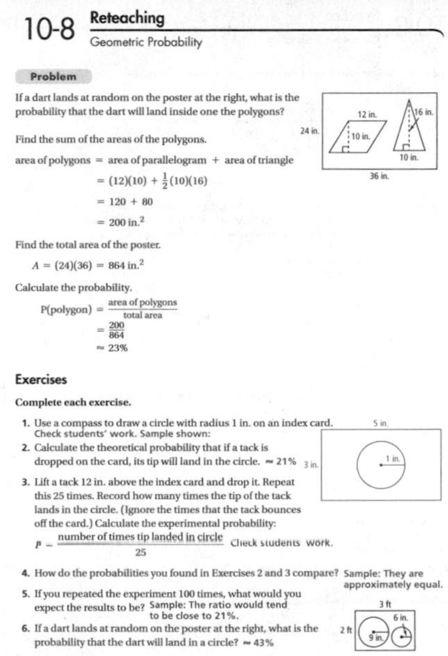
10-8 Reteaching
Geometric Probability

Problem

If a dart lands at random on the poster at the right, what is the probability that the dart will land inside one the polygons?

Find the sum of the areas of the polygons.

area of polygons = area of parallelogram + area of triangle
$= (12)(10) + \frac{1}{2}(10)(16)$
$= 120 + 80$
$= 200 \text{ in.}^2$

Find the total area of the poster.

$A = (24)(36) = 864 \text{ in.}^2$

Calculate the probability.

$P(\text{polygon}) = \frac{\text{area of polygons}}{\text{total area}}$
$= \frac{200}{864}$
$\approx 23\%$

Exercises

Complete each exercise.

1. Use a compass to draw a circle with radius 1 in. on an index card. Check students' work. Sample shown:
2. Calculate the theoretical probability that if a tack is dropped on the card, its tip will land in the circle. ≈ 21%
3. Lift a tack 12 in. above the index card and drop it. Repeat this 25 times. Record how many times the tip of the tack lands in the circle. (Ignore the times that the tack bounces off the card.) Calculate the experimental probability: $P = \frac{\text{number of times tip landed in circle}}{25}$ Check students' work.
4. How do the probabilities you found in Exercises 2 and 3 compare? Sample: They are approximately equal.
5. If you repeated the experiment 100 times, what would you expect the results to be? Sample: The ratio would tend to be close to 21%.
6. If a dart lands at random on the poster at the right, what is the probability that the dart will land in a circle? ≈ 43%

All-in-One Resources/Online

English Language Learner Support

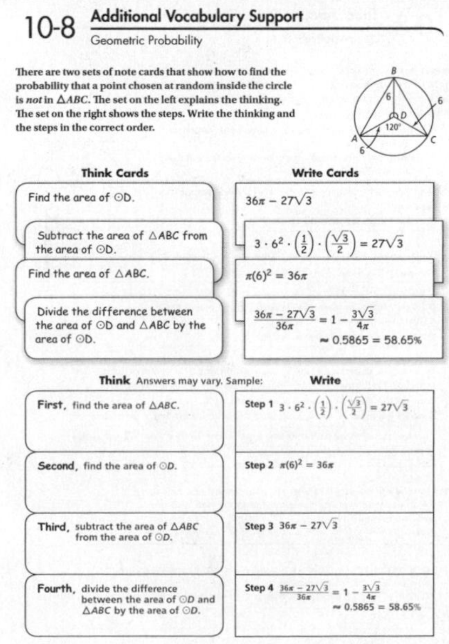
10-8 Additional Vocabulary Support
Geometric Probability

There are two sets of note cards that show how to find the probability that a point chosen at random inside the circle is *not* in △ABC. The set on the left explains the thinking. The set on the right shows the steps. Write the thinking and the steps in the correct order.

Think Cards

Find the area of ⊙D.
Subtract the area of △ABC from the area of ⊙D.
Find the area of △ABC.
Divide the difference between the area of ⊙D and △ABC by the area of ⊙D.

Write Cards

$36\pi - 27\sqrt{3}$
$3 \cdot 6^2 \cdot \left(\frac{1}{2}\right) \cdot \left(\frac{\sqrt{3}}{2}\right) = 27\sqrt{3}$
$\pi(6)^2 = 36\pi$
$\frac{36\pi - 27\sqrt{3}}{36\pi} = 1 - \frac{3\sqrt{3}}{4\pi} \approx 0.5865 = 58.65\%$

Think	Write
First, find the area of △ABC.	Step 1 $3 \cdot 6^2 \cdot \left(\frac{1}{2}\right) \cdot \left(\frac{\sqrt{3}}{2}\right) = 27\sqrt{3}$
Second, find the area of ⊙D.	Step 2 $\pi(6)^2 = 36\pi$
Third, subtract the area of △ABC from the area of ⊙D.	Step 3 $36\pi - 27\sqrt{3}$
Fourth, divide the difference between the area of ⊙D and △ABC by the area of ⊙D.	Step 4 $\frac{36\pi - 27\sqrt{3}}{36\pi} = 1 - \frac{3\sqrt{3}}{4\pi} \approx 0.5865 = 58.65\%$

Differentiated Remediation *continued*

On-Level

- **Practice** (2 pages) Provides extra practice for each lesson. For simpler practice exercises, use the Form K Practice pages found in the All-in-One Teaching Resources and online.
- **Think About a Plan** Helps students develop specific problem-solving skills and strategies by providing scaffolded guiding questions.
- **Standardized Test Prep** Focuses on all major exercises, all major question types, and helps students prepare for the high-stakes assessments.

Extension

- **Enrichment** Provides students with interesting problems and activities that extend the concepts of the lesson.
- **Activities, Games, and Puzzles** Worksheets that can be used for concepts development, enrichment, and for fun!

Practice and Problem Solving Wkbk/ All-in-One Resources/Online

Practice page 1

10-8 **Practice** *Form G*
Geometric Probability

Find the probability that a point chosen at random from $\overline{AK}$ is on the given segment.

A B C D E F G H I J K
0 2 4 6 8 10 12 14 16 18 20

1. $\overline{CF}$ $\frac{3}{10}$
2. $\overline{BI}$ $\frac{7}{10}$
3. $\overline{GK}$ $\frac{2}{5}$
4. $\overline{FG}$ $\frac{1}{10}$
5. $\overline{AK}$ 1
6. $\overline{AC}$ $\frac{1}{5}$
7. Roberto's trolley runs every 45 minutes. It he arrives at the trolley stop at a random time, what is the probability that he will *not* have to wait more than 10 minutes? Draw a geometric model to solve the problem. 22.2%
8. The state of Connecticut is approximated by a rectangle 100 mi by 50 mi. Hartford is approximately at the center of Connecticut. If a meteor hit the earth within 200 mi of Hartford, find the probability that the meteor landed in Connecticut. 4.0%
9. A stoplight at an intersection stays red for 60 second, changes to green for 45 seconds, and then turns yellow for 15 seconds. If Jamal arrives at the intersection at a random time, what is the probability that he will have to wait at a red light for more than 15 seconds? 37.5%

A point between *A* and *B* on each number line is chosen at random. What is the probability that the point is between *C* and *D*?

10. A C D B (−2, 0, 2, 4) 20%
11. A C D B (−10, 0, 10, 20) 50%
12. A C D B (−2, 0, 2, 4) 40%
13. A C D B (−3, 0, 3, 6) $33\frac{1}{3}$%

Use the dartboard at the right for Exercises 14–16. Assume that a dart you throw will land on the dartboard and is equally likely to land at any point on the board.

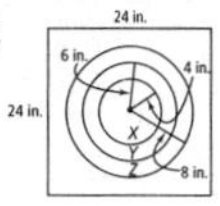

14. What is the probability of hitting region *X*? 8.7%
15. What is the probability of hitting region *Y*? 10.9%
16. What is the probability of hitting region *Z*? 15.3%

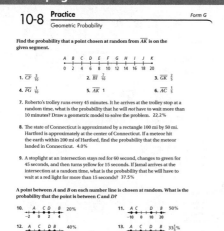

Practice and Problem Solving Wkbk/ All-in-One Resources/Online

Practice page 2

10-8 **Practice** (continued) *Form G*
Geometric Probability

A point in the figure is chosen at random. Find the probability that the point lies in the shaded region.

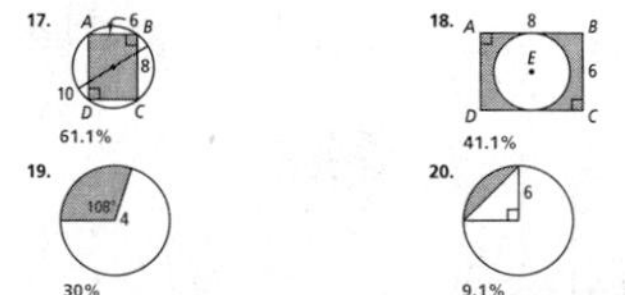

17. 61.1%
18. 41.1%
19. 30%
20. 9.1%

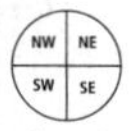

21. See the figure at the right. The wind comes out of one of the directions at random. What is the probability that the wind comes out of the southwest (SW)? $\frac{1}{4} = 25\%$

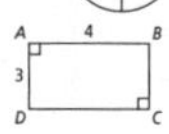

22. **Reasoning** Point *P* is chosen at random from the perimeter of rectangle *ABCD*.
    a. What is the probability that *P* lies on $\overline{DC}$? 28.6%
    b. Now *P* is chosen at random from the perimeter or the diagonals. Does this increase or decrease the probability that *P* lies on $\overline{DC}$? Explain. Find the new probability to support your conclusion.
    decreases it; because the total length is increased; $\frac{1}{6} \approx 16.7\%$

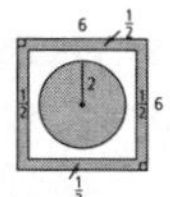

23. You and your friend are playing a target game based on the board at the right (not drawn to scale). You must hit the border to win a point. Your friend must hit the circle in the center.
    a. Is the game fair? That is, do you or your friend have an equal probability of hitting your target zones? Explain. If the game is not fair, find the radius of the circle that would make it fair.
    No; your friend has a better chance of winning because the area of the circle is greater than the area of the border; the radius required to make it fair is $r = \sqrt{\frac{11}{\pi}}$.
    b. Find the probability that you do not score a point. $1 - \frac{11}{36} = \frac{25}{36} \approx 69.4\%$
24. **Open-Ended** Make a game board using polygons and circles. Switch games with a partner and use geometric probability to find the likelihood of choosing each particular region of the game board.
    Check students' work.

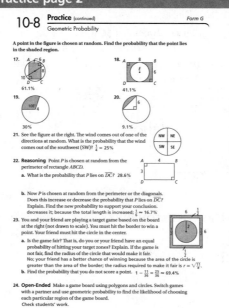

All-in-One Resources/Online

Enrichment

10-8 **Enrichment**
Geometric Probability

Points in Geometric Probability

There are various ways to extend the concept of geometric probability.

A B C D

1. Consider $\overline{AD}$ to the right.
    a. Point *P* on $\overline{AD}$ is chosen at random. What is the probability that *P* is on $\overline{BC}$? $\frac{BC}{AD}$
    b. Is the probability of choosing point *P* defined?
    No; probabilities are only defined for segments, not isolated points.
    c. Try to define the probability of choosing point *P*. What value should it have?
    Answers may vary. Sample: Define the probability to be zero because a point is one-dimensional.
    d. Let the probability of choosing any isolated point be zero. What paradox arises?
    Zero probability means the result cannot happen. So, no point could be chosen.
    e. Let the probability of choosing any isolated point be 0.001. What paradox arises?
    Answers may vary. Sample: The probability of choosing one of 1000 points would be 1. So, one of those 1000 points would have to be chosen. Yet we can always make a new point between any two points.
    f. Can these paradoxes be avoided? Explain.
    Yes; do not assign probability to individual points.

Geometric Probability in Three Dimensions

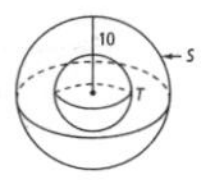

2. Suppose a bird moves randomly within 10 yd of her nest at all times. See the diagram at the right. Recall that the formula for the volume of a sphere is $V = \frac{4}{3}\pi r^3$.
    a. Point *P* in sphere *S* is chosen at random. How might you define the probability that *P* is in sphere *T*?
    Answers will vary. Sample: $P = \frac{\text{Volume of } T}{\text{Volume of } S}$
    b. What is the probability that the bird is within 1 yd of her nest? 0.1%
    c. There is a sphere *H* inside of *S* such that it is equally likely that the bird is inside sphere *H* as outside sphere *H*. What is the radius of sphere *H*? Round your answer to the nearest whole number. 8 yd
    d. What is the volume of sphere *S*? What is the volume of sphere *H*? Does this help explain your answer to part (c)? Explain.
    4189 yd^3; 2145 yd^3; yes; the volume of sphere *H* is about half the volume of sphere *S*.

Practice and Problem Solving Wkbk/ All-in-One Resources/Online

Think About a Plan

10-8 **Think About a Plan**
Geometric Probability

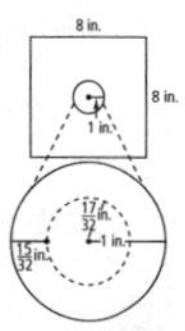

Games To win a prize at a carnival game, you must toss a quarter so that it lands entirely within a 1-in. circle as shown at the right. Assume that the center of a tossed quarter is equally likely to land at any point within the 8-in. square.

a. What is the probability that the quarter lands entirely in the circle in one toss?
b. **Reasoning** On average, how many coins must you toss to win a prize? Explain.

1. In this problem, what represents the favorable outcome? Be specific.
The quarter lands within the circle.
2. In this problem, what represents all the possible outcomes?
The quarter lands anywhere within the square.
3. If a section of the quarter is in the circle, does this count as a favorable outcome?
No; the entire quarter must be in the 1-in. circle.
4. How can you determine a smaller circle within which the center of the quarter must land for the quarter to be entirely within the 1-in. circle? What is the radius of this circle? The radius of a quarter is $\frac{15}{32}$ in.
Subtract the radius of the quarter from the radius of the 1-in. circle to find the radius of the circle within which the center of the quarter must fall; the radius is $\frac{17}{32}$ in., or $\frac{15}{32}$ in. less than the 1-in. circle.
5. Use words to write a probability ratio. Then rewrite the ratio using the appropriate formulas. Substitute the appropriate measures and find the probability.
$P(\text{event}) = \frac{\text{area of smaller circle}}{\text{area of square}} = \frac{\pi r^2}{s^2} = \frac{\pi \cdot 0.53125^2}{8^2} \approx \frac{0.89}{64} \approx 0.014$
6. Based on this, what is the average number of coins you must toss before you can expect to win a prize? Explain.
72; $72 > \frac{1}{0.014} > 71$, so you cannot expect to win a prize with the toss of 71 or fewer coins.

Practice and Problem Solving Wkbk/ All-in-One Resources/Online

Standardized Test Prep

10-8 **Standardized Test Prep**
Geometric Probability

Multiple Choice

For Exercises 1–4, choose the correct letter.

Q R S T

1. Point *X* on $\overline{QT}$ is chosen at random. What is the probability that *X* is on $\overline{ST}$? B
    Ⓐ $\frac{QT}{ST}$ Ⓑ $\frac{ST}{QT}$ Ⓒ $\frac{QS}{ST}$ Ⓓ $\frac{ST}{QS}$
2. Point *P* on $\overline{AD}$ is chosen at random. For which of the figures below is the probability that *P* is on $\overline{BC}$ 25%? Note: Diagrams not drawn to scale. I
    Ⓕ A 2, B 5, C 8, D 10
    Ⓗ A 2, B 3, C 4, D 5
    Ⓖ A 1, B 2, C 3, D 4
    Ⓘ A 1, B 2, C 3, D 5
3. Point *P* is chosen at random in a circle. If a square is inscribed in the circle, what is the probability that *P* lies outside the square? B
    Ⓐ $1 - \frac{1}{2\pi}$ Ⓑ $1 - \frac{2}{\pi}$ Ⓒ $1 - \frac{\pi}{2}$ Ⓓ $1 - \frac{1}{4\pi}$
4. You have a 7-cm straw and a 10-cm straw. You want to cut the 10-cm straw into two pieces so that the three pieces make a triangle. If you cut the straw at a random point, what is the probability that you can make a triangle? I
    Ⓕ 30% Ⓖ 40% Ⓗ 60% Ⓘ 70%

Short Response

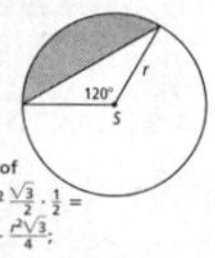

5. Point *P* is chosen at random in ⊙*S*. What is the probability that *P* lies in the shaded segment shown in the diagram at the right? Show your work.

Answers may vary. Sample: Area of sector $= \frac{120}{360} \cdot \pi r^2 = \frac{1}{3}\pi r^2$; area of triangle $= \frac{1}{2} \cdot bh = \frac{1}{2} \cdot 2r\sin(60) \cdot r\cos(60) = r^2 \sin 60 \cdot \cos 60 = r^2 \frac{\sqrt{3}}{2} \cdot \frac{1}{2} = \frac{r^2\sqrt{3}}{4}$; area of segment = area of sector − area of triangle $= \frac{1}{3}\pi r^2 - \frac{r^2\sqrt{3}}{4}$;
probability $= \frac{\text{Area of segment}}{\text{Area of circle}} = \frac{\frac{1}{3}\pi r^2 - \frac{r^2\sqrt{3}}{4}}{\pi r^2} = \frac{1}{3} - \frac{\sqrt{3}}{4\pi} \approx 0.1955 \approx 19.55\%$
[2] All answers correct [1] some answers correct [0] no answers correct

Online Teacher Resource Center

Activities, Games, and Puzzles

10-8 **Activity: Design a Dartboard**
Geometric Probability

Materials

- Suction-cup balls, one for each pair of students

Setup

Your teacher will divide the class into pairs. Suppose you and your partner work as designers for a company that manufactures dartboards.

Design Procedures

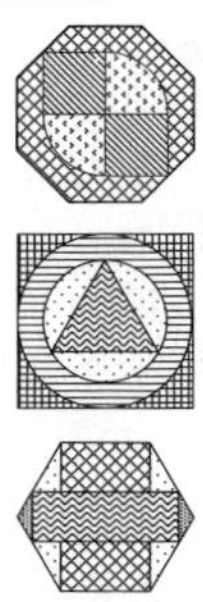

Use your knowledge of geometric probability and follow the steps below to produce a new design for a dartboard.

Step 1 Draw a dartboard using at least 3 different types of geometric figures. Measure and label any lengths you may need for calculating the area of each figure later.

Step 2 Color at least 3 distinct regions of your design using different colors.

Step 3 Determine the area of each region. Then, determine the geometric probability that a point chosen at random on the game board will be in that region.

Step 4 Assign point values to the various regions that you feel correspond to the difficulty a player may have landing in that region. In other words, regions with lower geometric probabilities should be worth more points.

Step 5 Write a short report to present your design to the class and justifying the point values you have assigned to each region.

Extension

Your teacher will select several of the best designs presented to the class in Step 5 above. He or she will make those designs into transparencies and display them.

Throw a suction-cup ball to the dartboard 10 times and record the region where the ball lands each time. Calculate the experimental geometric probability of each region of the dartboard.

How close are your results to the geometric probabilities the dartboard designer calculated in Step 3 above? Check students' work.

Variation

Instead of throwing a suction-cup ball in the Extension above, place stickers on the dartboard while blindfolded. Or make a game board out of paper, place it on the floor, and drop or flip chips onto it.

10 *Pull It* All Together

Completing the Performance Task

Look back at your results from the Apply What You've Learned sections in Lessons 10-1, 10-5, and 10-7. Use the work you did to complete the following.

To solve these problems you will pull together many concepts and skills that you have studied about area and perimeter of polygons and circles.

1. Solve the problem in the Task Description on page 613 by finding the probability that a person who plays the game wins a prize. Round your answer to the nearest whole percent. Assume the player's dart is equally likely to land at any point of the target. Show all your work and explain each step of your solution.
2. **Reflect** Choose one of the Mathematical Practices below and explain how you applied it in your work on the Performance Task.
   MP 1: Make sense of problems and persevere in solving them.
   MP 5: Use appropriate tools strategically.
   MP 7: Look for and make use of structure.

On Your Own

The diagram below shows a second target at the school fair. As before, a player throws one dart and wins a prize if the dart lands in any of the red or yellow regions of the target.

The target is composed of a square, two circles, and a regular hexagon that all have center *O*. One circle is inscribed in the hexagon, which is inscribed in the other circle. The length of one side of the regular hexagon in 7 in.

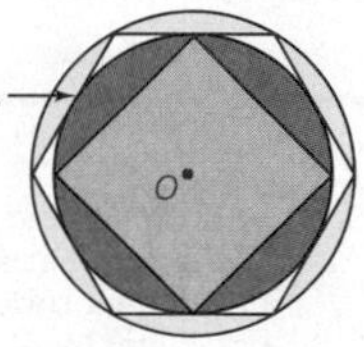

a. Find the probability that a person who plays the game wins a prize. Round your answer to the nearest whole percent. Assume a player's dart is equally likely to land at any point of the target.

b. Would your answer to part (a) change if the side length of the hexagon was 10 in. instead of 7 in.? Explain.

Completing the Performance Task

In the Apply What You've Learned sections in Lessons 10-1, 10-5, and 10-7, students found the areas of the red triangles, the regular octagon, the circle, and the yellow segments of the circle. Now students use some of their results to find a geometric probability to complete the Performance Task. Ask students the following questions as they work toward solving the problem.

Q How can you use the work you have done in the chapter to solve the problem? **[Sample: Find the total area of the red and yellow regions, and then compare that area to the area of the entire target.]**

Q What does the ratio of the area of the square to the area of the circle represent? How could you use this ratio to check your result for the probability of winning a prize? **[The probability of not winning a prize; sample: subtract the ratio from 1 and check that the result is the same as the calculated probability of winning a prize.]**

FOSTERING MATHEMATICAL DISCOURSE

Have students compare alternative methods of finding many of the lengths and areas involved in solving the problem in the Performance Task.

ANSWERS

1. about 36%
2. Check students' work.

On Your Own

This problem is similar to the problem posed on page 613, but now students must find areas in a more complex figure given only the side length of the regular hexagon. Students should strive to solve this problem independently.

ANSWERS

a. about 45%

b. No; all the corresponding shapes of the two targets are similar, so the ratios of the areas do not change.

Essential Questions

BIG idea Measurement

ESSENTIAL QUESTION How do you find the area of a polygon or find the circumference and area of a circle?

ANSWER You can find the area of a polygon, or the circumference or area of a circle, by first determining which formula to use. Then you can substitute the needed measures into the formula.

BIG idea Similarity

ESSENTIAL QUESTION How do perimeters and areas of similar polygons compare?

ANSWER The perimeters of similar polygons are proportional to the ratio of corresponding measures. The areas are proportional to the squares of the ratio of corresponding measures.

10 Chapter Review

Connecting BIG ideas and Answering the Essential Questions

1 Measurement
You can find the area of a polygon, or the circumference or area of a circle, by first determining which formula to use. Then you can substitute the needed measures into the formula.

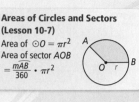

2 Similarity
The perimeters of similar polygons are proportional to the ratio of corresponding measures. The areas are proportional to the squares of corresponding measures.

Areas of Polygons (Lessons 10-1, 10-2, and 10-3)

Parallelogram	$A = bh$
Triangle	$A = \frac{1}{2}bh$
Trapezoid	$A = \frac{1}{2}h(b_1 + b_2)$
Rhombus or kite	$A = \frac{1}{2}d_1d_2$
Regular polygon	$A = \frac{1}{2}ap$

Area of a Triangle Given SAS (Lesson 10-5)

Area of $\triangle ABC = \frac{1}{2}bc(\sin A)$

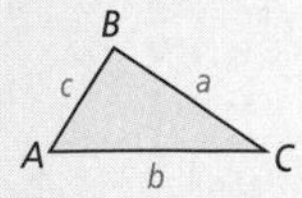

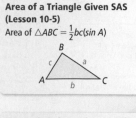

Circles and Arcs (Lesson 10-6)

$C = \pi d$ or $C = 2\pi r$

$m\widehat{ABC} = m\widehat{AB} + m\widehat{BC}$

length of $\widehat{AB} = \frac{m\widehat{AB}}{360} \cdot 2\pi r$

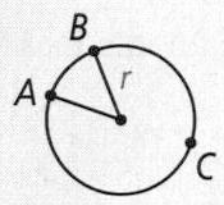

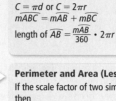

Areas of Circles and Sectors (Lesson 10-7)

Area of $\odot O = \pi r^2$

Area of sector $AOB = \frac{m\widehat{AB}}{360} \cdot \pi r^2$

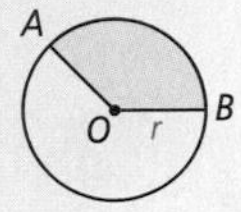

Perimeter and Area (Lesson 10-4)

If the scale factor of two similar figures is $\frac{a}{b}$, then

(1) the ratio of their perimeters is $\frac{a}{b}$ and

(2) the ratio of their areas is $\frac{a^2}{b^2}$.

Chapter Vocabulary

- adjacent arcs (p. 650)
- altitude (p. 616)
- apothem (p. 629)
- arc length (p. 653)
- base (pp. 616, 618)
- central angle (p. 649)
- circle (p. 649)
- circumference (p. 651)
- concentric circles (p. 651)
- congruent circles (p. 649)
- diameter (p. 649)
- geometric probability (p. 668)
- height (pp. 616, 618, 623)
- major arc (p. 649)
- minor arc (p. 649)
- radius (pp. 629, 649)
- sector of a circle (p. 661)
- segment of a circle (p. 662)
- semicircle (p. 649)

Choose the correct term to complete each sentence.

1. You can use any side as the __?__ of a triangle.

2. A(n) __?__ is a region bounded by an arc and the two radii to the arc's endpoints.

3. The distance from the center to a vertex is the __?__ of a regular polygon.

4. Two arcs of a circle with exactly one point in common are __?__.

Summative Questions

Use the following prompts as you review this chapter with your students. The prompts are designed to help you assess your students' understanding of the Big Ideas they have studied.

- What formulas for area do you know?
- How is arc length related to circumference of a circle?
- How do you find perimeters and areas of similar figures?

Answers

Chapter Review

1. base
2. sector
3. radius
4. adjacent arcs

10-1 Areas of Parallelograms and Triangles

Quick Review

You can find the area of a rectangle, a parallelogram, or a triangle if you know the **base** b and the **height** h.

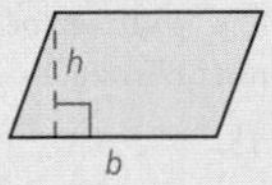

The area of a rectangle or parallelogram is $A = bh$.

The area of a triangle is $A = \frac{1}{2}bh$.

Example

What is area of the parallelogram?

$A = bh$ Use the area formula.

$= (12)(8) = 96$ Substitute and simplify.

The area of the parallelogram is 96 cm^2.

Exercises

Find the area of each figure.

5.

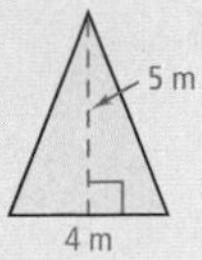

6.

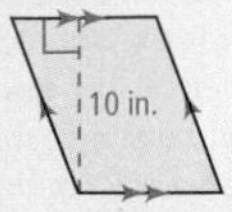

7.

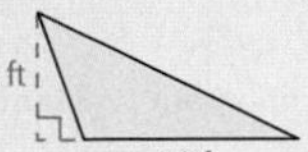

8.

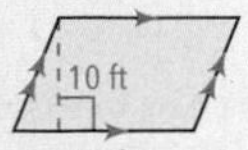

9. A right triangle has legs measuring 5 ft and 12 ft, and hypotenuse measuring 13 ft. What is its area?

10-2 Areas of Trapezoids, Rhombuses, and Kites

Quick Review

The **height of a trapezoid** h is the perpendicular distance between the bases, b_1 and b_2.

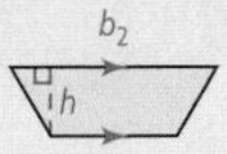

The area of a trapezoid is $A = \frac{1}{2}h(b_1 + b_2)$.

The area of a rhombus or a kite is $A = \frac{1}{2}d_1d_2$, where d_1 and d_2 are the lengths of its diagonals.

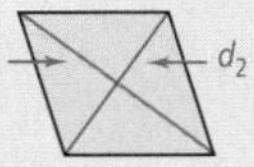

Example

What is the area of the trapezoid?

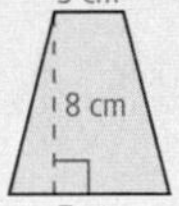

$A = \frac{1}{2}h(b_1 + b_2)$ Use the area formula.

$= \frac{1}{2}(8)(7 + 3)$ Substitute.

$= 40$ Simplify.

The area of the trapezoid is 40 cm^2.

Exercises

Find the area of each figure. If necessary, leave your answer in simplest radical form.

10.

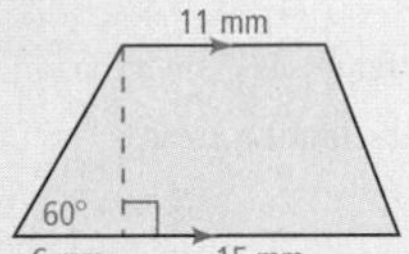

11.

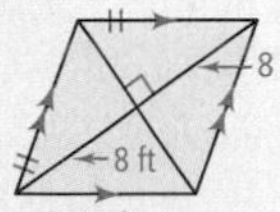

12.

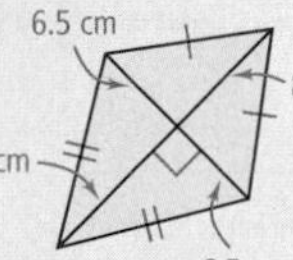

13.

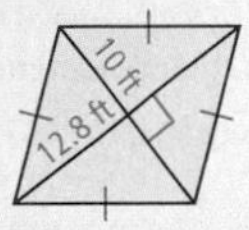

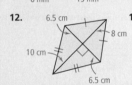

14. A trapezoid has a height of 6 m. The length of one base is three times the length of the other base. The sum of the base lengths is 18 m. What is the area of the trapezoid?

5. 10 m^2
6. 90 $in.^2$
7. 30 ft^2
8. 160 ft^2
9. 30 ft^2
10. $96\sqrt{3}$ mm^2
11. 96 ft^2
12. 117 cm^2
13. 256 ft^2
14. 54 m^2

Answers

Chapter Review (continued)

15. $9\sqrt{3}$ in.^{2}

16. 28 m^2

17. $2400\sqrt{3}$ cm^2

18. 112.5 m^2

19.

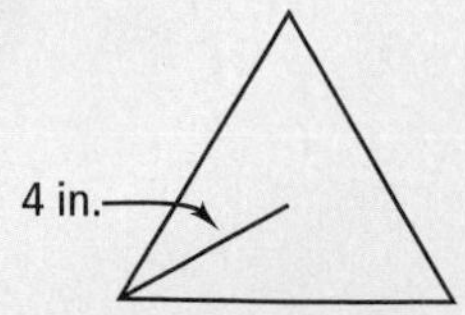

20.8 in.^{2}

20.

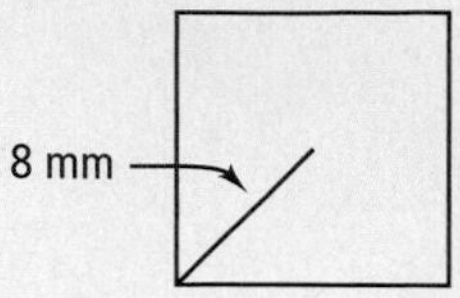

128 mm^2

21.

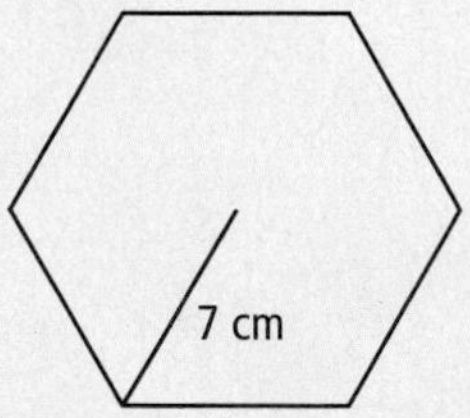

127.3 cm^2

22. 4 : 9

23. 9 : 4

24. 1 : 4

25. 4 : 1

26. $2\sqrt{2}$: 5

10-3 Areas of Regular Polygons

Quick Review

The **center of a regular polygon** C is the center of its circumscribed circle. The **radius** r is the distance from the center to a vertex. The **apothem** a is the perpendicular distance from the center to a side. The area of a regular polygon with apothem a and perimeter p is $A = \frac{1}{2}ap$.

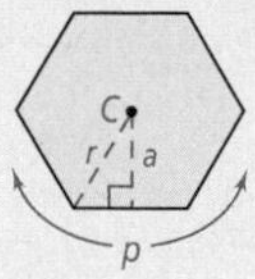

Example

What is the area of a hexagon with apothem 17.3 mm and perimeter 120 mm?

$A = \frac{1}{2}ap$ Use the area formula.

$= \frac{1}{2}(17.3)(120) = 1038$ Substitute and simplify.

The area of the hexagon is 1038 mm^2.

Exercises

Find the area of each regular polygon. If your answer is not an integer, leave it in simplest radical form.

15.

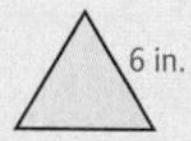

16.

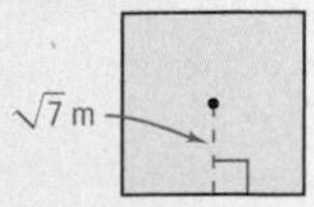

17. What is the area of a regular hexagon with a perimeter of 240 cm?

18. What is the area of a square with radius 7.5 m?

Sketch each regular polygon with the given radius. Then find its area to the nearest tenth.

19. triangle; radius 4 in.

20. square; radius 8 mm

21. hexagon; radius 7 cm

10-4 Perimeters and Areas of Similar Figures

Quick Review

If the scale factor of two similar figures is $\frac{a}{b}$, then the ratio of their perimeters is $\frac{a}{b}$, and the ratio of their areas is $\frac{a^2}{b^2}$.

Example

If the ratio of the areas of two similar figures is $\frac{4}{9}$, what is the ratio of their perimeters?

Find the scale factor.

$\frac{\sqrt{4}}{\sqrt{9}} = \frac{2}{3}$ Take the square root of the ratio of areas.

The ratio of the perimeters is the same as the ratio of corresponding sides, $\frac{2}{3}$.

Exercises

For each pair of similar figures, find the ratio of the area of the first figure to the area of the second.

22.

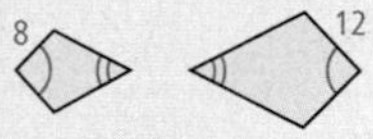

23.

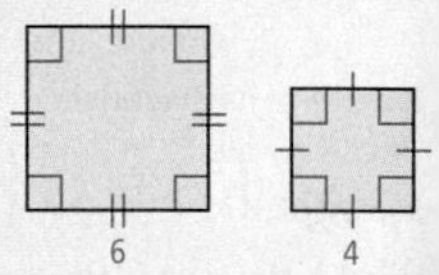

24.

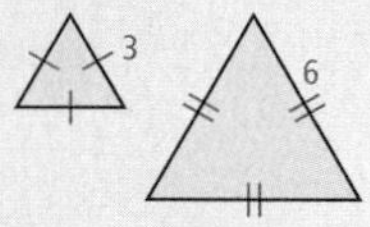

25.

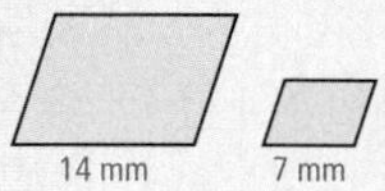

26. If the ratio of the areas of two similar hexagons is 8 : 25, what is the ratio of their perimeters?

10-5 Trigonometry and Area

Quick Review

You can use trigonometry to find the areas of regular polygons. You can also use trigonometry to find the area of a triangle when you know the lengths of two sides and the measure of the included angle.

Area of a $\triangle$
$= \frac{1}{2} \cdot$ side length $\cdot$ side length $\cdot$ sine of included angle

Example

What is the area of $\triangle XYZ$?

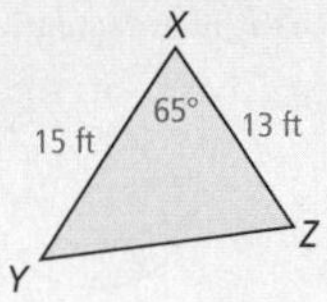

Area $= \frac{1}{2} \cdot XY \cdot XZ \cdot \sin X$
$= \frac{1}{2} \cdot 15 \cdot 13 \cdot \sin 65°$
≈ 88.36500924

The area of $\triangle XYZ$ is approximately 88 ft^2.

Exercises

Find the area of each polygon. Round your answers to the nearest tenth.

27. regular decagon with radius 5 ft

28. regular pentagon with apothem 8 cm

29. regular hexagon with apothem 6 in.

30. regular quadrilateral with radius 2 m

31. regular octagon with apothem 10 ft

32. regular heptagon with radius 3 ft

33.

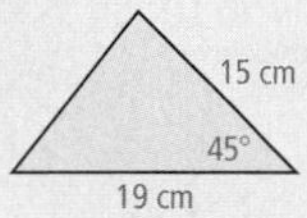

34.

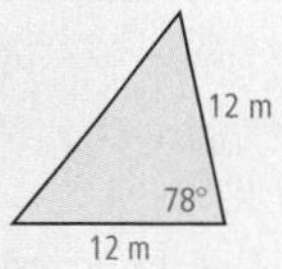

10-6 Circles and Arcs

Quick Review

A **circle** is the set of all points in a plane equidistant from a point called the **center**.

The **circumference** of a circle is $C = \pi d$ or $C = 2\pi r$.

Arc length is a fraction of a circle's circumference. The length of $\widehat{AB} = \frac{m\widehat{AB}}{360} \cdot 2\pi r$.

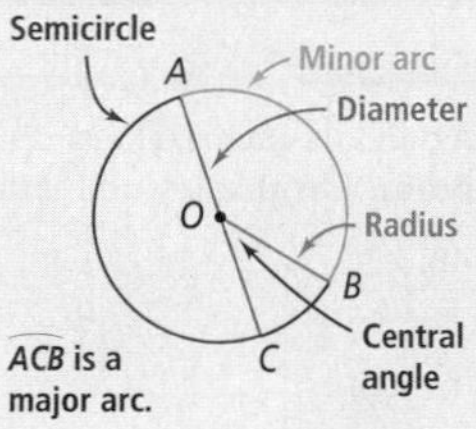

Example

A circle has a radius of 5 cm. What is the length of an arc measuring 80°?

length of $\widehat{AB} = \frac{m\widehat{AB}}{360} \cdot 2\pi r$ — Use the arc length formula.
$= \frac{80}{360} \cdot 2\pi(5)$ — Substitute.
$= \frac{20}{9}\pi$ — Simplify.

The length of the arc is $\frac{20}{9}\pi$ cm.

Exercises

Find each measure.

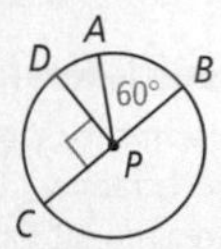

35. $m\angle APD$ **36.** $m\widehat{AC}$

37. $m\widehat{ABD}$ **38.** $m\angle CPA$

Find the length of each arc shown in red. Leave your answer in terms of π.

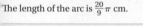

39.

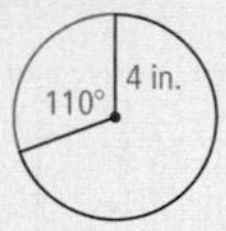

40.

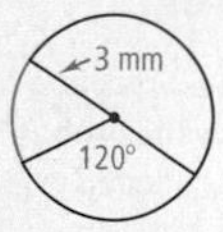

41.

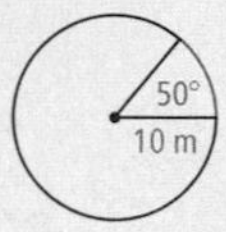

42.

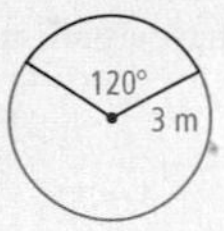

27. 73.5 ft^2
28. 232.5 cm^2
29. 124.7 in.^{2}
30. 8 m^2
31. 331.4 ft^2
32. 24.6 ft^2
33. 100.8 cm^2
34. 70.4 m^2
35. 30
36. 120
37. 330
38. 120
39. $\frac{22\pi}{9}$ in.
40. π mm
41. $\frac{25\pi}{9}$ m
42. 4π m

Answers

Chapter Review (continued)

43. 144π in.^{2}

44. $\frac{49\pi}{4}$ ft^2

45. 41.0 cm^2

46. 18.3 m^2

47. 36.2 cm^2

48. $\frac{1}{2}$, or 50%

49. $\frac{3}{8}$, or 37.5%

50. $\frac{1}{6}$, or about 16.7%

51. $\frac{1}{2}$, or 50%

52. $\frac{1}{2}$, or 50%

10-7 Areas of Circles and Sectors

Quick Review

The area of a circle is $A = \pi r^2$.

A **sector of a circle** is a region bounded by two radii and their intercepted arc. The area of sector $APB = \frac{m\overset{\frown}{AB}}{360} \cdot \pi r^2$.

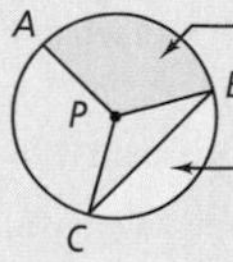

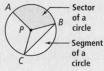

A **segment of a circle** is the part bounded by an arc and the segment joining its endpoints.

Example

What is the area of the shaded region?

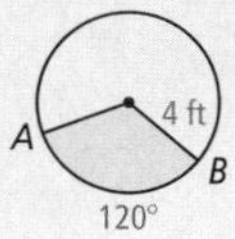

Area $= \frac{m\overset{\frown}{AB}}{360} \cdot \pi r^2$ Use the area formula.

$= \frac{120}{360} \cdot \pi(4)^2$ Substitute.

$= \frac{16\pi}{3}$ Simplify.

The area of the shaded region is $\frac{16\pi}{3}$ ft^2.

Exercises

What is the area of each circle? Leave your answer in terms of π.

43.

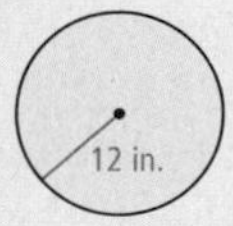

44.

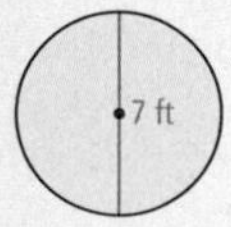

Find the area of each shaded region. Round your answer to the nearest tenth.

45.

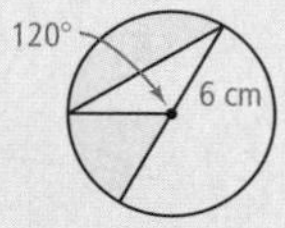

46.

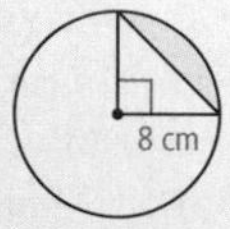

47. A circle has a radius of 20 cm. What is the area of the smaller segment of the circle formed by a 60° arc? Round to the nearest tenth.

10-8 Geometric Probability

Quick Review

Geometric probability uses geometric figures to represent occurrences of events. You can use a segment model or an area model. Compare the part that represents favorable outcomes to the whole, which represents all outcomes.

Example

A ball hits the target at a random point. What is the probability that it lands in the shaded region?

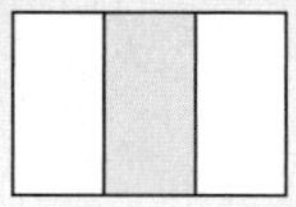

Since $\frac{1}{3}$ of the target is shaded, the probability that the ball hits the shaded region is $\frac{1}{3}$.

Exercises

A dart hits each dartboard at a random point. Find the probability that it lands in the shaded region.

48.

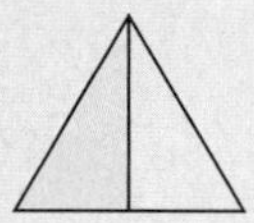

49.

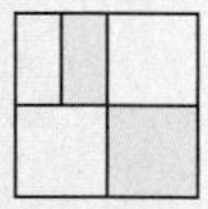

50.

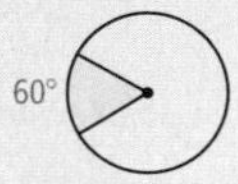

51.

52.

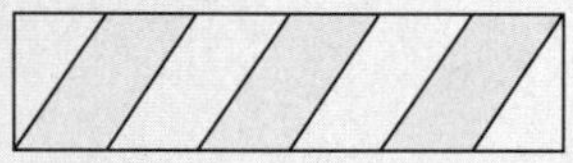

10 Chapter Test

MathXL® for School
Go to PowerGeometry.com

Do you know HOW?

Find the area of each figure. Round to the nearest tenth.

1.

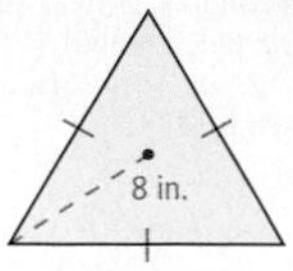

2.

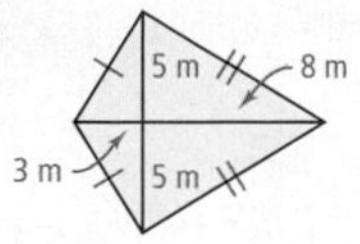

Find the area of each regular polygon. Round to the nearest tenth.

3.

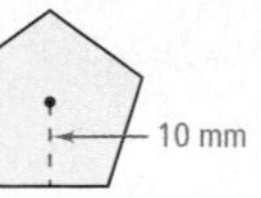

4.

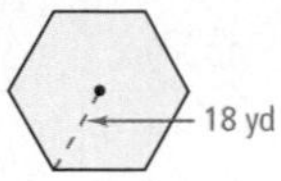

For each pair of similar figures, find the ratio of the area of the first figure to the area of the second.

5.

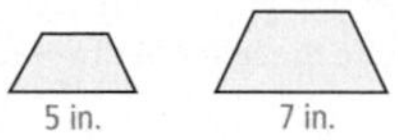

6.

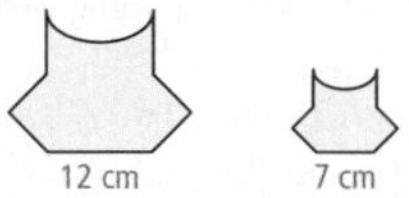

Find the area of each polygon to the nearest tenth.

7.

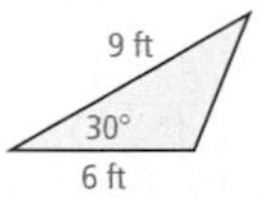

8.

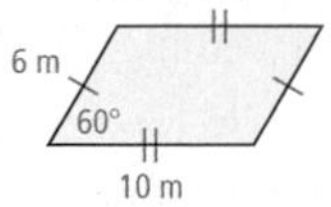

Find each measure for $\odot P$.

9. $m\angle BPC$ **10.** $m\overset{\frown}{AB}$

11. $m\overset{\frown}{ADC}$ **12.** $m\overset{\frown}{ADB}$

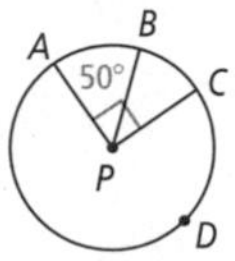

Find the length of each arc shown in red. Leave your answer in terms of π.

13.

14.

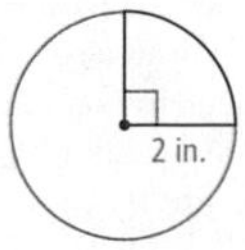

Find the area of each shaded region to the nearest hundredth.

15.

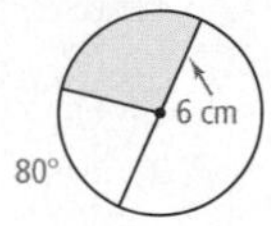

16.

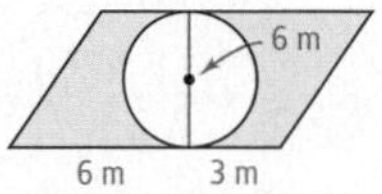

17. Probability Fly A lands on the edge of the ruler at a random point. Fly B lands on the surface of the target at a random point. Which fly is more likely to land in a yellow region? Explain.

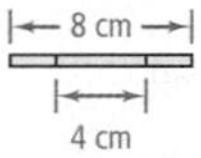

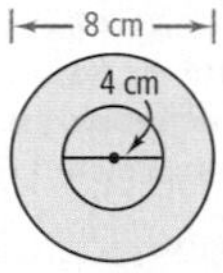

Do you UNDERSTAND?

18. Reasoning A right triangle has height 7 cm and base 4 cm. Find its area using the formulas $A = \frac{1}{2}bh$ and $A = \frac{1}{2}ab(\sin C)$. Are the results the same? Explain.

19. Mental Math Garden A has the shape of a quarter circle with radius 28 ft. Garden B has the shape of a circle with radius 14 ft. What is the ratio of the area of Garden A to the area of Garden B? Explain how you can find the answer without calculating the areas.

20. Reasoning Can a regular polygon have an apothem and a radius of the same length? Explain.

21. Open-Ended Use a compass to draw a circle. Shade a sector of the circle and find its area.

Answers

Chapter Test

1. 83.1 in.^{2}
2. 55 m^2
3. 363.3 mm^2
4. 841.8 yd^2
5. 25 : 49
6. 144 : 49
7. 13.5 ft^2
8. 52.0 m^2
9. 40
10. 50
11. 270
12. 310
13. 10π cm
14. 3π in.
15. 31.42 cm^2
16. 25.73 m^2
17. Fly A; the probability of landing on the yellow region of the ruler is $\frac{1}{2}$ and the probability of landing on the yellow region of the target is $\frac{1}{4}$.
18. Yes; $m\angle C = 90$ and sin $90° = 1$, so $\frac{1}{2}bh = 14$ cm^2 and $\frac{1}{2}ab$ (sin C) = 14 cm^2.
19. 1 : 1; explanations may vary. Sample: The radius of Garden A is twice the radius of Garden B, so the area of Garden A's complete circle is 4 times the area of Garden B. Garden A is a quarter circle, so its area is the same as the area of Garden B.
20. No; the apothem is a leg of a rt. △ and the radius is the hypotenuse of that △.
21. Check students' work.

PowerGeometry.com

MathXL for School

Prepare students for the Mid-Chapter Quiz and Chapter Test with online practice and review.

Item Number	Lesson	Content Standard
1	10-3	G-SRT.C.8
2	9-3	G-CO.A.5
3	9-3	G-CO.B.6
4	10-6	*G-GMD.A.1
5	8-2	G-SRT.C.8
6	8-1	G-SRT.C.8
7	10-3	G-SRT.C.8
8	10-7	G-C.B.5
9	9-1	G-CO.B.6
10	7-4	G-SRT.C.6
11	10-4	*G-GMD.A.3
12	10-3	G-MG.A.1
13	8-3	G-SRT.C.8
14	10-1	*G-GMD.A.3
15	7-2	G-SRT.B.5
16	10-7	G-C.B.5
17	10-4	*G-GMD.A.3
18	1-6	G-CO.A.1
19	6-7	G-SRT.C.8
20	4-6	G-CO.D.12
21	6-7	G-CO.C.11
22	7-2	G-SRT.B.5
23	10-6	G-C.A.2

*Prepares for standard

10 Common Core Cumulative Standards Review

ASSESSMENT

TIPS FOR SUCCESS

Some test questions require you to relate changes in lengths to changes in areas of similar figures. Read the sample question at the right. Then follow the tips to answer it.

TIP 1

The larger square must have a side length greater than that of the smaller square. You can eliminate any answer choices that are less than or equal to 6 in.

The ratio of the areas of the two squares shown below is 4 : 9. What is the length of a side of the larger square?

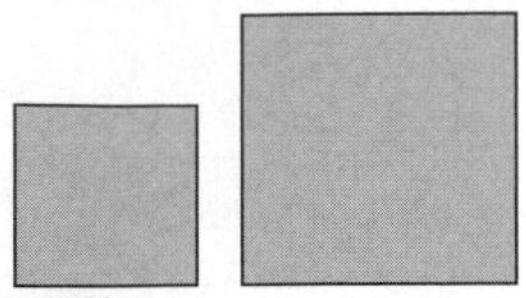

(A) 4 in.
(B) 9 in.
(C) 13.5 in.
(D) 36 in.

TIP 2

All squares are similar. The ratio of their areas is equal to the square of the ratio of their side lengths.

Think It Through

The ratio of the areas is 4 : 9. The ratio of the side lengths is $\sqrt{4} : \sqrt{9}$, or 2 : 3.

Use a proportion to find the length s of the larger square.

$\frac{2}{3} = \frac{6}{s}$

$s = 9$

The correct answer is B.

Vocabulary Builder

As you solve test items, you must understand the meanings of mathematical terms. Match each term with its mathematical meaning.

A. perimeter

B. segment of a circle

C. sector of a circle

D. area

I. the part of a circle bounded by an arc and the segment joining its endpoints

II. the region of a circle bounded by two radii and their intercepted arc

III. the number of square units enclosed by a figure

IV. the distance around a figure

Selected Response

Read each question. Then write the letter of the correct answer on your paper.

1. What is the exact area of an equilateral triangle with sides of length 10 m?

(A) $25\sqrt{3}$ m^2
(B) 25 m^2
(C) $10\sqrt{3}$ m^2
(D) $5\sqrt{3}$ m^2

2. If $\triangle CAT$ is rotated 90° around vertex C, what are the coordinates of A'?

(F) (1, 5)
(G) (3, 3)
(H) (−1, 3)
(I) (3, 5)

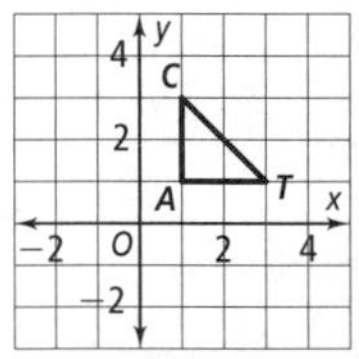

Answers

Common Core Cumulative Standards Review

A. IV
B. I
C. II
D. III
1. A
2. G

3. The vertices of $\triangle ABC$ have coordinates $A(-2, 3)$, $B(3, 4)$, and $C(1, -2)$. The vertices of $\triangle PQR$ have coordinates $P(-2, -3)$, $Q(3, -4)$, and $R(1, 2)$. Which transformation can you use to justify that $\triangle ABC$ is congruent to $\triangle PQR$?

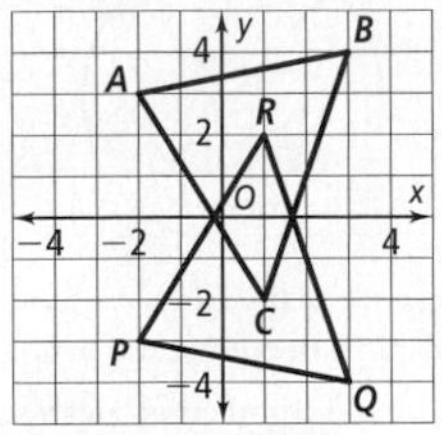

(A) a reflection across the y-axis

(B) a reflection across the x-axis

(C) a rotation 180° clockwise around the origin

(D) a translation 5 units left and 6 units down

4. If a truck's tire has a radius of 2 ft, what is the circumference of the tire in feet?

(F) 8π (H) 4π

(G) 6π (I) 2π

5. Your neighbor has a square garden, as shown below. He wants to install a sprinkler in the center of the garden so that the water sprays only as far as the corners of the garden. What is the approximate radius at which your neighbor should set the sprinkler?

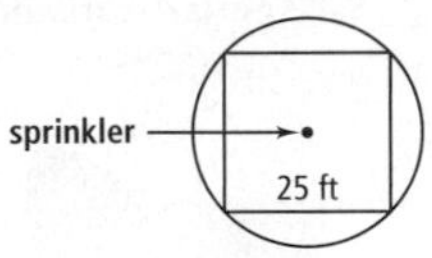

(A) 12.5 ft (C) 25 ft

(B) 18 ft (D) 35 ft

6. Which of the following could be the side lengths of a right triangle?

(F) 4.1, 6.2, 7.3 (H) 3.2, 5.4, 6.2

(G) 40, 60, 72 (I) 33, 56, 65

7. Every triangle in the figure at the right is an equilateral triangle. What is the total area of the shaded triangles?

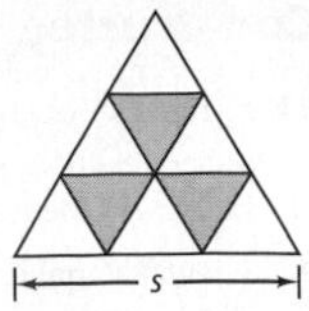

(A) $\frac{s^2\sqrt{3}}{36}$ (C) $\frac{s^2\sqrt{3}}{9}$

(B) $\frac{s^2\sqrt{3}}{12}$ (D) $\frac{s^2\sqrt{3}}{6}$

8. The shaded part of the circle below represents the portion of a garden where a landscaper planted rose bushes. What is the approximate area of the part of the garden used for rose bushes?

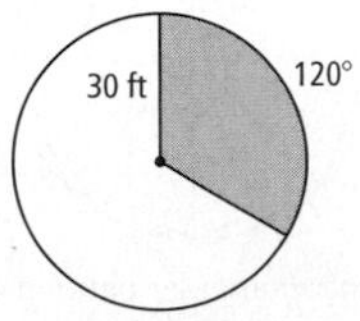

(F) 300 ft^2 (H) 3769 ft^2

(G) 942 ft^2 (I) 8482 ft^2

9. The vertices of $\triangle ART$ are $A(1, -2)$, $R(5, -1)$, and $T(4, -4)$. If $T_{<5, -2>}(\triangle ART) = A'R'T'$, what are the coordinates of the image triangle $A'R'T'$?

(A) $A'(5, -1)$, $R'(4, -4)$, $T'(1, -2)$

(B) $A'(6, -4)$, $R'(10, -3)$, $T'(9, -6)$

(C) $A'(3, 3)$, $R'(7, 4)$, $T'(6, 1)$

(D) $A'(-1, 3)$, $R'(3, 4)$, $T'(2, 1)$

10. In the figure below, what is the length of $\overline{AD}$?

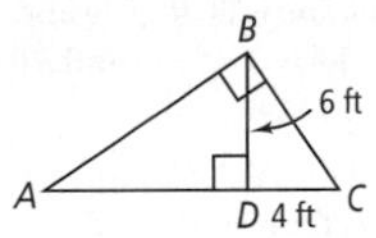

(F) $2\sqrt{13}$ ft (H) 10 ft

(G) 9 ft (I) $3\sqrt{13}$ ft

3. B
4. H
5. B
6. I
7. B
8. G
9. B
10. G

Answers

Common Core Cumulative Standards Review (continued)

11. 27

12. 67.5

13. 9891

14. 17,000

15. $\frac{1}{20}$

16. 84.8

17. 900

18. [2]

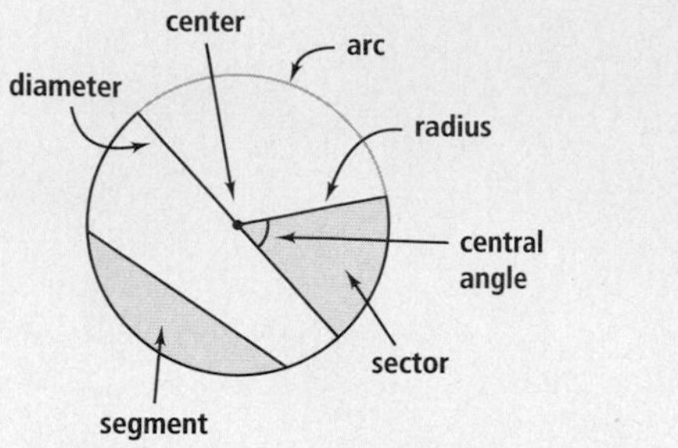

[1] incomplete diagram OR incorrect information

19. [2] No; using the Distance Formula,
$AB = \sqrt{(10-2)^2 + (9-3)^2} = 10$;
$BC = \sqrt{(10-10)^2 + (-3-9)^2} = 12$;
$AC = \sqrt{(10-2)^2 + (-3-3)^2} = 10$;
$\triangle ABC$ is isosc., but it is not equilateral.

[1] incomplete OR incorrect explanation

20. [2]

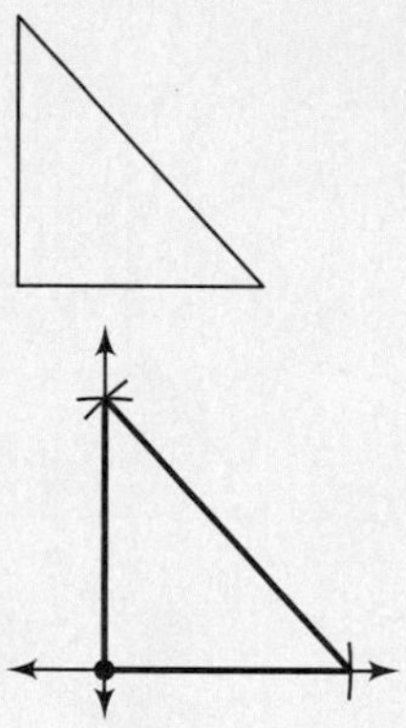

[1] sketch with no construction marks

21. [2] A rectangle; using the Distance Formula, $PR = \sqrt{(-2-1)^2 + (5-(-1))^2} = 3\sqrt{5}$ and $QS = \sqrt{(1-(-2))^2 + (5-(-1))^2} = 3\sqrt{5}$. So $\overline{PR} \cong \overline{QS}$. If the diagonals of a parallelogram are congruent, then the parallelogram is a rectangle.

[1] incomplete OR incorrect explanation

22. [4] Yes; by the Distance Formula, $AB = CD = 10$, $BC = 4$, $AD = 16$, $AF = GH = 5$, $FG = 2$, and $AH = 8$. Since $\frac{AB}{AF} = \frac{10}{5} = \frac{2}{1}$, $\frac{BC}{FG} = \frac{4}{2} = \frac{2}{1}$, $\frac{CD}{GH} = \frac{10}{5} = \frac{2}{1}$, $\frac{AD}{AH} = \frac{16}{8} = \frac{2}{1}$, corresp. sides are proportional. By the Refl. Prop. of $\cong$, $\angle A \cong \angle A$. The slopes of $\overline{AD}$, $\overline{BC}$, and $\overline{FG} = 0$ and the slopes of $\overline{CD}$ and $\overline{GH} = -\frac{4}{3}$, so $\overline{AD} \parallel \overline{BC} \parallel \overline{FG}$ and $\overline{CD} \parallel \overline{GH}$ because $\parallel$ lines have the same slope. If lines are $\parallel$, then corresp. $\angle$s are $\cong$, so $\angle AFG \cong \angle B$ and $\angle D \cong \angle GHA$. Also, $\angle C \cong \angle G$ by the Polygon-$\angle$-Sum Thm. Since corresp. sides are proportional and pairs of corresp. $\angle$s are $\cong$, $ABCD \sim AFGH$.

[3] one error OR incorrect statement

[2] two errors OR incomplete statements

[1] more than two errors OR incomplete statements

23. [4] **a.** The measure of the central $\angle$ for soccer is $(0.08)(360) = 28.8$. The measure of the central $\angle$ for cross-country is $(0.05)(360) = 18$. The difference is $28.8 - 18$ or 10.8.

b. 30; sample explanation: Since 8% of the students participating in fall sports play soccer and 5% run cross-country, you can set up a proportion: $\frac{\text{percentage playing soccer}}{\text{percentage running cross-country}} = \frac{8\%}{5\%}$. If x is the number of students who run cross-country, the proportion becomes $\frac{48}{x} = \frac{8}{5}$. Applying the Cross Products Property, $48(5) = 8x$. So, $x = \frac{48(5)}{8} = 30$.

[3] one error OR incorrect statement

[2] two errors OR incomplete statements

[1] more than two errors OR incomplete statements

Constructed Response

11. A triangle has a perimeter of 81 in. If you divide the length of each side by 3, what is the perimeter, in inches, of the new triangle?

12. A jewelry maker designed a pendant like the one shown below. It is a regular octagon set in a circle. Opposite vertices are connected by line segments. What is the measure of angle P in degrees?

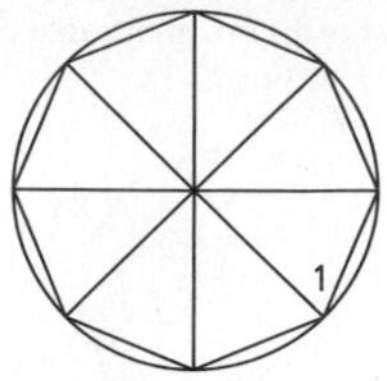

13. The diagonal of a rectangular patio makes a 70° angle with a side of the patio that is 60 ft long. What is the area, to the nearest square foot, of the patio?

14. What is the area, in square feet, of the unshaded part of the rectangle below?

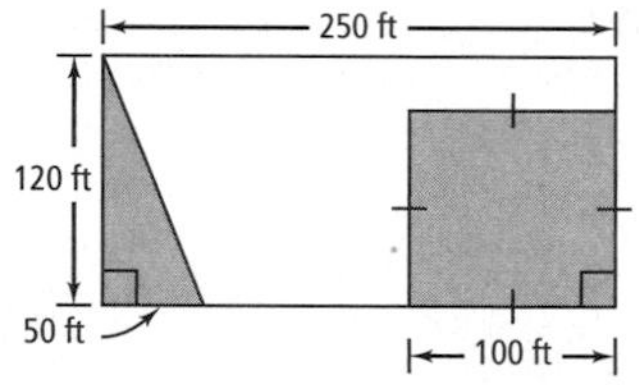

15. You are making a scale model of a building. The front of the actual building is 60 ft wide and 100 ft tall. The front of your model is 3 ft wide and 5 ft tall. What is the scale factor of the reduction?

16. The clock has a diameter of 18 in. At 8 o'clock, what is the area of the sector formed by the two hands of the clock? Round to the nearest tenth.

17. A square has an area of 225 cm^2. If you double the length of each side, what is the area, in square centimeters, of the new square?

18. Use a compass and straightedge to copy the diagram of the circle below. Label the center, radius, and diameter. Then label a central angle, an arc, a sector, and a segment.

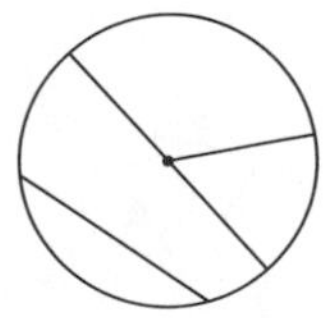

19. The coordinates of $\triangle ABC$ are $A(2, 3)$, $B(10, 9)$, and $C(10, -3)$. Is $\triangle ABC$ an equilateral triangle? Explain.

20. Draw a right triangle. Then construct a second triangle congruent to the first. Show your steps.

21. One diagonal of a parallelogram has endpoints at $P(-2, 5)$ and $R(1, -1)$. The other diagonal has endpoints at $Q(1, 5)$ and $S(-2, -1)$. What type of parallelogram is $PQRS$? Explain.

Extended Response

22. The coordinates of the vertices of isosceles trapezoid $ABCD$ are $A(0, 0)$, $B(6, 8)$, $C(10, 8)$, and $D(16, 0)$. The coordinates of the vertices of isosceles trapezoid $AFGH$ are $A(0, 0)$, $F(3, 4)$, $G(5, 4)$, and $H(8, 0)$. Are the two trapezoids similar? Justify your answer.

23. The circle graph below shows fall sports participation at one school.

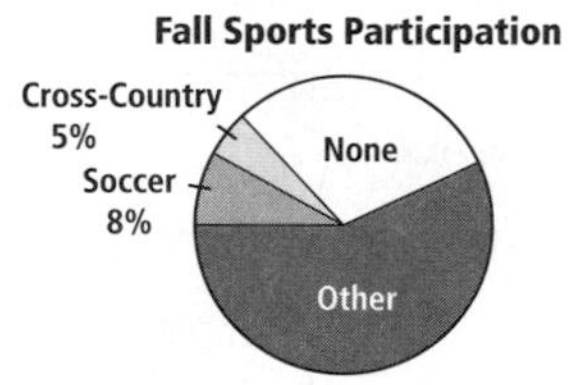

a. How many degrees greater is the measure of the angle that represents soccer than the measure of the angle that represents cross-country? Show your work.

b. If 48 students play soccer, how many run cross-country? Explain your reasoning.

Get Ready!

CHAPTER 11

Lesson 8-1 **The Pythagorean Theorem**

Algebra **Solve for *a, b,* or *c* in right $\triangle ABC$, where *a* and *b* are the lengths of the legs and *c* is the length of the hypotenuse.**

1. $a = 8, b = 15, c = ■$ **2.** $a = ■, b = 4, c = 12$ **3.** $a = 2\sqrt{3}, b = 2\sqrt{6}, c = ■$

Lesson 8-2 **Special Right Triangles**

4. Find the length of the shorter leg of a 30°-60°-90° triangle with hypotenuse $8\sqrt{5}$.

5. Find the length of the diagonal of a square whose perimeter is 24.

6. Find the height of an equilateral triangle with sides of length 8.

Lessons 10-1, 10-2, and 10-3 **Area**

Find the area of each figure. Leave your answer in simplest radical form.

7.
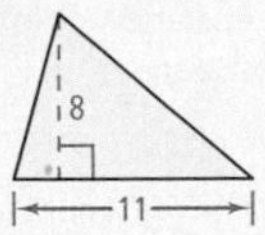

8.
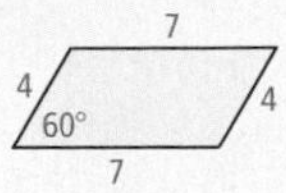

9.
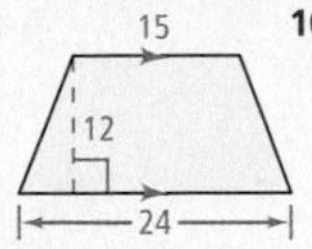

10.
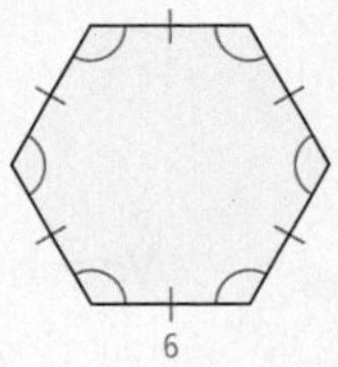

Lesson 10-4 **Perimeters and Areas of Similar Figures**

11. Two similar triangles have corresponding sides in a ratio of 3 : 5. Find the perimeter of the smaller triangle if the larger triangle has perimeter 40.

12. Two regular hexagons have areas of 8 and 25. Find the ratio of corresponding sides.

Looking Ahead Vocabulary

13. In Chapter 10, you learned about the *altitude* of a triangle. What do you think is the *altitude* of a three-dimensional figure?

14. You can turn a rock over in your hands to examine its *surface*. What do you think the *surface area* of a three-dimensional figure is?

15. What does an Egyptian *pyramid* look like?

Answers

Get Ready!

1. 17
2. $8\sqrt{2}$
3. 6
4. $4\sqrt{5}$
5. $6\sqrt{2}$
6. $4\sqrt{3}$
7. 44 units^2
8. $14\sqrt{3}$ units^2
9. 234 units^2
10. $54\sqrt{3}$ units^2
11. 24
12. $2\sqrt{2} : 5$
13. the ⊥ segment from one base to a parallel base or a vertex to the base
14. the sum of the areas of each side (face) of a figure
15. An Egyptian pyramid has 4 sides that are triangles and a bottom (base) that is a square.

Get Ready!

Assign this diagnostic assessment to determine if students have the prerequisite skills for Chapter 11.

Lesson	Skill
8-1	The Pythagorean Theorem
8-2	Special Right Triangles
10-1, 10-2, and 10-3	Area
10-4	Perimeters and Areas of Similar Figures

To remediate students, select from these resources (available for every lesson).

- Online Problems (PowerGeometry.com)
- Reteaching (All-in-One Teaching Resources)
- Practice (All-in-One Teaching Resources)

Why Students Need These Skills

THE PYTHAGOREAN THEOREM Students will use the Pythagorean Theorem to find missing heights and slant heights in pyramids.

SPECIAL RIGHT TRIANGLES Relationships in 30°-60°-90° and 45°-45°-90° triangles will provide shortcuts in problems involving missing measurements.

AREA Finding the surface area of space figures will require finding the area of plane figures.

PERIMETERS AND AREAS OF SIMILAR FIGURES Students will find surface areas of volumes of similar solids.

Looking Ahead Vocabulary

ALTITUDE Ask students what they already know about the term *altitude*. Have them give a real-world definition.

SURFACE AREA Show students a solid figure. Have students point to and identify the surfaces of the solid.

PYRAMID Show students photographs of pyramids used in architecture.

Chapter 11 Overview

In Chapter 11 students find surface areas and volumes of solid figures. Students will develop the answers to the Essential Questions as they learn the concepts and skills shown below.

BIG idea Visualization

ESSENTIAL QUESTION How can you determine the intersection of a solid and a plane?

- Students will examine cross sections.

BIG idea Measurement

ESSENTIAL QUESTION How do you find the surface area and volume of a solid?

- Students will use formulas to find surface areas and volumes of prisms and cylinders.
- Students will use formulas to find surface areas and volumes of pyramids and cones.
- Students will use formulas to find surface areas and volumes of spheres.

BIG idea Similarity

ESSENTIAL QUESTION How do the surface areas and volumes of similar solids compare?

- Students will examine ratios among similar solids.
- Given a figure and its surface area, students will be able to find the surface area of a solid similar to the original solid.
- Given a figure and its volume, students will be able to find the volume of a solid similar to the original solid.

Content Standards

Following are the standards covered in this chapter. Modeling standards are indicated by a star symbol (★).

CONCEPTUAL CATEGORY Geometry

Domain Geometric Measurement and Dimension G-GMD

Cluster Explain volume formulas and use them to solve problems (Standards G-GMD.A.1★, G-GMD.A.3★)
LESSONS 11-4, 11-5, 11-6

Cluster Visualize relationships between two-dimensional and three-dimensional objects (Standard G-GMD.B.4)
LESSON 11-1

Domain Modeling with Geometry G-MG

Cluster Apply geometric concepts in modeling situations (Standards G-MG.A.1★, G-MG.A.2★)
LESSONS 11-2, 11-3, 11-4, 11-5, 11-6, 11-7

CHAPTER 11

Surface Area and Volume

Download videos connecting math to your world. (VIDEO)

Interactive! Vary numbers, graphs, and figures to explore math concepts. (DYNAMIC ACTIVITIES)

The online Solve It will get you in gear for each lesson. (SOLVE IT!)

Math definitions in English and Spanish (VOCABULARY)

Online access to stepped-out problems aligned to Common Core (ONLINE PROBLEMS)

Get and view your assignments online. (ONLINE HOMEWORK)

Extra practice and review online (MathXL FOR SCHOOL)

Virtual Nerd™ tutorials with built-in support

Chapter Preview

- 11-1 Space Figures and Cross Sections
- 11-2 Surface Areas of Prisms and Cylinders
- 11-3 Surface Areas of Pyramids and Cones
- 11-4 Volumes of Prisms and Cylinders
- 11-5 Volumes of Pyramids and Cones
- 11-6 Surface Areas and Volumes of Spheres
- 11-7 Areas and Volumes of Similar Solids

BIG ideas

1 **Visualization**
Essential Question How can you determine the intersection of a solid and a plane?

2 **Measurement**
Essential Question How do you find the surface area and volume of a solid?

3 **Similarity**
Essential Question How do the surface areas and volumes of similar solids compare?

Vocabulary

English/Spanish Vocabulary Audio Online:

English	Spanish
cone, *p. 711*	cono
cross section, *p. 690*	sección de corte
cylinder, *p. 701*	cilindro
face, *p. 688*	cara
polyhedron, *p. 688*	poliedro
prism, *p. 699*	prisma
pyramid, *p. 708*	pirámide
similar solids, *p. 742*	cuerpos geométricos semejantes
sphere, *p. 733*	esfera
surface area, *pp. 700, 702, 709, 711*	área total
volume, *p. 717*	volumen

DOMAINS

- Geometric Measurement and Dimension
- Modeling with Geometry

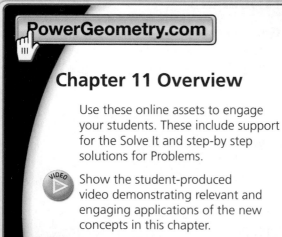

PowerGeometry.com

Chapter 11 Overview

Use these online assets to engage your students. These include support for the Solve It and step-by step solutions for Problems.

(VIDEO) Show the student-produced video demonstrating relevant and engaging applications of the new concepts in this chapter.

(VOCABULARY) Find online definitions for new terms in English and Spanish.

(SOLVE IT!) Start each lesson with an attention-getting Problem. View the Problem online with helpful hints.

Common Core Performance Task

Measuring the Immeasurable

Justin wants to measure the thickness of a paper towel, but his ruler is not the proper measuring device. Using a full roll, he takes the measurements shown in the diagram.

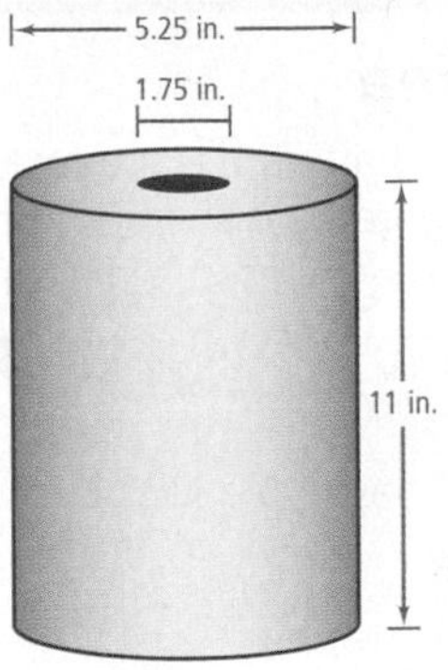

Not to scale

Justin reads the following additional information on the package.

70 SHEETS PER ROLL

EACH 9.0 IN. BY 11.0 IN.

Task Description

Find the thickness of a paper towel to the nearest thousandth of an inch.

Connecting the Task to the Math Practices

MATHEMATICAL PRACTICES

As you complete the task, you'll apply several Standards for Mathematical Practice.

- You'll describe the shape of the paper towel roll. (MP 1, MP 2)
- You'll use volume formulas to find the thickness of a paper towel. (MP 2, MP 4)

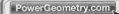

- Increase students' depth of knowledge with interactive online activities.
- Show Problems from each lesson solved step by step. Instant replay allows students to go at their own pace when studying online.
- Assign homework to individual students or to an entire class.
- Prepare students for the Mid-Chapter Quiz and Chapter Test with online practice and review.
- **Virtual Nerd™** Access Virtual Nerd student-centered math tutorials that directly relate to the content of the lesson.

Overview of the Performance Task

Students use formulas for the volume of a cylinder and the volume of a prism to find the thickness of a paper towel.

Students will work on the Performance Task in the following places in the chapter.

- Lesson 11-2 (p. 707)
- Lesson 11-4 (p. 724)
- Pull It All Together (p. 750)

Introducing the Performance Task

Tell students to read the problem on this page. Do not have them start work on the problem at this time, but ask them the following questions.

Q What is a strategy you could try in order to solve the problem? **[Sample: I can estimate the number of layers of paper towels there are in one roll, then divide the thickness of the roll by that number.]**

Q How could you find the volume of all the paper towels in the roll? **[Find the volume of the roll including the tube and subtract the volume of the tube.]**

PARCC CLAIMS

Sub-Claim A: Major Content with Connections to Practices

Sub-Claim D: Highlighted Practice MP 4 with Connections to Content

SBAC CLAIMS

Claim 2: Problem Solving

Claim 4: Modeling and Data Analysis

CHAPTER 11

SURFACE AREA AND VOLUME
Math Background

PROFESSIONAL DEVELOPMENT

The Understanding by Design® methodology was central to the development of the Big Ideas and the Essential Understandings. These will help your students build a structure on which to make connections to prior learning.

Visualization

BIG idea Visualization can help you connect properties of real objects with two-dimensional drawings of these objects.

ESSENTIAL UNDERSTANDINGS

11–1 A three-dimensional figure can be analyzed by describing the relationships among its vertices, edges, and faces.

11–2 to 11–3 The surface area of a three-dimensional figure is equal to the sum of the areas of each surface of the figure.

11–4 The volume of a prism and a cylinder can be found when its height and the area of its base are known.

11–5 The volume of a pyramid is related to the volume of a prism with the same base and height.

11–6 The surface area and the volume of a sphere can be found when its radius is known.

Measurement

BIG idea Some attributes of geometric figures, such as length, area, volume, and angle measure, are measurable. Units are used to describe these attributes.

ESSENTIAL UNDERSTANDINGS

11–2 to 11–3 The surface area of a three-dimensional figure is equal to the sum of the areas of each surface of the figure.

11–4 The volume of a prism and a cylinder can be found when its height and the area of its base are known.

11–5 The volume of a pyramid is related to the volume of a prism with the same base and height.

11–6 The surface area and the volume of a sphere can be found when its radius is known.

Similarity

BIG idea Two geometric figures are similar when corresponding lengths are proportional and corresponding angles are congruent. Areas of similar figures are proportional to the squares of their corresponding lengths. Volumes of similar figures are proportional to the cubes of their corresponding lengths.

ESSENTIAL UNDERSTANDING

11–7 Ratios can be used to compare the areas and volumes of similar solids.

Surface Area

The surface area of a figure is the total area of all the surfaces in the figure. This should include the lateral surfaces as well as any bases in the figure.

Prism

$SA = ph + 2B$

p = perimeter of base

B = area of base

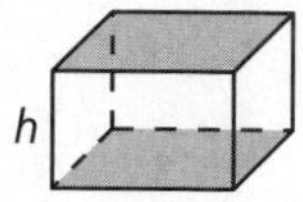

Cylinder

$SA = 2\pi rh + 2B$

B = area of base (πr^2)

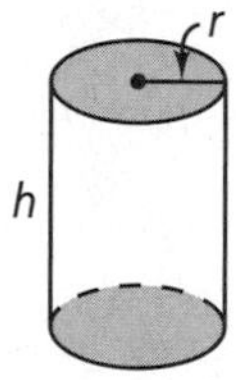

Pyramid

$SA = \frac{1}{2}p\ell + B$

p = perimeter of base

B = area of base

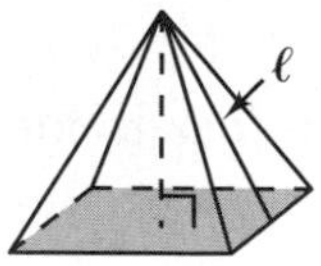

Cone

$SA = \pi r\ell + B$

B = area of base (πr^2)

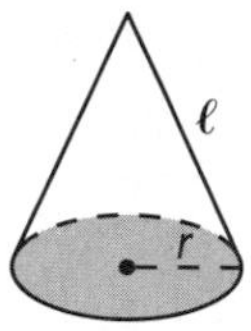

Sphere

$SA = 4\pi r^2$

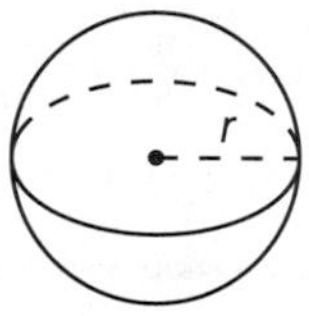

Common Errors With Surface Area

The surface areas of pyramids and cones are computed using slant height (ℓ), not height (h). If the problem gives the height and not the slant height, they need to use the Pythagorean Theorem to find the slant height.

Mathematical Practices

Look for and make use of structure Students develop formulas for surface area by building on their understanding of area formulas from the previous chapter. The surface area of a cone is found by extending understanding of a pyramid's surface area to an analogous circular object.

Volume

The volume of a figure is the amount of space it contains.

Prism

$V = Bh$

B = area of base

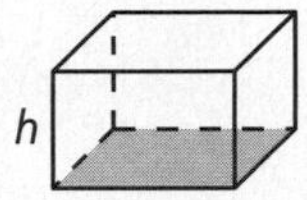

Cylinder

$V = Bh$

B = area of base (πr^2)

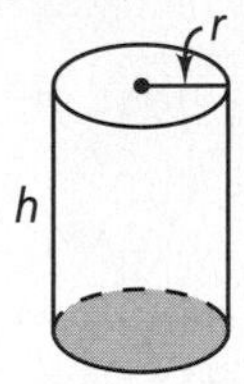

Pyramid

$V = \frac{1}{3}Bh$

B = area of base

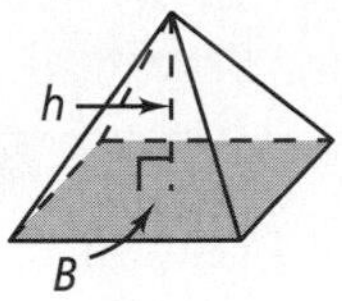

Cone

$V = \frac{1}{3}Bh$

B = area of base (πr^2)

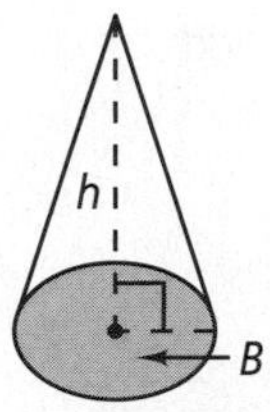

Sphere

$V = \frac{4}{3}\pi r^3$

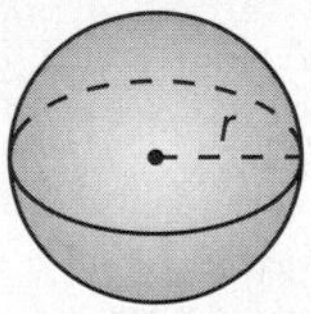

Common Errors With Volume

The surface area and volume formulas of spheres look similar.

$SA = 4\pi r^2$

$V = \frac{4}{3}\pi r^3$

Students who have worked to memorize the formulas of surface area and volume of spheres might accidentally transpose the 4 and $\frac{4}{3}$ or the r^2 and r^3.

Mathematical Practices

Construct viable arguments and critique the reasoning of others Students examine a derivation involving a large number of pyramids inside a sphere to find the sphere's volume. This builds on their knowledge of spherical surface area, volume of a pyramid and sophisticated algebraic skills.

Similar Solids

Lesson 7-2 introduced two properties of similar figures:

1. Corresponding angles are congruent.
2. Corresponding sides are proportional.

Lesson 10-4 introduced two more properties of similar figures:

3. The ratio of their perimeters is the scale factor.
4. The ratio of their areas is the square of the scale factor.

In Lesson 11-7, the topic of similarity is extended to three-dimensional solids. In particular, the following properties are introduced:

5. The ratio of the surface areas of similar solids is the square of the scale factor.
6. The ratio of the volumes of similar solids is the cube of the scale factor.

Consider the following similar solids:

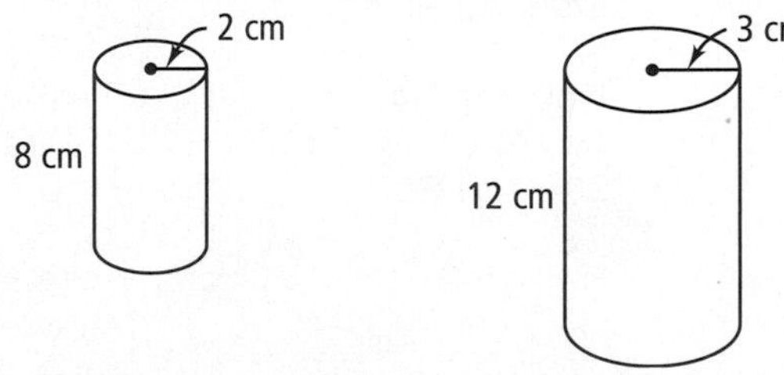

Ratio of corresponding sides $= \frac{8}{12} = \frac{2}{3}$

Ratio of surface areas

$= \left(\frac{2\pi \times 2 \times 8 + 2\pi \times 2^2}{2\pi \times 3 \times 12 + 2\pi \times 3^2}\right) = \frac{4}{9} = \left(\frac{2}{3}\right)^2$

Ratio of volumes $= \left(\frac{\pi \times 2^2 \times 8}{\pi \times 3^2 \times 12}\right) = \frac{8}{27} = \left(\frac{2}{3}\right)^3$

Common Errors With Similar Solids

Students have trouble remembering whether the scale factor should be squared or cubed. Connect this with the dimensions of the measurements. Perimeter is one-dimensional, so the perimeters are related by the scale factor. Area is two-dimensional, so the areas are related by the square of the scale factor. Volume is three-dimensional, so the volumes are related by the cube of the scale factor.

Mathematical Practices

Model with Mathematics Students work problems to see the ratios of similar solids in action and apply that understanding to varied situations in the lesson and exercises.

CHAPTER 11

SURFACE AREA AND VOLUME
Pacing and Assignment Guide

		TRADITIONAL			BLOCK
Lesson	**Teaching Day(s)**	**Basic**	**Average**	**Advanced**	**Block**
11-1	1	Problems 1–3 Exs. 6–17 all, 51–62	Problems 1–3 Exs. 7–17 odd, 51–62	Problems 1–3 Exs. 7–17 odd, 51–62	**Day 1** Problems 1–5 Exs. 7–23 odd, 24–40, 51–62
	2	Problems 4–5 Exs. 18–23 all, 24–34 even, 38	Problems 4–5 Exs. 19–23 odd, 24–40	Problems 4–5 Exs. 19–23 odd, 24–50	
11-2	1	Problems 1–2 Exs. 7–13 all, 44–47	Problems 1–2 Exs. 7–13 odd, 44–47	Problems 1–4 Exs. 7–19 odd, 21–47	**Day 2** Problems 1–4 Exs. 7–19 odd, 21–38, 44–47
	2	Problems 3–4 Exs. 14–20 all, 22–30 even, 37	Problems 3–4 Exs. 15–19 odd, 21–38		
11-3	1	Problems 1–4 Exs. 9–15 all, 44–53	Problems 1–4 Exs. 9–15 odd, 44–53	Problems 1–4 Exs. 9–21 odd, 22–53	**Day 3** Problems 1–4 Exs. 9–21 odd, 22–38, 44–53
	2	Problems 1–4 Exs. 16–21 all, 22, 25, 26–36 even	Problems 1–4 Exs. 17–21 odd, 22–38		
11-4	1	Problems 1–2 Exs. 6–13 all	Problems 1–2 Exs. 7–13 odd	Problems 1–4 Exs. 7–19 odd, 21–45	**Day 4** Problems 1–4 Exs. 7–19 odd, 21–42
	2	Problems 3–4 Exs. 14–21 all, 24, 30–32 all, 38	Problems 3–4 Exs. 15–19 odd, 21–42		
11-5	1	Problems 1–2 Exs. 5–14 all, 39–46	Problems 1–2 Exs. 5–13 odd, 39–46	Problems 1–4 Exs. 5–19 odd, 20–46	**Day 5** Problems 1–4 Exs. 5–19 odd, 20–34, 39–46
	2	Problems 3–4 Exs. 15–21 all, 24–32 even	Problems 3–4 Exs. 15–19 odd, 20–34		
11-6	1	Problems 1–2 Exs. 6–16 all, 60–71	Problems 1–2 Exs. 7–15 odd, 60–71	Problems 1–4 Exs. 7–25 odd, 26–71	**Day 6** Problems 1–4 Exs. 7–25 odd, 26–54, 60–71
	2	Problems 3–4 Exs. 17–26 all, 29–31, 34–42 even, 50	Problems 3–4 Exs. 17–25 odd, 26–54		
11-7	1	Problems 1–2 Exs. 5–14 all, 42–54	Problems 1–2 Exs. 5–13 odd, 42–54	Problems 1–4 Exs. 5–23 odd, 24–54	**Day 7** Problems 1–4 Exs. 5–23 odd, 24–38, 42–54
	2	Problems 3–4 Exs. 15–26 all, 28–29, 34–38 even	Problems 3–4 Exs. 15–23 odd, 24–38		
Review	1	Chapter 11 Review	Chapter 11 Review	Chapter 11 Review	**Day 8** Chapter 11 Review Chapter 11 Test
Assess	1	Chapter 11 Test	Chapter 11 Test	Chapter 11 Test	
Total		**16 Days**	**16 Days**	**10 Days**	**8 Days**

Note: Pacing does not include Concept Bytes and other feature pages.

Resources

KEY
I = Interactive asset at PowerGeometry.com
E = Editable master at PowerGeometry.com
P = Available in Print
M = Master at PowerGeometry.com
✓ = CD-ROM

	For the Chapter	11-1	11-2	11-3	11-4	11-5	11-6	11-7
Planning								
Teacher Center Online Planner & Grade Book	I	I	I	I	I	I	I	I
Interactive Learning & Guided Instruction								
My Math Video	I							
Solve It!		I M	I M	I M	I M	I M	I M	I M
Student Companion		P M	P M	P M	P M	P M	P M	
Vocabulary Support		I P M	I P M	I P M	I P M	I P M	I P M	I P M
Got It? Support		I P	I P	I P	I P	I P	I P	I P
Dynamic Activity	I							
Online Problems		I	I	I	I	I	I	I
Additional Problems		M	M	M	M	M	M	M
English Language Learner Support (TR)		E P M	E P M	E P M	E P M	E P M	E P M	E P M
Activities, Games, and Puzzles		E M	E M	E M	E M	E M	E M	E M
Teaching With TI Technology With CD-ROM								
TI-Nspire™ Support CD-ROM		✓	✓	✓	✓	✓	✓	✓
Lesson Check & Practice								
Student Companion		P M	P M	P M	P M	P M	P M	P M
Lesson Check Support		I P	I P	I P	I P	I P	I P	I P
Practice and Problem Solving Workbook		P	P	P	P	P	P	P
Think About a Plan (TR)		E P M	E P M	E P M	E P M	E P M	E P M	E P M
Practice Form G (TR)		E P M	E P M	E P M	E P M	E P M	E P M	E P M
Standardized Test Prep (TR)		P M	P M	P M	P M	P M	P M	P M
Practice *Form K* (TR)		E P M	E P M	E P M	E P M	E P M	E P M	E P M
Extra Practice	E M							
Find the Errors!	M							
Enrichment (TR)		E P M	E P M	E P M	E P M	E P M	E P M	E P M
Answers and Solutions CD-ROM	✓	✓	✓	✓	✓	✓	✓	✓
Assess & Remediate								
ExamView CD-ROM	✓	✓	✓	✓	✓	✓	✓	✓
Lesson Quiz		I M	I M	I M	I M	I M	I M	I M
Quizzes and Tests *Form G* (TR)	E P M			E P M				E P M
Quizzes and Tests *Form K* (TR)	E P M			E P M				E P M
Reteaching (TR)		E P M	E P M	E P M	E P M	E P M	E P M	E P M
Performance Tasks (TR)	P M							
Cumulative Review (TR)	P M							
Progress Monitoring Assessments	I P M							

(TR) Available in All-In-One Teaching Resources

1 Interactive Learning

Solve It!

PURPOSE To review drawing nets from Chapter 1 and develop an understanding of polyhedrons

PROCESS Students may sketch the different nets of the box and compare their drawings with students around them.

FACILITATE

Q How many corners does the tissue box have? How many flat surfaces? How many folded creases? **[The tissue box has 8 corners, 6 flat surfaces, and 12 folded creases.]**

ANSWER See Solve It in Answers on next page.

CONNECT THE MATH In the Solve It, students draw a net to represent the faces and edges of a polyhedron. In the lesson, students learn Euler's formula to relate the number of faces, edges, and vertices in a polyhedron.

2 Guided Instruction

Problem 1

Have students list each vertex, edge, and face, and shade each element using a different color.

Got It? VISUAL LEARNERS

It may help students to make a three-dimensional model of the figure.

11-1 Space Figures and Cross Sections

Common Core State Standards

G-GMD.B.4 Identify the shapes of two-dimensional cross-sections of three-dimensional objects, and identify three-dimensional objects generated by rotations of two-dimensional objects.

MP 1, MP 2, MP 3, MP 4, MP 5, MP 7

Objectives To recognize polyhedrons and their parts
To visualize cross sections of space figures

If you can shift your perspective by reflecting or rotating to get another net, then those nets are the same.

SOLVE IT! Getting Ready!

The tissue box at the right is a rectangular solid. Let x = the number of corners, y = the number of flat surfaces, and z = the number of folded creases. What is an equation that relates the quantities x, y, and z for a rectangular solid? Will your equation hold true for a cube? A solid with a triangular top and bottom? Explain.

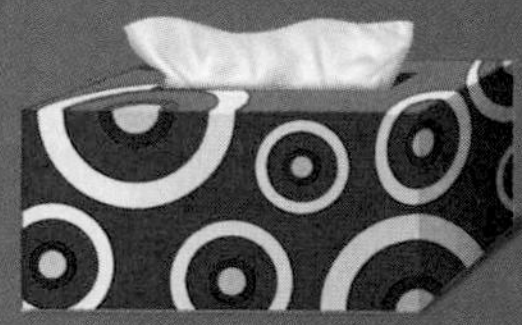

MATHEMATICAL PRACTICES In the Solve It, you used two-dimensional nets to represent a three-dimensional object.

Lesson Vocabulary
- polyhedron
- face
- edge
- vertex
- cross section

A **polyhedron** is a space figure, or three-dimensional figure, whose surfaces are polygons. Each polygon is a **face** of the polyhedron. An **edge** is a segment that is formed by the intersection of two faces. A **vertex** is a point where three or more edges intersect.

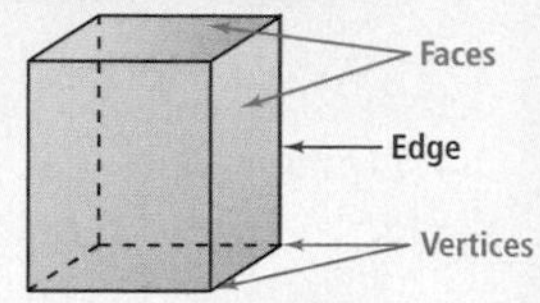

Essential Understanding You can analyze a three-dimensional figure by using the relationships among its vertices, edges, and faces.

Plan

Can you see the solid?
A dashed line indicates an edge that is hidden from view. This figure has one four-sided face and four triangular faces.

Problem 1 Identifying Vertices, Edges, and Faces

How many vertices, edges, and faces are in the polyhedron at the right? List them.

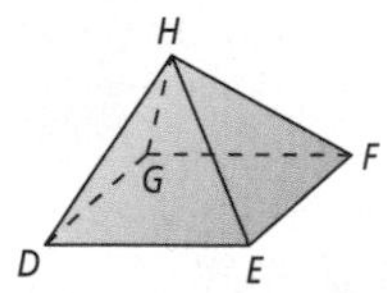

There are five vertices: D, E, F, G, and H.

There are eight edges: $\overline{DE}$, $\overline{EF}$, $\overline{FG}$, $\overline{GD}$, $\overline{DH}$, $\overline{EH}$, $\overline{FH}$, and $\overline{GH}$.

There are five faces: $\triangle DEH$, $\triangle EFH$, $\triangle FGH$, $\triangle GDH$, and quadrilateral $DEFG$.

Got It? 1. a. How many vertices, edges, and faces are in the polyhedron at the right? List them.

b. Reasoning Is $\overline{TV}$ an edge? Explain why or why not.

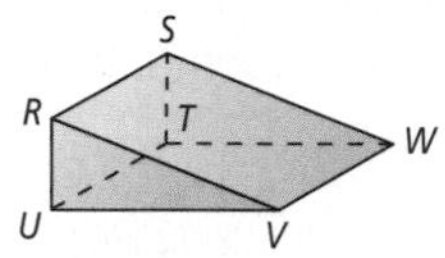

11-1 Preparing to Teach

BIG idea Visualization

ESSENTIAL UNDERSTANDINGS
- A three-dimensional figure can be analyzed by describing the relationships between its vertices, edges, and faces.
- A cross section is the intersection of a three-dimensional figure and a plane.

Math Background

Many students have difficulty interpreting diagrams of three-dimensional figures and drawing two-dimensional representations of them. This lesson focuses on the development of these visualization skills.

In order to study three-dimensional figures, it is necessary to define common aspects and characteristics of them. In this lesson, students will define faces, vertices, and edges of a polyhedron. They will learn about the numeric relationships among these parts of polyhedrons. Students will study Euler's Formula to see that the sum of the number of faces and the number of vertices is equal to the number of edges plus two. Students will also learn to visualize the cross sections of various solids. They will need to use visualization skills to determine how a plane will intersect a given solid. These skills will help students as they continue their study of three-dimensional figures.

Mathematical Practices

Reason abstractly and quantitatively. With Euler's Formula, students will decontextualize three-dimensional figures, reducing them to numeric values of their faces, edges, and vertices, as well as contextualize, pausing to consider the numeric values in terms of their original referents.

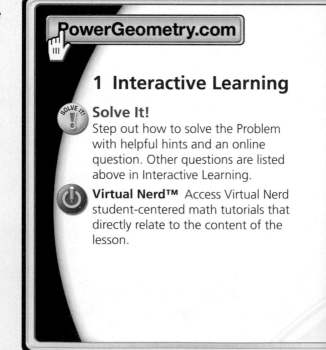

PowerGeometry.com

1 Interactive Learning

Solve It!
Step out how to solve the Problem with helpful hints and an online question. Other questions are listed above in Interactive Learning.

Virtual Nerd™ Access Virtual Nerd student-centered math tutorials that directly relate to the content of the lesson.

Leonhard Euler, a Swiss mathematician, discovered a relationship among the numbers of faces, vertices, and edges of any polyhedron. The result is known as Euler's Formula.

take note

Key Concept Euler's Formula

The sum of the number of faces (F) and vertices (V) of a polyhedron is two more than the number of its edges (E).

$F + V = E + 2$

Problem 2 Using Euler's Formula

Plan

How do you verify Euler's Formula?
Find the number of faces, vertices, and edges. Then substitute the values into Euler's Formula to make sure that the equation is true.

How many vertices, edges, and faces does the polyhedron at the right have? Use your results to verify Euler's Formula.

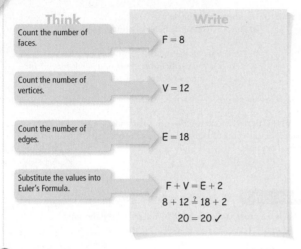

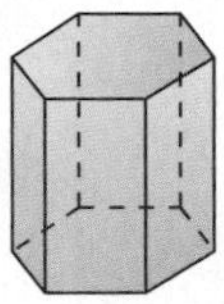

Think	Write
Count the number of faces.	$F = 8$
Count the number of vertices.	$V = 12$
Count the number of edges.	$E = 18$
Substitute the values into Euler's Formula.	$F + V = E + 2$ $8 + 12 \stackrel{?}{=} 18 + 2$ $20 = 20$ ✓

Got It? 2. For each polyhedron, use Euler's Formula to find the missing number.

a.

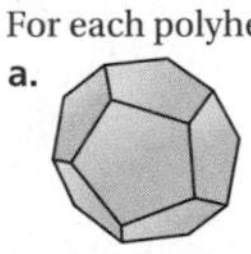

faces: ■
edges: 30
vertices: 20

b.

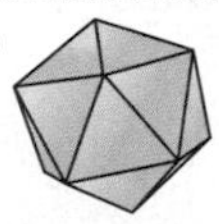

faces: 20
edges: ■
vertices: 12

Take Note

Have students verify Euler's Formula. Show them several examples of polyhedrons and have them count the number of faces, vertices, and edges and verify that they satisfy the formula.

Problem 2

Q How many faces does the polyhedron have? How many edges? **[8, 18]**

Q How many of those faces are rectangles that are on the sides of the polyhedron? **[6]**

Q How many edges are along the hexagonal faces of the polyhedron and its rectangular faces? Explain. **[12; 6 for the top face and 6 for the bottom face]**

Q Are there any other edges? Explain. **[Yes; there are edges where each of the rectangular faces meets another rectangular face.]**

Q How many vertices are around one of the hexagonal faces? **[6]**

Got It?

Q For 2a, what is Euler's formula with the known values substituted? **[$F + 20 = 30 + 2$]**

Q How do you solve the equation for F? **[Simplify the right side of the equation and then subtract 20 from each side.]**

Q For 2b, what is Euler's formula with the known values substituted? **[$20 + 12 = E + 2$]**

Q How do you solve the equation for E? **[Simplify the left side of the equation and then subtract 2 from each side.]**

2 Guided Instruction

Each Problem is worked out and supported online.

Problem 1
Identifying Vertices, Edges, and Faces
Animated

Problem 2
Using Euler's Formula
Animated

Problem 3
Verifying Euler's Formula in Two Dimensions
Animated

Problem 4
Describing a Cross Section

Problem 5
Drawing a Cross Section

Support in Geometry Companion
- Vocabulary
- Key Concepts
- Got It?

Answers

Solve It!

Equations may vary. Sample: One possible equation is $x + y = z + 2$, since $8 + 6 = 12 + 2$. Since a cube has the same number of corners, surfaces, and creases, the equation will hold true. A solid with a triangular bottom has 6 corners, 5 surfaces, and 9 creases, so the equation holds true.

Got It?

1–2. See back of book.

Problem 3

Q How are faces represented in the net of a solid? **[They are represented by the two-dimensional figures in the net.]**

Q How are edges represented in the net of a solid? **[They are represented by the segments that are the sides of the two-dimensional figures.]**

Q How many two-dimensional figures are in the net? **[8]**

Q How many vertices are in the net? **[22]**

Q How many segments are in the net? **[29]**

Got It?

Have students compare the nets they draw with the nets the students around them draw. There are several ways to draw the net of the solid.

Problem 4

Q What types of cross sections could be formed by a plane and the cylinder? **[a rectangle, a circle, an ellipse, or a truncated ellipse]**

Q Using the diagram shown, what is the position of the plane in relationship to the cylinder for each cross section named above? **[A rectangle corresponds to a vertical plane, as shown in the diagram; a circle corresponds to a horizontal cross section; an ellipse corresponds to a non-vertical plane that intersects the curved side of the cylinder. The ellipse will be truncated if it intersects one or both of the bases of the cylinder.]**

Got It? TACTILE LEARNERS

It may benefit students to hold and manipulate a bowl similar to the one in the diagram. Allow students to use a piece of paper to model a plane and draw the intersection of the plane on the bowl with a dry-erase marker.

In two dimensions, Euler's Formula reduces to $F + V = E + 1$, where F is the number of regions formed by V vertices linked by E segments.

Plan
What do you use for the variables?

In 3-D	In 2-D
F: Faces	→ Regions
V: Vertices	→ Vertices
E: Edges	→ Segments

Problem 3 Verifying Euler's Formula in Two Dimensions

How can you verify Euler's Formula for a net for the solid in Problem 2?

Draw a net for the solid.

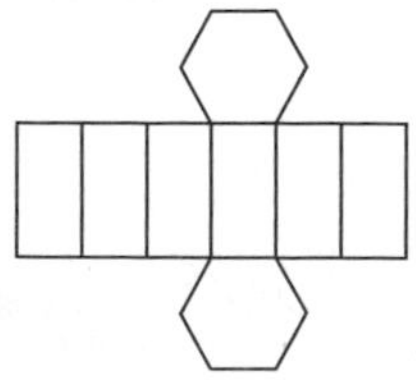

Number of regions: $F = 8$

Number of vertices: $V = 22$

Number of segments: $E = 29$

$F + V = E + 1$ Euler's Formula for two dimensions

$8 + 22 = 29 + 1$ Substitute.

$30 = 30$ ✔

Got It? 3. Use the solid at the right.

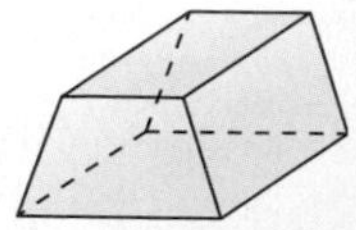

a. How can you verify Euler's Formula $F + V = E + 2$ for the solid?
b. Draw a net for the solid.
c. How can you verify Euler's Formula $F + V = E + 1$ for your two-dimensional net?

A **cross section** is the intersection of a solid and a plane. You can think of a cross section as a very thin slice of the solid.

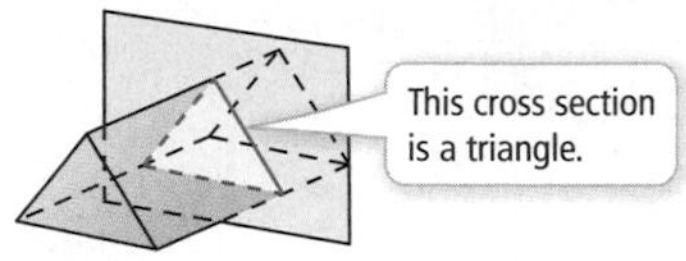

Think
How can you see the cross section?
Mentally rotate the solid so that the plane is parallel to your face.

Problem 4 Describing a Cross Section

What is the cross section formed by the plane and the solid at the right?

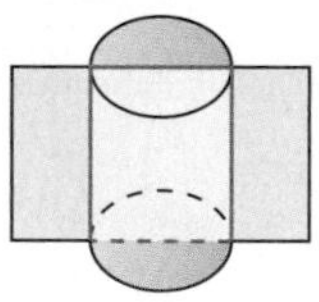

The cross section is a rectangle.

Got It? 4. For the solid at the right, what is the cross section formed by each of the following planes?

a. a horizontal plane
b. a vertical plane that divides the solid in half

Additional Problems

1. How many vertices, edges, and faces are in the polyhedron below? List them.

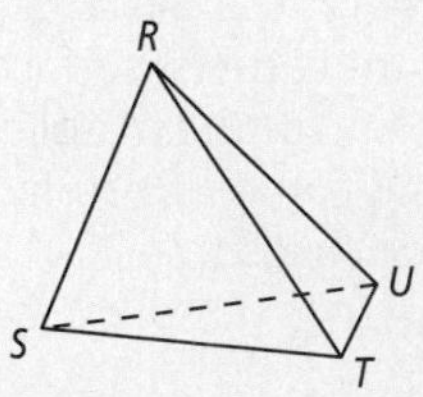

ANSWER There are four vertices: R, S, T, and U. There are six edges: $\overline{RS}$, $\overline{RT}$, $\overline{RU}$, $\overline{ST}$, $\overline{TU}$, and $\overline{US}$. There are four faces: $\triangle RST$, $\triangle RTU$, $\triangle RUS$, and $\triangle STU$.

2. How many edges does the polyhedron below have? Use Euler's Formula.

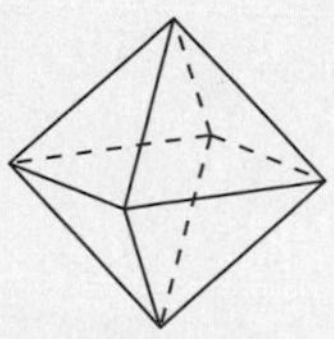

ANSWER 12

3. How can you verify Euler's Formula for a net of a cube?

ANSWER Draw a net for the cube. There are 6 regions ($F = 6$), 14 vertices ($V = 14$), and 19 segments ($E = 19$). So, $F + V = E + 1$, or $6 + 14 = 19 + 1$.

4. What is the cross section formed by the plane and the solid below?

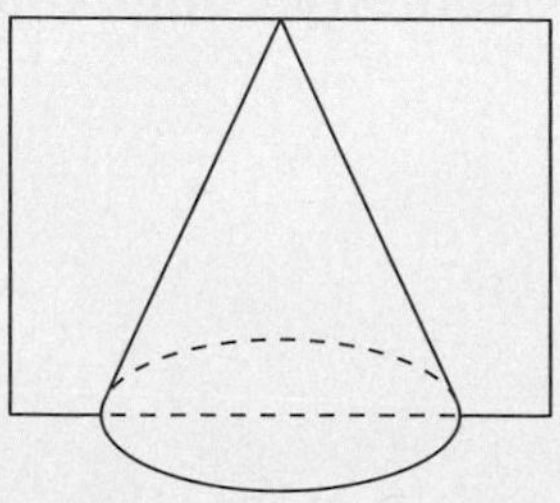

ANSWER The cross section is a triangle.

5. Draw a cross section formed by a vertical plane intersecting the right and left faces of the cube. What shape is the cross section?

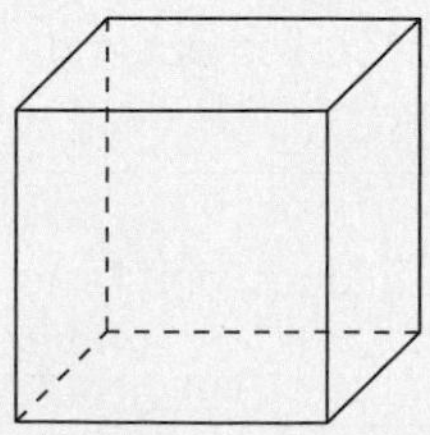

ANSWER The cross section is a square.

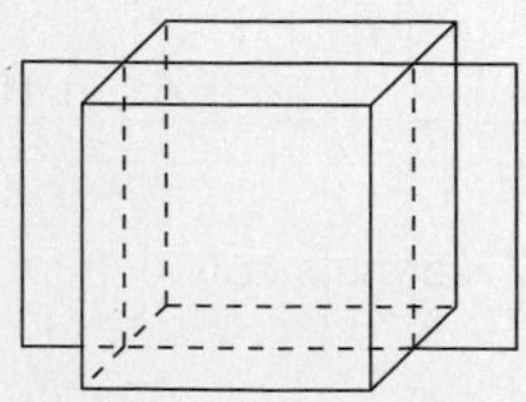

To draw a cross section, you can sometimes use the idea from Postulate 1-3 that the intersection of two planes is exactly one line.

Problem 5 Drawing a Cross Section

Visualization Draw a cross section formed by a vertical plane intersecting the front and right faces of the cube. What shape is the cross section?

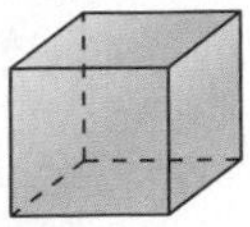

Think

How can you see parallel segments? Focus on the plane intersecting the front and right faces. The plane and both faces are vertical, so the intersections are vertical parallel lines.

Step 1 Visualize a vertical plane intersecting the vertical faces in parallel segments.

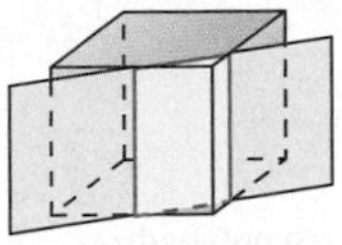

Step 2 Draw the parallel segments.

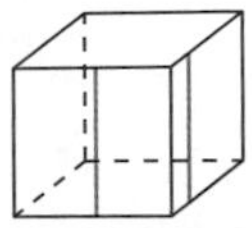

Step 3 Join their endpoints. Shade the cross section.

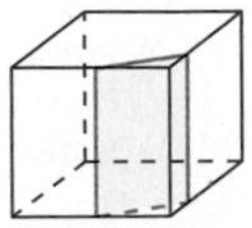

The cross section is a rectangle.

Got It? 5. Draw the cross section formed by a horizontal plane intersecting the left and right faces of the cube. What shape is the cross section?

Lesson Check

Do you know HOW?

1. How many faces, edges, and vertices are in the solid? List them.

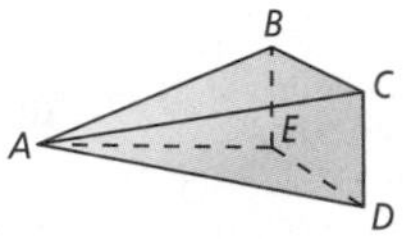

2. What is a net for the solid in Exercise 1? Verify Euler's Formula for the net.

3. What is the cross section formed by the cube and the plane containing the diagonals of a pair of opposite faces?

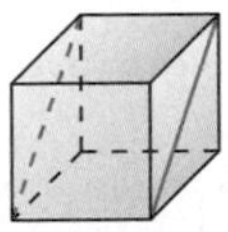

Do you UNDERSTAND? MATHEMATICAL PRACTICES

4. **Vocabulary** Suppose you build a polyhedron from two octagons and eight squares. Without using Euler's Formula, how many edges does the solid have? Explain.

5. **Error Analysis** Your math class is drawing polyhedrons. Which figure does not belong in the diagram below? Explain.

Problem 5

Q How does the location of where the plane intersects the faces affect the cross section? **[It will change the width of the rectangle.]**

Q For a narrower rectangle, should the plane be closer to the edge where the front and right faces meet or farther away? **[closer]**

Got It? TACTILE LEARNERS

Provide students with a cube and show them how to model a plane with their hands. This will help students visualize the cross section.

3 Lesson Check

Do you know HOW?

- If students have difficulty with Exercises 1-2, then have them review Problem 1 to see how to identify vertices, edges, and faces.

Do you UNDERSTAND?

- If students have difficulty with Exercise 5, then have them review the definition of polyhedron on the first page of the lesson.

Close

Q What are the three parts of a polyhedron that are related in Euler's Formula? **[faces, vertices, and edges]**

Q How are the number of faces, F, vertices, V, and edges, E, of a polyhedron related to each other? **[$F + V = E + 2$]**

Q What is a cross section? **[A cross section is the two-dimensional figure formed when a plane intersects a solid figure.]**

Answers

Got It? (continued)

3a. $6 + 8 = 12 + 2$

b.

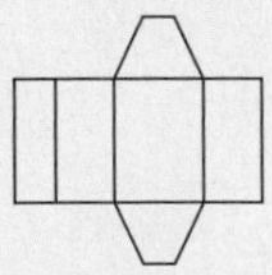

c. $6 + 14 = 19 + 1$

4a. a circle

b. an isosc. trapezoid

5.

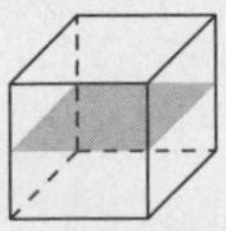

a square

Lesson Check

1. 5 faces: $\triangle ABC$, $\triangle ACD$, $\triangle ADE$, $\triangle AEB$, quadrilateral $BCDE$

 8 edges: $\overline{AB}$, $\overline{AC}$, $\overline{AD}$, $\overline{AE}$, $\overline{BC}$, $\overline{CD}$, $\overline{DE}$, $\overline{EB}$

 5 vertices: A, B, C, D, E

2. Sample:

 $F + V = 5 + 8$; $E + 1 = 12 + 1$; $5 + 8 = 12 + 1$

3. a rectangle
4. 24 edges: There are 8 edges on each of the two octagonal bases, and there are 8 edges that connect pairs of vertices of the bases.
5. A cylinder is not a polyhedron because its faces are not polygons.

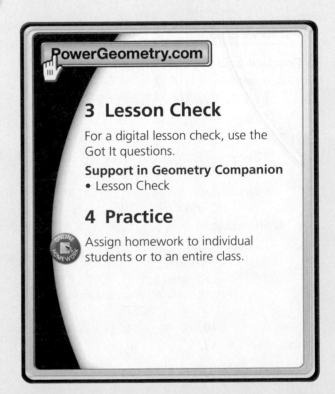

PowerGeometry.com

3 Lesson Check

For a digital lesson check, use the Got It questions.

Support in Geometry Companion

- Lesson Check

4 Practice

Assign homework to individual students or to an entire class.

4 Practice

ASSIGNMENT GUIDE
Basic: 6–23 all, 24–34 even, 38
Average: 7–23 odd, 24–40
Advanced: 7–23 odd, 24–50
Standardized Test Prep: 51–55
Mixed Review: 56–62

Mathematical Practices are supported by exercises with red headings. Here are the Practices supported in this lesson:

MP 1: Make Sense of Problems Ex. 34
MP 3: Construct Arguments Ex. 5, 27
MP 4: Model with Mathematics Ex. 40
MP 5: Use Appropriate Tools Ex. 24
MP 7: Shift Perspective Ex. 21–23, 26, 28–30, 31–33, 41, 42–50

Applications exercises have blue headings.

EXERCISE 28: Use the Think About a Plan worksheet in the **Practice and Problem Solving Workbook** (also available in the Teaching Resources in print and online) to further support students' development in becoming independent learners.

HOMEWORK QUICK CHECK
To check students' understanding of key skills and concepts, go over Exercises 11, 19, 26, 28, and 34.

Practice and Problem-Solving Exercises

MATHEMATICAL PRACTICES

A Practice

For each polyhedron, how many vertices, edges, and faces are there? List them. See Problem 1.

6.
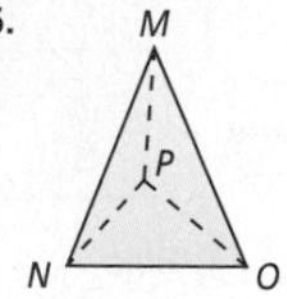

7.
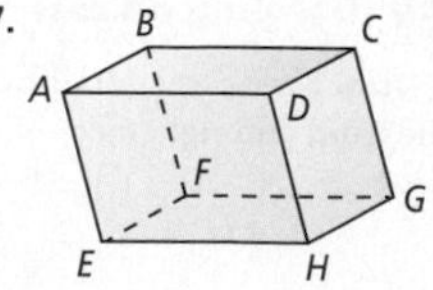

8.
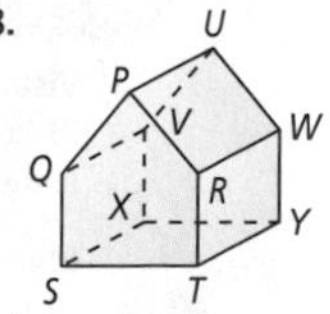

For each polyhedron, use Euler's Formula to find the missing number. See Problem 2.

9. faces: ■
edges: 15
vertices: 9

10. faces: 8
edges: ■
vertices: 6

11. faces: 20
edges: 30
vertices: ■

Use Euler's Formula to find the number of vertices in each polyhedron.

12. 6 square faces

13. 5 faces: 1 rectangle and 4 triangles

14. 9 faces: 1 octagon and 8 triangles

Verify Euler's Formula for each polyhedron. Then draw a net for the figure and verify Euler's Formula for the two-dimensional figure. See Problem 3.

15.
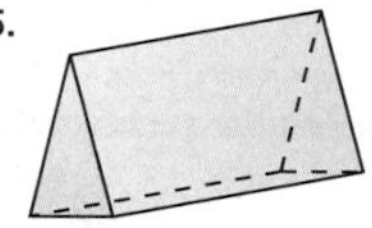

16.
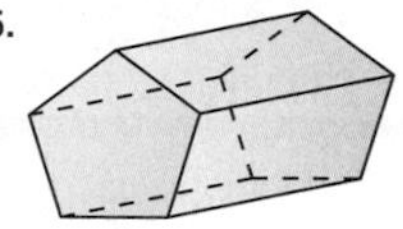

17.
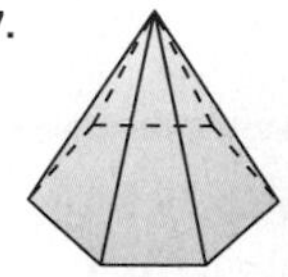

Describe each cross section. See Problem 4.

18.
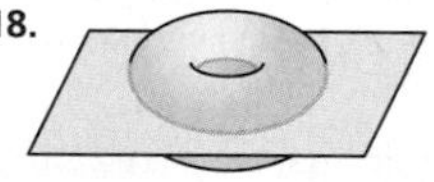

19.
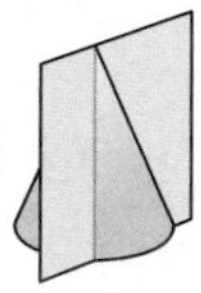

20.
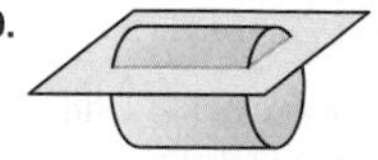

Visualization Draw and describe a cross section formed by a vertical plane intersecting the cube as follows. See Problem 5.

21. The vertical plane intersects the front and left faces of the cube.

22. The vertical plane intersects opposite faces of the cube.

23. The vertical plane contains the red edges of the cube.

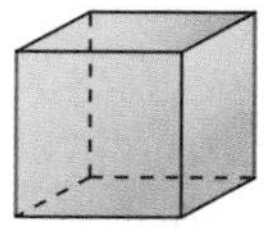

Answers

Practice and Problem-Solving Exercises

6. 4 vertices: *M*, *N*, *P*, *O*
6 edges: $\overline{MN}$, $\overline{MP}$, $\overline{MO}$, $\overline{NP}$, $\overline{PO}$, $\overline{ON}$
4 faces: $\triangle MNP$, $\triangle MPO$, $\triangle NPO$, $\triangle MNO$

7. 8 vertices: *A*, *B*, *C*, *D*, *E*, *F*, *G*, *H*
12 edges: $\overline{AB}$, $\overline{BC}$, $\overline{CD}$, $\overline{DA}$, $\overline{EF}$, $\overline{FG}$, $\overline{GH}$, $\overline{HE}$, $\overline{AE}$, $\overline{BF}$, $\overline{CG}$, $\overline{DH}$
6 faces: quadrilaterals *ABCD*, *EFGH*, *ABFE*, *BCGF*, *DCGH*, *ADHE*

8. 10 vertices: *P*, *Q*, *R*, *S*, *T*, *U*, *V*, *W*, *X*, *Y*
15 edges: $\overline{PQ}$, $\overline{QS}$, $\overline{ST}$, $\overline{TR}$, $\overline{RP}$, $\overline{UV}$, $\overline{VX}$, $\overline{XY}$, $\overline{YW}$, $\overline{WU}$, $\overline{PU}$, $\overline{QV}$, $\overline{SX}$, $\overline{TY}$, $\overline{RW}$
7 faces: quadrilaterals *STYX*, *RTYW*, *QSXV*, *PQVU*, *RWUP*, and pentagons *UVXYW* and *PQSTR*

9. 8 **10.** 12 **11.** 12
12. 8 **13.** 5 **14.** 9

15. $5 + 6 = 9 + 2$; answers may vary. Sample:

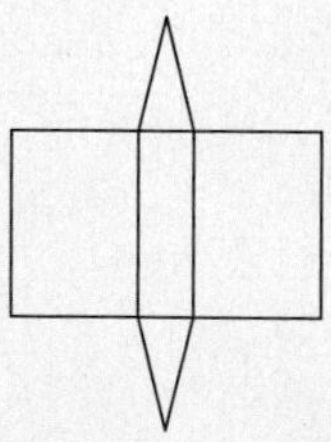

$5 + 10 = 14 + 1$

16. $7 + 10 = 15 + 2$; answers may vary. Sample:

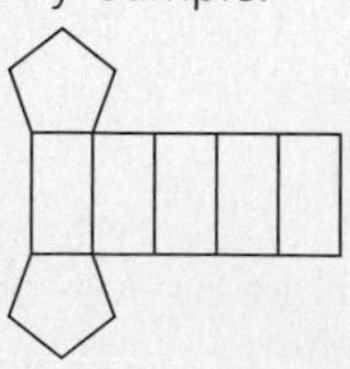

$7 + 18 = 24 + 1$

17. $7 + 7 = 12 + 2$; answers may vary. Sample:

$7 + 12 = 18 + 1$

18. two concentric circles
19. triangle
20. rectangle

21.
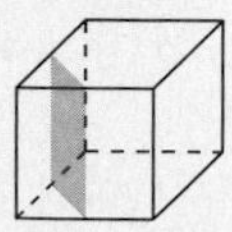
rectangle

22.
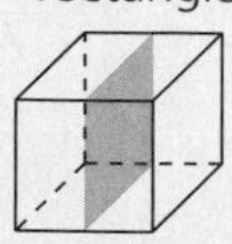
rectangle

23.
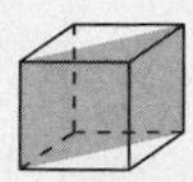
square

B Apply

24. a. Open-Ended Sketch a polyhedron whose faces are all rectangles. Label the lengths of its edges.
b. Use graph paper to draw two different nets for the polyhedron.

25. For the figure shown at the right, sketch each of following.
a. a horizontal cross section
b. a vertical cross section that contains the vertical line of symmetry

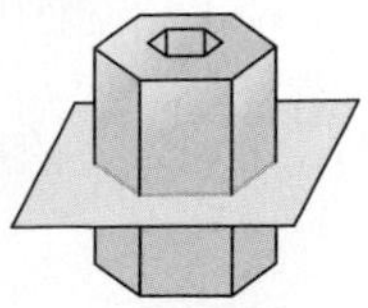

26. Reasoning Can you find a cross section of a cube that forms a triangle? Explain.

27. Reasoning Suppose the number of faces in a certain polyhedron is equal to the number of vertices. Can the polyhedron have nine edges? Explain.

Visualization Draw and describe a cross section formed by a plane intersecting the cube as follows.

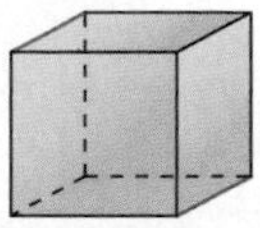

28. The plane is tilted and intersects the left and right faces of the cube.

29. The plane contains the red edges of the cube.

30. The plane cuts off a corner of the cube.

Visualization A plane region that revolves completely about a line sweeps out a solid of revolution. Use the sample to help you describe the *solid of revolution* you get by revolving each region about line ℓ.

Sample: Revolve the rectangular region about the line ℓ. You get a cylinder as the solid of revolution.

31.

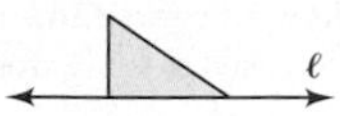

32.

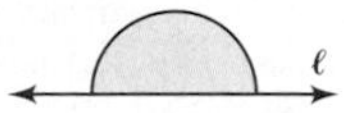

33.

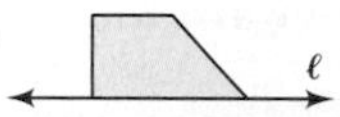

34. Think About a Plan Some balls are made from panels that suggest polygons. A soccer ball suggests a polyhedron with 20 regular hexagons and 12 regular pentagons. How many vertices does this polyhedron have?
- How can you determine the number of edges in a solid if you know the types of polygons that form the faces?
- What relationship can you use to find the number of vertices?

Euler's Formula $F + V = E + 1$ applies to any two-dimensional network where F is the number of regions formed by V vertices linked by E edges (or paths). Verify Euler's Formula for each network shown.

35.

36.

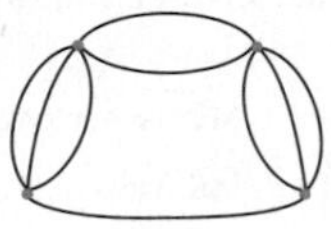

37.

31. cone
32. sphere
33. a cylinder attached to a cone
34. 60
35. $4 + 6 = 9 + 1$
36. $6 + 4 = 9 + 1$
37. $5 + 5 = 9 + 1$

24a. Sample:

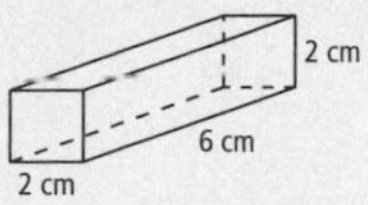

b. Sample:

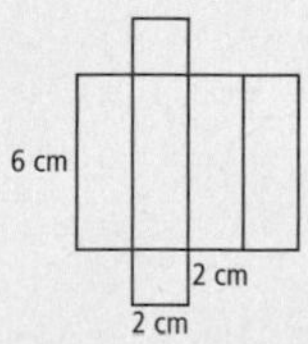

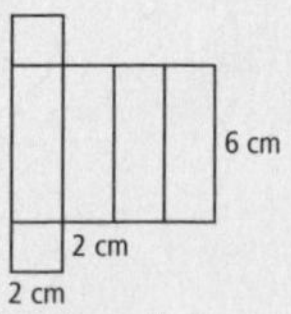

25a.

b.

26. Yes. Sample: The plane intersects a corner of the cube.

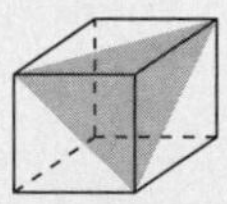

27. No; if $F = V$, then $F + V = 2F$, so $F + V$ is even. So $E \neq 9$ because $E + 2$ must be even.

28.

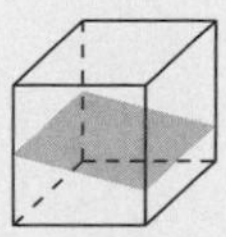

rectangle

29.

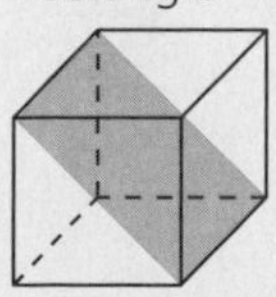

rectangle

30.

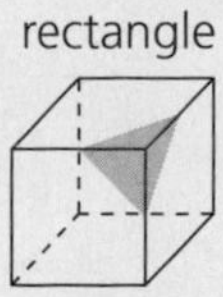

triangle

Answers

Practice and Problem-Solving Exercises (continued)

38a. A. icosahedron
B. octahedron
C. tetrahedron
D. hexahedron
E. dodecahedron

b. regular triangular pyramid, cube

c. $4 + 4 = 6 + 2$; $6 + 8 = 12 + 2$; $8 + 6 = 12 + 2$

39. 6 in.

40. Answers may vary. Check students' work.

41.

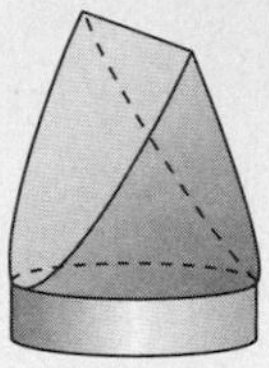

42.

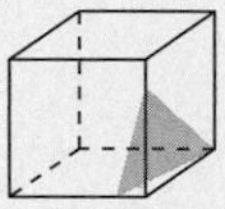

43.

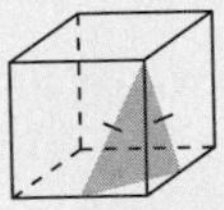

44.

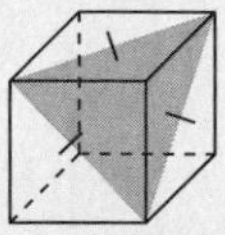

45.

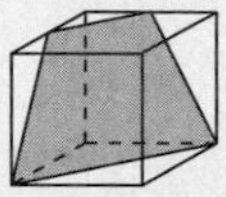

46.

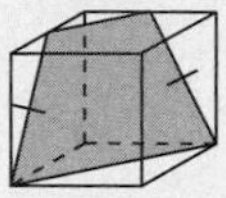

47.

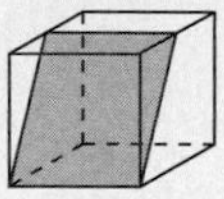

48.

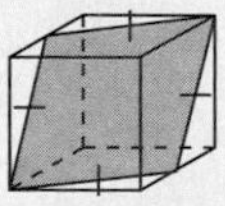

49.

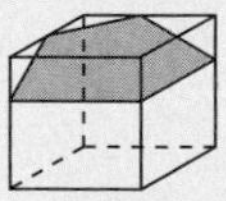

50.

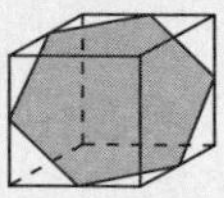

38. Platonic Solids There are five regular polyhedrons. They are called *regular* because all their faces are congruent regular polygons, and the same number of faces meet at each vertex. They are also called *Platonic solids* after the Greek philosopher Plato, who first described them in his work *Timaeus* (about 350 B.C.).

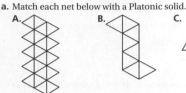

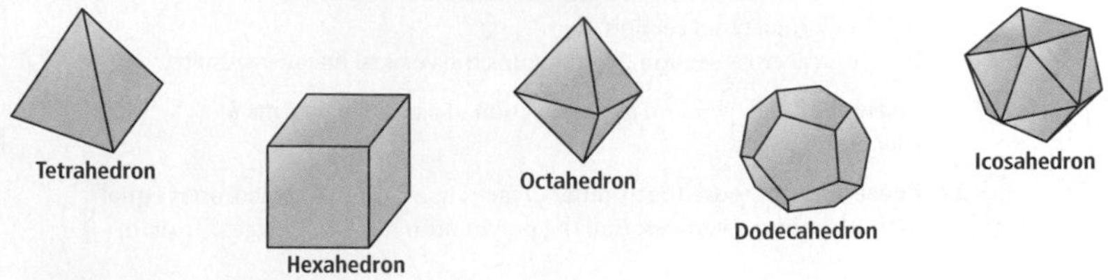

a. Match each net below with a Platonic solid.

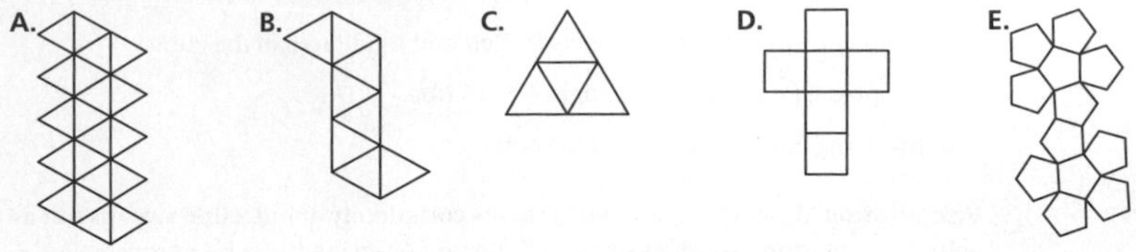

b. The first two Platonic solids have more familiar names. What are they?

c. Verify that Euler's Formula is true for the first three Platonic solids.

39. A cube has a net with area 216 in.^{2}. How long is an edge of the cube?

40. Writing Cross sections are used in medical training and research. Research and write a paragraph on how magnetic resonance imaging (MRI) is used to study cross sections of the brain.

Challenge

41. Open-Ended Draw a single solid that has the following three cross sections.

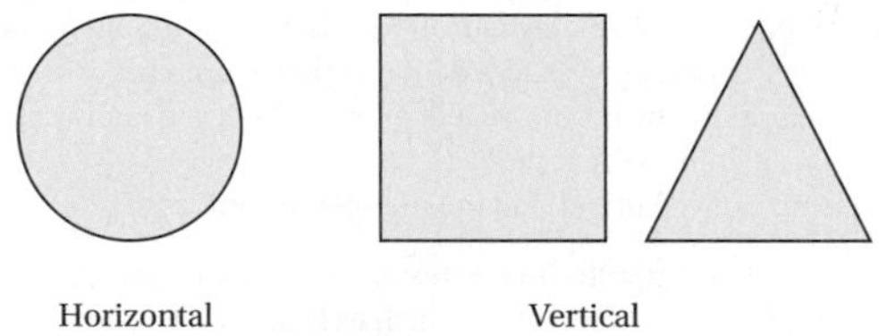

Visualization Draw a plane intersecting a cube to get the cross section indicated.

42. scalene triangle
43. isosceles triangle
44. equilateral triangle
45. trapezoid
46. isosceles trapezoid
47. parallelogram
48. rhombus
49. pentagon
50. hexagon

Standardized Test Prep

SAT/ACT

51. A polyhedron has four vertices and six edges. How many faces does it have?

(A) 2 (B) 4 (C) 5 (D) 10

52. Suppose the circumcenter of $\triangle ABC$ lies on one of its sides. What type of triangle must $\triangle ABC$ be?

(F) scalene (G) isosceles (H) equilateral (I) right

53. What is the area of a regular hexagon whose perimeter is 36 in.?

(A) $18\sqrt{3}$ in.^{2} (B) $27\sqrt{3}$ in.^{2} (C) $36\sqrt{3}$ in.^{2} (D) $54\sqrt{3}$ in.^{2}

54. What is the best description of the polygon at the right?

(F) concave decagon (H) regular pentagon

(G) convex decagon (I) regular decagon

Short Response

55. The coordinates of three vertices of a parallelogram are $A(2, 1)$, $B(1, -2)$ and $C(4, -1)$. What are the coordinates of the fourth vertex D? Explain.

Mixed Review

56. Probability Shuttle buses to an airport terminal leave every 20 min from a remote parking lot. Draw a geometric model and find the probability that a traveler who arrives at a random time will have to wait at least 8 min for the bus to leave the parking lot. See Lesson 10-8.

57. Games A dartboard is a circle with a 12-in. radius. What is the probability that you throw a dart that lands within 6 in. of the center of the dartboard?

Find the value of x to the nearest tenth. See Lesson 8-3.

58.

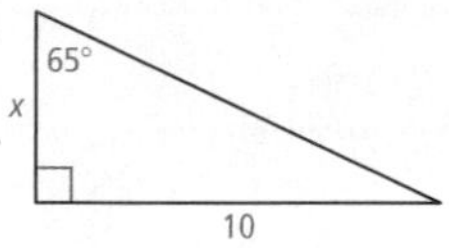

59.

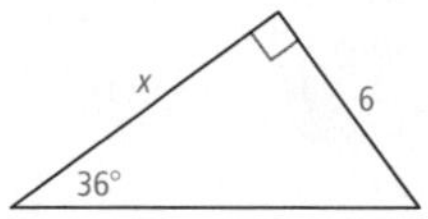

Get Ready! **To prepare for Lesson 11-2, do Exercises 60–62.**

Find the area of each net. See Lessons 1-8 and 10-3.

60.

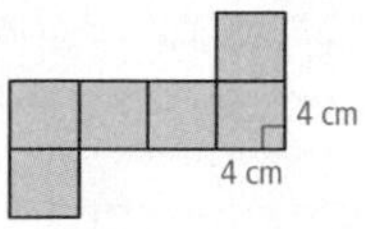

61.

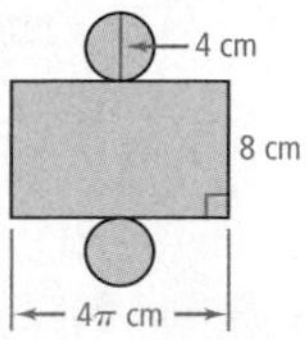

62.

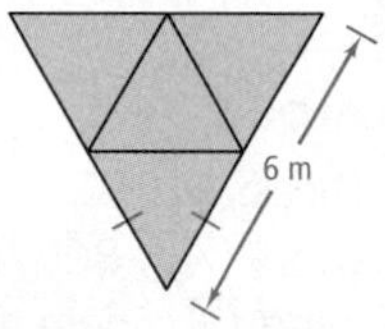

51. B

52. I

53. D

54. F

55. [2] (5, 2), (−1, 0), or (3, −4); the fourth vertex lies on a line parallel to an opposite side such that the length of the side is equal to the length of the opposite side.

[1] incomplete answer or explanation

56.

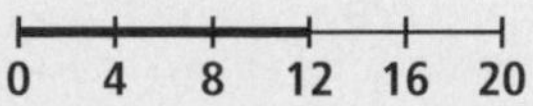

60%

57. 25%

58. 4.7

59. 8.3

60. 96 cm^2

61. 40π cm^2

62. $9\sqrt{3}$ m^2

11-1 Lesson Resources

Differentiated Remediation

Instructional Support

Geometry Companion

Students can use the **Geometry Companion** worktext (4 pages) . . .

- New Vocabulary
- Key Concepts
- Got It for each Problem
- Lesson Check

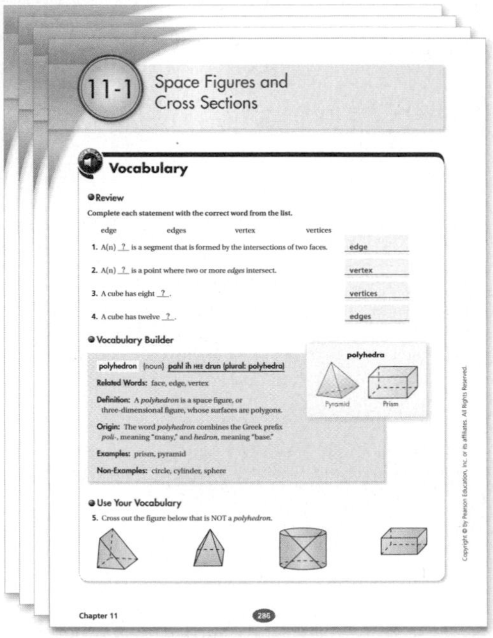
11-1 Space Figures and Cross Sections

Vocabulary

Review

Complete each statement with the correct word from the list.

edge edges vertex vertices

1. A(n) _?_ is a segment that is formed by the intersections of two faces. edge
2. A(n) _?_ is a point where two or more edges intersect. vertex
3. A cube has eight _?_. vertices
4. A cube has twelve _?_. edges

Vocabulary Builder

polyhedron (noun) pahl ih DREE drun (plural: polyhedra)

Related Words: face, edge, vertex

Definition: A *polyhedron* is a space figure, or three-dimensional figure, whose surfaces are polygons.

Origin: The word *polyhedron* combines the Greek prefix *poli-*, meaning "many," and *hedron*, meaning "base."

Examples: prism, pyramid

Non-Examples: circle, cylinder, sphere

Use Your Vocabulary

5. Cross out the figure below that is NOT a *polyhedron*.

Chapter 11 286

ELL Support

Using Manipulatives Using a shoebox or tissue box, move your hand over each face and say: *These are the faces of the prism.* Trace the edges with your finger as you say: *These are the edges of the prism where two faces intersect.* How many faces and edges are there?

5 Assess & Remediate

Lesson Quiz

1. How many vertices, edges, and faces are in the polyhedron below? List them.

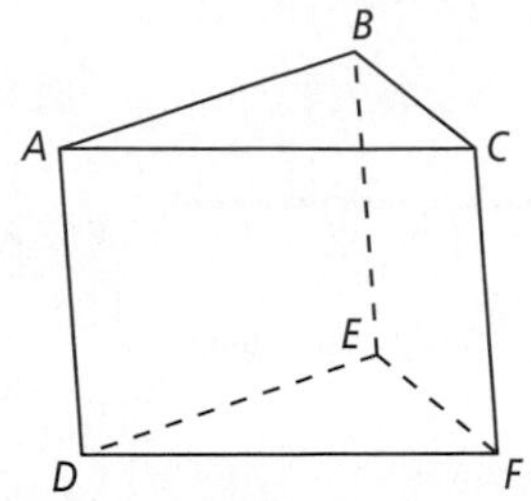

2. **Do You UNDERSTAND?** Use your answers to Exercise 1 to verify Euler's Theorem for a polyhedron.
3. What is the cross section formed by the plane and the solid figure below?

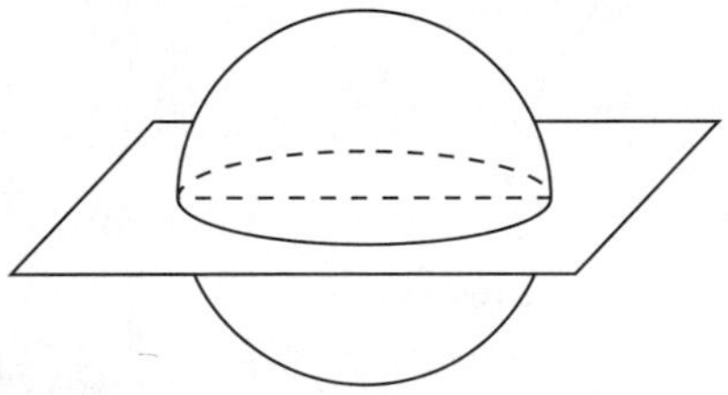

ANSWERS TO LESSON QUIZ

1. There are six vertices: *A, B, C, D, E,* and *F*. There are nine edges: $\overline{AB}$, $\overline{BC}$, $\overline{AC}$, $\overline{AD}$, $\overline{BE}$, $\overline{CF}$, $\overline{DE}$, $\overline{EF}$, and $\overline{DF}$. There are five faces: $\triangle ABC$, $\triangle DEF$, *ABED, BCFE,* and *ACFD*.
2. $F = 5$, $V = 6$, $E = 9$; $F + V = E + 2$; $5 + 6 = 9 + 2$
3. a circle

PRESCRIPTION FOR REMEDIATION

Use the student work on the Lesson Quiz to prescribe a differentiated review assignment.

Points	Differentiated Remediation
0–1	Intervention
2	On-level
3	Extension

PowerGeometry.com

5 Assess & Remediate

Assign the Lesson Quiz. Appropriate intervention, practice, or enrichment is automatically generated based on student performance.

Intervention

- **Reteaching** (2 pages) Provides reteaching and practice exercises for the key lesson concepts. Use with struggling students or absent students.
- **English Language Learner Support** Helps students develop and reinforce mathematical vocabulary and key concepts.

All-in-One Resources/Online

Reteaching

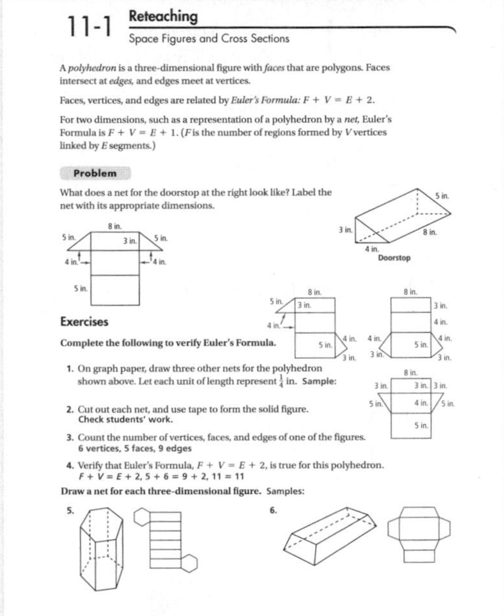
11-1 Reteaching

Space Figures and Cross Sections

A *polyhedron* is a three-dimensional figure with *faces* that are polygons. Faces intersect at *edges*, and edges meet at vertices.

Faces, vertices, and edges are related by *Euler's Formula*: $F + V = E + 2$.

For two dimensions, such as a representation of a polyhedron by a *net*, Euler's Formula is $F + V = E + 1$. (*F* is the number of regions formed by *V* vertices linked by *E* segments.)

Problem

What does a net for the doorstop at the right look like? Label the net with its appropriate dimensions.

Exercises

Complete the following to verify Euler's Formula.

1. On graph paper, draw three other nets for the polyhedron shown above. Let each unit of length represent $\frac{1}{2}$ in. Sample:
2. Cut out each net, and use tape to form the solid figure. Check students' work.
3. Count the number of vertices, faces, and edges of one of the figures. 6 vertices, 5 faces, 9 edges
4. Verify that Euler's Formula, $F + V = E + 2$, is true for this polyhedron. $F + V = E + 2$, $5 + 6 = 9 + 2$, $11 = 11$

Draw a net for each three-dimensional figure. Samples:

5.

6.

All-in-One Resources/Online

English Language Learner Support

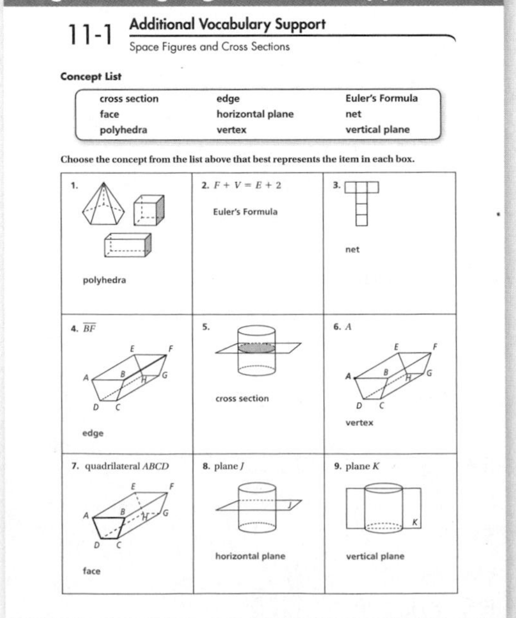
11-1 Additional Vocabulary Support

Space Figures and Cross Sections

Concept List

cross section	edge	Euler's Formula
face	horizontal plane	net
polyhedra	vertex	vertical plane

Choose the concept from the list above that best represents the item in each box.

1. polyhedra	2. $F + V = E + 2$ Euler's Formula	3. net
4. $\overline{BF}$ edge	5. cross section	6. *A* vertex
7. quadrilateral *ABCD* face	8. plane *J* horizontal plane	9. plane *K* vertical plane

Differentiated Remediation *continued*

On-Level

- **Practice** (2 pages) Provides extra practice for each lesson. For simpler practice exercises, use the Form K Practice pages found in the All-in-One Teaching Resources and online.
- **Think About a Plan** Helps students develop specific problem-solving skills and strategies by providing scaffolded guiding questions.
- **Standardized Test Prep** Focuses on all major exercises, all major question types, and helps students prepare for the high-stakes assessments.

Extension

- **Enrichment** Provides students with interesting problems and activities that extend the concepts of the lesson.
- **Activities, Games, and Puzzles** Worksheets that can be used for concepts development, enrichment, and for fun!

Practice and Problem Solving Wkbk/ All-in-One Resources/Online

Practice page 1

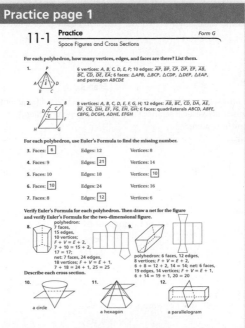

11-1 **Practice** *Form G*

Space Figures and Cross Sections

For each polyhedron, how many vertices, edges, and faces are there? List them.

1. 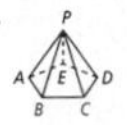

6 vertices: *A, B, C, D, E, P*; 10 edges: $\overline{AP}$, $\overline{BP}$, $\overline{CP}$, $\overline{DP}$, $\overline{EP}$, $\overline{AB}$, $\overline{BC}$, $\overline{CD}$, $\overline{DE}$, $\overline{EA}$; 6 faces: $\triangle APB$, $\triangle BCP$, $\triangle CDP$, $\triangle DEP$, $\triangle EAP$, and pentagon *ABCDE*

2. 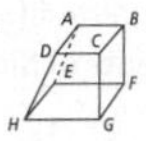

8 vertices: *A, B, C, D, E, F, G, H*; 12 edges: $\overline{AB}$, $\overline{BC}$, $\overline{CD}$, $\overline{DA}$, $\overline{AE}$, $\overline{BF}$, $\overline{CG}$, $\overline{DH}$, $\overline{EF}$, $\overline{FG}$, $\overline{EH}$, $\overline{GH}$; 6 faces: quadrilaterals *ABCD, ABFE, CBFG, DCGH, ADHE, EFGH*

For each polyhedron, use Euler's Formula to find the missing number.

3. Faces: 6 — Edges: 12 — Vertices: 8
4. Faces: 9 — Edges: 21 — Vertices: 14
5. Faces: 10 — Edges: 18 — Vertices: 10
6. Faces: 10 — Edges: 24 — Vertices: 16
7. Faces: 8 — Edges: 12 — Vertices: 6

Verify Euler's Formula for each polyhedron. Then draw a net for the figure and verify Euler's Formula for the two-dimensional figure.

8. 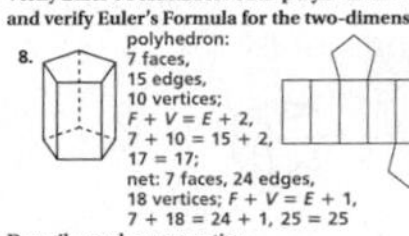
polyhedron: 7 faces, 15 edges, 10 vertices; $F + V = E + 2$, $7 + 10 = 15 + 2$, $17 = 17$; net: 7 faces, 24 edges, 18 vertices; $F + V = E + 1$, $7 + 18 = 24 + 1$, $25 = 25$

9. 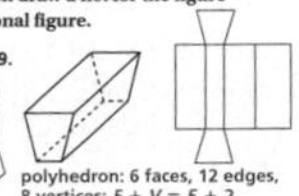
polyhedron: 6 faces, 12 edges, 8 vertices; $F + V = E + 2$, $6 + 8 = 12 + 2$, $14 = 14$; net: 6 faces, 19 edges, 14 vertices; $F + V = E + 1$, $6 + 14 = 19 + 1$, $20 = 20$

Describe each cross section.

10. 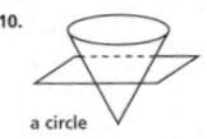
a circle

11. 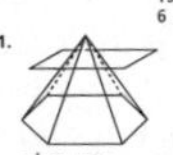
a hexagon

12. 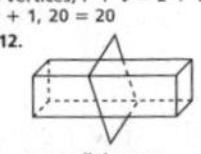
a parallelogram

Practice and Problem Solving Wkbk/ All-in-One Resources/Online

Practice page 2

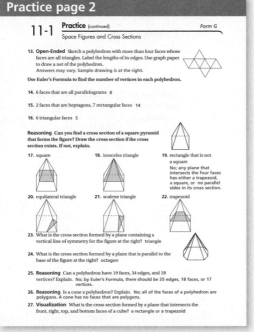

11-1 **Practice** (continued) *Form G*

Space Figures and Cross Sections

13. **Open-Ended** Sketch a polyhedron with more than four faces whose faces are all triangles. Label the lengths of its edges. Use graph paper to draw a net of the polyhedron.
Answers may vary. Sample drawing is at the right.

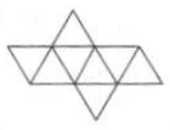

Use Euler's Formula to find the number of vertices in each polyhedron.

14. 6 faces that are all parallelograms 8
15. 2 faces that are heptagons, 7 rectangular faces 14
16. 6 triangular faces 5

Reasoning Can you find a cross section of a square pyramid that forms the figure? Draw the cross section if the cross section exists. If not, explain.

17. square

18. isosceles triangle
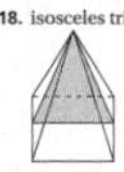

19. rectangle that is not a square
No; any plane that intersects the four faces has either a trapezoid, a square, or no parallel sides in its cross section.

20. equilateral triangle
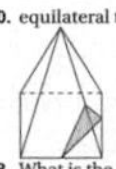

21. scalene triangle
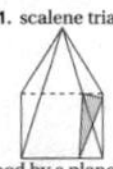

22. trapezoid
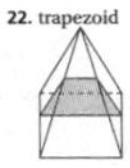

23. What is the cross section formed by a plane containing a vertical line of symmetry for the figure at the right? triangle
24. What is the cross section formed by a plane that is parallel to the base of the figure at the right? octagon
25. **Reasoning** Can a polyhedron have 19 faces, 34 edges, and 18 vertices? Explain. No; by Euler's Formula, there should be 35 edges, 18 faces, or 17 vertices.
26. **Reasoning** Is a cone a polyhedron? Explain. No; all of the faces of a polyhedron are polygons. A cone has no faces that are polygons.
27. **Visualization** What is the cross section formed by a plane that intersects the front, right, top, and bottom faces of a cube? a rectangle or a trapezoid

All-in-One Resources/Online

Enrichment

11-1 **Enrichment**

Space Figures and Cross Sections

Archimedean Solids

An Archimedean solid is a polyhedron that has a similar arrangement of polygons about each vertex. The faces of an Archimedean solid are made up of two or more types of regular polygons.

In total, there are 13 Archimedean solids. One example is the icosidodecahedron, shown in the drawings at the right.

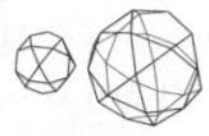

1. What are the regular polygons that make up the faces of an icosidodecahedron? Describe the polygons that meet at each vertex.
Pentagons and triangles; at each vertex two regular pentagons and two equilateral triangles meet.

Some of the Archimedean solids can be made by cutting off the corners of a regular polyhedron, or Platonic solid. Follow the steps below to make a truncated cube.

a. Draw a cube (or make a cube out of modeling clay).

b. Visualize a tilted plane that intersects the top, front, and right faces of the cube such that the cross section is an equilateral triangle. This plane should be closer to the corner of the cube than the midpoint of the edges. Draw this cross section (or cut off the corner of the cube of modeling clay).

c. Repeat step (b) with the other corners of the cube (or the clay). All cross sections should be congruent.

2. What are the regular polygons that make up the faces of a truncated cube? Describe the polygons that meet at each vertex.
Octagons and triangles; at each vertex two octagons and a triangle meet.
3. Suppose that the cross sections have been formed by planes that cut off the corners of the cube and intersect the midpoints of the edges. Draw or model the solid that would be formed. What are the regular polygons that make up the faces of this solid?
squares and triangles
4. Other Archimedean solids that can be made by cutting off the corners of a Platonic solid include the truncated tetrahedron and the truncated octahedron. Follow a set of steps similar to those above to create these solids. Then describe the faces of each.
truncated tetrahedron: hexagonal and triangular faces; truncated octahedron: hexagonal and square faces

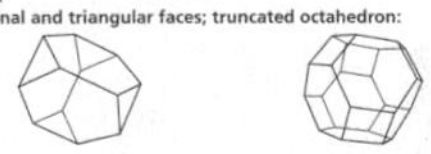

Practice and Problem Solving Wkbk/ All-in-One Resources/Online

Think About a Plan

11-1 **Think About a Plan**

Space Figures and Cross Sections

Visualization Draw and describe a cross section formed by a plane intersecting the cube as follows.

The plane is tilted and intersects the front and back faces of the cube perpendicular to the left and right faces.

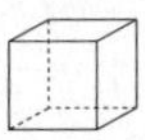

Understanding the Problem

1. What is a cube? Draw a typical view of a cube and describe a cube.
a rectangular prism with six square faces, all congruent
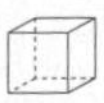
2. What is a plane? Draw a typical view of a plane and describe a plane.
a two-dimensional surface unbounded in all directions

3. What is a cross section? Draw a plane passing through your cube that is parallel to the top and bottom faces of your cube. Explain why the cross section is a square. A cross section is an intersection of a solid and a plane. It is a very thin slice of the solid. Because the plane is parallel to two opposing faces, the cross section is congruent to the faces. So, the cross section is a square congruent to the faces.
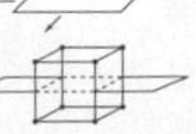

Planning the Solution

4. How can you use your understanding of cubes, planes, and cross sections to draw a plane that is tilted and intersects the front and back faces of the cube perpendicular to the left and right faces?
Look at the drawing of a parallel plane to understand how a tilted plane is different.
5. You showed that if a plane is not tilted, the cross section is a square. How can you use this knowledge to predict what the cross section will look like if the plane is tilted? Contrast the cross section you get when the plane is parallel to the cross section you will get when the plane is tilted.

Getting an Answer

6. Prepare to draw a plane that is tilted and intersects the front and back faces of the cube perpendicular to the left and right faces by studying your drawing of a plane parallel to the top and bottom faces. Imagine the thin slice created when this tilted plane passes through the cube. Describe this cross section.
The cross section is a rectangle. Two sides are the same length as the edges of the cube, and two sides are longer than the edges.

Practice and Problem Solving Wkbk/ All-in-One Resources/Online

Standardized Test Prep

11-1 **Standardized Test Prep**

Space Figures and Cross Sections

Multiple Choice

For Exercises 1–5, choose the correct letter.

1. A polyhedron has 6 vertices and 9 edges. How many faces does it have? B
Ⓐ 3 Ⓑ 5 Ⓒ 7 Ⓓ 9
2. A polyhedron has 25 faces and 36 edges. How many vertices does it have? H
Ⓕ 11 Ⓖ 12 Ⓗ 13 Ⓘ 14
3. Which of the following shows a net for a solid that has 8 faces, 12 vertices, and 18 edges? D
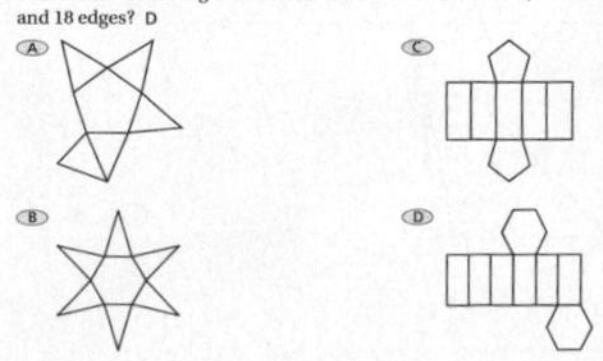

4. What is the cross section formed by a plane that contains a vertical line of symmetry for a tetrahedron? F
Ⓕ triangle Ⓖ square Ⓗ rectangle Ⓘ pentagon
5. What is the cross section formed by a plane that intersects three faces of a cube? A
Ⓐ triangle Ⓑ square Ⓒ rectangle Ⓓ pentagon

Short Response

6. How many edges and vertices are there for an octahedron, a polyhedron with eight congruent triangular faces?
[2] 12 edges AND 6 vertices
[1] 12 edges OR 6 vertices
[0] no correct response given

Online Teacher Resource Center

Activities, Games, and Puzzles

11-1 **Game: Card Link**

Space Figures and Cross Sections

Setup

Your teacher will divide the class into groups of 3 and give each group a set of cards. Give 5 cards to each player. Then place the remainder of the cards face down in a pile.

Game Play

The object of the game is to play all your cards so that you have none left. Flip over the first card and place it face up next to the pile of face-down cards. Decide who will go first. On your turn, play a card that "links" to the face-up card. Two cards can be linked if the number of vertices, edges, or faces on one card is the same as the number of vertices, edges, or faces on the other card.

For example, say the card shown at the right is face up. The figure on the card is a triangular pyramid with 4 vertices, 6 edges, and 4 faces.

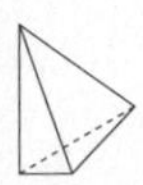

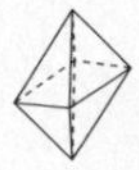

Suppose the card shown at the left is one of the five cards you have in your hand. The figure on the card is an octahedron with 6 vertices, 12 edges, and 8 faces.

You may link these cards. Place the card in your hand on top of the face-up card on the desk. When you play your card, you must state how the cards are linked. For the example shown, you should say, "the triangular pyramid has **6** edges and the octahedron has **6** vertices."

Your opponents should verify that the link you made is correct. If the link is incorrect, you must take your card back, and your turn is over. If you are unable to make a link, you must draw a card from the pile of face-down cards on the desk. If you can link the newly drawn card to the card on the pile, you may do so immediately. Otherwise, your turn is over.

Ending the Game

The game ends when one player has played all of his or her cards. That player wins the game.

Guided Instruction

PURPOSE To introduce perspective drawing
PROCESS Students will draw and recognize drawings in one-point perspective and two-point perspective.
DISCUSS Ask students to think of careers in which perspective drawings are used. **[Sample answers: architects, graphic designers]** Discuss the terms *vanishing point* and *horizon line* and their meanings.

Example 1

This Example has students draw a solid in one-point perspective.

Q In this one-point perspective drawing of a cube, from what perspective are you looking at the cube? **[straight at the front face]**

Example 2

This Example has students draw a solid in two-point perspective.

Q In this two-point perspective drawing of a box, from what perspective are you looking at the box? **[from one of its edges, sometimes referred to as a "corner view"]**

Q What is the difference between a one-point perspective drawing and a two-point perspective drawing? **[the number of vanishing points on the horizon line]**

Mathematical Practices This Concept Byte supports students in becoming proficient in shifting perspective, Mathematical Practice 7.

Concept Byte
Use With Lesson 11-1
EXTENSION

Perspective Drawing

Common Core State Standards
Extends G-GMD.B.4 Identify the shapes of two-dimensional cross-sections of three-dimensional objects, ...
MP 7

You can draw a three-dimensional space figure using a two-dimensional *perspective drawing*. Suppose two lines are parallel in three dimensions, but extend away from the viewer. You draw them—and create perspective—so that they meet at a *vanishing point* on a *horizon line*.

Example 1

Draw a cube in one-point perspective.

Step 1
Draw a square. Then draw a horizon line and a vanishing point on the line.

Step 2
Lightly draw segments from the vertices of the square to the vanishing point.

Step 3
Draw a square for the back of the cube. Each vertex should lie on a segment you drew in Step 2.

Step 4
Complete the figure by using dashes for the hidden edges of the cube. Erase unneeded lines.

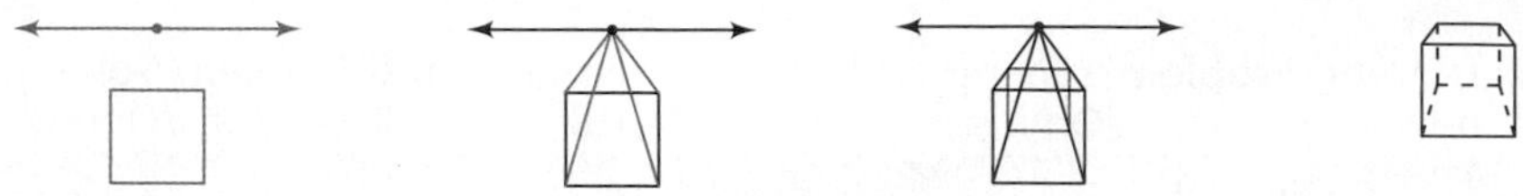

Example 2

Draw a box in two-point perspective.

Step 1
Draw a vertical segment. Then draw a horizon line and two vanishing points on the line.

Step 2
Lightly draw segments from the endpoints of the vertical segment to each vanishing point.

Step 3
Draw two vertical segments between the segments of Step 2.

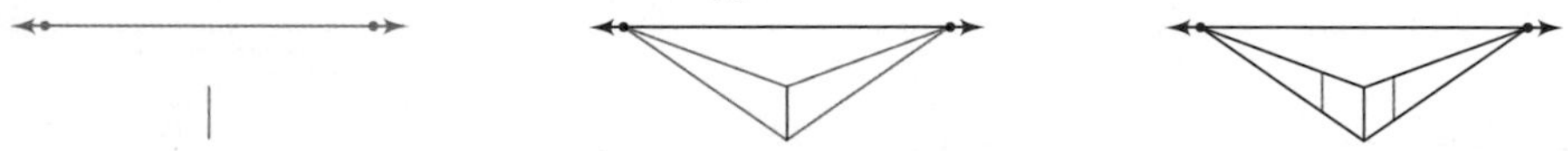

Step 4
Draw segments from the endpoints of the segments you drew in Step 3 to the vanishing points.

Step 5
Complete the figure by using dashes for the hidden edges of the figure. Erase unneeded lines.

Answers

Exercises

1. one-point
2. two-point
3. two-point
4. one-point
5.

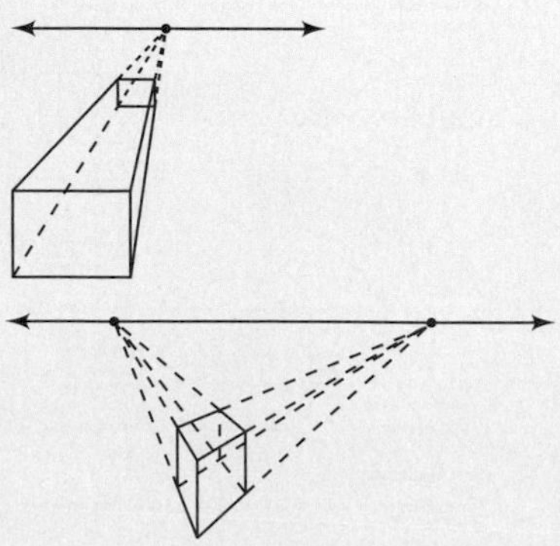

6.

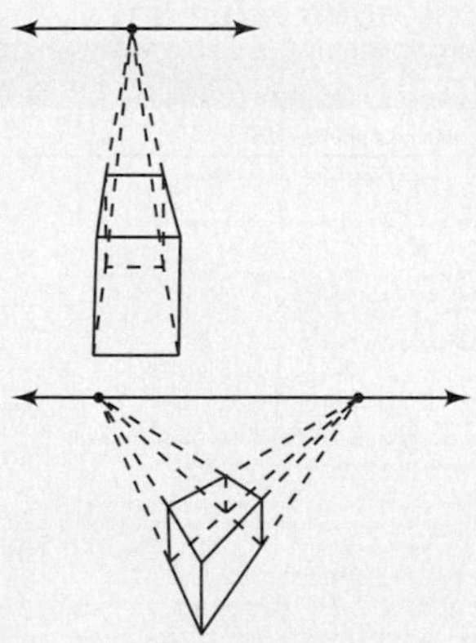

7.

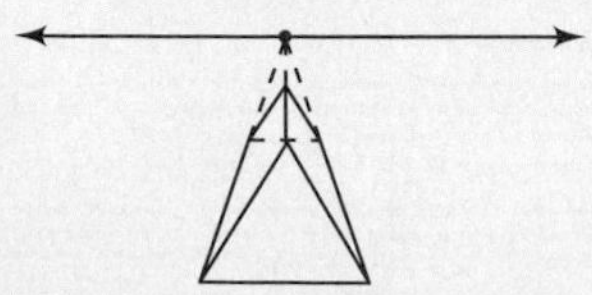

8.

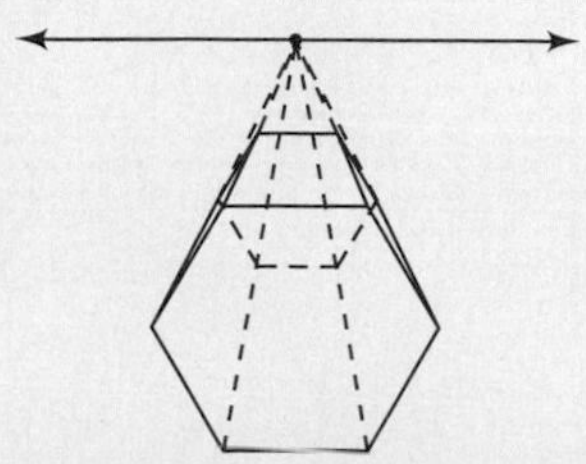

In one-point perspective, the front of the cube is parallel to the drawing surface. A two-point perspective drawing generally looks like a corner view. For either type of drawing, you should be able to envision each vanishing point.

Exercises

Is each object drawn in one-point or two-point perspective?

1. 

2. 

3. 

4. 

Draw each object in one-point perspective and then in two-point perspective.

5. a shoe box

6. a building in your town that sits on a street corner

Draw each container using one-point perspective. Show a base at the front.

7. a triangular carton

8. a hexagonal box

Copy each figure and locate the vanishing point(s).

9. 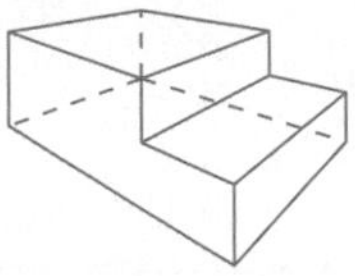

10. 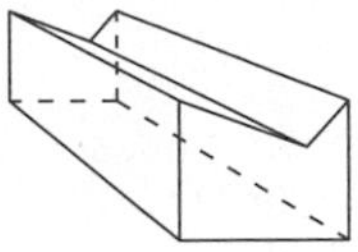

Optical Illusions **What is the optical illusion? Explain the role of perspective in each illusion.**

11. 

12. 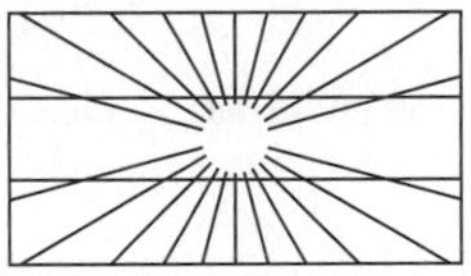

13. **Open-Ended** You can draw block letters in either one-point or two-point perspective. Write your initials in block letters using one-point perspective and two-point perspective.

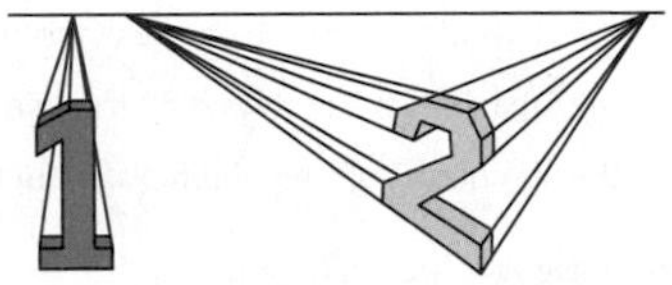

9. 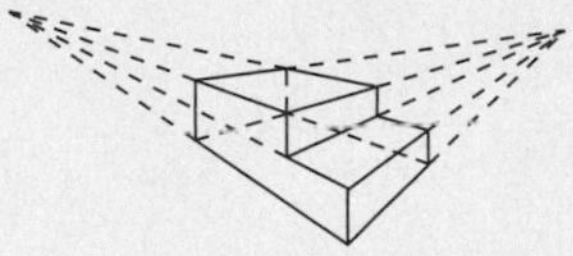

10. 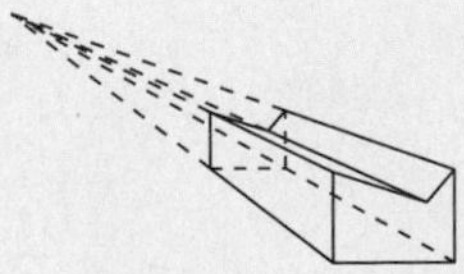

11–12. Answers may vary. Samples are given.

11. The horizontal segments appear to be different lengths, but they are the same length; the 4 nonhorizontal lines appear to converge at the vanishing point. So the upper horizontal line appears to be longer.

12. The horizontal lines appear to be curved, but they are straight; the slanted lines that would meet at the vanishing point create a cylinder effect.

13. Check students' work.

ERROR PREVENTION

Students will most likely be able to recognize a given drawing as a one-point or two-point perspective. However, it is often more challenging to take a given shape and recreate it. Therefore, when asked to do this in the exercises, be sure students model their work from the Examples and make their drawings step-by-step.

Guided Instruction

PURPOSE To review how to solve an equation for one variable in terms of another

PROCESS Students will manipulate a formula by solving for one variable of the formula in terms of the others.

DISCUSS Review the process of inverse operations and how to isolate a variable. Practice with one- and two-step equations with one variable for students who are having difficulty with the concepts.

Example

This Example demonstrates the step-by-step procedure for isolating a variable of a formula.

Q In part (a), which property of equality is used first? **[Division Property of Equality]**

Q Which property allows you to simplify $\frac{\pi r^2 h}{\pi r^2}$ as h? **[Identity Property of Multiplication $\left(\frac{\pi r^2}{\pi r^2} = 1\right)$]**

Q In part (b), why are two formulas used? **[The given information is written in terms of two formulas (area of a square and perimeter of the square). One formula is solved for the side length and then substituted into the other formula.]**

Algebra Review

Use With Lesson 11-2

Literal Equations

Common Core State Standards

Reviews A-CED.A.4 Rearrange formulas to highlight a quantity of interest, using the same reasoning as in solving equations.

A *literal equation* is an equation involving two or more variables. A formula is a special type of literal equation. You can transform a formula by solving for one variable in terms of the others.

Example

A The formula for the volume of a cylinder is $V = \pi r^2 h$. Find a formula for the height in terms of the radius and volume.

$V = \pi r^2 h$	Use the formula for the volume of a cylinder.
$\frac{V}{\pi r^2} = \frac{\pi r^2 h}{\pi r^2}$	Divide each side by πr^2, with $r \neq 0$.
$\frac{V}{\pi r^2} = h$	Simplify.

The formula for the height is $h = \frac{V}{\pi r^2}$.

B Find a formula for the area of a square in terms of its perimeter.

$P = 4s$	Use the formula for the perimeter of a square.
$\frac{P}{4} = s$	Solve for s in terms of P.
$A = s^2$	Use the formula for area.
$= \left(\frac{P}{4}\right)^2$	Substitute $\frac{P}{4}$ for s.
$= \frac{P^2}{16}$	Simplify.

The formula for the area is $A = \frac{P^2}{16}$.

Exercises

Algebra **Solve each equation for the variable in red.**

1. $C = 2\pi r$
2. $A = \frac{1}{2}bh$
3. $A = \pi r^2$

Algebra **Solve for the variable in red. Then solve for the variable in blue.**

4. $P = 2w + 2\ell$
5. $\tan A = \frac{y}{x}$
6. $A = \frac{1}{2}(b_1 + b_2)h$

Find a formula as described below.

7. the circumference C of a circle in terms of its area A
8. the area A of an isosceles right triangle in terms of the hypotenuse h
9. the apothem a of a regular hexagon in terms of the area A of the hexagon
10. Solve $A = \frac{1}{2}ab \sin C$ for $m\angle C$.

Answers

Exercises

1. $r = \frac{C}{2\pi}$
2. $b = \frac{2A}{h}$
3. $r = \sqrt{\frac{A}{\pi}}$
4. $w = \frac{P - 2\ell}{2}$; $\ell = \frac{P - 2w}{2}$
5. $y = x \tan A$; $x = \frac{y}{\tan A}$
6. $h = \frac{2A}{b_1 + b_2}$; $b_1 = \frac{2A}{h} - b_2$
7. $C = 2\sqrt{\pi A}$
8. $A = \frac{h^2}{4}$
9. $a = \frac{\sqrt{6A\sqrt{3}}}{6}$ or $a = \frac{\sqrt{6A} \cdot \sqrt[4]{3}}{6}$
10. $m\angle C = \sin^{-1}\left(\frac{2A}{ab}\right)$

11-2 Surface Areas of Prisms and Cylinders

Common Core State Standards
G-MG.A.1 Use geometric shapes, their measures, and their properties to describe objects.
MP 1, MP 3, MP 4, MP 6, MP 7, MP 8

Objective To find the surface area of a prism and a cylinder

You'll flatten this problem out in no time!

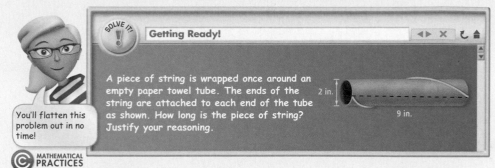

Solve It! Getting Ready!

A piece of string is wrapped once around an empty paper towel tube. The ends of the string are attached to each end of the tube as shown. How long is the piece of string? Justify your reasoning.

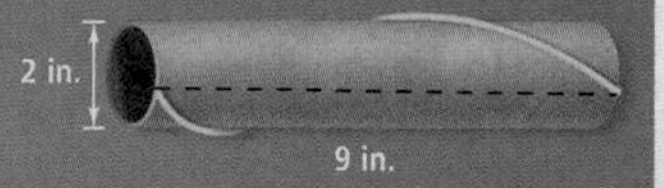

MATHEMATICAL PRACTICES

In the Solve It, you investigated the structure of a tube. In this lesson, you will learn properties of three-dimensional figures by investigating their surfaces.

Lesson Vocabulary
- prism (base, lateral face, altitude, height, lateral area, surface area)
- right prism
- oblique prism
- cylinder (base, altitude, height, lateral area, surface area)
- right cylinder
- oblique cylinder

Essential Understanding To find the surface area of a three-dimensional figure, find the sum of the areas of all the surfaces of the figure.

A **prism** is a polyhedron with two congruent, parallel faces, called **bases.** The other faces are **lateral faces.** You can name a prism using the shape of its bases.

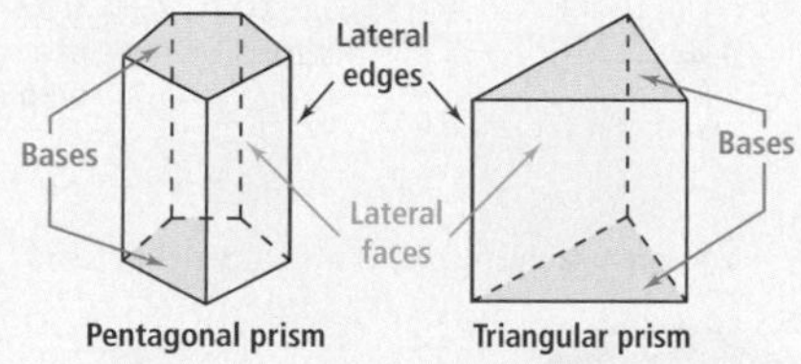

An **altitude** of a prism is a perpendicular segment that joins the planes of the bases. The **height** h of a prism is the length of an altitude. A prism may either be right or oblique.

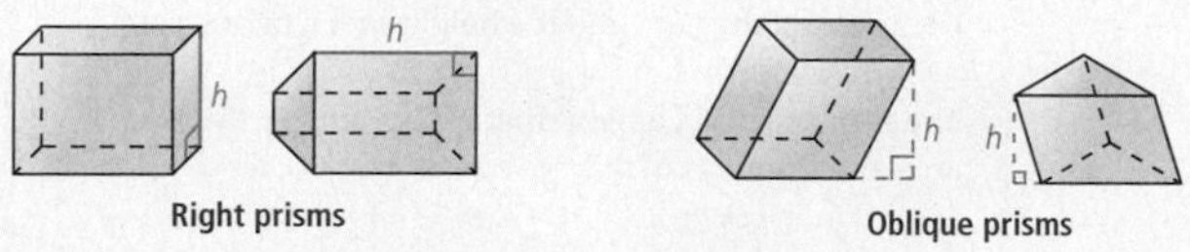

In a **right prism,** the lateral faces are rectangles and a lateral edge is an altitude. In an **oblique prism,** some or all of the lateral faces are nonrectangular. In this book, you may assume that a prism is a right prism unless stated or pictured otherwise.

1 Interactive Learning

Solve It!

PURPOSE To find the net of a cylinder

PROCESS Students may draw a net of the cylinder and find the diagonal of the rectangle.

FACILITATE

Q If you cut the paper towel roll along the dashed line and laid the cardboard flat, what shape would you have? **[rectangle]**

Q What are the dimensions of the rectangle? **[9 in. by 2π in.]**

Q What does the string represent on the rectangle? **[The string represents the diagonal of the rectangle.]**

Q How can you find the length of the diagonal? **[Use the Pythagorean Theorem to find the hypotenuse of a right triangle.]**

ANSWER See Solve It in Answers on next page.

CONNECT THE MATH Students can use a net to visualize and explore the situation in the Solve It. In the lesson, students use nets and formulas to find the surface area of prisms and cylinders.

2 Guided Instruction

Show students examples of different types of prisms. Ask them to identify the faces, bases, and heights of the prisms. Have them classify the prism as a right prism or oblique prism.

11-2 Preparing to Teach

BIG ideas Measurement
Visualization

ESSENTIAL UNDERSTANDING
- The area of a three-dimensional figure is equal to the sum of the areas of each surface of the figure.

Math Background

To continue their study of three-dimensional figures, students will learn to calculate the lateral and surface area of prisms and cylinders. The lateral area of a prism is the area of all of its lateral faces. The surface area is the sum of the areas of all the surfaces in the prism. Surface area is greater than lateral area because it includes the area of the figure's bases. Students will learn to use nets to calculate the surface area. They will discover how to derive a formula for the surface area using the perimeter of the base of the prism. Students will identify the differences between lateral area and surface area and apply the concepts in real-world problems.

Many students have difficulty with the concept of surface area and it may be helpful to relate the concept to a real-world situation, such as gift wrapping.

Mathematical Practices

Attend to precision. Students will define, explicitly use, and solve for the terms "lateral area" and "surface area" for cylinders and prisms.

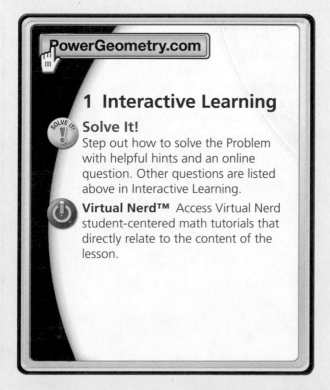

PowerGeometry.com

1 Interactive Learning

Solve It! Step out how to solve the Problem with helpful hints and an online question. Other questions are listed above in Interactive Learning.

Virtual Nerd™ Access Virtual Nerd student-centered math tutorials that directly relate to the content of the lesson.

Problem 1

Q How many different sized faces does the prism have? How many of each size? **[3; 2]**

Q What are the dimensions of the faces? **[5 cm by 4 cm; 4 cm by 3 cm; 5 cm by 3 cm]**

Q How can you find the surface area of the prism? **[Add the areas of all 6 faces.]**

Got It? VISUAL LEARNERS

Have students identify the faces of the prism. Ask them to calculate the area of each face separately, and then find the sum of the areas.

Q How many different sized faces does the triangular prism have? **[3]**

Q What shape are these faces? **[triangle, rectangle, rectangle]**

Take Note

Be sure that students understand the difference between lateral area and surface area. They should associate the lateral area with the lateral faces of the prism. Also, students should see that the surface area of a prism is greater than the lateral area because it includes the area of the bases and the lateral faces.

The **lateral area** (L.A.) of a prism is the sum of the areas of the lateral faces. The **surface area** (S.A.) is the sum of the lateral area and the area of the two bases.

Think

How do you know what units to use?
In the prism, the rectangle marked 5 on one side and 4 on the other has area 5 cm × 4 cm, or 20 cm^2. So use cm^2 as the unit for the surface area of the prism.

Problem 1 Using a Net to Find Surface Area of a Prism

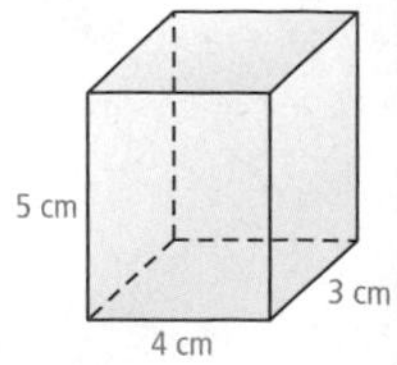

What is the surface area of the prism at the right? Use a net.

Draw a net for the prism. Then calculate the surface area.

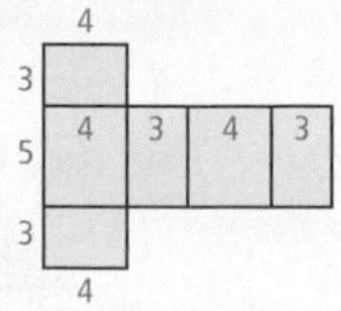

S.A. = sum of areas of all the faces

$= 5 \cdot 4 + 5 \cdot 3 + 5 \cdot 4 + 5 \cdot 3 + 3 \cdot 4 + 3 \cdot 4$

$= 20 + 15 + 20 + 15 + 12 + 12$

$= 94$

The surface area of the prism is 94 cm^2.

Got It? 1. What is the surface area of the triangular prism? Use a net.

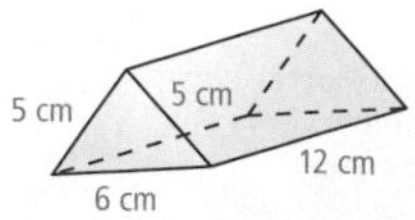

You can find formulas for lateral and surface areas of a prism by using a net.

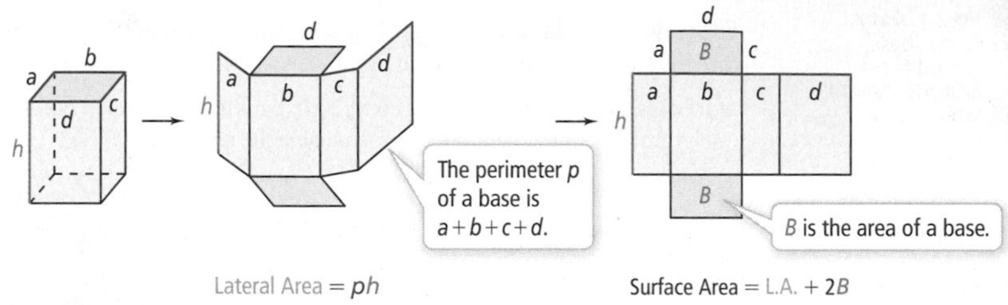

You can use the formulas with any right prism.

take note

Theorem 11-1 Lateral and Surface Areas of a Prism

The lateral area of a right prism is the product of the perimeter of the base and the height of the prism.

L.A. = ph

The surface area of a right prism is the sum of the lateral area and the areas of the two bases.

S.A. = L.A. + $2B$

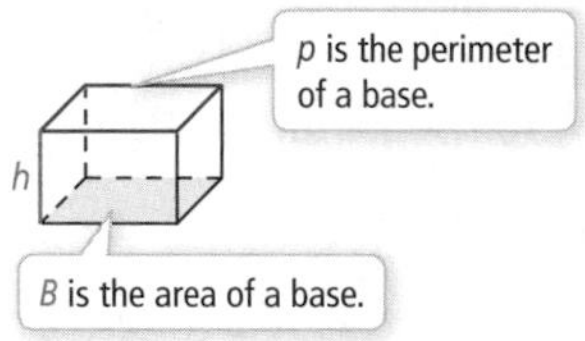

Answers

Solve It!

≈ 11 in.; the net of the tube is a rectangle 9 in. (length of the tube) by 2π in. (circumference). The string wraps around once, so it is a diagonal of the rectangle. Use the Pythag. Thm. to find the string's length: $\sqrt{9^2 + (2\pi)^2} \approx \sqrt{81 + 39.48} \approx 11$ in.

Got It?

1. 216 cm^2

PowerGeometry.com

2 Guided Instruction

Each Problem is worked out and supported online.

Problem 1
Using a Net to Find Surface Area of a Prism
Animated

Problem 2
Using Formulas to Find Surface Area of a Prism
Animated

Problem 3
Finding Surface Area of a Cylinder
Animated

Problem 4
Finding Lateral Area of a Cylinder

Support in Geometry Companion
- Vocabulary
- Key Concepts
- Got It?

Problem 2 Using Formulas to Find Surface Area of a Prism

Plan

What do you need to find first?
You first need to find the missing side length of a triangular base so that you can find the perimeter of a base.

What is the surface area of the prism at the right?

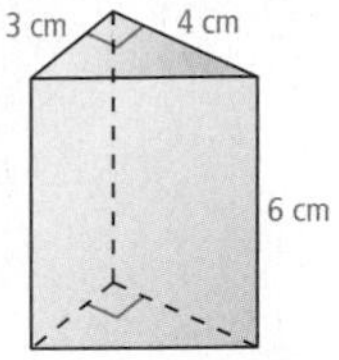

Step 1 Find the perimeter of a base.

The perimeter of the base is the sum of the side lengths of the triangle. Since the base is a right triangle, the hypotenuse is $\sqrt{3^2 + 4^2}$ cm, or 5 cm, by the Pythagorean Theorem.

$p = 3 + 4 + 5 = 12$

Step 2 Find the lateral area of the prism.

L.A. $= ph$	Use the formula for lateral area.
$= 12 \cdot 6$	Substitute 12 for p and 6 for h.
$= 72$	Simplify.

Think

Which height do you need?
For problems involving solids, make it a habit to note which height the formula requires. In Step 3, you need the height of the triangle, not the height of the prism.

Step 3 Find the area of a base.

$B = \frac{1}{2}bh$	Use the formula for the area of a triangle.
$= \frac{1}{2}(3 \cdot 4)$	Substitute 3 for b and 4 for h.
$= 6$	

Step 4 Find the surface area of the prism.

S.A. $=$ L.A. $+ 2B$	Use the formula for surface area.
$= 72 + 2(6)$	Substitute 72 for L.A. and 6 for B.
$= 84$	Simplify.

The surface area of the prism is 84 cm^2.

Got It? 2. **a.** What is the lateral area of the prism at the right?
b. What is the area of a base in simplest radical form?
c. What is the surface area of the prism rounded to a whole number?

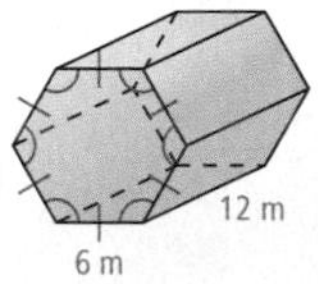

A **cylinder** is a solid that has two congruent parallel **bases** that are circles. An **altitude** of a cylinder is a perpendicular segment that joins the planes of the bases. The **height** h of a cylinder is the length of an altitude.

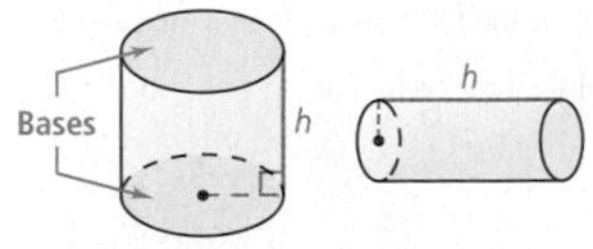

Right cylinders

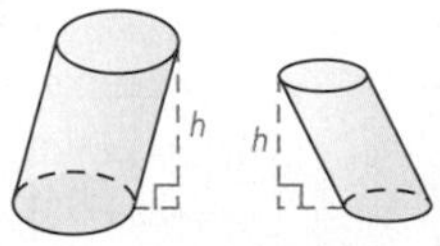

Oblique cylinders

In a **right cylinder,** the segment joining the centers of the bases is an altitude. In an **oblique cylinder,** the segment joining the centers is not perpendicular to the planes containing the bases. In this book, you may assume that a cylinder is a right cylinder unless stated or pictured otherwise.

Problem 2

Q Which faces of the prism are the bases? **[The triangular faces are the bases.]**

Q How can you find the unknown side length of the base? **[Use the Pythagorean Theorem.]**

Q Which sides of the base of the prism can be used as the base and height of the triangle to calculate the area? Explain. **[The legs of the right triangle are perpendicular so they can be used as the base and height of the triangle.]**

Got It?

Students will need to use trigonometry to find the length of the apothem. Have students draw a separate sketch of the right triangle created by the radius, apothem, and half the side.

Review the definition of an altitude. Emphasize that an altitude in an oblique prism may not be contained in the prism.

Additional Problems

1. What is the surface area of the prism below? Use a net.

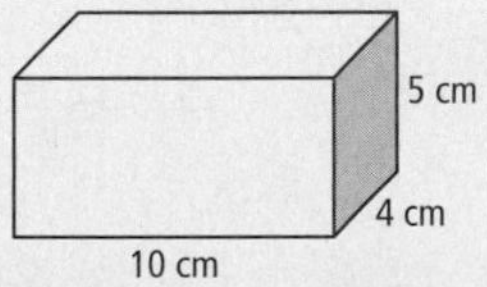

ANSWER 220 cm^2

2. What is the surface area of the prism below?

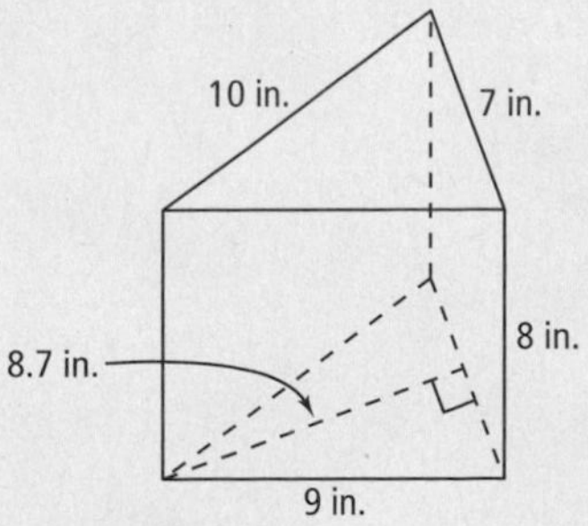

ANSWER 268.9 in.^{2}

3. The radius of the base of a cylinder is 6 cm and its height is 15 cm. What is the surface area of the cylinder in terms of π?

A. 176 π cm^2
B. 188 π cm^2
C. 234 π cm^2
D. 252 π cm^2

ANSWER D

4. A soup can is 4.5 in. high and has a diameter of 3 in. How much paper is needed to make a label that will completely cover the sides of the can without overlap?

ANSWER about 42.4 in.^{2}

Answers

Got It? (continued)

2a. 432 m^2
b. $54\sqrt{3}$ m^2
c. 619 m^2

Take Note

Have students review their work in the Solve It. They should use a similar method to derive the formula for the surface area of a cylinder. Be sure that students understand how to find the lateral area of a cylinder and add the area of the two circular bases to find the surface area.

Problem 3

Q What are the dimensions of the rectangle in the net of the cylinder? **[The rectangle is 8π in. by 6 in.]**

Q What additional information do you need to calculate the surface area? **[the area of the bases]**

Q How can you find the surface area of the cylinder? **[Add the area of the rectangle and two times the area of the base.]**

Got It?

Q Suppose a net of the cylinder was drawn. What would be the length of the rectangle in the net? **[20π cm]**

Q What else in the cylinder has the same length as the length of the rectangle? **[the circumference of each circular base]**

To find the area of the curved surface of a cylinder, visualize "unrolling" it. The area of the resulting rectangle is the **lateral area** of the cylinder. The **surface area** of a cylinder is the sum of the lateral area and the areas of the two circular bases. You can find formulas for these areas by looking at a net for a cylinder.

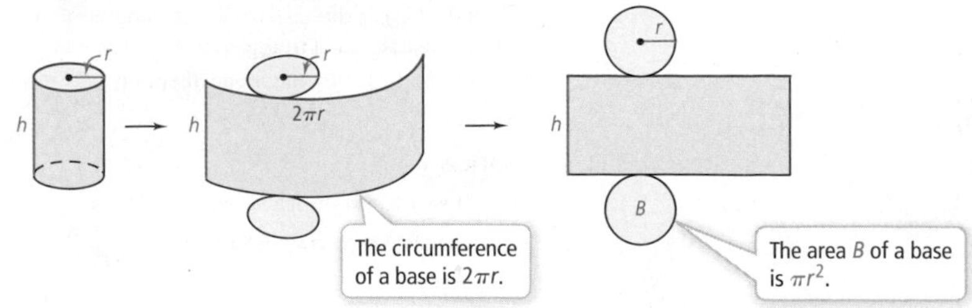

r
h
$2\pi r$
B
The circumference of a base is $2\pi r$.
The area B of a base is πr^2.

Lateral Area $= 2\pi rh$ Surface Area $=$ L.A. $+ 2\pi r^2$

take note

Theorem 11-2 Lateral and Surface Areas of a Cylinder

The lateral area of a right cylinder is the product of the circumference of the base and the height of the cylinder.

L.A. $= 2\pi r \cdot h$, or L.A. $= \pi dh$

The surface area of a right cylinder is the sum of the lateral area and the areas of the two bases.

S.A. $=$ L.A. $+ 2B$, or S.A. $= 2\pi rh + 2\pi r^2$

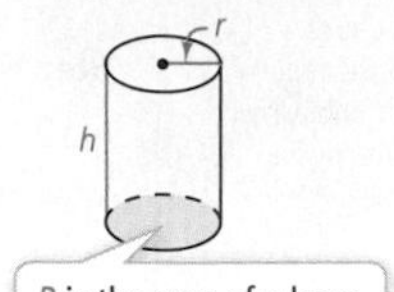

r
h

B is the area of a base.

Problem 3 Finding Surface Area of a Cylinder

Multiple Choice The radius of the base of a cylinder is 4 in. and its height is 6 in. What is the surface area of the cylinder in terms of π?

Ⓐ 32π in.^{2} Ⓑ 42π in.^{2} Ⓒ 80π in.^{2} Ⓓ 120π in.^{2}

Think

How is finding the surface area of a cylinder like finding the surface area of a prism?
For both, you need to find the L.A. and add it to twice the area of a base.

S.A. $=$ L.A. $+ 2B$	Use the formula for surface area of a cylinder.
$= 2\pi rh + 2(\pi r^2)$	Substitute the formulas for lateral area and area of a circle.
$= 2\pi(4)(6) + 2(\pi 4^2)$	Substitute 4 for r and 6 for h.
$= 48\pi + 32\pi$	Simplify.
$= 80\pi$	

The surface area of the cylinder is 80π in.^{2}. The correct choice is C.

Got It? 3. A cylinder has a height of 9 cm and a radius of 10 cm. What is the surface area of the cylinder in terms of π?

Answers

Got It? (continued)

3. 380π cm^2

Think

What is the problem asking you to find?
The area covered by one full turn of the roller is the rectangular part of the net of a cylinder. This is the lateral area.

Problem 4 Finding Lateral Area of a Cylinder

Interior Design **You are using the cylindrical stencil roller below to paint patterns on your floor. What area does the roller cover in one full turn?**

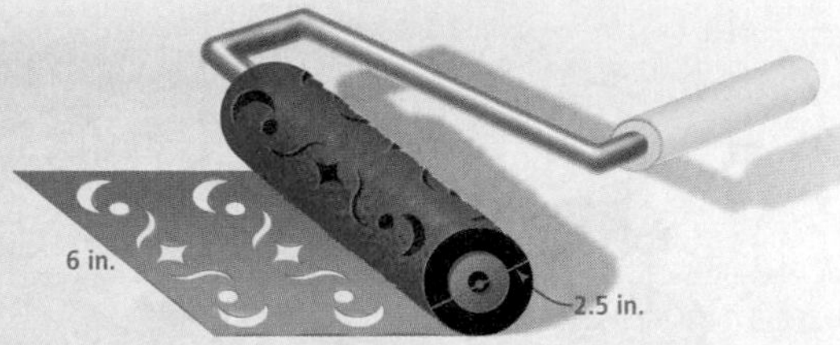

The area covered is the lateral area of a cylinder with height 6 in. and diameter 2.5 in.

L.A. $= \pi dh$	Use the formula for lateral area of a cylinder.
$= \pi(2.5)(6)$	Substitute 2.5 for d and 6 for h.
$= 15\pi \approx 47.1$	Simplify.

In one full turn, the stencil roller covers about 47.1 in.^{2}.

Got It? **4. a.** A smaller stencil roller has a height of 1.5 in. and the same diameter as the roller in Problem 4. What area does the smaller roller cover in one turn? Round your answer to the nearest tenth.

b. Reasoning What is the ratio of the smaller roller's height to the larger roller's height? What is the ratio of the areas the rollers can cover in one turn (smaller to larger)?

Lesson Check

Do you know HOW?

What is the surface area of each prism?

1.

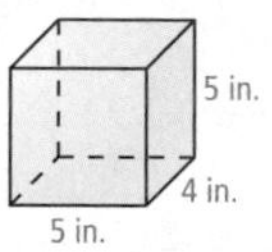

2.

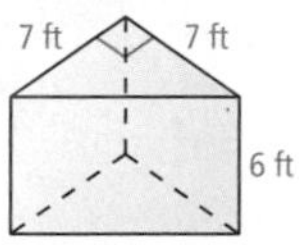

What is the surface area of each cylinder?

3.

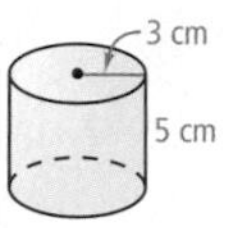

4.

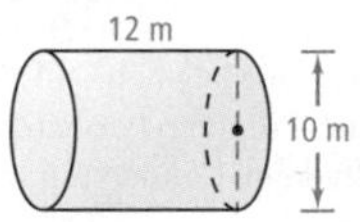

Do you UNDERSTAND? MATHEMATICAL PRACTICES

5. Vocabulary Name the lateral faces and the bases of the prism at the right.

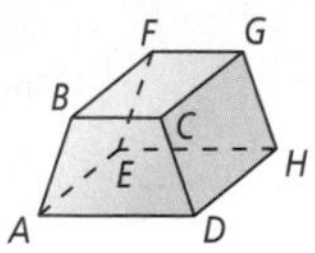

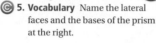

6. Error Analysis Your friend drew a net of a cylinder. What is your friend's error? Explain.

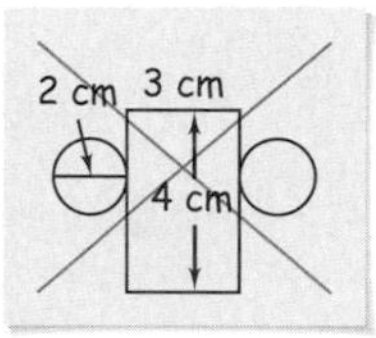

Problem 4

Q When a roller is used, what is the width of paint applied to the floor with one complete rotation of the roller? **[6 in.]**

Q What is the shape of the paint applied to the floor? **[rectangle]**

Q How can you use the diameter instead of the radius in the formula for lateral area? **[L.A. = πdh]**

Got It? TACTILE LEARNERS

Have students compare the areas of the rectangles that represent the areas covered by the rollers. The differences in the dimensions occur only in the width of the rectangles. Ask students to compare the areas in a ratio.

3 Lesson Check

Do you know HOW?

- If students have difficulty with Exercise 2, then have them review Problem 1 to draw a net for the prism.

Do you UNDERSTAND?

- If students have difficulty with Exercise 6, then have them review the Solve It to revisit the relationship of the figures in the net of a cylinder.

Close

Q What is the lateral area of a prism? **[The lateral area is the sum of the areas of all the lateral faces.]**

Q What is the surface area of a prism? **[The surface area is the lateral area plus the area of the bases of the prism.]**

4a. 11.8 in.^{2}

b. $\frac{1}{4}$; $\frac{1}{4}$

Lesson Check

1. 130 in.^{2}
2. $(133 + 42\sqrt{2})$ ft^2 or about 192.4 ft^2
3. 48π cm^2 or about 150.8 cm^2
4. 170π m^2 or about 534.1 m^2
5. lateral faces: *BFGC*, *DCGH*, *ADHE*, *EFBA*; bases: *ABCD*, *EFGH*
6. The diameter of the circular bases does not match the length of the rectangle. If the diameter is 2 cm, then the length must be 2π cm, or if the length is 4 cm, then the diameter should be $\frac{4}{\pi}$ cm, or about 1.3 cm.

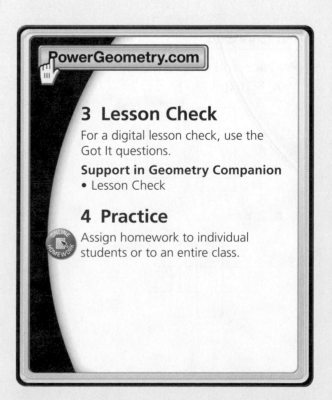

PowerGeometry.com

3 Lesson Check

For a digital lesson check, use the Got It questions.

Support in Geometry Companion
- Lesson Check

4 Practice

ONLINE HOMEWORK

Assign homework to individual students or to an entire class.

4 Practice

ASSIGNMENT GUIDE

Basic: 7–20 all, 22–30 even, 37

Average: 7–19 odd, 21–38

Advanced: 7–19 odd, 21–43

Standardized Test Prep: 44–47

Mathematical Practices are supported by exercises with red headings. Here are the Practices supported in this lesson:

MP 1: Make Sense of Problems Ex. 26

MP 3: Construct Arguments Ex. 25e, 37

MP 3: Communicate Ex. 22

MP 3: Critique the Reasoning of Others Ex. 6, 28

MP 7: Shift Perspective Ex. 24, 33–36

MP 8: Check for Reasonableness Ex. 27

Applications exercises have blue headings. Exercises 23, 29, and 30 support MP 4: Model.

STEM exercises focus on science or engineering applications.

EXERCISE 37: Use the Think About a Plan worksheet in the **Practice and Problem Solving Workbook** (also available in the Teaching Resources in print and online) to further support students' development in becoming independent learners.

HOMEWORK QUICK CHECK

To check students' understanding of key skills and concepts, go over Exercises 9, 17, 26, 28, and 37.

Practice and Problem-Solving Exercises

MATHEMATICAL PRACTICES

A Practice

Use a net to find the surface area of each prism.

See Problem 1.

7.

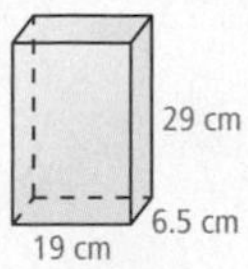

8.

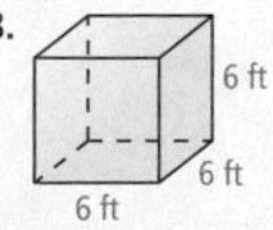

9.

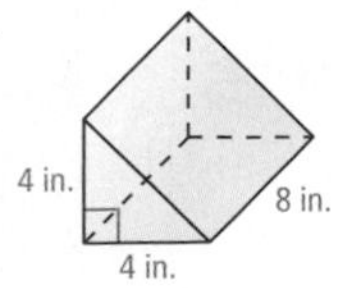

10. a. Classify the prism at the right.

b. Find the lateral area of the prism.

c. The bases are regular hexagons. Find the sum of their areas.

d. Find the surface area of the prism.

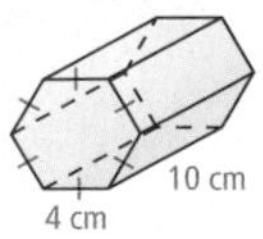

Use formulas to find the surface area of each prism. Round your answer to the nearest whole number.

See Problem 2.

11.

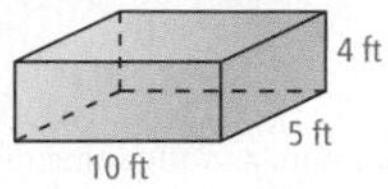

12.

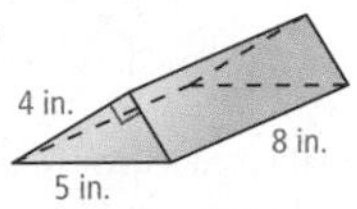

13. Regular octagon

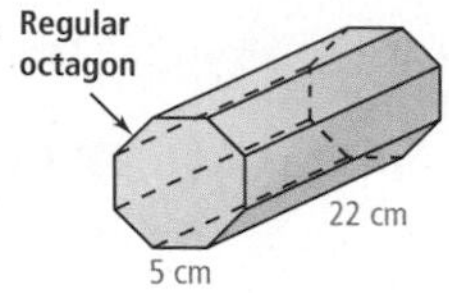

Find the lateral area of each cylinder to the nearest whole number.

See Problems 3 and 4.

14.

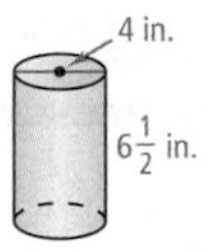

15.

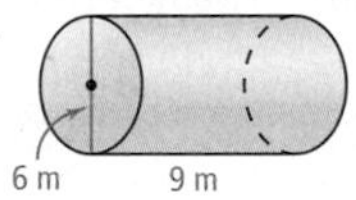

16.

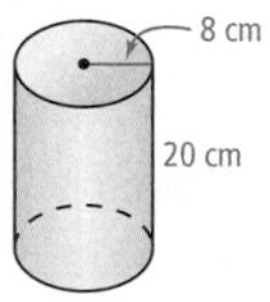

Find the surface area of each cylinder in terms of π.

17.

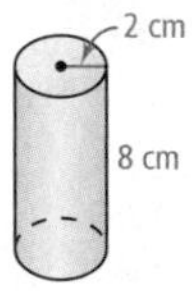

18.

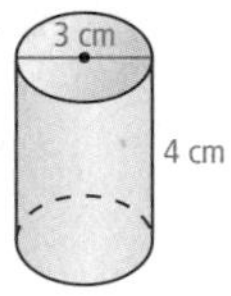

19.

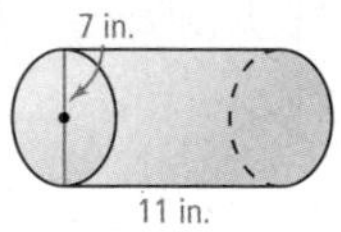

STEM **20. Packaging** A cylindrical carton of oatmeal with radius 3.5 in. is 9 in. tall. If all surfaces except the top are made of cardboard, how much cardboard is used to make the oatmeal carton? Assume no surfaces overlap. Round your answer to the nearest square inch.

Answers

Practice and Problem-Solving Exercises

7. 1726 cm^2

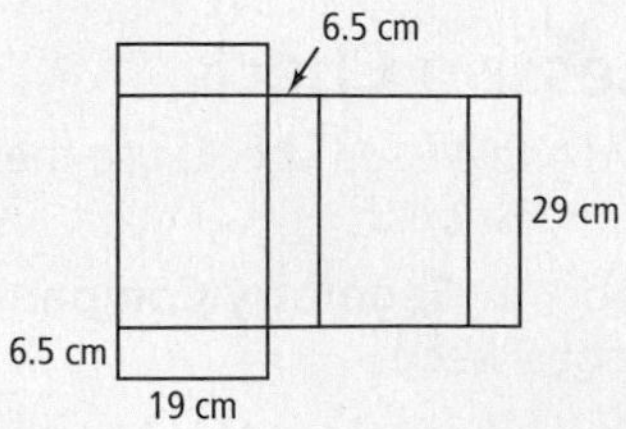

8. 216 ft^2

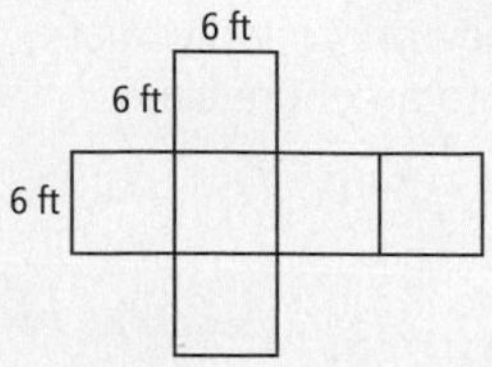

9. $(80 + 32\sqrt{2})$ in.^{2}, or about 125.3 in.^{2}

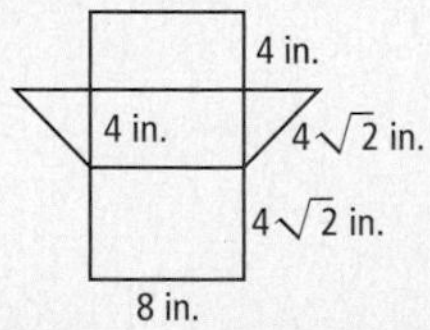

10a. right hexagonal prism

b. 240 cm^2

c. $48\sqrt{3}$ cm^2 or about 83.1 cm^2

d. $(240 + 48\sqrt{3})$ cm^2 or about 323.1 cm^2

11. 220 ft^2

12. 108 in.^{2}

13. 1121 cm^2

14. 82 in.^{2}

15. 170 m^2

16. 1005 cm^2

17. 40π cm^2

18. 16.5π cm^2

19. 101.5π in.^{2}

20. 236 in.^{2}

B Apply

21. A triangular prism has base edges 4 cm, 5 cm, and 6 cm long. Its lateral area is 300 cm^2. What is the height of the prism?

22. **Writing** Explain how a cylinder and a prism are alike and how they are different.

23. **Pencils** A hexagonal pencil is a regular hexagonal prism, as shown at the right. A base edge of the pencil has a length of 4 mm. The pencil (without eraser) has a height of 170 mm. What is the area of the surface of the pencil that gets painted?

24. **Open-Ended** Draw a net for a rectangular prism with a surface area of 220 cm^2.

25. Consider a box with dimensions 3, 4, and 5.
    a. Find its surface area.
    b. Double each dimension and then find the new surface area.
    c. Find the ratio of the new surface area to the original surface area.
    d. Repeat parts (a)–(c) for a box with dimensions 6, 9, and 11.
    e. **Make a Conjecture** How does doubling the dimensions of a rectangular prism affect the surface area?

26. **Think About a Plan** A cylindrical bead has a square hole, as shown at the right. Find the area of the surface of the entire bead.
    - Can you visualize the bead as a combination of familiar figures?
    - How do you find the area of the surface of the inner part of the bead?

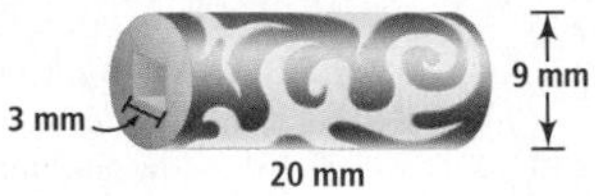

27. **Estimation** Estimate the surface area of a cube with edges 4.95 cm long.

28. **Error Analysis** Your class is drawing right triangular prisms. Your friend's paper is below. What is your friend's error?

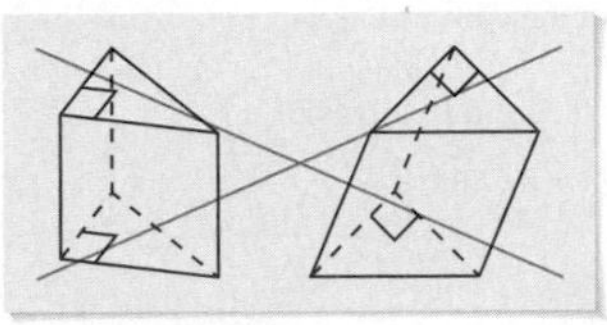

STEM 29. **Packaging** A cylindrical can of cocoa has the dimensions shown at the right. What is the approximate surface area available for the label? Round to the nearest square inch.

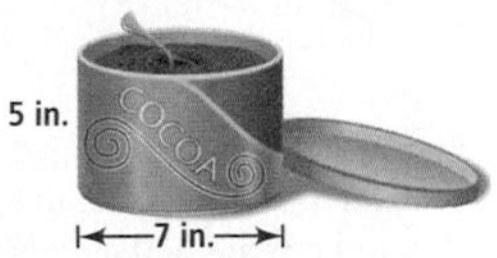

30. **Pest Control** A flour moth trap has the shape of a triangular prism that is open on both ends. An environmentally safe chemical draws the moth inside the prism, which is lined with an adhesive. What is the area of the trap that is lined with the adhesive?

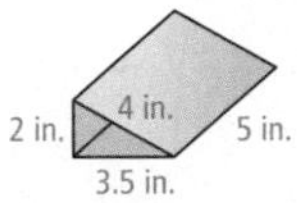

21. 20 cm
22. A cylinder and a prism both have two bases that are ∥ and ≅. The bases of a cylinder are circles, and the bases of a prism are polygons.
23. 4080 mm^2
24. Answers may vary. Sample:

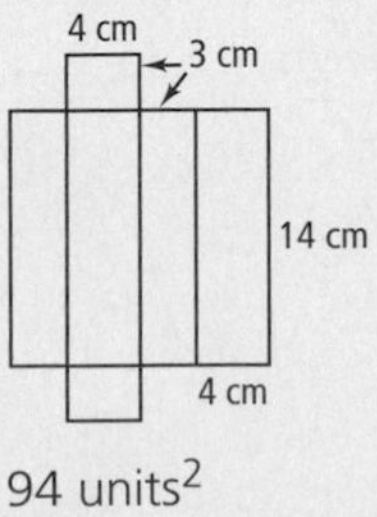

25a. 94 $units^2$
b. 376 $units^2$
c. 4 : 1
d. 438 $units^2$; 1752 $units^2$; 4 : 1
e. The surface area is multiplied by 4.

26. $(220.5\pi + 222)$ mm^2, or about 914.7 mm^2
27. just under 150 cm^2
28. The prism shown on the right is an oblique prism, not a right prism. Also, a right triangular prism does not have to have a rt. △ for a base. In a right triangular prism, each lateral edge must be ⊥ to the base.
29. 110 $in.^2$
30. 47.5 $in.^2$

Answers

Practice and Problem-Solving Exercises (continued)

31a. 7 units

b. 196π units^2 or about 615.8 units^2

32a. $A(3, 0, 0)$, $B(3, 5, 0)$, $C(0, 5, 0)$, $D(0, 5, 4)$

b. 5 units

c. 3 units

d. 4 units

e. 94 units^2

33. cylinder of radius 4 and height 2; 48π units^2

34. cylinder of radius 2 and height 4; 24π units^2

35. cylinder of radius 2 and height 4; 24π units^2

36. cylinder of radius 4 and height 2; 48π units^2

37a. The lateral area is doubled.

b. The surface area is more than doubled.

c. If r doubles, S.A. $= 2\pi(2r)^2 + 2\pi(2r)h = 8\pi r^2 + 4\pi rh = 2(4\pi r^2 + 2\pi rh)$. So the surface area $2\pi r^2 + 2\pi rh$ is more than doubled.

38a. $r \approx 1.2$ in.; $h = 6$ in.

b. about 54 in.^{2}

39. $(182\pi + 232)$ cm^2

40. $(84 + 20\pi)$ m^2

41. $(220 - 8\pi)$ in.^{2}

42a. 0, 8, 12, 6, 1

b. 1728 in.^{2}

43. $h = 6$ m; $r = 3$ m

31. Suppose that a cylinder has a radius of r units, and that the height of the cylinder is also r units. The lateral area of the cylinder is 98π square units.
a. **Algebra** Find the value of r.
b. Find the surface area of the cylinder.

32. Geometry in 3 Dimensions Use the diagram at the right.
a. Find the three coordinates of each vertex A, B, C, and D of the rectangular prism.
b. Find AB.
c. Find BC.
d. Find CD.
e. Find the surface area of the prism.

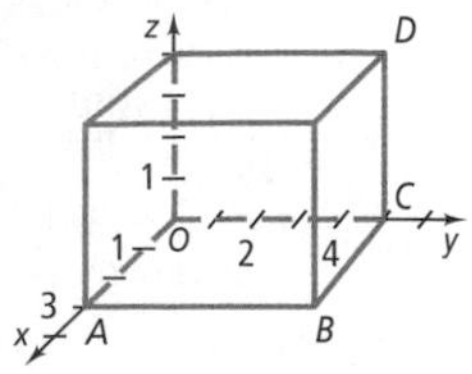

Visualization **Suppose you revolve the plane region completely about the given line to sweep out a solid of revolution. Describe the solid and find its surface area in terms of π.**

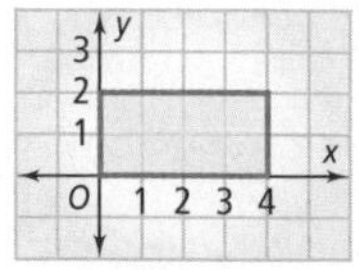

33. the y-axis

34. the x-axis

35. the line $y = 2$

36. the line $x = 4$

37. Reasoning Suppose you double the radius of a right cylinder.
a. How does that affect the lateral area?
b. How does that affect the surface area?
c. Use the formula for surface area of a right cylinder to explain why the surface area in part (b) was not doubled.

38. Packaging Some cylinders have wrappers with a spiral seam. Peeled off, the wrapper has the shape of a parallelogram. The wrapper for a biscuit container has base 7.5 in. and height 6 in.
a. Find the radius and height of the container.
b. Find the surface area of the container.

Challenge **What is the surface area of each solid in terms of π?**

39.

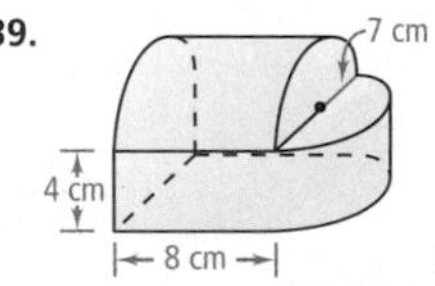

40.

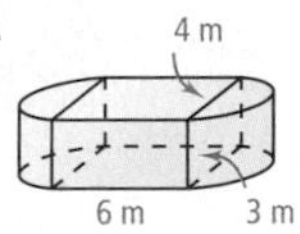

41.

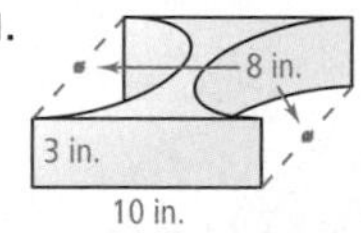

42. Each edge of the large cube at the right is 12 inches long. The cube is painted on the outside, and then cut into 27 smaller cubes. Answer these questions about the 27 cubes.
a. How many are painted on 4, 3, 2, 1, and 0 faces?
b. What is the total surface area that is unpainted?

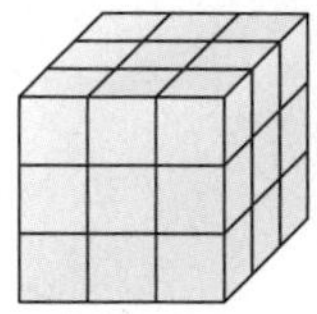

43. Algebra The sum of the height and radius of a cylinder is 9 m. The surface area of the cylinder is 54π m^2. Find the height and the radius.

Standardized Test Prep

GRIDDED RESPONSE

SAT/ACT

44. The height of a cylinder is twice the radius of the base. The surface area of the cylinder is 56π ft^2. What is the diameter of the base to the nearest tenth of a foot?

45. Two sides of a triangle measure 11 ft and 23 ft. What is the smallest possible whole number length, in feet, for the third side?

46. A polyhedron has one hexagonal face and six triangular faces. How many vertices does the polyhedron have?

47. The shortest shadow cast by a tree is 8 m long. The height of the tree is 20 m. To the nearest degree, what is the angle of elevation of the sun when the shortest shadow is cast?

Apply What You've Learned

MATHEMATICAL PRACTICES
MP 1, MP 2

Recall the paper towel roll from page 687, shown again below.

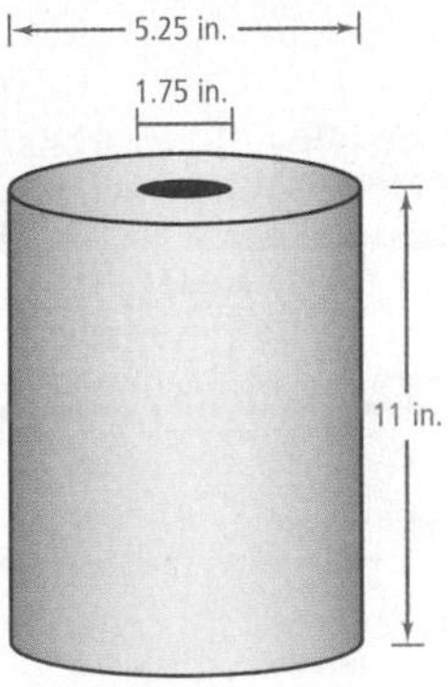

Not to scale

70 SHEETS PER ROLL
EACH 9.0 IN. BY 11.0 IN.

a. What three-dimensional shape is formed by the roll of paper towels?

b. What three-dimensional shape is formed by stretching the entire roll of paper towels flat on the floor? Explain.

c. What area does the stretched-out roll of paper towels cover?

44. 6.1
45. 13
46. 7
47. 68

Apply What You've Learned

Here students describe the shape of the roll of paper towels and find the area the paper towels from page 687 can cover. Later in the chapter, they will use volume formulas.

Mathematical Practices

Students **make sense of the problem** as they visualize unrolling the paper towels. (MP 1)
Students **reason quantitatively** to calculate the area covered by the unrolled paper towels. (MP 2)

ANSWERS

a. cylinder
b. Prism; the unrolled paper towels form a rectangle, with the thickness of an individual paper towel being the height of the prism.
c. 6930 $in.^2$

11-2 Lesson Resources

Differentiated Remediation

Instructional Support

Geometry Companion

Students can use the **Geometry Companion** worktext (4 pages) . . .

- New Vocabulary
- Key Concepts
- Got It for each Problem
- Lesson Check

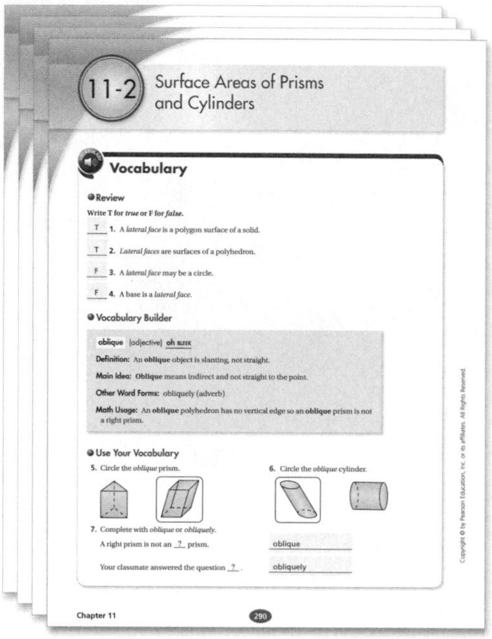
11-2 Surface Areas of Prisms and Cylinders

Vocabulary

Review

Write T for *true* or F for *false*.

T 1. A *lateral face* is a polygon surface of a solid.

T 2. *Lateral faces* are surfaces of a polyhedron.

F 3. A *lateral face* may be a circle.

F 4. A base is a *lateral face*.

Vocabulary Builder

oblique (adjective) oh BLEEK

Definition: An **oblique** object is slanting, not straight.

Main Idea: **Oblique** means indirect and not straight to the point.

Other Word Forms: obliquely (adverb)

Math Usage: An **oblique** polyhedron has no vertical edge so an **oblique** prism is not a right prism.

Use Your Vocabulary

5. Circle the *oblique* prism.

6. Circle the *oblique* cylinder.

7. Complete with *oblique* or *obliquely*.

A right prism is not an _?_ prism. oblique

Your classmate answered the question _?_ . obliquely

Copyright © by Pearson Education, Inc. or its affiliates. All Rights Reserved.

Chapter 11 290

ELL Support

Use Manipulatives Have students work in pairs. Hand out an empty card board tissue box, scissors, and a ruler to each pair. Tell students to measure and record the dimensions of the box. Students will first determine the surface area using a formula, then disassemble the box and find the area using the net formed by the disassembled box. Have student pairs present and discuss their methods and results. Repeat with cardboard oatmeal containers. Students can also make nets instead of taking apart the containers.

5 Assess & Remediate

Lesson Quiz

1. What is the surface area of the shoe box with the dimensions shown below? Use a net.

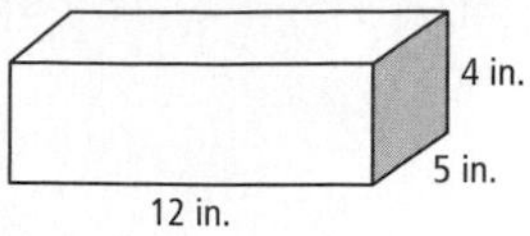

2. What is the surface area of a cube with 8 mm sides?

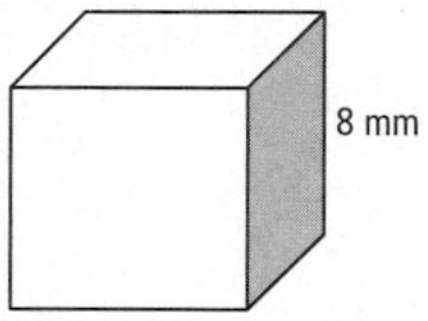

3. Do You UNDERSTAND? The pillars in front of Mr. Jefferson's home are shaped like cylinders with a height of 24 ft and a radius of 8 in. What is the lateral area of each pillar?

ANSWERS TO LESSON QUIZ

1. 256 $in.^2$

2. 384 mm^2

3. about 100.5 ft^2

PRESCRIPTION FOR REMEDIATION

Use the student work on the Lesson Quiz to prescribe a differentiated review assignment.

Points	Differentiated Remediation
0–1	Intervention
2	On-level
3	Extension

PowerGeometry.com

5 Assess & Remediate

Assign the Lesson Quiz. Appropriate intervention, practice, or enrichment is automatically generated based on student performance.

Intervention

- **Reteaching** (2 pages) Provides reteaching and practice exercises for the key lesson concepts. Use with struggling students or absent students.
- **English Language Learner Support** Helps students develop and reinforce mathematical vocabulary and key concepts.

All-in-One Resources/Online

Reteaching

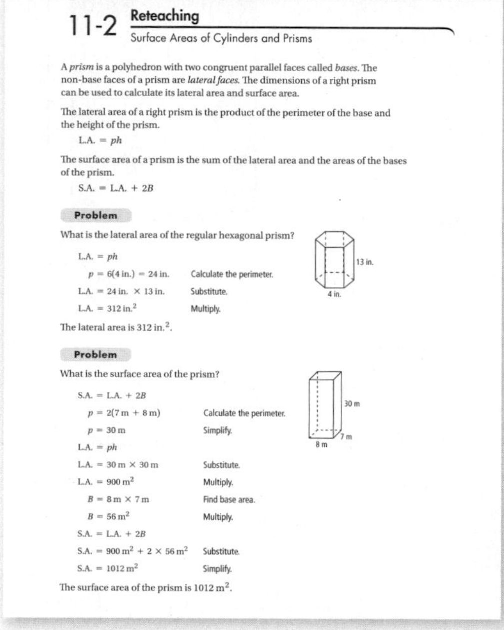
11-2 Reteaching
Surface Areas of Cylinders and Prisms

A *prism* is a polyhedron with two congruent parallel faces called *bases*. The non-base faces of a prism are *lateral faces*. The dimensions of a right prism can be used to calculate its lateral area and surface area.

The lateral area of a right prism is the product of the perimeter of the base and the height of the prism.

L.A. = ph

The surface area of a prism is the sum of the lateral area and the areas of the bases of the prism.

S.A. = L.A. + $2B$

Problem

What is the lateral area of the regular hexagonal prism?

L.A. = ph

$p = 6(4 \text{ in.}) = 24$ in. Calculate the perimeter.

L.A. = 24 in. × 13 in. Substitute.

L.A. = 312 $in.^2$ Multiply.

The lateral area is 312 $in.^2$.

Problem

What is the surface area of the prism?

S.A. = L.A. + $2B$

$p = 2(7\text{ m} + 8\text{ m})$ Calculate the perimeter.

$p = 30$ m Simplify.

L.A. = ph

L.A. = 30 m × 30 m Substitute.

L.A. = 900 m^2 Multiply.

$B = 8\text{ m} \times 7\text{ m}$ Find base area.

$B = 56\text{ m}^2$ Multiply.

S.A. = L.A. + $2B$

S.A. = 900 m^2 + 2 × 56 m^2 Substitute.

S.A. = 1012 m^2 Simplify.

The surface area of the prism is 1012 m^2.

All-in-One Resources/Online

English Language Learner Support

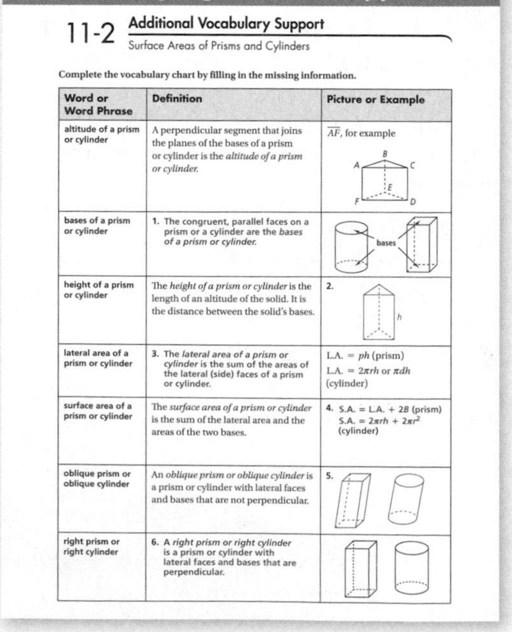
11-2 Additional Vocabulary Support
Surface Areas of Prisms and Cylinders

Complete the vocabulary chart by filling in the missing information.

Word or Word Phrase	Definition	Picture or Example
altitude of a prism or cylinder	A perpendicular segment that joins the planes of the bases of a prism or cylinder is the *altitude of a prism or cylinder.*	$\overline{AF}$, for example (A, B, C, D, E, F)
bases of a prism or cylinder	1. The congruent, parallel faces on a prism or a cylinder are the *bases of a prism or cylinder.*	bases
height of a prism or cylinder	The *height of a prism or cylinder* is the length of an altitude of the solid. It is the distance between the solid's bases.	2. h
lateral area of a prism or cylinder	3. The *lateral area of a prism or cylinder* is the sum of the areas of the lateral (side) faces of a prism or cylinder.	L.A. = ph (prism) L.A. = $2\pi rh$ or πdh (cylinder)
surface area of a prism or cylinder	The *surface area of a prism or cylinder* is the sum of the lateral area and the areas of the two bases.	4. S.A. = L.A. + $2B$ (prism) S.A. = $2\pi rh + 2\pi r^2$ (cylinder)
oblique prism or oblique cylinder	An *oblique prism or oblique cylinder* is a prism or cylinder with lateral faces and bases that are not perpendicular.	5.
right prism or right cylinder	6. A *right prism or right cylinder* is a prism or cylinder with lateral faces and bases that are perpendicular.	

Differentiated Remediation *continued*

On-Level

- **Practice** (2 pages) Provides extra practice for each lesson. For simpler practice exercises, use the Form K Practice pages found in the All-in-One Teaching Resources and online.
- **Think About a Plan** Helps students develop specific problem-solving skills and strategies by providing scaffolded guiding questions.
- **Standardized Test Prep** Focuses on all major exercises, all major question types, and helps students prepare for the high-stakes assessments.

Extension

- **Enrichment** Provides students with interesting problems and activities that extend the concepts of the lesson.
- **Activities, Games, and Puzzles** Worksheets that can be used for concepts development, enrichment, and for fun!

Practice and Problem Solving Wkbk/ All-in-One Resources/Online

Practice page 1

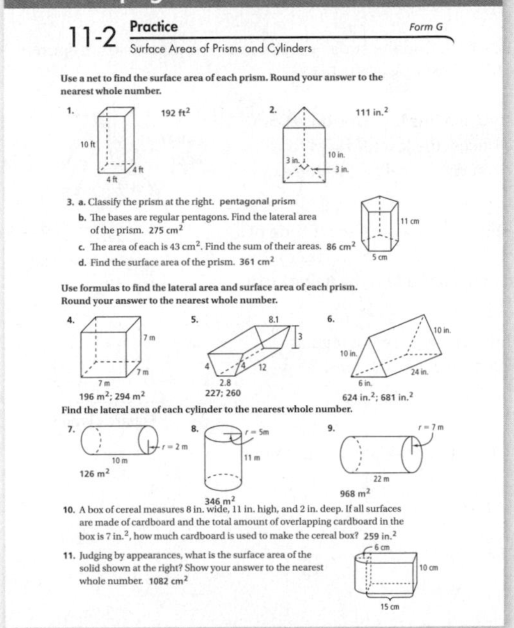
11-2 Practice — Surface Areas of Prisms and Cylinders — Form G

Use a net to find the surface area of each prism. Round your answer to the nearest whole number.

1. 192 ft² 2. 111 in.²

3. a. Classify the prism at the right. pentagonal prism
b. The bases are regular pentagons. Find the lateral area of the prism. 275 cm²
c. The area of each is 43 cm². Find the sum of their areas. 86 cm²
d. Find the surface area of the prism. 361 cm²

Use formulas to find the lateral area and surface area of each prism. Round your answer to the nearest whole number.

4. 196 m²; 294 m² 5. 227; 260 6. 624 in.²; 681 in.²

Find the lateral area of each cylinder to the nearest whole number.

7. 126 m² 8. 346 m² 9. 968 m²

10. A box of cereal measures 8 in. wide, 11 in. high, and 2 in. deep. If all surfaces are made of cardboard and the total amount of overlapping cardboard in the box is 7 in.², how much cardboard is used to make the cereal box? 259 in.²

11. Judging by appearances, what is the surface area of the solid shown at the right? Show your answer to the nearest whole number. 1082 cm²

Practice and Problem Solving Wkbk/ All-in-One Resources/Online

Practice page 2

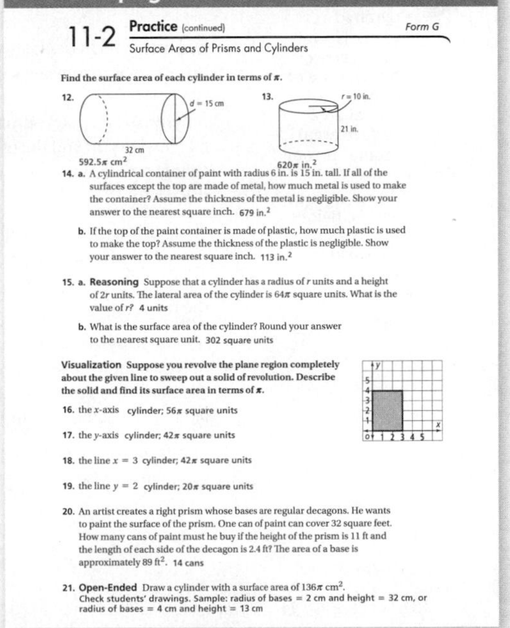
11-2 Practice (continued) — Surface Areas of Prisms and Cylinders — Form G

Find the surface area of each cylinder in terms of π.

12. 592.5π cm² 13. 620π in.²

14. a. A cylindrical container of paint with radius 6 in. is 15 in. tall. If all of the surfaces except the top are made of metal, how much metal is used to make the container? Assume the thickness of the metal is negligible. Show your answer to the nearest square inch. 679 in.²
b. If the top of the paint container is made of plastic, how much plastic is used to make the top? Assume the thickness of the plastic is negligible. Show your answer to the nearest square inch. 113 in.²

15. a. Reasoning Suppose that a cylinder has a radius of r units and a height of 2r units. The lateral area of the cylinder is 64π square units. What is the value of r? 4 units
b. What is the surface area of the cylinder? Round your answer to the nearest square unit. 302 square units

Visualization Suppose you revolve the plane region completely about the given line to sweep out a solid of revolution. Describe the solid and find its surface area in terms of π.

16. the x-axis cylinder; 56π square units
17. the y-axis cylinder; 42π square units
18. the line x = 3 cylinder; 42π square units
19. the line y = 2 cylinder; 20π square units

20. An artist creates a right prism whose bases are regular decagons. He wants to paint the surface of the prism. One can of paint can cover 32 square feet. How many cans of paint must he buy if the height of the prism is 11 ft and the length of each side of the decagon is 2.4 ft? The area of a base is approximately 89 ft². 14 cans

21. Open-Ended Draw a cylinder with a surface area of 136π cm². Check students' drawings. Sample: radius of bases = 2 cm and height = 32 cm, or radius of bases = 4 cm and height = 13 cm

All-in-One Resources/Online

Enrichment

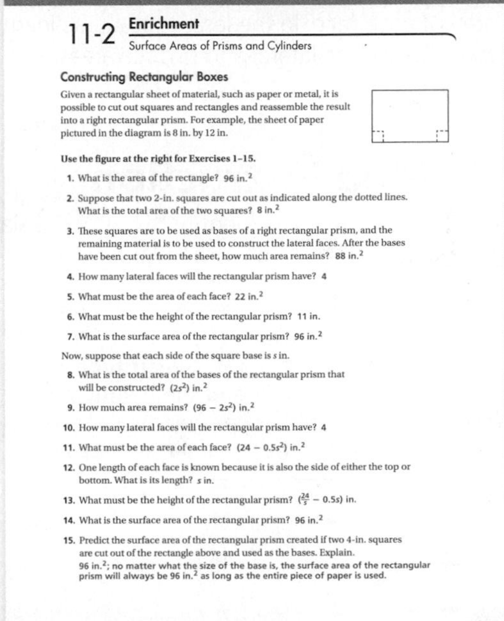
11-2 Enrichment — Surface Areas of Prisms and Cylinders

Constructing Rectangular Boxes

Given a rectangular sheet of material, such as paper or metal, it is possible to cut out squares and rectangles and reassemble the result into a right rectangular prism. For example, the sheet of paper pictured in the diagram is 8 in. by 12 in.

Use the figure at the right for Exercises 1–15.

1. What is the area of the rectangle? 96 in.²
2. Suppose that two 2-in. squares are cut out as indicated along the dotted lines. What is the total area of the two squares? 8 in.²
3. These squares are to be used as bases of a right rectangular prism, and the remaining material is to be used to construct the lateral faces. After the bases have been cut out from the sheet, how much area remains? 88 in.²
4. How many lateral faces will the rectangular prism have? 4
5. What must be the area of each face? 22 in.²
6. What must be the height of the rectangular prism? 11 in.
7. What is the surface area of the rectangular prism? 96 in.²

Now, suppose that each side of the square base is s in.

8. What is the total area of the bases of the rectangular prism that will be constructed? (2s²) in.²
9. How much area remains? (96 − 2s²) in.²
10. How many lateral faces will the rectangular prism have? 4
11. What must be the area of each face? (24 − 0.5s²) in.²
12. One length of each face is known because it is also the side of either the top or bottom. What is its length? s in.
13. What must be the height of the rectangular prism? (24/s − 0.5s) in.
14. What is the surface area of the rectangular prism? 96 in.²
15. Predict the surface area of the rectangular prism created if two 4-in. squares are cut out of the rectangle above and used as the bases. Explain. 96 in.²; no matter what the size of the base is, the surface area of the rectangular prism will always be 96 in.² as long as the entire piece of paper is used.

Practice and Problem Solving Wkbk/ All-in-One Resources/Online

Think About a Plan

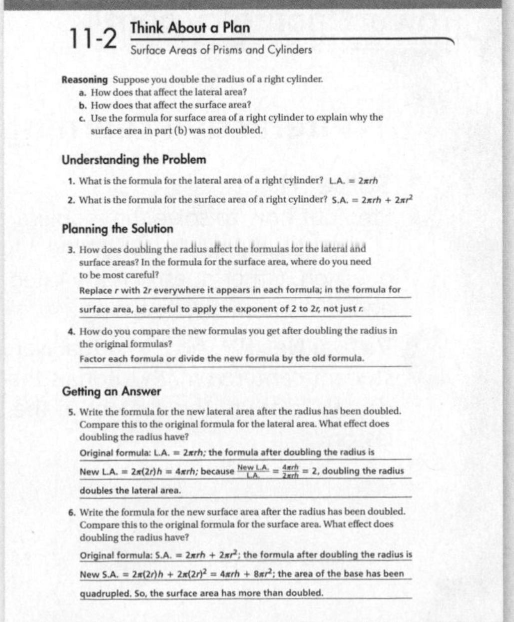
11-2 Think About a Plan — Surface Areas of Prisms and Cylinders

Reasoning Suppose you double the radius of a right cylinder.
a. How does that affect the lateral area?
b. How does that affect the surface area?
c. Use the formula for surface area of a right cylinder to explain why the surface area in part (b) was not doubled.

Understanding the Problem

1. What is the formula for the lateral area of a right cylinder? L.A. = 2πrh
2. What is the formula for the surface area of a right cylinder? S.A. = 2πrh + 2πr²

Planning the Solution

3. How does doubling the radius affect the formulas for the lateral and surface areas? In the formula for the surface area, where do you need to be most careful? Replace r with 2r everywhere it appears in each formula; in the formula for surface area, be careful to apply the exponent of 2 to 2r, not just r.
4. How do you compare the new formulas you get after doubling the radius in the original formulas? Factor each formula or divide the new formula by the old formula.

Getting an Answer

5. Write the formula for the new lateral area after the radius has been doubled. Compare this to the original formula for the lateral area. What effect does doubling the radius have? Original formula: L.A. = 2πrh; the formula after doubling the radius is New L.A. = 2π(2r)h = 4πrh; because New L.A./L.A. = 4πrh/2πrh = 2, doubling the radius doubles the lateral area.
6. Write the formula for the new surface area after the radius has been doubled. Compare this to the original formula for the surface area. What effect does doubling the radius have? Original formula: S.A. = 2πrh + 2πr²; the formula after doubling the radius is New S.A. = 2π(2r)h + 2π(2r)² = 4πrh + 8πr²; the area of the base has been quadrupled. So, the surface area has more than doubled.

Practice and Problem Solving Wkbk/ All-in-One Resources/Online

Standardized Test Prep

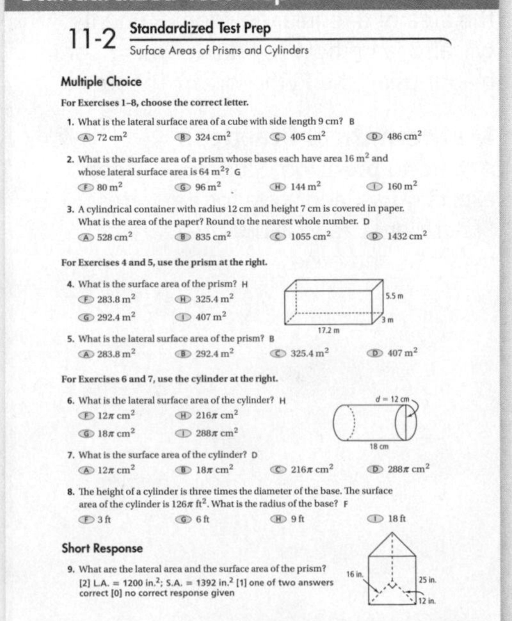
11-2 Standardized Test Prep — Surface Areas of Prisms and Cylinders

Multiple Choice

For Exercises 1–8, choose the correct letter.

1. What is the lateral surface area of a cube with side length 9 cm? B
A 72 cm² B 324 cm² C 405 cm² D 486 cm²
2. What is the surface area of a prism whose bases each have area 16 m² and whose lateral surface area is 64 m²? G
F 80 m² G 96 m² H 144 m² I 160 m²
3. A cylindrical container with radius 12 cm and height 7 cm is covered in paper. What is the area of the paper? Round to the nearest whole number. D
A 528 cm² B 835 cm² C 1055 cm² D 1432 cm²

For Exercises 4 and 5, use the prism at the right.

4. What is the surface area of the prism? H
F 283.8 m² H 325.4 m² G 292.4 m² I 407 m²
5. What is the lateral surface area of the prism? B
A 283.8 m² B 292.4 m² C 325.4 m² D 407 m²

For Exercises 6 and 7, use the cylinder at the right.

6. What is the lateral surface area of the cylinder? H
F 12π cm² H 216π cm² G 18π cm² I 288π cm²
7. What is the surface area of the cylinder? D
A 12π cm² B 18π cm² C 216π cm² D 288π cm²
8. The height of a cylinder is three times the diameter of the base. The surface area of the cylinder is 126π ft². What is the radius of the base? F
F 3 ft G 6 ft H 9 ft I 18 ft

Short Response

9. What are the lateral area and the surface area of the prism? [2] L.A. = 1200 in.²; S.A. = 1392 in.² [1] one of two answers correct [0] no correct response given

Online Teacher Resource Center

Activities, Games, and Puzzles

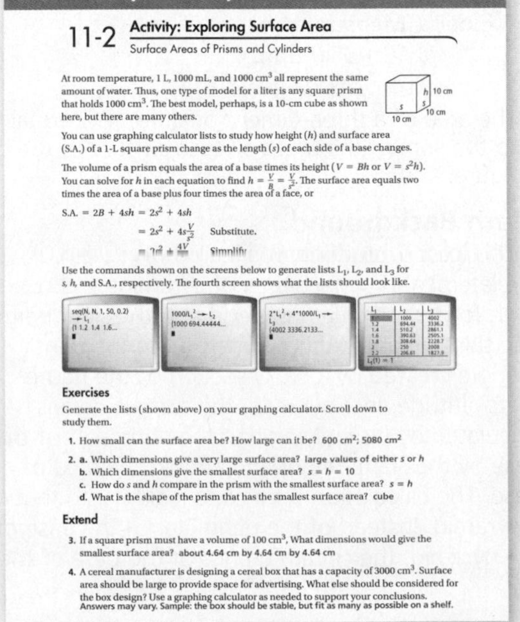
11-2 Activity: Exploring Surface Area — Surface Areas of Prisms and Cylinders

At room temperature, 1 L, 1000 mL, and 1000 cm³ all represent the same amount of water. Thus, one type of model for a liter is any square prism that holds 1000 cm³. The best model, perhaps, is a 10-cm cube as shown here, but there are many others.

You can use graphing calculator lists to study how height (h) and surface area (S.A.) of a 1-L square prism change as the length (s) of each side of a base changes.

The volume of a prism equals the area of a base times its height (V = Bh or V = s²h). You can solve for h in each equation to find h = V/B = V/s². The surface area equals two times the area of a base plus four times the area of a face, or

S.A. = 2B + 4sh = 2s² + 4sh
= 2s² + 4s(V/s²) Substitute.
= 2s² + 4V/s Simplify.

Use the commands shown on the screens below to generate lists L₁, L₂, and L₃ for s, h, and S.A., respectively. The fourth screen shows what the lists should look like.

Exercises

Generate the lists (shown above) on your graphing calculator. Scroll down to study them.

1. How small can the surface area be? How large can it be? 600 cm²; 5080 cm²
2. a. Which dimensions give a very large surface area? large values of either s or h
b. Which dimensions give the smallest surface area? s = h = 10
c. How do s and h compare in the prism with the smallest surface area? s = h
d. What is the shape of the prism that has the smallest surface area? cube

Extend

3. If a square prism must have a volume of 100 cm³, what dimensions would give the smallest surface area? about 4.64 cm by 4.64 cm by 4.64 cm
4. A cereal manufacturer is designing a cereal box that has a capacity of 3000 cm³. Surface area should be large to provide space for advertising. What else should be considered for the box design? Use a graphing calculator as needed to support your conclusions. Answers may vary. Sample: the box should be stable, but fit as many as possible on a shelf.

1 Interactive Learning

Solve It!

PURPOSE To find the lateral area of a pyramid
PROCESS Students may use the given information to determine the height of the pyramid and each triangular face.

FACILITATE

Q How can you find the height of the pyramid? **[Subtract the height of the rectangular prism from the height of the model.]**

Q What right triangle includes the height of one of the triangular faces of the pyramid? **[the triangle formed by the height of the pyramid and a segment from the center of the base to the midpoint of a base edge]**

ANSWER See Solve It in Answers on next page.
CONNECT THE MATH In the Solve It, students calculate the lateral area of a pyramid by finding the area of each face. In the lesson, students learn the formulas for the lateral and surface areas of pyramids and cones.

2 Guided Instruction

Be sure that students can identify the altitude, slant height, base, and vertex of pyramids.

Q Could the segment identified as the altitude in the hexagonal pyramid have also been identified as the height? Explain. **[yes; the altitude in the hexagonal pyramid and the height in the rectangular pyramid are perpendicular segments from the vertex of the pyramid to the base.]**

11-3 Surface Areas of Pyramids and Cones

Common Core State Standards
G-MG.A.1 Use geometric shapes, their measures, and their properties to describe objects.
MP 1, MP 3, MP 4, MP 6, MP 7

Objective To find the surface area of a pyramid and a cone

Think about what dimensions you need and how to get them from what you already know.

MATHEMATICAL PRACTICES

Getting Ready!

You are building a model of a clock tower. You have already constructed the basic structure of the tower at the right. Now you want to paint the roof. How much area does the paint need to cover? Give your answer in square inches. Explain your method.

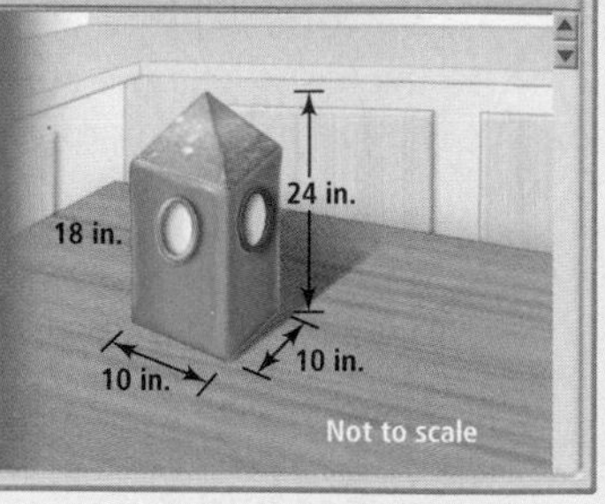

Lesson Vocabulary
- pyramid (base, lateral face, vertex, altitude, height, slant height, lateral area, surface area)
- regular pyramid
- cone (base, altitude, vertex, height, slant height, lateral area, surface area)
- right cone

The Solve It involves the triangular faces of a roof and the three-dimensional figures they form. In this lesson, you will learn to name such figures and to use formulas to find their areas.

Essential Understanding To find the surface area of a three-dimensional figure, find the sum of the areas of all the surfaces of the figure.

A **pyramid** is a polyhedron in which one face (the **base**) can be any polygon and the other faces (the **lateral faces**) are triangles that meet at a common vertex (called the **vertex** of the pyramid).

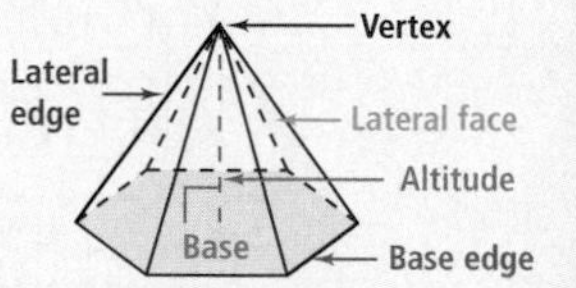

Hexagonal pyramid

You name a pyramid by the shape of its base. The **altitude** of a pyramid is the perpendicular segment from the vertex to the plane of the base. The length of the altitude is the **height** h of the pyramid.

A **regular pyramid** is a pyramid whose base is a regular polygon and whose lateral faces are congruent isosceles triangles. The **slant height** ℓ is the length of the altitude of a lateral face of the pyramid.

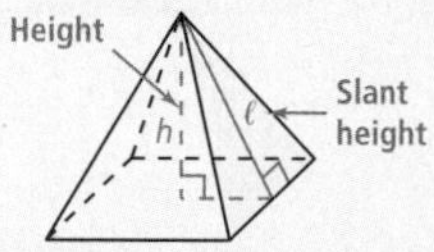

Square pyramid

In this book, you can assume that a pyramid is regular unless stated otherwise.

11-3 Preparing to Teach

BIG ideas **Measurement**
Visualization

ESSENTIAL UNDERSTANDING
- The area of a three-dimensional figure is equal to the sum of the areas of each surface of the figure.

Math Background

In this lesson, students will learn how to calculate the lateral and surface area of pyramids and cones. Both formulas use a new segment, the slant height. The slant height is the hypotenuse of the right triangle created by a cross-section of the figure at its altitude. In a pyramid, the surface area is calculated by multiplying half the perimeter of the base by the slant height and adding the area of the base. The calculation for a cone is similar to that of a pyramid. Instead of the perimeter of the base of the pyramid, the circumference of the base of the cone is used. The formula for the area of a circle is substituted for the formula for the area of a regular polygon. Students will also learn how to find a missing slant height using the Pythagorean Theorem.

Mathematical Practices

Attend to precision. Students will define, explicitly use, and solve for the terms "lateral area" and "surface area" for pyramids and cones.

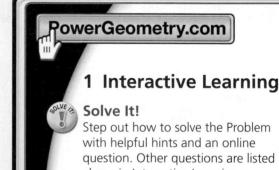

PowerGeometry.com

1 Interactive Learning

Solve It!
Step out how to solve the Problem with helpful hints and an online question. Other questions are listed above in Interactive Learning.

Virtual Nerd™ Access Virtual Nerd student-centered math tutorials that directly relate to the content of the lesson.

The **lateral area** of a pyramid is the sum of the areas of the congruent lateral faces. You can find a formula for the lateral area of a pyramid by looking at its net.

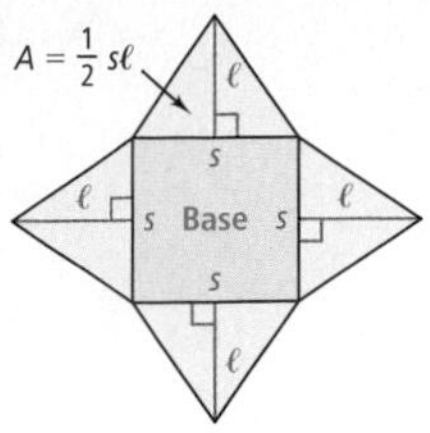

L.A. $= 4\left(\frac{1}{2}s\ell\right)$	The area of each lateral face is $\frac{1}{2}s\ell$.
$= \frac{1}{2}(4s)\ell$	Commutative and Associative Properties of Multiplication
$= \frac{1}{2}p\ell$	The perimeter p of the base is $4s$.

To find the **surface area** of a pyramid, add the area of its base to its lateral area.

take note

Theorem 11-3 Lateral and Surface Areas of a Pyramid

The lateral area of a regular pyramid is half the product of the perimeter p of the base and the slant height ℓ of the pyramid.

$$\text{L.A.} = \frac{1}{2}p\ell$$

The surface area of a regular pyramid is the sum of the lateral area and the area B of the base.

$$\text{S.A.} = \text{L.A.} + B$$

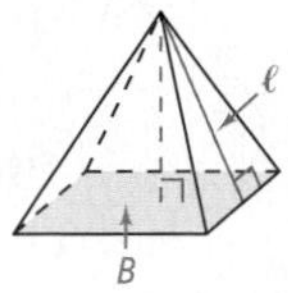

Problem 1 Finding the Surface Area of a Pyramid

Think

What is *B*?
B is the area of the base, which is a hexagon. You are given the apothem of the hexagon and the length of a side. Use them to find the area of the base.

What is the surface area of the hexagonal pyramid?

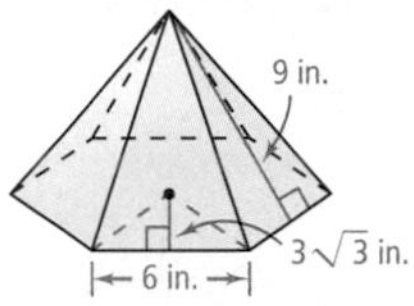

S.A. $=$ L.A. $+ B$	Use the formula for surface area.
$= \frac{1}{2}p\ell + \frac{1}{2}ap$	Substitute the formulas for L.A. and B.
$= \frac{1}{2}(36)(9) + \frac{1}{2}\left(3\sqrt{3}\right)(36)$	Substitute.
≈ 255.5307436	Use a calculator.

The surface area of the pyramid is about 256 in.^{2}.

Got It? 1. a. A square pyramid has base edges of 5 m and a slant height of 3 m. What is the surface area of the pyramid?

b. Reasoning Suppose the slant height of a pyramid is doubled. How does this affect the lateral area of the pyramid? Explain.

When the slant height of a pyramid is not given, you must calculate it before you can find the lateral area or surface area.

Point out to students that s represents the length of a side of the base of the pyramid. The variable ℓ represents the slant height of the pyramid. The slant height is the distance from the base of the pyramid to its vertex along a segment that is perpendicular to an edge of the base. Be certain that students can distinguish between the locations of s, ℓ, and h in diagrams of pyramids.

Take Note

Have students compare their work in the Solve It to the proof of the formula for the lateral area of a pyramid. Ask them to explain how their process was similar to the one shown at the top of the page. Emphasize the difference between lateral area and surface area.

Problem 1

Q What two measurements do you need to find the surface area of the pyramid? **[the lateral area and the area of the base]**

Q How do you find the area of a regular hexagon? **[Multiply half the perimeter by the length of the apothem.]**

Got It? VISUAL LEARNERS

Q In 1a, what is the area of the base of the pyramid? Show your work. **[25 m^2; $A = s^2 = 5^2 = 25$]**

Q In 1a, what is the perimeter of the base of the pyramid? Show your work. **[20 m; $P = 4s = 4 \cdot 5 = 20$]**

For 1b, ask students to make a conjecture. Have students check their answer by calculating the lateral area of the square pyramid in 1a with a slant height of 9 m and a slant height of 18 m.

2 Guided Instruction

Each Problem is worked out and supported online.

Problem 1
Finding the Surface Area of a Pyramid
Animated

Problem 2
Finding the Lateral Area of a Pyramid
Animated

Problem 3
Finding the Surface Area of a Cone

Problem 4
Finding the Lateral Area of a Cone
Animated

Support in Geometry Companion

- Vocabulary
- Key Concepts
- Got It?

Answers

Solve It!

About 156.2 in.^{2}; explanations may vary. Sample: The roof consists of 4 congruent triangles with base 10 in. and height $\sqrt{61}$ in. Find the area of one triangle and multiply by 4.

Got It?

1a. 55 m^2

b. The L.A. will double. Sample explanation: Since L.A. $= \frac{1}{2}p\ell$, then replacing ℓ with 2ℓ gives $\frac{1}{2}p(2\ell) = 2(\frac{1}{2}p\ell) = 2 \cdot$ L.A.

Problem 2

Q What measurement do you need that is not given in the diagram? **[the slant height of the pyramid]**
Q How can you find the slant height of the pyramid? **[Use the right triangle and the Pythagorean Theorem.]**
Q What are the leg lengths of the right triangle that you will use to find the slant height? **[The pyramid height is 36.4 m. The other leg of the triangle is half the side length or 15 m.]**

Got It?

Q What measures do you need to use the formula for the lateral area of the pyramid? **[the perimeter of the base and the slant height of the pyramid]**
Q Which of the given measures do you use to find the perimeter? Explain. **[36 ft; there are 6 sides, each with a length of 36 ft. Multiply 6 by 36.]**
Q What property or theorem justifies the relationship of the slant height and the height of the pyramid? **[Pythagorean Theorem]**
Q Which of the given measures do you use to find the slant height? Explain. **[42 ft and $18\sqrt{3}$ ft; these measures are the lengths of the legs of a right triangle that has the slant height as its hypotenuse.]**

Problem 2 Finding the Lateral Area of a Pyramid

Social Studies **The Pyramid of Cestius is located in Rome, Italy. Using the dimensions in the figure below, what is the lateral area of the Pyramid of Cestius? Round to the nearest square meter.**

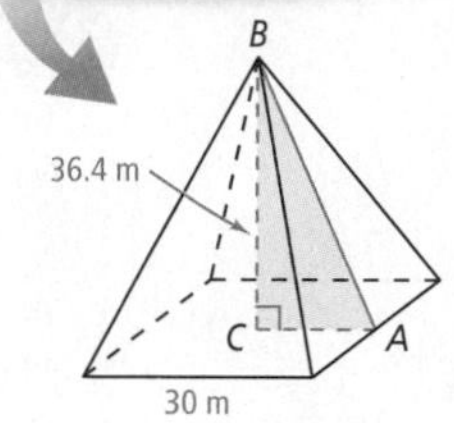

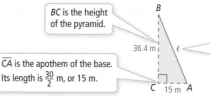

Know
- The height of the pyramid
- The base is a square with a side length of 30 m.
- $\triangle ABC$ is right, where AB is the slant height.

Need
The slant height of the pyramid

Plan
Find the perimeter of the base. Use the Pythagorean Theorem to find the slant height. Then use the formula for lateral area.

Think
Why is a new diagram helpful for finding the slant height?
The new diagram shows the information you need to use the Pythagorean Theorem.

Step 1 Find the perimeter of the base.

$p = 4s$	Use the formula for the perimeter of a square.
$= 4 \cdot 30$	Substitute 30 for s.
$= 120$	Simplify.

Step 2 Find the slant height of the pyramid.

BC is the height of the pyramid.

$\overline{CA}$ is the apothem of the base. Its length is $\frac{30}{2}$ m, or 15 m.

The slant height is the length of the hypotenuse of right $\triangle ABC$, or AB.

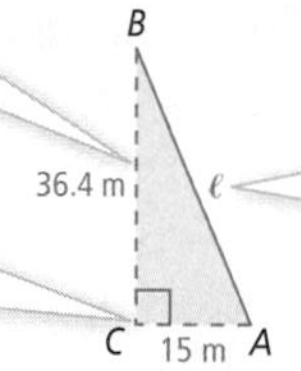

$\ell = \sqrt{CA^2 + BC^2}$	Use the Pythagorean Theorem.
$= \sqrt{15^2 + 36.4^2}$	Substitute 15 for CA and 36.4 for BC.
$= \sqrt{1549.96}$	Simplify.

Step 3 Find the lateral area of the pyramid.

$\text{L.A.} = \frac{1}{2}p\ell$	Use the formula for lateral area.
$= \frac{1}{2}(120)\sqrt{1549.96}$	Substitute 120 for p and $\sqrt{1549.96}$ for ℓ.
≈ 2362.171882	Use a calculator.

The lateral area of the Pyramid of Cestius is about 2362 m^2.

Got It? 2. a. What is the lateral area of the hexagonal pyramid at the right? Round to the nearest square foot.
b. Reasoning How does the slant height of a regular pyramid relate to its height? Explain.

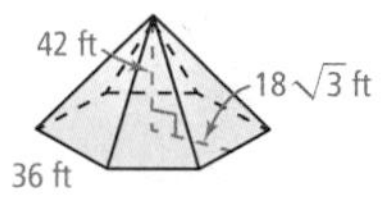

Additional Problems

1. What is the surface area of the square pyramid with base edges of 8 cm and a slant height of 12 cm?

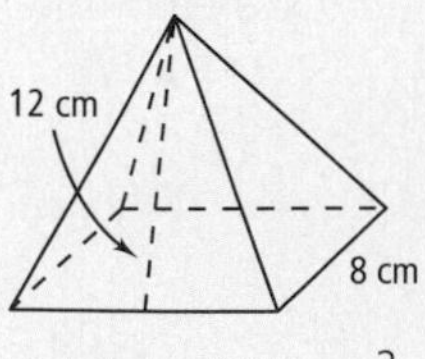

ANSWER 256 cm^2

2. What is the lateral area of a pyramid with a height of 8 in. and a square base that measures 9 in. on each side? Round to the nearest tenth.

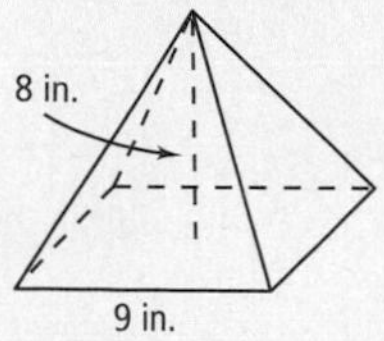

ANSWER about 165.2 in.^2

3. What is the lateral area of the cone in terms of π?

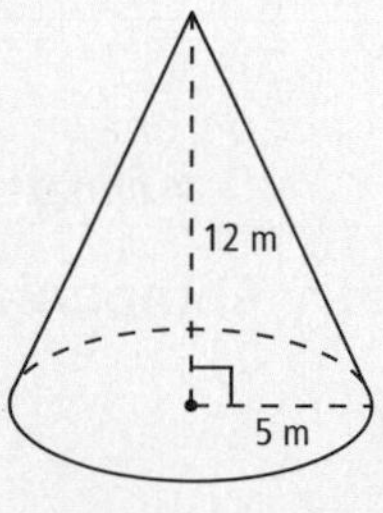

ANSWER 65π m^2

4. What is the lateral area of the ice cream cone shown below? Round to the nearest square centimeter.

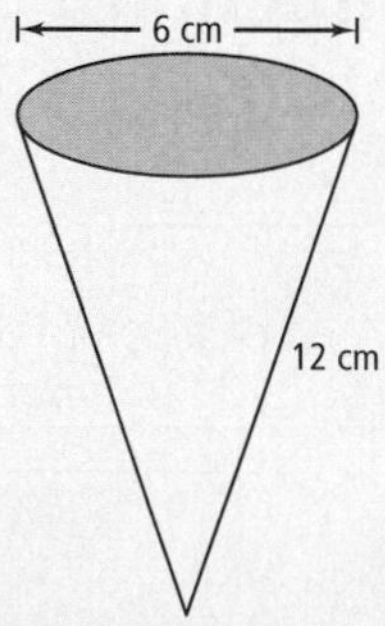

ANSWER about 113 cm^2

Like a pyramid, a **cone** is a solid that has one base and a vertex that is not in the same plane as the base. However, the **base** of a cone is a circle. In a **right cone,** the **altitude** is a perpendicular segment from the **vertex** to the center of the base. The **height** h is the length of the altitude. The **slant height** ℓ is the distance from the vertex to a point on the edge of the base. In this book, you can assume that a cone is a right cone unless stated or pictured otherwise.

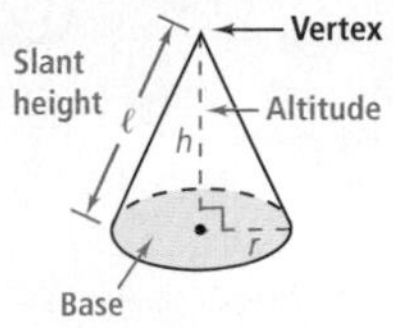

The **lateral area** is half the circumference of the base times the slant height. The formulas for the lateral area and **surface area** of a cone are similar to those for a pyramid.

take note

Theorem 11-4 Lateral and Surface Areas of a Cone

The lateral area of a right cone is half the product of the circumference of the base and the slant height of the cone.

$$\text{L.A.} = \frac{1}{2} \cdot 2\pi r \cdot \ell, \text{ or L.A.} = \pi r \ell$$

The surface area of a cone is the sum of the lateral area and the area of the base.

$$\text{S.A.} = \text{L.A.} + B$$

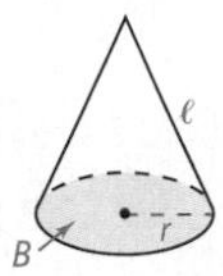

Problem 3 Finding the Surface Area of a Cone

Think

How is this different from finding the surface area of a pyramid?
For a pyramid, you need to find the perimeter of the base. For a cone, you need to find the circumference.

What is the surface area of the cone in terms of π?

$\text{S.A.} = \text{L.A.} + B$	Use the formula for surface area.
$= \pi r\ell + \pi r^2$	Substitute the formulas for L.A. and B.
$= \pi(15)(25) + \pi(15)^2$	Substitute 15 for r and 25 for ℓ.
$= 375\pi + 225\pi$	Simplify.
$= 600\pi$	

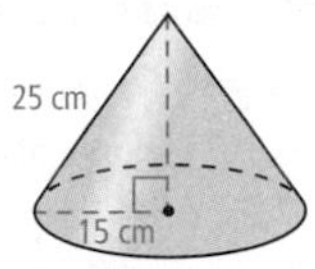

The surface area of the cone is 600π cm^2.

Got It? **3.** The radius of the base of a cone is 16 m. Its slant height is 28 m. What is the surface area in terms of π?

By cutting a cone and laying it out flat, you can see how the formula for the lateral area of a cone ($\text{L.A.} = \frac{1}{2} \cdot C_{\text{base}} \cdot \ell$) resembles that for the area of a triangle ($A = \frac{1}{2}bh$).

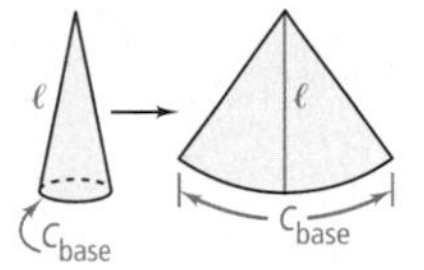

Take Note

Have students compare and contrast a pyramid and a cone. The main difference is the base of the figures. Therefore, the area and perimeter of the base will be calculated differently for each figure. Otherwise, the formulas are the same.

Problem 3

Q How do you find the circumference of a circle? **[Find the circumference of the circle using the formula $C = 2\pi r$.]**

Q What is the formula for the area of the base of a cone? **[$A = \pi r^2$]**

Q Do you need to find the height of the cone? Explain. **[no; the formulas have variables for the radius of the base and the slant height of the cone.]**

Got It?

Have students sketch a diagram of a cone with the given measurements.

Q What two numbers do you add in the final step to finding the surface area of this cone? **[448π and 256π]**

Answers

Got it! (continued)

2a. 5649 ft^2

b. The slant height is the hypotenuse of a rt. △ with a leg of length equal to the height of the pyramid, so the slant height is greater than the height.

3. 704π m^2

Problem 4

Q What measurement do you need to find the lateral area of the cone? **[the slant height]**

Q How can you find the slant height of the cone? **[Use the Pythagorean Theorem to find the hypotenuse of the right triangle formed by the height and radius of the cone.]**

Q What is the radius of the cone? Explain. **[40 mm; half the length of the diameter]**

Got It?

Have students draw a diagram of the cone and label the known measurements. Have them draw the right triangle created by the height, radius, and slant height of the cone.

3 Lesson Check

Do you know HOW?

- If students have difficulty with Exercise 2, then have them review Problem 1 to know how to find the values of the variables to substitute into the formulas.

Do you UNDERSTAND?

- If students have difficulty with Exercise 8, then have them draw a diagram like the one shown with Theorem 11-4.

Close

Q How do you find the surface area of a pyramid? **[Multiply half the perimeter by the slant height and add the area of the base.]**

Q What is the formula for the surface area of a cone? What do the variables represent? **[S.A. = L.A. + *B*, where L.A. is the lateral area and *B* is the area of the circular base.]**

Think

What is the problem asking you to find?
The problem is asking you to find the area that the filter paper covers. This is the lateral area of a cone.

Problem 4 Finding the Lateral Area of a Cone STEM

Chemistry **In a chemistry lab experiment, you use the conical filter funnel shown at the right. How much filter paper do you need to line the funnel?**

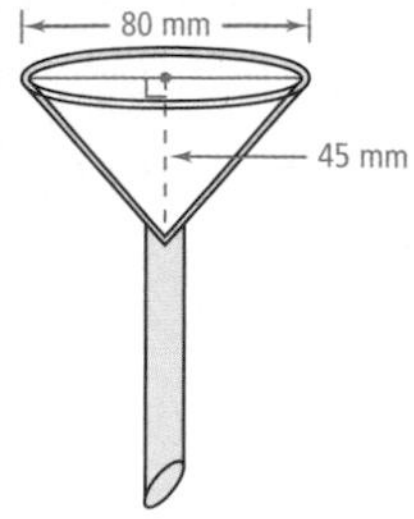

The top part of the funnel has the shape of a cone with a diameter of 80 mm and a height of 45 mm.

$\text{L.A.} = \pi r \ell$	Use the formula for lateral area of a cone.
$= \pi r\left(\sqrt{r^2 + h^2}\right)$	To find the slant height, use the Pythagorean Theorem.
$= \pi(40)\left(\sqrt{40^2 + 45^2}\right)$	Substitute $\frac{1}{2} \cdot 80$, or 40, for *r* and 45 for *h*.
≈ 7565.957013	Use a calculator.

You need about 7566 mm^2 of filter paper to line the funnel.

Got It? **4. a.** What is the lateral area of a traffic cone with radius 10 in. and height 28 in.? Round to the nearest whole number.

b. Reasoning Suppose the radius of a cone is halved, but the slant height remains the same. How does this affect the lateral area of the cone? Explain.

Lesson Check

Do you know HOW?

Use the diagram of the square pyramid at the right.

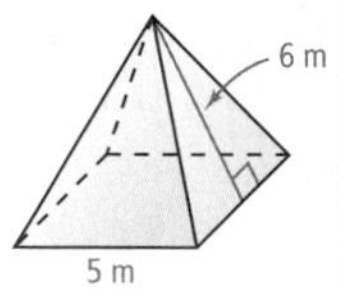

1. What is the lateral area of the pyramid?
2. What is the surface area of the pyramid?

Use the diagram of the cone at the right.

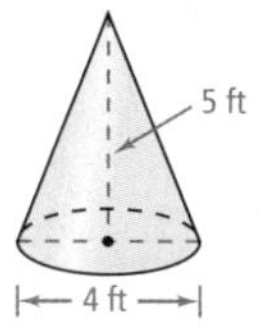

3. What is the lateral area of the cone?
4. What is the surface area of the cone?

Do you UNDERSTAND? MATHEMATICAL PRACTICES

5. **Vocabulary** How do the height and the slant height of a pyramid differ?
6. **Compare and Contrast** How are the formulas for the surface area of a prism and the surface area of a pyramid alike? How are they different?
7. **Vocabulary** How many lateral faces does a pyramid have if its base is pentagonal? Hexagonal? *n*-sided?
8. **Error Analysis** A cone has height 7 and radius 3. Your classmate calculates its lateral area. What is your classmate's error? Explain.

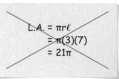

$\text{L.A.} = \pi r \ell$
$= \pi(3)(7)$
$= 21\pi$

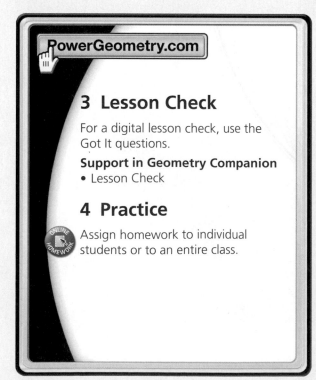

PowerGeometry.com

3 Lesson Check

For a digital lesson check, use the Got It questions.

Support in Geometry Companion

- Lesson Check

4 Practice

ONLINE HOMEWORK

Assign homework to individual students or to an entire class.

Answers

Got It! (continued)

4a. 934 in.^2

b. The L.A. will be halved. Sample explanation: Since $\text{L.A.} = \pi r \ell$, then replacing r with $\frac{r}{2}$ gives $\pi\left(\frac{r}{2}\right)\ell = \frac{1}{2}(\pi r \ell) = \frac{1}{2} \cdot \text{L.A.}$

Lesson Check

1. 60 m^2
2. 85 m^2
3. $2\pi\sqrt{29}$ ft^2, or about 33.8 ft^2
4. $(2\pi\sqrt{29} + 4\pi)$ ft^2, or about 46.4 ft^2
5. The height is the distance from the vertex to the center of the base, while the slant height is the distance from the vertex to the midpoint of an edge of the base.
6. Alike: Both are the sum of a lateral area and the areas of the bases. Different: For a prism the area includes two bases, while for a pyramid the surface area includes just one base.
7. 5; 6; *n*
8. The height 7 is not the slant height. The slant height is $\sqrt{7^2 + 3^2} = \sqrt{58}$, so $\text{L.A.} = \pi r \ell = \pi(3)(\sqrt{58}) = 3\pi\sqrt{58}$ units^2.

Practice and Problem-Solving Exercises

MATHEMATICAL PRACTICES

A Practice

Find the surface area of each pyramid to the nearest whole number.

See Problem 1.

9.

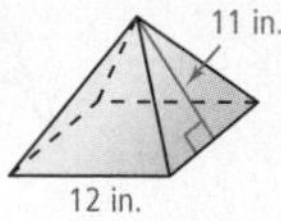

10.

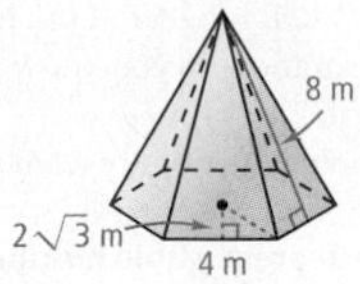

11.

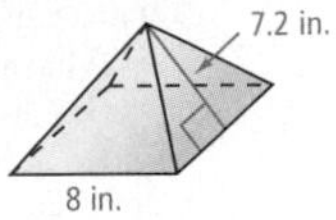

Find the lateral area of each pyramid to the nearest whole number.

See Problem 2.

12.

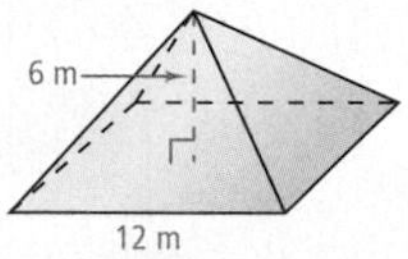

13.

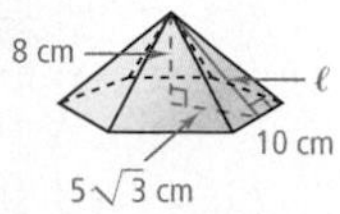

14.

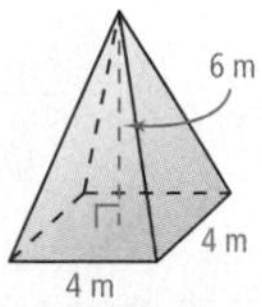

15. Social Studies The original height of the Pyramid of Khafre, located next to the Great Pyramid in Egypt, was about 471 ft. Each side of its square base was about 708 ft. What is the lateral area, to the nearest square foot, of a pyramid with those dimensions?

Find the lateral area of each cone to the nearest whole number.

See Problems 3 and 4.

16.

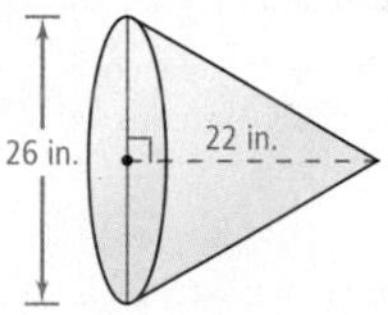

17.

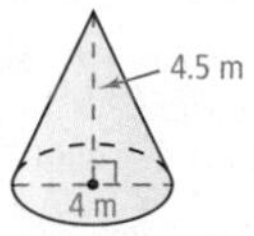

18.

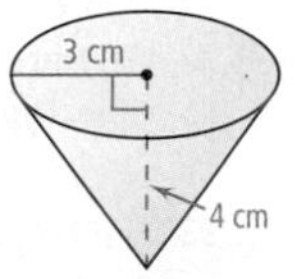

Find the surface area of each cone in terms of π.

19.

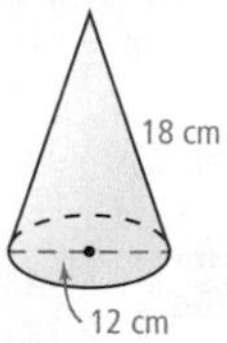

20.

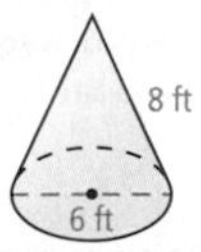

21.

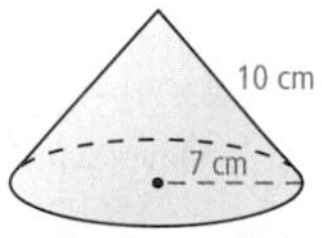

B Apply

22. Reasoning Suppose you could climb to the top of the Great Pyramid. Which route would be shorter, a route along a lateral edge or a route along the slant height of a side? Which of these routes is steeper? Explain your answers.

23. The lateral area of a cone is 4.8π in.^{2}. The radius is 1.2 in. Find the slant height.

4 Practice

ASSIGNMENT GUIDE

Basic: 9–21 all, 22, 25, 26–36 even
Average: 9–21 odd, 22–38
Advanced: 9–21 odd, 22–43
Standardized Test Prep: 44–48
Mixed Review: 49–53

Mathematical Practices are supported by exercises with red headings. Here are the Practices supported in this lesson:

MP 1: Make Sense of Problems Ex. 6, 25
MP 3: Construct Arguments Ex. 22, 24, 31
MP 3: Communicate Ex. 32
MP 3: Critique the Reasoning of Others Ex. 8
MP 6: Attend to Precision Ex. 29
MP 7: Shift Perspective Ex. 37–40

Applications exercises have blue headings. Exercises 15 and 30 support MP 4: Model.

STEM exercises focus on science or engineering applications.

EXERCISE 17: Use the Think About a Plan worksheet in the **Practice and Problem Solving Workbook** (also available in the Teaching Resources in print and online) to further support students' development in becoming independent learners.

HOMEWORK QUICK CHECK

To check students' understanding of key skills and concepts, go over Exercises 13, 17, 19, 22, and 25.

Practice and Problem-Solving Excercises

9. 408 in.^{2}
10. 138 m^2
11. 179 in.^{2}
12. 204 m^2
13. 354 cm^2
14. 51 m^2
15. 834,308 ft^2
16. 1044 in.^{2}
17. 31 m^2
18. 47 cm^2
19. 144π cm^2
20. 33π ft^2
21. 119π cm^2
22. Slant height; slant height; the slant height is shorter because it is one leg of a rt. △ with the lateral edge as the hypotenuse, and it is steeper because it rises the same vertical distance for a shorter horizontal distance.
23. 4 in.

Answers

Practice and Problem-Solving Excercises (continued)

24. $\overline{PT}$ is a leg in each of rt. △ *PTA*, *PTB*, *PTC*, and *PTD*. Since $\overline{PA}$, $\overline{PB}$, $\overline{PC}$, and $\overline{PD}$ are each the hypotenuse in those rt. △, $\overline{PT}$ must be shorter than $\overline{PA}$, $\overline{PB}$, $\overline{PC}$, and $\overline{PD}$.

25. 8 ft

26. 58 m^2

27. 471 ft^2

28. 41 m^2

29. Answers may vary. Sample:

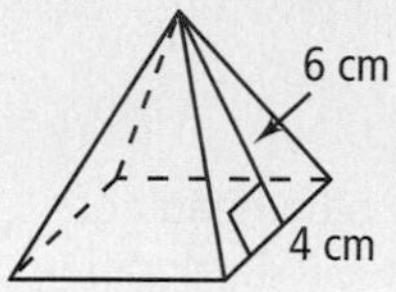

64 cm^2

30. 1580.6 ft^2

31. Cylinder; the L.A. of 2 cones is 30π in.^{2}, and the L.A. of the cylinder is 48π in.^{2}.

32. S.A. = L.A. + Base $= \pi r\ell + \pi r^2 = \pi r(\ell + r)$, or $(\ell + r)\pi r$. The formula S.A. $= \pi r\ell + \pi r^2$ involves 3 multiplications and one addition, while the formula S.A. $= (\ell + r)\pi r$ involves one addition and two multiplications.

33a. $\ell = \frac{\text{S.A.}}{\pi r} - r$

b. $r = \frac{-\pi\ell + \sqrt{\pi^2\ell^2 + 4\pi \cdot \text{S.A.}}}{2\pi}$

34. L.A. = 30 in.^{2}; $h \approx 4.8$ in.; $\ell = 5$ in.

35. $s = 12$ m, L.A. = 240 m^2; S.A. = 384 m^2

36. $s = 8$ cm; $\ell \approx 7.4$ cm; $h \approx 6.2$ cm

37. cone with $r = 4$ and $h = 3$; 36π units^2

38. cone with $r = 3$ and $h = 4$; 24π units^2

39. cylinder with cone-shaped hole; 60π units^2

40. cylinder with cone-shaped hole; 48π units^2

24. Writing Explain why the altitude $\overline{PT}$ in the pyramid at the right must be shorter than all of the lateral edges $\overline{PA}$, $\overline{PB}$, $\overline{PC}$, and $\overline{PD}$.

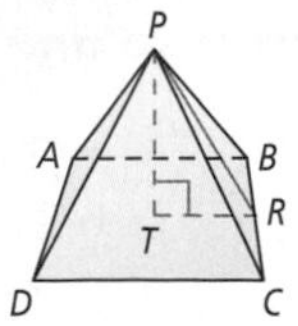

25. Think About a Plan The lateral area of a pyramid with a square base is 240 ft^2. Its base edges are 12 ft long. Find the height of the pyramid.
- What additional information do you know about the pyramid based on the given information?
- How can a diagram help you identify what you need to find?

Find the surface area to the nearest whole number.

26.

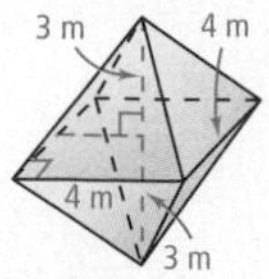

27.

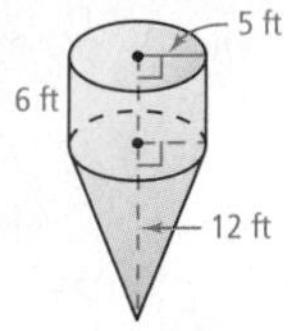

28.

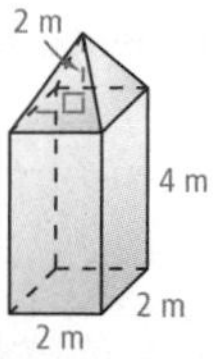

29. Open-Ended Draw a square pyramid with a lateral area of 48 cm^2. Label its dimensions. Then find its surface area.

STEM **30. Architecture** The roof of a tower in a castle is shaped like a cone. The height of the roof is 30 ft and the radius of the base is 15 ft. What is the lateral area of the roof? Round your answer to the nearest tenth.

31. Reasoning The figure at the right shows two glass cones inside a cylinder. Which has a greater surface area, the two cones or the cylinder? Explain.

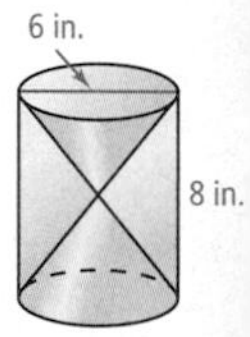

32. Writing You can use the formula S.A. $= (\ell + r)r\pi$ to find the surface area of a cone. Explain why this formula works. Also, explain why you may prefer to use this formula when finding surface area with a calculator.

33. Find a formula for each of the following.
a. the slant height of a cone in terms of the surface area and radius
b. the radius of a cone in terms of the surface area and slant height

The length of a side (s) of the base, slant height (ℓ), height (h), lateral area (L.A.), and surface area (S.A.) are measurements of a square pyramid. Given two of the measurements, find the other three to the nearest tenth.

34. $s = 3$ in., S.A. = 39 in.^{2} **35.** $h = 8$ m, $\ell = 10$ m **36.** L.A. = 118 cm^2, S.A. = 182 cm^2

Visualization **Suppose you revolve the plane region completely about the given line to sweep out a solid of revolution. Describe the solid. Then find its surface area in terms of π.**

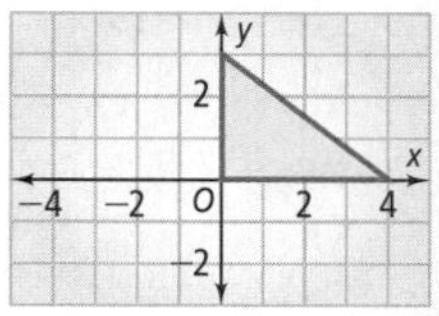

37. the y-axis **38.** the x-axis

Challenge **39.** the line $x = 4$ **40.** the line $y = 3$

41. A sector has been cut out of the disk at the right. The radii of the part that remains are taped together, without overlapping, to form the cone. The cone has a lateral area of 64π cm^2. Find the measure of the central angle of the cutout sector.

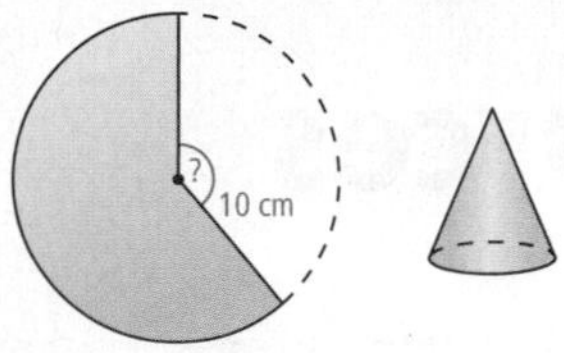

Each given figure fits inside a 10-cm cube. The figure's base is in one face of the cube and is as large as possible. The figure's vertex is in the opposite face of the cube. Draw a sketch and find the lateral and surface areas of the figure.

42. a square pyramid

43. a cone

Standardized Test Prep

SAT/ACT

44. To the nearest whole number, what is the surface area of a cone with diameter 27 m and slant height 19 m?

(A) 1378 m^2 (B) 1951 m^2 (C) 2757 m^2 (D) 3902 m^2

45. What is the hypotenuse of a right isosceles triangle with leg $2\sqrt{6}$?

(F) $4\sqrt{6}$ (G) $2\sqrt{3}$ (H) $4\sqrt{3}$ (I) $4\sqrt{2}$

46. Two angles in a triangle have measures 54 and 61. What is the measure of the smallest exterior angle?

(A) 119 (B) 115 (C) 112 (D) 126

Short Response

47. A diagonal divides a parallelogram into two isosceles triangles. Can the parallelogram be a rhombus? Explain.

Extended Response

48. *ABCD* has vertices at *A*(3, 4), *B*(7, 5), *C*(6, 1), and *D*(−2, −4). A dilation with center (0, 0) maps *A* to $A'\left(\frac{15}{2}, 10\right)$. What are the coordinates of *B*′, *C*′, and *D*′? Show your work.

Mixed Review

49. How much cardboard do you need to make a closed box 4 ft by 5 ft by 2 ft? See Lesson 11-2.

50. How much posterboard do you need to make a cylinder, open at each end, with height 9 in. and diameter $4\frac{1}{2}$ in.? Round your answer to the nearest square inch.

51. A kite with area 195 in.^{2} has a 15-in. diagonal. How long is the other diagonal? See Lesson 10-2.

Get Ready! **To prepare for Lesson 11-4, do Exercises 52 and 53.**

Find the area of each figure. If necessary, round to the nearest tenth. See Lessons 1-8 and 10-7.

52. a square with side length 2 cm

53. a circle with diameter 15 in.

41. 129.6

42. Check students' sketches;
L.A. = $100\sqrt{5}$ cm^2;
S.A. = $(100\sqrt{5} + 100)$ cm^2

43. Check students' sketches;
L.A. = $25\pi\sqrt{5}$ cm^2;
S.A. = $(25\pi\sqrt{5} + 25\pi)$ cm^2

44. A

45. H

46. B

47. [2] Yes; if the legs of each isosc. △ are two consecutive sides of the ▱, then the ▱ is a rhombus.
[1] incomplete

48. [4] A dilation with center (0, 0) takes *P*(*x*, *y*) to *P*′(*nx*, *ny*). If *A*(3, 4) → $A'\left(\frac{15}{2}, 10\right)$, then $n \cdot 4 = 10$, so $n = 2.5$. The other points will be *B*′(17.5, 12.5), *C*′(15, 2.5), and *D*′(−5, −10).
[3] correct explanation with one computational error
[2] correct answer with no explanation
[1] one or more computational errors or no explanation

49. 76 ft^2

50. 127 in.^{2}

51. 26 in.

52. 4 cm^2

53. 176.7 in.^{2}

11-3 Lesson Resources

Differentiated Remediation

Instructional Support

Geometry Companion

Students can use the **Geometry Companion** worktext (4 pages) . . .

- New Vocabulary
- Key Concepts
- Got It for each Problem
- Lesson Check

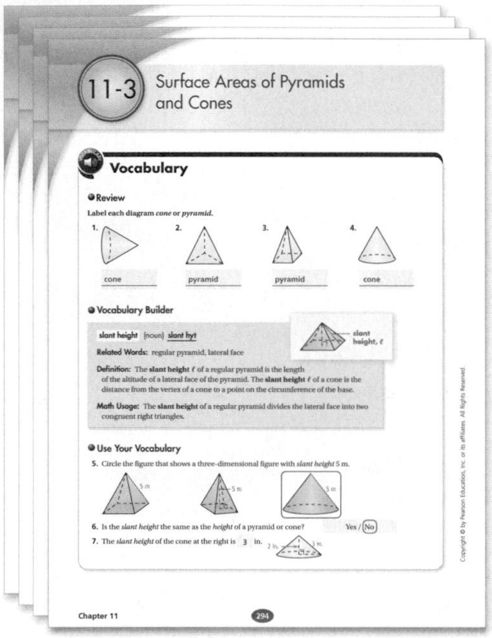

11-3 Surface Areas of Pyramids and Cones

Vocabulary

Review

Label each diagram *cone* or *pyramid*.

1. cone
2. pyramid
3. pyramid
4. cone

Vocabulary Builder

slant height (noun) **slant hyt**

Related Words: regular pyramid, lateral face

Definition: The **slant height** ℓ of a regular pyramid is the length of the altitude of a lateral face of the pyramid. The **slant height** ℓ of a cone is the distance from the vertex of a cone to a point on the circumference of the base.

Math Usage: The **slant height** of a regular pyramid divides the lateral face into two congruent right triangles.

Use Your Vocabulary

5. Circle the figure that shows a three-dimensional figure with *slant height* 5 m.
6. Is the *slant height* the same as the *height* of a pyramid or cone? Yes / No
7. The *slant height* of the cone at the right is 3 in.

Copyright © by Pearson Education, Inc. or its affiliates. All Rights Reserved.

Chapter 11 294

ELL Support

Focus on Language Project a transparency of the lesson on the board using an overhead projector. Read the lesson as students read along, pointing to the words as you read. Highlight or underline key words, such as pyramid, polygon, and vertex. Ask: What is a regular pyramid? Ask students to restate the meaning of each section in their own words, by asking, what is it saying here?

Assess Understanding Write the objective on the board and read it aloud as you point to each word. Ask students: What is surface area as you hold up a pyramid or cone? Have a student demonstrate visually using their hands and a solid figure. At the end of the lesson, ask students if the objective was met.

5 Assess & Remediate

Lesson Quiz

1. What is the surface area of a square pyramid with a height of 12 mm and a base that measures 10 mm on each side? Round to the nearest tenth if necessary.

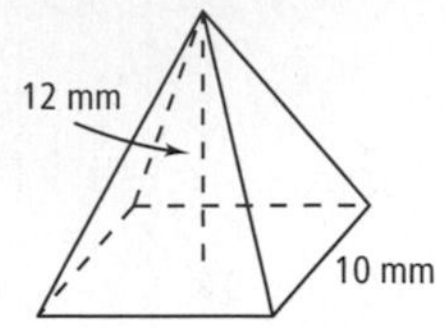

2. **Do you UNDERSTAND?** How much paper is needed to make the drinking cup below? Round to the nearest square inch.

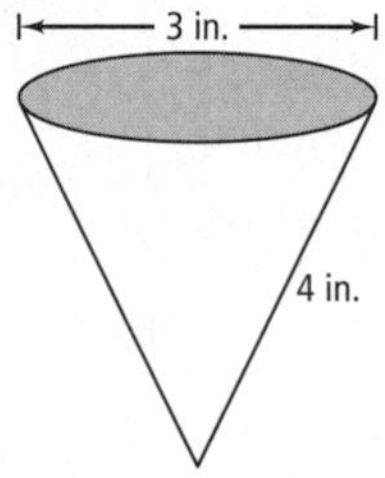

3. To the nearest square inch, what is the surface area of the cone in Question 2?

ANSWERS TO LESSON QUIZ

1. 360 mm^2
2. about 19 $in.^2$
3. 26 $in.^2$

PRESCRIPTION FOR REMEDIATION

Use the student work on the Lesson Quiz to prescribe a differentiated review assignment.

Points	Differentiated Remediation
0–1	Intervention
2	On-level
3	Extension

PowerGeometry.com

5 Assess & Remediate

Assign the Lesson Quiz. Appropriate intervention, practice, or enrichment is automatically generated based on student performance.

Intervention

- **Reteaching** (2 pages) Provides reteaching and practice exercises for the key lesson concepts. Use with struggling students or absent students.
- **English Language Learner Support** Helps students develop and reinforce mathematical vocabulary and key concepts.

All-in-One Resources/Online

Reteaching

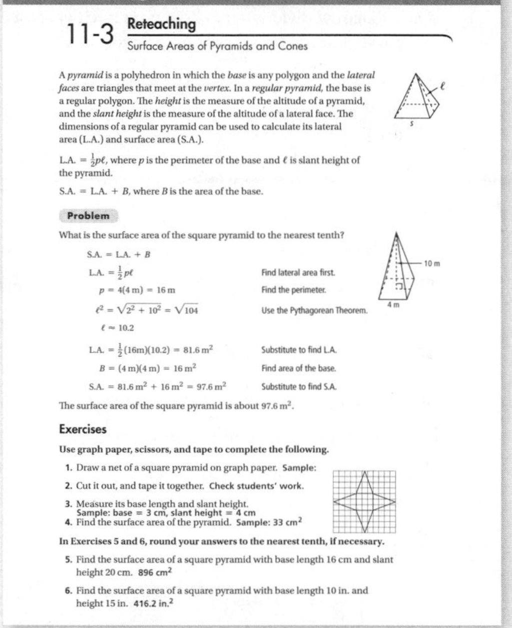

11-3 Reteaching

Surface Areas of Pyramids and Cones

A *pyramid* is a polyhedron in which the *base* is any polygon and the *lateral faces* are triangles that meet at the *vertex*. In a *regular pyramid*, the base is a regular polygon. The *height* is the measure of the altitude of a pyramid, and the *slant height* is the measure of the altitude of a lateral face. The dimensions of a regular pyramid can be used to calculate its lateral area (L.A.) and surface area (S.A.).

L.A. $= \frac{1}{2}p\ell$, where p is the perimeter of the base and ℓ is slant height of the pyramid.

S.A. = L.A. + B, where B is the area of the base.

Problem

What is the surface area of the square pyramid to the nearest tenth?

S.A. = L.A. + B	
L.A. $= \frac{1}{2}p\ell$	Find lateral area first.
$p = 4(4\text{ m}) = 16\text{ m}$	Find the perimeter.
$\ell^2 = \sqrt{2^2 + 10^2} = \sqrt{104}$	Use the Pythagorean Theorem.
$\ell \approx 10.2$	
L.A. $= \frac{1}{2}(16\text{m})(10.2) = 81.6\text{ m}^2$	Substitute to find L.A.
$B = (4\text{ m})(4\text{ m}) = 16\text{ m}^2$	Find area of the base.
S.A. $= 81.6\text{ m}^2 + 16\text{ m}^2 = 97.6\text{ m}^2$	Substitute to find S.A.

The surface area of the square pyramid is about 97.6 m^2.

Exercises

Use graph paper, scissors, and tape to complete the following.

1. Draw a net of a square pyramid on graph paper. **Sample:**
2. Cut it out, and tape it together. **Check students' work.**
3. Measure its base length and slant height. **Sample: base = 3 cm, slant height = 4 cm**
4. Find the surface area of the pyramid. **Sample: 33 cm^2**

In Exercises 5 and 6, round your answers to the nearest tenth, if necessary.

5. Find the surface area of a square pyramid with base length 16 cm and slant height 20 cm. **896 cm^2**
6. Find the surface area of a square pyramid with base length 10 in. and height 15 in. **416.2 $in.^2$**

All-in-One Resources/Online

English Language Learner Support

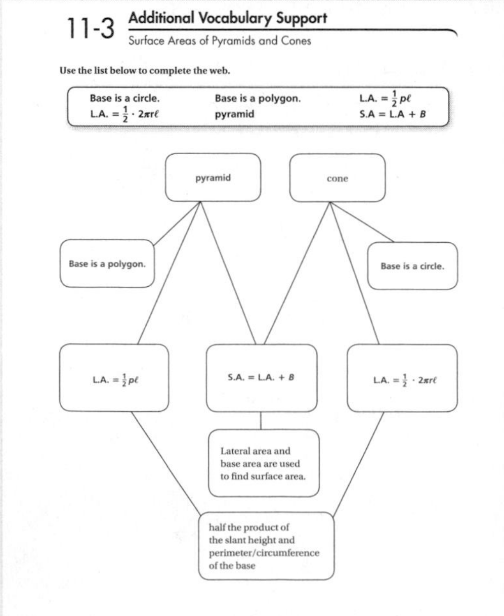

11-3 Additional Vocabulary Support

Surface Areas of Pyramids and Cones

Use the list below to complete the web.

Base is a circle.	Base is a polygon.	L.A. $= \frac{1}{2}p\ell$
L.A. $= \frac{1}{2} \cdot 2\pi r\ell$	pyramid	S.A = L.A + B

Differentiated Remediation *continued*

On-Level

- **Practice** (2 pages) Provides extra practice for each lesson. For simpler practice exercises, use the Form K Practice pages found in the All-in-One Teaching Resources and online.
- **Think About a Plan** Helps students develop specific problem-solving skills and strategies by providing scaffolded guiding questions.
- **Standardized Test Prep** Focuses on all major exercises, all major question types, and helps students prepare for the high-stakes assessments.

Extension

- **Enrichment** Provides students with interesting problems and activities that extend the concepts of the lesson.
- **Activities, Games, and Puzzles** Worksheets that can be used for concepts development, enrichment, and for fun!

Practice and Problem Solving Wkbk/ All-in-One Resources/Online

Practice page 1

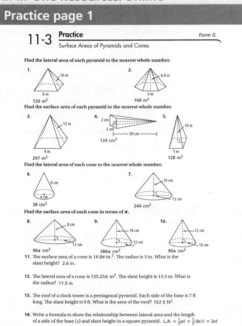

11-3 Practice *Form G*
Surface Areas of Pyramids and Cones

Find the lateral area of each pyramid to the nearest whole number.

1. 120 m²
2. 104 m²

Find the surface area of each pyramid to the nearest whole number.

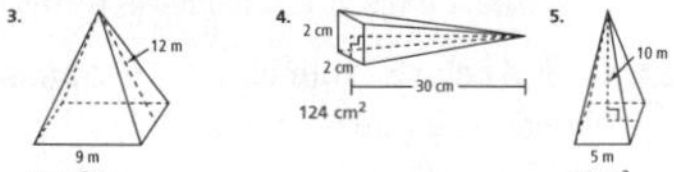

3. 297 m²
4. 124 cm²
5. 128 m²

Find the lateral area of each cone to the nearest whole number.

6. 38 cm²
7. 244 cm²

Find the surface area of each cone in terms of π.

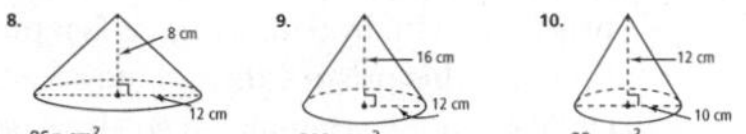

8. 96π cm²
9. 384π cm²
10. 90π cm²
11. The surface area of a cone is 16.8π in.². The radius is 3 in. What is the slant height? 2.6 in.
12. The lateral area of a cone is 155.25π m². The slant height is 13.5 m. What is the radius? 11.5 m
13. The roof of a clock tower is a pentagonal pyramid. Each side of the base is 7 ft long. The slant height is 9 ft. What is the area of the roof? 157.5 ft²
14. Write a formula to show the relationship between lateral area and the length of a side of the base (s) and slant height in a square pyramid. $L.A. = \frac{1}{2}p\ell = \frac{1}{2}(4s)\ell = 2s\ell$

Practice and Problem Solving Wkbk/ All-in-One Resources/Online

Practice page 2

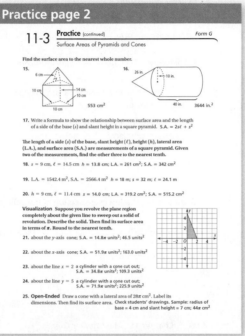

11-3 Practice (continued) *Form G*
Surface Areas of Pyramids and Cones

Find the surface area to the nearest whole number.

15. 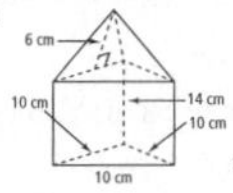

553 cm²

16. 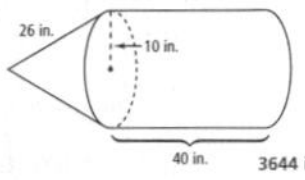

3644 in.²

17. Write a formula to show the relationship between surface area and the length of a side of the base (s) and slant height in a square pyramid. $S.A. = 2s\ell + s^2$

The length of a side (s) of the base, slant height (ℓ), height (h), lateral area (L.A.), and surface area (S.A.) are measurements of a square pyramid. Given two of the measurements, find the other three to the nearest tenth.

18. $s = 9$ cm, $\ell = 14.5$ cm $h = 13.8$ cm; L.A. = 261 cm²; S.A. = 342 cm²
19. L.A. = 1542.4 m², S.A. = 2566.4 m² $h = 18$ m; $s = 32$ m; $\ell = 24.1$ m
20. $h = 9$ cm, $\ell = 11.4$ cm $s = 14.0$ cm; L.A. = 319.2 cm²; S.A. = 515.2 cm²

Visualization **Suppose you revolve the plane region completely about the given line to sweep out a solid of revolution. Describe the solid. Then find its surface area in terms of π. Round to the nearest tenth.**

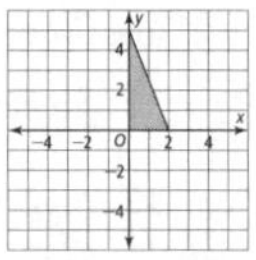

21. about the y-axis cone; S.A. = 14.8π units²; 46.5 units²
22. about the x-axis cone; S.A. = 51.9π units²; 163.0 units²
23. about the line $x = 2$ a cylinder with a cone cut out; S.A. = 34.8π units²; 109.3 units²
24. about the line $y = 5$ a cylinder with a cone cut out; S.A. = 71.9π units²; 225.9 units²
25. **Open-Ended** Draw a cone with a lateral area of 28π cm². Label its dimensions. Then find its surface area. Check students' drawings. Sample: radius of base = 4 cm and slant height = 7 cm; 44π cm²

All-in-One Resources/Online

Enrichment

11-3 Enrichment
Surface Areas of Pyramids and Cones

Frustum of a Solid

A *frustum* of a pyramid or cone is the figure made when the tip of the pyramid or cone is cut off by a cross section that is perpendicular to the height. In a frustum of a pyramid or cone, there are two bases, an upper base and a lower base.

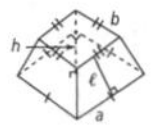

The figure at the upper right is a frustum of a square pyramid for which h is the height, a is the length of a side of the lower base, b is the length of a side of the upper base, and ℓ is the slant height.

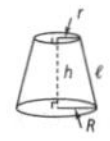

The figure at the lower right is a frustum of a cone for which h is the height, R is the radius of the lower base, r is the radius of the upper base, and ℓ is the slant height.

Use the frustum of the regular square pyramid above to derive a surface area formula.

1. What is the area of its lower base? a^2
2. What is the area of its upper base? b^2
3. What is the shape of each lateral face? trapezoid
4. What is the area of each lateral face? $\frac{1}{2}\ell(a + b)$
5. What is the lateral area of the figure? $2\ell(a + b)$
6. What is the formula for the surface area of a frustum of a square pyramid? $a^2 + b^2 + 2\ell(a + b)$

Use the frustum of the cone above to derive a surface area formula.

7. What is the area of its lower base? πR^2
8. What is the area of its upper base? πr^2
9. What is the lateral area of the figure? (*Hint:* Consider the formula for the lateral area of a cone.) $\pi R\ell + \pi r\ell$
10. What is the formula for the surface area of a frustum of a cone? $\pi(R^2 + r^2 + R\ell + r\ell)$

Find the surface area of each figure below. Round your answers to the nearest tenth.

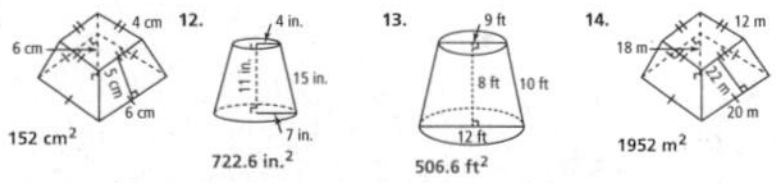

11. 152 cm²
12. 722.6 in.²
13. 506.6 ft²
14. 1952 m²

Practice and Problem Solving Wkbk/ All-in-One Resources/Online

Think About a Plan

11-3 Think About a Plan
Surface Areas of Pyramids and Cones

Find the lateral area of the cone to the nearest whole number.

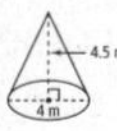

Understanding the Problem

1. What is the formula for the lateral area of a cone? $L.A. = \pi r\ell$
2. How are the two variables in this formula defined?
   The radius, r, is the distance from the center of the circle to a point on the circle. The slant height, ℓ, is the distance from the vertex of the cone to a point on the circle.
3. What two pieces of information are given in the figure of the cone?
   the diameter and the height

Planning the Solution

4. How can you use the given information to find the radius?
   Divide the length of the diameter by 2.
5. How can you use the given information and the radius to find the slant height?
   Substitute the height and the radius into the Pythagorean Theorem, $\ell = \sqrt{r^2 + h^2}$.

Getting an Answer

6. What is the radius? $r = \frac{1}{2} \cdot 4 \text{ m} = 2 \text{ m}$
7. What is the slant height of the cone? $\ell = \sqrt{2^2 + (4.5)^2} = \sqrt{24.25}$ m
8. What is the lateral area of the cone? $L.A. = \pi \cdot 2 \cdot \sqrt{24.25} = 2\sqrt{24.25}\pi \approx 31$ m²

Practice and Problem Solving Wkbk/ All-in-One Resources/Online

Standardized Test Prep

11-3 Standardized Test Prep
Surface Areas of Pyramids and Cones

Multiple Choice

For Exercises 1–5, choose the correct letter.

1. What is the lateral surface area of a square pyramid with side length 11.2 cm and slant height 20 cm? B
   (A) 224 cm² (B) 448 cm² (C) 896 cm² (D) 2508.8 cm²
2. What is the lateral surface area of a cone with radius 19 cm and slant height 11 cm? I
   (F) 19π cm² (G) 30π cm² (H) 200π cm² (I) 209π cm²
3. What is the lateral area of the square pyramid, to the nearest whole number? D

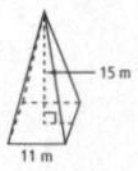

   (A) 165 m² (C) 330 m²
   (B) 176 m² (D) 351 m²
4. What is the surface area of the cone, to the nearest whole number? G

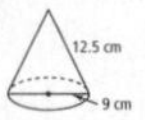

   (F) 221 cm² (H) 304 cm²
   (G) 240 cm² (I) 620 cm²
5. What is the surface area of a cone with diameter 28 cm and height 22 cm in terms of π? C
   (A) 196π cm² (B) 365π cm² (C) 561.1π cm² (D) 2202.8π cm²

Extended Response

6. What are the perimeter of the base, slant height, lateral area, and surface area for the square pyramid, to the nearest tenth of a meter or square meter?

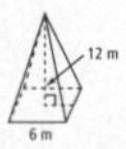

   [4] $p = 24$ m; $\ell = 12.4$ m; L.A. = 148.4 m²; S.A. = 184.4 m²
   [3] any three of the four values correctly given
   [2] any two of the four values correctly given
   [1] any one correct value given
   [0] no correct responses given

Online Teacher Resource Center

Activities, Games, and Puzzles

11-3 Puzzle: Unlock the Secret to Pyramids and Cones
Surface Areas of Pyramids and Cones

Below are three right triangle "keys" relating special measurements in cones and pyramids. Given that $S.A._{cone} = \pi r\ell + \pi r^2$ and $S.A._{pyramid} = \frac{1}{2}p\ell + \frac{1}{2}ap$, these keys provide the right triangle relationships you need in order to find the missing measurements in the equations. Cut out the 6 figures at the bottom of the page and place each beside the "key" you feel will unlock it. Then, find the surface area of each figure. You may assume that all the pyramids have regular polygons for bases.

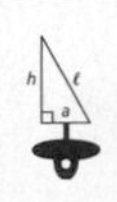	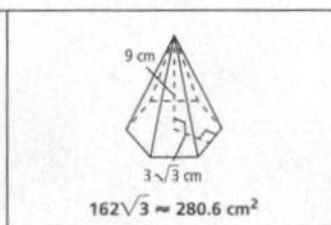 $162\sqrt{3} \approx 280.6$ cm²	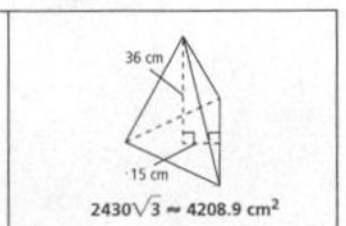 $2430\sqrt{3} \approx 4208.9$ cm²
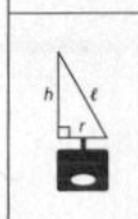	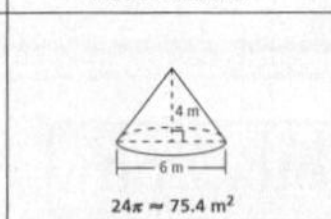 $24\pi \approx 75.4$ m²	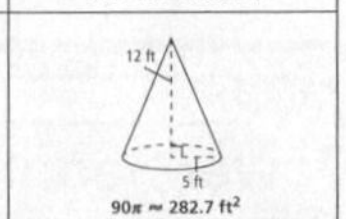 $90\pi \approx 282.7$ ft²
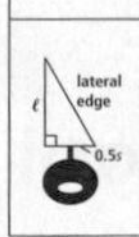	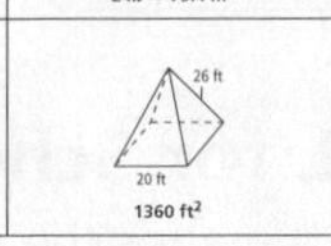 1360 ft²	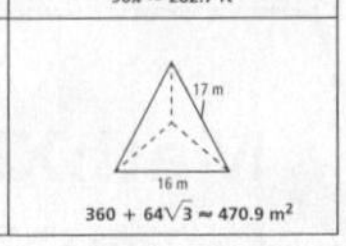 $360 + 64\sqrt{3} \approx 470.9$ m²

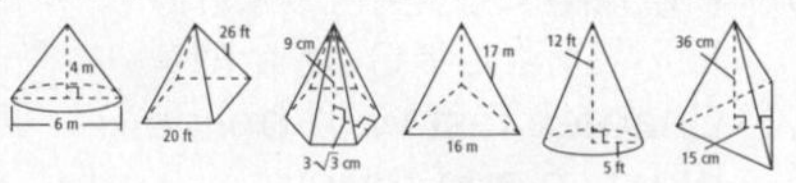

Answers

Mid-Chapter Quiz

1.
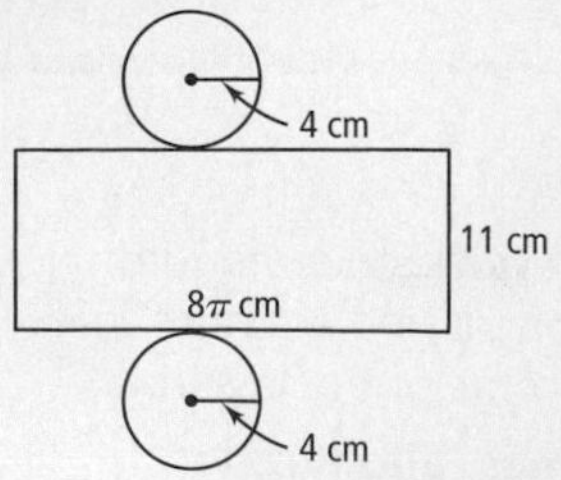

2.
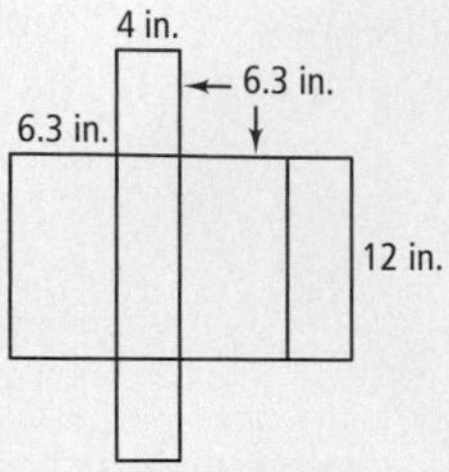

3.
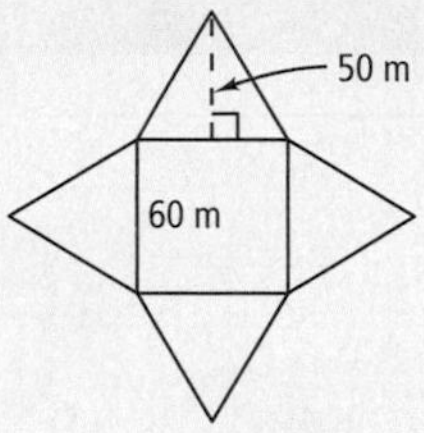

4.
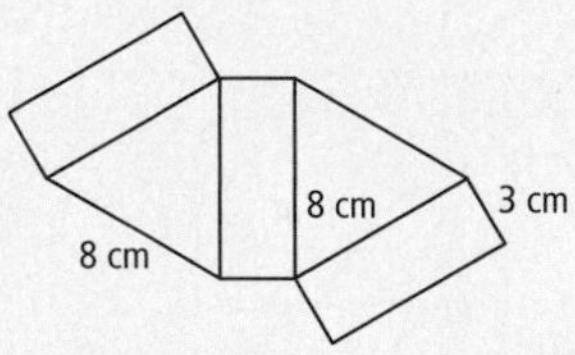

5. 377 cm^2 **6.** 298 in.^{2}
7. 9600 m^2 **8.** 75 ft^2
9–11. Check students' work.
9. 7 faces
10. 12 vertices
11. 10 vertices
12–14. Answers may vary. Samples are given.

12.
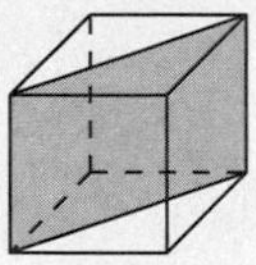

13.
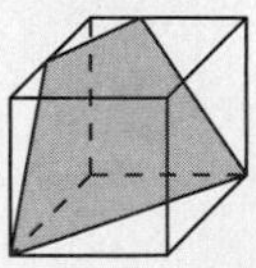

11 Mid-Chapter Quiz

MathXL® for School
Go to PowerGeometry.com

Do you know HOW?

Draw a net for each figure. Label the net with its dimensions.

1.
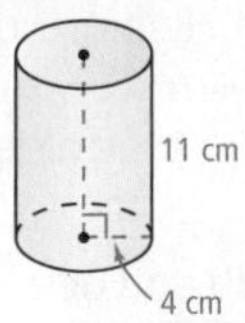

2.
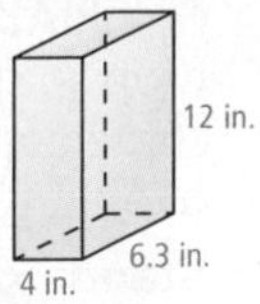

3.
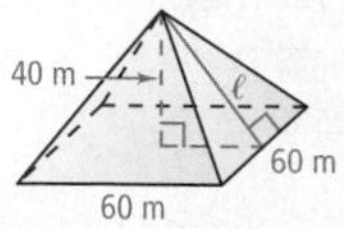

4.
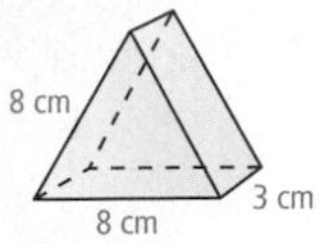

Find the surface area of each solid. Round to the nearest square unit.

5. cylinder in Exercise 1
6. prism in Exercise 2
7. pyramid in Exercise 3
8. cone in Exercise 4

For Exercises 9–11, use Euler's Formula. Show your work.

9. A polyhedron has 10 vertices and 15 edges. How many faces does it have?

10. A polyhedron has 2 hexagonal faces and 12 triangular faces. How many vertices does it have?

11. How many vertices does the net of a pentagonal pyramid have?

Draw a cube. Shade a cross section of the cube that forms each shape.

12. a rectangle **13.** a trapezoid **14.** a triangle

15. A square prism with base edges 2 in. has surface area 32 in.^{2}. What is its height?

Find the surface area of each figure to the nearest whole number.

16.
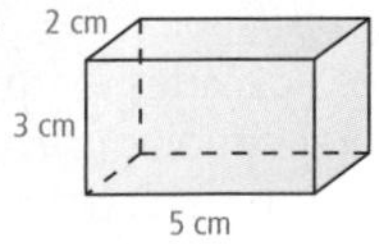

17.
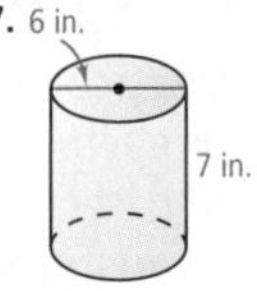

18.
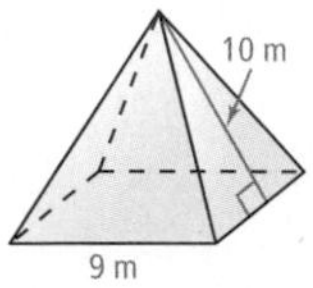

19.
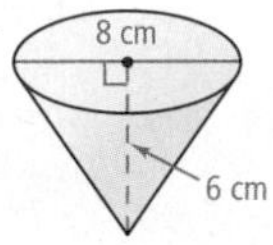

Do you UNDERSTAND?

20. Open-Ended Draw a net for a regular hexagonal prism.

21. Compare and Contrast How are a right prism and a regular pyramid alike? How are they different?

22. Algebra The height of a cylinder is twice the radius of the base. A cube has an edge length equal to the radius of the cylinder. Find the ratio between the surface area of the cylinder and the surface area of the cube. Leave your answer in terms of π.

23. Error Analysis Your class learned that the number of regions in the net for a solid is the same as the number of faces in that solid. Your friend concludes that a solid and its net must also have the same number of edges. What is your friend's error? Explain.

24. Visualization The dimensions of a rectangular prism are 3 in. by 5 in. by 10 in. Is it possible to intersect this prism with a plane so that the resulting cross section is a square? If so, draw and describe your cross section. Indicate the length of the edge of the square.

14.
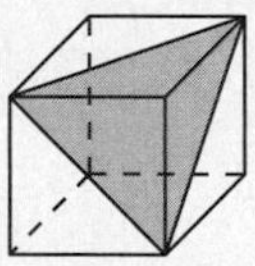

15. 3 in.
16. 62 cm^2 **17.** 188 in.^{2}
18. 261 m^2 **19.** 141 cm^2
20. Sample:
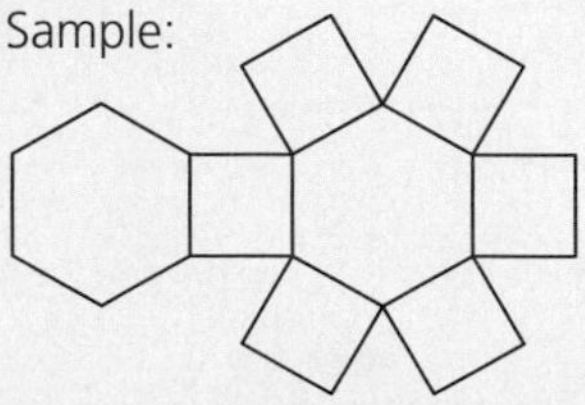

21. Answers may vary. Sample: Alike: Both have a polygon as a base. Different: A prism has two bases, a pyramid has only one base, and the lateral faces are rectangles for a prism and ≅ isosc. △ for a pyramid.

22. $\pi : 1$

23. Answers may vary. Sample: In a net, some edges appear twice; when the net is folded, those duplicate edges come together.

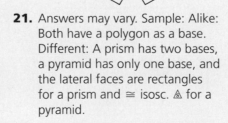

24. Answers may vary. Sample:
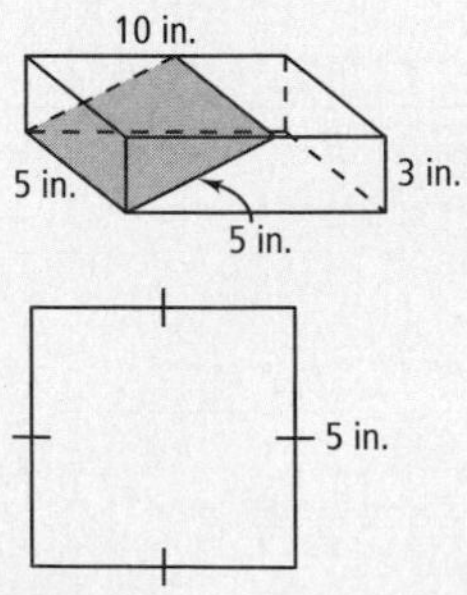

11-4 Volumes of Prisms and Cylinders

Common Core State Standards
G-GMD.A.1 Give an informal argument for the formulas for . . . volume of a cylinder . . . Use . . . Cavalieri's principle . . . **Also G-GMD.A.2, G-GMD.A.3, G-MG.A.1**
MP 1, MP 3, MP 4, MP 6, MP 7

Objective To find the volume of a prism and the volume of a cylinder

You can start by figuring out how many cubes will fit on the bottom of the box.

MATHEMATICAL PRACTICES

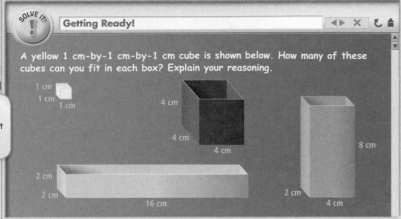

Getting Ready!

A yellow 1 cm-by-1 cm-by-1 cm cube is shown below. How many of these cubes can you fit in each box? Explain your reasoning.

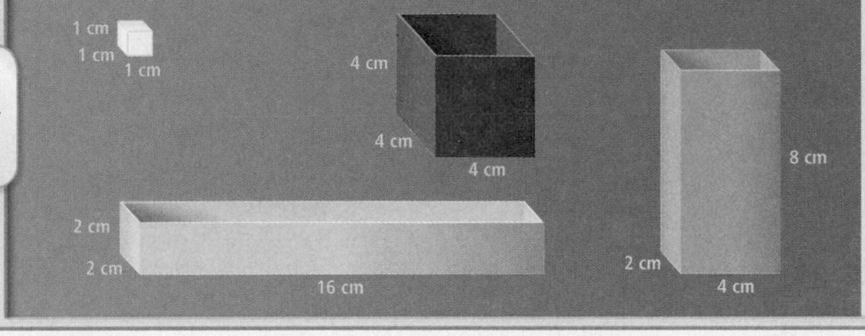

In the Solve It, you determined the volume of a box by finding how many 1 cm-by-1 cm-by-1 cm cubes the box holds.

Lesson Vocabulary
- volume
- composite space figure

Volume is the space that a figure occupies. It is measured in cubic units such as cubic inches (in.^{3}), cubic feet (ft^3), or cubic centimeters (cm^3). The volume V of a cube is the cube of the length of its edge e, or $V = e^3$.

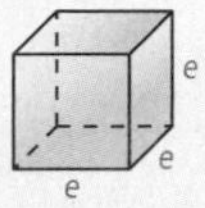

Essential Understanding You can find the volume of a prism or a cylinder when you know its height and the area of its base.

Both stacks of paper below contain the same number of sheets.

The first stack forms an oblique prism. The second forms a right prism. The stacks have the same height. The area of every cross section parallel to a base is the area of one sheet of paper. The stacks have the same volume. These stacks illustrate the following principle.

1 Interactive Learning

Solve It!

PURPOSE To find the volume of rectangular prisms
PROCESS Students may
- multiply the dimensions of the prisms.
- model the prisms with cubes.

FACILITATE

Q How many 1 cm cubes fit in the 4 cm square box? **[64]**

Q How can you express the dimensions of the other two boxes? **[8 cm-by-4 cm-by-2 cm and 2 cm-by-2 cm-by-16 cm]**

Q What is the relationship between the dimensions of a box and the number of cubic centimeters it holds? **[The number of cubic centimeters in a box is the product of the dimensions of the box.]**

ANSWER See Solve It in Answers on next page.
CONNECT THE MATH Students discover that the volume of a rectangular prism is equal to the product of its dimensions. In the lesson, students learn the formula $V = Bh$.

2 Guided Instruction

Have students identify the height and area of the base of each of the boxes in the Solve It. Ask them to verify that the volume of each box is the product of the area of its base and height.

11-4 Preparing to Teach

BIG ideas **Measurement**
Visualization

ESSENTIAL UNDERSTANDINGS
- The volume of a prism and a cylinder can be found when its height and the area of its base are known.
- The volume of a composite space figure is the sum of the volumes of the figures that are combined.

Math Background

Three-dimensional figures contain space, called volume. Both prisms and cylinders have the same cross sectional area at every distance from the base, therefore their volume formulas are similar. Students will learn how to calculate the volume of a prism and cylinder.

To calculate the volume of any prism or cylinder with a constant cross-sectional area, multiply the area of the base by the height of the solid. Students will relate this concept to the volume of a cylinder, which has a circular base. They will use the formula for the area of a circle to calculate the volume of a cylinder.

Placing disks on top of each other can help students see how the volume of a cylinder increases as the height of the cylinder increases.

Mathematical Practices

Attend to precision. In solving for volume of prisms and cylinders, students will specify cubic units of measure.

PowerGeometry.com

1 Interactive Learning

Solve It!
Step out how to solve the Problem with helpful hints and an online question. Other questions are listed above in Interactive Learning.

Virtual Nerd™ Access Virtual Nerd student-centered math tutorials that directly relate to the content of the lesson.

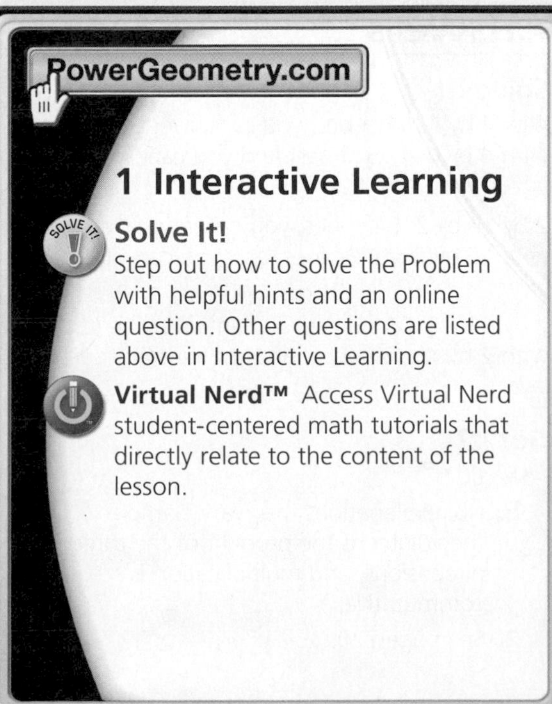

Take Note

Have students use Cavalieri's Principle to draw prisms with the same volume as each box in the Solve It. Students should start by drawing a figure with a base that has the same area.

Take Note

Have students calculate the volume of each figure in the diagram above Theorem 11-6. They can either choose a height for the figures, or let the height equal h. Students should be able to show that the volumes of all three solids are the same.

Problem 1

Q What is the base of the prism? **[The base is the rectangle with side lengths 24 cm and 20 cm.]**

Q How can you find the volume of the prism? **[Multiply the area of the base by the height.]**

Q What does a volume of 4800 cm^3 mean in terms of 1 cm cubes? **[It means that 4800 cubes fit inside of the box, or that 10 layers of 480 cubes fit inside the box.]**

Got It?

For 1b, have students make and support a conjecture before they try to calculate the volume of the prism turned on its side. Emphasize that the commutative and associative properties of multiplication imply that the results will be the same no matter which side is considered the base.

---

take note **Theorem 11-5 Cavalieri's Principle**

If two space figures have the same height and the same cross-sectional area at every level, then they have the same volume.

The area of each shaded cross section below is 6 cm^2. Since the prisms have the same height, their volumes must be the same by Cavalieri's Principle.

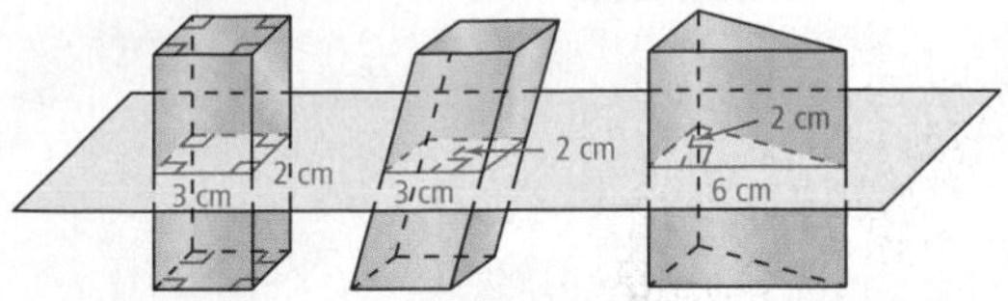

You can find the volume of a right prism by multiplying the area of the base by the height. Cavalieri's Principle lets you extend this idea to any prism.

take note **Theorem 11-6 Volume of a Prism**

The volume of a prism is the product of the area of the base and the height of the prism.

$$V = Bh$$

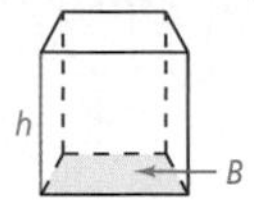

Problem 1 Finding the Volume of a Rectangular Prism

Plan

What do you need to use the formula? You need to find B, the area of the base. The prism has a rectangular base, so the area of the base is length × width.

What is the volume of the rectangular prism at the right?

$V = Bh$	Use the formula for the volume of a prism.
$= 480 \cdot 10$	The area of the base B is $24 \cdot 20$, or 480 cm^2, and the height is 10 cm.
$= 4800$	Simplify.

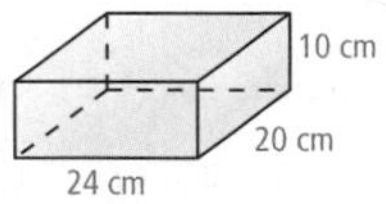

The volume of the rectangular prism is 4800 cm^3.

Got It? 1. a. What is the volume of the rectangular prism at the right?

b. Reasoning Suppose the prism at the right is turned so that the base is 4 ft by 5 ft and the height is 3 ft. Does the volume change? Explain.

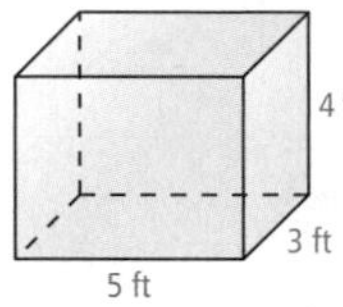

Answers

Solve It!

4 by 4 by 4: 64 cubes; you can cover the bottom with 4 by 4 = 16 cubes, and you can fit 4 such layers.

2 by 16 by 2: 64 cubes; you can cover the bottom with 16 by 2 = 32 cubes, and you can fit 2 such layers.

2 by 4 by 8: 64 cubes; you can cover the bottom with 2 by 4 = 8 cubes, and you can fit 8 such layers.

Got It?

1a. 60 ft^3

b. No; explanations may vary. Sample: the volume is the product of the three dimensions, and multiplication is commutative.

2. See page 720.

PowerGeometry.com

2 Guided Instruction

Each Problem is worked out and supported online.

Problem 1
Finding the Volume of a Rectangular Prism
Animated

Problem 2
Finding the Volume of a Triangular Prism
Animated

Problem 3
Finding the Volume of a Cylinder

Problem 4
Finding Volume of a Composite Figure
Animated

Support in Geometry Companion
- Vocabulary
- Key Concepts
- Got It?

Problem 2 Finding the Volume of a Triangular Prism

Multiple Choice What is the approximate volume of the triangular prism?

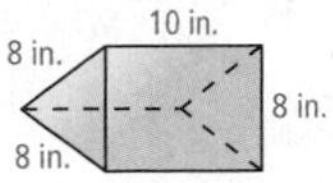

Ⓐ 188 in.^{3} Ⓒ 295 in.^{3}
Ⓑ 277 in.^{3} Ⓓ 554 in.^{3}

Step 1 Find the area of the base of the prism.

Each base of the triangular prism is an equilateral triangle, as shown at the right. An altitude of the triangle divides it into two 30°-60°-90° triangles. The height of the triangle is $\sqrt{3}$ • shorter leg, or $4\sqrt{3}$.

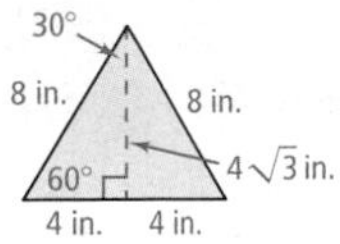

$B = \frac{1}{2}bh$ Use the formula for the area of a triangle.

$= \frac{1}{2}(8)(4\sqrt{3})$ Substitute 8 for b and $4\sqrt{3}$ for h.

$= 16\sqrt{3}$ Simplify.

Think

Which height do you use in the formula? Remember that the h in the formula for volume represents the height of the entire prism, not the height of the triangular base.

Step 2 Find the volume of the prism.

$V = Bh$ Use the formula for the volume of a prism.

$= 16\sqrt{3} \cdot 10$ Substitute $16\sqrt{3}$ for B and 10 for h.

$= 160\sqrt{3}$ Simplify.

≈ 277.1281292 Use a calculator.

The volume of the triangular prism is about 277 in.^{3}. The correct answer is B.

Got It? 2. a. What is the volume of the triangular prism at the right?
b. Reasoning Suppose the height of a prism is doubled. How does this affect the volume of the prism? Explain.

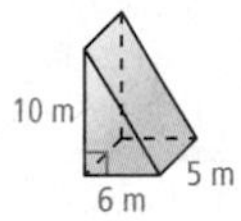

To find the volume of a cylinder, you use the same formula $V = Bh$ that you use to find the volume of a prism. Now, however, B is the area of the circle, so you use the formula $B = \pi r^2$ to find its value.

take note

Theorem 11-7 Volume of a Cylinder

The volume of a cylinder is the product of the area of the base and the height of the cylinder.

$$V = Bh, \text{ or } V = \pi r^2 h$$

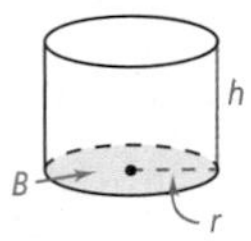

Problem 2

Q What figure is the base of the prism? **[an equilateral triangle]**

Q What additional measurement do you need to find the area of the triangular base? **[the height]**

Q What segment represents the height of an equilateral triangle? **[the perpendicular bisector of one side]**

Q What triangles are formed by the height of an equilateral triangle? **[30°-60°-90° triangles]**

Q How do you find the length of the longest leg of the triangle? **[Multiply $\sqrt{3}$ by the length of the shorter leg.]**

Got It?

For 2a, remind students that the height of a right triangle is the length of one of its legs. For 2b, have students change the height from 5 m to 10 m. Then write a conjecture about the result of doubling the height of a prism. They can verify their conjectures by substituting $2h$ into the formula for the volume.

Take Note

Have students compare and contrast the formula for the volume of a prism to the formula for the volume of a cylinder. In both formulas, the area of the base is multiplied by the height of the figure. In a prism, the base is a polygon. In a cylinder, the base is a circle.

Additional Problems

1. What is the volume of the prism below?

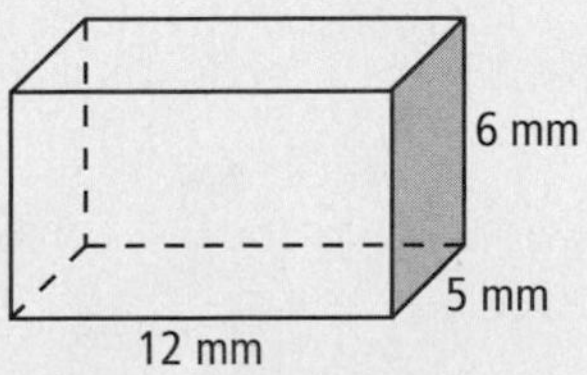

ANSWER 360 mm^3

2. What is the volume of the triangular prism?

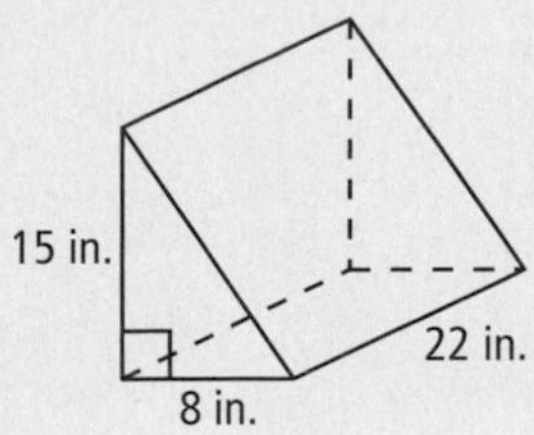

A. 1472 in.^{3}
B. 1320 in.^{3}
C. 1184 in.^{3}
D. 960 in.^{3}

ANSWER B

3. Find the volume of the cylinder in terms of π.

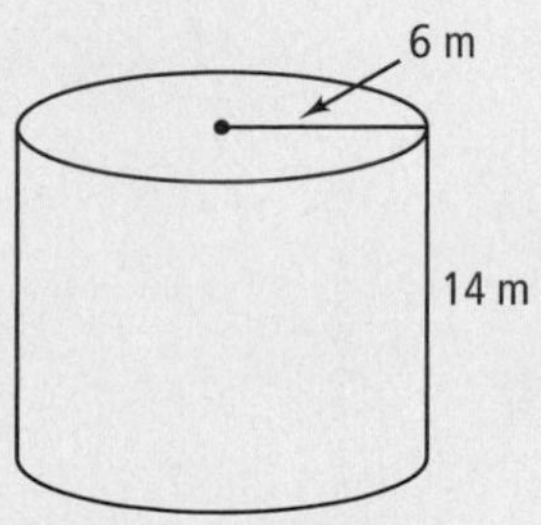

ANSWER 504π m^2

4. A lab technician made a 14 cm diameter hole through the middle of a cylinder that has a diameter of 20 cm and a height of 18 cm. What is the approximate volume of the finished cylinder?

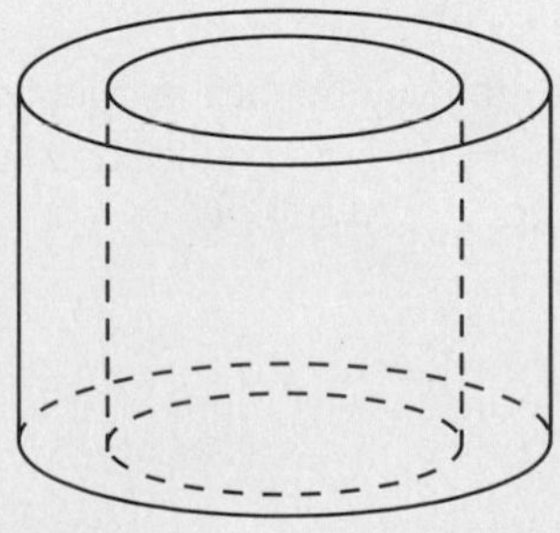

ANSWER about 2,884 cm^3

Problem 3

Q How do you find the area of the base? **[Calculate the area of a circle with a 3 cm radius.]**

Q How do you find the volume of the cylinder? **[Multiply the area of the base by the height of the cylinder.]**

Got It?

For 3a, remind students that the height of a cylinder must be perpendicular to both bases. They should realize that the segment marked is the height of the oblique cylinder.

For 3b, suggest that students calculate the volume of the cylinder and then calculate the volume again using a number for the height that is half of the height used to calculate the first volume.

Problem 4

Q What two solids form the entire aquarium? **[a rectangular prism and a half-cylinder]**

Q How can you find the volume of the half-cylinder? **[Divide the volume of the cylinder with radius 12 in. by 2.]**

Q What is the length of the rectangular prism without the half-cylinder on the end of the aquarium? Explain. **[The radius of the cylinder is 12 in. so the length of the rectangular prism is 48 in. − 12 in. = 36 in.]**

Got It? VISUAL LEARNERS

Have students sketch and label the rectangular prism and half-cylinder separately. Then they can calculate the volumes separately and find their sum.

Problem 3 Finding the Volume of a Cylinder

Plan

What do you know from the diagram?
You know that the radius r is 3 cm and the height h is 8 cm.

What is the volume of the cylinder in terms of π?

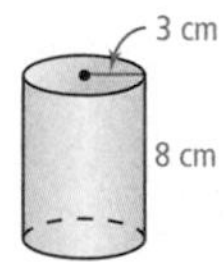

$V = \pi r^2 h$ Use the formula for the volume of a cylinder.

$= \pi(3)^2(8)$ Substitute 3 for r and 8 for h.

$= \pi(72)$ Simplify.

The volume of the cylinder is 72π cm^3.

Got It? 3. a. What is the volume of the cylinder at the right in terms of π?

b. Reasoning Suppose the radius of a cylinder is halved. How does this affect the volume of the cylinder? Explain.

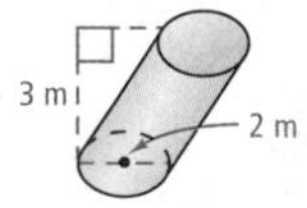

A **composite space figure** is a three-dimensional figure that is the combination of two or more simpler figures. You can find the volume of a composite space figure by adding the volumes of the figures that are combined.

Problem 4 Finding the Volume of a Composite Figure

What is the approximate volume of the bullnose aquarium to the nearest cubic inch?

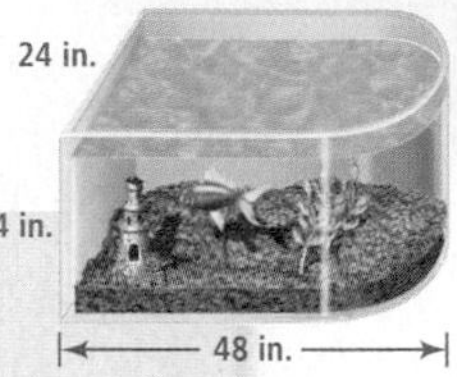

Plan

How can you find the volume by *solving a simpler problem*?
The aquarium is the combination of a rectangular prism and half of a cylinder. Find the volume of each figure.

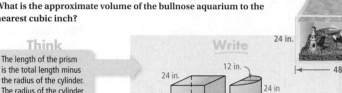

Think	Write
The length of the prism is the total length minus the radius of the cylinder. The radius of the cylinder is half the width of the prism.	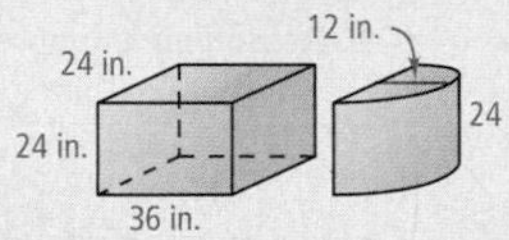
Find the volume of the prism and the half cylinder.	$V_1 = Bh$ $= (24 \cdot 36)(24)$ $= 20{,}736$ $V_2 = \frac{1}{2}\pi r^2 h$ $= \frac{1}{2}\pi(12)^2(24)$ ≈ 5429
Add the two volumes together.	$20{,}736 + 5429 = 26{,}165$ The approximate volume of the aquarium is 26,165 in.^{3}.

Got It? 4. What is the approximate volume of the lunchbox shown at the right? Round to the nearest cubic inch.

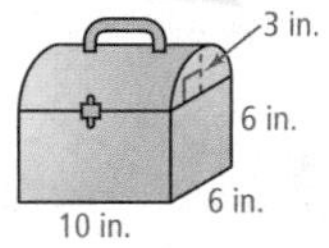

Answers

Got It? (continued)

2a. 150 m^3

b. The volume is doubled. Using $V = B \cdot h$ and replacing h with $2h$ gives $B \cdot (2h) = 2 \cdot B \cdot h = 2 \cdot V$.

3a. 3π m^3

b. The volume is $\frac{1}{4}$ the volume of the cylinder in part (a). Using $V = \pi r^2 h$ and replacing r with $\frac{r}{2}$ gives $\pi\left(\frac{r}{2}\right)^2 h = \frac{1}{4}\pi r^2 h = \frac{1}{4} \cdot V$.

4. 501 in.^{3}

Lesson Check

Do you know HOW?

What is the volume of each figure? If necessary, round to the nearest whole number.

1.

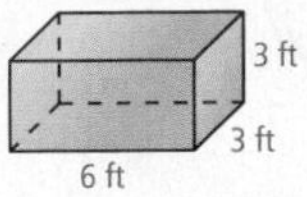

2.

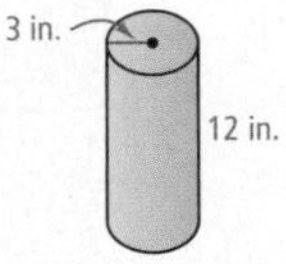

Do you UNDERSTAND? MATHEMATICAL PRACTICES

3. Vocabulary Is the figure at the right a composite space figure? Explain.

4. Compare and Contrast How are the formulas for the volume of a prism and the volume of a cylinder alike? How are they different?

5. Reasoning How is the volume of a rectangular prism with base 2 m by 3 m and height 4 m related to the volume of a rectangular prism with base 3 m by 4 m and height 2 m? Explain.

Practice and Problem-Solving Exercises MATHEMATICAL PRACTICES

A Practice

Find the volume of each rectangular prism. See Problem 1.

6.

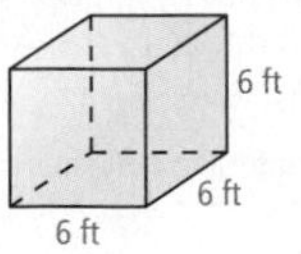

7.

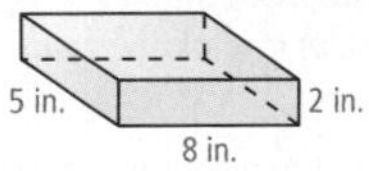

8.

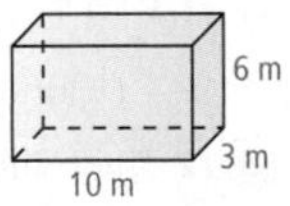

9. The base is a square with sides of 2 cm. The height is 3.5 cm.

Find the volume of each triangular prism. See Problem 2.

10.

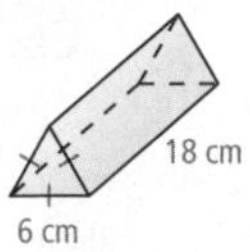

11.

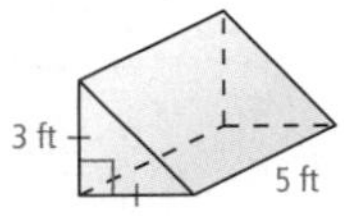

12.

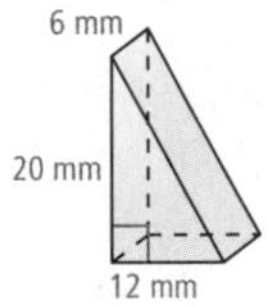

13. The base is a 45°-45°-90° triangle with a leg of 5 in. The height is 1.8 in.

Find the volume of each cylinder in terms of π and to the nearest tenth. See Problem 3.

14.

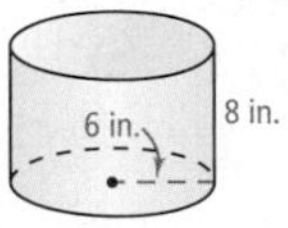

15.

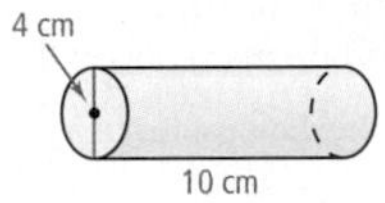

16.

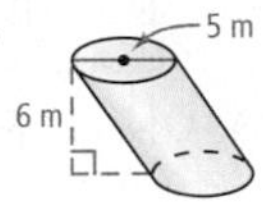

17. The diameter of the cylinder is 1 yd. The height is 4 yd.

3 Lesson Check

Do you know HOW?

- If students have difficulty with Exercise 2, then have them review Problem 3 to verify which measurements get substituted into the formula.

Do you UNDERSTAND?

- If students have difficulty with Exercise 3, then have them draw the figure as separate solids.

Close

Q How do you find the volume of a prism or a cylinder? **[Multiply the area of the base by the height of the solid.]**

Q The basic formulas for the volume of a cylinder and prism are the same. What is the difference between a cylinder and a prism? **[A cylinder has a circular base while the base of a prism is a polygon.]**

Q What is a composite figure? **[A composite figure is a geometric figure formed by two or more geometric figures.]**

Lesson Check

1. 54 ft^3 **2.** 339 $in.^3$

3. Yes; it is a combination of a cylinder and a cone.

4. Alike: Both are the product of the base area and the height. Different: For a prism the base is a polygon, while for a cylinder the base is a circle.

5. The volumes are the same, 24 m^3, because multiplication is commutative.

Practice and Problem-Solving Exercises

6. 216 ft^3
7. 80 $in.^3$
8. 180 m^3
9. 14 cm^3
10. about 280.6 cm^3
11. 22.5 ft^3
12. 720 mm^3
13. 22.5 $in.^3$
14. 288π $in.^3$; 904.8 $in.^3$
15. 40π cm^3; 125.7 cm^3
16. 37.5π m^3; 117.8 m^3
17. π yd^3; 3.1 yd^3

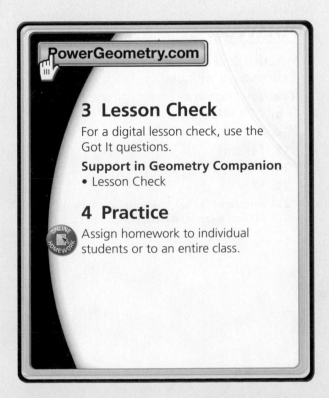

PowerGeometry.com

3 Lesson Check

For a digital lesson check, use the Got It questions.

Support in Geometry Companion

- Lesson Check

4 Practice

Assign homework to individual students or to an entire class.

4 Practice

ASSIGNMENT GUIDE
Basic: 6–21 all, 24, 30–32 all, 38
Average: 7–19 odd, 21–42
Advanced: 7–19 odd, 21–45

Mathematical Practices are supported by exercises with red headings. Here are the Practices supported in this lesson:

MP 1: Make Sense of Problems Ex. 4, 5, 21
MP 3: Construct Arguments Ex. 29, 31
MP 6: Attend to Precision Ex. 22
MP 7: Shift Perspective Ex. 39–42

Applications exercises have blue headings. Exercises 18, 26, 28, 30, 35, and 44 support MP 4: Model.

STEM exercises focus on science or engineering applications.

EXERCISE 30: Use the Think About a Plan worksheet in the **Practice and Problem Solving Workbook** (also available in the Teaching Resources in print and online) to further support students' development in becoming independent learners.

HOMEWORK QUICK CHECK
To check students' understanding of key skills and concepts, go over Exercises 15, 19, 21, 30, and 31.

See Problem 4.

18. Composite Figures Use the diagram of the backpack at the right.
  a. What two figures approximate the shape of the backpack?
  b. What is the volume of the backpack in terms of π?
  c. What is the volume of the backpack to the nearest cubic inch?

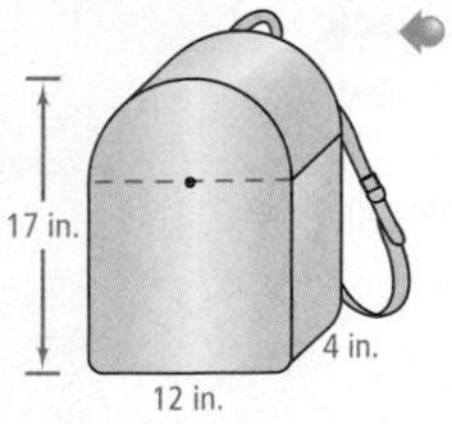

Find the volume of each composite space figure to the nearest whole number.

19.
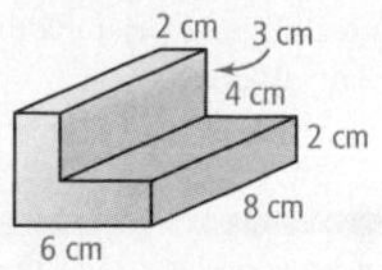

20.
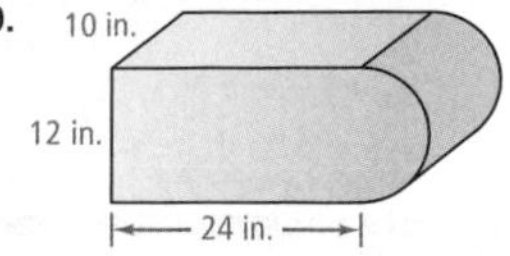

B Apply

21. Think About a Plan A full waterbed mattress is 7 ft by 4 ft by 1 ft. If water weighs 62.4 lb/ft^3, what is the weight of the water in the mattress to the nearest pound?
- How can you determine the amount of water the mattress can hold?
- The weight of the water is in pounds per cubic feet. How can you get an answer with a unit of pounds?

22. Open-Ended Give the dimensions of two rectangular prisms that have volumes of 80 cm^3 each but also have different surface areas.

Find the height of each figure with the given volume.

23.
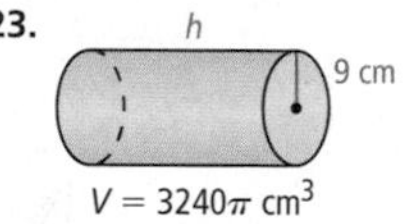

24.
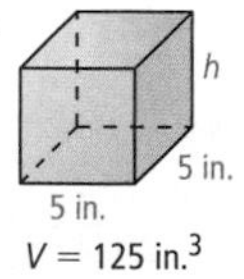

25.
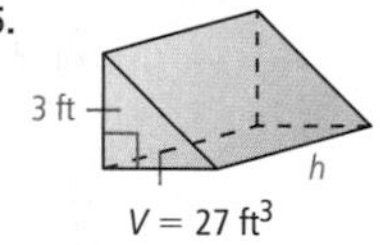

26. Sports A can of tennis balls has a diameter of 3 in. and a height of 8 in. Find the volume of the can to the nearest cubic inch.

27. What is the volume of the oblique prism shown at the right?

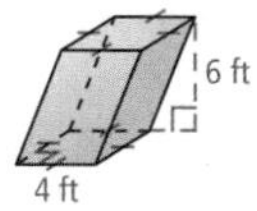

STEM **28. Environmental Engineering** A scientist suggests keeping indoor air relatively clean as follows: For a room with a ceiling 8 ft high, provide two or three pots of flowers for every 100 ft^2 of floor space. If your classroom has an 8-ft ceiling and measures 35 ft by 40 ft, how many pots of flowers should it have?

29. Reasoning Suppose the dimensions of a prism are tripled. How does this affect its volume? Explain.

Answers

Practice and Problem-Solving Exercises (continued)

18a. a rectangular prism and half a cylinder
b. $(528 + 72\pi)$ in.^{3}
c. 754 in.^{3}
19. 144 cm^3
20. 3445 in.^{3}
21. 1747 lb
22. Answers may vary. Sample: 2 cm by 4 cm by 10 cm; 4 cm by 4 cm by 5 cm
23. 40 cm
24. 5 in.
25. 6 ft
26. 57 in.^{3}
27. 96 ft^3
28. from 28 to 42 pots of flowers
29. Volume is 27 times greater. Using $V = B \cdot h = \ell \cdot w \cdot h$ for a rectangular prism, $(3\ell) \cdot (3w) \cdot (3h) = 27 \cdot \ell \cdot w \cdot h = 27 \cdot V$.

30. **Swimming Pool** The approximate dimensions of an Olympic-size swimming pool are 164 ft by 82 ft by 6.6 ft.
  a. Find the volume of the pool to the nearest cubic foot.
  b. If 1 $ft^3 \approx$ 7.48 gal, about how many gallons does the pool hold?

31. **Writing** The figures at the right can be covered by equal numbers of straws that are the same length. Describe how Cavalieri's Principle could be adapted to compare the areas of these figures.

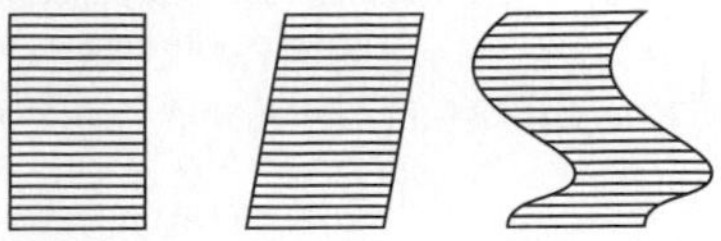

32. **Algebra** The volume of a cylinder is 600π cm^3. The radius of a base of the cylinder is 5 cm. What is the height of the cylinder?

33. **Coordinate Geometry** Find the volume of the rectangular prism at the right.

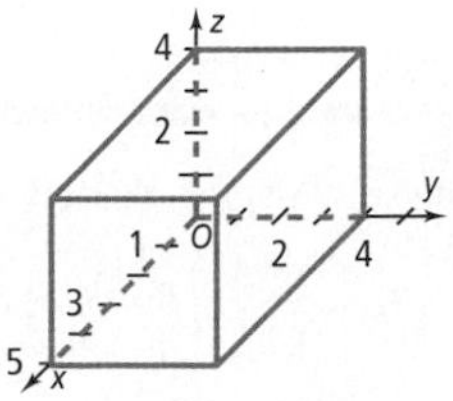

34. **Algebra** The volume of a cylinder is 135π cm^3. The height of the cylinder is 15 cm. What is the radius of a base of the cylinder?

35. **Landscaping** To landscape her 70 ft-by-60 ft rectangular backyard, your aunt is planning first to put down a 4-in. layer of topsoil. She can buy bags of topsoil at \$2.50 per 3-ft^3 bag, with free delivery. Or, she can buy bulk topsoil for \$22.00/yd^3, plus a \$20 delivery fee. Which option is less expensive? Explain.

36. The closed box at the right is shaped like a regular pentagonal prism. The exterior of the box has base edge 10 cm and height 14 cm. The interior has base edge 7 cm and height 11 cm. Find each measurement.
  a. the outside surface area
  b. the inside surface area
  c. the volume of the material needed to make the box

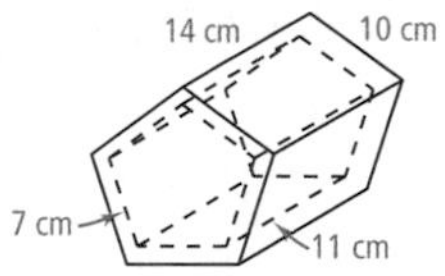

A cylinder has been cut out of each solid. Find the volume of the remaining solid. Round your answer to the nearest tenth.

37.

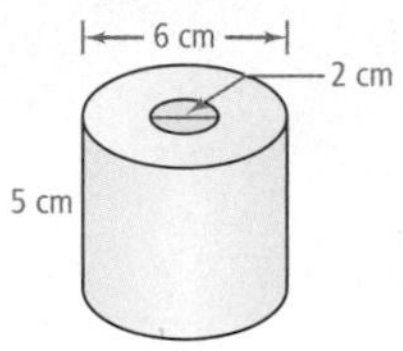

38.

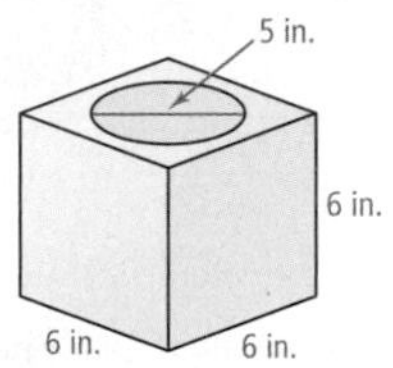

Visualization Suppose you revolve the plane region completely about the given line to sweep out a solid of revolution. Describe the solid and find its volume in terms of π.

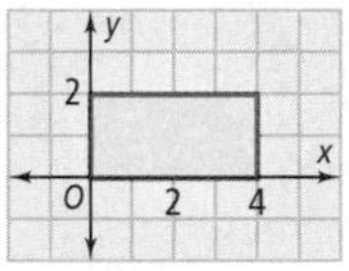

39. the x-axis
40. the y-axis
41. the line $y = 2$
42. the line $x = 5$

30a. 88,757 ft^3
b. 663,901 gal
31. Answers may vary. Sample: If two plane figures have the same height and the same width at every level, then they have the same area.
32. 24 cm
33. 80 $units^3$
34. 3 cm
35. bulk; cost of bags = \$1167.50, cost of bulk $\approx$ \$1164
36a. 1044.1 cm^2
b. 553.6 cm^2
c. 1481.3 cm^3
37. 125.7 cm^3
38. 98.2 $in.^3$
39. cylinder with $r = 2$ and $h = 4$; 16π $units^3$
40. cylinder with $r = 4$ and $h = 2$; 32π $units^3$
41. cylinder with $r = 2$ and $h = 4$; 16π $units^3$
42. cylinder with $r = 5$, $h = 2$, and a hole of radius 1; 48π $units^3$

Answers

Practice and Problem-Solving Exercises (continued)

43a. $C = 8.5$ in. and $h = 11$ in.: $V \approx 63.2$ in.^{3}; $C = 11$ in. and $h = 8.5$ in.: $V \approx 81.8$ in.^{3}; the cylinder with the greater circumference has the greater volume.

b. about 6.5 in. by 13.0 in.

44. 2827 cm^3

45. The volume of B is twice the volume of A.

Challenge

43. Paper Folding Any rectangular sheet of paper can be rolled into a right cylinder in two ways.

a. Use ordinary sheets of paper to model the two cylinders. Compute the volume of each cylinder. How do they compare?

b. Of all sheets of paper with perimeter 39 in., which size can be rolled into a right cylinder with greatest volume? (*Hint*: Try making a table.)

STEM **44. Plumbing** The outside diameter of a pipe is 5 cm. The inside diameter is 4 cm. The pipe is 4 m long. What is the volume of the material used for this length of pipe? Round your answer to the nearest cubic centimeter.

45. The radius of Cylinder B is twice the radius of Cylinder A. The height of Cylinder B is half the height of Cylinder A. Compare their volumes.

Apply What You've Learned

MATHEMATICAL PRACTICES MP 2, MP 4

Recall the paper towel roll from page 687, shown again below. Answer each of the following, leaving your answer in terms of π.

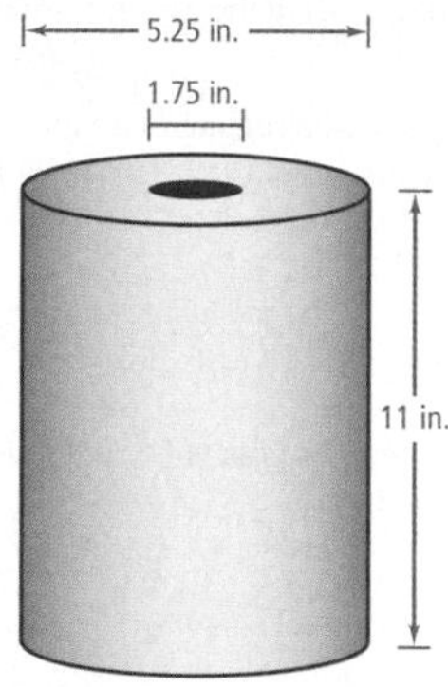

Not to scale

a. What is the volume of the entire package of paper towels?

b. What is the volume of the package that is not paper towels?

c. What is the volume of the paper towels in the package?

Apply What You've Learned

In the Apply What You've Learned for Lesson 11-2, students visualized unrolling the roll of paper towels shown on page 687 and found the area covered by the unrolled paper towels. Here, they answer questions about volume.

Mathematical Practices
Students **reason abstractly and quantitatively** by using the formula for the volume of a cylinder to find the volume of the paper towels in the package. (MP 2) Students use cylnders to **model** the roll of paper towels and its central tube.(MP 4)

ANSWERS

a. 75.796875π in.^{3}

b. 8.421875π in.^{3}

c. 67.375π in.^{3}

11-4 Lesson Resources

Differentiated Remediation

Instructional Support

Geometry Companion

Students can use the **Geometry Companion** worktext (4 pages) . . .

- New Vocabulary
- Key Concepts
- Got It for each Problem
- Lesson Check

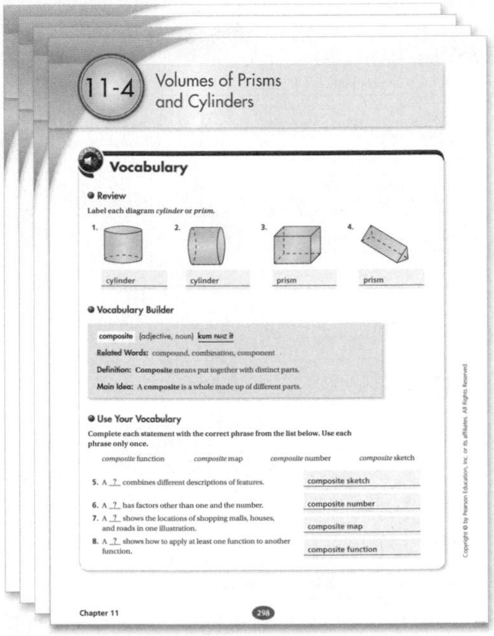

ELL Support

Connect to Prior Knowledge Connect the volume of a prism to the volume of a cylinder, by writing the formula for the volume of a prism on the board. Ask students where they have seen this formula before. Relate the formula, $V = Bh$, with the formulas for the volume of a cylinder by asking: What is the shape of the base of this cylinder? while holding up a cylinder. Substitute the area of the base, B, for the area of a circle, πr^2. Ask what B represents.

Use Graphic Organizers Model on the board how to write two-column notes. The left column will be for rephrasing key concepts, sketches, and examples. The right side is used for additional notes, questions for follow up, and for checks or stars to designate the most important information.

5 Assess & Remediate

Lesson Quiz

1. An aquarium has the dimensions shown below. What is the volume of the tank?

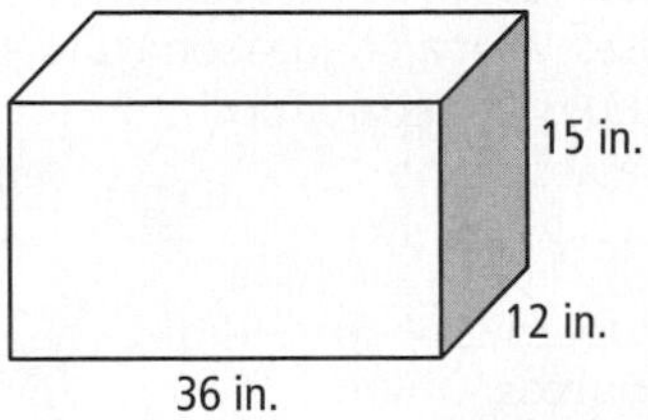

2. Find the volume of the cylinder in terms of π.

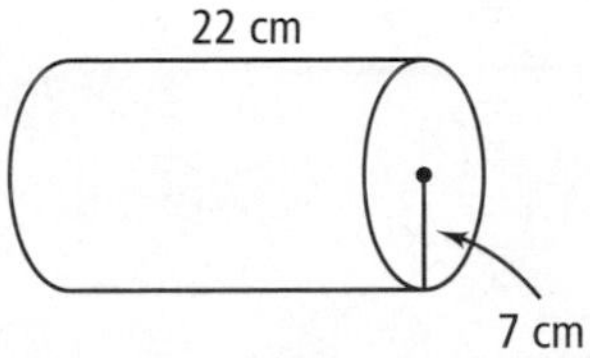

3. Do You UNDERSTAND? A machinist creates a spacer by drilling a 3 cm diameter hole through the middle of a cylinder that has a diameter of 5 cm and a height of 0.5 cm. What is the approximate volume of the finished spacer?

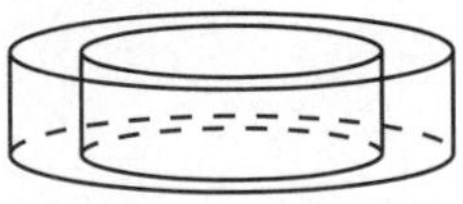

ANSWERS TO LESSON QUIZ

1. 6480 in.^{3}

2. 1078π cm^3

3. about 6.3 cm^3

PRESCRIPTION FOR REMEDIATION

Use the student work on the Lesson Quiz to prescribe a differentiated review assignment.

Points	Differentiated Remediation
0–1	Intervention
2	On-level
3	Extension

PowerGeometry.com

5 Assess & Remediate

Assign the Lesson Quiz. Appropriate intervention, practice, or enrichment is automatically generated based on student performance.

Intervention

- **Reteaching** (2 pages) Provides reteaching and practice exercises for the key lesson concepts. Use with struggling students or absent students.
- **English Language Learner Support** Helps students develop and reinforce mathematical vocabulary and key concepts.

All-in-One Resources/Online

Reteaching

11-4 **Reteaching**
Volumes of Prisms and Cylinders

Problem

Which is greater: the volume of the cylinder or the volume of the prism?

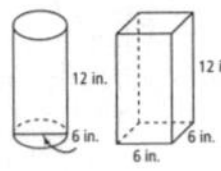

Volume of the cylinder: $V = Bh$

$= \pi r^2 \cdot h$

$= \pi(3)^2 \cdot 12$

≈ 339.3 in.^{3}

Volume of the prism: $V = Bh$

$= s^2 \cdot h$

$= 6^2 \cdot 12$

$= 432$ in.^{3}

The volume of the prism is greater.

Exercises

Find the volume of each object.

1. the rectangular prism part of the milk container
   490 cm^3
2. the cylindrical part of the measuring cup

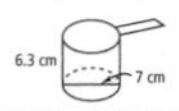

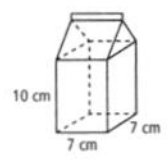

77.175π or about 242.5 cm^3

Find the volume of each of the following. Round your answers to the nearest tenth, if necessary.

3. a square prism with base length 7 m and height 15 m
   735 m^3
4. a cylinder with radius 9 in. and height 10 in.
   2544.7 in.^{3}
5. a triangular prism with height 14 ft and a right triangle base with legs measuring 9 ft and 12 ft
   756 ft^3
6. a cylinder with diameter 24 cm and height 5 cm
   2261.9 cm^3

All-in-One Resources/Online

English Language Learner Support

11-4 **Additional Vocabulary Support**
Volumes of Prisms and Cylinders

Problem

What is the volume of the prism at the right? Justify and explain your work.

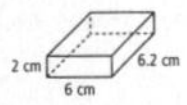

Explain	Work	Justify
First, write the formula for volume of a prism.	$V = Bh$	formula for volume of a prism
Second, substitute the value of the height.	$V = B(2)$	substitution, $h = 2$ cm
Next, substitute an expression for the area of the base.	$V = (6.2 \cdot 6)(2)$	The base is a rectangle with side lengths 6 cm and 6.2 cm.
Finally, find the product. The volume of the rectangular prism is 74.4 cm^3.	$V = 74.4$ Solution 74.4 cm^3	Multiply.

Exercise

What is the volume of the prism at the right? Justify and explain your work.

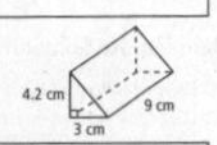

Explain	Work	Justify
First, write the formula for volume of a prism.	$V = Bh$	formula for volume of a prism
Second, substitute the value of the height.	$V = B(9)$	substitution, $h = 9$ cm
Next, substitute an expression for the area of the base.	$V = \frac{1}{2}(3 \cdot 4.2)(9)$	The base is a right triangle with legs 3 cm and 4.2 cm.
Finally, find the product. The volume of the rectangular prism is 56.7 cm^3.	$V = 56.7$ Solution 56.7 cm^3	Multiply.

Differentiated Remediation *continued*

On-Level

- **Practice** (2 pages) Provides extra practice for each lesson. For simpler practice exercises, use the Form K Practice pages found in the All-in-One Teaching Resources and online.
- **Think About a Plan** Helps students develop specific problem-solving skills and strategies by providing scaffolded guiding questions.
- **Standardized Test Prep** Focuses on all major exercises, all major question types, and helps students prepare for the high-stakes assessments.

Extension

- **Enrichment** Provides students with interesting problems and activities that extend the concepts of the lesson.
- **Activities, Games, and Puzzles** Worksheets that can be used for concepts development, enrichment, and for fun!

Practice and Problem Solving Wkbk/ All-in-One Resources/Online

Practice page 1

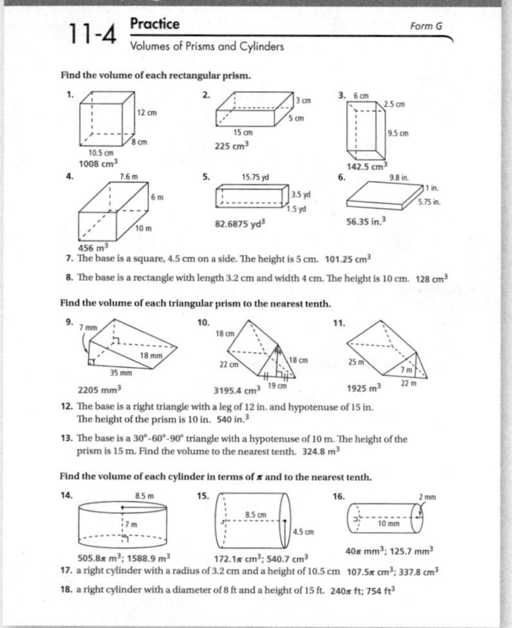

11-4 Practice — Volumes of Prisms and Cylinders — Form G

Find the volume of each rectangular prism.

1. 1008 cm³ 2. 225 cm³ 3. 142.5 cm³ 4. 456 m³ 5. 82.6875 yd³ 6. 56.35 in.³

7. The base is a square, 4.5 cm on a side. The height is 5 cm. 101.25 cm³

8. The base is a rectangle with length 3.2 cm and width 4 cm. The height is 10 cm. 128 cm³

Find the volume of each triangular prism to the nearest tenth.

9. 2205 mm³ 10. 3195.4 cm³ 11. 1925 m³

12. The base is a right triangle with a leg of 12 in. and hypotenuse of 15 in. The height of the prism is 10 in. 540 in.³

13. The base is a 30°-60°-90° triangle with a hypotenuse of 10 m. The height of the prism is 15 m. Find the volume to the nearest tenth. 324.8 m³

Find the volume of each cylinder in terms of π and to the nearest tenth.

14. 505.8π m³; 1588.9 m³ 15. 172.1π cm³; 540.7 cm³ 16. 40π mm³; 125.7 mm³

17. a right cylinder with a radius of 3.2 cm and a height of 10.5 cm 107.5π cm³; 337.8 cm³

18. a right cylinder with a diameter of 8 ft and a height of 15 ft. 240π ft; 754 ft³

Practice and Problem Solving Wkbk/ All-in-One Resources/Online

Practice page 2

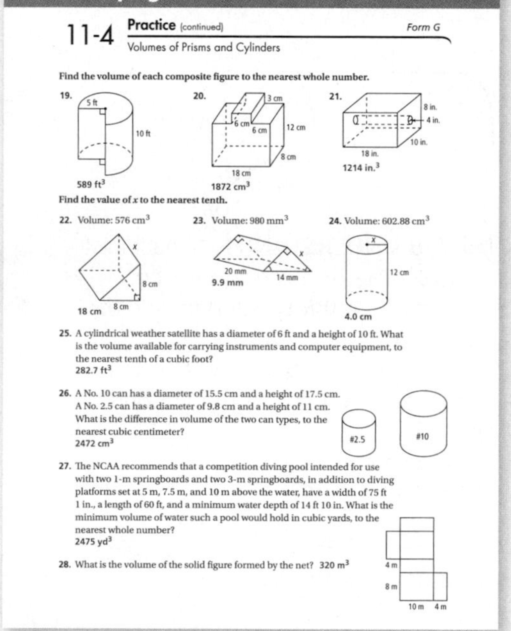

11-4 Practice (continued) — Volumes of Prisms and Cylinders — Form G

Find the volume of each composite figure to the nearest whole number.

19. 589 ft³ 20. 1872 cm³ 21. 1214 in.³

Find the value of x to the nearest tenth.

22. Volume: 576 cm³ 18 cm 23. Volume: 980 mm³ 9.9 mm 24. Volume: 602.88 cm³ 4.0 cm

25. A cylindrical weather satellite has a diameter of 6 ft and a height of 10 ft. What is the volume available for carrying instruments and computer equipment, to the nearest tenth of a cubic foot? 282.7 ft³

26. A No. 10 can has a diameter of 15.5 cm and a height of 17.5 cm. A No. 2.5 can has a diameter of 9.8 cm and a height of 11 cm. What is the difference in volume of the two can types, to the nearest cubic centimeter? 2472 cm³

27. The NCAA recommends that a competition diving pool intended for use with two 1-m springboards and two 3-m springboards, in addition to diving platforms set at 5 m, 7.5 m, and 10 m above the water, have a width of 75 ft 1 in., a length of 60 ft, and a minimum water depth of 14 ft 10 in. What is the minimum volume of water such a pool would hold in cubic yards, to the nearest whole number? 2475 yd³

28. What is the volume of the solid figure formed by the net? 320 m³

All-in-One Resources/Online

Enrichment

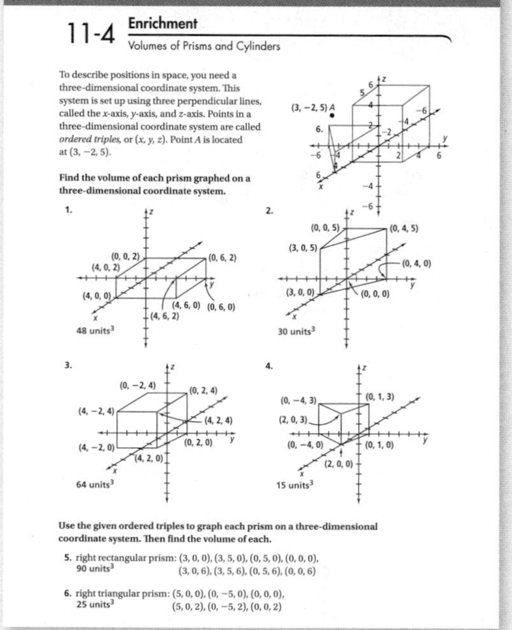

11-4 Enrichment — Volumes of Prisms and Cylinders

To describe positions in space, you need a three-dimensional coordinate system. This system is set up using three perpendicular lines, called the x-axis, y-axis, and z-axis. Points in a three-dimensional coordinate system are called *ordered triples*, or (x, y, z). Point A is located at (3, −2, 5).

Find the volume of each prism graphed on a three-dimensional coordinate system.

1. 48 units³ 2. 30 units³ 3. 64 units³ 4. 15 units³

Use the given ordered triples to graph each prism on a three-dimensional coordinate system. Then find the volume of each.

5. right rectangular prism: (3, 0, 0), (3, 5, 0), (0, 5, 0), (0, 0, 0), (3, 0, 6), (3, 5, 6), (0, 5, 6), (0, 0, 6) 90 units³

6. right triangular prism: (5, 0, 0), (0, −5, 0), (0, 0, 0), (5, 0, 2), (0, −5, 2), (0, 0, 2) 25 units³

Practice and Problem Solving Wkbk/ All-in-One Resources/Online

Think About a Plan

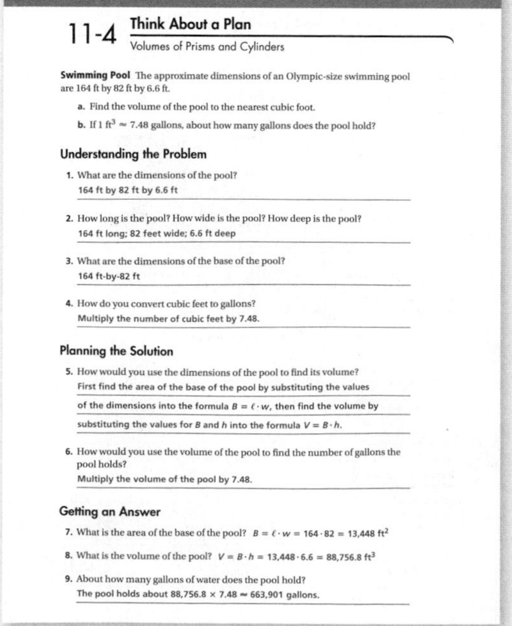

11-4 Think About a Plan — Volumes of Prisms and Cylinders

Swimming Pool The approximate dimensions of an Olympic-size swimming pool are 164 ft by 82 ft by 6.6 ft.

a. Find the volume of the pool to the nearest cubic foot.

b. If 1 ft³ ≈ 7.48 gallons, about how many gallons does the pool hold?

Understanding the Problem

1. What are the dimensions of the pool? 164 ft by 82 ft by 6.6 ft

2. How long is the pool? How wide is the pool? How deep is the pool? 164 ft long; 82 feet wide; 6.6 ft deep

3. What are the dimensions of the base of the pool? 164 ft-by-82 ft

4. How do you convert cubic feet to gallons? Multiply the number of cubic feet by 7.48.

Planning the Solution

5. How would you use the dimensions of the pool to find its volume? First find the area of the base of the pool by substituting the values of the dimensions into the formula $B = \ell \cdot w$, then find the volume by substituting the values for B and h into the formula $V = B \cdot h$.

6. How would you use the volume of the pool to find the number of gallons the pool holds? Multiply the volume of the pool by 7.48.

Getting an Answer

7. What is the area of the base of the pool? $B = \ell \cdot w = 164 \cdot 82 = 13{,}448$ ft²

8. What is the volume of the pool? $V = B \cdot h = 13{,}448 \cdot 6.6 = 88{,}756.8$ ft³

9. About how many gallons of water does the pool hold? The pool holds about 88,756.8 × 7.48 ≈ 663,901 gallons.

Practice and Problem Solving Wkbk/ All-in-One Resources/Online

Standardized Test Prep

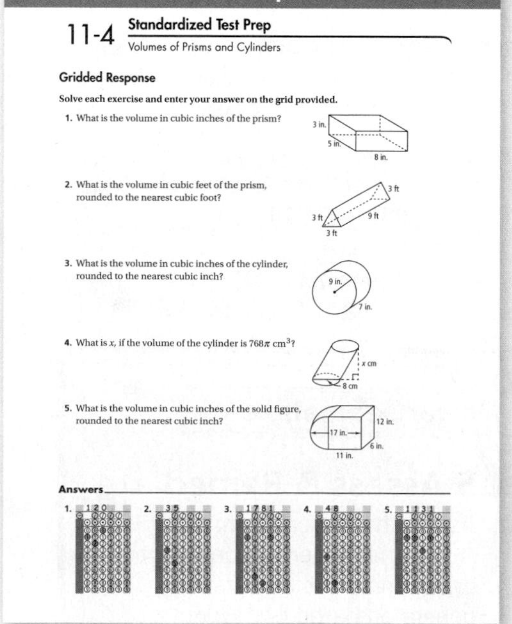

11-4 Standardized Test Prep — Volumes of Prisms and Cylinders

Gridded Response

Solve each exercise and enter your answer on the grid provided.

1. What is the volume in cubic inches of the prism?

2. What is the volume in cubic feet of the prism, rounded to the nearest cubic foot?

3. What is the volume in cubic inches of the cylinder, rounded to the nearest cubic inch?

4. What is x, if the volume of the cylinder is 768π cm³?

5. What is the volume in cubic inches of the solid figure, rounded to the nearest cubic inch?

Answers 1. 120 2. 35 3. 1781 4. 48 5. 1331

Online Teacher Resource Center

Activities, Games, and Puzzles

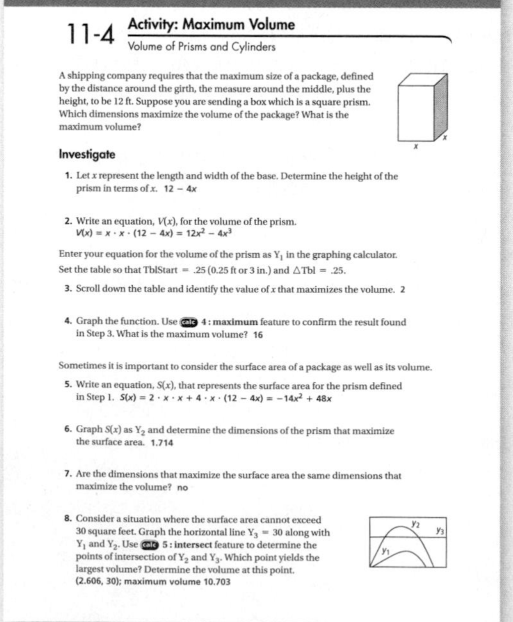

11-4 Activity: Maximum Volume — Volumes of Prisms and Cylinders

A shipping company requires that the maximum size of a package, defined by the distance around the girth, the measure around the middle, plus the height, to be 12 ft. Suppose you are sending a box which is a square prism. Which dimensions maximize the volume of the package? What is the maximum volume?

Investigate

1. Let x represent the length and width of the base. Determine the height of the prism in terms of x. $12 - 4x$

2. Write an equation, V(x), for the volume of the prism. $V(x) = x \cdot x \cdot (12 - 4x) = 12x^2 - 4x^3$

Enter your equation for the volume of the prism as Y_1 in the graphing calculator. Set the table so that TblStart = .25 (0.25 ft or 3 in.) and ΔTbl = .25.

3. Scroll down the table and identify the value of x that maximizes the volume. 2

4. Graph the function. Use calc 4 : maximum feature to confirm the result found in Step 3. What is the maximum volume? 16

Sometimes it is important to consider the surface area of a package as well as its volume.

5. Write an equation, S(x), that represents the surface area for the prism defined in Step 1. $S(x) = 2 \cdot x \cdot x + 4 \cdot x \cdot (12 - 4x) = -14x^2 + 48x$

6. Graph S(x) as Y_2 and determine the dimensions of the prism that maximize the surface area. 1.714

7. Are the dimensions that maximize the surface area the same dimensions that maximize the volume? no

8. Consider a situation where the surface area cannot exceed 30 square feet. Graph the horizontal line $Y_3 = 30$ along with Y_1 and Y_2. Use calc 5 : intersect feature to determine the points of intersection of Y_2 and Y_3. Which point yields the largest volume? Determine the volume at this point. (2.606, 30); maximum volume 10.703

Concept Byte

Use With Lesson 11-5

ACTIVITY

Finding Volume

Common Core State Standards

Prepares for G-GMD.A.1 Give an informal argument for the formulas for the circumference of a circle, area of a circle, volume of a cylinder, pyramid, and cone.

MP 7

You know how to find the volumes of a prism and of a cylinder. Use the following activities to explore finding the volumes of a pyramid and of a cone.

Activity 1

Step 1 Draw the nets shown at the right on heavy paper.

Step 2 Cut out the nets and tape them together to make a cube and a square pyramid. Each model will have one open face.

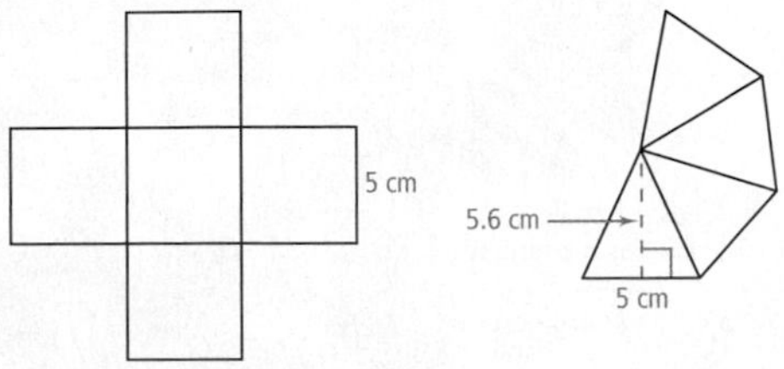

1. How do the areas of the bases of the cube and the pyramid compare?
2. How do the heights of the cube and the pyramid compare?
3. Fill the pyramid with rice or other material. Then pour the rice from the pyramid into the cube. How many pyramids full of rice does the cube hold?
4. The volume of the pyramid is what fractional part of the volume of the cube?
5. **Make a Conjecture** What do you think is the formula for the volume of a pyramid? Explain.

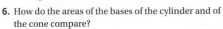

Activity 2

Step 1 Draw the nets shown at the right on heavy paper.

Step 2 Cut out the nets and tape them together to make a cylinder and a cone. Each model will have one open face.

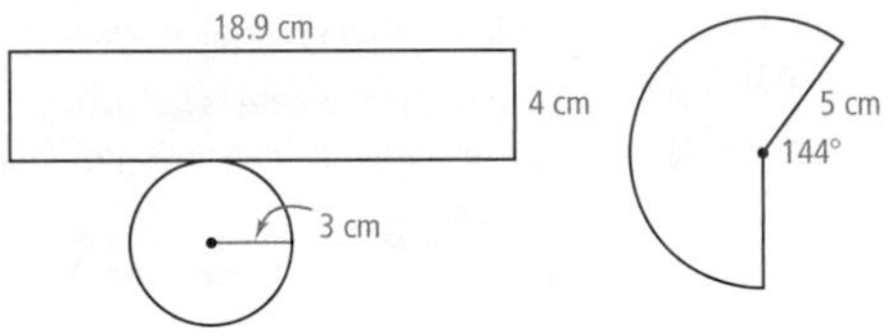

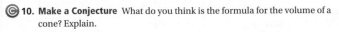

6. How do the areas of the bases of the cylinder and of the cone compare?
7. How do the heights of the cylinder and of the cone compare?
8. Fill the cone with rice or other material. Then pour the rice from the cone into the cylinder. How many cones full of rice does the cylinder hold?
9. What fractional part of the volume of the cylinder is the volume of the cone?
10. **Make a Conjecture** What do you think is the formula for the volume of a cone? Explain.

Answers

Activity 1

1. The areas are equal.
2. The heights are about equal.
3. about 3 pyramids
4. $\frac{1}{3}$
5. $V = \frac{1}{3}Bh$; the volume of a cube is Bh. The volume of the pyramid is $\frac{1}{3}$ the volume of the cube, or $\frac{1}{3}Bh$.

Activity 2

6. The areas are =.
7. The heights are =.
8. about 3 cones
9. $\frac{1}{3}$
10. $V = \frac{1}{3}Bh$; the volume of a cylinder is Bh. The volume of a cone is $\frac{1}{3}$ the volume of the cylinder, or $\frac{1}{3}Bh$.

Guided Instruction

PURPOSE To derive formulas for finding the volumes of pyramids and cones

PROCESS Students will

- make models of a cube, a square pyramid, a cylinder, and a cone from four provided net templates and compare the dimensions of their models.
- fill their open-faced pyramids with rice and transfer its contents to the cube or cylinder and make conjectures about the relationships between the volume of a prism and a pyramid and between a cylinder and a cone.

DISCUSS Review the formula for volume of a prism ($V = bh$) and a cylinder ($V = \pi r^2 h$).

Activity 1

In this Activity students compare the volume of a prism to the volume of a pyramid with a congruent base.

Q What is the perimeter of each base? **[20 cm]**

Q What is the area of each base? **[25 cm]**

Q Why is the height given on one face of the net of the pyramid different from the height of the prism? **[The height given is the slant height; the Pythagorean Theorem allows you to find a pyramid height of 5 cm.]**

Activity 2

In this Activity students compare the volume of a cylinder to the volume of a cone with a congruent base.

Q What is the height of the cylinder? **[4 cm]**

Q The length of the rectangle shown in the net of the cylinder is equal to what other measurement? Justify your answer. **[the circumference of the base; $C = 2\pi r = 2 \cdot 3 \cdot \pi \approx 18.85$]**

Q What does the segment labeled 5 cm in the net of the cone represent? **[the slant height]**

Q The cylinder and the cone formed by these nets are related. What is the radius of the base of the cone and the height of the cone? **[$r = 3$ cm; $h = 4$ cm]**

Q What formula can you use to verify the relationship of the radius, height, and slant height of the cone? Explain. **[Pythagorean Theorem; the slant height is the hypotenuse of a right triangle that has legs equal to the radius and the height. $5^2 = 3^2 + 4^2$]**

Mathematical Practices This Concept Byte supports students in becoming proficient in shifting perspective, Mathematical Practice 7.

1 Interactive Learning

Solve It!

PURPOSE To discover the formula for the volume of a pyramid

PROCESS Students may look for a pattern in the volumes of the pyramids in the diagram.

FACILITATE

Q How do you find the volume of a prism? **[Multiply the area of the base by the height.]**

Q What are the products of the areas of the bases and the heights for all the pyramids in the diagram? **[1 ft^3, 3 ft^3, 12 ft^3, 6 ft^3, 18 ft^3]**

Q What is the relationship between the volumes of the pyramids and the products you calculated? **[The volumes are one-third the products.]**

ANSWER See Solve It in Answers on next page.

CONNECT THE MATH Students discover the formula for the volume of a pyramid. In the lesson, students learn the formula for the volume of a pyramid and that a pyramid has $\frac{1}{3}$ the volume of a prism with the same dimensions.

2 Guided Instruction

Take Note

TACTILE LEARNERS

Challenge students to explain why the volume of a pyramid is one-third of its corresponding rectangular solid. It may help them to model the figures with clay. They can use string to cut a rectangular solid into a pyramid.

11-5 Volumes of Pyramids and Cones

Common Core State Standards

G-GMD.A.3 Use volume formulas for . . . pyramids, cones . . . to solve problems.

G-MG.A.1 Use geometric shapes, their measures, and their properties to describe objects.

MP 1, MP 3, MP 4, MP 7

Objective To find the volume of a pyramid and of a cone

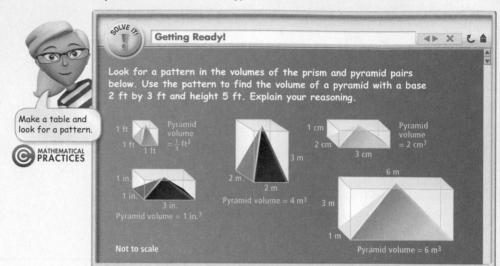

Getting Ready!

Look for a pattern in the volumes of the prism and pyramid pairs below. Use the pattern to find the volume of a pyramid with a base 2 ft by 3 ft and height 5 ft. Explain your reasoning.

MATHEMATICAL PRACTICES

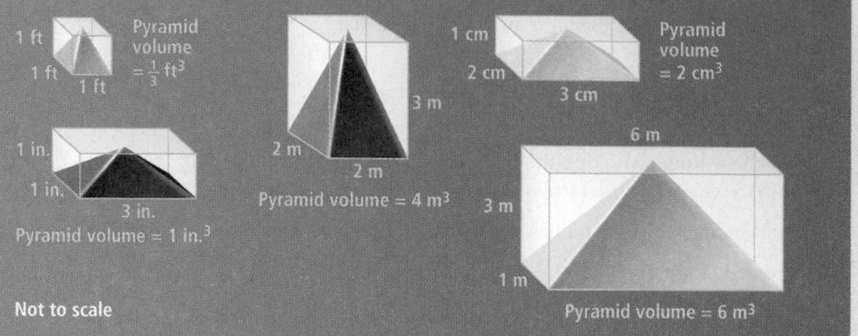

In the Solve It, you analyzed the relationship between the volume of a prism and the volume of an embedded pyramid.

Essential Understanding The volume of a pyramid is related to the volume of a prism with the same base and height.

take note

Theorem 11-8 Volume of a Pyramid

The volume of a pyramid is one third the product of the area of the base and the height of the pyramid.

$V = \frac{1}{3}Bh$

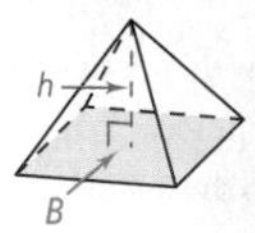

Because of Cavalieri's Principle, the volume formula is true for all pyramids. The height h of an oblique pyramid is the length of the perpendicular segment from its vertex to the plane of the base.

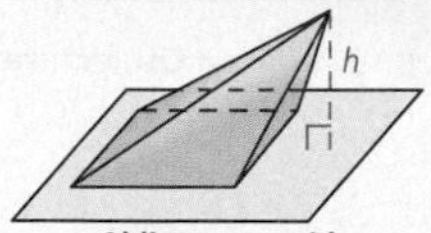

Oblique pyramid

11-5 Preparing to Teach

BIG ideas **Measurement**
Visualization

ESSENTIAL UNDERSTANDINGS

- The volume of a pyramid is related to the volume of a prism with the same base and height.
- The volume of a cone is related to the volume of a cylinder with the same base and height.

Math Background

Volume can be calculated for any solid figure. Pyramids and cones are related to analogous solids—the prism and the cylinder, respectively, so their volume formulas are similar.

Students will use their knowledge of volume to learn about the volumes of pyramids and cones. A pyramid has one-third the volume of the rectangular prism with the same base and height. A cone has one-third the volume of the cylinder with the same base and height. Cones and pyramids can also be a part of composite figures.

Cavalieri's Principle also applies to the volume formulas for these solids.

Mathematical Practices

Construct viable arguments and critique the reasoning of others. With a knowledge of Cavalieri's Principle with respect to prisms and cylinders, students will construct an argument about finding the volume of oblique cones and pyramids.

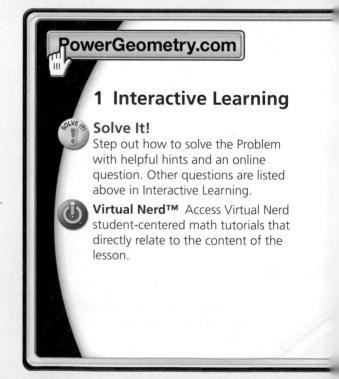

PowerGeometry.com

1 Interactive Learning

Solve It!
Step out how to solve the Problem with helpful hints and an online question. Other questions are listed above in Interactive Learning.

Virtual Nerd™ Access Virtual Nerd student-centered math tutorials that directly relate to the content of the lesson.

Problem 1 Finding Volume of a Pyramid STEM

Architecture **The entrance to the Louvre Museum in Paris, France, is a square pyramid with a height of 21.64 m. What is the approximate volume of the Louvre Pyramid?**

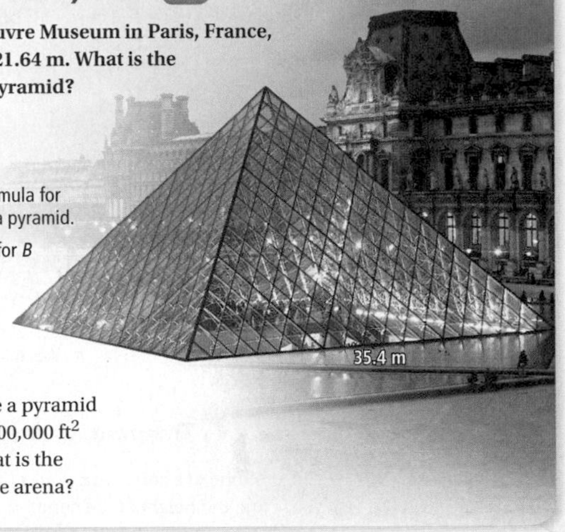

Think
How is this similar to finding the volume of a prism?
In both cases, you need the area of the base and the height.

The area of the base of the pyramid is 35.4 m • 35.4 m, or 1253.16 m^2.

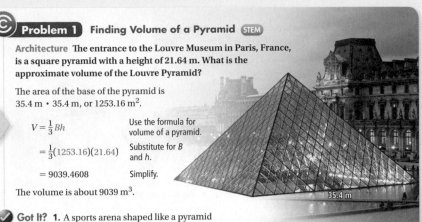

$V = \frac{1}{3}Bh$	Use the formula for volume of a pyramid.
$= \frac{1}{3}(1253.16)(21.64)$	Substitute for *B* and *h*.
$= 9039.4608$	Simplify.

The volume is about 9039 m^3.

Got It? 1. A sports arena shaped like a pyramid has a base area of about 300,000 ft^2 and a height of 321 ft. What is the approximate volume of the arena?

Problem 2 Finding the Volume of a Pyramid GRIDDED RESPONSE

What is the volume in cubic feet of a square pyramid with base edges 40 ft and slant height 25 ft?

Think
How do you use the slant height?
The slant height is the length of the hypotenuse of the right triangle. Use the slant height to find the height of the pyramid.

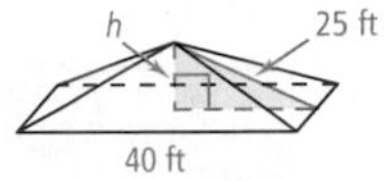

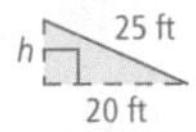

Step 1 Find the height of the pyramid.

$c^2 = a^2 + b^2$	Use the Pythagorean Theorem.
$25^2 = h^2 + 20^2$	Substitute 25 for *c*, *h* for *a*, and $\frac{40}{2}$, or 20, for *b*.
$625 = h^2 + 400$	Simplify.
$h^2 = 225$	Solve for h^2.
$h = 15$	Take the positive square root of both sides.

Step 2 Find the volume of the pyramid.

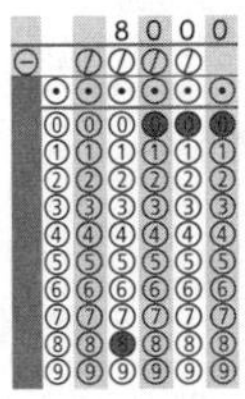

$V = \frac{1}{3}Bh$	Use the formula for volume of a pyramid.
$= \frac{1}{3}(40 \cdot 40)(15)$	Substitute 40 • 40 for *B* and 15 for *h*.
$= 8000$	Simplify.

The volume of the pyramid is 8000 ft^3.

Got It? 2. What is the volume of a square pyramid with base edges 24 m and slant height 13 m?

Problem 1

Q What is the base of the pyramid? **[The base is a square with side lengths 35.4 m.]**

Q How do you find the volume of the pyramid? **[Find one-third the product of the area of the base times the height.]**

Got It?

Have students sketch a diagram of the pyramid. Make sure that students label the correct measurements in their diagrams.

Problem 2

Q How is the height related to the slant height of the pyramid? **[It forms a right triangle with one leg half the length of a side of the base of the pyramid.]**

Q How can you find the height of the pyramid? **[Use the Pythagorean Theorem.]**

Q How do you find the volume of the pyramid? **[Multiply the area of the base times the height times $\frac{1}{3}$.]**

Got It?

Have students draw a diagram of the pyramid and label the known dimensions. Have them draw the right triangle formed by the height and the slant height. Perform the calculation using half the side length as one side of the right triangle.

2 Guided Instruction

Each Problem is worked out and supported online.

Problem 1
Finding Volume of a Pyramid
Animated

Problem 2
Finding the Volume of a Pyramid
Animated

Problem 3
Finding the Volume of a Cone

Problem 4
Finding the Volume of an Oblique Cone
Animated

Support in Geometry Companion

- Vocabulary
- Key Concepts
- Got It?

Answers

Solve It!

Pattern: From the examples shown, the pyramids and a prism have the same base and the same height, and the volume of each pyramid is one third the volume of its corresponding prism. Based on this pattern, the volume of the pyramid that fits in a prism with base 2 ft by 3 ft with height 5 ft will be $\frac{1}{3}(30)$ or 10 ft^3.

Got It?

1. 32,100,000 ft^3
2. 960 m^3

Take Note

Q What solid can have the same base and height as a cone, but a larger, different shape? **[cylinder]**

Q What solid can have the same base and height as a pyramid, but a larger, different shape? **[prism]**

Have students compare and contrast the formulas for the volume of a cone to the formula for the volume of a cylinder. For a cone, its volume equals the volume of its corresponding cylinder divided by 3.

Problem 3

Q How can you find the area of the base of the tepee? **[Find the area of the circular base.]**

Q What is the radius of the base of the tepee? **[$\frac{1}{2}$(14 ft) = 7 ft]**

Q How can you find the volume of the tepee? **[Multiply $\frac{1}{3}$ times 12 times the area of the circular base.]**

Got It?

Q For 3a, what is the height of the child's tepee? **[6 ft]**

Q For 3a, what is the radius of the base of the child's tepee? **[3.5 ft]**

For 3b, ask students what their expectations are about the relationship of the original tepee and the child's tepee. After they have performed the calculations, see if anyone can explain why the expectations were not accurate. For on-level and advanced classes, you can derive the ratio using the formula with the variables still in place. Do not make substitutions.

Essential Understanding The volume of a cone is related to the volume of a cylinder with the same base and height.

The cones and the cylinder have the same base and height. It takes three cones full of rice to fill the cylinder.

take note

Theorem 11-9 Volume of a Cone

The volume of a cone is one third the product of the area of the base and the height of the cone.

$V = \frac{1}{3}Bh$, or $V = \frac{1}{3}\pi r^2 h$

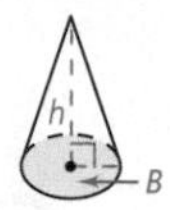

A cone-shaped structure can be particularly strong, as downward forces at the vertex are distributed to all points in its circular base.

Problem 3 Finding the Volume of a Cone STEM

Traditional Architecture **The covering on a tepee rests on poles that come together like concurrent lines. The resulting structure approximates a cone. If the tepee pictured is 12 ft high with a base diameter of 14 ft, what is its approximate volume?**

Think
How is this similar to finding the volume of a cylinder?
In both cases, you need to find the base area of a circle.

$V = \frac{1}{3}\pi r^2 h$	Use the formula for the volume of a cone.
$= \frac{1}{3}\pi(7)^2(12)$	Substitute $\frac{14}{2}$, or 7, for r and 12 for h.
≈ 615.7521601	Use a calculator.

The volume of the tepee is approximately 616 ft^3.

Got It? 3. a. The height and radius of a child's tepee are half those of the tepee in Problem 3. What is the volume of the child's tepee to the nearest cubic foot?

b. Reasoning What is the relationship between the volume of the original tepee and the child's tepee?

728 Chapter 11 Surface Area and Volume

Additional Problems

1. The Great Pyramid of Giza is the largest of the original Seven Wonders of the World. The pyramid originally had the dimensions shown below. What was the approximate volume of the Great Pyramid?

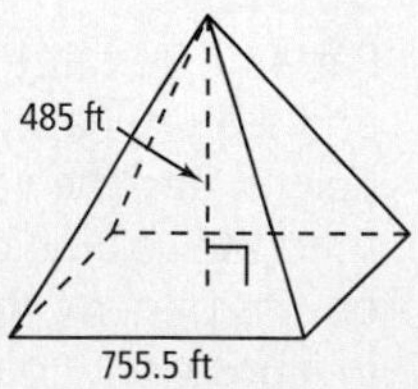

ANSWER about 92,276,140 ft^3

2. What is the volume in cubic yards of a square pyramid with base edges 18 yd and slant height 15 yd?

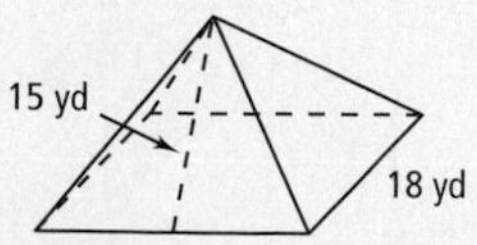

ANSWER 1296 yd^3

3. About how many cubic centimeters of water does the paper drinking cup hold?

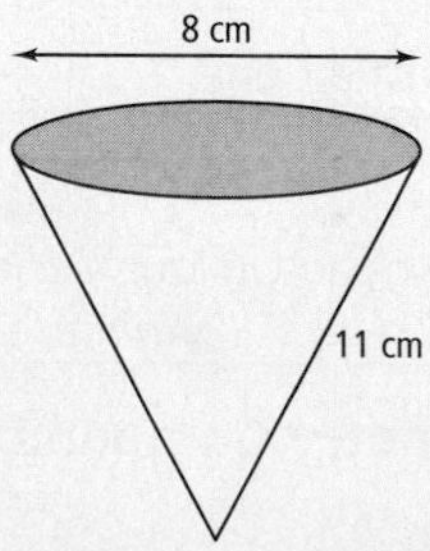

ANSWER about 172 cm^3

4. What is the volume of the oblique cone below? Give your answer in terms of π and also rounded to the nearest cubic inch.

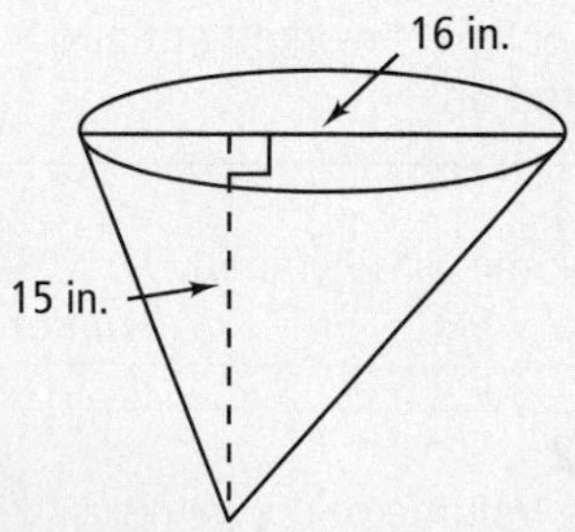

ANSWER 320π $in.^3$ or about 1005 $in.^3$

This volume formula applies to all cones, including oblique cones.

Problem 4 Finding the Volume of an Oblique Cone

What is the volume of the oblique cone at the right? Give your answer in terms of π and also rounded to the nearest cubic foot.

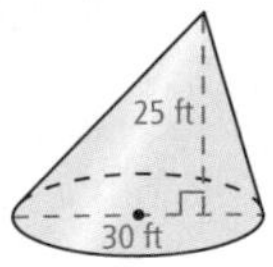

Think

What is the height of the oblique cone?
The height is the length of the perpendicular segment from the vertex of the cone to the base, which is 25 ft. In an oblique cone, the segment does not intersect the center of the base.

$V = \frac{1}{3}\pi r^2 h$	Use the formula for volume of a cone.
$= \frac{1}{3}\pi(15)^2(25)$	Substitute 15 for *r* and 25 for *h*.
$= 1875\pi$	Simplify.
≈ 5890.486225	Use a calculator.

The volume of the cone is 1875π ft^3, or about 5890 ft^3.

Got It? 4. a. What is the volume of the oblique cone at the right in terms of π and rounded to the nearest cubic meter?

b. Reasoning How does the volume of an oblique cone compare to the volume of a right cone with the same diameter and height? Explain.

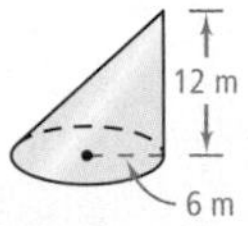

Lesson Check

Do you know HOW?

What is the volume of each figure? If necessary, round to the nearest tenth.

1.

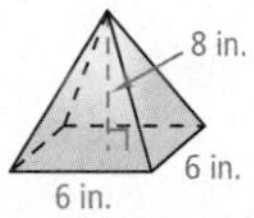

2.

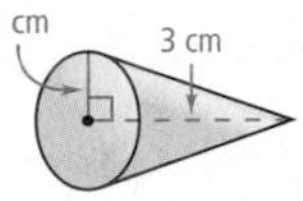

Do you UNDERSTAND? MATHEMATICAL PRACTICES

3. Compare and Contrast How are the formulas for the volume of a pyramid and the volume of a cone alike? How are they different?

4. Error Analysis A square pyramid has base edges 13 ft and height 10 ft. A cone has diameter 13 ft and height 10 ft. Your friend claims the figures have the same volume because the volume formulas for a pyramid and a cone are the same: $V = \frac{1}{3}Bh$. What is her error?

Practice and Problem-Solving Exercises MATHEMATICAL PRACTICES

A Practice Find the volume of each square pyramid. See Problem 1.

5. base edges 10 cm, height 6 cm

6. base edges 18 in., height 12 in.

7. base edges 5 m, height 6 m

8. Buildings The Transamerica Pyramid Building in San Francisco is 853 ft tall with a square base that is 149 ft on each side. To the nearest thousand cubic feet, what is the volume of the Transamerica Pyramid?

Problem 4
Remind students that the height of an object is the perpendicular segment from the vertex to the base.

Q What is the length of the radius of the base of the cone? **[15 ft]**

Got It? VISUAL LEARNERS
Have students review Cavalieri's Principle from Lesson 11-4. They should be able to see that the volume of an oblique cone is the same as a right cone with the same dimensions.

3 Lesson Check

Do you know HOW?
- If students have difficulty with Exercise 2, then have them review Problem 3 to verify which measurements are substituted into the formula for the volume.

Do you UNDERSTAND?
- If students have difficulty with Exercise 4, then have them draw diagrams of the two solids.

Close

Q How do you find the volume of a pyramid or a cone? **[Find one-third of the product of the area of the base times the height.]**

Q How do the volumes of right solids compare to the volume of oblique solids with the same dimensions? **[The volumes are the same if the dimensions are the same.]**

Answers

Got It? (continued)

3a. 77 ft^3

b. The volume of the original tepee is 8 times the volume of the child's tepee.

4a. 144π m^3; 452 m^3

b. They are equal because both cones have the same base and same height.

Lesson Check

1. 96 in.^{3}

2. 3.1 cm^3

3. Alike: Both formulas are $\frac{1}{3}$ the area of the base times height. Different: Because the bases are different figures, the base area will require different formulas.

4. The areas of the bases are not equal; the area of the base of the pyramid is $13^2 = 169$ ft^2, but the area of the base of the cone is $\pi(6.5)^2 \approx 132.7$ ft^2.

Practice and Problem-Solving Exercises

5. 200 cm^3

6. 1296 in.^{3}

7. 50 m^3

8. 6,312,000 ft^3

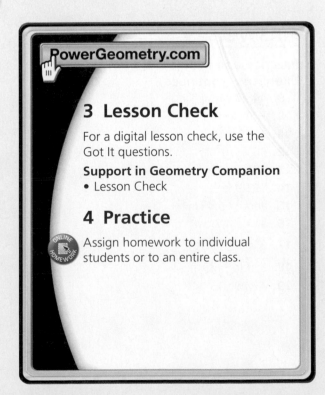

PowerGeometry.com

3 Lesson Check
For a digital lesson check, use the Got It questions.

Support in Geometry Companion
- Lesson Check

4 Practice
Assign homework to individual students or to an entire class.

4 Practice

ASSIGNMENT GUIDE

Basic: 5–21 all, 24–32 even
Average: 5–19 odd, 20–34
Advanced: 5–19 odd, 20–38
Standardized Test Prep: 39–42
Mixed Review: 43–46

Mathematical Practices are supported by exercises with red headings. Here are the Practices supported in this lesson:

MP 1: Make Sense of Problems Ex. 3, 20
MP 3: Construct Arguments Ex. 21, 29
MP 3: Communicate Ex. 22, 26
MP 3: Critique the Reasoning of Others Ex. 4
MP 7: Shift Perspective Ex. 33–36

Applications exercises have blue headings. Exercises 8, 15, 16, 27, 28, and 37b support MP 4: Model.

STEM exercises focus on science or engineering applications.

EXERCISE 26: Use the Think About a Plan worksheet in the **Practice and Problem Solving Workbook** (also available in the Teaching Resources in print and online) to further support students' development in becoming independent learners.

HOMEWORK QUICK CHECK

To check students' understanding of key skills and concepts, go over Exercises 11, 15, 20, 21, and 26.

Find the volume of each square pyramid. Round to the nearest tenth if necessary.

9.
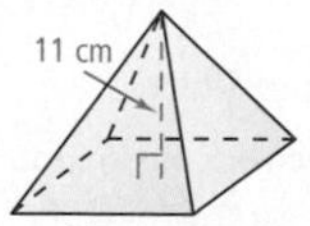

10.
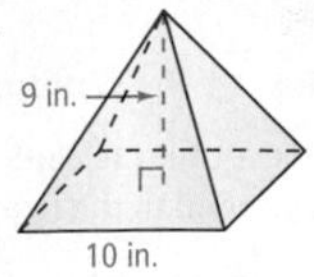

11.
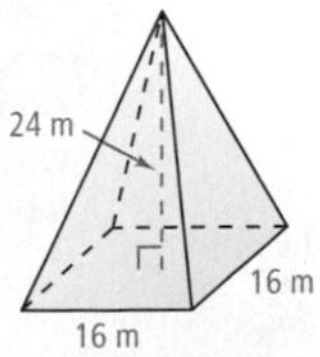

Find the volume of each square pyramid, given its slant height. Round to the nearest tenth. See Problem 2.

12.
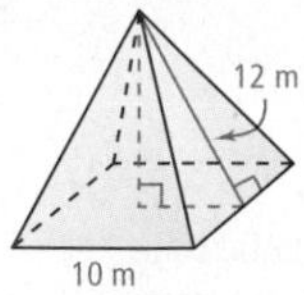

13.
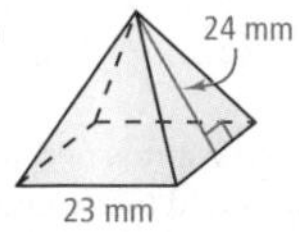

14.
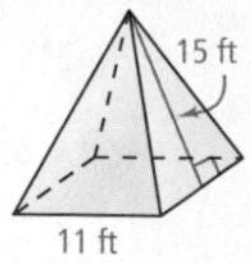

STEM **15. Chemistry** In a chemistry lab you use a filter paper cone to filter a liquid. The diameter of the cone is 6.5 cm and its height is 6 cm. How much liquid will the cone hold when it is full? See Problem 3.

STEM **16. Chemistry** This cone has a filter that was being used to remove impurities from a solution but became clogged and stopped draining. The remaining solution is represented by the shaded region. How many cubic centimeters of the solution remain in the cone?

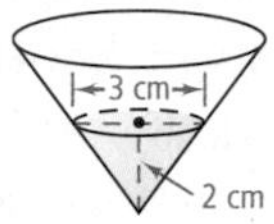

Find the volume of each cone in terms of π and also rounded as indicated. See Problem 4.

17. nearest cubic foot

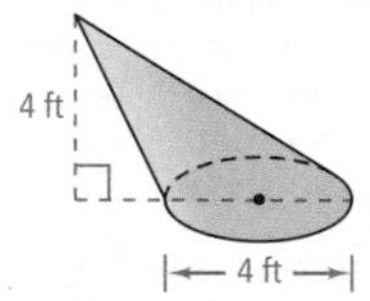

18. nearest cubic inch

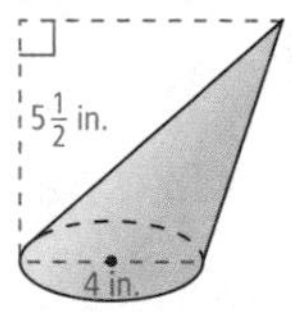

19. nearest cubic meter

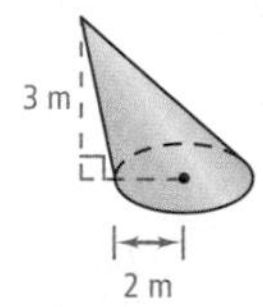

B Apply

20. Think About a Plan A cone with radius 1 fits snugly inside a square pyramid, which fits snugly inside a cube. What are the volumes of the three figures?
- How can you *draw a diagram* of the situation?
- What dimensions do the cone, pyramid, and cube have in common?

21. Reasoning Suppose the height of a pyramid is halved. How does this affect its volume? Explain.

22. Writing Without doing any calculations, explain how the volume of a cylinder with $B = 5\pi$ cm^2 and $h = 20$ cm compares to the volume of a cone with the same base area and height.

Answers

Practice and Problem-Solving Exercises (continued)

9. 443.7 cm^3
10. 300 in.^{3}
11. 2048 m^3
12. 363.6 m^3
13. 3714.5 mm^3
14. 562.9 ft^3
15. about 66.4 cm^3
16. about 4.7 cm^3
17. $\frac{16}{3}\pi$ ft^3; 17 ft^3
18. $\frac{22}{3}\pi$ in.^{3}; 23 in.^{3}
19. 4π m^3; 13 m^3
20. cube: 8 units^3; cone: $\frac{2}{3}\pi$ units^3; pyramid: $\frac{8}{3}$ units^3
21. Volume is halved; $V = \frac{1}{3}Bh$, so if h is replaced with $\frac{h}{2}$, then the volume is $\frac{1}{3}B\left(\frac{h}{2}\right) = \frac{1}{2}\left[\frac{1}{3}Bh\right]$.
22. The volume of the cylinder is 3 times the volume of the cone. (V of cylinder $= Bh$, V of cone $= \frac{1}{3}Bh$)

Find the volume to the nearest whole number.

23\.

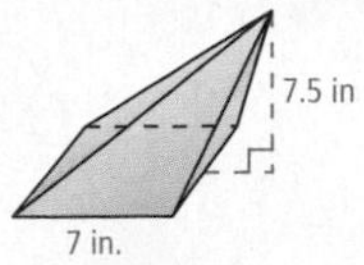

Square base

24\.

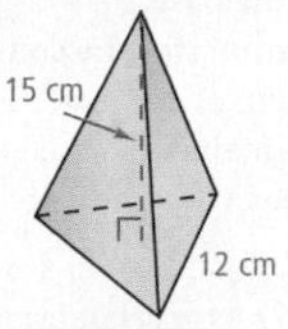

Equilateral base

25\.

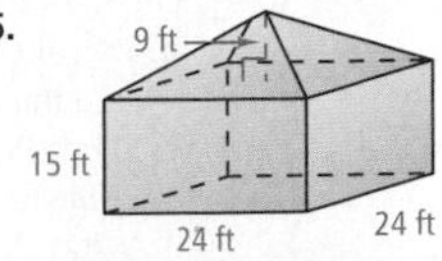

Square base

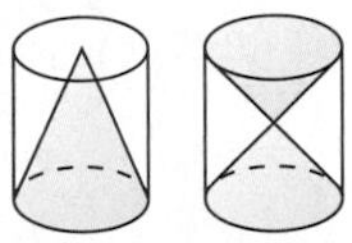

26\. **Writing** The two cylinders pictured at the right are congruent. How does the volume of the larger cone compare to the total volume of the two smaller cones? Explain.

STEM 27. **Architecture** The Pyramid of Peace is an opera house in Astana, Kazakhstan. The height of the pyramid is approximately 62 m and one side of its square base is approximately 62 m.

a. What is its volume to the nearest thousand cubic meters?

b. How tall would a prism-shaped building with the same square base as the Pyramid of Peace have to be to have the same volume as the pyramid?

STEM 28. **Hardware** Builders use a plumb bob to find a vertical line. The plumb bob shown at the right combines a regular hexagonal prism with a pyramid. Find its volume to the nearest cubic centimeter.

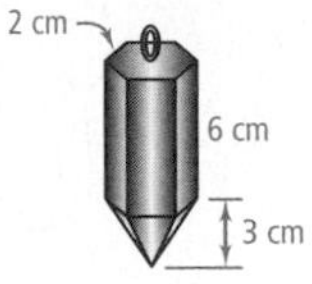

29\. **Reasoning** A cone with radius 3 ft and height 10 ft has a volume of 30π ft^3. What is the volume of the cone formed when the following happens to the original cone?

a. The radius is doubled. **b.** The height is doubled.

c. The radius and the height are both doubled.

Algebra Find the value of the variable in each figure. Leave answers in simplest radical form. The diagrams are not to scale.

30\.

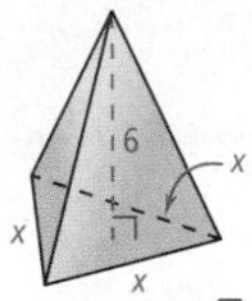

Volume $= 18\sqrt{3}$

31\.

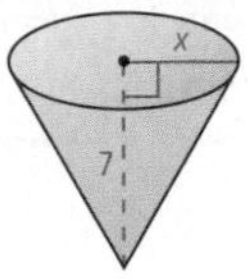

Volume $= 21\pi$

32\.

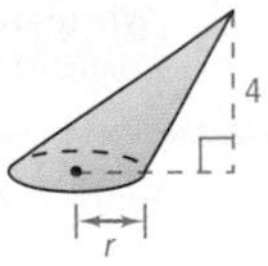

Volume $= 24\pi$

Visualization Suppose you revolve the plane region completely about the given line to sweep out a solid of revolution. Describe the solid. Then find its volume in terms of π.

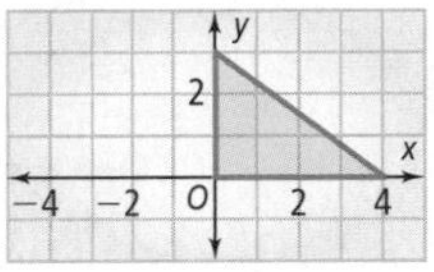

33\. the y-axis

34\. the x-axis

Challenge

35\. the line $x = 4$

36\. the line $y = -1$

23. 123 in.^{3}

24. 312 cm^3

25. 10,368 ft^3

26. They are =; both volumes are $\frac{1}{3}\pi r^2 h$.

27a. 79,000 m^3

b. $20\frac{2}{3}$ m, or about 20.7 m

28. 73 cm^3

29a. 120π ft^3

b. 60π ft^3

c. 240π ft^3

30. 6

31. 3

32. $3\sqrt{2}$

33. cone with $r = 4$ and $h = 3$; 16π

34. cone with $r = 3$ and $h = 4$; 12π

35. cylinder with $r = 4$, $h = 3$, with a cone of $r = 4$, $h = 3$ removed from it; 32π

36. cone with $r = 4$, $h = 5\frac{1}{3}$, with a cone of $r = 1$, $h = 1\frac{1}{3}$ cut off the top, and a cylinder of $r = 1$, $h = 4$ cut out of its center; 24π

Answers

Practice and Problem-Solving Exercises (continued)

37a. The frustum has volume that is the difference of the volumes of the entire cone and the small cone. The frustum has volume $V = \frac{1}{3}\pi R^2 H - \frac{1}{3}\pi r^2 h$ or $\frac{1}{3}\pi(R^2 H - r^2 h)$.

b. about 784.6 in.^{3}

38a. 47.1 m

b. 176.7 m^2

c. 389.6 m^3

39. A

40. G

41. B

42. [4]

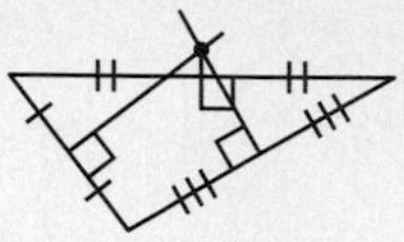

The circumcenter is outside.

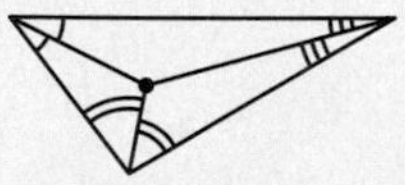

The incenter is inside.

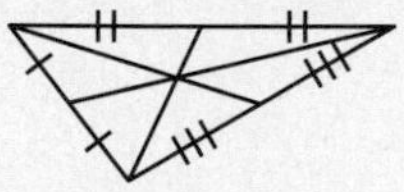

The centroid is inside.

[3] one part incorrect

[2] correct answers, diagrams incorrect

[1] correct answers, diagrams missing

43. 3600 cm^3

44. $JC > KN$

45. 7.1 in.^{2}

46. 13 cm

37. A *frustum* of a cone is the part that remains when the top of the cone is cut off by a plane parallel to the base.

a. Explain how to use the formula for the volume of a cone to find the volume of a frustum of a cone.

b. **Containers** A popcorn container 9 in. tall is the frustum of a cone. Its small radius is 4.5 in. and its large radius is 6 in. What is its volume?

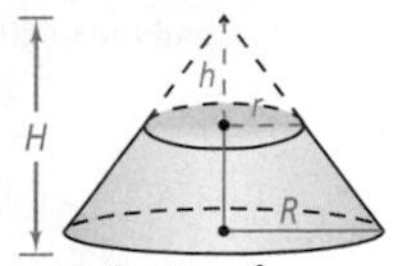

Frustum of cone

38. A disk has radius 10 m. A 90° sector is cut away, and a cone is formed.

a. What is the circumference of the base of the cone?

b. What is the area of the base of the cone?

c. What is the volume of the cone? (*Hint:* Use the slant height and the radius of the base to find the height.)

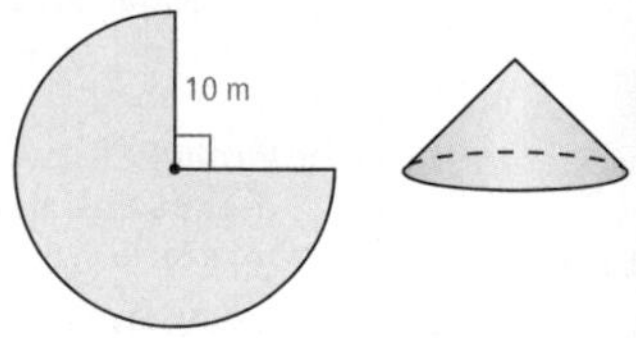

Standardized Test Prep

SAT/ACT

39. A cone has diameter 8 in. and height 14 in. A rectangular prism is 6 in. by 4 in. by 10 in. A square pyramid has base edge 8 in. and height 12 in. What are the volumes of the three figures in order from least to greatest?

(A) cone, prism, pyramid
(B) prism, cone, pyramid
(C) pyramid, cone, prism
(D) prism, pyramid, cone

40. One row of a truth table lists p as true and q as false. Which of the following statements is true?

(F) $p \rightarrow q$ (G) $p \vee q$ (H) $p \wedge q$ (I) $\sim p$

41. If a polyhedron has 8 vertices and 12 edges, how many faces does it have?

(A) 4 (B) 6 (C) 12 (D) 24

Extended Response

42. The point of concurrency of the three altitudes of a triangle lies outside the triangle. Where are its circumcenter, incenter, and centroid located in relation to the triangle? Draw diagrams to support your answers.

Mixed Review

43. A triangular prism has height 30 cm. Its base is a right triangle with legs 10 cm and 24 cm. What is the volume of the prism? See Lesson 11-4.

44. Given $\triangle JAC$ and $\triangle KIN$, you know $\overline{JA} \cong \overline{KI}$, $\overline{AC} \cong \overline{IN}$, and $m\angle A > m\angle I$. What can you conclude about JC and KN? See Lesson 5-7.

Get Ready! **To prepare for Lesson 11-6, do Exercises 45 and 46.**

45. Find the area of a circle with diameter 3 in. to the nearest tenth. See Lesson 1-8.

46. Find the circumference of a circle with radius 2 cm to the nearest centimeter.

11-5 Lesson Resources

Differentiated Remediation

Instructional Support

Geometry Companion

Students can use the **Geometry Companion** worktext (4 pages) . . .

- New Vocabulary
- Key Concepts
- Got It for each Problem
- Lesson Check

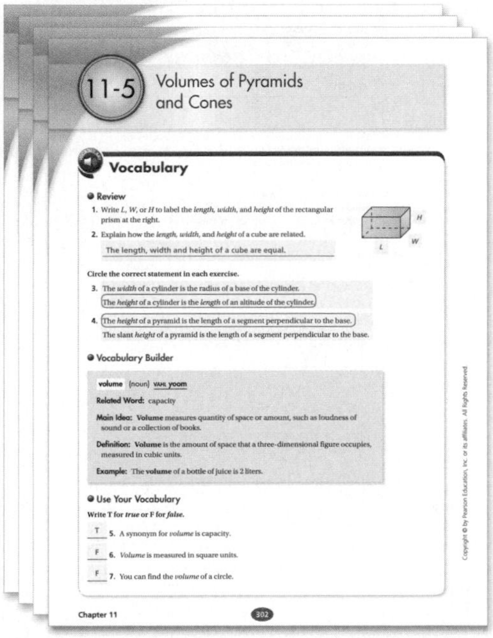
11-5 Volumes of Pyramids and Cones

Vocabulary

Review

1. Write *L*, *W*, or *H* to label the *length*, *width*, and *height* of the rectangular prism at the right.
2. Explain how the *length*, *width*, and *height* of a cube are related.
The length, width and height of a cube are equal.

Circle the correct statement in each exercise.

3. The width of a cylinder is the radius of a base of the cylinder. / The height of a cylinder is the length of an altitude of the cylinder.
4. The height of a pyramid is the length of a segment perpendicular to the base. / The slant height of a pyramid is the length of a segment perpendicular to the base.

Vocabulary Builder

volume (noun) VAHL yoom
Related Word: capacity
Main Idea: Volume measures quantity of space or amount, such as loudness of sound or a collection of books.
Definition: Volume is the amount of space that a three-dimensional figure occupies, measured in cubic units.
Example: The volume of a bottle of juice is 2 liters.

Use Your Vocabulary

Write T for *true* or F for *false*.

T 5. A synonym for *volume* is capacity.
F 6. *Volume* is measured in square units.
F 7. You can find the *volume* of a circle.

Chapter 11 302

ELL Support

Use Manipulatives If possible, use a volume kit to model the Solve It. Show students how a square pyramid fills one-third of a rectangular prism with the same base. As an alternative, use rice or water to demonstrate the volume is one-third the prism. Encourage students to make and share their observations.

Assess Understanding Have students work in pairs of mixed ability. Tell students to draw a net of a square pyramid and then use the net to make a solid figure. Tell them to make a section of a circle to make a cone. Have students write a paragraph that summarizes how to find the volume of each figure. Ask students to share their paragraphs by reading them to each other. Encourage students to question things they hear that they do not understand.

5 Assess & Remediate

Lesson Quiz

1. What is the volume of a square pyramid with sides that are 12.5 m long and a height of 16 m?

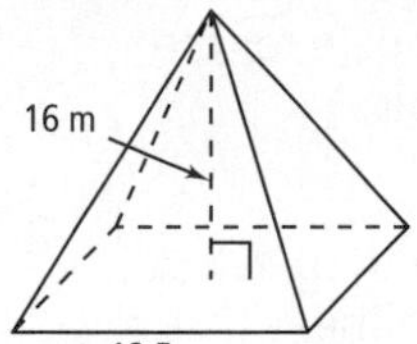

2. What is the volume of a traffic cone that has a height of 2.4 ft and a diameter of 1.25 feet?
3. **Do You UNDERSTAND?** What is the volume of the oblique cone below? Give your answer in terms of π.

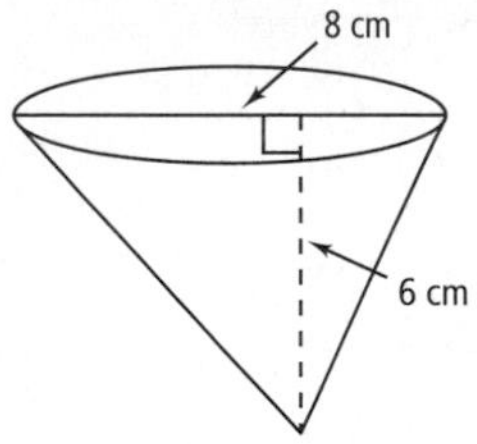

ANSWERS TO LESSON QUIZ

1. $833\frac{1}{3}$ m^3
2. about 1 ft^3
3. 32π cm^3

PRESCRIPTION FOR REMEDIATION

Use the student work on the Lesson Quiz to prescribe a differentiated review assignment.

Points	Differentiated Remediation
0–1	Intervention
2	On-level
3	Extension

PowerGeometry.com

5 Assess & Remediate

Assign the Lesson Quiz. Appropriate intervention, practice, or enrichment is automatically generated based on student performance.

Intervention

- **Reteaching** (2 pages) Provides reteaching and practice exercises for the key lesson concepts. Use with struggling students or absent students.
- **English Language Learner Support** Helps students develop and reinforce mathematical vocabulary and key concepts.

All-in-One Resources/Online

Reteaching

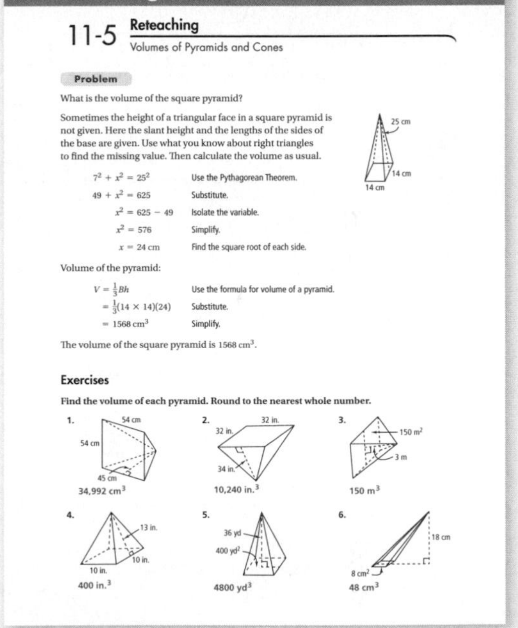
11-5 Reteaching
Volumes of Pyramids and Cones

Problem

What is the volume of the square pyramid?

Sometimes the height of a triangular face in a square pyramid is not given. Here the slant height and the lengths of the sides of the base are given. Use what you know about right triangles to find the missing value. Then calculate the volume as usual.

$7^2 + x^2 = 25^2$ Use the Pythagorean Theorem.
$49 + x^2 = 625$ Substitute.
$x^2 = 625 - 49$ Isolate the variable.
$x^2 = 576$ Simplify.
$x = 24$ cm Find the square root of each side.

Volume of the pyramid:

$V = \frac{1}{3}Bh$ Use the formula for volume of a pyramid.
$= \frac{1}{3}(14 \times 14)(24)$ Substitute.
$= 1568$ cm^3 Simplify.

The volume of the square pyramid is 1568 cm^3.

Exercises

Find the volume of each pyramid. Round to the nearest whole number.

1. 34,992 cm^3 2. 10,240 in.^{3} 3. 150 m^3
4. 400 in.^{3} 5. 4800 yd^3 6. 48 cm^3

All-in-One Resources/Online

English Language Learner Support

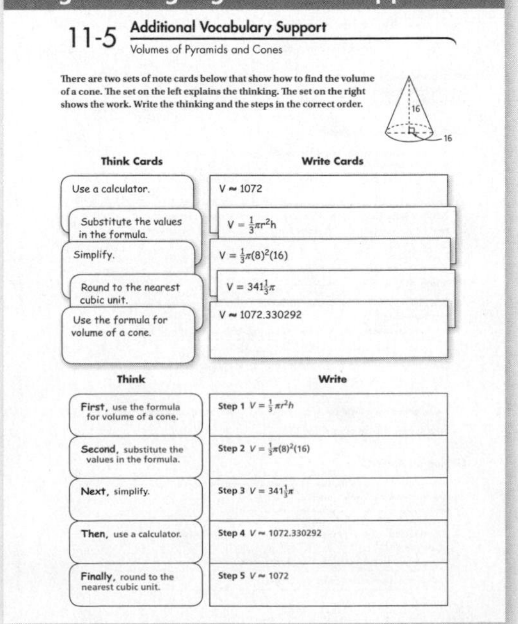
11-5 Additional Vocabulary Support
Volumes of Pyramids and Cones

There are two sets of note cards below that show how to find the volume of a cone. The set on the left explains the thinking. The set on the right shows the work. Write the thinking and the steps in the correct order.

Think Cards	Write Cards
Use a calculator.	$V \approx 1072$
Substitute the values in the formula.	$V = \frac{1}{3}\pi r^2 h$
Simplify.	$V = \frac{1}{3}\pi(8)^2(16)$
Round to the nearest cubic unit.	$V = 341\frac{1}{3}\pi$
Use the formula for volume of a cone.	$V \approx 1072.330292$

Think	Write
First, use the formula for volume of a cone.	Step 1 $V = \frac{1}{3}\pi r^2 h$
Second, substitute the values in the formula.	Step 2 $V = \frac{1}{3}\pi(8)^2(16)$
Next, simplify.	Step 3 $V = 341\frac{1}{3}\pi$
Then, use a calculator.	Step 4 $V \approx 1072.330292$
Finally, round to the nearest cubic unit.	Step 5 $V \approx 1072$

Differentiated Remediation *continued*

On-Level

- **Practice** (2 pages) Provides extra practice for each lesson. For simpler practice exercises, use the Form K Practice pages found in the All-in-One Teaching Resources and online.
- **Think About a Plan** Helps students develop specific problem-solving skills and strategies by providing scaffolded guiding questions.
- **Standardized Test Prep** Focuses on all major exercises, all major question types, and helps students prepare for the high-stakes assessments.

Extension

- **Enrichment** Provides students with interesting problems and activities that extend the concepts of the lesson.
- **Activities, Games, and Puzzles** Worksheets that can be used for concepts development, enrichment, and for fun!

Practice and Problem Solving Wkbk/ All-in-One Resources/Online

Practice page 1

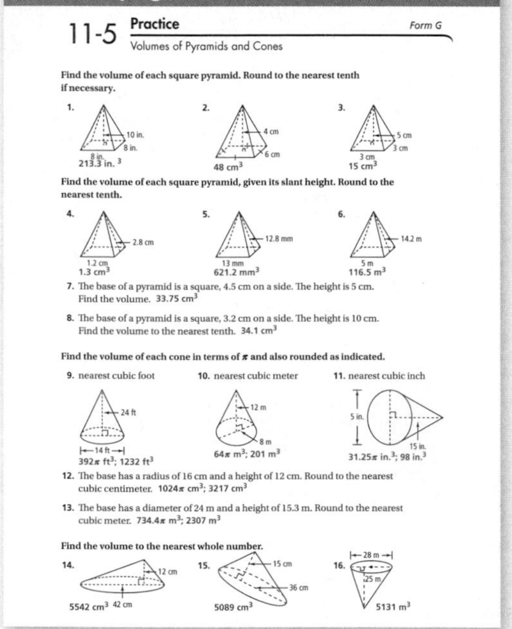

11-5 Practice — Form G
Volumes of Pyramids and Cones

Find the volume of each square pyramid. Round to the nearest tenth if necessary.

1. $213.\overline{3}$ in.^{3} 2. 48 cm^3 3. 15 cm^3

Find the volume of each square pyramid, given its slant height. Round to the nearest tenth.

4. 1.3 cm^3 5. 621.2 mm^3 6. 116.5 m^3

7. The base of a pyramid is a square, 4.5 cm on a side. The height is 5 cm. Find the volume. 33.75 cm^3
8. The base of a pyramid is a square, 3.2 cm on a side. The height is 10 cm. Find the volume to the nearest tenth. 34.1 cm^3

Find the volume of each cone in terms of π and also rounded as indicated.

9. nearest cubic foot — 392π ft^3; 1232 ft^3
10. nearest cubic meter — 64π m^3; 201 m^3
11. nearest cubic inch — 31.25π in.^{3}; 98 in.^{3}
12. The base has a radius of 16 cm and a height of 12 cm. Round to the nearest cubic centimeter. 1024π cm^3; 3217 cm^3
13. The base has a diameter of 24 m and a height of 15.3 m. Round to the nearest cubic meter. 734.4π m^3; 2307 m^3

Find the volume to the nearest whole number.

14. 5542 cm^3 15. 5089 cm^3 16. 5131 m^3

Practice page 2

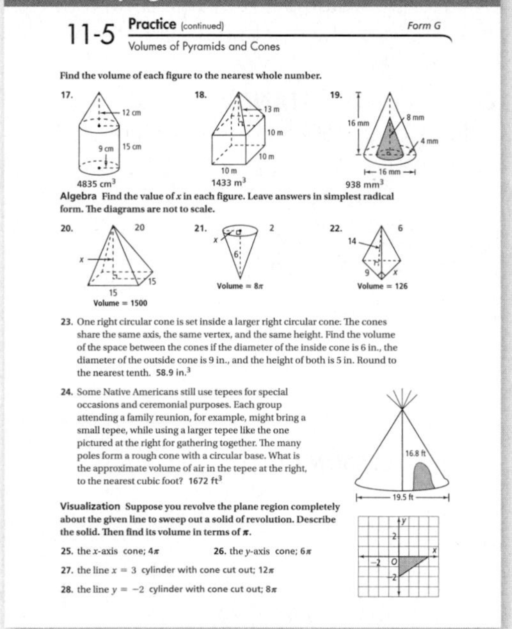

11-5 Practice (continued) — Form G
Volumes of Pyramids and Cones

Find the volume of each figure to the nearest whole number.

17. 4835 cm^3 18. 1433 m^3 19. 938 mm^3

Algebra Find the value of x in each figure. Leave answers in simplest radical form. The diagrams are not to scale.

20. Volume = 1500 21. Volume = 8π 22. Volume = 126

23. One right circular cone is set inside a larger right circular cone. The cones share the same axis, the same vertex, and the same height. Find the volume of the space between the cones if the diameter of the inside cone is 6 in., the diameter of the outside cone is 9 in., and the height of both is 5 in. Round to the nearest tenth. 58.9 in.^{3}
24. Some Native Americans still use tepees for special occasions and ceremonial purposes. Each group attending a family reunion, for example, might bring a small tepee, while using a larger tepee like the one pictured at the right for gathering together. The many poles form a rough cone with a circular base. What is the approximate volume of air in the tepee at the right, to the nearest cubic foot? 1672 ft^3

Visualization Suppose you revolve the plane region completely about the given line to sweep out a solid of revolution. Describe the solid. Then find its volume in terms of π.

25. the x-axis cone; 4π
26. the y-axis cone; 6π
27. the line $x = 3$ cylinder with cone cut out; 12π
28. the line $y = -2$ cylinder with cone cut out; 8π

All-in-One Resources/Online

Enrichment

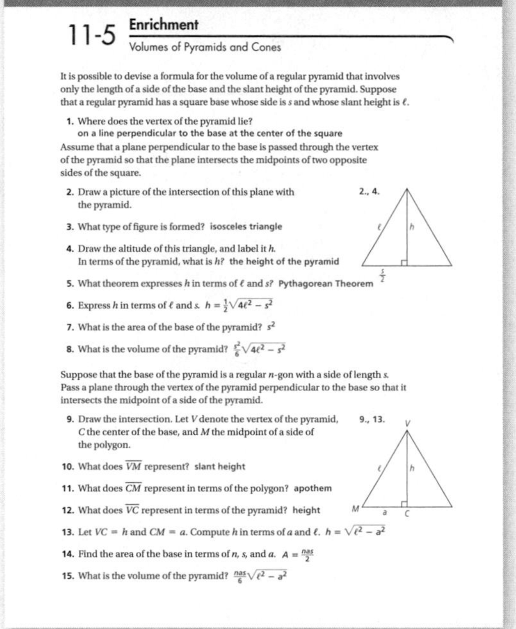

11-5 Enrichment
Volumes of Pyramids and Cones

It is possible to devise a formula for the volume of a regular pyramid that involves only the length of a side of the base and the slant height of the pyramid. Suppose that a regular pyramid has a square base whose side is s and whose slant height is ℓ.

1. Where does the vertex of the pyramid lie? on a line perpendicular to the base at the center of the square

Assume that a plane perpendicular to the base is passed through the vertex of the pyramid so that the plane intersects the midpoints of two opposite sides of the square.

2. Draw a picture of the intersection of this plane with the pyramid. 2., 4.
3. What type of figure is formed? isosceles triangle
4. Draw the altitude of this triangle, and label it h. In terms of the pyramid, what is h? the height of the pyramid
5. What theorem expresses h in terms of ℓ and s? Pythagorean Theorem
6. Express h in terms of ℓ and s. $h = \frac{1}{2}\sqrt{4\ell^2 - s^2}$
7. What is the area of the base of the pyramid? s^2
8. What is the volume of the pyramid? $\frac{s^2}{6}\sqrt{4\ell^2 - s^2}$

Suppose that the base of the pyramid is a regular n-gon with a side of length s. Pass a plane through the vertex of the pyramid perpendicular to the base so that it intersects the midpoint of a side of the pyramid.

9. Draw the intersection. Let V denote the vertex of the pyramid, C the center of the base, and M the midpoint of a side of the polygon. 9., 13.
10. What does $\overline{VM}$ represent? slant height
11. What does $\overline{CM}$ represent in terms of the polygon? apothem
12. What does $\overline{VC}$ represent in terms of the pyramid? height
13. Let $VC = h$ and $CM = a$. Compute h in terms of a and ℓ. $h = \sqrt{\ell^2 - a^2}$
14. Find the area of the base in terms of n, s, and a. $A = \frac{nas}{2}$
15. What is the volume of the pyramid? $\frac{nas}{6}\sqrt{\ell^2 - a^2}$

Practice and Problem Solving Wkbk/ All-in-One Resources/Online

Think About a Plan

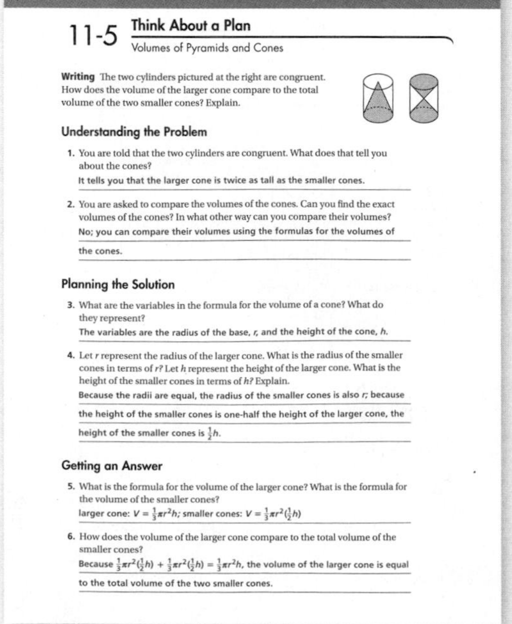

11-5 Think About a Plan
Volumes of Pyramids and Cones

Writing The two cylinders pictured at the right are congruent. How does the volume of the larger cone compare to the total volume of the two smaller cones? Explain.

Understanding the Problem

1. You are told that the two cylinders are congruent. What does that tell you about the cones?
It tells you that the larger cone is twice as tall as the smaller cones.
2. You are asked to compare the volumes of the cones. Can you find the exact volumes of the cones? In what other way can you compare their volumes?
No; you can compare their volumes using the formulas for the volumes of the cones.

Planning the Solution

3. What are the variables in the formula for the volume of a cone? What do they represent?
The variables are the radius of the base, r, and the height of the cone, h.
4. Let r represent the radius of the larger cone. What is the radius of the smaller cones in terms of r? Let h represent the height of the larger cone. What is the height of the smaller cones in terms of h? Explain.
Because the radii are equal, the radius of the smaller cones is also r; because the height of the smaller cones is one-half the height of the larger cone, the height of the smaller cones is $\frac{1}{2}h$.

Getting an Answer

5. What is the formula for the volume of the larger cone? What is the formula for the volume of the smaller cones?
larger cone: $V = \frac{1}{3}\pi r^2 h$; smaller cones: $V = \frac{1}{3}\pi r^2(\frac{1}{2}h)$
6. How does the volume of the larger cone compare to the total volume of the smaller cones?
Because $\frac{1}{3}\pi r^2(\frac{1}{2}h) + \frac{1}{3}\pi r^2(\frac{1}{2}h) = \frac{1}{3}\pi r^2 h$, the volume of the larger cone is equal to the total volume of the two smaller cones.

Practice and Problem Solving Wkbk/ All-in-One Resources/Online

Standardized Test Prep

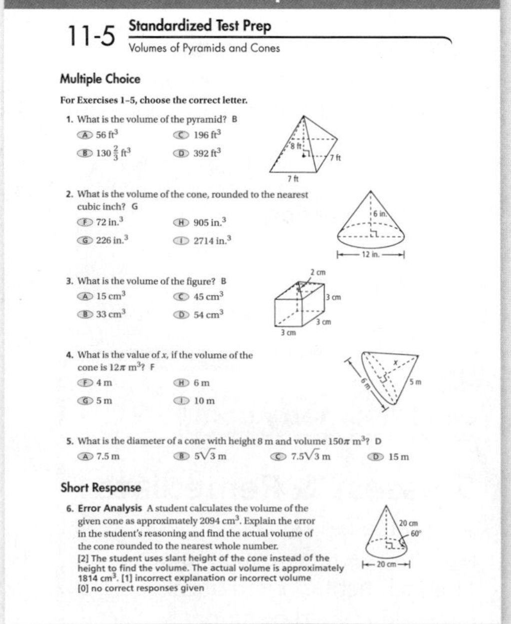

11-5 Standardized Test Prep
Volumes of Pyramids and Cones

Multiple Choice

For Exercises 1–5, choose the correct letter.

1. What is the volume of the pyramid? B
Ⓐ 56 ft^3 Ⓑ $130\frac{2}{3}$ ft^3 Ⓒ 196 ft^3 Ⓓ 392 ft^3
2. What is the volume of the cone, rounded to the nearest cubic inch? G
Ⓕ 72 in.^{3} Ⓖ 226 in.^{3} Ⓗ 905 in.^{3} Ⓘ 2714 in.^{3}
3. What is the volume of the figure? B
Ⓐ 15 cm^3 Ⓑ 33 cm^3 Ⓒ 45 cm^3 Ⓓ 54 cm^3
4. What is the value of x, if the volume of the cone is 12π m^3? F
Ⓕ 4 m Ⓖ 5 m Ⓗ 6 m Ⓘ 10 m
5. What is the diameter of a cone with height 8 m and volume 150π m^3? D
Ⓐ 7.5 m Ⓑ $5\sqrt{3}$ m Ⓒ $7.5\sqrt{3}$ m Ⓓ 15 m

Short Response

6. Error Analysis A student calculates the volume of the given cone as approximately 2094 cm^3. Explain the error in the student's reasoning and find the actual volume of the cone rounded to the nearest whole number.
[2] The student uses slant height of the cone instead of the height to find the volume. The actual volume is approximately 1814 cm^3. [1] incorrect explanation or incorrect volume [0] no correct responses given

Online Teacher Resource Center

Activities, Games, and Puzzles

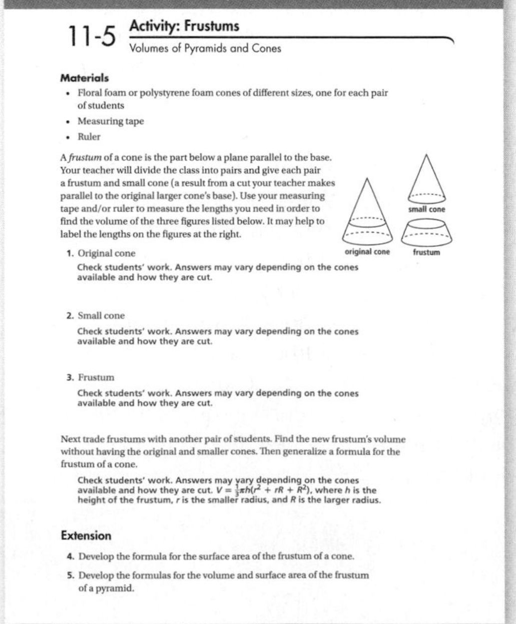

11-5 Activity: Frustums
Volumes of Pyramids and Cones

Materials
- Floral foam or polystyrene foam cones of different sizes, one for each pair of students
- Measuring tape
- Ruler

A *frustum* of a cone is the part below a plane parallel to the base. Your teacher will divide the class into pairs and give each pair a frustum and small cone (a result from a cut your teacher makes parallel to the original larger cone's base). Use your measuring tape and/or ruler to measure the lengths you need in order to find the volume of the three figures listed below. It may help to label the lengths on the figures at the right.

1. Original cone
Check students' work. Answers may vary depending on the cones available and how they are cut.
2. Small cone
Check students' work. Answers may vary depending on the cones available and how they are cut.
3. Frustum
Check students' work. Answers may vary depending on the cones available and how they are cut.

Next trade frustums with another pair of students. Find the new frustum's volume without having the original and smaller cones. Then generalize a formula for the frustum of a cone.

Check students' work. Answers may vary depending on the cones available and how they are cut. $V = \frac{1}{3}\pi h(r^2 + rR + R^2)$, where h is the height of the frustum, r is the smaller radius, and R is the larger radius.

Extension

4. Develop the formula for the surface area of the frustum of a cone.
5. Develop the formulas for the volume and surface area of the frustum of a pyramid.

11-6 Surface Areas and Volumes of Spheres

Common Core State Standards
G-GMD.A.3 Use volume formulas for . . . spheres to solve problems.
G-MG.A.1 Use geometric shapes, their measures, and their properties to describe objects.
MP 1, MP 3, MP 4, MP 6, MP 7, MP 8

Objective To find the surface area and volume of a sphere

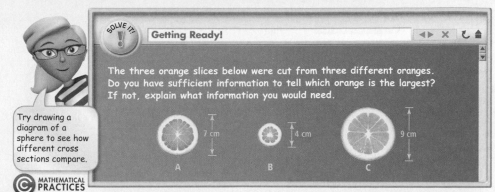

Solve It! Getting Ready!

The three orange slices below were cut from three different oranges. Do you have sufficient information to tell which orange is the largest? If not, explain what information you would need.

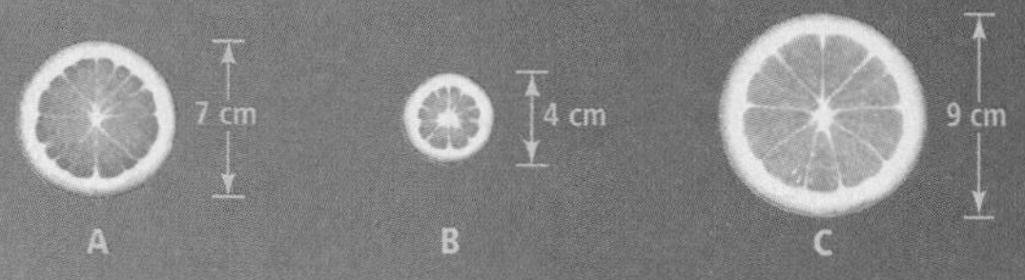

Try drawing a diagram of a sphere to see how different cross sections compare.

MATHEMATICAL PRACTICES

Lesson Vocabulary
- sphere
- center of a sphere
- radius of a sphere
- diameter of a sphere
- circumference of a sphere
- great circle
- hemisphere

In the Solve It, you considered the sizes of objects with circular cross sections.

A **sphere** is the set of all points in space equidistant from a given point called the **center.** A **radius** is a segment that has one endpoint at the center and the other endpoint on the sphere. A **diameter** is a segment passing through the center with endpoints on the sphere.

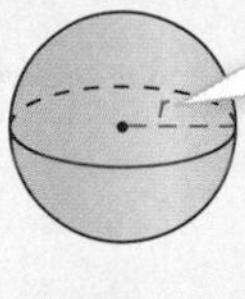

Essential Understanding You can find the surface area and the volume of a sphere when you know its radius.

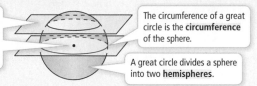

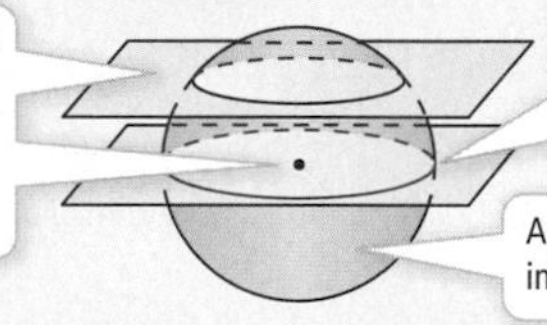

When a plane and a sphere intersect in more than one point, the intersection is a circle. If the center of the circle is also the center of the sphere, it is called a **great circle.**
The circumference of a great circle is the **circumference** of the sphere.
A great circle divides a sphere into two **hemispheres.**

A baseball can model a sphere. To approximate its surface area, you can take apart its covering. Each of the two pieces suggests a pair of circles with radius r, which is approximately the radius of the ball. The area of the four circles, $4\pi r^2$, suggests the surface area of the ball.

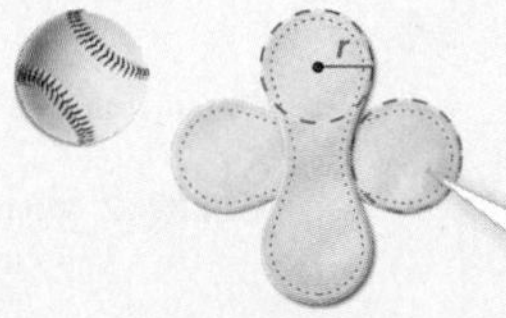

1 Interactive Learning

Solve It!

PURPOSE To realize that the cross section of a sphere does not determine the size of the sphere

PROCESS Students may use visual reasoning.

FACILITATE

Q Could orange B be larger than orange A? Why or why not? **[Yes, the slice from orange B could be from the top, where the slice from orange A could go through the center of the orange.]**

Q Can you predict the radius of the oranges from the given information? Why or why not? **[No, you do not know where the slices came from in the orange.]**

Q What additional information would you need to determine the radii of the oranges? **[You would need to know that the slices cut through the center of the oranges.]**

ANSWER See Solve It in Answers on next page.

CONNECT THE MATH Students should realize that they cannot tell which orange is the largest from the diagrams. In the lesson, students learn that the measurements of a sphere go through the center of the sphere.

2 Guided Instruction

Make sure that students can identify important segments and figures in a sphere. Practice by having them call out figures from a diagram you draw on the board.

11-6 Preparing to Teach

BIG ideas **Measurement**
Visualization

ESSENTIAL UNDERSTANDING
- The surface area and the volume of a sphere can be found when its radius is known.

Math Background

In contrast with prisms and cylinders, the surface area and volume of spheres is not obvious. To find the surface area of a sphere, students will learn the formula $4\pi r^2$. The volume of a sphere is $\frac{4}{3}\pi r^3$. The volume formula can be derived from what students know about the volume of pyramids. A set of congruent pyramids that share a common vertex approximates the shape of a sphere. The sum of the volumes of the pyramids approximates the volume of the sphere they represent. Students will use the formulas for surface area and volume of spheres in real-world problems.

Mathematical Practices

Model with mathematics. In finding the volume of a sphere, students will make an approximation of a sphere using pyramids and find the volume of the pyramids, realizing that the model is only approximate and needs revision.

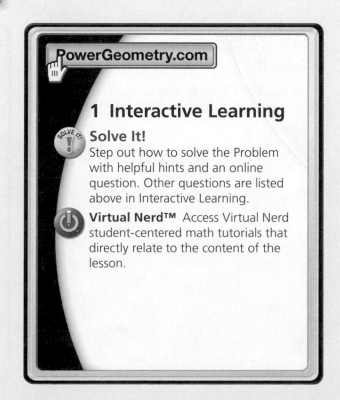

PowerGeometry.com

1 Interactive Learning

Solve It!
Step out how to solve the Problem with helpful hints and an online question. Other questions are listed above in Interactive Learning.

Virtual Nerd™ Access Virtual Nerd student-centered math tutorials that directly relate to the content of the lesson.

Take Note

Discuss if students have ever seen a diagram for the net of a sphere. It is not common for textbooks or other reference material to show this net. It is impossible to make a perfect sphere from a flat sheet of paper. The reason for this is that paper cannot curve in two directions at the same time. Any visual display for the net of a sphere can only be an approximation. There are methods for making approximated nets of spheres that include a polyhedron or pointed ellipses. You can have students reference cartographers' attempts to reduce distortions of the continents by creating a map of the globe represented by an elliptical object.

Problem 1

Q What is the radius of the sphere? Explain. **[The radius is half the diameter: 5 m.]**

Q How can you find the volume of the sphere? **[Multiply 4 times π and the radius squared.]**

Got It?

Have students sketch a diagram of the sphere and label the known measurement.

Problem 2

Q What measurement of the Earth is represented by the equator? **[the circumference]**

Q How can you find the radius of the Earth given the circumference? **[Solve the equation 24,902 = $2\pi r$ for r.]**

Q How can you find the surface area of the Earth? **[Multiply 4 times π and the radius squared.]**

Got It?

Have students draw a diagram of the melon. They will need to solve the formula of the circumference of a circle for the radius.

---

take note

Theorem 11-10 Surface Area of a Sphere

The surface area of a sphere is four times the product of π and the square of the radius of the sphere.

S.A. = $4\pi r^2$

Problem 1 Finding the Surface Area of a Sphere

Plan

What are you given? In sphere problems, make it a habit to note whether you are given the radius or the diameter. In this case, you are given the diameter.

What is the surface area of the sphere in terms of π?

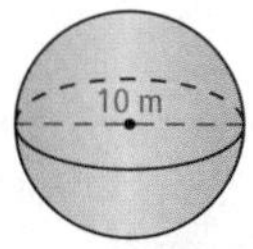

The diameter is 10 m, so the radius is $\frac{10}{2}$ m, or 5 m.

S.A. = $4\pi r^2$	Use the formula for surface area of a sphere.
= $4\pi(5)^2$	Substitute 5 for r.
= 100π	Simplify.

The surface area is 100π m^2.

Got It? **1.** What is the surface area of a sphere with a diameter of 14 in.? Give your answer in terms of π and rounded to the nearest square inch.

You can use spheres to approximate the surface areas of real-world objects.

Problem 2 Finding Surface Area

Plan

How can you use the length of Earth's equator? Earth's equator is a great circle that divides Earth into two hemispheres. Its length is Earth's circumference. Use it to find Earth's radius.

Geography **Earth's equator is about 24,902 mi long. What is the approximate surface area of Earth? Round to the nearest thousand square miles.**

Step 1 Find the radius of Earth.

$C = 2\pi r$	Use the formula for circumference.
$24{,}902 = 2\pi r$	Substitute 24,902 for C.
$\frac{24{,}902}{2\pi} = r$	Divide each side by 2π.
$r \approx 3963.276393$	Use a calculator.

Step 2 Use the radius to find the surface area of Earth.

S.A. = $4\pi r^2$	Use the formula for surface area.
= 4π ANS x^2 enter	Use a calculator. ANS uses the value of r from Step 1.
≈ 197387017.5	

The surface area of Earth is about 197,387,000 mi^2.

Got It? **2.** What is the surface area of a melon with circumference 18 in.? Round your answer to the nearest ten square inches.

Answers

Solve It!

No; you need to know how far from the center of the orange each slice was cut.

Got It?

1. 196π in.^{2}; 616 in.^{2}
2. 100 in.^{2}

PowerGeometry.com

2 Guided Instruction

Each Problem is worked out and supported online.

Problem 1
Finding the Surface Area of a Sphere
Animated

Problem 2
Finding Surface Area
Animated

Problem 3
Finding the Volume of a Sphere

Problem 4
Using Volume to Find Surface Area
Animated

Support in Geometry Companion
- Vocabulary
- Key Concepts
- Got It?

In the previous lesson, you learned that the volume of a cone is $\frac{1}{3}\pi r^3$. You can use this with Cavalieri's Principle to find the formula for the volume of a sphere.

Both figures at the right have a parallel plane x units above their centers that form circular cross sections.

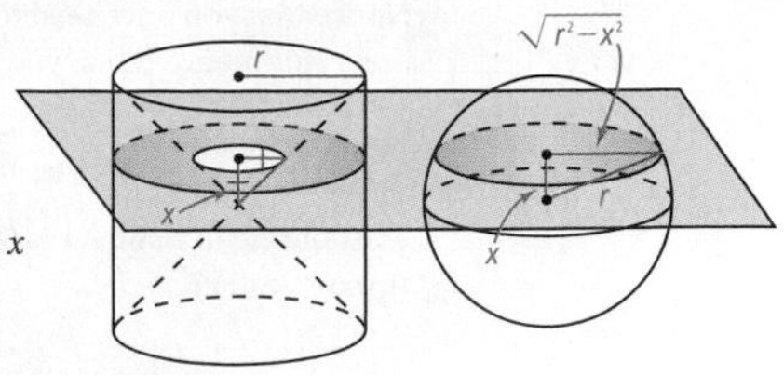

The area of the cross section of the cylinder minus the area of the cross section of the cone is the same as the area of the cross section of the sphere. Every horizontal plane will cut the figures into cross sections of equal area. By Cavalieri's Principle, the volume of the sphere = the volume of the cylinder = the volume of two cones.

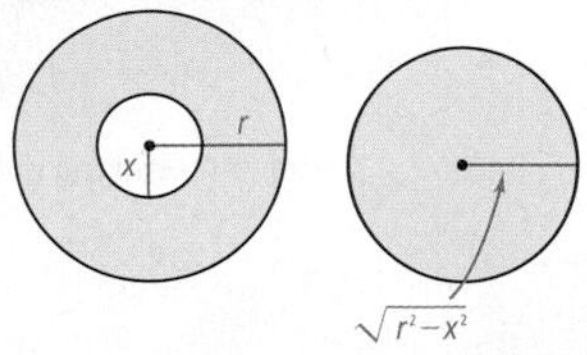

Volume of a sphere $= \pi r^2(2r) - 2(\frac{1}{3}\pi r^3)$

$= 2\pi r^3 - \frac{2}{3}\pi r^3$

$= \frac{4}{3}\pi r^3$

take note

Theorem 11-11 Volume of a Sphere

The volume of a sphere is four thirds the product of π and the cube of the radius of the sphere.

$$V = \frac{4}{3}\pi r^3$$

Problem 3 Finding the Volume of a Sphere

What is the volume of the sphere in terms of π?

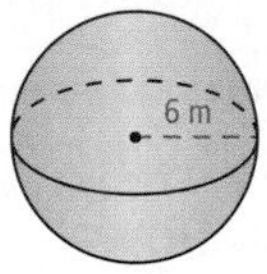

$V = \frac{4}{3}\pi r^3$ Use the formula for volume of a sphere.

$= \frac{4}{3}\pi(6)^3$ Substitute.

$= 288\pi$

The volume of the sphere is 288π m^3.

Think

What are the units of the answer?
You are cubing the radius, which is in meters (m), so your answer should be in cubic meters (m^3).

Got It? 3. a. A sphere has a diameter of 60 in. What is its volume to the nearest cubic inch?

b. Reasoning Suppose the radius of a sphere is halved. How does this affect the volume of the sphere? Explain.

The derivation of the formula for the volume of a sphere is based on the volume of a cone. Lead students through the logic of this argument.

Problem 3

Q How will you substitute to find the volume of the sphere? **[length of the radius, 6]**

Q How do you leave your answer in terms of π? **[Multiply all the constants in the equation and leave π like a variable at the end of the expression.]**

Got It?

Q What operation must you do to halve a value? **[Divide it by 2.]**

For 3b, have students review other questions in this chapter that asked how taking half of a measurement affected the volume. Then have them make a conjecture about the effect of using a radius that is half of a previous radius. Students should substitute $\frac{1}{2}r$ into the formula for volume and compare the results to the original formula.

Additional Problems

1. What is the surface area of the sphere in terms of π?

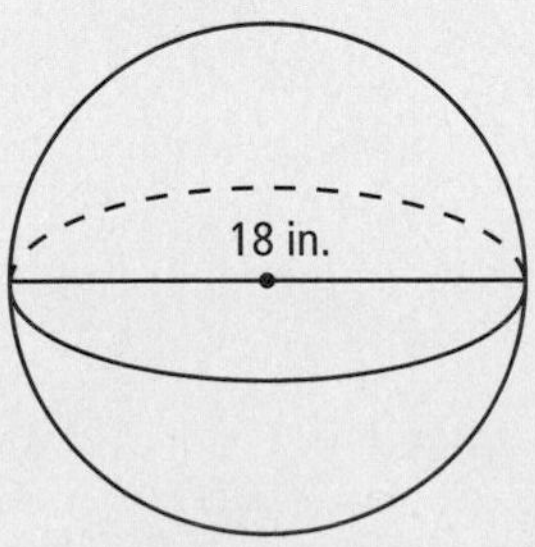

ANSWER 324π in.^{2}

2. What is the surface area of a ball with a circumference of about 25 in.? Round to the nearest ten square inches.

ANSWER about 200 in.^{2}

3. What is the volume of the sphere in terms of π?

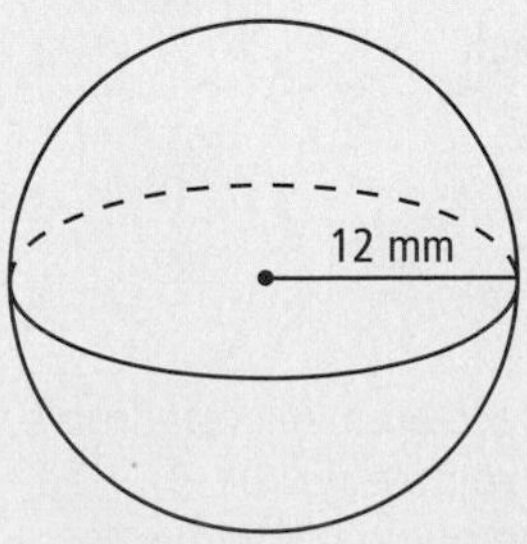

ANSWER 2304π mm^3

4. The volume of a sphere is 2200 yd^3. What is the surface area of the sphere?

ANSWER about 818 yd^2

Answers

Got It? (continued)

3a. 113,097 in.^{3}

b. The volume is $\left(\frac{1}{2}\right)^3 = \frac{1}{8}$ of the original volume. Using $V = \frac{4}{3}\pi r^3$, replacing r with $\frac{r}{2}$ gives $V = \frac{4}{3}\pi\left(\frac{r}{2}\right)^3 = \frac{1}{8}\left(\frac{4}{3}\pi r^3\right)$.

Problem 4

Q What measurement do you need to calculate the surface area of a sphere? **[the radius]**

Q How can you find the radius of the sphere from the given information? **[Set the formula for volume equal to the given number and solve for *r*.]**

Q What should you do after you find the value of *r*? **[Substitute it into the formula for the surface area of the sphere.]**

Got It? VISUAL LEARNERS

Make sure that students understand both steps involved in solving for the radius and finding the surface area of the sphere.

3 Lesson Check

Do you know HOW?

- If students have difficulty with Exercise 1, then have them review Problem 1 and follow how the variables were substituted into the formula and then the expression was simplified.

Do you UNDERSTAND?

- If students have difficulty with Exercise 5, then have them choose a value for the radius and calculate the surface area. Then calculate again with the radius doubled.

Close

Q How do you find the surface area of a sphere? **[Multiply 4π by the square of the radius.]**

Q How do you find the volume of a sphere? **[Multiply $\frac{4}{3}\pi$ by the cube of the radius.]**

Notice that you only need to know the radius of a sphere to find its volume and surface area. This means that if you know the volume of a sphere, you can find its surface area.

Problem 4 Using Volume to Find Surface Area

The volume of a sphere is 5000 m^3. What is its surface area to the nearest square meter?

Know	Need	Plan
The volume of a sphere	The radius of the sphere	*Work backward* by using the formula for volume and solving for *r*. Then use the radius to calculate surface area.

Step 1 Find the radius of the sphere.

$V = \frac{4}{3}\pi r^3$ Use the formula for volume of a sphere.

$5000 = \frac{4}{3}\pi r^3$ Substitute.

$5000\left(\frac{3}{4\pi}\right) = r^3$ Solve for r^3.

$\sqrt[3]{5000\left(\frac{3}{4\pi}\right)} = r$ Take the cube root of each side.

$r \approx 10.60784418$ Use a calculator.

Step 2 Find the surface area of the sphere.

$\text{S.A.} = 4\pi r^2$ Use the formula for surface area of a sphere.

$= 4\pi$ ANS x^2 enter Use a calculator.

≈ 1414.04792

The surface area of the sphere is about 1414 m^2.

Got It? 4. The volume of a sphere is 4200 ft^3. What is its surface area to the nearest tenth?

Lesson Check

Do you know HOW?

The diameter of a sphere is 12 ft.

1. What is its surface area in terms of π?

2. What is its volume to the nearest tenth?

3. The volume of a sphere is 80π cm^3. What is its surface area to the nearest whole number?

Do you UNDERSTAND? MATHEMATICAL PRACTICES

4. Vocabulary What is the ratio of the area of a great circle to the surface area of the sphere?

5. Error Analysis Your classmate claims that if you double the radius of a sphere, its surface area and volume will quadruple. What is your classmate's error? Explain.

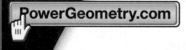

PowerGeometry.com

3 Lesson Check

For a digital lesson check, use the Got It questions.

Support in Geometry Companion

- Lesson Check

4 Practice

ONLINE HOMEWORK

Assign homework to individual students or to an entire class.

Answers

Got It? (continued)

4. 1258.9 ft^2

Lesson Check

1. 144π ft^2

2. 904.8 ft^3

3. 193 cm^2

4. 1 : 4

5. The surface area will quadruple, but the volume will be 8 times the original volume. $V = \frac{4}{3}\pi(2r)^3 = 8\left(\frac{4}{3}\pi r^3\right)$

Practice and Problem-Solving Exercises

MATHEMATICAL PRACTICES

A Practice

Find the surface area of the sphere with the given diameter or radius. Leave your answer in terms of π.

See Problem 1.

6. $d = 30$ m **7.** $r = 10$ in. **8.** $d = 32$ mm **9.** $r = 100$ yd

Sports **Find the surface area of each ball. Leave each answer in terms of π.**

10.

$d = 68$ mm

11.

$d = 21$ cm

12.

$d = 2\frac{1}{16}$ in.

Use the given circumference to find the surface area of each spherical object. Round your answer to the nearest whole number.

See Problem 2.

13. a grapefruit with $C = 14$ cm

14. a bowling ball with $C = 27$ in.

15. a pincushion with $C = 8$ cm

16. a head of lettuce with $C = 22$ in.

Find the volume of each sphere. Give each answer in terms of π and rounded to the nearest cubic unit.

See Problem 3.

17.

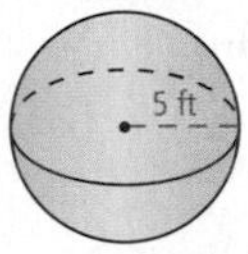

18.

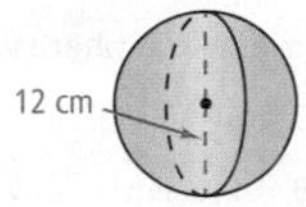

19.

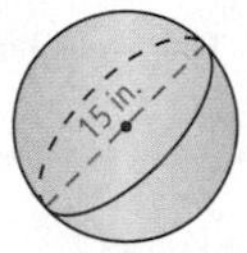

20.

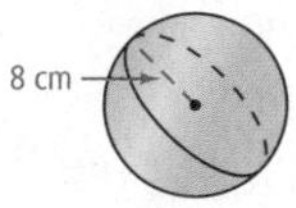

21.

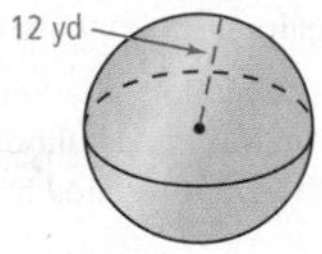

22.

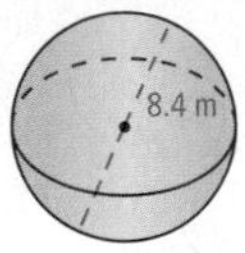

A sphere has the volume given. Find its surface area to the nearest whole number.

See Problem 4.

23. $V = 900$ in.^{3} **24.** $V = 3000$ m^3 **25.** $V = 140$ cm^3

B Apply

26. Mental Math Use $\pi \approx 3$ to estimate the surface area and volume of a sphere with radius 3 cm.

27. Open-Ended Give the dimensions of a cylinder and a sphere that have the same volume.

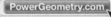

4 Practice

ASSIGNMENT GUIDE

Basic: 6–25 all, 26, 29–31 all, 34–42 even, 50
Average: 7–25 odd, 26–54
Advanced: 7–25 odd, 26–59
Standardized Test Prep: 60–64
Mixed Review: 65–71

Mathematical Practices are supported by exercises with red headings. Here are the Practices supported in this lesson:

MP 1: Make Sense of Problems Ex. 29
MP 3: Communicate Ex. 57b
MP 3: Critique the Reasoning of Others Ex. 5
MP 6: Attend to Precision Ex. 27
MP 7: Shift Perspective Ex. 28
MP 8: Check for Reasonableness Ex. 26

Applications exercises have blue headings. Exercises 31, 42, 43, 46, and 53 support MP 4: Model.

STEM exercises focus on science or engineering applications.

EXERCISE 31: Use the Think About a Plan worksheet in the **Practice and Problem Solving Workbook** (also available in the Teaching Resources in print and online) to further support students' development in becoming independent learners.

HOMEWORK QUICK CHECK

To check students' understanding of key skills and concepts, go over Exercises 11, 17, 26, 29, and 31.

Practice and Problem-Solving Exercises

6. 900π m^2
7. 400π in.^{2}
8. 1024π mm^2
9. $40{,}000\pi$ yd^2
10. 4624π mm^2
11. 441π cm^2
12. $\frac{1089}{256}\pi$ in.^{2}
13. 62 cm^2
14. 232 in.^{2}
15. 20 cm^2
16. 154 in.^{2}
17. $\frac{500}{3}\pi$ ft^3; 524 ft^3
18. 288π cm^3; 905 cm^3
19. $\frac{1125}{2}\pi$ in.^{3}; 1767 in.^{3}
20. $\frac{2048}{3}\pi$ cm^3; 2145 cm^3
21. 2304π yd^3; 7238 yd^3
22. $\frac{12{,}348}{125}\pi$ m^3; 310 m^3
23. 451 in.^{2}
24. 1006 m^2
25. 130 cm^2
26. S.A. $\approx$ 108 cm^2; $V \approx$ 108 cm^3
27. Answers may vary. Sample: sphere with $r = 3$ in., cylinder with $r = 3$ in. and $h = 4$ in.

Answers

Practice and Problem-Solving Exercises (continued)

28a. sphere with $r = 4$

b. $\frac{256}{3}\pi$ units^3

c. 64π units^2

29. 0.9 in.

30. C

31. 1.7 lb

32. 8 in. sphere; the volume of the three spheres is $3(4.5\pi)$ or 13.5π units^3, and of the large sphere is $85\frac{1}{3}\pi$ units^3.

33. An infinite number of planes pass through the center of a sphere, so there are an infinite number of great circles.

34. $\frac{4}{3}\pi$ m^3

35. 36π in.^{3}

36. $\frac{9}{2}\pi$ ft^3

37. $\frac{500}{3}\pi$ mm^3

38. $\frac{125}{6}\pi$ yd^3

39. 288π cm^3

40. $\frac{343}{6}\pi$ m^3

41. $\frac{1125}{2}\pi$ mi^3

42a. $457\frac{1}{3}\pi$ in.^{3}

b. $228\frac{2}{3}\pi$ in.^{3}

c. 11 in.

43a. about 8.9 in.^{2}

b. The answer is less than the actual surface area since the dimples on the golf ball add to the surface area.

44. Answers may vary. Sample: (5, 0, 0), (0, 5, 0), (0, 0, 5), (−5, 0, 0), (0, −5, 0), (0, 0, −5)

45a. on

b. inside

c. outside

28. Visualization The region enclosed by the semicircle at the right is revolved completely about the x-axis.

a. Describe the solid of revolution that is formed.
b. Find its volume in terms of π.
c. Find its surface area in terms of π.

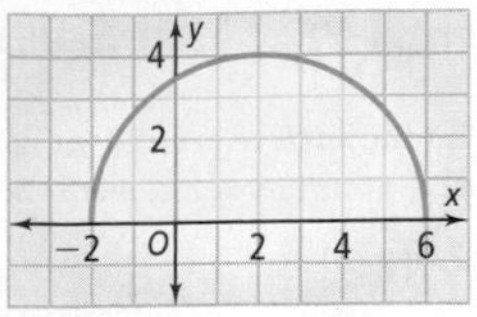

29. Think About a Plan A cylindrical tank with diameter 20 in. is half filled with water. How much will the water level in the tank rise if you place a metallic ball with radius 4 in. in the tank? Give your answer to the nearest tenth.

- What causes the water level in the tank to rise?
- Which volume formulas should you use?

30. The sphere at the right fits snugly inside a cube with 6-in. edges. What is the approximate volume of the space between the sphere and cube?

(A) 28.3 in.^{3} (B) 76.5 in.^{3} (C) 102.9 in.^{3} (D) 113.1 in.^{3}

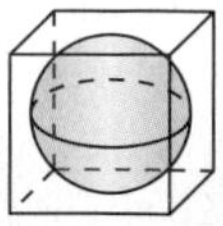

6 in.

31. Meteorology On September 3, 1970, a hailstone with diameter 5.6 in. fell at Coffeyville, Kansas. It weighed about 0.018 lb/in.^{3} compared to the normal 0.033 lb/in.^{3} for ice. About how heavy was this Kansas hailstone?

32. Reasoning Which is greater, the total volume of three spheres, each of which has diameter 3 in., or the volume of one sphere that has diameter 8 in.?

33. Reasoning How many great circles does a sphere have? Explain.

Find the volume in terms of π of each sphere with the given surface area.

34. 4π m^2	**35.** 36π in.^{2}	**36.** 9π ft^2	**37.** 100π mm^2
38. 25π yd^2	**39.** 144π cm^2	**40.** 49π m^2	**41.** 225π mi^2

42. Recreation A spherical balloon has a 14-in. diameter when it is fully inflated. Half of the air is let out of the balloon. Assume that the balloon remains a sphere.

a. Find the volume of the fully inflated balloon in terms of π.
b. Find the volume of the half-inflated balloon in terms of π.
c. What is the diameter of the half-inflated balloon to the nearest inch?

43. Sports Equipment The diameter of a golf ball is 1.68 in.

a. Approximate the surface area of the golf ball.
b. **Reasoning** Do you think that the value you found in part (a) is greater than or less than the actual surface area of the golf ball? Explain.

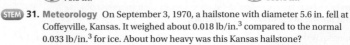

Geometry in 3 Dimensions **A sphere has center (0, 0, 0) and radius 5.**

44. Name the coordinates of six points on the sphere.

45. Tell whether each of the following points is *inside, outside,* or *on the sphere.*

a. $A(0, -3, 4)$ b. $B(1, -1, -1)$ c. $C(4, -6, -10)$

46. **Food** An ice cream vendor presses a sphere of frozen yogurt into a cone, as shown at the right. If the yogurt melts into the cone, will the cone overflow? Explain.

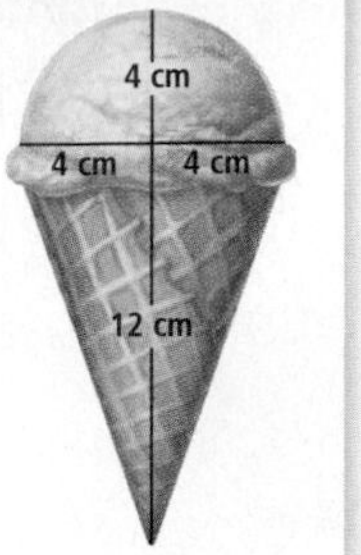

47. The surface area of a sphere is 5541.77 ft^2. What is its volume to the nearest tenth?

48. **Geography** The circumference of Earth at the equator is approximately 40,075 km. About 71% of Earth is covered by oceans and other bodies of water. To the nearest thousand square kilometers, how much of Earth's surface is land?

Find the surface area and volume of each figure.

49. 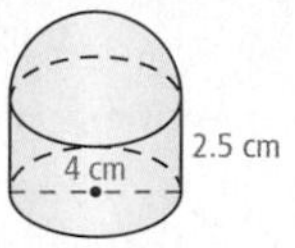

50. 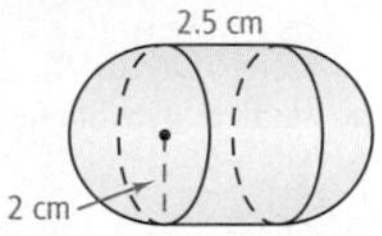

51. 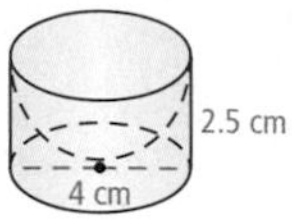

STEM 52. **Astronomy** The diameter of Earth is about 7926 mi. The diameter of the moon is about 27% of the diameter of Earth. What percent of the volume of Earth is the volume of the moon? Round your answer to the nearest whole percent.

STEM 53. **Science** The density of steel is about 0.28 lb/in.^{3}. Could you lift a solid steel ball with radius 4 in.? With radius 6 in.? Explain.

54. A cube with edges 6 in. long fits snugly inside a sphere as shown at the right. The diagonal of the cube is the diameter of the sphere.
   a. Find the length of the diagonal and the radius of the sphere. Leave your answer in simplest radical form.
   b. What is the volume of the space between the sphere and the cube to the nearest tenth?

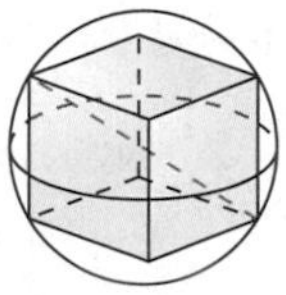

Challenge **Find the radius of a sphere with the given property.**

55. The number of square meters of surface area equals the number of cubic meters of volume.

56. The ratio of surface area in square meters to volume in cubic meters is 1 : 5.

57. Suppose a cube and a sphere have the same volume.
   a. Which has the greater surface area? Explain.
   b. Writing Explain why spheres are rarely used for packaging.

58. A plane intersects a sphere to form a circular cross section. The radius of the sphere is 17 cm and the plane comes to within 8 cm of the center. Draw a sketch and find the area of the cross section, to the nearest whole number.

59. **History** At the right, the sphere fits snugly inside the cylinder. Archimedes (about 287–212 B.C.) requested that such a figure be put on his gravestone along with the ratio of their volumes, a finding that he regarded as his greatest discovery. What is that ratio?

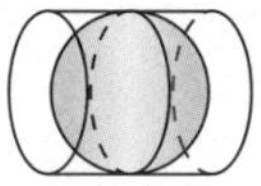

46. Yes; the volume of the frozen yogurt is $\frac{256}{3}\pi$ cm^3, and the volume of the cone is 64π cm^3.

47. 38,792.4 ft^3

48. about 148,250,000 km^2

49. 22π cm^2; $\frac{46}{3}\pi$ cm^3

50. 26π cm^2; $\frac{62}{3}\pi$ cm^3

51. 22π cm^2; $\frac{14}{3}\pi$ cm^3

52. 2%

53. Answers may vary. Sample: You could lift the small ball because it weighs about 75 lb. The big ball would be much harder to lift since it weighs about 253 lb.

54a. $6\sqrt{3}$ in.; $3\sqrt{3}$ in.

b. 371.7 in.^{3}

55. 3 m

56. 15 m

57a. Cube; explanations may vary. Sample: If $s^3 = \frac{4}{3}\pi r^3$, then $s = r \cdot \sqrt[3]{\frac{4\pi}{3}}$. So $6s^2 = 6\left(r \cdot \sqrt[3]{\frac{4\pi}{3}}\right)^2 \approx 15.6r^2 > 4\pi r^2$ (which is about $12.6r^2$).

b. Answers may vary. Sample: Spheres are difficult to stack in a display or on a shelf.

58.

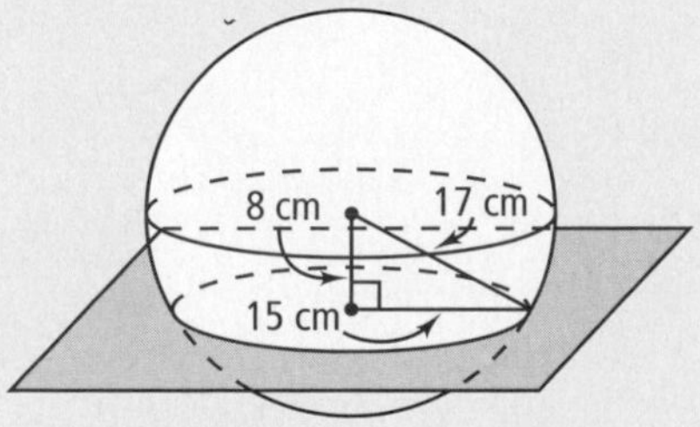

707 cm^2

59. 2 : 3

Answers

Practice and Problem-Solving Exercises (continued)

60. B

61. G

62. C

63. I

64. [2] $x^2 + 3^2 = 6^2$, $x^2 = 36 - 9 = 27$, $x = 3\sqrt{3}$. Use ~ $\triangle$s to set up and solve a proportion. $\frac{x}{6} = \frac{3}{y}$, $\frac{3\sqrt{3}}{6} = \frac{3}{y}$, $3\sqrt{3} \cdot y = 18$, $y = \frac{18}{3\sqrt{3}} = \frac{18\sqrt{3}}{9} = 2\sqrt{3}$.

[1] correct answer, no work shown

65. 16 m^3

66. 19 in.^3

67. $19{,}396 \text{ mm}^3$

68. 35; 55

69. 109, 71, 109, 71

70. yes; 3 : 1

71. yes; $3 : \sqrt{2}$ or $3\sqrt{2} : 2$

Standardized Test Prep

SAT/ACT

60. What is the diameter of a sphere whose surface area is $100\pi \text{ m}^2$?

(A) 5 m (B) 10 m (C) 5π m (D) 25π m

61. Which of the following statements contradict each other?

I. Opposite sides of $\square ABCD$ are parallel.
II. Diagonals of $\square ABCD$ are perpendicular.
III. $\square ABCD$ is not a rhombus.

(F) I and II (G) II and III (H) I and III (I) none

62. What is the reflection image of (3, 7) across the line $y = 4$?

(A) (3, 3) (B) (−7, 3) (C) (3, 1) (D) (3, −7)

63. The radius of a sphere is doubled. By what factor does the surface area of the sphere change?

(F) $\frac{1}{4}$ (G) $\frac{1}{2}$ (H) 2 (I) 4

Short Response

64. Find the values of x and y. Show your work.

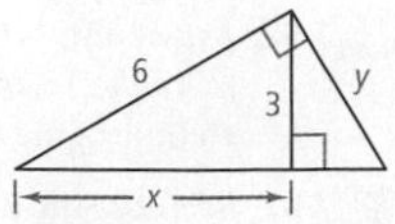

Mixed Review

Find the volume of each figure to the nearest cubic unit.

See Lesson 11-5.

65.

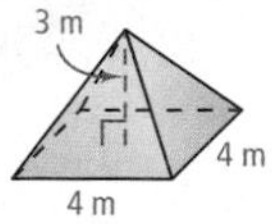

66.

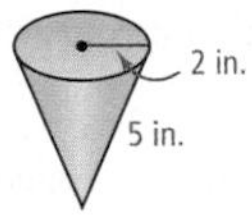

67.

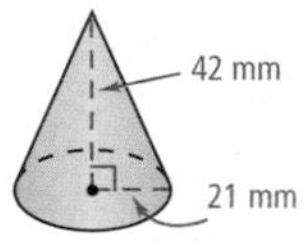

68. A leg of a right triangle has a length of 4 cm and the hypotenuse has a length of 7 cm. Find the measure of each acute angle of the triangle to the nearest degree.

See Lessons 6-4 and 8-3.

69. The length of each side of a rhombus is 16. The longer diagonal has length 26. Find the measures of the angles of the rhombus to the nearest degree.

Get Ready! **To prepare for Lesson 11-7, do Exercises 70 and 71.**

Are the figures similar? If so, give the scale factor.

See Lesson 7-2.

70. two squares, one with 3-in. sides and the other with 1-in. sides

71. two right isosceles triangles, one with a 3-cm hypotenuse and the other with a 1-cm leg

11-6 Lesson Resources

Differentiated Remediation

Instructional Support

Geometry Companion

Students can use the **Geometry Companion** worktext (4 pages) . . .

- New Vocabulary
- Key Concepts
- Got It for each Problem
- Lesson Check

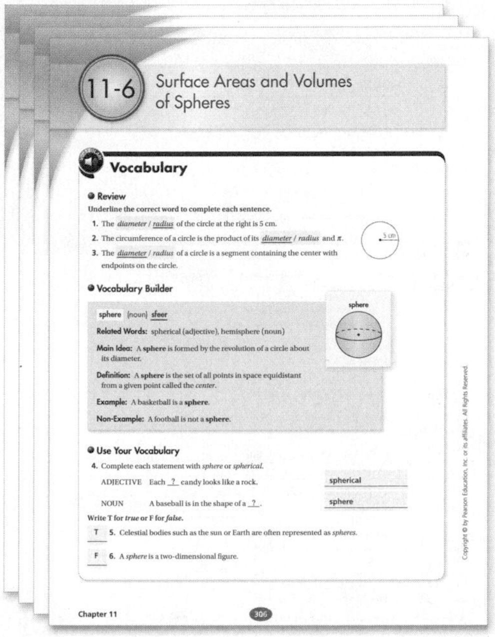

11-6 Surface Areas and Volumes of Spheres

Vocabulary

Review

Underline the correct word to complete each sentence.

1. The *diameter* / *radius* of the circle at the right is 5 cm.
2. The circumference of a circle is the product of its *diameter* / *radius* and π.
3. The *diameter* / *radius* of a circle is a segment containing the center with endpoints on the circle.

Vocabulary Builder

sphere (noun) sfeer

Related Words: spherical (adjective), hemisphere (noun)

Main Idea: A **sphere** is formed by the revolution of a circle about its diameter.

Definition: A **sphere** is the set of all points in space equidistant from a given point called the *center*.

Example: A basketball is a **sphere**.

Non-Example: A football is not a **sphere**.

Use Your Vocabulary

4. Complete each statement with *sphere* or *spherical*.

ADJECTIVE Each _?_ candy looks like a rock. spherical

NOUN A baseball is in the shape of a _?_. sphere

Write T for *true* or F for *false*.

T 5. Celestial bodies such as the sun or Earth are often represented as *spheres*.

F 6. A *sphere* is a two-dimensional figure.

Copyright © by Pearson Education, Inc. or its affiliates. All Rights Reserved.

Chapter 11 306

ELL Support

Focus on Language Examine the word *sphere*. Ask students for examples of a sphere, or ways that sphere may be used in language (for example, a sphere of influence). Examples may be food such as oranges or tomatoes, objects such as balls and globes, or environmental items such as a planet or certain kinds of seeds. Then ask students for the forms of the word sphere that mean in the shape of a sphere. Examples may be spherical or sphere-like.

Sphere may originate from the late Latin sphera, or sphaera, which means globe.

5 Assess & Remediate

Lesson Quiz

1. What is the surface area of the sphere in terms of π?

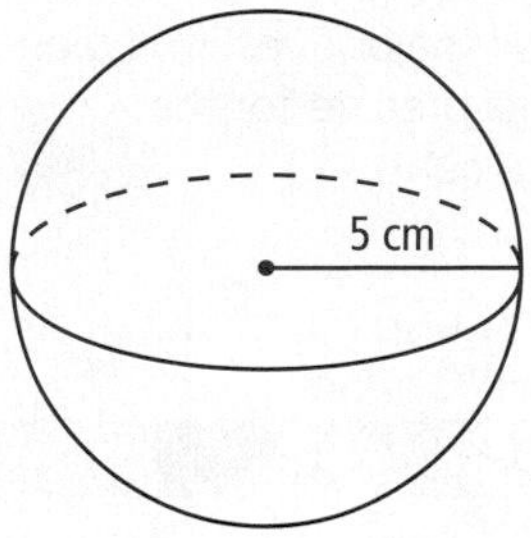

2. What is the volume of the sphere in terms of π?

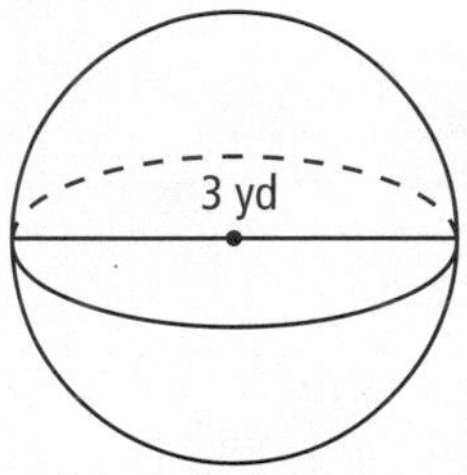

3. Do You UNDERSTAND? The surface area of the globe in a teacher's homeroom is about 1018 in.^{2}. What is the diameter of the globe? Round to the nearest whole number.

ANSWERS TO LESSON QUIZ

1. 100π cm^2

2. 4.5π yd^3

3. about 18 in.

PRESCRIPTION FOR REMEDIATION

Use the student work on the Lesson Quiz to prescribe a differentiated review assignment.

Points	Differentiated Remediation
0–1	Intervention
2	On-level
3	Extension

PowerGeometry.com

5 Assess & Remediate

Assign the Lesson Quiz. Appropriate intervention, practice, or enrichment is automatically generated based on student performance.

Intervention

- **Reteaching** (2 pages) Provides reteaching and practice exercises for the key lesson concepts. Use with struggling students or absent students.
- **English Language Learner Support** Helps students develop and reinforce mathematical vocabulary and key concepts.

All-in-One Resources/Online

Reteaching

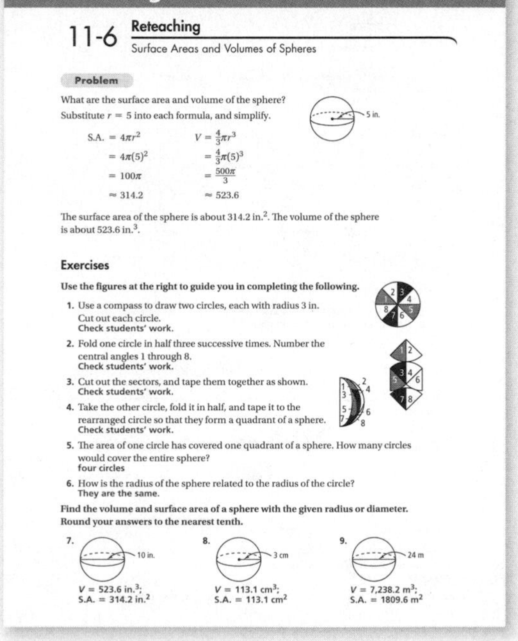

11-6 Reteaching

Surface Areas and Volumes of Spheres

Problem

What are the surface area and volume of the sphere?

Substitute $r = 5$ into each formula, and simplify.

$$\text{S.A.} = 4\pi r^2 = 4\pi(5)^2 = 100\pi \approx 314.2$$

$$V = \frac{4}{3}\pi r^3 = \frac{4}{3}\pi(5)^3 = \frac{500\pi}{3} \approx 523.6$$

The surface area of the sphere is about 314.2 in.^{2}. The volume of the sphere is about 523.6 in.^{3}.

Exercises

Use the figures at the right to guide you in completing the following.

1. Use a compass to draw two circles, each with radius 3 in. Cut out each circle. **Check students' work.**
2. Fold one circle in half three successive times. Number the central angles 1 through 8. **Check students' work.**
3. Cut out the sectors, and tape them together as shown. **Check students' work.**
4. Take the other circle, fold it in half, and tape it to the rearranged circle so that they form a quadrant of a sphere. **Check students' work.**
5. The area of one circle has covered one quadrant of a sphere. How many circles would cover the entire sphere? **four circles**
6. How is the radius of the sphere related to the radius of the circle? **They are the same.**

Find the volume and surface area of a sphere with the given radius or diameter. Round your answers to the nearest tenth.

7. 10 in. **$V = 523.6$ in.^{3}; S.A. $= 314.2$ in.^{2}**
8. 3 cm **$V = 113.1$ cm^3; S.A. $= 113.1$ cm^2**
9. 24 m **$V = 7{,}238.2$ m^3; S.A. $= 1809.6$ m^2**

All-in-One Resources/Online

English Language Learner Support

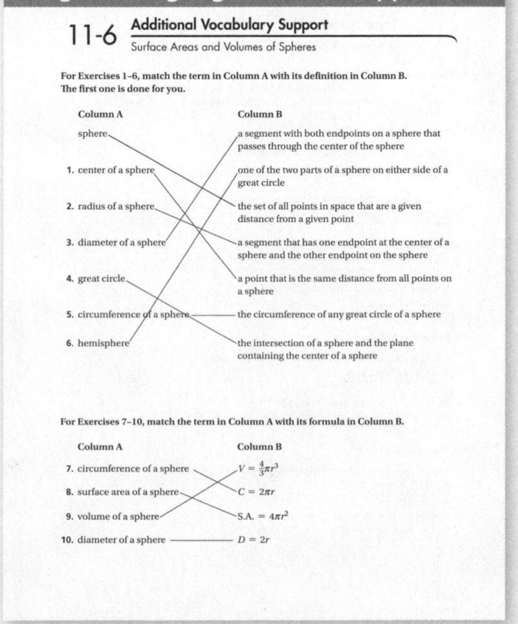

11-6 Additional Vocabulary Support

Surface Areas and Volumes of Spheres

For Exercises 1–6, match the term in Column A with its definition in Column B. The first one is done for you.

Column A	Column B
sphere	a segment with both endpoints on a sphere that passes through the center of the sphere
1. center of a sphere	one of the two parts of a sphere on either side of a great circle
2. radius of a sphere	the set of all points in space that are a given distance from a given point
3. diameter of a sphere	a segment that has one endpoint at the center of a sphere and the other endpoint on the sphere
4. great circle	a point that is the same distance from all points on a sphere
5. circumference of a sphere	the circumference of any great circle of a sphere
6. hemisphere	the intersection of a sphere and the plane containing the center of a sphere

For Exercises 7–10, match the term in Column A with its formula in Column B.

Column A	Column B
7. circumference of a sphere	$V = \frac{4}{3}\pi r^3$
8. surface area of a sphere	$C = 2\pi r$
9. volume of a sphere	S.A. $= 4\pi r^2$
10. diameter of a sphere	$D = 2r$

Differentiated Remediation *continued*

On-Level

- **Practice** (2 pages) Provides extra practice for each lesson. For simpler practice exercises, use the Form K Practice pages found in the All-in-One Teaching Resources and online.
- **Think About a Plan** Helps students develop specific problem-solving skills and strategies by providing scaffolded guiding questions.
- **Standardized Test Prep** Focuses on all major exercises, all major question types, and helps students prepare for the high-stakes assessments.

Extension

- **Enrichment** Provides students with interesting problems and activities that extend the concepts of the lesson.
- **Activities, Games, and Puzzles** Worksheets that can be used for concepts development, enrichment, and for fun!

Practice and Problem Solving Wkbk/ All-in-One Resources/Online

Practice page 1

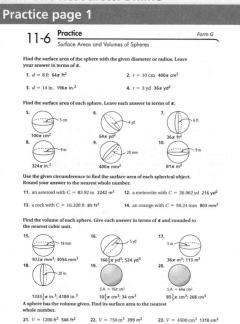

11-6 Practice — Form G
Surface Areas and Volumes of Spheres

Find the surface area of the sphere with the given diameter or radius. Leave your answer in terms of π.

1. $d = 8$ ft 64π ft^2
2. $r = 10$ cm 400π cm^2
3. $d = 14$ in. 196π in.^{2}
4. $r = 3$ yd 36π yd^2

Find the surface area of each sphere. Leave each answer in terms of π.

5. 100π cm^2 6. 64π yd^2 7. 36π ft^2 8. 324π in.^{2} 9. 400π mm^2 10. 81π m^2

Use the given circumference to find the surface area of each spherical object. Round your answer to the nearest whole number.

11. an asteroid with $C = 83.92$ m 2242 m^2
12. a meteorite with $C = 26.062$ yd 216 yd^2
13. a rock with $C = 16.328$ ft 85 ft^2
14. an orange with $C = 50.24$ mm 803 mm^2

Find the volume of each sphere. Give each answer in terms of π and rounded to the nearest cubic unit.

15. 972π mm^3; 3054 mm^3 16. $166\frac{2}{3}\pi$ yd^3; 524 yd^3 17. 36π m^3; 113 m^3
18. $1333\frac{1}{3}\pi$ in.^{3}; 4189 in.^{3} 19. S.A. = 16π cm^2 $10\frac{2}{3}\pi$ cm^3; 34 cm^3 20. S.A. = 64π cm^2 $85\frac{1}{3}\pi$ cm^3; 268 cm^3

A sphere has the volume given. Find its surface area to the nearest whole number.

21. $V = 1200$ ft^3 546 ft^2
22. $V = 750$ m^3 399 m^2
23. $V = 4500$ cm^3 1318 cm^2

Practice and Problem Solving Wkbk/ All-in-One Resources/Online

Practice page 2

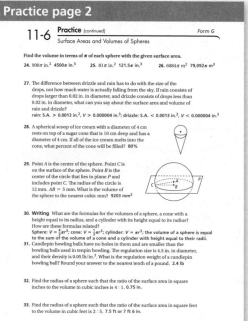

11-6 Practice (continued) — Form G
Surface Areas and Volumes of Spheres

Find the volume in terms of π of each sphere with the given surface area.

24. 900π in.^{2} 4500π in.^{3} 25. 81π in.^{2} 121.5π in.^{3} 26. 6084π m^2 $79{,}092\pi$ m^3

27. The difference between drizzle and rain has to do with the size of the drops, not how much water is actually falling from the sky. If rain consists of drops larger than 0.02 in. in diameter, and drizzle consists of drops less than 0.02 in. in diameter, what can you say about the surface area and volume of rain and drizzle? rain: S.A. > 0.0013 in.^{2}, V > 0.000004 in.^{3}; drizzle: S.A. < 0.0013 in.^{2}, V < 0.000004 in.^{3}

28. A spherical scoop of ice cream with a diameter of 4 cm rests on top of a sugar cone that is 10 cm deep and has a diameter of 4 cm. If all of the ice cream melts into the cone, what percent of the cone will be filled? 80%

29. Point A is the center of the sphere. Point C is on the surface of the sphere. Point B is the center of the circle that lies in plane P and includes point C. The radius of the circle is 12 mm. AB = 5 mm. What is the volume of the sphere to the nearest cubic mm? 9203 mm^3

30. Writing What are the formulas for the volumes of a sphere, a cone with a height equal to its radius, and a cylinder with its height equal to its radius? How are these formulas related? Sphere: $V = \frac{4}{3}\pi r^3$; cone: $V = \frac{1}{3}\pi r^3$; cylinder: $V = \pi r^3$; the volume of a sphere is equal to the sum of the volume of a cone and a cylinder with height equal to their radii.

31. Candlepin bowling balls have no holes in them and are smaller than the bowling balls used in tenpin bowling. The regulation size is 4.5 in. in diameter, and their density is 0.05 lb/in.^{3}. What is the regulation weight of a candlepin bowling ball? Round your answer to the nearest tenth of a pound. 2.4 lb

32. Find the radius of a sphere such that the ratio of the surface area in square inches to the volume in cubic inches is 4 : 1. 0.75 in.

33. Find the radius of a sphere such that the ratio of the surface area in square feet to the volume in cubic feet is 2 : 5. 7.5 ft or 7 ft 6 in.

All-in-One Resources/Online

Enrichment

11-6 Enrichment
Surface Areas and Volumes of Spheres

Two spheres of radius R and r are concentric if the centers of both spheres are the same point. Assume that $r > R$.

1. What is the volume of the outer (larger) sphere? $\frac{4}{3}\pi r^3$
2. What is the volume of the inner (smaller) sphere? $\frac{4}{3}\pi R^3$
3. What is the volume of the space between the smaller and the larger sphere? $\frac{4}{3}\pi(r^3 - R^3)$

The radius of Earth is approximately 4000 mi, and the radius of its core, which is a sphere of molten metals, is about 800 mi.

4. What is the volume of Earth that lies outside the core to the nearest billion mi^3? 266,000,000,000 mi^3
5. Use $(x^3 - y^3) = (x - y)(x^2 + xy + y^2)$ to factor the expression for the volume lying between two concentric spheres. $\frac{4}{3}\pi(r - R)(r^2 + rR + R^2)$

The thickness d of the thin spherical shell representing the difference between the two spheres is defined to be $d = r - R$.

6. Using $r \approx R$, compute the approximate volume of a thin spherical shell of radius R and thickness d. $\frac{4}{3}\pi d(3R^2) = 4\pi dR^2$
7. To see the accuracy of this approximation, fill in the table to two decimal places.

R	d	Exact Volume of Shell	Approximate Volume of Shell
10	0.1	126.92	125.66
50	1	32,048.43	31,415.93
100	3	388,413.95	376,991.12

Most golf balls consist of three concentric spheres: a liquid-filled center, elastic windings, and a dimpled cover. The size and materials of these three components vary for each type of golf ball. A specific type of golf ball has radius 0.84375 in. The center of the golf ball has radius 0.565 in., and the cover has thickness 0.05 in.

elastic windings / liquid-filled center / dimpled cover

8. What is the volume of the outside cover to the nearest thousandth? 0.421 in.^{3}
9. What is the volume of the elastic winding to the nearest hundredth? 1.34 in.^{3}

Practice and Problem Solving Wkbk/ All-in-One Resources/Online

Think About a Plan

11-6 Think About a Plan
Surface Areas and Volumes of Spheres

Meteorology On September 3, 1970, a hailstone with diameter 5.6 in. fell at Coffeyville, Kansas. It weighed about 0.018 lb/in.^{3} compared to the normal 0.033 lb/in.^{3} for ice. About how heavy was this Kansas hailstone?

Understanding the Problem

1. What is the situation described in the problem? A hailstone fell from the sky.
2. What key piece of information is unstated or implicit? The hailstone is spherical.
3. What question do you have to answer? How heavy was the hailstone?
4. Is there any unnecessary information? Explain. Yes; we don't need to know the normal lb/in.^{3} of ice.

Planning the Solution

5. What formula can you use to find the volume of the hailstone? You can use the formula $V = \frac{4}{3}\pi r^3$.
6. How can you use the volume of the hailstone to find its weight? Multiply the volume by the density, 0.018 lb/in.^{3}.

Getting an Answer

7. What is the volume of the hailstone? Show your work. $V = \frac{4}{3}\pi(\frac{5.6}{2})^3 \approx 92$ in.^{3}
8. How do you find the weight of the hailstone? The weight is 92 × 0.018 = 1.656 lb or 1 lb 10.5 oz.

Practice and Problem Solving Wkbk/ All-in-One Resources/Online

Standardized Test Prep

11-6 Standardized Test Prep
Surface Areas and Volumes of Spheres

Multiple Choice

For Exercises 1–5, choose the correct letter.

1. What is the approximate volume of the sphere? A (A) 524 m^3 (C) 1256 m^3 (B) 1000 m^3 (D) 1570 m^3
2. What is the approximate surface area of the sphere? G (F) 225 yd^2 (H) 1767 yd^2 (G) 707 yd^2 (I) 5301 yd^2
3. What is the approximate volume of the sphere if the surface area is 482.8 mm^2? A (A) 998 mm^3 (B) 1126 mm^3 (C) 2042 mm^3 (D) 2993 mm^3
4. What is the approximate surface area of the sphere? G (F) 342.3 km^2 (H) 903.4 km^2 (G) 451.9 km^2 (I) 2713 km^2 — C = 37.68 km
5. What is the approximate radius of a sphere whose volume is 1349 cm^3? B (A) 5.7 cm (B) 6.9 cm (C) 11 cm (D) 14.7 cm

Short Response

6. Suppose a wealthy entrepreneur commissions the design of a spherical spaceship to house a small group for a week in orbit around the Earth. The designer allocates 1000 ft^3 for each person, plus an additional 4073.5 ft^3 for various necessary machines. As in a recreational vehicle, the personal space is largely occupied by items such as beds, shower and toilet facilities, and a kitchenette. The diameter of the ship is 26.8 ft. What is the volume of the spaceship, and for approximately how many people is the ship designed?
[2] $V = 10{,}078.7$ ft^3; 6 people
[1] incorrect volume or incorrect number of people
[0] no response or incorrect response

Online Teacher Resource Center

Activities, Games, and Puzzles

11-6 Puzzle: Follow the Pi Train
Surface Areas and Volumes of Spheres

Find the volume and surface area of each figure below, writing your answers on the track pieces at the right as you go. Cut out those pieces along with the start and finish pieces, and try to make a continuous track by matching up identical values.

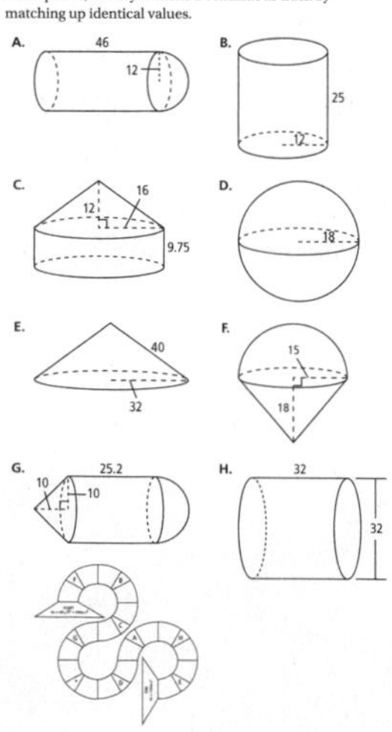

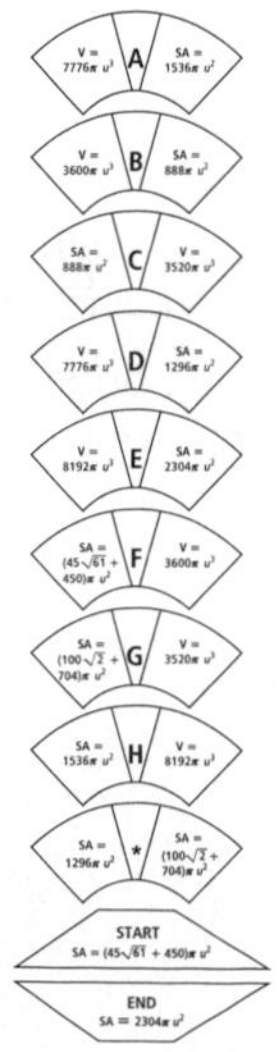

Concept Byte

Use With Lesson 11-7

TECHNOLOGY

Exploring Similar Solids

Common Core State Standards

Prepares for G-MG.A.2 Apply concepts of density based on area and volume in modeling situations . . .

MP 7

To explore surface areas and volumes of similar rectangular prisms, you can set up a spreadsheet like the one below. You choose the numbers for length, width, height, and scale factor. The computer uses formulas to calculate all the other numbers.

MATHEMATICAL PRACTICES

Activity

	A	B	C	D	E	F	G	H	I
1								Ratio of	Ratio of
2					Surface		Scale	Surface	Volumes
3		Length	Width	Height	Area	Volume	Factor (II : I)	Areas (II : I)	(II : I)
4	Rectangular Prism I	6	4	23	508	552	2	4	8
5	Similar Prism II	12	8	46	2032	4416			

In cell E4, enter the formula =2*(B4*C4+B4*D4+C4*D4). This will calculate the sum of the areas of the six faces of Prism I. In cell F4, enter the formula =B4*C4*D4. This will calculate the volume of Prism I.

In cells B5, C5, and D5 enter the formulas =G4*B4, =G4*C4, and =G4*D4, respectively. These will calculate the dimensions of similar Prism II. Copy the formulas from E4 and F4 into E5 and F5 to calculate the surface area and volume of Prism II.

In cell H4 enter the formula =E5/E4 and in cell I4 enter the formula =F5/F4. These will calculate the ratios of the surface areas and volumes.

Investigate In row 4, enter numbers for the length, width, height, and scale factor. Change the numbers to explore how the ratio of the surface areas and the ratio of the volumes are related to the scale factor.

Exercises

State a relationship that seems to be true about the scale factor and the given ratio.

1. the ratio of volumes
2. the ratio of surface areas

Set up spreadsheets that allow you to investigate the following ratios. State a conclusion from each investigation.

3. the volumes of similar cylinders
4. the lateral areas of similar cylinders
5. the surface areas of similar cylinders
6. the volumes of similar square pyramids

Guided Instruction

PURPOSE To use a spreadsheet to explore similar solids, and how ratios of surface areas and ratios of volumes are related to scale factor

PROCESS Students will use a spreadsheet program to draw conclusions about the scale factor and the ratio of volumes and the ratio of surface areas between similar solids.

DISCUSS For geometric solids to be similar, they must have the same shape. Their angles have equal measure, and the lengths of their edges are proportional.

For students who have had little or no experience with spreadsheets, it might be helpful to go through the steps of inserting a function, copying a formula from one cell to another, and widening a column to fit the given wording.

Activity

In this Activity students will examine the relationship between ratios of surface areas and ratios of volumes to the scale factor of rectangular prisms.

Q For this Activity, why do you think it is important to set-up your spreadsheet exactly as the one shown? **[Answers will vary. Sample: Because when you get to the part where you are inserting the formulas, the formulas given will only work if your cell entries mirror those shown.]**

Q When you copy the formulas from E4 and F4 into E5 and F5, do you have to go in and change the function to reflect the new row? **[No, when you copy and paste them, the functions get updated automatically.]**

Mathematical Practices This Concept Byte supports students in becoming proficient in looking for patterns, Mathematical Practice 7.

Answers

Exercises

1. The ratio of volumes is the scale factor cubed.
2. The ratio of surface areas is the scale factor squared.
3. The ratio of volumes is the scale factor cubed.
4. The ratio of lateral areas is the scale factor squared.
5. The ratio of surface areas is the scale factor squared.
6. The ratio of volumes is the scale factor cubed.

1 Interactive Learning

Solve It!

PURPOSE To determine the relationship between the volumes of similar figures

PROCESS Students may use visual reasoning.

FACILITATE

- **Q** What effect do you think taking half of all three dimensions will have on the volume of the middle layer? **[It will be $\left(\frac{1}{2}\right)^3 = \frac{1}{8}$ the volume of the lower layer.]**
- **Q** What is the relationship between the height of the bottom and top layers? **[The top layer is $\frac{1}{4}$ as high as the bottom layer.]**
- **Q** What will be the ratio between the volumes of the top and bottom layers? **[$\left(\frac{1}{4}\right)^3 = \frac{1}{64}$]**

ANSWER See Solve It in Answers on next page.

CONNECT THE MATH Students see that the ratio of volumes of similar solid figures will be equal to the cube of the scale factor. In the lesson, students study different similar solid figures to learn their scale factors.

2 Guided Instruction

Problem 1

- **Q** How can you tell if two figures are similar? **[The ratios of corresponding sides should be equal.]**
- **Q** Are the ratios of corresponding sides equal in both 1A and 1B? **[no]**

11-7 Areas and Volumes of Similar Solids

Common Core State Standards

G-MG.A.1 Use geometric shapes, their measures, and their properties to describe objects.

G-MG.A.2 Apply concepts of density based on area and volume in modeling situations . . .

MP 3, MP 4, MP 7, MP 8

Objective To compare and find the areas and volumes of similar solids

Solve It! Getting Ready!

A baker is making a three-layer wedding cake. Each layer has a square base. Each dimension of the middle layer is $\frac{1}{2}$ the corresponding dimension of the bottom layer. Each dimension of the top layer is $\frac{1}{2}$ the corresponding dimension of the middle layer. What conjecture can you make about the relationship between the volumes of the layers? Calculate the volumes to check your answer. Modify your conjecture if necessary.

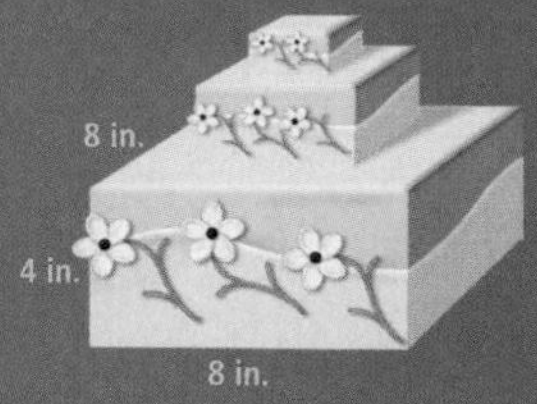

Will the bottom-to-middle ratio be the same as the middle-to-top ratio?

MATHEMATICAL PRACTICES

Lesson Vocabulary
- similar solids

Essential Understanding You can use ratios to compare the areas and volumes of similar solids.

Similar solids have the same shape, and all their corresponding dimensions are proportional. The ratio of corresponding linear dimensions of two similar solids is the scale factor. Any two cubes are similar, as are any two spheres.

Plan

How do you check for similarity?
Check that the ratios of the corresponding dimensions are the same. A rectangular prism has three dimensions (ℓ, w, h), so you must check three ratios.

Problem 1 Identifying Similar Solids

Are the two rectangular prisms similar? If so, what is the scale factor of the first figure to the second figure?

A

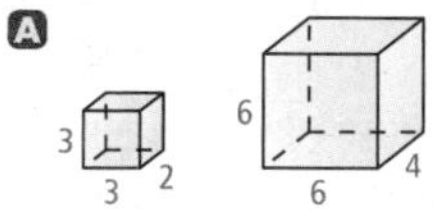

$\frac{3}{6} = \frac{2}{4} = \frac{3}{6}$

The prisms are similar because the corresponding linear dimensions are proportional.

The scale factor is $\frac{1}{2}$.

B

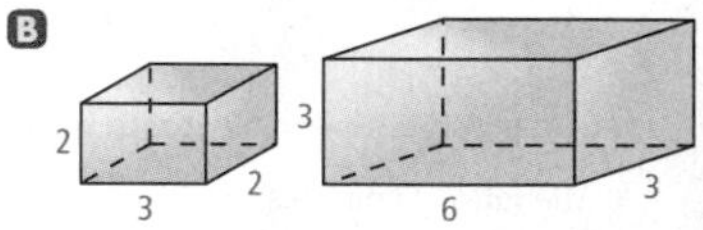

$\frac{2}{3} = \frac{2}{3} \neq \frac{3}{6}$

The prisms are not similar because the corresponding linear dimensions are not proportional.

11-7 Preparing to Teach

BIG ideas Similarity
Measurement

ESSENTIAL UNDERSTANDING
- Ratios can be used to compare the areas and volumes of similar solids.

Math Background

Students have learned how to calculate the volume and surface area of solid figures. In this lesson, they will use this knowledge to examine the relationship between the surface areas and volumes of similar figures. Similar solids have proportional sides. This knowledge allows predictions to be made about figures that are enlarged or reduced.

The surface areas of similar solids are related to the square of the scale factor for the two solids.

The volumes of similar solids are related by the cube of the scale factor for the two solids.

The relationships between linear, area, and volume measures of similar solids are nonintuitive but easily verified algebraically. It may help students who expect all to be linear relationships to make graphs or tables of the changes in diameter, surface area, and volume of a sphere as the radius increases in increments of 1.

Mathematical Practices

Construct viable arguments and critique the reasoning of others. Students will analyze the relationship between the scale factor, surface area, and volume of similar three-dimensional geometric figures by relating them to rectangular prisms.

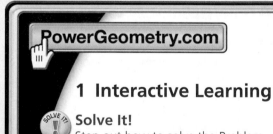

PowerGeometry.com

1 Interactive Learning

Solve It!
Step out how to solve the Problem with helpful hints and an online question. Other questions are listed above in Interactive Learning.

Virtual Nerd™ Access Virtual Nerd student-centered math tutorials that directly relate to the content of the lesson.

Got It? 1. Are the two cylinders similar? If so, what is the scale factor of the first figure to the second figure?

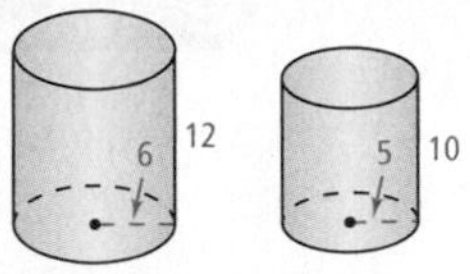

The two similar prisms shown here suggest two important relationships for similar solids.

The ratio of the side lengths is 1 : 2.
The ratio of the surface areas is 22 : 88, or 1 : 4.
The ratio of the volumes is 6 : 48, or 1 : 8.

The ratio of the surface areas is the square of the scale factor. The ratio of the volumes is the cube of the scale factor. These two facts apply to all similar solids.

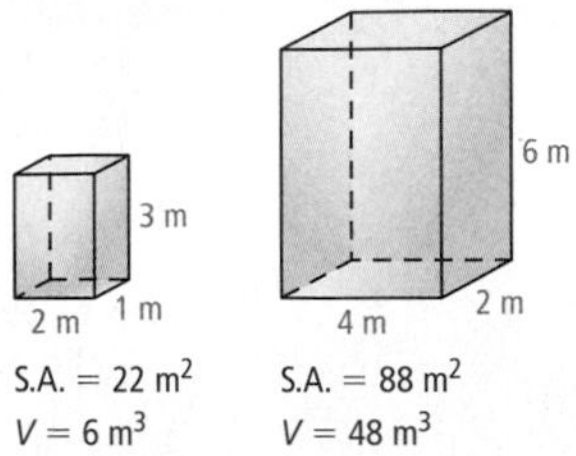

take note

Theorem 11-12 Areas and Volumes of Similar Solids

If the scale factor of two similar solids is $a : b$, then

- the ratio of their corresponding areas is $a^2 : b^2$
- the ratio of their volumes is $a^3 : b^3$

Problem 2 Finding the Scale Factor

Plan

How can you use the given information? You are given the volumes of two similar solids. Write a proportion using the ratio $a^3 : b^3$.

The square prisms at the right are similar. What is the scale factor of the smaller prism to the larger prism?

$\frac{a^3}{b^3} = \frac{729}{1331}$ The ratio of the volumes is $a^3 : b^3$.

$\frac{a}{b} = \frac{9}{11}$ Take the cube root of each side.

The scale factor is 9 : 11.

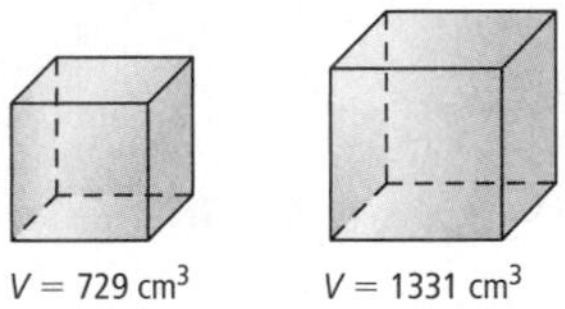

Got It? 2. a. What is the scale factor of two similar prisms with surface areas 144 m^2 and 324 m^2?

b. Reasoning Are any two square prisms similar? Explain.

Got It?

Q What two measurements should be proportional for the cylinders to be similar? **[the radius and the height]**

Q What is the ratio of the radii? $\frac{6}{5}$

Q What is the ratio of the heights? $\frac{6}{5}$

Take Note

Ask students to explain why the ratios are squared for area and cubed for volume. Area is a measurement of two dimensions: length and width. Volume is a measurement of three dimensions: length, width, and height.

Problem 2

Q How are the volumes of similar figures related? **[The ratio of volumes is the cube of the scale factor between the two figures.]**

Q How can you find the scale factor? **[Take the cube root of the ratio of volumes.]**

Got It?

For 2a, be sure that students realize that the measurements given are areas, not volumes.

If students struggle with 1b, have them draw diagrams of several square prisms and compare them with the drawings of students near them. They should realize that the ratios of sides are always equal.

2 Guided Instruction

Each Problem is worked out and supported online.

Problem 1
Identifying Similar Solids
Animated

Problem 2
Finding the Scale Factor

Problem 3
Using a Scale Factor
Animated

Problem 4
Using a Scale Factor to Find Capacity
Animated

Support in Geometry Companion
- Vocabulary
- Key Concepts
- Got It?

Answers

Solve It!

The volume of the middle layer is $\frac{1}{8}$ the volume of the bottom layer, and the volume of the top layer is $\frac{1}{8}$ the volume of the middle layer and $\frac{1}{64}$ the volume of the bottom layer; bottom layer $V = 256$ in.^{3}; middle layer $V = 32$ in.^{3}; top layer $V = 4$ in.^{3}

Got It?

1. yes; 6 : 5 or $\frac{6}{5}$

2a. 2 : 3

b. No; the bases are similar but the heights may not be in the same ratio as the edges of the bases.

Problem 3

Q How are the lateral areas of similar figures related? **[The ratio of the areas is the square of the scale factor.]**

Q How can you find the scale factor? **[Take the square root of the ratio of the lateral areas.]**

Q How are the volumes of similar figures related? **[The ratio of volumes is the cube of the scale factor.]**

Q How can you use the scale factor to find the volume of the larger can? **[Set up a proportion with the cube of the scale factor equal to the ratio of the known and unknown volumes.]**

Q What property of proportions will you use to solve? **[Cross-Products Property]**

Got It?

Students need to take the cube root of the ratio to find the scale factor. They need to square the scale factor to find the surface area of the smaller solid.

Problem 3 Using a Scale Factor

Painting **The lateral areas of two similar paint cans are 1019 cm² and 425 cm². The volume of the smaller can is 1157 cm³. What is the volume of the larger can?**

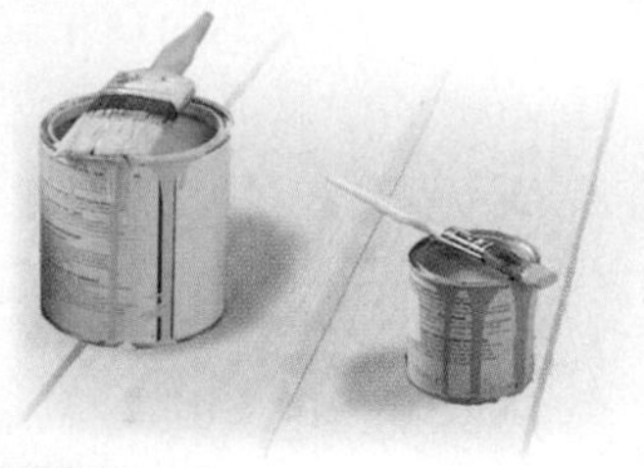

Know
- The lateral areas
- The volume of the smaller can

Need
The scale factor

Plan
Use the lateral areas to find the scale factor $a : b$. Then write and solve a proportion using the ratio $a^3 : b^3$ to find the volume of the larger can.

Think
Does it matter how you set up the proportion?
Yes. The numerators should refer to the same paint can, and the denominators should refer to the other can.

Step 1 Find the scale factor $a : b$.

$\frac{a^2}{b^2} = \frac{1019}{425}$ The ratio of the surface areas is $a^2 : b^2$.

$\frac{a}{b} = \frac{\sqrt{1019}}{\sqrt{425}}$ Take the positive square root of each side.

Step 2 Use the scale factor to find the volume.

$\frac{V_{large}}{V_{small}} = \frac{(\sqrt{1019})^3}{(\sqrt{425})^3}$ The ratio of the volumes is $a^3 : b^3$.

$\frac{V_{large}}{1157} = \frac{(\sqrt{1019})^3}{(\sqrt{425})^3}$ Substitute 1157 for V_{small}.

$V_{large} = 1157 \cdot \frac{(\sqrt{1019})^3}{(\sqrt{425})^3}$ Solve for V_{large}.

$V_{large} \approx 4295.475437$ Use a calculator.

The volume of the larger paint can is about 4295 cm³.

Got It? 3. The volumes of two similar solids are 128 m³ and 250 m³. The surface area of the larger solid is 250 m². What is the surface area of the smaller solid?

You can compare the capacities and weights of similar objects. The capacity of an object is the amount of fluid the object can hold. The capacities and weights of similar objects made of the same material are proportional to their volumes.

Additional Problems

1. Are the two rectangular prisms similar? If so, what is the scale factor?

a.

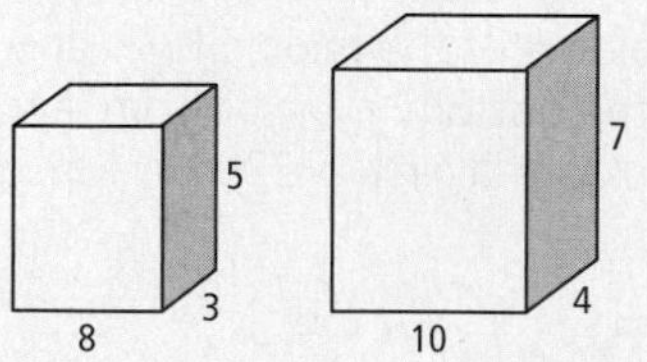

b.

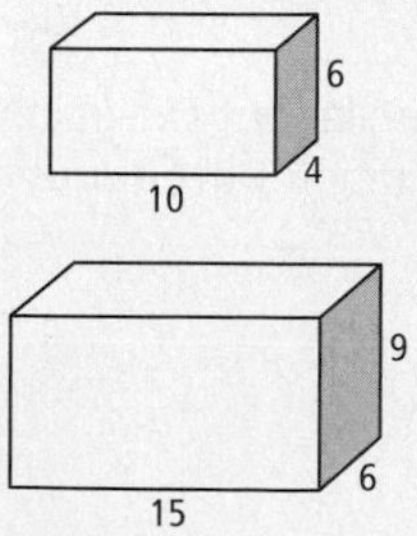

ANSWER **a.** no **b.** yes, 2 : 3

2. What is the scale factor of the similar rectangular prisms shown below?

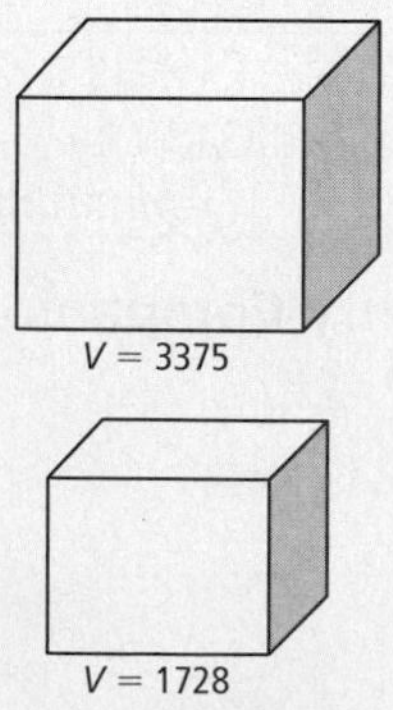

ANSWER 5 : 4

3. The volumes of two similar solids are 40 in.³ and 135 in.³. If the surface area of the smaller solid is 48 in.², what is the surface area of the larger solid?

ANSWER 108 in.²

4. An office supplies store sells paper clips in small and large boxes. A small box holds about 220 paper clips. The large box is formed by doubling the dimensions of the small box. About how many paper clips should fit in the large box?

ANSWER about 1760 paper clips

Problem 4 Using a Scale Factor to Find Capacity STEM

Think

How does capacity relate to volume? Since the capacities of similar objects are proportional to their volumes, the ratio of their capacities is equal to the ratio of their volumes.

Containers **A bottle that is 10 in. high holds 34 oz of milk. The sandwich shop shown at the right is shaped like a milk bottle. To the nearest thousand ounces how much milk could the building hold?**

The scale factor of the bottles is 1 : 48.

The ratio of their volumes, and hence the ratio of their capacities, is $1^3 : 48^3$, or 1 : 110,592.

$\frac{1}{110,592} = \frac{34}{x}$ Let x = the capacity of the milk-bottle building.

$x = 34 \cdot 110,592$ Use the Cross Products Property.

$x = 3,760,128$ Simplify.

The milk-bottle building could hold about 3,760,000 oz.

Got It? **4.** A marble paperweight shaped like a pyramid weighs 0.15 lb. How much does a similarly shaped marble paperweight weigh if each dimension is three times as large?

Lesson Check

Do you know HOW?

1. Which two of the following cones are similar? What is their scale factor?

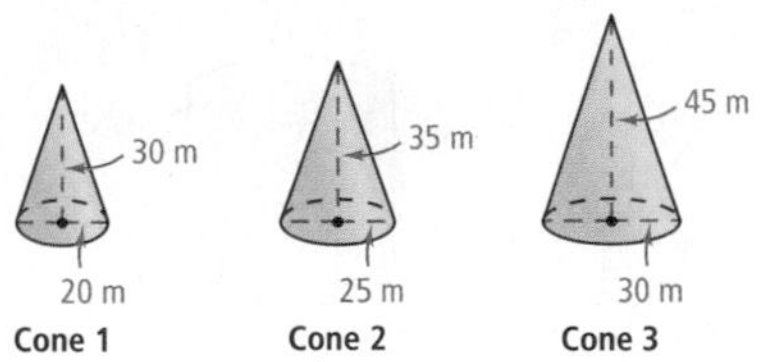

2. The volumes of two similar containers are 115 $in.^3$ and 67 $in.^3$. The surface area of the smaller container is 108 $in.^2$. What is the surface area of the larger container?

Do you UNDERSTAND? MATHEMATICAL PRACTICES

3. Vocabulary How are similar solids different from similar polygons? Explain.

4. Error Analysis Two cubes have surface areas 49 cm^2 and 64 cm^2. Your classmate tried to find the scale factor of the larger cube to the smaller cube. Explain and correct your classmate's error.

$\frac{a^2}{b^2} = \frac{49}{64}$

$\frac{a}{b} = \frac{7}{8}$

The scale factor of the larger cube to the smaller cube is 7 : 8.

Problem 4

Q What is the scale factor between the bottle and building? **[10 : 480 = 1 : 48]**

Q What is the ratio of the volumes of the two figures? **[$1^3 : 48^3$ = 1 : 110,592]**

Q How can you find the amount of milk that would be contained in the building? **[Multiply the volume of the bottle by the scale factor.]**

Got It? VISUAL LEARNERS

Have students identify the ratio of the pyramids.

3 Lesson Check

Do you know HOW?

- If students have difficulty with Exercise 1, then have them review Problem 1 to see how to set the ratios of dimensions.

Do you UNDERSTAND?

- If students have difficulty with Exercise 3, then have them draw diagrams of similar polygons and similar solids.

Close

Q How are the areas of similar solids related? **[The ratio of areas is equal to the square of the scale factor for the two solids.]**

Q How are the volumes of similar figures related? **[The ratio of volumes is equal to the cube of the scale factor for the two solids.]**

Answers

Got It? (continued)

3. 160 m^2

4. 4.05 lb

Lesson Check

1. Cone 1 and Cone 3 are similar; 2 : 3.

2. about 155 $in.^2$

3. Sample answer: There are many relationships that must be true for the solids to be similar: all corresponding angles must be ≅; the corresponding faces must be similar and all corresponding edges and heights proportional.

4. Your classmate found the scale factor of the smaller cube to the larger cube. The scale factor should be 8 : 7.

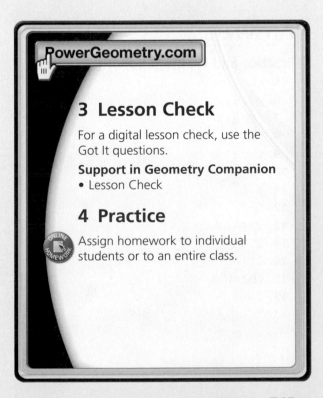

PowerGeometry.com

3 Lesson Check

For a digital lesson check, use the Got It questions.

Support in Geometry Companion

- Lesson Check

4 Practice

Assign homework to individual students or to an entire class.

4 Practice

ASSIGNMENT GUIDE

Basic: 5–26 all, 28–29, 34–38 even
Average: 5–23 odd, 24–38
Advanced: 5–23 odd, 24–41
Standardized Test Prep: 42–46
Mixed Review: 47–54

Mathematical Practices are supported by exercises with red headings. Here are the Practices supported in this lesson:

MP 3: Construct Arguments Ex. 26, 28, 29, 39
MP 3: Critique the Reasoning of Others Ex. 4
MP 7: Look for Patterns Ex. 23
MP 8: Check for Reasonableness Ex. 25

Applications exercises have blue headings. Exercises 21, 22, and 38 support MP 4: Model.

STEM exercises focus on science or engineering applications.

EXERCISE 29: Use the Think About a Plan worksheet in the **Practice and Problem Solving Workbook** (also available in the Teaching Resources in print and online) to further support students' development in becoming independent learners.

HOMEWORK QUICK CHECK

To check students' understanding of key skills and concepts, go over Exercises 11, 21, 25, 28, and 29.

Practice and Problem-Solving Exercises

MATHEMATICAL PRACTICES

A Practice

For Exercises 5–10, are the two figures similar? If so, give the scale factor of the first figure to the second figure. See Problem 1.

5.

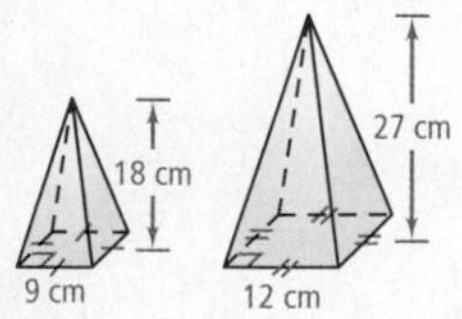

6.

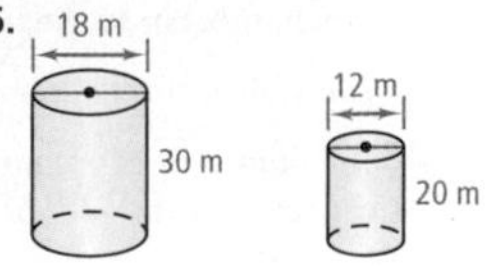

7.

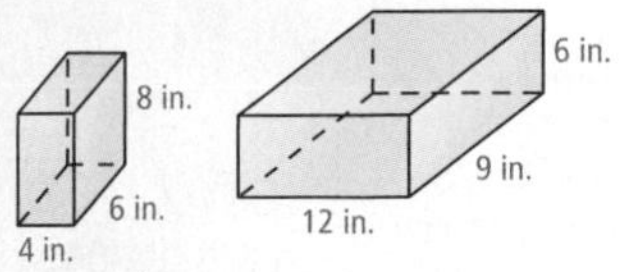

8.

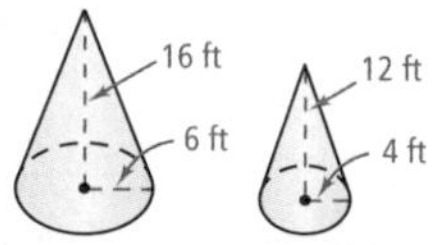

9. two cubes, one with 3-cm edges, the other with 4.5-cm edges

10. a cylinder and a square prism both with 3-in. radius and 1-in. height

Each pair of figures is similar. Use the given information to find the scale factor of the smaller figure to the larger figure. See Problem 2.

11.

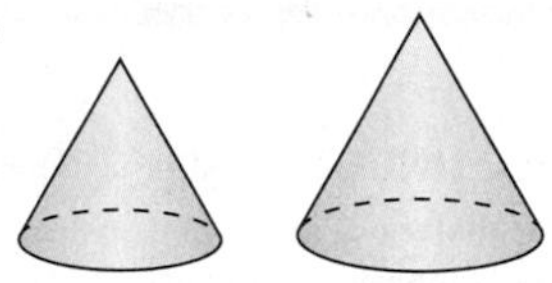

$V = 250\pi$ ft^3 $V = 432\pi$ ft^3

12.

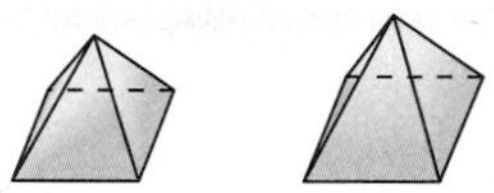

$V = 216$ in.^{3} $V = 343$ in.^{3}

13.

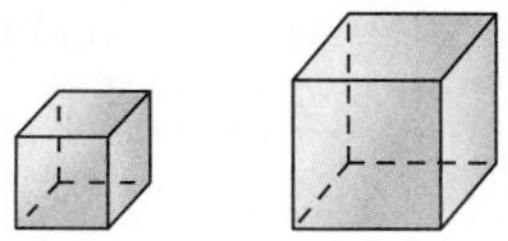

S.A. = 18 m^2 S.A. = 32 m^2

14.

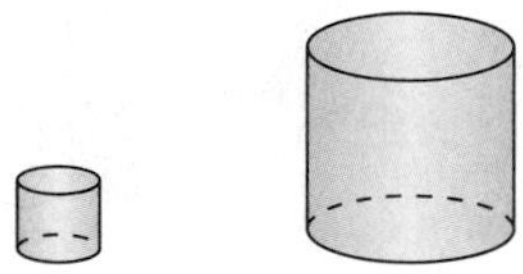

S.A. = 20π yd^2 S.A. = 125π yd^2

The surface areas of two similar figures are given. The volume of the larger figure is given. Find the volume of the smaller figure. See Problem 3.

15. S.A. = 248 in.^{2}
S.A. = 558 in.^{2}
$V = 810$ in.^{3}

16. S.A. = 192 m^2
S.A. = 1728 m^2
$V = 4860$ m^3

17. S.A. = 52 ft^2
S.A. = 208 ft^2
$V = 192$ ft^3

Answers

Practice and Problem-Solving Exercises

5. no
6. yes; 3 : 2
7. yes; 2 : 3
8. no
9. yes; 2 : 3
10. no
11. 5 : 6
12. 6 : 7
13. 3 : 4
14. 2 : 5
15. 240 in.^{3}
16. 180 m^3
17. 24 ft^3

The volumes of two similar figures are given. The surface area of the smaller figure is given. Find the surface area of the larger figure.

18. $V = 27$ in.^{3}
$V = 125$ in.^{3}
S.A. $= 63$ in.^{2}

19. $V = 27\text{ m}^3$
$V = 64\text{ m}^3$
S.A. $= 63\text{ m}^2$

20. $V = 2\text{ yd}^3$
$V = 250\text{ yd}^3$
S.A. $= 13\text{ yd}^2$

See Problem 4.

STEM **21. Packaging** There are 750 toothpicks in a regular-sized box. If a jumbo box is made by doubling all the dimensions of the regular-sized box, how many toothpicks will the jumbo box hold?

STEM **22. Packaging** A cylinder with a 4-in. diameter and a 6-in. height holds 1 lb of oatmeal. To the nearest ounce, how much oatmeal will a similar 10-in.-high cylinder hold? (*Hint:* 1 lb = 16 oz)

23. Compare and Contrast A regular pentagonal prism has 9-cm base edges. A larger, similar prism of the same material has 36-cm base edges. How does each indicated measurement for the larger prism compare to the same measurement for the smaller prism?

a. the volume
b. the weight

B Apply

24. Two similar prisms have heights 4 cm and 10 cm.
a. What is their scale factor?
b. What is the ratio of their surface areas?
c. What is the ratio of their volumes?

25. Think About a Plan A company announced that it had developed the technology to reduce the size of its atomic clock, which is used in electronic devices that transmit data. The company claims that the smaller clock will be similar to the existing clock made of the same material. The dimensions of the smaller clock will be $\frac{1}{10}$ the dimensions of the company's existing atomic clocks, and it will be $\frac{1}{100}$ the weight. Do these ratios make sense? Explain.

- What is the scale factor of the smaller clock to the larger clock?
- How are the weights of the two objects related to their scale factor?

26. Reasoning Is there a value of x for which the rectangular prisms at the right are similar? Explain.

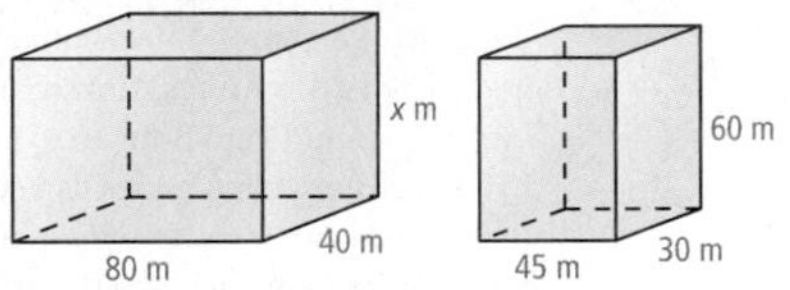

27. The volume of a spherical balloon with radius 3.1 cm is about 125 cm^3. Estimate the volume of a similar balloon with radius 6.2 cm.

28. Writing Are all spheres similar? Explain.

29. Reasoning A carpenter is making a blanket chest based on an antique chest. Both chests have the shape of a rectangular prism. The length, width, and height of the new chest will all be 4 in. greater than the respective dimensions of the antique. Will the chests be similar? Explain.

30. Two similar pyramids have lateral area 20 ft^2 and 45 ft^2. The volume of the smaller pyramid is 8 ft^3. Find the volume of the larger pyramid.

18. 175 in.^{2}
19. 112 m^2
20. 325 yd^2
21. 6000 toothpicks
22. 74 oz
23a. It is 64 times the volume of the smaller prism.
b. It is 64 times the weight of the smaller prism.
24a. 2 : 5
b. 4 : 25
c. 8 : 125
25. No; explanations may vary. Sample: If the scale factor is $\frac{1}{10}$, then the weight of the smaller clock should be $\frac{1}{1000}$ the weight of the existing clock.
26. yes; 60: $\frac{80}{60} = \frac{40}{30} = \frac{60}{45} = \frac{4}{3}$
27. about 1000 cm^3
28. Yes; explanations may vary. Sample: The ratio of the radii of the spheres equals the ratios of all other corresponding linear dimensions.
29. No; the same increase to all the dimensions does not result in proportional ratios unless the original prism is a cube.
30. 27 ft^3

Answers

Practice and Problem-Solving Exercises (continued)

31a. 3 : 1

b. 9 : 1

32a. 11 : 14

b. 121 : 196

33. 864 in.^{3}

34. 1 : 4; 1 : 8

35. 9 : 25; 27 : 125

36. 7 : 9; 343 : 729

37. 5 : 8; 25 : 64

38a. 144 coats

b. 1728 meals

39a. 100 times

b. 1000 times

c. His weight is 1000 times the weight of an average person, but his bones can support only 600 times the weight of an average person.

40a. 384 cm^3

b. 16 : 1

c. A: 384 cm^2; B: 24 cm^2

31. The volumes of two spheres are 729 in.^{3} and 27 in.^{3}.
  a. Find the ratio of their radii.
  b. Find the ratio of their surface areas.

32. The volumes of two similar pyramids are 1331 cm^3 and 2744 cm^3.
  a. Find the ratio of their heights.
  b. Find the ratio of their surface areas.

33. A clown's face on a balloon is 4 in. tall when the balloon holds 108 in.^{3} of air. How much air must the balloon hold for the face to be 8 in. tall?

Copy and complete the table for the similar solids.

	Similarity Ratio	Ratio of Surface Areas	Ratio of Volumes
34.	1 : 2	■ : ■	■ : ■
35.	3 : 5	■ : ■	■ : ■
36.	■ : ■	49 : 81	■ : ■
37.	■ : ■	■ : ■	125 : 512

38. Literature In *Gulliver's Travels,* by Jonathan Swift, Gulliver first traveled to Lilliput. The Lilliputian average height was one twelfth of Gulliver's height.
  a. How many Lilliputian coats could be made from the material in Gulliver's coat? (*Hint:* Use the ratio of surface areas.)
  b. How many Lilliputian meals would be needed to make a meal for Gulliver? (*Hint:* Use the ratio of volumes.)

Challenge

39. Indirect Reasoning Some stories say that Paul Bunyan was ten times as tall as the average human. Assume that Paul Bunyan's bone structure was proportional to that of ordinary people.
  a. Strength of bones is proportional to the area of their cross section. How many times as strong as the average person's bones would Paul Bunyan's bones be?
  b. Weights of objects made of like material are proportional to their volumes. How many times the average person's weight would Paul Bunyan's weight be?
  c. Human leg bones can support about 6 times the average person's weight. Use your answers to parts (a) and (b) to explain why Paul Bunyan could not exist with a bone structure that was proportional to that of ordinary people.

40. Square pyramids A and B are similar. In pyramid A, each base edge is 12 cm. In pyramid B, each base edge is 3 cm and the volume is 6 cm^3.
  a. Find the volume of pyramid A.
  b. Find the ratio of the surface area of A to the surface area of B.
  c. Find the surface area of each pyramid.

41. A cone is cut by a plane parallel to its base. The small cone on top is similar to the large cone. The ratio of the slant heights of the cones is 1 : 2. Find each ratio.
  a. the surface area of the large cone to the surface area of the small cone
  b. the volume of the large cone to the volume of the small cone
  c. the surface area of the frustum to the surface area of the large cone and to the surface area of the small cone
  d. the volume of the frustum to the volume of the large cone and to the volume of the small cone

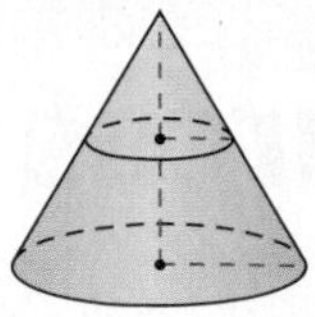

Standardized Test Prep

GRIDDED RESPONSE

SAT/ACT

42. The slant heights of two similar pyramids are in the ratio 1 : 5. The volume of the smaller pyramid is 60 m^3. What is the volume in cubic meters of the larger pyramid?

43. What is the value of x in the figure at the right?

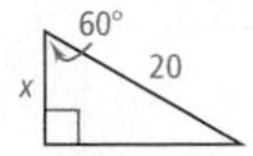

44. A dilation maps $\triangle JEN$ onto $\triangle J'E'N'$. If $JE = 4.5$ ft, $EN = 6$ ft, and $J'E' = 13.5$ ft, what is $E'N'$ in feet?

45. $\triangle CAR \cong \triangle BUS$, $m\angle C = 25$, and $m\angle R = 39$. What is $m\angle U$?

46. A regular pentagon has a radius of 5 in. What is the area of the pentagon to the nearest square inch?

Mixed Review

47. **Sports Equipment** The circumference of a regulation basketball is between 75 cm and 78 cm. What are the smallest and the largest surface areas that a basketball can have? Give your answers to the nearest whole unit. See Lesson 11-6.

Find the volume of each sphere to the nearest tenth.

48. diameter = 6 in. **49.** circumference = 2.5π m **50.** radius = 6 in.

51. The altitude to the hypotenuse of right $\triangle ABC$ divides the hypotenuse into 12-mm and 16-mm segments. Find the length of each of the following. See Lesson 7-4.
  a. the altitude to the hypotenuse
  b. the shorter leg of $\triangle ABC$ **c.** the longer leg of $\triangle ABC$

Get Ready! **To prepare for Lesson 12-1, do Exercises 52–54.**

Find the value of x. See Lesson 8-1.

52.

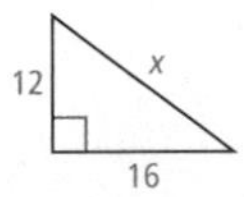

53.

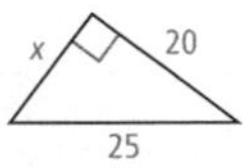

54.

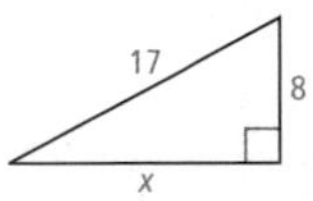

41a. 4 : 1
  b. 8 : 1
  c. $(3\ell + 5r) : (4\ell + 4r)$; $(3\ell + 5r) : (\ell + r)$, where r is the radius and ℓ is the slant height of the small cone.
  d. 7 : 8 and 7 : 1
42. 7500
43. 10
44. 18
45. 116
46. 59
47. about 1790 cm^2 and 1937 cm^2
48. 113.1 $in.^3$
49. 8.2 m^3
50. 904.8 $in.^3$
51a. $8\sqrt{3}$ mm, or about 13.9 mm
  b. $4\sqrt{21}$ mm, or about 18.3 mm
  c. $8\sqrt{7}$ mm, or about 21.2 mm
52. 20
53. 15
54. 15

11-7 Lesson Resources

Differentiated Remediation

Instructional Support

Geometry Companion

Students can use the **Geometry Companion** worktext (4 pages) . . .

- New Vocabulary
- Key Concepts
- Got It for each Problem
- Lesson Check

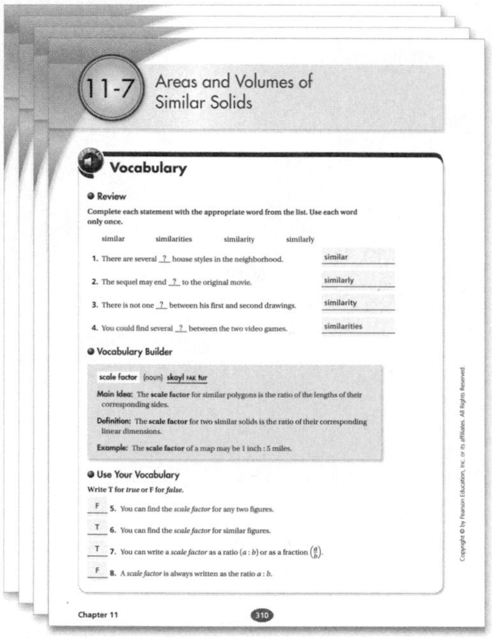

11-7 Areas and Volumes of Similar Solids

Vocabulary

Review

Complete each statement with the appropriate word from the list. Use each word only once.

similar similarities similarity similarly

1. There are several _?_ house styles in the neighborhood. similar
2. The sequel may end _?_ to the original movie. similarly
3. There is not one _?_ between his first and second drawings. similarity
4. You could find several _?_ between the two video games. similarities

Vocabulary Builder

scale factor (noun) skayl FAK tur

Main Idea: The **scale factor** for similar polygons is the ratio of the lengths of their corresponding sides.

Definition: The **scale factor** for two similar solids is the ratio of their corresponding linear dimensions.

Example: The **scale factor** of a map may be 1 inch : 5 miles.

Use Your Vocabulary

Write T for *true* or F for *false*.

F 5. You can find the *scale factor* for any two figures.

T 6. You can find the *scale factor* for similar figures.

T 7. You can write a *scale factor* as a ratio $(a : b)$ or as a fraction $\left(\frac{a}{b}\right)$.

F 8. A *scale factor* is always written as the ratio $a : b$.

Chapter 11 310

ELL Support

Focus on Language Use flashcards for vocabulary review. Arrange the class into 7 mixed groups and assign each a lesson from the chapter. Have each group write the vocabulary words and their definitions on a sheet of paper. Invite volunteers to write their words on the board, using the opportunity to discuss and critique their definitions as a class. Ask: What does that definition mean? Can you give an example? Have students write the vocabulary words on one side of an index card with the definition on the other. Students can use the cards to quiz each other and for review later.

5 Assess & Remediate

Lesson Quiz

1. Are the two rectangular prisms similar? If so, what is the scale factor?

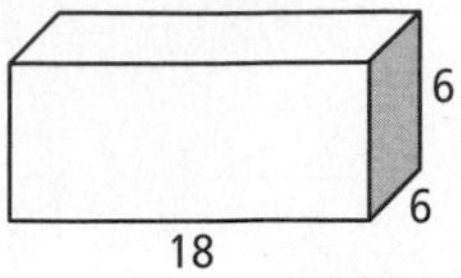

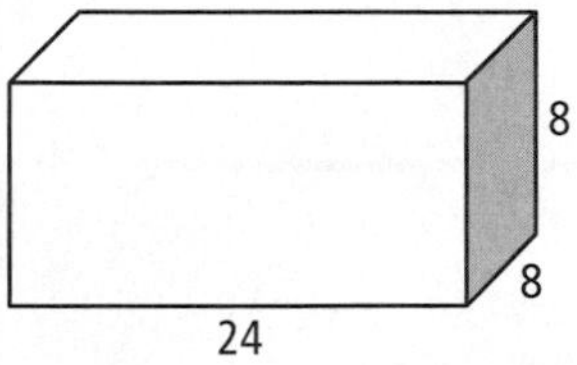

2. The surface areas of two similar solids are 441 cm^2 and 225 cm^2. If the volume of the smaller solid is 250 cm^3, what is the volume of the larger solid?
3. **Do You UNDERSTAND?** A small box holds 18.5 oz of cereal. The family size box is formed by scaling the dimensions of the small box by a factor of 1.15. How much cereal does the family size box hold? Round to the nearest tenth if necessary.

ANSWERS TO LESSON QUIZ

1. yes, 3 : 4
2. 686 cm^3
3. about 28.1 oz

PRESCRIPTION FOR REMEDIATION

Use the student work on the Lesson Quiz to prescribe a differentiated review assignment.

Points	Differentiated Remediation
0–1	Intervention
2	On-level
3	Extension

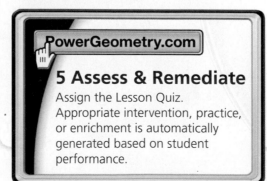

PowerGeometry.com

5 Assess & Remediate

Assign the Lesson Quiz. Appropriate intervention, practice, or enrichment is automatically generated based on student performance.

Intervention

- **Reteaching** (2 pages) Provides reteaching and practice exercises for the key lesson concepts. Use with struggling students or absent students.
- **English Language Learner Support** Helps students develop and reinforce mathematical vocabulary and key concepts.

All-in-One Resources/Online

Reteaching

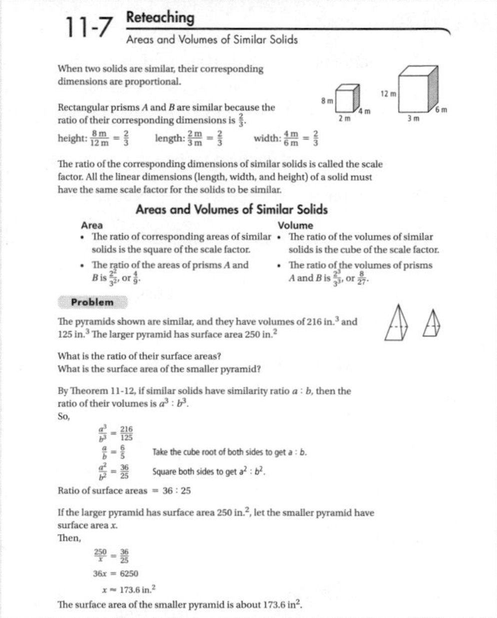

11-7 Reteaching
Areas and Volumes of Similar Solids

When two solids are similar, their corresponding dimensions are proportional.

Rectangular prisms A and B are similar because the ratio of their corresponding dimensions is $\frac{2}{3}$.

height: $\frac{8\text{ m}}{12\text{ m}} = \frac{2}{3}$ length: $\frac{2\text{ m}}{3\text{ m}} = \frac{2}{3}$ width: $\frac{4\text{ m}}{6\text{ m}} = \frac{2}{3}$

The ratio of the corresponding dimensions of similar solids is called the scale factor. All the linear dimensions (length, width, and height) of a solid must have the same scale factor for the solids to be similar.

Areas and Volumes of Similar Solids

Area
- The ratio of corresponding areas of similar solids is the square of the scale factor.
- The ratio of the areas of prisms A and B is $\frac{2^2}{3^2}$, or $\frac{4}{9}$.

Volume
- The ratio of the volumes of similar solids is the cube of the scale factor.
- The ratio of the volumes of prisms A and B is $\frac{2^3}{3^3}$, or $\frac{8}{27}$.

Problem

The pyramids shown are similar, and they have volumes of 216 $in.^3$ and 125 $in.^3$ The larger pyramid has surface area 250 $in.^2$

What is the ratio of their surface areas?
What is the surface area of the smaller pyramid?

By Theorem 11-12, if similar solids have similarity ratio $a : b$, then the ratio of their volumes is $a^3 : b^3$.
So,

$\frac{a^3}{b^3} = \frac{216}{125}$

$\frac{a}{b} = \frac{6}{5}$ Take the cube root of both sides to get $a : b$.

$\frac{a^2}{b^2} = \frac{36}{25}$ Square both sides to get $a^2 : b^2$.

Ratio of surface areas = 36 : 25

If the larger pyramid has surface area 250 $in.^2$, let the smaller pyramid have surface area x.
Then,

$\frac{250}{x} = \frac{36}{25}$

$36x = 6250$

$x \approx 173.6\ in.^2$

The surface area of the smaller pyramid is about 173.6 in^2.

All-in-One Resources/Online

English Language Learner Support

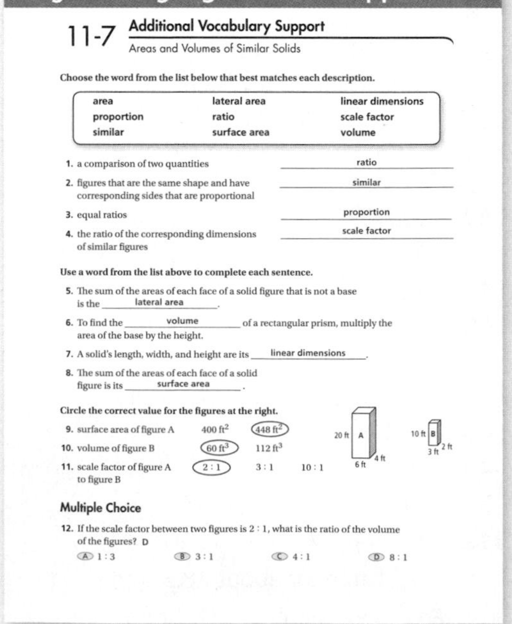

11-7 Additional Vocabulary Support
Areas and Volumes of Similar Solids

Choose the word from the list below that best matches each description.

area	lateral area	linear dimensions
proportion	ratio	scale factor
similar	surface area	volume

1. a comparison of two quantities ratio
2. figures that are the same shape and have corresponding sides that are proportional similar
3. equal ratios proportion
4. the ratio of the corresponding dimensions of similar figures scale factor

Use a word from the list above to complete each sentence.

5. The sum of the areas of each face of a solid figure that is not a base is the lateral area.
6. To find the volume of a rectangular prism, multiply the area of the base by the height.
7. A solid's length, width, and height are its linear dimensions.
8. The sum of the areas of each face of a solid figure is its surface area.

Circle the correct value for the figures at the right.

9. surface area of figure A 400 ft^2 (448 ft^2)
10. volume of figure B (60 ft^3) 112 ft^3
11. scale factor of figure A to figure B (2 : 1) 3 : 1 10 : 1

Multiple Choice

12. If the scale factor between two figures is 2 : 1, what is the ratio of the volume of the figures? D

Ⓐ 1 : 3 Ⓑ 3 : 1 Ⓒ 4 : 1 Ⓓ 8 : 1

Differentiated Remediation *continued*

On-Level

- **Practice** (2 pages) Provides extra practice for each lesson. For simpler practice exercises, use the Form K Practice pages found in the All-in-One Teaching Resources and online.
- **Think About a Plan** Helps students develop specific problem-solving skills and strategies by providing scaffolded guiding questions.
- **Standardized Test Prep** Focuses on all major exercises, all major question types, and helps students prepare for the high-stakes assessments.

Extension

- **Enrichment** Provides students with interesting problems and activities that extend the concepts of the lesson.
- **Activities, Games, and Puzzles** Worksheets that can be used for concepts development, enrichment, and for fun!

Practice and Problem Solving Wkbk/ All-in-One Resources/Online

Practice page 1

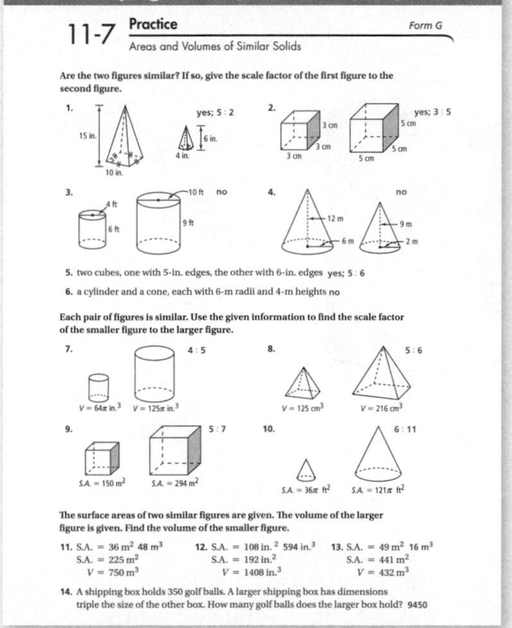
11-7 Practice — Form G
Areas and Volumes of Similar Solids

Are the two figures similar? If so, give the scale factor of the first figure to the second figure.

1. yes; 5 : 2
2. yes; 3 : 5
3. no
4. no
5. two cubes, one with 5-in. edges, the other with 6-in. edges yes; 5 : 6
6. a cylinder and a cone, each with 6-m radii and 4-m heights no

Each pair of figures is similar. Use the given information to find the scale factor of the smaller figure to the larger figure.

7. 4 : 5
8. 5 : 6
9. 5 : 7
10. 6 : 11

The surface areas of two similar figures are given. The volume of the larger figure is given. Find the volume of the smaller figure.

11. S.A. = 36 m² 48 m³; S.A. = 225 m²; V = 750 m³
12. S.A. = 108 in.² 594 in.³; S.A. = 192 in.²; V = 1408 in.³
13. S.A. = 49 m² 16 m³; S.A. = 441 m²; V = 432 m³
14. A shipping box holds 350 golf balls. A larger shipping box has dimensions triple the size of the other box. How many golf balls does the larger box hold? 9450

Practice and Problem Solving Wkbk/ All-in-One Resources/Online

Practice page 2

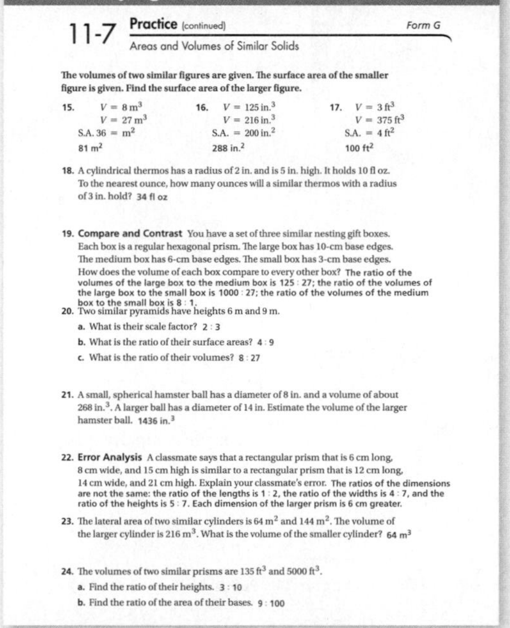
11-7 Practice (continued) — Form G
Areas and Volumes of Similar Solids

The volumes of two similar figures are given. The surface area of the smaller figure is given. Find the surface area of the larger figure.

15. V = 8 m³; V = 27 m³; S.A. 36 = m²; 81 m²
16. V = 125 in.³; V = 216 in.³; S.A. = 200 in.²; 288 in.²
17. V = 3 ft³; V = 375 ft³; S.A. = 4 ft²; 100 ft²

18. A cylindrical thermos has a radius of 2 in. and is 5 in. high. It holds 10 fl oz. To the nearest ounce, how many ounces will a similar thermos with a radius of 3 in. hold? 34 fl oz

19. Compare and Contrast You have a set of three similar nesting gift boxes. Each box is a regular hexagonal prism. The large box has 10-cm base edges. The medium box has 6-cm base edges. The small box has 3-cm base edges. How does the volume of each box compare to every other box? The ratio of the volumes of the large box to the medium box is 125 : 27; the ratio of the volumes of the large box to the small box is 1000 : 27; the ratio of the volumes of the medium box to the small box is 8 : 1.

20. Two similar pyramids have heights 6 m and 9 m.
a. What is their scale factor? 2 : 3
b. What is the ratio of their surface areas? 4 : 9
c. What is the ratio of their volumes? 8 : 27

21. A small, spherical hamster ball has a diameter of 8 in. and a volume of about 268 in.³. A larger ball has a diameter of 14 in. Estimate the volume of the larger hamster ball. 1436 in.³

22. Error Analysis A classmate says that a rectangular prism that is 6 cm long, 8 cm wide, and 15 cm high is similar to a rectangular prism that is 12 cm long, 14 cm wide, and 21 cm high. Explain your classmate's error. The ratios of the dimensions are not the same: the ratio of the lengths is 1 : 2, the ratio of the widths is 4 : 7, and the ratio of the heights is 5 : 7. Each dimension of the larger prism is 6 cm greater.

23. The lateral area of two similar cylinders is 64 m² and 144 m². The volume of the larger cylinder is 216 m³. What is the volume of the smaller cylinder? 64 m³

24. The volumes of two similar prisms are 135 ft³ and 5000 ft³.
a. Find the ratio of their heights. 3 : 10
b. Find the ratio of the area of their bases. 9 : 100

All-in-One Resources/Online

Enrichment

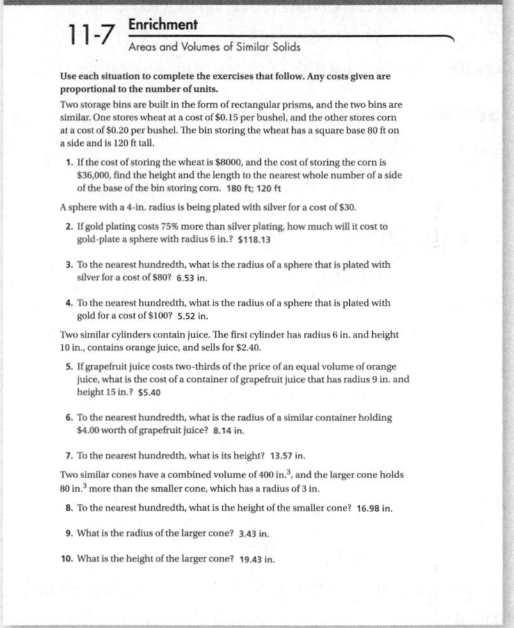
11-7 Enrichment
Areas and Volumes of Similar Solids

Use each situation to complete the exercises that follow. Any costs given are proportional to the number of units.

Two storage bins are built in the form of rectangular prisms, and the two bins are similar. One stores wheat at a cost of $0.15 per bushel, and the other stores corn at a cost of $0.20 per bushel. The bin storing the wheat has a square base 80 ft on a side and is 120 ft tall.

1. If the cost of storing the wheat is $8000, and the cost of storing the corn is $36,000, find the height and the length to the nearest whole number of a side of the base of the bin storing corn. 180 ft; 120 ft

A sphere with a 4-in. radius is being plated with silver for a cost of $30.

2. If gold plating costs 75% more than silver plating, how much will it cost to gold-plate a sphere with radius 6 in.? $118.13
3. To the nearest hundredth, what is the radius of a sphere that is plated with silver for a cost of $80? 6.53 in.
4. To the nearest hundredth, what is the radius of a sphere that is plated with gold for a cost of $100? 5.52 in.

Two similar cylinders contain juice. The first cylinder has radius 6 in. and height 10 in., contains orange juice, and sells for $2.40.

5. If grapefruit juice costs two-thirds of the price of an equal volume of orange juice, what is the cost of a container of grapefruit juice that has radius 9 in. and height 15 in.? $5.40
6. To the nearest hundredth, what is the radius of a similar container holding $4.00 worth of grapefruit juice? 8.14 in.
7. To the nearest hundredth, what is its height? 13.57 in.

Two similar cones have a combined volume of 400 in.³, and the larger cone holds 80 in.³ more than the smaller cone, which has a radius of 3 in.

8. To the nearest hundredth, what is the height of the smaller cone? 16.98 in.
9. What is the radius of the larger cone? 3.43 in.
10. What is the height of the larger cone? 19.43 in.

Practice and Problem Solving Wkbk/ All-in-One Resources/Online

Think About a Plan

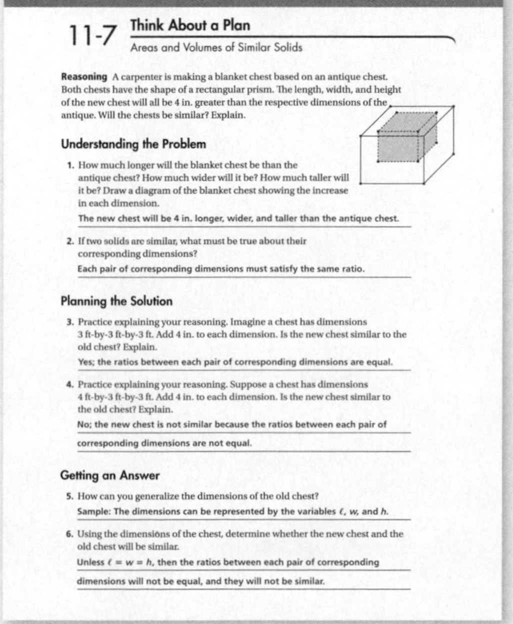
11-7 Think About a Plan
Areas and Volumes of Similar Solids

Reasoning A carpenter is making a blanket chest based on an antique chest. Both chests have the shape of a rectangular prism. The length, width, and height of the new chest will all be 4 in. greater than the respective dimensions of the antique. Will the chests be similar? Explain.

Understanding the Problem

1. How much longer will the blanket chest be than the antique chest? How much wider will it be? How much taller will it be? Draw a diagram of the blanket chest showing the increase in each dimension.
The new chest will be 4 in. longer, wider, and taller than the antique chest.
2. If two solids are similar, what must be true about their corresponding dimensions?
Each pair of corresponding dimensions must satisfy the same ratio.

Planning the Solution

3. Practice explaining your reasoning. Imagine a chest has dimensions 3 ft-by-3 ft-by-3 ft. Add 4 in. to each dimension. Is the new chest similar to the old chest? Explain.
Yes; the ratios between each pair of corresponding dimensions are equal.
4. Practice explaining your reasoning. Suppose a chest has dimensions 4 ft-by-3 ft-by-3 ft. Add 4 in. to each dimension. Is the new chest similar to the old chest? Explain.
No; the new chest is not similar because the ratios between each pair of corresponding dimensions are not equal.

Getting an Answer

5. How can you generalize the dimensions of the old chest?
Sample: The dimensions can be represented by the variables ℓ, w, and h.
6. Using the dimensions of the chest, determine whether the new chest and the old chest will be similar.
Unless ℓ = w = h, then the ratios between each pair of corresponding dimensions will not be equal, and they will not be similar.

Practice and Problem Solving Wkbk/ All-in-One Resources/Online

Standardized Test Prep

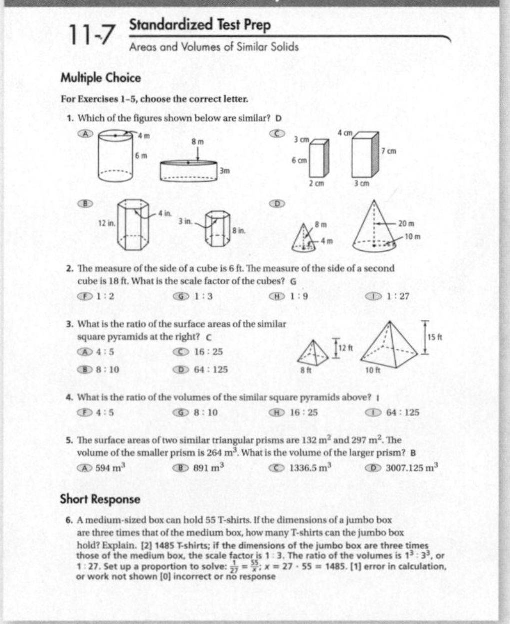
11-7 Standardized Test Prep
Areas and Volumes of Similar Solids

Multiple Choice

For Exercises 1–5, choose the correct letter.

1. Which of the figures shown below are similar? D
2. The measure of the side of a cube is 6 ft. The measure of the side of a second cube is 18 ft. What is the scale factor of the cubes? G
F 1 : 2 G 1 : 3 H 1 : 9 I 1 : 27
3. What is the ratio of the surface areas of the similar square pyramids at the right? C
A 4 : 5 B 8 : 10 C 16 : 25 D 64 : 125
4. What is the ratio of the volumes of the similar square pyramids above? I
F 4 : 5 G 8 : 10 H 16 : 25 I 64 : 125
5. The surface areas of two similar triangular prisms are 132 m² and 297 m². The volume of the smaller prism is 264 m³. What is the volume of the larger prism? B
A 594 m³ B 891 m³ C 1336.5 m³ D 3007.125 m³

Short Response

6. A medium-sized box can hold 55 T-shirts. If the dimensions of a jumbo box are three times that of the medium box, how many T-shirts can the jumbo box hold? Explain. [2] 1485 T-shirts; if the dimensions of the jumbo box are three times those of the medium box, the scale factor is 1 : 3. The ratio of the volumes is 1³ : 3³, or 1 : 27. Set up a proportion to solve: $\frac{1}{27} = \frac{55}{x}$; x = 27 · 55 = 1485. [1] error in calculation, or work not shown [0] incorrect or no response

Online Teacher Resource Center

Activities, Games, and Puzzles

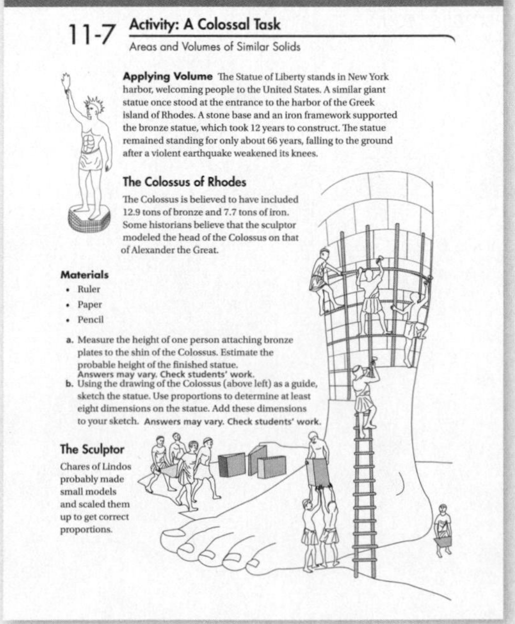
11-7 Activity: A Colossal Task
Areas and Volumes of Similar Solids

Applying Volume The Statue of Liberty stands in New York harbor, welcoming people to the United States. A similar giant statue once stood at the entrance to the harbor of the Greek island of Rhodes. A stone base and an iron framework supported the bronze statue, which took 12 years to construct. The statue remained standing for only about 66 years, falling to the ground after a violent earthquake weakened its knees.

The Colossus of Rhodes
The Colossus is believed to have included 12.9 tons of bronze and 7.7 tons of iron. Some historians believe that the sculptor modeled the head of the Colossus on that of Alexander the Great.

Materials
- Ruler
- Paper
- Pencil

a. Measure the height of one person attaching bronze plates to the shin of the Colossus. Estimate the probable height of the finished statue. Answers may vary. Check students' work.
b. Using the drawing of the Colossus (above left) as a guide, sketch the statue. Use proportions to determine at least eight dimensions on the statue. Add these dimensions to your sketch. Answers may vary. Check students' work.

The Sculptor
Chares of Lindos probably made small models and scaled them up to get correct proportions.

Completing the Performance Task

In the Apply What You've Learned sections in Lessons 11-2 and 11-4, students answered questions about the roll of paper towels shown on page 687. Here, students use their findings to complete the Performance Task. Ask students the following questions as they work toward solving the problem.

Q How can you use the work you have done in the chapter to solve the problem? **[Sample: I can divide the volume of the paper towels by the area they cover.]**

Q How can you check that your answer to the problem is reasonable? **[Sample: I can multiply my answer by 9 • 11 to find the volume of a single paper towel. Then I can multiply by 70 and compare the result to the volume I found in Lesson 11-4 for all the paper towels.]**

FOSTERING MATHEMATICAL DISCOURSE

Have students discuss what would happen if certain measures were changed on the paper towel roll.

ANSWERS

1. about 0.031 in.
2. Check students' work.

On Your Own

This problem is similar to the problem posed on page 687, but now students must determine the number of paper towels in the roll, instead of the thickness of a paper towel. Students should strive to solve this problem independently.

ANSWER

No; explanations may vary. Sample: Solve the equation

$$11\pi\left(\frac{6}{2}\right)^2 - 11\pi\left(\frac{1}{2}\right)^2 = 9 \cdot 11 \cdot t \cdot x$$

for x, where t is the thickness of one paper towel found in the original performance task (about 0.031 in.), and x is the number of paper towels in the larger roll. The number of paper towels in the larger roll is approximately 99, which is only about 41% greater than the number of sheets in the smaller roll.

11 Pull It All Together

To solve these problems, you will pull together many concepts and skills that you have studied about surface area and volume.

Completing the Performance Task

Look back at your results from the Apply What You've Learned sections in Lessons 11-2 and 11-4. Use the work you did to complete the following.

1. Solve the problem in the Task Description on page 687 by finding the thickness of a paper towel to the nearest thousandth of an inch. Show all your work and explain each step of your solution.
2. **Reflect** Choose one of the Mathematical Practices below and explain how you applied it in your work on the Performance Task.
   - MP 1: Make sense of problems and persevere in solving them.
   - MP 2: Reason abstractly and quantitatively.
   - MP 4: Model with mathematics.

On Your Own

The same company that makes the paper towel roll shown on page 687 sells a larger roll, shown below, that they claim has 50% more individual paper towels than the roll Justin measured. The dimensions of each individual paper towel are the same for both rolls.

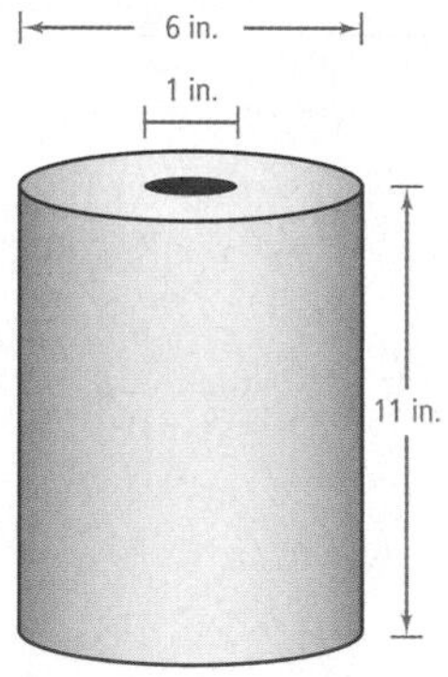

Not to scale

Is the company's claim true? Explain.

11 Chapter Review

Connecting BIG ideas and Answering the Essential Questions

1 Visualization
You can determine the intersection of a solid and a plane by visualizing how the plane slices the solid to form a two-dimensional cross section.

Space Figures and Cross Sections (Lesson 11-1)
This vertical plane intersects the cylinder in a rectangular cross section.

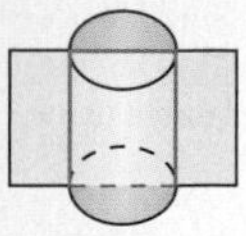

2 Measurement
You can find the surface area or volume of a solid by first choosing a formula to use and then substituting the needed dimensions into the formula.

Surface Areas and Volumes of Prisms, Cylinders, Pyramids, and Cones (Lessons 11-2 through 11-5)

	Surface Area (S.A.)	Volume (*V*)
Prism	$ph + 2B$	Bh
Cylinder	$2\pi rh + 2B$	Bh
Pyramid	$\frac{1}{2}p\ell + B$	$\frac{1}{3}Bh$
Cone	$\pi r\ell + B$	$\frac{1}{3}Bh$

Surface Areas and Volumes of Spheres (Lesson 11-6)
S.A. $= 4\pi r^2$
$V = \frac{4}{3}\pi r^3$

3 Similarity
The surface areas of similar solids are proportional to the squares of their corresponding dimensions. The volumes are proportional to the cubes of their corresponding dimensions.

Areas and Volumes of Similar Solids (Lesson 11-7)
If the scale factor of two similar solids is $a : b$, then
- the ratio of their areas is $a^2 : b^2$
- the ratio of their volumes is $a^3 : b^3$

Chapter Vocabulary

- altitude (pp. 699, 701, 708, 711)
- center of a sphere (p. 733)
- cone (p. 711)
- cross section (p. 690)
- cylinder (p. 701)
- edge (p. 688)
- face (p. 688)
- great circle (p. 733)
- hemisphere (p. 733)
- lateral area (pp. 700, 702, 709, 711)
- lateral face (pp. 699, 708)
- polyhedron (p. 688)
- prism (p. 699)
- pyramid (p. 708)
- right cone (p. 711)
- right cylinder (p. 701)
- right prism (p. 699)
- slant height (pp. 708, 711)
- sphere (p. 733)
- surface area (pp. 700, 702, 709, 711)
- volume (p. 717)

Choose the correct term to complete each sentence.

1. A set of points in space equidistant from a given point is called a(n) __?__.
2. A(n) __?__ is a polyhedron in which one face can be any polygon and the lateral faces are triangles that meet at a common vertex.
3. If you slice a prism with a plane, the intersection of the prism and the plane is a(n) __?__ of the prism.

Essential Questions

BIG idea Visualization

ESSENTIAL QUESTION How can you determine the intersection of a solid and a plane?

ANSWER You can determine the intersection of a solid and a plane by visualizing how the plane slices the solid to form a two-dimensional cross section.

BIG idea Measurement

ESSENTIAL QUESTION How do you find the surface area and volume of a solid?

ANSWER You can find the surface area or volume of a solid by first choosing a formula to use and then substituting the needed dimensions into the formula.

BIG idea Similarity

ESSENTIAL QUESTION How do the surface areas and volumes of similar solids compare?

ANSWER The surface areas of similar solids are proportional to the squares of their corresponding dimensions. The volumes are proportional to the cubes of their corresponding dimensions.

Answers

Chapter Review

1. sphere
2. pyramid
3. cross section

Summative Questions

Use the following prompts as you review this chapter with your students. The prompts are designed to help you assess your students' understanding of the Big Ideas they have studied.

- What formulas for surface area do you know?
- What formulas for volume do you know?
- How do you find surface areas and volumes of similar solids?

Answers

Chapter Review (continued)

4–5. Answers may vary. Samples are given.

4.

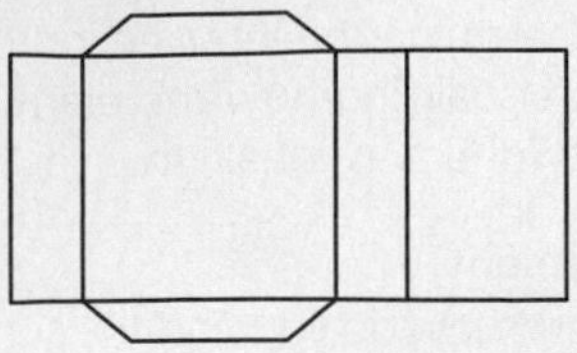

5.

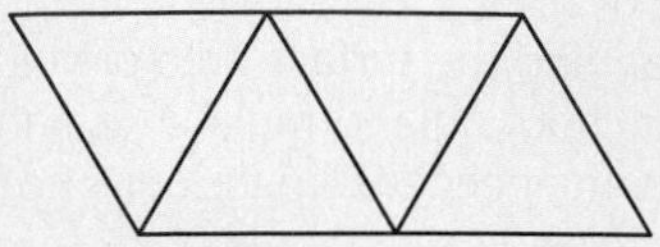

6. 8

7. 8

8. 5

9. a circle

10.

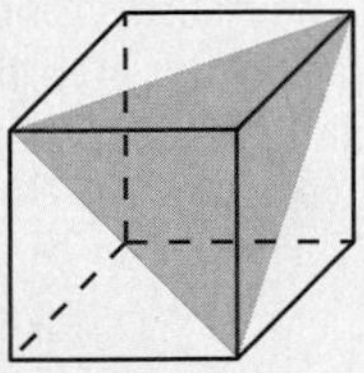

11. 36 cm^2

12. 66π m^2

13. 208 $in.^2$

14. 36π cm^2

15. 32.5π cm^2

11-1 Space Figures and Cross Sections

Quick Review

A **polyhedron** is a three-dimensional figure whose surfaces are polygons. The polygons are **faces** of the polyhedron. An **edge** is a segment that is the intersection of two faces. A **vertex** is a point where three or more edges intersect. A **cross section** is the intersection of a solid and a plane.

Example

How many faces and edges does the polyhedron have?

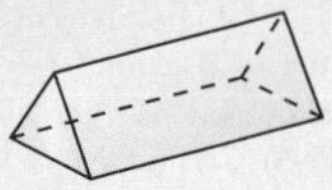

The polyhedron has 2 triangular bases and 3 rectangular faces for a total of 5 faces.

The 2 triangles have a total of 6 edges. The 3 rectangles have a total of 12 edges. The total number of edges in the polyhedron is one half the total of 18 edges, or 9.

Exercises

Draw a net for each three-dimensional figure.

4.

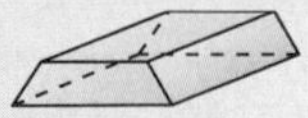

5.

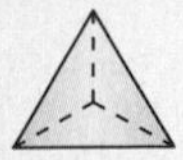

Use Euler's Formula to find the missing number.

6. $F = 5, V = 5, E = ■$

7. $F = 6, V = ■, E = 12$

8. How many vertices are there in a solid with 4 triangular faces and 1 square base?

9. Describe the cross section in the figure at the right.

10. Sketch a cube with an equilateral triangle cross section.

11-2 Surface Areas of Prisms and Cylinders

Quick Review

The **lateral area of a right prism** is the product of the perimeter of the base and the height. The **lateral area of a right cylinder** is the product of the circumference of the base and the height of the cylinder. The **surface area** of each solid is the sum of the lateral area and the areas of the bases.

Example

What is the surface area of a cylinder with radius 3 m and height 6 m? Leave your answer in terms of π.

S.A. = L.A. + 2*B*	Use the formula for surface area of a cylinder.
$= 2\pi rh + 2(\pi r^2)$	Substitute formulas for lateral area and area of a circle.
$= 2\pi(3)(6) + 2\pi(3)^2$	Substitute 3 for *r* and 6 for *h*.
$= 36\pi + 18\pi$	Simplify.
$= 54\pi$	

The surface area of the cylinder is 54π m^2.

Exercises

Find the surface area of each figure. Leave your answers in terms of π where applicable.

11.

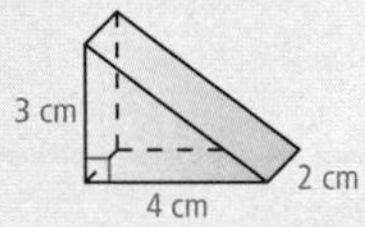

12.

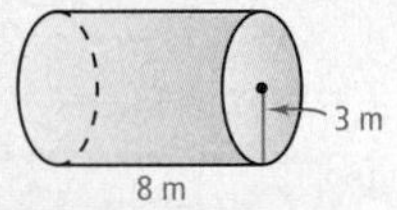

13.

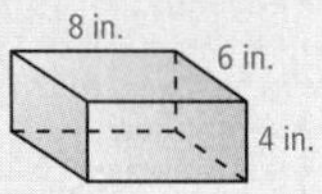

14.

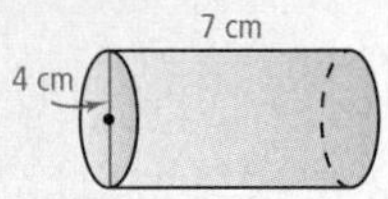

15. A cylinder has radius 2.5 cm and lateral area 20π cm^2. What is the surface area of the cylinder in terms of π?

11-3 Surface Areas of Pyramids and Cones

Quick Review

The **lateral area of a regular pyramid** is half the product of the perimeter of the base and the slant height. The **lateral area of a right cone** is half the product of the circumference of the base and the slant height. The **surface area** of each solid is the sum of the lateral area and the area of the base.

Example

What is the surface area of a cone with radius 3 in. and slant height 10 in.? Leave your answer in terms of π.

$\text{S.A.} = \text{L.A.} + B$	Use the formula for surface area of a cone.
$= \pi r\ell + \pi r^2$	Substitute formulas for lateral area and area of a circle.
$= \pi(3)(10) + \pi(3)^2$	Substitute 3 for r and 10 for ℓ.
$= 30\pi + 9\pi$	Simplify.
$= 39\pi$	

The surface area of the cone is 39π in.^{2}.

Exercises

Find the surface area of each figure. Round your answers to the nearest tenth.

16.

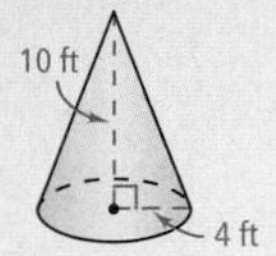

17.

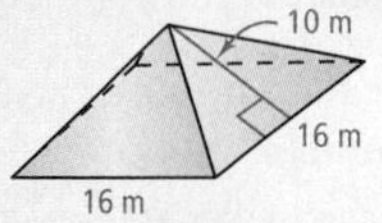

18.

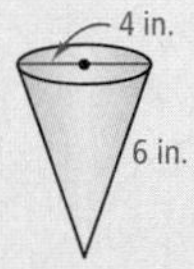

19.

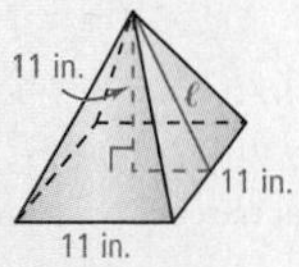

20. Find the formula for the base area of a prism in terms of surface area and lateral area.

11-4 and 11-5 Volumes of Prisms, Cylinders, Pyramids, and Cones

Quick Review

The **volume** of a space figure is the space that the figure occupies. Volume is measured in cubic units. The **volume of a prism** and the **volume of a cylinder** are the product of the area of a base and the height of the solid. The **volume of a pyramid** and the **volume of a cone** are one third the product of the area of the base and the height of the solid.

Example

What is the volume of a rectangular prism with base 3 cm by 4 cm and height 8 cm?

$V = Bh$	Use the formula for the volume of a prism.
$= (3 \cdot 4)(8)$	Substitute.
$= 96$	Simplify.

The volume of the prism is 96 cm^3.

Exercises

Find the volume of each figure. If necessary, round to the nearest tenth.

21.

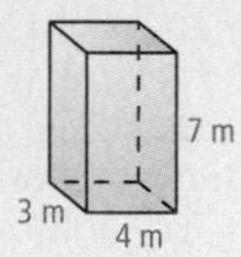

22.

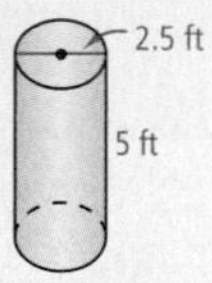

23.

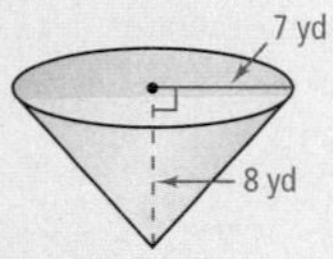

24.

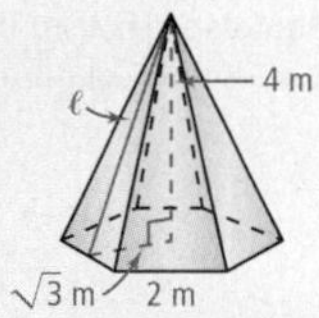

16. about 185.6 ft^2
17. 576 m^2
18. about 50.3 in.^{2}
19. about 391.6 in.^{2}
20. $B = \frac{\text{S.A.} - \text{L.A.}}{2}$
21. 84 m^3
22. 24.5 ft^3
23. 410.5 yd^3
24. 13.9 m^3

Answers

Chapter Review (continued)

25. S.A. = 314.2 in.^{2}; V = 523.6 in.^{3}
26. S.A. = 153.9 cm^2; V = 179.6 cm^3
27. S.A. = 50.3 ft^2; V = 33.5 ft^3
28. S.A. = 8.0 ft^2; V = 2.1 ft^3
29. 904.78 cm^3
30. 314 m^2
31. 8.6 in.^{3}
32. Answers may vary. Sample:

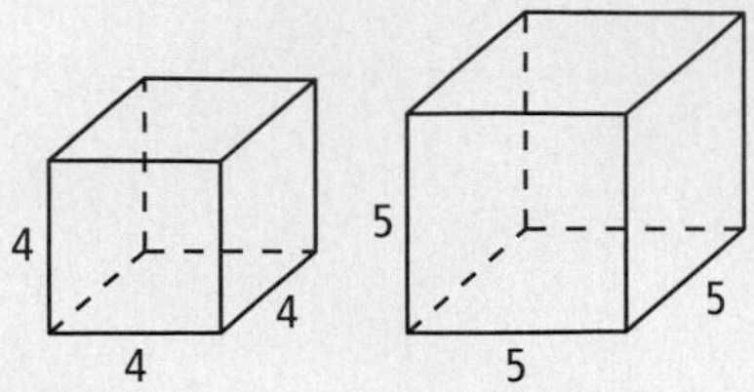

33. 27 : 64
34. 64 : 27
35. 324 pencils

11-6 Surface Areas and Volumes of Spheres

Quick Review

The **surface area of a sphere** is four times the product of π and the square of the radius of the sphere. The **volume of a sphere** is $\frac{4}{3}$ the product of π and the cube of the radius of the sphere.

Example

What is the surface area of a sphere with radius 7 ft? Round your answer to the nearest tenth.

S.A. $= 4\pi r^2$	Use the formula for surface area of a sphere.
$= 4\pi(7)^2$	Substitute.
≈ 615.8	Simplify.

The surface area of the sphere is about 615.8 ft^2.

Exercises

Find the surface area and volume of a sphere with the given radius or diameter. Round your answers to the nearest tenth.

25. r = 5 in.
26. d = 7 cm
27. d = 4 ft
28. r = 0.8 ft

29. What is the volume of a sphere with a surface area of 452.39 cm^2? Round your answer to the nearest hundredth.

30. What is the surface area of a sphere with a volume of 523.6 m^3? Round your answer to the nearest square meter.

31. Sports Equipment The circumference of a lacrosse ball is 8 in. Find its volume to the nearest tenth of a cubic inch.

11-7 Areas and Volumes of Similar Solids

Quick Review

Similar solids have the same shape and all their corresponding dimensions are proportional.

If the scale factor of two similar solids is $a : b$, then the ratio of their corresponding surface areas is $a^2 : b^2$, and the ratio of their volumes is $a^3 : b^3$.

Example

Is a cylinder with radius 4 in. and height 12 in. similar to a cylinder with radius 14 in. and height 35 in.? If so, give the scale factor.

$$\frac{4}{14} \neq \frac{12}{35}$$

The cylinders are not similar because the ratios of corresponding linear dimensions are not equal.

Exercises

32. Open-Ended Sketch two similar solids whose surface areas are in the ratio 16 : 25. Include dimensions.

For each pair of similar solids, find the ratio of the volume of the first figure to the volume of the second.

33.

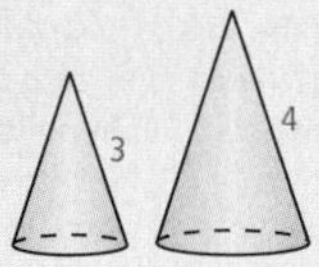

34.

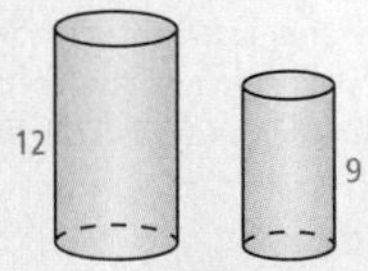

35. Packaging There are 12 pencils in a regular-sized box. If a jumbo box is made by tripling all the dimensions of the regular-sized box, how many pencils will the jumbo box hold?

11 Chapter Test

Do you know HOW?

Draw a net for each figure. Label the net with appropriate dimensions.

1. 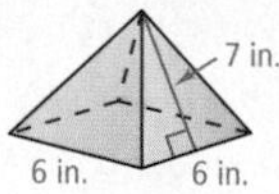

2. 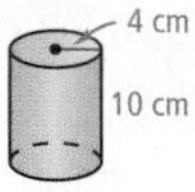

Use the polyhedron at the right for Exercises 3 and 4.

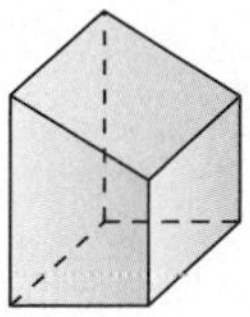

3. Verify Euler's Formula for the polyhedron.
4. Draw a net for the polyhedron. Verify $F + V = E + 1$ for the net.
5. What is the number of edges in a pyramid with seven faces?

Describe the cross section formed in each diagram.

6. 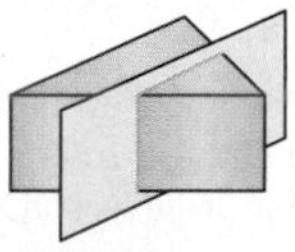

7. 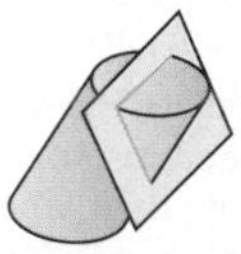

8. **Aviation** The flight data recorders on commercial airlines are rectangular prisms. The base of a recorder is 15 in. by 8 in. Its height ranges from 15 in. to 22 in. What are the largest and smallest possible volumes for the recorder?

Find the volume and surface area of each figure to the nearest tenth.

9. 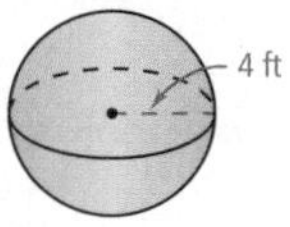

10. 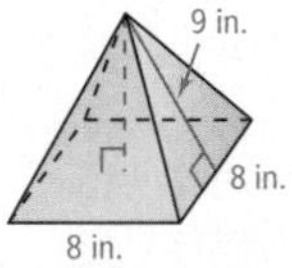

11. 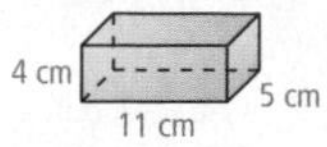

12. 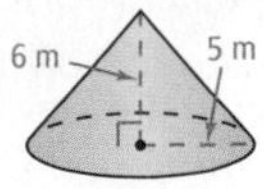

13. 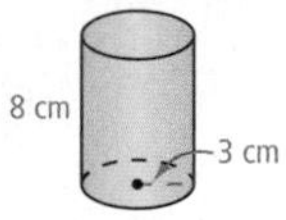

14. 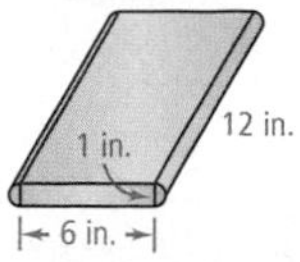

15. List these solids in order from the one with least volume to the one with the greatest volume.
    A. a cube with edge 5 cm
    B. a cylinder with radius 4 cm and height 4 cm
    C. a square pyramid with base edges 6 cm and height 6 cm
    D. a cone with radius 4 cm and height 9 cm
    E. a rectangular prism with a 5 cm-by-5 cm base and height 6 cm
16. **Painting** The floor of a bedroom is 12 ft by 15 ft and the walls are 7 ft high. One gallon of paint covers about 450 ft^2. How many gallons of paint do you need to paint the walls of the bedroom?

Do you UNDERSTAND?

17. **Reasoning** What solid has a cross section that could either be a circle or a rectangle?
18. **Visualization** The triangle is revolved completely about the y-axis.

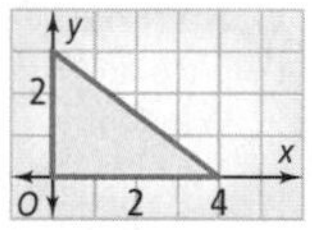

    a. Describe the solid of revolution that is formed.
    b. Find its lateral area and volume in terms of π.
19. **Open-Ended** Draw two different solids that have volume 100 in.^{3}. Label the dimensions of each solid.

Answers

Chapter Test

1. 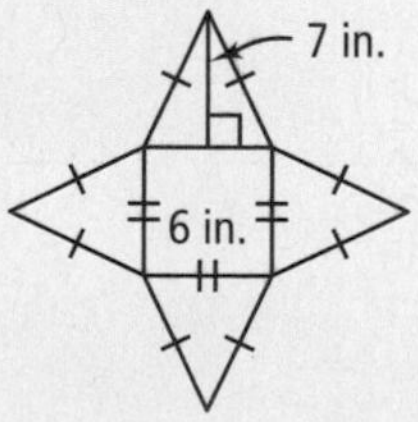

2. 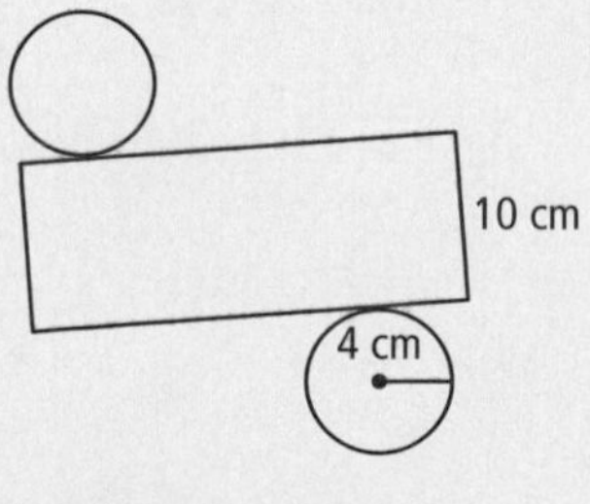

3. $6 + 8 = 12 + 2$

4. 

$6 + 14 = 19 + 1$

5. 12 edges
6. rectangle
7. an arc of an ellipse with a straight segment joining the two ends
8. 2640 in.^{3}; 1800 in.^{3}
9. 268.1 ft^3; 201.1 ft^2
10. 172.0 in.^{3}; 208 in.^{2}
11. 220 cm^3; 238 cm^2
12. 157.1 m^3; 201.2 m^2
13. 226.2 cm^3; 207.3 cm^2
14. 81.4 in.^{3}; 195.3 in.^{2}
15. C, A, E, D, B
16. 1 gal
17. cylinder

18a. cone with $r = 4$, $h = 3$

b. 20π units^2; 16π units^3

19. Answers may vary. Sample:

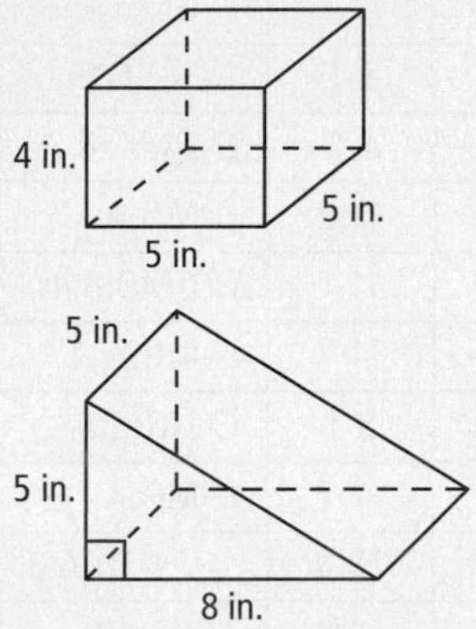

Item Number	Lesson	Content Standard
1	11-7	G-GMD.A.3
2	11-4	*G-GMD.A.3
3	6-6	G-CO.C.11
4	10-6	G-C.B.5
5	11-3	G-GMD.B.4
6	10-6	G-C.B.5
7	4-7	G-CO.C.9
8	11-5	G-GMD.A.3
9	8-1	G-SRT.C.8
10	8-2	G-SRT.C.8
11	11-2	G-MG.A.1
12	11-1	G-GMD.B.4
13	11-7	G-MG.A.2
14	8-1	G-SRT.C.8
15	11-2	G-MG.A.1
16	10-4	*G-GMD.A.3
17	10-2	*G-GMD.A.3
18	11-2	*G-GMD.A.3
19	11-6	G-GMD.A.3
20	11-4	G-MG.A.1
21	10-6	G-C.B.5
22	11-2	G-MG.A.1
23	8-2	G-SRT.C.8

*Prepares for standard

11 Common Core Cumulative Standards Review

ASSESSMENT

TIPS FOR SUCCESS

Some questions ask you to use a formula to estimate volume. Read the sample question at the right. Then follow the tips to answer it.

The tank below is filled with gasoline. Which is closest to the volume of the tank in cubic feet?

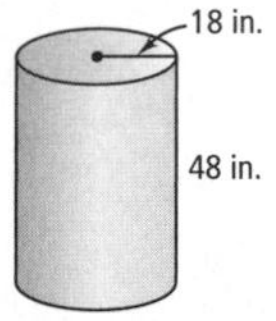

A $28\ ft^3$ C $4072\ ft^3$
B $864\ ft^3$ D $48{,}858\ ft^3$

TIP 1
Check whether the units in the problem match the units in the answer choices. You may need to convert measurements.

TIP 2
Identify the given solid and what you want to find so you can select the correct formula.

Think It Through
Convert the radius and the height given in the figure to feet: $r = 1.5$ ft and $h = 4$ ft. Substitute the values for r and h into the formula for the volume of a cylinder and simplify: $V = \pi r^2 h = \pi(1.5)^2(4) \approx 28$. The correct answer is A.

Vocabulary Builder

As you solve problems, you must understand the meanings of mathematical terms. Choose the correct term to complete each sentence.

A. The bases of a cylinder are (*circles, polygons*).

B. An arc of a circle that is larger than a semicircle is a (*major arc, minor arc*).

C. Each polygon of a polyhedron is called a(n) (*edge, face*).

D. A (*hemisphere, great circle*) is the intersection of a plane and a sphere through the center of the sphere.

E. The length of the altitude of a pyramid is the (*height, slant height*) of the pyramid.

F. The area of a net of a polyhedron is equal to the (*lateral area, surface area*) of the polyhedron.

G. In a parallelogram, one diagonal is always the (*perpendicular bisector, segment bisector*) of the other diagonal.

Selected Response

Read each question. Then write the letter of the correct answer on your paper.

1. A pyramid has a volume of $108\ m^3$. A similar pyramid has base edges and a height that are $\frac{1}{3}$ those of the original pyramid. What is the volume of the second pyramid?

A $3\ m^3$ C $12\ m^3$
B $4\ m^3$ D $36\ m^3$

2. What is the volume of the triangular prism at the right?

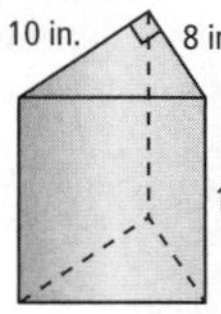

F $520\ in.^3$
G $560\ in.^3$
H $600\ in.^3$
I $1120\ in.^3$

3. Which quadrilateral CANNOT have one diagonal that bisects the other?

A square C kite
B trapezoid D parallelogram

Answers

Common Core Cumulative Standards Review

A. circles
B. major arc
C. face
D. great circle
E. height
F. surface area
G. segment bisector
1. B
2. G
3. B

4. The diameter of the circle is 12 units. What is the minor arc length from point C to point D?

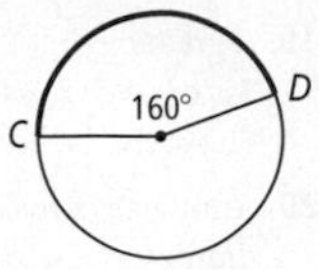

Ⓕ $\frac{16}{3}\pi$ units
Ⓖ 8π units
Ⓗ 16π units
Ⓘ $\frac{160}{3}\pi$ units

5. Which is a net for a pentagonal pyramid?

Ⓐ
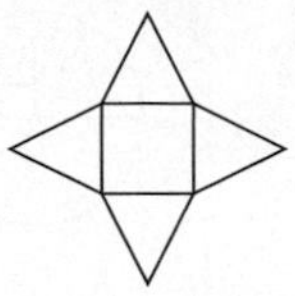

Ⓑ
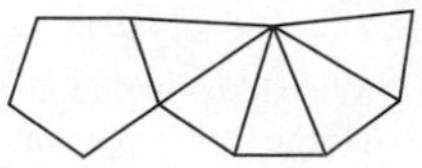

Ⓒ
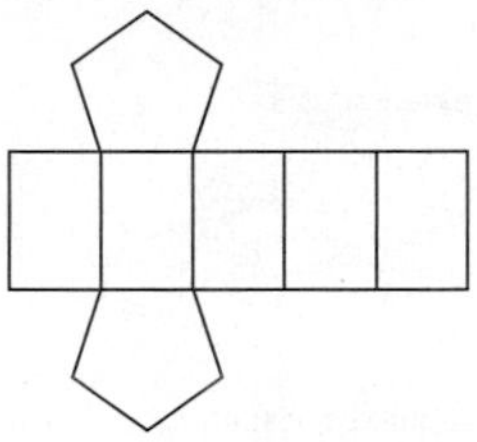

Ⓓ
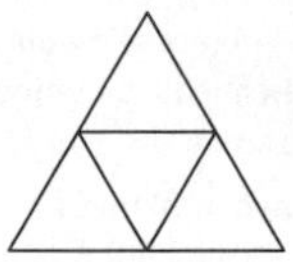

6. A frozen dinner is divided into three sections on a circular plate with a 12-in. diameter. What is the approximate arc length of the section containing green beans?

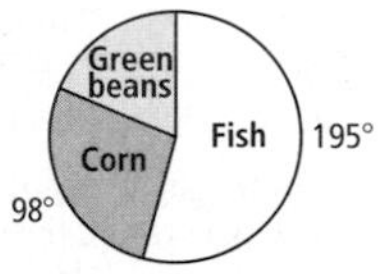

Ⓕ 7 in.
Ⓗ 16 in.
Ⓖ 10 in.
Ⓘ 20 in.

7. Suppose $\overline{AD} \cong \overline{AE}$, $\overline{DB} \cong \overline{EC}$, and $\angle ABC \cong \angle ACB$. Which statement must be proved first to prove $\triangle ABE \cong \triangle ACD$ by SSS?

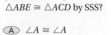

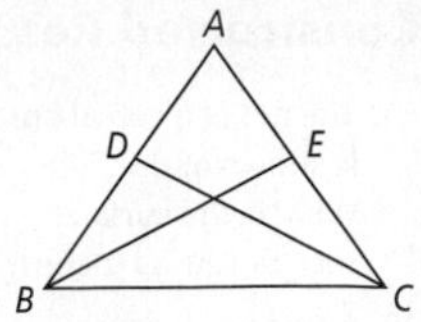

Ⓐ $\angle A \cong \angle A$
Ⓒ $\overline{AC} \cong \overline{AC}$
Ⓑ $\overline{AB} \cong \overline{BC}$
Ⓓ $\overline{DC} \cong \overline{EB}$

8. The formula for the volume of a cone is $V = \frac{1}{3}\pi r^2 h$. Which statement is true?

Ⓕ The volume depends on the mean of π, the radius, and the height.
Ⓖ The volume depends only on the square of the radius.
Ⓗ The radius depends on the product of the height and π.
Ⓘ The volume depends on both the height and the radius.

9. A wooden pole was broken during a windstorm. Before the pole broke, the height of the pole above the ground was 18 ft. When it broke, the top of the pole hit the ground 12 ft from the base.

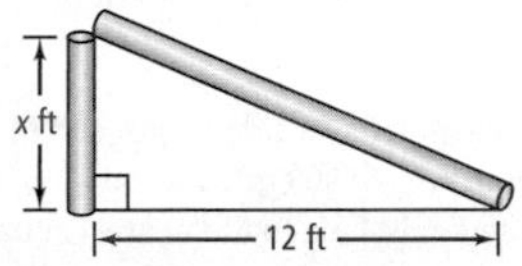

How tall is the part of the pole that remained standing?

Ⓐ 5 ft
Ⓒ 13 ft
Ⓑ 10 ft
Ⓓ 15 ft

10. $\triangle ABC$ is a right isosceles triangle. Which statement CANNOT be true?

Ⓕ The length of the hypotenuse is $\sqrt{2} \cdot$ the length of a leg.
Ⓖ $AB = BC$
Ⓗ The hypotenuse is the shortest side of the triangle.
Ⓘ $m\angle B = 90$

4. F
5. B
6. F
7. D
8. I
9. A
10. H

Answers

Common Core Cumulative Standards Review (continued)

11. 234.25

12. 15

13. 128

14. 124

15. 484

16. 77

17. 72

18. 484

19. [2] $C = 2\pi r$ so $3\pi = 2\pi r$ and $r = \frac{3}{2}$. Using $V = \frac{4}{3}\pi r^3$, then $V = \frac{4}{3}\pi\left(\frac{3}{2}\right)^3 = \frac{9}{2}\pi$ cm^3.

[1] incomplete OR incorrect explanation

20. [2] Answers may vary. Sample: 6 in. by 7 in. by 8 in.; cylinder: $V = 108\pi$ in.^{3} ≈ 339.3 in.^{3}; rectangular prism: $V = 336$ in.^{3}.

[1] one computational error OR correct answer without work

21. [2] Since the measure of the central $\angle$ is 90, $m\widehat{AB} = 90$. So length of $\widehat{AB} = \frac{90}{360} \cdot$ (circumference of $\odot$) $= \frac{1}{4} \cdot 12\pi = 3\pi$.

[1] one computational error OR correct answer without work

22. [4] The surface area equals $\frac{1}{2}$ the surface area of the log plus the area of the rectangular surface of the cut through the center. Using inches:

- $\frac{1}{2}$ (S.A. of log) $=$ $\frac{1}{2}[2\pi(9)(36) + 2\pi(9^2)] =$ $\frac{1}{2}(810\pi) = 405\pi$
- area of rectangular face $= 18 \cdot 36 = 648$
- $\frac{1}{2}$ (S.A. of log) + area of rectangular face $= 405\pi + 648 \approx 1920.345025$

The surface area of one of the halves is about 1920 in.^{2}.

[3] one computational error

[2] two errors OR missing statements

[1] more than two errors OR missing statements

23. [4]

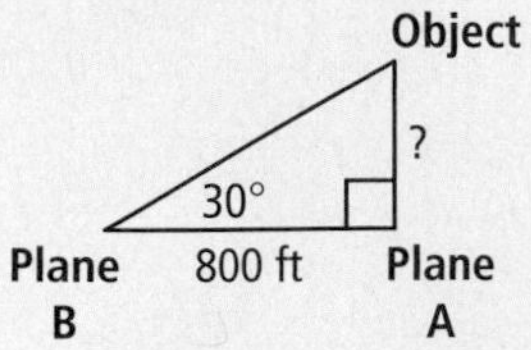

There is a 30° angle between each hour number on a dial clock, so the two planes and the object form a 30°-60°-90° rt. △. The object is $\frac{800}{\sqrt{3}} \approx 462$ ft from Plane A.

[3] one error OR missing statement

[2] two errors OR missing statements

[1] more than two errors OR missing statements

Constructed Response

11. The net of a cereal box is shown at the right. What is the surface area in square inches?

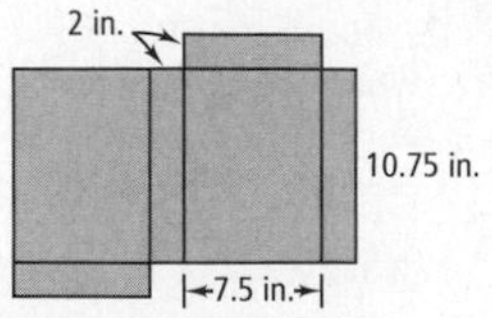

12. A polyhedron has 10 vertices and 7 faces. How many edges does the polyhedron have?

13. The radius of a sphere with a volume of 2 m^3 is quadrupled. What is the volume of the new sphere in cubic meters?

14. Molly stands at point *A* directly below a kite that Kat is flying from point *B*. Kat has let out 130 ft of the string. Molly stands 40 ft from Kat. To the nearest foot, how many feet from the ground is the kite?

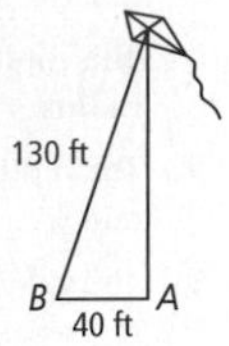

15. The figure below shows a cylindrical tin. To the nearest square centimeter, what is the surface area of the tin?

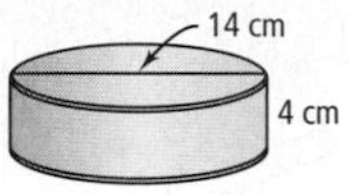

16. A regular octagon has a side length of 8 m and an approximate area of 309 m^2. The side length of another regular octagon is 4 m. To the nearest square meter, what is the approximate area of the smaller octagon?

17. The lengths of the bases of an isosceles trapezoid are shown below. If the perimeter of this trapezoid is 44 units, what is its area in square units?

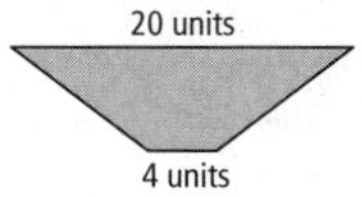

18. What is the surface area, in square centimeters, of a rectangular prism that is 9 cm by 8 cm by 10 cm?

19. A great circle of a sphere has a circumference of 3π cm. What is the volume of the sphere in terms of π? Show your work.

20. Anna has cereal in a cylindrical container that has a diameter of 6 in. and a height of 12 in. She wants to store the cereal in a rectangular prism container. What is one possible set of dimensions for a rectangular prism container that would be close to the volume of the cylindrical container? Show your work.

21. What is the length of $\widehat{AB}$ in terms of π? Show your work.

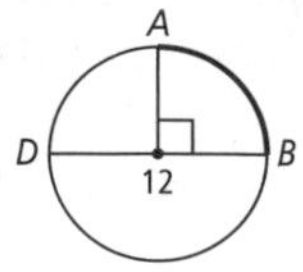

Extended Response

22. A log is cut lengthwise through its center. To the nearest square inch, what is the surface area of one of the halves? Explain your reasoning.

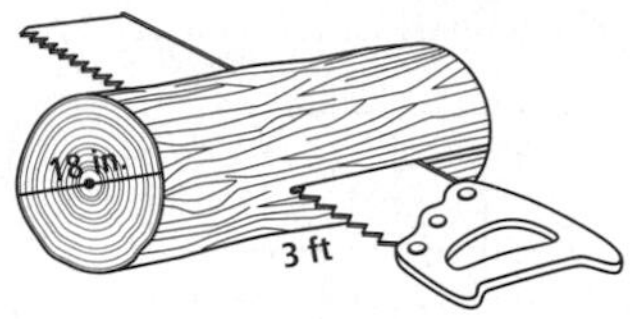

23. When airplane pilots make a visual sighting of an object outside the airplane, they often refer to the dial of a clock to help locate the object. For example, an object at 12 o'clock is straight ahead, an object at 3 o'clock is 90° to the right, and so on.

Suppose that two pilots flying two airplanes in the same direction spot the same object. Pilot A reports the object at 12 o'clock, and Pilot B reports the object at 2 o'clock. At the same time, Pilot A reports seeing the other airplane at 9 o'clock.

Draw a diagram showing the possible locations of the two planes and the object. If the planes are 800 ft apart, how far is Pilot A from the object? Show your work.

CHAPTER 12

Get Ready!

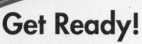

Skills Handbook, p. 893

Solving Equations

Algebra Solve for x.

1. $\frac{1}{2}(x + 42) = 62$ **2.** $(5 + 3)8 = (4 + x)6$ **3.** $(9 + x)2 = (12 + 4)3$

Lesson 1-7

Distance Formula

Find the distance between each pair of points.

4. $(13, 7), (6, 31)$ **5.** $(-4, 2,), (2, -4)$ **6.** $(-3, -1), (0, 3)$ **7.** $(2\sqrt{3}, 5), (-\sqrt{3}, 2)$

Lesson 4-5

Isosceles and Equilateral Triangles

Algebra Find the value of x.

8.

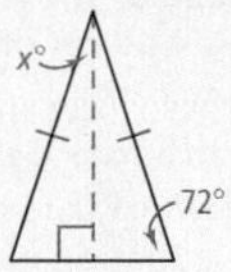

9.

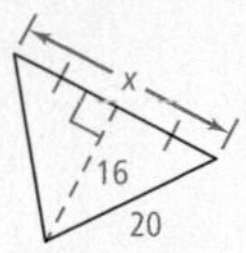

10.

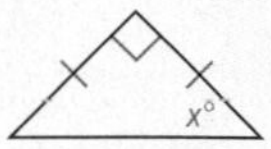

11.

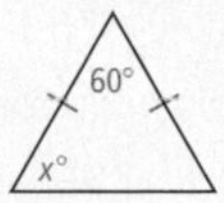

Lesson 8-1

The Pythagorean Theorem

Algebra Find the value of x. Leave your answer in simplest radical form.

12.

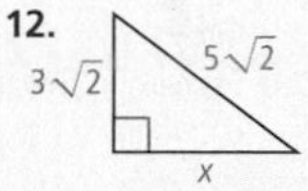

13.

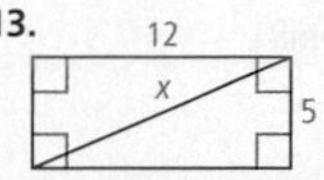

14.

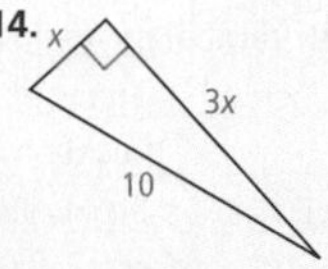

15.

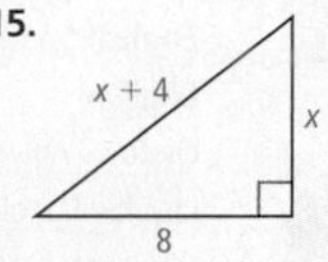

Looking Ahead Vocabulary

16. When you are in a conversation and you go off on a *tangent,* you are leading the conversation away from the main topic. What do you think a line that is *tangent* to a circle might look like?

17. You learned how to *inscribe* a triangle in a circle in Chapter 5. What do you think an *inscribed* angle is?

18. A defensive player *intercepts* a pass when he catches the football before it reaches the intended receiver. On a circle, what might an *intercepted* arc of an angle be?

Answers

Get Ready!

1. 82
2. $6\frac{2}{3}$
3. 15
4. 25
5. $6\sqrt{2}$
6. 5
7. 6
8. 18
9. 24
10. 45
11. 60
12. $4\sqrt{2}$
13. 13
14. $\sqrt{10}$
15. 6
16. Answers may vary. Sample: A tangent touches a circle at one point.
17. Answers may vary. Sample: An inscribed ∠ has its vertex on a circle and its sides are inside the circle.
18. Answers may vary. Sample: An intercepted arc is the part of a circle that lies in the interior of an ∠.

Get Ready!

Assign this diagnostic assessment to determine if students have the prerequisite skills for Chapter 12.

Lesson	Skill
Skills Handbook, p. T894	Solving Equations
1-7	Distance Formula
4-5	Isosceles and Equilateral Triangles
8-1	The Pythagorean Theorem

To remediate students, select from these resources (available for every lesson).

- Online Problems (PowerGeometry.com)
- Reteaching (All-in-One Teaching Resources)
- Practice (All-in-One Teaching Resources)

Why Students Need These Skills

SOLVING EQUATIONS Students will solve equations to find unknown quantities in circles and the special segments related to circles.

DISTANCE FORMULA The Distance Formula will be used to generate the equations of circles on the coordinate plane.

ISOSCELES AND EQUILATERAL TRIANGLES Students will use triangles created by tangent lines, secant lines and chords to solve for unknown measures.

THE PYTHAGOREAN THEOREM Right triangles are formed when a radius of a circle intersects a tangent line. Students will use the Pythagorean Theorem to solve for unknown side lengths in these right triangles.

Looking Ahead Vocabulary

TANGENT Discuss what it means to "go off on a tangent."

INSCRIBED Draw an image of a polygon inscribed in a circle.

INTERCEPTED Have students give other examples of objects that may be intercepted.

Chapter 12 Overview

In Chapter 12 students explore concepts related to circles. Students will develop the answers to the Essential Questions as they learn the concepts and skills listed below.

BIG idea Reasoning and Proof

ESSENTIAL QUESTION How can you prove relationships between angles and arcs in a circle?

- Students will examine angles formed by lines that intersect inside and outside a circle.
- Students will relate arcs and angles.

BIG idea Measurement

ESSENTIAL QUESTION When lines intersect a circle, or within a circle, how do you find the measures of resulting angles, arcs, and segments?

- Students will use properties of tangent lines.
- Students will use the relationships among chords, arcs, and central angles.
- Students will solve problems with angles formed by secants and tangents.

BIG idea Coordinate Geometry

ESSENTIAL QUESTION How do you find the equation of a circle in the coordinate plane?

- The center and radius of a circle in a coordinate plane can be used to find the equation of a circle.

Content Standards

Following are the standards covered in this chapter.

CONCEPTUAL CATEGORY Geometry

Domain Circles G-C

Cluster Understand and apply theorems about circles. (Standards G-C.A.2, G-C.A.3, G-C.A.4)
LESSONS 12-2, 12-3

Domain Expressing Geometric Properties with Equations G-GPE

Cluster Translate between the geometric description and the equation for a conic section. (Standard G-GPE.A.1)
LESSON 12-5

Domain Geometric Measurement and Dimension G-GMD

Cluster Visualize relationships between two-dimensional and three-dimensional objects. (Standard G-GMD.B.4)
LESSON 12-6

CHAPTER 12

Circles

Download videos connecting math to your world.

Interactive! Vary numbers, graphs, and figures to explore math concepts.

The online Solve It will get you in gear for each lesson.

Math definitions in English and Spanish

Online access to stepped-out problems aligned to Common Core

Get and view your assignments online.

Extra practice and review online

Virtual Nerd™ tutorials with built-in support

Chapter Preview

BIG ideas

1 **Reasoning and Proof**
Essential Question How can you pro relationships between angles and arcs a circle?

2 **Measurement**
Essential Question When lines inters a circle or within a circle, how do you the measures of resulting angles, arcs, segments?

3 **Coordinate Geometry**
Essential Question How do you find the equation of a circle in the coordin plane?

DOMAINS
- Circles
- Expressing Geometric Properties with Equations

Vocabulary

English/Spanish Vocabulary Audio Online:

English	Spanish
chord, *p. 771*	cuerda
inscribed angle, *p. 780*	ángulo inscrito
intercepted arc, *p. 780*	arco interceptor
locus, *p. 804*	lugar geométrico
point of tangency, *p. 762*	punto de tangencia
secant, *p. 791*	secante
standard form of an equation of a circle, *p. 799*	forma normal de una ecuación de un círculo
tangent to a circle, *p. 762*	tagente de un círculo

PowerGeometry.com

Chapter 12 Overview

Use these online assets to engage your students. These include support for the Solve It and step-by step solutions for Problems.

Show the student-produced video demonstrating relevant and engaging applications of the new concepts in this chapter.

Find online definitions for new terms in English and Spanish.

Start each lesson with an attention-getting Problem. View the Problem online with helpful hints.

Common Core Performance Task

Determining the Dimensions of a Logo

The logo of the Sunshine Sailboat Company features several triangular sails in front of a circular sun. The company plans to make a large version of the logo for a showroom display, as shown in the diagram below. The red sail will be made of copper. In the diagram, $\overline{CD}$ is tangent to ⊙*O* at point *B*.

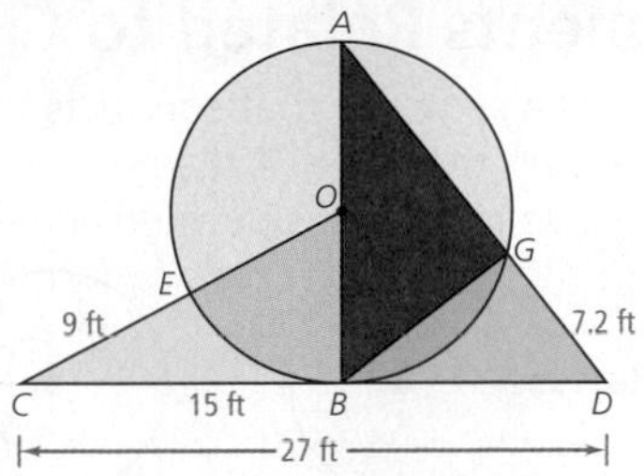

Task Description

Determine the area of the copper needed for the red sail in the logo for the showroom display.

Connecting the Task to the Math Practices

MATHEMATICAL PRACTICES

As you complete the task, you'll apply several Standards for Mathematical Practice.

- You'll write an equation to represent a relationship among lengths in the logo. (MP 2)
- You'll analyze the diagram to find relationships among angles and arcs in the logo design. (MP 1)
- You'll interpret mathematical results in the context of a real-world problem and decide whether your results make sense. (MP 4)

Increase students' depth of knowledge with interactive online activities.

Show Problems from each lesson solved step by step. Instant replay allows students to go at their own pace when studying online.

Assign homework to individual students or to an entire class.

Prepare students for the Mid-Chapter Quiz and Chapter Test with online practice and review.

Virtual Nerd™ Access Virtual Nerd student-centered math tutorials that directly relate to the content of the lesson.

Overview of the Performance Task

Students will use theorems about tangents and secants, as well as the Pythagorean Theorem, to find lengths of segments in a logo design. They will then use these lengths to find the area of the red sail in the logo.

Students will work on the Performance Task in the following places in the chapter.

- Lesson 12-1 (p. 769)
- Lesson 12-3 (p. 787)
- Lesson 12-4 (p. 797)
- Pull It All Together (p. 812)

Introducing the Performance Task

Tell students to read the problem on this page. Do not have them start work on the problem at this time, but ask them the following questions.

Q What is a strategy you could try in order to solve the problem? **[Sample: I could look for relationships that will allow me to use the given measurements to find the lengths of a base and height of the red triangle.]**

Q What relationships do you know exist between some of the segments in the diagram? Explain. **[Sample: *OE*, *OA*, and *OB* are all equal because they are radii of the circle.]**

PARCC CLAIMS

Sub-Claim B: Additional and Supporting Content With Connections to Practices
Sub-Claim D: Highlighted Practice MP 4 With Connections to Content

SBAC CLAIMS

Claim 2: Problem Solving
Claim 4: Modeling and Data Analysis

CHAPTER 12

CIRCLES
Math Background

PROFESSIONAL DEVELOPMENT

The Understanding by Design® methodology was central to the development of the Big Ideas and the Essential Understandings. These will help your students build a structure on which to make connections to prior learning.

Reasoning and Proof

BIG idea Definitions establish meanings and remove possible misunderstanding. Other truths are more complex and difficult to see. It is often possible to verify complex truths by reasoning from simpler ones by using deductive reasoning.

ESSENTIAL UNDERSTANDINGS

12-1 A radius of a circle and the tangent that intersects the endpoint of the radius on the circle have a special relationship.

12-2 Information about congruent parts of a circle (or congruent circles) can be used to find information about other parts of the circle (or circles).

12-6 The description of a locus can be used to sketch a geometric relationship.

Measurement

BIG idea Some attributes of geometric figures, such as length, area, volume, and angle measure, are measurable. Units are used to describe these attributes.

ESSENTIAL UNDERSTANDINGS

12-3 to 12-4 Angles formed by intersecting lines have a special relationship to the arcs the intersecting lines intercept. This includes (1) arcs formed by chords that inscribe angles, (2) angles and arcs formed by lines intersecting either within a circle or outside a circle, and (3) intersecting chords, intersecting secants, or a secant that intersects a tangent.

Coordinate Geometry

BIG idea It is possible to verify some complex truths on the coordinate plane using deductive reasoning in combination with Distance, Midpoint, and Slope formulas.

ESSENTIAL UNDERSTANDING

12-5 The information in the equation of a circle allows the circle to be graphed. The equation of a circle can be written if its center and radius are known.

Segments Related to Circles

A radius is a segment that connects the center of a circle to a point on the circle. A diameter connects two points on the circle and passes through the center of the circle.

Radius

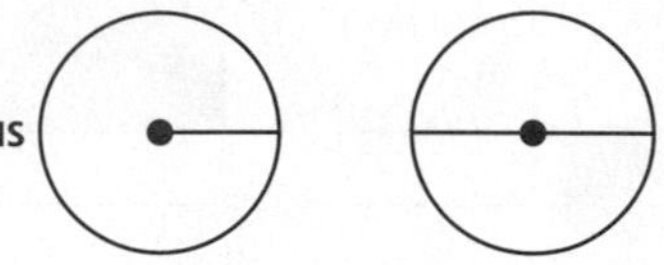

Diameter

A tangent touches the outside of a circle at exactly one point. A chord intersects a circle in two points.

Tangent

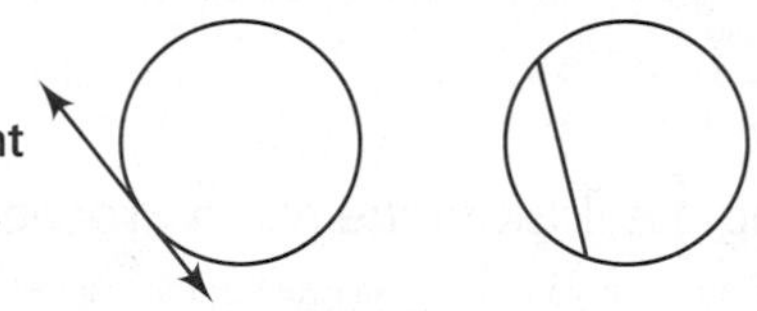

Chord

Tangents to a circle are congruent if they share a common endpoint outside of the circle.

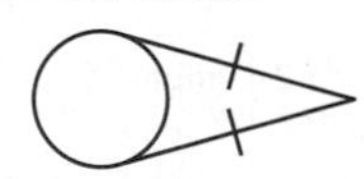

Chords in a circle are congruent if:
- The central angles that create them are congruent.
- Their arcs are congruent.
- They are the same distance from the center.

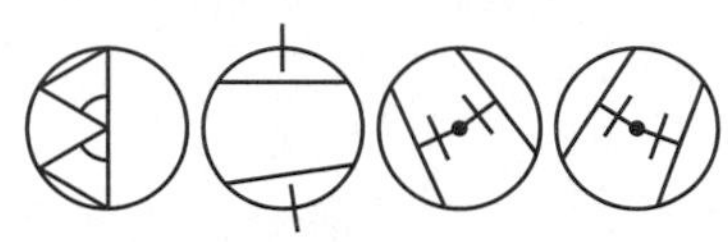

Common Errors With Segments Related to Circles

Students sometimes get confused identifying segments of a circle. Have students create a vocabulary sheet that includes definitions and diagrams of each type of segment. Discuss the relationship between the segments as well. For example, is a diameter a chord? [Yes, but a chord is not necessarily a diameter.]

Mathematical Practices

Attend to Precision Students acquire a large collection of terms relating to circles and must be precise in differentiating between them.

Angles Formed by Special Segments of Circles

The measure of angles formed by chords, tangents and secants can be determined from arc lengths.

Using

Two chords that intersect on a circle:

$m\angle B = \frac{1}{2}m\widehat{AC}$

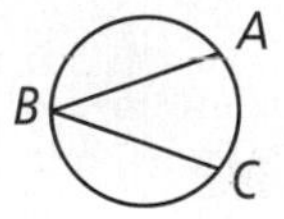

A chord and a tangent:

$m\angle C = \frac{1}{2}m\widehat{BDC}$

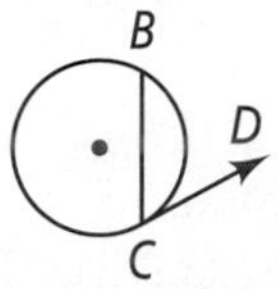

Two secants:

$m\angle ACE = \frac{1}{2}(m\widehat{AE} - m\widehat{BD})$

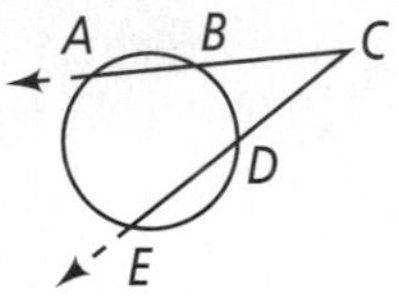

Two chords or secants that intersect inside a circle:

$m\angle ABC = \frac{1}{2}(m\widehat{AC} + m\widehat{DE})$

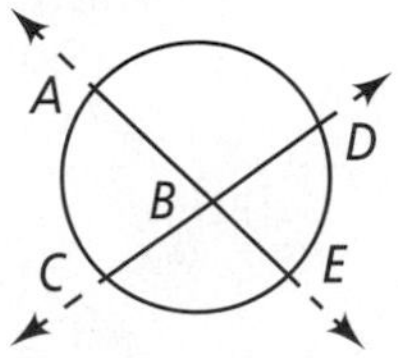

Common Errors With Angles Formed by Special Segments of Circles

Students might try to apply the Inscribed Angle Theorem, $m\angle B = \frac{1}{2}m\widehat{AC}$, when the vertex is not on the circle.

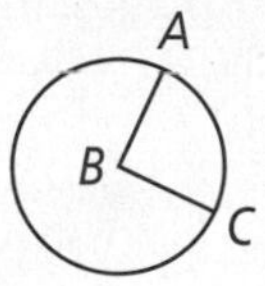

$m\angle B \neq \frac{1}{2}m\widehat{AC}$

Mathematical Practices

Construct viable arguments and critique the reasoning of others Students develop relationships between angles in and around circles by building on exterior angles, using isosceles triangles, and algebraic substitution. They use inscribed angles to expand their knowledge of angle-arc relationships.

Equations of Circles

A circle is the set of all points equidistant from the center of the circle.

The equation of a circle on a coordinate plane is:

$(x - h)^2 + (y - k)^2 = r^2$,

where (h, k) is the center of the circle and r is the radius of the circle.

Given the equation $(x - 2)^2 + (y + 3)^2 = 4$, you can find the center and radius of the circle.

$(h, k) = (2, -3)$

$r^2 = 4$ so $r = 2$

To graph the circle, plot the center of the circle and four points which are 2 units from the center. Then draw the circle.

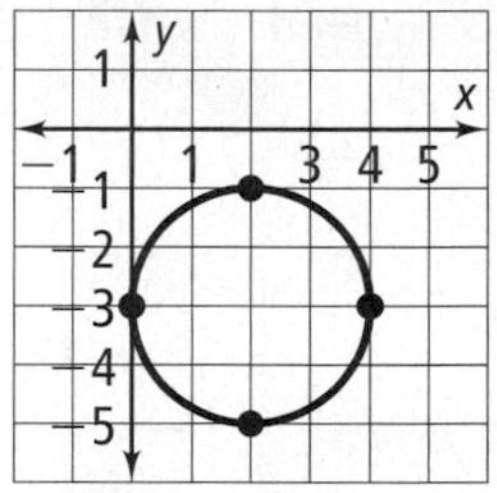

Common Errors With Equations of Circles

When asked to give the center of the circle represented by the equation below, students might not take into account the subtraction inside the parentheses.

For the equation $(x - 3)^2 + (y + 7)^2 = 64$, students may write the center as $(-3, 7)$. The center is actually $(3, -7)$.

Mathematical Practices

Look for and make use of structure Students develop algebraic equations for circles from the Pythagorean Theorem. They apply their understanding of radius to vary the problems they can solve.

CHAPTER 12

CIRCLES

Pacing and Assignment Guide

Lesson	Teaching Day(s)	TRADITIONAL: Basic	Average	Advanced	BLOCK: Block
12-1	1	Problems 1–3 Exs. 6–14	Problems 1–3 Exs. 7–13 odd	Problems 1–3 Exs. 7–13 odd	**Day 1** Problems 1–5 Exs. 7–19 odd, 20–29
	2	Problems 4–5 Exs. 15–22, 26	Problems 4–5 Exs. 15–19 odd, 20–29	Problems 4–5 Exs. 15–19 odd, 20–31	
12-2	1	Problems 1–2 Exs. 6–10, 40–52	Problems 1–2 Exs. 7–9 odd, 40–52	Problems 1–2 Exs. 7–9 odd, 40–52	**Day 2** Problems 1–4 Exs. 7–15 odd, 16–34, 40–52
	2	Problems 3–4 Exs. 11–16, 18, 23–25, 29	Problems 3–4 Exs. 11–15 odd, 16–34	Problems 3–4 Exs. 11–15 odd, 16–39	
12-3	1	Problems 1–3 Exs. 6–19 all, 20–24 even, 28–29	Problems 1–3 Exs. 7–17 odd, 19–34	Problems 1–3 Exs. 7–17 odd, 19–39	**Day 3** Problems 1–3 Exs. 7–17 odd, 19–34
12-4	1	Problems 1–3 Exs. 8–20 all, 22–26 even, 27–31 odd	Problems 1–3 Exs. 9–19 odd, 21–39	Problems 1–3 Exs. 9–19 odd, 21–43	Problems 1–3 Exs. 9–19 odd, 21–39
12-5	1	Problems 1–3 Exs. 8–30 all, 31–52 even, 58–65	Problems 1–3 Exs. 9–29 odd, 31–56, 58–65	Problems 1–3 Exs. 9–29 odd, 31–65	**Day 4** Problems 1–3 Exs. 9–29 odd, 31–56, 58–65
12-6	1	Problems 1–3 Exs. 7–19 all, 20–24 even, 25, 32–40 even, 46, 55–65	Problems 1–3 Exs. 7–19 odd, 20–48, 55–65	Problems 1–3 Exs. 7–19 odd, 20–65	Problems 1–3 Exs. 7–19 odd, 20–48, 55–65
Review	1	Chapter 12 Review	Chapter 12 Review	Chapter 12 Review	**Day 5** Chapter 12 Review
Assess	1	Chapter 12 Test	Chapter 12 Test	Chapter 12 Test	Chapter 12 Test
Total		**10 Days**	**10 Days**	**10 Days**	**5 Days**

Note: Pacing does not include Concept Bytes and other feature pages.

Resources

KEY
I = Interactive asset at PowerGeometry.com
E = Editable master at PowerGeometry.com
P = Available in Print
M = Master at PowerGeometry.com
✓ = CD-ROM

	For the Chapter	12-1	12-2	12-3	12-4	12-5	12-6
Planning							
Teacher Center Online Planner & Grade Book	I	I	I	I	I	I	I
Interactive Learning & Guided Instruction							
My Math Video	I						
Solve It!		I M	I M	I M	I M	I M	I M
Student Companion		P M	P M	P M	P M	P M	
Vocabulary Support		I P M	I P M	I P M	I P M	I P M	I P M
Got It? Support		I P	I P	I P	I P	I P	I P
Online Problems		I	I	I	I	I	I
Additional Problems		M	M	M	M	M	M
English Language Learner Support (TR)		E P M	E P M	E P M	E P M	E P M	E P M
Activities, Games, and Puzzles		E M	E M	E M	E M	E M	E M
Teaching With TI Technology With CD-ROM				✓ P			
TI-Nspire™ Support CD-ROM		✓	✓	✓	✓	✓	✓
Lesson Check & Practice							
Student Companion		P M	P M	P M	P M	P M	P M
Lesson Check Support		I P	I P	I P	I P	I P	I P
Practice and Problem Solving Workbook		P	P	P	P	P	P
Think About a Plan (TR)		E P M	E P M	E P M	E P M	E P M	E P M
Practice Form G (TR)		E P M	E P M	E P M	E P M	E P M	E P M
Standardized Test Prep (TR)		P M	P M	P M	P M	P M	P M
Practice *Form K* (TR)		E P M	E P M	E P M	E P M	E P M	E P M
Extra Practice	E M						
Find the Errors!	M						
Enrichment (TR)		E P M	E P M	E P M	E P M	E P M	E P M
Answers and Solutions CD-ROM	✓	✓	✓	✓	✓	✓	✓
Assess & Remediate							
ExamView CD-ROM	✓	✓	✓	✓	✓	✓	✓
Lesson Quiz		I M	I M	I M	I M	I M	I M
Quizzes and Tests *Form G* (TR)	E P M			E P M			E P M
Quizzes and Tests *Form K* (TR)	E P M			E P M			E P M
Reteaching (TR)		E P M	E P M	E P M	E P M	E P M	E P M
Performance Tasks (TR)	P M						
Cumulative Review (TR)	P M						
Progress Monitoring Assessments	I P M						

(TR) Available in All-In-One Teaching Resources

1 Interactive Learning

Solve It!

PURPOSE To discover characteristics of tangent segments

PROCESS Students measure the lengths of tangent segments with a common endpoint and use inductive reasoning to make a conjecture.

FACILITATE

As students draw circles with different tangent segments, be sure that they are drawing them correctly. Model the process on the board. They should align a straight edge so that it touches the circle in only one point.

Q Label the points of intersection of the lines with the circle *B* and *C*. What triangles could be formed? **[$\triangle ABO$, $\triangle ACO$, $\triangle ABC$, and $\triangle BCO$]**

Q How can you classify the segments $\overline{BO}$ and $\overline{CO}$? **[They are radii of the circle.]**

Q What is the relationship between $\triangle ABO$ and $\triangle ACO$? Justify your answer. **[They are congruent by SSS.]**

ANSWER See Solve It in Answers on next page.

CONNECT THE MATH Students should see that two tangent segments drawn from the same point to a circle are congruent. In the lesson, students will examine this relationship and others regarding tangent lines and circles.

2 Guided Instruction

Take Note

Have students identify the hypothesis and conclusion of Theorem 12-1. Ask them to name the right angles in the diagram.

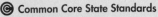

12-1 Tangent Lines

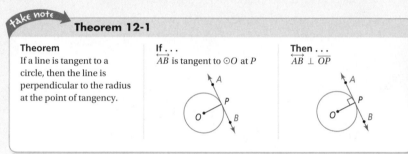

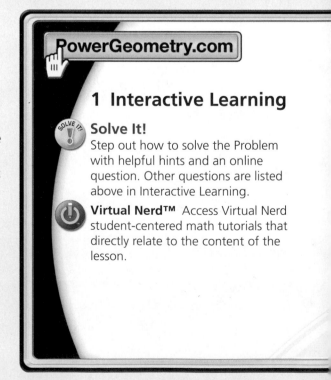

Common Core State Standards

G-C.A.2 Identify and describe relationships among inscribed angles, radii, and chords . . . the radius of a circle is perpendicular to the tangent where the radius intersects the circle.

MP 1, MP 3, MP 4

Objective To use properties of a tangent to a circle

MATHEMATICAL PRACTICES

SOLVE IT! Getting Ready!

Draw a diagram like the one at the right. Each ray from Point A touches the circle in only one place no matter how far it extends. Measure AB and AC. Repeat the procedure with a point farther away from the circle. Consider any two rays with a common endpoint outside the circle. Make a conjecture about the lengths of the two segments formed when the rays touch the circle.

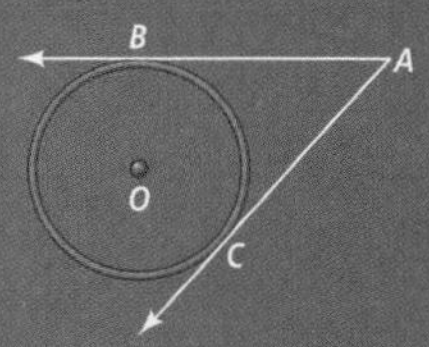

In the Solve It, you drew lines that touch a circle at only one point. These lines are called tangents. This use of the word *tangent* is related to, but different from, the tangent ratio in right triangles that you studied in Chapter 8.

Lesson Vocabulary
- tangent to a circle
- point of tangency

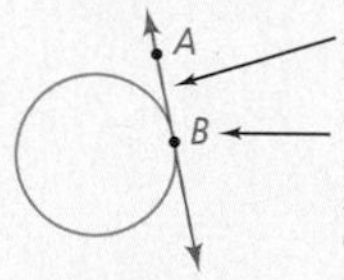

A **tangent to a circle** is a line in the plane of the circle that intersects the circle in exactly one point.

The point where a circle and a tangent intersect is the **point of tangency.**

$\overrightarrow{BA}$ is a tangent ray and $\overline{BA}$ is a tangent segment.

Essential Understanding A radius of a circle and the tangent that intersects the endpoint of the radius on the circle have a special relationship.

take note

Theorem 12-1

Theorem	If . . .	Then . . .
If a line is tangent to a circle, then the line is perpendicular to the radius at the point of tangency.	$\overleftrightarrow{AB}$ is tangent to $\odot O$ at P 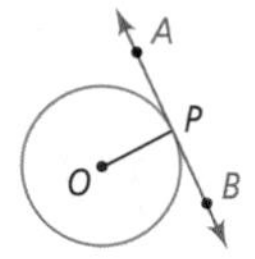	$\overleftrightarrow{AB} \perp \overline{OP}$ 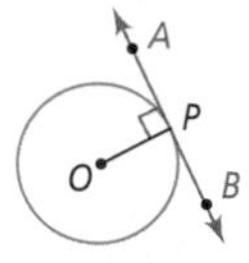

12-1 Preparing to Teach

BIG ideas **Reasoning and Proof**
Measurement

ESSENTIAL UNDERSTANDINGS

- A radius of a circle and the tangent that intersects the endpoint of the radius on the circle have a special relationship.
- A circle has a special relationship to a triangle whose sides are tangent to the circle.

Math Background

Students will use their understanding of congruent triangles to prove statements about tangent lines. Tangent lines have two important characteristics: they only touch the circle at one point, and they are perpendicular to the radius at the point of tangency. These characteristics are illustrated by real-world relationships between the Earth and lines of sight. Students can determine characteristics of circumscribed figures using characteristics of tangent lines.

Have students imagine and describe twirling and then releasing a ball attached to a string. The released ball will follow a path (easy to see because of the string) influenced by gravity, but its direction at the point of release is along the line tangent to its original orbit at the point of release. Point out that the law of physics that the ball obeys is related to the geometry theorems in this lesson.

In future lessons, students will learn about other special segments in circles. Tangent lines will be used again in the study of the slope of a nonlinear function.

Mathematical Practices

Make sense of problems and persevere in solving them. Students will draw diagrams of circles and tangent lines and identify important relationships, such as segment lengths and angle measurements.

PowerGeometry.com

1 Interactive Learning

Solve It!
Step out how to solve the Problem with helpful hints and an online question. Other questions are listed above in Interactive Learning.

Virtual Nerd™ Access Virtual Nerd student-centered math tutorials that directly relate to the content of the lesson.

Proof Indirect Proof of Theorem 12-1

Given: n is tangent to $\odot O$ at P.

Prove: $n \perp \overline{OP}$

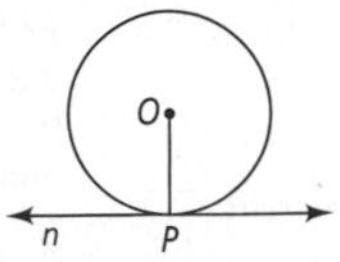

Step 1 Assume that n is not perpendicular to $\overline{OP}$.

Step 2 If line n is not perpendicular to $\overline{OP}$, then, for some other point L on n, $\overline{OL}$ must be perpendicular to n. Also there is a point K on n such that $\overline{LK} \cong \overline{LP}$. $\angle OLK \cong \angle OLP$ because perpendicular lines form congruent adjacent angles. $\overline{OL} \cong \overline{OL}$. So, $\triangle OLK \cong \triangle OLP$ by SAS.

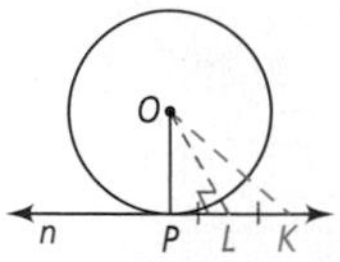

Since corresponding parts of congruent triangles are congruent, $\overline{OK} \cong \overline{OP}$. So K and P are both on $\odot O$ by the definition of a circle. For two points on n to also be on $\odot O$ contradicts the given fact that n is tangent to $\odot O$ at P. So the assumption that n is not perpendicular to $\overline{OP}$ must be false.

Step 3 Therefore, $n \perp \overline{OP}$ must be true.

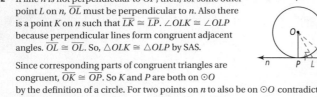

Problem 1 Finding Angle Measures

Multiple Choice $\overline{ML}$ **and** $\overline{MN}$ **are tangent to** $\odot O$**. What is the value of** x**?**

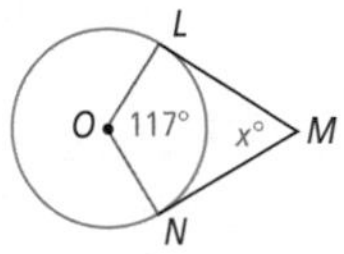

A 58
C 90
B 63
D 117

Think
What kind of angle is formed by a radius and a tangent?
The angle formed is a right angle, so the measure is 90.

Since $\overline{ML}$ and $\overline{MN}$ are tangent to $\odot O$, $\angle L$ and $\angle N$ are right angles. $LMNO$ is a quadrilateral. So the sum of the angle measures is 360.

$m\angle L + m\angle M + m\angle N + m\angle O = 360$

$90 + m\angle M + 90 + 117 = 360$ Substitute.

$297 + m\angle M = 360$ Simplify.

$m\angle M = 63$ Solve.

The correct answer is B.

Got It? 1. **a.** $\overline{ED}$ is tangent to $\odot O$. What is the value of x?

b. Reasoning Consider a quadrilateral like the one in Problem 1. Write a formula you could use to find the measure of any angle x formed by two tangents when you know the measure of the central angle c whose radii intersect the tangents.

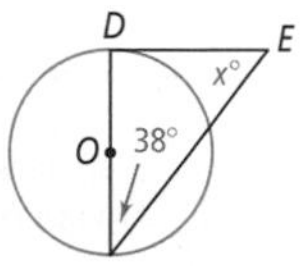

Students may need to review the definition of circles to understand how Theorem 12-1 proves that K and P must both be on the circle. Because $\overline{OK} \cong \overline{OP}$, K and P are equidistant from O. Students may benefit from working backwards to prove the theorem. Ask them to identify the statement that contradicts the given information. They must use the definition of a tangent.

Problem 1

Q How can you classify $\overline{LO}$ and $\overline{NO}$? **[They are radii of the circle.]**

Q What is $m\angle L$ and $m\angle N$? **[Both angles measure 90.]**

Q What type of geometric figure do the tangents and radii form? **[a quadrilateral]**

Q How can you find the missing angle measure in the quadrilateral? **[Subtract the known angle measures from 360.]**

Got It? ERROR PREVENTION

Have students label the known angle measures in the diagram. Emphasize that the third angle, call it $\angle F$, is not 90 because $\overline{EF}$ is not tangent to the circle. Ask them to write an equation using the angle measures.

Students may need to be reminded that the angles of a triangle have a sum of 180.

2 Guided Instruction

Each Problem is worked out and supported online

Problem 1
Finding Angle Measures

Problem 2
Finding Distance

Problem 3
Finding a Radius
Animated

Problem 4
Identifying a Tangent

Problem 5
Circles Inscribed in Polygons
Animated

Support in Geometry Companion
- Vocabulary
- Key Concepts
- Got It?

Answers

Solve It!

The two segments have the same length.

Got It?

1a. 52

b. $x = 180 - c$

Problem 2

Encourage students to draw a diagram of the situation. Discuss which segments and angles can be labeled with their measurements.

Q How can you classify the line that represents the line of sight from the top of the tower? **[The line of sight is tangent to the Earth.]**

Q What does the horizon represent in relation to the circle that is Earth? **[The horizon is the point of tangency of the line of sight.]**

Q What type of angle do the line of sight and the radius of the Earth form? **[a right angle]**

Q How can you find the distance between the top of the tower and the center of the Earth? **[Add the height of the tower and the radius of the Earth.]**

Got It? VISUAL LEARNERS

Have students draw a diagram of the situation. Ask them to identify the tangent line and point of tangency. Have them label the known lengths.

Take Note

Be sure that students can identify Theorems 12-1 and 12-2 as converses. Have them identify the hypothesis and conclusion of each theorem. Then ask students to combine the two theorems to form a biconditional statement.

Problem 2 Finding Distance STEM

Plan

How does knowing Earth's radius help?
The radius forms a right angle with a tangent line from the observation deck to the horizon. So, you can use two radii, the tower's height, and the tangent to form a right triangle.

Earth Science **The CN Tower in Toronto, Canada, has an observation deck 447 m above ground level. About how far is it from the observation deck to the horizon? Earth's radius is about 6400 km.**

Step 1 Make a sketch. The length 447 m is about 0.45 km.

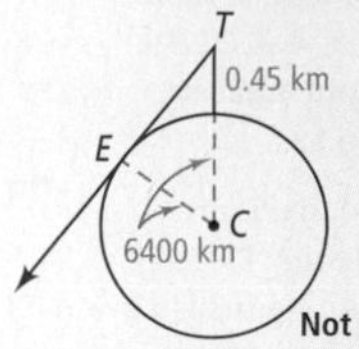

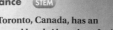

Step 2 Use the Pythagorean Theorem.

$CT^2 = TE^2 + CE^2$	
$(6400 + 0.45)^2 = TE^2 + 6400^2$	Substitute.
$(6400.45)^2 = TE^2 + 6400^2$	Simplify.
$40{,}965{,}760.2025 = TE^2 + 40{,}960{,}000$	Use a calculator.
$5760.2025 = TE^2$	Subtract 40,960,000 from each side.
$76 \approx TE$	Take the positive square root of each side.

The distance from the CN Tower to the horizon is about 76 km.

Got It? 2. What is the distance to the horizon that a person can see on a clear day from an airplane 2 mi above Earth? Earth's radius is about 4000 mi.

Theorem 12-2 is the converse of Theorem 12-1. You can use it to prove that a line or segment is tangent to a circle. You can also use it to construct a tangent to a circle.

take note

Theorem 12-2

Theorem	If . . .	Then . . .
If a line in the plane of a circle is perpendicular to a radius at its endpoint on the circle, then the line is tangent to the circle.	$\overleftrightarrow{AB} \perp \overline{OP}$ at P 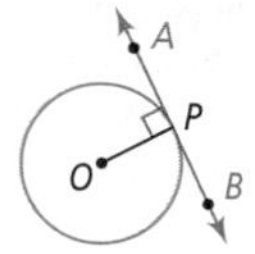	$\overleftrightarrow{AB}$ is tangent to $\odot O$

You will prove Theorem 12-2 in Exercise 30.

Additional Problems

1. $\overline{AD}$ and $\overline{AG}$ are tangent to $\odot O$. What is the value of x?

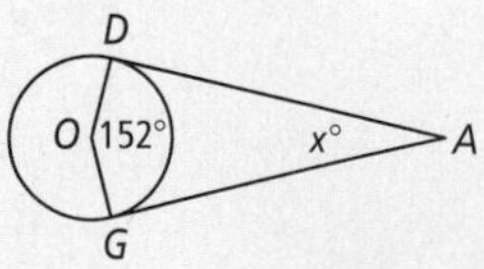

A. 28
B. 56
C. 76
D. 152

ANSWER A

2. Jasmine is riding in an airplane at an altitude of about 6.5 mi above the Earth. How far on the Earth can she see if the Earth's radius is about 4000 mi? Round to the nearest mile.

ANSWER 228 mi

3. What is the value of x?

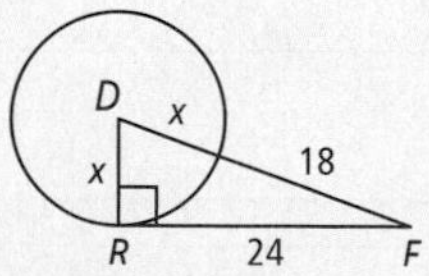

ANSWER 7

4. Is $\overline{MN}$ tangent to $\odot P$ at N? Explain.

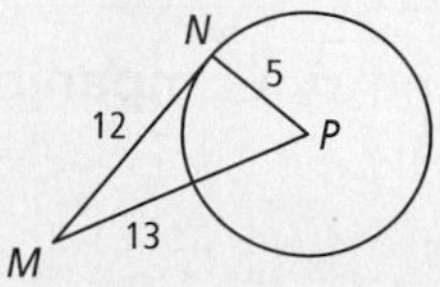

ANSWER Yes. Triangle MNP is a right triangle by the Converse of the Pythagorean Theorem. So, $\overline{MN}$ is tangent to $\odot P$ at N by Theorem 12-2.

5. $\odot S$ is inscribed in $\triangle LMN$. What is the perimeter of $\triangle LMN$?

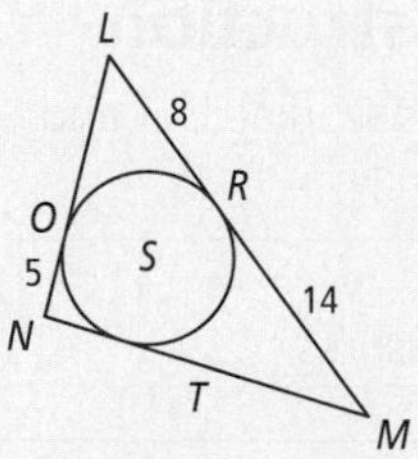

ANSWER 54

Problem 3 Finding a Radius

Think

Why does the value x appear on each side of the equation?
The length of $\overline{AC}$, the hypotenuse, is the radius plus 8, which is on the left side of the equation. On the right side of the equation, the radius is one side of the triangle.

What is the radius of $\odot C$?

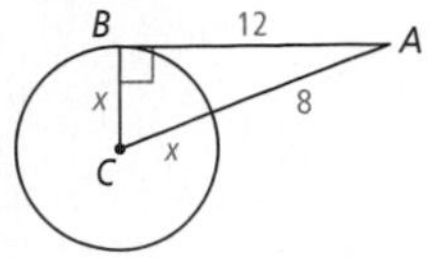

$AC^2 = AB^2 + BC^2$	Pythagorean Theorem
$(x + 8)^2 = 12^2 + x^2$	Substitute.
$x^2 + 16x + 64 = 144 + x^2$	Simplify.
$16x = 80$	Subtract x^2 and 64 from each side.
$x = 5$	Divide each side by 16.

The radius is 5.

Got It? 3. What is the radius of $\odot O$?

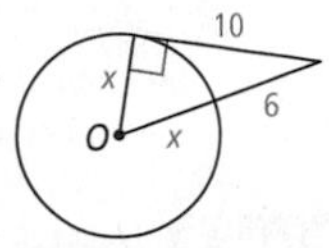

Problem 4 Identifying a Tangent

Think

What information does the diagram give you?
- LMN is a triangle.
- $NM = 25$, $LM = 24$, $NL = 7$
- $\overline{NL}$ is a radius.

Is $\overline{ML}$ tangent to $\odot N$ at L? Explain.

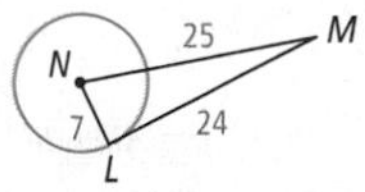

Know
The lengths of the sides of $\triangle LMN$

Need
To determine whether $\overline{ML}$ is tangent to $\odot O$

Plan
$\overline{ML}$ is a tangent if $\overline{ML} \perp \overline{NL}$. Use the Converse of the Pythagorean Theorem to determine whether $\triangle LMN$ is a right triangle.

$NL^2 + ML^2 \stackrel{?}{=} NM^2$	
$7^2 + 24^2 \stackrel{?}{=} 25^2$	Substitute.
$625 = 625$	Simplify.

By the Converse of the Pythagorean Theorem, $\triangle LMN$ is a right triangle with $\overline{ML} \perp \overline{NL}$. So $\overline{ML}$ is tangent to $\odot N$ at L because it is perpendicular to the radius at the point of tangency (Theorem 12-2).

Got It? 4. Use the diagram in Problem 4. If $NL = 4$, $ML = 7$, and $NM = 8$, is $\overline{ML}$ tangent to $\odot N$ at L? Explain.

In the Solve It, you made a conjecture about the lengths of two tangents from a common endpoint outside a circle. Your conjecture may be confirmed by the following theorem.

Problem 3

Q Which segment is tangent to the circle? **[$\overline{AB}$]**
Q What is the length of the hypotenuse in the right triangle? **[$x + 8$]**
Q How can you check your answer? **[Verify that the side lengths satisfy the Pythagorean Theorem.]**

Got It? VISUAL LEARNERS

Have students label each side of the right triangle *a*, *b*, or *c* making sure that *c* is the hypotenuse. Have them write out the Pythagorean Theorem and substitute the values given for *a*, *b*, and *c*. Check that the equation they write contains a quadratic expression.

Problem 4

Q If $\overline{ML}$ is tangent to the circle, then what type of triangle is $\triangle LMN$? **[a right triangle]**
Q Which segment is the hypotenuse of $\triangle LMN$? **[$\overline{MN}$]**
Q How can you verify that $\triangle LMN$ is a right triangle? **[Verify that the side lengths satisfy the Pythagorean Theorem.]**

Got It? ERROR PREVENTION

Have students redraw the diagram and label it with the new lengths. Ask them to write an equation to relate the lengths of the segments using the Pythagorean Theorem. Once they have come to a conclusion about $\overline{ML}$, have them write an indirect proof to justify their answer.

Answers

Got It? (continued)

2. about 127 mi
3. $5\frac{1}{3}$
4. no; $4^2 + 7^2 = 65 \neq 8^2$

Problem 5

Q How can you classify the sides of the triangle? Justify your answer. **[Because the circle is inscribed in the triangle, the sides are tangent to the circle.]**

Q Which pairs of segments are congruent in $\triangle ABC$? **[$\overline{AD} \cong \overline{AF}$, $\overline{BD} \cong \overline{BE}$, and $\overline{CE} \cong \overline{CF}$]**

Got It?

Have students write an equation relating the segment lengths to the perimeter of the triangle.

3 Lesson Check

Do you know HOW?

- If students have difficulty with Exercise 1, then remind them that the measures of the angles of a triangle have a sum of 180°.
- If students have difficulty with Exercise 2, then point out that there are two radius lengths in the diagram.

Do you UNDERSTAND?

- If students have difficulty with Exercise 5, then ask them if a triangle can have more than one right angle. Then review Theorem 12-1.

Close

Q What type of angle is formed at the intersection of a tangent and the radius of a circle? **[a right angle]**

Q What is the relationship between two segments drawn from the same point, tangent to a circle? **[The segments are congruent.]**

take note

Theorem 12-3

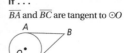

Theorem	If . . .	Then . . .
If two tangent segments to a circle share a common endpoint outside the circle, then the two segments are congruent.	$\overline{BA}$ and $\overline{BC}$ are tangent to $\odot O$ 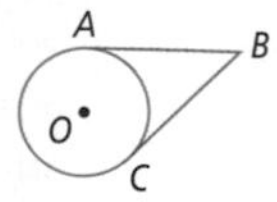	$\overline{BA} \cong \overline{BC}$

You will prove Theorem 12-3 in Exercise 23.

In the figure at the right, the sides of the triangle are tangent to the circle. The circle is *inscribed in* the triangle. The triangle is *circumscribed about* the circle.

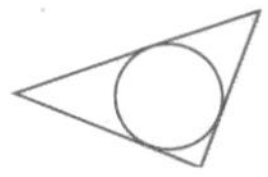

Problem 5 Circles Inscribed in Polygons

Plan

How can you find the length of $\overline{BC}$?
Find the segments congruent to $\overline{BE}$ and $\overline{EC}$. Then use segment addition.

$\odot O$ is inscribed in $\triangle ABC$. What is the perimeter of $\triangle ABC$?

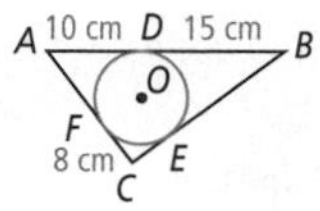

$AD = AF = 10$ cm
$BD = BE = 15$ cm
$CF = CE = 8$ cm

Two segments tangent to a circle from a point outside the circle are congruent, so they have the same length.

$p = AB + BC + CA$	Definition of perimeter p
$= AD + DB + BE + EC + CF + FA$	Segment Addition Postulate
$= 10 + 15 + 15 + 8 + 8 + 10$	Substitute.
$= 66$	

The perimeter is 66 cm.

Got It? 5. $\odot O$ is inscribed in $\triangle PQR$, which has a perimeter of 88 cm. What is the length of $\overline{QY}$?

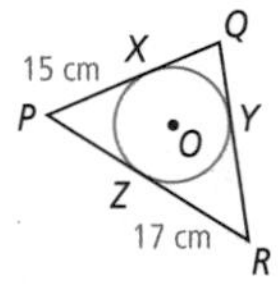

Lesson Check

Do you know HOW?

1. If $m\angle A = 58$, what is $m\angle ACB$?
2. If $BC = 8$ and $DC = 4$, what is the radius?
3. If $AC = 12$ and $BC = 9$, what is the radius?

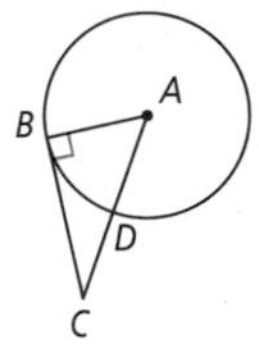

Do you UNDERSTAND? MATHEMATICAL PRACTICES

4. **Vocabulary** How are the phrases *tangent ratio* and *tangent of a circle* used differently?
5. **Error Analysis** A classmate insists that $\overline{DF}$ is a tangent to $\odot E$. Explain how to show that your classmate is wrong.

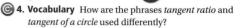

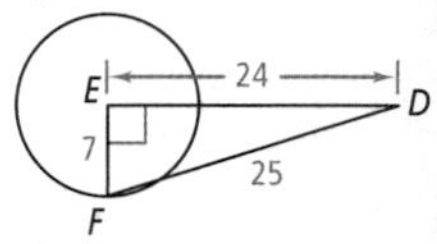

PowerGeometry.com

3 Lesson Check

For a digital lesson check, use the Got It questions.

Support in Geometry Companion

- Lesson Check

4 Practice

ONLINE HOMEWORK

Assign homework to individual students or to an entire class.

Answers

Got It? (continued)

5. 12 cm

Lesson Check

1. 32
2. 6 units
3. $\sqrt{63} \approx 7.9$ units
4. Answers may vary. Sample: *Tangent ratio* refers to a ratio of the lengths of two sides of a rt. △, while *tangent to a circle* refers to a line or a part of a line that is in the plane of a circle and touches the circle in exactly one point.
5. If $\overline{DF}$ is tangent to $\odot E$, then $\overline{DF} \perp \overline{EF}$. That would mean that $\triangle DEF$ contains two rt. ∠s, which is impossible. So $\overline{DF}$ is not a tangent to $\odot E$.

Practice and Problem-Solving Exercises

MATHEMATICAL PRACTICES

A Practice

Algebra Lines that appear to be tangent are tangent. O is the center of each circle. What is the value of x?

See Problem 1.

6.

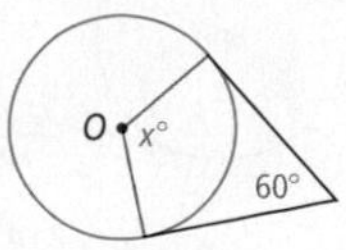

7.

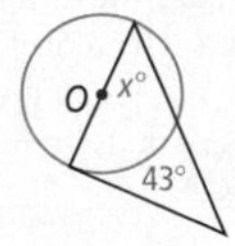

8.

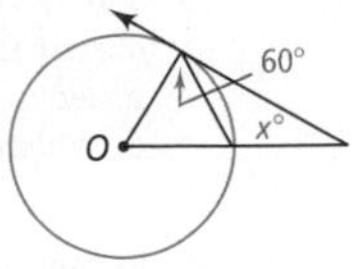

STEM Earth Science The circle at the right represents Earth. The radius of Earth is about 6400 km. Find the distance d to the horizon that a person can see on a clear day from each of the following heights h above Earth. Round your answer to the nearest tenth of a kilometer.

See Problem 2.

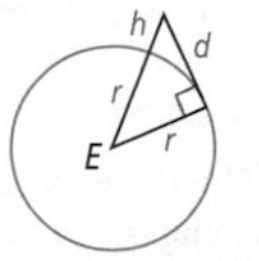

9. 5 km

10. 1 km

11. 2500 m

Algebra In each circle, what is the value of x, to the nearest tenth?

See Problem 3.

12.

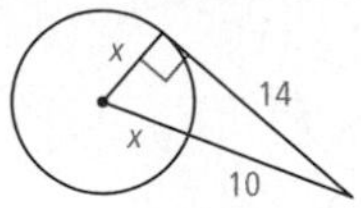

13.

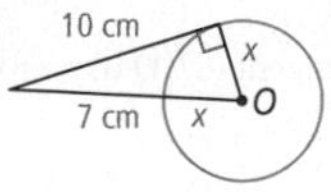

14.

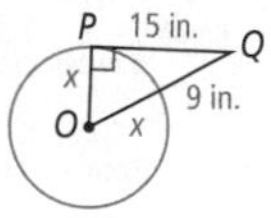

Determine whether a tangent is shown in each diagram. Explain.

See Problem 4.

15.

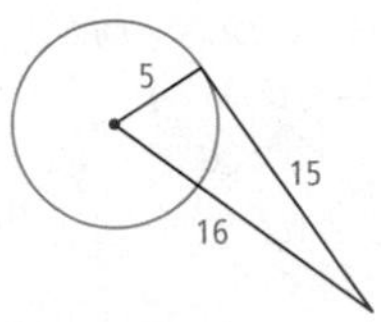

16.

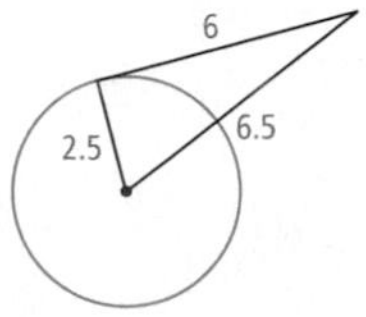

17.

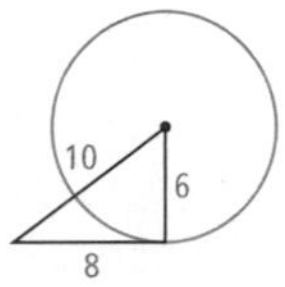

Each polygon circumscribes a circle. What is the perimeter of each polygon?

See Problem 5.

18.

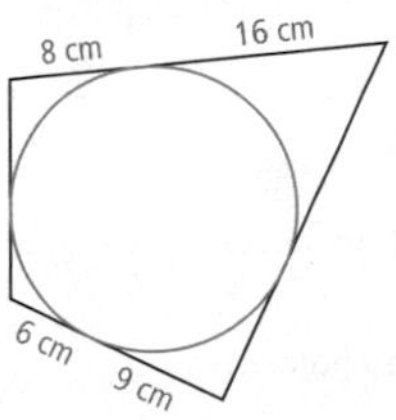

19.

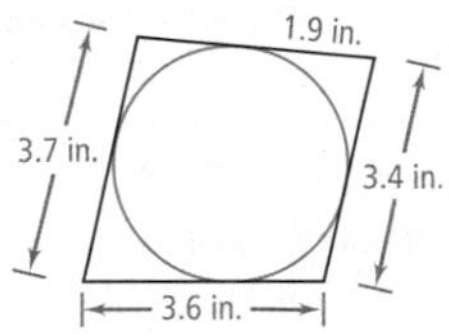

4 Practice

ASSIGNMENT GUIDE

Basic: 6–22, 26
Average: 7–19 odd, 20–29
Advanced: 7–19 odd, 20–31

Mathematical Practices are supported by exercises with red headings. Here are the Practices supported in this lesson:

MP 1: Make Sense of Problems Ex. 22
MP 3: Construct Arguments Ex. 21
MP 3: Critique the Reasoning of Others Ex. 5

Applications exercises have blue headings. Exercises 9–11 and 20 support MP 4: Model.

STEM exercises focus on science or engineering applications.

EXERCISE 26: Use the Think About a Plan worksheet in the **Practice and Problem Solving Workbook** (also available in the Teaching Resources in print and online) to further support students' development in becoming independent learners.

HOMEWORK QUICK CHECK

To check students' understanding of key skills and concepts, go over Exercises 7, 17, 21, 22, and 26.

Practice and Problem-Solving Exercises

6. 120
7. 47
8. 30
9. 253.0 km
10. 113.1 km
11. 178.9 km
12. 4.8
13. 3.6 cm
14. 8 in.
15. no; $5^2 + 15^2 \neq 16^2$
16. yes; $2.5^2 + 6^2 = 6.5^2$
17. yes; $6^2 + 8^2 = 10^2$
18. 78 cm
19. 14.2 in.

Answers

Practice and Problem-Solving Exercises (continued)

20a. external
b. external
c. internal
d. blue segments; green segments

21. All 4 are ≅; the two tangents to each coin from A are ≅, so by the Transitive Prop. of ≅, all the tangents are ≅.

22. Answers may vary. Sample: One square is inscribed in the circle and the other square circumscribes the circle. If the circle has radius a, each side of the smaller square has length $a\sqrt{2}$ and the area of the square is $2a^2$. Each side of the larger square has length $2a$ and the area of the square is $4a^2$. So the larger square has double the area of the smaller square.

23. 1. $\overline{BA}$ and $\overline{BC}$ are tangent to ⊙O at A and C. (Given) 2. $\overline{AB} \perp \overline{OA}$ and $\overline{BC} \perp \overline{OC}$ (If a line is tan. to a ⊙, it is ⊥ to the radius.) 3. $\triangle BAO$ and $\triangle BCO$ are rt. △s. (Def. of rt. △) 4. $\overline{AO} \cong \overline{OC}$ (Radii of a circle are ≅.) 5. $\overline{BO} \cong \overline{BO}$ (Refl. Prop. of ≅) 6. $\triangle BAO \cong \triangle BCO$ (HL) 7. $\overline{BA} \cong \overline{BC}$ (Corresp. parts of ≅ △s are ≅.)

24. 1. $\overline{BC}$ is tangent to ⊙A at D. (Given) 2. $\overline{DB} \cong \overline{DC}$ (Given) 3. $\overline{AD} \perp \overline{BC}$ (If a line is tan. to a ⊙, it is ⊥ to the radius.) 4. $\angle ADB$ and $\angle ADC$ are rt. ∠s (Def. of ⊥) 5. $\angle ADB \cong \angle ADC$ (Rt. ∠s are ≅.) 6. $\overline{AD} \cong \overline{AD}$ (Refl. Prop. of ≅) 7. $\triangle ADB \cong \triangle ADC$ (SAS) 8. $\overline{AB} \cong \overline{AC}$ (Corresp. parts of ≅ △s are ≅.)

25. 1. ⊙A and ⊙B with common tangents $\overline{DF}$ and $\overline{CE}$ (Given) 2. $GD = GC$ and $GE = GF$ (Two tan. segments from a pt. to a ⊙ are ≅.)
3. $\frac{GD}{GC} = 1$, $\frac{GF}{GE} = 1$ (Div. Prop. of =)
4. $\frac{GD}{GC} = \frac{GF}{GE}$ (Trans. Prop. of =)
5. $\angle DGC \cong \angle EGF$ (Vert. ∠s are ≅.)
6. $\triangle GDC \sim \triangle GFE$ (SAS ~ Thm.)

26a. Rectangle; $\overline{AB}$ is tangent to ⊙D and ⊙E; $\overline{DB} \perp \overline{AB}$ and $\overline{AE} \perp \overline{AB}$ (A line tangent to a ⊙ is ⊥ to the radius.); $\overline{BC} \parallel \overline{AE}$ (Two coplanar lines ⊥ to the same line are ∥.) So, $ABCE$ is a ▱ with two rt. ∠s. Therefore, $ABCE$ is a rectangle.
b. 35 in.
c. 35.5 in.

27. 57.5

28.

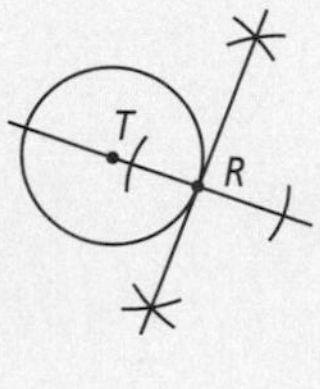

29.

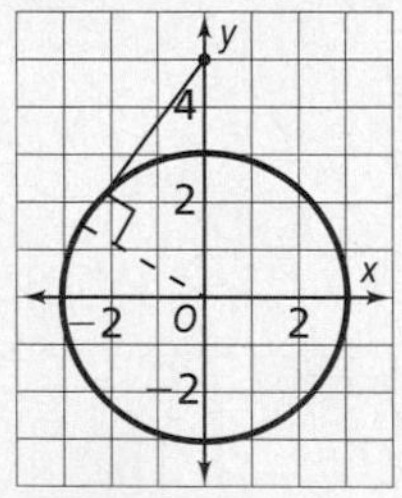

4 units

B Apply STEM **20. Solar Eclipse** Common tangents to two circles may be *internal* or *external*. If you draw a segment joining the centers of the circles, a common internal tangent will intersect the segment. A common external tangent will not. For this cross-sectional diagram of the sun, moon, and Earth during a solar eclipse, use the terms above to describe the types of tangents of each color.

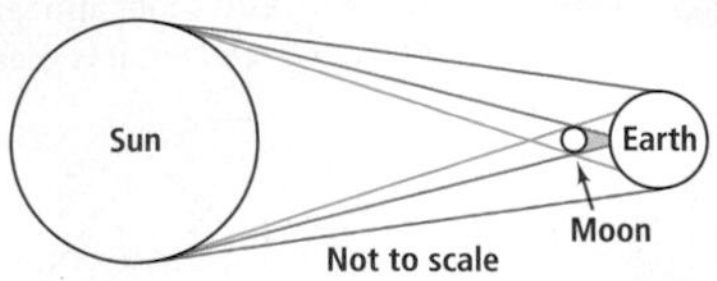

a. red **b.** blue **c.** green
d. Which tangents show the extent on Earth's surface of total eclipse? Of partial eclipse?

21. Reasoning A nickel, a dime, and a quarter are touching as shown. Tangents are drawn from point A to both sides of each coin. What can you conclude about the four tangent segments? Explain.

22. Think About a Plan Leonardo da Vinci wrote, "When each of two squares touch the same circle at four points, one is double the other." Explain why the statement is true.
- How will drawing a sketch help?
- Are both squares inside the circle?

23. Prove Theorem 12-3.
Proof
Given: $\overline{BA}$ and $\overline{BC}$ are tangent to ⊙O at A and C, respectively.
Prove: $\overline{BA} \cong \overline{BC}$

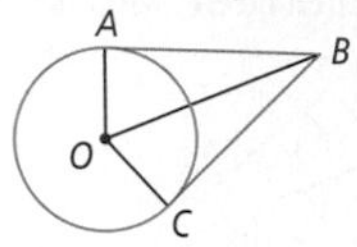

24. Given: $\overline{BC}$ is tangent to ⊙A at D.
Proof
$\overline{DB} \cong \overline{DC}$
Prove: $\overline{AB} \cong \overline{AC}$

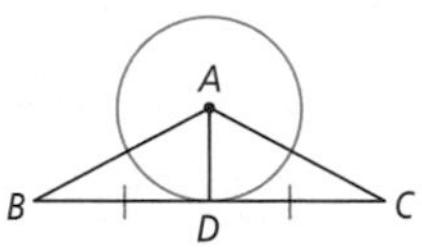

25. Given: ⊙A and ⊙B with common tangents $\overline{DF}$ and $\overline{CE}$
Proof
Prove: $\triangle GDC \sim \triangle GFE$

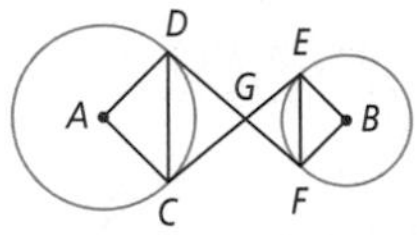

26. a. A belt fits snugly around the two circular pulleys. $\overline{CE}$ is an auxiliary line from E to $\overline{BD}$. $\overline{CE} \parallel \overline{BA}$. What type of quadrilateral is $ABCE$? Explain.
b. What is the length of $\overline{CE}$?
c. What is the distance between the centers of the pulleys to the nearest tenth?

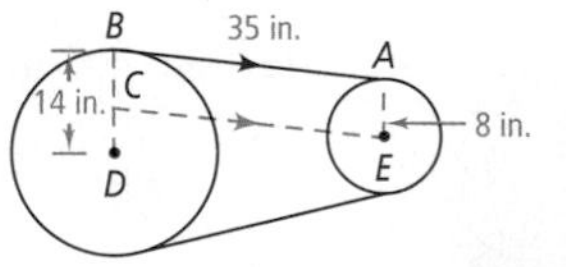

27. $\overline{BD}$ and $\overline{CK}$ at the right are diameters of ⊙A. $\overline{BP}$ and $\overline{QP}$ are tangents to ⊙A. What is $m\angle CDA$?

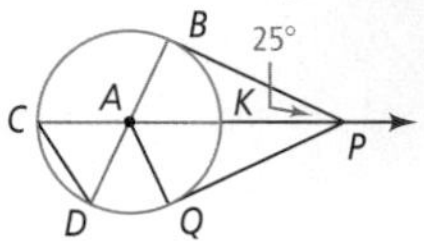

28. Constructions Draw a circle. Label the center T. Locate a point on the circle and label it R. Construct a tangent to ⊙T at R.

29. Coordinate Geometry Graph the equation $x^2 + y^2 = 9$. Then draw a segment from (0, 5) tangent to the circle. Find the length of the segment.

Challenge

30\. Write an indirect proof of Theorem 12-2.
Proof
Given: $\overline{AB} \perp \overline{OP}$ at P.
Prove: $\overline{AB}$ is tangent to $\odot O$.

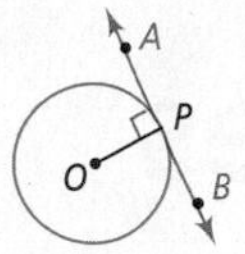

31\. Two circles that have one point in common are *tangent circles*. Given any triangle, explain how to draw three circles that are centered at each vertex of the triangle and are tangent to each other.

Apply What You've Learned

MATHEMATICAL PRACTICES
MP 2, MP 4

Look back at the information given on page 761 about the logo for the showroom display. The diagram of the logo is shown again below.

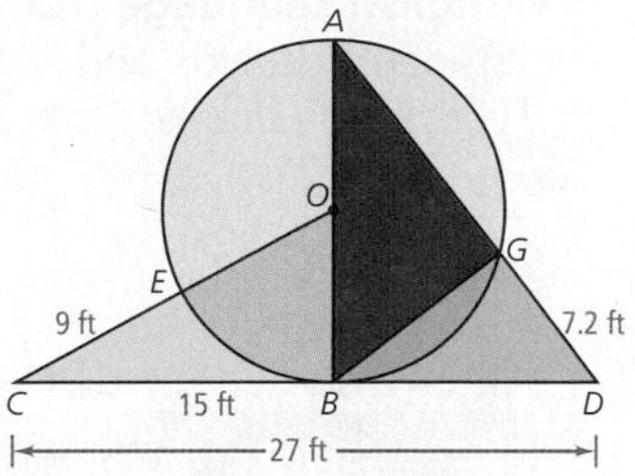

a. What can you conclude about $\angle OBC$? Justify your answer.

b. Write and solve an equation to find the length of $\overline{OE}$.

c. Explain how you know your answer to part (b) is reasonable.

d. How can you use the length you found in part (b) to find the length of one side of the red sail in the logo for the showroom display? What is that length?

30. Assume $\overleftrightarrow{AB}$ is not tangent to $\odot O$. Then either $\overleftrightarrow{AB}$ does not intersect $\odot O$ or $\overleftrightarrow{AB}$ intersects $\odot O$ at two pts. If $\overleftrightarrow{AB}$ does not intersect $\odot O$, then P is not on $\odot O$, which contradicts $\overline{OP}$ being a radius. If $\overleftrightarrow{AB}$ intersects $\odot O$ at two pts., P and Q, then $\overline{OP} \cong \overline{OQ}$ ($\cong$ radii), $\triangle OPQ$ is isosc., and $\angle OPQ \cong \angle OQP$. But $\angle OPQ$ is a rt. $\angle$ since $\overleftrightarrow{AB} \perp \overline{OP}$, and $\triangle OPQ$ has two rt. $\triangle$s. This is a contradiction also, so $\overleftrightarrow{AB}$ is tangent to $\odot O$.

31. At each vertex, let the radius of a circle be the distance from the vertex to either point of tangency of the inscribed circle.

Apply What You've Learned

Here students begin to make sense of the diagram of the logo on page 761. To do so, they focus on the tangent $\overline{CD}$ and on $\triangle OBC$, and find the length of one side of the red triangular sail. Later in the chapter, they will find lengths of the other sides of the red sail.

Mathematical Practices

Students **reason abstractly and quantitatively** to write and solve an equation and interpret the result in terms of the diagram. (MP 2)

ANSWERS

a. $\angle OBC$ is a right angle because a tangent to a circle is perpendicular to the radius at the point of tangency.

b. $(9 + x)^2 = x^2 + 15^2$, where x is the length of $\overline{OE}$; $OE = 8$ ft

c. Sample: The answer is reasonable because it results in a right triangle with legs of lengths 8 ft and 15 ft, and a hypotenuse of length 17 ft. The hypotenuse is the longest side of the right triangle, and the sum of the lengths of any two sides of the triangle is greater than the length of the third side.

d. The length of $\overline{OE}$ (8 ft) is the radius of the circle, so OA and OB are also 8 ft. You can find the length of side $\overline{AB}$ in $\triangle AGB$ as the sum of OA and OB (16 ft).

12-1 Lesson Resources

Differentiated Remediation

Instructional Support

Geometry Companion

Students can use the **Geometry Companion** worktext (4 pages) . . .

- New Vocabulary
- Key Concepts
- Got It for each Problem
- Lesson Check

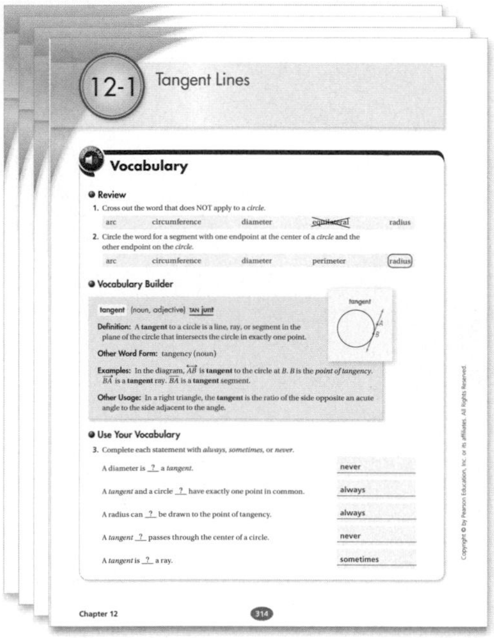

12-1 Tangent Lines

Vocabulary

Review

1. Cross out the word that does NOT apply to a *circle*.

arc circumference diameter ~~equilateral~~ radius

2. Circle the word for a segment with one endpoint at the center of a *circle* and the other endpoint on the *circle*.

arc circumference diameter perimeter (radius)

Vocabulary Builder

tangent (noun, adjective) TAN junt

Definition: A **tangent** to a circle is a line, ray, or segment in the plane of the circle that intersects the circle in exactly one point.

Other Word Form: tangency (noun)

Examples: In the diagram, $\overleftrightarrow{AB}$ is **tangent** to the circle at *B*. *B* is the *point of tangency*. $\overrightarrow{BA}$ is a **tangent** ray. $\overline{BA}$ is a **tangent** segment.

Other Usage: In a right triangle, the **tangent** is the ratio of the side opposite an acute angle to the side adjacent to the angle.

Use Your Vocabulary

3. Complete each statement with *always*, *sometimes*, or *never*.

A diameter is _?_ a *tangent*. never

A *tangent* and a circle _?_ have exactly one point in common. always

A radius can _?_ be drawn to the point of tangency. always

A *tangent* _?_ passes through the center of a circle. never

A *tangent* is _?_ a ray. sometimes

Chapter 12 314

ELL Support

Focus on Language Project the lesson on the board and read Theorem 12-1 as you point to each word. Invite students to define key words such as *tangent* and *perpendicular*. Model an example of a line that is tangent to a circle. Trace the tangent line, the point of tangency, and the angles formed by the perpendicular lines as you restate the theorem and identify each part. Now have students rewrite the theorem in their own words and draw their own examples. Invite students to share their work. Repeat with Theorem 12-2.

5 Assess & Remediate

Lesson Quiz

1. **Do you UNDERSTAND?** $\overline{AD}$ and $\overline{AG}$ are tangent to ⊙*O*. What is the value of *x*?

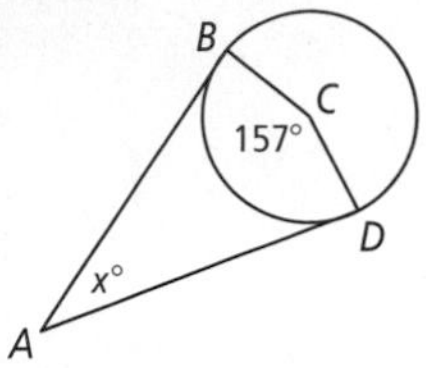

2. What is the radius of ⊙*F*?

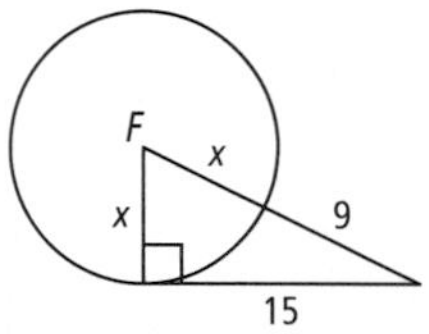

3. $\overline{FT}$ is tangent to ⊙*P* at *T*. What is *PT*?

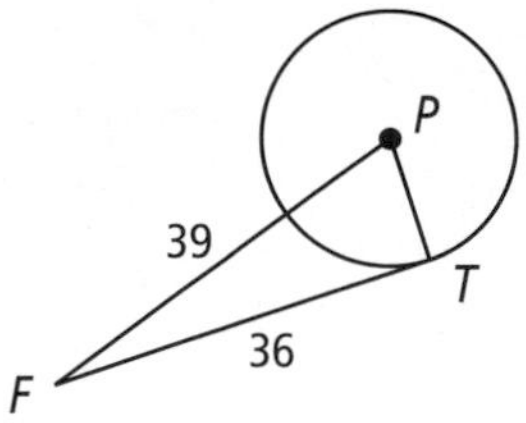

ANSWERS TO LESSON QUIZ

1. 23
2. 8
3. 15

PRESCRIPTION FOR REMEDIATION

Use the student work on the Lesson Quiz to prescribe a differentiated review assignment.

Points	Differentiated Remediation
0–1	Intervention
2	On-level
3	Extension

PowerGeometry.com

5 Assess & Remediate

Assign the Lesson Quiz. Appropriate intervention, practice, or enrichment is automatically generated based on student performance.

Intervention

- **Reteaching** (2 pages) Provides reteaching and practice exercises for the key lesson concepts. Use with struggling students or absent students.
- **English Language Learner Support** Helps students develop and reinforce mathematical vocabulary and key concepts.

All-in-One Resources/Online

Reteaching

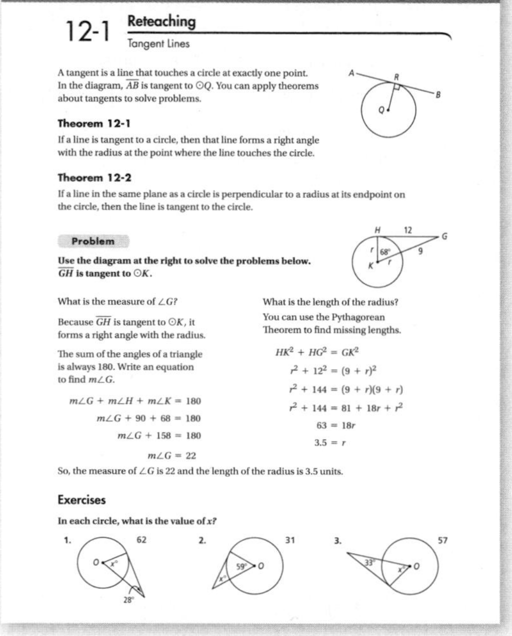

12-1 Reteaching

Tangent Lines

A tangent is a line that touches a circle at exactly one point. In the diagram, $\overline{AB}$ is tangent to ⊙*Q*. You can apply theorems about tangents to solve problems.

Theorem 12-1

If a line is tangent to a circle, then that line forms a right angle with the radius at the point where the line touches the circle.

Theorem 12-2

If a line in the same plane as a circle is perpendicular to a radius at its endpoint on the circle, then the line is tangent to the circle.

Problem

Use the diagram at the right to solve the problems below. $\overline{GH}$ is tangent to ⊙*K*.

What is the measure of ∠*G*?

Because $\overline{GH}$ is tangent to ⊙*K*, it forms a right angle with the radius.

The sum of the angles of a triangle is always 180. Write an equation to find $m\angle G$.

$$m\angle G + m\angle H + m\angle K = 180$$
$$m\angle G + 90 + 68 = 180$$
$$m\angle G + 158 = 180$$
$$m\angle G = 22$$

What is the length of the radius?

You can use the Pythagorean Theorem to find missing lengths.

$$HK^2 + HG^2 = GK^2$$
$$r^2 + 12^2 = (9 + r)^2$$
$$r^2 + 144 = (9 + r)(9 + r)$$
$$r^2 + 144 = 81 + 18r + r^2$$
$$63 = 18r$$
$$3.5 = r$$

So, the measure of ∠*G* is 22 and the length of the radius is 3.5 units.

Exercises

In each circle, what is the value of *x*?

1. 62
2. 31
3. 57

All-in-One Resources/Online

English Language Learner Support

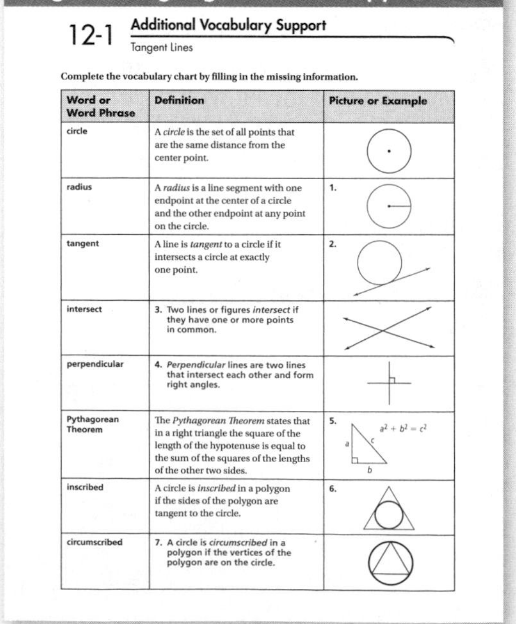

12-1 Additional Vocabulary Support

Tangent Lines

Complete the vocabulary chart by filling in the missing information.

Word or Word Phrase	Definition	Picture or Example
circle	A *circle* is the set of all points that are the same distance from the center point.	
radius	A *radius* is a line segment with one endpoint at the center of a circle and the other endpoint at any point on the circle.	1.
tangent	A line is *tangent* to a circle if it intersects a circle at exactly one point.	2.
intersect	3. Two lines or figures *intersect* if they have one or more points in common.	
perpendicular	4. *Perpendicular* lines are two lines that intersect each other and form right angles.	
Pythagorean Theorem	The *Pythagorean Theorem* states that in a right triangle the square of the length of the hypotenuse is equal to the sum of the squares of the lengths of the other two sides.	5. $a^2 + b^2 = c^2$
inscribed	A circle is *inscribed* in a polygon if the sides of the polygon are tangent to the circle.	6.
circumscribed	7. A circle is *circumscribed* in a polygon if the vertices of the polygon are on the circle.	

Differentiated Remediation *continued*

On-Level

- **Practice** (2 pages) Provides extra practice for each lesson. For simpler practice exercises, use the Form K Practice pages found in the All-in-One Teaching Resources and online.
- **Think About a Plan** Helps students develop specific problem-solving skills and strategies by providing scaffolded guiding questions.
- **Standardized Test Prep** Focuses on all major exercises, all major question types, and helps students prepare for the high-stakes assessments.

Extension

- **Enrichment** Provides students with interesting problems and activities that extend the concepts of the lesson.
- **Activities, Games, and Puzzles** Worksheets that can be used for concepts development, enrichment, and for fun!

Practice and Problem Solving Wkbk/ All-in-One Resources/Online

Practice page 1

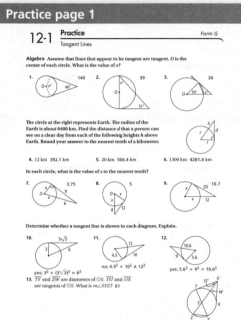

12-1 **Practice** — Tangent Lines — *Form G*

Algebra Assume that lines that appear to be tangent are tangent. *O* is the center of each circle. What is the value of *x*?

1. 140 2. 39 3. 20

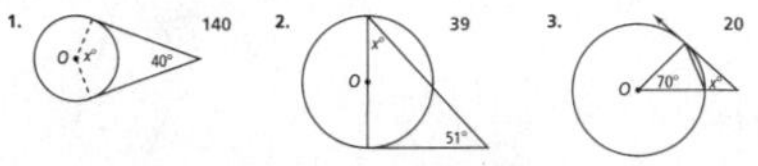

The circle at the right represents Earth. The radius of the Earth is about 6400 km. Find the distance *d* that a person can see on a clear day from each of the following heights *h* above Earth. Round your answer to the nearest tenth of a kilometer.

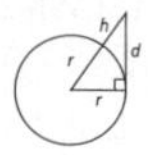

4. 12 km 392.1 km 5. 20 km 506.4 km 6. 1300 km 4281.4 km

In each circle, what is the value of *x* to the nearest tenth?

7. 3.75 8. 5 9. 10.7

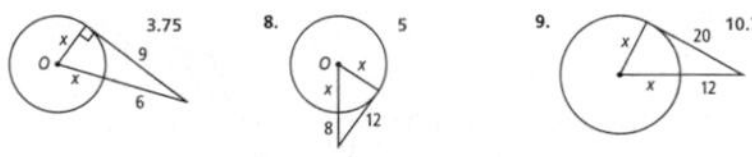

Determine whether a tangent line is shown in each diagram. Explain.

10. 11. 12.

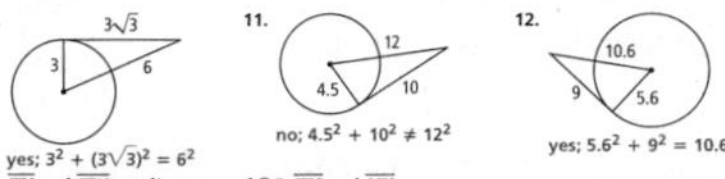

10. yes; $3^2 + (3\sqrt{3})^2 = 6^2$
11. no; $4.5^2 + 10^2 \neq 12^2$
12. yes; $5.6^2 + 9^2 = 10.6^2$

13. $\overline{TY}$ and $\overline{ZW}$ are diameters of ⊙*S*. $\overline{TU}$ and $\overline{UX}$ are tangents of ⊙*S*. What is $m\angle SYZ$? 61

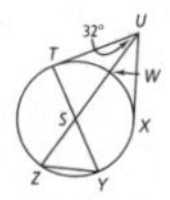

Practice and Problem Solving Wkbk/ All-in-One Resources/Online

Practice page 2

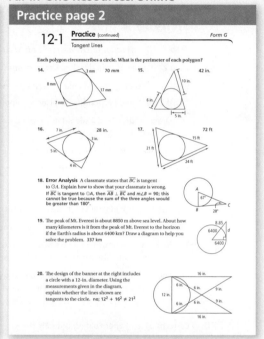

12-1 **Practice** (continued) — Tangent Lines — *Form G*

Each polygon circumscribes a circle. What is the perimeter of each polygon?

14. 70 mm 15. 42 in.

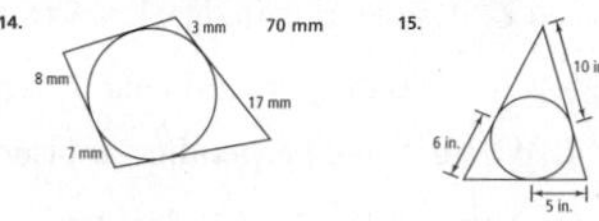

16. 28 in. 17. 72 ft

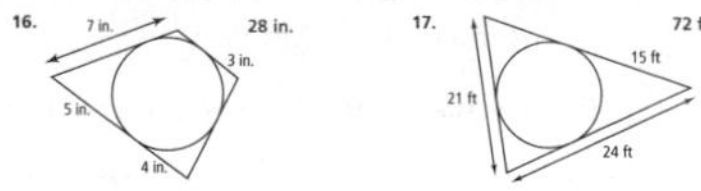

18. **Error Analysis** A classmate states that $\overline{BC}$ is tangent to ⊙*A*. Explain how to show that your classmate is wrong. If $\overline{BC}$ is tangent to ⊙*A*, then $\overline{AB} \perp \overline{BC}$ and $m\angle B = 90$; this cannot be true because the sum of the three angles would be greater than 180°.

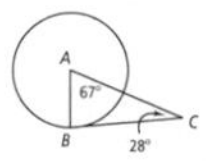

19. The peak of Mt. Everest is about 8850 m above sea level. About how many kilometers is it from the peak of Mt. Everest to the horizon if the Earth's radius is about 6400 km? Draw a diagram to help you solve the problem. 337 km

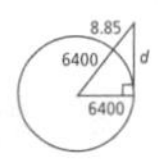

20. The design of the banner at the right includes a circle with a 12-in. diameter. Using the measurements given in the diagram, explain whether the lines shown are tangents to the circle. no; $12^2 + 16^2 \neq 21^2$

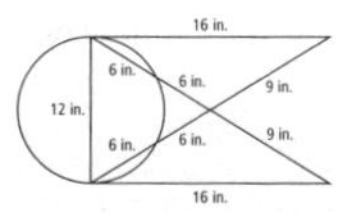

All-in-One Resources/Online

Enrichment

12-1 **Enrichment** — Tangent Lines

Inscribed Circles and Right Triangles

The following theorem is about a relationship between the radius of the inscribed circle of a right triangle and the lengths of the triangle's sides.

Complete the proof.

Given: Right △*ABC* with right angle at *C*. ⊙*O* is inscribed in △*ABC*. *M*, *N*, and *P* are the points of tangency. Radius $\overline{ON}$ is drawn.

Prove: $2ON = AC + BC - AB$

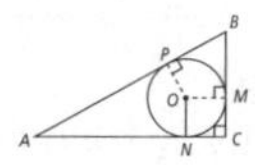

Statements	Reasons
1) In right △*ABC*, ∠*C* is a right angle. ⊙*O* is inscribed in *ABC*. *M*, *N*, and *P* are points of tangency. Radius $\overline{ON}$ is drawn.	1) _?_ Given
2) Draw radii $\overline{OM}$ and $\overline{OP}$.	2) _?_ Two points determine a line segment.
3) $\overline{BP} \cong \overline{BM}$; $\overline{CM} \cong \overline{CN}$; $\overline{AP} \cong \overline{AN}$	3) _?_ Theorem 12-3 (Two tangents drawn to a circle from a point outside the circle are congruent.)
4) $\overline{ON} \cong \overline{OM}$	4) _?_ Radii of a circle are congruent.
5) ∠*OMC* is a right angle.	5) _?_ Theorem 12-1 (A radius and a tangent line drawn to the same point of contact form a right angle.)
6) ∠*ONC* is a right angle.	6) _?_ Theorem 12-1 (A radius and a tangent line drawn to the same point of contact form a right angle.)
7) Quad. *OMCN* is a square.	7) _?_ Definition of a square
8) $\overline{ON} \cong \overline{OM} \cong \overline{CM} \cong \overline{CN}$	8) _?_ The sides of a square are congruent.
9) $BP + CM = BM + CN$ $AP + CM = AN + CN$	9) _?_ Addition Property
10) $AN + CN = AC$ $BM + CM = BC$ $BP + AP = AB$	10) _?_ Segment Addition Postulate
11) $BP + CM = BC$	11) _?_ Substitution Property
12) $(BP + AP) + 2CM = BC + AC$	12) _?_ Addition Property and Substitution Property
13) $AB + 2ON = BC + AC$	13) _?_ Substitution Property
14) $2ON = AC + BC - AB$	14) _?_ Subtraction Property

Practice and Problem Solving Wkbk/ All-in-One Resources/Online

Think About a Plan

12-1 **Think About a Plan** — Tangent Lines

a. A belt fits snugly around the two circular pulleys. $\overline{CE}$ is an auxiliary line from *E* to $\overline{BD}$. $\overline{CE} \parallel \overline{AB}$. What type of quadrilateral is *ABCE*? Explain.

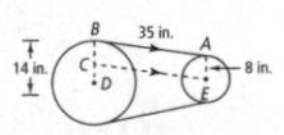

b. What is the length of $\overline{CE}$?

c. What is the distance between the centers of the pulleys to the nearest tenth?

1. What is the definition of a tangent line? a line that touches a circle at only one point
2. What is the relationship between a line tangent to a circle and the radius at the point of tangency (Theorem 12-1)? They are perpendicular.
3. Where is the point of tangency for $\overline{AB}$ on ⊙*D*? On ⊙*E*? B; A
4. What is the measure of ∠*CBA*? What is the measure of ∠*BAE*? Explain. 90°; 90°; they are formed by perpendicular line segments
5. How can you use parallel lines to find the measure of ∠*CEA*? Because $\overline{CE} \parallel \overline{AB}$, ∠*BAE* and ∠*CEA* are supplementary angles, so $m\angle CEA = 90°$.
6. How can you use parallel lines or the Polygon Angle-Sum Theorem to find the measure of ∠*BCE*? Because $\overline{CE} \parallel \overline{AB}$, ∠*CBA* and ∠*BCE* are supplementary angles, so $m\angle BCE = 90°$. Or, ∠*BCE* is the fourth angle in a quadrilateral in which the other angles sum to 270°, so its measure is 90°.
7. What type of quadrilateral has four right angles? a rectangle
8. What is the length of $\overline{BA}$? 35 in.
9. What is the length of $\overline{CE}$? 35 in.
10. What are the center points of the pulleys? D; E
11. How can you use the Segment Addition Postulate to find $\overline{CD}$? $BD - BC = CD$, $BD = 14$ in., and $BC = 8$ in.
12. What is the measure of $\overline{CD}$? 6 in.
13. How can you use the Pythagorean Theorem to find the length of $\overline{DE}$? $a^2 + b^2 = c^2$; if $CD = a$, $CE = b$, and $DE = c$, then $DE = \sqrt{(6^2) + (35^2)} = \sqrt{1261} \approx 35.5$ in.

Practice and Problem Solving Wkbk/ All-in-One Resources/Online

Standardized Test Prep

12-1 **Standardized Test Prep** — Tangent Lines

Multiple Choice

For Exercises 1–5, choose the correct letter.

1. $\overline{AB}$ and $\overline{BC}$ are tangents to ⊙*P*. What is the value of *x*? B
   (A) 73 (B) 107 (C) 117 (D) 146

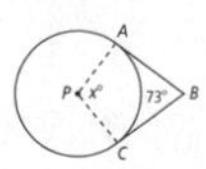

2. Earth's radius is about 4000 mi. To the nearest mile, what is the distance a person can see on a clear day from an airplane 5 mi above Earth? G
   (F) 63 mi (G) 200 mi (H) 4000 mi (I) 5660 mi

3. $\overline{YZ}$ is a tangent to ⊙*X*, and *X* is the center of the circle. What is the length of the radius of the circle? B
   (A) 4 (B) 6 (C) 12 (D) 12.8

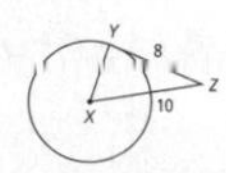

4. The radius of ⊙*G* is 4 cm. Which is a tangent of ⊙*G*? I
   (F) $\overline{AB}$ (G) $\overline{CD}$ (H) $\overline{BF}$ (I) $\overline{FE}$

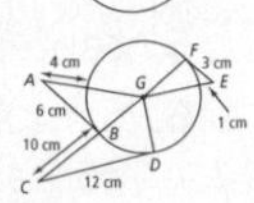

5. ⊙*A* is inscribed in a quadrilateral. What is the perimeter of the quadrilateral? B
   (A) 25 mm (B) 50 mm (C) 60 mm (D) 150 mm

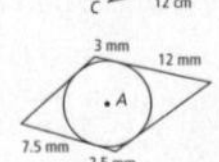

Short Response

6. **Given:** $\overline{GI}$ is a tangent to ⊙*J*.
   Prove: △*FGH* ≅ △*FIH* [2] Statements: 1) $\overline{GI}$ is a tangent to ⊙*J*; 2) $\overline{FH} \perp \overline{GI}$; 3) ∠*FHG* and ∠*FHI* are right △; 4) ∠*FHG* ≅ ∠*FHI*; 5) $\overline{FH} \cong \overline{FH}$; 6) $m\angle GFH = 42$, $m\angle FIH = 48$; 7) $m\angle HFI = 42$; 8) ∠*HFI* ≅ ∠*HFG*; 9) △*FGH* ≅ △*FIH*; Reasons: 1) Given; 2) Thm. 12-1; 3) Def. of ⊥; 4) Def. of ≅; 5) Refl. Prop.; 6) Given; 7) Triangle Angle-Sum Thm.; 8) Def. of ≅; 9) ASA Post. [1] proof missing steps [0] incorrect or missing proof

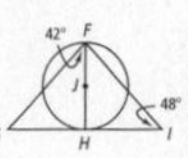

Online Teacher Resource Center

Activities, Games, and Puzzles

12-1 **Activity: Tangent Segments and Angle Measure** — Tangent Lines

Materials
- Compass
- Ruler
- Protractor

Work with a partner, and follow these steps to create angles whose measures may have a special relationship.

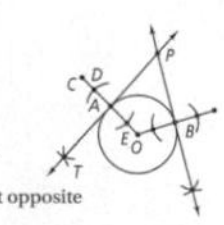

Step 1 Using a compass, draw a large circle, and mark its center *O*.

Step 2 Draw two radii, $\overline{OA}$ and $\overline{OB}$. Make sure that the radii are not opposite segments forming a diameter.

Step 3 Construct a tangent line perpendicular to radius $\overline{OA}$ at *A*, as follows:
- a. Extend $\overline{OA}$ through *A* to a point *C*.
- b. Place the tip of the compass on point *A*. With a small radius on the compass, draw an arc on one side of $\overline{OC}$ that intersects $\overline{OC}$ in two points, *D* and *E*.
- c. Set the compass with a radius that is greater than *AD*.
- d. Put the point of the compass on *D*, and draw a small arc on the same side of $\overline{OC}$ that contains the larger arc drawn in step 3b.
- e. Keeping the same compass setting, put the point of the compass on *E*, and draw a small arc to intersect the arc drawn in step 3d. Call this intersection point *T*.
- f. Draw $\overline{TA}$, which is perpendicular to radius $\overline{OA}$ at *A* and therefore tangent to circle *O* at *A*.

Step 4 Construct a tangent line perpendicular to radius $\overline{OB}$ at *B* by following the procedure given in steps 3a–f.

Step 5 Extend the tangent lines constructed in steps 3 and 4 until they intersect. Call this point of intersection *P*.

1. Use a protractor to measure ∠*APB*, the angle formed by the two tangent lines. Check students' work.
2. Use a protractor to measure ∠*AOB*, the angle formed by the two radii. Check students' work.
3. What special relationship appears to be true about these angles? The angles appear to be supplementary.
4. Repeat steps 1–5, using a different circle and radii that form an angle of a different measure. Use a protractor to measure the angle formed by the two tangents and the angle formed by the two radii. Does the same relationship hold? Yes; the angle formed by the two tangents and the angle formed by the two radii are supplementary.

Guided Instruction

PURPOSE To use paper-folding activities to explore properties of chords

PROCESS Students will

- form chords in a circle and make conjectures.
- form the perpendicular bisector of a chord and make conjectures.

DISCUSS Explain the conditions that a segment must meet to be a chord of a circle.

Activity 1

This Activity focuses on the relationships of the lengths of chords that are equidistant from the center of a circle.

Q What is the longest chord of a circle? **[diameter]**

Q What can you conclude about the distance from the center of a circle to chords that are congruent? **[The distance from the center is the same.]**

Q Suppose you have a circle with two chords that are not congruent. Is the shorter chord farther from the center or closer to the center than the longer chord? **[farther]**

Activity 2

This Activity focuses on the relationship between perpendicular bisectors to chords.

Q Where do perpendicular bisectors intersect the intercepted arcs? **[in the middle]**

Q What can you say about perpendicular bisectors of parallel chords? **[They coincide.]**

Q Because all perpendicular bisectors of chords pass through the center of the circle, what is another term you can use to describe each perpendicular bisector? **[diameter]**

Mathematical Practices This Concept Byte supports students in becoming proficient in using appropriate tools, Mathematical Practice 5.

Concept Byte

Use With Lesson 12-2

ACTIVITY

Paper Folding With Circles

Common Core State Standards

Prepares for G-C.A.2 Identify and describe relationships among inscribed angles, radii, and chords . . .

MP 5

A *chord* is a segment with endpoints on a circle. In these activities, you will explore some of the properties of chords.

Activity 1

Step 1 Use a compass. Draw a circle on tracing paper.

Step 2 Use a straightedge. Draw two radii.

Step 3 Set your compass to a distance shorter than the radii. Place its point at the center of the circle. Mark two congruent segments, one on each radius.

Step 4 Fold a line perpendicular to each radius at the point marked on the radius.

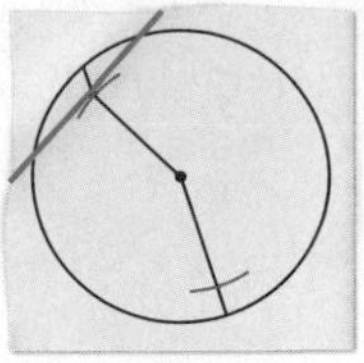

1. How do you measure the distance between a point and a line?
2. Each perpendicular contains a chord. Compare the lengths of the chords.
3. **Make a Conjecture** What is the relationship among the lengths of the chords that are equidistant from the center of a circle?

Activity 2

Step 1 Use a compass. Draw a circle on tracing paper.

Step 2 Use a straightedge. Draw two chords that are not diameters.

Step 3 Fold the perpendicular bisector for each chord.

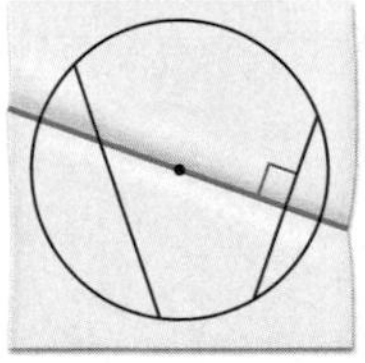

4. Where do the perpendicular bisectors appear to intersect?
5. Draw a third chord and fold its perpendicular bisector. Where does it appear to intersect the other two?
6. **Make a Conjecture** What is true about the perpendicular bisector of a chord?

Exercises

7. Write a proof of your conjecture from Exercise 3 or give a counterexample.
8. What theorem provides a quick proof of your conjecture from Exercise 6?
9. **Make a Conjecture** Suppose two chords have different lengths. How do their distances from the center of the circle compare?
10. You are building a circular patio table. You have to drill a hole through the center of the tabletop for an umbrella. How can you find the center?

Answers

Activity 1

1. Measure the ⊥ segment from the pt. to the line.
2. The lengths are =.
3. Sample: Chords equidistant from the center of a circle are ≅.

Activity 2

4. center of the circle
5. center of the circle
6. Sample: The ⊥ bis. of a chord contains the circle's center.

Exercises

7. 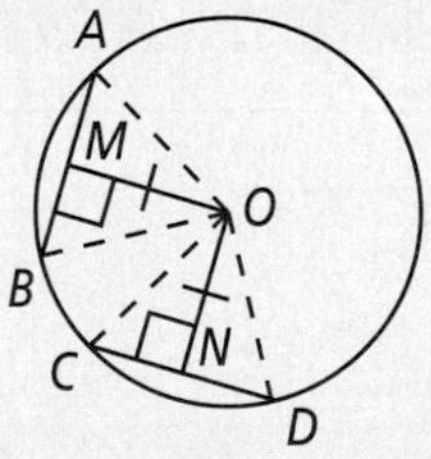

Given: $\overline{OM} \perp \overline{AB}$, $\overline{ON} \perp \overline{CD}$, $OM = ON$

Prove: $\overline{AB} \cong \overline{CD}$

Proof: $\triangle AOM \cong \triangle BOM \cong \triangle CON \cong \triangle DON$ by HL. Therefore, $AM = BM = CN = DN$. $AB = AM + MB = CN + ND = CD$, so $\overline{AB} \cong \overline{CD}$.

8. Converse of ⊥ Bis. Theorem
9. Sample: The shorter (longer) chord is farther from (closer to) the center.
10. Answers may vary. Sample: Use the carpenter's square to locate two chords and the ⊥ bis. of each. The two ⊥ bis. will intersect at the center of the circle.

12-2 Chords and Arcs

Common Core State Standards
G-C.A.2 Identify and describe relationships among inscribed angles, radii, and chords.
MP 1, MP 3

Objectives To use congruent chords, arcs, and central angles
To use perpendicular bisectors to chords

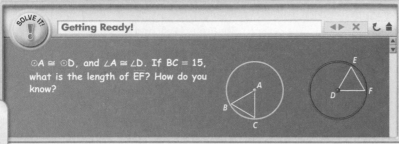

Getting Ready!

$\odot A \cong \odot D$, and $\angle A \cong \angle D$. If $BC = 15$, what is the length of EF? How do you know?

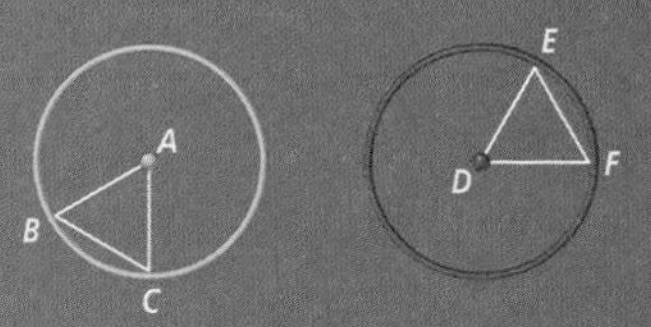

How can you use congruent triangles to help with this one?

MATHEMATICAL PRACTICES In the Solve It, you found the leng th of a **chord,** which is a segment whose endpoints are on a circle. The diagram shows the chord $\overline{PQ}$ and its related arc, $\widehat{PQ}$.

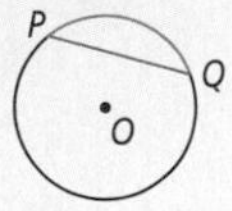

Essential Understanding You can use information about congruent parts of a circle (or congruent circles) to find information about other parts of the circle (or circles).

The following theorems and their converses confirm that if you know that chords, arcs, or central angles in a circle are congruent, then you know the other two parts are congruent.

Lesson Vocabulary
chord

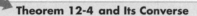

take note **Theorem 12-4 and Its Converse**

Theorem
Within a circle or in congruent circles, congruent central angles have congruent arcs.

Converse
Within a circle or in congruent circles, congruent arcs have congruent central angles.

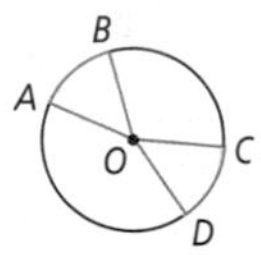

If $\angle AOB \cong \angle COD$, then $\widehat{AB} \cong \widehat{CD}$.
If $\widehat{AB} \cong \widehat{CD}$, then $\angle AOB \cong \angle COD$.

You will prove Theorem 12-4 and its converse in Exercises 19 and 35.

1 Interactive Learning

Solve It!

PURPOSE To show that in congruent circles chords formed by congruent central angles are congruent
PROCESS Students use congruent triangles to show that the chords are congruent.

FACILITATE

Q What does it mean for two circles to be congruent? **[Their radii are equal.]**

Q Which four segments in the diagram are congruent? **[$\overline{AB} \cong \overline{AC} \cong \overline{DE} \cong \overline{DF}$]**

Q How can you prove $\triangle ABC \cong \triangle DEF$? **[The triangles are congruent by SAS.]**

Q What theorem allows you to conclude that $\overline{BC} \cong \overline{EF}$? **[Corresponding Parts of Congruent Triangles]**

ANSWER See Solve It in Answers on next page.
CONNECT THE MATH In the Solve It, students explored congruent circles and segments to prove triangles were congruent. In the lesson, students will use congruent triangles to identify congruent chords.

2 Guided Instruction

Take Note

Have students list the congruent parts in the diagram. They should list congruent radii, central angles, and arcs.

12-2 Preparing to Teach

BIG ideas **Reasoning and Proof**
Measurement

ESSENTIAL UNDERSTANDING

- Information about congruent parts of a circle (or congruent circles) can be used to find information about other parts of the circle (or circles).

Math Background

In this lesson, students will broaden their understanding of special segments in circles. They will learn that congruent chords and arcs are formed by congruent central angles. Congruent chords are also equidistant from the center of the circle. Additionally, students will learn that a diameter that is perpendicular to a chord bisects the chord and its related arc. These properties of segments in circles can be used to determine characteristics of circles.

Paper folding activities offer students a good way to develop key concepts related to central angles, chords, and arcs.

Mathematical Practices

Construct viable arguments and critique the reasoning of others. With a knowledge of proving triangles congruent, students will prove that chords that are equidistant from the center of a circle are congruent.

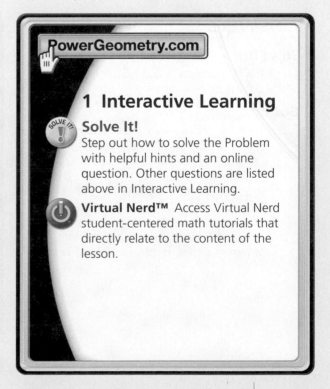

PowerGeometry.com

1 Interactive Learning

Solve It!
Step out how to solve the Problem with helpful hints and an online question. Other questions are listed above in Interactive Learning.

Virtual Nerd™ Access Virtual Nerd student-centered math tutorials that directly relate to the content of the lesson.

Take Note
Discuss and summarize the information in Theorems 12-5 and 12-6. Review the definition of a central angle and arc. Be sure that students can explain the connection between the measures of these figures.

Problem 1

Q Which arcs are related to $\overline{BC}$ and $\overline{DF}$? **[$\widehat{BC}$ and $\widehat{DF}$]**

Q Which central angles are related to $\overline{BC}$ and $\overline{DF}$? **[$\angle O$ and $\angle P$]**

Got It?
ERROR PREVENTION

Have students identify the central angles and chords that are related to the given arcs. Challenge students to prove the chords and angles congruent using congruent triangles instead of the theorems.

Take Note
Review the definition of the distance from a point to a line. Emphasize that the segments that represent the distances must be perpendicular to the chords.

take note

Theorem 12-5 and Its Converse

Theorem
Within a circle or in congruent circles, congruent central angles have congruent chords.

Converse
Within a circle or in congruent circles, congruent chords have congruent central angles.

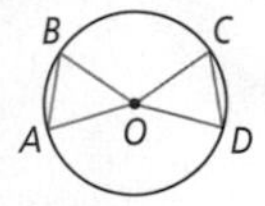

If $\angle AOB \cong \angle COD$, then $\overline{AB} \cong \overline{CD}$.
If $\overline{AB} \cong \overline{CD}$, then $\angle AOB \cong \angle COD$.

You will prove Theorem 12-5 and its converse in Exercises 20 and 36.

Theorem 12-6 and Its Converse

Theorem
Within a circle or in congruent circles, congruent chords have congruent arcs.

Converse
Within a circle or in congruent circles, congruent arcs have congruent chords.

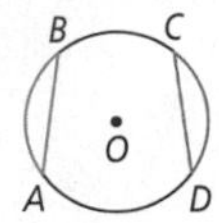

If $\overline{AB} \cong \overline{CD}$, then $\widehat{AB} \cong \widehat{CD}$.
If $\widehat{AB} \cong \widehat{CD}$, then $\overline{AB} \cong \overline{CD}$.

You will prove Theorem 12-6 and its converse in Exercises 21 and 37.

Problem 1 Using Congruent Chords

Think
Why is it important that the circles are congruent?
Two circles may have central angles with congruent chords, but the central angles will not be congruent unless the circles are congruent.

In the diagram, $\odot O \cong \odot P$. Given that $\overline{BC} \cong \overline{DF}$, what can you conclude?

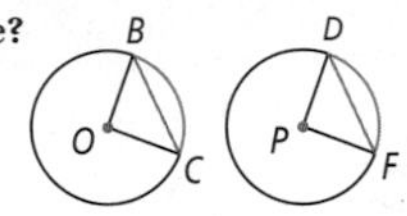

$\angle O \cong \angle P$ because, within congruent circles, congruent chords have congruent central angles (conv. of Thm. 12-5). $\widehat{BC} \cong \widehat{DF}$ because, within congruent circles, congruent chords have congruent arcs (Thm. 12-6).

Got It? 1. Reasoning Use the diagram in Problem 1. Suppose you are given $\odot O \cong \odot P$ and $\angle OBC \cong \angle PDF$. How can you show $\angle O \cong \angle P$? From this, what else can you conclude?

take note

Theorem 12-7 and Its Converse

Theorem
Within a circle or in congruent circles, chords equidistant from the center or centers are congruent.

Converse
Within a circle or in congruent circles, congruent chords are equidistant from the center (or centers).

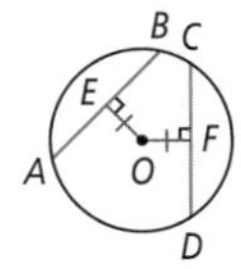

If $OE = OF$, then $\overline{AB} \cong \overline{CD}$.
If $\overline{AB} \cong \overline{CD}$, then $OE = OF$.

You will prove the converse of Theorem 12-7 in Exercise 38.

Answers

Solve It!
15; $\triangle ABC \cong \triangle DEF$ by SAS, so $\overline{EF} \cong \overline{BC}$ because corresp. parts of $\cong$ △ are $\cong$.

Got It?
1. Since the circles are $\cong$, their radii are $=$ and △ BOC and DPF are isosceles. So $\overline{OB} \cong \overline{OC} \cong \overline{PD} \cong \overline{DF}$. Since $\angle B \cong \angle D$ and the △ are isosceles, $\angle B \cong \angle C \cong \angle D \cong \angle F$. So $\triangle BOC \cong \triangle DPF$ by AAS. So $\angle O \cong \angle P$. Therefore, $\overline{BC} \cong \overline{DF}$ (either by corresp. parts of $\cong$ △ are $\cong$ or by within $\cong$ circles, $\cong$ central ∠s have $\cong$ chords) and $\widehat{BC} \cong \widehat{DF}$ (within $\cong$ circles, $\cong$ central ∠s have $\cong$ arcs).

PowerGeometry.com

2 Guided Instruction

Each Problem is worked out and supported online.

Problem 1
Using Congruent Chords

Problem 2
Finding the Length of a Chord
Animated

Problem 3
Using Diameters and Chords

Problem 4
Finding Measures in a Circle
Animated

Support in Geometry Companion
- Vocabulary
- Key Concepts
- Got It?

Proof of Theorem 12-7

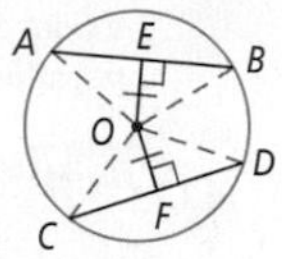

Given: $\odot O$, $\overline{OE} \cong \overline{OF}$, $\overline{OE} \perp \overline{AB}$, $\overline{OF} \perp \overline{CD}$

Prove: $\overline{AB} \cong \overline{CD}$

Statements	Reason
1) $\overline{OA} \cong \overline{OB} \cong \overline{OC} \cong \overline{OD}$	1) Radii of a circle are congruent.
2) $\overline{OE} \cong \overline{OF}$, $\overline{OE} \perp \overline{AB}$, $\overline{OF} \perp \overline{CD}$	2) Given
3) $\angle AEO$ and $\angle CFO$ are right angles.	3) Def. of perpendicular segments
4) $\triangle AEO \cong \triangle CFO$	4) HL Theorem
5) $\angle A \cong \angle C$	5) Corres. parts of $\cong$ $\triangle$s are $\cong$.
6) $\angle B \cong \angle A$, $\angle C \cong \angle D$	6) Isosceles Triangle Theorem
7) $\angle B \cong \angle D$	7) Transitive Property of Congruence
8) $\angle AOB \cong \angle COD$	8) If two $\angle$s of a $\triangle$ are $\cong$ to two $\angle$s of another $\triangle$, then the third $\angle$s are $\cong$.
9) $\overline{AB} \cong \overline{CD}$	9) $\cong$ central angles have $\cong$ chords.

Problem 2 Finding the Length of a Chord

GRIDDED RESPONSE

What is the length of $\overline{RS}$ in $\odot O$?

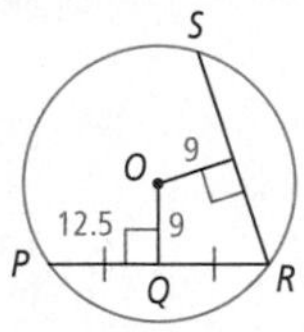

Know
The diagram indicates that $PQ = QR = 12.5$ and $\overline{PR}$ and $\overline{RS}$ are both 9 units from the center.

Need
The length of chord $\overline{RS}$

Plan
$\overline{PR} \cong \overline{RS}$, since they are the same distance from the center of the circle. So finding PR gives the length of $\overline{RS}$.

$PQ = QR = 12.5$ Given in the diagram

$PQ + QR = PR$ Segment Addition Postulate

$12.5 + 12.5 = PR$ Substitute.

$25 = PR$ Add.

$RS = PR$ Chords equidistant from the center of a circle are congruent.

$RS = 25$ Substitute.

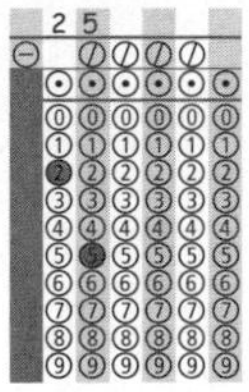

Got It? 2. What is the value of x? Justify your answer.

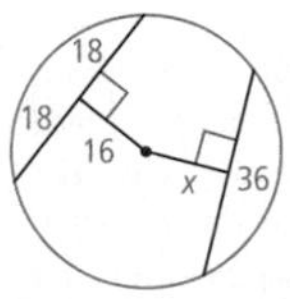

It may benefit students to work backwards through the proof. Ask them to identify a theorem that will help them show the chords are congruent. **[Converse of Theorem 12-5]**

Using the converse of Theorem 12-5, students can identify triangles that they need to prove congruent to show the central angles are congruent.

Problem 2

Q What type of segment is $\overline{OQ}$? **[It is the perpendicular bisector of $\overline{PR}$.]**

Q How is PQ related to PR? **[$PQ + QR = PR$ or $2PQ = PR$]**

Q How are the chords in the circle related? Justify your answer. **[They are congruent because they are equidistant from the center of the circle.]**

Got It? VISUAL LEARNERS

Ask students to describe the relationship between the chords in the circle. Have them identify the theorem that allows them to determine the distance from the center of the circle to the chord.

Got It? (continued)

2. 16; $\cong$ chords are equidistant from the center.

Take Note

Review the logic of these theorems with students. They should identify the congruent triangles created by the endpoints of the chord, the center of the circle, and the midpoint of the segment.

For Theorem 12-8: $\triangle OCD$ is an isosceles triangle, so $\angle C \cong \angle D$. $\triangle COE \cong \triangle DOE$ by AAS. $\overline{CE} \cong \overline{DE}$ because they are corresponding parts. $\angle COE \cong \angle DOE$ because they are corresponding parts. Arcs AC and AD are congruent because they are formed by congruent central angles.

For Theorem 12-9: $\triangle COE \cong \triangle DOE$ by SAS. $\overline{CE} \cong \overline{DE}$ because they are corresponding parts. $\angle COE \cong \angle DOE$ because they are corresponding parts. Arcs AC and AD are congruent because they are formed by congruent central angles.

For Theorem 12-10: Review the properties of perpendicular bisectors and lines. Have students connect the definitions to classify $\overline{AB}$.

The Converse of the Perpendicular Bisector Theorem from Lesson 5-2 has special applications to a circle and its diameters, chords, and arcs.

take note

Theorem 12-8

Theorem	If . . .	Then . . .
In a circle, if a diameter is perpendicular to a chord, then it bisects the chord and its arc.	$\overline{AB}$ is a diameter and $\overline{AB} \perp \overline{CD}$ 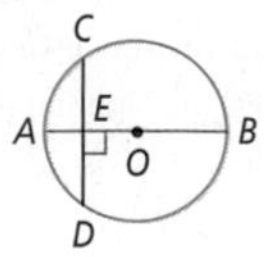	$\overline{CE} \cong \overline{ED}$ and $\overset{\frown}{CA} \cong \overset{\frown}{AD}$ 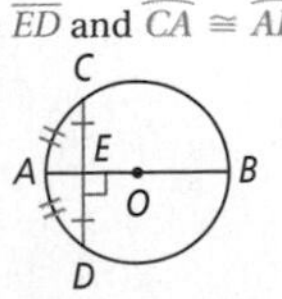

You will prove Theorem 12-8 in Exercise 22.

Theorem 12-9

Theorem	If . . .	Then . . .
In a circle, if a diameter bisects a chord (that is not a diameter), then it is perpendicular to the chord.	$\overline{AB}$ is a diameter and $\overline{CE} \cong \overline{ED}$ 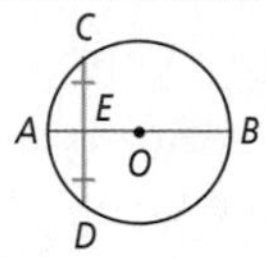	$\overline{AB} \perp \overline{CD}$ 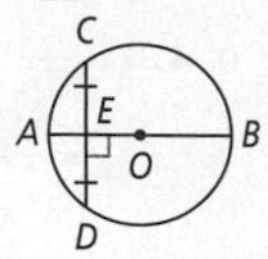

Theorem 12-10

Theorem	If . . .	Then . . .
In a circle, the perpendicular bisector of a chord contains the center of the circle.	$\overline{AB}$ is the perpendicular bisector of chord $\overline{CD}$ 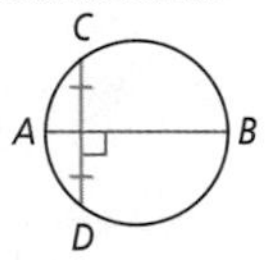	$\overline{AB}$ contains the center of $\odot O$ 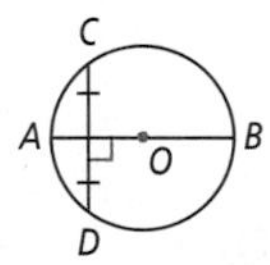

You will prove Theorem 12-10 in Exercise 33.

Proof Proof of Theorem 12-9

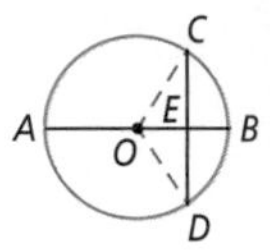

Given: $\odot O$ with diameter $\overline{AB}$ bisecting $\overline{CD}$ at E

Prove: $\overline{AB} \perp \overline{CD}$

Proof: $OC = OD$ because the radii of a circle are congruent. $CE = ED$ by the definition of bisect. Thus, O and E are both equidistant from C and D. By the Converse of the Perpendicular Bisector Theorem, both O and E are on the perpendicular bisector of $\overline{CD}$. Two points determine one line or segment, so $\overline{OE}$ is the perpendicular bisector of $\overline{CD}$. Since $\overline{OE}$ is part of $\overline{AB}$, $\overline{AB} \perp \overline{CD}$.

Additional Problems

1. In the diagram, $\overset{\frown}{VT} \cong \overset{\frown}{RP}$. What can you conclude?

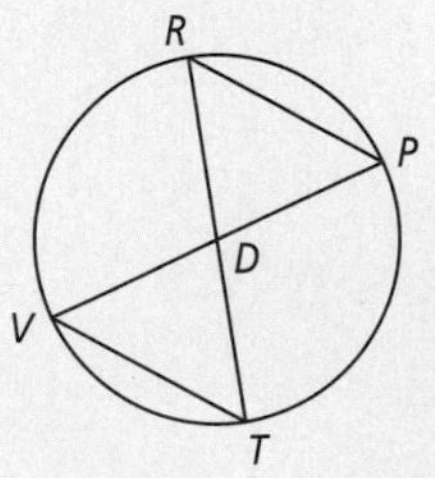

ANSWER $\overline{VT} \cong \overline{RP}$, $\angle VDT \cong \angle RDP$

2. What is the value of a in the circle?

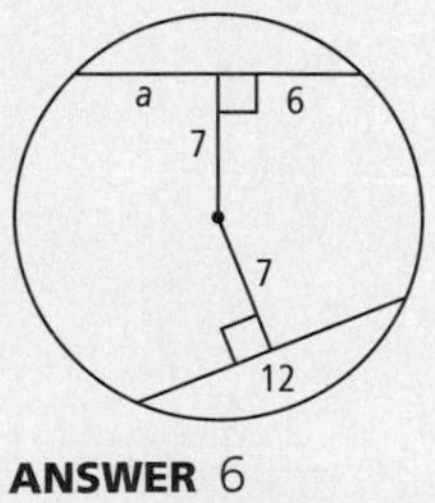

ANSWER 6

3. Given an arc like the one below, how can you find the center of the circle that contains the arc?

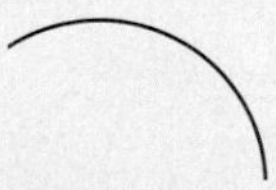

ANSWER

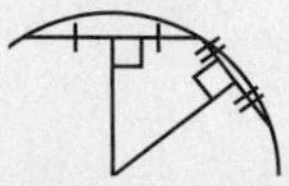

Draw two chords. Then construct the perpendicular bisector to each chord. By Theorem 12-10, the perpendicular bisectors are diameters of the circle that contains the arc. Therefore they intersect at the center of the circle.

4. What is the missing length to the nearest tenth?

a.

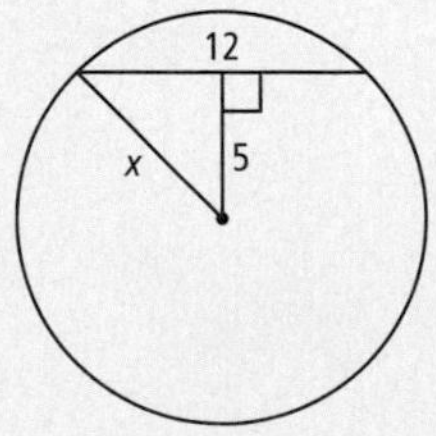

b.

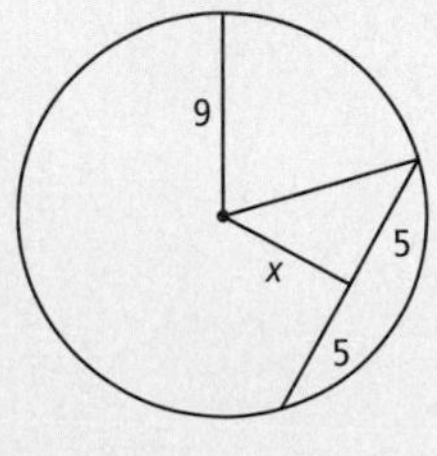

ANSWER a. 7.8 **b.** 7.5

Problem 3 Using Diameters and Chords

Archaeology **An archaeologist found pieces of a jar. She wants to find the radius of the rim of the jar to help guide her as she reassembles the pieces. What is the radius of the rim?**

Think

How does the construction help find the center?
The perpendicular bisectors contain diameters of the circle. Two diameters intersect at the circle's center.

Step 1 Trace a piece of the rim. Draw two chords and construct perpendicular bisectors.

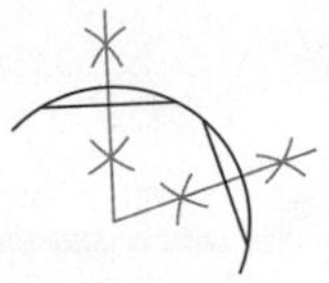

Step 2 The center is the intersection of the perpendicular bisectors. Use the center to find the radius.

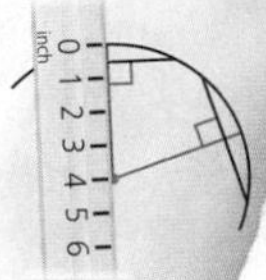

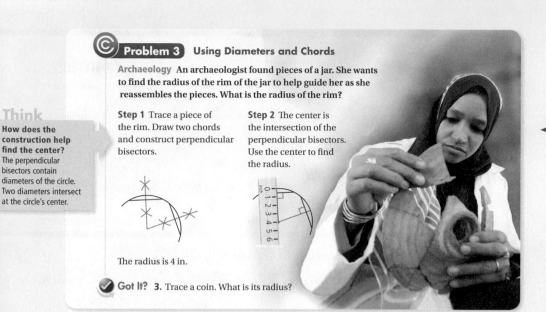

The radius is 4 in.

Got It? 3. Trace a coin. What is its radius?

Problem 4 Finding Measures in a Circle

Algebra **What is the value of each variable to the nearest tenth?**

Plan

Find two sides of a right triangle. The third side is either the answer or leads to an answer.

A

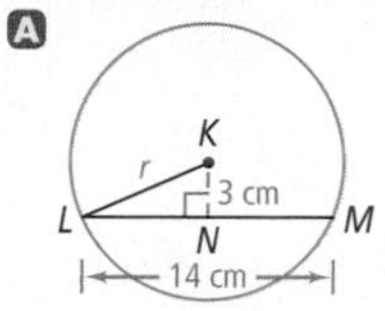

$LN = \frac{1}{2}(14) = 7$ A diameter ⊥ to a chord bisects the chord.

$r^2 = 3^2 + 7^2$ Use the Pythagorean Theorem.

$r \approx 7.6$ Find the positive square root of each side.

B

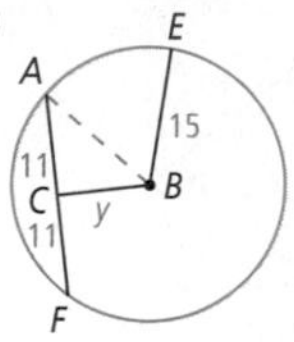

$\overline{BC} \perp \overline{AF}$ A diameter that bisects a chord that is not a diameter is ⊥ to the chord.

$BA = BE = 15$ Draw an auxiliary $\overline{BA}$. The auxiliary $\overline{BA} \cong \overline{BE}$ because they are radii of the same circle.

$y^2 + 11^2 = 15^2$ Use the Pythagorean Theorem.

$y^2 = 104$ Solve for y^2.

$y \approx 10.2$ Find the positive square root of each side.

Got It? 4. **Reasoning** In part (b), how does the auxiliary $\overline{BA}$ make the problem simpler to solve?

Problem 3

Q How is the perpendicular bisector of a chord related to a circle? **[It contains a diameter of the circle.]**

Q How many diameters do you need to draw to locate the center of a circle? **[at least two]**

Q How can you find the perpendicular bisectors of chords? **[Construct perpendicular bisectors using a compass and straightedge.]**

Got It? VISUAL LEARNERS

Review the steps involved in constructing the perpendicular bisector of a segment. Be sure that students are constructing perpendicular bisectors correctly before they measure the radius of the circle.

Problem 4

Q In 4A, what is the relationship between $\overline{KN}$ and $\overline{LM}$? **[They are perpendicular, so $\overline{KN}$ bisects $\overline{LM}$ by Theorem 12-8.]**

Q What type of triangle is formed by the chord, the radius, and the perpendicular bisector? **[a right triangle]**

Q In 4B, what is the relationship between $\overline{BC}$ and $\overline{AF}$? **[$\overline{BC}$ bisects $\overline{AF}$, so they are perpendicular by Theorem 12-9.]**

Q How are $\overline{BA}$ and $\overline{BE}$ related? **[They are radii of the same circle, so they are congruent.]**

Got It? ERROR PREVENTION

Review the definition of an auxiliary line. Ask students why the segment is called auxiliary.

Answers

Got It? (continued)

3. Check students' work.
4. $\overline{BA}$ is the hypotenuse of rt. $\triangle BAC$, so the Pythagorean Theorem can be used.

3 Lesson Check

Do you know HOW?

- If students have difficulty with Exercise 1, then have them classify $\angle AOB$ and $\angle COD$.

Do you UNDERSTAND?

- If students have difficulty with Exercise 5, then have them review the diagram in Problem 2.

Close

Q If two central angles are congruent, what can you say about the arcs and chords that they create? **[The arcs and chords are congruent.]**

Q What is true about two congruent chords in a circle? **[They are equidistant from the center of the circle.]**

Q If a diameter is perpendicular to a chord, what is true about the chord and its related arc? **[The diameter bisects the chord and arc.]**

Lesson Check

Do you know HOW?

In $\odot O$, $m\overset{\frown}{CD} = 50$ and $\overline{CA} \cong \overline{BD}$.

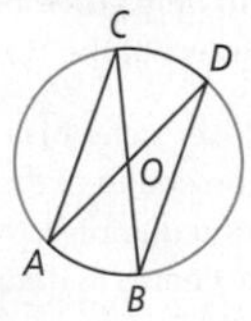

1. What is $m\overset{\frown}{AB}$? How do you know?
2. What is true of $\overset{\frown}{CA}$ and $\overset{\frown}{BD}$? Why?
3. Since $CA = BD$, what do you know about the distance of $\overline{CA}$ and $\overline{BD}$ from the center of $\odot O$?

Do you UNDERSTAND? MATHEMATICAL PRACTICES

4. **Vocabulary** Is a radius a chord? Is a diameter a chord? Explain your answers.
5. **Error Analysis** What is the error in the diagram?

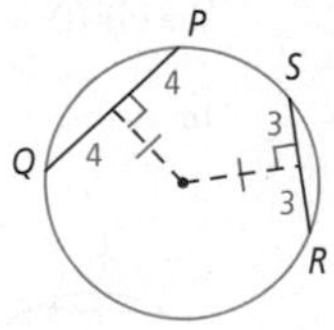

Practice and Problem-Solving Exercises MATHEMATICAL PRACTICES

A Practice In Exercises 6 and 7, the circles are congruent. What can you conclude? See Problem 1.

6. 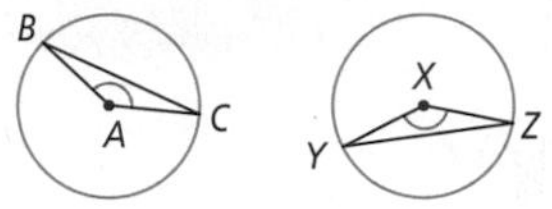

7. 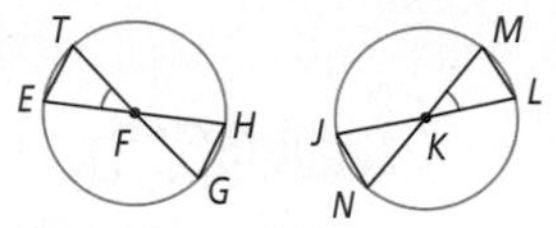

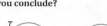

Find the value of *x*. See Problem 2.

8. 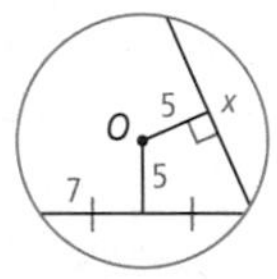

9. 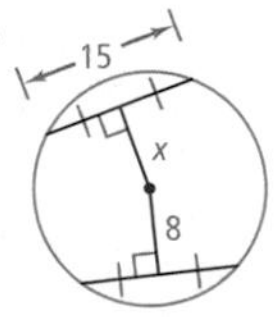

10. 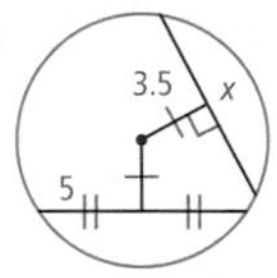

See Problems 3 and 4.

11. In the diagram at the right, $\overline{GH}$ and $\overline{KM}$ are perpendicular bisectors of the chords they intersect. What can you conclude about the center of the circle? Justify your answer.

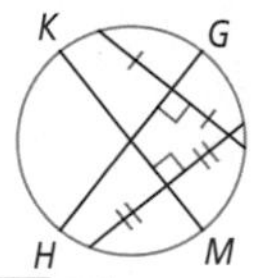

12. In $\odot O$, $\overline{AB}$ is a diameter of the circle and $\overline{AB} \perp \overline{CD}$. What conclusions can you make?

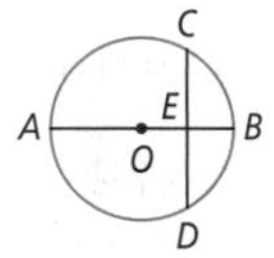

PowerGeometry.com

3 Lesson Check

For a digital lesson check, use the Got It questions.

Support in Geometry Companion

- Lesson Check

4 Practice

Assign homework to individual students or to an entire class.

Answers

Lesson Check

1. 50; $\angle COD \cong \angle AOB$ (Vert. ∠s are ≅), so $\overset{\frown}{CD} \cong \overset{\frown}{AB}$ because ≅ central ∠s have ≅ arcs. Therefore, $m\overset{\frown}{CD} = m\overset{\frown}{AB}$.
2. $\overset{\frown}{CA} \cong \overset{\frown}{BD}$ because in a circle ≅ chords have ≅ arcs.
3. The distances are equal because in a circle ≅ chords are equidistant from the center.
4. A radius is *not* a chord because one of its endpoints is not on the circle. A diameter *is* a chord because both of its endpoints are on the circle.
5. Chords $\overline{SR}$ and $\overline{QP}$ are equidistant from the center, so their lengths must be equal.

Practice and Problem-Solving Exercises

6. $\overset{\frown}{BC} \cong \overset{\frown}{YZ}$, $\overline{BC} \cong \overline{YZ}$
7. Answers may vary. Sample: $\overset{\frown}{ET} \cong \overset{\frown}{GH} \cong \overset{\frown}{JN} \cong \overset{\frown}{ML}$; $\overline{ET} \cong \overline{GH} \cong \overline{JN} \cong \overline{ML}$; $\angle TFE \cong \angle HFG$; $\angle JKN \cong \angle MKL$
8. 14
9. 8
10. 10
11. The center is at the intersection of $\overline{GH}$ and $\overline{KM}$, because if a chord is the ⊥ bis. of another chord, then the first chord is a diameter; two diameters intersect at the center of a circle.
12. $CE = ED$, $\overset{\frown}{BC} \cong \overset{\frown}{BD}$

Algebra Find the value of x to the nearest tenth.

13. 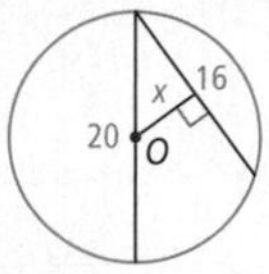

14. 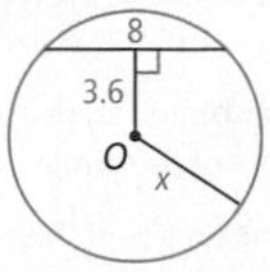

15. 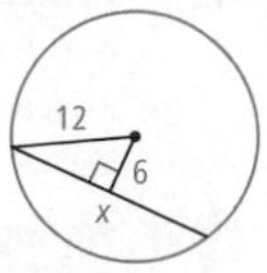

B Apply

16. Geometry in 3 Dimensions In the figure at the right, sphere O with radius 13 cm is intersected by a plane 5 cm from center O. Find the radius of the cross section $\odot A$.

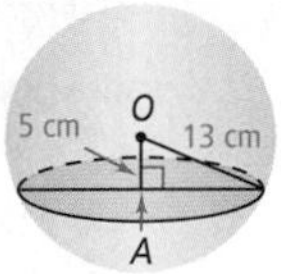

17. Geometry in 3 Dimensions A plane intersects a sphere that has radius 10 in., forming the cross section $\odot B$ with radius 8 in. How far is the plane from the center of the sphere?

18. Think About a Plan Two concentric circles have radii of 4 cm and 8 cm. A segment tangent to the smaller circle is a chord of the larger circle. What is the length of the segment to the nearest tenth?
- How will you start the diagram?
- Where is the best place to position the radius of each circle?

19. Prove Theorem 12-4.
Proof
Given: $\odot O$ with $\angle AOB \cong \angle COD$
Prove: $\overparen{AB} \cong \overparen{CD}$

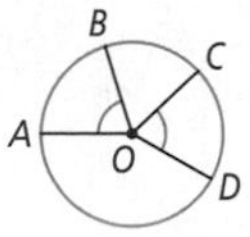

20. Prove Theorem 12-5.
Proof
Given: $\odot O$ with $\angle AOB \cong \angle COD$
Prove: $\overline{AB} \cong \overline{CD}$

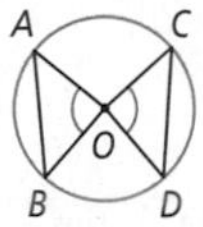

21. Prove Theorem 12-6.
Proof
Given: $\odot O$ with $\overline{AB} \cong \overline{CD}$
Prove: $\overparen{AB} \cong \overparen{CD}$

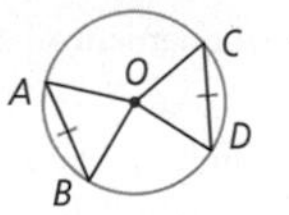

22. Prove Theorem 12-8.
Proof
Given: $\odot O$ with diameter $\overline{ED} \perp \overline{AB}$ at C
Prove: $\overline{AC} \cong \overline{BC}$, $\overparen{AD} \cong \overparen{BD}$

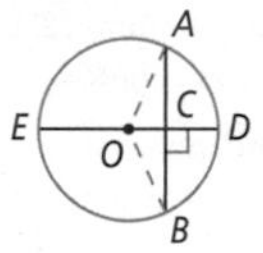

$\odot A$ and $\odot B$ are congruent. $\overline{CD}$ is a chord of both circles.

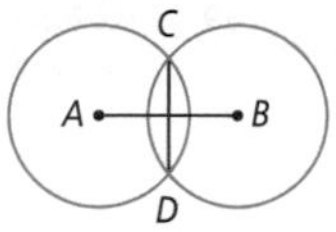

23. If $AB = 8$ in. and $CD = 6$ in., how long is a radius?

24. If $AB = 24$ cm and a radius $= 13$ cm, how long is $\overline{CD}$?

25. If a radius $= 13$ ft and $CD = 24$ ft, how long is $\overline{AB}$?

26. Construction Use Theorem 12-5 to construct a regular octagon.

4 Practice

ASSIGNMENT GUIDE
Basic: 6–16, 18, 23–25, 29
Average: 7–15 odd, 16–34
Advanced: 7–15 odd, 16–39
Standardized Test Prep: 40–43
Mixed Review: 44–52

Mathematical Practices are supported by exercises with red headings. Here are the Practices supported in this lesson:

MP 1: Make Sense of Problems Ex. 18
MP 3: Communicate Ex. 29
MP 3: Critique the Reasoning of Others Ex. 5

Applications exercises have blue headings.

EXERCISE 23: Use the Think About a Plan worksheet in the **Practice and Problem Solving Workbook** (also available in the Teaching Resources in print and online) to further support students' development in becoming independent learners.

HOMEWORK QUICK CHECK
To check students' understanding of key skills and concepts, go over Exercises 7, 11, 18, 23, and 29.

13. 6
14. 5.4
15. 20.8
16. 12 cm
17. 6 in.
18. 13.9 cm
19. Since $\angle AOB \cong \angle COD$, it follows that $m\angle AOB = m\angle COD$. Now $m\angle AOB = m\overparen{AB}$ and $m\angle COD = m\overparen{CD}$ (definition of arc measure). So $m\overparen{AB} = m\overparen{CD}$ (Substitution). Therefore, $\overparen{AB} \cong \overparen{CD}$ (definition of $\cong$ arcs).
20. $\odot O$ with $\angle AOB \cong \angle COD$ (given); $\overline{AO} \cong \overline{BO} \cong \overline{CO} \cong \overline{DO}$ (all radii of a $\odot$ are $\cong$). $\triangle AOB \cong \triangle COD$ (SAS); $\overline{AB} \cong \overline{CD}$ (corresp. parts of $\cong$ △s are $\cong$).
21. $\odot O$ with $\overline{AB} \cong \overline{CD}$ (given); $\overline{AO} \cong \overline{BO} \cong \overline{CO} \cong \overline{DO}$ (all radii of a $\odot$ are $\cong$); $\triangle AOB \cong \triangle COD$ (SSS); $\angle AOB \cong \angle COD$ (corresp. parts of $\cong$ $\triangle$s are $\cong$.); $\overparen{AB} \cong \overparen{CD}$ ($\cong$ central ∠s have $\cong$ arcs).
22. $\odot O$ with diameter $\overline{ED} \perp \overline{AB}$ at C (given). Draw $\overline{OA}$ and $\overline{OB}$ (2 pts. determine a line). $\angle ACO$ and $\angle BCO$ are rt. ∠s ($\perp$ lines form rt. ∠s). $\triangle ACO$ and $\triangle BCO$ are rt. △s (Def. of a rt. $\triangle$). $\overline{OA} \cong \overline{OB}$ (all radii of a $\odot$ are $\cong$); $\overline{OC} \cong \overline{OC}$ (Refl. Prop. of $\cong$); $\triangle ACO \cong \triangle BCO$ (HL); $\overline{AC} \cong \overline{BC}$ (corresp. parts of $\cong$ △s are $\cong$); $\angle AOC \cong \angle BOC$ (corresp. parts of $\cong$ △s are $\cong$); $\overparen{AD} \cong \overparen{BD}$ ($\cong$ central ∠s have $\cong$ arcs).
23. 5 in.
24. 10 cm
25. 10 ft
26. 

Answers

Practice and Problem-Solving Exercises (continued)

27. 9.2 units

28.

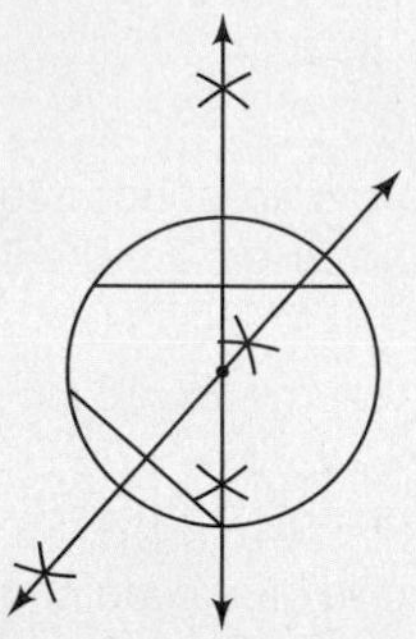

29. The length of a chord or an arc is determined not only by the measure of the central ∠s, but also by the radius of the ⊙.

30. 108

31. 90

32. about 123.9

33. $\overline{XW} \cong \overline{XY}$ (all radii of a circle are ≅); X is on the ⊥ bis. of $\overline{WY}$ (Converse of ⊥ Bis. Thm.); ℓ is the ⊥ bis. of $\overline{WY}$ (given); X is on ℓ (Subst. Prop.), so ℓ contains the center of ⊙X.

34. ⊙A with $\overline{CE} \perp \overline{BD}$ (given); $\overline{CF} \cong \overline{CF}$ (Refl. Prop. of ≅); $\overline{BF} \cong \overline{DF}$ (a diameter ⊥ to a chord bisects the chord); ∠BFC and ∠DFC are rt. ∠s (⊥ lines form rt. ∠s); ∠BFC ≅ ∠DFC (Rt. ∠s are ≅); △BFC ≅ △DFC (SAS); $\overline{BC} \cong \overline{CD}$ (corresp. parts of ≅ △s are ≅); $\widehat{BC} \cong \widehat{DC}$ (≅ chords have ≅ arcs).

35.

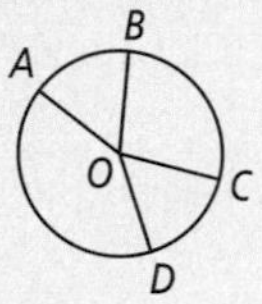

Given: ⊙O with $\widehat{AB} \cong \widehat{CD}$

Prove: ∠AOB ≅ ∠COD

Proof: $m\angle AOB = m\widehat{AB}$ and $m\angle COD = m\widehat{CD}$ (definition of arc measure). $\widehat{AB} \cong \widehat{CD}$ (given), so $m\widehat{AB} = m\widehat{CD}$ (Def. of ≅ arcs). Therefore, $m\angle AOB = m\angle COD$ (Substitution). Hence ∠AOB ≅ ∠COD (Def. of ≅ ∠s).

36.

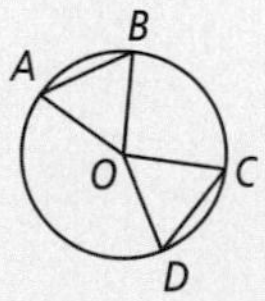

Given: ⊙O with $\overline{AB} \cong \overline{CD}$

Prove: ∠AOB ≅ ∠COD

Proof: In circle O, $AO = BO = CO = DO$ (radii of a ⊙ are ≅) and $\overline{AB} \cong \overline{CD}$ (given). So △AOB ≅ △COD (SSS) and ∠AOB ≅ ∠COD (corresp. parts of ≅ △s are ≅).

37.

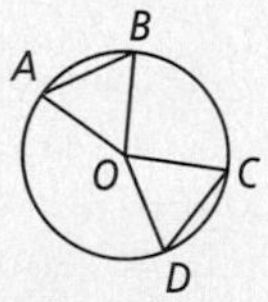

Given: ⊙O with $\widehat{AB} \cong \widehat{CD}$

Prove: $\overline{AB} \cong \overline{CD}$

Proof: It is given that $\widehat{AB} \cong \widehat{CD}$, so ∠$AOB$ ≅ ∠COD (if arcs are ≅ then their central ∠s are ≅). Also, $AO = BO = CO = DO$ (radii of a ⊙ are ≅), so △AOB ≅ △COD (SAS), and $\overline{AB} \cong \overline{CD}$ (corresp. parts of ≅ △s are ≅).

38.

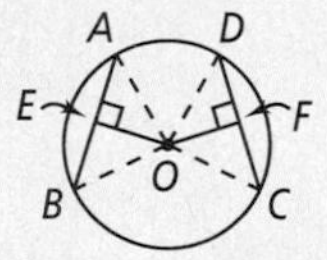

Given: ⊙O with $\overline{AB} \cong \overline{CD}$, $\overline{OE} \perp \overline{AB}$, $\overline{OF} \perp \overline{CD}$

Prove: $\overline{OE} \cong \overline{OF}$

Proof: All radii of ⊙O are ≅ and it is given that $\overline{AB} \cong \overline{CD}$, so △$AOB$ ≅ △COD by SSS. ∠A ≅ ∠C (corresp. parts of ≅ △s are ≅). ∠OEA and ∠OFC are rt. ∠s (⊥ lines form rt. ∠s). So, ∠OEA ≅ ∠OFC (Rt. ∠s are ≅). Thus, △OEA ≅ △OFC by AAS, and $\overline{OE} \cong \overline{OF}$ (corresp. parts of ≅ △s are ≅).

---

27. In the diagram at the right, the endpoints of the chord are the points where the line $x = 2$ intersects the circle $x^2 + y^2 = 25$. What is the length of the chord? Round your answer to the nearest tenth.

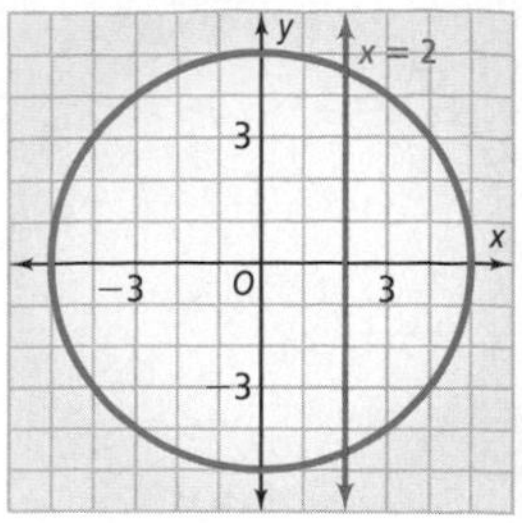

28. Construction Use a circular object such as a can or a saucer to draw a circle. Construct the center of the circle.

29. Writing Theorems 12-4 and 12-5 both begin with the phrase, "within a circle or in congruent circles." Explain why the word *congruent* is essential for both theorems.

Find $m\widehat{AB}$. (*Hint:* You will need to use trigonometry in Exercise 32.)

30.

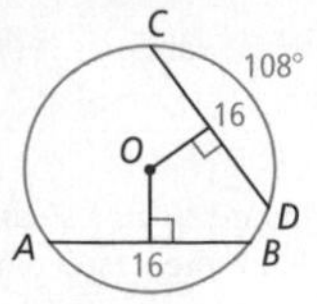

31.

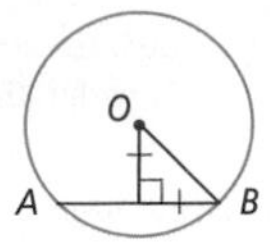

32.

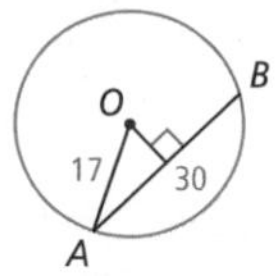

33. Proof Prove Theorem 12-10.

Given: ℓ is the ⊥ bisector of $\overline{WY}$.

Prove: ℓ contains the center of ⊙X.

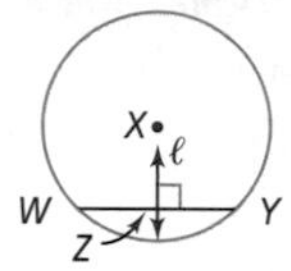

34. Proof **Given:** ⊙A with $\overline{CE} \perp \overline{BD}$

Prove: $\widehat{BC} \cong \widehat{DC}$

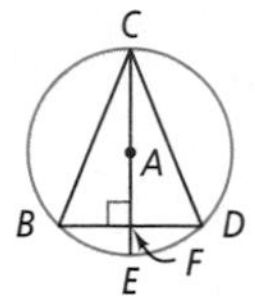

Challenge **Prove each of the following.**

Proof

35. Converse of Theorem 12-4: Within a circle or in congruent circles, congruent arcs have congruent central angles.

36. Converse of Theorem 12-5: Within a circle or in congruent circles, congruent chords have congruent central angles.

37. Converse of Theorem 12-6: Within a circle or in congruent circles, congruent arcs have congruent chords.

38. Converse of Theorem 12-7: Within a circle or congruent circles, congruent chords are equidistant from the center (or centers).

39. Proof If two circles are concentric and a chord of the larger circle is tangent to the smaller circle, prove that the point of tangency is the midpoint of the chord.

Standardized Test Prep

SAT/ACT

40. The diameter of a circle is 25 cm and a chord of the same circle is 16 cm. To the nearest tenth, what is the distance of the chord from the center of the circle?

Ⓐ 9.0 cm　Ⓒ 18.0 cm
Ⓑ 9.6 cm　Ⓓ 19.2 cm

41. The Smart Ball Company makes plastic balls for small children. The diameter of a ball is 8 cm. The cost for creating a ball is 2 cents per square centimeter. Which value is the most reasonable estimate for the cost of making 1000 balls?

Ⓕ \$2010　Ⓗ \$16,080
Ⓖ \$4021　Ⓘ \$42,900

42. From the top of a building you look down at an object on the ground. Your eyes are 50 ft above the ground and the angle of depression is 50°. Which distance is the best estimate of how far the object is from the base of the building?

Ⓐ 42 ft　Ⓒ 65 ft
Ⓑ 60 ft　Ⓓ 78 ft

Short Response

43. A bicycle tire has a diameter of 17 in. How many revolutions of the tire are necessary to travel 800 ft? Show your work.

Mixed Review

Assume that the lines that appear to be tangent are tangent. *O* is the center of each circle. Find the value of *x* to the nearest tenth.

See Lesson 12-1.

44.

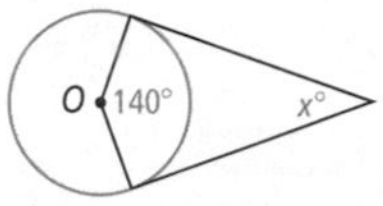

45.

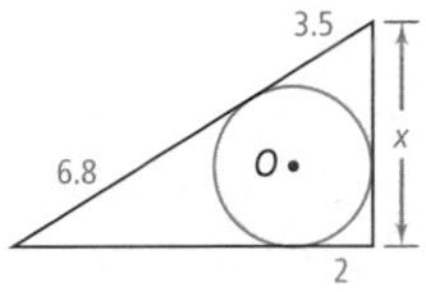

46. The legs of a right triangle are 10 in. and 24 in. long. The bisector of the right angle cuts the hypotenuse into two segments. What is the length of each segment, rounded to the nearest tenth?

See Lesson 7-5.

Get Ready! **To prepare for Lesson 12-3, do Exercises 47–52.**

See Lesson 10-6.

Identify the following in ⊙*P* at the right.

47. a semicircle　**48.** a minor arc　**49.** a major arc

Find the measure of each arc in ⊙*P*.

50. $\overset{\frown}{ST}$　**51.** $\overset{\frown}{STQ}$　**52.** $\overset{\frown}{RT}$

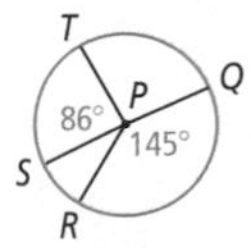

39.

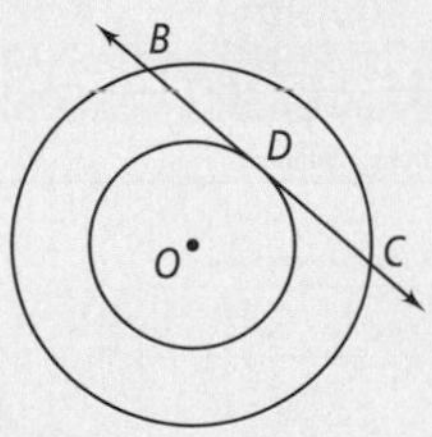

Given: Concentric circles, $\overline{BC}$ is tangent to the smaller circle at *D*
Prove: *D* is the midpt. of $\overline{BC}$
Proof: It is given that $\overline{BC}$ is tangent to the smaller circle, so $\overline{BC} \perp \overline{OD}$ (a tangent is ⊥ to a radius at the point of tangency). $\overline{OD}$ is part of a diameter of the larger circle, so $\overline{BD} \cong \overline{CD}$ (if a diameter is ⊥ to a chord, it bisects the chord). *D* is the midpt. of $\overline{BC}$ (Def. of midpt.)

40. B
41. G
42. A
43. [2] During one revolution the bicycle moves $C = \pi d = \pi(17) = 53.4$ in., or about 4.45 ft. So the number of revolutions needed to travel 800 ft is $\frac{800}{4.45} \approx 180$ revolutions.
[1] correct method, but inches not converted to feet
44. 40
45. 5.5
46. 7.6 in. and 18.4 in.
47–49. Answers may vary. Samples are given.
47. $\overset{\frown}{STQ}$
48. $\overset{\frown}{ST}$
49. $\overset{\frown}{STR}$
50. 86
51. 180
52. 121

12-2 Lesson Resources

Differentiated Remediation

Instructional Support

Geometry Companion

Students can use the **Geometry Companion** worktext (4 pages) . . .

- New Vocabulary
- Key Concepts
- Got It for each Problem
- Lesson Check

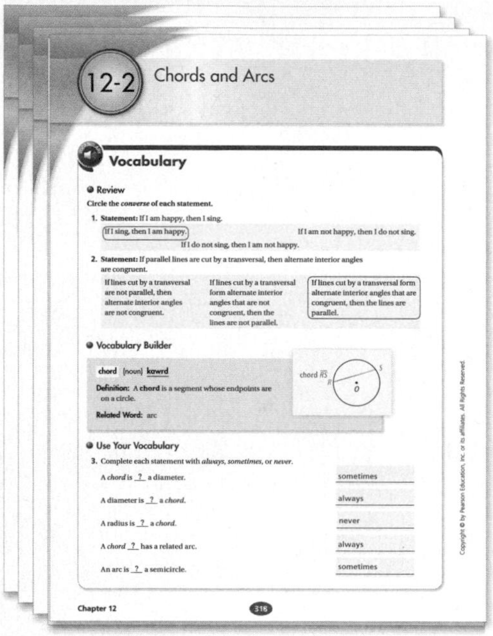

12-2 Chords and Arcs

Vocabulary

Review

Circle the *converse* of each statement.

1. **Statement:** If I am happy, then I sing.
   If I sing, then I am happy. — If I am not happy, then I do not sing. — If I do not sing, then I am not happy.
2. **Statement:** If parallel lines are cut by a transversal, then alternate interior angles are congruent.
   If lines cut by a transversal are not parallel, then alternate interior angles are not congruent. — If lines cut by a transversal form alternate interior angles that are not congruent, then the lines are not parallel. — If lines cut by a transversal form alternate interior angles that are congruent, then the lines are parallel.

Vocabulary Builder

chord (noun) **kawrd**

Definition: A **chord** is a segment whose endpoints are on a circle.

Related Word: arc

Use Your Vocabulary

3. Complete each statement with *always*, *sometimes*, or *never*.

A *chord* is _?_ a diameter. sometimes

A diameter is _?_ a *chord*. always

A radius is _?_ a *chord*. never

A *chord* _?_ has a related arc. always

An arc is _?_ a semicircle. sometimes

Chapter 12 318

ELL Support

Use Manipulatives Have students work in pairs. Hand out different sizes of circles that have been cut from construction paper. Tell students to use a protractor and a ruler to draw the center and two congruent chords on one circle, and two congruent central angles on the other. Students can trade their circles and prove the chords and central angles are congruent. Discuss the results.

Assess Understanding Draw three different sized circles on the board, each with a central angle of the same measure. Ask whether the central angles are congruent and whether the arcs are congruent. Ask students to explain what is different and the same.

5 Assess & Remediate

Lesson Quiz

1. **Do you UNDERSTAND?** In the diagram, $\angle GHF \cong \angle KHJ$. What can you conclude?

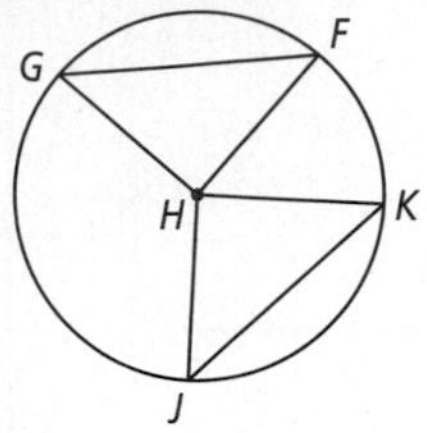

2. In the above diagram, $JK = 8$. The perimeter of $\triangle JHK = 18$. What is HK?
3. What is the missing length?

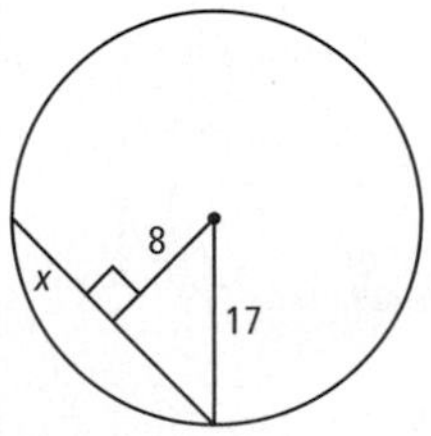

ANSWERS TO LESSON QUIZ

1. $\overline{GF} \cong \overline{KJ}$, $\overparen{GF} \cong \overparen{KJ}$
2. 5
3. 15

PRESCRIPTION FOR REMEDIATION

Use the student work on the Lesson Quiz to prescribe a differentiated review assignment.

Points	Differentiated Remediation
0–1	Intervention
2	On-level
3	Extension

PowerGeometry.com

5 Assess & Remediate

Assign the Lesson Quiz. Appropriate intervention, practice, or enrichment is automatically generated based on student performance.

Intervention

- **Reteaching** (2 pages) Provides reteaching and practice exercises for the key lesson concepts. Use with struggling students or absent students.
- **English Language Learner Support** Helps students develop and reinforce mathematical vocabulary and key concepts.

All-in-One Resources/Online

Reteaching

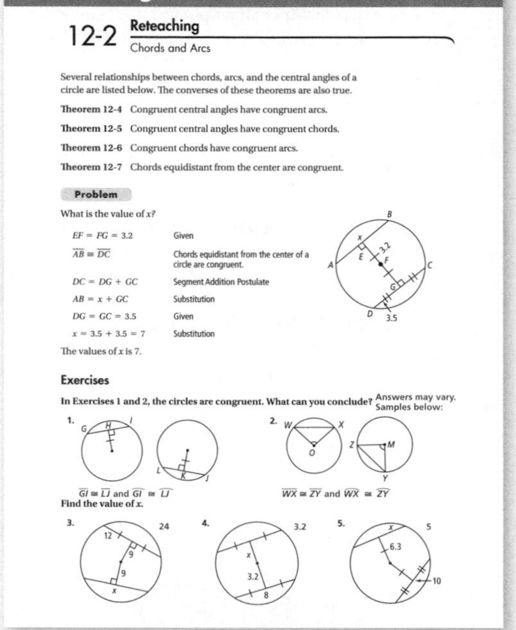

12-2 Reteaching
Chords and Arcs

Several relationships between chords, arcs, and the central angles of a circle are listed below. The converses of these theorems are also true.

Theorem 12-4 Congruent central angles have congruent arcs.

Theorem 12-5 Congruent central angles have congruent chords.

Theorem 12-6 Congruent chords have congruent arcs.

Theorem 12-7 Chords equidistant from the center are congruent.

Problem

What is the value of x?

$EF = FG = 3.2$	Given
$\overline{AB} \cong \overline{DC}$	Chords equidistant from the center of a circle are congruent.
$DC = DG + GC$	Segment Addition Postulate
$AB = x + GC$	Substitution
$DG = GC = 3.5$	Given
$x = 3.5 + 3.5 = 7$	Substitution

The values of x is 7.

Exercises

In Exercises 1 and 2, the circles are congruent. What can you conclude? Answers may vary. Samples below:

1. $\overline{GI} \cong \overline{LJ}$ and $\overparen{GI} \cong \overparen{LJ}$
2. $\overline{WX} \cong \overline{ZY}$ and $\overparen{WX} \cong \overparen{ZY}$

Find the value of x.

3. 24
4. 3.2
5. 5

All-in-One Resources/Online

English Language Learner Support

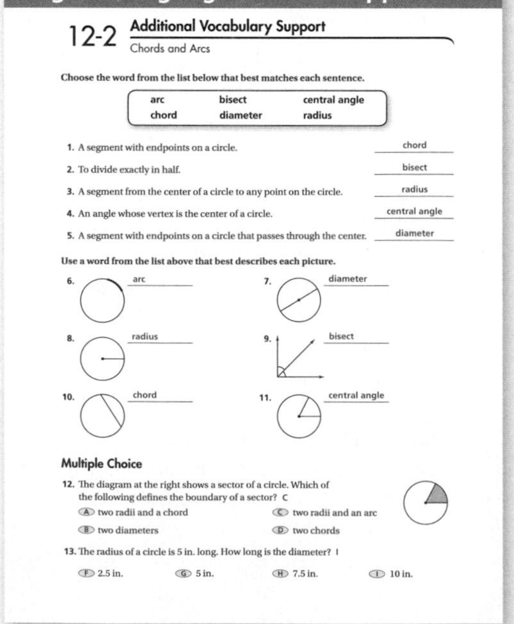

12-2 Additional Vocabulary Support
Chords and Arcs

Choose the word from the list below that best matches each sentence.

arc	bisect	central angle
chord	diameter	radius

1. A segment with endpoints on a circle. chord
2. To divide exactly in half. bisect
3. A segment from the center of a circle to any point on the circle. radius
4. An angle whose vertex is the center of a circle. central angle
5. A segment with endpoints on a circle that passes through the center. diameter

Use a word from the list above that best describes each picture.

6. arc
7. diameter
8. radius
9. bisect
10. chord
11. central angle

Multiple Choice

12. The diagram at the right shows a sector of a circle. Which of the following defines the boundary of a sector? C
    A two radii and a chord
    B two diameters
    C two radii and an arc
    D two chords
13. The radius of a circle is 5 in. long. How long is the diameter? I
    F 2.5 in.
    G 5 in.
    H 7.5 in.
    I 10 in.

Differentiated Remediation *continued*

On-Level

- **Practice** (2 pages) Provides extra practice for each lesson. For simpler practice exercises, use the Form K Practice pages found in the All-in-One Teaching Resources and online.
- **Think About a Plan** Helps students develop specific problem-solving skills and strategies by providing scaffolded guiding questions.
- **Standardized Test Prep** Focuses on all major exercises, all major question types, and helps students prepare for the high-stakes assessments.

Extension

- **Enrichment** Provides students with interesting problems and activities that extend the concepts of the lesson.
- **Activities, Games, and Puzzles** Worksheets that can be used for concepts development, enrichment, and for fun!

Practice and Problem Solving Wkbk/ All-in-One Resources/Online

Practice page 1

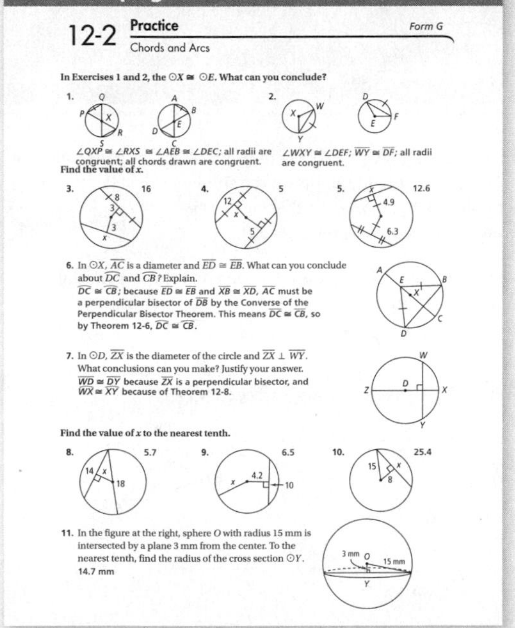
12-2 **Practice** *Form G*
Chords and Arcs

In Exercises 1 and 2, the $\odot X \cong \odot E$. What can you conclude?

1. $\angle QXP \cong \angle RXS \cong \angle AEB \cong \angle DEC$; all radii are congruent; all chords drawn are congruent.
2. $\angle WXY \cong \angle DEF$; $\overline{WY} \cong \overline{DF}$; all radii are congruent.

Find the value of *x*.

3. 16
4. 5
5. 12.6
6. In $\odot X$, $\overline{AC}$ is a diameter and $\overline{ED} \cong \overline{EB}$. What can you conclude about $\widehat{DC}$ and $\widehat{CB}$? Explain. $\widehat{DC} \cong \widehat{CB}$; because $\overline{ED} \cong \overline{EB}$ and $\overline{XB} \cong \overline{XD}$, $\overline{AC}$ must be a perpendicular bisector of $\overline{DB}$ by the Converse of the Perpendicular Bisector Theorem. This means $\overline{DC} \cong \overline{CB}$, so by Theorem 12-6, $\widehat{DC} \cong \widehat{CB}$.
7. In $\odot D$, $\overline{ZX}$ is the diameter of the circle and $\overline{ZX} \perp \overline{WY}$. What conclusions can you make? Justify your answer. $\overline{WD} \cong \overline{DY}$ because $\overline{ZX}$ is a perpendicular bisector, and $\widehat{WX} \cong \widehat{XY}$ because of Theorem 12-8.

Find the value of *x* to the nearest tenth.

8. 5.7
9. 6.5
10. 25.4
11. In the figure at the right, sphere *O* with radius 15 mm is intersected by a plane 3 mm from the center. To the nearest tenth, find the radius of the cross section $\odot Y$. 14.7 mm

Practice and Problem Solving Wkbk/ All-in-One Resources/Online

Practice page 2

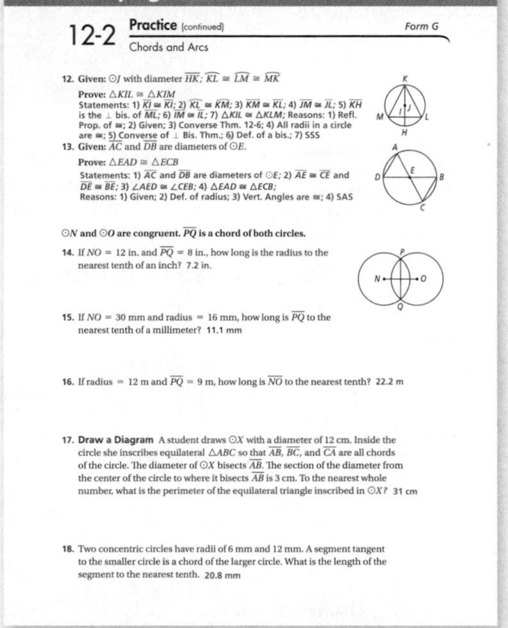
12-2 **Practice** (continued) *Form G*
Chords and Arcs

12. **Given:** $\odot J$ with diameter $\overline{HK}$; $\widehat{KL} \cong \widehat{LM} \cong \widehat{MK}$
Prove: $\triangle KIL \cong \triangle KIM$
Statements: 1) $\overline{KI} \cong \overline{KI}$; 2) $\overline{KL} \cong \overline{KM}$; 3) $\overline{KM} \cong \overline{KL}$; 4) $\overline{JM} \cong \overline{JL}$; 5) $\overline{KH}$ is the $\perp$ bis. of $\overline{ML}$; 6) $\overline{IM} \cong \overline{IL}$; 7) $\triangle KIL \cong \triangle KLM$; Reasons: 1) Refl. Prop. of $\cong$; 2) Given; 3) Converse Thm. 12-6; 4) All radii in a circle are $\cong$; 5) Converse of $\perp$ Bis. Thm.; 6) Def. of a bis.; 7) SSS
13. **Given:** $\overline{AC}$ and $\overline{DB}$ are diameters of $\odot E$.
Prove: $\triangle EAD \cong \triangle ECB$
Statements: 1) $\overline{AC}$ and $\overline{DB}$ are diameters of $\odot E$; 2) $\overline{AE} \cong \overline{CE}$ and $\overline{DE} \cong \overline{BE}$; 3) $\angle AED \cong \angle CEB$; 4) $\triangle EAD \cong \triangle ECB$;
Reasons: 1) Given; 2) Def. of radius; 3) Vert. Angles are $\cong$; 4) SAS

$\odot N$ and $\odot O$ are congruent. $\overline{PQ}$ is a chord of both circles.

14. If $NO = 12$ in. and $PQ = 8$ in., how long is the radius to the nearest tenth of an inch? 7.2 in.
15. If $NO = 30$ mm and radius $= 16$ mm, how long is $\overline{PQ}$ to the nearest tenth of a millimeter? 11.1 mm
16. If radius $= 12$ m and $\overline{PQ} = 9$ m, how long is $\overline{NO}$ to the nearest tenth? 22.2 m
17. **Draw a Diagram** A student draws $\odot X$ with a diameter of 12 cm. Inside the circle she inscribes equilateral $\triangle ABC$ so that $\overline{AB}$, $\overline{BC}$, and $\overline{CA}$ are all chords of the circle. The diameter of $\odot X$ bisects $\overline{AB}$. The section of the diameter from the center of the circle to where it bisects $\overline{AB}$ is 3 cm. To the nearest whole number, what is the perimeter of the equilateral triangle inscribed in $\odot X$? 31 cm
18. Two concentric circles have radii of 6 mm and 12 mm. A segment tangent to the smaller circle is a chord of the larger circle. What is the length of the segment to the nearest tenth. 20.8 mm

All-in-One Resources/Online

Enrichment

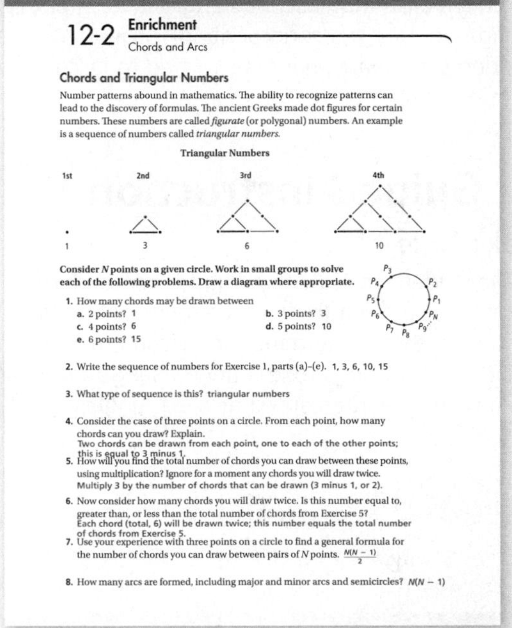
12-2 **Enrichment**
Chords and Arcs

Chords and Triangular Numbers

Number patterns abound in mathematics. The ability to recognize patterns can lead to the discovery of formulas. The ancient Greeks made dot figures for certain numbers. These numbers are called *figurate* (or polygonal) numbers. An example is a sequence of numbers called *triangular numbers.*

Triangular Numbers

1st 1; 2nd 3; 3rd 6; 4th 10

Consider *N* points on a given circle. Work in small groups to solve each of the following problems. Draw a diagram where appropriate.

1. How many chords may be drawn between
   a. 2 points? 1
   b. 3 points? 3
   c. 4 points? 6
   d. 5 points? 10
   e. 6 points? 15
2. Write the sequence of numbers for Exercise 1, parts (a)–(e). 1, 3, 6, 10, 15
3. What type of sequence is this? triangular numbers
4. Consider the case of three points on a circle. From each point, how many chords can you draw? Explain. Two chords can be drawn from each point, one to each of the other points; this is equal to 3 minus 1.
5. How will you find the total number of chords you can draw between these points, using multiplication? Ignore for a moment any chords you will draw twice. Multiply 3 by the number of chords that can be drawn (3 minus 1, or 2).
6. Now consider how many chords you will draw twice. Is this number equal to, greater than, or less than the total number of chords from Exercise 5? Each chord (total, 6) will be drawn twice; this number equals the total number of chords from Exercise 5.
7. Use your experience with three points on a circle to find a general formula for the number of chords you can draw between pairs of *N* points. $\frac{N(N-1)}{2}$
8. How many arcs are formed, including major and minor arcs and semicircles? $N(N-1)$

Practice and Problem Solving Wkbk/ All-in-One Resources/Online

Think About a Plan

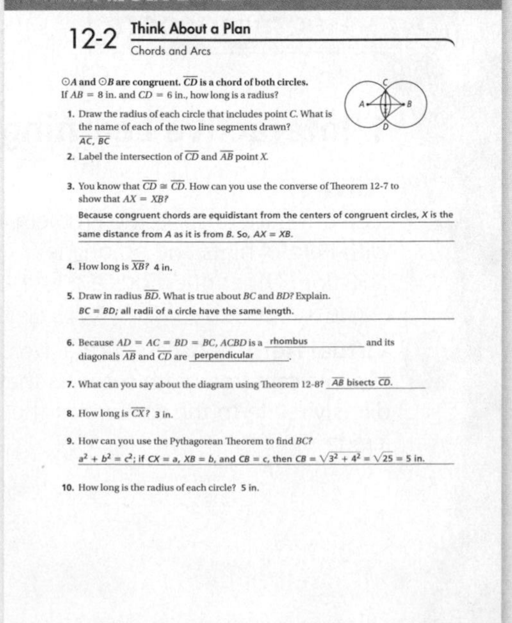
12-2 **Think About a Plan**
Chords and Arcs

$\odot A$ and $\odot B$ are congruent. $\overline{CD}$ is a chord of both circles.
If $AB = 8$ in. and $CD = 6$ in., how long is a radius?

1. Draw the radius of each circle that includes point *C*. What is the name of each of the two line segments drawn? $\overline{AC}$, $\overline{BC}$
2. Label the intersection of $\overline{CD}$ and $\overline{AB}$ point *X*.
3. You know that $\overline{CD} \cong \overline{CD}$. How can you use the converse of Theorem 12-7 to show that $AX = XB$? Because congruent chords are equidistant from the centers of congruent circles, *X* is the same distance from *A* as it is from *B*. So, $AX = XB$.
4. How long is $\overline{XB}$? 4 in.
5. Draw in radius $\overline{BD}$. What is true about *BC* and *BD*? Explain. $BC = BD$; all radii of a circle have the same length.
6. Because $AD = AC = BD = BC$, *ACBD* is a rhombus and its diagonals $\overline{AB}$ and $\overline{CD}$ are perpendicular.
7. What can you say about the diagram using Theorem 12-8? $\overline{AB}$ bisects $\overline{CD}$.
8. How long is $\overline{CX}$? 3 in.
9. How can you use the Pythagorean Theorem to find *BC*? $a^2 + b^2 = c^2$; if $CX = a$, $XB = b$, and $CB = c$, then $CB = \sqrt{3^2 + 4^2} = \sqrt{25} = 5$ in.
10. How long is the radius of each circle? 5 in.

Practice and Problem Solving Wkbk/ All-in-One Resources/Online

Standardized Test Prep

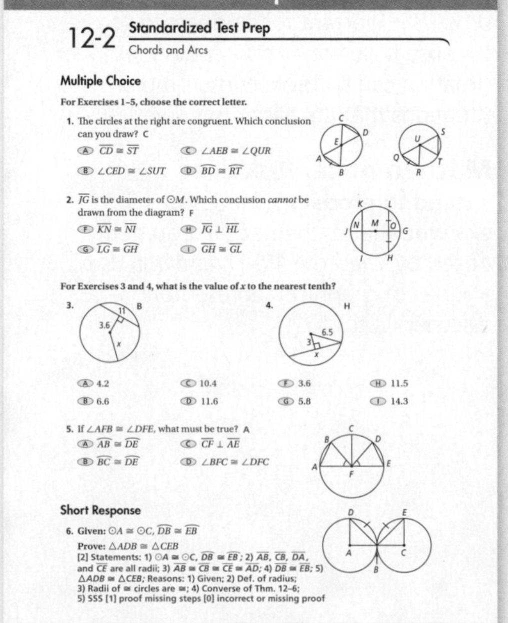
12-2 **Standardized Test Prep**
Chords and Arcs

Multiple Choice

For Exercises 1–5, choose the correct letter.

1. The circles at the right are congruent. Which conclusion can you draw? C
   A. $\overline{CD} \cong \overline{ST}$
   B. $\angle CED \cong \angle SUT$
   C. $\angle AEB \cong \angle QUR$
   D. $\widehat{BD} \cong \widehat{RT}$
2. $\overline{JG}$ is the diameter of $\odot M$. Which conclusion *cannot* be drawn from the diagram? F
   F. $\overline{KN} \cong \overline{NI}$
   G. $\widehat{LG} \cong \widehat{GH}$
   H. $\overline{JG} \perp \overline{HL}$
   I. $\overline{GH} \cong \overline{GL}$

For Exercises 3 and 4, what is the value of *x* to the nearest tenth?

3. A. 4.2 B. 6.6 C. 10.4 D. 11.6
4. F. 3.6 G. 5.8 H. 11.5 I. 14.3
5. If $\angle AFB \cong \angle DFE$, what must be true? A
   A. $\widehat{AB} \cong \widehat{DE}$
   B. $\widehat{BC} \cong \widehat{DE}$
   C. $\overline{CF} \perp \overline{AE}$
   D. $\angle BFC \cong \angle DFC$

Short Response

6. **Given:** $\odot A \cong \odot C$, $\widehat{DB} \cong \widehat{EB}$
Prove: $\triangle ADB \cong \triangle CEB$
[2] Statements: 1) $\odot A \cong \odot C$, $\widehat{DB} \cong \widehat{EB}$; 2) $\overline{AB}$, $\overline{CB}$, $\overline{DA}$, and $\overline{CE}$ are all radii; 3) $\overline{AB} \cong \overline{CB} \cong \overline{CE} \cong \overline{AD}$; 4) $\overline{DB} \cong \overline{EB}$; 5) $\triangle ADB \cong \triangle CEB$; Reasons: 1) Given; 2) Def. of radius; 3) Radii of $\cong$ circles are $\cong$; 4) Converse of Thm. 12-6; 5) SSS [1] proof missing steps [0] incorrect or missing proof

Online Teacher Resource Center

Activities, Games, and Puzzles

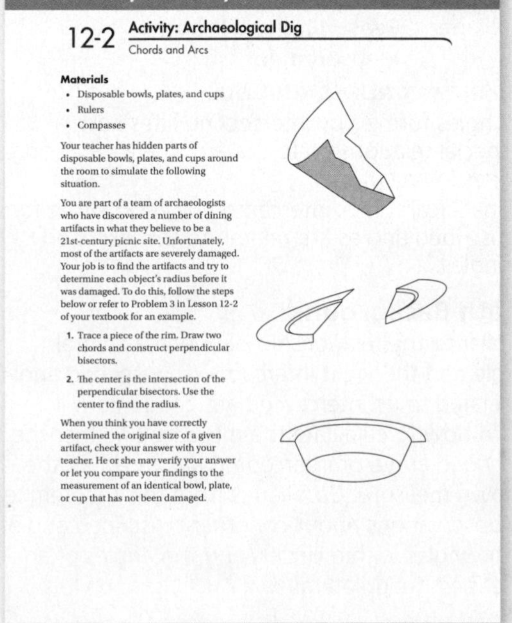
12-2 **Activity: Archaeological Dig**
Chords and Arcs

Materials
- Disposable bowls, plates, and cups
- Rulers
- Compasses

Your teacher has hidden parts of disposable bowls, plates, and cups around the room to simulate the following situation.

You are part of a team of archaeologists who have discovered a number of dining artifacts in what they believe to be a 21st-century picnic site. Unfortunately, most of the artifacts are severely damaged. Your job is to find the artifacts and try to determine each object's radius before it was damaged. To do this, follow the steps below or refer to Problem 3 in Lesson 12-2 of your textbook for an example.

1. Trace a piece of the rim. Draw two chords and construct perpendicular bisectors.
2. The center is the intersection of the perpendicular bisectors. Use the center to find the radius.

When you think you have correctly determined the original size of a given artifact, check your answer with your teacher. He or she may verify your answer or let you compare your findings to the measurement of an identical bowl, plate, or cup that has not been damaged.

1 Interactive Learning

Solve It!

PURPOSE To discover that inscribed angles that intercept the same arc are congruent

PROCESS Students may measure the angles using a protractor and draw a conclusion based on their findings.

FACILITATE

Q For each player, where does the angle of shots intersect with the circle? **[Each angle intersects the sides of the goal.]**

Q How are the measures of the angles related? **[They are the same.]**

Q Which player has the widest angle in which to shoot? **[They all have the same angle.]**

ANSWER See Solve It in Answers on next page.

CONNECT THE MATH In the Solve It, students should realize that inscribed angles that intersect congruent arcs are congruent. In this lesson, students will learn theorems related to inscribed angles.

2 Guided Instruction

Take Note

Have students practice finding the measure of $\overset{\frown}{AC}$ or $\angle B$. Give them the measurement of one figure and have them determine the measurement of the other. Ask students to identify the length of an arc that would be intercepted by a 90° angle.

12-3 Inscribed Angles

Common Core State Standards
G-C.A.2 Identify and describe relationships among inscribed angles, radii, and chords. **Also G-C.A.3, G-C.A.4**
MP 1, MP 3, MP 4, MP 6

Objectives To find the measure of an inscribed angle
To find the measure of an angle formed by a tangent and a chord

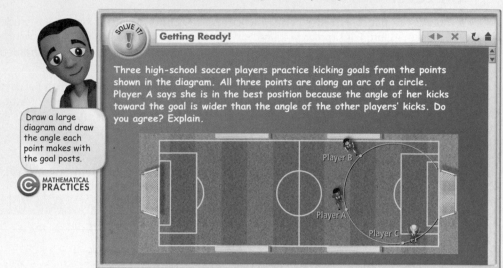

Draw a large diagram and draw the angle each point makes with the goal posts.

MATHEMATICAL PRACTICES

Getting Ready!

Three high-school soccer players practice kicking goals from the points shown in the diagram. All three points are along an arc of a circle. Player A says she is in the best position because the angle of her kicks toward the goal is wider than the angle of the other players' kicks. Do you agree? Explain.

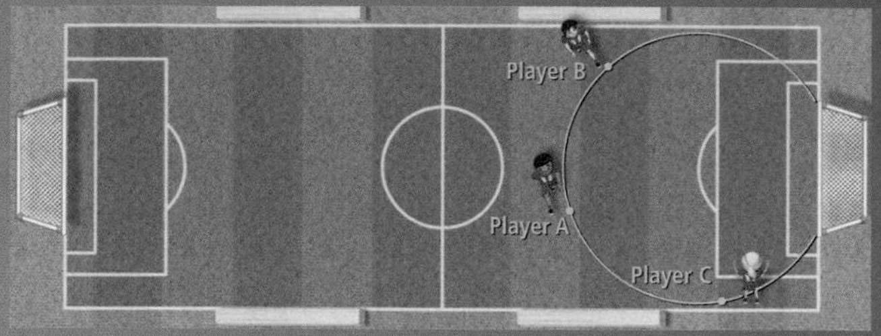

Lesson Vocabulary
- inscribed angle
- intercepted arc

An angle whose vertex is on the circle and whose sides are chords of the circle is an **inscribed angle.** An arc with endpoints on the sides of an inscribed angle, and its other points in the interior of the angle is an **intercepted arc.** In the diagram, inscribed $\angle C$ intercepts $\overset{\frown}{AB}$.

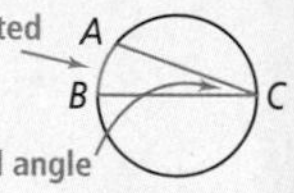

Essential Understanding Angles formed by intersecting lines have a special relationship to the arcs the intersecting lines intercept. In this lesson, you will study arcs formed by inscribed angles.

take note **Theorem 12-11** Inscribed Angle Theorem

The measure of an inscribed angle is half the measure of its intercepted arc.

$$m\angle B = \frac{1}{2} m\overset{\frown}{AC}$$

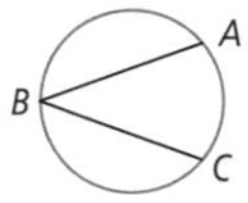

12-3 Preparing to Teach

BIG ideas **Reasoning and Proof**
Measurement

ESSENTIAL UNDERSTANDINGS

- Angles formed by intersecting lines have a special relationship to the arcs the intersecting lines intercept.
- Specifically, arcs intercepted by chords that form inscribed angles are related to the inscribed angles.

Math Background

Similar to the relationship between a central angle and the arc it intercepts, an inscribed angle is related to its intercepted arc. Students will learn how to calculate the measure of either the inscribed angle or intercepted arc based on the known measure. Corollaries from this theorem lead to observations about congruent inscribed angles, right angles within circles, and the angles of an inscribed quadrilateral.

The proof of the Inscribed Angle Theorem uses a divide-and-conquer strategy. The proof is important for students to see, because it illustrates how a complex situation can be broken into simpler situations that are easier to prove.

Mathematical Practices

Attend to precision. Students will examine the mathematical argument posed by Theorem 12-12 and will use a series of graphical representations to discover its truth.

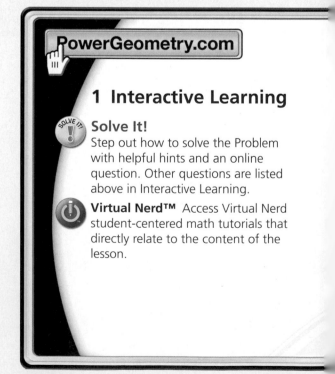

PowerGeometry.com

1 Interactive Learning

Solve It!
Step out how to solve the Problem with helpful hints and an online question. Other questions are listed above in Interactive Learning.

Virtual Nerd™ Access Virtual Nerd student-centered math tutorials that directly relate to the content of the lesson.

To prove Theorem 12-11, there are three cases to consider.

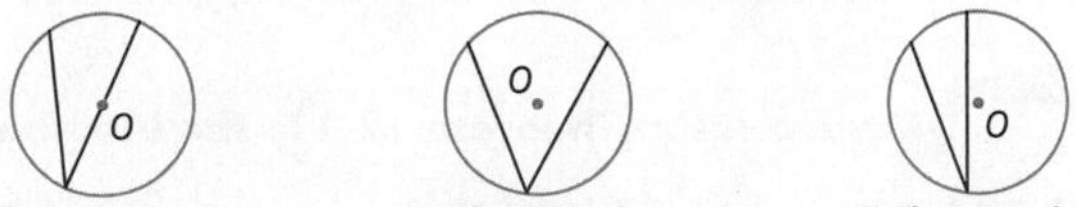

I: The center is on a side of the angle. II: The center is inside the angle. III: The center is outside the angle.

Below is a proof of Case I. You will prove Case II and Case III in Exercises 26 and 27.

Proof **Proof of Theorem 12-11, Case I**

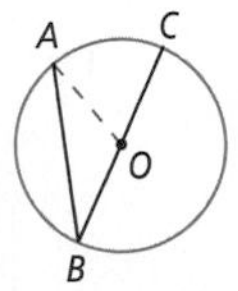

Given: $\odot O$ with inscribed $\angle B$ and diameter $\overline{BC}$

Prove: $m\angle B = \frac{1}{2} m\widehat{AC}$

Draw radius $\overline{OA}$ to form isosceles $\triangle AOB$ with $OA = OB$ and, hence, $m\angle A = m\angle B$ (Isosceles Triangle Theorem).

$m\angle AOC = m\angle A + m\angle B$	Triangle Exterior Angle Theorem
$m\widehat{AC} = m\angle AOC$	Definition of measure of an arc
$m\widehat{AC} = m\angle A + m\angle B$	Substitute.
$m\widehat{AC} = 2m\angle B$	Substitute and simplify.
$\frac{1}{2} m\widehat{AC} = m\angle B$	Divide each side by 2.

Problem 1 **Using the Inscribed Angle Theorem**

Plan

Which variable should you solve for first? You know the inscribed angle that intercepts $\widehat{PT}$, which has the measure *a*. You need *a* to find *b*. So find *a* first.

What are the values of *a* and *b*?

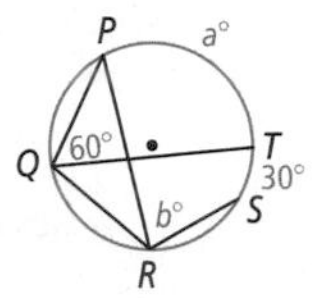

$m\angle PQT = \frac{1}{2} m\widehat{PT}$	Inscribed Angle Theorem
$60 = \frac{1}{2} a$	Substitute.
$120 = a$	Multiply each side by 2.
$m\angle PRS = \frac{1}{2} m\widehat{PS}$	Inscribed Angle Theorem
$m\angle PRS = \frac{1}{2}(m\widehat{PT} + m\widehat{TS})$	Arc Addition Postulate
$b = \frac{1}{2}(120 + 30)$	Substitute.
$b = 75$	Simplify.

Got It? 1. a. In $\odot O$, what is $m\angle A$?

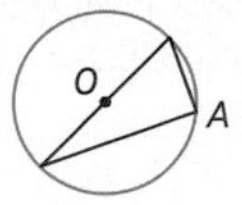

b. What are $m\angle A$, $m\angle B$, $m\angle C$, and $m\angle D$?

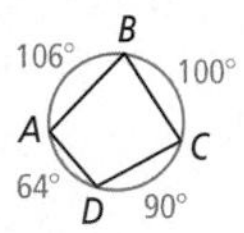

c. What do you notice about the sums of the measures of the opposite angles in the quadrilateral in part (b)?

Have students compare the three cases. Ask them why each case must be considered separately. Emphasize to students that properties of equality must be used to show that two quantities are equal.

Problem 1

Q Which arc is related to $\angle PQT$? **[$\widehat{PT}$]**

Q How can you find the measure of the arc? **[Multiply the measure of the inscribed angle by 2.]**

Q Which angle is related to $\widehat{PT}$? **[$\angle PRS$]**

Q How can you find the measure of the angle? **[Divide the associated arc measure by 2.]**

Got It? ERROR PREVENTION

Have students identify the arc that is intercepted by the angle. They must find the sum of the smaller arcs to find the measure of the associated arc.

2 Guided Instruction

Each Problem is worked out and supported online.

Problem 1
Using the Inscribed Angle Theorem

Problem 2
Using Corollaries to Find Angle Measures

Problem 3
Using Arc Measure

Support in Geometry Companion
- Vocabulary
- Key Concepts
- Got It?

Answers

Solve It!

No. Note to teacher: Through some method, students must determine that all three △ are ≅.

Got It?

1a. 90

b. $m\angle A = 95$, $m\angle B = 77$, $m\angle C = 85$, and $m\angle D = 103$

c. The sum of the measures of opposite △ is 180.

Take Note

Have students review their answers for the Solve It at the beginning of this lesson. Then, challenge students to explain the logic of each corollary. They should use Theorem 12-11 for each Corollary. For Corollary 3, ask students to write each angle measure as an expression involving its corresponding arc measure.

Problem 2

Q In 2A, what information do you not need? **[the measures of the small arcs]**

Q What type of arc does $\angle 1$ intercept? **[a semicircle]**

Q In 2B, which arc does $\angle 2$ intercept? **[the same arc as the angle marked 38°]**

Got It? **VISUAL LEARNERS**

Students may benefit from drawing each inscribed angle separately. Once students identify the measure of each angle, have them identify the theorem or corollary that justifies their answers.

You will use three corollaries to the Inscribed Angle Theorem to find measures of angles in circles. The first corollary may confirm an observation you made in the Solve It.

take note **Corollaries to Theorem 12-11: The Inscribed Angle Theorem**

Corollary 1
Two inscribed angles that intercept the same arc are congruent.

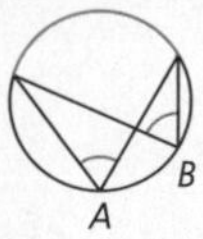

Corollary 2
An angle inscribed in a semicircle is a right angle.

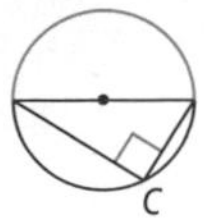

Corollary 3
The opposite angles of a quadrilateral inscribed in a circle are supplementary.

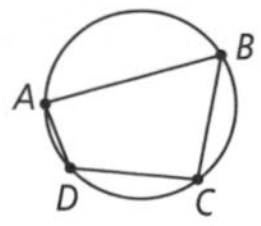

You will prove these corollaries in Exercises 31–33.

Problem 2 **Using Corollaries to Find Angle Measures**

Think
Is there too much information?
Each diagram has more information than you need. Focus on what you need to find.

What is the measure of each numbered angle?

A

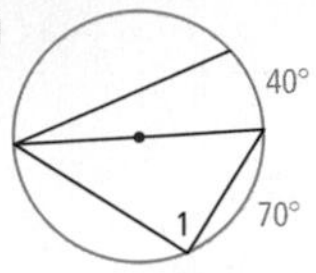

$\angle 1$ is inscribed in a semicircle.
By Corollary 2, $\angle 1$ is a right angle, so $m\angle 1 = 90$.

B

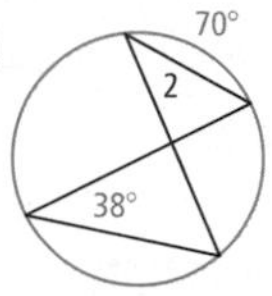

$\angle 2$ and the 38° angle intercept the same arc. By Corollary 1, the angles are congruent, so $m\angle 2 = 38$.

Got It? **2.** In the diagram at the right, what is the measure of each numbered angle?

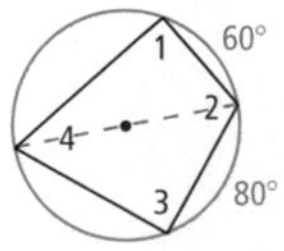

The following diagram shows point *A* moving along the circle until a tangent is formed. From the Inscribed Angle Theorem, you know that in the first three diagrams $m\angle A$ is $\frac{1}{2}m\widehat{BC}$. As the last diagram suggests, this is also true when *A* and *C* coincide.

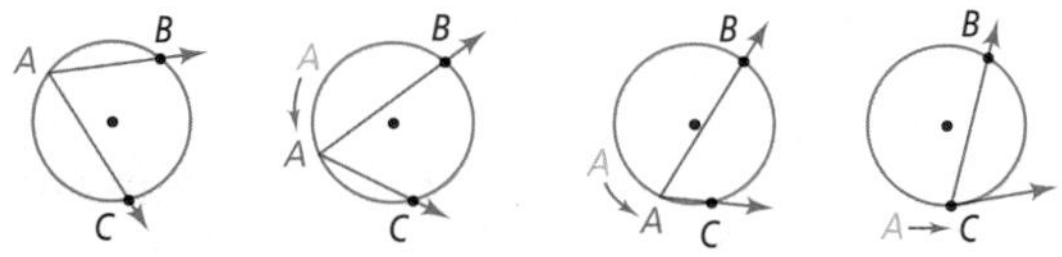

Additional Problems

1. What are the values of *a* and *b*?

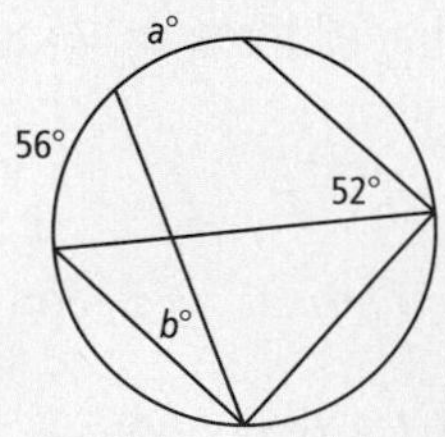

ANSWER $a = 48$, $b = 28$

2. What is the measure of each numbered angle?

a.

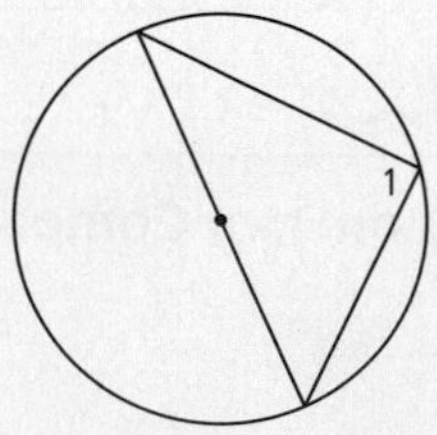

b.

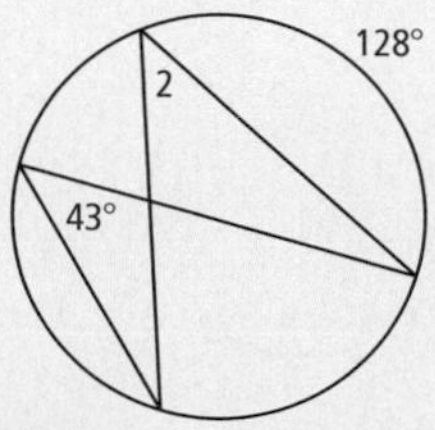

ANSWER a. 90 **b.** 43

3. In the diagram, $\overleftrightarrow{AC}$ is a tangent to the circle at *B*. If the measure of $\widehat{BED}$ is 214, what is $m\angle DBC$?

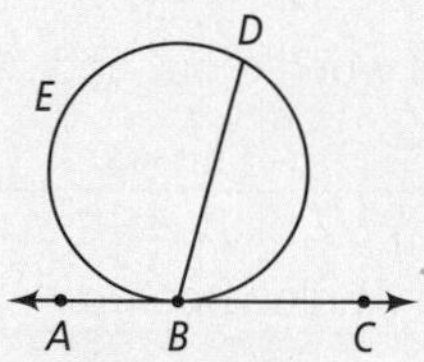

ANSWER 73

take note

Theorem 12-12

The measure of an angle formed by a tangent and a chord is half the measure of the intercepted arc.

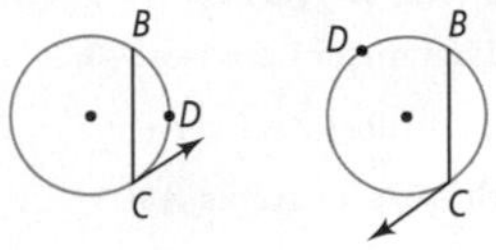

$m\angle C = \frac{1}{2} m\widehat{BDC}$

You will prove Theorem 12-12 in Exercise 34.

Problem 3 Using Arc Measure

In the diagram, $\overleftrightarrow{SR}$ is a tangent to the circle at Q. If $m\widehat{PMQ} = 212$, what is $m\angle PQR$?

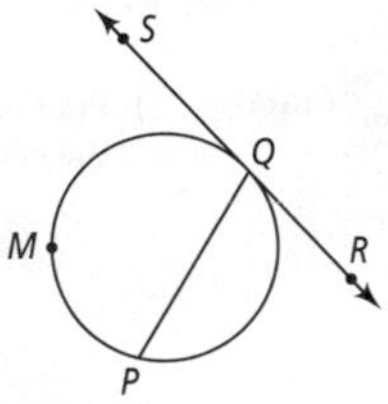

Know

- $\overleftrightarrow{SR}$ is tangent to the circle at Q
- $m\widehat{PMQ} = 212$

Need

$m\angle PQR$

Plan

$m\angle PQS + m\angle PQR = 180$. So first find $m\angle PQS$ using $\widehat{PMQ}$.

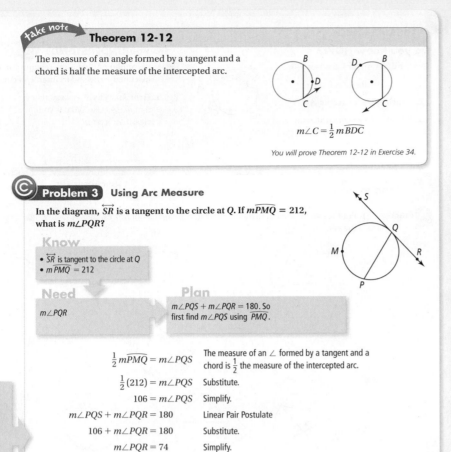

$\frac{1}{2} m\widehat{PMQ} = m\angle PQS$	The measure of an $\angle$ formed by a tangent and a chord is $\frac{1}{2}$ the measure of the intercepted arc.
$\frac{1}{2}(212) = m\angle PQS$	Substitute.
$106 = m\angle PQS$	Simplify.
$m\angle PQS + m\angle PQR = 180$	Linear Pair Postulate
$106 + m\angle PQR = 180$	Substitute.
$m\angle PQR = 74$	Simplify.

Think

How can you check the answer?
One way is to use $m\angle PQR$ to find $m\widehat{PQ}$. Confirm that $m\widehat{PQ} + m\widehat{PMQ} = 360$.

Got It? 3. a. In the diagram at the right, $\overline{KJ}$ is tangent to $\odot O$. What are the values of x and y?

b. Reasoning In part (a), an inscribed angle ($\angle Q$) and an angle formed by a tangent and chord ($\angle KJL$) intercept the same arc. What is always true of these angles? Explain.

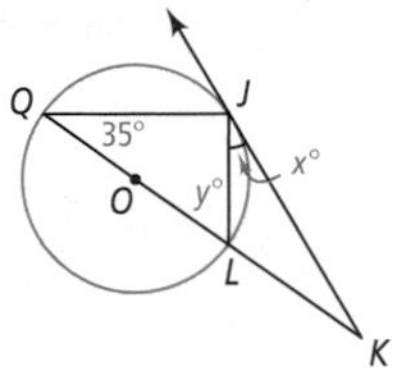

Take Note

Ask students to identify the tangent and the chord in the diagram. Be sure that students connect this theorem with Theorem 12-11.

Problem 3

Q What angle is related to $\widehat{PMQ}$? **[$\angle PQS$]**

Q How is $\angle PQS$ related to $\angle PQR$? **[They are supplementary.]**

Q How can you find the measure of this angle? **[Subtract the measure of $\angle PQS$ from 180°.]**

Got It? VISUAL LEARNERS

Q What type of triangle is $\triangle JLQ$? Justify your answer. **[$\triangle JLQ$ is a right triangle because $\angle LJQ$ is an inscribed angle that intercepts a semicircle.]**

Q How can you find y? **[Subtract 35 from 90.]**

Q What arc does $\angle KJL$ intercept? **[$\widehat{JL}$]**

Q Which other inscribed angle intercepts the same arc? **[$\angle JQL$]**

Answers

Got It? (continued)

2. $m\angle 1 = 90$, $m\angle 2 = 110$, $m\angle 3 = 90$, $m\angle 4 = 70$

3a. $x = 35$, $y = 55$

b. An inscribed $\angle$, and an $\angle$ formed by a tangent and chord, are both equal to half the measure of the intercepted arc. Since the $\angle$s intercept the same arc, their measures are = and they are $\cong$.

3 Lesson Check

Do you know HOW?

- If students have difficulty with Exercise 1, then have them draw the angle separately.

Do you UNDERSTAND?

- If students have difficulty with Exercise 5, then have them review Problem 2 and compare the drawings.

Close

Q How can you find the measure of an inscribed angle? **[Divide the measure of its intercepted arc by 2.]**

Q What is the relationship between an angle formed by a tangent and a chord and the arc it intercepts? **[The measure of the angle is half the measure of the intercepted arc.]**

Lesson Check

Do you know HOW?

Use the diagram for Exercises 1–3.

1. Which arc does $\angle A$ intercept?
2. Which angle intercepts $\overset{\frown}{ABC}$?
3. Which angles of quadrilateral $ABCD$ are supplementary?

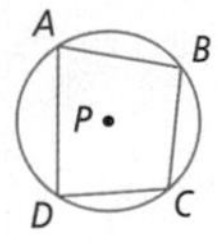

Do you UNDERSTAND? MATHEMATICAL PRACTICES

4. **Vocabulary** What is the relationship between an inscribed angle and its intercepted arc?
5. **Error Analysis** A classmate says that $m\angle A = 90$. What is your classmate's error?

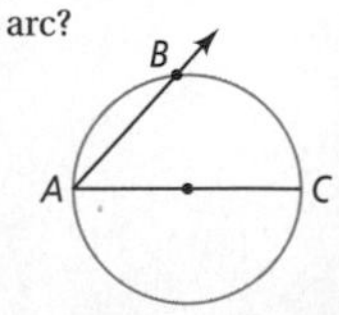

Practice and Problem-Solving Exercises MATHEMATICAL PRACTICES

A Practice

Find the value of each variable. For each circle, the dot represents the center.

See Problems 1 and 2.

6.
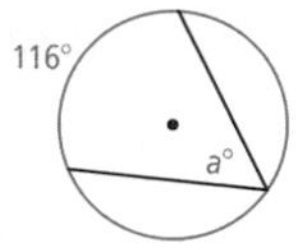

7.
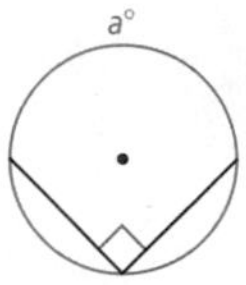

8.
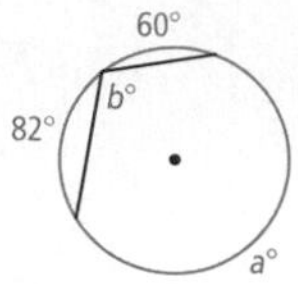

9.
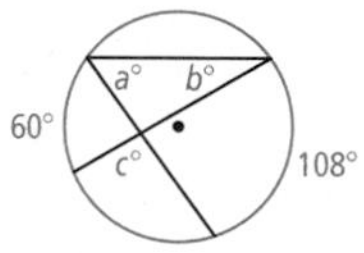

10.
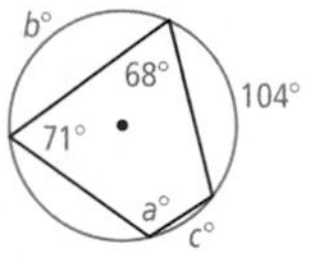

11.
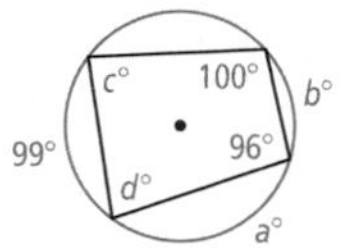

12.
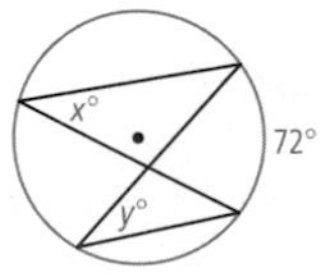

13.
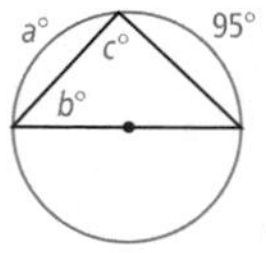

14.
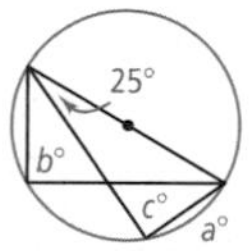

15.
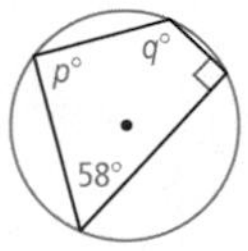

Find the value of each variable. Lines that appear to be tangent are tangent.

See Problem 3.

16.
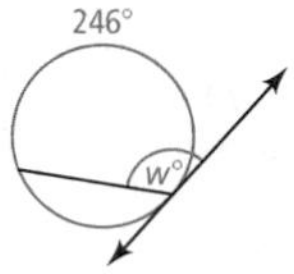

17.
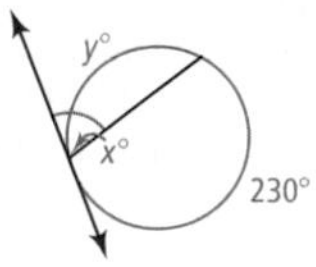

18.
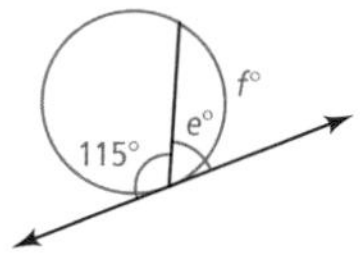

B Apply

19. **Writing** A parallelogram inscribed in a circle must be what kind of parallelogram? Explain.

PowerGeometry.com

3 Lesson Check

For a digital lesson check, use the Got It questions.

Support in Geometry Companion

- Lesson Check

4 Practice

Assign homework to individual students or to an entire class.

Answers

Lesson Check

1. $\overset{\frown}{BD}$
2. $\angle D$
3. $\angle A$ and $\angle C$ are suppl., and $\angle B$ and $\angle D$ are suppl.
4. Sample answer: For inscribed $\angle ABC$, B is the vertex and A, B, and C are points on the circle. The intercepted arc of $\angle ABC$ consists of points A, C, and all the points on the circle in the interior of $\angle ABC$.
5. $\angle A$ is not inscribed in a semicircle.

Practice and Problem-Solving Exercises

6. 58
7. 180
8. $a = 218$, $b = 109$
9. $a = 54$, $b = 30$, $c = 96$
10. $a = 112$, $b = 120$, $c = 38$
11. $a = 101$, $b = 67$, $c = 84$, $d = 80$
12. $x = 36$, $y = 36$
13. $a = 85$, $b = 47.5$, $c = 90$
14. $a = 50$, $b = 90$, $c = 90$
15. $p = 90$, $q = 122$
16. 123
17. $x = 65$, $y = 130$
18. $e = 65$, $f = 130$
19. Rectangle; opposite ∠s are ≅ (because figure is ▱) and suppl. (because opp. ∠s intercept arcs whose measures sum to 360). ≅ suppl. ∠s are rt. ∠s, so the inscribed ▱ must be a rectangle.

Find each indicated measure for $\odot O$.

20. a. $m\widehat{BC}$
b. $m\angle B$
c. $m\angle C$
d. $m\widehat{AB}$

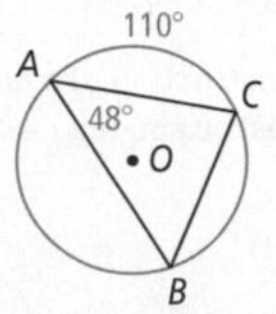

21. a. $m\angle A$
b. $m\widehat{CE}$
c. $m\angle C$
d. $m\angle D$
e. $m\angle ABE$

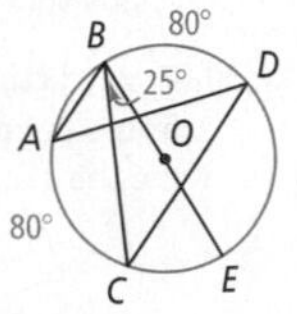

22. Think About a Plan What kind of trapezoid can be inscribed in a circle? Justify your response.
- Draw several diagrams to make a conjecture.
- How can parallel lines help?

Find the value of each variable. For each circle, the dot represents the center.

23.

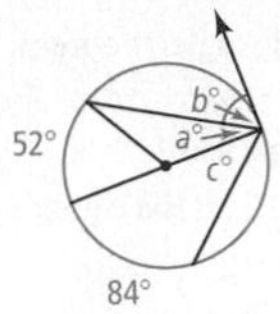

24.

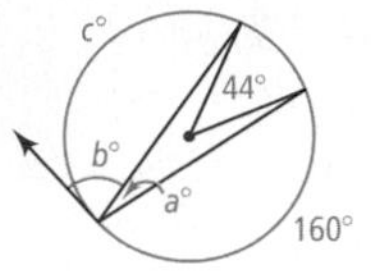

25.

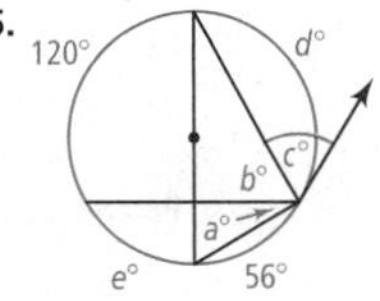

Write a proof for Exercises 26 and 27.

26. Inscribed Angle Theorem, Case II
Proof **Given:** $\odot O$ with inscribed $\angle ABC$
Prove: $m\angle ABC = \frac{1}{2}m\widehat{AC}$

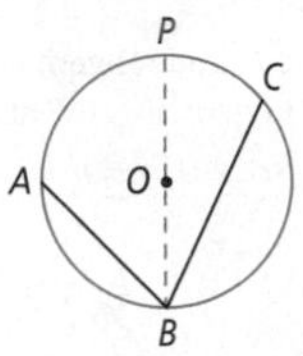

(*Hint:* Use the Inscribed Angle Theorem, Case I.)

27. Inscribed Angle Theorem, Case III
Proof **Given:** $\odot S$ with inscribed $\angle PQR$
Prove: $m\angle PQR = \frac{1}{2}m\widehat{PR}$

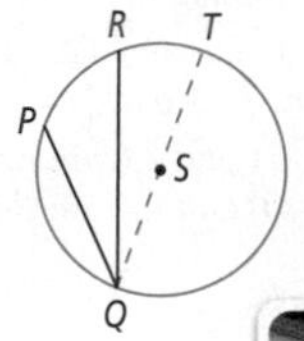

(*Hint:* Use the Inscribed Angle Theorem, Case I.)

28. Television The director of a telecast wants the option of showing the same scene from three different views.
a. Explain why cameras in the positions shown in the diagram will transmit the same scene.
b. Reasoning Will the scenes look the same when the director views them on the control room monitors? Explain.

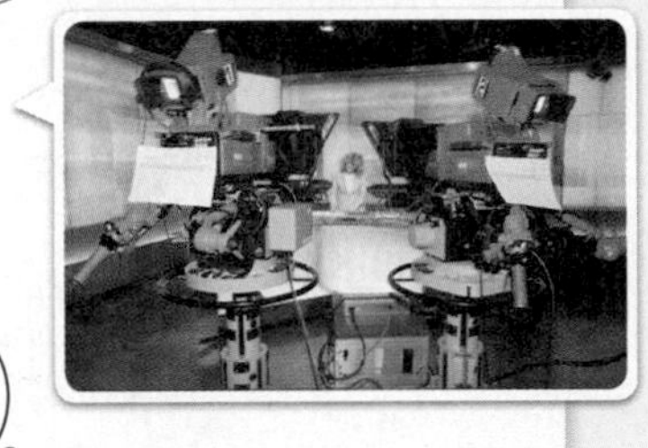

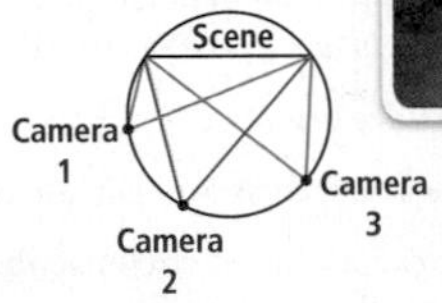

4 Practice

ASSIGNMENT GUIDE
Basic: 6–18 all, 19, 20–24 even, 28–29
Average: 7–17 odd, 19–34
Advanced: 7–17 odd, 19–39

Mathematical Practices are supported by exercises with red headings. Here are the Practices supported in this lesson:

MP 1: Persevere in Solving Problems Ex. 22
MP 3: Construct Arguments Ex. 28b, 29, 35–37
MP 3: Critique the Reasoning of Others Ex. 5

Applications exercises have blue headings. Exercise 28 supports MP 4: Model.

EXERCISE 24: Use the Think About a Plan worksheet in the **Practice and Problem Solving Workbook** (also available in the Teaching Resources in print and online) to further support students' development in becoming independent learners.

HOMEWORK QUICK CHECK
To check students' understanding of key skills and concepts, go over Exercises 9, 17, 22, 24, and 29.

20a. 96
b. 55
c. 77
d. 154
21a. 40
b. 50
c. 40
d. 40
e. 65
22. Isosc. trapezoid; answers may vary. Sample: For inscribed trapezoid $ABCD$, $\angle A$ must be suppl. to $\angle C$ (Corollary 3 to Thm. 12-11), and $\angle C$ must be suppl. to $\angle B$ (same-side int. ∡ of parallel lines are suppl). So $\angle A \cong \angle B$, and the trapezoid must be isosc.
23. $a = 26$, $b = 64$, $c = 42$
24. $a = 22$, $b = 78$, $c = 156$
25. $a = 30$, $b = 60$, $c = 62$, $d = 124$, $e = 60$

26. $\odot O$ with inscribed $\angle ABC$ (given); $m\angle ABO = \frac{1}{2}m\widehat{AP}$ and $m\angle OBC = \frac{1}{2}m\widehat{PC}$ (Inscribed $\angle$ Thm., Case I); $m\angle ABO + m\angle OBC = m\angle ABC$ ($\angle$ Add. Post.); $\frac{1}{2}m\widehat{AP} + \frac{1}{2}m\widehat{PC} = m\angle ABC$ (Subst. Prop.); $\frac{1}{2}(m\widehat{AP} + m\widehat{PC}) = m\angle ABC$ (Distr. Prop.); $\frac{1}{2}m\widehat{AC} = m\angle ABC$ (Arc Add. Post.)

27. $\odot S$ with inscribed $\angle PQR$ (given); $m\angle PQT = \frac{1}{2}m\widehat{PT}$ (Inscribed $\angle$ Thm., Case I); $m\angle RQT = \frac{1}{2}m\widehat{RT}$ (Inscribed $\angle$ Thm., Case I); $m\widehat{PR} = m\widehat{PT} - m\widehat{RT}$ (Arc Add. Post.); $m\angle PQR = m\angle PQT - m\angle RQT$ ($\angle$ Add. Post.); $m\angle PQR = \frac{1}{2}m\widehat{PT} - \frac{1}{2}m\widehat{RT}$ (Subst. Prop.); $m\angle PQR = \frac{1}{2}m\widehat{PR}$ (Subst. Prop.)

28. Answers may vary. Sample:
a. If the cameras' lenses open at ≅ ∡, then in the positions shown they share the same arc of the scene.
b. No; the distances from each position of the scene to each camera affect the look of the scene.

Answers

Practice and Problem-Solving Exercises (continued)

29. No; since opposite $\angle$s of a quadrilateral inscribed in a circle must be supplementary, the only rhombus that meets the criteria is a square.

30. $\angle ACB$ is a rt. $\angle$ because it is inscribed in semicircle $\overset{\frown}{ACB}$, so $\overline{AC} \perp \overleftrightarrow{BC}$. If a line is $\perp$ to a radius at its endpoint, it is tangent to the circle.

31. $\odot O$, $\angle A$ intercepts $\overset{\frown}{BC}$, and $\angle D$ intercepts $\overset{\frown}{BC}$ (Given); $m\angle A = \frac{1}{2}m\overset{\frown}{BC}$ and $m\angle D = \frac{1}{2}m\overset{\frown}{BC}$ (Inscribed $\angle$ Thm.); $m\angle A = m\angle D$ (Subst. Prop.); $\angle A \cong \angle D$ (Def. of $\cong$ $\angle$s).

32. $\odot O$ with $\angle CAB$ inscribed in a semicircle (Given); $m\angle CAB = \frac{1}{2}m\overset{\frown}{BDC}$ (Inscribed $\angle$ Thm.); $m\overset{\frown}{BDC} = 180$ (A semicircle has a measure of 180.); $m\angle CAB = 90$ (Subst. Prop.); $\angle CAB$ is a rt. $\angle$ (Def. of rt. $\angle$).

33. Quadrilateral ABCD inscribed in $\odot O$ (Given); $m\angle A = \frac{1}{2}m\overset{\frown}{BCD}$ and $m\angle C = \frac{1}{2}m\overset{\frown}{BAD}$ (Inscribed $\angle$ Thm.); $m\angle A + m\angle C = \frac{1}{2}m\overset{\frown}{BCD} + \frac{1}{2}m\overset{\frown}{BAD}$ (Add. Prop.); $m\overset{\frown}{BCD} + m\overset{\frown}{BAD} = 360$ (Arc measure of circle is 360.); $\frac{1}{2}m\overset{\frown}{BCD} + \frac{1}{2}m\overset{\frown}{BAD} = 180$ (Mult. Prop.) $m\angle A + m\angle C = 180$ (Subst. Prop.); $\angle A$ and $\angle C$ are suppl. (Def. of suppl.); $m\angle B = \frac{1}{2}m\overset{\frown}{ADC}$ and $m\angle D = \frac{1}{2}m\overset{\frown}{ABC}$ (Inscribed $\angle$ Thm.); $m\angle B + m\angle D = \frac{1}{2}m\overset{\frown}{ADC} + \frac{1}{2}m\overset{\frown}{ABC}$ (Add. Prop.); $m\overset{\frown}{ADC} + m\overset{\frown}{ABC} = 360$ (Arc measure of circle is 360.); $\frac{1}{2}m\overset{\frown}{ADC} + \frac{1}{2}m\overset{\frown}{ABC} = 180$ (Mult. Prop.) $m\angle B + m\angle D = 180$ (Subst. Prop.); $\angle B$ and $\angle D$ are suppl. (Def. of suppl. $\angle$s).

34. $\overline{GH}$ and tangent ℓ intersecting $\odot E$ at H (Given); draw $\overleftrightarrow{HE}$ intersecting $\odot E$ at D so $\overline{HD}$ is a diameter (2 pts. determine a line.); $\angle DHI$ is a rt. $\angle$ (A tangent line is $\perp$ to radius at pt. of tangency.); $m\angle DHI = 90$ (Def. of rt. $\angle$); $m\overset{\frown}{DGH} = 180$ (A semicircle has a measure of 180.); $m\angle DHG + m\angle GHI = m\angle DHI$ ($\angle$ Add. Post.); $m\overset{\frown}{DG} + m\overset{\frown}{GFH} = m\overset{\frown}{DGH}$ (Arc Add. Post.); $m\angle DHG + m\angle GHI = 90$ (Subst. Prop.); $m\overset{\frown}{DG} + m\overset{\frown}{GFH} = 180$ (Subst. Prop.); $\frac{1}{2}(m\overset{\frown}{DG} + m\overset{\frown}{GFH}) = 90$ (Mult. Prop. of =); $m\angle DHG + m\angle GHI = \frac{1}{2}(m\overset{\frown}{DG} + m\overset{\frown}{GFH})$ (Subst. Prop.); $m\angle DHG = \frac{1}{2}m\overset{\frown}{DG}$ (Inscribed $\angle$ Thm.); $\frac{1}{2}m\overset{\frown}{DG} + m\angle GHI = \frac{1}{2}m\overset{\frown}{DG} + \frac{1}{2}m\overset{\frown}{GFH}$ (Subst. Prop. and Distr. Prop.); $m\angle GHI = \frac{1}{2}m\overset{\frown}{GFH}$ (Subtr. Prop. of =).

35. false

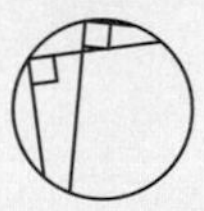

36. True; the measure of the intercepted arc must be 2 • 90 or 180, so the intercepted arc is a semicircle.

37. True; opposite $\angle$s in an inscribed quadrilateral intercept nonoverlapping arcs totaling 360 and inscribed $\angle$s have half the measure of the intercepted arcs, so the opposite $\angle$s are suppl.

29. Reasoning Can a rhombus that is not a square be inscribed in a circle? Justify your answer.

30. Constructions The diagrams below show the construction of a tangent to a circle from a point outside the circle. Explain why $\overleftrightarrow{BC}$ must be tangent to $\odot A$. (*Hint:* Copy the third diagram and draw $\overline{AC}$.)

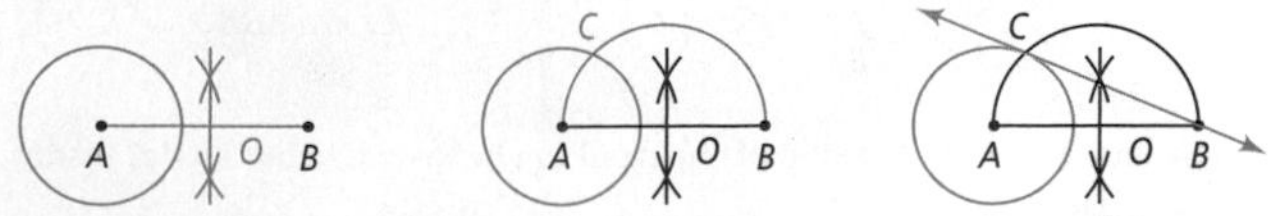

Given: $\odot A$ and point B Construct the midpoint of $\overline{AB}$. Label the point O.

Construct a semicircle with radius OA and center O. Label its intersection with $\odot A$ as C.

Draw $\overleftrightarrow{BC}$.

Write a proof for Exercises 31–34.

31. Inscribed Angle Theorem, Corollary 1

Proof **Given:** $\odot O$, $\angle A$ intercepts $\overset{\frown}{BC}$, $\angle D$ intercepts $\overset{\frown}{BC}$.

Prove: $\angle A \cong \angle D$

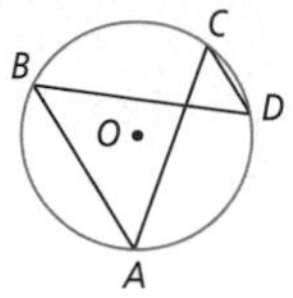

32. Inscribed Angle Theorem, Corollary 2

Proof **Given:** $\odot O$ with $\angle CAB$ inscribed in a semicircle

Prove: $\angle CAB$ is a right angle.

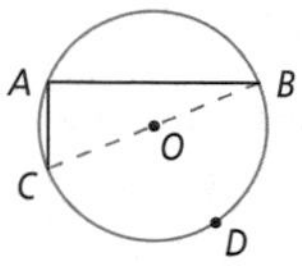

33. Inscribed Angle Theorem, Corollary 3

Proof **Given:** Quadrilateral $ABCD$ inscribed in $\odot O$

Prove: $\angle A$ and $\angle C$ are supplementary. $\angle B$ and $\angle D$ are supplementary.

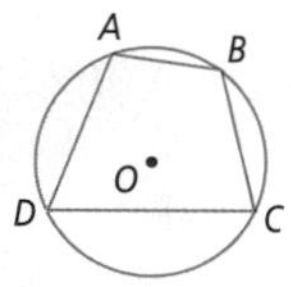

34. Theorem 12-12

Proof **Given:** $\overline{GH}$ and tangent ℓ intersecting $\odot E$ at H

Prove: $m\angle GHI = \frac{1}{2}m\overset{\frown}{GFH}$

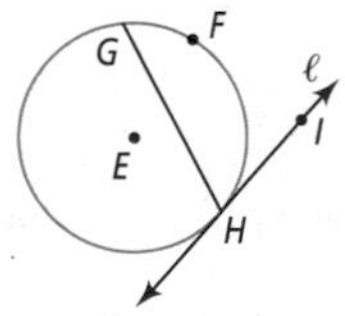

Challenge

Reasoning Is the statement *true* or *false*? If it is true, give a convincing argument. If it is false, give a counterexample.

35. If two angles inscribed in a circle are congruent, then they intercept the same arc.

36. If an inscribed angle is a right angle, then it is inscribed in a semicircle.

37. A circle can always be circumscribed about a quadrilateral whose opposite angles are supplementary.

38. Prove that if two arcs of a circle are included between parallel chords, then the arcs are congruent.
Proof

39. Constructions Draw two segments. Label their lengths *x* and *y*. Construct the geometric mean of *x* and *y*. (*Hint:* Recall a theorem about a geometric mean.)

Apply What You've Learned

MATHEMATICAL PRACTICES MP 1

Look back at the information given on page 761 about the logo for the showroom display. The diagram of the logo is shown again below.

Consider relationships of angles and arcs in the diagram. Select all of the following that are true. Explain your reasoning.

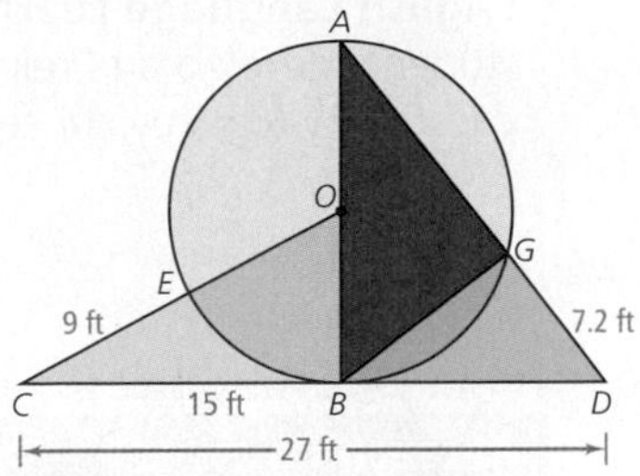

A. $\angle ADB$ is an inscribed angle in $\odot O$.

B. $\angle AGB$ is an inscribed angle in $\odot O$.

C. $\angle AGB$ intercepts $\overset{\frown}{GB}$.

D. $\angle AGB$ intercepts $\overset{\frown}{AEB}$.

E. The measure of $\overset{\frown}{AG}$ is half the measure of $\angle AGB$.

F. $\triangle AGB$ is a right triangle.

G. $\triangle ABG \sim \triangle BDG$

38.

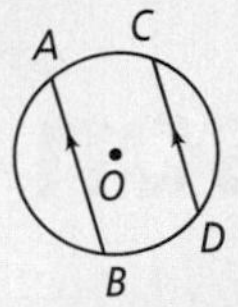

Given: $\overline{AB} \parallel \overline{CD}$

Prove: $m\overset{\frown}{AC} \cong m\overset{\frown}{BD}$

Proof: $\odot O$ with $\overline{AB} \parallel \overline{CD}$ (given); draw $\overline{AD}$ (2 pts. determine a line); $\angle CDA \cong \angle DAB$ (|| lines have $\cong$ alt. int. ∠s); $m\overset{\frown}{AC} \cong m\overset{\frown}{BD}$ ($\cong$ inscribed ∠s intercept $\cong$ arcs).

39.

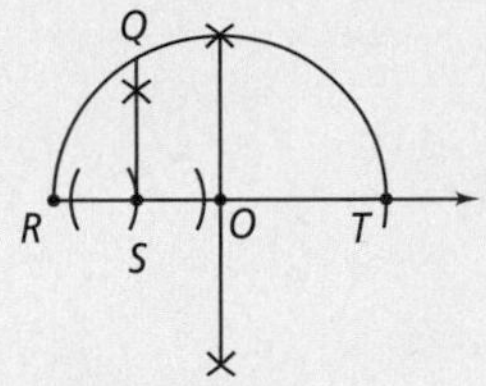

Construct $\overline{RT}$ so $RS = x$ and $ST = y$. Find O, the midpt. of $\overline{RT}$, and draw a semicircle with diameter $\overline{RT}$. Construct $\overline{SQ} \perp \overline{RT}$. Then $\triangle RQT$ is a rt. $\triangle$ and QS is the geometric mean of RS and ST.

Apply What You've Learned

In the Apply What You've Learned for Lesson 12-1, students worked with $\triangle OBC$ in the diagram of the logo on page 761. Here, they turn their attention to $\triangle AGB$ and relationships determined by the inscribed angles in the circle. Later in the chapter, they will use the conclusion that $\triangle AGB$ is a right triangle to find one of its side lengths.

Mathematical Practices

Students continue to **make sense of the problem** as they analyze the relationships of the angles and arcs in the diagram. (MP 1)

ANSWERS

Choices B, D, F, and G are all true.

12-3 Lesson Resources

Differentiated Remediation

Instructional Support

Geometry Companion

Students can use the **Geometry Companion** worktext (4 pages) . . .

- New Vocabulary
- Key Concepts
- Got It for each Problem
- Lesson Check

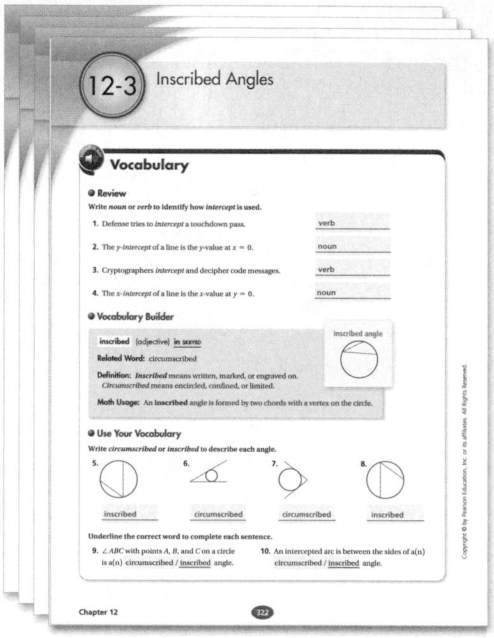

12-3 Inscribed Angles

Vocabulary

Review

Write *noun* or *verb* to identify how *intercept* is used.

1. Defense tries to *intercept* a touchdown pass. verb
2. The *y-intercept* of a line is the *y*-value at $x = 0$. noun
3. Cryptographers *intercept* and decipher code messages. verb
4. The *x-intercept* of a line is the *x*-value at $y = 0$. noun

Vocabulary Builder

inscribed (adjective) in SKRYBD

Related Word: circumscribed

Definition: ***Inscribed*** means written, marked, or engraved on. *Circumscribed* means encircled, confined, or limited.

Math Usage: An **inscribed** angle is formed by two chords with a vertex on the circle.

Use Your Vocabulary

Write *circumscribed* or *inscribed* to describe each angle.

5. inscribed
6. circumscribed
7. circumscribed
8. inscribed

Underline the correct word to complete each sentence.

9. $\angle ABC$ with points *A*, *B*, and *C* on a circle is a(n) circumscribed / inscribed angle.
10. An intercepted arc is between the sides of a(n) circumscribed / inscribed angle.

Chapter 12 322

ELL Support

Use Role Playing Arrange students into heterogeneous groups of four. Assign a student to act as the instructor. Have a student from each group form a temporary group with one student from each of the other groups. Assign each group a concept: Case I, II, or III from Theorem 12-11 or Theorem 12-12. Students in the temporary groups will thoroughly learn about their concepts and be prepared to provide an oral explanation, a drawn example, and a demonstration of how to solve a problem. The problems students can model should be from the lesson, or ones you have specifically assigned. Students will rejoin their groups and peer-teach their concept to the rest of their group members.

5 Assess & Remediate

Lesson Quiz

1. What are the values of *x* and *b*?

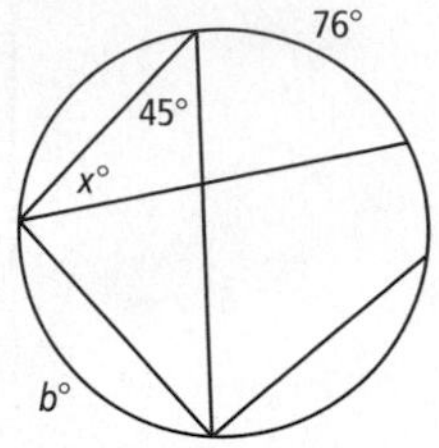

2. Do you UNDERSTAND? In the diagram, $\overleftrightarrow{RT}$ is tangent to the circle at *S*. If the measure of $\overparen{SVU}$ is 138°, what is $m\angle TSU$?

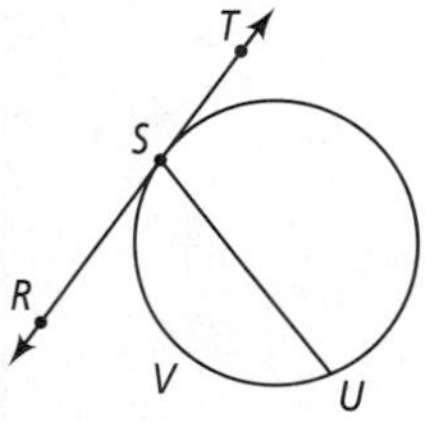

3. What is $m\angle C$?

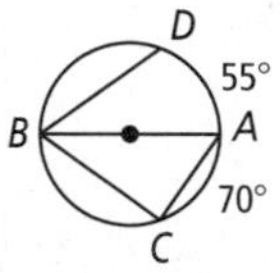

ANSWERS TO LESSON QUIZ

1. $x = 38$, $b = 90$

2. 111

3. 90

PRESCRIPTION FOR REMEDIATION

Use the student work on the Lesson Quiz to prescribe a differentiated review assignment.

Points	Differentiated Remediation
0–1	Intervention
2	On-level
3	Extension

PowerGeometry.com

5 Assess & Remediate

Assign the Lesson Quiz. Appropriate intervention, practice, or enrichment is automatically generated based on student performance.

Intervention

- **Reteaching** (2 pages) Provides reteaching and practice exercises for the key lesson concepts. Use with struggling students or absent students.
- **English Language Learner Support** Helps students develop and reinforce mathematical vocabulary and key concepts.

All-in-One Resources/Online

Reteaching

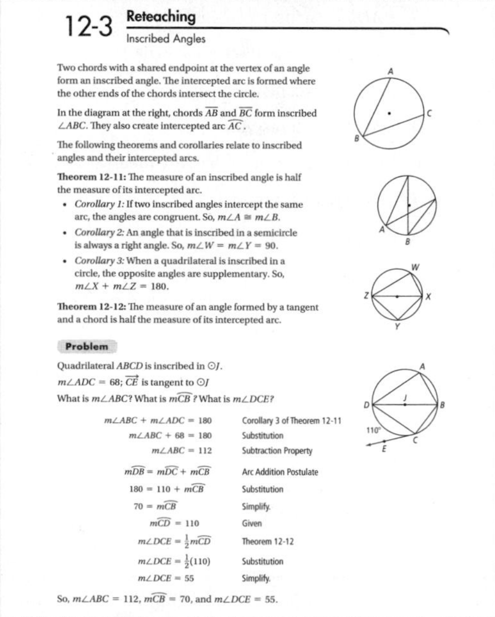

12-3 Reteaching
Inscribed Angles

Two chords with a shared endpoint at the vertex of an angle form an inscribed angle. The intercepted arc is formed where the other ends of the chords intersect the circle.

In the diagram at the right, chords $\overline{AB}$ and $\overline{BC}$ form inscribed $\angle ABC$. They also create intercepted arc $\overparen{AC}$.

The following theorems and corollaries relate to inscribed angles and their intercepted arcs.

Theorem 12-11: The measure of an inscribed angle is half the measure of its intercepted arc.

- *Corollary 1:* If two inscribed angles intercept the same arc, the angles are congruent. So, $m\angle A \cong m\angle B$.
- *Corollary 2:* An angle that is inscribed in a semicircle is always a right angle. So, $m\angle W = m\angle Y = 90$.
- *Corollary 3:* When a quadrilateral is inscribed in a circle, the opposite angles are supplementary. So, $m\angle X + m\angle Z = 180$.

Theorem 12-12: The measure of an angle formed by a tangent and a chord is half the measure of its intercepted arc.

Problem

Quadrilateral *ABCD* is inscribed in $\odot J$.
$m\angle ADC = 68$; $\overrightarrow{CE}$ is tangent to $\odot J$
What is $m\angle ABC$? What is $m\overparen{CB}$? What is $m\angle DCE$?

$m\angle ABC + m\angle ADC = 180$	Corollary 3 of Theorem 12-11
$m\angle ABC + 68 = 180$	Substitution
$m\angle ABC = 112$	Subtraction Property
$m\overparen{DB} = m\overparen{DC} + m\overparen{CB}$	Arc Addition Postulate
$180 = 110 + m\overparen{CB}$	Substitution
$70 = m\overparen{CB}$	Simplify.
$m\overparen{CD} = 110$	Given
$m\angle DCE = \frac{1}{2}m\overparen{CD}$	Theorem 12-12
$m\angle DCE = \frac{1}{2}(110)$	Substitution
$m\angle DCE = 55$	Simplify.

So, $m\angle ABC = 112$, $m\overparen{CB} = 70$, and $m\angle DCE = 55$.

All-in-One Resources/Online

English Language Learner Support

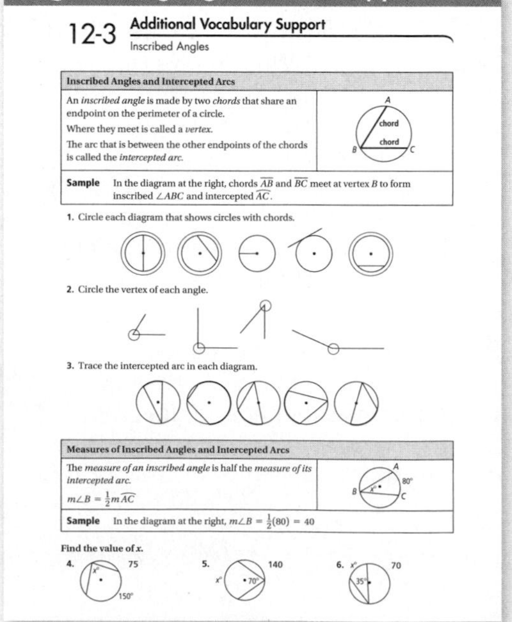

12-3 Additional Vocabulary Support
Inscribed Angles

Inscribed Angles and Intercepted Arcs

An *inscribed angle* is made by two *chords* that share an endpoint on the perimeter of a circle.
Where they meet is called a *vertex*.
The arc that is between the other endpoints of the chords is called the *intercepted arc*.

Sample In the diagram at the right, chords $\overline{AB}$ and $\overline{BC}$ meet at vertex *B* to form inscribed $\angle ABC$ and intercepted $\overparen{AC}$.

1. Circle each diagram that shows circles with chords.
2. Circle the vertex of each angle.
3. Trace the intercepted arc in each diagram.

Measures of Inscribed Angles and Intercepted Arcs

The *measure of an inscribed angle* is half the *measure of its intercepted arc*.
$m\angle B = \frac{1}{2}m\overparen{AC}$

Sample In the diagram at the right, $m\angle B = \frac{1}{2}(80) = 40$

Find the value of *x*.

4. 75
5. 140
6. 70

Differentiated Remediation *continued*

On-Level

- **Practice** (2 pages) Provides extra practice for each lesson. For simpler practice exercises, use the Form K Practice pages found in the All-in-One Teaching Resources and online.
- **Think About a Plan** Helps students develop specific problem-solving skills and strategies by providing scaffolded guiding questions.
- **Standardized Test Prep** Focuses on all major exercises, all major question types, and helps students prepare for the high-stakes assessments.

Extension

- **Enrichment** Provides students with interesting problems and activities that extend the concepts of the lesson.
- **Activities, Games, and Puzzles** Worksheets that can be used for concepts development, enrichment, and for fun!

Practice and Problem Solving Wkbk/ All-in-One Resources/Online

Practice page 1

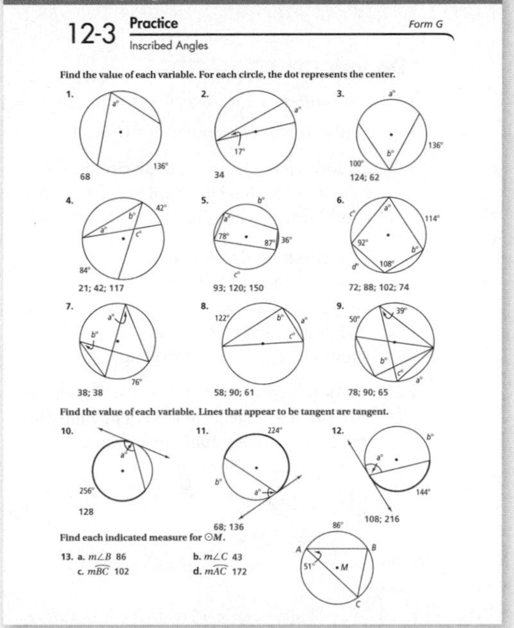

12-3 **Practice** — Inscribed Angles — *Form G*

Find the value of each variable. For each circle, the dot represents the center.

1. 68
2. 34
3. 124; 62
4. 21; 42; 117
5. 93; 120; 150
6. 72; 88; 102; 74
7. 38; 38
8. 58; 90; 61
9. 78; 90; 65

Find the value of each variable. Lines that appear to be tangent are tangent.

10. 128
11. 68; 136
12. 108; 216

Find each indicated measure for $\odot M$.

13. a. $m\angle B$ 86 b. $m\angle C$ 43 c. $m\widehat{BC}$ 102 d. $m\widehat{AC}$ 172

Practice and Problem Solving Wkbk/ All-in-One Resources/Online

Practice page 2

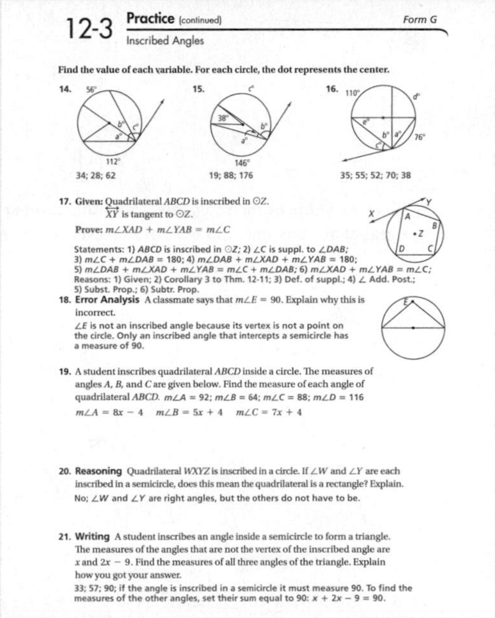

12-3 **Practice** (continued) — Inscribed Angles — *Form G*

Find the value of each variable. For each circle, the dot represents the center.

14. 34; 28; 62
15. 19; 88; 176
16. 35; 55; 52; 70; 38

17. **Given:** Quadrilateral $ABCD$ is inscribed in $\odot Z$. $\overleftrightarrow{XY}$ is tangent to $\odot Z$.
Prove: $m\angle XAD + m\angle YAB = m\angle C$
Statements: 1) $ABCD$ is inscribed in $\odot Z$; 2) $\angle C$ is suppl. to $\angle DAB$; 3) $m\angle C + m\angle DAB = 180$; 4) $m\angle DAB + m\angle XAD + m\angle YAB = 180$; 5) $m\angle DAB + m\angle XAD + m\angle YAB = m\angle C + m\angle DAB$; 6) $m\angle XAD + m\angle YAB = m\angle C$; Reasons: 1) Given; 2) Corollary 3 to Thm. 12-11; 3) Def. of suppl.; 4) $\angle$ Add. Post.; 5) Subst. Prop.; 6) Subtr. Prop.

18. **Error Analysis** A classmate says that $m\angle E = 90$. Explain why this is incorrect.
$\angle E$ is not an inscribed angle because its vertex is not a point on the circle. Only an inscribed angle that intercepts a semicircle has a measure of 90.

19. A student inscribes quadrilateral $ABCD$ inside a circle. The measures of angles A, B, and C are given below. Find the measure of each angle of quadrilateral $ABCD$. $m\angle A = 92$; $m\angle B = 64$; $m\angle C = 88$; $m\angle D = 116$
$m\angle A = 8x - 4$ $m\angle B = 5x + 4$ $m\angle C = 7x + 4$

20. **Reasoning** Quadrilateral $WXYZ$ is inscribed in a circle. If $\angle W$ and $\angle Y$ are each inscribed in a semicircle, does this mean the quadrilateral is a rectangle? Explain.
No; $\angle W$ and $\angle Y$ are right angles, but the others do not have to be.

21. **Writing** A student inscribes an angle inside a semicircle to form a triangle. The measures of the angles that are not the vertex of the inscribed angle are x and $2x - 9$. Find the measures of all three angles of the triangle. Explain how you got your answer.
33; 57; 90; if the angle is inscribed in a semicircle it must measure 90. To find the measures of the other angles, set their sum equal to 90: $x + 2x - 9 = 90$.

All-in-One Resources/Online

Enrichment

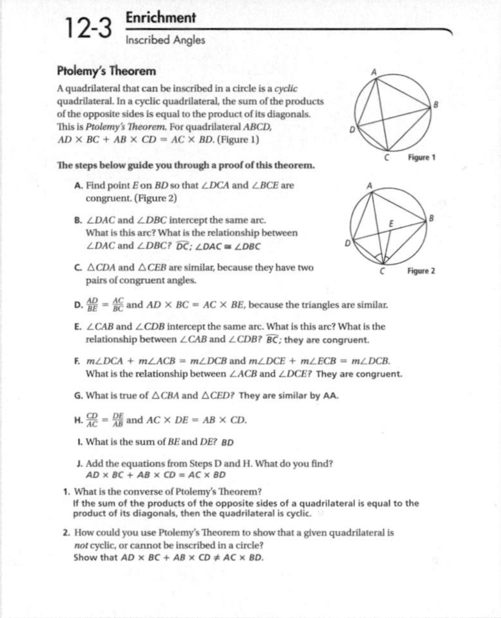

12-3 **Enrichment** — Inscribed Angles

Ptolemy's Theorem

A quadrilateral that can be inscribed in a circle is a *cyclic* quadrilateral. In a cyclic quadrilateral, the sum of the products of the opposite sides is equal to the product of its diagonals. This is *Ptolemy's Theorem.* For quadrilateral $ABCD$, $AD \times BC + AB \times CD = AC \times BD$. (Figure 1)

The steps below guide you through a proof of this theorem.

A. Find point E on BD so that $\angle DCA$ and $\angle BCE$ are congruent. (Figure 2)
B. $\angle DAC$ and $\angle DBC$ intercept the same arc. What is this arc? What is the relationship between $\angle DAC$ and $\angle DBC$? $\widehat{DC}$; $\angle DAC \cong \angle DBC$
C. $\triangle CDA$ and $\triangle CEB$ are similar, because they have two pairs of congruent angles.
D. $\frac{AD}{BE} = \frac{AC}{BC}$ and $AD \times BC = AC \times BE$, because the triangles are similar.
E. $\angle CAB$ and $\angle CDB$ intercept the same arc. What is this arc? What is the relationship between $\angle CAB$ and $\angle CDB$? $\widehat{BC}$; they are congruent.
F. $m\angle DCA + m\angle ACB = m\angle DCB$ and $m\angle DCE + m\angle ECB = m\angle DCB$. What is the relationship between $\angle ACB$ and $\angle DCE$? They are congruent.
G. What is true of $\triangle CBA$ and $\triangle CED$? They are similar by AA.
H. $\frac{CD}{AC} = \frac{DE}{AB}$ and $AC \times DE = AB \times CD$.
I. What is the sum of BE and DE? BD
J. Add the equations from Steps D and H. What do you find? $AD \times BC + AB \times CD = AC \times BD$

1. What is the converse of Ptolemy's Theorem?
If the sum of the products of the opposite sides of a quadrilateral is equal to the product of its diagonals, then the quadrilateral is cyclic.
2. How could you use Ptolemy's Theorem to show that a given quadrilateral is *not* cyclic, or cannot be inscribed in a circle?
Show that $AD \times BC + AB \times CD \neq AC \times BD$.

Figure 1

Figure 2

Practice and Problem Solving Wkbk/ All-in-One Resources/Online

Think About a Plan

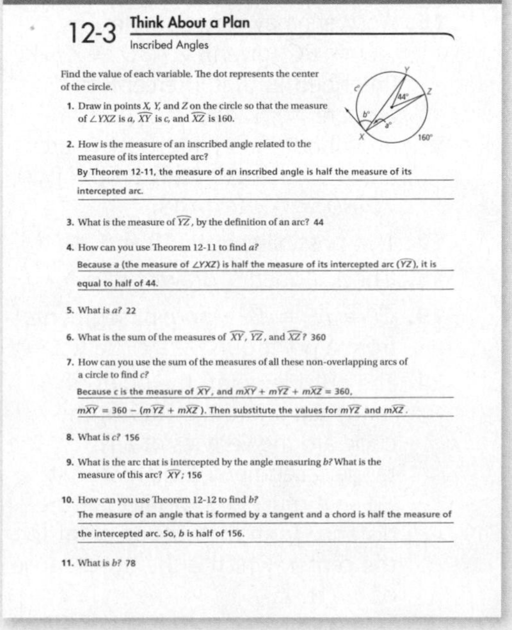

12-3 **Think About a Plan** — Inscribed Angles

Find the value of each variable. The dot represents the center of the circle.

1. Draw in points X, Y, and Z on the circle so that the measure of $\angle YXZ$ is a, $\widehat{XY}$ is c, and $\widehat{XZ}$ is 160.
2. How is the measure of an inscribed angle related to the measure of its intercepted arc?
By Theorem 12-11, the measure of an inscribed angle is half the measure of its intercepted arc.
3. What is the measure of $\widehat{YZ}$, by the definition of an arc? 44
4. How can you use Theorem 12-11 to find a?
Because a (the measure of $\angle YXZ$) is half the measure of its intercepted arc ($\widehat{YZ}$), it is equal to half of 44.
5. What is a? 22
6. What is the sum of the measures of $\widehat{XY}$, $\widehat{YZ}$, and $\widehat{XZ}$? 360
7. How can you use the sum of the measures of all these non-overlapping arcs of a circle to find c?
Because c is the measure of $\widehat{XY}$, and $m\widehat{XY} + m\widehat{YZ} + m\widehat{XZ} = 360$, $m\widehat{XY} = 360 - (m\widehat{YZ} + m\widehat{XZ})$. Then substitute the values for $m\widehat{YZ}$ and $m\widehat{XZ}$.
8. What is c? 156
9. What is the arc that is intercepted by the angle measuring b? What is the measure of this arc? $\widehat{XY}$; 156
10. How can you use Theorem 12-12 to find b?
The measure of an angle that is formed by a tangent and a chord is half the measure of the intercepted arc. So, b is half of 156.
11. What is b? 78

Practice and Problem Solving Wkbk/ All-in-One Resources/Online

Standardized Test Prep

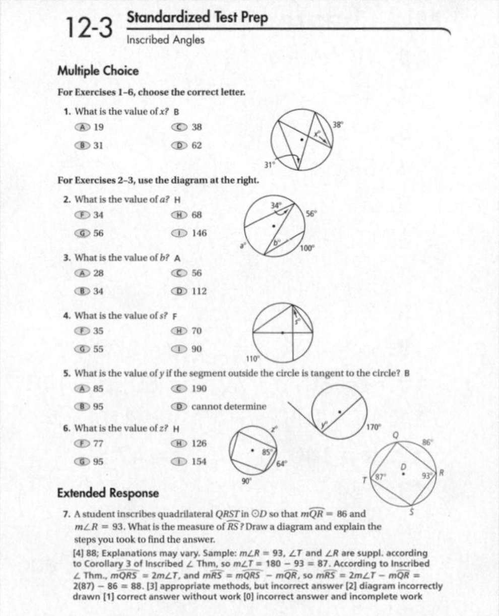

12-3 **Standardized Test Prep** — Inscribed Angles

Multiple Choice

For Exercises 1–6, choose the correct letter.

1. What is the value of x? B
(A) 19 (B) 31 (C) 38 (D) 62

For Exercises 2–3, use the diagram at the right.

2. What is the value of a? H
(F) 34 (G) 56 (H) 68 (I) 146
3. What is the value of b? A
(A) 28 (B) 34 (C) 56 (D) 112
4. What is the value of s? F
(F) 35 (G) 55 (H) 70 (I) 90
5. What is the value of y if the segment outside the circle is tangent to the circle? B
(A) 85 (B) 95 (C) 190 (D) cannot determine
6. What is the value of z? H
(F) 77 (G) 95 (H) 126 (I) 154

Extended Response

7. A student inscribes quadrilateral $QRST$ in $\odot D$ so that $m\widehat{QR} = 86$ and $m\angle R = 93$. What is the measure of $\widehat{RS}$? Draw a diagram and explain the steps you took to find the answer.
[4] 88; Explanations may vary. Sample: $m\angle R = 93$, $\angle T$ and $\angle R$ are suppl. according to Corollary 3 of Inscribed $\angle$ Thm, so $m\angle T = 180 - 93 = 87$. According to Inscribed $\angle$ Thm., $m\widehat{QRS} = 2m\angle T$, and $m\widehat{RS} = m\widehat{QRS} - m\widehat{QR}$, so $m\widehat{RS} = 2m\angle T - m\widehat{QR} = 2(87) - 86 = 88$. [3] appropriate methods, but incorrect answer [2] diagram incorrectly drawn [1] correct answer without work [0] incorrect answer and incomplete work

Online Teacher Resource Center

Activities, Games, and Puzzles

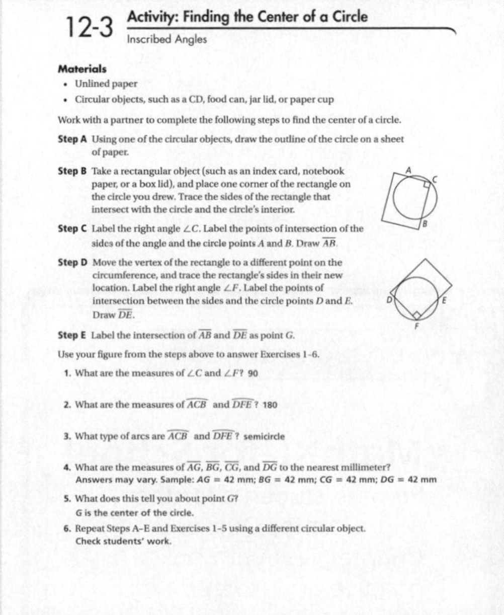

12-3 **Activity: Finding the Center of a Circle** — Inscribed Angles

Materials

- Unlined paper
- Circular objects, such as a CD, food can, jar lid, or paper cup

Work with a partner to complete the following steps to find the center of a circle.

Step A Using one of the circular objects, draw the outline of the circle on a sheet of paper.
Step B Take a rectangular object (such as an index card, notebook paper, or a box lid), and place one corner of the rectangle on the circle you drew. Trace the sides of the rectangle that intersect with the circle and the circle's interior.
Step C Label the right angle $\angle C$. Label the points of intersection of the sides of the angle and the circle points A and B. Draw $\overline{AB}$.
Step D Move the vertex of the rectangle to a different point on the circumference, and trace the rectangle's sides in their new location. Label the right angle $\angle F$. Label the points of intersection between the sides and the circle points D and E. Draw $\overline{DE}$.
Step E Label the intersection of $\overline{AB}$ and $\overline{DE}$ as point G.

Use your figure from the steps above to answer Exercises 1–6.

1. What are the measures of $\angle C$ and $\angle F$? 90
2. What are the measures of $\widehat{ACB}$ and $\widehat{DFE}$? 180
3. What type of arcs are $\widehat{ACB}$ and $\widehat{DFE}$? semicircle
4. What are the measures of $\overline{AG}$, $\overline{BG}$, $\overline{CG}$, and $\overline{DG}$ to the nearest millimeter?
Answers may vary. Sample: $AG = 42$ mm; $BG = 42$ mm; $CG = 42$ mm; $DG = 42$ mm
5. What does this tell you about point G?
G is the center of the circle.
6. Repeat Steps A–E and Exercises 1–5 using a different circular object.
Check students' work.

12 Mid-Chapter Quiz

MathXL® for School
Go to PowerGeometry.com

Do you know HOW?

Each polygon below circumscribes the circle. Find the perimeter of the polygon.

1.
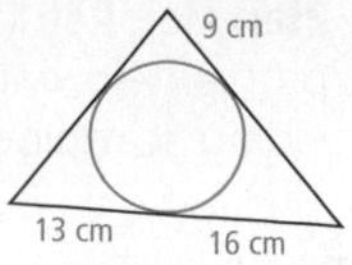

2.
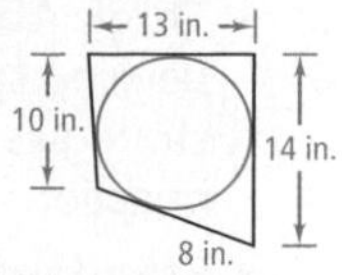

3.
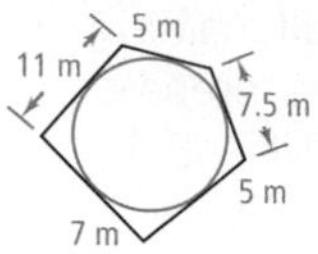

4.
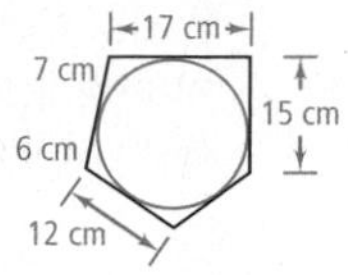

Algebra **Find the value of x in $\odot O$.**

5.
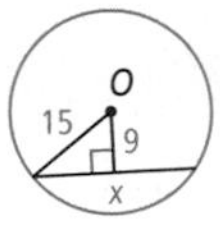

6.
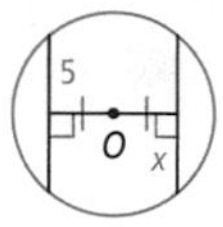

7.
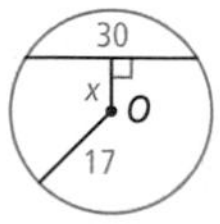

8.
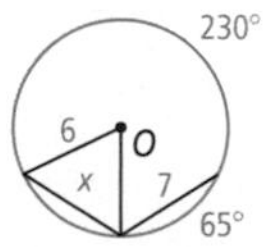

Find the value of each variable. Lines that appear to be tangent are tangent, and the dot represents the center.

9.
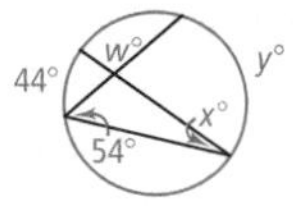

10.
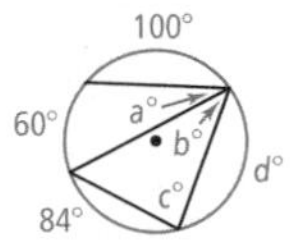

11.
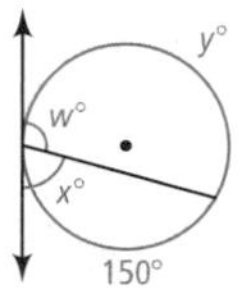

12.
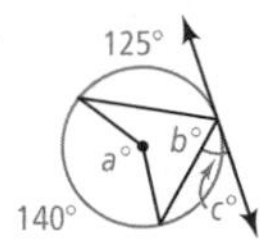

Find $m\widehat{AB}$.

13.
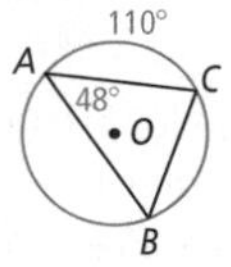

14.
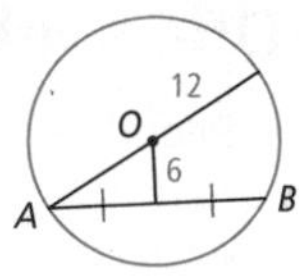

Write a two-column proof, paragraph proof, or flow proof.

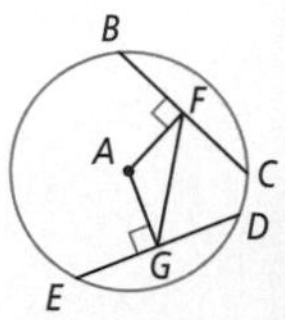

15. Given: $\odot A$ with $\overline{BC} \cong \overline{DE}$, $\overline{AF} \perp \overline{BC}$, $\overline{AG} \perp \overline{DE}$
Proof
Prove: $\angle AFG \cong \angle AGF$

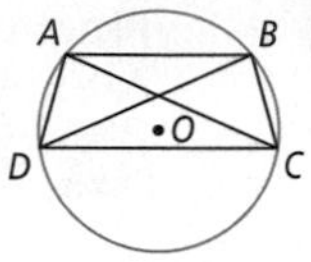

16. Given: $\odot O$ with $\widehat{AD} \cong \widehat{BC}$
Proof
Prove: $\triangle ABD \cong \triangle BAC$

Do you UNDERSTAND?

17. Reasoning In $\odot C$, $m\widehat{PQ} = 50$ and $m\widehat{QR} = 20$. Find two possible values for $m\widehat{PR}$.

18. Open-Ended Draw a triangle circumscribed about a circle. Then draw the radii to each tangent. How many convex quadrilaterals are in your figure?

19. Reasoning $\overline{EF}$ is tangent to both $\odot A$ and $\odot B$ at F. $\overline{CD}$ is tangent to $\odot A$ at C and to $\odot B$ at D. What can you conclude about $\overline{CE}$, $\overline{DE}$, and $\overline{FE}$? Explain.

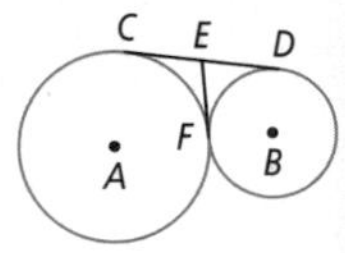

20. Writing Explain why the length of a segment tangent to a circle from a point outside the circle will always be less than the distance from the point to the center of the circle.

Answers

Mid-Chapter Quiz

1. 76 cm
2. 48 in.
3. 51 m
4. 68 cm
5. 24
6. 5
7. 8
8. 7
9. $w = 104$, $x = 22$, $y = 108$
10. $a = 30$, $b = 42$, $c = 80$, $d = 116$
11. $w = 105$, $x = 75$, $y = 210$
12. $a = 140$, $b = 70$, $c = 47.5$
13. 154
14. 120
15. $\odot A$ with $\overline{BC} \cong \overline{DE}$, $\overline{AF} \perp \overline{BC}$ and $\overline{AG} \perp \overline{DE}$ (given); $\overline{AF} \cong \overline{AG}$ ($\cong$ chords in a $\odot$ are equidistant from the center); $\angle AFG \cong \angle AGF$ (an isosc. $\triangle$ has $\cong$ base $\angle$s).
16. Answers may vary. Sample: $\widehat{AD} \cong \widehat{BC}$ (given); $\angle ABD \cong \angle BAC$ (Inscribed $\angle$s that intercept $\cong$ arcs are $\cong$); $\angle ADB \cong \angle BCA$ (Inscribed $\angle$s that intercept $\cong$ arcs are $\cong$); $\overline{AB} \cong \overline{AB}$ (Refl. Prop. of $\cong$); $\triangle ABD \cong \triangle BAC$ (AAS).
17. Two possibilities are 70 and 30.
18. Check students' drawings; 3
19. $\overline{CE} \cong \overline{FE} \cong \overline{DE}$ (tangent segments from a point outside a circle are $\cong$)
20. The outside point, the point of tangency, and the center of the circle are the vertices of a rt. $\triangle$. The tangent segment is a leg of that $\triangle$, so it must be shorter than the distance from the outside point to the center, which is the hypotenuse of the rt. $\triangle$.

PowerGeometry.com

MathXL for School
Prepare students for the Mid-Chapter Quiz and Chapter Test with online practice and review.

Concept Byte

Use With Lesson 12-4

TECHNOLOGY

Exploring Chords and Secants

Common Core State Standards

Extends G-C.A.2 Identify and describe relationships among inscribed angles, radii, and chords.

MP 7

Activity 1

Construct ⊙A and two chords $\overline{BC}$ and $\overline{DE}$ that intersect at F.

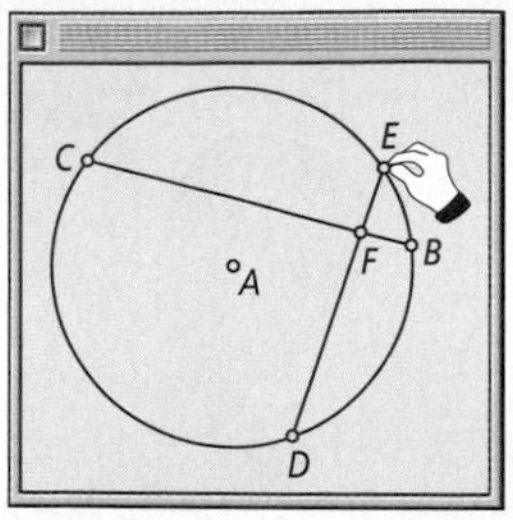

1. Measure $\overline{BF}$, $\overline{FC}$, $\overline{EF}$, and $\overline{FD}$.
2. Use the calculator program of your software to find $BF \cdot FC$ and $EF \cdot FD$.
3. Manipulate the lines. What pattern do you observe in the products?
4. **Make a Conjecture** What appears to be true for two intersecting chords?

Activity 2

A *secant* is a line that intersects a circle in two points. A *secant segment* is a segment that contains a chord of the circle and has only one endpoint outside the circle. Construct a new circle and two secants $\overleftrightarrow{DG}$ and $\overleftrightarrow{DE}$ that intersect outside the circle at point D. Label the intersections with the circle as shown.

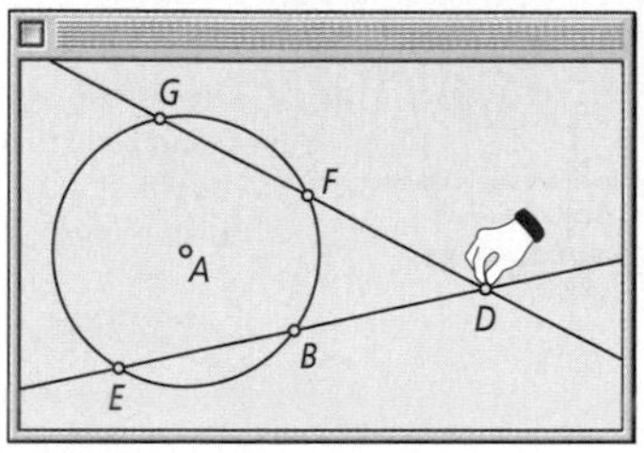

5. Measure $\overline{DG}$, $\overline{DF}$, $\overline{DE}$, and $\overline{DB}$.
6. Calculate the products $DG \cdot DF$ and $DE \cdot DB$.
7. Manipulate the lines. What pattern do you observe in the products?
8. **Make a Conjecture** What appears to be true for two intersecting secants?

Activity 3

Construct ⊙A with tangent $\overline{DG}$ perpendicular to radius $\overline{AG}$ and secant $\overline{DE}$ that intersects the circle at B and E.

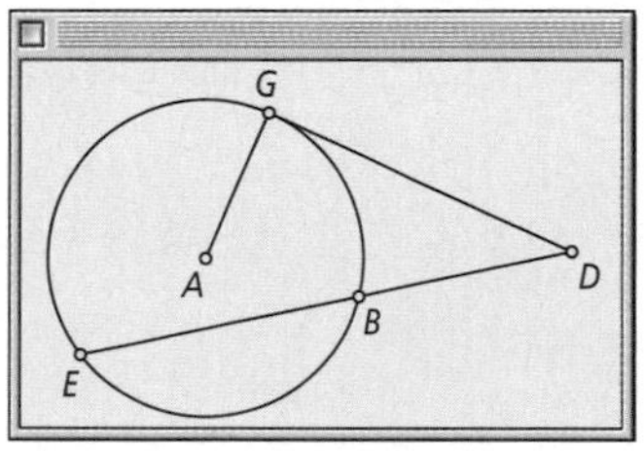

9. Measure $\overline{DG}$, $\overline{DE}$, and $\overline{DB}$.
10. Calculate the products $(DG)^2$ and $DE \cdot DB$.
11. Manipulate the lines. What pattern do you observe in the products?
12. **Make a Conjecture** What appears to be true for the tangent segment and secant segment?

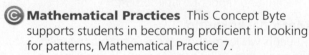

Answers

Activity 1

1. Check students' work.
2. Check students' work.
3. $BF \cdot FC = EF \cdot FD$
4. The products of the lengths of the segments of the chords are equal.

Activity 2

5. Check students' work.
6. Check students' work.
7. $DG \cdot DF = DE \cdot DB$
8. The product of the lengths of one secant seg. and its external seg. = the product of the lengths of the other secant seg. and its external seg.

Activity 3

9. Check students' work.
10. Check students' work.
11. $(DG)^2 = DE \cdot DB$
12. The square of the length of the tangent seg. = the product of the lengths of the secant seg. and its external seg.

Guided Instruction

PURPOSE To explore various properties of chords and secants

PROCESS Students will

- construct and manipulate chords.
- construct and manipulate secants.
- construct and manipulate a tangent and secant.
- measure and calculate products of the segments formed by chords, secants, and tangents, and make conjectures about these calculations.

DISCUSS Review how to use geometric software to construct, manipulate, and calculate lengths of specific segments that are formed between various chords, secants, and tangents of circles.

Activity 1

In this Activity students make conjectures about partial segments of chords.

Q How many partial segments are made from chord $\overline{BC}$? **[2]**

Q What conclusion can you make regarding the products of the partial segments of the chords? **[They are equal.]**

Q If you are given the lengths of three of the partial segments, how can you find the length of the fourth partial segment? **[Use an equation or solve a proportion.]**

Activity 2

In this Activity students make conjectures about secants.

Q Which segments are external segments? **[$\overline{DF}$ and $\overline{DB}$]**

Q Is the relationship between segments formed by secants similar to the relationship between segments formed by chords? **[Yes; for each line, you multiply the distances from the given point to the two points of intersection with the circle.]**

Activity 3

In this Activity students make conjectures about a secant and a tangent.

Q Can you say that the situation of tangent and secant is the special case of the situation with two secants? **[Yes; when the two points of intersection with the circle for one of the secants coincide, the secant becomes a tangent.]**

Q Why isn't one more case, with two tangents, considered? **[Because two tangent segments drawn from one point are congruent by Theorem 12-3.]**

Show students that they can redo Activity 2 and manipulate the secant so that G coincides with F. Thus, they can derive the relationship between a secant and a tangent.

Mathematical Practices This Concept Byte supports students in becoming proficient in looking for patterns, Mathematical Practice 7.

1 Interactive Learning

Solve It!

PURPOSE To discover the measure of an angle formed by two chords in a circle

PROCESS Students must use their knowledge of inscribed angles and exterior angles of triangles to write an equation for $m\angle 1$.

FACILITATE

Q How is $\angle 1$ related to the triangles formed by the chords? **[It is an exterior angle to the triangles.]**

Q What do you know about an exterior angle in a triangle? **[Its measure is equal to the sum of the remote interior angles.]**

Q What is $m\angle C$ and $m\angle D$? **[$m\angle C = \frac{1}{2}\widehat{AD}$, $m\angle D = \frac{1}{2}\widehat{BC}$]**

ANSWER See Solve It in Answers on next page.

CONNECT THE MATH In the Solve It, students determine the relationship of an angle inside a circle formed by two chords and the measures of the arcs that the chords form. In this lesson, students use inscribed angles to see that the measure of an interior angle is equal to half the sum of the intercepted arcs.

2 Guided Instruction

Take Note

Have students compare and contrast the theorems presented in the Take Note. They should see that the theorems discuss intersecting lines. Inside a circle, the sum of the intercepted arcs is used. Outside, the difference is used.

12-4 Angle Measures and Segment Lengths

Common Core State Standards
Extends G-C.A.2 Identify and describe relationships among inscribed angles, radii, and chords.
MP 1, MP 3, MP 4

Objectives To find measures of angles formed by chords, secants, and tangents
To find the lengths of segments associated with circles

Think about how inscribed angles will help you out.

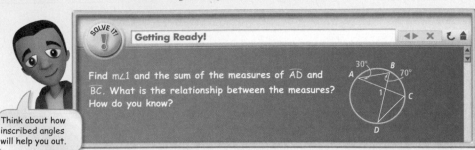

SOLVE IT! Getting Ready!

Find $m\angle 1$ and the sum of the measures of $\widehat{AD}$ and $\widehat{BC}$. What is the relationship between the measures? How do you know?

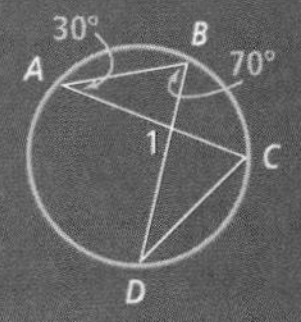

MATHEMATICAL PRACTICES

Essential Understanding Angles formed by intersecting lines have a special relationship to the related arcs formed when the lines intersect a circle. In this lesson, you will study angles and arcs formed by lines intersecting either within a circle or outside a circle.

Lesson Vocabulary
- secant

take note

Theorem 12-13

The measure of an angle formed by two lines that intersect inside a circle is half the sum of the measures of the intercepted arcs.

$$m\angle 1 = \frac{1}{2}(x + y)$$

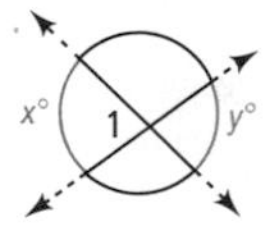

Theorem 12-14

The measure of an angle formed by two lines that intersect outside a circle is half the difference of the measures of the intercepted arcs.

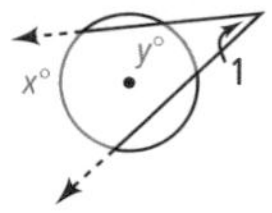
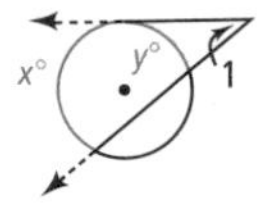
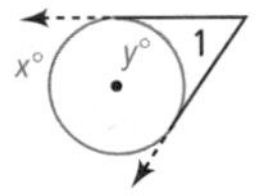

$$m\angle 1 = \frac{1}{2}(x - y)$$

You will prove Theorem 12-14 in Exercises 35 and 36.

12-4 Preparing to Teach

BIG ideas **Reasoning and Proof**
Measurement

ESSENTIAL UNDERSTANDINGS
- Angles formed by intersecting lines have a special relationship to the arcs the intersecting lines intercept.
- Arcs formed by lines intersecting either within a circle or outside a circle are related to the angles formed by the lines.
- There are special relationships between intersecting chords, intersecting secants, or a secant and tangent that intersect.

Math Background

Students will learn about the relationship between the angle formed by two segments that intersect a circle and the intercepted arcs. They will also learn that the segment lengths created by intersecting lines are proportional. Students may benefit from creating a table that contains a summary of all the information they have learned about circles so far. Have them organize the information by the figures involved. Students should have a section for arc length, inscribed angles, tangents, and chords. Some information may appear in more than one section. Help students to organize all the theorems that they have learned about circles.

Dynamic geometry software can provide students with an excellent tool to investigate conjectures about angle measures and segment lengths related to circles.

Mathematical Practices

Model with mathematics. In Problem 2, students will apply their knowledge of circles and tangent lines to model the view from a satellite in orbit.

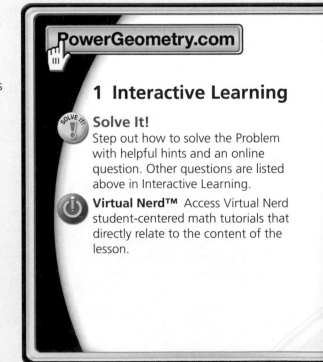

PowerGeometry.com

1 Interactive Learning

Solve It!
Step out how to solve the Problem with helpful hints and an online question. Other questions are listed above in Interactive Learning.

Virtual Nerd™ Access Virtual Nerd student-centered math tutorials that directly relate to the content of the lesson.

In Theorem 12-13, the lines from a point outside the circle going through the circle are called secants. A **secant** is a line that intersects a circle at two points. $\overleftrightarrow{AB}$ is a secant, $\overrightarrow{AB}$ is a secant ray, and $\overline{AB}$ is a secant segment. A chord is part of a secant.

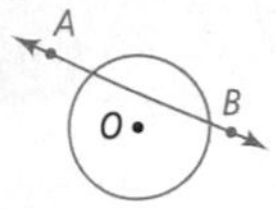

Proof **Proof of Theorem 12-13**

Given: $\odot O$ with intersecting chords $\overline{AC}$ and $\overline{BD}$

Prove: $m\angle 1 = \frac{1}{2}(m\widehat{AB} + m\widehat{CD})$

Begin by drawing auxiliary $\overline{AD}$ as shown in the diagram.

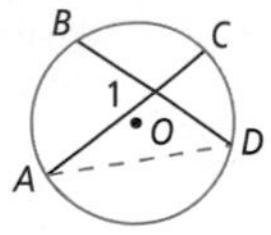

$m\angle BDA = \frac{1}{2}m\widehat{AB}$, and $m\angle CAD = \frac{1}{2}m\widehat{CD}$
Inscribed Angle Theorem

$m\angle 1 = m\angle BDA + m\angle CAD$
△Exterior Angle Theorem

$m\angle 1 = \frac{1}{2}m\widehat{AB} + \frac{1}{2}m\widehat{CD}$
Substitute.

$m\angle 1 = \frac{1}{2}(m\widehat{AB} + m\widehat{CD})$
Distributive Property

Problem 1 Finding Angle Measures

Algebra What is the value of each variable?

Think
Remember to add arc measures for arcs intercepted by lines that intersect inside a circle and subtract arc measures for arcs intercepted by lines that intersect outside a circle.

A

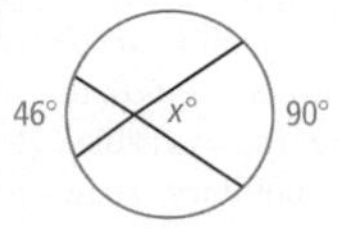

$x = \frac{1}{2}(46 + 90)$ Theorem 12-13
$x = 68$ Simplify.

B

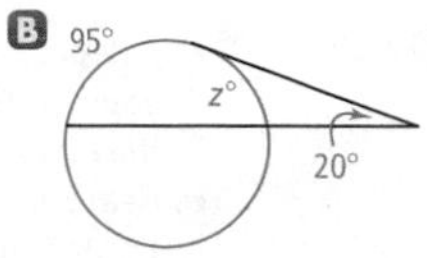

$20 = \frac{1}{2}(95 - z)$ Theorem 12-14
$40 = 95 - z$ Multiply each side by 2.
$z = 55$ Solve for z.

Got It? **1.** What is the value of each variable?

a.

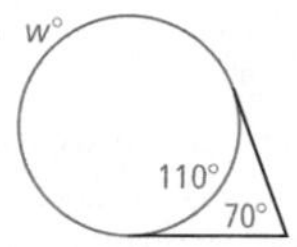

b.

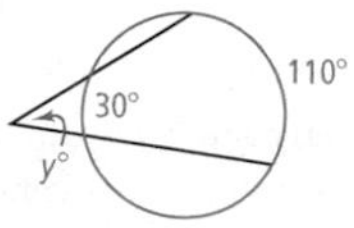

c.

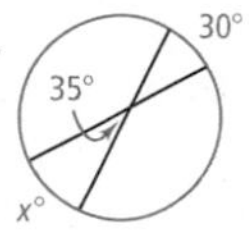

2 Guided Instruction

Each Problem is worked out and supported online.

Problem 1
Finding Angle Measures
Animated

Problem 2
Finding an Arc Measure

Problem 3
Finding Segment Lengths
Animated

Support in Geometry Companion
- Vocabulary
- Key Concepts
- Got It?

Students should be able to produce this proof given their work in the Solve It. Ask them to identify which segment needs to be drawn to create a triangle within the circle. Then, students can use the same argument they developed in the Solve It.

Problem 1

Q In 1A, does the vertex of the angle fall inside or outside the circle? **[inside]**

Q Should you use the sum or difference of the arc measures to calculate the measure of the angle? **[the sum]**

Q In 1B, what measure is unknown? **[one of the intercepted arcs]**

Q Should you use addition or subtraction to calculate the measure of the arc? Why? **[Use the difference because the vertex of the angle is outside the circle.]**

Got It? ERROR PREVENTION

Have students describe the process for finding each type of figure. In 1a, the measure of an arc is unknown so there will be a variable in the equation. In 1b, students should be able to write an expression that can be simplified to find the value of y.

Answers

Solve It!

$m\angle 1 = \frac{1}{2}(m\widehat{AD} + m\widehat{BC})$. Sample explanation: $m\angle 1 = 100$ (Exterior ∠ Thm.); $m\angle B = \frac{1}{2}m\widehat{AD}$, so $m\widehat{AD} = 2m\angle B = 140$; $m\angle A = \frac{1}{2}m\widehat{BC}$, so $m\widehat{BC} = 2m\angle A = 60$; $m\widehat{AD} + m\widehat{BC} = 140 + 60 = 200$; $m\angle 1 = \frac{1}{2}(m\widehat{AD} + m\widehat{BC})$

Got It?

1a. 250
b. 40
c. 40

Problem 2

Q What arcs are intercepted by the tangent lines? **[$\overset{\frown}{AB}$ and $\overset{\frown}{AEB}$]**

Q What is the relationship between the intercepted arcs? **[They form the entire circle, so their sum is 360°.]**

Q Which theorem can you use to find the measure of the arcs? **[Theorem 12-14]**

Got It? VISUAL LEARNERS

For 2a, have students draw a diagram with the new angle measure. The equation they write should be similar to the one in Problem 2. For 2b, have students act out the problem. They can hold their arms in an angle and view the chalkboard. Ask whether moving closer to the chalkboard increases or decreases the area of the chalkboard they can see between their arms.

Have students draw a diagram similar to the one at the bottom of the page. They can use a ruler to measure the segments created and find the product of those lengths. Have students write a conjecture about why they think the product remains constant.

Problem 2 Finding an Arc Measure

Satellite **A satellite in a geostationary orbit above Earth's equator has a viewing angle of Earth formed by the two tangents to the equator. The viewing angle is about 17.5°. What is the measure of the arc of Earth that is viewed from the satellite?**

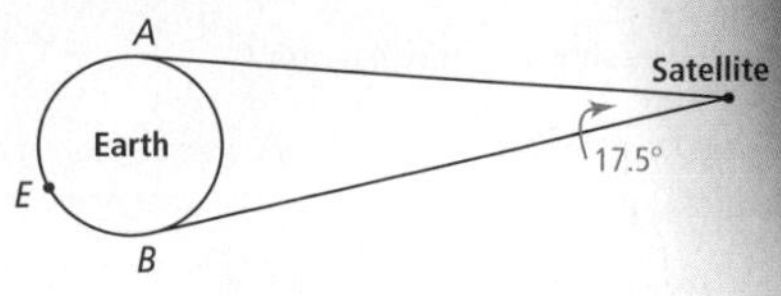

Think

How can you represent the measures of the arcs?
The sum of the measures of the arcs is 360°. If the measure of one arc is x, the measure of the other is $360 - x$.

Let $m\overset{\frown}{AB} = x$.

Then $m\overset{\frown}{AEB} = 360 - x$.

$17.5 = \frac{1}{2}(m\overset{\frown}{AEB} - m\overset{\frown}{AB})$	Theorem 12-14
$17.5 = \frac{1}{2}[(360 - x) - x]$	Substitute.
$17.5 = \frac{1}{2}(360 - 2x)$	Simplify.
$17.5 = 180 - x$	Distributive Property
$x = 162.5$	Solve for x.

A 162.5° arc can be viewed from the satellite.

Got It? **2. a.** A departing space probe sends back a picture of Earth as it crosses Earth's equator. The angle formed by the two tangents to the equator is 20°. What is the measure of the arc of the equator that is visible to the space probe?

b. Reasoning Is the probe or the geostationary satellite in Problem 2 closer to Earth? Explain.

Essential Understanding There is a special relationship between two intersecting chords, two intersecting secants, or a secant that intersects a tangent. This relationship allows you to find the lengths of unknown segments.

From a given point P, you can draw two segments to a circle along infinitely many lines. For example, $\overline{PA_1}$ and $\overline{PB_1}$ lie along one such line. Theorem 12-15 states the surprising result that no matter which line you use, the product of the lengths $PA \cdot PB$ remains constant.

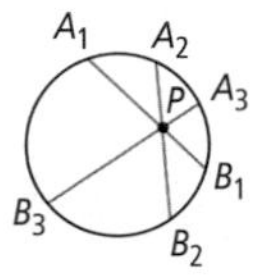

Additional Problems

1. What is the value of each variable?

a.

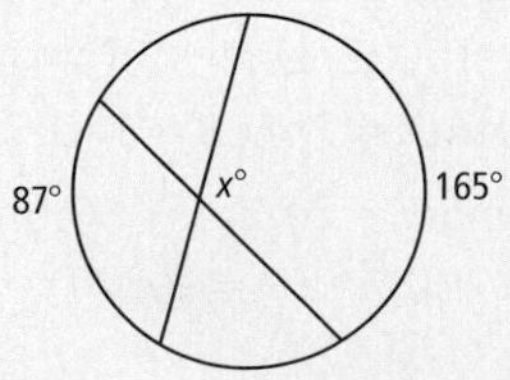

b.

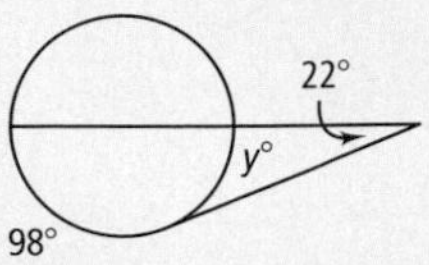

ANSWER a. 126 **b.** 54

2. The viewing angle of the moon from a spacecraft is formed by the two tangents to the moon. The viewing angle is 18°. What is the measure, in degrees, of the arc of the moon that is viewed from the spacecraft?

ANSWER 162

3. Find the value of the variable. If the answer is not a whole number, round to the nearest tenth.

a.

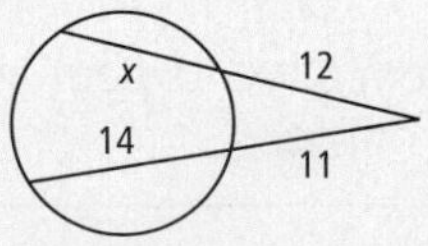

b.

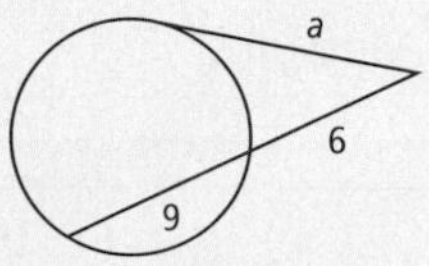

ANSWER a. 10.9 **b.** 9.5

take note

Theorem 12-15

For a given point and circle, the product of the lengths of the two segments from the point to the circle is constant along any line through the point and circle.

I.

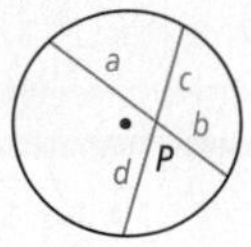

$a \cdot b = c \cdot d$

II.

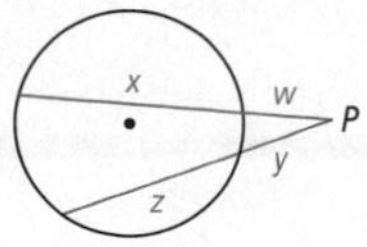

$(w + x)w = (y + z)y$

III.

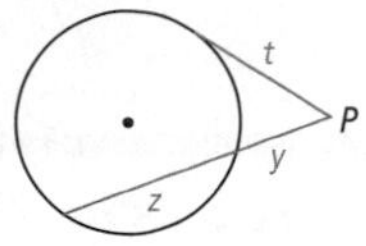

$(y + z)y = t^2$

As you use Theorem 12-15, remember the following.

- **Case I:** The products of the chord segments are equal.
- **Case II:** The products of the secants and their outer segments are equal.
- **Case III:** The product of a secant and its outer segment equals the square of the tangent.

Here is a proof for Case I. You will prove Case II and Case III in Exercises 37 and 38.

Proof Proof of Theorem 12-15, Case I

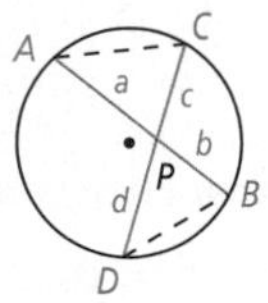

Given: A circle with chords $\overline{AB}$ and $\overline{CD}$ intersecting at P

Prove: $a \cdot b = c \cdot d$

Draw $\overline{AC}$ and $\overline{BD}$. $\angle A \cong \angle D$ and $\angle C \cong \angle B$ because each pair intercepts the same arc, and angles that intercept the same arc are congruent. $\triangle APC \sim \triangle DPB$ by the Angle-Angle Similarity Postulate. The lengths of corresponding sides of similar triangles are proportional, so $\frac{a}{d} = \frac{c}{b}$. Therefore, $a \cdot b = c \cdot d$.

Problem 3 Finding Segment Lengths

Plan

How can you identify the segments needed to use Theorem 12-15? Find where segments intersect each other relative to the circle. The lengths of segments that are part of one line will be on the same side of an equation.

Algebra Find the value of the variable in $\odot N$.

A

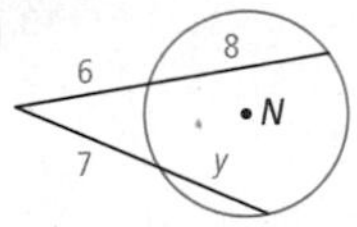

$(6 + 8)6 = (7 + y)7$	Thm. 12-15, Case II
$84 = 49 + 7y$	Distributive Property
$35 = 7y$	
$5 = y$	Solve for y.

B

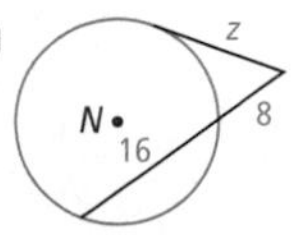

$(8 + 16)8 = z^2$	Thm. 12-15, Case III
$192 = z^2$	Simplify.
$13.9 \approx z$	Solve for z.

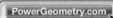

Take Note

Emphasize that the products must use the entire segment length. Ask students to explain how this is shown in Case II and Case III.

Q What type of equation can lead to a set of equal products? **[a proportion]**

Q What type of triangles uses proportions? **[similar triangles]**

Problem 3

Q Which case of Theorem 12-15 is represented in 3A? **[Case II]**

Q What segment lengths should you use in the equation? **[6, 14, 7, and 7 + y]**

Q Which case of Theorem 12-15 is represented in 3B? **[Case III]**

Q Which segment length will be repeated in the equation? **[8]**

Answers

Got It? (continued)

2a. 160

b. The probe is closer; as an observer moves away from Earth, the viewing angle decreases and the measure of the arc of Earth that is viewed gets larger and approaches 180°.

Got It? VISUAL LEARNERS
Have students visually identify the case of Theorem 12-15 that is represented in each part. Make sure that students are using the sum of segment lengths to write their equations for 3a.

3 Lesson Check

Do you know HOW?

- If students have difficulty with Exercise 2, then have them review Problem 1 and write an equation for this diagram.

Do you UNDERSTAND?

- If students have difficulty with Exercise 7, then have them review Problem 3 and write an equation that models the one written for 3B.

Close

Q When two lines intersect inside a circle, how is the angle measure related to the measure of the intercepted arcs? **[The angle measure is equal to half the sum of the intercepted arcs.]**

Q When two lines intersect inside a circle, how are the segment lengths related? **[The products of the lengths of the segments from the intersection to the circle are equal.]**

Got It? 3. What is the value of the variable to the nearest tenth?

a.

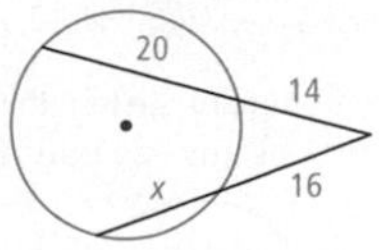

b.

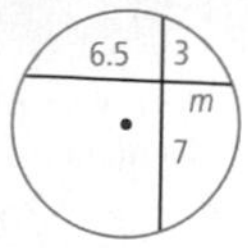

Lesson Check

Do you know HOW?

1. What is the value of x?
2. What is the value of y?
3. What is the value of z, to the nearest tenth?

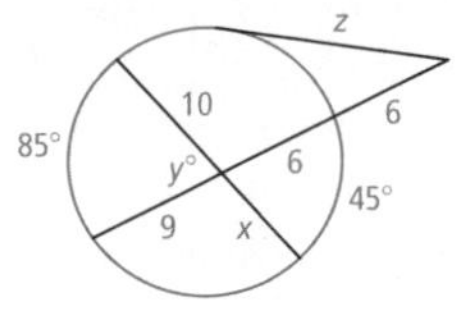

4. The measure of the angle formed by two tangents to a circle is 80. What are the measures of the intercepted arcs?

Do you UNDERSTAND? MATHEMATICAL PRACTICES

5. **Vocabulary** Describe the difference between a *secant* and a *tangent*.
6. In the diagram for Exercises 1–3, is it possible to find the measures of the unmarked arcs? Explain.
7. **Error Analysis** To find the value of x, a student wrote the equation $(7.5)6 = x^2$. What error did the student make?

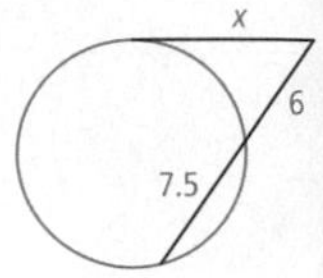

Practice and Problem-Solving Exercises MATHEMATICAL PRACTICES

A Practice

Algebra Find the value of each variable.

See Problems 1 and 2.

8.

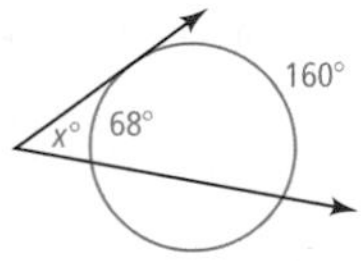

9.

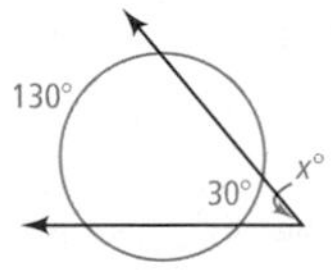

10.

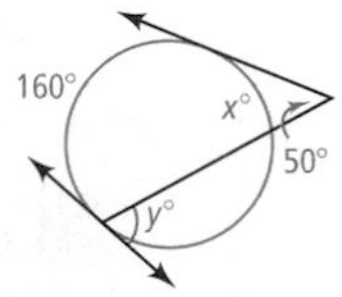

11.

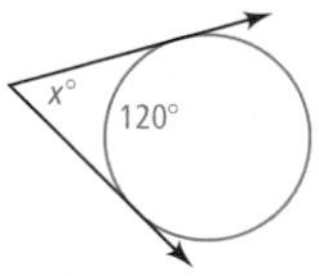

12.

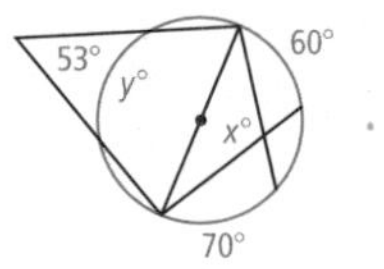

13.

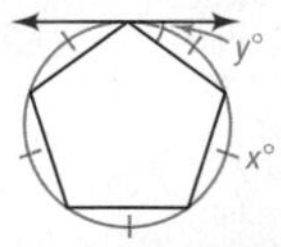

14. **Photography** You focus your camera on a circular fountain. Your camera is at the vertex of the angle formed by tangents to the fountain. You estimate that this angle is 40°. What is the measure of the arc of the circular basin of the fountain that will be in the photograph?

PowerGeometry.com

3 Lesson Check

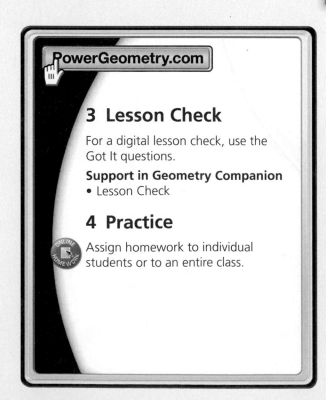

For a digital lesson check, use the Got It questions.

Support in Geometry Companion

- Lesson Check

4 Practice

Assign homework to individual students or to an entire class.

Answers

Got It? (continued)

3a. 13.8
b. 3.2

Lesson Check

1. 5.4
2. 65
3. 11.2
4. 100, 260
5. A secant is a line that intersects a circle at two points; a tangent is a line that intersects a circle at one point.
6. No; we can find the sum of the measures of the two arcs (in this situation, that sum is 230), but there is not enough information to find the measure of each arc.
7. The student forgot to multiply by the length of the entire secant seg.; the equation should be $(13.5)(6) = x^2$.

Practice and Problem-Solving Exercises

8. 46
9. 50
10. $x = 60$, $y = 70$
11. 60
12. $x = 115$, $y = 74$
13. $x = 72$, $y = 36$
14. 140

Algebra Find the value of each variable using the given chord, secant, and tangent lengths. If the answer is not a whole number, round to the nearest tenth.

See Problem 3.

15.

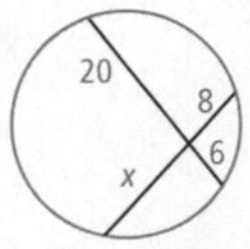

16.

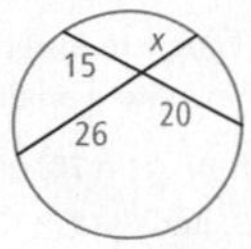

17.

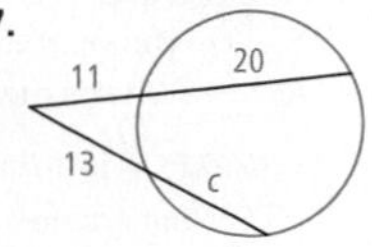

18.

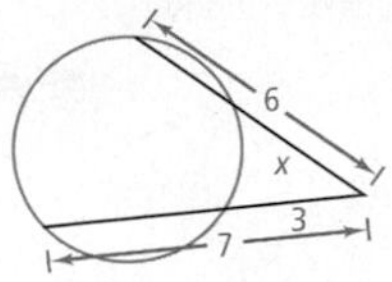

19.

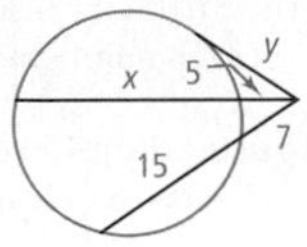

20.

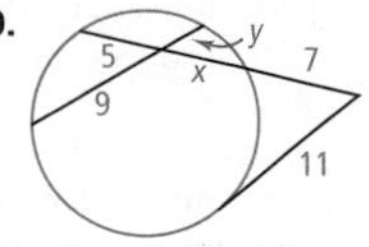

B Apply

Algebra $\overline{CA}$ and $\overline{CB}$ are tangents to $\odot O$ Write an expression for each arc or angle in terms of the given variable.

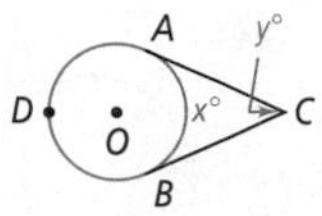

21. $m\widehat{ADB}$ using x **22.** $m\angle C$ using x **23.** $m\widehat{AB}$ using y

Find the diameter of $\odot O$. A line that appears to be tangent is tangent. If your answer is not a whole number, round it to the nearest tenth.

24.

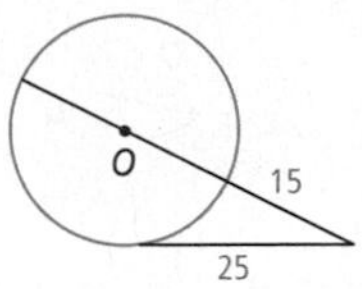

25.

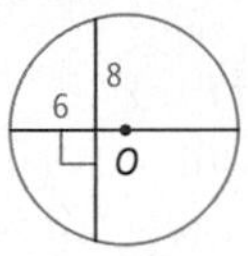

26.

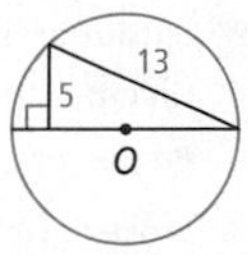

27. A circle is inscribed in a quadrilateral whose four angles have measures 85, 76, 94, and 105. Find the measures of the four arcs between consecutive points of tangency.

Wankel engine

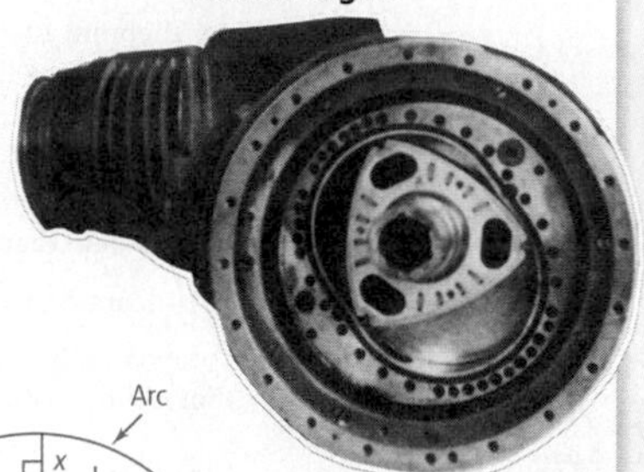

STEM 28. Engineering The basis for the design of the Wankel rotary engine is an equilateral triangle. Each side of the triangle is a chord to an arc of a circle. The opposite vertex of the triangle is the center of the circle that forms the arc. In the diagram below, each side of the equilateral triangle is 8 in. long.

a. Use what you know about equilateral triangles and find the value of x.

b. Reasoning Copy the diagram and complete the circle with the given center. Then use Theorem 12-15 to find the value of x. Show that your answers to parts (a) and (b) are equal.

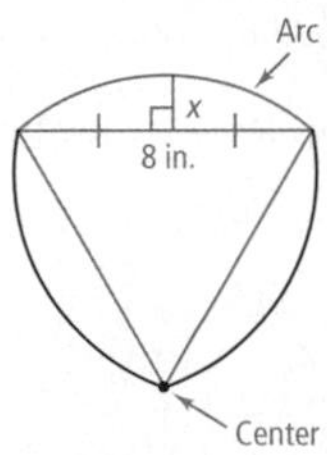

4 Practice

ASSIGNMENT GUIDE

Basic: 8–20 all, 22–26 even, 27, 29, 31
Average: 9–19 odd, 21–39
Advanced: 9–19 odd, 21–43

Mathematical Practices are supported by exercises with red headings. Here are the Practices supported in this lesson:

MP 1: Make Sense of Problems Ex. 29, 31
MP 3: Construct Arguments Ex. 28b
MP 3: Critique the Reasoning of Others Ex. 7

Applications exercises have blue headings. Exercises 14 and 28 support MP 4: Model.

STEM exercises focus on science or engineering applications.

EXERCISE 27: Use the Think About a Plan worksheet in the **Practice and Problem Solving Workbook** (also available in the Teaching Resources in print and online) to further support students' development in becoming independent learners.

HOMEWORK QUICK CHECK

To check students' understanding of key skills and concepts, go over Exercises 11, 15, 27, 29, and 31.

15. 15
16. 11.5
17. 13.2
18. 3.5
19. $x = 25.8, y \approx 12.4$
20. $x \approx 5.3, y \approx 2.9$
21. $360 - x$
22. $180 - x$
23. $180 - y$
24. 26.7
25. 16.7
26. 14.1
27. 95, 104, 86, 75
28a. $(8 - 4\sqrt{3})$ in.
b. $\frac{4}{2+\sqrt{3}}$ in.; $\frac{4}{2+\sqrt{3}} \cdot \frac{2-\sqrt{3}}{2-\sqrt{3}} = \frac{8-4\sqrt{3}}{4-3} = 8 - 4\sqrt{3}$

29. **Think About a Plan** In the diagram, the circles are concentric. What is a formula you could use to find the value of c in terms of a and b?
- How can you use the inscribed angle to find the value of c?
- What is the relationship of the inscribed angle to a and b?

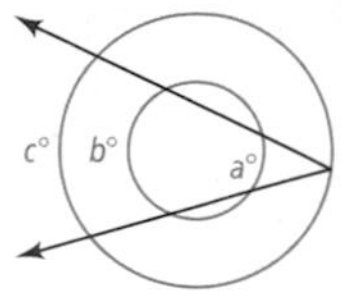

30. $\triangle PQR$ is inscribed in a circle with $m\angle P = 70$, $m\angle Q = 50$, and $m\angle R = 60$. What are the measures of $\overparen{PQ}$, $\overparen{QR}$, and $\overparen{PR}$?

31. **Reasoning** Use the diagram at the right. If you know the values of x and y, how can you find the measure of each numbered angle?

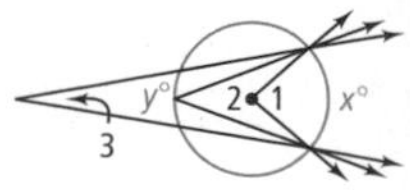

Algebra Find the values of x and y using the given chord, secant, and tangent lengths. If your answer is not a whole number, round it to the nearest tenth.

32.

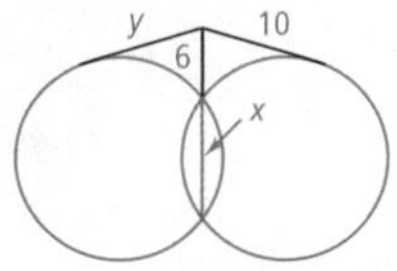

33.

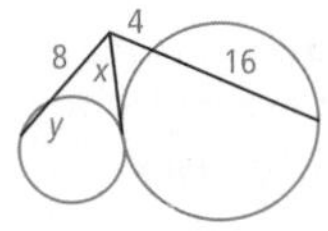

34.

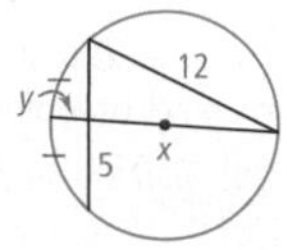

35. Prove Theorem 12-14 as it applies to two secants that intersect outside a circle.
Proof

Given: $\odot O$ with secants $\overline{CA}$ and $\overline{CE}$

Prove: $m\angle ACE = \frac{1}{2}(m\overparen{AE} - m\overparen{BD})$

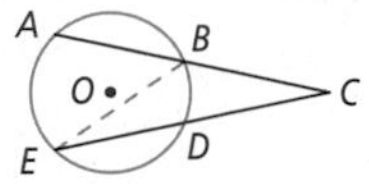

36. Prove the other two cases of Theorem 12-14. (See Exercise 35.)
Proof

For Exercises 37 and 38, write proofs that use similar triangles.

37. Prove Theorem 12-15, Case II.
Proof

38. Prove Theorem 12-15, Case III.
Proof

39. The diagram at the right shows a *unit circle*, a circle with radius 1.
   a. What triangle is similar to $\triangle ABE$?
   b. Describe the connection between the ratio for the tangent of $\angle A$ and the segment that is tangent to $\odot A$.
   c. The secant ratio is $\frac{\text{hypotenuse}}{\text{length of leg adjacent to an angle}}$. Describe the connection between the ratio for the secant of $\angle A$ and the segment that is the secant in the unit circle.

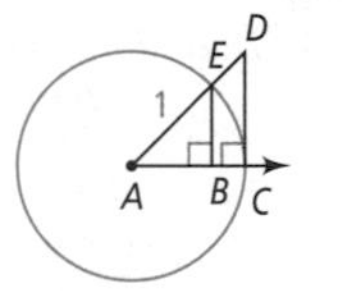

Challenge **For Exercises 40 and 41, use the diagram at the right. Prove each statement.**

40. $m\angle 1 + m\overparen{PQ} = 180$
Proof

41. $m\angle 1 + m\angle 2 = m\overparen{QR}$
Proof

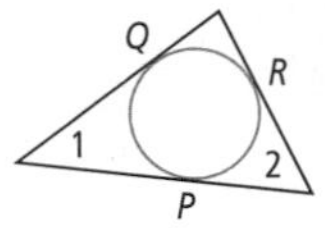

Answers

Practice and Problem-Solving Exercises (continued)

29. $c = b - a$

30. $m\overparen{PQ} = 120$, $m\overparen{QR} = 140$, $m\overparen{PR} = 100$

31. $\angle 1$ is a central $\angle$, so $m\angle 1 = x$; $\angle 2$ is an inscribed $\angle$, so $m\angle 2 = \frac{1}{2}x$; $\angle 3$ is formed by the secants, so $m\angle 3 = \frac{1}{2}(x - y)$.

32. $x \approx 10.7$, $y = 10$

33. $x \approx 8.9$, $y = 2$

34. $x \approx 10.9$, $y \approx 2.3$

35. 1. $\odot O$ with secants $\overline{CA}$ and $\overline{CE}$. (Given)
2. Draw $\overline{BE}$ (2 pts. determine a line.)
3. $m\angle BEC = \frac{1}{2}m\overparen{BD}$ and $m\angle ABE = \frac{1}{2}m\overparen{AE}$ (The measure of an inscribed $\angle$ is half the measure of its intercepted arc.) 4. $m\angle BEC + m\angle BCE = m\angle ABE$ (Ext. $\angle$ Thm.) 5. $\frac{1}{2}m\overparen{BD} + m\angle BCE = \frac{1}{2}m\overparen{AE}$ (Subst. Prop. of =)
6. $m\angle BCE = \frac{1}{2}m\overparen{AE} - \frac{1}{2}m\overparen{BD}$ (Subtr. Prop. of =) 7. $m\angle BCE = \frac{1}{2}(m\overparen{AE} - m\overparen{BD})$ (Distr. Prop.) 8. $\angle BCE \cong \angle ACE$ (Refl. Prop. of $\cong$)
9. $m\angle ACE = \frac{1}{2}(m\overparen{AE} - m\overparen{BD})$ (Subst. Prop. of =)

36.

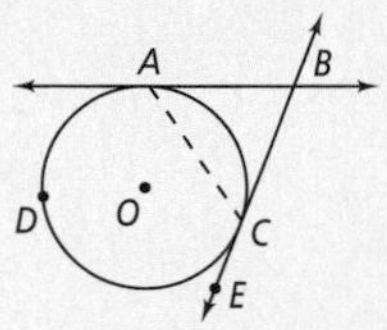

Given: $\overleftrightarrow{BA}$ is tangent to $\odot O$ at A, $\overleftrightarrow{BC}$ is tangent to $\odot O$ at C
Prove: $m\angle ABC = \frac{1}{2}(m\overparen{ADC} - m\overparen{AC})$
Proof: 1. Draw $\overline{AC}$ (2 pts. determine a line)
2. $m\angle ACE = \frac{1}{2}m\overparen{ADC}$ (The measure of an $\angle$ formed by a tangent and a chord is half the measure of the intercepted arc.)
3. $m\angle ACE = m\angle ABC + m\angle BAC$ (Ext. $\angle$ Thm.) 4. $m\angle BAC = \frac{1}{2}m\overparen{AC}$ (The measure of an $\angle$ formed by a tangent and a chord is half the measure of the intercepted arc.)
5. $\frac{1}{2}m\overparen{ADC} = m\angle ABC + m\angle BAC$ (Subst. Prop. of =) 6. $\frac{1}{2}m\overparen{ADC} = m\angle ABC + \frac{1}{2}m\overparen{AC}$ (Subst. Prop. of =) 7. $m\angle ABC = \frac{1}{2}m\overparen{ADC} - \frac{1}{2}m\overparen{AC}$ (Subtr. Prop. of =)
8. $m\angle ABC = \frac{1}{2}(m\overparen{ADC} - m\overparen{AC})$ (Distr. Prop.)

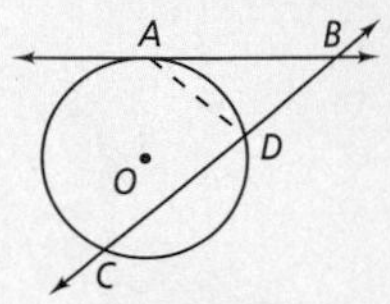

Given: $\odot O$ with tangent $\overleftrightarrow{BA}$ and secant $\overleftrightarrow{BC}$
Prove: $m\angle ABC = \frac{1}{2}(m\overparen{AC} - m\overparen{DA})$
Proof: 1. Draw $\overline{AD}$. (2 pts. determine a line.)
2. $m\angle DAB = \frac{1}{2}m\overparen{AD}$ (The measure of an $\angle$ formed by a tangent and a chord is half the measure of the intercepted arc.)
3. $m\angle ADC = \frac{1}{2}m\overparen{AC}$ (The measure of an inscribed $\angle$ is half the measure of its intercepted arc.) 4. $m\angle ABC = m\angle ADC - m\angle DAB$ (Subtr. Prop. of = and Ext. $\angle$ Thm.) 5. $m\angle ABC = \frac{1}{2}m\overparen{AC} - \frac{1}{2}m\overparen{AD}$ (Subst. Prop. of =) 6. $m\angle B = \frac{1}{2}(m\overparen{AC} - m\overparen{AD})$ (Distr. Prop.)

37.

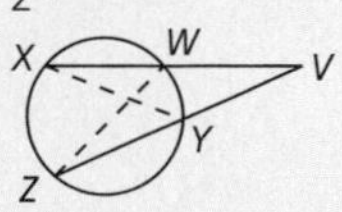

Given: A $\odot$ with secant segments $\overline{XV}$ and $\overline{ZV}$
Prove: $XV \cdot WV = ZV \cdot YV$.
Proof: Draw $\overline{XY}$ and $\overline{ZW}$. (2 pts. determine a line.); $\angle XVY \cong \angle ZVW$ (Refl. Prop. of $\cong$); $\angle VXY \cong \angle WZV$ (2 inscribed $\measuredangle$ that intercept the same arc are $\cong$.); $\triangle XVY \sim \triangle ZVW$ (AA~); $\frac{XV}{ZV} = \frac{YV}{WV}$ (In similar figures, corresp. sides are proportional.); $XV \cdot WV = ZV \cdot YV$ (Prop. of Proportion).

38.

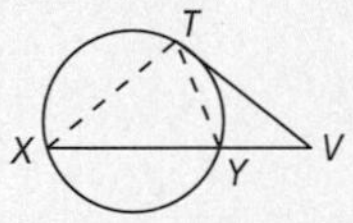

Given: A circle with tangent $\overline{TV}$ and secant $\overline{XV}$ Prove: $XV \cdot YV = (TV)^2$.
Proof: 1. Draw $\overline{TX}$ and $\overline{TY}$. (2 pts. determine a line.)
2. $m\angle TXV = \frac{1}{2}m\overparen{TY}$ (The measure of an inscribed $\angle$ is half the measure of the intercepted arc.)
3. $m\angle VTY = \frac{1}{2}m\overparen{TY}$ (The measure of an $\angle$ formed by a chord and a tangent is half the measure of the intercepted arc.) 4. $m\angle TXV = m\angle VTY$ (Trans. Prop. of =)
5. $\angle TVY \cong \angle TVX$ (Reflexive Prop. of $\cong$) 6. $\triangle TVY \sim \triangle XVT$ (AA ~)
7. $\frac{YV}{TV} = \frac{TV}{XV}$ (In similar figures, corresp. sides are proportional.)
8. $XV \cdot YV = (TV)^2$ (Prop. of Proportion)

42. Use the diagram at the right and the theorems of this lesson to prove the Proof Pythagorean Theorem.

43. If an equilateral triangle is inscribed in a circle, prove that the tangents to the Proof circle at the vertices form an equilateral triangle.

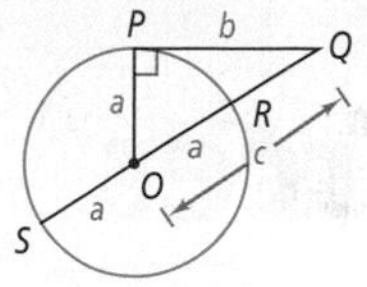

Apply What You've Learned

MATHEMATICAL PRACTICES
MP 2, MP 4

Look back at the information given on page 761 about the logo for the showroom display. The diagram of the logo is shown again below.

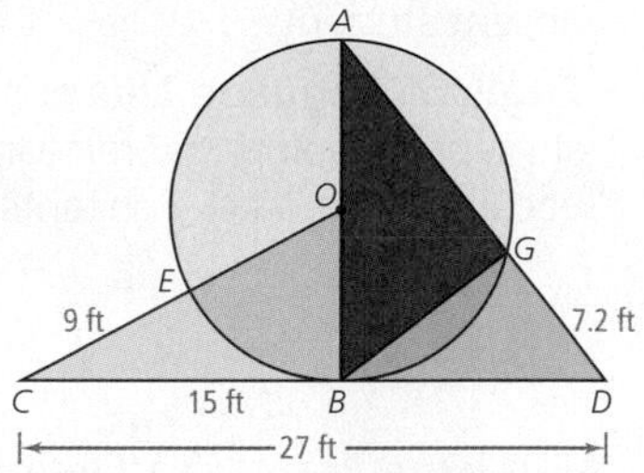

a. Name each segment in the diagram that is neither a chord nor a radius of ⊙O. Which of these segments is a secant?

b. In the Apply What You've Learned in Lesson 12-1, you found the length of one side of the red sail in the logo. Use the secant you identified in part (a) and a tangent to write and solve an equation to find the length of another side of the red sail in the logo.

c. Does the length you found in part (b) make sense given everything you know so far about the figure in the diagram? Explain.

Apply What You've Learned

In the Apply What You've Learned section for Lesson 12-3, students identified $\triangle AGB$ as a right triangle. Now students calculate the length of $\overline{AG}$ in this triangle. Later in the chapter, they will find the length of $\overline{BG}$ and the area of the red sail.

Mathematical Practices

Students interpret their mathematical **model** in the context of a real-world problem and decide whether the results make sense. (MP 4)

ANSWERS

a. $\overline{CE}$, $\overline{CO}$, $\overline{CB}$, $\overline{CD}$, $\overline{BD}$, $\overline{DG}$, $\overline{DA}$; $\overline{DA}$

b. $7.2(7.2 + x) = 12^2$, where x is the length of $\overline{AG}$; $AG = 12.8$ ft

c. Yes; sample: The length makes sense since $\overline{AG}$ is a chord of the circle and its length, 12.8 ft, is less than the diameter of the circle, 16 ft.

39a. $\triangle ACD$

b. $\tan A = \frac{DC}{AC} = \frac{DC}{1} = DC$, length of tangent seg.

c. $\text{secant } A = \frac{AD}{AC} = \frac{AD}{1} = AD$, length of secant seg.

40. $m\angle 1 = \frac{1}{2}m\widehat{QRP} - \frac{1}{2}m\widehat{PQ}$ (vertex outside ⊙, $m\angle = \frac{1}{2}$ difference of intercepted arcs); $m\angle 1 + m\widehat{PQ} = \frac{1}{2}m\widehat{QRP} + \frac{1}{2}m\widehat{PQ}$ (Add. Prop. of =); $m\angle 1 + m\widehat{PQ} = \frac{1}{2}(m\widehat{QRP} + m\widehat{PQ})$ (Distr. Prop.); $m\angle 1 + m\widehat{PQ} = \frac{1}{2}(360)$ (arc measure of ⊙ is 360); $m\angle 1 + m\widehat{PQ} = 180$ (Simplify.)

41. $m\angle 1 = \frac{1}{2}m\widehat{QRP} - \frac{1}{2}m\widehat{PQ}$ and $m\angle 2 = \frac{1}{2}m\widehat{RQP} - \frac{1}{2}m\widehat{RP}$ (vertex outside ⊙, $m\angle$ = half difference of intercepted arcs); $m\angle 1 + m\angle 2 = \frac{1}{2}m\widehat{QRP} + \frac{1}{2}m\widehat{RQP} - \frac{1}{2}m\widehat{PQ} - \frac{1}{2}m\widehat{RP}$ (Subst. Prop. of =); $m\angle 1 + m\angle 2 = \frac{1}{2}m\widehat{QR} + \frac{1}{2}m\widehat{RP} + \frac{1}{2}m\widehat{QR} + \frac{1}{2}m\widehat{PQ} - \frac{1}{2}m\widehat{PQ} - \frac{1}{2}m\widehat{RP}$ (Arc Add. Postulate and Distr. Prop.); $m\angle 1 + m\angle 2 = m\widehat{QR}$ (Distr. Prop.).

42. 1. $(PQ)^2 = (QS)(QR)$ (Square of the tangent equals the product of the secant times the external segment.)
2. $b^2 = (c + a)(c - a)$ (Substitution)
3. $b^2 = c^2 - a^2$ (Distributive Prop.)
4. $b^2 + a^2 = c^2$ (Addition Prop. of Equality)

43.

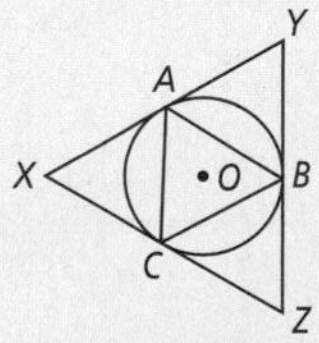

Given: Equilateral $\triangle ABC$ is inscribed in ⊙O; $\overline{XY}$, $\overline{YZ}$, and $\overline{XZ}$ are tangents to ⊙O
Prove: $\triangle XYZ$ is equilateral.
Proof: $m\widehat{AB} = m\widehat{BC} = m\widehat{AC} = 120$, since chords $\overline{AB}$, $\overline{BC}$, and $\overline{CA}$ are all ≅. So the measures of $\angle X$, $\angle Y$, and $\angle Z$ are $\frac{1}{2}(240 - 120) = 60$, and $\triangle XYZ$ is equiangular, so it is also equilateral.

12-4 Lesson Resources

Instructional Support

Geometry Companion

Students can use the **Geometry Companion** worktext (4 pages) . . .

- New Vocabulary
- Key Concepts
- Got It for each Problem
- Lesson Check

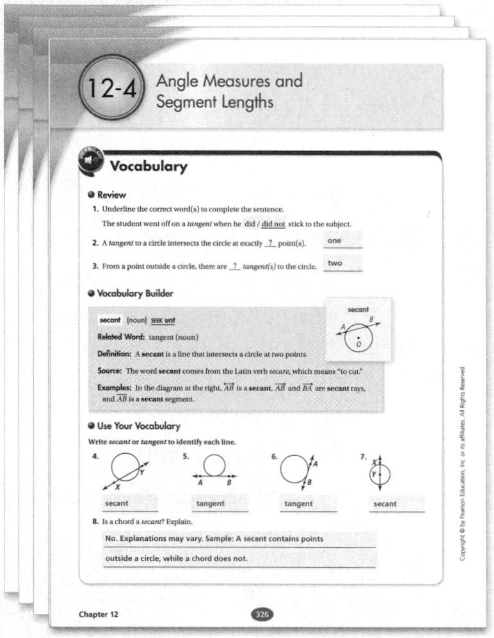
12-4 Angle Measures and Segment Lengths

Vocabulary

Review

1. Underline the correct word(s) to complete the sentence.

The student went off on a *tangent* when he did / did not stick to the subject.

2. A *tangent* to a circle intersects the circle at exactly _?_ point(s). one

3. From a point outside a circle, there are _?_ tangent(s) to the circle. two

Vocabulary Builder

secant (noun) SEEK unt

Related Word: tangent (noun)

Definition: A **secant** is a line that intersects a circle at two points.

Source: The word **secant** comes from the Latin verb *secare*, which means "to cut."

Examples: In the diagram at the right, $\overleftrightarrow{AB}$ is a **secant**, $\overrightarrow{AB}$ and $\overrightarrow{BA}$ are **secant** rays, and $\overline{AB}$ is a **secant** segment.

Use Your Vocabulary

Write *secant* or *tangent* to identify each line.

4. secant 5. tangent 6. tangent 7. secant

8. Is a chord a secant? Explain.

No. Explanations may vary. Sample: A secant contains points outside a circle, while a chord does not.

Chapter 12 326

ELL Support

Focus on Language Place students in small groups. Students can write the vocabulary words from the chapter on one side of individual index cards. Invite students to discuss the meaning of the word and restate a definition in their own words. Students can write the definition on the back of the card, along with sketches or examples. Have students use the cards to quiz each other and then keep the cards for later review.

Use Multiple Representation Have other texts and workbooks of different instructional levels that explore tangents, secants, chords, angles, and the other chapter concepts.

5 Assess & Remediate

Lesson Quiz

1. What is the value of the variable?

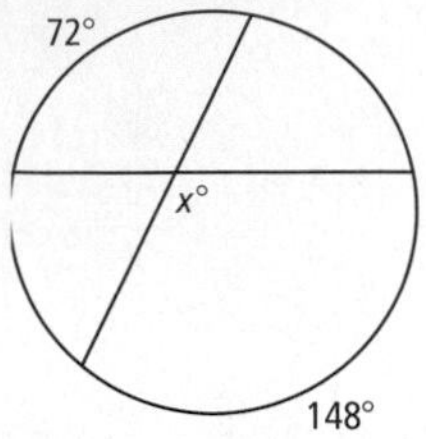

2. Do you UNDERSTAND? Find the value of the variable. If the answer is not a whole number, round to the nearest tenth.

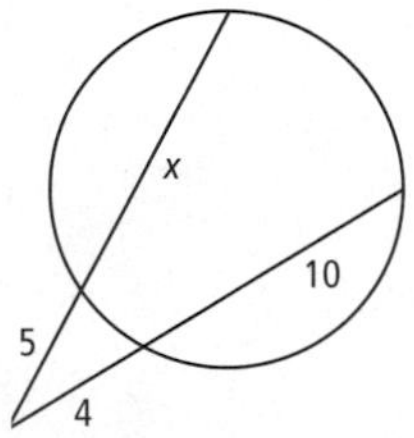

3. What is the value of g?

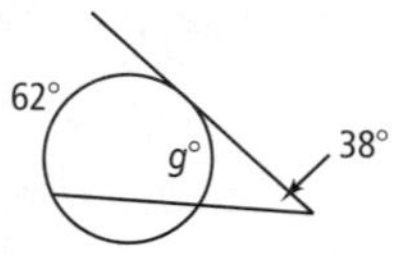

ANSWERS TO LESSON QUIZ

1. 110

2. 6.2

3. 43

PRESCRIPTION FOR REMEDIATION

Use the student work on the Lesson Quiz to prescribe a differentiated review assignment.

Points	Differentiated Remediation
0–1	Intervention
2	On-level
3	Extension

PowerGeometry.com

5 Assess & Remediate

Assign the Lesson Quiz. Appropriate intervention, practice, or enrichment is automatically generated based on student performance.

Differentiated Remediation

Intervention

- **Reteaching** (2 pages) Provides reteaching and practice exercises for the key lesson concepts. Use with struggling students or absent students.
- **English Language Learner Support** Helps students develop and reinforce mathematical vocabulary and key concepts.

All-in-One Resources/Online

Reteaching

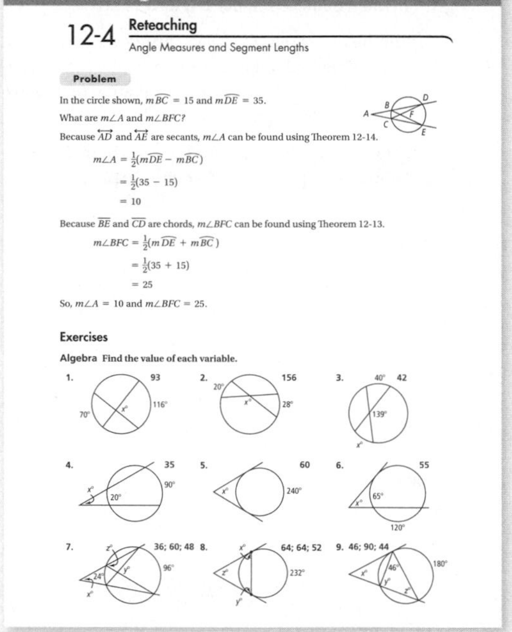
12-4 Reteaching
Angle Measures and Segment Lengths

Problem

In the circle shown, $m\widehat{BC} = 15$ and $m\widehat{DE} = 35$. What are $m\angle A$ and $m\angle BFC$?

Because $\overline{AD}$ and $\overline{AE}$ are secants, $m\angle A$ can be found using Theorem 12-14.

$m\angle A = \frac{1}{2}(m\widehat{DE} - m\widehat{BC})$
$= \frac{1}{2}(35 - 15)$
$= 10$

Because $\overline{BE}$ and $\overline{CD}$ are chords, $m\angle BFC$ can be found using Theorem 12-13.

$m\angle BFC = \frac{1}{2}(m\widehat{DE} + m\widehat{BC})$
$= \frac{1}{2}(35 + 15)$
$= 25$

So, $m\angle A = 10$ and $m\angle BFC = 25$.

Exercises

Algebra Find the value of each variable.

1. 93 2. 156 3. 40; 42 4. 35 5. 60 6. 55 7. 36; 60; 48 8. 64; 64; 52 9. 46; 90; 44

All-in-One Resources/Online

English Language Learner Support

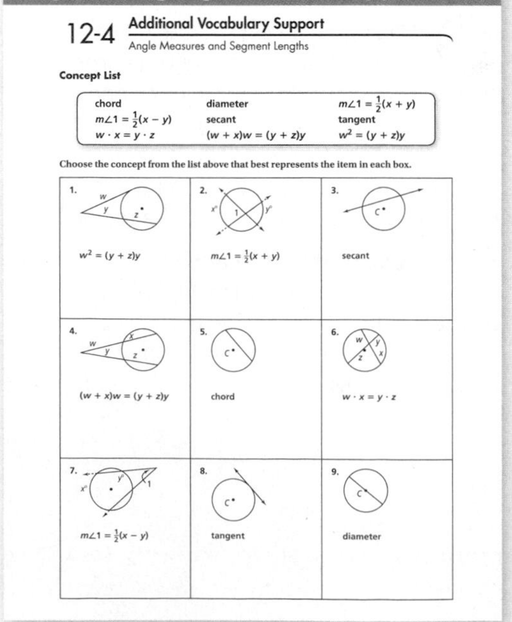
12-4 Additional Vocabulary Support
Angle Measures and Segment Lengths

Concept List

chord	diameter	$m\angle 1 = \frac{1}{2}(x + y)$
$m\angle 1 = \frac{1}{2}(x - y)$	secant	tangent
$w \cdot x = y \cdot z$	$(w + x)w = (y + z)y$	$w^2 = (y + z)y$

Choose the concept from the list above that best represents the item in each box.

1. $w^2 = (y + z)y$ 2. $m\angle 1 = \frac{1}{2}(x + y)$ 3. secant 4. $(w + x)w = (y + z)y$ 5. chord 6. $w \cdot x = y \cdot z$ 7. $m\angle 1 = \frac{1}{2}(x - y)$ 8. tangent 9. diameter

Differentiated Remediation *continued*

On-Level

- **Practice** (2 pages) Provides extra practice for each lesson. For simpler practice exercises, use the Form K Practice pages found in the All-in-One Teaching Resources and online.
- **Think About a Plan** Helps students develop specific problem-solving skills and strategies by providing scaffolded guiding questions.
- **Standardized Test Prep** Focuses on all major exercises, all major question types, and helps students prepare for the high-stakes assessments.

Extension

- **Enrichment** Provides students with interesting problems and activities that extend the concepts of the lesson.
- **Activities, Games, and Puzzles** Worksheets that can be used for concepts development, enrichment, and for fun!

Practice and Problem Solving Wkbk/ All-in-One Resources/Online

Practice page 1

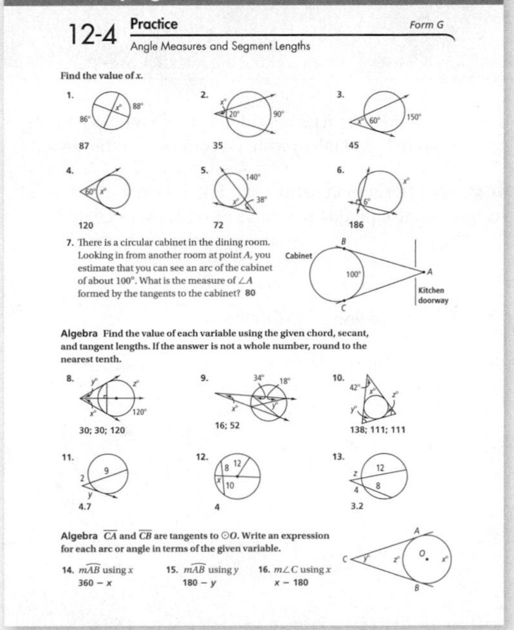

12-4 **Practice** *Form G*
Angle Measures and Segment Lengths

Find the value of x.

1. 87
2. 35
3. 45
4. 120
5. 72
6. 186
7. There is a circular cabinet in the dining room. Looking in from another room at point A, you estimate that you can see an arc of the cabinet of about 100°. What is the measure of $\angle A$ formed by the tangents to the cabinet? **80**

Algebra Find the value of each variable using the given chord, secant, and tangent lengths. If the answer is not a whole number, round to the nearest tenth.

8. 30; 30; 120
9. 16; 52
10. 138; 111; 111
11. 4.7
12. 4
13. 3.2

Algebra $\overline{CA}$ and $\overline{CB}$ are tangents to $\odot O$. Write an expression for each arc or angle in terms of the given variable.

14. $m\widehat{AB}$ using x **$360 - x$**
15. $m\widehat{AB}$ using y **$180 - y$**
16. $m\angle C$ using x **$x - 180$**

Practice and Problem Solving Wkbk/ All-in-One Resources/Online

Practice page 2

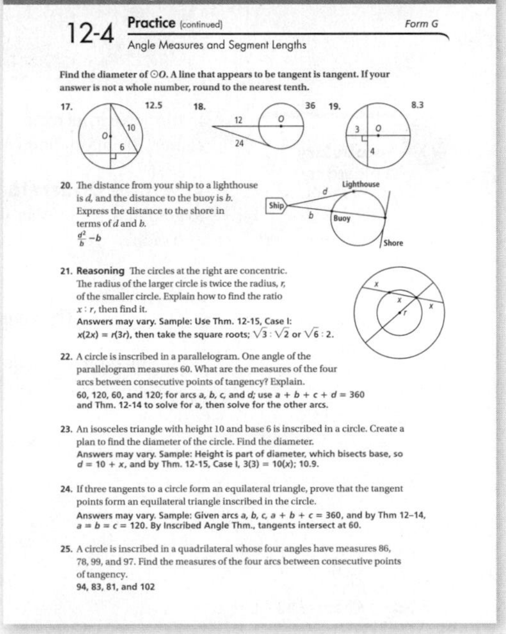

12-4 **Practice** (continued) *Form G*
Angle Measures and Segment Lengths

Find the diameter of $\odot O$. A line that appears to be tangent is tangent. If your answer is not a whole number, round to the nearest tenth.

17. 12.5
18. 36
19. 8.3
20. The distance from your ship to a lighthouse is d, and the distance to the buoy is b. Express the distance to the shore in terms of d and b.
$\frac{d^2}{b} - b$

21. **Reasoning** The circles at the right are concentric. The radius of the larger circle is twice the radius, r, of the smaller circle. Explain how to find the ratio $x : r$, then find it.
Answers may vary. Sample: Use Thm. 12-15, Case I: $x(2x) = r(3r)$, then take the square roots; $\sqrt{3} : \sqrt{2}$ or $\sqrt{6} : 2$.
22. A circle is inscribed in a parallelogram. One angle of the parallelogram measures 60. What are the measures of the four arcs between consecutive points of tangency? Explain.
60, 120, 60, and 120; for arcs a, b, c, and d; use $a + b + c + d = 360$ and Thm. 12-14 to solve for a, then solve for the other arcs.
23. An isosceles triangle with height 10 and base 6 is inscribed in a circle. Create a plan to find the diameter of the circle. Find the diameter.
Answers may vary. Sample: Height is part of diameter, which bisects base, so $d = 10 + x$, and by Thm. 12-15, Case I, $3(3) = 10(x)$; 10.9.
24. If three tangents to a circle form an equilateral triangle, prove that the tangent points form an equilateral triangle inscribed in the circle.
Answers may vary. Sample: Given arcs a, b, c, $a + b + c = 360$, and by Thm 12–14, $a = b = c = 120$. By Inscribed Angle Thm., tangents intersect at 60.
25. A circle is inscribed in a quadrilateral whose four angles have measures 86, 78, 99, and 97. Find the measures of the four arcs between consecutive points of tangency.
94, 83, 81, and 102

All-in-One Resources/Online

Enrichment

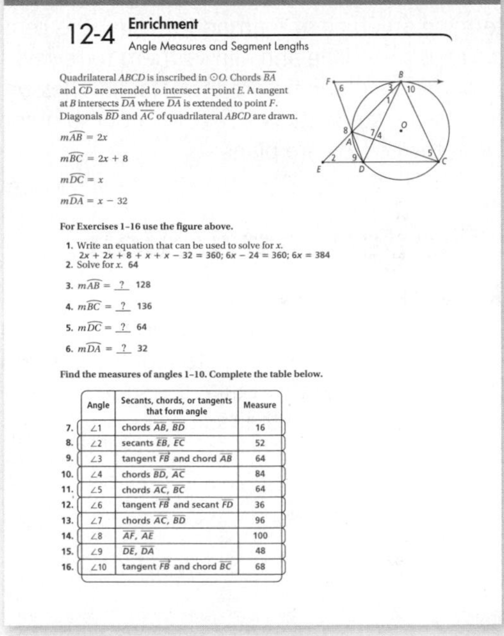

12-4 **Enrichment**
Angle Measures and Segment Lengths

Quadrilateral $ABCD$ is inscribed in $\odot O$. Chords $\overline{BA}$ and $\overline{CD}$ are extended to intersect at point E. A tangent at B intersects $\overline{DA}$ where $\overline{DA}$ is extended to point F. Diagonals $\overline{BD}$ and $\overline{AC}$ of quadrilateral $ABCD$ are drawn.

$m\widehat{AB} = 2x$

$m\widehat{BC} = 2x + 8$

$m\widehat{DC} = x$

$m\widehat{DA} = x - 32$

For Exercises 1–16 use the figure above.

1. Write an equation that can be used to solve for x.
$2x + 2x + 8 + x + x - 32 = 360$; $6x - 24 = 360$; $6x = 384$
2. Solve for x. **64**
3. $m\widehat{AB} =$ __?__ **128**
4. $m\widehat{BC} =$ __?__ **136**
5. $m\widehat{DC} =$ __?__ **64**
6. $m\widehat{DA} =$ __?__ **32**

Find the measures of angles 1–10. Complete the table below.

	Angle	Secants, chords, or tangents that form angle	Measure
7.	$\angle 1$	chords $\overline{AB}$, $\overline{BD}$	16
8.	$\angle 2$	secants $\overline{EB}$, $\overline{EC}$	52
9.	$\angle 3$	tangent $\overrightarrow{FB}$ and chord $\overline{AB}$	64
10.	$\angle 4$	chords $\overline{BD}$, $\overline{AC}$	84
11.	$\angle 5$	chords $\overline{AC}$, $\overline{BC}$	64
12.	$\angle 6$	tangent $\overrightarrow{FB}$ and secant $\overline{FD}$	36
13.	$\angle 7$	chords $\overline{AC}$, $\overline{BD}$	96
14.	$\angle 8$	$\overline{AF}$, $\overline{AE}$	100
15.	$\angle 9$	$\overline{DE}$, $\overline{DA}$	48
16.	$\angle 10$	tangent $\overrightarrow{FB}$ and chord $\overline{BC}$	68

Practice and Problem Solving Wkbk/ All-in-One Resources/Online

Think About a Plan

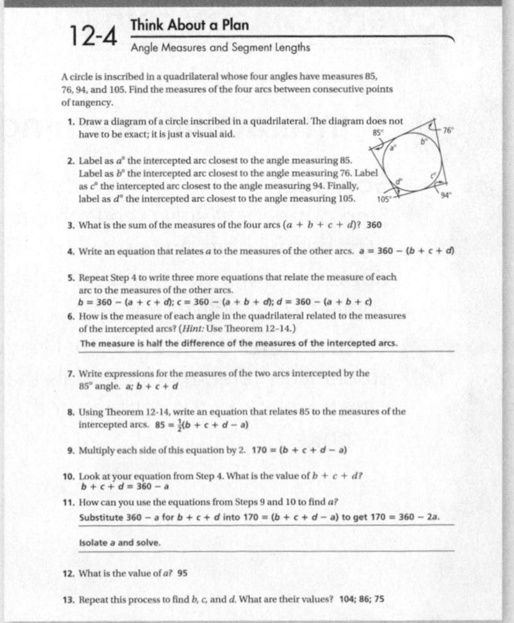

12-4 **Think About a Plan**
Angle Measures and Segment Lengths

A circle is inscribed in a quadrilateral whose four angles have measures 85, 76, 94, and 105. Find the measures of the four arcs between consecutive points of tangency.

1. Draw a diagram of a circle inscribed in a quadrilateral. The diagram does not have to be exact; it is just a visual aid.
2. Label as $a°$ the intercepted arc closest to the angle measuring 85. Label as $b°$ the intercepted arc closest to the angle measuring 76. Label as $c°$ the intercepted arc closest to the angle measuring 94. Finally, label as $d°$ the intercepted arc closest to the angle measuring 105.
3. What is the sum of the measures of the four arcs $(a + b + c + d)$? **360**
4. Write an equation that relates a to the measures of the other arcs. **$a = 360 - (b + c + d)$**
5. Repeat Step 4 to write three more equations that relate the measure of each arc to the measures of the other arcs.
$b = 360 - (a + c + d)$; $c = 360 - (a + b + d)$; $d = 360 - (a + b + c)$
6. How is the measure of each angle in the quadrilateral related to the measures of the intercepted arcs? (*Hint:* Use Theorem 12-14.)
The measure is half the difference of the measures of the intercepted arcs.
7. Write expressions for the measures of the two arcs intercepted by the 85° angle. **a; $b + c + d$**
8. Using Theorem 12-14, write an equation that relates 85 to the measures of the intercepted arcs. **$85 = \frac{1}{2}(b + c + d - a)$**
9. Multiply each side of this equation by 2. **$170 = (b + c + d - a)$**
10. Look at your equation from Step 4. What is the value of $b + c + d$?
$b + c + d = 360 - a$
11. How can you use the equations from Steps 9 and 10 to find a?
Substitute $360 - a$ for $b + c + d$ into $170 = (b + c + d - a)$ to get $170 = 360 - 2a$.
Isolate a and solve.
12. What is the value of a? **95**
13. Repeat this process to find b, c, and d. What are their values? **104; 86; 75**

Practice and Problem Solving Wkbk/ All-in-One Resources/Online

Standardized Test Prep

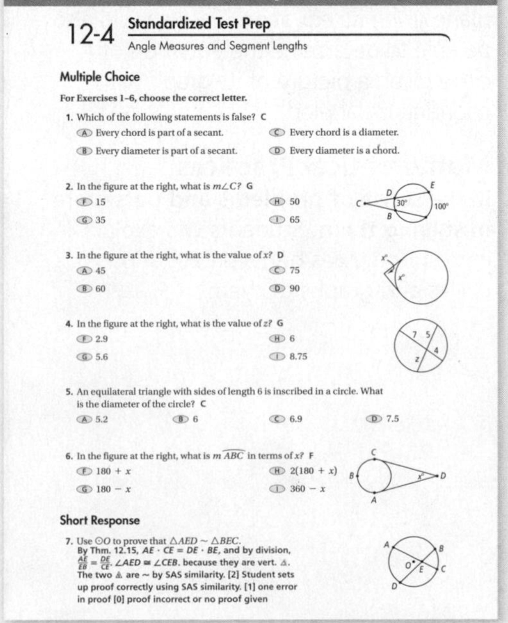

12-4 **Standardized Test Prep**
Angle Measures and Segment Lengths

Multiple Choice

For Exercises 1–6, choose the correct letter.

1. Which of the following statements is false? **C**
A. Every chord is part of a secant.
B. Every diameter is part of a secant.
C. Every chord is a diameter.
D. Every diameter is a chord.
2. In the figure at the right, what is $m\angle C$? **G**
F. 15
G. 35
H. 50
I. 65
3. In the figure at the right, what is the value of x? **D**
A. 45
B. 60
C. 75
D. 90
4. In the figure at the right, what is the value of z? **G**
F. 2.9
G. 5.6
H. 6
I. 8.75
5. An equilateral triangle with sides of length 6 is inscribed in a circle. What is the diameter of the circle? **C**
A. 5.2
B. 6
C. 6.9
D. 7.5
6. In the figure at the right, what is $m\widehat{ABC}$ in terms of x? **F**
F. $180 + x$
G. $180 - x$
H. $2(180 + x)$
I. $360 - x$

Short Response

7. Use $\odot O$ to prove that $\triangle AED \sim \triangle BEC$.
By Thm. 12.15, $AE \cdot CE = DE \cdot BE$, and by division, $\frac{AE}{EB} = \frac{DE}{CE}$. $\angle AED \cong \angle CEB$. because they are vert. △. The two △ are ~ by SAS similarity. [2] Student sets up proof correctly using SAS similarity. [1] one error in proof [0] proof incorrect or no proof given

Online Teacher Resource Center

Activities, Games, and Puzzles

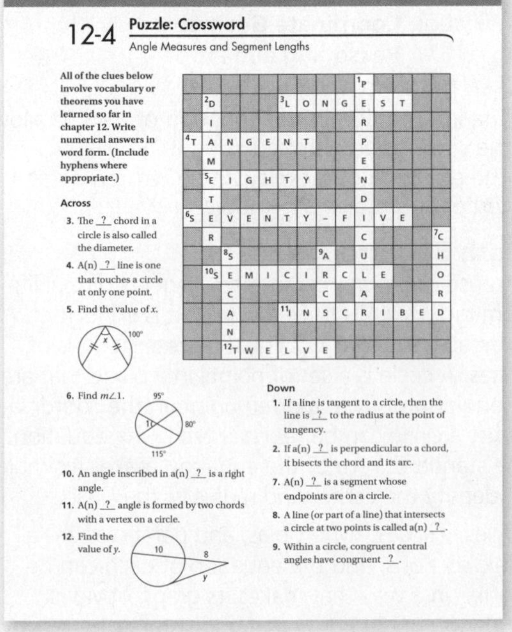

12-4 **Puzzle: Crossword**
Angle Measures and Segment Lengths

All of the clues below involve vocabulary or theorems you have learned so far in chapter 12. Write numerical answers in word form. (Include hyphens where appropriate.)

									1 P				
	2 D				3 L	O	N	G	E	S	T		
	I								R				
4 T	A	N	G	E	N	T			P				
	M								E				
	5 E	I	G	H	T	Y			N				
	T								D				
6 S	E	V	E	N	T	Y	-	F	I	V	E		
	R								C				7 C
		8 S					9 A		U				H
	10 S	E	M	I	C	I	R	C	L	E			O
		C					C		A				R
		A			11 I	N	S	C	R	I	B	E	D
		N											
		12 T	W	E	L	V	E						

Across

3. The __?__ chord in a circle is also called the diameter.
4. A(n) __?__ line is one that touches a circle at only one point.
5. Find the value of x.
6. Find $m\angle 1$.
10. An angle inscribed in a(n) __?__ is a right angle.
11. A(n) __?__ angle is formed by two chords with a vertex on a circle.
12. Find the value of y.

Down

1. If a line is tangent to a circle, then the line is __?__ to the radius at the point of tangency.
2. If a(n) __?__ is perpendicular to a chord, it bisects the chord and its arc.
7. A(n) __?__ is a segment whose endpoints are on a circle.
8. A line (or part of a line) that intersects a circle at two points is called a(n) __?__.
9. Within a circle, congruent central angles have congruent __?__.

1 Interactive Learning

Solve It!

PURPOSE To use the center and radius of a circle in a real-world setting

PROCESS Students may

- draw a circle with radius $\frac{1}{2}$ on the grid and determine if all points fall inside the circle.
- find the distance from base station to each point and determine if all points are less than $\frac{1}{2}$ mi from it.

FACILITATE

Q What type of figure can represent the range of the walkie-talkies? **[a circle]**

Q What is the center and radius of the circle? **[The center is at (2, 4) and the radius is $\frac{1}{2}$ mi.]**

ANSWER See Solve It in Answers on next page.

CONNECT THE MATH The Solve It helps students determine an efficient method for using the center and radius of a circle and allows them to review the coordinate plane before they start this lesson. In the lesson, students will discover how to graph a circle in the coordinate plane.

2 Guided Instruction

Take Note

Emphasize that writing the equation for a circle in standard form makes it easier to identify the center (h, k). Remind students to take the square root of the constant r^2 to find the radius.

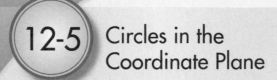

12-5 Circles in the Coordinate Plane

Common Core State Standards

G-GPE.A.1 Derive the equation of a circle given center and radius using the Pythagorean Theorem . . .

MP 1, MP 3, MP 4, MP 7

Objectives To write the equation of a circle

To find the center and radius of a circle

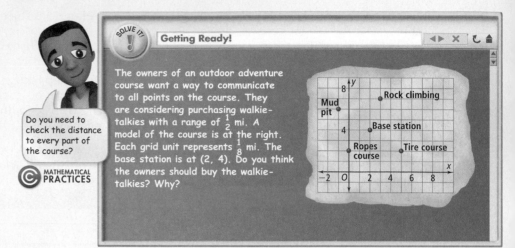

Do you need to check the distance to every part of the course?

MATHEMATICAL PRACTICES

Solve It! Getting Ready!

The owners of an outdoor adventure course want a way to communicate to all points on the course. They are considering purchasing walkie-talkies with a range of $\frac{1}{2}$ mi. A model of the course is at the right. Each grid unit represents $\frac{1}{8}$ mi. The base station is at (2, 4). Do you think the owners should buy the walkie-talkies? Why?

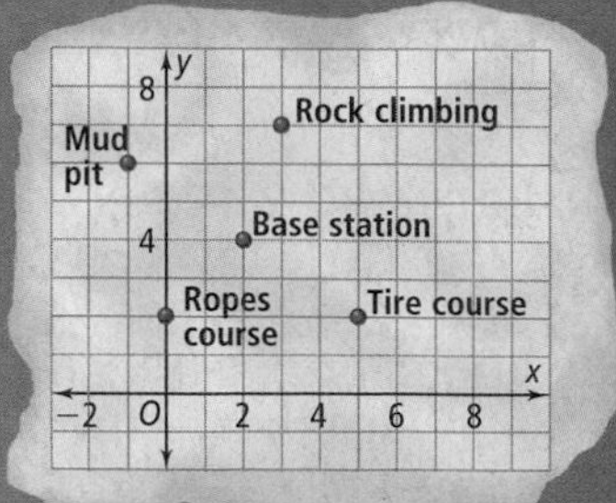

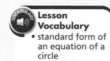

Lesson Vocabulary

- standard form of an equation of a circle

In the Solve It, all of the obstacles lie within or on a circle with the base station as the center. The information from the diagram is enough to write an equation for the circle.

Essential Understanding The information in the equation of a circle allows you to graph the circle. Also, you can write the equation of a circle if you know its center and radius.

Take note

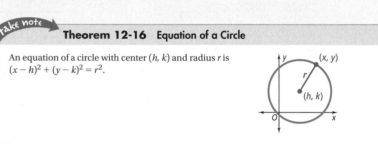

Theorem 12-16 Equation of a Circle

An equation of a circle with center (h, k) and radius r is $(x - h)^2 + (y - k)^2 = r^2$.

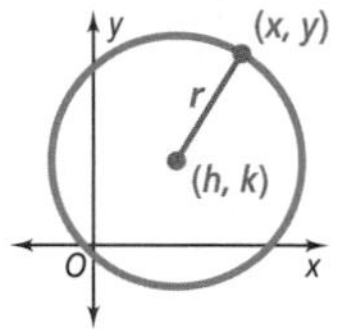

12-5 Preparing to Teach

BIG ideas **Coordinate Geometry**
Reasoning and Proof

ESSENTIAL UNDERSTANDINGS

- The information in the equation of a circle allows the circle to be graphed.
- The equation of a circle can be written if its center and radius are known.

Math Background

Because circles appear in many aspects of real life from radio signals to tree trunks, it is important to be able to construct visual representations of circles. A circle is a set of points in a plane that are all equidistant from a given point. In the coordinate plane, a circle can be represented by an equation. The standard form of the equation makes it simple to identify the center and radius of the circle.

Circles, ellipses, hyperbolas, and parabolas are all conic sections, and the equation of each can be written in a way that makes its graph obvious. Students should know and understand the equation of a circle and be able to sketch a circle in its proper location in the coordinate plane given its equation. They should also be able to determine the equation of a circle given a picture of its graph with coordinates labeled.

Mathematical Practices

Make sense of problems and persevere in solving them. Students will explain correspondences between equations of circles and graphs of them.

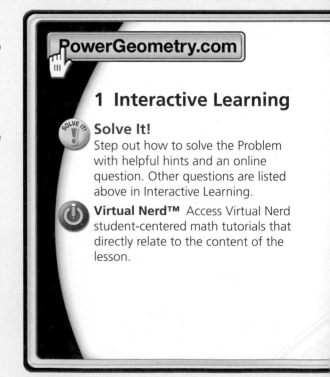

PowerGeometry.com

1 Interactive Learning

Solve It!

Step out how to solve the Problem with helpful hints and an online question. Other questions are listed above in Interactive Learning.

Virtual Nerd™ Access Virtual Nerd student-centered math tutorials that directly relate to the content of the lesson.

Here's Why It Works You can use the Distance Formula to find an equation of a circle with center (h, k) and radius r, which proves Theorem 12-16. Let (x, y) be any point on the circle. Then the radius r is the distance from (h, k) to (x, y).

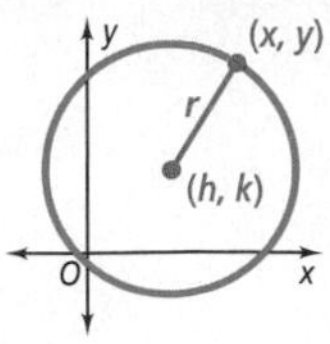

$d = \sqrt{(x_2 - x_1)^2 + (y_2 - y_1)^2}$ Distance Formula

$r = \sqrt{(x - h)^2 + (y - k)^2}$ Substitute (x, y) for (x_2, y_2) and (h, k) for (x_1, y_1).

$r^2 = (x - h)^2 + (y - k)^2$ Square both sides.

The equation $(x - h)^2 + (y - k)^2 = r^2$ is the **standard form of an equation of a circle.** You may also call it the *standard equation* of a circle.

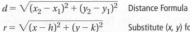

Plan

What do you need to know to write the equation of a circle?
You need to know the values of h, k, and r; h is the x-coordinate of the center, k is the y-coordinate of the center, and r is the radius.

Problem 1 Writing the Equation of a Circle

What is the standard equation of the circle with center $(5, -2)$ and radius 7?

$(x - h)^2 + (y - k)^2 = r^2$ Use the standard form of an equation of a circle.

$(x - 5)^2 + [y - (-2)]^2 = 7^2$ Substitute $(5, -2)$ for (h, k) and 7 for r.

$(x - 5)^2 + (y + 2)^2 = 49$ Simplify.

Got It? **1.** What is the standard equation of each circle?
a. center $(3, 5)$; radius 6 **b.** center $(-2, -1)$; radius $\sqrt{2}$

Think

How is this problem different from Problem 1?
In this problem, you don't know r. So the first step is to find r.

Problem 2 Using the Center and a Point on a Circle

What is the standard equation of the circle with center $(1, -3)$ that passes through the point $(2, 2)$?

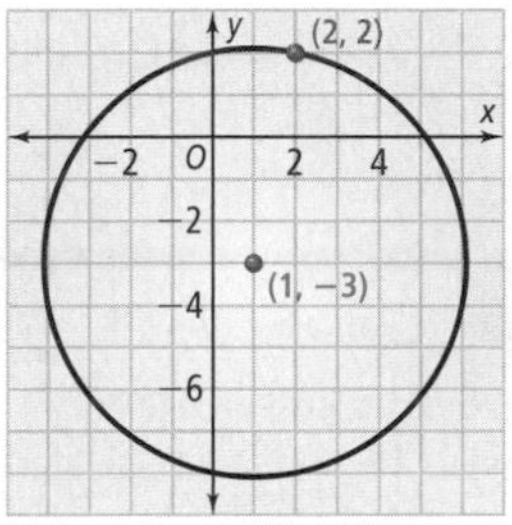

Step 1 Use the Distance Formula to find the radius.

$r = \sqrt{(x_2 - x_1)^2 + (y_2 - y_1)^2}$ Use the Distance Formula.

$= \sqrt{(1 - 2)^2 + (-3 - 2)^2}$ Substitute $(1, -3)$ for (x_2, y_2) and $(2, 2)$ for (x_1, y_1).

$= \sqrt{(-1)^2 + (-5)^2}$ Simplify.

$= \sqrt{26}$

Step 2 Use the radius and the center to write an equation.

$(x - h)^2 + (y - k)^2 = r^2$ Use the standard form of an equation of a circle.

$(x - 1)^2 + [y - (-3)]^2 = (\sqrt{26})^2$ Substitute $(1, -3)$ for (h, k) and $\sqrt{26}$ for r.

$(x - 1)^2 + (y + 3)^2 = 26$ Simplify.

Got It? **2.** What is the standard equation of the circle with center $(4, 3)$ that passes through the point $(-1, 1)$?

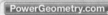

Here's Why It Works

Review the definition of a circle with students. Emphasize that a circle is the set of points that are equidistant from the center. Then, ask students to write an equation using the Distance Formula to show this relationship.

Problem 1

Q The center $(5, -2)$ represents what variables from the standard equation of a circle? **[(h, k)]**

Q How will the circle change if the value of k is changed from -2 to 2? **[The circle will be reflected about the x-axis.]**

Q How does the radius appear in the equation? **[The radius appears on one side of the equal sign and it is squared.]**

Got It? ERROR PREVENTION

Have students label each center as (h, k). Then have them find the square of the radius. Finally, have students substitute the values into the equation.

Problem 2

Q How can you find the radius of the circle? **[Find the distance between the center of the circle and the point on the circle.]**

Q How do you find this distance? **[Use the ordered pair of the center and the ordered pair of a point on the circle in the Distance Formula.]**

Got It? VISUAL LEARNERS

Be sure that students understand why the distance between the center and a point on the circle is equal to the radius.

2 Guided Instruction

Each Problem is worked out and supported online.

Problem 1
Writing the Equation of a Circle

Problem 2
Using the Center and a Point on a Circle

Problem 3
Graphing a Circle Given Its Equation

Support in Geometry Companion

- Vocabulary
- Key Concepts
- Got It?

Answers

Solve It!

Yes; the base station is located less than $\frac{1}{2}$ mi from all of the obstacles.

Got It?

1a. $(x - 3)^2 + (y - 5)^2 = 36$
b. $(x + 2)^2 + (y + 1)^2 = 2$
2. $(x - 4)^2 + (y - 3)^2 = 29$

Problem 3

Q What values in the equation give you the center of the circle? **[The values subtracted from x (7) and from y (−2) are the coordinates of the center (7, −2).]**

Q How do you find the radius of the circle? **[Take the square root of 64.]**

Q What two steps are needed to graph the circle? **[Plot the center and draw a circle with radius 8.]**

Got It? VISUAL LEARNERS

Remind students that the equation for a circle includes subtraction signs in the quantities. So in this case, the elements of the ordered pair that represents the center of the circle are positive.

3 Lesson Check

Do you know HOW?

- If students have difficulty with Exercises 1-2, then have them review Problem 1 and write the equation of a circle before substituting the values into the equation.

Do you UNDERSTAND?

- If students have difficulty with Exercise 5, then have them review the Take Note on page 798 and identify the meaning of each variable in the equation of a circle.

Close

Q What is the equation of a circle? **[$(x - h)^2 + (y - k)^2 = r^2$]**

Q How can you identify the center and radius of the circle? **[The center is (h, k) and the radius $= r$.]**

If you know the standard equation of a circle, you can describe the circle by naming its center and radius. Then you can use this information to graph the circle.

Problem 3 Graphing a Circle Given Its Equation STEM

Communications **When you make a call on a cell phone, a tower receives and transmits the call. A way to monitor the range of a cell tower system is to use equations of circles. Suppose the equation $(x - 7)^2 + (y + 2)^2 = 64$ represents the position and the transmission range of a cell tower. What is the graph that shows the position and range of the tower?**

Know	Need	Plan
The equation representing the cell tower's position and range	To draw a graph	Determine the values of (h, k) and r in the equation. Then draw a graph.

$(x - 7)^2 + (y + 2)^2 = 64$ Use the standard equation of a circle.

$(x - 7)^2 + [y - (-2)] = 8^2$ Rewrite to find h, k, and r.

($h = 7$, $k = -2$, $r = 8$)

The center is (7, −2) and the radius is 8.

To graph the circle, place the compass point at the center (7, −2) and draw a circle with radius 8.

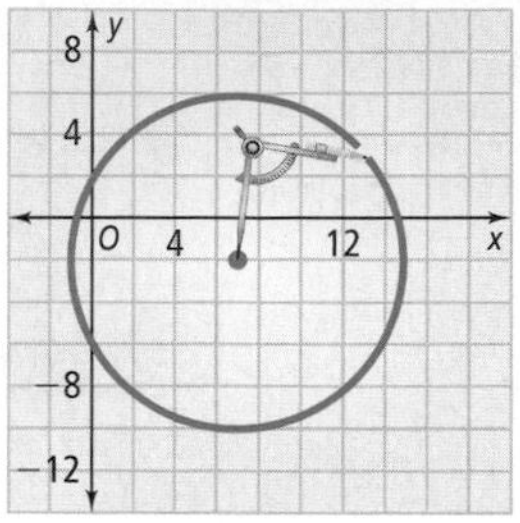

Got It? 3. a. In Problem 3, what does the center of the circle represent? What does the radius represent?

b. What is the center and radius of the circle with equation $(x - 2)^2 + (y - 3)^2 = 100$? Graph the circle.

Lesson Check

Do you know HOW?

What is the standard equation of each circle?

1. center (0, 0); $r = 4$
2. center (1, −1); $r = \sqrt{5}$

What is the center and radius of each circle?

3. $(x - 8)^2 + y^2 = 9$
4. $(x + 2)^2 + (y - 4)^2 = 7$

Do you UNDERSTAND? MATHEMATICAL PRACTICES

5. What is the least amount of information that you need to graph a circle? To write the equation of a circle?
6. Suppose you know the center of a circle and a point on the circle. How do you determine the equation of the circle?
7. **Error Analysis** A student says that the center of a circle with equation $(x - 2)^2 + (y + 3)^2 = 16$ is (−2, 3). What is the student's error?

PowerGeometry.com

3 Lesson Check

For a digital lesson check, use the Got It questions.

Support in Geometry Companion

- Lesson Check

4 Practice

Assign homework to individual students or to an entire class.

Additional Problems

1. Write the standard equation of the circle with center (−6, 8) and radius 5.

ANSWER $(x + 6)^2 + (y - 8)^2 = 25$

2. What is the equation of the circle with center (4, 2) that passes through the point (2, −3)?

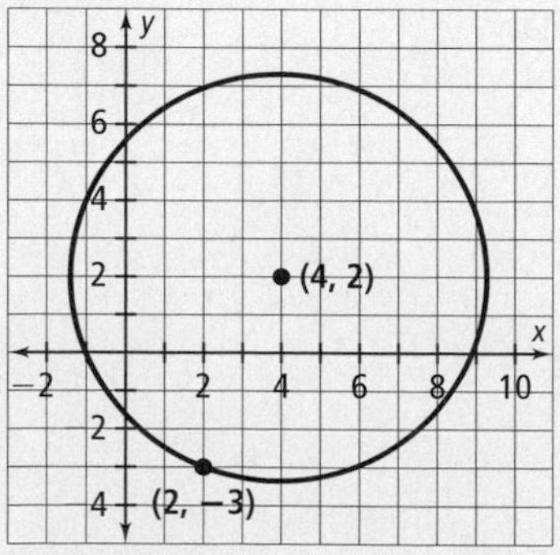

ANSWER $(x - 4)^2 + (y - 2)^2 = 29$

3. What is the center and radius of the circle with equation $(x - 3)^2 + (y + 2)^2 = 36$? Graph the circle.

ANSWER center (3, −2); radius 6

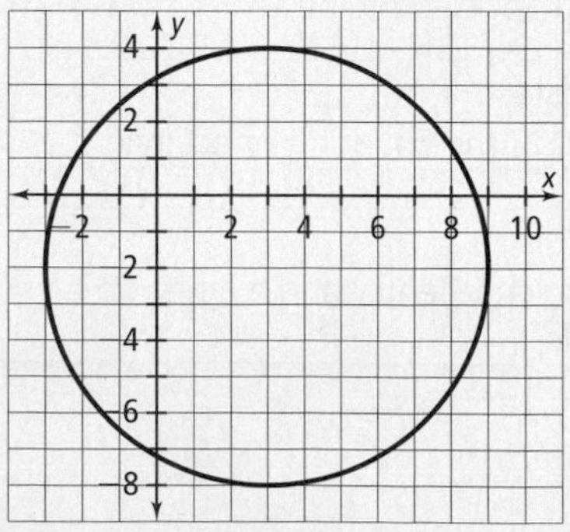

Practice and Problem-Solving Exercises

MATHEMATICAL PRACTICES

A Practice

Write the standard equation of each circle. See Problem 1.

8. center $(2, -8)$; $r = 9$
9. center $(0, 3)$; $r = 7$
10. center $(0.2, 1.1)$; $r = 0.4$
11. center $(5, -1)$; $r = 12$
12. center $(-6, 3)$; $r = 8$
13. center $(-9, -4)$; $r = \sqrt{5}$
14. center $(0, 0)$; $r = 4$
15. center $(-4, 0)$; $r = 3$
16. center $(-1, -1)$; $r = 1$

Write a standard equation for each circle in the diagram at the right. See Problem 2.

17. $\odot P$
18. $\odot Q$

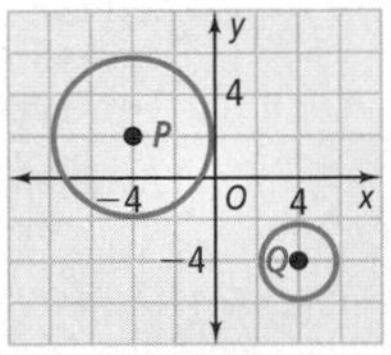

Write the standard equation of the circle with the given center that passes through the given point.

19. center $(-2, 6)$; point $(-2, 10)$
20. center $(1, 2)$; point $(0, 6)$
21. center $(7, -2)$; point $(1, -6)$
22. center $(-10, -5)$; point $(-5, 5)$
23. center $(6, 5)$; point $(0, 0)$
24. center $(-1, -4)$; point $(-4, 0)$

Find the center and radius of each circle. Then graph the circle. See Problem 3.

25. $(x + 7)^2 + (y - 5)^2 = 16$
26. $(x - 3)^2 + (y + 8)^2 = 100$
27. $(x + 4)^2 + (y - 1)^2 = 25$
28. $x^2 + y^2 = 36$

Public Safety Each equation models the position and range of a tornado alert siren. Describe the position and range of each.

29. $(x - 5)^2 + (y - 7)^2 = 81$
30. $(x + 4)^2 + (y - 9)^2 = 144$

B Apply

Write the standard equation of each circle.

31. 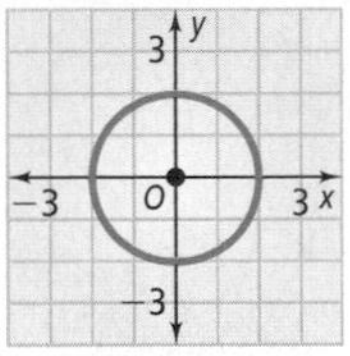

32. 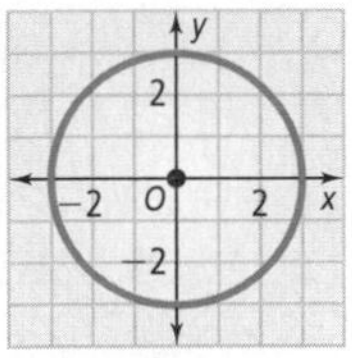

33. 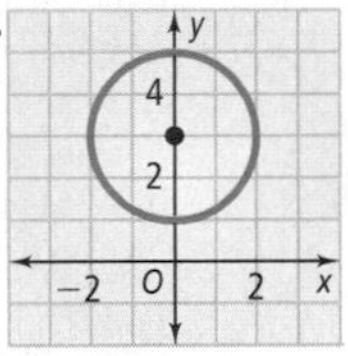

34. 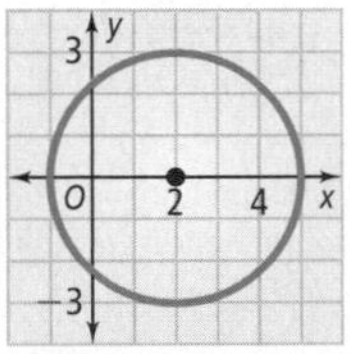

35. 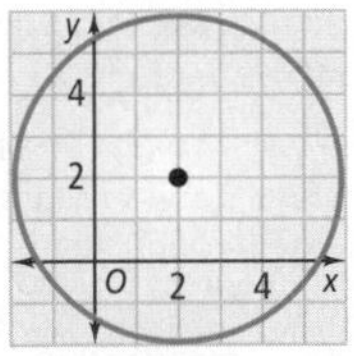

36. 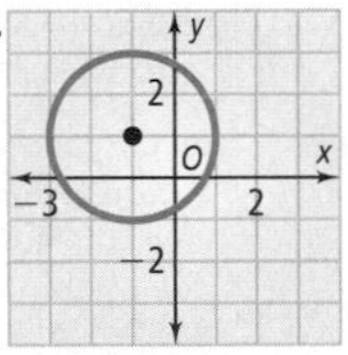

4 Practice

ASSIGNMENT GUIDE

Basic: 8–30 all, 31–52 even
Average: 9–29 odd, 31–56
Advanced: 9–29 odd, 31–57
Standardized Test Prep: 58–60
Mixed Review: 61–65

Mathematical Practices are supported by exercises with red headings. Here are the Practices supported in this lesson:

MP 1: Make Sense of Problems Ex. 44
MP 3: Communicate Ex. 56
MP 3: Critique the Reasoning of Others Ex. 7
MP 7: Shift Perspective Ex. 40

Applications exercises have blue headings. Exercise 57 supports MP 4: Model.

EXERCISE 46: Use the Think About a Plan worksheet in the **Practice and Problem Solving Workbook** (also available in the Teaching Resources in print and online) to further support students' development in becoming independent learners.

HOMEWORK QUICK CHECK

To check students' understanding of key skills and concepts, go over Exercises 11, 21, 40, 44, and 46.

Answers

Got It? (continued)

3a. The center of the circle represents the cell tower's position. The radius represents the cell tower's transmission range.

b. center (2, 3); radius 10

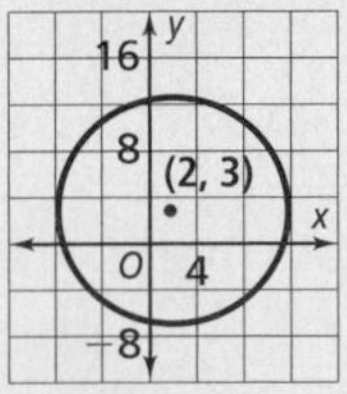

Lesson Check

1. $x^2 + y^2 = 16$
2. $(x - 1)^2 + (y + 1)^2 = 5$
3. center (8, 0); radius 3
4. center $(-2, 4)$; radius $\sqrt{7}$
5. its center and its radius; its center and its radius
6. Using the two known points, use the Distance Formula to find the distance between them; that is the radius. Then use the center and the radius to write the standard equation for the circle.
7. Sample explanation: The student should have rewritten the equation as $(x - 2)^2 + (y - (-3))^2 = 16$ to realize that the center is $(2, -3)$.

Practice and Problem-Solving Exercises

8. $(x - 2)^2 + (y + 8)^2 = 81$
9. $x^2 + (y - 3)^2 = 49$
10. $(x - 0.2)^2 + (y - 1.1)^2 = 0.16$
11. $(x - 5)^2 + (y + 1)^2 = 144$
12. $(x + 6)^2 + (y - 3)^2 = 64$
13. $(x + 9)^2 + (y + 4)^2 = 5$
14. $x^2 + y^2 = 16$
15. $(x + 4)^2 + y^2 = 9$
16. $(x + 1)^2 + (y + 1)^2 = 1$
17. $(x + 4)^2 + (y - 2)^2 = 16$
18. $(x - 4)^2 + (y + 4)^2 = 4$
19. $(x + 2)^2 + (y - 6)^2 = 16$
20. $(x - 1)^2 + (y - 2)^2 = 17$
21. $(x - 7)^2 + (y + 2)^2 = 52$
22. $(x + 10)^2 + (y + 5)^2 = 125$
23. $(x - 6)^2 + (y - 5)^2 = 61$
24. $(x + 1)^2 + (y + 4)^2 = 25$
25. center $(-7, 5)$; radius 4

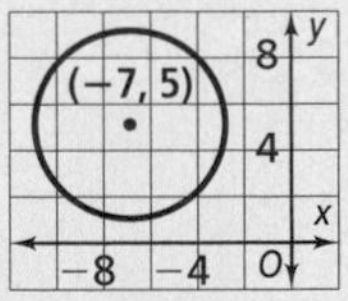

26–36. See next page.

Answers

Practice and Problem-Solving Exercises (continued)

26. center (3, −8); radius 10

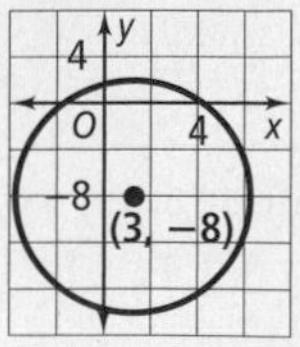

27. center (−4, 1); radius 5

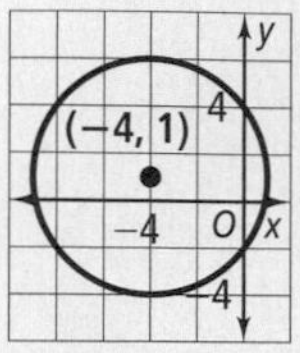

28. center (0, 0); radius 6

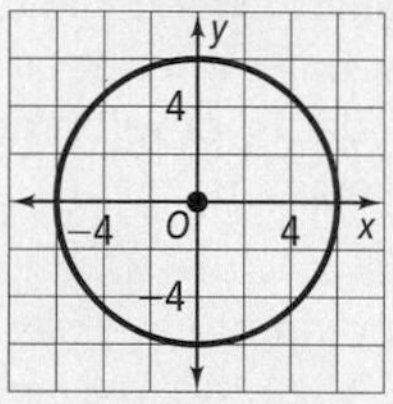

29. position (5, 7); range 9

30. position (−4, 9); range 12

31. $x^2 + y^2 = 4$

32. $x^2 + y^2 = 9$

33. $x^2 + (y - 3)^2 = 4$

34. $(x - 2)^2 + y^2 = 9$

35. $(x - 2)^2 + (y - 2)^2 = 16$

36. $(x + 1)^2 + (y - 1)^2 = 4$

37. $(x - 4)^2 + (y - 3)^2 = 25$

38. $(x - 5)^2 + (y - 3)^2 = 13$

39. $(x - 3)^2 + (y - 3)^2 = 8$

40. The graph is the point (0, 0).

41. Yes; it is a circle with center (1, −2) and radius 3.

42. No; the *x* and *y* terms are not squared.

43. No; the *x* term is not squared.

44. circumference 16π units; area 64π units^2

45. $(x - 4)^2 + (y - 7)^2 = 36$

46. *x*-int. 13; *y*-int. 9.75

47. $(x - h)^2 + (y - k)^2 = r^2$

$(y - k)^2 = r^2 - (x - h)^2$

$y - k = \pm\sqrt{r^2 - (x - h)^2}$

$y = \pm\sqrt{r^2 - (x - h)^2} + k$

48.

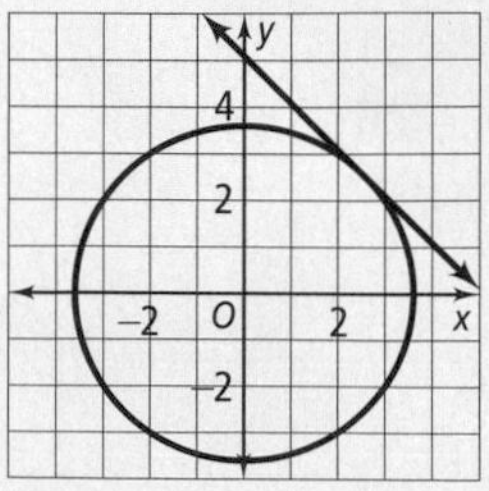

(3, 2), (2, 3)

Write an equation of a circle with diameter $\overline{AB}$.

37. *A*(0, 0), *B*(8, 6) **38.** *A*(3, 0), *B*(7, 6) **39.** *A*(1, 1), *B*(5, 5)

40. Reasoning Describe the graph of $x^2 + y^2 = r^2$ when $r = 0$.

Determine whether each equation is the equation of a circle. Justify your answer.

41. $(x - 1)^2 + (y + 2)^2 = 9$ **42.** $x + y = 9$ **43.** $x + (y - 3)^2 = 9$

44. Think About a Plan Find the circumference and area of the circle whose equation is $(x - 9)^2 + (y - 3)^2 = 64$. Leave your answers in terms of π.
- What essential information do you need?
- What formulas will you use?

45. Write an equation of a circle with area 36π and center (4, 7).

46. What are the *x*- and *y*-intercepts of the line tangent to the circle $(x - 2)^2 + (y - 2)^2 = 5^2$ at the point (5, 6)?

47. For $(x - h)^2 + (y - k)^2 = r^2$, show that $y = \sqrt{r^2 - (x - h)^2} + k$ or $y = -\sqrt{r^2 - (x - h)^2} + k$.

Sketch the graphs of each equation. Find all points of intersection of each pair of graphs.

48. $x^2 + y^2 = 13$, $y = -x + 5$ **49.** $x^2 + y^2 = 17$, $y = -\frac{1}{4}x$ **50.** $x^2 + y^2 = 8$, $y = 2$

51. $x^2 + y^2 = 20$, $y = -\frac{1}{2}x + 5$ **52.** $(x + 1)^2 + (y - 1)^2 = 18$, $y = x + 8$ **53.** $(x - 2)^2 + (y - 2)^2 = 10$, $y = -\frac{1}{3}x + 6$

54. You can use completing the square and factoring to find the center and radius of a circle.

a. What number *c* do you need to add to each side of the equation $x^2 + 6x + y^2 - 4y = -4$ so that $x^2 + 6x + c$ can be factored into a perfect square binomial?

b. What number *d* do you need to add to each side of the equation $x^2 + 6x + y^2 - 4y = -4$ so that $y^2 - 4y + d$ can be factored into a perfect square binomial?

c. Rewrite $x^2 + 6x + y^2 - 4y = -4$ using your results from parts (a) and (b).

d. What are the center and radius of $x^2 + 6x + y^2 - 4y = -4$?

e. What are the center and radius of $x^2 + 4x + y^2 - 20y + 100 = 0$?

Challenge

55. The concentric circles $(x - 3)^2 + (y - 5)^2 = 64$ and $(x - 3)^2 + (y - 5)^2 = 25$ form a ring. The lines $y = \frac{2}{3}x + 3$ and $y = 5$ intersect the ring, making four sections. Find the area of each section. Round your answers to the nearest tenth of a square unit.

49.

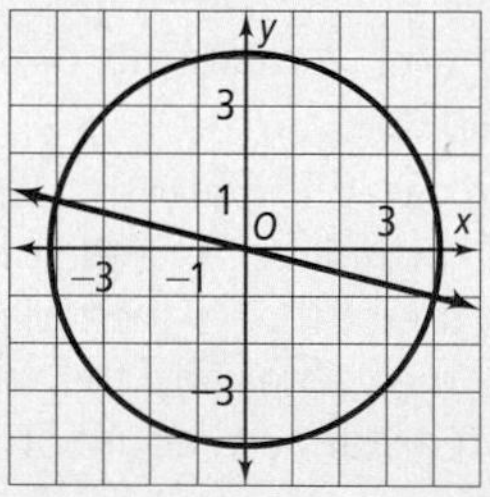

(4, −1), (−4, 1)

50.

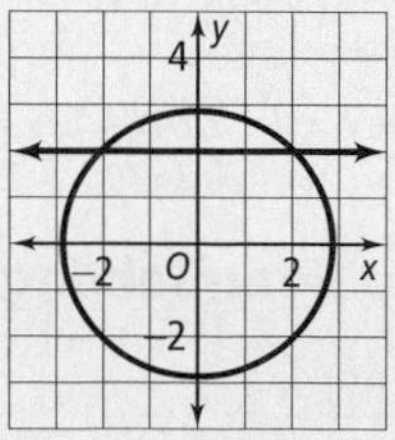

(2, 2), (−2, 2)

51.

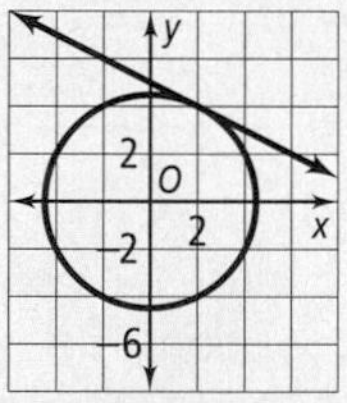

(2, 4)

52.

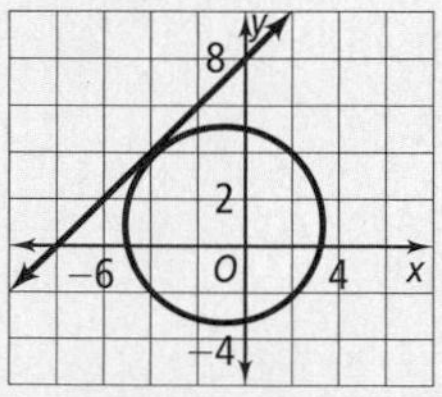

(−4, 4)

53.

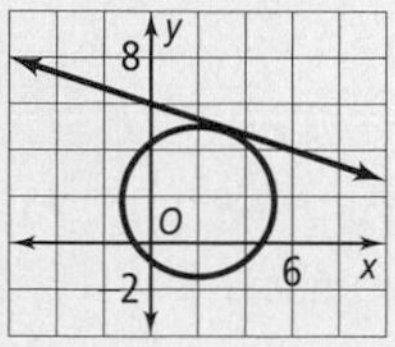

(3, 5)

56. **Geometry in 3 Dimensions** The equation of a sphere is similar to the equation of a circle. The equation of a sphere with center (h, j, k) and radius r is $(x-h)^2 + (y-j)^2 + (z-k)^2 = r^2$. $M(-1, 3, 2)$ is the center of a sphere passing through $T(0, 5, 1)$. What is the radius of the sphere? What is the equation of the sphere?

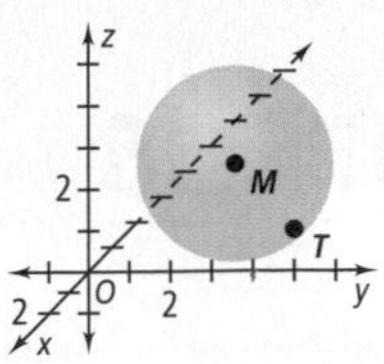

57. **Nautical Distance** A close estimate of the radius of Earth's equator is 3960 mi.
   a. Write the equation of the equator with the center of Earth as the origin.
   b. Find the length of a 1° arc on the equator to the nearest tenth of a mile.
   c. **History** Columbus planned his trip to the East by going west. He thought each 1° arc was 45 mi long. He estimated that the trip would take 21 days. Use your answer to part (b) to find a better estimate.

Standardized Test Prep

SAT/ACT

58. What is an equation of a circle with radius 16 and center $(2, -5)$?

Ⓐ $(x-2)^2 + (y+5)^2 = 16$
Ⓑ $(x+2)^2 + (y-5)^2 = 4$
Ⓒ $(x+2)^2 + (y-5)^2 = 256$
Ⓓ $(x-2)^2 + (y+5)^2 = 256$

59. What can you NOT conclude from the diagram at the right?

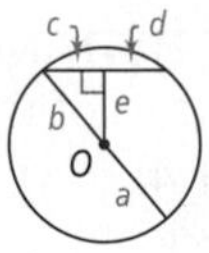

Ⓕ $c = d$
Ⓖ $c^2 + e^2 = b^2$
Ⓗ $a = b$
Ⓘ $e = d$

Short Response

60. Are the following statements equivalent?
- In a circle, if two central angles are congruent, then they have congruent arcs.
- In a circle, if two arcs are congruent, then they have congruent central angles.

Mixed Review

Find the value of each variable.

See Lesson 12-4.

61.

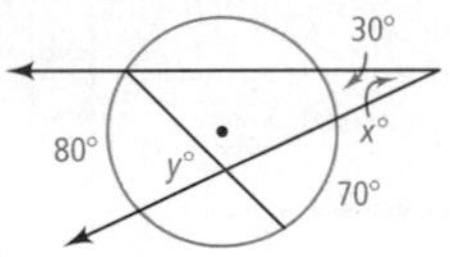

62.

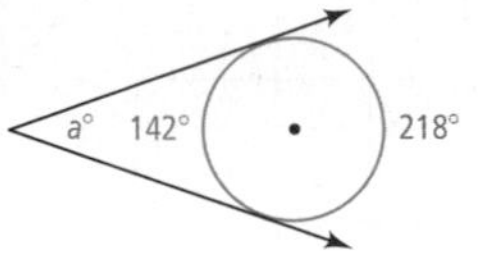

Get Ready! **To prepare for Lesson 12-6, do Exercises 63–65.**

See Lessons 1-2 and 1-5.

Sketch each of the following.

63. the perpendicular bisector of $\overline{BC}$

64. line k parallel to line m and perpendicular to line w, all in plane N

65. $\angle EFG$ bisected by $\overrightarrow{FH}$

54a. 9
b. 4
c. $(x+3)^2 + (y-2)^2 = 9$
d. center $(-3, 2)$; radius 3
e. center $(-2, 10)$; radius 2
55. about 11.5, 11.5, 49.8, and 49.8 units^2
56. $r = \sqrt{6}$; $(x+1)^2 + (y-3)^2 + (z-2)^2 = 6$
57a. $x^2 + y^2 = 15{,}681{,}600$
b. 69.1 mi
c. about 32 days
58. D
59. I
60. [2] No; the second statement is the converse of the first statement, and a conditional and its converse are not equivalent statements.
[1] correct answer without explanation
61. $x = 25$, $y = 75$
62. 38
63.

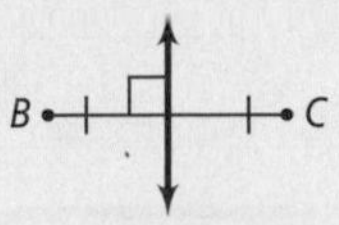

64.

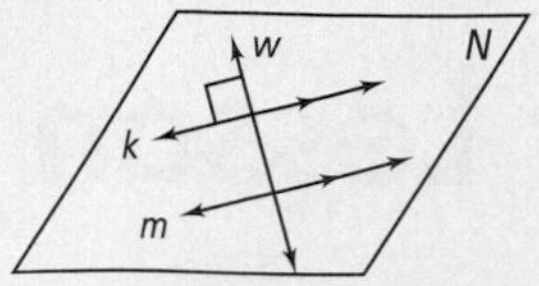

65.

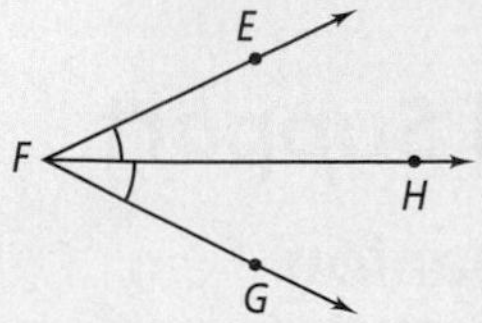

12-5 Lesson Resources

Differentiated Remediation

Instructional Support

Geometry Companion

Students can use the **Geometry Companion** worktext (4 pages) . . .

- New Vocabulary
- Key Concepts
- Got It for each Problem
- Lesson Check

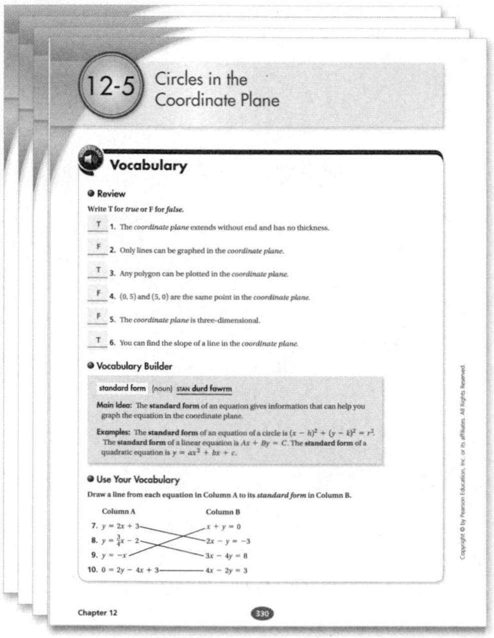
12-5 Circles in the Coordinate Plane

Vocabulary

Review

Write T for *true* or F for *false*.

T 1. The *coordinate plane* extends without end and has no thickness.

F 2. Only lines can be graphed in the *coordinate plane*.

T 3. Any polygon can be plotted in the *coordinate plane*.

F 4. (0, 5) and (5, 0) are the same point in the *coordinate plane*.

F 5. The *coordinate plane* is three-dimensional.

T 6. You can find the slope of a line in the *coordinate plane*.

Vocabulary Builder

standard form (noun) STAN durd fawrm

Main Idea: The **standard form** of an equation gives information that can help you graph the equation in the coordinate plane.

Examples: The **standard form** of an equation of a circle is $(x-h)^2+(y-k)^2=r^2$. The **standard form** of a linear equation is $Ax+By=C$. The **standard form** of a quadratic equation is $y=ax^2+bx+c$.

Use Your Vocabulary

Draw a line from each equation in Column A to its *standard form* in Column B.

Column A	Column B
7. $y = 2x + 3$	$x + y = 0$
8. $y = \frac{1}{2}x - 2$	$2x - y = -3$
9. $y = -x$	$3x - 4y = 8$
10. $0 = 2y - 4x + 3$	$4x - 2y = 3$

Chapter 12 330

ELL Support

Assess Understanding Arrange students into pairs of mixed abilities. On the board, draw a circle on a coordinate plane. One student will write an equation of the circle using the center and the radius, and the other student will use the center and one point. Tell them to share their equations and discuss any discrepancies. Vary this activity by having one student draw a circle on a coordinate plane and the other write the equation. The drawings can be done on graph paper in a page protector so that the paper can be clean and reused.

5 Assess & Remediate

Lesson Quiz

1. Write the standard equation of the circle with center $(5, -4)$ and radius 3.

2. What is the equation of the circle with center $(1, -6)$ that passes through the point $(-4, 3)$?

3. Do you UNDERSTAND? What is the center and radius of the circle with equation $(x+4)^2+(y+1)^2=49$? Graph the circle.

ANSWERS TO LESSON QUIZ

1. $(x-5)^2+(y+4)^2=9$

2. $(x-1)^2+(y+6)^2=106$

3. center $(-4, -1)$; radius 7

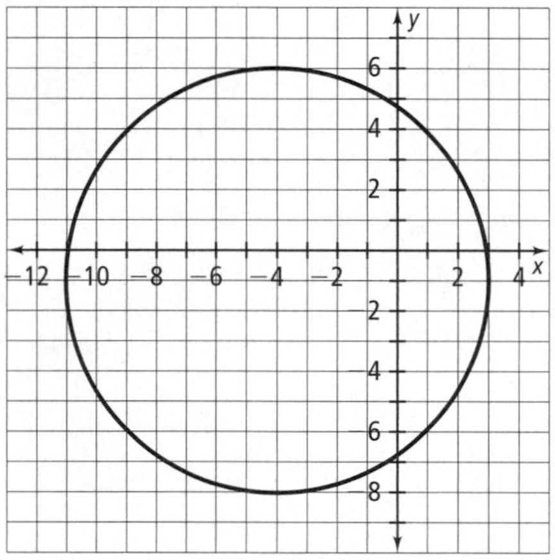

PRESCRIPTION FOR REMEDIATION

Use the student work on the Lesson Quiz to prescribe a differentiated review assignment.

Points	Differentiated Remediation
0–1	Intervention
2	On-level
3	Extension

PowerGeometry.com

5 Assess & Remediate

Assign the Lesson Quiz. Appropriate intervention, practice, or enrichment is automatically generated based on student performance.

Intervention

- **Reteaching** (2 pages) Provides reteaching and practice exercises for the key lesson concepts. Use with struggling students or absent students.
- **English Language Learner Support** Helps students develop and reinforce mathematical vocabulary and key concepts.

All-in-One Resources/Online

Reteaching

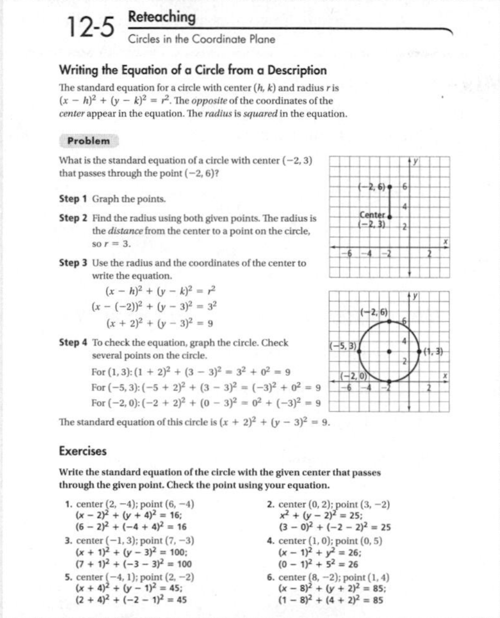
12-5 Reteaching

Circles in the Coordinate Plane

Writing the Equation of a Circle from a Description

The standard equation for a circle with center (h, k) and radius r is $(x-h)^2+(y-k)^2=r^2$. The *opposite* of the coordinates of the *center* appear in the equation. The *radius* is *squared* in the equation.

Problem

What is the standard equation of a circle with center $(-2, 3)$ that passes through the point $(-2, 6)$?

Step 1 Graph the points.

Step 2 Find the radius using both given points. The radius is the *distance* from the center to a point on the circle, so $r = 3$.

Step 3 Use the radius and the coordinates of the center to write the equation.

$(x-h)^2+(y-k)^2=r^2$

$(x-(-2))^2+(y-3)^2=3^2$

$(x+2)^2+(y-3)^2=9$

Step 4 To check the equation, graph the circle. Check several points on the circle.

For $(1, 3)$: $(1+2)^2+(3-3)^2=3^2+0^2=9$

For $(-5, 3)$: $(-5+2)^2+(3-3)^2=(-3)^2+0^2=9$

For $(-2, 0)$: $(-2+2)^2+(0-3)^2=0^2+(-3)^2=9$

The standard equation of this circle is $(x+2)^2+(y-3)^2=9$.

Exercises

Write the standard equation of the circle with the given center that passes through the given point. Check the point using your equation.

1. center $(2, -4)$; point $(6, -4)$ $(x-2)^2+(y+4)^2=16$; $(6-2)^2+(-4+4)^2=16$
2. center $(0, 2)$; point $(3, -2)$ $x^2+(y-2)^2=25$; $(3-0)^2+(-2-2)^2=25$
3. center $(-1, 3)$; point $(7, -3)$ $(x+1)^2+(y-3)^2=100$; $(7+1)^2+(-3-3)^2=100$
4. center $(1, 0)$; point $(0, 5)$ $(x-1)^2+y^2=26$; $(0-1)^2+5^2=26$
5. center $(-4, 1)$; point $(2, -2)$ $(x+4)^2+(y-1)^2=45$; $(2+4)^2+(-2-1)^2=45$
6. center $(8, -2)$; point $(1, 4)$ $(x-8)^2+(y+2)^2=85$; $(1-8)^2+(4+2)^2=85$

All-in-One Resources/Online

English Language Learner Support

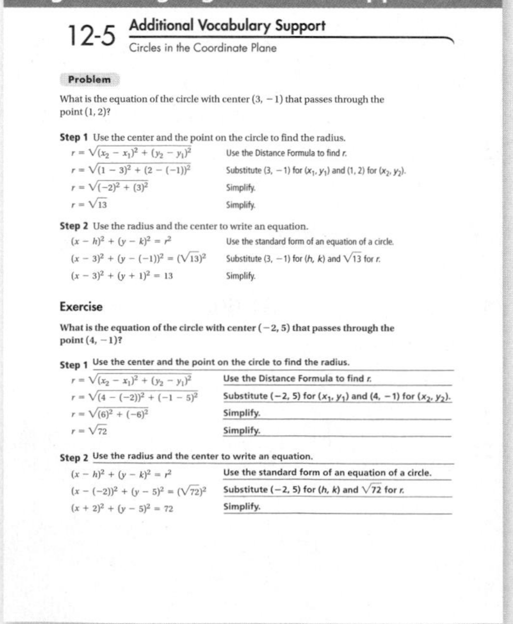
12-5 Additional Vocabulary Support

Circles in the Coordinate Plane

Problem

What is the equation of the circle with center $(3, -1)$ that passes through the point $(1, 2)$?

Step 1 Use the center and the point on the circle to find the radius.

$r=\sqrt{(x_2-x_1)^2+(y_2-y_1)^2}$	Use the Distance Formula to find r.
$r=\sqrt{(1-3)^2+(2-(-1))^2}$	Substitute $(3, -1)$ for (x_1, y_1) and $(1, 2)$ for (x_2, y_2).
$r=\sqrt{(-2)^2+(3)^2}$	Simplify.
$r=\sqrt{13}$	Simplify.

Step 2 Use the radius and the center to write an equation.

$(x-h)^2+(y-k)^2=r^2$	Use the standard form of an equation of a circle.
$(x-3)^2+(y-(-1))^2=(\sqrt{13})^2$	Substitute $(3, -1)$ for (h, k) and $\sqrt{13}$ for r.
$(x-3)^2+(y+1)^2=13$	Simplify.

Exercise

What is the equation of the circle with center $(-2, 5)$ that passes through the point $(4, -1)$?

Step 1 Use the center and the point on the circle to find the radius.

$r=\sqrt{(x_2-x_1)^2+(y_2-y_1)^2}$	Use the Distance Formula to find r.
$r=\sqrt{(4-(-2))^2+(-1-5)^2}$	Substitute $(-2, 5)$ for (x_1, y_1) and $(4, -1)$ for (x_2, y_2).
$r=\sqrt{(6)^2+(-6)^2}$	Simplify.
$r=\sqrt{72}$	Simplify.

Step 2 Use the radius and the center to write an equation.

$(x-h)^2+(y-k)^2=r^2$	Use the standard form of an equation of a circle.
$(x-(-2))^2+(y-5)^2=(\sqrt{72})^2$	Substitute $(-2, 5)$ for (h, k) and $\sqrt{72}$ for r.
$(x+2)^2+(y-5)^2=72$	Simplify.

Differentiated Remediation *continued*

On-Level

- **Practice** (2 pages) Provides extra practice for each lesson. For simpler practice exercises, use the Form K Practice pages found in the All-in-One Teaching Resources and online.
- **Think About a Plan** Helps students develop specific problem-solving skills and strategies by providing scaffolded guiding questions.
- **Standardized Test Prep** Focuses on all major exercises, all major question types, and helps students prepare for the high-stakes assessments.

Extension

- **Enrichment** Provides students with interesting problems and activities that extend the concepts of the lesson.
- **Activities, Games, and Puzzles** Worksheets that can be used for concepts development, enrichment, and for fun!

Practice and Problem Solving Wkbk/ All-in-One Resources/Online

Practice page 1

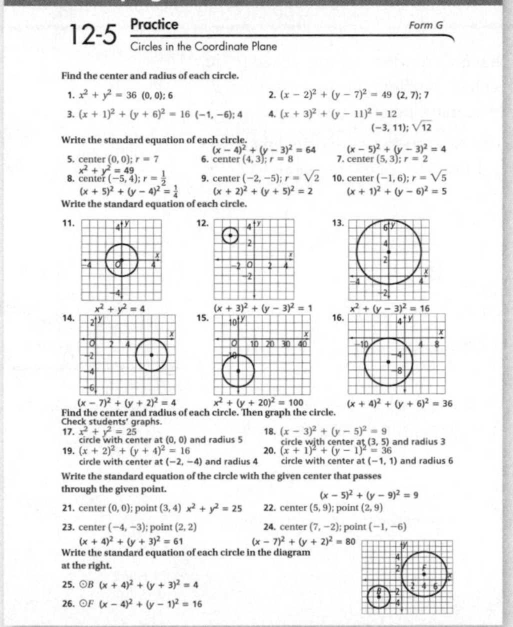

12-5 Practice — Form G
Circles in the Coordinate Plane

Find the center and radius of each circle.

1. $x^2 + y^2 = 36$ (0, 0); 6
2. $(x-2)^2 + (y-7)^2 = 49$ (2, 7); 7
3. $(x+1)^2 + (y+6)^2 = 16$ (−1, −6); 4
4. $(x+3)^2 + (y-11)^2 = 12$ (−3, 11); $\sqrt{12}$

Write the standard equation of each circle.

5. center (0, 0); $r = 7$ $x^2 + y^2 = 49$
6. center (4, 3); $r = 8$ $(x-4)^2 + (y-3)^2 = 64$
7. center (5, 3); $r = 2$ $(x-5)^2 + (y-3)^2 = 4$
8. center (−5, 4); $r = \frac{1}{2}$ $(x+5)^2 + (y-4)^2 = \frac{1}{4}$
9. center (−2, −5); $r = \sqrt{2}$ $(x+2)^2 + (y+5)^2 = 2$
10. center (−1, 6); $r = \sqrt{5}$ $(x+1)^2 + (y-6)^2 = 5$

Write the standard equation of each circle.

11. $x^2 + y^2 = 4$
12. $(x+3)^2 + (y-3)^2 = 1$
13. $x^2 + (y-3)^2 = 16$
14. $(x-7)^2 + (y+2)^2 = 4$
15. $x^2 + (y+20)^2 = 100$
16. $(x+4)^2 + (y+6)^2 = 36$

Find the center and radius of each circle. Then graph the circle. Check students' graphs.

17. $x^2 + y^2 = 25$ circle with center at (0, 0) and radius 5
18. $(x-3)^2 + (y-5)^2 = 9$ circle with center at (3, 5) and radius 3
19. $(x+2)^2 + (y+4)^2 = 16$ circle with center at (−2, −4) and radius 4
20. $(x+1)^2 + (y-1)^2 = 36$ circle with center at (−1, 1) and radius 6

Write the standard equation of the circle with the given center that passes through the given point.

21. center (0, 0); point (3, 4) $x^2 + y^2 = 25$
22. center (5, 9); point (2, 9) $(x-5)^2 + (y-9)^2 = 9$
23. center (−4, −3); point (2, 2) $(x+4)^2 + (y+3)^2 = 61$
24. center (7, −2); point (−1, −6) $(x-7)^2 + (y+2)^2 = 80$

Write the standard equation of each circle in the diagram at the right.

25. ⊙B $(x+4)^2 + (y+3)^2 = 4$
26. ⊙F $(x-4)^2 + (y-1)^2 = 16$

Practice and Problem Solving Wkbk/ All-in-One Resources/Online

Practice page 2

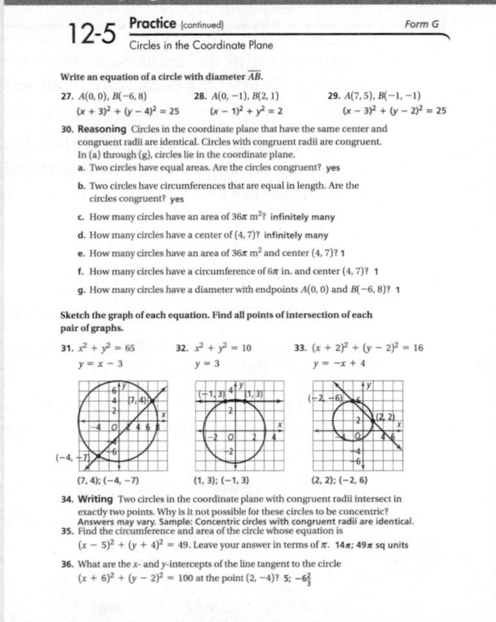

12-5 Practice (continued) — Form G
Circles in the Coordinate Plane

Write an equation of a circle with diameter $\overline{AB}$.

27. A(0, 0), B(−6, 8) $(x+3)^2 + (y-4)^2 = 25$
28. A(0, −1), B(2, 1) $(x-1)^2 + y^2 = 2$
29. A(7, 5), B(−1, −1) $(x-3)^2 + (y-2)^2 = 25$

30. **Reasoning** Circles in the coordinate plane that have the same center and congruent radii are identical. Circles with congruent radii are congruent. In (a) through (g), circles lie in the coordinate plane.
a. Two circles have equal areas. Are the circles congruent? yes
b. Two circles have circumferences that are equal in length. Are the circles congruent? yes
c. How many circles have an area of 36π m²? infinitely many
d. How many circles have a center of (4, 7)? infinitely many
e. How many circles have an area of 36π m² and center (4, 7)? 1
f. How many circles have a circumference of 6π in. and center (4, 7)? 1
g. How many circles have a diameter with endpoints A(0, 0) and B(−6, 8)? 1

Sketch the graph of each equation. Find all points of intersection of each pair of graphs.

31. $x^2 + y^2 = 65$, $y = x - 3$ (7, 4); (−4, −7)
32. $x^2 + y^2 = 10$, $y = 3$ (1, 3); (−1, 3)
33. $(x+2)^2 + (y-2)^2 = 16$, $y = -x + 4$ (2, 2); (−2, 6)

34. **Writing** Two circles in the coordinate plane with congruent radii intersect in exactly two points. Why is it not possible for these circles to be concentric? Answers may vary. Sample: Concentric circles with congruent radii are identical.
35. Find the circumference and area of the circle whose equation is $(x-5)^2 + (y+4)^2 = 49$. Leave your answer in terms of π. 14π; 49π sq units
36. What are the x- and y-intercepts of the line tangent to the circle $(x+6)^2 + (y-2)^2 = 100$ at the point (2, −4)? 5; $-6\frac{2}{3}$

All-in-One Resources/Online

Enrichment

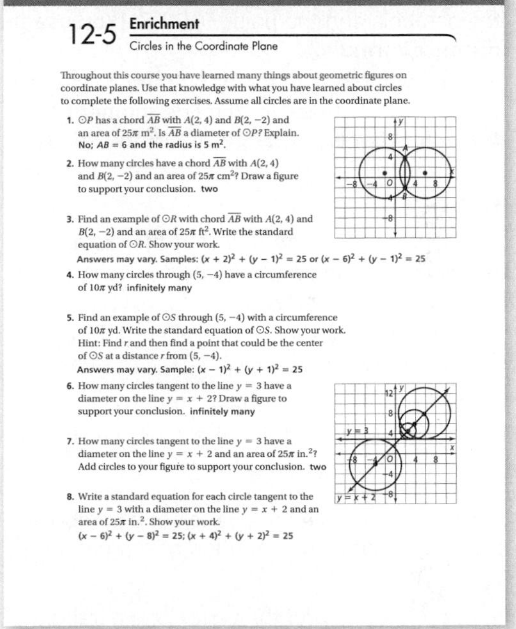

12-5 Enrichment
Circles in the Coordinate Plane

Throughout this course you have learned many things about geometric figures on coordinate planes. Use that knowledge with what you have learned about circles to complete the following exercises. Assume all circles are in the coordinate plane.

1. ⊙P has a chord $\overline{AB}$ with A(2, 4) and B(2, −2) and an area of 25π m². Is $\overline{AB}$ a diameter of ⊙P? Explain. No; AB = 6 and the radius is 5 m².
2. How many circles have a chord $\overline{AB}$ with A(2, 4) and B(2, −2) and an area of 25π cm²? Draw a figure to support your conclusion. two
3. Find an example of ⊙R with chord $\overline{AB}$ with A(2, 4) and B(2, −2) and an area of 25π ft². Write the standard equation of ⊙R. Show your work. Answers may vary. Samples: $(x+2)^2 + (y-1)^2 = 25$ or $(x-6)^2 + (y-1)^2 = 25$
4. How many circles through (5, −4) have a circumference of 10π yd? infinitely many
5. Find an example of ⊙S through (5, −4) with a circumference of 10π yd. Write the standard equation of ⊙S. Show your work. Hint: Find r and then find a point that could be the center of ⊙S at a distance r from (5, −4). Answers may vary. Sample: $(x-1)^2 + (y+1)^2 = 25$
6. How many circles tangent to the line $y = 3$ have a diameter on the line $y = x + 2$? Draw a figure to support your conclusion. infinitely many
7. How many circles tangent to the line $y = 3$ have a diameter on the line $y = x + 2$ and an area of 25π in.²? Add circles to your figure to support your conclusion. two
8. Write a standard equation for each circle tangent to the line $y = 3$ with a diameter on the line $y = x + 2$ and an area of 25π in.². Show your work. $(x-6)^2 + (y-8)^2 = 25$; $(x+4)^2 + (y+2)^2 = 25$

Practice and Problem Solving Wkbk/ All-in-One Resources/Online

Think About a Plan

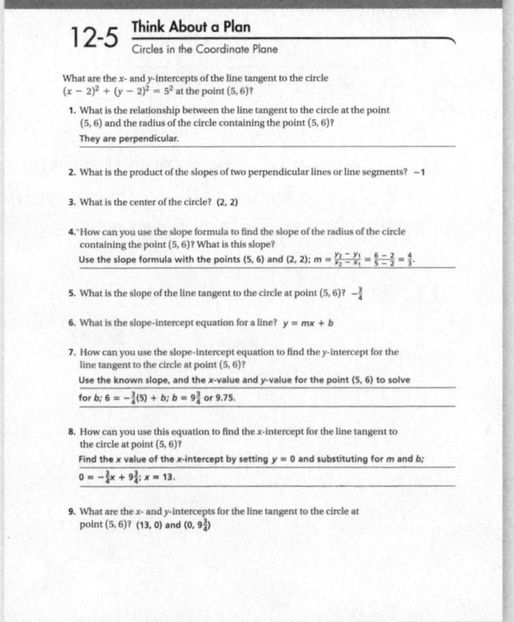

12-5 Think About a Plan
Circles in the Coordinate Plane

What are the x- and y-intercepts of the line tangent to the circle $(x-2)^2 + (y-2)^2 = 5^2$ at the point (5, 6)?

1. What is the relationship between the line tangent to the circle at the point (5, 6) and the radius of the circle containing the point (5, 6)? They are perpendicular.
2. What is the product of the slopes of two perpendicular lines or line segments? −1
3. What is the center of the circle? (2, 2)
4. How can you use the slope formula to find the slope of the radius of the circle containing the point (5, 6)? What is this slope? Use the slope formula with the points (5, 6) and (2, 2); $m = \frac{y_2 - y_1}{x_2 - x_1} = \frac{6-2}{5-2} = \frac{4}{3}$.
5. What is the slope of the line tangent to the circle at point (5, 6)? $-\frac{3}{4}$
6. What is the slope-intercept equation for a line? $y = mx + b$
7. How can you use the slope-intercept equation to find the y-intercept for the line tangent to the circle at point (5, 6)? Use the known slope, and the x-value and y-value for the point (5, 6) to solve for b; $6 = -\frac{3}{4}(5) + b$; $b = 9\frac{3}{4}$ or 9.75.
8. How can you use this equation to find the x-intercept for the line tangent to the circle at point (5, 6)? Find the x value of the x-intercept by setting $y = 0$ and substituting for m and b; $0 = -\frac{3}{4}x + 9\frac{3}{4}$; $x = 13$.
9. What are the x- and y-intercepts for the line tangent to the circle at point (5, 6)? (13, 0) and (0, $9\frac{3}{4}$)

Practice and Problem Solving Wkbk/ All-in-One Resources/Online

Standardized Test Prep

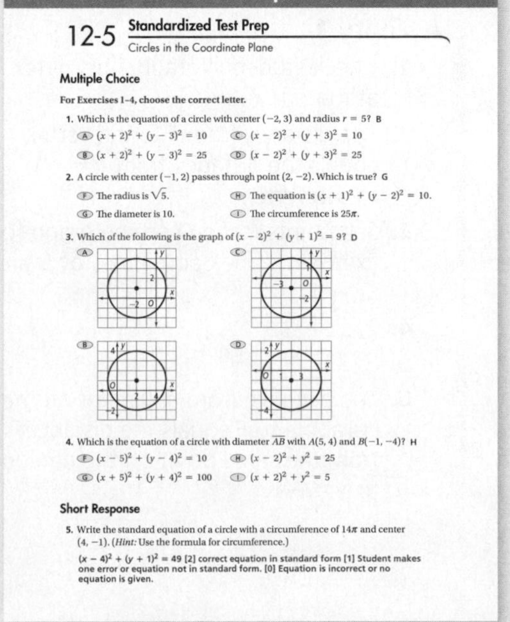

12-5 Standardized Test Prep
Circles in the Coordinate Plane

Multiple Choice

For Exercises 1–4, choose the correct letter.

1. Which is the equation of a circle with center (−2, 3) and radius $r = 5$? B
A $(x+2)^2 + (y-3)^2 = 10$
B $(x+2)^2 + (y-3)^2 = 25$
C $(x-2)^2 + (y+3)^2 = 10$
D $(x-2)^2 + (y+3)^2 = 25$
2. A circle with center (−1, 2) passes through point (2, −2). Which is true? G
F The radius is $\sqrt{5}$.
G The diameter is 10.
H The equation is $(x+1)^2 + (y-2)^2 = 10$.
I The circumference is 25π.
3. Which of the following is the graph of $(x-2)^2 + (y+1)^2 = 9$? D
4. Which is the equation of a circle with diameter $\overline{AB}$ with A(5, 4) and B(−1, −4)? H
F $(x-5)^2 + (y-4)^2 = 10$
G $(x+5)^2 + (y+4)^2 = 100$
H $(x-2)^2 + y^2 = 25$
I $(x+2)^2 + y^2 = 5$

Short Response

5. Write the standard equation of a circle with a circumference of 14π and center (4, −1). (*Hint:* Use the formula for circumference.) $(x-4)^2 + (y+1)^2 = 49$ [2] correct equation in standard form [1] Student makes one error or equation not in standard form. [0] Equation is incorrect or no equation is given.

Online Teacher Resource Center

Activities, Games, and Puzzles

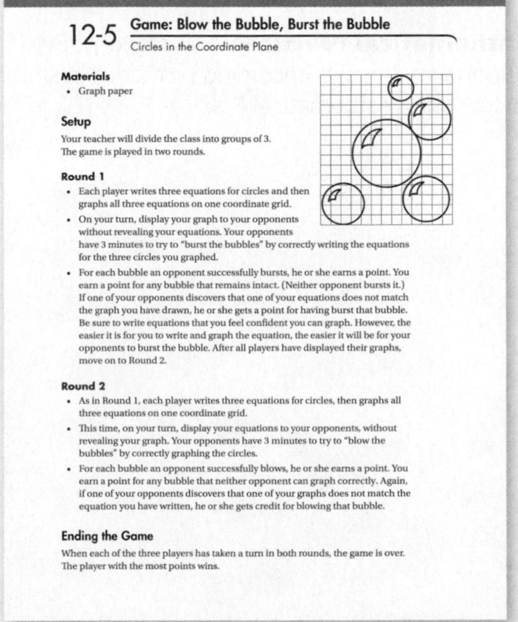

12-5 Game: Blow the Bubble, Burst the Bubble
Circles in the Coordinate Plane

Materials
- Graph paper

Setup
Your teacher will divide the class into groups of 3. The game is played in two rounds.

Round 1
- Each player writes three equations for circles and then graphs all three equations on one coordinate grid.
- On your turn, display your graph to your opponents without revealing your equations. Your opponents have 3 minutes to try to "burst the bubbles" by correctly writing the equations for the three circles you graphed.
- For each bubble an opponent successfully bursts, he or she earns a point. You earn a point for any bubble that remains intact. (Neither opponent bursts it.) If one of your opponents discovers that one of your equations does not match the graph you have drawn, he or she gets a point for having burst that bubble. Be sure to write equations that you feel confident you can graph. However, the easier it is for you to write and graph the equation, the easier it will be for your opponents to burst the bubble. After all players have displayed their graphs, move on to Round 2.

Round 2
- As in Round 1, each player writes three equations for circles, then graphs all three equations on one coordinate grid.
- This time, on your turn, display your equations to your opponents, without revealing your graph. Your opponents have 3 minutes to try to "blow the bubbles" by correctly graphing the circles.
- For each bubble an opponent successfully blows, he or she earns a point. You earn a point for any bubble that neither opponent can graph correctly. Again, if one of your opponents discovers that one of your graphs does not match the equation you have written, he or she gets credit for blowing that bubble.

Ending the Game
When each of the three players has taken a turn in both rounds, the game is over. The player with the most points wins.

Guided Instruction

PURPOSE To write the equation of a parabola using the focus and directrix

PROCESS In the first Activity, students determine that the points on a parabola are the same distance from a fixed point and a fixed line. Then they develop the vocabulary terms *focus* and *directrix* and learn how to use these elements to write an equation.

DISCUSS Before students begin, you may want to review the equation of a parabola with students. In Algebra 1, they generally used the standard form $y = ax^2 + bx + c$.

Activity 1

In this Activity, students explore the relationship of a given point and line to a parabola. They establish that any point on the parabola is equidistant from the point and the line.

Q How do you find the distance between a point on the parabola and the point $F(0, 1)$? **[You use the distance formula.]**

Q How do you measure the distance from the point to a line? **[You measure the line segment through the point that is perpendicular to the line.]**

Activity 2

In this Activity, using the fact that the distance from a point on the parabola to the focus equals the distance from the point on the parabola to the directrix, students write a general equation for a parabola with vertex at (0, 0).

After students find the equation in Exercise 9, challenge students to think about how to use this equation.

Q Suppose $c = 0.5$ and the vertex of a parabola is (0, 0). What is the equation of the parabola? **[$y = \frac{1}{2}x^2$]**

Q Suppose $c = 0.5$ and the vertex of a parabola is (1, 1). What is the equation of the parabola? **[$y = \frac{1}{2}(x - 1)^2 + 1$, or in standard form, $y = \frac{1}{2}x^2 - x + \frac{3}{2}$.]**

Mathematical Practices This Concept Byte supports students in becoming proficient in shifting perspective, Mathematical Practice 7.

Concept Byte

Use With Lesson 12-5

EXTENSION

Equation of a Parabola

Common Core State Standards

G-GPE.A.2 Derive the equation of a parabola given a focus and directrix.

MP 7

Recall from Algebra that a *parabola* is the graph of a quadratic function. In the following activities you will explore other properties of parabolas and how to translate between the geometric description of a parabola and its equation.

Activity 1

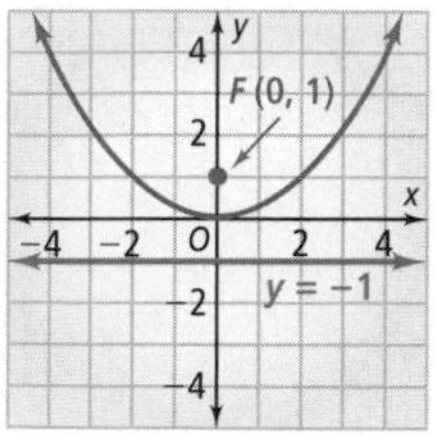

1. Copy the graph of $y = \frac{1}{4}x^2$. What is the vertex of this parabola?
2. Look at the point $F(0, 1)$ and the line $y = -1$ on the coordinate plane. What do you notice about the distance from F to the vertex and the distance from the line $y = -1$ to the vertex?
3. Mark the point $P(2, 1)$ on the parabola. Mark the point D directly below point P on the line $y = -1$. What are the coordinates of point D?
4. What are the lengths of $\overline{FP}$ and $\overline{PD}$?
5. Mark the point $Q(4, 4)$ on the parabola. What is the distance from this point to F? What is the vertical distance from (4, 4) to the line $y = -1$?
6. Make a conjecture about any point on the parabola, the point $F(0, 1)$ and the line $y = -1$. Test your conjecture by using other points on the parabola.

A parabola can be defined as the set of all points that are equidistant from a fixed point, called the **focus of the parabola**, and a fixed line called the **directrix**. That means that any point on a parabola must be the same distance from its focus and directrix.

A focus and directrix uniquely determine a parabola. So, you can derive the equation for a parabola using the definition above. Call an arbitrary point on the parabola (x, y). Then use the distance formula to write an equation.

Activity 2

Refer to the graph to complete this activity.

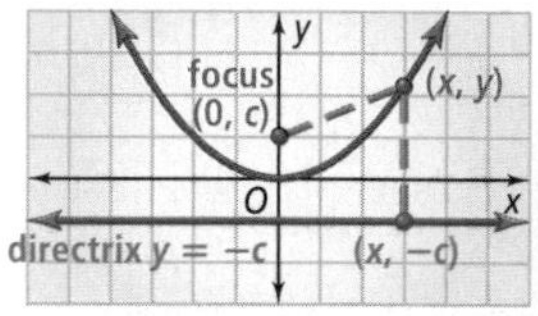

7. What is an expression for the distance between $(0, c)$ and (x, y)?
8. How can you write an expression for the distance between (x, y) and $(x, -c)$? What is the expression?
9. Write an equation for a parabola by setting the expressions you wrote in Exercises 7 and 8 equal to each other. Simplify the equation to get a quadratic equation in standard form.
10. What does the value of c represent in the equation?
11. What is an expression that represents the distance between the focus and the directrix?

Answers

Activity 1

1. Check students' graph. The vertex is at the origin.
2. The distance from F to the vertex equals the distance from $y = -1$ to the vertex.
3. See points P and D on the graph for Exercise 1. The coordinates of D are $(2, -1)$.
4. 2; 2
5. 5; 5
6. The distance from any point on the parabola to F equals the distance from the same point on the parabola to $y = -1$.

Activity 2

7. $\sqrt{(x - 0)^2 + (y - c)^2}$
8. $y - (-c)$ or $\sqrt{(x - x)^2 + [y - (-c)]^2}$
9. $y = \frac{1}{4c}x^2$
10. The value of c represents the distance from the focus or the distance from the directrix to any point on the parabola.
11. $2c$

Activity 3

In this Activity, you will derive the equation of a parabola with focus $F(2, 4)$ and directrix $y = -2$.

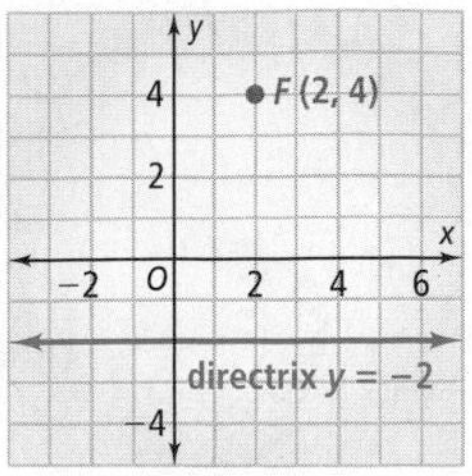

12. What are the coordinates of the vertex of the parabola? Explain how you know.

13. How can you use what you know to make a sketch of the parabola on the coordinate plane?

14. What do you know about the distance from any point (x, y) on the parabola to point $F(2, 4)$ to and the distance from (x, y) to the line $y = -2$?

15. Write an equation that represents the relationship described in Exercise 15.

16. What is the equation for the parabola in standard form?

Exercises

Find the equation of the parabola with the focus and directrix shown.

17.

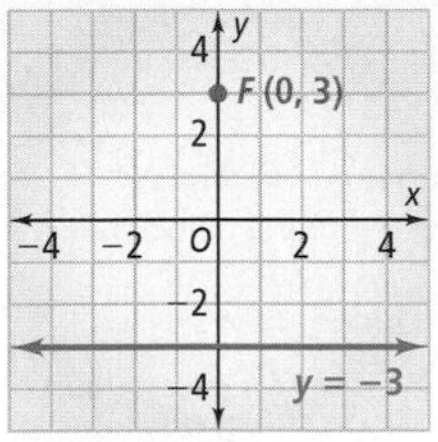

18.

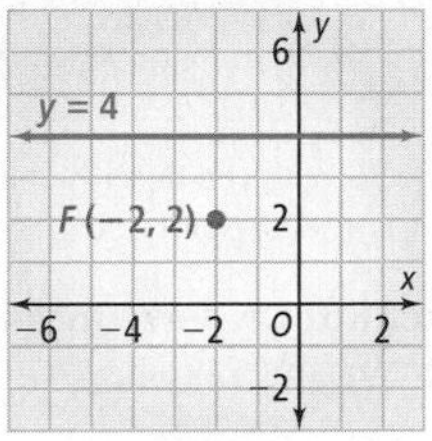

19.

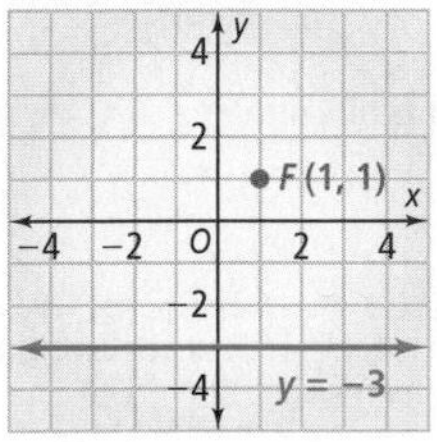

20.

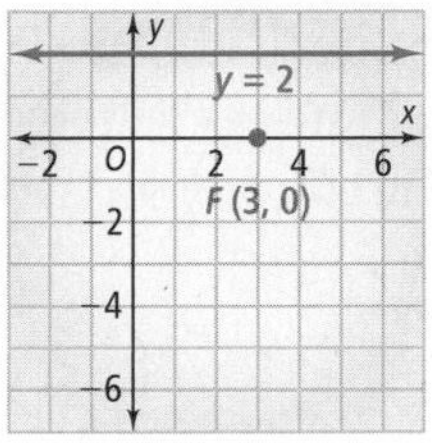

21. What is the equation of a parabola with focus $F(3, -2)$ and directrix $y = -4$?

22. The vertex of a parabola is at $(3, -2)$ and its focus is at $(3, 8)$. What is the equation of its directrix?

STEM **23.** **Solar Energy** Solar energy can be captured by using a solar disk. A cross section of the disk is the shape if a parabola. If you can model the cross section so that the vertex is at the origin, and the receiver (the focus) is at $(0, 7)$, what is the equation of this parabola?

Activity 3

In this Activity, using the focus and directrix, students find the equation of a specific parabola.

Q Is it easy to draw the parabola based on knowing the distance of any given point to the focus and directrix? **[Answers may vary. Sample: Because one distance is at a slant and the other is vertical, it is not easy to determine exact points.]**

Q What is the advantage of knowing what the equation is in standard form? **[Answers may vary. Sample: It is easy to graph the equation using a graphing calculator.]**

If students have graphing calculators, have them graph the function they find, determine a point on the graph using the TRACE function, and then use the point to confirm that their equation is correct and that no error occurred in going to standard form.

Activity 3

12. (2, 1); the vertex is halfway between the focus and the directrix.

13. Find points that are the same distance from $y = -2$ and $(2, 4)$.

14. They are the same.

15. $\sqrt{(x-2)^2 + (y-4)^2} = y - (-2)$

16. $y = \frac{1}{12}x^2 - \frac{1}{3}x + \frac{4}{3}$

Exercises

17. $y = \frac{1}{12}x^2$

18. $y = -\frac{1}{4}x^2 - x + 2$

19. $y = \frac{1}{8}x^2 - \frac{1}{4}x - \frac{7}{8}$

20. $y = -\frac{1}{4}x^2 + \frac{3}{2}x - \frac{5}{4}$

21. $y = \frac{1}{4}x^2 - \frac{3}{2}x - \frac{3}{4}$

22. $y = -12$

23. $y = \frac{1}{28}x^2$

1 Interactive Learning

Solve It!

PURPOSE To write a description of a set of points satisfying a condition about a real-world situation
PROCESS Students may identify a set of points that satisfies the condition or a street that satisfies the condition.

FACILITATE

- **Q** What is one point that is equidistant from both offices? **[the intersection of B St. and 2nd St.]**
- **Q** Is there a street that is always equidistant from the two offices? If so, what is it? **[Yes; B St.]**
- **Q** How could you classify B St. in comparison to 2nd St.? **[It is the perpendicular bisector of 2nd St.]**

ANSWER See Solve It in Answers on next page.
CONNECT THE MATH Students should recognize that the perpendicular bisector of a segment is the set of all points that are equidistant from the endpoints of the segment. In this lesson, students will use the mathematical language they have learned in this course to describe situations given a condition.

2 Guided Instruction

Problem 1

- **Q** What type of figure represents the set of all points in a plane that are equidistant from a given point? **[a circle]**

12-6 Locus: A Set of Points

Common Core State Standards
G-GMD.B.4 . . . Identify three-dimensional objects generated by rotations of two-dimensional objects.
MP 1, MP 3, MP 4, MP 6

Objective To draw and describe a locus

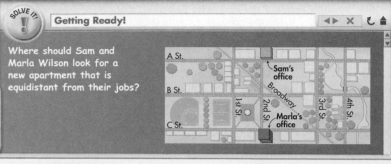

Solve It! Getting Ready!

Where should Sam and Marla Wilson look for a new apartment that is equidistant from their jobs?

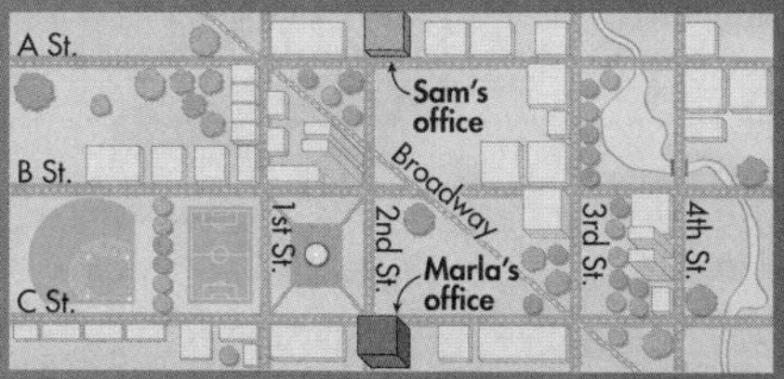

You studied the distance between two points in Lesson 5-2. If you need help, look back.

MATHEMATICAL PRACTICES In the Solve It, you described the possible locations based on a certain condition. A **locus** is a set of points, all of which meet a stated condition. *Loci* is the plural of locus.

Lesson Vocabulary
- locus

Essential Understanding You can use the description of a locus to sketch a geometric relationship.

Problem 1 Describing a Locus in a Plane

What is a sketch and description for each locus of points in a plane?

A **the points 1 cm from a given point C**

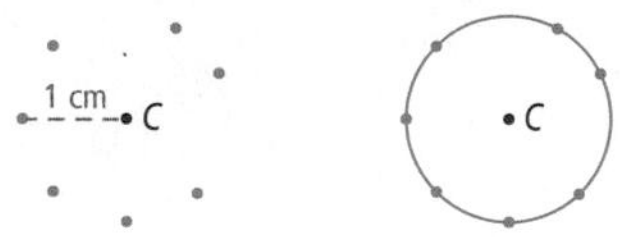

Draw a point C. Sketch several points 1 cm from C. Keep doing so until you see a pattern. Draw the figure the pattern suggests.

The locus is a circle with center C and radius 1 cm.

B **the points 1 cm from $\overline{AB}$**

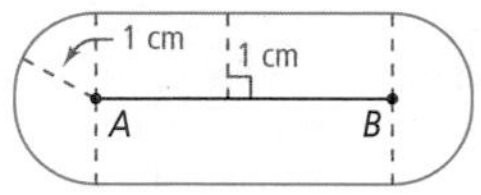

Draw $\overline{AB}$. Sketch several points on either side of $\overline{AB}$. Also sketch points 1 cm from point A and point B. Keep doing so until you see a pattern. Draw the figure the pattern suggests.

The locus is a pair of parallel segments, each 1 cm from $\overline{AB}$, and two semicircles with centers at A and B.

Think
Have you considered all possibilities?
Make sure that the endpoints as well as the segment are included in the sketch.

12-6 Preparing to Teach

BIG ideas **Reasoning and Proof**
Measurement

ESSENTIAL UNDERSTANDING
- The description of a locus can be used to sketch a geometric relationship.

Math Background

The topic of loci, traditionally seen as counterintuitive and visually difficult for students, can become a rich illustrative tool for use in exploring the meaning of a definition and in developing mathematical reasoning.

Loci are used to find a set of points that satisfy a given condition. Students have already studied several examples of loci: perpendicular bisectors, angle bisectors, and circles. In this lesson students will see how these figures can be defined by the points they contain rather than by the method by which they can be drawn.

Mathematical Practices

Attend to precision. Students will define and make explicit use of the term "locus."

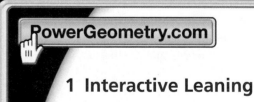

PowerGeometry.com

1 Interactive Leaning

Solve It!
Step out how to solve the Problem with helpful hints and an online question. Other questions are listed above in Interactive Learning.

Virtual Nerd™ Access Virtual Nerd student-centered math tutorials that directly relate to the content of the lesson.

Got It? 1. Reasoning If the question for part (b) asked for the locus of points in a plane 1 cm from $\overleftrightarrow{AB}$, how would the sketch change?

You can use locus descriptions for geometric terms.

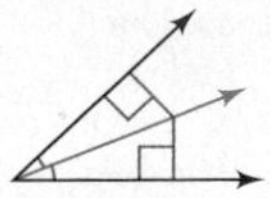

The locus of points in the interior of an angle that are equidistant from the sides of the angle is an angle bisector.

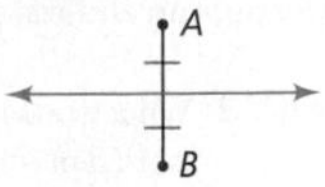

In a plane, the locus of points that are equidistant from a segment's endpoints is the perpendicular bisector of the segment.

Sometimes a locus is described by two conditions. You can draw the locus by first drawing the points that satisfy each condition. Then find their intersection.

Problem 2 Drawing a Locus for Two Conditions

What is a sketch of the locus of points in a plane that satisfy these conditions?

- **the points equidistant from intersecting lines *k* and *m***
- **the points 5 cm from the point where *k* and *m* intersect**

Know: Lines *k* and *m* intersect.

Need: Sketch that satisfies the given conditions

Plan: Make a sketch to satisfy the first condition. Then sketch the second condition. Look for the points in common.

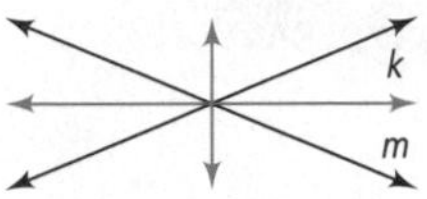

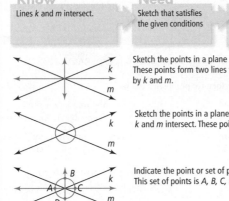

Sketch the points in a plane equidistant from lines *k* and *m*. These points form two lines that bisect the vertical angles formed by *k* and *m*.

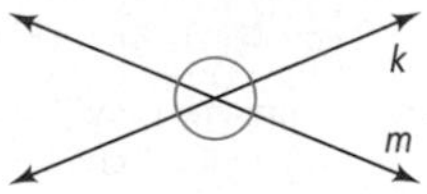

Sketch the points in a plane 5 cm from the point where *k* and *m* intersect. These points form a circle.

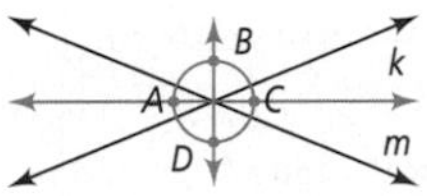

Indicate the point or set of points that satisfies both conditions. This set of points is *A*, *B*, *C*, and *D*.

Got It? 2. What is a sketch of the locus of points in a plane that satisfy these conditions?

- the points equidistant from two points *X* and *Y*
- the points 2 cm from the midpoint of $\overline{XY}$

Got It? ERROR PREVENTION

Have students draw $\overleftrightarrow{AB}$. The locus will be two lines parallel to $\overleftrightarrow{AB}$ on opposite sides of $\overleftrightarrow{AB}$.

Problem 2

Q What is the locus of points that are equidistant from two intersecting lines? **[The locus is a set of perpendicular lines that bisect the angles created by line *k* and *m*.]**

Q What is the locus of points that are a given distance from one point? **[a circle]**

Q What is the intersection of the two loci in Q2? **[The angle bisectors and circle intersect at four points.]**

Got It? VISUAL LEARNERS

Q What is the intersection of the two loci? **[The line and circle intersect at two points.]**

2 Guided Instruction

Each Problem is worked out and supported online.

Problem 1
Describing a Locus in a Plane

Problem 2
Drawing a Locus for Two Conditions

Problem 3
Describing a Locus in Space

Support in Geometry Companion

- Vocabulary
- Key Concepts
- Got It?

Answers

Solve It!

anywhere on B Street

Got It?

1. a pair of ∥ lines, each 1 cm from $\overleftrightarrow{AB}$
2. 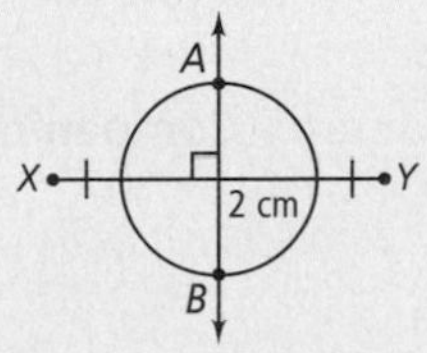

Points *A* and *B* satisfy both conditions.

Problem 3

Q In space, what figure is related to a circle? **[a sphere]**

Q How would you describe the set of all points in space that are equidistant from a line? **[The points are equidistant from the line in every direction.]**

Q What figure do these points form? **[a cylinder or tube]**

Got It? VISUAL LEARNERS

Have students visualize each locus in space. If necessary, model the loci with paper.

3 Lesson Check

Do you know HOW?

- If students have difficulty with Exercises 1-4, then have them review Problems 1 and 2 and model their sketches after those in the problem.

Do you UNDERSTAND?

- If students have difficulty with Exercise 6, then have them model the situation in space.

Close

Q What do you call a set of points that satisfy a given condition? **[a locus]**

Q What are some examples of geometric figures you have studied that represent loci? **[Answers will vary. Sample: perpendicular bisector, angle bisector, circle]**

Problem 3 Describing a Locus in Space

Think

How can making a sketch help?
Make a sketch of the points in a plane and then visualize what the figure would look like in three dimensions.

A **What is the locus of points in space that are *c* units from a point *D*?**

The locus is a sphere with center at point *D* and radius *c*.

B **What is the locus of points in space that are 3 cm from a line ℓ?**

The locus is an endless cylinder with radius 3 cm and centerline ℓ.

Got It? **3.** What is each locus of points?

a. in a plane, the points that are equidistant from two parallel lines

b. in space, the points that are equidistant from two parallel planes

Lesson Check

Do you know HOW?

What is a sketch and description for each locus of points in a plane?

1. points 4 cm from a point *X*

2. points 2 in. from $\overline{UV}$

3. points 3 mm from $\overleftrightarrow{LM}$

4. points 1 in. from a circle with radius 3 in.

Do you UNDERSTAND? MATHEMATICAL PRACTICES

5. Vocabulary How are the words *locus* and *location* related?

6. Compare and Contrast How are the descriptions of the locus of points for each situation alike? How are they different?

- in a plane, the points equidistant from points *J* and *K*
- in space, the points equidistant from points *J* and *K*

Practice and Problem-Solving Exercises MATHEMATICAL PRACTICES

A Practice

Sketch and describe each locus of points in a plane. See Problem 1.

7. points equidistant from the endpoints of $\overline{PQ}$

8. points in the interior of $\angle ABC$ and equidistant from the sides of $\angle ABC$

9. points equidistant from two perpendicular lines

10. midpoints of radii of a circle with radius 2 cm

For Exercises 11–15, sketch the locus of points in a plane that satisfy the given conditions. See Problem 2.

11. equidistant from points *M* and *N* and on a circle with center *M* and radius $= \frac{1}{2}MN$

12. 3 cm from $\overline{GH}$ and 5 cm from *G*, where $GH = 4.5$ cm

13. equidistant from the sides of $\angle PQR$ and on a circle with center *P* and radius *PQ*

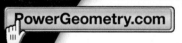

PowerGeometry.com

3 Lesson Check

For a digital lesson check, use the Got It questions.

Support in Geometry Companion

- Lesson Check

4 Practice

Assign homework to individual students or to an entire class.

Answers

Got It? (continued)

3a. The locus is the line ‖ to and equidistant from the given ‖ lines (midway between them).

b. The locus is a plane ‖ to and equidistant from the given ‖ planes (midway between them).

Lesson Check

1–6. See back of book.

Practice and Problem-Solving Exercises

7–29. See back of book.

14. equidistant from both points A and B and points C and D

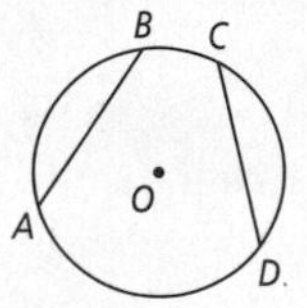

15. equidistant from the sides of $\angle JKL$ and on $\odot C$

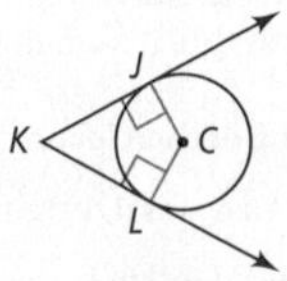

Describe each locus of points in space.

See Problem 3.

16. points 3 cm from a point F

17. points 4 cm from $\overleftrightarrow{DE}$

18. points 1 in. from plane M

19. points 5 mm from $\overrightarrow{PQ}$

B Apply

Describe the locus that each blue figure represents.

20.

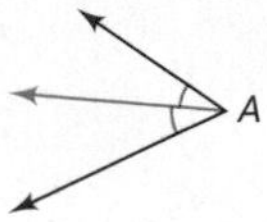

21.

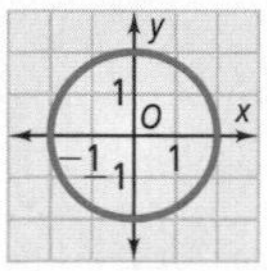

22.

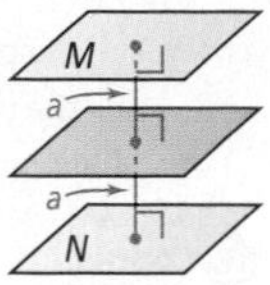

23. Open-Ended Give two examples of loci from everyday life, one in a plane and one in space.

24. Writing A classmate says that it is impossible to find a point equidistant from three collinear points. Is she correct? Explain.

25. Think About a Plan Write a locus description of the points highlighted in blue on the coordinate plane.

- How many conditions will be involved?
- What is the condition with respect to the origin?
- What are the conditions with respect to the x- and y-axes?

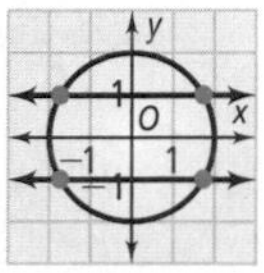

Coordinate Geometry Write an equation for the locus of points in a plane equidistant from the two given points.

26. $A(0, 2)$ and $B(2, 0)$

27. $P(1, 3)$ and $Q(5, 1)$

28. $T(2, -3)$ and $V(6, 1)$

STEM **29. Meteorology** An anemometer measures wind speed and wind direction. In an anemometer, there are three cups mounted on an axis. Consider a point on the edge of one of the cups.

a. Describe the locus that this point traces as the cup spins in the wind.

b. Suppose the distance of the point from the axis of the anemometer is 2 in. Write an equation for the locus of part (a). Use the axis as the origin.

4 Practice

ASSIGNMENT GUIDE

Basic: 7–19 all, 20–24 even, 25, 32–40 even, 46

Average: 7–19 odd, 20–48

Advanced: 7–19 odd, 20–54

Standardized Test Prep: 55–57

Mixed Review: 58–65

Mathematical Practices are supported by exercises with red headings. Here are the Practices supported in this lesson:

MP 1: Make Sense of Problems Ex. 6, 25

MP 3: Construct Arguments Ex. 24, 45c

MP 4: Model with Mathematics Ex. 23

Applications exercises have blue headings. Exercises 29, 30, and 50–54 support MP 4: Model.

STEM exercises focus on science or engineering applications.

EXERCISE 46: Use the Think About a Plan worksheet in the **Practice and Problem Solving Workbook** (also available in the Teaching Resources in print and online) to further support students' development in becoming independent learners.

HOMEWORK QUICK CHECK

To check students' understanding of key skills and concepts, go over Exercises 7, 15, 24, 25, and 46.

Additional Problems

1. What is a sketch and description for each locus (or loci) of points in a plane?

a. the points 1 cm from the endpoints of $\overline{CD}$

b. the points 2 cm from a given point P

ANSWER a. The loci are circles with a radius of 1 cm centered at C and D.

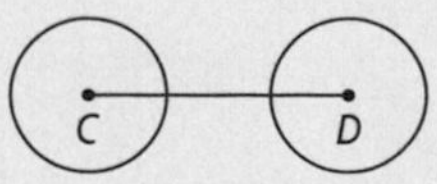

b. The locus is a circle with center P and radius 2 cm.

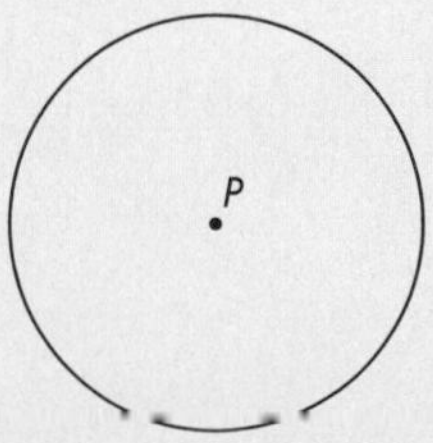

2. What is a sketch of the locus of points in a plane that satisfy these conditions?

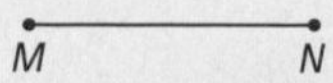

- the points equidistant from the endpoints of segment MN
- the points less than or equal to 2 cm from the midpoint of segment MN

ANSWER The locus is a segment of the perpendicular bisector of $\overline{MN}$ with a length of 2 cm on each side of MN.

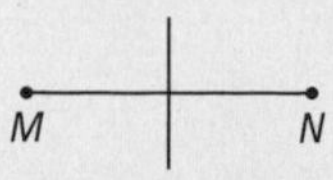

3. a. Describe the locus (or loci) of points in space that are 5 cm from plane P.

b. Describe the locus of points in space that are 3 in. from point Q.

ANSWER a. two parallel planes in space that are 5 cm from the original plane **b.** a sphere in space with a radius of 3 in. and center Q.

Answers

Practice and Problem-Solving Exercises (continued)

30. 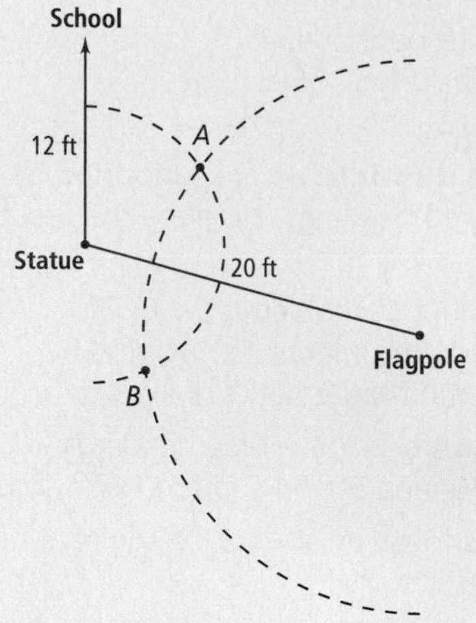

The radius of the arc from the statue represents 8ft. The arc from the flagpole represents 16ft. Points *A* and *B* are the two possible positions for the fountain.

31–35. Answers may vary. Samples are given.

31. top view

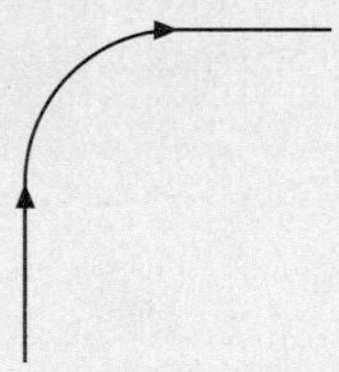

32. top view

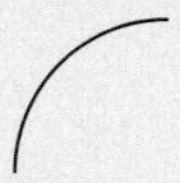

33. side view

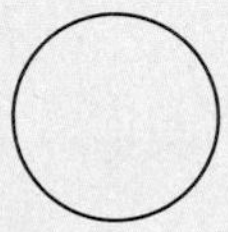

34. top view

35. side view

36. No; the loci do not intersect.

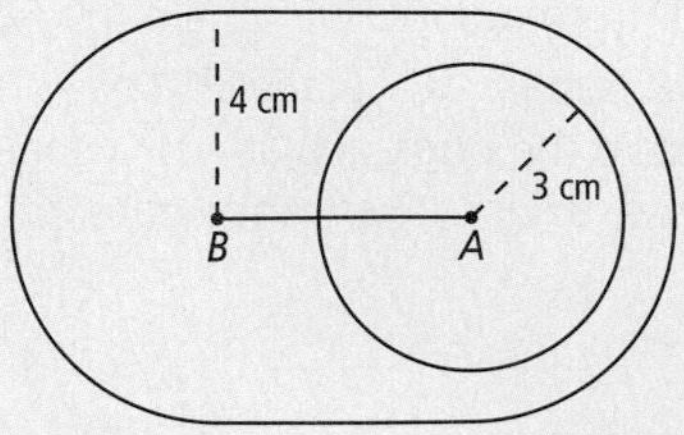

37. 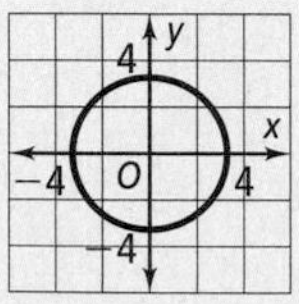

38. 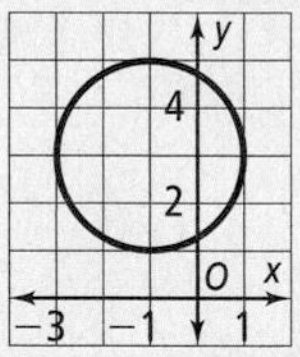

39. 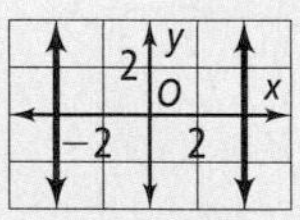

40. 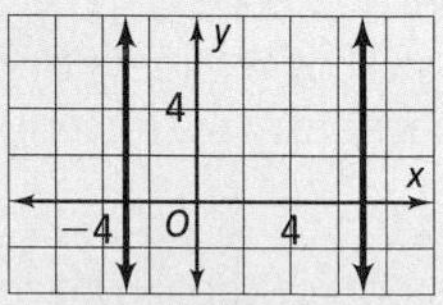

41. 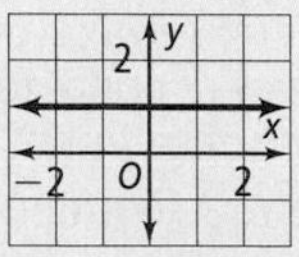

42. 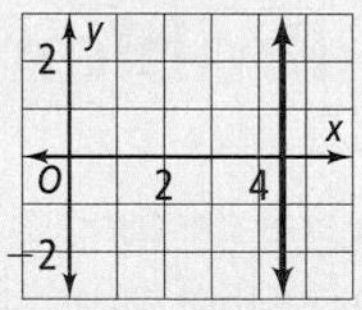

43. 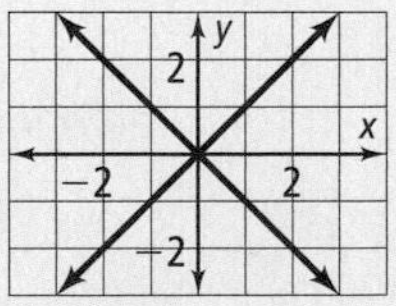

44. 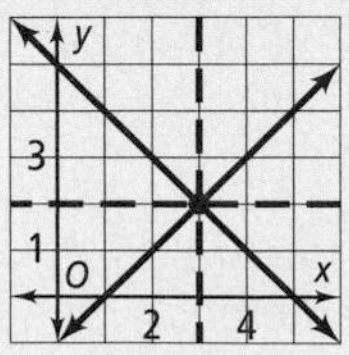

30. Landscaping The school board plans to construct a fountain in front of the school. What are all the possible locations for a fountain such that the fountain is 8 ft from the statue and 16 ft from the flagpole?

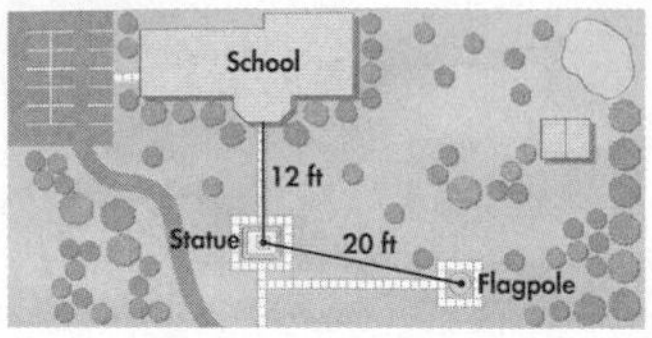

Make a drawing of each locus.

31. the path of a car as it turns to the right

32. the path of a doorknob as a door opens

33. the path of a knot in the middle of a jump-rope as it is being used

34. the path of the tip of your nose as you turn your head

35. the path of a fast-pitched softball

36. Reasoning Points *A* and *B* are 5 cm apart. Do the following loci in a plane have any points in common?

the points 3 cm from *A*

the points 4 cm from $\overline{AB}$

Illustrate your answer with a sketch.

Coordinate Geometry **Draw each locus on the coordinate plane.**

37. all points 3 units from the origin

38. all points 2 units from $(-1, 3)$

39. all points 4 units from the *y*-axis

40. all points 5 units from $x = 2$

41. all points equidistant from $y = 3$ and $y = -1$

42. all points equidistant from $x = 4$ and $x = 5$

43. all points equidistant from the *x*- and *y*-axes

44. all points equidistant from $x = 3$ and $y = 2$

45. a. Draw a segment to represent the base of an isosceles triangle. Locate three points that could be the vertex of the isosceles triangle.

b. Describe the locus of possible vertices for the isosceles triangle.

c. Writing Explain why points in the locus you described are the only possibilities for the vertex of the isosceles triangle.

46. Describe the locus of points in a plane 3 cm from the points on a circle with radius 8 cm.

47. Describe the locus of points in a plane 8 cm from the points on a circle with radius 3 cm.

48. Sketch the locus of points for the air valve on the tire of a bicycle as the bicycle moves down a straight path.

Challenge

49. In the diagram, Moesha, Jan, and Leandra are seated at uniform distances around a circular table. Copy the diagram. Shade the points on the table that are closer to Moesha than to Jan or Leandra.

Playground Equipment **Think about the path of a child on each piece of playground equipment. Draw the path from (a) a top view, (b) a front view, and (c) a side view.**

50. a swing
51. a straight slide
52. a corkscrew slide
53. a merry-go-round
54. a firefighters' pole

Standardized Test Prep

SAT/ACT

55. What are the coordinates of the center of the circle whose equation is $(x-9)^2+(y+4)^2=1$?

Ⓐ $(3, -2)$ Ⓑ $(-3, 2)$ Ⓒ $(-9, 4)$ Ⓓ $(9, -4)$

56. A plane passes through two adjacent faces of a rectangular prism. The plane is perpendicular to the base of the prism. Which term is the most specific name for a figure formed by the cross section of the plane and the prism?

Ⓕ square Ⓖ rectangle Ⓗ parallelogram Ⓘ kite

Short Response

57. Margie's cordless telephone can transmit up to 0.5 mi from her home. Carol's cordless telephone can transmit up to 0.25 mi from her home. Carol and Margie live 0.25 mi from each other. Can Carol's telephone work in a region that Margie's cannot? Sketch and label your diagram.

Mixed Review

Write an equation of the circle with center *C* and radius *r*. See Lesson 12-5.

58. $C(6, -10), r = 5$ **59.** $C(1, 7), r = 6$ **60.** $C(-8, -1), r = \sqrt{13}$

Find the surface area of each figure to the nearest tenth. See Lesson 11-2.

61.

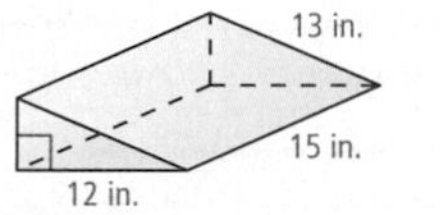

62.

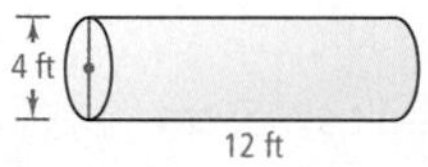

In ⊙*O*, find the area of sector *AOB*. Leave your answer in terms of π. See Lesson 10-7.

63. $OA = 4, m\widehat{AB} = 90$ **64.** $OA = 8, m\widehat{AB} = 72$ **65.** $OA = 10, m\widehat{AB} = 36$

45a. Sample:

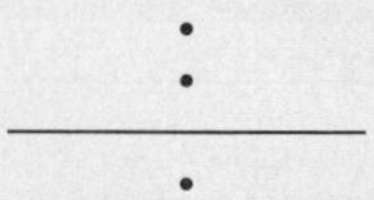

b. The locus is the ⊥ bis. of the base except for the midpt of the base.

c. Sample explanation: The vertex of the isosc. △ must be equidistant from the endpoints of the base, and all the points (in a plane) that are equidistant from two points lie on the ⊥ bis. of the segment whose endpoints are the two given points.

46. The locus is two circles concentric with the original, one with radius 5 cm and one with radius 11 cm.

47. The locus is a circle of radius 11 cm, concentric with the original.

48.

49.

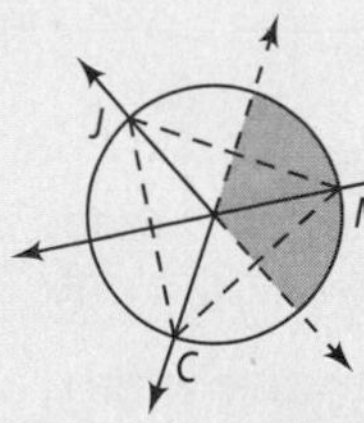

50.

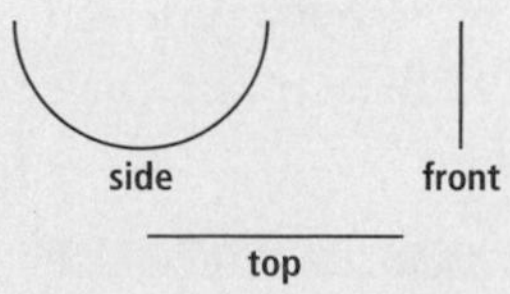

51.

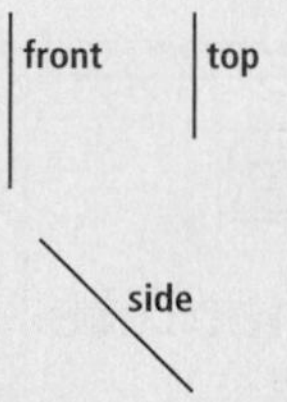

52.

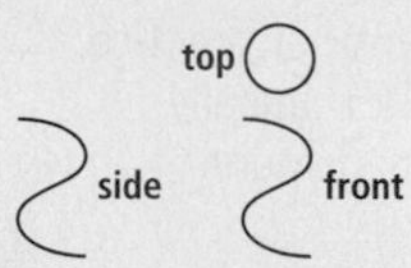

53.

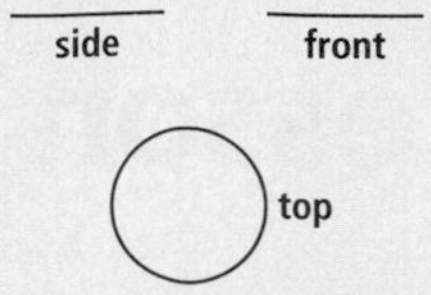

54.

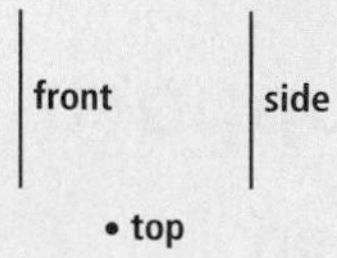

55. D

56. G

57. [2] no

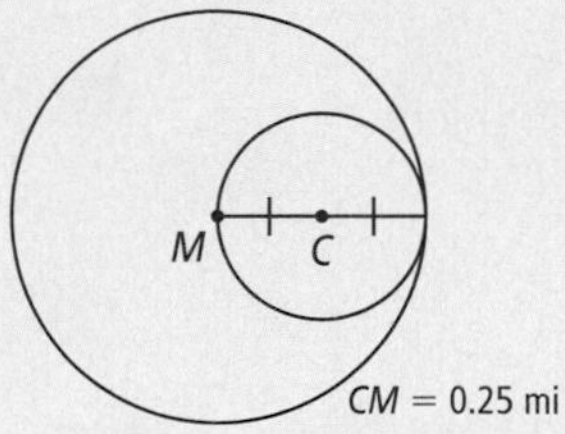

[1] incorrect answer OR incorrect diagram/ explanation

58. $(x-6)^2+(y+10)^2=25$

59. $(x-1)^2+(y-7)^2=36$

60. $(x+8)^2+(y+1)^2=13$

61. 510 in.^{2}

62. 175.9 ft^2

63. 4π units^2

64. $\frac{64\pi}{5}$ units^2

65. 10π units^2

12-6 Lesson Resources

Differentiated Remediation

Instructional Support

Geometry Companion

Students can use the **Geometry Companion** worktext (4 pages) . . .

- New Vocabulary
- Key Concepts
- Got It for each Problem
- Lesson Check

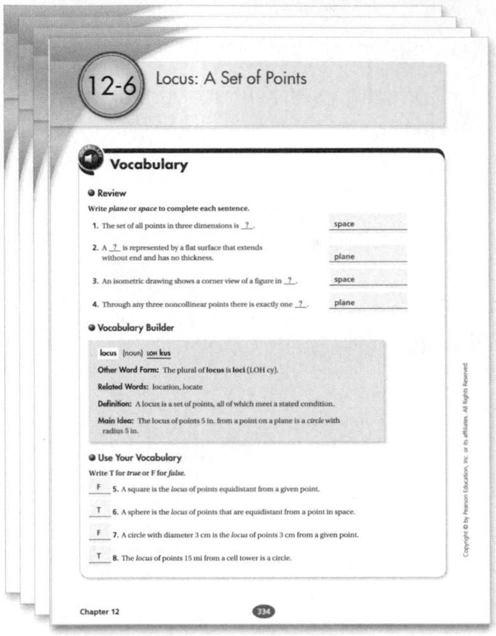

12-6 Locus: A Set of Points

Vocabulary

Review

Write *plane* or *space* to complete each sentence.

1. The set of all points in three dimensions is _?_. space
2. A _?_ is represented by a flat surface that extends without end and has no thickness. plane
3. An isometric drawing shows a corner view of a figure in _?_. space
4. Through any three noncollinear points there is exactly one _?_. plane

Vocabulary Builder

locus (noun) LOH kus

Other Word Form: The plural of **locus** is **loci** (LOH cy).

Related Words: location, locate

Definition: A locus is a set of points, all of which meet a stated condition.

Main Idea: The locus of points 5 in. from a point on a plane is a *circle* with radius 5 in.

Use Your Vocabulary

Write T for *true* or F for *false*.

F 5. A square is the *locus* of points equidistant from a given point.

T 6. A sphere is the *locus* of points that are equidistant from a point in space.

F 7. A circle with diameter 3 cm is the *locus* of points 3 cm from a given point.

T 8. The *locus* of points 15 mi from a cell tower is a circle.

Copyright © by Pearson Education, Inc. or its affiliates. All Rights Reserved.

Chapter 12 334

ELL Support

Focus on Language Examine the word *locus*. Ask students if there are any words that they have heard that sound like *locus* [*location, local, locust*]. What are synonyms for *locus*? Examples may include *place*, *site*, *spot*, or *position*. The plural of *locus* is *loci*.

Locus is a noun meaning a place, a center, or a source. In mathematics it is the set of points that satisfy some condition. It comes from the Latin word *locus*, which means a place.

5 Assess & Remediate

Lesson Quiz

1. What is a description for the midpoints of the radii of a circle with radius 3 cm? Draw a sketch.

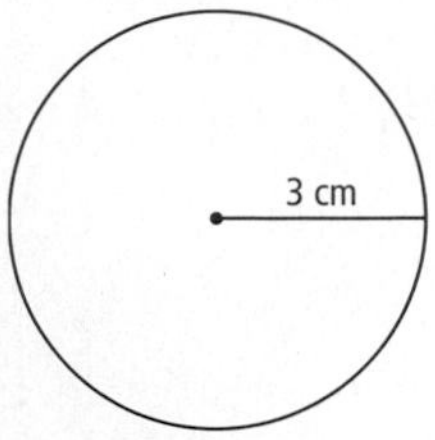

2. **Do you UNDERSTAND?** What is a sketch for the points in the interior of $\angle DEF$ and equidistant from the sides of $\angle DEF$? What is a name for the locus?

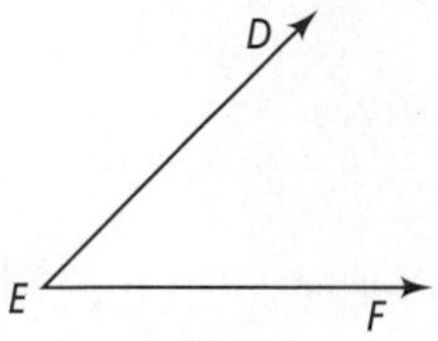

ANSWERS TO LESSON QUIZ

1. a concentric circle with radius 1.5 cm

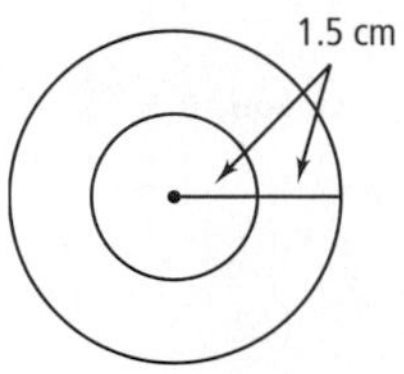

2.

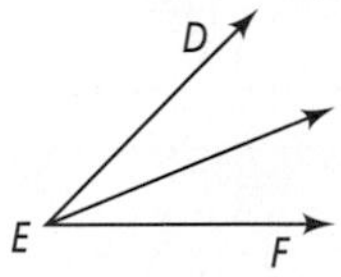

the angle bisector of $\angle DEF$

PRESCRIPTION FOR REMEDIATION

Use the student work on the Lesson Quiz to prescribe a differentiated review assignment.

Points	Differentiated Remediation
0	Intervention
1	On-level
2	Extension

PowerGeometry.com

5 Assess & Remediate

Assign the Lesson Quiz. Appropriate intervention, practice, or enrichment is automatically generated based on student performance.

Intervention

- **Reteaching** (2 pages) Provides reteaching and practice exercises for the key lesson concepts. Use with struggling students or absent students.
- **English Language Learner Support** Helps students develop and reinforce mathematical vocabulary and key concepts.

All-in-One Resources/Online

Reteaching

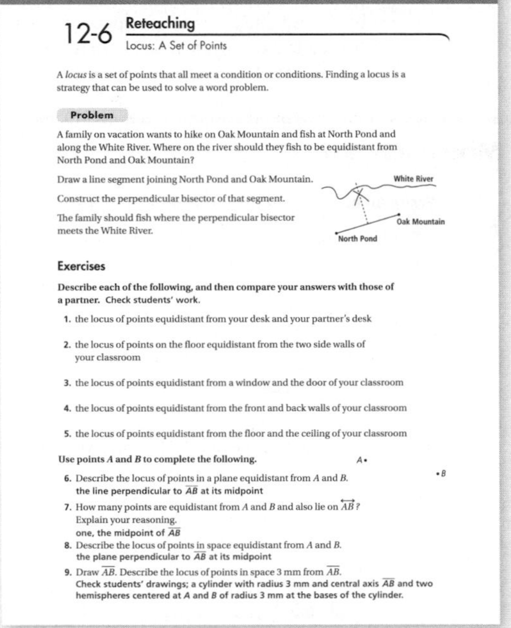

12-6 Reteaching

Locus: A Set of Points

A *locus* is a set of points that all meet a condition or conditions. Finding a locus is a strategy that can be used to solve a word problem.

Problem

A family on vacation wants to hike on Oak Mountain and fish at North Pond and along the White River. Where on the river should they fish to be equidistant from North Pond and Oak Mountain?

Draw a line segment joining North Pond and Oak Mountain.

Construct the perpendicular bisector of that segment.

The family should fish where the perpendicular bisector meets the White River.

Exercises

Describe each of the following, and then compare your answers with those of a partner. Check students' work.

1. the locus of points equidistant from your desk and your partner's desk
2. the locus of points on the floor equidistant from the two side walls of your classroom
3. the locus of points equidistant from a window and the door of your classroom
4. the locus of points equidistant from the front and back walls of your classroom
5. the locus of points equidistant from the floor and the ceiling of your classroom

Use points *A* and *B* to complete the following.

6. Describe the locus of points in a plane equidistant from *A* and *B*. the line perpendicular to $\overline{AB}$ at its midpoint
7. How many points are equidistant from *A* and *B* and also lie on $\overleftrightarrow{AB}$? Explain your reasoning. one, the midpoint of $\overline{AB}$
8. Describe the locus of points in space equidistant from *A* and *B*. the plane perpendicular to $\overline{AB}$ at its midpoint
9. Draw $\overline{AB}$. Describe the locus of points in space 3 mm from $\overline{AB}$. Check students' drawings; a cylinder with radius 3 mm and central axis $\overline{AB}$ and two hemispheres centered at *A* and *B* of radius 3 mm at the bases of the cylinder.

All-in-One Resources/Online

English Language Learner Support

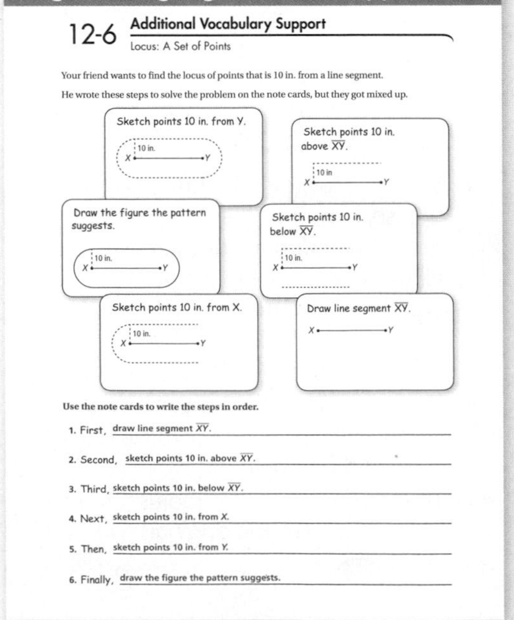

12-6 Additional Vocabulary Support

Locus: A Set of Points

Your friend wants to find the locus of points that is 10 in. from a line segment.

He wrote these steps to solve the problem on the note cards, but they got mixed up.

Use the note cards to write the steps in order.

1. First, draw line segment $\overline{XY}$.
2. Second, sketch points 10 in. above $\overline{XY}$.
3. Third, sketch points 10 in. below $\overline{XY}$.
4. Next, sketch points 10 in. from *X*.
5. Then, sketch points 10 in. from *Y*.
6. Finally, draw the figure the pattern suggests.

Differentiated Remediation *continued*

On-Level

- **Practice** (2 pages) Provides extra practice for each lesson. For simpler practice exercises, use the Form K Practice pages found in the All-in-One Teaching Resources and online.
- **Think About a Plan** Helps students develop specific problem-solving skills and strategies by providing scaffolded guiding questions.
- **Standardized Test Prep** Focuses on all major exercises, all major question types, and helps students prepare for the high-stakes assessments.

Extension

- **Enrichment** Provides students with interesting problems and activities that extend the concepts of the lesson.
- **Activities, Games, and Puzzles** Worksheets that can be used for concepts development, enrichment, and for fun!

Practice and Problem Solving Wkbk/ All-in-One Resources/Online

Practice page 1

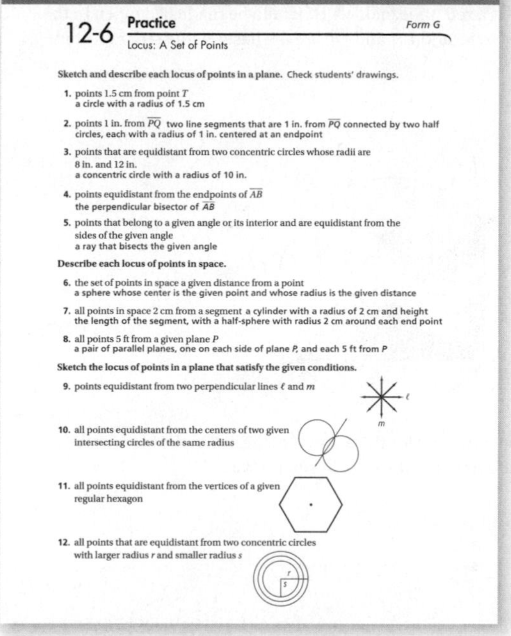

12-6 **Practice** *Form G*
Locus: A Set of Points

Sketch and describe each locus of points in a plane. Check students' drawings.

1. points 1.5 cm from point T
   a circle with a radius of 1.5 cm
2. points 1 in. from $\overline{PQ}$ two line segments that are 1 in. from $\overline{PQ}$ connected by two half circles, each with a radius of 1 in. centered at an endpoint
3. points that are equidistant from two concentric circles whose radii are 8 in. and 12 in.
   a concentric circle with a radius of 10 in.
4. points equidistant from the endpoints of $\overline{AB}$
   the perpendicular bisector of $\overline{AB}$
5. points that belong to a given angle or its interior and are equidistant from the sides of the given angle
   a ray that bisects the given angle

Describe each locus of points in space.

6. the set of points in space a given distance from a point
   a sphere whose center is the given point and whose radius is the given distance
7. all points in space 2 cm from a segment a cylinder with a radius of 2 cm and height the length of the segment, with a half-sphere with radius 2 cm around each end point
8. all points 5 ft from a given plane P
   a pair of parallel planes, one on each side of plane P, and each 5 ft from P

Sketch the locus of points in a plane that satisfy the given conditions.

9. points equidistant from two perpendicular lines ℓ and m
10. all points equidistant from the centers of two given intersecting circles of the same radius
11. all points equidistant from the vertices of a given regular hexagon
12. all points that are equidistant from two concentric circles with larger radius r and smaller radius s

Practice and Problem Solving Wkbk/ All-in-One Resources/Online

Practice page 2

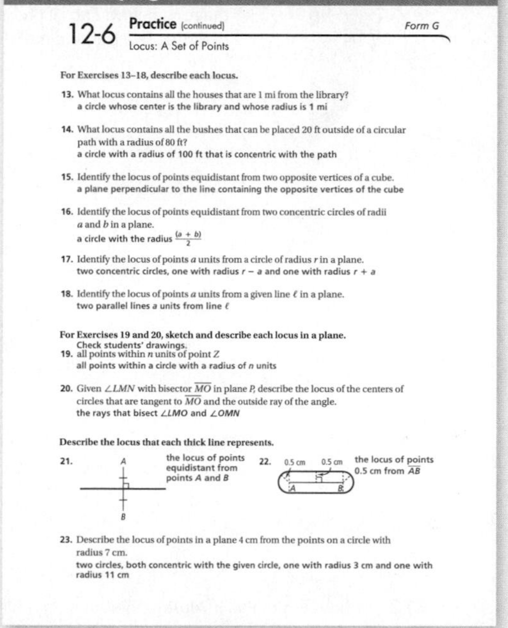

12-6 **Practice** (continued) *Form G*
Locus: A Set of Points

For Exercises 13–18, describe each locus.

13. What locus contains all the houses that are 1 mi from the library?
    a circle whose center is the library and whose radius is 1 mi
14. What locus contains all the bushes that can be placed 20 ft outside of a circular path with a radius of 80 ft?
    a circle with a radius of 100 ft that is concentric with the path
15. Identify the locus of points equidistant from two opposite vertices of a cube.
    a plane perpendicular to the line containing the opposite vertices of the cube
16. Identify the locus of points equidistant from two concentric circles of radii a and b in a plane.
    a circle with the radius $\frac{(a + b)}{2}$
17. Identify the locus of points a units from a circle of radius r in a plane.
    two concentric circles, one with radius $r - a$ and one with radius $r + a$
18. Identify the locus of points a units from a given line ℓ in a plane.
    two parallel lines a units from line ℓ

For Exercises 19 and 20, sketch and describe each locus in a plane.
Check students' drawings.

19. all points within n units of point Z
    all points within a circle with a radius of n units
20. Given $\angle LMN$ with bisector $\overline{MO}$ in plane P, describe the locus of the centers of circles that are tangent to $\overline{MO}$ and the outside ray of the angle.
    the rays that bisect $\angle LMO$ and $\angle OMN$

Describe the locus that each thick line represents.

21. the locus of points equidistant from points A and B
22. the locus of points 0.5 cm from $\overline{AB}$
23. Describe the locus of points in a plane 4 cm from the points on a circle with radius 7 cm.
    two circles, both concentric with the given circle, one with radius 3 cm and one with radius 11 cm

All-in-One Resources/Online

Enrichment

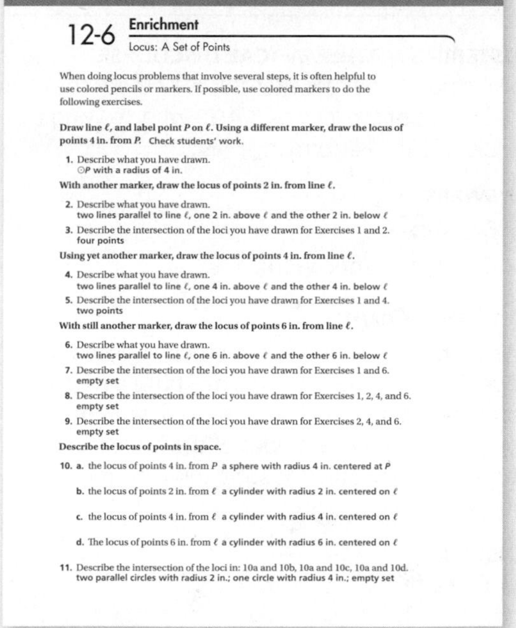

12-6 **Enrichment**
Locus: A Set of Points

When doing locus problems that involve several steps, it is often helpful to use colored pencils or markers. If possible, use colored markers to do the following exercises.

Draw line ℓ, and label point P on ℓ. Using a different marker, draw the locus of points 4 in. from P. Check students' work.

1. Describe what you have drawn.
   $\odot P$ with a radius of 4 in.

With another marker, draw the locus of points 2 in. from line ℓ.

2. Describe what you have drawn.
   two lines parallel to line ℓ, one 2 in. above ℓ and the other 2 in. below ℓ
3. Describe the intersection of the loci you have drawn for Exercises 1 and 2.
   four points

Using yet another marker, draw the locus of points 4 in. from line ℓ.

4. Describe what you have drawn.
   two lines parallel to line ℓ, one 4 in. above ℓ and the other 4 in. below ℓ
5. Describe the intersection of the loci you have drawn for Exercises 1 and 4.
   two points

With still another marker, draw the locus of points 6 in. from line ℓ.

6. Describe what you have drawn.
   two lines parallel to line ℓ, one 6 in. above ℓ and the other 6 in. below ℓ
7. Describe the intersection of the loci you have drawn for Exercises 1 and 6.
   empty set
8. Describe the intersection of the loci you have drawn for Exercises 1, 2, 4, and 6.
   empty set
9. Describe the intersection of the loci you have drawn for Exercises 2, 4, and 6.
   empty set

Describe the locus of points in space.

10. a. the locus of points 4 in. from P a sphere with radius 4 in. centered at P
    b. the locus of points 2 in. from ℓ a cylinder with radius 2 in. centered on ℓ
    c. the locus of points 4 in. from ℓ a cylinder with radius 4 in. centered on ℓ
    d. The locus of points 6 in. from ℓ a cylinder with radius 6 in. centered on ℓ
11. Describe the intersection of the loci in: 10a and 10b, 10a and 10c, 10a and 10d.
    two parallel circles with radius 2 in.; one circle with radius 4 in.; empty set

Practice and Problem Solving Wkbk/ All-in-One Resources/Online

Think About a Plan

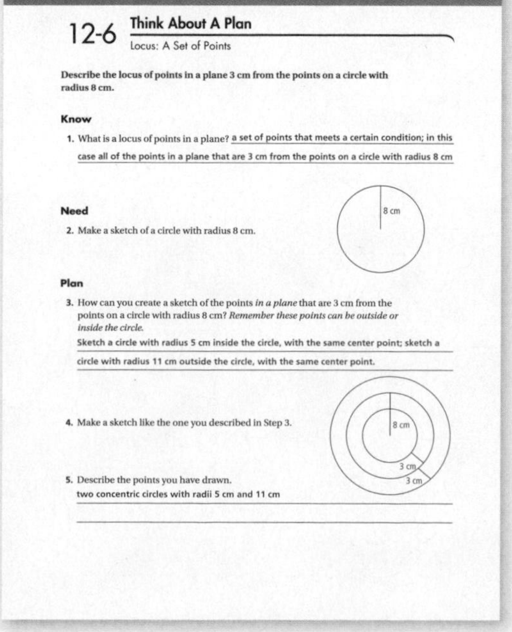

12-6 **Think About A Plan**
Locus: A Set of Points

Describe the locus of points in a plane 3 cm from the points on a circle with radius 8 cm.

Know

1. What is a locus of points in a plane? a set of points that meets a certain condition; in this case all of the points in a plane that are 3 cm from the points on a circle with radius 8 cm

Need

2. Make a sketch of a circle with radius 8 cm.

Plan

3. How can you create a sketch of the points *in a plane* that are 3 cm from the points on a circle with radius 8 cm? *Remember these points can be outside or inside the circle.*
   Sketch a circle with radius 5 cm inside the circle, with the same center point; sketch a circle with radius 11 cm outside the circle, with the same center point.
4. Make a sketch like the one you described in Step 3.
5. Describe the points you have drawn.
   two concentric circles with radii 5 cm and 11 cm

Practice and Problem Solving Wkbk/ All-in-One Resources/Online

Standardized Test Prep

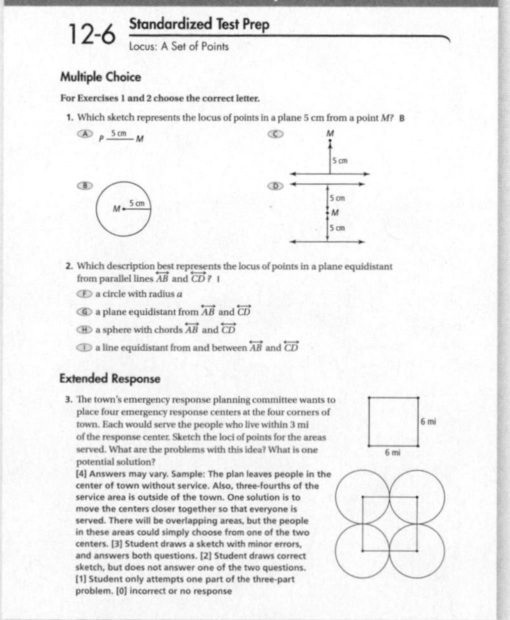

12-6 **Standardized Test Prep**
Locus: A Set of Points

Multiple Choice

For Exercises 1 and 2 choose the correct letter.

1. Which sketch represents the locus of points in a plane 5 cm from a point M? B
2. Which description best represents the locus of points in a plane equidistant from parallel lines $\overleftrightarrow{AB}$ and $\overleftrightarrow{CD}$? I
   F. a circle with radius a
   G. a plane equidistant from $\overleftrightarrow{AB}$ and $\overleftrightarrow{CD}$
   H. a sphere with chords $\overleftrightarrow{AB}$ and $\overleftrightarrow{CD}$
   I. a line equidistant from and between $\overleftrightarrow{AB}$ and $\overleftrightarrow{CD}$

Extended Response

3. The town's emergency response planning committee wants to place four emergency response centers at the four corners of town. Each would serve the people who live within 3 mi of the response center. Sketch the loci of points for the areas served. What are the problems with this idea? What is one potential solution?
   [4] Answers may vary. Sample: The plan leaves people in the center of town without service. Also, three-fourths of the service area is outside of the town. One solution is to move the centers closer together so that everyone is served. There will be overlapping areas, but the people in these areas could simply choose from one of the two centers. [3] Student draws a sketch with minor errors, and answers both questions. [2] Student draws correct sketch, but does not answer one of the two questions. [1] Student only attempts one part of the three-part problem. [0] incorrect or no response

Online Teacher Resource Center

Activities, Games, and Puzzles

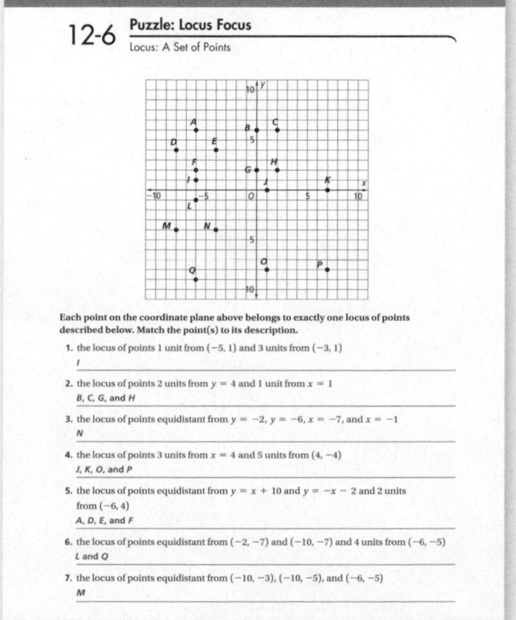

12-6 **Puzzle: Locus Focus**
Locus: A Set of Points

Each point on the coordinate plane above belongs to exactly one locus of points described below. Match the point(s) to its description.

1. the locus of points 1 unit from $(-5, 1)$ and 3 units from $(-3, 1)$
   I
2. the locus of points 2 units from $y = 4$ and 1 unit from $x = 1$
   B, C, G, and H
3. the locus of points equidistant from $y = -2$, $y = -6$, $x = -7$, and $x = -1$
   N
4. the locus of points 3 units from $x = 4$ and 5 units from $(4, -4)$
   J, K, O, and P
5. the locus of points equidistant from $y = x + 10$ and $y = -x - 2$ and 2 units from $(-6, 4)$
   A, D, E, and F
6. the locus of points equidistant from $(-2, -7)$ and $(-10, -7)$ and 4 units from $(-6, -5)$
   L and Q
7. the locus of points equidistant from $(-10, -3)$, $(-10, -5)$, and $(-6, -5)$
   M

Completing the Performance Task

In the Apply What You've Learned sections in Lessons 12-1, 12-3, and 12-4, students determined that $\triangle AGB$ is a right triangle and found the length of the hypotenuse and of one leg of $\triangle AGB$. Here, students use these lengths to help them complete the Performance Task. Ask students the following questions as they work toward solving the problem.

Q How can you use the work you have done in the chapter to solve the problem? **[I have found the lengths of all the line segments in the diagram except *BG*. I can use the fact that $\triangle AGB$ is a right triangle to find the length of $\overline{BG}$ using the Pythagorean Theorem. Then I will have a base and height for $\triangle AGB$.]**

Q What are two different ways to determine the length of $\overline{BG}$? **[Use the Pythagorean Theorem and use Corollary 1 to Theorem 7-3 since $\overline{BG}$ is the altitude to the hypotenuse of $\triangle AGB$.]**

FOSTERING MATHEMATICAL DISCOURSE

Have students discuss alternative methods for finding the lengths of the legs of $\triangle AGB$ when solving the problem in the Performance Task.

ANSWERS

1. 61.44 ft^2

2. Check students' work.

On Your Own

This problem is similar to the problem posed on page 761, but now students must use a different combination of circle theorems to determine the lengths of the relevant sides of the trapezoid. Students should strive to solve this problem independently.

ANSWER

about 72 ft^2

12 Pull It All Together

Completing the Performance Task

To solve these problems you will pull together many concepts and skills that you have studied about relationships within circles.

Look back at your results from the Apply What You've Learned sections in Lessons 12-1, 12-3, and 12-4. Use the work you did to complete the following.

1. Solve the problem in the Task Description on page 761 by determining the area of the copper needed for the red sail in the logo for the showroom display. Show all your work and explain each step of your solution.

2. Reflect Choose one of the Mathematical Practices below and explain how you applied it in your work on the Performance Task.

MP 1: Make sense of problems and persevere in solving them.

MP 2: Reason abstractly and quantitatively.

MP 4: Model with mathematics.

On Your Own

The Sunshine Sailboat Company is considering an alternate logo for their showroom display. This logo consists of one triangular sail against a circular sun, as shown in the diagram below. Part of the sail is a red trapezoid, which would be made of copper. In the diagram, *Q* is the center of the circle, and $\overline{PK}$ and $\overline{QL}$ are the bases of the red trapezoid.

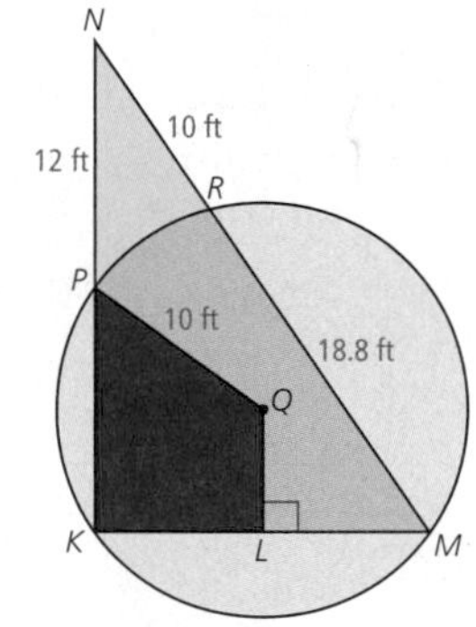

Determine the area of the copper needed for the red part of the sail in the logo for the showroom display. Round your answer to the nearest square foot.

12 Chapter Review

Connecting BIG ideas and Answering the Essential Questions

1 Reasoning and Proof
The measure of an arc equals the measure of its central angle. You can use this angle and arc to prove relationships of other angles and arcs.

Tangents, Chords, and Arcs (Lessons 12-1, 12-2)

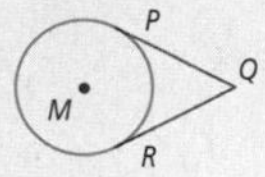

$\overline{PQ}$ and $\overline{RQ}$ are tangents to ⊙M. So $\overline{PQ} \cong \overline{RQ}$.

In a circle or congruent circles, if one is true, all are true: $\angle AOB \cong \angle COD$, $\overline{AB} \cong \overline{CD}$, $\overset{\frown}{AB} \cong \overset{\frown}{CD}$

Locus: A Set of Points (Lesson 12-6)
A locus is a set of points, all of which meet a stated condition.

2 Measurement
Segments intersecting circles form angles and intercepted arcs. You can find some missing measures using given information and appropriate formulas.

Angles and Circles (Lessons 12-3, 12-4)

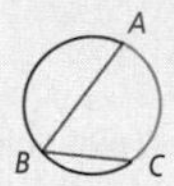

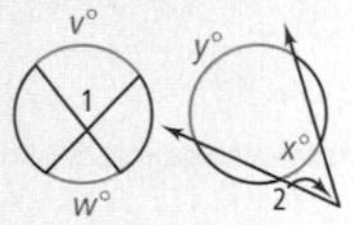

$m\angle B = \frac{1}{2}m\overset{\frown}{AC}$

$m\angle 1 = \frac{1}{2}(v + w)$

$m\angle 2 = \frac{1}{2}(y - x)$

Segment Lengths (Lesson 12-4)

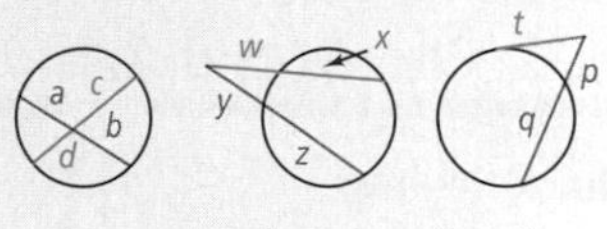

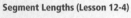

$a \cdot b = c \cdot d$ $(w + x)w = (y + z)y$

$(p + q)p = t^2$

3 Coordinate Geometry
You can use the center and the radius to write an equation of a circle.

Circles in the Coordinate Plane (Lesson 12-5)
$(x - h)^2 + (y - k)^2 = r^2$

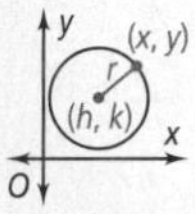

Chapter Vocabulary

- chord (p. 771)
- inscribed angle (p. 780)
- intercepted arc (p. 780)
- locus (p. 804)
- point of tangency (p. 762)
- secant (p. 791)
- standard form of an equation of a circle (p. 799)
- tangent to a circle (p. 762)

Use the figure to choose the correct term to complete each sentence.

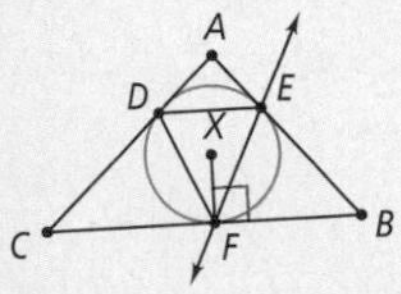

1. $\overleftrightarrow{EF}$ is (*a secant of, tangent to*) ⊙X.

2. $\overline{DF}$ is a (*chord, locus*) of ⊙X.

3. △ABC is made of (*chords in, tangents to*) ⊙X.

4. ∠DEF is an (*intercepted arc, inscribed angle*) of ⊙X.

5. The set of all points equidistant from the endpoints of $\overline{CB}$ is a (*locus, tangent*).

Essential Questions

BIG idea Reasoning and Proof

ESSENTIAL QUESTION How can you prove relationships between angles and arcs in a circle?
ANSWER The measure of an arc equals the measure of its central angle. You can use this angle and arc to prove relationships of other angles and arcs.

BIG idea Measurement

ESSENTIAL QUESTION When lines intersect a circle, or within a circle, how do you find the measures of resulting angles, arcs, and segments?
ANSWER Segments intersecting circles form angles and intercepted arcs. You can find some missing measures using given information and appropriate formulas.

BIG idea Coordinate Geometry

ESSENTIAL QUESTION How do you find the equation of a circle in the coordinate plane?
ANSWER You can use the center and radius to write an equation of a circle.

Answers

Chapter Review

1. secant of
2. chord
3. tangents to
4. inscribed ∠
5. locus

Summative Questions

Use the following prompts as you review this chapter with your students. The prompts are designed to help you assess your students' understanding of the BIG Ideas they have studied.

- What is a tangent?
- What is a segment?
- How can you find the measure of an angle formed in the interior of a circle?
- How can you find the measure of an angle formed on the exterior of a circle?
- How do you write the equation of a circle?

Answers

Chapter Review (continued)

6. 20 units
7. $\sqrt{3}$
8. 120
9. 90
10. $2:1$ or $\frac{2}{1}$
11. $\overline{AB}$ is a diameter of the circle.
12. 4.5
13. $\frac{\sqrt{181}}{2} \approx 6.7$

12-1 Tangent Lines

Quick Review

A **tangent** to a circle is a line that intersects the circle at exactly one point. The radius to that point is perpendicular to the tangent. From any point outside a circle, you can draw two segments tangent to a circle. Those segments are congruent.

Example

$\overrightarrow{PA}$ and $\overrightarrow{PB}$ are tangents. Find x.

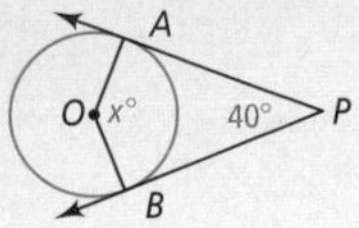

The radii are perpendicular to the tangents. Add the angle measures of the quadrilateral:

$$x + 90 + 90 + 40 = 360$$
$$x + 220 = 360$$
$$x = 140$$

Exercises

Use $\odot O$ for Exercises 6–8.

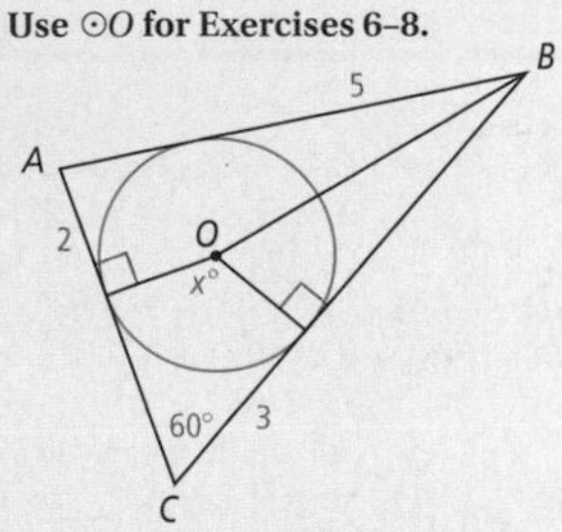

6. What is the perimeter of $\triangle ABC$?
7. $OB = \sqrt{28}$. What is the radius?
8. What is the value of x?

12-2 Chords and Arcs

Quick Review

A **chord** is a segment whose endpoints are on a circle. Congruent chords are equidistant from the center. A diameter that bisects a chord that is not a diameter is perpendicular to the chord. The perpendicular bisector of a chord contains the center of the circle.

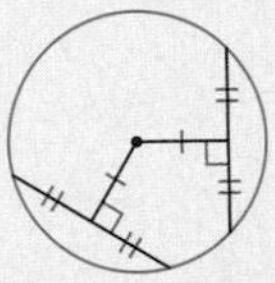

Example

What is the value of d?

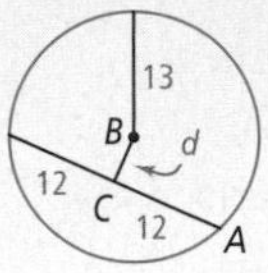

Since the chord is bisected, $m\angle ACB = 90$. The radius is 13 units. So an auxiliary segment from A to B is 13 units. Use the Pythagorean Theorem.

$$d^2 + 12^2 = 13^2$$
$$d^2 = 25$$
$$d = 5$$

Exercises

Use the figure at the right for Exercises 9–11.

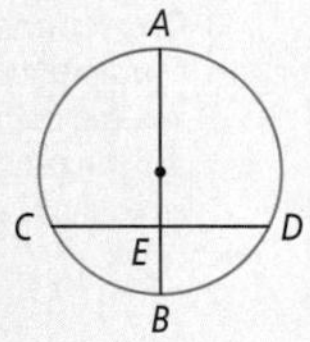

9. If $\overline{AB}$ is a diameter and $CE = ED$, then $m\angle AEC =$ __?__ .
10. If $\overline{AB}$ is a diameter and is at right angles to $\overline{CD}$, what is the ratio of CD to DE?
11. If $CE = \frac{1}{2}CD$ and $m\angle DEB = 90$, what is true of $\overline{AB}$?

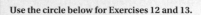

Use the circle below for Exercises 12 and 13.

12. What is the value of x?
13. What is the value of y?

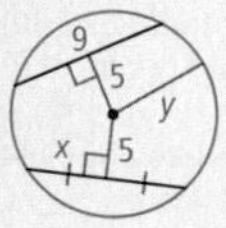

12-3 Inscribed Angles

Quick Review

An **inscribed angle** has its vertex on a circle and its sides are chords. An **intercepted arc** has its endpoints on the sides of an inscribed angle, and its other points in the interior of the angle. The measure of an inscribed angle is half the measure of its intercepted arc.

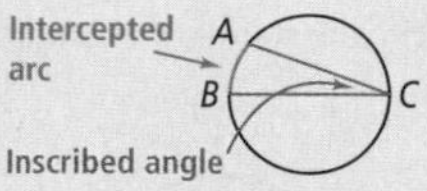

Example

What is $m\widehat{PS}$? What is $m\angle R$?

The $m\angle Q = 60$ is half of $m\widehat{PS}$, so $m\widehat{PS} = 120$. $\angle R$ intercepts the same arc as $\angle Q$, so $m\angle R = 60$.

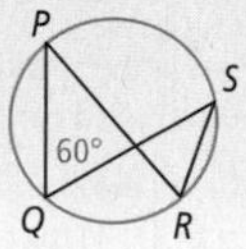

Exercises

Find the value of each variable. Line ℓ is a tangent.

14.

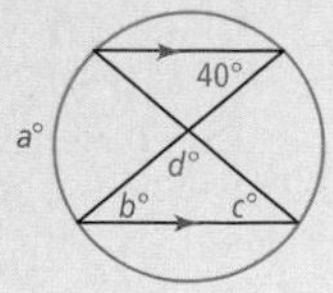

15.

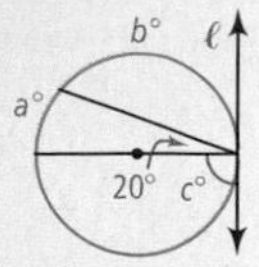

16.

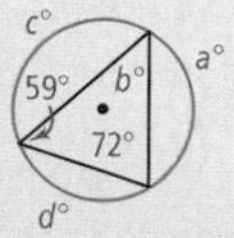

17.

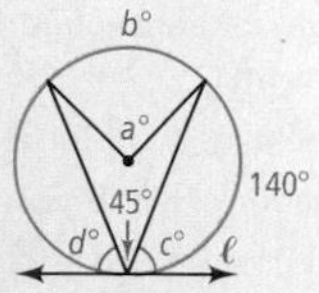

12-4 Angle Measures and Segment Lengths

Quick Review

A **secant** is a line that intersects a circle at two points. The following relationships are true:

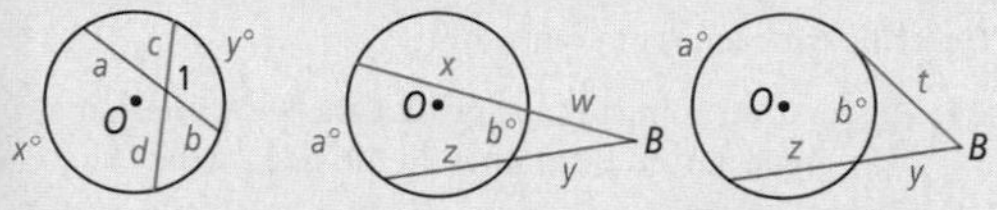

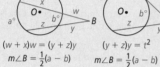

$a \cdot b = c \cdot d$ $\quad (w + x)w = (y + z)y \quad (y + z)y = t^2$

$m\angle 1 = \frac{1}{2}(x + y) \quad m\angle B = \frac{1}{2}(a - b) \quad m\angle B = \frac{1}{2}(a - b)$

Example

What is the value of x?

$(x + 10)10 = (19 + 9)9$

$10x + 100 = 252$

$x = 15.2$

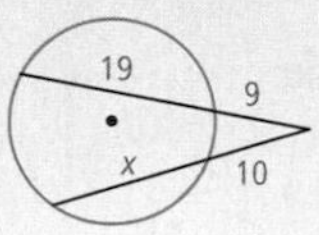

Exercises

Find the value of each variable.

18.

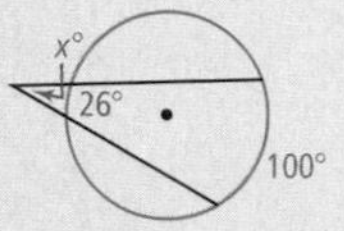

19.

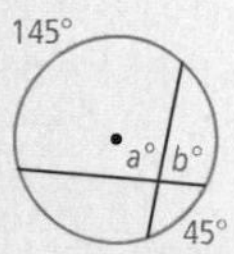

20.

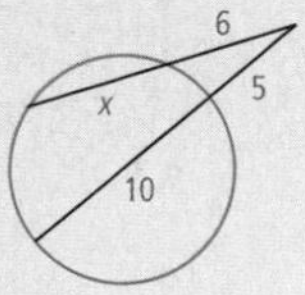

21.

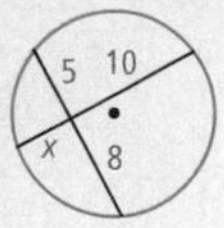

14. $a - 80$, $b = 40$, $c = 40$, $d = 100$
15. $a = 40$, $b = 140$, $c = 90$
16. $a = 118$, $b = 49$, $c = 144$, $d = 98$
17. $a = 90$, $b = 90$, $c = 70$, $d = 65$
18. 37
19. $a = 95$, $b = 85$
20. 6.5
21. 4

Answers

Chapter Review (continued)

22. $x^2 + (y + 2)^2 = 9$

23. $(x - 3)^2 + (y - 2)^2 = 4$

24. $(x + 3)^2 + (y + 4)^2 = 25$

25. $(x - 1)^2 + (y - 4)^2 = 9$

26. center $(7, -5)$; radius 6

27. The locus is the ray that bisects the $\angle$.

28. The locus is a circle, concentric with the given circle, with radius 7 cm.

29. The locus is two lines, one on each side of the given line and $\parallel$ to it, each at a distance of 8 in. from the given line.

30. The locus consists of a cylinder with radius 6 in. that has $\overline{AB}$ as its centerline, along with two hemispheres with centers A and B, each with radius 6 in.

12-5 Circles in the Coordinate Plane

Quick Review

The **standard form of an equation of a circle** with center (h, k) and radius r is

$(x - h)^2 + (y - k)^2 = r^2$.

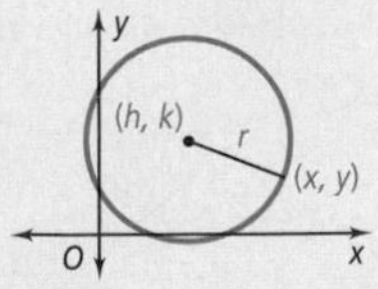

Example

Write the standard equation of the circle shown.

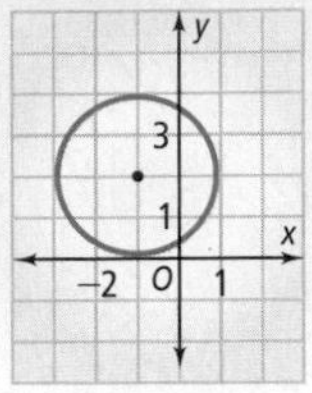

The center is $(-1, 2)$. The radius is 2.

The equation of the circle is

$(x - (-1))^2 + (y - 2)^2 = 2^2$

or

$(x + 1)^2 + (y - 2)^2 = 4$.

Exercises

Write the standard equation of each circle below.

22.

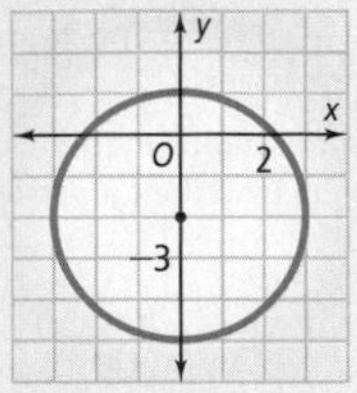

23.

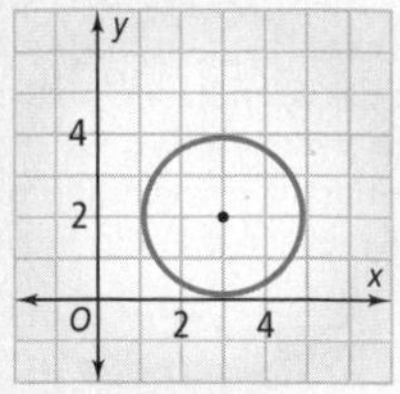

24. What is the standard equation of the circle with radius 5 and center $(-3, -4)$?

25. What is the standard equation of the circle with center $(1, 4)$ that passes through $(-2, 4)$?

26. What are the center and radius of the circle with equation $(x - 7)^2 + (y + 5)^2 = 36$?

12-6 Locus: A Set of Points

Quick Review

A **locus** is a set of points that satisfies a stated condition.

Example

Sketch and describe the locus of points in a plane equidistant from points *A* and *B*.

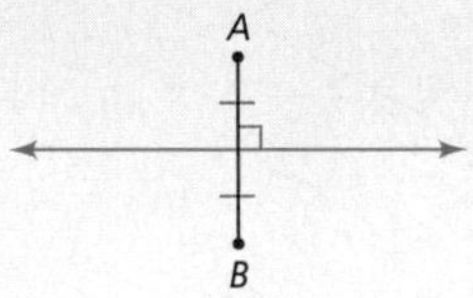

The locus is the perpendicular bisector of $\overline{AB}$.

Exercises

Describe each locus of points.

27. The set of all points in a plane that are in the interior of an angle and equidistant from the sides of the angle.

28. The set of all points in a plane that are 5 cm from a circle with radius 2 cm.

29. The set of all points in a plane at a distance 8 in. from a given line.

30. The set of all points in space that are a distance 6 in. from $\overline{AB}$.

12 Chapter Test

Do you know HOW?

Algebra **For Exercises 1–8, lines that appear tangent are tangent. Find the value of each variable. Round decimals to the nearest tenth.**

1. 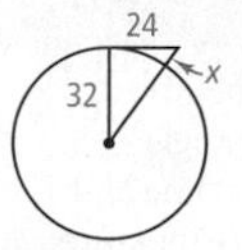

2. 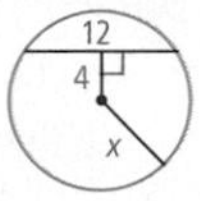

3. 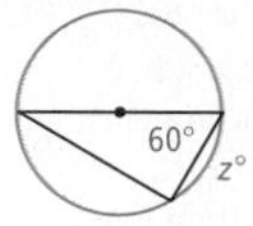

4. 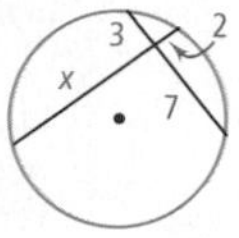

5. 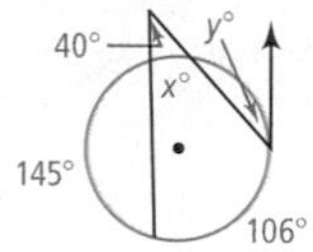

6. 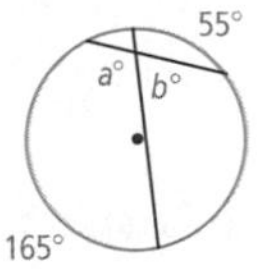

7. 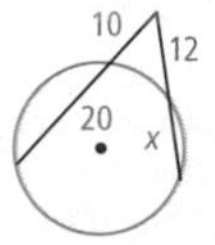

8. 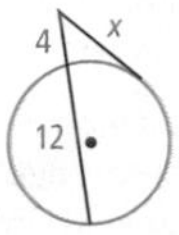

Find $m\widehat{AB}$.

9. 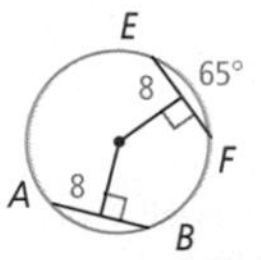

10. 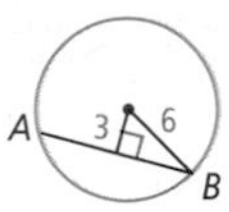

11. Graph $(x + 3)^2 + (y - 2)^2 = 9$. Then label the center and radius.

12. Write an equation of the circle with cente r (3, 0) that passes through point (−2, −4).

13. What is the graph of $x^2 + y^2 = 0$?

14. Write an equation for the locus of points in the coordinate plane that are 4 units from (−5, 2).

Write the standard equation of each circle.

15. 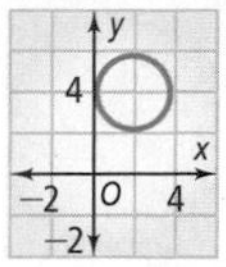

16. 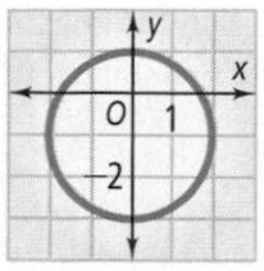

Sketch each locus on a coordinate plane.

17. the set of all points 3 units from the line $y = -2$

18. the set of all points equidistant from the axes

Do you UNDERSTAND?

19. **Writing** What is special about a rhombus inscribed in a circle? Justify your answer.

20. **Reasoning** $\overleftrightarrow{EF}$ is the perpendicular bisector of chord $\overline{AB}$, and $\overline{CD} \parallel \overline{AB}$. Show that $\overleftrightarrow{EF}$ is the perpendicular bisector of chord $\overline{CD}$.

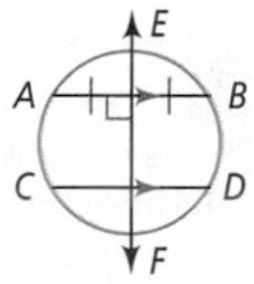

21. **Error Analysis** A student says that $\angle RPO \cong \angle NPG$ in this circle, since they are vertical angles, and thus $\overline{RO} \cong \overline{NG}$. Why is this incorrect?

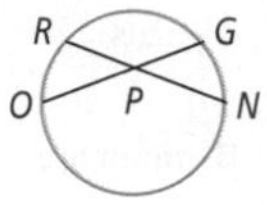

22. **Reasoning** $\overleftrightarrow{PA}$ and $\overleftrightarrow{PB}$ are tangent to ⊙O. PA is equal to the radius of the circle. What kind of quadrilateral is PAOB? Explain.

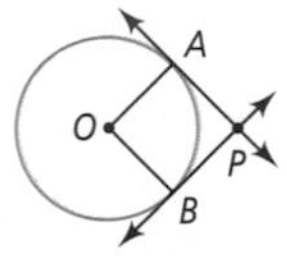

23. **Reasoning** A secant line passes through a circle at points A and B. Point C is also on the circle. Describe the locus of points P that satisfy these conditions: P and C are on the same side of the secant, and $m\angle APB = m\angle ACB$.

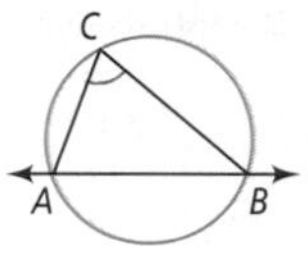

Answers

Chapter Test

1. 8
2. 7.2
3. 60
4. 10.5
5. $x = 26$, $y = 41.5$
6. $a = 110$, $b = 70$
7. 13
8. 8
9. 65
10. 120
11. 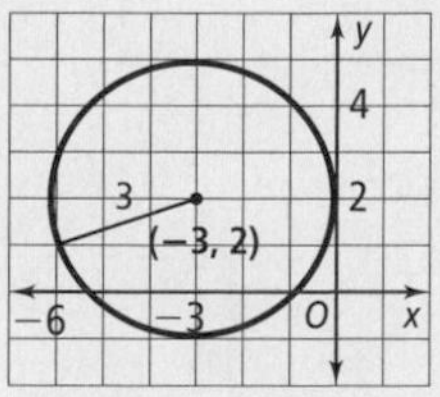

12. $(x - 3)^2 + y^2 = 41$
13. The graph is the point (0, 0).
14. $(x + 5)^2 + (y - 2)^2 = 16$
15. $(x - 2)^2 + (y - 4)^2 = 4$
16. $x^2 + (y + 1)^2 = 4$
17. 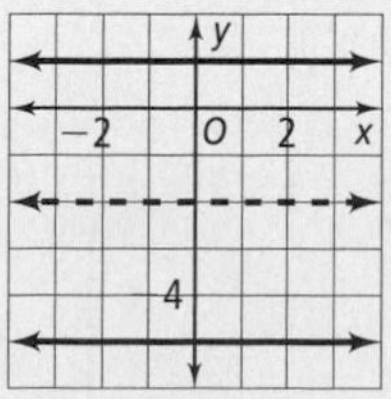

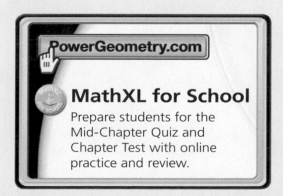

18. 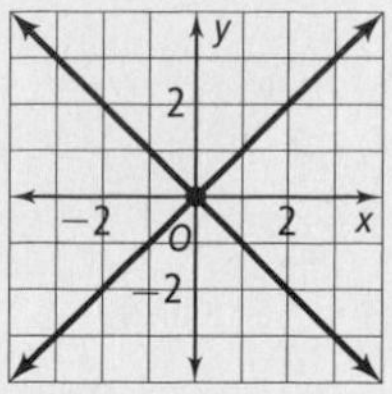

19. It is a square. Sample justification: Since opp. ∡ of an inscribed quadrilateral are suppl., and opp. ∡ of a rhombus are ≅, all four ∡ are rt. ∡.
20. $\overleftrightarrow{EF}$ is the ⊥ bis. of $\overline{AB}$ (given), so $\overleftrightarrow{EF}$ contains the center of the circle. $\overline{CD} \parallel \overline{AB}$ (given), so $\overleftrightarrow{EF} \perp \overline{CD}$ (if a line is ⊥ to one of two ∥ lines, it is ⊥ to the other), and $\overleftrightarrow{EF}$ is the ⊥ bis. of $\overline{CD}$ (if a diameter is ⊥ to a chord, it bisects the chord).
21. ∠RPO and ∠NPG are not central ∡. Nothing can be assumed about the measure of the intercepted arcs.
22. A square; $\overline{PB} \cong \overline{PA}$ (tangent segments are ≅) and $PA = OA = OB$ (given), so $\overline{PB} \cong \overline{PA} \cong \overline{OA} \cong \overline{OB}$ (= segments are ≅). Hence PAOB is a rhombus (all sides are ≅). Now $m\angle OAP = m\angle OBP = 90$ (tangent is ⊥ to radius), so $m\angle AOB = m\angle APB = 90$ (consecutive ∡ in a ▱ are suppl.). Since all sides are ≅ and all ∡ are rt. ∡, PAOB is a square.
23. The locus is all points P on $\widehat{APB}$, excluding A and B.

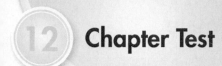

Item Number	Lesson	Content Standard
1	4-6	G-SRT.C.8
2	8-1	G-SRT.C.8
3	9-3	G-CO.A.4
4	12-3	G-C.A.2
5	11-7	*G-GMD.A.3
6	3-2	G-CO.C.9
7	10-6	*G-GMD.A.3
8	12-4	G-C.A.2
9	10-3	G-MG.A.1
10	12-5	G-GPE.A.1
11	8-3	G-SRT.C.8
12	3-2	G-CO.C.10
13	7-2	G-SRT.B.5
14	12-1	G-C.A.2
15	11-2	G-MG.A.1
16	6-7	G-GPE.B.7
17	12-2	G-C.A.2
18	6-1	G-CO.A.3
19	3-8	G-GPE.B.5
20	12-4	G-C.A.2

*Prepares for standard

12 Common Core Cumulative Standards Review

ASSESSMENT

TIPS FOR SUCCESS

Some test questions require you to use the relationships between lines, segments, and circles. Read the sample question at the right. Then follow the tips to answer it.

In the figure below, $\overleftrightarrow{AC}$ is tangent to the circle at point B. If $m\widehat{BDE} = 156°$, what is $m\angle ABE$?

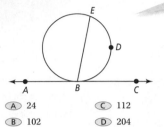

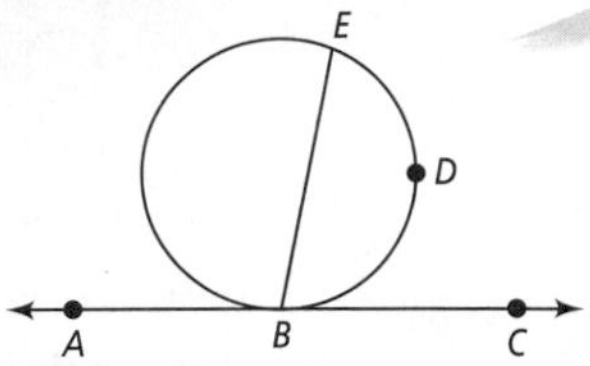

Ⓐ 24 Ⓒ 112

Ⓑ 102 Ⓓ 204

TIP 1

Because $\angle ABE$ is an obtuse angle, you can eliminate choice A. Also, because the measure of a straight angle is 180° and choice D is greater than 180°, choice D can be eliminated.

TIP 2

Use the relationship between tangents, chords, angles, and intercepted arcs.

Think It Through

The measure of an angle formed by a tangent line and a chord is one half the measure of the intercepted arc. So, $m\angle EBC = \frac{1}{2} m\widehat{BDE} = 78°$. Because $\angle ABC$ is a straight angle, $\angle ABE$ and $\angle EBC$ form a linear pair. Therefore, $m\angle ABE = 180 - 78 = 102$. The correct answer is B.

Vocabulary Builder

As you solve test items, you must understand the meanings of mathematical terms. Match each term with its mathematical meaning.

A. locus

B. minor arc

C. major arc

D. cross section

I. the intersection of a solid and a plane

II. part of a circle that is smaller than a semicircle

III. a set of points, all of which meet a stated condition

IV. a part of a circle that is larger than a semicircle

Selected Response

Read each question. Then write the letter of the correct answer on your paper.

1. A vertical mast is on top of building and is positioned 6 ft from the front edge. The mast casts a shadow perpendicular to the front of the building, and the tip of the shadow is 90 ft from the front of the building. At the same time, the 24-ft building casts a 64-ft shadow. What is the height of the mast?

Ⓐ 7 ft 6 in. Ⓒ 12 ft

Ⓑ 9 ft 9 in. Ⓓ 33 ft

2. Javier leans a 20-ft long ladder against a wall. If the base of the ladder is positioned 5 feet from the wall, how high up the wall does the ladder reach? Round to the nearest tenth.

Ⓕ 19.4 ft Ⓗ 17.5 ft

Ⓖ 18.8 ft Ⓘ 15 ft

Answers

Common Core Cumulative Standards Review

A. III
B. II
C. IV
D. I
1. D
2. F

3. $\triangle FGH$ has the vertices shown below. If the triangle is rotated 90° counterclockwise about the origin, what are the coordinates of the rotated point F'?

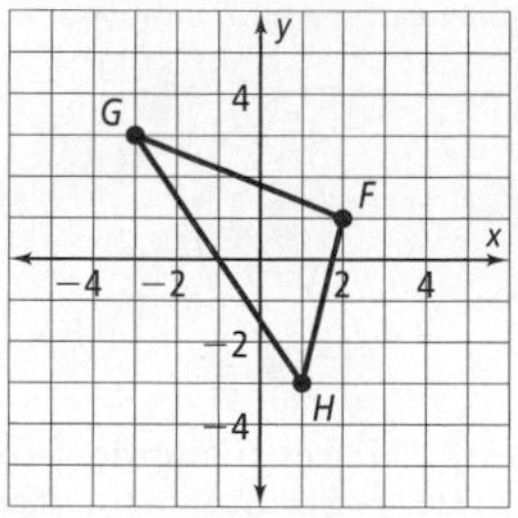

Ⓐ $(1, -3)$ Ⓒ $(-1, 2)$

Ⓑ $(3, -1)$ Ⓓ $(1, -2)$

4. What is $m\overset{\frown}{RT}$ in the figure at the right?

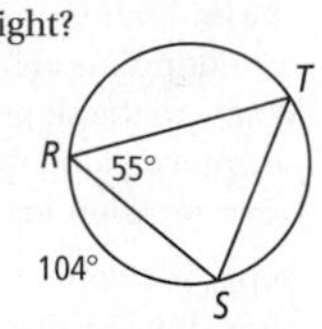

Ⓕ 162°

Ⓖ 146°

Ⓗ 110°

Ⓘ 73°

5. A rectangular prism has a volume of 204 cubic inches. A similar prism is created scaling each dimension by a factor of 2. What is the volume of the larger prism?

Ⓐ 102 in.^3 Ⓒ 816 in.^3

Ⓑ 408 in.^3 Ⓓ 1632 in.^3

6. What is the value of x?

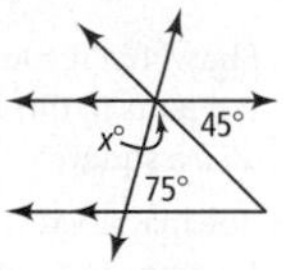

Ⓕ 45°

Ⓖ 75°

Ⓗ 60°

Ⓘ 105°

7. A bicycle tire has a radius of 14.5 inches. About how far does the bicycle travel if the tire makes 15 complete revolutions? Use 3.14 for π.

Ⓐ 57 ft Ⓒ 825 ft

Ⓑ 114 ft Ⓓ 1366 ft

8. What is the measure of x in the figure shown below? What is the value of x?

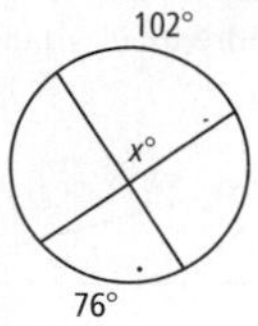

Ⓕ 112° Ⓗ 102°

Ⓖ 104° Ⓘ 89°

9. A stop sign is shaped like a regular octagon with the dimensions shown. What is the area of the stop sign?

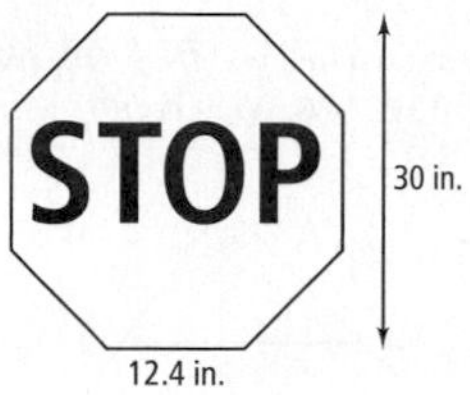

Ⓐ 99.2 in.^2 Ⓒ 744 in.^2

Ⓑ 268.6 in.^2 Ⓓ 1488 in.^2

10. What is the equation of the circle shown on the coordinate grid below?

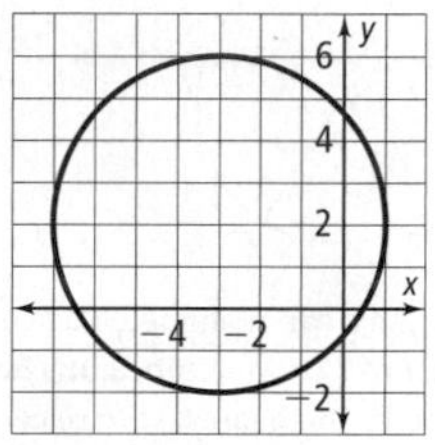

Ⓕ $(x + 3)^2 + (y - 2)^2 = 4$

Ⓖ $(x + 3)^2 + (y - 2)^2 = 16$

Ⓗ $(x - 3)^2 + (y + 2)^2 = 4$

Ⓘ $(x - 3)^2 + (y + 2)^2 = 16$

3. C
4. G
5. D
6. H
7. B
8. I
9. C
10. G

Answers

Common Core Cumulative Standards Review (continued)

11. 1.24

12. 5

13. 22.5

14. 130

15. 341

16. [2] $\overline{DE} = \sqrt{(1+2)^2 + (1-4)^2} \approx 4.2$

$\overline{EF} = \sqrt{(-2-4)^2 + (4-7)^2} \approx 6.7$

$\overline{FD} = \sqrt{(4-1)^2 + (7-1)^2} \approx 6.7$

Perimeter is about 17.7 units.

[1] one minor computational error

17. [2] 11.2

[1] one minor computational error

18. [2] 30

[1] one minor computational error

19. [4] slope of $\overline{HK} = \frac{7-1}{-3-6} = \frac{6}{-9} = -\frac{2}{3}$;

slope of $\overline{MN} = \frac{-8-10}{-5-7} = \frac{-18}{-12} = \frac{3}{2}$;

$\overline{HK}$ and $\overline{MN}$ are perpendicular because the slopes are opposite reciprocals.

[3] one minor computational error

[2] partial correct answer OR no explanation

[1] minor computational error AND no explanation

20. [4] **a.**

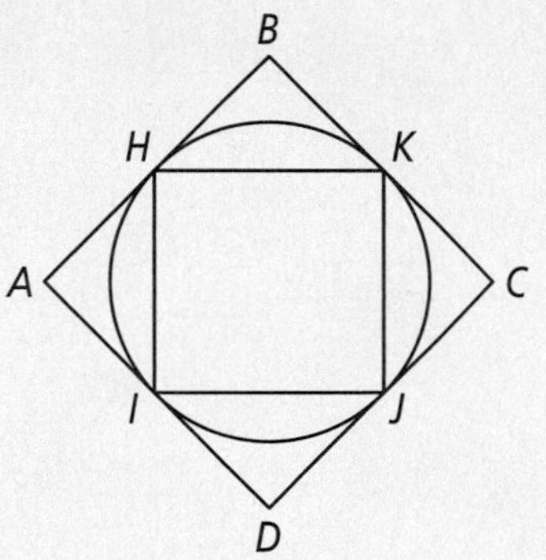

$BH = BK$, $CK = CJ$, $DJ = DI$, and $AH = AI$ by Theorem 12-3. Also $AH = HB$ since $\widehat{HI} = \widehat{HK}$, and likewise $BK = BC$, etc. Therefore $ABCD$ is a rhombus, since all its sides are congruent. Finally, since $m\widehat{HK} = 90$ and $m\angle B = 90$, $ABCD$ is a square. Therefore the new figure is a square.

b. 2

c. Yes, because the tangent lines at the vertices of the inscribed polygon form a new polygon with equivalent side lengths.

[3] one minor computational error

[2] partial correct answer OR no explanation

[1] minor computational error AND no explanation

Constructed Response

11. A ski ramp on a lake has the dimensions shown below. To the nearest hundredth of a meter, what is the height *h* of the ramp?

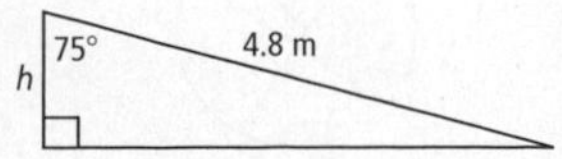

12. What is the value of *n* in the trapezoid shown below?

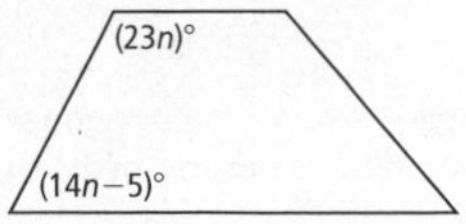

13. In the figure shown below, $\overline{MN} \parallel \overline{OP}$, $LM = 12$, $MN = 15$, and $MO = 6$. What is OP?

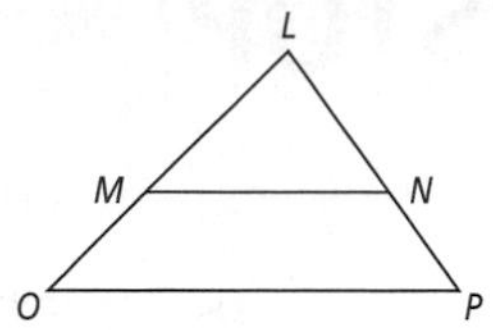

14. In the figure below, $\overline{XY}$ and $\overline{XZ}$ are tangent to ⊙*O* at points *Y* and *Z*, respectively. What is $m\angle YOZ$?

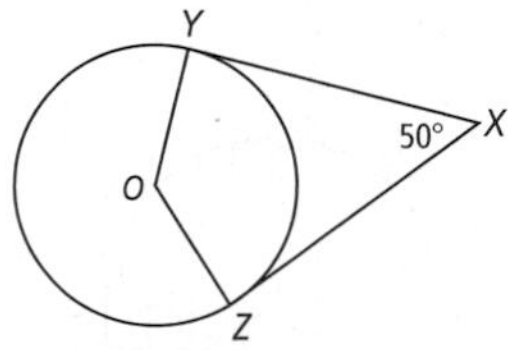

15. An aluminum juice can is shaped like a cylinder with a diameter of 7 centimeters and a height of 12 centimeters. To the nearest square centimeter, how much material was needed to make the can? Use 3.14 for π.

16. $\triangle DEF$ has vertices $D(1, 1)$, $E(-2, 4)$, and $F(4, 7)$. What is the perimeter of $\triangle DEF$? Show your work.

17. In ⊙*A* below, $AE = 13.1$ and $\overline{AC} \perp \overline{BD}$. If $BC = 6.8$, what is AC, to the nearest tenth? Show your work.

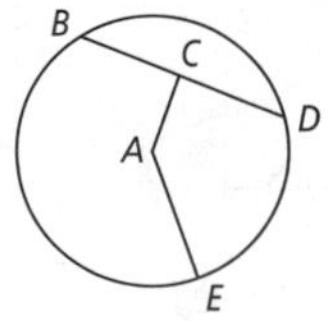

18. What is the measure of an exterior angle of a regular dodecagon (12-sided polygon)? Show your work.

Extended Response

19. The endpoints of $\overline{HK}$ are $H(-3, 7)$ and $K(6, 1)$, and the endpoints of $\overline{MN}$ are $M(-5, -8)$ and $N(7, 10)$. What are the slopes of the line segments? Are the line segments parallel, perpendicular, or neither? Explain how you know and show your work.

20. Suppose a square is inscribed in a circle such as square *HIJK* shown below.

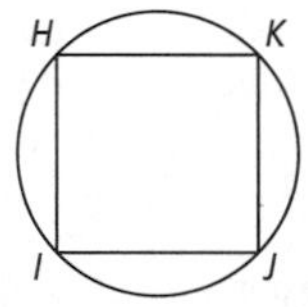

a. Show that if you form a new figure by connecting the tangents to the circle at *H*, *I*, *J*, and *K*, the new figure is also a square.

b. The inscribed square and the square formed by the tangents are similar. What is the scale factor of the similar figures?

c. Let a regular polygon with *n*-sides be inscribed in a circle. Do the tangent lines at the vertices of the polygon form another regular polygon with *n*-sides? Explain.

CHAPTER 13

Get Ready!

Skills Handbook, Page 895

Converting Between Percents, Fractions, and Decimals

Write two equivalent forms for each of the following.

1. 0.7875 **2.** $\frac{7}{8}$ **3.** 0.95 **4.** 60%

Skills Handbook, Page 891

Simplifying Ratios

Simplify each ratio.

5. $\frac{45}{60}$ **6.** $\frac{24}{36}$ **7.** $\frac{7 \cdot 6}{4 \cdot 3}$

Skills Handbook, Page 890

Evaluating Expressions

Evaluate each expression for $x = 6$ and $y = -2$.

8. $9x - 5y + 8$ **9.** $\frac{x}{y}$ **10.** $\frac{2x + y}{y^2}$

Skills Handbook, Page 894

Solving Equations

Solve each equation.

11. $2n + 7 = 23$ **12.** $3(n - 2) = 8$ **13.** $\frac{5}{7}n - 12 = 15$

Lesson 10-8

Finding Geometric Probability

14. What is the probability that a point chosen at random is in the inner circle of the figure at the right?

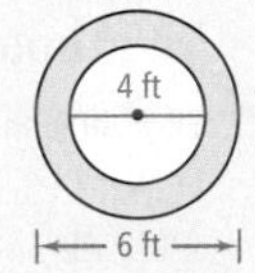

Looking Ahead Vocabulary

15. When you describe the likelihood that a baseball player will get a hit, you are finding an *experimental probability*. What do you think is the experiment in this situation?

16. People training to be pilots sometimes use flight simulators, machines that create a simulation of actual flight. What do you suppose are some ways *simulations* could be used in a math class?

17. When the occurrence of one event does not affect the occurrence of another event, they are called *independent events*. Do you think the events *tossing heads with a penny* and *tossing tails with a quarter* are independent events? Explain.

Answers

Get Ready!

1. $\frac{63}{80}$, 78.75%
2. 0.875, 87.5%
3. $\frac{19}{20}$, 95%
4. $\frac{3}{5}$, 0.6
5. $\frac{3}{4}$
6. $\frac{2}{3}$
7. $\frac{7}{2}$
8. 72
9. −3
10. $\frac{5}{2}$
11. 8
12. $\frac{14}{3}$
13. $\frac{189}{5}$
14. $\frac{4}{9}$
15. the batter's at-bats
16. Answers may vary. Sample answer: A simulation can be used to model situations that are difficult or impractical to conduct.
17. Yes, how the penny lands does not affect how the quarter lands.

Get Ready!

Assign this diagnostic assessment to determine if students have the prerequisite skills for Chapter 13.

Lesson	Skill
Skills Handbook, Page T895	Converting Between Percents, Fractions, and Decimals
Skills Handbook, Page T891	Simplifying Ratios
Skills Handbook, Page T890	Evaluating Expressions
Skills Handbook, Page T894	Solving Equations
10-8	Geometric Probability

To remediate students, select from these resources (available for every lesson).

- Online Problems (PowerGeometry.com)
- Reteaching (All-in-One Teaching Resources)
- Practice (All-in-One Teaching Resources)

Why Students Need These Skills

CONVERTING BETWEEN PERCENTS, FRACTIONS, AND DECIMALS Converting between percents, fractions, and decimals is essential to expressing probabilities in various ways.

SIMPLIFYING RATIOS Simplifying ratios is essential for working with probability.

EVALUATING EXPRESSIONS Evaluating expressions is essential for working with probability formulas.

SOLVING EQUATIONS The thinking through steps that students must do in solving equations is helpful in thinking through the steps required in probability.

FINDING GEOMETRIC PROBABILITY Students prior experience with geometric probability prepares them for the concepts involved in experimental probability.

Looking Ahead Vocabulary

EXPERIMENTAL PROBABILITY Ask students what they think of when they think of an experiment.

SIMULATION Discuss with students what they already know about the term *simulate*. Have them give a real-world definition.

INDEPENDENT EVENTS Ask students to identify other events that cannot occur at the same time.

Chapter 13 Overview

In Chapter 13, students expand on their knowledge of probability. Students will develop the answers to the Essential Questions as they learn the concepts and skills shown below.

BIG idea Probability

ESSENTIAL QUESTION What is the difference between experimental probability and theoretical probability?

- Students will find probabilities based on real-world observations as well as probabilities based strictly on mathematics.

BIG idea Data Representation

ESSENTIAL QUESTION What is a frequency table?

- Students will use frequency tables to find relative frequency.
- Students will use two-way frequency tables to calculate conditional probability.

BIG idea Probability

ESSENTIAL QUESTION What does it mean for an event to be random?

- Students will learn different ways to model randomness and make fair decisions.

Content Standards

Following are the standards covered in this chapter.

CONCEPTUAL CATEGORY Statistics and Probability

Domain Conditional Probability and the Rules of Probability S-CP

Cluster Understand independence and conditional probability and use them to interpret data. (Standards S-CP.A.1, S-CP.A.2, S-CP.A.3, S-CP.A.4, S-CP.A.5)

LESSONS 13-1, 13-2, 13-5, 13-6

Cluster Use the rules of probability to compute probabilities of compound events in a uniform probability model. (Standards S-CP.B.6, S-CP.B.7, S-CP.B.8, S-CP.B.9)

LESSONS 13-4, 13-6

Domain Using Probability to Make Decisions S-MD

Cluster Calculate expected values and use them to solve problems. (Standards S-MD.B.6, S-MD.B.7)

LESSON 13-7

CHAPTER 13

Probability

Download videos connecting math to your world.

Interactive! Vary numbers, graphs, and figures to explore math concepts.

The online Solve It will get you in gear for each lesson.

Math definitions in English and Spanish

Online access to stepped-out problems aligned to Common Core

Get and view your assignments online.

Extra practice and review online

Virtual Nerd™ tutorials with built-in support

Chapter Preview

BIG ideas

1 **Probability**
Essential Question What is the difference between experimental probability and theoretical probability

2 **Data Representation**
Essential Question What is a freque table?

3 **Probability**
Essential Question What does it me for an event to be random?

Vocabulary

English/Spanish Vocabulary Audio Online:

English	Spanish
combination, *p. 838*	combinación
conditional probability, *p. 851*	probabilidad condicional
dependent events, *p. 844*	sucesos dependientes
experimental probability, *p. 825*	probabilidad experimental
independent events, *p. 844*	sucesos independientes
mutually exclusive events, *p. 845*	sucesos mutuamente excluyentes
permutation, *p. 837*	permutación
sample space, *p. 824*	espacio de muestral
theoretical probability, *p. 825*	probabilidad teórica

DOMAINS

- Conditional Probability and Rules of Probability
- Using Probability to Make Decisions

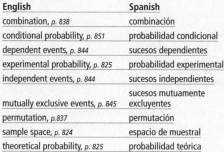

PowerGeometry.com

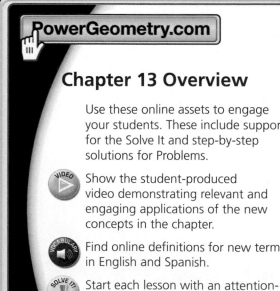

Chapter 13 Overview

Use these online assets to engage your students. These include support for the Solve It and step-by-step solutions for Problems.

Show the student-produced video demonstrating relevant and engaging applications of the new concepts in the chapter.

Find online definitions for new terms in English and Spanish.

Start each lesson with an attention-getting problem. View the Problem online with helpful hints.

Common Core Performance Task

Analyzing a Survey

Loretta's Ice Cream tested two new flavors, caramel and raspberry. After analyzing a random survey of 300 customers, Loretta's assistant presents the following information about the customers' responses to the new flavors.

- Seventy-two customers like only raspberry.
- Twice as many customers like only caramel as like both flavors.
- The number of customers that like both new flavors is equal to the number of customers that do not like either flavor.

Loretta asks her assistant to use the survey results to tell her which of the following is more likely.

- A customer who likes raspberry also likes caramel.
- A customer who likes caramel also likes raspberry.

Task Description

Determine whether it is more likely that a customer who likes raspberry also likes caramel, or that a customer who likes caramel also likes raspberry.

Connecting the Task to the Math Practices

MATHEMATICAL PRACTICES

As you complete the task, you'll apply these Standards for Mathematical Practice.

- You'll analyze the survey data and calculate experimental probabilities. (MP 1, MP 4)
- You'll use quantitative reasoning to calculate conditional probabilities. (MP 2)

Show Problems from each lesson solved step by step. Instant replay allows students to go at their own pace when studying online.

Assign homework to individual students or to an entire class.

Prepare students for the Mid-Chapter Quiz and Chapter Test with online practice and review.

Virtual Nerd™ Access Virtual Nerd student-centered math tutorials that directly relate to the content of the lesson.

Overview of the Performance Task

Students will analyze the results of a survey and use the results to find experimental probabilities. They will also calculate and compare conditional probabilities.

Students will work on the Performance Task in the following places in the chapter.

- Lesson 13-1 (p. 829)
- Lesson 13-6 (p. 861)
- Pull It All Together (p. 869)

Introducing the Performance Task

Tell students to read the problem on this page. Do not have them start work on the problem at this time, but ask them the following question.

Q What is a strategy you could try in order to solve the problem? **[Sample: I could make a table or Venn diagram to organize the information about the numbers of customers who liked only raspberry, only caramel, both flavors, or neither flavor.]**

PARCC CLAIMS

Sub-Claim B: Additional and Supporting Content with Connections to Practices
Sub-Claim D: Highlighted Practice MP 4 with Connections to Content

SBAC CLAIMS

Claim 3: Communicating Reasoning
Claim 4: Modeling and Data Analysis

CHAPTER 13

PROBABILITY
Math Background

PROFESSIONAL DEVELOPMENT

Understanding by Design principles were central to the development of the Big Ideas and the Essential Understandings. These will help your students build a structure on which to make connections to prior learning.

PROGRAM ORGANIZATION · BIG IDEA · ESSENTIAL UNDERSTANDING · PROGRAM ORGANIZATION

Probability

BIG idea Probability describes the likelihood that an event will occur. The probability of an event can range from 0 (impossible) to 1 (certain). Experimental probability is based on observations or trials of an experiment, while theoretical probability is based on what should happen mathematically. Combinations and permutations can be used to count the number of possible outcomes in a sample space.

ESSENTIAL UNDERSTANDINGS

13–1 Probability is a measure of the likelihood that an event will occur.

13–3 Counting techniques can be used to find all of the possible ways to complete different tasks or choose items from a list.

13–4 The probability of compound events can be found by using the probability of each part of the compound event.

Data Representation

BIG idea A frequency table is a data display that shows how often an item appears in a particular category. Frequency tables can be used to calculate the relative frequencies of each item. A two-way frequency table, or contingency table, displays the frequencies of data in two different categories. Contingency tables can be used to find conditional probabilities.

ESSENTIAL UNDERSTANDINGS

13–2 Tables can be used to organize data by frequency and find probabilities.

13–5 Two-way frequency tables can be used to organize data and identify sample spaces to approximate probabilities.

13–6 Tables, tree diagrams, and formulas can be used to find conditional probability.

Probability

BIG idea A random event has no bias or inclination toward any particular outcome. Random number tables and electronic random number generators can be used to model random events. In order to reach a fair decision, each possible choice must have the same probability of being selected. Expected value uses theoretical probability to tell you what you can expect in the long run, which can help you make more informed decisions.

ESSENTIAL UNDERSTANDING

13–7 Probability can be used to make fair decisions based on prior experience.

Probability

The probability of an event is a number from 0 to 1 that describes how likely it is that an event to occur. Probabilities closer to 1 are more likely to occur, and probabilities closer to 0 are less likely to occur.

- Experimental Probability:

$$P(\text{event}) = \frac{\text{number of times the event occurs}}{\text{number of times the experiment is done}}$$

- Theoretical Probability:

$$P(\text{event}) = \frac{\text{number of favorable outcomes}}{\text{number of possible outcomes}}$$

Counting Methods

Counting methods can be used to find the number of ways to choose objects from different sets. You can use counting methods to find the total number of outcomes in a sample space.

- **Fundamental Counting Principle** If event M can occur in m ways and event N can occur in n ways, then event M followed by event N can occur in $m \cdot n$ ways.
- A **permutation** is an arrangement of items in which the order of the objects is important.

$${}_nP_r = \frac{n!}{(n-r)!};\ 0 \leq r \leq n$$

- A **combination** is a selection of items in which order is not important.

$${}_nC_r = \frac{n!}{r!(n-r)!};\ 0 \leq r \leq n$$

Compound Probability

A **compound event** is made up of two or more events. The outcomes of **independent events** do not affect each other. If the outcome of an event affects the outcome of another event, they are **dependent events. Mutually exclusive events** cannot occur at the same time.

- For independent events A and B, $P(A \text{ and } B) = P(A) \cdot P(B)$.
- For any two events A and B, $P(A \text{ or } B) = P(A) + P(B) - P(A \text{ and } B)$.
- For mutually exclusive events A and B, $P(A \text{ or } B) = P(A) + P(B)$.

Mathematical Practices

Model with mathematics. Make sense of problems and persevere in solving them. Basic probability concepts are introduced and then analyzed in real-world contexts, giving students a quick overview of how the laws of probability predictably govern random behavior.

Using Data Displays

Frequency Tables

A **frequency table** is a data display that shows how often an item appears in a particular category.

- The **relative frequency** of an item is the ratio of the number of times the item occurs to the total number of items in the sample space.
- A probability distribution can be shown in a frequency table.

Satisfaction Level	Very Satisfied	Somewhat Satisfied	Somewhat Unsatisfied	Very Unsatisfied
Frequency	8	4	3	1
Probability	0.5	0.25	0.1875	0.0625

Two-Way Frequency Tables

A **two-way frequency table,** or contingency table, is a data display that shows the frequencies of data in two different categories.

	Plan to Attend College	Do not Plan to Attend College	Totals
Juniors	29	3	32
Seniors	33	5	38
Totals	62	8	70

Conditional Probability

The conditional probability of an event *A* given that event *B* has occurred is written $P(A \mid B)$.

- $P(A \mid B) = \frac{P(A \text{ and } B)}{P(B)}$, where $P(B) \neq 0$
- In $P(A \mid B)$, the sample space for *A* is limited to those situations that also involve *B* occurring.
- You can use contingency tables to help you find conditional probabilities.

Mathematical Practices

Reason abstractly and quantitatively. Students relate relative frequency to probability and use two-way frequency tables to find probability. These experiences lead to an understanding of conditional probability and the use of formulas to find conditional probability.

Random Events

A random event has no predetermined pattern or bias toward one outcome or the other. Two useful tools for modeling random processes are:

- tables of random numbers
- electronic random number generators

Simulations

Simulations can be used to model real world situations and make predictions based on probability. This is useful for situations that are difficult or impractical to actually perform. To conduct a simulation, follow these steps:

- Step 1: Identify all possible outcomes and determine the theoretical probability of each outcome.
- Step 2: Determine a probability model with probabilities that match those of the actual outcomes.
- Step 3: Determine what each trial of the simulation represents, how many trials to conduct, and how the results will be recorded.

Expected Value

Expected value uses theoretical probability to predict what should happen in a situation. If you have an idea of what *should* happen in the long run, you can make better decisions in problem situations.

- If *A* is an event that includes outcomes $A_1, A_2, A_3, \ldots$ and Value(A_n) is a quantitative value associated with each outcome, the expected value of *A* is given by: $\text{Value}(A) = P(A_1) \cdot \text{Value}(A_1) + P(A_2) \cdot \text{Value}(A_2) + \ldots$
- As the number of trials of an experiment increases, the results should get closer and closer to the expected value.

Common Errors When Modeling Random Events

Simulations Errors can occur when making a prediction based on too few trials of a simulation. In order to have an accurate model, many trials of the simulation should be conducted and an average value taken.

Expected Value Remind students that they must include all of the possible outcomes when calculating expected value. Expected value can only be calculated when the possible outcomes have a quantitative value.

Mathematical Practices

Construct viable arguments and critique the reasoning of others. Students develop simulations and justify the appropriateness of the simulation. Students also find expected value and assess decisions based on expected results.

CHAPTER 13

PROBABILITY

Pacing and Assignment Guide

Lesson	Teaching Day(s)	Basic (Traditional)	Average (Traditional)	Advanced (Traditional)	Block
13-1	1	Problems 1–3 Exs. 7–18, 20, 23–26	Problems 1–3 Exs. 7–15 odd, 17–31	Problems 1–3 Exs. 7–15 odd, 17–19, 28–33	**Day 1** Problems 1–3 Exs. 7–15 odd, 17–31
13-2	1	Problems 1–3 Exs. 8–21, 25–36	Problems 1–3 Exs. 9–15 odd, 16–23, 25–36	Problems 1–3 Exs. 9–15 odd, 16–36	Problems 1–3 Exs. 9–15 odd, 16–23, 25–36
13-3	1	Problems 1–3 Exs. 10–15, 22, 23, 31, 32	Problems 1–3 Exs. 11–15 odd, 31	Problems 1–3 Exs. 11–17 odd, 28–31	**Day 2** Problems 1–6 Exs. 11–27 odd, 28–36
	2	Problems 4–6 Exs. 16–21, 33–36	Problems 4–6 Exs. 17–27 odd, 28–30, 32–36	Problems 4–6 Exs. 19–27 odd, 33–36	
13-4	1	Problems 1–4 Exs. 6–20, 22, 24, 30–38	Problems 1–4 Exs. 7–21 odd, 23–28, 30–38	Problems 1–4 Exs. 7–21 odd, 23–29, 30–38	**Day 3** Problems 1–4 Exs. 7–21 odd, 23–28, 30–38
13-5	1	Problems 1–3 Exs. 6–13, 19, 20–22 even, 24–36	Problems 1–3 Exs. 7–13 odd, 14–22, 24–36	Problems 1–3 Exs. 7–13 odd, 14–36	**Day 4** Problems 1–3 Exs. 7–13 odd, 14–22, 24–36
13-6	1	Problems 1–4 Exs. 6–10, 12, 14, 16, 19–29	Problems 1–4 Exs. 7, 9–17, 19–29	Problems 1–4 Exs. 7, 9–29	**Day 5** Problems 1–4 Exs. 7, 9–17, 19–29
13-7	1	Problems 1–2 Exs. 6–9, 12, 14, 19–21	Problems 1–2 Exs. 7, 9, 12–14, 19–21	Problems 1–2 Exs. 7, 9, 12–14, 19–21	**Day 6** Problems 1–4 Exs. 7–17 odd, 18–21
	2	Problems 3–4 Exs. 10, 11, 15–17	Problems 3–4 Exs. 11, 15–18	Problems 3–4 Exs. 11, 15–18	
Review	1	Chapter 13 Review	Chapter 13 Review	Chapter 13 Review	**Day 7** Chapter 1 Review Chapter 1 Test
Assess	1	Chapter 1 Test	Chapter 1 Test	Chapter 1 Test	
Total		**11 Days**	**11 Days**	**11 Days**	**7 Days**

Note: Pacing does not include Concept Bytes and other feature pages.

Resources

KEY

I = Interactive asset at PowerGeometry.com
E = Editable master at PowerGeometry.com
P = Available in Print
M = Master at PowerGeometry.com
✓ = CD-ROM

	For the Chapter	13-1	13-2	13-3	13-4	13-5	13-6	13-7
Planning								
Teacher Center Online Planner & Grade Book	I	I	I	I	I	I	I	I
Interactive Learning & Guided Instruction								
My Math Video	I							
Solve It!		I M	I M	I M	I M	I M	I M	I M
Student Companion		P M	P M	P M	P M	P M	P M	P M
Vocabulary Support		I P M	I P M	I P M	I P M	I P M	I P M	I P M
Got It? Support		I P	I P	I P	I P	I P	I P	I P
Dynamic Activity	I							
Online Problems		I	I	I	I	I	I	I
Additional Problems		M	M	M	M	M	M	M
English Language Learner Support (TR)		E P M	E P M	E P M	E P M	E P M	E P M	E P M
Activities, Games, and Puzzles		E M	E M	E M	E M	E M	E M	E M
Teaching With TI Technology With CD-ROM								
TI-Nspire™ Support CD-ROM		✓	✓	✓	✓	✓	✓	✓
Lesson Check & Practice								
Student Companion		P M	P M	P M	P M	P M	P M	P M
Lesson Check Support		I P	I P	I P	I P	I P	I P	I P
Practice and Problem Solving Workbook		P	P	P	P	P	P	P
Think About a Plan (TR)		E P M	E P M	E P M	E P M	E P M	E P M	E P M
Practice Form G (TR)		E P M	E P M	E P M	E P M	E P M	E P M	E P M
Standardized Test Prep (TR)		P M	P M	P M	P M	P M	P M	P M
Practice *Form K* (TR)		E P M	E P M	E P M	E P M	E P M	E P M	E P M
Extra Practice	E M							
Find the Errors!	M							
Enrichment (TR)		E P M	E P M	E P M	E P M	E P M	E P M	E P M
Answers and Solutions CD-ROM	✓	✓	✓	✓	✓	✓	✓	✓
Assess & Remediate								
ExamView CD-ROM	✓	✓	✓	✓	✓	✓	✓	✓
Lesson Quiz		I M	I M	I M	I M	I M	I M	I M
Quizzes and Tests *Form G* (TR)	E P M					E P M		
Quizzes and Tests *Form K* (TR)	E P M					E P M		
Reteaching (TR)		E P M	E P M	E P M	E P M	E P M	E P M	E P M
Performance Tasks (TR)	P M							
Cumulative Review (TR)	P M							
Progress Monitoring Assessments	I P M							

(TR) Available in All-In-One Teaching Resources

1 Interactive Learning

Solve It!

PURPOSE To use different strategies to describe a theoretical probability.

PROCESS Students may recognize that they need to compare the number of hours between 12:00 and 2:00 with the total number of hours on the clock face.

FACILITATE

Q How does the minute hand help you determine the number of possible positions for the hour hand? **[You can use the placement of the minute hand to determine that the hour hand is about midway between a pair of numbers on the clock.]**

ANSWER See Solve It in Answers on next page.

CONNECT THE MATH In the Solve It, students examine possible outcomes to describe the probability of an event. In the lesson, students will compare the favorable outcomes of an event to the total number of possible outcomes to find the probability of the event.

2 Guided Instruction

Take Note — VISUAL LEARNERS

Use the number line to reinforce students' understanding of probability. The probability of an event can be any value between (and including) 0 and 1. Impossible events have a probability of 0, and certain events have a probability of 1.

13-1 Experimental and Theoretical Probability

Common Core State Standards

S-CP.A.1 Describe events as subsets of a sample space using characteristics of the outcomes, or as unions, intersections, or complements of other events. **Also S-CP.A.4**

MP 1, MP 2, MP 3, MP 4, MP 6

Objectives To calculate experimental and theoretical probability

Visualize where the hour hand could be.

Solve It! Getting Ready!

You find an old clock in your attic. The hour hand of the clock is broken off. Between 12:00 and 2:00, how many positions are possible for the hour hand? For a 12-hour period, how many positions are possible for the hour hand? What is the probability that the clock stopped some time between 12:00 and 2:00?

MATHEMATICAL PRACTICES

Lesson Vocabulary
- outcome
- event
- sample space
- probability
- experimental probability
- theoretical probability
- complement of an event

In the Solve It, you probably considered where the hour hand would be based on where the minute hand is. In the language of probability, this position would be a *favorable outcome*. An **outcome** is the possible result of a situation or experiment. An **event** may be a single outcome or a group of outcomes. For the clock, the hour hand being about halfway between 12 and 1 or about halfway between 1 and 2 are two favorable outcomes for the event of where the hour hand may stop between 12:00 and 2:00. The set of all possible outcomes is the **sample space**.

Essential Understanding Probability is a measure of the likelihood that an event will occur.

take note — Key Concept Probability

Definition

If the outcomes in a sample space are equally likely to occur, the **probability** of an event P(event) is a numerical value from 0 to 1 that measures the likelihood of an event.

$$P(\text{event}) = \frac{\text{number of favorable outcomes}}{\text{number of possible outcomes}}$$

You can write the probability of an event as a ratio, decimal, or percent.

impossible — equally likely to occur or not occur — certain

0 ← less likely — 0.5 — more likely → 1

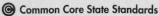

13-1 Preparing to Teach

BIG idea Probability

ESSENTIAL UNDERSTANDING
- Probability is a measure of the likelihood that an event will occur.

Math Background

Whether calculating experimental or theoretical probabilities, the process of comparing favorable outcomes to total possible outcomes or observations is the same.

- Experimental probability compares the number of favorable observations to the number of trials of an experiment. This describes the likelihood of an event based on observations or the results of an experiment.
- Theoretical probability compares the number of favorable outcomes to the number of outcomes in the sample space. This describes what *should* happen based on mathematical reasoning.

As the number of trials of an experiment increases, the experimental probability should approach the theoretical probability. This is known as the Law of Large Numbers.

Mathematical Practices

Attend to precision. Students will define and use theoretical and experimental probability and will find complements of events.

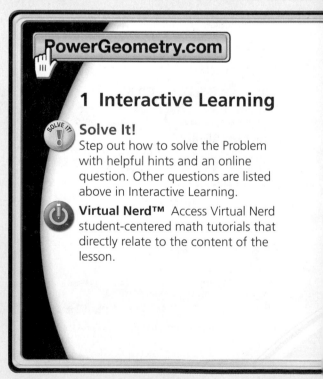

PowerGeometry.com

1 Interactive Learning

Solve It! Step out how to solve the Problem with helpful hints and an online question. Other questions are listed above in Interactive Learning.

Virtual Nerd™ Access Virtual Nerd student-centered math tutorials that directly relate to the content of the lesson.

You can find probabilities by using the results of an experiment or by reasoning mathematically.

take note

Key Concept Experimental Probability

Experimental probability of an event measures the likelihood that the event occurs based on the actual results of an experiment.

$$P(\text{event}) = \frac{\text{number of times the event occurs}}{\text{number of times the experiment is done}}$$

Problem 1 Calculating Experimental Probability

Quality Control **A quality control inspector samples 500 LCD monitors and finds defects in three of them.**

A What is the experimental probability that a monitor selected at random will have a defect?

$$P(\text{defect}) = \frac{\text{number of monitors with a defect}}{\text{number of monitors inspected}}$$

$$= \frac{3}{500}$$

$$= 0.006 \text{ or } 0.6\%$$

Think
You can report the result as a fraction, decimal, or percent. Use the form that will communicate most clearly.

The experimental probability that a monitor selected at random is defective is 0.6%.

B If the company manufactures 15,240 monitors in a month, how many are likely to have a defect based on the quality inspector's results?

number of defective monitors = $P(\text{defect})$ • total number of monitors

= 0.006 • 15,240

= 91.44

It is likely that approximately 91 monitors are defective.

Got It? 1. A park has 538 trees. You choose 40 at random and determine that 25 are maple trees. What is the experimental probability that a tree chosen at random is a maple tree? About how many trees in the park are likely to be maple trees?

When a sample space consists of real data, you can find the experimental probability. **Theoretical probability** describes the likelihood of an event based on mathematical reasoning.

This chapter uses *standard number cubes* to illustrate probability. A standard number cube has 6 faces with a number from 1 to 6 on each face. No number is used twice.

Take Note

Experimental probability is based on observations or the results of trials, so it likely will not match the theoretical probability exactly.

Problem 1

Q Given that the probability of a monitor being defective is 0.6%, if the manufacturer surveys 100 previous customers, how many of the customers will likely report that they bought a defective monitor? Explain. **[either 0 or 1; 100 × 0.006 = 0.6]**

Got It?

Q Can you determine the probability that a tree selected at random is not a maple tree? Explain. **[Yes, if 25 of the 40 randomly selected trees are maple then 15 trees in the sample are not maple.]**

2 Guided Instruction

Each Problem is worked out and supported online.

Problem 1
Calculating Experimental Probability

Problem 2
Calculating Theoretical Probability

Problem 3
Using Probabilities of Events and Their Complements

Support in Geometry Companion

- Vocabulary
- Key Concepts
- Got It?

Answers

Solve It!

There are 2 possible positions the hour hand could have stopped between 12:00 and 2:00 for the given position of the minute hand. There are 12 possible positions for a 12-hour period. The probability is $\frac{2}{12}$ or $\frac{1}{6}$.

Got It?

1. 0.625 or 62.5%; about 336

Problem 2

In this problem, students compare the number of favorable outcomes to the number of possible outcomes in a sample space to find theoretical probability.

Q How can you be certain that each possible outcome in the sample space is equally likely to occur? **[For fair number cubes, each number is equally likely to be occur.]**

Got It?

Q How many different ways can you roll a sum of 9 with two number cubes? a sum of 2? a sum of 13? **[4; 1; 0]**

Take Note

SYNTHESIZING

Show students how they can check the formula for finding the probability of a complementary event. By comparing the number of unfavorable outcomes to the total number of outcomes, they will arrive at the same solution.

Problem 3

In this problem, students will find the probability of the complement of an event.

Q What is the sum of P(green) and P(not green)? Explain. **[1; An event and its complement make up the entire sample space.]**

Got It?

Q What is another way to write P(not red)? **[P(green, blue, or white)]**

Q Show how finding P(green, blue, or white) results in the same answer. **[$\frac{(8 + 5 + 6)}{29} = \frac{19}{29}$]**

Think

What results should you be looking for?
Any two cubes that result in the sum of 7, like a 1 and a 6.

Problem 2 Calculating Theoretical Probability

What is the probability of rolling numbers that add to 7 when rolling two standard number cubes?

Step 1 Make a table of the possible results for the rolls of two number cubes. Circle the ones that sum to 7.

	1	2	3	4	5	6
1	1,1	2,1	3,1	4,1	5,1	6,1
2	1,2	2,2	3,2	4,2	5,2	6,2
3	1,3	2,3	3,3	4,3	5,3	6,3
4	1,4	2,4	3,4	4,4	5,4	6,4
5	1,5	2,5	3,5	4,5	5,5	6,5
6	1,6	2,6	3,6	4,6	5,6	6,6

Step 2 Find the number of possible outcomes for the event that the sum of two cubes is 7.

Step 3 Find the probability.

$P(\text{rolling a sum of } 7) = \frac{6}{36}$

The probability of rolling numbers that add to 7 is $\frac{6}{36}$, or $\frac{1}{6}$.

Got It? 2. What is the probability of getting each sum when rolling two standard number cubes?

a. 9 **b.** 2 **c.** 13

The **complement of an event** consists of all of the possible outcomes in the sample space that are not part of the event. For example, if you roll a standard number cube, the probability of rolling a number less than 3 is $P(\text{rolling} < 3) = \frac{2}{6}$, or $\frac{1}{3}$. The probability of *not* rolling a number less than 3 is $P(\text{not} < 3) = \frac{4}{6}$, or $\frac{2}{3}$.

take note

Key Concept Probability of a Complement

The sum of the probability of an event and the probability of its complement is 1.

$P(\text{event}) + P(\text{not event}) = 1$ $\qquad$ $P(\text{not event}) = 1 - P(\text{event})$

Think

Can you find P(not green) another way?
You can find the total number of marbles that are not green, and then divide by the total number of marbles: $\frac{(10 + 5 + 6)}{29}$.

Problem 3 Using Probabilities of Events and Their Complements

A jar contains 10 red marbles, 8 green marbles, 5 blue marbles, and 6 white marbles. What is the probability that a randomly selected marble is not green?

$P(\text{not green}) = 1 - P(\text{green})$ Probability of the complement

$= 1 - \frac{8}{29}$ Find P(green).

$= \frac{21}{29}$ Simplify.

The probability that the chosen marble is not green is $\frac{21}{29}$.

Got It? 3. What is the probability that a randomly chosen marble is not red?

Additional Problems

1. Of the students in Carlos' homeroom, 11 are studying Spanish, 6 are studying German, and 8 are studying French. If a student is selected at random, what is the probability that he or she is studying German?

ANSWER 0.24

2. What is the probability of rolling a sum of 6 with two standard number cubes?

ANSWER $\frac{5}{36}$

3. There are 43 freshmen, 28 sophomores, 35 juniors, and 26 seniors in the school auditorium. If a student is selected at random, what is the probability that the student is not a junior?

ANSWER $\frac{97}{132}$

Answers

Got It? (continued)

2a. $\frac{4}{36}$ or $\frac{1}{9}$ **b.** $\frac{1}{36}$ **c.** 0

3. $\frac{19}{29}$

Lesson Check

Do you know HOW?

Use the spinner to find each theoretical probability.

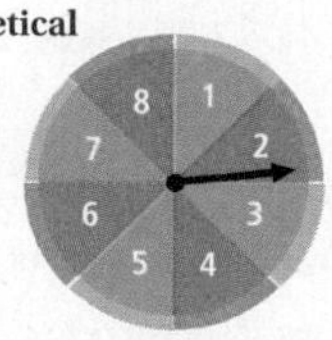

1. *P*(an even number)
2. *P*(a number greater than 5)
3. *P*(a prime number)

Do you UNDERSTAND? MATHEMATICAL PRACTICES

4. **Vocabulary** How are experimental and theoretical probability similar? How are they different?
5. **Open-Ended** Give an example of an impossible event.
6. **Error Analysis** Your friend says that the probability of rolling a number less than 7 on a standard number cube is 100. Explain your friend's error and find the correct probability.

Practice and Problem-Solving Exercises MATHEMATICAL PRACTICES

A Practice

See Problem 1.

7. A baseball player got a hit 19 times of his last 64 times at bat.
   a. What is the experimental probability that the player got a hit?
   b. If the player comes up to bat 200 times in a season, about how many hits is he likely to get?
8. A medical study tests a new cough medicine on 4250 people. It is effective for 3982 people. What is the experimental probability that the medicine is effective? For a group of 9000 people, predict the approximate number of people for whom the medicine will be effective.

A bag contains letter tiles that spell the name of the state MISSISSIPPI. Find the theoretical probability of drawing one tile at random for each of the following.

See Problems 2 and 3.

9. *P*(M)
10. *P*(I)
11. *P*(S)
12. *P*(P)
13. *P*(not M)
14. *P*(not I)
15. *P*(not S)
16. *P*(not P)

B Apply

17. **Think About a Plan** Suppose that you flip 3 coins. What is the theoretical probability of getting at least 2 heads?
    - What is the sample space of possible outcomes?
    - What are the favorable outcomes?
18. **Music** A music collection includes 10 rock CDs, 8 country CDs, 5 classical CDs, and 7 hip hop CDs.
    a. What is the probability that a CD randomly selected from the collection is a classical CD?
    b. What is the probability that a CD randomly selected from the collection is not a classical CD?

3 Lesson Check

Do you know HOW? ERROR INTERVENTION

- If students have trouble finding theoretical probabilities, then have them look at Problem 2.
- If students have difficulty solving Exercise 2, then ask them how many possible outcomes there are that result in a number greater than 5.

Do you UNDERSTAND?

- For Exercise 4, have students give examples of a situation that involves experimental probability and a situation that involves theoretical probability. Ask them to determine what distinguishes one situation from the other.

Close

Q For what kinds of situations would you choose to use experimental probability rather than theoretical probability? Explain. **[Situations based on observations or experiments or situations in which the entire sample space is unknown.]**

Q Give a possible probability for an event that is:
impossible [0]
very unlikely [sample: 0.1]
somewhat unlikely [sample: 0.3]
equally likely as not [0.5]
somewhat likely [sample: 0.7]
very likely [sample: 0.9]
certain [1]

Lesson Check

1. $\frac{4}{8}$ or $\frac{1}{2}$
2. $\frac{3}{8}$
3. $\frac{4}{8}$ or $\frac{1}{2}$
4. Answers may vary. Sample: The computation of each is similar, but experimental probability is based on the results of trials. Theoretical probability is based on what should happen mathematically.
5. Answers may vary. Sample: Rolling a 9 with a standard number cube.
6. Probabilities are expressed as numbers from 0 to 1 or from 0% to 100%. The probability is 1 or 100%.

Practice and Problem-Solving Exercises

7a. about 0.297 or 29.7%
b. about 59 times
8. about 0.937 or 93.7%; about 8432 people
9. $\frac{1}{11}$
10. $\frac{4}{11}$
11. $\frac{4}{11}$
12. $\frac{2}{11}$
13. $\frac{10}{11}$
14. $\frac{7}{11}$
15. $\frac{7}{11}$
16. $\frac{9}{11}$
17. $\frac{4}{8}$ or $\frac{1}{2}$
18a. $\frac{5}{30}$ or $\frac{1}{6}$
b. $\frac{25}{30}$ or $\frac{5}{6}$

4 Practice

ASSIGNMENT GUIDE
Basic: 7–18, 20, 23–26
Average: 7–15 odd, 17–31
Advanced: 7–15 odd, 17–19, 28–33

Mathematical Practices are supported by exercises with red headings. Here are the Practices supported in this lesson:

MP 1: Make Sense of Problems Ex. 17
MP 2: Reason Quantitatively Ex. 27–31
MP 3: Communicate Ex. 4–5
MP 3: Critique the Reasoning of Others Ex. 6

Applications exercises have blue headings. Exercises 18 and 21 support MP 4: Model.

STEM exercises focus on science or engineering applications.

EXERCISE 18: Use the Think About a Plan worksheet in the **Practice and Problem Solving Workbook** (also available in the Teaching Resources in print and online) to further support students' development in becoming independent learners.

HOMEWORK QUICK CHECK
To check students' understanding of key skills and concepts, go over Exercises 9, 11, 15, 17, and 18.

19. You are playing a board game with a standard number cube. It is your last turn and if you roll a number greater than 2, you will win the game. What is the probability that you will not win the game?

STEM **20. Weather** If there is a 70% chance of snow this weekend, what is the probability that it will not snow?

21. Quality Control From 15,000 graphing calculators produced by a manufacturer, an inspector selects a random sample of 450 calculators and finds 4 defective calculators. Estimate the total number of defective calculators out of the 15,000.

22. Suppose you choose a letter at random from the word shown below. What is the probability that you will not choose a B?

PROBABILITY

A student randomly selected 65 vehicles in the student parking lot and noted the color of each. She found that 9 were black, 10 were blue, 13 were brown, 7 were green, 12 were red, and 14 were a variety of other colors. What is each experimental probability?

23. *P*(red)
24. *P*(black)
25. *P*(not blue)
26. *P*(not green)

STEM **27. Genetics** Genetics was first studied by Gregor Mendel, who experimented with pea plants. He crossed pea plants having yellow, round seeds with pea plants having green, wrinkled seeds. The following are the probabilities for each type of new seed.

yellow, round: 56.25% yellow, wrinkled: 18.75%
green, round: 18.75% green, wrinkled: 6.25%

If 2014 seeds were produced, how many of each variety would you expect?

Reasoning For Exercises 28–31, describe each of the following situations using one of the following probabilities. Explain your answer.

I. 0 **II.** between 0 and 0.5 **III.** between 0.5 and 1 **IV.** 1

28. having school on Tuesday
29. two elephants in the city zoo having the same weight
30. getting your driver's license at the age of 10
31. turning on the TV while a commercial is playing

Challenge

32. The students in a math class took turns rolling a standard number cube. The results are shown in the table at the right.
a. What is the theoretical probability of rolling the number 1 with the number cube?
b. What was the experimental probability of rolling the number 1 for the experiment in class?

Number Cube Experiment

Outcome	1	2	3	4	5	6
Times Rolled	39	40	47	42	38	44

Answers

Practice and Problem-Solving Exercises (continued)

19. $\frac{2}{6}$ or $\frac{1}{3}$
20. 30%
21. about 133
22. $\frac{9}{11}$
23. $\frac{12}{65}$
24. $\frac{9}{65}$
25. $\frac{55}{65}$ or $\frac{11}{13}$
26. $\frac{58}{65}$
27. yellow, round: about 1133
yellow, wrinkled: about 378
green, round: about 378
green, wrinkled: about 126
28–31. Answers will vary. Check students' work.
32a. $\frac{1}{6}$
b. 0.156 or 15.6%

33. Another way to express probability is with *odds*. Odds compare the number of favorable outcomes to the number of unfavorable outcomes. Odds in favor of an event are usually written as

number of favorable outcomes : number of unfavorable outcomes.

Suppose the probability of drawing a red marble from a bag of marbles is $\frac{3}{10}$.

a. What are the odds in favor of drawing a red marble?

b. What are the odds against drawing a red marble?

Apply What You've Learned

MATHEMATICAL PRACTICES
MP 1, MP 4

Look back at the information on page 823 about the survey Lorreta's Ice Cream conducted. Copy and complete the table below.

Flavors Liked	Number of Responses
Only Raspberry	■
Only Caramel	■
Both	■
Neither	■

Select all of the following that are true. Explain your reasoning.

A. An equation that can be used to find the number of customers in the survey that like both flavors is $72 + 3x = 300$, where x is the number who like both.

B. The number of customers in the survey that like both flavors is 57.

C. The number of customers in the survey that like both flavors is 76.

D. The experimental probability that a customer who tries both new flavors will like only raspberry is $\frac{6}{25}$.

E. The experimental probability that a customer who tries both new flavors will like only raspberry is 0.72.

F. The experimental probability that a customer who tries both new flavors will like only caramel is 57%.

G. The experimental probability that a customer who tries both new flavors will like only caramel is 0.38.

33a. 3 : 7
b. 7 : 3

Apply What You've Learned

Here students work with the survey information given on page 823. They use the information to find the number of customers who gave each response shown in the table, and then find two experimental probabilities. Later in the chapter, they will find conditional probabilities.

Mathematical Practices

Students begin to **make sense of the problem** as they analyze the survey information to complete the table. (MP 1) Students **model with mathematics** as they calculate experimental probabilities. (MP 4)

ANSWERS

Choices B, D, and G are all true.

13-1 Lesson Resources

Additional Instructional Support

Geometry Companion

Students can use the **Geometry Companion** worktext (4 pages) as you teach the lesson. Use the Companion to support

- Solve It!
- New Vocabulary
- Key Concepts

13-1 Experimental and Theoretical Probability

Vocabulary

Review

Use the following sets to complete Exercises 1–3.

$A = \{1, 2, 3\}$ $B = \{1, 2\}$ $C = \{3, 4, 5\}$

For Execises 1–3, fill in the blanks to complete the statement.

1. What is the *union* of *sets A* and *C*?
   $A \cup C =$ {1, 2, 3, 4, 5}
2. What is the *intersection* of sets *A* and *B*?
   $A \cap B =$ {1, 2}
3. Which set is a *subset* of set *A*? Set *B*

For Exercises 4–6, rewrite each fraction as a percentage to one decimal place.

4. $\frac{9}{15}$ 60% 5. $\frac{18}{22}$ 81.8% 6. $\frac{118}{150}$ 78.7%

Vocabulary Builder

Probability (noun) prä-bə-bi-lə-tē

Related Words: outcome, event, sample space, experimental probability, theoretical probability

Definition: **Probability** is a measure of the likelihood that an event will occur.

Example: You can write the **probability** of an event as a ratio, decimal, or percent.

Use Your Vocabulary

The words *probability*, *probable*, and *probably* sound alike but have different meanings. In Exercises 7–9, underline the correct choice to complete each sentence.

7. The probability / probable / probably of rain today is 60%.
8. It will probability / probable / probably rain today.
9. It is probability / probable / probably that it will rain today.

Chapter 13 338

ELL Support

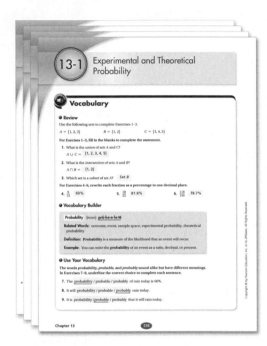

Use Real Objects Let students practice finding theoretical and experimental probabilities by using real objects. For example, divide students into groups and give each group a bowl of different colored candies or marbles. Have each group find the theoretical probability of selecting each color when a candy or marble is chosen at random. Then have them conduct an experiment of 50 trials (with replacement) to find the experimental probabilities of selecting each color at random. Discuss with students the similarities and differences between the experimental and theoretical probabilities. Combine the experimental results to show that the larger number of trials results in experimental probabilities that are closer to the theoretical probabilities.

5 Assess & Remediate

Lesson Quiz

Use the Lesson Quiz to assess students' mastery of the skills and concepts of this lesson.

1. **Do you UNDERSTAND?** In the first round of a golf tournament there were 2 eagles, 16 birdies, 39 pars, 16 bogies, and 8 double bogies on the 18th hole. If a player is selected at random, what is the probability that he had a birdie on the 18th hole during the first round?
2. What is the probability of tossing at least 1 tail if you toss 3 coins at once?
3. A bag has 42 red marbles, 15 blue marbles, 8 green marbles, and 3 orange marbles in it. If a marble from the bag is selected at random, what is the probability that the marble is not orange?

ANSWERS TO LESSON QUIZ

1. $\frac{16}{81}$
2. $\frac{7}{8}$
3. $\frac{65}{68}$

PRESCRIPTION FOR REMEDIATION

Use the student work on the Lesson Quiz to prescribe a differentiated review assignment.

Points	Differentiated Remediation
0–1	Intervention
2	On-level
3	Extension

PowerGeometry.com

5 Assess & Remediate

Assign the Lesson Quiz. Appropriate intervention, practice, or enrichment is automatically generated based on student performance.

Differentiated Remediation

Intervention

- **Reteaching** (2 pages) Provides reteaching and practice exercises for the key lesson concepts. Use with struggling students or absent students.
- **English Language Learner Support** Helps students develop and reinforce mathematical vocabulary and key concepts.

All-in-One Resources/Online

Reteaching

13-1 Reteaching
Experimental and Theoretical Probability

Probability is a measure of the likelihood that something will occur.

An *outcome* is one possible result of an experiment. The *sample space* is the set of all possible outcomes. For example, one possible outcome of flipping a coin is "heads". The sample space is all possible outcomes, or {heads, tails}.

An *event* is an outcome or a collection of outcomes.

The probability of tossing a coin and having "heads" show can be expressed mathematically with a value from 0 and 1:

$$P(\text{event}) = \frac{\text{number of favorable outcomes}}{\text{number of possible outcomes}}$$

$$P(\text{heads}) = \frac{1\text{ (heads)}}{2\text{ (heads or tails)}}$$

The probability of a coin landing on "heads" is $\frac{1}{2}$.

The situation used above is an example of *theoretical probability*. Theoretical probability is determined based on the number of possible outcomes. An experiment is not necessary in order to determine theoretical probability.

Problem

Suppose a card is randomly chosen from a deck of 52 cards. What is the theoretical probability of choosing a king?

$P(\text{event}) = \frac{\text{number of favorable outcomes}}{\text{number of possible outcomes}}$ Use probability notation.

$P(\text{king}) = \frac{4\text{ kings}}{52\text{ cards}}$ Substitute.

$P(\text{king}) = \frac{4}{52} = \frac{1}{13}$ Simplify.

The theoretical probability of randomly choosing a king from a deck of 52 cards is $\frac{1}{13}$.

All-in-One Resources/Online

English Language Learner Support

13-1 Additional Vocabulary Support
Experimental and Theoretical Probability

Complete the vocabulary chart by filling in the missing information.

Word or Word Phrase	Definition	Picture or Example
probability	1. Probability is a numerical value between 0 and 1 that measures the likelihood of an event.	Actuaries use *probability* to assess risk and make tables for insurance companies.
experimental probability	2. Experimental probability of an event is a measure of the likelihood that an event occurs based on the actual results of an experiment.	A team won 6 of their 10 games. The experimental probability of the team winning is $\frac{6}{10}$.
theoretical probability	Theoretical probability of an event is a measure of the likelihood that the event occurs based on mathematical reasoning.	3. Answers may vary. Sample: The theoretical probability of a penny landing on heads is $\frac{1}{2}$.
outcome	4. An outcome is the possible result of a situation or experiment.	One *outcome* of rolling a number cube is rolling a 5.
event	5. An event is an outcome or a group of outcomes.	A possible *event* of choosing two cards from a deck is a king of diamonds and a 2 of clubs.
sample space	The sample space is all possible outcomes.	6. The sample space for rolling one number cube is {1, 2, 3, 4, 5, 6}.
complement of an event	7. The complement of an event consists of all possible outcomes in the sample space that are not part of the event.	The complement of rolling an even number on a number cube is {1, 3, 5}.

Differentiated Remediation *continued*

On-Level

- **Practice** (2 pages) Provides extra practice for each lesson. For simpler practice exercises, use the Form K Practice pages found in the All-in-One Teaching Resources and online.
- **Think About a Plan** Helps students develop specific problem-solving skills and strategies by providing scaffolded guiding questions.
- **Standardized Test Prep** Focuses on all major exercises, all major question types, and helps students prepare for the high-stakes assessments.

Extension

- **Enrichment** Provides students with interesting problems and activities that extend the concepts of the lesson.
- **Activities, Games, and Puzzles** Worksheets that can be used for concepts development, enrichment, and for fun!

Practice and Problem Solving Wkbk/ All-in-One Resources/Online

Practice page 1

13-1 **Practice** *Form G*

Experimental and Theoretical Probability

You roll a standard number cube 10 times. The results are shown below.

6, 4, 6, 1, 5, 2, 4, 2, 4, 3

Find the experimental probability of each outcome.

1. P(rolling a 5) $\frac{1}{10}$
2. P(rolling a 6) $\frac{1}{5}$
3. P(rolling an even number) $\frac{7}{10}$
4. P(rolling a 1) $\frac{1}{10}$
5. What is the experimental probability of rolling an odd number on a standard number cube? For 50 rolls of the number cube, predict the number of rolls that will result in an odd number. 0.3; 15

Find the theoretical probability of each outcome.

6. P(rolling a 5) $\frac{1}{6}$
7. P(rolling a 6) $\frac{1}{6}$
8. P(rolling an even number) $\frac{1}{2}$
9. P(rolling a 1) $\frac{1}{6}$
10. P(rolling an odd number) $\frac{1}{2}$
11. P(rolling a multiple of 3) $\frac{1}{3}$

A bag contains 2 red ping-pong balls, 3 green ping-pong balls, 3 blue ping-pong balls, and 1 yellow ping-pong ball. Find the probability of randomly selecting each outcome.

12. P(not red) $\frac{7}{9}$
13. P(not green) $\frac{2}{3}$
14. P(not blue) $\frac{2}{3}$
15. P(not yellow) $\frac{8}{9}$

Practice and Problem Solving Wkbk/ All-in-One Resources/Online

Practice page 2

13-1 **Practice** (continued) *Form G*

Experimental and Theoretical Probability

16. A game is played where students throw beanbags at the target shown to the right. Each region of the target is the same size and every beanbag hits the target. For one game, section A was hit 6 times, section B 3 times, section C 8 times, and section D 5 times.

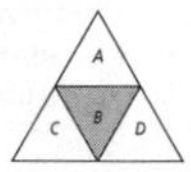

   a. What is the experimental probability of hitting section D? $\frac{5}{22}$

   b. What is the theoretical probability of hitting section D? $\frac{1}{4}$

17. **Reasoning** How are the probability of an event and the probability of its complement related mathematically? Their sum is equal to 1.

Two standard number cubes are rolled. Find each probability.

18. P(a sum equal to 2) $\frac{1}{36}$
19. P(sum not equal to 2) $\frac{35}{36}$
20. P(a product equal to 15) $\frac{1}{18}$
21. P(a sum greater than 6) $\frac{7}{12}$
22. P(a product less than or equal to 2) $\frac{1}{9}$
23. P(a sum equal to 12) $\frac{1}{36}$
24. **Open-Ended** Is it possible for an event to have a probability of 1? Explain your answer. Yes; Answers may vary. Sample: The probability of rolling a number cube and rolling a number between 1 and 6, inclusive, is 1.
25. **Error Analysis** Out of 20 coin flips, your classmate gets heads 14 times. She determines that the experimental probability of getting heads is $\frac{1}{2}$. What error did your classmate make? What is the correct value for experimental probability? Explain. Your classmate confused theoretical probability with experimental probability. The experimental probability of getting heads is the number of number of heads divided by the total number of flips, or $\frac{14}{20} = \frac{7}{10}$.

All-in-One Resources/Online

Enrichment

13-1 **Enrichment**

Experimental and Theoretical Probability

Geometric probability uses area ratios to determine probability.

The target to the right has bands that are 2 ft wide, surrounding a central circle with diameter 2 ft. The probability of an object, such as an arrow, hitting the target and landing in the gray band is the ratio of the area of the gray band to the area of the entire target.

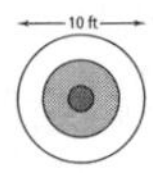

1. How can you calculate area of the gray band? Subtract the area of the central circle from the area of the gray circle.
2. What is the area of the gray circle? 28.27 ft^2
3. What is the area of the central circle? 3.14 ft^2
4. What equation can you use to determine the area of the gray band? $28.27 - 3.14 = 25.13\ ft^2$
5. What is the area of the entire target? 78.54 ft^2

   The probability of hitting the target and landing in the gray band is the ratio of the area of the gray band to the area of the entire target.

6. What is the probability of hitting the target and landing in the gray band? $\frac{25.13}{78.54}$ OR 0.32 OR 32%
7. What must happen for this probability to be reliable? the arrow has to hit the target

Practice and Problem Solving Wkbk/ All-in-One Resources/Online

Think About a Plan

13-1 **Think About a Plan**

Experimental and Theoretical Probability

Music A music collection includes 10 rock CDs, 8 country CDs, 5 classical CDs, and 7 hip hop CDs.

a. What is the probability that a CD randomly selected from the collection is a classical CD?

b. What is the probability that a CD randomly selected from the collection is not a classical CD?

Understanding the Problem

1. In what form is probability usually expressed? as a fraction, decimal, or percent
2. What operation must you use to determine the number of outcomes for the [illegible]

Planning the Solution

3. How can you express a probability as a fraction?

   The numerator is the number of favorable outcomes of the desired event.

   The denominator is the number of possible outcomes.

4. How can you find the number of possible outcomes in this scenario?

   Find the total number of CDs in the collection (all types).

5. Describe how to find the number of favorable outcomes.

   Count the number of the desired type of CD.

Getting an Answer

6. Find the number of possible outcomes. Show your work. $10 + 8 + 5 + 7 = 30$ possible outcomes
7. How many CDs in the collection are classical? 5
8. What is the probability that a CD randomly selected from the collection is a classical CD? Write your answer as a fraction in simplest form and as a decimal. $\frac{5}{30} = \frac{1}{6}$ or about 0.167
9. Use the probability that a CD randomly selected from the collection is a classical CD to find the probability that a CD randomly selected from the collection is *not* a classical CD. $1 - \frac{1}{6} = \frac{5}{6}$ or about 0.833

Practice and Problem Solving Wkbk/ All-in-One Resources/Online

Standardized Test Prep

13-1 **Standardized Test Prep**

Experimental and Theoretical Probability

Multiple Choice

For Exercises 1–4, choose the correct letter.

1. What is the probability of rolling a number greater than 2 on a number cube? D

   (A) $\frac{1}{6}$ (B) $\frac{1}{3}$ (C) $\frac{1}{2}$ (D) $\frac{2}{3}$

2. A coin is tossed 30 times and lands on heads 17 times. What is the experimental probability of the coin landing on *tails*? H

   (F) $\frac{17}{30}$ (G) $\frac{1}{2}$ (H) $\frac{13}{30}$ (I) $\frac{8}{15}$

3. What is the theoretical probability of randomly choosing a science book from a shelf that holds 3 mystery books, 5 science books, and 4 nature books? C

   (A) $\frac{1}{4}$ (B) $\frac{1}{3}$ (C) $\frac{5}{12}$ (D) $\frac{7}{12}$

4. What is the complement of rolling a 2 or 5 on a number cube? G

   (F) {5, 2} (G) {1, 3, 4, 6} (H) {1, 3, 4, 5} (I) {7}

Short Response

5. A spinner has four equal sections labeled 2, 4, 6 and 8. Suppose you spin the spinner twice. What is the theoretical probability that the sum of the outcomes is 10? Show the favorable outcomes.

   [2] $\frac{1}{4}$ (OR equivalent fraction); (2, 8), (4, 6), (6, 4), (8, 2)

   [1] Student provides a correct probability based on a partial list of favorable outcomes.

   [0] Student's response shows no effort OR does not correctly use sample space to determine probability.

Online Teacher Resource Center

Activities, Games, and Puzzles

13-1 **Game: A Game of Chance**

Theoretical and Experimental Probability

What You Need

2 players
2 six-sided numbered cubes

How to Play

Choose one player to be the roller and one to be the caller. The roller tosses the number cubes for each turn. The caller says a number (prior to the toss) when the number cubes are tossed, attempting to predict the sum of the roll. If the caller says the correct sum, he or she wins; otherwise the roller wins.

Before You Begin

Work together to determine which number should be called so that the caller has the best chance of winning. Then determine which number should be called to give the roller the best chance of winning. Show your work in the area provided below.

The best number for the caller is 7 because it has a 1 in 6 chance of being rolled. The best number(s) for the roller is either 2 or 12 because each has a 1 in 36 chance of being rolled.

Playing the Game

Now play the game 24 times. The first 12 times, the caller should call the number determined to give the caller the best chance of winning. The second 12 times, the caller should call the number determined to give the roller the best chance of winning. Record your results in the table below.

	Number of Times Caller Wins	Number of Times Roller Wins
Games Where Caller Has Best Chance		
Games Where Roller Has best Chance		

1 Interactive Learning

Solve It!

PURPOSE To calculate probability using data from a frequency table.

PROCESS Students may recognize the situation as representing experimental probability and use the methods that they learned in lesson 13-1.

FACILITATE

Q How many cars in all were observed passing the mile marker? **[50]**

Q How many of the cars were traveling faster than 65 mph? **[13]**

ANSWER See Solve It in Answers on next page.

CONNECT THE MATH In the Solve It, students use data that is presented in a frequency table to calculate an experimental probability. In this lesson, they will learn that this is also known as a relative frequency of a probability distribution.

2 Guided Instruction

Take Note

EXTENSION

Point out to students that the relative frequencies of different categories in a frequency table are the same as the experimental probabilities of the categories.

Problem 1

In this problem, students find a relative frequency using a frequency table.

Q What are the possible numerical values of a relative frequency? **[between 0 and 1]**

Q What is the sum of the relative frequencies in a frequency table? **[1]**

13-2 Probability Distributions and Frequency Tables

Common Core State Standards

S-CP.A.4 Construct and interpret two-way frequency tables of data when two categories are associated with each object being classified . . . **Also S-CP.A.5**

MP 1, MP 2, MP 3

Objective To make and use frequency tables and probability distributions

Think about how you can use data in tables to find probabilities.

MATHEMATICAL PRACTICES

Getting Ready!

The table at the right shows the speeds of cars as they pass a certain mile marker on highway 66. The speed limit is 65 mph. What is the total number of cars that passed the marker? What is the probability that a car stopped at random will be traveling faster than the speed limit?

Speed (mph)	Number of Cars
< 55	2
55–60	12
60–65	23
> 65	13

In the Solve It, you used information from the table to calculate probability. A **frequency table** is a data display that shows how often an item appears in a category.

Essential Understanding You can use data organized in tables that show frequencies to find probabilities.

Lesson Vocabulary
- frequency table
- relative frequency
- probability distribution

take note

Key Concept Relative Frequency

Relative frequency is the ratio of the frequency of the category to the total frequency.

Problem 1 Finding Relative Frequencies

Surveys **The results of a survey of students' music preferences are organized in this frequency table. What is the relative frequency of preference for rock music?**

Type of Music Preferred	Frequency
Rock	10
Hip Hop	7
Country	8
Classical	5
Alternative	6
Other	4

Plan

How do you find the denominator for a relative frequency? Find the sum of the frequencies in the frequency table.

Use the frequency table to find the number of times rock music is chosen as the preference, and the total number of survey results.

$$\text{relative frequency} = \frac{\text{frequency of rock music preference}}{\text{total frequency}}$$

$$= \frac{10}{10 + 7 + 8 + 5 + 6 + 4} = \frac{10}{40} = \frac{1}{4}$$

The relative frequency of preference for rock music is $\frac{1}{4}$.

13-2 Preparing to Teach

BIG idea **Data Representation**

ESSENTIAL UNDERSTANDING
- Tables can be used to organize data by frequency and find probabilities.

Math Background

Students should recognize that the processes of calculating relative frequencies and experimental probabilities are the same. Explain to students that a frequency table is one way of showing a probability distribution. Students can also use the relative frequencies from a table to create a histogram and see how the probabilities are distributed.

In a probability distribution,
- the value of each relative frequency is greater than or equal to 0 and less than or equal to 1, and
- the sum of the relative frequencies is 1.

Mathematical Practices

Reason abstractly and quantitatively. Students will use frequency tables to find relative frequency and connect relative frequency to probability, moving between data to analysis of data by using tools of probability.

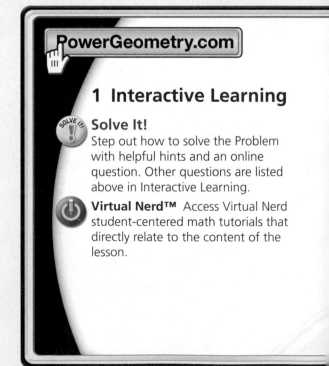

PowerGeometry.com

1 Interactive Learning

Solve It! Step out how to solve the Problem with helpful hints and an online question. Other questions are listed above in Interactive Learning.

Virtual Nerd™ Access Virtual Nerd student-centered math tutorials that directly relate to the content of the lesson.

Got It? 1. What is the relative frequency of preference for each type of music?

a. classical

b. hip hop

c. country

d. **Critical Thinking** Without calculating the rest of the relative frequencies, what is the sum of the relative frequencies for all the types of music? Explain how you know.

You can use relative frequency to approximate the probabilities of events.

Problem 2 Calculating Probability by Using Relative Frequencies

A student conducts a probability experiment by tossing 3 coins one after the other. Using the results below, what is the probability that exactly two heads will occur in the next three tosses?

Coin Toss Result	HHH	HHT	HTT	HTH	THH	THT	TTT	TTH
Frequency	5	7	9	6	2	9	10	2

Step 1 **Find the number of times a trial results in exactly two heads.**

The possible results that show exactly two heads are HHT, HTH, and THH.

The frequency of these results is $7 + 6 + 2 = 15$.

Step 2 **Find the total of all the frequencies.**

$5 + 7 + 9 + 6 + 2 + 9 + 10 + 2 = 50$

Step 3 **Find the relative frequency of a trial with exactly two heads.**

$$\text{relative frequency} = \frac{\text{frequency of exactly two heads}}{\text{total of the frequencies}} = \frac{15}{50} = \frac{3}{10}$$

Based on the data collected, the probability that the next toss will be exactly two heads is $\frac{3}{10}$.

Think

How can you use the frequency table to find the probability?
The relative frequency is an approximation of the overall probability of the result.

Got It? 2. A student conducts a probability experiment by spinning the spinner shown. Using the results in the frequency table, what is the probability of the spinner pointing at 4 on the next spin?

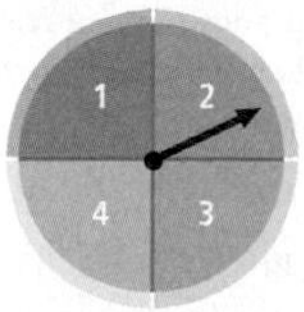

Spinner Result	1	2	3	4
Frequency	29	32	21	18

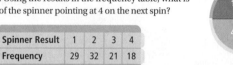

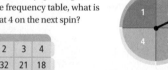

A **probability distribution** shows the probability of each possible outcome. A probability distribution can be shown in a frequency table.

Got It?

Q Once you calculate one relative frequency from a frequency table, how can you use this result to help you find the other relative frequencies? **[You know the total frequency, which will be the same for each relative frequency.]**

Problem 2 SYNTHESIZING

In this problem, the connection between relative frequencies and experimental probabilities is explicitly made.

Q What is the theoretical probability that exactly two heads will show when you toss three coins? Is this the same as the result of Problem 2? Explain. **[$\frac{3}{8}$; No; this is different from the results of the experiment.]**

Got It?

Q How would you find the probability of each other possible outcome? **[divide the frequency of each spinner result by 100]**

2 Guided Instruction

Each Problem is worked out and supported online.

Problem 1
Finding Relative Frequencies

Problem 2
Calculating Probability by Using Relative Frequencies

Problem 3
Finding a Probability Distribution

Support in Geometry Companion

- Vocabulary
- Key Concepts
- Got It?

Answers

Solve It!

$\frac{13}{50}$; find the total number of cars that were observed passing the mile marker, 50. Then divide the number of cars going faster than 65 mph by this number.

Got It?

1a. $\frac{1}{8}$

b. $\frac{7}{40}$

c. $\frac{1}{5}$

d. $\frac{40}{40}$ or 1; the sum of the frequencies includes all of the events in the sample space.

2. $\frac{18}{100}$, or $\frac{9}{50}$

Problem 3

In this problem, students create a frequency table and find a probability distribution.

Q What are the possible outcomes described in the problem? **[the number of bull's eyes hit by each archer]**

Q What is the total frequency? What does this number represent? **[50; the total number of bull's eyes hit by all of the archers]**

Got It?

Q What do the probabilities in the distribution represent? **[The probabilities represent the likelihood that a randomly selected test score fall in a certain range.]**

3 Lesson Check

Do you know HOW? ERROR INTERVENTION

- If students have trouble finding the relative frequencies in Exercises 1–4, then ask them how they would find the experimental probabilities of each music format.

Do you UNDERSTAND?

- If students do not know how to describe the sample space, then ask them how many different outcomes are possible when you flip a coin.
- For Exercise 7, ask students how many people have a chance to win the sweepstakes and how many winners are possible.

Close

Q How do you calculate the relative frequencies of data given in a frequency table? **[divide the frequency of each category by the total number of items]**

Problem 3 **Finding a Probability Distribution**

Archery **In a recent competition, 50 archers shot 6 arrows each at a target. Three archers hit no bull's eyes; 5 hit one bull's eye; 7 hit two bull's eyes; 7 hit three bull's eyes; 11 hit four bull's eyes; 10 hit five bull's eyes; and 7 hit six bull's eyes. What is the probability distribution for the number of bull's eyes each archer hit?**

Know	Need	Plan
The possible outcomes and the frequency of each outcome.	The probabilities of each outcome.	Make a frequency table and use relative frequencies to complete the probability distribution.

First, create a frequency table showing all of the possible outcomes: 0, 1, 2, 3, 4, 5, or 6 bull's eyes and the frequency for each.

Next, use the table to find the relative frequencies for each number of bull's eyes. The relative frequencies are the probability distribution.

Probability Distribution of Bull's Eyes Hits

Number of Bull's Eyes Hit	0	1	2	3	4	5	6
Frequency	3	5	7	7	11	10	7
Probability	$\frac{3}{50}$	$\frac{5}{50}$	$\frac{7}{50}$	$\frac{7}{50}$	$\frac{11}{50}$	$\frac{10}{50}$	$\frac{7}{50}$

Got It? **3.** On a math test, there were 10 scores between 90 and 100, 12 scores between 80 and 89, 15 scores between 70 and 79, 8 scores between 60 and 69, and 2 scores below 60. What is the probability distribution for the test scores?

Lesson Check

Do you know HOW?

Consumer Research **The results of a survey show the frequency of responses for preferred music formats. What are the relative frequencies of the following formats?**

1. CD **2.** MP3

3. Blu-ray **4.** Radio

5. What is the probability distribution for the music formats?

Preferred Format	Frequency
CD	54
Radio	50
Blu-ray	10
MP3	28

Do you UNDERSTAND? MATHEMATICAL PRACTICES

6. Describe the sample space of possible outcomes for tossing a coin once.

7. Error Analysis A friend says that her sweepstakes ticket will either win or not win, so the theoretical probability of it winning must be 50%. Explain your friend's error.

Additional Problems

1. The table shows the number of athletes from different sports at a rewards ceremony. What is the relative frequency of lacrosse players at the ceremony?

Sport	Frequency
Soccer	29
Cross Country	18
Football	35
Lacrosse	22

ANSWER $\frac{22}{104}$ or $\frac{11}{52}$

2. In Additional Problem 1, what is the probability that a randomly selected athlete is a cross country runner?

ANSWER $\frac{18}{104}$ or $\frac{9}{52}$

3. Terrance has had 33 hits, 12 walks, 21 strike outs, 22 ground outs, and 27 fly outs this season. What is the probability distribution for his batting performance this season?

ANSWER

Batting Results

Result	Hit	Walk	K	GO	FO
Frequency	33	12	21	22	27
Probability	$\frac{33}{115}$	$\frac{12}{115}$	$\frac{21}{115}$	$\frac{22}{115}$	$\frac{27}{115}$

Answers

Got It? (continued)

3.

Math Test Scores

Score	90–100	80–89	70–79	60–69	0–59
Frequency	10	12	15	8	2
Probability	$\frac{10}{47}$	$\frac{12}{47}$	$\frac{15}{47}$	$\frac{8}{47}$	$\frac{2}{47}$

Lesson Check

1. $\frac{54}{142}$ or $\frac{37}{71}$

2. $\frac{28}{142}$ or $\frac{14}{71}$

3. $\frac{10}{142}$ or $\frac{5}{71}$

4. $\frac{50}{142}$ or $\frac{25}{71}$

5.

Preferred Music Format

Result	CD	Radio	Blu-ray	MP3
Frequency	54	50	10	28
Probability	$\frac{37}{71}$	$\frac{25}{71}$	$\frac{5}{71}$	$\frac{14}{71}$

Practice and Problem-Solving Exercises

MATHEMATICAL PRACTICES

A Practice

Blood Drive **The honor society at a local high school sponsors a blood drive. High school juniors and seniors who weigh over 110 pounds may donate. The table at the right indicates the frequency of each donor blood type.**

See Problem 1.

Blood Drive Results

Blood Type	Frequency
O	30
A	25
B	6
AB	2

8. What is the relative frequency of blood type AB?

9. What is the relative frequency of blood type A?

10. Which blood type has the highest relative frequency? What is the relative frequency for this blood type?

11. The blood drive is extended for a second day, and the frequency doubles for each blood type. Does the relative frequency for each blood type change? Explain.

See Problems 2 and 3.

12. The data collected for new students enrolled at a community college show that 26 are under the age of eighteen, 395 are between the ages of eighteen and twenty-two, 253 are between the ages of twenty-three and twenty-seven, 139 are between the ages of twenty-eight and thirty-two, and 187 are over the age of thirty-two. What is the probability distribution for these data?

Twenty-three preschoolers were asked what their favorite snacks are. The results are shown in the bar graph at the right.

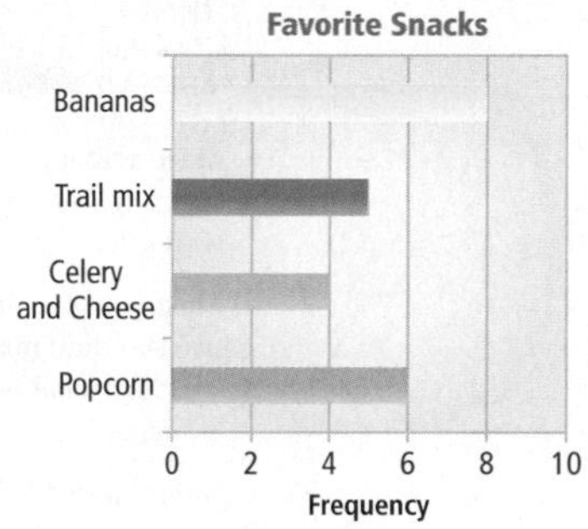

13. What is the probability that a preschooler chosen at random chose popcorn as his or her favorite snack?

14. What is the probability that a preschooler chosen at random did not choose bananas as his or her favorite snack?

15. What is the probability distribution for the data in the graph?

B Apply

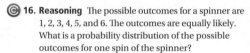

16. Reasoning The possible outcomes for a spinner are 1, 2, 3, 4, 5, and 6. The outcomes are equally likely. What is a probability distribution of the possible outcomes for one spin of the spinner?

17. Text Messages The table at the right shows numbers of text messages sent in one month by students at Metro High School.

a. If a student is chosen at random, what is the probability that the student sends 1500 or fewer text messages in one month?

b. If a student is chosen at random, what is the probability that the student sends more than 1500 messages in a month?

Number of Text Messages t	Number of Students
$t \leq 500$	25
$500 < t \leq 1500$	120
$1500 < t \leq 2500$	300
$t > 2500$	538

4 Practice

ASSIGNMENT GUIDE

Basic: 8–21
Average: 9–15 odd, 16–23
Advanced: 9–15 odd, 16–24
Standardized Test Prep: 25–28
Mixed Review: 29–36

Mathematical Practices are supported by exercises with red headings. Here are the Practices supported in this lesson:

MP 1: Make Sense of Problems Ex. 19
MP 2: Reason Abstractly Ex. 16
MP 3: Critique the Reasoning of Others Ex. 7, 23

Applications exercises have blue headings.

EXERCISE 18: Use the Think About a Plan worksheet in the **Practice and Problem Solving Workbook** (also available in the Teaching Resources in print and online) to further support students' development in becoming independent learners.

HOMEWORK QUICK CHECK

To check students' understanding of key skills and concepts, go over Exercises 9, 12, 19, 22, and 23.

6. The possible outcomes are heads or tails.

7. There are many ways she will not win but only one way she will win, so the probability of winning is much smaller.

Practice and Problem-Solving Exercises

8. $\frac{2}{63}$

9. $\frac{25}{63}$

10. O; $\frac{30}{63}$ or $\frac{10}{21}$

11. The relative frequencies do not change because both the numerator and denominator are multiplied by 2.

12.

Community College Enrollment

Age	Under 18	18–22	23–27	28–32	Over 32
Frequency	26	395	253	139	187
Probability	$\frac{26}{1000}$	$\frac{395}{1000}$	$\frac{253}{1000}$	$\frac{139}{1000}$	$\frac{187}{1000}$

13. $\frac{6}{23}$

14. $\frac{15}{23}$

15.

Favorite Snacks

Snack	Bananas	Trail Mix	C&C	Popcorn
Frequency	8	5	4	6
Probability	$\frac{8}{23}$	$\frac{5}{23}$	$\frac{4}{23}$	$\frac{6}{23}$

16.

Spinner Results

Result	1	2	3	4	5	6
Frequency	1	1	1	1	1	1
Probability	$\frac{1}{6}$	$\frac{1}{6}$	$\frac{1}{6}$	$\frac{1}{6}$	$\frac{1}{6}$	$\frac{1}{6}$

17a. $\frac{145}{983}$

b. $\frac{838}{983}$

Answers

Practice and Problem-Solving Exercises (continued)

18. a.

Computer Survey

Number of Computers	0	1	2	3	More than 3
Frequency	12	29	31	6	2
Probability	$\frac{12}{80}$	$\frac{29}{80}$	$\frac{31}{80}$	$\frac{6}{80}$	$\frac{2}{80}$

b. $\approx$ 1140

c. $\approx$ 7030

19. Answers may vary. Sample answer:

Coin Toss

Result	4 Heads	3 Heads 1 Tail	2 Heads 1 Tail	1 Head 3 Tails	4 Tails
Frequency	1	4	6	4	1
Probability	$\frac{1}{16}$	$\frac{1}{4}$	$\frac{3}{8}$	$\frac{1}{4}$	$\frac{1}{16}$

20–22. See below.

18. Computers The results of a survey of 80 households in Westville are shown at the right.

a. What is the probability distribution of the number of computers in Westville households?

b. If Westville has 15,200 households, predict the number of households that will have exactly 3 computers.

c. How many households will have either two or three computers?

Computer Survey

Number of Computers	Frequency
0	12
1	29
2	31
3	6
More than 3	2

19. Think About a Plan You can make a probability distribution table for theoretical probabilities. What is a probability distribution for four tosses of a single coin?

- What are all of the possible outcomes?
- What is the probability of each outcome?
- How can you display this in a table?

20. Employment The table below shows the number of people (in thousands) working in each occupational category, according to the U.S. Bureau of Labor Statistics.

U.S. Occupational Categories of Employed Workers

Occupational Category	Management, professional, and related	Service	Sales and office	Natural resources, construction, maintenance	Production, transportation, and material moving
Number (thousands)	9773	2271	1456	289	177

a. Make a table showing the relative frequency for each occupational category.

b. If an employed person is randomly selected, what is the likelihood the person works in the Service category? Explain how you know.

21. a. Make a probability distribution for the sum of the faces after rolling two standard number cubes.

b. Are the probabilities theoretical or based on experimental results? Explain.

22. A student chose 100 letters at random from a page of a textbook. Partial results are in the table at the right. Make a probability distribution for the data. Include a category for all other letters not represented by the table.

Letter	Tally
a	𝍸 III
e	𝍸 𝍸 IIII
i	𝍸 II
n	𝍸 I
o	𝍸 III
r	IIII
s	𝍸 I
t	𝍸 IIII

23. Error Analysis The results of a survey about students' favorite type of pet are shown in the probability distribution below. What is wrong with this probability distribution? What information do you need to correct the probability distribution?

Favorite Pet	Dogs	Cats	Birds	Fish	Hamsters
Probability	0.36	0.34	0.15	0.08	0.12

20. a.

U.S. Occupational Categories of Employed Workers

Occupational Category	Management, professional, and related	Service	Sales and office	Natural resources, construction, maintenance	Production, transportation, and material moving
Number (thousands)	9773	2271	1456	289	177
Relative Frequency	$\frac{9773}{13,966}$	$\frac{2271}{13,966}$	$\frac{1456}{13,966}$	$\frac{289}{13,966}$	$\frac{177}{13,966}$

b. $\approx$ 16.3% because this is the relative frequency

21. a.

Sum	2	3	4	5	6	7	8	9	10	11	12
Frequency	1	1	2	4	5	6	5	4	3	2	1
Probability	$\frac{1}{36}$	$\frac{1}{18}$	$\frac{1}{12}$	$\frac{1}{9}$	$\frac{5}{36}$	$\frac{1}{6}$	$\frac{5}{36}$	$\frac{1}{9}$	$\frac{1}{12}$	$\frac{1}{18}$	$\frac{1}{36}$

b. Theoretical the results are based on each face of one number cube having a probability of $\frac{1}{6}$.

22.

Letter	a	e	i	n	o	r	s	t	other	Total
Frequency	8	14	7	6	8	4	6	9	38	100
Probability	0.08	0.14	0.07	0.06	0.08	0.04	0.06	0.09	0.38	1

Challenge 24. A spinner has an equal number of even and odd numbers. The probability of getting an outcome of an even number on any spin is 50%. Make a probability distribution for even numbers for two spins, for three spins, and for four spins. How does the probability distribution change for each additional spin?

Standardized Test Prep

SAT/ACT

25. One hundred fifty students were asked the number of hours they spend on homework each night. Twenty students responded that they spend less than 1 hour on homework each night. What is the relative frequency of this response?

Ⓐ 0.13 Ⓑ 0.87 Ⓒ 20 Ⓓ 130

26. Which pair of angles are supplementary?

Ⓐ 110° and 70° Ⓑ 25° and 65°
Ⓒ 90° and 80° Ⓓ 160° and 200°

27. What is the maximum number of right angles possible in a pentagon?

Ⓐ 1 Ⓑ 2 Ⓒ 3 Ⓓ 4

Short Response

28. What is the measure of an interior angle of a regular octagon?

Mixed Review

Determine whether each of the following is an experimental or theoretical probability. Explain your choice. See Lesson 13-1.

29. the probability of rolling two 5's when rolling two standard number cubes
30. the probability that a poll respondent chooses red as their favorite color
31. the probability of a number being even when randomly chosen from the whole numbers less than 11

The pairs of figures are similar. Find *x*. See Lesson 7-2.

32.
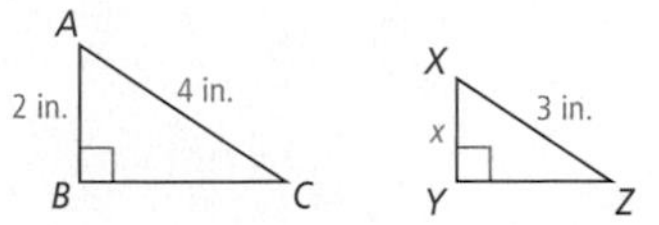

33.
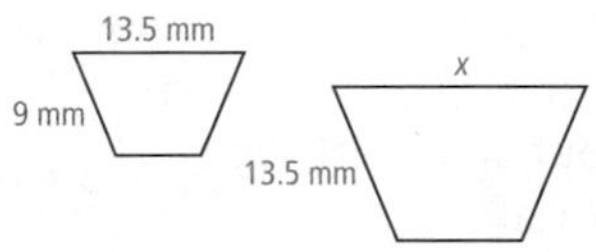

Get Ready! **To Prepare for Lesson 13-3, do Exercises 34–36.** See p. 891.

Simplify each ratio.

34. $\frac{ab}{bc}$ 35. $\frac{24 \cdot 15}{8 \cdot 3}$ 36. $\frac{10 \cdot 9 \cdot 8 \cdot 7 \cdot 6 \cdot 5 \cdot 4 \cdot 3 \cdot 2 \cdot 1}{8 \cdot 7 \cdot 6 \cdot 5 \cdot 4 \cdot 3 \cdot 2 \cdot 1}$

23. The probabilities do not add to 1; they add to 1.05. You need to know the frequency distribution of the favorite pets.

24.

Evens in Two Spins

Number of Evens	0	1	2
Probability	$\frac{1}{4}$	$\frac{1}{2}$	$\frac{1}{4}$

Evens in Three Spins

Number of Evens	0	1	2	3
Probability	$\frac{1}{8}$	$\frac{3}{8}$	$\frac{3}{8}$	$\frac{1}{8}$

Evens in Four Spins

Number of Evens	0	1	2	3	4
Probability	$\frac{1}{16}$	$\frac{4}{16}$	$\frac{6}{16}$	$\frac{4}{16}$	$\frac{1}{16}$

The denominator doubles with each additional spin.

25. A
26. A
27. C
28. 135°
29. theoretical; though experimental results can be found, the more you do the experiment, the closer the results will be to the theoretical value.
30. experimental; no information is available to find results until an experiment is done.
31. theoretical; all of the data is available to find the probability without an experiment.
32. 1.5 in.
33. 20.25 mm
34. $\frac{a}{c}$
35. 15
36. 90

13-2 Lesson Resources

Additional Instructional Support

Geometry Companion

Students can use the Geometry Companion worktext (4 pages) as you teach the lesson. Use the Companion to support

- Solve It!
- New Vocabulary
- Key Concepts

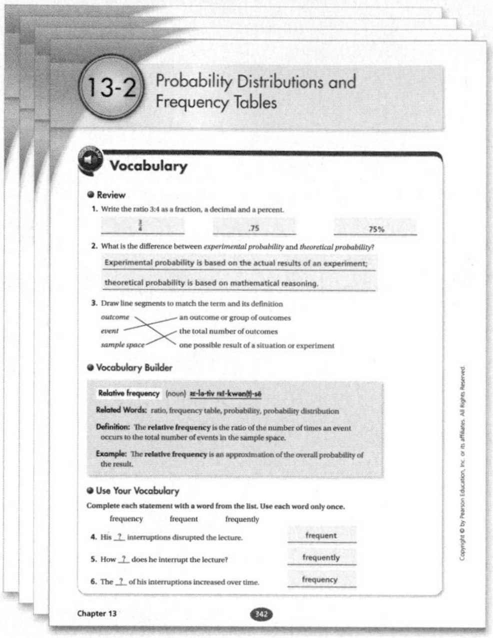

13-2 Probability Distributions and Frequency Tables

Vocabulary

Review

1. Write the ratio 3:4 as a fraction, a decimal and a percent. $\frac{3}{4}$ .75 75%
2. What is the difference between *experimental probability* and *theoretical probability*? Experimental probability is based on the actual results of an experiment; theoretical probability is based on mathematical reasoning.
3. Draw line segments to match the term and its definition

outcome	an outcome or group of outcomes
event	the total number of outcomes
sample space	one possible result of a situation or experiment

Vocabulary Builder

Relative frequency (noun) rel-a-tiv frē-kwan(t)-sē

Related Words: ratio, frequency table, probability, probability distribution

Definition: The **relative frequency** is the ratio of the number of times an event occurs to the total number of events in the sample space.

Example: The **relative frequency** is an approximation of the overall probability of the result.

Use Your Vocabulary

Complete each statement with a word from the list. Use each word only once.

frequency frequent frequently

4. His _?_ interruptions disrupted the lecture. frequent
5. How _?_ does he interrupt the lecture? frequently
6. The _?_ of his interruptions increased over time. frequency

Chapter 13 342

ELL Support

Use Small-Group Interactions Organize students into small groups and have them conduct a survey to gather data about their fellow students. Have them be as creative as possible and ask a survey question that is interesting to them. Possible topics might include:

- favorite school subject
- favorite sport
- favorite kind of music
- favorite after school activity
- favorite cafeteria food

Each group should organize their data in a frequency table. Then use the data to create a probability distribution. Ask students to use the probability distributions to make predictions about the entire class.

5 Assess & Remediate

Lesson Quiz

Use the Lesson Quiz to assess students' mastery of the skills and concepts of this lesson.

1. The table below shows the number of members from each class on the student council. What is the relative frequency of freshmen on the student council?

Class	Frequency
Freshmen	2
Sophomores	3
Juniors	5
Seniors	4

2. What is the probability that a randomly selected member of the student council is a senior?

3. Deanne has 8 jazz, 5 rock, 4 blues, and 15 pop songs on a playlist. What is a probability distribution for the types of music on the playlist?

ANSWERS TO LESSON QUIZ

1. $\frac{2}{14}$ or $\frac{1}{7}$

2. $\frac{4}{14}$

3. Check students' work.

jazz: $\frac{8}{32}$; rock: $\frac{5}{32}$; blues: $\frac{4}{32}$; pop: $\frac{15}{32}$

PRESCRIPTION FOR REMEDIATION

Use the student work on the Lesson Quiz to prescribe a differentiated review assignment.

Points	Differentiated Remediation
0–1	Intervention
2	On-level
3	Extension

PowerGeometry.com

5 Assess & Remediate

Assign the Lesson Quiz. Appropriate intervention, practice, or enrichment is automatically generated based on student performance.

Differentiated Remediation

Intervention

- **Reteaching** (2 pages) Provides reteaching and practice exercises for the key lesson concepts. Use with struggling students or absent students.
- **English Language Learner Support** Helps students develop and reinforce mathematical vocabulary and key concepts.

All-in-One Resources/Online

Reteaching

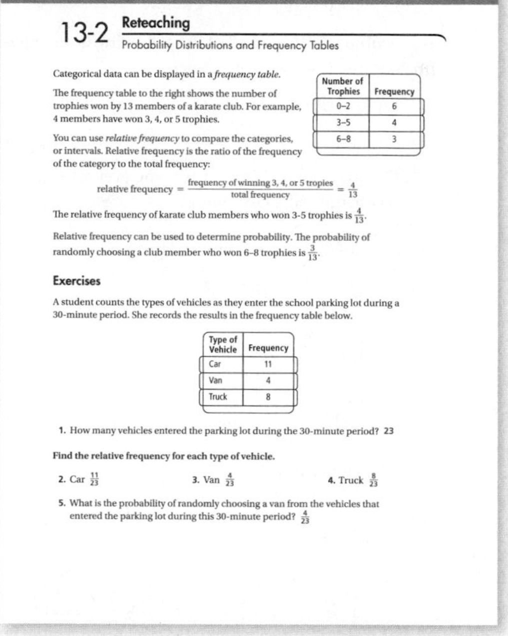

13-2 Reteaching

Probability Distributions and Frequency Tables

Categorical data can be displayed in a *frequency table.*

The frequency table to the right shows the number of trophies won by 13 members of a karate club. For example, 4 members have won 3, 4, or 5 trophies.

Number of Trophies	Frequency
0–2	6
3–5	4
6–8	3

You can use *relative frequency* to compare the categories, or intervals. Relative frequency is the ratio of the frequency of the category to the total frequency:

$$\text{relative frequency} = \frac{\text{frequency of winning 3, 4, or 5 tropies}}{\text{total frequency}} = \frac{4}{13}$$

The relative frequency of karate club members who won 3-5 trophies is $\frac{4}{13}$.

Relative frequency can be used to determine probability. The probability of randomly choosing a club member who won 6–8 trophies is $\frac{3}{13}$.

Exercises

A student counts the types of vehicles as they enter the school parking lot during a 30-minute period. She records the results in the frequency table below.

Type of Vehicle	Frequency
Car	11
Van	4
Truck	8

1. How many vehicles entered the parking lot during the 30-minute period? 23

Find the relative frequency for each type of vehicle.

2. Car $\frac{11}{23}$ 3. Van $\frac{4}{23}$ 4. Truck $\frac{8}{23}$

5. What is the probability of randomly choosing a van from the vehicles that entered the parking lot during this 30-minute period? $\frac{4}{23}$

All-in-One Resources/Online

English Language Learner Support

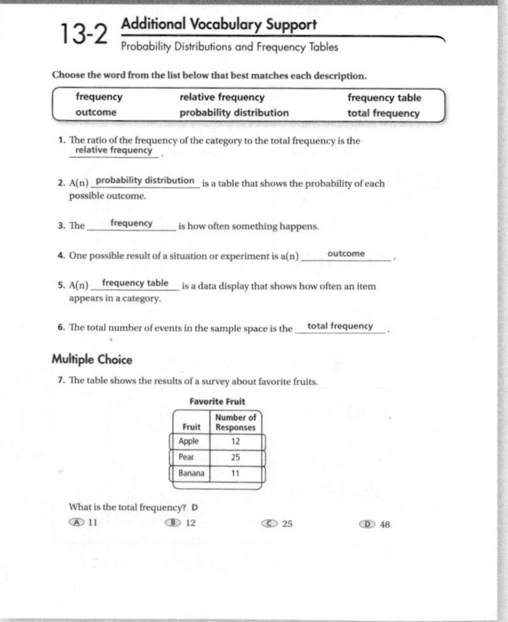

13-2 Additional Vocabulary Support

Probability Distributions and Frequency Tables

Choose the word from the list below that best matches each description.

frequency	relative frequency	frequency table
outcome	probability distribution	total frequency

1. The ratio of the frequency of the category to the total frequency is the relative frequency.
2. A(n) probability distribution is a table that shows the probability of each possible outcome.
3. The frequency is how often something happens.
4. One possible result of a situation or experiment is a(n) outcome.
5. A(n) frequency table is a data display that shows how often an item appears in a category.
6. The total number of events in the sample space is the total frequency.

Multiple Choice

7. The table shows the results of a survey about favorite fruits.

Favorite Fruit

Fruit	Number of Responses
Apple	12
Pear	25
Banana	11

What is the total frequency? D

Ⓐ 11 Ⓑ 12 Ⓒ 25 Ⓓ 48

Differentiated Remediation *continued*

On-Level

- **Practice** (2 pages) Provides extra practice for each lesson. For simpler practice exercises, use the Form K Practice pages found in the All-in-One Teaching Resources and online.
- **Think About a Plan** Helps students develop specific problem-solving skills and strategies by providing scaffolded guiding questions.
- **Standardized Test Prep** Focuses on all major exercises, all major question types, and helps students prepare for the high-stakes assessments.

Practice and Problem Solving Wkbk/ All-in-One Resources/Online

Practice page 1

13-2 Practice *Form G*
Probability Distributions and Frequency Tables

A camp counselor records the number of camp attendees who participate in the daily activities. The results are shown in the table below.

Camp Activities

Activity	Number of People
Waterskiing	12
Hiking	18
Canoeing	13

Find the relative frequency of each activity.

1. Waterskiing $\frac{12}{43}$
2. Hiking $\frac{18}{43}$
3. Canoeing $\frac{13}{43}$

A spinner has 3 equal sections colored red, blue, and green. A student conducts an experiment where she spins the spinner twice and records the results. The results are shown in the frequency table below.

Colors	RR	RB	RG	BB	BR	BG	GG	GR	GB
Frequency	3	5	2	1	4	2	3	2	1

4. What is the probability of spinning red *exactly* once on the next two spins? $\frac{13}{23}$
5. What is the probability of spinning blue twice on the next two spins? $\frac{1}{23}$

At a mini-golf course, 20 friends each take 5 shots to try and get a hole-in-one. The results are shown in the probability distribution below. Complete the table.

Number of Holes-in-one	0	1	2	3	4	5
Frequency	2	6	5	4	2	1
Probability	6. $\frac{1}{10}$	7. $\frac{3}{10}$	8. $\frac{1}{4}$	9. $\frac{1}{5}$	10. $\frac{1}{10}$	11. $\frac{1}{20}$

Practice and Problem Solving Wkbk/ All-in-One Resources/Online

Practice page 2

13-2 Practice (continued) *Form G*
Probability Distributions and Frequency Tables

12. A student records the favorite season for 50 students. The results are shown in the table at the right.
    a. What is the relative frequency of spring? $\frac{13}{50}$
    b. What is the relative frequency of winter? $\frac{7}{50}$
    c. If the table included the number of responses for only three of the seasons, could you determine the relative frequency of the remaining season? Explain. Yes. Since there are 50 respondents, the number of responses for the remaining season is the difference between 50 and the sum of the other three seasons. This number of responses is the numerator in the relative frequency, and 50 is the denominator.

Favorite Season

Season	Number of Responses
Winter	7
Spring	13
Summer	19
Fall	11

13. **Reasoning** How are the relative frequencies in a frequency table mathematically related? The sum of the relative frequencies is equal to 1.
14. A student randomly chooses songs on her MP3 player. Out of 30 different choices, she chooses 8 hip-hops songs, 4 country songs, 11 rock songs, and 7 classical songs. What is the probability that the student chooses a country song? $\frac{2}{15}$
15. **Writing** A certain probability distribution includes simplified fractions for some of the probabilities. Will the sum of the numerators be equal to the total frequency? Explain. No. The total frequency is the sum of all the probability numerators before they are simplified. Once a probability fraction is simplified, the sum will be less than the total frequency.
16. **Error Analysis** A cross-country coach makes the probability distribution below for the number of wins for some of the team members.

Number of Wins	0	1	2	3	4
Frequency	7	4	6	5	3
Probability	$\frac{7}{10}$	$\frac{4}{10}$	$\frac{6}{10}$	$\frac{5}{10}$	$\frac{3}{10}$

Explain the error the coach made when making the table. The denominator of the probabilities is wrong; it should be 25 (the sum of the frequencies), not 10 (the sum of the number of wins).

Extension

- **Enrichment** Provides students with interesting problems and activities that extend the concepts of the lesson.
- **Activities, Games, and Puzzles** Worksheets that can be used for concepts development, enrichment, and for fun!

All-in-One Resources/Online

Enrichment

13-2 Enrichment
Probability Distributions and Frequency Tables

The frequency table below compares the number of hours spent studying each week by students in one class.

Hours Spent Studying

Number of Hours	0–3	4–7	8–11	12–15
Frequency	3	5	6	4

1. How many students are in the class? 18
2. What is the relative frequency of students who study 4 – 7 hours per week? $\frac{5}{18}$
3. What is the same for each category? The range is 3 OR the number of values is the same, 4.

A frequency table can be used to make a *histogram*, a type of bar graph. A histogram using the table "Hours Spent Studying" is started below.

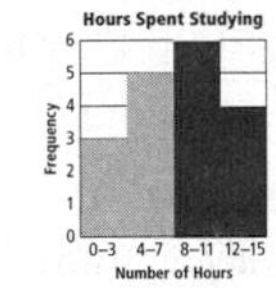

4. Why do you think the bars touch? The values are continuous, with no gaps.
5. Complete the histogram. See graph.
6. How would the visual comparison of categories be affected if one of the categories included six values instead of four? That category would appear proportionally larger than it is.

Practice and Problem Solving Wkbk/ All-in-One Resources/Online

Think About a Plan

13-2 Think About a Plan
Probability Distributions and Frequency Tables

Survey The results of a survey of 80 households in Westville are shown at the right.

a. What is the probability distribution of the number of computers in Westville households?
b. If Westville has 15,200 households, predict the number of households that will have exactly 3 computers.
c. How many households will have either two or three computers?

Computer Survey

Number of Computers	Frequency
0	12
1	29
2	31
3	6
More than 3	2

Understanding the Problem

1. How can you find the number of households in parts (b) and (c) if the survey did not poll all of the households in Westville? Use the survey results as a model for the entire population and extrapolate from the data.

Planning the Solution

2. How can a probability distribution for the surveyed households assist in calculating the answers?
   A probability table will determine the probability of households with varying numbers of computers for the survey of 80 households. Then the probabilities can be applied to all 15,200 households.

Getting an Answer

3. Complete the probability distribution based on the given information.

Number of Computers	0	1	2	3	>3
Frequency	12	29	31	6	2
Probability	$\frac{12}{80}$	$\frac{29}{80}$	$\frac{31}{80}$	$\frac{6}{80}$	$\frac{2}{80}$

$P(0) = \frac{12}{80} = 0.15$, $P(1) = \frac{29}{80} = 0.3625$, $P(2) = \frac{31}{80} = 0.3875$, $P(3) = \frac{6}{80} = 0.075$, $P(>3) = \frac{2}{80} = 0.025$

4. Predict the number of households in Westville that will have exactly 3 computers. Show your work. $15{,}200 \cdot P(3) = 15{,}200 \cdot 0.075 = 1140$ households
5. Use your answer to Step 4 to find the number of households that will have two or three computers. Show your work. $15{,}200 \cdot P(2) + 1140 = 15{,}200 \cdot 0.3875 + 1140 = 5890 + 1140 = 7030$ households

Practice and Problem Solving Wkbk/ All-in-One Resources/Online

Standardized Test Prep

13-2 Standardized Test Prep
Probability Distributions and Frequency Tables

Multiple Choice

Use the table below for Exercises 1–4. Choose the correct letter.

The table below shows the results of a soccer team's scores for games played this season.

Goals	0	1	2	3
Frequency	4	7	5	3

1. How many games did the team play? C
   (A) 3 (B) 6 (C) 19 (D) 25
2. What is the relative frequency of games with 1 goal scored? H
   (F) $\frac{3}{19}$ (G) $\frac{7}{25}$ (H) $\frac{7}{19}$ (I) $\frac{8}{15}$
3. What is the probability that the team scored 2 or more goals? B
   (A) $\frac{5}{19}$ (B) $\frac{8}{19}$ (C) $\frac{5}{8}$ (D) $\frac{8}{11}$
4. Which expression can be used to determine the probability of scoring fewer than 3 goals? F
   (F) $1 - \frac{3}{19}$ (G) $\frac{3}{19} - 1$ (H) $1 - \frac{16}{19}$ (I) $1 + \frac{3}{19}$

Short Response

5. The relative frequencies of two of three possible outcomes are $\frac{1}{2}$ and $\frac{1}{3}$. What is the relative frequency of the third outcome?
   [2] $\frac{1}{6}$ (or equivalent fraction)
   [1] Student provides evidence showing an understanding that the sum of the frequencies is 1.
   [0] Student's response shows no effort OR does not attempt to make the frequencies add to 1.

Online Teacher Resource Center

Activities, Games, and Puzzles

13-2 Puzzle: Internet Search Engines
Probability Distributions and Frequency Tables

Use the information in the table to calculate the relative frequencies for each type of search engine listed. The total frequency is 234 search engines. The frequency for each engine is given. Use the results for the relative frequencies to solve the riddle:

What do the Internet and spiders have in common?

Type of Engine	Frequency	Relative Frequency	Key	Type of Engine	Frequency	Relative Frequency	Key
All Purpose	15	15/234	A	Medical	10	5/117	N
Bit Torrent	8	4/117	B	MetaSearch	18	1/13	O
Blog	7	7/234	C	MultiMedia	12	2/39	P
Books	2	1/117	D	News	6	1/39	R
Business	7	7/234	E	Open Source	20	10/117	S
Email	3	1/78	F	People	9	1/26	T
Enterprise	18	1/13	G	Question & Answer	12	2/39	U
Forum	1	1/234	H	Real Estate	8	4/117	V
Games	3	1/78	I	School	3	1/78	W
International	20	10/117	J	Shopping	12	2/39	X
Job	19	19/234	K	Source Code	6	1/39	Y
Legal	7	7/234	L	Visual Search	2	1/117	Z
Maps	6	1/39	M				

Some frequencies are repeated in the table. Therefore, some spaces in the answer below may have more than one letter that matches its assigned value. You must determine which letters to use to make a word that answers the riddle.

W	E	B	S	I	T	E	S
$\frac{1}{78}$	$\frac{7}{234}$	$\frac{4}{117}$	$\frac{10}{117}$	$\frac{1}{78}$	$\frac{1}{26}$	$\frac{7}{234}$	$\frac{10}{117}$

1 Interactive Learning

Solve It!

PURPOSE To use counting techniques to determine the number of possible arrangements.
PROCESS Students may make an organized list or act out the situation to determine the solution.

FACILITATE

Q How could you act out the situation to determine the number of possible arrangements? **[Let 4 items represents the samples. Choose items in different orders.]**

Q Begin by solving a simpler problem. How many arrangements are there for 1 sample? for 2 samples? for 3 samples? **[1; 2; 6]**

ANSWER See Solve It in Answers on next page.
CONNECT THE MATH In the Solve It, students explore counting techniques to find the number of possible arrangements of samples. In this lesson, students will use the Fundamental Counting Principle, permutations, and combinations to count the number of ways events can occur and find probabilities.

2 Guided Instruction

Take Note

The Fundamental Counting Principle is also known as the Multiplication Counting Principle.

Here's Why It Works — VISUAL LEARNERS

The tree diagram visually demonstrates the Fundamental Counting Principle. Ask students why it might not always be feasible to draw a tree diagram to count possible outcomes.

13-3 Permutations and Combinations

Common Core State Standards
Prepares for S-CP.B.9 Use permutations and combinations to compute probabilities of compound events and solve problems.
MP 1, MP 3, MP 6

Objectives To use permutations and combinations to solve problems

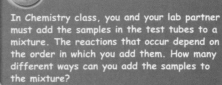

SOLVE IT! Getting Ready!

In Chemistry class, you and your lab partner must add the samples in the test tubes to a mixture. The reactions that occur depend on the order in which you add them. How many different ways can you add the samples to the mixture?

MATHEMATICAL PRACTICES In the Solve It, you may have drawn a diagram or listed all of the different possible ways to add the samples. Sometimes there are so many possibilities that listing them all is not practical.

Essential Understanding You can use counting techniques to find all of the possible ways to complete different tasks or choose items from a list.

Lesson Vocabulary
- Fundamental Counting Principle
- permutation
- *n* factorial
- combination

take note — Key Concept Fundamental Counting Principle

The **Fundamental Counting Principle** says that if event M occurs in m ways and event N occurs in n ways, then event M followed by event N can occur in $m \cdot n$ ways.

Example 4 entrees and 6 drinks gives $4 \cdot 6 = 24$ possible lunch specials

Here's Why It Works Suppose that you are choosing an outfit by first selecting blue jeans or black pants and then selecting a red, green, or blue shirt. The tree diagram shows the possible outfits you can choose. For m pant choices and n shirt choices, there are $m \cdot n$ or $2 \cdot 3 = 6$ possible outfits.

Pants	Shirts	Outfits
blue jeans	red	jeans, red shirt
	green	jeans, green shirt
	blue	jeans, blue shirt
black pants	red	pants, red shirt
	green	pants, green shirt
	blue	pants, blue shirt

13-3 Preparing to Teach

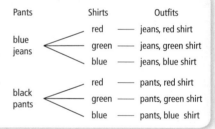

BIG idea Probability

ESSENTIAL UNDERSTANDING
- Counting techniques can be used to find all of the possible ways to complete different tasks or choose items from a list.

Math Background

This lesson introduces students to different techniques that can be used to count items and find probabilities.
- Fundamental Counting Principle

To find the number of ways multiple events can occur, multiply the numbers of ways each event can occur individually.
- Permutations

With permutations, the order in which items are selected is important.

$_nP_r = \frac{n!}{(n-r)!}$ for $0 \leq r \leq n$

- Combinations

With combinations, the order in which items are selected is not important.

$_nC_r = \frac{n!}{r!(n-r)!}$ for $0 \leq r \leq n$

Mathematical Practices

Attend to precision. Students will utilize the factorial, combination, and permutation symbols and determine when to use each.

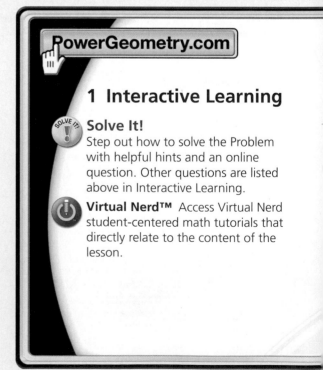

PowerGeometry.com

1 Interactive Learning

Solve It!
Step out how to solve the Problem with helpful hints and an online question. Other questions are listed above in Interactive Learning.

Virtual Nerd™ Access Virtual Nerd student-centered math tutorials that directly relate to the content of the lesson.

You can use the Fundamental Counting Principle for situations that involve more than two events.

Plan
How can you find the number of possible lunch specials without listing each one?
Use the Fundamental Counting Principle. Multiply the choices for each category.

Problem 1 Using the Fundamental Counting Principle

Menus **A deli offers a lunch special if you choose one from each of the following types of sandwiches, side items, and drink choices. How many different lunch specials are possible?**

Deli Menu

Sandwiches	Side Items	Drinks
ham & turkey	chips	juice
salami	potato salad	iced tea
tuna	fruit salad	lemonade
club	garden salad	milk
veggie		water
meatball		

There are 6 possible sandwiches, 4 different side items, and 5 drink choices.

$6 \cdot 4 \cdot 5 = 120$ Use the Fundamental Counting Principle.

There are 120 different possible lunch specials.

Got It? **1.** Suppose that a computer generates passwords that begin with a letter followed by 2 digits, like R38. The same digit can be used more than once. How many different passwords can the computer generate?

A **permutation** is an arrangement of items in which the order of the objects is important. You can use the Fundamental Counting Principle to find the total number of permutations. Suppose you want to find the number of ways to line up 4 friends. There are 4 ways to choose the first person, 3 ways to choose the second person, 2 ways to choose the third, and only 1 way to choose the last. So, there are $4 \cdot 3 \cdot 2 \cdot 1 = 24$ permutations, or ways to arrange your friends.

You can use *factorial notation* to write $4 \cdot 3 \cdot 2 \cdot 1$ as 4! (You say this as "4 factorial.") For any positive integer *n*, ***n* factorial** is $n! = n(n-1)(n-2) \cdot \ldots \cdot 3 \cdot 2 \cdot 1$. Zero factorial is defined to be equal to 1.

Think
Why use the Fundamental Counting Principle?
There are too many possibilities to make a list.

Problem 2 Finding the Number of Permutations

Music **You download 8 songs on your music player. If you play the songs using the random shuffle option, how many different ways can the sequence of songs be played?**

$8! = 8 \cdot 7 \cdot 6 \cdot \ldots \cdot 2 \cdot 1 = 40{,}320$ Use the Fundamental Counting Principle.

There are 40,320 different ways to randomly play the 8 songs.

Got It? **2.** In how many ways can you arrange 12 books on a shelf?

2 Guided Instruction

Each Problem is worked out and supported online.

Problem 1
Using the Fundamental Counting Principle

Problem 2
Finding the Number of Permutations

Problem 3
Finding a Permutation

Problem 4
Using the Combination Formula

Problem 5
Identifying Combinations and Permutations

Problem 6
Finding Probabilities

Support in Geometry Companion
- Vocabulary
- Key Concepts
- Got It?

Problem 1
In this problem, students apply the Fundamental Counting Principle.

Q How many decisions are there to make? What are the decisions? **[three decisions: sandwich, side item, and drink]**

Q What would be the advantages or disadvantages to solving the problem with a tree diagram? **[Making a tree diagram would show each possible lunch option, but it would be very time consuming.]**

Got It?

Q Suppose a different password has 2 letters followed by 1 digit. Would there be more or fewer possibilities for this type of password? Explain. **[more; there are 26 letters but only 10 digits.]**

Problem 2 SYNTHESIZING
Have students discuss how the methods used in Problem 2 could be applied to the Solve It problem.

Q If the first and last songs are interchanged, would this represent a different permutation? Explain. **[Yes, the order would be different so it is a different permutation.]**

Got It?

Q Suppose 2 of the books are taken off the shelf. How can you use your result to find the number of permutations of 10 books on the shelf? How many permutations are there? **[Divide the result by $12 \cdot 11 = 132$; 3,628,800]**

Answers

Solve It!
There are $4 \cdot 3 \cdot 2 \cdot 1 = 24$ different ways to add the samples to the mixture.

Got It?
1. 2600
2. 479,001,600

Take Note

Point out to students that *r* must be less than or equal to *n* because you cannot have an ordering of more items than you actually have. For example, if you only have 6 singers, you cannot have a permutation of 7 singers.

Problem 3 SYNTHESIZING

Show students that they can also apply the Fundamental Counting Principle to solve Problem 3. There are 10 ways to elect the president, 9 ways to elect the vice president, and 8 ways to elect the treasurer. So, there are $10 \cdot 9 \cdot 8 = 720$ ways to elect the officers.

Q What is the result when you cancel the common factors in the expression $\frac{10 \cdot 9 \cdot 8 \cdot 7 \cdot 6 \cdot 5 \cdot 4 \cdot 3 \cdot 2 \cdot 1}{7 \cdot 6 \cdot 5 \cdot 4 \cdot 3 \cdot 2 \cdot 1}$? **[10 · 9 · 8]**

Q Which counting technique does this simplified expression represent? **[the Fundamental Counting Principle]**

Got It?

Q Why do you find the number of permutations? **[The order in which the swimmers finish is important.]**

Q If trophies were only awarded to first and second places, how many ways could the trophies be won? **[132]**

Take Note

Discuss with students the differences between permutations and combinations. Include in the discussion the differences in the applications as well as the differences in the formulas.

Suppose you want 3 songs to play at a time from the 8 songs you have downloaded. There are 8 ways to select the first song, 7 ways to select the second song, and 6 ways to select the third song. By the Fundamental Counting Principle, this is $8 \cdot 7 \cdot 6$ ways, or 336 ways. You can express the result using factorials, as shown below.

take note

Key Concept Permutation Notation

The number of permutations of *n* items of a set arranged *r* items at a time is

$${}_nP_r = \frac{n!}{(n-r)!} \text{ for } 0 \le r \le n.$$

Example $${}_8P_3 = \frac{8!}{(8-3)!} = \frac{8!}{5!} = \frac{8 \cdot 7 \cdot 6 \cdot 5 \cdot 4 \cdot 3 \cdot 2 \cdot 1}{5 \cdot 4 \cdot 3 \cdot 2 \cdot 1} = 8 \cdot 7 \cdot 6 = 336$$

Problem 3 Finding a Permutation

The environmental club is electing a president, vice president, and treasurer. How many different ways can the officers be chosen from the 10 members?

Think

How do you know that the way the officers are chosen is a permutation?
This situation uses permutations because a different arrangement of the same 3 members is a different result. So the order of the choices matters.

Method 1 Use the formula for permutations.

There are 10 members, arranged 3 at a time. So $n = 10$ and $r = 3$.

$${}_{10}P_3 = \frac{10!}{(10-3)!} \quad \text{Substitute 10 for } n \text{ and 3 for } r.$$

$$= \frac{3{,}628{,}800}{5040} = 720 \quad \text{Simplify.}$$

Method 2 Use a graphing calculator.

Press 1 0 math ◁ 2 3 enter

${}_{10}P_3 = 720$

There are 720 ways that three of the ten members can be chosen as officers.

Got It? 3. Twelve swimmers compete in a race. In how many possible ways can the swimmers finish first, second, and third?

A **combination** is a selection of items in which order is *not* important. Suppose you select 3 different fruits to make a fruit salad. The order you select the fruits does not matter.

take note

Key Concept Combination Notation

The number of combinations of *n* items chosen *r* at a time is

$${}_nC_r = \frac{n!}{r!(n-r)!} \text{ for } 0 \le r \le n.$$

Example $${}_9C_4 = \frac{9!}{4!(9-4)!} = \frac{9!}{4!\,5!} = \frac{9 \cdot 8 \cdot 7 \cdot 6 \cdot 5 \cdot 4 \cdot 3 \cdot 2 \cdot 1}{(4 \cdot 3 \cdot 2 \cdot 1)(5 \cdot 4 \cdot 3 \cdot 2 \cdot 1)} = \frac{9 \cdot 8 \cdot 7 \cdot 6}{4 \cdot 3 \cdot 2 \cdot 1} = 126$$

Additional Problems

1. A restaurant offers 6 different appetizers, 11 entrées, and 5 desserts. How many dinners are possible if a customers orders 1 appetizer, 1 entrée, and 1 dessert?
   ANSWER 330
2. Seven runners are competing in a race. In how many ways can the runners finish the race?
   ANSWER 5040
3. A baseball team manager has 9 batters. How many ways can he choose the first 4 batters of the lineup if the order of the batters is important?
   ANSWER 3024
4. A 5-person committee is to be made up from 12 teachers. How many different committees are possible?
   ANSWER 792
5. Mrs. Winthrop is choosing 5 singers from a group of 16 to sing in an ensemble during a choir production. How many ways can she choose the singers?
   ANSWER 4368
6. A 4-digit combination lock is made up using the digits from 1 through 9. Each digit can be used only once. What is the probability that a combination chosen at random contains all odd digits?
   ANSWER $\frac{5}{126}$

Answers

Got It? (continued)

3. 1320

Problem 4 Using the Combination Formula

Reading **Suppose that you choose 4 books to read on summer vacation from a reading list of 12 books. How many different combinations of the books are possible?**

Method 1 Use the formula for finding combinations.

There are 12 books chosen 4 at a time.

$$_{12}C_4 = \frac{12!}{4!(12-4)!} = \frac{12!}{4!\,8!} = \frac{12 \cdot 11 \cdot 10 \cdot 9 \cdot 8 \cdot 7 \cdot 6 \cdot 5 \cdot 4 \cdot 3 \cdot 2 \cdot 1}{(4 \cdot 3 \cdot 2 \cdot 1)(8 \cdot 7 \cdot 6 \cdot 5 \cdot 4 \cdot 3 \cdot 2 \cdot 1)} = 495$$

Think
If you are not using a calculator, how can you simplify the calculation?
Divide all common factors before multiplying.

Method 2 Use a calculator.

Press 1 2 math ◁ 3 4 enter

$_{12}C_4 = 495$

There are 495 ways to choose 4 books from a reading list of 12 books.

Got It? 4. A service club has 8 freshmen. Five of the freshmen are to be on the clean-up crew for the town's annual picnic. How many different ways are there to choose the 5 member clean-up crew?

To determine whether to use the permutation formula or the combination formula, you must decide whether order is important.

Problem 5 Identifying Combinations and Permutations

A **A college student is choosing 3 classes to take during first, second, and third semseter from the 5 elective classes offered in his major. How many possible ways can the student schedule the three classes?**

The order in which the classes are chosen does matter. Use a permutation.

$$_5P_3 = \frac{5!}{(5-3)!} = \frac{5!}{2!} = 60$$

There are 60 ways that the student can schedule the three classes.

Think
Why does order not matter?
After the selection takes place, the same 12 people are on the jury, regardless of the order in which they were chosen.

B **A jury of 12 people is chosen from a pool of 35 potential jurors. How many different juries can be chosen?**

The order in which the jurors are chosen is not important. Use a combination.

$$_{35}C_{12} = \frac{35!}{12!(35-12)!} = \frac{35!}{12!\,23!} = 834{,}451{,}800$$

There are 834,451,800 possible juries of 12 people.

Got It? 5. A yogurt shop allows you to choose any 3 of the 10 possible mix-ins for a Just Right Smoothie. How many different Just Right Smoothies are possible?

Problem 4 ERROR PREVENTION

In this problem, students use the combination formula to solve a problem.

Q When simplifying the expression $\frac{12!}{4!8!}$, are $\frac{12!}{4!}$ and 3! equivalent? Explain. **[No, $\frac{12!}{4!}$ simplifies to $12 \cdot 11 \cdot \ldots \cdot 6 \cdot 5$, but $3! = 3 \cdot 2 \cdot 1$.]**

Got It?

Q Why is order not important when choosing the members for the clean-up crew? **[The groups of members will not change depending on the order they are selected. For example, set ABCDE contains the same members as set CBADE or set BACED.]**

Problem 5

When determining whether a situation represents a permutation or a combination, students must consider whether or not the order of selection matters.

Q In part A, why is the order in which the classes are chosen important? **[The order the classes are chosen will result in different schedules. For example, schedule ABC is different than schedule BCA.]**

Got It?

Q Does the order in which the mix-ins are chosen matter? Explain. **[No, the smoothie will be the same regardless of the order of the ingredients.]**

4. 56
5. 120

Problem 6 **ERROR PREVENTION**
Students must be able to determine whether or not order is important when finding the number of favorable outcomes as well as when finding the number of possible outcomes. Using the wrong formula will result in an incorrect answer.

Q Does the solution in Step 2 use a permutation or combination to find the number of ways to choose 1, 2, and 3? Explain. **[permutation; the order in which the balls are selected is important.]**

Got It?

Q How many ways can you select the 1 ball first, the 2 ball second, and the 3 ball third? Which counting technique does this represent? Explain. **[1; Fundamental Counting Principle; There is only 1 way to select each ball.]**

3 Lesson Check

Do you know HOW? ERROR INTERVENTION

- If students have trouble solving Exercises 1 through 6, review with them how to use their calculators to find permutations and combinations.

Do you UNDERSTAND?

- In Exercise 8, ask students to give an example of a situation that is modeled by a permutation and a situation that is modeled by a combination.

Close

Q How is the Fundamental Counting Principle applied? **[Multiply the number of ways each event can happen individually to find the total number of ways all events can happen.]**

Q How do you determine whether to use a permutation or a combination? **[permutation: order matters, combination: order does not matter]**

You can use permutations and combinations to help you solve probability problems.

Problem 6 Finding Probabilities

Three pool balls are randomly chosen from a set numbered from 1 to 15. What is the probability that the pool balls chosen are numbered 5, 7, and 9?

Think
What is the total number of outcomes? Because the problem requires 3 pool balls chosen at random, the total number of outcomes is all the ways that you can choose 3 pool balls from a set of 15 pool balls.

Step 1 Use the probability formula.

$$P(\text{choosing 5, 7, and 9}) = \frac{\text{number of possible ways to choose 5, 7, and 9}}{\text{number of ways to choose 3 pool balls}}$$

Step 2 Find the numerator. Use the Fundamental Counting Principle to find the number of possible ways to choose 5, 7, and 9.

There are 3 ways to pick the first ball, 2 ways to pick the second ball, and 1 way to pick the last ball.

$$3 \cdot 2 \cdot 1 = 6$$

Step 3 Find the sample space. Because choosing pool balls numbered 5, 7, and 9 is the same outcome as choosing pool balls numbered 9, 5, and 7, the order does not matter. Use the combination formula to find the total number of ways to choose 3 pool balls from 15 pool balls.

$$_{15}C_3 = \frac{15}{3!(15-3)!} = 455$$

Step 4 Find the probability.

$$P(\text{choosing 5, 7, and 9}) = \frac{6}{455} \approx 0.013$$

The probability of choosing the pool balls numbered 5, 7, and 9 is about 0.013, or 1.3%.

Got It? 6. What is the probability of choosing first the number 1 ball, then the number 2 ball, and then the number 3 ball?

Lesson Check

Do you know HOW?

Evaluate each expression.

1. 3! **2.** 0! **3.** $_6P_2$

4. $_6P_3$ **5.** $_6C_2$ **6.** $_6C_3$

7. Sports How many ways can you choose 6 people to form a volleyball team out of a group of 10 players?

Do you UNDERSTAND? MATHEMATICAL PRACTICES

8. Compare and Contrast How are combinations and permutations similar? How are they different?

9. Reasoning Your friend says that she can calculate any probability if she knows how many successful outcomes there are. Is there something else needed? Explain.

Answers

Got It? (continued)

6. $\frac{1}{2730}$

Lesson Check

1. 6
2. 1
3. 30
4. 120
5. 15
6. 20
7. 210
8. Both are methods of counting. With permutations order is important, but with combinations order is not important.
9. Yes, she also needs to know the total number of possible outcomes in the sample space.

Practice and Problem-Solving Exercises

MATHEMATICAL PRACTICES

A Practice

See Problem 1.

10. Telephones International calls require the use of a country code. Many country codes are 3-digit numbers. Country codes do not begin with a 0 or 1. There are no restrictions on the second and third digits. How many different 3-digit country codes are possible?

11. Security To make an entry code, you need to first choose a single-digit number and then two letters, which can repeat. How many entry codes can you make?

Find the value of each expression.

See Problems 2–4.

12. 6!   **13.** $\frac{15!}{(15-10)!}$ **14.** ${}_{10}P_6$ **15.** ${}_{10}C_6$

16. Linguistics The Hawaiian alphabet has 12 letters. How many permutations are possible for each number of letters?
a. 3 letters   **b.** 5 letters

17. A class has 30 students. In how many ways can committees be formed using the following numbers of students?
a. 3 students   **b.** 5 students

For Exercises 18–19, determine whether to use a permutation or a combination. Then solve the problem.

See Problem 5.

18. You and your friends pick up seven movies to watch over a holiday. You have time to watch only two. In how many ways can you select the two to watch?

19. Suppose that the math team at your school competes in a regional tournament. The math team has 12 members. Regional teams are made up of 4 people. How many different regional teams are possible?

See Problem 6.

20. You have a stack of 8 cards numbered 1–8. What is the probability that the first cards selected are 5 and 6?

21. To win a lottery, 6 numbers are drawn at random from a pool of 50 numbers. Numbers cannot repeat. You have one lottery ticket. What is the probability you hold the winning ticket?

B Apply

22. Entertainment Suppose that you and 4 friends go to a popular movie. You arrive late and cannot sit together, but you find 3 available seats in a row. How many possible ways can you and your friends sit in these seats?

23. Think About a Plan There are 8 online songs that you want to download. If you only have enough money to download 3 of the songs, how many different groups of songs can you buy?
- Does the order in which you select the songs matter? Explain.
- Should you use the permutation formula or combination formula?
- What are the values of n and r?

4 Practice

ASSIGNMENT GUIDE
Basic: 10–21, 30, 32
Average: 11–29 odd, 28, 30
Advanced: 23–29 odd, 28, 30
Standardized Test Prep: 31–32
Mixed Review: 33–36

Mathematical Practices are supported by exercises with red headings. Here are the Practices supported in this lesson:

MP 1: Make Sense of Problems Ex. 23
MP 3: Understand Definitions Ex. 8, 28
MP 3: Critique the Reasoning of Others Ex. 9, 27
MP 3: Communicate Ex. 24

Applications exercises have blue headings.

EXERCISE 29: Use the Think About a Plan worksheet in the **Practice and Problem Solving Workbook** (also available in the Teaching Resources in print and online) to further support students' development in becoming independent learners.

HOMEWORK QUICK CHECK
To check students' understanding of key skills and concepts, go over Exercises 16, 17, 20, 23, and 27–29.

Practice and Problem-Solving Exercises

10. 800
11. 6760
12. 720
13. 10,897,286,400
14. 151,200
15. 210
16a. 1320
b. 95,040
17a. 4060
b. 142,506
18. 21
19. 495
20. $\frac{2}{28}$
21. $\frac{1}{15{,}890{,}700}$
22. 24 ways
23. 56

24. **Government** The are 24 members of the U.S. Senate Committee on Finance. How many possible ways are there to choose a 13-member subcommittee to review current energy legislation?

25. **Music** What is the probability of the youngest four members of a 15-member choir being randomly selected for a quartet (a group of four singers)?

26. What is the probability of randomly choosing a specific set of 7 books off a bookshelf holding 12 books?

27. **Error Analysis** A friend says that there are 6720 different ways to combine 5 out of 8 ingredients to make a stew. Explain the error and find the correct answer.

28. **Reasoning** Can ${}_nC_r$ ever be equal to ${}_nP_r$? Explain.

29. A 4-digit code is needed to unlock a bicycle lock. The digits 0 through 9 can be used only once in the code.
   a. What is the probability that all of the digits are even?
   b. Writing These types of locks are usually called combination locks. What name might a mathematician prefer? Why?

Challenge

30. *Circular permutations* are arrangements of objects in a circle or a loop. For example, the number of different ways a group of friends can sit around a table represent circular permutations. Permutations that are equivalent after a rotation of the circle are considered the same. Make a table for the regular and circular permutations of 2 out of 2, 3 out of 3, and 4 out of 4. What formula do you think could be used to find the number of circular permutations for n out of n items?

Standardized Test Prep

GRIDDED RESPONSE

SAT/ACT

31. A cube has a volume of 64 cm^3. What is the total surface area of this cube in square centimeters?

32. The major arc of a circle measures 210°. What is the measure of the corresponding minor arc?

Mixed Review

33. The results of a survey on favorite movie genres are shown below in the frequency table below. What is the probability distribution of the data? See Lesson 13-2.

Favorite Movie Genres

Genre	Action	Comedy	Drama	Horror	Other
Frequency	9	8	3	6	4

Get Ready! **To Prepare for Lesson 13-4, do Exercises 34–36.**

Two standard number cubes are tossed. Find the following probabilities. See Lesson 13-1.

34. P(two 5s) 35. P(sum of 10) 36. P(sum less than 9)

Answers

Practice and Problem-Solving Exercises (continued)

24. 2,496,144
25. $\frac{1}{1365}$
26. $\frac{1}{792}$
27. The friend used a permutation instead of a combination. There are 56 ways.
28. Yes, for example when $r = 0$ or 1.
29a. $\frac{1}{42}$
   b. Sample answer: The number of possible codes is actually a permutation.
30. $(n - 1)!$
31. 96
32. 150
33.

Favorite Movie Genres

Genres	Action	Comedy	Drama	Horror	Other
Frequency	9	8	3	6	4
Probability	$\frac{3}{10}$	$\frac{4}{15}$	$\frac{1}{10}$	$\frac{1}{5}$	$\frac{2}{15}$

34. $\frac{1}{36}$
35. $\frac{3}{36}$
36. $\frac{26}{36}$

13-3 Lesson Resources

Differentiated Remediation

Additional Instructional Support

Geometry Companion

Students can use the **Geometry Companion** worktext (4 pages) as you teach the lesson. Use the Companion to support

- Solve It!
- New Vocabulary
- Key Concepts

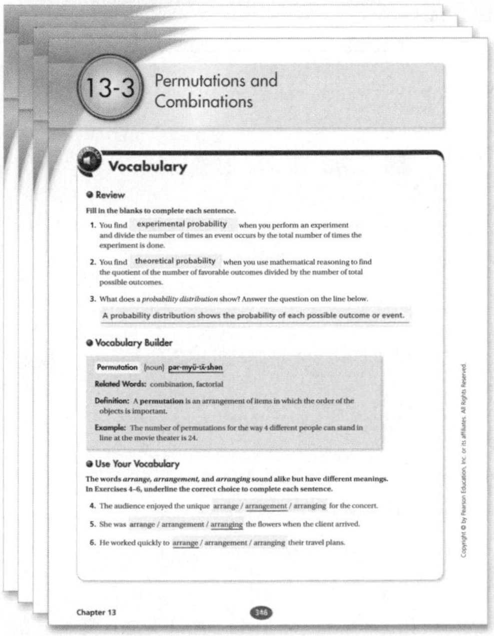

13-3 Permutations and Combinations

Vocabulary

Review

Fill in the blanks to complete each sentence.

1. You find experimental probability when you perform an experiment and divide the number of times an event occurs by the total number of times the experiment is done.
2. You find theoretical probability when you use mathematical reasoning to find the quotient of the number of favorable outcomes divided by the number of total possible outcomes.
3. What does a *probability distribution* show? Answer the question on the line below.

A probability distribution shows the probability of each possible outcome or event.

Vocabulary Builder

Permutation (noun) per-myū-tā-shən

Related Words: combination, factorial

Definition: A **permutation** is an arrangement of items in which the order of the objects is important.

Example: The number of permutations for the way 4 different people can stand in line at the movie theater is 24.

Use Your Vocabulary

The words *arrange, arrangement,* and *arranging* sound alike but have different meanings. In Exercises 4–6, underline the correct choice to complete each sentence.

4. The audience enjoyed the unique arrange / arrangement / arranging for the concert.
5. She was arrange / arrangement / arranging the flowers when the client arrived.
6. He worked quickly to arrange / arrangement / arranging their travel plans.

Copyright © by Pearson Education, Inc. or its affiliates. All Rights Reserved.

Chapter 13 346

ELL Support

Connect to Prior Knowledge of Math

Some students will struggle with finding probabilities for situations that involve permutations and combinations. Remind them that in order to calculate the probability of an event, you need to divide the number of favorable outcomes by the total number of possible outcomes in the sample space. In both of these computations, remind students that they need to think about whether or not the order is important. If order is important, use permutations. If order is not important, use combinations. Also, be sure students are aware that some probability problems will require using both a combination and a permutation.

5 Assess & Remediate

Lesson Quiz

Use the Lesson Quiz to assess students' mastery of the skills and concepts of this lesson.

1. Megan has 4 different skirts and 8 different blouses to choose from. How many outfits are possible if she chooses 1 skirt and 1 blouse?
2. How many different ways can 5 students stand in line in the cafeteria?
3. How many ways can Elena choose 4 songs from a list of 15 if the order of the songs is important?
4. A restaurant employs 8 cooks. How many ways can the manager choose 4 of the cooks to attend a training seminar?
5. **Do you UNDERSTAND?** The top 4 finishers in a race will go on to run in the finals. If there are 9 runners in the race, how many ways can the runners for the finals be decided? Does this represent a permutation or a combination?
6. If Gene rolls 5 number cubes at the same time, what is the probability that all 5 number cubes show the same number?

ANSWERS TO LESSON QUIZ

1. 32
2. 120
3. 32,760
4. 70
5. 126; combination
6. $\frac{1}{1296}$

PRESCRIPTION FOR REMEDIATION

Use the student work on the Lesson Quiz to prescribe a differentiated review assignment.

Points	Differentiated Remediation
0–2	Intervention
3–4	On-level
5–6	Extension

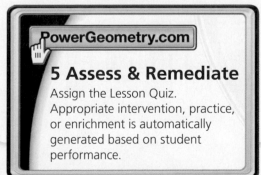

PowerGeometry.com

5 Assess & Remediate

Assign the Lesson Quiz. Appropriate intervention, practice, or enrichment is automatically generated based on student performance.

Intervention

- **Reteaching** (2 pages) Provides reteaching and practice exercises for the key lesson concepts. Use with struggling students or absent students.
- **English Language Learner Support** Helps students develop and reinforce mathematical vocabulary and key concepts.

All-in-One Resources/Online

Reteaching

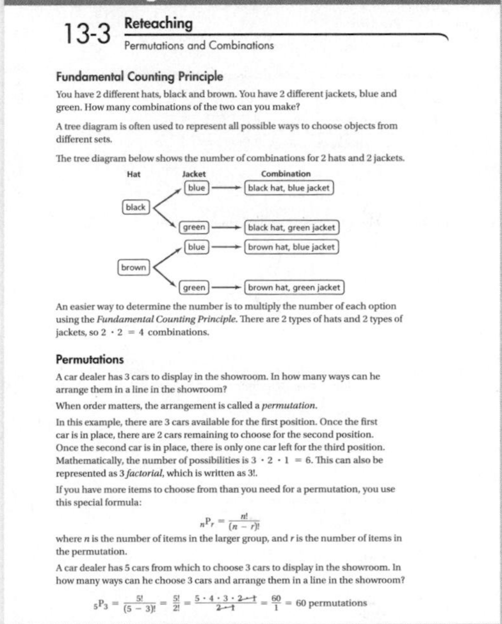

13-3 Reteaching
Permutations and Combinations

Fundamental Counting Principle

You have 2 different hats, black and brown. You have 2 different jackets, blue and green. How many combinations of the two can you make?

A tree diagram is often used to represent all possible ways to choose objects from different sets.

The tree diagram below shows the number of combinations for 2 hats and 2 jackets.

An easier way to determine the number is to multiply the number of each option using the *Fundamental Counting Principle*. There are 2 types of hats and 2 types of jackets, so $2 \cdot 2 = 4$ combinations.

Permutations

A car dealer has 3 cars to display in the showroom. In how many ways can he arrange them in a line in the showroom?

When order matters, the arrangement is called a *permutation*.

In this example, there are 3 cars available for the first position. Once the first car is in place, there are 2 cars remaining to choose for the second position. Once the second car is in place, there is only one car left for the third position. Mathematically, the number of possibilities is $3 \cdot 2 \cdot 1 = 6$. This can also be represented as 3 *factorial*, which is written as 3!.

If you have more items to choose from than you need for a permutation, you use this special formula:

$${}_nP_r = \frac{n!}{(n-r)!}$$

where n is the number of items in the larger group, and r is the number of items in the permutation.

A car dealer has 5 cars from which to choose 3 cars to display in the showroom. In how many ways can he choose 3 cars and arrange them in a line in the showroom?

$${}_5P_3 = \frac{5!}{(5-3)!} = \frac{5!}{2!} = \frac{5 \cdot 4 \cdot 3 \cdot 2 \cdot 1}{2 \cdot 1} = \frac{60}{1} = 60 \text{ permutations}$$

All-in-One Resources/Online

English Language Learner Support

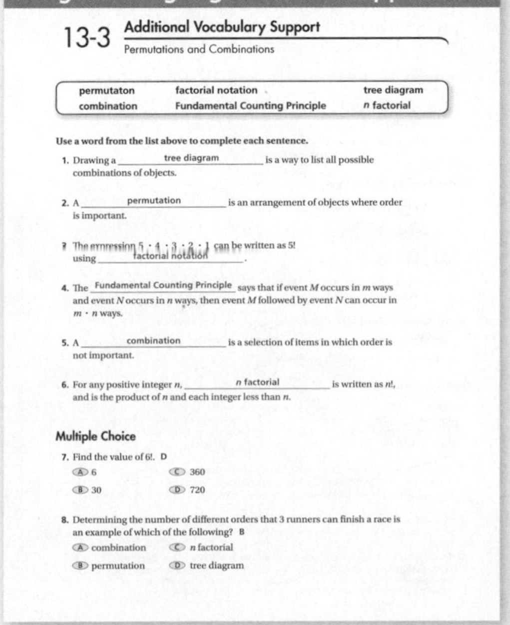

13-3 Additional Vocabulary Support
Permutations and Combinations

permutaton	factorial notation	tree diagram
combination	Fundamental Counting Principle	*n* factorial

Use a word from the list above to complete each sentence.

1. Drawing a tree diagram is a way to list all possible combinations of objects.
2. A permutation is an arrangement of objects where order is important.
3. The expression $5 \cdot 4 \cdot 3 \cdot 2 \cdot 1$ can be written as 5! using factorial notation.
4. The Fundamental Counting Principle says that if event M occurs in m ways and event N occurs in n ways, then event M followed by event N can occur in $m \cdot n$ ways.
5. A combination is a selection of items in which order is not important.
6. For any positive integer n, n factorial is written as $n!$, and is the product of n and each integer less than n.

Multiple Choice

7. Find the value of 6!. D
   - A 6
   - B 30
   - C 360
   - D 720
8. Determining the number of different orders that 3 runners can finish a race is an example of which of the following? B
   - A combination
   - B permutation
   - C n factorial
   - D tree diagram

Differentiated Remediation *continued*

On-Level

- **Practice** (2 pages) Provides extra practice for each lesson. For simpler practice exercises, use the Form K Practice pages found in the All-in-One Teaching Resources and online.
- **Think About a Plan** Helps students develop specific problem-solving skills and strategies by providing scaffolded guiding questions.
- **Standardized Test Prep** Focuses on all major exercises, all major question types, and helps students prepare for the high-stakes assessments.

Extension

- **Enrichment** Provides students with interesting problems and activities that extend the concepts of the lesson.
- **Activities, Games, and Puzzles** Worksheets that can be used for concepts development, enrichment, and for fun!

Practice and Problem Solving Wkbk/ All-in-One Resources/Online

Practice page 1

13-3 **Practice** *Form G*
Permutations and Combinations

1. A band sells t-shirts in 3 sizes and 2 different colors. How many different t-shirts are there to choose from? **6**

2. Each player on the baseball team can order a baseball bat using the table to the right. How many choices does each player have? **12 choices**

Finish	Length	Wood Type
Natural	32"	Ash
Black	33"	Maple
	34"	

3. In how many different orders can 5 runners finish a race? **120 orders**

4. Evaluate 7!. **5040**

5. What is the value of $\frac{25!}{24!}$? **25**

6. How many possible combinations of 3 items from a group of 5 are possible? **10 combinations**

7. Evaluate $_6P_3$. **120**

8. A basketball coach will choose 5 players from a group of 8 players to start the next game. How many different groups of starting players are possible? **56**

9. What is the value of $_nC_r$ when $n = 7$ and $r = 4$? **35**

10. What is the probability of randomly choosing a penny and a nickel from a cup of coins that contains a penny, a nickel, a dime, and a quarter? **$\frac{2}{6}$ OR equivalent fraction, decimal, or percent.**

11. Three playing cards are randomly chosen from a set numbered from 1 to 7. What is the probability that the chosen cards are numbered 1, 2 and 3? **approximately 0.17**

Practice and Problem Solving Wkbk/ All-in-One Resources/Online

Practice page 2

13-3 **Practice** (continued) *Form G*
Permutations and Combinations

12. **Recreation** When renting a bike from a local bike shop, you can choose from the types, sizes, and colors in the table shown below?

Type	Size	Color
Mountain	Small	Green
Cruising	Medium	Red
Road	Large	Blue

How many different choices do you have? **27**

13. **Open-Ended** Use an example to explain why you can use *n*! to find the number of possible orders for *n* objects. **Answers will vary. Sample: There are 4 people in a race. There are 4 possible first-place finishers. Once the first place finisher is determined, there are 3 possible 2nd-place, etc. So the number of orders is $4 \cdot 3 \cdot 2 \cdot 1 = 4! = 24$. For a race with *n* people, there are *n*! possible orders of finishing.**

14. **Reasoning** A hiker has 2 pairs of hiking shoes, 3 different shirts, and 2 different pairs of shorts to choose from. How does the number of combinations of shoes, shirts, and shorts change as the hiker adds shirts to his collection? Explain. **Each time the hiker adds a shirt to his collection, 4 additional combinations are possible. The initial number of combinations is $2 \cdot 3 \cdot 2 = 12$. When another shirt is added, the number of combinations becomes $2 \cdot 4 \cdot 2 = 16$.**

15. **Business** For each weekly meeting of a group of business leaders, members take turns being the note-taker, the facilitator, and the speaker. In how many different ways can these positions be chosen from the 9 members? **504 ways**

16. **Writing** Explain how the Fundamental Counting Principle is related to a tree diagram. **A tree diagram lists options of each type in columns, and links each type of option to show outcomes. The Fundamental Counting Principle uses numbers only. The numbers of options in each column of a tree diagram are multiplied, yielding the number of possible outcomes.**

17. A game at the fair involves ping-pong balls numbered 1 to 18. You can win a prize if you correctly choose the 5 numbers that are randomly drawn. What are your chances of winning? **approximately 0.014**

All-in-One Resources/Online

Enrichment

13-3 **Enrichment**
Permutations and Combinations

You can determine the possible number of permutations for a problem even when there are special conditions.

You are asked to arrange the following 6 shapes in a line.

The special conditions are that the triangles and quadrilaterals must alternate, and the first shape must be a triangle. How many 6-shape arrangements, with no repetitions, satisfy these conditions? Fill in the blanks in the equation below by answering the questions.

__3__ · __3__ · __2__ · __2__ · __1__ · __1__ = __36__ arrangements

1. How many possibilities are there for the first shape? Fill in the first blank with your answer. Explain. **There are 3 possibilities, since the shape needs to be a triangle and there are 3 triangles.**

2. How many possibilities are there for the second shape? Fill in the second blank with your answer. Explain. **There are 3 possibilities, since the shape needs to be a quadrilateral and there are 3 quadrilaterals.**

3. How many possibilities are there for the third shape? Fill in the third blank with your answer. Explain. **There are 2 possibilities, since one triangle has already been used in the first position, and the shape needs to be a triangle.**

4. How many possibilities are there for the fourth shape? Fill in the fourth blank with your answer. Explain. **There are 2 possibilities, since one quadrilateral has already been used in the second position, and the shape needs to be a quadrilateral.**

5. Fill in the remaining blanks. How many possible arrangements are there? **36**

6. How many possible arrangements are there *without* the special conditions? **720**

Practice and Problem Solving Wkbk/ All-in-One Resources/Online

Think About a Plan

13-3 **Think About a Plan**
Permutations and Combinations

A 4-digit code is needed to unlock a bicycle lock. The digits 0 through 9 can be used only once in the code.

a. What is the probability that all of the digits are even? (Hint: 0 is not considered an even number.)

b. **Writing** These types of locks are usually called combination locks. What name might a mathematician prefer? Why?

Understanding the Problem

1. There are [10] possible numbers that can appear in a code.

2. Of those, a total of [4] digit(s) are even.

Planning the Solution

3. To solve the problem, I need to find __the probability that the numbers chosen for the lock are 2, 4, 6 and 8__.

4. Use the probability formula: P(choosing [2, 4, 6, and 8]) equals the number of possible ways to choose __2, 4, 6, and 8__ divided by the total number of ways to choose __4__ numbers.

Getting an Answer

5. Find the number of possible ways to choose 2, 4, 6, and 8. There are __4__ way(s) to pick the first digit, __3__ way(s) to pick the second digit, __2__ way(s) to pick the third digit, and __1__ way(s) to pick the fourth digit, so there are __4__ × __3__ × __2__ × __1__ = __24__ ways.

6. Because choosing the code 2-4-6-8 is not the same as choosing the code 8-6-4-2, the order does matter. Use the __permutation__ formula to find the total number of ways to choose __4__ numbers from __10__ numbers. $_{10}P_4 = 5040$

7. Find probability that all of the digits are even by finding the quotient of the answers to Step 7 and Step 8. Write your answer in simplest form. $\frac{24}{5040} = \frac{1}{210}$

8. These types of locks are usually called combination locks. What name might a mathematician prefer? Why? **Order does not matter for combinations, but order does matter in the code for a lock. Therefore, the number of possible codes is actually a permutation, so a better name might be a "permutation lock."**

Practice and Problem Solving Wkbk/ All-in-One Resources/Online

Standardized Test Prep

13-3 **Standardized Test Prep**
Permutations and Combinations

Gridded response

Solve each exercise and enter your answer on the grid provided.

1. In how many different ways can 6 books be arranged on a bookshelf?

2. The options for a college's science classes are shown in the table.

Title	Grade Types	Times
Science 101	Pass/Fail	9:00 am
Science 105	Letter Grade	10:30 am

How many combinations of class, type, and times are available?

3. How many combinations of 4 fish can you choose from a tank containing 8 fish?

4. A bag contains 7 marbles: one each of red, orange, yellow, green, blue, violet, and white. A child randomly pulls 4 marbles from the bag. What is the probability that the marbles chosen are green, blue, red, and yellow? Round your answer to the nearest hundredth.

5. A teacher wants to choose one student to take attendance, one student to hand out papers, and one student to collect homework. If there are 16 students in the class, in how many different ways can the students be chosen?

Answers

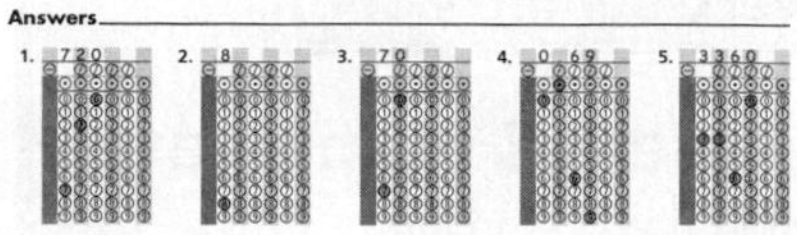

Online Teacher Resource Center

Activities, Games, and Puzzles

13-3 **Game: Combination Bingo**
Combinations

Setup

Choose numbers from those listed below to make a game board. You will only use 25 of the numbers. Place them randomly in the squares.

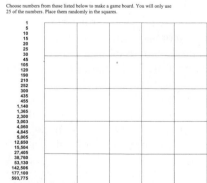

1
5
10
15
20
25
30
45
105
120
190
210
252
300
435
455
1,140
1,365
2,300
3,003
4,060
4,845
5,005
12,650
15,504
27,405
38,760
53,130
142,506
177,100
593,775

Game Play

The game should be played as a class or in groups of 8 to 10.

Take turns rolling a pair of number cubes. Use the greatest number rolled as *n* and the other number rolled as *r*.

Each player calculates $_nC_r$. Each player with the calculated value on his or her game board marks the square. Pass the number cubes to another player for the next roll. Continue to roll and calculate the value of the expression until one player marks off five squares in a row—vertical, horizontal, or diagonal. That player is the winner.

13 Mid-Chapter Quiz

MathXL® for School
Go to PowerGeometry.com

Do you know HOW?

A bag contains colored tiles. The tiles are randomly selected, and the results are recorded in the table below. Find the experimental probabilities.

Colored Tiles				
Tile	red	blue	green	yellow
Frequency	8	10	9	12

1. *P*(red) **2.** *P*(green) **3.** *P*(yellow)

4. You roll two standard number cubes. What is the probability that you roll two odd numbers?

Suppose that you choose a number at random from the set {2, 5, 8, 10, 12, 19, 35}.

5. What is the probability that the number is even?

6. What is the probability that the number is not less than 10?

Students in the sophomore class at a local high school were asked whether they were left-handed or right-handed. The results of the survey are below.

Handedness	Frequency
Left-Handed	72
Right-Handed	478

7. What is the relative frequency of sophomores that are left-handed?

8. What is the probability distribution for this data?

Evaluate each expression.

9. 8! **10.** 3!

11. $\frac{9!}{5!}$ **12.** $\frac{12!}{5!\,7!}$

13. ${}_8C_5$ **14.** ${}_{15}P_5$

15. ${}_4C_3$ **16.** ${}_{10}C_9$

In Exercises 17 and 18, determine whether each situation involves a permutation or a combination.

17. How many ways can presentation groups of 6 students be formed from a class of 24 students?

18. How many ways can you arrange 5 books on a shelf from a group of 9 books?

19. How many different security codes similar to A12 can you form using the digits 1–4 and the letters A, B, and N?

20. Elections A club has 16 members. All of the members are eligible for the offices of president, vice president, secretary, and treasurer. In how many different ways can the officers be elected?

21. The names of all 19 students in a class are written on slips of paper and placed in a basket. How many different ways can 5 names be randomly chosen from the basket?

Do you UNDERSTAND?

22. Reasoning The probability of an event is $\frac{4}{5}$. Is this event likely or unlikely to occur? Explain.

23. Compare and Contrast What is the same about experimental and theoretical probability? What is different?

24. Open-Ended Give an example of a situation that can be described using permutations and another situation that can be described using combinations.

25. Error Analysis The frequency table below shows the results of an experiment where a coin is tossed 100 times. Your friend says that the results show that the experimental probability of tossing heads is $\frac{42}{58}$. Explain your friend's error.

Coin Shows	Frequency
Heads	42
Tails	58

Answers

Mid-Chapter Quiz

1. $\frac{8}{39}$

2. $\frac{9}{39}$

3. $\frac{12}{39}$

4. $\frac{1}{4}$

5. $\frac{4}{7}$

6. $\frac{4}{7}$

7. $\frac{36}{275}$

8. left-handed: $\frac{36}{275}$, right-handed: $\frac{239}{275}$

9. 40,320

10. 6

11. 3024

12. 792

13. 56

14. 360,360

15. 4

16. 10

17. combination

18. permutation

19. 48

20. 43,680

21. 11,628

22. The event is likely to occur because the probability is close to 1.

23. Both experimental and theoretical probability are computed by finding the ratio of the number of favorable outcomes to the total number of outcomes possible. However, theoretical probability is based on what should happen mathematically and experimental probability is based on what is observed.

24. Answers may vary. Check students' work.

25. The friend found the ratio heads : tails instead of heads : total tosses.

PowerGeometry.com

MathXL for School

Prepare students for the Mid-Chapter Quiz and Chapter Test with online practice and review.

1 Interactive Learning

Solve It!

PURPOSE To explore how to find the probability of a compound event.

PROCESS Students may conclude that the weather forecasts on the same day in Philadelphia and San Diego are independent because of the great distance between the two cities.

FACILITATE

Q Do you think you should add the probabilities for the chance of rain in each city? Why or why not? **[No, this would result in a probability greater than 100% which is not possible.]**

ANSWER See Solve It in Answers on next page.

CONNECT THE MATH In the Solve It, students explore a situation where they need to find the probability that two different events will both occur. In this lesson, students will learn how to find the probability of compound events including independent, mutually exclusive, and overlapping events.

2 Guided Instruction

Problem 1

ELL SUPPORT

Ask students to give examples of how they use the words *dependent* and *independent* in a context outside of mathematics. Then discuss how these concepts apply to the mathematical definitions of dependent and independent events.

13-4 Compound Probability

Common Core State Standards
S-CP.B.7 Apply the Addition Rule, $P(A \text{ or } B) = P(A) + P(B) - P(A \text{ and } B)$ . . . **Also S-CP.B.8, S-CP.B.9**
MP 3, MP 4, MP 6

Objective To identify independent and dependent events
To find compound probabilities

MATHEMATICAL PRACTICES

Solve It! Getting Ready!

Suppose you are traveling from Philadelphia, PA, to San Diego, CA. Do you think the probability of rain in Philadelphia affects the probability of rain in San Diego? Justify your reasoning.

Philadelphia Today	Philadelphia Tomorrow	San Diego Today	San Diego Tomorrow
Rain	Sunny	Cloudy	Partly Cloudy
43° F	58° F	70° F	75° F
Chance of rain: 80%	Chance of rain: 0%	Chance of rain: 50%	Chance of rain: 10%

Lesson Vocabulary
- compound event
- independent events
- dependent events
- mutually exclusive events
- overlapping events

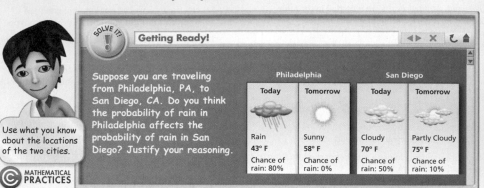

If you were to find the probability of rain in both cities in the Solve It, you would be finding the probability of a *compound event.* A **compound event** is an event that is made up of two or more events.

Essential Understanding You can find the probability of compound events by using the probability of each part of the compound event.

If the occurrence of an event does not affect how another event occurs, the events are called **independent events**. If the occurence of an event does affect how another event occurs, the events are called **dependent events**. To calculate the probability of a compound event, first determine whether the events are independent or dependent.

Think
How can you tell that two events are independent?
Two events are independent if one does not affect the other.

Problem 1 Identifying Independent and Dependent Events

Are the outcomes of each trial independent or dependent events?

A Choose a number tile from 12 tiles. Then spin a spinner.

The choice of number tile does not affect the spinner result. The events are independent.

B Pick one card from a set of 15 sequentially numbered cards. Then, without replacing the card, pick another card.

The first card chosen affects the possible outcomes of the second pick, so the events are dependent.

13-4 Preparing to Teach

BIG idea Probability

ESSENTIAL UNDERSTANDING
- The probability of compound events can be found by using the probability of each part of the compound event.

Math Background

When the outcome of one event affects the outcome of another event, the events are *dependent* events. When the outcomes do not affect each other, the events are *independent*.

- For independent events A and B, the probability that both A and B will occur is given by $P(A \text{ or } B) = P(A) \cdot P(B)$.
- *Mutually exclusive* events cannot happen at the same time. For mutually exclusive events A and B, the probability that either A or B occurs is given by $P(A \text{ or } B) = P(A) + P(B)$.
- *Overlapping* events have some common outcomes. For overlapping events A and B, the probability that either A or B occurs is given by $P(A \text{ or } B) = P(A) + P(B) - P(A \text{ and } B)$.

Mathematical Practices

Construct viable arguments. Students must analyze situations and determine which formula for compound probability is appropriate for the given situation.

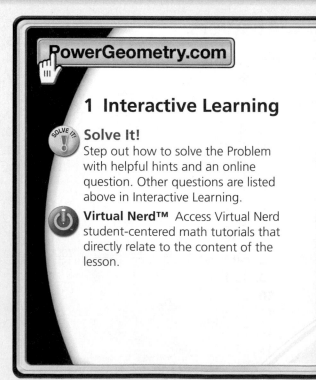

PowerGeometry.com

1 Interactive Learning

Solve It!
Step out how to solve the Problem with helpful hints and an online question. Other questions are listed above in Interactive Learning.

Virtual Nerd™ Access Virtual Nerd student-centered math tutorials that directly relate to the content of the lesson.

Got It? 1. You roll a standard number cube. Then you flip a coin. Are the outcomes independent or dependent events? Explain.

You can find the probability that two independent events will both occur by multiplying the probabilities of each event.

take note

Key Concept Probability of A and B

If A and B are independent events, then $P(A \text{ and } B) = P(A) \cdot P(B)$.

Problem 2 Finding the Probability of Independent Events

A desk drawer contains 5 red pens, 6 blue pens, 3 black pens, 24 silver paper clips, and 16 white paper clips. If you select a pen and a paper clip from the drawer without looking, what is the probability that you select a blue pen and a white paper clip?

Plan

Why are the events independent?
Selecting a blue pen has no affect on selecting a white paper clip.

Step 1 Let A = selecting a blue pen. Find the probability of A.

$P(A) = \frac{6}{14} = \frac{3}{7}$ 6 blue pens out of 14 pens

Step 2 Let B = selecting a white paper clip. Find the probability of B.

$P(B) = \frac{16}{40} = \frac{2}{5}$ 16 white paper clips out of 40 clips

Step 3 Find $P(A \text{ and } B)$. Use the formula for the probability of independent events.

$P(A \text{ and } B) = P(A) \cdot P(B) = \frac{3}{7} \cdot \frac{2}{5} = \frac{6}{35} \approx 0.171$, or 17.1%

The probability that you select a blue pen and a white paper clip is about 17.1%.

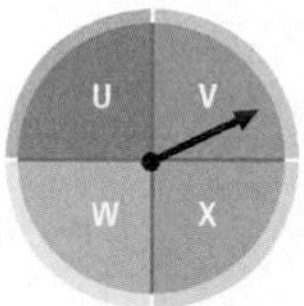

Got It? 2. You roll a standard number cube and spin the spinner at the right. What is the probability that you roll a number less than 3 and the spinner lands on a vowel?

Events that cannot happen at the same time are called **mutually exclusive events**. For example, you cannot roll a 2 and a 5 on a standard number cube at the same time, so the events are mutually exclusive. If events A and B are mutually exclusive, then the probability of both A and B occurring is 0. The probability that either A or B occurs is the sum of the probability of A occurring and the probability of B occurring.

take note

Key Concept Probability of Mutually Exclusive Events

If A and B are mutually exclusive events, then $P(A \text{ and } B) = 0$, and $P(A \text{ or } B) = P(A) + P(B)$.

Q Will picking a number tile have any impact on the spinner result? Explain. **[No, the spinner can land anywhere regardless of what tile is picked.]**

Take Note

ERROR PREVENTION

Point out to students that the multiplication rule only applies to independent events, so it is very important to first determine whether or not two events are independent.

Problem 2

In this problem, students will find the probability of two independent events.

Q After you have selected a pen and a paper clip, will the probability for the next trial be the same? Explain. **[No, if the first pen and paper clip are not replaced in the desk, the sample space of possible outcomes will change.]**

Got It?

Q Are the events dependent or independent? Explain. **[The events are independent because the number rolled will not affect the letter that is spun.]**

2 Guided Instruction

Each Problem is worked out and supported online.

Problem 1
Identifying Independent and Dependent Events

Problem 2
Finding the Probability of Independent Events

Problem 3
Finding the Probability of Mutually Exclusive Events

Problem 4
Finding Probabilities of Overlapping Events

Support in Geometry Companion
- Vocabulary
- Key Concepts
- Got It?

Answers

Solve It!

The probability of rain in one city does not affect the probability of rain in the other city because the cities are very far apart.

Got It?

1. independent; the outcomes do not affect each other
2. $\frac{1}{12}$

Problem 3

In this problem, students will find a compound probability for two events that cannot occur at the same time.

Q Why are the probabilities added instead of multiplied in Problem 3? **[because the events are mutually exclusive and you are finding *P*(*A* or *B*).]**

Q How would a frequency table help you find the probability of mutually exclusive events? **[You can add the frequencies of two categories *A* and *B* and divide by the total number to find *P*(*A* or *B*).]**

Got It?

Q What assumption do you need to make in order to solve this problem? Explain. **[That an athlete cannot be on both the baseball team and the track team.]**

Take Note

VISUAL LEARNERS

Use Venn diagrams to show the relationships for overlapping and mutually exclusive events.

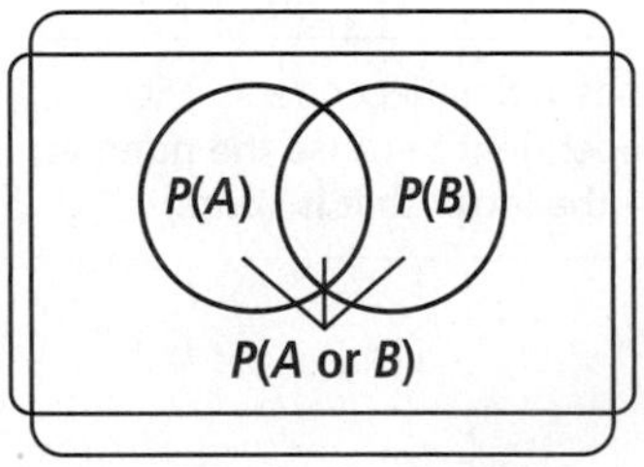

Overlapping Events

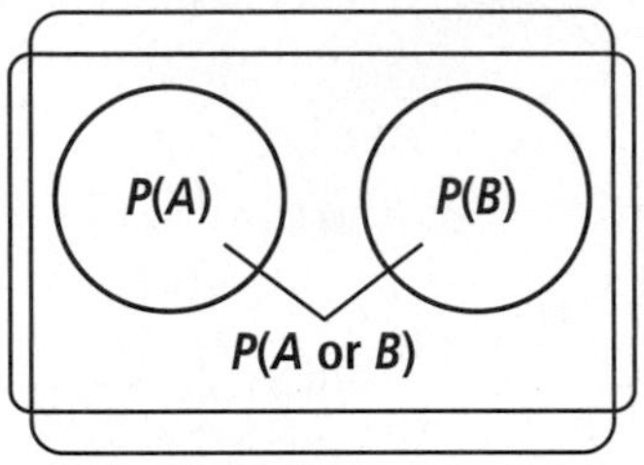

Mutually Exclusive Events

Plan

Is there a way to simplify this problem? You can model the probabilities with a simpler problem. Suppose there are 100 athletes. In the model 28 athletes will play basketball, and 24 will be on the swim team.

Problem 3 Finding the Probability of Mutually Exclusive Events

Athletics **Student athletes at a local high school may participate in only one sport each season. During the fall season, 28% of student athletes play basketball and 24% are on the swim team. What is the probability that a randomly selected student athlete plays basketball or is on the swim team?**

Because athletes participate in only one sport each season, the events are mutually exclusive. Use the formula $P(A \text{ or } B) = P(A) + P(B)$.

$P(\text{basketball or swim team}) = P(\text{basketball}) + P(\text{swim team})$

$= 28\% + 24\% = 52\%$ Substitute and Simplify.

The probability of an athlete either playing basketball or being on the swim team is 52%.

Got It? **3.** In the spring season, 15% of the athletes play baseball and 23% are on the track team. What is the probability of an athlete either playing baseball or being on the track team?

Overlapping events have outcomes in common. For example, for a standard number cube, the event of rolling an even number and the event of rolling a multiple of 3 overlap because a roll of 6 is a favorable outcome for both events.

take note

Key Concept **Probability of Overlapping Events**

If *A* and *B* are overlapping events, then $P(A \text{ or } B) = P(A) + P(B) - P(A \text{ and } B)$.

Here's Why It Works Suppose you have 7 index cards, each having one of the following letters written on it:

A B C D E F G

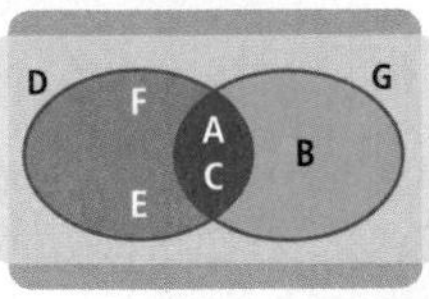

P(FACE), the probability of selecting a letter from the word FACE, is $\frac{4}{7}$. *P*(CAB), the probability of selecting a letter from the word CAB, is $\frac{3}{7}$.

Consider *P*(FACE or CAB), the probability of choosing a letter from either the word FACE or the word CAB. These events overlap since the words have two letters in common. If you simply add *P*(FACE) and *P*(CAB), you get $\frac{4}{7} + \frac{3}{7} = \frac{4+3}{7}$. The value of the numerator should be the number of favorable outcomes, but there are only 5 distinct letters in the words FACE and CAB. The problem is that when you simply add, the letters A and C are counted twice, once in the favorable outcomes for the word FACE, and once for the favorable outcomes for the word CAB. You must subtract the number of letters that the two words have in common so they are only counted once.

$$P(\text{FACE or CAB}) = \frac{4+3-2}{7} = \frac{4}{7} + \frac{3}{7} - \frac{2}{7} = P(\text{FACE}) + P(\text{CAB}) - P(\text{AC})$$

Additional Problems

1. Cameron selects a letter tile, does not replace it, and then selects another letter tile. Are these events independent or dependent?
ANSWER dependent

2. Tyrone selects a letter of the alphabet and chooses a digit from 0 through 9. What is the probability that he selects a consonant and a number greater than 3?
ANSWER $\frac{63}{130}$

3. At Danielle's school, 34% of the students take the bus to school and 12% walk. What is the probability that a student at Danielle's school rides the bus or walks?
ANSWER 46%

4. Suppose Caleb rolls 2 standard number cubes. What is the probability that the sum is greater than 9 or even?
ANSWER $\frac{5}{9}$

Answers

Got It? (continued)

3. 38%

Problem 4 Finding Probabilities of Overlapping Events

What is the probability of rolling either an even number or a multiple of 3 when rolling a standard number cube?

Know	Need	Plan
You are rolling a standard number cube. The events are overlapping events because 6 is both even and a multiple of 3.	You need the probability of rolling an even number and the probability of rolling a multiple of 3.	Find the probabilities and use the formula for probabilities of overlapping events.

Think

Why do you need to subtract the overlapping probability?
If the overlapping probability is not subtracted, it is counted twice. This would introduce an error.

$$P(\text{even or multiple of } 3) = P(\text{even}) + P(\text{multiple of } 3) - P(\text{even and multiple of } 3)$$
$$= \frac{3}{6} + \frac{2}{6} - \frac{1}{6}$$
$$= \frac{4}{6}, \text{ or } \frac{2}{3}$$

The probability of rolling an even or a multiple of 3 is $\frac{2}{3}$.

Got It? **4.** What is the probability of rolling either an odd number or a number less than 4 when rolling a standard number cube?

Lesson Check

Do you know HOW?

1. Suppose A and B are independent events. What is $P(A \text{ and } B)$ if $P(A) = 50\%$ and $P(B) = 25\%$?
2. Suppose A and B are mutually exclusive events. What is $P(A \text{ or } B)$ if $P(A) = 0.6$ and $P(B) = 0.25$?
3. Suppose A and B are overlapping events. What is $P(A \text{ and } B)$ if $P(A) = \frac{1}{3}$; $P(B) = \frac{1}{2}$ and $P(A \text{ and } B) = \frac{1}{5}$?

Do you Understand? MATHEMATICAL PRACTICES

4. **Open-Ended** Give an example of independent events, and an example of dependent events. Describe how the examples differ.
5. **Error Analysis** Your brother says that being cloudy tomorrow and raining tomorrow are independent events. Explain your brother's error.

Practice and Problem-Solving Exercises MATHEMATICAL PRACTICES

A Practice **Determine whether the outcomes of the two actions are *independent* or *dependent* events.**

See Problem 1.

6. You toss a coin and roll a number cube.
7. You draw a marble from a bag without looking. You do not replace it. You draw another marble from the bag.
8. Choose a card at random from a standard deck of cards and replace it. Then choose another card.
9. Ask a student's age and ask what year the student expects to graduate.

Problem 4

In this problem, students will find $P(A \text{ or } B)$ for overlapping events.

Q How would you find P(even *and* multiple of 3)? **[Find the number of outcomes that are both even and a multiple of 3. Then divide by the size of the sample space.]**

Got It?

Q How many possible outcomes are there when the number cube is rolled? How many of them are either odd or less than 4? **[6; 4]**

3 Lesson Check

Do you know HOW? ERROR INTERVENTION

- If students have difficulty remembering to subtract $P(A \text{ and } B)$ in Exercise 3, then show them a Venn diagram to illustrate why this must be subtracted for overlapping events.

Do you UNDERSTAND?

- If students have difficulty identifying the error in Exercise 5, then ask them what conditions need to be met in order for it to rain.

Close

Q How are independent and dependent events different? **[The outcomes of dependent events affect each other, but the outcomes of independent events do not.]**

Q How do you find $P(A \text{ and } B)$ for two independent events? **[Multiply $P(A)$ by $P(B)$.]**

Q How do you find $P(A \text{ or } B)$ for any two events? **[Add $P(A)$ to $P(B)$ and subtract $P(A \text{ and } B)$.]**

Answers

Got It? (continued)

4. $\frac{2}{3}$

Lesson Check

1. 12.5%
2. 0.85
3. $\frac{19}{30}$
4. Answers will vary. When the outcomes do not affect each other, the events are independent. If the outcome of one event affects the outcome of another event, they are dependent events.
5. Sample answer: The events are dependent because it needs to be cloudy in order for it to rain.

Practice and Problem Solving Exercises

6. independent
7. dependent
8. independent
9. dependent

4 Practice

ASSIGNMENT GUIDE

Basic: 6–20, 22, 24
Average: 7–21 odd, 23–28
Advanced: 7–21 odd, 23–29
Standardized Test Prep: 30–32
Mixed Review: 33–38

Mathematical Practices are supported by exercises with red headings. Here are the Practices supported in this lesson:

MP 3: Communicate Ex. 4, 25
MP 3: Critique the Reasoning of Others Ex. 5
MP 3: Construct Arguments Ex. 27–28
MP 4: Model with Mathematics Ex. 22
MP 6: Attend to Precision Ex. 29

Applications exercises have blue headings.

EXERCISE 21: Use the Think About a Plan worksheet in the **Practice and Problem Solving Workbook** (also available in the Teaching Resources in print and online) to further support students' development in becoming independent learners.

HOMEWORK QUICK CHECK

To check students' understanding of key skills and concepts, go over Exercises 6–9, 21, 22, and 24.

You spin the spinner at the right and without looking, you choose a tile from a set of tiles numbered from 1 to 10. Find each probability. See Problem 2.

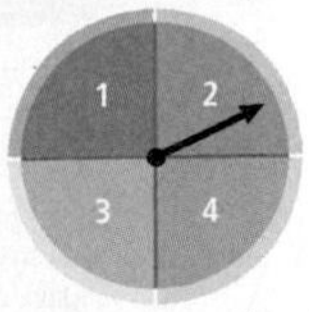

10. P(spinner lands on 2 and choose a 3)

11. P(spinner lands on an odd number and choose an even number)

12. P(spinner lands a number less than 4 and choose a 9 or 10)

A bag contains 3 blue chips, 6 black chips, 2 green chips, and 4 red chips. Use this information to find each probability if a chip is selected at random. See Problem 3.

13. P(blue chip or black chip)

14. P(green chip or red chip)

15. P(green chip or black chip)

16. P(blue, black, or red chip)

A set of cards contains four suits (red, blue, green, and yellow). In each suit there are cards numbered from 1 to 10. Calculate the following probabilities for one card selected at random. See Problem 4.

17. P(blue card or card numbered 10)

18. P(green or yellow card, or card numbered 1)

19. P(red card or card greater than 5)

20. P(red or blue card, or card less than 6)

B Apply

21. Pets In a litter of 8 kittens, there are 2 brown females, 1 brown male, 3 spotted females, and 2 spotted males. If a kitten is selected at random, what is the probability that the kitten will be female or brown?

22. Think About A Plan Suppose you are taking a test and there are three multiple-choice questions that you do not know the answers to. Each has four answer choices. Rather than leave the answers blank, you decide to guess. What is the probability that you answer all three questions correctly?
- Is each guess independent or dependent?
- What is the probability that a random guess answers a question when there are four answer choices?

23. Vacation In a math class, 75% of the students have visited the ocean and 50% have visited the mountains on vacation before. If 45% of the students have visited the ocean and the mountains on vacation before, what is the probability that a randomly selected student has visited the ocean or the mountains?

24. What is the probability that a standard number cube rolled three times will roll first even, then odd, and then even?

25. Writing Describe the difference between mutually exclusive and overlapping events. Give examples of each.

26. When you draw a marble out of a bag and then draw another without replacing the first, the probability of the second event is different from the probability of the first.
a. What is the probability of drawing a red marble out of a bag containing 3 red and 7 blue marbles?
b. What is the probability of drawing a second red marble if a red marble is drawn the first time and not replaced?
c. What is the probability of drawing two red marbles in a row?

Answers

Practice and Problem-Solving Exercises (continued)

10. $\frac{1}{40}$
11. $\frac{1}{4}$
12. $\frac{3}{20}$
13. $\frac{3}{5}$
14. $\frac{2}{5}$
15. $\frac{8}{15}$
16. $\frac{13}{15}$
17. $\frac{13}{40}$
18. $\frac{11}{20}$
19. $\frac{5}{8}$
20. $\frac{3}{4}$
21. $\frac{3}{4}$
22. $\frac{1}{64}$
23. 80%
24. $\frac{1}{8}$
25. Mutually exclusive events cannot happen at the same time, but overlapping events can occur at the same time.
26a. $\frac{3}{10}$
b. $\frac{2}{9}$
c. $\frac{1}{15}$

Challenge

Reasoning For each set of probabilities, determine if the events A and B are mutually exclusive. Explain.

27. $P(A) = \frac{1}{2}$, $P(B) = \frac{1}{3}$, $P(A \text{ or } B) = \frac{2}{3}$

28. $P(A) = \frac{1}{6}$, $P(B) = \frac{3}{8}$, $P(A \text{ and } B) = 0$

29. Reasoning Are mutually exclusive events dependent or independent? Explain.

Standardized Test Prep

SAT/ACT

30. Which of the following statements is NOT true?

Ⓐ The side lengths of an isosceles right triangle can be all whole numbers.

Ⓑ The side lengths of a right triangle can form a Pythagorean triple.

Ⓒ The side lengths of an equilateral triangle can be all whole numbers.

Ⓓ The angle measures of an equilateral triangle can be all whole numbers.

Short Response

31. An arc of a circle measures 90° and is 10 cm long. How long is the circle's diameter?

32. You roll a standard number cube and then spin the spinner shown at the right. What is the probability that you will roll a 5 and spin a 3?

Mixed Review

Calculate the following permutations and combinations. See Lesson 13-3.

33. The number of 3 letter sequences that can be made without reusing any letter.

34. The number of ways that 8 runners can finish a race, if there are no ties.

35. The number of ways a 5-member subcommittee can be formed from a 12-member student government.

Get Ready! To Prepare for Lesson 13-5, do Exercises 36–38.

See Lesson 13-2.

Students were asked about the number of siblings they have. The results of the survey are shown in the frequency table at the right. Find the following probabilities if a student is chosen at random from the respondents.

36. $P(\text{2 siblings})$

37. $P(\text{fewer than 3 siblings})$

38. $P(\text{more than 1 sibling})$

Number of Siblings	Frequency
0	5
1	12
2	15
3	7

27. No; Answers may vary. Sample: $\frac{1}{2} + \frac{1}{3} \neq \frac{2}{3}$

28. Yes; events A and B are mutually exclusive when $P(A \text{ and } B) = 0$.

29. dependent events; If one event occurs, then the other event cannot occur because they are mutually exclusive. So, the outcome of the second event depends on the outcome of the first event.

30. A

31. $\frac{40}{\pi}$ cm

32. $\frac{1}{36}$

33. 15,600

34. 40,320

35. 792

36. $\frac{5}{13}$

37. $\frac{32}{39}$

38. $\frac{22}{39}$

13-4 Lesson Resources

Differentiated Remediation

Additional Instructional Support

Geometry Companion

Students can use the **Geometry Companion** worktext (4 pages) as you teach the lesson. Use the Companion to support

- Solve It!
- New Vocabulary
- Key Concepts

13-4 Compound Probability

Vocabulary

Review

Fill in the blanks to complete each sentence.

1. For the spinner on the right, what is the probability of the spinner landing on each number?

   $P(1) = \frac{2}{5}$. $P(2) = \frac{2}{5}$ $P(3) = \frac{1}{5}$

Fill in the blank to complete each sentence.

2. An event is an outcome or a group of outcomes
3. The probability of an *impossible* event is 0 .
   The probability of a *certain* event is 1 .

Vocabulary Builder

Independent events (noun) in-da-PEN-dant i-VENTS

Related Words: compound event, dependent events, mutually exclusive events

Definition: If the occurrence of an event does not affect how another event occurs, the events are called **independent events.**

Example: Being odd and being even are independent events.

Use Your Vocabulary

Complete each statement with a word from the list. Use each word only once.

independence independent independently

4. The contest rules required that each participant complete the tasks _?_. independently
5. During the American Revolution, the Thirteen Colonies fought for their _?_ from Great Britain. independence
6. The pharmaceutical company hired an _?_ research group to verify its latest results. independent

ELL Support

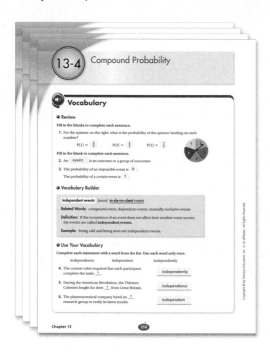

Have Students Create Test Items In order to deepen students' understanding of compound events and how to find the probabilities of compound events, have students create their own test items. Instruct students to write a test item that involves finding the probability of independent events, mutually exclusive events, or overlapping events. Each author should solve their own problem and keep the solution to themselves. Have students exchange their test items with each other and solve. Then have each pair of students compare whether or not they used the same approach and came up with the same solution.

5 Assess & Remediate

Lesson Quiz

Use the Lesson Quiz to assess students' mastery of the skills and concepts of this lesson.

1. **Do you UNDERSTAND?** Regina eats a piece of fruit from a basket of apples, bananas, pears, and melons. Later her brother eats a piece of fruit from the basket. Do these represent independent or dependent events?
2. What is $P(A \text{ and } B)$ if A and B are independent events, $P(A) = 0.24$, and $P(B) = 0.15$?
3. At a grocery store, there is a 62% chance that a customer will pay with a credit card and a 25% chance that a customer will pay with cash. What is the probability that a randomly selected customer will pay with a credit card or with cash?
4. Angelo chooses a number from 1 to 10. What is the probability that he chooses an odd number or a multiple of 5?

ANSWERS TO LESSON QUIZ

1. dependent
2. 0.036
3. 87%
4. $\frac{3}{5}$

PRESCRIPTION FOR REMEDIATION

Use the student work on the Lesson Quiz to prescribe a differentiated review assignment.

Points	Differentiated Remediation
0–2	Intervention
3	On-level
4	Extension

PowerGeometry.com

5 Assess & Remediate

Assign the Lesson Quiz. Appropriate intervention, practice, or enrichment is automatically generated based on student performance.

Intervention

- **Reteaching** (2 pages) Provides reteaching and practice exercises for the key lesson concepts. Use with struggling students or absent students.
- **English Language Learner Support** Helps students develop and reinforce mathematical vocabulary and key concepts.

All-in-One Resources/Online

Reteaching

13-4 **Reteaching**
Compound Probability

Events that cannot happen at the same time are *mutually exclusive events.*

For mutually exclusive events: $P(A \text{ and } B) = 0$
$P(A \text{ or } B) = P(A) + P(B)$

Problem

You put 8 tiles, numbered 3 through 10, into a bag.

a. What is the probability of randomly choosing a tile that is both an even number and a prime number?

   The two events are: Event A: randomly choose a tile that is an even number
   Event B: randomly choose a tile that is a prime number

   It is not possible to choose a tile that is an even number *and* a prime number, since all even numbers between 3 and 10 are divisible by 2. Since the events cannot happen at the same time, or during the same choice, the events are mutually exclusive.

   $$P(A \text{ and } B) = 0$$

b. What is the probability of randomly choosing a tile that is an even number or a prime number?

   There are 4 even tiles: $P(A) = \frac{4}{8} = \frac{1}{2}$

   There are 3 prime number tiles: $P(B) = \frac{3}{8}$

   Using substitution, $P(\text{even or prime})$ is

   $$P(A \text{ or } B) = P(A) + P(B) = \frac{1}{2} + \frac{3}{8} = \frac{4}{8} + \frac{3}{8} = \frac{7}{8}$$

Exercises

1. Event A and Event B are mutually exclusive events.
   a. Evaluate $P(A \text{ or } B)$ when $P(A) = \frac{1}{3}$ and $P(B) = \frac{2}{9}$. $\frac{5}{9}$
   b. Evaluate $P(A \text{ or } B)$ when $P(A) = \frac{2}{5}$ and $P(B) = \frac{3}{7}$. $\frac{29}{35}$
2. A tank at the aquarium holds a variety of fish, of which 24% are catfish, 13% are black tetras, and 8% are angel fish. What is the probability of randomly choosing a fish that is a black tetra or an angel fish? **0.21 or 21%**
3. A teacher surveyed students to determine where to go on a field trip. The results were 43% preferred a water park, 33% preferred the beach, and 24% preferred the zoo. What is the probability of randomly choosing a student who prefers the beach or the zoo? **0.57 or 57%**

All-in-One Resources/Online

English Language Learner Support

13-4 **Additional Vocabulary Support**
Compound Probability

Use the chart to review vocabulary. These vocabulary words will help you complete this page.

Word Phrase	Definition	Example
compound events	A compound event is an event that is made up of two or more events.	Get heads in a coin toss and a 3 in a roll of a fair number cube.
independent events	If the occurrence of an event does not affect how another event occurs, then the events are called **independent events.**	Choose a green apple from a bowl of fruit, then put the apple back in the bowl. Then choose an orange from the bowl.
dependent events	If the occurrence of an event does affect how another event occurs, then the events are called **dependent events.**	Choose a red marble from a bag. Without replacing the first marble, choose a blue marble from the bag.
mutually exclusive events	Events that cannot happen at the same time are called mutually exclusive events.	Spin a 4 and a 6 in one spin of a spinner.
overlapping events	Overlapping events have outcomes in common.	Roll a 4 and a multiple of 2 on a number cube.

A student has a standard deck of 52 cards. Choose the word phrase from the table above that best matches each situation.

1. Draw the Jack of Spades, keep it, then draw the Queen of Diamonds. dependent events
2. Draw one card and get a King or a Club. overlapping events
3. Draw two cards and get the 2 of Hearts and the 3 of Hearts. compound events
4. Draw one card and get a Queen or an Ace. mutually exclusive events
5. Draw the 10 of Hearts, replace it, then choose the 5 of Clubs. independent events

Differentiated Remediation *continued*

On-Level

- **Practice** (2 pages) Provides extra practice for each lesson. For simpler practice exercises, use the Form K Practice pages found in the All-in-One Teaching Resources and online.
- **Think About a Plan** Helps students develop specific problem-solving skills and strategies by providing scaffolded guiding questions.
- **Standardized Test Prep** Focuses on all major exercises, all major question types, and helps students prepare for the high-stakes assessments.

Practice and Problem Solving Wkbk/ All-in-One Resources/Online

Practice page 1

13-4 **Practice** — Compound Probability — *Form G*

For Exercises 1–3, determine whether the events are *independent* or *dependent*.

1. You roll a 2 on a number cube and spin a 3 on a spinner. independent
2. You choose a King from a deck of cards and get heads in a coin toss. independent
3. You roll a number cube and get a 6, and roll again if the first roll is a 6. dependent
4. What is $P(A \text{ and } B)$ if $P(A) = \frac{1}{2}$ and $P(B) = \frac{2}{3}$, where A and B are independent events? $\frac{1}{3}$
5. What is the probability of rolling a 4 on a fair number cube and getting "tails" when tossing a coin? $\frac{1}{12}$
6. What is $P(A \text{ or } B)$ if $P(A) = 32\%$ and $P(B) = 17\%$, where A and B are mutually exclusive events? 0.49 or 49%
7. At a local high school, 34% of the students take a bus to school and 56% of the students walk to school. What is the probability of randomly selecting a student that takes a bus or walks to school? 0.9 or 90%
8. What is $P(A \text{ or } B)$ if $P(A) = \frac{1}{4}$ and $P(B) = \frac{1}{2}$, where A and B are overlapping events? $\frac{5}{8}$
9. A spinner has 8 equal sections numbered 1 to 8. What is the probability of the spinner stopping on a number that is a multiple of 3 or is greater than 5? $\frac{1}{2}$

Practice and Problem Solving Wkbk/ All-in-One Resources/Online

Practice page 2

13-4 **Practice** (continued) — Compound Probability — *Form G*

10. A local aquarium has 6 turtles, 12 penguins, and 8 sharks. You randomly select 1 animal to watch. What is the probability that you select a turtle or a shark? $\frac{7}{13}$
11. **Writing** A bag contains red, green, and blue golf balls and golf tees. You reach into a bag to randomly select one golf ball and one golf tee. Describe how to calculate the probability that you select a red golf ball and a red golf tee. First count the total number of balls and tees, and the number of red balls and red tees. The probability of choosing a red ball is the number of red balls divided by the total number of balls. The probability of choosing a red tee is the number of red tees divided by the total number of tees. Since randomly choosing a ball and a tee are independent events, the probability of randomly selecting both a red ball and a red tee is the product of these probabilities.
12. In a local town, 55% of the residents drive to work, 23% of the residents own a dog, and 6% of the residents walk to work. Find the probability that a randomly chosen resident owns a dog or walks to work. 0.28 or 28%
13. You donate 8 baseballs to a local baseball team. Your uncle donates 12 baseballs. If a total of 50 baseballs are donated, what is the probability that the first pitch of the season uses one of your baseballs or one of your uncle's baseballs? Write your answer as a percent. 0.4 or 40%

Use the spinner at the right for Exercises 14–17.

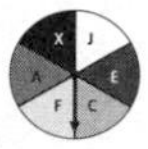

14. What is the probability of the arrow stopping on a consonant or one of the first 4 letters of the alphabet? $\frac{5}{6}$
15. What is the probability of the arrow stopping on "X" on the first spin and "F" on the second spin? $\frac{1}{36}$
16. What is the probability of the arrow stopping on "J" or "A" on one spin? $\frac{1}{3}$
17. **Reasoning** What is the probability of the arrow stopping on "J" and "A" on one spin? Explain. 0; Explanations may vary. Sample: The spinner would have to stop on both the "J" and the "A" some of the time, which is impossible. The events are mutually exclusive.

Practice and Problem Solving Wkbk/ All-in-One Resources/Online

Think About a Plan

13-4 **Think About a Plan** — Compound Probability

Pets In a litter of 8 kittens, there are 2 brown females, 1 brown male, 3 spotted females, and 2 spotted males. If a kitten is selected at random, what is the probability that the kitten will be female or brown?

Understanding the Problem

1. Can a kitten be both female and brown? Yes.
2. Can a kitten be both female and male? No.

Planning the Solution

3. What are the characteristics of the selected kitten? It is brown or it is female.
4. How many brown kittens are there? 3
5. How many female kittens are there? 5
6. How many kittens are brown and female? 2

Getting an Answer

7. Determine the probabilities for choosing a brown kitten, choosing a female kitten, and choosing a brown female kitten. Brown: $\frac{3}{8}$ Female: $\frac{5}{8}$ Brown female: $\frac{2}{8}$
8. Since this is an overlapping probability, use your values in the following formula: $P(\text{female or brown}) = P(\text{female}) + P(\text{brown}) - P(\text{female and brown})$ $\frac{3}{8} + \frac{5}{8} - \frac{2}{8} = \frac{6}{8} = \frac{3}{4}$
9. What is the probability that the kitten will be female or brown? Write your answer as a fraction and as a decimal. $\frac{3}{4}$ or 0.75

Practice and Problem Solving Wkbk/ All-in-One Resources/Online

Standardized Test Prep

13-4 **Standardized Test Prep** — Compound Probability

Multiple Choice

For Exercises 1–4, choose the correct letter.

1. What is the probability of rolling a 5 on a number cube and randomly drawing the 2 of Clubs from a deck of cards? A
   (A) $\frac{1}{312}$ (B) $\frac{1}{260}$ (C) $\frac{1}{24}$ (D) $\frac{1}{2}$
2. In one class, 19% of the students received an A on the last test and 13% of the students received a C. What is the probability that a randomly chosen student received an A or a C? I
   (F) 0.06 (G) 0.13 (H) 0.16 (I) 0.32
3. What is the probability of rolling a 3 or a number less than 4 on a number cube? B
   (A) $\frac{5}{19}$ (B) $\frac{1}{2}$ (C) $\frac{2}{3}$ (D) $\frac{3}{4}$
4. You win 6 out of every 10 races you run. Your friend wins 7 out of every 9 dancing competitions she enters. What is the probability of you both winning your next events? G
   (F) $\frac{7}{16}$ (G) $\frac{7}{15}$ (H) $\frac{21}{40}$ (I) $\frac{13}{19}$

Short Response

5. The results of a survey revealed that 26% of the students read fiction in their spare time, 21% of the students read non-fiction, and 7% don't read in their spare time. What is the probability that a randomly chosen student reads fiction or doesn't read in her spare time?
   [2] 0.33 or 33%
   [1] Student provides evidence of combining 2 of the percentages.
   [0] Student's response shows no effort OR does not attempt to add 2 of the percentages.

Extension

- **Enrichment** Provides students with interesting problems and activities that extend the concepts of the lesson.
- **Activities, Games, and Puzzles** Worksheets that can be used for concepts development, enrichment, and for fun!

All-in-One Resources/Online

Enrichment

13-4 **Enrichment** — Compound Probability

Venn Diagrams

You can represent a set of numbers or objects in a *Venn diagram*. A Venn diagram can show mutually exclusive events or overlapping events.

Example 1

You roll a number cube. Event *A* is rolling an odd number. Event *B* is rolling an even number. This is represented in the Venn diagram at the right.

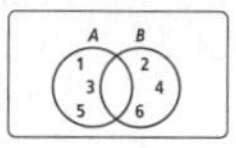

Because a number cannot be both odd and even, Event *A* and Event *B* are mutually exclusive. Therefore, the overlapping area of the circles is empty.

Example 2

You roll a number cube. Event *A* is rolling an even number. Event *B* is rolling a number greater than 3. This is represented in the Venn diagram at the right.

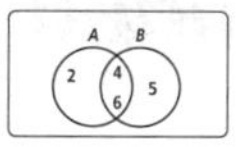

Because 4 and 6 belong to both events, they are shown in the overlapping area of the circles.

Exercises

For Exercises 1 and 2, determine whether the overlapping area of the Venn diagram for the situation would be *empty* or *not empty*.

1. You randomly choose a student from drama club that is a boy or has played a lead role in a play. not empty
2. You randomly choose a lettered tile or a numbered tile from a hat. empty
3. Make a Venn diagram for the following:
   Event A: Randomly choosing a letter from the word HOCKEY.
   Event B: Randomly choosing a letter from the word TRACK. See diagram to the right.

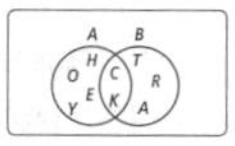

Online Teacher Resource Center

Activities, Games, and Puzzles

13-4 **Game: The Probability Path** — Compound Events

Provide the host with the following questions and answers (shown in brackets).

What is the probability of the following?

1. rolling a 2 on your first turn and a 6 on your second turn [1/36]
2. rolling an even number on your first turn and a 2 on your second turn [1/12]
3. landing on a space with an even denominator [1/2]
4. landing on a $\frac{1}{3}$ or a space with an identical neighbor on each side [7/15]
5. from the $\frac{1}{12}$ in the middle, the next move landing on another $\frac{1}{12}$ [1/3]
6. landing on $\frac{1}{2}$ on your second move [1/12]
7. landing on a $\frac{1}{6}$ or a $\frac{1}{3}$ [7/30]
8. the finishing roll being a 3 [0]
9. landing on a $\frac{1}{12}$ or "forward two spaces" [7/30]
10. landing on a space with an even denominator or an odd denominator [9/10]
11. landing on a space with an odd denominator or requiring the piece to move two spaces [1/2]
12. landing on a $\frac{1}{4}$ after being on a corner space and then landing on a $\frac{1}{12}$ [1/18]
13. landing on a $\frac{1}{6}$ on your first two moves [1/36]
14. landing on two $\frac{1}{2}$ s in a row [0]
15. rolling a 4 and landing on a $\frac{1}{6}$ at some point in the game [1/10]
16. rolling a 3 on your first turn and a 3 on your second turn [1/36]
17. rolling a 1 on your first turn and a 3 or 5 on your second turn [1/18]
18. landing on a space with an odd denominator [2/5]
19. landing on a $\frac{1}{12}$ or a space with an identical neighbor on each side [1/3]
20. moving "forward two spaces" after being on a $\frac{1}{3}$ [1/6]
21. landing on $\frac{1}{3}$ on your second move [1/6]
22. landing on a $\frac{1}{4}$ or a $\frac{1}{2}$ [2/15]
23. landing on $\frac{1}{12}$ on your second move [1/9]
24. landing on a $\frac{2}{3}$ or "back two spaces" [2/5]
25. landing on a space with an even denominator or requiring the piece to move two spaces [3/5]
26. landing on a space with a 1 or a 2 in the numerator [9/10]
27. landing on a "forward two spaces" after being on a $\frac{1}{3}$ and then landing on a $\frac{1}{6}$ on the next turn [1/18]
28. landing on a $\frac{1}{3}$ on your last two moves [0]
29. from the $\frac{1}{12}$ in the right corner landing on $\frac{1}{12}$ [1/2]
30. rolling a 3 and landing on a $\frac{1}{6}$ at some point in the game [1/6]

1 Interactive Learning

Solve It!

PURPOSE To analyze data presented in a table.
PROCESS Students may recognize that they can break the table down into 2 columns (passed and failed) and two rows (took the class and did not take the class.)

FACILITATE

Q How many students took the driving test? How many of them passed the driving test? **[80; 50]**

Q How many students took the driver's education class? What percentage of them passed the driving test? **[39; about 82%]**

Q How many students did not take the driver's education class? What percentage of them passed the driving test? **[41; about 44%]**

ANSWER See Solve It in Answers on next page.
CONNECT THE MATH In the Solve It, students explore and analyze a set of data from two different categories. In this lesson, students will learn how to use two-way frequency tables to find conditional probabilities, and they will use probability models to analyze data.

2 Guided Instruction

Problem 1

In this problem, students will use a two-way frequency table to find a probability.

Q Can you calculate the relative frequencies of the values in the Totals row and Totals column? What would these represent? **[Yes, these would represent the probabilities of each category: male, female, activities, no activities.]**

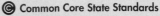

13-5 Probability Models

Common Core State Standards
S-CP.A.4 Construct and interpret two-way frequency tables of data . . . Use the two-way table as a sample space to decide if events are independent and to approximate conditional probabilities.
MP 1, MP 3

Objective To construct and use probability models

Support your answer based on the data!

MATHEMATICAL PRACTICES

SOLVE IT! Getting Ready!

The table at the right shows the number of students who passed their driving test as well as whether they took a driver's education class to prepare. What effect, if any, does taking the driver's education class have?

	Passed	Failed	Totals
Took the class	32	7	39
Do not take the class	18	23	41
Totals	50	30	80

In the Solve It, the data is displayed in a *two-way frequency table*. A **two-way frequency table**, or *contingency table*, displays the frequencies of data in two different categories.

Lesson Vocabulary
- two-way frequency table
- conditional probability

Essential Understanding You can use two-way frequency tables to organize data and identify sample spaces to approximate probabilities.

Think
What is the connection between relative frequency and probability?
You can use relative frequency to approximate a probability.

Problem 1 Using a Two-way Frequency Table

Activities **The table shows data about student involvement in extracurricular activities at a local high school. What is the probability that a randomly chosen student is a female who is not involved in extracurricular activities?**

Extracurricular Activities

	Involved in Activities	Not Involved in Activities	Totals
Male	112	145	257
Female	139	120	259
Totals	251	265	516

To find the probability, calculate the relative frequency.

relative frequency $= \frac{\text{females not involved}}{\text{total number of students}} = \frac{120}{516} \approx 0.233$

The probability that a student is a female who is involved in extracurricular activities is about 23.3%.

13-5 Preparing to Teach

BIG idea Conditional Probability

ESSENTIAL UNDERSTANDINGS
- Two-way frequency tables can be used to organize data, identify sample spaces, and approximate probabilities.
- In real world situations frequency tables can be used to find conditional probabilities and determine if treatments are effective.

Math Background

When discussing how to use two-way frequency tables to find conditional probabilities, tell students that they can isolate a particular row or column and then simply calculate the relative frequencies using the total of that row or column. This may simplify the process by allowing them to focus in on smaller portions of the table.

By their design, two-way frequency tables lend themselves to finding conditional probabilities. These conditional probabilities can be used to analyze data in a number of real world applications by showing patterns in the data. A two-way table containing relative frequencies is a form of a probability distribution.

Mathematical Practices

Make sense of problems. Students use two-way frequency tables to find probabilities based on real-world data.

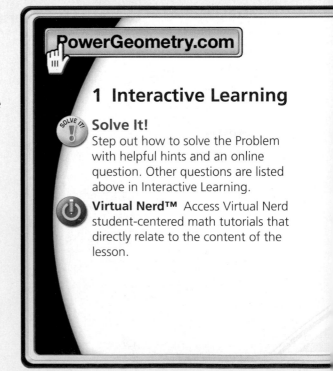

PowerGeometry.com

1 Interactive Learning

Solve It!
Step out how to solve the Problem with helpful hints and an online question. Other questions are listed above in Interactive Learning.

Virtual Nerd™ Access Virtual Nerd student-centered math tutorials that directly relate to the content of the lesson.

Got It? 1. The two-way frequency table at the right shows the number of male and female students by grade level on the prom committee. What is the probability that a member of the prom committee is a male who is a junior?

	Male	Female	Totals
Juniors	3	4	7
Seniors	3	2	5
Totals	6	6	12

The probability that an event will occur, given that another event has already occurred is called a **conditional probability**. You can write the conditional probability of event B, given that event A has already occurred as $P(B \mid A)$. You read $P(B \mid A)$ as "the probability of event B, given event A."

Problem 2 **Finding Probability**

Opinion Polls **Respondents of a poll were asked whether they were for, against, or had no opinion about a bill before the state legislature that would increase the minimum wage. What is the probability that a randomly selected person is over 60 years old, given that the person had no opinion on the state bill?**

Age Group	For	Against	No Opinion	Totals
18–29	310	50	20	380
30–45	200	30	10	240
45–60	120	20	30	170
Over 60	150	20	40	210
Totals	780	120	100	1000

Plan

What part of the table do you need to use?
Since the group you are interested in is the one with no opinion, you only need to look at that column.

The condition that the person selected has no opinion on the minimum-wage bill limits the total outcomes to the 100 people who had no opinion. Of those 100 people, 40 respondents were over 60 years old.

$P(\text{over } 60 \mid \text{no opinion}) = \frac{40}{100} = 0.4$

Got It? 2. a. What is the probability that a randomly selected person is 30–45 years old, given that the person is in favor of the minimum-wage bill?

b. **Reasoning** What is the probability that a randomly selected person is not 18–29, given that the person is in favor of the minimum wage bill?

Got It?

Q How many students are on the prom committee altogether? Of these, how many are male juniors? **[12; 3]**

Problem 2 ERROR PREVENTION

In order to find the correct conditional probability, be sure students divide by the total in the No Opinion column.

Q How would you use the table to find $P(\text{For} \mid \text{age } 18\text{–}29)$? **[Divide the number of respondents age 18–29 who are for the bill (310) by the total number of respondents 18–29 years old (380).]**

Got It?

Q Is this the same as finding the probability that a respondent is for the state bill given that the person is 30–45 years old? Explain. **[No, this probability is much higher at about 0.83.]**

Q Which column represents the condition in the problem? **[not 18–29]**

2 Guided Instruction

Each Problem is worked out and supported online.

Problem 1
Using a Two-Way Frequency Table

Problem 2
Finding Probability

Problem 3
Using Relative Frequencies

Support in Geometry Companion

- Vocabulary
- Key Concepts
- Got It?

Answers

Solve It!

Students who took the class were more likely to pass the test.; Divide the number of students who passed the test (50) by the total number of students (80). To analyze the effectiveness of the class, compare how likely a student was to pass the test based on whether or not he or she took the class. The probability for students who took the class and passed is 0.88. The probability for students who did not take the class and passed was 0.44.

Got It?

1. $\frac{1}{4}$

2a. about 0.26

b. about 0.60

Problem 3

In this problem, students use a two-way relative frequency table to find conditional probability.

Q Method 1: How do you find the number of people who attended the seminar? **[Find how many sales representatives the company has and multiply by the relative frequency of those that attended, 0.8]**

Q Method 2: Why is the denominator of the fraction "total frequency of increased sales"? **[Because it is *P*(*A*) and *A* is increased sales.]**

Got It?

Q Which column is needed for solving this problem? **[Did not attend the seminar.]**

3 Lesson Check

Do you know HOW? ERROR INTERVENTION

- If students have difficulty solving Exercise 2, then ask them how the expression *P*(democrat | supports the issue) is read.

Do you UNDERSTAND?

- In Exercise 3, ask students what the sample space is when finding the relative frequency of Republicans who do not support the issue.

Close

Q How can a two-way frequency table help you calculate conditional probabilities? **[Use the total from a given row or column as the sample space. Divide a value in that row or column by the total.]**

Q What information is shown in a probability distribution? **[the probabilities or relative frequencies of each outcome]**

Problem 3 Using Relative Frequencies

Business A company has 150 sales representatives. Two months after a sales seminar, the company vice-president made the table of relative frequencies based on sales results. What is the probability that someone who attended the seminar had an increase in sales?

	Attended Seminar	Did not Attend Seminar	Totals
Increased Sales	0.48	0.02	0.5
No Increase in Sales	0.32	0.18	0.5
Totals	0.8	0.2	1

Think
Given the relative frequencies, how do you find the frequencies?
Multiply the number of sales representatives times the relative frequency for a category.

Method 1 Find frequencies first.

Find the number of people who attended the seminar and had increased sales: $0.48 \cdot 150 = 72$

Find the number of people who attended the seminar: $0.8 \cdot 150 = 120$

Find *P*(increased sales | sales seminar): $\frac{72}{120} = 0.6$, or 60%

Method 2 Use relative frequencies.

P(increased sales | sales seminar):

$$= \frac{\text{relative frequency of attend seminar and increased sales}}{\text{relative frequency of attended seminar}} = \frac{0.48}{0.8} = 0.6$$

The probability that some who attended the seminar had an increase in sales is 0.6, or 60%.

Got It? 3. What is the probability that a randomly selected sales representative, who did not attend the seminar, did not see an increase in sales?

Lesson Check

Do you know HOW?

Use the two-way frequency table to find the probabilities.

	Supports the Issue	Does not supports the issue	Totals
Democrat	24	36	60
Republican	27	33	60
Totals	51	69	120

1. *P*(democrat and supports the issue)
2. *P*(democrat | supports the issue)

Do you UNDERSTAND? MATHEMATICAL PRACTICES

3. **Error Analysis** Using the table at the left, a student calculated the relative frequency of those who do not support the issue, given that they are Republican, as $\frac{33}{33+36} \approx 0.478$. What error did the student make?
4. **Vocabulary** What is a two-way frequency table?
5. Suppose *A* is a female student and *B* is a student who plays sports. What does $P(B \mid A)$ mean?

Additional Problems

Use the table below to solve Additional Problems 1 and 2.

	Right Handed	Left Handed	Totals
Males	22	5	27
Females	19	4	23
Totals	41	9	50

1. What is the probability that a randomly chosen person from the survey is a right handed male?
ANSWER $\frac{11}{25}$
2. What is the probability that a randomly chosen person from the survey is left handed, given that the person is a female?
ANSWER $\frac{4}{23}$

3. Find the missing relative frequencies in the probability distribution.

	Part time job	No part time job	Totals
11th graders	0.19	0.31	■
12th graders	0.44	■	0.5
Totals	■	0.37	1

ANSWER 0.5; 0.06; 0.63

4. Use your answers from Problem 3 to answer the following questions.
   a. If the total number of students surveyed is 350 students, how many students were in the 12th grade and had part time jobs?
   b. Find *P*(12th graders | Part time job)
   c. Find *P*(No part time job | 11th graders)

ANSWER a. 154 students
b. about 0.70
c. 0.62

Practice and Problem-Solving Exercises

MATHEMATICAL PRACTICES

A Practice

For Exercises 6–11, use the two-way frequency table below to find the probability of each event.

See Problems 1 and 2.

6. P(7th-grade girl)
7. P(8th-grade boy)
8. P(6th-grade girl)
9. P(girl | 7th-grade)
10. P(6th-grade | boy)
11. P(8th-grade | girl)

Attendance at Soccer Camp

	6th Graders	7th Graders	8th Graders	Totals
Boys	7	6	10	23
Girls	8	7	12	27
Totals	15	13	22	50

Veterinary Medicine **For Exercises 12–13, use the table at the right. It shows relative frequencies of treatments that a veterinarian gave dogs and cats during one week.**

See Problem 3.

12. When you find the probability that a treatment consists of shots only, given that the vet is treating a cat, how do you limit the sample space?
13. What is the probability that an animal chosen at random is a dog, given that the treatment was shots only?

	Shots only	Shots and Checkup	Totals
Dogs	0.31	0.23	0.54
Cats	0.26	0.20	0.46
Totals	0.57	0.43	1

B Apply

Academic Competition **The table at the right shows numbers of participants in an academic competition. Use this information for Exercises 14–18.**

14. What is P(female)?
15. What is P(freshman)?
16. What is P(female freshman)?
17. What is P(female | freshman)?
18. What is P(freshman | female)?

	Male	Female	Totals
Freshmen	3	5	8
Sophomores	6	4	10
Juniors	7	5	12
Seniors	4	6	10
Totals	20	20	40

19. **Think About a Plan** The two-way frequency table at the right shows the number of employees at a factory who have attended a safety workshop and the number who have been injured on the job. What is the probability that a worker from the factory has attended the safety workshop, given that the employee has not been injured on the job?
    - Which column is needed for solving this problem?
    - Do you need the total number of workers or the total not injured?

	Injured	Not injured	Totals
Safety workshop	3	36	39
No safety workshop	12	24	36
Totals	15	60	75

Answers

Got It? (continued)

3. 0.90

Lesson Check

1. 0.2
2. about 0.47
3. He divided by the number of people who do not support the issue instead of by the total number of Republicans.
4. A two way frequency table displays the frequencies of data in two different categories.
5. The probability that a student plays sports, given that the student is a female.

Practice and Problem Solving Exercises

6. 0.14
7. 0.2
8. 0.16
9. $\frac{7}{13}$
10. $\frac{7}{23}$
11. about 0.44
12. the sample space is limited to cats only
13. about 0.54
14. $\frac{1}{2}$
15. $\frac{1}{5}$
16. $\frac{1}{8}$
17. $\frac{5}{8}$
18. $\frac{1}{4}$
19. $\frac{3}{5}$

4 Practice

ASSIGNMENT GUIDE

Basic: 6–13, 19, 20–22 even
Average: 7–13 odd, 14–22
Advanced: 7–13 odd, 14–23
Standardized Test Prep: 24–27
Mixed Review: 28–36

Mathematical Practices are supported by exercises with red headings. Here are the Practices supported in this lesson:

MP 1: Make Sense of Problems Ex. 19
MP 3: Critique the Reasoning of Others Ex. 3
MP 3: Understand Definitions Ex. 4
MP 3: Construct Arguments Ex. 20
MP 3: Communicate Ex. 21

Applications exercises have blue headings.

EXERCISE 22: Use the Think About a Plan worksheet in the **Practice and Problem Solving Workbook** (also available in the Teaching Resources in print and online) to further support students' development in becoming independent learners.

HOMEWORK QUICK CHECK

To check students' understanding of key skills and concepts, go over Exercises 15, 19, 20, 27, and 31.

20. Reasoning Recall that two events are independent when the occurrence of one has no effect the other. The table at the right is a frequency table.

	B	*D*	*F*	Total
A	7	5	4	16
C	3	4	5	12
E	11	7	2	28
Total	21	16	11	48

a. Calculate $P(B)$.
b. Calculate $P(B|A)$.
c. Does the occurrence of *A* have any effect on the probability of *B*? What can you say about events *A* and *B*?
d. Are events *C* and *D* independent? Explain.
e. Are events *C* and *F* independent? Explain.

21. Writing What is the sum of the probabilities of all possible outcomes in a probability experiment? How does this relate to the relative frequencies in a contingency table? Explain and give an example.

22. Healthcare The table at the right is a relative frequency distribution for healthy people under the age of 65.
a. Copy and complete the table.
b. What is the probability of getting the flu, given you are vaccinated?
c. What is the probability of getting the flu, given you have not been vaccinated?

Flu Vaccines

	Got the Flu	Did not Get the Flu	Totals
Vaccinated	■	54%	60%
Not Vaccinated	■	■	■
Totals	15%	■	100%

Challenge

23. When you construct a two-way frequency table, you add across rows and down columns to find values for the totals row and column. The values in the totals column and totals row of the table are called *marginal frequencies*, and the values in the interior of the table are called *joint frequencies*. In the table below, the values 9, 3, 5, and 8 represent the joint frequencies, and the values 14, 11, 12, 13, and 25 represent the marginal frequencies.

	Exam Score ≥ 85%	Exam Score < 85%	Totals
Studied more than 4 hours	9	3	12
Studied less than 4 hours	5	8	13
Totals	14	11	25

a. Why do you think the values in the interior of the table are called joint frequencies?
b. What do the marginal frequencies of the table represent?
c. If you replace the joint and marginal frequencies in a two-way frequency table with the respective relative frequencies, what do the values in the totals row and column represent? Use the given table to provide an example.

Answers

Practice and Problem-Solving Exercises (continued)

20a. $\frac{7}{16}$
b. $\frac{7}{16}$
c. Since *P*(*B*) and *P*(*B* | *A*) are the same, event *A* occurring has no effect on the probability of event *B* occurring. *A* and *B* are independent events.
d. Since *P*(*D*) and *P*(*D* | *C*) are the same, event *C* occurring has no effect on the probability of event *D* occurring. *D* and *C* are independent events.
e. Since *P*(*F*) and *P*(*F* | *C*) are not equal, event *C* occurring does have an effect on the probability of event *F* occurring. *F* and *C* are dependent events.
21. Answers may vary. Sample: The sum of the probabilities is 1. The sum of the relative frequencies is also 1 because they represent the experimental probabilities of each possible outcome. Check students' examples.
22a. 6%, 9%, 31%, 40%, 85%
b. $\frac{1}{10}$
c. $\frac{9}{40}$
23a. Each cell contains the frequency of the joint event described by its row and column.
b. The marginal frequencies are the total frequencies for each category in the two-way table.
c. The sum of the joint probabilities in any row or column of a joint probability distribution is the same as the marginal probability associated with that row or column. Check students' work.

Standardized Test Prep

GRIDDED RESPONSE

SAT/ACT

24. The two-way frequency table shows the number of males and females that either support or are against the building of a new mall. What is the probability that a randomly selected person is a female, given that the person supports the new mall? Round to the nearest hundredth.

	For	Against	Totals
Male	62	48	110
Female	78	32	110
Totals	140	80	220

25. The area of a kite is 150 in.^{2}. The length of one diagonal is 50 in. What is the length, in inches, of the other diagonal?

26. What is the x-coordinate of the midpoint of $\overline{AB}$ for $A(-3, 9)$ and $B(-5, -3)$?

27. The segment with endpoints $A(1, 5)$ and $B(2,1)$ is reflected over $x = 3$ and translated up 2 units and to the right 3 units. What is the x-coordinate of the midpoint of $\overline{A'B'}$?

Mixed Review

Find the sum of the interior angle measures of each polygon.

See Lesson 6-1.

28. quadrilateral

29. dodecagon

30. 20-gon

31. A cylindrical carton with a radius of 2 in. is 7 in. tall. Assuming no surfaces overlap, what is the surface area of the carton? Round your answer to the nearest square inch.

See Lesson 11-2.

32. A cube with edges 8 cm long fits within a sphere, as shown at the right. The diagonal of the cube is the diameter of the sphere.

a. Find the radius of the sphere. Leave your answer in simplest radical form.

b. What is the volume of the sphere?

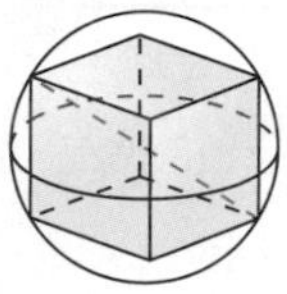

See Lesson 11-6.

Get Ready! **To prepare for Lesson 13-6, do Exercises 33–36.**

Find the probability of each event. Use the two-way frequency table.

33. $P(\text{large} \mid \text{red})$

34. $P(\text{red} \mid \text{large})$

35. $P(\text{small} \mid \text{blue})$

36. $P(\text{large} \mid \text{blue})$

	Large	Small	Totals
Blue	17	3	20
Red	8	12	20
Totals	25	15	40

24. 0.56
25. 6
26. −4
27. 7.5
28. 360°
29. 1800°
30. 3240°
31. 113 in.^{2}
32. a. $4\sqrt{3}$ cm
b. about 1393 cm^3

33. $\frac{2}{5}$
34. $\frac{8}{25}$
35. $\frac{3}{20}$
36. $\frac{17}{20}$

13-5 Lesson Resources

Differentiated Remediation

Additional Instructional Support

Geometry Companion

Students can use the **Geometry Companion** worktext (4 pages) as you teach the lesson. Use the Companion to support

- Solve It!
- New Vocabulary
- Key Concepts

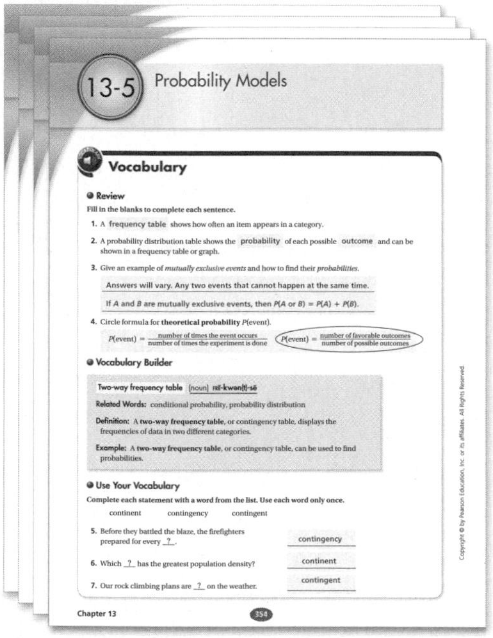

13-5 Probability Models

Vocabulary

Review

Fill in the blanks to complete each sentence.

1. A frequency table shows how often an item appears in a category.
2. A probability distribution table shows the probability of each possible outcome and can be shown in a frequency table or graph.
3. Give an example of *mutually exclusive events* and how to find their *probabilities*.
   Answers will vary. Any two events that cannot happen at the same time.
   If A and B are mutually exclusive events, then $P(A \text{ or } B) = P(A) + P(B)$.
4. Circle formula for **theoretical probability** P(event).
   $P(\text{event}) = \frac{\text{number of times the event occurs}}{\text{number of times the experiment is done}}$ $\quad P(\text{event}) = \frac{\text{number of favorable outcomes}}{\text{number of possible outcomes}}$

Vocabulary Builder

Two-way frequency table (noun) rē-kwen(t)-sē

Related Words: conditional probability, probability distribution

Definition: A **two-way frequency table,** or contingency table, displays the frequencies of data in two different categories.

Example: A **two-way frequency table,** or contingency table, can be used to find probabilities.

Use Your Vocabulary

Complete each statement with a word from the list. Use each word only once.

continent contingency contingent

5. Before they battled the blaze, the firefighters prepared for every _?_. contingency
6. Which _?_ has the greatest population density? continent
7. Our rock climbing plans are _?_ on the weather. contingent

ELL Support

Use Small-Group Interactions Divide students into small groups and ask each group to think of two different categories that their classmates would fall into based on a single characteristic.

Sample categories:

- play sports/does not play sports
- has a part time job/does not have a part time job
- right handed/left handed
- plays a musical instrument/does not play an instrument
- plans to attend college/does not plan to attend college

Use Males and Females as the other category headings. Have students gather data and complete a two-way frequency table and a probability distribution for their data. Then have them use the tables to calculate simple probabilities such as *P*(female and plays a musical instrument) as well as conditional probabilities such as *P*(male | has a part time job).

5 Assess & Remediate

Lesson Quiz

Use the Lesson Quiz to assess students' mastery of the skills and concepts of this lesson.

Use the two-way frequency table below to solve Exercises 1 and 2.

	Varsity	Non varsity	Totals
Sophomores	3	14	17
Juniors	8	5	13
Totals	11	19	30

1. **Do you UNDERSTAND?** What is *P*(sophomore and varsity)?
2. What is *P*(sophomore | non varsity)?
3. Use the table below to find *P*(Male | Against).

	Female	Male	Totals
For	0.22	0.26	0.48
Against	0.32	0.20	0.52
Totals	0.54	0.46	1

ANSWERS TO LESSON QUIZ

1. 0.1
2. about 0.74
3. about 0.38

PRESCRIPTION FOR REMEDIATION

Use the student work on the Lesson Quiz to prescribe a differentiated review assignment.

Points	Differentiated Remediation
0–1	Intervention
2	On-level
3	Extension

PowerGeometry.com

5 Assess & Remediate

Assign the Lesson Quiz. Appropriate intervention, practice, or enrichment is automatically generated based on student performance.

Intervention

- **Reteaching** (2 pages) Provides reteaching and practice exercises for the key lesson concepts. Use with struggling students or absent students.
- **English Language Learner Support** Helps students develop and reinforce mathematical vocabulary and key concepts.

All-in-One Resources/Online

Reteaching

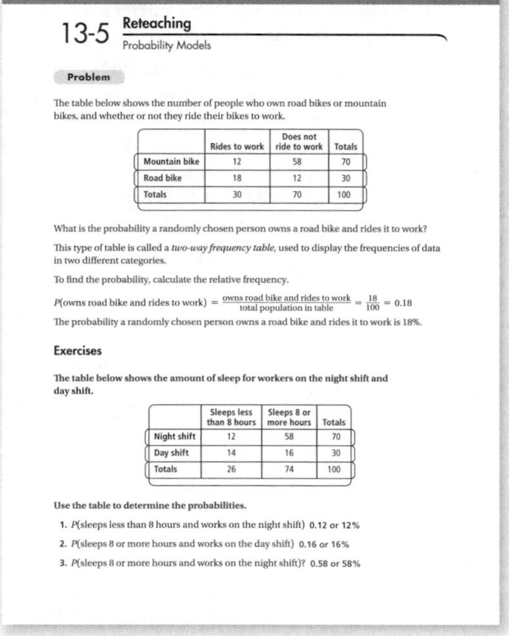

13-5 Reteaching
Probability Models

Problem

The table below shows the number of people who own road bikes or mountain bikes, and whether or not they ride their bikes to work.

	Rides to work	Does not ride to work	Totals
Mountain bike	12	58	70
Road bike	18	12	30
Totals	30	70	100

What is the probability a randomly chosen person owns a road bike and rides it to work?

This type of table is called a *two-way frequency table,* used to display the frequencies of data in two different categories.

To find the probability, calculate the relative frequency.

$P(\text{owns road bike and rides to work}) = \frac{\text{owns road bike and rides to work}}{\text{total population in table}} = \frac{18}{100} = 0.18$

The probability a randomly chosen person owns a road bike and rides it to work is 18%.

Exercises

The table below shows the amount of sleep for workers on the night shift and day shift.

	Sleeps less than 8 hours	Sleeps 8 or more hours	Totals
Night shift	12	58	70
Day shift	14	16	30
Totals	26	74	100

Use the table to determine the probabilities.

1. *P*(sleeps less than 8 hours and works on the night shift) 0.12 or 12%
2. *P*(sleeps 8 or more hours and works on the day shift) 0.16 or 16%
3. *P*(sleeps 8 or more hours and works on the night shift)? 0.58 or 58%

All-in-One Resources/Online

English Language Learner Support

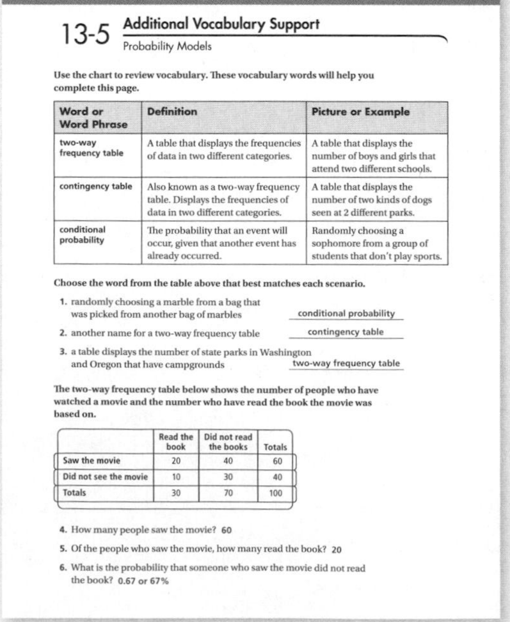

13-5 Additional Vocabulary Support
Probability Models

Use the chart to review vocabulary. These vocabulary words will help you complete this page.

Word or Word Phrase	Definition	Picture or Example
two-way frequency table	A table that displays the frequencies of data in two different categories.	A table that displays the number of boys and girls that attend two different schools.
contingency table	Also known as a two-way frequency table. Displays the frequencies of data in two different categories.	A table that displays the number of two kinds of dogs seen at 2 different parks.
conditional probability	The probability that an event will occur, given that another event has already occurred.	Randomly choosing a sophomore from a group of students that don't play sports.

Choose the word from the table above that best matches each scenario.

1. randomly choosing a marble from a bag that was picked from another bag of marbles conditional probability
2. another name for a two-way frequency table contingency table
3. a table displays the number of state parks in Washington and Oregon that have campgrounds two-way frequency table

The two-way frequency table below shows the number of people who have watched a movie and the number who have read the book the movie was based on.

	Read the book	Did not read the books	Totals
Saw the movie	20	40	60
Did not see the movie	10	30	40
Totals	30	70	100

4. How many people saw the movie? 60
5. Of the people who saw the movie, how many read the book? 20
6. What is the probability that someone who saw the movie did not read the book? 0.67 or 67%

Differentiated Remediation *continued*

On-Level

- **Practice** (2 pages) Provides extra practice for each lesson. For simpler practice exercises, use the Form K Practice pages found in the All-in-One Teaching Resources and online.
- **Think About a Plan** Helps students develop specific problem-solving skills and strategies by providing scaffolded guiding questions.
- **Standardized Test Prep** Focuses on all major exercises, all major question types, and helps students prepare for the high-stakes assessments.

Extension

- **Enrichment** Provides students with interesting problems and activities that extend the concepts of the lesson.
- **Activities, Games, and Puzzles** Worksheets that can be used for concepts development, enrichment, and for fun!

Practice and Problem Solving Wkbk/ All-in-One Resources/Online

Practice page 1

13-5 **Practice** *Form G*
Probability Models

For Exercises 1–4, use the two-way frequency table below. It shows the number of one doctor's female patients who caught a cold one winter and whether or not they exercised regularly.

	Caught a cold	Did not catch a cold	Totals
Exercised	8	30	38
Did not exercise	10	2	12
Totals	18	32	50

1. How many patients exercised? **38**
2. What is the probability that a randomly chosen patient caught a cold and did not exercise? **0.2**
3. What is the probability that a randomly chosen patient exercised and did not catch a cold? **0.6**
4. What is *P*(did not exercise | did not catch a cold)? **0.0625**

The table below shows the students in a physical education class. Use this information for Exercises 5–7.

	Has played tennis	Has not played tennis	Totals
Boys	10	6	16
Girls	10	4	14
Totals	20	10	30

5. What is *P*(girl)? **0.467**
6. What is *P*(has not played tennis)? **0.33**
7. What is the probability that a randomly chosen student has played tennis given he is a boy? **0.625**

Practice and Problem Solving Wkbk/ All-in-One Resources/Online

Practice page 2

13-5 **Practice** (continued) *Form G*
Probability Models

For Exercises 8–10, use the table below. It shows the relative frequencies of students in a science club who have pets, and whether or not they have a yard.

	Pets	No pets	Totals
Yard	0.60	0.05	0.65
No yard	0.25	0.10	0.35
Totals	0.85	0.15	1

8. What is the probability that a randomly selected student has a yard given that they have pets? **0.71**
9. What is *P*(does not have a yard | have no pets)? **0.67**
10. **Error Analysis** Your friend determines that *P*(has a yard | has no pets) is 0.08. What error did your friend make? What is the correct probability? **The friend found the probability *P*(has no pets | has a yard); 0.33**

A biologist surveyed one type of plant growing on a wooded acre. Use his results, shown in the table below, for Exercises 11 and 12.

	Lobed Leaves	Non-lobed Leaves	Totals
Red Berries	12	48	60
No Red Berries	40	0	40
Totals	52	48	100

11. What is *P*(has red berries | has lobed leaves)? **0.23**
12. What is *P*(has lobed leaves | has red berries)? **0.2**

All-in-One Resources/Online

Enrichment

13-5 **Enrichment**
Probability Models

Relative frequencies can be used to determine actual values.

	Singles	Doubles	Triples	Homeruns	Totals
Outfielders	.45	.17	.09	.05	.76
Infielders	.13	.06	.02	.03	.24
Totals	.58	.23	.11	.08	1

1. Complete the table. **See table.**
2. What is *P*(a given hit was a triple | that hit was made by an infielder)? **0.083**
3. If there were 720 hits last year, how many homeruns were hit by infielders? **22**
4. What is the probability that a hit was a double given it was hit by an infielder? **0.25**
5. Complete the table below with the actual number of hits. Round where necessary. **See table.**

	Singles	Doubles	Triples	Homeruns	Totals
Outfielders	324	122	65	36	547
Infielders	94	43	14	22	173
Totals	418	165	79	58	720

6. Using your new table, what fraction should you use to determine *P*(player gets a double | player is an outfielder)? $\frac{122}{547}$

Practice and Problem Solving Wkbk/ All-in-One Resources/Online

Think About a Plan

13-5 **Think About a Plan**
Probability Models

Healthcare The table at the right is a relative frequency distribution for healthy people under the age of 65.

a. Copy and complete the table.
b. What is the probability of getting the flu given you are vaccinated?
c. What is the probability of getting the flu given you have not been vaccinated?

	Got the Flu	Did not Get the Flu	Totals
Vaccinated	■	54%	60%
Not Vaccinated	■	■	■
Totals	15%	■	100%

Understanding the Problem

1. [illegible] mutually exclusive, what operation(s) can you use to find missing probabilities in the table? **Addition and subtraction**
2. How is the two-way frequency table helpful in finding the probabilities? **The table organizes and helps determine the outcome of one event "given" the outcome of another event.**

Planning the Solution

3. Write the formula for conditional probability.
$P(A \mid B) = \frac{P(A \text{ and } B)}{P(B)}$

Getting an Answer

4. Complete the table. Remember that the sum of each pair of values in the row or column should equal the amount in the Total row or column, respectively.

	Got the Flu	Did not Get the Flu	Totals
Vaccinated	6%	54%	60%
Not Vaccinated	9%	31%	40%
Totals	15%	85%	100%

5. Use the formula from Step 3 to find the probability of getting the flu, given you are vaccinated.
$P(\text{got the flu} \mid \text{vaccinated}) = \frac{P(\text{got the flu and vaccinated})}{P(\text{vaccinated})} = \frac{6\%}{60\%} = 10\%$
6. Use the formula from Step 3 to find the probability of getting the flu, given you have not been vaccinated.
$P(\text{got the flu} \mid \text{not vaccinated}) = \frac{9\%}{40\%} = 22.5\%$

Practice and Problem Solving Wkbk/ All-in-One Resources/Online

Standardized Test Prep

13-5 **Standardized Test Prep**
Probability Models

Multiple Choice

For Exercises 1–4, choose the correct letter.

The table below shows the number of participants at a charity event who walked or ran, and who wore a red t-shirt or a blue t-shirt. Use the table for Exercises 1–4.

	Blue t-shirt	Red t-shirt	Totals
Walk	80	30	110
Run	20	30	50
Totals	100	60	160

1. What is the probability that a randomly chosen person ran and wore a blue t-shirt? **A**
(A) 0.125 (B) 0.25 (C) 0.4 (D) 25
2. What is the probability that a randomly chosen person walked and wore a red t-shirt? **G**
(F) 0.18 (G) 0.1875 (H) 0.3525 (I) 0.5
3. What is *P*(ran | wore a blue t-shirt)? **A**
(A) 0.2 (B) 0.25 (C) 0.4 (D) 0.8
4. What is the probability that a randomly chosen runner wore a blue t-shirt? **G**
(F) 0.3 (G) 0.4 (H) 0.5 (I) 0.6

Short Response

5. When calculating $P(B \mid A)$, why is $P(A)$ in the denominator?
[2] *P*(*A*) is the denominator because it is the event that has already happened, and represents the total population for the probability.
[1] Student provides evidence showing an understanding of conditional probability calculations.
[0] Student's response shows no effort OR does not attempt to explain the probability in terms of a fraction.

Online Teacher Resource Center

Activities, Games, and Puzzles

13-5 **Activity: Birthday Records**
Probability Models

Use the table below to record the birth month and date of each student in your class. Then total the number of birthdays in each month and in each date range. Use the two-way frequency table you have created to answer the questions below.

Month/ Date Range	1–8	9–16	17–24	25–31	Total
January					
February					
March					
April					
May					
[illegible]					
July					
August					
September					
October					
November					
December					
Total					

Answers will vary based on data collected.

1. What is *P*(May)? ____
2. What is *P*(9 to 16)? ____
3. What is *P*(October and 17 to 24)? ____
4. What is *P*(January and 1 to 8)? ____
5. What is *P*(June, July, or August)? ____
6. What is *P*(1 to 8 | February)? ____
7. What is *P*(17 to 24 | December)? ____
8. What is *P*(25 to 31 | April)? ____
9. Write your own probability question. Include an answer. ____

1 Interactive Learning

Solve It!

PURPOSE To analyze a situation involving conditional probability.

PROCESS Students may recognize that there are two scenarios in which they would clean the garage: it rains and they clean the garage; it does not rain and they clean the garage.

FACILITATE

Q Would a tree diagram help you to analyze the situation? Explain. **[Yes, a tree diagram would show all of the possible outcomes.]**

Q If you can find the probability that it rains and you clean and the probability that it does not rain and you clean, how could you find the probability that you clean the garage? **[Add the two probabilities.]**

ANSWER See Solve It in Answers on next page.

CONNECT THE MATH In the Solve It, students explore a situation involving conditional probability. In this lesson, students will find conditional probabilities using probability models, tree diagrams, and formulas.

2 Guided Instruction

Take Note

Q In the expression $P(B \mid A)$, which event represents the conditional event? Explain. **[*A*; the expression is read the probability of event *B*, given event *A*.]**

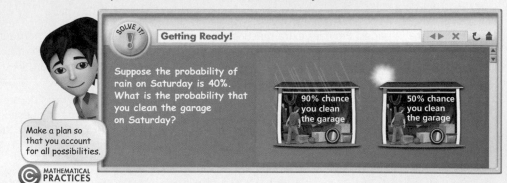

13-6 Conditional Probability Formulas

Common Core State Standards

S-CP.A.5 Recognize and explain the concepts of conditional probability and independence in everyday language and everyday situations. **Also S-CP.A.2, S-CP.A.3, S-CP.B.6**

MP 1, MP 2, MP 3, MP 4

Objective To understand and calculate conditional probabilities

Getting Ready!

Suppose the probability of rain on Saturday is 40%. What is the probability that you clean the garage on Saturday?

Make a plan so that you account for all possibilities.

MATHEMATICAL PRACTICES

In the Solve It, you may have calculated the probability that it rains and you clean the garage and the probability that it does not rain and you clean the garage and then added the probabilities together.

Essential Understanding You can find conditional probabilities using a formula.

In the previous lesson, you found probabilities using frequency tables. For example, in the relative frequency table at the right, the probability that a sales representative had an increase in sales, given that he attended the training seminar, is $\frac{0.48}{0.8} = 0.6$.

	Attended Seminar	Did not Attend Seminar	Totals
Increased Sales	0.48	0.02	0.5
No Increase in Sales	0.32	0.18	0.5
Totals	0.8	0.2	1

With respect to probability, this would be

$$P(\text{increased sales} \mid \text{attend seminar}) = \frac{P(\text{attend seminar and had increased sales})}{P(\text{attended seminar})}.$$

This suggests an algebraic formula for finding conditional probability.

Take note **Conditional Probability Formula**

For any two events A and B, the probability of B occurring, given that event A has occurred, is $P(B \mid A) = \frac{P(A \text{ and } B)}{P(A)}$, where $P(A) \neq 0$.

13-6 Preparing to Teach

BIG idea Data Representation

ESSENTIAL UNDERSTANDING

- Tables, tree diagrams, and formulas can be used to find conditional probability.

Math Background

Conditional probability is the probability that an event will occur given that another event has already occurred. For example, if A and B are dependent events, then $P(B \mid A)$ is the conditional probability that B will occur, given that A has occurred. After event A has occurred, the sample space for event B is reduced to those outcomes that include A.

In this lesson, students use the following methods to calculate conditional probabilities:

- frequency tables,
- the conditional probability formula,
- and tree diagrams.

Mathematical Practices

Reason abstractly and quantitatively. Students use probability formulas and tree diagrams to find conditional probability.

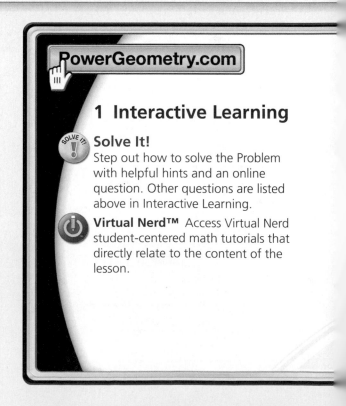

PowerGeometry.com

1 Interactive Learning

Solve It! Step out how to solve the Problem with helpful hints and an online question. Other questions are listed above in Interactive Learning.

Virtual Nerd™ Access Virtual Nerd student-centered math tutorials that directly relate to the content of the lesson.

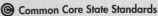

Problem 1 Using Conditional Probabilities

Think
How can you find $P(A)$?
A is the event that represents the sample space. In this case, it is the volunteers that received the drug. Since this is half of the volunteers, $P(A)$ is $\frac{1}{2}$.

Pharmaceutical Testing In a study designed to test the effectiveness of a new drug, half of the volunteers received the drug. The other half of the volunteers received a placebo, a tablet or pill containing no medication. The probability of a volunteer receiving the drug and getting well was 45%. What is the probability of someone getting well, given that he receives the drug?

Step 1 Identify the probabilities.

$P(B \mid A) = P(\text{getting well, given taking the new drug})$

$P(A) = P(\text{taking the new drug}) = \frac{1}{2} = 0.5$

$P(A \text{ and } B) = P(\text{taking the new drug and getting well}) = 45\%, \text{ or } 0.45$

Step 2 Find $P(B \mid A)$.

$P(B \mid A) = \frac{0.45}{0.5} = 0.9, \text{ or } 90\%$ Use the conditional probability formula.

The probability of getting well if someone received the drug is 90%.

Got It? 1. The probability of a volunteer receiving the placebo and having his or her health improve was 20%. What is the conditional probability of a volunteer's health improving, given that they received the placebo?

Conditional probabilities are usually not reversible. $P(A \mid B) \neq P(B \mid A)$.

Problem 2 Comparing Conditional Probabilities

Think
How are the sample spaces limited in each of the conditional probabilities?
In $P(\text{cat} \mid \text{dog})$, the sample space is limited to the probability that a pet owner owns a dog. Similarly, $P(\text{dog} \mid \text{cat})$, the sample space is the probability that an owner owns a cat.

Pets In a survey of pet owners, 45% own a dog, 27% own a cat, and 12% own both a dog and a cat. What is the conditional probability that a dog owner also owns a cat? What is the conditional probability that a cat owner also owns a dog?

$P(\text{cat} \mid \text{dog}) = \frac{P(\text{owns cat and dog})}{P(\text{owns dog})}$	Definition	$P(\text{dog} \mid \text{cat}) = \frac{P(\text{owns cat and dog})}{P(\text{owns cat})}$
$= \frac{0.12}{0.45}$	Substitute.	$= \frac{0.12}{0.27}$
$\approx 0.267, \text{ or } 26.7\%$	Simplify.	$\approx 0.444, \text{ or } 44.4\%$

The conditional probability that a dog owner also owns a cat is about 26.7%.
The conditional probability that a cat owner also owns a dog is about 44.4%.

Got It? 2. The same survey showed that 5% of the pet owners own a dog, a cat, and at least one other type of pet.

a. What is the conditional probability that a pet owner owns a cat and some other type of pet, given that they own a dog?

b. What is the conditional probability that a pet owner owns a dog and some other type of pet, given that they own a cat?

c. Critical Thinking Can you calculate the conditional probability of owning another pet for a pet owner owning a cat and no dogs? Explain.

Problem 1
In this problem, students use the formula for conditional probability.

Q Draw a tree diagram to model the situation. **[Check students' drawings.]**

Q What is the conditional probability that a volunteer did not get better, given that he or she received the drug? **[5%]**

Got It?

Q What is the probability that a volunteer's health improved regardless of whether he or she received the drug or the placebo? Explain. **[65%; *P*(drug and improve) = 45% and *P*(placebo and improve) = 20%]**

Problem 2
ERROR PREVENTION

Suggest to students that they label one event A and the other event B and write out the probabilities $P(A)$, $P(B)$, and $P(A \text{ and } B)$. This way they will be sure to substitute the correct values in the formula for conditional probability.

Q What would have to be true in order for $P(B \mid A)$ to be equal to $P(A \mid B)$? Explain. **[$P(A) = P(B)$, so that the denominators are the same in both formulas.]**

Got It?
Have students draw a Venn diagram that shows the relationship between the probabilities. This will help them see that they do not have enough information to solve the question in part (c).

2 Guided Instruction

Each Problem is worked out and supported online.

Problem 1
Using Conditional Probabilities

Problem 2
Comparing Conditional Probabilities

Problem 3
Using a Tree Diagram

Support in Geometry Companion
- Vocabulary
- Key Concepts
- Got It?

Answers

Solve It!
66%; Multiply the chance of rain by the probability that you clean the garage if it rains. Then multiply the chance of no rain by the probability that you clean the garage if it doesn't rain. Finally, add the results.

Got It?
1. 40%

2a. $\frac{1}{9}$

b. $\frac{5}{27}$

c. No, you don't know *P*(other pet) or *P*(cat and no dogs).

Problem 3 **VISUAL LEARNERS**

In this problem, students use a tree diagram to help them combine conditional probabilities.

Q Multiply to find the probabilities of each of the 4 paths in the tree diagram. What is the sum of these probabilities? **[0.42, 0.28, 0.24, 0.06; 1]**

Got It?

Q How can you use complements to find the probability that the team will not win their next game? What is this probability? **[Subtract the result from part (a) from 1.; 45.5%]**

Q In part (b), what does the increasing chance of a muddy field represent in terms of how the probabilities are weighted? Explain. **[As the chance of a muddy field increases, the higher probability of the team winning on a muddy field is weighted heavier.]**

Because $P(B|A) = \frac{P(A \text{ and } B)}{P(A)}$, then $P(A \text{ and } B) = P(A) \cdot P(B|A)$.

You can use this form of the conditional rule when you know the conditional probability. You can also combine conditional probabilities to find the probability of an event that can happen in more than one way.

Problem 3 Using a Tree Diagram

Graduation Rate **A college reported the following based on their graduation data.**

- **70% of freshmen had attended public schools**
- **60% of freshmen who had attended public schools graduated within 5 years**
- **80% of other freshmen graduated within 5 years**

What percent of freshmen graduated within 5 years?

You can use a tree diagram to organize the information.

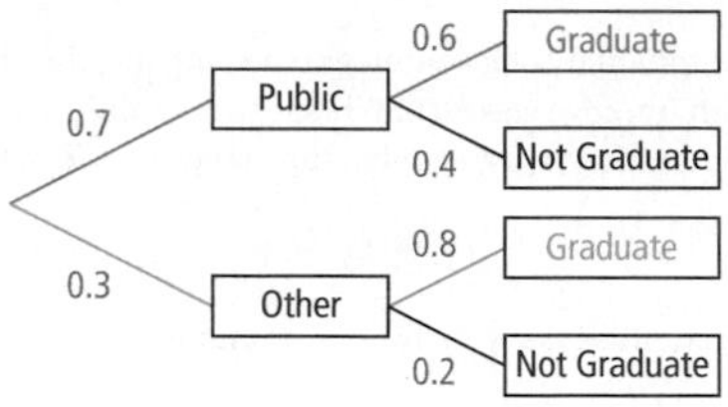

$P(\text{Public and Graduate}) = P(\text{Graduate, given Public}) \cdot P(\text{Public})$
$= 0.6 \cdot 0.7$
$= 0.42$

$P(\text{Other and Graduate}) = P(\text{Graduate, given Other}) \cdot P(\text{Other})$
$= 0.8 \cdot 0.3$
$= 0.24$

Plan
Why do you add the probabilities here?
You add the probabilities because the events are mutually exclusive.

$P(\text{Graduate}) = P(\text{Public and Graduate}) + P(\text{Other and Graduate})$
$= 0.42 + 0.24$
$= 0.66$

66% of the freshmen graduate within 5 years.

Got It? 3. a. A soccer team wins 65% of its games on muddy fields and 30% of their games on dry fields. The probability of the field being muddy for their next game is 70%. What is the probability that the team will win their next game?

b. Critical Thinking If the probability of the field being muddy increases, how will that influence the probability of the soccer team winning their next game? Explain.

Additional Problems

1. Half of the volunteers in a study received a test drug and half received a placebo. If the probability that a volunteer received the placebo and got better is 12%, what is the conditional probability of getting better given that a volunteer received the placebo?
ANSWER 24%

2. In a survey of shoppers, 54% use coupons, 36% belong to shopper clubs, and 18% use coupons and belong to shopper clubs. What is the conditional probability that a shopper uses coupons given that he or she belongs to shoppers clubs? What is the conditional probability that a shopper belongs to shoppers clubs given that he or she uses coupons?
ANSWER $\frac{1}{2}$; $\frac{1}{3}$

3. Suppose that a basketball team has a 75% chance of winning a tournament if their star player is able to play. Otherwise, the team has a 40% chance of winning the tournament. The doctor says that the star player has a 60% chance of being able to play. What is the probability that the team will win the tournament?
ANSWER 61%

Lesson Check

Do you know HOW?

A jar contains 10 large red marbles, 4 small red marbles, 6 large blue marbles, and 5 small blue marbles. Calculate the following conditional probabilities for choosing a marble at random.

1. $P(\text{large} \mid \text{red})$
2. $P(\text{large} \mid \text{blue})$
3. For the same jar of marbles, which of the conditional probabilities is larger, $P(\text{red} \mid \text{small})$ or $P(\text{small} \mid \text{red})$? Explain.

Do you UNDERSTAND? MATHEMATICAL PRACTICES

4. **Error Analysis** Your friend says that the conditional probability of one event is 0% if it is independent of another given event. Explain your friend's error.
5. **Compare and Contrast** How is finding a conditional probability like finding a compound probability? How is it different?

Practice and Problem-Solving Exercises MATHEMATICAL PRACTICES

A Practice

6. **Allowance** Suppose that 62% of children are given a weekly allowance, and 38% of children do household chores to earn an allowance. What is the probability that a child does household chores, given that the child gets an allowance?

See Problems 1 and 2.

7. You roll two standard number cubes. What is the probability that the sum is even, given that one number cube shows a 2?

Softball Suppose that your softball team has a 75% chance of making the playoffs. Your cross-town rivals have an 80% chance of making the playoffs. Teams that make the playoffs have a 25% chance of making the finals. Use this information to find the following probabilities.

See Problem 3.

8. P(your team makes the playoffs and the finals)
9. P(cross-town rivals make the playoffs and the finals)

B Apply

10. **Think About a Plan** Suppose there are two stop lights between your home and school. On the many times you have taken this route, you have determined that 70% of the time you are stopped on the first light and 40% of the time you are stopped on both lights. If you are not stopped on the first light, there is a 50% chance you are stopped on the second light. What is the probability that you make it to school without having to stop at a stoplight?
    - What conditional probability are you looking for?
    - How can a tree diagram help?

3 Lesson Check

Do you know HOW? ERROR INTERVENTION

- If students have trouble solving Exercises 1 and 2, then have them identify the sample space for each conditional probability.
- If students think that the conditional probabilities in Exercise 3 are equal, then remind them that $P(B \mid A)$ is rarely the same as $P(A \mid B)$ because the sample spaces are usually different.

Do you UNDERSTAND?

- In Exercise 4, point out to students that $P(B \mid A) = P(B)$ if A and B are independent because the outcome of B does not depend on A. Show students how the formula for conditional probability is reduced because $P(A \text{ and } B) = P(A) \cdot P(B)$ for independent events.

Close

Q What are three tools you can use to help you find conditional probabilities? **[probability distributions, tree diagrams, and the conditional probability formula]**

Q How do you calculate $P(B \mid A)$? **[Divide $P(A \text{ and } B)$ by $P(A)$.]**

Q In general, is $P(B \mid A)$ equal to $P(A \mid B)$? Explain. **[No, the sample spaces are usually different.]**

Answers

Got It? (continued)

3a. 54.5%

b. Their chances of winning will increase because they have a better chance of winning on a muddy field.

Lesson Check

1. $\frac{5}{7}$
2. $\frac{6}{11}$
3. $P(\text{red} \mid \text{small})$; $P(\text{red} \mid \text{small}) = \frac{4}{9}$, $P(\text{small} \mid \text{red}) = \frac{2}{7}$
4. No. If two events are independent, then $P(A \text{ and } B) = P(A) \cdot P(B)$, which is not necessarily zero. The conditional probability would be 0 for mutually exclusive events.
5. Sample answer: When finding a conditional probability you divide, but when you find a compound probability you often multiply.

Practice and Problem Solving Exercises

6. $\frac{19}{31}$
7. $\frac{1}{2}$
8. 18.75%
9. 20%
10. 15%

4 Practice

ASSIGNMENT GUIDE

Basic: 6–10, 12, 14, 16
Average: 7, 9–17
Advanced: 7, 9–18
Standardized Test Prep: 19–21

Mathematical Practices are supported by exercises with red headings. Here are the Practices supported in this lesson:

MP 1: Make Sense of Problems Ex. 12
MP 3: Compare Arguments Ex. 5
MP 4: Model with Mathematics Ex. 22

Applications exercises have blue headings. Exercises 8, 9, and 11 support MP 4: Model.

STEM exercises focus on science or engineering applications.

EXERCISE 16: Use the Think About a Plan worksheet in the **Practice and Problem Solving Workbook** (also available in the Teaching Resources in print and online) to further support students' development in becoming independent learners.

HOMEWORK QUICK CHECK

To check students' understanding of key skills and concepts, go over Exercises 7, 9, 10, 12, and 16.

11. Sports There is a 40% chance that a school's basketball team will make the playoffs this year. If they make the playoffs, there is a 15% chance that they will win the championship. There is also a 30% chance that the same school's volleyball team will make the playoffs this year. If the volleyball team makes the playoffs, there is a 30% chance that they will win the championship. What is the probability that at least one of these teams will win a championship this year?

STEM **Science** **In a research study, one third of the volunteers received drug A, one third received drug B, and one third received a placebo. Out of all the volunteers, 10% received drug A and got better, 8% received drug B and got better, and 12% received the placebo and got better.**

12. What is the conditional probability of a volunteer getting better if they were given drug A?

13. What is the conditional probability of a volunteer getting better if they were given drug B?

14. What is the conditional probability of a volunteer getting better if they were given the placebo?

STEM **15. Chemistry** A scientist discovered that a certain element was present in 35% of the samples she studied. In 15% of all samples, the element was found in a special compound. What is the probability that the compound is in a sample that contains the element?

16. Music Three students compared the music on their MP3 players. Find the probability that a randomly selected song is country, given that it is not on Student A's MP3 player. Round to the nearest hundredth.

Student	Rock	Country	R & B
A	0.10	0.02	0.03
B	0.13	0.05	0.23
C	0.10	0.01	0.33

17. Fire Drill A fire drill will begin at a randomly-chosen time between 8:30 A.M. and 3:30 P.M. You have a math test planned for 2:05 P.M. to 3:00 P.M. If the fire drill is in the afternoon, what is the probability that it will start during the test?

Challenge

18. The table below shows the average standardized test scores for a group of students. If 35% of the students are seniors, 40% are juniors, and 25% are freshman, what is the probability that a student chosen at random will have a score at least 125?

	Average Score < 125	Average Score between 125 and 145	Average Score > 145
Freshman	48%	37%	15%
Junior	36%	52%	12%
Senior	13%	69%	18%

Answers

Practice and Problem-Solving Exercises (continued)

11. 14.46%
12. 30%
13. 24%
14. 36%
15. $\frac{3}{7}$
16. 0.07
17. $\frac{11}{42}$
18. 69.05%

Standardized Test Prep

SAT/ACT

19. Which of the following statements is true?

(A) A kite has two pairs of congruent angles.

(B) The measure of an inscribed angle equals the measure of the arc it intersects.

(C) A rhombus has two consecutive angles congruent if and only if the rhombus is a square.

(D) Two equilateral triangles can be combined to form a square.

20. If $CD = 3$ and $AD = 12$, what is BD?

(A) 2 (C) 6

(B) 4 (D) 8

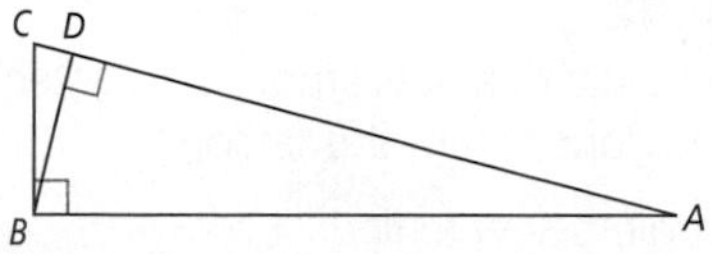

Short Response

21. The sides of a rectangle are 5 cm and 12 cm long. What is the sum of the lengths of the rectangle's diagonals?

Apply What You've Learned

MATHEMATICAL PRACTICES MP 2

Review the information on page 823 about the survey Loretta's Ice Cream conducted and the question Loretta asked her assistant. To answer Loretta's question, her assistant needs to find these two probabilities:

- the probability that a customer who likes raspberry also likes caramel
- the probability that a customer who likes caramel also likes raspberry

a. The probabilities listed above are conditional probabilities. Write the conditional probability formula for the probability that a customer who likes raspberry also likes caramel.

b. What two probabilities do you need to find in order to use the conditional probablity formula to find the probablity that a customer who likes raspberry also likes caramel? Use the the information in the table you completed in the Apply What You've Learned in Lesson 13-1 to find these probabilties.

c. What is the probability that a customer who likes raspberry also likes caramel? Round your answer to the nearest whole percent.

19. C
20. C
21. 26 cm

Apply What You've Learned

In the Apply What You've Learned for Lesson 13-1, students completed the table and calculated experimental probabilities related to the survey information given on page 823. Here, students will work with conditional probabilities.

Mathematical Practices

Students **reason abstractly and quantitatively** as they use the conditional probability formula to represent and calculate a probability. (MP 2)

ANSWERS

a. $P(\text{likes caramel} \mid \text{likes raspberry}) = \dfrac{P(\text{likes raspberry and caramel})}{P(\text{likes raspberry})}$

b. the probability that a customer likes both flavors, and the probability that a customer likes raspberry; $P(\text{likes both}) = 19\%$, $P(\text{likes raspberry}) = 43\%$

c. about 44%

13-6 Lesson Resources

Differentiated Remediation

Additional Instructional Support

Geometry Companion

Students can use the **Geometry Companion** worktext (4 pages) as you teach the lesson. Use the Companion to support

- Solve It!
- New Vocabulary
- Key Concepts

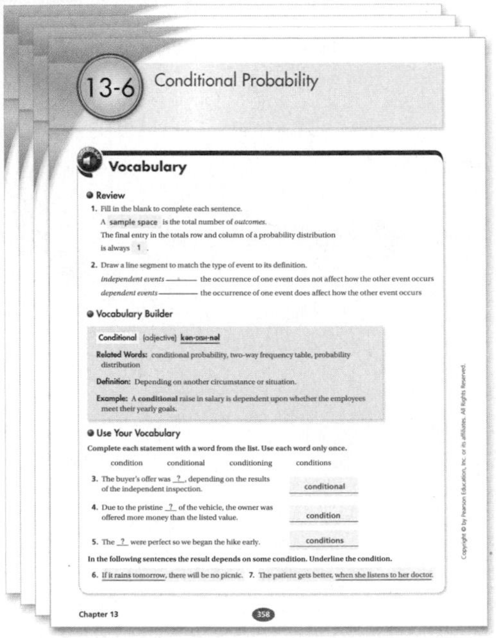

13-6 Conditional Probability

Vocabulary

Review

1. Fill in the blank to complete each sentence.
   A **sample space** is the total number of *outcomes*.
   The final entry in the totals row and column of a probability distribution is always **1**.
2. Draw a line segment to match the type of event to its definition.
   independent events — the occurrence of one event does not affect how the other event occurs
   dependent events — the occurrence of one event does affect how the other event occurs

Vocabulary Builder

Conditional (adjective) **kən-DISH-nal**

Related Words: conditional probability, two-way frequency table, probability distribution

Definition: Depending on another circumstance or situation.

Example: A **conditional** raise in salary is dependent upon whether the employees meet their yearly goals.

Use Your Vocabulary

Complete each statement with a word from the list. Use each word only once.

condition conditional conditioning conditions

3. The buyer's offer was _?_, depending on the results of the independent inspection. **conditional**
4. Due to the pristine _?_ of the vehicle, the owner was offered more money than the listed value. **condition**
5. The _?_ were perfect so we began the hike early. **conditions**

In the following sentences the result depends on some condition. Underline the condition.

6. If it rains tomorrow, there will be no picnic. 7. The patient gets better, when she listens to her doctor.

Copyright © by Pearson Education, Inc. or its affiliates. All Rights Reserved.

Chapter 13 358

ELL Support

Have Students Create Test Items Organize students in small groups or pairs. Have each group create a sample contingency table that uses "Male" and "Female" to describe one characteristic of the students in the classroom. Then have each group decide on a second characteristic that can be used to form the table. Have them complete the table and find the relative frequencies.

Next, have each group write 2–3 test items using the data in the contingency table. Sample questions might include:

- What is *P*(wears glasses | female)?
- What is *P*(male | wearing black)?

Ask for groups to volunteer to share their tables and test items with the class and work together to solve them.

5 Assess & Remediate

Lesson Quiz

Use the Lesson Quiz to assess students' mastery of the skills and concepts of this lesson.

1. Use the table below to find *P*(4-wheel drive | pickup truck).

	2-wheel drive	4-wheel drive	Totals
Pickup Trucks	0.15	0.33	0.48
SUVs	0.28	0.24	0.52
Totals	0.43	0.57	1

2. The probability that a student is a senior and has a part-time job is 8%. The probability that a student is a senior is 30%. What is the probability that a student has a part-time job, given that the student is a senior?

3. In a survey of computer owners, 80% use an anti-virus program, 25% use an anti-spyware program, and 15% use both. What is *P*(anti-virus | anti-spyware)? What is *P*(anti-spyware | anti-virus)?

ANSWERS TO LESSON QUIZ

1. 68.75%
2. $\frac{4}{15}$
3. 60%; 18.75%

PRESCRIPTION FOR REMEDIATION

Use the student work on the Lesson Quiz to prescribe a differentiated review assignment.

Points	Differentiated Remediation
0–1	Intervention
2	On-level
3	Extension

PowerGeometry.com

5 Assess & Remediate

Assign the Lesson Quiz. Appropriate intervention, practice, or enrichment is automatically generated based on student performance.

Intervention

- **Reteaching** (2 pages) Provides reteaching and practice exercises for the key lesson concepts. Use with struggling students or absent students.
- **English Language Learner Support** Helps students develop and reinforce mathematical vocabulary and key concepts.

All-in-One Resources/Online

Reteaching

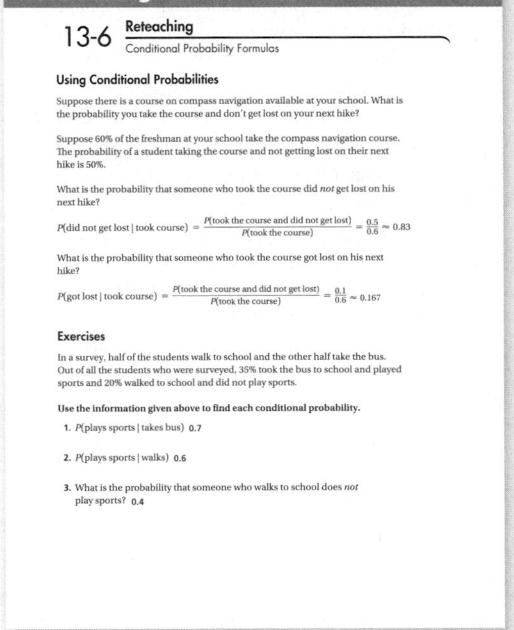

13-6 **Reteaching**
Conditional Probability Formulas

Using Conditional Probabilities

Suppose there is a course on compass navigation available at your school. What is the probability you take the course and don't get lost on your next hike?

Suppose 60% of the freshman at your school take the compass navigation course. The probability of a student taking the course and not getting lost on their next hike is 50%.

What is the probability that someone who took the course did *not* get lost on his next hike?

$P(\text{did not get lost} \mid \text{took course}) = \frac{P(\text{took the course and did not get lost})}{P(\text{took the course})} = \frac{0.5}{0.6} \approx 0.83$

What is the probability that someone who took the course got lost on his next hike?

$P(\text{got lost} \mid \text{took course}) = \frac{P(\text{took the course and did not get lost})}{P(\text{took the course})} = \frac{0.1}{0.6} \approx 0.167$

Exercises

In a survey, half of the students walk to school and the other half take the bus. Out of all the students who were surveyed, 35% took the bus to school and played sports and 20% walked to school and did not play sports.

Use the information given above to find each conditional probability.

1. *P*(plays sports | takes bus) **0.7**
2. *P*(plays sports | walks) **0.6**
3. What is the probability that someone who walks to school does *not* play sports? **0.4**

All-in-One Resources/Online

English Language Learner Support

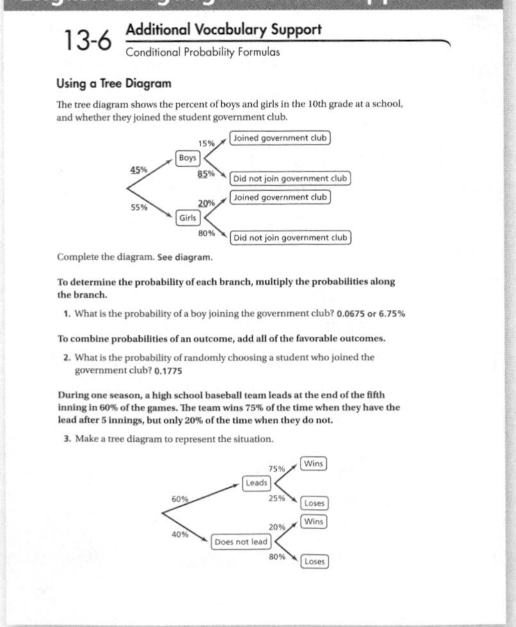

13-6 **Additional Vocabulary Support**
Conditional Probability Formulas

Using a Tree Diagram

The tree diagram shows the percent of boys and girls in the 10th grade at a school, and whether they joined the student government club.

Complete the diagram. **See diagram.**

To determine the probability of each branch, multiply the probabilities along the branch.

1. What is the probability of a boy joining the government club? **0.0675 or 6.75%**

To combine probabilities of an outcome, add all of the favorable outcomes.

2. What is the probability of randomly choosing a student who joined the government club? **0.1775**

During one season, a high school baseball team leads at the end of the fifth inning in 60% of the games. The team wins 75% of the time when they have the lead after 5 innings, but only 20% of the time when they do not.

3. Make a tree diagram to represent the situation.

Differentiated Remediation *continued*

On-Level

- **Practice** (2 pages) Provides extra practice for each lesson. For simpler practice exercises, use the Form K Practice pages found in the All-in-One Teaching Resources and online.
- **Think About a Plan** Helps students develop specific problem-solving skills and strategies by providing scaffolded guiding questions.
- **Standardized Test Prep** Focuses on all major exercises, all major question types, and helps students prepare for the high-stakes assessments.

Extension

- **Enrichment** Provides students with interesting problems and activities that extend the concepts of the lesson.
- **Activities, Games, and Puzzles** Worksheets that can be used for concepts development, enrichment, and for fun!

Practice and Problem Solving Wkbk/ All-in-One Resources/Online

Practice page 1

13-6 **Practice** *Form G*
Conditional Probability Formulas

At a recent swim meet, half of the swim club members experienced an improvement in their race times over a previous swim meet. The probability of a swim club member experiencing an improvement in their race time and training the week before the meet was 30%.

1. What is the probability that a swimmer trained the week before the meet given that his or her race time improved? **0.6 or 60%**
2. The probability that a swimmer did not experience an improvement in his or her race times and trained the week before the meet was 10%. What is *P*(trained | did not improve)? **0.2 or 20%**
3. Half of a class took Form A of a test, and half took Form B. Of the students who took Form B, 39% passed. What is the conditional probability that a randomly chosen student took Form B and passed? **0.195 or 19.5%**
4. Three-fourths of a research team worked in a lab while one-fourth of the team worked near a pond. Of the researchers who worked near the pond, 14% collected insects. What is the probability that a randomly chosen researcher worked near the pond and collected insects? **0.035 or 3.5%**
5. In the senior class, 24% of the students play softball, 32% of the students play field hockey, and 14% play both. What are the probabilities that a softball player also plays field hockey, and a field hockey player also plays softball? **0.5833 and 0.4375**

Use the diagram at the right for Exercises 6 and 7.

The tree diagram shows the percentages of plants that received sunlight and whether or not they grew.

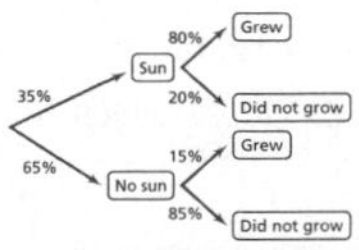

6. What is the combined probability that a plant grew? **0.3775**
7. What is the combined probability that a plant did not grow? **0.6225**

Practice and Problem Solving Wkbk/ All-in-One Resources/Online

Practice page 2

13-6 **Practice** (continued) *Form G*
Conditional Probability Formulas

Of the people who went to an amusement park last week, 85% rode a rollercoaster, 45% attended a musical review show, and 18% did both.

8. What is the conditional probability that a person who rode a rollercoaster also attended a musical review show? **0.212**
9. **Writing** Explain the meaning of *P*(rode a rollercoaster | attended musical review). Then calculate the probability. **P(rode a rollercoaster | attended musical review) is the number of people who rode a rollercoaster given that they attended a musical review show; 0.4**
10. **Writing** Half of your 200 classmates went to the zoo. Of the students who went to the zoo, 25% saw the dolphin show. Explain how to calculate the number of students that attended the dolphin show. **The percent of those who went to the zoo and saw the dolphin show is 50% times 25%, or 12.5%. To find the number of students who saw the dolphin show, multiply the total number of students by this percentage. $200 \cdot 0.125 = 25$ students**

The diagram at the right shows the percent of blue-eyed voters and brown-eyed voters that voted for 2 candidates. Use the table for Exercises 11 and 12.

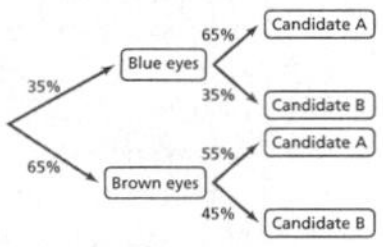

11. What is the combined probability that Candidate A won? **0.585**
12. **Error Analysis** Your friend says the combined probability of Candidate B winning is 80%. What error did she make? What is the correct combined probability? **She added the probabilities beneath each candidate instead of finding the combined probabilities of each branch and adding them; 0.415**
13. Of a group of friends, 28% take dance lessons, 32% take singing lessons, and 8% take both. What is the probability that a dancer takes singing lessons? What is the probability that a singer takes dance lessons? **0.286 and 0.25**

All-in-One Resources/Online

Enrichment

13-6 **Enrichment**
Conditional Probability Formulas

You can use a tree diagram to organize several levels of probabilities.

The tree diagram below shows the percentage of a school's basketball players who live within 2 miles of school, whether or not they walk to practice, and whether or not they are one of the 5 players in the first string (players who start the game).

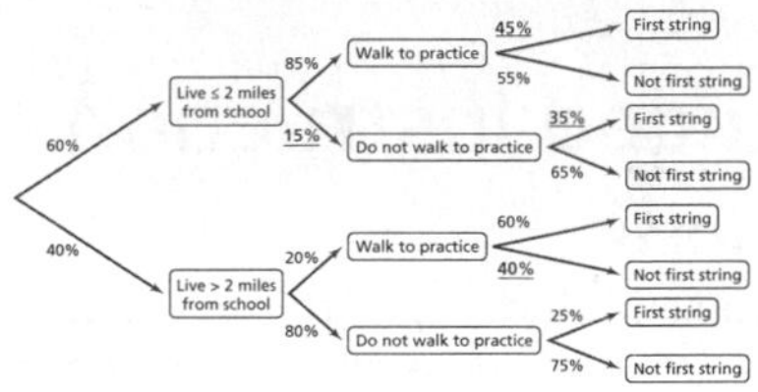

1. Complete the diagram. **See above.**
2. What is the probability that a player lives within 2 miles of school and walks to practice? **0.51**
3. What is the probability that a player walks to practice? **0.59**
4. What is the probability that a player lives within 2 miles of school, walks to practice, and is on the first string? **0.2295**
5. What is the probability that a player is in the first string? **0.389**

Practice and Problem Solving Wkbk/ All-in-One Resources/Online

Think About a Plan

13-6 **Think About a Plan**
Conditional Probability Formulas

Chemistry A scientist discovered that a certain element was present in 35% of the samples she studied. In 15% of those samples, the element was found in a special compound. What is the probability that the compound is in a sample that contains the element?

Understanding the Problem

1. What percent of the samples contain the element? **0.35 or 35%**
2. What percent of the samples contain the special compound *and* contain the element? **0.15 or 15%**
3. Will you use the percent of the samples that did *not* contain the element? **No.**

Planning the Solution

4. Write the formula for conditional probability.
   $P(A \mid B) = \frac{P(A \text{ and } B)}{P(B)}$
5. What is the value of *P*(*A* and *B*)? **0.15 or 15%**
6. What is the value of *P*(*B*)? **0.35 or 35%**

Getting an Answer

7. Evaluate the conditional probability formula using the values for *A* and *B*.
   $P(A \mid B) = \frac{0.15}{0.35} = 0.\overline{428571}$
8. What is the probability that the compound is in a sample that contains the element? **About 0.429 or 42.9%**

Practice and Problem Solving Wkbk/ All-in-One Resources/Online

Standardized Test Prep

13-6 **Standardized Test Prep**
Conditional Probability Formulas

Multiple Choice

For Exercises 1–4, choose the correct letter.

A physician determined that on average, 40% of his patients get the flu each year. Of this group, 10% received the flu vaccine. Of the patients who do not get the flu, 20% received the flu vaccine. Use this information for Exercises 1 and 2.

1. What is the probability that someone who did not receive the vaccine got the flu? **B**
   A. 0.25　B. 0.43　C. 0.75　D. 0.75
2. What is the probability that someone who received the vaccine got the flu? **F**
   F. 0.25　G. 0.33　H. 0.5　I. 0.66
3. Of the 85% of the students in a class who studied for a test, 75% passed the test. Of the 15% of the students who did not study, 30% passed. What is the combined probability of passing? **C**
   A. 0.3　B. 0.6375　C. 0.6825　D. 0.75
4. In a survey, 60% of the people own a laptop computer, 80% own a desktop computer, and 30% own both. What is the conditional probability that a laptop computer owner also owns a desktop computer? **H**
   F. 0.3　G. 0.4　H. 0.5　I. 0.6

Short Response

5. When calculating *P*(*B* | *A*), what does *A* represent?
   [2] *A* represents an event that has already happened, and is the population you are finding the probability within.
   [1] Student provides evidence showing an understanding of conditional probability calculations.
   [0] Student's response shows no effort OR does not attempt to explain the probability in terms of an event that has already happened.

Online Teacher Resource Center

Activities, Games, and Puzzles

13-6 **Puzzle: Planning Errands**
Conditional Probability Formulas

Mrs. Whitt and her son Graham have four errands to run on Saturday morning. Graham is learning about conditional probability and he proposes that they use the concepts he is learning to decide the order they will complete the errands. Mrs. Whitt shares with Graham that she has recently read an article that states that 65% of middle school students own cell phones, 40% have a TV in their bedrooms, and 25% have their own laptops. The article also stated that 18% of middle school students had both a cell phone and a laptop and 6% have a cell phone, a TV in their bedroom, and a laptop. She writes the probability questions below and provides the map for Graham to complete.

Directions: Calculate the probabilities. Match your answer to a location on the map and fill in the blank with the name of that errand. When all answers are complete you will have the order that Graham and his mom will complete the errands.

1. What is the conditional probability that a student has all three devices given he or she has a cell phone? 9.2% (House)
2. What is the conditional probability that a student has a cell phone given he or she has a laptop? 72% (Bank)
3. What is the conditional probability that a student has all three electronic devices given he or she has a TV? 15% (Gas Station)
4. What is the conditional probability that a student has a laptop given he or she has a cell phone? 27.7% (Dry Cleaners)
5. What is the conditional probability that a student has all three devices given he or she has a laptop? 24% (Grocery)

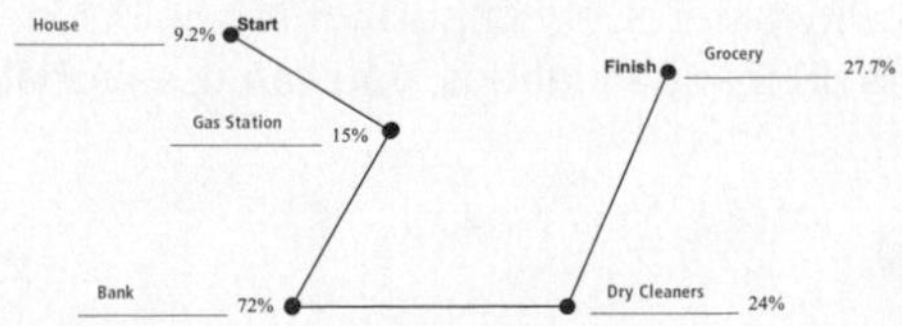

1 Interactive Learning

Solve It!

PURPOSE To explore ways of modeling a random selection.

PROCESS Ask students how they have made fair decisions among their friends or families in the past.

FACILITATE

Q What must be true about the selection in order for a fair decision to be reached? Explain. **[Each group of four should have the same chance of being selected.]**

ANSWER See Solve It in Answers on next page.

CONNECT THE MATH In the Solve It, students try to come up with ways of modeling a random selection and making a fair decision. In this lesson, students will learn some ways to model randomness and make decisions based on expected values.

2 Guided Instruction

Problem 1

In this problem, students learn how to model a random situation using a table of random numbers.

Q Do you think it matters which line of a table you use to make a random selection? Explain. **[No. Since each line of a random number table features random numbers, the decision will be fair no matter which line you choose.]**

13-7 Modeling Randomness

Common Core State Standards

S-MD.B.6 Use probabilities to make fair decisions (e.g., drawing by lots, using a random number generator). **Also S-MD.B.7**

MP 1, MP 3, MP 4

Objective To understand random numbers
To use probabilities in decision-making

Think of as many ways as you can.

MATHEMATICAL PRACTICES

Solve It! Getting Ready!

A class of 25 students wants to choose 4 students at random to bring food for a class party to be used in the platter shown. Any set of 4 students should have an equal chance of being chosen. What are some ways that the class could determine a random selection of 4 students?

In the Solve It, you thought of ways to select 4 people. In this lesson, you will learn ways to make a random selection and use probabilities to make decisions and solve real world problems.

Lesson Vocabulary
- expected value

Essential Understanding You can use probability to make choices and to help make decisions based on prior experience.

A random event has no predetermined pattern or bias toward one outcome or another. You can use random number tables or randomly generated numbers using graphing calculators or computer software to help you make fair decisions. In Reference 5 of this book, you will find a random number table to use with this lesson.

Think

How can a single-digit number be represented by two digits?
A zero and a number between one and nine can represent a single-digit number. For example, 09 can represent 9.

Problem 1 Making Random Selections

There are 28 students in a homeroom. Four students are chosen at random to represent the homeroom on a student committee. How can a random number table be used to fairly choose the students?

Step 1 Select a line from a random number table.

18823 18160 93593 67294 09632 62617 86779

Step 2 Group the line from the table into two digit numbers.

18 82 31 81 60 93 59 36 72 94 09 63 26 26 17 86 77 9

13-7 Preparing to Teach

BIG idea Probability

ESSENTIAL UNDERSTANDING
- Probability and random selection can be used in making appropriate and fair decisions.

Math Background

Students probably model random situations in many areas of their lives and do not even realize it. Whether it's flipping a coin to see who has to wash the dishes or playing rock-paper-scissors to determine who serves first in a volleyball game, these are all ways of using probability to make fair decisions.

Probability plays a very important role in decision making. In many situations, you can use probability to help you predict what is likely or unlikely to happen. Expected value uses theoretical probability to help you predict what will happen in the long run. Having a better idea of what is *likely* to happen mathematically can help you make better decisions in problem situations.

Mathematical Practices

Model with mathematics. Reason abstractly and quantitatively. Students model situations through simulations using random number tables and random number generators. They find expected value and use this tool to make decisions.

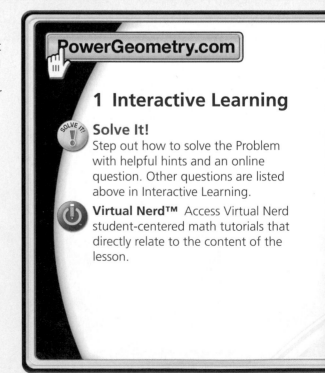

PowerGeometry.com

1 Interactive Learning

Solve It!
Step out how to solve the Problem with helpful hints and an online question. Other questions are listed above in Interactive Learning.

Virtual Nerd™ Access Virtual Nerd student-centered math tutorials that directly relate to the content of the lesson.

Step 3 Match the first four numbers less than 28 with the position of the students' names on a list. Duplicates and numbers greater than 28 are discarded because they don't correspond to any student on the list.

18 82 31 81 60 93 59 36 72 94 09 63 26 26 17 86 77 9

The students listed 18th, 9th, 26th, and 17th on the list are chosen fairly.

Got It? 1. A teacher wishes to choose three students from a class of 25 students to raise the school's flag. What numbers should the teacher choose based on this line from a random number table?

65358 70469 87149 89509 72176 18103 55169 79954 72002 20582

Problem 2 Making a Simulation

A cereal company is having a promotion in which 1 of 6 different prizes is given away with each box. The prizes are equally and randomly distributed in the boxes of cereal. On average, how many boxes of cereal will a customer need to buy in order to get all 6 prizes?

Plan

How do you start to model a situation? The first step in modeling a situation is determining the possible outcomes and assigning a number to each outcome.

Step 1 Let the digits from 1 to 6 represent the six prizes.

Step 2 Use a graphing calculator and enter the function randInt(1,6) to generate integers from 1 to 6 to simulate getting each prize. One trial is completed when all 6 digits have appeared. The circled numbers represent the first appearance of each number.

(2)(5)(3) 2 5 2 2 (4) 5 (1) 2 2 4 4 5 (6)

Step 3 Count how many boxes of cereal will be bought before all the digits 1 through 6 have appeared.

In this trial, all six prizes were collected after the purchase of the 16th box of cereal.

Step 4 Conduct additional trials. For 19 more simulations the results were: 17, 12, 21, 17, 8, 11, 14, 10, 16, 23, 19, 15, 9, 27, 20, 10, 18, 12, 13. For the 20 simulations, it will take, on average, $\frac{308}{20} = 15.4$ boxes of cereal to get all 6 prizes.

Got It? 2. Suppose that to win a game, you must roll a standard number cube until all of the sides show at least one time. On average, how many times will you have to roll the number cube before each side shows at least one time? Use a random number table or a random number generator.

Got It?

Q Will different lines from a random number table always result in different groups of students? **[Different lines will likely result in different groups being selected, but it is possible to get the same group.]**

Problem 2 EXTENSION

In Problem 2, students are introduced to using technology to model a random situation.

Q What is another method that you could have used to model how many tries it would take to get all 6 prizes? **[Sample answer: Use a number cube and let each number represent a different prize.]**

Got It?

Q Does the result imply that it will always take about 15 rolls to get all 6 numbers? Explain. **[No. This is an average value. Sometimes it will take more rolls, sometimes fewer.]**

Take Note EXTENSION

Remind students that a probability distribution shows the probability of each possible outcome. Discuss with students how a probability distribution can assist them in finding expected value.

2 Guided Instruction

Each Problem is worked out and supported online.

Problem 1 Making Random Selections

Problem 2 Making a Simulation

Problem 3 Calculating an Expected Value

Problem 4 Making Decisions Based on Expected Values

Support in Geometry Companion

- Vocabulary
- Key Concepts
- Got It?

Answers

Solve It!

Answers will vary.; Students should discuss different ways of modeling a random selection such as writing names on slips of paper, placing them in a box and shaking well, and then drawing 3 names without looking.

Got It?

1. 04, 14, 09
2. Answers may vary. Sample: about 15

In Problem 2, you figured out how many boxes of cereal on average you would expect to have to buy to collect all 6 prizes. You can use this information to make a decision about whether it is worth trying to collect all 6 prizes. Similarly, you can use the *expected value* of a situation that involves uncertainty to make a decision. *Expected value* uses theoretical probability to tell you what you can expect in the long run. If you know what *should* happen mathematically, you make better decisions in problem situations. The **expected value** is the sum of each outcome's value multiplied by its probability.

take note

Key Concept Calculating Expected Value

If A is an action that includes outcomes $A_1, A_2, A_3, \ldots$ and Value(A_n) is a quantitative value associated with each outcome, the expected value of A is given by

$$\text{Value}(A) = P(A_1) \cdot \text{Value}(A_1) + P(A_2) \cdot \text{Value}(A_2) + \ldots$$

Problem 3 Calculating an Expected Value

Suppose you are at a carnival and are throwing darts at a board like the one at the right. There is an equally likely chance that your dart lands anywhere on the board. You receive 20 points if your dart lands in the white area, 10 points if it lands in the red area, and -5 points if it lands in the blue area. How many points can you expect to get given that the areas for each region are white, 36 in.^{2}; red, 108 in.^{2}; and blue, 432 in.^{2}? The total area is 576 in.^{2}.

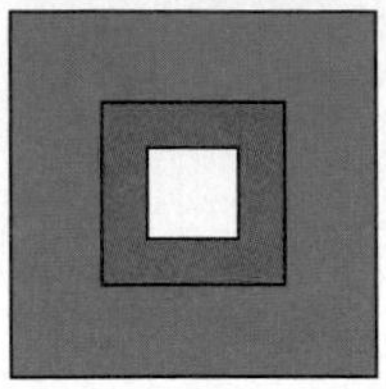

Value(throw)

$= P(\text{white area}) \cdot (\text{white points}) + P(\text{red area}) \cdot (\text{red points}) + P(\text{blue area}) \cdot (\text{blue points})$

$= \frac{36}{576} \cdot 20 + \frac{108}{576} \cdot 10 + \frac{432}{576} \cdot (-5)$ Substitute the ratio of each area to the total area and the points for each section.

$= 1.25 + 1.875 + (-3.75)$ Multiply.

$= -0.625$ Simplify.

You can expect to get -0.625 points.

Got It? **3. a.** How many points can you expect to get for the dart board shown in Example 3 if you receive these points: 30 points for white, 15 points for red, and -10 points for blue?

b. For part (a), what number of points could be set for the blue area that would result in an expected value of 0?

Expected values can be used to make data-driven decisions. You can calculate the expected values of situations of interest, and compare them to decide which is most favorable.

Problem 3 ELL SUPPORT

Discuss with students how the meanings of the words *expected* and *value* are used in everyday life and how these uses relate to the mathematical meaning of expected value.

Q How would you find the expected value if the board had additional areas worth various point values? **[You would find the expected value of each area and add.]**

Got It?

Q If a player wins the game only when he or she ends with a positive score, is it likely that there will be many winners? Explain. **[No. If a player can expect to get −6.25 points per game, this game is not in the player's favor.]**

Additional Problems

1. Use the random numbers below to find the first two singers from a group of 15 if they are numbered 01, 02, . . . , 15.
95985 41520 68320 78623 59658
ANSWER 15, 07

2. On average Regina wins 60% of the races that she enters. Describe how you could use random numbers to model how many of her next 10 races Regina will win.
ANSWER Sample answer: Use a calculator to generate random numbers from 1 to 5 and let 1, 2, and 3 represent a win and 4 and 5 represent a loss.

3. A game ticket for a local charity has a 70% chance of winning \$1, a 25% chance of winning \$5, and a 5% chance of winning \$50. What is the expected value of the game ticket?
ANSWER \$4.45

4. An investment has a 30% chance of earning \$4000 and a 60% chance of losing \$1000, and a 10% chance of losing \$1000. Based on expected value, would you recommend the investment? Explain.
ANSWER Yes, the expected value of the investment is \$1700.

Answers

Got It? (continued)

3a. −2.8125

b. −6.25

Think

How can a coach know the probability for his team's kicking a field goal?
He can know the experimental probability based on how many successful field goals divided by the number of attempts in similar situations.

Problem 4 Making Decisions Based on Expected Values

Football On the opening drive, a A football coach must decide whether to kick a field goal (FG) or go for a touchdown (TD). The probabilities for each choice based on his team's experience are shown on the page from his playbook at the right. A field goal will give his team 3 points. His team will get 7 points if the touchdown is successful. Which play should he choose?

	Pts.	Prob.
FG	3	90%
TD	7	35%

Step 1 Calculate the expected value of both plays.

Field Goal: $90\% \cdot 3 = 2.7$ points

Touchdown: $35\% \cdot 7 = 2.45$ points

Step 2 Choose the play with the greater expected value.

$2.7 > 2.45$

The coach should choose the field goal.

Got It? **4. a.** Suppose the probability for the field goal was 80% and the probability for a touchdown was 30%. Which play should the coach choose?

b. Are there situations where the coach should choose a play that doesn't have the greatest expected value? Explain.

Lesson Check

Do you know HOW?

1. What are the first four numbers between 1 and 45 which would be chosen on the basis of the following line from a random number table?

   81638 36566 42709 33717 59943 12027 46547

2. A basketball player can either attempt a 3-point shot (with a 25% probability of scoring) or pass to a teammate with a 50% probability of scoring 2 points. What are the expected values for each choice? Which choice should he make?

Do you UNDERSTAND? MATHEMATICAL PRACTICES

3. **Reasoning** A friend says that using a random number table to pick a person at random isn't as fair as throwing a dart at the list of names. Explain your friend's error.

4. **Vocabulary** Explain the meaning of expected value. Include an example.

5. **Writing** Describe how you can use a random number generator to model the results of tossing a coin a hundred times.

Problem 4

In this problem, students will use expected value to make a decision based on probability.

Q Would the expected value analysis likely be more accurate toward the beginning of the season or the end of the season? Explain. **[end of the season; The probabilities are experimental, so they will be more accurate as more data is collected.]**

Got It?

Q If the coach makes the correct decision based on expected value, is he guaranteed of any particular outcome? Explain. **[No, the decision is based on what is likely to happen, but there is no guarantee.]**

3 Lesson Check

Do you know HOW? ERROR INTERVENTION

- If students have trouble solving Exercise 1, then remind them to treat 1 as 01, 2 as 02, and so on.

Do you UNDERSTAND?

- In Exercise 3, ask students if they have any control over what numbers appear in a random number table. Then ask them if they have any control over where the dart lands.

Close

Q How do you find the expected value of a situation that involves outcomes with quantitative values? **[Multiply each value by the corresponding probability of the outcome and find the sum.]**

Q How can expected value help you make decisions? **[By giving you some insight into what should happen mathematically.]**

Got It? (continued)

4a. field goal

b. Answers may vary. Sample: Yes. If it is late in the game and they are down by more than 3 points, the touchdown might be a better option.

Lesson Check

1. 42, 37, 17, 31
2. 0.75; 1; pass to the teammate
3. With the dart, there is some control over which name will be chosen. The table is more random.
4. Expected value is the average value of a trial based on the probabilities and the values of the outcomes. Check students' examples.
5. Answers may vary. Sample: Let 0–4 represent heads and 5–9 represent tails.

4 Practice

ASSIGNMENT GUIDE

Basic: 6–17
Average: 7–11 odd, 12–18
Advanced: 7–11 odd, 12–18
Standardized Test Prep: 19–21
Mixed Review: 22–25

Mathematical Practices are supported by exercises with red headings. Here are the Practices supported in this lesson:

MP 1: Make Sense of Problems Ex. 12
MP 3: Understand Definitions Ex. 4
MP 3: Critique the Reasoning of Others Ex. 3
MP 3: Communicate Ex. 5

Applications exercises have blue headings. Exercises 13 and 14 support MP 4: Model.

EXERCISE 16: Use the Think About a Plan worksheet in the **Practice and Problem Solving Workbook** (also available in the Teaching Resources in print and online) to further support students' development in becoming independent learners.

HOMEWORK QUICK CHECK

To check students' understanding of key skills and concepts, go over Exercises 7, 14, 16, 20, and 21.

Practice and Problem-Solving Exercises

MATHEMATICAL PRACTICES

A Practice

Use the lines from a random number table to select numbers to use in each problem. See Problem 1.

6. Choose 3 students from a list of 45 volunteers.

72749 13347 65030 26128 49067 27904 49953 74674 94617 13317

7. Choose 5 families to survey from a phone directory page with 950 names.

11873 57196 32209 67663 07990 12288 59245 83638 23642 61715

For Exercises 8 and 9, describe how to use random numbers to do a simulation for each situation. See Problem 2.

8. A teacher assigns students new seats randomly each week. On average, how long will it be before a student is assigned the same seat for two weeks in a row?

9. Sports A basketball player makes 80% of her free throw attempts. Find the average number of free throw attempts needed in order to make 3 free throws in a row.

10. Games In a game show, a contestant receives a prize that has a 5% probability of being worth \$1000 and a 95% probability of being worth \$1. What is the expected value of winning a prize? See Problems 3 and 4.

11. Video Games In a video game, a player has a 80% probability of winning 1000 points if he attacks a certain monster. There is a 20% probability that he will lose 25,000 points. What is the expected value of points earned? Should the player attack the monster? Explain.

B Apply

12. Think About a Plan A business owner is choosing which of two potential products to develop. Developing Product A will cost \$10,000, and the product has a 60% probability of being successful. Product B will cost \$15,000 to develop, and it has a 30% probability of being successful. If Product A is successful, the business will gain \$200,000. If product B is successful, the business will gain \$450,000. Which product should the business owner choose?

- How can the expected values of Product A and Product B help the business owner make the decision?
- Compare the expected values. Recommend the business owner choose the product with the greater expected value.

13. Business A company is deciding whether to invest in a new business opportunity. There is a 40% chance that the company will lose \$25,000, a 25% chance that the company will break even, and a 35% chance that the company will make \$40,000. Should company executives decide to invest in the opportunity? Explain.

14. Finance A stock has a 25% probability of increasing by \$10 per share over the next month. The stock has a 75% probability of decreasing by \$5 per share over the same period. What is the expected value of the stock's increase or decrease? Based on this information, should you buy the stock? Explain.

Answers

Practice and Problem Solving Exercises

6. 33, 03, 02

7. 118, 735, 719, 632, 209

8. Answers may vary. Sample: If there are *n* students in the class, assign each student a unique number from 1 through *n*. Use a random number table to assign each student to a seat each week. Count how many new seating assignments it takes until at least one student is in the same seat as he or she was in the previous week. Conduct several simulations and find the average value.

9. Answers may vary. Sample: Use a random number generator to produce numbers from 1 to 5. Count how many trials it takes before producing 3 numbers in a row that are not 1. Conduct the simulations many times and find the average value.

10. \$50.95

11. −4200; No, the player should not continue because the expected value is negative.

12. After taking into account startup costs, product A has an expected value of \$110,000, and product B has an expected value of \$120,000. The business owner should choose to develop product B.

13. Yes, the expected value is \$4000.

14. Answers may vary. Sample: −\$1.25; No, you should not buy the stock based on the expected value being negative.

15. **Test Taking** You earn 1 point for each correct response on a multiple choice test and lose 0.5 point for an incorrect response. Each question has four answer choices. If a student does not know the correct answer and guesses, what is the expected value of the guess?

16. **Games** A bag contains 10 marbles; 4 are red, 5 are yellow, and 1 is blue. You draw one marble from the bag without looking. If you draw a blue marble, you win \$10. If you draw a red marble, you win \$5, and if you draw a yellow marble, you win \$3. What is the expected value of drawing one marble? Would you play this game for \$5? Explain.

17. **Writing** A polling company has been hired to survey 500 households out of a possible 2500 households. How can the 500 households be randomly selected for the survey?

Challenge

18. **Stocks** An investor is choosing between three stocks she might purchase. The potential outcomes are listed in the table.
    a. Calculate the expected value of the gain or loss for each stock.
    b. Set up a model and use a random number table to model the investment in each stock.
    c. How do the results of the random number table model compare with the expected values?
    d. Does the expected value tell you anything about the riskiness of each stock?

Potential Outcomes

	Lose \$25 per share	Gain \$5 per share	Gain \$45 per share
Stock ABC	40%	15%	45%
Stock JKL	15%	65%	20%
Stock MNO	5%	80%	15%

Standardized Test Prep

SAT/ACT

19. What is the sum of the interior angles of an octagon?

(A) 135 (B) 360 (C) 480 (D) 1080

20. A kite has an 80° angle and a 50° angle. Which of the following might be the measure of the remaining angles?

(A) 80 and 50 (B) 50 and 180 (C) 80 and 150 (D) 50 and 150

Short Response

21. What are the lengths of the two legs of a right triangle with a 60° angle and a hypotenuse 1 meter long? Round your answer to the nearest centimeter.

Mixed Review

Calculate the following conditional probabilities. See Lesson 13-5.

22. The conditional probability $P(A \mid B)$ when $P(B) = 30\%$ and $P(A \text{ and } B) = 20\%$

23. The conditional probability $P(A \mid B)$ when $P(B) = 75\%$ and $P(A \text{ and } B) = 50\%$

What is the standard equation of each circle? See Lesson 12-5.

24. center $(3, -5)$; radius, 9

25. center $(-2, 8)$; radius, 5

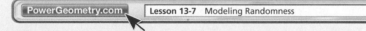

15. −0.125
16. \$4.50; No, the expected value is less than \$5.
17. Answers may vary. Sample: Assign numbers from 1 to 2500 to each household. Use a calculator to generate random numbers between 1 and 2500. Select the first 500 unique numbers that the calculator generates.
18a. ABC: \$11; JKL: \$8.50; MNO: \$9.50
b. Check students' models.
c. Answers will vary.
d. The expected values tell you that stock ABC will likely be the safest investment.
19. D
20. C
21. 50 cm and 87 cm
22. $\frac{2}{3}$
23. $\frac{2}{3}$
24. $(x-3)^2 + (y+5)^2 = 81$
25. $(x+2)^2 + (y-8)^2 = 25$

13-7 Lesson Resources

Differentiated Remediation

Additional Instructional Support

Geometry Companion

Students can use the **Geometry Companion** worktext (4 pages) as you teach the lesson. Use the Companion to support

- Solve It!
- New Vocabulary
- Key Concepts

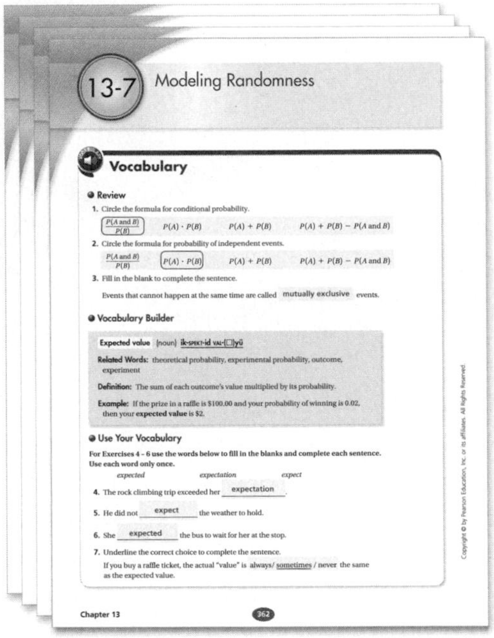

13-7 Modeling Randomness

Vocabulary

Review

1. Circle the formula for conditional probability.
2. Circle the formula for probability of independent events.
3. Fill in the blank to complete the sentence.

Events that cannot happen at the same time are called mutually exclusive events.

Vocabulary Builder

Expected value (noun)

Related Words: theoretical probability, experimental probability, outcome, experiment

Definition: The sum of each outcome's value multiplied by its probability.

Example: If the prize in a raffle is $100.00 and your probability of winning is 0.02, then your expected value is $2.

Use Your Vocabulary

For Exercises 4–6 use the words below to fill in the blanks and complete each sentence. Use each word only once.

expected expectation expect

4. The rock climbing trip exceeded her expectation.
5. He did not expect the weather to hold.
6. She expected the bus to wait for her at the stop.
7. Underline the correct choice to complete the sentence.

If you buy a raffle ticket, the actual "value" is always/ sometimes / never the same as the expected value.

Chapter 13 362

ELL Support

Use Brainstorming Divide students into small groups and have each group think of a situation in which a fair decision is needed. Some sample situations might include:

- deciding which student gets to read the announcements
- deciding which family member has to take the garbage out
- deciding which players get to be team captains

Have each group of students brainstorm different ideas of how a fair decision could best be reached. To end the activity, have each group report their situations and decision methods to the rest of the class.

5 Assess & Remediate

Lesson Quiz

Use the Lesson Quiz to assess students' mastery of the skills and concepts of this lesson.

1. Use the random numbers listed below to choose a winner from among 3000 raffle tickets. Assume the first ticket is number 0001, the second 0002, and so on.
   83205 92014 03463 79473 45013
2. Based on past records, a salesperson has a 15% chance of closing a sale with new customers. How could you model the number of sales he can expect with his next 5 customers?
3. **Do you UNDERSTAND?** On your next turn in a board game there is a 60% chance that you will move ahead 4 spaces and a 40% chance that you will move back 2 spaces. What is the expected value for the number of spaces moved?
4. An antique painting has a 40% chance of increasing in value by $10,000 and a 60% chance of decreasing in value by $5000. Based on this information, would you recommend investing in the painting?

ANSWERS TO LESSON QUIZ

1. The winning raffle ticket is numbered 1403.
2. Answers will vary. Check students' answers.
3. 1.6 spaces forward
4. Yes, the expected value of the painting's worth is an increase of $1000.

PRESCRIPTION FOR REMEDIATION

Use the student work on the Lesson Quiz to prescribe a differentiated review assignment.

Points	Differentiated Remediation
0–2	Intervention
3	On-level
4	Extension

PowerGeometry.com

5 Assess & Remediate

Assign the Lesson Quiz. Appropriate intervention, practice, or enrichment is automatically generated based on student performance.

Intervention

- **Reteaching** (2 pages) Provides reteaching and practice exercises for the key lesson concepts. Use with struggling students or absent students.
- **English Language Learner Support** Helps students develop and reinforce mathematical vocabulary and key concepts.

All-in-One Resources/Online

Reteaching

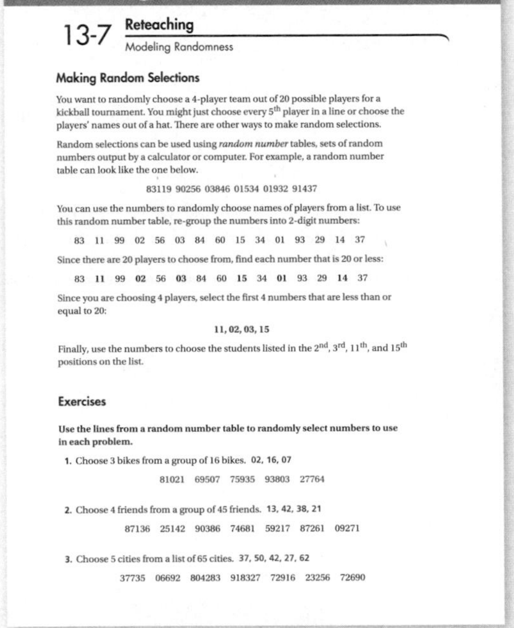

13-7 Reteaching

Modeling Randomness

Making Random Selections

You want to randomly choose a 4-player team out of 20 possible players for a kickball tournament. You might just choose every 5th player in a line or choose the players' names out of a hat. There are other ways to make random selections.

Random selections can be used using *random number* tables, sets of random numbers output by a calculator or computer. For example, a random number table can look like the one below.

83119 90256 03846 01534 01932 91437

You can use the numbers to randomly choose names of players from a list. To use this random number table, re-group the numbers into 2-digit numbers:

83 11 99 02 56 03 84 60 15 34 01 93 29 14 37

Since there are 20 players to choose from, find each number that is 20 or less:

83 **11** 99 **02** 56 **03** 84 60 **15** 34 **01** 93 29 **14** 37

Since you are choosing 4 players, select the first 4 numbers that are less than or equal to 20:

11, 02, 03, 15

Finally, use the numbers to choose the students listed in the 2nd, 3rd, 11th, and 15th positions on the list.

Exercises

Use the lines from a random number table to randomly select numbers to use in each problem.

1. Choose 3 bikes from a group of 16 bikes. 02, 16, 07
   81021 69507 75935 93803 27764
2. Choose 4 friends from a group of 45 friends. 13, 42, 38, 21
   87136 25142 90386 74681 59217 87261 09271
3. Choose 5 cities from a list of 65 cities. 37, 50, 42, 27, 62
   37735 06692 804283 918327 72916 23256 72690

All-in-One Resources/Online

English Language Learner Support

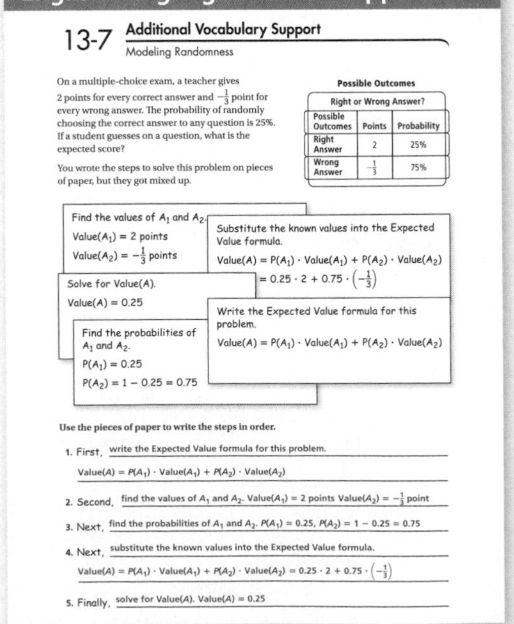

13-7 Additional Vocabulary Support

Modeling Randomness

On a multiple-choice exam, a teacher gives 2 points for every correct answer and $-\frac{1}{3}$ point for every wrong answer. The probability of randomly choosing the correct answer to any question is 25%. If a student guesses on a question, what is the expected score?

You wrote the steps to solve this problem on pieces of paper, but they got mixed up.

Possible Outcomes

Right or Wrong Answer?		
Possible Outcomes	Points	Probability
Right Answer	2	25%
Wrong Answer	$-\frac{1}{3}$	75%

Find the values of A_1 and A_2. Value(A_1) = 2 points Value(A_2) = $-\frac{1}{3}$ points

Substitute the known values into the Expected Value formula. Value(A) = P(A_1) · Value(A_1) + P(A_2) · Value(A_2) = 0.25 · 2 + 0.75 · $\left(-\frac{1}{3}\right)$

Solve for Value(A). Value(A) = 0.25

Write the Expected Value formula for this problem. Value(A) = P(A_1) · Value(A_1) + P(A_2) · Value(A_2)

Find the probabilities of A_1 and A_2. P(A_1) = 0.25 P(A_2) = 1 − 0.25 = 0.75

Use the pieces of paper to write the steps in order.

1. First, write the Expected Value formula for this problem. Value(A) = P(A_1) · Value(A_1) + P(A_2) · Value(A_2)
2. Second, find the values of A_1 and A_2. Value(A_1) = 2 points Value(A_2) = $-\frac{1}{3}$ point
3. Next, find the probabilities of A_1 and A_2. P(A_1) = 0.25, P(A_2) = 1 − 0.25 = 0.75
4. Next, substitute the known values into the Expected Value formula. Value(A) = P(A_1) · Value(A_1) + P(A_2) · Value(A_2) = 0.25 · 2 + 0.75 · $\left(-\frac{1}{3}\right)$
5. Finally, solve for Value(A). Value(A) = 0.25

Differentiated Remediation *continued*

On-Level

- **Practice** (2 pages) Provides extra practice for each lesson. For simpler practice exercises, use the Form K Practice pages found in the All-in-One Teaching Resources and online.
- **Think About a Plan** Helps students develop specific problem-solving skills and strategies by providing scaffolded guiding questions.
- **Standardized Test Prep** Focuses on all major exercises, all major question types, and helps students prepare for the high-stakes assessments.

Extension

- **Enrichment** Provides students with interesting problems and activities that extend the concepts of the lesson.
- **Activities, Games, and Puzzles** Worksheets that can be used for concepts development, enrichment, and for fun!

Practice and Problem Solving Wkbk/ All-in-One Resources/Online

Practice page 1

13-7 **Practice** *Form G*
Modeling Randomness

For Exercises 1 and 2, use the lines from a random number table to select numbers to use in each problem.

1. Choose 4 relatives from a group of 24 relatives. **15, 9, 18, 20**

   15287 69109 65187 09154 48712 07934 63825

2. Choose 5 customers from a group of 52 customers. **13, 36, 45, 34**

   73546 87013 67823 68454 76984 57934 90853

3. A manufacturer makes five types of car, represented by the numbers 1 through 5. Using the random number table below, how many randomly chosen cars are picked before all 4 types have been chosen? **18**

   15132 53521 35123 35441 14243

4. At halftime of a basketball game, 8 different types of t-shirts are thrown into the crowd. Using the random number table below, how many randomly chosen t-shirts are thrown before all 8 types have been distributed? **25**

   53636 45227 45424 31115 21428 53737 82763 61732 72181

5. In each hand of a card game, there is a 54% chance of winning 3 points and a 46% chance of losing 4 points. What is the expected gain or loss on each hand? **−0.22**
6. On a certain test, 2 points are awarded for a correct answer and 1 point is deducted for each incorrect answer. The last time the test was given, the students answered correctly 68% of the time. What is the expected value for each question? **1.04**
7. A certain basketball player has success rates of 39% for her 3-point shots and 51% for her 2-point shots. Find the expected values and compare them using <, >, or =.

   3-point **1.17** [>] **1.02** 2-point

8. In a trivia game, you answer correctly on 26% of the 5-point bonus questions, and you answer correctly on 68% of the 2-point questions. Find the expected values and compare them using <, >, or =.

   5-point **1.3** [<] **1.36** 2-point

Practice and Problem Solving Wkbk/ All-in-One Resources/Online

Practice page 2

13-7 **Practice** (continued) *Form G*
Modeling Randomness

9. **Reasoning** Two strategic moves in a board game have expected values of 0.9 and 0.95. Are you guaranteed to perform better by continuously using the move with the greater expected value? Why or why not? **No. Expected value is a probability, and events won't always follow the probability.**
10. The directors of 7 different community choirs are asked to provide 30 vocalists to perform in a special holiday concert. The choirs are represented by the numbers 1 through 7 in the random number table below. How many random choices of vocalists are made until all 7 choirs are represented in the concert by one vocalist each? **22**

    71762 65624 22261 42512 73141 64572

11. In a carnival game, contestants can throw at 2 targets to win prizes. There is a 12% success rate of hitting target A, which yields a $7 prize. There is a 48% success rate of hitting target B, which yields a $3 prize. Compare the expected values using <, >, or =.

    $7 prize **0.84** [<] **1.44** $3 prize

12. **Error Analysis** Your friend uses the random number table below to choose 4 baseball players from a group of 13 baseball players.

    04855 91291 08956 51342 48291 18275 19172

    He picks the numbers 4, 12, 8, and 11. What error did he make? **When the numbers are grouped in twos, 13 was skipped - it appears before 11**

13. A theater coach is able to send only 5 of his 26 students to a special acting workshop. He wants to be fair and choose them randomly. Using the random number table below, which numbers will he use to help choose the students? **9, 2, 12, 26, 7**

    09364 97140 29376 30287 45008 43875 83926 31248 26075

14. A board game uses a spinner with equal-sized sections numbered 1, 2, 3, 4, 5, 6, and 7. Spinning an even number enables a player to move 2 spaces forward. Spinning an odd number makes a player move 1 space backward. What is the expected value, in fraction form, of each spin? $\frac{2}{7}$

All-in-One Resources/Online

Enrichment

13-7 **Enrichment**
Modeling Randomness

Game Time

Design your own game based on rolling one number cube.

1. Assign scores for 3 possible events for each roll of the cube. Use one negative value for one of the events. Write the point value for each roll below.
   Answers will vary. Sample answer:
   Roll a 1, 2, or 5; 1 point.
   Roll a 3; − 4 points.
   Roll a 4 or 6; 2 points.
2. Determine the expected value for each roll. Show your work.
   Answers will vary. Sample answer:
   Roll a 1, 2, or 5; $\frac{1}{2} \cdot 1 = \frac{1}{2}$
   Roll a 3;−4 points. $\frac{1}{6} \cdot (-4) = -\frac{2}{3}$
   Roll a 4 or 6; 2 points. $\frac{1}{3} \cdot 2 = \frac{2}{3}$
   $\frac{1}{2} + \left(-\frac{2}{3}\right) + \frac{2}{3} = \frac{1}{2}$ **or 0.5. The expected value is 0.5.**
3. Test your game by rolling the number cube 4 times. What total score did you get for each roll? Write the results of your rolls below.
   Answers will vary. Students should have the outcome and point value for each roll.
4. How would you change your point system to make the expected value of each roll closer to 0? Show your work.
   Answers will vary. The results of the new point system should have an expected value closer to 0 than the initial expected value.
5. Explain how a player of this game could have a positive point score even if the expected value of each roll is negative.
   Although the expected value is negative, some of the outcomes have positive point values. There is a chance the positive rolls will be greater in value than the negative.

Practice and Problem Solving Wkbk/ All-in-One Resources/Online

Think About a Plan

13-7 **Think About a Plan**
Modeling Randomness

Games A bag contains 10 marbles: 4 are red, 5 are yellow, and 1 is blue. You draw one marble from the bag without looking. If you draw a blue marble, you win $10. If you draw a red marble, you win $5, and if you draw a yellow marble, you win $3. What is the expected value you will win or lose? Would you play this game for $5? Explain.

Understanding the Problem

1. How many possibilities are there? **3**
2. What is the combined probability of the events? **1 or 100%**

Planning the Solution

3. Write a formula for the expected value of these events.
   $\text{Value}(A) = P(A_1) \cdot \text{Value}(A_1) + P(A_2) \cdot \text{Value}(A_2) + P(A_3) \cdot \text{Value}(A_3)$
4. What is the probability of drawing a blue marble? How much will you win if you draw a blue marble? $\frac{1}{10} = 10\%$; **$10**
5. What is the probability of drawing a red marble? How much will you win if you draw a red marble? $\frac{4}{10} = 40\%$; **$5**
6. What is the probability of drawing a yellow marble? How much will you win if you draw a yellow marble? $\frac{5}{10} = 50\%$; **$3**

Getting an Answer

7. Use the expected value formula from Step 3 to find the expected value you will win or lose in the game. Show your work.
   $\text{Value}(A) = 10\%(10) + 40\%(5) + 50\%(3) = 1 + 2 + 1.5 = 4.5$
   The expected value is a **gain (or win)** of **$4.50**.
8. Would you play this game for $5? Why or why not? **No; the expected value of $4.50 is less than $5, so the player is not expected to win.**

Practice and Problem Solving Wkbk/ All-in-One Resources/Online

Standardized Test Prep

13-7 **Standardized Test Prep**
Modeling Randomness

Multiple Choice

For Exercises 1–4, choose the correct letter.

For Exercises 1 and 2, use the lines from a random number table to select numbers to use in each problem.

1. Using the table below, which numbers would you use to choose 3 students from a group of 50 students? **B**

   36674 86790 98265 42947 20763

   (A) 36, 48, 42 (B) 36, 48, 26 (C) 36, 48, 9 (D) 48, 26, 42

2. There are students from 4 different towns at a conference. Using the random number table below, how many students would be randomly chosen until one from each town is chosen? **H**

   13231 23121 42314 13423

   (F) 3 (G) 10 (H) 11 (I) 15

3. In each roll of a game piece, there is a 64% chance of losing 2 points and a 36% chance of winning 7 points. What is the expected value for each roll? **B**

   (A) -1.28 (B) 1.24 (C) 2.52 (D) 3.8

4. In a game at the fair, a player has a 43% chance of making a 3-point shot and a 32% chance of making a 4-point shot. Which shows the greater probability shot and difference between expected values? **H**

   (F) 4-point by 0.01 (G) 4-point by 0.08 (H) 3-point by 0.01 (I) 3-point by 0.08

Extended Response

5. In a game at a fundraiser, Choice A has a 12% of winning 8 prize tickets and Choice B has a 46% chance of winning 2 prize tickets. Describe how you would choose between playing Choice A or Choice B.
   [4] I would play choose Choice A because it has an expected value 0.04 greater than Choice B. Choice A: $0.12 \times 8 = 0.96$; Choice B: $0.46 \times 2 = 0.92$; $0.96 - 0.92 = 0.04$
   [3] Appropriate methods with one computational error.
   [2] Appropriate methods with multiple computational errors.
   [1] Correct answers without work shown.
   [0] Student's response shows no effort OR does not attempt to explain the probability in terms of an event that has already happened.

Online Teacher Resource Center

Activities, Games, and Puzzles

13-7 **Game: Who Does More?**
Modeling Randomness

Deanna and her brother Gary have been given a list of 12 household chores by their parents. Typically, they would split the chores evenly and get to work. Gary does not want to do 6 chores; he wants to do less than Deanna. Help Gary develop a game for choosing chores that will result in him having to complete fewer chores. Complete each category below for your game. Your game must have at least 3 rules.

Supplies Needed

Answers will vary. Sample: spinner with 12 sections, numbered 1-12

Rules

Answers will vary. Sample: Deanna and Gary will take turns spinning the spinner. If the spinner lands on a number that is not a multiple of 3, Deanna chooses a chore. If the spinner lands on a multiple of 3, Gary chooses a chore.

1. What is the probability that Deanna chooses a chore?
   Sample answer based on rules above: 2/3
2. What is the probability that Gary chooses a chore?
   Sample answer based on rules above: 1/3
3. What change to the rules could Deanna suggest that would make the game fair?
   Answers will vary. Sample based on rules above: If the spinner lands on a factor of 12, Deanna choose a chore, otherwise Gary chooses a chore.
4. What change to the rules could Deanna suggest that would make the game favor her?
   Answers will vary. Sample based on rules above: If the spinner lands on a prime number, Deanna choose a chore, otherwise Gary chooses a chore.

Guided Instruction

PURPOSE To use probability concepts in decision making.

PROCESS Students will
- calculate theoretical probabilities of events in order to decide the best strategy in a game.
- use experimental probability to make decisions about product testing and development.

DISCUSS Before having students work on the Activities, ask them to think of ways they have used probability concepts to make decisions. If they have difficulty coming up with examples, ask them if they have seen other people use probability concepts to make decisions in contexts like sports and television game shows. Before students complete Activity 2, remind them that a placebo is a treatment with no active ingredients. Ask students why they think a treatment group might be given a placebo.

Activity 1

In this activity, students use theoretical probability and number theory to find the best strategy in a game.

Q What is the sample space for each probability? **[1, 2, 3, 4, 5, 6]**

Q What are the favorable outcomes for Player 1? What are the favorable outcomes for Player 2? **[Player 1: 2, 4, 5, 6; Player 2: 1, 2, 3, 4, 6]**

Activity 2

In this activity, students use a two-way frequency table to determine whether a product is effective.

Q What is the sample space for each probability? What are the sizes of each sample space? **[all of the participants in Group 1, 80; all of the participants in Group 2, 80]**

Mathematical Practices This Concept Byte supports students in becoming proficient in reasoning abstractly and quantitatively, Mathematical Practice 2.

Concept Byte

For Use With Lesson 13-7

ACTIVITY

Probability and Decision Making

Common Core State Standards

S-MD.B.7 Analyze decisions and strategies using probability concepts (e.g., product testing, medical testing, pulling a hockey goalie at the end of a game).

MP 2

Understanding probability is important in areas from game playing to testing new pharmaceutical drugs.

Activity 1

Your older sister has devised a game to play with your younger brother. She says that your brother can choose who goes first. Here are the rules:

Roll a standard number cube.

Player 1 wins a point if the number cube shows an even number or a 5.

Player 2 wins a point if the number cube is a factor of 24.

1. a. In each turn, what is the probability that Player 1 scores a point? What is the probability that Player 2 scores a point?

b. What advice would you give your brother? Should he choose to be Player 1 or Player 2? Explain.

Activity 2

A pharmaceutical company tested a new formulation for eye drops on 160 people who suffered from very dry eyes. Two groups of test subjects were randomly selected. The first group received a placebo and the second group received the new eye drops. Below are the results of the testing.

Eye Drop Test Results

	Improvement	No Improvement
Group 1	48	32
Group 2	39	41

2. What is the probability that someone in Group 1 showed improvement?

3. What is the probability that someone in Group 2 showed improvement?

4. Would you advise further development on the new formulation for eye drops or for the company to drop this new medication? Justify your answer using probabilities.

Answers

Activity 1

1a. $\frac{2}{3}, \frac{5}{6}$

b. Your brother should choose to be Player 2. The probability that Player 2 scores a point is greater than the probability that Player 1 scores a point.

Activity 2

2. 60%

3. 48.75%

4. Answers may vary. Sample: No. The probability of success for the new eye drops is not sufficiently higher than the probability of success for the placebo to warrant the investment.

13 *Pull It* All Together

Completing the Performance Task

Look back at your results from the Apply What You've Learned sections in Lessons 13-1 and 13-6. Use the work you did to complete the following.

To solve these problems, you will pull together concepts and skills related to probability.

1. Solve the problem in the Task Description on page 823 by determining whether it is more likely that a customer who likes raspberry also likes caramel, or that a customer who likes caramel also likes raspberry. Show all your work and explain each step of your solution.
2. **Reflect** Choose one of the Mathematical Practices below and explain how you applied it in your work on the Performance Task.
   MP 1: Make sense of problems and persevere in solving them.
   MP 2: Reason abstractly and quantitatively.

On Your Own

Tyson's Frozen Yogurt tested two new flavors, cinnamon and mango. From a random survey of 150 customers, the store received the following information about the customers' responses to the new flavors:

- Twenty-four customers like only mango.
- Half as many customers like both flavors as like only cinnamon.
- The number of customers that like only cinnamon is four times the number of customers that do not like either flavor.

Tyson says he will start serving both new flavors if there is at least a 50% chance that a customer who likes mango also likes cinnamon.

Will Tyson start serving the two new flavors? Explain.

Completing the Performance Task

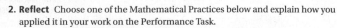

In the Apply What You've Learned sections in Lessons 13-1 and 13-6, students found experimental probabilities and a conditional probability related to the survey information given on page 823. Now, they can use their results to complete the Performance Task. Ask students the following question as they work toward solving the problem.

Q How can you use the work you have done in the chapter to solve the problem? **[Sample: I can use the information in the table I completed in Lesson 13-1 to calculate the probability that a customer who likes caramel also likes raspberry, and compare the result to the conditional probability that I found in Lesson 13-6.]**

FOSTERING MATHEMATICAL DISCOURSE

Have students read and critique each other's solutions of the Performance Task.

ANSWERS

1. $P(\text{likes caramel} \mid \text{likes raspberry}) \approx 0.44$, or 44% and $P(\text{likes raspberry} \mid \text{likes caramel}) \approx 0.33$, or 33%, so it is more likely that a customer who likes raspberry also likes caramel, than a customer who likes caramel also likes raspberry.
2. Check students' work.

On Your Own

This problem is similar to the problem posed on page 823, but now the student must use a conditional probability to decide if the shop owner will serve the new flavors.

ANSWER

Yes; explanations may vary. Sample:

$$P(\text{likes cinnamon} \mid \text{likes mango}) = \frac{P(\text{likes both})}{P(\text{likes mango})} = \frac{0.36}{0.60} = 0.60\text{, or } 60\%$$

Since 60% > 50%, Tyson will serve the two new flavors.

Essential Questions

BIG idea Probability

ESSENTIAL QUESTION What is the difference between experimental probability and theoretical probability?

ANSWER Theoretical probability is based on what should happen mathematically. Experimental probability is based on observations.

BIG idea Data Representation

ESSENTIAL QUESTION What is a frequency table?

ANSWER A frequency table represents data from one category. A two-way frequency table represents data from two categories.

BIG idea Probability

ESSENTIAL QUESTION What does it mean for an event to be random?

ANSWER A random event has no bias or inclination toward any outcome.

13 Chapter Review

Connecting BIG ideas and Answering the Essential Questions

1 Probability
Theoretical probability is based on what should happen mathematically. Experimental probability is based on observations.

Experimental and Theoretical Probability (Lesson 13-1)

Experimental Probability:

$P(\text{event}) = \dfrac{\text{number of times the event occurs}}{\text{number of times the experiment is done}}$

Theoretical Probability:

$P(\text{event}) = \dfrac{\text{number of favorable outcomes}}{\text{number of possible outcomes}}$

Compound Probability (Lesson 13–4)

For independent events *A* and *B*, $P(A \text{ and } B) = P(A) \cdot P(B)$.

For mutually exclusive events *A* and *B*, $P(A \text{ or } B) = P(A) + P(B)$.

For overlapping events *A* and *B*, $P(A \text{ or } B) = P(A) + P(B) - P(A \text{ and } B)$.

2 Data Representation
A frequency table represents data from one category. A two-way frequency table represents data from two categories.

Probability Distributions and Frequency Tables and Probability Models (Lessons 13-2 and 13-5)

Tables are used to organize data, and can be helpful when finding probabilities.

Permutations and Combinations (Lesson 13–3)

${}_nP_r = \dfrac{n!}{(n-r)!}$ for $0 \le r \le n$

${}_nC_r = \dfrac{n!}{r!(n-r)!}$ for $0 \le r \le n$

Conditional Probability Formulas (Lesson 13–6)

You can use two-way frequency tables to calculate conditional probability.
For dependent events *A* and *B*,
$P(A \text{ and } B) = P(A) \cdot P(B \mid A)$.

3 Probability
A random event has no bias or inclination toward any outcome.

Modeling Randomness (Lesson 13–7)

You can use random number tables and graphing calculators to model random events and make fair decisions.

Chapter Vocabulary

- combination (p. 838)
- complement of an event (p. 826)
- conditional probability (p. 851)
- dependent events (p. 844)
- event (p. 824)
- experimental probability (p. 825)
- expected value (p. 864)
- frequency table (p. 830)
- independent events (p. 844)
- mutually exclusive events (p. 845)
- permutation (p. 837)
- probability (p. 824)
- sample space (p. 824)
- theoretical probability (p. 825)
- two-way frequency table (p. 850)

Choose the correct term to complete each sentence.

1. A __?__ is a grouping of items in which order is important.

2. __?__ is based on what should happen mathematically.

3. The outcomes of __?__ events affect each other.

4. A __?__ should be used to show the results of a survey on students' favorite teacher.

870 Chapter 13 Chapter Review

Summative Questions

Use the following prompts as you review this chapter with your students. The prompts are designed to help you assess your students' understand of the BIG ideas they have studied.

- Compare and contrast theoretical and experimental probability.
- What kind of data can be displayed in a frequency table? What kind of data can be displayed in a two-way frequency table?
- Compare and contrast combinations and permutations. Give an example of a situation that could be modeled by a combination and one that could be modeled by a permutation.
- How can a simulation help you make predictions about what is likely to happen in future situations?
- What is expected value? How can you use expected value to help you make informed decisions?

Answers

Chapter Review

1. permutation
2. Theoretical probability
3. dependent
4. frequency table

13-1 Experimental and Theoretical Probability

Quick Review

Experimental probability is based on how many successes are observed in repeated trials of an experiment. **Theoretical probability** describes the likelihood of an event based on what should happen mathematically. The **sample space** is the set of all possible outcomes of an experiment.

Example

There are 14 boys and 11 girls in a homeroom. If a student is selected at random, what is the probability that the student is a girl?

There are 25 possible outcomes in the sample space. Of these, 11 are girls. The probability that the student is a girl is $P(\text{girl}) = \frac{11}{25} = 0.44$.

Exercises

There are 12 math teachers, 9 science teachers, 3 music teachers, and 6 social studies teachers at a conference. If a teacher is selected at random, find each probability.

5. $P(\text{science})$

6. $P(\text{not music})$

7. $P(\text{reading})$

8. $P(\text{not math})$

9. Bowling Suppose you rolled 5 strikes and 7 non-strikes in a bowling game. Based on these results, what is the experimental probability that you will roll a strike?

13-2 Probability Distributions and Frequency Tables

Quick Review

A **frequency table** shows how often an item appears in a category. The **relative frequency** of an item is the ratio of the frequency of the category to the total frequency. A **probability distribution** shows the probabilities of every possible outcome of an experiment.

Example

What is the relative frequency of sedans?

There are 30 cars in all, and 12 of them are sedans. So, the relative frequency of sedans is $\frac{12}{30} = 0.4$.

Type of Car	Frequency
Truck	3
Minivan	6
Sedan	12
Sports	2
Other	7

Exercises

Use the frequency table below to estimate each probability.

Test Grade	Frequency
A	7
B	11
C	6
D	3
F	1

10. $P(\text{A})$

11. $P(\text{B})$

12. $P(\text{B or C})$

13. $P(\text{grade below C})$

14. $P(\text{grade above C})$

15. $P(\text{grade above D})$

5. $\frac{3}{10}$
6. $\frac{9}{10}$
7. 0
8. $\frac{3}{5}$
9. $\frac{5}{12}$
10. $\frac{1}{4}$
11. $\frac{11}{28}$
12. $\frac{17}{28}$
13. $\frac{1}{7}$
14. $\frac{9}{14}$
15. $\frac{6}{7}$

Answers

Chapter Review (continued)

16. 120
17. 40,320
18. 840
19. 220
20. 6720
21. 90
22. 36
23. 1140
24. 78,624,000
25. 2300
26. 0.115
27. 0.595
28. $\frac{7}{16}$
29. Answers may very. Sample: If the events are mutually exclusive, then *P*(*A* and *B*) = 0 and the formula becomes *P*(*A* or *B*) = *P*(*A*) + *P*(*B*). Otherwise, the formula is used as stated.

13-3 Permutations and Combinations

Quick Review

According to the **Fundamental Counting Principle**, if event M can occur in m ways and event N can occur in n ways, then event M and event N can occur in $m \cdot n$ ways.

***n* factorial:** $n! = n \cdot (n-1) \cdot (n-2) \cdot \ldots \cdot 3 \cdot 2 \cdot 1$

Permutation formula (order matters):

$${}_nP_r = \frac{n!}{(n-r)!}$$

Combination formula (order is not important):

$${}_nC_r = \frac{n!}{r!(n-r)!}$$

Example

How many ways can a teacher choose a group of 4 students from 15 if order does not matter?

$${}_{15}C_4 = \frac{15!}{4!(15-4)!} = 1365 \text{ ways}$$

Exercises

Evaluate each of the following.

16. $5!$
17. $8!$
18. $\frac{7!}{3!}$
19. $\frac{12!}{9!3!}$
20. ${}_8P_5$
21. ${}_{10}P_2$
22. ${}_9C_7$
23. ${}_{20}C_3$

24. License Plate A license plate is made up of 3 letters followed by 4 digits. How many different license plates are possible if none of the letters or digits can be repeated?

25. Music A student wants to put 3 songs on her MP3 player from a collection of 25 songs. How many different ways can she do this?

13-4 Compound Probability

Quick Review

For independent events A and B, $P(A \text{ and } B) = P(A) \cdot P(B)$. For mutually exclusive events A and B, $P(A \text{ or } B) = P(A) + P(B)$. For overlapping events A and B, $P(A \text{ or } B) = P(A) + P(B) - P(A \text{ and } B)$.

Example

If a classmate rolls a standard number cube, what is the probability that she rolls a prime number or an even number?

$P(\text{prime}) = \frac{3}{6} = \frac{1}{2}$ — 2, 3, and 5 are prime.

$P(\text{even}) = \frac{3}{6} = \frac{1}{2}$ — 2, 4, and 6 are even.

$P(\text{even and prime}) = \frac{1}{6}$ — 2 is even and prime.

$P(\text{even or prime}) = \frac{1}{2} + \frac{1}{2} - \frac{1}{6} = \frac{5}{6}$

Exercises

Suppose A and B are independent events, $P(A) = 0.46$, and $P(B) = 0.25$. Find each probability.

26. $P(A \text{ and } B)$
27. $P(A \text{ or } B)$

28. A bag contains 3 red marbles, 4 green marbles, 5 blue marbles, and 4 yellow marbles. If a marble is selected at random, what is the probability that it is red or green?

29. Writing Explain why the formula $P(A \text{ or } B) = P(A) + P(B) - P(A \text{ and } B)$ can be used for any two events A and B.

13-5 Probability Models

Quick Review

A **two-way frequency** table shows data that fall into two different categories. You can use two-way frequency tables to find relative frequencies of items in different categories.

Example

What is the relative frequency of female juniors on the student council?

	Male	Female	Totals
Juniors	2	2	4
Seniors	1	3	4
Totals	3	5	8

There are 8 members and 2 of them are female juniors. The relative frequency is $\frac{2}{8} = 0.25$.

Exercises

Studying Use the two-way frequency table to answer each question.

	Studied	Did not study	Totals
Passed	19	7	26
Failed	1	4	5
Totals	20	11	31

30. Did more students choose to study or to not study for the quiz?

31. What is the relative frequency of students who studied for the quiz and passed?

32. What is the relative frequency of students who did not study and failed?

13-6 Conditional Probability Formulas

Quick Review

To find the conditional probability that dependent events A and B occur, multiply the probability that event A occurs by the probability that event B occurs, given that event A has already occurred.

$P(A \text{ and } B) = P(A) \cdot P(B|A)$

For any two events A and B with $P(A) \neq 0$,

$P(B|A) = \frac{P(A \text{ and } B)}{P(A)}$.

Example

A coach's desk drawer contains 3 blue pens, 5 black pens, and 4 red pens. What is the probability that he selects a black pen followed by a red pen, if the first pen is not replaced?

$P(\text{black, then red}) = P(\text{black}) \cdot P(\text{red, given black})$

$= \frac{5}{12} \cdot \frac{4}{11} = \frac{5}{33}$

Exercises

Solve each problem.

33. Recreation The ball bin in the gym contains 8 soccer balls, 12 basketballs, and 10 kickballs. If Coach Meyers selects 2 balls at random and does not replace the first, what is the probability that she selects 2 basketballs?

34. Tests Suppose your teacher has given 2 pop quizzes this week. Seventy percent of her class passed the first quiz, and 50% passed both quizzes. If a student is selected at random, what is the probability that he or she passed the second quiz, given that the student passed the first quiz?

35. Shopping At a shoe store yesterday, 74% of customers bought shoes, 38% bought accessories, and 55% bought both shoes and accessories. If a customer is selected at random, what is the probability that he bought accessories, given that he bought shoes?

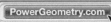

30. More students chose to study.
31. 0.95
32. about 0.364
33. $\frac{22}{145}$
34. $\frac{5}{7}$ or about 71.4%
35. $\frac{55}{74}$ or about 74.3%

Answers

Chapter Review (continued)

36. 20

37. 13

38. $\frac{13}{20}$

13-7 Modeling Randomness

Quick Review

Random events have no predetermined pattern or bias toward one outcome or the other. You can use random number tables and graphing calculators to help you model random events and make fair decisions. A simulation is a probability model that can be used to make predictions about real world situations.

Example

Twenty students are assigned 2-digit numbers from 01 to 20. Use the excerpt from the random number table below to randomly select the first 3 students.

80261 53267 85048 53091 18479 06286 38323

Reading from left to right, the 2-digit numbers are:

80, 26, 15, 32, 67, 85, 04, 85, 30, 91, 18, . . .

The first 3 students are indicated by the numbers 15, 04, and 18.

Exercises

Golf On average, Raymond shoots lower than 80 in 3 out of 4 rounds of golf. He conducts a simulation to predict how many times he will shoot lower than 80 in his next 5 rounds. Let 1, 2, and 3 represent shooting lower than 80, and let 4 represent shooting 80 or higher.

22113	22413	34141	12131	34432
34112	34142	24241	44231	41122
23132	13314	14133	11432	41214
34423	13332	12422	11422	33312

36. How many trials of the simulation did Raymond conduct?

37. In how many trials of the simulation did Raymond shoot lower than 80 in at least 4 of the 5 rounds?

38. According to this simulation, what is the experimental probability that Raymond will shoot lower than 80 in at least 4 of his next 5 rounds?

13 Chapter Test

Do you know HOW?

You select a number from 1 to 10 at random. Find each probability.

1. P(odd number)
2. P(1 or 4)
3. P(less than 8)
4. P(greater than 10)

For Exercises 5–8, use the frequency table to find each relative frequency.

Trees Planted	Frequency
Oak	7
Maple	12
Birch	9
Pear	17

5. oak trees
6. maple trees
7. birch trees
8. pear trees

Evaluate each expression.

9. 8!
10. $\frac{9!}{5!}$
11. $\frac{6!}{4!2!}$
12. ${}_7C_4$
13. ${}_8P_5$
14. ${}_6C_6$

If A and B are independent events, find $P(A$ and $B)$ and $P(A$ or $B)$.

15. $P(A) = 0.2$, $P(B) = 0.4$

16. $P(A) = \frac{2}{3}$, $P(B) = \frac{1}{4}$

For Exercises 17 and 18, find each conditional probability, given that $P(A) = \frac{11}{20}$, $P(B) = \frac{17}{25}$, and $P(A$ and $B) = \frac{3}{25}$.

17. $P(A \mid B)$
18. $P(B \mid A)$

Use the excerpt from a random number table below to find each of the following in Exercises 19 and 20.

59717 37738 07632 84854 38223 65202 50822

19. the first three 2-digit numbers from 01 to 30

20. the first two 3-digit numbers from 100 to 299

Do you UNDERSTAND?

For Exercises 21 and 22, determine whether each pair of events is independent or dependent.

21. playing a stringed instrument and playing a percussion instrument

22. selecting a letter tile, then selecting a second letter tile without replacing the first

23. Writing If four friends are lining up for a photo, does the situation describe a combination or a permutation? Explain. How many different ways are there for four friends to line up for a photo?

24. Weather The probability of school being cancelled because of snow on any given day in January is 7.5%. Describe a simulation you could conduct to predict how many snow days you will have in January if there are 21 school days during the month.

25. A and B are independent events. Find two probabilities, $P(A)$ and $P(B)$, such that $P(A$ and $B) = \frac{1}{8}$ and $P(A$ or $B) = \frac{5}{8}$.

26. Writing Describe how to find relative frequencies of items in different categories in a contingency table.

Answers

Chapter Test

1. 0.5
2. 0.2
3. 0.7
4. 0
5. $\frac{7}{45}$
6. $\frac{4}{15}$
7. $\frac{1}{5}$
8. $\frac{17}{45}$
9. 40,320
10. 3024
11. 15
12. 35
13. 6720
14. 1
15. 0.08; 0.52
16. $\frac{1}{6}$; $\frac{3}{4}$
17. $\frac{3}{17}$
18. $\frac{12}{55}$
19. 07, 28, 22
20. 173, 202
21. independent
22. dependent
23. Sample answer: permutation because order is important, 24 ways
24. Sample answer: Use a graphing calculator to generate random numbers from 1 to 40. Let the numbers 1, 2, and 3 represent a snow day, and let the other numbers represent a non-snow day. Generate a number 21 times. This is one trial. Repeat the trial many times, and find the average number of snow days.
25. $P(A) = \frac{1}{2}$, $P(B) = \frac{1}{4}$ or $P(A) = \frac{1}{4}$, $P(B) = \frac{1}{2}$
26. Sample answer: Divide the value by the total in the given category.

Item Number	Lesson	Content Standard
1	10-6	G-CO.A.1
2	13-1	S-CP.A.1
3	12-1	G-C.A.2
4	8-3	G-SRT.C.7
5	10-1	G-SRT.C.8
6	12-5	G-GPE.A.1
7	10-7	G-C.B.5
8	9-3	G-CO.B.6
9	12-3	G-C.A.2
10	12-6	G-CO.A.5
11	9-3	G-CO.A.5
12	13-4	S-CP.A.2
13	12-4	G-SRT.B.5
14	6-4	G-CO.C.11
15	3-5	G-CO.C.9
16	9-2	G-SRT.A.1a
17	8-1	G-SRT.C.8
18	7-5	G-SRT.B.5
19	9-2	G-CO.A.5
20	4-3	G-CO.C.10
21	8-3	G-SRT.C.8
22	8-1	G-SRT.C.8
23	5-4	G-CO.C.10
24	6-5	G-CO.C.11
25	3-5	G-CO.C.10
26	5-3	G-CO.C.9
27	13-5	S-CP.A.4
28	11-5	G-GMD.A.3
29	4-2	G-SRT.B.5
30	7-4	G-SRT.B.5
31	8-5	G-SRT.D.10
32	7-2	G-SRT.B.5
33	8-1	G-SRT.C.8
34	13-6	S-CP.B.7
35	6-3	G-CO.C.10
36	10-6	G-SRT.C.8
37	9-4	G-CO.A.5
38	8-3	G-SRT.C.8
39	13-7	S-MD.B.6
40	8-3	G-SRT.D.10
41	12-4	G-C.A.2
42	8-1	G-SRT.C.8
43	7-4	G-SRT.B.5
44	11-4	G-GMD.A.3
45	4-5	G-CO.C.10
46	3-5	G-CO.C.10
47	11-4	G-GMD.A.3
48	11-7	G-GMD.A.3
49	12-1	G-C.A.2
50	10-4	G-SRT.B.5
51	10-5	G-SRT.C.8
52	12-1	G-C.A.2
53	7-5	G-SRT.B.5
54	4-7	G-CO.C.10
55	5-4	G-CO.C.12
56	9-3	G-CO.A.3
57	5-7	G-C.A.3
58	11-6	G-GMD.A.3
59	11-1	G-GMD.B.4
60	12-1	G-C.A.2
61	7-4	G-SRT.B.5
62	3-6	G-CO.C.12
63	6-3	G-CO.C.11
64	12-3	G-C.A.2

Common Core End-of-Course Assessment

ASSESSMENT

Selected Response

Read each question. Then write the letter of the correct answer on your paper.

1. Which term is defined by the following phrase?

a set of points in a plane a given distance from a point

(A) segment (B) circle (C) sector (D) angle

2. The United States has a land area of about 3,536,278 mi². Illinois has a land area of about 57,918 mi². What is the probability that a location in the United States chosen at random is not in Illinois?

(F) 1.6% (G) 9.8% (H) 16.3% (I) 98.4%

3. $\overleftrightarrow{AB}$ is tangent to ⊙C at point B. Which of the following can you NOT conclude is true?

(A) $m\angle CAB < m\angle ACB$

(B) $AB^2 + BC^2 = AC^2$

(C) $\angle CAB$ and $\angle ACB$ are complements.

(D) $\overleftrightarrow{AB} \perp \overleftrightarrow{BC}$

4. What is the relationship between the sine of an angle and the cosine of the angle's complement?

(F) They are inverses. (G) Their sum is 1. (H) They are the same. (I) They are opposites.

5. What is the area of $\triangle RST$ in square feet?

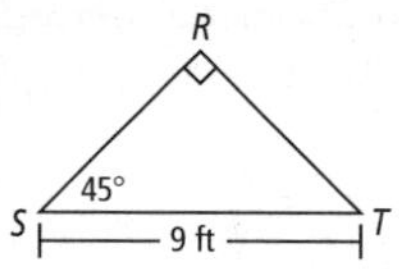

(A) $9\sqrt{2}$ (B) 20.25 (C) 40.5 (D) $81\sqrt{2}$

6. Which is an equation of the circle below?

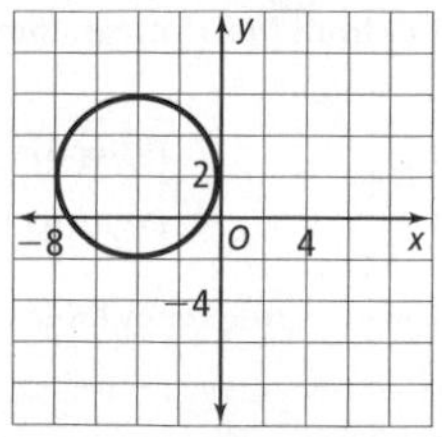

(F) $(x + 4)^2 + (y - 2)^2 = 16$

(G) $(x - 4)^2 + (y - 2)^2 = 4$

(H) $(x - 2)^2 + (y + 1)^2 = 4$

(I) $(x + 2)^2 + (y - 1)^2 = 2$

7. What is the area of the shaded sector? Use 3.14 for π.

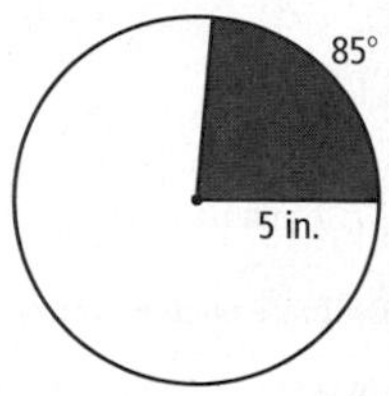

(A) about 18.5 in.²

(B) about 66.8.5 in.²

(C) about 78.5 in.²

(D) about 85 in.²

8. Which describes the transformation of $\triangle ABC$ to $\triangle A'B'C'$?

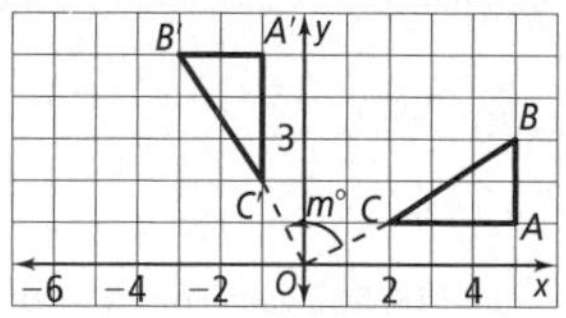

(F) $R_{y=x}(\triangle ABC) = \triangle A'B'C'$

(G) $T_{<2,5>}(\triangle ABC) = \triangle A'B'C'$

(H) $r_{(m°, O)}(\triangle ABC) = \triangle A'B'C'$

(I) $R_{y=x} + T_{<2,0>}(\triangle ABC) = \triangle A'B'C'$

Answers

Common Core End-of-Course Assessment

1. B
2. I
3. A
4. H
5. B
6. F
7. A
8. H

9. What are the values of a and b in the figure below?

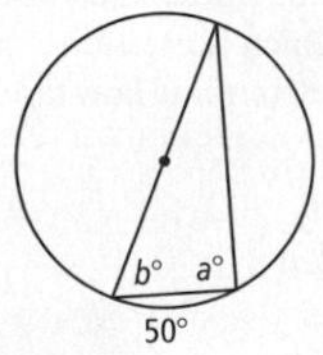

Ⓐ $a = 90$ and $b = 130$

Ⓑ $a = 50$ and $b = 130$

Ⓒ $a = 90$ and $b = 65$

Ⓓ $a = 65$ and $b = 65$

10. What is the locus of points in space that are 5 cm from segment m?

Ⓕ a cylinder with radius 5 cm capped at each end with a half sphere

Ⓖ a cylinder with radius 5 cm

Ⓗ a sphere with radius 5 cm

Ⓘ a cylinder with radius 5 cm capped at each end with a filled circle

11. Figure $QRST$ is shown in the coordinate plane. Which transformation creates an image with a vertex at the point $(-2, 1)$?

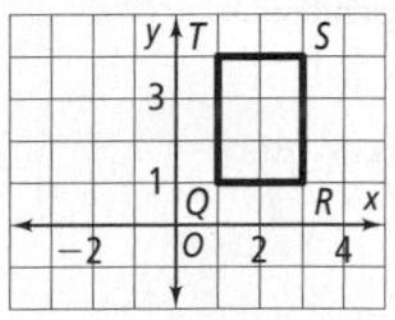

Ⓐ Rotate figure $QRST$ by 90° about R.

Ⓑ Reflect figure $QRST$ across the line $y = 1$.

Ⓒ Reflect figure $QRST$ across the line $x = 1$.

Ⓓ Rotate figure $QRST$ by 90° about Q.

12. Suppose you roll two number cubes, one red and one blue. What is the probability that you will roll 3 on the red cube and an even number on the blue cube?

Ⓕ $\frac{1}{4}$　Ⓗ $\frac{1}{6}$

Ⓖ $\frac{1}{3}$　Ⓘ $\frac{1}{12}$

13. Amy is trying to prove that $a \cdot b = c \cdot d$ using $\odot O$ at the right by first proving that $\triangle APC \sim \triangle DPB$. Which similarity theorem or postulate can Amy use?

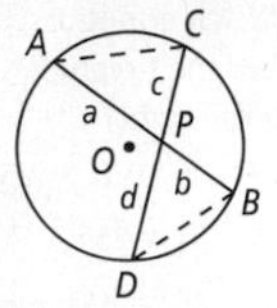

Ⓐ Side-Side-Side Similarity Theorem

Ⓑ Side-Angle-Side Similarity Theorem

Ⓒ Angle-Angle Similarity Postulate

Ⓓ None of these

14. All four angles of a quadrilateral have the same measure. Which statement is true?

Ⓕ All four sides of the quadrilateral must have the same length.

Ⓖ All four angles of the quadrilateral are acute.

Ⓗ Opposite sides of the quadrilateral are parallel.

Ⓘ The quadrilateral must be a square.

15. A triangular park is bordered by three streets, as shown in the map below.

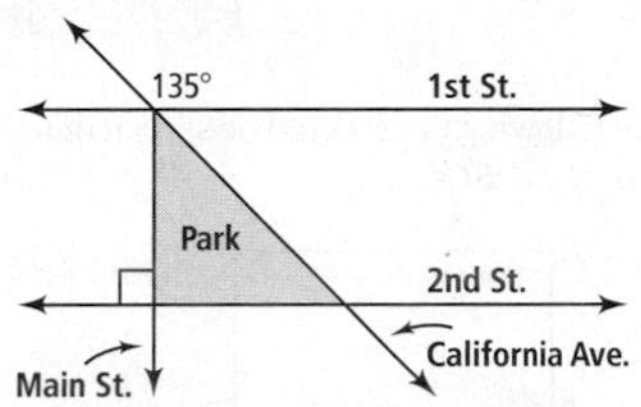

If 1st Street and 2nd Street are parallel, what are the measures of the three angles of the park?

Ⓐ 90, 45, 45　Ⓒ 90, 25, 65

Ⓑ 90, 35, 55　Ⓓ 135, 25, 10

16. $\overline{AB}$ is dilated by a scale factor of 2 with a center of dilation C, which is not on $\overline{AB}$. Which best describes $\overline{A'B'}$ and its length?

I. $A'B' = 2AB$

II. $\overline{A'B'} \parallel \overline{AB}$

Ⓕ I only　Ⓗ I and II

Ⓖ II only　Ⓘ neither I nor II

9. C
10. F
11. D
12. I
13. C
14. H
15. A
16. H

Answers

Common Core End-of-Course Assessment (continued)

17. C
18. I
19. D
20. I
21. B
22. I
23. C
24. H
25. B

17. Which equation can be used to find the height h of the triangle at the right?

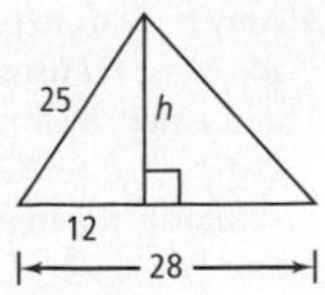

Ⓐ $h^2 = 28^2 - 25^2$
Ⓑ $h^2 = 25^2 + 12^2$
Ⓒ $25^2 = 12^2 + h^2$
Ⓓ $25^2 = 12^2 - h^2$

18. In the figure at the right, what is the length of $\overline{PS}$?

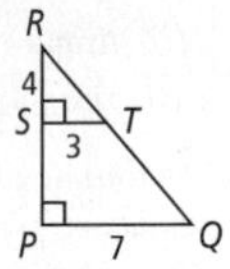

Ⓕ 3
Ⓖ $3\sqrt{2}$
Ⓗ 4
Ⓘ $5\frac{1}{3}$

19. Quadrilateral $PQRS$ is reflected across the line $x = 1$. Its image is $P'Q'R'S'$. What are the coordinates of P'?

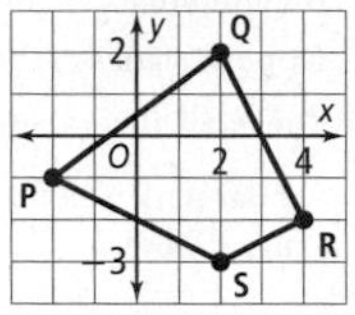

Ⓐ $(2, -1)$
Ⓑ $(4, 2)$
Ⓒ $(-2, 3)$
Ⓓ $(4, -1)$

20. Which of the following facts would be sufficient to prove $\triangle ACB \cong \triangle DBC$?

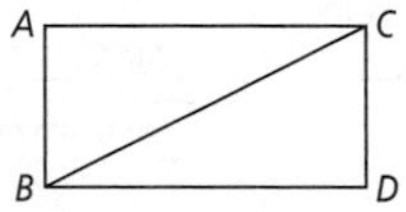

Ⓕ $\angle A$ is a right angle.
Ⓖ $\overline{BC} \cong \overline{BC}$
Ⓗ $\angle ABC$ and $\angle DCB$ are acute.
Ⓘ $\overline{AB} \parallel \overline{CD}$ and $\overline{AC} \parallel \overline{BD}$.

21. To the nearest tenth of a foot, what is the value of x in the figure below?

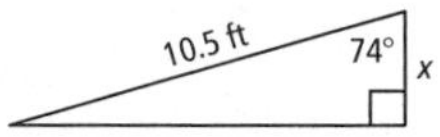

Ⓐ 2.8
Ⓑ 2.9
Ⓒ 10.1
Ⓓ 36.6

22. Bob walked diagonally across a rectangular field that measured 240 ft by 320 ft. Which expression could be used to determine how far Bob walked?

Ⓕ $2(240 + 320)$
Ⓖ $\sqrt{240} + \sqrt{320}$
Ⓗ $\frac{240 + 320}{2}$
Ⓘ $\sqrt{240^2 + 320^2}$

23. Which triangle is drawn with its medians?

Ⓐ
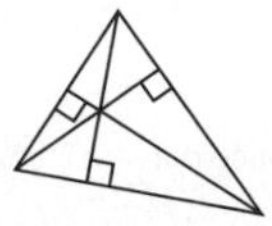

Ⓒ
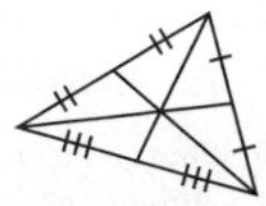

Ⓑ
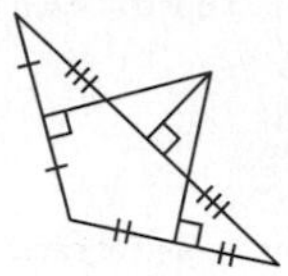

Ⓓ
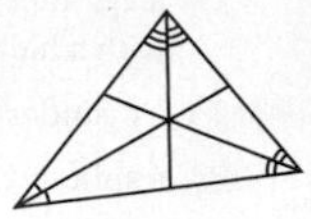

24. Which statement is true for both a rhombus and a kite?

Ⓕ The diagonals are congruent.
Ⓖ Opposite sides are congruent.
Ⓗ The diagonals are perpendicular.
Ⓘ Opposite sides are parallel.

25. Given $\triangle ABC$ below, what is $m\angle B$?

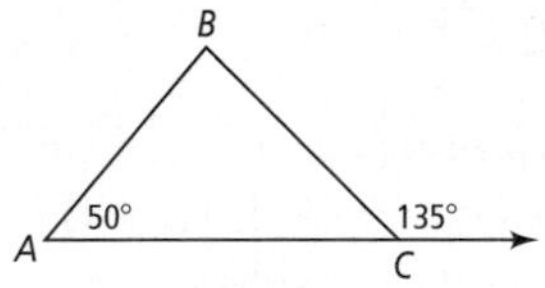

Ⓐ 45
Ⓑ 85
Ⓒ 95
Ⓓ 100

26. How can you determine that a point lies on the perpendicular bisector of $\overline{PQ}$ with endpoints $P(-3, -6)$ and $Q(-3, 4)$?

Ⓕ The point has x-coordinate -3.

Ⓖ The point has y-coordinate -3.

Ⓗ The point lies on the line $x = -1$.

Ⓘ The point lies on the line $y = -1$.

27. The table shows the number of degree recipients in a recent year. What is the probability that the recipient was a female getting a bachelor's degree?

Number of Degree Recipients (thousands)

Degree	Male	Female
Associates	245	433
Bachelors	598	853

Ⓐ 36%   Ⓒ 59%

Ⓑ 40%   Ⓓ 64%

28. What is the volume of the square pyramid?

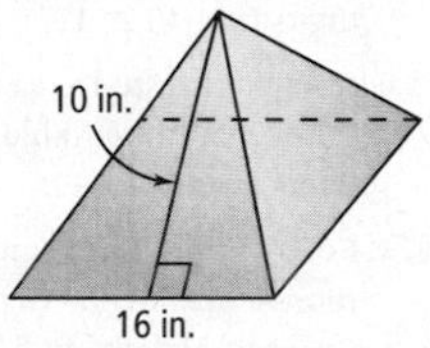

Ⓕ 512 in.^{3}

Ⓖ 256 in.^{3}

Ⓗ $106\frac{2}{3}$ in.^{3}

Ⓘ 64 in.^{3}

29. Which postulate or theorem justifies the statement $\triangle JLV \cong \triangle PMK$?

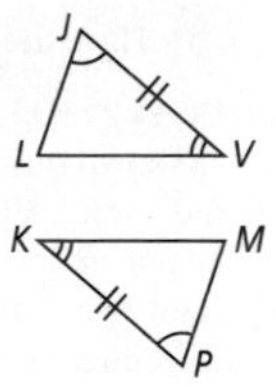

Ⓐ ASA

Ⓑ SAS

Ⓒ AAS

Ⓓ SSS

30. What is the value of $\frac{y}{x}$?

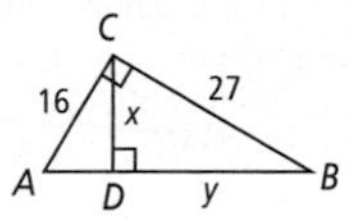

Ⓕ $\frac{16}{27}$

Ⓖ $\frac{\sqrt{985}}{27}$

Ⓗ $\frac{27}{16}$

Ⓘ $\frac{\sqrt{985}}{26}$

31. For the triangle at the right, what is $m\angle E$?

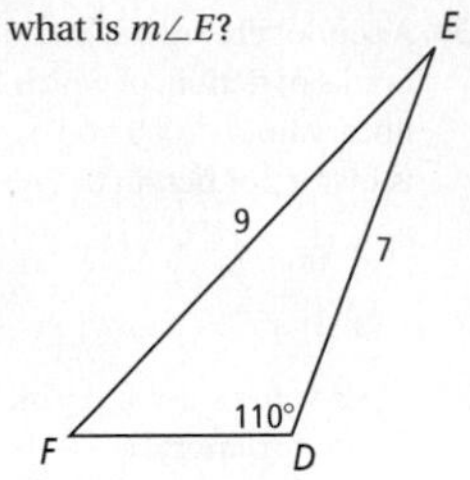

Ⓐ 17°

Ⓑ 23°

Ⓒ 47°

Ⓓ 53°

32. In the figure below, $\overline{PQ}$ is parallel to $\overline{RS}$, and $\overline{PS}$ and $\overline{QR}$ intersect at point T. What is the length of $\overline{PS}$?

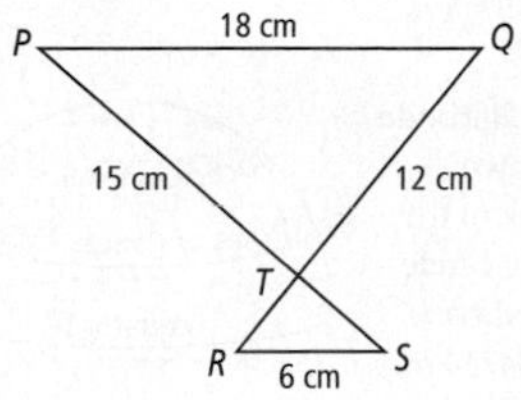

Ⓕ 4 cm   Ⓗ 19 cm

Ⓖ 5 cm   Ⓘ 20 cm

33. A hiker is traveling north through Anza Borrego State Park. When he is 0.8 mi from his destination, he veers off course for 1 mi. Use the diagram below to determine how many miles x the hiker is from his destination.

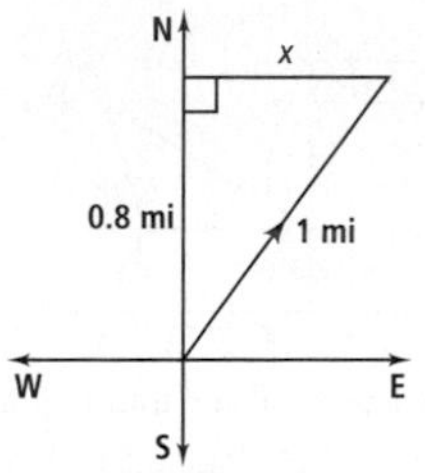

Ⓐ 1.8   Ⓒ 0.36

Ⓑ 0.6   Ⓓ 0.2

26. I
27. B
28. F
29. A
30. H
31. B
32. I
33. B

Answers

Common Core End-of-Course Assessment (continued)

34. H
35. B
36. G
37. B
38. I
39. B
40. I

34. A school library has 85 books on history. It also has 240 books of fiction, of which 25 are historical fiction. If the library holds 1000 books, what is the probability that it is history or fiction but not historical fiction?

F $\frac{1}{14}$ H $\frac{3}{10}$

G $\frac{2}{5}$ I $\frac{13}{14}$

35. What values of x and y make the quadrilateral a parallelogram?

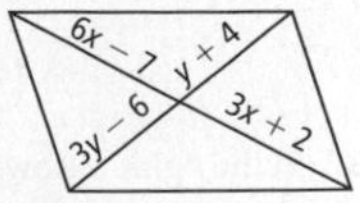

A $x = 3,\ y = 3$ C $x = 5,\ y = 3$

B $x = 3,\ y = 5$ D $x = 5,\ y = 7$

36. On a globe, lines of latitude form circles, as shown at the right. The arc from the equator to the Artic Circle is 66.5°. The approximate radius of Earth is 6471 km. What is the circumference of the Artic Circle?

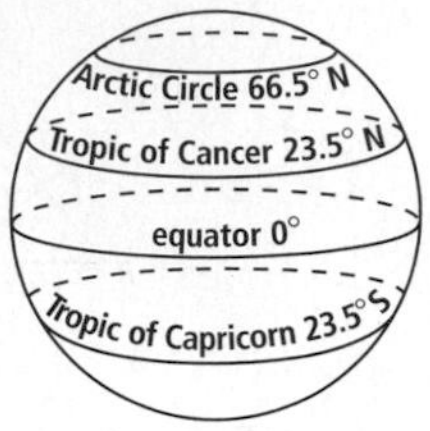

F about 2693 km

G about 16,213 km

H about 2,662,006 km

I about 22,775,168 km

37. Which sequence of transformations maps $\triangle ABC$ onto $\triangle DEF$?

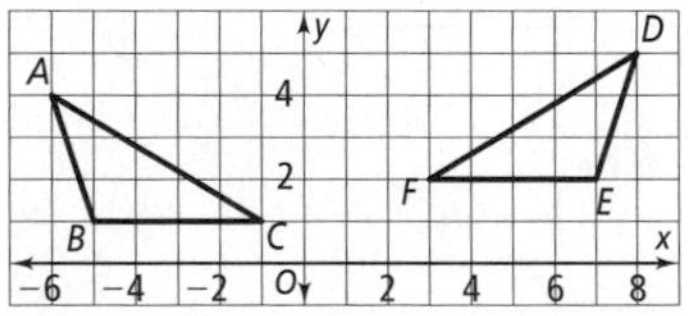

A reflection across the line $x = 0$ and translation right 2 units

B reflection across the line $x = 1$ and translation up 1 unit

C 180° rotation about the origin and reflection over the line $x = \frac{1}{2}$

D 180° rotation about the origin and translation right 2 units up 1 unit

38. Read this excerpt from a news article.

The Casco Bay Bridge, a double-leaf drawbridge in Maine, opened in 1997. The bridge replaced the old Million Dollar Bridge over the Fore River. The old bridge had a clearance of 24 feet between the water and the closed bridge. The new bridge has a clearance of 65 feet when closed, so it does not need to be opened as often as the old bridge. Each leaf of the new bridge is approximately 143 feet long and opens up to an angle of 78°. The new bridge may need to be opened less often, but it takes about 6 minutes longer to open and close than the old bridge.

How high off the water must the tip of each leaf be when the Casco Bay Bridge is open?

F 65 ft H 140 ft

G 95 ft I 205 ft

39. Which method is a fair way to randomly select 12 students from a class of 30 students?

A Assign each student a number from 01 to 30. Use a random number table, selecting the first 12 pairs of digits from 01 to 12.

B Assign each student a number from 01 to 30. Use a random number table, selecting the first 12 pairs of digits from 01 to 30.

C Roll red and blue number cubes. The red cube represents the first digit and the second represents the second digit. Roll 12 times.

D Have the 30 students elect the 12 students.

40. Derek lives close to both his school and the library, as shown below. After school, Derek walked 5 min to the library. He then walked 8 min to his home when he realized that he left a book at school. Which time is the best estimate of how long it will take Derek to walk directly from his home to his school if the angle formed by the School-Library-Derek's home is 87°? (Assume Derek walks at the same rate each time.)

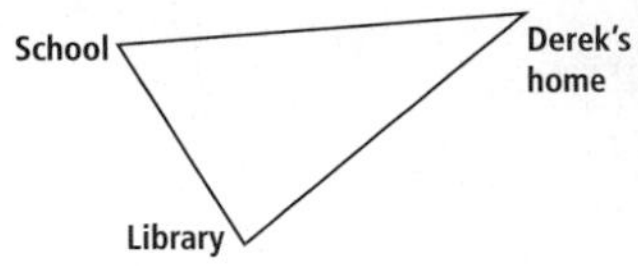

F 9 min H 13 min

G 11 min I 15 min

41. What is the value of x in the circle at the right?

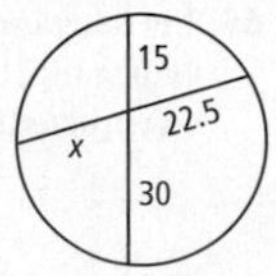

Ⓐ 25 Ⓒ 11.25

Ⓑ 20 Ⓓ 7.5

42. Which of the following could be the side lengths of a right triangle?

Ⓕ 4.1, 6.2, 7.3

Ⓖ 40, 60, 72

Ⓗ 3.2, 5.4, 6.2

Ⓘ 33, 56, 65

43. In the figure at the right, what is the length of $\overline{CD}$?

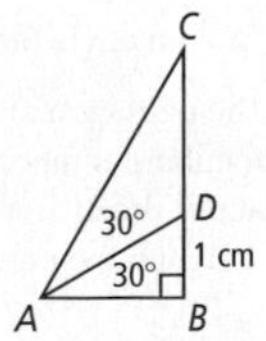

Ⓐ $2\sqrt{3}$ cm

Ⓑ 3 cm

Ⓒ 2 cm

Ⓓ $\sqrt{3}$ cm

44. A manufacturer is comparing two packages for a new product. Package A is a rectangular prism that is 9 in. by 5 in. by 8 in. Package B is a triangular prism with height 15 in. Its bases are right triangles with 6-in. and 8-in. legs. Which statement best describes the relationship between the two prisms?

Ⓕ The triangular prism has $\frac{1}{2}$ the volume of the rectangular prism.

Ⓖ The rectangular prism has the greater volume.

Ⓗ The volumes are equal.

Ⓘ not enough information

45. Which of the following must be true?

I. $\angle BAC \cong \angle B$

II. $\angle B \cong \angle C$

III. $\overline{AD} \cong \overline{AB}$

IV. $\overline{BD} \cong \overline{CD}$

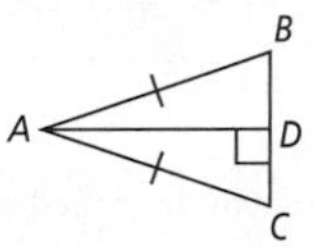

Ⓐ I and II only Ⓒ II and IV only

Ⓑ I and III only Ⓓ III and IV only

46. What are the values of x and y?

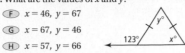

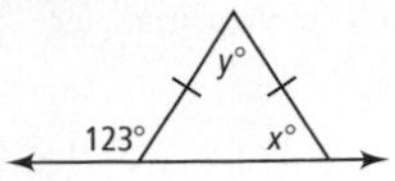

Ⓕ $x = 46,\ y = 67$

Ⓖ $x = 67,\ y = 46$

Ⓗ $x = 57,\ y = 66$

Ⓘ $x = 66,\ y = 57$

47. What is the volume of the cylinder to the nearest cubic inch?

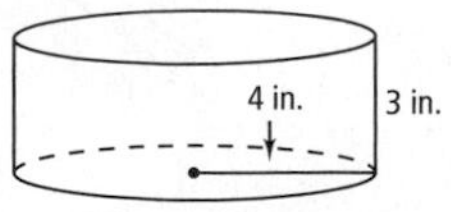

Ⓐ 603 in.^{3} Ⓒ 151 in.^{3}

Ⓑ 226 in.^{3} Ⓓ 113 in.^{3}

48. A large box of laundry detergent has the shape of a rectangular prism. A similar box has length, width, and height that are one half of the large box. How many times the volume of the small box is the volume of the large box?

Ⓕ 4 Ⓗ 64

Ⓖ 8 Ⓘ 512

49. In the figure at the right, $\overline{AB}$ is tangent to $\odot O$, $AB = 15$ cm, and $BC = 9$ cm. What is the radius of $\odot O$?

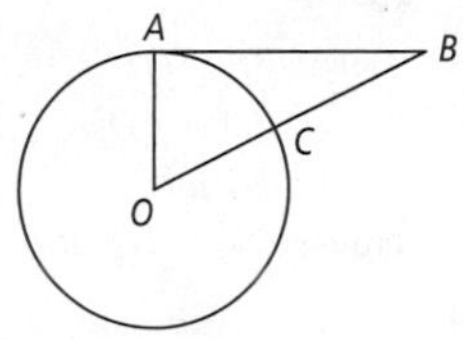

Ⓐ 7 cm Ⓒ 9 cm

Ⓑ 8 cm Ⓓ 16 cm

50. Myra multiplies the length of each side of a triangle by $\frac{1}{5}$. By what factor can she multiply the perimeter of the original triangle to find the perimeter of the new triangle?

Ⓕ 0.008 Ⓗ 0.2

Ⓖ 0.04 Ⓘ 5

51. What is the area of a regular pentagon with side length 4 cm? Round your answer to the nearest tenth.

Ⓐ 16.2 cm^2 Ⓒ 41.5 cm^2

Ⓑ 27.5 cm^2 Ⓓ 55.0 cm^2

41. B
42. I
43. C
44. H
45. C
46. H
47. C
48. G
49. B
50. H
51. B

Answers

Common Core End-of-Course Assessment (continued)

52. G
53. C
54. G
55. D
56. H
57. B

52. The triangle circumscribes the circle. What is the perimeter of the triangle?

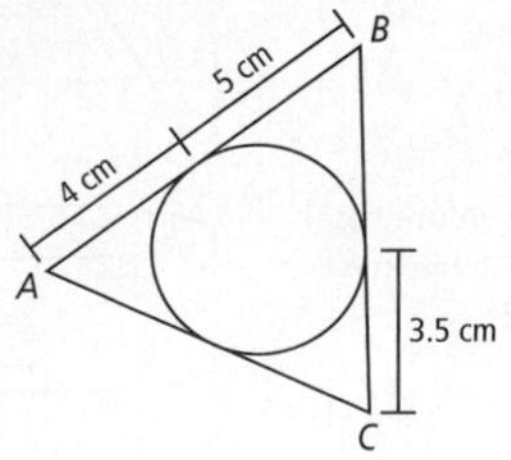

(F) 24 cm
(G) 25 cm
(H) 25.5 cm
(I) 37.5 cm

53. What is the value of x in the figure at the right?

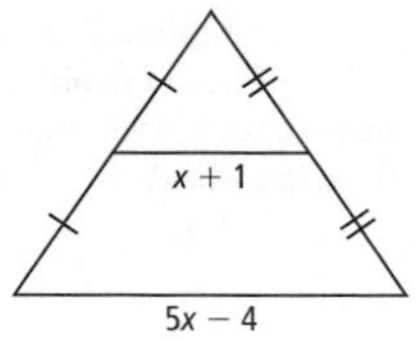

(A) 1
(B) $\frac{5}{4}$
(C) 2
(D) $\frac{7}{2}$

54. Given: $\overline{AE} \cong \overline{FD}$
$\angle A \cong \angle D$
$\overline{FG} \cong \overline{GE}$
Prove: $\triangle ABE \cong \triangle DCF$

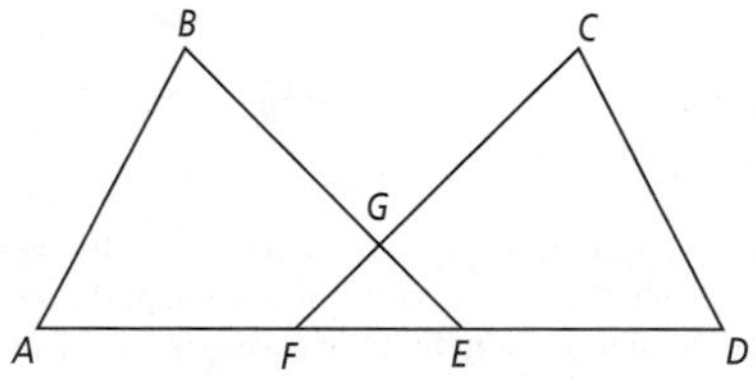

Which next step in the proof would be most helpful in proving $\triangle ABE \cong \triangle DCF$?

(F) $\angle BGF \cong \angle CGE$ (Vertical Angles Theorem)
(G) $\angle GFE \cong \angle GEF$ (Isosceles Triangle Theorem)
(H) $AF + FE = DE + FE$ (Segment Addition Postulate)
(I) $\angle B \cong \angle C$ (Corresp. Parts of $\cong$ $\triangle$s are $\cong$.)

55. The diagram below shows a standard construction with straightedge and compass. What has been constructed?

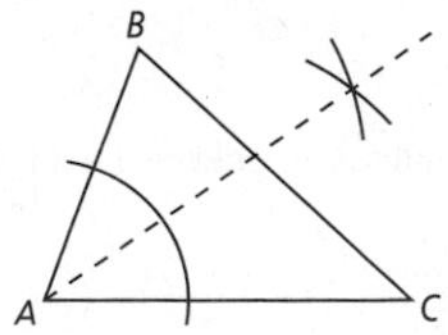

(A) a median
(B) an altitude
(C) a perpendicular bisector
(D) an angle bisector

56. The pentagon at the right is a regular pentagon. Which rotation about the center of the pentagon will map the pentagon onto itself?

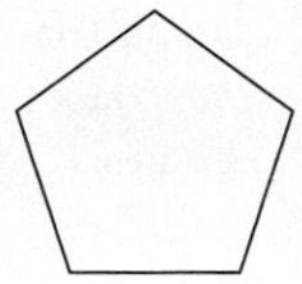

(F) 58°
(G) 90°
(H) 144°
(I) 180°

57. In $\triangle ABC$ below, point O has been constructed.

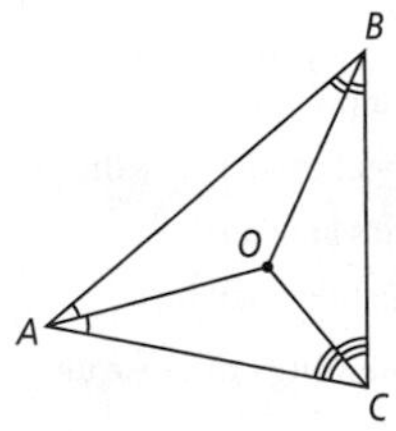

What is point O?

(A) centroid
(B) center of the inscribed circle
(C) center of the circumscribed circle
(D) none of the above

882 Common Core End-of-Course Assessment

58. What is the surface area of the sphere?

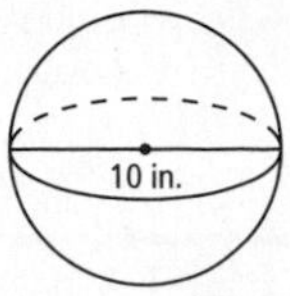

Ⓕ 100π in.^{2}

Ⓖ 100π in.^{3}

Ⓗ $\frac{500}{3}\pi$ in.^{2}

Ⓘ 400π in.^{2}

59. Which is the cross section of a cylinder that is intersected by a plane that is perpendicular to the two bases?

Ⓐ circle

Ⓑ rectangle

Ⓒ trapezoid

Ⓓ triangle

60. If *O* is the center of the circle, what can you conclude from the diagram?

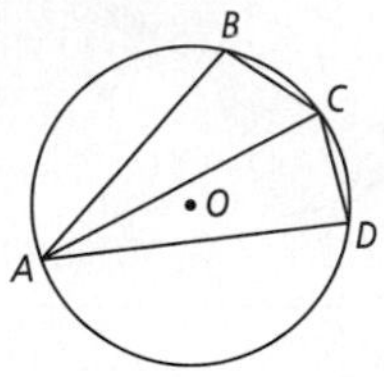

Ⓕ $AB > AD$

Ⓖ $AB = AD$

Ⓗ $AB < AD$

Ⓘ There is not enough information to compare *AB* and *AD*.

61. The height of the cone below is 5.

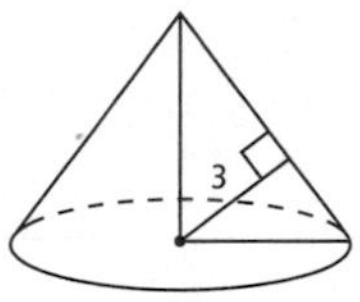

What is the radius of the cone?

Ⓐ 3

Ⓑ $\sqrt{10}$

Ⓒ 3.75

Ⓓ 4

Constructed Response

62. How do you construct a $\triangle DEF$ congruent to $\triangle ACB$?

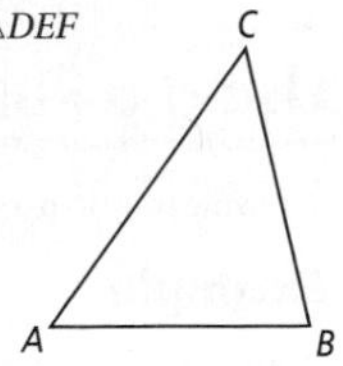

63. Rieko is trying to prove the following theorem:

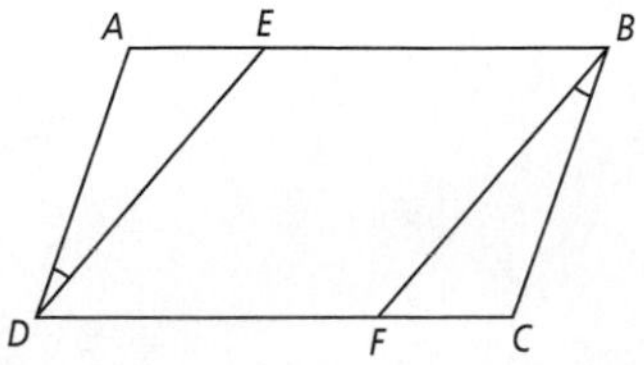

If *ABCD* is a parallelogram, and $\angle ADE \cong \angle CBF$, then *DEBF* is a parallelogram.

One strategy is to show that both pairs of opposite angles are congruent. How would you show that $\angle EDF \cong \angle EBF$?

Extended Response

64. In one game on a reality TV show, players must dig for a prize hidden somewhere within a sand-filled circular area. There are 18 posts equally spaced around the circle and clues are given so that players narrow the location by crossing two ropes between the posts. What is the angle measure *w* formed by the ropes in the diagram? Explain your reasoning.

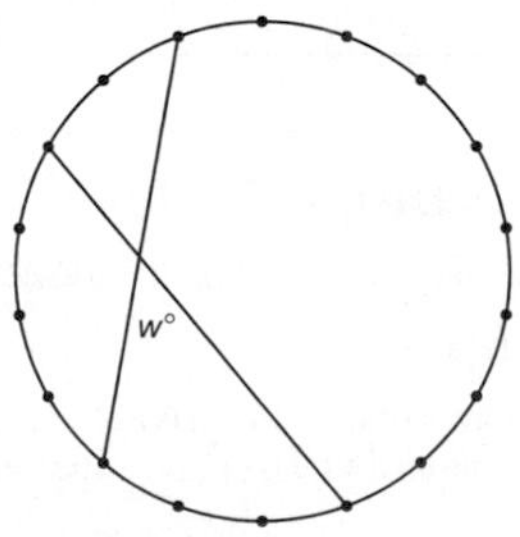

58. F

59. B

60. I

61. C

62. [2] Answers may vary. Sample: Draw a line, mark off length *AB* on that line, and label the endpoints of the segment as *D* and *F*. Using *D* as center and *AC* as radius, draw an arc. Using *F* as center and *BC* as radius, draw an arc that intersects the arc with center *D*. Label the intersection of the two arcs as point *E*. [This construction uses SSS ≅; other valid constructions use SAS or ASA ≅.]

[1] incomplete OR incorrect description

63. [2] Because opposite pairs of ∠s of a ▱ are ≅, $\angle ADF \cong \angle CBE$. By the ∠ Add. Post., $m\angle ADF = m\angle ADE + m\angle EDF$ and $m\angle CBE = m\angle CBF + m\angle EBF$. So, $m\angle EDF = m\angle ADF - m\angle ADE$ and $m\angle EBF = m\angle CBE - m\angle CBF$. It is given that $\angle ADE \cong \angle CBF$. So, by the Subst. Prop., $m\angle EBF = m\angle ADF - m\angle ADE$. Therefore, $m\angle EDF = m\angle EBF$ and $\angle EDF \cong \angle EBF$.

[1] incomplete explanation

64. [4] 50; the 18 posts divide the circumference of the circle into 18 ≅ arcs, so the measure of each arc is $\frac{360}{18} = 20$. The ∠ whose measure is *w* intercepts two arcs, one with measure $2 \cdot 20 = 40$ and one with measure $3 \cdot 20 = 60$. Using the formula for the measure of an ∠ formed by two chords, $w = \frac{1}{2}(40 + 60) = \frac{1}{2}(100) = 50$; $w = 50$.

[3] correct method, one computational error

[2] correct answer, incomplete explanation

[1] correct answer, no work

Skills Handbook

Using a Ruler and Protractor

Knowing how to use a ruler and protractor is crucial for success in geometry.

Example

Draw a triangle that has a 28° angle between sides of length 5.2 cm and 3.0 cm.

Step 1 Use a ruler to draw a segment 5.2 cm long.

Step 2 Place the hole of a protractor at one endpoint of the segment. Make a small mark at the 28° position along the protractor.

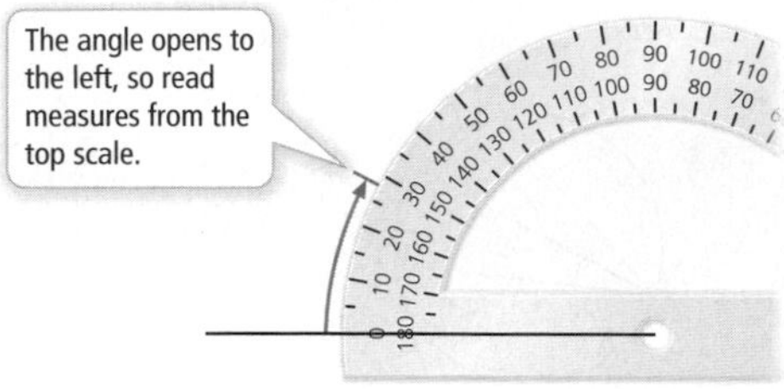

The angle opens to the left, so read measures from the top scale.

Step 3 Align the ruler along the small mark and the same endpoint. Place the zero point of the ruler at the endpoint. Draw a segment 3.0 cm long.

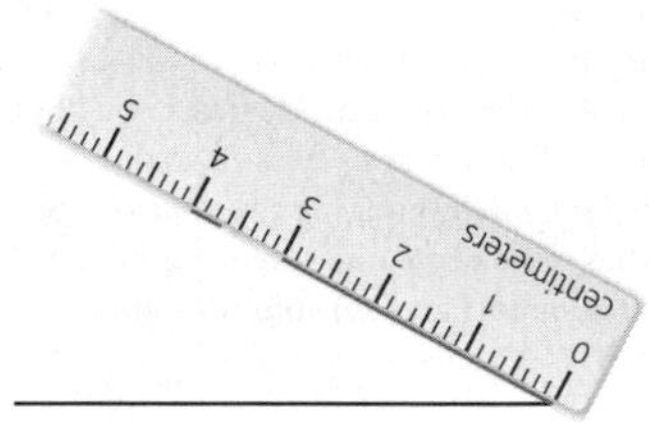

Step 4 Complete the triangle by connecting the endpoints of the first and second segments.

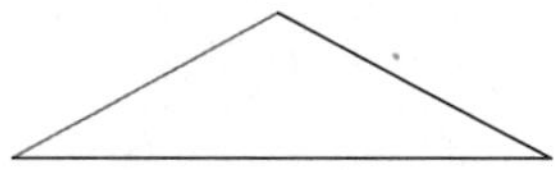

Exercises

1. Measure sides $\overline{AB}$ and $\overline{BC}$ of $\triangle ABC$ to the nearest millimeter.
2. Measure each angle of $\triangle ABC$ to the nearest degree.
3. Draw a triangle that has a side of length 2.4 cm between a 43° angle and a 102° angle.

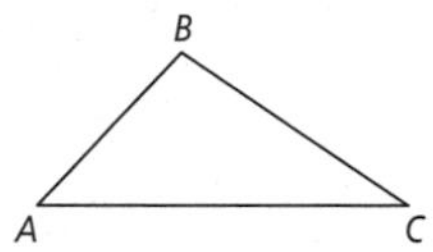

Answers

Using a Ruler and Protractor

1–2. Answers may vary slightly due to measuring method.

1. 20 mm; 25 mm

2. $m\angle A = 43$; $m\angle B = 103$; $m\angle C = 34$

3.

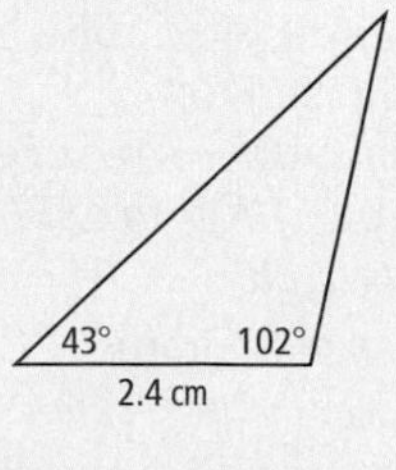

Classifying Triangles

You can classify a triangle by its angles and sides.

Equiangular
all angles congruent

Acute
all angles acute

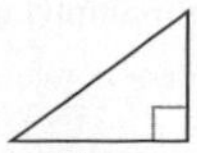

Right
one right angle

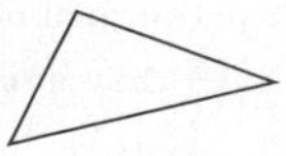

Obtuse
one obtuse angle

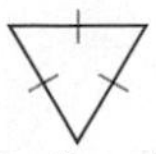

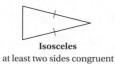

Equilateral
all sides congruent

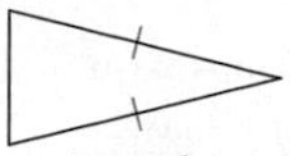

Isosceles
at least two sides congruent

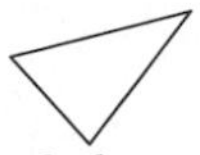

Scalene
no sides congruent

Example

What type of triangle is shown below?

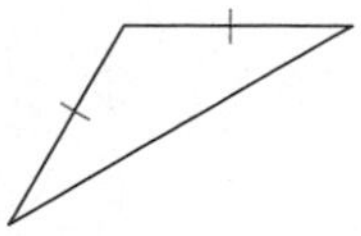

At least two sides are congruent, so the triangle is isosceles. One angle is obtuse, so the triangle is obtuse. The triangle is an obtuse isosceles triangle.

Exercises

Classify each triangle by its sides and angles.

1. 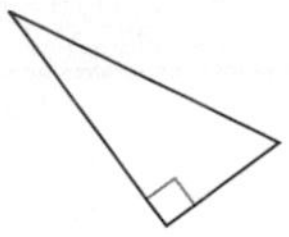

2. 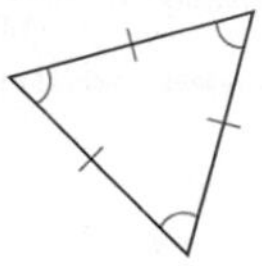

3. 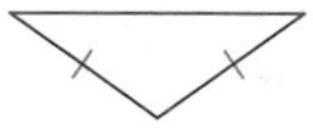

If possible, draw a triangle to fit each description. Mark the triangle to show known information. If you cannot draw the triangle, write *not possible* and explain why.

4. acute equilateral

5. right equilateral

6. obtuse scalene

7. acute isosceles

8. right isosceles

9. acute scalene

Skills Handbook

Skills Handbook

Classifying Triangles

1. right, scalene

2. acute equiangular, equilateral

3. obtuse, isosceles

4.

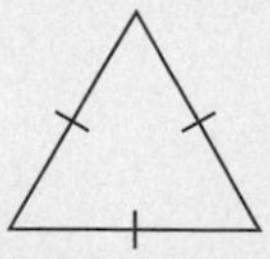

5. Not possible; a rt. △ will always have one longest side opposite the rt. ∠.

6.

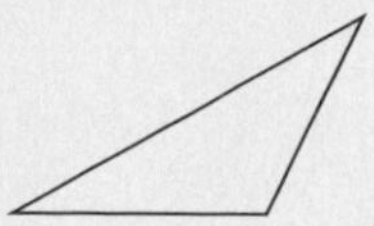

7.

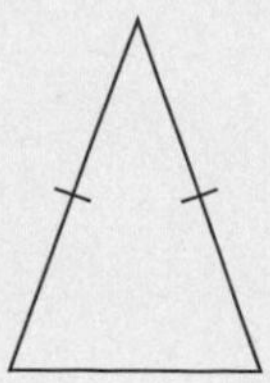

8.

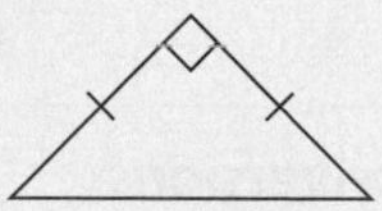

9.

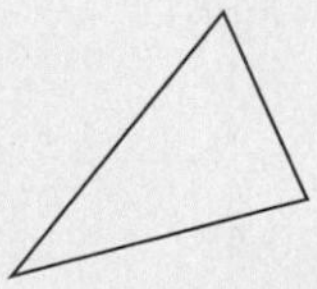

Skills Handbook

Measurement Conversions

To convert from one unit of measure to another, you multiply by a conversion factor in the form of a fraction. The numerator and denominator are in different units, but they represent the same amount. So you can think of this as multiplying by 1.

An example of a conversion factor is $\frac{1\text{ ft}}{12\text{ in.}}$. You can create other conversion factors using the table on page 837.

Example 1

Complete each statement.

a. 88 in. = ■ ft

$88\text{ in.} \cdot \frac{1\text{ ft}}{12\text{ in.}} = \frac{88}{12}\text{ ft} = 7\frac{1}{3}\text{ ft}$

b. 5.3 m = ■ cm

$5.3\text{ m} \cdot \frac{100\text{ cm}}{1\text{ m}} = 5.3(100)\text{ cm} = 530\text{ cm}$

Area is always in square units, and volume is always in cubic units.

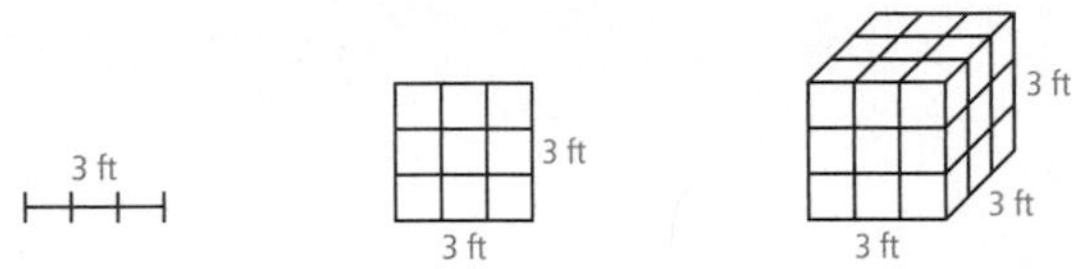

1 yd = 3 ft | **1 yd² = 9 ft²** | **1 yd³ = 27 ft³**

Example 2

Complete each statement.

a. 300 in.^2 = ■ ft^2

$1\text{ ft} = 12\text{ in.}$, so $1\text{ ft}^2 = (12\text{ in.})^2 = 144\text{ in.}^2$.

$300\text{ in.}^2 \cdot \frac{1\text{ ft}^2}{144\text{ in.}^2} = 2\frac{1}{12}\text{ ft}^2$

b. $200{,}000\text{ cm}^3$ = ■ m^3

$1\text{ m} = 100\text{ cm}$, so $1\text{ m}^3 = (100\text{ cm})^3 = 1{,}000{,}000\text{ cm}^3$.

$200{,}000\text{ cm}^3 \cdot \frac{1\text{ m}^3}{1{,}000{,}000\text{ cm}^3} = 0.2\text{ m}^3$

Exercises

Complete each statement.

1. 40 cm = ■ m
2. 1.5 kg = ■ g
3. 60 cm = ■ mm
4. 200 in. = ■ ft
5. 28 yd = ■ in.
6. 1.5 mi = ■ ft
7. 15 g = ■ mg
8. 430 mg = ■ g
9. 34 L = ■ mL
10. 1.2 m = ■ cm
11. 43 mm = ■ cm
12. 3600 s = ■ min
13. 14 gal = ■ qt
14. 4500 lb = ■ t
15. 234 min = ■ h
16. 3 ft^2 = ■ in.^2
17. 108 m^2 = ■ cm^2
18. 21 cm^2 = ■ mm^2
19. 1.4 yd^2 = ■ ft^2
20. 0.45 km^2 = ■ m^2
21. 1300 ft^2 = ■ yd^2
22. 1030 in.^2 = ■ ft^2
23. $20{,}000{,}000\text{ ft}^2$ = ■ mi^2
24. 1000 cm^3 = ■ m^3

Answers

Measurement Conversions

1. 0.4
2. 1500
3. 600
4. $16\frac{2}{3}$
5. 1008
6. 7920
7. 15,000
8. 0.43
9. 34,000
10. 120
11. 4.3
12. 60
13. 56
14. $2\frac{1}{4}$
15. 3.9
16. 432
17. 1,080,000
18. 2100
19. 12.6
20. 450,000
21. $144\frac{4}{9}$
22. $7\frac{11}{72}$
23. $\frac{3125}{4356}$
24. 0.001

Measurement, Rounding Error, and Reasonableness

There is no such thing as an *exact* measurement. Measurements are always approximate. No matter how precise it is, a measurement actually represents a range of values.

Example 1

Chris's height, to the nearest inch, is 5 ft 8 in. What range of values does this measurement represent?

The height is given to the nearest inch, so the error is $\frac{1}{2}$ in. Chris's height, then, is between 5 ft $7\frac{1}{2}$ in. and 5 ft $8\frac{1}{2}$ in., or 5 ft 8 in. $\pm$ $\frac{1}{2}$ in. Within this range are all the measures that, when rounded to the nearest inch, equal 5 ft 8 in.

As you calculate with measurements, errors can accumulate.

Example 2

Jean drives 18 km to work each day. This distance is given to the nearest kilometer. What is the range of values for the round-trip distance?

The driving distance is between 17.5 and 18.5 km, or 18 $\pm$ 0.5 km. Double the lower limit, 17.5, and the upper limit, 18.5. Thus, the round trip can be anywhere between 35 and 37 km, or 36 $\pm$ 1 km. Notice that the error for the round trip is double the error for a single leg of the trip.

So that your answers will be reasonable, keep precision and error in mind as you calculate. For example, in finding AB, the length of the hypotenuse of $\triangle ABC$, it would be inappropriate to give the answer as 8.6533 if the sides are given to the nearest tenth. Round your answer to 8.7.

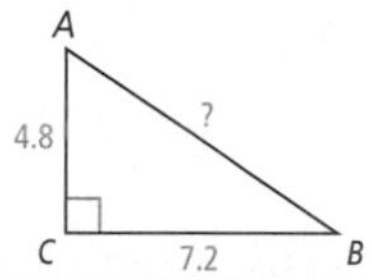

Exercises

Each measurement is followed by its unit of greatest precision. Find the range of values that each measurement represents.

1. 24 ft (ft)
2. 124 cm (cm)
3. 340 mL (mL)
4. $5\frac{1}{2}$ mi. $\left(\frac{1}{2}\text{ mi}\right)$
5. 73.2 mm (0.1 mm)
6. 34 yd^2 (yd^2)
7. The lengths of the sides of $TJCM$ are given to the nearest tenth of a centimeter. What is the range of values for the figure's perimeter?
8. To the nearest degree, two angles of a triangle are 49° and 73°. What is the range of values for the measure of the third angle?
9. The lengths of the legs of a right triangle are measured as 131 m and 162 m. You use a calculator to find the length of the hypotenuse. The calculator display reads 208.33867. What should your answer be?

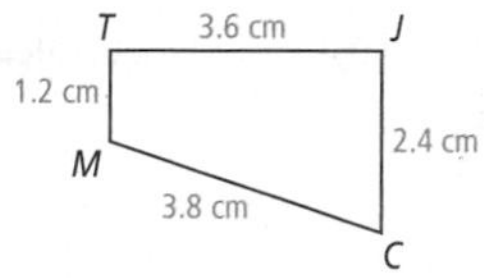

Skills Handbook

Measurement, Rounding Error, and Reasonableness

1. $23\frac{1}{2}$ ft to $24\frac{1}{2}$ ft
2. $123\frac{1}{2}$ cm to $124\frac{1}{2}$ cm
3. $339\frac{1}{2}$ mL to $340\frac{1}{2}$ mL
4. $5\frac{1}{4}$ mi to $5\frac{3}{4}$ mi
5. 73.15 mm to 73.25 mm
6. $33\frac{1}{2}$ yd^2 to $34\frac{1}{2}$ yd^2
7. 10.8 cm to 11.2 cm
8. 57 to 59
9. 208 cm

Skills Handbook

The Effect of Measurement Errors on Calculations

Measurements are always approximate, and calculations with these measurements produce error. Percent error is a measure of accuracy of a measurement or calculation. It is the ratio of the greatest possible error to the measurement.

$$\text{percent error} = \frac{\text{greatest possible error}}{\text{measurement}}$$

Example

The dimensions of a box are measured as 18 in., 12 in., and 9 in. What is the percent error in calculating the box's volume?

The measurements are to the nearest inch, so the greatest possible error is 0.5 in.

Volume:

as measured	maximum value	minimum value
$V = \ell \cdot w \cdot h$	$V = \ell \cdot w \cdot h$	$V = \ell \cdot w \cdot h$
$= 18 \cdot 12 \cdot 9$	$= 18.5 \cdot 12.5 \cdot 9.5$	$= 17.5 \cdot 11.5 \cdot 8.5$
$= 1944 \text{ in.}^3$	$\approx 2196.9 \text{ in.}^3$	$\approx 1710.6 \text{ in.}^3$

Possible Error:

maximum value − measured	measured − minimum value
$2196.9 - 1944 = 252.9$	$1944 - 1710.6 = 233.4$

$$\text{percent error} = \frac{\text{greatest possible error}}{\text{measurement}}$$

$$= \frac{252.9}{1944}$$

$$\approx 0.1300926$$

The percent error is about 13%.

Exercises

Find the percent error in calculating the volume of each box given its dimensions. Round to the nearest percent.

1. 10 cm by 5 cm by 20 cm

2. 1.2 mm by 5.7 mm by 2.0 mm

3. 1.24 cm by 4.45 cm by 5.58 cm

4. $8\frac{1}{4}$ in. by $17\frac{1}{2}$ in. by 5 in.

Find the percent error in calculating the perimeter of each figure.

5.

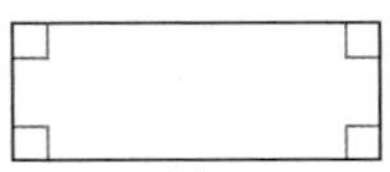

6.

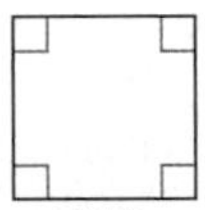

7.

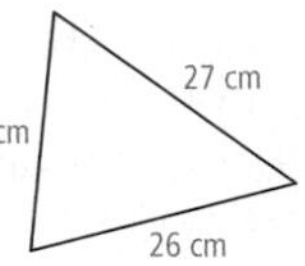

Answers

The Effect of Measurement Errors on Calculations

1. 18%
2. 8%
3. 1%
4. 5%
5. ≈ 7%
6. ≈ 2%
7. ≈ 2%

Squaring Numbers and Finding Square Roots

The square of a number is found by multiplying the number by itself. An exponent of 2 is used to indicate that a number is being squared.

Example 1

Simplify.

a. 5^2

$5^2 = 5 \cdot 5$
$= 25$

b. $(-3.5)^2$

$(-3.5)^2 = -3.5) \cdot (-3.5)$
$= 12.25$

c. $\left(\frac{2}{7}\right)^2$

$\left(\frac{2}{7}\right)^2 = \frac{2}{7} \cdot \frac{2}{7}$
$= \frac{4}{49}$

The square root of a number is itself a number that, when squared, results in the original number. A radical symbol ($\sqrt{\ }$) is used to represent the positive square root of a number.

Example 2

Simplify. Round to the nearest tenth if necessary.

a. $\sqrt{36}$

$\sqrt{36} = 6$, since $6^2 = 36$.

b. $\sqrt{174}$

$\sqrt{174} \approx 13.2$, since $13.2^2 \approx 174$.

You can solve equations that include squared numbers.

Example 3

Algebra Solve.

a. $x^2 = 144$

$x = 12$ or -12

b. $a^2 + 3^2 = 5^2$

$a^2 + 9 = 25$
$a^2 = 16$
$a = 4$ or -4

Exercises

Simplify.

1. 11^2 **2.** $(-14)^2$ **3.** 5.1^2 **4.** $\left(\frac{8}{5}\right)^2$ **5.** -6^2 **6.** $\left(-\frac{3}{7}\right)^2$

Simplify. Round to the nearest tenth if necessary.

7. $\sqrt{100}$ **8.** $\sqrt{169}$ **9.** $\sqrt{74}$ **10.** $\sqrt{50}$ **11.** $\sqrt{\frac{4}{9}}$ **12.** $\sqrt{\frac{49}{81}}$

Algebra Solve. Round to the nearest tenth if necessary.

13. $x^2 = 49$ **14.** $a^2 = 9$ **15.** $y^2 + 7 = 8$ **16.** $5 + x^2 = 11$

17. $8^2 + b^2 = 10^2$ **18.** $5^2 + 4^2 = c^2$ **19.** $p^2 + 12^2 = 13^2$ **20.** $20^2 = 15^2 + a^2$

Skills Handbook

Squaring Numbers and Finding Square Roots

1. 121 **2.** 196
3. 26.01 **4.** $\frac{64}{25}$
5. -36 **6.** $\frac{9}{49}$
7. 10 **8.** 13
9. 8.6 **10.** 7.1
11. $\frac{2}{3}$ **12.** $\frac{7}{9}$
13. ± 7 **14.** ± 3
15. ± 1 **16.** ± 2.4
17. ± 6 **18.** ± 6.4
19. ± 5 **20.** ± 13.2

Evaluating and Simplifying Expressions

To evaluate an expression with variables, substitute a number for each variable. Then simplify the expression using the order of operations. Be especially careful with exponents and negative signs. For example, the expression $-x^2$ always yields a negative or zero value, and $(-x)^2$ is always positive or zero.

Order of Operations

1. Perform any operation(s) inside grouping symbols.
2. Simplify any term with exponents.
3. Multiply and divide in order from left to right.
4. Add and subtract in order from left to right.

Example 1

Algebra **Evaluate each expression for $r = 4$.**

a. $-r^2$

$-r^2 = -4^2 = -16$

b. $-3r^2$

$-3r^2 = -3(4^2) = -3(16) = -48$

c. $(r + 2)^2$

$(r + 2)^2 = (4 + 2)^2 = (6)^2 = 36$

To simplify an expression, you eliminate any parentheses and combine like terms.

Example 2

Algebra **Simplify each expression.**

a. $5r - 2r + 1$

Combine like terms.
$5r - 2r + 1 = 3r + 1$

b. $\pi(3r - 1)$

Use the Distributive Property.
$\pi(3r - 1) = 3\pi r - \pi$

c. $(r + \pi)(r - \pi)$

Multiply polynomials.
$(r + \pi)(r - \pi) = r^2 - \pi^2$

Exercises

Algebra **Evaluate each expression for $x = 5$ and $y = -3$.**

1. $-2x^2$ **2.** $-y + x$ **3.** $-xy$ **4.** $(x + 5y) \div x$

5. $x + 5y \div x$ **6.** $(-2y)^2$ **7.** $(2y)^2$ **8.** $(x - y)^2$

9. $\frac{x+1}{y}$ **10.** $y - (x - y)$ **11.** $-y^x$ **12.** $\frac{2(1-x)}{y-x}$

13. $x \cdot y - x$ **14.** $x - y \cdot x$ **15.** $\frac{y^3 - x}{x - y}$ **16.** $-y(x - 3)^2$

Algebra **Simplify.**

17. $6x - 4x + 8 - 5$ **18.** $2(\ell + w)$ **19.** $-(4x + 7)$ **20.** $y(4 - y)$

21. $-4x(x - 2)$ **22.** $3x - (5 + 2x)$ **23.** $2t^2 + 4t - 5t^2$ **24.** $(r - 1)^2$

25. $(1 - r)^2$ **26.** $(y + 1)(y - 3)$ **27.** $4h + 3h - 4 + 3$ **28.** $\pi r - (1 + \pi r)$

29. $(x + 4)(2x - 1)$ **30.** $2\pi h(1 - r)^2$ **31.** $3y^2 - (y^2 + 3y)$ **32.** $-(x + 4)^2$

Answers

Evaluating and Simplifying Expressions

1. -50 **2.** 8
3. 15 **4.** -2
5. 2 **6.** 36
7. 36 **8.** 64
9. -2 **10.** -11
11. 243 **12.** 1
13. -20 **14.** 20
15. -4 **16.** 12
17. $2x + 3$ **18.** $2\ell + 2w$
19. $-4x - 7$ **20.** $4y - y^2$
21. $-4x^2 + 8x$ **22.** $x - 5$
23. $-3t^2 + 4t$ **24.** $r^2 - 2r + 1$
25. $1 - 2r + r^2$ **26.** $y^2 - 2y - 3$
27. $7h - 1$ **28.** -1
29. $2x^2 + 7x - 4$
30. $2\pi hr^2 - 4\pi hr + 2\pi h$
31. $2y^2 - 3y$ **32.** $-x^2 - 8x - 16$

Simplifying Ratios

The ratio of the length of the shorter leg to the length of the longer leg for this right triangle is 4 to 6. This ratio can be written in three ways.

4 to 6 $\qquad \frac{4}{6} \qquad$ 4 : 6

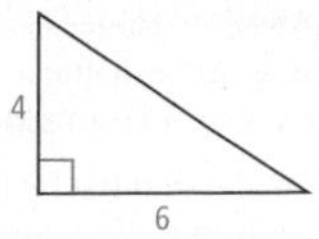

Example

Algebra **Simplify each ratio.**

a. **4 to 6**

$4 \text{ to } 6 = \frac{4}{6}$

$= \frac{2 \cdot 2}{2 \cdot 3}$ Find and remove the common factor.

$= \frac{2}{3}$

b. $3ab : 27ab$

$3ab : 27ab = \frac{3ab}{27ab}$

$= \frac{3ab \cdot 1}{3ab \cdot 9}$

$= \frac{1}{9}$

c. $\frac{4a + 4b}{a + b}$

$\frac{4a + 4b}{a + b} = \frac{4(a + b)}{a + b}$ Factor the numerator. The denominator cannot be factored. Remove the common factor $(a + b)$.

$= 4$

Exercises

Algebra **Simplify each ratio.**

1. 25 to 15 **2.** 6 : 9 **3.** $\frac{36}{54}$

4. 0.8 to 2.4 **5.** $\frac{7}{14x}$ **6.** $\frac{12c}{14c}$

7. $22x^2$ to $35x$ **8.** $0.5ab : 8ab$ **9.** $\frac{4xy}{0.25x}$

10. $1\frac{1}{2}x$ to $5x$ **11.** $\frac{x^2 + x}{2x}$ **12.** $\frac{1}{4}r^2$ to $6r$

13. $0.72t : 7.2t^2$ **14.** $(2x - 6) : (6x - 4)$ **15.** $12xy : 8x$

16. $(9x - 9y)$ to $(x - y)$ **17.** $\frac{\pi r}{r^2 + \pi r}$ **18.** $\frac{8ab}{32xy}$

Express each ratio in simplest form.

19. shorter leg : longer leg **20.** hypotenuse to shorter leg

21. $\frac{\text{shorter leg}}{\text{hypotenuse}}$ **22.** $\frac{\text{longer leg}}{\text{hypotenuse}}$

23. longer leg to shorter leg **24.** hypotenuse : longer leg

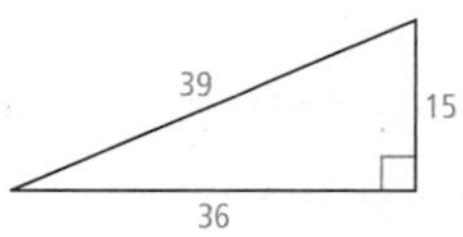

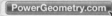

Simplifying Ratios

1. $\frac{5}{3}$ **2.** $\frac{2}{3}$

3. $\frac{2}{3}$ **4.** $\frac{1}{3}$

5. $\frac{1}{2x}$ **6.** $\frac{6}{7}$

7. $\frac{22x}{35}$ **8.** $\frac{1}{16}$

9. $16y$ **10.** $\frac{3}{10}$

11. $\frac{x + 1}{2}$ **12.** $\frac{r}{24}$

13. $\frac{1}{10t}$ **14.** $\frac{x - 3}{3x - 2}$

15. $\frac{3}{2}$ **16.** 9

17. $\frac{\pi}{r + \pi}$ **18.** $\frac{ab}{4xy}$

19. $\frac{5}{12}$ **20.** $\frac{13}{5}$

21. $\frac{5}{13}$ **22.** $\frac{12}{13}$

23. $\frac{12}{5}$ **24.** $\frac{13}{12}$

Absolute Value

Absolute value is used to represent the distance of a number from 0 on a number line. Since distance is always referred to as a nonnegative number, the absolute value of an expression is nonnegative.

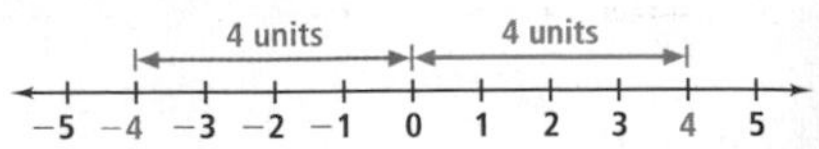

On the number line at the right, both 4 and −4 are four units from zero. Therefore, $|4|$ and $|-4|$ are both equal to four.

When working with more complicated expressions, always remember to simplify within absolute value symbols first.

Example 1

Simplify each expression.

a. $|4| + |-19|$

$|4| + |-19| = 4 + 19$

$= 23$

b. $|4 - 8|$

$|4 - 8| = |-4|$

$= 4$

c. $-3|-7-4|$

$-3|-7-4| = -3|-11|$

$= -3 \cdot 11$

$= -33$

To solve the absolute value equation $|x| = a$, find all the values x that are a units from 0 on a number line.

Example 2

Algebra Solve.

a. $|x| = 7$

$x = 7$ or -7

b. $|x| - 3 = 22$

$|x| - 3 = 22$

$|x| = 25$

$x = 25$ or -25

Exercises

Simplify each expression.

1. $|-8|$ **2.** $|11|$ **3.** $|-7| + |15|$ **4.** $|-12| - |-12|$

5. $|-5| - |10|$ **6.** $|-4| + |-2|$ **7.** $10 - |-20|$ **8.** $|-9| - 15$

9. $|4 - 17|$ **10.** $|-9 - 11|$ **11.** $2|-21 + 16|$ **12.** $-8|-9 + 4|$

Algebra Solve.

13. $|x| = 16$ **14.** $1 = |x|$ **15.** $|x| + 7 = 27$ **16.** $|x| - 9 = 15$

Answers

Absolute Value

1. 8 **2.** 11
3. 22 **4.** 0
5. −5 **6.** 6
7. −10 **8.** −6
9. 13 **10.** 20
11. 10 **12.** −40
13. −16 or 16 **14.** −1 or 1
15. −20 or 20 **16.** −24 or 24

The Coordinate Plane

Two number lines that intersect at right angles form a coordinate plane. The horizontal axis is the x-axis and the vertical axis is the y-axis. The axes intersect at the origin and divide the coordinate plane into four sections called quadrants.

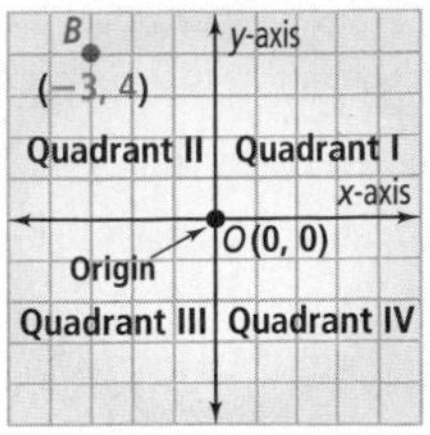

An ordered pair of numbers names the location of a point in the plane. These numbers are the coordinates of the point. Point B has coordinates $(-3, 4)$.

The first coordinate is the x-coordinate. $(-3, 4)$ The second coordinate is the y-coordinate.

You use the x-coordinate to tell how far to move right (positive) or left (negative) from the origin. You then use the y-coordinate to tell how far to move up (positive) or down (negative) to reach the point (x, y).

Example 1

Graph each point in the coordinate plane. In which quadrant or on which axis would you find each point?

a. Graph point $A(-2, 3)$ in the coordinate plane.

To graph $A(-2, 3)$, move 2 units to the left of the origin. Then move 3 units up. Since the x-coordinate is negative and the y-coordinate is positive, point A is in Quadrant II.

b. Graph point $B(2, 0)$ in the coordinate plane.

To graph $B(2, 0)$, move 2 units to the right of the origin. Since the y-coordinate is 0, point B is on the x-axis.

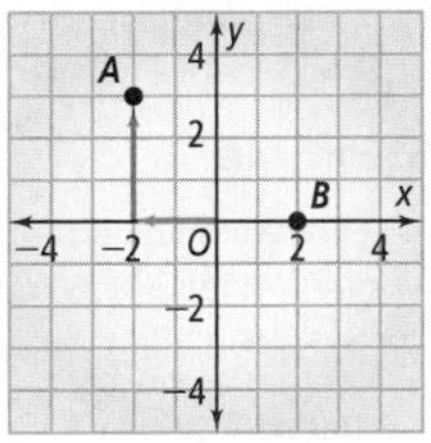

Exercises

Name the coordinates of each point in the coordinate plane at the right.

1. S **2.** T **3.** U **4.** V

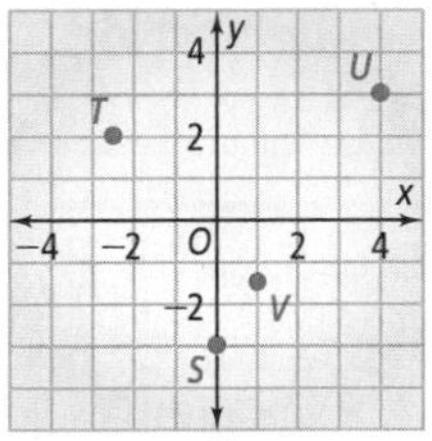

Graph each ordered pair in the same coordinate plane.

5. $(0, -5)$ **6.** $(4, -1)$ **7.** $(-2, -2)$ **8.** $\left(-1\frac{1}{2}, 4\right)$

In which quadrant or on which axis would you find each point?

9. $(0, 10)$ **10.** $\left(1\frac{1}{2}, -3\right)$ **11.** $(-5, 0)$ **12.** $(-9, -2)$

The Coordinate Plane

1. $(0, -3)$

2. $\left(-2\frac{1}{2}, 2\right)$

3. $(4, 3)$

4. $\left(1, -1\frac{1}{2}\right)$

5–8.

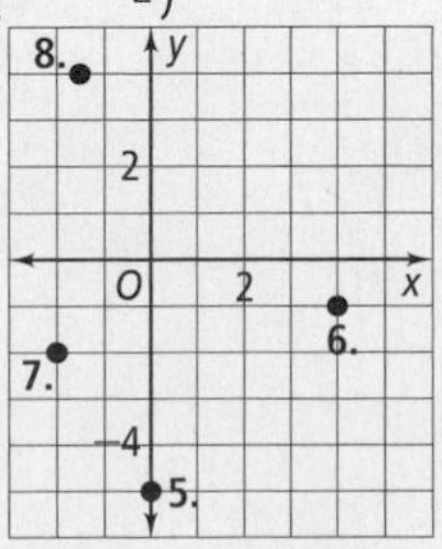

9. y-axis **10.** IV

11. x-axis **12.** III

Skills Handbook

Solving and Writing Linear Equations

To solve a linear equation, use the properties of equality and properties of real numbers to find the value of the variable that satisfies the equation.

Example 1

Algebra **Solve each equation.**

a. $5x - 3 = 2$

$5x - 3 = 2$

$5x = 5$ Add 3 to each side.

$x = 1$ Divide each side by 5.

b. $1 - 2(x + 1) = x$

$1 - 2(x + 1) = x$

$1 - 2x - 2 = x$ Use the Distributive Property.

$-1 - 2x = x$ Simplify the left side.

$-1 = 3x$ Add $2x$ to each side.

$-\frac{1}{3} = x$ Divide each side by 3.

You will sometimes need to translate word problems into equations. Look for words that suggest a relationship or some type of mathematical operation.

Example 2

Algebra **A student has grades of 80, 65, 78, and 92 on four tests. What is the minimum grade she must earn on her next test to ensure an average of 80?**

Relate average of 80, 65, 78, 92, and next test is 80 — Pull out the key words and numbers.

Define Let x = the grade on the next test. — Let a variable represent what you are looking for.

Write $\frac{80 + 65 + 78 + 92 + x}{5} = 80$ Write an equation.

$\frac{315 + x}{5} = 80$ Combine like terms.

$315 + x = 400$ Multiply each side by 5.

$x = 85$ Subtract 315 from each side.

The student must earn 85 on the next test for an average of 80.

Exercises

Algebra **Solve each equation.**

1. $3n + 2 = 17$ **2.** $5a - 2 = -12$ **3.** $2x + 4 = 10$

4. $3(n - 4) = 15$ **5.** $4 + 2y = 8y$ **6.** $-6z + 1 = 13 - 3z$

7. $6 - (3t + 4) = t$ **8.** $7 = -2(4n - 4.5)$ **9.** $(w + 5) - 5 = (2w + 5)$

10. $\frac{5}{7}p - 10 = 30$ **11.** $\frac{m}{-3} - 3 = 1$ **12.** $5k + 2(k + 1) = 23$

13. Twice a number subtracted from 35 is 9. What is the number?

14. The Johnsons pay $9.95 a month plus $.035 per min for local phone service. Last month, they paid $12.75. How many minutes of local calls did they make?

Answers

Solving and Writing Linear Equations

1. 5 **2.** -2

3. 3 **4.** 9

5. $\frac{2}{3}$ **6.** -4

7. $\frac{1}{2}$ **8.** $\frac{1}{4}$

9. -5 **10.** 56

11. -12 **12.** 3

13. $35 - 2x = 9$; 13

14. $\$9.95 + \$.035m = \$12.75$; 80 min

Percents

A percent is a ratio in which a number is compared to 100. For example, the expression *60 percent* means "60 out of 100." The symbol % stands for "percent."

A percent can be written in decimal form by first writing it in ratio form, and then writing the ratio as a decimal. For example, 25% is equal to the ratio $\frac{25}{100}$ or $\frac{1}{4}$. As a decimal, $\frac{1}{4}$ is equal to 0.25. Note that 25% can also be written directly as a decimal by moving the decimal point two places to the left.

Example 1

Convert each percent to a decimal.

a. 42%
42% = 0.42

b. 157%
157% = 1.57

c. 12.4%
12.4% = 0.124

d. 4%
4% = 0.04

To calculate a percent of a number, write the percent as a decimal and multiply.

Example 2

Simplify. Where necessary, round to the nearest tenth.

a. 30% of 242

$30\% \text{ of } 242 = 0.3 \cdot 242$
$= 72.6$

b. 7% of 38

$7\% \text{ of } 38 = 0.07 \cdot 38$
$= 2.66 \approx 2.7$

For a percent problem, it is a good idea to check that your answer is reasonable by estimating it.

Example 3

Estimate 23% of 96.

$23\% \approx 25\%$ and $96 \approx 100$. So 25% $\left(\text{or } \frac{1}{4}\right)$ of 100 = 25.
A reasonable estimate is 25.

Exercises

Convert each percent to a decimal.

1. 50% **2.** 27% **3.** 6% **4.** 84.6% **5.** 109% **6.** 2.5%

Simplify. Where necessary, round to the nearest tenth.

7. 21% of 40 **8.** 45% of 200 **9.** 6% of 120 **10.** 23.8% of 176

Estimate.

11. 12% of 70 **12.** 48% of 87 **13.** 73% of 64 **14.** 77% of 42

Skills Handbook

Percents

1. 0.5 **2.** 0.27
3. 0.06 **4.** 0.846
5. 1.09 **6.** 0.025
7. 8.4 **8.** 90
9. 7.2 **10.** 41.9
11–14. Answers may vary. Samples are given.
11. 7 **12.** 43
13. 45 **14.** 32

Reference

Table 1 Measures

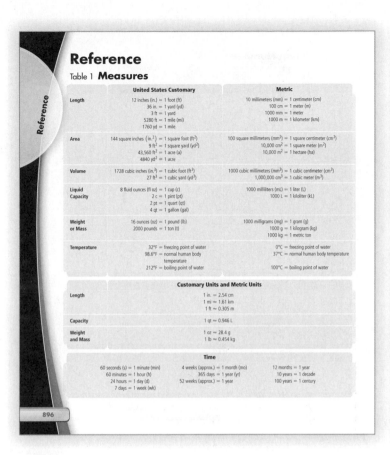

	United States Customary	Metric
Length	12 inches (in.) = 1 foot (ft) 36 in. = 1 yard (yd) 3 ft = 1 yard 5280 ft = 1 mile (mi) 1760 yd = 1 mile	10 millimeters (mm) = 1 centimeter (cm) 100 cm = 1 meter (m) 1000 mm = 1 meter 1000 m = 1 kilometer (km)
Area	144 square inches $(in.^2)$ = 1 square foot (ft^2) 9 ft^2 = 1 square yard (yd^2) 43,560 ft^2 = 1 acre (a) 4840 yd^2 = 1 acre	100 square millimeters (mm^2) = 1 square centimeter (cm^2) 10,000 cm^2 = 1 square meter (m^2) 10,000 m^2 = 1 hectare (ha)
Volume	1728 cubic inches $(in.^3)$ = 1 cubic foot (ft^3) 27 ft^3 = 1 cubic yard (yd^3)	1000 cubic millimeters (mm^3) = 1 cubic centimeter (cm^3) 1,000,000 cm^3 = 1 cubic meter (m^3)
Liquid Capacity	8 fluid ounces (fl oz) = 1 cup (c) 2 c = 1 pint (pt) 2 pt = 1 quart (qt) 4 qt = 1 gallon (gal)	1000 milliliters (mL) = 1 liter (L) 1000 L = 1 kiloliter (kL)
Weight or Mass	16 ounces (oz) = 1 pound (lb) 2000 pounds = 1 ton (t)	1000 milligrams (mg) = 1 gram (g) 1000 g = 1 kilogram (kg) 1000 kg = 1 metric ton
Temperature	32°F = freezing point of water 98.6°F = normal human body temperature 212°F = boiling point of water	0°C = freezing point of water 37°C = normal human body temperature 100°C = boiling point of water

	Customary Units and Metric Units
Length	1 in. = 2.54 cm 1 mi ≈ 1.61 km 1 ft ≈ 0.305 m
Capacity	1 qt ≈ 0.946 L
Weight and Mass	1 oz ≈ 28.4 g 1 lb ≈ 0.454 kg

Time		
60 seconds (s) = 1 minute (min)	4 weeks (approx.) = 1 month (mo)	12 months = 1 year
60 minutes = 1 hour (h)	365 days = 1 year (yr)	10 years = 1 decade
24 hours = 1 day (d)	52 weeks (approx.) = 1 year	100 years = 1 century
7 days = 1 week (wk)		

Table 2 Formulas

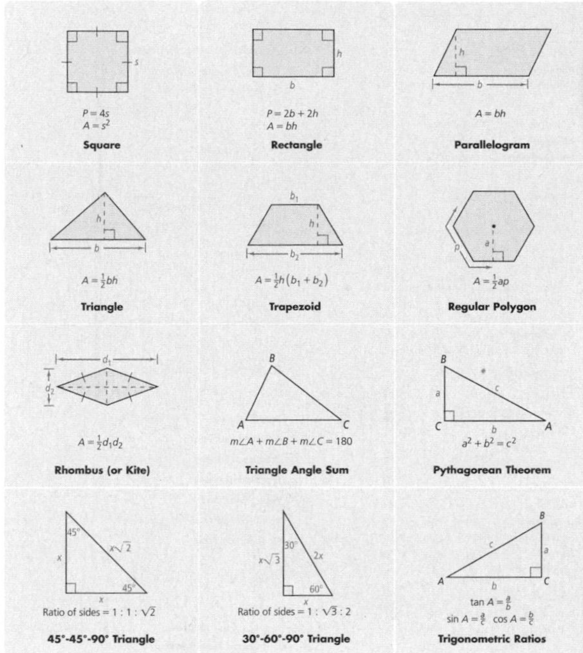

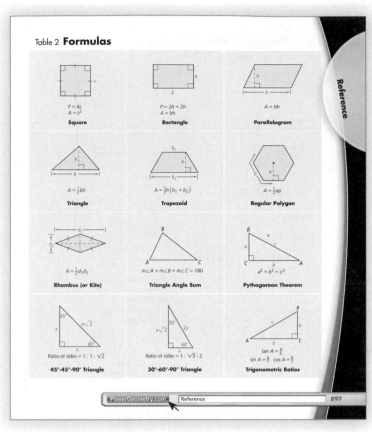

Square: $P = 4s$, $A = s^2$

Rectangle: $P = 2b + 2h$, $A = bh$

Parallelogram: $A = bh$

Triangle: $A = \frac{1}{2}bh$

Trapezoid: $A = \frac{1}{2}h(b_1 + b_2)$

Regular Polygon: $A = \frac{1}{2}ap$

Rhombus (or Kite): $A = \frac{1}{2}d_1d_2$

Triangle Angle Sum: $m\angle A + m\angle B + m\angle C = 180$

Pythagorean Theorem: $a^2 + b^2 = c^2$

45°-45°-90° Triangle: Ratio of sides = $1 : 1 : \sqrt{2}$

30°-60°-90° Triangle: Ratio of sides = $1 : \sqrt{3} : 2$

Trigonometric Ratios: $\tan A = \frac{a}{b}$, $\sin A = \frac{a}{c}$, $\cos A = \frac{b}{c}$

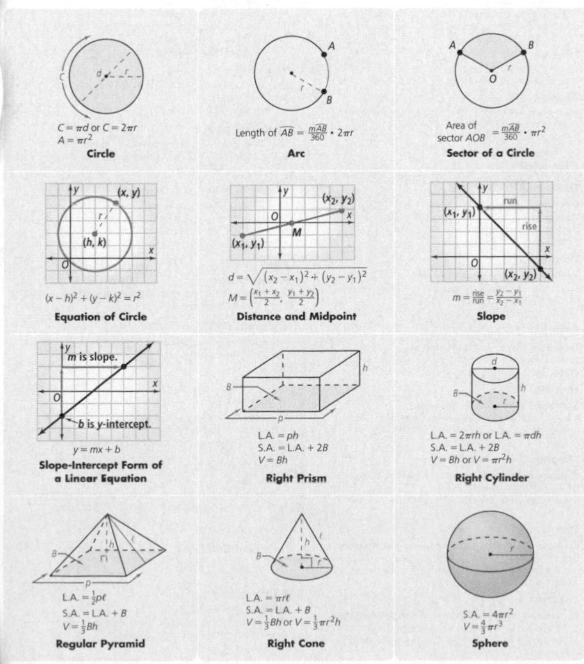

Circle: $C = \pi d$ or $C = 2\pi r$, $A = \pi r^2$

Arc: Length of $\widehat{AB} = \frac{m\widehat{AB}}{360} \cdot 2\pi r$

Sector of a Circle: Area of sector $AOB = \frac{m\widehat{AB}}{360} \cdot \pi r^2$

Equation of Circle: $(x - h)^2 + (y - k)^2 = r^2$

Distance and Midpoint: $d = \sqrt{(x_2 - x_1)^2 + (y_2 - y_1)^2}$, $M = \left(\frac{x_1 + x_2}{2}, \frac{y_1 + y_2}{2}\right)$

Slope: $m = \frac{\text{rise}}{\text{run}} = \frac{y_2 - y_1}{x_2 - x_1}$

Slope-Intercept Form of a Linear Equation: $y = mx + b$

Right Prism: L.A. $= ph$, S.A. $=$ L.A. $+ 2B$, $V = Bh$

Right Cylinder: L.A. $= 2\pi rh$ or L.A. $= \pi dh$, S.A. $=$ L.A. $+ 2B$, $V = Bh$ or $V = \pi r^2 h$

Regular Pyramid: L.A. $= \frac{1}{2}p\ell$, S.A. $=$ L.A. $+ B$, $V = \frac{1}{3}Bh$

Right Cone: L.A. $= \pi r\ell$, S.A. $=$ L.A. $+ B$, $V = \frac{1}{3}Bh$ or $V = \frac{1}{3}\pi r^2 h$

Sphere: S.A. $= 4\pi r^2$, $V = \frac{4}{3}\pi r^3$

Table 3 Reading Math Symbols

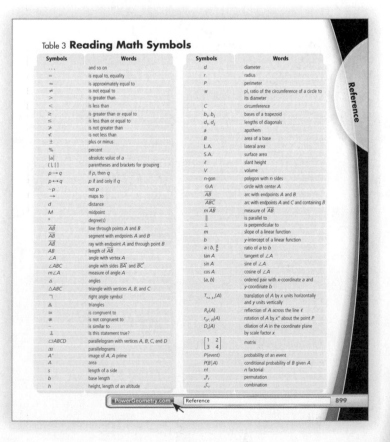

Symbols	Words
. . .	and so on
=	is equal to, equality
≈	is approximately equal to
≠	is not equal to
>	is greater than
<	is less than
≥	is greater than or equal to
≤	is less than or equal to
≯	is not greater than
≮	is not less than
±	plus or minus
%	percent
$\|a\|$	absolute value of a
( ), [ ]	parentheses and brackets for grouping
$p \rightarrow q$	if p, then q
$p \leftrightarrow q$	p if and only if q
$\sim p$	not p
→	maps to
d	distance
M	midpoint
°	degree(s)
$\overleftrightarrow{AB}$	line through points A and B
$\overline{AB}$	segment with endpoints A and B
$\overrightarrow{AB}$	ray with endpoint A and through point B
AB	length of $\overline{AB}$
$\angle A$	angle with vertex A
$\angle ABC$	angle with sides $\overrightarrow{BA}$ and $\overrightarrow{BC}$
$m\angle A$	measure of angle A
∡s	angles
$\triangle ABC$	triangle with vertices A, B, and C
⌝	right angle symbol
△s	triangles
≅	is congruent to
≇	is not congruent to
~	is similar to
≟	Is this statement true?
▱$ABCD$	parallelogram with vertices A, B, C, and D
▱s	parallelograms
A'	image of A, A prime
A	area
s	length of a side
b	base length
h	height, length of an altitude
d	diameter
r	radius
P	perimeter
π	pi, ratio of the circumference of a circle to its diameter
C	circumference
b_1, b_2	bases of a trapezoid
d_1, d_2	lengths of diagonals
a	apothem
B	area of a base
L.A.	lateral area
S.A.	surface area
ℓ	slant height
V	volume
n-gon	polygon with n sides
$\odot A$	circle with center A
$\widehat{AB}$	arc with endpoints A and B
$\widehat{ABC}$	arc with endpoints A and C and containing B
$m\widehat{AB}$	measure of $\widehat{AB}$
∥	is parallel to
⊥	is perpendicular to
m	slope of a linear function
b	y-intercept of a linear function
$a : b$, $\frac{a}{b}$	ratio of a to b
$\tan A$	tangent of $\angle A$
$\sin A$	sine of $\angle A$
$\cos A$	cosine of $\angle A$
(a, b)	ordered pair with x-coordinate a and y-coordinate b
$T_{\langle x, y \rangle}(A)$	translation of A by x units horizontally and y units vertically
$R_\ell(A)$	reflection of A across the line ℓ
$r_{(x°, P)}(A)$	rotation of A by $x°$ about the point P
$D_x(A)$	dilation of A in the coordinate plane by scale factor x
$\begin{bmatrix} 1 & 2 \\ 3 & 4 \end{bmatrix}$	matrix
P(event)	probability of an event
$P(B \mid A)$	conditional probability of B given A
$n!$	n factorial
$_nP_r$	permutation
$_nC_r$	combination

Table 4 **Properties of Real Numbers**

Unless otherwise stated, *a*, *b*, *c*, and *d* represent real numbers.

Identity Properties

Addition $a + 0 = a$ and $0 + a = a$

Multiplication $a \cdot 1 = a$ and $1 \cdot a = a$

Commutative Properties

Addition $a + b = b + a$

Multiplication $a \cdot b = b \cdot a$

Associative Properties

Addition $(a + b) + c = a + (b + c)$

Multiplication $(a \cdot b) \cdot c = a \cdot (b \cdot c)$

Inverse Properties

Addition The sum of a number and its *opposite*, or *additive inverse*, is zero.

$a + (-a) = 0$ and $-a + a = 0$

Multiplication The *reciprocal*, or *multiplicative inverse*, of a rational number $\frac{a}{b}$ is $\frac{b}{a}$ $(a, b \neq 0)$.

$a \cdot \frac{1}{a} = 1$ and $\frac{1}{a} \cdot a = 1$ $(a \neq 0)$

Distributive Properties

$a(b + c) = ab + ac$ $(b + c)a = ba + ca$

$a(b - c) = ab - ac$ $(b - c)a = ba - ca$

Properties of Equality

Addition If $a = b$, then $a + c = b + c$.

Subtraction If $a = b$, then $a - c = b - c$.

Multiplication If $a = b$, then $a \cdot c = b \cdot c$.

Division If $a = b$ and $c \neq 0$, then $\frac{a}{c} = \frac{b}{c}$.

Substitution If $a = b$, then b can replace a in any expression.

Reflexive $a = a$

Symmetric If $a = b$, then $b = a$.

Transitive If $a = b$ and $b = c$, then $a = c$.

Properties of Proportions

$\frac{a}{b} = \frac{c}{d}$ $(a, b, c, d \neq 0)$ is equivalent to

(1) $ad = bc$ (2) $\frac{b}{a} = \frac{d}{c}$

(3) $\frac{a}{c} = \frac{b}{d}$ (4) $\frac{a+b}{b} = \frac{c+d}{d}$

Zero-Product Property

If $ab = 0$, then $a = 0$ or $b = 0$.

Properties of Inequality

Addition If $a > b$ and $c \geq d$, then $a + c > b + d$.

Multiplication If $a > b$ and $c > 0$, then $ac > bc$. If $a > b$ and $c < 0$, then $ac < bc$.

Transitive If $a > b$ and $b > c$, then $a > c$.

Comparison If $a = b + c$, and $c > 0$, then $a > b$.

Properties of Exponents

For any nonzero numbers *a* and *b*, any positive number *c*, and any integers *m* and *n*,

Zero Exponent $a^0 = 1$

Negative Exponent $a^{-n} = \frac{1}{a^n}$

Product of Powers $a^m \cdot a^n = a^{m+n}$

Quotient of Powers $\frac{a^m}{a^n} = a^{m-n}$

Power to a Power $(c^m)^n = c^{mn}$

Product to a Power $(ab)^n = a^n b^n$

Quotient to a Power $\left(\frac{a}{b}\right)^n = \frac{a^n}{b^n}$

Properties of Square Roots

For any nonnegative numbers *a* and *b*, and any positive number *c*,

Product of Square Roots $\sqrt{a} \cdot \sqrt{b} = \sqrt{ab}$

Quotient of Square Roots $\frac{\sqrt{a}}{\sqrt{c}} = \sqrt{\frac{a}{c}}$

Table 5 **Random Numbers**

18823	18160	93593	67294	19632	62617	86779	74024
65358	70469	87149	89509	72176	18103	55169	79954
72002	20582	25325	22451	22445	62132	81638	36566
42709	33717	59943	12027	46547	72749	13347	65030
26128	49067	27904	49953	74674	94617	13317	58697
31973	06303	94202	62287	56164	79157	98375	24558
99241	38449	46438	91579	01907	72146	05764	22400
94490	49833	09258	11873	57196	32209	67663	07990
12288	59245	83638	23642	61715	35483	84563	79956
88618	54619	24853	59783	47537	88822	47227	09262
25041	57862	19203	86103	02800	23198	70639	43757
52064	75820	50994	31050	67304	16730	29373	96700
07845	69584	70548	52973	72302	97594	92241	15204
42665	29990	57260	75846	01152	30141	35982	96088
04003	36893	51639	65625	28426	90634	32979	05449
32959	06776	72420	55622	81422	67587	93193	67479
29041	35939	80920	31801	38638	87905	37617	53135
63364	20495	50868	54130	32625	30799	94255	03514
27838	19139	82031	46143	93922	32001	05378	42457
94248	29387	32682	86235	35805	66529	00886	25875
40156	92636	95648	79767	16307	71133	15714	44142
44293	19195	30569	41277	01417	34656	80207	33362
71878	31767	40056	52582	30766	70264	86253	07179
24757	57502	51033	16551	66731	87844	41420	10084
55529	68560	50069	50652	76104	42086	48720	96632
39724	50318	91370	68016	06222	26806	86726	52832
80950	27135	14110	92292	17049	60257	01638	04460
21694	79570	74409	95087	75424	57042	27349	16229
06930	85441	37191	75134	12845	67868	51500	97761
18740	35448	56096	37910	35485	19640	07689	31027
40657	14875	70695	92569	40703	69318	95070	01541
52249	56515	59058	34509	35791	22150	56558	75286
86570	07303	40560	57856	22009	67712	19435	90250
62962	66253	93288	01838	68388	55481	00336	19271
78066	09117	62350	58972	80778	46458	83677	16125
89106	30219	30068	54030	49295	48985	01624	72881
88310	18172	89450	04987	02781	37935	76222	93595
20942	90911	57643	34009	20728	88785	81212	08214
93926	66687	58252	18674	18501	22362	37319	33201
88294	55814	67443	77285	36229	26886	66782	89931
29751	08485	49910	83844	56013	26596	20875	34568
11169	15529	33241	83594	01727	86595	65723	82322
06062	54400	80649	70749	50395	48993	77447	24862
87445	17139	43278	55031	79971	18515	61850	49101
39283	22821	44330	82225	53534	77235	42973	60190

Postulates, Theorems, and Constructions

Chapter 1 Tools of Geometry

Postulate 1-1

Through any two points there is exactly one line. (p. 13)

Postulate 1-2

If two distinct lines intersect, then they intersect in exactly one point. (p. 13)

Postulate 1-3

If two distinct planes intersect, then they intersect in exactly one line. (p. 14)

Postulate 1-4

Through any three noncollinear points there is exactly one plane. (p. 15)

Postulate 1-5

Ruler Postulate

Every point on a line can be paired with a real number. This makes a one-to-one correspondence between the points on the line and the real numbers. (p. 20)

Postulate 1-6

Segment Addition Postulate

If three points *A*, *B*, and *C* are collinear and *B* is between *A* and *C*, then $AB + BC = AC$. (p. 21)

Postulate 1-7

Protractor Postulate

Consider $\overrightarrow{OB}$ and a point *A* on one side of $\overrightarrow{OB}$. Every ray of the form $\overrightarrow{OA}$ can be paired one to one with a real number from 0 to 180. (p. 28)

Postulate 1-8

Angle Addition Postulate

If point *B* is in the interior of $\angle AOC$, then $m\angle AOB + m\angle BOC = m\angle AOC$. (p. 30)

Postulate 1-9

Linear Pair Postulate

If two angles form a linear pair, then they are supplementary. (p. 36)

The Midpoint Formulas

On a Number Line

The coordinate of the midpoint *M* of $\overline{AB}$ is $\frac{a+b}{2}$.

In the Coordinate Plane

Given $\overline{AB}$ where $A(x_1, y_1)$ and $B(x_2, y_2)$, the coordinates of the midpoint of $\overline{AB}$ are $M\left(\frac{x_1 + x_2}{2}, \frac{y_1 + y_2}{2}\right)$. (p. 50)

The Distance Formula

The distance between two points $A(x_1, y_1)$ and $B(x_2, y_2)$ is $d = \sqrt{(x_2 - x_1)^2 + (y_2 - y_1)^2}$. (p. 52)

- Proof on p. 497, Exercise 35

The Distance Formula (Three Dimensions)

In a three-dimensional coordinate system, the distance between two points (x_1, y_1, z_1) and (x_2, y_2, z_2) can be found with this extension of the Distance Formula.

$d = \sqrt{(x_2 - x_1)^2 + (y_2 - y_1)^2 + (z_2 - z_1)^2}$ (p. 56)

Postulate 1-10

Area Addition Postulate

The area of a region is the sum of the areas of its nonoverlapping parts. (p. 63)

Chapter 2 Reasoning and Proof

Law of Detachment

If the hypothesis of a true conditional is true, then the conclusion is true. In symbolic form:
If $p \rightarrow q$ is true and p is true, then q is true. (p. 106)

Law of Syllogism

If $p \rightarrow q$ is true and $q \rightarrow r$ is true, then $p \rightarrow r$ is true. (p. 108)

Properties of Congruence

Reflexive Property

$\overline{AB} \cong \overline{AB}$ and $\angle A \cong \angle A$

Symmetric Property

If $\overline{AB} \cong \overline{CD}$, then $\overline{CD} \cong \overline{AB}$.
If $\angle A \cong \angle B$, then $\angle B \cong \angle A$.

Transitive Property

If $\overline{AB} \cong \overline{CD}$, and $\overline{CD} \cong \overline{EF}$, then $\overline{AB} \cong \overline{EF}$.
If $\angle A \cong \angle B$, and $\angle B \cong \angle C$, then $\angle A \cong \angle C$.
If $\angle B \cong \angle A$, and $\angle B \cong \angle C$, then $\angle A \cong \angle C$. (p. 114)

Theorem 2-1

Vertical Angles Theorem

Vertical angles are congruent. (p. 120)

- Proof on p. 121

Theorem 2-2

Congruent Supplements Theorem

If two angles are supplements of the same angle (or of congruent angles), then the two angles are congruent. (p. 122)

- Proof on p. 123, Problem 3

Theorem 2-3

Congruent Complements Theorem

If two angles are complements of the same angle (or of congruent angles), then the two angles are congruent. (p. 123)

- Proof on p. 125, Exercise 13

Theorem 2-4

All right angles are congruent. (p. 123)

- Proof on p. 125, Exercise 18

Theorem 2-5

If two angles are congruent and supplementary, then each is a right angle. (p. 123)

- Proof on p. 126, Exercise 23

Chapter 3 Parallel and Perpendicular Lines

Theorem 3-2

Corresponding Angles Theorem

If a transversal intersects two parallel lines, then corresponding angles are congruent. (p. 149)

- Proof on p. 155, Exercise 25

Theorem 3-1

Alternate Interior Angles Theorem

If a transversal intersects two parallel lines, then alternate interior angles are congruent. (p. 149)

- Proof on p. 150

Postulate 3-1

Same-Side Interior Angles Postulate

If a transversal intersects two parallel lines, then same-side interior angles are supplementary. (p. 148)

Theorem 3-3

Alternate Exterior Angles Theorem

If a transversal intersects two parallel lines, then alternate exterior angles are congruent. (p. 151)

- Proof on p. 150, Got It 2

Theorem 3-4

Converse of the Corresponding Angles Theorem

If two lines and a transversal form corresponding angles that are congruent, then the lines are parallel. (p. 156)

- Proof on p. 161, Exercise 29

Theorem 3-5

Converse of the Alternate Interior Angles Theorem

If two lines and a transversal form alternate interior angles that are congruent, then the two lines are parallel. (p. 157)

- Proof on p. 158

Theorem 3-6

Converse of the Same-Side Interior Angles Postulate

If two lines and a transversal form same-side interior angles that are supplementary, then the two lines are parallel. (p. 157)

- Proof on p. 158, Got It 2

Theorem 3-7

Converse of the Alternate Exterior Angles Theorem

If two lines and a transversal form alternate exterior angles that are congruent, then the two lines are parallel. (p. 157)

- Proof on p. 158, Problem 2

Theorem 3-8

If two lines are parallel to the same line, then they are parallel to each other. (p. 164)

- Proof on p. 167, Exercise 7

Theorem 3-9

In a plane, if two lines are perpendicular to the same line, then they are parallel to each other. (p. 165)

- Proof on p. 165

Theorem 3-10

Perpendicular Transversal Theorem

In a plane, if a line is perpendicular to one of two parallel lines, then it is perpendicular to the other. (p. 166)

- Proof on p. 168, Exercise 10

Postulate 3-2

Parallel Postulate

Through a point not on a line, there is one and only one line parallel to the given line. (p. 171)

Theorem 3-11

Triangle Angle-Sum Theorem

The sum of the measures of the angles of a triangle is 180. (p. 172)

- Proof on p. 172

Theorem 3-12

Triangle Exterior Angle Theorem

The measure of each exterior angle of a triangle equals the sum of the measures of its two remote interior angles. (p. 173)

- Proof on p. 177, Exercise 33

Corollary

The measure of an exterior angle of a triangle is greater than the measure of each of its remote interior angles. (p. 325)

- Proof on p. 325

Spherical Geometry Parallel Postulate

Through a point not on a line, there is no line parallel to the given line. (p. 179)

Postulate 3-3

Perpendicular Postulate

Through a point not on a line, there is one and only one line perpendicular to the given line. (p. 184)

Slopes of Parallel Lines
If two nonvertical lines are parallel, then their slopes are equal. If the slopes of two distinct nonvertical lines are equal, then the lines are parallel. Any two vertical lines or horizontal lines are parallel. (p. 197)
- Proofs on p. 457, Exercises 33, 34

Slopes of Perpendicular Lines
If two nonvertical lines are perpendicular, then the product of their slopes is -1. If the slopes of two lines have a product of -1, then the lines are perpendicular. Any horizontal line and vertical line are perpendicular. (p. 198)
- Proofs on p. 418, Exercise 28; p. 497, Exercise 51; p. 466, Exercise 44

Chapter 4 Congruent Triangles

Theorem 4-1
Third Angles Theorem
If the two angles of one triangle are congruent to two angles of another triangle, then the third angles are congruent. (p. 220)
- Proof on p. 220

Postulate 4-1
Side-Side-Side (SSS) Postulate
If the three sides of one triangle are congruent to the three sides of another triangle, then the two triangles are congruent. (p. 227)

Postulate 4-2
Side-Angle-Side (SAS) Postulate
If two sides and the included angle of one triangle are congruent to two sides and the included angle of another triangle, then the two triangles are congruent. (p. 228)

Postulate 4-3
Angle-Side-Angle (ASA) Postulate
If two angles and the included side of one triangle are congruent to two angles and the included side of another triangle, then the two triangles are congruent. (p. 234)

Theorem 4-2
Angle-Angle-Side (AAS) Theorem
If two angles and a nonincluded side of one triangle are congruent to two angles and the corresponding nonincluded side of another triangle, then the triangles are congruent. (p. 236)
- Proof on p. 236

Theorem 4-3
Isosceles Triangle Theorem
If two sides of a triangle are congruent, then the angles opposite those sides are congruent. (p. 250)
- Proofs on p. 251; p. 255, Exercise 22

Corollary
If a triangle is equilateral, then the triangle is equiangular. (p. 252)
- Proof on p. 255, Exercise 24

Theorem 4-4
Converse of the Isosceles Triangle Theorem
If two angles of a triangle are congruent, then the sides opposite the angles are congruent. (p. 251)
- Proof on p. 255, Exercise 23

Corollary
If a triangle is equiangular, then the triangle is equilateral. (p. 252)
- Proof on p. 255, Exercise 24

Theorem 4-5
If a line bisects the vertex angle of an isosceles triangle, then the line is also the perpendicular bisector of the base. (p. 252)
- Proof on p. 255, Exercise 26

Theorem 4-6
Hypotenuse-Leg (HL) Theorem
If the hypotenuse and a leg of one right triangle are congruent to the hypotenuse and a leg of another right triangle, then the triangles are congruent. (p. 259)
- Proof on p. 259

Chapter 5 Relationships Within Triangles

Theorem 5-1
Triangle Midsegment Theorem
If a segment joins the midpoints of two sides of a triangle, then the segment is parallel to the third side and is half as long. (p. 285)
- Proof on p. 415, Got It 2

Theorem 5-2
Perpendicular Bisector Theorem
If a point is on the perpendicular bisector of a segment, then it is equidistant from the endpoints of the segment. (p. 293)
- Proof on p. 298, Exercise 32

Theorem 5-3
Converse of the Perpendicular Bisector Theorem
If a point is equidistant from the endpoints of a segment, then it is on the perpendicular bisector of the segment. (p. 293)
- Proof on p. 298, Exercise 33

Theorem 5-4
Angle Bisector Theorem
If a point is on the bisector of an angle, then the point is equidistant from the sides of the angle. (p. 295)
- Proof on p. 298, Exercise 34

Theorem 5-5
Converse of the Angle Bisector Theorem
If a point in the interior of an angle is equidistant from the sides of the angle, then the point is on the angle bisector. (p. 295)
- Proof on p. 298, Exercise 35

Theorem 5-6
Concurrency of Perpendicular Bisectors Theorem
The perpendicular bisectors of the sides of a triangle are concurrent at a point equidistant from the vertices. (p. 301)
- Proof on p. 302

Theorem 5-7
Concurrency of Angle Bisectors Theorem
The bisectors of the angles of a triangle are concurrent at a point equidistant from the sides of the triangle. (p. 303)
- Proof on p. 306, Exercise 24

Theorem 5-8
Concurrency of Medians Theorem
The medians of a triangle are concurrent at a point that is two-thirds the distance from each vertex to the midpoint of the opposite side. (p. 309)
- Proof on p. 417, Exercise 25

Theorem 5-9
Concurrency of Altitudes Theorem
The lines that contain the altitudes of a triangle are concurrent. (p. 310)
- Proof on p. 417, Exercise 26

Comparison Property of Inequality
If $a = b + c$ and $c > 0$, then $a > b$. (p. 324)
- Proof on p. 324

Theorem 5-10
If two sides of a triangle are not congruent, then the larger angle lies opposite the longer side. (p. 325)
- Proof on p. 330, Exercise 40

Theorem 5-11
If two angles of a triangle are not congruent, then the longer side lies opposite the larger angle. (p. 326)
- Proof on p. 326

Theorem 5-12
Triangle Inequality Theorem
The sum of the lengths of any two sides of a triangle is greater than the length of the third side. (p. 327)
- Proof on p. 331, Exercise 45

Theorem 5-13
The Hinge Theorem (SAS Inequality Theorem)
If two sides of one triangle are congruent to two sides of another triangle and the included angles are not congruent, then the longer third side is opposite the larger included angle. (p. 332)
- Proof on p. 338, Exercise 25

Theorem 5-14
Converse of the Hinge Theorem (SSS Inequality)
If two sides of one triangle are congruent to two sides of another triangle and the third sides are not congruent, then the larger included angle is opposite the longer third side. (p. 334)
- Proof on p. 334

Chapter 6 Polygons and Quadrilaterals

Theorem 6-1
Polygon Angle-Sum Theorem
The sum of the measures of the angles of an n-gon is $(n - 2)180$. (p. 353)
- Proof on p. 357, Exercise 40

Corollary
The measure of each angle of a regular n-gon is $\frac{(n-2)180}{n}$. (p. 354)
- Proof on p. 358, Exercise 43

Theorem 6-2
Polygon Exterior Angle-Sum Theorem
The sum of the measures of the exterior angles of a polygon, one at each vertex, is 360. (p. 355)
- Proofs on p. 352 (using a computer); p. 357, Exercise 39

Theorem 6-3
If a quadrilateral is a parallelogram, then its opposite sides are congruent. (p. 359)
- Proof on p. 360

Theorem 6-4
If a quadrilateral is a parallelogram, then its consecutive angles are supplementary. (p. 360)
- Proof on p. 365, Exercise 32

Theorem 6-5
If a quadrilateral is a parallelogram, then its opposite angles are congruent. (p. 361)
- Proof on p. 361, Problem 2

Theorem 6-6
If a quadrilateral is a parallelogram, then its diagonals bisect each other. (p. 362)
- Proof on p. 364, Exercise 13

Theorem 6-7
If three (or more) parallel lines cut off congruent segments on one transversal, then they cut off congruent segments on every transversal. (p. 363)
- Proof on p. 366, Exercise 43

Theorem 6-8
If both pairs of opposite sides of a quadrilateral are congruent, then the quadrilateral is a parallelogram. (p. 367)
- Proof on p. 373, Exercise 20

Theorem 6-9
If an angle of a quadrilateral is supplementary to both of its consecutive angles, then the quadrilateral is a parallelogram. (p. 368)
- Proof on p. 373, Exercise 21

Theorem 6-10
If both pairs of opposite angles of a quadrilateral are congruent, then the quadrilateral is a parallelogram. (p. 368)
- Proof on p. 373, Exercise 18

Theorem 6-11
If the diagonals of a quadrilateral bisect each other, then the quadrilateral is a parallelogram. (p. 369)
- Proof on p. 369

Theorem 6-12
If one pair of opposite sides of a quadrilateral is both congruent and parallel, then the quadrilateral is a parallelogram. (p. 370)
- Proof on p. 373, Exercise 19

Theorem 6-13
If a parallelogram is a rhombus, then its diagonals are perpendicular. (p. 376)
- Proof on p. 377

Theorem 6-14
If a parallelogram is a rhombus, then each diagonal bisects a pair of opposite angles. (p. 376)
- Proof on p. 381, Exercise 45

Theorem 6-15
If a parallelogram is a rectangle, then its diagonals are congruent. (p. 378)
- Proof on p. 381, Exercise 41

Theorem 6-16
If the diagonals of a parallelogram are perpendicular, then the parallelogram is a rhombus. (p. 383)
- Proof on p. 383

Theorem 6-17
If one diagonal of a parallelogram bisects a pair of opposite angles, then the parallelogram is a rhombus. (p. 384)
- Proof on p. 387, Exercise 23

Theorem 6-18
If the diagonals of a parallelogram are congruent, then the parallelogram is a rectangle. (p. 384)
- Proof on p. 387, Exercise 24

Theorem 6-19
If a quadrilateral is an isosceles trapezoid, then each pair of base angles is congruent. (p. 389)
- Proof on p. 396, Exercise 45

Theorem 6-20
If a quadrilateral is an isosceles trapezoid, then its diagonals are congruent. (p. 391)
- Proof on p. 396, Exercise 53

Theorem 6-21
Trapezoid Midsegment Theorem
If a quadrilateral is a trapezoid, then
(1) the midsegment is parallel to the bases, and
(2) the length of the midsegment is half the sum of the lengths of the bases. (p. 391)
- Proofs on p. 409, Problem 3; p. 415, Problem 2

Theorem 6-22
If a quadrilateral is a kite, then its diagonals are perpendicular. (p. 392)
- Proof on p. 392

Chapter 7 Similarity

Postulate 7-1
Angle-Angle Similarity (AA ~) Postulate
If two angles of one triangle are congruent to two angles of another triangle, then the triangles are similar. (p. 450)

Theorem 7-1
Side-Angle-Side Similarity (SAS ~) Theorem
If an angle of one triangle is congruent to an angle of a second triangle, and the sides that include the two angles are proportional, then the triangles are similar. (p. 451)
- Proof on p. 457, Exercise 35

Theorem 7-2
Side-Side-Side Similarity (SSS ~) Theorem
If the corresponding sides of two triangles are proportional, then the triangles are similar. (p. 451)
- Proof on p. 458, Exercise 36

Theorem 7-3
The altitude to the hypotenuse of a right triangle divides the triangle into two triangles that are similar to the original triangle and to each other. (p. 460)
- Proof on p. 461

Corollary 1
The length of the altitude to the hypotenuse of a right triangle is the geometric mean of the lengths of the segments of the hypotenuse. (p. 462)
- Proof on p. 466, Exercise 42

Corollary 2
The altitude to the hypotenuse of a right triangle separates the hypotenuse so that the length of each leg of the triangle is the geometric mean of the length of the hypotenuse and the length of the segment of the hypotenuse adjacent to the leg. (p. 463)
- Proof on p. 466, Exercise 43

Theorem 7-4
Side-Splitter Theorem
If a line is parallel to one side of a triangle and intersects the other two sides, then it divides those sides proportionally. (p. 471)
- Proof on p. 472

Converse
If a line divides two sides of a triangle proportionally, then it is parallel to the third side.
- Proof on p. 476, Exercise 37

Corollary
If three parallel lines intersect two transversals, then the segments intercepted on the transversals are proportional. (p. 473)
- Proof on p. 477, Exercise 46

Theorem 7-5
Triangle-Angle-Bisector Theorem
If a ray bisects an angle of a triangle, then it divides the opposite side into two segments that are proportional to the other two sides of the triangle. (p. 473)
- Proof on p. 477, Exercise 47

Chapter 8 Right Triangles and Trigonometry

Theorem 8-1
Pythagorean Theorem
If a triangle is a right triangle, then the sum of the squares of the lengths of the legs is equal to the square of the length of the hypotenuse.
$a^2 + b^2 = c^2$ (p. 491)
- Proof on p. 497, Exercise 49

Theorem 8-2
Converse of the Pythagorean Theorem
If the sum of the squares of the lengths of two sides of a triangle is equal to the square of the length of the third side, then the triangle is a right triangle. (p. 493)
- Proof on p. 498, Exercise 52

Theorem 8-3
If the square of the length of the longest side of a triangle is greater than the sum of the squares of the lengths of the other two sides, then the triangle is obtuse. (p. 494)
- Proof on p. 498, Exercise 53

Theorem 8-4
If the square of the length of the longest side of a triangle is less than the sum of the squares of the lengths of the other two sides, then the triangle is acute. (p. 494)
- Proof on p. 498, Exercise 54

Theorem 8-5
45°-45°-90° Triangle Theorem
In a 45°-45°-90° triangle, both legs are congruent and the length of the hypotenuse is $\sqrt{2}$ times the length of a leg.
hypotenuse $= \sqrt{2} \cdot$ leg (p. 499)
- Proof on p. 499

Theorem 8-6
30°-60°-90° Triangle Theorem
In a 30°-60°-90° triangle, the length of the hypotenuse is twice the length of the shorter leg. The length of the longer leg is $\sqrt{3}$ times the length of the shorter leg.
hypotenuse $= 2 \cdot$ shorter leg
longer leg $= \sqrt{3} \cdot$ shorter leg (p. 501)
- Proof on p. 501

Law of Sines
$\frac{\sin A}{a} = \frac{\sin B}{b} = \frac{\sin C}{c}$ (p. 522)
- Proof on p. 522

Law of Cosines
$a^2 = b^2 + c^2 - 2bc \cos A$
$b^2 = a^2 + c^2 - 2ac \cos B$
$c^2 = a^2 + b^2 - 2ab \cos C$ (p. 527)
- Proof on p. 527

Chapter 9 Transformations

Theorem 9-1
The composition of two or more isometries is an isometry. (p. 570)

Theorem 9-2
Reflections Across Parallel Lines
A composition of reflections across two parallel lines is a translation. (p. 571)

Theorem 9-3
Reflections Across Intersecting Lines
A composition of reflections across two intersecting lines is a rotation. (p. 572)

Chapter 10 Area

Theorem 10-1
Area of a Rectangle
The area of a rectangle is the product of its base and height.
$A = bh$ (p. 616)

Theorem 10-2
Area of a Parallelogram
The area of a parallelogram is the product of a base and the corresponding height.
$A = bh$ (p. 616)

Theorem 10-3
Area of a Triangle
The area of a triangle is half the product of a base and the corresponding height.
$A = \frac{1}{2}bh$ (p. 618)

Theorem 10-4
Area of a Trapezoid
The area of a trapezoid is half the product of the height and the sum of the bases.
$A = \frac{1}{2}h(b_1 + b_2)$ (p. 623)

Theorem 10-5
Area of a Rhombus or a Kite
The area of a rhombus or a kite is half the product of the lengths of its diagonals.
$A = \frac{1}{2}d_1d_2$ (p. 624)

Postulate 10-1
If two figures are congruent, then their areas are equal. (p. 630)

Theorem 10-6
Area of a Regular Polygon
The area of a regular polygon is half the product of the apothem and the perimeter.
$A = \frac{1}{2}ap$ (p. 630)
- Proof on p. 630

Theorem 10-7
Perimeters and Areas of Similar Figures
If the scale factor of two similar figures is $\frac{a}{b}$, then
(1) the ratio of their perimeters is $\frac{a}{b}$ and
(2) the ratio of their areas is $\frac{a^2}{b^2}$. (p. 635)

Theorem 10-8
Area of a Triangle Given SAS
The area of a triangle is half the product of the lengths of two sides and the sine of the included angle.
Area of $\triangle ABC = \frac{1}{2}bc(\sin A)$ (p. 645)
- Proof on p. 645

Postulate 10-2
Arc Addition Postulate
The measure of the arc formed by two adjacent arcs is the sum of the measures of the two arcs.
$m\overset{\frown}{ABC} = m\overset{\frown}{AB} + m\overset{\frown}{BC}$ (p. 650)

Theorem 10-9
Circumference of a Circle
The circumference of a circle is π times the diameter.
$C = \pi d$ or $C = 2\pi r$ (p. 651)

Theorem 10-10
Arc Length
The length of an arc of a circle is the product of the ratio $\frac{\text{measure of the arc}}{360}$ and the circumference of the circle.
length of $\overset{\frown}{AB} = \frac{m\overset{\frown}{AB}}{360} \cdot 2\pi r$ or
length of $\overset{\frown}{AB} = \frac{m\overset{\frown}{AB}}{360} \cdot \pi d$ (p. 653)

Theorem 10-11
Area of a Circle
The area of a circle is the product of π and the square of the radius.
$A = \pi r^2$ (p. 660)

Theorem 10-12
Area of a Sector of a Circle
The area of a sector of a circle is the product of the ratio $\frac{\text{measure of the arc}}{360}$ and the area of the circle.
Area of sector $AOB = \frac{m\overset{\frown}{AB}}{360} \cdot \pi r^2$ (p. 661)

Chapter 11 Surface Area and Volume

Theorem 11-1
Lateral and Surface Areas of a Prism
The lateral area of a right prism is the product of the perimeter of the base and the height of the prism.
L.A. $= ph$
The surface area of a right prism is the sum of the lateral area and the areas of the two bases.
S.A. = L.A. $+ 2B$ (p. 700)

Theorem 11-2
Lateral and Surface Areas of a Cylinder
The lateral area of a right cylinder is the product of the circumference of the base and the height of the cylinder.
L.A. $= 2\pi rh$, or L.A. $= \pi dh$
The surface area of a right cylinder is the sum of the lateral area and areas of the two bases.
S.A. = L.A. $+ 2B$, or S.A. $= 2\pi rh + 2\pi r^2$ (p. 702)

Theorem 11-3
Lateral and Surface Areas of a Pyramid
The lateral area of a regular pyramid is half the product of the perimeter p of the base and the slant height ℓ of the pyramid.
L.A. $= \frac{1}{2}p\ell$
The surface area of a regular pyramid is the sum of the lateral area and the area B of the base.
S.A. = L.A. $+ B$ (p. 709)

Theorem 11-4
Lateral and Surface Areas of a Cone
The lateral area of a right cone is half the product of the circumference of the base and the slant height of the cone.
L.A. $= \frac{1}{2} \cdot 2\pi r\ell$, or L.A. $= \pi r\ell$
The surface area of a right cone is the sum of the lateral area and the area of the base.
S.A. = L.A. $+ B$ (p. 711)

Theorem 11-5
Cavalieri's Principle
If two space figures have the same height and the same cross-sectional area at every level, then they have the same volume. (p. 718)

Theorem 11-6
Volume of a Prism
The volume of a prism is the product of the area of the base and the height of the prism.
$V = Bh$ (p. 718)

Theorem 11-7
Volume of a Cylinder
The volume of a cylinder is the product of the area of the base and the height of the cylinder.
$V = Bh$, or $V = \pi r^2 h$ (p. 719)

Theorem 11-8
Volume of a Pyramid
The volume of a pyramid is one third the product of the area of the base and the height of the pyramid.
$V = \frac{1}{3}Bh$ (p. 726)

Theorem 11-9
Volume of a Cone
The volume of a cone is one third the product of the area of the base and the height of the cone.
$V = \frac{1}{3}Bh$, or $V = \frac{1}{3}\pi r^2 h$ (p. 728)

Theorem 11-10
Surface Area of a Sphere
The surface area of a sphere is four times the product of π and the square of the radius of the sphere.
S.A. $= 4\pi r^2$ (p. 734)

Theorem 11-11
Volume of a Sphere
The volume of a sphere is four thirds the product of π and the cube of the radius of the sphere.
$V = \frac{4}{3}\pi r^3$ (p. 735)

Theorem 11-12
Areas and Volumes of Similar Solids
If the scale factor of two similar solids is $a : b$, then
- the ratio of their corresponding areas is $a^2 : b^2$, and
- the ratio of their volumes is $a^3 : b^3$. (p. 743)

Chapter 12 Circles

Theorem 12-1
If a line is tangent to a circle, then the line is perpendicular to the radius at the point of tangency. (p. 762)
- Proof on p. 763

Theorem 12-2
If a line in the plane of a circle is perpendicular to a radius at its endpoint on the circle, then the line is tangent to the circle. (p. 764)
- Proof on p. 769, Exercise 30

Theorem 12-3
If two segments are tangent to a circle from a point outside the circle, then the two segments are congruent. (p. 766)
- Proof on p. 768, Exercise 23

Theorem 12-4
Within a circle or in congruent circles, congruent central angles have congruent arcs. (p. 771)
- Proof on p. 777, Exercise 19

Converse
Within a circle or in congruent circles, congruent arcs have congruent central angles. (p. 771)
- Proof on p. 778, Exercise 35

Theorem 12-5
Within a circle or in congruent circles, congruent central angles have congruent chords. (p. 772)
- Proof on p. 777, Exercise 20

Converse
Within a circle or in congruent circles, congruent chords have congruent central angles. (p. 772)
- Proof on p. 778, Exercise 36

Theorem 12-6
Within a circle or in congruent circles, congruent chords have congruent arcs. (p. 772)
- Proof on p. 777, Exercise 21

Converse
Within a circle or in congruent circles, congruent arcs have congruent chords. (p. 772)
- Proof on p. 778, Exercise 37

Theorem 12-7
Within a circle or in congruent circles, chords equidistant from the center (or centers) are congruent. (p. 772)
- Proof on p. 773

Converse
Within a circle or in congruent circles, congruent chords are equidistant from the center (or centers). (p. 772)
- Proof on p. 778, Exercise 38

Theorem 12-8
In a circle, if a diameter is perpendicular to a chord, it bisects the chord and its arc. (p. 774)
- Proof on p. 777, Exercise 22

Theorem 12-9
In a circle, if a diameter bisects a chord (that is not a diameter), it is perpendicular to the chord. (p. 774)
- Proof on p. 774

Theorem 12-10
In a circle, the perpendicular bisector of a chord contains the center of the circle. (p. 774)
- Proof on p. 778, Exercise 33

Theorem 12-11
Inscribed Angle Theorem
The measure of an inscribed angle is half the measure of its intercepted arc. (p. 780)
- Proofs on p. 781; p. 785, Exercises 26, 27

Corollary 1
Two inscribed angles that intercept the same arc are congruent. (p. 782)
- Proof on p. 786, Exercise 31

Corollary 2
An angle inscribed in a semicircle is a right angle. (p. 782)
- Proof on p. 786, Exercise 32

Corollary 3
The opposite angles of a quadrilateral inscribed in a circle are supplementary. (p. 782)
- Proof on p. 786, Exercise 33

Theorem 12-12
The measure of an angle formed by a tangent and a chord is half the measure of the intercepted arc. (p. 783)
- Proof on p. 786, Exercise 34

Theorem 12-13
The measure of an angle formed by two lines that intersect inside a circle is half the sum of the measures of the intercepted arcs. (p. 790)
- Proof on p. 791

Theorem 12-14
The measure of an angle formed by two lines that intersect outside a circle is half the difference of the measures of the intercepted arcs. (p. 790)
- Proofs on p. 796, Exercises 35, 36

Theorem 12-15
For a given point and circle, the product of the lengths of the two segments from the point to the circle is constant along any line through the point and circle. (p. 793)
- Proofs on p. 793; p. 796, Exercises 37, 38

Theorem 12-16
An equation of a circle with center (h, k) and radius r is $(x - h)^2 + (y - k)^2 = r^2$. (p. 798)
- Proof on p. 799

Constructions

Construction 1
Congruent Segments
Construct a segment congruent to a given segment. (p. 43)

Construction 2
Congruent Angles
Construct an angle congruent to a given angle. (p. 44)

Construction 3
Perpendicular Bisector
Construct the perpendicular bisector of a segment. (p. 45)

Construction 4
Angle Bisector
Construct the bisector of an angle. (p. 45)

Construction 5
Parallel Through a Point Not on a Line
Construct the line parallel to a given line and through a given point that is not on the line. (p. 182)

Construction 6
Quadrilateral With Parallel Sides
Construct a quadrilateral with one pair of parallel sides of lengths a and b. (p. 183)

Construction 7
Perpendicular Through a Point on a Line
Construct the perpendicular to a given line at a given point on the line. (p. 184)

Construction 8
Perpendicular Through a Point Not on a Line
Construct the perpendicular to a given line through a given point not on the line. (p. 185)

Visual Glossary

English	Spanish

A

Acute angle (p. 29) An acute angle is an angle whose measure is between 0 and 90.

Ángulo agudo (p. 29) Un ángulo agudo es un ángulo que mide entre 0 y 90 grados.

Example

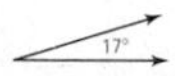

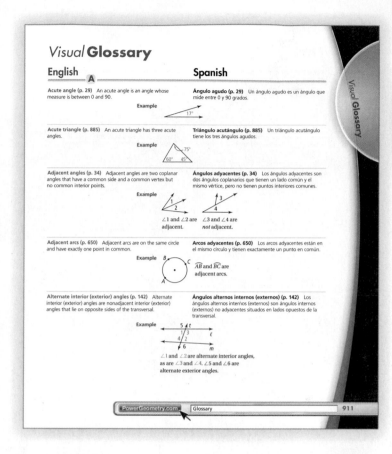

Acute triangle (p. 885) An acute triangle has three acute angles.

Triángulo acutángulo (p. 885) Un triángulo acutángulo tiene los tres ángulos agudos.

Example

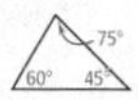

Adjacent angles (p. 34) Adjacent angles are two coplanar angles that have a common side and a common vertex but no common interior points.

Ángulos adyacentes (p. 34) Los ángulos adyacentes son dos ángulos coplanarios que tienen un lado común y el mismo vértice, pero no tienen puntos interiores comunes.

Example

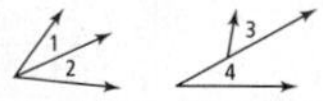

$\angle 1$ and $\angle 2$ are adjacent. $\angle 3$ and $\angle 4$ are *not* adjacent.

Adjacent arcs (p. 650) Adjacent arcs are on the same circle and have exactly one point in common.

Arcos adyacentes (p. 650) Los arcos adyacentes están en el mismo círculo y tienen exactamente un punto en común.

Example

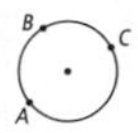

$\widehat{AB}$ and $\widehat{BC}$ are adjacent arcs.

Alternate interior (exterior) angles (p. 142) Alternate interior (exterior) angles are nonadjacent interior (exterior) angles that lie on opposite sides of the transversal.

Ángulos alternos internos (externos) (p. 142) Los ángulos alternos internos (externos) son ángulos internos (externos) no adyacentes situados en lados opuestos de la transversal.

Example

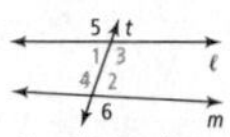

$\angle 1$ and $\angle 2$ are alternate interior angles, as are $\angle 3$ and $\angle 4$. $\angle 5$ and $\angle 6$ are alternate exterior angles.

Visual Glossary

English	Spanish

Altitude *See* **cone; cylinder; parallelogram; prism; pyramid; trapezoid; triangle.**

Altura *Ver* **cone; cylinder; parallelogram; prism; pyramid; trapezoid; triangle.**

Altitude of a triangle (p. 310) An altitude of a triangle is the perpendicular segment from a vertex to the line containing the side opposite that vertex.

Altura de un triángulo (p. 310) Una altura de un triángulo es el segmento perpendicular que va desde un vértice hasta la recta que contiene el lado opuesto a ese vértice.

Example

Angle (p. 27) An angle is formed by two rays with the same endpoint. The rays are the *sides* of the angle and the common endpoint is the *vertex* of the angle.

Ángulo (p. 27) Un ángulo está formado por dos semirrectas que convergen en un mismo extremo. Las semirrectas son los *lados* del ángulo y los extremos en común son el *vértice*.

Example

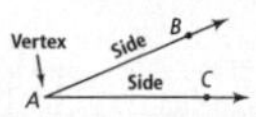

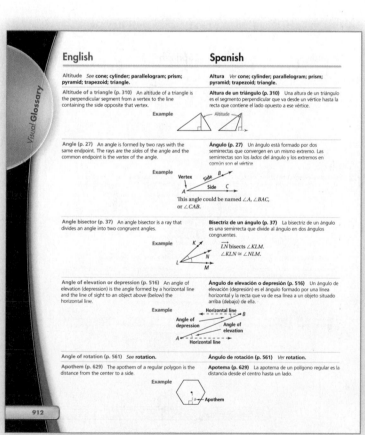

This angle could be named $\angle A$, $\angle BAC$, or $\angle CAB$.

Angle bisector (p. 37) An angle bisector is a ray that divides an angle into two congruent angles.

Bisectriz de un ángulo (p. 37) La bisectriz de un ángulo es una semirrecta que divide al ángulo en dos ángulos congruentes.

Example

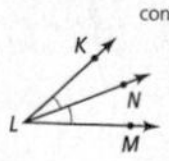

$\overrightarrow{LN}$ bisects $\angle KLM$. $\angle KLN \cong \angle NLM$.

Angle of elevation or depression (p. 516) An angle of elevation (depression) is the angle formed by a horizontal line and the line of sight to an object above (below) the horizontal line.

Ángulo de elevación o depresión (p. 516) Un ángulo de elevación (depresión) es el ángulo formado por una línea horizontal y la recta que va de esa línea a un objeto situado arriba (debajo) de ella.

Example

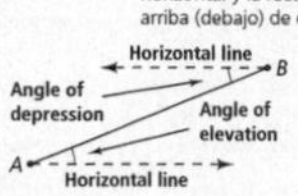

Angle of rotation (p. 561) *See* **rotation.**

Ángulo de rotación (p. 561) *Ver* **rotation.**

Apothem (p. 629) The apothem of a regular polygon is the distance from the center to a side.

Apotema (p. 629) La apotema de un polígono regular es la distancia desde el centro hasta un lado.

Example

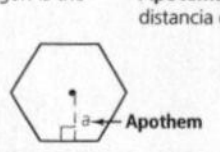

Visual Glossary

English	Spanish

Arc *See* **major arc; minor arc.** *See also* **arc length; measure of an arc.**

Arco *Ver* **major arc; minor arc.** *Ver también* **arc length; measure of an arc.**

Arc length (p. 653) The length of an arc of a circle is the product of the ratio $\frac{\text{measure of the arc}}{360}$ and the circumference of the circle.

Longitud de un arco (p. 653) La longitud del arco de un círculo es el producto del cociente $\frac{\text{medida del arco}}{360}$ por la circunferencia del círculo.

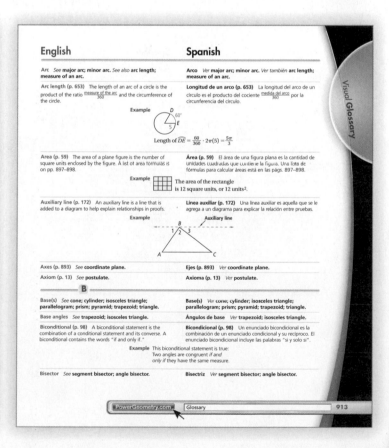

Example

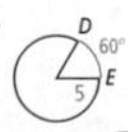

Length of $\widehat{DE} = \frac{60}{360} \cdot 2\pi(5) = \frac{5\pi}{3}$

Area (p. 59) The area of a plane figure is the number of square units enclosed by the figure. A list of area formulas is on pp. 897–898.

Área (p. 59) El área de una figura plana es la cantidad de unidades cuadradas que contiene la figura. Una lista de fórmulas para calcular áreas está en las págs. 897–898.

Example

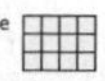

The area of the rectangle is 12 square units, or 12 units^2.

Auxiliary line (p. 172) An auxiliary line is a line that is added to a diagram to help explain relationships in proofs.

Línea auxiliar (p. 172) Una línea auxiliar es aquella que se le agrega a un diagrama para explicar la relación entre pruebas.

Example

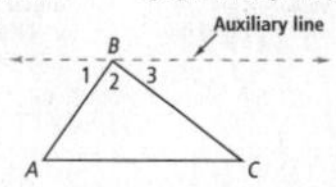

Axes (p. 893) *See* **coordinate plane.**

Ejes (p. 893) *Ver* **coordinate plane.**

Axiom (p. 13) *See* **postulate.**

Axioma (p. 13) *Ver* **postulate.**

B

Base(s) *See* **cone; cylinder; isosceles triangle; parallelogram; prism; pyramid; trapezoid; triangle.**

Base(s) *Ver* **cone; cylinder; isosceles triangle; parallelogram; prism; pyramid; trapezoid; triangle.**

Base angles *See* **trapezoid; isosceles triangle.**

Ángulos de base *Ver* **trapezoid; isosceles triangle.**

Biconditional (p. 98) A biconditional statement is the combination of a conditional statement and its converse. A biconditional contains the words "if and only if."

Bicondicional (p. 98) Un enunciado bicondicional es la combinación de un enunciado condicional y su recíproco. El enunciado bicondicional incluye las palabras "si y solo si".

Example This biconditional statement is true: Two angles are congruent *if and only if* they have the same measure.

Bisector *See* **segment bisector; angle bisector.**

Bisectriz *Ver* **segment bisector; angle bisector.**

Visual Glossary

Visual **Glossary**

English — C — Spanish

Center *See* **circle; dilation; regular polygon; rotation; sphere.**

Centro *Ver* **circle; dilation; regular polygon; rotation; sphere.**

Central angle of a circle (p. 649) A central angle of a circle is an angle whose vertex is the center of the circle.

Ángulo central de un círculo (p. 649) Un ángulo central de un círculo es un ángulo cuyo vértice es el centro del círculo.

Example

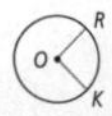

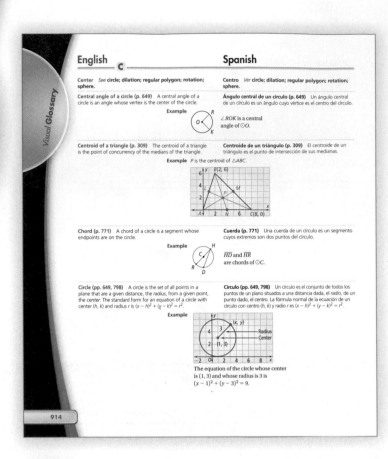

$\angle ROK$ is a central angle of $\odot O$.

Centroid of a triangle (p. 309) The centroid of a triangle is the point of concurrency of the medians of the triangle.

Centroide de un triángulo (p. 309) El centroide de un triángulo es el punto de intersección de sus medianas.

Example *P* is the centroid of $\triangle ABC$.

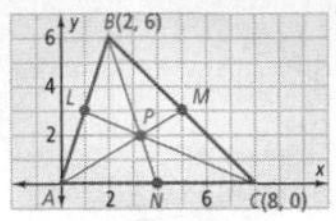

Chord (p. 771) A chord of a circle is a segment whose endpoints are on the circle.

Cuerda (p. 771) Una cuerda de un círculo es un segmento cuyos extremos son dos puntos del círculo.

Example

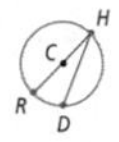

$\overline{HD}$ and $\overline{HR}$ are chords of $\odot C$.

Circle (pp. 649, 798) A circle is the set of all points in a plane that are a given distance, the *radius*, from a given point, the *center*. The standard form for an equation of a circle with center (h, k) and radius r is $(x - h)^2 + (y - k)^2 = r^2$.

Círculo (pp. 649, 798) Un círculo es el conjunto de todos los puntos de un plano situados a una distancia dada, el *radio*, de un punto dado, el *centro*. La fórmula normal de la ecuación de un círculo con centro (h, k) y radio r es $(x - h)^2 + (y - k)^2 = r^2$.

Example

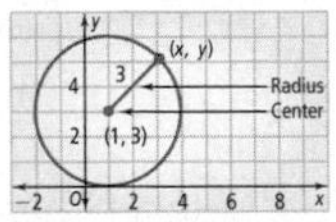

The equation of the circle whose center is (1, 3) and whose radius is 3 is $(x - 1)^2 + (y - 3)^2 = 9$.

English — Spanish

Circumcenter of a triangle (p. 301) The circumcenter of a triangle is the point of concurrency of the perpendicular bisectors of the sides of the triangle.

Circuncentro de un triángulo (p. 301) El circuncentro de un triángulo es el punto de intersección de las bisectrices perpendiculares de los lados del triángulo.

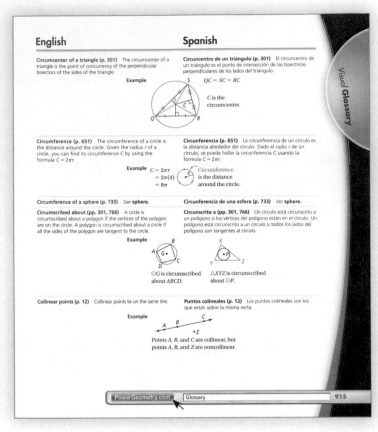

Example

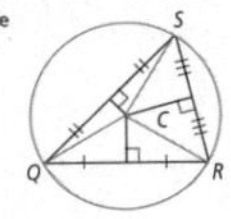

$QC = SC = RC$

C is the circumcenter.

Circumference (p. 651) The circumference of a circle is the distance around the circle. Given the radius *r* of a circle, you can find its circumference *C* by using the formula $C = 2\pi r$.

Circunferencia (p. 651) La circunferencia de un círculo es la distancia alrededor del círculo. Dado el radio *r* de un círculo, se puede hallar la circunferencia *C* usando la fórmula $C = 2\pi r$.

Example $C = 2\pi r$
$= 2\pi(4)$
$= 8\pi$

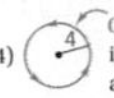

Circumference is the distance around the circle.

Circumference of a sphere (p. 733) *See* **sphere.**

Circunferencia de una esfera (p. 733) *Ver* **sphere.**

Circumscribed about (pp. 301, 766) A circle is circumscribed about a polygon if the vertices of the polygon are on the circle. A polygon is circumscribed about a circle if all the sides of the polygon are tangent to the circle.

Circunscrito a (pp. 301, 766) Un círculo está circunscrito a un polígono si los vértices del polígono están en el círculo. Un polígono está circunscrito a un círculo si todos los lados del polígono son tangentes al círculo.

Example

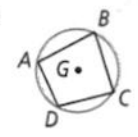

$\odot G$ is circumscribed about *ABCD*.

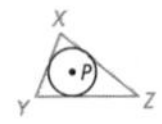

$\triangle XYZ$ is circumscribed about $\odot P$.

Collinear points (p. 12) Collinear points lie on the same line.

Puntos colineales (p. 12) Los puntos colineales son los que están sobre la misma recta.

Example

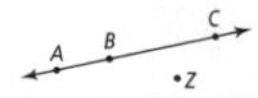

Points *A*, *B*, and *C* are collinear, but points *A*, *B*, and *Z* are noncollinear.

English — Spanish

Combination (p. 838) Any unordered selection of *r* objects from a set of *n* objects is a combination. The number of combinations of *n* objects taken *r* at a time is ${}_nC_r = \frac{n!}{r!(n - r)!}$ for $0 \le r \le n$.

Combinación (p. 838) Cualquier selección no ordenada de *r* objetos tomados de un conjunto de *n* objetos es una combinación. El número de combinaciones de *n* objetos, cuando se toman *r* objetos cada vez, es ${}_nC_r = \frac{n!}{r!(n - r)!}$ para $0 \le r \le n$.

Example The number of combinations of seven items taken four at a time is ${}_7C_4 = \frac{7!}{4!(7 - 4)!} = 35$. There are 35 ways to choose four items from seven items without regard to order.

Compass (p. 43) A compass is a geometric tool used to draw circles and parts of circles, called arcs.

Compás (p. 43) El compás es un instrumento usado para dibujar círculos y partes de círculos, llamados arcos.

Complement of an event (p. 826) All possible outcomes that are not in the event.
P(complement of event) = 1 − *P*(event)

Complemento de un suceso (p. 826) Todos los resultados posibles que no se dan en el suceso.
P(complemento de un suceso) = 1 − *P*(suceso)

Example The complement of rolling a 1 or a 2 on a standard number cube is rolling a 3, 4, 5, or 6.

Complementary angles (p. 34) Two angles are complementary angles if the sum of their measures is 90.

Ángulos complementarios (p. 34) Dos ángulos son complementarios si la suma de sus medidas es igual a 90 grados.

Example

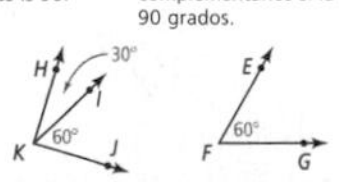

$\angle HKI$ and $\angle IKJ$ are complementary angles, as are $\angle HKI$ and $\angle EFG$.

Composite space figures (p. 720) A composite space figure is the combination of two or more figures into one object.

Figuras geométricas compuestas (p. 720) Una figura geométrica compuesta es la combinación de dos o más figuras en un mismo objeto.

Example

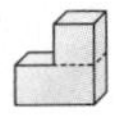

English — Spanish

Composition of transformations (p. 548) A composition of two transformations is a transformation in which a second transformation is performed on the image of a first transformation.

Composición de transformaciones (p. 548) Una composición de dos transformaciones es una transformación en la cual una segunda transformación se realiza a partir de la imagen de la primera.

Example

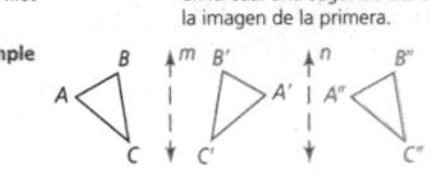

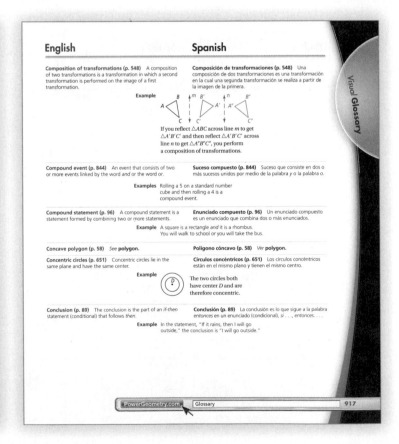

If you reflect $\triangle ABC$ across line *m* to get $\triangle A'B'C'$ and then reflect $\triangle A'B'C'$ across line *n* to get $\triangle A''B''C''$, you perform a composition of transformations.

Compound event (p. 844) An event that consists of two or more events linked by the word *and* or the word *or*.

Suceso compuesto (p. 844) Suceso que consiste en dos o más sucesos unidos por medio de la palabra *y* o la palabra *o*.

Examples Rolling a 5 on a standard number cube and then rolling a 4 is a compound event.

Compound statement (p. 96) A compound statement is a statement formed by combining two or more statements.

Enunciado compuesto (p. 96) Un enunciado compuesto es un enunciado que combina dos o más enunciados.

Example A square is a rectangle *and* it is a rhombus.
You will walk to school *or* you will take the bus.

Concave polygon (p. 58) *See* **polygon.**

Polígono cóncavo (p. 58) *Ver* **polygon.**

Concentric circles (p. 651) Concentric circles lie in the same plane and have the same center.

Círculos concéntricos (p. 651) Los círculos concéntricos están en el mismo plano y tienen el mismo centro.

Example

The two circles both have center *D* and are therefore concentric.

Conclusion (p. 89) The conclusion is the part of an *if-then* statement (conditional) that follows *then*.

Conclusión (p. 89) La conclusión es lo que sigue a la palabra *entonces* en un enunciado (condicional), *si* . . ., *entonces*. . . .

Example In the statement, "If it rains, then I will go outside," the conclusion is "I will go outside."

English | Spanish

Concurrent lines (p. 301) Concurrent lines are three or more lines that meet in one point. The point at which they meet is the *point of concurrency.*

Rectas concurrentes (p. 301) Las rectas concurrentes son tres o más rectas que se unen en un punto. El punto en que se unen es el *punto de concurrencia.*

Example

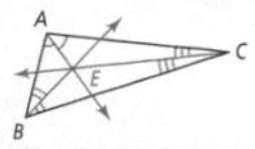

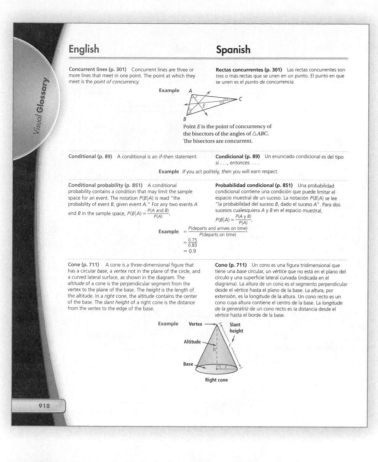

Point E is the point of concurrency of the bisectors of the angles of $\triangle ABC$. The bisectors are concurrent.

Conditional (p. 89) A conditional is an *if-then* statement.

Condicional (p. 89) Un enunciado condicional es del tipo *si . . ., entonces. . . .*

Example *If* you act politely, *then* you will earn respect.

Conditional probability (p. 851) A conditional probability contains a condition that may limit the sample space for an event. The notation $P(B|A)$ is read "the probability of event B, given event A." For any two events A and B in the sample space, $P(B|A) = \frac{P(A \text{ and } B)}{P(A)}$.

Probabilidad condicional (p. 851) Una probabilidad condicional contiene una condición que puede limitar el espacio muestral de un suceso. La notación $P(B|A)$ se lee "la probabilidad del suceso B, dado el suceso A". Para dos sucesos cualesquiera A y B en el espacio muestral, $P(B|A) = \frac{P(A \text{ y } B)}{P(A)}$.

Example $= \frac{P(\text{departs and arrives on time})}{P(\text{departs on time})}$

$= \frac{0.75}{0.83}$

≈ 0.9

Cone (p. 711) A cone is a three-dimensional figure that has a circular *base*, a *vertex* not in the plane of the circle, and a curved lateral surface, as shown in the diagram. The *altitude* of a cone is the perpendicular segment from the vertex to the plane of the base. The *height* is the length of the altitude. In a *right cone*, the altitude contains the center of the base. The *slant height* of a right cone is the distance from the vertex to the edge of the base.

Cono (p. 711) Un cono es una figura tridimensional que tiene una *base* circular, un *vértice* que no está en el plano del círculo y una superficie lateral curvada (indicada en el diagrama). La *altura* de un cono es el segmento perpendicular desde el vértice hasta el plano de la base. La *altura*, por extensión, es la longitude de la altura. Un *cono recto* es un cono cuya altura contiene el centro de la base. La *longitude de la generatriz* de un cono recto es la distancia desde el vértice hasta el borde de la base.

Example

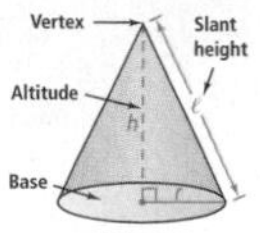

English | Spanish

Congruence transformation (p. 580) *See* **isometry.**

Transformación de congruencia (p. 580) *Ver* **isometry.**

Congruent angles (p. 29) Congruent angles are angles that have the same measure.

Ángulos congruentes (p. 29) Los ángulos congruentes son ángulos que tienen la misma medida.

Example

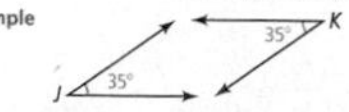

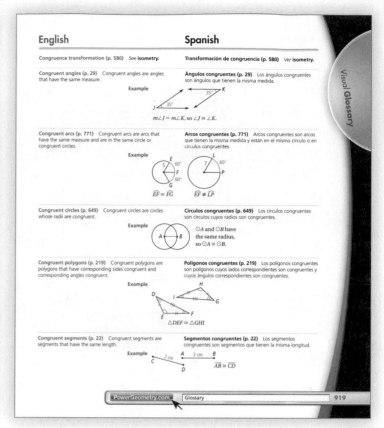

$m\angle J = m\angle K$, so $\angle J \cong \angle K$.

Congruent arcs (p. 771) Congruent arcs are arcs that have the same measure and are in the same circle or congruent circles.

Arcos congruentes (p. 771) Arcos congruentes son arcos que tienen la misma medida y están en el mismo círculo o en círculos congruentes.

Example

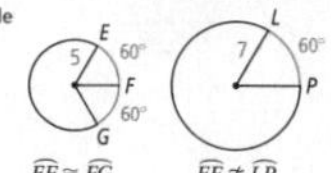

$\overset{\frown}{EF} \cong \overset{\frown}{FG}$ $\overset{\frown}{EF} \ncong \overset{\frown}{LP}$

Congruent circles (p. 649) Congruent circles are circles whose radii are congruent.

Círculos congruentes (p. 649) Los círculos congruentes son círculos cuyos radios son congruentes.

Example

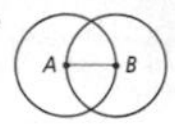

$\odot A$ and $\odot B$ have the same radius, so $\odot A \cong \odot B$.

Congruent polygons (p. 219) Congruent polygons are polygons that have corresponding sides congruent and corresponding angles congruent.

Polígonos congruentes (p. 219) Los polígonos congruentes son polígonos cuyos lados correspondientes son congruentes y cuyos ángulos correspondientes son congruentes.

Example

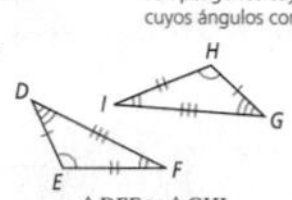

$\triangle DEF \cong \triangle GHI$

Congruent segments (p. 22) Congruent segments are segments that have the same length.

Segmentos congruentes (p. 22) Los segmentos congruentes son segmentos que tienen la misma longitud.

Example

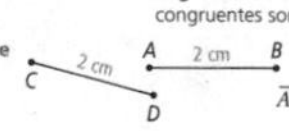

$\overline{AB} \cong \overline{CD}$

English | Spanish

Conjecture (p. 83) A conjecture is a conclusion reached by using inductive reasoning.

Conjetura (p. 83) Una conjetura es una conclusión obtenida usando el razonamiento inductivo.

Example As you walk down the street, you see many people holding unopened umbrellas. You make the conjecture that the forecast must call for rain.

Conjunction (p. 96) A conjunction is a compound statement formed by connecting two or more statements with the word *and*.

Conjunción (p. 96) Una conjunción es un enunciado compuesto que conecta dos o más enunciados por medio de la palabra *y*.

Example The sky is blue *and* the grass is green.

Consecutive angles (p. 360) Consecutive angles of a polygon share a common side.

Ángulos consecutivos (p. 360) Los ángulos consecutivos de un polígono tienen un lado común.

Example

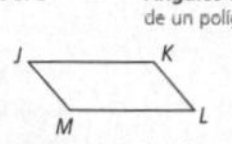

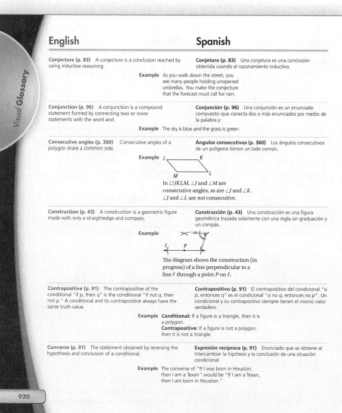

In $\square JKLM$, $\angle J$ and $\angle M$ are consecutive angles, as are $\angle J$ and $\angle K$. $\angle J$ and $\angle L$ are *not* consecutive.

Construction (p. 43) A construction is a geometric figure made with only a straightedge and compass.

Construcción (p. 43) Una construcción es una figura geométrica trazada solamente con una regla sin graduación y un compás.

Example

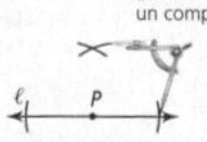

The diagram shows the construction (in progress) of a line perpendicular to a line ℓ through a point P on ℓ.

Contrapositive (p. 91) The contrapositive of the conditional "if p, then q" is the conditional "if not q, then not p." A conditional and its contrapositive always have the same truth value.

Contrapositivo (p. 91) El contrapositivo del condicional "si p, entonces q" es el condicional "si no q, entonces no p". Un condicional y su contrapositivo siempre tienen el mismo valor verdadero.

Example **Conditional:** If a figure is a triangle, then it is a polygon.
Contrapositive: If a figure is not a polygon, then it is not a triangle.

Converse (p. 91) The statement obtained by reversing the hypothesis and conclusion of a conditional.

Expresión recíproca (p. 91) Enunciado que se obtiene al intercambiar la hipótesis y la conclusión de una situación condicional.

Example The converse of "If I was born in Houston, then I am a Texan" would be "If I am a Texan, then I am born in Houston."

English | Spanish

Convex polygon (p. 58) *See* **polygon.**

Polígono convexo (p. 58) *Ver* **polygon.**

Coordinate(s) of a point (pp. 20, 893) The coordinate of a point is its distance and direction from the origin of a number line. The coordinates of a point on a coordinate plane are in the form (x, y), where x is the x-coordinate and y is the y-coordinate.

Coordenada(s) de un punto (pp. 20, 893) La coordenada de un punto es su distancia y dirección desde el origen en una recta numérica. Las coordenadas de un punto en un plano de coordenadas se expresan como (x, y), donde x es la coordenada x, e y es la coordenada y.

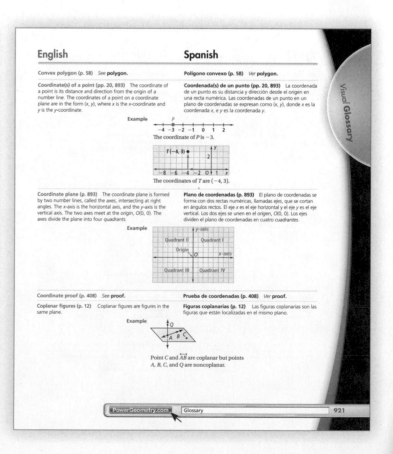

Example

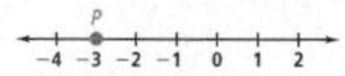

The coordinate of P is -3.

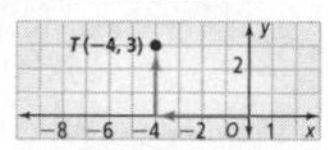

The coordinates of T are $(-4, 3)$.

Coordinate plane (p. 893) The coordinate plane is formed by two number lines, called the *axes*, intersecting at right angles. The x-axis is the horizontal axis, and the y-axis is the vertical axis. The two axes meet at the *origin*, $O(0, 0)$. The axes divide the plane into four *quadrants*.

Plano de coordenadas (p. 893) El plano de coordenadas se forma con dos rectas numéricas, llamadas *ejes*, que se cortan en ángulos rectos. El eje x es el eje horizontal y el eje y es el eje vertical. Los dos ejes se unen en el *origen*, $O(0, 0)$. Los ejes dividen el plano de coordenadas en cuatro *cuadrantes*.

Example

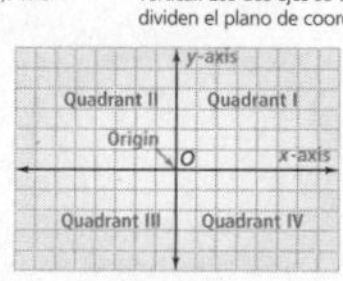

Coordinate proof (p. 408) *See* **proof.**

Prueba de coordenadas (p. 408) *Ver* **proof.**

Coplanar figures (p. 12) Coplanar figures are figures in the same plane.

Figuras coplanarias (p. 12) Las figuras coplanarias son las figuras que están localizadas en el mismo plano.

Example

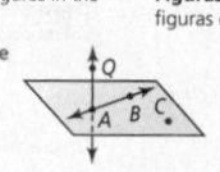

Point C and $\overleftrightarrow{AB}$ are coplanar but points A, B, C, and Q are noncoplanar.

English / Spanish

Corollary (p. 252) A corollary is a theorem that can be proved easily using another theorem.

Corolario (p. 252) Un corolario es un teorema que se puede probar fácilmente usando otro teorema.

Example **Theorem:** If two sides of a triangle are congruent, then the angles opposite those sides are congruent.
Corollary: If a triangle is equilateral, then it is equiangular.

Corresponding angles (p. 142) Corresponding angles lie on the same side of the transversal *t* and in corresponding positions relative to ℓ and *m*.

Ángulos correspondientes (p. 142) Los ángulos correspondientes están en el mismo lado de la transversal *t* y en las correspondientes posiciones relativas a ℓ y *m*.

Example

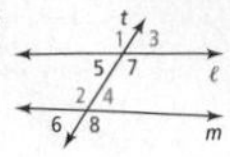

$\angle 1$ and $\angle 2$ are corresponding angles, as are $\angle 3$ and $\angle 4$, $\angle 5$ and $\angle 6$, and $\angle 7$ and $\angle 8$.

Cosine ratio (p. 507) *See* **trigonometric ratios.**

Razón coseno (p. 507) *Ver* **trigonometric ratios.**

Counterexample (pp. 84, 90) An example showing that a statement is false.

Contraejemplo (pp. 84, 90) Ejemplo que demuestra que un enunciado es falso.

Example **Statement:** All apples are red.
Counterexample: A Granny Smith Apple is green.

Cross Products Property (p. 434) The product of the extremes of a proportion is equal to the product of the means.

Propiedad de los productos cruzados (p. 434) El producto de los extremos de una proporción es igual al producto de los medios.

Example If $\frac{x}{3} = \frac{12}{21}$, then $21x = 3 \cdot 12$.

Cross section (p. 690) A cross section is the intersection of a solid and a plane.

Sección de corte (p. 690) Una sección de corte es la intersección de un plano y un cuerpo geométrico.

Example

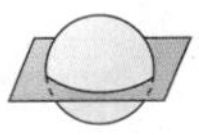

The cross section is a circle.

Cube (p. 691) A cube is a polyhedron with six faces, each of which is a square.

Cubo (p. 691) Un cubo es un poliedro de seis caras, cada una de las caras es un cuadrado.

Example

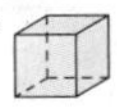

English / Spanish

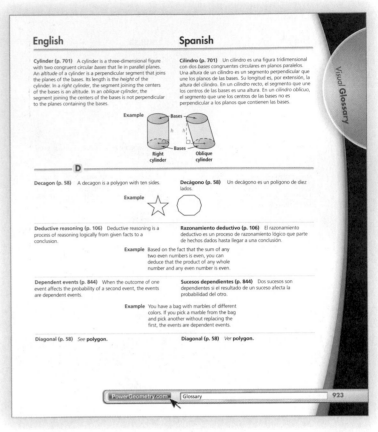

Cylinder (p. 701) A cylinder is a three-dimensional figure with two congruent circular *bases* that lie in parallel planes. An *altitude* of a cylinder is a perpendicular segment that joins the planes of the bases. Its length is the *height* of the cylinder. In a *right cylinder*, the segment joining the centers of the bases is an altitude. In an *oblique cylinder*, the segment joining the centers of the bases is not perpendicular to the planes containing the bases.

Cilindro (p. 701) Un cilindro es una figura tridimensional con dos *bases* congruentes circulares en planos paralelos. Una *altura* de un cilindro es un segmento perpendicular que une los planos de las bases. Su longitud es, por extensión, la *altura* del cilindro. En un *cilindro recto*, el segmento que une los centros de las bases es una altura. En un *cilindro oblicuo*, el segmento que une los centros de las bases no es perpendicular a los planos que contienen las bases.

Example

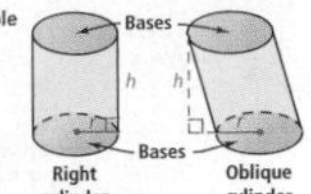

D

Decagon (p. 58) A decagon is a polygon with ten sides.

Decágono (p. 58) Un decágono es un polígono de diez lados.

Example

Deductive reasoning (p. 106) Deductive reasoning is a process of reasoning logically from given facts to a conclusion.

Razonamiento deductivo (p. 106) El razonamiento deductivo es un proceso de razonamiento lógico que parte de hechos dados hasta llegar a una conclusión.

Example Based on the fact that the sum of any two even numbers is even, you can deduce that the product of any whole number and any even number is even.

Dependent events (p. 844) When the outcome of one event affects the probability of a second event, the events are dependent events.

Sucesos dependientes (p. 844) Dos sucesos son dependientes si el resultado de un suceso afecta la probabilidad del otro.

Example You have a bag with marbles of different colors. If you pick a marble from the bag and pick another without replacing the first, the events are dependent events.

Diagonal (p. 58) *See* **polygon.**

Diagonal (p. 58) *Ver* **polygon.**

English / Spanish

Diameter of a circle (p. 649) A diameter of a circle is a segment that contains the center of the circle and whose endpoints are on the circle. The term *diameter* can also mean the length of this segment.

Diámetro de un círculo (p. 649) Un diámetro de un círculo es un segmento que contiene el centro del círculo y cuyos extremos están en el círculo. El término *diámetro* también puede referirse a la longitud de este segmento.

Example

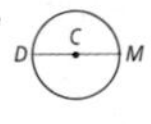

$\overline{DM}$ is a diameter of $\odot C$.

Diameter of a sphere (p. 733) The diameter of a sphere is a segment passing through the center, with endpoints on the sphere.

Diámetro de una esfera (p. 733) El diámetro de una esfera es un segmento que contiene el centro de la esfera y cuyos extremos están en la esfera.

Example

Dilation (p. 587) A dilation is a transformation that has *center C* and *scale factor n*, where $n > 0$, and maps a point R to R' in such a way that R' is on $\overrightarrow{CR}$ and $CR' = n \cdot CR$. The center of a dilation is its own image. If $n > 1$, the dilation is an *enlargement*, and if $0 < n < 1$, the dilation is a *reduction*.

Dilatación (p. 587) Una dilatación, o *transformación de semejanza*, tiene *centro C* y *factor de escala n* para $n > 0$, y asocia un punto R a R' de tal modo que R' está en $\overrightarrow{CR}$ y $CR' = n \cdot CR$. El centro de una dilatación es su propia imagen. Si $n > 1$, la dilatación es un *aumento*, y si $0 < n < 1$, la dilatación es una *reducción*.

Example

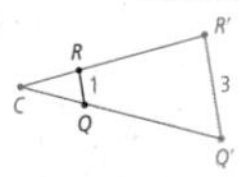

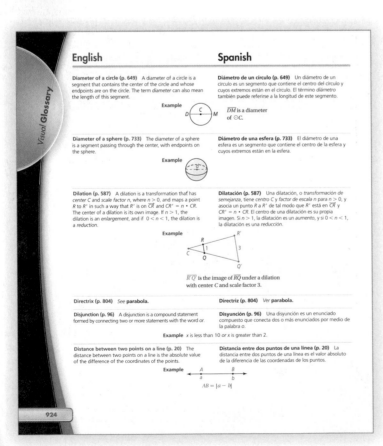

$\overline{R'Q'}$ is the image of $\overline{RQ}$ under a dilation with center C and scale factor 3.

Directrix (p. 804) *See* **parabola.**

Directriz (p. 804) *Ver* **parabola.**

Disjunction (p. 96) A disjunction is a compound statement formed by connecting two or more statements with the word *or*.

Disyunción (p. 96) Una disyunción es un enunciado compuesto que conecta dos o más enunciados por medio de la palabra *o*.

Example x is less than 10 *or* x is greater than 2.

Distance between two points on a line (p. 20) The distance between two points on a line is the absolute value of the difference of the coordinates of the points.

Distancia entre dos puntos de una línea (p. 20) La distancia entre dos puntos de una línea es el valor absoluto de la diferencia de las coordenadas de los puntos.

Example

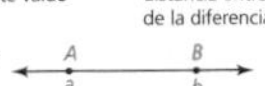

$AB = |a - b|$

English / Spanish

Distance from a point to a line (p. 294) The distance from a point to a line is the length of the perpendicular segment from the point to the line.

Distancia desde un punto hasta una recta (p. 294) La distancia desde un punto hasta una recta es la longitud del segmento perpendicular que va desde el punto hasta la recta.

Example

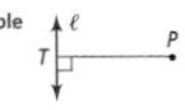

The distance from point P to a line ℓ is PT.

E

Edge (p. 688) *See* **polyhedron.**

Arista (p. 688) *Ver* **polyhedron.**

Endpoint (p. 12) *See* **ray; segment.**

Extremo (p. 12) *Ver* **ray; segment.**

Enlargement (p. 588) *See* **dilation.**

Aumento (p. 588) *Ver* **dilation.**

Equiangular triangle or polygon (pp. 354, 885) An equiangular triangle (polygon) is a triangle (polygon) whose angles are all congruent.

Triángulo o polígono equiángulo (pp. 354, 885) Un triángulo (polígono) equiángulo es un triángulo (polígono) cuyos ángulos son todos congruentes.

Example

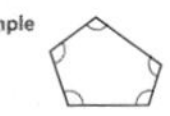

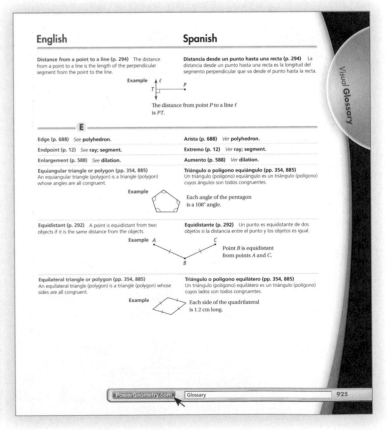

Each angle of the pentagon is a 108° angle.

Equidistant (p. 292) A point is equidistant from two objects if it is the same distance from the objects.

Equidistante (p. 292) Un punto es equidistante de dos objetos si la distancia entre el punto y los objetos es igual.

Example

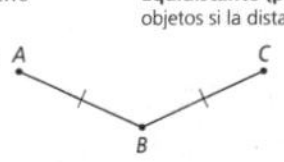

Point B is equidistant from points A and C.

Equilateral triangle or polygon (pp. 354, 885) An equilateral triangle (polygon) is a triangle (polygon) whose sides are all congruent.

Triángulo o polígono equilátero (pp. 354, 885) Un triángulo (polígono) equilátero es un triángulo (polígono) cuyos lados son todos congruentes.

Example

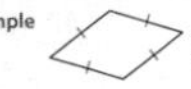

Each side of the quadrilateral is 1.2 cm long.

English	Spanish

Equivalent statements (p. 91) Equivalent statements are statements with the same truth value.

Enunciados equivalentes (p. 91) Los enunciados equivalentes son enunciados con el mismo valor verdadero.

Example The following statements are equivalent:
If a figure is a square, then it is a rectangle.
If a figure is not a rectangle, then it is not a square.

Euclidean geometry (p. 179) Euclidean geometry is a geometry of the plane in which Euclid's Parallel Postulate is accepted as true.

Geometría euclidiana (p.179) La geometría euclidiana es una geometría del plano en donde el postulado paralelo de Euclides es verdadero.

Example

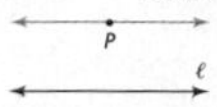

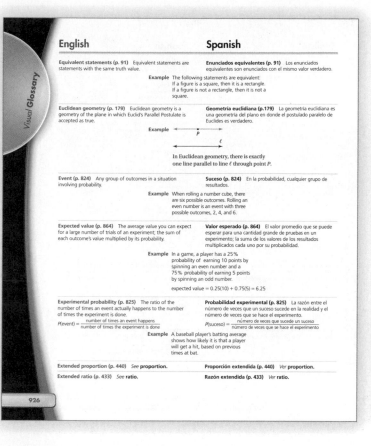

In Euclidean geometry, there is exactly one line parallel to line ℓ through point P.

Event (p. 824) Any group of outcomes in a situation involving probability.

Suceso (p. 824) En la probabilidad, cualquier grupo de resultados.

Example When rolling a number cube, there are six possible outcomes. Rolling an even number is an event with three possible outcomes, 2, 4, and 6.

Expected value (p. 864) The average value you can expect for a large number of trials of an experiment; the sum of each outcome's value multiplied by its probability.

Valor esperado (p. 864) El valor promedio que se puede esperar para una cantidad grande de pruebas en un experimento; la suma de los valores de los resultados multiplicados cada uno por su probabilidad.

Example In a game, a player has a 25% probability of earning 10 points by spinning an even number and a 75% probability of earning 5 points by spinning an odd number.

expected value = 0.25(10) + 0.75(5) = 6.25

Experimental probability (p. 825) The ratio of the number of times an event actually happens to the number of times the experiment is done.

$$P(\text{event}) = \frac{\text{number of times an event happens}}{\text{number of times the experiment is done}}$$

Probabilidad experimental (p. 825) La razón entre el número de veces que un suceso sucede en la realidad y el número de veces que se hace el experimento.

$$P(\text{suceso}) = \frac{\text{número de veces que sucede un suceso}}{\text{número de veces que se hace el experimento}}$$

Example A baseball player's batting average shows how likely it is that a player will get a hit, based on previous times at bat.

Extended proportion (p. 440) *See* **proportion.**

Proporción extendida (p. 440) *Ver* **proportion.**

Extended ratio (p. 433) *See* **ratio.**

Razón extendida (p. 433) *Ver* **ratio.**

English	Spanish

Exterior angle of a polygon (p. 173) An exterior angle of a polygon is an angle formed by a side and an extension of an adjacent side.

Ángulo exterior de un polígono (p. 173) El ángulo exterior de un polígono es un ángulo formado por un lado y una extensión de un lado adyacente.

Example

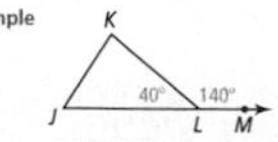

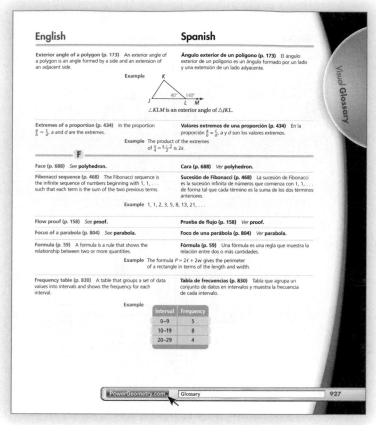

$\angle KLM$ is an exterior angle of $\triangle JKL$.

Extremes of a proportion (p. 434) In the proportion $\frac{a}{b} = \frac{c}{d}$, a and d are the extremes.

Valores extremos de una proporción (p. 434) En la proporción $\frac{a}{b} = \frac{c}{d}$, a y d son los valores extremos.

Example The product of the extremes of $\frac{x}{4} = \frac{x+3}{2}$ is $2x$.

F

Face (p. 688) *See* **polyhedron.**

Cara (p. 688) *Ver* **polyhedron.**

Fibonacci sequence (p. 468) The Fibonacci sequence is the infinite sequence of numbers beginning with 1, 1, . . . such that each term is the sum of the two previous terms.

Sucesión de Fibonacci (p. 468) La sucesión de Fibonacci es la sucesión infinita de números que comienza con 1, 1, . . . de forma tal que cada término es la suma de los dos términos anteriores.

Example 1, 1, 2, 3, 5, 8, 13, 21, . . .

Flow proof (p. 158) *See* **proof.**

Prueba de flujo (p. 158) *Ver* **proof.**

Focus of a parabola (p. 804) *See* **parabola.**

Foco de una parábola (p. 804) *Ver* **parabola.**

Formula (p. 59) A formula is a rule that shows the relationship between two or more quantities.

Fórmula (p. 59) Una fórmula es una regla que muestra la relación entre dos o más cantidades.

Example The formula $P = 2\ell + 2w$ gives the perimeter of a rectangle in terms of the length and width.

Frequency table (p. 830) A table that groups a set of data values into intervals and shows the frequency for each interval.

Tabla de frecuencias (p. 830) Tabla que agrupa un conjunto de datos en intervalos y muestra la frecuencia de cada intervalo.

Example

Interval	Frequency
0–9	5
10–19	8
20–29	4

English	Spanish

Fundamental Counting Principle (p. 836) If there are m ways to make the first selection and n ways to make the second selection, then there are $m \cdot n$ ways to make the two selections.

Principio fundamental de Conteo (p. 836) Si hay m maneras de hacer la primera selección y n maneras de hacer la segunda selección, quiere decir que hay $m \cdot n$ maneras de hacer las dos selecciones.

Example For 5 shirts and 8 pairs of shorts, the number of possible outfits is $5 \cdot 8 = 40$.

G

Geometric mean (p. 462) The geometric mean is the number x such that $\frac{a}{x} = \frac{x}{b}$, where a, b, and x are positive numbers.

Media geométrica (p. 462) La media geométrica es el número x tanto que $\frac{a}{x} = \frac{x}{b}$, donde a, b y x son números positivos.

Example The geometric mean of 6 and 24 is 12.

$$\frac{6}{x} = \frac{x}{24}$$
$$x^2 = 144$$
$$x = 12$$

Geometric probability (p. 668) Geometric probability is a probability that uses a geometric model in which points represent outcomes.

Probabilidad geométrica (p. 668) La probabilidad geométrica es una probabilidad que utiliza un modelo geométrico donde se usan puntos para representar resultados.

Example

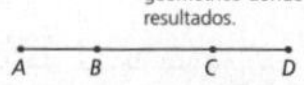

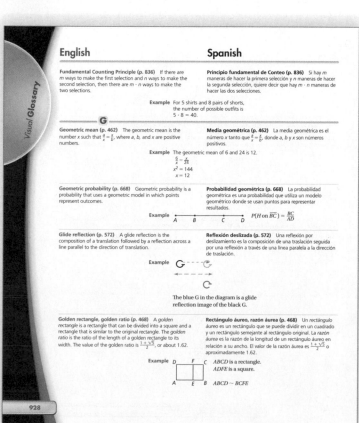

$P(H \text{ on } \overline{BC}) = \frac{BC}{AD}$

Glide reflection (p. 572) A glide reflection is the composition of a translation followed by a reflection across a line parallel to the direction of translation.

Reflexión deslizada (p. 572) Una reflexión por deslizamiento es la composición de una traslación seguida por una reflexión a través de una línea paralela a la dirección de traslación.

Example

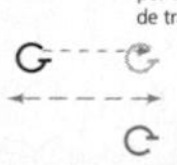

The blue G in the diagram is a glide reflection image of the black G.

Golden rectangle, golden ratio (p. 468) A *golden rectangle* is a rectangle that can be divided into a square and a rectangle that is similar to the original rectangle. The *golden ratio* is the ratio of the length of a golden rectangle to its width. The value of the golden ratio is $\frac{1+\sqrt{5}}{2}$, or about 1.62.

Rectángulo áureo, razón áurea (p. 468) Un *rectángulo áureo* es un rectángulo que se puede dividir en un cuadrado y un rectángulo semejante al rectángulo original. La *razón áurea* es la razón de la longitud de un rectángulo áureo en relación a su ancho. El valor de la razón áurea es $\frac{1+\sqrt{5}}{2}$ o aproximadamente 1.62.

Example

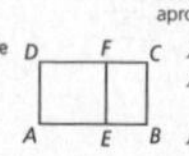

$ABCD$ is a rectangle.
$ADFE$ is a square.

$ABCD \sim BCFE$

English	Spanish

Great circle (p. 733) A great circle is the intersection of a sphere and a plane containing the center of the sphere. A great circle divides a sphere into two *hemispheres.*

Círculo máximo (p. 733) Un círculo máximo es la intersección de una esfera y un plano que contiene el centro de la esfera. Un círculo máximo divide una esfera en dos *hemisferios.*

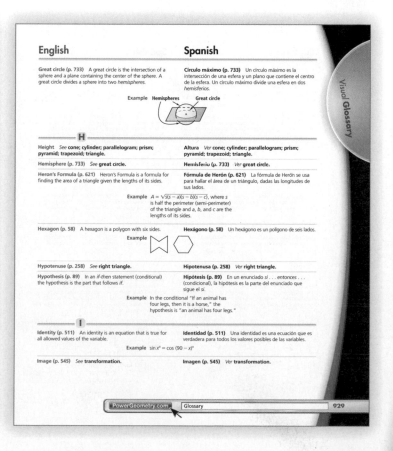

Example

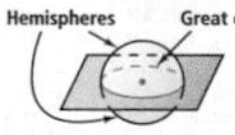

H

Height *See* **cone; cylinder; parallelogram; prism; pyramid; trapezoid; triangle.**

Altura *Ver* **cone; cylinder; parallelogram; prism; pyramid; trapezoid; triangle.**

Hemisphere (p. 733) *See* **great circle.**

Hemisferio (p. 733) *Ver* **great circle.**

Heron's Formula (p. 621) Heron's Formula is a formula for finding the area of a triangle given the lengths of its sides.

Fórmula de Herón (p. 621) La fórmula de Herón se usa para hallar el área de un triángulo, dadas las longitudes de sus lados.

Example $A = \sqrt{s(s-a)(s-b)(s-c)}$, where s is half the perimeter (semi-perimeter) of the triangle and a, b, and c are the lengths of its sides.

Hexagon (p. 58) A hexagon is a polygon with six sides.

Hexágono (p. 58) Un hexágono es un polígono de seis lados.

Example

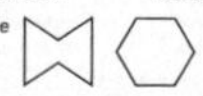

Hypotenuse (p. 258) *See* **right triangle.**

Hipotenusa (p. 258) *Ver* **right triangle.**

Hypothesis (p. 89) In an *if-then* statement (conditional) the hypothesis is the part that follows *if.*

Hipótesis (p. 89) En un enunciado *si . . . entonces . . .* (condicional), la hipótesis es la parte del enunciado que sigue el *si.*

Example In the conditional "If an animal has four legs, then it is a horse," the hypothesis is "an animal has four legs."

I

Identity (p. 511) An identity is an equation that is true for all allowed values of the variable.

Identidad (p. 511) Una identidad es una ecuación que es verdadera para todos los valores posibles de las variables.

Example $\sin x° = \cos (90 - x)°$

Image (p. 545) *See* **transformation.**

Imagen (p. 545) *Ver* **transformation.**

English | Spanish

Incenter of a triangle (p. 303) The incenter of a triangle is the point of concurrency of the angle bisectors of the triangle.

Incentro de un triángulo (p. 303) El incentro de un triángulo es el punto donde concurren las tres bisectrices de los ángulos del triángulo.

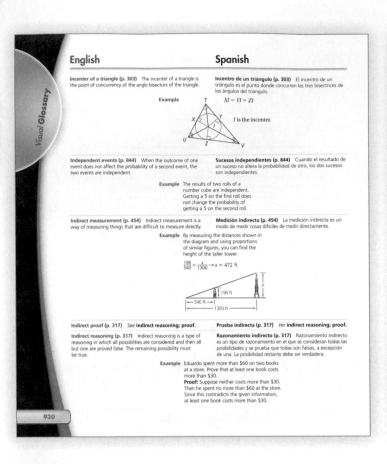

Example

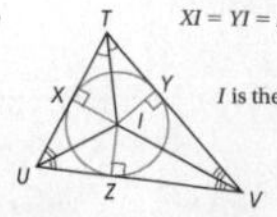

$XI = YI = ZI$

I is the incenter.

Independent events (p. 844) When the outcome of one event does not affect the probability of a second event, the two events are independent.

Sucesos independientes (p. 844) Cuando el resultado de un suceso no altera la probabilidad de otro, los dos sucesos son independientes.

Example The results of two rolls of a number cube are independent. Getting a 5 on the first roll does not change the probability of getting a 5 on the second roll.

Indirect measurement (p. 454) Indirect measurement is a way of measuring things that are difficult to measure directly.

Medición indirecta (p. 454) La medición indirecta es un modo de medir cosas difíciles de medir directamente.

Example By measuring the distances shown in the diagram and using proportions of similar figures, you can find the height of the taller tower.

$\frac{196}{540} = \frac{x}{1300} \rightarrow x \approx 472$ ft

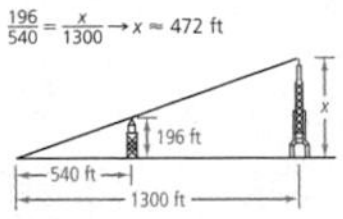

Indirect proof (p. 317) *See* **indirect reasoning; proof.**

Prueba indirecta (p. 317) *Ver* **indirect reasoning; proof.**

Indirect reasoning (p. 317) Indirect reasoning is a type of reasoning in which all possiblities are considered and then all but one are proved false. The remaining possibility must be true.

Razonamiento indirecto (p. 317) Razonamiento indirecto es un tipo de razonamiento en el que se consideran todas las posibilidades y se prueba que todas son falsas, a excepción de una. La posibilidad restante debe ser verdadera.

Example Eduardo spent more than $60 on two books at a store. Prove that at least one book costs more than $30.
Proof: Suppose neither costs more than $30. Then he spent no more than $60 at the store. Since this contradicts the given information, at least one book costs more than $30.

English | Spanish

Inductive reasoning (p. 82) Inductive reasoning is a type of reasoning that reaches conclusions based on a pattern of specific examples or past events.

Razonamiento inductivo (p. 82) El razonamiento inductivo es un tipo de razonamiento en el cual se llega a conclusiones con base en un patrón de ejemplos específicos o sucesos pasados.

Example You see four people walk into a building. Each person emerges with a small bag containing food. You use inductive reasoning to conclude that this building contains a restaurant.

Inscribed angle (p. 780) An angle is inscribed in a circle if the vertex of the angle is on the circle and the sides of the angle are chords of the circle.

Ángulo inscrito (p. 780) Un ángulo está inscrito en un círculo si el vértice del ángulo está en el círculo y los lados del ángulo son cuerdas del círculo.

Example

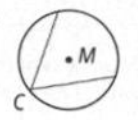

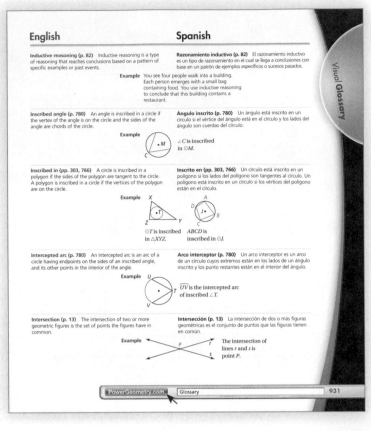

$\angle C$ is inscribed in $\odot M$.

Inscribed in (pp. 303, 766) A circle is inscribed in a polygon if the sides of the polygon are tangent to the circle. A polygon is inscribed in a circle if the vertices of the polygon are on the circle.

Inscrito en (pp. 303, 766) Un círculo está inscrito en un polígono si los lados del polígono son tangentes al círculo. Un polígono está inscrito en un círculo si los vértices del polígono están en el círculo.

Example

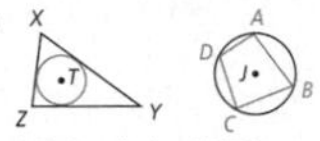

$\odot T$ is inscribed in $\triangle XYZ$.

$ABCD$ is inscribed in $\odot J$.

Intercepted arc (p. 780) An intercepted arc is an arc of a circle having endpoints on the sides of an inscribed angle, and its other points in the interior of the angle.

Arco interceptor (p. 780) Un arco interceptor es un arco de un círculo cuyos extremos están en los lados de un ángulo inscrito y los punto restantes están en el interior del ángulo.

Example

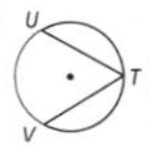

$\widehat{UV}$ is the intercepted arc of inscribed $\angle T$.

Intersection (p. 13) The intersection of two or more geometric figures is the set of points the figures have in common.

Intersección (p. 13) La intersección de dos o más figuras geométricas es el conjunto de puntos que las figuras tienen en común.

Example

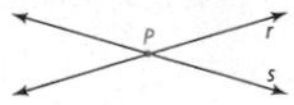

The intersection of lines r and s is point P.

English | Spanish

Inverse (p. 91) The inverse of the conditional "if p, then q" is the conditional "if not p, then not q."

Inverso (p. 91) El inverso del condicional "si p, entonces q" es el condicional "si no p, entonces no q".

Example **Conditional:** If a figure is a square, then it is a parallelogram.
Inverse: If a figure is not a square, then it is not a parallelogram.

Isometric drawing (p. 5) An isometric drawing shows a corner view of a three-dimensional figure. It is usually drawn on isometric dot paper. An isometric drawing allows you to see the top, front, and side of an object in the same drawing.

Dibujo isométrico (p. 5) Un dibujo isométrico muestra la perspectiva de una esquina de una figura tridimensional. Generalmente se dibuja en papel punteado isométrico. Un dibujo isométrico permite ver la cima, el frente, y el lado de un objeto en el mismo dibujo.

Example

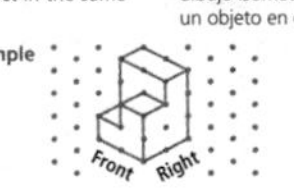

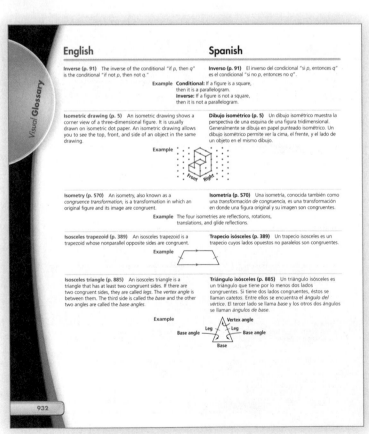

Isometry (p. 570) An isometry, also known as a *congruence transformation,* is a transformation in which an original figure and its image are congruent.

Isometría (p. 570) Una isometría, conocida también como una *transformación de congruencia,* es una transformación en donde una figura original y su imagen son congruentes.

Example The four isometries are reflections, rotations, translations, and glide reflections.

Isosceles trapezoid (p. 389) An isosceles trapezoid is a trapezoid whose nonparallel opposite sides are congruent.

Trapecio isósceles (p. 389) Un trapecio isosceles es un trapecio cuyos lados opuestos no paralelos son congruentes.

Example

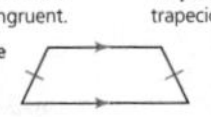

Isosceles triangle (p. 885) An isosceles triangle is a triangle that has at least two congruent sides. If there are two congruent sides, they are called *legs*. The *vertex angle* is between them. The third side is called the *base* and the other two angles are called the *base angles*.

Triángulo isósceles (p. 885) Un triángulo isósceles es un triángulo que tiene por lo menos dos lados congruentes. Si tiene dos lados congruentes, éstos se llaman *catetos*. Entre ellos se encuentra el *ángulo del vértice*. El tercer lado se llama *base* y los otros dos ángulos se llaman *ángulos de base*.

Example

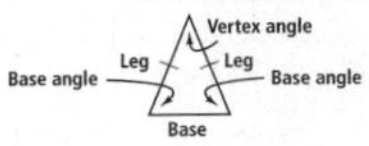

English | Spanish

K

Kite (p. 392) A kite is a quadrilateral with two pairs of consecutive sides congruent and no opposite sides congruent.

Cometa (p. 392) Una cometa es un cuadrilátero con dos pares de lados congruentes consecutivos y sin lados opuestos congruentes.

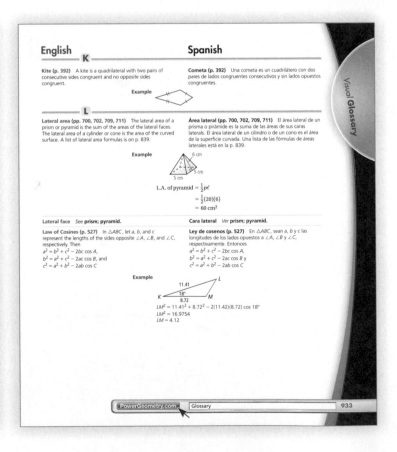

Example

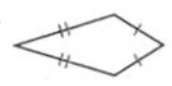

L

Lateral area (pp. 700, 702, 709, 711) The lateral area of a prism or pyramid is the sum of the areas of the lateral faces. The lateral area of a cylinder or cone is the area of the curved surface. A list of lateral area formulas is on p. 839.

Área lateral (pp. 700, 702, 709, 711) El área lateral de un prisma o pirámide es la suma de las áreas de sus caras laterals. El área lateral de un cilindro o de un cono es el área de la superficie curvada. Una lista de las fórmulas de áreas laterales está en la p. 839.

Example

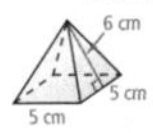

$$\text{L.A. of pyramid} = \tfrac{1}{2}p\ell = \tfrac{1}{2}(20)(6) = 60\ \text{cm}^2$$

Lateral face *See* **prism; pyramid.**

Cara lateral *Ver* **prism; pyramid.**

Law of Cosines (p. 527) In $\triangle ABC$, let a, b, and c represent the lengths of the sides opposite $\angle A$, $\angle B$, and $\angle C$, respectively. Then
$a^2 = b^2 + c^2 - 2bc \cos A$,
$b^2 = a^2 + c^2 - 2ac \cos B$, and
$c^2 = a^2 + b^2 - 2ab \cos C$

Ley de cosenos (p. 527) En $\triangle ABC$, sean a, b y c las longitudes de los lados opuestos a $\angle A$, $\angle B$ y $\angle C$, respectivamente. Entonces
$a^2 = b^2 + c^2 - 2bc \cos A$,
$b^2 = a^2 + c^2 - 2ac \cos B$ y
$c^2 = a^2 + b^2 - 2ab \cos C$

Example

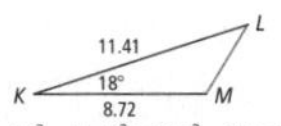

$LM^2 = 11.41^2 + 8.72^2 - 2(11.42)(8.72) \cos 18°$
$LM^2 = 16.9754$
$LM = 4.12$

English | Spanish

Law of Sines (p. 522) In $\triangle ABC$, let a, b, and c represent the lengths of the sides opposite $\angle A$, $\angle B$, and $\angle C$, respectively. Then $\frac{\sin A}{a} = \frac{\sin B}{b} = \frac{\sin C}{c}$.

Ley de senos (p. 522) En $\triangle ABC$, sean a, b y c las longitudes de los lados opuestos a $\angle A$, $\angle B$ y $\angle C$, respectivamente. Entonces $\frac{\text{sen } A}{a} = \frac{\text{sen } B}{b} = \frac{\text{sen } C}{c}$.

Example

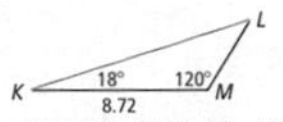

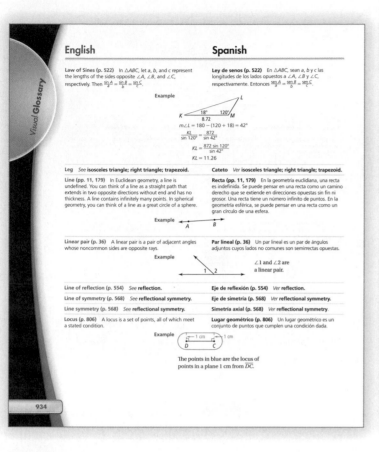

$m\angle L = 180 - (120 + 18) = 42°$

$\frac{KL}{\sin 120°} = \frac{872}{\sin 42°}$

$KL = \frac{872 \sin 120°}{\sin 42°}$

$KL = 11.26$

Leg *See* **isosceles triangle; right triangle; trapezoid.**

Cateto *Ver* **isosceles triangle; right triangle; trapezoid.**

Line (pp. 11, 179) In Euclidean geometry, a line is undefined. You can think of a line as a straight path that extends in two opposite directions without end and has no thickness. A line contains infinitely many points. In spherical geometry, you can think of a line as a great circle of a sphere.

Recta (pp. 11, 179) En la geometría euclidiana, una recta es indefinida. Se puede pensar en una recta como un camino derecho que se extiende en direcciones opuestas sin fin ni grosor. Una recta tiene un número infinito de puntos. En la geometría esférica, se puede pensar en una recta como un gran círculo de una esfera.

Example

Linear pair (p. 36) A linear pair is a pair of adjacent angles whose noncommon sides are opposite rays.

Par lineal (p. 36) Un par lineal es un par de ángulos adjuntos cuyos lados no comunes son semirrectas opuestas.

Example

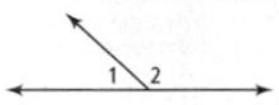

$\angle 1$ and $\angle 2$ are a linear pair.

Line of reflection (p. 554) *See* **reflection.**

Eje de reflexión (p. 554) *Ver* **reflection.**

Line of symmetry (p. 568) *See* **reflectional symmetry.**

Eje de simetría (p. 568) *Ver* **reflectional symmetry.**

Line symmetry (p. 568) *See* **reflectional symmetry.**

Simetría axial (p. 568) *Ver* **reflectional symmetry**.

Locus (p. 806) A locus is a set of points, all of which meet a stated condition.

Lugar geométrico (p. 806) Un lugar geométrico es un conjunto de puntos que cumplen una condición dada.

Example

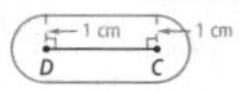

The points in blue are the locus of points in a plane 1 cm from $\overline{DC}$.

English | Spanish

M

Major arc (p. 649) A major arc of a circle is an arc that is larger than a semicircle.

Arco mayor (p. 649) Un arco mayor de un círculo es cualquier arco más grande que un semicírculo.

Example

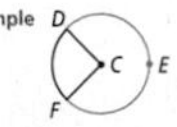

$\widehat{DEF}$ is a major arc of $\odot C$.

Map (p. 545) *See* **transformation.**

Trazar (p. 545) *Ver* **transformation.**

Means of a proportion (p. 434) In the proportion $\frac{a}{b} = \frac{c}{d}$, b and c are the means.

Valores medios de una proporción (p. 434) En la proporción $\frac{a}{b} = \frac{c}{d}$, b y c son los valores medios.

Example The product of the means of $\frac{x}{4} = \frac{x+3}{2}$ is $4(x + 3)$ or $4x + 12$.

Measure of an angle (p. 28) Consider $\overrightarrow{OD}$ and a point C on one side of $\overleftrightarrow{OD}$. Every ray of the form $\overrightarrow{OC}$ can be paired one to one with a real number from 0 to 180. The measure of $\angle COD$ is the absolute value of the difference of the real numbers paired with $\overrightarrow{OC}$ and $\overrightarrow{OD}$.

Medida de un ángulo (p. 28) Toma en cuenta $\overrightarrow{OD}$ y un punto C a un lado de $\overleftrightarrow{OD}$. Cada semirrecta de la forma $\overrightarrow{OC}$ puede ser emparejada exactamente con un número real de 0 a 180. La medida de $\angle COD$ es el valor absoluto de la diferencia de los números reales emparejados con $\overrightarrow{OC}$ y $\overrightarrow{OD}$.

Example

$m\angle COD = 105$

English | Spanish

Measure of an arc (p. 650) The measure of a minor arc is the measure of its central angle. The measure of a major arc is 360 minus the measure of its related minor arc.

Medida de un arco (p. 650) La medida de un arco menor es la medida de su ángulo central. La medida de un arco mayor es 360 menos la medida en grados de su arco menor correspondiente.

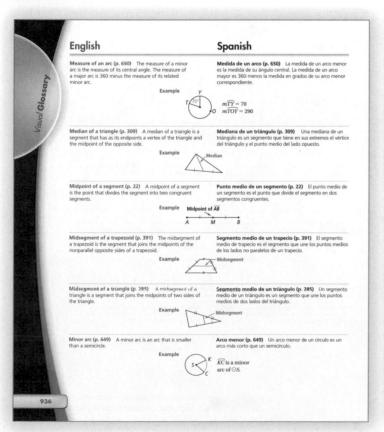

Example

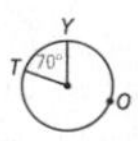

$m\widehat{TY} = 70$

$m\widehat{TOY} = 290$

Median of a triangle (p. 309) A median of a triangle is a segment that has as its endpoints a vertex of the triangle and the midpoint of the opposite side.

Mediana de un triángulo (p. 309) Una mediana de un triángulo es un segmento que tiene en sus extremos el vértice del triángulo y el punto medio del lado opuesto.

Example

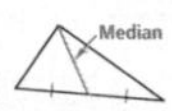

Midpoint of a segment (p. 22) A midpoint of a segment is the point that divides the segment into two congruent segments.

Punto medio de un segmento (p. 22) El punto medio de un segmento es el punto que divide el segmento en dos segmentos congruentes.

Example

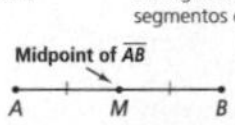

Midsegment of a trapezoid (p. 391) The midsegment of a trapezoid is the segment that joins the midpoints of the nonparallel opposite sides of a trapezoid.

Segmento medio de un trapecio (p. 391) El segmento medio de trapecio es el segmento que une los puntos medios de los lados no paralelos de un trapecio.

Example

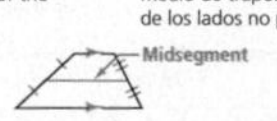

Midsegment of a triangle (p. 285) A midsegment of a triangle is a segment that joins the midpoints of two sides of the triangle.

Segmento medio de un triángulo (p. 285) Un segmento medio de un triángulo es un segmento que une los puntos medios de dos lados del triángulo.

Example

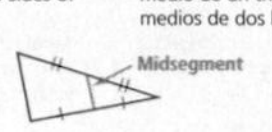

Minor arc (p. 649) A minor arc is an arc that is smaller than a semicircle.

Arco menor (p. 649) Un arco menor de un círculo es un arco más corto que un semicírculo.

Example

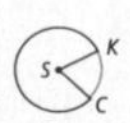

$\widehat{KC}$ is a minor arc of $\odot S$.

English | Spanish

Mutually exclusive events (p. 845) When two events cannot happen at the same time, the events are mutually exclusive. If A and B are mutually exclusive events, then $P(A \text{ or } B) = P(A) + P(B)$.

Sucesos mutuamente excluyentes (p. 845) Cuando dos sucesos no pueden ocurrir al mismo tiempo, son mutuamente excluyentes. Si A y B son sucesos mutuamente excluyentes, entonces $P(A \text{ o } B) = P(A) + P(B)$.

Example Rolling an even number E and rolling a multiple of five M on a standard number cube are mutually exclusive events.

$P(E \text{ or } M) = P(E) + P(M)$

$= \frac{3}{6} + \frac{1}{6}$

$= \frac{4}{6}$

$= \frac{2}{3}$

N

***n* factorial (p. 837)** The product of the integers from n down to 1, for any positive integer n. You write n factorial as $n!$. The value of $0!$ is defined to be 1.

***n* factorial (p. 837)** Producto de todos los enteros desde n hasta 1, de cualquier entero positivo n. El factorial de n se escribe $n!$. El valor de $0!$ se define como 1.

Example $4! = 4 \cdot 3 \cdot 2 \cdot 1 = 24$

Negation (p. 91) The negation of a statement has the opposite meaning of the original statement.

Negación (p. 91) La negación de un enunciado tiene el sentido opuesto del enunciado original.

Example **Statement:** The angle is obtuse.
Negation: The angle is not obtuse.

Net (p. 4) A net is a two-dimensional pattern that you can fold to form a three-dimensional figure.

Plantilla (p. 4) Una plantilla es una figura bidimensional que se puede doblar para formar una figura tridimensional.

Example

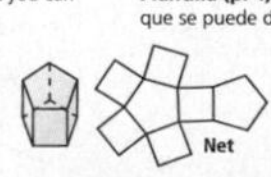

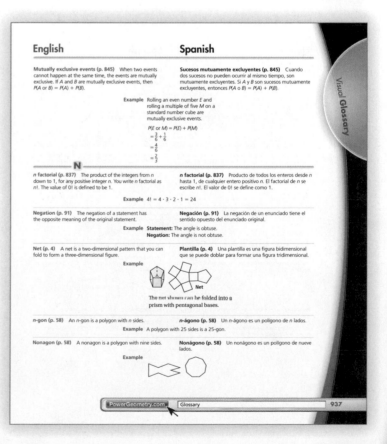

The net shown can be folded into a prism with pentagonal bases.

***n*-gon (p. 58)** An n-gon is a polygon with n sides.

***n*-ágono (p. 58)** Un n-ágono es un polígono de n lados.

Example A polygon with 25 sides is a 25-gon.

Nonagon (p. 58) A nonagon is a polygon with nine sides.

Nonágono (p. 58) Un nonágono es un polígono de nueve lados.

Example

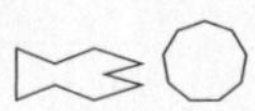

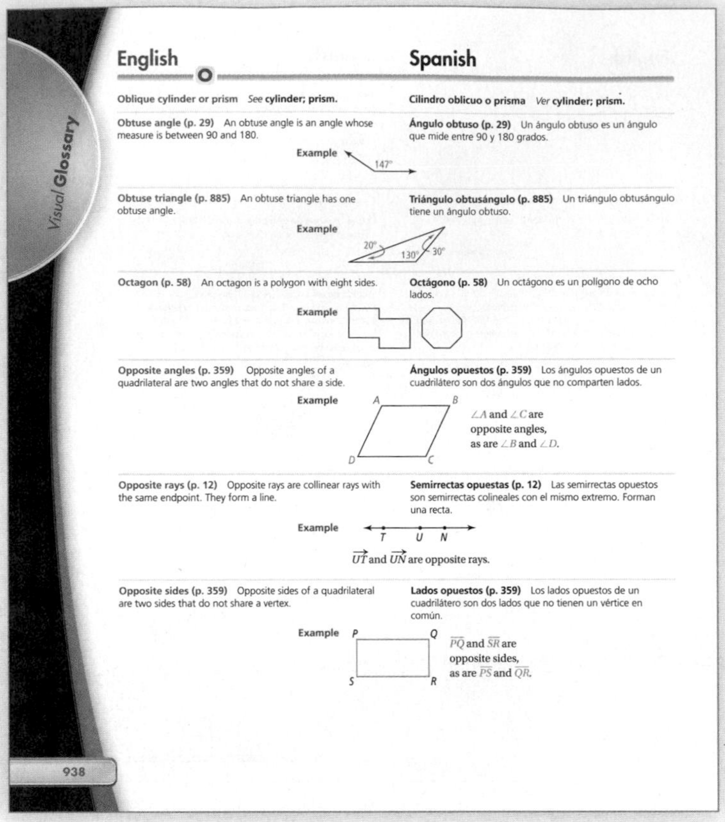

English / Spanish

O

Oblique cylinder or prism *See* **cylinder; prism.**
Cilindro oblicuo o prisma *Ver* **cylinder; prism.**

Obtuse angle (p. 29) An obtuse angle is an angle whose measure is between 90 and 180.
Ángulo obtuso (p. 29) Un ángulo obtuso es un ángulo que mide entre 90 y 180 grados.

Example 147°

Obtuse triangle (p. 885) An obtuse triangle has one obtuse angle.
Triángulo obtusángulo (p. 885) Un triángulo obtusángulo tiene un ángulo obtuso.

Example 20°, 130°, 30°

Octagon (p. 58) An octagon is a polygon with eight sides.
Octágono (p. 58) Un octágono es un polígono de ocho lados.

Example

Opposite angles (p. 359) Opposite angles of a quadrilateral are two angles that do not share a side.
Ángulos opuestos (p. 359) Los ángulos opuestos de un cuadrilátero son dos ángulos que no comparten lados.

Example $\angle A$ and $\angle C$ are opposite angles, as are $\angle B$ and $\angle D$.

Opposite rays (p. 12) Opposite rays are collinear rays with the same endpoint. They form a line.
Semirrectas opuestas (p. 12) Las semirrectas opuestos son semirrectas colineales con el mismo extremo. Forman una recta.

Example $\overrightarrow{UT}$ and $\overrightarrow{UN}$ are opposite rays.

Opposite sides (p. 359) Opposite sides of a quadrilateral are two sides that do not share a vertex.
Lados opuestos (p. 359) Los lados opuestos de un cuadrilátero son dos lados que no tienen un vértice en común.

Example $\overline{PQ}$ and $\overline{SR}$ are opposite sides, as are $\overline{PS}$ and $\overline{QR}$.

938

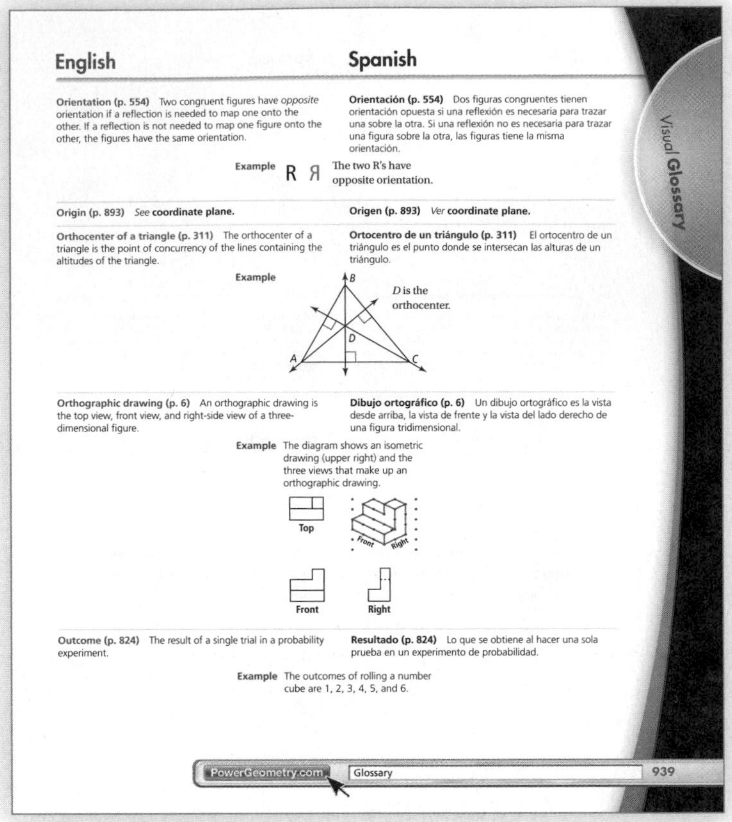

English / Spanish

Orientation (p. 554) Two congruent figures have *opposite* orientation if a reflection is needed to map one onto the other. If a reflection is not needed to map one figure onto the other, the figures have the same orientation.
Orientación (p. 554) Dos figuras congruentes tienen orientación opuesta si una reflexión es necesaria para trazar una sobre la otra. Si una reflexión no es necesaria para trazar una figura sobre la otra, las figuras tiene la misma orientación.

Example R Я The two R's have opposite orientation.

Origin (p. 893) *See* **coordinate plane.**
Origen (p. 893) *Ver* **coordinate plane.**

Orthocenter of a triangle (p. 311) The orthocenter of a triangle is the point of concurrency of the lines containing the altitudes of the triangle.
Ortocentro de un triángulo (p. 311) El ortocentro de un triángulo es el punto donde se intersecan las alturas de un triángulo.

Example *D* is the orthocenter.

Orthographic drawing (p. 6) An orthographic drawing is the top view, front view, and right-side view of a three-dimensional figure.
Dibujo ortográfico (p. 6) Un dibujo ortográfico es la vista desde arriba, la vista de frente y la vista del lado derecho de una figura tridimensional.

Example The diagram shows an isometric drawing (upper right) and the three views that make up an orthographic drawing.

Top, Front, Right

Outcome (p. 824) The result of a single trial in a probability experiment.
Resultado (p. 824) Lo que se obtiene al hacer una sola prueba en un experimento de probabilidad.

Example The outcomes of rolling a number cube are 1, 2, 3, 4, 5, and 6.

PowerGeometry.com Glossary 939

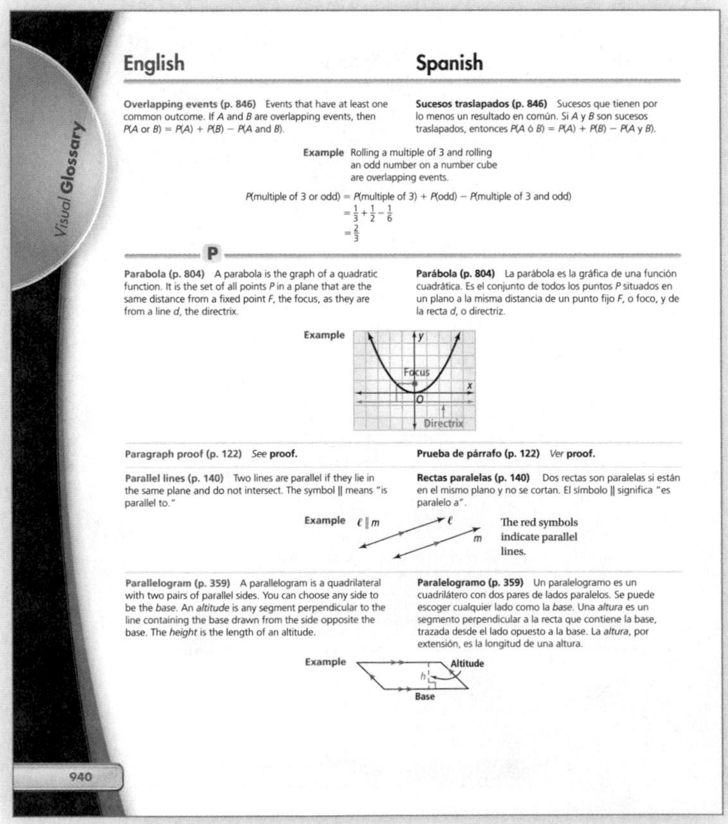

English / Spanish

Overlapping events (p. 846) Events that have at least one common outcome. If *A* and *B* are overlapping events, then $P(A \text{ or } B) = P(A) + P(B) - P(A \text{ and } B)$.
Sucesos traslapados (p. 846) Sucesos que tienen por lo menos un resultado en común. Si *A* y *B* son sucesos traslapados, entonces $P(A \text{ ó } B) = P(A) + P(B) - P(A \text{ y } B)$.

Example Rolling a multiple of 3 and rolling an odd number on a number cube are overlapping events.

$$P(\text{multiple of 3 or odd}) = P(\text{multiple of 3}) + P(\text{odd}) - P(\text{multiple of 3 and odd}) = \frac{1}{3} + \frac{1}{2} - \frac{1}{6} = \frac{2}{3}$$

P

Parabola (p. 804) A parabola is the graph of a quadratic function. It is the set of all points *P* in a plane that are the same distance from a fixed point *F*, the focus, as they are from a line *d*, the directrix.
Parábola (p. 804) La parábola es la gráfica de una función cuadrática. Es el conjunto de todos los puntos *P* situados en un plano a la misma distancia de un punto fijo *F*, o foco, y de la recta *d*, o directriz.

Example Focus, Directrix

Paragraph proof (p. 122) *See* **proof.**
Prueba de párrafo (p. 122) *Ver* **proof.**

Parallel lines (p. 140) Two lines are parallel if they lie in the same plane and do not intersect. The symbol ∥ means "is parallel to."
Rectas paralelas (p. 140) Dos rectas son paralelas si están en el mismo plano y no se cortan. El símbolo ∥ significa "es paralelo a".

Example $\ell \parallel m$ The red symbols indicate parallel lines.

Parallelogram (p. 359) A parallelogram is a quadrilateral with two pairs of parallel sides. You can choose any side to be the *base*. An *altitude* is any segment perpendicular to the line containing the base drawn from the side opposite the base. The *height* is the length of an altitude.
Paralelogramo (p. 359) Un paralelogramo es un cuadrilátero con dos pares de lados paralelos. Se puede escoger cualquier lado como la *base*. Una *altura* es un segmento perpendicular a la recta que contiene la base, trazada desde el lado opuesto a la base. La *altura*, por extensión, es la longitud de una altura.

Example Altitude, *h*, Base

940

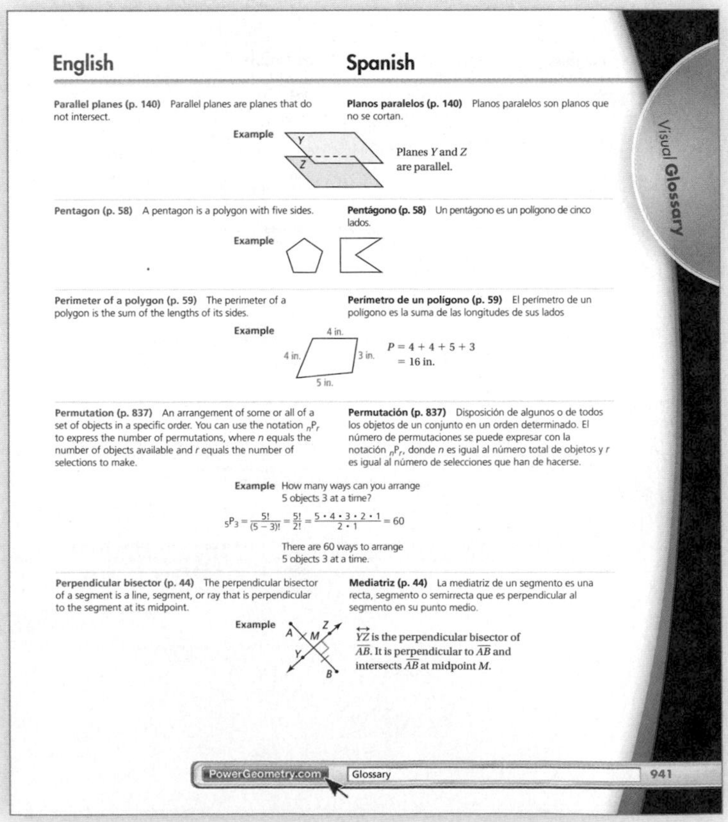

English / Spanish

Parallel planes (p. 140) Parallel planes are planes that do not intersect.
Planos paralelos (p. 140) Planos paralelos son planos que no se cortan.

Example Planes *Y* and *Z* are parallel.

Pentagon (p. 58) A pentagon is a polygon with five sides.
Pentágono (p. 58) Un pentágono es un polígono de cinco lados.

Example

Perimeter of a polygon (p. 59) The perimeter of a polygon is the sum of the lengths of its sides.
Perímetro de un polígono (p. 59) El perímetro de un polígono es la suma de las longitudes de sus lados

Example 4 in., 4 in., 3 in., 5 in.

$P = 4 + 4 + 5 + 3 = 16$ in.

Permutation (p. 837) An arrangement of some or all of a set of objects in a specific order. You can use the notation $_nP_r$ to express the number of permutations, where *n* equals the number of objects available and *r* equals the number of selections to make.
Permutación (p. 837) Disposición de algunos o de todos los objetos de un conjunto en un orden determinado. El número de permutaciones se puede expresar con la notación $_nP_r$, donde *n* es igual al número total de objetos y *r* es igual al número de selecciones que han de hacerse.

Example How many ways can you arrange 5 objects 3 at a time?

$$_5P_3 = \frac{5!}{(5-3)!} = \frac{5!}{2!} = \frac{5 \cdot 4 \cdot 3 \cdot 2 \cdot 1}{2 \cdot 1} = 60$$

There are 60 ways to arrange 5 objects 3 at a time.

Perpendicular bisector (p. 44) The perpendicular bisector of a segment is a line, segment, or ray that is perpendicular to the segment at its midpoint.
Mediatriz (p. 44) La mediatriz de un segmento es una recta, segmento o semirrecta que es perpendicular al segmento en su punto medio.

Example $\overleftrightarrow{YZ}$ is the perpendicular bisector of $\overline{AB}$. It is perpendicular to $\overline{AB}$ and intersects $\overline{AB}$ at midpoint *M*.

PowerGeometry.com Glossary 941

English | Spanish

Perpendicular lines (p. 44) Perpendicular lines are lines that intersect and form right angles. The symbol ⊥ means "is perpendicular to."

Rectas perpendiculares (p. 44) Las rectas perpendiculares son rectas que se cortan y forman ángulos rectos. El símbolo ⊥ significa "es perpendicular a".

Example

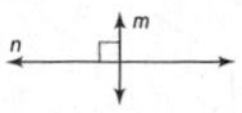

$m \perp n$

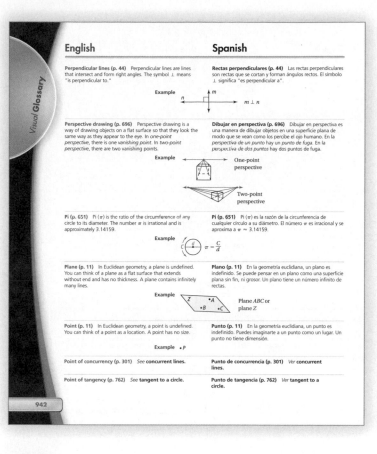

Perspective drawing (p. 696) Perspective drawing is a way of drawing objects on a flat surface so that they look the same way as they appear to the eye. In *one-point perspective*, there is one *vanishing point*. In *two-point perspective*, there are two vanishing points.

Dibujar en perspectiva (p. 696) Dibujar en perspectiva es una manera de dibujar objetos en una superficie plana de modo que se vean como los percibe el ojo humano. En la *perspectiva de un punto* hay un *punto de fuga*. En la *perspectiva de dos puntos* hay dos puntos de fuga.

Example

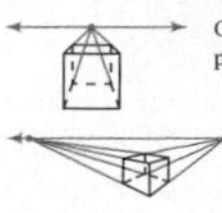

Pi (p. 651) Pi (π) is the ratio of the circumference of any circle to its diameter. The number π is irrational and is approximately 3.14159.

Pi (p. 651) Pi (π) es la razón de la circunferencia de cualquier círculo a su diámetro. El número π es irracional y se aproxima a $\pi \approx 3.14159$.

Example

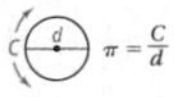

$\pi = \frac{C}{d}$

Plane (p. 11) In Euclidean geometry, a plane is undefined. You can think of a plane as a flat surface that extends without end and has no thickness. A plane contains infinitely many lines.

Plano (p. 11) En la geometría euclidiana, un plano es indefinido. Se puede pensar en un plano como una superficie plana sin fin, ni grosor. Un plano tiene un número infinito de rectas.

Example

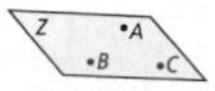

Plane *ABC* or plane *Z*

Point (p. 11) In Euclidean geometry, a point is undefined. You can think of a point as a location. A point has no size.

Punto (p. 11) En la geometría euclidiana, un punto es indefinido. Puedes imaginarte a un punto como un lugar. Un punto no tiene dimensión.

Example • *P*

Point of concurrency (p. 301) *See* **concurrent lines.**

Punto de concurrencia (p. 301) *Ver* **concurrent lines.**

Point of tangency (p. 762) *See* **tangent to a circle.**

Punto de tangencia (p. 762) *Ver* **tangent to a circle.**

English | Spanish

Point-slope form (p. 190) The point-slope form for a nonvertical line with slope *m* and through point (x_1, y_1) is $y - y_1 = m(x - x_1)$.

Forma punto-pendiente (p. 190) La forma punto-pendiente para una recta no vertical con pendiente *m* y que pasa por el punto (x_1, y_1) es $y - y_1 = m(x - x_1)$.

Example $y + 1 = 3(x - 4)$

In this equation, the slope is 3 and (x_1, y_1) is $(4, -1)$.

Point symmetry (p. 568) Point symmetry is the type of symmetry for which there is a rotation of 180° that maps a figure onto itself.

Simetría central (p. 568) La simetría central es un tipo de simetría en la que una figura se ha rotado 180° sobre sí misma.

Example

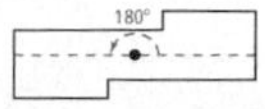

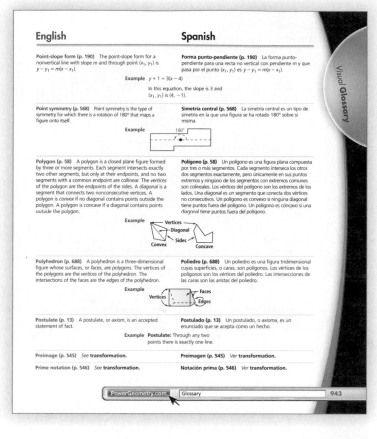

Polygon (p. 58) A polygon is a closed plane figure formed by three or more segments. Each segment intersects exactly two other segments, but only at their endpoints, and no two segments with a common endpoint are collinear. The *vertices* of the polygon are the endpoints of the sides. A *diagonal* is a segment that connects two nonconsecutive vertices. A polygon is *convex* if no diagonal contains points outside the polygon. A polygon is *concave* if a diagonal contains points outside the polygon.

Polígono (p. 58) Un polígono es una figura plana compuesta por tres o más segmentos. Cada segmento interseca los otros dos segmentos exactamente, pero únicamente en sus puntos extremos y ningúno de los segmentos con extremos comunes son colineales. Los *vértices* del polígono son los extremos de los lados. Una *diagonal* es un segmento que conecta dos vértices no consecutivos. Un polígono es *convexo* si ninguna diagonal tiene puntos fuera del polígono. Un polígono es *cóncavo* si una diagonal tiene puntos fuera del polígono.

Example

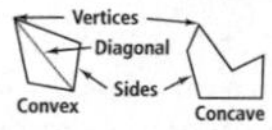

Polyhedron (p. 688) A polyhedron is a three-dimensional figure whose surfaces, or *faces*, are polygons. The vertices of the polygons are the *vertices* of the polyhedron. The intersections of the faces are the *edges* of the polyhedron.

Poliedro (p. 688) Un poliedro es una figura tridimensional cuyas superficies, o *caras*, son polígonos. Los vértices de los polígonos son los *vértices* del poliedro. Las intersecciones de las caras son las *aristas* del poliedro.

Example

Postulate (p. 13) A postulate, or *axiom*, is an accepted statement of fact.

Postulado (p. 13) Un postulado, o *axioma*, es un enunciado que se acepta como un hecho.

Example **Postulate:** Through any two points there is exactly one line.

Preimage (p. 545) *See* **transformation.**

Preimagen (p. 545) *Ver* **transformation.**

Prime notation (p. 546) *See* **transformation.**

Notación prima (p. 546) *Ver* **transformation.**

English | Spanish

Prism (p. 699) A prism is a polyhedron with two congruent and parallel faces, which are called the *bases*. The other faces, which are parallelograms, are called the *lateral faces*. An *altitude* of a prism is a perpendicular segment that joins the planes of the bases. Its length is the *height* of the prism. A *right prism* is one whose lateral faces are rectangular regions and a lateral edge is an altitude. In an *oblique prism*, some or all of the lateral faces are nonrectangular.

Prisma (p. 699) Un prisma es un poliedro con dos caras congruentes paralelas llamadas *bases*. Las otras caras son paralelogramos llamados *caras laterales*. La *altura* de un prisma es un segmento perpendicular que une los planos de las bases. Su longitud es también la *altura* del prisma. En un *prisma rectangular*, las caras laterales son rectangulares y una de las aristas laterales es la altura. En un *prisma oblicuo*, algunas o todas las caras laterales no son rectangulares.

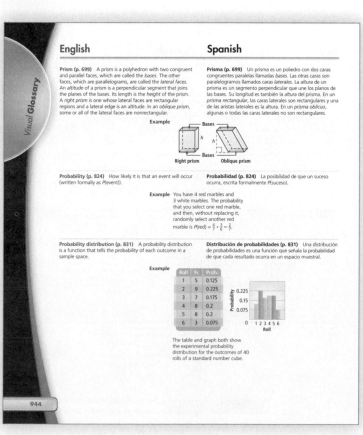

Example

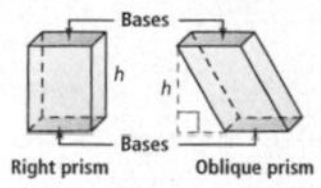

Probability (p. 824) How likely it is that an event will occur (written formally as *P*(event)).

Probabilidad (p. 824) La posibilidad de que un suceso ocurra, escrita formalmente *P*(suceso).

Example You have 4 red marbles and 3 white marbles. The probability that you select one red marble, and then, without replacing it, randomly select another red marble is $P(\text{red}) = \frac{4}{7} \cdot \frac{3}{6} = \frac{2}{7}$.

Probability distribution (p. 831) A probability distribution is a function that tells the probability of each outcome in a sample space.

Distribución de probabilidades (p. 831) Una distribución de probabilidades es una función que señala la probabilidad de que cada resultado ocurra en un espacio muestral.

Example

Roll	Fr.	Prob.
1	5	0.125
2	9	0.225
3	7	0.175
4	8	0.2
5	8	0.2
6	3	0.075

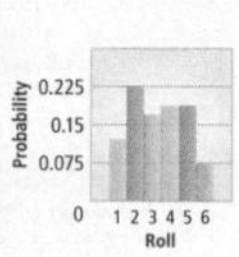

The table and graph both show the experimental probability distribution for the outcomes of 40 rolls of a standard number cube.

English | Spanish

Proof (pp. 115, 122, 158, 317, 408) A proof is a convincing argument that uses deductive reasoning. A proof can be written in many forms. In a two-column proof, the statements and reasons are aligned in columns. In a paragraph proof, the statements and reasons are connected in sentences. In a flow proof, arrows show the logical connections between the statements. In a coordinate proof, a figure is drawn on a coordinate plane and the formulas for slope, midpoint, and distance are used to prove properties of the figure. An indirect proof involves the use of indirect reasoning.

Prueba (pp. 115, 122, 158, 317, 408) Una prueba es un argumento convincente en el cual se usa el razonamiento deductivo. Una prueba se puede escribir de varias maneras. En una *prueba de dos columnas*, los enunciados y las razones se alinean en columnas. En una *prueba de párrafo*, los enunciados y razones están unidos en oraciones. En una *prueba de flujo*, hay flechas que indican las conexiones lógicas entre enunciados. En una *prueba de coordenadas*, se dibuja una figura en un plano de coordenadas y se usan las fórmulas de la pendiente, punto medio y distancia para probar las propiedades de la figura. Una *prueba indirecta* incluye el uso de razonamiento indirecto.

Example

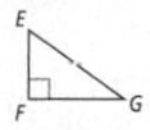

Given: $\triangle EFG$, with right angle $\angle F$
Prove: $\angle E$ and $\angle G$ are complementary.

Paragraph Proof: Because $\angle F$ is a right angle, $m\angle F = 90$. By the Triangle Angle-Sum Theorem, $m\angle E + m\angle F + m\angle G = 180$. By substitution, $m\angle E + 90 + m\angle G = 180$. Subtracting 90 from each side yields $m\angle E + m\angle G = 90$. $\angle E$ and $\angle G$ are complementary by definition.

Proportion (p. 434) A proportion is a statement that two ratios are equal. An *extended proportion* is a statement that three or more ratios are equal.

Proporción (p. 434) Una proporción es un enunciado en el cual dos razones son iguales. Una *proporción extendida* es un enunciado que dice que tres razones o más son iguales.

Example $\frac{x}{5} = \frac{3}{4}$ is a proportion.
$\frac{9}{27} = \frac{3}{9} = \frac{1}{3}$ is an extended proportion.

English	Spanish

Pyramid (p. 708) A pyramid is a polyhedron in which one face, the *base*, is a polygon and the other faces, the *lateral faces*, are triangles with a common vertex, called the *vertex* of the pyramid. An *altitude* of a pyramid is the perpendicular segment from the vertex to the plane of the base. Its length is the *height* of the pyramid. A *regular pyramid* is a pyramid whose base is a regular polygon and whose lateral faces are congruent isosceles triangles. The *slant height* of a regular pyramid is the length of an altitude of a lateral face.

Pirámide (p. 708) Una pirámide es un poliedro en donde una cara, la *base*, es un polígono y las otras caras, las *caras laterales*, son triángulos con un vértice común, llamado el *vértice* de la pirámide. Una *altura* de una pirámide es el segmento perpendicular que va del vértice hasta el plano de la base. Su longitude es, por extensión, la *altura* de la pirámide. Una *pirámide regular* es una pirámide cuya base es un polígono regular y cuyas caras laterales son triángulos isósceles congruentes. La *apotema* de una pirámide regular es la longitude de la altura de la cara lateral.

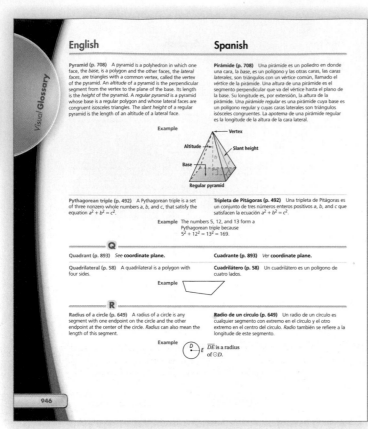

Example

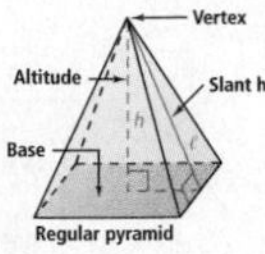

Pythagorean triple (p. 492) A Pythagorean triple is a set of three nonzero whole numbers a, b, and c, that satisfy the equation $a^2 + b^2 = c^2$.

Tripleta de Pitágoras (p. 492) Una tripleta de Pitágoras es un conjunto de tres números enteros positivos a, b, and c que satisfacen la ecuación $a^2 + b^2 = c^2$.

Example The numbers 5, 12, and 13 form a Pythagorean triple because $5^2 + 12^2 = 13^2 = 169$.

Q

Quadrant (p. 893) *See* **coordinate plane.**

Cuadrante (p. 893) *Ver* **coordinate plane.**

Quadrilateral (p. 58) A quadrilateral is a polygon with four sides.

Cuadrilátero (p. 58) Un cuadrilátero es un polígono de cuatro lados.

Example

R

Radius of a circle (p. 649) A radius of a circle is any segment with one endpoint on the circle and the other endpoint at the center of the circle. *Radius* can also mean the length of this segment.

Radio de un círculo (p. 649) Un radio de un círculo es cualquier segmento con extremo en el círculo y el otro extremo en el centro del círculo. *Radio* también se refiere a la longitude de este segmento.

Example

$\overline{DE}$ is a radius of $\odot D$.

English	Spanish

Radius of a regular polygon (p. 629) The radius of a regular polygon is the distance from the center to a vertex.

Radio de un polígono regular (p. 629) El radio de un polígono regular es la distancia desde el centro hasta un vértice.

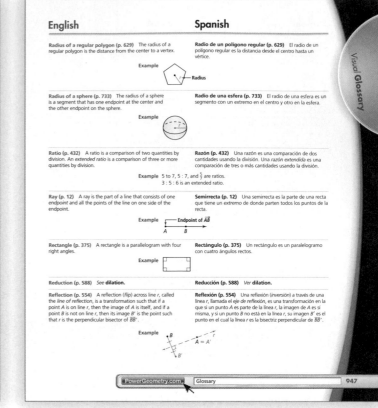

Example

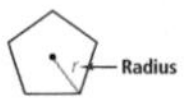

Radius of a sphere (p. 733) The radius of a sphere is a segment that has one endpoint at the center and the other endpoint on the sphere.

Radio de una esfera (p. 733) El radio de una esfera es un segmento con un extremo en el centro y otro en la esfera.

Example

Ratio (p. 432) A ratio is a comparison of two quantities by division. An *extended ratio* is a comparison of three or more quantities by division.

Razón (p. 432) Una razón es una comparación de dos cantidades usando la división. Una *razón extendida* es una comparación de tres o más cantidades usando la división.

Example 5 to 7, 5 : 7, and $\frac{5}{7}$ are ratios.
3 : 5 : 6 is an extended ratio.

Ray (p. 12) A ray is the part of a line that consists of one *endpoint* and all the points of the line on one side of the endpoint.

Semirrecta (p. 12) Una semirrecta es la parte de una recta que tiene un *extremo* de donde parten todos los puntos de la recta.

Example

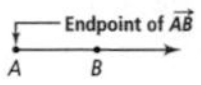

Rectangle (p. 375) A rectangle is a parallelogram with four right angles.

Rectángulo (p. 375) Un rectángulo es un paralelogramo con cuatro ángulos rectos.

Example

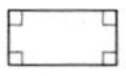

Reduction (p. 588) *See* **dilation.**

Reducción (p. 588) *Ver* **dilation.**

Reflection (p. 554) A reflection (*flip*) across line r, called the *line of reflection*, is a transformation such that if a point A is on line r, then the image of A is itself, and if a point B is not on line r, then its image B' is the point such that r is the perpendicular bisector of $\overline{BB'}$.

Reflexión (p. 554) Una reflexión (*inversión*) a través de una línea r, llamada el *eje de reflexión*, es una transformación en la que si un punto A es parte de la línea r, la imagen de A es sí misma, y si un punto B no está en la línea r, su imagen B' es el punto en el cual la línea r es la bisectriz perpendicular de $\overline{BB'}$.

Example

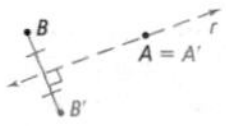

English	Spanish

Reflectional symmetry (p. 568) Reflectional symmetry, or *line symmetry*, is the type of symmetry for which there is a reflection that maps a figure onto itself. The reflection line is the *line of symmetry*. The line of symmetry divides a figure with reflectional symmetry into two congruent halves.

Simetría reflexiva (p. 568) Simetría reflexiva, o *simetría lineal*, es el tipo de simetría donde hay una reflexión que ubica una figura en sí misma. El eje de reflexión es el *eje de simetría*. El eje de simetría divide una figura con simetría reflexiva en dos mitades congruentes.

Example

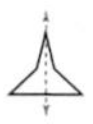

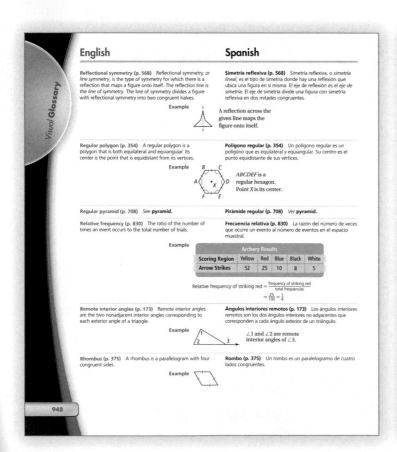

A reflection across the given line maps the figure onto itself.

Regular polygon (p. 354) A regular polygon is a polygon that is both equilateral and equiangular. Its *center* is the point that is equidistant from its vertices.

Polígono regular (p. 354) Un polígono regular es un polígono que es equilateral y equiangular. Su *centro* es el punto equidistante de sus vértices.

Example

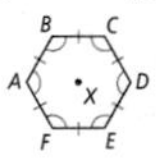

$ABCDEF$ is a regular hexagon. Point X is its center.

Regular pyramid (p. 708) *See* **pyramid.**

Pirámide regular (p. 708) *Ver* **pyramid.**

Relative frequency (p. 830) The ratio of the number of times an event occurs to the total number of trials.

Frecuencia relativa (p. 830) La razón del número de veces que ocurre un evento al número de eventos en el espacio muestral.

Example

Archery Results

Scoring Region	Yellow	Red	Blue	Black	White
Arrow Strikes	52	25	10	8	5

$$\text{Relative frequency of striking red} = \frac{\text{frequency of striking red}}{\text{total frequencies}} = \frac{25}{100} = \frac{1}{4}$$

Remote interior angles (p. 173) Remote interior angles are the two nonadjacent interior angles corresponding to each exterior angle of a triangle.

Ángulos interiores remotos (p. 173) Los ángulos interiores remotos son los dos ángulos interiores no adyacentes que corresponden a cada ángulo exterior de un triángulo.

Example

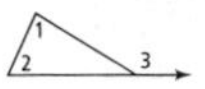

$\angle 1$ and $\angle 2$ are remote interior angles of $\angle 3$.

Rhombus (p. 375) A rhombus is a parallelogram with four congruent sides.

Rombo (p. 375) Un rombo es un paralelogramo de cuatro lados congruentes.

Example

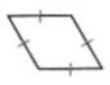

English	Spanish

Right angle (p. 29) A right angle is an angle whose measure is 90.

Ángulo recto (p. 29) Un ángulo recto es un ángulo que mide 90.

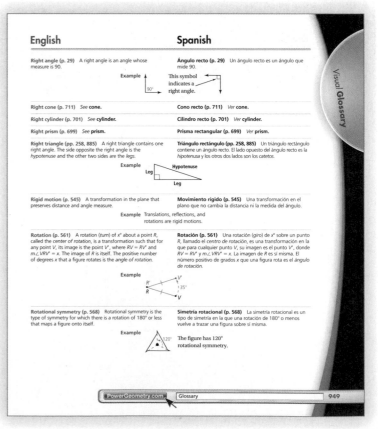

Example

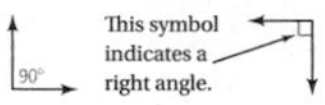

Right cone (p. 711) *See* **cone.**

Cono recto (p. 711) *Ver* **cone.**

Right cylinder (p. 701) *See* **cylinder.**

Cilindro recto (p. 701) *Ver* **cylinder.**

Right prism (p. 699) *See* **prism.**

Prisma rectangular (p. 699) *Ver* **prism.**

Right triangle (pp. 258, 885) A right triangle contains one right angle. The side opposite the right angle is the *hypotenuse* and the other two sides are the *legs*.

Triángulo rectángulo (pp. 258, 885) Un triángulo rectángulo contiene un ángulo recto. El lado opuesto del ángulo recto es la *hipotenusa* y los otros dos lados son los *catetos*.

Example

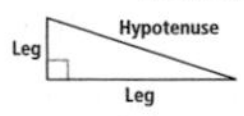

Rigid motion (p. 545) A transformation in the plane that preserves distance and angle measure.

Movimiento rígido (p. 545) Una transformación en el plano que no cambia la distancia ni la medida del ángulo.

Example Translations, reflections, and rotations are rigid motions.

Rotation (p. 561) A rotation (*turn*) of $x°$ about a point R, called the *center of rotation*, is a transformation such that for any point V, its image is the point V', where $RV = RV'$ and $m\angle VRV' = x$. The image of R is itself. The positive number of degrees x that a figure rotates is the *angle of rotation*.

Rotación (p. 561) Una rotación (*giro*) de $x°$ sobre un punto R, llamado el *centro de rotación*, es una transformación en la que para cualquier punto V, su imagen es el punto V', donde $RV = RV'$ y $m\angle VRV' = x$. La imagen de R es sí misma. El número positivo de grados x que una figura rota es el *ángulo de rotación*.

Example

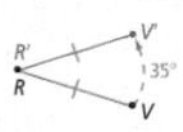

Rotational symmetry (p. 568) Rotational symmetry is the type of symmetry for which there is a rotation of 180° or less that maps a figure onto itself.

Simetría rotacional (p. 568) La simetría rotacional es un tipo de simetría en la que una rotación de 180° o menos vuelve a trazar una figura sobre sí misma.

Example

The figure has 120° rotational symmetry.

Visual **Glossary**

Visual **Glossary**

English

Spanish

S

Same-side interior angles (p. 142) Same-side interior angles lie on the same side of the transversal t and between ℓ and m.

Ángulos internos del mismo lado (p. 142) Los ángulos internos del mismo lado están situados en el mismo lado de la transversal t y dentro de ℓ y m.

Example

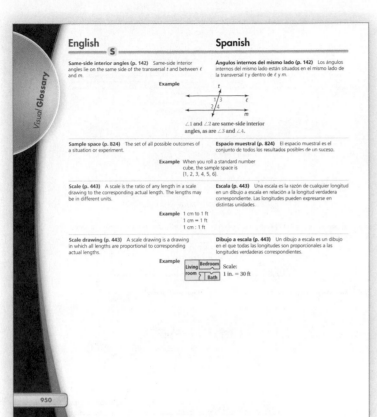

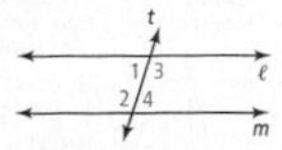

$\angle 1$ and $\angle 2$ are same-side interior angles, as are $\angle 3$ and $\angle 4$.

Sample space (p. 824) The set of all possible outcomes of a situation or experiment.

Espacio muestral (p. 824) El espacio muestral es el conjunto de todos los resultados posibles de un suceso.

Example When you roll a standard number cube, the sample space is {1, 2, 3, 4, 5, 6}.

Scale (p. 443) A scale is the ratio of any length in a scale drawing to the corresponding actual length. The lengths may be in different units.

Escala (p. 443) Una escala es la razón de cualquier longitud en un dibujo a escala en relación a la longitud verdadera correspondiente. Las longitudes pueden expresarse en distintas unidades.

Example 1 cm to 1 ft
1 cm = 1 ft
1 cm : 1 ft

Scale drawing (p. 443) A scale drawing is a drawing in which all lengths are proportional to corresponding actual lengths.

Dibujo a escala (p. 443) Un dibujo a escala es un dibujo en el que todas las longitudes son proporcionales a las longitudes verdaderas correspondientes.

Example

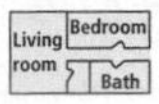

Scale:
1 in. = 30 ft

English

Spanish

Scale factor (pp. 440, 742) A scale factor is the ratio of corresponding linear measurements of two similar figures.

Factor de escala (pp. 440, 742) El factor de escala es la razón de las medidas lineales correspondientes de dos figuras semejantes.

Example

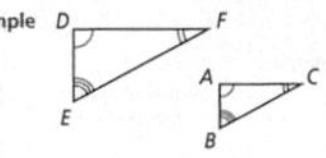

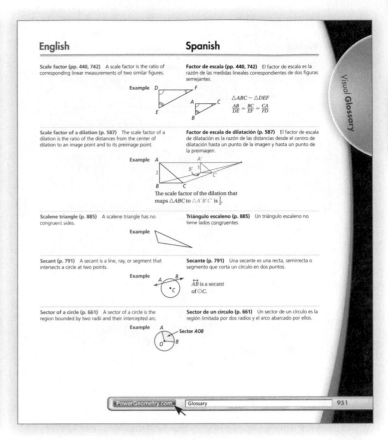

$\triangle ABC \sim \triangle DEF$

$\frac{AB}{DE} = \frac{BC}{EF} = \frac{CA}{FD}$

Scale factor of a dilation (p. 587) The scale factor of a dilation is the ratio of the distances from the center of dilation to an image point and to its preimage point.

Factor de escala de dilatación (p. 587) El factor de escala de dilatación es la razón de las distancias desde el centro de dilatación hasta un punto de la imagen y hasta un punto de la preimagen.

Example

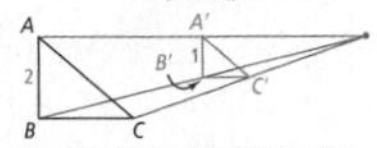

The scale factor of the dilation that maps $\triangle ABC$ to $\triangle A'B'C'$ is $\frac{1}{2}$.

Scalene triangle (p. 885) A scalene triangle has no congruent sides.

Triángulo escaleno (p. 885) Un triángulo escaleno no tiene lados congruentes.

Example

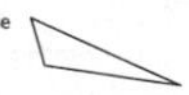

Secant (p. 791) A secant is a line, ray, or segment that intersects a circle at two points.

Secante (p. 791) Una secante es una recta, semirrecta o segmento que corta un círculo en dos puntos.

Example

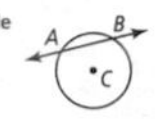

$\overleftrightarrow{AB}$ is a secant of $\odot C$.

Sector of a circle (p. 661) A sector of a circle is the region bounded by two radii and their intercepted arc.

Sector de un círculo (p. 661) Un sector de un círculo es la región limitada por dos radios y el arco abarcado por ellos.

Example

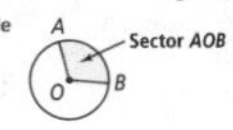

English

Spanish

Segment (p. 12) A segment is the part of a line that consists of two points, called *endpoints*, and all points between them.

Segmento (p. 12) Un segmento es la parte de una recta que tiene dos puntos, llamados *extremos*, entre los cuales están todos los puntos de esa recta.

Example

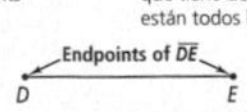

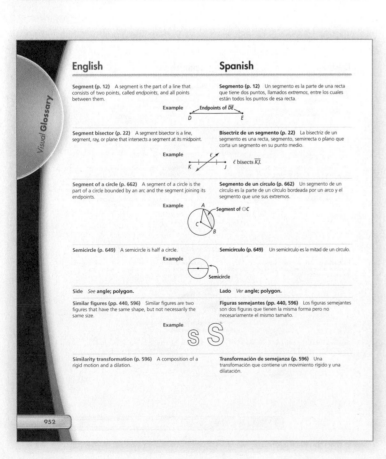

Segment bisector (p. 22) A segment bisector is a line, segment, ray, or plane that intersects a segment at its midpoint.

Bisectriz de un segmento (p. 22) La bisectriz de un segmento es una recta, segmento, semirrecta o plano que corta un segmento en su punto medio.

Example

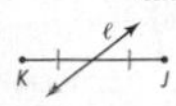

ℓ bisects $\overline{KJ}$.

Segment of a circle (p. 662) A segment of a circle is the part of a circle bounded by an arc and the segment joining its endpoints.

Segmento de un círculo (p. 662) Un segmento de un círculo es la parte de un círculo bordeada por un arco y el segmento que une sus extremos.

Example

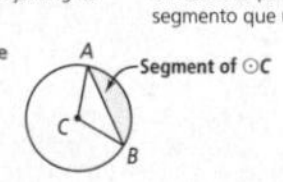

Semicircle (p. 649) A semicircle is half a circle.

Semicírculo (p. 649) Un semicírculo es la mitad de un círculo.

Example

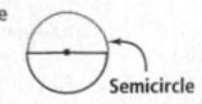

Side *See* **angle; polygon.**

Lado *Ver* **angle; polygon.**

Similar figures (pp. 440, 596) Similar figures are two figures that have the same shape, but not necessarily the same size.

Figuras semejantes (pp. 440, 596) Los figuras semejantes son dos figuras que tienen la misma forma pero no necesariamente el mismo tamaño.

Example

Similarity transformation (p. 596) A composition of a rigid motion and a dilation.

Transformación de semejanza (p. 596) Una transfomación que contiene un movimiento rígido y una dilatación.

English

Spanish

Similar polygons (p. 440) Similar polygons are polygons having corresponding angles congruent and the lengths of corresponding sides proportional. You denote similarity by ~.

Polígonos semejantes (p. 440) Los polígonos semejantes son polígonos cuyos ángulos correspondientes son congruentes y las longitudes de los lados correspondientes son proporcionales. El símbolo ~ significa "es semejante a".

Example

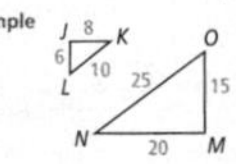

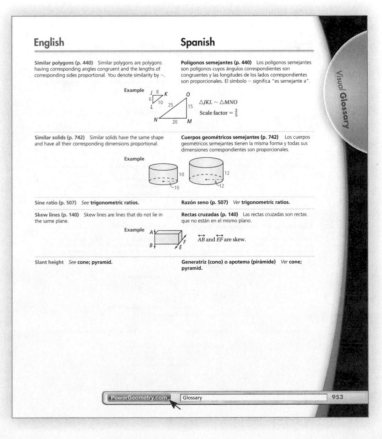

$\triangle JKL \sim \triangle MNO$

Scale factor = $\frac{2}{5}$

Similar solids (p. 742) Similar solids have the same shape and have all their corresponding dimensions proportional.

Cuerpos geométricos semejantes (p. 742) Los cuerpos geométricos semejantes tienen la misma forma y todas sus dimensiones correspondientes son proporcionales.

Example

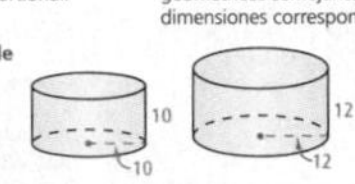

Sine ratio (p. 507) *See* **trigonometric ratios.**

Razón seno (p. 507) *Ver* **trigonometric ratios.**

Skew lines (p. 140) Skew lines are lines that do not lie in the same plane.

Rectas cruzadas (p. 140) Las rectas cruzadas son rectas que no están en el mismo plano.

Example

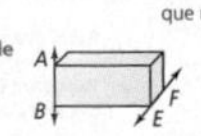

$\overleftrightarrow{AB}$ and $\overleftrightarrow{EF}$ are skew.

Slant height *See* **cone; pyramid.**

Generatriz (cono) o apotema (pirámide) *Ver* **cone; pyramid.**

English | Spanish

Slope-intercept form (p. 190) The slope-intercept form of a linear equation is $y = mx + b$, where m is the slope of the line and b is the y-intercept.

Forma pendiente-intercepto (p. 190) La forma pendiente-intercepto es la ecuación lineal $y = mx + b$, en la que m es la pendiente de la recta y b es el punto de intersección de esa recta con el eje y.

Example $y = \frac{1}{2}x - 3$
In this equation, the slope is $\frac{1}{2}$ and the y-intercept is -3.

Slope of a line (p. 189) The slope of a line is the ratio of its vertical change in the coordinate plane to the corresponding horizontal change. If (x_1, y_1) and (x_2, y_2) are points on a nonvertical line, then the slope is $\frac{y_2 - y_1}{x_2 - x_1}$. The slope of a horizontal line is 0 and the slope of a vertical line is undefined.

Pendiente de una recta (p. 189) La pendiente de una recta es la razón del cambio vertical en el plano de coordenadas en relación al cambio horizontal correspondiente. Si (x_1, y_1) y (x_2, y_2) son puntos en una recta no vertical, entonces la pendiente es $\frac{y_2 - y_1}{x_2 - x_1}$. La pendiente de una recta horizontal es 0, y la pendiente de una recta vertical es indefinida.

Example

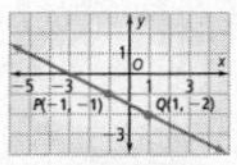

The line containing $P(-1, -1)$ and $Q(1, -2)$ has slope $\frac{-2-(-1)}{1-(-1)} = \frac{-1}{2} = -\frac{1}{2}$.

Space (p. 12) Space is the set of all points.

Espacio (p. 12) El espacio es el conjunto de todos los puntos.

Sphere (p. 733) A sphere is the set of all points in space that are a given distance r, the *radius*, from a given point C, the *center*. A *great circle* is the intersection of a sphere with a plane containing the center of the sphere. The *circumference* of a sphere is the circumference of any great circle of the sphere.

Esfera (p. 733) Una esfera es el conjunto de los puntos del espacio que están a una distancia dada r, el *radio*, de un punto dado C, el *centro*. Un *círculo máximo* es la intersección de una esfera y un plano que contiene el centro de la esfera. La *circunferencia* de una esfera es la circunferencia de cualquier círculo máximo de la esfera.

Example

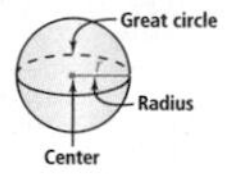

English | Spanish

Spherical geometry (p. 179) In spherical geometry, a plane is considered to be the surface of a sphere and a line is considered to be a great circle of the sphere. In spherical geometry, through a point not on a given line there is no line parallel to the given line.

Geometría esférica (p. 179) En la geometría esférica, un plano es la superficie de una esfera y una recta es un círculo máximo de la esfera. En la geometría esférica, a través de un punto que no está en una recta dada, no hay recta paralela a la recta dada.

Example

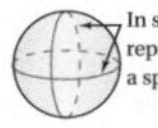

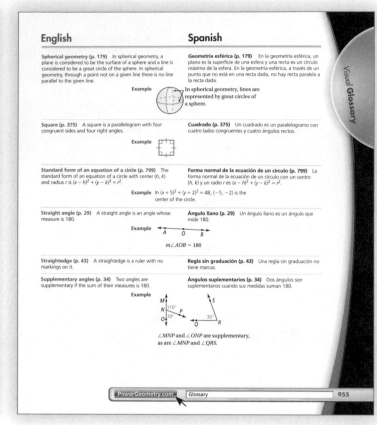

In spherical geometry, lines are represented by great circles of a sphere.

Square (p. 375) A square is a parallelogram with four congruent sides and four right angles.

Cuadrado (p. 375) Un cuadrado es un paralelogramo con cuatro lados congruentes y cuatro ángulos rectos.

Example

Standard form of an equation of a circle (p. 799) The standard form of an equation of a circle with center (h, k) and radius r is $(x - h)^2 + (y - k)^2 = r^2$.

Forma normal de la ecuación de un círculo (p. 799) La forma normal de la ecuación de un círculo con un centro (h, k) y un radio r es $(x - h)^2 + (y - k)^2 = r^2$.

Example In $(x + 5)^2 + (y + 2)^2 = 48$, $(-5, -2)$ is the center of the circle.

Straight angle (p. 29) A straight angle is an angle whose measure is 180.

Ángulo llano (p. 29) Un ángulo llano es un ángulo que mide 180.

Example

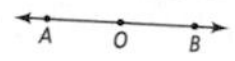

$m\angle AOB = 180$

Straightedge (p. 43) A straightedge is a ruler with no markings on it.

Regla sin graduación (p. 43) Una regla sin graduación no tiene marcas.

Supplementary angles (p. 34) Two angles are supplementary if the sum of their measures is 180.

Ángulos suplementarios (p. 34) Dos ángulos son suplementarios cuando sus medidas suman 180.

Example

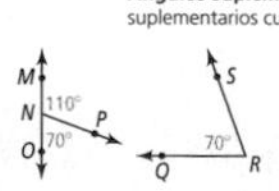

$\angle MNP$ and $\angle ONP$ are supplementary, as are $\angle MNP$ and $\angle QRS$.

English | Spanish

Surface area (pp. 700, 702, 709, 711, 734) The surface area of a prism, cylinder, pyramid, or cone is the sum of the lateral area and the areas of the bases. The surface area of a sphere is four times the area of a great circle. A list of surface area formulas is on p. 839.

Área (pp. 700, 702, 709, 711, 734) El área de un prisma, pirámide, cilindro o cono es la suma del área lateral y las áreas de las bases. El área de una esfera es igual a cuatro veces el área de un círculo máximo. Una lista de fórmulas de áreas está en la p. 839.

Example

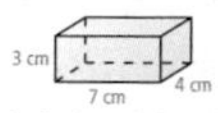

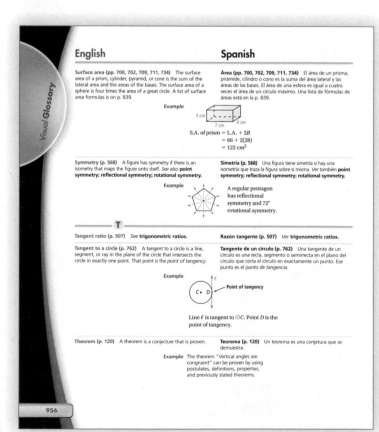

S.A. of prism = L.A. + 2B
= 66 + 2(28)
= 122 cm²

Symmetry (p. 568) A figure has symmetry if there is an isometry that maps the figure onto itself. *See also* **point symmetry; reflectional symmetry; rotational symmetry.**

Simetría (p. 568) Una figura tiene simetría si hay una isometría que traza la figura sobre sí misma. *Ver también* **point symmetry; reflectional symmetry; rotational symmetry.**

Example

A regular pentagon has reflectional symmetry and 72° rotational symmetry.

T

Tangent ratio (p. 507) *See* **trigonometric ratios.**

Razón tangente (p. 507) *Ver* **trigonometric ratios.**

Tangent to a circle (p. 762) A tangent to a circle is a line, segment, or ray in the plane of the circle that intersects the circle in exactly one point. That point is the *point of tangency*.

Tangente de un círculo (p. 762) Una tangente de un círculo es una recta, segmento o semirrecta en el plano del círculo que corta el círculo en exactamente un punto. Ese punto es el *punto de tangencia*.

Example

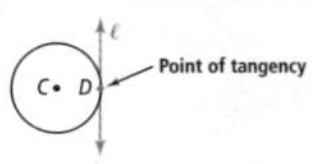

Line ℓ is tangent to $\odot C$. Point D is the point of tangency.

Theorem (p. 120) A theorem is a conjecture that is proven.

Teorema (p. 120) Un teorema es una conjetura que se demuestra.

Example The theorem "Vertical angles are congruent" can be proven by using postulates, definitions, properties, and previously stated theorems.

English | Spanish

Theoretical probability (p. 825) The ratio of the number of favorable outcomes to the number of possible outcomes if all outcomes have the same chance of happening.

$P(\text{event}) = \frac{\text{number of favorable outcomes}}{\text{number of possible outcomes}}$

Probabilidad teórica (p. 825) Si cada resultado tiene la misma probabilidad de darse, la probabilidad teórica de un suceso se calcula como la razón del número de resultados favorables al número de resultados posibles.

$P(\text{suceso}) = \frac{\text{numero de resultados favorables}}{\text{numero de resultados posibles}}$

Example In tossing a coin, the events of getting heads or tails are equally likely. The likelihood of getting heads is $P(\text{heads}) = \frac{1}{2}$.

Transformation (p. 545) A transformation is a change in the position, size, or shape of a geometric figure. The given figure is called the *preimage* and the resulting figure is called the *image*. A transformation *maps* a figure onto its image. *Prime notation* is sometimes used to identify image points. In the diagram, X' (read "X prime") is the image of X.

Transformación (p. 545) Una transformación es un cambio en la posición, tamaño o forma de una figura. La figura dada se llama la *preimagen* y la figura resultante se llama la *imagen*. Una transformación *traza* la figura sobre su propia imagen. La *notación prima* a veces se utilize para identificar los puntos de la imagen. En el diagrama de la derecha, X' (leído X prima) es la imagen de X.

Example

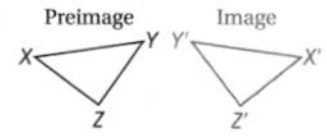

$\triangle XYZ \rightarrow \triangle X'Y'Z'$

Translation (p. 547) A translation (*slide*) is a transformation that moves points the same distance and in the same direction.

Traslación (p. 547) Una traslación (*desplazamiento*) es una transformación en la que se mueven puntos la misma distancia en la misma dirección.

Example

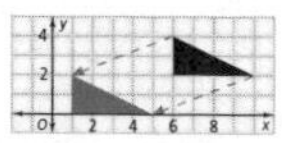

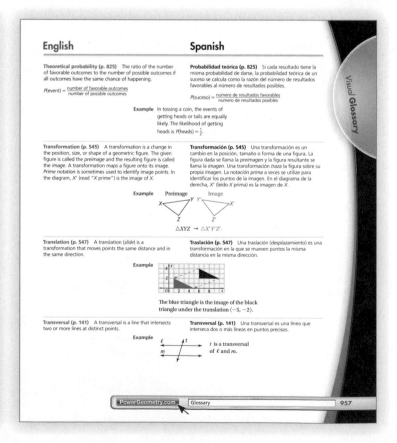

The blue triangle is the image of the black triangle under the translation $\langle -5, -2 \rangle$.

Transversal (p. 141) A transversal is a line that intersects two or more lines at distinct points.

Transversal (p. 141) Una transversal es una línea que interseca dos o más líneas en puntos precisos.

Example

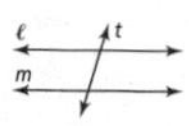

t is a transversal of ℓ and m.

English | Spanish

Trapezoid (p. 389) A trapezoid is a quadrilateral with exactly one pair of parallel sides, the *bases*. The nonparallel sides are called the *legs* of the trapezoid. Each pair of angles adjacent to a base are *base angles* of the trapezoid. An *altitude* of a trapezoid is a perpendicular segment from one base to the line containing the other base. Its length is called the *height* of the trapezoid.

Trapecio (p. 389) Un trapecio es un cuadrilátero con exactamente un par de lados paralelos, las *bases*. Los lados no paralelos se llaman los *catetos* del trapecio. Cada par de ángulos adyacentes a la base son los *ángulos de base* del trapecio. Una *altura* del trapecio es un segmento perpendicular que va de una base a la recta que contiene la otra base. Su longitud se llama, por extensión, la *altura* del trapecio.

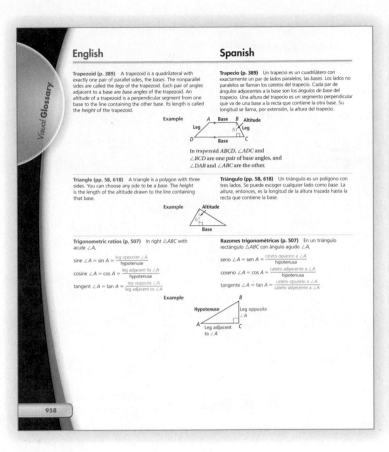

Example

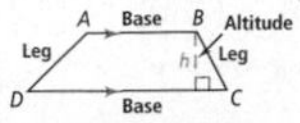

In trapezoid $ABCD$, $\angle ADC$ and $\angle BCD$ are one pair of base angles, and $\angle DAB$ and $\angle ABC$ are the other.

Triangle (pp. 58, 618) A triangle is a polygon with three sides. You can choose any side to be a *base*. The *height* is the length of the altitude drawn to the line containing that base.

Triángulo (pp. 58, 618) Un triángulo es un polígono con tres lados. Se puede escoger cualquier lado como *base*. La *altura*, entonces, es la longitud de la altura trazada hasta la recta que contiene la base.

Example

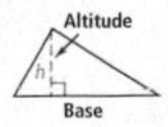

Trigonometric ratios (p. 507) In right $\triangle ABC$ with acute $\angle A$,

sine $\angle A = \sin A = \dfrac{\text{leg opposite } \angle A}{\text{hypotenuse}}$

cosine $\angle A = \cos A = \dfrac{\text{leg adjacent to } \angle A}{\text{hypotenuse}}$

tangent $\angle A = \tan A = \dfrac{\text{leg opposite } \angle A}{\text{leg adjacent to } \angle A}$

Razones trigonométricas (p. 507) En un triángulo rectángulo $\triangle ABC$ con ángulo agudo $\angle A$,

seno $\angle A = \text{sen } A = \dfrac{\text{cateto opuesto a } \angle A}{\text{hipotenusa}}$

coseno $\angle A = \cos A = \dfrac{\text{cateto adyecente a } \angle A}{\text{hipotenusa}}$

tangente $\angle A = \tan A = \dfrac{\text{cateto opuesto a } \angle A}{\text{cateto adyecente a } \angle A}$

Example

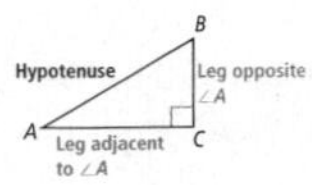

English | Spanish

Truth table (p. 97) A truth table is a table that lists all the possible combinations of truth values for two or more statements.

Tabla de verdad (p. 97) Una tabla de verdad es una tabla que muestra todas las combinaciones posibles de valores de verdad de dos o más enunciados.

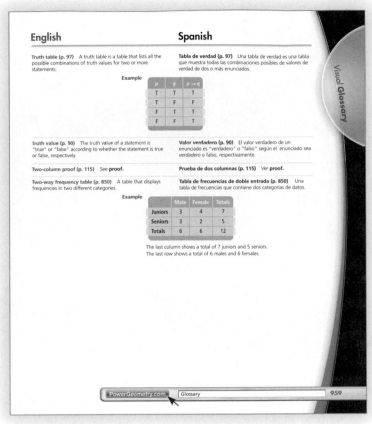

Example

p	q	$p \to q$
T	T	T
T	F	F
F	T	T
F	F	T

Truth value (p. 90) The truth value of a statement is "true" or "false" according to whether the statement is true or false, respectively.

Valor verdadero (p. 90) El valor verdadero de un enunciado es "verdadero" o "falso" según el enunciado sea *verdadero* o falso, respectivamente.

Two-column proof (p. 115) *See* **proof.**

Prueba de dos columnas (p. 115) *Ver* **proof.**

Two-way frequency table (p. 850) A table that displays frequencies in two different categories.

Tabla de frecuencias de doble entrada (p. 850) Una tabla de frecuencias que contiene dos categorías de datos.

Example

	Male	Female	Totals
Juniors	3	4	7
Seniors	3	2	5
Totals	6	6	12

The last column shows a total of 7 juniors and 5 seniors.
The last row shows a total of 6 males and 6 females.

English | Spanish

V

Vertex *See* **angle; cone; polygon; polyhedron; pyramid.** The plural form of *vertex* is *vertices*.

Vértice *Ver* **angle; cone; polygon; polyhedron; pyramid.**

Vertex angle (p. 250) *See* **isosceles triangle.**

Ángulo del vértice (p. 250) *Ver* **isosceles triangle.**

Vertical angles (p. 34) Vertical angles are two angles whose sides form two pairs of opposite rays.

Ángulos opuestos por el vértice (p. 34) Dos ángulos son ángulos opuestos por el vértice si sus lados son semirrectas opuestas.

Example

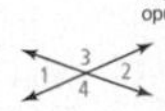

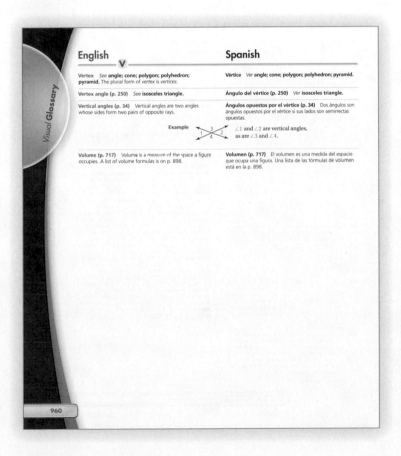

$\angle 1$ and $\angle 2$ are vertical angles, as are $\angle 3$ and $\angle 4$.

Volume (p. 717) Volume is a measure of the space a figure occupies. A list of volume formulas is on p. 898.

Volumen (p. 717) El volumen es una medida del espacio que ocupa una figura. Una lista de las fórmulas de volumen está en la p. 898.

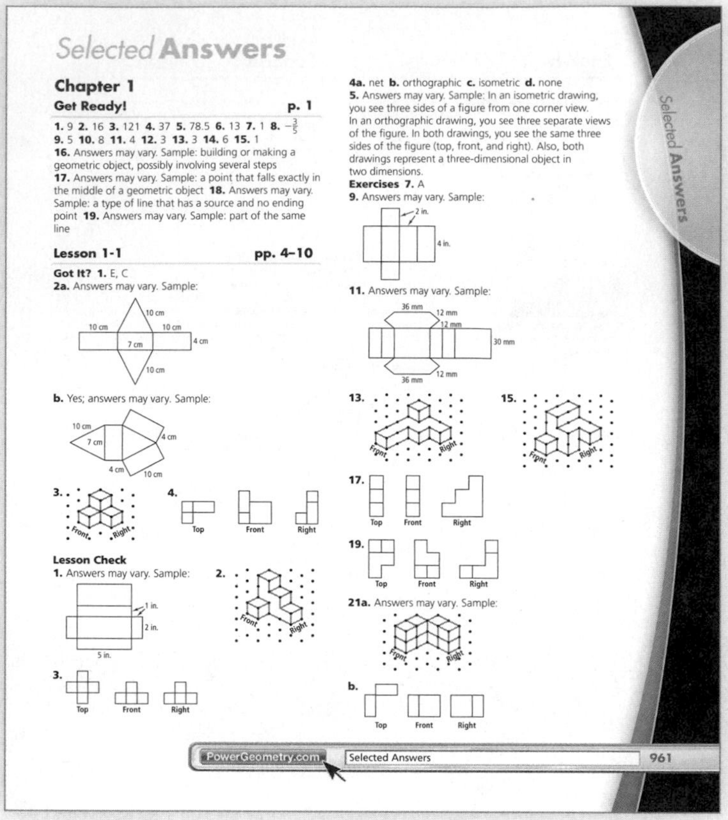

Selected Answers

Chapter 1

Get Ready! p. 1

1. 9 **2.** 16 **3.** 121 **4.** 37 **5.** 78.5 **6.** 13 **7.** 1 **8.** $-\frac{3}{5}$ **9.** 5 **10.** 8 **11.** 4 **12.** 3 **13.** 3 **14.** 6 **15.** 1 **16.** Answers may vary. Sample: building or making a geometric object, possibly involving several steps **17.** Answers may vary. Sample: a point that falls exactly in the middle of a geometric object **18.** Answers may vary. Sample: a type of line that has a source and no ending point **19.** Answers may vary. Sample: part of the same line

Lesson 1-1 pp. 4–10

Got It? 1. E, C
2a. Answers may vary. Sample:
b. Yes; answers may vary. Sample:
3. **4.**

Lesson Check
1. Answers may vary. Sample: **2.**
3.

4a. net **b.** orthographic **c.** isometric **d.** none
5. Answers may vary. Sample: In an isometric drawing, you see three sides of a figure from one corner view. In an orthographic drawing, you see three separate views of the figure. In both drawings, you see the same three sides of the figure (top, front, and right). Also, both drawings represent a three-dimensional object in two dimensions.
Exercises 7. A
9. Answers may vary. Sample:
11. Answers may vary. Sample:
13. **15.**
17.
19.
21a. Answers may vary. Sample:
b.

PowerGeometry.com | Selected Answers | 961

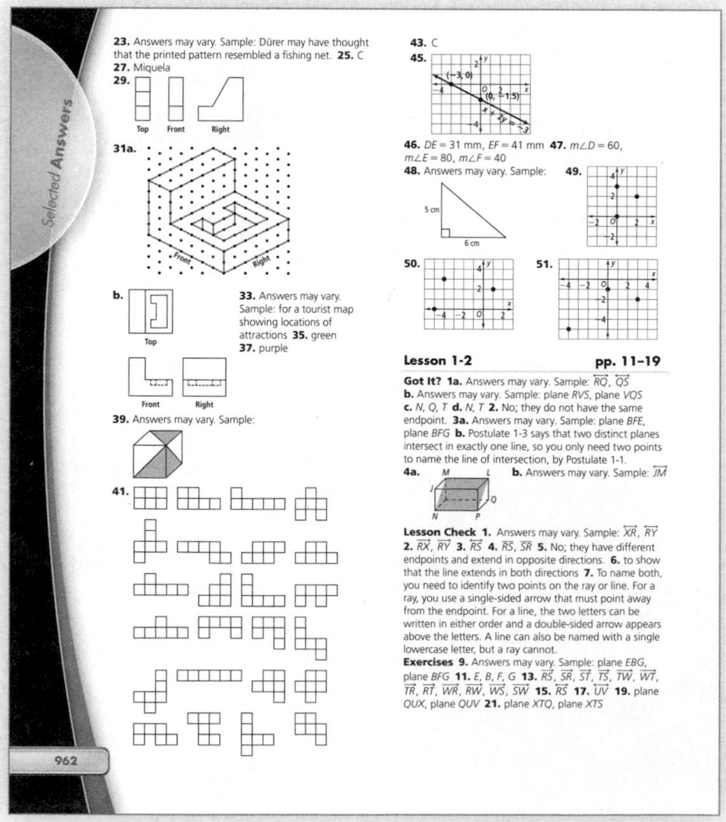

23. Answers may vary. Sample: Dürer may have thought that the printed pattern resembled a fishing net. **25.** C **27.** Miquela
29.
31a.
b.
33. Answers may vary. Sample: for a tourist map showing locations of attractions **35.** green **37.** purple
39. Answers may vary. Sample:
41.

43. C
45.
46. $DE = 31$ mm, $EF = 41$ mm **47.** $m\angle D = 60$, $m\angle E = 80$, $m\angle F = 40$
48. Answers may vary. Sample: **49.**
50. **51.**

Lesson 1-2 pp. 11–19

Got It? 1a. Answers may vary. Sample: $\overleftrightarrow{RQ}$, $\overleftrightarrow{QS}$ **b.** Answers may vary. Sample: plane *RVS*, plane *VQS* **c.** *N, Q, T* **d.** *N, T* **2.** No; they do not have the same endpoint. **3a.** Answers may vary. Sample: plane *BFE*, plane *BFG* **b.** Postulate 1-3 says that two distinct planes intersect in exactly one line, so you only need two points to name the line of intersection, by Postulate 1-1.
4a. **b.** Answers may vary. Sample: $\overleftrightarrow{JM}$

Lesson Check 1. Answers may vary. Sample: $\overleftrightarrow{XR}$, $\overleftrightarrow{RY}$ **2.** $\overrightarrow{RX}$, $\overrightarrow{RY}$ **3.** $\overleftrightarrow{RS}$ **4.** $\overline{RS}$, $\overline{SR}$ **5.** No; they have different endpoints and extend in opposite directions. **6.** to show that the line extends in both directions **7.** To name both, you need to identify two points on the ray or line. For a ray, you use a single-sided arrow that must point away from the endpoint. For a line, the two letters can be written in either order and a double-sided arrow appears above the letters. A line can also be named with a single lowercase letter, but a ray cannot.
Exercises 9. Answers may vary. Sample: plane *EBG*, plane *BFG* **11.** *E, B, F, G* **13.** $\overrightarrow{RS}$, $\overrightarrow{SR}$, $\overrightarrow{ST}$, $\overrightarrow{TS}$, $\overrightarrow{TW}$, $\overrightarrow{WT}$, $\overrightarrow{TR}$, $\overrightarrow{RT}$, $\overrightarrow{WR}$, $\overrightarrow{RW}$, $\overrightarrow{WS}$, $\overrightarrow{SW}$ **15.** $\overleftrightarrow{RS}$ **17.** $\overleftrightarrow{UV}$ **19.** plane *QUX*, plane *QUV* **21.** plane *XTQ*, plane *XTS*

962

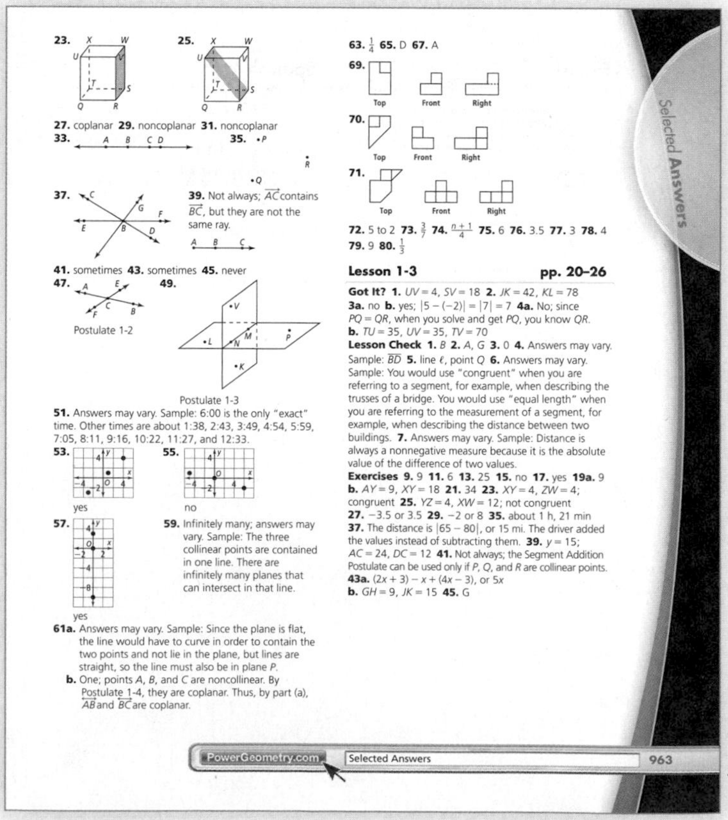

23. **25.**
27. coplanar **29.** noncoplanar **31.** noncoplanar
33. **35.**
37.
39. Not always; $\overrightarrow{AC}$ contains $\overrightarrow{BC}$, but they are not the same ray.
41. sometimes **43.** sometimes **45.** never
47. Postulate 1-2 **49.** Postulate 1-3
51. Answers may vary. Sample: 6:00 is the only "exact" time. Other times are about 1:38, 2:43, 3:49, 4:54, 5:59, 7:05, 8:11, 9:16, 10:22, 11:27, and 12:33.
53. yes **55.** no
57. yes
59. Infinitely many; answers may vary. Sample: The three collinear points are contained in one line. There are infinitely many planes that can intersect in that line.
61a. Answers may vary. Sample: Since the plane is flat, the line would have to curve in order to contain the two points and not lie in the plane, but lines are straight, so the line must also be in plane *P*.
b. One; points *A*, *B*, and *C* are noncollinear. By Postulate 1-4, they are coplanar. Thus, by part (a), $\overleftrightarrow{AB}$ and $\overleftrightarrow{BC}$ are coplanar.

63. $\frac{1}{4}$ **65.** D **67.** A
69.
70.
71.
72. 5 to 2 **73.** $\frac{3}{7}$ **74.** $\frac{n+1}{4}$ **75.** 6 **76.** 3.5 **77.** 3 **78.** 4 **79.** 9 **80.** $\frac{1}{3}$

Lesson 1-3 pp. 20–26

Got It? 1. $UV = 4$, $SV = 18$ **2.** $JK = 42$, $KL = 78$ **3a.** no **b.** yes; $|5 - (-2)| = |7| = 7$ **4a.** No; since $PQ = QR$, when you solve and get PQ, you know QR. **b.** $TU = 35$, $UV = 35$, $TV = 70$
Lesson Check 1. *B* **2.** *A*, *G* **3.** 0 **4.** Answers may vary. Sample: $\overline{BD}$ **5.** line ℓ, point *Q* **6.** Answers may vary. Sample: You would use "congruent" when you are referring to a segment, for example, when describing the trusses of a bridge. You would use "equal length" when you are referring to the measurement of a segment, for example, when describing the distance between two buildings. **7.** Answers may vary. Sample: Distance is always a nonnegative measure because it is the absolute value of the difference of two values.
Exercises 9. 9 **11.** 6 **13.** 25 **15.** no **17.** yes **19a.** 9 **b.** $AY = 9$, $XY = 18$ **21.** 34 **23.** $XY = 4$, $ZW = 4$; congruent **25.** $YZ = 4$, $XW = 12$; not congruent **27.** −3.5 or 3.5 **29.** −2 or 8 **35.** about 1 h, 21 min **37.** The distance is $|65 - 80|$, or 15 mi. The driver added the values instead of subtracting them. **39.** $y = 15$; $AC = 24$, $DC = 12$ **41.** Not always; the Segment Addition Postulate can be used only if *P*, *Q*, and *R* are collinear points. **43a.** $(2x + 3) - x + (4x - 3)$, or $5x$ **b.** $GH = 9$, $JK = 15$ **45.** G

PowerGeometry.com | Selected Answers | 963

Lesson 1-4 pp. 27–33

Got It? 1a. $\angle LMK$, $\angle 2$ **b.** No; since there are three ∡ that have vertex M, it would not be clear which one you intended. **2.** $m\angle LKH = 35$, acute; $m\angle HKN = 180$, straight; $m\angle MKH = 145$, obtuse **3.** 49 **4.** $m\angle DEC = 142$, $m\angle CEF = 38$
Lesson Check 1. $\angle ABC$, $\angle CBA$ **2.** $85 - x$ **3.** acute **4.** 0 or 1; congruent ∡ may be two separate angles, or they may have the same vertex and share one side. **5.** No; the diagram is not marked with ≅ ∡.
Exercises 7. $\angle ABC$, $\angle CBA$, $\angle B$, or $\angle 1$ **9.** 70, acute **11.** 110, obtuse **13.** 85, acute **15.** Answers may vary. Sample:

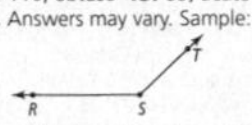

17.

19. $\angle BJA$ **21.** 130 **23.** $m\angle RQS = 43$, $m\angle TQS = 137$ **25.** about 90°; right **27.** about 88°; acute **29.** $x = 8$; $m\angle AOB = 30$, $m\angle BOC = 50$, $m\angle COD = 30$ **31.** A **33.** 180 **35.** 30 **37.** 40 **39a.** yes;

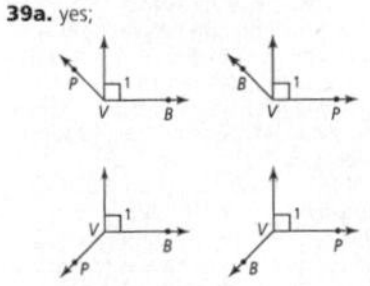

Lesson 1-5 pp. 34–40

Got It? 1a. Yes; $\angle AFE$ and $\angle CFD$ are formed by opposite rays $\overrightarrow{FA}$, $\overrightarrow{FD}$, $\overrightarrow{FC}$, and $\overrightarrow{FE}$. **b.** No; $m\angle BFC = 28$ and $m\angle DFE = 118$, so $28 + 118 \neq 180$. **c.** Yes; $\angle BFD$ and $\angle AFB$ share $\overrightarrow{FB}$, and they have no common interior points. **2a.** Yes; they have corresponding ≅ tick marks. **b.** No; they do not have corresponding ≅ tick marks. **c.** No; it (or its supplements) do not have a right angle symbol. **d.** No; $\overline{PW}$ and $\overline{WQ}$ do not have corresponding ≅ tick marks. **3a.** Adding the measures of both ∡ should give 180. **b.** $m\angle ADB = 77$, $m\angle BDC = 103$ **4.** 36

Lesson Check 1–3. Answers may vary. Samples are given. **1.** $\angle AFE$ and $\angle CFD$ (or $\angle AFC$ and $\angle EFD$) **2.** $\angle AEF$ and $\angle DEF$ (or $\angle AEC$ and $\angle DEC$) **3.** $\angle BCE$ and $\angle ECD$ (or any two adjacent ∡ with common vertex F) **4.** 20 **5.** Answers may vary. Sample: The angles combine to form a line. **6.** Since the ∡ are complementary, the sum of the two measures should be 90, not 180. So, $x = 15$.
Exercises 7. Yes; the angles share a common side and vertex, and have no interior points in common. **9.** No; they are supplementary. **11.** $\angle DOC$, $\angle AOB$ **13.** $\angle EOC$ **15.** Answers may vary. Sample: $\angle AOB$, $\angle DOC$ **17.** No; they are not marked as ≅. **19.** Yes. Answers may vary. Sample: The two ∡ form a linear pair. **21.** No; $\overline{JC}$ and $\overline{CD}$ are not marked as ≅. **23.** Yes; they are formed by $\overleftrightarrow{JF}$ and $\overleftrightarrow{ED}$. **25.** $m\angle EFG = 69$, $m\angle GFH = 111$ **27.** $x = 5$, $m\angle ABC = 50$ **29.** $x = 11$, $m\angle ABC = 56$ **31.** 120; 60 **33.** 90 **35.** 155 **37a.** 19.5 **b.** $m\angle RQS = 43$, $m\angle TQS = 137$ **c.** Answers may vary. Sample: $43 + 137 = 180$ **39.** Both are correct; if you multiply both sides of the equation $m\angle ABX = \frac{1}{2}m\angle ABC$ by 2, you get $2m\angle ABX = m\angle ABC$. **41.** The four vertical angles are all right angles. **43.** $\angle KML$ **45.** $\angle PMR$, $\angle KML$, $\angle KMQ$, $\angle MQP$ **47.** 30

Lesson 1-6 pp. 43–48

Got It?

1.

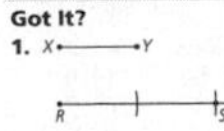

2a.

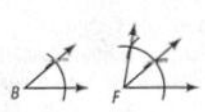

b. Answers may vary. Sample: You use a compass setting to copy a distance.

3.

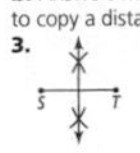

4.

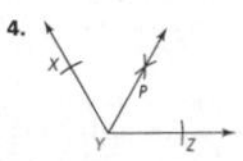

Lesson Check

1.

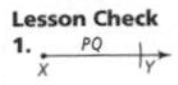

2.

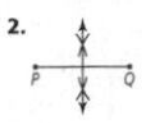

3.

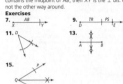

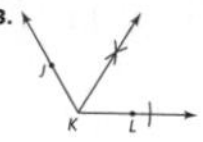

4. compass, straightedge **5.** Answers may vary. Sample: When you sketch a figure, it does not require accurate measurements for angles and sides. When you draw a figure with a ruler and protractor, you use measurements to determine the lengths of sides or the sizes of angles. When you construct a figure, the only tools you use are a compass and straightedge. **6.** Since $\overleftrightarrow{XY}$ is ⊥ to and contains the midpoint of $\overline{AB}$, then $\overleftrightarrow{XY}$ is the ⊥ bis. of $\overline{AB}$, not the other way around.
Exercises

7.

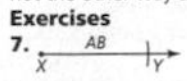

9.

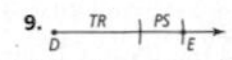

11.

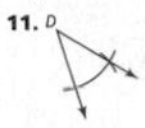

13.

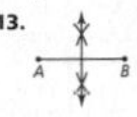

15.

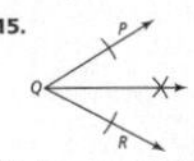

17. Answers may vary. Sample:

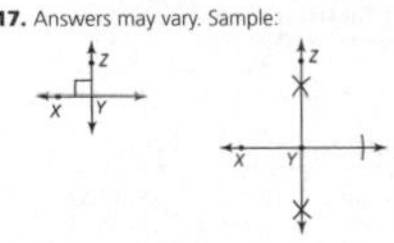

Find a segment on $\overleftrightarrow{XY}$ so that you can construct $\overleftrightarrow{YZ}$ as its perpendicular bisector.
19. Answers may vary. Sample: Both constructions involve drawing arcs with the same radius from two different points, and using the point(s) of intersection of those arcs. Arcs must intersect at two points for the ⊥ bis., but only one point for the ∠ bis. **21a.** A segment has exactly one midpoint; using the Ruler Postulate (Post. 1-5), each point corresponds with exactly one number, and exactly one number represents half the length of a segment. **b.** A segment has infinitely many bisectors because many lines can be drawn through the midpoint. **c.** In the plane with the segment, there is one ⊥ bis. because only one line in that plane can be drawn through the midpoint so that it forms a right angle with the given segment. **d.** Consider the plane that is the ⊥ bis. of the segment. Any line in that plane that contains the midpoint of the segment is a ⊥ bis. of the segment, and there are infinitely many such lines.

23.

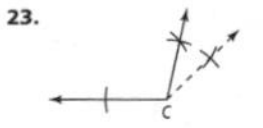

25a. With P as center, draw an arc with radius slightly more than $\frac{1}{2}PQ$. Keeping that radius, draw an arc with Q as center. Those two arcs meet at 2 points; the line through those 2 points intersects $\overline{PQ}$ at its midpoint. **b.** Follow the steps in part (a) to find the midpoint C of $\overline{PQ}$. Then repeat the process for segments $\overline{PC}$ and $\overline{CQ}$.
27. possible

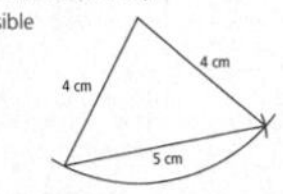

29. Not possible; the two 2-cm sides do not meet.
31a.

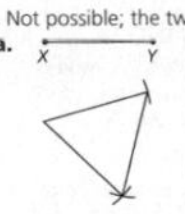

b. The measure of each angle is 60°. **c.** Draw an angle congruent to one of the angles of the triangle from part (a) to get a 60° ∠. Then construct its angle bisector to get two 30° ∡. **33.** ⊥; contains the intersection of that line with the plane

35.

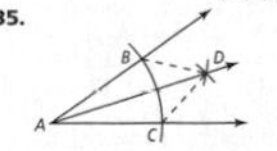

In the angle bisector construction, $\overline{AB} \cong \overline{AC}$, $\overline{BD} \cong \overline{CD}$, and $\overline{AD} \cong \overline{AD}$. Using the statement that two triangles are ≅ if three pairs of sides are ≅, then $\triangle ABD \cong \triangle ACD$. Since the △s are ≅, each ∠ of one △ is ≅ to an ∠ of the other △. So, $\angle BAD \cong \angle CAD$ and $\overrightarrow{AD}$ is the ∠ bisector of $\angle BAC$.
37. I **39.** 116 **40.** yes; $m\angle TUV + m\angle VUW = 180$ **41.** 6 **42.** 10 **43.** 4 **44.** 3 **45.** 196 **46.** 10 **47.** −1

Lesson 1-7 pp. 50–56

Got It? 1a. −4 **b.** (4, −2) **2.** (11, −13) **3a.** 15.8 **b.** Yes; the diff. of the coordinates are opposite, but their squares are the same. $VU = \sqrt{(-11)^2 + 8^2} = \sqrt{185} = 13.6$ **4.** $\sqrt{1325}$ m, or about 36.4 m
Lesson Check 1. (0.5, 5.5), or $\left(\frac{1}{2}, \frac{11}{2}\right)$ **2.** (7, −8) **3.** $\sqrt{73}$, or about 8.5 units **4.** Answers may vary. Sample: For two different points, the expression $(x_2 - x_1)^2 + (y_2 - y_1)^2$ in the Distance Formula is always positive. So the positive square root of a positive number is

positive. **5.** He did not keep the x-value and y-values together; so, $d = \sqrt{(1-3)^2 + (5-8)^2} = \sqrt{4+9} = \sqrt{13}$ units.
Exercises 7. −1.5, or $-\frac{3}{2}$ **9.** −10 **11.** (3, 1) **13.** (6, 1) **15.** $\left(3\frac{7}{8}, -3\right)$ **17.** (5, −1) **19.** (12, −24) **21.** (5.5, −13.5) **23.** 18 **25.** 9 **27.** 10 **29.** 12.2 **31.** 8.2 **33.** 8.5 **35.** Everett, Charleston, Brookline, Fairfield, Davenport **37a.** 5.8 **b.** $\left(\frac{3}{2}, \frac{1}{2}\right)$, or (1.5, 0.5) **39a.** 5.4 **b.** $\left(-\frac{5}{2}, 3\right)$, or (−2.5, 3) **41a.** 2.8 **b.** (−4, −4) **43a.** 5.4 **b.** $\left(3, \frac{1}{2}\right)$, or (3, 0.5) **45.** 165 units; flying T to V then to U is shortest distance. **47a.** Answers may vary. Sample: Distance Formula (Find KP, then divide it by 2.) **b.** Answers may vary. Sample: Distance Formula (If M is the given midpoint, find KM and then multiply it by 2.) **49a.** 10.7 **b.** (3, −4)
51a.

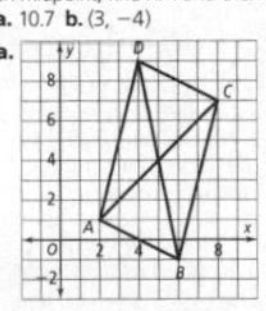

The midpoints are the same, (5, 4).
b. Answers may vary. Sample: The diagonals bisect each other. **53.** 7 mi **55.** 3.2 mi **57a.** Answers may vary. Sample: (0, 2) and (4, 2); (2, 0) and (2, 4); (0, 4) and (4, 0); (0, 0) and (4, 4) **b.** Infinitely many; draw a circle with center (2, 2) and radius 4. Any diameter of that circle has length 8 and midpoint (2, 2).
59. $A(0, 0, 0)$, $B(6, 0, 0)$, $C(6, -3, 0)$, $D(0, -3, 0)$, $E(0, 0, 9)$, $F(6, 0, 9)$, $G(0, -3, 9)$ **61.** 11.7 units **63.** F

65.

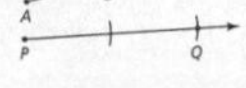

66.

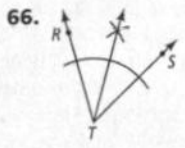

67. $\angle PQR$, $\angle RQP$ **68.** 150 **69.** $10\frac{5}{6}$ **70.** 504 **71.** 9 **72.** 10,560

Review p. 58

1. yes **3.** no; not a plane figure **5.** Sample: $FBWMX$; sides are $\overline{FB}$, $\overline{BW}$, $\overline{WM}$, $\overline{MX}$, $\overline{XF}$; angles are $\angle F$, $\angle B$, $\angle W$, $\angle M$, $\angle X$. **7.** Sample: $AGNHEPT$; sides are $\overline{AG}$, $\overline{GN}$, $\overline{NH}$, $\overline{HE}$, $\overline{EP}$, $\overline{PT}$, $\overline{TA}$; angles are $\angle A$, $\angle G$, $\angle N$, $\angle H$, $\angle E$, $\angle P$, $\angle T$. **9.** nonagon or enneagon, convex

Lesson 1-8 pp. 59–67

Got It? 1a. 24 in. **b.** 32 in. **2a.** 48π m **b.** 75.4 m
3.

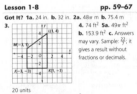

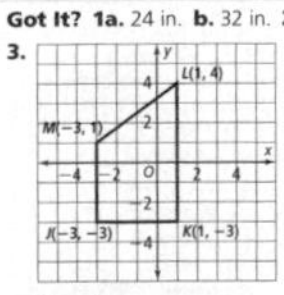

20 units

4. 74 ft² **5a.** 49π ft² **b.** 153.9 ft² **c.** Answers may vary. Sample: $\frac{22}{7}$; it gives a result without fractions or decimals.

6a.

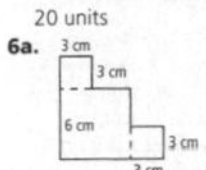

b. 64 ft²

Lesson Check 1. 20 in.; 21 in.² **2a.** 56.5 in.; 254.5 in.² **b.** 22.9 m; 41.9 m² **3.** $(12 + 2\sqrt{2})$ units; 10 square units **4.** Answers may vary. Sample: To fence a garden you would find the perimeter; to determine the material needed to make a tablecloth you would find the area. **5.** Answers may vary. Sample: Remind your friend that $2\pi r$ has only one variable, so it must compute the circumference. πr^2 has one variable squared, and square units indicate area. **6.** The classmate seems to have forgotten to multiply r^2 by π. The correct answer is $A = \pi r^2 = \pi(30)^2 = 900\pi \approx 2827.4$ in.².
Exercises 7. 22 in. **9.** 38 ft **11.** 10π ft **13.** $\frac{\pi}{2}$ m
15.

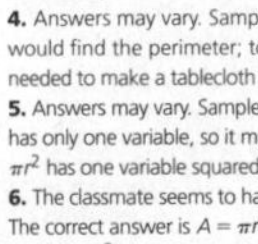

$(17 + \sqrt{65})$ units
17.

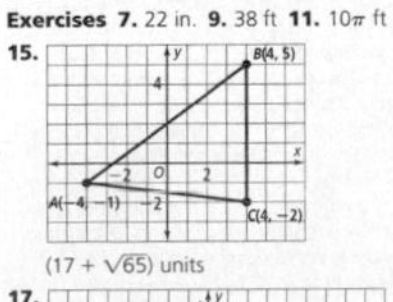

38 units

19. 4320 in.², or $3\frac{1}{3}$ yd² **21.** 8000 cm², or 0.8 m² **23.** 400π m² **25.** $\frac{3969}{400}\pi$ ft² **27.** 153.9 ft² **29.** 452.4 cm² **31.** 310 m² **33.** 208 ft² **35.** Perimeter; the crown molding must fit the edges of the ceiling. **37.** Area; the floor is a surface. **39a.** 144 in.²; 1 ft² **b.** 144 **41.** 16 cm **43.** 96 cm² **45.** 27 in.² **47a.** Yes; substitute s for each of a and b to get perimeter, $P = 2s + 2s$ or $P = 4s$. **b.** No; we need to know the length and width of a rectangle to find its perimeter. **c.** $A = \frac{P^2}{16}$ **49.** $\frac{25}{4}\pi$ units²
51.

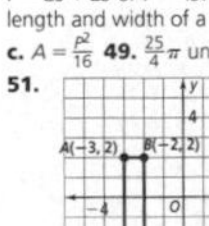

10 units, 4 square units
53a. Answers may vary. Sample:

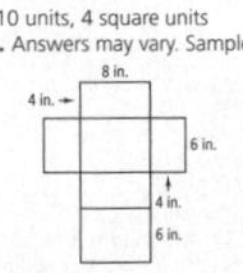

b. 208 in.²; 208 in.² **55.** $35.70 **57.** $\frac{3a}{20}$ square units **59.** $(10x^2 + \frac{7}{2}xy - 3y^2)$ square units **61.** 1104 ft² **63.** 27.9 **64a.** 8.5 units **b.** $\left(\frac{11}{2}, 5\right)$, or (5.5, 5) **65a.** 5.8 units **b.** $\left(-\frac{3}{2}, \frac{11}{2}\right)$, or (−1.5, 5.5) **66a.** 6.7 units **b.** $\left(-\frac{5}{2}, -2\right)$, or (−2.5, −2) **67.** 90 **68.** $\overline{WK}$, $\overline{KR}$ **69a.** $1^2 = 1$, $2^2 = 4$, $3^2 = 9$, $4^2 = 16$, $5^2 = 25$, $6^2 = 36$, $7^2 = 49$, $8^2 = 64$, $9^2 = 81$, $10^2 = 100$ **b.** It is odd.

Chapter Review pp. 70–74

1. angle bisector **2.** perpendicular lines **3.** net **4.** complementary angles **5.** 4, 6, 11
6.

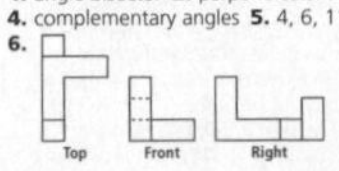

7. Answers may vary. Sample: $\overleftrightarrow{QA}$ and $\overleftrightarrow{AB}$ **8.** $\overleftrightarrow{QR}$ **9.** Answers may vary. Sample: A, B, C **10.** True; Postulate 1-1 states, "Through any two points, there is exactly one line." **11.** False; they have different endpoints. **12.** −7, 3 **13.** $\frac{1}{2}$ or 0.5 **14.** 15 **15.** $XY = 21$, $YZ = 29$ **16.** acute **17.** right **18.** 36 **19.** 14 **20–23.** Answers may vary. Samples are given. **20.** $\angle ADB$ and $\angle BDC$ **21.** $\angle ADB$ and $\angle BDF$ **22.** $\angle ADC$ and $\angle EDF$ **23.** $\angle ADC$ and $\angle ADE$ **24.** 31 **25.** 15

26.

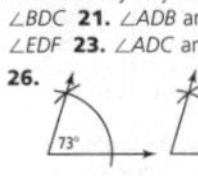

27.

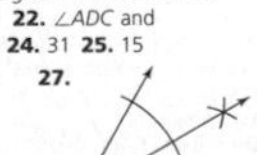

28.

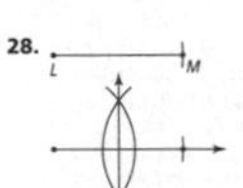

29a–b.

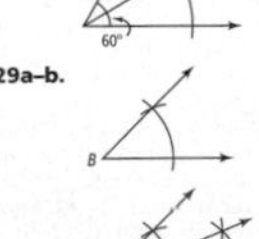

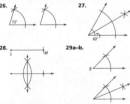

30. 1.4 units **31.** 7.6 units **32.** 14.4 units **33.** (0, 0) **34.** 7.2 units **35.** (6, −2) **36.** (1, 1) **37.** (−6, −7) **38.** 32 cm; 64 cm² **39.** 32 in.; 40 in.² **40.** 6π in.; 9π in.² **41.** 15π m; $\frac{225}{4}\pi$ m²

Chapter 2

Get Ready! p. 79

1. 50 **2.** −3 **3.** 25.5 **4.** 10.5 **5.** 15 **6.** 11 **7.** 7 **8.** 5 **9.** 6 **10.** 20 **11.** 18 **12.** $\angle ACD$, $\angle DCA$ **13.** 3 **14.** $m\angle 1 = 48$, $m\angle 2 = 42$ **15.** $\angle ADC$ and $\angle CDB$ **16.** $\angle 1$ and $\angle 2$ **17.** $\angle ADB$ or $\angle BDA$ **18.** Answers may vary. Sample: Similar: They are both statements you start with. Different: In geometry you do not try to prove the hypothesis of a statement. **19.** Answers may vary. Sample: A conclusion in geometry answers questions raised by the hypothesis. **20.** Answers may vary. Sample: In geometry you use deductive reasoning to draw conclusions from other information.

Lesson 2-1 pp. 82–88

Got It? 1a. 25, 20 **b.**

2. Every 3rd term is B, so the 21st term will be B. **3.** The sum of the first 30 odd numbers is 30^2, or 900. **4a.** Sales will be about 500 fewer than 8000, or 7500. **b.** No; sales may increase because students may want backpacks for school. **5a–c.** Answers may vary. Samples are given. **a.** A

carnation can be red, and it is not a rose. **b.** When three points are collinear, the number of planes that can be drawn through them is infinite. **c.** When you multiply 5 (or any odd number) by 3, the product is not divisible by 6.
Lesson Check **1.** 31, 37 **2.** 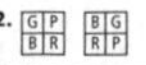

3. Answers may vary. Sample: any nonsquare rectangle **4.** One meaning of *counter* is "against," so a counterexample is an example that goes against a statement. **5.** In the pattern 2, 4, . . . , the next term is 6 if the rule is "add 2"; the next term is 8 if the rule is "double the previous term"; and the next term is 7 if the rule is "add 2, then add 3, then add 4, . . ." Just giving the first 2 terms does not give enough information to describe the pattern.
Exercises **7.** Find the next square; 36, 49. **9.** Multiply the previous term by $\frac{1}{2}$; $\frac{1}{16}$, $\frac{1}{32}$. **11.** Subtract 3 from the previous number; 3, 0. **13.** the first letter of the months; J, J **15.** the Presidents of the U.S.; Madison, Monroe **17.** state postal abbreviations in alphabetical order; CO, CT **19.** 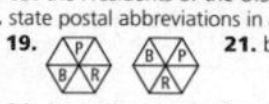

21. blue **23.** blue
25–30. Answers may vary. Samples are given. **25.** The sum of the first 100 positive odd numbers is 100^2, or 10,000. **27.** The sum of two odd numbers is even. **29.** The product of two even numbers is even. **31.** 1 mi **33–37.** Answers may vary. Samples are given. **33.** two right angles **35.** −2 and −3 **37.** −2 and −3 **39.** Add 1 then add 3; add 1 then add 3; . . . ; 10, 13. **41.** Multiply by 3, add 1; multiply by 3, add 1; . . . ; 201, 202. **43.** Add $\frac{1}{2}$, add $\frac{1}{4}$, add $\frac{1}{8}$ . . . ; $\frac{31}{32}$, $\frac{63}{64}$.
45. 123,454,321 **47.** 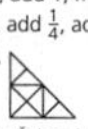
49. 102 cm
51a. sì-shí-sān; lìu-shí-qī; bā-shí-sì **b.** Yes; the second part of the number repeats each ten numbers. **53.** His conjecture is probably false because most people's growth slows by 18 until they stop growing sometime between 18 and 22 years. **55.** Answers may vary.
57. 1 × 1: 64 squares; 2 × 2: 49 squares; 3 × 3: 36 squares; 4 × 4: 25 squares; 5 × 5: 16 squares; 6 × 6: 9 squares; 7 × 7: 4 squares; 8 × 8: 1 square; total number of squares: 204

Lesson 2-2 pp. 89–95

Got It? **1.** Hypothesis: An angle measures 130. Conclusion: The angle is obtuse. **2.** If an animal is a dolphin, then it is a mammal. **3a.** False; January has 28 days, plus 3 more. **b.** True; the sum of the measures of two angles that form a linear pair is 180.
4. Counterexamples may vary. Samples are given. Converse: If a vegetable contains beta carotene, then it is a carrot. Inverse: If a vegetable is not a carrot, then it does not contain beta carotene. Contrapositive: If a vegetable does not contain beta carotene, then it is not a carrot. The conditional and the contrapositive are true. The converse and inverse are false; counterexample: any vegetable, such as spinach, that contains beta carotene.
Lesson Check **1.** Hypothesis: Someone is a resident of Key West.Conclusion: The person lives in Florida. Conditional: If someone is a resident of Key West, then that person lives in Florida. **2.** Converse: If a figure has a perimeter of 10 cm, then it is a rectangle with sides 2 cm and 3 cm. Inverse: If a figure is not a rectangle with sides 2 cm and 3 cm, then it does not have a perimeter of 10 cm. Contrapositive: If a figure does not have a perimeter of 10 cm, then it is not a rectangle with sides 2 cm and 3 cm. The original conditional and the contrapositive are true. **3.** The hypothesis and conclusion were exchanged. The conditional should be "If it is Sunday, then you jog." **4.** Both are true because a conditional and its contrapositive have the same truth value, and a converse and an inverse have the same truth value.
Exercises **5.** Hypothesis: You are an American citizen. Conclusion: You have the right to vote. **7.** Hypothesis: You want to be healthy. Conclusion: You should eat vegetables. **9.** If $3x - 7 = 14$, then $3x = 21$. **11.** If an object or example is a counterexample for a conjecture, then the object or example shows that the conjecture is false. **13.** If something is blue, then it has a color.
15. If something is wheat, then it is a grain. **17.** false; Mexico **19.** true **21.** Conditional: If a person is a pianist, then that person is a musician. Converse: If a person is a musician, then that person is a pianist. Inverse: If a person is not a pianist, then that person is not a musician. Contrapositive: If a person is not a musician, then that person is not a pianist. The conditional and contrapositive are true. The converse and inverse are false; counterexample: a percussionist is a musician.
23. Conditional: If a number is an odd natural number

less than 8, then the number is prime. Converse: If a number is prime, then it is an odd natural number less than 8. Inverse: If a number is not an odd natural number less than 8, then the number is not prime. Contrapositive: If a number is not prime, then it is not an odd natural number less than 8. All four statements are false; counterexamples: 1 and 11. **25.** If a group is half the people, then that group should make up half the Congress. **27.** If an event has a probability of 1, then that event is certain to occur. **29.** Answers may vary. Sample: If an angle is acute, its measure is less than 90; if the measure of an angle is 85, then it is acute. **31.** Natalie is correct because a conditional statement and its contrapositive have the same truth value.
33. 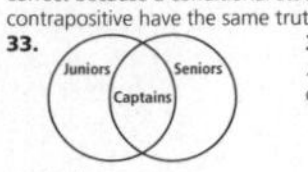

35. If $|x| = 6$, then $x = -6$; false, $x = 6$ is a counterexample.
37. If $x^3 < 0$, then $x < 0$; true. **39.** If you wear Snazzy sneakers, then you will look cool. **41.** If two figures are congruent, then they have equal areas. **43.** All integers divisible by 8 are divisible by 2. **45.** Some musicians are students.

Lesson 2-3 pp. 98–104

Got It? **1.** Converse: If two angles are congruent, then the angles have equal measure; true. Biconditional: Two angles have equal measure if and only if the angles are congruent. **2.** If two numbers are reciprocals, then their product is 1. If the product of two numbers is 1, then the numbers are reciprocals. **3.** Yes, it is reversible; an angle is a straight angle if and only if its measure is 180. **4a.** No, it is not precise; a rectangle is also a figure with four right angles. **b.** Answers may vary. Sample: Obtuse angles have measures between 90 and 180.
Lesson Check **1.** If points are collinear, then they lie on the same line. If points lie on the same line, then they are collinear. **2.** This month is June if and only if next month is July. **3.** Two angles are vertical angles if and only if their sides are opposite rays. **4.** The prefix *bi-* means "two." **5.** The word *gigantic* is not precise. **6.** The second statement is a better definition. A counterexample for the first statement is any two nonadjacent right angles.
Exercises **7.** Converse: If two segments are congruent, then they have the same length; true. Biconditional: Two segments have the same length if and only if they are congruent. **9.** Converse: If a number is even, then it is divisible by 20; false. **11.** Converse: If it is Independence Day in the United States, then it is July 4; true. Biconditional: In the United States, it is July 4 if and only if it is Independence Day. **13.** If a line bisects a segment, then it intersects the segment only at its midpoint. If a line intersects a segment only at its midpoint, then the line bisects the segment. **15.** If you live in Washington, D.C., then you live in the capital of the United States. If you live in the capital of the United States, then you live in Washington, D.C. **17.** If an angle is a right angle, then it measures 90. If an angle measures 90, then it is a right angle. **19.** A line, segment, or ray is a perpendicular bisector of a segment if and only if it is perpendicular to the segment at its midpoint. **21.** A person is a Tarheel if and only if the person was born in North Carolina.
23. not reversible **25.** No, it is not reversible; some endangered animals are not red wolves. **27.** No, it is not precise; straightedges and protractors are geometric tools. **29.** yes **31.** No; a straight angle has a measure greater than 90, but it is not an obtuse angle. **33.** That statement, as a biconditional, is "an angle is a right angle if and only if it is greater than an acute angle." Counterexamples to that statement are obtuse angles and straight angles. **35.** A point is in Quadrant III if and only if it has two negative coordinates. **37.** A number is a whole number if and only if it is a nonnegative integer. **39.** good definition **41.** good definition **43.** If $\angle A$ and $\angle B$ are a linear pair, then they are supplementary. **45.** If $\angle A$ and $\angle B$ are a linear pair, then they are adjacent, supplementary angles. **47.** Answers may vary. Sample: A line is a circle on the sphere formed by the intersection of the sphere and a plane containing the center of the sphere. **49.** D **51a.** If you go to the store, then you want to buy milk; false. **b.** Answers may vary. Sample: A counterexample is going to the store because you want to buy juice. **52.** If your grades suffer, then you do not get enough sleep. **53.** If you have a good voice, then you are in the school chorus.
54. true **55.** 60, 50 **56.** 4, $\frac{4}{5}$ **57.** 4, −2

Lesson 2-4 pp. 106–112

Got It? **1a.** Marla is not safe out in the open. **b.** No conclusion is possible. **2a.** If a whole number ends in 0, then it is divisible by 5; Law of Syllogism. **b.** No conclusion is possible. **3a.** The Nile is the longest river in the world; Law of Syllogism and Law of Detachment.
b. Yes; if you use the Law of Detachment first, then you

must use it again to reach the same conclusion. The Law of Syllogism is not used.
Lesson Check **1.** No conclusion is possible. **2.** Figure *ABC* is a triangle; Law of Detachment. **3.** If it is Saturday, then you wear sneakers; Law of Syllogism. **4.** The Law of Detachment cannot be applied because the hypothesis is not satisfied. **5.** Answers may vary. Sample: Deductive reasoning uses logic to reach conclusions, while inductive reasoning bases conclusions on unproved (but possibly true) conjectures.
Exercises **7.** No conclusion is possible; the hypothesis has not been satisfied. **9.** No conclusion is possible; the hypothesis has not been satisfied. **11.** If an animal is a Florida panther, then it is endangered. **13.** If a line intersects a segment at its midpoint, then it divides the segment into two congruent segments. **15.** Alaska's Mount McKinley is the highest mountain in the U.S.
17. If you are studying botany, then you are studying a science. (Law of Syllogism only) No conclusion can be made about Shanti. **19.** Must be true; by E and A, it is breakfast time; by D, Julio is drinking juice. **21.** May be true; by E and A, it is breakfast time. You don't know what Kira drinks at breakfast. **23.** May be true; by E, Maria is drinking juice. You don't know if she also drinks water. **25.** strange **27.** If a figure is a square, then it is a rectangle; *ABCD* is a rectangle. **29.** If a person is a high school student, then the person likes art; no conclusion is possible because the hypothesis is not satisfied.
31a. The result is two more than the chosen integer.
b. $x + 2$
c. The expression in part (b) is equivalent to the conjecture in part (a). In part (a) inductive reasoning was used to make a conjecture based on a pattern. In part (b) deductive reasoning was used in order to write and simplify an expression.

Lesson 2-5 pp. 113–119

Got It? **1.** 75; $x = 2x - 75$ (Def. of an $\angle$ bis.); $x + 75 = 2x$ (Add. Prop. of Eq.); $75 = 2x - x$ (Subtr. Prop. of Eq.); $75 = x$ **2a.** Sym. Prop. of $\cong$ **b.** Distr. Prop. **c.** Mult. Prop. of Eq. **d.** Refl. Prop. of Eq.
3a. Answers may vary. Sample: $\overline{AB} \cong \overline{CD}$ (Given); $AB = CD$ ($\cong$ segments have = length.); $BC = BC$ (Refl. Prop. of Eq.); $AB + BC = BC + CD$ (Add. Prop. of =); $AB + BC = AC$, $BC + CD = BD$ (Seg. Add. Post.); $AC = BD$ (Trans. Prop. of Eq.); $\overline{AC} \cong \overline{BD}$ (Segments with = length are $\cong$.) **b.** Answers may vary. Sample: You need to establish equality in order to add the same quantity ($m\angle 2$) to each side of the equation in Statement 3.
Lesson Check **1.** Trans. Prop. of Eq. **2.** Distr. Prop. **3.** Subtr. Prop of Eq. **4a.** Given **b.** Subtr. Prop. of Eq. **c.** Div. Prop. of Eq.
Exercises **5a.** Mult. Prop. of Eq. **b.** Distr. Prop. **c.** Add. Prop. of Eq. **7a.** def. of suppl. $\angle$s **b.** Subst. Prop. **c.** Distr. Prop. **d.** Subtr. Prop. of Eq. **e.** Div. Prop. of Eq. **9.** Distr. Prop. **11.** Sym. Prop. of $\cong$
13a. Given **b.** A midpt. divides a seg. into two $\cong$ segments. **c.** Substitution **d.** $2x = 12$ **e.** Div. Prop. of Eq. **15.** $\angle K$ **17.** 3 **19.** $\angle XYZ \cong \angle WYT$
21. Since $\overline{LR}$ and $\overline{RL}$ are two ways to name the same segment and $\angle CBA$ and $\angle ABC$ are two ways to name the same $\angle$, then both statements are examples of saying that something is $\cong$ to itself. **23.** $KM = 35$ (Given); $KL + LM = KM$ (Seg. Add. Post.); $(2x - 5) + 2x = 35$ (Subst. Prop.); $4x - 5 = 35$ (Distr. Prop.); $4x = 40$ (Add. Prop. of Eq.); $x = 10$ (Div. Prop. of Eq.); $KL = 2x - 5$ (Given); $KL = 2(10) - 5$ (Subst. Prop.); $KL = 15$ (Simplify) **25.** The error is in the 5th step when both sides of the equation are divided by $b - a$, which is 0, and division by 0 is not defined. **27.** Transitive only; A cannot be taller than A; if A is taller than B, then B is not taller than A. **29.** 58.5 **31.** 153.86 **33.** 58 **34.** Walt's science teacher is concerned. **35.** 80 **36.** 65 **37.** 125 **38.** 90 **39.** 50 **40.** 90 **41.** 35

Lesson 2-6 pp. 120–127

Got It? **1.** 40 **2a.** $\angle 1 \cong \angle 2$ (Given); $\angle 1 \cong \angle 3$, $\angle 2 \cong \angle 4$ (Vert. $\angle$s are $\cong$); $\angle 1 \cong \angle 4$, $\angle 2 \cong \angle 3$ (Trans. Prop. of $\cong$); $\angle 1 \cong \angle 2 \cong \angle 3 \cong 4$ (Trans. Prop. of $\cong$)
b. Answers may vary. Sample: $m\angle 1 + m\angle 2 = 180$ because they form a linear pair. So $m\angle 1 = 90$ and $m\angle 2 = 90$ because $\angle 1 \cong \angle 2$. Then, using the relationship that $m\angle 2 + m\angle 3 = 180$ and $m\angle 1 + m\angle 4 = 180$, you can show that $m\angle 3 + m\angle 4 = 180$ by the Subtr. Prop. of Eq. Then $\angle 1 \cong \angle 2 \cong \angle 3 \cong \angle 4$ because their measures are =. **3.** Answers may vary. Sample: $\angle 1$ and $\angle 3$ are vert. $\angle$s because it is given. $\angle 1$ and $\angle 2$ are suppl. and $\angle 2$ and $\angle 3$ are suppl. because $\angle$s that form a linear pair are suppl. So, $m\angle 1 + m\angle 2 = 180$ and $m\angle 2 + m\angle 3 = 180$ by the def. of suppl. $\angle$s. By the Trans. Prop. of Eq., $m\angle 1 + m\angle 2 = m\angle 2 + m\angle 3$. By the Subtr. Prop. of Eq., $m\angle 1 = m\angle 3$. So, $\angle 1 \cong \angle 3$ because $\angle$s with the same measure are $\cong$.
Lesson Check **1.** $m\angle 1 = 90$, $m\angle 2 = 50$, $m\angle 3 = 40$ **2.** B **3.** $\angle B \cong \angle C$ because both are suppl. to $\angle A$ and if two $\angle$s are suppl. to the same $\angle$, then they are $\cong$. **4.** He used the Trans. Prop. of $\cong$, which does not apply here. $\angle 2$ and $\angle 3$ are $\cong$, not compl. If two $\angle$s are compl. to the same $\angle$, then they are $\cong$ to each other. **5.** Answers may vary. Sample: A postulate is a statement that is assumed to be true, while a theorem is a statement that is proved to be true.
Exercises **7.** $x = 38$, $y = 104$ **9.** 60, 60 **11.** 120, 120 **13a.** 90 **b.** 90 **c.** $m\angle 3$ **d.** $\cong$ **15.** Answers may vary. Sample: scissors **17.** $x = 14$, $y = 15$; $3x + 8 = 50$, $5x - 20 = 50$, $5x + 4y = 130$ **19.** $x = 50$, $y = 50$
21. $\angle EIG \cong \angle FIH$ because all rt. $\angle$s are $\cong$; $\angle EIF \cong \angle HIG$ because each one is compl. to $\angle FIG$ and compl. of the same $\angle$ are $\cong$. **23a.** It is given. **b.** $m\angle V$ **c.** 180 **d.** Division **e.** right **25.** By Theorem 2-5: If two $\angle$s are $\cong$ and suppl., then each is a right $\angle$. **27.** $m\angle A = 30$, $m\angle B = 60$ **29.** $m\angle A = 90$, $m\angle B = 90$ **31.** Answers may vary. Sample: (−5, −1) **33.** $x = 30$, $y = 90$; 60, 120, 60 **34.** $x = 35$, $y = 70$; 70, 110, 70 **37.** 20 **39.** 60 **40.** Subtr. Prop. of Eq. **41.** Div. Prop. of Eq. **42.** Trans. Prop. of $\cong$ **43.** points *F*, *I*, *H*, *B* **44.** no **45.** yes **46.** line *r* (or $\overleftrightarrow{EG}$, $\overleftrightarrow{GH}$, $\overleftrightarrow{HC}$, and so on) **47.** any three of $\overleftrightarrow{FI}$ (or $\overleftrightarrow{IF}$), $\overleftrightarrow{FH}$, $\overleftrightarrow{FB}$, $\overleftrightarrow{IH}$, $\overleftrightarrow{IB}$, $\overleftrightarrow{HB}$ **48.** point *H*

Chapter Review pp. 129–132

1. conclusion **2.** deductive reasoning **3.** truth value **4.** converse **5.** biconditional **6.** theorem **7.** hypothesis **8.** Divide the previous term by 10; 1, $\frac{1}{10}$. **9.** Multiply the previous term by −1; 5, −5. **10.** Subtract 7 from the previous term; 6, −1. **11.** Multiply the previous term by 4; 1536, 6144. **12.** Answers may vary. Sample: $-1 \cdot 2 = -2$, and −2 is not greater than 2 **13.** Answers may vary. Sample: Portland, Maine **14.** If a person is a motorcyclist, then that person wears a helmet. **15.** If two nonparallel lines intersect, then they intersect in one point. **16.** If two $\angle$s form a linear pair, then the $\angle$s are supplementary. **17.** If today is one of a certain group of holidays, then school is closed. **18.** Converse: If the measure of an $\angle$ is greater than 90 and less than 180, then the $\angle$ is obtuse. Inverse: If an angle is not obtuse, then it is not true that its measure is greater than 90 and less than 180. Contrapositive: If it is not true that the measure of an $\angle$ is greater than 90 and less than 180, then the $\angle$ is not obtuse. All four statements are true.
19. Converse: If a figure has four sides, then the figure is a square. Inverse: If a figure is not a square, then it does not have four sides. Contrapositive: If a figure does not have four sides, then it is not a square. The conditional and the contrapositive are true. The converse and inverse are false. **20.** Converse: If you play an instrument, then you play the tuba. Inverse: If you do not play the tuba, then you do not play an instrument. Contrapositive: If you do not play an instrument, then you do not play the tuba. The conditional and the contrapositive are true. The converse and inverse are false. **21.** Converse: If you are busy on Saturday night, then you baby-sit. Inverse: If you do not baby-sit, then you are not busy on Saturday night. Contrapositive: If you are not busy on Saturday night, then you do not baby-sit. The conditional and the contrapositive are true. The converse and inverse are false. **22.** No; it is not reversible; a magazine is a counterexample. **23.** yes **24.** No; it is not reversible; a line is a counterexample. **25.** A phrase is an oxymoron if and only if it contains contradictory terms. **26.** If two $\angle$s are complementary, then the sum of their measures is 90; if the sum of the measures of two $\angle$s is 90, then the $\angle$s are complementary. **27.** Colin will become a better player. **28.** $m\angle 1 + m\angle 2 = 180$ **29.** If two angles are vertical, then their measures are equal. **30.** If your father buys new gardening gloves, then he will plant tomatoes.
31a. Given **b.** Seg. Add. Post. **c.** Subst. Prop. **d.** Distr. Prop. **e.** Subtr. Prop. of Eq. **f.** Div. Prop. of Eq.
32. *BY* **33.** $p - 2q$ **34.** 18 **35.** 74 **36.** 74 **37.** 106
38. $\angle 1$ is compl. to $\angle 2$, $\angle 3$ is compl. to $\angle 4$, and $\angle 2 \cong \angle 4$ are all given. $m\angle 2 = m\angle 4$ by the def. of $\cong$. $\angle 1$ and $\angle 4$ are compl. by the Subst. Post. $\angle 1 \cong \angle 3$ by the $\cong$ Compl. Thm.

Chapter 3

Get Ready! p. 137

1. $\angle 1$ and $\angle 5$, $\angle 5$ and $\angle 2$ **2.** $\angle 3$ and $\angle 4$ **3.** $\angle 1$ and $\angle 2$ **4.** $\angle 1$ and $\angle 5$, $\angle 5$ and $\angle 2$ **5.** Div. Prop. of = **6.** Trans. Prop. of $\cong$ **7.** 4 **8.** 61 **9.** 15 **10.** 5 **11.** $2\sqrt{17}$
12. $\sqrt{17}$ **13.** Answers may vary. Sample: A figure divides a plane or space into three parts: the figure itself, the region inside the figure—called its interior—and the region outside the figure—called its exterior.
14. Answers may vary. Sample: *Trans-* means "cross"; a transversal crosses other lines. **15.** Answers may vary. Sample: A flow proof shows the individual steps of the proof and how each step is related to the other steps.

Lesson 3-1 pp. 140–146

Got It? **1a.** $\overline{EH}$, $\overline{BC}$, $\overline{FG}$ **b.** Sample: They are both in plane *FEDC*, so they are coplanar. **c.** plane *BCG* ∥ plane *ADH* **d.** any two of $\overline{AB}$, $\overline{BF}$, $\overline{EF}$, and $\overline{AE}$ **2.** any three of $\angle 1$ and $\angle 3$, $\angle 2$ and $\angle 4$, $\angle 8$ and $\angle 6$, $\angle 7$ and $\angle 5$
3. corresp. $\angle$s
Lesson Check **1–7.** Answers may vary. Samples are given. **1.** $\overline{EF}$ and $\overline{HG}$ **2.** $\overline{EF}$ and $\overline{GC}$ **3.** plane *ABF* ∥ plane *DCG* **4.** $\angle 8$ and $\angle 6$ **5.** $\angle 3$ and $\angle 8$ **6.** $\angle 1$ and $\angle 3$ **7.** $\angle 1$ and $\angle 4$ **8.** Although lines that are not coplanar do not intersect, they are not parallel. **9.** Alt. int. $\angle$s are $\angle$s between two lines on opposite sides of a transversal. **10.** Carly; the lines are coplanar since they are both in plane *ABH*, so $\overline{AB} \parallel \overline{HG}$.
Exercises **11.** plane *JCD* ∥ plane *ELH* **13.** $\overleftrightarrow{GB}$, $\overleftrightarrow{JE}$, $\overleftrightarrow{CL}$, $\overleftrightarrow{FA}$ **15.** $\overleftrightarrow{GB}$, $\overleftrightarrow{DH}$, $\overleftrightarrow{CL}$ **17.** $\angle 7$ and $\angle 6$ (lines *a* and *b* with transversal *d*), $\angle 2$ and $\angle 5$ (lines *b* and *c* with transversal *e*) **19.** $\angle 5$ and $\angle 6$ (lines *d* and *e* with transversal *b*); $\angle 2$ and $\angle 4$ (lines *b* and *e* with transversal *c*) **21.** $\angle 1$ and $\angle 2$ are corresp. $\angle$s; $\angle 3$ and $\angle 4$ are alt. int. $\angle$s; $\angle 5$ and $\angle 6$ are corresp. $\angle$s. **23.** $\angle 1$ and $\angle 2$ are corresp. $\angle$s; $\angle 3$ and $\angle 4$ are same-side int. $\angle$s; $\angle 5$ and $\angle 6$ are alt. int. $\angle$s.

25. 2 pairs **27.** 2 pairs **29.** Skew; answers may vary. Sample: Since the paths are not coplanar, they are skew. **31.** False; $\overleftrightarrow{ED}$ and $\overleftrightarrow{HG}$ are skew. **33.** False; the planes intersect. **35.** False; both lines are in plane *ABC*. **37.** always **39.** always **41.** sometimes **43a.** Lines may be intersecting, parallel, or skew. **b.** Answers may vary. Sample: In a classroom, two adjacent edges of the floor are intersecting, two opposite edges of the floor are parallel, and one edge of the floor is skew to each of the vertical edges of the opposite wall. **45a.** The lines of intersection are ∥. **b.** Sample: The lines of intersection of a wall with the ceiling and floor (or the lines of intersection of any of the 6 planes with two different, opposite faces)

47. Yes;

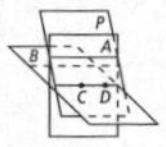

Lesson 3-2 pp. 148–155

Got It? 1a. Yes, if you have the measure of at least one angle. **2.** (1) $a \parallel b$ (Given) (2) $\angle 1 \cong \angle 5$ (If lines are ∥, then corresp. ∠s are ≅.) (3) $\angle 5 \cong \angle 7$ (Vert. ∠s are ≅.) (4) $\angle 1 \cong \angle 7$ (Trans. Prop. of ≅) **3a.** 75; $m\angle 1 = m\angle 4$ by the Alt. Int. ∠s Thm. **b.** 75; $m\angle 2 = m\angle 4$ by the Corresp. ∠s Thm. **c.** 105; $m\angle 5 = 105$ by the Corresp. ∠s Thm. **d.** 105; Alt. Int. ∠s Thm. **e.** 105; Vert. ∠s. Thm. **f.** 105; $\angle 8 \cong \angle 6$ by the Corresp. ∠s Thm. **4a.** $x = 64$, $y = 40$ **b.** Clockwise from the bottom left, the measures are 52, 128, 120, 60.

Lesson Check 1–2. Answer may vary. Samples are given. **1.** ∠4 and ∠5, ∠2 and ∠6, ∠3 and ∠7, ∠4 and ∠8 **2.** ∠2 and ∠5, ∠4 and ∠7 **3.** 70 **4.** 55 **5.** Alike: Two parallel lines are cut by a tra nsversal and the angles are congruent; different: The int. ∠s are between the two parallel lines, while the ext. ∠s are not between the two parallel lines. **6.** same-side ext. ∠s, because they are ext. ∠s on the same side of the transversal

Exercises 7. ∠1 (vert. ∠s), ∠7 (alt. int. ∠s), ∠4 (corresp. ∠s) **9.** ∠3 (alt. int. ∠s), ∠1 (corresp. ∠s) **11.** (1) $a \parallel b$; $c \parallel d$ (Given) (2) $\angle 1 \cong \angle 4$ (Alt. int. ∠s are ≅.) (3) $\angle 4 \cong \angle 3$ (Corresp. ∠s are ≅.) (4) $\angle 1 \cong \angle 3$ (Trans. Prop. of ≅) **13.** $m\angle 1 = 120$ because corresp. ∠s are ≅; $m\angle 2 = 60$ because ∠2 forms a linear pair with the given ∠. **15.** $x = 115$, $x - 50 = 65$ **17.** 20; $5x = 100$, $4x = 80$ **19.** $x = 135$, $y = 45$ **21.** 90; all the ∠s are ≅ because each pair form vert. ∠s, corresp. ∠s, or suppl. ∠s.

23a. 117 **b.** same-side int. ∠s

25. (1) $\ell \parallel m$ (Given)
(2) $m\angle 2 + m\angle 3 = 180$ (∠s that form a linear pair are suppl.)
(3) $m\angle 3 + m\angle 6 = 180$ (Same-side interior angles are suppl.)
(4) $m\angle 2 + m\angle 3 = m\angle 3 + m\angle 6$ (Substitution)
(5) $m\angle 2 = m\angle 6$ (Subtraction)
(6) $\angle 2 \cong \angle 6$ (Def. of Congruence)

27. $m\angle 1 = 48$, $m\angle 2 = 132$

Lesson 3-3 pp. 156–163

Got It? 1. $\ell \parallel m$ by the Converse of the Corresp. ∠s Thm.
2. Answers may vary. Sample:

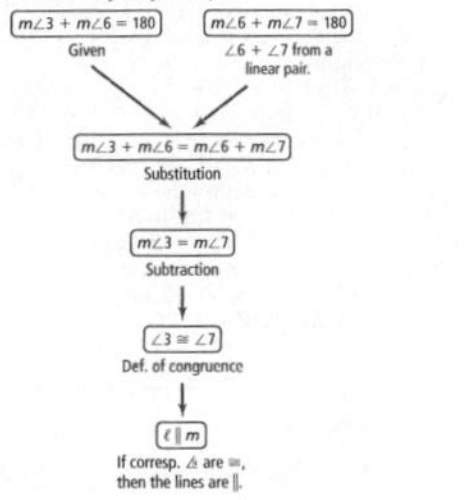

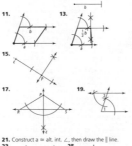

3. $\angle 2 \cong \angle 3$ (Vert. ∠s are ≅.), so $\angle 1 \cong \angle 3$ (Trans. Prop. of ≅). So $r \parallel s$ by the Converse of the Corresp. ∠s Thm. **4.** 19

Lesson Check 1. Conv. of Corresp. ∠s Thm. **2.** Conv. of Alt. Int. ∠s Thm. **3.** 115 **4.** If you want to prove that alt. int. ∠s are ≅, use the Alt. Int. ∠s Thm.; if you want to prove that two lines are parallel, use the Converse of the Alt. Int. ∠s Thm. **5.** Alike: Both give statements and reasons; different: The proofs use different formats. **6.** $\overleftrightarrow{DC}$ is the transversal, so the two same-side int. ∠s show that $\overleftrightarrow{AD}$ and $\overleftrightarrow{BC}$ are parallel.

Exercises 7. $\overleftrightarrow{BE} \parallel \overleftrightarrow{CG}$ by the Converse of the Corresp. ∠s Thm. **9.** $\overleftrightarrow{CA} \parallel \overleftrightarrow{HR}$ by the Converse of the Corresp. ∠s Thm. **11a.** Given **b.** ∠1 and ∠2 form a linear pair. **c.** ∠s that form a linear pair are suppl. **d.** $\angle 2 \cong \angle 3$ **e.** If corresp. ∠s are ≅, then lines are ∥. **13.** 30 **15.** 59

17. $a \parallel b$; if same-side int. ∠s are suppl., then the lines are ∥. **19.** $a \parallel b$; if same-side int. ∠s are suppl., then the lines are ∥. **21.** none **23.** $a \parallel b$ (Conv. of the Alt. Ext. ∠s Thm.) **25.** none **27.** 5 **29.** ∠2 and ∠3 are suppl. (Linear Pair Post.), so $\angle 3 \cong \angle 7$ (Congr. Suppl. Thm.). Therefore, $\ell \parallel m$ (Conv. of the Corresp. ∠s Thm.). **31.** $x = 10$; $m\angle 1 = m\angle 2 = 70$ **33.** $x = 2.5$; $m\angle 1 = m\angle 2 = 30$ **35.** Answers may vary. Sample: If $\angle 3 \cong \angle 5$, then $\ell \parallel n$ by the Converse of Corresp. ∠s Thm. **37.** Answers may vary. Sample: If $\angle 5 \cong \angle 3$, then $j \parallel k$ by the Converse of the Corresp. ∠s Thm. **39.** If alt. ext. ∠s are ≅, then the lines are ∥.

41. Answers may vary. Sample:

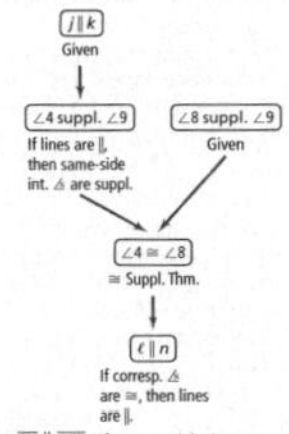

43. $\overline{PL} \parallel \overline{NA}$; if same-side int. ∠s are suppl., then the lines are ∥. **45.** $\overline{PN} \parallel \overline{LA}$; if same-side int. ∠s are suppl., then the lines are ∥. **47.** A **49.** C **51.** A sketch of a closed plane figure consisting of 5 segments (sides); the pentagon is convex because all diagonals are inside the pentagon OR the pentagon is concave because at least one diagonal has points outside the pentagon. **52.** $m\angle 1 = 70$ (∠1 is suppl. to a 110° ∠.); $m\angle 2 = 110$ (∠2 is suppl. to ∠1, which is a 70° angle.) **53.** $m\angle 1 = 66$ (Alt. int. ∠s are ≅.); $m\angle 2 = 86$ (∠2 is suppl. to a 94° angle.) **54.** always **55.** sometimes **56.** sometimes **57.** never

Lesson 3-4 pp. 164–169

Got It? 1. Yes; place the pieces with 60° ∠s opposite each other and place the pieces with 30° ∠s opposite each other. All four corners will be 90°, so opposite sides will be ∥. **2.** Yes; $a \parallel b$ because they are both ⊥ to *d*, and in a plane, two lines ⊥ to the same line are ∥.

Lesson Check 1. They are ⊥; using Main Street as a transversal, Avenue B ⊥ Main Street by Thm. 3-10. **2.** $a \parallel b$; in a plane, if two lines are ⊥ to the same line, then they are ∥. **3.** Sample: Even if the 3 lines are not in the same plane, each line is parallel to the other 2 lines. **4.** Thm. 3-9 uses the Converse of the Corresp. ∠s Thm.; the ⊥ Trans. Thm. uses the Corresp. ∠s Thm. **5.** The diagram should show that *m* and *r* are ⊥.

Exercises 7a. corresp. ∠s **b.** ∠1 **c.** ∠3 **d.** Converse of Corresp. ∠s Thm. **9.** Measure any three int. ∠s to be rt. ∠s and opp. walls will be ∥ because two walls ⊥ to the same wall are ∥. **11.** The rungs are ∥ to each other because they are all ⊥ to the same side. **13.** The rungs are ⊥ to both sides. The rungs are ⊥ to one of two ∥ sides, so they are ⊥ to both sides. **15.** The rungs are ∥ because they are all ⊥ to one side. **17.** Sample: Using the diagram underneath Thm. 3-10, $\overleftrightarrow{EC}$ and $\overleftrightarrow{AB}$ are both ⊥ to $\overleftrightarrow{AC}$, but $\overleftrightarrow{EC}$ and $\overleftrightarrow{AB}$ are skew, so they cannot be ∥. **19.** $a \parallel d$ by Thm. 3-8 **21.** $a \perp d$ by Thm. 3-10 **23.** $a \parallel d$ by Thms. 3-8 and 3-9 **25.** Reflexive: $a \parallel a$; false; every line intersects itself. Symmetric: If $a \parallel b$ then $b \parallel a$; true; lines *a* and *b* are coplanar and do not meet. Transitive: If $a \parallel b$ and $b \parallel c$, then $a \parallel c$; true; that is Thm. 3-8. **27.** A **29.** C **31.** 53 **32.** 46 **33.** right **34.** obtuse **35.** acute **36.** 60 **37.** 20 **38.** 40 **39.** 58

Lesson 3-5 pp. 171–178

Got It? 1. 29 **2.** 127, 127, 106 **3.** Yes; answers may vary. Sample: $m\angle ACB$ must = 100, so by the △ ∠-Sum Thm., $m\angle A + 30 + 100 = 180$, and $m\angle A = 50$.

Lesson Check 1. 58 **2.** 45 **3.** 68 **4.** $130 - x$ **5.** $m\angle 1 = 130$ **6.** $m\angle 3 = 38$ **7.** Answers may vary. Sample: Consider the int. ∠A of △ABC. By the △ ∠-Sum Thm., the sum of the measures of angles *A*, *B*, and *C* is 180°. ∠A is suppl. to its ext. ∠. So the sum of the measures of angles *B* and *C* is equal to the measure of the ext. ∠ of ∠A. **8.** A; all 3 ∠s are int. ∠s, so the solution should use the △ ∠-Sum Thm.

Exercises 9. 30 **11.** 90 **13.** $x = y = 80$ **15a.** ∠5, ∠6, ∠8 **b.** For ∠5: ∠1 and ∠3; for ∠6: ∠1 and ∠2; for ∠8: ∠1 and ∠2 **c.** $\angle 6 \cong \angle 8$ **17.** 123 **19.** $m\angle 3 = 92$, $m\angle 4 = 88$ **21.** 114 **23.** 60, 80 **25.** 102, 65, 13 **27.** 60; answers may vary. Sample: $180 \div 3 = 60$, so each ∠ is 60. **29.** $x = 37$; $m\angle P = 65$, $m\angle Q = 78$, $m\angle R = 37$ **31.** $a = 67$, $b = 58$, $c = 125$, $d = 23$, $e = 90$ **33.** ∠1 is an ext. ∠ of the △. (Given); ∠1 and ∠4 are suppl. (∠s that form a straight ∠ are suppl.); $m\angle 1 + m\angle 4 = 180$ (Def. of suppl.); $m\angle 2 + m\angle 3 + m\angle 4 = 180$ (△ ∠-Sum Thm.); $m\angle 1 + m\angle 4 = m\angle 2 + m\angle 3 + m\angle 4$ (Subst. Prop.); $m\angle 1 = m\angle 2 + m\angle 3$ (Subtr. Prop. of =) **35.** 40, 50 **37.** $\frac{1}{7}$ **39.** $\frac{1}{19}$ **41.** 115 **43.** A **45.** C

Lesson 3-6 pp. 182–188

Got It? 1. ∠1 and ∠NHJ are corresp. ∠s for lines *m* and ℓ. Since $\angle 1 \cong \angle NHJ$, then $m \parallel \ell$.
2a. Answers may vary. Sample:

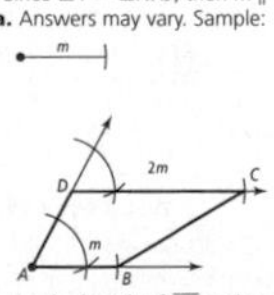

b. No; the length of $\overline{AB}$ and $m\angle A$ are not determined.

3.

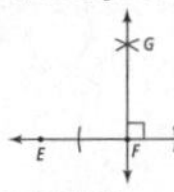

4.

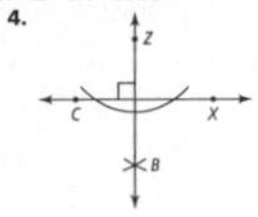

Lesson Check

1.

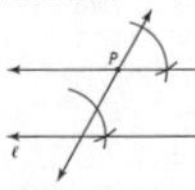

2.

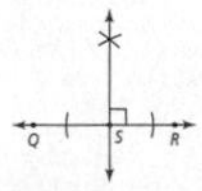

3.

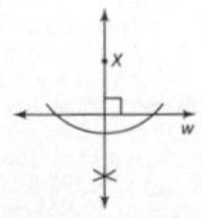

4. Yes; the same compass opening is used to draw the arcs at *C*. **5.** No; points *E* and *F* would have been further apart, but the new point *G* would determine the same line $\overleftrightarrow{RG}$ as in Step 4. **6.** Similar: You are constructing a line ⊥ to a given line through a given point. Different: The given point is on the given line in Problem 3 and is not on the given line in Problem 4.

Exercises

7.

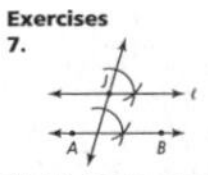

9.

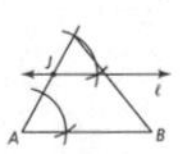

11–13. Constructions may vary. Samples using the following segments are given.

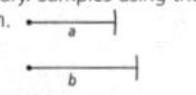

11.

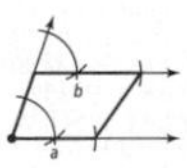

13.

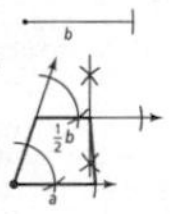

15.

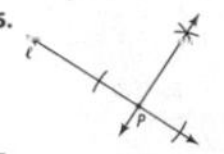

17.

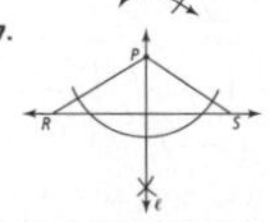

19.

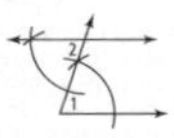

21. Construct a ≅ alt. int. ∠, then draw the ∥ line.

23.

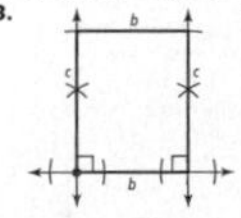

25.

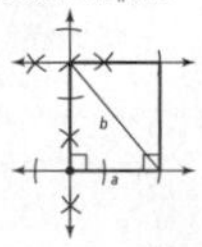

27.

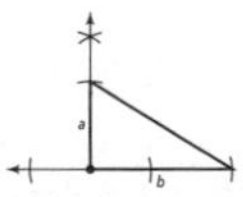

29a. II, IV, III, I **b.** III: points *C* and *G*; I: the intersection of $\overleftrightarrow{GC}$ with the arcs from Step III

31–39. Constructions may vary. Samples are given.

31.

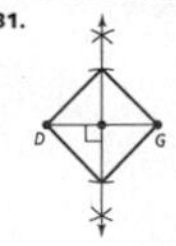

square

33.

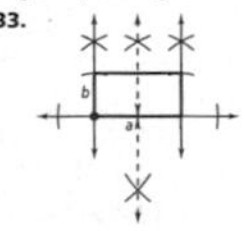

35.

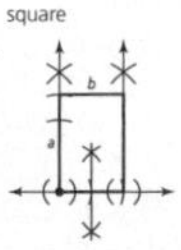

37.

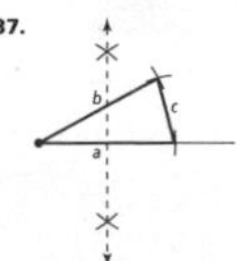

39. Not possible; the shorter sides would meet at a point on the longer side, forming a segment. **41.** I **43.** $3y = 120$, $(y - 15) = 25$ **44.** $x = 104$, $(x - 28) = 76$, $y = 35$, $(2y - 1) = 69$ **45.** $\frac{1}{2}$ **46.** 1 **47.** -2

Lesson 3-7 pp. 189–196

Got It? 1a. $\frac{4}{3}$ **b.** 0

2a.

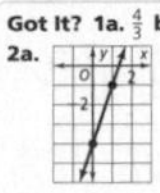

b.

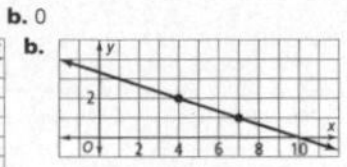

3a. $y = -\frac{1}{2}x + 2$ **b.** $y - 4 = -3(x + 1)$ **4a.** $y + 1 = \frac{6}{5}(x + 2)$ **b.** $y = \frac{6}{5}x + \frac{7}{5}$; $y = \frac{6}{5}x + \frac{7}{5}$; they represent the same line. **5a.** horizontal: $y = -3$; vertical: $x = 4$ **b.** No; the slope is undefined for a vertical line, so you cannot use the slope-intercept form because that requires a value for the slope.

Lesson Check 1. 5 **2.** -2 **3.** $y = 8x + 10$ **4.** $y - 3 = 4(x - 3)$ or $y - 7 = 4(x - 4)$ **5.** Answers may vary. Sample: The slope-intercept form $y = mx + b$ uses the slope *m* and the *y*-intercept *b*; the point-slope form $y - y_1 = m(x - x_1)$ uses a point (x_1, y_1) on the line and the slope *m*. **6.** The lines have the same *y*-int., but one line has a steep positive slope and the other has a less steep negative slope. **7.** Your classmate switched the *x*- and *y*-values in the formula for slope. The slope of the line is undefined.

Exercises 9. $-\frac{5}{6}$ **11.** $-\frac{3}{2}$ **13.** -8 **15.** undefined

17.

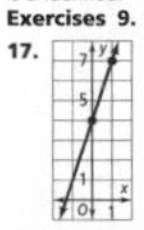

19.

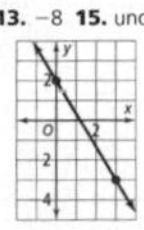

21.

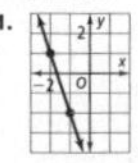

23.

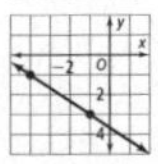

25. $y = \frac{1}{2}x - 5$ **27.** $y + 1 = -3(x - 4)$ **29.** $y - 6 = -(x + 2)$ or $y - 3 = -(x - 1)$ **31.** $y - 2 = -\frac{1}{2}(x - 6)$ or $y - 4 = -\frac{1}{2}(x - 2)$ **33.** $y = \frac{1}{2}(x + 1)$ or $y + 1 = \frac{1}{2}(x + 3)$ **35.** horizontal: $y = -2$; vertical: $x = 3$ **37.** horizontal: $y = 4$; vertical: $x = 6$ **39.**

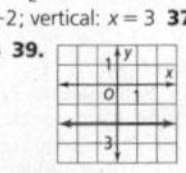

41.

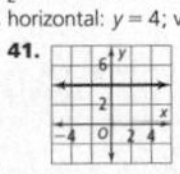

43. Yes; if the ramp is 24 in. high and 72 in. long, the slope will be $\frac{24}{72} = 0.\overline{3}$, which is less than the maximum slope of $\frac{4}{11} = 0.\overline{36}$. **45.** $y = -x + 2$ **47.** $y = -\frac{3}{2}x + 5$

49. (6, −4)

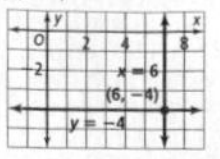

51. (−1, 3)

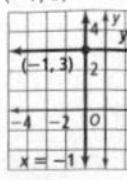

53. No; answers may vary. Sample: $\frac{1}{12} < \frac{3}{10}$ so the ramp would need to zigzag to comply with the law.
55a. Undefined; the y-axis is a vertical line, and the slope of a vertical line is undefined. **b.** $x = 0$ **57a.** $y = \frac{5}{2}x$
b. $y - 5 = -\frac{5}{2}(x - 2)$ or $y = -\frac{5}{2}x + 10$ **c.** The abs. value of the slopes is the same, but one slope is pos. and the other is neg. One y-int. is 0 and the other is 10.
59. No; the slope of the line through the first two points is $-\frac{1}{3}$ and the slope of the line through the last two points is −1, so the points do not lie on the same line.
61. 9 **63.** $\frac{1}{6}$ **65.** G **67.** G
69.

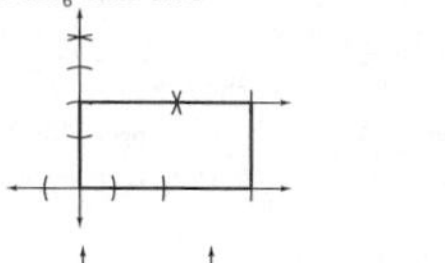

70.

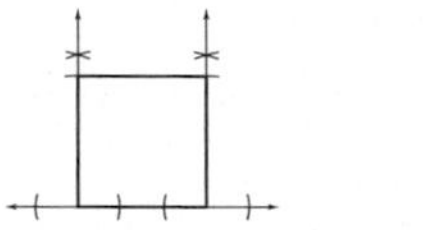

71. Distr. Prop. **72.** Substitution **73.** Reflexive Prop. of ≅ **74.** Symmetric Prop. of ≅ **75.** $\frac{1}{2}$ **76.** $\frac{5}{2}$ **77.** −5

Lesson 3-8 pp. 197–204

Got It? 1. No; the slope of ℓ_3 is $\frac{6-2}{-13-(-1)} = \frac{4}{-12} = -\frac{1}{3}$ And the slope of ℓ_4 is $\frac{6-7}{3-6} = \frac{-1}{-3} = \frac{1}{3}$. The slopes are not equal. **2.** $y - 3 = -(x + 5)$ **3.** No; the slope of ℓ_3 is $\frac{7-(-1)}{2-3} = \frac{8}{-1} = -8$ and the slope of ℓ_4 is $\frac{6-7}{-2-8} = \frac{-1}{-10} = \frac{1}{10}$. Since the product of the slopes is not −1, the Lines are not ⊥. **4.** $y - 7 = \frac{1}{3}(x + 3)$ **5.** $y - 40 = \frac{1}{3}(x - 90)$
Lesson Check 1. ⊥; the slope of $\overleftrightarrow{AB}$ is 2 and the slope of $\overleftrightarrow{CD}$ is $-\frac{1}{2}$. Since $(2)\left(-\frac{1}{2}\right) = -1$, the lines are ⊥. **2.** ∥; the slope of $\overleftrightarrow{AB}$ is 6 and the slope of $\overleftrightarrow{CD}$ is 6. Since the slopes are equal, the lines are ∥. **3.** Neither; the slope of $\overleftrightarrow{AB}$ is 0 and the slope of $\overleftrightarrow{CD}$ is 1. Since the slopes are not equal and their product is not −1, the lines are neither ∥ nor ⊥. **4.** Answers may vary. Sample: $y + 3 = \frac{1}{4}(x - 2)$
5. The second line should say "slope of parallel line = 3" because ∥ lines have equal slopes. **6.** Sample: ∥ line equations have equal slopes. ⊥ line equations have slopes with product −1.
Exercises
7. Yes; the slope of ℓ_1 is $-\frac{1}{2}$ and the slope of ℓ_2 is $-\frac{1}{2}$, and two lines with the same slope are ∥. **9.** No; the slope of ℓ_1 is $\frac{3}{2}$ and the slope of ℓ_2 is 2. Since the slopes are not equal the lines are not ∥. **11.** $y = -2x + 3$
13. $y - 4 = \frac{1}{2}(x + 2)$ **15.** Yes; the slope of ℓ_1 is $-\frac{1}{2}$ and the slope of ℓ_2 is 2. Since the product of the slopes is −1, the lines are ⊥. **17.** No; the slope of ℓ_1 is −1 and the slope of ℓ_2 is $\frac{4}{5}$. Since the product of the slopes is not −1, the lines are not ⊥. **19.** $y - 6 = -\frac{3}{2}(x - 6)$
21. $y - 4 = \frac{1}{2}(x - 4)$ **23.** Yes; both slopes are −1 so the lines are ∥. **25.** No; the slope of the first line is $-\frac{3}{4}$ and the slope of the second line is −3. Since the slopes are not equal, the lines are not ∥. **27.** −4 **29.** No; if two equations represent lines with the same slope and the same y-intercept, the equations must represent the same line. **31.** slope of $\overline{AB}$ = slope of $\overline{CD} = -\frac{3}{4}$, $\overline{AB} \parallel \overline{CD}$; slope of $\overline{BC}$ = slope of $\overline{AD} = 1$, $\overline{BC} \parallel \overline{AD}$ **33.** slope of $\overline{AB}$ = slope of $\overline{CD} = 0$, $\overline{AB} \parallel \overline{CD}$; slope of $\overline{BC} = 3$, slope of $\overline{AD} = \frac{3}{2}$, $\overline{BC} \nparallel \overline{AD}$ **35.** A **37.** Yes; the equations represent a horizontal line and a vertical line, and every horizontal line is ⊥ to every vertical line. **39.** Answers may vary. Sample: The three lines must have the same slope or undefined slope, so all three lines are ∥.
41a. $y = -\frac{1}{2}x + 100$ **b.** (100, 50) **c.** 112 yd
43. Slope of $\overline{AB}$ is $-\frac{1}{8}$; slope of $\overline{CD}$ is 8; the lines are ⊥. **45.** The slope of $\overline{AC}$ is $\frac{10}{-2} = -5$ and the slope of $\overline{BD}$ is $\frac{2}{10} = \frac{1}{5}$. Since the product of the slopes is −1, the diagonals are ⊥. The midpoint of $\overline{AC}$ is $\left(\frac{7+9}{2}, \frac{11+1}{2}\right) = (8, 6)$ and the midpoint of $\overline{BD}$ is $\left(\frac{13+3}{2}, \frac{7+5}{2}\right) = (8, 6)$. Since the two diagonals have the same midpoint, they bisect each other.
47. $y - 5 = \frac{1}{3}(x - 4)$ **49.** 6 **51.** 7
53. $y = -\frac{1}{2}x + 3$ **54.** $y - 2 = \frac{5}{3}(x + 4)$ or $y - 7 = \frac{5}{3}(x + 1)$ **55.** $y + 2 = \frac{3}{4}(x - 3)$ or $y + 8 = \frac{3}{4}(x + 5)$ **56.** Reflexive Prop. of ≅ **57.** Mult. Prop. of Equality **58.** Distr. Prop. **59.** Symmetric Prop. of ≅
60. Yes; ∠1 and ∠2 are vert. $\angle$s, and vert. $\angle$s are ≅.
61. Yes; ∠1 and ∠2 are both rt. $\angle$s, and all rt. $\angle$s are ≅.

62. No; $m\angle 1 = 54$ (Given) and $m\angle 2 = 90 - 54 = 36$ (because ∠1 and ∠2 are compl.).

Chapter Review pp. 206–210

1. transversal **2.** ext. ∠ **3.** point-slope **4.** alt. int. $\angle$s
5. skew lines **6.** slope-intercept **7.** ∠2 and ∠7, *a* and *b*, transversal *d*; ∠3 and ∠6, *c* and *d*, transversal *e*; ∠3 and ∠8; *b* and *e*; transversal *c* **8.** ∠5 and ∠8, lines *a* and *b*, transversal *c*; ∠2 and ∠6; *a* and *e*; transversal *d* **9.** ∠1 and ∠4, lines *c* and *d*, transversal *b*; ∠2 and ∠4, lines *a* and *b*, transversal *d*; ∠2 and ∠5, lines *c* and *d*, transversal *a*; ∠1 and ∠5, lines *a* and *b*, transversal *c*; ∠3 and ∠4; *b* and *c*; transversal *e* **10.** ∠1 and ∠7, lines *c* and *d*, transversal *b* **11.** corresp. $\angle$s **12.** alt. int. $\angle$s
13. $m\angle 1 = 120$ because corresp. $\angle$s are ≅; $m\angle 2 = 120$ because ∠1 and ∠2 are vert. $\angle$s. **14.** $m\angle 1 = 75$ because same-side int. $\angle$s are suppl.; $m\angle 2 = 105$ because alt. int. $\angle$s are ≅. **15.** $x = 118, y = 37$ **16.** 20
17. 20 **18.** $n \parallel p$; if corresp. $\angle$s are ≅, then the lines are ∥. **19.** none; ∠3 and ∠6 form a linear pair. **20.** $\ell \parallel m$; if same-side int. $\angle$s are suppl., then the lines are ∥.
21. $n \parallel p$; if alt. int. $\angle$s are ≅, then the lines are ∥.
22. ∥ **23.** *a* **24.** 1st Street and 3rd Street are ∥ because they are both ∥ to Morris Avenue. Since 1st Street and 5th street are both ∥ to 3rd Street, 1st Street and 5th Street are ∥ to each other. **25.** $x = 60, y = 60$
26. $x = 45, y = 45$ **27.** 30 **28.** 55 **29.** 3
30.

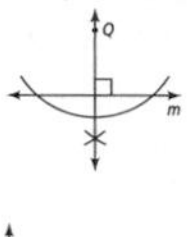

31.

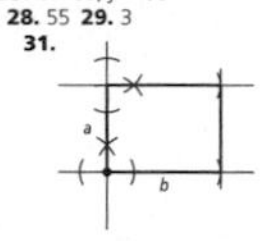

32.

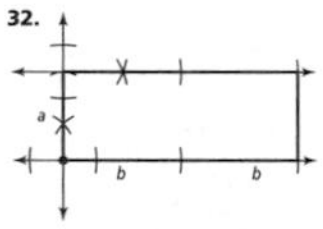

33.

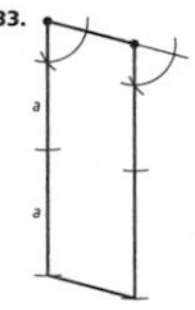

34. −1 **35.** undefined

36. slope: 2; y-intercept: −1

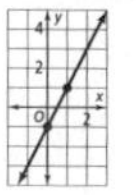

37. slope: −2; point: (−5, 3)

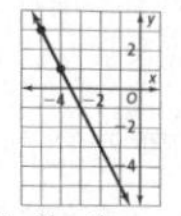

38. $y = -\frac{1}{2}x + 12$ **39.** $y + 9 = 3(x - 1)$
40. $y - 2 = 4(x - 4)$ or $y + 2 = 4(x - 3)$
41. neither **42.** ∥ **43.** ⊥ **44.** ∥
45. $y - 2 = 8(x + 6)$ **46.** $y + 3 = -6(x - 3)$

Chapter 4

Get Ready! p. 215

1. $AB = 4$, $BC = 3$, $AC = 5$ **2.** $AB = 8$, $BC = \sqrt{265}$, $AC = \sqrt{137}$ **3.** $AB = \sqrt{58}$, $BC = \sqrt{32}$, $AC = \sqrt{58}$
4. $\angle J \cong \angle L$ **5.** $m\angle M = m\angle N = 90$ **6.** ∠B is a rt. ∠.
7. $\angle AFB \cong \angle CFD$ **8.** $\angle B \cong \angle C$, $\angle A \cong \angle D$, $\angle AEB \cong \angle CED$ **9.** $\angle DAC \cong \angle BCA$, $\angle DCA \cong \angle BAC$, $\angle DAB \cong \angle BCD$, $\angle B \cong \angle D$ **10.** $m\angle A = 21$, $m\angle B = 71$, $m\angle C = 88$ **11-13.** Answers may vary. Sample: **11.** The base is the side that meets each of the two ≅ sides of the △. **12.** The legs are the ≅ sides of an isosc. △.
13. Corresp. parts are the sides or $\angle$s that are in the same relative position in each figure.

Lesson 4-1 pp. 218–224

Got It? 1. $\overline{WY} \cong \overline{MK}$, $\overline{YS} \cong \overline{KV}$, $\overline{WS} \cong \overline{MV}$, $\angle W \cong \angle M$, $\angle Y \cong \angle K$, $\angle S \cong \angle V$ **2.** $m\angle V = 83$; $\angle W \cong \angle M$ and $\angle Y \cong \angle K$ because they are corresp. parts of ≅ △s. By the Triangle Angle-Sum Theorem, $m\angle M + m\angle K + m\angle V = 180$. By substitution, $62 + 35 + m\angle V = 180$. So by subtraction, $m\angle V = 83$. **3.** Answers may vary. Sample: You know that $\overline{AD} \cong \overline{CD}$ (Given) and $\overline{BD} \cong \overline{BD}$ (Reflexive Prop. of ≅), but you have no other information about the sides and $\angle$s of the △, so you cannot conclude that $\triangle ABD \cong \triangle CBD$. **4.** $\angle A \cong \angle D$ (Given), and $\angle ABE \cong \angle DBC$ because vertical $\angle$s are ≅. Also, $\angle AEB \cong \angle DCB$ (Third $\angle$s Theorem). The three pairs of sides are ≅ (Given), so $\triangle AEB \cong \triangle DCB$ by the def. of ≅ △s.
Lesson Check 1a. $\overline{NY}$ **b.** ∠X **2a.** $\overline{RO}$
b. ∠T **3a.** ∠A **b.** $\overline{KL}$ **c.** CKLU **4a.** $\angle M \cong \angle T$
b. 92 **5.** Answers may vary. Sample: finding the correct top for a food container **6.** No; the △s could be the same shape but not necessarily the same size. **7.** He has not shown that corresp. $\angle$s are ≅.

Exercises 9. $\overline{EF} \cong \overline{HI}$, $\overline{FG} \cong \overline{IJ}$, $\overline{EG} \cong \overline{HJ}$, $\angle EFG \cong \angle HIJ$, $\angle FGE \cong \angle IJH$, $\angle FEG \cong \angle IHJ$ **11.** $\overline{CM}$ **13.** ∠B **15.** ∠J
17. △CLM **19.** △MCL **21.** $\angle P \cong \angle S$, $\angle O \cong \angle I$, $\angle L \cong \angle D$, $\angle Y \cong \angle E$ **23.** 45 ft **25.** 52 **27.** 280 ft
29. 128 **31.** No; there are not three pairs of ≅ corresp. sides. **33.** C **35.** $m\angle A = m\angle D = 20$ **37.** $BC = EF = 8$
39. 43 **41.** 5 **43.** Answers may vary. Sample: If $\triangle PQR \cong \triangle XYZ$, then $\overline{PQ} \cong \overline{XY}$, $\overline{QR} \cong \overline{YZ}$, $\overline{PR} \cong \overline{XZ}$, $\angle P \cong \angle X$, $\angle Q \cong \angle Y$, and $\angle R \cong \angle Z$. **45.** Two pairs of ≅ sides are given, and the third pair of sides are ≅ because $\overline{PQ}$ bisects $\overline{RT}$, so $\overline{TS} \cong \overline{RS}$. $\overline{PR} \parallel \overline{TQ}$, so $\angle P \cong \angle Q$ and $\angle R \cong \angle T$ because they are alt. int. $\angle$s; the third pair of $\angle$s are vertical $\angle$s, so they are ≅. Thus $\triangle PRS \cong \triangle QTS$ by the def. of ≅ △s. **47.** $KL = 4$, $LM = 3$, $KM = 5$
49a. 15 quadrilaterals

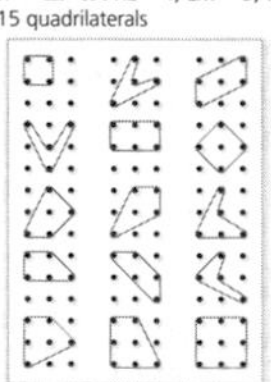

b. 11 convex, 4 concave

Lesson 4-2 pp. 226–233

Got It? 1. Two pairs of sides are given as ≅, and $\overline{BD} \cong \overline{BD}$ by the Refl. Prop. of ≅. So $\triangle BCD \cong \triangle BFD$ by SSS. **2.** $\overline{LE} \cong \overline{BN}$ **3.** SSS; three pairs of corresp. sides are ≅.
Lesson Check 1a. ∠PEN (or ∠E) **b.** ∠NPE (or ∠P)
2a. $\overline{HA}$ and $\overline{HT}$ **b.** $\overline{TH}$ and $\overline{TA}$ **3.** SAS **4.** SSS
5. Answers may vary. Sample: Alike: Both use three pairs of ≅ parts to prove △s ≅. Different: SSS uses three pairs of ≅ sides, while SAS uses two pairs of ≅ sides and their ≅ included $\angle$s. **6.** No; the ≅ $\angle$s are not included between the pairs of ≅ sides. **7.** No; the △s have the same perimeter, but the three side lengths of one △ are not necessarily = to the three side lengths of the other △, so you cannot use SSS. There is no information about the $\angle$s of the △s, so you cannot use SAS.
Exercises
9. *F* is the midpt. of $\overline{GI}$ (Given), so $\overline{IF} \cong \overline{GF}$ because a midpt. divides a segment into two ≅ segments. The other two pairs of sides are given as ≅, so $\triangle EFI \cong \triangle HFG$ by SSS. **11.** You need to know $\overline{LG} \cong \overline{MN}$; the diagram shows that $\overline{LT} \cong \overline{MQ}$ and $\angle L \cong \angle M$. ∠L is included between $\overline{LG}$ and $\overline{LT}$, and ∠M is included between $\overline{MN}$ and $\overline{MQ}$. **13.** Not enough information; the congruent vertical angles TQP and RQS are not included by the pairs of ≅ sides. **15.** If the 40° ∠ is *always* included between the two 5-in. sides, then all the △s will be ≅ by SAS. If the 40° ∠ is *never* included between the two 5-in. sides, then the angles of the △ will be 40°, 40°, and 100°, with the 100° angle included between the 5-in. sides, so all the △s will be ≅ by SAS. But a △ with the 40° angle included between the 5-in. sides will not be ≅ to a △ with the 40° angle not included between the 5-in. sides. **17.** *X* is the midpt. of $\overline{AG}$ and $\overline{NR}$ (Given), so $\overline{AX} \cong \overline{GX}$ and $\overline{NX} \cong \overline{RX}$ by the def. of midpt. Also, $\angle AXN \cong \angle GXR$ because they are vertical $\angle$s, so $\triangle ANX \cong \triangle GRX$ by SAS. **19.** $AB = \sqrt{25 + 16} = \sqrt{41}$ and $DE = \sqrt{16 + 36} = \sqrt{52}$, so $\triangle ABC \ncong \triangle DEF$.
21. Answers may vary. Sample: roof trusses for a house, sections of a ferris wheel, sawhorses used by a carpenter; explanations will vary.
23a. Answers may vary. Sample:

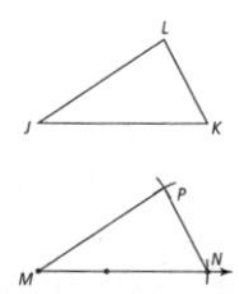

b. Answers may vary. Sample:

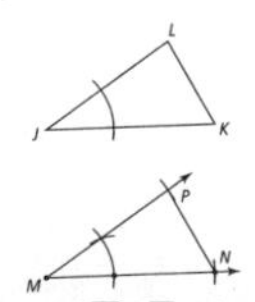

25. Not enough information; you need $\overline{DY} \cong \overline{TK}$ to show the △s are ≅ by SSS, or you need $\angle H \cong \angle P$ to show the △s are ≅ by SAS. **27.** Not necessarily; the ≅ $\angle$s are not included between the pairs of ≅ sides. **29.** $\overline{AE}$ and $\overline{BD}$ bisect each other (Given), so $\overline{AC} \cong \overline{EC}$ and $\overline{DC} \cong \overline{BC}$ (Def. of bisector). $\angle ACB \cong \angle ECD$ (Vert. $\angle$s are ≅.), so $\triangle ACB \cong \triangle ECD$ by SAS. **31.** Given the ⊥ segments, $\angle B \cong \angle CMA$ because all rt. $\angle$s are ≅. *M* is the midpt. of $\overline{AB}$ (Given), so $\overline{AM} \cong \overline{MB}$ by the def. of midpt. Since $\overline{DB} \cong \overline{CM}$ (Given), then $\triangle AMC \cong \triangle MBD$ by SAS.
33. Answers may vary. Sample: $\angle N \cong \angle L$, $\overline{MN} \cong \overline{OL}$, and $\overline{NO} \cong \overline{LM}$ (Given), so $\triangle MNO \cong \triangle OLM$ by SAS. $\angle NMO \cong \angle LOM$ (Corresp. parts of ≅ △s are ≅.) So $\overline{MN} \parallel \overline{OL}$ because if alt. int. $\angle$s are ≅, then the lines are ∥.
35. A **37.** D **39.** ∠E **40.** $\overline{AB}$ **41.** $\overline{FG}$ **42.** ∠C **43.** If $2x = 6$, then $x = 3$; both are true. **44.** If $x^2 = 9$, then $x = 3$; the statement is true and its converse is false.
45. $\overline{JH}$ **46.** ∠MNL (or ∠N)

Lesson 4-3 pp. 234–241

Got It? 1. $\triangle HGO \cong \triangle ACT$ because $\overline{HG} \cong \overline{AC}$ and the ≅ segments are included between two pairs of ≅ $\angle$s.
2. $\angle B \cong \angle E$ because all rt. $\angle$s are ≅. $\overline{AB} \cong \overline{AE}$ and $\angle CAB \cong \angle DAE$ (Given), so $\triangle ABC \cong \triangle AED$ by ASA.
3a. $\overline{RP}$ bisects ∠SRQ (Given), so $\angle SRP \cong \angle QRP$ by the def. of ∠ bisector. $\angle S \cong \angle Q$ (Given) and $\overline{RP} \cong \overline{RP}$ (Refl. Prop. of ≅), so $\triangle SRP \cong \triangle QRP$ by AAS. **b.** After Step 3 in the proof, state that $\angle MRW \cong \angle KWR$ by the Third $\angle$s Theorem and write Step 4, so $\triangle WMR \cong \triangle RKW$ by ASA. **4.** Yes; $\overline{PR} \cong \overline{SR}$ and $\angle A \cong \angle I$ (Given). $\angle ARP \cong \angle IRS$ (Vert. $\angle$s are ≅.), so $\triangle PAR \cong \triangle SIR$ by AAS.
Lesson Check 1. $\overline{RS}$ **2.** ∠N, ∠O **3.** ASA
4. AAS **5.** Answers may vary. Sample: Alike: Both postulates use three pairs of ≅ corresp. parts. Different: To use the ASA Postulate, the sides must be included between the pairs of corresp. $\angle$s, while to use the SAS Postulate, the $\angle$s must be included between the pairs of corresp. sides. **6.** $\overline{LM}$ is not included between the pairs of ≅ corresp. $\angle$s. **7.** $\angle F \cong \angle G$; $\angle D \cong \angle H$
Exercises
9. $\triangle ABC \cong \triangle EDF$ **11.** $\overline{AC} \perp \overline{BD}$ (Given), so $\angle ACB \cong \angle ACD$ because ⊥ lines form rt. $\angle$s, and all rt. $\angle$s are ≅. $\angle BAC \cong \angle DAC$ (Given) and $\overline{AC} \cong \overline{AC}$ (Refl. Prop. of ≅), so $\triangle ABC \cong \triangle ADC$ by ASA. **13a.** Vert. $\angle$s are ≅. **b.** Given **c.** $\overline{TQ} \cong \overline{RQ}$ **d.** AAS **15.** Given the ⊥ segments, $\angle Q \cong \angle S$ because ⊥ lines form rt. $\angle$s, and all rt. $\angle$s are ≅. It is given that *T* is the midpt. of $\overline{PR}$, so $\overline{PT} \cong \overline{RT}$ by the def. of midpt. $\angle PTQ \cong \angle RTS$ because vert. $\angle$s are ≅, so $\triangle PQT \cong \triangle RST$ by AAS.
17. $\triangle UST \cong \triangle RTS$ by AAS. **19.** It is given that $\angle N \cong \angle P$ and $\overline{MO} \cong \overline{QO}$. Also, $\angle MON \cong \angle QOP$ because vert. $\angle$s are ≅. So $\triangle MON \cong \triangle QOP$ by AAS.
21. Answers may vary. Sample: Yes; ASA guarantees a unique triangle with vertices at the oak tree, the maple tree, and the time capsule. **23.** No; the common side is included between the two ≅ $\angle$s in one △, but it is not included between the ≅ $\angle$s in the other △. **25.** $\overline{AE} \parallel \overline{BD}$ (Given), so $\angle A \cong \angle DBC$ (If ∥ lines, corresp. $\angle$s are ≅.). Since $\angle E \cong \angle D$ and $\overline{AE} \cong \overline{BD}$ (Given), then $\triangle AEB \cong \triangle BDC$ by ASA.

27. Answers may vary. Sample:

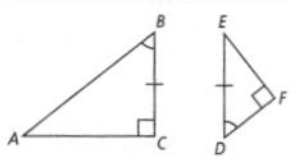

29. $\triangle EAB \cong \triangle ECD$, $\triangle EBC \cong \triangle EDA$, $\triangle ABD \cong \triangle CDB$, $\triangle ABC \cong \triangle CDA$
31. $\frac{13}{20}$ **33.** I
35. Converse: If you are too young to vote in the United States, then you are less than 18 years old; true.

Lesson 4-4 pp. 244–248

Got It? 1. $\overline{BA} \cong \overline{DA}$ and $\overline{CA} \cong \overline{ED}$ (Given). $\angle CAB \cong \angle EAD$ (Vert. $\angle$s are ≅.) So $\triangle ABC \cong \triangle ADE$ by SAS and $\angle C \cong \angle E$ because corresp. parts of ≅ △s are ≅.
2a. It is given that *M* is the midpt. of $\overline{BC}$, so $\overline{BM} \cong \overline{CM}$ by the def. of midpt. $\overline{AB} \cong \overline{AC}$ (Given) and $\overline{AM} \cong \overline{AM}$ (Refl. Prop. of ≅), so $\triangle AMB \cong \triangle AMC$ by SSS. Thus $\angle AMB \cong \angle AMC$ because corresp. parts of ≅ △s are ≅.
b. No; while $\overline{TR} \perp \overline{RS}$, if point *L* is not at sea level, then $\overline{TR}$ would not be ⊥ to $\overline{RL}$.
Lesson Check 1. SAS; so $\overline{EA} \cong \overline{MA}$ because corresp. parts of ≅ △s are ≅. **2.** SSS; so $\angle U \cong \angle E$ because corresp. parts of ≅ △s are ≅. **3.** "Corresp. parts of ≅ △s are ≅" is a short version of the def. of ≅ △s.
4. $\triangle KHL \cong \triangle NHM$ by AAS Thm.
Exercises 5. $\triangle KLJ \cong \triangle OMN$ by SAS; $\overline{KJ} \cong \overline{ON}$, $\angle K \cong \angle O$, $\angle J \cong \angle N$. **7.** $\overline{OM} \cong \overline{EB}$ and $\overline{ME} \cong \overline{RO}$ (Given). $\overline{OE} \cong \overline{OE}$ by the Refl. Prop. of ≅. $\triangle MOE \cong \triangle REO$ by SSS, so $\angle M \cong \angle R$ because corresp. parts of ≅ △s are ≅.
9. A pair of ≅ sides and a pair of ≅ $\angle$s are given. Since $\overline{PT} \cong \overline{PT}$ (Refl. Prop. of ≅), then $\triangle STP \cong \triangle OTP$ by SAS. $\angle S \cong \angle O$ because corresp. parts of ≅ △s are ≅. **11.** $\overline{KL}$ bisects ∠PKQ, so $\angle PKL \cong \angle QKL$. $\overline{KL} \cong \overline{KL}$ by Refl. Prop. of ≅. $\triangle PKL \cong \triangle QKL$ by SAS, so $\angle P \cong \angle Q$ because corresp. parts of ≅ △s are ≅. **13.** $\angle PLK \cong \angle QLK$ because ⊥ lines form rt. $\angle$s, and all rt. $\angle$s are ≅. From the def. of ∠ bisector, $\angle PKL \cong \angle QKL$. So with $\overline{KL} \cong \overline{KL}$ by the Refl. Prop. of ≅, $\triangle PKL \cong \triangle QKL$ by ASA and $\angle P \cong \angle Q$ because corresp. parts of ≅ △s are ≅.
15. $\overline{BA} \cong \overline{BC}$ (Given) and $\overline{BD}$ bisects ∠ABC (Given) $\angle ABD \cong \angle CBD$ (Def. of ∠ bisector). $\overline{BD} \cong \overline{BD}$ (Refl. Prop. of ≅), so $\triangle ABD \cong \triangle CBD$ by SAS. $\angle ADB \cong \angle CDB$

Selected Answers

(Corresp. parts of $\cong$ $\triangle$s are $\cong$.) and $\angle ADB$ and $\angle CDB$ are suppl. so they must be rt. $\angle$s. By def. of $\perp$ lines, $\overline{BD} \perp \overline{AC}$. $\overline{AD} \cong \overline{CD}$ (Corresp. parts of $\cong$ $\triangle$s are $\cong$.), so $\overline{BD}$ bisects $\overline{AC}$ (Def. of seg. bisector). **17.** The construction makes $\overline{AC} \cong \overline{BE}$, $\overline{AD} \cong \overline{BF}$, and $\overline{CD} \cong \overline{EF}$. So $\triangle ACD \cong \triangle BEF$ by SSS. Thus $\angle A \cong \angle B$ because corresp. parts of $\cong$ $\triangle$s are $\cong$. **19.** It is given that $\overline{JK} \parallel \overline{QP}$, so $\angle K \cong \angle Q$ and $\angle J \cong \angle P$ because they are alt. int. $\angle$s. With $\overline{JK} \cong \overline{PQ}$ (Given), $\triangle KJM \cong \triangle QPM$ by ASA and then $\overline{JM} \cong \overline{PM}$ because corresp. parts of $\cong$ $\triangle$s are $\cong$. Thus M is the midpt. of $\overline{JP}$ by def. of midpt. So $\overline{KQ}$, which contains point M, bisects $\overline{JP}$ by the def. of segment bisector. **21.** Using the given information and $\overline{AE} \cong \overline{AE}$ (Refl. Prop. of $\cong$), $\triangle AKE \cong \triangle ABE$ by SSS. Thus $\angle KAS \cong \angle BAS$ because corresp. parts of $\cong$ $\triangle$s are $\cong$. In $\triangle KAS$ and $\triangle BAS$, $\overline{AK} \cong \overline{AB}$ (Given) and $\overline{AS} \cong \overline{AS}$ (Refl. Prop. of $\cong$), so $\triangle KAS \cong \triangle BAS$ by SAS. Thus $\overline{KS} \cong \overline{BS}$ because corresp. parts of $\cong$ $\triangle$s are $\cong$, and S is the midpt. of $\overline{BK}$ by the def. of midpt.

Lesson 4-5 pp. 250–256

Got It? 1a. Yes; since $\overline{WV} \cong \overline{WS}$, $\angle WVS \cong \angle S$ by the Isosc. $\triangle$ Thm.; yes; since $\angle WVS \cong \angle S$, and $\angle R \cong \angle WVS$ (Given), $\angle R \cong \angle S$ (Trans. Prop. of $\cong$). Therefore, $\overline{TR} \cong \overline{TS}$ by the Converse of isosc. $\triangle$ Thm. **b.** No; there is not enough information about the sides or $\angle$s of $\triangle RUV$. **2.** 63 **3.** $m\angle A = 61$, $m\angle BCD = 119$
Lesson Check 1a. 70 **b.** 53 **2a.** 75 **b.** 48 **3.** 23, 134 **4a.** The $\angle$s opposite the $\cong$ sides are $\cong$. **b.** All three $\angle$s have measure 60, and all three sides are $\cong$. **5.** The $\cong$ $\angle$s should be opposite the $\cong$ sides.
Exercises 7. $\overline{UW}$; Converse of Isosc. $\triangle$ Thm. **9.** Answers may vary. Sample: $\angle VUY$; Isosc. $\triangle$ Thm. **11.** $x = 38$, $y = 4$ **13.** 108 **15.** 45 and 45; the sum of the measures of the acute $\angle$s must be 90, so the measure of each acute $\angle$ must be half of 90. **17.** 2.5 **19.** 35 **21.** 20, 80, 80 or 50, 50, 80 **23a.** $\overline{RS}$ **b.** $\overline{RS}$; Proof: $\overline{RS} \cong \overline{RS}$ (Refl. Prop. of $\cong$) and $\angle PRS \cong \angle QRS$ (def. of $\angle$ bisector). Also, $\angle P \cong \angle Q$ (Given). So $\triangle PRS \cong \triangle QRS$ by AAS. $\overline{PR} \cong \overline{QR}$ because corresp. parts of $\cong$ $\triangle$s are $\cong$. **25.** $\overline{AE} \cong \overline{DE}$ (Given), so $\angle A \cong \angle D$ by the Isosc. $\triangle$ Thm. Since $\overline{AB} \cong \overline{DC}$ (Given), then $\triangle ABE \cong \triangle DCE$ by SAS. **27a.** isosc. $\triangle$ **b.** 900 ft; 1100 ft **c.** The tower is the $\perp$ bisector of each base.
29.

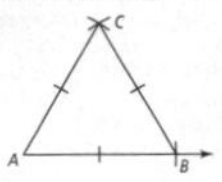

Draw $\overline{AB}$. Using AB as a radius, draw arcs with centers A and B. The intersection of these arcs is C. **31.** $m = 36$, $n = 27$

33. (−4, 0), (0, 0), (0, −4), (4, 4), (4, 8), (8, 4) **35.** (−1, 6), (2, 6), (2, 9), (5, 0), (5, 3), (8, 3) **37.** B **39.** C **41.** $RC = GV$; there are three pairs of $\cong$ $\angle$s and one pair of $\cong$ sides, so $\triangle TRC \cong \triangle HGV$ by AAS or ASA, and $\overline{RC} \cong \overline{GV}$ because corresp. parts of $\cong$ $\triangle$s are $\cong$. **42.** The letters are the first letters of the days of the week; S, S. **43.** Yes; the $\triangle$s share a common side, so they are $\cong$ by SAS. **44.** Yes; the vertical $\angle$s are $\cong$, so the $\triangle$s are $\cong$ by SAS.

Algebra Review p. 257

1. (−3, −7) **3.** no solution **5.** infinitely many solutions **7.** infinitely many solutions

Lesson 4-6 pp. 258–264

Got It? 1a. $\triangle PRS$ and $\triangle RPQ$ are rt. $\triangle$s with $\cong$ hypotenuses ($\overline{SP} \cong \overline{QR}$) and $\cong$ legs ($\overline{PR} \cong \overline{PR}$). So $\triangle PRS \cong \triangle RPQ$ by HL. **b.** Yes; the two $\triangle$s satisfy the three conditions of the HL Thm., so they are $\cong$. **2.** It is given that $\overline{AD}$ is the $\perp$ bisector of $\overline{CE}$, so $\triangle CBD$ and $\triangle EBA$ are rt. $\triangle$s and $\overline{CB} \cong \overline{EB}$ by the def. of $\perp$ bisector. Also, $\overline{CD} \cong \overline{EA}$ (Given), so $\triangle CBD \cong \triangle EBA$ by HL.
Lesson Check 1. yes; $\triangle BCA \cong \triangle EFD$ **2.** yes; $\triangle MPL \cong \triangle MNO$ **3.** no **4.** yes; $\triangle XVR \cong \triangle TVR$ **5.** 13 cm; the hypotenuse is the longest side of a rt. $\triangle$. **6.** Answers may vary. Sample: Alike: They both require information on two pairs of sides and one pair of $\angle$s. Different: For HL, the rt. $\angle$s are NOT included between the two pairs of $\cong$ sides, while for SAS the $\angle$s ARE included between the two pairs of $\cong$ sides. **7.** No; $\triangle LMJ$ and $\triangle JKL$ are rt. $\triangle$s with $\cong$ hypotenuses ($\overline{MJ} \cong \overline{KL}$) and $\cong$ legs ($\overline{LJ} \cong \overline{LJ}$), so $\triangle LMJ \cong \triangle JKL$ by HL.
Exercises 9a. $\triangle ABE$ and $\triangle DEB$ are rt. $\triangle$s. **b.** $\overline{BE} \cong \overline{EB}$ **c.** $\overline{AB} \cong \overline{DE}$ **d.** HL **11.** From the given information about $\perp$ segments, $\triangle PTM$ and $\triangle RMJ$ are rt. $\triangle$s $\overline{PM} \cong \overline{RJ}$ (Given), and since M is the midpt. of $\overline{TJ}$, $\overline{TM} \cong \overline{JM}$. Thus $\triangle PTM \cong \triangle RMJ$ by HL. **13.** $x = -1$, $y = 3$ **15.** Yes; the two $\triangle$s are rt. $\triangle$s with $\cong$ hypotenuses and one pair of $\cong$ legs, so the two $\triangle$s are $\cong$ by HL. Then $\overline{RQ} \cong \overline{CB}$ because corresp. parts of $\cong$ $\triangle$s are $\cong$. **17.** Using the information about $\perp$ segments, $\triangle RST$ and $\triangle TUV$ are rt. $\triangle$s $\overline{RS} \cong \overline{TU}$ (Given), and T is the midpt. of $\overline{RV}$ (Given), so $\overline{RT} \cong \overline{TV}$ (Def. of midpt.). Thus $\triangle RST \cong \triangle TUV$ by HL.
19.

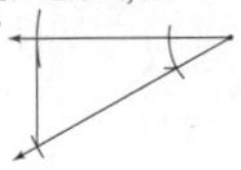

21.

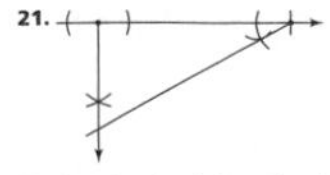

23. From the given information about an isosc. $\triangle$, rt. $\angle$s, and midpt., you can conclude that $\overline{KG} \cong \overline{KE}$ (Def. of isosc. $\triangle$), $\triangle LKG$ and $\triangle DKE$ are rt. $\triangle$s (Def. of rt. $\triangle$), and $\overline{LK} \cong \overline{DK}$ (Def. of midpt.). So $\triangle LKG \cong \triangle DKE$ by HL, and $\overline{LG} \cong \overline{DE}$ because corresp. parts of $\cong$ $\triangle$s are $\cong$. **25.** No, the triangles are not $\cong$. Explanations may vary. Sample: $\overline{DF}$ is the hypotenuse of $\triangle DEF$, so it is the longest side of the triangle. Therefore, it is greater than 5 and greater than 13 because it is longer than either of the legs. So $\overline{DF}$ cannot be congruent to $\overline{AC}$, which is the hypotenuse of $\triangle ABC$ and has length 13. **27.** $\triangle AEB$ and $\triangle CEB$ are rt. $\triangle$s because the given information includes $\overline{BE} \perp \overline{EA}$ and $\overline{BE} \perp \overline{EC}$. $\triangle ABC$ is equilateral (Given), so $\overline{AB} \cong \overline{CB}$ by the def. of equilateral. Also, $\overline{BE} \cong \overline{BE}$ by the Refl. Prop. of $\cong$. So $\triangle AEB \cong \triangle CEB$ by HL. **29.** C **31.** $\triangle XRY \cong \triangle XRZ$ by HL, $\triangle YQX \cong \triangle YQZ$ by HL, $\triangle ZPX \cong \triangle ZPY$ by HL, and $\triangle XPS \cong \triangle YPS$ by SAS OR other correct pairs and explanations. **32.** $\triangle STU$ is isosceles. $\overline{ST} \cong \overline{UT}$ because corresp. parts of $\cong$ $\triangle$s are $\cong$. **33.** $\triangle STU$ is equilateral. $\overline{ST} \cong \overline{US}$, $\overline{TU} \cong \overline{ST}$, and $\overline{US} \cong \overline{TU}$ because corresp. parts of $\cong$ $\triangle$s are $\cong$. **34.** Yes; $\triangle ABC \cong \triangle LMN$ by HL. **35.** No; $\triangle LMN$ and $\triangle HJK$ have one pair of $\cong$ sides and one pair of $\cong$ $\angle$s, but that is not enough to conclude that they are $\cong$. **36.** No; the hypotenuse of rt. $\triangle ABC$ is $\cong$ to a leg of rt. $\triangle RST$, so the $\triangle$s cannot be $\cong$.

Lesson 4-7 pp. 265–271

Got It? 1a. $\overline{AD}$ **b.** $\overline{AB}$ **2.** It is given that $\triangle ACD \cong \triangle BDC$, so $\angle ADC \cong \angle BCD$ because corresp. parts of $\cong$ $\triangle$s are $\cong$. Therefore, $\overline{CE} \cong \overline{DE}$ by the Converse of the Isosc. $\triangle$ Thm. **3.** $\triangle PSQ \cong \triangle RSQ$ by SAS because $\overline{PS} \cong \overline{RS}$ (Given), $\angle PSQ \cong \angle RSQ$ (Given), and $\overline{SQ} \cong \overline{SQ}$ (Refl. Prop. of $\cong$). So $\overline{PQ} \cong \overline{RQ}$ and $\angle PQT \cong \angle RQT$ (Corresp. parts of $\cong$ $\triangle$s are $\cong$.). Also, $\overline{QT} \cong \overline{QT}$ (Refl. Prop. of $\cong$), so $\triangle QPT \cong \triangle QRT$ by SAS. **4.** Using $\overline{AD} \cong \overline{AD}$ (Refl. Prop. of $\cong$) and the two given pairs of $\cong$ $\angle$s, $\triangle ACD \cong \triangle AED$ by AAS. Then $\overline{CD} \cong \overline{ED}$ (Corresp. parts of $\cong$ $\triangle$s are $\cong$.) and $\angle BDC \cong \angle FDE$ (Vert. $\angle$s are $\cong$.). Therefore, $\triangle BDC \cong \triangle FDE$ by ASA, and $\overline{BD} \cong \overline{FD}$ because corresp. parts of $\cong$ $\triangle$s are $\cong$.

Lesson Check 1. $\overline{JK}$ **2.** $\angle D$
3.

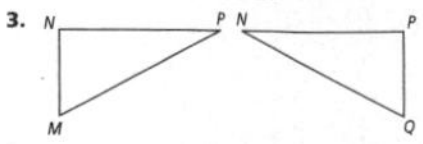

4.

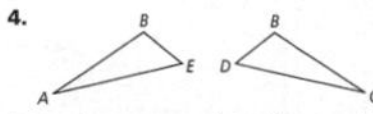

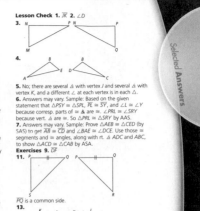

5. No; there are several $\angle$s with vertex J and several $\angle$s with vertex K, and a different $\angle$ at each vertex is in each $\triangle$. **6.** Answers may vary. Sample: Based on the given statement that $\triangle PSY \cong \triangle SPL$, $\overline{PL} \cong \overline{SY}$, and $\angle L \cong \angle Y$ because corresp. parts of $\cong$ $\triangle$s are $\cong$. $\angle PRL \cong \angle SRY$ because vert. $\angle$s are $\cong$. So $\triangle PRL \cong \triangle SRY$ by AAS. **7.** Answers may vary. Sample: Prove $\triangle AEB \cong \triangle CED$ (by SAS) to get $\overline{AB} \cong \overline{CD}$ and $\angle BAE \cong \angle DCE$. Use those $\cong$ segments and $\cong$ angles, along with rt. $\angle$s ADC and ABC, to show $\triangle ACD \cong \triangle CAB$ by ASA.
Exercises 9. $\overline{DF}$
11.

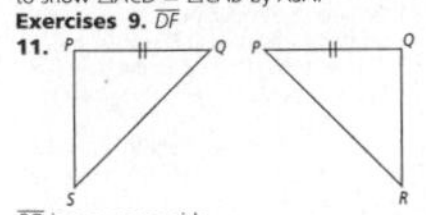

$\overline{PQ}$ is a common side.
13.

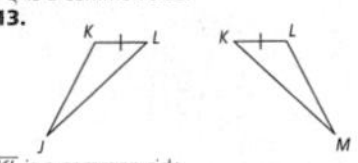

$\overline{KL}$ is a common side.
15. $\overline{RS} \cong \overline{UT}$ and $\overline{RT} \cong \overline{US}$ (Given), and $\overline{ST} \cong \overline{ST}$ (Refl. Prop. of $\cong$), so $\triangle RST \cong \triangle UTS$ by SSS. **17.** $\angle 1 \cong \angle 2$ and $\angle 3 \cong \angle 4$ (Given), and $\overline{QB} \cong \overline{QB}$ by the Refl. Prop. of $\cong$. So $\triangle QTB \cong \triangle QUB$ by ASA. Thus $\overline{QT} \cong \overline{QU}$ (Corresp. parts of $\cong$ $\triangle$s are $\cong$.). $\overline{QE} \cong \overline{QE}$ (Refl. Prop. of $\cong$), so $\triangle QET \cong \triangle QEU$ by SAS. **19.** Since $VT = VU + UT = UT + TS = US$, $\overline{VT} \cong \overline{US}$. Therefore, $\triangle QVT \cong \triangle PSU$ by SAS. **21.** It is given that $\overline{AC} \cong \overline{EC}$ and $\overline{CD} \cong \overline{CB}$, and $\angle C \cong \angle C$ by the Refl. Prop. of $\cong$. So $\triangle ACD \cong \triangle ECB$ by SAS, and $\angle A \cong \angle E$ because corresp. parts of $\cong$ $\triangle$s are $\cong$. **23.** Answers may vary. Sample:

Selected Answers

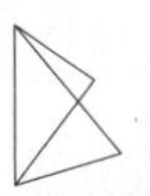

25. $\overline{TE} \cong \overline{RI}$ and $\overline{TI} \cong \overline{RE}$ (Given) and $\overline{EI} \cong \overline{EI}$ (Refl. Prop. of $\cong$), so $\triangle TEI \cong \triangle RIE$ by SSS. Thus $\angle TIE \cong \angle REI$ because corresp. parts of $\cong$ $\triangle$s are $\cong$. Also, $\angle TDI \cong \angle ROE$ because $\angle TDI$ and $\angle ROE$ are rt. $\angle$s (Given) and all rt. $\angle$s are $\cong$. So $\triangle TDI \cong \triangle ROE$ by AAS and $\overline{TD} \cong \overline{RO}$ because corresp. parts of $\cong$ $\triangle$s are $\cong$. **27.** The overlapping $\triangle$s are $\triangle CAE$ and $\triangle CBD$. It is given that $\overline{AC} \cong \overline{BC}$ and $\angle A \cong \angle B$. Also, $\angle C \cong \angle C$ by the Refl. Prop. of $\cong$. So $\triangle CAE \cong \triangle CBD$ by ASA. **29.** D **31.** C **33a.** right **b.** right **c.** Reflexive **d.** HL
34.

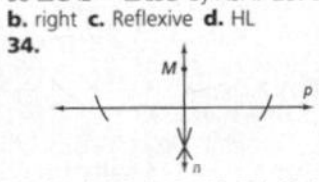

35. (1, 2) **36.** (1.5, 5.5) **37.** (1, 1)

Chapter Review pp. 273–276

1. legs **2.** hypotenuse **3.** corollary **4.** congruent polygons **5.** $\overline{ML}$ **6.** $\angle U$ **7.** $\overline{ST}$ **8.** $ONMLK$ **9.** 80 **10.** 3 **11.** 5 **12.** 35 **13.** 100 **14.** 145 **15.** $\angle D$ **16.** $\overline{MR}$ **17.** not enough information **18.** not enough information **19.** SAS **20.** AAS or ASA **21.** $\triangle TVY \cong \triangle YWX$ by AAS, so $\overline{TV} \cong \overline{YW}$ because corresp. parts of $\cong$ $\triangle$s are $\cong$. **22.** $\triangle BEC \cong \triangle DEC$ by ASA, so $\overline{BE} \cong \overline{DE}$ because corresp. parts of $\cong$ $\triangle$s are $\cong$. **23.** $\triangle BEC \cong \triangle DEC$ by SSS, so $\angle B \cong \angle D$ because corresp. parts of $\cong$ $\triangle$s are $\cong$. **24.** If $\parallel$ lines, alt. int. $\angle$s are $\cong$, so $\angle LKM \cong \angle NMK$. Then $\triangle LKM \cong \triangle NMK$ by SAS, and $\overline{KN} \cong \overline{ML}$ because corresp. parts of $\cong$ $\triangle$s are $\cong$. **25.** $x = 4$, $y = 65$ **26.** $x = 55$, $y = 62.5$ **27.** $x = 65$, $y = 90$ **28.** $x = 7$, $y = 60$ **29.** $\overline{LN} \perp \overline{KM}$ (Given), so $\triangle KLN$ and $\triangle MLN$ are rt. $\triangle$s. $\overline{KL} \cong \overline{ML}$ (Given) and $\overline{LN} \cong \overline{LN}$ (Refl. Prop. of $\cong$), so $\triangle KLN \cong \triangle MLN$ by HL. **30.** The given information on $\perp$ segments means $\triangle PSQ$ and $\triangle RQS$ are rt. $\triangle$s. You know $\overline{PQ} \cong \overline{RS}$ (Given) and $\overline{QS} \cong \overline{QS}$ (Refl. Prop. of $\cong$). So $\triangle PSQ \cong \triangle RQS$ by HL. **31.** $\triangle AEC \cong \triangle ABD$ by SAS or ASA or AAS. **32.** $\triangle FIH \cong \triangle GHI$ by SAS. **33.** $\triangle TAR \cong \triangle TSP$ by ASA.

Chapter 5

Get Ready! p. 281

1.

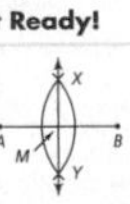

2.

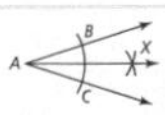

3. midpt. of $\overline{AB}$: (1, 2); midpt. of $\overline{BC}$: (−1, −2); midpt. of $\overline{AC}$: (3, −3); $AB = 2\sqrt{17}$; $BC = 2\sqrt{29}$; $AC = 4\sqrt{5}$ **4.** midpt. of $\overline{AB}$: (4, 2); midpt. of $\overline{BC}$: (4, 5); midpt. of $\overline{AC}$: (−1, 5); $AB = 10$; $BC = 2\sqrt{34}$; $AC = 6$ **5.** midpt. of $\overline{AB}$: (0, −3); midpt. of $\overline{BC}$: (1, 0); midpt. of $\overline{AC}$: (−1, 0); $AB = 4$; $BC = 2\sqrt{10}$; $AC = 2\sqrt{10}$ **6.** The team did not win. **7.** It is too late. **8.** $m\angle R \le 60$ **9.** −6 **10.** $-\frac{8}{3}$ **11.** undefined **12.** the length of a segment from a vertex to the opposite side **13.** the length of a $\perp$ segment from the point to the line **14.** a segment that connects the midpts. of 2 sides of the $\triangle$ **15.** The lines intersect at one point, or the lines have exactly one point in common.

Lesson 5-1 pp. 285–291

Got It? 1a. $\overline{AC} \parallel \overline{YZ}$, $\overline{CB} \parallel \overline{XY}$, $\overline{AB} \parallel \overline{XZ}$ **b.** 65; $\overline{UV}$ is a midsegment of $\triangle NOM$, so by the $\triangle$ Midseg. Thm., $\overline{UV} \parallel \overline{NM}$. Then $m\angle VUO = m\angle N = 65$ because corresp. $\angle$s of $\parallel$ lines are $\cong$. **2.** $DC = 6$; $AC = 12$; $EF = 6$; $AB = 15$ **3.** 1320 ft
Lesson Check 1. $\overline{NO}$ **2.** 23 **3.** 4 **4.** A midsegment is a segment whose endpoints are the midpts. of two sides of a triangle. **5.** The segments are $\parallel$. **6.** The student is assuming that L is the midpt. of $\overline{OT}$, which is not given.
Exercises 7. $\overline{UY} \parallel \overline{XV}$, $\overline{UW} \parallel \overline{TX}$, $\overline{YW} \parallel \overline{TV}$ **9.** $\overline{FE}$ **11.** $\overline{AB}$ **13.** $\overline{AC}$ **15.** 40 **17.** 160 **19.** 13 **21.** 6 **23.** 17 **25.** 156 m **27.** 114 ft 9 in.; because the red segments divide the legs into four $\cong$ parts, the white segment divides each leg into two $\cong$ parts. The white segment is a midsegment of the triangular face of the building, so its length is one half the length of the base. **29.** 40; $\overline{ST}$ is a midsegment of $\triangle PQR$, so by the $\triangle$ Midseg. Thm., $\overline{ST} \parallel \overline{PR}$. Then $m\angle QPR = m\angle QST$ because corresp. $\angle$s of $\parallel$ lines are $\cong$. **31.** 60 **33.** 100 **35.** 18.5 **37.** C **39.** 50 **41.** $x = 6$; $y = 6.5$ **43.** 24 **45.** Draw $\overrightarrow{CA}$. Find P on $\overrightarrow{CA}$ such that $CA = AP$. Draw $\overline{PD}$. Construct the $\perp$ bisector of $\overline{PD}$. Label the intersection point B. Draw $\overline{AB}$. This is a midsegment of $\triangle CPD$. According to the $\triangle$ Midsegment Thm., $\overline{AB} \parallel \overline{CD}$ and $AB = \frac{1}{2}CD$. **47.** $G(4, 4)$; $H(0, 2)$; $J(8, 0)$

Lesson 5-2 pp. 292–299

Got It? 1. 8 **2a.** Any point on the $\perp$ bis. of $\overline{PS}$ **b.** At the intersection point of ℓ and the perpendicular bisector of $\overline{PS}$; let X be the intersection point of ℓ and the perpendicular bisector of $\overline{PS}$. By the $\perp$ Bis. Thm., $XR = XS$ and $XS = XP$, so $XR = XS = XP$. Thus, X is equidistant from R, S, and P. **3.** 21
Lesson Check 1. $\overline{AC}$ is the $\perp$ bisector of $\overline{DB}$. **2.** 15 **3.** 18 **4.** Answers may vary. Sample:

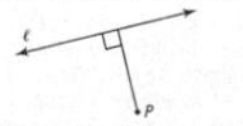

5. Draw the $\perp$ segments that join P to $\overrightarrow{OL}$ and $\overrightarrow{OX}$. Use a ruler to determine if $OL = OX$. If $OL = OX$, then P is on the bisector of $\angle LOX$.
Exercises 7. 3 **9.** Coleman School; it is on 6th Ave., which is (approximately) the $\perp$ bisector of 14th St. between 8th Ave. and Union Square. **11.** Draw $\overline{HS}$ and find its midpt., M. Through M, construct the line $\perp$ to $\overline{HS}$. Any point on this line will be equidistant from H and S. **13.** $\overrightarrow{HL}$ bisects $\angle KHF$; point L is equidistant from the sides of the $\angle$, so L is on the bisector of $\angle KHF$ by the Converse of the $\angle$ Bisector Thm. **15.** 54; 54 **17.** $y = 3$, $ST = 15$, $TU = 15$ **19.** 10 **21.** isosc., because $TW = ZW$ **23.** At the point on $\overline{XY}$ that lies on the bisector of $\angle GPL$; the goalie does not know to which side of her the player will aim his shot, so she should keep herself equidistant from the sides of $\angle GPL$. Points on the bisector of $\angle GPL$ are equidistant from $\overline{PG}$ and $\overline{PL}$. If she moves to a point on the $\perp$ bisector of $\overline{GL}$, she will be closer to $\overline{PL}$ than to $\overline{PG}$.
25a.

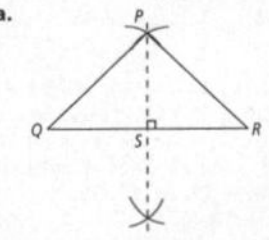

b. Answers may vary. Sample: Since P is on the perpendicular bisector of QR, it is equidistant from Q and R by the perpendicular bisector theorem. **27.** A pt. is on the bisector of an $\angle$ if and only if it is equidistant from the sides of the $\angle$. **29.** No; A is not equidistant from the sides of $\angle TXR$. **31.** Yes; A is equidistant from the sides of $\angle TXR$. **33.** $\overline{PA} \cong \overline{PB}$ (Given) and $\angle AMP \cong \angle BMP$ because all rt. $\angle$s are $\cong$. Also, $\overline{PM} \cong \overline{PM}$ by the Refl. Prop. of $\cong$. So rt. $\triangle PMA \cong$ rt. $\triangle PAB$ by HL and $\overline{AM} \cong \overline{BM}$ because corresp. parts of $\cong$ $\triangle$s are $\cong$. Therefore $\overleftrightarrow{PM}$ is the $\perp$ bisector of $\overline{AB}$, by the def. of $\perp$ bisector. **35.** In rt. $\triangle SPQ$ and rt. $\triangle SRQ$, $\overline{SP} \cong \overline{SR}$ (Given) and $\overline{QS} \cong \overline{QS}$ (Refl. Prop. of $\cong$), so $\triangle SPQ \cong \triangle SRQ$ by HL. Thus $\angle PQS \cong \angle RQS$ because corresp. parts of $\cong$ $\triangle$s are $\cong$, and $\overrightarrow{QS}$ bisects $\angle PQR$ by the def. of $\angle$ bisector. **37.** Line ℓ through the midpts. of two sides of $\triangle ABC$ is equidistant from A, B, and C. This is because $\triangle 1 \cong \triangle 2$ and $\triangle 3 \cong \triangle 4$ by ASA. $\overline{AD} \cong \overline{BE}$ and $\overline{BE} \cong \overline{CF}$ because corresp. parts of $\cong$ $\triangle$s are $\cong$. By the Trans. Prop. of $\cong$,

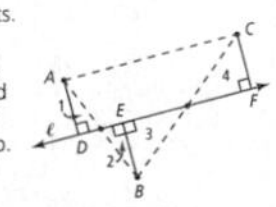

$\overline{AD} \cong \overline{BE} \cong \overline{CF}$. By the def. of $\cong$, $AD = BE = CF$, so points A, B, and C are equidistant from line ℓ. **39.** C **41.** C **43.** 6 **44.** 120 **45.** $\frac{1}{3}$ **46.** undefined **47.** Answers may vary. Sample: It is a vertical line that contains the point (5, 0).

Lesson 5-3 pp. 301–307

Got It? 1. (6, 5) **2.** at the circumcenter of the $\triangle$ whose vertices are the three trees

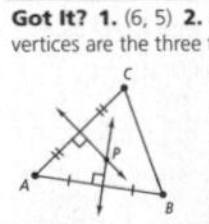

3a. 61 **b.** No; answers may vary. Sample: The distance from Q to $\overline{KL}$ is QN, the length of the shortest segment from Q to $\overline{KL}$. From part (a), $QN = 61$, so $QP > 61$.
Lesson Check 1. (3, 2.5) **2.** 6 **3.** obtuse $\triangle$ **4.** Since the three $\perp$ bisectors of a $\triangle$ are concurrent, the third $\perp$ bisector goes through the pt. of intersection of the other two $\perp$ bisectors. **5.** Answers may vary. Sample: The diagram does not show that $\overline{QC}$ bisects $\angle SQR$, so you cannot conclude that point C is equidistant from the sides of $\angle SQR$. **6.** Each one is a point of concurrency of bisectors of parts of a $\triangle$, each is equidistant from three parts of the $\triangle$, and each is the center of a $\odot$ that contains three points of the $\triangle$. The circumcenter is equidistant from three points, while the incenter is equidistant from three segments. The $\triangle$ is inside the $\odot$ centered at the circumcenter and outside the $\odot$ centered at the incenter.
Exercises 7. (−2, −3) **9.** (1.5, 1) **11.** (−3, 1.5) **13.** (3.5, 3) **15.** C **17.** 2 **19.** Isosceles; $SR = ST$,

so $\angle SRT \cong \angle STR$ (Isosc. $\triangle$ Thm.). Since P is the incenter of $\triangle RST$, $\overline{PR}$ and $\overline{PT}$ are $\angle$ bisectors. So $m\angle PRT = \frac{1}{2}m\angle SRT = \frac{1}{2}m\angle STR = m\angle PTR$. Thus $PR = PT$ by the Converse of the Isosc. $\triangle$ Thm.
21. Same method as for Exercise 20.

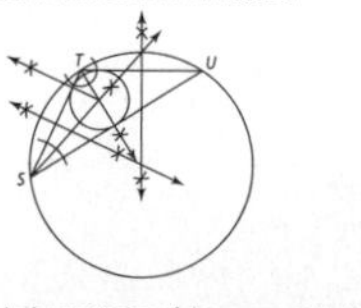

23. An interpretation of the passage is that the treasure is equidistant from three Norway pines. To find the treasure, Karl can find the circumcenter of the $\triangle$ whose vertices are the three pines. **25.** P; the markings in the diagram show that P is the incenter of the triangular station and C is the circumcenter. If you stand at P, you will be equidistant from the three sides along which the buses are parked. If you move away from P, you will move closer to some of the buses. **27.** true
29.

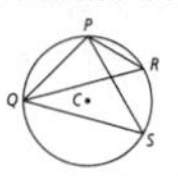

As the diagram shows, circle C is circumscribed about both $\triangle PQR$ and $\triangle PQS$, so points R and S do not have to coincide.

31. Never; if you have three $\parallel$ lines ℓ, m, and n, with m in between ℓ and n, then a point equidistant from ℓ and m would be (midway) between them. A point equidistant from m and n would be (midway) between those two lines. A point equidistant from all three would therefore have to be on both sides of m! This is impossible.
33. B **35.** C **37.** 4 **38.** 17 **39.** (3, 8) **40.** (5, 3.5)

Lesson 5-4 pp. 309–315

Got It? 1a. 13.5 **b.** 2 : 1; $ZA = \frac{2}{3}CZ$ and $AC = \frac{1}{3}CZ$, so $ZA : AC = \frac{2}{3} : \frac{1}{3} = 2 : 1$. **2a.** A median; it connects a vertex of $\triangle ABC$ and the midpt. of the opposite side.
b. Neither; E is a midpt. of $\triangle ABC$, but G is not a vertex of $\triangle ABC$. **c.** An altitude; it extends from a vertex of $\triangle ABC$ and is $\perp$ to the opposite side. **3.** (1, 2)
Lesson Check 1. median **2.** 6 **3.** 7.5 **4.** $\overline{AB}$, $\overline{AC}$
5. $\overline{HJ}$ does not contain a vertex of $\triangle ABC$, so it is not an altitude of $\triangle ABC$. **6.** No; any pair of altitudes meet at the orthocenter of the $\triangle$. **7.** They are $\perp$; since A is the orthocenter of $\triangle ABC$, A lies on the altitude from B to $\overline{AC}$. B also lies on this altitude, so the altitude from B to $\overline{AC}$ must be $\overline{BA}$. Therefore, $\overline{BC} \perp \overline{AC}$.
Exercises 9. $ZY = 4.5$, $ZU = 13.5$ **11.** Median; it connects a vertex of $\triangle ABC$ and the midpt. of the opposite side. **13.** Altitude; it extends from a vertex of $\triangle ABC$ and is $\perp$ to the opposite side. **15.** (6, 4)
17. H **19.** J **21.** 125
23.

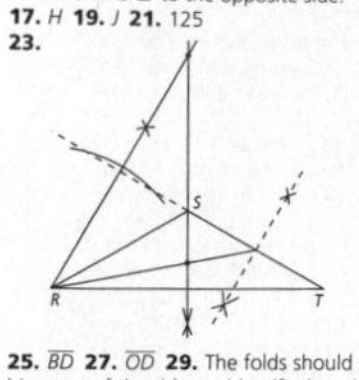

25. $\overline{BD}$ **27.** $\overline{OD}$ **29.** The folds should show the $\perp$ bisectors of the sides to identify the midpt. of each side, and also show the fold through each vertex and the midpt. of the opposite side. **31.** C **33.** Answers may vary. Sample: The $\angle$ bisector of the vertex $\angle$ forms two $\triangle$s that are $\cong$ by SAS. Therefore the 2 segments formed on the base are $\cong$ (so the $\angle$ bisector contains a median), and the two $\angle$s formed by the $\angle$ bisector and the base are rt. $\angle$s (so the $\angle$ bisector contains an altitude). Thus the median and the altitude are the same.
35.

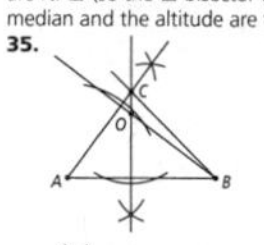

Draw $\overleftrightarrow{AB}$. Construct the $\perp$ to $\overleftrightarrow{AB}$ through O. Draw $\overleftrightarrow{BO}$. Construct the $\perp$ to $\overleftrightarrow{BO}$ through A. The two $\perp$s intersect at C. Draw $\overline{BC}$.
37. A is the intersection of the altitudes, so it is the orthocenter; B is the intersection of the $\angle$ bisectors, so it is the incenter; C is the intersection of the medians, so it is the centroid; D is the intersection of the $\perp$ bisectors of the sides, so it is the circumcenter. **39.** incenter **41.** H
43. Both; the markings show directly that $\overline{XY}$ is a $\perp$ bisector. The two $\triangle$s formed are congruent by SAS, so the two $\angle$s at top are $\cong$. Therefore, $\overline{XY}$ is also an $\angle$ bisector. **44.** Neither; $\overline{XY}$ connects vertex X and the midpt., Y, of side $\overline{PQ}$, so $\overline{XY}$ is a median. **45.** Two angles are not congruent. **46.** You are 16 years old.
47. $m\angle A \geq 90$

Lesson 5-5 pp. 317–322

Got It? 1a. Assume temporarily that $\triangle BOX$ is acute.
b. Assume temporarily that no pair of shoes you bought cost more than \$25. **2a.** II and III **b.** No; if $\triangle ABC$ is an isosc., nonequilateral $\triangle$, then Statement III is true but Statement II is not true. Therefore, Statements II and III are not equivalent. **3.** Assume temporarily that $y = 6$. Then $7(x + 6) = 70$; divide each side by 7 to get $x + 6 = 10$ and so $x = 4$. But this contradicts the given statement that $x \neq 4$. The temporary assumption that $y = 6$ led to a contradiction, so we can conclude that $y \neq 6$.
Lesson Check 1. Assume temporarily that at least one $\angle$ in quadrilateral $ABCD$ is not a rt. $\angle$.
2. Lines a and b meet at P.

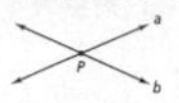

3. The negation of "$\angle A$ is obtuse" is "$\angle A$ is an acute or a rt. $\angle$."
Exercises 5. Assume temporarily that $\angle J$ is a rt. $\angle$.
7. Assume temporarily that no $\angle$ is obtuse. **9.** Assume temporarily that $m\angle 2 \leq 90$. **11.** I and III **13.** I and II
15a. rt. $\angle$ **b.** rt. $\angle$s **c.** 90 **d.** 180 **e.** 90 **f.** 90 **g.** 0
h. more than one rt. $\angle$ **i.** at most one rt. $\angle$ **17.** Assume temporarily $\ell \parallel p$. Then $\angle 1 \cong \angle 2$ because if lines are $\parallel$ then corresp. $\angle$s are $\cong$. But this contradicts the given statement that $\angle 1 \not\cong \angle 2$. Therefore the temporary assumption is false, and we can conclude that $\ell \nparallel p$.
19. Assume temporarily that $\overline{XB} \not\cong \overline{XA}$. **21.** I and III
23. Assume temporarily that at least one base $\angle$ is a rt. $\angle$. Then both base $\angle$s must be rt. $\angle$s, by the Isosc. $\triangle$ Thm. But this contradicts the fact that a $\triangle$ is formed, because in a plane two lines $\perp$ to the same line are $\parallel$. Therefore the temporary assumption is false that at least one base $\angle$ is a rt. $\angle$, and we can conclude that neither base $\angle$ is a rt. $\angle$.
25. Assume temporarily that an obtuse $\triangle$ can contain a rt. $\angle$. Then the measure of the obtuse $\angle$ plus the measure of the rt. $\angle$ must be greater than $90 + 90 = 180$. This contradicts the $\triangle$ Angle-Sum Thm., so the temporary assumption that an obtuse $\triangle$ can contain a rt. $\angle$ is incorrect. We can conclude that an obtuse $\triangle$ cannot contain a rt. $\angle$. **27.** The culprit entered the room through a hole in the roof; all the other possibilities were ruled out. **29.** Assume temporarily $\overline{XB} \perp \overline{AC}$. Then $\angle BXA \cong \angle BXC$ (All rt. $\angle$s are $\cong$.), $\angle ABX \cong \angle CBX$ (Given), and $\overline{BX} \cong \overline{BX}$ (Reflexive Prop. of $\cong$), so $\triangle BXA \cong \triangle BXC$ by ASA and $BA = BC$ because corresp. parts of $\cong$ $\triangle$s are $\cong$. But this contradicts the given statement that $\triangle ABC$ is scalene. Therefore the temporary assumption that $\overline{XB} \perp \overline{AC}$ is wrong, and we can conclude that $\overline{XB}$ is not $\perp$ to $\overline{AC}$. **31.** D **33.** C **35.** 24 cm
36. 30 and 120 **37.** Law of Syllogism
38. $\overline{AC}$, $\overline{BC}$, $\overline{AB}$ **39.** $\overline{CA}$, $\overline{BC}$, $\overline{BA}$

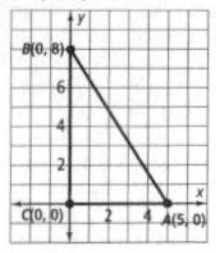

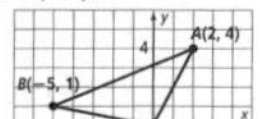

40. $\overline{AB}$, $\overline{AC} \cong \overline{BC}$

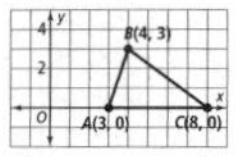

Algebra Review p. 323

1. $x \leq -1$ **3.** $x > -10.5$ **5.** $a \leq 90$ **7.** $z > 0.5$
9. $n \leq -\frac{7}{2}$ **11.** $x > -5$ **13.** $x \leq -1$ **15.** $x > -8$
17. $x \geq -362$

Lesson 5-6 pp. 324–331

Got It? 1. $\angle 5$ is an ext. $\angle$ of $\triangle ACD$, so by the Corollary to the $\triangle$ Ext. $\angle$ Thm., $m\angle 5 > m\angle C$.
2. Holingsworth Rd. and MLK Blvd. **3.** $\overline{OX}$; $m\angle X = 180 - (130 + 24) = 26$, so $m\angle O > m\angle X > m\angle S$. By Theorem 5-11, $SX > OS > OX$. **4a.** No; $2 + 6 \ngtr 9$. **b.** Yes; the sum of the lengths of any two sides is greater than the length of the third side. **5.** 3 in. $< x <$ 11 in.
Lesson Check 1. $\overline{BC}$ **2.** $\angle C$ **3.** No; $5 + 4 \ngtr 10$.
4. If the perimeter is 16 and the length of one side is 8, then the sum of the lengths of the other two sides is $16 - 8 = 8$. However, the $\triangle$ Inequality Thm. tells you that if the length of one side is 8, then the sum of the lengths of the other two sides is greater than 8. So the friend is incorrect. **5.** No; the adjacent interior $\angle$ would measure 92. Then, because a second $\angle$ of the $\triangle$ measures 90, the sum of the $\angle$ measures would exceed 180, which contradicts the $\triangle$ Angle-Sum Thm.
Exercises 7. This is true by the Corollary to the $\triangle$ Ext. $\angle$ Thm. **9.** $\angle M$, $\angle L$, $\angle K$ **11.** $\angle G$, $\angle H$, $\angle J$ **13.** $\angle E$, $\angle F$, $\angle D$ **15.** $\overline{MN}$, $\overline{NO}$, $\overline{MO}$ **17.** $\overline{TU}$, $\overline{UV}$, $\overline{TV}$ **19.** $\overline{EF}$, $\overline{DE}$, $\overline{DF}$
21. No; $2 + 3 \ngtr 6$. **23.** No; $8 + 10 \ngtr 19$. **25.** Yes; $2 + 9 > 10$, $9 + 10 > 2$, and $2 + 10 > 9$.
27. 4 ft $< x <$ 20 ft **29.** 0 cm $< x <$ 12 cm
31. 3 yd $< x <$ 11 yd **33.** Place the computer at the corner that forms a rt. $\angle$; place the bookshelf along the wall opposite the rt. $\angle$. In a rt. $\triangle$ the rt. $\angle$ is the largest $\angle$, and the longest side of a $\triangle$ is opposite the largest $\angle$.

35. The dashed red line and the courtyard walkway determine three sides of a $\triangle$, so by the $\triangle$ Inequality Thm., the path that follows the dashed red line is longer than the courtyard walkway. **37.** $\overline{RS}$ **39.** $\overline{XY}$
41. Answers may vary. Sample: The sum of the $\angle$ measures of a $\triangle$ is 180, so $m\angle T + m\angle P + m\angle A = 180$. Since $m\angle T = 90$, $m\angle P + m\angle A = 90$ and so $m\angle T > m\angle A$ (Comparison Prop. of Inequality). Therefore $PA > PT$ by Thm. 5-11. **43.** (2, 4), (2, 5), (2, 6), (3, 3), (3, 4), (3, 5), (3, 6), (3, 7), (4, 3), (4, 4), (4, 5), (4, 6), (4, 7), (4, 8)
45. Answers may vary. Sample:

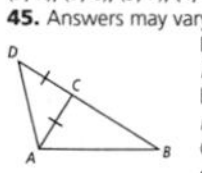

Find point D on $\overrightarrow{BC}$ such that $DC = AC$. $m\angle D = m\angle CAD$ by the Isosc. $\triangle$ Thm. Now $m\angle DAB > m\angle DAC$ by the Comparison Prop. of Inequality, and so $m\angle DAB > m\angle D$ by substitution. Thus $DB > AB$ by Thm. 5-11. We know $DC + CB = DB$ by the Segment Add. Post., so $DC + CB > AB$ (Substitution) and $AC + CB > AB$ (Substitution).

Lesson 5-7 pp. 332–339

Got It? 1a. $LN > OQ$ **b.** Assume temporarily that $m\angle P \ngtr m\angle A$. If $m\angle P = m\angle A$, then $\triangle ABC \cong \triangle PQR$ (SAS), but this contradicts the fact that $BC \neq QR$. If $m\angle P < m\angle A$, then by the Hinge Thm., $QR < BC$. This contradicts the fact that $QR > BC$. Therefore, $m\angle P > m\angle A$. **2.** The 40° opening; the lengths of the blades do not change as the scissors open. The included angle between the blades of the 40° opening is greater than the included angle of the 35° opening, so by the Hinge Thm., the distance between the blades is greater for the 40° opening. **3.** $-6 < x < 24$ **4.** From the given information, $LO = ON$ (def. of midpt.) and $m\angle MOL = 100$ (suppl. $\angle$ to $\angle MON$). Since $\overline{MO} \cong \overline{MO}$, and $m\angle MOL > m\angle MON$, the Hinge Thm. yields $LM > MN$.
Lesson Check 1. $FD > BC$ **2.** $m\angle UST > m\angle VST$
3. Answers may vary. Sample: As a door opens, and the angle between the door and doorway increases, the distance between the door jamb and the nonhinge vertical edge of the door increases. **4.** The two $\angle$s that are formed by $\cong$ sides are $\angle ABD$ and $\angle CDB$. Since the side opposite $\angle ABD$ is longer than the side opposite $\angle CDB$, the correct conclusion is $m\angle ABD > m\angle CDB$. **5.** Answers may vary. Sample: Both deal with a pair of $\triangle$s that have two pairs of $\cong$ corresponding sides along with a relationship between the $\angle$s formed by those sides.
Exercises 7. $PR < RT$ **9.** no conclusion
11. $6 < x < 38$ **13.** $3.5 < y < 17.5$ **15a.** Converse of Isosc. $\triangle$ Thm. **b.** Given **c.** Def. of midpt. **d.** $BC = CD$
e. Given **f.** Hinge Theorem **17.** $m\angle QTR > m\angle RTS$; $m\angle PTQ + m\angle QTR + m\angle RTS = 180$, so $m\angle PTQ + m\angle RTS = 88$. Thus $m\angle RTS < 88$ by the Comparison Prop. of Inequality, so $m\angle QTR > m\angle RTS$ by the Transitive Prop. of Inequality. **19a.** The two labeled $\angle$s are formed by $\cong$ corresp. sides of the two $\triangle$s, so the side opposite the 94° $\angle$ should be longer than the side opposite the 91° $\angle$, by the Hinge Thm. Thus the side labeled "13" must be longer than the side labeled "14."
b. Answers may vary. Sample: Switch the angle labels 91° and 94°. **21.** A **23.** $\triangle ABE \cong \triangle CBD$ (Given) so $\triangle ABE$ and $\triangle CBD$ are isosc. with $AB = EB = DB = CB$. Since $m\angle EBD > m\angle ABE$ (Given), $ED > AE$ by the Hinge Thm.
25. A Using the diagram in the Plan for Proof, $BC = YZ$, $BD = YX$, and $m\angle ZYX = m\angle CBD$, so $\triangle DBC \cong \triangle XYZ$ by SAS. $\angle FBA \cong \angle FBD$ (Def. of $\angle$ bisector), $\overline{BD} \cong \overline{BA}$ (because each is $\cong$ to $\overline{XY}$), and $\overline{BF} \cong \overline{BF}$, so $\triangle ABF \cong \triangle DBF$ by SAS. $\overline{AF} \cong \overline{DF}$, because corresp. parts of $\cong$ $\triangle$s are $\cong$. $AF + FC = AC$ (Segment Addition Post.), so $DF + FC = AC$. Using the $\triangle$ Inequality Thm. in $\triangle FDC$, $DF + FC > DC$. Now $AC > DC$ by substitution. Since $DC = XZ$ (Corresp. parts of $\cong$ $\triangle$s are $\cong$.), it follows that $AC > XZ$ by substitution.

Chapter Review pp. 341–344

1. median **2.** distance from a point to a line **3.** incenter
4. 15 **5.** 11 **6.** $L\left(\frac{5}{2}, -\frac{1}{2}\right)$; $M\left(\frac{7}{2}, \frac{1}{2}\right)$; slope of $\overline{AB} = 1$ and slope of $\overline{LM} = 1$, so $\overline{LM} \parallel \overline{AB}$; $AB = 2\sqrt{2}$ and $LM = \sqrt{2}$, so $LM = \frac{1}{2}AB$. **7.** Let point S be second base and point T be third base. Find the midpt. M of $\overline{ST}$ and then through M construct the line $\ell \perp$ to $\overline{ST}$. Points of the baseball field that are on line ℓ are equidistant from second and third base. **8.** 40 **9.** 40 **10.** 6 **11.** 11 **12.** 33
13. 33 **14.** (0, 0) **15.** (3, 2) **16.** (4, 4) **17.** (5, 1)
18. 45 **19.** 40 **20.** 25 **21.** $\overline{AB}$ is an altitude; it is a segment from a vertex that is $\perp$ to the opposite side.
22. $\overline{AB}$ is a median; it is a segment from a vertex to the midpt. of the opposite side. **23.** $QZ = 8$, $QM = 12$
24. (0, −1) **25.** (2, −3) **26.** Assume temporarily that neither of the two numbers is even. That means each number is odd, so the product of the two numbers must be odd. That contradicts the statement that the product of the two numbers is even. Thus the temporary assumption is false, and we can conclude that at least one of the numbers must be even. **27.** Assume temporarily that the third line intersects neither of the first two. Then it is $\parallel$ to both of them. Since the first two lines are $\parallel$ to the same line, they are $\parallel$ to each other. This contradicts the given information. Therefore the temporary assumption is false, and the third line must intersect at least one of the two others. **28.** Assume temporarily that there is a $\triangle$ with two obtuse $\angle$s. Then the sum of the measures of those two $\angle$s is greater than 180, which contradicts the $\triangle$ Angle-Sum Thm. Therefore the temporary assumption is false, and a $\triangle$ can have at most one obtuse $\angle$. **29.** Assume temporarily that an equilateral $\triangle$ has an obtuse $\angle$. Since all the $\angle$s are $\cong$ in an equilateral $\triangle$, then all three $\angle$s must be obtuse. But we showed in Ex. 28 that a $\triangle$ can have at most one obtuse $\angle$. Therefore the temporary assumption is false, and an equilateral $\triangle$ cannot have an obtuse $\angle$.
30. Assume temporarily that each of the three integers is less than or equal to 3. Then the sum of the three integers must be less than or equal to $3 \cdot 3$, or 9. This contradicts the given statement that the sum of the three integers is greater than 9. Therefore the temporary assumption is false, and you can conclude that one of the integers must be greater than 3. **31.** $\overline{RS}$, $\overline{ST}$, $\overline{RT}$ **32.** No; $5 + 8 \ngtr 15$.
33. Yes; $10 + 12 > 20$, $10 + 20 > 12$, and $12 + 20 > 10$. **34.** 1 ft $< x <$ 25 ft **35.** < **36.** >
37. <

Chapter 6

Get Ready! p. 349

1. 30 **2.** 42 **3.** 22 **4.** yes **5.** no **6.** yes **7.** $\parallel$ **8.** $\perp$
9. neither **10.** ASA **11.** SAS **12.** AAS **13.** Answers may vary. Sample: polygon in which all the $\angle$s are $\cong$
14. Answers may vary. Sample: four-sided figure formed by joining two isosc. $\triangle$s **15.** Answers may vary. Sample: Angles that follow one right after the other.

Lesson 6-1 pp. 353–358

Got It? 1a. 2700 **b.** Answers may vary. Sample: Divide 1980 by 180, and then add 2. **2.** 140 **3.** 102 **4.** 40
Lesson Check 1. 1620 **2.** 360 **3.** 144, 36 **4.** Yes; explanations may vary. Sample: rectangle that is not square **5.** $\angle 2$ and $\angle 4$; their measures are equal; answers may vary. Sample: Two $\angle$s suppl. to the same $\angle$ must be $\cong$. **6.** Answers may vary. Sample: ext. $\angle$ would measure 50, which is not a factor of 360.
Exercises 7. 900 **9.** 2160 **11.** 180,000 **13.** 150
15. 60, 120, 120, 60 **17.** 145 **19.** 10 **21.** 3.6 **23.** 8
25. 18 **27.** octagon; $m\angle 1 = 135$, $m\angle 2 = 45$
29. $y = 103$, $z = 70$ **31.** 36 **33.** 144; 10 **35.** 150; 12
37. 45, 45, 90 **39a.** $180n$ **b.** $(n - 2) \cdot 180$
c. $180n - [(n - 2) \cdot 180] = 360$ **d.** Polygon Ext. $\angle$ Sum Theorem **41.** octagon **43a.** Answers may vary. Sample: The sum of the interior $\angle$ measures $= (n - 2)180$. All $\angle$s of a regular n-gon are $\cong$. So each interior $\angle$ measure $= \frac{180(n-2)}{n}$, and $\frac{180(n-2)}{n} = \frac{180n - 360}{n} = 180 - \frac{360}{n}$. **b.** As n gets larger, $\frac{360}{n}$ gets smaller. The interior angle measure gets closer to 180. The polygon becomes more like a circle. **45.** 225 **47.** 79 **49.** $\overline{CD}$; the longer side is opposite the larger $\angle$. **50.** Distr. Prop. **51.** Refl. Prop. of $\cong$ **52.** Sym. Prop. of $\cong$ **53.** ASA **54a.** $\angle HGE$
b. $\angle GHE$ **c.** $\angle HEG$ **d.** $\overline{GH}$ **e.** $\overline{HE}$ **f.** $\overline{EG}$

Lesson 6-2 pp. 359–366

Got It? 1. 94
2. 1. $ABCD$ is a ▱ and $\overline{AK} \cong \overline{MK}$. (Given)
2. $\angle A \cong \angle BCD$ (Opp. $\angle$s of a ▱ are $\cong$.)
3. $\angle A \cong \angle CMD$ (Isosc. $\triangle$ Theorem)
4. $\angle BCD \cong \angle CMD$ (Transitive Prop. of $\cong$)

3a. $x = 4$, $y = 5$, $PR = 16$, $SQ = 10$ **b.** No; answers may vary. Sample: Solutions to a system of equations do not depend on the method used to solve it. **4.** 5
Lesson Check 1. 53 **2.** 127 **3.** 5 **4.** 7 **5.** $ED = 12$, $FD = 24$ **6.** Answers may vary. Sample: The $\angle$ opposite the given $\angle$ is congruent to it. The other two $\angle$s and the given $\angle$ are consecutive $\angle$s, so they are supplements of the given $\angle$. **7.** A quad. and a ▱ both have four sides, but if both pairs of opp. sides are $\parallel$, then the figure is a ▱. **8.** It is not given that $\overleftrightarrow{PQ}$, $\overleftrightarrow{RS}$, and $\overleftrightarrow{TV}$ are $\parallel$.
Exercises 9. 127 **11.** 100 **13a.** Def. of ▱ **b.** If lines are $\parallel$, then alt. int. $\angle$s are $\cong$. **c.** Opp. sides of a ▱ are $\cong$.
d. $\triangle ABE \cong \triangle CDE$ **e.** Corresp. parts of $\cong$ $\triangle$s are $\cong$.
f. $\overline{AC}$ and $\overline{BD}$ bisect each other at E. **15.** $x = 5$, $y = 7$
17. 3 **19.** 9 **21.** 2.25 **23.** 4.5 **25.** 20 **27.** $x = 12$, $y = 4$
29. 22, $AB = 23.6$, $BC = 18.5$, $CD = 23.6$, $AD = 18.5$
31a. 2.5 ft **b.** 129 **c.** Answers may vary. Sample: As $m\angle E$ increases, $m\angle D$ decreases. $\angle E$ and $\angle D$ are suppl.
33. Answers may vary. Sample:
1. ▱s $LENS$ and $NGTH$ (Given)
2. $\angle L \cong \angle ENS$ and $\angle GNH \cong \angle T$. (Opp. $\angle$s of a ▱ are $\cong$.)
3. $\angle ENS \cong \angle GNH$ (Vert. $\angle$s are $\cong$.)
4. $\angle L \cong \angle T$ (Transitive Prop. of $\cong$)

35. Answers may vary. Sample:
1. ▱s $LENS$ and $NGTH$ (Given)
2. $\angle E$ is suppl. to $\angle ENS$. (Consecutive $\angle$s in a ▱ are suppl.)
3. $\angle GNH \cong \angle ENS$ (Vert. $\angle$s are $\cong$.)
4. $\angle GNH \cong \angle T$ (Opp. $\angle$s of a ▱ are $\cong$.)
5. $\angle ENS \cong \angle T$ (Transitive Prop. of $\cong$)
6. $\angle E$ is suppl. to $\angle T$. (Substitution Prop.)

37. 1. ▱s $RSTW$ and $XYTZ$ (Given)
2. $\overline{XY} \parallel \overline{TZ}$ and $\overline{TZ} \parallel \overline{RS}$. (Def. of ▱)
3. $\overline{XY} \parallel \overline{RS}$ (If two lines are $\parallel$ to the same line, then they are $\parallel$ to each other.)

39. $m\angle 1 = 71$, $m\angle 2 = 28$, $m\angle 3 = 81$
41. $AB = CD = 13$, $BC = AD = 33$
43. Answers may vary. Sample:
1. $\overline{AB} \parallel \overline{CD}$, $\overline{CD} \parallel \overline{EF}$ (Given)
2. $\overline{BG} \parallel \overline{AC}$, $\overline{DH} \parallel \overline{CE}$ (Construction)
3. $ABGC$ and $CDHE$ are ▱s. (Def. of ▱)
4. $\overline{AC} \cong \overline{BG}$, $\overline{CE} \cong \overline{DH}$ (Opp. sides of a ▱ are ≅.)
5. $\overline{AC} \cong \overline{CE}$ (Given)
6. $\overline{BG} \cong \overline{DH}$ (Trans. Prop. of ≅)
7. $\overline{BG} \parallel \overline{DH}$ (If two lines are ∥ to the same line, then they are ∥ to each other.)
8. $\angle 3 \cong \angle 6$ and $\angle GBD \cong \angle HDF$. (If lines are ∥, corresp. ∠s are ≅.)
9. $\triangle GBD \cong \triangle HDF$ (AAS)
10. $\overline{BD} \cong \overline{DF}$ (Corresp. parts of ≅ △s are ≅.)

45. D **47.** C **49.** 1440 **50.** 2520
51. 4140 **52.** 6840 **53.** $\overline{AC} \perp \overline{DB}$ (or $\angle ACD$ and $\angle ACB$ are rt. ∠s) **54.** 42

Lesson 6-3 pp. 367–374

Got It? 1. $x = 10$, $y = 43$ **2a.** No; $DEFG$ could be an isosc. trapezoid. (One pair of sides must be both ≅ and ∥.)
b. yes
1. $\angle ALN \cong \angle DNL$; $\angle ANL \cong \angle DLN$ (Given)
2. $\overline{AN} \parallel \overline{LD}$ and $\overline{AL} \parallel \overline{ND}$. (If alt. int. ∠s are ≅, then lines are ∥.)
3. $LAND$ is a ▱. (Def. of ▱)

3. 6 ft; explanations may vary. Sample: The maximum height occurs when $\overline{QP}$ is vertical.
Lesson Check 1. 112 **2.** Yes; opp. ∠s are ≅. **3.** No; the diagonals may not bisect each other. **4.** because Thm. 6-3 and its converse are both true **5.** Thm. 6-11 and Thm. 6-6 are converses of each other. Use Thm. 6-11 if you need to show the figure is a ▱. Use Thm. 6-6 if it is given that the figure is a ▱. **6.** It is a ▱ only if the same pair of opp. sides are ≅ and ∥.
Exercises 7. 5 **9.** $x = 21$, $y = 39$ **11.** 5 **13.** Yes; both pairs of opp. sides are ≅. **15.** Yes; both pairs of opp. ∠s are ≅. **17.** A quad. is a ▱ if and only if its opp. sides are ≅; a quad. is a ▱ if and only if its consecutive ∠s are suppl.; a quad. is a ▱ if and only if its opp. ∠s are ≅; a quad. is a ▱ if and only if its diagonals bisect each other.
19. Answers may vary. Sample:
1. Draw $\overline{BD}$. (Construction)
2. $\angle CBD \cong \angle ADB$ (Alt. int. ∠s are ≅.)
3. $\overline{BC} \cong \overline{DA}$ (Given)
4. $\overline{BD} \cong \overline{BD}$ (Refl. Prop. of ≅)
5. $\triangle BCD \cong \triangle DAB$ (SAS)
6. $\angle BDC \cong \angle DBA$ (Corresp. parts of ≅ △s are ≅.)
7. $\overline{AB} \parallel \overline{CD}$ (If alt. int. ∠s are ≅, then lines are ∥.)
8. $ABCD$ is a ▱. (Def. of ▱)

21. Answers may vary. Sample:
1. $\angle A$ is suppl. to $\angle B$. (Given)
2. $\overline{BC} \parallel \overline{AD}$ (Converse of Corresp. ∠s Postulate)
3. $\angle A$ is suppl. to $\angle D$. (Given)
4. $\overline{AB} \parallel \overline{DC}$ (Converse of Corresp. ∠s Postulate)
5. $ABCD$ is a ▱. (Def. of ▱)

23. $x = 3$, $y = 11$
25. Answers may vary. Sample:
1. $\triangle TRS \cong \triangle RTW$ (Given)
2. $\overline{SR} \cong \overline{WT}$ and $\overline{ST} \cong \overline{WR}$. (Corresp. parts of ≅ △s are ≅.)
3. $RSTW$ is a ▱. (If both pairs of opp. sides of a quad. are ≅, then the quad. is a ▱.)

27. Answers may vary. Sample:
1. $\overline{AB} \cong \overline{CD}$, $\overline{AC} \cong \overline{BD}$ (Construction)
2. ▱$ABCD$ (Opp. sides of a ▱ are ≅.)
3. M is the midpt. of $\overline{BC}$. (Diagonals of a ▱ bisect each other.)
4. $\overline{AM}$ is a median. (Def. of median)

29. D
31a. $7x - 11 = 6x$, $x = 11$ **b.** Yes; $m\angle F = 66$ and $m\angle FED = 114$. So $m\angle F + m\angle FED = 66 + 114 = 180$, and $\overline{AF} \parallel \overline{DE}$. (Converse of Corresp. ∠s Postulate)
c. Yes; $\overline{BD} \parallel \overline{FE}$ (Given) and $\overline{BF} \parallel \overline{DE}$ from part (b). So $BDEF$ is a ▱. (Def. of ▱)
32. $a = 8$, $h = 30$, $k = 120$ **33.** $m = 9.5$, $x = 15$
34. $c = 204$, $e = 13$, $f = 11$
35.
1. $\overline{AD} \cong \overline{BC}$, $\angle DAB \cong \angle CBA$ (Given)
2. $\overline{AB} \cong \overline{AB}$ (Refl. Prop. of ≅)
3. $\triangle ACB \cong \triangle BDA$ (SAS)
4. $\overline{AC} \cong \overline{BD}$ (Corresp. parts of ≅ △s are ≅.)

36. 7.47 **37.** 7.47 **38.** 7.47 **39.** 3.5 **40.** 13.2 **41.** 124
42. 56 **43.** 56 **44.** 28

Lesson 6-4 pp. 375–382

Got It? 1. Rhombus; opp. sides of a ▱ are ≅, so all sides of $EFGH$ are ≅, and there are no rt. ∠s.
2. $m\angle 1 = m\angle 2 = m\angle 3 = m\angle 4 = 38$ **3a.** 43
b. Isosc.; diagonals of a rectangle are ≅ and bisect each other.
Lesson Check 1. Square; it is a rectangle because of the rt. ∠, and a rhombus because it has 4 ≅ sides.
2. Rhombus; it has 4 ≅ sides, and no rt. ∠s. **3.** $m\angle 1 = 40$, $m\angle 2 = 90$, $m\angle 3 = 50$ **4.** 4, 4 **5.** rectangle and square; rhombus and square **6.** The first step should be $2x + 8 + 9x - 6 = 90$.
Exercises 7. Rectangle; the ▱ has 4 rt. ∠s and does not have 4 ≅ sides. **9.** $m\angle 1 = m\angle 2 = m\angle 3 = m\angle 4 = 37$
11. $m\angle 1 = 118$, $m\angle 2 = m\angle 3 = 31$ **13.** $m\angle 1 = 32$, $m\angle 2 = 90$, $m\angle 3 = 58$, $m\angle 4 = 32$ **15.** $m\angle 1 = 55$, $m\angle 2 = 35$, $m\angle 3 = 55$, $m\angle 4 = 90$ **17.** $m\angle 1 = 90$,

$m\angle 2 = 55$, $m\angle 3 = 90$ **19.** $x = 3$; $LN = MP = 7$
21. $x = 9$; $LN = MP = 67$ **23.** $x = 2.5$; $LN = MP = 12.5$ **25.** ▱ **27.** rectangle **29.** ▱, rhombus, rectangle, square **31.** ▱, rhombus, rectangle, square
33. ▱, rhombus, rectangle, square **35.** rectangle, square **37.** rhombus, square **39.** $x = 5$, $y = 4$; all sides are 3. **41a.** Given **b.** Def. of rectangle **c.** Refl. Prop. of ≅ **d.** Def. of rectangle **e.** $\overline{AB} \cong \overline{DC}$ **f.** $\triangle ABC \cong \triangle DCB$ **g.** All rt. ∠s are ≅. **h.** Corresp. parts of ≅ △s are ≅.
43. $x = 5$, $y = 32$, $z = 7.5$
45. Answers may vary. Sample:
1. $ABCD$ is a rhombus. (Given)
2. $\overline{AB} \cong \overline{AD}$ and $\overline{CB} \cong \overline{CD}$. (Def. of rhombus)
3. $\overline{AC} \cong \overline{AC}$ (Refl. Prop. of ≅)
4. $\triangle ABC \cong \triangle ADC$ (SSS)
5. $\angle 3 \cong \angle 4$ and $\angle 2 \cong \angle 1$. (Corresp. parts of ≅ △s are ≅.)
6. $\overline{AC}$ bisects $\angle BAD$ and $\angle BCD$. (Def. of ∠ bisector)

47. $m\angle H = m\angle J = 58$, $m\angle K = m\angle G = 122$, $HK = KJ = JG = GH = 6$ **49.** $AC = BD = 16$
51. $AC = BD = 1$ **53.** 2 **55.** D **57.** A **59.** Yes; both pairs of opp. sides of the quad. are ≅. **60.** No; two opp. sides are ≅ and two opp. sides are ∥, but not the same pair of opp. sides.
61. Yes; diagonals of the quad. bisect each other. **62.** 6
63. 16 **64.** 5 **65.** $\overline{RQ}$ **66.** $\overline{PR}$ **67.** $\overline{ST}$
68. Answers may vary. Sample:

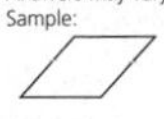

69. Answers may vary. Sample:

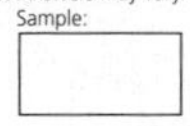

Lesson 6-5 pp. 383–388

Got It? 1a. The ▱ is not a rectangle or a square because ∠s are not rt. ∠s. It might be a rhombus. **b.** No; the fact that the diagonals bisect each other is true of all ▱s. **2.** 4 **3.** Yes; make diagonals ⊥. The result will be a rectangle and a rhombus, so it is square.
Lesson Check 1. Rectangle; diagonals are ≅.
2. Rhombus; diagonals are ⊥. **3.** 2 **4.** 3 **5a.** rhombus, square **b.** rectangle, square **c.** rhombus, square **d.** rectangle, rhombus, square **e.** rhombus, square
6. The only ▱s with ⊥ diagonals are rhombuses and squares. **7.** Rectangle; diagonals are ≅.
Exercises 9. Rhombus; diagonals are ⊥. **11.** 12 **13.** 10
15. Answers may vary. Sample: Measure the lengths of the frame's diagonals. If they are ≅, then the frame has the shape of a rectangle, and therefore a parallelogram; measure the two pairs of alt. int. ∠s formed by the turnbuckle (the transversal). If both pairs of ∠s are ≅, then both pairs of opposite sides of the frame are ∥.

17. 11 **19.** 16
21. Rhombus; answers may vary. Sample:

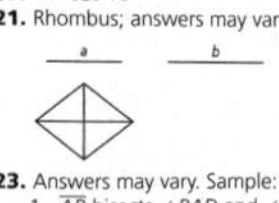

23. Answers may vary. Sample:
1. $\overline{AB}$ bisects $\angle BAD$ and $\angle BCD$. (Given)
2. $\angle 1 \cong \angle 2$ and $\angle 3 \cong \angle 4$. (Def. of bisect)
3. $\overline{AC} \cong \overline{AC}$ (Refl. Prop. of ≅)
4. $\triangle ABC \cong \triangle ADC$ (ASA)
5. $\overline{AB} \cong \overline{AD}$ and $\overline{BC} \cong \overline{CD}$. (Corresp. parts of ≅ △s are ≅.)
6. $\overline{AB} \cong \overline{CD}$ and $\overline{BC} \cong \overline{AD}$. (Opp. sides of a ▱ are ≅.)
7. $\overline{AB} \cong \overline{AD} \cong \overline{BC} \cong \overline{CD}$ (Trans. Prop. of ≅)
8. $ABCD$ is a rhombus. (Def. of rhombus)

25. Construct the midpt. of each diagonal. Copy the diagonals so the two midpts. coincide. Connect the endpoints of the diagonals. **27.** Construct the midpts. of each diagonal. Construct two ⊥ lines, and mark off diagonal lengths on the ⊥ lines. Connect the endpoints of the diagonals. **29.** Yes; ≅ diagonals in a ▱ mean it can be a rectangle with 2 opp. sides 2 cm long. **31.** "If one diagonal of a ▱ bisects one ∠, then the ▱ is a rhombus." The new statement is true. If $\overline{AC}$ bisects $\angle BCD$, then $\angle BCA \cong \angle DCA$. (Def. of ∠ bisector) $ABCD$ is a ▱, so $\angle B \cong \angle D$. (Opp. ∠s of a ▱ are ≅.) $\overline{AC} \cong \overline{AC}$ (Reflexive Prop. of ≅), so $\triangle BCA \cong \triangle DCA$ (AAS) and $\overline{BC} \cong \overline{DC}$. (Corresp. parts of ≅ △s are ≅.) Since opp. sides of a ▱ are ≅, $\overline{AB} \cong \overline{CD}$ and $\overline{BC} \cong \overline{DA}$. So $\overline{AB} \cong \overline{BC} \cong \overline{CD} \cong \overline{DA}$, and ▱$ABCD$ is a rhombus. (Def. of rhombus) **33.** I
35. $\left(\frac{-7 + x}{2}, \frac{10 + y}{2}\right) = (-1, 4)$. $-7 + x = -2$, $x = 5$, and $10 + y = 8$, $y = -2$, so $Q(5, -2)$.
36. $m\angle 1 = 128$, $m\angle 2 = 26$, $m\angle 3 = 26$
37. $m\angle 1 = 57$, $m\angle 2 = 57$, $m\angle 3 = 66$ **38.** $m\angle 1 = 90$, $m\angle 2 = 58$, $m\angle 3 = 90$ **39.** A ▱ is a rhombus if and only if its diagonals are ⊥. **40.** A ▱ is a rectangle if and only if its diagonals are ≅. **41.** $a = 5.6$, $b = 6.8$; 4.5, 4.5, 4.2, 4.2 **42.** 3; 18, 4.8, 18, 16.4 **43.** $m = 5$, $n = 15$; 15, 15, 21, 21

Lesson 6-6 pp. 389–397

Got It? 1a. $m\angle P = m\angle Q = 74$, $m\angle S = 106$ **b.** Yes; $\overline{DE} \parallel \overline{CF}$ so same-side int. ∠s are suppl. **2.** obtuse ∠ measure: 102; acute ∠ measure: 78 **3a.** 6; 23 **b.** 3; 1; A △ has 3 midsegments joining any pair of side midpts. A trapezoid has 1 midsegment joining the midpts. of the two legs. **4.** $m\angle 1 = 90$, $m\angle 2 = 54$, $m\angle 3 = 36$

Lesson Check 1. $m\angle 1 = 78$, $m\angle 2 = 90$, $m\angle 3 = 12$
2. $m\angle 1 = 94$, $m\angle 2 = 132$ **3.** 20 **4.** No; a kite's opp. sides are not ≅ or ∥. **5.** Answers may vary. Sample: Similar: diagonals are ⊥, consecutive sides ≅. Different: one diagonal of a kite bisects opp. ∠s but the other diagonal does not; all sides of a rhombus are ≅. **6.** Def. of trapezoid is a quad. with exactly one pair of ∥ sides. A ▱ has two pairs of ∥ sides, so a ▱ is not a trapezoid.
Exercises 7. $m\angle 1 = 77$, $m\angle 2 = 103$, $m\angle 3 = 103$
9. $m\angle 1 = 49$, $m\angle 2 = 131$, $m\angle 3 = 131$
11. $m\angle 1 = m\angle 2 = 115$, $m\angle 3 = 65$ **13.** 9 **15.** 9
17. $m\angle 1 = 90$, $m\angle 2 = 45$, $m\angle 3 = 45$ **19.** $m\angle 1 = 90$, $m\angle 2 = 26$, $m\angle 3 = 90$ **21.** $m\angle 1 = 90$, $m\angle 2 = 55$, $m\angle 3 = 90$, $m\angle 4 = 55$, $m\angle 5 = 35$ **23.** $m\angle 1 = 90$, $m\angle 2 = 90$, $m\angle 3 = 90$, $m\angle 4 = 90$, $m\angle 5 = 46$, $m\angle 6 = 34$, $m\angle 7 = 56$, $m\angle 8 = 44$, $m\angle 9 = 56$, $m\angle 10 = 44$
25. Answers may vary. Sample:

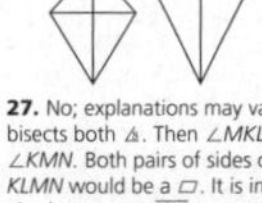

27. No; explanations may vary. Sample: Assume $\overline{KM}$ bisects both ∠s. Then $\angle MKL \cong \angle MKN \cong \angle KML \cong \angle KMN$. Both pairs of sides of $KLMN$ would be ∥, and $KLMN$ would be a ▱. It is impossible for an isosc. trap. to also be a ▱, so $\overline{KM}$ cannot bis. $\angle LMN$ and $\angle LKN$.
29. 15 **31.** $AD = 4$, $EF = 9$, $BC = 14$ **33.** $HG = 2$, $CD = 5$, $EF = 8$ **35.** $x = 35$, $y = 30$ **37.** Isosc. trapezoid; $\overline{AB} \parallel \overline{DC}$ (If alt. int. ∠s are ≅, then lines are ∥) and $\overline{AD} \cong \overline{BC}$. (Corresp. parts of ≅ △s are ≅.) **39.** Yes; the ≅ ∠s can be obtuse. **41.** Yes; if two ≅ ∠s are rt. ∠s, they are suppl. The other two ∠s are also suppl. **43.** Yes; the ≅ ∠s each have measure 45.
45. Answers may vary. Sample:
1. Draw $\overline{AE} \parallel \overline{DC}$. (Construction)
2. $AECD$ is a ▱. (Def. of ▱)
3. $\overline{AE} \cong \overline{DC}$ (Opp. sides of a ▱ are ≅.)
4. $\angle 1 \cong \angle C$ (If ∥ lines, corresp. ∠s are ≅.)
5. $\angle B \cong \angle 1$ (Isosc. △ Thm.)
6. $\angle B \cong \angle C$ (Transitive Prop. of ≅)
7. $\angle D$ and $\angle C$ are suppl. (If ∥ lines, same-side int. ∠s are suppl.)
8. $\angle BAD$ and $\angle B$ are suppl. (If ∥ lines, same-side int. ∠s are suppl.)
9. $\angle BAD \cong \angle D$ (∠s suppl. to ≅ ∠s are ≅.)

47. Isosc. trapezoid; answers may vary. Sample:

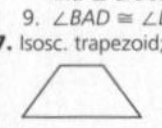

49. Rectangle, square; answers may vary. Sample:

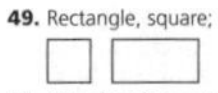

51. Kite, rhombus, square; answers may vary. Sample:

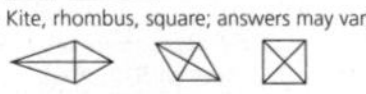

53. Answers may vary. Sample:
1. $\overline{AB} \cong \overline{DC}$ (Given)
2. $\angle BAD \cong \angle CDA$ (Base ∠s of an isosc. trapezoid are ≅.)
3. $\overline{AD} \cong \overline{AD}$ (Refl. Prop. of ≅)
4. $\triangle BAD \cong \triangle CDA$ (SAS)
5. $\overline{BD} \cong \overline{CA}$ (Corresp. parts of ≅ △s are ≅.)

55. Answers may vary. Sample:
1. Draw $\overline{TA}$ and $\overline{PR}$. (Construction)
2. $\overline{TR} \cong \overline{PA}$ (Given)
3. $\angle TRA \cong \angle PAR$ (Base ∠s of an isosc. trapezoid are ≅.)
4. $\overline{RA} \cong \overline{RA}$ (Refl. Prop. of ≅)
5. $\triangle TRA \cong \triangle PAR$ (SAS)
6. $\angle RTA \cong \angle APR$ (Corresp. parts of ≅ △s are ≅.)

57. True; a square is a ▱ with 4 rt. ∠s. **59.** False; a rhombus has 4 ≅ sides, and a kite does not. **61.** False; counterexample: kites and trapezoids are not ▱s.
63. Answers may vary. Sample:
1. $\overline{RT}$ and $\overline{PA}$ are not ∥. (Def. of trapezoid)
2. Extend $\overline{RT}$ and $\overline{PA}$ to meet at M. (Construction)
3. $\angle MTP \cong \angle R$ and $\angle MPT \cong \angle A$. (If ∥ lines, then corresp. ∠s are ≅.)
4. $\angle MTP \cong \angle MPT$ (Trans. Prop. of ≅)
5. $\overline{MT} \cong \overline{MP}$ (Converse of Isosc. △ Thm.)
6. $\angle MIT$ and $\angle MIP$ are rt. ∠s. (A line ⊥ to one of two ∥ lines is also ⊥ to the other line.)
7. $\overline{MI} \cong \overline{MI}$ (Refl. Prop. of ≅)
8. $\triangle MIT \cong \triangle MIP$ (HL)
9. $\overline{TI} \cong \overline{PI}$ (Corresp. parts of ≅ △s are ≅.)
10. $\overline{BI}$ is the ⊥ bis. of $\overline{TP}$. (Def. of ⊥ bis.)

65. half the difference of the bases; △ Midsegment Thm.

Algebra Review p. 399

1. $5\sqrt{2}$ **3.** 8 **5.** $4\sqrt{3}$ **7.** $6\sqrt{2}$ **9.** 6 **11.** $7\sqrt{2}$ **13.** $2\sqrt{6}$
15. $\frac{3\sqrt{10}}{2}$

Lesson 6-7 pp. 400–405

Got It? 1. scalene **2a.** Yes; slope of $\overline{MN}$ = slope of $\overline{PQ} = -3$ and slope of $\overline{NP}$ = slope of $\overline{MQ} = \frac{1}{3}$, so opp. sides are ∥. The product of slopes is −1, so sides are ⊥.

b. Yes; $MN = PQ = NP = MQ = \sqrt{10}$. **c.** Yes; slope of $\overline{AB} = \frac{3}{4}$ and slope of $\overline{BC} = -\frac{4}{3}$, so the product of their slopes is −1. Therefore, $\overline{AB} \perp \overline{BC}$ and $\angle B$ is a rt. ∠. So $\triangle ABC$ is a rt. △ by def. of rt. △. **3.** rhombus (The length of each side is $\sqrt{13}$.)
Lesson Check 1. isosceles **2.** No; explanations may vary. Sample: The diagonal lengths ($\sqrt{29}$ and 5) are not equal. **3.** Find the coordinates and use the Distance Formula to compare lengths. **4.** Answers may vary. Sample: $DEFG$ is not a ▱.
Exercises 5. Scalene; side lengths are 4, 5, and $\sqrt{17}$.
7. Isosceles; side lengths are $2\sqrt{2}$, $\sqrt{34}$, and $\sqrt{34}$.
9. Rhombus; explanations may vary. Sample: All four sides are ≅ (with length $\sqrt{5}$), and diagonals are not ≅ (with lengths 2 and 4). **11.** None; explanations may vary. Sample: Consecutive sides are not ≅ or ⊥. **13.** Rhombus; explanations may vary. Sample: All sides are ≅ and consecutive sides are not ⊥.
15. rhombus
17.

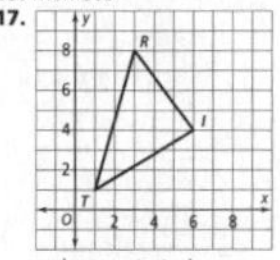

scalene; not rt. △

19.

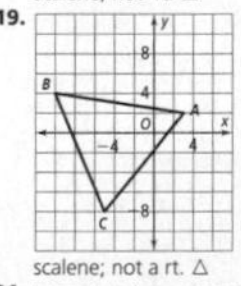

scalene; not a rt. △

21.

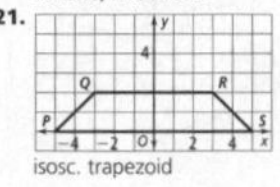

isosc. trapezoid

23.

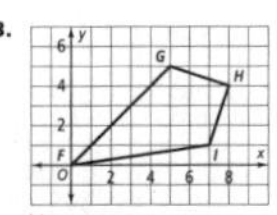

kite

25.

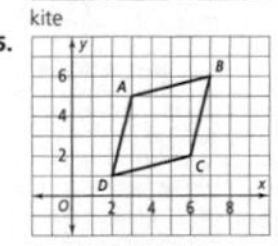

rhombus

27.

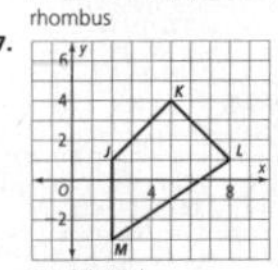

quadrilateral

29.

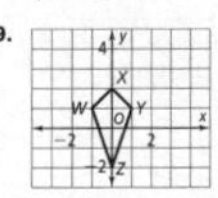

kite

31. Yes; $PR = SW = 4$, $PQ = ST = \sqrt{10}$, $QR = TW = 3\sqrt{2}$, so $\triangle PQR \cong \triangle STW$ by SSS. **33.** ▱; 24 units^2
35. slope of $\overline{DE} = 2$; slope of $\overline{AB} = 2$; $DE = \frac{1}{2}\sqrt{5}$; $AB = \sqrt{5}$. So $\overline{DE} \parallel \overline{AB}$ and $DE = \frac{1}{2}AB$. **37.** Answers may vary. Sample: Chairs are not at vertices of a ▱. Move right-most chair down by 1 grid unit. **39.** $G(-4, 1)$, $H(1, 3)$
41. $\left(-1, 6\frac{2}{3}\right)$, $\left(1, 8\frac{1}{3}\right)$, $(3, 10)$, $\left(5, 11\frac{2}{3}\right)$, $\left(7, 13\frac{1}{3}\right)$
43. $(-2.76, 5.2)$, $(-2.52, 5.4)$, $(-2.28, 5.6)$, . . . , $(8.52, 14.6)$, $(8.76, 14.8)$ **45.** D **47.** A **49.** $m\angle 1 = 62$, $m\angle 2 = m\angle 3 = 118$, $x = 2.5$ **50.** $(3, 2)$ **51.** $(-3, -4)$
52. −1 **53.** 0 **54.** $\frac{b}{c + d - a}$

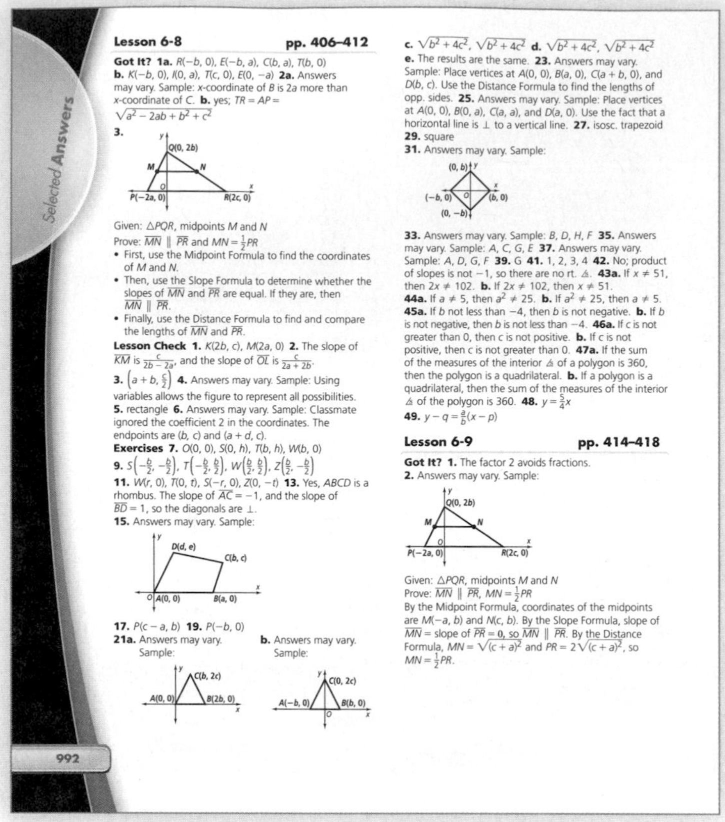

Lesson 6-8 pp. 406–412

Got It? 1a. $R(-b, 0)$, $E(-b, a)$, $C(b, a)$, $T(b, 0)$ **b.** $K(-b, 0)$, $I(0, a)$, $T(c, 0)$, $E(0, -a)$ **2a.** Answers may vary. Sample: x-coordinate of B is $2a$ more than x-coordinate of C. **b.** yes; $TR = AP = \sqrt{a^2 - 2ab + b^2 + c^2}$

3.

Given: $\triangle PQR$, midpoints M and N
Prove: $\overline{MN} \parallel \overline{PR}$ and $MN = \frac{1}{2}PR$
- First, use the Midpoint Formula to find the coordinates of M and N.
- Then, use the Slope Formula to determine whether the slopes of $\overline{MN}$ and $\overline{PR}$ are equal. If they are, then $\overline{MN} \parallel \overline{PR}$.
- Finally, use the Distance Formula to find and compare the lengths of $\overline{MN}$ and $\overline{PR}$.

Lesson Check 1. $K(2b, c)$, $M(2a, 0)$ **2.** The slope of $\overline{KM}$ is $\frac{c}{2b - 2a}$, and the slope of $\overline{OL}$ is $\frac{c}{2a + 2b}$. **3.** $\left(a + b, \frac{c}{2}\right)$ **4.** Answers may vary. Sample: Using variables allows the figure to represent all possibilities. **5.** rectangle **6.** Answers may vary. Sample: Classmate ignored the coefficient 2 in the coordinates. The endpoints are (b, c) and $(a + d, c)$.
Exercises 7. $O(0, 0)$, $S(0, h)$, $T(b, h)$, $W(b, 0)$ **9.** $S\left(-\frac{b}{2}, -\frac{b}{2}\right)$, $T\left(-\frac{b}{2}, \frac{b}{2}\right)$, $W\left(\frac{b}{2}, \frac{b}{2}\right)$, $Z\left(\frac{b}{2}, -\frac{b}{2}\right)$ **11.** $W(r, 0)$, $T(0, t)$, $S(-r, 0)$, $Z(0, -t)$ **13.** Yes, $ABCD$ is a rhombus. The slope of $\overline{AC} = -1$, and the slope of $\overline{BD} = 1$, so the diagonals are $\perp$.
15. Answers may vary. Sample:

17. $P(c - a, b)$ **19.** $P(-b, 0)$
21a. Answers may vary. Sample: **b.** Answers may vary. Sample:

c. $\sqrt{b^2 + 4c^2}$, $\sqrt{b^2 + 4c^2}$ **d.** $\sqrt{b^2 + 4c^2}$, $\sqrt{b^2 + 4c^2}$ **e.** The results are the same. **23.** Answers may vary. Sample: Place vertices at $A(0, 0)$, $B(a, 0)$, $C(a + b, 0)$, and $D(b, c)$. Use the Distance Formula to find the lengths of opp. sides. **25.** Answers may vary. Sample: Place vertices at $A(0, 0)$, $B(0, a)$, $C(a, a)$, and $D(a, 0)$. Use the fact that a horizontal line is $\perp$ to a vertical line. **27.** isosc. trapezoid **29.** square
31. Answers may vary. Sample:

33. Answers may vary. Sample: B, D, H, F **35.** Answers may vary. Sample: A, C, G, E **37.** Answers may vary. Sample: A, D, G, F **39.** G **41.** 1, 2, 3, 4 **42.** No; product of slopes is not -1, so there are no rt. $\triangle$s. **43a.** If $x \neq 51$, then $2x \neq 102$. **b.** If $2x \neq 102$, then $x \neq 51$. **44a.** If $a \neq 5$, then $a^2 \neq 25$. **b.** If $a^2 \neq 25$, then $a \neq 5$. **45a.** If b not less than -4, then b is not negative. **b.** If b is not negative, then b is not less than -4. **46a.** If c is not greater than 0, then c is not positive. **b.** If c is not positive, then c is not greater than 0. **47a.** If the sum of the measures of the interior $\angle$s of a polygon is 360, then the polygon is a quadrilateral. **b.** If a polygon is a quadrilateral, then the sum of the measures of the interior $\angle$s of the polygon is 360. **48.** $y = \frac{5}{4}x$ **49.** $y - q = \frac{a}{b}(x - p)$

Lesson 6-9 pp. 414–418

Got It? 1. The factor 2 avoids fractions.
2. Answers may vary. Sample:

Given: $\triangle PQR$, midpoints M and N
Prove: $\overline{MN} \parallel \overline{PR}$, $MN = \frac{1}{2}PR$
By the Midpoint Formula, coordinates of the midpoints are $M(-a, b)$ and $N(c, b)$. By the Slope Formula, slope of $\overline{MN}$ = slope of $\overline{PR} = 0$, so $\overline{MN} \parallel \overline{PR}$. By the Distance Formula, $MN = \sqrt{(c + a)^2}$ and $PR = 2\sqrt{(c + a)^2}$, so $MN = \frac{1}{2}PR$.

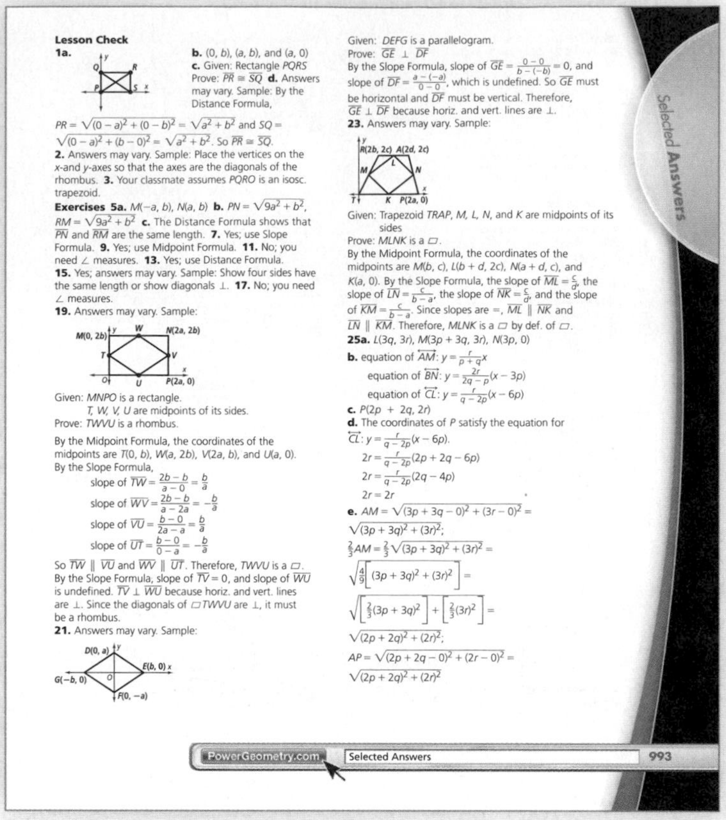

Lesson Check
1a. **b.** $(0, b)$, (a, b), and $(a, 0)$ **c.** Given: Rectangle $PQRS$ Prove: $\overline{PR} \cong \overline{SQ}$ **d.** Answers may vary. Sample: By the Distance Formula, $PR = \sqrt{(0 - a)^2 + (0 - b)^2} = \sqrt{a^2 + b^2}$ and $SQ = \sqrt{(0 - a)^2 + (b - 0)^2} = \sqrt{a^2 + b^2}$. So $\overline{PR} \cong \overline{SQ}$.
2. Answers may vary. Sample: Place the vertices on the x-and y-axes so that the axes are the diagonals of the rhombus. **3.** Your classmate assumes $PQRO$ is an isosc. trapezoid.
Exercises 5a. $M(-a, b)$, $N(a, b)$ **b.** $PN = \sqrt{9a^2 + b^2}$, $RM = \sqrt{9a^2 + b^2}$ **c.** The Distance Formula shows that $\overline{PN}$ and $\overline{RM}$ are the same length. **7.** Yes; use Slope Formula. **9.** Yes; use Midpoint Formula. **11.** No; you need $\angle$ measures. **13.** Yes; use Distance Formula. **15.** Yes; answers may vary. Sample: Show four sides have the same length or show diagonals $\perp$. **17.** No; you need $\angle$ measures.
19. Answers may vary. Sample:

Given: $MNPO$ is a rectangle.
T, W, V, U are midpoints of its sides.
Prove: $TWVU$ is a rhombus.

By the Midpoint Formula, the coordinates of the midpoints are $T(0, b)$, $W(a, 2b)$, $V(2a, b)$, and $U(a, 0)$. By the Slope Formula,

$$\text{slope of } \overline{TW} = \frac{2b - b}{a - 0} = \frac{b}{a}$$
$$\text{slope of } \overline{WV} = \frac{2b - b}{a - 2a} = -\frac{b}{a}$$
$$\text{slope of } \overline{VU} = \frac{b - 0}{2a - a} = \frac{b}{a}$$
$$\text{slope of } \overline{UT} = \frac{b - 0}{0 - a} = -\frac{b}{a}$$

So $\overline{TW} \parallel \overline{VU}$ and $\overline{WV} \parallel \overline{UT}$. Therefore, $TWVU$ is a $\square$. By the Slope Formula, slope of $\overline{TV} = 0$, and slope of $\overline{WU}$ is undefined. $\overline{TV} \perp \overline{WU}$ because horiz. and vert. lines are $\perp$. Since the diagonals of $\square TWVU$ are $\perp$, it must be a rhombus.
21. Answers may vary. Sample:

Given: $DEFG$ is a parallelogram.
Prove: $\overline{GE} \perp \overline{DF}$
By the Slope Formula, slope of $\overline{GE} = \frac{0 - 0}{b - (-b)} = 0$, and slope of $\overline{DF} = \frac{a - (-a)}{0 - 0}$, which is undefined. So $\overline{GE}$ must be horizontal and $\overline{DF}$ must be vertical. Therefore, $\overline{GE} \perp \overline{DF}$ because horiz. and vert. lines are $\perp$.
23. Answers may vary. Sample:

Given: Trapezoid $TRAP$, M, L, N, and K are midpoints of its sides
Prove: $MLNK$ is a $\square$.
By the Midpoint Formula, the coordinates of the midpoints are $M(b, c)$, $L(b + d, 2c)$, $N(a + d, c)$, and $K(a, 0)$. By the Slope Formula, the slope of $\overline{ML} = \frac{c}{d}$, the slope of $\overline{LN} = \frac{c}{b - a}$, the slope of $\overline{NK} = \frac{c}{d}$, and the slope of $\overline{KM} = \frac{c}{b - a}$. Since slopes are =, $\overline{ML} \parallel \overline{NK}$ and $\overline{LN} \parallel \overline{KM}$. Therefore, $MLNK$ is a $\square$ by def. of $\square$.
25a. $L(3q, 3r)$, $M(3p + 3q, 3r)$, $N(3p, 0)$
b. equation of $\overleftrightarrow{AM}$: $y = \frac{r}{p + q}x$
equation of $\overleftrightarrow{BN}$: $y = \frac{2r}{2q - p}(x - 3p)$
equation of $\overleftrightarrow{CL}$: $y = \frac{r}{q - 2p}(x - 6p)$
c. $P(2p + 2q, 2r)$
d. The coordinates of P satisfy the equation for $\overleftrightarrow{CL}$: $y = \frac{r}{q - 2p}(x - 6p)$.

$$2r = \frac{r}{q - 2p}(2p + 2q - 6p)$$
$$2r = \frac{r}{q - 2p}(2q - 4p)$$
$$2r = 2r$$

e. $AM = \sqrt{(3p + 3q - 0)^2 + (3r - 0)^2} = \sqrt{(3p + 3q)^2 + (3r)^2}$;
$\frac{2}{3}AM = \frac{2}{3}\sqrt{(3p + 3q)^2 + (3r)^2} =$
$\sqrt{\frac{4}{9}\left[(3p + 3q)^2 + (3r)^2\right]} =$
$\sqrt{\left[\frac{2}{3}(3p + 3q)^2\right] + \left[\frac{2}{3}(3r)^2\right]} =$
$\sqrt{(2p + 2q)^2 + (2r)^2}$;
$AP = \sqrt{(2p + 2q - 0)^2 + (2r - 0)^2} = \sqrt{(2p + 2q)^2 + (2r)^2}$

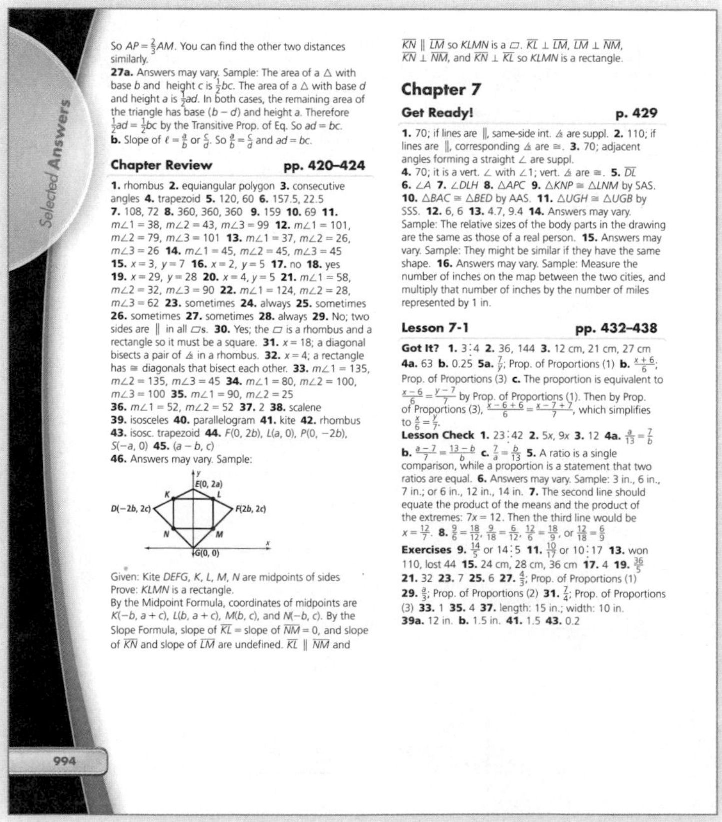

So $AP = \frac{2}{3}AM$. You can find the other two distances similarly.
27a. Answers may vary. Sample: The area of a $\triangle$ with base b and height c is $\frac{1}{2}bc$. The area of a $\triangle$ with base d and height a is $\frac{1}{2}ad$. In both cases, the remaining area of the triangle has base $(b - d)$ and height a. Therefore $\frac{1}{2}ad = \frac{1}{2}bc$ by the Transitive Prop. of Eq. So $ad = bc$. **b.** Slope of $\ell = \frac{a}{b}$ or $\frac{c}{d}$. So $\frac{a}{b} = \frac{c}{d}$ and $ad = bc$.

Chapter Review pp. 420–424

1. rhombus **2.** equiangular polygon **3.** consecutive angles **4.** trapezoid **5.** 120, 60 **6.** 157.5, 22.5 **7.** 108, 72 **8.** 360, 360, 360 **9.** 159 **10.** 69 **11.** $m\angle 1 = 38$, $m\angle 2 = 43$, $m\angle 3 = 99$ **12.** $m\angle 1 = 101$, $m\angle 2 = 79$, $m\angle 3 = 101$ **13.** $m\angle 1 = 37$, $m\angle 2 = 26$, $m\angle 3 = 26$ **14.** $m\angle 1 = 45$, $m\angle 2 = 45$, $m\angle 3 = 45$ **15.** $x = 3$, $y = 7$ **16.** $x = 2$, $y = 5$ **17.** no **18.** yes **19.** $x = 29$, $y = 28$ **20.** $x = 4$, $y = 5$ **21.** $m\angle 1 = 58$, $m\angle 2 = 32$, $m\angle 3 = 90$ **22.** $m\angle 1 = 124$, $m\angle 2 = 28$, $m\angle 3 = 62$ **23.** sometimes **24.** always **25.** sometimes **26.** sometimes **27.** sometimes **28.** always **29.** No; two sides are $\parallel$ in all $\square$s. **30.** Yes; the $\square$ is a rhombus and a rectangle so it must be a square. **31.** $x = 18$; a diagonal bisects a pair of $\angle$s in a rhombus. **32.** $x = 4$; a rectangle has $\cong$ diagonals that bisect each other. **33.** $m\angle 1 = 135$, $m\angle 2 = 135$, $m\angle 3 = 45$ **34.** $m\angle 1 = 80$, $m\angle 2 = 100$, $m\angle 3 = 100$ **35.** $m\angle 1 = 90$, $m\angle 2 = 25$ **36.** $m\angle 1 = 52$, $m\angle 2 = 52$ **37.** 2 **38.** scalene **39.** isosceles **40.** parallelogram **41.** kite **42.** rhombus **43.** isosc. trapezoid **44.** $F(0, 2b)$, $L(a, 0)$, $P(0, -2b)$, $S(-a, 0)$ **45.** $(a - b, c)$
46. Answers may vary. Sample:

Given: Kite $DEFG$, K, L, M, N are midpoints of sides
Prove: $KLMN$ is a rectangle.
By the Midpoint Formula, coordinates of midpoints are $K(-b, a + c)$, $L(b, a + c)$, $M(b, c)$, and $N(-b, c)$. By the Slope Formula, slope of $\overline{KL}$ = slope of $\overline{NM} = 0$, and slope of $\overline{KN}$ and slope of $\overline{LM}$ are undefined. $\overline{KL} \parallel \overline{NM}$ and $\overline{KN} \parallel \overline{LM}$ so $KLMN$ is a $\square$. $\overline{KL} \perp \overline{LM}$, $\overline{LM} \perp \overline{NM}$, $\overline{KN} \perp \overline{NM}$, and $\overline{KN} \perp \overline{KL}$ so $KLMN$ is a rectangle.

Chapter 7

Get Ready! p. 429

1. 70; if lines are $\parallel$, same-side int. $\angle$s are suppl. **2.** 110; if lines are $\parallel$, corresponding $\angle$s are $\cong$. **3.** 70; adjacent angles forming a straight $\angle$ are suppl. **4.** 70; it is a vert. $\angle$ with $\angle 1$; vert. $\angle$s are $\cong$. **5.** $\overline{DL}$ **6.** $\angle A$ **7.** $\angle DLH$ **8.** $\triangle APC$ **9.** $\triangle KNP \cong \triangle LNM$ by SAS. **10.** $\triangle BAC \cong \triangle BED$ by AAS. **11.** $\triangle UGH \cong \triangle UGB$ by SSS. **12.** 6, 6 **13.** 4.7, 9.4 **14.** Answers may vary. Sample: The relative sizes of the body parts in the drawing are the same as those of a real person. **15.** Answers may vary. Sample: They might be similar if they have the same shape. **16.** Answers may vary. Sample: Measure the number of inches on the map between the two cities, and multiply that number of inches by the number of miles represented by 1 in.

Lesson 7-1 pp. 432–438

Got It? 1. 3 : 4 **2.** 36, 144 **3.** 12 cm, 21 cm, 27 cm **4a.** 63 **b.** 0.25 **5a.** $\frac{7}{y}$; Prop. of Proportions (1) **b.** $\frac{x + 6}{6}$; Prop. of Proportions (3) **c.** The proportion is equivalent to $\frac{x - 6}{6} = \frac{y - 7}{7}$ by Prop. of Proportions (1). Then by Prop. of Proportions (3), $\frac{x - 6 + 6}{6} = \frac{x - 7 + 7}{7}$, which simplifies to $\frac{x}{6} = \frac{y}{7}$.
Lesson Check 1. 23 : 42 **2.** $5x$, $9x$ **3.** 12 **4a.** $\frac{a}{13} = \frac{7}{b}$ **b.** $\frac{a - 7}{7} = \frac{13 - b}{b}$ **c.** $\frac{7}{a} = \frac{b}{13}$ **5.** A ratio is a single comparison, while a proportion is a statement that two ratios are equal. **6.** Answers may vary. Sample: 3 in., 6 in., 7 in.; or 6 in., 12 in., 14 in. **7.** The second line should equate the product of the means and the product of the extremes: $7x = 12$. Then the third line would be $x = \frac{12}{7}$. **8.** $\frac{9}{6} = \frac{18}{12}$, $\frac{9}{18} = \frac{6}{12}$, $\frac{12}{6} = \frac{18}{9}$, or $\frac{12}{18} = \frac{6}{9}$
Exercises 9. $\frac{14}{5}$ or 14 : 5 **11.** $\frac{10}{17}$ or 10 : 17 **13.** won 110, lost 44 **15.** 24 cm, 28 cm, 36 cm **17.** 4 **19.** $\frac{36}{5}$ **21.** 32 **23.** 7 **25.** 6 **27.** $\frac{4}{3}$; Prop. of Proportions (1) **29.** $\frac{a}{3}$; Prop. of Proportions (2) **31.** $\frac{7}{4}$; Prop. of Proportions (3) **33.** 1 **35.** 4 **37.** length: 15 in.; width: 10 in. **39a.** 12 in. **b.** 1.5 in. **41.** 1.5 **43.** 0.2

45. **47.** The product of the means is $26 \cdot 16 = 416$, and the product of the extremes is $10 \cdot 42 = 420$. Since $416 \neq 420$, it is not a valid proportion. **49.** $\frac{9}{4}$; divide each side by $4n$. **51.** $\frac{b}{2}$; Prop. of Proportions (3) **53.** $\frac{c}{d}$; Prop. of Proportions (2), then (3), then (2) **55.** $\frac{a}{b} = \frac{c}{d}$ (given); $\frac{a}{b}(bd) = \frac{c}{d}(bd)$ (Mult. Prop. of =); $ad = bc$ (simplify and Commutative Prop. of Mult.); $bc = ad$ (Sym. Prop. of =); $\frac{bc}{ac} = \frac{ad}{ac}$ (Div. Prop. of =); $\frac{b}{a} = \frac{d}{c}$ (simplify) **57.** $\frac{a}{b} = \frac{c}{d}$ (given); $\frac{a}{b} + 1 = \frac{c}{d} + 1$ (Add. Prop. of Eq.); $\frac{a}{b} + \frac{b}{b} = \frac{c}{d} + \frac{d}{d}$ (Subst. Prop. of Eq.); $\frac{a + b}{b} = \frac{c + d}{d}$ (simplify) **59.** $-\frac{5}{6}$, $\frac{1}{2}$

Algebra Review p. 439

1. -7, 2 **3.** -3, $-\frac{1}{2}$ **5.** $\frac{5 + \sqrt{3}}{2}$, $\frac{5 - \sqrt{3}}{2}$; 3.37, 1.63 **7.** -4, $\frac{5}{2}$ **9.** 0, 4 **11.** $\frac{-5 + \sqrt{55}}{6}$, $\frac{-5 - \sqrt{55}}{6}$; 0.40, -2.07

Lesson 7-2 pp. 440–447

Got It? 1a. $\angle D \cong \angle H$, $\angle E \cong \angle J$, $\angle F \cong \angle K$, $\angle G \cong \angle L$ **b.** $\frac{DE}{HJ} = \frac{EF}{JK} = \frac{FG}{KL} = \frac{GD}{LH}$ **2a.** not similar **b.** $ABCDE \sim SRVUT$ or $ABCDE \sim UVRST$; 2 : 1 **3.** $\frac{10}{3}$ **4.** 28.8 in. high by 48 in. wide **5a.** Using 0.8 cm as the height of the towers, then $\frac{1}{200} = \frac{0.8}{h}$ and $h = 160$ m. **b.** No; using a scale of 1 in. = 50 ft, the paper must be more than 12 in. long.
Lesson Check 1. $\angle H$ **2.** JT **3.** yes; $DEGH \sim PLQR$; 3 : 2 **4.** 6 **5.** Answers may vary. Sample: The scale indicates how many units of length of the actual object are represented by each unit of length in the drawing. **6.** A is incorrect. Sample explanation: In the diagram, $\angle T$ corresp. to $\angle P$ (or to $\angle U$), but in the similarity statement $TRUV \sim NPQV$, $\angle T$ corresp. to $\angle N$. **7.** Every figure is $\sim$ to itself, so similarity is reflexive. If figure 1 $\sim$ figure 2 and figure 2 $\sim$ figure 3, then figure 1 $\sim$ figure 3, so similarity is transitive. If figure 1 $\sim$ figure 2, then figure 2 $\sim$ figure 1, so the similarity is symmetric. **8.** any three of the following: $\triangle ABS \sim \triangle PRS$, $\triangle ASB \sim \triangle PSR$, $\triangle SAB \sim \triangle SPR$, $\triangle SBA \sim \triangle SRP$, $\triangle BAS \sim \triangle RPS$, $\triangle BSA \sim \triangle RSP$
Exercises 9. $\angle R \cong \angle D$, $\angle S \cong \angle E$, $\angle T \cong \angle F$, $\angle V \cong \angle G$; $\frac{RS}{DE} = \frac{ST}{EF} = \frac{TV}{FG} = \frac{VR}{GD}$ **11.** $\angle K \cong \angle H$, $\angle L \cong \angle G$, $\angle M \cong \angle F$, $\angle N \cong \angle D$, $\angle P \cong \angle C$; $\frac{KL}{HG} = \frac{LM}{GF} = \frac{MN}{FD} = \frac{NP}{DC} = \frac{PK}{CH}$ **13.** $ABDC \sim FEDG$ (or $ABDC \sim FGDE$, $ABDC \sim DEFG$, $ABDC \sim DGFE$); scale factor is 2 : 3. **15.** Not similar; sample explanation: The ratio of the longer sides is $\frac{12}{9}$ or $\frac{4}{3}$, and the ratio of the shorter sides is $\frac{10}{8}$ or $\frac{5}{4}$. Since $\frac{4}{3} \neq \frac{5}{4}$, the corresp. sides are not proportional and the figures are not $\sim$. **17.** Not similar; sample explanation: The $\angle$ measures are not the same. **19.** $x = 8$, $y = 9$, $z = 5.25$ **21.** 120 pixels wide by 90 pixels high **23.** 5 in. **25.** 3 : 5 **27.** 5 : 3 **29.** 25 **31a.** The slope of $\overline{AB}$, $\overline{CD}$, $\overline{AE}$, and $\overline{FG}$ is -2. The slope of $\overline{BC}$, $\overline{AD}$, $\overline{EF}$, and $\overline{AG}$ is $\frac{1}{2}$. For each pair of consecutive sides of $ABCD$, the slopes are negative reciprocals, so $ABCD$ has four rt. $\angle$s. Similarly, $AEFG$ has four rt. $\angle$s. The measure of $\angle A$, $\angle ABC$, $\angle BCD$, $\angle CDA$, $\angle E$, $\angle F$, and $\angle G$, is 90. **b.** By the Distance Formula, $AB = BC = CD = AD = \sqrt{5}$ and $AE = EF = FG = AG = 2\sqrt{5}$. **c.** All the angles of $AEFG$ and $ABCD$ are $\cong$. $\frac{AB}{AE} = \frac{BC}{EF} = \frac{CD}{FG} = \frac{AD}{AG} = \frac{\sqrt{5}}{2\sqrt{5}} = \frac{1}{2}$ The corresp. sides are proportional, so $AEFG \sim ABCD$. **33.** No; for polygons with more than 3 sides, you also need to know that corresp. $\angle$s are $\cong$ in order to state that the polygons are $\sim$. **35.** 1 : 3 **37.** $x = 10$; 2 : 1 **43.** always **45.** sometimes **47.** 21 ft by 40 ft **49.** All $\angle$s in any rectangle are right $\angle$s, so all corresp. $\angle$s are $\cong$. The ratio of two pair of consecutive sides for each rectangle is the same. Since opposite sides of a parallelogram are equal, the other two pair of sides will also have the same ratio. So corresp. sides form equal ratios and are proportional. So $BCEG \sim LJAW$.

Lesson 7-3 pp. 450–458

Got It? 1a. The measures of the two acute $\angle$s in each $\triangle$ are 39 and 51, so the $\triangle$s are $\sim$ by the AA $\sim$ Post. **b.** Each of the base $\angle$s in the $\triangle$ at the left measures 68, while each of the base $\angle$s in the $\triangle$ at the right measures $\frac{1}{2}(180 - 62) = 59$; the $\triangle$s are not $\sim$. **2a.** The ratio for each of the three pairs of corresp. sides is 3 : 4, so $\triangle ABC \sim \triangle EFG$ by SSS $\sim$. **b.** $\angle A$ is in each $\triangle$ and $\frac{AL}{AC} = \frac{AW}{AE} = \frac{1}{2}$, so $\triangle ALW \sim \triangle ACE$ by SAS$\sim$.
3a. $\overline{MP} \parallel \overline{AC}$ (given), so $\angle A \cong \angle P$ and $\angle C \cong \angle M$ because if two lines are $\parallel$, then alt. int. $\angle$s are $\cong$. So $\triangle ABC \sim \triangle PBM$ by AA$\sim$. **b.** No; the $\cong$ vertical angles are not included by the proportional sides, so it is not possible to prove that the triangles are similar. **4.** The

triangles formed will not be similar unless both Darius and the cliff form right angles with the ground.
Lesson Check **1.** Yes; $m\angle R = 180 - (35 + 45) = 100$, and $\angle AEZ \cong \angle REB$ (Vert. $\angle$s are $\cong$.), so $\triangle AEZ \sim \triangle REB$ by AA~. **2.** Yes; the ratios of corresp. sides are all 2 : 3, so $\triangle ABC \sim \triangle FED$ by SSS~. **3.** Yes; $\angle G \cong \angle E$ and $\frac{UG}{FE} = \frac{AG}{BE} = \frac{4}{5}$, so $\triangle GUA \sim \triangle EFB$ by SAS ~. **4.** Answers may vary. Sample: Measure your shadow and the flagpole's shadow. Use the proportion $\frac{\text{your shadow}}{\text{flagpole's shadow}} = \frac{\text{your height}}{\text{flagpole's height}}$
5. Method A is not correct because the ratio, $\frac{4}{8}$ does not use corresp. sides. **6a.** Answers may vary. Sample: Both use two pairs of corresp. sides and the $\angle$s included by those sides, but SAS~ uses pairs of equal ratios, while SAS $\cong$ uses pairs of $\cong$ sides. **b.** Both involve all three sides of a $\triangle$, but corresp. sides are proportional for SSS ~ and $\cong$ for SSS $\cong$.
Exercises **7.** $\triangle FGH \sim \triangle KJH$; AA~ **9.** $\triangle RST \sim \triangle PSQ$; SAS~ **11.** Not ~; $m\angle U = 180 - (25 + 35) = 120$, while $m\angle A = 110$. **13.** $\angle A \cong \angle A$ (Refl. Prop. of $\cong$) and $\angle ABC \cong \angle ACD$ (given), so $\triangle ABC \sim \triangle ACD$ by AA~. **15.** There are a pair of $\cong$ vert. $\angle$s and a pair of $\cong$ rt. $\angle$s, so the $\triangle$s are ~ by AA~; 180 ft **17.** about 169.2 m **19.** $\triangle LMN \sim \triangle SMT$ by AA~. **21a.** No; the ratios of the sides that form the vertex $\angle$s are =, but the vertex $\angle$s may not be $\cong$. **b.** Yes; sample explanation: An isosc. rt. $\triangle$ has two $\angle$s 45°, so any two isosc. rt. $\triangle$s are ~ by AA~. **23.** 180 ft **25.** 20 **27.** In $\triangle PQR$ and $\triangle STV$, $\angle Q \cong \angle T$ because $\perp$ lines form rt. $\angle$s, which are $\cong$. The sides that contain the $\angle$s are proportional (given). So $\triangle PQR \sim \triangle STV$ by SAS~, and $\angle KRV \cong \angle KVR$ because corresp. $\angle$s of ~ $\triangle$s are $\cong$. Thus $\triangle VKR$ is isosc. by the Converse of Isosc. $\triangle$ Thm. **29.** Yes; the two $\parallel$ lines and the two sides determine two pairs of $\cong$ corr. $\angle$s, so the two $\triangle$s are ~ by AA~. **31.** 4 : 3; sample explanation: Since $\angle P \cong \angle S$ and $\angle PQM \cong \angle STR$, $\triangle PQM \sim \triangle STR$ by AA~. So the ratio $\frac{MQ}{RT} = \frac{PM}{SR}$ = the ratio of corresp. sides in $\triangle PMN$ and $\triangle SRW$ namely, 4 : 3. **33.** It is given that $\ell_1 \parallel \ell_2$, so $\angle BAC \cong \angle EDF$ because if lines are $\parallel$, then corresponding $\angle$s are $\cong$. The given $\perp$ lines mean $\angle ACB \cong \angle DFE$ because $\perp$ lines form rt. $\angle$s, which are $\cong$. So $\triangle ABC \sim \triangle DEF$ by AA~, and $\frac{BC}{EF} = \frac{AC}{DF}$ because corresp. sides of ~ $\triangle$s are proportional. Then Prop. of Proportions (2) lets us conclude that $\frac{BC}{AC} = \frac{EF}{DF}$.
35.

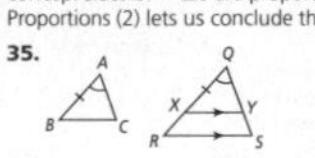

Choose point X on $\overline{QR}$ so that $QX = AB$. Then draw $\overline{XY} \parallel \overline{RS}$ (Through a point not on a line, there is exactly one line $\parallel$ to the given line.). $\angle A \cong \angle Q$ (Given) and $\angle QXY \cong \angle R$ (If two lines are $\parallel$, then corresp. $\angle$s are $\cong$.), so $\triangle QXY \sim \triangle QRS$ by AA~. Therefore, $\frac{QX}{QR} = \frac{XY}{RS} = \frac{QY}{QS}$ because corresp. sides of ~ $\triangle$s are proportional. Since $QX = AB$, substitute QX for AB in the given proportion $\frac{AB}{QR} = \frac{AC}{QS}$ to get $\frac{QX}{QR} = \frac{AC}{QS}$. Therefore, $\frac{QX}{QR} = \frac{QY}{QS} = \frac{AC}{QS}$, and $QY = AC$. So $\triangle ABC \cong \triangle QXY$ by SAS. $\angle B \cong \angle QXY$ (Corresp. parts of $\cong$ $\triangle$s are $\cong$.) and $\angle B \cong \angle R$ by the Transitive Prop. of $\cong$. Therefore, $\triangle ABC \sim \triangle QRS$ by AA~.
37. C **39.** C **41.** 2 : 3 **42.** 135 **43.** 12 **44.** $\frac{3}{2}$ **45.** 125; obtuse **46.** 88; acute **47.** 180; straight **48.** 110; obtuse **49.** 8, 18; x, 24; 6 **50.** m, 18; 12, 20; $\frac{40}{3}$ or $13\frac{1}{3}$ **51.** $x + 2$, 9; 15, x; 3 **52.** $x + 4$, 5; $x - 3$, 9; $\frac{47}{4}$ or 11.75

Lesson 7-4 pp. 460–467

Got It? **1a.** $\triangle PRQ \sim \triangle SPQ \sim \triangle SRP$ **b.** $\frac{SR}{SP} = \frac{SP}{SQ}$, $\frac{SR}{SP} = \frac{PR}{QP}$ **2.** $6\sqrt{2}$ **3.** $x = 6$, $y = 2\sqrt{5}$ **4.** 12 in.
Lesson Check **1.** 6 **2.** $\sqrt{48}$ or $4\sqrt{3}$ **3.** h **4.** g **5.** j, h or h, j **6.** d, d **7a.** $\overline{RT}$ **b.** $\overline{RP}$, $\overline{PT}$ **c.** $\overline{PT}$ **8.** The length 8 is the entire hypotenuse, so the segments of the hypotenuse have lengths 3 and 5. The correct proportion is $\frac{3}{x} = \frac{x}{5}$.
Exercises **9.** Answers may vary. Sample: $\triangle KJL \sim \triangle NJK \sim \triangle NKL$ **11.** Answers may vary. Sample: $\triangle OMN \sim \triangle PMO \sim \triangle PON$ **13.** 12 **15.** $\sqrt{63}$ or $3\sqrt{7}$ **17.** 14 **19.** $x = 20$, $y = 10\sqrt{5}$ **21.** $x = 3\sqrt{7}$, $y = 12$ **23a.** 4 cm
b.

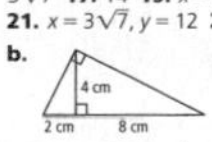

c. Answers may vary. Sample: Draw a 10-cm segment. Construct a $\perp$ of length 4 cm that is 2 cm from one endpoint; connect to form a $\triangle$. **25.** 2.5 **27.** 1 **29.** Yes; the proportion $\frac{a}{\sqrt{ab}} = \frac{\sqrt{ab}}{b}$ is true by the Cross Products Prop. and satisfies the definition of the geometric mean.
31. 8.50 m **33.** $\ell_1 = \sqrt{2}$, $\ell_2 = \sqrt{2}$, $a = 1$, $s_2 = 1$ **35.** $\ell_2 = 2\sqrt{3}$, $h = 4$, $a = \sqrt{3}$, $s_1 = 1$ **37.** (−2, 6), (10, 6) **39.** 4 **41.** 5 **43.** $\triangle ABC \sim \triangle ACD$ and $\triangle ABC \sim \triangle CBD$ by Thm. 7-3. Then $\frac{AB}{AC} = \frac{AC}{AD}$ and $\frac{AB}{CB} = \frac{BC}{BD}$ because corresp. sides of ~ $\triangle$s are proportional.
45a.

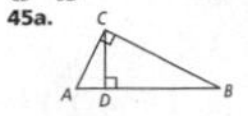

Given: $\overline{AC} \perp \overline{BC}$, $\overline{AB} \perp \overline{CD}$
Prove: $AC \cdot BC = AB \cdot CD$
b. The conjecture is true. You can express the area of $\triangle ABC$ as $\frac{1}{2}(AC)(BC)$ or as $\frac{1}{2}(AB)(CD)$, so $AC \cdot BC = AB \cdot CD$.
47. Answers may vary. Sample: $\triangle ABC \sim \triangle DEC$ (AA ~ Post.), so $\frac{AC}{DC} = \frac{BC}{EC}$ (Corr. sides of ~ $\triangle$s are in prop.). By the Subtraction Property of =, $\frac{AC - DC}{DC} = \frac{BC - EC}{EC}$, or $\frac{AD}{DC} = \frac{BE}{EC}$. **49.** H **51.** $\angle R \cong \angle P$ (given) and $\angle RNM \cong \angle PNQ$ (Vert. $\angle$s are $\cong$.), so $\triangle NRM \sim \triangle NPQ$

by AA ~. **52.** $x = 5$, $y = 8$ **53.** $x = 3$, $y = 4$ **54.** 28 cm **55.** 9.8 in. **56.** $\frac{24}{7}$ mm or $3\frac{3}{7}$ mm

Lesson 7-5 pp. 471–478

Got It? **1a.** 8 **b.** $RS = \frac{1}{2}XZ$ (Midsegment Thm.) **2.** 5.76 yd **3.** 14.4
Lesson Check **1.** d **2.** c **3.** d **4.** 5 **5.** 15 **6.** Answers may vary. Sample: The corollary to the Side-Splitter Thm. takes the same three (or more) $\parallel$ lines as in Thm. 6-7, but instead of cutting off $\cong$ segments it allows the segments to be proportional. **7.** Answers may vary. Sample: Alike: Both involve a $\triangle$ and a seg. from one vertex to the opposite side of the $\triangle$. Different: In Corollary 1 to Thm. 7-3, the $\triangle$ is a rt. $\triangle$ and the seg. is an alt., while in the $\triangle$-$\angle$-Bis. Thm. the $\triangle$ does not have to be a rt. $\triangle$ and the seg. is an $\angle$ bis. **8.** The Side-Splitter Thm. involves only the segments formed on the two sides intersected by the $\parallel$ line. (To find x, you can use a proportion involving the two ~ $\triangle$s.)
Exercises **9.** 7.5 **11.** 10 **13.** 8 mm **15.** 7.5 **17.** $3\frac{5}{13}$ **19.** 6 **21.** 35 **23.** Use the Side-Splitter Thm. to write the proportion $\frac{AB}{BD} = \frac{AC}{CE}$, then find the values of BD, AC, and CE to calculate the unknown length AB. **25.** KS by the $\triangle$-$\angle$-Bis. Thm. **27.** JP by the Side-Splitter Thm. **29.** KM by the $\triangle$-$\angle$-Bis. Thm. **31.** 575 ft **33.** 20 **35.** $\frac{2}{7}$ or 3 **37.** $\frac{XR}{RQ} = \frac{YS}{SQ}$ (Given); $\frac{XR + RQ}{RQ} = \frac{YS + SQ}{SQ}$ (Prop. of Proportions (3)); $XQ = XR + RQ$, $YQ = YS + SQ$ (Seg. Add. Post.); $\frac{XQ}{RQ} = \frac{YQ}{SQ}$ (Subst.); $\angle Q \cong \angle Q$ (Refl. Prop. of $\cong$); $\triangle XQY \sim \triangle RQS$ (SAS ~ Post.); $\angle 1 \cong \angle 2$ (Corresp. $\angle$s of ~ $\triangle$s are $\cong$.); $\overline{RS} \parallel \overline{XY}$ (If corresp. $\angle$s are $\cong$, the lines are $\parallel$.) **39.** no; $\frac{28}{12} \neq \frac{24}{10}$ **41.** 12.5 cm or 4.5 cm **43.** Isosc.; $AC : BC$ is 1 : 1 by the $\triangle$-$\angle$-Bis. Thm. **45.** 5.2 **47.** By the Side-Splitter Thm., $\frac{CD}{DB} = \frac{CA}{AF}$. By the Corresp. $\angle$s Post., $\angle 3 \cong \angle 1$. Since $\overrightarrow{AD}$ bisects $\angle CAB$, $\angle 1 \cong \angle 2$. By the Alt. Int. $\angle$s Thm., $\angle 2 \cong \angle 4$. So, $\angle 3 \cong \angle 4$ by the Trans. Prop. of $\cong$. By the Converse of the Isosc. $\triangle$ Thm., $BA = AF$. Substituting BA for AF, $\frac{CD}{DB} = \frac{CA}{BA}$.
49. Use the diagram with Ex. 47, with $\overline{AD} \parallel \overline{EB}$. It is given that $\frac{CD}{DB} = \frac{CA}{BA}$, and you want to prove that $\angle 1 \cong \angle 2$. By the Side-Splitter Thm., $\frac{CA}{AF} = \frac{CD}{DB}$. So $\frac{CA}{BA} = \frac{CA}{AF}$ by the Transitive Prop. of =, and $BA = AF$. Therefore, $\angle 3 \cong \angle 4$ by the Isosc. $\triangle$Thm. Using properties of $\parallel$ lines, $\angle 1 \cong \angle 3$ and $\angle 2 \cong \angle 4$. So $\angle 1 \cong \angle 2$ by the Transitive Prop. of $\cong$, and $\overline{AD}$ bisects $\angle CAB$ by the def. of $\angle$ bis.
51. 20 **53.** 118 **55.** m **56.** m **57.** c **58.** h **59.** (3, −3) **60.** (0, 2) **61.** (1.5, 2.5) **62.** $(3\text{ m})^2 = 9\text{ m}^2$, $(4\text{ m})^2 = 16\text{ m}^2$, $(5\text{ m})^2 = 25\text{ m}^2$ **63.** $(5\text{ in.})^2 = 25\text{ in.}^2$, $(12\text{ in.})^2 = 144\text{ in.}^2$, $(13\text{ in.})^2 = 169\text{ in.}^2$ **64.** $(4\text{ m})^2 = 16\text{ m}^2$, $(4\sqrt{2}\text{ m})^2 = 32\text{ m}^2$

Chapter Review pp. 480–482

1. similar **2.** proportion **3.** scale factor **4.** means, extremes (in either order) **5.** 1 : 116 or $\frac{1}{116}$ **6.** 36 **7.** 6 **8.** $\frac{55}{4}$ or $13\frac{3}{4}$ **9.** 6 **10.** 7 **11.** $JEHN \sim JKLP$; 3 : 4 **12.** $\triangle PQR \sim \triangle XYZ$; 3 : 2 **13.** 120 ft **14.** 45 ft **15.** The ratios of each pair of corresp. sides is 2 : 1, so $\triangle AMY \sim \triangle ECD$ by SSS~. **16.** If lines are $\parallel$, then corresp. $\angle$s are $\cong$, so $\triangle RPT \sim \triangle SGT$ by AA~.
17. 12 **18.** $2\sqrt{15}$ **19.** $x = 6\sqrt{2}$, $y = 6\sqrt{6}$ **20.** $\sqrt{35}$ **21.** $x = 2\sqrt{21}$; $y = 4\sqrt{3}$ **22.** $x = 12$, $y = 4\sqrt{5}$ **23.** 7.5 **24.** 3.6 **25.** 22.5 **26.** 12 **27.** 17.5 **28.** 77

Chapter 8

Get Ready! p. 487

1. 4.648 **2.** 40.970 **3.** 6149.090 **4.** −5 **5.** AA ~ **6.** SSS ~ **7.** SAS ~ **8.** 12 **9.** 8 **10.** $2\sqrt{13}$ **11.** 9 **12.** Answers may vary. Sample: When something is "elevated" you look up to see it, so an $\angle$ of elevation is formed by a horizontal line and the line of sight. **13.** Answers may vary. Sample: The prefix *tri-* means 3; triangles are associated with trigonometric ratios. **14.** Answers may vary. Sample: Because the Law of Cosines can be derived from the Pythagorean Theorem, and you can use the Law of Cosines to find angle measures, the Law of Cosines can probably be used to find side lengths and angle measures.

Lesson 8-1 pp. 491–498

Got It? **1a.** 26 **b.** Yes; 10, 24, and 26 are whole numbers that satisfy $a^2 + b^2 = c^2$. **2.** $6\sqrt{3}$ **3.** 15.5 in. **4a.** No; $16^2 + 48^2 \neq 50^2$. **b.** No; $a^2 + b^2 = b^2 + a^2$ for any values of a and b. **5.** acute
Lesson Check **1.** 37 **2.** $\sqrt{130}$ **3.** 4 **4.** $4\sqrt{3}$ **5.** The three numbers a, b, and c must be whole numbers that satisfy $a^2 + b^2 = c^2$. **6.** The longest side is 34, so the student should have tested $16^2 + 30^2 \stackrel{?}{=} 34^2$.
Exercises **7.** 10 **9.** 34 **11.** 97 **13.** no; $4^2 + 5^2 \neq 6^2$ **15.** yes; $15^2 + 20^2 = 25^2$ **17.** $\sqrt{33}$ **19.** $\sqrt{105}$ **21.** $5\sqrt{3}$ **23.** 17 m **25.** no; $8^2 + 24^2 \neq 25^2$ **27.** acute **29.** acute **31.** right **33.** 4.2 in. **35a.** $|x_2 - x_1|$; $|y_2 - y_1|$ **b.** $PQ^2 = (x_2 - x_1)^2 + (y_2 - y_1)^2$ **c.** $PQ = \sqrt{(x_2 - x_1)^2 + (y_2 - y_1)^2}$ **37.** $8\sqrt{5}$ **39.** 29 **41.** 84 **43–48.** Answers may vary. Samples are given. **43a.** 6 **b.** 7 **45a.** 8 **b.** 11 **47a.** 8 **b.** 10 **49.** $\frac{q}{b} = \frac{b}{c}$ and $\frac{r}{a} = \frac{a}{c}$ because each leg is the geometric mean of the adj. hypotenuse segment and the hypotenuse. By the Cross Products Property, $b^2 = qc$ and $a^2 = rc$. Then $a^2 + b^2 = qc + rc = c(q + r)$. Substituting c for $q + r$

gives $a^2 + b^2 = c^2$. **51a.** Horiz. lines have slope 0, and vert. lines have undef. slope. Neither could be mult. to get −1. **b.** Assume the lines do not intersect. Then they have the same slope m. Then $m \cdot m = m^2 = -1$, which is impossible. So the lines must intersect. **c.** Let ℓ_1 be $y = \frac{b}{a}x$ and ℓ_2 be $y = -\frac{a}{b}x$. Define $C(a,b)$, $A(0,0)$, and $B(a, -\frac{a^2}{b})$.

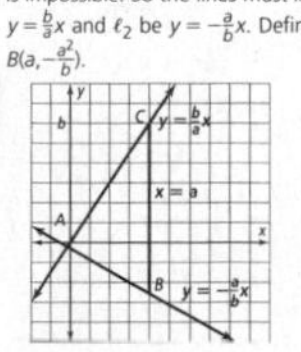

Using the Distance Formula, $AC = \sqrt{a^2 + b^2}$, $BA = \sqrt{a^2 + \frac{a^4}{b^2}}$, and $CB = b + \frac{a^2}{b}$. Then $AC^2 + BA^2 = CB^2$ and $m\angle A = 90$ by the Conv. of the Pythagorean Thm. So $\ell_1 \perp \ell_2$.
53. Draw right $\triangle FDE$ with legs $\overline{DE}$ of length a and $\overline{EF}$ of length b, and hypotenuse of length x. By the Pythagorean Thm., $a^2 + b^2 = x^2$. $\triangle ABC$ has sides of length a, b, and c, where $c^2 > a^2 + b^2 \cdot c^2 > x^2$ and $c > x$ by Prop. of Inequalities. If $c > x$, then $m\angle C > m\angle E$ by the Converse of the Hinge Thm. An angle with measure > 90 is obtuse, so $\triangle ABC$ is an obtuse $\triangle$. **55.** 4 **57.** 61 **59.** 4, 5 **60.** $\sqrt{3}$ **61.** $15\sqrt{2}$ **62.** $\frac{16\sqrt{3}}{3}$

Lesson 8-2 pp. 499–505

Got It? **1.** $5\sqrt{6}$ **2a.** $5\sqrt{2}$ **b.** $\frac{\sqrt{2}}{\sqrt{2}} = 1$, so multiplying by $\frac{\sqrt{2}}{\sqrt{2}}$ is the same as multiplying by 1. **3.** 141 ft **4.** $\frac{10\sqrt{3}}{3}$ **5.** 15.6 mm
Lesson Check **1.** $7\sqrt{2}$ **2.** 3 **3.** $4\sqrt{2}$ **4.** $6\sqrt{3}$ **5.** Rika; 5 should be opposite the 30° $\angle$ and $5\sqrt{3}$ should be opposite the 60° $\angle$. **6.** Answers may vary. Sample: The $\triangle$ is isosc. The length of each leg is the same. Use the Pythagorean Thm. to find the hypotenuse; 6, $6\sqrt{2}$.
Exercises **7.** $x = 8$, $y = 8\sqrt{2}$ **9.** $60\sqrt{2}$ **11.** $5\sqrt{2}$ **13.** 14.1 cm **15.** $x = 20$, $y = 20\sqrt{3}$ **17.** $x = 5$, $y = 5\sqrt{3}$ **19.** $x = 4$, $y = 2$ **21.** 50 ft **23.** $a = 7$, $b = 14$, $c = 7$, $d = 7\sqrt{3}$ **25.** $a = 10\sqrt{3}$, $b = 5\sqrt{3}$, $c = 15$, $d = 5$ **27.** $a = 3$, $b = 7$ **29.** 14.4 s
31. Answers may vary. Sample: A ramp up to a door is 12 ft long. The ramp forms a 30° $\angle$ with the ground. How high off the ground is the door? 6 ft
33a. $\sqrt{3}$ units **b.** $2\sqrt{3}$ units **c.** $s\sqrt{3}$ units **35.** I
37. $AC = 6$; $\frac{3}{AC} = \frac{AC}{12}$, $AC^2 = 36$ **38.** $\sqrt{11}$ in. **39.** $4\sqrt{21}$ cm **40.** $\frac{12}{7}$ **41.** $\frac{54}{11}$ **42.** $\frac{15}{2}$ **43.** $\frac{60}{7}$

Lesson 8-3 pp. 507–513

Got It? **1.** $\frac{15}{17}$; $\frac{8}{17}$; $\frac{15}{8}$ **2a.** 13.8 **b.** 1.9 **c.** 3.8 **d.** 44 ft **3a.** 68 **b.** No; you can use any of the three trigonometric ratios as long as you identify the appropriate leg that is opp. or adj. to each acute $\angle$.
Lesson Check **1.** $\frac{8}{10}$ or $\frac{4}{5}$ **2.** $\frac{6}{10}$ or $\frac{3}{5}$ **3.** $\frac{8}{6}$ or $\frac{4}{3}$ **4.** $\frac{6}{10}$ or $\frac{3}{5}$ **5.** $\frac{8}{10}$ or $\frac{4}{5}$ **6.** $\frac{6}{8}$ or $\frac{3}{4}$ **7.** 12.1 **8.** 57.5
9. The word is made up of the first letters of each ratio: $S = \frac{O}{H}$, $C = \frac{A}{H}$, and $T = \frac{O}{A}$. **10.** No; $\sin X = \frac{YZ}{YX}$, $\sin A = \frac{BC}{BA}$, and $\triangle XYZ \sim \triangle ABC$ by AA ~, so $\frac{YZ}{YX} = \frac{BC}{BA}$ because corresp. sides of ~ $\triangle$s are proportional. Therefore, $\sin X = \sin A$.
Exercises **11.** $\frac{7}{25}$; $\frac{24}{25}$; $\frac{7}{24}$ **13.** $\frac{\sqrt{3}}{2}$; $\frac{1}{2}$; $\sqrt{3}$ **15.** 8.3 **17.** 17.0 **19.** 21.4 **21.** 1085 ft **23.** 58 **25.** 59 **27.** 66 **29.** about 17 ft 8 in.
31. $\cos X \cdot \tan X = \frac{\text{adjacent}}{\text{hypotenuse}} \cdot \frac{\text{opposite}}{\text{adjacent}} = \frac{\text{opposite}}{\text{hypotenuse}} = \sin X$
33. $w = 3$, $x \approx 41$ **35.** $w \approx 68.3$, $x \approx 151.6$
37a. They are equal; yes; sine and cosine of compl. $\angle$s are =. **b.** $\angle B$; $\angle A$ **c.** Sample: The cosine is the complement's sine.
39a.

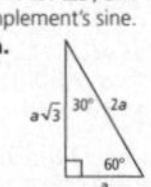

Using the ratio of sides $1 : \sqrt{3} : 2$ for a 30°-60°-90° $\triangle$, $\tan 60° = \frac{\sqrt{3}}{1} = \sqrt{3}$.
b. Answers may vary. Sample: $\sin 60° = \sqrt{3} \cdot \cos 60°$
41. $\frac{15}{9}$ or $\frac{5}{3}$ **43.** $\frac{15}{9}$ or $\frac{5}{3}$ **45.** $\frac{12}{9}$ or $\frac{4}{3}$
47a. No; answers may vary. Sample: $\tan 45° + \tan 30° = 1 + \frac{\sqrt{3}}{3} \approx 1.6$, but $\tan 75° \approx 3.7$.
b. No; assume $\tan A - \tan B = \tan (A - B)$; $\tan A = \tan B + \tan (A - B)$ by the Add. Prop. of =; let $A = B + C$, then $\tan (B + C) = \tan B + \tan C$ by the Subst. Prop.; part (a) proved this false; this contradicts the assumption, so $\tan A - \tan B \neq \tan (A - B)$.
49. $(\sin B)^2 + (\cos B)^2 = \left(\frac{b}{c}\right)^2 + \left(\frac{a}{c}\right)^2 = \frac{b^2}{c^2} + \frac{a^2}{c^2} = \frac{b^2 + a^2}{c^2} = \frac{c^2}{c^2} = 1$
51. $\frac{1}{(\sin A)^2} - \frac{1}{(\tan A)^2} = \frac{1}{\left(\frac{a}{c}\right)^2} - \frac{1}{\left(\frac{a}{b}\right)^2} = \frac{c^2}{a^2} - \frac{b^2}{a^2} = \frac{c^2 - b^2}{a^2} = \frac{a^2}{a^2} = 1$

53a. 1.5 AU **b.** 5.2 AU

Lesson 8-4 pp. 516–521

Got It? **1a.** $\angle$ of elevation from the person in the hot-air balloon to bird **b.** $\angle$ of depression from the person in the hot-air balloon to base of mountain **2.** about 631 ft **3.** about 6.2 km
Lesson Check **1.** $\angle$ of elevation from C to A **2.** $\angle$ of depression from A to C **3.** $\angle$ of elevation from A to D **4.** $\angle$ of elevation from A to B **5.** $\angle$ of depression from B to A **6.** $\angle 1 \cong \angle 2$ (alt. int. $\angle$s); $\angle 4 \cong \angle 5$ (alt. int. $\angle$s) **7.** Answers may vary. Sample: An $\angle$ of elevation is formed by two rays with a common endpoint when one ray is horizontal and the other ray is above the horizontal ray. **8.** Answers may vary. Sample: The $\angle$ labeled in the sketch is the complement of the $\angle$ of depression.
Exercises **9.** $\angle$ of elevation from sub to boat **11.** $\angle$ of elevation from boat to tree **13.** $\angle$ of elevation from Max to top of waterfall **15.** $\angle$ of depression from top of waterfall to Max **17.** 34.2 ft **19.** 986 m **21.** 0.6 km **23.** 64° **25.** 72, 72 **27.** 27, 27 **29a.** length of any guy wire = distance on the ground from the tower to the guy wire div. by the cosine of the $\angle$ formed by the guy wire and the ground **b.** height of attachment = distance on the ground from the tower to the guy wire times the tangent of the $\angle$ formed by the guy wire and the ground **31.** about 2.8 **33.** 3300 m **35.** Answers may vary. **37.** G **39.** 85.2 m **40.** 38.2 ft **41.** 45 **42.** $2\sqrt{17} \approx 8.2$ **43.** $\sqrt{229} \approx 15.1$ **44.** $2\sqrt{37} \approx 12.2$

Lesson 8-5 pp. 522–526

Got It? **1.** 9.5 units **2.** 38.4 **3.** 40.6 ft
Lesson Check **1.** 43.6 **2.** about 18.1 **3.** about 9.1 **4.** No, you need to know at least one angle measure to use the Law of Sines. **5.** $\angle R$ is opposite side PQ, so the proportion should be written as $\frac{\sin 75°}{4} = \frac{\sin P}{3}$.
Exercises **7.** 54.2 **9.** $x \approx 2.1$, $y \approx 3.6$ **11.** $x \approx 12.7$, $y \approx 9.4$ **13.** 259.4 yd **15.** about 97.4 mi **17.** 36.9°

Lesson 8-6 pp. 527–532

Got It? **1.** 61.8 **2.** 76.4° **3.** 2.1 mi
Lesson Check **1.** 11.2 **2.** 81.5 **3.** 26.0 sq units **4.** $m\angle X \approx 86.4$, $m\angle Y \approx 58.8$, $m\angle Z \approx 34.8$ **5.** The variable c should be squared, so the actual value of c is about 7.98. **6.** Use the Law of Cosines to solve for the measure of the angle that is opposite the longest side of the triangle.
Exercises **7.** 10.0 **9.** 3.9 **11.** $x \approx 46.8$, $y \approx 35.0$ **13.** $x \approx 54.1$, $y \approx 72.0$ **15.** 54.7 **17.** 115 ft **19.** Law of Sines; 59.0 **21.** Law of Sines; 17.5 **23.** 18.4 ft **25.** 26.9 **27.** 73.7, 53.1, 53.1

Chapter Review pp. 534–536

1. Trigonometric ratios **2.** $\angle$ of elevation **3.** Pythagorean triple **4.** $2\sqrt{113}$ **5.** 17 **6.** $12\sqrt{2}$ **7.** $9\sqrt{3}$ **8.** $x = 7$, $y = 7\sqrt{2}$ **9.** $5\sqrt{2}$ **10.** $x = 6\sqrt{3}$, $y = 12$ **11.** $x = 7$, $y = 7\sqrt{3}$ **12.** 70.7 ft **13.** $\frac{2\sqrt{19}}{20}$ or $\frac{\sqrt{19}}{10}$; $\frac{18}{20}$ or $\frac{9}{10}$; $\frac{2\sqrt{19}}{18}$ or $\frac{\sqrt{19}}{9}$ **14.** $\frac{16}{20}$ or $\frac{4}{5}$; $\frac{12}{20}$ or $\frac{3}{5}$; $\frac{16}{12}$ or $\frac{4}{3}$ **15.** 16.5 **16.** 33.1 **17.** 38.2 ft **18.** $x = 14.7$ cm **19.** 29.9° **20.** $m\angle D = 82.1$ **21.** $m\angle N = 86.0$

Chapter 9

Get Ready! p. 541

1. $\triangle ADC$ **2.** $\triangle LJK$ **3.** $\triangle RTS$ **4.** $\triangle LHC$ **5.** 108 **6.** 135 **7.** 144 **8.** 160 **9.** always **10.** never **11.** sometimes **12.** always **13.** 50 ft; Because 1 in. = 20 ft, 2.5 in. = 2.5(20 ft) = 50 ft. **14.** 0.25 in.; Because 20 ft = 1 in. and 5 ft = $\frac{1}{4}$(20 ft), by substitution, 5 ft = $\frac{1}{4}$(1 in.) = 0.25 in. **15.** left hand; 4 ft **16.** the point at the center of the clock **17.** Answers may vary. Sample: When you dilate a geometric figure, you change its size.

Lesson 9-1 pp. 545–552

Got It? **1a.** Yes; the distance between the vertices and the angle measures of the image are the same as the preimage. **b.** Yes; the distances between the vertices and the angle measures of the image are the same as the preimage.
2a. $\angle U$; P **b.** $\overline{NI}$ and $\overline{SU}$, $\overline{ID}$ and $\overline{UP}$, $\overline{DN}$ and $\overline{PS}$
3a–b. Graph:

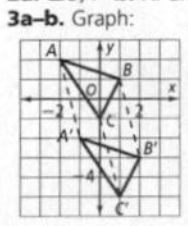

a. $A'(-1, -2)$, $B'(2, -3)$, $C'(1, -5)$ **b.** $AA' = BB' = CC'$ and $\overline{AA'} \parallel \overline{BB'} \parallel \overline{CC'}$; because rigid motions preserve distance and the slope of each segment is −4.
4. $T_{<7, -1>}(\triangle LMN)$
5. 3 squares right and 5 squares down

Lesson Check 1. P'; $\overline{T'J'}$

2.

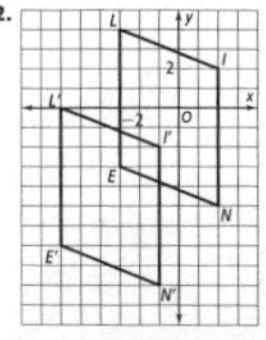

3. $T_{<-12, 4>}(H)$

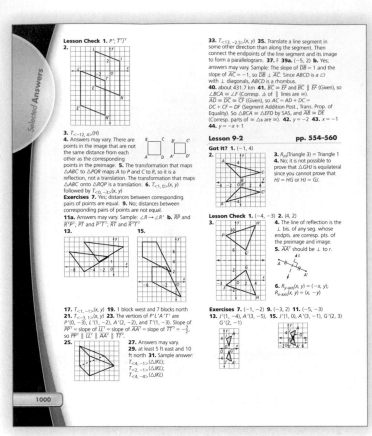

4. Answers may vary. There are points in the image that are not the same distance from each other as the corresponding points in the preimage.

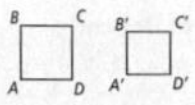

5. The transformation that maps $\triangle ABC$ to $\triangle PQR$ maps A to P and C to R, so it is a reflection, not a translation. The transformation that maps $\triangle ABC$ onto $\triangle RQP$ is a translation. **6.** $T_{<1, 0>}(x, y)$ followed by $T_{<0, -3>}(x, y)$

Exercises 7. Yes; distances between corresponding pairs of points are equal. **9.** No; distances between corresponding pairs of points are not equal.

11a. Answers may vary. Sample: $\angle R \to \angle R'$ **b.** $\overline{RP}$ and $\overline{R'P'}$; $\overline{PT}$ and $\overline{P'T'}$; $\overline{RT}$ and $\overline{R'T'}$

13.

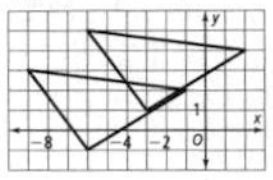

15.

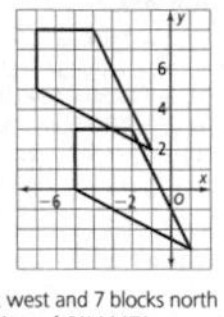

17. $T_{<1, -1>}(x, y)$ **19.** 1 block west and 7 blocks north **21.** $T_{<-3, 1>}(x, y)$ **23.** The vertices of $P'L'A'T'$ are $P'(0, -3)$, $L'(1, -2)$, $A'(2, -2)$, and $T'(1, -3)$. Slope of $\overline{PP'}$ = slope of $\overline{LL'}$ = slope of $\overline{AA'}$ = slope of $\overline{TT'} = -\frac{3}{2}$, so $\overline{PP'} \parallel \overline{LL'} \parallel \overline{AA'} \parallel \overline{TT'}$.

25.

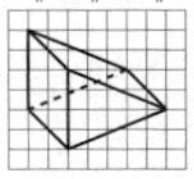

27. Answers may vary.
29. at least 5 ft east and 10 ft north **31.** Sample answer: $T_{<4, -1>}(\triangle JKL)$; $T_{<2, -1>}(\triangle JKL)$; $T_{<4, -4>}(\triangle JKL)$

33. $T_{<13, -2.5>}(x, y)$ **35.** Translate a line segment in some other direction than along the segment. Then connect the endpoints of the line segment and its image to form a parallelogram. **37.** F **39a.** $(-5, 2)$ **b.** Yes; answers may vary. Sample: The slope of $\overline{DB} = 1$ and the slope of $\overline{AC} = -1$, so $\overline{DB} \perp \overline{AC}$. Since $ABCD$ is a ▱ with ⊥ diagonals, $ABCD$ is a rhombus.
40. about 431.7 km **41.** $\overline{BC} \cong \overline{EF}$ and $\overline{BC} \parallel \overline{EF}$ (Given), so $\angle BCA \cong \angle F$ (Corresp. ∠s of ∥ lines are ≅). $\overline{AD} \cong \overline{DC} \cong \overline{CF}$ (Given), so $AC = AD + DC = DC + CF = DF$ (Segment Addition Post., Trans. Prop. of Equality). So $\triangle BCA \cong \triangle EFD$ by SAS, and $\overline{AB} \cong \overline{DE}$ (Corresp. parts of ≅ △s are ≅). **42.** $y = -2$ **43.** $x = -1$
44. $y = -x + 1$

Lesson 9-2 — pp. 554–560

Got It? 1. $(-1, 4)$

2.

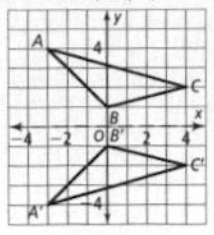

3. R_m(Triangle 3) = Triangle 1
4. No; it is not possible to prove that $\triangle GHJ$ is equilateral since you cannot prove that $HJ = HG$ or $HJ = GJ$.

Lesson Check 1. $(-4, -3)$ **2.** $(4, 2)$

3.

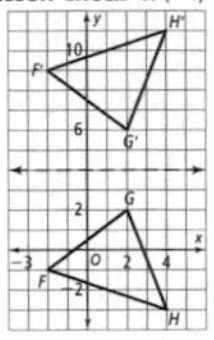

4. The line of reflection is the ⊥ bis. of any seg. whose endpts. are corresp. pts. of the preimage and image.
5. $\overline{AA'}$ should be ⊥ to r.

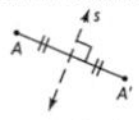

6. $R_{y\text{-axis}}(x, y) = (-x, y)$; $R_{x\text{-axis}}(x, y) = (x, -y)$

Exercises 7. $(-1, -2)$ **9.** $(-3, 2)$ **11.** $(-5, -3)$
13. $J'(1, -4)$, $A'(3, -5)$, $G'(2, -1)$ **15.** $J'(1, 0)$, $A'(3, -1)$, $G'(2, 3)$

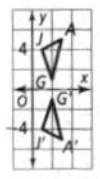

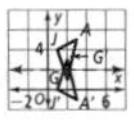

17. $J'(-3, 4)$, $A'(-5, 5)$, $G'(-4, 1)$

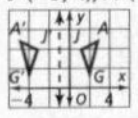

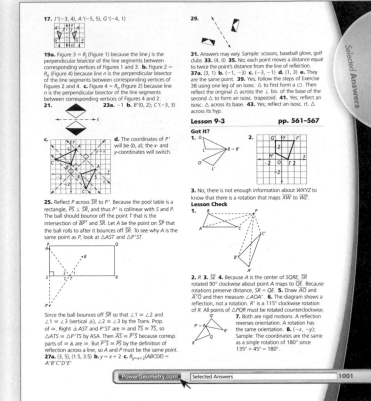

19a. Figure 3 = R_j (Figure 1) because the line j is the perpendicular bisector of the line segments between corresponding vertices of Figures 1 and 3. **b.** Figure 2 = R_n (Figure 4) because line n is the perpendicular bisector of the line segments between corresponding vertices of Figures 2 and 4. **c.** Figure 4 = R_n (Figure 2) because line n is the perpendicular bisector of the line segments between corresponding vertices of Figures 4 and 2.

21.

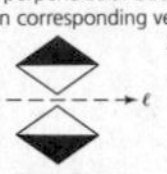

23a. -1 **b.** $B'(0, 2)$; $C'(-3, 3)$

c.

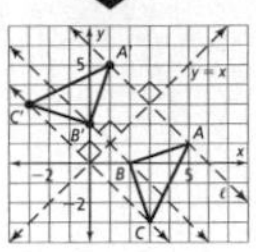

d. The coordinates of P' will be (b, a); the x- and y-coordinates will switch.

25. Reflect P across $\overline{SR}$ to P'. Because the pool table is a rectangle, $\overline{PS} \perp \overline{SR}$, and thus P' is collinear with S and P. The ball should bounce off the point T that is the intersection of $\overline{BP'}$ and $\overline{SR}$. Let A be the point on $\overline{SP}$ that the ball rolls to after it bounces off $\overline{SR}$. To see why A is the same point as P, look at $\triangle AST$ and $\triangle P'ST$.

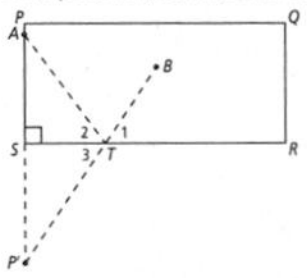

Since the ball bounces off $\overline{SR}$ so that $\angle 1 \cong \angle 2$ and $\angle 1 \cong \angle 3$ (vertical ∠s), $\angle 2 \cong \angle 3$ by the Trans. Prop. of ≅. Right △s AST and $P'ST$ are ≅ and $\overline{TS} \cong \overline{TS}$, so $\triangle ATS \cong \triangle P'TS$ by ASA. Then $\overline{AS} \cong \overline{P'S}$ because corresp. parts of ≅ △s are ≅. But $\overline{P'S} \cong \overline{PS}$ by the definition of reflection across a line, so A and P must be the same point.
27a. (3, 5), (1.5, 3.5) **b.** $y = x + 2$ **c.** $R_{y=x+2}(ABCDE) = A'B'C'D'E'$

29.

31. Answers may vary. Sample: scissors, baseball glove, golf clubs **33.** $(4, 0)$ **35.** No; each point moves a distance equal to twice the point's distance from the line of reflection.
37a. $(3, 1)$ **b.** $(-1, -3)$ **c.** $(-3, -1)$ **d.** $(1, 3)$ **e.** They are the same point. **39.** Yes; follow the steps of Exercise 38 using one leg of an isosc. △ to first form a ▱. Then reflect the original △ across the ⊥ bis. of the base of the second △ to form an isosc. trapezoid. **41.** Yes; reflect an isosc. △ across its base. **43.** Yes; reflect an isosc. rt. △ across its hyp.

Lesson 9-3 — pp. 561–567

Got It?

1.

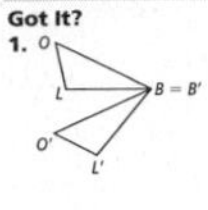

2.

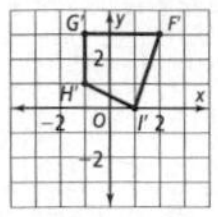

3. No; there is not enough information about $WXYZ$ to know that there is a rotation that maps $\overline{XW}$ to $\overline{WZ}$.

Lesson Check

1.

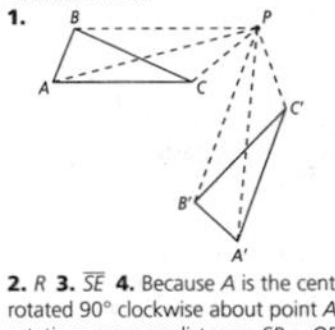

2. R **3.** $\overline{SE}$ **4.** Because A is the center of $SQRE$, $\overline{SR}$ rotated 90° clockwise about point A maps to $\overline{QE}$. Because rotations preserve distance, $SR = QE$. **5.** Draw $\overline{AO}$ and $\overline{A'O}$ and then measure $\angle AOA'$. **6.** The diagram shows a reflection, not a rotation. R' is a 115° clockwise rotation of R. All points of $\triangle PQR$ must be rotated counterclockwise.

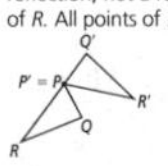

7. Both are rigid motions. A reflection reverses orientation. A rotation has the same orientation. **8.** $(-x, -y)$; Sample: The coordinates are the same as a single rotation of 180° since $135° + 45° = 180°$.

Exercises

9.

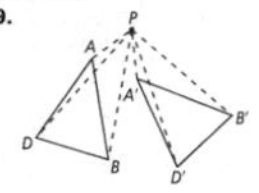

11.

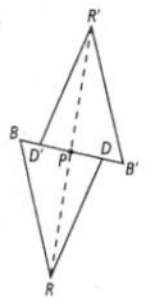

13.

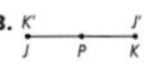

15.

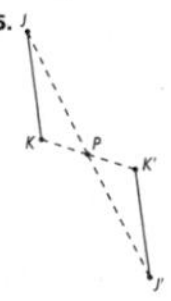

17.

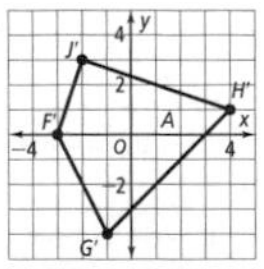

19.

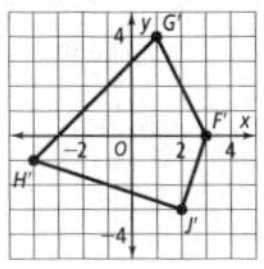

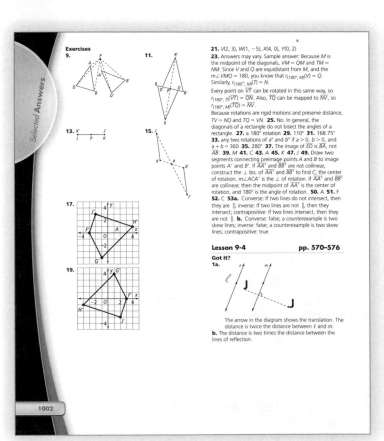

21. $V(2, 3)$, $W(1, -5)$, $X(4, 0)$, $Y(0, 2)$
23. Answers may vary. Sample answer: Because M is the midpoint of the diagonals, $VM = QM$ and $TM = NM$. Since V and Q are equidistant from M, and the $m\angle VMQ = 180$, you know that $r_{(180°, M)}(V) = Q$. Similarly, $r_{(180°, M)}(T) = N$.
Every point on $\overline{VT}$ can be rotated in this same way, so $r_{(180°, T)}(\overline{VT}) = \overline{QN}$. Also, $\overline{TQ}$ can be mapped to $\overline{NV}$, so $r_{(180°, M)}(\overline{TQ}) = \overline{NV}$.
Because rotations are rigid motions and preserve distance, $TV = NQ$ and $TQ = VN$. **25.** No. In general, the diagonals of a rectangle do not bisect the angles of a rectangle. **27.** a 180° rotation **29.** 110° **31.** 168.75° **33.** any two rotations of $a°$ and $b°$ if $a > 0$, $b > 0$, and $a + b = 360$ **35.** 280° **37.** The image of $\overline{ED}$ is $\overline{BA}$, not $\overline{AB}$. **39.** M **41.** C **43.** A **45.** K **47.** J **49.** Draw two segments connecting preimage points A and B to image points A' and B'. If $\overline{AA'}$ and $\overline{BB'}$ are not collinear, construct the ⊥ bis. of $\overline{AA'}$ and $\overline{BB'}$ to find C, the center of rotation. $m\angle ACA'$ is the ∠ of rotation. If $\overline{AA'}$ and $\overline{BB'}$ are collinear, then the midpoint of $\overline{AA'}$ is the center of rotation, and 180° is the angle of rotation. **50.** A **51.** F **52.** C **53a.** Converse: If two lines do not intersect, then they are ∥; inverse: If two lines are not ∥, then they intersect; contrapositive: If two lines intersect, then they are not ∥. **b.** Converse: false; a counterexample is two skew lines; inverse: false; a counterexample is two skew lines; contrapositive: true.

Lesson 9-4 — pp. 570–576

Got It?

1a.

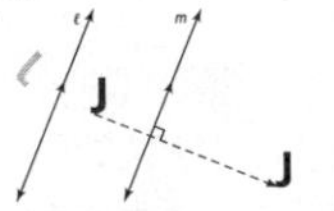

The arrow in the diagram shows the translation. The distance is twice the distance between ℓ and m.
b. The distance is two times the distance between the lines of reflection.

2a.

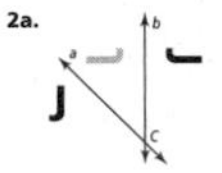

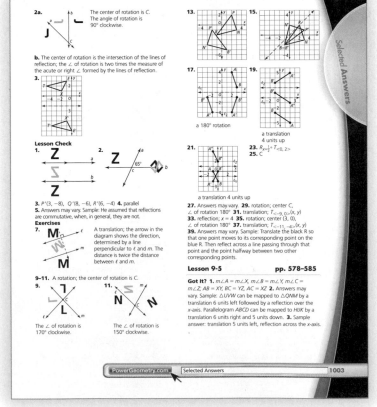

The center of rotation is C. The angle of rotation is 90° clockwise.

b. The center of rotation is the intersection of the lines of reflection; the ∠ of rotation is two times the measure of the acute or right ∠ formed by the lines of reflection.

3.

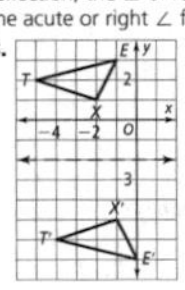

Lesson Check

1.

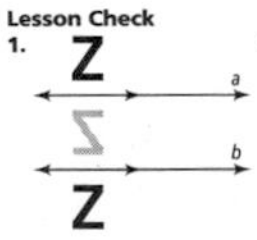

2.

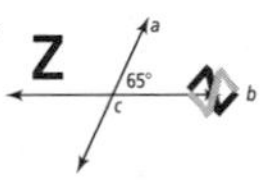

3. $P'(3, -8)$, $Q'(8, -6)$, $R'(6, -4)$ **4.** parallel
5. Answers may vary. Sample: He assumed that reflections are commutative, when, in general, they are not.

Exercises

7.

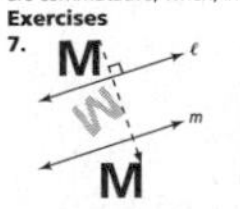

A translation; the arrow in the diagram shows the direction, determined by a line perpendicular to ℓ and m. The distance is twice the distance between ℓ and m.

9–11. A rotation; the center of rotation is C.

9.

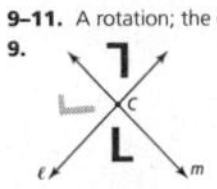

The ∠ of rotation is 170° clockwise.

11.

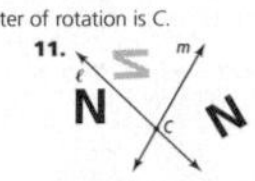

The ∠ of rotation is 150° clockwise.

13.

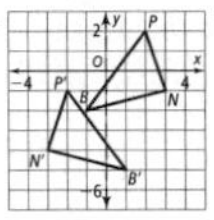

15.

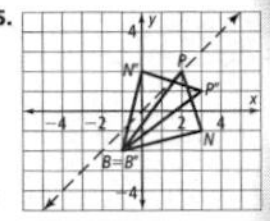

17.

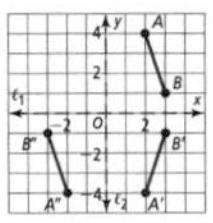

a 180° rotation

19.

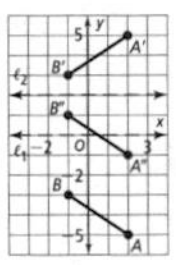

a translation 4 units up

21.

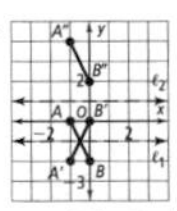

a translation 4 units up

23. $R_{x=\frac{1}{2}} \circ T_{<0, 2>}$
25. C

27. Answers may vary. **29.** rotation; center C, ∠ of rotation 180° **31.** translation; $T_{<-9, 0>}(x, y)$
33. reflection; $x = 4$ **35.** rotation; center (3, 0), ∠ of rotation 180° **37.** translation; $T_{<-11, -4>}(x, y)$
39. Answers may vary. Sample: Translate the black R so that one point moves to its corresponding point on the blue R. Then reflect across a line passing through that point and the point halfway between two other corresponding points.

Lesson 9-5 — pp. 578–585

Got It? 1. $m\angle A = m\angle X$, $m\angle B = m\angle Y$, $m\angle C = m\angle Z$; $AB = XY$, $BC = YZ$, $AC = XZ$ **2.** Answers may vary. Sample: $\triangle UVW$ can be mapped to $\triangle QNM$ by a translation 6 units left followed by a reflection over the x-axis. Parallelogram $ABCD$ can be mapped to $HIJK$ by a translation 6 units right and 5 units down. **3.** Sample answer: translation 5 units left, reflection across the x-axis.

4. Answers may vary. Sample: Translate $\triangle YDT$ so that point D and point N coincide. Since $\overline{TD} \cong \overline{EN}$, you can rotate $\triangle YDT$ so that $\overline{TD}$ and $\overline{EN}$ coincide. Since rotations preserve angle measure and distance, the other two pairs of sides will also coincide. Therefore, this composition of a translation followed by a rotation maps $\triangle YDT$ to $\triangle SNE$, and $\triangle YDT \cong \triangle SNE$. **5.** No, there is no congruence transformation from one figure onto the other.
Lesson Check 1. $\triangle RAV \cong \triangle QSI$ **2.** Sample answer: Translate triangle RAV 5 units right and 1 unit down; then reflect across the x-axis. **3.** Sample answer: Using transformations, you can define congruence of figures other than polygons. **4.** Yes, because a rotation is a rigid motion and a glide reflection is a composition of a translation and reflection, so a rotation followed by a glide reflection is a congruence transformation. **5.** Sample answer: The game of chess requires that the chess pieces move on the board by using congruence transformations.
Exercises 7. $\overline{GC} \cong \overline{FD}$; Sample answer: Reflect segment GC over the y-axis; then translate 1 unit right and 2 units down. **9.** Sample answer: Rotate $\triangle LMN$ 180° about the origin. **11.** Sample answer: Rotate $\triangle LMN$ 180° about the origin. **13.** Answers may vary. Sample answer: Translate $\triangle IQC$ so that points Q and Z coincide. Rotate $\triangle IQC$ so that sides $\overline{QC}$ and $\overline{NZ}$ coincide. Reflect $\triangle IQC$ across side $\overline{NZ}$ so that the triangles overlap. $\triangle IQC \cong \triangle VZN$. **15.** no, there is no congruence transformation **17.** translation **19.** reflection
21. Answers may vary. Sample: Translate the top triangle down 6 units; reflect across the x-axis; rotate the bottom triangle 180° about the point $(-3, 0)$, then reflect across the line $x = -3$; reflect the bottom triangle across the line $x = -4$, then rotate 180° about the point $(-4, 0)$.
23a. rotations and glide reflections **b.** translations and glide reflections **25a.** Congruence transformations preserve distances and angle measures. **b.** Use SAS, proven in Problem 4. **27.** Sample answer: Draw and label the midpoint of $\overline{GH}$ as point M. Draw $\overline{FM}$. Using the identity mapping, $\overline{FM} \cong \overline{FM}$. It is given that $\overline{FG} \cong \overline{FH}$, and $\overline{GM} \cong \overline{MH}$ by the definition of midpoint. Therefore, if triangle FHM is reflected across $\overline{FM}$, it will overlap triangle FGM. Because there is an isometry mapping triangle FHM onto triangle FGM, $\triangle FHM \cong \triangle FGM$. Therefore, $\angle G \cong \angle H$ because corresp. parts of $\cong$ ▲ are $\cong$. **29.** 14.14 **31.** 6.4 **33.** $A'(-3, 3)$, $B'(-4, -2)$, $C'(-4, 4)$ **34.** greater than 10 in. and less than 42 in. **35.** greater than 1 ft and less than 40 ft **36.** greater than 0 m and less than 18 m **37.** greater than 3.5 yd and less than 12.5 yd **38.** 3 in. by 4 in. **39.** 2 in. by 2.5 in. **40.** 1.5 in. by 2.25 in.

Lesson 9-6 pp. 587–593

Got It? 1. reduction; $n = \frac{1}{2}$ **2a.** $P'(-0.5, 0)$, $Z'(-1, 0.5)$, $G'(0, -1)$ **b.** Answers may vary. Sample: Use the Distance Formula to find the lengths of the sides of $\triangle P'Z'G'$ and $\triangle PZG$. Then show that the corresp. sides are proportional, so the ▲ are ~ by SSS ~ Thm. **3.** 5.1 cm
Lesson Check 1. enlargement; 1.5 **2.** $D'(2, -10)$ **3.** $T'(0, 2)$ **4.** $M'(0, 0)$ **5.** a number between 0 and 1 **6a.** The student used 6, instead of $2 + 6 = 8$, as the preimage length in the denominator; the correct scale factor is $n = \frac{2}{2+6} = \frac{1}{4}$. **b.** The student did not write the scale factor with the image length in the numerator; the correct scale factor is $n = \frac{1}{4}$.
Exercises 7. enlargement; $\frac{3}{2}$ **9.** enlargement; $\frac{3}{2}$ **11.** reduction; $\frac{1}{3}$ **13.** reduction; $\frac{1}{2}$ **15.** enlargement; $\frac{3}{2}$ **17.** $P'(-50, 10)$, $Q'(-30, 30)$, $R'(10, -30)$

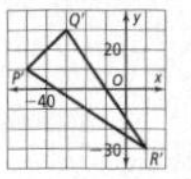

19. 1.2 cm **21.** 0.2 cm **23.** $L'(-15, 0)$ **25.** $A'(-9, 3)$ **27.** $B'\left(\frac{1}{8}, -\frac{3}{20}\right)$ **29.** $Q'\left(-\frac{3}{4}, 1\right)$, $R'\left(-\frac{1}{2}, -\frac{1}{4}\right)$, $T'\left(\frac{3}{4}, \frac{1}{4}\right)$, $W'\left(\frac{3}{4}, \frac{5}{4}\right)$ **31.** $Q'(-2.7, 3.6)$, $R'(-1.8, -0.9)$, $T'(2.7, 0.9)$, $W'(2.7, 4.5)$ **33.** $Q'(-300, 400)$, $R'(-200, -100)$, $T'(300, 100)$, $W'(300, 500)$ **35.** $x = 3$, $y = 60$; the image of a dilation is similar to the preimage, so $\triangle L'N'M' \sim \triangle LNM$. The ratio of the corresp. sides is the same as the scale factor of the dilation, which is 4 : 2, or 2 : 1. To find x, solve the proportion $\frac{x+3}{x} = \frac{2}{1}$. $y = 60$ because corresponding angles of ~ figures are ≅.
37.

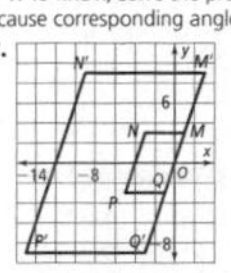

41. $HI = 32$ ft; $I'J' = 7.5$ ft
43. a. Since $\overleftrightarrow{AB}$ does not pass through the origin, it has x- and y-intercepts $(a, 0)$ and $(0, b)$ for some a, $b \neq 0$. So $D_k(a, 0) = (ka, 0)$ and $D_k(0, b) = (0, kb)$. So $\overleftrightarrow{A'B'}$ has different x- and y- intercepts than $\overleftrightarrow{AB}$ and thus $\overleftrightarrow{AB} \neq \overleftrightarrow{A'B'}$.
b. The slope of $\overleftrightarrow{AB}$ is $\frac{b_2 - a_2}{b_1 - a_1}$. Since $A' = (ka_1, ka_2)$ and $B' = (kb_1, kb_2)$, the slope of $\overleftrightarrow{A'B'}$ is $\frac{kb_2 - ka_2}{kb_1 - ka_1} = \frac{b_2 - a_2}{b_1 - a_1}$.
c. If $a_1 = b_1$ then $\overleftrightarrow{AB}$ is vertical. Also, $a_1 = b_1$ implies $ka_1 = kb_1$, so $\overleftrightarrow{A'B'}$ is also vertical.
45. False; a dilation does not map a segment to a ≅ segment unless the scale factor is 1.

47. True; the image and preimage are ~, so the corresp. ∠ are ≅.
49.

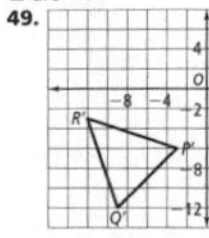

51. 1 ft **53.** G **55a.** no **b.** yes; 1
56. $J'(25, 1)$, $K'(6, 2)$, $L'(3, -20)$ **57.** 4 **58.** 7.5 **59.** 40

Lesson 9-7 pp. 594–600

Got It? 1. $L''(0, 2)$, $M''(0.5, -0.5)$, $N''(1.5, 1.5)$ **2.** $D_{0.5} \circ R_{y\text{-axis}}$ **3.** No, there is no similarity transformation that maps one triangle to the other. The side lengths are not all proportional. **4.** Yes, there is a similarity transformation: rotation, translation, and then dilation.
Lesson Check 1. Answers may vary. Sample: $D_{1.5} \circ R_{y\text{-axis}}$ **2.** $R''(\frac{3}{4}, -\frac{1}{2})$, $S''(\frac{1}{4}, -\frac{1}{4})$, $T''(\frac{1}{2}, -1)$
3. Sample answer: The pupils of your eyes dilate when you go from dark to bright locations or from bright to dark. The pupils are reduced or enlarged proportionally to form similar pupils. **4.** Answers may vary.
Exercises
5.

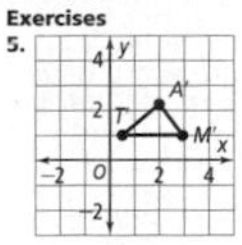

7.

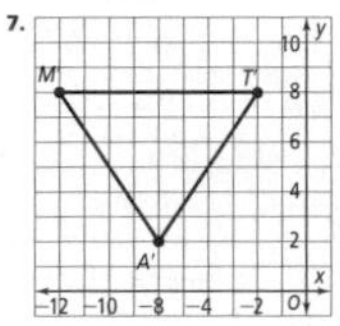

9. $D_{1.5} \circ r_{(180°, O)}$ **11.** Sample answer: $\triangle AVS$ is similar to $\triangle RGI$. Translate $\triangle RGI$ so that points R and A coincide. Rotate by 180°. Then dilate with center A and scale factor 1.5. **13.** Sample answer: The similarity transformation is a rotation about point A followed by a dilation with respect to point A. $\triangle SCA$ is similar to $\triangle ELA$. **15.** Sample answer: Yes, there is a similarity transformation between the two figures: a translation and a rotation followed by a dilation. **17.** Yes, the triangles are similar because there is a similarity transformation that maps $\triangle AJL$ to $\triangle ABC$.
19. sometimes **21.** always **23a.** Yes, the triangles are similar because there is a similarity transformation between them: rotation followed by a dilation. **b.** 330 m
25. Sample answer: Yes, a rigid motion is a similarity transformation with a scale factor of 1. The preimage and image of a rigid motion are congruent, so they are also similar. **27.** Sample answer: No, there is not a similarity transformation. To create $\triangle NOP$, $\triangle ABC$ is reflected across the x-axis. Then the x-coordinates are scaled by a factor of 5 and the y-coordinates are scaled by a factor of 4. Because the reflected triangle is 5 times as wide as $\triangle ABC$ but only 4 times as tall, the figures are not similar. Therefore, there is no similarity transformation between them. **29a.** true **b.** true **c.** false **31.** 6 **33.** 7.5
34. H, 180°; I, 180°; N, 180°; O, any rotation; S, 180°; X, 180°; Z, 180° **35.** (–2, –2) or (7, 1) **36.** 25 cm^2
37. 28 in.^{2} **38.** 11.5 m^2 **39.** 1.5 ft^2

Chapter Review pp. 602–606

1. transformation **2.** similarity transformation
3. translation **4.** image **5a.** No; the distances between corresponding points in the image and preimage are not the same. **b.** $\overline{LA}$, W
6. $R'(-4, 3)$, $S'(-6, 6)$, $T'(-10, 8)$

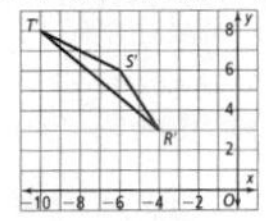

7. $T_{<-5, 10>}(x, y)$
9. $A'(6, -4)$, $B'(-2, -1)$, $C'(5, 0)$

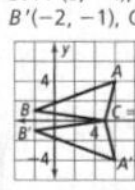

8. $T_{<-2, 7>}(x, y)$
10. $A'(2, 4)$, $B'(10, 1)$, $C'(3, 0)$

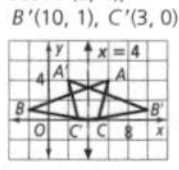

11. $A'(4, 6)$, $B'(1, -2)$, $C'(0, 5)$

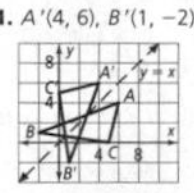

12.

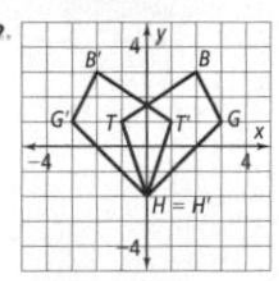

13.

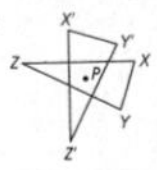

14. $P'(4, -1)$

15. $W'(-1, -3)$, $X'(2, -5)$, $Y'(8, 0)$, and $Z'(-1, -2)$
16.

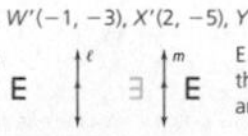

E is translated right, twice the distance between ℓ and m.

17. same; rotation **18.** same; translation **19.** opposite; glide reflection **20.** $\triangle T'A'M'$ with vertices $T'(-4, -9)$, $A'(0, -5)$, $M'(-1, -10)$ **21.** $(r_{(90°, O)} \circ R_{x\text{-axis}})(\triangle XYZ)$
22. Answers may vary. Sample: Yes, the letters are congruent. The p can be mapped to the d with a composition of a translation followed by a rotation.
23. enlargement; 2 **24.** $M'(-15, 20)$, $A'(-30, -5)$, $T'(0, 0)$, $H'(15, 10)$
25.

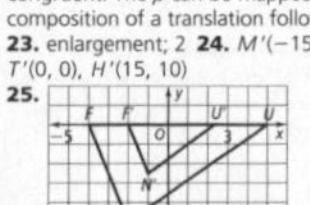

26. $L'N' = 6.5$ ft, $M'N' = 11.25$ ft
27.

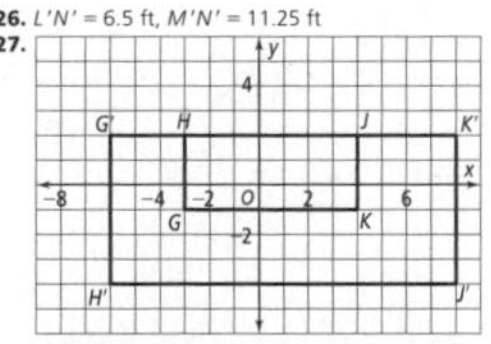

28. No. The side lengths are not proportional. **29.** No, because all of the dimensions of the airplane must dilate by the same scale factor for the figures to be similar.
30. Answers may vary. Sample answer: The figures are similar because a composition of a translation, rotation, and a dilation maps p to d.

Chapter 10

Get Ready! p. 611

1. 9 **2.** 64 **3.** 144 **4.** 225 **5.** 4 **6.** 8 **7.** 10 **8.** 13 **9.** ±8 **10.** ±15 **11.** ±12 **12.** $2\sqrt{2}$ **13.** $3\sqrt{3}$ **14.** $5\sqrt{3}$ **15.** $24\sqrt{2}$ **16.** $\frac{2}{7}$ **17.** 5 bags **18.** rhombus
19. parallelogram **20.** rhombus **21–23.** Answers may vary. Samples are given. **21.** half of a circle **22.** more than half a circle **23.** arcs that are next to each other

Lesson 10-1 pp. 616–622

Got It? 1. 108 m^2 **2.** 7.5 cm **3.** 30 in.^{2} or $\frac{5}{24}$ ft^2
4. The area is doubled.
Lesson Check 1. 200 m^2 **2.** 64 ft^2 **3.** 96 cm^2
4. 36 in.^{2} **5.** No; two altitudes of an obtuse △ lie outside the △. The legs of a right △ are two altitudes of the △.
6. Answers may vary. Sample: You can cut and paste a section of the ▱ to make a rectangle that is ≅ to the given rectangle. **7.** The area of $\triangle ABC$ is half the area of the ▱.
Exercises 9. 26.79 in.^{2} **11.** 11.2 units **13.** $16\frac{8}{13}$ units **15.** 13.5 yd^2 **17a.** 1390 ft^2 **b.** Find the entire area and subtract the areas for flowers.
c. $(50)(31) - 2\left[\frac{1}{2}(10)(16)\right] = 1550 - 160 = 1390$ ft^2
19. B **21.** 18 in.; 12 in. **25.** 6 units^2 **27.** 12 units^2
29. 3 units^2 **31.** The area is tripled; explanations may vary. Sample: If $A = \frac{1}{2}b \cdot h$, then $\frac{1}{2}(b \cdot 3h) = 3 \cdot \frac{1}{2}(b \cdot h) = 3A$.

33a.

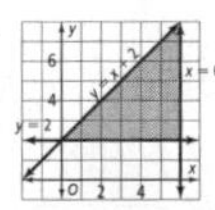

b. 18 units^2

35a.

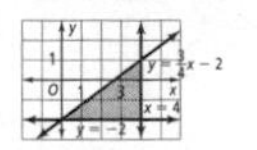

b. 6 units^2

37. 60 units^2 **39.** 20 units^2 **41.** 312.5 ft^2
43. 12,800 m^2 **45.** 126 m^2 **47.** B
49. No; the sum of the two shorter legs is 6 + 4. By the △ Inequality Thm., that sum must be greater than the length of the third side of the △. Since $6 + 4 < 11$, a △ with sides 6, 4, and 11 is not possible.

Lesson 10-2 pp. 623–628

Got It? 1. 94.5 cm^2 **2.** 12 m^2 **3.** 54 in.^{2} **4.** 96 cm^2
Lesson Check 1. 42 m^2 **2.** 378 in.^{2} **3.** 30 ft^2
4. 288 in.^{2} **5.** 300 m^2 **6.** 8 cm^2 **7.** No; in the formula for the area of a trapezoid, half the sum of the bases would have to equal the length of the base of the parallelogram in order for the areas to be the same. This is not possible since the other base of the trapezoid will be longer or shorter than the given base. **8.** No; if you know the height, then you need only the lengths of the bases, but not the legs, to find the area. **9.** No; unless the rhombus is a square, you cannot calculate the area without knowing the lengths of the diagonals. **10.** No; you can calculate the area of a kite from the lengths of the diagonals, without knowing the lengths of the sides.
Exercises 11. 472 in.^{2} **13.** 108 ft^2 **15.** $\frac{5}{6}$ ft^2
17. 30 ft^2 **19.** 72 m^2 **21.** 18 m^2 **23.** 1200 ft^2
25. 24 m^2 **27.** about 35.4 cm^2 **29.** 11.3 cm^2
31. 1.8 m^2 **33.** 15 units^2 **35.** C **37.** 18 cm^2
39. $\frac{128\sqrt{3}}{3}$ in.^{2} **41a.** $A = \frac{1}{2}b_1h$, $A = \frac{1}{2}b_2h$ **b.** Add the areas of the ▲ to get the area of the trapezoid: Area of trapezoid $= \frac{1}{2}b_1h + \frac{1}{2}b_2h = \frac{1}{2}h(b_1 + b_2)$.
43. 1.5m^2 **45.** A

47.

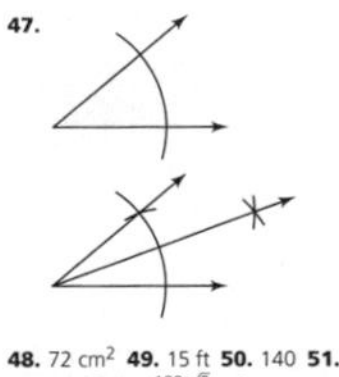

48. 72 cm^2 **49.** 15 ft **50.** 140 **51.** $25\sqrt{3}$ cm^2
52. 50 ft^2 **53.** $\frac{100\sqrt{3}}{3}$ m^2

Lesson 10-3 pp. 629–634

Got It? 1. $m\angle 1 = 45$, $m\angle 2 = 22.5$, $m\angle 3 = 67.5$
2a. 232 cm^2 **b.** It is reduced by half; explanations may vary. Sample: The perimeter of the original polygon is $n \cdot s$. If the side is reduced to half its length, the new perimeter is $n \cdot \frac{1}{2}s$, or $\frac{1}{2}ns$. **3.** 665 ft^2
Lesson Check 1. 100.0 in^2 **2.** 23.4 ft^2 **3.** 5.2 m^2
4. 166.3 units^2 **5.** A radius is the distance from the center to a vertex, while the apothem is the perpendicular distance from the center to a side. **6a.** $s = 2a$
b. $s = \frac{2\sqrt{3}}{3}a$ **c.** $s = 2\sqrt{3}a$ **7.** Special ▲ have ∠ of 30°, 60°, 90° or 45°, 45°, 90° and are found in equilateral ▲, squares, and regular hexagons.
Exercises 9. $m\angle 4 = 90$, $m\angle 5 = 45$, $m\angle 6 = 45$
11. 2144.475 cm^2 **13.** 12,080 in.^{2} **15.** 1168.5 m^2
17. 841.8 ft^2 **19.** 93.5 m^2 **21.** 72 cm^2 **23.** $162\sqrt{3}$ m^2
25. $12\sqrt{3}$ in.^{2} **27a.** 45 **b.** 67.5 **29a.** 30 **b.** 75
31. 9.7 ft **33a.** 9.1 in. **b.** 6 in. **c.** 3.7 in. **d.** Answers may vary. Sample: About 4 in.; the length of a side of a pentagon should be between 3.7 in. and 6 in.
35. The apothem is one leg of a rt. △ and the radius is the hypotenuse. **37.** 17.0; 18 **39.** 51.0; 187.1 **41.** The apothem is ⊥ to a side of the pentagon. Two right ▲ are formed with the radii of the pentagon. The ▲ are ≅ by HL. So, the ∠ formed by the apothem and radii are ≅ because corresp. parts of ≅ ▲ are ≅. Therefore, the apothem bisects the vertex ∠. **43a.** (2.8, 2.8)
b. 5.6 units^2 **c.** 45 units^2 **45.** F **47.** $(2, 2\sqrt{3})$ and $(2, -2\sqrt{3})$, or equivalent decimal approximations (2, 3.464) and (2, −3.464); the length of each side of the △ is 4 units. The third vertex must lie on the altitude of the triangle, which is a point on the line $x = 2$ and has x-coordinate 2. Using the Distance Formula, $\sqrt{(2-0)^2 + (y-0)^2} = 4$; $\sqrt{4 + y^2} = 4$; $4 + y^2 = 16$; $y^2 = 12$; $y = \pm\sqrt{12}$; $y = \pm 2\sqrt{3}$

48. $46\ m^2$ **49.** 8 m **50.** $P = 28$ in.; $A = 49\ in.^2$
51. $P = 24$ m; $A = 32\ m^2$ **52.** $P = 24$ cm; $A = 24\ cm^2$

Lesson 10-4 pp. 635–641

Got It? 1a. 7 : 5 **b.** 49 : 25 **2.** $54\ in.^2$ **3a.** \$6.94 **b.** In order for the two plots to be ~, the pairs of corresp. sides must have the same ratio. **4.** $5\sqrt{5} : 3$
Lesson Check 1. 2 : 3; 4 : 9 **2.** 4 : 3; 16 : 9 **3.** $69.3\ ft^2$ **4.** $\sqrt{6} : 4$ **5.** For two ~ figures, the ratio of their areas is the square of the ratio of the perimeters. **6.** $\sqrt{2} : 1$; the ratio of the areas is 2 : 1, so the ratio of the perimeters is the square root of that ratio, which is $\sqrt{2} : 1$. **7.** Answers may vary. Sample: The ratios of perimeters and areas of ~ figures are not = (unless the figures are ≅, in which case each ratio is 1). **8.** The ratio of the areas of two ≅ figures is 1, while the ratio of the areas of two ~ figures is the square of the scale factor.
Exercises 9. 1 : 2; 1 : 4 **11.** 2 : 3; 4 : 9 **13.** $24\ in.^2$ **15.** $59\ ft^2$ **17.** \$384 **19.** 1 : 2; 1 : 2 **21.** 7 : 3; 7 : 3 **23.** 4 : 1; 4 : 1 **25.** 3 : 1; 9 : 1 **27.** 2 : 3; 4 : 9 **29.** 6 : 1; 36 : 1 **31.** While the ratio of lengths is 2 : 1, the ratio of areas is 4 : 1. **33.** $252\ m^2$ **35.** $x = 2\sqrt{2}$ cm, $y = 3\sqrt{2}$ cm **37.** $x = \frac{8\sqrt{3}}{3}$ cm, $y = 4\sqrt{3}$ cm **39.** $x = 8$ cm, $y = 12$ cm **43a.** 8 : 3 **b.** 64 : 9 **45a.** $6\sqrt{3}\ cm^2$ **b.** $54\sqrt{3}\ cm^2$; $13.5\sqrt{3}\ cm^2$; $96\sqrt{3}\ cm^2$ **47a–c.** Answers may vary. Samples are given.
a.

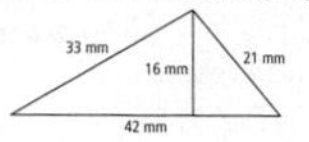

b. 96 mm; $336\ mm^2$ **c.** 457 yd; $7619\ yd^2$ **49.** Sometimes; a 1 unit-by-8 unit rectangle and a 2 unit-by-4 unit rectangle have the same area, but they are not ~. **51.** Sometimes; if they are ≅, they are ~ and have = areas. **53.** $\frac{26}{3}$ **55.** 155 **56.** $50\ cm^2$ **57.** $690\ units^2$ **58.** $480\ units^2$ **59.** $5\frac{1}{3}$ cm, 12 cm **60.** $36\ m^2$ **61.** $4536\ in.^2$ **62.** $168\ ft^2$

Lesson 10-5 pp. 643–648

Got It? 1. $28\ in.^2$ **2a.** $265\ in.^2$ **b.** The area is quadrupled; explanations may vary. Sample: Both the apothem and the side length are doubled if the radius is doubled. **3.** $45\ in.^2$
Lesson Check 1. $41.6\ m^2$ **2.** $277.0\ cm^2$ **3.** $22\ in.^2$ **4.** Yes; the diagonal of a regular hexagon is two times the side, and you have several ways to find the area of a regular hexagon with 6-cm sides. **5.** He set up the wrong ratio. The correct ratio is $\frac{4}{a} = \tan 36°$.
Exercises 7. $123.1\ yd^2$ **9.** $141.7\ in.^2$ **11.** $12.4\ mm^2$ **13.** $2540.5\ cm^2$ **15.** $18.0\ ft^2$ **17.** $311.3\ km^2$ **19.** $0.8\ ft^2$ **21.** Multiply the formula for the area of an equilateral △, $A = \frac{s^2\sqrt{3}}{4}$, by 6 to get $\frac{3s^2\sqrt{3}}{2}$; use a 30°-60°-90° △ to find the height of one equilateral △ with side s, then multiply the area of that △ by 6; or use the tangent ratio to find the apothem and then use the formula $A = \frac{1}{2}ap$. **23.** 20.8 m, $20.8\ m^2$ **25.** 61.2 m, $282.8\ m^2$ **27.** $1{,}459{,}000\ ft^2$ **29.** about $925.8\ cm^2$ **31.** area of Pentagon A ≈ 1.53 • (area of Pentagon B) **33.** area of Octagon B ≈ 1.17 • (area of Octagon A) **35.** $162\sqrt{3}\ ft^2$ or about $280.6\ ft^2$ **37.** about $48.2\ cm^2$ **39.** 320 ft

Lesson 10-6 pp. 649–657

Got It? 1a. $\overset{\frown}{SP}$, $\overset{\frown}{SQ}$, $\overset{\frown}{PQ}$, $\overset{\frown}{QR}$, $\overset{\frown}{RS}$ **b.** $\overset{\frown}{RSP}$, $\overset{\frown}{RQP}$ **c.** $\overset{\frown}{PQS}$, $\overset{\frown}{PSQ}$, $\overset{\frown}{SPR}$, $\overset{\frown}{QRS}$, $\overset{\frown}{RSQ}$ **2a.** 77 **b.** 103 **c.** 208 **d.** 283 **3a.** about 29.5 ft **b.** 2 : 1; if the radius of ⊙A is r, then its circumference is $2\pi r$. ⊙B will have a circumference of πr. The ratio of their circumferences is $\frac{2\pi r}{\pi r} = \frac{2}{1}$, or 2 : 1. **4.** 1.3π m
Lesson Check 1–3. Answers may vary. Samples are given. **1.** $\overset{\frown}{AB}$ **2.** $\overset{\frown}{DAB}$ **3.** $\overset{\frown}{CAB}$ **4.** 81 **5.** 18π cm **6.** $\frac{23\pi}{4}$ cm **7.** The measure of an arc corresponds to the measure of a central angle; an arc length is a fraction of the circle's circumference. **8.** The student substituted the diameter into the formula that requires the radius.
Exercises 9. $\overset{\frown}{BC}$, $\overset{\frown}{BD}$, $\overset{\frown}{CD}$, $\overset{\frown}{CE}$, $\overset{\frown}{DE}$, $\overset{\frown}{DF}$, $\overset{\frown}{EF}$, $\overset{\frown}{FB}$ **11.** $\overset{\frown}{BCE}$, $\overset{\frown}{BFE}$, $\overset{\frown}{CBF}$, $\overset{\frown}{CDF}$ **13.** 180 **15.** 270 **17.** 308 **19.** 90 **21.** 90 **23.** 270 **25.** 6π ft **27.** 14π in. **29.** 19 in. **31.** 8π ft **33.** 33π in. **35.** $\frac{5\pi}{4}$ m **37.** 70 **39.** 110 **41.** 235 **43.** about 183.3 ft **45.** Find the measure of the major arc, then use Thm. 10-10; or find the length of the minor arc using Thm. 10-10, then subtract that length from the circumference of the circle. **47.** 38 **49.** 31 m **51.** 3 : 4 **53.** 2.6π in. **55.** 7.9 units **57.** Since $\overline{AR} \cong \overline{RW}$ and $AR + RW = AW$ by the Seg. Add. Post., $AW = 2 \cdot AR$. So the radius of the outer circle is twice the radius of the inner circle. Because ∠QAR and ∠SAU are vertical ∡, and $m\angle SAT = \frac{1}{2}\, m\angle SAU$, $m\angle QAR = 2 \cdot m\angle SAT$. The length of $\overset{\frown}{ST} = \frac{m\angle SAT}{360} \cdot 2\pi(2 \cdot AR) = \frac{m\angle SAT}{90} \cdot \pi(AR)$ and the length of $\overset{\frown}{QR} = \frac{m\angle QAR}{360} \cdot 2\pi(AR) = \frac{2 \cdot m\angle SAT}{360} \cdot 2\pi(AR) = \frac{m\angle SAT}{90} \cdot \pi(AR)$. Therefore the length of $\overset{\frown}{ST}$ = the length of $\overset{\frown}{QR}$ by the Trans. Prop. of Eq. **59.** 325.7 yd, 333.5 yd, 341.4 yd, 349.2 yd, 357.1 yd, 365.0 yd, 372.8 yd, 380.6 yd **61.** F **63.** Using the Distance Formula, $AB = CD = 3$, $BC = AD = 5$ and $RS = TV = 6$, $ST = RV = 10$. The slopes of $\overline{AB}$ and $\overline{CD} = 0$ and the slopes of $\overline{BC}$ and $\overline{AD}$ are undefined. So both $\overline{AB}$ and $\overline{CD}$ are ⊥ to $\overline{BC}$ and $\overline{AD}$. Therefore, ABCD is a rectangle and ∡ A, B, C, and D are rt. ∡. The slopes of $\overline{RS}$ and $\overline{TV}$ are undefined and the slopes of $\overline{ST}$ and $\overline{RV} = 0$. So, RSTV is a rectangle and ∡ R, S, T, and V are rt. ∡. Since all rt. ∡ are =, the pairs of corresponding ∡ are ≅. The short sides of the two rectangles are 3 and 6, and the long sides are 5 and 10. Since $\frac{3}{6} = \frac{5}{10} = \frac{1}{2}$, the corresp. sides are proportional. Therefore, ABCD ~ RSTV by the def. of ~ polygons. **64.** $m\angle 1 = 30$, $m\angle 2 = 15$, $m\angle 3 = 75$, $m\angle 4 = 30$ **65.** 18.6 mm **66.** Answers may vary slightly. Samples: 120 mm; $1116\ mm^2$ **67.** No; it could be an isosc. trapezoid. **68.** Yes; the diagonals bis. each other, so it is a ▱. **69.** Yes; one pair of sides is both ≅ and ∥, so it is a ▱. **70.** 17π in. or about 53.4 in. **71.** 3π cm or about 9.4 cm

Lesson 10-7 pp. 660–666

Got It? 1a. about $1385\ ft^2$ **b.** The area is $\frac{1}{4}$ the original area; explanations may vary. Sample: half the radius is $\frac{r}{2}$. So, if $A = \pi r^2$, then $\pi\left(\frac{r}{2}\right)^2 = \frac{1}{4}\pi r^2 = \frac{1}{4}A$. **2.** $2\pi\ in.^2$ **3.** $4.6\ m^2$
Lesson Check 1. $64\pi\ in.^2$ **2.** $\frac{135}{8}\pi\ in.^2$, or $16.875\pi\ in.^2$ **3.** $\left(\frac{4}{3}\pi - \sqrt{3}\right) m^2$ **4.** A sector of a circle is a region bounded by an arc and the two radii to the endpoints of the arc. A segment is a part of a circle bounded by an arc and the seg. joining the arc's endpoints. **5.** No; the central ∡ corresponding to the arcs and the radii of the circles may be different. Circles with different radii do not have the same area. **6.** 6^2 was incorrectly evaluated as 6 • 2.
Exercises 7. $9\pi\ m^2$ **9.** $0.7225\pi\ ft^2$ **11.** about $282{,}743\ ft^2$ **13.** $40.5\pi\ yd^2$ **15.** $\frac{169\pi}{6}\ m^2$ **17.** $12\pi\ ft^2$ **19.** $\frac{25\pi}{4}\ m^2$ **21.** $24\pi\ in.^2$ **23.** $22.1\ cm^2$ **25.** $3.3\ m^2$ **27.** $(54\pi + 20.25\sqrt{3})\ cm^2$ **29.** $(4 - \pi)\ ft^2$ **31.** $(784 - 196\pi)\ in.^2$ **33.** $314\ ft^2$ **35.** $116\ mm^2$ **37.** $22.6\ mm^2$ **39.** 12 in. **41a.** Answers may vary. Sample: Subtract the minor arc segment from the area of the circle; or add the areas of the major sector and the △ that is part of the minor arc sector. **b.** $(25\pi - 50)\ units^2$; $(75\pi + 50)\ units^2$ **43.** $4.4\ m^2$ **45.** $\left(\frac{5\pi}{6} - 2 \cdot \sin 75°\right) ft^2$, or $\left[\frac{5\pi}{6} - 4(\sin 37.5°)(\cos 37.5°)\right]$ **47.** $(200 - 50\pi)\ m^2$ **49.** Blue region: Let $AB = 2$. Area of blue $= 4 - \pi$; area of yellow $= \pi - 2$, and $4 - \pi < \pi - 2$.

Lesson 10-8 pp. 668–674

Got It? 1. $\frac{1}{2}$ or 50% **2.** $\frac{1}{5}$ or 20% **3.** $\frac{1}{2}$ or 50% **4a.** 0.04, or 4% **b.** The black zone; the area of the black zone is greater than the area of the red zone, so P(black zone) > P(red zone).
Lesson Check 1. $\frac{3}{7}$ **2.** $\frac{6}{7}$ **3.** $\frac{4}{7}$ **4.** $\frac{3}{7}$ **5.** about 0.09, or 9% **6.** $\frac{2}{3}$; explanations may vary. Sample: Since $\frac{SQ}{QT} = \frac{1}{2}$, you can let $SQ = x$ and $QT = 2x$, where x is not 0. Then $ST = 3x$ and the ratio $\frac{QT}{ST} = \frac{2x}{3x} = \frac{2}{3}$. **7.** The numerator should be (area of square − area of semicircles); the favorable region is the shaded region and its area is the area left when the areas of the semicircles are subtracted from the area of the square.
Exercises 9. $\frac{1}{10}$ **11.** $\frac{2}{5}$ **13.** $\frac{2}{5}$ **15.** $\frac{2}{5}$, or 40% **17.** $\frac{2}{9}$, or about 22% **19.** $\frac{5}{9}$, or about 56% **21.** $\frac{1}{49}$, or about 2% **23.** $\frac{24}{49}$, or about 49% **25.** $\frac{3}{10}$, or 30% **27.** $\frac{3}{20}$, or 15% **29.** $\frac{9}{19}$, or about 47% **31.** $\frac{1}{4}$; $m\overset{\frown}{AB} = 90$, so the length of $\overset{\frown}{AB} = \frac{90}{360} \cdot 2\pi r = \frac{1}{4} \cdot 2\pi r$. The ratio of the length of $\overset{\frown}{AB}$ to the circumference is $\frac{1}{4}$. **33.** $\frac{3}{5}$ **35.** $\frac{1}{10}$ **37.** $\frac{1}{40}$ **39.** $\frac{\pi}{20}$, or about 16% **41.** 36 s **43a.** about 8.7% **b.** about 19.6% **45.** 50% **47.** G **49.** G **51.** $100\pi\ ft^2$ **52.** $12\pi\ cm^2$ **53.** $A'''(-2, -2)$; $B'''(1, -1)$; $C'''(-1, 1)$ **54.** $A'''(-5, 6)$; $B'''(-4, 3)$; $C'''(-2, 5)$ **55.** $A'''(-2, 1)$; $B'''(2, -1)$; $C'''(1, -2)$

56.

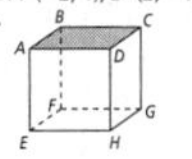

57.

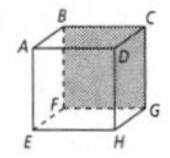

58.

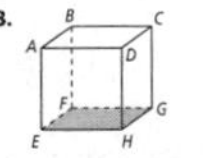

59.

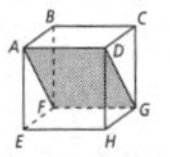

60.

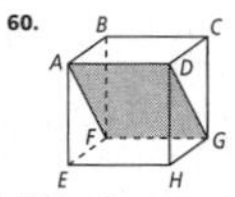

61.

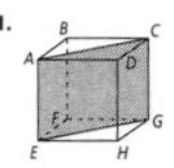

62. Sample:

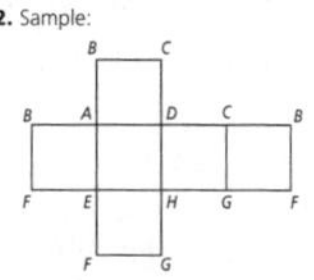

Chapter Review pp. 676–680

1. base **2.** sector **3.** radius **4.** adjacent arcs **5.** $10\ m^2$ **6.** $90\ in.^2$ **7.** $30\ ft^2$ **8.** $160\ ft^2$ **9.** $30\ ft^2$ **10.** $96\sqrt{3}\ mm^2$ **11.** $96\ ft^2$ **12.** $117\ cm^2$ **13.** $256\ ft^2$ **14.** $54\ m^2$ **15.** $9\sqrt{3}\ in.^2$ **16.** $28\ m^2$ **17.** $2400\sqrt{3}\ cm^2$ **18.** $112.5\ m^2$

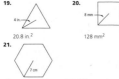

19.

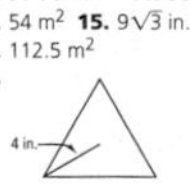

$20.8\ in.^2$

20.

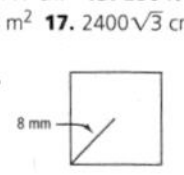

$128\ mm^2$

21.

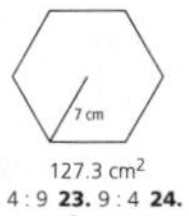

$127.3\ cm^2$

22. 4 : 9 **23.** 9 : 4 **24.** 1 : 4 **25.** 4 : 1 **26.** $2\sqrt{2} : 5$ **27.** $73.5\ ft^2$ **28.** $232.5\ cm^2$ **29.** $124.7\ in.^2$ **30.** $8\ m^2$ **31.** $331.4\ ft^2$ **32.** $24.6\ ft^2$ **33.** $100.8\ cm^2$ **34.** $70.4\ m^2$ **35.** 30 **36.** 120 **37.** 330 **38.** 120 **39.** $\frac{22\pi}{9}$ in. **40.** π mm **41.** $\frac{25\pi}{9}$ m **42.** 4π m **43.** $144\pi\ in.^2$ **44.** $\frac{49\pi}{4}\ ft^2$ **45.** $41.0\ cm^2$ **46.** $18.3\ m^2$ **47.** $36.2\ cm^2$ **48.** $\frac{1}{2}$, or 50% **49.** $\frac{3}{8}$, or 37.5% **50.** $\frac{1}{6}$, or about 16.7% **51.** $\frac{1}{2}$, or 50% **52.** $\frac{1}{2}$, or 50%

Chapter 11

Get Ready! p. 685

1. 17 **2.** $8\sqrt{2}$ **3.** 6 **4.** $4\sqrt{5}$ **5.** $6\sqrt{2}$ **6.** $4\sqrt{3}$ **7.** $44\ units^2$ **8.** $14\sqrt{3}\ units^2$ **9.** $234\ units^2$ **10.** $54\sqrt{3}\ units^2$ **11.** 24 **12.** $2\sqrt{2} : 5$ **13.** the ⊥ segment from one base to a parallel base or a vertex to the base **14.** the sum of the areas of each side (face) of a figure **15.** An Egyptian pyramid has 4 sides that are triangles and a bottom (base) that is a square.

Lesson 11-1 pp. 688–695

Got It? 1a. 6 vertices: R, S, T, U, V, W; 9 edges: $\overline{SR}$, $\overline{ST}$, $\overline{UR}$, $\overline{UV}$, $\overline{UT}$, $\overline{RV}$, $\overline{SW}$, $\overline{VW}$, $\overline{TW}$; 5 faces: △URV, △STW, quadrilateral RSTU, quadrilateral RSWV, quadrilateral TWVU **b.** No; an edge is a segment formed by the intersection of two faces. $\overline{TV}$ is a segment that is contained in only one face, so it is not an edge. **2a.** 12 **b.** 30 **3a.** $6 + 8 = 12 + 2$
b.

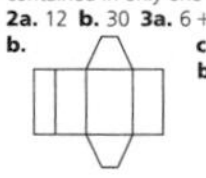

c. $6 + 14 = 19 + 1$ **4a.** a circle **b.** an isosc. trapezoid

5.

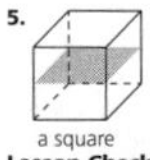

a square

Lesson Check
1. 5 faces: △ABC, △ACD, △ADE, △AEB, quadrilateral BCDE; 8 edges: $\overline{AB}$, $\overline{AC}$, $\overline{AD}$, $\overline{AE}$, $\overline{BC}$, $\overline{CD}$, $\overline{DE}$, $\overline{EB}$; 5 vertices: A, B, C, D, E
2. Sample:

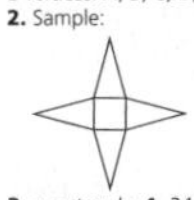

$F + V = 5 + 8$;
$E + 1 = 12 + 1$;
$5 + 8 = 12 + 1$

3. a rectangle **4.** 24 edges: There are 8 edges on each of the two octagonal bases, and there are 8 edges that connect pairs of vertices of the bases. **5.** A cylinder is not a polyhedron because its faces are not polygons.
Exercises
7. 8 vertices: A, B, C, D, E, F, G, H; 12 edges: $\overline{AB}$, $\overline{BC}$, $\overline{CD}$, $\overline{DA}$, $\overline{EF}$, $\overline{FG}$, $\overline{GH}$, $\overline{HE}$, $\overline{AE}$, $\overline{BF}$, $\overline{CG}$, $\overline{DH}$; 6 faces: quadrilaterals ABCD, EFGH, ABFE, BCGF, DCGH, ADHE **9.** 8 **11.** 12 **13.** 5
15. $5 + 6 = 9 + 2$; answers may vary. Sample:

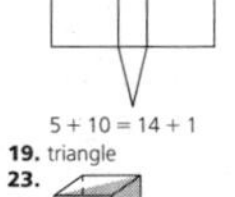

$5 + 10 = 14 + 1$

17. $7 + 7 = 12 + 2$; answers may vary. Sample:

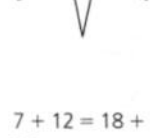

$7 + 12 = 18 + 1$

19. triangle
21.

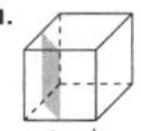

rectangle

23.

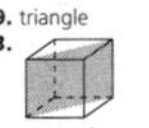

rectangle

25a.

b.

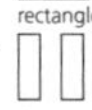

27. No; if $F = V$, then $F + V = 2F$, so $F + V$ is even. So $E \neq 9$ because $E + 2$ must be even.
29.

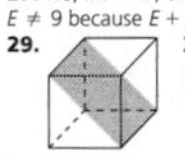

rectangle

31. cone **33.** a cylinder attached to a cone **35.** $4 + 6 = 9 + 1$ **37.** $5 + 5 = 9 + 1$ **39.** 6 in.

41.

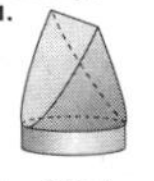

43.

45.

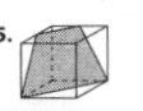

47.

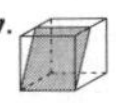

49.

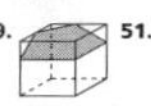

51. B **53.** D

55. (5, 2), (−1, 0), or (3, −4); the fourth vertex lies on a line parallel to an opposite side such that the length of the side is equal to the length of the opposite side.
56.

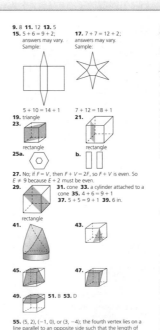

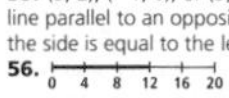

60%

57. 25% **58.** 4.7 **59.** 8.3 **60.** $96\ cm^2$ **61.** $40\pi\ cm^2$ **62.** $9\sqrt{3}\ m^2$

Review p. 698

1. $r = \frac{C}{2\pi}$ **3.** $r = \sqrt{\frac{A}{\pi}}$ **5.** $y = x \tan A$; $x = \frac{y}{\tan A}$ **7.** $C = 2\sqrt{\pi A}$ **9.** $a = \frac{\sqrt{6A\sqrt{3}}}{6}$ or $a = \frac{\sqrt{6A} \cdot \sqrt[4]{3}}{6}$

Lesson 11-2 pp. 699–707

Got It? 1. $216\ cm^2$ **2a.** $432\ m^2$ **b.** $54\sqrt{3}\ m^2$ **c.** $619\ m^2$ **3.** $380\pi\ cm^2$ **4a.** $11.8\ in.^2$ **b.** $\frac{1}{4}$; $\frac{1}{4}$
Lesson Check 1. $130\ in.^2$ **2.** $(133 + 42\sqrt{2})\ ft^2$ or about $192.4\ ft^2$ **3.** $48\pi\ cm^2$ or about $150.8\ cm^2$ **4.** $170\pi\ m^2$ or about $534.1\ m^2$ **5.** lateral faces: BFGC, DCGH, ADHE, EFBA; bases: ABCD, EFGH **6.** The diameter of the circular bases does not match the length of the rectangle. If the diameter is 2 cm, then the length must be 2π cm, or if the length is 4 cm, then the diameter should be $\frac{4}{\pi}$ cm, or about 1.3 cm.
Exercises
7. $1726\ cm^2$

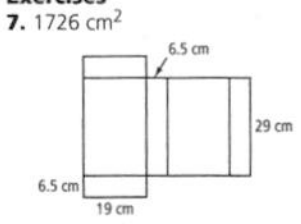

9. $(80 + 32\sqrt{2})\ in.^2$, or about $125.3\ in.^2$

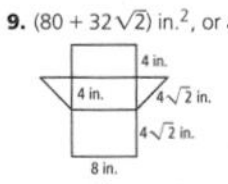

11. $220\ ft^2$ **13.** $1121\ cm^2$ **15.** $170\ m^2$ **17.** $40\pi\ cm^2$ **19.** $101.5\pi\ in.^2$ **21.** 20 cm **23.** $4080\ mm^2$ **25a.** $94\ units^2$ **b.** $376\ units^2$ **c.** 4 : 1 **d.** $438\ units^2$; $1752\ units^2$; 4 : 1 **e.** The surface area is multiplied by 4. **27.** just under $150\ cm^2$ **29.** $110\ in.^2$ **31a.** 7 units **b.** $196\pi\ units^2$ **33.** cylinder of radius 4 and height 2; $48\pi\ units^2$ **35.** cylinder of radius 2 and height 4; $24\pi\ units^2$ **37a.** The lateral area is doubled. **b.** The surface area is more than doubled. **c.** If r doubles, $S.A. = 2\pi(2r)^2 + 2\pi(2r)h = 8\pi r^2 + 4\pi rh = 2(4\pi r^2 + 2\pi rh)$. So the surface area $2\pi r^2 + 2\pi rh$ is more than doubled. **39.** $(182\pi + 232)\ cm^2$ **41.** $(220 - 8\pi)\ in.^2$ **43.** $h = 6$ m; $r = 3$ m **45.** 13 **47.** 68

Lesson 11-3 pp. 708–715

Got It? 1a. 55 m^2 **b.** The L.A. will double. Sample explanation: Since L.A. $= \frac{1}{2}p\ell$, then replacing ℓ with 2ℓ gives $\frac{1}{2}p(2\ell) = 2(\frac{1}{2}p\ell) = 2 \cdot$ L.A. **2a.** 5649 ft^2 **b.** The slant height is the hypotenuse of a rt. △ with a leg of length equal to the height of the pyramid, so the slant height is greater than the height. **3.** 704π m^2 **4a.** 934 $in.^2$ **b.** The L.A. will be halved. Sample explanation: Since L.A. $= \pi r\ell$, then replacing r with $\frac{r}{2}$ gives $\pi\left(\frac{r}{2}\right)\ell = \frac{1}{2}(\pi r\ell) = \frac{1}{2} \cdot$ L.A.
Lesson Check 1. 60 m^2 **2.** 85 m^2 **3.** $2\pi\sqrt{29}$ ft^2, or about 33.8 ft^2 **4.** $(2\pi\sqrt{29} + 4\pi)$ ft^2, or about 46.4 ft^2 **5.** The height is the distance from the vertex to the center of the base, while the slant height is the distance from the vertex to the midpoint of an edge of the base. **6.** Alike: Both are the sum of a lateral area and the areas of the bases. Different: For a prism the area includes two bases, while for a pyramid the surface area includes just one base. **7.** 5; 6; n **8.** The height 7 is not the slant height. The slant height is $\sqrt{7^2 + 3^2} = \sqrt{58}$, so L.A. $= \pi r\ell = \pi(3)(\sqrt{58}) = 3\pi\sqrt{58}$ $units^2$.
Exercises 9. 408 $in.^2$ **11.** 179 $in.^2$ **13.** 354 cm^2 **15.** 834,308 ft^2 **17.** 31 m^2 **19.** 144π cm^2 **21.** 119π **23.** 4 in. **25.** 8 ft **27.** 471 ft^2
29. Answers may vary. Sample:

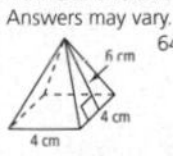

31. Cylinder; the L.A. of 2 cones is 30π $in.^2$, and the L.A. of the cylinder is 48π $in.^2$. **33a.** $\ell = \frac{S.A.}{\pi r} - r$ **b.** $r = \frac{-\pi\ell + \sqrt{\pi^2\ell^2 + 4\pi \cdot S.A.}}{2\pi}$ **35.** $s = 12$ m, L.A. = 240 m^2; S.A. = 384 m^2 **37.** cone with $r = 4$ and $h = 3$; 36π **39.** cylinder with cone-shaped hole; 60π $units^2$ **41.** 129.6 **43.** L.A. $= 25\pi\sqrt{5}$ cm^2; S.A. $= (25\pi\sqrt{5} + 25\pi)$ cm^2 **45.** H **47.** Yes; if the legs of each isosc. △ are two consecutive sides of the ▱, then the ▱ is a rhombus. **49.** 76 ft^2 **50.** 127 $in.^2$ **51.** 26 in. **52.** 4 cm^2 **53.** 176.7 $in.^2$

Lesson 11-4 pp. 717–724

Got It? 1a. 60 ft^3 **b.** No; explanations may vary. Sample: The volume is the product of the three dimensions, and multiplication is commutative. **2a.** 150 m^3 **b.** The volume is doubled. Using $V = B \cdot h$ and replacing h with $2h$ gives $B \cdot (2h) = 2 \cdot B \cdot h = 2 \cdot V$. **3a.** 3π m^3 **b.** The volume is $\frac{1}{4}$ the volume of the cylinder in part (a). Using $V = \pi r^2 h$ and replacing r with $\frac{r}{2}$ gives $\pi\left(\frac{r}{2}\right)^2 h = \frac{1}{4}\pi r^2 h = \frac{1}{4} \cdot V$. **4.** 501 $in.^3$
Lesson Check 1. 54 ft^3 **2.** 339 $in.^3$ **3.** Yes; it is a combination of a cylinder and a cone. **4.** Alike: Both are the product of the base area and the height. Different: For a prism the base is a polygon, while for a cylinder the base is a circle. **5.** The volumes are the same, 24 m^3, because multiplication is commutative.
Exercises 7. 80 $in.^3$ **9.** 14 cm^3 **11.** 22.5 ft^3 **13.** 22.5 $in.^3$ **15.** 40π cm^3; 125.7 cm^3 **17.** π yd^3; 3.1 yd^3 **19.** 144 cm^3 **21.** 1747 lb **23.** 40 cm **25.** 6 ft **27.** 96 ft^3 **29.** Volume is 27 times greater. Using $V = B \cdot h = \ell \cdot w \cdot h$ for a rectangular prism, $(3\ell) \cdot (3w) \cdot (3h) = 27 \cdot \ell \cdot w \cdot h = 27 \cdot V$. **31.** Answers may vary. Sample: If two plane figures have the same height and the same width at every level, then they have the same area. **33.** 80 $units^3$ **35.** bulk; cost of bags = \$1167.50, cost of bulk ≈ \$1164 **37.** 125.7 cm^3 **39.** cylinder with $r = 2$ and $h = 4$; 16π $units^3$ **41.** cylinder with $r = 2$ and $h = 4$; 16π $units^3$ **43a.** $C = 8.5$ in. and $h = 11$ in.: $V \approx 63.2$ $in.^3$; $C = 11$ in. and $h = 8.5$ in.: $V \approx 81.8$ $in.^3$; the cylinder with the greater circumference has the greater volume. **45.** The volume of B is twice the volume of A.

Lesson 11-5 pp. 726–732

Got It? 1. 32,100,000 ft^3 **2.** 960 m^3 **3a.** 77 ft^3 **b.** The volume of the original tepee is 8 times the volume of the child's tepee. **4a.** 144π m^3; 452 m^3 **b.** They are equal because both cones have the same base and same height.
Lesson Check 1. 96 $in.^3$ **2.** 3.1 cm^3 **3.** Alike: Both formulas are $\frac{1}{3}$ the area of the base times height. Different: Because the bases are different figures, the base area will require different formulas. **4.** The areas of the bases are not equal; the area of the base of the pyramid is $13^2 = 169$ ft^2, but the area of the base of the cone is $\pi(6.5)^2 \approx 132.7$ ft^2.

Exercises 5. 200 cm^3 **7.** 50 m^3 **9.** 443.7 cm^3 **11.** 2048 m^3 **13.** 3714.5 mm^3 **15.** about 66.4 cm^3 **17.** $\frac{16}{3}\pi$ ft^3; 17 ft^3 **19.** 4π m^3; 13 m^3 **21.** Volume is halved; $V = \frac{1}{3}Bh$, so if h is replaced with $\frac{h}{2}$, then the volume is $\frac{1}{3}B\left(\frac{h}{2}\right) = \frac{1}{2}\left[\frac{1}{3}Bh\right]$. **23.** 123 $in.^3$ **25.** 10,368 ft^3 **27a.** 79,000 m^3 **b.** $20\frac{2}{3}$ m, or about 20.7 m **29a.** 120π ft^3 **b.** 60π ft^3 **c.** 240π ft^3 **31.** 3 **33.** cone with $r = 4$ and $h = 3$; 16π **35.** cylinder with $r = 4$, $h = 3$, with a cone of $r = 4$, $h = 3$ removed from it; 32π **37a.** The frustum has volume that is the difference of the volumes of the entire cone and the small cone. The frustum has volume $V = \frac{1}{3}\pi R^2 H - \frac{1}{3}\pi r^2 h$ or $\frac{1}{3}\pi(R^2 H - r^2 h)$. **b.** about 784.6 $in.^3$ **39.** A **41.** B **43.** 3600 cm^3 **44.** $JC > KN$ **45.** 7.1 $in.^2$ **46.** 13 cm

Lesson 11-6 pp. 733–740

Got It? 1. 196π $in.^2$; 616 $in.^2$ **2.** 100 $in.^2$ **3a.** 113,097 $in.^3$ **b.** The volume is $\left(\frac{1}{2}\right)^3 = \frac{1}{8}$ of the original volume. Using $V = \frac{4}{3}\pi r^3$, replacing r with $\frac{r}{2}$ gives $V = \frac{4}{3}\pi\left(\frac{r}{2}\right)^3 = \frac{1}{8}\left(\frac{4}{3}\pi r^3\right)$. **4.** 1258.9 ft^2
Lesson Check 1. 144π ft^2 **2.** 904.8 ft^3 **3.** 193 cm^2 **4.** 1 : 4 **5.** The surface area will quadruple, but the volume will be 8 times the original volume. $V = \frac{4}{3}\pi(2r)^3 = 8\left(\frac{4}{3}\pi r^3\right)$
Exercises 7. 400π $in.^2$ **9.** $40{,}000\pi$ yd^2 **11.** 441π cm^2 **13.** 62 cm^2 **15.** 20 cm^2 **17.** $\frac{500}{3}\pi$ ft^3; 524 ft^3 **19.** $\frac{1125}{2}\pi$ $in.^3$; 1767 $in.^3$ **21.** 2304π yd^3; 7238 yd^3 **23.** 451 $in.^2$ **25.** 130 cm^2 **27.** Answers may vary. Sample: sphere with $r = 3$ in., cylinder with $r = 3$ in. and $h = 4$ in. **29.** 0.9 in. **31.** 1.7 lb **33.** An infinite number of planes pass through the center of a sphere, so there are an infinite number of great circles. **35.** 36π $in.^3$ **37.** $\frac{500}{3}\pi$ mm^3 **39.** 288π cm^3 **41.** $\frac{1125}{2}\pi$ mi^3 **43a.** about 8.9 $in.^2$ **b.** The answer is less than the actual surface area since the dimples on the golf ball add to the surface area. **45a.** on **b.** inside **c.** outside **47.** 38,792.4 ft^3 **49.** 22π cm^2; $\frac{46}{3}\pi$ cm^3 **51.** 22π cm^2; $\frac{14}{3}\pi$ cm^3 **53.** Answers may vary. Sample: You could lift the small ball because it weighs about 75 lb. The big ball would be much harder to lift since it weighs about 253 lb. **55.** 3 m **57a.** Cube; explanations may vary. Sample: If $s^3 = \frac{4}{3}\pi r^3$, then $s = r \cdot \sqrt[3]{\frac{4\pi}{3}}$. So $6s^2 = 6\left(r \cdot \sqrt[3]{\frac{4\pi}{3}}\right)^2 \approx 15.6r^2 > 4\pi r^2$ (which is about $12.6r^2$). **b.** Answers may vary. Sample: Spheres are difficult to stack in a display or on a shelf. **59.** 2 : 3 **61.** G **63.** I **65.** 16 m^3 **66.** 19 $in.^3$ **67.** 19,396 mm^3 **68.** 35; 55 **69.** 109, 71, 109, 71 **70.** yes; 3 : 1 **71.** yes; $3 : \sqrt{2}$ or $3\sqrt{2} : 2$

Lesson 11-7 pp. 742–749

Got It? 1. yes; 6 : 5 or $\frac{6}{5}$ **2a.** 2 : 3 **b.** No; the bases are similar but the heights may not be in the same ratio as the edges of the bases. **3.** 160 m^2 **4.** 4.05 lb
Lesson Check 1. Cone 1 and Cone 3 are similar; 2 : 3. **2.** about 155 $in.^2$ **3.** Answers may vary. Sample: There are many relationships that must be true for the solids to be similar: all corresponding angles must be ≅; the corresponding faces must be similar and all corresponding edges and heights proportional. **4.** Your classmate found the scale factor of the smaller cube to the larger cube. The scale factor should be 8 : 7.
Exercises 5. no **7.** yes; 2 : 3 **9.** yes; 2 : 3 **11.** 5 : 6 **13.** 3 : 4 **15.** 240 $in.^3$ **17.** 24 ft^3 **19.** 112 m^2 **21.** 6000 toothpicks **23a.** It is 64 times the volume of the smaller prism. **b.** It is 64 times the weight of the smaller prism. **25.** No; explanations may vary. Sample: If the scale factor is $\frac{1}{10}$, then the weight of the smaller clock should be $\frac{1}{1000}$ the weight of the existing clock. **27.** about 1000 cm^3 **29.** No; the same increase to all the dimensions does not result in proportional ratios unless the original prism is a cube. **31a.** 3 : 1 **b.** 9 : 1 **33.** 864 $in.^3$ **35.** 9 : 25; 27 : 125 **37.** 5 : 8; 25 : 64 **39a.** 100 times **b.** 1000 times **c.** His weight is 1000 times the weight of an average person, but his bones can support only 600 times the weight of an average person. **41a.** 4 : 1 **b.** 8 : 1 **c.** $(3\ell + 5r) : (4\ell + 4r)$; $(3\ell + 5r) : (\ell + r)$, where r is the radius and ℓ is the slant height of the small cone. **d.** 7 : 8 and 7 : 1 **43.** 10 **45.** 116 **47.** about 1790 cm^2 and 1937 cm^2 **48.** 113.1 $in.^3$ **49.** 8.2 m^3 **50.** 904.8 $in.^3$ **51a.** $8\sqrt{3}$ mm, or about 13.9 mm **b.** $4\sqrt{21}$ mm, or about 18.3 mm **c.** $8\sqrt{7}$ mm, or about 21.2 mm **52.** 20 **53.** 15 **54.** 15

Chapter Review pp. 751–754

1. sphere **2.** pyramid **3.** cross section **4–5.** Answers may vary. Samples are given.
4.

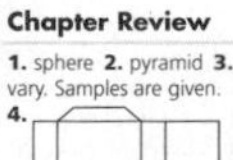

5.

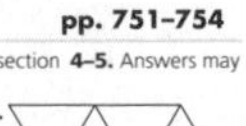

6. 8 **7.** 8 **8.** 5 **9.** a circle
10.

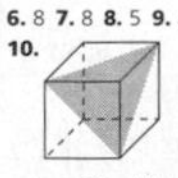

11. 36 cm^2 **12.** 66π m^2 **13.** 208 $in.^2$ **14.** 36π cm^2 **15.** 32.5π cm^2 **16.** 185.6 ft^2 **17.** 576 m^2 **18.** 50.3 $in.^2$ **19.** 391.6 $in.^2$ **20.** $B = \frac{S.A. - L.A.}{2}$ **21.** 84 m^3 **22.** 24.5 ft^3 **23.** 410.5 yd^3 **24.** 13.9 m^3 **25.** S.A. = 314.2 $in.^2$; $V = 523.6$ $in.^3$ **26.** S.A. = 153.9 cm^2; $V = 179.6$ cm^3 **27.** S.A. = 50.3 ft^2; $V = 33.5$ ft^3 **28.** S.A. = 8.0 ft^2; $V = 2.1$ ft^3 **29.** 904.78 cm^3 **30.** 314 m^2 **31.** 8.6 $in.^3$
32. Answers may vary. Sample:

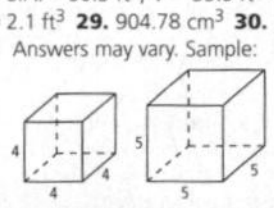

33. 27 : 64 **34.** 64 : 27 **35.** 324 pencils

Chapter 12

Get Ready! p. 759

1. 82 **2.** $6\frac{2}{3}$ **3.** 15 **4.** 25 **5.** $6\sqrt{2}$ **6.** 5 **7.** 6 **8.** 18 **9.** 24 **10.** 45 **11.** 60 **12.** $4\sqrt{2}$ **13.** 13 **14.** $\sqrt{10}$ **15.** 6 **16.** Answers may vary. Sample: A tangent touches a circle at one point. **17.** Answers may vary. Sample: An inscribed ∠ has its vertex on a circle and its sides are inside the circle. **18.** Answers may vary. Sample: An intercepted arc is the part of a circle that lies in the interior of an ∠.

Lesson 12-1 pp. 762–769

Got It? 1a. 52 **b.** $x = 180 - c$ **2.** about 127 mi **3.** $5\frac{1}{3}$ **4.** no; $4^2 + 7^2 = 65 \neq 8^2$ **5.** 12 cm
Lesson Check 1. 32 **2.** 6 units **3.** $\sqrt{63} \approx 7.9$ units **4.** Answers may vary. Sample: *Tangent ratio* refers to a ratio of the lengths of two sides of a rt. △, while *tangent to a circle* refers to a line or part of a line that is in the plane of a circle and touches the circle in exactly one point. **5.** If $\overline{DF}$ is tangent to ⊙E, then $\overline{DF} \perp \overline{EF}$. That would mean that △$DEF$ contains two rt. ∆s, which is impossible. So $\overline{DF}$ is not a tangent to ⊙E.
Exercises 7. 47 **9.** 253.0 km **11.** 178.9 km **13.** 3.6 cm **15.** no; $5^2 + 15^2 \neq 16^2$ **17.** yes; $6^2 + 8^2 = 10^2$ **19.** 14.2 in. **21.** All 4 are ≅; the two tangents to each coin from A are ≅, so by the Transitive Prop. of ≅, all the tangents are ≅. **23.** 1. $\overline{BA}$ and $\overline{BC}$ are tangent to ⊙O at A and C. (Given) 2. $\overline{AB} \perp \overline{OA}$ and $\overline{BC} \perp \overline{OC}$ (If a line is tan. to a ⊙, it is ⊥ to the radius.) 3. △BAO and △BCO are rt. ∆s. (Def. of rt. △) 4. $\overline{AO} \cong \overline{OC}$ (Radii of a circle are ≅.) 5. $\overline{BO} \cong \overline{BO}$ (Refl. Prop. of ≅) 6. △BAO ≅ △BCO (HL) 7. $\overline{BA} \cong \overline{BC}$ (Corresp. parts of ≅ ∆s are ≅.) **25.** 1. ⊙A and ⊙B with common tangents $\overline{DF}$ and $\overline{CE}$ (Given) 2. $GD = GC$ and $GE = GF$ (Two tan. segments from a pt. to a ⊙ are ≅.) 3. $\frac{GD}{GC} = 1$, $\frac{GF}{GE} = 1$ (Div. Prop. of =) 4. $\frac{GD}{GC} = \frac{GF}{GE}$ (Trans. Prop. of =) 5. ∠DGC ≅ ∠EGF (Vert. ∆s are ≅.) 6. △GDC ~ △GFE (SAS ~ Thm.) **27.** 57.5
29.

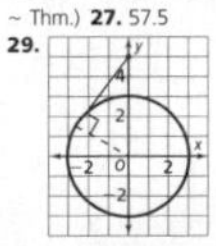

4 units

Lesson 12-2 pp. 771–779

Got It? 1. Since the circles are ≅, their radii are = and ∆BOC and DPF are isosceles. So $\overline{OB} \cong \overline{OC} \cong \overline{PD} \cong \overline{DF}$. Since ∠$B$ ≅ ∠D and the ∆s are isosceles, ∠B ≅ ∠C ≅ ∠D ≅ ∠F. So △BOC ≅ △DPF by AAS. So ∠O ≅ ∠P. Therefore, $\overline{BC} \cong \overline{DF}$ (either by corresp. parts of ≅ ∆s are ≅ or by within ≅ circles, ≅ central ∆s have ≅ chords) and $\overset{\frown}{BC} \cong \overset{\frown}{DF}$ (within ≅ circles, ≅ central ∆s have ≅ arcs). **2.** 16; ≅ chords are equidistant from the center **4.** $\overline{BA}$ is the hypotenuse of rt. △BAC, so the Pythagorean Theorem can be used.
Lesson Check 1. 50; ∠COD ≅ ∠AOB (Vert. ∆s are ≅), so $\overset{\frown}{CD} \cong \overset{\frown}{AB}$ because ≅ central ∆s have ≅ arcs. Therefore, $m\overset{\frown}{CD} = m\overset{\frown}{AB}$. **2.** $\overset{\frown}{CA} \cong \overset{\frown}{BD}$ because in a circle ≅ chords have ≅ arcs. **3.** The distances are equal because in a circle ≅ chords are equidistant from the center. **4.** A radius is *not* a chord because one of its endpoints is not on the circle. A diameter *is* a chord because both of its endpoints are on the circle. **5.** Chords $\overline{SR}$ and $\overline{QP}$ are equidistant from the center, so their lengths must be equal.
Exercises 7. Answers may vary. Sample: $\overset{\frown}{ET} \cong \overset{\frown}{GH} \cong \overset{\frown}{JN} \cong \overset{\frown}{ML}$; $\overline{ET} \cong \overline{GH} \cong \overline{JN} \cong \overline{ML}$; ∠$TFE$ ≅ ∠HFG; ∠JKN ≅ ∠MKL **9.** 8 **11.** The center is at the intersection of $\overline{GH}$ and $\overline{KM}$, because if a chord is the ⊥ bis. of another chord, then the first chord is a diameter; two diameters intersect at the center of a circle. **13.** 6 **15.** 20.8 **17.** 6 in. **19.** Since ∠AOB ≅ ∠COD, it follows that $m∠AOB = m∠COD$. Now $m∠AOB = m\overset{\frown}{AB}$ and $m∠COD = m\overset{\frown}{CD}$ (Definition of arc measure). So $m\overset{\frown}{AB} = m\overset{\frown}{CD}$ (Substitution). Therefore, $\overset{\frown}{AB} \cong \overset{\frown}{CD}$ (Definition of ≅ arcs). **21.** ⊙O with $\overline{AB} \cong \overline{CD}$ (Given); $\overline{AO} \cong \overline{BO} \cong \overline{CO} \cong \overline{DO}$ (All radii of a ⊙ are ≅); △AOB ≅ △COD (SSS); ∠AOB ≅ ∠COD (Corresp. parts of ≅ ∆s are ≅.); $\overset{\frown}{AB} \cong \overset{\frown}{CD}$ (≅ central ∆s have ≅ arcs.). **23.** 5 in. **25.** 10 ft **27.** 9.2 units **29.** The length of a chord or an arc is determined not only by the measure of the central ∆, but also by the radius of the ⊙. **31.** 90 **33.** $\overline{XW} \cong \overline{XY}$ (All radii of a circle are ≅); X is on the ⊥ bis. of $\overline{WY}$ (Converse of ⊥ Bis. Thm.); ℓ is the ⊥ bis. of $\overline{WY}$ (Given); X is on ℓ (Subst. Prop.), so ℓ contains the center of ⊙X.
35.

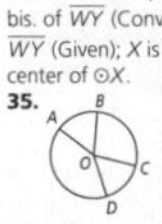

Given: ⊙O with $\overset{\frown}{AB} \cong \overset{\frown}{CD}$
Prove: ∠AOB ≅ ∠COD
Proof: $m∠AOB = m\overset{\frown}{AB}$ and $m∠COD = m\overset{\frown}{CD}$ (definition of arc measure). $\overset{\frown}{AB} \cong \overset{\frown}{CD}$ (given), so $m\overset{\frown}{AB} = m\overset{\frown}{CD}$ (Def. of ≅ arcs). Therefore, $m∠AOB = m∠COD$ (Substitution). Hence ∠AOB ≅ ∠COD (Def. of ≅ ∆s).
37.

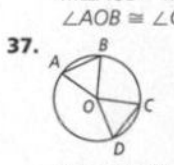

Given: ⊙O with $\overset{\frown}{AB} \cong \overset{\frown}{CD}$
Prove: $\overline{AB} \cong \overline{CD}$
Proof: It is given that $\overset{\frown}{AB} \cong \overset{\frown}{CD}$, so ∠$AOB$ ≅ ∠COD (if arcs are ≅ then their central ∆s are ≅). Also, $AO = BO = CO = DO$ (radii of a ⊙ are ≅), so △AOB ≅ △COD (SAS), and $\overline{AB} \cong \overline{CD}$ (corresp. parts of ≅ ∆s are ≅).
39.

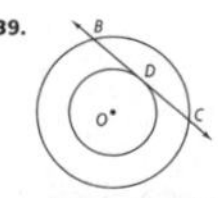

Given: Concentric circles, $\overline{BC}$ is tangent to the smaller circle at D
Prove: D is the midpt. of $\overline{BC}$
Proof: It is given that $\overline{BC}$ is tangent to the smaller circle, so $\overline{BC} \perp \overline{OD}$ (a tangent is ⊥ to a radius at the point of tangency). $\overline{OD}$ is part of a diameter of the larger circle, so $\overline{BD} \cong \overline{CD}$ (if a diameter is ⊥ to a chord, it bisects the chord). D is the midpt. of $\overline{BC}$ (Def. of midpt.)
41. G
43. During one revolution the bicycle moves $C = \pi d = \pi(17) = 53.4$ in., or about 4.45 ft. So the number of revolutions needed to travel 800 ft is $\frac{800}{4.45} \approx 180$ revolutions. **44.** 40 **45.** 5.5 **46.** 7.6 in. and 18.4 in. **47-49.** Answers may vary. Samples are given. **47.** $\overset{\frown}{STQ}$ **48.** $\overset{\frown}{ST}$ **49.** $\overset{\frown}{STR}$ **50.** 86 **51.** 180 **52.** 121

Lesson 12-3 pp. 780–787

Got It? 1a. 90 **b.** $m∠A = 95$, $m∠B = 77$, $m∠C = 85$, and $m∠D = 103$ **c.** The sum of the measures of opposite ∆s is 180. **2.** $m∠1 = 90$, $m∠2 = 110$, $m∠3 = 90$, $m∠4 = 70$ **3a.** $x = 35$, $y = 55$ **b.** An inscribed ∠, and an ∠ formed by a tangent and chord, are both equal to half the measure of the intercepted arc. Since the ∆s intercept the same arc, their measures are = and they are ≅.
Lesson Check 1. $\overset{\frown}{BD}$ **2.** ∠D **3.** ∠A and ∠C are suppl., and ∠B and ∠D are suppl. **4.** Sample answer: For inscribed ∠ABC, B is the vertex and A, B, and C are points on the circle. The intercepted arc of ∠ABC consists of points A, C, and all the points on the circle in the interior of ∠ABC. **5.** ∠A is not inscribed in a semicircle.
Exercises 7. 180 **9.** $a = 54$, $b = 30$, $c = 96$ **11.** $a = 101$, $b = 67$, $c = 84$, $d = 80$ **13.** $a = 85$, $b = 47.5$, $c = 90$ **15.** $p = 90$, $q = 122$ **17.** $x = 65$, $y = 130$ **19.** Rectangle; opposite ∆s are ≅ (because figure is ▱) and suppl. (because opp. ∆s intercept arcs whose measures sum to 360). ≅ suppl. ∆s are rt. ∆s, so the inscribed ▱ must be a rectangle. **21a.** 40 **b.** 50 **c.** 40 **d.** 40 **e.** 65 **23.** $a = 26$, $b = 64$, $c = 42$

25. $a = 30, b = 60, c = 62, d = 124, e = 60$ **27.** $\odot S$ with inscribed $\angle PQR$ (Given); $m\angle PQT = \frac{1}{2}m\widehat{PT}$ (Inscribed $\angle$ Thm., Case I); $m\angle RQT = \frac{1}{2}m\widehat{RT}$ (Inscribed $\angle$ Thm., Case I); $m\widehat{PR} = m\widehat{PT} - m\widehat{RT}$ (Arc Add. Post.); $m\angle PQR = m\angle PQT - m\angle RQT$ ($\angle$ Add. Post.); $m\angle PQR = \frac{1}{2}m\widehat{PT} - \frac{1}{2}m\widehat{RT}$ (Subst. Prop.); $m\angle PQR = \frac{1}{2}m\widehat{PR}$ (Subst. Prop.) **29.** No; since opposite $\angle$s of a quadrilateral inscribed in a circle must be supplementary, the only rhombus that meets the criteria is a square. **31.** $\odot O$, $\angle A$ intercepts $\widehat{BC}$, and $\angle D$ intercepts $\widehat{BC}$ (Given); $m\angle A = \frac{1}{2}m\widehat{BC}$ and $m\angle D = \frac{1}{2}m\widehat{BC}$ (Inscribed $\angle$ Thm.); $m\angle A = m\angle D$ (Subst. Prop.); $\angle A \cong \angle D$ (Def. of $\cong$ $\angle$s). **33.** Quadrilateral $ABCD$ inscribed in $\odot O$ (Given); $m\angle A = \frac{1}{2}m\widehat{BCD}$ and $m\angle C = \frac{1}{2}m\widehat{BAD}$ (Inscribed $\angle$ Thm.); $m\angle A + m\angle C = \frac{1}{2}m\widehat{BCD} + \frac{1}{2}m\widehat{BAD}$ (Add. Prop.); $m\widehat{BCD} + m\widehat{BAD} = 360$ (Arc measure of circle is 360); $\frac{1}{2}m\widehat{BCD} + \frac{1}{2}m\widehat{BAD} = 180$ (Mult. Prop.) $m\angle A + m\angle C = 180$ (Subst. Prop.); $\angle A$ and $\angle C$ are suppl. (Def. of suppl.); $m\angle B = \frac{1}{2}m\widehat{ADC}$ and $m\angle D = \frac{1}{2}m\widehat{ABC}$ (Inscribed $\angle$ Thm.); $m\angle B + m\angle D = \frac{1}{2}m\widehat{ADC} + \frac{1}{2}m\widehat{ABC}$ (Add. Prop.); $m\widehat{ADC} + m\widehat{ABC} = 360$ (Arc measure of circle is 360); $\frac{1}{2}m\widehat{ADC} + \frac{1}{2}m\widehat{ABC} = 180$ (Mult. Prop.) $m\angle B + m\angle D = 180$ (Subst. Prop.); $\angle B$ and $\angle D$ are suppl. (Def. of suppl. $\angle$s).

35. false **37.** True; opposite $\angle$s in an inscribed quadrilateral intercept nonoverlapping arcs totaling 360 and inscribed $\angle$s have half the measure of the intercepted arcs, so the opposite $\angle$s are suppl.

Lesson 12-4 pp. 790–797

Got It? 1a. 250 **b.** 40 **c.** 40 **2a.** 160 **b.** The probe is closer; as an observer moves away from Earth, the viewing angle decreases and the measure of the arc of Earth that is viewed gets larger and approaches 180. **3a.** 13.8 **b.** 3.2
Lesson Check 1. 5.4 **2.** 65 **3.** 11.2 **4.** 100, 260
5. A secant is a line that intersects a circle at two points; a tangent is a line that intersects a circle at one point.
6. No; we can find the sum of the measures of the two arcs (in this situation, that sum is 230), but there is not enough information to find the measure of each arc.
7. The student forgot to multiply by the length of the entire secant seg.; the equation should be $(13.5)(6) = x^2$.
Exercises 9. 50 **11.** 60 **13.** $x = 72, y = 36$ **15.** 15 **17.** 13.2 **19.** $x = 25.8, y \approx 12.4$ **21.** $360 - x$ **23.** $180 - y$ **25.** 16.7 **27.** 95, 104, 86, 75 **29.** $c = b - a$ **31.** $\angle 1$ is a central $\angle$, so $m\angle 1 = x$; $\angle 2$ is an inscribed $\angle$, so $m\angle 2 = \frac{1}{2}x$; $\angle 3$ is formed by the secants, so $m\angle 3 = \frac{1}{2}(x - y)$. **33.** $x \approx 8.9, y = 2$ **35.** 1. $\odot O$ with secants $\overline{CA}$ and $\overline{CE}$ (Given) 2. Draw $\overline{BE}$ (2 pts. determine a line.) 3. $m\angle BEC = \frac{1}{2}m\widehat{BD}$ and $m\angle ABE = \frac{1}{2}m\widehat{AE}$ (The measure of an inscribed $\angle$ is half the measure of its intercepted arc.) 4. $m\angle BEC + m\angle BCE = m\angle ABE$ (Ext. $\angle$ Thm.) 5. $\frac{1}{2}m\widehat{BD} + m\angle BCE = \frac{1}{2}m\widehat{AE}$ (Subst. Prop. of =) 6. $m\angle BCE = \frac{1}{2}m\widehat{AE} - \frac{1}{2}m\widehat{BD}$ (Subst. Prop. of =) 7. $m\angle BCE = \frac{1}{2}(m\widehat{AE} - m\widehat{BD})$ (Distr. Prop.) 8. $\angle BCE \cong \angle ACE$ (Refl. Prop. of $\cong$) 9. $m\angle ACE = \frac{1}{2}(m\widehat{AE} - m\widehat{BD})$ (Subst. Prop. of =)
37.

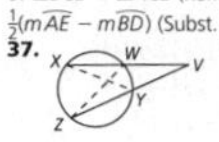

Given: A $\odot$ with secant segments $\overline{XV}$ and $\overline{ZV}$
Prove: $XV \cdot WV = ZV \cdot YV$.
Proof: Draw $\overline{XY}$ and $\overline{ZW}$ (2 pts. determine a line); $\angle XVY \cong \angle ZVW$ (Refl. Prop. of $\cong$); $\angle VXY \cong \angle WZV$ (2 inscribed $\angle$s that intercept the same arc are $\cong$); $\triangle XVY \sim \triangle ZVW$ (AA~); $\frac{XV}{ZV} = \frac{YV}{WV}$ (In similar figures, corresp. sides are proportional); $XV \cdot WV = ZV \cdot YV$ (Prop. of Proportion) **39a.** $\triangle ACD$ **b.** $\tan A = \frac{DC}{AC} = \frac{DC}{1} = DC$, length of tangent seg. **c.** secant $A = \frac{AD}{AC} = \frac{AD}{1} = AD$, length of secant seg. **41.** $m\angle 1 = \frac{1}{2}m\widehat{QRP} - \frac{1}{2}m\widehat{PQ}$ and $m\angle 2 = \frac{1}{2}m\widehat{RQP} - \frac{1}{2}m\widehat{RP}$ (vertex outside $\odot$, $m\angle$ = half difference of intercepted arcs); $m\angle 1 + m\angle 2 = \frac{1}{2}m\widehat{QRP} + \frac{1}{2}m\widehat{RQP} - \frac{1}{2}m\widehat{PQ} - \frac{1}{2}m\widehat{RP}$ (Subst. Prop. of =); $m\angle 1 + m\angle 2 = \frac{1}{2}m\widehat{QR} + \frac{1}{2}m\widehat{RP} + \frac{1}{2}m\widehat{QR} + \frac{1}{2}m\widehat{PQ} - \frac{1}{2}m\widehat{PQ} - \frac{1}{2}m\widehat{RP}$ (Arc Add. Postulate and Distr. Prop.); $m\angle 1 + m\angle 2 = m\widehat{QR}$ (Distr. Prop.).

43.

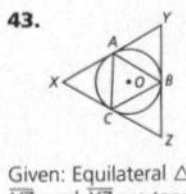

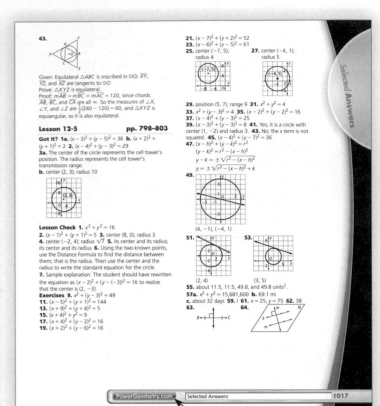

Given: Equilateral $\triangle ABC$ is inscribed in $\odot O$; $\overline{XY}$, $\overline{YZ}$, and $\overline{XZ}$ are tangents to $\odot O$
Prove: $\triangle XYZ$ is equilateral.
Proof: $m\widehat{AB} = m\widehat{BC} = m\widehat{AC} = 120$, since chords $\overline{AB}$, $\overline{BC}$, and $\overline{CA}$ are all $\cong$. So the measures of $\angle X$, $\angle Y$, and $\angle Z$ are $\frac{1}{2}(240 - 120) = 60$, and $\triangle XYZ$ is equiangular, so it is also equilateral.

Lesson 12-5 pp. 798–803

Got It? 1a. $(x - 3)^2 + (y - 5)^2 = 36$ **b.** $(x + 2)^2 + (y + 1)^2 = 2$ **2.** $(x - 4)^2 + (y - 3)^2 = 29$
3a. The center of the circle represents the cell tower's position. The radius represents the cell tower's transmission range.
b. center (2, 3); radius 10

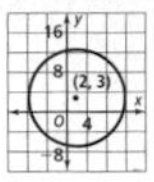

Lesson Check 1. $x^2 + y^2 = 16$
2. $(x - 1)^2 + (y + 1)^2 = 5$ **3.** center (8, 0); radius 3
4. center (−2, 4); radius $\sqrt{7}$ **5.** its center and its radius; its center and its radius **6.** Using the two known points, use the Distance Formula to find the distance between them; that is the radius. Then use the center and the radius to write the standard equation for the circle.
7. Sample explanation: The student should have rewritten the equation as $(x - 2)^2 + (y - (-3))^2 = 16$ to realize that the center is (2, −3).
Exercises 9. $x^2 + (y - 3)^2 = 49$
11. $(x - 5)^2 + (y + 1)^2 = 144$
13. $(x + 9)^2 + (y + 4)^2 = 5$
15. $(x + 4)^2 + y^2 = 9$
17. $(x + 4)^2 + (y - 2)^2 = 16$
19. $(x + 2)^2 + (y - 6)^2 = 16$
21. $(x - 7)^2 + (y + 2)^2 = 52$
23. $(x - 6)^2 + (y - 5)^2 = 61$
25. center (−7, 5); radius 4

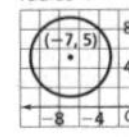

27. center (−4, 1); radius 5

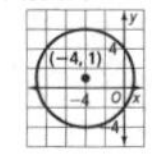

29. position (5, 7); range 9 **31.** $x^2 + y^2 = 4$
33. $x^2 + (y - 3)^2 = 4$ **35.** $(x - 2)^2 + (y - 2)^2 = 16$
37. $(x - 4)^2 + (y - 3)^2 = 25$
39. $(x - 3)^2 + (y - 3)^2 = 8$ **41.** Yes; it is a circle with center (1, −2) and radius 3. **43.** No; the x term is not squared. **45.** $(x - 4)^2 + (y - 7)^2 = 36$
47. $(x - h)^2 + (y - k)^2 = r^2$
$(y - k)^2 = r^2 - (x - h)^2$
$y - k = \pm\sqrt{r^2 - (x - h)^2}$
$y = \pm\sqrt{r^2 - (x - h)^2} + k$
49.

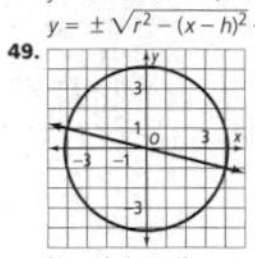

(4, −1), (−4, 1)

51.

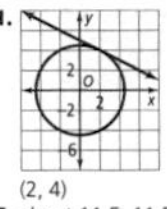

(2, 4)

53.

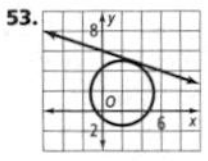

(3, 5)

55. about 11.5, 11.5, 49.8, and 49.8 units^2.
57a. $x^2 + y^2 = 15{,}681{,}600$ **b.** 69.1 mi
c. about 32 days **59.** I **61.** $x = 25, y = 75$ **62.** 38
63.

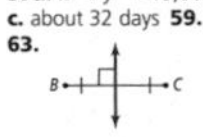

64.

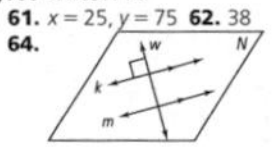

65.

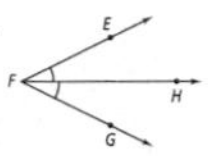

Lesson 12-6 pp. 806–811

Got It? 1. a pair of ∥ lines, each 1 cm from $\overleftrightarrow{AB}$
2.

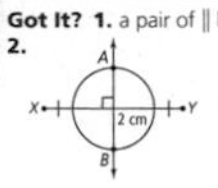

Points A and B satisfy both conditions.

3a. The locus is the line ∥ to and equidistant from the given ∥ lines (midway between them). **b.** The locus is a plane ∥ to and equidistant from the given ∥ planes (midway between them).
Lesson Check
1.

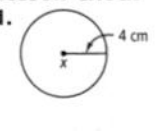

The locus is a circle with center x and radius 4 cm.

2.

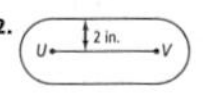

The locus is a pair of ∥ segments, each segment 2 in. from $\overline{UV}$, and two semicircles with radius 2 in. and centers U and V.

3.

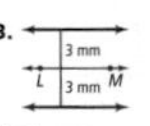

The locus is a pair of ∥ lines, each 3 mm from $\overleftrightarrow{LM}$.

4.

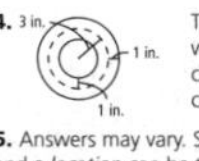

The locus is two circles concentric with the original circle; the smaller circle has radius 2 in. and the larger circle has radius 4 in.

5. Answers may vary. Sample: A *locus* is a set of points, and a *location* can be thought of as a description of a single point. **6.** The locus in a plane is a line (the ⊥ bis. of $\overline{JK}$) and the locus in space is a plane (it contains the midpt. of $\overline{JK}$ and is ⊥ to it).
Exercises
7. The locus is the ⊥ bis. of $\overline{PQ}$.

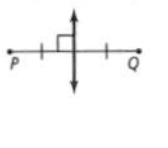

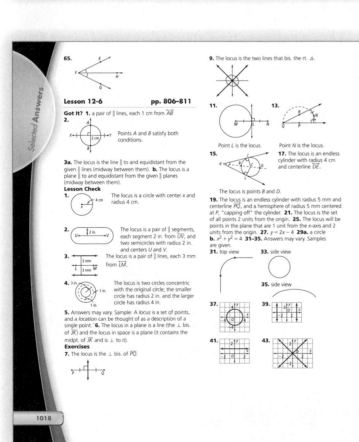

9. The locus is the two lines that bis. the rt. $\angle$s.

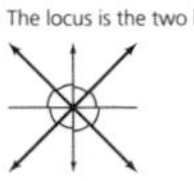

11.

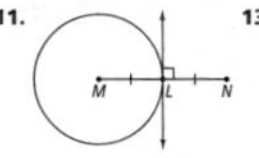

Point L is the locus.

13.

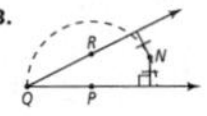

Point N is the locus.

15.

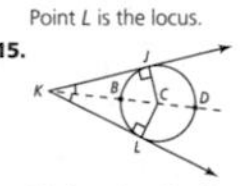

The locus is points B and D.

17. The locus is an endless cylinder with radius 4 cm and centerline $\overleftrightarrow{DE}$.

19. The locus is an endless cylinder with radius 5 mm and centerline $\overrightarrow{PQ}$, and a hemisphere of radius 5 mm centered at P, "capping off" the cylinder. **21.** The locus is the set of all points 2 units from the origin. **25.** The locus will be points in the plane that are 1 unit from the x-axis and 2 units from the origin. **27.** $y = 2x - 4$ **29a.** a circle **b.** $x^2 + y^2 = 4$ **31–35.** Answers may vary. Samples are given.
31. top view

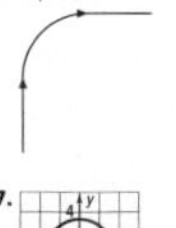

33. side view

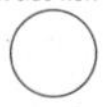

35. side view

37.

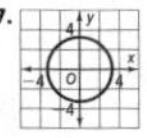

39.

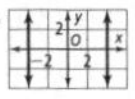

41.

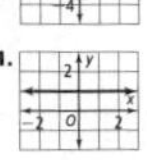

43.

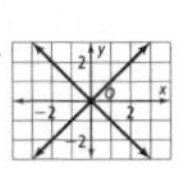

45a. Sample:

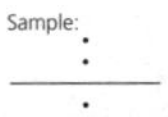

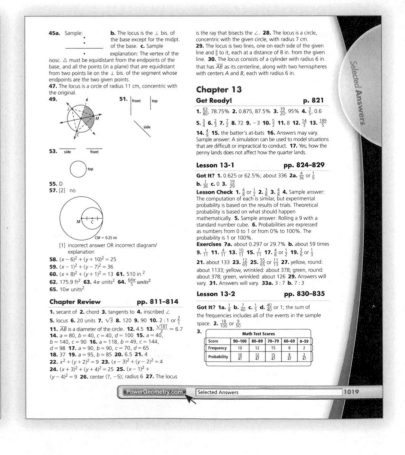

b. The locus is the ⊥ bis. of the base except for the midpt. of the base. **c.** Sample explanation: The vertex of the isosc. △ must be equidistant from the endpoints of the base, and all the points (in a plane) that are equidistant from two points lie on the ⊥ bis. of the segment whose endpoints are the two given points.
47. The locus is a circle of radius 11 cm, concentric with the original.
49.

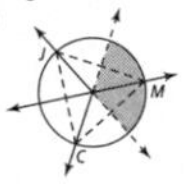

51.

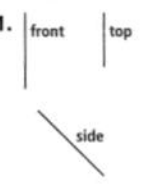

53.

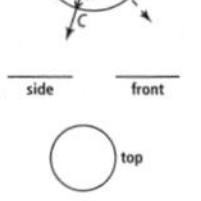

55. D
57. [2] no

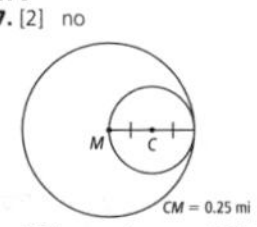

[1] incorrect answer OR incorrect diagram/explanation
58. $(x - 6)^2 + (y + 10)^2 = 25$
59. $(x - 1)^2 + (y - 7)^2 = 36$
60. $(x + 8)^2 + (y + 1)^2 = 13$ **61.** 510 in.^{2}
62. 175.9 ft^2 **63.** 4π units^2 **64.** $\frac{64\pi}{5}$ *units*^{2}
65. 10π units^2

Chapter Review pp. 811–814

1. secant of **2.** chord **3.** tangents to **4.** inscribed $\angle$
5. locus **6.** 20 units **7.** $\sqrt{3}$ **8.** 120 **9.** 90 **10.** 2 : 1 or $\frac{2}{1}$
11. $\overline{AB}$ is a diameter of the circle. **12.** 4.5 **13.** $\frac{\sqrt{181}}{2} \approx 6.7$
14. $a = 80, b = 40, c = 40, d = 100$ **15.** $a = 40, b = 140, c = 90$ **16.** $a = 118, b = 49, c = 144, d = 98$ **17.** $a = 90, b = 90, c = 70, d = 65$
18. 37 **19.** $a = 95, b = 85$ **20.** 6.5 **21.** 4
22. $x^2 + (y + 2)^2 = 9$ **23.** $(x - 3)^2 + (y - 2)^2 = 4$
24. $(x + 3)^2 + (y + 4)^2 = 25$ **25.** $(x - 1)^2 + (y - 4)^2 = 9$ **26.** center (7, −5); radius 6 **27.** The locus is the ray that bisects the $\angle$. **28.** The locus is a circle, concentric with the given circle, with radius 7 cm.
29. The locus is two lines, one on each side of the given line and ∥ to it, each at a distance of 8 in. from the given line. **30.** The locus consists of a cylinder with radius 6 in. that has $\overline{AB}$ as its centerline, along with two hemispheres with centers A and B, each with radius 6 in.

Chapter 13

Get Ready! p. 821

1. $\frac{63}{80}$, 78.75% **2.** 0.875, 87.5% **3.** $\frac{19}{20}$, 95% **4.** $\frac{3}{5}$, 0.6
5. $\frac{3}{4}$ **6.** $\frac{2}{3}$ **7.** $\frac{7}{2}$ **8.** 72 **9.** −3 **10.** $\frac{5}{2}$ **11.** 8 **12.** $\frac{14}{3}$ **13.** $\frac{189}{5}$
14. $\frac{4}{9}$ **15.** the batter's at-bats **16.** Answers may vary. Sample answer: A simulation can be used to model situations that are difficult or impractical to conduct. **17.** Yes; how the penny lands does not affect how the quarter lands.

Lesson 13-1 pp. 824–829

Got It? 1. 0.625 or 62.5%; about 336 **2a.** $\frac{4}{36}$ or $\frac{1}{9}$
b. $\frac{1}{36}$ **c.** 0 **3.** $\frac{19}{29}$
Lesson Check 1. $\frac{4}{8}$ or $\frac{1}{2}$ **2.** $\frac{3}{8}$ **3.** $\frac{4}{8}$ **4.** Sample answer: The computation of each is similar, but experimental probability is based on the results of trials. Theoretical probability is based on what should happen mathematically. **5.** Sample answer: Rolling a 9 with a standard number cube. **6.** Probabilities are expressed as numbers from 0 to 1 or from 0% to 100%. The probability is 1 or 100%.
Exercises 7a. about 0.297 or 29.7% **b.** about 59 times
9. $\frac{1}{11}$ **11.** $\frac{4}{11}$ **13.** $\frac{10}{11}$ **15.** $\frac{7}{11}$ **17.** $\frac{4}{8}$ or $\frac{1}{2}$ **19.** $\frac{2}{6}$ or $\frac{1}{3}$
21. about 133 **23.** $\frac{12}{65}$ **25.** $\frac{55}{65}$ or $\frac{11}{13}$ **27.** yellow, round: about 1133; yellow, wrinkled: about 378; green, round: about 378; green, wrinkled: about 126 **29.** Answers will vary. **31.** Answers will vary. **33a.** 3 : 7 **b.** 7 : 3

Lesson 13-2 pp. 830–835

Got It? 1a. $\frac{1}{8}$ **b.** $\frac{7}{40}$ **c.** $\frac{1}{5}$ **d.** $\frac{40}{40}$ or 1; the sum of the frequencies includes all of the events in the sample space. **2.** $\frac{18}{100}$ or $\frac{9}{50}$
3.

Math Test Scores

Score	90–100	80–89	70–79	60–69	0–59
Frequency	10	12	15	8	2
Probability	$\frac{10}{47}$	$\frac{12}{47}$	$\frac{15}{47}$	$\frac{8}{47}$	$\frac{2}{47}$

Lesson Check **1.** $\frac{54}{142}$ or $\frac{37}{71}$ **2.** $\frac{28}{142}$ or $\frac{14}{71}$ **3.** $\frac{10}{142}$ or $\frac{5}{71}$ **4.** $\frac{50}{142}$ or $\frac{25}{71}$

5.

Preferred Music Format				
Result	CD	Radio	Blu-ray	MP3
Frequency	54	50	10	28
Probability	$\frac{37}{71}$	$\frac{25}{71}$	$\frac{5}{71}$	$\frac{14}{71}$

6. The possible outcomes are heads or tails. **7.** There are many ways she will not win but only one way she will win, so both the probability of winning is much smaller.

Exercises **9.** $\frac{25}{63}$ **11.** The relative frequencies do not change because both the numerator and denominator are multiplied by 2. **13.** $\frac{6}{23}$

15.

Favorite Snacks				
Snack	Bananas	Trail Mix	C&C	Popcorn
Frequency	8	5	4	6
Probability	$\frac{8}{23}$	$\frac{5}{23}$	$\frac{4}{23}$	$\frac{6}{23}$

17a. $\frac{145}{983}$ **b.** $\frac{838}{983}$

19. Answers may vary. Sample answer:

Coin Toss					
Result	4 Heads	3 Heads 1 Tail	2 Heads 1 Tail	1 Head 3 Tails	4 Tails
Frequency	8	5	4	6	1
Probability	$\frac{1}{16}$	$\frac{1}{4}$	$\frac{3}{8}$	$\frac{1}{4}$	$\frac{1}{16}$

21a.

Sum	2	3	4	5	6	7	8	9	10	11	12
Frequency	1	1	2	4	5	6	5	4	3	2	1
Probability	$\frac{1}{36}$	$\frac{1}{18}$	$\frac{1}{12}$	$\frac{1}{9}$	$\frac{5}{36}$	$\frac{1}{6}$	$\frac{5}{36}$	$\frac{1}{9}$	$\frac{1}{12}$	$\frac{1}{18}$	$\frac{1}{36}$

b. theoretical; the results are based on each face of one number cube having a probability of $\frac{1}{6}$.

23. The probabilities do not add to 1; they add to 1.05. You need to know the frequency distribution of the favorite pets. **25.** A **27.** C **29.** theoretical; though experimental results can be found, the more you do the experiment, the closer the results will be to the theoretical value.

30. experimental; no information is available to find results until an experiment is done. **31.** theoretical; all of the data are available to find the probability without an experiment.

32. 1.5 in. **33.** 20.25 mm **34.** $\frac{a}{c}$ **35.** 15 **36.** 90

Lesson 13-3 pp. 836–842

Got It? **1.** 2600 **2.** 479,001,600 **3.** 1320 **4.** 56 **5.** 120 **6.** $\frac{1}{2730}$

Lesson Check **1.** 6 **2.** 1 **3.** 30 **4.** 120 **5.** 15 **6.** 20 **7.** 210 **8.** Both are methods of counting. With permutations order is important, but with combinations order is not important. **9.** Yes, she also needs to know the total number of possible outcomes in the sample space.

Exercises **11.** 6760 **13.** 10,897,286,400 **15.** 210 **17a.** 4060 **b.** 142,506 **19.** 495 **21.** $\frac{1}{15{,}890{,}700}$ **23.** 56 **25.** $\frac{1}{1365}$ **27.** The friend used a permutation instead of a combination. There are 56 ways. **29a.** $\frac{1}{42}$ **b.** Sample: The number of possible codes is actually a permutation.

31. 96 cm^2

33.

Favorite Movie Genres					
Genres	Action	Comedy	Drama	Horror	Other
Frequency	9	8	3	6	4
Probability	$\frac{3}{10}$	$\frac{4}{15}$	$\frac{1}{10}$	$\frac{1}{5}$	$\frac{2}{15}$

34. $\frac{1}{36}$ **35.** $\frac{3}{36}$ **36.** $\frac{26}{36}$

Lesson 13-4 pp. 844–849

Got It? **1.** independent; the outcomes do not affect each other **2.** $\frac{1}{12}$ **3.** 38% **4.** $\frac{2}{3}$

Lesson Check **1.** 12.5% **2.** 0.85 **3.** $\frac{19}{30}$ **4.** Answers will vary. When the outcomes do not affect each other, the events are independent. If the outcome of one event affects the outcome of another event, they are dependent events. **5.** Sample answer: The events are dependent because it needs to be cloudy in order for it to rain.

Exercises **7.** dependent **9.** dependent **11.** $\frac{1}{4}$ **13.** $\frac{3}{5}$ **15.** $\frac{8}{15}$ **17.** $\frac{13}{40}$ **19.** $\frac{5}{8}$ **21.** $\frac{3}{4}$ **23.** 80% **25.** Mutually exclusive events cannot happen at the same time, but overlapping events can occur at the same time.

27. No; Answers may vary. Sample: $\frac{1}{2} + \frac{1}{3} \neq \frac{2}{3}$

29. dependent events; If one event occurs, then the other event cannot occur because they are mutually exclusive. So, the outcome of the second event depends on the outcome of the first event. **31.** $\frac{41}{\pi}$ cm **33.** 15,600 **34.** 40,320 **35.** 792 **36.** $\frac{5}{13}$ **37.** $\frac{32}{39}$ **38.** $\frac{22}{39}$

Lesson 13-5 pp. 850–855

Got It? **1.** $\frac{1}{4}$ **2a.** about 0.26 **b.** about 0.76 **3.** 0.90

Lesson Check **1.** 0.2 **2.** about 0.47 **3.** He divided by the number of people who do not support the issue instead of by the total number of Republicans.

4. A two-way frequency table displays the frequencies of data in two different categories. **5.** The probability that a student plays sports, given that the student is a female.

Exercises **7.** 0.2 **9.** $\frac{7}{13}$ **11.** about 0.444 **13.** about 0.54 **15.** $\frac{1}{5}$ **17.** $\frac{5}{8}$ **19.** $\frac{3}{5}$ **21.** Answers may vary: Sample: The sum of the probabilities is 1. The sum of the relative frequencies is also 1 because they represent the experimental probabilities of each possible outcome. **23a.** Each cell contains the frequency of the joint event described by its row and column. **b.** The marginal frequencies are the total frequencies for each category in the two-way table. **c.** The sum of the joint probabilities in any row or column of a joint probability distribution is the same as the marginal probability associated with that row or column.

25. 6 **27.** 7.5 **28.** 360° **29.** 1800° **30.** 3240°

31. 113 $in.^2$ **32a.** $4\sqrt{3}$ cm **b.** about 1393 cm^3 **33.** $\frac{2}{5}$ **34.** $\frac{8}{25}$ **35.** $\frac{3}{20}$ **36.** $\frac{17}{20}$

Lesson 13-6 pp. 856–861

Got It? **1.** 40% **2a.** $\frac{1}{9}$ **b.** $\frac{5}{27}$ **c.** No, you don't know *P*(other pet) or *P*(cat and no dogs).

3a. 54.5% **b.** Their chances of winning will increase because they have a better chance of winning on a muddy field.

Lesson Check **1.** $\frac{5}{7}$ **2.** $\frac{6}{11}$ **3.** $P(\text{red} \mid \text{small})$; $P(\text{red} \mid \text{small}) = \frac{4}{9}$, $P(\text{small} \mid \text{red}) = \frac{2}{7}$ **4.** No. If two events are independent, then $P(A \text{ and } B) = P(A) \cdot P(B)$, which is not necessarily zero. The conditional probability would be 0 for mutually exclusive events. **5.** Sample answer: When finding a conditional probability you divide, but when you find a compound probability you often multiply.

Exercises **7.** $\frac{1}{2}$ **9.** 20% **11.** 14.46% **13.** 24% **15.** $\frac{3}{7}$ **17.** $\frac{11}{42}$

Lesson 13-7 pp. 862–867

Got It? **1.** 04, 14, 09 **2.** Answers may vary. Sample: about 15 **3a.** −2.8125 **b.** −6.25 **4a.** field goal **b.** Sample answer: Yes. If it is late in the game and they are down by more than 3 points, the touchdown might be a better option.

Lesson Check **1.** 42, 37, 17, 31 **2.** 0.75; 1; pass to the teammate **3.** With the dart, there is some control over which name will be chosen. The table is more random. **4.** Expected value is the average value of a trial based on the probabilities and the values of the outcomes. **5.** Sample answer: Let 0–4 represent heads and 5–9 represent tails.

Exercises **7.** 118, 735, 719, 632, 209

9. Answers may vary. Sample: Use a random number generator to produce numbers from 1 to 5. Count how many trials it takes before producing 8 numbers in a row that are not 1. Conduct the simulations many times and find the average value. **11.** −4200; No, the player should not continue because the expected value is negative.

13. Yes, the expected value is \$4000. **15.** −0.125

17. Answers may vary. Sample: Assign numbers from 1 to 2500 to each household. Use a calculator to generate random numbers between 1 and 2500. Select the first 500 unique numbers that the calculator generates.

19. D **21.** 50 cm and 87 cm **22.** $\frac{2}{3}$ **23.** $\frac{2}{3}$

24. $(x-3)^2 + (y+5)^2 = 81$

25. $(x+2)^2 + (y-8)^2 = 25$

Chapter Review pp. 870–874

1. permutation **2.** Theoretical probability **3.** dependent **4.** frequency table **5.** $\frac{3}{10}$ **6.** $\frac{9}{10}$ **7.** 0 **8.** $\frac{3}{5}$ **9.** $\frac{5}{12}$ **10.** $\frac{1}{4}$ **11.** $\frac{11}{28}$ **12.** $\frac{17}{28}$ **13.** $\frac{1}{7}$ **14.** $\frac{9}{14}$ **15.** $\frac{6}{7}$ **16.** 120 **17.** 40,320 **18.** 840 **19.** 220 **20.** 6720 **21.** 90 **22.** 36 **23.** 1140 **24.** 78,624,000 **25.** 2300 **26.** 0.115 **27.** 0.595 **28.** $\frac{7}{16}$ **29.** Sample answer: If the events are mutually exclusive, then $P(A \text{ and } B) = 0$ and the formula becomes $P(A \text{ or } B) = P(A) + P(B)$. Otherwise, the formula is used as stated. **30.** More students chose to study.

31. 0.95 **32.** about 0.364 **33.** $\frac{22}{145}$ **34.** $\frac{5}{7}$ or about 71.4% **35.** $\frac{55}{74}$ or about 74.3% **36.** 20 **37.** 13 **38.** $\frac{13}{20}$

Skills Handbook

p. 884 **1.** Answer may vary slightly due to measuring method. Sample: 20 mm; 25 mm

3.

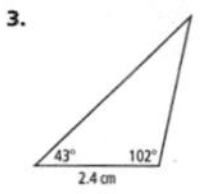

p. 885 **1.** right, scalene **3.** obtuse, isosceles **5.** Not possible; a rt. △ will always have one longest side opposite the rt. ∠.

7.

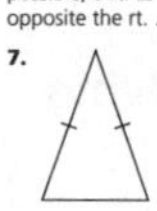

9.

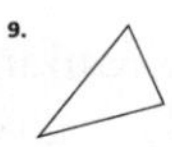

p. 886 **1.** 0.4 **3.** 600 **5.** 1008 **7.** 15,000 **9.** 34,000 **11.** 4.3 **13.** 56 **15.** 3.9 **17.** 1,080,000 **19.** 12.6 **21.** $144\frac{4}{9}$ **23.** $\frac{3125}{4356}$

p. 887 **1.** $23\frac{1}{2}$ ft to $24\frac{1}{2}$ ft **3.** $339\frac{1}{2}$ mL to $340\frac{1}{2}$ mL **5.** 73.15 mm to 73.25 mm **7.** 10.8 cm to 11.2 cm **9.** 208 cm

p. 888 **1.** 18% **3.** 1% **5.** ≈ 9% **7.** ≈ 2%

p. 889 **1.** 121 **3.** 26.01 **5.** −36 **7.** 10 **9.** 8.6 **11.** $\frac{2}{3}$ **13.** ±7 **15.** ±1 **17.** ±6 **19.** ±5

p. 890 **1.** −50 **3.** 15 **5.** 2 **7.** 36 **9.** −2 **11.** 243 **13.** −20 **15.** −4 **17.** $2x + 3$ **19.** $-4x - 7$ **21.** $-4x^2 + 8x$ **23.** $-3t^2 + 4t$ **25.** $1 - 2r + r^2$ **27.** $7h - 1$ **29.** $2x^2 + 7x - 4$ **31.** $2y^2 - 3y$

p. 891 **1.** $\frac{5}{3}$ **3.** $\frac{2}{3}$ **5.** $\frac{1}{2x}$ **7.** $\frac{22x}{35}$ **9.** $16y$ **11.** $\frac{x+1}{2}$ **13.** $\frac{1}{10r}$ **15.** $\frac{3y}{2}$ **17.** $\frac{\pi}{r+\pi}$ **19.** $\frac{5}{12}$ **21.** $\frac{5}{13}$ **23.** $\frac{12}{5}$

p. 892 **1.** 8 **3.** 22 **5.** −5 **7.** −10 **9.** 13 **11.** 10 **13.** −16 or 16 **15.** −20 or 20

p. 893 **1.** (0, −3) **3.** (4, 3)

5–8.

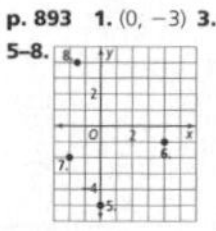

9. *y*-axis **11.** *x*-axis

p. 894 **1.** 5 **3.** 3 **5.** $\frac{2}{3}$ **7.** $\frac{1}{2}$ **9.** −5 **11.** −12 **13.** $35 - 2x = 9$; 13

p. 895 **1.** 0.5 **3.** 0.06 **5.** 1.09 **7.** 8.4 **9.** 7.2 **11–14.** Answers may vary. Samples are given. **11.** 7 **13.** 45

Additional Answers

Chapter 1

Lesson 1-1

Got It? page 5

2a. Answers may vary. Sample:

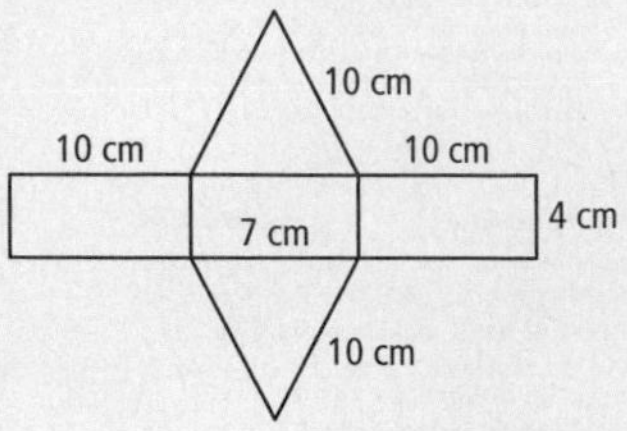

b. Yes; answers may vary. Sample:

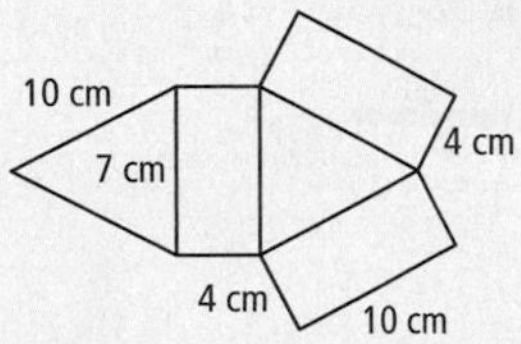

Lesson 1-6

Practice and Problem-Solving Exercises page 48

35.

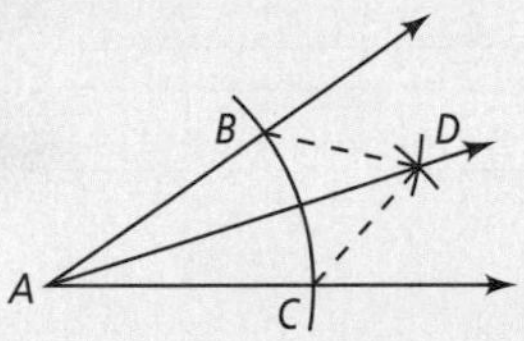

In the angle bisector construction, $\overline{AB} \cong \overline{AC}$, $\overline{BD} \cong \overline{CD}$, and $\overline{AD} \cong \overline{AD}$. Using the statement that two triangles are ≅ if three pairs of sides are ≅, then $\triangle ABD \cong \triangle ACD$. Since the ⚠ are ≅, each ∠ of one △ is ≅ to an ∠ of the other △. So, $\angle BAD \cong \angle CAD$ and $\overrightarrow{AD}$ is the ∠ bisector of $\angle BAC$.

36. D

37. I

38. [2] $x^2 - 2 = x$
$x^2 - x - 2 = 0$
$(x - 2)(x + 1) = 0$
$x - 2 = 0$ or $x + 1 = 0$
$x = 2$ or $x = -1$ (not possible)
$x = 2$

[1] incomplete steps OR both values for x OR incorrect factoring of equation OR incorrect equation

39. 116

40. yes; $m\angle TUV + m\angle VUW = 180$

41. 6

42. 10

43. 4

44. 3

45. 196

46. 10

47. −1

Chapter 3

Lesson 3-6

Practice and Problem-Solving Exercises page 186

7.

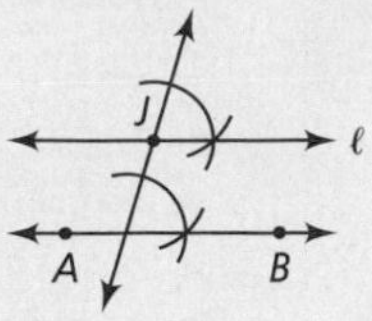

8.

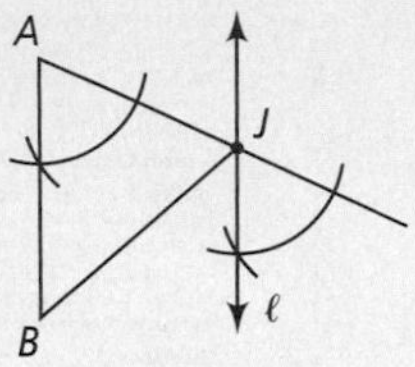

9.

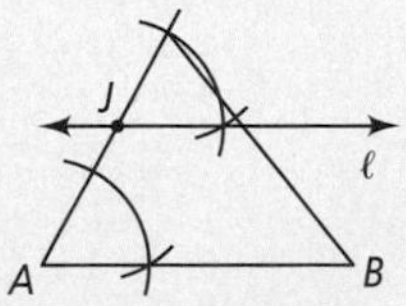

10.

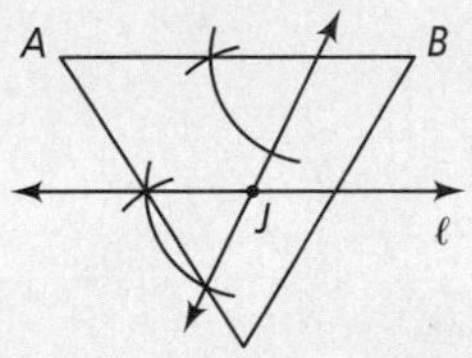

11–13. Constructions may vary. Samples using the following segments are given.

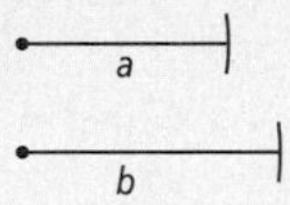

11.

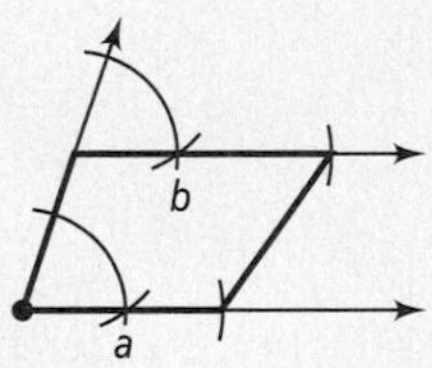

12.

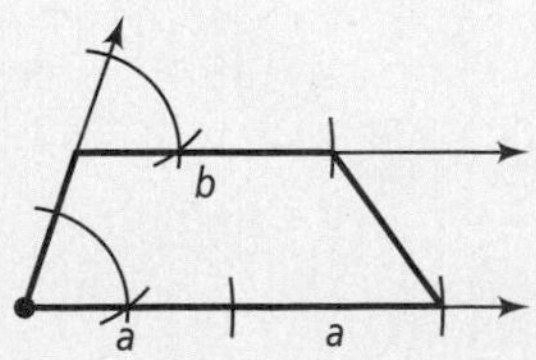

13.

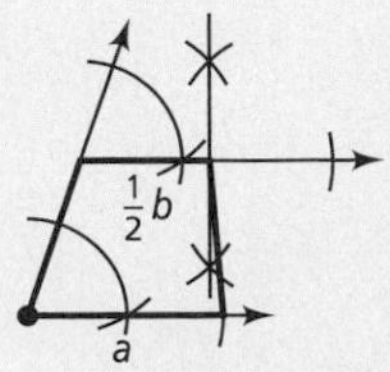

14.

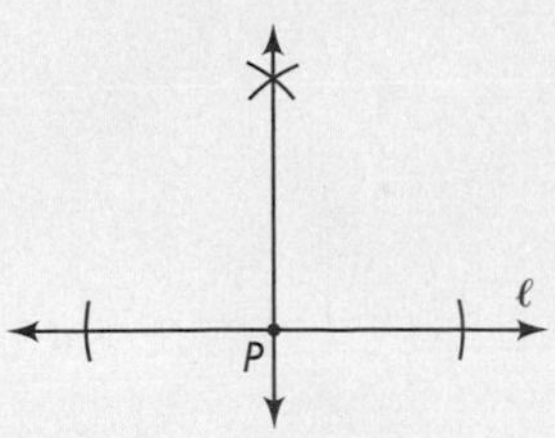

15.

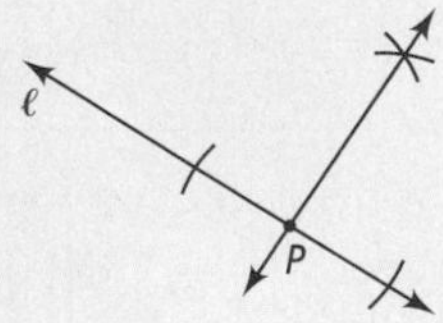

16.

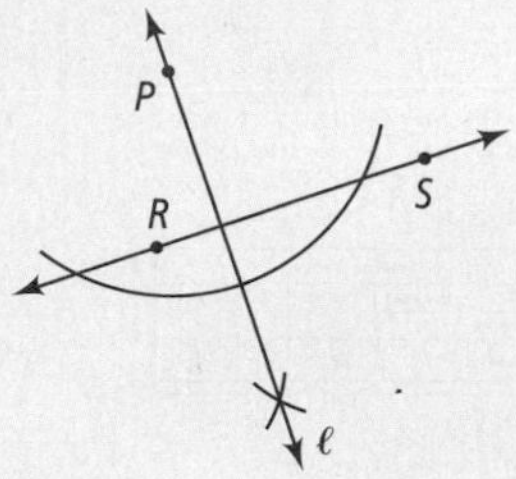

17.

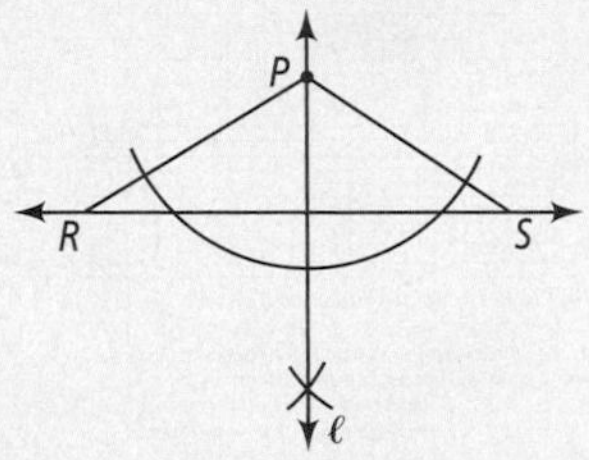

18.

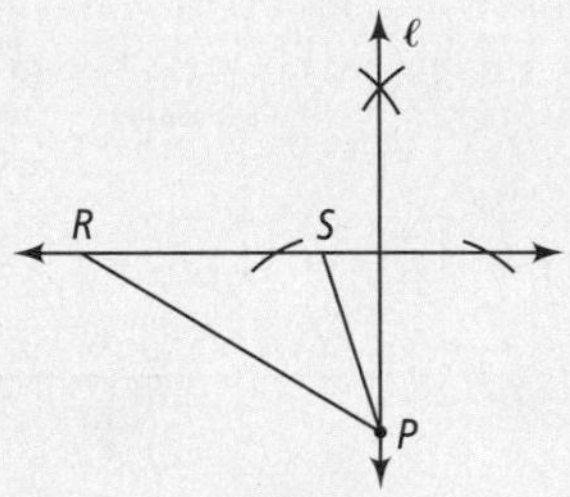

19.

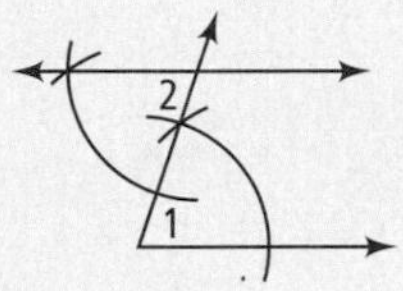

20.

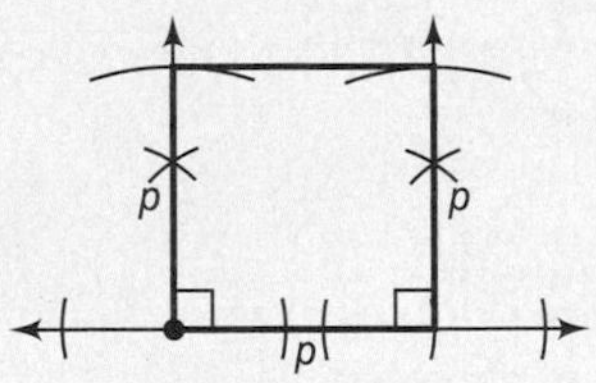

Chapter 4

Lesson 4-7

Got It? page 267

4. Using $\overline{AD} \cong \overline{AD}$ (Refl. Prop. of ≅) and the two given pairs of ≅ ∠s, $\triangle ACD \cong \triangle AED$ by AAS. Then $\overline{CD} \cong \overline{ED}$ (Corresp. parts of ≅ ⚠ are ≅.) and $\angle BDC \cong \angle FDE$ (Vert. ∠s are ≅.). Therefore, $\triangle BDC \cong \triangle FDE$ by ASA, and $\overline{BD} \cong \overline{FD}$ because corresp. parts of ≅ ⚠ are ≅.

Chapter 6

Lesson 6-7

Got It? page 401

1. scalene

2a. Yes; slope of $\overline{MN}$ = slope of $\overline{PQ} = -3$ and slope of $\overline{NP}$ = slope of $\overline{MQ} = \frac{1}{3}$, so opp. sides are ∥. The product of slopes is −1, so sides are ⊥.

b. Yes; $MN = PQ = NP = MQ = \sqrt{10}$.

c. Yes; slope of $\overline{AB} = \frac{3}{4}$ and slope of $\overline{BC} = -\frac{4}{3}$, so the product of their slopes is −1. Therefore, $\overline{AB} \perp \overline{BC}$ and ∠B is a rt. ∠. So △ABC is a rt. △ by def. of rt. △.

Lesson 6-7

Practice and Problem-Solving Exercises page 405

37. Answers may vary. Sample: Chairs are not at vertices of a ▱. Move right-most chair down by 1 grid unit.

38a. rectangle

b. rectangle

c. Yes; corresp. sides are ≅ and corresp. ∠s are ≅ (rt. ∠s), so ABCD ≅ EFGH.

39. G(−4, 1), H(1, 3)

40. (0, 7.5), (3, 10), (6, 12.5)

41. $\left(-1, 6\frac{2}{3}\right)$, $\left(1, 8\frac{1}{3}\right)$, (3, 10), $\left(5, 11\frac{2}{3}\right)$, $\left(7, 13\frac{1}{3}\right)$

42. (−1.8, 6), (−0.6, 7), (0.6, 8), (1.8, 9), (3, 10), (4.2, 11), (5.4, 12), (6.6, 13), (7.8, 14)

43. (−2.76, 5.2), (−2.52, 5.4), (−2.28, 5.6), . . . , (8.52, 14.6), (8.76, 14.8)

44. $\left(-3 + a\left(\frac{12}{n}\right), 5 + a\left(\frac{10}{n}\right)\right)$ for $a = 1, 2, 3, \ldots, n - 1$

45. D

46. G

47. A

48. [2] No; the slope of $\overline{AC}$ is 0, the slope of $\overline{AB}$ is $-\frac{3}{2}$, and the slope of $\overline{BC}$ is 1. No slopes have product of −1, so the sides are not ⊥.

[1] correct answer with no explanation

49. $m\angle 1 = 62$, $m\angle 2 = m\angle 3 = 118$, $x = 2.5$

50. (3, 2)

51. (−3, −4)

52. −1

53. 0

54. $\frac{b}{c + d - a}$

Lesson 6-8

Got It? page 409

3.

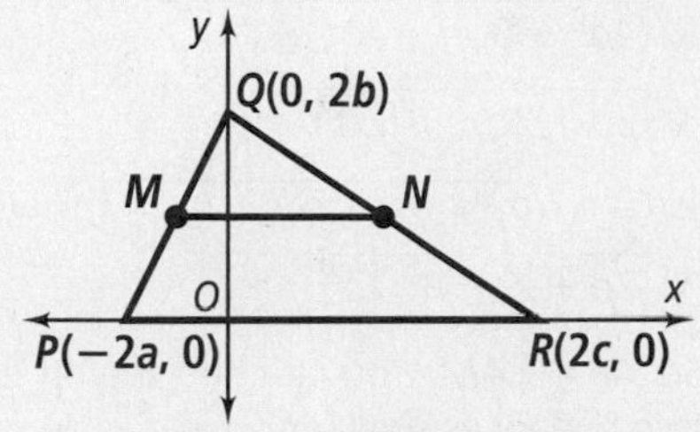

Given: △PQR, midpoints M and N

Prove: $\overline{MN} \parallel \overline{PR}$ and $MN = \frac{1}{2}PR$

- First, use the Midpoint Formula to find the coordinates of M and N.
- Then, use the Slope Formula to determine whether the slopes of $\overline{MN}$ and $\overline{PR}$ are equal. If they are, then $\overline{MN} \parallel \overline{PR}$.
- Finally, use the Distance Formula to find and compare the lengths of $\overline{MN}$ and $\overline{PR}$.

Lesson 6-9

Got It? page 415

2. Answers may vary. Sample:

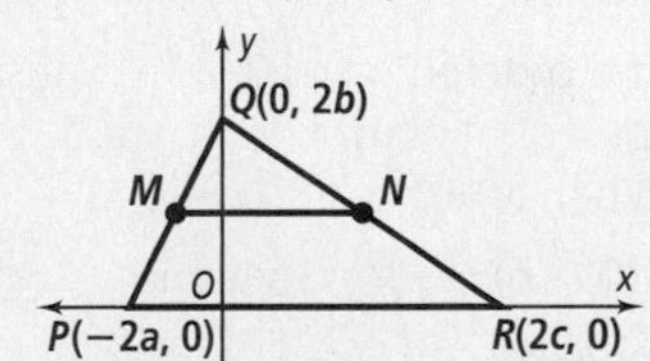

Given: △PQR, midpoints M and N

Prove: $\overline{MN} \parallel \overline{PR}$, $MN = \frac{1}{2}PR$

By the Midpoint Formula, coordinates of the midpoints are M(−a, b) and N(c, b). By the Slope Formula, slope of $\overline{MN}$ = slope of $\overline{PR} = 0$, so $\overline{MN} \parallel \overline{PR}$.

By the Distance Formula, $MN = \sqrt{(c + a)^2}$ and $PR = 2\sqrt{(c + a)^2}$, so $MN = \frac{1}{2}PR$.

Lesson 6-9

Practice and Problem-Solving Exercises page 417

19. Answers may vary. Sample:

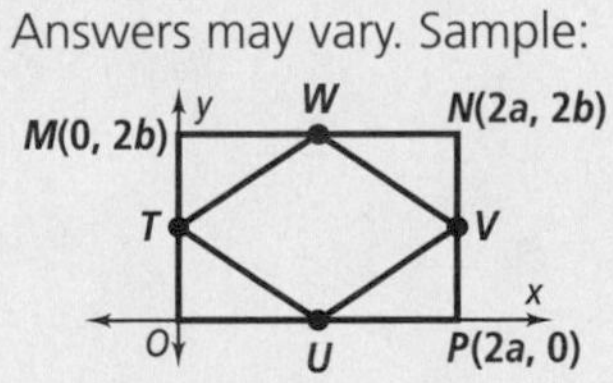

Given: MNPO is a rectangle.
T, W, V, U are midpoints of its sides.

Prove: TWVU is a rhombus.

By the Midpoint Formula, the coordinates of the midpoints are T(0, b), W(a, 2b), V(2a, b), and U(a, 0). By the Slope Formula,

slope of $\overline{TW} = \frac{2b - b}{a - 0} = \frac{b}{a}$

slope of $\overline{WV} = \frac{2b - b}{a - 2a} = -\frac{b}{a}$

slope of $\overline{VU} = \frac{b - 0}{2a - a} = \frac{b}{a}$

slope of $\overline{UT} = \frac{b - 0}{0 - a} = -\frac{b}{a}$

So $\overline{TW} \parallel \overline{VU}$ and $\overline{WV} \parallel \overline{UT}$. Therefore, TWVU is a ▱.

By the Slope Formula, slope of $\overline{TV} = 0$, and slope of $\overline{WU}$ is undefined. $\overline{TV} \perp \overline{WU}$ because horiz. and vert. lines are ⊥. Since the diagonals of ▱TWVU are ⊥, it must be a rhombus.

20. Answers may vary. Sample: Lines are ⊥ when product of their slopes is −1; it is difficult to find the product without using coordinate methods.

21. Answers may vary. Sample:

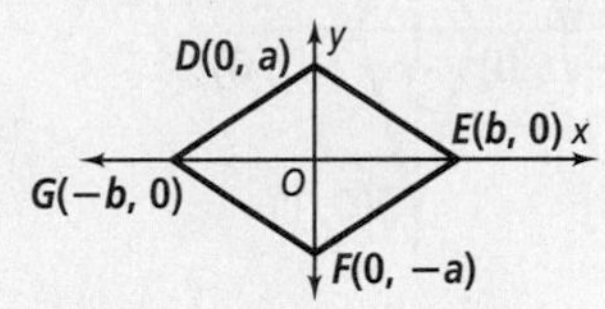

Given: DEFG is a rhombus.

Prove: $\overline{GE} \perp \overline{DF}$

By the Slope Formula, slope of $\overline{GE} = \frac{0 - 0}{b - (-b)} = 0$, and slope of $\overline{DF} = \frac{a - (-a)}{0 - 0}$, which is undefined. So $\overline{GE}$ must be horizontal and $\overline{DF}$ must be vertical. Therefore, $\overline{GE} \perp \overline{DF}$ because horiz. and vert. lines are ⊥.

22. Answers may vary. Sample:

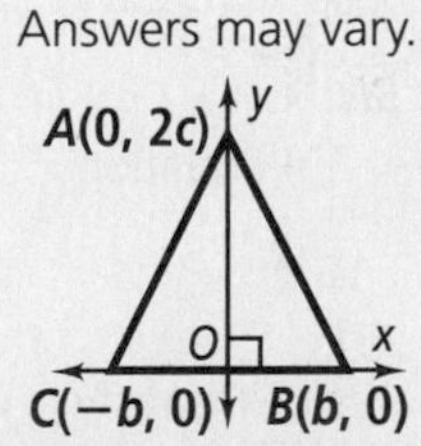

Given: Isosc. △ABC with base $\overline{BC}$ and altitude $\overline{AO}$

Prove: $\overline{AO}$ bisects $\overline{BC}$.

By the Distance Formula, $CO = \sqrt{[0 - (-b)]^2 + (0 - 0)^2} = b$ and $BO = \sqrt{(b - 0)^2 + (0 - 0)^2} = b$. Since $CO = BO$, $\overline{CO} \cong \overline{BO}$, so $\overline{AO}$ bisects $\overline{BC}$ by def. of seg. bisect.

23. Answers may vary. Sample:

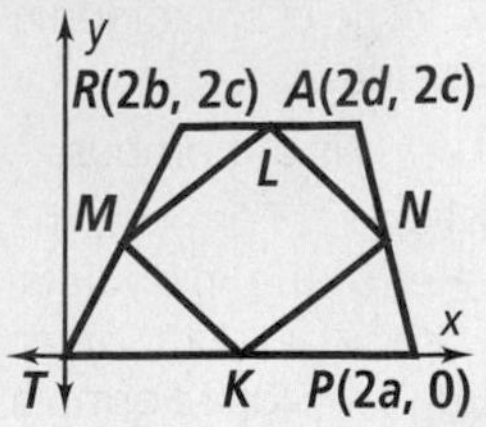

Given: Trapezoid *TRAP*, *M*, *L*, *N*, and *K* are midpoints of its sides
Prove: *MLNK* is a ▱.
By the Midpoint Formula, the coordinates of the midpoints are $M(b, c)$, $L(b + d, 2c)$, $N(a + d, c)$, and $K(a, 0)$. By the Slope Formula, the slope of $\overline{ML} = \frac{c}{d}$, the slope of $\overline{LN} = \frac{c}{b-a}$, the slope of $\overline{NK} = \frac{c}{d}$, and the slope of $\overline{KM} = \frac{c}{b-a}$. Since slopes are =, $\overline{ML} \parallel \overline{NK}$ and $\overline{LN} \parallel \overline{KM}$. Therefore, *MLNK* is a ▱ by def. of ▱.

24. Answers may vary. Sample:

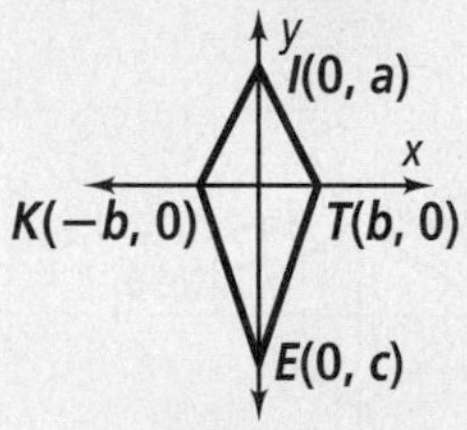

Given: Kite *KITE*
Prove: $\triangle KIE \cong \triangle TIE$
By the Distance Formula, $KI = IT = \sqrt{a^2 + b^2}$ and $KE = TE = \sqrt{b^2 + c^2}$, and $IE = \sqrt{(a - c)^2}$. $\overline{IE} \cong \overline{IE}$ by the Refl. Prop. of ≅. So $\triangle KIE \cong \triangle TIE$ by SSS.

25a. $L(3q, 3r)$, $M(3p + 3q, 3r)$, $N(3p, 0)$

b. equation of $\overleftrightarrow{AM}$: $y = \frac{r}{p + q}x$
equation of $\overleftrightarrow{BN}$:
$y = \frac{2r}{2q - p}(x - 3p)$ equation of $\overleftrightarrow{CL}$: $y = \frac{r}{q - 2p}(x - 6p)$

c. $P(2p + 2q, 2r)$

d. The coordinates of *P* satisfy the equation for $\overleftrightarrow{CL}$:
$y = \frac{r}{q - 2p}(x - 6p)$.
$2r = \frac{r}{q - 2p}(2p + 2q - 6p)$
$2r = \frac{r}{q - 2p}(2q - 4p)$
$2r = 2r$

e.
$AM = \sqrt{(3p + 3q - 0)^2 + (3r - 0)^2} =$
$\sqrt{(3p + 3q)^2 + (3r)^2}$;
$\frac{2}{3}AM = \frac{2}{3}\sqrt{(3p + 3q)^2 + (3r)^2} =$
$\sqrt{\frac{4}{9}\left[(3p + 3q)^2 + (3r)^2\right]} =$
$\sqrt{\left[\frac{4}{9}(3p + 3q)^2\right] + \left[\frac{4}{9}(3r)^2\right]} =$
$\sqrt{\left[\frac{4}{9}(9p^2 + 18pq + 9q^2)\right] + \left[\frac{4}{9}(9r^2)\right]} =$
$\sqrt{(4p^2 + 8pq + 4q^2) + (4r^2)} =$
$\sqrt{(2p + 2q)^2 + (2r)^2}$;
$AP = \sqrt{(2p + 2q - 0)^2 + (2r - 0)^2} =$
$\sqrt{(2p + 2q)^2 + (2r)^2}$
So $AP = \frac{2}{3}AM$. You can find the other two distances similarly.

26a. $\frac{b}{c}$

b. The point-slope formula for point $(a, 0)$ and $m = \frac{b}{c}$ is $y - 0 = \frac{b}{c}(x - a)$ or $y = \frac{b}{c}(x - a)$.

c. $x = 0$

d. The ordered pair $\left(0, \frac{-ab}{c}\right)$ satisfies the equation of line *q*, $x = 0$. When $x = 0$, $y = \frac{b}{c}(x - a)$ $= \frac{b}{c}(0 - a) = \frac{-ab}{c}$. So *p* and *q* intersect at $\left(0, -\frac{ab}{c}\right)$.

e. $\frac{a}{c}$

f. The point-slope formula for point $(b, 0)$ and $m = \frac{a}{c}$ is $y - 0 = \frac{a}{c}(x - b)$ or $y = \frac{a}{c}(x - b)$.

g. The ordered pair $\left(0, \frac{-ab}{c}\right)$ satisfies the equation of line *q*, $x = 0$. When $x = 0$, $y = \frac{a}{c}(x - b) =$ $\frac{a}{c}(0 - b) = \frac{-ab}{c}$. So *q* and *r* intersect at $\left(0, -\frac{ab}{c}\right)$.

h. $\left(0, \frac{-ab}{c}\right)$

27a. Answers may vary. Sample:
The area of a △ with base *b* and height *c* is $\frac{1}{2}bc$. The area of a △ with base *d* and height *a* is $\frac{1}{2}ad$. In both cases, the remaining area of the triangle has base $(b - d)$ and height *a*. Therefore $\frac{1}{2}ad = \frac{1}{2}bc$ by the Transitive Prop. of Eq. So $ad = bc$.

b. Slope of $\ell = \frac{a}{b}$ or $\frac{c}{d}$. So $\frac{a}{b} = \frac{c}{d}$ and $ad = bc$.

Chapter 9

Lesson 9-4

Got It? **page 571**

1a.

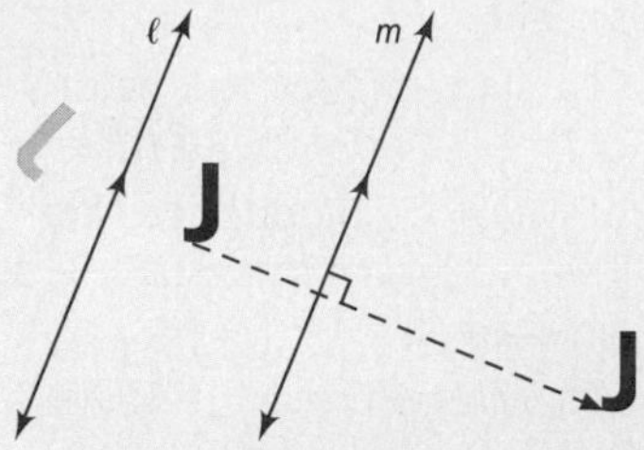

The arrow shows the translation. The distance is twice the distance between ℓ and *m*.

b. The distance is two times the distance between the lines of reflection.

Chapter 11

Lesson 11-1

Got It? **page 689**

1a. 6 vertices: *R*, *S*, *T*, *U*, *V*, *W*;
9 edges: $\overline{SR}$, $\overline{ST}$, $\overline{UR}$, $\overline{UV}$, $\overline{UT}$, $\overline{RV}$, $\overline{SW}$, $\overline{VW}$, $\overline{TW}$;
5 faces: △*URV*, △*STW*, quadrilateral *RSTU*, quadrilateral *RSWV*, quadrilateral *TWVU*

b. No; an edge is a segment formed by the intersection of two faces. $\overline{TV}$ is a segment that is contained in only one face, so it is not an edge.

2a. 12

b. 30

Chapter 12

Lesson 12-6

Lesson Check **page 808**

1.

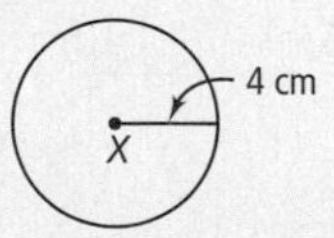

The locus is a circle with center *X* and radius 4 cm.

2.

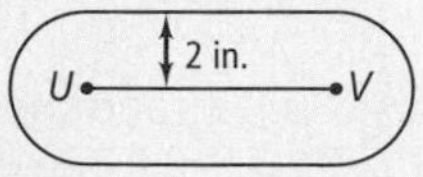

The locus is a pair of ∥ segments, each segment 2 in. from $\overline{UV}$, and two semicircles with radius 2 in. and centers *U* and *V*.

3.

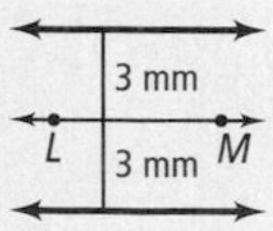

The locus is a pair of $\parallel$ lines, each 3 mm from $\overleftrightarrow{LM}$.

4. 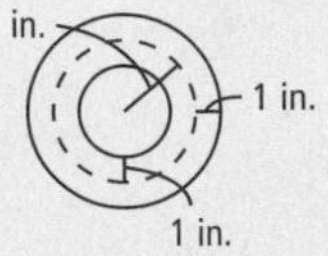

The locus is two circles concentric with the original circle; the smaller circle has radius 2 in. and the larger circle has radius 4 in.

5. Answers may vary. Sample: A *locus* is a set of points, and a *location* can be thought of as a description of a single point.

6. Alike: Both contain the midpt. of $\overline{JK}$ and are $\perp$ to $\overline{JK}$. Differences: The locus in a plane is a line and the locus in space is a plane.

Practice and Problem-Solving Exercises **page 808**

7. The locus is the $\perp$ bis. of $\overline{PQ}$.

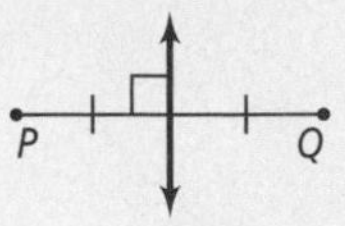

8. The locus is the ray that bis. $\angle ABC$.

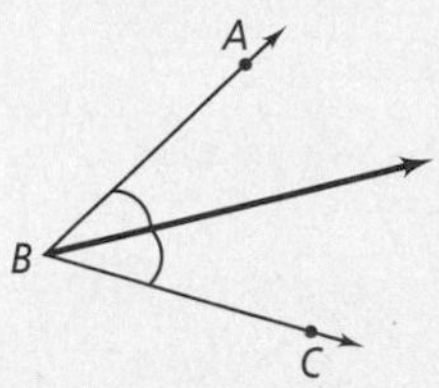

9. The locus is the two lines that bis. the rt. $\angle$s.

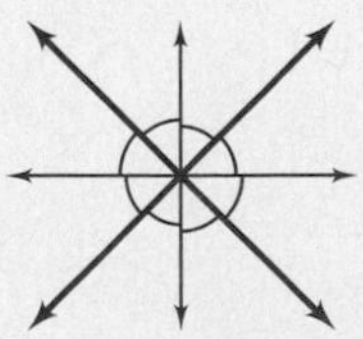

10. The locus is a circle, concentric with the given circle, with radius 1 cm.

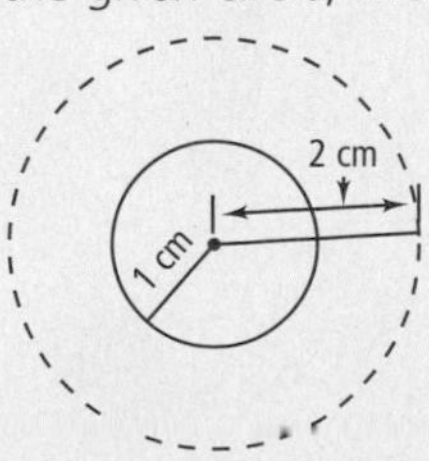

11. 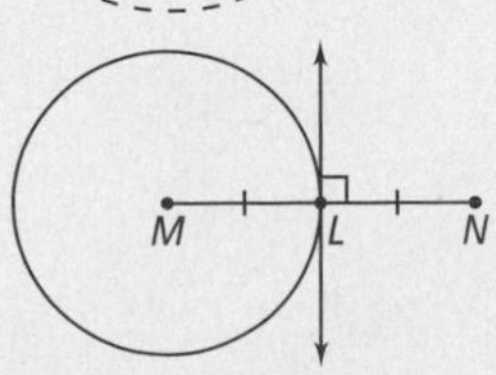

The locus is point L.

12. 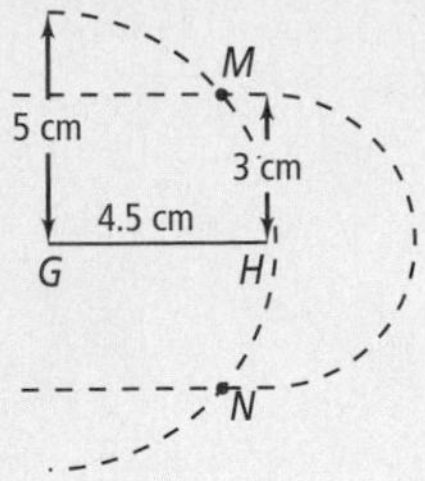

The locus is points M and N.

13. 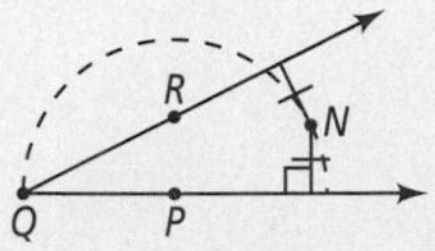

The locus is point N.

14. 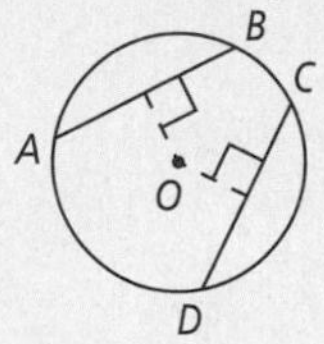

The locus is the center O.

15. 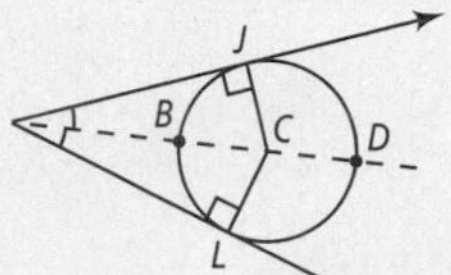

The locus is points B and D.

16. The locus is a sphere with center F and radius 3 cm.

17. The locus is an endless cylinder with radius 4 cm and centerline $\overleftrightarrow{DE}$.

18. The locus is two planes, each $\parallel$ to plane M, and each 1 in. from M.

19. The locus is an endless cylinder with radius 5 mm and centerline $\overrightarrow{PQ}$, and a hemisphere of radius 5 mm centered at P, "capping off" the cylinder.

20. The locus is the set of all points in the interior of $\angle A$ and equidistant from the sides of $\angle A$.

21. The locus is the set of all points 2 units from the origin.

22. The locus is the set of all points equidistant from two $\parallel$ planes M and N.

23. Check students' work.

24. Yes; if the collinear pts. are A, B, and C, then the locus of pts. equidist. from A and B is a plane M, $\perp$ to $\overline{AB}$ at its midpt. Similarly, pts. equidist. from B and C are on plane N, $\perp$ at the midpt. of $\overline{BC}$. But $M \parallel N$.

25. The locus is the set of points in the plane that are 1 unit from the *x*-axis and 2 units from the origin.

26. $y = x$

27. $y = 2x - 4$

28. $y = -x + 3$

29a. a circle

b. $x^2 + y^2 = 4$

Index

A

Index

D

E

H

I

K

Index

Index

Q

R

Index

U

V

W

X

Y

Z

Acknowledgments

Staff Credits

The people who made up the High School Mathematics team—representing composition services, core design digital and multimedia production services, digital product development, editorial, editorial services, manufacturing, marketing, and production management—are listed below.

Emily Allman, Dan Anderson, Scott Andrews, Christopher Anton, Carolyn Artin, Michael Avidon, Margaret Banker, Charlie Bink, Niki Birbilis, Suzanne Biron, Beth Blumberg, Tim Breeze-Thorndike, Kyla Brown, Rebekah Brown, Judith Buice, Sylvia Bullock, Stacie Cartwright, Carolyn Chappo, Christia Clarke, Mary Ellen Cole, Tom Columbus, Andrew Coppola, AnnMarie Coyne, Bob Craton, Nicholas Cronin, Patrick Culleton, Damaris Curran, Steven Cushing, Sheila DeFazio, Cathie Dillender, Emily Dumas, Patty Fagan, Frederick Fellows, Jorgensen Fernandez, Mandy Figueroa, Suzanne Finn, Sara Freund, Matt Frueh, Jon Fuhrer, Andy Gaus, Mark Geyer, Mircea Goia, Andrew Gorlin, Shelby Gragg, Ellen Granter, Gerard Grasso, Lisa Gustafson, Toni Haluga, Greg Ham, Marc Hamilton, Chris Handorf, Angie Hanks, Scott Harris, Cynthia Harvey, Phil Hazur, Thane Heninger, Aun Holland, Amanda House, Chuck Jann, Linda Johnson, Blair Jones, Marian Jones, Tim Jones, Gillian Kahn, Matthew Keefer, Brian Keegan, Jim Kelly, Jonathan Kier, Jennifer King, Tamara King, Elizabeth Krieble, Meytal Kotik, Brian Kubota, Roshni Kutty, Mary Landry, Christopher Langley, Christine Lee, Sara Levendusky, Lisa Lin, Wendy Marberry, Dominique Mariano, Clay Martin, Rich McMahon, Eve Melnechuk, Cynthia Metallides, Hope Morley, Christine Nevola, Michael O'Donnell, Michael Oster, Ameer Padshah, Stephen Patrias, Jeffrey Paulhus, Jonathan Penyack, Valerie Perkins, Brian Reardon, Wendy Rock, Marcy Rose, Carol Roy, Irene Rubin, Hugh Rutledge, Vicky Shen, Jewel Simmons, Ted Smykal, Emily Soltanoff, William Speiser, Jayne Stevenson, Richard Sullivan, Dan Tanguay, Dennis Tarwood, Susan Tauer, Tiffany Taylor-Sullivan, Catherine Terwilliger, Maria Torti, Mark Tricca, Leonid Tunik, Ilana Van Veen, Lauren Van Wart, John Vaughan, Laura Vivenzio, Samuel Voigt, Kathy Warfel, Don Weide, Laura Wheel, Eric Whitfield, Sequoia Wild, Joseph Will, Kristin Winters, Allison Wyss, Dina Zolotusky

Additional Credits: Michele Cardin, Robert Carlson, Kate Dalton-Hoffman, Dana Guterman, Narae Maybeth, Carolyn McGuire, Manjula Nair, Rachel Terino, Steve Thomas

Illustration

Stephen Durke: 4, 5, 8, 9, 11, 18, 20, 25, 27, 32, 43, 50, 53, 54, 59, 60, 61, 66, 67, 140, 148, 156, 161, 164, 168, 171, 174, 179, 182,189, 195, 197, 203, 218, 219, 222, 226, 228, 232, 234, 235, 240, 244, 245, 248, 250, 255, 265, 270, 285, 290, 292, 294, 296, 297, 301, 303, 305, 306, 309, 314, 317, 324, 326, 330, 332, 333, 334, 336, 338, 466, 504, 511, 516, 519, 520, 521, 548, 620, 641, 655, 656, 657, 665, 811; **Jeff Grunwald represented by Wilkinson Studios, Inc.:** 82, 83, 89, 94, 98, 106, 109, 111, 113, 120, 125, 128, 200, 357, 359, 371, 375, 383, 385, 389, 400, 404, 406, 417, 437, 440, 443, 446, 450, 454, 456, 464, 465, 466, 471, 473, 475, 476, 499, 516, 639; **Phil Guzy:** 34, 321, 330, 491, 496, 507, 513, 544, 549, 554, 559, 569, 570, 616, 623, 643, 649, 652, 655, 672, 673, 688, 697, 699, 703, 705, 706, 708, 717, 720, 726, 728, 733, 739, 742, 748, 758, 762, 770, 771, 780, 790, 792, 798; **Rob Schuster** 522; **XNR Productions:** 527, 623, 806, 810.

Technical Illustration

Aptara, Inc.; Datagrafix, Inc.

Photographs

Every effort has been made to secure permission and provide appropriate credit for photographic material. The publisher deeply regrets any omission and pledges to correct errors called to its attention in subsequent editions.

Unless otherwise acknowledged, all photographs are the property of Pearson Education, Inc.

Photo locators denoted as follows: Top (T), Center (C), Bottom (B), Left (L), Right (R), Background (Bkgd)

Cover
Bon Appetit/Marc O. Finley/Alamy Images

Front Matter
ix, **x** Hoge Noorden/epa/Corbis

15 (TR) Kelly Redinger/Alamy; **30** (T) Pete Saloutos/Corbis; **32** (BCR) Stuart Melvin/Alamy Images; **39** (BR) Richard Megna/Fundamental Photographs; **62** (TR) Charles O. Cecil /Alamy Images; **103** (T) Material courtesy of Bill Vicars and Lifeprint; **126** (CR) Jenny Thompson/Fotolia; **143** (T) Bill Brooks/Alamy Images; **144** (BC) Kevin Fleming/Corbis; **154** (BR) Frank Adelstein; **159** (TR) Robert Slade/Manor Photography/Alamy Images; **162** (TR) Robert Llewellyn/Corbis; **176** (TCR) Peter Cade/Iconica/Getty Images; **227** (B) Stan Honda/Staff/AFP/Getty Images; **246** (BR) Viktor Kitaykin/iStockphoto; **253** (TR) John Wells/Photo Researchers, Inc.; **258** (CR) Tony Freeman/PhotoEdit, Inc.; **260** (C) Paul Jones/Iconica/Getty Images, (TL) Veer/Corbis; **263** (TR) Image Source Black/Jupiter Images; **287** (BR) SeanPavonePhoto/Fotolia; **289** (CR) Joseph Sohm/Visions of America, LLC/Alamy Images; **333** (BR) Gunter Marx/Corbis; **354** (CR) Anthony Bannister/Gallo Images/Corbis; Robert Harding Picture Library/Alamy Images; **356** (CR) Christiana Subekti/iStockphoto, (CL) Inger Hogstrom/Danita Delimont, Agent, (CC) Laurie Strachan/Alamy; **360** (BR) Eric Hood/iStockphoto; **365** (TR) Esa Hiltula/Alamy Images; **367** (TCR) Victor Fraile/Reuters Media; **376** (TR) Kirsty McLaren/Alamy Images; **379** (CL) Claro Cortes IV/Reuters/Landov LLC, (CR) Michael Jenner/Alamy Images; **387** (TR, TCR) Rodney Raschke/Active Photo Service; **395** (BR) Colin Underhill/Alamy Images; **432** (BR) Chuck Eckert/Alamy Images; **443** (C) Ron Watts/Corbis; **466** (TR) ©James L. Amos/Corbis; **469** (BC) Corey Hochachka/Design Pics, (BR) Getty Images, (BL) Mark A. Johnson/Corbis; **475** (TCR) Victor R. Boswell Jr./Contributor National Geographic/Getty Images; **493** (T) Petra Wegner/Alamy Images; **502** (BR) PhotoObjects/Thinkstock; **508** (BR) Steve Vidler/Imagestate Media; **517** (BCR) Dave Reede/All Canada Photos/Alamy Images; **545** (BR, BL, BC) RubberBall/Alamy; **559** (BCR) North Wind Picture Archives/Alamy Images; **566** (TR) Alan Copson/City Pictures/Alamy Images; **584** (TL) M.C. Escher's "Symmetry E56" © 2009 The M.C. Escher Company-Holland. All rights reserved. www.mcescher.com (TR) M.C. Escher's "Symmetry E18" © 2009 The M.C. Escher Company-Holland. All rights reserved. www.mcescher.com; **587** (C) Martin William Allen/Alamy; **589** (BR) Keith Leighton/Alamy, (BCR) Reven T.C. Wurman/Alamy Images; **598** (BL) D. Hurst/Alamy, (BR) Owaki – Kulla/Corbis; **601** (CR) Douglas Kirkland/Corbis; **618** (TR) Bob Gates/Alamy Images; **625** (TR) Joe Sohm/Visions of America/Alamy; **627** (TR) artpartner-images/Alamy Images; **633** (TR) Dennis Marsico/Corbis; **644** (TR) Alan Schein Photography/Corbis; **661** (TR) Matthias Tunger/Photonica/Getty Images; **668** (TCR) Clive Streeter/©DK Images; **670** (BR) amana images inc./Alamy; **693** (BCR) Sports Bokeh/Alamy; **705** (TR) Ron Chapple Stock/Alamy; **710** (TR) Adam Eastland/Alamy Images; **727** (TR) age fotostock/SuperStock; **728** (BCR) John E Marriott/Alamy Images; **733** (BCR) D. Hurst/Alamy; **737** (TL) Andrew Paterson/Alamy, (TC) Getty Images, (TR) Image Source/Getty Images; **738** (BR) Stephen Sweet/Alamy; **744** (TR) Jupiterimages/BananaStock/Alamy; **745** (TR) Andre Jenny/Alamy Images; **764** (TCR) Marshall Space Flight Center Collection/NASA, (TR) T. Pohling/Alamy Images; **768** (TR, TCR, TC) Clive Streeter/©DK Images; **775** (TR) Cris Bouroncle/Staff/AFP/Getty Images; **785** (BR) Vario Images GmbH & Co.KG/Alamy Images; **792** (TR) Marshall Space Flight Center Collection/NASA; **794** (BR) Melvyn Longhurst/Alamy; **795** (BR) dpa/Corbis; **809** (BR) ©matthiasengelien/Alamy.

This page intentionally left blank.